Lisa A. Urry
MILLS COLLEGE, OAKLAND, CALIFORNIA

Noel Meyers
LA TROBE UNIVERSITY, VICTORIA

Michael L. Cain
NEW MEXICO STATE UNIVERSITY

Steven A. Wasserman
UNIVERSITY OF CALIFORNIA, SAN DIEGO

Peter V. Minorsky
MERCY COLLEGE, DOBBS FERRY, NEW YORK

Rebecca B. Orr
COLLIN COLLEGE, PLANO, TEXAS

Karen Burke da Silva
FLINDERS UNIVERSITY, SOUTH AUSTRALIA

Ann Parkinson
UNIVERSITY OF THE SUNSHINE COAST, QUEENSLAND

Lesley Lluka
UNIVERSITY OF QUEENSLAND, QUEENSLAND

Prasad Chunduri
UNIVERSITY OF QUEENSLAND, QUEENSLAND

Pearson Australia
707 Collins Street
Melbourne VIC 3008
Ph: 03 9811 2400

www.pearson.com.au

Project Management Team Leader: Jill Gillies
Senior Production Manager: Lisa D'Cruz
Portfolio Associate: Jessica Darnell

Printed and bound in Australia by The SOS Print + Media Group

ISBN: 9780655707721

About this Custom Book

Welcome to the fourth edition of *LFS100 Cell Biology.*

This custom book is a compilation of chapters from *Campbell Biology: Australian and New Zealand Version,* 12th Edition by Urry *et al*. The chapters were selected to meet the specific requirements of your course. As only relevant content has been included, you may come across references to chapters that are not in this custom book.

We wish you success in your studies.

This page is intentionally blank.

Brief Contents

About the Authors

Lisa A. Urry (Chapter 1 and Units 1–3) is Professor of Biology at Mills College. After earning a BA at Tufts University, she completed her PhD at the Massachusetts Institute of Technology (MIT). Lisa has conducted research on gene expression during embryonic and larval development in sea urchins. Deeply committed to promoting opportunities in science for women and underrepresented minorities, she has taught courses ranging from introductory and developmental biology to an immersive course on the US/Mexico border.

Noel Meyers completed his PhD in plant pollination biology at the University of Queensland. He has completed two postdoctoral research fellowships with the CSIRO Division of Plant Industry. For his teaching, Noel won an Australian Award for University Teaching and a Pearson Uniserve Award for his contributions to science students' learning. He has also earned a Fellowship of the Higher Education Research and Development Society of Australasia (FHERDSA). Noel dedicates his life to science education.

Michael L. Cain (Units 4, 5, and 8) is an ecologist and evolutionary biologist who is now writing full-time. Michael earned an AB from Bowdoin College, an MSc from Brown University, and a PhD from Cornell University. As a faculty member at New Mexico State University, he taught introductory biology, ecology, evolution, botany, and conservation biology. Michael is the author of dozens of scientific papers on topics that include foraging behaviour in insects and plants, long-distance seed dispersal, and speciation in crickets. He is also a coauthor of an ecology textbook.

Steven A. Wasserman (Unit 7) is Professor of Biology at the University of California, San Diego (UCSD). He earned an AB from Harvard University and a PhD from MIT. Working on the fruit fly *Drosophila*, Steve has done research on developmental biology, reproduction, and immunity. Having taught genetics, development, and physiology to undergraduate, graduate, and medical students, he now focuses on introductory biology, for which he has been honoured with UCSD's Distinguished Teaching Award.

Peter V. Minorsky (Unit 6) is Professor of Biology at Mercy College in New York, where he teaches introductory biology, ecology, and botany. He received his AB from Vassar College and his PhD from Cornell University. Peter taught at Kenyon College, Union College, Western Connecticut State University, and Vassar College; he is also the science writer for the journal *Plant Physiology*. His research interests concern how plants sense environmental change. Peter received the 2008 Award for Teaching Excellence at Mercy College.

Rebecca B. Orr (Ready-to-Go Teaching Modules, Interactive Visual Activities, eText Media Integration) is Professor of Biology at Collin College in Plano, Texas, where she teaches introductory biology. She earned her BS from Texas A&M University and her PhD from University of Texas Southwestern Medical Center at Dallas. Rebecca has a passion for investigating strategies that result in more effective learning and retention, and she is a certified Team-Based Learning Collaborative Trainer Consultant. She enjoys focusing on the creation of learning opportunities that both engage and challenge students.

Karen Burke da Silva is the Dean (Education) in the College of Science and Engineering at Flinders University. Karen is recognised as one of Australia's most influential science educators; she was awarded the Australian University Teacher of the Year in 2016 and the South Australian STEM tertiary educator of the year in 2015. She has published numerous journal articles, book chapters, and curricula in STEM education. Karen is a conservation biologist and is the Founder of the Saving Nemo Conservation and Citizen Science Program. She is actively involved in social media campaigns to raise awareness around conservation and environmental issues building greater understanding of science in the public arena.

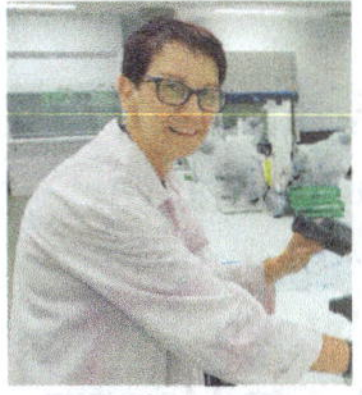

Ann Parkinson completed her PhD in muscle physiology at the University of New South Wales. Ann is a Senior Lecturer in Physiology and Anatomy at the University of the Sunshine Coast and has over 20 years' experience in developing and delivering curriculum in cell biology and physiology. She has earned recognition for excellence in learning and teaching, including a ALTC Citation and the Senior Fellow level of the Higher Education Academy. Ann's research is primarily in higher education practice in teaching biology and physiology, with particular emphasis on the use of visualisation technologies to enhance learning. This includes the novel CAVE2TM environment using 3D immersive visualisation to view cells.

Lesley Lluka is a pharmacist and neuroscientist, and completed a PhD in Pharmacology at the University of Queensland. Her research has focused on the molecular targets for the action of antidepressant drugs, with research collaborations in Europe and the United States, and on science education. After teaching pharmacology for many years, Lesley is now dedicated to teaching large first-year biology classes for students with a broad range of career trajectories. Since 2005, she has worked at the University of Queensland, where she has held teaching and learning leadership positions at school and faculty level.

Prasad Chunduri completed his PhD in the field of cardiovascular biology at The University of Queensland. He currently teaches physiology to many undergraduate students, especially those in their first year of university studies. He has received several awards for his teaching, including a University Commendation for Outstanding Contribution to Student Learning and an Award for Teaching Excellence from the Faculty of Medicine at The University of Queensland. His disciplinary research interests continue in the field of cardiovascular biology and stroke. In addition, he also researches in Scholarship of Teaching and Learning, predominantly in the design and evaluation of better teaching methods for first-year biology students.

Neil A. Campbell (1946–2004) earned his MA from the University of California, Los Angeles, and his PhD from the University of California, Riverside. His research focused on desert and coastal plants. Neil's 30 years of teaching included introductory biology courses at Cornell University, Pomona College, and San Bernardino Valley College, where he received the college's first Outstanding Professor Award in 1986. For many years he was also a visiting scholar at UC Riverside. Neil was the founding author of *Campbell Biology*.

Preface

From the Australian and New Zealand authors

Biology has always been important, although never as important as it is now. There has also never been a more exciting time to make biology your lifelong passion and career.

The author team and those who brought this text to life worked to convey to you the challenges, the wonders, and the foundations you will need as a practicing biologist specialising in the Southern Hemisphere, and in Australia and New Zealand particularly.

To help build your foundational knowledge, we share with you a suite of unifying themes. The evolution of unique biological legacies in both New Zealand and Australia represents a central tenet of this book. Starting with Gondwana, the supercontinent centred on modern Antarctica, our countries shared a plethora of dinosaurs and plants—we have the fossils to prove it. New Zealand, isolated from Gondwana over 90 million years ago, was long thought to have evolved its biological uniqueness from a random selection of Gondwana plants and animals. However, Hamish Campbell's work (combined with that of countless talented New Zealand and Australian researchers) suggests this is unlikely. The waves that rolled over New Zealand for millions of years may have erased this distinctive biological experiment. As the oceans rolled back from New Zealand, say 23 million years ago, a slow process of colonisation began. Today, roughly 95% of the plants and animals found in New Zealand evolved from immigrants who arrived from Australia, probably by sea and air.

As history has already taught the authors of this text, perhaps a third of the information we share with you now will ultimately prove misleading or, worse, flat out wrong. As authors, we face an inevitable, unresolvable conundrum: We don't know which third will change. If the authors of the last several editions had taken bets on what new discoveries would sweep away understandings we shared with previous readers, few, if any, of us would have wagered that the confluence of genetics, biology, and geology would show us the close and comparatively recent affinities between the Australian and New Zealand biologies. We seek solutions based on pragmatism and determination.

Here, we chronicle the best of our understandings of the biological world—as we understand it today. It is likely that you or one of your classmates will reset our direction and guide us on a new course, as we continually seek to improve our knowledge of the biological world. Your discoveries, through dedicated works and committed actions, may prove instrumental in shifting humanity's biological perspectives and imperatives.

Don't believe us? Take Dr Abigail Allwood's work in the Pilbara region of Western Australia (see her interview at the beginning of Unit 1). Her work reveals evidence of some of the earliest life and bacterial reefs on Earth. Abby currently works for NASA, where she searches Jezero Crater on Mars for evidence of ancient life.

Human-induced climate change—another theme linking the understandings contained in this book—has in the last several years produced unprecedented fires and floods, along with more intense and increasingly frequent extreme weather events. Collectively, the resilience of ecosystems to these events will become intimately tied to the management and stewardship biologists can bring to issues. Biologists can and do shift policies that profoundly change how we think and act. While challenges abound, so too do innovative and new ways of thinking: New Zealand's campaign to become free from introduced predators by 2050 represents the kind of call to action that sets the standard for other nations. Your ingenuity is all the more important in facing down these unprecedented challenges.

Biologists from Australia and New Zealand played an oversized role in the development of vaccines and in enriching our understandings to help combat the sudden acute respiratory syndrome coronavirus 2 (SARS-CoV-2) that causes the coronavirus (COVID-19) disease. A disease that appeared as if from nowhere—except, it didn't. The consequences of the disease altered how we learnt, changed the ways we as individuals studied, and transformed the ways we lived. During lockdown, we may have reflected on the human-centric biosphere gone haywire—except, it hadn't.

Along with parts of the rest of the world, Australia and New Zealand experienced climate-induced shifts in fires, droughts, and floods. While apparently unrelated, your study of biology will show the subtle yet insidious link between new viruses and habitat degradation, decays in biodiversity, the spread of introduced species, and pollution.

We bookend Dr Allwood's ancient bacterial reefs with the work of Ove Hoegh-Guldberg, who seeks to understand the most biologically diverse modern coral reefs on the planet. His work, and that of his colleagues, seeks to lay the foundations for conservation and the restoration of these biological wonders. The work of conservation biologists to preserve intact ecosystems has never been more important. Professor Michael Archer's Lazarus Project seeks to return the Tasmanian tiger from the clutches of extinction. If it comes to fruition, perhaps other species will follow. Cloning may represent a hedge against the unthinkable, the erosion of biological diversity, despite our most thorough conservation efforts. Successful cloning of ferrets brings new hope for endangered species—provided we have conserved the unique Australian and New Zealand ecosystems on which they will again rely.

Ultimately, the authors of this book invest their energy and lives to enrich your capacity to build on the contributions of those that have gone before, so that you can contribute more, with greater sustainability and intergenerational equity than at any other time in history. What an exciting time to be alive, to be sentient, and to have the opportunity to curiously wander the world equipped with the underpinnings of biological wisdom.

From the US Author Team

We are honoured to present the Twelfth Edition of *Campbell Biology*. For the last three decades, *Campbell Biology* has been the leading college text in the biological sciences. It has been translated into 19 languages and has provided millions of students with a solid foundation in college-level biology. This success is a testament not only to Neil Campbell's original vision but also to the dedication of hundreds of reviewers (listed on pages xxv-xxviii), who, together with editors, artists, and contributors, have shaped and inspired this work.

Our goals for the Twelfth Edition include:

- **supporting students** with new visual presentations of content and new study tools
- **supporting instructors** by providing new teaching modules with tools and materials for introducing, teaching, and assessing important and often challenging topics
- **integrating text and media** to engage, guide, and inform students in an active process of inquiry and learning.

Our starting point, as always, is our commitment to crafting text and visuals that are accurate, are current, and reflect our passion for teaching biology.

New to This Edition

Here we provide an overview of the new features that we have developed for the Twelfth Edition; we invite you to explore pages xii–xix for more information and examples.

- **NEW! Chapter Openers Re-envisioned.** Catalysed by feedback from students and instructors, informed by data analytics, and building on the results of science education research, we have redesigned the opening of every chapter of the text. The result is more visual, more interactive, and more engaging. In place of an opening narrative, the first page of each chapter is organised around three new elements that provide students with the specific tools and approaches needed to achieve the learning objectives of that chapter:
  - **NEW! Visual Overview.** Centered on a basic biological question related to the opening photo and legend, the Visual Overview illustrates a core idea of the chapter with straightforward art and text. Students get an immediate sense of what the chapter is about and what kinds of thinking will underlie its exploration.
  - **NEW! Study Tip.** Just as the Visual Overview introduces students to *what* they will learn, the study tip offers guidance in *how* to learn. It encourages students to learn actively through such proven strategies as drawing a flow chart, labelling a diagram, or making a table. Each tip provides an effective strategy for tackling important content in the chapter.
- **NEW! Updated Content.** As in each new edition of *Campbell Biology*, the Twelfth Edition incorporates new content, summarised on pages ix–xi. Content updates reflect rapid, ongoing changes in knowledge about climate change, genomics, gene-editing technology (CRISPR), evolutionary biology, microbiome-based therapies, and more. In addition, Unit 7 includes a new section on "Biological Sex, Gender Identity, and Sexual Orientation in Human Sexuality," which provides instructors and students with a thoughtful, clear, and current introduction to topics of tremendous relevance to biology, to student lives, and to current public discourse and events.
- **5 NEW! Ready-to-Go Teaching Modules.** The Ready-to-Go Teaching Modules provide instructors with active learning exercises and questions to use in class, plus Mastering Biology assignments that can be assigned before and after class. A total of 15 modules are now available in the Instructor Resources area of Mastering Biology.

Mastering Biology

Mastering Biology provides valuable resources for instructors to assign homework and for students to study on their own:

- **Assignments.** Mastering Biology is the most widely used online assessment and tutorial program for biology, providing an extensive library of thousands of tutorials and questions that are graded automatically.
  - **NEW! Early Alerts** give instructors a quick way to monitor students' progress and provide feedback, even before the first test.

- Hundreds of self-paced **tutorials** provide individualised coaching with specific hints and feedback on the most difficult topics in the course.
- Optional **Adaptive Follow-up Assignments** provide additional questions tailored to each student's needs.

- **Pearson eText.** The Pearson eText can be directly accessed from Mastering Biology.
- **Dynamic Study Modules.** These popular review tools can be assigned, or students can use them for self-study.
- **Study Area.** The Study Area provides students with access to high-quality media, including BioFlix animations, as well as Figure Walkthroughs, BBC Videos and HHMI BioInteractive short films.

For more information, see pages xvii–xix and visit www.masteringbiology.com.

Our Hallmark Features

Teachers of general biology face a daunting challenge: to help students acquire a conceptual framework for organising an ever-expanding amount of information. The hallmark features of *Campbell Biology* provide such a framework, while promoting a deeper understanding of biology and the process of science. As such, they are well-aligned with the core competencies outlined by the **Vision and Change** national conferences. Furthermore, the core concepts defined by Vision and Change have close parallels in the unifying themes that are introduced in Chapter 1 and integrated throughout the book.

Chief among the themes of both Vision and Change and *Campbell Biology* is **evolution.** Each chapter of this text includes at least one Evolution section that explicitly focuses on evolutionary aspects of the chapter material, and each chapter ends with an Evolution Connection Question and a Write About a Theme Question.

To help students distinguish "the forest from the trees," each chapter is organised around a framework of three to seven carefully chosen **Key Concepts**. The text, Concept Check Questions, Summary of Key Concepts, and Mastering Biology resources all reinforce these main ideas and essential facts.

Because text and illustrations are equally important for learning biology, **integration of text and figures** has been a hallmark of *Campbell Biology* since the First Edition. The new Visual Overviews, together with our popular Visualising Figures, Exploring Figures, and Make Connections Figures, epitomise this approach.

To encourage **active reading** of the text, *Campbell Biology* includes numerous opportunities for students to stop and think about what they are reading, often by putting pencil to paper to draw a sketch, annotate a figure, or graph data. Answering these questions requires students to write or draw as well as think and thus helps develop the core competency of communicating science.

Finally, *Campbell Biology* has always featured **scientific inquiry**. The inquiry activities provide students practice in applying the process of science and using quantitative reasoning, addressing core competencies from Vision and Change.

Our Partnership with Instructors and Students

The real test of any text is how well it helps instructors teach and students learn. We welcome comments from both students and instructors. Please address your suggestions to customer.service@pearson.com.au.

Highlights of New Content

This section highlights selected new content in *Campbell Biology*, Twelfth Edition. In addition to the content updates noted here, every chapter has a **new Visual Overview** on the chapter opening page.

Unit 1 THE CHEMISTRY OF LIFE

In Unit 1, new content engages students in learning foundational chemistry. Chapter 2 includes a new micrograph of the tiny hairs on a gecko's foot that allow it to walk up a wall. The opening photo for Chapter 3 features a ringed seal, a species endangered by the melting of Arctic sea ice due to climate change. Chapter 3 also has added coverage on the discovery of a large subsurface reservoir of liquid water on Mars and the first CO_2 enhancement study done on an unconfined natural coral reef (both reported in 2018). Chapter 4 now includes the discovery of carbon-based compounds on Mars reported by NASA in 2018. In Chapter 5, the technique of cryo-electron microscopy is introduced, due to its increasing importance in the determination of molecular structure.

Unit 2 THE CELL

Our main goal for this unit was to make the material more accessible, inviting, and exciting to students. Chapter 6 includes a new text description of cryo-electron microscopy (cryo-EM) and a new cryo-EM image in Figure 6.3. Art has been added to Figure 6.17 to illustrate the dynamic nature of mitochondrial networks. Chapter 7 begins with a new chapter-opening image showing neurotransmitter release during exocytosis. Figure 8.1 includes a new photo of bioluminescent click beetle larvae on the outside of a termite mound and a new Visual Overview that illustrates how the laws of thermodynamics apply to metabolic reactions like bioluminescence.

▼ Figure 8.1

Figure 8.1 The green glowing spots on the outside of this Brazilian termite mound are larvae of the click beetle, *Pyrophorus nyctophanus*. These larvae convert the energy stored in organic molecules to light, a process called bioluminescence, which attracts termites that the larvae eat. Bioluminescence and other metabolic activities in a cell are energy transformations that are subject to physical laws.

How do the laws of thermodynamics relate to biological processes?

The first law of thermodynamics: *Energy can be transferred and transformed, but it cannot be created or destroyed.* For example:

Light energy from the sun
Transformed by plants to
Chemical energy in organic molecules in plants
Transformed by termites to
Chemical energy in organic molecules in termites
Transformed by click beetle larvae to
Light energy produced by bioluminescence

The second law of thermodynamics: *Every energy transfer or transformation increases the entropy (disorder) of the universe.* For example, during every energy transformation, some energy is converted to thermal energy and released as heat:

Heat
Plant using light to make organic molecules
Heat
Termite digesting plant and making new molecules
Heat
Click beetle larva digesting termite and emitting light

145

Chapter 9 includes new information on human brown fat usage, the role of fermentation during the production of chocolate, and recent research on the role of lactate in mammalian metabolism. Chapter 10 begins with a new concept that puts photosynthesis into a big-picture ecological context. Chapter 10 also includes a discussion of the 2018 discovery of a new form of chlorophyll found in cyanobacteria that can carry out photosynthesis using far-red light. In Chapter 11, the relevance of synaptic signaling is underscored by mentioning that it is a target for treatment of depression, anxiety, and PTSD. In Chapter 12, the cell cycle figure (Figure 12.6) now includes cell images and labels describing the events of each phase.

Unit 3 GENETICS

Chapters 13–17 incorporate changes that help students grasp the more abstract concepts of genetics and their chromosomal and molecular underpinnings. For example, a new Concept Check 13.2 question asks students about shoes as an analogy for chromosomes. In Chapter 14, the classic idea of a single gene determining hair or eye colour, or even earlobe attachment, is discussed as an oversimplification. Also, the "Fetal Testing" section has been updated to reflect current practices in obstetrics. Chapter 15 now includes new information on "three-parent" babies. In Concept 16.3, the text and Figure 16.23 have been extensively revised to reflect recent models of the structure and organisation of interphase chromatin, as well as how chromosomes condense during preparation for mitosis. Chapter 17 now describes the mutation responsible for the albino phenotype of the Asinara donkeys featured in the chapter-opening photo. To make it easier to cover CRISPR, a new section has been added to Concept 17.5 describing the CRISPR-Cas9 system, including Figure 17.28, "Gene editing using the CRISPR-Cas9 system".

Chapters 18–21 are extensively updated, driven by exciting new discoveries based on DNA sequencing and gene-editing technology. In Chapter 18, the coverage of epigenetic inheritance has been enhanced and updated, including the new Figure 18.8. Also in Chapter 18, a description of topologically associated domains has been added, along with an update on the 4D Nucleome Network. In Chapter 19, the topic of emerging viral diseases has been updated extensively and reorganised to

▼ Figure 18.8 Examples of epigenetic inheritance.

(a) Effects of maternal diet on genetically identical mice.

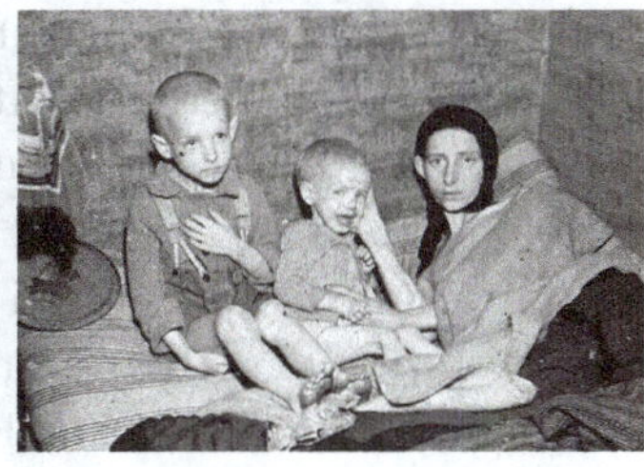

(b) The Dutch Hunger Winter.

clearly differentiate influenza viruses that are emerging from those that cause seasonal flu. Other Chapter 19 updates include information on vaccine programs, mentioning a large measles outbreak in 2019 that correlated with lower vaccination rates in that region. Information has also been added on improvement of treatment regimes for HIV. Chapter 20 has been extensively updated, including addition of two new subsections, "Personal Genome Analysis" and "Personalised Medicine," with new information on direct-to-consumer genome analysis. Other updates include the first cloning of a primate, stem cell treatment of age-related macular degeneration, CRISPR correction of the sickle-cell disease allele in mice, and a report of gene editing of fertilised human eggs that resulted in live births. Chapter 21 updates include results of the Cancer Genome Atlas Project, a newly discovered function of retrotransposon transcription, and new information on the *FOXP2* gene.

Unit 4 MECHANISMS OF EVOLUTION

The revision of Unit 4 uses an evidence-based approach to strengthen how we help students understand key evolutionary concepts. For example, new text in Concept 24.3 describes how hybrids can become reproductively isolated from both parent species, leading to the formation of a new species. Evidence supporting this new material comes from a 2018 study on the descendants of hybrids between two species of Galápagos finches and provides an example of how scientists can observe the formation of a new species in nature. In Concept 25.2, the discussion of fossils as a form of scientific evidence is supported by a new figure (Figure 25.5) that highlights five different types of fossils and how they are formed. The unit also features new material that connects evolutionary concepts and societal issues. A new figure (Figure 25.11) provides fossil evidence of an enormous change in the evolutionary history of life: the first appearance of large, multicellular eukaryotes.

Unit 5 THE EVOLUTIONARY HISTORY OF BIOLOGICAL DIVERSITY

Chapter 34 has been updated with recent genomic data and fossil discoveries indicating that Neanderthals and Denisovans are more closely related to each other than to humans and that they interbred with each other (and with humans). In Chapter 29, a new figure (Figure 29.1) provides a visual overview of major steps in the colonisation of land by plants, and revisions to text in Concept 29.1 strengthen our description of derived traits of plants that facilitated life on land. Chapter 27 includes a new section of text that describes the rise of antibiotic resistance and multidrug resistance and discusses novel approaches in the search for new antibiotics. This new material is supported by two new figures, Figure 27.22 and Figure 27.23. Other updates include the revision of many phylogenies to reflect recent phylogenomic data; a new Inquiry Figure (Figure 28.26) on the root of the eukaryotic tree; and new text describing the 2017 discovery of 315,000-year-old fossils of a hominin that had facial features like those of humans, while the back of its skull was elongated, as in earlier species.

▶ Figure 27.22 The rise of antibiotic resistance.

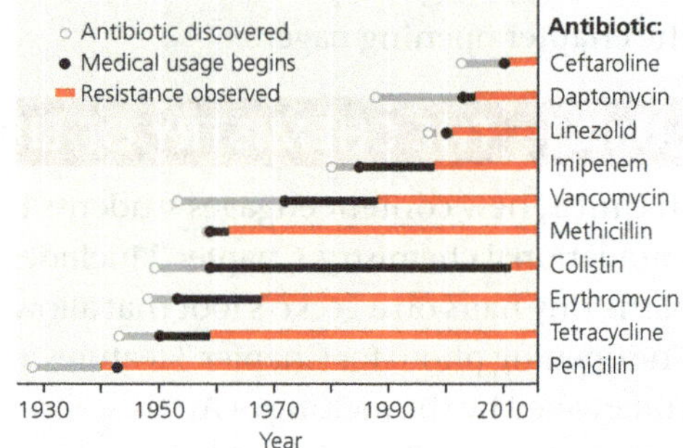

Unit 6 PLANT FORM AND FUNCTION

In Chapter 35, greater emphasis is placed on how structure fits function in vascular plants by way of a new Visual Overview. In Chapter 36, a new Visual Skills Question provides a quantitative exercise in estimating stomatal density. Chapter 37 begins with an emphasis on the importance of crop fertilisation in feeding the world. To increase student engagement, renewed emphasis is placed on the link between the nutrition of plants and the nutrition of the organisms, including humans, that feed on them. Table 37.1 concerning plant essential elements has been expanded to include micronutrients as well as macronutrients. In Concept 37.2, a new subsection titled "Global Climate Change and Food Quality" discusses new evidence that global climate change may be negatively impacting the nutritional mineral content of crops. In Chapter 38, the discussion of genetic engineering and agriculture has been enhanced by a discussion of biofortification and by updates concerning "Golden Rice." Chapter 39 includes new updates on the location of the IAA receptor in plant cells and the role of abscisic acid in bud dormancy. The introduction to Concept 39.2 has been revised to emphasise that plants use many classes of chemicals in addition to the classic hormones to communicate information.

Unit 7 ANIMAL FORM AND FUNCTION

The Unit 7 revisions feature pedagogical innovations coupled with updates for currency. A striking new underwater image of Emperor penguins (Figure 40.1) opens the unit and highlights the contributions of form, function, and behaviour to homeostasis in general as well as to the specific topic of thermoregulation. The artwork used to introduce and explore homeostasis throughout the unit (Figures 40.8, 40.17, 41.23, 42.28, 44.20, 44.22, and 45.18) has been improved and refined to provide a clear and consistent presentation of the role of perturbation in triggering a response. In Chapter 43, the introduction of the adaptive immune response has been shifted to later in the chapter, allowing students to build on the features of innate immunity before tackling the more demanding topic of the adaptive response. In Chapter 48, the structural overview of neurons is now completed before the introduction of information processing. A new illustration, Figure 49.8, provides a concise visual comparison of sympathetic and parasympathetic neurons with each other and with motor neurons of the CNS. In addition, in-depth consideration of glia is now provided in Concept 49.1, where it is more logically integrated into the overview of nervous systems. At the end of the unit, an eye-catching photograph of the male frigatebird's courtship display (Figure 51.1) introduces the topic of animal behaviour. Among the content updates that enhance currency and student engagement throughout the unit are discussions of phage therapy and faecal transplantation, state-of-the-art treatments that both rely on microbiome data, and chronic traumatic encephalopathy (CTE), as well as the latest findings on dinosaur locomotion (Concept 40.1), the awarding of a Nobel Prize in 2017 in the field of circadian rhythms (Concept 40.2), and reference to the ongoing public health crisis of opioid addiction in the context of considering the brain's reward system (Concept 49.5).

▶ **Figure 49.8 Comparison of pathways in the motor and autonomic nervous systems.**

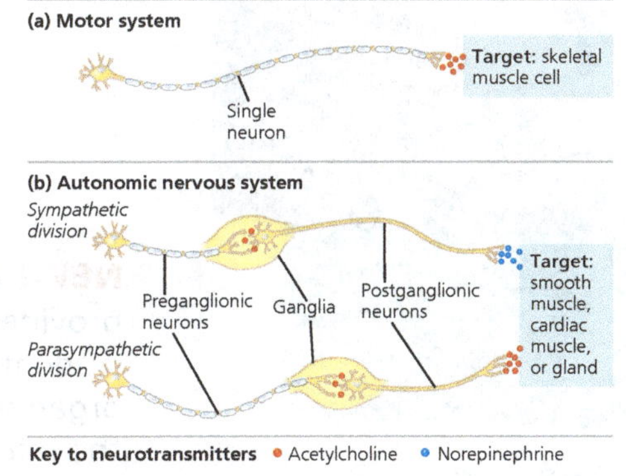

Unit 8 ECOLOGY

Complementary goals of the Unit 8 revision were to strengthen our coverage of core concepts while also increasing our coverage of how human actions affect ecological communities. Revisions include a new section of text and a new figure (Figure 52.7) on how plants (and deforestation) can affect the local or regional climate; a new section of text in Concept 55.1 that summarises how ecosystems work; new text and a new figure (Figure 52.31) illustrating how rapid evolution can cause rapid ecological change; new material in Concept 55.2 on eutrophication and how it can cause the formation of large "dead zones" in aquatic ecosystems; and new text and a new figure (Figure 54.22) on how the abundance of organisms at each trophic level can be controlled by bottom-up or top-down control. A new figure (Figure 56.28) shows the extent of the record-breaking 2017 dead zone in the Gulf of Mexico and the watershed that contributes to its nutrient load. In addition, Concept 56.1 includes a new section that describes attempts to use cloning to resurrect species lost to extinction, while Concept 56.4 includes a new section of text and two new figures (Figure 56.32 and 56.33) on plastic waste, a major and growing environmental problem. In keeping with our book-wide goal of expanding our coverage of climate change, Chapter 56 has a new Scientific Skills Exercise in which students interpret changes in atmospheric CO_2 concentrations; a new figure (Figure 56.37) describes human and natural factors that contribute to rising global temperatures; and a new section of text in Concept 56.4 describes how global climate change models are developed and why they are valuable.

▼ **Figure 56.28 A dead zone arising from nitrogen pollution in the Mississippi basin.**

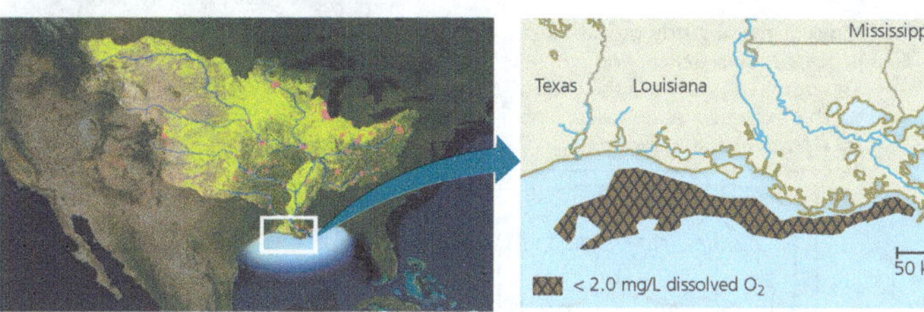

(a) Nutrients drain from agricultural land (green) and cities (red) through the vast Mississippi watershed to the Gulf of Mexico.

(b) The 2017 dead zone, represented here, was the largest yet measured. It occupied 22,730 km^2, an area slightly larger than New Jersey.

A New Visual Experience for Every Chapter

NEW! Chapter Openers introduce each chapter and feature a question answered with a clear, simple image to help students visualise and remember concepts as they move through each chapter.

17 Gene Expression: From Gene to Protein

KEY CONCEPTS

Figure 17.1 An albino joey (the name for a young koala) with its regularly coloured mother. The joey's albino colouration arose because a recessive mutation occurred in the DNA of an ancestor koala and passed through several generations until it manifested in this mother. The recessive mutation disables pigment synthesis. When a male with the same heterozygous gene mates with a similarly genetically endowed mother, there remains a small chance that their offspring will become a homozygous albino koala.

Study Tip

Make a visual study guide: Sketch the process shown below, and add labels and details as you read the chapter. (In this exercise, assume all processes take place in a eukaryotic cell.)

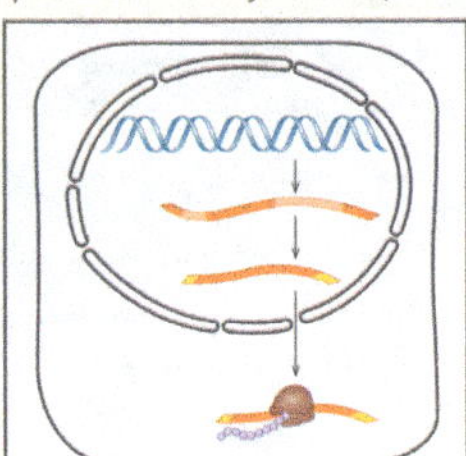

Go to Mastering Biology to access Dynamic Study Modules for revision, 3D BioFlix® animations and high-quality videos, and your interactive Pearson eText.

How can one change in DNA result in such a dramatic change in appearance?

Proteins are the link between genotype and phenotype. Gene expression is the process by which DNA directs the synthesis of proteins:

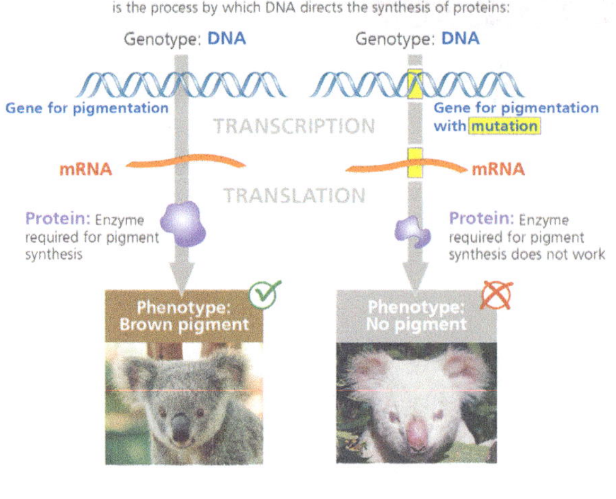

337

NEW! A Study Tip provides an activity for students to help them organise and learn the information in the chapter.

NEW! A Visual Overview helps students start with the big picture.

Develop Scientific Skills

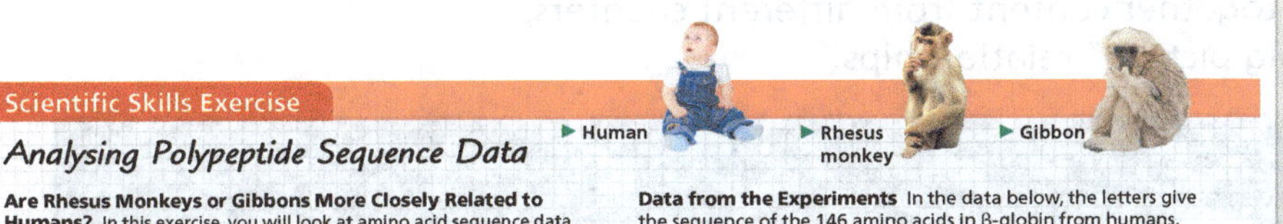

Analysing Polypeptide Sequence Data

Are Rhesus Monkeys or Gibbons More Closely Related to Humans? In this exercise, you will look at amino acid sequence data for the β polypeptide chain of haemoglobin, often called β-globin. You will then interpret the data to hypothesise whether the monkey or the gibbon is more closely related to humans.

How Such Experiments Are Done Researchers can isolate the polypeptide of interest from an organism and then determine the amino acid sequence. More frequently, the DNA of the relevant gene is sequenced, and the amino acid sequence of the polypeptide is deduced from the DNA sequence of its gene.

Data from the Experiments In the data below, the letters give the sequence of the 146 amino acids in β-globin from humans, rhesus monkeys, and gibbons. Because a complete sequence would not fit on one line here, the sequences are divided into three segments: amino acids 1–50, 51–100, and 101–146. The sequences for the three different species are aligned so that you can compare them easily. For example, you can see that for all three species, the first amino acid is V (valine) and the 146th amino acid is H (histidine).

Species	Alignment of Amino Acid Sequences of β-globin
Human	1 VHLTPEEKSA VTALWGKVNV DEVGGEALGR LLVVYPWTQR FFESFGDLST
Monkey	1 VHLTPEEKNA VTTLWGKVNV DEVGGEALGR LLLVYPWTQR FFESFGDLSS
Gibbon	1 VHLTPEEKSA VTALWGKVNV DEVGGEALGR LLVVYPWTQR FFESFGDLST
Human	51 PDAVMGNPKV KAHGKKVLGA FSDGLAHLDN LKGTFATLSE LHCDKLHVDP
Monkey	51 PDAVMGNPKV KAHGKKVLGA FSDGLNHLDN LKGTFAQLSE LHCDKLHVDP
Gibbon	51 PDAVMGNPKV KAHGKKVLGA FSDGLAHLDN LKGTFAQLSE LHCDKLHVDP
Human	101 ENFRLLGNVL VCVLAHHFGK EFTPPVQAAY QKVVAGVANA LAHKYH
Monkey	101 ENFRLLGNVL VCVLAHHFGK EFTPPVQAAY QKVVAGVANA LAHKYH
Gibbon	101 ENFRLLGNVL VCVLAHHFGK EFTPPVQAAY QKVVAGVANA LAHKYH

Data from Human: http://www.ncbi.nlm.nih.gov/protein/AAA21113.1; rhesus monkey: http://www.ncbi.nlm.nih.gov/protein/122634; gibbon: http://www.ncbi.nlm.nih.gov/protein/122616

INTERPRET THE DATA

1. Scan the monkey and gibbon sequences, letter by letter, circling any amino acids that do not match the human sequence. **(a)** How **many amino acids differ between the monkey** and the human sequences? **(b)** Between the gibbon and human?
2. For each nonhuman species, what percent of its amino acids are identical to the human sequence of β-globin?
3. Based on these data alone, state a **hypothesis for which of these two species is** more closely related to humans. What is your reasoning?
4. What other evidence could you use to support your hypothesis?

Scientific Skills Exercises in every chapter of the text use real data to build key skills needed for biology, including data analysis, graphing, experimental design, and maths skills. Each exercise is also available as an automatically graded assignment in Mastering Biology with answer-specific feedback for students.

Problem-Solving Exercises guide students in applying scientific skills and interpreting real data in the context of solving a real-world problem. A version of each Problem-Solving Exercise can also be assigned in Mastering Biology.

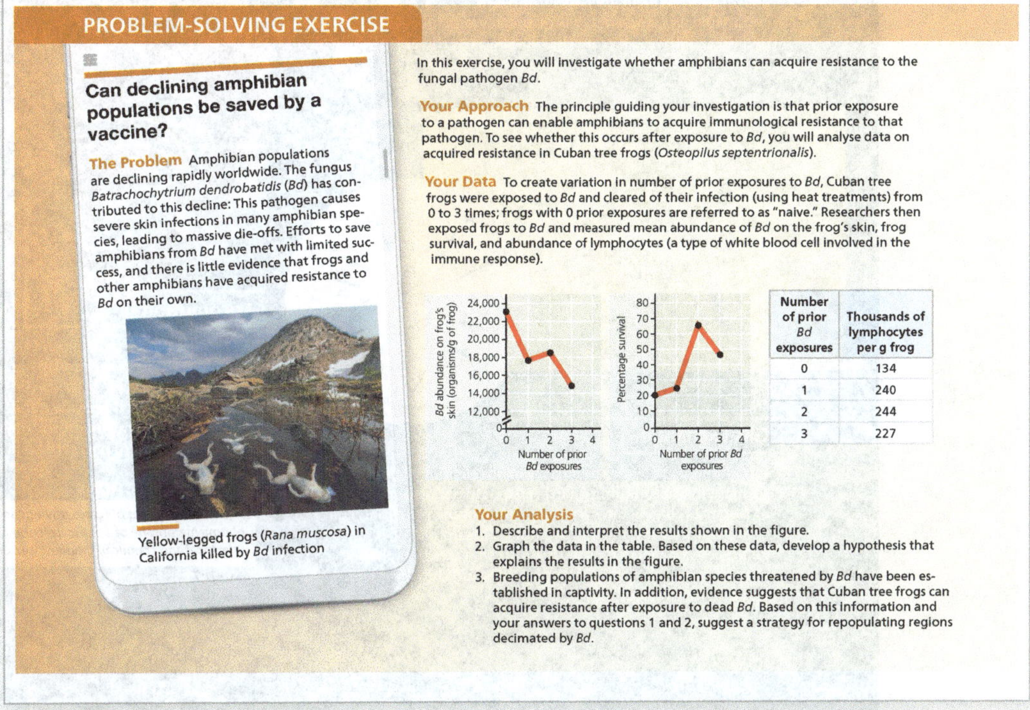

PROBLEM-SOLVING EXERCISE

Can declining amphibian populations be saved by a vaccine?

The Problem Amphibian populations are declining rapidly worldwide. The fungus *Batrachochytrium dendrobatidis* (*Bd*) has contributed to this decline: This pathogen causes severe skin infections in many amphibian species, leading to massive die-offs. Efforts to save amphibians from *Bd* have met with limited success, and there is little evidence that frogs and other amphibians have acquired resistance to *Bd* on their own.

Yellow-legged frogs (*Rana muscosa*) in California killed by *Bd* infection

In this exercise, you will investigate whether amphibians can acquire resistance to the fungal pathogen *Bd*.

Your Approach The principle guiding your investigation is that prior exposure to a pathogen can enable amphibians to acquire immunological resistance to that pathogen. To see whether this occurs after exposure to *Bd*, you will analyse data on acquired resistance in Cuban tree frogs (*Osteopilus septentrionalis*).

Your Data To create variation in number of prior exposures to *Bd*, Cuban tree frogs were exposed to *Bd* and cleared of their infection (using heat treatments) from 0 to 3 times; frogs with 0 prior exposures are referred to as "naive." Researchers then exposed frogs to *Bd* and measured mean abundance of *Bd* on the frog's skin, frog survival, and abundance of lymphocytes (a type of white blood cell involved in the immune response).

Number of prior *Bd* exposures	Thousands of lymphocytes per g frog
0	134
1	240
2	244
3	227

Your Analysis

1. Describe and interpret the results shown in the figure.
2. Graph the data in the table. Based on these data, develop a hypothesis that explains the results in the figure.
3. Breeding populations of amphibian species threatened by *Bd* have been established in captivity. In addition, evidence suggests that Cuban tree frogs can acquire resistance after exposure to dead *Bd*. Based on this information and your answers to questions 1 and 2, suggest a strategy for repopulating regions decimated by *Bd*.

Make Connections Across Multiple Concepts

Make Connections Figures pull together content from different chapters, providing a visual representation of "big picture" relationships.

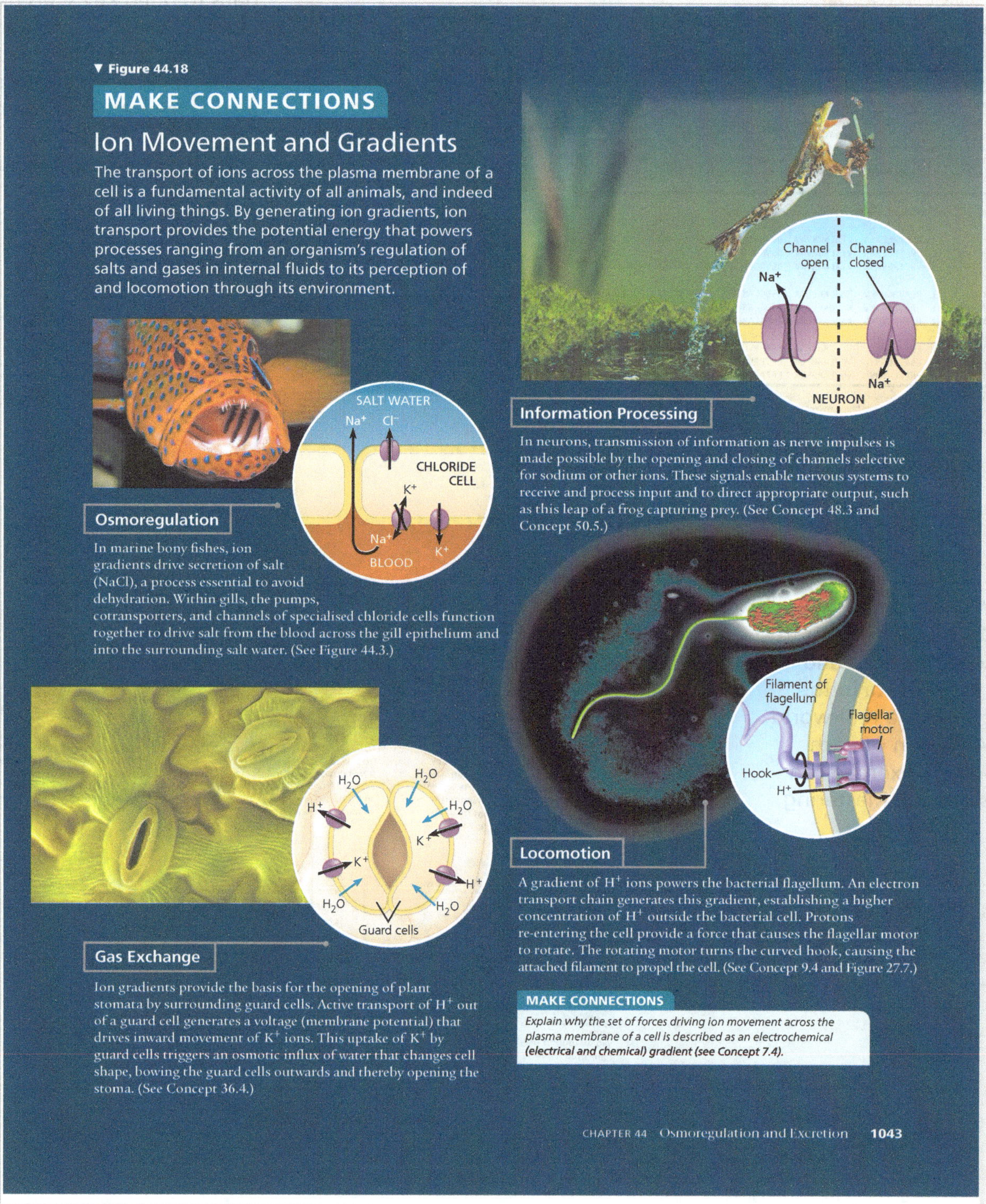

▼ **Figure 44.18**

MAKE CONNECTIONS

Ion Movement and Gradients

The transport of ions across the plasma membrane of a cell is a fundamental activity of all animals, and indeed of all living things. By generating ion gradients, ion transport provides the potential energy that powers processes ranging from an organism's regulation of salts and gases in internal fluids to its perception of and locomotion through its environment.

Osmoregulation

In marine bony fishes, ion gradients drive secretion of salt (NaCl), a process essential to avoid dehydration. Within gills, the pumps, cotransporters, and channels of specialised chloride cells function together to drive salt from the blood across the gill epithelium and into the surrounding salt water. (See Figure 44.3.)

Gas Exchange

Ion gradients provide the basis for the opening of plant stomata by surrounding guard cells. Active transport of H^+ out of a guard cell generates a voltage (membrane potential) that drives inward movement of K^+ ions. This uptake of K^+ by guard cells triggers an osmotic influx of water that changes cell shape, bowing the guard cells outwards and thereby opening the stoma. (See Concept 36.4.)

Information Processing

In neurons, transmission of information as nerve impulses is made possible by the opening and closing of channels selective for sodium or other ions. These signals enable nervous systems to receive and process input and to direct appropriate output, such as this leap of a frog capturing prey. (See Concept 48.3 and Concept 50.5.)

Locomotion

A gradient of H^+ ions powers the bacterial flagellum. An electron transport chain generates this gradient, establishing a higher concentration of H^+ outside the bacterial cell. Protons re-entering the cell provide a force that causes the flagellar motor to rotate. The rotating motor turns the curved hook, causing the attached filament to propel the cell. (See Concept 9.4 and Figure 27.7.)

MAKE CONNECTIONS

Explain why the set of forces driving ion movement across the plasma membrane of a cell is described as an electrochemical (electrical and chemical) gradient (see Concept 7.4).

CHAPTER 44 Osmoregulation and Excretion **1043**

CONCEPT CHECK 24.2

1. Summarise key differences between allopatric and sympatric speciation. Which type of speciation is more common, and why?
2. Describe two mechanisms that can decrease gene flow in sympatric populations, thereby making sympatric speciation more likely to occur.
3. **WHAT IF?** Is allopatric speciation more likely to occur on an island close to a mainland or on a more isolated island of the same size? Explain your prediction.
4. **MAKE CONNECTIONS** Review the process of meiosis in Figure 13.8. Describe how an error during meiosis could lead to polyploidy.

For suggested answers, see Appendix A.

Make Connections Questions in every chapter ask students to relate content to material presented earlier in the course.

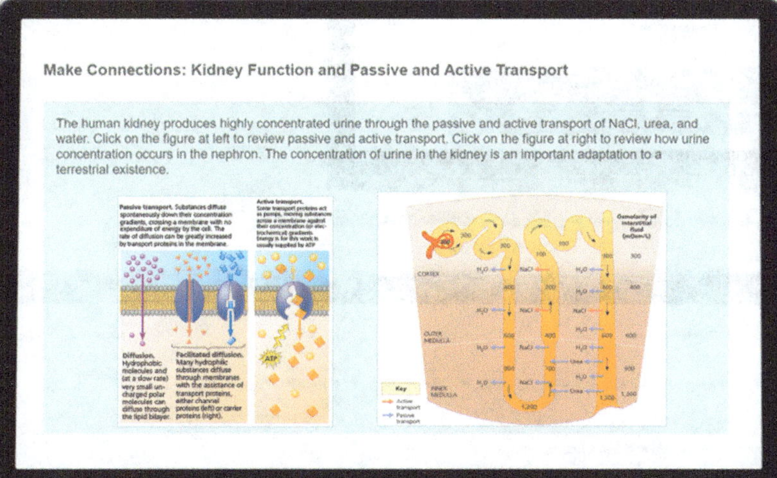

Make Connections Tutorials connect content from two different chapters using art from the book. Make Connections Tutorials are assignable and automatically graded in Mastering Biology and include answer-specific feedback for students.

NEW! An expanded collection of Figure Walkthroughs guide students through key figures with narrated explanations and figure mark-ups that reinforce important points. **These are available for assignment in Mastering Biology**.

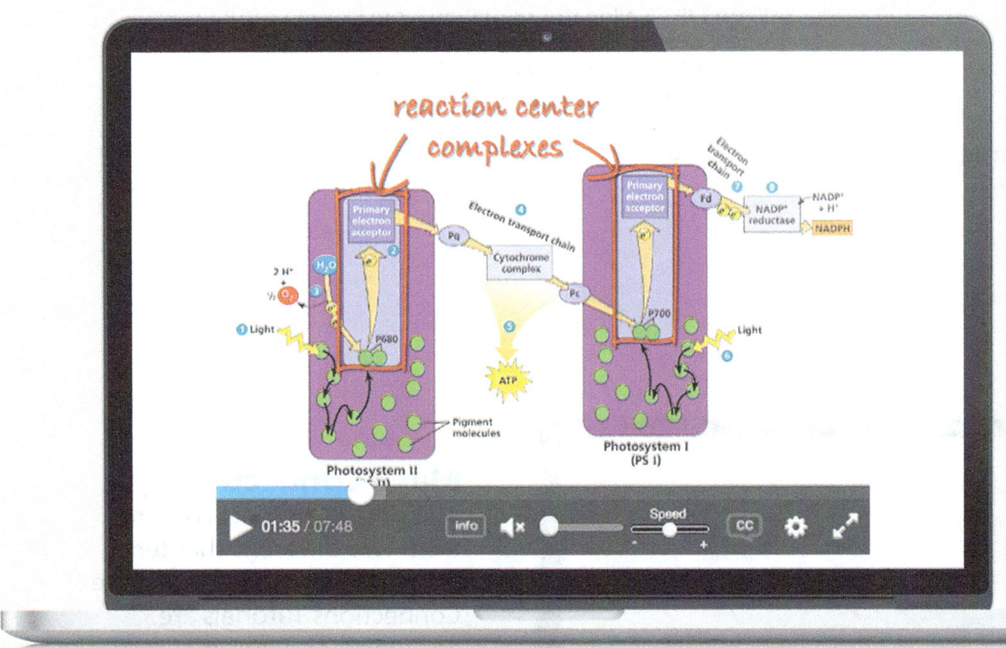

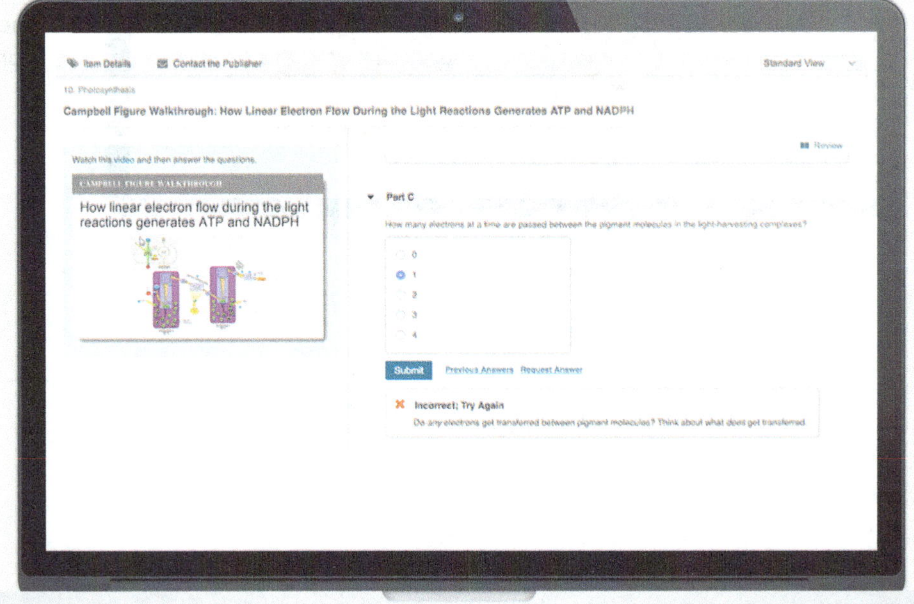

Innovation in Assessment

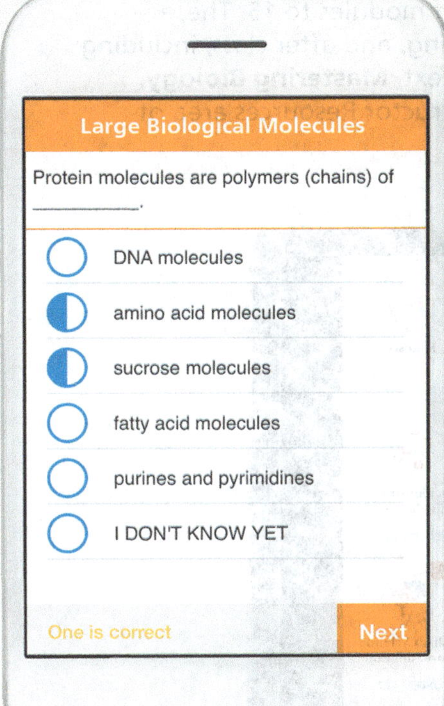

Dynamic Study Modules use the latest developments in cognitive science to help students study by adapting to their performance in real time. Students build confidence and understanding, enabling them to participate and perform better, both in and out of class. Available on smartphones, tablets, and computers.

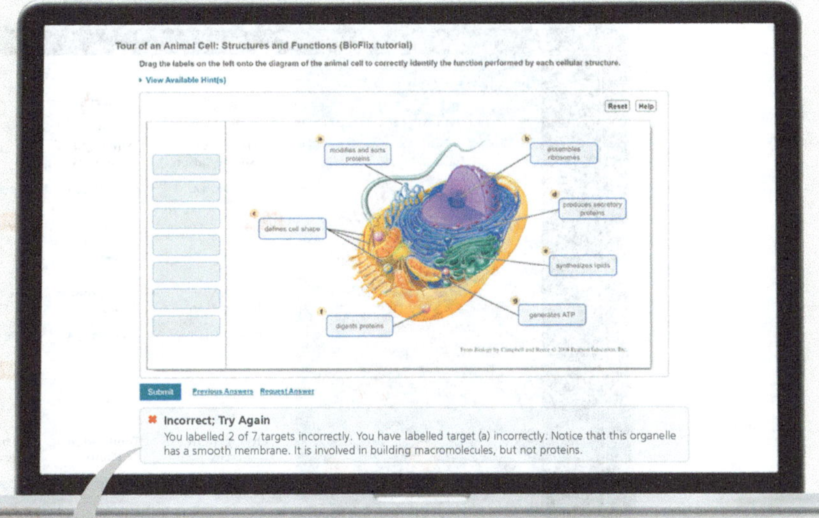

✖ **Incorrect; Try Again**

You labelled 2 of 7 targets incorrectly. You have labelled target (a) incorrectly. Notice that this organelle has a smooth membrane. It is involved in building macromolecules, but not proteins.

Wrong-Answer Feedback Using data gathered from all of the students using the program, **Mastering Biology** offers wrong-answer feedback that is specific to each student. Rather than simply providing feedback of the "right/wrong/try again" variety, Mastering Biology guides students towards the correct final answer without giving the answer away.

UPDATED! Test Bank questions have been analysed and revised with student success in mind. Revisions account for how students read, analyse, and engage with the content.

Innovation in Instructor Resources

NEW! 5 new Ready-to-Go Teaching Modules expand the number of modules to 15. These instructor resources are designed to make use of teaching tools before, during, and after class, including new ideas for in-class activities. The modules incorporate the best that the text, **Mastering Biology**, and **Learning Catalytics** have to offer and can be accessed through the Instructor Resources area of Mastering Biology.

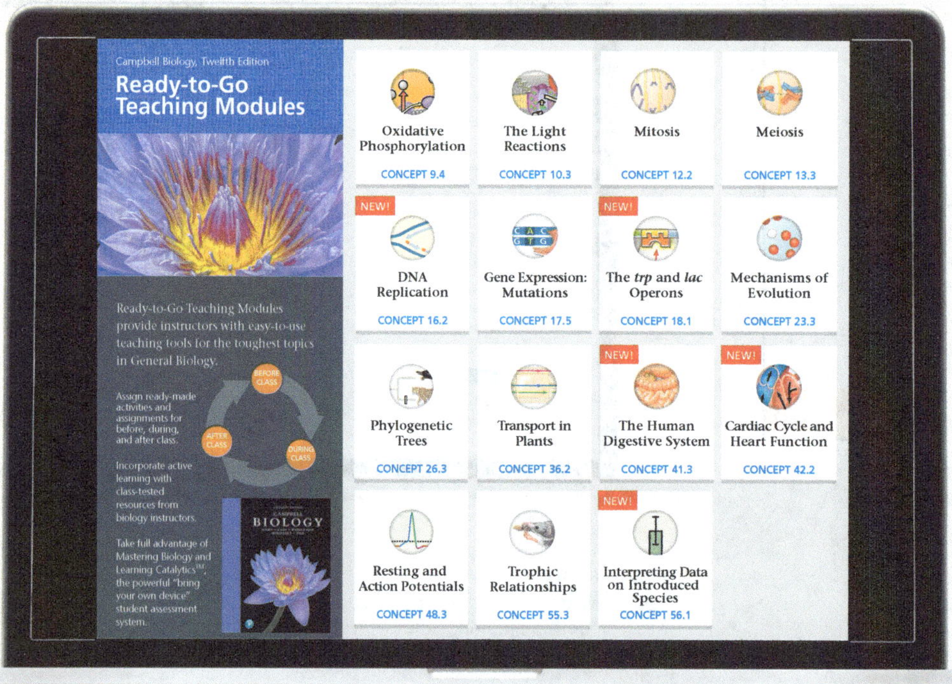

NEW! Early Alerts in **Mastering Biology** help instructors know when students may be struggling in the course. This insight enables instructors to provide personalised communication and support at the moment students need it so they can stay—and succeed—in the course.

Test Bank

This extensively revised resource—available in Word as well as Blackboard, Canvas and Moodle-compatible formats—contains over 4,500 questions, including scenario-based questions and art, graph, and data interpretation questions.

Digital Image PowerPoints

Labelled and unlabelled versions of the art, graphs, and photos from the text are provided in a PowerPoint template for easy lecture presentation or customisation.

Skills Exercises

Scientific Skills Exercises

*Available only in Mastering Biology. All other Scientific Skills Exercises are in the print book, eText, and Mastering Biology.

Problem-Solving Exercises

Featured Figures

Visualising Figures

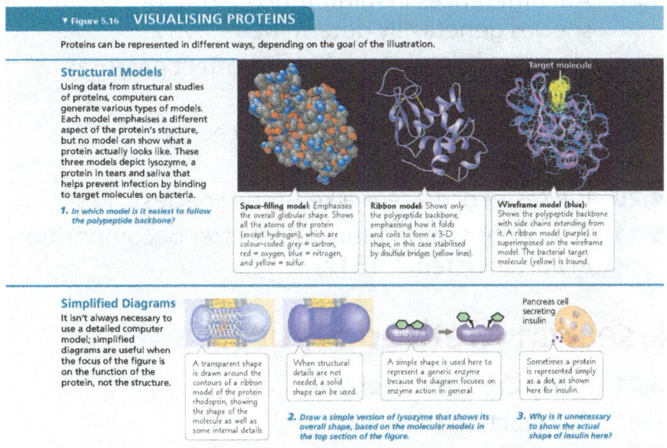

Make Connections Figures

Exploring Figures

Inquiry Figures

Research Method Figures

*The Inquiry Figure, original research paper, and a worksheet to guide you through the paper are provided in *Inquiry in Action: Interpreting Scientific Papers*, Fourth Edition.
†A related Experimental Inquiry Tutorial can be assigned in Mastering Biology.

Interviews

Abigail Allwood
Jet Propulsion Laboratory, NASA

Diana Bautista
University of California, Berkeley

Francisco Mojica
University of Alicante, Spain

Ary Hoffmann
University of Melbourne

Hamish Campbell
GNS Science

Dennis Gonsalves
Agricultural Research Center, Hilo, Hawaii

Steffanie Strathdee
University of California, San Diego

Ove Hoegh-Guldberg
University of Queensland

Acknowledgments

Australian and New Zealand Edition

We all thank our reviewers who asked us to think harder, clarify better, and distil more fully the wonder and wisdom of your ancestors in biological research so that you can achieve all you can, and more. Without the dedication and unstinting commitment of the entire Pearson team, the authors could not have formed the public face of a team of hundreds that brought this work to you. Thanks in particular to Anna Carter, Julie Ganner, Bernadette Chang, Lisa Woodland, Madeleine Roberts, and of course Mandy Sheppard. Although their names do not appear on the cover of the text, their unstinting dedication to perfection worked to enrich this text in innumerable ways. We exhort you to repay our work simply: Engage with passion in your unstinting quest to reveal a little more of the awesome biological workings of the world in which you will live, love, and learn.

US Edition

The authors wish to express their gratitude to the global community of instructors, researchers, students, and publishing professionals who have contributed to the Twelfth Edition of *Campbell Biology*.

As authors of this text, we are mindful of the daunting challenge of keeping up to date in all areas of our rapidly expanding subject. We are grateful to the many scientists who helped shape this text by discussing their research fields with us, answering specific questions in their areas of expertise, and sharing their ideas about biology education. We are especially grateful to the following, listed alphabetically: Graham Alexander, Elizabeth Atkinson, Kristian Axelsen, Ron Bassar, Christopher Benz, David Booth, George Brooks, Abby Dernberg, Jean DeSaix, Alex Engel, Rachel Kramer Green, Fred Holtzclaw, Theresa Holtzclaw, Tim James, Kathy Jones, Azarias Karamanlidis, Gary Karpen, Joe Montoya, Laurie Nemzer, Kevin Peterson, T. K. Reddy, David Reznick, Thomas Schneider, Alastair Simpson, Martin Smith, Steven Swoap, and John Taylor. In addition, the biologists listed on pages xxviii–xxxi provided detailed reviews, helping us ensure the text's scientific accuracy and improve its pedagogical effectiveness. Thanks also to Mary Camuso and Ann Sinclair for contributing a creative Study Tip for their fellow students.

Thanks also to the other professors and students, from all over the world, who contacted the authors directly with useful suggestions. We alone bear the responsibility for any errors that remain, but the dedication of our consultants, reviewers, and other correspondents makes us confident in the accuracy and effectiveness of this text.

Interviews with prominent scientists have been a hallmark of *Campbell Biology* since its inception, and conducting these interviews was again one of the great pleasures of revising the book.

Mastering Biology and the other electronic accompaniments for this text are invaluable teaching and learning aids. We are grateful to the contributors for the Ready-to-Go Teaching Modules: Chad Brassil, Ruth Buskirk, Eileen Gregory, Angela Hodgson, Molly Jacobs, Bridgette Kirkpatrick, Maureen Leupold, Jennifer Metzler, Karen Resendes, Justin Shaffer, Allison Silveus, Jered Studinski, Cynthia Surmacz, Sara Tallarovic, and Carole Twichell. We would also like to extend our sincere appreciation to Carolyn Wetzel for her hard work on the Figure Walkthroughs. And our gratitude goes to Bryan Jennings and Roberta Batorsky for their work on the Reading Questions. Thanks also to Ann Brokaw and Bob Cooper for their contributions to the AAAS Science in the Classroom activities; we also appreciate the support of Beth Reudi, Shelby Lake, and Lydia Kaprelian from AAAS.

The value of *Campbell Biology* as a learning tool is greatly enhanced by the supplementary materials that have been created for instructors and students. We recognise that the dedicated authors of these materials are essentially writing mini (and not so mini) books. We appreciate the hard work and creativity of all the authors listed, with their creations, on pages xxiv–xxv. We are also grateful to Kathleen Fitzpatrick and Nicole Tunbridge (PowerPoint® Lecture Presentations); Roberta Batorsky, Douglas Darnowski, James Langeland, and David Knochel (Clicker Questions); Sonish Azam, Ford Lux, Karen Bernd, Janet Lanza, Chris Romero, Marshall Sundberg, Justin Shaffer, Ed Zalisko, and David Knochel (Test Bank).

Campbell Biology results from an unusually strong synergy between a team of scientists and a team of publishing professionals.

Our editorial team at Pearson Education again demonstrated unmatched talents, commitment, and pedagogical insights. Josh Frost, our Manager of Higher Ed Global Content Strategy for Life Sciences, brought publishing savvy, intelligence, and a much-appreciated level head to leading the whole team. The clarity and effectiveness of every page owe much to our extraordinary Supervising Editors Beth Winickoff and Pat Burner, who worked with a top-notch team of Senior Developmental Editors in John Burner, Mary Ann Murray, Hilair Chism, Andrew Recher, and Mary Hill. Our unsurpassed Director of Content Development Ginnie Simione Jutson and Courseware Portfolio Management Director Beth Wilbur were indispensable in moving the project in the right direction. We also want to thank Robin Heyden for organising the annual Biology Leadership Conferences and keeping us in touch with the world of AP Biology. We also extend our thanks to Ashley Fallon, Editorial Assistant, Chelsea Noack, Associate Content Analyst, and Rebecca Berardy Schwartz, Product Manager.

You would not have this beautiful text if not for the work of the production team: Director, Content Production & Digital Studio Erin Gregg; Managing Producer Michael Early; Senior Content Producer Lori Newman; Photo Researcher Maureen Spuhler; Copy Editor Joanna Dinsmore; Proofreader Pete Shanks; Rights & Permissions Manager Ben Ferrini; Rights & Permissions Project Manager Matt Perry; Senior Project Manager Margaret McConnell and the rest of the staff at Integra Software Services, Inc.; Art Production Manager Rebecca Marshall, Artist Kitty Auble, and the rest of the staff at Lachina Creative; Design Manager Mark Ong; Text and Cover Designer Jeff Puda; and Manufacturing Buyer Stacey Weinberger. We also thank those who worked on the text's supplements: Project Manager Shiny Rajesh and her team at Integra Software Services.

For creating the wonderful package of electronic media that accompanies the text, we are grateful to Senior Content Developer Sarah Jensen; Content Producers Kaitlin Smith and Ashley Gordon; Director, Production & Digital Studio Katie Foley; Director, Production & Digital Studio Laura Tommasi; Supervising Media Producer Tod Regan; Specialist, Instructional Design and Development Sarah Young-Dualan; Digital Program Manager, Science, Caroline Ayres; Project Manager Katie Cook; Media Producer Ziki Dekel; Manager, Creative Technology Greg Davis; and Senior Learning Tools Strategist Kassi Foley.

For their enthusiasm, encouragement, and support, we are grateful to Jeanne Zalesky, Director, Global Higher Ed Content Management and Strategy, Science & Health Sciences; Michael Gillespie, Director, Higher Ed Product Management, Life Sciences; Adam Jaworski, VP Product Management Higher Ed, Science; and Paul Corey, SVP Global Content Strategy, Higher Ed.

For representing our text to our international audience, we thank our sales and marketing partners throughout the world. They are all strong allies in biology education.

Finally, we wish to thank our families and friends for their encouragement and patience throughout this long project. Our special thanks to Lily and Alex (L.A.U.); Debra and Hannah (M.L.C.); Aaron, Sophie, Noah, and Gabriele (S.A.W.); Natalie (P.V.M.); and Jim, Abby, Dan, and Emily (R.B.O). Thanks to Jane Reece, now retired, for her generosity and thoughtfulness throughout her many years as a Campbell author. And, as always, thanks to Rochelle, Allison, Jason, McKay, and Gus.

Lisa A. Urry, Michael L. Cain, Steven A. Wasserman,
Peter V. Minorsky, and Rebecca B. Orr

Reviewers

Twelfth Australian and New Zealand Edition Reviewers

Michael Calver, *Murdoch University*; Natalie Colson, *Griffith University*; Michelle Coulson, *University of Adelaide*; John Dearnaley, *University of Southern Queensland*; Christopher Jones, *Western Sydney University*; Saw Hoon Lim, *University of Melbourne*; Leigh Martin, *University of Technology, Sydney*; Steve Melvin, *Griffith University*; Tasmin Rymer, *James Cook University*; Masha Smallhorn, *Flinders University*; Huseyin Sumer, *Swinburne University of Technology*; Brydget Tulloch, *University of Waikato*.

Twelfth US Edition Reviewers

Sheena Abernathy, *College of the Mainland*; James Arnone, *William Paterson University*; Josh Auld, *West Chester University*; Gemma Bartha, *Springfield College*; Louise Beard, *University of Essex*; Marin Beaupre, *Massasoit Community College*; Kevin Bennett, *University of Hawai'i*; Kelsie Bernot, *North Carolina A&T State University*; Christine Bezotte, *Elmira College*; Chris Bloch, *Bridgewater State University*; Aiwei Borengasser, *University of Arkansas – Pulaski Tech*; Robert Borgon, *University of Central Florida*; Nicole Bournias-Vardiabasis, *California State University, San Bernardino*; George Brooks, *University of California, Berkeley*; Michael Buoni, *Delaware Technical Community College*; Kelcey Burris, *Union High School*; Elena Cainas, *Broward College*; Mickael Cariveau, *Mount Olive College*; Billy Carver, *Lees-McRae College*; Anne Casper, *Eastern Michigan University*; Bruce Chase, *University of Nebraska, Omaha*; Amanda Chau, *Blinn College*; Katie Clark, *University of California, Riverside*; Catharina Coenen, *Allegheny College*; Curt Coffman, *Vincennes University*; Juliet Collins, *University of Wisconsin, Madison*; Bob Cooper, *Pennsbury High School*; Robin Cotter, *Phoenix College*; Marilyn Cruz-Alvarez, *Florida Gulf Coast University*; Noelle Cutter, *Molloy College*; Deborah Dardis, *Southeastern Louisiana University*; Farahad Dastoor, *University of Maine, Orono*; Andrew David, *Clarkson University*; Jeremiah Davie, *D'Youville College*; Brian Deis, *University of Hawai'i* Jean DeSaix, *University of North Carolina at Chapel Hill* Kelly Dubois, *Calvin College*; Cynthia Eayre, *Fresno City College*; Arri Eisen, *Emory University*; Lisa Elfring, *University of Arizona*; Kurt Elliott, *Northwest Vista College*; Ana Esther Escandon, *Los Angeles Harbor College*; Linda Fergusson-Kolmes, *Portland Community College*; April Fong, *Portland Community College*; Robert Fowler, *San Jose State University*; Brittany Gasper, *Florida Southern College*; Carri Gerber, *The Ohio State University*; Marina Gerson, *California State University, Stanislaus*; Brian Gibbens, *University of Minnesota*; Sara Gremillion, *Georgia Southern University*; Ron Gross, *Community College of Allegheny County*; Melissa Gutierrez, *University of Southern Mississippi*; Gokhan Hacisalihoglu, *Florida A&M University*; Monica Hall-Woods, *St. Charles Community College*; Catherine Hartkorn, *New Mexico State University*; Valerie Haywood, *Case Western Reserve University*; Maryann Herman, *St. John Fisher College*; Alexander Heyl, *Adelphi University*; Laura Hill, *University of Vermont*; Anne-Marie Hoskinson, *South Dakota State University*; Katriina Ilves, *Pace University*; James Jacob, *Tompkins Cortland Community College*; Darrel James, *Fred C. Beyer High School*; Jerry Johnson, *Corban University*; Greg Jones, *Santa Fe College*; Kathryn Jones, *Howard Community College*; Seth Jones, *University of Kentucky*; Steven Karafit, *University of Central Arkansas*; Lori Kayes, *Oregon State University*; Ben Kolber, *Duquesne University*; Catherine Konopka, *John Carroll University*; Bill Kroll, *Loyola University, Chicago*; MaryLynne LaMantia, *Golden West College*; Neil Lamb, *HudsonAlpha Institute for Biotechnology*; Michelle LaPorte, *St. Louis Community College*; Neil Lax, *Duquesne University*; John Lepri, *Carrboro High School*; Jani Lewis, *State University of New York at Geneseo*; Eddie Lunsford, *Southwestern Community College*; Alyssa MacDonald, *Leeward Community College*; Charles Mallery, *University of Miami*; Marlee Marsh, *Columbia College*; Nicole McDaniels, *Herkimer College*; Mike Meighan, *University of California, Berkeley*; Jennifer Metzler, *Ball State University*; Grace Ju Miller, *Indiana Wesleyan University*; Terry Miller, *Central Carolina Community College*; Shamone Mizenmayer, *Central High School*; Cam Muir, *University of Hawai'i at Hilo*; Heather Murdock, *San Francisco State University*; Madhavan Narayanan, *Mercy College*; Jennifer Nauen, *University of Delaware*; Karen Neal, *J. Sargeant Reynolds Community College, Richmond*; Leonore Neary, *Joliet Junior College*; Shanna Nifoussi, *University of Superior*; Jennifer Ortiz, *North Hills School District*; Fernanda Oyarzun, *Universidad Católica de la Santísima Concepción, Chile*; Stephanie Pandolfi, *Wayne State University*; John Plunket, *Horry-Georgetown Technical College*; Elena Pravosudova, *University of Nevada, Reno*; Pushpa Ramakrishna, *Chandler-Gilbert Community College*; Sami Raut, *University of Alabama, Birmingham*; Robert Reavis, *Glendale Community College*; Linda Rehfuss, *Bucks County Community College*; Deborah Rhoden, *Snead State Community College*; Linda Richardson, *Blinn College*; Brian Ring, *Valdosta State University*; Rob Ruliffson, *Minneapolis Community and Technical College*; Judy Schoonmaker, *Colorado School of Mines*; David Schultz, *University of Louisville*; David Schwartz, *Houston Community College*; Duane Sears, *University of California, Santa Barbara*; J. Michael Sellers, *University of Southern Mississippi*; Pramila Sen, *Houston Community College*; Jyotsna Sharma, *University of Texas San Antonio*; Joan Sharp, *Simon Fraser University*; Marcia Shofner, *University of Maryland, College Park*; Linda Sigismondi, *University of Rio Grande*; Davida Smyth, *Mercy College*; Helen Snodgrass, *YES Prep North Forest*; Ayodotun Sodipe, *Texas Southern University*; Kathy Sparace, *Tri-County Technical College*; Patricia Steinke, *San Jacinto College Central*; Elizabeth Sudduth, *Georgia Gwinnett College*; Aaron Sullivan, *Houghton College*; Yvonne Sun, *University of Dayton*; Andrea Swei, *San Francisco State University*; Greg Thurmon, *Central Methodist University*; Stephanie Toering-Peters, *Wartburg College*; Monica Togna, *Drexel University*; Gail Tompkins, *Wake Technical Community College*; Tara Turley-Stoulig, *Southeastern Louisiana University*; Bishnu Twanabasu, *Weatherford College*; Erin Ventresca, *Albright College*; Wei Wan, *Texas A&M University*; Alan Wasmoen, *Metropolitan Community College, Nebraska*; Fred Wasserman, *Boston University*; Vicki Watson, *University of Montana*; Bill Wesley, *Mars Area High School*; Clay White, *Lone Star College*; Lisa Whitenack, *Allegheny College*; Larry Wimmers, *Towson University*; Heather Woodson, *Gaston College*; Shelly Wu, *Texas Christian University*; Mary Wuerth, *Tamalpais High School*; John Yoder, *University of Alabama*; Alyson Zeamer, *University of Texas San Antonio*.

Reviewers of Previous Editions

Steve Abedon, *Ohio State University*; Kenneth Able, *State University of New York, Albany*; Thomas Adams, *Michigan State University*; Martin Adamson, *University of British Columbia*; Dominique Adriaens, *Ghent University*; Ann Aguanno, *Marymount Manhattan College*; Shylaja Akkaraju, *Bronx Community College of CUNY*; Marc Albrecht, *University of Nebraska*; John Alcock, *Arizona State University*; Eric Alcorn, *Acadia University*; George R. Aliaga, *Tarrant County College*; Philip Allman, *Florida Gulf Coast College*; Rodney Allrich, *Purdue University*; Richard Almon, *State University of New York, Buffalo*; Bonnie Amos, *Angelo State University*; Katherine Anderson, *University of California, Berkeley*; Richard J. Andren, *Montgomery County Community College*; Estry Ang, *University of Pittsburgh, Greensburg*; Jeff Appling, *Clemson University*; J. David Archibald, *San Diego State University*; David Armstrong, *University of Colorado, Boulder*; Howard J. Arnott, *University of Texas, Arlington*; Mary Ashley, *University of Illinois, Chicago*; Angela S. Aspbury, *Texas State University*; Robert Atherton, *University of Wyoming*; Karl Aufderheide, *Texas A&M University*; Leigh Auleb, *San Francisco State University*; Terry Austin, *Temple College*; P. Stephen Baenziger, *University of Nebraska*; Brian Bagatto, *University of Akron*; Ellen Baker, *Santa Monica College*; Katherine Baker, *Millersville University*; Virginia Baker, *Chipola College*; Teri Balser, *University of Wisconsin, Madison*; William Barklow, *Framingham State College*; Susan Barman, *Michigan State University*; Steven Barnhart, *Santa Rosa Junior College*; Jim Barron, *Montana State University Billings*; Andrew Barton, *University of Maine Farmington*; Rebecca A. Bartow, *Western Kentucky University*; Ron Basmajian, *Merced College*; David Bass, *University of Central Oklahoma*; Stephen Bauer, *Belmont Abbey College*; Bonnie Baxter, *Westminster College*; Tim Beagley, *Salt Lake Community College*; Margaret E. Beard, *College of the Holy Cross*; Tom Beatty, *University of British Columbia*; Chris Beck, *Emory University*; Wayne Becker, *University of Wisconsin, Madison*; Patricia Bedinger, *Colorado State University*; Jane Beiswenger, *University of Wyoming*; Anne Bekoff, *University of Colorado, Boulder*; Marc Bekoff, *University of Colorado, Boulder*; Tania Beliz, *College of San Mateo*; Adrianne Bendich, *Hoffman-La Roche, Inc.*; Marilee Benore, *University of Michigan, Dearborn*; Barbara Bentley, *State University of New York, Stony Brook*; Darwin Berg, *University of California, San Diego*; Werner Bergen, *Michigan State University*; Gerald Bergstrom, *University of Wisconsin, Milwaukee*; Anna W. Berkovitz, *Purdue University*; Aimee Bernard, *University of Colorado Denver*; Dorothy Berner, *Temple University*; Annalisa Berta, *San Diego State University*; Paulette Bierzychudek, *Pomona College*; Charles Biggers, *Memphis State University*; Teresa Bilinski, *St. Edward's University*; Kenneth Birnbaum, *New York University*; Sarah Bissonnette, *University of California, Berkeley*; Catherine Black, *Idaho State University*; Michael W. Black, *California Polytechnic State University, San Luis Obispo*; William Blaker, *Furman University*; Robert Blanchard, *University of New Hampshire*; Andrew R. Blaustein, *Oregon State University*; Judy Bluemer, *Morton College*; Edward Blumenthal, *Marquette University*; Robert Blystone, *Trinity University*; Robert Boley, *University of Texas, Arlington*; Jason E. Bond, *East Carolina University*; Eric Bonde, *University of Colorado, Boulder*; Cornelius Bondzi, *Hampton University*; Richard Boohar, *University of Nebraska, Omaha*; Carey L. Booth, *Reed College*; Allan Bornstein, *Southeast Missouri State University*; David Bos, *Purdue University*; Oliver Bossdorf, *State University of New York, Stony Book*; James L. Botsford, *New Mexico State University*; Lisa Boucher, *University of Nebraska, Omaha*; Jeffery Bowen, *Bridgewater State University*; J. Michael Bowes, *Humboldt State University*; Richard Bowker, *Alma College*; Robert Bowker, *Glendale Community College, Arizona*; Scott Bowling, *Auburn University*; Barbara Bowman, *Mills College*; Barry Bowman, *University of California, Santa Cruz*; Deric Bownds, *University of Wisconsin, Madison*; Robert Boyd, *Auburn University*; Sunny Boyd, *University of Notre Dame*; Jerry Brand, *University of Texas, Austin*; Edward Braun, *Iowa State University*; Theodore A. Bremner, *Howard University*; James Brenneman, *University of Evansville*; Charles H. Brenner, *Berkeley, California*; Lawrence Brewer, *University of Kentucky*; Donald P. Briskin, *University of Illinois, Urbana*; Paul Broady, *University of Canterbury*; Chad Brommer, *Emory University*; Judith L. Bronstein, *University of Arizona*; David Broussard, *Lycoming College*; Danny Brower, *University of Arizona*; Carole Browne, *Wake Forest University*; Beverly Brown, *Nazareth College*; Mark Browning, *Purdue University*; David Bruck, *San Jose State University*; Robb T. Brumfield, *Louisiana State University*; Herbert Bruneau, *Oklahoma State University*; Gary Brusca, *Humboldt State University*; Richard C. Brusca, *University of Arizona, Arizona-Sonora Desert Museum*; Alan H. Brush, *University of Connecticut, Storrs*; Howard Buhse, *University of Illinois, Chicago*; Arthur Buikema, *Virginia Tech*; Beth Burch, *Huntington University*; Tessa Burch, *University of Tennessee*; Al Burchsted, *College of Staten Island*; Warren Burggren, *University of North Texas*; Meg Burke, *University of North Dakota*; Edwin Burling, *De Anza College*; Dale Burnside, *Lenoir-Rhyne University*; William Busa, *Johns Hopkins University*; Jorge Busciglio, *University of California, Irvine*; John Bushnell, *University of Colorado*; Linda Butler, *University of Texas, Austin*; David Byres, *Florida Community College*,

Jacksonville; Patrick Cafferty, *Emory University*; Guy A. Caldwell, *University of Alabama*; Jane Caldwell, *West Virginia University*; Kim A. Caldwell, *University of Alabama*; Ragan Callaway, *The University of Montana*; Kenneth M. Cameron, *University of Wisconsin, Madison*; R. Andrew Cameron, *California Institute of Technology*; Alison Campbell, *University of Waikato*; Iain Campbell, *University of Pittsburgh*; Michael Campbell, *Penn State University*; Patrick Canary, *Northland Pioneer College*; W. Zacheus Cande, *University of California, Berkeley*; Deborah Canington, *University of California, Davis*; Robert E. Cannon, *University of North Carolina, Greensboro*; Frank Cantelmo, *St. John's University*; John Capeheart, *University of Houston, Downtown*; Gregory Capelli, *College of William and Mary*; Cheryl Keller Capone, *Pennsylvania State University*; Richard Cardullo, *University of California, Riverside*; Nina Caris, *Texas A&M University*; Mickael Cariveau, *Mount Olive College*; Jeffrey Carmichael, *University of North Dakota*; Robert Carroll, *East Carolina University*; Laura L. Carruth, *Georgia State University*; J. Aaron Cassill, *University of Texas, San Antonio*; Karen I. Champ, *Central Florida Community College*; David Champlin, *University of Southern Maine*; Brad Chandler, *Palo Alto College*; Wei-Jen Chang, *Hamilton College*; Bruce Chase, *University of Nebraska, Omaha*; P. Bryant Chase, *Florida State University*; Doug Cheeseman, *De Anza College*; Shepley Chen, *University of Illinois, Chicago*; Giovina Chinchar, *Tougaloo College*; Joseph P. Chinnici, *Virginia Commonwealth University*; Jung H. Choi, *Georgia Institute of Technology*; Steve Christensen, *Brigham Young University, Idaho*; Geoffrey Church, *Fairfield University*; Henry Claman, *University of Colorado Health Science Center*; Anne Clark, *Binghamton University*; Greg Clark, *University of Texas*; Patricia J. Clark, *Indiana University-Purdue University, Indianapolis*; Ross C. Clark, *Eastern Kentucky University*; Lynwood Clemens, *Michigan State University*; Janice J. Clymer, *San Diego Mesa College*; Reggie Cobb, *Nashville Community College*; William P. Coffman, *University of Pittsburgh*; Austin Randy Cohen, *California State University, Northridge*; Bill Cohen, *University of Kentucky*; J. John Cohen, *University of Colorado Health Science Center*; James T. Colbert, *Iowa State University*; Sean Coleman, *University of the Ozarks*; Jan Colpaert, *Hasselt University*; Robert Colvin, *Ohio University*; Jay Comeaux, *McNeese State University*; David Cone, *Saint Mary's University*; Erin Connolly, *University of South Carolina*; Elizabeth Connor, *University of Massachusetts*; Joanne Conover, *University of Connecticut*; Ron Cooper, *University of California, Los Angeles;* Gregory Copenhaver, *University of North Carolina, Chapel Hill*; John Corliss, *University of Maryland*; James T. Costa, *Western Carolina University*; Stuart J. Coward, *University of Georgia*; Charles Creutz, *University of Toledo*; Bruce Criley, *Illinois Wesleyan University*; Norma Criley, *Illinois Wesleyan University*; Joe W. Crim, *University of Georgia*; Greg Crowther, *University of Washington*; Karen Curto, *University of Pittsburgh*; William Cushwa, *Clark College*; Anne Cusic, *University of Alabama, Birmingham*; Richard Cyr, *Pennsylvania State University*; Curtis Daehler, *University of Hawaii at Manoa*; Marymegan Daly, *The Ohio State University*; W. Marshall Darley, *University of Georgia*; Douglas Darnowski, *Indiana University Southeast*; Cynthia Dassler, *The Ohio State University*; Shannon Datwyler, *California State University, Sacramento*; Marianne Dauwalder, *University of Texas, Austin*; Larry Davenport, *Samford University*; Bonnie J. Davis, *San Francisco State University*; Jerry Davis, *University of Wisconsin, La Crosse*; Michael A. Davis, *Central Connecticut State University*; Thomas Davis, *University of New Hampshire*; Melissa Deadmond, *Truckee Meadows Community College*; John Dearn, *University of Canberra*; Maria E. de Bellard, *California State University, Northridge*; Teresa DeGolier, *Bethel College*; James Dekloe, *University of California, Santa Cruz*; Eugene Delay, *University of Vermont*; Patricia A. DeLeon, *University of Delaware*; Veronique Delesalle, *Gettysburg College*; T. Delevoryas, *University of Texas, Austin*; Roger Del Moral, *University of Washington*; Charles F. Delwiche, *University of Maryland*; Diane C. DeNagel, *Northwestern University*; William L. Dentler, *University of Kansas*; Jennifer Derkits, *J. Sergeant Reynolds Community College*; Daniel DerVartanian, *University of Georgia*; Jean DeSaix, *University of North Carolina, Chapel Hill*; Janet De Souza-Hart, *Massachusetts College of Pharmacy & Health Sciences*; Biao Ding, *Ohio State University*; Michael Dini, *Texas Tech University*; Kevin Dixon, *Florida State University*; Andrew Dobson, *Princeton University*; Stanley Dodson, *University of Wisconsin, Madison*; Jason Douglas, *Angelina College*; Mark Drapeau, *University of California, Irvine*; John Drees, *Temple University School of Medicine*; Charles Drewes, *Iowa State University*; Marvin Druger, *Syracuse University*; Gary Dudley, *University of Georgia*; David Dunbar, *Cabrini College*; Susan Dunford, *University of Cincinnati*; Kathryn A. Durham, *Lorain Community College*; Betsey Dyer, *Wheaton College*; Robert Eaton, *University of Colorado*; Robert S. Edgar, *University of California, Santa Cruz*; Anna Edlund, *Lafayette College*; Douglas J. Eernisse, *California State University, Fullerton*; Betty J. Eidemiller, *Lamar University*; Brad Elder, *Doane College*; Curt Elderkin, *College of New Jersey*; William D. Eldred, *Boston University*; Michelle Elekonich, *University of Nevada, Las Vegas*; George Ellmore, *Tufts University*; Mary Ellard-Ivey, *Pacific Lutheran University*; Kurt Elliott, *North West Vista College*; Norman Ellstrand, *University of California, Riverside*; Johnny El-Rady, *University of South Florida*; Bert Ely, *University of South Carolina*; Dennis Emery, *Iowa State University*; John Endler, *University of California, Santa Barbara*; Rob Erdman, *Florida Gulf Coast College*; Dale Erskine, *Lebanon Valley College*; Margaret T. Erskine, *Lansing Community College*; Susan Erster, *Stony Brook University*; Gerald Esch, *Wake Forest University*; Frederick B. Essig, *University of South Florida*; Mary Eubanks, *Duke University*; David Evans, *University of Florida*; Robert C. Evans, *Rutgers University, Camden*; Sharon Eversman, *Montana State University*; Olukemi Fadayomi, *Ferris State University*; Lincoln Fairchild, *Ohio State University*; Peter Fajer, *Florida State University*; Bruce Fall, *University of Minnesota*; Sam Fan, *Bradley University*; Lynn Fancher, *College of DuPage*; Ellen H. Fanning, *Vanderbilt University*; Paul Farnsworth, *University of New Mexico*; Larry Farrell, *Idaho State University*; Jerry F. Feldman, *University of California, Santa Cruz*; Lewis Feldman, *University of California, Berkeley*; Myriam Alhadeff Feldman, *Cascadia Community College*; Eugene Fenster, *Longview Community College*; Linda Fergusson-Kolmes, *Portland Community College, Sylvania Campus*; Russell Fernald, *University of Oregon*; Danilo Fernando, *SUNY College of Environmental Science and Forestry, Syracuse*; Rebecca Ferrell, *Metropolitan State College of Denver*; Christina Fieber, *Horry-Georgetown Technical College*; Melissa Fierke, *SUNY College of Environmental Science and Forestry*; Kim Finer, *Kent State University*; Milton Fingerman, *Tulane University*; Barbara Finney, *Regis College*; Teresa Fischer, *Indian River Community College*; Frank Fish, *West Chester University*; David Fisher, *University of Hawaii, Manoa*; Jonathan S. Fisher, *St. Louis University*; Steven Fisher, *University of California, Santa Barbara*; David Fitch, *New York University*; Kirk Fitzhugh, *Natural History Museum of Los Angeles County*; Lloyd Fitzpatrick, *University of North Texas*; William Fixsen, *Harvard University*; T. Fleming, *Bradley University*; Abraham Flexer, *Manuscript Consultant, Boulder, Colorado*; Mark Flood, *Fairmont State University*; Margaret Folsom, *Methodist College*; Kerry Foresman, *University of Montana*; Norma Fowler, *University of Texas, Austin*; Robert G. Fowler, *San Jose State University*; David Fox, *University of Tennessee, Knoxville*; Carl Frankel, *Pennsylvania State University, Hazleton*; Stewart Frankel, *University of Hartford*; Robert Franklin, *College of Charleston*; James Franzen, *University of Pittsburgh*; Art Fredeen, *University of Northern British Columbia*; Kim Fredericks, *Viterbo University*; Bill Freedman, *Dalhousie University*; Matt Friedman, *University of Chicago*; Otto Friesen, *University of Virginia*; Frank Frisch, *Chapman University*; Virginia Fry, *Monterey Peninsula College*; Bernard Frye, *University of Texas, Arlington*; Jed Fuhrman, *University of Southern California*; Alice Fulton, *University of Iowa*; Chandler Fulton, *Brandeis University*; Sara Fultz, *Stanford University*; Berdell Funke, *North Dakota State University*; Anne Funkhouser, *University of the Pacific*; Zofia E. Gagnon, *Marist College*; Michael Gaines, *University of Miami*; Cynthia M. Galloway, *Texas A&M University, Kingsville*; Arthur W. Galston, *Yale University*; Stephen Gammie, *University of Wisconsin, Madison*; Carl Gans, *University of Michigan*; John Gapter, *University of Northern Colorado*; Andrea Gargas, *University of Wisconsin, Madison*; Lauren Garner, *California Polytechnic State University, San Luis Obispo*; Reginald Garrett, *University of Virginia*; Craig Gatto, *Illinois State University*; Kristen Genet, *Anoka Ramsey Community College*; Patricia Gensel, *University of North Carolina*; Chris George, *California Polytechnic State University, San Luis Obispo*; Robert George, *University of Wyoming*; J. Whitfield Gibbons, *University of Georgia*; J. Phil Gibson, *University of Oklahoma*; Frank Gilliam, *Marshall University*; Eric Gillock, *Fort Hayes State University*; Simon Gilroy, *University of Wisconsin, Madison*; Edwin Ginés-Candelaria, *Miami Dade College*; Alan D. Gishlick, *Gustavus Adolphus College*; Todd Gleeson, *University of Colorado*; Jessica Gleffe, *University of California, Irvine*; John Glendinning, *Barnard College*; David Glenn-Lewin, *Wichita State University*; William Glider, *University of Nebraska*; Tricia Glidewell, *Marist School*; Elizabeth A. Godrick, *Boston University*; Jim Goetze, *Laredo Community College*; Lynda Goff, *University of California, Santa Cruz*; Elliott Goldstein, *Arizona State University*; Paul Goldstein, *University of Texas, El Paso*; Sandra Gollnick, *State University of New York, Buffalo*; Roy Golsteyn, *University of Lethbridge*; Anne Good, *University of California, Berkeley*; Judith Goodenough, *University of Massachusetts, Amherst*; Wayne Goodey, *University of British Columbia*; Barbara E. Goodman, *University of South Dakota*; Robert Goodman, *University of Wisconsin, Madison*; Ester Goudsmit, *Oakland University*; Linda Graham, *University of Wisconsin, Madison*; Robert Grammer, *Belmont University*; Joseph Graves, *Arizona State University*; Eileen Gregory, *Rollins College*; Phyllis Griffard, *University of Houston, Downtown*; A. J. F. Griffiths, *University of British Columbia*; Bradley Griggs, *Piedmont Technical College*; William Grimes, *University of Arizona*; David Grise, *Texas A&M University, Corpus Christi*; Mark Gromko, *Bowling Green State University*; Serine Gropper, *Auburn University*; Katherine L. Gross, *Ohio State University*; Gary Gussin, *University of Iowa*; Edward Gruberg, *Temple University*; Carla Guthridge, *Cameron University*; Mark Guyer, *National Human Genome Research Institute*; Ruth Levy Guyer, *Bethesda, Maryland*; Carla Haas, *Pennsylvania State University*; R. Wayne Habermehl, *Montgomery County Community College*; Pryce Pete Haddix, *Auburn University*; Mac Hadley, *University of Arizona*; Joel Hagen, *Radford University*; Jack P. Hailman, *University of Wisconsin*; Leah Haimo, *University of California, Riverside*; Ken Halanych, *Auburn University*; Jody Hall, *Brown University*; Heather Hallen-Adams, *University of Nebraska, Lincoln*; Douglas Hallett, *Northern Arizona University*; Rebecca Halyard, *Clayton State College*; Devney Hamilton, *Stanford University* (student); E. William Hamilton, *Washington and Lee University*; Matthew B. Hamilton, *Georgetown University*; Sam Hammer, *Boston University*; Penny Hanchey-Bauer, *Colorado State University*; William F. Hanna, *Massasoit Community College*; Dennis Haney, *Furman University*; Laszlo Hanzely, *Northern Illinois University*; Jeff Hardin, *University of Wisconsin, Madison*; Jean Hardwick, *Ithaca College*; Luke Harmon, *University of Idaho*; Lisa Harper, *University of California, Berkeley*; Deborah Harris, *Case Western Reserve University*; Jeanne M. Harris, *University of Vermont*; Richard Harrison, *Cornell University*; Stephanie Harvey, *Georgia Southwestern State University*; Carla Hass, *Pennsylvania State University*; Chris Haufler, *University of Kansas*; Bernard A. Hauser, *University of Florida*; Chris Haynes, *Shelton State Community College*; Evan B. Hazard, *Bemidji State University* (emeritus); H. D. Heath, *California State University, East Bay*; George Hechtel, *State University of New York, Stony Brook*; S. Blair Hedges, *Pennsylvania State University*; Brian Hedlund, *University of Nevada, Las Vegas*; David Heins, *Tulane University*; Jean Heitz, *University of Wisconsin, Madison*; Andreas Hejnol, *Sars International Centre for Marine Molecular Biology*; John D. Helmann, *Cornell University*; Colin Henderson, *University of Montana*; Susan Hengeveld, *Indiana University*; Michelle Henricks, *University of California, Los Angeles*; Caroll Henry, *Chicago State University*; Frank Heppner, *University of Rhode Island*; Albert Herrera, *University of Southern California*; Scott Herrick, *Missouri Western State College*; Ira Herskowitz, *University of California, San Francisco*; Paul E. Hertz, *Barnard College*; Chris Hess, *Butler University*; David Hibbett, *Clark University*; R. James Hickey, *Miami University*; Karen Hicks, *Kenyon College*; Kendra Hill, *San Diego State University*; William Hillenius, *College of Charleston*; Kenneth Hillers, *California Polytechnic State University, San Luis Obispo*; Ralph Hinegardner, *University of California, Santa Cruz*; William Hines, *Foothill College*; Robert Hinrichsen, *Indiana University of Pennsylvania*; Helmut Hirsch, *State University of New York, Albany*; Tuan-hua David Ho, *Washington University*; Carl Hoagstrom, *Ohio Northern University*; Elizabeth Hobson, *New Mexico State University*; Jason Hodin, *Stanford University*; James Hoffman, *University of Vermont*; A. Scott Holaday, *Texas Tech University*; Mark Holbrook, *University of Iowa*; N. Michele Holbrook, *Harvard University*; James Holland, *Indiana State University, Bloomington*; Charles Holliday, *Lafayette College*; Lubbock Karl Holte, *Idaho State University*; Alan R. Holyoak, *Brigham Young University, Idaho*; Laura Hoopes, *Occidental College*; Nancy Hopkins, *Massachusetts Institute of*

Technology; Sandra Horikami, *Daytona Beach Community College*; Kathy Hornberger, *Widener University*; Pius F. Horner, *San Bernardino Valley College*; Becky Houck, *University of Portland*; Margaret Houk, *Ripon College*; Laura Houston, *Northeast Lakeview College*; Daniel J. Howard, *New Mexico State University*; Ronald R. Hoy, *Cornell University*; Sandra Hsu, *Skyline College*; Sara Huang, *Los Angeles Valley College*; Cristin Hulslander, *University of Oregon*; Donald Humphrey, *Emory University School of Medicine*; Catherine Hurlbut, *Florida State College, Jacksonville*; Diane Husic, *Moravian College*; Robert J. Huskey, *University of Virginia*; Steven Hutcheson, *University of Maryland, College Park*; Linda L. Hyde, *Gordon College*; Bradley Hyman, *University of California, Riverside*; Jeffrey Ihara, *Mira Costa College*; Mark Iked, *San Bernardino Valley College*; Cheryl Ingram-Smith, *Clemson University*; Erin Irish, *University of Iowa*; Sally Irwin, *University of Hawaii, Maui College*; Harry Itagaki, *Kenyon College*; Alice Jacklet, *State University of New York, Albany*; John Jackson, *North Hennepin Community College*; Thomas Jacobs, *University of Illinois*; Kathy Jacobson, *Grinnell College*; Mark Jaffe, *Nova Southeastern University*; John C. Jahoda, *Bridgewater State College*; Douglas Jensen, *Converse College*; Jamie Jensen, *Brigham Young University*; Dan Johnson, *East Tennessee State University*; Lance Johnson, *Midland Lutheran College*; Lee Johnson, *The Ohio State University*; Randall Johnson, *University of California, San Diego*; Roishene Johnson, *Bossier Parish Community College*; Stephen Johnson, *William Penn University*; Wayne Johnson, *Ohio State University*; Kenneth C. Jones, *California State University, Northridge*; Russell Jones, *University of California, Berkeley*; Cheryl Jorcyk, *Boise State University*; Chad Jordan, *North Carolina State University*; Ann Jorgensen, *University of Hawaii*; Alan Journet, *Southeast Missouri State University*; Walter Judd, *University of Florida*; Ari Jumpponen, *Kansas State University*; Thomas W. Jurik, *Iowa State University*; Caroline M. Kane, *University of California, Berkeley*; Doug Kane, *Defiance College*; Thomas C. Kane, *University of Cincinnati*; The-Hui Kao, *Pennsylvania State University*; Tamos Kapros, *University of Missouri*; Kasey Karen, *Georgia College & State University*; E. L. Karlstrom, *University of Puget Sound*; David Kass, *Eastern Michigan University*; Jennifer Katcher, *Pima Community College*; Laura A. Katz, *Smith College*; Judy Kaufman, *Monroe Community College*; Maureen Kearney, *Field Museum of Natural History*; Eric G. Keeling, *Cary Institute of Ecosystem Studies*; Patrick Keeling, *University of British Columbia*; Thomas Keller, *Florida State University*; Elizabeth A. Kellogg, *University of Missouri, St. Louis*; Paul Kenrick, *Natural History Museum, London*; Norm Kenkel, *University of Manitoba*; Chris Kennedy, *Simon Fraser University*; George Khoury, *National Cancer Institute*; Stephen T. Kilpatrick, *University of Pittsburgh at Johnstown*; Rebecca T. Kimball, *University of Florida*; Shannon King, *North Dakota State University*; Mark Kirk, *University of Missouri, Columbia*; Robert Kitchin, *University of Wyoming*; Hillar Klandorf, *West Virginia University*; Attila O. Klein, *Brandeis University*; Karen M. Klein, *Northampton Community College*; Daniel Klionsky, *University of Michigan*; Mark Knauss, *Georgia Highlands College*; Janice Knepper, *Villanova University*; Charles Knight, *California Polytechnic State University*; Jennifer Knight, *University of Colorado*; Ned Knight, *Linfield College*; Roger Koeppe, *University of Arkansas*; David Kohl, *University of California, Santa Barbara*; Greg Kopf, *University of Pennsylvania School of Medicine*; Thomas Koppenheffer, *Trinity University*; Peter Kourtev, *Central Michigan University*; Margareta Krabbe, *Uppsala University*; Jacob Krans, *Western New England University*; Anselm Kratochwil, *Universität Osnabrück*; Eliot Krause, *Seton Hall University*; Deborah M. Kristan, *California State University, San Marcos*; Steven Kristoff, *Ivy Tech Community College*; Dubear Kroening, *University of Wisconsin*; William Kroll, *Loyola University, Chicago*; Janis Kuby, *San Francisco State University*; Barbara Kuemerle, *Case Western Reserve University*; Justin P. Kumar, *Indiana University*; Rukmani Kuppuswami, *Laredo Community College*; David Kurijaka, *Ohio University*; Lee Kurtz, *Georgia Gwinnett College*; Michael P. Labare, *United States Military Academy, West Point*; Marc-André Lachance, *University of Western Ontario*; J. A. Lackey, *State University of New York, Oswego*; Elaine Lai, *Brandeis University*; Mohamed Lakrim, *Kingsborough Community College*; Ellen Lamb, *University of North Carolina, Greensboro*; William Lamberts, *College of St Benedict and St John's University*; William L'Amoreaux, *College of Staten Island*; Lynn Lamoreux, *Texas A&M University*; Carmine A. Lanciani, *University of Florida*; Kenneth Lang, *Humboldt State University*; Jim Langeland, *Kalamazoo College*; Dominic Lannutti, *El Paso Community College*; Allan Larson, *Washington University*; Grace Lasker, *Lake Washington Institute of Technology*; John Latto, *University of California, Santa Barbara*; Diane K. Lavett, *State University of New York, Cortland, and Emory University*; Charles Leavell, *Fullerton College*; C. S. Lee, *University of Texas*; Daewoo Lee, *Ohio University*; Tali D. Lee, *University of Wisconsin, Eau Claire*; Hugh Lefcort, *Gonzaga University*; Robert Leonard, *University of California, Riverside*; Michael R. Leonardo, *Coe College*; John Lepri, *University of North Carolina, Greensboro*; Donald Levin, *University of Texas, Austin*; Joseph Levine, *Boston College*; Mike Levine, *University of California, Berkeley*; Alcinda Lewis, *University of Colorado, Boulder*; Bill Lewis, *Shoreline Community College*; Jani Lewis, *State University of New York*; John Lewis, *Loma Linda University*; Lorraine Lica, *California State University, East Bay*; Harvey Liftin, *Broward Community College*; Harvey Lillywhite, *University of Florida, Gainesville*; Graeme Lindbeck, *Valencia Community College*; Clark Lindgren, *Grinnell College*; Eric W. Linton, *Central Michigan University*; Diana Lipscomb, *George Washington University*; Christopher Little, *The University of Texas, Pan American*; Kevin D. Livingstone, *Trinity University*; Andrea Lloyd, *Middlebury College*; Tatyana Lobova, *Old Dominion University*; Sam Loker, *University of New Mexico*; David Longstreth, *Louisiana State University*; Christopher A. Loretz, *State University of New York, Buffalo*; Donald Lovett, *College of New Jersey*; Jane Lubchenco, *Oregon State University*; Douglas B. Luckie, *Michigan State University*; Hannah Lui, *University of California, Irvine*; Margaret A. Lynch, *Tufts University*; Steven Lynch, *Louisiana State University, Shreveport*; Lisa Lyons, *Florida State University*; Richard Machemer Jr., *St. John Fisher College*; Elizabeth Machunis-Masuoka, *University of Virginia*; James MacMahon, *Utah State University*; Nancy Magill, *Indiana University*; Christine R. Maher, *University of Southern Maine*; Linda Maier, *University of Alabama, Huntsville*; Jose Maldonado, *El Paso Community College*; Richard Malkin, *University of California, Berkeley*; Charles Mallery, *University of Miami*; Keith Malmos, *Valencia Community College, East Campus*; Cindy Malone, *California State University, Northridge*; Mark Maloney, *University of South Mississippi*; Carol Mapes, *Kutztown University of Pennsylvania*; William Margolin, *University of Texas Medical School*; Lynn Margulis, *Boston University*; Julia Marrs, *Barnard College* (student); Kathleen A. Marrs, *Indiana University-Purdue University, Indianapolis*; Edith Marsh, *Angelo State University*; Diane L. Marshall, *University of New Mexico*; Mary Martin, *Northern Michigan University*; Karl Mattox, *Miami University of Ohio*; Joyce Maxwell, *California State University, Northridge*; Jeffrey D. May, *Marshall University*; Mike Mayfield, *Ball State University*; Kamau Mbuthia, *Bowling Green State University*; Lee McClenaghan, *San Diego State University*; Richard McCracken, *Purdue University*; Andrew McCubbin, *Washington State University*; Kerry McDonald, *University of Missouri, Columbia*; Tanya McGhee, *Craven Community College*; Jacqueline McLaughlin, *Pennsylvania State University, Lehigh Valley*; Neal McReynolds, *Texas A&M International*; Darcy Medica, *Pennsylvania State University*; Lisa Marie Meffert, *Rice University*; Susan Meiers, *Western Illinois University*; Michael Meighan, *University of California, Berkeley*; Scott Meissner, *Cornell University*; Paul Melchior, *North Hennepin Community College*; Phillip Meneely, *Haverford College*; John Merrill, *Michigan State University*; Brian Metscher, *University of California, Irvine*; Jenny Metzler, *Ball State University*; Ralph Meyer, *University of Cincinnati*; James Mickle, *North Carolina State University*; Jan Mikesell, *Gettysburg College*; Roger Milkman, *University of Iowa*; Grace Miller, *Indiana Wesleyan University*; Helen Miller, *Oklahoma State University*; John Miller, *University of California, Berkeley*; Jonathan Miller, *Edmonds Community College*; Kenneth R. Miller, *Brown University*; Mill Miller, *Wright State University*; Alex Mills, *University of Windsor*; Sarah Milton, *Florida Atlantic University*; Eli Minkoff, *Bates College*; John E. Minnich, *University of Wisconsin, Milwaukee*; Subhash Minocha, *University of New Hampshire*; Michael J. Misamore, *Texas Christian University*; Kenneth Mitchell, *Tulane University School of Medicine*; Ivona Mladenovic, *Simon Fraser University*; Alan Molumby, *University of Illinois, Chicago*; Nicholas Money, *Miami University*; Russell Monson, *University of Colorado, Boulder*; Joseph P. Montoya, *Georgia Institute of Technology*; Frank Moore, *Oregon State University*; Janice Moore, *Colorado State University*; Linda Moore, *Georgia Military College*; Randy Moore, *Wright State University*; William Moore, *Wayne State University*; Carl Moos, *Veterans Administration Hospital, Albany, New York*; Linda Martin Morris, *University of Washington*; Michael Mote, *Temple University*; Alex Motten, *Duke University*; Jeanette Mowery, *Madison Area Technical College*; Deborah Mowshowitz, *Columbia University*; Rita Moyes, *Texas A&M, College Station*; Darrel L. Murray, *University of Illinois, Chicago*; Courtney Murren, *College of Charleston*; John Mutchmor, *Iowa State University*; Elliot Myerowitz, *California Institute of Technology*; Barbara Nash, *Mercy College*; Gavin Naylor, *Iowa State University*; John Neess, *University of Wisconsin, Madison*; Ross Nehm, *Ohio State University*; Tom Neils, *Grand Rapids Community College*; Kimberlyn Nelson, *Pennsylvania State University*; Raymond Neubauer, *University of Texas, Austin*; Todd Newbury, *University of California, Santa Cruz*; James Newcomb, *New England College*; Jacalyn Newman, *University of Pittsburgh*; Harvey Nichols, *University of Colorado, Boulder*; Deborah Nickerson, *University of South Florida*; Bette Nicotri, *University of Washington*; Caroline Niederman, *Tomball College*; Eric Nielsen, *University of Michigan*; Maria Nieto, *California State University, East Bay*; Anders Nilsson, *University of Umeå*; Greg Nishiyama, *College of the Canyons*; Charles R. Noback, *College of Physicians and Surgeons, Columbia University*; Jane Noble-Harvey, *Delaware University*; Mary C. Nolan, *Irvine Valley College*; Kathleen Nolta, *University of Michigan*; Peter Nonacs, *University of California, Los Angeles*; Mohamed A. F. Noor, *Duke University*; Shawn Nordell, *St. Louis University*; Richard S. Norman, *University of Michigan, Dearborn* (emeritus); David O. Norris, *University of Colorado, Boulder*; Steven Norris, *California State University, Channel Islands*; Gretchen North, *Occidental College*; Cynthia Norton, *University of Maine, Augusta*; Steve Norton, *East Carolina University*; Steve Nowicki, *Duke University*; Bette H. Nybakken, *Hartnell College*; Brian O'Conner, *University of Massachusetts, Amherst*; Gerard O'Donovan, *University of North Texas*; Eugene Odum, *University of Georgia*; Mark P. Oemke, *Alma College*; Linda Ogren, *University of California, Santa Cruz*; Patricia O'Hern, *Emory University*; Olabisi Ojo, *Southern University at New Orleans*; Nathan O. Okia, *Auburn University, Montgomery*; Jeanette Oliver, *St. Louis Community College, Florissant Valley*; Gary P. Olivetti, *University of Vermont*; Margaret Olney, *St. Martin's College*; John Olsen, *Rhodes College*; Laura J. Olsen, *University of Michigan*; Sharman O'Neill, *University of California, Davis*; Wan Ooi, *Houston Community College*; Aharon Oren, *The Hebrew University*; John Oross, *University of California, Riverside*; Rebecca Orr, *Collin College*; Catherine Ortega, *Fort Lewis College*; Charissa Osborne, *Butler University*; Gay Ostarello, *Diablo Valley College*; Henry R. Owen, *Eastern Illinois University*; Thomas G. Owens, *Cornell University*; Penny Padgett, *University of North Carolina, Chapel Hill*; Kevin Padian, *University of California, Berkeley*; Dianna Padilla, *State University of New York, Stony Brook*; Anthony T. Paganini, *Michigan State University*; Fatimata Pale, *Thiel College*; Barry Palevitz, *University of Georgia*; Michael A. Palladino, *Monmouth University*; Matt Palmtag, *Florida Gulf Coast University*; Stephanie Pandolfi, *Michigan State University*; Daniel Papaj, *University of Arizona*; Peter Pappas, *County College of Morris*; Nathalie Pardigon, *Institut Pasteur*; Bulah Parker, *North Carolina State University*; Stanton Parmeter, *Chemeketa Community College*; Susan Parrish, *McDaniel College*; Cindy Paszkowski, *University of Alberta*; Robert Patterson, *San Francisco State University*; Ronald Patterson, *Michigan State University*; Crellin Pauling, *San Francisco State University*; Kay Pauling, *Foothill Community College*; Daniel Pavuk, *Bowling Green State University*; Debra Pearce, *Northern Kentucky University*; Patricia Pearson, *Western Kentucky University*; Andrew Pease, *Stevenson University*; Nancy Pelaez, *Purdue University*; Shelley Penrod, *North Harris College*; Imara Y. Perera, *North Carolina State University*; Beverly Perry, *Houston Community College*; Irene Perry, *University of Texas of the Permian Basin*; Roger Persell, *Hunter College*; Eric Peters, *Chicago State University*; Larry Peterson, *University of Guelph*; David Pfennig, *University of North Carolina, Chapel Hill*; Mark Pilgrim, *College of Coastal Georgia*; David S. Pilliod, *California Polytechnic State University, San Luis Obispo*; Vera M. Piper, *Shenandoah University*; Deb Pires, *University of California, Los Angeles*; J. Chris Pires, *University of Missouri, Columbia*; Jarmila Pittermann, *University of California, Santa Cruz*; Bob Pittman, *Michigan State University*; James Platt, *University of Denver*; Martin Poenie, *University of Texas, Austin*; Scott Poethig, *University of Pennsylvania*; Crima Pogge, *San Francisco Community College*; Michael Pollock, *Mount Royal University*;

Roberta Pollock, *Occidental College*; Jeffrey Pommerville, *Texas A&M University*; Therese M. Poole, *Georgia State University*; Angela R. Porta, *Kean University*; Jason Porter, *University of the Sciences, Philadelphia*; Warren Porter, *University of Wisconsin*; Daniel Potter, *University of California, Davis*; Donald Potts, *University of California, Santa Cruz*; Robert Powell, *Avila University*; Andy Pratt, *University of Canterbury*; David Pratt, *University of California, Davis*; Halina Presley, *University of Illinois, Chicago*; Eileen Preston, *Tarrant Community College Northwest*; Mary V. Price, *University of California, Riverside*; Mitch Price, *Pennsylvania State University*; Steven Price, *Virginia Commonwealth University*; Terrell Pritts, *University of Arkansas, Little Rock*; Rong Sun Pu, *Kean University*; Rebecca Pyles, *East Tennessee State University*; Scott Quackenbush, *Florida International University*; Ralph Quatrano, *Oregon State University*; Peter Quinby, *University of Pittsburgh*; Val Raghavan, *Ohio State University*; Deanna Raineri, *University of Illinois, Champaign-Urbana*; David Randall, *City University Hong Kong*; Talitha Rajah, *Indiana University Southeast*; Charles Ralph, *Colorado State University*; Pushpa Ramakrishna, *Chandler-Gilbert Community College*; Thomas Rand, *Saint Mary's University*; Monica Ranes-Goldberg, *University of California, Berkeley*; Samiksha Raut, *University of Alabama at Birmingham*; Robert S. Rawding, *Gannon University*; Robert H. Reavis, *Glendale Community College*; Kurt Redborg, *Coe College*; Ahnya Redman, *Pennsylvania State University*; Brian Reeder, *Morehead State University*; Bruce Reid, *Kean University*; David Reid, *Blackburn College*; C. Gary Reiness, *Lewis & Clark College*; Charles Remington, *Yale University*; Erin Rempala, *San Diego Mesa College*; David Reznick, *University of California, Riverside*; Fred Rhoades, *Western Washington State University*; Douglas Rhoads, *University of Arkansas*; Eric Ribbens, *Western Illinois University*; Christina Richards, *New York University*; Sarah Richart, *Azusa Pacific University*; Wayne Rickoll, *University of Puget Sound*; Christopher Riegle, *Irvine Valley College*; Loren Rieseberg, *University of British Columbia*; Bruce B. Riley, *Texas A&M University*; Todd Rimkus, *Marymount University*; John Rinehart, *Eastern Oregon University*; Donna Ritch, *Pennsylvania State University*; Carol Rivin, *Oregon State University East*; Laurel Roberts, *University of Pittsburgh*; Diane Robins, *University of Michigan*; Kenneth Robinson, *Purdue University*; Thomas Rodella, *Merced College*; Luis Rodriguez, *San Antonio College*; Deb Roess, *Colorado State University*; Heather Roffey, *Marianopolis College*; Rodney Rogers, *Drake University*; Suzanne Rogers, *Seton Hill University*; William Roosenburg, *Ohio University*; Kara Rosch, *Blinn College*; Mike Rosenzweig, *Virginia Polytechnic Institute and State University*; Wayne Rosing, *Middle Tennessee State University*; Thomas Rost, *University of California, Davis*; Stephen I. Rothstein, *University of California, Santa Barbara*; John Ruben, *Oregon State University*; Albert Ruesink, *Indiana University*; Patricia Rugaber, *College of Coastal Georgia*; Scott Russell, *University of Oklahoma*; Jodi Rymer, *College of the Holy Cross*; Neil Sabine, *Indiana University*; Tyson Sacco, *Cornell University*; Glenn-Peter Saetre, *University of Oslo*; Rowan F. Sage, *University of Toronto*; Tammy Lynn Sage, *University of Toronto*; Sanga Saha, *Harold Washington College*; Don Sakaguchi, *Iowa State University*; Walter Sakai, *Santa Monica College*; Per Salvesen, *University of Bergen*; Mark F. Sanders, *University of California, Davis*; Kathleen Sandman, *Ohio State University*; Davison Sangweme, *University of North Georgia*; Louis Santiago, *University of California, Riverside*; Ted Sargent, *University of Massachusetts, Amherst*; K. Sathasivan, *University of Texas, Austin*; Gary Saunders, *University of New Brunswick*; Thomas R. Sawicki, *Spartanburg Community College*; Inder Saxena, *University of Texas, Austin*; Karin Scarpinato, *Georgia Southern University*; Carl Schaefer, *University of Connecticut*; Andrew Schaffner, *Cal Poly San Luis Obispo*; Maynard H. Schaus, *Virginia Wesleyan College*; Renate Scheibe, *University of Osnabrück*; Cara Schillington, *Eastern Michigan University*; David Schimpf, *University of Minnesota, Duluth*; William H. Schlesinger, *Duke University*; Mark Schlissel, *University of California, Berkeley*; Christopher J. Schneider, *Boston University*; Thomas W. Schoener, *University of California, Davis*; Robert Schorr, *Colorado State University*; Patricia M. Schulte, *University of British Columbia*; Karen S. Schumaker, *University of Arizona*; Brenda Schumpert, *Valencia Community College*; David J. Schwartz, *Houston Community College*; Carrie Schwarz, *Western Washington University*; Christa Schwintzer, *University of Maine*; Erik P. Scully, *Towson State University*; Robert W. Seagull, *Hofstra University*; Edna Seaman, *Northeastern University*; Duane Sears, *University of California, Santa Barbara*; Brent Selinger, *University of Lethbridge*; Orono Shukdeb Sen, *Bethune-Cookman College*; Wendy Sera, *Seton Hill University*; Alison M. Shakarian, *Salve Regina University*; Timothy E. Shannon, *Francis Marion University*; Victoria C. Sharpe, *Blinn College*; Elaine Shea, *Loyola College, Maryland*; Stephen Sheckler, *Virginia Polytechnic Institute and State University*; Robin L. Sherman, *Nova Southeastern University*; Richard Sherwin, *University of Pittsburgh*; Alison Sherwood, *University of Hawaii at Manoa*; Lisa Shimeld, *Crafton Hills College*; James Shinkle, *Trinity University*; Barbara Shipes, *Hampton University*; Brian Shmaefsky, *Lone Star College*; Richard M. Showman, *University of South Carolina*; Eric Shows, *Jones County Junior College*; Peter Shugarman, *University of Southern California*; Alice Shuttey, *DeKalb Community College*; James Sidie, *Ursinus College*; Daniel Simberloff, *Florida State University*; Rebecca Simmons, *University of North Dakota*; Anne Simon, *University of Maryland, College Park*; Robert Simons, *University of California, Los Angeles*; Alastair Simpson, *Dalhousie University*; Susan Singer, *Carleton College*; Sedonia Sipes, *Southern Illinois University, Carbondale*; John Skillman, *California State University, San Bernardino*; Roger Sloboda, *Dartmouth University*; John Smarrelli, *Le Moyne College*; Andrew T. Smith, *Arizona State University*; Kelly Smith, *University of North Florida*; Nancy Smith-Huerta, *Miami Ohio University*; John Smol, *Queen's University*; Andrew J. Snope, *Essex Community College*; Mitchell Sogin, *Woods Hole Marine Biological Laboratory*; Doug Soltis, *University of Florida, Gainesville*; Julio G. Soto, *San Jose State University*; Susan Sovonick-Dunford, *University of Cincinnati*; Rebecca Sperry, *Salt Lake Community College*; Frederick W. Spiegel, *University of Arkansas*; Clint Springer, *Saint Joseph's University*; John Stachowicz, *University of California, Davis*; Joel Stafstrom, *Northern Illinois University*; Alam Stam, *Capital University*; Amanda Starnes, *Emory University*; Karen Steudel, *University of Wisconsin*; Barbara Stewart, *Swarthmore College*; Gail A. Stewart, *Camden County College*; Cecil Still, *Rutgers University, New Brunswick*; Margery Stinson, *Southwestern College*; James Stockand, *University of Texas Health Science Center, San Antonio*; John Stolz, *California Institute of Technology*; Judy Stone, *Colby College*; Richard D. Storey, *Colorado College*; Stephen Strand, *University of California, Los Angeles*; Eric Strauss, *University of Massachusetts, Boston*; Antony Stretton, *University of Wisconsin, Madison*; Russell Stullken, *Augusta College*; Mark Sturtevant, *Oakland University, Flint*; John Sullivan, *Southern Oregon State University*; Gerald Summers, *University of Missouri*; Judith Sumner, *Assumption College*; Marshall D. Sundberg, *Emporia State University*; Cynthia Surmacz, *Bloomsburg University*; Lucinda Swatzell, *Southeast Missouri State University*; Daryl Sweeney, *University of Illinois, Champaign-Urbana*; Diane Sweeney, *Punahou School*; Samuel S. Sweet, *University of California, Santa Barbara*; Janice Swenson, *University of North Florida*; Michael A. Sypes, *Pennsylvania State University*; Lincoln Taiz, *University of California, Santa Cruz*; David Tam, *University of North Texas*; Yves Tan, *Cabrillo College*; Samuel Tarsitano, *Southwest Texas State University*; David Tauck, *Santa Clara University*; Emily Taylor, *California Polytechnic State University, San Luis Obispo*; James Taylor, *University of New Hampshire*; John W. Taylor, *University of California, Berkeley*; Kristen Taylor, *Salt Lake Community College*; Martha R. Taylor, *Cornell University*; Franklyn Tan Te, *Miami Dade College*; Thomas Terry, *University of Connecticut*; Roger Thibault, *Bowling Green State University*; Kent Thomas, *Wichita State University*; Rebecca Thomas, *College of St. Joseph*; William Thomas, *Colby-Sawyer College*; Cyril Thong, *Simon Fraser University*; John Thornton, *Oklahoma State University*; Robert Thornton, *University of California, Davis*; William Thwaites, *Tillamook Bay Community College*; Stephen Timme, *Pittsburg State University*; Mike Toliver, *Eureka College*; Eric Toolson, *University of New Mexico*; Leslie Towill, *Arizona State University*; James Traniello, *Boston University*; Paul Q. Trombley, *Florida State University*; Nancy J. Trun, *Duquesne University*; Constantine Tsoukas, *San Diego State University*; Marsha Turell, *Houston Community College*; Victoria Turgeon, *Furman University*; Robert Tuveson, *University of Illinois, Urbana*; Maura G. Tyrrell, *Stonehill College*; Catherine Uekert, *Northern Arizona University*; Claudia Uhde-Stone, *California State University, East Bay*; Gordon Uno, *University of Oklahoma*; Lisa A. Urry, *Mills College*; Saba Valadkhan, *Center for RNA Molecular Biology*; James W. Valentine, *University of California, Santa Barbara*; Joseph Vanable, *Purdue University*; Theodore Van Bruggen, *University of South Dakota*; Kathryn VandenBosch, *Texas A&M University*; Gerald Van Dyke, *North Carolina State University*; Brandi Van Roo, *Framingham State College*; Moira Van Staaden, *Bowling Green State University*; Martin Vaughan, *Indiana University-Purdue University Indianapolis*; Sarah VanVickle-Chavez, *Washington University, St. Louis*; William Velhagen, *New York University*; Steven D. Verhey, *Central Washington University*; Kathleen Verville, *Washington College*; Sara Via, *University of Maryland*; Meena Vijayaraghavan, *Tulane University*; Frank Visco, *Orange Coast College*; Laurie Vitt, *University of California, Los Angeles*; Neal Voelz, *St. Cloud State University*; Thomas J. Volk, *University of Wisconsin, La Crosse*; Leif Asbjørn Vøllestad, *University of Oslo*; Amy Volmer, *Swarthmore College*; Janice Voltzow, *University of Scranton*; Margaret Voss, *Penn State Erie*; Susan D. Waaland, *University of Washington*; Charles Wade, *C.S. Mott Community College*; William Wade, *Dartmouth Medical College*; John Waggoner, *Loyola Marymount University*; Jyoti Wagle, *Houston Community College*; Edward Wagner, *University of California, Irvine*; D. Alexander Wait, *Southwest Missouri State University*; Claire Walczak, *Indiana University*; Jerry Waldvogel, *Clemson University*; Dan Walker, *San Jose State University*; Robert Lee Wallace, *Ripon College*; Jeffrey Walters, *North Carolina State University*; Linda Walters, *University of Central Florida*; James Wandersee, *Louisiana State University*; James T. Warren Jr., *Pennsylvania State University*; Nickolas M. Waser, *University of California, Riverside*; Fred Wasserman, *Boston University*; Margaret Waterman, *University of Pittsburgh*; Charles Webber, *Loyola University of Chicago*; Peter Webster, *University of Massachusetts, Amherst*; Terry Webster, *University of Connecticut, Storrs*; Beth Wee, *Tulane University*; James Wee, *Loyola University, New Orleans*; Andrea Weeks, *George Mason University*; John Weishampel, *University of Central Florida*; Peter Wejksnora, *University of Wisconsin, Milwaukee*; Charles Wellman, *Sheffield University*; Kentwood Wells, *University of Connecticut*; David J. Westenberg, *University of Missouri, Rolla*; Richard Wetts, *University of California, Irvine*; Christopher Whipps, *State University of New York College of Environmental Science and Forestry*; Jessica White-Phillip, *Our Lady of the Lake University*; Matt White, *Ohio University*; Philip White, *James Hutton Institute*; Susan Whittemore, *Keene State College*; Murray Wiegand, *University of Winnipeg*; Ernest H. Williams, *Hamilton College*; Kathy Williams, *San Diego State University*; Kimberly Williams, *Kansas State University*; Stephen Williams, *Glendale Community College*; Elizabeth Willott, *University of Arizona*; Christopher Wills, *University of California, San Diego*; Paul Wilson, *California State University, Northridge*; Fred Wilt, *University of California, Berkeley*; Peter Wimberger, *University of Puget Sound*; Robert Winning, *Eastern Michigan University*; E. William Wischusen, *Louisiana State University*; Clarence Wolfe, *Northern Virginia Community College*; Vickie L. Wolfe, *Marshall University*; Janet Wolkenstein, *Hudson Valley Community College*; Robert T. Woodland, *University of Massachusetts Medical School*; Joseph Woodring, *Louisiana State University*; Denise Woodward, *Pennsylvania State University*; Patrick Woolley, *East Central College*; Sarah E. Wyatt, *Ohio University*; Grace Wyngaard, *James Madison University*; Shuhai Xiao, *Virginia Polytechnic Institute*, Ramin Yadegari, *University of Arizona*; Paul Yancey, *Whitman College*; Philip Yant, *University of Michigan*; Linda Yasui, *Northern Illinois University*; Anne D. Yoder, *Duke University*; Hideo Yonenaka, *San Francisco State University*; Robert Yost, *Indiana University-Purdue University Indianapolis*; Tia Young, *Pennsylvania State University*; Gina M. Zainelli, *Loyola University, Chicago*; Edward Zalisko, *Blackburn College*; Nina Zanetti, *Siena College*; Sam Zeveloff, *Weber State University*; Zai Ming Zhao, *University of Texas, Austin*; John Zimmerman, *Kansas State University*; Miriam Zolan, *Indiana University*; Theresa Zucchero, *Methodist University*; Uko Zylstra, *Calvin College*.

Detailed Contents

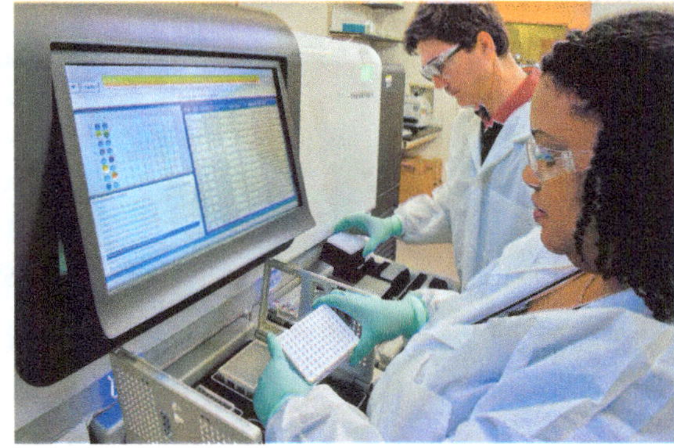

8 An Introduction to Metabolism 145

9 Cellular Respiration and Fermentation 166

10 Photosynthesis 189

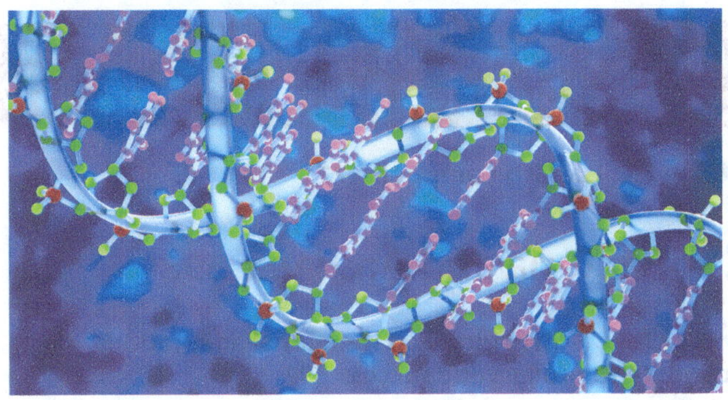

16 The Molecular Basis of Inheritance 316

17 Gene Expression: From Gene to Protein 337

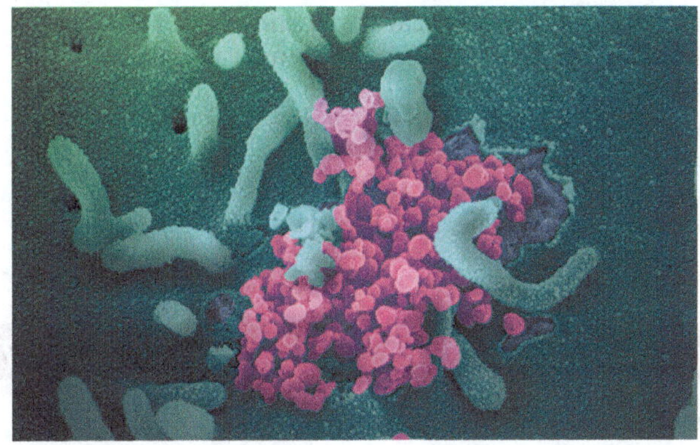

19 Viruses 400

20 DNA Tools and Biotechnology 419

Unit 4 Mechanisms of Evolution 471

25 The History of Life on Earth 533

Unit 5 The Evolutionary History of Biological Diversity 577

27 Bacteria and Archaea 598

This page is intentionally blank.

1 Evolution, the Themes of Biology, and Scientific Inquiry

KEY CONCEPTS

Study Tip

Make a table: List the five unifying themes of biology across the top. Enter at least three examples of each theme as you read this chapter. One example is filled in for you. To help you focus on these big ideas, continue adding examples throughout your study of biology.

Evolution	Organisation			
The species *Banksia serrata* has evolved to survive more frequent fires.				

Go to Mastering Biology

to access Dynamic Study Modules for revision, 3D BioFlix® animations and high-quality videos, and your interactive Pearson eText.

Figure 1.1 Over millions of years, Australia's climate became hotter and drier. Plants like this banksia (*Banksia serrata*) evolved to survive more frequent fires. New branches and leaves grow from undamaged cells and tissues beneath the protective and burned bark. Stored resources allow the banksias to reestablish their canopy, which shades potential competitors. Maintaining a place in the landscape, in spite of frequent fires, maximises the potential for the species to survive and flourish.

How does this banksia illustrate the unifying themes of biology?

As a result of **evolution** through natural selection over long periods of time, banksias developed traits that allowed them to survive frequent fires.

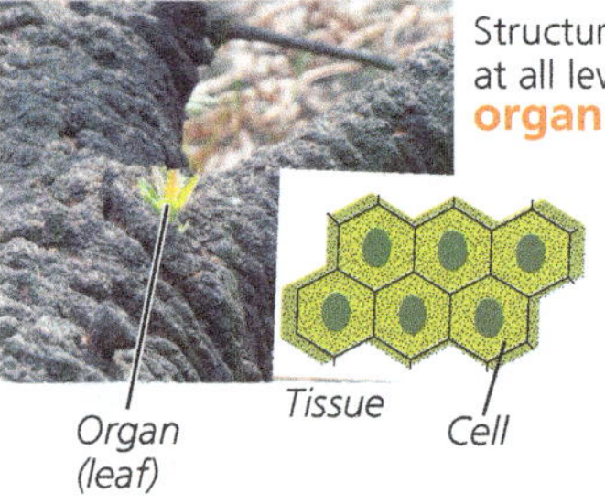

Structure fits function at all levels of a banksia's **organisation.**

Genetic **information** encoded in DNA determines how the banksia grows.

Energy flows one way from the sun to plants to a grasshopper; **matter** cycles between a grasshopper and its environment.

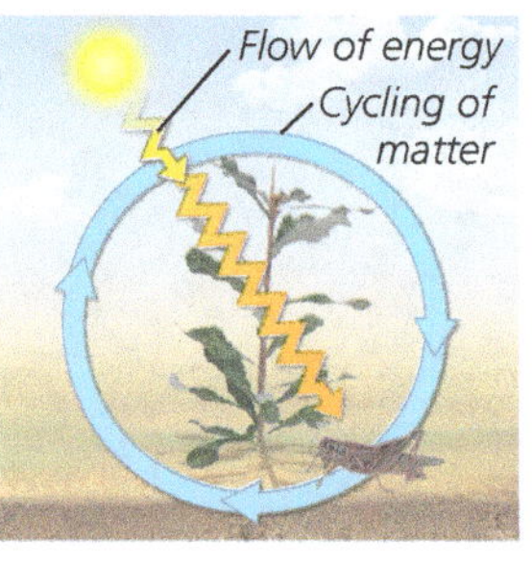

A plant being eaten by a grasshopper and a mouse being preyed upon by a swamp harrier are **interactions** within a system.

CONCEPT 1.1

The study of life reveals unifying themes

At the most fundamental level, we may ask: What is life? Even a child realises that a dog or a plant is alive, while a rock or a car is not. Yet the phenomenon we call life defies a simple definition. We recognise life by what living things do. **Figure 1.2** highlights some of the properties and processes associated with life.

Biology, the scientific study of life, is a subject of enormous scope, and exciting new biological discoveries are being made every day. How can you organise all the information you'll encounter as you study biology into a comprehensible framework? Focusing on a few big ideas will help. Five unifying themes—timeless ways to think about life—will serve you well and remain useful decades from now.

- Organisation
- Information
- Energy and Matter
- Interactions
- Evolution

In this section and the next, we'll briefly explore each theme.

▼ **Figure 1.2 Some properties of life.**

▼ **Order.** This close-up of a sunflower illustrates the highly ordered structure that characterises life.

▲ **Evolutionary adaptation.** The overall appearance of this pygmy sea horse camouflages the animal in its environment. Such adaptations evolve over countless generations by the reproductive success of those individuals with heritable traits that are best suited to their environments.

▲ **Regulation.** The regulation of blood flow through the blood vessels of this bilby's ears helps maintain a constant body temperature by adjusting heat exchange with the surrounding air.

▼ **Reproduction.** Organisms (living things) reproduce their own kind.

▲ **Energy processing.** This butterfly obtains fuel in the form of nectar from flowers. The butterfly will use chemical energy stored in its food to power flight and other work.

▲ **Growth and development.** Inherited information carried by genes controls the pattern of growth and development of organisms, such as this oak seedling.

▲ **Response to the environment.** The Venus flytrap on the left closed its trap rapidly in response to the environmental stimulus of a grasshopper landing on the open trap.

Theme: New Properties Emerge at Successive Levels of Biological Organisation

ORGANISATION The study of life on Earth extends from the microscopic scale of the molecules and cells that make up organisms to the global scale of the entire living planet. As biologists, we can divide this enormous range into different levels of biological organisation. In **Figure 1.3**, we look down on the Earth from space. We zoom in and narrow our focus until we can examine life in an alpine meadow. Our journey, depicted as a series of numbered steps, illustrates one example of the hierarchy of biological organisation.

Zooming in to progressively-finer resolution illustrates the principle that underlies *reductionism*, an approach that reduces complex systems to simpler components that are more manageable to study. Reductionism is a powerful strategy in biology. For example, by studying the molecular structure of DNA that had been extracted from cells, James Watson and Francis Crick inferred the chemical basis of biological inheritance. Despite its importance, reductionism provides an incomplete view of life on Earth, as you'll see next.

Emergent Properties

Let's reexamine Figure 1.3, beginning this time at the molecular level and then zooming out. This approach allows us to see novel properties emerge at each level that are absent

▼ Figure 1.3 Exploring Levels of Biological Organisation

◄ 1 The Biosphere

Even from space, we can see signs of Earth's life—in the mosaic of greens indicating forests, for example. We can also see the **biosphere**, which consists of all life on Earth and all the places where life exists: most regions of land, most bodies of water, the atmosphere to an altitude of several kilometres and even sediments far below the ocean floor.

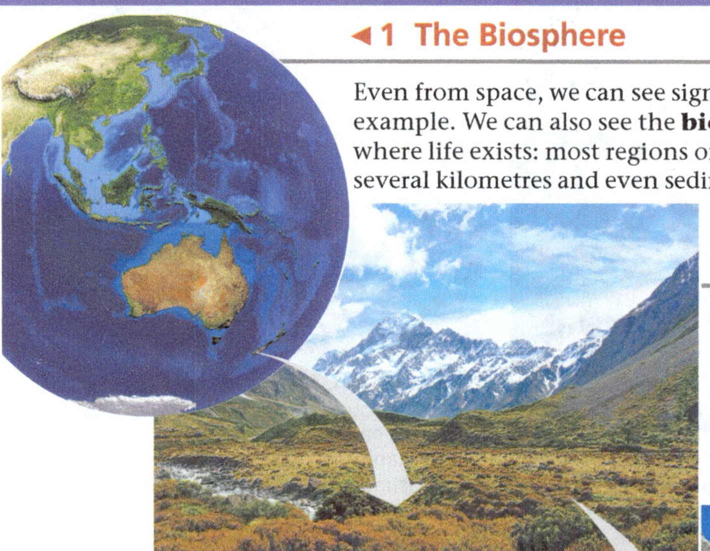

◄ 2 Ecosystems

Our first scale change brings us to a New Zealand alpine meadow, which is an example of an ecosystem, as are a tropical forest, grassland, desert, and coral reef. An **ecosystem** consists of all the living things in a particular area, along with all the nonliving components of the environment with which life interacts, such as soil, water, atmospheric gases, and light.

► 3 Communities

The array of organisms inhabiting a particular ecosystem is called a biological **community**. The community in our meadow ecosystem includes many kinds of plants, various animals, mushrooms and other fungi, and enormous numbers of diverse microorganisms, such as bacteria, that are too small to see without a microscope. Each of these forms of life belongs to a *species*—a group whose members can only reproduce with other members of the group.

► 4 Populations

A **population** consists of all the individuals of a species living within the bounds of a specified area that interbreed with each other. For example, our meadow includes a population of Mount Cook lily (some of which are shown here), a tussock butterfly population, and a population of keas. A community is therefore the set of populations that inhabit a particular area.

▲ 5 Organisms

Individual living things are called **organisms**. Each plant in the meadow is an organism, and so is each animal, fungus, and bacterium.

from the preceding one. These **emergent properties** are due to the arrangement and interactions of parts as complexity increases. For example, although photosynthesis occurs in an intact chloroplast, it will not take place if chlorophyll and other chloroplast molecules are simply mixed in a test tube. The coordinated processes of photosynthesis require a specific organisation of these molecules in the chloroplast. Isolated components of living systems—the objects of study in a reductionist approach—lack a number of significant properties that emerge at higher levels of organisation.

Emergent properties are not unique to life. A box of bicycle parts won't transport you anywhere, but if they are arranged in a certain way, you can pedal to your chosen destination. Compared with such nonliving examples, however, biological systems are far more complex, making the emergent properties of life especially challenging to study.

To fully explore emergent properties, biologists today complement reductionism with **systems biology**, the exploration of a biological system by analysing the interactions among its parts. In this context, a single leaf cell can be considered a system, as can a frog, an ant colony, or a desert ecosystem. By examining and modelling the dynamic behaviour of an integrated network of components, systems biology enables us to pose new kinds of questions. For example, how do networks of molecular interactions in our bodies generate our 24-hour cycle of wakefulness and sleep? At a larger scale,

▼ 6 Organs

The structural hierarchy of life continues to unfold as we explore the architecture of a complex organism. This lily leaf is an example of an **organ**, a body part that is made up of multiple tissues and completes specific bodily functions. Leaves, stems, and roots are the major organs of plants. Within an organ, each tissue has a distinct arrangement and contributes particular properties to organ function.

▼ 7 Tissues

Viewing the tissues of a leaf requires a microscope. Each **tissue** is a group of cells that work together, performing a specialised function. The leaf shown here has been cut on an angle. The honeycombed tissue in the interior of the leaf (left side of photo) is the main location of photosynthesis, the process that converts light energy to the chemical energy of sugar. The jigsaw puzzle–like "skin" on the surface of the leaf (right side of photo) is a tissue called epidermis. The pores through the epidermis allow entry of the gas CO_2, a raw material for sugar production.

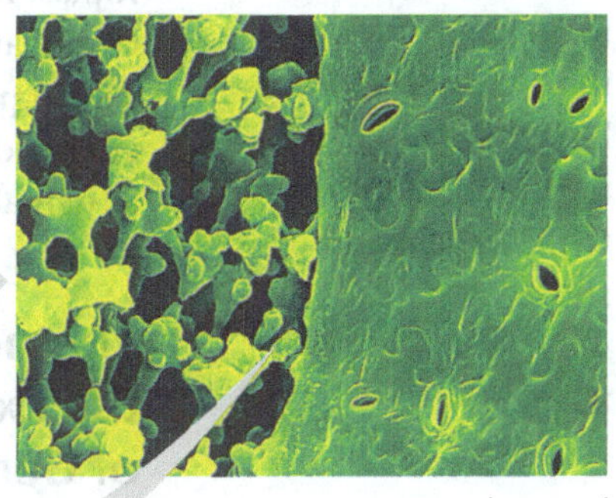

► 8 Cells

The **cell** is life's fundamental unit of structure and function. Some organisms consist of a single cell, which performs all the functions of life. Other organisms are multicellular and feature a division of labour among specialised cells. Here we see a magnified view of a cell in a leaf tissue. This cell is about 40 micrometres (μm) across—about 500 of them would reach across a small coin. Within these tiny cells are even smaller green structures called chloroplasts, which are responsible for photosynthesis.

▼ 9 Organelles

Chloroplasts are examples of **organelles**, the various functional components present in cells. The image below, taken by a powerful microscope, shows a single chloroplast.

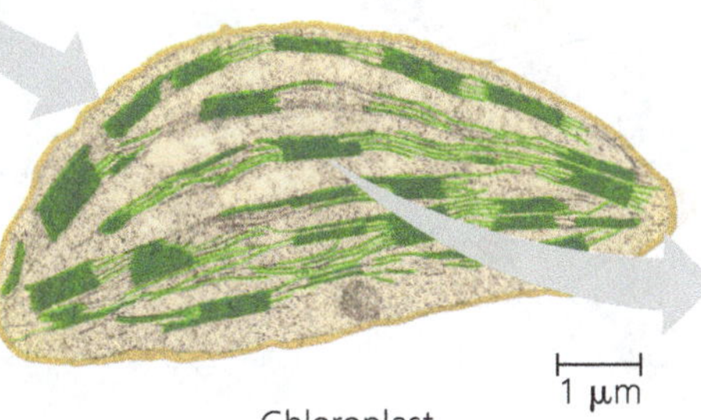

▼ 10 Molecules

Our last scale change drops us into a chloroplast for a view of life at the molecular level. A **molecule** is a chemical structure consisting of two or more units called atoms, represented as balls in this computer graphic of a chlorophyll molecule. Chlorophyll is the pigment that makes a leaf green, and it absorbs sunlight during photosynthesis. Within each chloroplast, millions of chlorophyll molecules are organised into systems that convert light energy to the chemical energy of food.

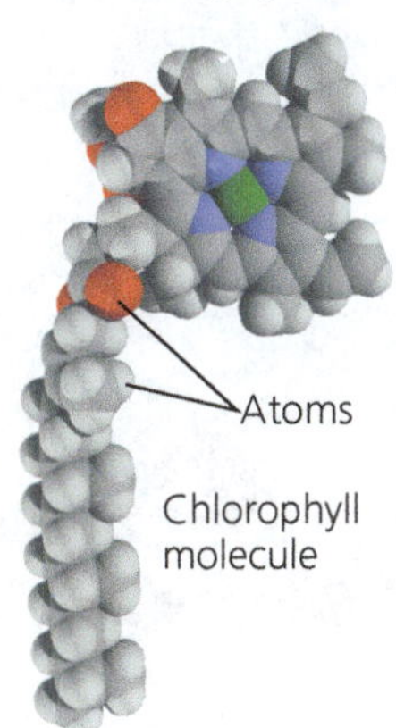

how does a gradual increase in atmospheric carbon dioxide alter ecosystems and the entire biosphere? Systems biology can be used to study life at all levels.

Structure and Function

At each level of the biological hierarchy, we find a correlation between structure and function. Consider the leaf in Figure 1.3: Its broad shape maximises the capture of sunlight by chloroplasts. Because such correlations of structure and function are common in all living things, analysing a biological structure gives us clues about what it does and how it works. For example, the hummingbird's anatomy allows its wings to rotate at the shoulder, so hummingbirds have the ability, unique among birds, to fly backward or hover in place. While hovering, the birds can extend their long, slender beaks into flowers and feed on nectar. The elegant match of form and function in the structures of life is explained by natural selection, which we'll explore shortly.

The Cell: An Organism's Basic Unit of Structure and Function

The cell is the smallest unit of organisation that can perform all activities required for life. The so-called Cell Theory was first developed in the 1800s, based on the observations of many scientists. The theory states that all living organisms are made of cells, which are the basic unit of life. In fact, the actions of organisms are all based on the activities of cells. For instance, the movement of your eyes as you read this sentence results from the activities of muscle and nerve cells. Even a process that occurs on a global scale, such as the recycling of carbon atoms, is the product of cellular functions, including the photosynthetic activity of chloroplasts in leaf cells.

All cells share certain characteristics. For instance, every cell is enclosed by a membrane that regulates the passage of materials between the cell and its surroundings. Nevertheless, we distinguish two main forms of cells: prokaryotic and eukaryotic. Prokaryotic cells are found in two groups of single-celled microorganisms, bacteria (singular, *bacterium*) and archaea (singular, *archaean*). All other forms of life, including plants and animals, are composed of eukaryotic cells.

A **eukaryotic cell** contains membrane-enclosed organelles **(Figure 1.4)**. Some organelles, such as the DNA-containing nucleus, are found in the cells of all eukaryotes; other organelles are specific to particular cell types. For example, the chloroplast in Figure 1.3 is an organelle found only in eukaryotic cells that carry out photosynthesis. In contrast to eukaryotic cells, a **prokaryotic cell** lacks a nucleus or other membrane-enclosed organelles. Furthermore, prokaryotic cells are generally smaller than eukaryotic cells, as shown in Figure 1.4.

Theme: Life's Processes Involve the Expression and Transmission of Genetic Information

INFORMATION Within cells, structures called chromosomes contain genetic material in the form of **DNA (deoxyribonucleic acid)**. In cells that are preparing to

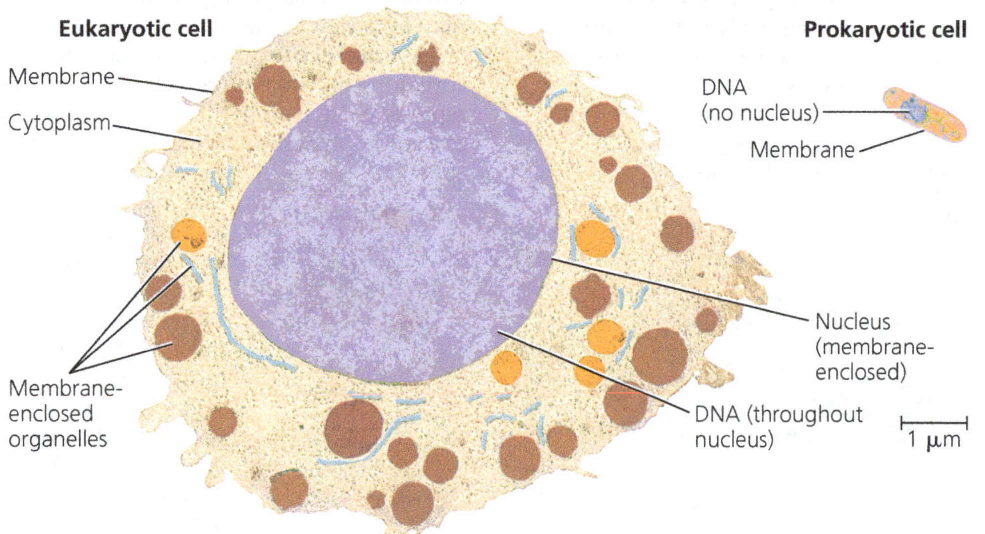

◄ **Figure 1.4 Contrasting eukaryotic and prokaryotic cells in size and complexity.** The cells are shown to scale here; to see a larger magnification of a prokaryotic cell, see Figure 6.5.

VISUAL SKILLS *Measure the scale bar, the length of the prokaryotic cell, and the diameter of the eukaryotic cell. Knowing that this scale bar represents 1 μm, calculate the length of the prokaryotic cell and the diameter of the eukaryotic cell in μm.*

▼ **Figure 1.5 A lung cell from a newt divides into two smaller cells that will grow and divide again.**

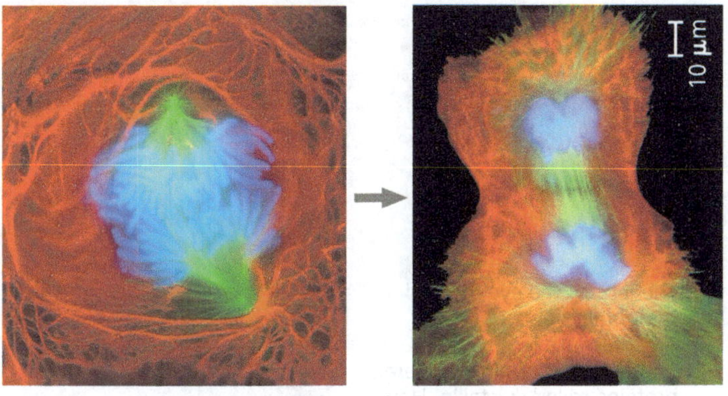

divide, the chromosomes may be made visible using a dye that appears blue when bound to the DNA **(Figure 1.5)**.

DNA, the Genetic Material

Each chromosome contains one very long DNA molecule with hundreds or thousands of **genes**, each a section of the DNA of the chromosome. Transmitted from parents to offspring, genes are the units of inheritance. They encode the information necessary to build all of the molecules synthesised within a cell, which in turn establish that cell's identity and function. You began as a single cell stocked with DNA inherited from your parents. The replication of that DNA prior to each cell division transmitted copies of the DNA to what eventually became the trillions of cells of your body. As the cells grew and divided, the genetic information encoded by the DNA directed your development **(Figure 1.6)**.

The molecular structure of DNA accounts for its ability to store information. A DNA molecule is made up of two long chains, called strands, arranged in a double helix. Each chain is made up of four kinds of chemical building blocks called nucleotides, abbreviated A, T, C, and G **(Figure 1.7)**. Specific sequences of these four nucleotides encode the information in genes. The way DNA encodes information is analogous to how we arrange the letters of the alphabet into words and phrases with specific meanings. The word *rat*, for example, evokes a rodent; the words *tar* and *art*, which contain the same letters, mean very different things. We can think of nucleotides as a four-letter alphabet.

For many genes, the sequence provides the blueprint for making a protein. For instance, a given bacterial gene may specify a particular protein (such as an enzyme) required to break down a certain sugar molecule, while one particular human gene may denote an enzyme, and another gene a different protein (an antibody, perhaps) that helps fight off infection. Overall, proteins are major players in building and maintaining the cell and carrying out its activities.

▼ **Figure 1.6 Inherited DNA directs the development of an organism.**

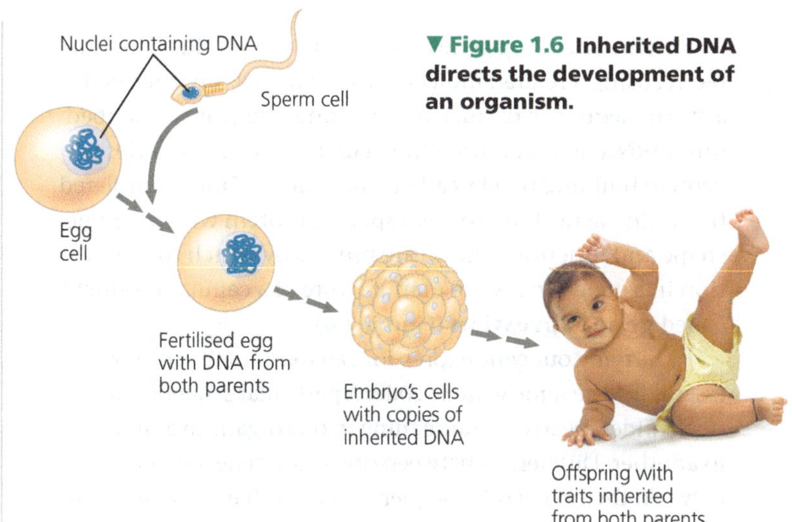

▼ **Figure 1.7 DNA: the genetic material.**

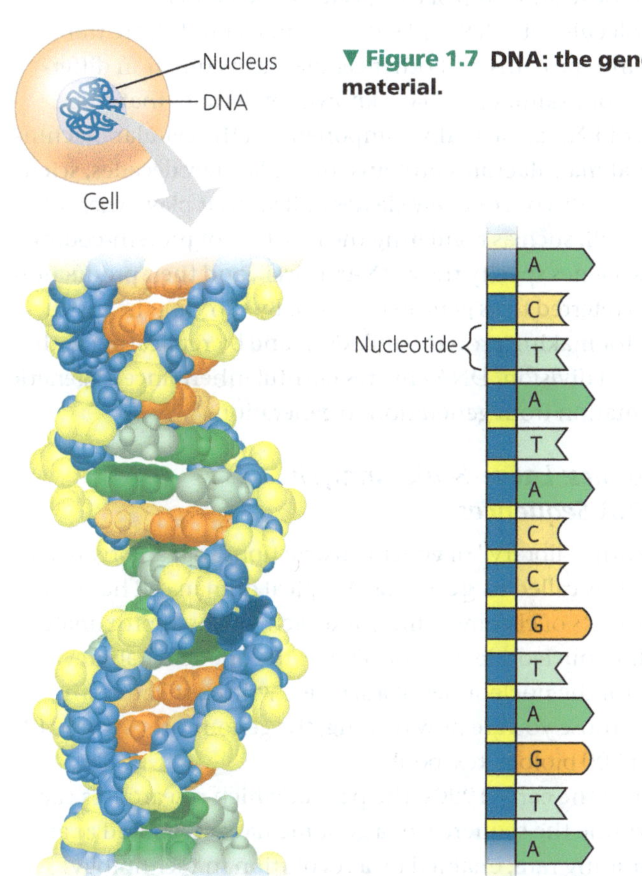

(a) DNA double helix. This model shows the atoms in a segment of DNA. Made up of two long chains (strands) of building blocks called nucleotides, a DNA molecule takes the three-dimensional form of a double helix.

(b) Single strand of DNA. These geometric shapes and letters are simple symbols for the nucleotides in a small section of one strand of a DNA molecule. Genetic information is encoded in specific sequences of the four types of nucleotides. Their names are abbreviated A, T, C, and G.

Protein-encoding genes control protein production indirectly, using a related molecule called RNA as an intermediary. The sequence of nucleotides along a gene is transcribed into mRNA, which is then translated into a linked series of protein building blocks called amino acids. Once completed, the amino acid chain forms a specific protein with a unique shape and function. The entire process by which the information in a gene directs the manufacture of a cellular product is called **gene expression** (**Figure 1.8**).

In carrying out gene expression, all forms of life employ essentially the same genetic code: A particular sequence of nucleotides means the same thing in one organism as it does in another. Differences between organisms reflect differences between their nucleotide sequences rather than between their genetic codes. This universality of the genetic code is a strong piece of evidence that all life is related. Comparing the sequences in several species for a gene that codes for a particular protein can provide valuable information both about the protein and about the relationship of the species to each other.

Molecules of mRNA, like the one in Figure 1.8, are translated into proteins, but other cellular RNAs function differently. For example, we have known for decades that some types of RNA are actually components of the cellular machinery that manufactures proteins. In the last few decades, scientists have discovered new classes of RNA that play other roles in the cell, such as regulating the function of protein-coding genes. Genes specify these RNAs as well, and their production is also referred to as gene expression. By carrying the instructions for making proteins and RNAs and by replicating with each cell division, DNA ensures faithful inheritance of genetic information from generation to generation.

Genomics: Large-Scale Analysis of DNA Sequences

The entire "library" of genetic instructions that an organism inherits is called its **genome**. A typical human cell has two similar sets of chromosomes, and each set has approximately 3 billion nucleotide pairs of DNA. If the one-letter abbreviations for the nucleotides of a set were written in letters the size of those you are now reading, the genomic text would fill about 700 biology textbooks.

Since the early 1990s, the pace at which researchers can determine the sequence of a genome has accelerated at an astounding rate, enabled by a revolution in technology. The genome sequence—the entire sequence of nucleotides for a representative member of a species—is now known for humans and many other animals, as well as numerous plants, fungi, bacteria, and archaea. To make sense of the deluge of data from genome-sequencing projects and the growing catalog of known gene functions, scientists are applying a systems biology approach at the cellular and molecular levels. Rather than investigating a single gene at

▼ Figure 1.8 Gene expression: Cells use information encoded in a gene to synthesise a functional protein.

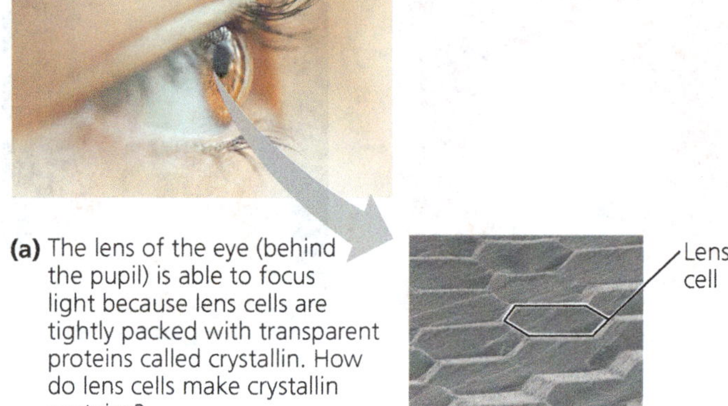

(a) The lens of the eye (behind the pupil) is able to focus light because lens cells are tightly packed with transparent proteins called crystallin. How do lens cells make crystallin proteins?

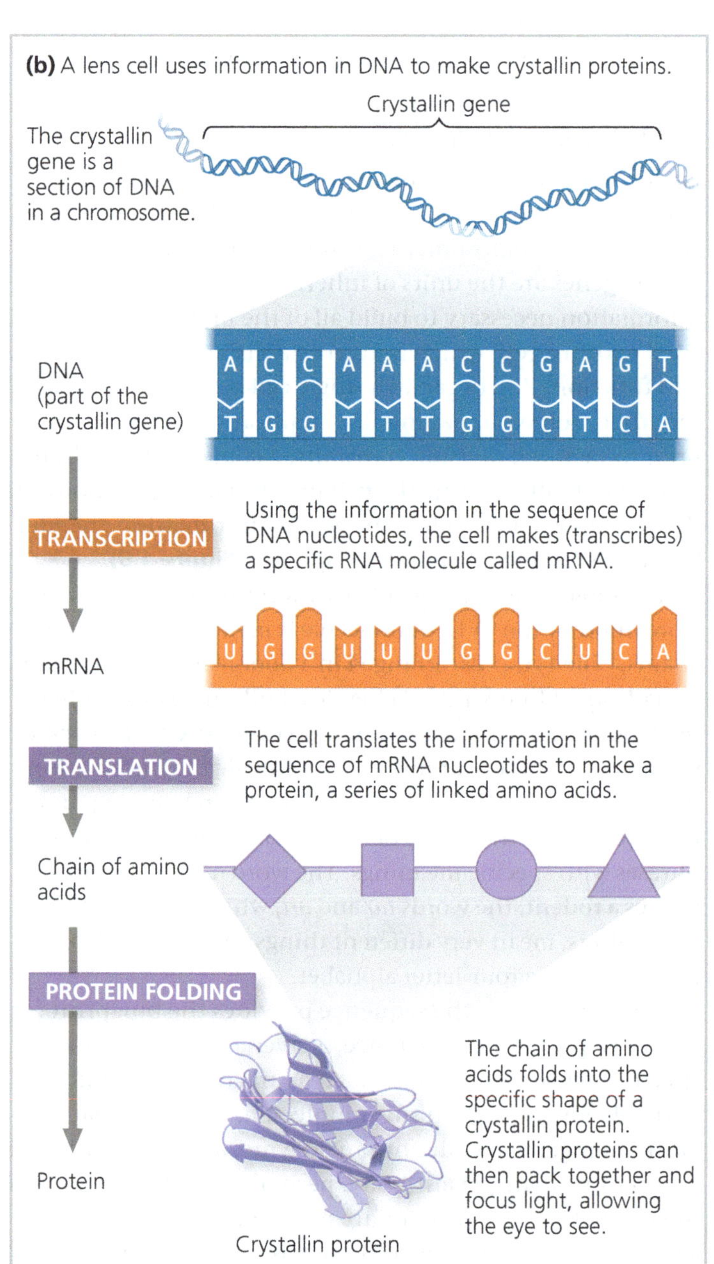

a time, researchers study whole sets of genes (or other DNA) in one or more species—an approach called **genomics**. Likewise, the term **proteomics** refers to the study of sets of proteins and their properties. (The entire set of proteins expressed by a given cell, tissue, or organism is called a **proteome**.)

Three important research developments have made the genomic and proteomic approaches possible. One is "high-throughput" technology, tools that can analyse many biological samples very rapidly. The second major development is **bioinformatics**, the use of computational tools to store, organise, and analyse the huge volume of data that results from high-throughput methods. The third development is the formation of interdisciplinary research teams—groups of diverse specialists that may include computer scientists, mathematicians, engineers, chemists, physicists, and, of course, biologists from a variety of fields. Researchers in such teams aim to learn how the activities of all the proteins and RNAs encoded by the DNA are coordinated in cells and in whole organisms.

Theme: Life Requires the Transfer and Transformation of Energy and Matter

ENERGY AND MATTER Moving, growing, reproducing, and the various cellular activities of life are work, and work requires energy. The input of energy, primarily from the sun, and the transformation of energy from one form to another make life possible **(Figure 1.9)**. When a plant's leaves absorb sunlight in the process of photosynthesis, molecules within the leaves convert the energy of sunlight to the chemical energy of food, such as sugars. The chemical energy in the food molecules is then passed along from plants and other photosynthetic organisms (**producers**) to consumers. A **consumer** is an organism that feeds on other organisms or their remains.

When an organism uses chemical energy to perform work, such as muscle contraction or cell division, some of that energy is lost to the surroundings as heat. As a result, energy *flows through* an ecosystem in one direction, usually entering as light and exiting as heat. In contrast, chemicals *cycle within* an ecosystem, where they are used and then recycled (see Figure 1.9). Chemicals that a plant absorbs from the air or soil may be incorporated into the plant's body and then passed to an animal that eats the plant. Eventually, these chemicals will be returned to the environment by decomposers such as bacteria and fungi that break down waste products, leaf litter, and the bodies of dead organisms. The chemicals are then available to be taken up by plants again, thereby completing the cycle.

Theme: From Molecules to Ecosystems, Interactions Are Important in Biological Systems

INTERACTIONS At any level of the biological hierarchy, interactions between the components of the system ensure smooth integration of all the parts, such that they function as a whole. This holds true equally well for molecules in a cell and the components of an ecosystem; we'll look at both as examples.

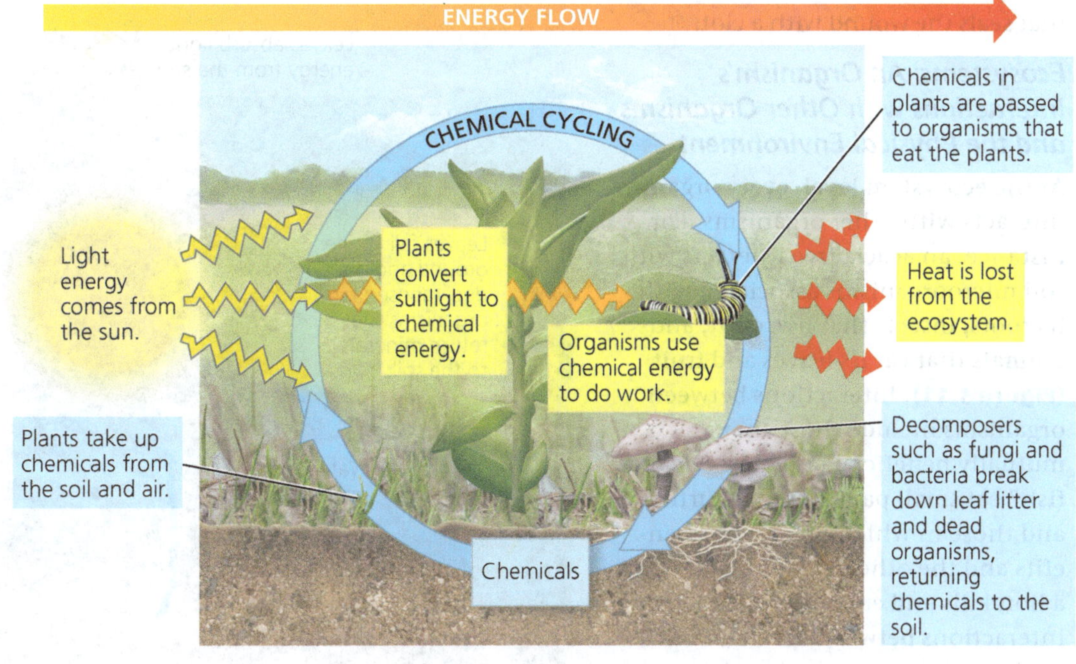

▶ **Figure 1.9 Energy flow and chemical cycling.** There is a one-way flow of energy in an ecosystem: During photosynthesis, plants convert energy from sunlight to chemical energy (stored in food molecules such as sugars), which is used by plants and other organisms to do work and is eventually lost from the ecosystem as heat. In contrast, chemicals cycle between organisms and the physical environment.

Molecules: Interactions Within Organisms

At lower levels of organisation, the interactions between components that make up living organisms—organs, tissues, cells, and molecules—are crucial to their smooth operation. Consider the regulation of blood sugar level, for instance. Cells in the body must match the supply of fuel (sugar) to demand, regulating the opposing processes of sugar breakdown and storage. The key is the ability of many biological processes to self-regulate by a mechanism called feedback.

In **feedback regulation**, the output or product of a process regulates that very process. The most common form of regulation in living systems is *negative feedback*, a loop in which the response reduces the initial stimulus. As seen in the example of insulin signalling **(Figure 1.10)**, after a meal the level of the sugar glucose in your blood rises, which stimulates cells of the pancreas to secrete insulin. Insulin, in turn, causes body cells to take up glucose and liver cells to store it, thus decreasing the blood glucose level. This eliminates the stimulus for insulin secretion, shutting off the pathway. Thus, the output of the process (insulin) negatively regulates that process.

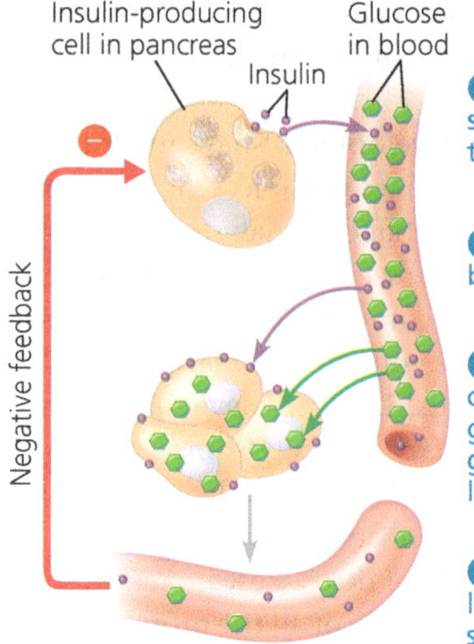

▼ **Figure 1.10 Feedback regulation.** The human body regulates the use and storage of glucose, a major cellular fuel. This figure shows negative feedback: The response to insulin reduces the initial stimulus.

VISUAL SKILLS *In this example, what is the response to insulin? What is the initial stimulus that is reduced by the response?*

Though less common than processes regulated by negative feedback, there are also many biological processes regulated by *positive feedback*, in which an end product *speeds up* its own production. The clotting of your blood in response to injury is an example. When a blood vessel is damaged, structures in the blood called platelets begin to aggregate at the site. Positive feedback occurs as chemicals released by the platelets attract *more* platelets. The platelet pileup then initiates a complex process that seals the wound with a clot.

Ecosystems: An Organism's Interactions with Other Organisms and the Physical Environment

At the ecosystem level, every organism interacts with other organisms. For instance, an acacia tree interacts with soil microorganisms associated with its roots, insects that live on it, and animals that eat its leaves and fruit **(Figure 1.11)**. Interactions between organisms include those that are mutually beneficial (as when "cleaner fish" eat small parasites on a turtle) and those in which one species benefits and the other is harmed (as when a lion kills and eats a zebra). In some interactions between species, both are harmed—for example, when two plants compete for a soil resource that is in short supply. Interactions among organisms help regulate the functioning of the ecosystem as a whole.

Each organism also interacts continuously with physical factors in its environment. The leaves of a tree, for example, absorb light from the sun, take in carbon dioxide from the air,

▼ **Figure 1.11 Interactions of an African acacia tree with other organisms and the physical environment.**

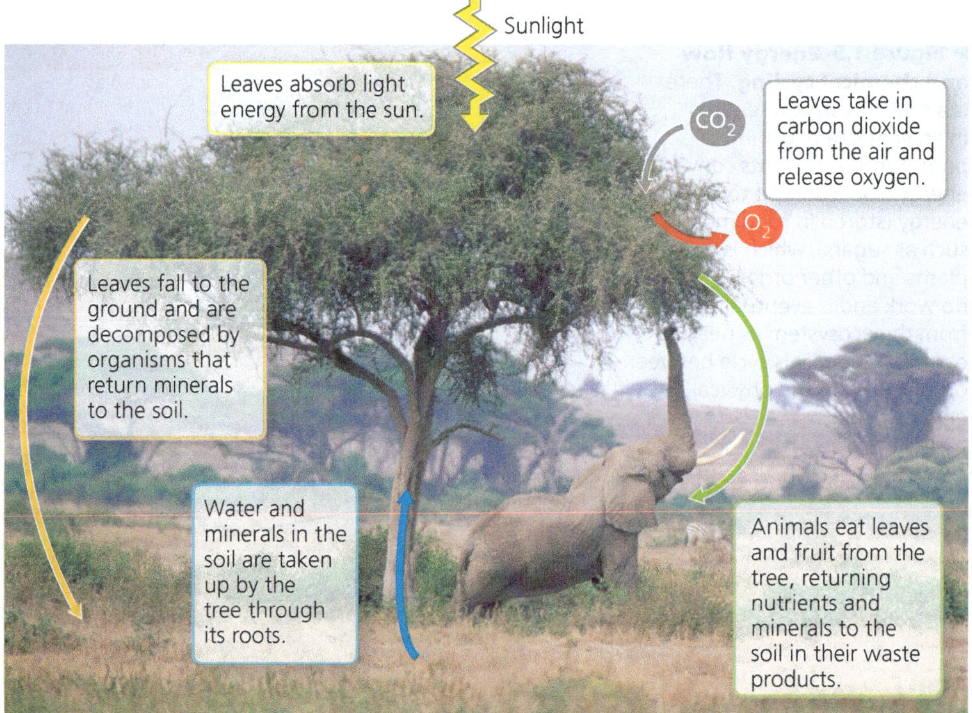

and release oxygen to the air (see Figure 1.11). The environment is also affected by organisms. For instance, in addition to taking up water and minerals from the soil, the roots of a plant break up rocks as they grow, contributing to the formation of soil. On a global scale, plants and other photosynthetic organisms have generated all the oxygen in the atmosphere.

Like other organisms, we humans interact with our environment. Our interactions sometimes have dire consequences: For example, over the past 150 years, humans have greatly increased the burning of fossil fuels (coal, oil, and gas). This practice releases large amounts of carbon dioxide (CO_2) and other gases into the atmosphere, causing heat to be trapped close to Earth's surface (see Figure 56.29). Scientists calculate that the CO_2 added to the atmosphere by human activities has increased the average temperature of the planet by about 1°C since 1900. At the current rates that CO_2 and other gases are being added to the atmosphere, global models predict an additional rise of between 3°C and 7°C before the end of this century.

This ongoing global warming is a major aspect of **climate change**, a directional change to the global climate that lasts for three decades or more (as opposed to short-term changes in the weather). But global warming is not the only way the climate is changing: Wind and precipitation patterns are also shifting, and extreme weather events such as storms and droughts are occurring more often. Climate change has already affected organisms and their habitats all over the planet. For example, polar bears have lost much of the ice platform from which they hunt, leading to food shortages and increased mortality rates. As habitats deteriorate, hundreds of plant and animal species are shifting their ranges to more suitable locations—but for some, there is insufficient suitable habitat, or they may not be able to migrate quickly enough. As a result, the populations of many species are shrinking in size or even disappearing **(Figure 1.12)**. (For more examples of how climate change is affecting life on Earth, see Make Connections Figure 56.30.)

The loss of populations due to climate change can ultimately result in extinction, the permanent loss of a species. As we'll explore in greater detail in Concept 56.4, the consequences of these changes for humans and other organisms may be profound. While the climate crisis is dire, the loss of biological diversity evolved over billions of years may represent an even greater challenge to life on Earth.

▶ **Figure 1.12 Threatened by global warming.** A warmer environment causes lizards in the genus *Sceloporus* to spend more time in refuges from the heat, reducing time for foraging. Their food intake drops, decreasing reproductive success. Surveys of 200 *Sceloporus* populations in Mexico show that 12% of these populations have disappeared since 1975.

Having considered four of the unifying themes (organisation, information, energy and matter, and interactions), let's now turn to evolution. There is consensus among biologists that evolution is the core theme of biology, and it is discussed in detail in the next section.

CONCEPT CHECK 1.1

1. Starting with the molecular level in Figure 1.3, write a sentence that includes components from the previous (lower) level of biological organisation, for example: "A molecule consists of *atoms* bonded together." Continue with organelles, moving up the biological hierarchy.
2. Identify the theme or themes exemplified by (a) the sharp quills of an echidna, (b) the development of a multicellular organism from a single fertilised egg, and (c) a hummingbird using sugar to power its flight.
3. **WHAT IF?** For each theme discussed in this section, give an example not mentioned in the text.

For suggested answers, see Appendix A.

CONCEPT 1.2

The Core Theme: Evolution accounts for the unity and diversity of life

EVOLUTION An understanding of evolution helps us to make sense of everything we know about life on Earth. As the fossil record clearly shows, life has been evolving for billions of years, resulting in a vast diversity of past and present organisms. But along with the diversity there is also unity, in the form of shared features. For example, while sea horses, bilbies, hummingbirds, and giraffes all look very different, their skeletons are organised in the same basic way.

The scientific explanation for the unity and diversity of organisms is **evolution**: a process of biological change in which species accumulate differences from their ancestors as they adapt to different environments over time. Thus, we can account for differences between two species (diversity) with the idea that certain heritable changes occurred after the two species diverged from their common ancestor. However, they also share certain traits (unity) simply because they have descended from a common ancestor. An abundance of evidence of different types supports the occurrence of evolution and the mechanisms that describe how it takes place, which we'll explore in detail in Chapters 22–25. To quote one of the founders of modern evolutionary theory, Theodosius Dobzhansky, "Nothing in biology makes sense except in the light of evolution." To understand this statement, we need to examine how biologists think about the vast diversity of life on the planet.

Classifying the Diversity of Life

Diversity is a hallmark of life. Biologists have so far identified and named about 1.8 million species of organisms. Each species is given a two-part name: The first part is the name of the genus (plural, *genera*) to which the species belongs, and the second part is unique to the species within the genus. (For example, *Homo sapiens* is the name of our species.)

To date, known species include at least 100,000 species of fungi, 290,000 plant species, 57,000 vertebrate species (animals with backbones), and 1 million insect species (more than half of all known forms of life)—not to mention the myriad types of single-celled organisms. Researchers identify thousands of additional species each year. Estimates of the total number of species range from about 10 million to over 100 million. Whatever the actual number, the enormous variety of life gives biology a very broad scope. Biologists face a major challenge in attempting to make sense of this variety.

The Three Domains of Life

Humans tend to group diverse items according to their similarities and relationships to each other. Consequently, biologists have long used careful comparisons of structure, function, and other obvious features to classify forms of life into groups. In the last few decades, new methods of assessing species relationships, such as comparisons of DNA sequences, have led to a reevaluation of the classification of life. Although this reevaluation is ongoing, biologists currently place all organisms into three groups called domains: Bacteria, Archaea, and Eukarya **(Figure 1.13)**.

Two of the three domains—**Bacteria** and **Archaea**—consist of single-celled prokaryotic organisms. All the eukaryotes (organisms with eukaryotic cells) are in domain **Eukarya**. This domain includes four subgroups: kingdom Plantae, kingdom Fungi, kingdom Animalia, and the protists. The three kingdoms are distinguished partly by their modes of nutrition: Plants produce their own sugars and other food

▼ **Figure 1.13 The three domains of life.**

(a) Domain Bacteria

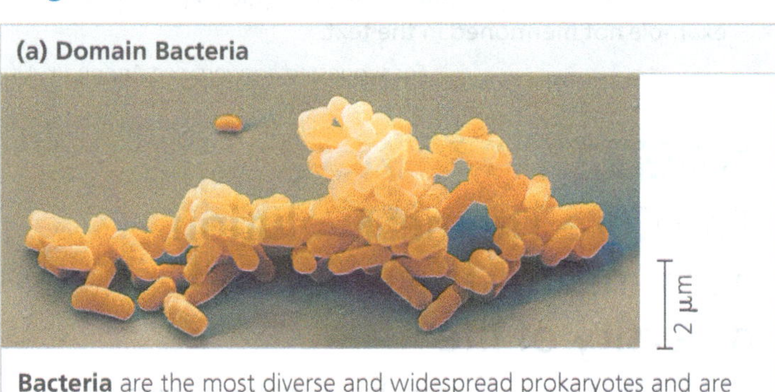

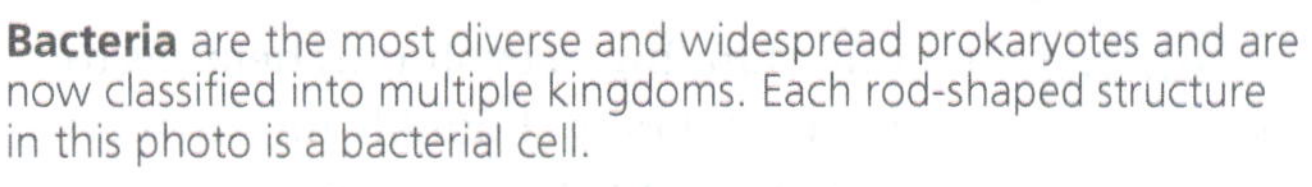
Bacteria are the most diverse and widespread prokaryotes and are now classified into multiple kingdoms. Each rod-shaped structure in this photo is a bacterial cell.

(b) Domain Archaea

Domain Archaea includes multiple kingdoms. Some of the prokaryotes known as **archaea** live in Earth's extreme environments, such as salty lakes and boiling hot springs. Each round structure in this photo is an archaeal cell.

(c) Domain Eukarya

▲ **Kingdom Plantae** (plants) consists of multicellular eukaryotes that carry out photosynthesis, the conversion of light energy to the chemical energy in food. Most plant species live on land.

▶ **Kingdom Fungi** is characterised in part by the nutritional mode of its members (such as this mushroom), which absorb nutrients from outside their bodies.

◀ **Kingdom Animalia** consists of multicellular eukaryotes that ingest other organisms.

▶ **Protists** are mostly unicellular eukaryotes and some relatively simple multicellular relatives. Pictured here is an assortment of protists inhabiting pond water. Scientists are currently debating how to classify protists in a way that accurately reflects their evolutionary relationships.

molecules by photosynthesis, fungi absorb nutrients in dissolved form from their surroundings, and animals obtain food by eating and digesting other organisms. Animalia is, of course, the kingdom to which we belong.

The most numerous and diverse eukaryotes are the protists, which are mostly single-celled organisms. Although protists were once placed in a single kingdom, they are now classified into several groups. One major reason for this change is the recent DNA evidence showing that some protists are less closely related to other protists than they are to plants, animals, or fungi.

Unity in the Diversity of Life

As diverse as life is, there is also remarkable unity among forms of life. Consider, for example, the similar skeletons of different animals and the universal genetic language of DNA (the genetic code), both mentioned earlier. In fact, similarities between organisms are evident at all levels of the biological hierarchy. For example, unity is obvious in many features of cell structure, even among distantly related organisms **(Figure 1.14)**.

How can we account for life's dual nature of unity and diversity? The process of evolution, explained next, illuminates both the similarities and differences in the world of life. It also introduces another important dimension of biology: the passage of time. The history of life, as documented by fossils and other evidence, is the saga of an ever-changing Earth billions of years old, inhabited by an evolving cast of living forms **(Figure 1.15)**.

▼ **Figure 1.15 Studying the history of life.** Researchers in South Africa reconstruct skeletons of *Homo naledi*, an extinct relative of *Homo sapiens*. The fossils were discovered in an underground cave that may have been a burial chamber.

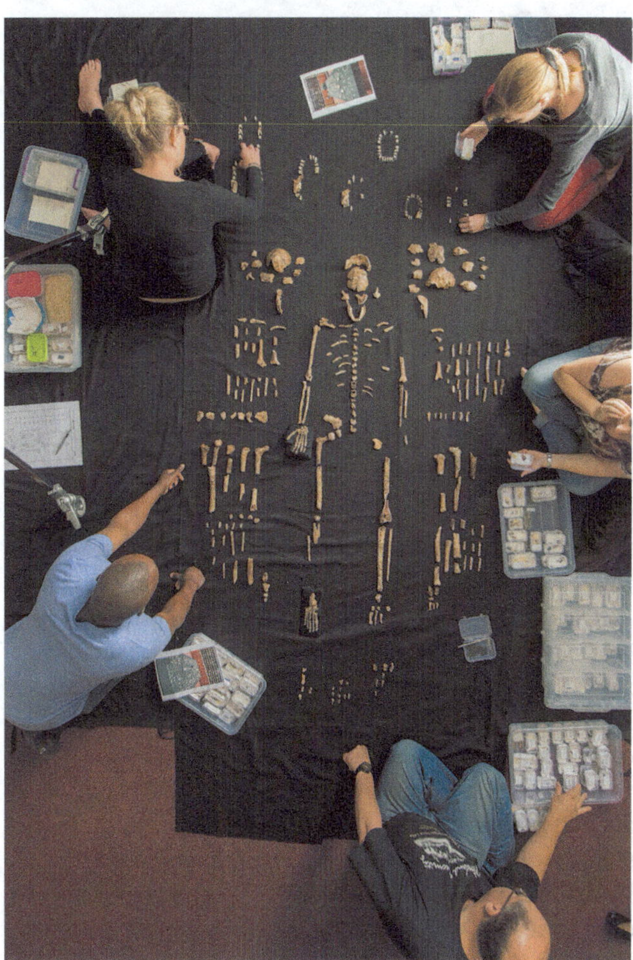

▼ **Figure 1.14 An example of unity underlying the diversity of life: the architecture of cilia in eukaryotes.** Cilia (singular, *cilium*) are extensions of cells that function in locomotion. They occur in eukaryotes as diverse as *Paramecium* (found in pond water) and humans. Even organisms so different share a common architecture for their cilia, which have an elaborate system of tubules that is striking in cross-sectional views.

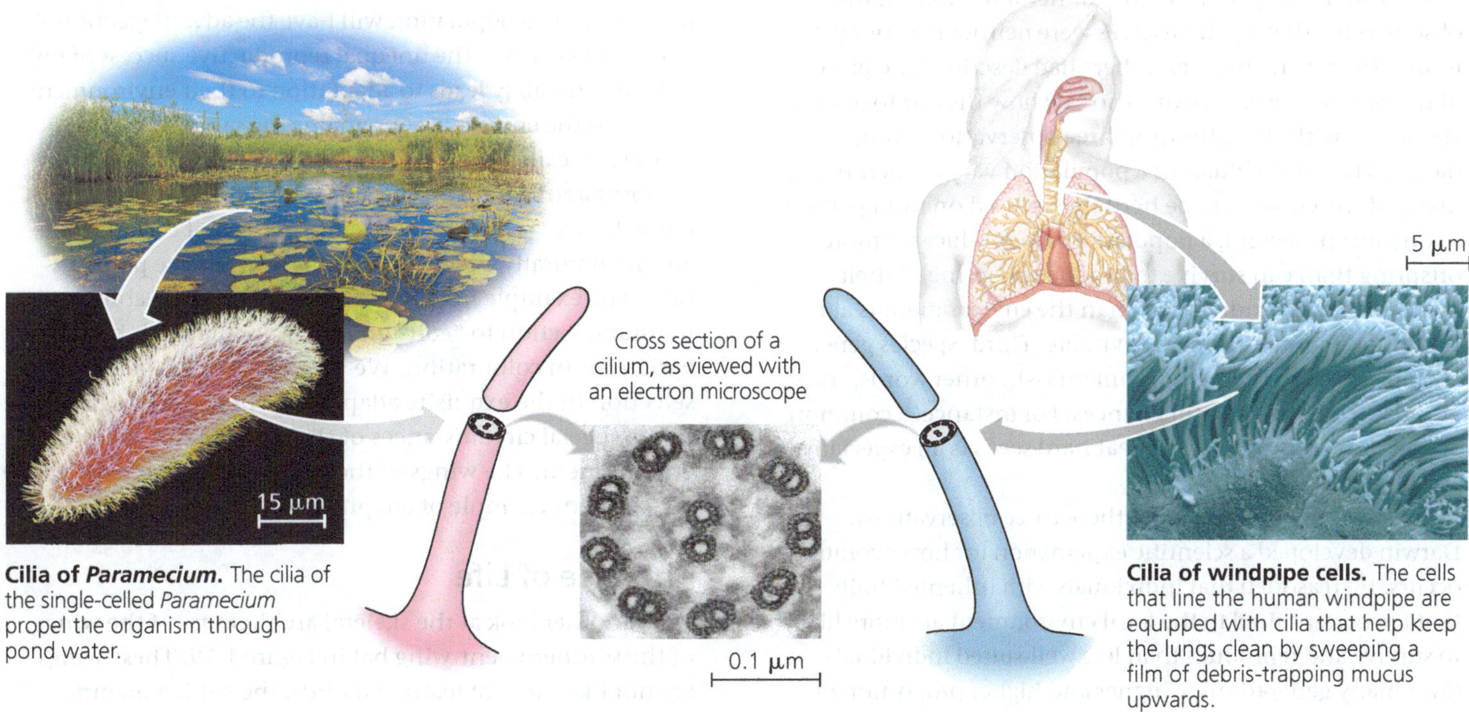

Cilia of *Paramecium*. The cilia of the single-celled *Paramecium* propel the organism through pond water.

Cilia of windpipe cells. The cells that line the human windpipe are equipped with cilia that help keep the lungs clean by sweeping a film of debris-trapping mucus upwards.

▼ Figure 1.16 Charles Darwin. The portrait shows Darwin in about 1840, well before the 1859 publication of his revolutionary book, commonly referred to as *The Origin of Species*.

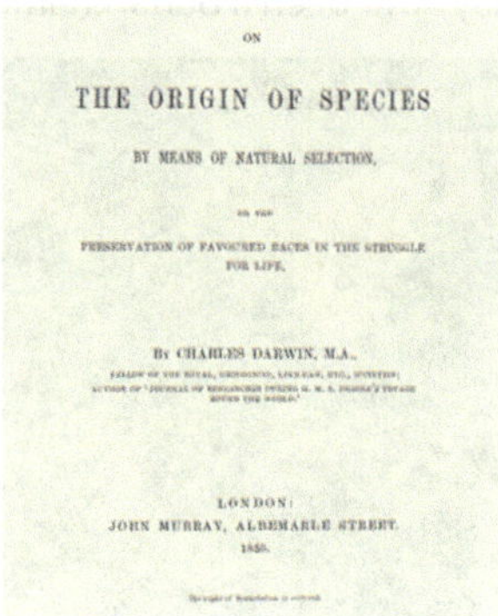

ON

THE ORIGIN OF SPECIES

BY MEANS OF NATURAL SELECTION,

PRESERVATION OF FAVOURED RACES IN THE STRUGGLE FOR LIFE.

BY CHARLES DARWIN, M.A.,

LONDON:

JOHN MURRAY, ALBEMARLE STREET.

Charles Darwin and the Theory of Natural Selection

An evolutionary view of life came into sharp focus in November 1859, when Charles Darwin published one of the most important and influential books ever written, *On the Origin of Species by Means of Natural Selection* **(Figure 1.16)**. *The Origin of Species* articulated two main points. The first point was that, as species adapt to different environments over time, they accumulate differences from their ancestors. Darwin called this process "descent with modification." This insightful phrase captured the duality of life's unity and diversity—unity in the kinship among species that descended from common ancestors and diversity in the modifications that evolved as species branched from their common ancestors **(Figure 1.17)**. Darwin's second main point was his proposal that "natural selection" is a primary cause of descent with modification.

Darwin developed his theory of natural selection from observations that by themselves were neither new nor profound. However, although others had described the pieces of the puzzle, it was Darwin who saw how they fit together. He started with the following three observations from nature: First, individuals in a population vary in their traits, many of which seem to be heritable (passed on from parents to offspring). Second, a population can produce far more offspring than can survive to produce offspring of their own. With more individuals than the environment is able to support, competition is inevitable. Third, species generally are suited to their environments—in other words, they are adapted to their circumstances. For instance, a common adaptation among birds that eat hard seeds is an especially strong beak.

By making inferences from these three observations, Darwin developed a scientific explanation for how evolution occurs. He reasoned that individuals with inherited traits that are better suited to the local environment are more likely to survive and reproduce than less well-suited individuals. Over many generations, a higher and higher proportion of individuals in a population will have the advantageous traits. Evolution occurs as the unequal reproductive success of individuals ultimately leads to adaptation to their environment, as long as the environment remains the same.

▼ Figure 1.17 Unity and diversity among birds. These four birds are variations on a common body plan. For example, each has feathers, a beak, and wings. However, these common features are highly specialised for the birds' diverse lifestyles.

▼ Red-tailed hawk (*Buteo borealis*)

▼ Cassowary (*Casuarius casuarius johnsonii*)

▲ Chatham Islands black robin (*Petroica traversi*)

▲ Gentoo penguin (*Pygoscelis papua*)

Darwin called this mechanism of evolutionary adaptation **natural selection** because the natural environment consistently "selects" for the propagation of certain traits among naturally occurring variant traits in the population. The example in **Figure 1.18** illustrates the ability of natural selection to "edit" an insect population's heritable variations in colouration. We see the products of natural selection in the exquisite adaptations of various organisms to the special circumstances of their way of life and their environment. The wings of the bat shown in **Figure 1.19** are an excellent example of adaptation.

The Tree of Life

Take another look at the skeletal architecture of the wings of the southern bent-wing bat in Figure 1.19. These wings are not like those of feathered birds; the bat is a mammal.

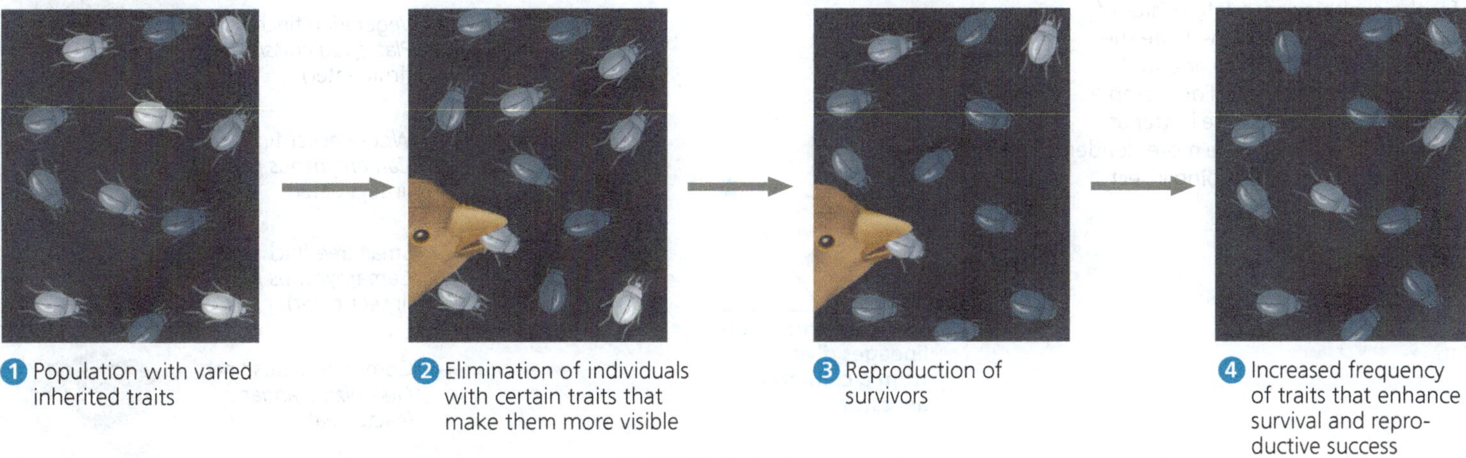

▼ **Figure 1.18 Natural selection.** This imaginary beetle population has colonised a locale where the soil has been blackened by a recent brush fire. Initially, the population varies extensively in the inherited colouration of the individuals, from very light grey to charcoal. For hungry birds that prey on the beetles, it is easiest to spot the beetles that are lightest in colour.

DRAW IT *Over time, the soil will gradually become lighter in colour. Draw another step to show how the soil, when lightened to medium colour, would affect natural selection. Write a caption for this new step 5. Then explain how the population would change over time as the soil becomes lighter.*

▲ **Figure 1.19 Evolutionary adaptation.** Bats, the only mammals capable of active flight, have wings with webbing between extended "fingers." Darwin proposed that such adaptations are refined over time by natural selection.

The bones, joints, nerves, and blood vessels in the bat's forelimbs, though adapted for flight, are very similar to those in the human arm, the foreleg of a horse, and the flipper of a whale. Indeed, all mammalian forelimbs are anatomical variations of a common architecture. According to the Darwinian concept of descent with modification, the shared anatomy of mammalian limbs reflects inheritance of the limb structure from a common ancestor—the "prototype" mammal from which all other mammals descended. The diversity of mammalian forelimbs results from modification by natural selection operating over millions of years in different environmental contexts. Fossils and other evidence corroborate anatomical unity in supporting this view of mammalian descent from a common ancestor.

Darwin proposed that natural selection, by its cumulative effects over long periods of time, could cause an ancestral species to give rise to two or more descendant species. This could occur, for example, if one population of organisms became fragmented into several subpopulations isolated in different environments. In these separate arenas of natural selection, one species could gradually radiate into multiple species as the geographically isolated populations adapted over many generations to different environmental conditions.

The Galápagos finches are a famous example of the process of radiation of new species from a common ancestor. Darwin collected specimens of these birds during his 1835 visit to the remote Galápagos Islands, 900 kilometres (km) off the Pacific coast of South America. These relatively young volcanic islands are home to many species of plants and animals found nowhere else in the world, though many Galápagos organisms are clearly related to species on the South American mainland. The Galápagos finches are thought to have descended from an ancestral finch species that reached the archipelago from South America or the Caribbean. Over time, the Galápagos finches diversified from their ancestor as populations became adapted to different food sources on their particular islands. Years after Darwin collected the finches, researchers began to sort out their evolutionary relationships, first from anatomical and geographic data and more recently with the help of DNA sequence comparisons.

Biologists' diagrams of evolutionary relationships generally take treelike forms, though the trees are often turned sideways

▶ **Figure 1.20 Descent with modification: adaptive radiation of finches on the Galápagos Islands.** This "tree" illustrates a current hypothesis for the evolutionary relationships of finches on the Galápagos. Note the various beaks, which are adapted to particular food sources. For example, heavier, thicker beaks are better at cracking seeds, while the more slender beaks are better at grasping insects.

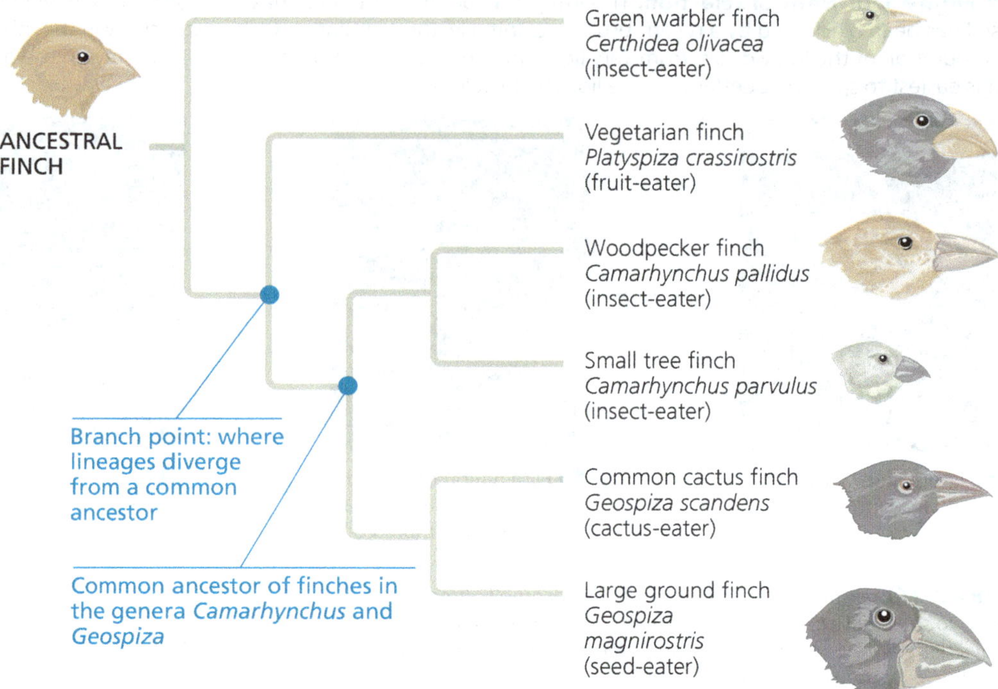

as in **Figure 1.20**. Tree diagrams make sense: Just as an individual has a genealogy that can be diagrammed as a family tree, each species is one twig of a branching tree of life extending back in time through ancestral species more and more remote. Species that are very similar, such as the Galápagos finches, share a relatively recent common ancestor. Through an ancestor that lived much further back in time, finches are related to sparrows, hawks, penguins, and all other birds. Furthermore, finches and other birds are related to us through a common ancestor even more ancient. Trace life back far enough, and we reach the early prokaryotes that inhabited Earth over 3.5 billion years ago. We can recognise their vestiges in our own cells—in the universal genetic code, for example. Indeed, all of life is connected through its long evolutionary history.

CONCEPT CHECK 1.2

1. Explain why "editing" is a metaphor for how natural selection acts on a population's heritable variation.
2. Referring to Figure 1.20, provide a possible explanation for how, over a very long time, the green warbler finch came to have a slender beak.
3. **DRAW IT** The three domains you learned about in Concept 1.2 can be represented in the tree of life as the three main branches, with three subbranches on the eukaryotic branch being the kingdoms Plantae, Fungi, and Animalia. What if fungi and animals are more closely related to each other than either of these kingdoms is to plants—as recent evidence strongly suggests? Draw a simple branching pattern that symbolises the proposed relationship between these three eukaryotic kingdoms.

For suggested answers, see Appendix A.

CONCEPT 1.3

In studying nature, scientists form and test hypotheses

Science is a way of knowing—an approach to understanding the natural world. It developed out of our curiosity about ourselves, other life-forms, our planet, and the universe. The word *science* is derived from a Latin verb meaning "to know." Striving to understand seems to be one of our basic urges.

At the heart of science is **inquiry**, the search for information and explanations of natural phenomena. There is no formula for successful scientific inquiry, no single scientific method that researchers must rigidly follow. As in all quests, science includes elements of challenge, adventure, and luck, along with careful planning, reasoning, creativity, patience, and the persistence to overcome setbacks. Such diverse elements of inquiry make science far less structured than most people realise. That said, it is possible to highlight certain characteristics that help to distinguish science from other ways of describing and explaining nature.

Scientists use a process of inquiry that includes making observations, forming logical, testable explanations (*hypotheses*), and testing them. The process is necessarily repetitive: In testing a hypothesis, more observations may inspire revision of the original hypothesis or formation of a new one, thus leading to further testing. In this way, scientists circle closer and closer to their best estimation of the laws governing nature.

Exploration and Observation

Biology, like other sciences, begins with careful observations. In gathering information, biologists often use tools such as microscopes, precision thermometers, or high-speed cameras that extend their senses or facilitate careful measurement. Observations can reveal valuable information about the natural world. For example, a series of detailed observations have shaped our understanding of cell structure, and another set of observations is currently expanding our databases of genome sequences from diverse species and databases of genes whose expression is altered in various diseases.

In exploring nature, biologists also rely heavily on the scientific literature, the published contributions of fellow scientists. By reading about and understanding past studies, scientists can build on the foundation of existing knowledge, focusing their investigations on observations that are original and on hypotheses that are consistent with previous findings. Identifying publications relevant to a new line of research is now easier than at any point in the past, thanks to indexed and searchable electronic databases.

Gathering and Analysing Data

Recorded observations are called **data**. Put another way, data are items of information on which scientific inquiry is based. The term *data* implies numbers to many people. But some data are *qualitative*, often in the form of recorded descriptions rather than numerical measurements. For example, Jane Goodall spent decades recording her observations of chimpanzee behaviour during field research in a Tanzanian jungle **(Figure 1.21)**. In her studies, Goodall also enriched the field of animal behaviour with volumes of *quantitative* data, such as the frequency and duration of specific behaviours for different members of a group of chimpanzees in a variety of situations. Quantitative data are generally expressed as numerical measurements and often organised into tables and graphs. Scientists analyse their data using a type of mathematics called *statistics* to test whether their results are significant or merely due to random fluctuations. All results presented in this text have been shown to be statistically significant.

Collecting and analysing observations can lead to important conclusions based on a type of logic called **inductive reasoning**. Through induction, we derive generalisations from a large number of specific observations. "The sun always rises in the east" is one example. Another biological example is the generalisation "All organisms are made of cells," which was based on two centuries of microscopic observations made by biologists examining cells in diverse biological specimens. Careful observations and data analyses, along with

▼ **Figure 1.21 Jane Goodall collecting qualitative data on chimpanzee behaviour.** Goodall recorded her observations in field notebooks, often with sketches of the animals' behaviour.

generaliations reached by induction, are fundamental to our understanding of nature.

Forming and Testing Hypotheses

Our innate curiosity often stimulates us to pose questions about the natural basis for the phenomena we observe in the world. What caused the different chimpanzee behaviours observed in the wild? Answering such questions usually involves forming and testing logical explanations—that is, hypotheses.

In science, a **hypothesis** is an explanation, based on observations and assumptions, that leads to a testable prediction. Said another way, a hypothesis is an explanation on trial. The hypothesis is usually a rational accounting for a set of observations, based on the available data and guided by inductive reasoning. A scientific hypothesis must lead to predictions that can be tested by making additional observations or by performing experiments. An **experiment** is a scientific test, carried out under controlled conditions.

We all make observations and develop questions and hypotheses in solving everyday problems. Let's say, for example, that your desk lamp is plugged in and turned on but the bulb isn't lit. That's an observation. The question is obvious: Why doesn't the lamp work? Two reasonable hypotheses based on your experience are that (1) the bulb is burnt out or (2) the bulb is not screwed in properly. Each of these alternative hypotheses leads to predictions you can test with experiments. For example, the burnt-out bulb hypothesis predicts that replacing the bulb will fix the problem. **Figure 1.22** diagrams this informal inquiry. Figuring things out in this way by trial and error is a hypothesis-based approach.

▼ **Figure 1.22 A simplified view of the scientific process.** The idealised process sometimes called the "scientific method" is shown in this flow chart, which illustrates hypothesis testing for a desk lamp that doesn't work.

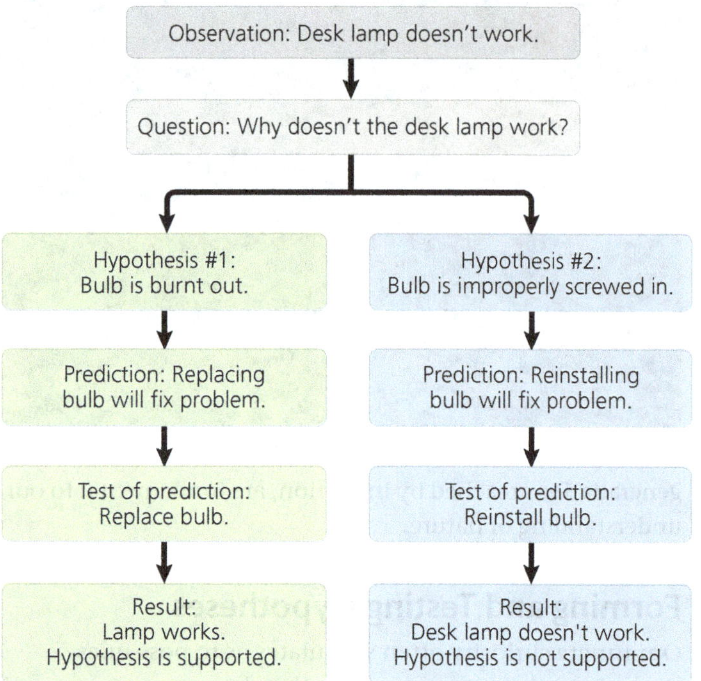

Deductive Reasoning

A type of logic called deduction is also built into the use of hypotheses in science. While induction entails reasoning from a set of specific observations to reach a general conclusion, **deductive reasoning** involves logic that flows in the opposite direction, from the general to the specific. From general premises, we extrapolate to the specific results we should expect if the premises are true. In the scientific process, deductions usually take the form of predictions of results that will be found if a particular hypothesis (premise) is correct. We then test the hypothesis by carrying out experiments or observations to see whether or not the results are as predicted. This deductive testing takes the form of "*If . . . then*" logic. In the case of the desk lamp example: *If* the burnt-out bulb hypothesis is correct, *then* the lamp should work if you replace the bulb with a new one.

We can use the desk lamp example to illustrate two other key points about the use of hypotheses in science. First, one can always devise additional hypotheses to explain a set of observations. For instance, another hypothesis to explain our nonworking desk lamp is that the wall socket is faulty. Although you could design an experiment to test this hypothesis, you can never test all possible hypotheses. Second, we can never *prove* that a hypothesis is true. Suppose that replacing the bulb fixed the lamp. The burnt-out bulb hypothesis would be the most likely explanation, but testing supports that hypothesis *not* by proving that it is correct, but rather by failing to prove it incorrect. For example, even if replacing the bulb fixed the desk lamp, it might have been because there was a temporary power outage that just happened to end while the bulb was being changed.

Although a hypothesis can never be proved beyond all doubt, testing it in various ways can significantly increase our confidence in its validity. Often, rounds of hypothesis formulation and testing lead to a scientific consensus—the shared conclusion of many scientists that a particular hypothesis explains the known data well and stands up to experimental testing.

Questions That Can and Cannot Be Addressed by Science

Scientific inquiry is a powerful way to learn about nature, but there are limitations to the kinds of questions it can answer. A scientific hypothesis must be *testable*; there must be some observation or experiment that could reveal if such an idea is likely to be true or false. The hypothesis that a burnt-out bulb is the sole reason the lamp doesn't work would not be supported if replacing the bulb with a new one didn't fix the lamp.

Not all hypotheses meet the criteria of science: You wouldn't be able to test the hypothesis that invisible ghosts are fooling with your desk lamp! Because science only deals with natural, testable explanations for natural phenomena, it can neither support nor contradict the invisible ghost hypothesis, nor whether spirits or elves cause storms, rainbows, or illnesses. Such supernatural explanations are simply outside the bounds of science, as are religious matters, which are issues of personal faith. Science and religion are not mutually exclusive or contradictory; they are simply concerned with different issues.

The Flexibility of the Scientific Process

The way that researchers answer questions about the natural and physical world is often idealised as the *scientific method*. However, very few scientific inquiries adhere rigidly to the sequence of steps that are typically used to describe this approach. For example, a scientist may start to design an experiment, but then backtrack after realising that more preliminary observations are necessary. In other cases, observations remain too puzzling to prompt well-defined questions until further study provides a new context in which to view those observations. For example, scientists could not unravel the details of how genes encode proteins until *after* the discovery of the structure of DNA (an event that took place in 1953).

A more realistic model of the scientific process is shown in **Figure 1.23**. The focus of this model, shown in the central circle in the figure, is the forming and testing of hypotheses. This core set of activities is the reason that science does so well in explaining phenomena in the natural world. These activities, however, are shaped by exploration and discovery (the upper circle in Figure 1.23) and influenced by interactions with other scientists and with society more generally (lower circles). For example, the community of scientists influences which hypotheses are tested, how test results are interpreted, and what value is placed on the findings. Similarly, societal needs—such as the push to cure cancer or understand the process of climate change—may help shape what research projects are funded and how extensively the results are discussed.

Now that we have highlighted the key features of scientific inquiry—making observations and forming and testing hypotheses—you should be able to recognise these features in a case study of actual scientific research.

▼ **Figure 1.23 The process of science: a realistic model.** In reality, the process of science is not linear, but instead involves backtracking, repetitions, and feedback between different parts of the process. This illustration is based on a model (How Science Works) from the website Understanding Science (www.understandingscience.org).

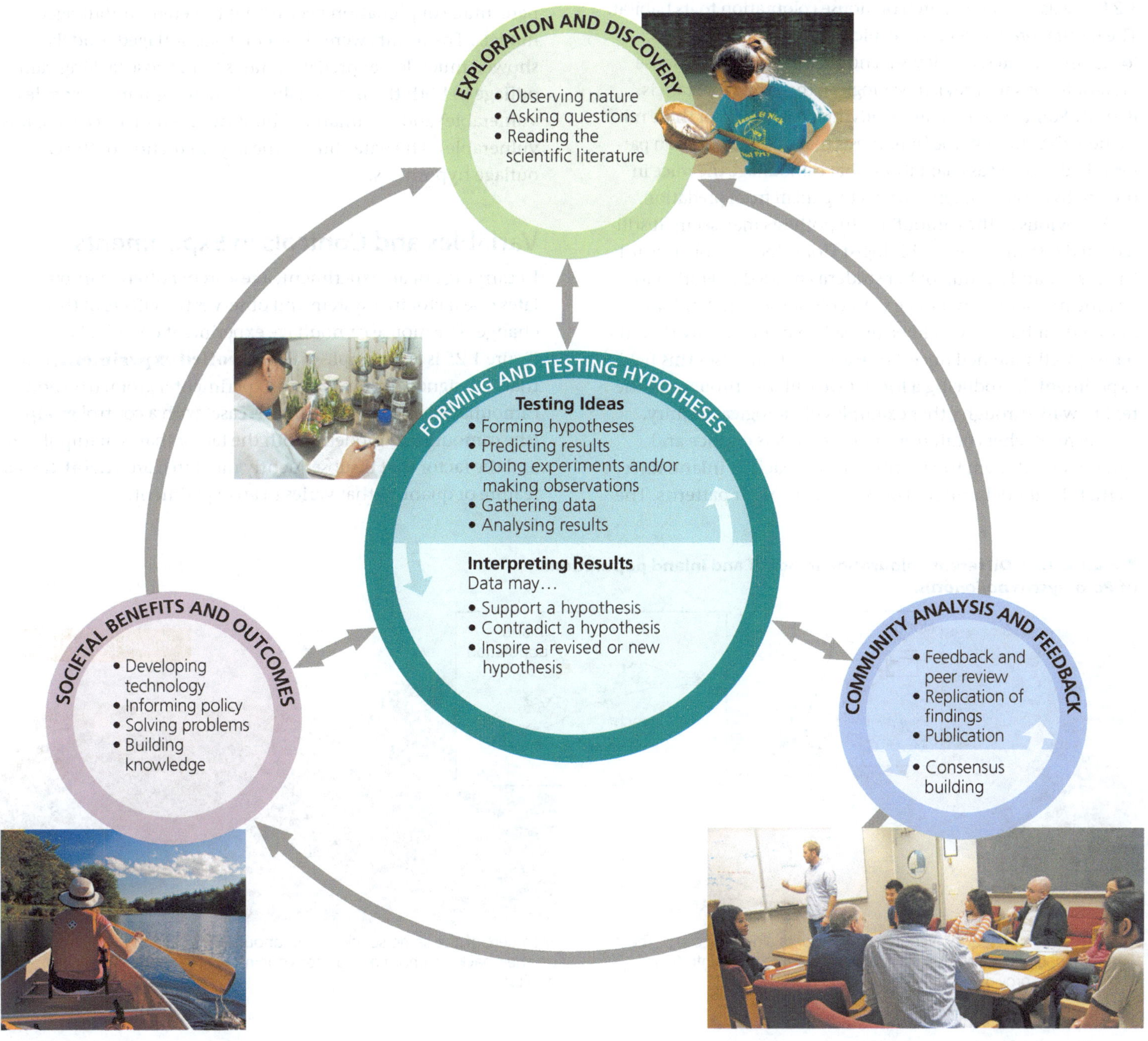

A Case Study in Scientific Inquiry: Investigating Coat Colouration in Mouse Populations

Our case study begins with a set of observations and inductive generalisations. Colour patterns of animals vary widely in nature, sometimes even among members of the same species. What accounts for such variation? As you may recall, the two mice depicted at the beginning of this chapter are members of the same species (*Peromyscus polionotus*), but they have different coat (fur) colour patterns and reside in different environments. The beach mouse lives along the Florida seashore, a habitat of brilliant white sand dunes with sparse clumps of beach grass. The inland mouse lives on darker, more fertile soil further inland **(Figure 1.24)**. Even a brief glance at the photographs in Figure 1.24 reveals a striking match of mouse colouration to its habitat. The natural predators of these mice, including hawks, owls, foxes, and coyotes, are all visual hunters (they use their sense of sight to look for prey). It was logical, therefore, for Francis Bertody Sumner, a naturalist studying populations of these mice in the 1920s, to form the hypothesis that their colouration patterns had evolved as adaptations that camouflage the mice in their native environments, protecting them from predation.

As obvious as the camouflage hypothesis may seem, it still required testing. In 2010, biologist Hopi Hoekstra of Harvard University and a group of her students headed to Florida to test the prediction that mice with colouration that did not match their habitat would be preyed on more heavily than the native, well-matched mice. **Figure 1.25** summarises this field experiment, introducing a format we will use throughout the text to walk through other examples of biological inquiry.

The researchers built hundreds of models of mice and spray-painted them to resemble either beach or inland mice, so that the models differed only in their colour patterns. The researchers placed equal numbers of these model mice randomly in both habitats and left them overnight. The mouse models resembling the native mice in the habitat were the *control* group (for instance, light-coloured mouse models in the beach habitat), while the mouse models with the non-native colouration were the *experimental* group (for example, darker models in the beach habitat). The following morning, the team counted and recorded signs of predation events, which ranged from bites and gouge marks on some models to the outright disappearance of others. Judging by the shape of the predators' bites and the tracks surrounding the experimental sites, the predators appeared to be split fairly evenly between mammals (such as foxes and coyotes) and birds (such as owls, herons, and hawks).

For each environment, the researchers then calculated the percentage of predation events that targeted camouflaged models. The results were clear-cut: Camouflaged models showed much lower predation rates than those lacking camouflage in both the beach habitat (where light mice were less vulnerable) and the inland habitat (where dark mice were less vulnerable). The data thus fit the key prediction of the camouflage hypothesis.

Variables and Controls in Experiments

In carrying out an experiment, a researcher often manipulates one factor in a system and observes the effects of this change. The mouse camouflage experiment described in Figure 1.25 is an example of a **controlled experiment**, one that is designed to compare an experimental group (the non-camouflaged mice models, in this case) with a control group (the camouflaged models). Both the factor that is manipulated and the factor that is subsequently measured are **variables**—a feature or quantity that varies in an experiment.

▼ Figure 1.24 Different colouration in beach and inland populations of *Peromyscus polionotus*.

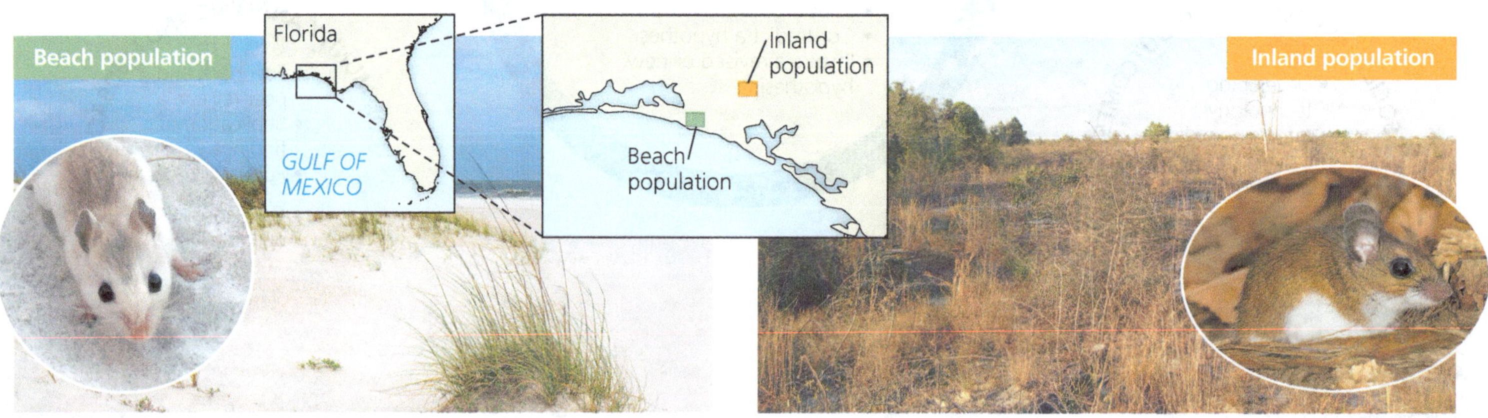

Beach mice live on sparsely vegetated sand dunes along the coast. The light tan, dappled fur on their backs causes them to blend into their surroundings, providing camouflage.

Members of the same species living about 30 km inland have dark fur on their backs, camouflaging them against the dark ground of their habitat.

▼ Figure 1.25 Inquiry

Does camouflage affect predation rates on two populations of mice?

Experiment Hopi Hoekstra and colleagues tested the hypothesis that coat colouration provides camouflage that protects beach and inland populations of *Peromyscus polionotus* mice from predation in their habitats. The researchers spray-painted mouse models with light or dark colour patterns that matched those of the beach and inland mice and placed models with each of the patterns in both habitats. The next morning, they counted damaged or missing models.

Results For each habitat, the researchers calculated the percentage of attacked models that were camouflaged or non-camouflaged. In both habitats, the models whose pattern did not match their surroundings suffered much higher "predation" than did the camouflaged models.

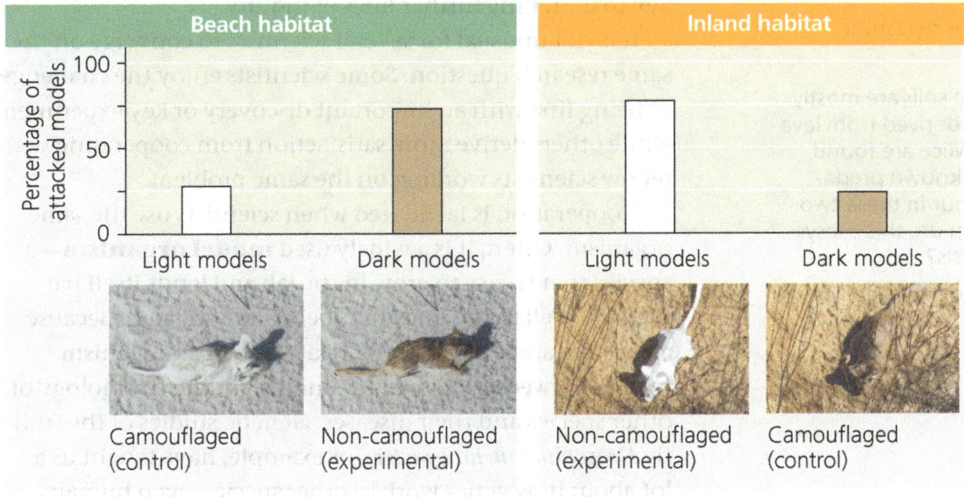

Conclusion The results are consistent with the researchers' prediction: that mouse models with camouflage colouration would be attacked less often than non-camouflaged mouse models. Thus, the experiment supports the camouflage hypothesis.

Data from S. N. Vignieri, J. G. Larson, and H. E. Hoekstra, The selective advantage of crypsis in mice, *Evolution* 64:2153–2158 (2010).

INTERPRET THE DATA *The bars indicate the percentage of the attacked models that were either light or dark. Assume 100 mouse models were attacked in each habitat. For the beach habitat, how many were light models? Dark models? Answer the same questions for the inland habitat. Do the results of the experiment support the camouflage hypothesis? Explain.*

In our example, the colour of the mouse model was the **independent variable**—the factor being manipulated by the researchers. The **dependent variable** is the factor being measured that is predicted to be affected by the independent variable; in this case, the researchers measured the amount of predation in response to variation in colour of the mouse model. Note also that the experimental and control groups differ in only one independent variable: colour.

As a result, the researchers could rule out other factors as causes of the more frequent attacks on the non-camouflaged mice—such as different numbers of predators or different temperatures in the different test areas. The clever experimental design left colouration as the only factor that could account for the low predation rate on models camouflaged with respect to the surrounding environment.

A common misconception is that the term *controlled experiment* means that scientists control all features of the experimental environment. But that's impossible in field research and can be very difficult even in highly regulated laboratory environments. Researchers usually "control" unwanted variables not by *eliminating* them through environmental regulation, but by *cancelling out* their effects by using control groups.

Theories in Science

"It's just a theory!" Our everyday use of the term *theory* often implies an untested speculation. But the term *theory* has a different meaning in science. What is a scientific theory, and how is it different from a hypothesis or from mere speculation?

First, a scientific **theory** is much broader in scope than a hypothesis. *This* is a hypothesis: "Coat colouration well-matched to their habitat is an adaptation that protects mice from predators." But *this* is a theory: "Evolutionary adaptations arise by natural selection." This theory proposes that natural selection is the evolutionary mechanism that accounts for an enormous variety of adaptations, of which coat colour in mice is but one example.

Second, a theory is general enough to spin off many new, testable hypotheses. For example, the theory of natural selection motivated two researchers at Princeton University, Peter and Rosemary Grant, to test the specific hypothesis that the beaks of Galápagos finches evolve in response to changes in the types of available food. (Their results supported their hypothesis; see Figure 23.2.)

And third, compared to any one hypothesis, a theory is generally supported by a much greater body of evidence. The theory of natural selection has been supported by a vast quantity of evidence, with more being found every day, and has not been contradicted by any scientific data. Those theories that become widely adopted in science (such as the theory of natural selection and the theory of gravity) explain a great diversity of observations and are supported by a vast accumulation of evidence.

Finally, scientists will sometimes modify or even reject a previously supported theory if new research consistently produces results that don't fit. For example, biologists once lumped bacteria and archaea together as a kingdom of prokaryotes. When new methods for comparing cells and molecules could be used to test such relationships, the evidence led scientists to reject the theory that bacteria and archaea are members of the same kingdom. If there is "truth" in science, it is at best conditional, based on the weight of available evidence.

CONCEPT CHECK **1.3**

1. What qualitative observation led to the quantitative study in Figure 1.25?
2. Contrast inductive reasoning with deductive reasoning.
3. Why is natural selection called a theory?
4. **WHAT IF?** In the deserts of New Mexico, the soils are mostly sandy, with occasional regions of black rock derived from lava flows that occurred about 1,000 years ago. Mice are found in both sandy and rocky areas, and owls are known predators. What might you expect about coat colour in these two mouse populations? Explain. How would you use this ecosystem to further test the camouflage hypothesis?

For suggested answers, see Appendix A.

CONCEPT **1.4**

Science benefits from a cooperative approach and diverse viewpoints

Movies and cartoons sometimes portray scientists as loners in white lab coats, working in isolated labs. In reality, science is an intensely social activity. Most scientists work in teams, which often include both graduate and undergraduate students. And to succeed in science, it helps to be a good communicator. Research results have no impact until shared with a community of peers through seminars, publications, and websites. And, in fact, research papers aren't published until they are vetted by colleagues in what is called the "peer review" process. The examples of scientific inquiry described in this text, for instance, have all been published in peer-reviewed journals.

Building on the Work of Others

The great scientist Isaac Newton once said: "To explain all nature is too difficult a task for any one man or even for any one age. 'Tis much better to do a little with certainty, and leave the rest for others that come after you." Anyone who becomes a scientist, driven by curiosity about how nature works, is sure to benefit greatly from the rich storehouse of discoveries by others who have come before. In fact, Hopi Hoekstra's experiment benefited from the work of another researcher, D. W. Kaufman, 40 years earlier. You can study the design of Kaufman's experiment and interpret the results in the **Scientific Skills Exercise**.

Scientific results are continually scrutinised through the repetition of observations and experiments. Scientists working in the same research field often check one another's claims by attempting to confirm observations or repeat experiments. If scientific colleagues cannot repeat experimental findings, this failure may reflect some underlying weakness in the original claim, which will then have to be revised. In this sense, science polices itself. Integrity and adherence to high professional standards in reporting results are central to the scientific endeavour, since the validity of experimental data is key to designing further lines of inquiry.

It is not unusual for several scientists to converge on the same research question. Some scientists enjoy the challenge of being first with an important discovery or key experiment, while others derive more satisfaction from cooperating with fellow scientists working on the same problem.

Cooperation is facilitated when scientists use the same organism. Often, it is a widely used **model organism**—a species that is easy to grow in the lab and lends itself particularly well to the questions being investigated. Because all species are evolutionarily related, such an organism may be viewed as a model for understanding the biology of other species and their diseases. Genetic studies of the fruit fly *Drosophila melanogaster*, for example, have taught us a lot about how genes work in other species, even humans. Some other popular model organisms are the mustard plant *Arabidopsis thaliana*, the soil worm *Caenorhabditis elegans*, the zebrafish *Danio rerio*, the mouse *Mus musculus*, and the bacterium *Escherichia coli*. As you read through this text, note the many contributions that these and other model organisms have made to the study of life.

Biologists may approach interesting questions from different angles. Some biologists focus on ecosystems, while others study natural phenomena at the level of organisms or cells. This text is divided into units that look at biology at different levels and investigate problems through different approaches. Yet any given problem can be addressed from many perspectives, which in fact complement each other. For example, Hoekstra not only carried out field studies showing that coat colouration can affect predation rates but also did lab studies that uncovered at least one genetic mutation that underlies the differences between beach and inland mouse colouration. Her lab includes biologists specialising in different biological levels, allowing links to be made between the evolutionary adaptations she focuses on and their molecular basis in DNA sequences.

As a biology student, you can benefit from making connections between the different levels of biology. You can develop this skill by noticing when certain topics crop up again and again in different units. One such topic is sickle-cell disease, a well-understood genetic condition that is prevalent among

Scientific Skills Exercise

Interpreting a Pair of Bar Graphs

How much does camouflage affect predation on mice by owls with and without moonlight? D. W. Kaufman hypothesised that the extent to which the coat colour of a mouse contrasted with the colour of its surroundings would affect the rate of nighttime predation by owls. He also hypothesised that the contrast would be affected by the amount of moonlight. In this exercise, you will analyse data from his studies of owl predation on mice that tested these hypotheses.

How the Experiment Was Done Pairs of mice (*Peromyscus polionotus*) with different coat colour, one light brown and one dark brown, were released simultaneously into an enclosure that contained a hungry owl. The researcher recorded the colour of the mouse that was first caught by the owl. If the owl did not catch either mouse within 15 minutes, the test was recorded as a zero. The release trials were repeated multiple times in enclosures with either a dark-coloured soil surface or a light-coloured soil surface. The presence or absence of moonlight during each assay was recorded.

Data from the Experiment

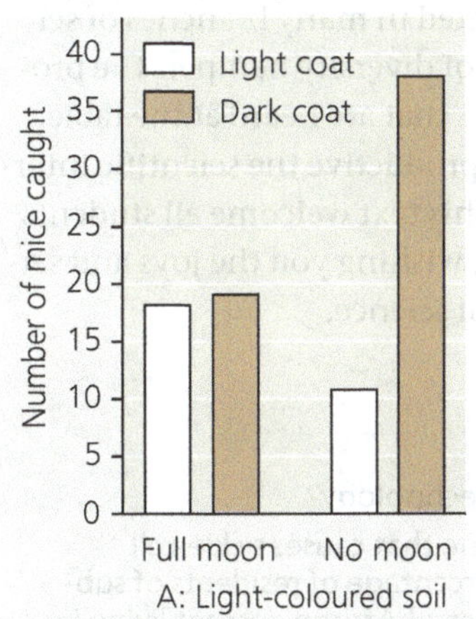

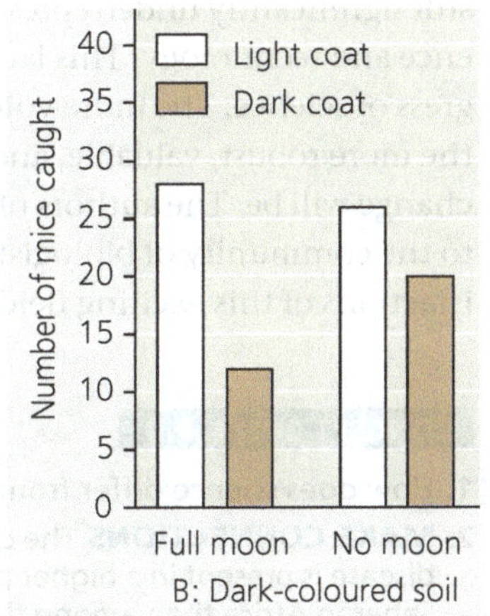

Data from D. W. Kaufman, Adaptive coloration in *Peromyscus polionotus*: Experimental selection by owls, *Journal of Mammalogy* 55:271–283 (1974).

INTERPRET THE DATA

1. First, make sure you understand how the graphs are set up. Graph A shows data from the light-coloured soil enclosure and graph B from the dark-coloured enclosure, but in all other respects the graphs are the same. **(a)** There is more than one independent variable in these graphs. What are the independent variables, the variables that were tested by the researcher? Which axis of the graphs has the independent variables? **(b)** What is the dependent variable, the response to the variables being tested? Which axis of the graphs has the dependent variable?
2. **(a)** How many dark brown mice were caught in the light-coloured soil enclosure on a moonlit night? **(b)** How many dark brown mice were caught in the dark-coloured soil enclosure on a moonlit night? **(c)** On a moonlit night, would a dark brown mouse be more likely to escape predation by owls on dark- or light-coloured soil? Explain your answer.
3. **(a)** Is a dark brown mouse on dark-coloured soil more likely to escape predation under a full moon or with no moon? **(b)** What about a light brown mouse on light-coloured soil? Explain.
4. **(a)** Under which conditions would a dark brown mouse be most likely to escape predation at night? **(b)** A light brown mouse?
5. **(a)** What combination of independent variables led to the highest predation level in enclosures with light-coloured soil? **(b)** What combination of independent variables led to the highest predation level in enclosures with dark-coloured soil?
6. Thinking about your answers to question 5, provide a simple statement describing conditions that are especially deadly for either colour of mouse.
7. Combining the data from both graphs, estimate the number of mice caught in moonlight versus no-moonlight conditions. Which condition is optimal for predation by the owl? Explain.

native inhabitants of Africa and other warm regions and their descendants. Sickle-cell disease will appear in several units of the text, each time addressed at a new level. In addition, Make Connections figures connect the content in different chapters, and Make Connections questions ask you to make the connections yourselves. We hope these features will help you integrate the material you're learning and enhance your enjoyment of biology by encouraging you to keep the big picture in mind.

Science, Technology, and Society

The research community is part of society at large, and the relationship of science to society becomes clearer when we add technology to the picture (see Figure 1.23). Though science and technology sometimes employ similar inquiry patterns, their basic goals differ. The goal of science is to understand natural phenomena, while that of **technology** is to *apply* scientific knowledge for some specific purpose. Because scientists put new technology to work in their research, science and technology are interdependent.

The potent combination of science and technology can have dramatic effects on society. Sometimes, the applications of basic research that turn out to be the most beneficial come out of the blue, from completely unanticipated observations in the course of scientific exploration. For example, discovery of the structure of DNA by Watson and Crick in 1953 and subsequent achievements in DNA science led to the technologies of DNA manipulation that are transforming applied fields such as medicine, agriculture, and forensics. Perhaps Watson and Crick envisioned that their discovery would someday lead to important applications, but it is unlikely that they could have predicted exactly what all those applications would be.

The directions that technology takes depend less on the curiosity that drives basic science than on the current needs and wants of people and on the social environment of the times. Debates about technology centre more on "*should* we do it" than "*can* we do it." With advances in technology come difficult choices. For example, under what circumstances is it acceptable to use DNA technology to find out if particular people have genes for hereditary diseases? Should such tests always be voluntary, or are there circumstances when genetic testing should be mandatory? Should insurance companies or employers have access to the information, as they do for many other types of personal health data? These questions are becoming much more urgent as the sequencing of individual genomes becomes quicker and cheaper.

Ethical issues raised by such questions have as much to do with politics, economics, and cultural values as with science and technology. All citizens—not only professional scientists—have a responsibility to be informed about how science works and about the potential benefits and risks of technology. The relationship between science, technology, and society increases the significance and value of any biology course.

The Value of Diverse Viewpoints in Science

Many of the technological innovations with the most profound impact on human society originated in settlements along trade routes, where a rich mix of different cultures ignited new ideas. For example, the printing press, which helped spread knowledge to all social classes, was invented by the German Johannes Gutenberg around 1440. This invention relied on several innovations from China, including paper and ink. Paper travelled along trade routes from China to Baghdad, where technology was developed for its mass production. This technology then migrated to Europe, as did water-based ink from China, which was modified by Gutenberg to become oil-based ink. We have the cross-fertilisation of diverse cultures to thank for the printing press, and the same can be said for other important inventions.

Along similar lines, science benefits from a diversity of backgrounds and viewpoints among its practitioners. But just how diverse a population are scientists in relation to gender, race, ethnicity, and other attributes?

The scientific community reflects the cultural standards and behaviours of the society around it. It is therefore not surprising that until recently, women, people of colour, and other underrepresented groups have faced huge obstacles in their pursuit to become professional scientists in many countries around the world. Over the past 50 years, changing attitudes about career choices have increased the proportion of women in biology and some other sciences, so that now women constitute roughly half of undergraduate biology majors and biology PhD students.

The pace has been slow at higher levels in the profession, however, and women and many racial and ethnic groups are still significantly underrepresented in many branches of science and technology. This lack of diversity hampers the progress of science. The more voices that are heard at the table, the more robust, valuable, and productive the scientific interchange will be. The authors of this text welcome all students to the community of biologists, wishing you the joys and satisfactions of this exciting field of science.

CONCEPT CHECK 1.4

1. How does science differ from technology?
2. **MAKE CONNECTIONS** The gene that causes sickle-cell disease is present in a higher percentage of residents of sub-Saharan Africa than among those of African descent living in the New Zealand or Australia. Even though this gene causes sickle-cell disease, it also provides some protection from malaria, a serious disease that is widespread in sub-Saharan Africa but absent in the Australia or New Zealand. Discuss an evolutionary process that could account for the different percentages of the sickle-cell gene among residents of the two regions. (See Concept 1.2.)

For suggested answers, see Appendix A.

1 Chapter Review

SUMMARY OF KEY CONCEPTS

CONCEPT 1.1

The study of life reveals unifying themes *(pp. 3–11)*

Organisation Theme: New Properties Emerge at Successive Levels of Biological Organisation

- The hierarchy of life unfolds as follows: biosphere > ecosystem > community > population > organism > organ system > organ > tissue > cell > organelle > molecule > atom. With each step from atoms to the biosphere, new **emergent properties** result from interactions among components at the lower levels. In an approach called reductionism, complex systems are broken down to simpler components that are more manageable to study. In **systems biology**, scientists attempt to model the dynamic behaviour of whole biological systems by studying the interactions among the system's parts.
- Structure and function are correlated at all levels of biological organisation. The cell, an organism's basic unit of structure and function, is the lowest level that can perform all activities required for life. Cells are either prokaryotic or eukaryotic. **Eukaryotic cells** have a DNA-containing nucleus and other membrane-enclosed organelles. **Prokaryotic cells** lack such organelles.

Information Theme: Life's Processes Involve the Expression and Transmission of Genetic Information

- Genetic information is encoded in the nucleotide sequences of **DNA**. It is DNA that transmits heritable information from parents to offspring. DNA sequences (called **genes**) program a cell's protein production by being transcribed into mRNA and then translated into specific proteins, a process called **gene expression**. Gene expression also produces RNAs that are not translated into protein but serve other important functions. **Genomics** is the large-scale analysis of the DNA sequences of a species (its **genome**) as well as the comparison of genomes between species. **Bioinformatics** uses computational tools to deal with huge volumes of sequence data.

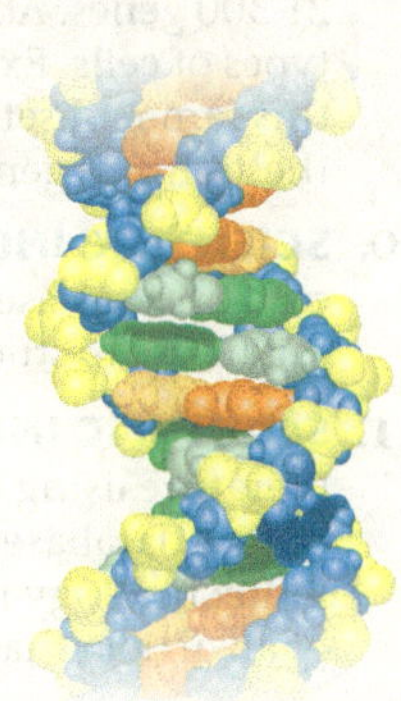

Energy and Matter Theme: Life Requires the Transfer and Transformation of Energy and Matter

- Energy flows through an ecosystem. All organisms must perform work, which requires energy. **Producers** convert energy from sunlight to chemical energy, some of which is used by them and by **consumers** to do work, and is eventually lost from the ecosystem as heat. Chemicals cycle between organisms and the environment.

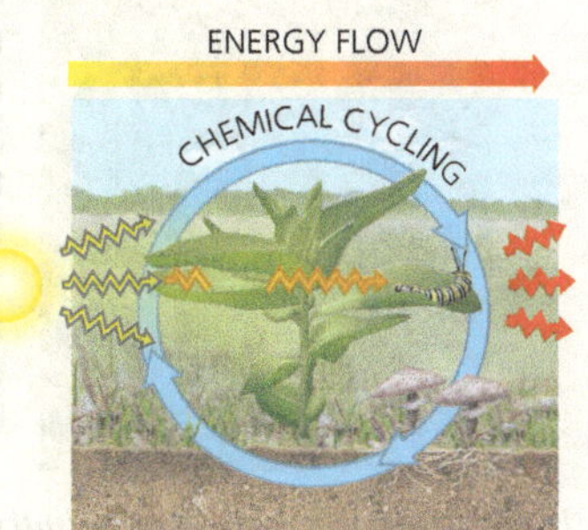

Interactions Theme: From Molecules to Ecosystems, Interactions Are Important in Biological Systems

- In **feedback regulation**, a process is regulated by its output or end product. In negative feedback, accumulation of the end product slows its production. In positive feedback, an end product speeds up its own production.
- Organisms interact continuously with physical factors. Plants take up nutrients from the soil and chemicals from the air and use energy from the sun.

? *Thinking about the muscles and nerves in your hand, how does the activity of text messaging reflect the four unifying themes of biology described in this section?*

CONCEPT 1.2

The Core Theme: Evolution accounts for the unity and diversity of life *(pp. 11–16)*

- **Evolution**, the process of change that has transformed life on Earth, accounts for the unity and diversity of life. It also explains evolutionary adaptation—the match of organisms to their environments.
- Biologists classify species according to a system of broader and broader groups. Domain **Bacteria** and domain **Archaea** consist of prokaryotes. Domain **Eukarya**, the eukaryotes, includes various groups of protists and the kingdoms Plantae, Fungi, and Animalia. As diverse as life is, there is also evidence of remarkable unity, revealed in the similarities between different species.
- Darwin proposed **natural selection** as the mechanism for evolutionary adaptation of populations to their environments. Natural selection is the evolutionary process that occurs when a population is exposed to environmental factors that consistently cause individuals with certain heritable traits to have greater reproductive success than do individuals with other heritable traits.

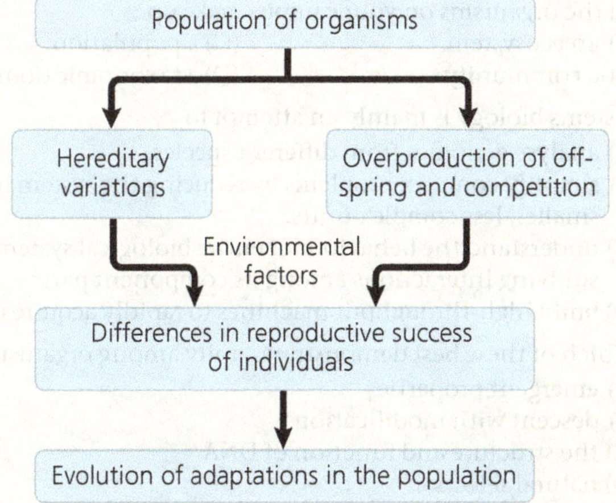

- Each species is one twig of a branching tree of life extending back in time through more and more remote ancestral species. All of life is connected through its long evolutionary history.

? *How could natural selection have led to the evolution of adaptations such as camouflaging coat colour in beach mice?*

CONCEPT 1.3

In studying nature, scientists form and test hypotheses *(pp. 16–22)*

- In scientific **inquiry**, scientists collect **data** and use **inductive reasoning** to draw a general conclusion, which can be developed into a testable **hypothesis**. **Deductive reasoning** uses predictions to test hypotheses. Hypotheses must be testable; science can address neither the possibility of supernatural phenomena nor religious beliefs. Hypotheses can be tested by conducting **experiments** or, when that is not possible, by making observations. In the process of science, the core activity is testing ideas. This endeavour is influenced by exploration and discovery, community analysis and feedback, and societal outcomes.
- **Controlled experiments** are designed to demonstrate the effect of one **variable** by testing control groups and experimental groups that differ in only that one variable.
- A scientific **theory** is broad in scope, generates new hypotheses, and is supported by a large body of evidence.

? *What are the roles of gathering and interpreting data?*

CONCEPT 1.4

Science benefits from a cooperative approach and diverse viewpoints *(pp. 22–24)*

- Science is a social activity. The work of each scientist builds on the work of others who have come before. Scientists must be able to repeat each other's results, and integrity is key. Biologists approach questions at different levels; their approaches complement each other.
- **Technology** consists of any method or device that applies scientific knowledge for some specific purpose that affects society. The impact of basic research is not always immediately obvious.
- Diversity among scientists promotes progress in science.

? *Explain why different approaches and diverse backgrounds among scientists are important.*

TEST YOUR UNDERSTANDING

Levels 1-2: Remembering/Understanding

1. All the organisms on your campus make up
(A) an ecosystem.
(B) a community.
(C) a population.
(D) a taxonomic domain.

2. Systems biology is mainly an attempt to
(A) analyse genomes from different species.
(B) simplify complex problems by reducing the system into smaller, less complex units.
(C) understand the behaviour of entire biological systems by studying interactions among its component parts.
(D) build high-throughput machines to rapidly acquire data.

3. Which of these best demonstrates unity among organisms?
(A) emergent properties
(B) descent with modification
(C) the structure and function of DNA
(D) natural selection

4. A controlled experiment is one that
(A) proceeds slowly so a scientist can make careful records.
(B) tests experimental and control groups in parallel.
(C) is repeated many times to make sure the results are accurate.
(D) keeps all variables constant.

5. Which of the following statements best distinguishes hypotheses from theories in science?
(A) Theories are hypotheses that have been proved.
(B) Hypotheses are guesses; theories are correct answers.
(C) Hypotheses usually are relatively narrow in scope; theories have broad explanatory power.
(D) Theories are proved true; hypotheses are often contradicted by experimental results.

Levels 3-4: Applying/Analysing

6. Which of the following is an example of qualitative data?
(A) The fish swam in a zigzag motion.
(B) The contents of the stomach are mixed every 20 seconds.
(C) The temperature decreased from 20°C to 15°C.
(D) The six pairs of robins hatched an average of three chicks each.

7. Which sentence best describes the logic of scientific inquiry?
(A) If I generate a testable hypothesis, tests and observations will support it.
(B) If my prediction is correct, it will lead to a testable hypothesis.
(C) If my observations are accurate, they will support my hypothesis.
(D) If my prediction turns out to be correct, my hypothesis is supported.

8. **DRAW IT** Draw a biological hierarchy similar to the one in Figure 1.3 but using a coral reef as the ecosystem, a fish as the organism, its stomach as the organ, and DNA as the molecule. Include all levels in the hierarchy.

Levels 5-6: Evaluating/Creating

9. **EVOLUTION CONNECTION** A typical prokaryotic cell has about 3,000 genes in its DNA, while a human cell has about 21,300 genes. About 1,000 of these genes are present in both types of cells. Explain how such different organisms could have this same subset of 1,000 genes. What sorts of functions might these shared genes have?

10. **SCIENTIFIC INQUIRY** Based on the results of the mouse colouration case study, suggest another hypothesis researchers might use to study the role of predators in natural selection.

11. **SCIENTIFIC INQUIRY** Scientists search the scientific literature using electronic databases such as PubMed, a free online database maintained by the National Center for Biotechnology Information. Use PubMed to find the abstract of an article that Hopi Hoekstra published in 2017 or later.

12. **WRITE ABOUT A THEME: EVOLUTION** In a short essay (100–150 words), discuss Darwin's view of how natural selection resulted in both unity and diversity of life. Include in your discussion some of his evidence.

13. **SYNTHESISE YOUR KNOWLEDGE**

Can you pick out the mossy leaf-tailed gecko lying against the tree trunk in this photo? How is the appearance of the gecko a benefit in terms of survival? Given what you learned about evolution, natural selection, and genetic information in this chapter, describe how the gecko's colouration might have evolved.

For selected answers, see Appendix A.

Unit 1 THE CHEMISTRY OF LIFE

Abigail Allwood is Principal Investigator at the Jet Propulsion Laboratory of the US National Aeronautics and Space Administration (NASA) in Pasadena, California. She has a passion for biology, and used her expertise in geology and chemistry to study Earth's earliest life. Now she looks for evidence of the formation of life preserved in the rocks of another planet. Abby was the first Australian to become a NASA principal investigator, for NASA's 2020 Mars Rover mission.

AN INTERVIEW WITH

Abigail Allwood

It must be enormously gratifying to know that a piece of equipment you and your team designed and built is on Mars.

It's still hard to believe. I have to pinch myself to remember it is real, and the culmination of a dream.

Tell us about PIXL and how you are using it.

The PIXL (Planetary Instrument for X-ray Lithochemistry) uses X-rays to scan areas of rock the size of a postage stamp. We use it to look for traces of organic materials that represent the chemical signatures of microbial life or microfossils. It searches for fossil evidence of microbial colonies that lived in ancient oceans, or formed in lakes during periods of sediment deposition, like those we know occurred on Earth and Mars. The stromatolite colonies you can see today in Western Australia's Shark Bay closely resemble their ancient cousins, and perhaps those that once lived on Mars.

How did your work on fossils in the Pilbara (Western Australia) inform your early and subsequent work?

Earth's oldest stromatolites [formed by ancient communities of microbes—see the chapter, "The History of Life on Earth"] allow us to study Earth's earliest biosphere, because they help us identify fossil signatures of emergent microbial life. Using the knowledge gained, we could then search for similar signatures of life beyond Earth.

What made you interested in studying the Pilbara region, and what was the study site like?

My PhD supervisor, Professor Malcolm Walter, and I knew that the Pilbara Craton (1,250 km north of Perth) provides one area where you can find parts of Earth's crust that are of similar age to Martian rocks.

Part of my study site preserved the ripples in the sand that formed on the margins of the ocean. Near that shore, I found evidence of different kinds of stromatolite fossils. In our *Nature* paper, we showed that the stromatolite-infested rock outcrop formed part of a 10-kilometre-long fossil reef that could only have developed from biological processes. That work provided evidence of the earliest known life on Earth, around 3.4 billion years ago.

We know from our studies of Earth and Mars that both planets shared characteristics that could have supported life a long time ago.

"I'd found my passion, and I built the tools I needed to pursue it."

We don't know whether life ever began on Mars. I wanted to find out if similar fossils to the ones we found in the Pilbara during the earliest times on Earth also occurred on Mars.

How did you come to work at NASA?

Even after we published our *Nature* paper, sceptics wanted more microbial-scale evidence to support my work. I was happy to provide it. I went to the Jet Propulsion Laboratory to get irrefutable evidence.

I also realised that I had the opportunity to develop an X-ray tool like the one used in the Pilbara to scan rocks on Mars. Over five years, my team and I miniaturised a device so it could fit on a Martian rover. Now I can't wait to find out what it will show us.

What led you to pursue science?

I remember in Grade 8, a teacher asked: "Who wants to be a scientist?" One person put their hand up, and it wasn't me. Yet, here I am.

I found my passion late, in my first year at university. I had to learn the physics and maths I'd never studied in school. I built those skills in my undergraduate studies.

You can do it too.

2 The Chemical Context of Life

KEY CONCEPTS

Study Tip

Make a table: As you read the chapter, make a summary table like the following. Add more rows as you proceed.

Property	Element (atom) C	H	O	N
Atomic number				
# Electrons				
# Neutrons				
Mass number				
Electron distribution diagram				
# Valence electrons				

Go to Mastering Biology

to access Dynamic Study Modules for revision, 3D BioFlix® animations and high-quality videos, and your interactive Pearson eText.

Figure 2.1 Wood ants (*Formica rufa*) use chemistry to ward off enemies. When threatened from above, they shoot volleys of formic acid from their abdomens into the air. The acid bombards and stings potential predators, such as hungry birds.

What determines the properties of a compound such as formic acid?

A compound is made of atoms joined by bonds. Formic acid (CH_2O_2) consists of carbon (C), hydrogen (H), and oxygen (O).

The number of protons (⊕) determines an atom's identity. Oxygen has 8 protons.

An atom's electron (⊖) distribution determines its ability to form bonds. Oxygen has space for 2 more electrons, so it can form 2 bonds.

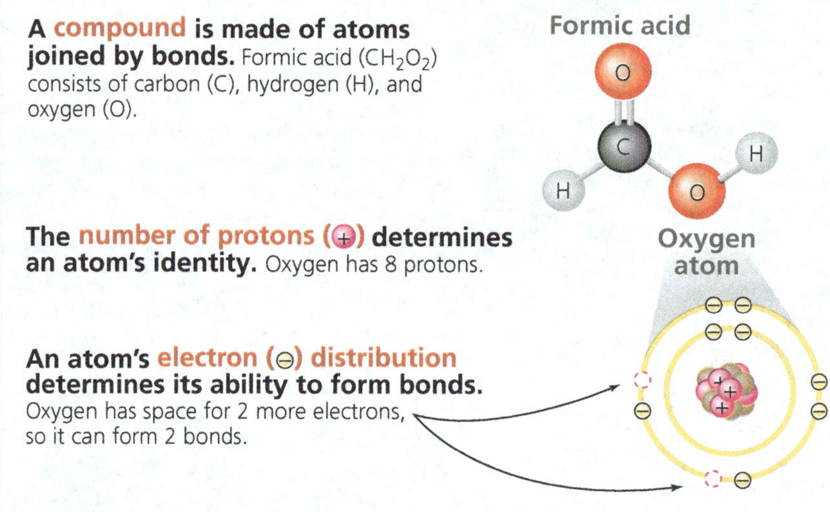

A compound's properties depend on its atoms and how they are bonded together.

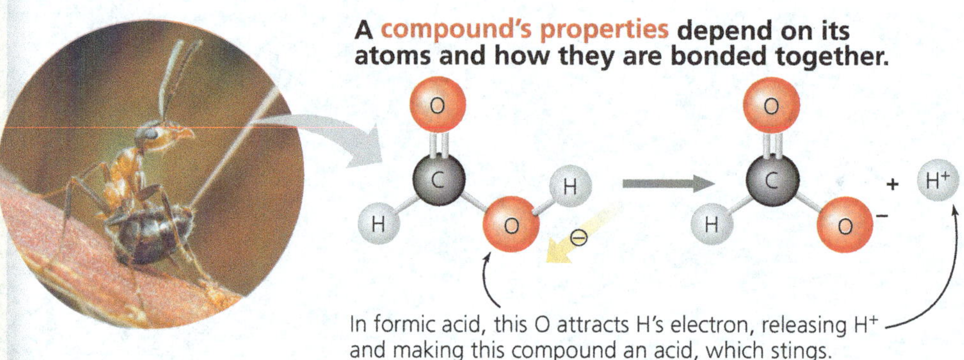

CONCEPT 2.1

Matter consists of chemical elements in pure form and in combinations called compounds

Organisms are composed of **matter**, which is anything that takes up space and has mass. Matter exists in many forms. Rocks, metals, oils, gases, and living organisms are a few examples of what seems to be an endless assortment of matter.

Elements and Compounds

Matter is made up of elements. An **element** is a substance that cannot be broken down to other substances by chemical reactions. Today, chemists recognise 92 elements occurring in nature; gold, copper, carbon, and oxygen are examples. Each element has a symbol, usually the first letter or two of its name. Some symbols are derived from Latin or German; for instance, the symbol for sodium is Na, from the Latin word *natrium*.

A **compound** is a substance consisting of two or more different elements combined in a fixed ratio. Table salt, for example, is sodium chloride (NaCl), a compound composed of the elements sodium (Na) and chlorine (Cl) in a 1:1 ratio. Pure sodium is a metal, and pure chlorine is a poisonous gas. When chemically combined, however, sodium and chlorine form an edible compound. Water (H_2O), another compound, consists of the elements hydrogen (H) and oxygen (O) in a 2:1 ratio. These are simple examples of organised matter having emergent properties: A compound has characteristics different from those of its elements **(Figure 2.2)**.

The Elements of Life

Of the 92 natural elements, about 20–25% are **essential elements** that an organism needs to live a healthy life and reproduce. The essential elements are similar among organisms, but there is some variation—for example, humans need 25 elements, but plants need only 17.

Just four elements—oxygen (O), carbon (C), hydrogen (H), and nitrogen (N)—make up approximately 96% of living matter. Calcium (Ca), phosphorus (P), potassium (K), sulfur (S), and a few other elements account for most of the remaining 4% or so of an organism's mass. **Trace elements** are required by an organism in only minute quantities. Some trace elements, such as iron (Fe), are needed by all forms of life; others are required only by certain species. For example, in vertebrates (animals with backbones), the element iodine (I) is an essential ingredient of a hormone produced by the thyroid gland. A daily intake of only 0.15 milligram (mg) of iodine is adequate for normal activity of the human thyroid. An iodine deficiency in the diet causes the thyroid gland to grow to abnormal size, a condition called goitre. Consuming seafood or iodised salt reduces the incidence of goitre. Relative amounts of all the elements in the human body are listed in **Table 2.1**.

Some naturally occurring elements are toxic to organisms. In humans, for instance, the element arsenic has been linked to numerous diseases and can be lethal. In some areas of the world, arsenic occurs naturally and can make its way into the groundwater. As a result of using water from drilled wells in southern Asia, millions of people have been inadvertently exposed to arsenic-laden water. Efforts are under way to reduce arsenic levels in their water supply.

▼ Figure 2.2 The emergent properties of a compound. The metal sodium combines with the poisonous gas chlorine, forming the edible compound sodium chloride, or table salt.

Na Sodium + **Cl** Chlorine (gas) → **NaCl** Sodium chloride

Table 2.1 Elements in the Human Body

Element	Symbol	Percentage of Body Mass (including water)	
Oxygen	O	65.0%	96.3%
Carbon	C	18.5%	
Hydrogen	H	9.5%	
Nitrogen	N	3.3%	
Calcium	Ca	1.5%	3.7%
Phosphorus	P	1.0%	
Potassium	K	0.4%	
Sulfur	S	0.3%	
Sodium	Na	0.2%	
Chlorine	Cl	0.2%	
Magnesium	Mg	0.1%	

Trace elements (less than 0.01% of mass): boron (B), chromium (Cr), cobalt (Co), copper (Cu), fluorine (F), iodine (I), iron (Fe), manganese (Mn), molybdenum (Mo), selenium (Se), silicon (Si), tin (Sn), vanadium (V), zinc (Zn)

INTERPRET THE DATA *Given the makeup of the human body, what compound do you think accounts for the high percentage of oxygen?*

Case Study: Evolution of Tolerance to Toxic Elements

EVOLUTION Some species have become adapted to environments containing elements that occur in toxic concentrations. Some plant species have adapted to soils with high concentrations of compounds that are toxic to the majority of plants. For example, in parts of Western Australia, Tasmania, and some locations in New Zealand, soils are so acidic, saline, and nutrient-poor that few plants can survive there. The foundation for these soils developed millions of years ago. Quartzite soils formed when sandstone was exposed to intense heat. When quartzite breaks down, it releases few nutrients and high concentrations of salts, and produces highly acidic soils. Although many plants cannot survive in soils derived from quartzite, a small number of plant species have adaptations that allow them to do so **(Figure 2.3)**. Presumably, variants of ancestral, nonquartzite species arose that could survive in acidic, salty soils, and subsequent natural selection resulted in the distinctive array of species we see in these areas today. Plants adapted to acidic and salty soil are of great interest to researchers because studying them can teach us so much about natural selection and evolutionary adaptations on a local scale.

▼ **Figure 2.3 Plant community established on soils with potentially toxic concentrations of elements.** These plants grow on soils eroded from quartzite-rich rocks. The inset shows a close-up of a quartz rock. Over very long periods, the quartz will erode to produce highly acidic and salt-rich soils. Members of this plant community have evolved to tolerate the high concentrations of these elements that would kill many other plants.

CONCEPT CHECK 2.1

1. **MAKE CONNECTIONS** Explain how table salt has emergent properties. (See Concept 1.1.)
2. Is a trace element an essential element? Explain.
3. **WHAT IF?** In humans, iron is a trace element required for the proper functioning of haemoglobin, the molecule that carries oxygen in red blood cells. What might be the effects of an iron deficiency?
4. **MAKE CONNECTIONS** Explain how natural selection might have played a role in the evolution of species that are tolerant of serpentine soils. (Review Concept 1.2.)

For suggested answers, see Appendix A.

CONCEPT 2.2

An element's properties depend on the structure of its atoms

Each element consists of a certain type of atom that is different from the atoms of any other element. An **atom** is the smallest unit of matter that still retains the properties of an element. Atoms are so small that it would take about a million of them to stretch across the period printed at the end of this sentence. We symbolise atoms with the same abbreviation used for the element that is made up of those atoms. For example, the symbol C stands for both the element carbon and a single carbon atom.

Subatomic Particles

Although the atom is the smallest unit having the properties of an element, these tiny bits of matter are composed of even smaller parts, called *subatomic particles*. Using high-energy collisions, physicists have produced more than 100 types of particles from the atom, but only three kinds of particles are relevant here: **neutrons**, **protons**, and **electrons**. Protons and electrons are electrically charged. Each proton has one unit of positive charge, and each electron has one unit of negative charge. A neutron, as its name implies, is electrically neutral.

Protons and neutrons are packed together tightly in a dense core, or **atomic nucleus**, at the centre of an atom; protons give the nucleus a positive charge. The rapidly moving electrons form a "cloud" of negative charge around the nucleus, and it is the attraction between opposite charges that keeps the electrons in the vicinity of the nucleus. **Figure 2.4** shows two commonly used models of the structure of the helium atom as an example.

▼ **Figure 2.4 Simplified models of a helium (He) atom.** The helium nucleus consists of 2 neutrons (brown) and 2 protons (pink). Two electrons (yellow) exist outside the nucleus. These models are not to scale; they greatly overestimate the size of the nucleus in relation to the electron cloud.

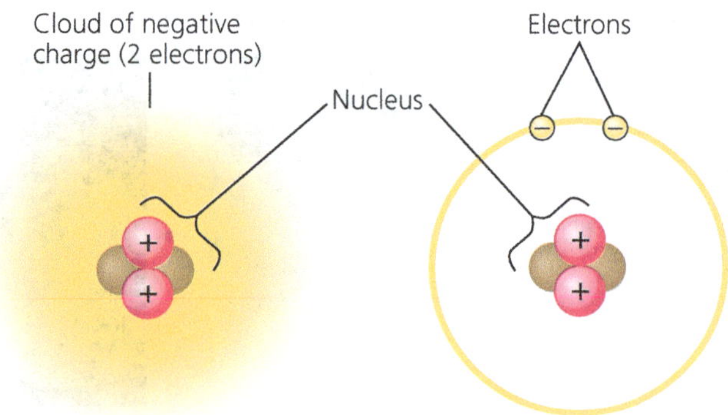

(a) This model represents the two electrons as a cloud of negative charge, a result of their motion around the nucleus.

(b) In this more simplified model, the electrons are shown as two small yellow spheres on a circle around the nucleus.

The neutron and proton are almost identical in mass, each about 1.7×10^{-24} gram (g). Grams and other conventional units are not very useful for describing the mass of objects that are so minuscule. Thus, for atoms and subatomic particles (and for molecules, too), we use a unit of measurement called the **dalton**, in honor of John Dalton, the British scientist who helped develop atomic theory around 1800. (The dalton is the same as the *atomic mass unit*, or *amu*, a unit you may have encountered elsewhere.) Neutrons and protons have masses close to 1 dalton. Because the mass of an electron is only about 1/2,000 that of a neutron or proton, we can ignore electrons when computing the total mass of an atom.

Atomic Number and Atomic Mass

Atoms of the various elements differ in their number of subatomic particles. All atoms of a particular element have the same number of protons in their nuclei. This number of protons, which is unique to that element, is called the **atomic number** and is written as a subscript to the left of the symbol for the element. The abbreviation $_2He$, for example, tells us that an atom of the element helium has 2 protons in its nucleus. Unless otherwise indicated, an atom is neutral in electrical charge, which means that its protons must be balanced by an equal number of electrons. Therefore, the atomic number tells us the number of protons and also the number of electrons in an electrically neutral atom.

We can deduce the number of neutrons from a second quantity, the **mass number**, which is the total number of protons and neutrons in the nucleus of an atom. The mass number is written as a superscript to the left of an element's symbol. For example, we can use this shorthand to write an atom of helium as ^4_2He. Because the atomic number indicates how many protons there are, we can determine the number of neutrons by subtracting the atomic number from the mass number. In our example, the helium atom ^4_2He has 2 neutrons. For sodium (Na):

$^{23}_{11}Na$

Mass number = number of protons + neutrons
= 23 for sodium

Atomic number = number of protons
= number of electrons in a neutral atom
= 11 for sodium

Number of neutrons = mass number − atomic number
= 23 − 11 = 12 for sodium

The simplest atom is hydrogen ^1_1H, which has no neutrons; it consists of a single proton with a single electron.

Because the contribution of electrons to mass is negligible, almost all of an atom's mass is concentrated in its nucleus. Neutrons and protons each have a mass very close to 1 dalton, so the mass number is close to, but slightly different from, the total mass of an atom, called its **atomic mass**. For example, the mass number of sodium ($^{23}_{11}Na$) is 23, but its atomic mass is 22.9898 daltons; the difference is explained below.

Isotopes

All atoms of a given element have the same number of protons, but some atoms have more neutrons than other atoms of the same element and therefore have greater mass. These different atomic forms of the same element are called **isotopes** of the element. In nature, an element may occur as a mixture of its isotopes. As an example, the element carbon, which has the atomic number 6, has three naturally occurring isotopes. The most common isotope is carbon-12, $^{12}_6C$, which accounts for about 99% of the carbon in nature. The isotope $^{12}_6C$ has 6 neutrons. Most of the remaining 1% of carbon consists of atoms of the isotope $^{13}_6C$, with 7 neutrons. A third, even rarer isotope, $^{14}_6C$, has 8 neutrons. Notice that all three isotopes of carbon have 6 protons; otherwise, they would not be carbon. Although the isotopes of an element have slightly different masses, they behave identically in chemical reactions. (For an element with more than one naturally occurring isotope, the atomic mass is an average of those isotopes, weighted by their abundance. Thus, carbon has an atomic mass of 12.01 daltons.)

Both ^{12}C and ^{13}C are stable isotopes, meaning that their nuclei do not have a tendency to lose subatomic particles, a process called decay. The isotope ^{14}C, however, is unstable, or radioactive. A **radioactive isotope** is one in which the nucleus decays spontaneously, giving off particles and energy. When the radioactive decay leads to a change in the number of protons, it transforms the atom to an atom of a different element. For example, when a carbon-14 (^{14}C) atom decays, a neutron decays into a proton, transforming the atom into a nitrogen (^{14}N) atom. Radioactive isotopes have many useful applications in biology.

Radioactive Tracers

Radioactive isotopes are often used as diagnostic tools in medicine. Cells can use radioactive atoms just as they would use nonradioactive isotopes of the same element. The radioactive isotopes are incorporated into biologically active molecules, which are then used as tracers to track atoms during metabolism, the chemical processes of an organism. For example, certain kidney disorders are diagnosed by injecting small doses of radioactively labelled substances into the blood and then analysing the tracer molecules excreted in the urine. Radioactive tracers are also used in combination with sophisticated imaging instruments, such as PET scanners that

▶ **Figure 2.5 A PET scan, a medical use for radioactive isotopes.** PET, an acronym for positron-emission tomography, detects locations of intense chemical activity in the body. The bright yellow spot marks an area with an elevated level of radioactively labelled glucose, which in turn indicates high metabolic activity, a hallmark of cancerous tissue.

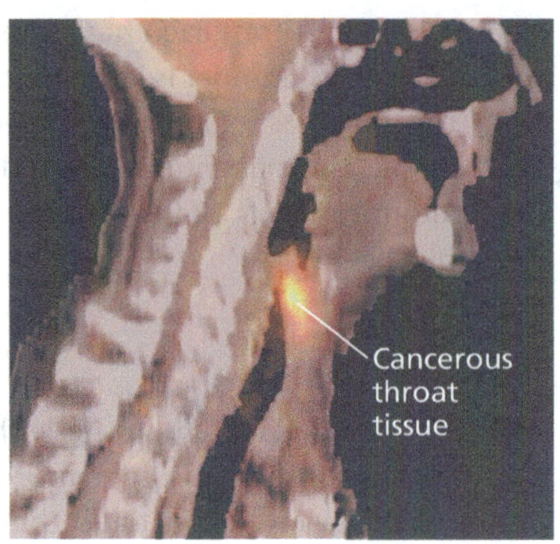

can monitor growth and metabolism of cancers in the body **(Figure 2.5)**.

Although radioactive isotopes are very useful in biological research and medicine, radiation from decaying isotopes also poses a hazard to life by damaging cellular molecules. The severity of this damage depends on the type and amount of radiation an organism absorbs. One of the most serious environmental threats is radioactive fallout from nuclear accidents. The doses of most isotopes used in medical diagnosis, however, are relatively safe.

Radiometric Dating

EVOLUTION Researchers measure radioactive decay in fossils to date these relics of past life. Fossils provide a large body of evidence for evolution, documenting differences between organisms from the past and those living at present and giving us insight into species that have disappeared over time. While the layering of fossil beds establishes that deeper fossils are older than more shallow ones, the actual age (in years) of the fossils in each layer cannot be determined by position alone. This is where radioactive isotopes come in.

A "parent" isotope decays into its "daughter" isotope at a fixed rate, expressed as the **half-life** of the isotope—the time it takes for 50% of the parent isotope to decay. Each radioactive isotope has a characteristic half-life that is not affected by temperature, pressure, or any other environmental variable. Using a process called **radiometric dating**, scientists measure the ratio of different isotopes and calculate how many half-lives (in years) have passed since an organism was fossilised or a rock was formed. Half-life values range from very short for some isotopes, measured in seconds or days, to extremely long—uranium-238 has a half-life of 4.5 billion years! Each isotope can best "measure" a particular range of years: Uranium-238 was used to determine that moon rocks are approximately 4.5 billion years old, similar to the estimated age of Earth. In the **Scientific Skills Exercise**, you can work with data from an experiment that used carbon-14 to determine the age of an important fossil. (Figure 25.6 explains more about radiometric dating of fossils.)

The Energy Levels of Electrons

The simplified models of the atom in Figure 2.4 greatly exaggerate the size of the nucleus relative to that of the whole atom. If an atom of helium were the size of a typical football stadium, the nucleus would be the size of a pencil eraser in the centre of the field. Moreover, the electrons would be like two tiny gnats buzzing around the stadium. Atoms are mostly empty space. When two atoms approach each other during a chemical reaction, their nuclei do not come close enough to interact. Of the three subatomic particles we have discussed, only electrons are directly involved in chemical reactions.

An atom's electrons vary in the amount of energy they possess. **Energy** is defined as the capacity to cause change—for instance, by doing work. **Potential energy** is the energy that matter possesses because of its location or structure. For example, water in a reservoir on a hill has potential energy because of its altitude. When the gates of the reservoir's dam are opened and the water runs downhill, the energy can be used to do work, such as moving the blades of turbines to generate electricity. Because energy has been expended, the water has less energy at the bottom of the hill than it did in the reservoir. Matter has a natural tendency to move towards the lowest possible state of potential energy; in our example, the water runs downhill. To restore the potential energy of a reservoir, work must be done to elevate the water against gravity.

The electrons of an atom have potential energy due to their distance from the nucleus **(Figure 2.6)**. The negatively charged electrons are attracted to the positively charged nucleus.

▼ **Figure 2.6 Energy levels of an atom's electrons.** Electrons exist only at fixed levels of potential energy called electron shells.

(a) A ball bouncing down a flight of stairs can come to rest only on each step, not between steps. Similarly, an electron can exist only at certain energy levels, not between levels.

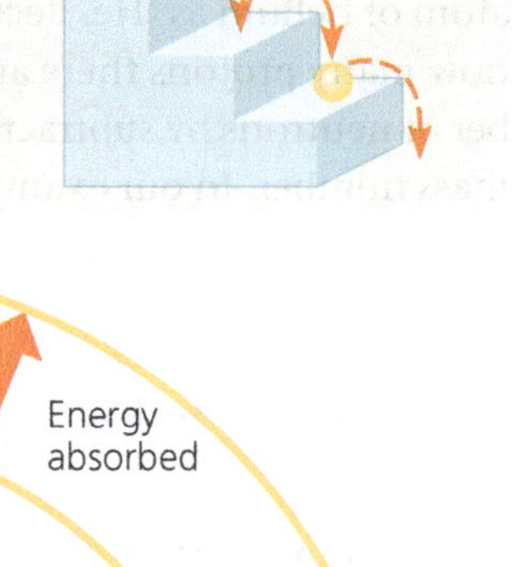

(b) An electron can move from one shell to another only if the energy it gains or loses is exactly equal to the difference in energy between the energy levels of the two shells. Arrows in this model indicate some of the stepwise changes in potential energy that are possible.

Scientific Skills Exercise

Calibrating a Standard Radioactive Isotope Decay Curve and Interpreting Data

How Long Might Neanderthals Have Co-Existed with Modern Humans (*Homo sapiens*)? Neanderthals (*Homo neanderthalensis*) were living in Europe by 350,000 years ago and may have coexisted with early *Homo sapiens* in parts of Eurasia for hundreds or thousands of years before Neanderthals became extinct. Researchers sought to more accurately determine the extent of their overlap by pinning down the latest date Neanderthals still lived in the area. They used carbon-14 dating to determine the age of a Neanderthal fossil from the most recent (uppermost) archeological layer containing Neanderthal bones. In this exercise you will calibrate a standard carbon-14 decay curve and use it to determine the age of this Neanderthal fossil. The age will help you approximate the last time the two species may have coexisted at the site where this fossil was collected.

How the Experiment Was Done Carbon-14 (^{14}C) is a radioactive isotope of carbon that decays to ^{14} at a constant rate. ^{14}C is present in the atmosphere in small amounts at a constant ratio with both ^{13}C and ^{12}C, two other isotopes of carbon. When carbon is taken up from the atmosphere by a plant during photosynthesis, ^{12}C, ^{13}C, and ^{14}C isotopes are incorporated into the plant in the same proportions in which they were present in the atmosphere. These proportions remain the same in the tissues of an animal that eats the plant. While an organism is alive, the ^{14}C in its body constantly decays to ^{14}N but is constantly replaced by new carbon from the environment. Once an organism dies, it stops taking in new ^{14}C but the ^{14}C in its tissues continues to decay, while the ^{12}C in its tissues remains the same because it is not radioactive and does not decay. Thus, scientists can calculate how long the pool of original ^{14}C has been decaying in a fossil by measuring the ratio of ^{14}C to ^{12}C and comparing it to the ratio of ^{14}C to ^{12}C present originally in the atmosphere. The fraction of ^{14}C in a fossil compared to the original fraction of ^{14}C can be converted to years because we know that the half-life of ^{14}C is 5,730 years—in other words, half of the ^{14}C in a fossil decays every 5,730 years.

Data from the Experiment The researchers found that the Neanderthal fossil had approximately 0.0078 (or, in scientific notation, 7.8×10^{-3}) as much ^{14}C as the atmosphere. The following questions will guide you through translating this fraction into the age of the fossil.

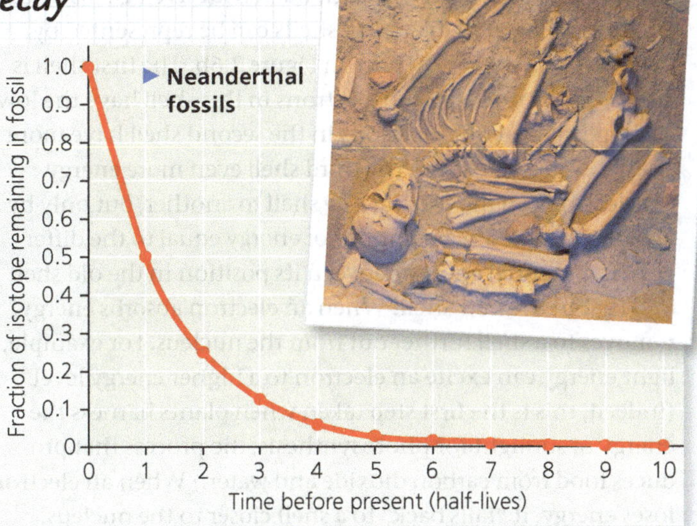

Data from R. Pinhasi et al., Revised age of late Neanderthal occupation and the end of the Middle Paleolithic in the northern Caucasus, *Proceedings of the National Academy of Sciences USA* 147:8611–8616 (2011). doi 10.1073/pnas.1018938108

INTERPRET THE DATA

1. The graph shows a standard curve of radioactive isotope decay. The line shows the fraction of the radioactive isotope over time (before the present) in units of half-lives. Recall that a half-life is the amount of time it takes for half of the radioactive isotope to decay. Labelling each data point with the corresponding fractions will help orient you to this graph. Draw an arrow to the data point for half-life = 1 and write the fraction of ^{14}C that will remain after one half-life. Calculate the fraction of ^{14}C remaining at each half-life and write the fractions on the graph near arrows pointing to the data points. Convert each fraction to a decimal number and round off to a maximum of three significant digits (zeros at the beginning of the number do not count as significant digits). Also write each decimal number in scientific notation.
2. Recall that ^{14}C has a half-life of 5,730 years. To calibrate the *x*-axis for ^{14}C decay, write the time before present in years below each half-life.
3. The researchers found that the Neanderthal fossil had approximately 0.0078 as much ^{14}C as found originally in the atmosphere. **(a)** Using the numbers on your graph, determine how many half-lives have passed since the Neanderthal died. **(b)** Using your ^{14}C calibration on the *x*-axis, what is the approximate age of the Neanderthal fossil in years (round off to the nearest thousand)? **(c)** Approximately when did Neanderthals become extinct according to this study? **(d)** The researchers cite evidence that modern humans (*H. sapiens*) became established in the same region as the last Neanderthals approximately 39,000–42,000 years ago. What does this suggest about possible overlap of Neanderthals and modern humans?
4. Carbon-14 dating works for fossils up to about 75,000 years old; fossils older than that contain too little ^{14}C to be detected. Most dinosaurs went extinct 65.5 million years ago. **(a)** Can ^{14}C be used to date dinosaur bones? Explain. **(b)** Radioactive uranium-235 has a half-life of 704 million years. If it was incorporated into dinosaur bones, could it be used to date the dinosaur fossils? Explain.

It takes work to move a given electron further away from the nucleus, so the more distant an electron is from the nucleus, the greater its potential energy. Unlike the continuous flow of water downhill, changes in the potential energy of electrons can occur only in steps of fixed amounts. An electron having a certain amount of energy is something like a ball on a staircase (see Figure 2.6a). The ball can have different amounts of potential energy, depending on which step it is on, but it cannot spend much time between the steps. Similarly, an electron's potential energy is determined by its energy level. An electron can exist only at certain energy levels, not between them.

An electron's energy level is correlated with its average distance from the nucleus. Electrons are found in different **electron shells**, each with a characteristic average distance and energy level. In diagrams, shells can be represented by concentric circles, as they are in Figure 2.6b. The first shell is closest to the nucleus, and electrons in this shell have the lowest potential energy. Electrons in the second shell have more energy, and electrons in the third shell even more energy. An electron can move from one shell to another, but only by absorbing or losing an amount of energy equal to the difference in potential energy between its position in the old shell and that in the new shell. When an electron absorbs energy, it moves to a shell further out from the nucleus. For example, light energy can excite an electron to a higher energy level. (Indeed, this is the first step taken when plants harness the energy of sunlight for photosynthesis, the process that produces food from carbon dioxide and water.) When an electron loses energy, it "falls back" to a shell closer to the nucleus, and the lost energy is usually released to the environment as visible light or ultraviolet radiation.

Electron Distribution and Chemical Properties

The chemical behaviour of an atom is determined by the distribution of electrons in the atom's electron shells. Beginning with hydrogen, the simplest atom, we can imagine building the atoms of the other elements by adding 1 proton and 1 electron at a time (along with an appropriate number of neutrons). **Figure 2.7**, a modified version of what is called the *periodic table of the elements*, shows this distribution of electrons for the first 18 elements, from hydrogen ($_1H$) to argon ($_{18}Ar$). The elements are arranged in three rows, or *periods*, corresponding to the number of electron shells in their atoms. The left-to-right sequence of elements in each row corresponds to the sequential addition of electrons and protons.

Hydrogen's 1 electron and helium's 2 electrons are located in the first shell. Electrons, like all matter, tend to exist in the lowest available state of potential energy. In an atom, this state is in the first shell. However, the first shell can hold no more than 2 electrons; thus, hydrogen and helium are the

▼ Figure 2.7 Electron distribution diagrams for the first 18 elements in the periodic table. In a standard periodic table, information for each element is presented as shown for helium in the inset. In the diagrams in this table, electrons are represented as yellow dots and electron shells as concentric circles. These diagrams are a convenient way to picture the distribution of an atom's electrons among its electron shells, but these simplified models do not accurately represent the shape of the atom or the location of its electrons. The elements are arranged in rows, each representing the filling of an electron shell. As electrons are added, they occupy the lowest available shell.

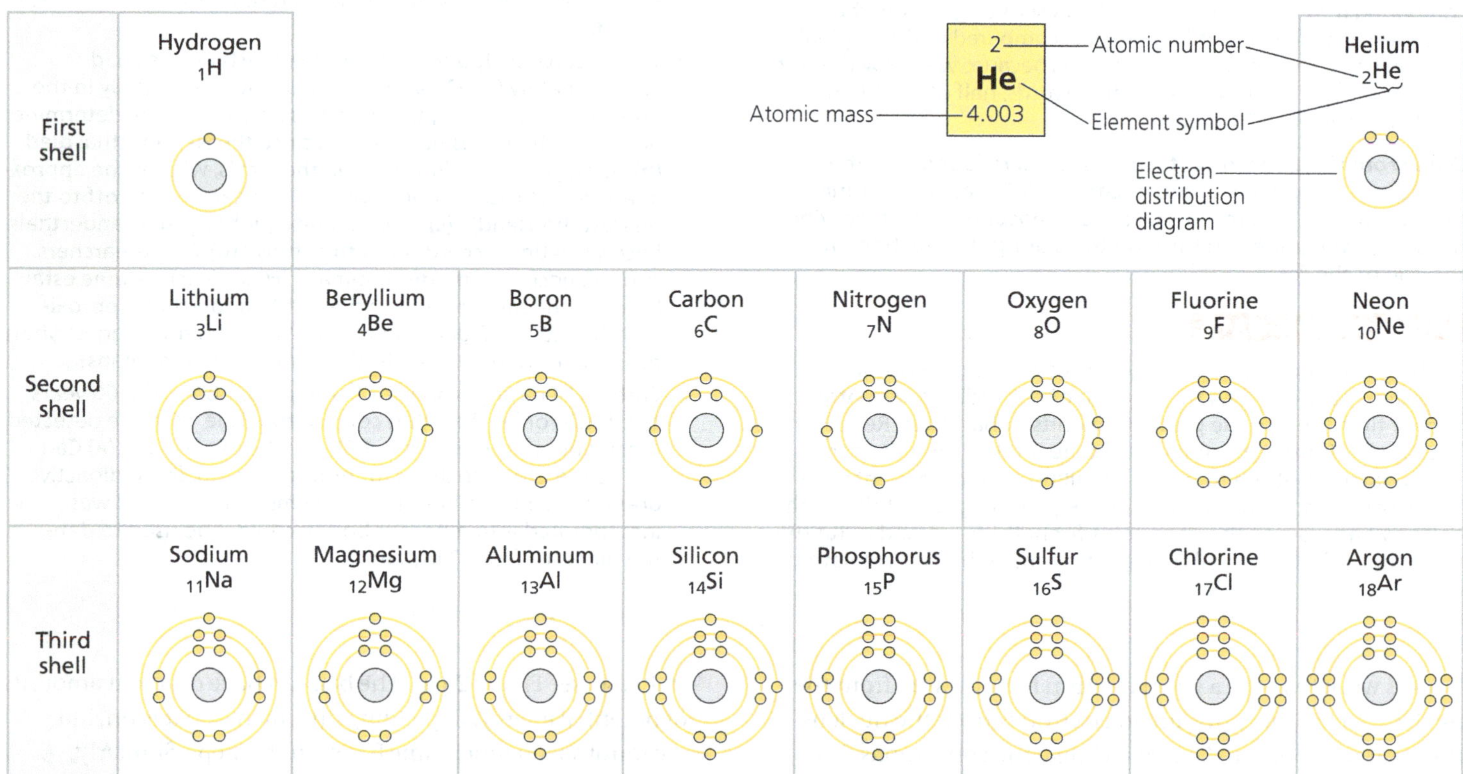

VISUAL SKILLS *What is the atomic number of magnesium? How many protons and electrons does it have? How many electron shells? How many valence electrons?*

only elements in the first row of the table. In an atom with more than 2 electrons, the additional electrons must occupy higher shells because the first shell is full. The next element, lithium, has 3 electrons. Two of these electrons fill the first shell, while the third electron occupies the second shell. The second shell holds a maximum of 8 electrons. Neon, at the end of the second row, has 8 electrons in the second shell, giving it a total of 10 electrons.

The chemical behaviour of an atom depends mostly on the number of electrons in its *outermost* shell. We call those outer electrons **valence electrons** and the outermost electron shell the **valence shell**. In the case of lithium, there is only 1 valence electron, and the second shell is the valence shell. Atoms with the same number of electrons in their valence shells exhibit similar chemical behaviour. For example, fluorine (F) and chlorine (Cl) both have 7 valence electrons, and both form compounds when combined with the element sodium (Na): Sodium fluoride (NaF) is commonly added to toothpaste to prevent tooth decay, and, as described earlier, NaCl is table salt (see Figure 2.2). An atom with a completed valence shell is unreactive; that is, it will not interact readily with other atoms. At the far right of the periodic table are helium, neon, and argon, the only three elements shown in Figure 2.7 that have full valence shells. These elements are said to be *inert*, meaning chemically unreactive. All the other atoms in Figure 2.7 are chemically reactive because they have incomplete valence shells.

Electron Orbitals

In the early 1900s, the electron shells of an atom were visualised as concentric paths of electrons orbiting the nucleus, somewhat like planets orbiting the sun. It is still convenient to use two-dimensional concentric-circle diagrams, as in Figure 2.7, to symbolise three-dimensional electron shells. However, you need to remember that each concentric circle represents only the *average* distance between an electron in that shell and the nucleus. Accordingly, the concentric-circle diagrams do not give a real picture of an atom. In reality, we can never know the exact location of an electron. What we can do instead is describe the space in which an electron spends most of its time. The three-dimensional space where an electron is found 90% of the time is called an **orbital**.

Each electron shell contains electrons at a particular energy level, distributed among a specific number of orbitals of distinctive shapes and orientations. **Figure 2.8** shows the orbitals of neon as an example, with its electron distribution diagram for reference. You can think of an orbital as a component of an electron shell. The first electron shell has only one spherical *s* orbital (called 1*s*), but the second shell has four orbitals: one large spherical *s* orbital (called 2*s*) and three dumbbell-shaped *p* orbitals (called 2*p* orbitals). (The third shell and other higher electron shells also have *s* and *p* orbitals, as well as orbitals of more complex shapes.)

▼ **Figure 2.8 Electron orbitals.**

Neon, with two filled shells (10 electrons)

First shell

Second shell

(a) Electron distribution diagram. An electron distribution diagram is shown here for a neon atom, which has a total of 10 electrons. Each concentric circle represents an electron shell, which can be subdivided into electron orbitals.

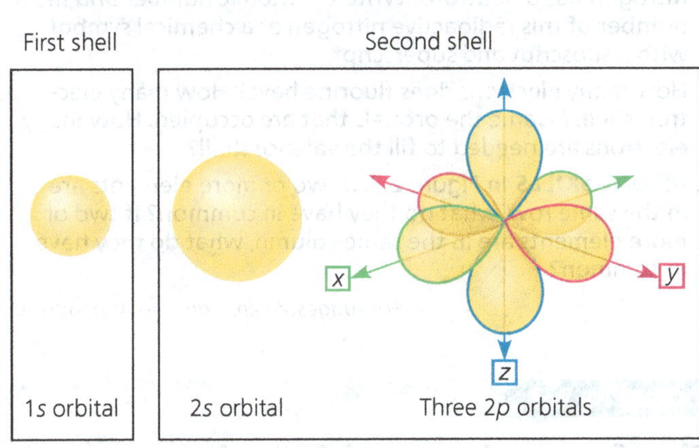

(b) Separate electron orbitals. The three-dimensional shapes represent electron orbitals—the volumes of space where the electrons of an atom are most likely to be found. Each orbital holds a maximum of 2 electrons. The first electron shell, on the left, has one spherical (*s*) orbital, designated 1*s*. The second shell, on the right, has one larger *s* orbital (designated 2*s* for the second shell) plus three dumbbell-shaped orbitals called *p* orbitals (2*p* for the second shell). The three 2*p* orbitals lie at right angles to one another along imaginary *x*-, *y*-, and *z*-axes of the atom. Each 2*p* orbital is outlined here in a different colour.

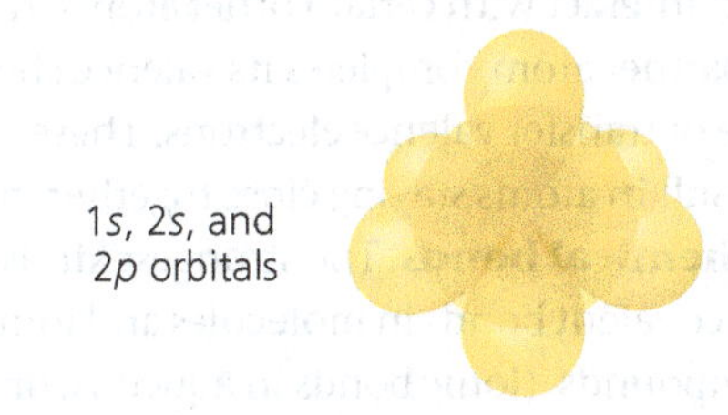

(c) Superimposed electron orbitals. To reveal the complete picture of the electron orbitals of neon, we superimpose the 1*s* orbital of the first shell and the 2*s* and three 2*p* orbitals of the second shell.

No more than 2 electrons can occupy a single orbital. The first electron shell can therefore accommodate up to 2 electrons in its *s* orbital. The lone electron of a hydrogen atom occupies the 1*s* orbital, as do the 2 electrons of a helium atom. The four orbitals of the second electron shell can hold up to 8 electrons, 2 in each orbital. Electrons in each of the four orbitals in the second shell have nearly the same energy, but they move in different volumes of space.

The reactivity of an atom arises from the presence of unpaired electrons in one or more orbitals of the atom's valence shell. As you will see in the next section, atoms interact in a way that completes their valence shells. When they do so, it is the *unpaired* electrons that are involved.

CONCEPT CHECK 2.2

1. A lithium atom has 3 protons and 4 neutrons. What is its mass number?
2. A nitrogen atom has 7 protons, and the most common isotope of nitrogen has 7 neutrons. A radioactive isotope of nitrogen has 8 neutrons. Write the atomic number and mass number of this radioactive nitrogen as a chemical symbol with a subscript and superscript.
3. How many electrons does fluorine have? How many electron shells? Name the orbitals that are occupied. How many electrons are needed to fill the valence shell?
4. **VISUAL SKILLS** In Figure 2.7, if two or more elements are in the same row, what do they have in common? If two or more elements are in the same column, what do they have in common?

For suggested answers, see Appendix A.

CONCEPT 2.3

The formation and function of molecules and ionic compounds depend on chemical bonding between atoms

Now that we have looked at the structure of atoms, we can move up the hierarchy of organisation and see how atoms combine to form molecules and ionic compounds. Atoms with incomplete valence shells can interact with certain other atoms in such a way that each partner atom completes its valence shell: The atoms either share or transfer valence electrons. These interactions usually result in atoms staying close together, held by attractions called **chemical bonds**. The strongest kinds of chemical bonds are covalent bonds in molecules and ionic bonds in dry ionic compounds. (Ionic bonds in aqueous, or water-based, solutions are weak interactions, as we will see later.)

Covalent Bonds

A **covalent bond** is the sharing of a pair of valence electrons by two atoms. For example, let's consider what happens when two hydrogen atoms approach each other. Recall that hydrogen has 1 valence electron in the first shell, but the shell's capacity is 2 electrons. When the two hydrogen atoms come close enough for their 1*s* orbitals to overlap, they can share their electrons **(Figure 2.9)**. Each hydrogen atom is now associated with 2 electrons in what amounts to a completed valence shell. Two or more atoms held together by covalent bonds constitute a **molecule**, in this case a hydrogen molecule.

Figure 2.9 Formation of a covalent bond.

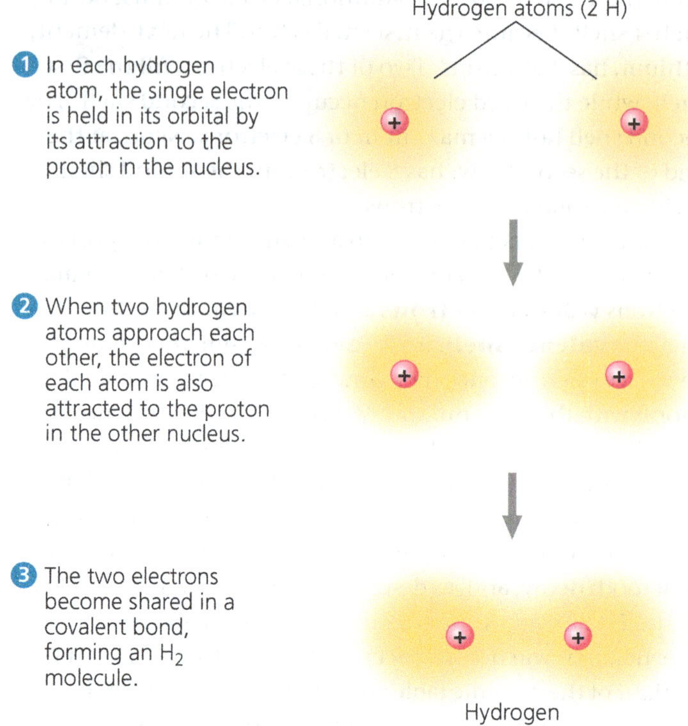

Figure 2.10a shows several ways of representing a hydrogen molecule. Its *molecular formula*, H_2, simply indicates that the molecule consists of two atoms of hydrogen. Electron sharing can be depicted by an electron distribution diagram or by a *Lewis dot structure*, in which element symbols are surrounded by dots that represent the valence electrons (H:H). We can also use a *structural formula*, H—H, where the line represents a **single bond**, a pair of shared electrons. A *space-filling model* comes closest to representing the actual shape of the molecule. (You may also be familiar with ball-and-stick models, which are shown in Figure 2.15.)

Oxygen has 6 electrons in its second electron shell and therefore needs 2 more electrons to complete its valence shell. Two oxygen atoms form a molecule by sharing *two* pairs of valence electrons **(Figure 2.10b)**. The atoms are thus joined by what is called a **double bond** (O═O).

Each atom that can share valence electrons has a bonding capacity corresponding to the number of covalent bonds the atom can form. When the bonds form, they give the atom a full complement of electrons in the valence shell. The bonding capacity of oxygen, for example, is 2. This bonding capacity is called the atom's **valence** and usually equals the number of electrons required to complete the atom's outermost (valence) shell. See if you can determine the valences of hydrogen, oxygen, nitrogen, and carbon by studying the electron distribution diagrams in Figure 2.7. You can see that the valence of hydrogen is 1; oxygen, 2; nitrogen, 3; and carbon, 4. The situation is more complicated for phosphorus, in the third row of the periodic table, which can have a valence of 3 or 5 depending on the combination of single and double bonds it makes.

The molecules H_2 and O_2 are pure elements rather than compounds because a compound is a combination of two or

▼ Figure 2.10 Covalent bonding in four molecules. The number of electrons required to complete an atom's valence shell generally determines how many covalent bonds that atom will form. This figure shows several ways of indicating covalent bonds.

Name and Molecular Formula	Electron Distribution Diagram	Lewis Dot Structure and Structural Formula	Space-Filling Model
(a) Hydrogen (H_2). Two hydrogen atoms share one pair of electrons, forming a single bond.	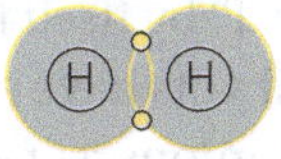 H H	H:H H—H	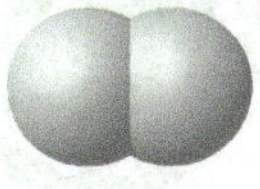
(b) Oxygen (O_2). Two oxygen atoms share two pairs of electrons, forming a double bond.	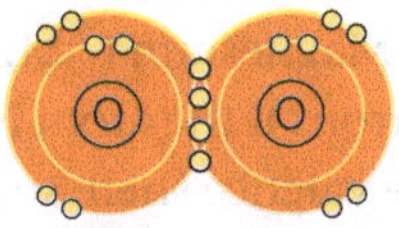 O O	Ö::Ö O=O	
(c) Water (H_2O). Two hydrogen atoms and one oxygen atom are joined by single bonds, forming a molecule of water.	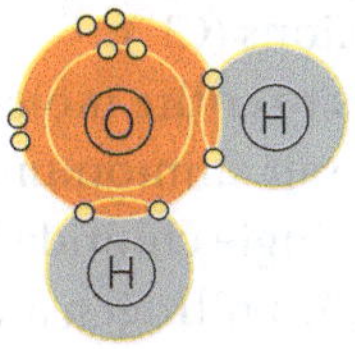 O H H	:Ö:H H O—H with H below	
(d) Methane (CH_4). Four hydrogen atoms can satisfy the valence of one carbon atom, forming methane.	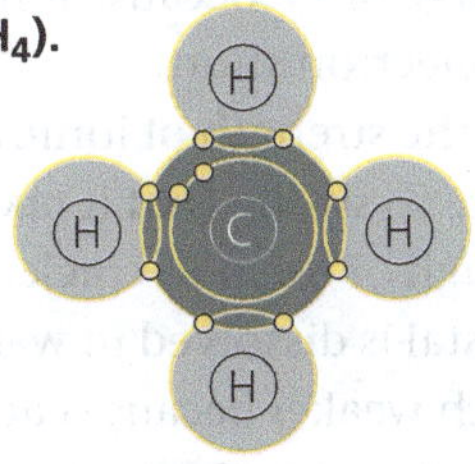 H H C H H	H:C:H with H above and below H—C—H with H above and below	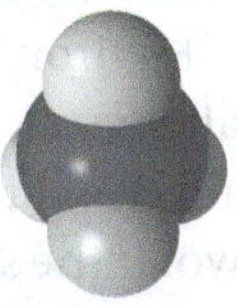

more *different* elements. Water, with the molecular formula H_2O, is a compound. Two atoms of hydrogen are needed to satisfy the valence of one oxygen atom. **Figure 2.10c** shows the structure of a water molecule. (Water is so important to life that the chapter, "Water and Life" is devoted to its structure and behaviour.)

Methane, the main component of natural gas, is a compound with the molecular formula CH_4. It takes four hydrogen atoms, each with a valence of 1, to complete the valence shell of a carbon atom, with its valence of 4 **(Figure 2.10d)**. (We'll look at other carbon compounds in the chapter, "Carbon and the Molecular Diversity of Life.")

Atoms in a molecule attract shared bonding electrons to varying degrees, depending on the element. The attraction of a particular atom for the electrons of a covalent bond is called its **electronegativity**. The more electronegative an atom is, the more strongly it pulls shared electrons towards itself. In a covalent bond between two atoms of the same element, the electrons are shared equally because the two atoms have the

▼ Figure 2.11 Polar covalent bonds in a water molecule.

Because oxygen (O) is more electronegative than hydrogen (H), shared electrons are pulled more towards oxygen. Thus, the covalent bonds in water are polar.

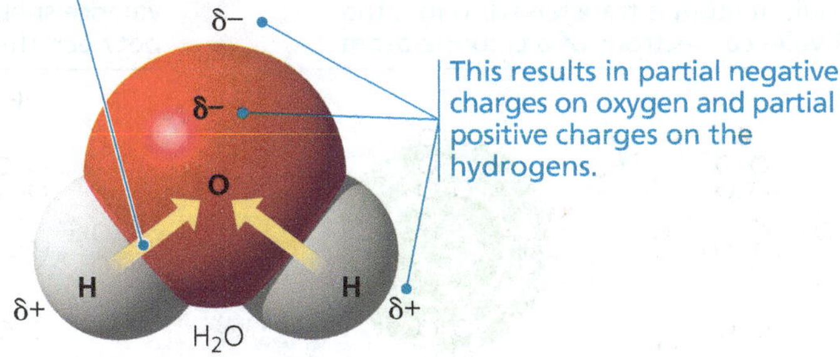

same electronegativity—the tug-of-war is at a standoff. Such a bond is called a **nonpolar covalent bond**. For example, the single bond of H_2 is nonpolar, as is the double bond of O_2. However, when an atom is bonded to a more electronegative atom, the electrons of the bond are not shared equally. This type of bond is called a **polar covalent bond**. Such bonds vary in their polarity, depending on the relative electronegativity of the two atoms. For example, the bonds between the oxygen and hydrogen atoms of a water molecule are quite polar **(Figure 2.11)**.

Oxygen is one of the most electronegative elements, attracting shared electrons much more strongly than hydrogen does. In a covalent bond between oxygen and hydrogen, the electrons spend more time near the oxygen nucleus than near the hydrogen nucleus. Because electrons have a negative charge and are pulled towards oxygen in a water molecule, the oxygen atom has partial negative charges (indicated by the Greek letter δ with a minus sign, δ−, or "delta minus"), and the hydrogen atoms have partial positive charges (δ+, or "delta plus"). In contrast, the individual bonds of methane (CH_4) are much less polar because the electronegativities of carbon and hydrogen are quite similar.

Ionic Bonds

In some cases, two atoms are so unequal in their attraction for valence electrons that the more electronegative atom strips an electron completely away from its partner. The two resulting oppositely charged atoms (or molecules) are called **ions**. A positively charged ion is called a **cation**, while a negatively charged ion is called an **anion**. (It may help you to think of the *t* in *cation* as a plus sign, and of *anion* as "a negative ion.") Because of their opposite charges, cations and anions attract each other; this attraction is called an **ionic bond**. Note that the transfer of an electron is not, by itself, the formation of a bond; rather, it allows a bond to form because it results in two ions of opposite charge. Any two ions of opposite charge can form an ionic bond. The ions do not need to have acquired their charge by an electron transfer with each other.

▼ **Figure 2.12 Electron transfer and ionic bonding.** The attraction between oppositely charged atoms, or ions, is an ionic bond. An ionic bond can form between any two oppositely charged ions, even if they have not been formed by transfer of an electron from one to the other.

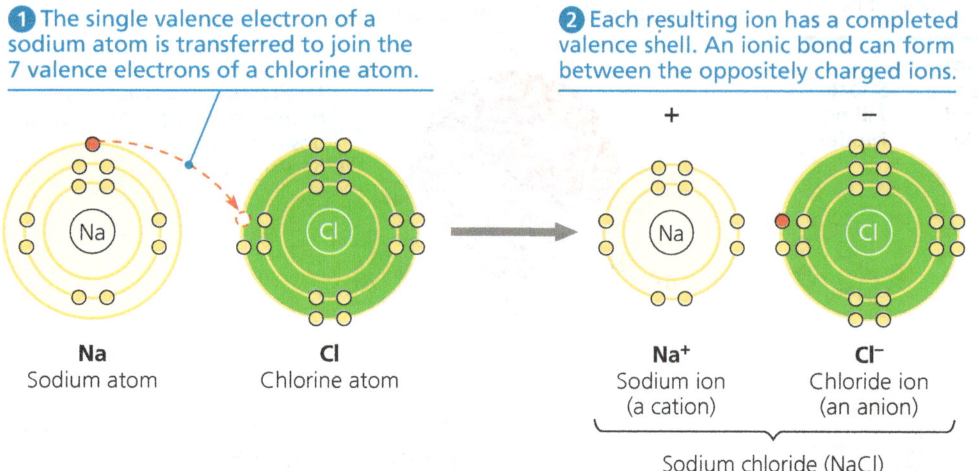

This is what happens when an atom of sodium ($_{11}Na$) encounters an atom of chlorine ($_{17}Cl$) **(Figure 2.12)**. A sodium atom has a total of 11 electrons, with its single valence electron in the third electron shell. A chlorine atom has a total of 17 electrons, with 7 electrons in its valence shell. When these two atoms meet, the lone valence electron of sodium is transferred to the chlorine atom, and both atoms end up with their valence shells complete. (Because sodium no longer has an electron in the third shell, the second shell is now the valence shell.) The electron transfer between the two atoms moves one unit of negative charge from sodium to chlorine. Sodium, now with 11 protons but only 10 electrons, has a net electrical charge of 1+; the sodium atom has become a cation. Conversely, the chlorine atom, having gained an extra electron, now has 17 protons and 18 electrons, giving it a net electrical charge of 1–; it has become a chloride ion—an anion.

Compounds formed by ionic bonds are called **ionic compounds**, or **salts**. We know the ionic compound sodium chloride (NaCl) as table salt **(Figure 2.13)**. Salts are often found in nature as crystals of various sizes and shapes. Each salt crystal is an aggregate of vast numbers of cations and anions bonded by their electrical attraction and arranged in a three-dimensional lattice. Unlike a covalent compound, which consists of molecules having a definite size and number of atoms, an ionic compound does not consist of molecules. The formula for an ionic compound, such as NaCl, indicates only the ratio of elements in a crystal of the salt. "NaCl" by itself is not a molecule.

▼ **Figure 2.13 A sodium chloride (NaCl) crystal.** The sodium ions (Na^+) and chloride ions (Cl^-) are held together by ionic bonds. The formula NaCl tells us that the ratio of Na^+ to Cl^- is 1:1.

Not all salts have equal numbers of cations and anions. For example, the ionic compound magnesium chloride ($MgCl_2$) has two chloride ions for each magnesium ion. Magnesium ($_{12}Mg$) must lose 2 outer electrons if the atom is to have a complete valence shell, so it has a tendency to become a cation with a net charge of 2+(Mg^{2+}). One magnesium cation can therefore form ionic bonds with two chloride anions (Cl^-).

The term *ion* also applies to entire molecules that are electrically charged. In the salt ammonium chloride (NH_4Cl), for instance, the anion is a single chloride ion (Cl^-), but the cation is ammonium (NH_4^+), a nitrogen atom covalently bonded to four hydrogen atoms. The whole ammonium ion has an electrical charge of 1+ because it has given up 1 electron and thus is 1 electron short.

Environment affects the strength of ionic bonds. In a dry salt crystal, the bonds are so strong that it takes a hammer and chisel to break enough of them to crack the crystal in two. If the same salt crystal is dissolved in water, however, the ionic bonds are much weaker because each ion is partially shielded by its interactions with water molecules. Most drugs are manufactured as salts because they are quite stable when dry but can dissociate (come apart) easily in water. (In Concept 3.2, you will learn how water dissolves salts.)

Weak Chemical Interactions

In organisms, most of the strongest chemical bonds are covalent bonds, which link atoms to form a cell's molecules. But weaker interactions within and between molecules are also indispensable, contributing greatly to the emergent properties of life. Many large biological molecules are held in their functional form by weak interactions. In addition, when two molecules in the cell make contact, they may adhere temporarily by weak interactions. The reversibility of weak interactions can be an advantage: Two molecules can come together, affect one another in some way, and then separate.

Several types of weak chemical interactions are important in organisms. One is the ionic bond as it exists between ions dissociated in water, which we just discussed. Hydrogen bonds and van der Waals interactions are also crucial to life.

Hydrogen Bonds

Among weak chemical interactions, hydrogen bonds are so central to the chemistry of life that they deserve special attention. When a hydrogen atom is covalently bonded to an electronegative atom, the hydrogen atom has a partial positive charge that allows it to be attracted to a different electronegative atom with a partial negative charge nearby. This noncovalent attraction between a hydrogen and an electronegative atom is called a **hydrogen bond**. In living cells, the electronegative partners are usually oxygen or nitrogen atoms. **Figure 2.14** shows hydrogen bonding between water (H_2O) and ammonia (NH_3).

▼ Figure 2.14 A hydrogen bond.

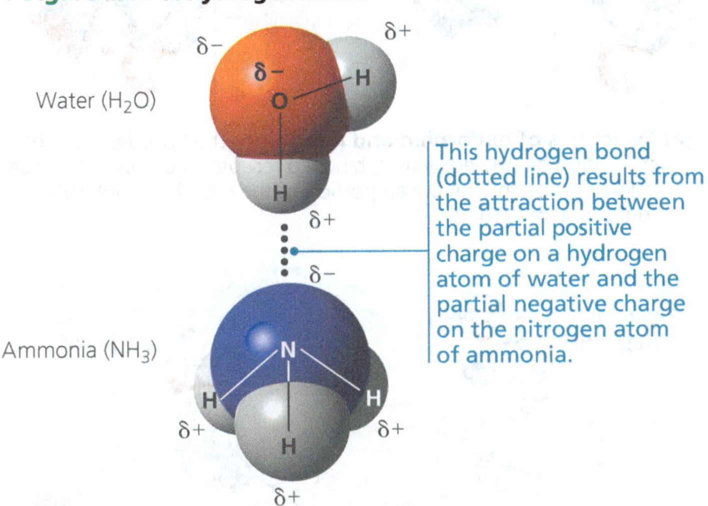

DRAW IT *Draw one water molecule hydrogen bonded to four other water molecules around it. Use simple outlines of space-filling models. Draw the partial charges on the water molecules and use dots for the hydrogen bonds.*

Van der Waals Interactions

Even a molecule with nonpolar covalent bonds may have positively and negatively charged regions. Electrons are not always evenly distributed; at any instant, they may accumulate by chance in one part of a molecule or another. The results are ever-changing regions of positive and negative charge that enable all atoms and molecules to stick to one another. These **van der Waals interactions** are individually weak and occur only when atoms and molecules are very close together. When many such interactions occur simultaneously, however, they can be powerful: Van der Waals interactions allow the gecko lizard shown here to walk straight up a wall! The anatomy of the gecko's foot—including toes with hundreds of thousands of tiny hairs, each with multiple projections—maximises surface contact with the wall. The van der Waals interactions between the foot molecules and the molecules of the wall's surface are so numerous that despite their individual weakness, together they can support the gecko's body weight.

Van der Waals interactions, hydrogen bonds, ionic bonds in water, and other weak interactions may form not only between molecules but also between parts of a large molecule, such as a protein or nucleic acid. The cumulative effect of weak interactions is to reinforce the three-dimensional shape of the molecule. (You will learn more about the very important biological roles of weak interactions in Figures 5.18 and 5.24.)

Molecular Shape and Function

A molecule has a characteristic size and shape, which are key to its function in the living cell. A molecule consisting of two atoms, such as H_2 or O_2, is always linear, but most molecules with more than two atoms have more complicated shapes. These shapes are determined by the positions of the atoms' orbitals **(Figure 2.15)**. When an atom forms covalent bonds, the orbitals in its valence shell undergo rearrangement.

▼ Figure 2.15 Molecular shapes due to hybrid orbitals.

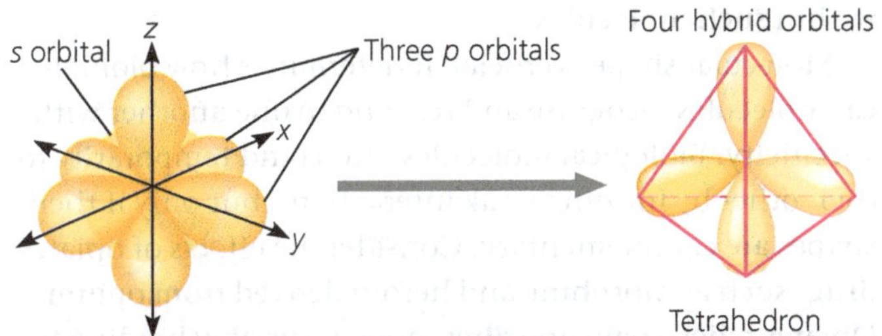

(a) Hybradisation of orbitals. The single *s* and three *p* orbitals of a valence shell involved in covalent bonding combine to form four teardrop-shaped hybrid orbitals. These orbitals extend to the four corners of an imaginary tetrahedron (outlined in pink).

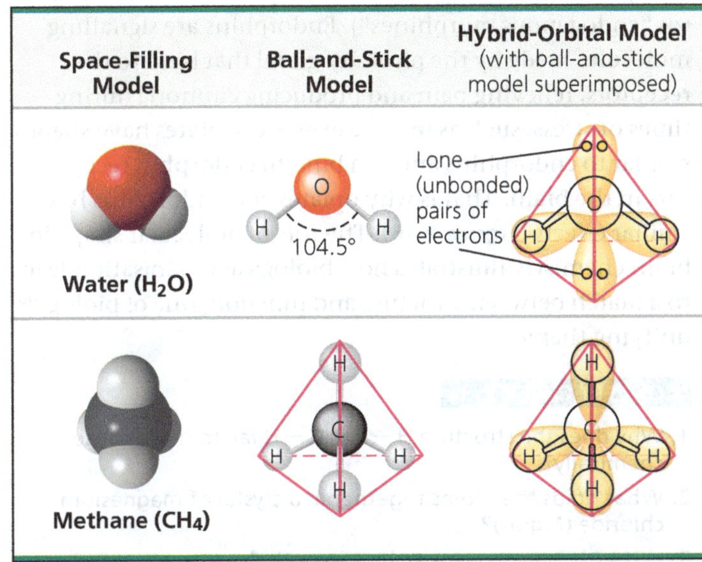

(b) Molecular-shape models. Three models representing molecular shape are shown for water and methane. The positions of the hybrid orbitals determine the shapes of the molecules.

For atoms with valence electrons in both *s* and *p* orbitals (review Figure 2.8), the single *s* and three *p* orbitals form four new hybrid orbitals shaped like identical teardrops extending from the region of the atomic nucleus, as shown in Figure 2.15a. If we connect the larger ends of the teardrops with lines, we have the outline of a geometric shape called a tetrahedron, a pyramid with a triangular base.

For water molecules (H_2O), two of the hybrid orbitals in the oxygen's valence shell are shared with hydrogens. The other two hybrid orbitals are occupied by lone (unbonded) pairs of electrons (see Figure 2.15b). The result is a molecule shaped roughly like a V, with its two covalent bonds at an angle of 104.5°.

The methane molecule (CH_4) has the shape of a completed tetrahedron because all four hybrid orbitals of the carbon atom are shared with hydrogen atoms (see Figure 2.15b). The carbon nucleus is at the centre, with its four covalent bonds radiating to hydrogen nuclei at the corners of the tetrahedron. Larger molecules containing multiple carbon atoms, including many of the molecules that make up living matter, have more complex overall shapes. However, the tetrahedral shape of a carbon atom bonded to four other atoms is often a repeating motif within such molecules.

Molecular shape is crucial: It determines how biological molecules recognise and respond to one another with specificity. Biological molecules often bind temporarily to each other by forming weak interactions, but only if their shapes are complementary. Consider the effects of opiates, drugs such as morphine and heroin derived from opium. Opiates relieve pain and alter mood by weakly binding to specific receptor molecules on the surfaces of brain cells. Why would brain cells carry receptors for opiates, compounds that are not *endogenous*, made by the body? In 1975, this question was answered by the discovery of endorphins (or "endogenous morphines"). Endorphins are signalling molecules made by the pituitary gland that bind to the receptors, relieving pain and producing euphoria during times of stress, such as intense exercise. Opiates have shapes similar to endorphins and can bind to endorphin receptors in the brain. That is why opiates and endorphins have similar effects **(Figure 2.16)**. The role of molecular shape in brain chemistry illustrates how biological organisation leads to a match between structure and function, one of biology's unifying themes.

CONCEPT CHECK 2.3

1. Why does the structure H—C=C—H fail to make sense chemically?
2. What holds the atoms together in a crystal of magnesium chloride ($MgCl_2$)?
3. **WHAT IF?** If you were a pharmaceutical researcher, why would you want to learn the three-dimensional shapes of naturally occurring signalling molecules?

For suggested answers, see Appendix A.

▼ **Figure 2.16 A molecular mimic.** Morphine affects pain perception and emotional state by mimicking the brain's natural endorphins.

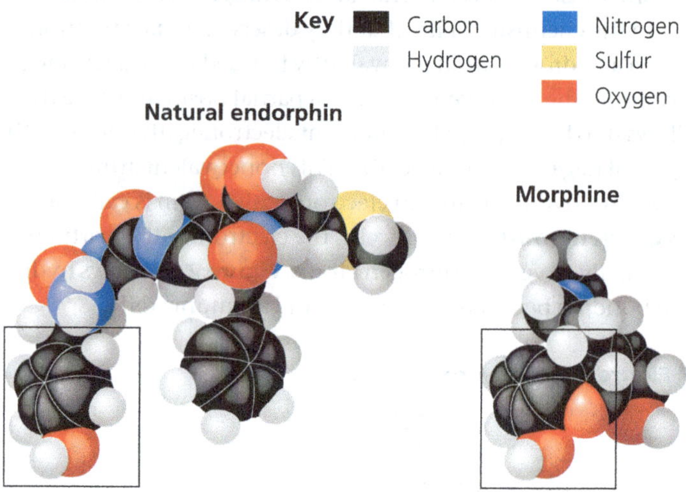

(a) Structures of endorphin and morphine. The boxed portion of the endorphin molecule (left) binds to receptor molecules on target cells in the brain. The boxed portion of the morphine molecule (right) is a close match.

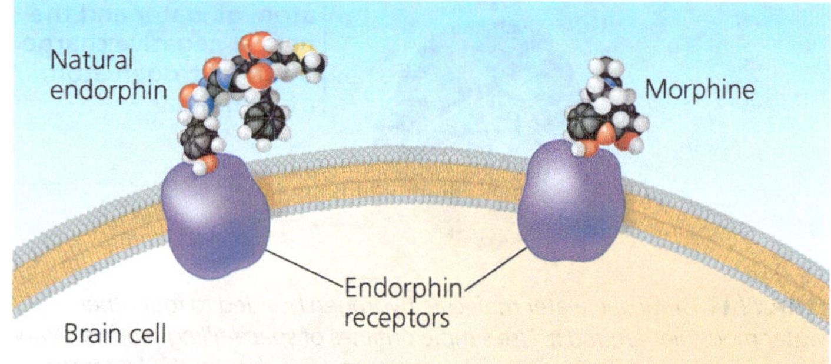

(b) Binding to endorphin receptors. Both endorphin and morphine can bind to endorphin receptors on the surface of a brain cell.

CONCEPT 2.4

Chemical reactions make and break chemical bonds

The making and breaking of chemical bonds, leading to changes in the composition of matter, are called **chemical reactions**. An example is the reaction between hydrogen and oxygen molecules that forms water:

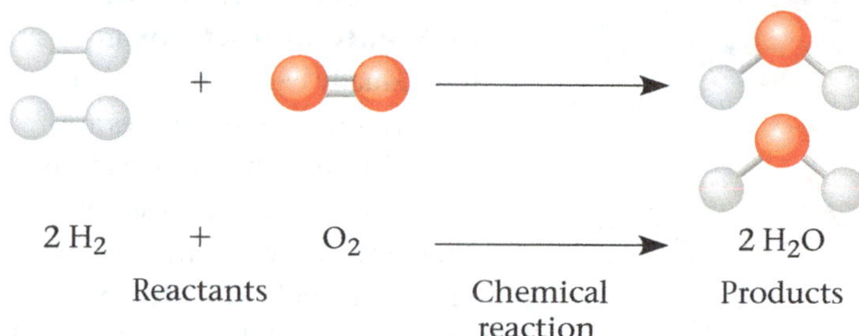

$$2\,H_2 + O_2 \longrightarrow 2\,H_2O$$

Reactants — Chemical reaction — Products

This reaction breaks the covalent bonds of H_2 and O_2 and forms the new bonds of H_2O. When we write the equation for a chemical reaction, we use an arrow to indicate the conversion of the starting materials, called the **reactants**, to the resulting materials, or **products**. The coefficients indicate the number of molecules involved; for example, the coefficient 2 before the H_2 means that the reaction starts with two molecules of hydrogen. Notice that all atoms of the reactants must be accounted for in the products. Matter is conserved in a chemical reaction: Reactions cannot create or destroy atoms but can only rearrange (redistribute) the electrons among them.

Photosynthesis, which takes place within the cells of green plant tissues, is an important biological example of how chemical reactions rearrange matter. Humans and other animals ultimately depend on photosynthesis for food and oxygen, and this process is at the foundation of life in almost all ecosystems.

The following summarises photosynthesis:

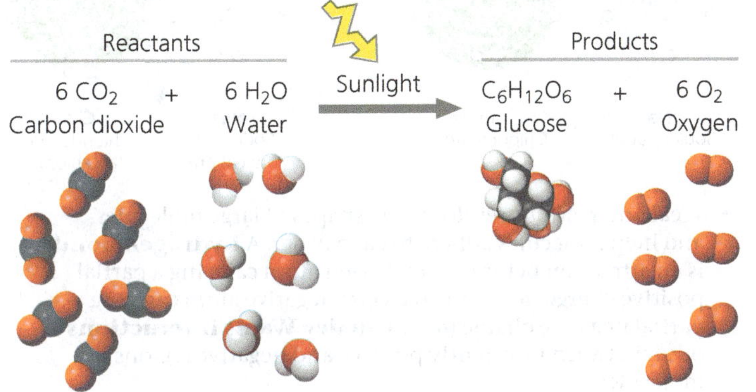

The raw materials of photosynthesis are carbon dioxide (CO_2) and water (H_2O), which land plants absorb from the air and soil, respectively. Within the plant cells, sunlight powers the conversion of these ingredients to a sugar called glucose ($C_6H_{12}O_6$) and oxygen molecules (O_2), a by-product that can be seen when released by a water plant **(Figure 2.17)**. Although photosynthesis is actually a sequence of many chemical reactions, we still end up with the same number and types of atoms that we had when we started. Matter has simply been rearranged, with an input of energy provided by sunlight.

All chemical reactions are theoretically reversible, with the products of the forward reaction becoming the reactants for the reverse reaction. For example, hydrogen and nitrogen molecules can combine to form ammonia, but ammonia can also decompose to regenerate hydrogen and nitrogen:

$$3\,H_2 + N_2 \rightleftharpoons 2\,NH_3$$

The two opposite-headed arrows indicate that the reaction is reversible.

One of the factors affecting the rate of a reaction is the concentration of reactants. The greater the concentration of reactant molecules, the more frequently they collide with one another and have an opportunity to react and form products. The same holds true for products. As

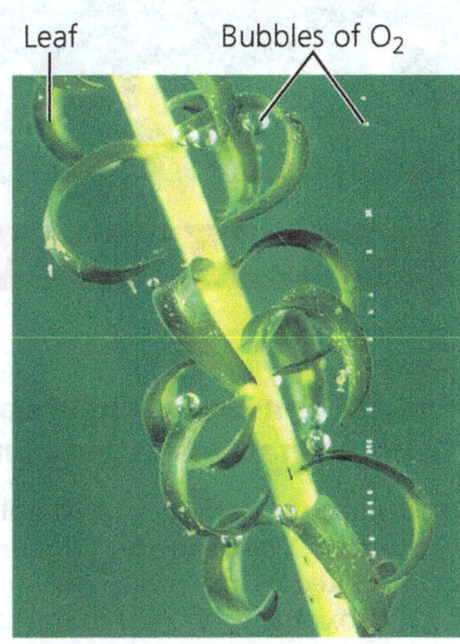

▶ **Figure 2.17 Photosynthesis: a solar-powered rearrangement of matter.** *Elodea*, a freshwater plant, produces sugar by rearranging the atoms of carbon dioxide and water in the chemical process known as photosynthesis, which is powered by sunlight. Much of the sugar is then converted to other food molecules. Oxygen gas (O_2) is a by-product of photosynthesis; notice the bubbles of O_2 gas escaping from the leaves submerged in water.

DRAW IT *Add labels and arrows on the photo showing the reactants and products of photosynthesis as it takes place in a leaf.*

products accumulate, collisions resulting in the reverse reaction become more frequent. Eventually, the forward and reverse reactions occur at the same rate, and the relative concentrations of products and reactants stop changing. The point at which the reactions offset one another exactly is called **chemical equilibrium**. This is a dynamic equilibrium; reactions are still going on in both directions, but with no net effect on the concentrations of reactants and products. Equilibrium does *not* mean that the reactants and products are equal in concentration, but only that their concentrations have stabilised at a particular ratio. The reaction involving ammonia reaches equilibrium when ammonia decomposes as rapidly as it forms. In some chemical reactions, the equilibrium point may lie so far to the right that these reactions go essentially to completion; that is, virtually all the reactants are converted to products.

We will return to the subject of chemical reactions after more detailed study of the various types of molecules that are important to life. In the next chapter, we focus on water, the substance in which all the chemical processes of organisms occur.

CONCEPT CHECK 2.4

1. **MAKE CONNECTIONS** Consider the reaction between hydrogen and oxygen that forms water, shown with ball-and-stick models at the beginning of Concept 2.4. After studying Figure 2.10, draw and label the Lewis dot structures representing this reaction.
2. Which type of chemical reaction, if any, occurs faster at equilibrium: the formation of products from reactants or that of reactants from products?
3. **WHAT IF?** Write an equation that uses the products of photosynthesis as reactants and the reactants of photosynthesis as products. Add energy as another product. This new equation describes a process that occurs in your cells. Describe this equation in words. How does this equation relate to breathing?

For suggested answers, see Appendix A.

2 Chapter Review

SUMMARY OF KEY CONCEPTS

CONCEPT 2.1

Matter consists of chemical elements in pure form and in combinations called compounds *(pp. 29–30)*

- **Elements** cannot be broken down chemically to other substances. A **compound** contains two or more different elements in a fixed ratio. Oxygen, carbon, hydrogen, and nitrogen make up approximately 96% of living matter.

? *Compare an element and a compound.*

CONCEPT 2.2

An element's properties depend on the structure of its atoms *(pp. 30–36)*

- An **atom**, the smallest unit of an element, has the following components:

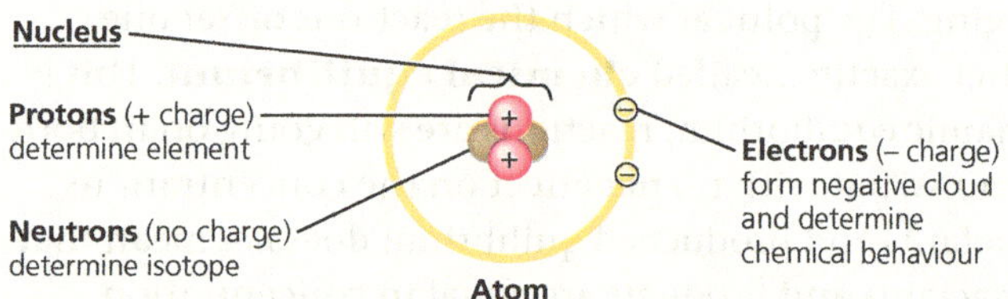

- An electrically neutral atom has equal numbers of electrons and protons; the number of protons determines the **atomic number**. The **atomic mass** is measured in **daltons** and is roughly equal to the **mass number**, the sum of protons plus neutrons. **Isotopes** of an element differ from each other in neutron number and therefore mass. Unstable isotopes give off particles and energy as radioactivity.
- In an atom, electrons occupy specific **electron shells**; the electrons in a shell have a characteristic energy level. Electron distribution in shells determines the chemical behaviour of an atom. An atom that has an incomplete outer shell, the **valence shell**, is reactive.
- Electrons exist in **orbitals**, three-dimensional spaces with specific shapes that are components of electron shells.

DRAW IT *Draw the electron distribution diagrams for neon ($_{10}$Ne) and argon ($_{18}$Ar). Use these diagrams to explain why these elements are chemically unreactive.*

CONCEPT 2.3

The formation and function of molecules and ionic compounds depend on chemical bonding between atoms *(pp. 36–40)*

- **Chemical bonds** form when atoms interact and complete their valence shells. **Covalent bonds** form when pairs of electrons are shared:

H· + H· ⟶ H:H (**Single covalent bond**)

:Ö· + ·Ö: ⟶ O::O (**Double covalent bond**)

- **Molecules** consist of two or more covalently bonded atoms. The attraction of an atom for the electrons of a covalent bond is its **electronegativity**. If both atoms are the same, they have the same electronegativity and share a **nonpolar covalent bond**. Electrons of a **polar covalent bond** are pulled closer to the more electronegative atom, such as the oxygen in H_2O.
- An **ion** forms when an atom or molecule gains or loses an electron and becomes charged. An **ionic bond** is the attraction between two oppositely charged ions:

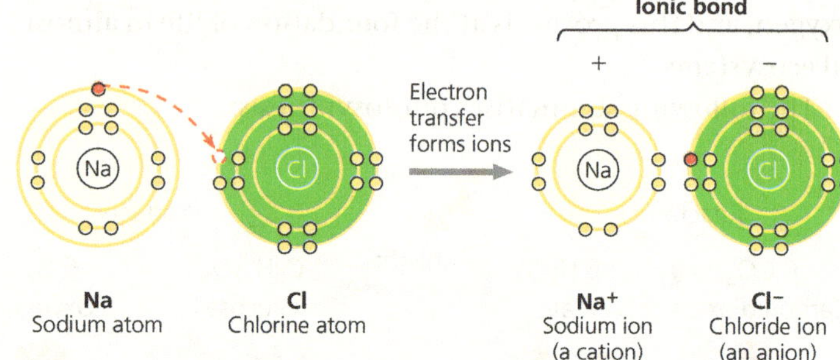

- Weak interactions reinforce the shapes of large molecules and help molecules adhere to each other. A **hydrogen bond** is an attraction between a hydrogen atom carrying a partial positive charge (δ+) and an electronegative atom carrying a partial negative charge (δ–). **Van der Waals interactions** occur between transiently positive and negative regions of molecules.
- A molecule's shape is determined by the positions of its atoms' valence orbitals. Covalent bonds result in hybrid orbitals, which are responsible for the shapes of H_2O, CH_4, and many more complex biological molecules. Molecular shape is usually the basis for the recognition of one biological molecule by another.

? *In terms of electron sharing between atoms, compare nonpolar covalent bonds, polar covalent bonds, and the formation of ions.*

CONCEPT 2.4

Chemical reactions make and break chemical bonds *(pp. 40–41)*

- **Chemical reactions** change **reactants** into **products** while conserving matter. All chemical reactions are theoretically reversible. **Chemical equilibrium** is reached when the forward and reverse reaction rates are equal.

? *What would happen to the concentration of products if more reactants were added to a reaction that was in chemical equilibrium? How would this addition affect the equilibrium?*

TEST YOUR UNDERSTANDING

Levels 1-2: Remembering/Understanding

1. Compared with ^{31}P, the radioactive isotope ^{32}P has
   (A) a different atomic number.
   (B) one more proton.
   (C) one more electron.
   (D) one more neutron.

2. In the term *trace element*, the adjective *trace* means that
   (A) the element is required in very small amounts.
   (B) the element can be used as a label to trace atoms through an organism's metabolism.
   (C) the element is very rare on Earth.
   (D) the element enhances health but is not essential for the organism's long-term survival.
3. The reactivity of an atom arises from
   (A) the average distance of the outermost electron shell from the nucleus.
   (B) the existence of unpaired electrons in the valence shell.
   (C) the sum of the potential energies of all the electron shells.
   (D) the potential energy of the valence shell.
4. Which statement is true of all atoms that are anions?
   (A) The atom has more electrons than protons.
   (B) The atom has more protons than electrons.
   (C) The atom has fewer protons than does a neutral atom of the same element.
   (D) The atom has more neutrons than protons.
5. Which of the following statements correctly describes any chemical reaction that has reached equilibrium?
   (A) The concentrations of products and reactants are equal.
   (B) The reaction is now irreversible.
   (C) Both forward and reverse reactions have halted.
   (D) The rates of the forward and reverse reactions are equal.

Levels 3-4: Applying/Analysing

6. We can represent atoms by listing the number of protons, neutrons, and electrons—for example, $2p^+$, $2n^0$, $2e^-$ for helium. Which of the following represents the ^{18}O isotope of oxygen?
   (A) $7p^+$, $2n^0$, $9e^-$
   (B) $8p^+$, $10n^0$, $8e^-$
   (C) $9p^+$, $9n^0$, $9e^-$
   (D) $10p^+$, $8n^0$, $9e^-$
7. The atomic number of sulfur is 16. Sulfur combines with hydrogen by covalent bonding to form a compound, hydrogen sulfide. Based on the number of valence electrons in a sulfur atom, predict the molecular formula of the compound.
   (A) HS (C) H_2S
   (B) HS_2 (D) H_4S
8. What coefficients must be placed in the following blanks so that all atoms are accounted for in the products?

$$C_6H_{12}O_6 \rightarrow \_\_\_\_\_ C_2H_6O + \_\_\_\_\_ CO_2$$

   (A) 2; 1 (C) 1; 3
   (B) 3; 1 (D) 2; 2
9. **DRAW IT** Draw Lewis dot structures for each hypothetical molecule shown below, using the correct number of valence electrons for each atom. Determine which molecule makes sense because each atom has a complete valence shell and each bond has the correct number of electrons. Explain what makes the other molecule nonsensical, considering the number of bonds each type of atom can make.

```
       H   H                 H           H
       |   |                 |           |
   H—O—C—C=O             H—C—H—C=O
       |                     |
(a)    H             (b)     H
```

Levels 5-6: Evaluating/Creating

10. **EVOLUTION CONNECTION** The percentages of naturally occurring elements making up the human body (see Table 2.1) are similar to the percentages of these elements found in other organisms. How could you account for this similarity among organisms?
11. **SCIENTIFIC INQUIRY** A north Australian orchid (*Bulbophyllum baileyi*) emits chemical signals (blossom odours) that attract male fruit flies (*Bactrocera* sp.). Male fruit flies detect the odour molecules up to several hundred metres from the plant. The fruit fly's antenna (shown in this picture) have hundreds of receptor cells that allow the insect to discern these signals. Once detected, the insect searches for and finds the orchid. Based on what you learned in this chapter, propose a hypothesis to account for the ability of the male fruit fly to detect a specific molecule in the presence of many other molecules in the air. What predictions does your hypothesis make? Design an experiment to test one of these predictions.

12. **WRITE ABOUT A THEME: ORGANISATION** While waiting at an airport, Neil Campbell once overheard this claim: "It's paranoid and ignorant to worry about industry or agriculture contaminating the environment with their chemical wastes. After all, this stuff is just made of the same atoms that were already present in our environment." Drawing on your knowledge of electron distribution, bonding, and emergent properties (see Concept 1.1), write a short essay (100–150 words) countering this argument.
13. **SYNTHESISE YOUR KNOWLEDGE**

This bombardier beetle is spraying a boiling hot liquid that contains irritating chemicals, used as a defence mechanism against its enemies. The beetle stores two sets of chemicals separately in its glands. Using what you learned about chemistry in this chapter, propose a possible explanation for why the beetle is not harmed by the chemicals it stores and what causes the explosive discharge.

For selected answers, see Appendix A.

3 Water and Life

KEY CONCEPTS

3.1 **Polar covalent bonds in water molecules result in hydrogen bonding** *p. 45*

3.2 **Four emergent properties of water contribute to Earth's suitability for life** *p. 45*

3.3 **Acidic and basic conditions affect living organisms** *p. 52*

Study Tip

Make a visual study guide: Draw a diagram and write a caption that explains how the structure of water supports life for each of the following properties of water:

The Properties of Water	
Cohesion of water molecules	Moderation of temperature
Ice floats	The solvent of life

Go to Mastering Biology

to access Dynamic Study Modules for revision, 3D BioFlix® animations and high-quality videos, and your interactive Pearson eText.

Figure 3.1 Ringed seals (*Phoca hispida*) depend on Arctic sea ice as a platform from which to hunt for fish in the water below. As Earth warms from climate change, the melting of sea ice is a threat to species that live on, under, and around the floating ice.

How does water's structure allow its solid form (ice) to float on liquid water?

Water (H_2O) is a polar molecule: At one end, the O has partial negative charges (δ−) because O pulls electrons towards itself. At the other end, the H atoms have partial positive charges (δ+).

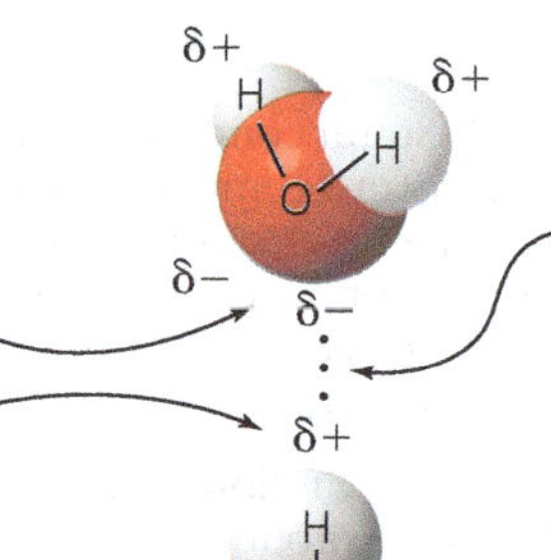

Weak attractions between oppositely charged regions of water molecules, called hydrogen bonds, allow water molecules to bond to each other.

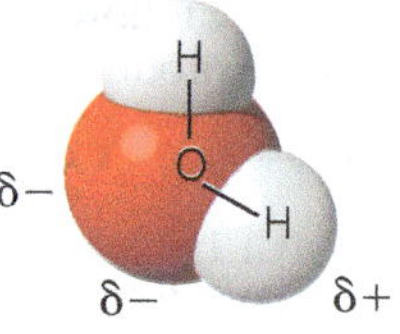

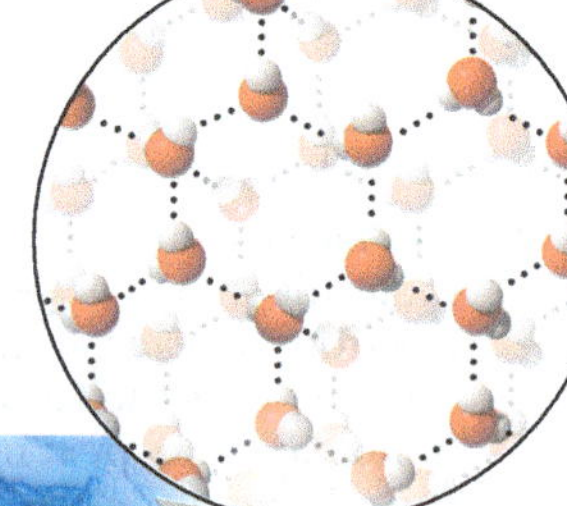

In liquid water, the hydrogen bonds constantly break and re-form. As a result, **the water molecules can slip closer together.**

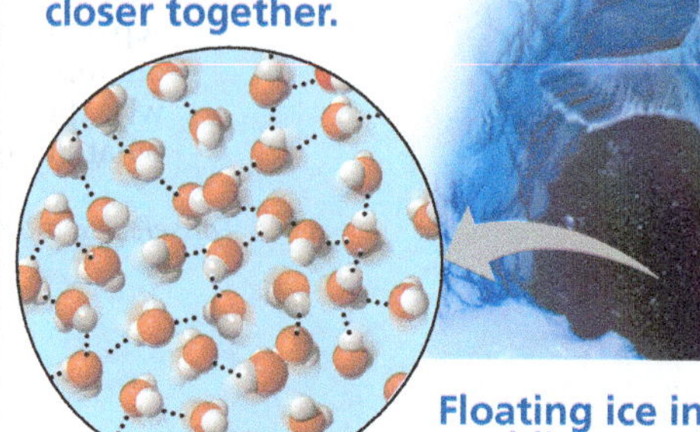

In ice, the hydrogen bonds are stable and **the water molecules are further apart.** Therefore, ice is less dense than liquid water, so it floats.

Floating ice insulates the water below, enabling survival of aquatic life. Water also has other life-supporting properties, as you'll see.

CONCEPT 3.1

Polar covalent bonds in water molecules result in hydrogen bonding

Water is so familiar to us that it is easy to overlook its many extraordinary qualities. Following the theme of emergent properties, we can trace water's unique behaviour to the structure and interactions of its molecules.

Studied on its own, the water molecule is deceptively simple. It is shaped like a wide V, with its two hydrogen atoms joined to the oxygen atom by single covalent bonds. Oxygen is more electronegative than hydrogen, so the electrons of the covalent bonds spend more time closer to oxygen than to hydrogen; these are **polar covalent bonds** (see Figure 2.11). This unequal sharing of electrons and water's V-like shape make it a **polar molecule**, meaning that its overall charge is unevenly distributed. In water, the oxygen of the molecule has partial negative charges ($\delta-$), and the hydrogens have partial positive charges ($\delta+$).

The properties of water arise from attractions between oppositely charged atoms of different water molecules: The partially positive hydrogen of one molecule is attracted to the partially negative oxygen of a nearby molecule. The two molecules are thus held together by a hydrogen bond **(Figure 3.2)**. When water is in its liquid form, its hydrogen bonds are very fragile, each only about 1/20 as strong as a covalent bond. The hydrogen bonds form, break, and re-form with great frequency. Each lasts only a few trillionths of a second, but the molecules are constantly forming new hydrogen bonds with a succession of partners. Therefore, at any instant, most of the water molecules are hydrogen-bonded to their neighbours. The extraordinary properties of water emerge from this hydrogen bonding, which organises water molecules into a higher level of structural order.

▼ **Figure 3.2 Hydrogen bonds between water molecules.**

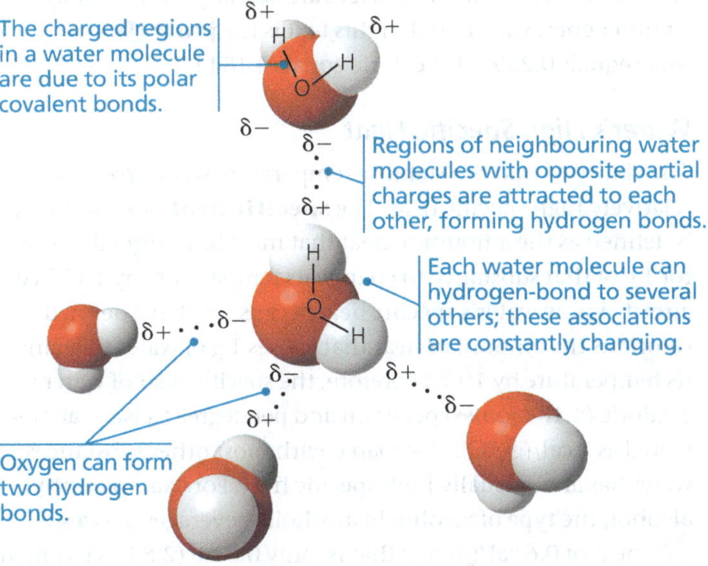

DRAW IT *Draw partial charges on the water molecule at the far left, and draw three more water molecules hydrogen-bonded to it.*

CONCEPT CHECK 3.1

1. **MAKE CONNECTIONS** What is electronegativity, and how does it affect interactions between water molecules? (Review Figure 2.11.)
2. **VISUAL SKILLS** Look at Figure 3.2 and explain why the central water molecule can hydrogen-bond to other water molecules.
3. Why is it unlikely that two neighbouring water molecules would be arranged like this? (structure: O with two H's facing H's of another O)
4. **WHAT IF?** What would be the effect on the properties of the water molecule if oxygen and hydrogen had equal electronegativity?

For suggested answers, see Appendix A.

CONCEPT 3.2

Four emergent properties of water contribute to Earth's suitability for life

We will examine four emergent properties of water that contribute to Earth's suitability as an environment for life: cohesive behaviour, ability to moderate temperature, expansion upon freezing, and versatility as a solvent.

Cohesion of Water Molecules

Water molecules stay close to each other as a result of hydrogen bonding. Although the arrangement of molecules in a sample of liquid water is constantly changing, at any given moment many of the molecules are linked by multiple hydrogen bonds. These linkages make water more structured than most other liquids. Collectively, the hydrogen bonds hold the substance together, a phenomenon called **cohesion**.

One result of cohesion due to hydrogen bonding is high **surface tension**, a measure of how difficult it is to stretch or break the surface of a liquid. At the air-water interface is an ordered arrangement of water molecules, hydrogen-bonded to one another and to the water below, but not to the air above. This asymmetry gives water an unusually high surface tension, making it behave as though it were coated with an invisible film. The spider in **Figure 3.3** takes advantage of the surface tension of water to walk across a pond

► **Figure 3.3 Walking on water.** The high surface tension of water, resulting from the collective strength of its hydrogen bonds, allows this raft spider to walk on the surface of a pond.

without breaking the surface, and some plants can float on water as well. You can observe the surface tension of water by slightly overfilling a drinking glass; the water will stand above the rim.

Cohesion also contributes to the transport of water and dissolved nutrients against gravity in plants **(Figure 3.4)**. Water from the roots reaches the leaves through a network of water-conducting cells. As water evaporates from a leaf, hydrogen bonds cause water molecules leaving the veins to tug on molecules further down, and the upward pull is transmitted through the water-conducting cells all the way to the roots. **Adhesion**, the clinging of one substance to another, also plays a role. Adhesion of water by hydrogen bonds to the molecules of cell walls helps counter the downward pull of gravity (see Figure 3.4).

▼ **Figure 3.4 Water transport in plants.** Evaporation from leaves pulls water upwards from the roots through water-conducting cells. Because of the properties of cohesion and adhesion, the tallest trees can transport water more than 100 m upwards—approximately one-quarter the height of the Empire State Building in New York City.

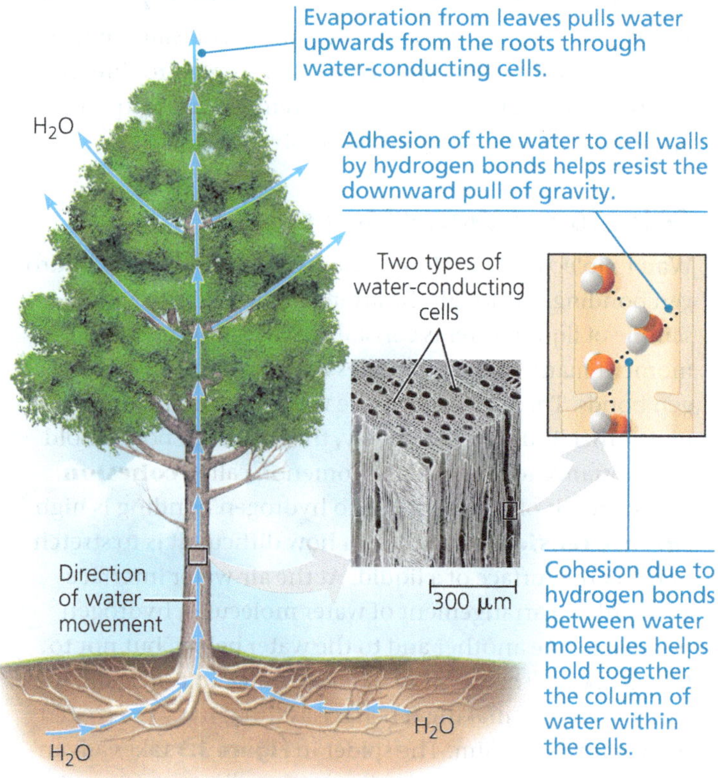

Moderation of Temperature by Water

Water moderates air temperature by absorbing heat from air that is warmer and releasing stored heat to air that is cooler. Water is effective as a heat bank because it can absorb or release a relatively large amount of heat with only a slight change in its own temperature. To understand this capability of water, let's first look at temperature and heat.

Temperature and Heat

Anything that moves has **kinetic energy**, the energy of motion. Atoms and molecules have kinetic energy because they are always moving, although not necessarily in any particular direction. The faster a molecule moves, the greater its kinetic energy. The kinetic energy associated with the random movement of atoms or molecules is called **thermal energy**. Thermal energy is related to temperature, but they are not the same thing. **Temperature** represents the *average* kinetic energy of the molecules in a body of matter, regardless of volume, whereas the thermal energy of a body of matter reflects the *total* kinetic energy, and thus depends on the matter's volume. When water is heated in a coffeemaker, the average speed of the molecules increases, and the thermometer records this as a rise in temperature of the liquid. The total amount of thermal energy also increases in this case. Note, however, that although the pot of coffee has a much higher temperature than, say, the water in a swimming pool, the swimming pool contains more thermal energy because of its much greater volume.

Whenever two objects of different temperature are brought together, thermal energy passes from the warmer to the cooler object until the two are the same temperature. Molecules in the cooler object speed up at the expense of the thermal energy of the warmer object. An ice cube cools a drink not by adding coldness to the liquid but by absorbing thermal energy from the liquid as the ice itself melts. Thermal energy in transfer from one body of matter to another is defined as **heat**.

One convenient unit of heat is the **calorie (cal)**. A calorie is the amount of heat it takes to raise the temperature of 1 g of water by 1°C. Conversely, a calorie is also the amount of heat that 1 g of water releases when it cools by 1°C. A **kilocalorie (kcal)**, 1,000 cal, is the quantity of heat required to raise the temperature of 1 kilogram (kg) of water by 1°C. (The "Calories" on food packages are actually kilocalories.) Another energy unit used in this text is the **joule (J)**. One joule equals 0.239 cal; 1 calorie equals 4.184 J.

Water's High Specific Heat

The ability of water to stabilise temperature stems from its relatively high specific heat. The **specific heat** of a substance is defined as the amount of heat that must be absorbed or lost for 1 g of that substance to change its temperature by 1°C. We already know water's specific heat because we have defined a calorie as the amount of heat that causes 1 g of water to change its temperature by 1°C. Therefore, the specific heat of water is 1 calorie (4.184 joules) per gram and per degree Celsius, abbreviated as 1 cal/(g · °C). Compared with most other substances, water has an unusually high specific heat. For example, ethyl alcohol, the type of alcohol in alcoholic beverages, has a specific heat of 0.6 cal/(g · °C); that is, only 0.6 cal (2.5 J) is required to raise the temperature of 1 g of ethyl alcohol by 1°C.

Because of the high specific heat of water relative to other materials, water will change its temperature less than other liquids when it absorbs or loses a given amount of heat. The reason you can burn your fingers by touching the side of an iron pot on the stove when the water in the pot is still lukewarm is that the specific heat of water is ten times greater than that of iron. In other words, the same amount of heat will raise the temperature of 1 g of the iron much faster than it will raise the temperature of 1 g of the water. Specific heat can be thought of as a measure of how well a substance resists changing its temperature when it absorbs or releases heat. Water resists changing its temperature; when it does change its temperature, it absorbs or loses a relatively large quantity of heat for each degree of change.

We can trace water's high specific heat, like many of its other properties, to hydrogen bonding. Heat must be absorbed in order to break hydrogen bonds; by the same token, heat is released when hydrogen bonds form. A calorie of heat causes a relatively small change in the temperature of water because much of the heat is used to disrupt hydrogen bonds before the water molecules can begin moving faster. And when the temperature of water drops slightly, many additional hydrogen bonds form, releasing a considerable amount of energy in the form of heat.

What is the relevance of water's high specific heat to life on Earth? A large body of water can absorb and store a huge amount of heat from the sun in the daytime and during summer while warming up only a few degrees. At night and during winter, the gradually cooling water can warm the air. This capability of water serves to moderate air temperatures in coastal areas **(Figure 3.5)**. The high specific heat of water also tends to stabilise ocean temperatures, creating a favourable environment for marine life. Thus, because of its high specific heat, the water that covers most of Earth keeps temperature fluctuations on land and in water within limits that permit life. Also, because organisms are made primarily of water, they are better able to resist changes in their own temperature than if they were made of a liquid with a lower specific heat.

▼ Figure 3.5 The influence of a large body of water on climate. Large bodies of water moderate coastal climates. The diagram records maximum temperatures on one September day in southern Western Australia.

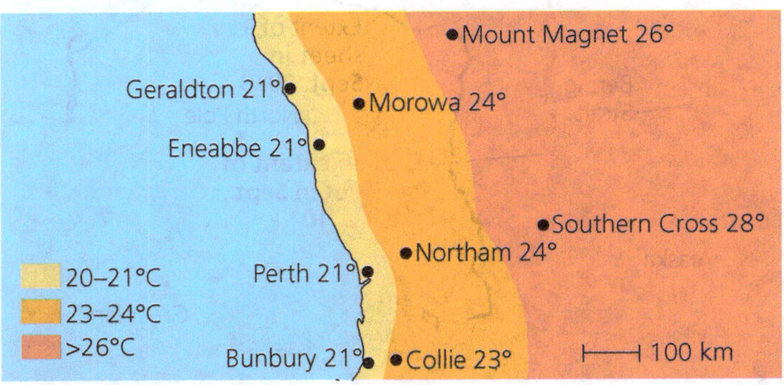

INTERPRET THE DATA *Explain the pattern of temperatures shown in this diagram.*

Evaporative Cooling

Molecules of any liquid stay close together because they are attracted to one another. Molecules moving fast enough to overcome these attractions can depart the liquid and enter the air as a gas (vapour). This transformation from a liquid to a gas is called vaporisation, or evaporation. Recall that the speed of molecular movement varies and that temperature is the *average* kinetic energy of molecules. Even at low temperatures, the speediest molecules can escape into the air. Some evaporation occurs at any temperature; a glass of water at room temperature, for example, will eventually evaporate completely. If a liquid is heated, the average kinetic energy of molecules increases and the liquid evaporates more rapidly.

Heat of vaporisation is the quantity of heat a liquid must absorb for 1 g of it to be converted from the liquid to the gaseous state. For the same reason that water has a high specific heat, it also has a high heat of vaporisation relative to most other liquids. To evaporate 1 g of water at 25°C, about 580 cal (2427 J) of heat is needed—nearly double the amount needed to vaporise a gram of alcohol or ammonia. Water's high heat of vaporisation is another emergent property resulting from the strength of its hydrogen bonds, which must be broken before the molecules can exit from the liquid in the form of water vapour.

The high amount of energy required to vaporise water has a wide range of effects. On a global scale, for example, it helps moderate Earth's climate. A considerable amount of solar heat absorbed by tropical seas is consumed during the evaporation of surface water. Then, as moist tropical air circulates polewards, it releases heat as it condenses and forms rain. On an organismal level, water's high heat of vaporisation accounts for the severity of steam burns. These burns are caused by the heat energy released (during formation of hydrogen bonds) when steam condenses into liquid on the skin.

As a liquid evaporates, the surface of the liquid that remains behind cools down (its temperature decreases). This **evaporative cooling** occurs because the "hottest" molecules, those with the greatest kinetic energy, are the most likely to leave as gas. It is as if the 100 fastest runners at a college transferred to another school; the average speed of the remaining students would decline.

Evaporative cooling of water contributes to the stability of temperature in lakes and ponds and also provides a mechanism that prevents terrestrial organisms from overheating. For example, evaporation of water from the leaves of a plant helps keep the tissues in the leaves from becoming too warm in the sunlight. Evaporation of sweat from human skin dissipates body heat and helps prevent overheating on a hot day or when excess heat is generated by strenuous activity. High humidity on a hot day increases discomfort because the high concentration of water vapour in the air inhibits the evaporation of sweat from the body. Animals without sweat glands, such as elephants, may spray water on themselves to cool down **(Figure 3.6)**.

▼ **Figure 3.6 Evaporative cooling.** In hot weather, an elephant sprays water from its trunk onto its head. Evaporation of this water cools the elephant down.

Ice Floats on Liquid Water

Water is one of the few substances that are less dense as a solid than as a liquid. Consequently, ice floats on liquid water. While other materials contract and become denser when they solidify, water expands. Hydrogen bonding causes this exotic behaviour. At temperatures above 4°C, water behaves like other liquids, expanding as it warms and contracting as it cools. As the temperature falls from 4°C to 0°C, water begins to freeze because more and more of its molecules are moving too slowly to break hydrogen bonds. At 0°C, the molecules form a crystalline lattice, where each water molecule hydrogen-bonds to four partners (see Figure 3.1). The hydrogen bonds keep the molecules at "arm's length," far enough apart to make ice about 10% less dense (10% fewer molecules in the same volume) than liquid water at 4°C. When ice absorbs enough heat for its temperature to rise above 0°C, hydrogen bonds between molecules become disrupted. As the crystal collapses, the ice melts and molecules have fewer hydrogen bonds, allowing them to slip closer together. Water reaches its greatest density at 4°C and then begins to expand as the molecules move faster. Even in liquid water, many of the molecules become connected by hydrogen bonds, though only transiently: The hydrogen bonds constantly break and re-form.

The ability of ice to float due to its lower density is an important factor in the suitability of the environment for life. If ice sank, then eventually ponds, lakes, and even oceans could freeze solid, making life as we know it impossible on Earth. During summer, only the upper few inches of the ocean would thaw. Instead, when a deep body of water cools, the ice floats, insulating the liquid water below. This prevents it from freezing and allows life to exist under the frozen surface, as shown in Figure 3.1. Besides insulating the water below, ice also provides a solid habitat for some animals, such as penguins and seals.

Many scientists have raised the alarm because of the rapid disappearance of these bodies of ice. Global warming, which is caused by carbon dioxide and other "greenhouse" gases in the atmosphere (see Figure 56.30), continues to profoundly influence icy environments around the globe. Antarctica's average air temperatures increased 3°C over the past 50 years. In addition to melting surface ice and snow, increased sea temperatures erode ice from beneath. In 2019, these factors contributed to a 1,683 km^2 iceberg (about the size of New Zealand's Stewart Island) carving away from Antarctica's Amery Ice Shelf. An obvious consequence was the reduction of Antarctica's animal habitat. A second consequence was

▼ **Figure 3.7 Effects of climate change on the Arctic.** Warmer temperatures in the Arctic cause more sea ice to melt in the summer. The loss of ice disrupts the ecosystem, affecting many species. (Map data is from the National Snow and Ice Data Center.)

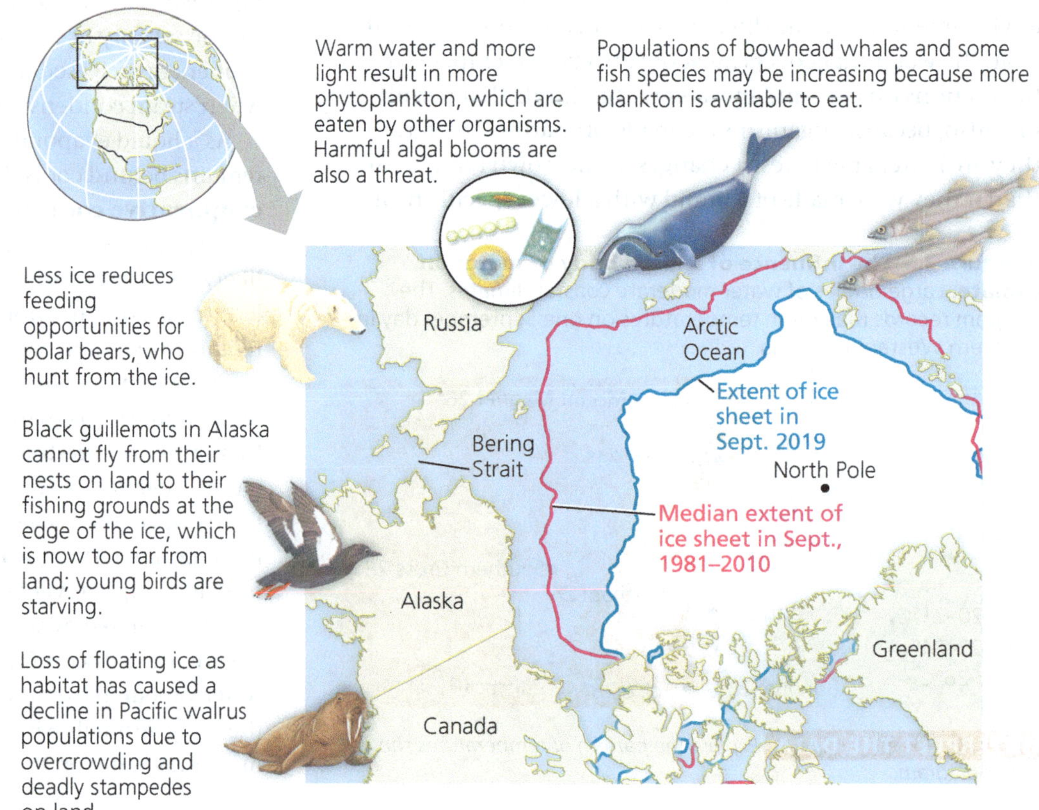

perhaps less obvious though: the melting of the 315 billion tonne iceberg contributed to raised sea levels.

Meanwhile, the Arctic's average air temperature has increased 2.2°C since 1961. Temperature increases impact seasonal balances between Arctic sea ice and liquid water, causing ice to form later in the year, to melt earlier, and to cover a smaller area. The rate at which coastal Antarctic ice and sea ice are disappearing, combined with declines in glaciers and Arctic sea ice, poses an extreme challenge to animals that depend on ice for their survival **(Figure 3.7)**.

Water: The Solvent of Life

A sugar cube placed in a glass of water will dissolve. In time, the glass will contain a uniform mixture of sugar and water; the concentration of dissolved sugar will be the same everywhere in the mixture. A liquid that is a completely homogeneous mixture of two or more substances is called a **solution**. The dissolving agent of a solution is the **solvent**, and the substance that is dissolved is the **solute**. In this case, water is the solvent and sugar is the solute. An **aqueous solution** is one in which the solute is dissolved in water; water is the solvent.

Water is a very versatile solvent, a quality we can trace to the polarity of the water molecule. Suppose, for example, that a spoonful of table salt, the ionic compound sodium chloride (NaCl), is placed in water **(Figure 3.8)**. At the surface of each crystal (grain) of salt, the sodium and chloride ions are exposed to the solvent. These ions and regions of the water molecules are attracted to each other due to their opposite charges. The oxygens of the water molecules have regions of partial negative charge that are attracted to sodium cations. The hydrogen regions are partially positively charged and are attracted to chloride anions. As a result, water molecules surround the individual sodium and chloride ions, separating and shielding them from one another. The sphere of water molecules around each dissolved ion is called a **hydration shell**. Working inwards from the surface of each salt crystal, water eventually dissolves all the ions. The result is a solution of two solutes, sodium cations and chloride anions, mixed homogeneously with water, the solvent. Other ionic compounds also dissolve in water. Seawater, for instance, contains a great variety of dissolved ions, as do living cells.

▼ Figure 3.8 Table salt dissolving in water. A sphere of water molecules, called a hydration shell, surrounds each solute ion.

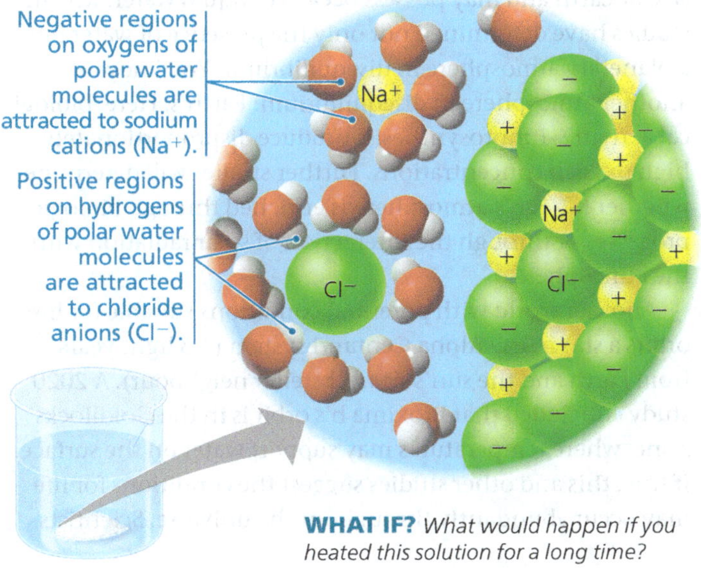

▼ Figure 3.9 A water-soluble protein. Human lysozyme is a protein found in tears and saliva that has antibacterial action (see Figure 5.16). This model shows the lysozyme molecule (purple) in an aqueous environment. Ionic and polar regions on the protein's surface attract the partially charged regions on water molecules.

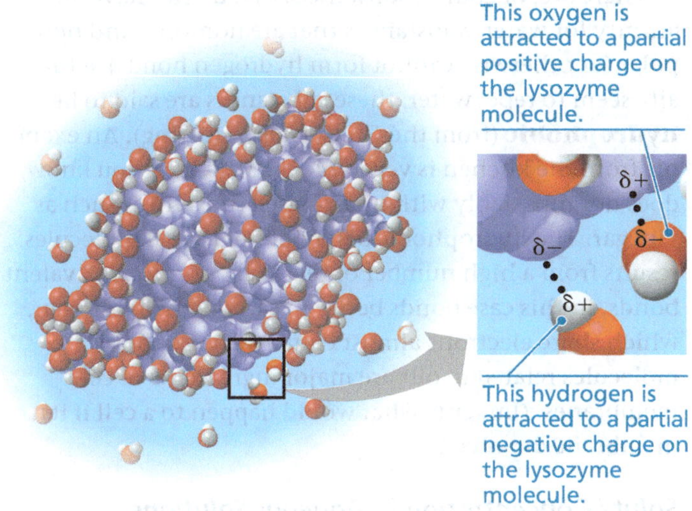

A compound does not need to be ionic to dissolve in water; many compounds made up of nonionic polar molecules, such as the sugar in the sugar cube mentioned earlier, are also water-soluble. Such compounds dissolve when water molecules surround each of the solute molecules, forming hydrogen bonds with them. Even molecules as large as proteins can dissolve in water if they have ionic and polar regions on their surface **(Figure 3.9)**. Many different kinds of polar compounds are dissolved (along with ions) in the water of such biological fluids as blood, the sap of plants, and the liquid within all cells. Water is the solvent of life.

Hydrophilic and Hydrophobic Substances

Any substance that has an affinity for water is said to be **hydrophilic** (from the Greek *hydro*, water, and *philos*, loving). In some cases, substances can be hydrophilic without actually dissolving. For example, some molecules in cells are so large that they do not dissolve. Another example of a hydrophilic substance that does not dissolve is cotton, a plant product. Cotton consists of giant molecules of cellulose, a compound with numerous regions of partial positive and partial negative charges that can form hydrogen bonds with water. Water adheres to the cellulose fibres. Thus, a cotton towel does a great job of drying the body, yet it does not dissolve in the washing machine. Cellulose is also present in the walls of water-conducting cells in a plant; you read earlier how the adhesion of water to these hydrophilic walls helps water move up the plant against gravity.

There are, of course, substances that do not have an affinity for water. Substances that are nonionic and nonpolar (or otherwise cannot form hydrogen bonds) actually seem to repel water; these substances are said to be **hydrophobic** (from the Greek *phobos*, fearing). An example from the kitchen is vegetable oil, which, as you know, does not mix stably with water-based substances such as vinegar. The hydrophobic behaviour of the oil molecules results from a high number of relatively nonpolar covalent bonds, in this case bonds between carbon and hydrogen, which share electrons almost equally. Hydrophobic molecules related to oils are major ingredients of cell membranes. (Imagine what would happen to a cell if its membrane dissolved!)

Solute Concentration in Aqueous Solutions

Most of the chemical reactions in organisms involve solutes dissolved in water. To understand such reactions, we must know how many atoms and molecules are involved and calculate the concentration of solutes in an aqueous solution (the number of solute molecules in a volume of solution).

When carrying out experiments, we use mass to calculate the number of molecules. We must first calculate the **molecular mass**, which is the sum of the masses of all the atoms in a molecule. As an example, let's calculate the molecular mass of table sugar (sucrose), $C_{12}H_{22}O_{11}$, by multiplying the number of atoms by the atomic mass of each element. In round numbers of daltons, the mass of a carbon atom is 12, the mass of a hydrogen atom is 1, and the mass of an oxygen atom is 16. Thus, sucrose has a molecular mass of $(12 \times 12) + (22 \times 1) + (11 \times 16) = 342$ daltons. Because we can't weigh out small numbers of molecules, we usually measure substances in units called moles. Just as a dozen always means 12 objects, a **mole (mol)** represents an exact number of objects: 6.02×10^{23}, which is called Avogadro's number. Because of the way in which Avogadro's number and the unit *dalton* were originally defined, there are 6.02×10^{23} daltons in 1 g. Once we determine the molecular mass of a molecule such as sucrose, we can use the same number (342), but with the unit *gram*, to represent the mass of 6.02×10^{23} molecules of sucrose, or 1 mol of sucrose (sometimes called the *molar mass*). To obtain 1 mol of sucrose in the lab, therefore, we weigh out 342 g.

The practical advantage of measuring a quantity of chemicals in moles is that a mole of one substance has exactly the same number of molecules as a mole of any other substance. If the molecular mass of substance A is 342 daltons and that of substance B is 10 daltons, then 342 g of A will have the same number of molecules as 10 g of B. A mole of ethyl alcohol (C_2H_6O) also contains 6.02×10^{23} molecules, but its mass is only 46 g because the mass of a molecule of ethyl alcohol is less than that of a molecule of sucrose. Measuring in moles makes it convenient for scientists working in the laboratory to combine substances in fixed ratios of molecules.

How would we make a litre (L) of solution consisting of 1 mol of sucrose dissolved in water? We would measure out 342 g of sucrose and then gradually add water, while stirring, until the sugar was completely dissolved. We would then add enough water to bring the total volume of the solution up to 1 L. At that point, we would have a 1-molar (1 *M*) solution of sucrose. **Molarity**—the number of moles of solute per litre of solution—is the unit of concentration most often used by biologists for aqueous solutions.

Water's capacity as a versatile solvent complements the other properties discussed in this chapter. Since these remarkable properties allow water to support life on Earth so well, scientists who seek life elsewhere in the universe look for water as a sign that a celestial body might sustain life.

Possible Evolution of Life on Other Planets

EVOLUTION Astrobiologists search for signs of the origins, evolution, and distribution of life in the universe. Initial work focused on celestial bodies that might possess water because they orbit in the circumstellar habitable zone or, more metaphorically, the "Goldilocks Zone"; that is, objects orbiting their star at distances where surface conditions remain "just right" for liquid water—not too cold (ice), and not too warm (water vapour is found only in the atmosphere). One branch of astrobiology continues this focus with studies of exoplanets: worlds orbiting stars outside our solar system. These worlds abound. Since 1992, astronomers have catalogued over 4,000 of them. Quite a few possess water vapour in their atmospheres. Here, we consider two exoplanets and their potential to support life.

K2-18b orbits a star 124 light years from Earth (located in the constellation of Leo). K2-18b is about 8.6 times the size of Earth and may possess oceans of liquid water. Recent studies have determined not only the presence of water but a planetary atmosphere in disequilibrium. You know of another atmosphere in disequilibrium: Earth's. Here, biological activities (photosynthesis) produce disproportionately high oxygen concentrations. Further studies will determine whether K2-18b's atmosphere is enriched through biological processes or through the interaction of solar radiation with the atmosphere.

Another world with potential conditions suitable for life orbits a star in the Alpha Centauri system (4.3 light years from Earth and the sun's nearest stellar neighbour). A 2020 study concluded that Proxima b's orbit is in the Goldilocks Zone, where temperatures may support water on the surface. If true, this and other studies suggest the conditions for life may occur abundantly throughout the universe. Scientists acknowledge that they collect this kind of data at the edge of our ability to discern. While ground- and space-based studies continue, we will eventually need to send robots and maybe humans to find out for ourselves whether life exists on those distance exoplanets. Concurrently, a second branch of astrobiology searches for water and life on the planets and moons of our solar system.

Mars has long fascinated our species as a world where life might have evolved. Water was once plentiful on the planet's surface and remains underground now. During the Martian Hesperian period (3.7 to 3 billion years before the present), one of Mars's extensive river systems carried millions of tonnes of water per minute over the Echus Chasma **(Figure 3.10a)**. The 4,000 metre-tall waterfall was probably the largest and highest in the solar system—ever. From there, water drained into the salty ocean that covered much of Mars's northern hemisphere. In the intervening aeons, Mars lost much of its atmosphere, and its oceans, although planetary geologists believe much of the water remains. For example, just beneath Mars's surface, we find small quantities of water ice. However, to find liquid water we need to look beyond the icy polar caps and descend up to 2.5 km through Mars's cryosphere. Scientists conjecture that the cryosphere caps a layer of liquid water. In the dark, Mars's salty waters await future explorers. Perhaps with increased access to interplanetary travel, you will one day search for life on Mars's surface and deep underground as part of your day job.

In the past 25 years, astrobiologists reconsidered definitions of the Goldilocks Zone, extending it to beyond and beneath planetary surfaces. Take Venus, for example: at first glance, it presents an environment hostile to life. Movies from the surface would show a Venus-scape rippling in the superheated atmosphere (nearly 500°C—hot enough to melt aluminium and lead), with pressures equivalent to those you would experience 1000 m below the surface of Earth's oceans. To find more hospitable environments, explorers need to ascend 57 km straight up into Venus's temperate atmospheric zone, where temperatures of 30°C and atmospheric pressures equivalent to those on Earth at sea level allow liquid water droplets to form. In 2020, scientists exploring the Venusian temperate atmospheric zones may have discovered a potential signature of life: the gas phosphine (PH_3). Scientists ruled out known forms of atmospheric chemistry and potential sources of the gas from geological sources. On Earth, two sources of phosphine exist: through human-mediated industrial processes, and as the by-product of bacterial metabolism. Astrobiologists suggest the gas may arise from biological activities. Such an extraordinary claim requires extraordinary evidence. Scientists have engaged with the process of uncovering the truth through analysis of old and new data, and have begun lobbying for spacecraft to visit the Venusian atmosphere. Other scientists propose alternative interpretations of the biological basis of these gas traces—gathered at the limits of our observational capacity. Scientists hope to determine whether phosphine gas on Venus represents a "biosignature" or the by-product of some yet-unknown chemical process. Scientists will likely find answers to this question within the duration of their undergraduate degree.

The Goldilocks Zone required further redefinition when we learned that water abounds further out into the darkness of our solar system. An abundance of water occurs on Jupiter's moons: Europa, Ganymede, and Callisto. However, we will turn our attention to one of Saturn's 82 moons, and examine Enceladus (pronounced En-cell-a-dus) **(Figure 3.10b)**. NASA's Cassini probe flew through geysers ejected from the

▼ **Figure 3.10a Echus Chasma.** Volcanic activity, cometary impacts, and breaches in sub-surface aquifers produced Echus Chasma, which carried water into the oceans of Mars.

▼ **Figure 3.10b NASA's Cassini probe image of Enceladus (one of Saturn's moons).** This enhanced image shows the moon's southern pole, and geysers of frozen saltwater. The probe discerned the presence of complex organic molecules in the plumes you see here.

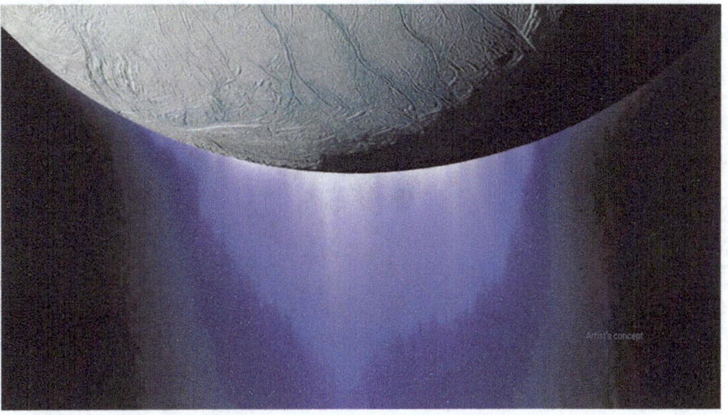

moon's southern pole. Aside from evidence of saltwater particles in the plumes, Cassini detected complex biochemical molecules.

Planetary scientists believe Saturn's strong gravitational fields squeeze Enceladus's liquid molten core to produce oceans, and deep-sea hydrothermal vents (Figure 25.3). Similar vents give rise to oases of life in Earth's deep oceans, where chemosynthetic bacteria, worms, and crustaceans abound. Many biologists conjecture that the earliest life on Earth evolved near hydrothermal vents. Astrobiologists wonder if vents on Enceladus, and on other icy moons throughout our solar system, create conditions that harbour life.

From our experiences on Earth, we can conclusively say that the evolution of life requires water. Unless, of course, it doesn't. Some studies observe that methane (CH_4), under certain conditions, possesses the same properties as water. Such conditions occur on another of Saturn's moons, Titan. Located in the cold darkness of the outer solar system, Titan receives less than 1% of the sun's illumination on Earth. Water freezes so hard that it forms the "bedrock" of that moon. At a chilly −180°C, liquid methane rains on Titan's surface, where it pools and flows into methane oceans. In liquid form, methane possesses the chemical properties essential for life on Earth. The presence of complex organic compounds in Titan's atmosphere, not readily explained by geological processes, offer tantalising clues about the potential of a very different form of biochemistry.

Further out into the outer solar system, our robot probes have found evidence of water ice on Triton (Neptune's largest moon) and the twin dwarf planets of Pluto and Charon. If water and compounds with similar chemical properties occur throughout the solar system and beyond, our search for biology beyond Earth may prove more fruitful than any of us can imagine.

▼ **Figure 3.11 Evidence for liquid water on Mars.** Water appears to have helped form these dark streaks that run downhill on Mars during the summer. NASA scientists also found evidence of hydrated salts, indicating water is present. (This digitally treated photograph was taken by the Mars Reconnaissance Orbiter.)

CONCEPT CHECK 3.2

1. Describe how properties of water contribute to the upward movement of water in a tree.
2. Explain the saying "It's not the heat; it's the humidity."
3. How can the freezing of water crack boulders?
4. **WHAT IF?** A water strider (an insect that can walk on water) has legs that are coated with a hydrophobic substance. What might be the benefit? What would happen if the substance were hydrophilic?
5. **INTERPRET THE DATA** The concentration of the appetite-regulating hormone ghrelin is about 1.3×10^{-10} *M* in the blood of a fasting person. How many molecules of ghrelin are in 1 L of blood?

For selected answers, see Appendix A.

CONCEPT 3.3

Acidic and basic conditions affect living organisms

Occasionally, a hydrogen atom participating in a hydrogen bond between two water molecules shifts from one molecule to the other. When this happens, the hydrogen atom leaves its electron behind, and what is actually transferred is a **hydrogen ion** (H^+), a single proton with a charge of 1+. The water molecule that lost a proton is now a **hydroxide ion** (OH^-), which has a charge of 1−. The proton binds to the other water molecule, making that molecule a **hydronium ion** (H_3O^+). We can picture the chemical reaction as follows:

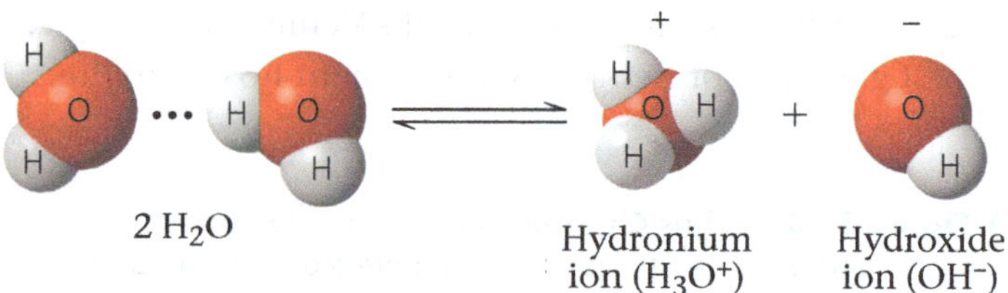

By convention, H^+ (the hydrogen ion) is used to represent H_3O^+ (the hydronium ion), and we follow that practice in this text. Keep in mind, though, that H^+ does not exist on its own in an aqueous solution. It is always associated with a water molecule in the form of H_3O^+.

As indicated by the double arrows, this is a reversible reaction that reaches a state of dynamic equilibrium when water molecules dissociate at the same rate that they are being reformed from H^+ and OH^-. At this equilibrium point, the concentration of water molecules greatly exceeds the concentrations of H^+ and OH^-. In pure water, only one water molecule in every 554 million is dissociated; the concentration of H^+ and of OH^- in pure water is therefore 10^{-7} *M* (at 25°C). This means there is only one ten-millionth of a mole of hydrogen ions per litre of pure water and an equal number of hydroxide ions. (Even so, this is a huge number—over 60,000 *trillion*—of each ion in a litre of pure water.)

Although the dissociation of water is reversible and statistically rare, it is exceedingly important in the chemistry of life. H^+ and OH^- are very reactive. Changes in their concentrations can drastically affect a cell's proteins and other complex molecules. As we have seen, the concentrations of H^+ and OH^- are equal in pure water, but adding certain kinds of solutes, called acids and bases, disrupts this balance. Biologists use something called the pH scale to describe how acidic or basic (the opposite of acidic) a solution is. In the remainder of this chapter, you will learn about acids, bases, and pH and why changes in pH can adversely affect organisms.

Acids and Bases

What would cause an aqueous solution to have an imbalance in H^+ and OH^- concentrations? When acids dissolve in water, they donate additional H^+ to the solution. An **acid** is a substance that increases the hydrogen ion concentration of a solution. For example, when hydrochloric acid (HCl) is added to water, hydrogen ions dissociate from chloride ions:

$$HCl \rightarrow H^+ + Cl^-$$

This source of H^+ (dissociation of water is the other source) results in an acidic solution—one having more H^+ than OH^-.

A substance that reduces the hydrogen ion concentration of a solution is called a **base**. Some bases reduce the H^+ concentration directly by accepting hydrogen ions. Ammonia (NH_3), for instance, acts as a base when the unshared electron pair in nitrogen's valence shell attracts a hydrogen ion from the solution, resulting in an ammonium ion (NH_4^+):

$$NH_3 + H^+ \rightleftharpoons NH_4^+$$

Other bases reduce the H^+ concentration indirectly by dissociating to form hydroxide ions, which combine with hydrogen ions and form water. One such base is sodium hydroxide (NaOH), which in water dissociates into its ions:

$$NaOH \rightarrow Na^+ + OH^-$$

In either case, the base reduces the H^+ concentration. Solutions with a higher concentration of OH^- than H^+ are known as basic solutions. A solution in which the H^+ and OH^- concentrations are equal is said to be neutral.

Notice that single arrows were used in the reactions for HCl and NaOH. These compounds dissociate completely when mixed with water, so hydrochloric acid is called a strong acid and sodium hydroxide a strong base. In contrast, ammonia is a weak base. The double arrows in the reaction for ammonia indicate that the binding and release of hydrogen ions are reversible reactions, although at equilibrium there will be a fixed ratio of NH_4^+ to NH_3.

Weak acids are acids that reversibly release and accept back hydrogen ions. An example is carbonic acid:

$$\underset{\text{Carbonic acid}}{H_2CO_3} \rightleftharpoons \underset{\text{Bicarbonate ion}}{HCO_3^-} + \underset{\text{Hydrogen ion}}{H^+}$$

Here the equilibrium so favours the reaction in the left direction that when carbonic acid is added to pure water, only 1% of the molecules are dissociated at any particular time. Still, that is enough to shift the balance of H^+ and OH^- from neutrality.

The pH Scale

In any aqueous solution at 25°C, the product of the H^+ and OH^- concentrations is constant at 10^{-14}. This can be written

$$[H^+][OH^-] = 10^{-14}$$

(The brackets indicate molar concentration.) As previously mentioned, in a neutral solution at 25°C, $[H^+] = 10^{-7}$ and $[OH^-] = 10^{-7}$. Therefore, the product of $[H^+]$ and $[OH^-]$ in a neutral solution at 25°C is 10^{-14}. If enough acid is added to a solution to increase $[H^+]$ to 10^{-5} *M*, then $[OH^-]$ will decline by an equivalent factor to 10^{-9} *M* (note that $10^{-5} \times 10^{-9} = 10^{-14}$). This constant relationship expresses the behaviour of acids and bases in an aqueous solution. An acid not only adds hydrogen ions to a solution, but also removes hydroxide ions because of the tendency for H^+ to combine with OH^-, forming water. A base has the opposite effect, increasing OH^- concentration but also reducing H^+ concentration by the formation of water. If enough of a base is added to raise the OH^- concentration to 10^{-4} *M*, it will cause the H^+ concentration to drop to 10^{-10} *M*. Whenever we know the concentration of either H^+ or OH^- in an aqueous solution, we can deduce the concentration of the other ion.

The pH scale **(Figure 3.12)** is a simple numerical method for expressing the range of H^+ concentrations. The H^+ concentrations of solutions can vary by a factor of 100 trillion or more. Instead of using moles per litre, the pH scale compresses the range of H^+ concentrations by employing logarithms. The **pH** of a solution is defined as the negative logarithm (base 10) of the H^+ concentration:

$$pH = -\log [H^+]$$

For a neutral aqueous solution, $[H^+]$ is 10^{-7} *M*, giving us

$$-\log 10^{-7} = -(-7) = 7$$

Notice that pH *decreases* as H^+ concentration *increases* (see Figure 3.11). Notice, too, that although the pH scale is based on H^+ concentration, it also implies OH^-

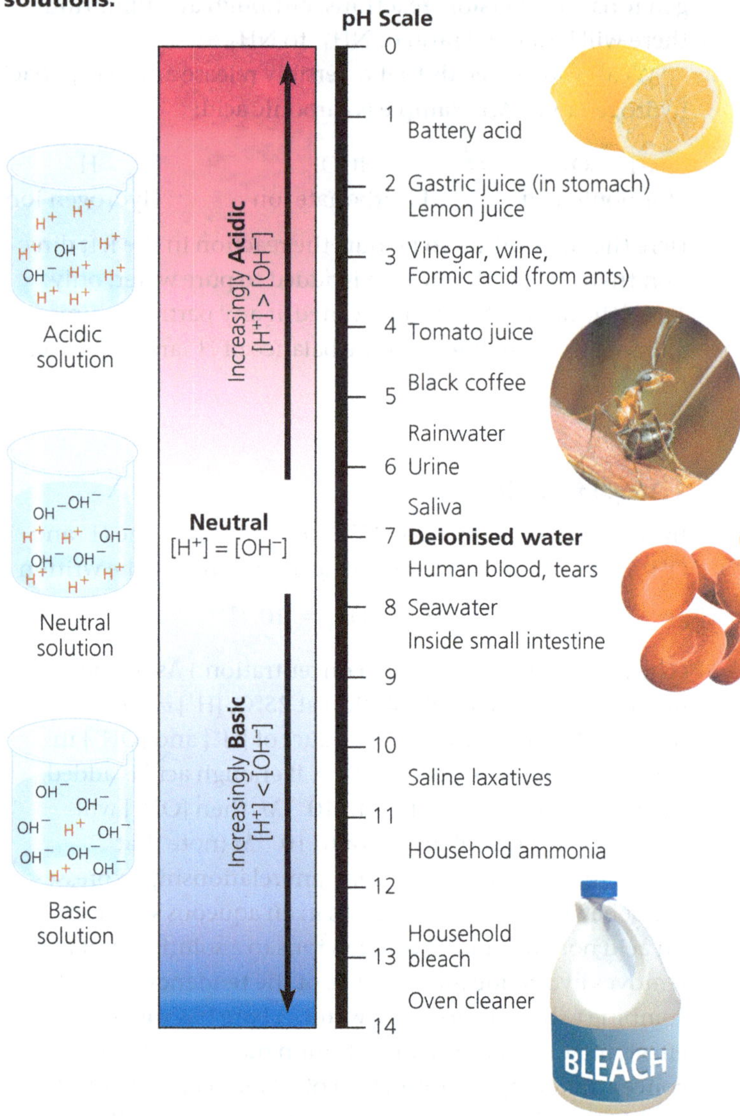

▼ **Figure 3.12 The pH scale and pH values of some aqueous solutions.**

concentration. A solution of pH 10 has a hydrogen ion concentration of 10^{-10} *M* and a hydroxide ion concentration of 10^{-4} *M*.

The pH of a neutral aqueous solution at 25°C is 7, the midpoint of the pH scale. A pH value less than 7 denotes an acidic solution; the lower the number, the more acidic the solution. The pH for basic solutions is above 7. Most biological fluids, such as blood and saliva, are within the range of pH 6–8. There are a few exceptions, however, including the strongly acidic digestive juice of the human stomach (gastric juice), which has a pH of about 2.

Remember that each pH unit represents a tenfold difference in H^+ and OH^- concentrations. It is this mathematical feature that makes the pH scale so compact. A solution of pH 3 is not twice as acidic as a solution of pH 6, but 1,000 times (10 × 10 × 10) more acidic. When the pH of a solution changes slightly, the actual concentrations of H^+ and OH^- in the solution change substantially.

Buffers

The internal pH of most living cells is close to 7. Even a slight change in pH can be harmful because the chemical processes of the cell are very sensitive to the concentrations of hydrogen and hydroxide ions. The pH of human blood is very close to 7.4, which is slightly basic. A person cannot survive for more than a few minutes if the blood pH drops to 7 or rises to 7.8, and a chemical system exists in the blood that maintains a stable pH. If 0.01 mol of a strong acid is added to a litre of pure water, the pH drops from 7.0 to 2.0. If the same amount of acid is added to a litre of blood, however, the pH decrease is only from 7.4 to 7.3. Why does the addition of acid have so much less of an effect on the pH of blood than it does on the pH of water?

The presence of substances called buffers allows biological fluids to maintain a relatively constant pH despite the addition of acids or bases. A **buffer** is a substance that minimises changes in the concentrations of H^+ and OH^- in a solution. It does so by accepting hydrogen ions from the solution when they are in excess and donating hydrogen ions to the solution when they have been depleted. Most buffer solutions contain a weak acid and its corresponding base, which combine reversibly with hydrogen ions.

Several buffers contribute to pH stability in human blood and many other biological solutions. One of these is carbonic acid (H_2CO_3), which is formed when CO_2 reacts with water in blood plasma. As mentioned earlier, carbonic acid dissociates to yield a bicarbonate ion (HCO_3^-) and a hydrogen ion (H^+):

$$\underset{\text{(acid)}}{\underset{H^+\text{ donor}}{H_2CO_3}} \underset{\text{Response to a drop in pH}}{\overset{\text{Response to a rise in pH}}{\rightleftharpoons}} \underset{\text{(base)}}{\underset{H^+\text{ acceptor}}{HCO_3^-}} + \underset{\text{ion}}{\underset{\text{Hydrogen}}{H^+}}$$

The chemical equilibrium between carbonic acid and bicarbonate acts as a pH regulator, the reaction shifting left or right as other processes in the solution add or remove hydrogen ions. If the H^+ concentration in blood begins to fall (that is, if pH rises), the reaction proceeds to the right and more carbonic acid dissociates, replenishing hydrogen ions. But when the H^+ concentration in blood begins to rise (when pH drops), the reaction proceeds to the left, with HCO_3^- (the base) removing the hydrogen ions from the solution and forming H_2CO_3. Thus, the carbonic acid–bicarbonate buffering system consists of an acid and a base in equilibrium with each other. Most other buffers are also acid-base pairs.

Acidification: A Threat to Our Oceans

Among the many threats to water quality posed by human activities is the burning of fossil fuels, which releases CO_2 into the atmosphere. The resulting increase in atmospheric CO_2 levels has caused global warming and other aspects of climate change (see Concept 56.4). In addition, about 25% of human-generated CO_2 is absorbed by the oceans. In spite of the huge volume of water in the oceans, scientists worry that the absorption of so much CO_2 will harm marine ecosystems.

Recent data have shown that such fears are well founded. When CO_2 dissolves in seawater, it reacts with water to form carbonic acid, which lowers ocean pH. This process, known as **ocean acidification**, alters the delicate balance of conditions for life in the oceans **(Figure 3.13)**. Based on measurements of the CO_2 level in air bubbles trapped in ice over thousands of years, scientists calculate that the pH of the oceans is 0.1 pH unit lower (more acidic) now than at any time in the past 420,000 years. Recent studies predict that it will drop another 0.3–0.5 pH unit by the end of this century.

As seawater acidifies, the extra hydrogen ions combine with carbonate ions (CO_3^{2-}) to form bicarbonate ions (HCO_3^-), thereby reducing the carbonate ion concentration (see Figure 3.12). Scientists predict that ocean acidification will cause the carbonate ion concentration to decrease by 40% by the year 2100. This is of great concern because carbonate ions are required for calcification, the production of calcium carbonate ($CaCO_3$) by many marine organisms, including reef-building corals and animals that build shells. The **Scientific Skills Exercise** allows you to work with data from an experiment examining the effect of carbonate ion concentration on coral reefs, using an artificial system. In 2018, researchers carried out the first CO_2 enhancement study on an unconfined natural coral reef, observing that addition of CO_2 suppressed calcification and concluding that ocean acidification is likely to cause "profound, ecosystem-wide changes in coral reefs." Coral reefs are sensitive ecosystems that act as havens for a great diversity of marine life. The disappearance of coral reef ecosystems would be a tragic loss of biological diversity.

If there is any reason for optimism about the future quality of water resources on our planet, it is that we have made progress in learning about the delicate chemical balances in oceans, lakes, and rivers. Continued progress can come only from the actions of informed individuals, like yourselves, who are concerned about environmental quality. This requires understanding the crucial role that water plays in the suitability of the environment for continued life on Earth.

▼ **Figure 3.13 Atmospheric CO_2 from human activities and its fate in the ocean.**

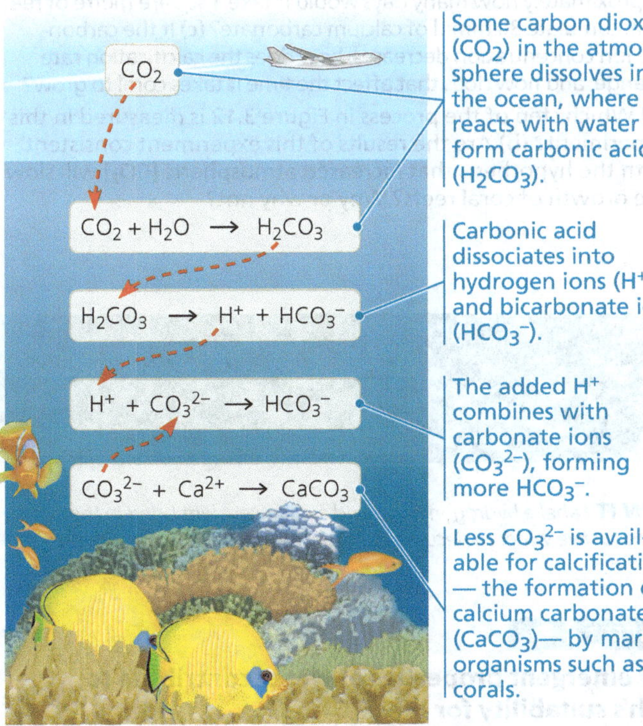

VISUAL SKILLS *Summarise the effect of adding excess CO_2 to the oceans on the calcification process in the final equation.*

CONCEPT CHECK 3.3

1. Compared with a basic solution at pH 9, the same volume of an acidic solution at pH 4 has ________ times as many hydrogen ions (H^+).
2. HCl is a strong acid that dissociates in water: $HCl \rightarrow H^+ + Cl^-$. What is the pH of 0.01 *M* HCl?
3. Acetic acid (CH_3COOH) can be a buffer, similar to carbonic acid. Write the dissociation reaction, identifying the acid, base, H^+ acceptor, and H^+ donor.
4. **WHAT IF?** Given a litre of pure water and a litre solution of acetic acid, what would happen to the pH, in general, if you added 0.01 mol of a strong acid to each? Use the reaction from question 3 to explain the result.

For suggested answers, see Appendix A.

Scientific Skills Exercise

Interpreting a Scatter Plot with a Regression Line

How Does the Carbonate Ion Concentration of Seawater Affect the Calcification Rate of a Coral Reef? Scientists predict that acidification of the ocean due to higher levels of atmospheric CO_2 will lower the concentration of dissolved carbonate ions, which living corals use to build calcium carbonate reef structures. In this exercise, you will analyse data from a controlled experiment that examined the effect of carbonate ion concentration ($[CO_3^{2-}]$) on calcium carbonate deposition, a process called calcification.

How the Experiment Was Done For several years, scientists conducted research on ocean acidification using a large coral reef aquarium at Biosphere 2 in Arizona. They measured the rate of calcification by the reef organisms and examined how the calcification rate changed with differing amounts of dissolved carbonate ions in the seawater.

Data from the Experiment The black data points in the graph form a scatter plot. The red line, known as a linear regression line, is the best-fitting straight line for these points.

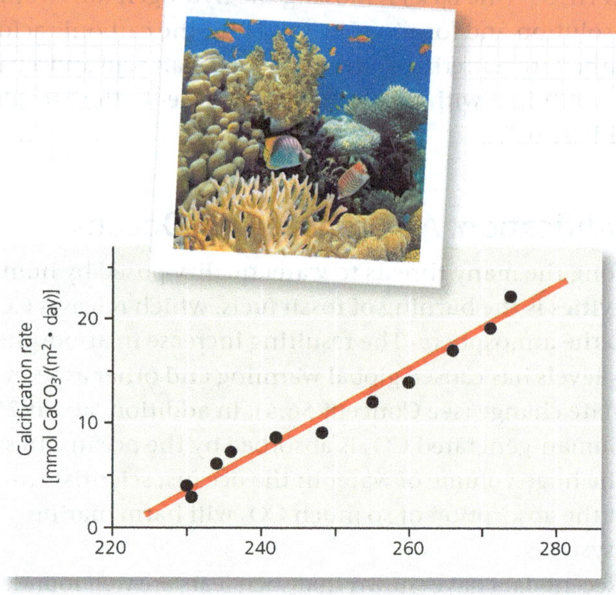

Data from C. Langdon et al., Effect of calcium carbonate saturation state on the calcification rate of an experimental coral reef, *Global Biogeochemical Cycles* 14:639–654 (2000).

INTERPRET THE DATA

1. When presented with a graph of experimental data, the first step in analysis is to determine what each axis represents. **(a)** In words, what is shown on the *x*-axis? (Include the units.) **(b)** What is on the *y*-axis? **(c)** Which variable is the independent variable—the one that was *manipulated* by the researchers? **(d)** Which is the dependent variable—the one that responded to or depended on the treatment, which was *measured* by the researchers? (For additional information about graphs, see the Scientific Skills Review in Appendix D.)
2. Based on the data shown in the graph, describe in words the relationship between carbonate ion concentration and calcification rate.
3. **(a)** If the seawater carbonate ion concentration is 270 μmol/kg, estimate the rate of calcification and how many days it would take 1 square metre of reef to accumulate 30 mmol of calcium carbonate ($CaCO_3$). **(b)** If the seawater carbonate ion concentration is 250 μmol/kg, what is the approximate rate of calcification, and approximately how many days would it take 1 square metre of reef to accumulate 30 mmol of calcium carbonate? **(c)** If the carbonate ion concentration decreases, how does the calcification rate change, and how does that affect the time it takes coral to grow?
4. **(a)** Which step of the process in Figure 3.12 is measured in this experiment? **(b)** Are the results of this experiment consistent with the hypothesis that increased atmospheric $[CO_2]$ will slow the growth of coral reefs? Why or why not?

3 Chapter Review

SUMMARY OF KEY CONCEPTS

CONCEPT 3.1

Polar covalent bonds in water molecules result in hydrogen bonding *(p. 45)*

- Water is a **polar molecule**. A hydrogen bond forms when a partially negatively charged region on the oxygen of one water molecule is attracted to the partially positively charged hydrogen of a nearby water molecule. Hydrogen bonding between water molecules is the basis for water's properties.

DRAW IT *Label a hydrogen bond and a polar covalent bond in the diagram of five water molecules. Is a hydrogen bond a covalent bond? Explain.*

CONCEPT 3.2

Four emergent properties of water contribute to Earth's suitability for life *(pp. 45–52)*

- Hydrogen bonding keeps water molecules close to each other, giving water **cohesion**. Hydrogen bonding is also responsible for water's **surface tension**.
- Water has a high **specific heat**: Heat is absorbed when hydrogen bonds break and is released when hydrogen bonds form. This helps keep **temperatures** relatively steady, within limits that permit life. **Evaporative cooling** is based on water's high **heat**

of vaporisation. The evaporative loss of the most energetic water molecules cools a surface.

- Ice floats because it is less dense than liquid water. This property allows life to exist under the frozen surfaces of lakes and polar seas.
- Water is an unusually versatile **solvent** because its polar molecules are attracted to ions and polar substances that can form hydrogen bonds. **Hydrophilic** substances have an affinity for water; **hydrophobic** substances do not. **Molarity**, the number of moles of **solute** per litre of **solution**, is used as a measure of solute concentration in solutions. A **mole** is a certain number of molecules of a substance. The mass of a mole of a substance in grams is the same as the **molecular mass** in daltons.
- The emergent properties of water support life on Earth and may contribute to the potential for life to have evolved on other planets.

? *Describe how different types of solutes dissolve in water. Explain what a solution is.*

CONCEPT 3.3

Acidic and basic conditions affect living organisms *(pp. 52–56)*

- A water molecule can transfer an H^+ to another water molecule to form H_3O^+ (represented simply by H^+) and OH^-.
- The concentration of H^+ is expressed as **pH**; $pH = -\log [H^+]$. A **buffer** consists of an acid-base pair that combines reversibly with hydrogen ions, allowing it to resist pH changes.
- The burning of fossil fuels increases the amount of CO_2 in the atmosphere. Some CO_2 dissolves in the oceans, causing **ocean acidification**, which has potentially grave consequences for marine organisms that rely on calcification.

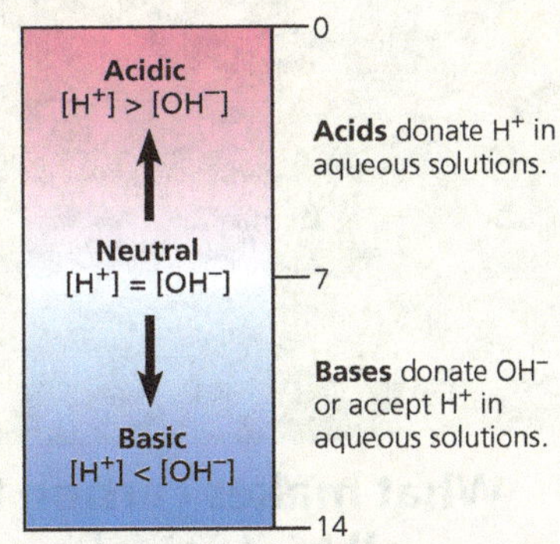

? *Explain what happens to the concentration of hydrogen ions in an aqueous solution when you add a base and cause the concentration of OH^- to rise to 10^{-3}. What is the pH of this solution?*

TEST YOUR UNDERSTANDING

Levels 1-2: Remembering/Understanding

1. Which of the following is a hydrophobic material?
   (A) paper (C) wax
   (B) table salt (D) sugar
2. We can be sure that a mole of table sugar and a mole of vitamin C are equal in their
   (A) mass. (C) number of atoms.
   (B) volume. (D) number of molecules.
3. Measurements show that the pH of a particular lake is 4.0. What is the hydrogen ion concentration of the lake?
   (A) 4.0 *M* (C) 10^{-4} *M*
   (B) 10^{-10} *M* (D) 10^{4} *M*
4. What is the *hydroxide* ion concentration of the lake described in question 3?
   (A) 10^{-10} *M* (C) 10^{-7} *M*
   (B) 10^{-4} *M* (D) 10.0 *M*

Levels 3-4: Applying/Analysing

5. A slice of pizza has 500 kcal (2092 kJ). If we could burn the pizza and use all the heat to warm a 50-L container of cold water, what would be the approximate increase in the temperature of the water? (Note: A litre of cold water weighs about 1 kg.)
   (A) 50°C (C) 100°C
   (B) 5°C (D) 10°C
6. **DRAW IT** Draw the hydration shells that form around a potassium ion and a chloride ion when potassium chloride (KCl) dissolves. Label the positive, negative, and partial charges.

Levels 5-6: Evaluating/Creating

7. Right before a predicted overnight freeze, farmers spray water on crops to protect the plants. Use the properties of water to explain how this method works. Be sure to mention why hydrogen bonds are responsible for this phenomenon.
8. **MAKE CONNECTIONS** What do climate change (see Concepts 1.1 and 3.2) and ocean acidification have in common?
9. **EVOLUTION CONNECTION** This chapter explains how the emergent properties of water contribute to the suitability of the environment for life. Until fairly recently, scientists assumed that other physical requirements for life included a moderate range of temperature, pH, atmospheric pressure, and salinity, as well as low levels of toxic chemicals. That view has changed with the discovery of organisms known as extremophiles, which flourish in hot, acidic sulfur springs, around hydrothermal vents deep in the ocean, and in soils with high levels of toxic metals. Why would astrobiologists study extremophiles? What does the existence of life in such extreme environments say about the possibility of life on other planets?
10. **SCIENTIFIC INQUIRY** Design a controlled experiment to test the hypothesis that water acidification caused by acidic rain would inhibit the growth of *Elodea*, a freshwater plant (see Figure 2.17).
11. **WRITE ABOUT A THEME: ORGANISATION** Several emergent properties of water contribute to the suitability of the environment for life. In a short essay (100–150 words), describe how the ability of water to function as a versatile solvent arises from the structure of water molecules.
12. **SYNTHESISE YOUR KNOWLEDGE**

How do cats drink? Scientists using high-speed video have shown that cats use an interesting technique to drink aqueous substances like water and milk. Four times a second, the cat touches the tip of its tongue to the water and draws a column of water up into its mouth (as you can see in the photo), which then shuts before gravity can pull the water back down. Describe how the properties of water allow cats to drink in this fashion, including how water's molecular structure contributes to the process.

For selected answers, see Appendix A.

4 Carbon and the Molecular Diversity of Life

KEY CONCEPTS

4.1 **Organic chemistry is key to the origin of life** *p. 59*

4.2 **Carbon atoms can form diverse molecules by bonding to four other atoms** *p. 60*

4.3 **A few chemical groups are key to molecular function** *p. 64*

Study Tip

Label chemical groups: After you have read through Figure 4.9, look through the chapters, "Carbon and the Molecular Diversity of Life" and "The Structure and Function of Large Biological Molecules" for molecules that have the chemical groups shown in that figure. Circle and label the chemical groups you find, as in the following example:

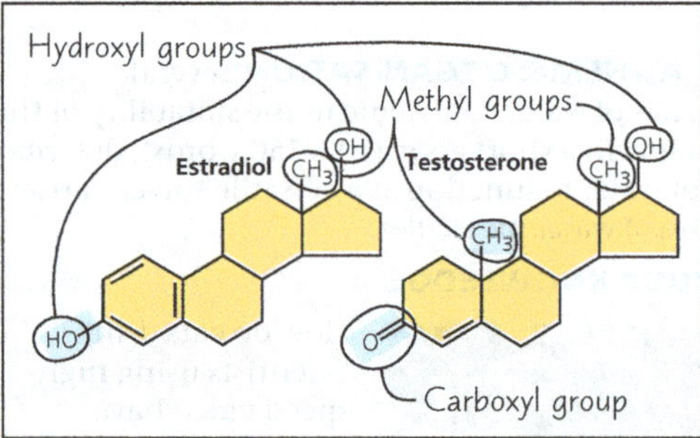

Go to Mastering Biology

to access Dynamic Study Modules for revision, 3D BioFlix® animations and high-quality videos, and your interactive Pearson eText.

Figure 4.1 Qinling golden snub-nosed monkeys in southwest China. These monkeys and other living organisms in this mountainous forest are made up of chemicals based mostly on the element carbon. Of all chemical elements, carbon is unparalleled in its ability to form molecules that are large, complex, and varied, making possible the diversity of organisms that have evolved on Earth.

What makes carbon the basis for all biological molecules?

Carbon can form four bonds, and therefore can bond to up to four other atoms or groups of atoms.

Carbon can bond to other carbons, resulting in carbon skeletons. Carbon also commonly bonds to

H **hydrogen,**

O **oxygen,** and

N **nitrogen.**

The properties of a carbon-containing molecule depend on the arrangement of its **carbon skeleton** and on its **chemical groups.**

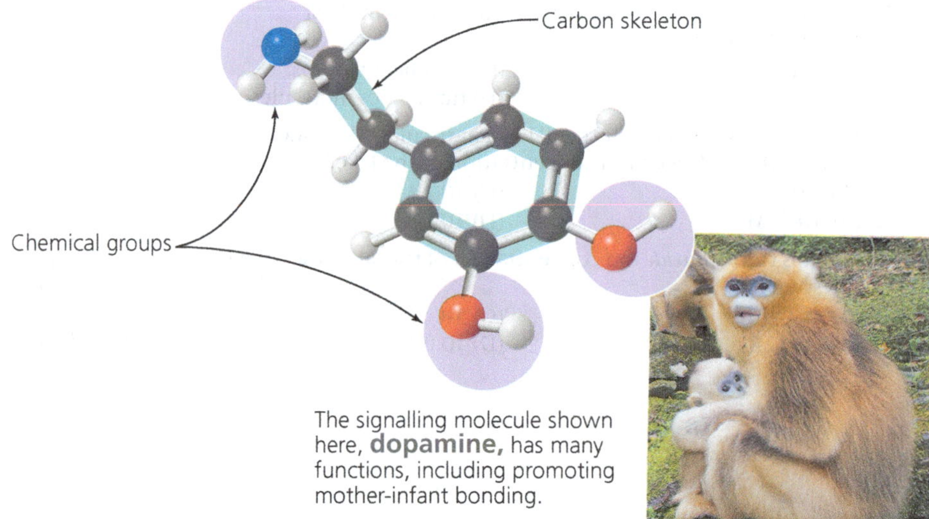

The signalling molecule shown here, **dopamine,** has many functions, including promoting mother-infant bonding.

CONCEPT 4.1

Organic chemistry is key to the origin of life

For historical reasons, compounds containing carbon are said to be organic, and their study is called **organic chemistry**. Organic compounds range from simple molecules, such as methane (CH_4), to colossal ones, such as proteins, with thousands of atoms.

EVOLUTION In 1953, Stanley Miller, a graduate student of Harold Urey at the University of Chicago, designed an experiment on the abiotic (nonliving) synthesis of organic compounds to investigate the origin of life. Study **Figure 4.2** to learn about his classic experiment. From his results, Miller concluded that complex organic molecules could arise spontaneously under conditions thought at that time to have existed on early Earth. You can work with the data from a related experiment in the **Scientific Skills Exercise**. These experiments support the idea that abiotic synthesis of organic compounds, perhaps near volcanoes, could have been an early stage in the origin of life (see Figure 25.2).

In Concept 3.2, you learned about evidence for the presence of water on Mars. Even more exciting, in 2018, NASA reported that the rover *Curiosity* had found carbon-based compounds on Mars in a crater where a lake once existed. While these compounds might have been brought to Mars on a meteorite or formed by geologic processes, an intriguing possibility is that they might have been the relics of life-forms that once existed on that planet.

The overall percentages of the major elements of life—C, H, O, N, S, and P—are quite uniform from one organism to another, reflecting the common evolutionary origin of all life. Because of carbon's ability to form four bonds, however, this limited assortment of atomic building blocks can be used to build an inexhaustible variety of organic molecules. Different species of organisms, and different individuals within a species, are distinguished by variations in the types of organic molecules they make. In a sense, the great diversity of living organisms we see on the planet (and in fossil remains) is made possible by the unique chemical versatility of the carbon atom.

CONCEPT CHECK 4.1

1. **VISUAL SKILLS** See Figure 4.2. Miller carried out a control experiment without discharging sparks and found no organic compounds. What might explain this result?

For suggested answers, see Appendix A.

▼ **Figure 4.2 Inquiry**

Can organic molecules form under conditions estimated to simulate those on the early Earth?

Experiment In 1953, Stanley Miller set up a closed system to mimic conditions thought at that time to have existed on the early Earth. A flask of water simulated the primeval sea. The water was heated so that some vaporised and moved into a second, higher flask containing the "atmosphere"—a mixture of gases. Sparks were discharged in the synthetic atmosphere to mimic lightning.

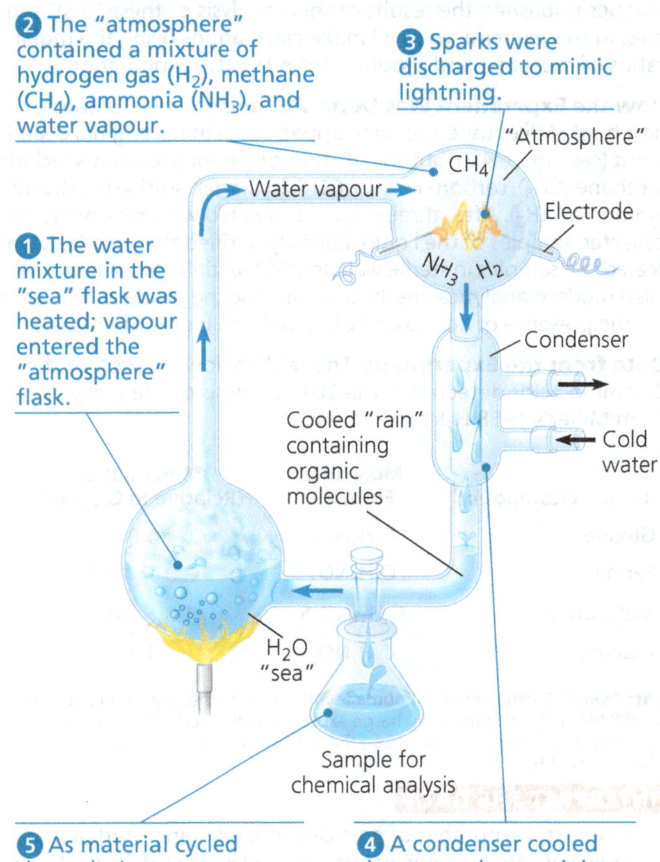

Results Miller identified a variety of organic molecules that are common in organisms. These included simple compounds, such as formaldehyde (CH_2O) and hydrogen cyanide (HCN), and more complex molecules, such as amino acids and long chains of carbon and hydrogen known as hydrocarbons.

Conclusion Organic molecules, a first step in the origin of life, may have been synthesised abiotically on the early Earth. Although later evidence indicated that the early-Earth atmosphere was different from the "atmosphere" used by Miller in this experiment, recent experiments using the revised list of chemicals also produced organic molecules. (We will explore this hypothesis in more detail in Concept 25.1.)

Data from S. L. Miller, A production of amino acids under possible primitive Earth conditions, *Science* 117:528–529 (1953).

WHAT IF? *If Miller had increased the concentration of NH_3 in his experiment, how might the relative amounts of the products HCN and CH_2O have differed?*

Scientific Skills Exercise

Working with Moles and Molar Ratios

Could the First Biological Molecules Have Formed Near Volcanoes on Early Earth? In 2007, Jeffrey Bada, a former graduate student of Stanley Miller, discovered some vials of samples that had never been analysed from an experiment performed by Miller in 1958. In that experiment, Miller used hydrogen sulfide gas (H_2S) as one of the gases in the reactant mixture. Since H_2S is released by volcanoes, the H_2S experiment was designed to mimic conditions near volcanoes on early Earth. In 2011, Bada and colleagues published the results of their analysis of these "lost" samples. In this exercise, you will make calculations using the molar ratios of reactants and products from the H_2S experiment.

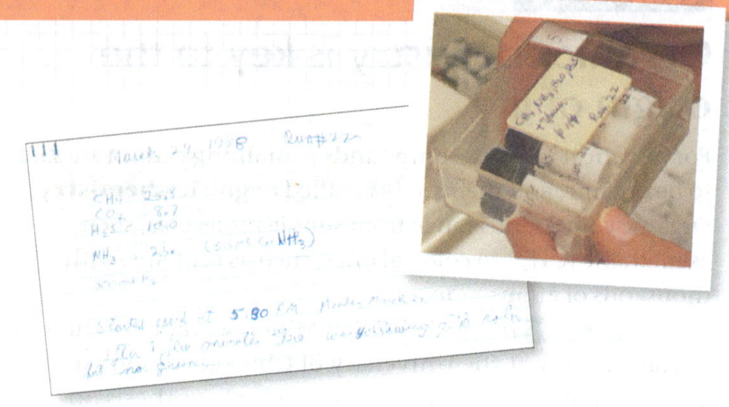

▲ **Some of Stanley Miller's notes from his 1958 hydrogen sulfide (H_2S) experiment along with his original vials.**

How the Experiment Was Done According to his laboratory notebook, Miller used the same apparatus as in his original experiment (see Figure 4.2), but the mixture of gaseous reactants included methane (CH_4), carbon dioxide (CO_2), hydrogen sulfide (H_2S), and ammonia (NH_3). After three days of simulated volcanic activity, he collected samples of the liquid, partially purified the chemicals, and sealed the samples in sterile vials. In 2011, Bada's research team used modern analytical methods to analyse the products in the vials for the presence of amino acids, the building blocks of proteins.

Data from the Experiment The table below shows 4 of the 23 amino acids detected in the 2011 analysis of the samples from Miller's 1958 H_2S experiment.

Product Compound	Molecular Formula	Molar Ratio (Relative to Glycine)
Glycine	$C_2H_5NO_2$	1.0
Serine	$C_3H_7NO_3$	3.0×10^{-2}
Methionine	$C_5H_{11}NO_2S$	1.8×10^{-3}
Alanine	$C_3H_7NO_2$	1.1

Data from E. T. Parker et al., Primordial synthesis of amines and amino acids in a 1958 Miller H_2S-rich spark discharge experiment, *Proceedings of the National Academy of Sciences USA* 108:5526-5531 (2011). www.pnas.org/cgi/doi/10.1073/pnas.1019191108.

INTERPRET THE DATA

1. A *mole* is the number of particles of a substance with a mass equivalent to its molecular (or atomic) mass in daltons. There are 6.02×10^{23} molecules (or atoms) in 1.0 mole (Avogadro's number; see Concept 3.2). The data table shows the "molar ratios" of some of the products from the Miller H_2S experiment. In a molar ratio, each unitless value is expressed relative to a standard for that experiment. Here, the standard is the number of moles of the amino acid glycine, which is set to a value of 1.0. For instance, serine has a molar ratio of 3.0×10^{-2}, meaning that for every mole of glycine, there is 3.0×10^{-2} mole of serine. **(a)** Give the molar ratio of methionine to glycine and explain what it means. **(b)** How many molecules of glycine are present in 1.0 mole? **(c)** For every 1.0 mole of glycine in the sample, how many molecules of methionine are present? (Recall that to multiply two numbers with exponents, you add their exponents; to divide them, you subtract the exponent in the denominator from that in the numerator.)
2. **(a)** Which amino acid is present in higher amounts than glycine? **(b)** How many more molecules of that amino acid are present than the number of molecules in 1.0 mole of glycine?
3. The synthesis of products is limited by the amount of reactants. **(a)** If one mole each of CH_4, NH_3, H_2S, and CO_2 is added to 1 litre of water (= 55.5 moles of H_2O) in a flask, how many moles of hydrogen, carbon, oxygen, nitrogen, and sulfur are in the flask? **(b)** Looking at the molecular formula in the table, how many moles of each element would be needed to make 1.0 mole of glycine? **(c)** What is the maximum number of moles of glycine that could be made in that flask, with the specified ingredients, if no other molecules were made? Explain. **(d)** If serine or methionine were made individually, which element(s) would be used up first for each? How much of each product could be made?
4. The earlier published experiment carried out by Miller did not include H_2S in the reactants (see Figure 4.2). Which of the compounds shown in the data table can be made in the H_2S experiment but could not be made in the earlier experiment?

CONCEPT 4.2

Carbon atoms can form diverse molecules by bonding to four other atoms

The key to an atom's chemical characteristics is its electron configuration. This configuration determines the kinds and number of bonds an atom will form with other atoms. Recall that it is the valence electrons, those in the outermost shell, that are available to form bonds with other atoms.

The Formation of Bonds with Carbon

Carbon has 6 electrons, with 2 in the first electron shell and 4 in the second shell; thus, it has 4 valence electrons in a shell that can hold up to 8 electrons. A carbon atom usually completes its valence shell by sharing its 4 electrons with other atoms so that 8 electrons are present. Each pair of shared electrons constitutes a covalent bond (see Figure 2.10d). In organic molecules, carbon usually forms single or double covalent bonds. Each carbon atom acts as an intersection point from which a molecule can branch off in as many as four directions. This enables carbon to form large, complex molecules.

▼ **Figure 4.3 The shapes of three simple organic molecules.**

Molecule and Molecular Shape	Molecular Formula	Structural Formula	Ball-and-Stick Model (molecular shape in pink)	Space-Filling Model
(a) Methane. When a carbon atom has four single bonds to other atoms, the molecule is tetrahedral.	CH_4	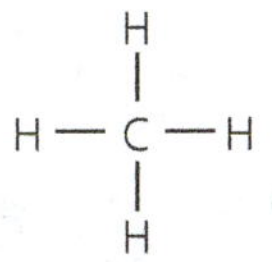 H–C–H with H above and below C	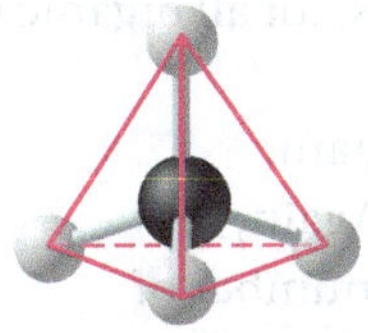	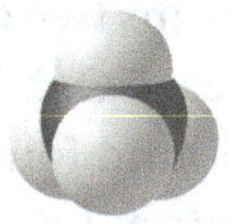
(b) Ethane. A molecule may have more than one tetrahedral group of single-bonded atoms. (Ethane consists of two such groups.)	C_2H_6	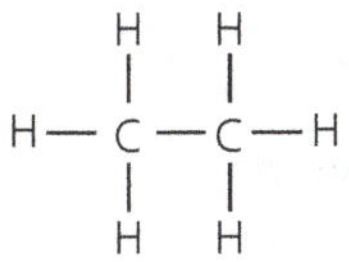 H–C–C–H with H above and below each C	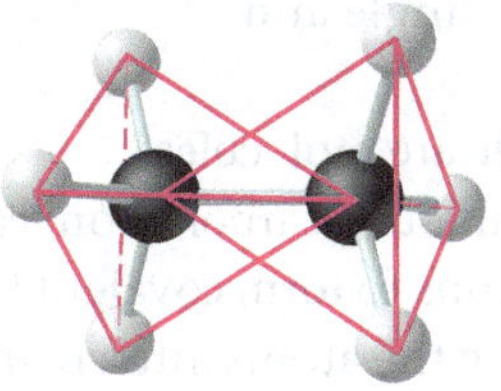	
(c) Ethene (ethylene). When two carbon atoms are joined by a double bond, all atoms attached to those carbons are in the same plane, and the molecule is flat.	C_2H_4	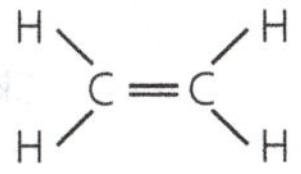 H₂C=CH₂ (each C bonded to two H)	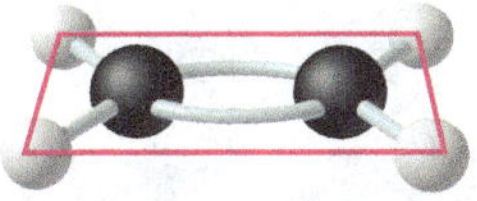	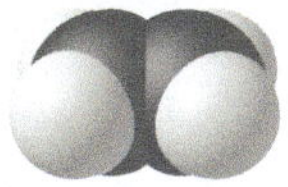

When a carbon atom forms four single covalent bonds, the arrangement of its four hybrid orbitals causes the bonds to angle towards the corners of an imaginary tetrahedron. The bond angles in methane (CH_4) are 109.5° **(Figure 4.3a)**, and they are roughly the same in any group of atoms where carbon has four single bonds. For example, ethane (C_2H_6) is shaped like two overlapping tetrahedrons **(Figure 4.3b)**. In molecules with more carbons, every grouping of a carbon bonded to four other atoms has a tetrahedral shape. But when two carbon atoms are joined by a double bond, as in ethene (C_2H_4), the bonds from both carbons are all in the same plane, so the atoms joined to those carbons are in the same plane as well **(Figure 4.3c)**. We find it convenient to write molecules as structural formulas, as if the molecules being represented are two-dimensional, but keep in mind that molecules are three-dimensional and that the shape of a molecule is central to its function.

The number of electrons required to fill the valence shell of an atom is generally equal to the atom's **valence**, the number of covalent bonds it can form. **Figure 4.4** shows the valences of carbon and its most frequent bonding partners—hydrogen, oxygen, and nitrogen. These are the four main atoms in organic molecules.

The electron configuration of carbon gives it covalent compatibility with many different elements. Let's consider how valence and the rules of covalent bonding apply to carbon atoms with partners other than hydrogen. We'll look at two examples, the simple molecules carbon dioxide and urea.

▼ **Figure 4.4 Valences of the major elements of organic molecules.** Valence, the number of covalent bonds an atom can form, is generally equal to the number of electrons required to fill the valence shell. (Sodium, phosphorus, and chlorine are exceptions.)

	Hydrogen	Oxygen	Nitrogen	Carbon
Lewis dot structure showing existing valence electrons	H•	•Ö:	•N̈•	•Ċ•
Electron distribution diagram with red circles showing electrons needed to fill the valence shell	 H	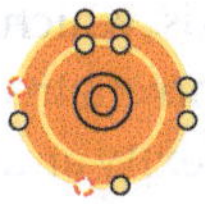 O	 N	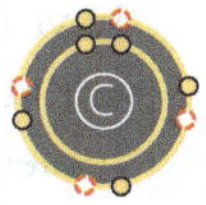 C
Number of electrons needed to fill the valence shell	1	2	3	4
Valence: Number of bonds the element can form	1	2	3	4

MAKE CONNECTIONS *Draw the Lewis dot structures for sodium, silicon, phosphorus, sulfur, and chlorine. (Refer to Figure 2.7.)*

In the carbon dioxide molecule (CO_2), a single carbon atom is joined to two atoms of oxygen by double covalent bonds. The structural formula for CO_2 is shown here:

$$O{=}C{=}O$$

Each line in a structural formula represents a pair of shared electrons. Thus, the two double bonds in CO_2 have the same number of shared electrons as four single bonds. The arrangement completes the valence shells of all atoms in the molecule:

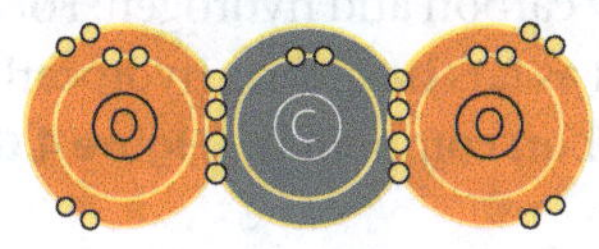

Because CO_2 is a very simple molecule and lacks hydrogen, it is often considered inorganic, even though it contains carbon. Whether we call CO_2 organic or inorganic, however, it is clearly important to the living world as the source of carbon, via photosynthetic organisms, for all organic molecules in organisms (see Concept 2.4).

Urea, $CO(NH_2)_2$, is an organic compound found in urine. Again, each atom has the required number of covalent bonds. In this case, one carbon atom participates in both single and double bonds.

Urea

Urea and carbon dioxide are molecules with only one carbon atom. But a carbon atom can also use one or more valence electrons to form covalent bonds to other carbon atoms, linking the atoms into chains, as shown here for C_3H_8:

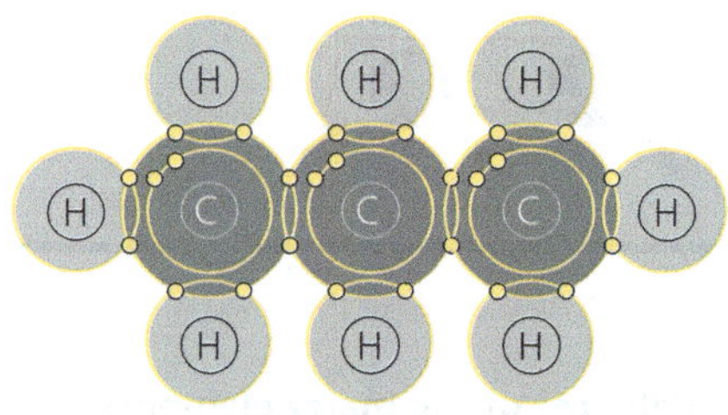

Molecular Diversity Arising from Variation in Carbon Skeletons

Carbon chains form the basis of most organic molecules. Carbon skeletons vary in length and may be straight, branched, or arranged in closed rings **(Figure 4.5)**. Some carbon chains have double bonds, which vary in number and location. Such variation in carbon chains is one important source of the molecular complexity and diversity that characterise living matter. In addition, the skeletons of biological molecules often include atoms of other elements, like oxygen and phosphorus; such atoms can also be bonded to carbons of the skeleton.

Hydrocarbons

All of the molecules that are shown in Figures 4.3 and 4.5 are **hydrocarbons**, organic molecules consisting of only carbon and hydrogen. Atoms of hydrogen are attached to the carbon skeleton wherever electrons are available for covalent bonding. Hydrocarbons are the major components of petroleum, which is called a fossil fuel because it consists of the partially decomposed remains of organisms that lived millions of years ago.

Although hydrocarbons are not prevalent in most living organisms, some of a cell's organic molecules have regions consisting of only carbon and hydrogen. For example, the molecules known as fats have long hydrocarbon tails attached to a nonhydrocarbon component **(Figure 4.6)**. Neither

▼ **Figure 4.5 Four ways that carbon skeletons can vary.**

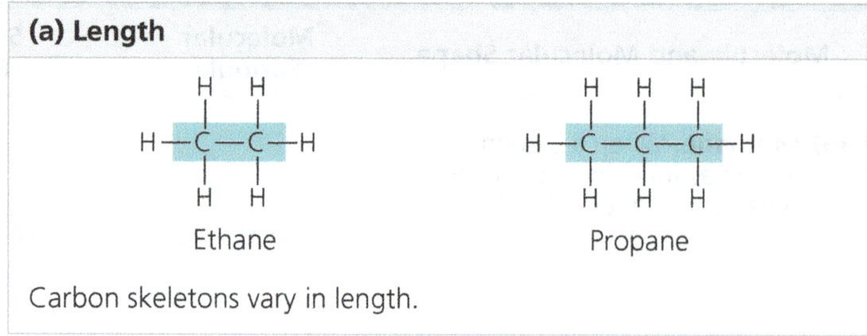

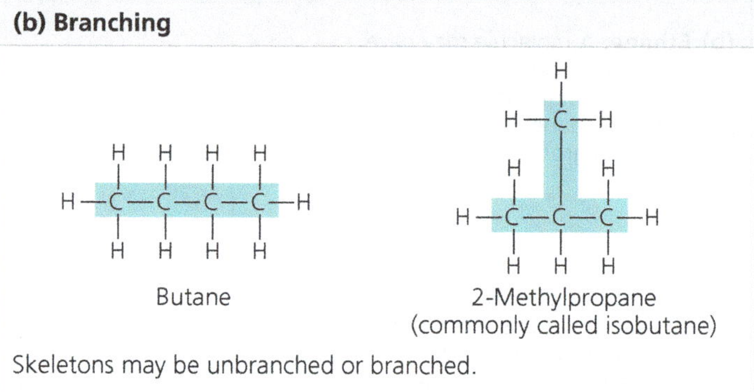

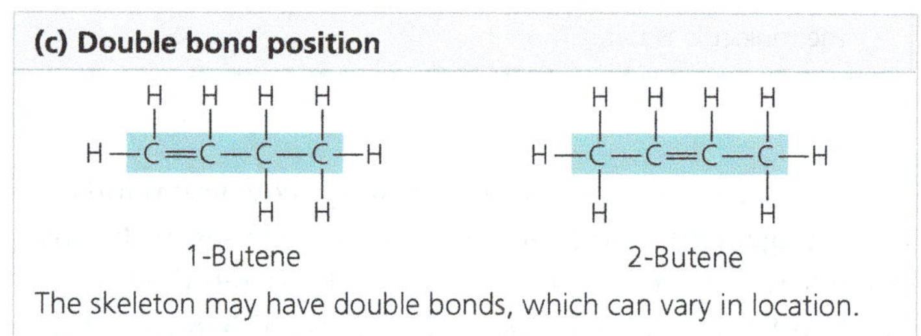

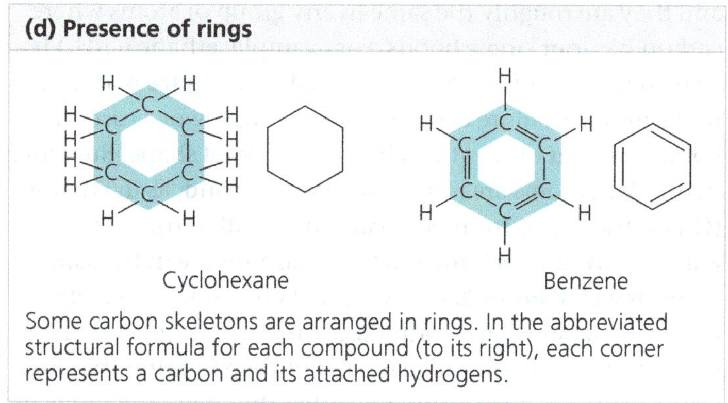

petroleum nor fat dissolves in water; both are hydrophobic compounds because the great majority of their bonds are relatively nonpolar carbon-to-hydrogen linkages. Another characteristic of hydrocarbons is that they can undergo reactions that release a relatively large amount of energy. The gasoline that fuels a car consists of hydrocarbons, and the hydrocarbon tails of fats serve as stored fuel for plant embryos (seeds) and animals.

Isomers

Variation in the architecture of organic molecules can be seen in **isomers**, compounds that have the same numbers of

▼ **Figure 4.6 The role of hydrocarbons in fats.**
(a) Mammalian adipose cells stockpile fat molecules as a fuel reserve. This colourised micrograph shows part of a human adipose cell with many fat droplets, each containing a large number of fat molecules. **(b)** A fat molecule consists of a small, nonhydrocarbon component joined to three hydrocarbon tails that account for the hydrophobic behaviour of fats. The tails can be broken down to provide energy. (Black = carbon; grey = hydrogen; red = oxygen.)

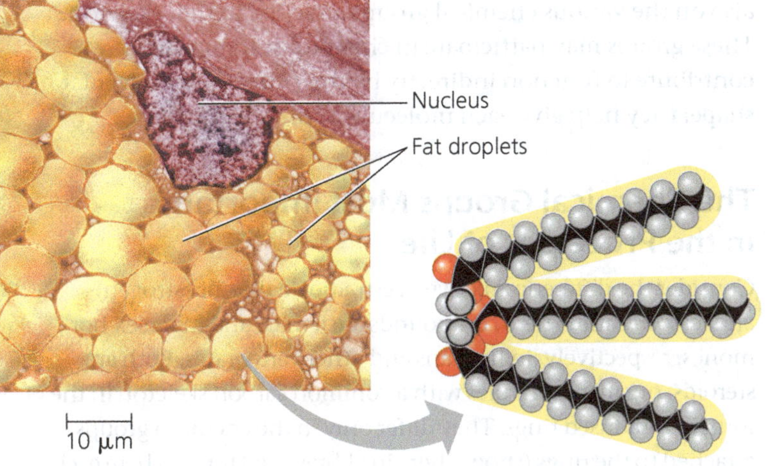

(a) Part of a human adipose cell **(b) A fat molecule**

MAKE CONNECTIONS *How do the tails account for the hydrophobic nature of fats? (See Concept 3.2.)*

▼ **Figure 4.7 Three types of isomers.** Isomers are compounds that have the same molecular formula but different structures.

(a) Structural isomers

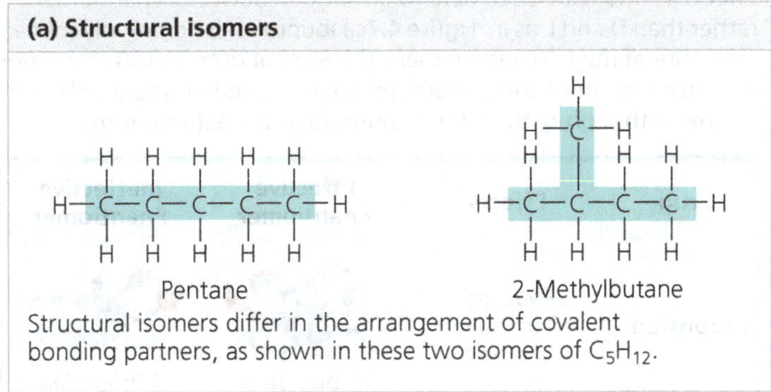

Structural isomers differ in the arrangement of covalent bonding partners, as shown in these two isomers of C_5H_{12}.

(b) *Cis-trans* isomers (also known as geometric isomers)

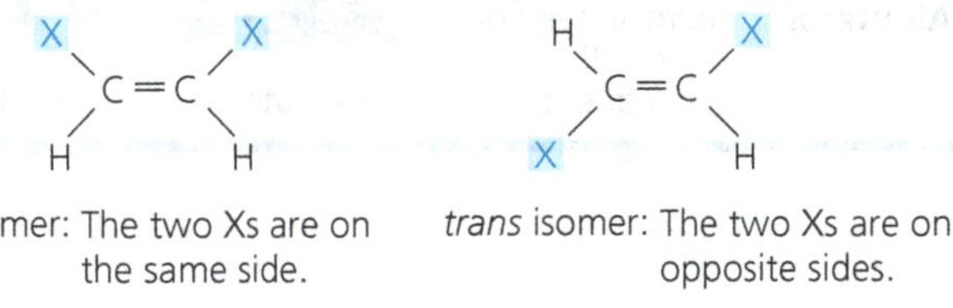

cis isomer: The two Xs are on the same side.

trans isomer: The two Xs are on opposite sides.

Cis-trans isomers differ in arrangement about a double bond. In these diagrams, X represents an atom or group of atoms attached to a double-bonded carbon.

(c) Enantiomers

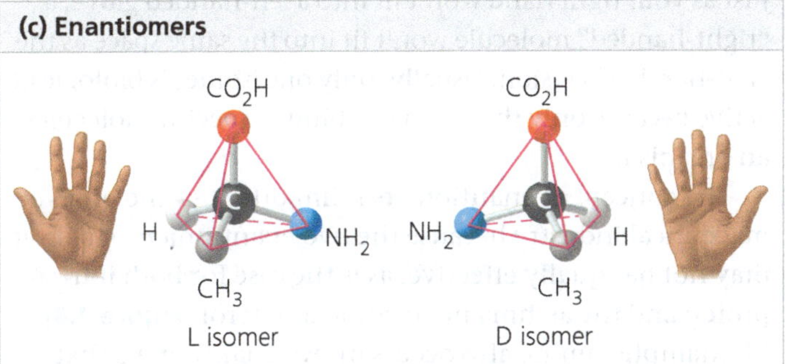

Enantiomers differ in spatial arrangement around an asymmetric carbon, resulting in molecules that are mirror images, like left and right hands. The two isomers here are designated the L and D isomers from the Latin for "left" and "right" (*levo* and *dextro*). Enantiomers cannot be superimposed on each other.

DRAW IT *There are three structural isomers of C_5H_{12}; draw the one not shown in (a).*

atoms of the same elements but different structures and hence different properties. We will examine three types of isomers: structural isomers, *cis-trans* isomers, and enantiomers.

Structural isomers differ in the covalent arrangements of their atoms. Compare, for example, the two five-carbon compounds in **Figure 4.7a**. Both have the molecular formula C_5H_{12}, but they differ in the covalent arrangement of their carbon skeletons. The skeleton is straight in one compound but branched in the other. The number of possible isomers increases tremendously as carbon skeletons increase in size. There are only three forms of C_5H_{12} (two of which are shown in Figure 4.7a), but there are 18 variants of C_8H_{18} and 366,319 possible structural isomers of $C_{20}H_{42}$. Structural isomers may also differ in the location of double bonds.

In ***cis-trans* isomers** (also known as *geometric isomers*), carbons have covalent bonds to the same atoms, but these atoms differ in their spatial arrangements due to the inflexibility of double bonds. Single bonds allow the atoms they join to rotate freely about the bond axis without changing the compound. In contrast, double bonds do not permit such rotation. If a double bond joins two carbon atoms, and each C also has two different atoms (or groups of atoms) attached to it, then two distinct *cis-trans* isomers are possible. Consider a simple molecule with two double-bonded carbons, each of which has an H and an X attached to it **(Figure 4.7b)**. The arrangement with both Xs on the same side of the double bond is called a *cis isomer*, and that with the Xs on opposite sides is called a *trans isomer*. The subtle difference in shape between such isomers can have a dramatic effect on the biological activities of organic molecules. For example, the biochemistry of vision involves a light-induced change of retinal, a chemical compound in the eye, from the *cis* isomer to the *trans* isomer (see Figure 50.17). Another example involves *trans* fats, harmful fats formed during food processing that are discussed in Concept 5.3.

Enantiomers are isomers that are mirror images of each other and that differ in shape due to the presence of an *asymmetric carbon*, one that is attached to four different atoms or groups of atoms. (See the middle carbon in the ball-and-stick models shown in **Figure 4.7c**.) The four groups can

▼ Figure 4.8 The pharmacological importance of enantiomers. Ibuprofen and albuterol are drugs whose enantiomers have different effects. (S and R are used here to distinguish between enantiomers, rather than D and L as in Figure 4.7c.) Ibuprofen is commonly sold as a mixture of the two enantiomers; the S enantiomer is 100 times more effective than the R form. Albuterol is synthesised and sold only as the R form of that drug; the S form counteracts the active R form.

Drug	Effects	Effective Enantiomer	Ineffective Enantiomer
Ibuprofen	Reduces inflammation and pain	S-Ibuprofen	R-Ibuprofen
Albuterol	Relaxes bronchial (airway) muscles, improving airflow in asthma patients	R-Albuterol	S-Albuterol

be arranged in space around the asymmetric carbon in two different ways that are mirror images. Enantiomers are, in a way, left-handed and right-handed versions of the molecule. Just as your right hand won't fit into a left-handed glove, a "right-handed" molecule won't fit into the same space as the "left-handed" version. Usually, only one isomer is biologically active because only that form can bind to specific molecules in an organism.

The concept of enantiomers is important in the pharmaceutical industry because the two enantiomers of a drug may not be equally effective, as is the case for both ibuprofen and the asthma medication albuterol **(Figure 4.8)**. Methamphetamine also occurs in two enantiomers that have very different effects. One enantiomer is the highly addictive stimulant drug known as "crank," sold illegally in the street drug trade. The other has a much weaker effect and is the active ingredient in an over-the-counter vapour inhaler for treatment of nasal congestion. The differing effects of enantiomers in the body demonstrate that organisms are sensitive to even the subtlest variations in molecular architecture. Once again, we see that molecules have emergent properties that depend on the specific arrangement of their atoms.

CONCEPT CHECK **4.2**

1. **DRAW IT** (a) Draw a structural formula for C_2H_4. (b) Draw the *trans* isomer of $C_2H_2Cl_2$.
2. **VISUAL SKILLS** Which two pairs of molecules in Figure 4.5 are isomers? For each pair, identify the type of isomer.
3. How are gasoline and fat chemically similar?
4. **VISUAL SKILLS** See Figures 4.5a and 4.7. Can propane (C_3H_8) form isomers? Explain.

For suggested answers, see Appendix A.

CONCEPT 4.3

A few chemical groups are key to molecular function

The distinctive properties of an organic molecule depend not only on the arrangement of its mostly carbon skeleton but also on the various chemical groups attached to that skeleton. These groups may participate in chemical reactions or may contribute to function indirectly by their effects on molecular shape; they help give each molecule its unique properties.

The Chemical Groups Most Important in the Processes of Life

Consider the differences between estradiol (a type of estrogen) and testosterone. These compounds are female and male sex hormones, respectively, in humans and other vertebrates. Both are steroids, organic molecules with a common carbon skeleton in the form of four fused rings. They differ only in the chemical groups attached to the rings (shown here in abbreviated form, where each corner represents a carbon and its attached hydrogens); the distinctions in molecular architecture are shaded in blue:

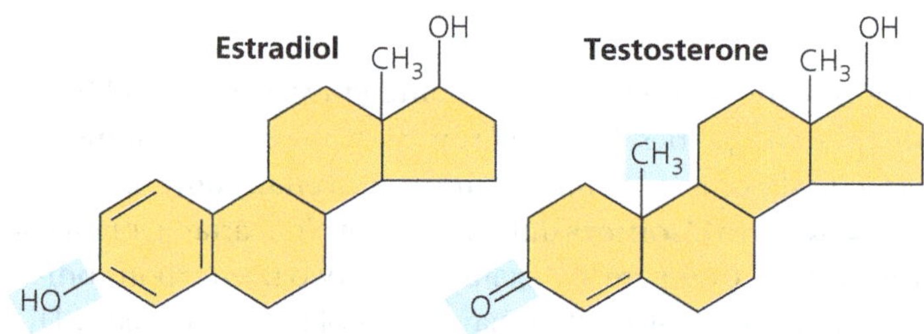

The different actions of these two molecules on many targets throughout the body are the basis of sexual characteristics, producing the contrasting features of male and female vertebrates. In this case, the chemical groups are important because they affect molecular shape, contributing to function.

In other cases, chemical groups are directly involved in chemical reactions; such groups are known as **functional groups**. Each has certain properties, such as shape and charge, that cause it to participate in chemical reactions in a characteristic way.

The seven chemical groups most important in biological processes are the hydroxyl, carbonyl, carboxyl, amino, sulfhydryl, phosphate, and methyl groups. The first six groups can be chemically reactive; of these six, all except the sulfhydryl group are also hydrophilic and thus increase the solubility of organic compounds in water. The methyl group is not reactive, but instead often serves as a recognisable tag on biological molecules. Study **Figure 4.9** to become familiar with these biologically important chemical groups. As shown at the right of the figure, the carboxyl group and the amino group are ionised at normal cellular pH.

Figure 4.9 Some biologically important chemical groups.

Chemical Group	Group Properties and Compound Name	Examples
Hydroxyl group (—OH) —OH 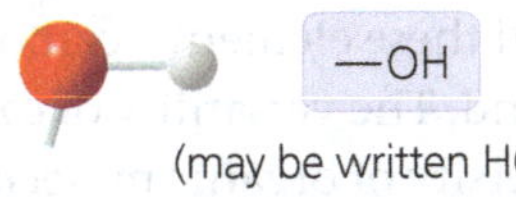 (may be written HO—)	Is polar due to electronegative oxygen. Forms hydrogen bonds with water, helping dissolve compounds such as sugars. Compound name: **Alcohol** (specific name usually ends in *-ol*)	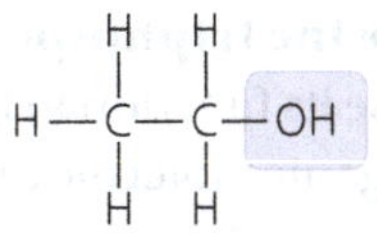 **Ethanol**, the alcohol present in alcoholic beverages
Carbonyl group (>C=O) 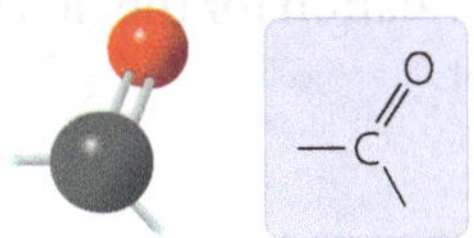	Sugars with ketone groups are called ketoses; those with aldehydes are called aldoses. Compound name: **Ketone** (carbonyl group is within a carbon skeleton) or **aldehyde** (carbonyl group is at the end of a carbon skeleton)	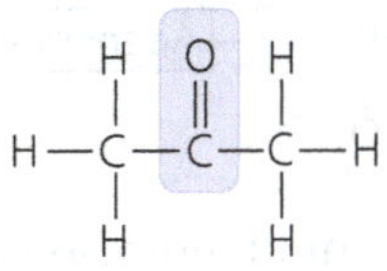 **Acetone**, the simplest ketone 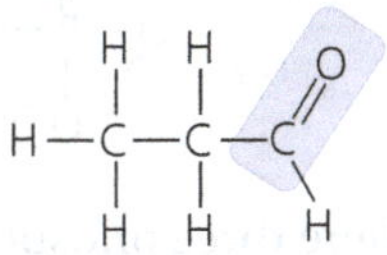 **Propanal**, an aldehyde
Carboxyl group (—COOH) 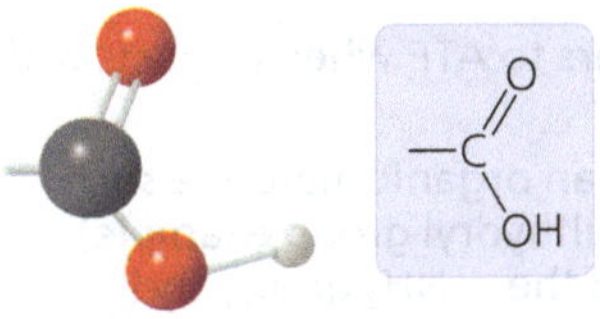	Acts as an acid (can donate H^+) because the covalent bond between oxygen and hydrogen is so polar. Compound name: **Carboxylic acid**, or **organic acid**	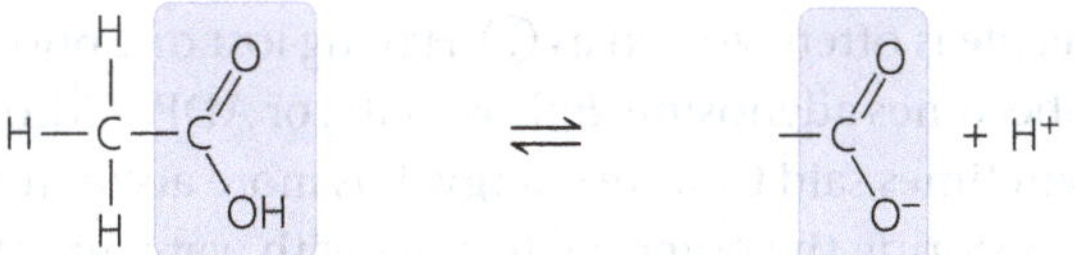 **Acetic acid**, which gives vinegar its sour taste ⇌ $+ H^+$ ionised form of —COOH (carboxylate ion), found in cells
Amino group (—NH_2) 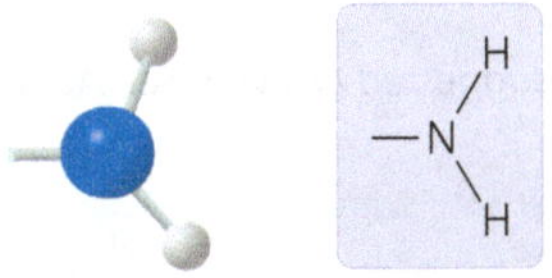	Acts as a base; can pick up an H^+ from the surrounding solution (water, in living organisms). Compound name: **Amine**	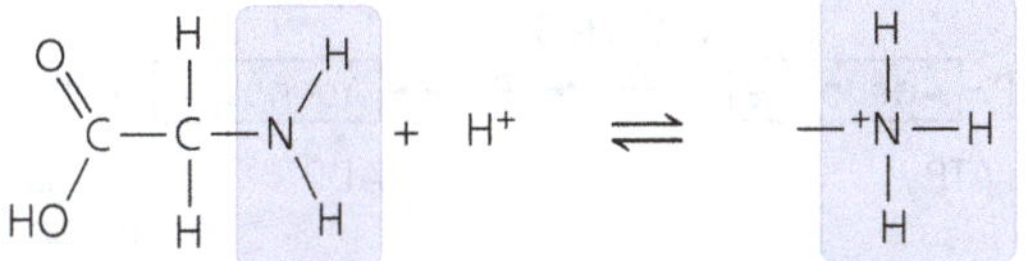 **Glycine**, an amino acid (note its carboxyl group) $+ H^+$ ⇌ ionised form of —NH_2, found in cells
Sulfhydryl group (—SH) —SH 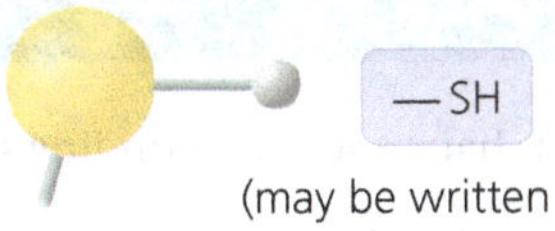 (may be written HS —)	Two —SH groups can react, forming a "cross-link" that helps stabilise protein structure. Hair protein cross-links maintain the straightness or curliness of hair; in hair salons, "permanent" treatments break cross-links, then re-form them while the hair is in the desired shape. Compound name: **Thiol**	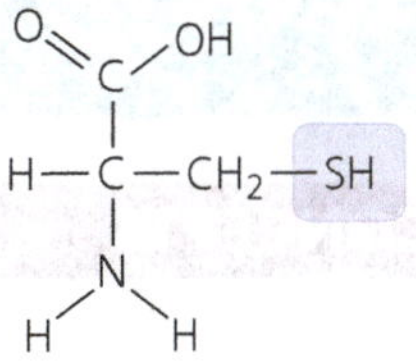 **Cysteine**, a sulfur-containing amino acid
Phosphate group (—OPO_3^{2-}) 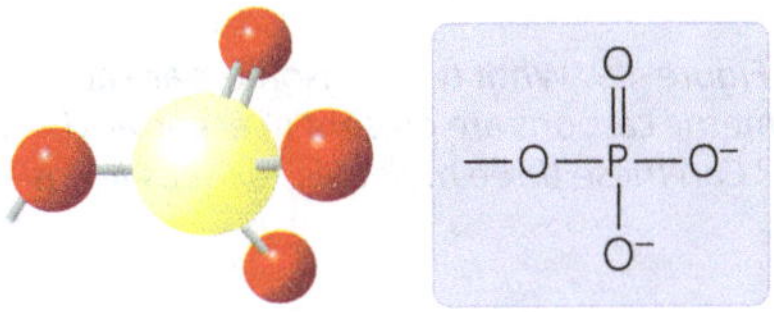	Contributes negative charge (1– when positioned inside a chain of phosphates; 2– when at the end). When attached, confers on a molecule the ability to react with water, releasing energy. Compound name: **Organic phosphate**	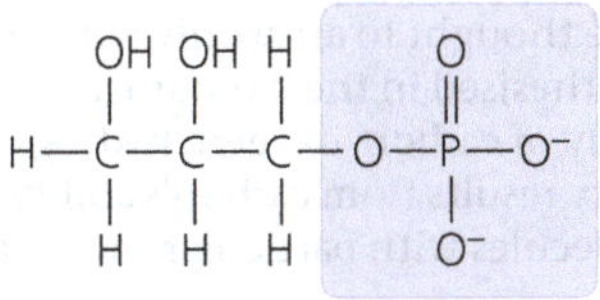 **Glycerol phosphate**, which takes part in many important chemical reactions in cells
Methyl group (—CH_3) 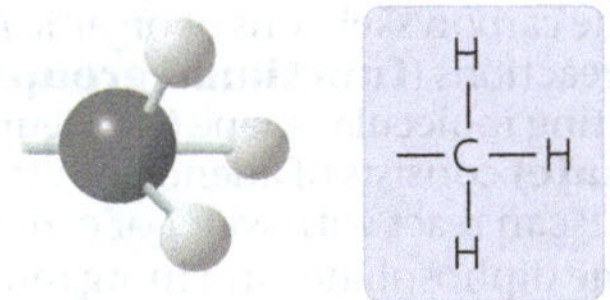	Affects the expression of genes when bonded to DNA or to proteins that bind to DNA. Affects the shape and function of male and female sex hormones. Compound name: **Methylated compound**	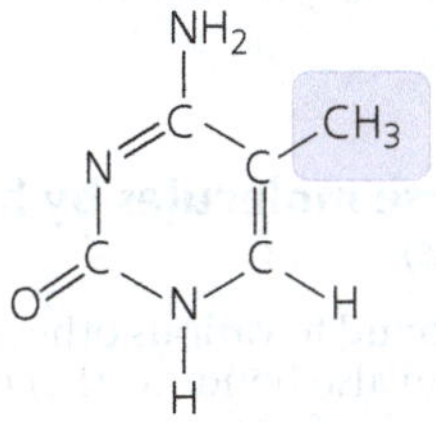 **5-Methylcytosine**: Cytosine, a component of DNA, has been modified by addition of a methyl group.

ATP: An Important Source of Energy for Cellular Processes

The "Phosphate group" row in Figure 4.9 shows a simple example of an organic phosphate molecule. A more complicated organic phosphate, **adenosine triphosphate**, or **ATP**, is worth mentioning here because its function in the cell is so important. ATP consists of an organic molecule called adenosine attached to a string of three phosphate groups:

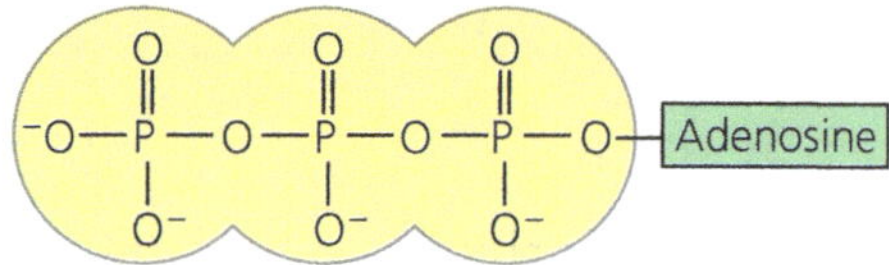

Where three phosphates are present in series, as in ATP, one phosphate may be split off as a result of a reaction with water. This inorganic phosphate ion, $HOPO_3^{2-}$, is often abbreviated Ⓟ$_i$ in this text, and a phosphate group in an organic molecule is often written as Ⓟ. Having lost one phosphate, ATP becomes adenosine *di*phosphate, or ADP. Although ATP is sometimes said to store energy, it is more accurate to think of it as storing the potential to react with water or other molecules. Overall, the process releases energy that can be used by the cell. You'll learn more about this in Concept 8.3.

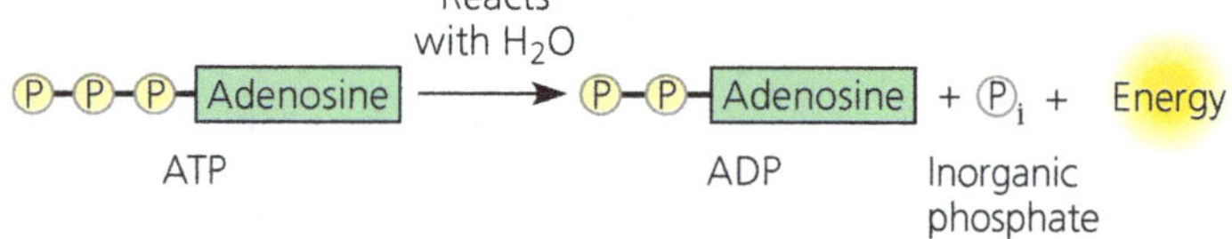

The Chemical Elements of Life: *A Review*

Living matter, as you have learned, consists mainly of carbon, oxygen, hydrogen, and nitrogen, with smaller amounts of sulfur and phosphorus. These elements all form strong covalent bonds, an essential characteristic in the architecture of complex organic molecules. Of all these elements, carbon is the virtuoso of the covalent bond. The versatility of carbon makes possible the great diversity of organic molecules, each with particular properties that emerge from the unique arrangement of its mostly carbon skeleton and the chemical groups attached to that skeleton. This variation at the molecular level provides the foundation for the rich biological diversity found on our planet.

CONCEPT CHECK 4.3

1. **VISUAL SKILLS** What does the term *amino acid* signify about the structure of such a molecule? See Figure 4.9.
2. What chemical change occurs to ATP when it reacts with water and releases energy?
3. **DRAW IT** Suppose you had an organic molecule such as cysteine (see Figure 4.9, sulfhydryl group example), and you chemically removed the $-NH_2$ group and replaced it with $-COOH$. Draw this structure. How would this change the chemical properties of the molecule? Is the central carbon asymmetric before the change? After?

For suggested answers, see Appendix A.

4 Chapter Review

SUMMARY OF KEY CONCEPTS

CONCEPT 4.1

Organic chemistry is key to the origin of life *(pp. 59–60)*

- **Organic** compounds, once thought to arise only within living organisms, were finally synthesised in the laboratory.
- Living matter is made mostly of carbon, oxygen, hydrogen, and nitrogen. Biological diversity results from carbon's ability to form a huge number of molecules with particular shapes and properties.

? *How did Stanley Miller's experiments support the idea that, even at life's origins, physical and chemical laws govern the processes of life?*

CONCEPT 4.2

Carbon atoms can form diverse molecules by bonding to four other atoms *(pp. 60–64)*

- Carbon, with a valence of 4, can bond to various other atoms, including O, H, and N. Carbon can also bond to other carbon atoms, forming the carbon skeletons of organic compounds. These skeletons vary in length and shape and have bonding sites for atoms of other elements.
- **Hydrocarbons** consist of carbon and hydrogen.
- **Isomers** are compounds that have the same molecular formula but different structures and therefore different properties. Three types of isomers are **structural isomers, *cis-trans* isomers**, and **enantiomers**.

VISUAL SKILLS *Refer back to Figure 4.9. What type of isomers are acetone and propanal? How many asymmetric carbons are present in acetic acid, glycine, and glycerol phosphate? Can these three molecules exist as forms that are enantiomers?*

CONCEPT 4.3

A few chemical groups are key to molecular function *(pp. 64–66)*

- Chemical groups attached to the carbon skeletons of organic molecules participate in chemical reactions (**functional groups**) or contribute to function by affecting molecular shape (see Figure 4.9).
- **ATP (adenosine triphosphate)** consists of adenosine attached to three phosphate groups. ATP can react with water or other molecules, forming ADP (adenosine diphosphate) and inorganic phosphate. This reaction releases energy that can be used by the cell.

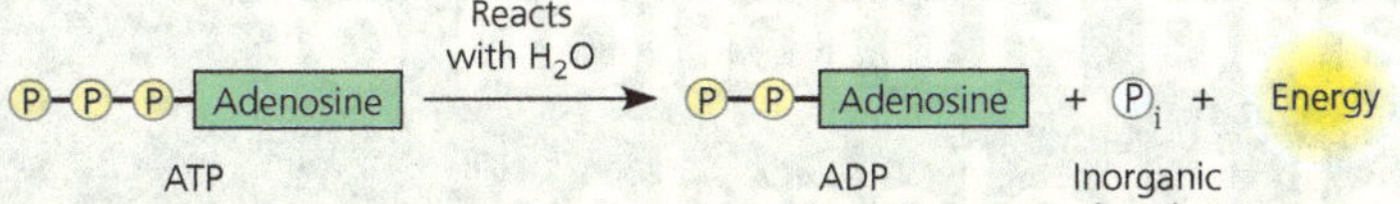

? *In what ways does a methyl group differ chemically from the other six important chemical groups shown in Figure 4.9?*

TEST YOUR UNDERSTANDING

Levels 1-2: Remembering/Understanding

1. Organic chemistry is currently defined as
(A) the study of compounds made only by living cells.
(B) the study of carbon compounds.
(C) the study of natural (as opposed to synthetic) compounds.
(D) the study of hydrocarbons.

2. VISUAL SKILLS Which functional group is present in this molecule?
(A) sulfhydryl
(B) carboxyl
(C) methyl
(D) phosphate

```
HO   O
  \ //
   C   H
   |   |
H—C—C—OH
   |   |
   N   H
  / \
 H   H
```

3. MAKE CONNECTIONS Which chemical group is most likely to be responsible for an organic molecule behaving as a base (see Concept 3.3)?
(A) hydroxyl
(B) carbonyl
(C) amino
(D) phosphate

Levels 3-4: Applying/Analysing

4. VISUAL SKILLS Visualise the structural formula of each of the following hydrocarbons. Which hydrocarbon has a double bond in its carbon skeleton?
(A) C_3H_8
(B) C_2H_6
(C) C_2H_4
(D) C_2H_2

5. VISUAL SKILLS Choose the term that correctly describes the relationship between these two sugar molecules.
(A) structural isomers
(B) *cis-trans* isomers
(C) enantiomers
(D) isotopes

```
    H            H   O
    |             \ //
H—C—OH            C
    |             |
    C=O       H—C—OH
    |             |
H—C—OH        H—C—OH
    |             |
    H             H
```

6. VISUAL SKILLS Identify the asymmetric carbon in this molecule.

```
O    OH H  H  H
 \\ A |B |C |  |D
  C—C—C—C—C—H
 /   |  |  |  |
H    H  H  H  H
```

7. Which action could produce a carbonyl group?
(A) the replacement of the —OH of a carboxyl group with hydrogen
(B) the addition of a thiol to a hydroxyl
(C) the addition of a hydroxyl to a phosphate
(D) the replacement of the nitrogen of an amine with oxygen

8. VISUAL SKILLS Which of the molecules shown in question 5 has an asymmetric carbon? Which carbon is asymmetric?

Levels 5-6: Evaluating/Creating

9. EVOLUTION CONNECTION • DRAW IT Some scientists think that life elsewhere in the universe might be based on the element silicon, rather than on carbon, as on Earth. Look at the electron distribution diagram for silicon in Figure 2.7 and draw the Lewis dot structure for silicon. What properties does silicon share with carbon that would make silicon-based life more likely than, say, neon-based life or aluminum-based life?

10. SCIENTIFIC INQUIRY Fifty years ago, pregnant women who were prescribed thalidomide for morning sickness gave birth to children with birth defects. Thalidomide is a mixture of two enantiomers; one reduces morning sickness, but the other causes severe birth defects. Today, the Advisory Committee on Medicines (ACM) has approved this drug for non-pregnant individuals with leprosy (Hansen's disease) or newly diagnosed multiple myeloma, a blood and bone marrow cancer. The beneficial enantiomer can be synthesised and given to patients, but over time, both the beneficial *and* the harmful enantiomer can be detected in the body. Propose a possible explanation for the presence of the harmful enantiomer.

11. WRITE ABOUT A THEME: ORGANISATION In 1918, an epidemic of sleeping sickness caused an unusual rigid paralysis in some survivors, similar to symptoms of advanced Parkinson's disease. Years later, L-dopa (below, left), a chemical used to treat Parkinson's disease, was given to some of these patients. L-dopa was remarkably effective at eliminating the paralysis, at least temporarily. However, its enantiomer, D-dopa (right), was subsequently shown to have no effect at all, as is the case for Parkinson's disease. In a short essay (100–150 words), discuss how the effectiveness of one enantiomer and not the other illustrates the theme of structure and function.

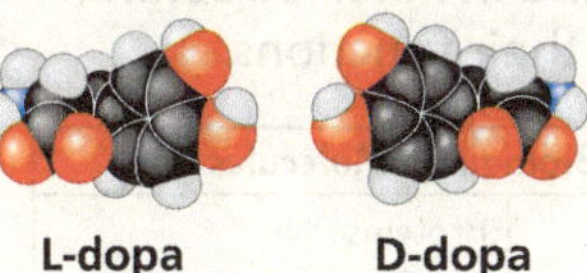

12. SYNTHESISE YOUR KNOWLEDGE

Explain how the chemical structure of the carbon atom accounts for the differences between the male and female lions seen in the photo.

For selected answers, see Appendix A.

5 The Structure and Function of Large Biological Molecules

KEY CONCEPTS

Study Tip

Make a visual study guide: For each class of biological molecules, draw two examples and list their structural similarities and their functions.

Important Biological Molecules	
Carbohydrates	Proteins
Nucleic acids	Lipids

Go to Mastering Biology

to access Dynamic Study Modules for revision, 3D BioFlix® animations and high-quality videos, and your interactive Pearson eText.

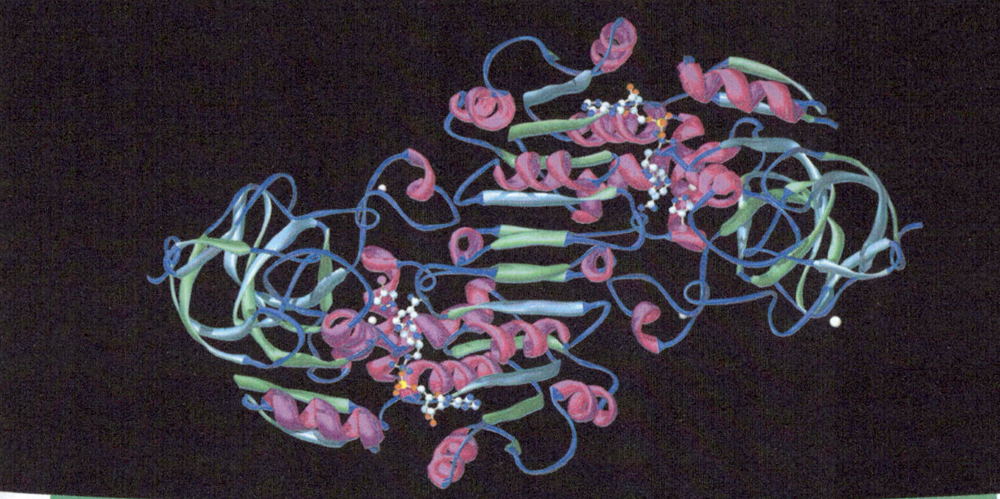

Figure 5.1 Alcohol dehydrogenase, a protein that breaks down alcohol in the body, is shown here as a molecular model. The form of this protein that an individual possesses affects how well that person tolerates drinking alcohol. Proteins are one class of large molecules, or macromolecules.

What are the structures and functions of the four important classes of biological molecules?

Three classes are macromolecules that are polymers (long chains of monomer subunits).

Monomer

Polymer

Carbohydrates are a source of energy and provide structural support.

Carbohydrate (starch)

Proteins have a wide range of functions, such as catalysing reactions and transporting substances into and out of cells.

Amino acid

Protein (alcohol dehydrogenase)

Nucleic acids store genetic information and function in gene expression.

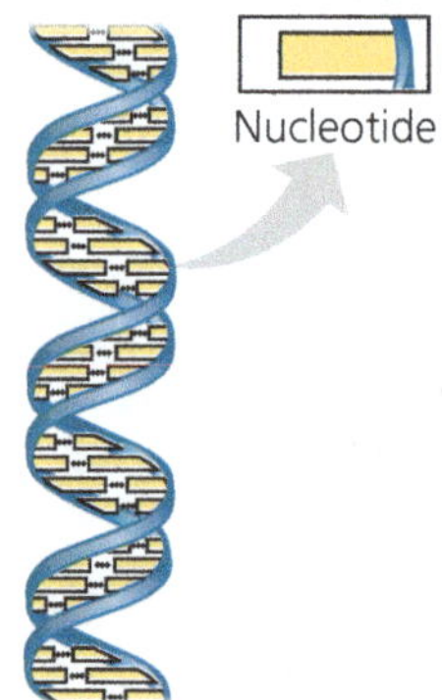

Nucleic acid (DNA)

The fourth class, lipids, are not polymers or macromolecules.

Lipids are a group of diverse molecules that do not mix well with water. Key functions include providing energy, making up cell membranes, and acting as hormones.

Lipid (phospholipid)

CONCEPT 5.1

Macromolecules are polymers, built from monomers

Large carbohydrates, proteins, and nucleic acids, also known as **macromolecules** for their huge size, are chain-like molecules called polymers (from the Greek *polys*, many, and *meros*, part). A **polymer** is a long molecule consisting of many similar or identical building blocks linked by covalent bonds, much as a train consists of a chain of boxcars. The repeating units that serve as the building blocks of a polymer are smaller molecules called **monomers** (from the Greek *monos*, single). In addition to forming polymers, some monomers have functions of their own.

The Synthesis and Breakdown of Polymers

Although each class of polymer is made up of a different type of monomer, the chemical mechanisms by which cells make polymers (polymerisation) and break them down are similar for all classes of large biological molecules. In cells, these processes are facilitated by **enzymes**, specialised macromolecules (usually proteins) that speed up chemical reactions. The reaction that connects a monomer to another monomer or a polymer is a *condensation reaction*, a reaction in which two molecules are covalently bonded to each other with the loss of a small molecule. If a water molecule is lost, it is known as a **dehydration reaction**. For example, carbohydrate and protein polymers are synthesised by dehydration reactions. Each reactant contributes part of the water molecule that is released during the reaction: One provides a hydroxyl group (—OH), while the other provides a hydrogen (—H) **(Figure 5.2a)**. This reaction is repeated as monomers are added to the chain one by one, lengthening the polymer.

Polymers are disassembled to monomers by **hydrolysis**, a process that is essentially the reverse of the dehydration reaction **(Figure 5.2b)**. Hydrolysis means water breakage (from the Greek *hydro*, water, and *lysis*, break). The bond between monomers is broken by the addition of a water molecule, with a hydrogen from water attaching to one monomer and the hydroxyl group attaching to the other. An example of hydrolysis within our bodies is the process of digestion. The bulk of the organic material in our food is in the form of polymers that are much too large to enter our cells. Within the digestive tract, various enzymes attack the polymers, speeding up hydrolysis. Released monomers are then absorbed into the bloodstream for distribution to all body cells. Those cells can then use dehydration reactions to assemble the monomers into new, different polymers that can perform specific functions required by the cell. (Dehydration reactions and hydrolysis can also be involved in the formation and breakdown of molecules that are not polymers, such as some lipids.)

▼ Figure 5.2 The synthesis and breakdown of carbohydrate and protein polymers.

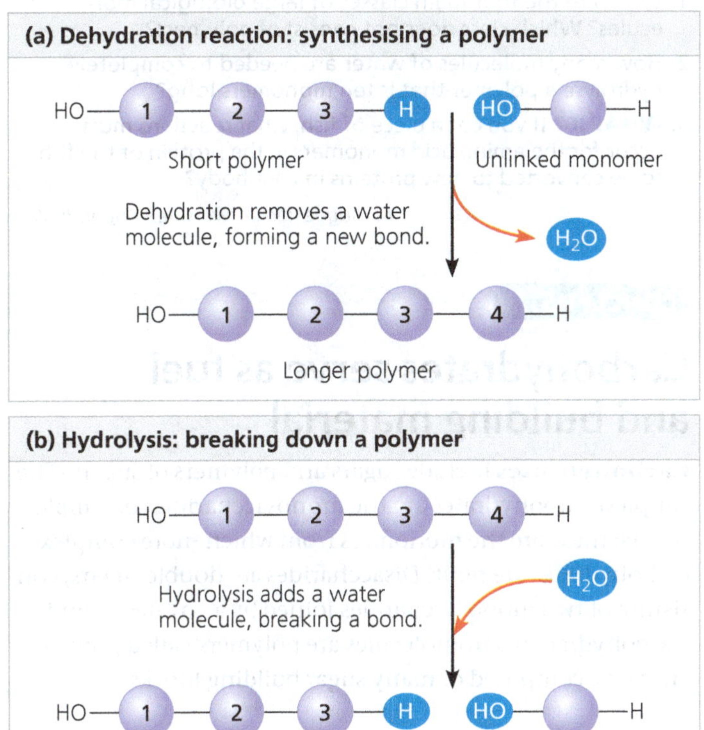

The Diversity of Polymers

A cell has thousands of different macromolecules; the collection varies from one type of cell to another. The inherited differences between close relatives, such as human siblings, reflect small variations in polymers, particularly DNA and proteins. Molecular differences between unrelated individuals are more extensive, and those between species greater still. The diversity of macromolecules in the living world is vast, and the possible variety is effectively limitless.

What is the basis for such diversity in life's polymers? These molecules are constructed from only 40 to 50 common monomers and some others that occur rarely. Building a huge variety of polymers from such a limited number of monomers is analogous to constructing hundreds of thousands of words from only 26 letters of the alphabet. The key is arrangement—the particular linear sequence that the units follow. However, this analogy falls far short of describing the great diversity of macromolecules because most biological polymers have many more monomers than the number of letters in even the longest word. Proteins, for example, are built from 20 kinds of amino acids arranged in chains that are typically hundreds of amino acids long. The molecular logic of life is simple but elegant: Small molecules common to all organisms act as building blocks that are ordered into unique macromolecules.

Despite this immense diversity, molecular structure and function can still be grouped roughly by class. Let's examine each of the four major classes of large biological molecules. For each class, the large molecules have emergent properties not found in their individual components.

CONCEPT CHECK 5.1

1. What are the four main classes of large biological molecules? Which class does not consist of polymers?
2. How many molecules of water are needed to completely hydrolyse a polymer that is ten monomers long?
3. **WHAT IF?** If you eat a piece of fish, what reactions must occur for the amino acid monomers in the protein of the fish to be converted to new proteins in your body?

For suggested answers, see Appendix A.

CONCEPT 5.2

Carbohydrates serve as fuel and building material

Carbohydrates include sugars and polymers of sugars. The simplest carbohydrates are the monosaccharides, or simple sugars; these are the monomers from which more complex carbohydrates are built. Disaccharides are double sugars, consisting of two monosaccharides joined by a covalent bond. Carbohydrate macromolecules are polymers called polysaccharides, composed of many sugar building blocks.

Sugars

Monosaccharides (from the Greek *monos*, single, and *sacchar*, sugar) generally have molecular formulas that are some multiple of the unit CH_2O. Glucose ($C_6H_{12}O_6$), the most common monosaccharide, is of central importance in the chemistry of life. In the structure of glucose, we can see the trademarks of a monosaccharide: The molecule has a carbonyl group, $>C=O$, and multiple hydroxyl groups, —OH **(Figure 5.3)**. Depending on the location of the carbonyl group, a monosaccharide is either an aldose (aldehyde sugar) or a ketose (ketone sugar). Glucose, for example, is an aldose; fructose, an isomer of glucose, is a ketose. (Most names for sugars end in *-ose*.) Another criterion for classifying monosaccharides is the size of the carbon skeleton, which ranges from three to seven carbons long. Glucose, fructose, and other sugars that have six carbons are called hexoses. Trioses (three-carbon sugars) and pentoses (five-carbon sugars) are also common.

Still another source of diversity for simple sugars is in the way their parts are arranged spatially around asymmetric carbons. (Recall that an asymmetric carbon is a carbon attached to four different atoms or groups of atoms.) Glucose and galactose, for example, differ only in the placement of parts around one asymmetric carbon (see the purple boxes in Figure 5.3). What seems like a small difference is significant enough to give the two sugars distinctive shapes and binding activities, thus different behaviours.

Although it is convenient to draw glucose with a linear carbon skeleton, this representation is not completely accurate. In aqueous solutions, glucose molecules, as well as most other

▼ **Figure 5.3 The structure and classification of some monosaccharides.** Sugars vary in the location of their carbonyl groups (orange), the length of their carbon skeletons, and the way their parts are arranged spatially around asymmetric carbons (compare, for example, the purple portions of glucose and galactose).

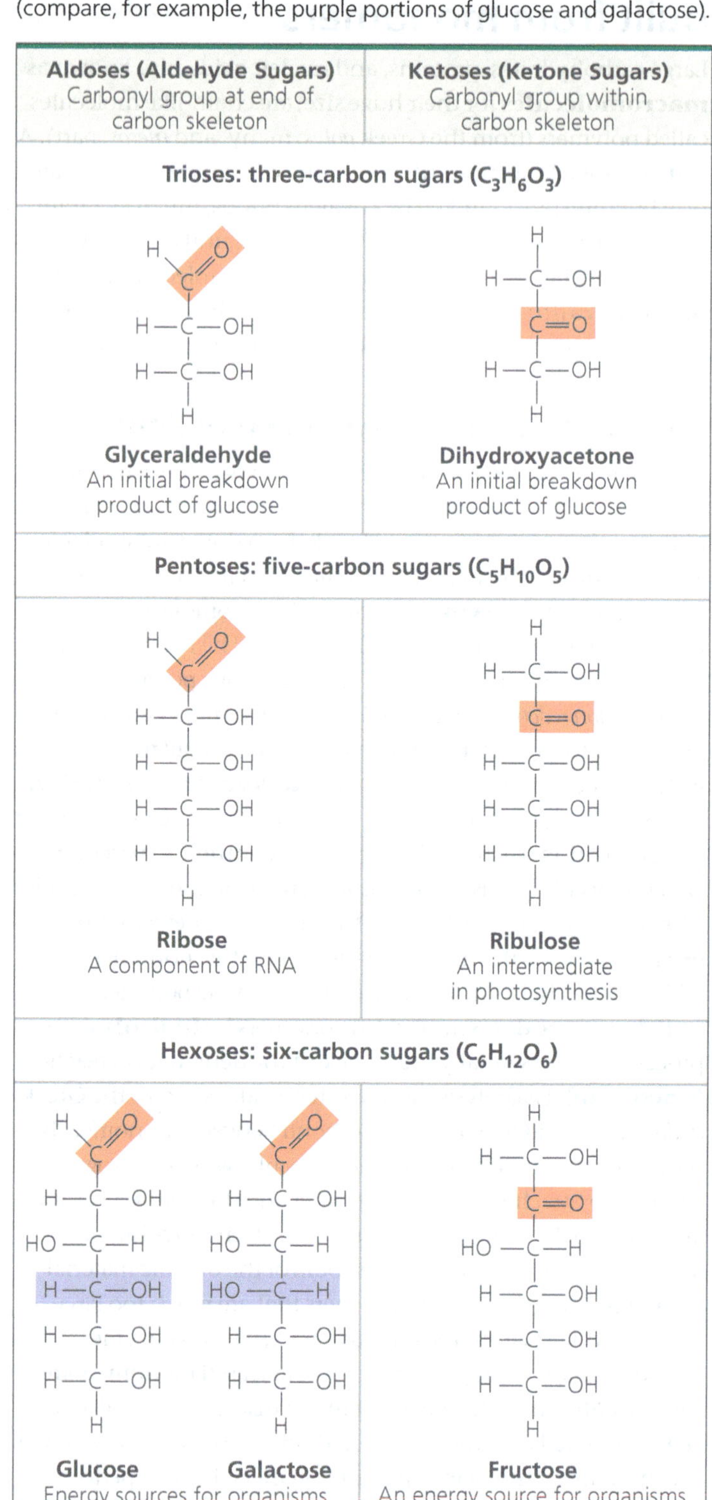

MAKE CONNECTIONS *In the 1970s, a process was developed that converts the glucose in corn syrup to its sweeter-tasting isomer, fructose. High-fructose corn syrup, a common ingredient in soft drinks and processed food, is a mixture of glucose and fructose. What type of isomers are glucose and fructose? (See Figure 4.7.)*

▼ Figure 5.4 Linear and ring forms of glucose.

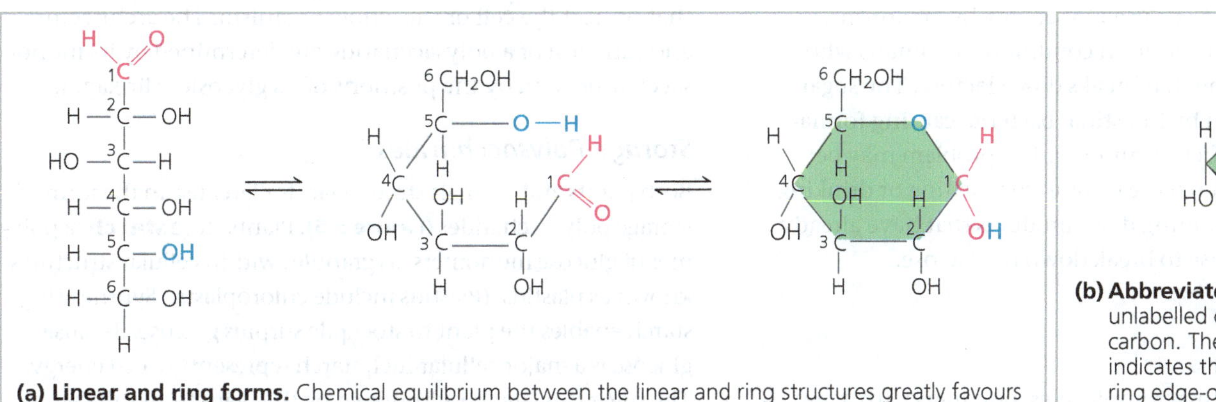

(a) Linear and ring forms. Chemical equilibrium between the linear and ring structures greatly favours the formation of rings. The carbons of the sugar are numbered 1 to 6, as shown. To form the glucose ring, carbon 1 (magenta) bonds to the oxygen (blue) attached to carbon 5.

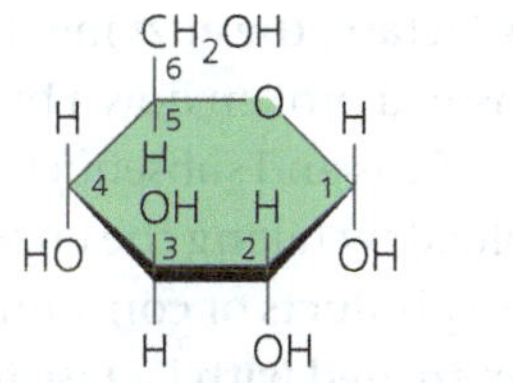

(b) Abbreviated ring structure. Each unlabelled corner represents a carbon. The ring's thicker edge indicates that you are looking at the ring edge-on; the components attached to the ring lie above or below the plane of the ring.

DRAW IT *Start with the linear form of fructose (see Figure 5.3) and draw the formation of the fructose ring in two steps, as shown in (a). First, number the carbons starting at the top of the linear structure. Then draw the molecule in a ringlike orientation, attaching carbon 5 via its oxygen to carbon 2. Compare the number of carbons in the ring portions of fructose and glucose.*

five- and six-carbon sugars, form rings, because they are the most stable form of these sugars under physiological conditions **(Figure 5.4)**.

Monosaccharides, particularly glucose, are major nutrients for cells. In the process known as cellular respiration, cells extract energy from glucose molecules by breaking them down in a series of reactions. Not only are monosaccharides a major fuel for cellular work, but their carbon skeletons also serve as raw material for the synthesis of other types of small organic molecules, such as amino acids and fatty acids. Monosaccharides that are not immediately used in these ways are generally incorporated as monomers into disaccharides or polysaccharides, discussed next.

A **disaccharide** consists of two monosaccharides joined by a **glycosidic linkage**, a covalent bond formed between two monosaccharides by a dehydration reaction (*glyco* refers to carbohydrate). For example, maltose is a disaccharide formed by the linking of two molecules of glucose **(Figure 5.5a)**. Also known as malt sugar, maltose is an ingredient used in brewing beer. The most prevalent disaccharide is sucrose, or table sugar. Its two monomers are glucose and fructose **(Figure 5.5b)**. Plants generally transport carbohydrates from leaves to roots and other nonphotosynthetic organs in the form of sucrose. Lactose, the sugar present in milk, is another disaccharide, in this case a glucose molecule joined

▼ Figure 5.5 Examples of disaccharide synthesis.

(a) Dehydration reaction in the synthesis of maltose. The bonding of two glucose units forms maltose. The 1–4 glycosidic linkage joins the number 1 carbon of one glucose to the number 4 carbon of the second glucose. Joining the glucose monomers in a different way would result in a different disaccharide.

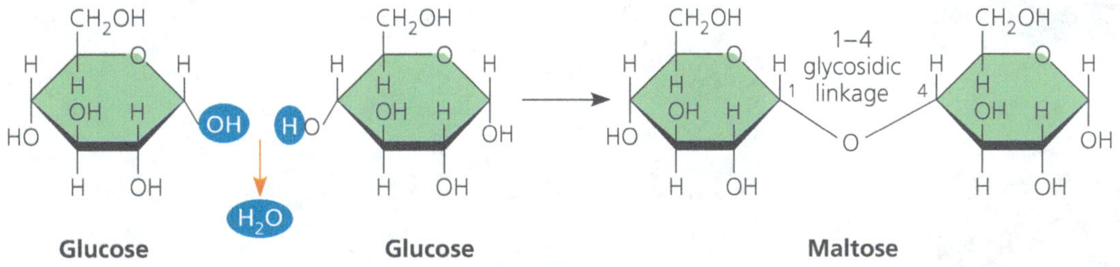

Glucose + Glucose → Maltose + H_2O (1–4 glycosidic linkage)

(b) Dehydration reaction in the synthesis of sucrose. Sucrose is a disaccharide formed from glucose and fructose. Notice that fructose forms a five-sided ring, though it is a hexose like glucose.

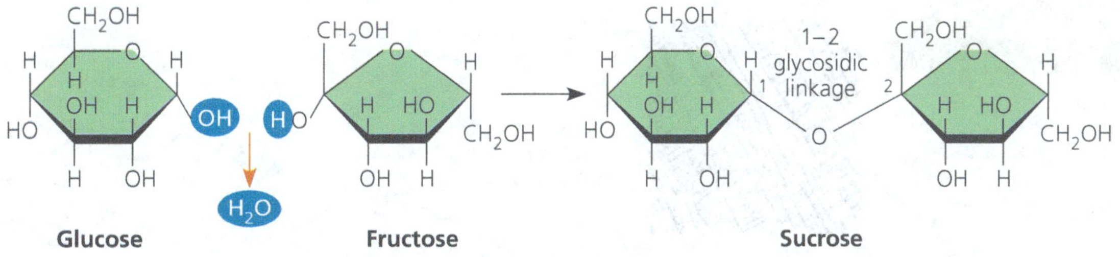

Glucose + Fructose → Sucrose + H_2O (1–2 glycosidic linkage)

DRAW IT *Referring to Figures 5.3 and 5.4, number the carbons in each sugar in this figure. How does the name of each linkage relate to the numbers?*

to a galactose molecule. Disaccharides must be broken down into monosaccharides to be used for energy by organisms. Lactose intolerance is a common condition in humans who lack lactase, the enzyme that breaks down lactose. The sugar is instead broken down by intestinal bacteria, causing formation of gas and subsequent cramping. The problem may be avoided by taking the enzyme lactase when eating or drinking dairy products or consuming dairy products that have already been treated with lactase to break down the lactose.

Polysaccharides

Polysaccharides are macromolecules, polymers with a few hundred to a few thousand monosaccharides joined by glycosidic linkages. Some polysaccharides serve as storage material, hydrolysed as needed to provide monosaccharides for cells. Other polysaccharides serve as building material for structures that protect the cell or the whole organism. The architecture and function of a polysaccharide are determined by its monosaccharides and by the positions of its glycosidic linkages.

Storage Polysaccharides

Both plants and animals store sugars for later use in the form of storage polysaccharides **(Figure 5.6)**. Plants store **starch**, a polymer of glucose monomers, as granules within cellular structures known as plastids. (Plastids include chloroplasts.) Synthesising starch enables the plant to stockpile surplus glucose. Because glucose is a major cellular fuel, starch represents stored energy. The sugar can later be withdrawn by the plant from this carbohydrate "bank" by hydrolysis, which breaks the bonds between the glucose monomers. Most animals, including humans, also have enzymes that can hydrolyse plant starch, making glucose

▼ Figure 5.6 Polysaccharides of plants and animals. (a) Starch stored in plant cells, **(b)** glycogen stored in muscle cells, and **(c)** structural cellulose fibres in plant cell walls are all polysaccharides composed entirely of glucose monomers (green hexagons). In starch and glycogen, the polymer chains tend to form helices in unbranched regions because of the angle of the 1–4 linkages between glucose molecules. There are two kinds of starch: amylose and amylopectin. Cellulose, with a different kind of glucose linkage, is always unbranched.

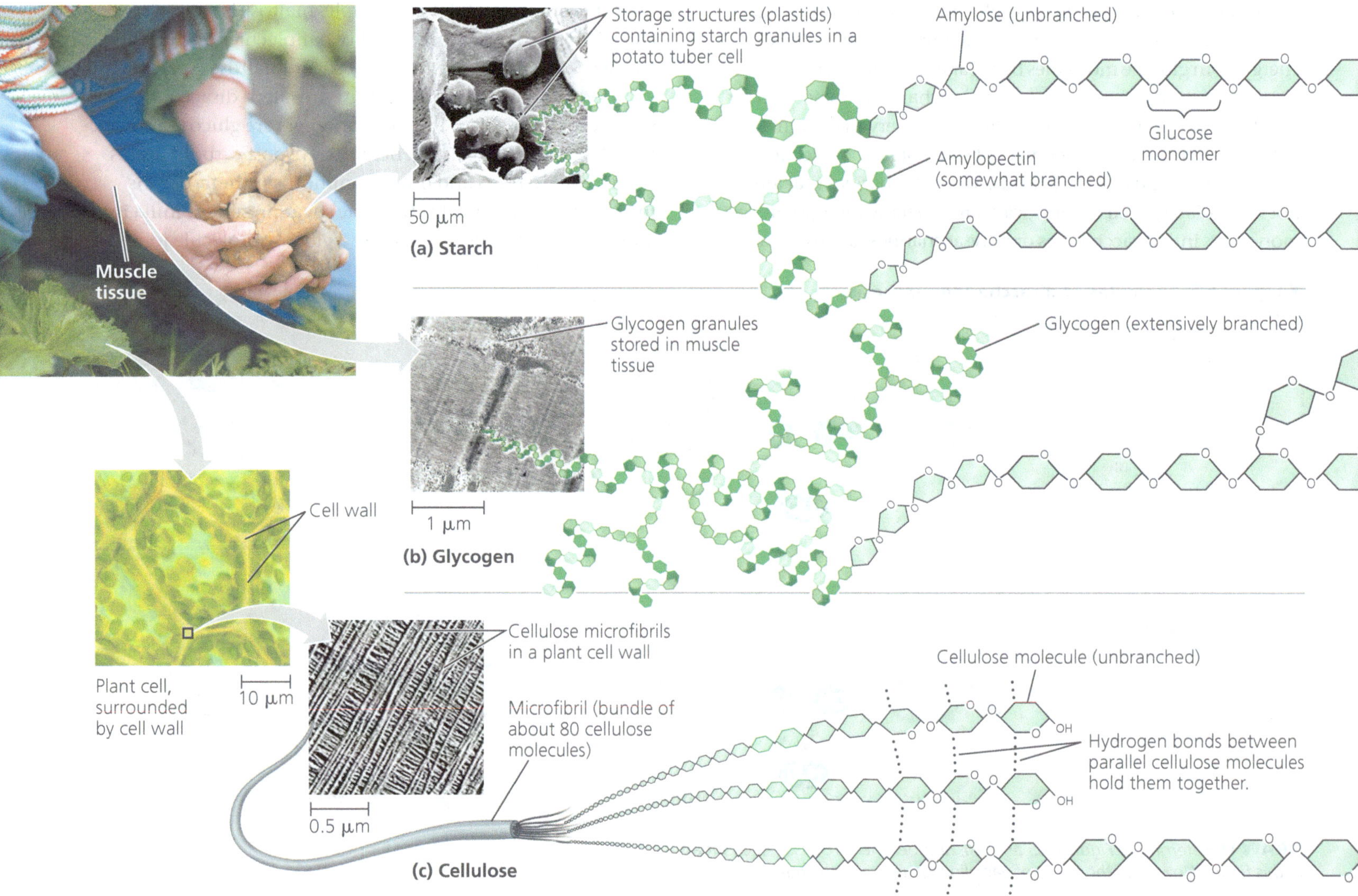

available as a nutrient for cells. Potato tubers and grains—the fruits of wheat, corn, rice, and other grasses—are the major sources of starch in the human diet.

Most of the glucose monomers in starch are joined by 1–4 linkages (number 1 carbon to number 4 carbon), like the glucose units in maltose (see Figure 5.5a). The simplest form of starch, amylose, is unbranched. Amylopectin, a more complex starch, is a branched polymer with 1–6 linkages at the branch points. Both of these starches are shown in **Figure 5.6a**.

Animals store a polysaccharide called **glycogen**, a polymer of glucose that is like amylopectin but more extensively branched **(Figure 5.6b)**. Vertebrates store glycogen mainly in liver and muscle cells. Breakdown of glycogen in these cells releases glucose when the demand for energy increases. (The extensively branched structure of glycogen fits its function: More free ends are available for breakdown.) This stored fuel cannot sustain an animal for long, however. In humans, for example, glycogen stores are depleted in about a day unless they are replenished by eating. This is an issue of concern in low-carbohydrate diets, which can result in weakness and fatigue.

Structural Polysaccharides

Organisms build strong materials from structural polysaccharides. For example, the polysaccharide called **cellulose** is a major component of the tough walls that enclose plant cells **(Figure 5.6c)**. Globally, plants produce almost 10^{14} kg (100 billion tons) of cellulose per year; it is the most abundant organic compound on Earth.

Like starch, cellulose is a polymer of glucose with 1–4 glycosidic linkages, but the linkages in these two polymers differ. The difference is based on the fact that there are actually two slightly different ring structures for glucose **(Figure 5.7a)**. When glucose forms a ring, the hydroxyl group attached to the number 1 carbon is positioned either below or above the plane of the ring. These two ring forms for glucose are called alpha (α) and beta (β), respectively. (Greek letters are often used as a "numbering" system for different versions of biological structures, much as we use the letters a, b, c, and so on for the parts of a question or a figure.) In starch, all the glucose monomers are in the α configuration **(Figure 5.7b)**, the arrangement we saw in Figures 5.4 and 5.5. In contrast, the glucose monomers of cellulose are all in the β configuration, making every glucose monomer "upside down" with respect to its neighbours (**Figure 5.7c**; see also Figure 5.6c).

The differing glycosidic linkages in starch and cellulose give the two molecules distinct three-dimensional shapes. Certain starch molecules are largely helical, fitting their function of efficiently storing glucose units. Conversely, a cellulose molecule is straight. Cellulose is never branched, and some hydroxyl groups on its glucose monomers are free to hydrogen-bond with the hydroxyls of other cellulose molecules lying parallel to it. In plant cell walls, parallel cellulose molecules held together in this way are grouped into units called microfibrils (see Figure 5.6c). These cable-like microfibrils are a strong building material for plants and an important substance for humans because cellulose is the major constituent of paper and the only component of cotton. The unbranched structure of cellulose thus fits its function: imparting strength to parts of the plant.

Enzymes that digest starch by hydrolysing its α linkages are unable to hydrolyse the β linkages of cellulose due to the different shapes of these two molecules. In fact,

▼ **Figure 5.7 Starch and cellulose structures.**

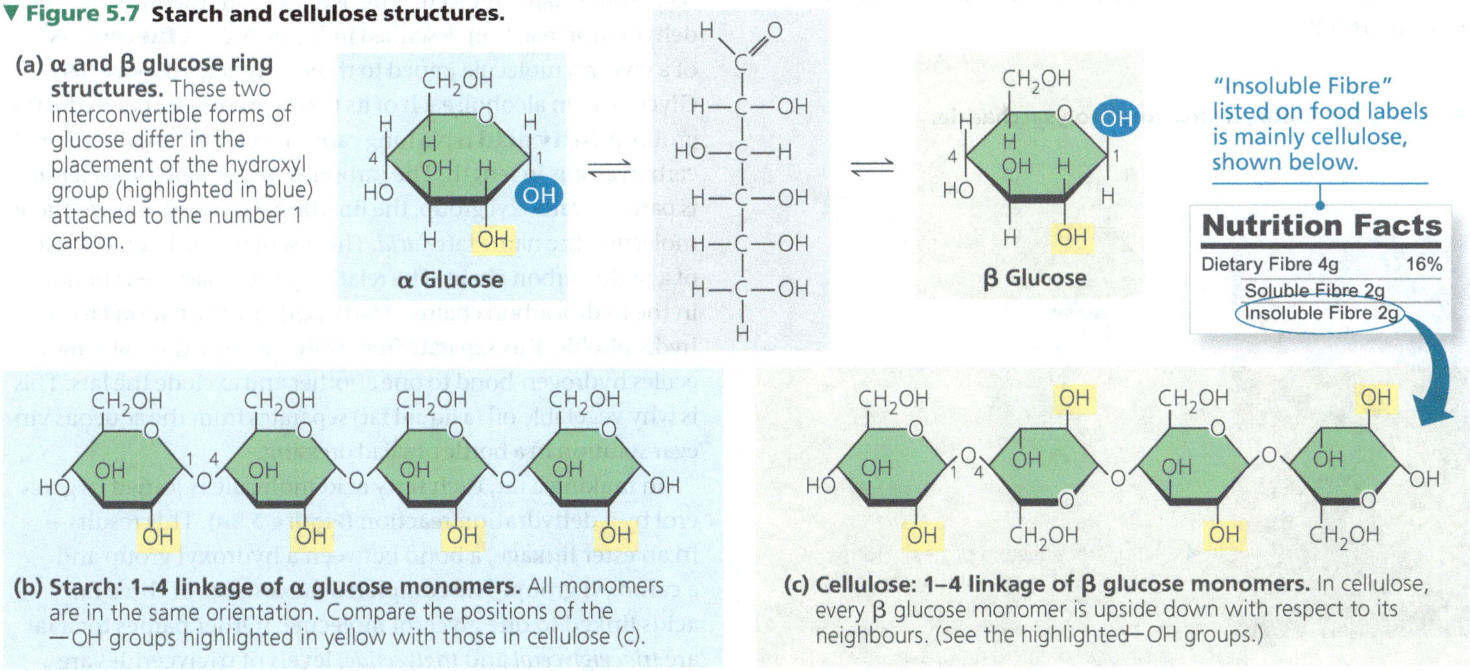

(a) α and β glucose ring structures. These two interconvertible forms of glucose differ in the placement of the hydroxyl group (highlighted in blue) attached to the number 1 carbon.

"Insoluble Fibre" listed on food labels is mainly cellulose, shown below.

Nutrition Facts

Dietary Fibre 4g	16%
Soluble Fibre 2g	
Insoluble Fibre 2g	

(b) Starch: 1–4 linkage of α glucose monomers. All monomers are in the same orientation. Compare the positions of the —OH groups highlighted in yellow with those in cellulose (c).

(c) Cellulose: 1–4 linkage of β glucose monomers. In cellulose, every β glucose monomer is upside down with respect to its neighbours. (See the highlighted—OH groups.)

few organisms possess enzymes that can digest cellulose. Almost all animals, including humans, do not; the cellulose in our food passes through the digestive tract and is eliminated with the feces. Along the way, the cellulose abrades the wall of the digestive tract and stimulates the lining to secrete mucus, which aids in the smooth passage of food through the tract. Thus, although cellulose is not a nutrient for humans, it is an important part of a healthy diet. Most fruits, vegetables, and whole grains are rich in cellulose. On food packages, "insoluble fibre" refers mainly to cellulose (see Figure 5.7).

Some microorganisms can digest cellulose, breaking it down into glucose monomers. A cow harbours cellulose-digesting prokaryotes and protists in its gut. These microbes hydrolyse the cellulose of hay and grass and convert the glucose to other compounds that nourish the cow. Similarly, a termite, which is unable to digest cellulose by itself, has prokaryotes or protists living in its gut that can make a meal of wood. Some fungi can also digest cellulose in soil and elsewhere, thereby helping recycle chemical elements within Earth's ecosystems.

Another important structural polysaccharide is **chitin**, the carbohydrate used by arthropods (insects, spiders, crustaceans, and related animals) to build their exoskeletons—hard cases that surround the soft parts of an animal **(Figure 5.8)**. Made up of chitin embedded in a layer of proteins, the case is leathery and flexible at first, but becomes hardened when the proteins are chemically linked to each other (as in insects) or encrusted with calcium carbonate (as in crabs). Chitin is also found in fungi, which use this polysaccharide rather than cellulose as the building material for their cell walls. Chitin is similar to cellulose, with β linkages, except that the glucose monomer of chitin has a nitrogen-containing attachment (see Figure 5.8).

▼ **Figure 5.8 Chitin, a structural polysaccharide.**

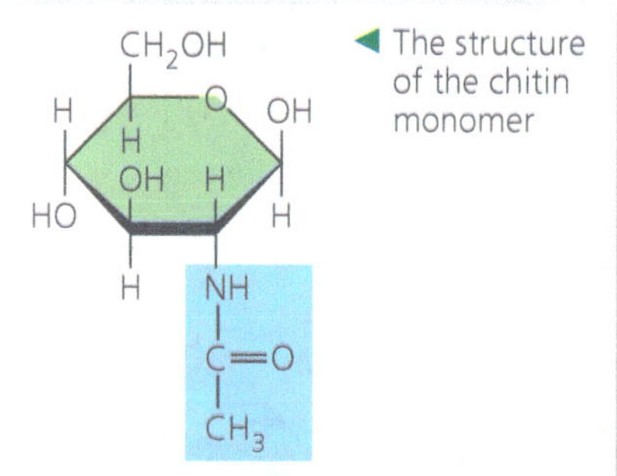

◀ The structure of the chitin monomer

◀ Chitin, embedded in proteins, forms the exoskeleton of arthropods. This emperor dragonfly (*Anax imperator*) is molting—shedding its old exoskeleton (brown) and emerging upside down in adult form.

CONCEPT CHECK 5.2

1. Write the formula for a monosaccharide that has three carbons.
2. A dehydration reaction joins two glucose molecules to form maltose. The formula for glucose is $C_6H_{12}O_6$. What is the formula for maltose?
3. **WHAT IF?** After a cow is given antibiotics to treat an infection, a vet gives the animal a drink of "gut culture" containing various prokaryotes. Why is this necessary?

For suggested answers, see Appendix A.

CONCEPT 5.3

Lipids are a diverse group of hydrophobic molecules

Lipids are the one class of large biological molecules that does not include true polymers, and they are generally not big enough to be considered macromolecules. The compounds called **lipids** are grouped with each other because they share one important trait: They are hydrophobic: They mix poorly, if at all, with water. This behaviour of lipids is based on their molecular structure. Although they may have some polar bonds associated with oxygen, lipids consist mostly of hydrocarbon regions with relatively non-polar C—H bonds. Lipids are varied in form and function. They include waxes and certain pigments, but we will focus on the types of lipids that are most important biologically: fats, phospholipids, and steroids.

Fats

Although fats are not polymers, they are large molecules assembled from smaller molecules by dehydration reactions, like the dehydration reaction described in Figure 5.2a. A **fat** consists of a glycerol molecule joined to three fatty acids **(Figure 5.9)**. Glycerol is an alcohol; each of its three carbons bears a hydroxyl group. A **fatty acid** has a long carbon skeleton, usually 16 or 18 carbon atoms in length. The carbon at one end of the skeleton is part of a carboxyl group, the functional group that gives these molecules the name fatty *acid*. The rest of the skeleton consists of a hydrocarbon chain. The relatively nonpolar C—H bonds in the hydrocarbon chains of fatty acids are the reason fats are hydrophobic. Fats separate from water because the water molecules hydrogen-bond to one another and exclude the fats. This is why vegetable oil (a liquid fat) separates from the aqueous vinegar solution in a bottle of salad dressing.

In making a fat, each fatty acid molecule is joined to glycerol by a dehydration reaction **(Figure 5.9a)**. This results in an ester linkage, a bond between a hydroxyl group and a carboxyl group. The completed fat consists of three fatty acids linked to one glycerol molecule. (Other names for a fat are *triacylglycerol* and *triglyceride*; levels of triglycerides are reported when blood is tested for lipids.) The fatty acids in a

▼ Figure 5.9 The synthesis and structure of a fat, or triacylglycerol. The molecular building blocks of a fat are one molecule of glycerol and three molecules of fatty acids. The carbons of the fatty acids are arranged zigzag to suggest the actual orientations of the four single bonds extending from each carbon (see Figures 4.3a and 4.6b).

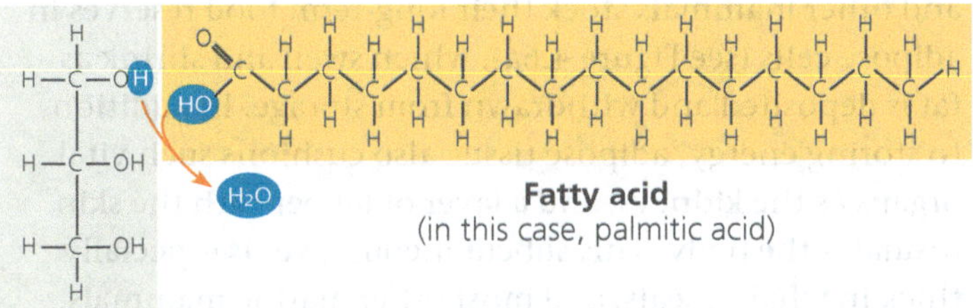

(a) One of three dehydration reactions in the synthesis of a fat. One water molecule is removed for each fatty acid joined to the glycerol.

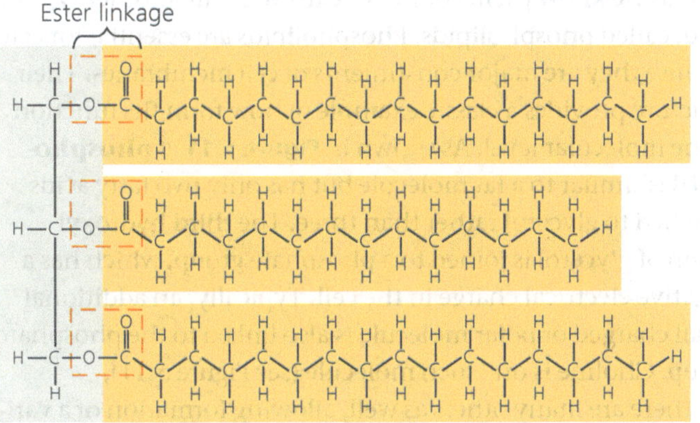

(b) A fat molecule (triacylglycerol) with three fatty acid units. In this example, two of the fatty acid units are identical.

▼ Figure 5.10 Saturated and unsaturated fats and fatty acids.

(a) Saturated fat

At room temperature, the molecules of a saturated fat, such as the fat in butter, are packed closely together, forming a solid.

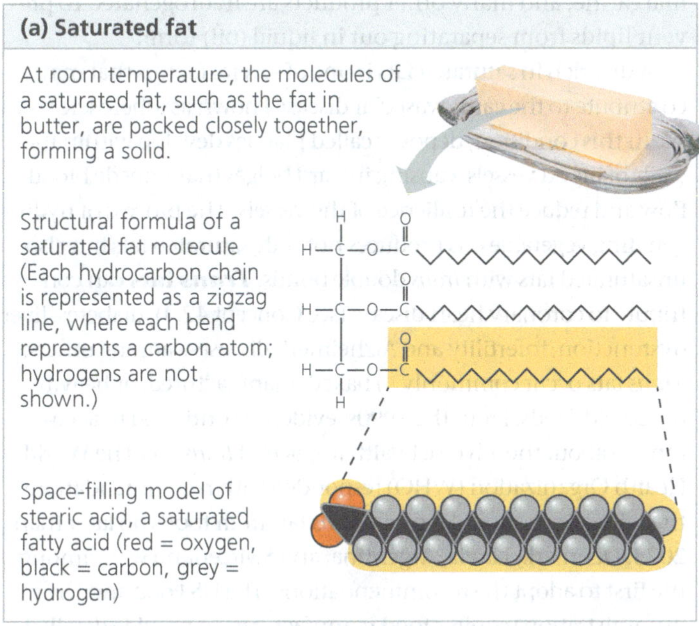

Structural formula of a saturated fat molecule (Each hydrocarbon chain is represented as a zigzag line, where each bend represents a carbon atom; hydrogens are not shown.)

Space-filling model of stearic acid, a saturated fatty acid (red = oxygen, black = carbon, grey = hydrogen)

(b) Unsaturated fat

At room temperature, the molecules of an unsaturated fat such as olive oil cannot pack together closely enough to solidify because of the kinks in some of their fatty acid hydrocarbon chains.

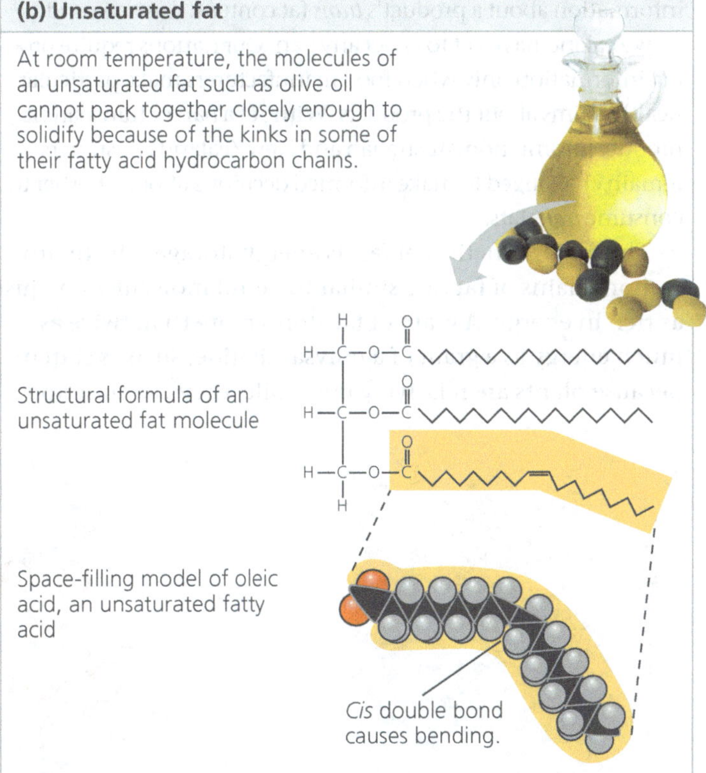

Structural formula of an unsaturated fat molecule

Space-filling model of oleic acid, an unsaturated fatty acid

fat can all be the same, or they can be of two or three different kinds, as in **Figure 5.9b**.

The terms *saturated* fats and *unsaturated* fats are commonly used in the context of nutrition **(Figure 5.10)**. These terms refer to the structure of the hydrocarbon chains of the fatty acids. If there are no double bonds between carbon atoms composing a chain, then as many hydrogen atoms as possible are bonded to the carbon skeleton. Such a structure is said to be *saturated* with hydrogen, and the resulting fatty acid is therefore called a **saturated fatty acid (Figure 5.10a)**. An **unsaturated fatty acid** has one or more double bonds, with one fewer hydrogen atom on each double-bonded carbon. Nearly every double bond in naturally occurring fatty acids is a *cis* double bond, which creates a kink in the hydrocarbon chain wherever it occurs **(Figure 5.10b)**. (See Figure 4.7b to remind yourself about *cis* and *trans* double bonds.)

A fat made from saturated fatty acids is called a saturated fat. Most animal fats are saturated: The hydrocarbon chains of their fatty acids—the "tails" of the fat molecules—lack double bonds (see Figure 5.10a), and their flexibility allows the fat molecules to pack together tightly. Saturated animal fats, such as lard and butter, are solid at room temperature. In contrast, the fats of plants and fishes are generally unsaturated, meaning that they are composed of one or more types of unsaturated fatty acids. Usually liquid at room temperature, plant and fish fats are referred to as oils—olive oil and cod liver oil are examples. The kinks where the *cis* double bonds are located (see Figure 5.10b) prevent the molecules from packing together closely enough to solidify at room temperature. The phrase *hydrogenated vegetable oils* on food labels means that unsaturated fats have been synthetically converted to saturated fats by adding hydrogen, allowing them to solidify. Peanut butter, margarine, and many other products are hydrogenated to prevent lipids from separating out in liquid (oil) form.

A diet rich in saturated fats is one of several factors that may contribute to the cardiovascular disease known as atherosclerosis. In this condition, deposits called plaques develop within the walls of blood vessels, causing inward bulges that impede blood flow and reduce the resilience of the vessels. The process of hydrogenating vegetable oils produces not only saturated fats but also unsaturated fats with *trans* double bonds. ***Trans* fats** can contribute to coronary heart disease (see Concept 42.4), diabetes, liver dysfunction, infertility and Alzheimer's disease—among others. *Trans* fats occur commonly in baked, manfuactured, or heavily processed foods. From the 1990s, evidence continued to accumulate about the adverse health impacts of *trans* fats.The World Health Organization (WHO) responded with recommendations that all nations should remove *trans* fats from foods no later than 2023. Denmark, Latvia, Argentina, and Switzerland were among the first to adopt the recommendations. The US Food and Drug Administration requires food manufacturers to provide detailed information about a product's *trans* fat content. Australia and New Zealand have yet to enact any ban. Our nations require product information only when food manufacturers make particular health claims about the products. Without a ban or transparent product information, Australian and New Zealand consumers remain challenged to make informed decisions about whether to consume *trans* fats.

The major function of fats is energy storage. The hydrocarbon chains of fats are similar to petrol molecules and just as rich in energy. A gram of fat stores more than twice as much energy as a gram of a polysaccharide, such as starch. Because plants are relatively immobile, they can function with bulky energy storage in the form of starch. (Vegetable oils are generally obtained from seeds, where more compact storage is an asset to the plant.) Animals, however, must carry their energy stores with them, so there is an advantage to having a more compact reservoir of fuel—fat. Humans and other mammals stock their long-term food reserves in adipose cells (see Figure 4.6a), which swell and shrink as fat is deposited and withdrawn from storage. In addition to storing energy, adipose tissue also cushions such vital organs as the kidneys, and a layer of fat beneath the skin insulates the body. This subcutaneous layer is especially thick in whales, seals, and most other marine mammals, insulating their bodies in cold ocean water.

Phospholipids

Cells as we know them could not exist without another type of lipid, called phospholipids. Phospholipids are essential for cells because they are major constituents of cell membranes. Their structure provides a classic example of how form fits function at the molecular level. As shown in **Figure 5.11**, a **phospholipid** is similar to a fat molecule but has only two fatty acids attached to glycerol rather than three. The third hydroxyl group of glycerol is joined to a phosphate group, which has a negative electrical charge in the cell. Typically, an additional small charged or polar molecule is also linked to the phosphate group. Choline is one such molecule (see Figure 5.11), but there are many others as well, allowing formation of a variety of phospholipids that differ from each other.

The two ends of phospholipids show different behaviours with respect to water. The hydrocarbon tails are hydrophobic and are excluded from water. However, the

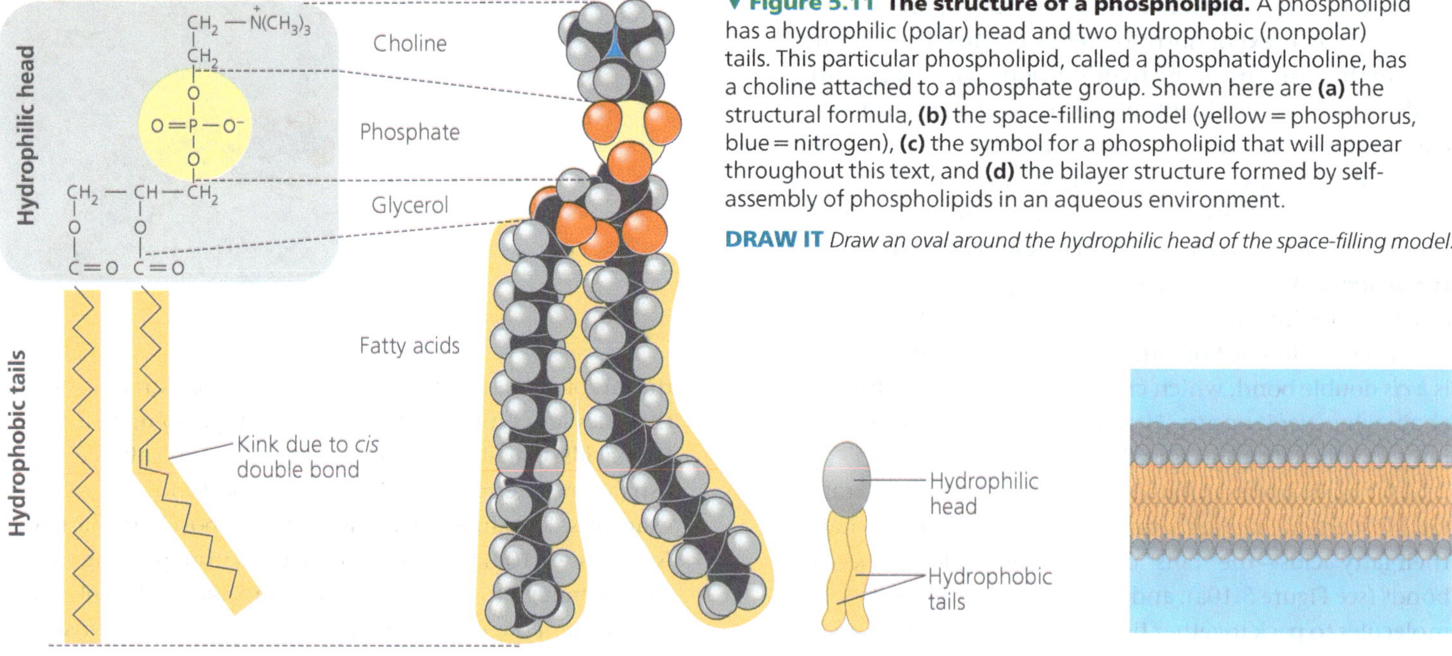

▼ **Figure 5.11 The structure of a phospholipid.** A phospholipid has a hydrophilic (polar) head and two hydrophobic (nonpolar) tails. This particular phospholipid, called a phosphatidylcholine, has a choline attached to a phosphate group. Shown here are **(a)** the structural formula, **(b)** the space-filling model (yellow = phosphorus, blue = nitrogen), **(c)** the symbol for a phospholipid that will appear throughout this text, and **(d)** the bilayer structure formed by self-assembly of phospholipids in an aqueous environment.

DRAW IT *Draw an oval around the hydrophilic head of the space-filling model.*

phosphate group and its attachments form a hydrophilic head that has an affinity for water. When phospholipids are added to water, they self-assemble into a double-layered sheet called a "bilayer" that shields their hydrophobic fatty acid tails from water (Figure 5.11d).

At the surface of a cell, phospholipids are arranged in a similar bilayer. The hydrophilic heads of the molecules are on the outside of the bilayer, in contact with the aqueous solutions inside and outside of the cell. The hydrophobic tails point towards the interior of the bilayer, away from the water. The phospholipid bilayer forms a boundary between the cell and its external environment and establishes separate compartments within eukaryotic cells; in fact, the existence of cells depends on the properties of phospholipids.

Steroids

Steroids are lipids characterised by a carbon skeleton consisting of four fused rings. Different steroids are distinguished by the particular chemical groups attached to this ensemble of rings. **Cholesterol**, a type of steroid, is a crucial molecule in animals **(Figure 5.12)**. It is a common component of animal cell membranes and is also the precursor from which other steroids, such as the vertebrate sex hormones, are synthesised. In vertebrates, cholesterol is synthesised in the liver and is also obtained from the diet. A high level of cholesterol in the blood may contribute to atherosclerosis, although some researchers are questioning the roles of cholesterol and saturated fats in the development of this condition.

▼ **Figure 5.12 Cholesterol, a steroid.** Cholesterol is the molecule from which other steroids, including the sex hormones, are synthesised. Steroids vary in the chemical groups attached to their four interconnected rings (shown in gold).

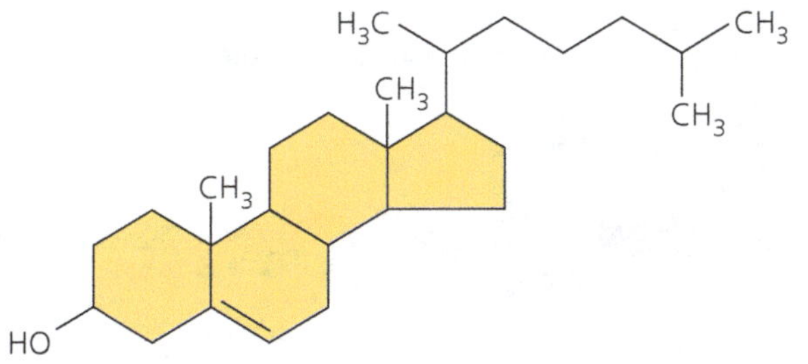

MAKE CONNECTIONS *Compare cholesterol with the sex hormones shown in the figure at the beginning of Concept 4.3. Circle the chemical groups that cholesterol has in common with estradiol; put a square around the chemical groups that cholesterol has in common with testosterone.*

CONCEPT CHECK 5.3

1. Compare the structure of a fat (triglyceride) with that of a phospholipid.
2. Why are human sex hormones considered lipids?
3. **WHAT IF?** Suppose a membrane surrounded an oil droplet, as it does in the cells of plant seeds and in some animal cells. Describe and explain the form it might take.

For suggested answers, see Appendix A.

CONCEPT 5.4

Proteins include a diversity of structures, resulting in a wide range of functions

Nearly every dynamic function of a living being depends on proteins. In fact, the importance of proteins is underscored by their name, which comes from the Greek word *proteios*, meaning "first," or "primary." Proteins account for more than 50% of the dry mass of most cells, and they are instrumental in almost everything organisms do. Some proteins speed up chemical reactions, while others play a role in defence, storage, transport, cellular communication, movement, or structural support. **Figure 5.13** shows examples of proteins with these functions, which you'll learn more about in later chapters.

Life would not be possible without enzymes, most of which are proteins. Enzymatic proteins regulate metabolism by acting as **catalysts**, chemical agents that selectively speed up chemical reactions without being consumed in the reaction. Because an enzyme can perform its function over and over again, these molecules can be thought of as workhorses that keep cells running by carrying out the processes of life.

A human has tens of thousands of different proteins, each with a specific structure and function; proteins, in fact, are the most structurally sophisticated molecules known. Consistent with their diverse functions, they vary extensively in structure, each type of protein having a unique three-dimensional shape.

Proteins are all constructed from the same set of 20 amino acids, linked in unbranched polymers. The bond between amino acids is called a *peptide bond*, so a polymer of amino acids is called a **polypeptide**. A **protein** is a biologically functional molecule made up of one or more polypeptides, each folded and coiled into a specific three-dimensional structure.

Amino Acids (Monomers)

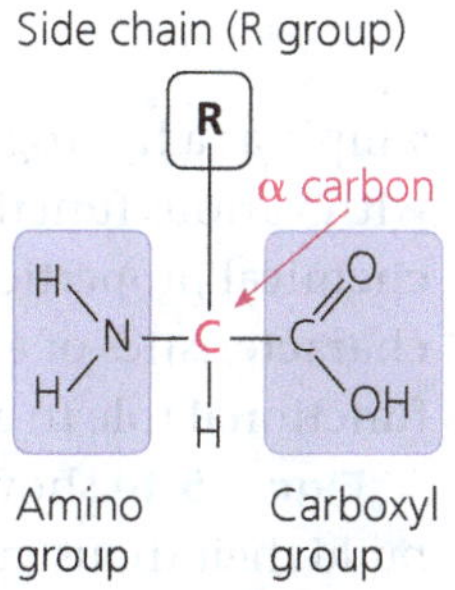

All amino acids share a common structure. An **amino acid** is an organic molecule with both an amino group and a carboxyl group (see Figure 4.9); the small figure shows the general formula for an amino acid. At the centre of the amino acid is an asymmetric carbon atom called the *alpha* (α) *carbon*. Its four different partners are an amino group, a carboxyl group, a hydrogen atom, and a variable group symbolised by R. The R group, also called the side chain, differs with each amino acid. The R group may be as

▼ Figure 5.13 An overview of protein functions.

Enzymatic proteins

Function: Selective acceleration of chemical reactions

Example: Digestive enzymes catalyse the hydrolysis of bonds in food molecules.

Enzyme

Defensive proteins

Function: Protection against disease

Example: Antibodies inactivate and help destroy viruses and bacteria.

Antibodies
Virus
Bacterium

Storage proteins

Function: Storage of amino acids

Examples: Casein, the protein of milk, is the major source of amino acids for baby mammals. Plants have storage proteins in their seeds. Ovalbumin is the protein of egg white, used as an amino acid source for the developing embryo.

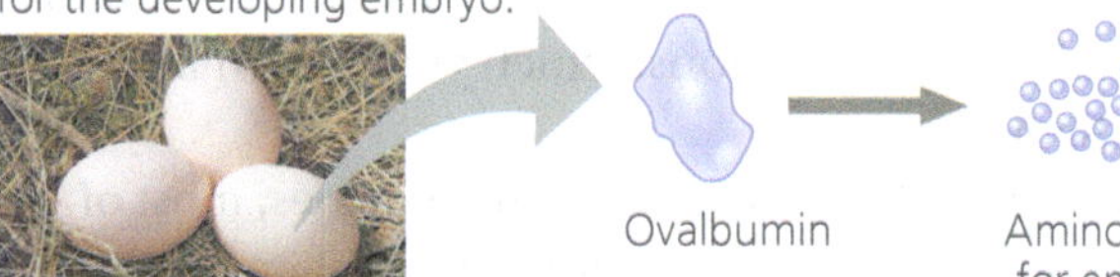

Transport proteins

Function: Transport of substances

Examples: Haemoglobin, the iron-containing protein of vertebrate blood, transports oxygen from the lungs to other parts of the body. Other proteins transport molecules across membranes, as shown here.

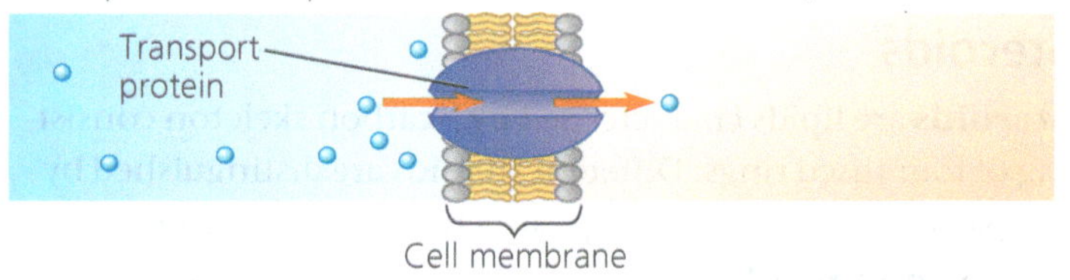

Hormonal proteins

Function: Coordination of an organism's activities

Example: Insulin, a hormone secreted by the pancreas, causes other tissues to take up glucose, thus regulating blood sugar concentration.

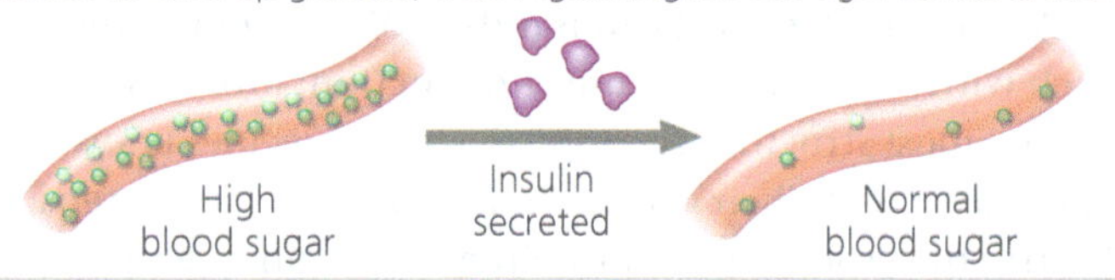

Receptor proteins

Function: Response of cell to chemical stimuli

Example: Receptors built into the membrane of a nerve cell detect signalling molecules released by other nerve cells.

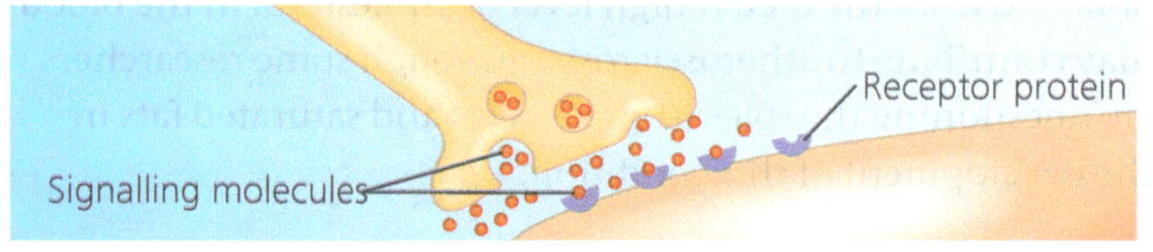

Contractile and motor proteins

Function: Movement

Examples: Motor proteins are responsible for the undulations of cilia and flagella. Actin and myosin proteins are responsible for the contraction of muscles.

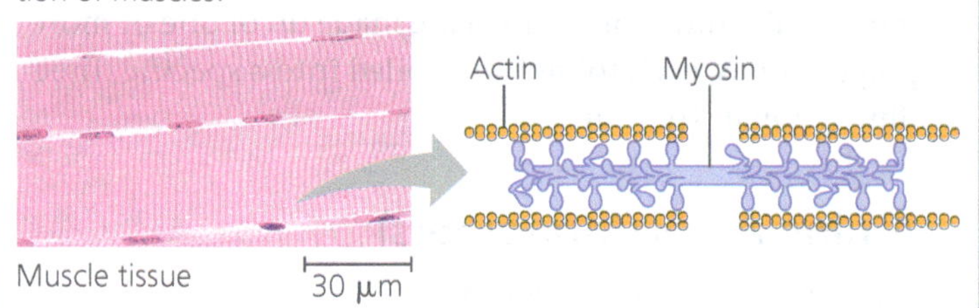

Structural proteins

Function: Support

Examples: Keratin is the protein of hair, horns, feathers, and other skin appendages. Insects and spiders use silk fibres to make their cocoons and webs, respectively. Collagen and elastin proteins provide a fibrous framework in animal connective tissues.

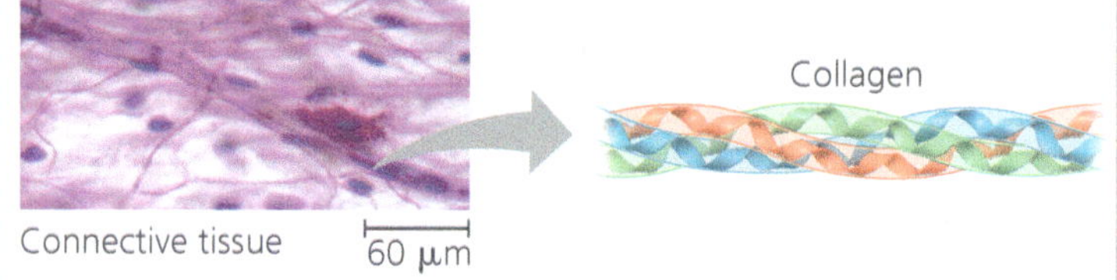

simple as a hydrogen atom, or it may be a carbon skeleton with various functional groups attached. The physical and chemical properties of the side chain determine the unique characteristics of a particular amino acid, thus affecting its functional role in a polypeptide.

Figure 5.14 shows the 20 amino acids that cells use to build their thousands of proteins. Here the amino groups and carboxyl groups are all depicted in ionised form, the way they usually exist at the pH found in a cell. The amino acids are grouped according to the properties of their side chains. One group consists of amino acids with nonpolar side chains, which are hydrophobic. Another group consists of amino acids with polar side chains, which are hydrophilic. Acidic amino acids have side chains that are generally negative in charge due to the presence of a carboxyl group, which is usually dissociated (ionised) at cellular pH. Basic amino acids have amino groups in their side chains that are generally positive in charge. (The terms *acidic* and *basic* in this context refer only to groups in the side chains because *all* amino acids—as monomers—have carboxyl groups and amino groups.) Because they are charged, acidic and basic side chains are also hydrophilic.

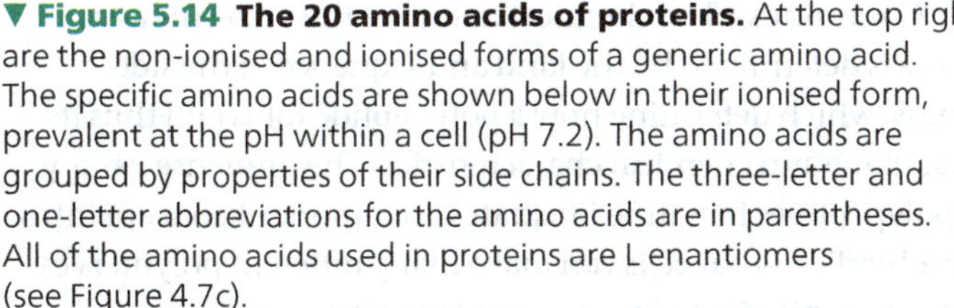

▼ Figure 5.14 The 20 amino acids of proteins. At the top right are the non-ionised and ionised forms of a generic amino acid. The specific amino acids are shown below in their ionised form, prevalent at the pH within a cell (pH 7.2). The amino acids are grouped by properties of their side chains. The three-letter and one-letter abbreviations for the amino acids are in parentheses. All of the amino acids used in proteins are L enantiomers (see Figure 4.7c).

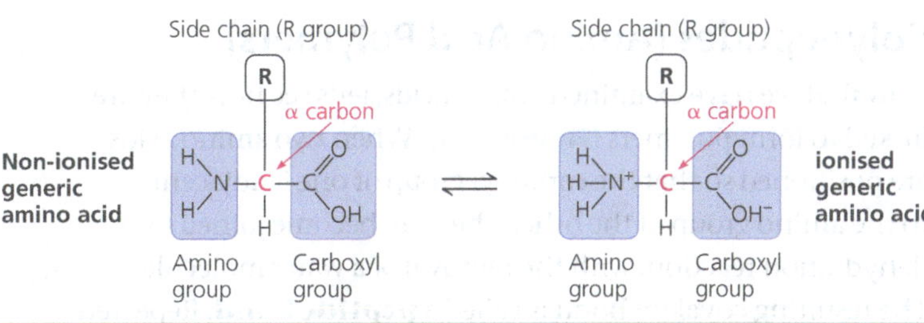

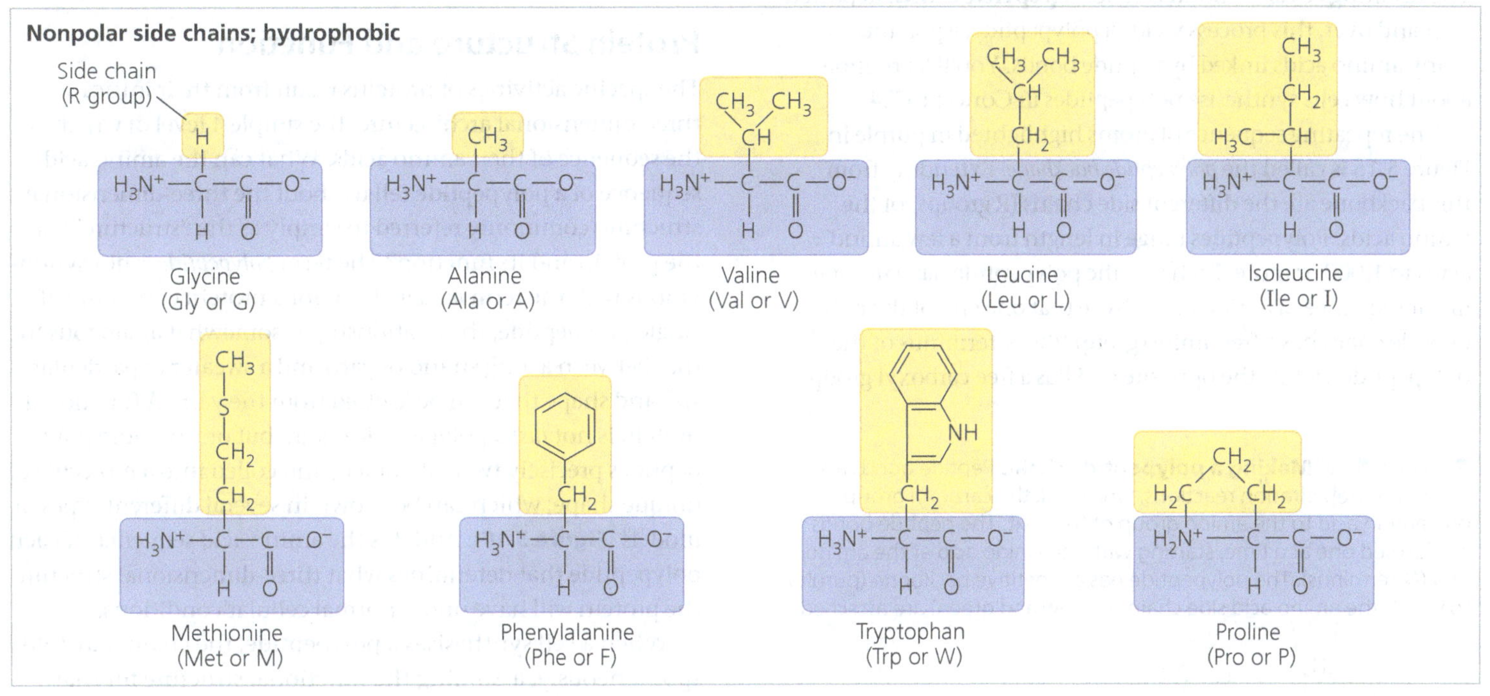

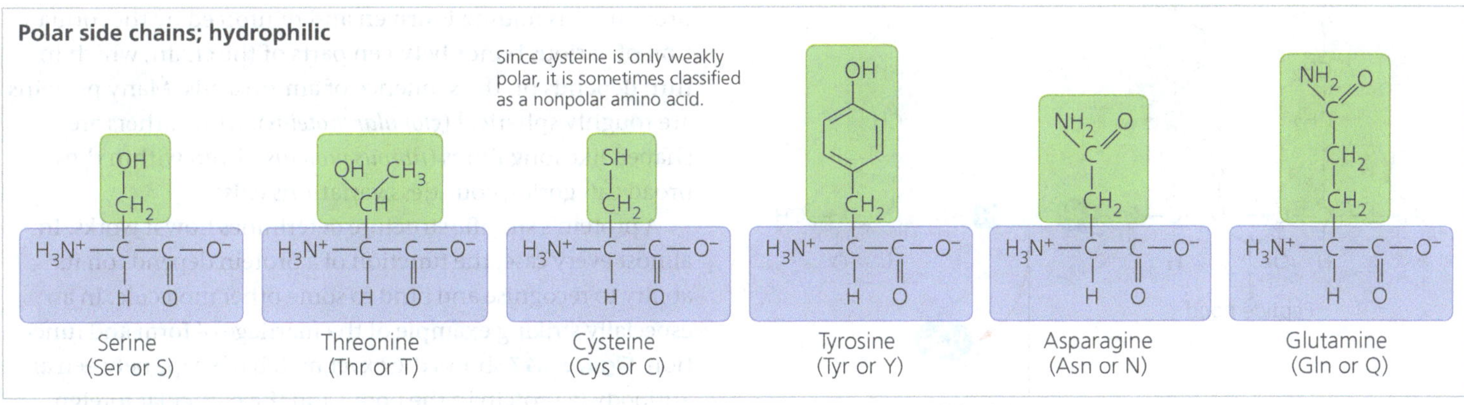

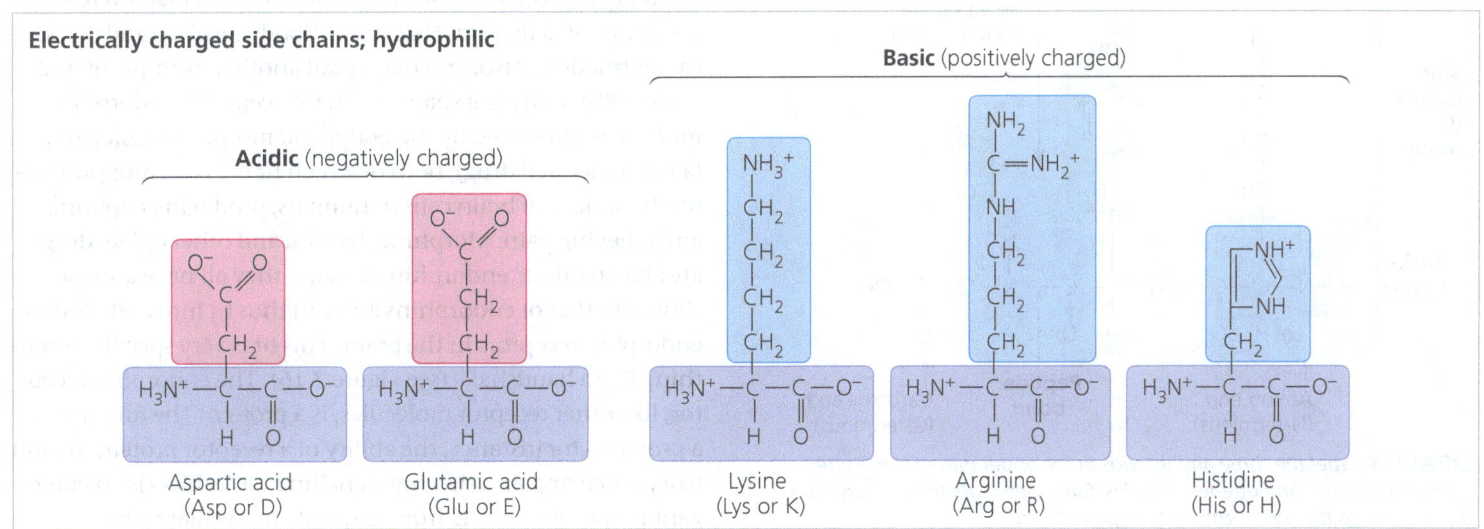

Polypeptides (Amino Acid Polymers)

Now that we have examined amino acids, let's see how they are linked to form polymers **(Figure 5.15)**. When two amino acids are positioned so that the carboxyl group of one is adjacent to the amino group of the other, they can become joined by a dehydration reaction, with the removal of a water molecule. The resulting covalent bond is called a **peptide bond**. Repeated over and over, this process yields a polypeptide, a polymer of many amino acids linked by peptide bonds. You'll learn more about how cells synthesise polypeptides in Concept 17.4.

The repeating sequence of atoms highlighted in purple in Figure 5.15 is called the *polypeptide backbone*. Extending from this backbone are the different side chains (R groups) of the amino acids. Polypeptides range in length from a few amino acids to 1,000 or more. Each specific polypeptide has a unique linear sequence of amino acids. Note that one end of the polypeptide chain has a free amino group (the N-terminus of the polypeptide), while the opposite end has a free carboxyl group (the C-terminus). The chemical nature of the molecule as a whole is determined by the kind and sequence of the side chains, which determine how a polypeptide folds and thus its final shape and chemical characteristics. The immense variety of polypeptides in nature illustrates an important concept introduced earlier—that cells can make many different polymers by linking a limited set of monomers into diverse sequences.

▼ Figure 5.15 Making a polypeptide chain. Peptide bonds are formed by dehydration reactions, which link the carboxyl group of one amino acid to the amino group of the next. The peptide bonds are formed one at a time, starting with the amino acid at the amino end (N-terminus). The polypeptide has a repetitive backbone (purple) to which the amino acid side chains (yellow and green) are attached.

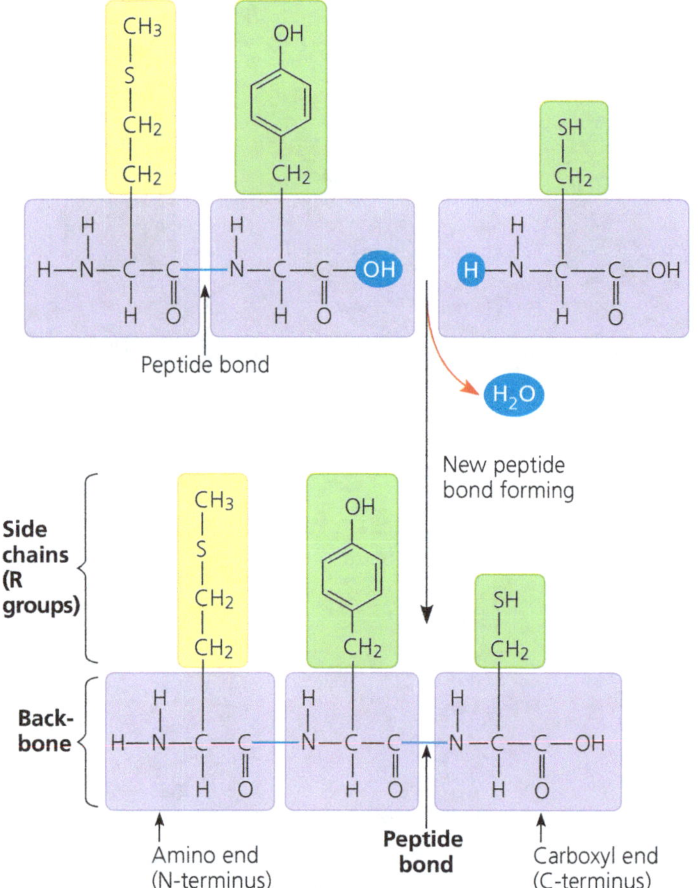

DRAW IT *Label the three amino acids in the upper part of the figure using three-letter and one-letter codes. Circle and label the carboxyl and amino groups that will form the new peptide bond.*

Protein Structure and Function

The specific activities of proteins result from their intricate three-dimensional architecture, the simplest level of which is the sequence of their amino acids. What can the amino acid sequence of a polypeptide tell us about the three-dimensional structure (commonly referred to simply as the "structure") of the protein and its function? The term *polypeptide* is not synonymous with the term *protein*. Even for a protein consisting of a single polypeptide, the relationship is somewhat analogous to that between a long strand of yarn and a sweater of particular size and shape that can be knitted from the yarn. A functional protein is not *just* a polypeptide chain, but one or more polypeptides precisely twisted, folded, and coiled into a molecule of unique shape, which can be shown in several different types of models **(Figure 5.16)**. And it is the amino acid sequence of each polypeptide that determines what three-dimensional structure the protein will have under normal cellular conditions.

When a cell synthesises a polypeptide, the chain may fold spontaneously, assuming the functional structure for that protein. This folding is driven and reinforced by the formation of various bonds between parts of the chain, which in turn depends on the sequence of amino acids. Many proteins are roughly spherical (*globular proteins*), while others are shaped like long fibres (*fibrous proteins*). Even within these broad categories, countless variations exist.

A protein's specific structure determines how it works. In almost every case, the function of a protein depends on its ability to recognise and bind to some other molecule. In an especially striking example of the marriage of form and function, **Figure 5.17** shows the exact match of shape between an antibody (a protein in the body) and the particular foreign substance on a flu virus that the antibody binds to and marks for destruction. Also, you may recall another example of molecules with matching shapes from Concept 2.3: endorphin molecules (produced by the body) and morphine molecules (a manufactured drug), both of which fit into receptor proteins on the surface of brain cells in humans, producing euphoria and relieving pain. Morphine, heroin, and other opiate drugs are able to mimic endorphins because they all have a shape similar to that of endorphins and can thus fit into and bind to endorphin receptors in the brain. This fit is very specific, something like a handshake (see Figure 2.16). The endorphin receptor, like other receptor molecules, is a protein. The function of a protein—for instance, the ability of a receptor protein to bind to a particular pain-relieving signalling molecule—is an emergent property resulting from exquisite molecular order.

▼ Figure 5.16 VISUALISING PROTEINS

Proteins can be represented in different ways, depending on the goal of the illustration.

Structural Models

Using data from structural studies of proteins, computers can generate various types of models. Each model emphasises a different aspect of the protein's structure, but no model can show what a protein actually looks like. These three models depict lysozyme, a protein in tears and saliva that helps prevent infection by binding to target molecules on bacteria.

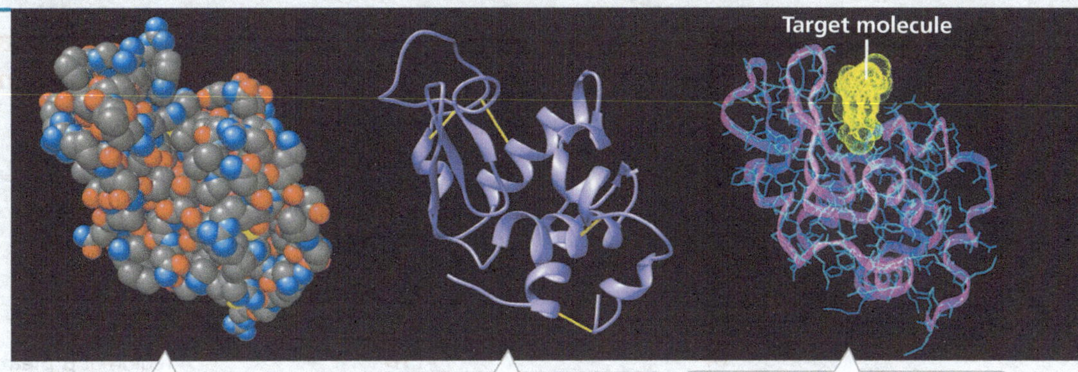

Space-filling model: Emphasises the overall globular shape. Shows all the atoms of the protein (except hydrogen), which are colour-coded: grey = carbon, red = oxygen, blue = nitrogen, and yellow = sulfur.

Ribbon model: Shows only the polypeptide backbone, emphasising how it folds and coils to form a 3-D shape, in this case stabilised by disulfide bridges (yellow lines).

Wireframe model (blue): Shows the polypeptide backbone with side chains extending from it. A ribbon model (purple) is superimposed on the wireframe model. The bacterial target molecule (yellow) is bound.

***1.** In which model is it easiest to follow the polypeptide backbone?*

Simplified Diagrams

It isn't always necessary to use a detailed computer model; simplified diagrams are useful when the focus of the figure is on the function of the protein, not the structure.

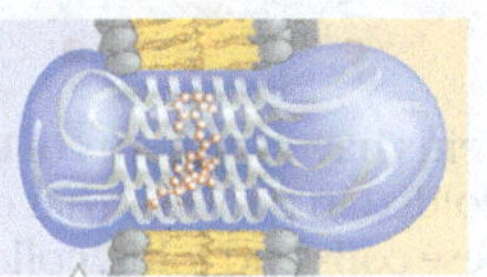

A transparent shape is drawn around the contours of a ribbon model of the protein rhodopsin, showing the shape of the molecule as well as some internal details.

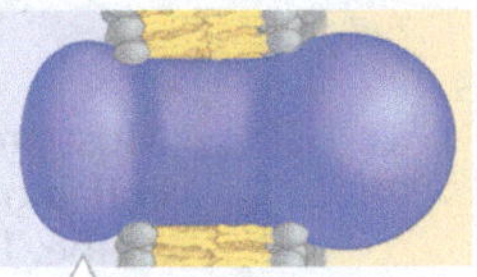

When structural details are not needed, a solid shape can be used.

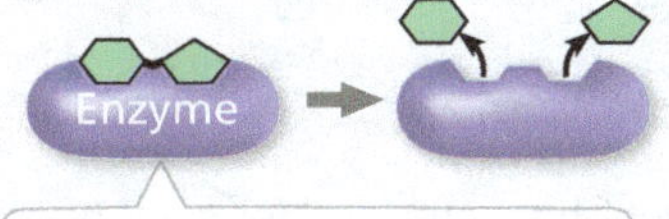

A simple shape is used here to represent a generic enzyme because the diagram focuses on enzyme action in general.

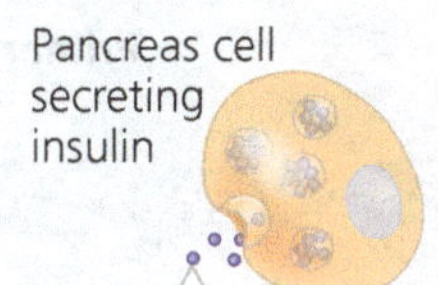

Sometimes a protein is represented simply as a dot, as shown here for insulin.

***2.** Draw a simple version of lysozyme that shows its overall shape, based on the molecular models in the top section of the figure.*

***3.** Why is it unnecessary to show the actual shape of insulin here?*

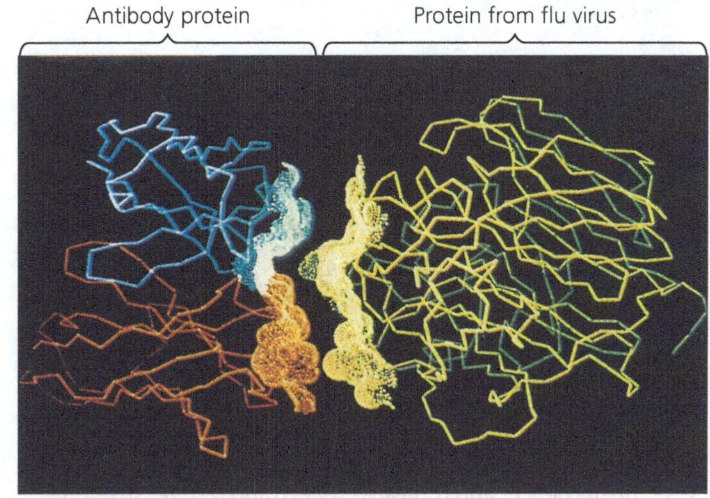

▶ Figure 5.17 Complementarity of shape between two protein surfaces. A technique called X-ray crystallography was used to generate a computer model of an antibody protein (blue and orange, left) bound to a flu virus protein (yellow and green, right). This is a wireframe model modified by adding an "electron density map" in the region where the two proteins meet. Computer software was then used to back the images away from each other slightly.

VISUAL SKILLS *What do these computer models allow you to see about the two proteins?*

Four Levels of Protein Structure

In spite of their great diversity, proteins share three superimposed levels of structure, known as primary, secondary, and tertiary structure. A fourth level, quaternary structure, arises when a protein consists of two or more polypeptide chains. **Figure 5.18** describes these four levels of protein structure. Be sure to study this figure thoroughly before going on to the next section.

▼ Figure 5.18 Exploring Levels of Protein Structure

Primary Structure

Linear chain of amino acids

The **primary structure** of a protein is its sequence of amino acids. As an example, let's consider transthyretin, a globular blood protein that transports vitamin A and one of the thyroid hormones. Transthyretin is made up of four identical polypeptide chains, each composed of 127 amino acids. Shown here is one of these chains unraveled for a closer look at its primary structure. Each of the 127 positions along the chain is occupied by one of the 20 amino acids, indicated here by its three-letter abbreviation.

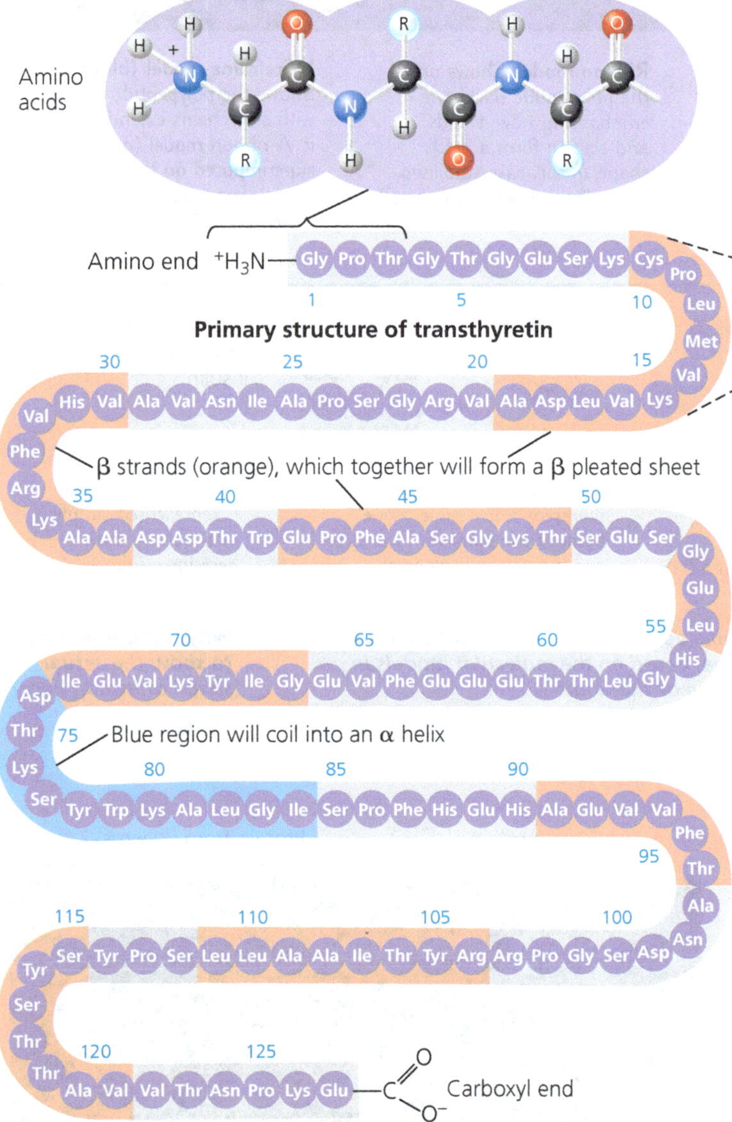

The primary structure is like the order of letters in a very long word. If left to chance, there would be 20^{127} different ways of making a polypeptide chain 127 amino acids long. However, the precise primary structure of a protein is determined not by the random linking of amino acids, but by inherited genetic information. The primary structure in turn dictates secondary structure (α helices and β pleated sheets) and tertiary structure, due to the chemical nature of the backbone and the side chains (R groups) of the amino acids along the polypeptide.

Secondary Structure

Regions stabilised by hydrogen bonds between atoms of the polypeptide backbone

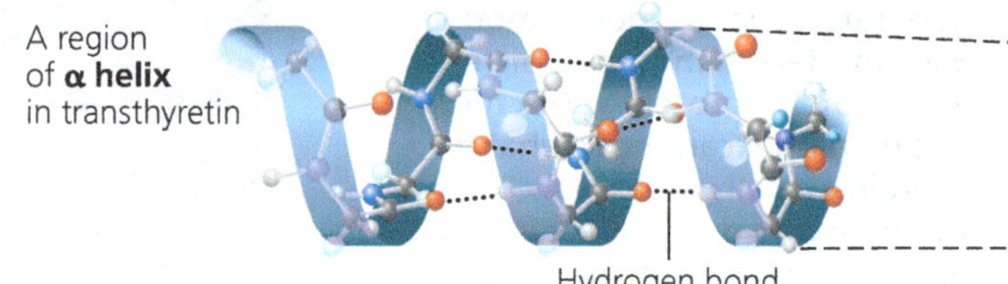

A region of **β pleated sheet** (made up of adjacent β strands) in transthyretin

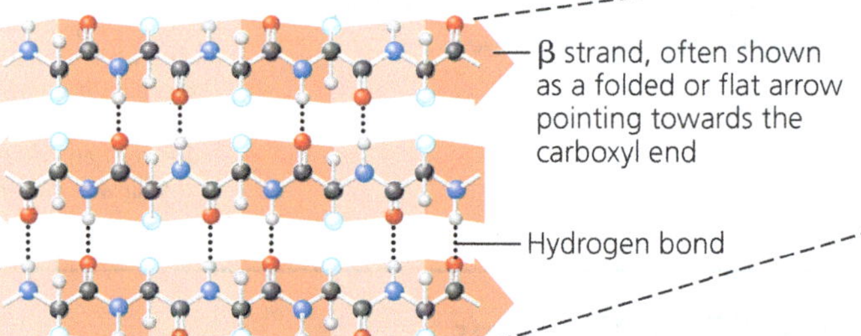

Most proteins have segments of their polypeptide chains repeatedly coiled or folded in patterns that contribute to the protein's overall shape. These coils and folds, collectively referred to as **secondary structure**, are the result of hydrogen bonds between the repeating constituents of the polypeptide backbone (not the amino acid side chains). Within the backbone, the oxygen atoms have a partial negative charge, and the hydrogen atoms attached to the nitrogens have a partial positive charge (see Figure 2.14); therefore, hydrogen bonds can form between these atoms. Individually, these hydrogen bonds are weak, but because they are repeated many times over a relatively long region of the polypeptide chain, they can support a particular shape for that part of the protein.

One such secondary structure is the **α helix**, a delicate coil held together by hydrogen bonding between every fourth amino acid, as shown above. Although each transthyretin polypeptide has only one α helix region (see the Primary and Tertiary Structure sections), other globular proteins have multiple stretches of α helix separated by nonhelical regions (see haemoglobin in the Quaternary Structure section). Some fibrous proteins, such as α-keratin, the structural protein of hair, have the α helix formation over most of their length.

The other secondary structure is the **β pleated sheet**. As shown above, two or more segments of the polypeptide chain lying side by side (called β strands) are connected by hydrogen bonds between parts of the two parallel segments. β pleated sheets make up the core of many globular proteins, as is the case for transthyretin (see Tertiary Structure), and dominate some fibrous proteins, including the silk protein of a spider's web. The teamwork of so many hydrogen bonds makes each spider silk fibre stronger than a steel strand.

► Spiders secrete silk fibres made of a structural protein containing β pleated sheets, which allow the spiderweb to stretch and recoil.

Tertiary Structure

Three-dimensional shape stabilised by interactions between side chains

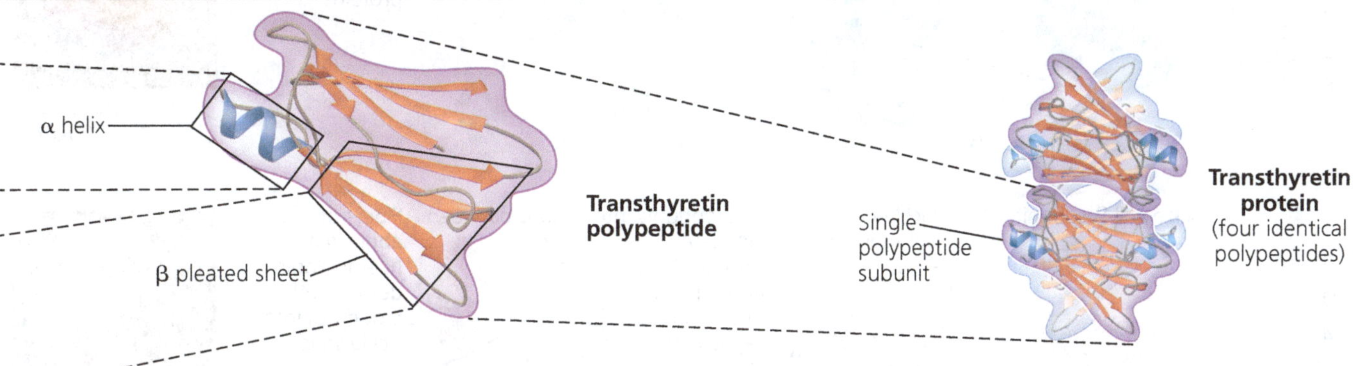

Superimposed on the patterns of secondary structure is a protein's tertiary structure, shown here in a ribbon model of the transthyretin polypeptide. While secondary structure involves interactions between backbone constituents, **tertiary structure** is the overall shape of a polypeptide resulting from interactions between the side chains (R groups) of the various amino acids. One type of interaction that contributes to tertiary structure is called—somewhat misleadingly—a **hydrophobic interaction**. As a polypeptide folds into its functional shape, amino acids with hydrophobic (nonpolar) side chains usually end up in clusters at the core of the protein, out of contact with water. Thus, a "hydrophobic interaction" is actually caused by the exclusion of nonpolar substances by water molecules. Once nonpolar amino acid side chains are close together, van der Waals interactions help hold them together. Meanwhile, hydrogen bonds between polar side chains and ionic bonds between positively and negatively charged side chains also help stabilise tertiary structure. These are all weak interactions in the aqueous cellular environment, but their cumulative effect helps give the protein a unique shape.

Covalent bonds called **disulfide bridges** may further reinforce the shape of a protein. Disulfide bridges form where two cysteine monomers, which have sulfhydryl groups (—SH) on their side chains (see Figure 4.9), are brought close together by the folding of the protein. The sulfur of one cysteine bonds to the sulfur of the second, and the disulfide bridge (—S—S—) rivets parts of the protein together (see yellow lines in Figure 5.16 ribbon model). All of these different kinds of interactions can contribute to the tertiary structure of a protein, as shown here in a small part of a hypothetical protein:

Hydrogen bond
Hydrophobic interactions and van der Waals interactions
Ionic bond
Disulfide bridge
Polypeptide backbone of small part of a protein

R groups are coloured as in Figure 5.14.

Quaternary Structure

Association of two or more polypeptides (some proteins only)

Some proteins consist of two or more polypeptide chains aggregated into one functional macromolecule. **Quaternary structure** is the overall protein structure that results from the aggregation of these polypeptide subunits. For example, shown above is the complete globular transthyretin protein, made up of its four polypeptides.

Another example is collagen, which is a fibrous protein that has three identical helical polypeptides intertwined into a larger triple helix, giving the long fibres great strength. This suits collagen fibres to their function as the girders of connective tissue in skin, bone, tendons, ligaments, and other body parts. (Collagen accounts for 40% of the protein in a human body.)

Collagen

Haemoglobin, the oxygen-binding protein of red blood cells, is another example of a globular protein with quaternary structure. It consists of four polypeptide subunits, two of one kind (α) and two of another kind (β). Both α and β subunits consist primarily of α-helical secondary structure. Each subunit has a nonpolypeptide component, called haem, with an iron atom that binds oxygen.

Haem
Iron
β subunit
α subunit
α subunit
β subunit
Haemoglobin

▼ Figure 5.19 A single amino acid substitution in a protein causes sickle-cell disease.

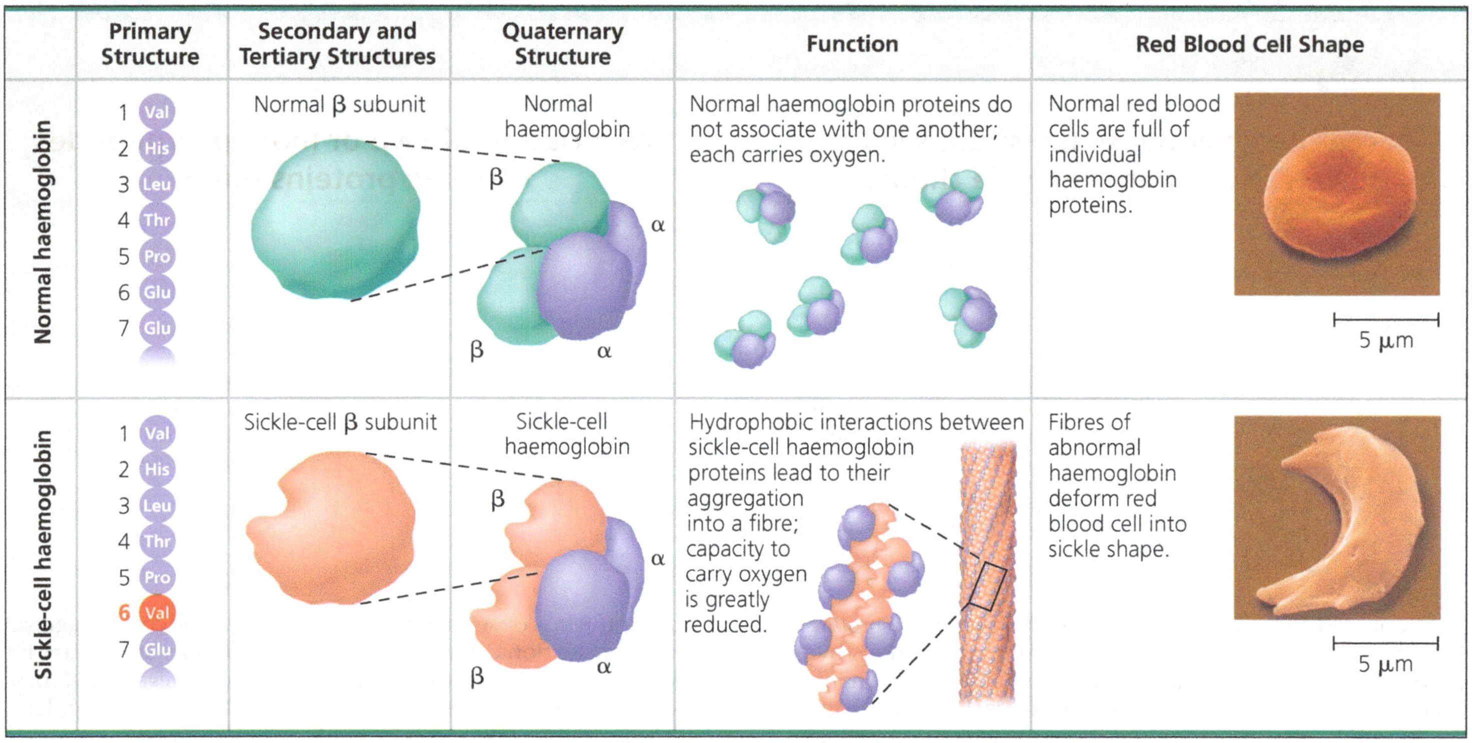

	Primary Structure	Secondary and Tertiary Structures	Quaternary Structure	Function	Red Blood Cell Shape
Normal haemoglobin	1 Val 2 His 3 Leu 4 Thr 5 Pro 6 Glu 7 Glu	Normal β subunit	Normal haemoglobin β α β α	Normal haemoglobin proteins do not associate with one another; each carries oxygen.	Normal red blood cells are full of individual haemoglobin proteins. 5 μm
Sickle-cell haemoglobin	1 Val 2 His 3 Leu 4 Thr 5 Pro 6 Val 7 Glu	Sickle-cell β subunit	Sickle-cell haemoglobin β α β α	Hydrophobic interactions between sickle-cell haemoglobin proteins lead to their aggregation into a fibre; capacity to carry oxygen is greatly reduced.	Fibres of abnormal haemoglobin deform red blood cell into sickle shape. 5 μm

MAKE CONNECTIONS *Considering the chemical characteristics of the amino acids valine and glutamic acid (see Figure 5.14), propose a possible explanation for the dramatic effect on protein function that occurs when valine is substituted for glutamic acid.*

Sickle-Cell Disease: A Change in Primary Structure

Even a slight change in primary structure can affect a protein's shape and ability to function. For instance, **sickle-cell disease**, an inherited blood disorder, is caused by the substitution of one amino acid (valine) for the normal one (glutamic acid) at the position of the sixth amino acid in the primary structure of haemoglobin, the protein that carries oxygen in red blood cells. Normal red blood cells are disk-shaped, but in sickle-cell disease, the abnormal haemoglobin molecules tend to aggregate into chains, deforming some of the cells into a sickle shape **(Figure 5.19)**. A person with the disease has periodic "sickle-cell crises" when the angular cells clog tiny blood vessels, impeding blood flow. The toll taken on such patients is a dramatic example of how a simple change in protein structure can have devastating effects on protein function.

What Determines Protein Structure?

You've learned that a unique shape endows each protein with a specific function. But what are the key factors determining protein structure? You already know most of the answer: A polypeptide chain of a given amino acid sequence can be arranged into a three-dimensional shape determined by the interactions responsible for secondary and tertiary structure. This folding normally occurs as the protein is being synthesised in the crowded environment within a cell, aided by other proteins. However, protein structure also depends on the physical and chemical conditions of the protein's environment. If the pH, salt concentration, temperature, or other aspects of its environment are altered, the weak chemical bonds and interactions within a protein may be destroyed, causing the protein to unravel and lose its native shape, a change called **denaturation (Figure 5.20)**. Because it is misshapen, the denatured protein is biologically inactive.

Most proteins become denatured if they are transferred from an aqueous environment to a nonpolar solvent, such as ether or chloroform; the polypeptide chain refolds so that

▼ Figure 5.20 Denaturation and renaturation of a protein. High temperatures or various chemical treatments will denature a protein, causing it to lose its shape and hence its ability to function. If the denatured protein remains dissolved, it may renature when the chemical and physical aspects of its environment are restored to normal.

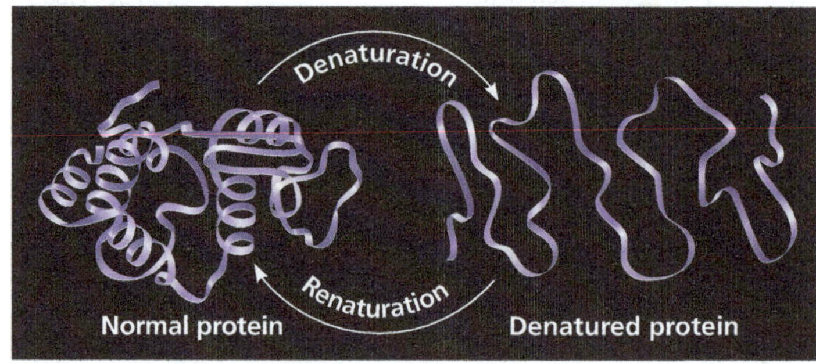

its hydrophobic regions face outwards towards the solvent. Other denaturation agents include chemicals that disrupt the hydrogen bonds, ionic bonds, and disulfide bridges that maintain a protein's shape. Denaturation can also result from excessive heat, which agitates the polypeptide chain enough to overpower the weak interactions that stabilise the structure. The white of an egg becomes opaque during cooking because the denatured proteins are insoluble and solidify. This also explains why excessively high fevers can be fatal: Proteins in the blood tend to denature at very high body temperatures.

When a protein in a test-tube solution has been denatured by heat or chemicals, it can sometimes return to its functional shape when the denaturing agent is removed. (Sometimes this is not possible: For example, a fried egg will not become liquefied when placed back into the refrigerator!) We can conclude that the information for building specific shape is intrinsic to the protein's primary structure; this is often the case for small proteins. The sequence of amino acids determines the protein's shape—where an α helix can form, where β pleated sheets can exist, where disulfide bridges are located, where ionic bonds can form, and so on. But how does protein folding occur in the cell?

Protein Folding in the Cell

Biochemists now know the amino acid sequence for about 160 million proteins, with about 4.5–5 million added each month, and the three-dimensional shape for about 40,000. Researchers have tried to correlate the primary structure of many proteins with their three-dimensional structure to discover the rules of protein folding. Unfortunately, however, the protein-folding process is not that simple. Most proteins probably go through several intermediate structures on their way to a stable shape, and looking at the mature structure does not reveal the stages of folding required to achieve that form. However, biochemists have developed methods for tracking a protein through such stages and learning more about this important process.

Misfolding of polypeptides in cells is a serious problem that has come under increasing scrutiny by medical researchers. Many diseases—such as cystic fibrosis, Alzheimer's, Parkinson's, and mad cow disease—are associated with an accumulation of misfolded proteins. In fact, misfolded versions of the transthyretin protein featured in Figure 5.18 have been implicated in several diseases, including one form of senile dementia.

Even when scientists have a correctly folded protein in hand, determining its exact three-dimensional structure is not simple, for a single protein has thousands of atoms. The method most commonly used to determine the 3-D structure of a protein is **X-ray crystallography**, which depends on the diffraction of an X-ray beam by the atoms of a crystallised molecule. Using this technique, scientists can build a 3-D model that shows the exact position of every atom in a protein molecule **(Figure 5.21)**. Nuclear magnetic resonance (NMR) spectroscopy, cryo-electron microscopy (cryo-EM; see Concept 6.1) and bioinformatics (see Concept 5.6) are complementary approaches to understanding protein structure and function.

▼ Figure 5.21 Research Method

X-Ray Crystallography

Application Scientists use X-ray crystallography to determine the three-dimensional (3-D) structure of macromolecules such as nucleic acids and proteins.

Technique Researchers aim an X-ray beam through a crystallised protein or nucleic acid. The atoms of the crystal diffract (bend) the X-rays into an orderly array that a digital detector records as a pattern of spots called an X-ray diffraction pattern, an example of which is shown here.

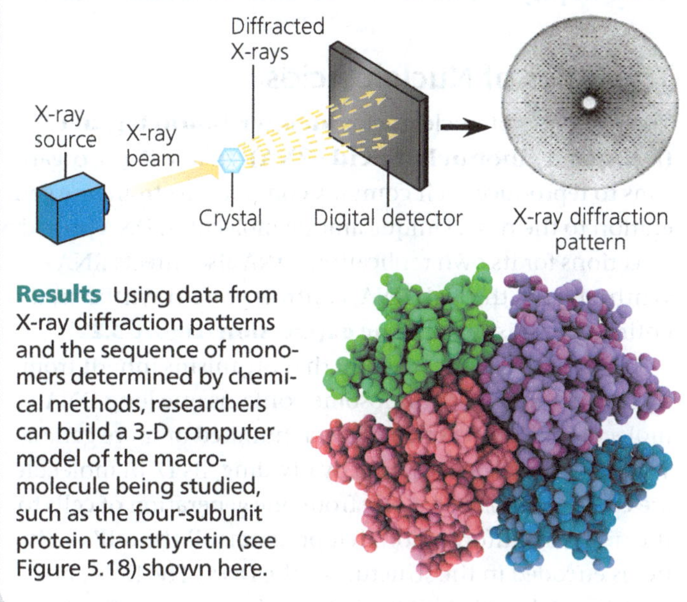

Results Using data from X-ray diffraction patterns and the sequence of monomers determined by chemical methods, researchers can build a 3-D computer model of the macromolecule being studied, such as the four-subunit protein transthyretin (see Figure 5.18) shown here.

The structure of some proteins is difficult to determine for a simple reason: A growing body of biochemical research has revealed that a significant number of proteins, or regions of proteins, do not have a distinct 3-D structure until they interact with a target protein or other molecule. Their flexibility and indefinite structure are important for their function, which may require binding with different targets at different times. These proteins, which may account for 20–30% of mammalian proteins, are called *intrinsically disordered proteins* and are the focus of current research.

CONCEPT CHECK 5.4

1. What parts of a polypeptide participate in the bonds that hold together secondary structure? Tertiary structure?
2. Thus far in the chapter, the Greek letters α and β have been used to specify at least three different pairs of structures. Name and briefly describe them.
3. Each amino acid has a carboxyl group and an amino group. Are these groups present in a polypeptide? Explain.
4. **WHAT IF?** Where would you expect a polypeptide region rich in the amino acids valine, leucine, and isoleucine to be located in a folded polypeptide? Explain.

For suggested answers, see Appendix A.

CONCEPT 5.5

Nucleic acids store, transmit, and help express hereditary information

If the primary structure of polypeptides determines a protein's shape, what determines primary structure? The amino acid sequence of a polypeptide is programmed by a discrete unit of inheritance known as a **gene**. Genes consist of DNA, which belongs to the class of compounds called nucleic acids. **Nucleic acids** are polymers made of monomers called nucleotides.

The Roles of Nucleic Acids

The two types of nucleic acids, **deoxyribonucleic acid (DNA)** and **ribonucleic acid (RNA)**, enable living organisms to reproduce their complex components from one generation to the next. Unique among molecules, DNA provides directions for its own replication. DNA also directs RNA synthesis and, through RNA, controls protein synthesis; this entire process is called **gene expression** (Figure 5.22).

DNA is the genetic material that organisms inherit from their parents. Each chromosome contains one long DNA molecule, usually carrying several hundred or more genes. When a cell reproduces itself by dividing, its DNA molecules are copied and passed along from one generation of cells to the next. The information that programs all the cell's activities is encoded in the structure of the DNA. The DNA, however, is not directly involved in running the operations of the cell, any more than computer software by itself can read the bar code on a box of cereal. Just as a scanner is needed to read a bar code, proteins are required to implement genetic programs. The molecular hardware of the cell—the tools that carry out biological functions—consists mostly of proteins. For example, the oxygen carrier in red blood cells is the protein haemoglobin that you saw earlier (see Figure 5.18), not the DNA that specifies its structure.

How does RNA, the other type of nucleic acid, fit into gene expression, the flow of genetic information from DNA to proteins? A given gene along a DNA molecule can direct synthesis of a type of RNA called *messenger RNA* (*mRNA*). The mRNA molecule interacts with the cell's protein-synthesising machinery to direct production of a polypeptide, which folds into all or part of a protein. We can summarise the flow of genetic information as DNA → RNA → protein (see Figure 5.22). The sites of protein synthesis are cellular structures called ribosomes. In a eukaryotic cell, ribosomes are in the cytoplasm—the region between the nucleus and the cell's outer boundary, the plasma membrane—but DNA resides in the nucleus. Messenger RNA conveys genetic instructions for building proteins from the nucleus to the cytoplasm. Prokaryotic cells lack nuclei but still use mRNA to convey a message from the DNA to ribosomes and other cellular equipment that translate the coded information into amino acid sequences. Later, you'll read about other functions of some recently discovered RNA molecules; the stretches of DNA that direct synthesis of these RNAs are also considered genes (see Concept 18.3).

▼ **Figure 5.22 Gene expression: DNA → RNA → protein.** In a eukaryotic cell, DNA in the nucleus programs protein production in the cytoplasm by dictating synthesis of messenger RNA (mRNA).

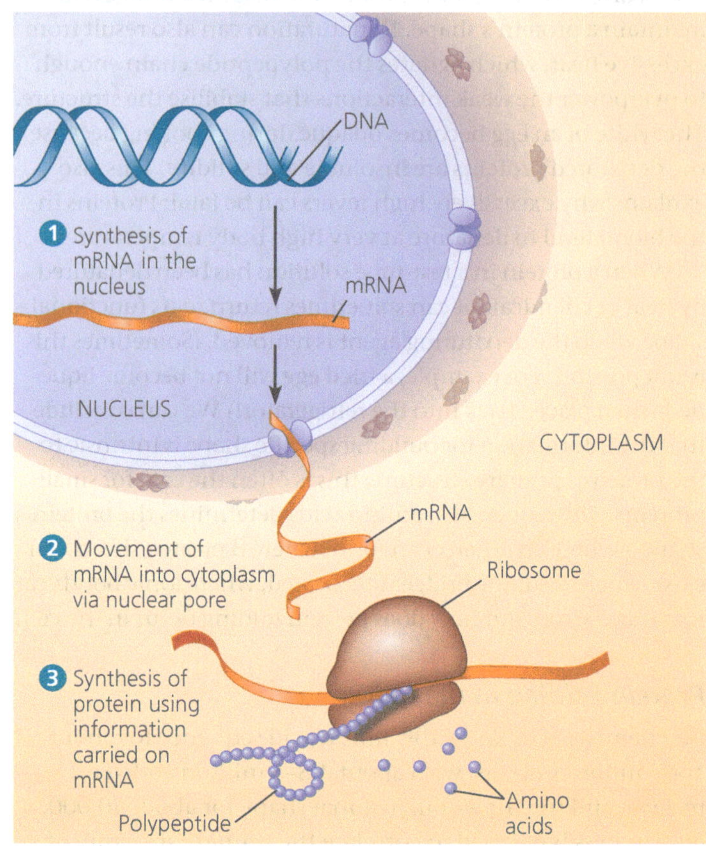

The Components of Nucleic Acids

Nucleic acids are macromolecules that exist as polymers called **polynucleotides** (Figure 5.23a). As indicated by the name, each polynucleotide consists of monomers called **nucleotides**. A nucleotide, in general, is composed of three parts: a five-carbon sugar (a pentose), a nitrogen-containing (nitrogenous) base, and one to three phosphate groups (Figure 5.23b). The beginning monomer used to build a polynucleotide has three phosphate groups, but two are lost during the polymerisation process. The portion of a nucleotide without any phosphate groups is called a *nucleoside*.

To understand the structure of a single nucleotide, let's first consider the nitrogenous bases (Figure 5.23c). Each nitrogenous base has one or two rings that include nitrogen atoms. (They are called nitrogenous *bases* because the nitrogen atoms tend to take up H^+ from solution, thus acting as bases.) There are two families of nitrogenous bases: pyrimidines and purines. A **pyrimidine** has one six-membered ring of carbon and nitrogen atoms. The members of the pyrimidine family are cytosine (C), thymine (T), and uracil (U).

Purines are larger, with a six-membered ring fused to a five-membered ring. The purines are adenine (A) and guanine (G). The specific pyrimidines and purines differ in the chemical groups attached to the rings. Adenine, guanine, and cytosine are found in both DNA and RNA; thymine is found only in DNA and uracil only in RNA.

Now let's add the sugar to which the nitrogenous base is attached. In DNA the sugar is **deoxyribose**; in RNA it is **ribose** (see Figure 5.23c). The only difference between these two sugars is that deoxyribose lacks an oxygen atom on the second carbon in the ring, hence the name *deoxy*ribose.

So far, we have built a nucleoside (base plus sugar). To complete the construction of a nucleotide, we attach one to three phosphate groups to the 5′ carbon of the sugar (the carbon numbers in the sugar include ′, the prime symbol; see Figure 5.23b). With one phosphate, this is a nucleoside monophosphate, more often called a nucleotide.

Nucleotide Polymers

The linkage of nucleotides into a polynucleotide involves a condensation reaction. (You will learn the details in Concept 16.2.) In the polynucleotide, adjacent nucleotides are joined by a phosphodiester linkage, which consists of a phosphate group that covalently links the sugars of two nucleotides. This bonding results in a repeating pattern of sugar-phosphate units called the *sugar-phosphate backbone* (see Figure 5.23a). (Note that the nitrogenous bases are not part of the backbone.) The two free ends of the polymer are distinctly different from each other. One end has a phosphate attached to a 5′ carbon, and the other end has a hydroxyl group on a 3′ carbon; we refer to these as the *5′ end* and the *3′ end*, respectively. We can say that a polynucleotide has a built-in directionality along its sugar-phosphate backbone, from 5′ to 3′, somewhat like a one-way street. The bases are attached all along the sugar-phosphate backbone.

The sequence of bases along a DNA (or mRNA) polymer is unique for each gene and provides very specific information to the cell. Because genes are hundreds to thousands of nucleotides long, the number of possible base sequences is effectively limitless. The information carried by the gene is encoded in its specific sequence of the four DNA bases. For example, the sequence 5′-AGGTAACTT-3′ means one thing, whereas the sequence 5′-CGCTTTAAC-3′ has a different meaning. (Entire genes, of course, are much longer.) The linear order of bases in a gene specifies the amino acid sequence—the primary structure—of a protein, which in turn specifies that protein's 3-D structure, thus enabling its function in the cell.

▼ **Figure 5.23 Components of nucleic acids. (a)** A polynucleotide has a sugar-phosphate backbone with variable appendages, the nitrogenous bases. **(b)** In a polynucleotide, each nucleotide monomer includes a nitrogenous base, a sugar, and a phosphate group. Note that carbon numbers in the sugar include primes (′). **(c)** A nucleoside includes a nitrogenous base (purine or pyrimidine) and a five-carbon sugar (deoxyribose or ribose).

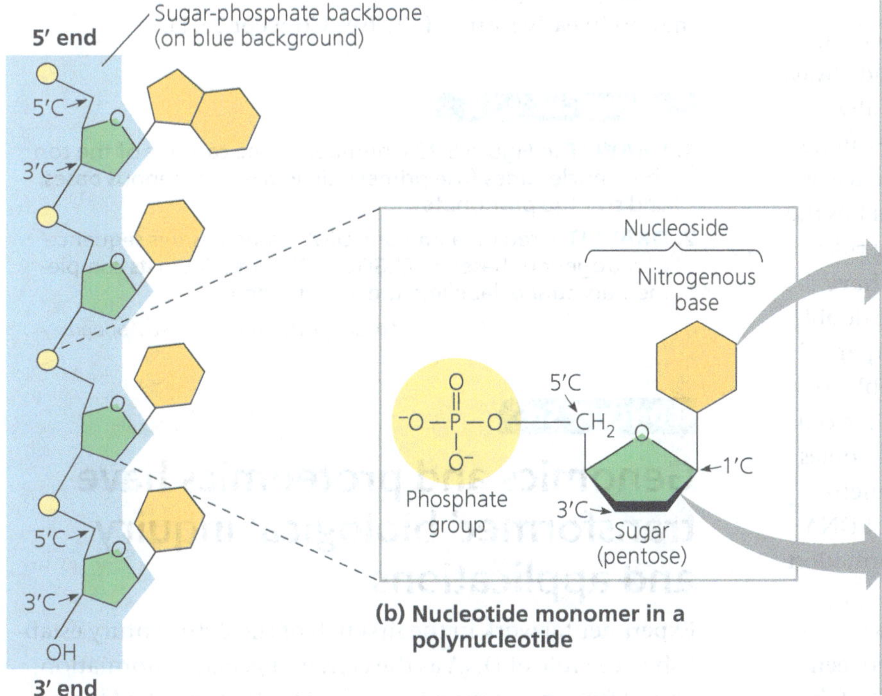

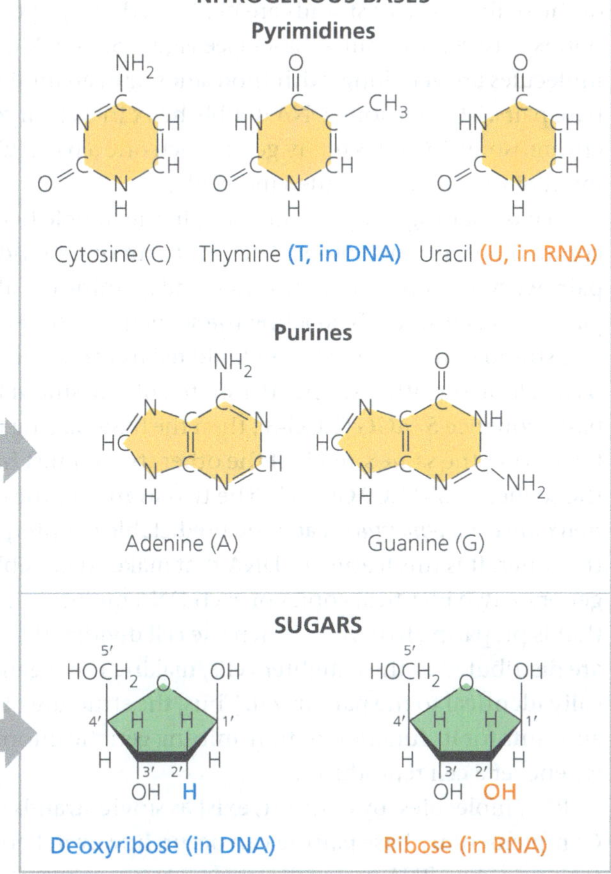

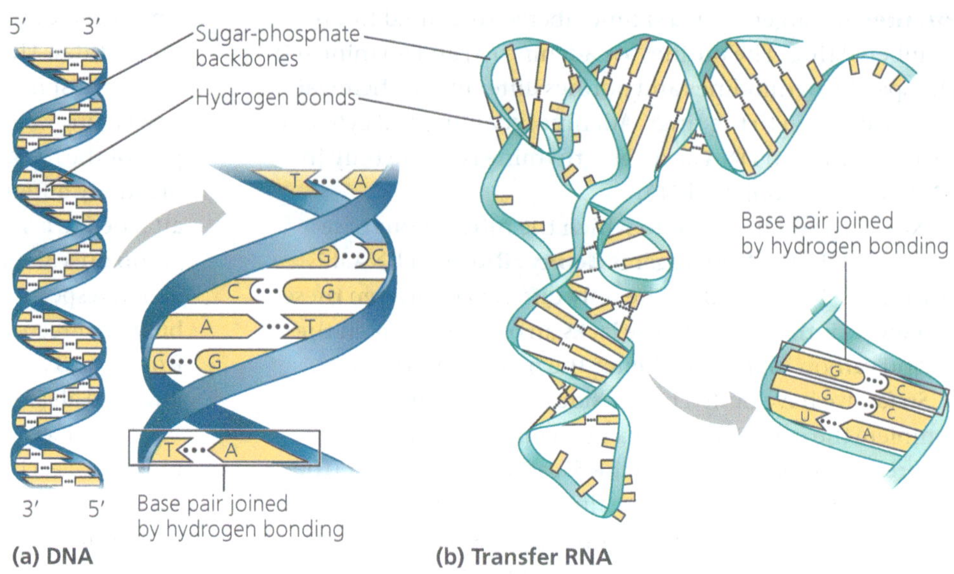

▶ **Figure 5.24 The structures of DNA and tRNA molecules. (a)** The DNA molecule is usually a double helix, with the sugar-phosphate backbones of the antiparallel polynucleotide strands (symbolised here by blue ribbons) on the outside of the helix. Hydrogen bonds between pairs of nitrogenous bases hold the two strands together. As illustrated here with symbolic shapes for the bases, adenine (A) can pair only with thymine (T), and guanine (G) can pair only with cytosine (C). Each DNA strand in this figure is the structural equivalent of the polynucleotide diagrammed in Figure 5.23a. **(b)** A tRNA molecule has a roughly L-shaped structure due to complementary base pairing of antiparallel stretches of RNA. In RNA, A pairs with U.

The Structures of DNA and RNA Molecules

A DNA molecule has two polynucleotides, or "strands," that wind around an imaginary axis, forming a **double helix** **(Figure 5.24a)**. The two sugar-phosphate backbones run in opposite 5′ → 3′ directions from each other; this arrangement is referred to as **antiparallel**, somewhat like a divided highway. The sugar-phosphate backbones are on the outside of the helix, and the nitrogenous bases are paired in the interior of the helix. The two strands are held together by hydrogen bonds between the paired bases (see Figure 5.24a). Most DNA molecules are very long, with thousands or even millions of base pairs. The one long DNA double helix in a eukaryotic chromosome includes many genes, each one a particular segment of the double-stranded molecule.

In base pairing, only certain bases in the double helix are compatible with each other. Adenine (A) in one strand always pairs with thymine (T) in the other, and guanine (G) always pairs with cytosine (C). Reading the sequence of bases along one strand of the double helix would tell us the sequence of bases along the other strand. If a stretch of one strand has the base sequence 5′-AGGTCCG-3′, then the base-pairing rules tell us that the same stretch of the other strand must have the sequence 3′-TCCAGGC-5′. The two strands of the double helix are *complementary*, each the predictable counterpart of the other. It is this feature of DNA that makes it possible to generate two identical copies of each DNA molecule in a cell that is preparing to divide. When the cell divides, the copies are distributed to the daughter cells, making them genetically identical to the parent cell. Thus, the structure of DNA accounts for its function of transmitting genetic information whenever a cell reproduces.

RNA molecules, by contrast, exist as single strands. Complementary base pairing can occur, however, between regions of two RNA molecules or even between two stretches of nucleotides in the *same* RNA molecule. In fact, base pairing within an RNA molecule allows it to take on the particular three-dimensional shape necessary for its function. Consider, for example, the type of RNA called *transfer RNA (tRNA)*, which brings amino acids to the ribosome during the synthesis of a polypeptide. A tRNA molecule is about 80 nucleotides in length. Its functional shape results from base pairing between nucleotides where complementary stretches of the molecule can run antiparallel to each other **(Figure 5.24b)**.

Note that in RNA, adenine (A) pairs with uracil (U); thymine (T) is not present in RNA. Another difference between RNA and DNA is that DNA almost always exists as a double helix, whereas RNA molecules are more variable in shape. RNAs are versatile molecules, and many biologists believe RNA may have preceded DNA as the carrier of genetic information in early forms of life (see Concept 25.1).

CONCEPT CHECK 5.5

1. **DRAW IT** In Figure 5.23a, number all the carbons of the top three nucleotides (use primes), circle the nitrogenous bases, and star the phosphates.
2. **DRAW IT** A region along one DNA strand has this sequence of nitrogenous bases: 5′-TAGGCCT-3′. Write down its complementary strand, labelling the 5′ and 3′ ends.

For suggested answers, see Appendix A.

CONCEPT 5.6

Genomics and proteomics have transformed biological inquiry and applications

Experimental work in the first half of the 20th century established the role of DNA as the carrier of genetic information, passed from generation to generation, that specified the functioning of living cells and organisms. Once the structure of the DNA molecule was described in 1953, and the linear sequence of nucleotide bases was understood to specify the amino acid sequence of proteins, biologists sought to "decode" genes by learning their nucleotide sequences (often called "base sequences").

The first chemical techniques for *DNA sequencing*, or determining the sequence of nucleotides along a DNA strand, one by one, were developed in the 1970s. Researchers began to study gene sequences, gene by gene, and the more they learned, the more questions they had: How was expression of genes regulated? Genes and their protein products clearly interacted with each other, but how? What was the function, if any, of the DNA that is not part of genes? To fully understand the genetic functioning of a living organism, the entire sequence of the full complement of DNA, the organism's *genome*, would be most enlightening. In spite of the apparent impracticality of this idea, in the late 1980s several prominent biologists put forth an audacious proposal to launch a project that would sequence the entire human genome—all 3 billion bases of it! This endeavor began in 1990 and was effectively completed in the early 2000s.

An unplanned but profound side benefit of this project—the Human Genome Project—was the rapid development of faster and less expensive methods of sequencing. This trend has continued: The cost for sequencing 1 million bases in 2001, well over $5,000, has decreased to less than $0.02 in 2017. And a human genome, the first of which took over 10 years to sequence, could be completed at today's pace in day or less **(Figure 5.25)**. The number of genomes that have been fully sequenced has exploded, generating reams of data and prompting development of **bioinformatics**, the use of computer software and other computational tools that can handle and analyse these large data sets.

The reverberations of these developments have transformed the study of biology and related fields. Biologists often look at problems by analysing large sets of genes or even comparing whole genomes of different species, an approach called **genomics**. A similar analysis of large sets of proteins, including their sequences, is called **proteomics**. (Protein sequences can be determined either by using biochemical techniques or by translating the DNA sequences that code for them.) These approaches permeate all fields of biology, some examples of which are shown in **Figure 5.26**.

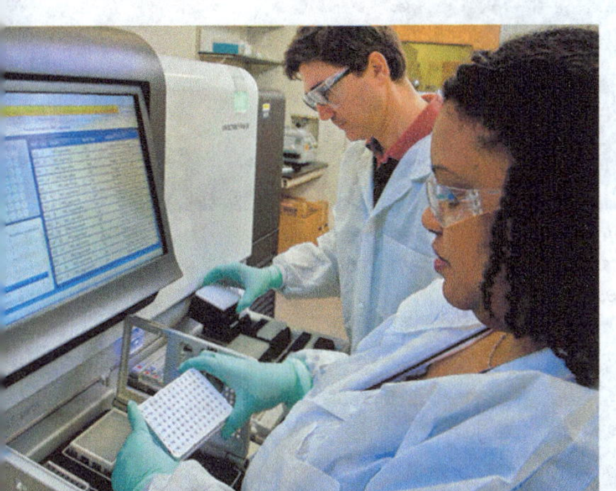

◀**Figure 5.25 Automatic DNA-sequencing machines and abundant computing power enable rapid sequencing of genes and genomes.**

Perhaps the most significant impact of genomics and proteomics on the field of biology as a whole has been their contributions to our understanding of evolution. In addition to confirming evidence for evolution from the study of fossils and characteristics of currently existing species, genomics has helped us tease out relationships among different groups of organisms that had not been resolved by previous types of evidence, and thus infer evolutionary history.

DNA and Proteins as Tape Measures of Evolution

EVOLUTION We are accustomed to thinking of shared traits, such as hair and milk production in mammals, as evidence of shared ancestry. Because DNA carries heritable information in the form of genes, sequences of genes and their protein products document the hereditary background of an organism. The linear sequences of nucleotides in DNA molecules are passed from parents to offspring; these sequences determine the amino acid sequences of proteins. As a result, siblings have greater similarity in their DNA and proteins than do unrelated individuals of the same species.

Given our evolutionary view of life, we can extend this concept of "molecular genealogy" to relationships between species: We would expect two species that appear to be closely related based on anatomical evidence (and possibly fossil evidence) to also share a greater proportion of their DNA and protein sequences than do less closely related species. In fact, that is the case. An example is the comparison of the β polypeptide chain of human haemoglobin with the corresponding haemoglobin polypeptide in other vertebrates. In this chain of 146 amino acids, humans and gorillas differ in just 1 amino acid, while humans and frogs, more distantly related, differ in 67 amino acids. In the **Scientific Skills Exercise**, you can apply this sort of reasoning to additional species. And this conclusion holds true as well when comparing whole genomes: The human genome is 95–98% identical to that of the chimpanzee, but only roughly 85% identical to that of the mouse, a more distant evolutionary relative. Molecular biology has added a new tape measure to the toolkit biologists use to assess evolutionary kinship.

Comparing genomic sequences has practical applications as well. In the **Problem-Solving Exercise**, you can see how this type of genomic analysis can help you detect consumer fraud.

CONCEPT CHECK 5.6

1. How would sequencing the entire genome of an organism help scientists to understand how that organism functioned?
2. Given the function of DNA, why would you expect two species with very similar traits to also have very similar genomes?

For suggested answers, see Appendix A.

▼ **Figure 5.26**

MAKE CONNECTIONS

Contributions of Genomics and Proteomics to Biology

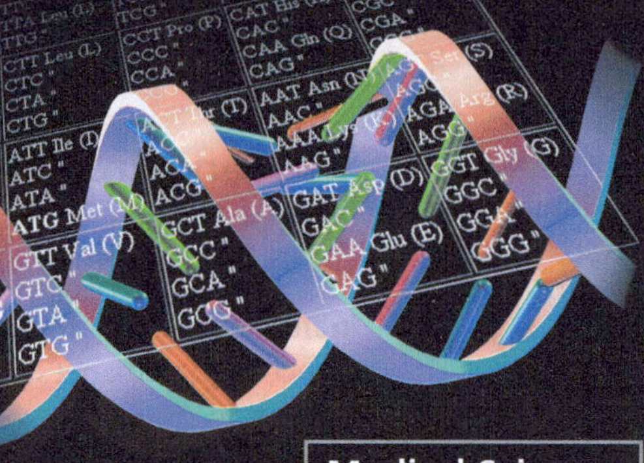

Nucleotide sequencing and the analysis of large sets of genes and proteins can be done rapidly and inexpensively due to advances in technology and information processing. Taken together, genomics and proteomics have advanced our understanding of biology across many different fields.

Paleontology

New DNA sequencing techniques have allowed decoding of minute quantities of DNA found in ancient tissues from our extinct relatives, the Neanderthals (*Homo neanderthalensis*). Sequencing the Neanderthal genome has informed our understanding of their physical appearance (as in this reconstruction), as well as their relationship with modern humans. (See Figures 34.51 and 34.52.)

Medical Science

Identifying the genetic basis for human diseases like cancer helps researchers focus their search for potential future treatments. Currently, sequencing the sets of genes expressed in an individual's tumour can allow a morea targeted approach to treating the cancer, a type of "personalised medicine." (See Figures 12.20 and 18.27.)

Evolution

A major aim of evolutionary biology is to understand the relationships among species, both living and extinct. For example, genome sequence comparisons have identified the hippopotamus as the land mammal sharing the most recent common ancestor with whales. (See Figure 22.20.)

Hippopotamus

Short-finned pilot whale

Species Interactions

Most plant species exist in a mutually beneficial partnership with fungi (right) and bacteria associated with the plants' roots; these interactions improve plant growth. Genome sequencing and analysis of gene expression have allowed characterisation of plant-associated communities. Such studies will help advance our understanding of such interactions and may improve agricultural practices. (See the Chapter 31 Scientific Skills Exercise and Figure 37.10.)

Conservation Biology

The tools of molecular genetics and genomics are increasingly used by forensic ecologists to identify which species of animals and plants are killed illegally. In one case, genomic sequences of DNA from illegal shipments of elephant tusks were used to track down poachers and pinpoint the territory where they were operating. (See Figure 56.8.)

MAKE CONNECTIONS

Considering the examples provided here, describe how the approaches of genomics and proteomics help us to address a variety of biological questions.

Scientific Skills Exercise

Analysing Polypeptide Sequence Data

▶ Human

▶ Rhesus monkey

▶ Gibbon

Are Rhesus Monkeys or Gibbons More Closely Related to Humans? In this exercise, you will look at amino acid sequence data for the β polypeptide chain of haemoglobin, often called β-globin. You will then interpret the data to hypothesise whether the monkey or the gibbon is more closely related to humans.

How Such Experiments Are Done Researchers can isolate the polypeptide of interest from an organism and then determine the amino acid sequence. More frequently, the DNA of the relevant gene is sequenced, and the amino acid sequence of the polypeptide is deduced from the DNA sequence of its gene.

Data from the Experiments In the data below, the letters give the sequence of the 146 amino acids in β-globin from humans, rhesus monkeys, and gibbons. Because a complete sequence would not fit on one line here, the sequences are divided into three segments: amino acids 1–50, 51–100, and 101–146. The sequences for the three different species are aligned so that you can compare them easily. For example, you can see that for all three species, the first amino acid is V (valine) and the 146th amino acid is H (histidine).

Species	Alignment of Amino Acid Sequences of β-globin					
Human	1	VHLTPEEKSA	VTALWGKVNV	DEVGGEALGR	LLVVYPWTQR	FFESFGDLST
Monkey	1	VHLTPEEKNA	VTTLWGKVNV	DEVGGEALGR	LLLVYPWTQR	FFESFGDLSS
Gibbon	1	VHLTPEEKSA	VTALWGKVNV	DEVGGEALGR	LLVVYPWTQR	FFESFGDLST
Human	51	PDAVMGNPKV	KAHGKKVLGA	FSDGLAHLDN	LKGTFATLSE	LHCDKLHVDP
Monkey	51	PDAVMGNPKV	KAHGKKVLGA	FSDGLNHLDN	LKGTFAQLSE	LHCDKLHVDP
Gibbon	51	PDAVMGNPKV	KAHGKKVLGA	FSDGLAHLDN	LKGTFAQLSE	LHCDKLHVDP
Human	101	ENFRLLGNVL	VCVLAHHFGK	EFTPPVQAAY	QKVVAGVANA	LAHKYH
Monkey	101	ENFRLLGNVL	VCVLAHHFGK	EFTPPVQAAY	QKVVAGVANA	LAHKYH
Gibbon	101	ENFRLLGNVL	VCVLAHHFGK	EFTPPVQAAY	QKVVAGVANA	LAHKYH

Data from Human: http://www.ncbi.nlm.nih.gov/protein/AAA21113.1; rhesus monkey: http://www.ncbi.nlm.nih.gov/protein/122634; gibbon: http://www.ncbi.nlm.nih.gov/protein/122616

INTERPRET THE DATA

1. Scan the monkey and gibbon sequences, letter by letter, circling any amino acids that do not match the human sequence. **(a)** How many amino acids differ between the monkey and the human sequences? **(b)** Between the gibbon and human?
2. For each nonhuman species, what percent of its amino acids are identical to the human sequence of β-globin?
3. Based on these data alone, state a hypothesis for which of these two species is more closely related to humans. What is your reasoning?
4. What other evidence could you use to support your hypothesis?

PROBLEM-SOLVING EXERCISE

Are you a victim of fish fraud?

When buying salmon, perhaps you prefer the more expensive wild-caught Pacific salmon (*Oncorhynchus* species) over farmed Atlantic salmon (*Salmo salar*). But studies reveal that about 40% of the time, you aren't getting the fish you paid for!

In this exercise, you will investigate whether a piece of salmon has been fraudulently labelled.

Your Approach The principle guiding your investigation is that DNA sequences from within a species or from closely related species are more similar to each other than are sequences from more distantly related species.

Your Data You've been sold a piece of salmon labelled as coho salmon (*Oncorhynchus kisutch*). To see whether your fish was labelled correctly, you will compare a short DNA sequence from your sample with known sequences from the same gene for three salmon species. The sequences are as follows:

	Sample labelled as *O. kisutch* (coho salmon)	5 ′-CGGCACCGCCCTAAGTCTCT-3 ′
Known sequences	*O. kisutch* (coho salmon)	5 ′-AGGCACCGCCCTAAGTCTAC-3 ′
	O. keta (chum salmon)	5 ′-AGGCACCGCCCTGAGCCTAC-3 ′
	Salmo salar (Atlantic salmon)	5 ′-CGGCACCGCCCTAAGTCTCT-3 ′

Data from the International Barcode of Life.

Your Analysis

1. Circle any bases in the known sequences that do not match the sequence from your fish sample.
2. How many bases differ between **(a)** *O. kisutch* and your fish sample? **(b)** *O. keta* and the sample? **(c)** *S. salar* and the sample?
3. For each known sequence, what percentage of its bases are identical to your sample?
4. Based on these data alone, state a hypothesis for the species identity of your sample. What is your reasoning?

5 Chapter Review

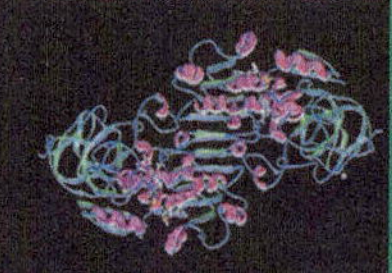

SUMMARY OF KEY CONCEPTS

CONCEPT 5.1

Macromolecules are polymers, built from monomers *(pp. 69–70)*

- Large carbohydrates (polysaccharides), proteins, and nucleic acids are **polymers,** chains of **monomers**. Components of lipids vary. Many monomers can form larger molecules by **dehydration reactions**, in which water molecules are released. Polymers can disassemble by the reverse process, **hydrolysis**. An immense variety of polymers can be built from a small set of monomers.

? *What is the fundamental basis for the differences between large carbohydrates, proteins, and nucleic acids?*

Large Biological Molecules	Components	Examples	Functions
CONCEPT 5.2 **Carbohydrates serve as fuel and building material** *(pp. 70–74)* ? *Compare starch and cellulose. What role does each play in the human body?*	CH_2OH, H, O, H, H, OH, H, HO, OH, H, OH Monosaccharide monomer	**Monosaccharides:** glucose, fructose **Disaccharides:** lactose, sucrose	Fuel; carbon sources that can be converted to other molecules or combined into polymers
		Polysaccharides: • Cellulose (plants) • Starch (plants) • Glycogen (animals) • Chitin (animals and fungi)	• Cellulose strengthens plant cell walls • Starch stores glucose for energy in plants • Glycogen stores glucose for energy in animals • Chitin strengthens animal exoskeletons and fungal cell walls
CONCEPT 5.3 **Lipids are a diverse group of hydrophobic molecules** *(pp. 74–77)* ? *Why are lipids not considered to be polymers or macromolecules?*	Glycerol — 3 fatty acids	**Triacylglycerols** (fats or oils): glycerol + three fatty acids	Important energy source
	Head with (P) 2 fatty acids	**Phospholipids:** glycerol + phosphate group + two fatty acids	Lipid bilayers of membranes Hydrophilic heads / Hydrophobic tails
	Steroid backbone	**Steroids:** four fused rings with attached chemical groups	• Component of cell membranes (cholesterol) • Signalling molecules that travel through the body (hormones)
CONCEPT 5.4 **Proteins include a diversity of structures, resulting in a wide range of functions** *(pp. 77–85)* ? *Explain the basis for the great diversity of proteins.*	R, H, N, C, C, O, H, OH, H Amino acid monomer (20 types)	• Enzymes • Defensive proteins • Storage proteins • Transport proteins • Hormones • Receptor proteins • Motor proteins • Structural proteins	• Catalyse chemical reactions • Protect against disease • Store amino acids • Transport substances • Coordinate organismal responses • Receive signals from outside cell • Function in cell movement • Provide structural support
CONCEPT 5.5 **Nucleic acids store, transmit, and help express hereditary information** *(pp. 86–88)* ? *What role does complementary base pairing play in nucleic acids?*	Nitrogenous base Phosphate group (P)—CH_2 O Sugar Nucleotide (monomer of a polynucleotide)	**DNA:** • Sugar = deoxyribose • Nitrogenous bases = C, G, A, T • Usually double-stranded	Stores hereditary information
		RNA: • Sugar = ribose • Nitrogenous bases = C, G, A, U • Usually single-stranded	Various functions in gene expression, including carrying instructions from DNA to ribosomes

CONCEPT 5.6

Genomics and proteomics have transformed biological inquiry and applications *(pp. 88–91)*

- Recent technological advances in DNA sequencing have given rise to **genomics**, an approach that analyses large sets of genes or whole genomes, and **proteomics**, a similar approach for large sets of proteins. **Bioinformatics** is the use of computational tools and computer software to analyse these large data sets.
- The more closely two species are related evolutionarily, the more similar their DNA sequences are. DNA sequence data confirm models of evolution based on fossils and anatomical evidence.

? *Given the sequences of a particular gene in fruit flies, fish, mice, and humans, predict the relative similarity of the human sequence to that of each of the other species.*

TEST YOUR UNDERSTANDING

Levels 1-2: Remembering/Understanding

1. Which of the following categories includes all others in the list?
(A) disaccharide
(B) polysaccharide
(C) starch
(D) carbohydrate

2. The enzyme amylase can break glycosidic linkages between glucose monomers only if the monomers are in the α form. Which of the following could amylase break down?
(A) glycogen, starch, and amylopectin
(B) glycogen and cellulose
(C) cellulose and chitin
(D) starch, chitin, and cellulose

3. Which statement about *unsaturated* fats is true?
(A) They are more common in animals than in plants.
(B) They have double bonds in their fatty acid chains.
(C) They generally solidify at room temperature.
(D) They contain more hydrogen than do saturated fats having the same number of carbon atoms.

4. The structural level of a protein *least* affected by a disruption in hydrogen bonding is the
(A) primary level.
(B) secondary level.
(C) tertiary level.
(D) quaternary level.

5. Enzymes that break down DNA catalyse the hydrolysis of the covalent bonds that join nucleotides together. What would happen to DNA molecules treated with these enzymes?
(A) The two strands of the double helix would separate.
(B) The phosphodiester linkages of the polynucleotide backbone would be broken.
(C) The pyrimidines would be separated from the deoxyribose sugars.
(D) All bases would be separated from the deoxyribose sugars.

Levels 3-4: Applying/Analysing

6. The molecular formula for glucose is $C_6H_{12}O_6$. What would be the molecular formula for a polymer made by linking ten glucose molecules together by dehydration reactions?
(A) $C_{60}H_{120}O_{60}$
(B) $C_{60}H_{102}O_{51}$
(C) $C_{60}H_{100}O_{50}$
(D) $C_{60}H_{111}O_{51}$

7. Which of the following pairs of base sequences could form a short stretch of a normal double helix of DNA?
(A) 5′-AGCT-3′ with 5′-TCGA-3′
(B) 5′-GCGC-3′ with 5′-TATA-3′
(C) 5′-ATGC-3′ with 5′-GCAT-3′
(D) All of these pairs are correct.

8. Construct a table that organises the following terms and label the columns and rows.

Monosaccharides	Polypeptides	Phosphodiester linkages
Fatty acids	Triacylglycerols	Peptide bonds
Amino acids	Polynucleotides	Glycosidic linkages
Nucleotides	Polysaccharides	Ester linkages

9. DRAW IT Copy the polynucleotide strand in Figure 5.23a and label the bases G, T, C, and T, starting from the 5′ end. Assuming this is a DNA polynucleotide, now draw the complementary strand, using the same symbols for phosphates (circles), sugars (pentagons), and bases. Label the bases. Draw arrows showing the 5′ → 3′ direction of each strand. Use the arrows to make sure the second strand is antiparallel to the first. Hint: After you draw the first strand vertically, turn the paper upside down; it is easier to draw the second strand from the 5′ towards the 3′ direction as you go from top to bottom.

Levels 5-6: Evaluating/Creating

10. EVOLUTION CONNECTION Comparisons of amino acid sequences can shed light on the evolutionary divergence of related species. If you were comparing two living species, would you expect all proteins to show the same degree of divergence? Why or why not? Justify your answer.

11. SCIENTIFIC INQUIRY Suppose you are a research assistant in a lab studying DNA-binding proteins. You have been given the amino acid sequences of all the proteins encoded by the genome of a certain species and have been asked to find candidate proteins that could bind DNA. What type of amino acids would you expect to see in the DNA-binding regions of such proteins? Explain your thinking.

12. WRITE ABOUT A THEME: ORGANISATION Proteins, which have diverse functions in a cell, are all polymers of the same kinds of monomers—amino acids. Write a short essay (100–150 words) that discusses how the structure of amino acids allows this one type of polymer to perform so many functions.

13. SYNTHESISE YOUR KNOWLEDGE

Given that the function of egg yolk is to nourish and support the developing chick, explain why egg yolks are so high in fat, protein, and cholesterol.

For selected answers, see Appendix A.

Dr Diana Bautista is an Associate Professor of Molecular and Cell Biology at the University of California, Berkeley, in the United States, and a Howard Hughes Medical Institute Faculty Scholar. Recipient of the Gill Transformative Investigator Award and the Society for Neuroscience Young Investigator Award, she received a BS in Biology from the University of Oregon and a PhD in Neuroscience from Stanford University. Dr Bautista grew up in Chicago, the first in her family to attend university and to pursue a career in science. She and her laboratory members investigate the cell-signalling pathways associated with pain and itch and how their misregulation can result in chronic pain and chronic itch. Her research is interdisciplinary: Dr Bautista collaborates with computational biologists, immunologists, and physiologists. She believes that science benefits most from a diverse group of colleagues, each bringing a different perspective.

AN INTERVIEW WITH

Diana M. Bautista

What got you interested in biology, and in neuroscience in particular?

I always liked science as a kid, but it was never a big focus in my life. I went to college interested in art, then took some time off because I was a terrible artist. I ended up working with an environmental group helping low-income communities oppose a hazardous waste incinerator. That got me excited about science because it was at the interface of the environment and community empowerment, so I decided to return to college and major in biology. Once I returned to school, at the University of Oregon, I found out you could get a work-study job in a lab. I ended up working in Peter O'Day's lab studying vision in fruit flies by recording electrical signals in response to light. For me, shining light on a fly eye and recording an electrical signal from their nervous system in real time was just the coolest thing I had ever seen. The lab was a really inclusive, warm environment, and Peter was an amazing mentor—if it weren't for him, I wouldn't be in science. I went to graduate school at Stanford in neuroscience, where I worked with Dr Richard Lewis on a newly discovered calcium channel and how calcium signalling pathways drive cellular behaviours.

▼ Dr Bautista studies cell responses to stimuli using equipment that measures electrical signals.

Why did you focus your research on pain?

After my PhD I wanted to continue working on cell signalling but from a more organismal viewpoint, so I did my postdoctoral research with Dr David Julius at the University of California, San Francisco. His lab had recently identified the cellular protein (TRPV1) that is responsive to capsaicin—the molecule that gives chili peppers their "hotness"—and showed that this protein is also activated by painful heat. That was really exciting to me because it involved thinking about proteins involved in cell signaling, but in the bigger context of pain. Proteins like TRPV1 serve important protective functions—for example, by stopping us from grabbing a hot frying pan that could cause a burn. One of my projects was to identify the wasabi receptor, a protein on the surface of nerve cells that plays a key role in inflammatory pain hypersensitivity. Once you activate this receptor protein—with irritants such as wasabi, mustard oil, or inflammatory mediators produced in the body by tissue injury such as a burn—it activates and makes pain cells more sensitive: Warmth triggers sensations of burning heat, and light touch triggers pain. Normally, pain hypersensitivity dies down after a few days as the tissue heals, but in some patients, it becomes a chronic debilitating disease.

"It doesn't matter what you did before or where you come from—anybody can be a scientist."

What does your lab work on now?

We are interested in the sense of touch. Touch-sensitive neurons innervate the skin and mediate gentle touch sensations and texture preferences: Do you like wool sweaters, or do you find them itchy? Touch-sensitive neurons also innervate our musculature, allowing us to detect limb position and mediating coordinated movements, such as our ability to text or to walk in a straight line without staring at our feet. In someone suffering from chronic pain, a gentle touch feels incredibly painful, while a chronic itch sufferer experiences light touch as itchy and pain as pleasure. When you scratch an itch, you're scratching really hard—if you didn't have an itch it would hurt, but when you do have an itch it feels good. We have no idea how our body processes the same type of stimulus under different injury or disease conditions, and that's what really excited me when I started my own lab.

What is your advice to an undergraduate considering a career in biology?

I'm an advisor for undergraduates, so I talk to a lot of students. I like to tell them about my path and let them know that it doesn't matter what you did before or where you come from—anybody can be a scientist. I think it's sometimes tough early on in biology, because at first you're taught facts, but as you progress, biology is about figuring out the unknown. And if you can get into a lab early on, then you really see that side of science—that it's about learning what we know, and then going beyond that, doing experiments to identify new mechanisms.

6 A Tour of the Cell

KEY CONCEPTS

Study Tip

Draw animal and plant cells: Draw an outline of an animal cell and add structures, labels, and functions. Draw a plant cell labelled with structures unique to plant cells.

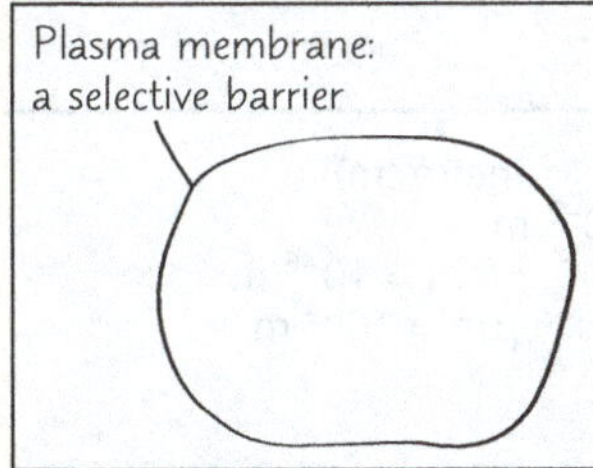

Go to Mastering Biology

to access Dynamic Study Modules for revision, 3D BioFlix® animations and high-quality videos, and your interactive Pearson eText.

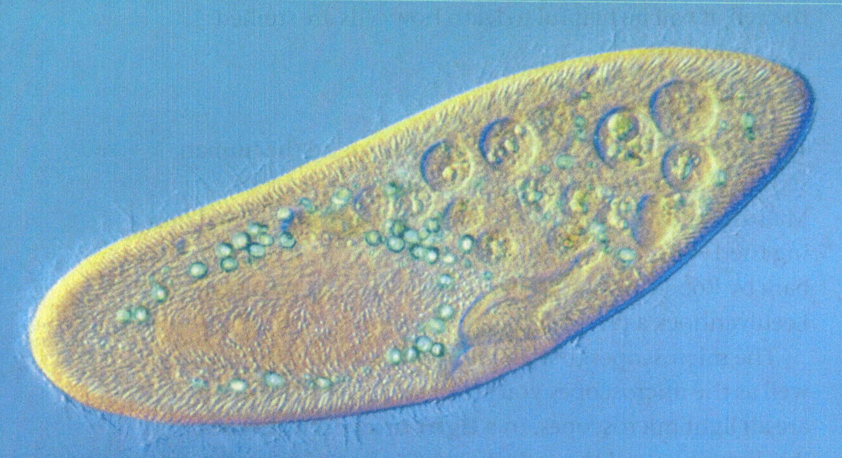

Figure 6.1 The cell is an organism's basic unit of structure and function. Many forms of life exist as single-celled organisms, such as the *Paramecium* shown here. Larger, more complex organisms, including plants and animals, are multicellular. In this chapter, we focus mainly on eukaryotic cells—cells with a nucleus.

How does the internal organisation of eukaryotic cells allow them to perform the functions of life?

Internal membranes divide a cell, such as this plant cell, into compartments where specific chemical reactions occur.

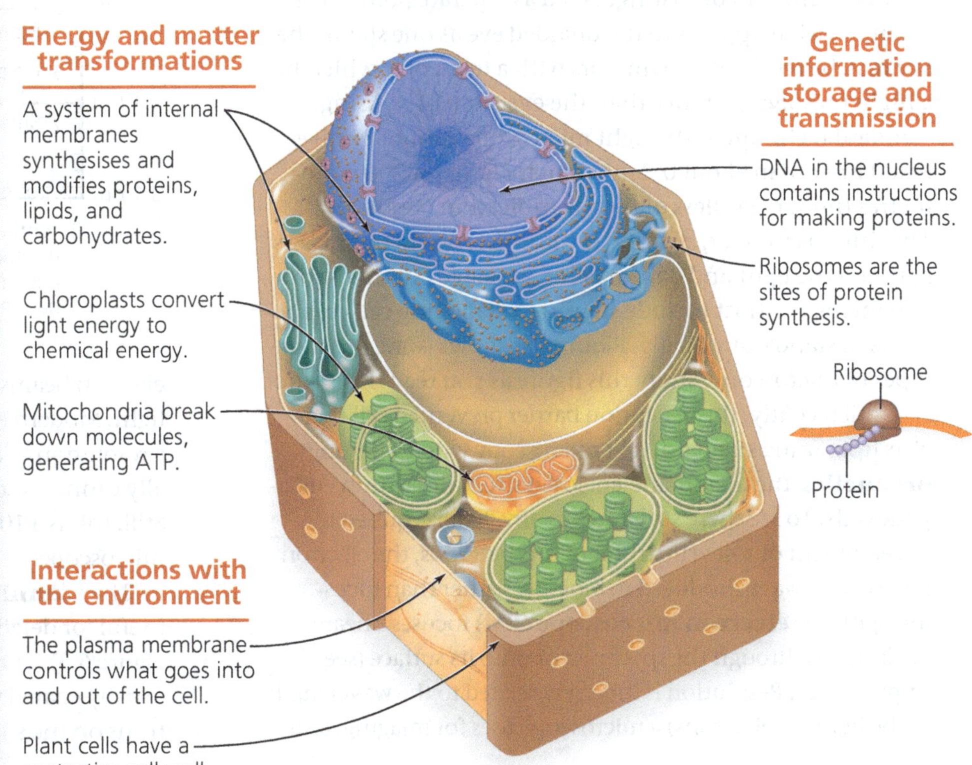

CONCEPT 6.1

Biologists use microscopes and biochemistry to study cells

How can cell biologists investigate the inner workings of a cell, usually too small to be seen by the unaided eye? Before we tour the cell, it will be helpful to learn how cells are studied.

Microscopy

The development of instruments that extend the human senses allowed the discovery and early study of cells. Microscopes were invented in 1590 and further refined during the 1600s. Cell walls were first seen on dead cells of oak bark by Robert Hooke in 1665 and living cells by Antoni van Leeuwenhoek a few years later.

The microscopes first used by Renaissance scientists, as well as the microscopes you are likely to use in the laboratory, are all light microscopes. In a **light microscope (LM)**, visible light is passed through the specimen and then through glass lenses. The lenses refract (bend) the light in such a way that the image of the specimen is magnified as it is projected into the eye or into a camera (see Appendix C).

Three important parameters in microscopy are magnification, resolution, and contrast. *Magnification* is the ratio of an object's image size to its real size. Light microscopes can magnify effectively to about 1,000 times the actual size of the specimen; at greater magnifications, additional details cannot be seen clearly. *Resolution* is a measure of the clarity of the image; it is the minimum distance two points can be separated and still be distinguished as separate points. For example, what appears to the unaided eye as one star in the sky may be resolved as twin stars with a telescope, which has a higher resolving ability than the eye. Similarly, using standard techniques, the light microscope cannot resolve detail finer than about 0.2 micrometre (μm), or 200 nanometres (nm), regardless of the magnification **(Figure 6.2)**. The third parameter, *contrast*, is the difference in brightness between the light and dark areas of an image. Methods for enhancing contrast include staining or labelling cell components to stand out visually. **Figure 6.3** shows some different types of microscopy; study this figure as you read this section.

Until recently, the resolution barrier prevented cell biologists from using standard light microscopy when studying **organelles**, the membrane-enclosed structures within eukaryotic cells. To see these structures in any detail required the development of a new instrument. In the 1950s, the electron microscope was introduced to biology. Rather than focusing light, the **electron microscope (EM)** focuses a beam of electrons through the specimen or onto its surface (see Appendix C). Resolution is inversely related to the wavelength of the light (or electrons) a microscope uses for imaging, and

▼ **Figure 6.2 The size range of cells.** Most cells are between 1 and 100 μm in diameter (yellow region of chart), and their components are even smaller (see Figure 6.32), as are viruses. Notice that the scale along the left side is logarithmic, to accommodate the range of sizes shown. Starting at the top of the scale with 10 m and going down, each reference measurement marks a tenfold decrease in diameter or length.

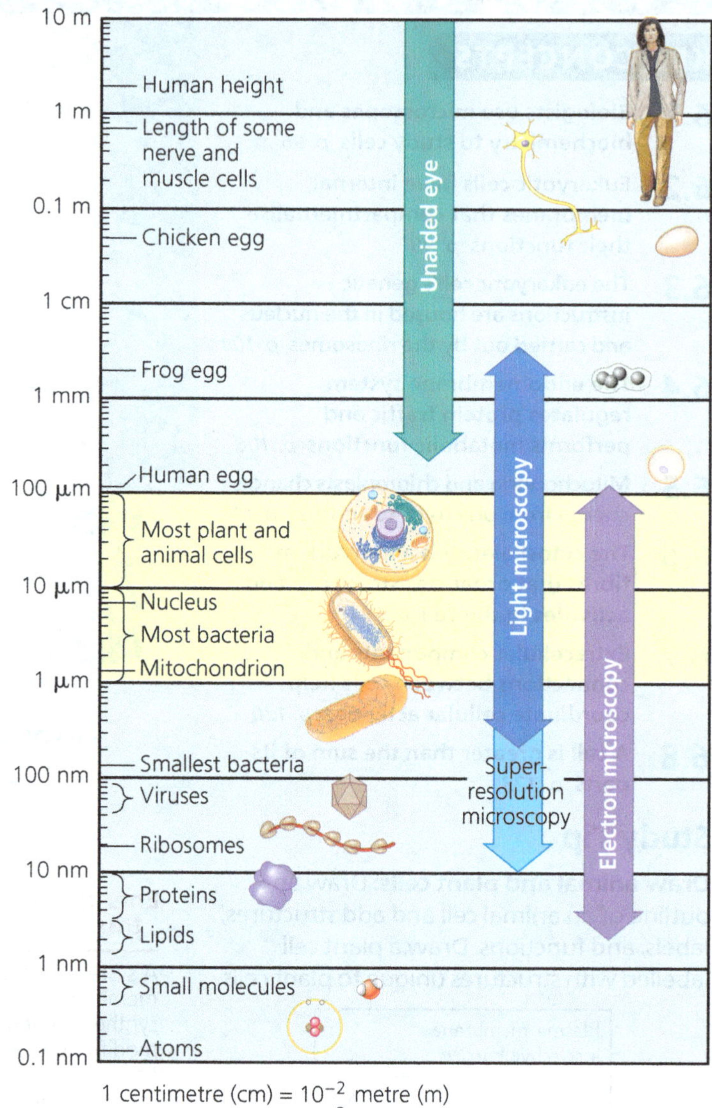

1 centimetre (cm) = 10^{-2} metre (m)
1 millimetre (mm) = 10^{-3} m
1 micrometre (μm) = 10^{-3} mm = 10^{-6} m
1 nanometre (nm) = 10^{-3} μm = 10^{-9} m

electron beams have much shorter wavelengths than visible light. Modern electron microscopes can theoretically achieve a resolution of about 0.002 nm, though in practice they usually cannot resolve structures smaller than about 2 nm across. Still, this is a 100-fold improvement over the standard light microscope.

The **scanning electron microscope (SEM)** is especially useful for detailed study of the topography of a specimen (see Figure 6.3). The electron beam scans the surface of the sample, usually coated with a thin film of gold. The beam excites electrons on the surface, and these secondary electrons are detected

▼ Figure 6.3 Exploring Microscopy

Light Microscopy (LM)

Brightfield. Light passes directly through the specimen. Unstained (left), the image has little contrast. Staining with dyes (right) enhances contrast. Most stains require cells to be preserved, which kills them.

Phase-contrast. Variations in density within the specimen are amplified to enhance contrast in unstained cells; this is especially useful for examining living, unstained cells.

Differential interference contrast (Nomarski). As in phase-contrast microscopy, optical modifications are used to exaggerate differences in density; the image appears almost 3-D.

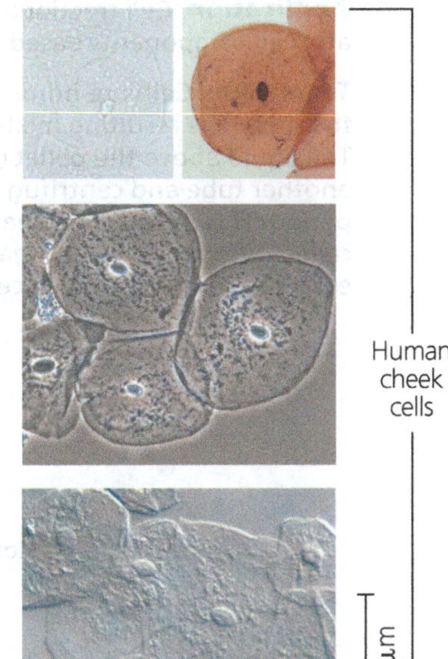

Fluorescence. Locations of specific molecules are revealed by labelling the molecules with fluorescent dyes or antibodies, which absorb ultraviolet radiation and emit visible light. In this fluorescently labelled uterine cell, DNA is blue, organelles called mitochondria are orange, and part of the cell's "skeleton" (called the *cytoskeleton*) is green.

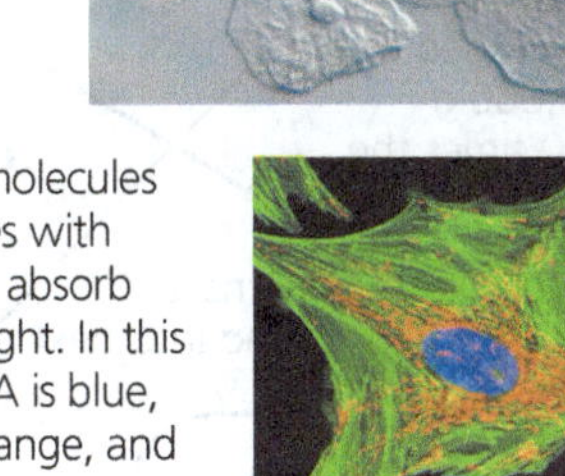

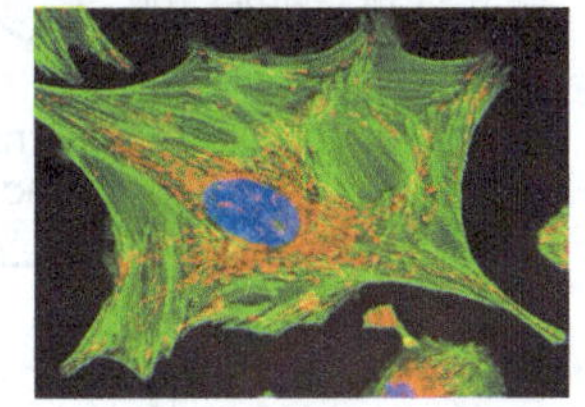

Confocal. This image shows two types of fluorescence micrographs: confocal (top) and standard (bottom). (Nerve cells are green, support cells orange, areas of overlap yellow.) In confocal microscopy, a laser is used to create a single plane of fluorescence; out-of-focus light from other planes is eliminated. By capturing sharp images at many different planes, a 3-D reconstruction can be created. A standard fluorescence micrograph is blurry because out-of-focus light is not excluded.

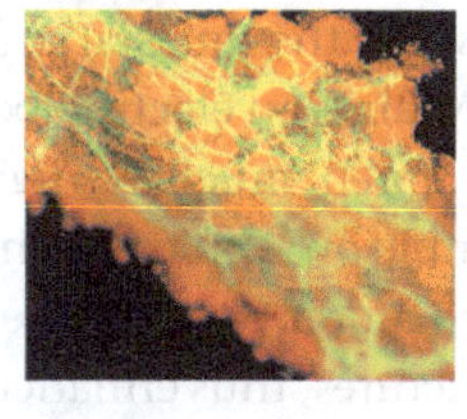

Deconvolution. The top of this image of a white blood cell was reconstructed from many blurry fluorescence images at different planes, each processed using deconvolution software. This process digitally removes out-of-focus light and reassigns it to its source, creating a much sharper 3-D image. The bottom is a compilation of standard fluorescent micrographs through the same cell.

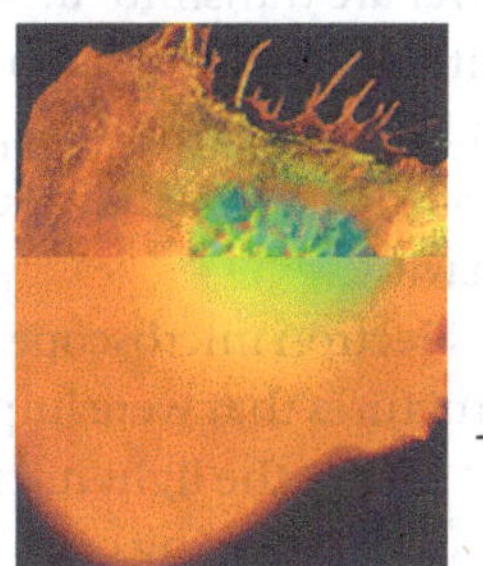

Super-resolution. To make this super-resolution image of a cow aorta cell (top), individual fluorescent molecules were excited by UV light and their position recorded. (DNA is blue, mitochondria red, and part of the cytoskeleton green.) Combining information from many molecules in different places "breaks" the resolution limit, resulting in the sharp image on top. The size of each dot is well below the 200-nm resolution of a standard light microscope, as seen in the confocal image (bottom) of the same cell.

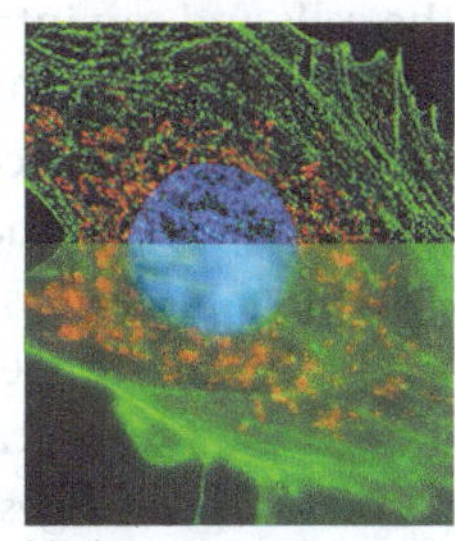

Electron Microscopy (EM)

Scanning electron microscopy (SEM). Micrographs taken with a scanning electron microscope show a 3-D image of the surface of a specimen. This SEM shows the surface of a cell from a trachea (windpipe) covered with cell projections called cilia. Electron micrographs are black and white but are often artificially colourised to highlight particular structures, as has been done with all three electron micrographs shown here.

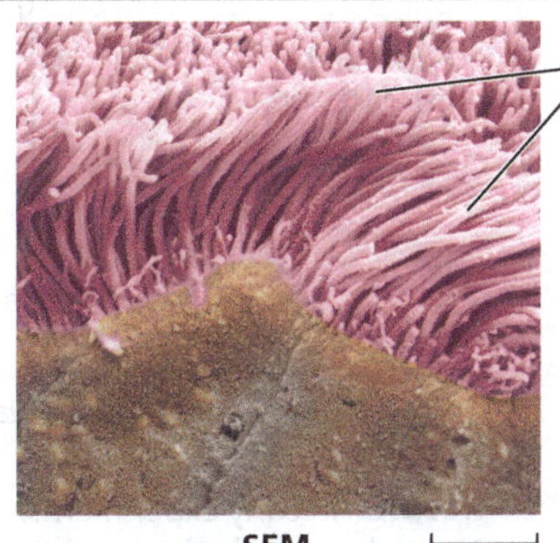

Transmission electron microscopy (TEM). A transmission electron microscope profiles a thin section of a specimen. This TEM shows a section through a tracheal cell, revealing its internal structure. In preparing the specimen, some cilia were cut along their lengths, creating longitudinal sections, while other cilia were cut straight across, creating cross sections.

Cryo-electron microscopy (cryo-EM). Specimens of tissue or aqueous solutions of proteins are frozen rapidly at temperatures less than −160°C, locking the molecules into a rigid state. A beam of electrons is passed through the sample to visualise the molecules by electron microscopy, and software is used to merge a series of such micrographs, creating a 3-D image like the one below.

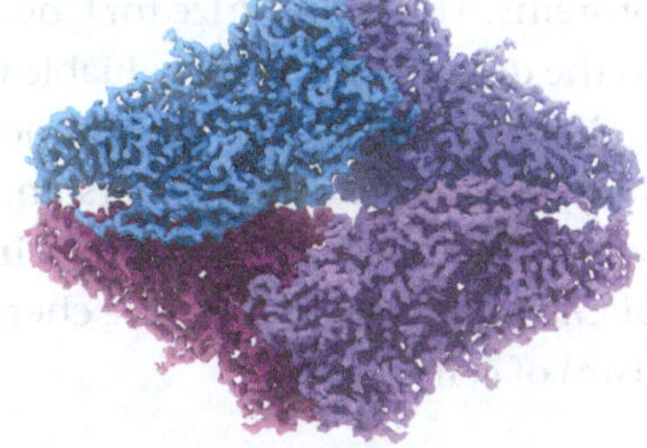

Computer-generated image of the bacterial enzyme β-galactosidase, which breaks down lactose. This image was compiled from more than 90,000 cryo-EM images.

Abbreviations used in figure legends in this text:
LM = Light Micrograph
SEM = Scanning Electron Micrograph
TEM = Transmission Electron Micrograph

VISUAL SKILLS *When the tissue was sliced, what was the orientation of the cilia in the lower portion of the TEM? The upper portion? Explain how the orientation of the cilia determined the type of sections we see.*

by a device that translates the pattern of electrons into an electronic signal sent to a video screen. The result is an image of the specimen's surface that appears three-dimensional.

The **transmission electron microscope (TEM)** is used to study the internal structure of cells (see Figure 6.3). The TEM aims an electron beam through a very thin section of the specimen, much as a light microscope aims light through a sample on a slide. For the TEM, the specimen has been stained with atoms of heavy metals, which attach to certain cellular structures, thus enhancing the electron density of some parts of the cell more than others. The electrons passing through the specimen are scattered more in the denser regions, so fewer are transmitted. The image displays the pattern of transmitted electrons. Instead of using glass lenses, both the SEM and TEM use electromagnets as lenses to bend the paths of the electrons, ultimately focusing the image onto a monitor for viewing.

Electron microscopes have revealed many subcellular structures that were impossible to resolve with the light microscope. But the light microscope offers advantages, especially in studying living cells. A disadvantage of electron microscopy is that the methods customarily used to prepare the specimen kill the cells and can introduce artifacts, structural features seen in micrographs that do not exist in the living cell.

In the past several decades, light microscopy has been revitalised by major technical advances (see Figure 6.3). Labelling individual cellular molecules or structures with fluorescent markers has made it possible to see such structures with increasing detail. In addition, both confocal and deconvolution microscopy have produced sharper images of three-dimensional tissues and cells. Finally, a group of new techniques and labelling molecules developed in recent years, called *super-resolution microscopy*, has allowed researchers to "break" the resolution barrier and distinguish subcellular structures as small as 10–20 nm across.

A recently developed new type of TEM called cryo-electron microscopy (cryo-EM) (see Figure 6.3) allows specimens to be preserved at extremely low temperatures. This avoids the use of preservatives, allowing visualisation of structures in their cellular environment. This method is increasingly used to complement X-ray crystallography in revealing protein complexes and subcellular structures like ribosomes, described later. Cryo-EM has even been used to resolve some individual proteins. The Nobel Prize for Chemistry was awarded in 2017 to the developers of this valuable technique.

Microscopes are the most important tools of *cytology*, the study of cell structure. Understanding the function of each structure, however, required the integration of cytology and *biochemistry*, the study of the chemical processes (metabolism) of cells.

Cell Fractionation

A useful technique for studying cell structure and function is **cell fractionation (Figure 6.4)**, which takes cells

▼ Figure 6.4 Research Method

Cell Fractionation

Application Cell fractionation is used to separate (fractionate) cell components based on size and density.

Technique Cells are homogenised in a blender to break them up. The resulting mixture (*homogenate*) is centrifuged. The liquid above the pellet (*supernatant*) is poured into another tube and centrifuged at a higher speed for a longer period. This process is repeated several times. This process, called *differential centrifugation*, results in a series of pellets, each containing different cell components.

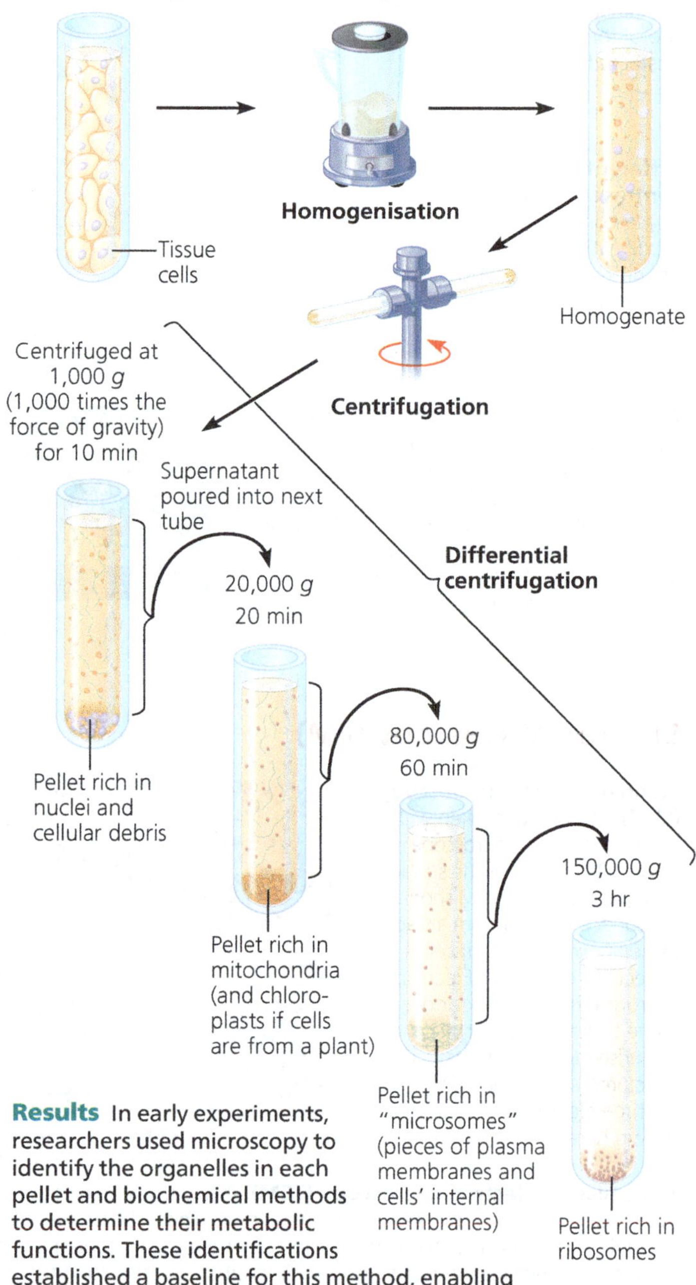

Results In early experiments, researchers used microscopy to identify the organelles in each pellet and biochemical methods to determine their metabolic functions. These identifications established a baseline for this method, enabling today's researchers to know which cell fraction they should collect in order to isolate and study particular organelles.

MAKE CONNECTIONS *If you wanted to study the process of translation of proteins from mRNA, which part of which fraction would you use? (See Figure 5.22.)*

apart and separates major organelles and other subcellular structures from one another. The piece of equipment that is used for this task is the centrifuge, which spins test tubes holding mixtures of disrupted cells at a series of increasing speeds, a process called *differential centrifugation*. At each speed, the resulting force causes a subset of the cell components to settle to the bottom of the tube, forming a pellet. At lower speeds, the pellet consists of larger components, and higher speeds result in a pellet with smaller components.

Cell fractionation enables researchers to prepare specific cell components in bulk and identify their functions, a task not usually possible with intact cells. For example, on one of the cell fractions, biochemical tests showed the presence of enzymes involved in cellular respiration, while electron microscopy revealed large numbers of the organelles called mitochondria. Together, these data helped biologists determine that mitochondria are the sites of cellular respiration. Biochemistry and cytology thus complement each other in correlating cell function with structure.

CONCEPT CHECK 6.1

1. How do stains used for light microscopy compare with those used for electron microscopy?
2. **WHAT IF?** Which type of microscope would you use to study (a) the changes in shape of a living white blood cell? (b) the details of surface texture of a hair?

For suggested answers, see Appendix A.

CONCEPT 6.2

Eukaryotic cells have internal membranes that compartmentalise their functions

Cells—the basic structural and functional units of every organism—are of two distinct types: prokaryotic and eukaryotic. Organisms of the domains Bacteria and Archaea consist of prokaryotic cells. Organisms of the domain Eukarya—protists, fungi, animals, and plants—all consist of eukaryotic cells. ("Protist" is an informal term referring to a diverse group of mostly unicellular eukaryotes.)

Comparing Prokaryotic and Eukaryotic Cells

All cells share certain basic features: They are all bounded by a selective barrier, called the *plasma membrane* (or the cell membrane). Inside all cells is a semifluid, jellylike substance called **cytosol**, in which subcellular components are suspended. All cells contain *chromosomes*, which carry genes in the form of DNA. And all cells have *ribosomes*, tiny complexes that make proteins according to instructions from the genes.

A major difference between prokaryotic and eukaryotic cells is the location of their DNA. In a **eukaryotic cell**, most of the DNA is in an organelle called the *nucleus*, which is bounded by a double membrane (see Figure 6.8). In a **prokaryotic cell**, the DNA is concentrated in a region that is not membrane-enclosed, called the **nucleoid** (**Figure 6.5**).

▼ **Figure 6.5 A prokaryotic cell.** Lacking a true nucleus and the other membrane-enclosed organelles of the eukaryotic cell, the prokaryotic cell appears much simpler in internal structure. Prokaryotes include bacteria and archaea; the general cell structure of these two domains is quite similar.

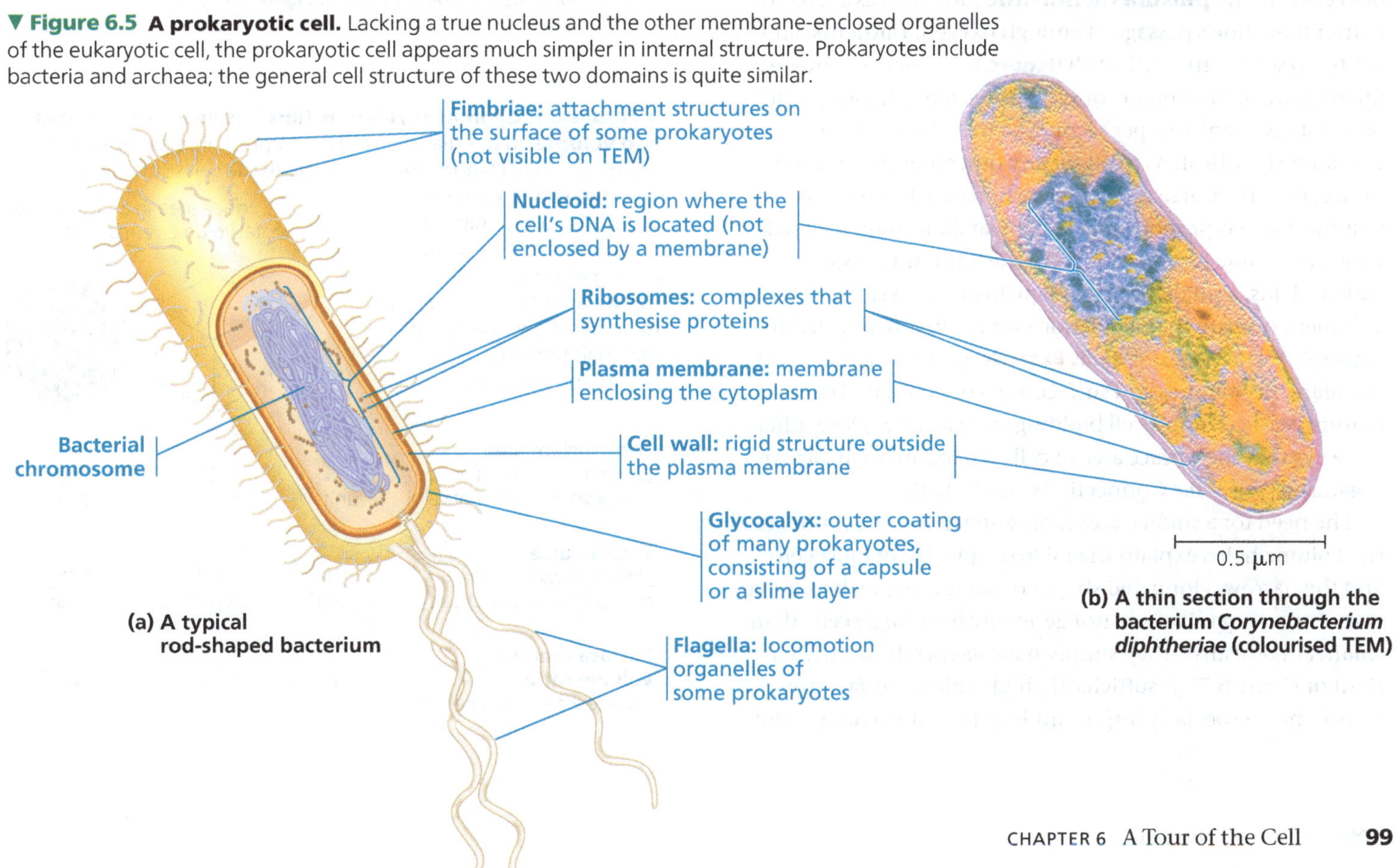

(a) A typical rod-shaped bacterium

(b) A thin section through the bacterium *Corynebacterium diphtheriae* (colourised TEM)

Eukaryotic means "true nucleus" (from the Greek *eu*, true, and *karyon*, kernel, referring to the nucleus), and *prokaryotic* means "before nucleus" (from the Greek *pro*, before), reflecting the earlier evolution of prokaryotic cells.

The interior of either type of cell is called the **cytoplasm**; in eukaryotic cells, this term refers only to the region between the nucleus and the plasma membrane. Within the cytoplasm of a eukaryotic cell, suspended in cytosol, are a variety of organelles of specialised form and function. These membrane-bounded structures are absent in almost all prokaryotic cells, another distinction between prokaryotic and eukaryotic cells. In spite of the absence of organelles, though, the prokaryotic cytoplasm is not a formless soup. For example, some prokaryotes contain regions surrounded by proteins (not membranes), within which specific reactions take place.

Eukaryotic cells are generally much larger than prokaryotic cells (see Figure 6.2). Size is a general feature of cell structure that relates to function. The logistics of carrying out cellular metabolism sets limits on cell size. At the lower limit, the smallest cells known are bacteria called mycoplasmas, which have diameters between 0.1 and 1.0 μm. These are perhaps the smallest packages with enough DNA to program metabolism and enough enzymes and other cellular equipment to carry out the activities necessary for a cell to sustain itself and reproduce. Typical bacteria are 1–5 μm in diameter, about ten times the size of mycoplasmas. Eukaryotic cells are typically 10–100 μm in diameter.

Metabolic requirements also impose theoretical upper limits on the size that is practical for a single cell. At the boundary of every cell, the **plasma membrane** functions as a selective barrier that allows passage of enough oxygen, nutrients, and wastes to service the entire cell **(Figure 6.6)**. For each square micrometre of membrane, only a limited amount of a particular substance can cross per second, so the ratio of surface area to volume is critical. As a cell (or any other object) increases in size, its surface area grows proportionately less than its volume. (Area is proportional to a linear dimension squared, whereas volume is proportional to the linear dimension cubed.) Thus, a smaller cell has a greater ratio of surface area to volume: Compare the calculations for the first two "cells" in **Figure 6.7**. The **Scientific Skills Exercise** gives you a chance to calculate the volumes and surface areas of two actual cells—a mature yeast cell and a cell budding from it. To explore different ways that the surface area of cells is maximised in various organisms, see Make Connections Figure 33.8.

The need for a surface area large enough to accommodate the volume helps explain the microscopic size of most cells and the narrow, elongated shapes of some cells, such as nerve cells. Larger organisms do not generally have *larger* cells than smaller organisms—they simply have *more* cells (see the far right of Figure 6.7). A sufficiently high ratio of surface area to volume is especially important in cells that exchange a lot

▼ Figure 6.6 The plasma membrane. The plasma membrane and the membranes of organelles consist of a double layer (bilayer) of phospholipids with various proteins attached to or embedded in it. The hydrophobic parts of phospholipids and membrane proteins are found in the interior of the membrane, while the hydrophilic parts are in contact with aqueous solutions on either side. Carbohydrate side chains may be attached to proteins or lipids on the outer surface of the plasma membrane.

(a) Colourised TEM of a plasma membrane. The plasma membrane appears as a pair of dark bands separated by a gold band.

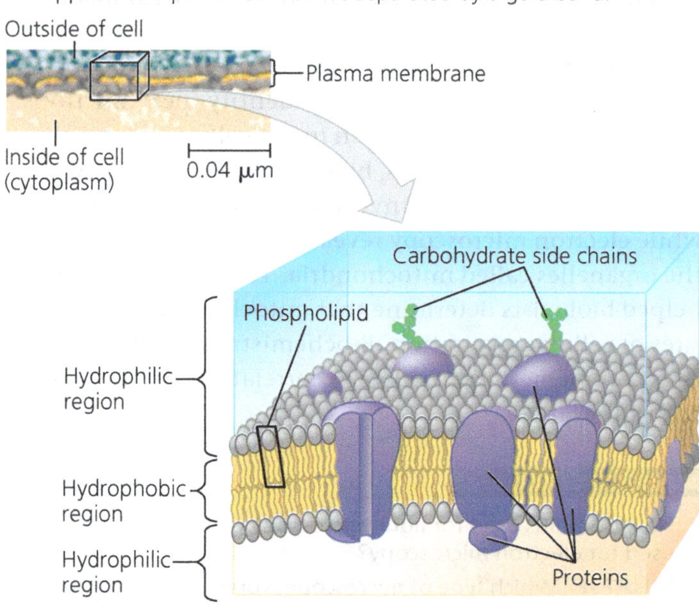

(b) Structure of the plasma membrane

VISUAL SKILLS *What parts of the membrane diagram in (b) correspond to the dark bands in the TEM in (a)? What parts correspond to the gold band? (Review Figure 5.11.)*

▼ Figure 6.7 Geometric relationships between surface area and volume. In this diagram, cells are represented as cubes. Using arbitrary units of length, we can calculate the cell's surface area (in square units, or units^2), volume (in cubic units, or units^3), and ratio of surface area to volume. A high surface area-to-volume ratio facilitates the exchange of materials between a cell and its environment.

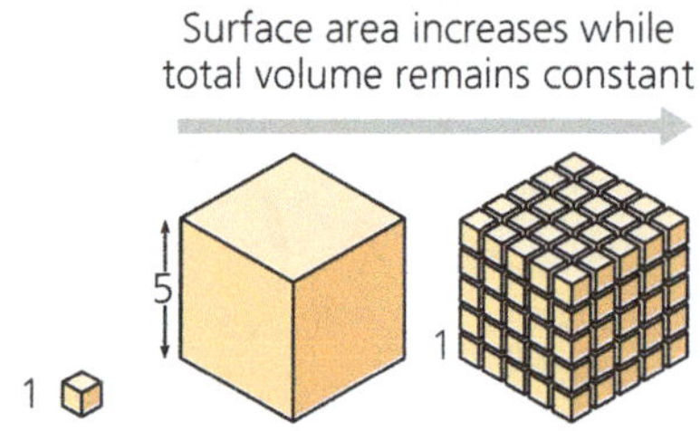

Surface area increases while total volume remains constant

1 5 1

Total surface area [(height × width of 1 side) × 6 sides × number of cells]	6 units^2	150 units^2	750 units^2
Total volume [(height × width × length of 1 cell) × number of cells]	1 unit^3	125 units^3	125 units^3
Surface area-to-volume ratio [surface area ÷ volume]	6	1.2	6

Scientific Skills Exercise

Using a Scale Bar to Calculate Volume and Surface Area of a Cell

How Much New Cytoplasm and Plasma Membrane Are Made by a Growing Yeast Cell? The unicellular yeast *Saccharomyces cerevisiae* divides by budding off a small new cell that then grows to full size (see the yeast cells at the bottom of Figure 6.8). During its growth, the new cell synthesises new cytoplasm, which increases its volume, and new plasma membrane, which increases its surface area. In this exercise, you will use a scale bar to determine the sizes of a mature parent yeast cell and a cell budding from it. You will then calculate the volume and surface area of each cell. You will use your calculations to determine how much cytoplasm and plasma membrane the new cell needs to synthesise to grow to full size.

How the Experiment Was Done Yeast cells were grown under conditions that promoted division by budding. The cells were then viewed with a differential interference contrast light microscope and photographed.

Data from the Experiment This light micrograph shows a budding yeast cell about to be released from the mature parent cell:

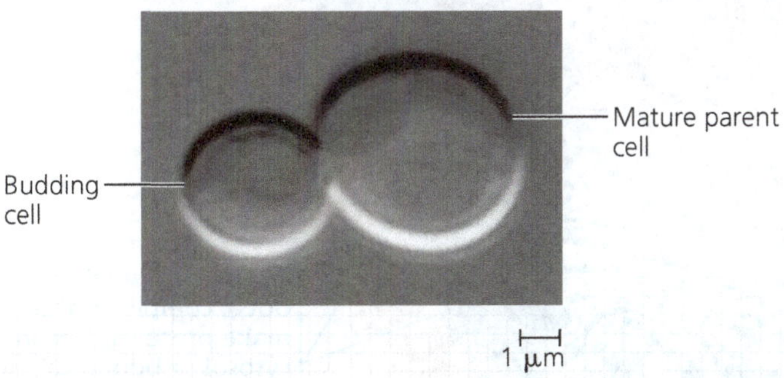

Micrograph from K. Tatchell, using yeast cells grown for experiments described in L. Kozubowski et al., Role of the septin ring in the asymmetric localisation of proteins at the mother-bud neck in *Saccharomyces cerevisiae, Molecular Biology of the Cell* 16:3455–3466 (2005).

INTERPRET THE DATA

1. Examine the micrograph of the yeast cells. The scale bar under the photo is labelled 1 μm. The scale bar works in the same way as a scale on a map, where, for example, 1 centimetre equals 1 kilometre. In this case the bar represents one thousandth of a millimetre. Using the scale bar as a basic unit, determine the diameter of the mature parent cell and the new cell. Start by measuring the scale bar and the diameter of each cell. The units you use are irrelevant, but working in millimetres is convenient. Divide each diameter by the length of the scale bar and then multiply by the scale bar's length value to give you the diameter in micrometres.
2. The shape of a yeast cell can be approximated by a sphere. **(a)** Calculate the volume of each cell using the formula for the volume of a sphere:

$$V = \frac{4}{3}\pi r^3$$

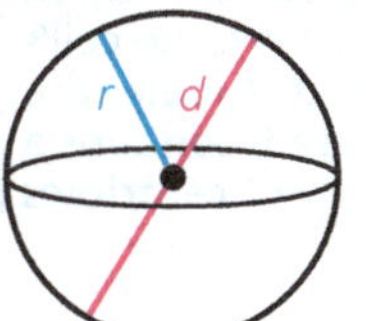

Note that π (the Greek letter pi) is a constant with an approximate value of 3.14, *d* stands for diameter, and *r* stands for radius, which is half the diameter. **(b)** What volume of new cytoplasm will the new cell have to synthesise as it matures? To determine this, calculate the difference between the volume of the full-sized cell and the volume of the new cell.
3. As the new cell grows, its plasma membrane needs to expand to contain the increased volume of the cell. **(a)** Calculate the surface area of each cell using the formula for the surface area of a sphere: $A = 4\pi r^2$. **(b)** How much area of new plasma membrane will the new cell have to synthesise as it matures?
4. When the new cell matures, it will be approximately how many times greater in volume and how many times greater in surface area than its current size?

of material with their surroundings, such as intestinal cells. Such cells may have many long, thin projections from their surface called *microvilli*, which increase surface area without an appreciable increase in volume.

The evolutionary relationships between prokaryotic and eukaryotic cells will be discussed later in this chapter, and prokaryotic cells will be described in detail elsewhere (see the chapter, "Bacteria and Archaea"). Most of the discussion of cell structure that follows in this chapter applies to eukaryotic cells.

A Panoramic View of the Eukaryotic Cell

In addition to the plasma membrane at its outer surface, a eukaryotic cell has extensive, elaborately arranged internal membranes that divide the cell into compartments—the organelles mentioned earlier. The cell's compartments provide different local environments that support specific metabolic functions, so incompatible processes can occur simultaneously in a single cell. The plasma membrane and organelle membranes also participate directly in the cell's metabolism because many enzymes are built right into the membranes.

The basic fabric of most biological membranes is a double layer of phospholipids and other lipids. Embedded in this lipid bilayer or attached to its surfaces are diverse proteins (see Figure 6.6). However, each type of membrane has a unique composition of lipids and proteins suited to that membrane's specific functions. For example, enzymes embedded in the membranes of the organelles called mitochondria function in cellular respiration. Because membranes are so fundamental to the organisation of the cell, the chapter, "Membrane Structure and Function" will discuss them in detail.

Before continuing with this chapter, examine the eukaryotic cells in **Figure 6.8**. The generalised diagrams of an animal cell and a plant cell introduce the various organelles and show the key differences between animal and plant cells. The micrographs at the bottom of the figure give you a glimpse of cells from different types of eukaryotic organisms.

▼ Figure 6.8 Exploring Eukaryotic Cells

Animal Cell (cutaway view of generalised cell)

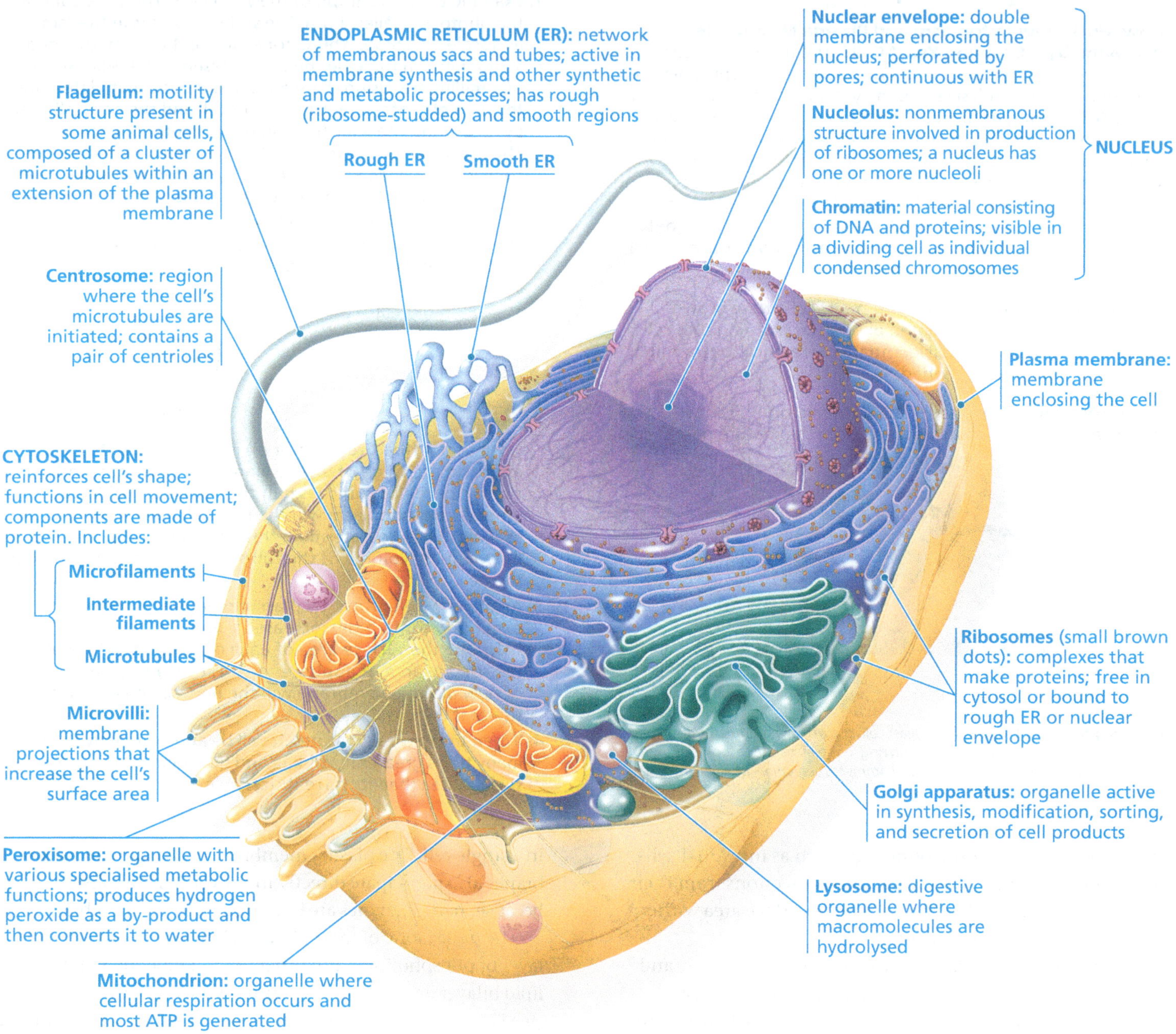

ENDOPLASMIC RETICULUM (ER): network of membranous sacs and tubes; active in membrane synthesis and other synthetic and metabolic processes; has rough (ribosome-studded) and smooth regions

Rough ER

Smooth ER

Flagellum: motility structure present in some animal cells, composed of a cluster of microtubules within an extension of the plasma membrane

Centrosome: region where the cell's microtubules are initiated; contains a pair of centrioles

CYTOSKELETON: reinforces cell's shape; functions in cell movement; components are made of protein. Includes:

- **Microfilaments**
- **Intermediate filaments**
- **Microtubules**

Microvilli: membrane projections that increase the cell's surface area

Peroxisome: organelle with various specialised metabolic functions; produces hydrogen peroxide as a by-product and then converts it to water

Mitochondrion: organelle where cellular respiration occurs and most ATP is generated

NUCLEUS

- **Nuclear envelope:** double membrane enclosing the nucleus; perforated by pores; continuous with ER
- **Nucleolus:** nonmembranous structure involved in production of ribosomes; a nucleus has one or more nucleoli
- **Chromatin:** material consisting of DNA and proteins; visible in a dividing cell as individual condensed chromosomes

Plasma membrane: membrane enclosing the cell

Ribosomes (small brown dots): complexes that make proteins; free in cytosol or bound to rough ER or nuclear envelope

Golgi apparatus: organelle active in synthesis, modification, sorting, and secretion of cell products

Lysosome: digestive organelle where macromolecules are hydrolysed

Animal Cells

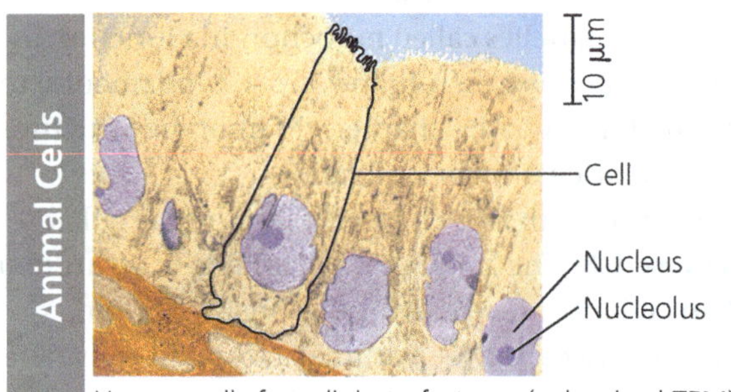

Human cells from lining of uterus (colourised TEM)

Unicellular Fungi

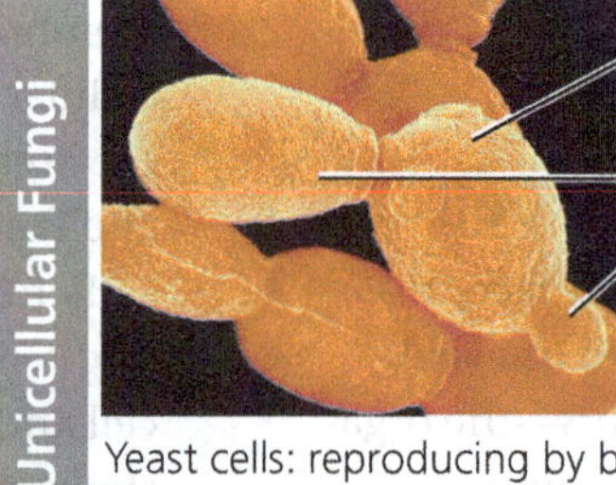

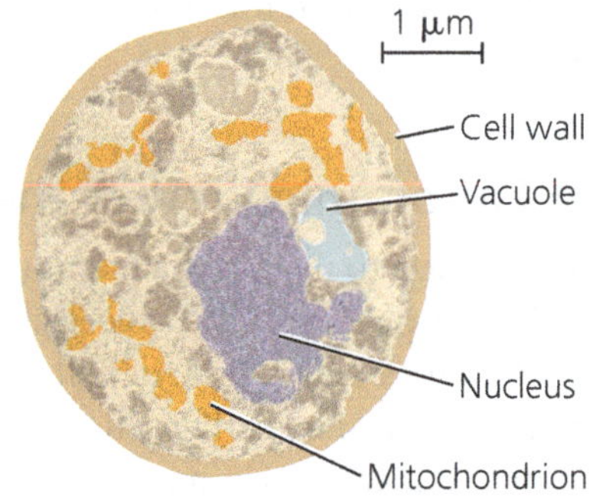

Yeast cells: reproducing by budding (above, colourised SEM) and a single cell (right, colourised TEM)

Plant Cell (cutaway view of generalised cell)

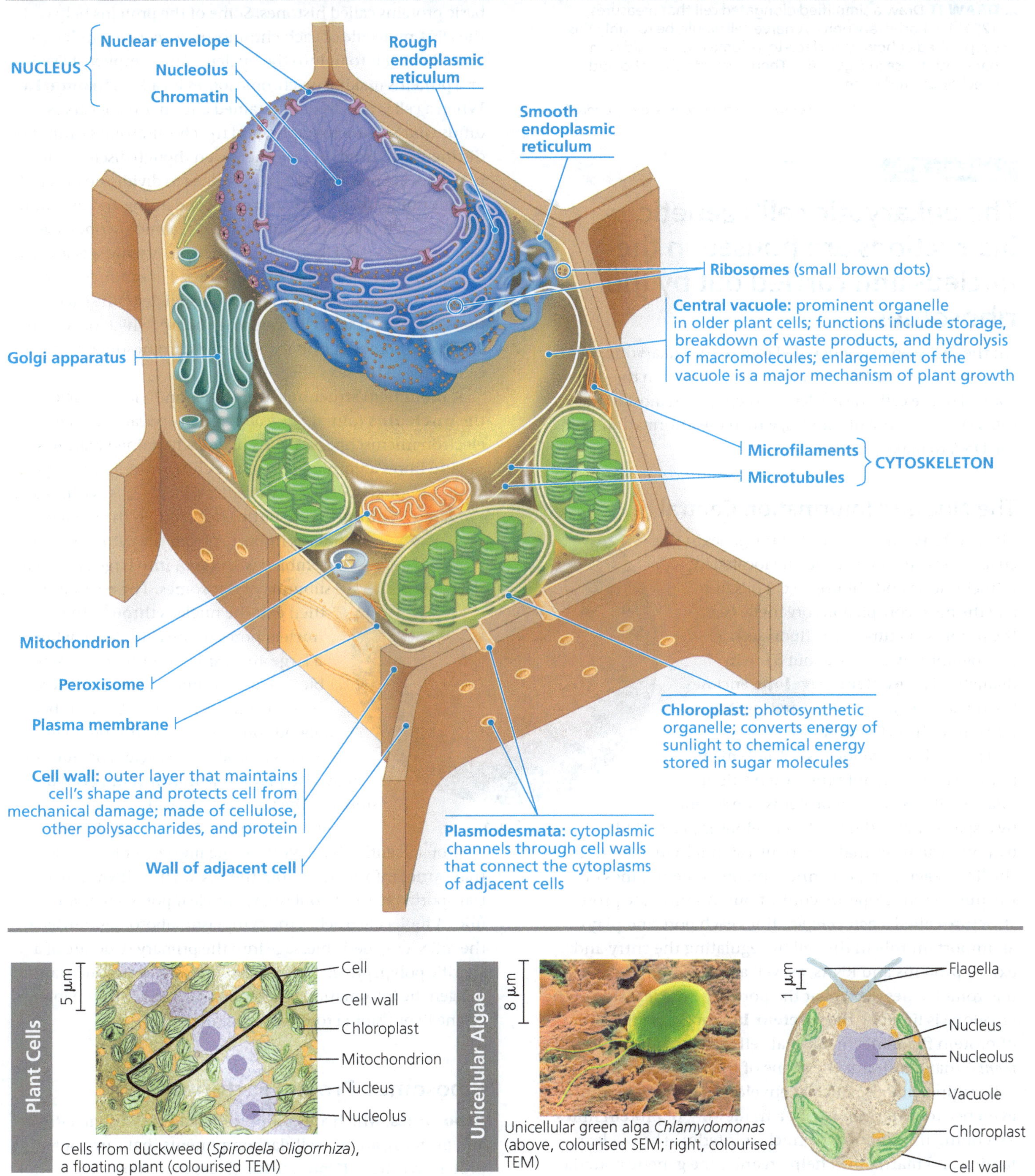

Cells from duckweed (*Spirodela oligorrhiza*), a floating plant (colourised TEM)

Unicellular green alga *Chlamydomonas* (above, colourised SEM; right, colourised TEM)

CONCEPT CHECK 6.2

1. Briefly describe the structure and function of the nucleus, the mitochondrion, the chloroplast, and the endoplasmic reticulum.
2. **DRAW IT** Draw a simplified elongated cell that measures $125 \times 1 \times 1$ arbitrary units. A nerve cell would be roughly this shape. Predict how its surface-to-volume ratio would compare with those in Figure 6.7. Then calculate the ratio and check your prediction.

For suggested answers, see Appendix A.

CONCEPT 6.3

The eukaryotic cell's genetic instructions are housed in the nucleus and carried out by the ribosomes

On the first stop of our detailed tour of the eukaryotic cell, let's look at two cellular components involved in the genetic control of the cell: the nucleus, which houses most of the cell's DNA, and the ribosomes, which use information from the DNA to make proteins.

The Nucleus: Information Central

The **nucleus** contains most of the genes in the eukaryotic cell. (Some genes are located in mitochondria and chloroplasts.) It is usually the most conspicuous organelle (see the purple structure in the fluorescence micrograph), averaging about 5 μm in diameter. The **nuclear envelope** encloses the nucleus **(Figure 6.9)**, separating its contents from the cytoplasm.

Nucleus

5 μm

The nuclear envelope is a *double* membrane. The two membranes, each a lipid bilayer with associated proteins, are separated by a space of 20–40 nm. The envelope is perforated by pore structures that are about 100 nm in diameter. At the lip of each pore, the inner and outer membranes of the nuclear envelope are continuous. An intricate protein structure called a *pore complex* lines each pore and plays an important role in the cell by regulating the entry and exit of proteins and RNAs, as well as large complexes of macromolecules. Except at the pores, the nuclear side of the envelope is lined by the **nuclear lamina**, a netlike array of protein filaments (in animal cells, called *intermediate filaments*) that maintains the shape of the nucleus by mechanically supporting the nuclear envelope. There is also much evidence for a *nuclear matrix*, a framework of protein fibres extending throughout the nuclear interior. The nuclear lamina and matrix may help organise the genetic material so it functions efficiently.

Within the nucleus, the DNA is organised into discrete units called **chromosomes**, structures that carry the genetic information. Each chromosome contains one long DNA molecule associated with many proteins, including small basic proteins called histones. Some of the proteins help coil the DNA molecule of each chromosome, reducing its length and allowing it to fit into the nucleus. The complex of DNA and proteins making up chromosomes is called **chromatin**. When a cell is not dividing, stained chromatin appears as a diffuse mass in micrographs, and the chromosomes cannot be distinguished from one another, even though discrete chromosomes are present. As a cell prepares to divide, however, the chromosomes form loops and coil, condensing and becoming thick enough to be distinguished under a microscope as separate structures (see Figure 16.23). Each eukaryotic species has a characteristic number of chromosomes. For example, a typical human cell has 46 chromosomes in its nucleus; the exceptions are human sex cells (eggs and sperm), which have only 23 chromosomes. A fruit fly cell has 8 chromosomes in most cells and 4 in the sex cells.

A prominent structure within the nondividing nucleus is the **nucleolus** (plural, *nucleoli*), which appears through the electron microscope as a mass of densely stained granules and fibres adjoining part of the chromatin. Here a type of RNA called *ribosomal RNA* (rRNA) is synthesised from genes in the DNA. Also in the nucleolus, proteins imported from the cytoplasm are assembled with rRNA into large and small subunits of ribosomes. These subunits then exit the nucleus through the nuclear pores to the cytoplasm, where a large and a small subunit can assemble into a ribosome. Sometimes there are two or more nucleoli; the number depends on the species and the stage in the cell's reproductive cycle. The nucleoli may also play a role in controlling cell division and the life span of a cell.

As we saw in Figure 5.22, the nucleus directs protein synthesis by synthesising messenger RNA (mRNA) that carries information from the DNA. The mRNA is then transported to the cytoplasm via nuclear pores. Once an mRNA molecule reaches the cytoplasm, ribosomes translate the mRNA's genetic message into the primary structure of a specific polypeptide. (This process of transcribing and translating genetic information is described in detail in the chapter, "Gene Expression: From Gene to Protein.")

Ribosomes: Protein Factories

Ribosomes, which are complexes made of ribosomal RNAs and proteins, are the cellular components that carry out protein synthesis **(Figure 6.10)**. (Note that ribosomes are not membrane bounded and thus are not considered organelles.)

▼ Figure 6.9 The nucleus and its envelope. Within the nucleus are the chromosomes, which appear as a mass of chromatin (DNA and associated proteins) and one or more nucleoli (singular, *nucleolus*), which function in ribosome synthesis. The nuclear envelope, which consists of two membranes separated by a narrow space, is perforated with pores and lined by the nuclear lamina.

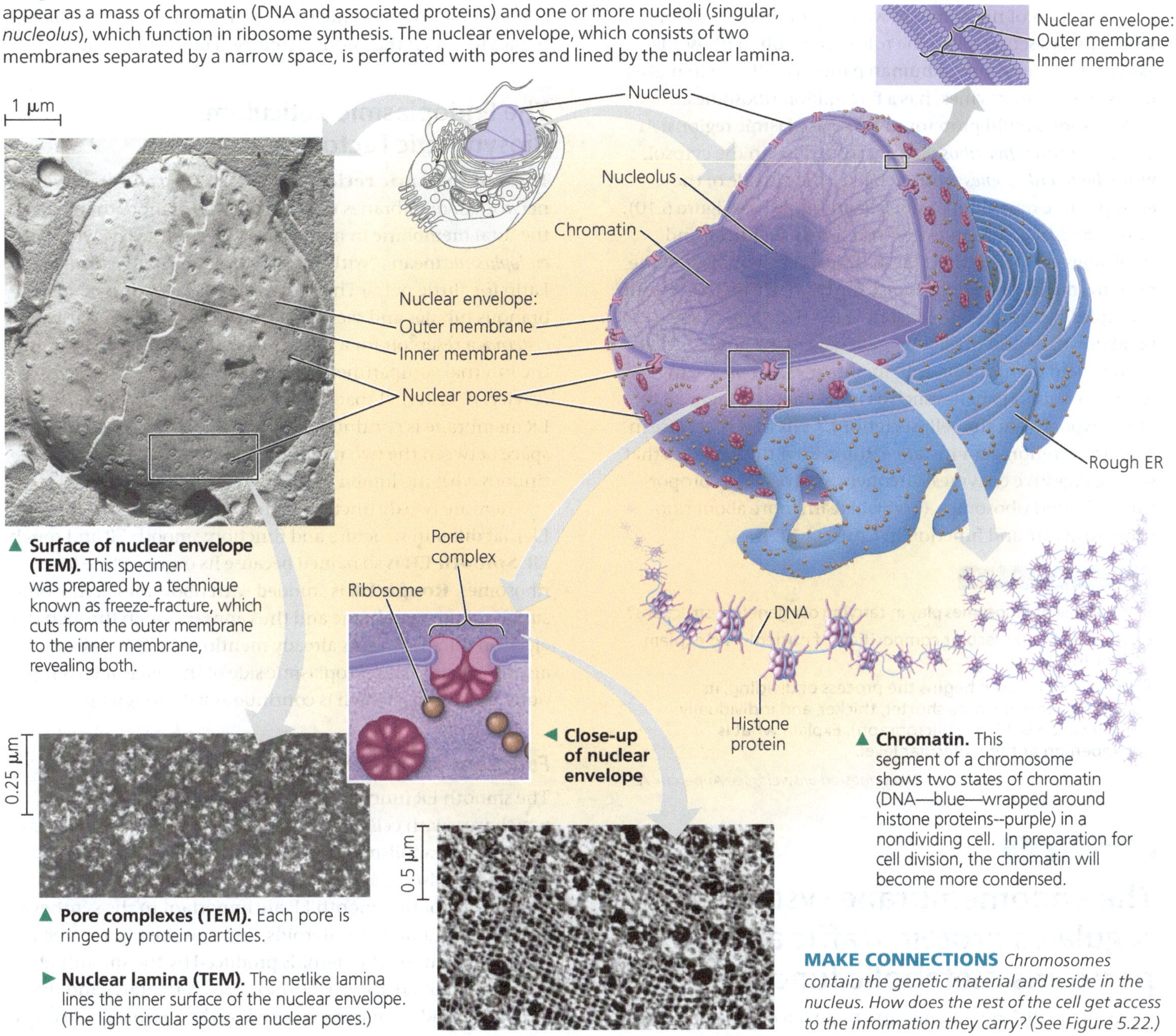

1 µm

Nucleus

Nucleolus

Chromatin

Nuclear envelope:
Outer membrane
Inner membrane

Nuclear pores

Rough ER

▲ Surface of nuclear envelope (TEM). This specimen was prepared by a technique known as freeze-fracture, which cuts from the outer membrane to the inner membrane, revealing both.

Ribosome

Pore complex

◄ Close-up of nuclear envelope

DNA

Histone protein

▲ Chromatin. This segment of a chromosome shows two states of chromatin (DNA—blue—wrapped around histone proteins--purple) in a nondividing cell. In preparation for cell division, the chromatin will become more condensed.

0.25 µm

▲ Pore complexes (TEM). Each pore is ringed by protein particles.

0.5 µm

► Nuclear lamina (TEM). The netlike lamina lines the inner surface of the nuclear envelope. (The light circular spots are nuclear pores.)

MAKE CONNECTIONS *Chromosomes contain the genetic material and reside in the nucleus. How does the rest of the cell get access to the information they carry? (See Figure 5.22.)*

▼ Figure 6.10 Ribosomes. This electron micrograph of a pancreas cell shows both free and bound ribosomes. The simplified diagram and computer model show the two subunits of a ribosome.

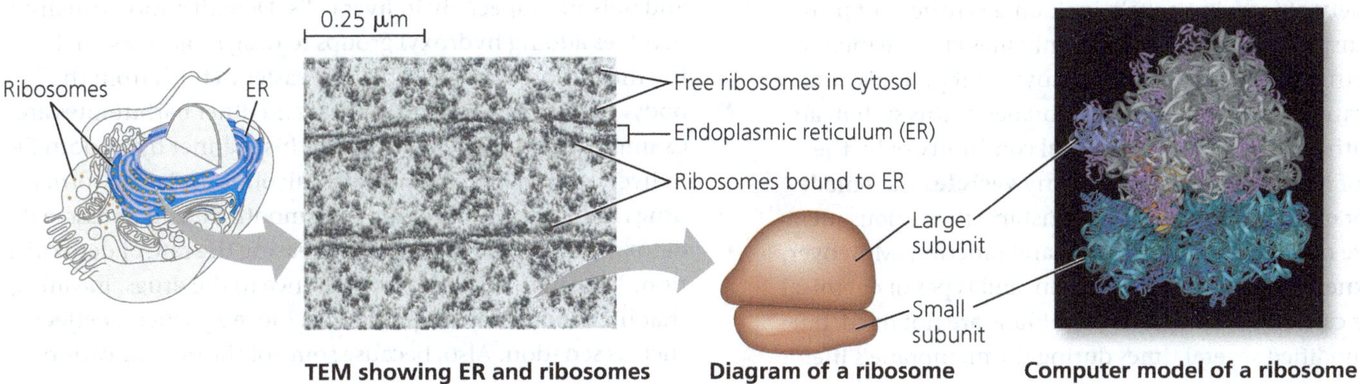

TEM showing ER and ribosomes **Diagram of a ribosome** **Computer model of a ribosome**

DRAW IT *After you have read the section on ribosomes, circle a ribosome in the micrograph that might be making a protein that will be secreted.*

Cells with high rates of protein synthesis have particularly large numbers of ribosomes as well as prominent nucleoli, which makes sense, given the role of nucleoli in ribosome assembly. For example, a human pancreas cell, which makes many digestive enzymes, has a few million ribosomes.

Ribosomes build proteins in two cytoplasmic regions: At any given time, *free ribosomes* are suspended in the cytosol, while *bound ribosomes* are attached to the outside of the endoplasmic reticulum or nuclear envelope (see Figure 6.10). Bound and free ribosomes are structurally identical, and ribosomes can play either role at different times. Most of the proteins made on free ribosomes function within the cytosol; examples are enzymes that catalyse the first steps of sugar breakdown. Bound ribosomes generally make proteins that are destined for insertion into membranes, for packaging within certain organelles such as lysosomes (see Figure 6.8), or for export from the cell (secretion). Cells that specialise in protein secretion—for instance, the cells of the pancreas that secrete digestive enzymes—frequently have a high proportion of bound ribosomes. (You will learn more about ribosome structure and function in Concept 17.4.)

CONCEPT CHECK **6.3**

1. What role do ribosomes play in carrying out genetic instructions?
2. Describe the molecular composition of nucleoli and explain their function.
3. **WHAT IF?** As a cell begins the process of dividing, its chromosomes become shorter, thicker, and individually visible in an LM (light micrograph). Explain what is happening at the molecular level.

For suggested answers, see Appendix A.

CONCEPT **6.4**

The endomembrane system regulates protein traffic and performs metabolic functions

Many of the different membrane-bounded organelles of the eukaryotic cell are part of the **endomembrane system**, which includes the nuclear envelope, the endoplasmic reticulum, the Golgi apparatus, lysosomes, various kinds of vesicles and vacuoles, and the plasma membrane. This system carries out a variety of tasks in the cell, including synthesis of proteins, transport of proteins into membranes and organelles or out of the cell, metabolism and movement of lipids, and detoxification of poisons. The membranes of this system are related either through direct physical continuity or by the transfer of membrane segments as tiny **vesicles** (sacs made of membrane). Despite these relationships, the various membranes are not identical in structure and function. Moreover, the thickness, molecular composition, and types of chemical reactions carried out in a given membrane are not fixed, but may be modified several times during the membrane's life.

Having already looked at the nuclear envelope, we will now focus on the endoplasmic reticulum and the other endomembranes to which the endoplasmic reticulum gives rise.

The Endoplasmic Reticulum: Biosynthetic Factory

The **endoplasmic reticulum (ER)** is such an extensive network of membranes that it accounts for more than half the total membrane in many eukaryotic cells. (The word *endoplasmic* means "within the cytoplasm," and *reticulum* is Latin for "little net.") The ER consists of a network of membranous tubules and sacs called cisternae (from the Latin *cisterna*, a reservoir for a liquid). The ER membrane separates the internal compartment of the ER, called the *ER lumen* (cavity) or cisternal space, from the cytosol. And because the ER membrane is continuous with the nuclear envelope, the space between the two membranes of the envelope is continuous with the lumen of the ER **(Figure 6.11)**.

There are two distinct, though connected, regions of the ER that differ in structure and function: smooth ER and rough ER. **Smooth ER** is so named because its outer surface lacks ribosomes. **Rough ER** is studded with ribosomes on the outer surface of the membrane and thus appears rough through the electron microscope. As already mentioned, ribosomes are also attached to the cytoplasmic side of the nuclear envelope's outer membrane, which is continuous with rough ER.

Functions of Smooth ER

The smooth ER functions in diverse metabolic processes, which vary with cell type. These processes include synthesis of lipids, metabolism of carbohydrates, detoxification of drugs and poisons, and storage of calcium ions.

Enzymes of the smooth ER are important in the synthesis of lipids, including oils, steroids, and new membrane phospholipids. Among the steroids produced by the smooth ER in animal cells are the sex hormones of vertebrates and the various steroid hormones secreted by the adrenal glands. The cells that synthesise and secrete these hormones—in the testes and ovaries, for example—are rich in smooth ER, a structural feature that fits the function of these cells.

Other enzymes of the smooth ER help detoxify drugs and poisons, especially in liver cells. Detoxification usually involves adding hydroxyl groups to drug molecules, making them more water-soluble and easier to flush from the body. The sedative phenobarbital and other barbiturates are examples of drugs metabolised in this manner by smooth ER in liver cells. In fact, barbiturates, alcohol, and many other drugs induce the proliferation of smooth ER and its associated detoxification enzymes, thus increasing the rate of detoxification. This, in turn, increases tolerance to the drugs, meaning that higher doses are required to achieve a particular effect, such as sedation. Also, because some of the detoxification enzymes have relatively broad action, the proliferation of smooth ER in response to one drug can increase the need for higher dosages of other drugs as well. Barbiturate abuse, for instance, can decrease the effectiveness of certain antibiotics and other useful drugs.

▼ **Figure 6.11 Endoplasmic reticulum (ER).** A membranous system of interconnected tubules and flattened sacs called cisternae, the ER is also continuous with the nuclear envelope, as shown in the cutaway diagram at the top. The membrane of the ER encloses a continuous compartment called the ER lumen (or cisternal space). Rough ER, which is studded on its outer surface with ribosomes, can be distinguished from smooth ER in the electron micrograph (TEM). Transport vesicles bud off from a region of the rough ER called transitional ER and travel to the Golgi apparatus and other destinations.

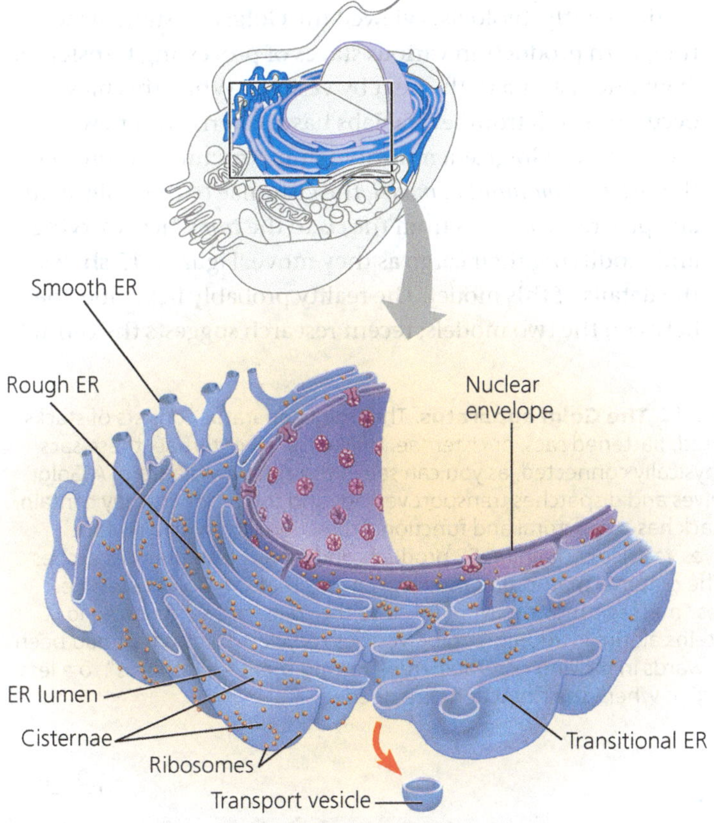

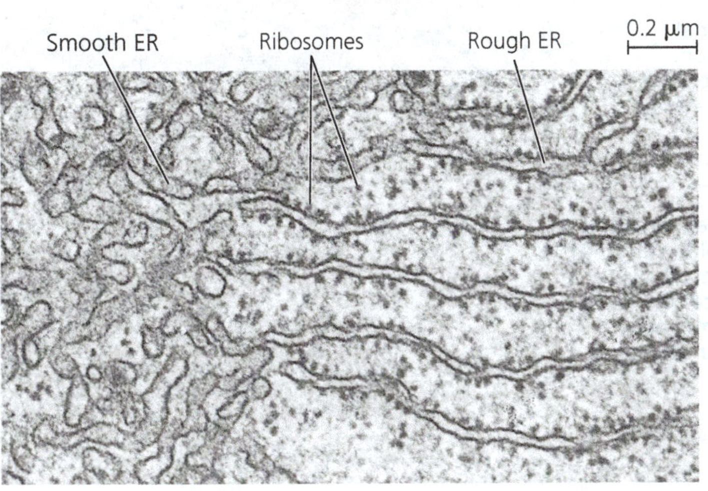

The smooth ER also stores calcium ions. In muscle cells, for example, the smooth ER membrane pumps calcium ions from the cytosol into the ER lumen. When a muscle cell is stimulated by a nerve impulse, calcium ions rush back across the ER membrane into the cytosol and trigger contraction of the muscle cell. In other cell types, release of calcium ions from the smooth ER triggers different responses, such as secretion of vesicles carrying newly synthesised proteins.

Functions of Rough ER

Many cells secrete proteins that are produced by ribosomes attached to rough ER. For instance, certain pancreatic cells synthesise the protein insulin in the ER and secrete this hormone into the bloodstream. As a polypeptide chain grows from a bound ribosome, the chain is threaded into the ER lumen through a pore formed by a protein complex in the ER membrane. The new polypeptide folds into its functional shape as it enters the ER lumen. Most secretory proteins are **glycoproteins**, proteins with carbohydrates covalently bonded to them. The carbohydrates are attached to the proteins in the ER lumen by enzymes built into the ER membrane.

After secretory proteins are formed, the ER membrane keeps them separate from proteins in the cytosol, which are produced by free ribosomes. Secretory proteins depart from the ER wrapped in the membranes of vesicles that bud like bubbles from a specialised region called transitional ER (see Figure 6.11). Vesicles in transit from one part of the cell to another are called **transport vesicles**; we will examine their fate shortly.

In addition to making secretory proteins, rough ER is a membrane factory for the cell; it grows in place by adding membrane proteins and phospholipids to its own membrane. As polypeptides destined to be membrane proteins grow from the ribosomes, they are inserted into the ER membrane itself and anchored there by their hydrophobic portions. Like the smooth ER, the rough ER also makes membrane phospholipids; enzymes built into the ER membrane assemble phospholipids from precursors in the cytosol. The ER membrane expands, and portions of it are transferred in the form of transport vesicles to other components of the endomembrane system.

The Golgi Apparatus: Shipping and Receiving Centre

After leaving the ER, many transport vesicles travel to the **Golgi apparatus**. We can think of the Golgi as a warehouse for receiving, sorting, shipping, and even some manufacturing. Here, products of the ER, such as proteins, are modified and stored and then sent to other destinations. Not surprisingly, the Golgi apparatus is especially extensive in cells specialised for secretion.

The Golgi apparatus consists of a group of associated, flattened membranous sacs—cisternae—looking like a stack of pita bread **(Figure 6.12)**. A cell may have many, even hundreds, of these stacks. The membrane of each cisterna in a stack separates its internal space from the cytosol. Vesicles concentrated in the vicinity of the Golgi apparatus are engaged in the transfer of material between parts of the Golgi and other structures.

A Golgi stack has a distinct structural directionality, with the membranes of cisternae on opposite sides of the stack differing in thickness and molecular composition. The two sides of a Golgi stack are referred to as the *cis* face and the *trans* face; these act, respectively, as the receiving and shipping departments of the Golgi apparatus. The term *cis* means "on the same side," and the *cis* face is usually located near the ER. Transport vesicles move material from the ER to the Golgi apparatus. A vesicle that buds from the ER can add its membrane and the contents of its lumen to the *cis* face by fusing with a Golgi membrane on that side. The *trans* face ("on the opposite side") gives rise to vesicles that pinch off and travel to other sites.

Products of the endoplasmic reticulum are usually modified during their transit from the *cis* region to the *trans* region of the Golgi apparatus. For example, glycoproteins formed in the ER have their carbohydrates modified, first in the ER itself, and then as they pass through the Golgi. The Golgi removes some sugar monomers and substitutes others, producing a large variety of carbohydrates. Membrane phospholipids may also be altered in the Golgi.

In addition to its finishing work, the Golgi apparatus also manufactures some macromolecules. Many polysaccharides secreted by cells are Golgi products. For example, pectins and certain other noncellulose polysaccharides are made in the Golgi of plant cells and then incorporated along with cellulose into their cell walls. Like secretory proteins, nonprotein Golgi products that will be secreted depart from the *trans* face of the Golgi inside transport vesicles that eventually fuse with the plasma membrane. The contents are released and the vesicle membrane is incorporated into the plasma membrane, adding to the surface area.

The Golgi manufactures and refines its products in stages, with different cisternae containing unique teams of enzymes. Until recently, biologists viewed the Golgi as a static structure, with products in various stages of processing transferred from one cisterna to the next by vesicles. While this may occur, research from several labs has given rise to a new model of the Golgi as a more dynamic structure. According to the *cisternal maturation model*, the cisternae of the Golgi actually progress forwards from the *cis* to the *trans* face, carrying and modifying their cargo as they move. Figure 6.12 shows the details of this model. The reality probably lies somewhere between the two models; recent research suggests the central

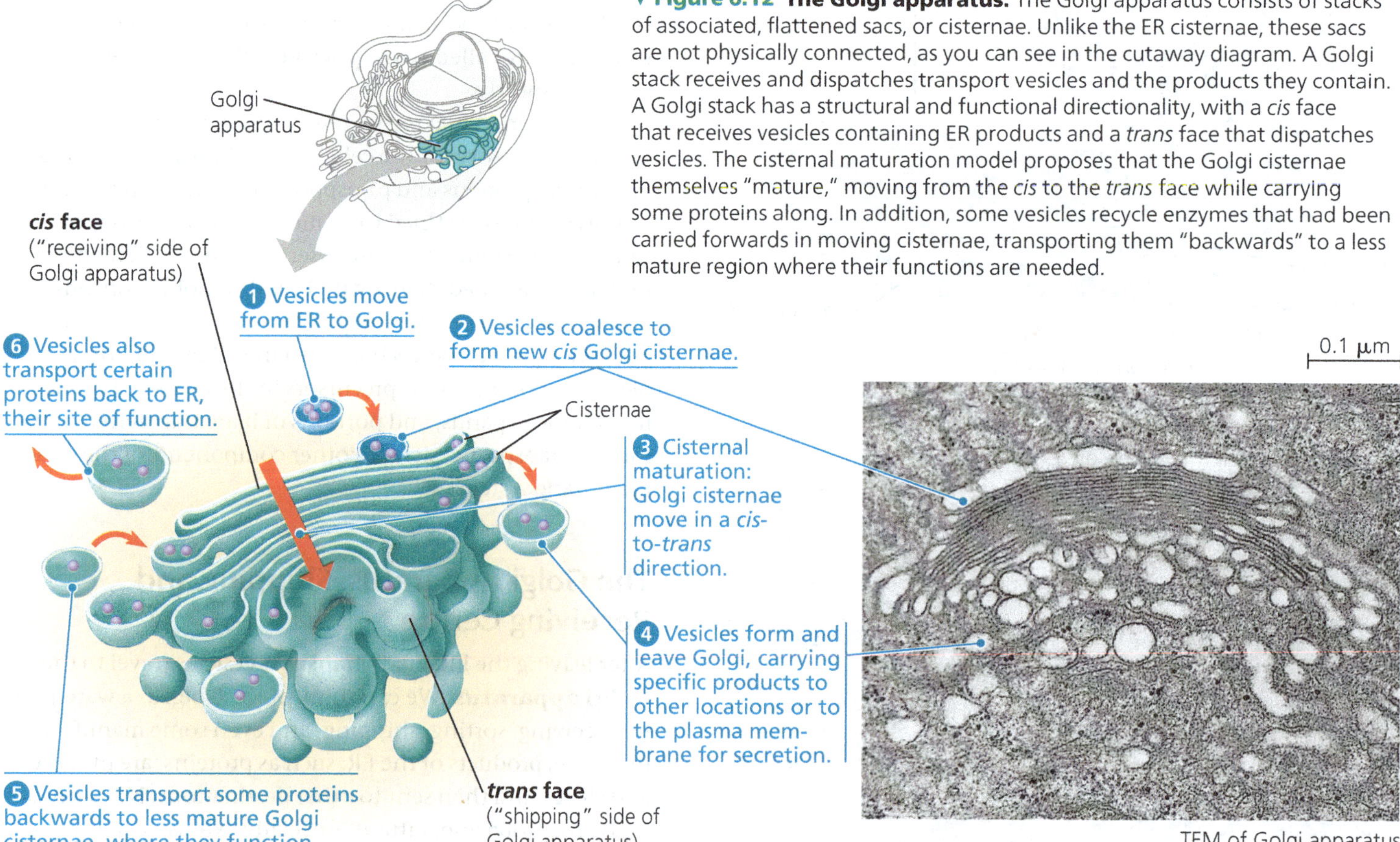

▼ **Figure 6.12 The Golgi apparatus.** The Golgi apparatus consists of stacks of associated, flattened sacs, or cisternae. Unlike the ER cisternae, these sacs are not physically connected, as you can see in the cutaway diagram. A Golgi stack receives and dispatches transport vesicles and the products they contain. A Golgi stack has a structural and functional directionality, with a *cis* face that receives vesicles containing ER products and a *trans* face that dispatches vesicles. The cisternal maturation model proposes that the Golgi cisternae themselves "mature," moving from the *cis* to the *trans* face while carrying some proteins along. In addition, some vesicles recycle enzymes that had been carried forwards in moving cisternae, transporting them "backwards" to a less mature region where their functions are needed.

regions of the cisternae may remain in place, while the outer ends are more dynamic.

Before a Golgi stack dispatches its products by budding vesicles from the *trans* face, it sorts these products and targets them for various parts of the cell. Molecular identification tags, such as phosphate groups added to the Golgi products, aid in sorting by acting like postcodes on mailing labels. Finally, transport vesicles budded from the Golgi may have external molecules on their membranes that recognise "docking sites" on the surface of specific organelles or on the plasma membrane, thus targeting the vesicles appropriately.

Lysosomes: Digestive Compartments

A **lysosome** is a membranous sac of hydrolytic enzymes that many eukaryotic cells use to digest (hydrolyse) macromolecules. Lysosomal enzymes work best in the acidic environment found in lysosomes. If a lysosome breaks open or leaks its contents, the released enzymes are not very active because the cytosol has a near-neutral pH. However, excessive leakage from a large number of lysosomes can destroy a cell by self-digestion.

Hydrolytic enzymes and lysosomal membrane are made by rough ER and then transferred to the Golgi apparatus for further processing. At least some lysosomes probably arise by budding from the *trans* face of the Golgi apparatus (see Figure 6.12). How are the proteins of the inner surface of the lysosomal membrane and the digestive enzymes themselves spared from destruction? Apparently, the three-dimensional shapes of these proteins protect vulnerable bonds from enzymatic attack.

Lysosomes carry out intracellular digestion in a variety of circumstances. Amoebas and many other unicellular protists eat by engulfing smaller organisms or food particles, a process called **phagocytosis** (from the Greek *phagein*, to eat, and *kytos*, vessel, referring here to the cell). The *food vacuole* formed in this way then fuses with a lysosome, whose enzymes digest the food (**Figure 6.13a**, top). Digestion products, including simple sugars, amino acids, and other monomers, pass into the cytosol and become nutrients for the cell. Some human cells also carry out phagocytosis. Among them are macrophages, a type of white blood cell that helps defend the body by engulfing and destroying bacteria and other invaders (see Figure 6.13a, bottom, and Figure 6.31).

▼ **Figure 6.13 Lysosomes.**

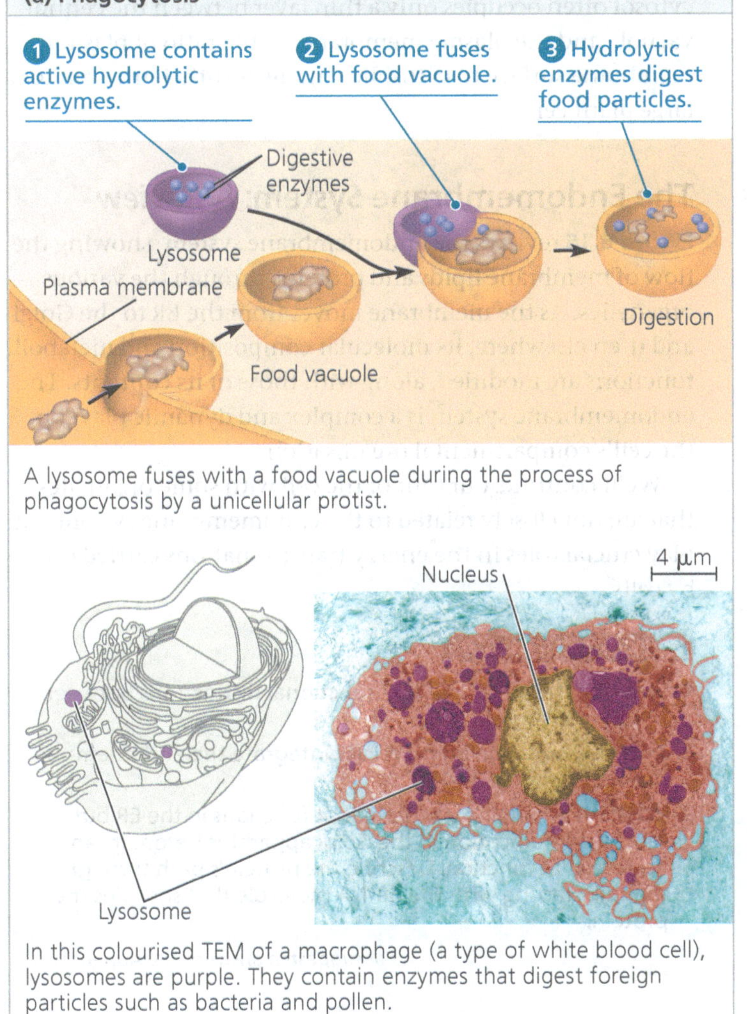

A lysosome fuses with a food vacuole during the process of phagocytosis by a unicellular protist.

In this colourised TEM of a macrophage (a type of white blood cell), lysosomes are purple. They contain enzymes that digest foreign particles such as bacteria and pollen.

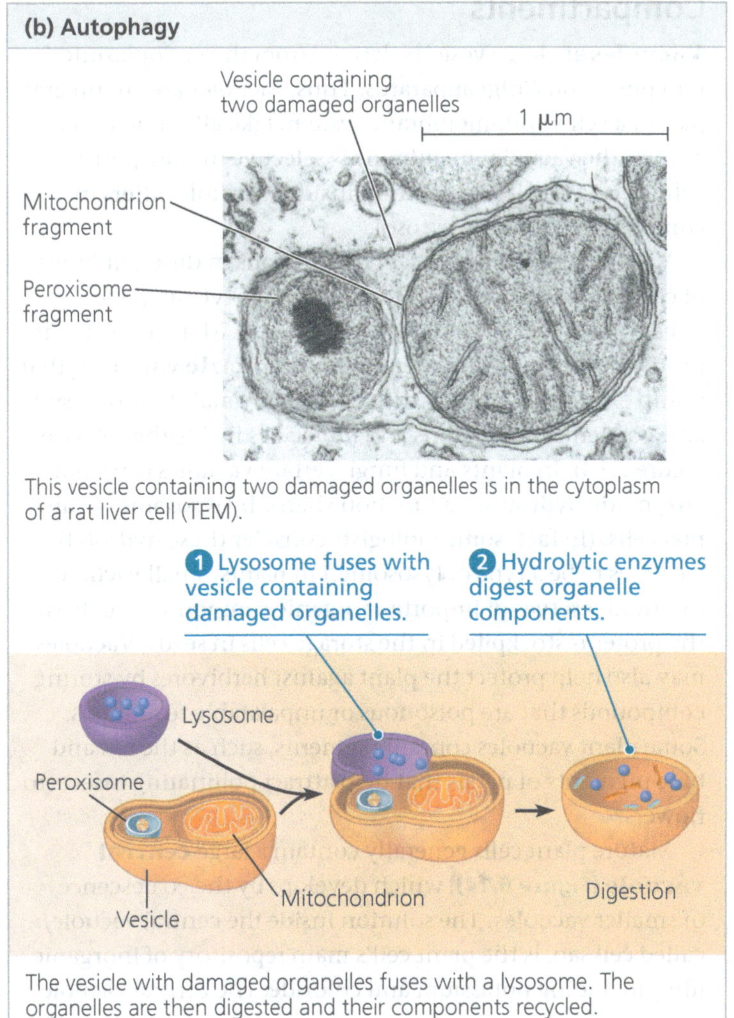

This vesicle containing two damaged organelles is in the cytoplasm of a rat liver cell (TEM).

The vesicle with damaged organelles fuses with a lysosome. The organelles are then digested and their components recycled.

Lysosomes also use their hydrolytic enzymes to recycle the cell's own organic material, a process called *autophagy*. During autophagy, a damaged organelle or small amount of cytosol becomes surrounded by a double membrane (of unknown origin), and a lysosome fuses with the outer membrane of this vesicle **(Figure 6.13b)**. The lysosomal enzymes dismantle the inner membrane and the enclosed material, and the resulting small organic compounds are released to the cytosol for reuse. With the help of lysosomes, the cell continually renews itself. A human liver cell, for example, recycles half of its macromolecules each week.

The cells of people with inherited lysosomal storage diseases lack a functioning hydrolytic enzyme normally present in lysosomes. The lysosomes become engorged with indigestible material, which begins to interfere with other cellular activities. In Tay-Sachs disease, for example, a lipid-digesting enzyme is missing or inactive, and the brain becomes impaired by an accumulation of lipids in the cells. Fortunately, lysosomal storage diseases are rare in the general population.

Vacuoles: Diverse Maintenance Compartments

Vacuoles are large vesicles derived from the endoplasmic reticulum and Golgi apparatus. Thus, vacuoles are an integral part of a cell's endomembrane system. Like all cellular membranes, the vacuolar membrane is selective in transporting solutes; as a result, the solution inside a vacuole differs in composition from the cytosol.

Vacuoles perform a variety of functions in different kinds of cells. **Food vacuoles**, formed by phagocytosis, have already been mentioned (see Figure 6.13a). Many unicellular protists living in fresh water have **contractile vacuoles** that pump excess water out of the cell, thereby maintaining a suitable concentration of ions and molecules inside the cell (see Figure 7.13). In plants and fungi, certain vacuoles carry out enzymatic hydrolysis, a function shared by lysosomes in animal cells. (In fact, some biologists consider these hydrolytic vacuoles to be a type of lysosome.) In plants, small vacuoles can hold reserves of important organic compounds, such as the proteins stockpiled in the storage cells in seeds. Vacuoles may also help protect the plant against herbivores by storing compounds that are poisonous or unpalatable to animals. Some plant vacuoles contain pigments, such as the red and blue pigments of petals that help attract pollinating insects to flowers.

Mature plant cells generally contain a large **central vacuole (Figure 6.14)**, which develops by the coalescence of smaller vacuoles. The solution inside the central vacuole, called cell sap, is the plant cell's main repository of inorganic ions, including potassium and chloride. The central vacuole

▼ Figure 6.14 The plant cell vacuole. The central vacuole is usually the largest compartment in a plant cell; the rest of the cytoplasm is often confined to a narrow zone between the vacuolar membrane and the plasma membrane (TEM).

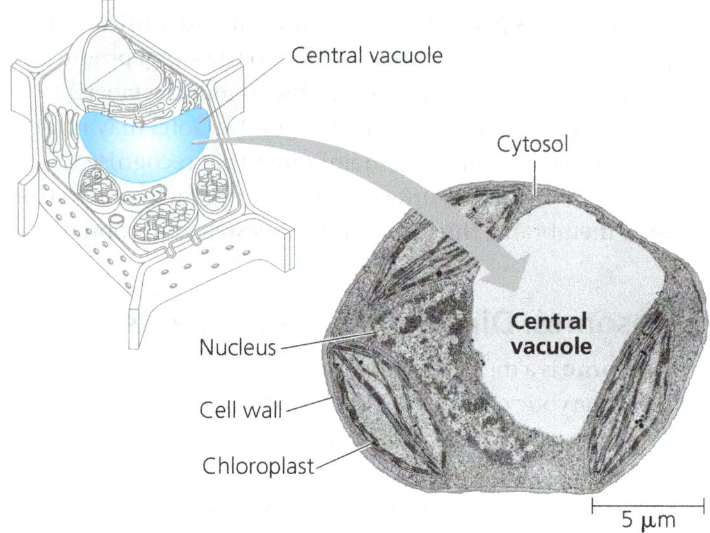

plays a major role in the growth of plant cells, which enlarge as the vacuole absorbs water, enabling the cell to become larger with a minimal investment in new cytoplasm. The cytosol often occupies only a thin layer between the central vacuole and the plasma membrane, so the ratio of plasma membrane surface to cytosolic volume is sufficient, even for a large plant cell.

The Endomembrane System: *A Review*

Figure 6.15 reviews the endomembrane system, showing the flow of membrane lipids and proteins through the various organelles. As the membrane moves from the ER to the Golgi and then elsewhere, its molecular composition and metabolic functions are modified, along with those of its contents. The endomembrane system is a complex and dynamic player in the cell's compartmental organisation.

We'll continue our tour of the cell with some organelles that are not closely related to the endomembrane system but play crucial roles in the energy transformations carried out by cells.

CONCEPT CHECK 6.4

1. Describe the structural and functional distinctions between rough and smooth ER.
2. Describe how transport vesicles integrate the endomembrane system.
3. **WHAT IF?** Imagine a protein that functions in the ER but requires modification in the Golgi apparatus before it can achieve that function. Describe the protein's path through the cell, starting with the mRNA molecule that specifies the protein.

For suggested answers, see Appendix A.

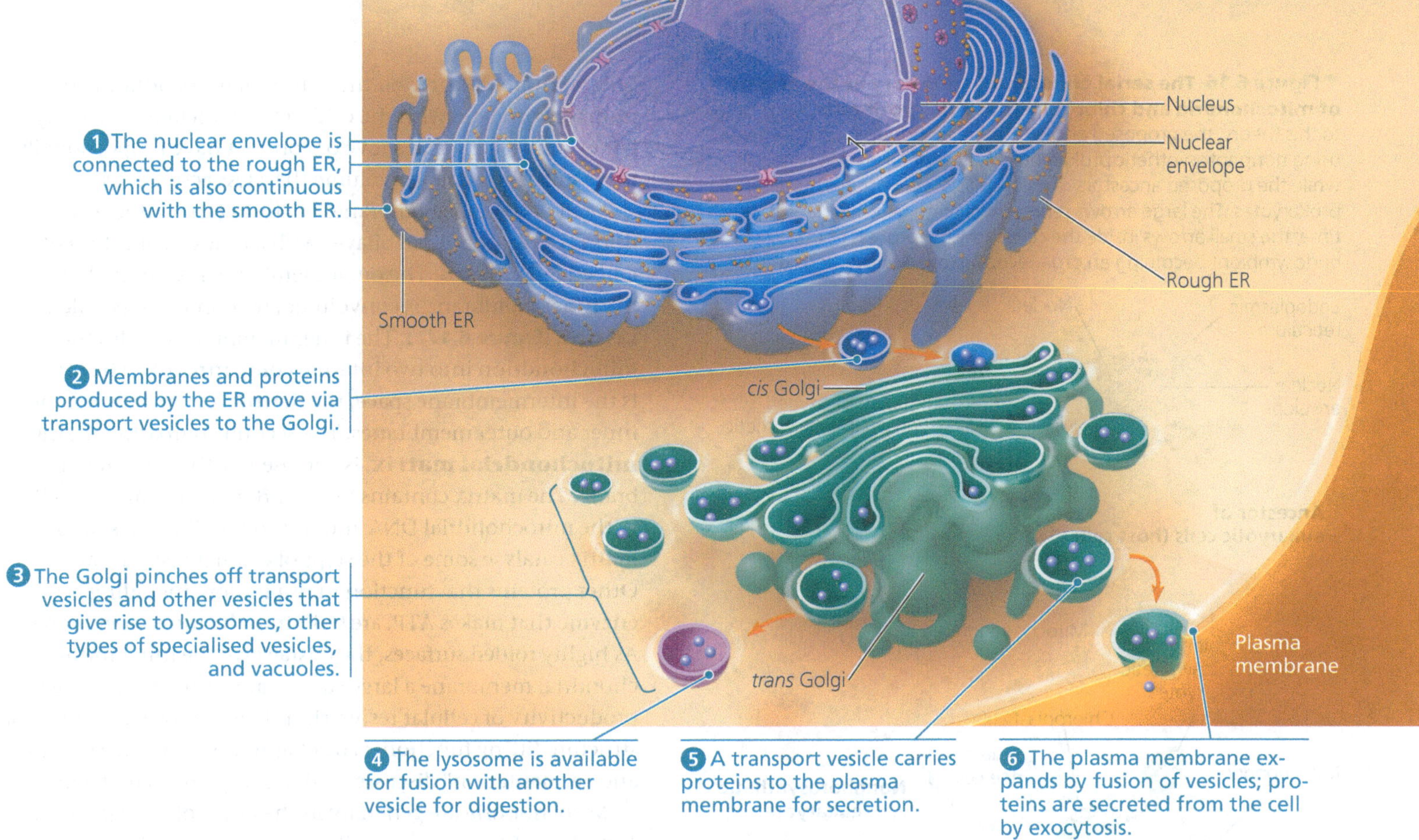

▲ **Figure 6.15 Review: Relationships among organelles of the endomembrane system.** The red arrows show some of the migration pathways for membranes and the materials they enclose.

CONCEPT 6.5

Mitochondria and chloroplasts change energy from one form to another

Organisms transform the energy they acquire from their surroundings. In eukaryotic cells, mitochondria and chloroplasts are the organelles that convert energy to forms that cells can use for work. **Mitochondria** (singular, *mitochondrion*) are the sites of cellular respiration, the metabolic process that uses oxygen to drive the generation of ATP by extracting energy from sugars, fats, and other fuels. **Chloroplasts**, found in plants and algae, are the sites of photosynthesis. This process in chloroplasts converts solar energy to chemical energy by absorbing sunlight and using it to drive the synthesis of organic compounds such as sugars from carbon dioxide and water.

In addition to having related functions, mitochondria and chloroplasts share similar evolutionary origins, which we'll look at briefly before examining their structures. In this section, we'll also consider the peroxisome, an oxidative organelle. The evolutionary origin of the peroxisome, as well as its relation to other organelles, is still a matter of some debate.

The Evolutionary Origins of Mitochondria and Chloroplasts

EVOLUTION Mitochondria and chloroplasts display similarities with bacteria that led to the **serial endosymbiont theory**, illustrated in Figure 6.16. This theory states that an early ancestor of eukaryotic cells (a *host* cell) engulfed an oxygen-using nonphotosynthetic prokaryotic cell. Eventually, the engulfed cell formed a relationship with the host cell in which it was enclosed, becoming an *endosymbiont* (a cell living within another cell). Indeed, over the course of evolution, the host cell and its endosymbiont merged into a single organism, a eukaryotic cell with the endosymbiont having become a mitochondrion. At least one of these cells may have then taken up a photosynthetic prokaryote, becoming the ancestor of eukaryotic cells that contain chloroplasts.

This is a widely accepted theory, which we will examine in more detail in Concept 25.3. This theory is consistent with many structural features of mitochondria and chloroplasts. First, rather than being bounded by a single membrane like organelles of the endomembrane system, mitochondria and typical chloroplasts have two membranes surrounding them. (Chloroplasts also have an internal system of membranous sacs.) There is evidence that the ancestral engulfed

▼ **Figure 6.16 The serial endosymbiont theory of the origins of mitochondria and chloroplasts in eukaryotic cells.** According to this theory, the proposed ancestors of mitochondria were oxygen-using nonphotosynthetic prokaryotes that were taken into host cells, while the proposed ancestors of chloroplasts were photosynthetic prokaryotes. The large arrows represent change over evolutionary time; the small arrows inside the cells show the process of the endosymbiont becoming an organelle, also over long periods of time.

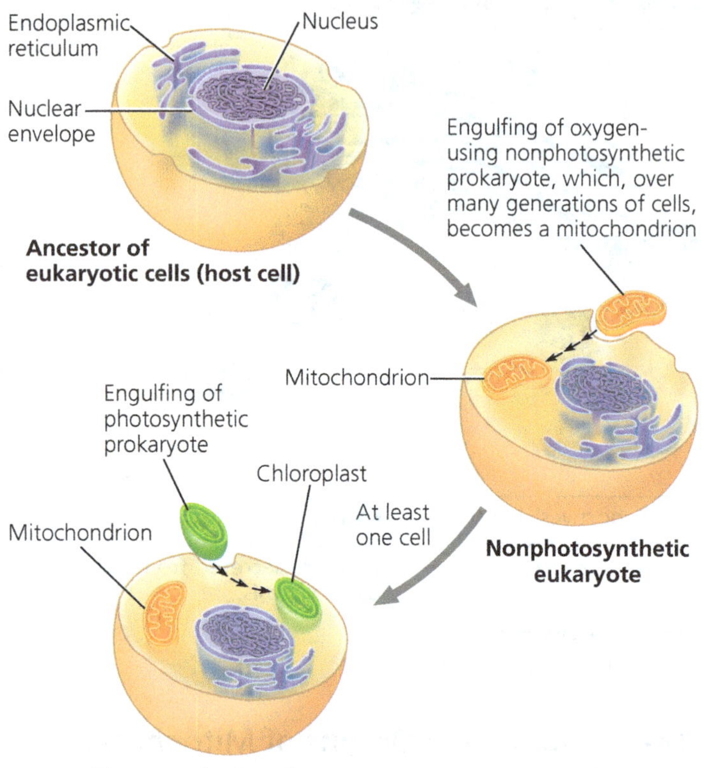

prokaryotes had two outer membranes, which became the double membranes of mitochondria and chloroplasts. Second, like prokaryotes, mitochondria and chloroplasts contain ribosomes, as well as circular DNA molecules—like bacterial chromosomes—associated with their inner membranes. The DNA in these organelles programs the synthesis of some organelle proteins on ribosomes that have been synthesised and assembled there as well. Third, also consistent with their probable evolutionary origins as cells, mitochondria and chloroplasts are autonomous (somewhat independent) organelles that grow and reproduce within the cell.

Next, we focus on the structures of mitochondria and chloroplasts, while providing an overview of their structures and functions. (In the chapters, "Cellular Respiration and Fermentation" and "Photosynthesis," we will examine their roles as energy transformers.)

Mitochondria: Chemical Energy Conversion

Mitochondria are found in nearly all eukaryotic cells, including those of plants, animals, fungi, and most protists. Some cells have a single large mitochondrion, but more often a cell has hundreds or even thousands of mitochondria; the number correlates with the cell's level of metabolic activity. For example, cells that move or contract have proportionally more mitochondria per volume than less active cells.

Each of the two membranes enclosing the mitochondrion is a phospholipid bilayer with a unique collection of embedded proteins. The outer membrane is smooth, but the inner membrane is convoluted, with infoldings called **cristae** (**Figure 6.17a**). The inner membrane divides the mitochondrion into two internal compartments. The first is the intermembrane space, the narrow region between the inner and outer membranes. The second compartment, the **mitochondrial matrix**, is enclosed by the inner membrane. The matrix contains many different enzymes as well as the mitochondrial DNA and ribosomes. Enzymes in the matrix catalyse some of the steps of cellular respiration. Other proteins that function in respiration, including the enzyme that makes ATP, are built into the inner membrane. As highly folded surfaces, the cristae give the inner mitochondrial membrane a large surface area, thus enhancing the productivity of cellular respiration. This is another example of structure fitting function. (The chapter, "Cellular Respiration and Fermentation" discusses cellular respiration in detail.)

Mitochondria are generally in the range of 1–10 μm long. Time-lapse films of living cells reveal mitochondria moving around, changing their shapes, and fusing or dividing into separate fragments, unlike the static structures seen in most diagrams and electron micrographs. These studies helped cell biologists understand that mitochondria in a living cell form a branched tubular network that is in a dynamic state of flux (see **Figure 6.17b** and **c**). In skeletal muscle, this network has been referred to by researchers as a "power grid."

Chloroplasts: Capture of Light Energy

Chloroplasts contain the green pigment chlorophyll, along with enzymes and other molecules that function in the photosynthetic production of sugar. These lens-shaped organelles, about 3–6 μm in length, are found in leaves and other green organs of plants and in algae (**Figure 6.18**; see also Figure 6.26c).

The contents of a chloroplast are partitioned from the cytosol by an envelope consisting of two membranes separated by a very narrow intermembrane space. Inside the chloroplast is another membranous system in the form of flattened, interconnected sacs called **thylakoids**. In some regions, thylakoids are stacked like poker chips; each stack is called a **granum** (plural, *grana*). The fluid outside the thylakoids is the **stroma**, which contains the chloroplast DNA and ribosomes as well as many enzymes. The membranes of the chloroplast divide the chloroplast space into three compartments: the intermembrane space, the stroma, and the thylakoid space. This compartmental organisation enables the chloroplast to convert light energy to chemical energy

▼ **Figure 6.17 The mitochondrion, site of cellular respiration. (a)** The inner and outer membranes seen in the drawing and TEM establish two compartments: the intermembrane space and the mitochondrial matrix. The cristae increase the surface area of the inner membrane. Circular DNA molecules are associated with the inner mitochondrial membrane. **(b)** Mitochondria are dynamic, at times separating into fragments or fusing. **(c)** The LM shows one protist (*Euglena gracilis*) at a much lower magnification than the TEM. The mitochondria form a branched tubular network; the mitochondrial matrix is stained green. The nuclear DNA is stained red; molecules of mitochondrial DNA appear as bright yellow spots.

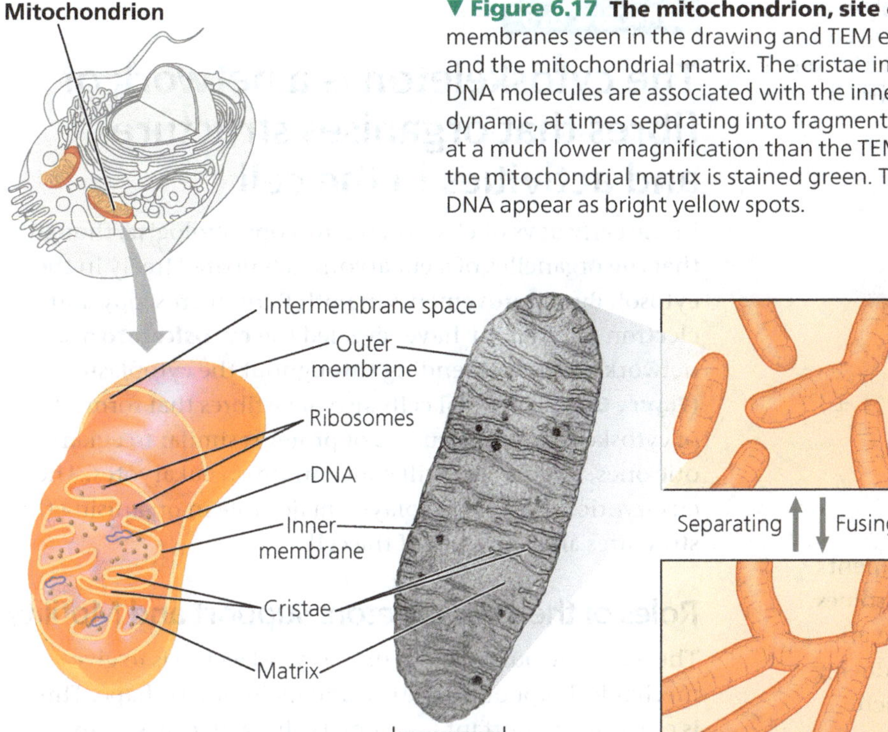

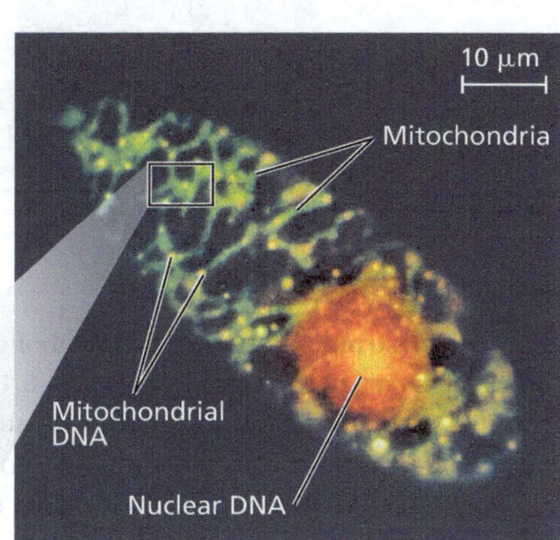

(a) Diagram and TEM of mitochondrion

(b) Dynamic nature of mitochondrial networks

(c) Network of mitochondria in *Euglena* (LM)

during photosynthesis. (You will learn more about photosynthesis in the chapter, "Photosynthesis.")

As with mitochondria, the static and rigid appearance of chloroplasts in micrographs or schematic diagrams cannot accurately depict their dynamic behaviour in the living cell. Their shape is changeable, and they grow and occasionally pinch in two, reproducing themselves. They are mobile and, with mitochondria and other organelles, move around the cell along tracks of the cytoskeleton, a structural network we will consider in Concept 6.6.

The chloroplast is a specialised member of a family of closely related plant organelles called **plastids**. One type of plastid, the *amyloplast*, is a colourless organelle that stores starch (amylose), particularly in roots and tubers. Another is the *chromoplast*, which has pigments that give fruits and flowers their orange and yellow hues.

▼ **Figure 6.18 The chloroplast, site of photosynthesis. (a)** Many plants have lens-shaped chloroplasts, as shown here in a diagram and a TEM. A chloroplast has three compartments: the intermembrane space, the stroma, and the thylakoid space. Free ribosomes are present in the stroma, as are copies of chloroplast DNA molecules. **(b)** This light micrograph, at a much lower magnification than the TEM, shows a whole cell of the green alga *Spirogyra crassa*, which is named for the spiral arrangement of each cell's chloroplasts.

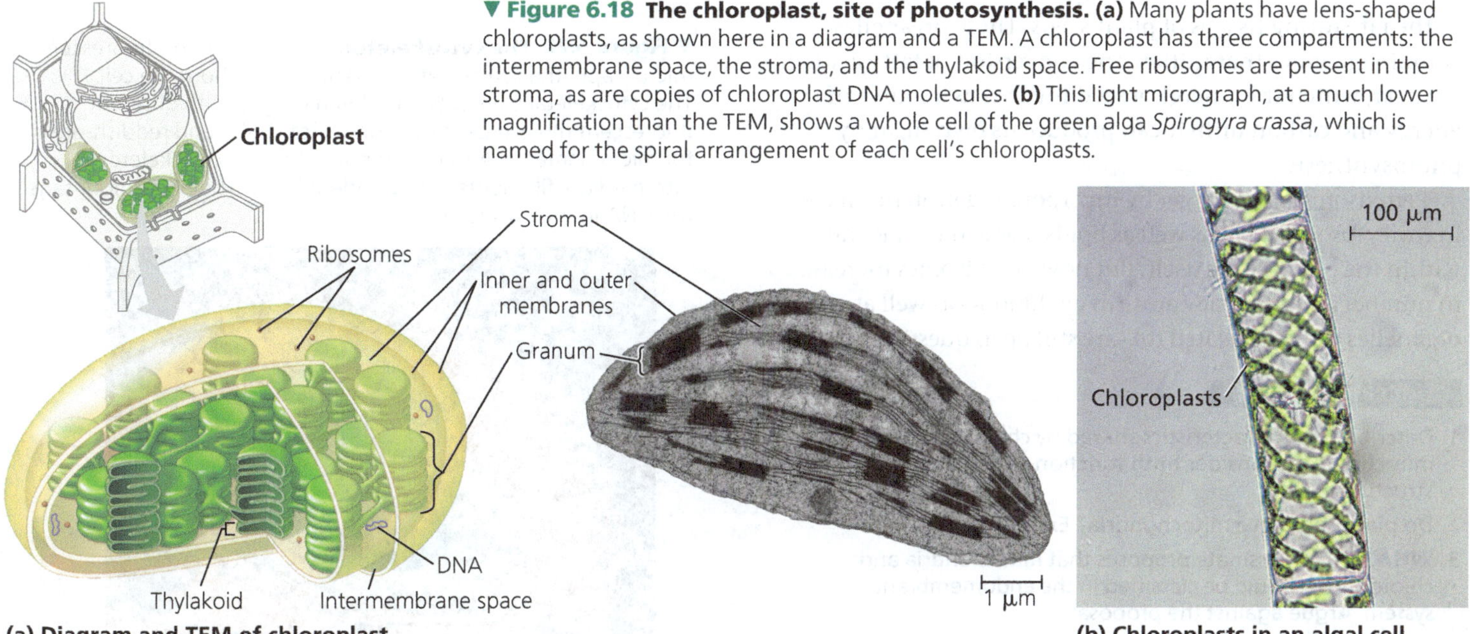

(a) Diagram and TEM of chloroplast

(b) Chloroplasts in an algal cell

▼ **Figure 6.19 A peroxisome.** Peroxisomes are roughly spherical and often have a granular or crystalline core that is thought to be a dense collection of enzyme molecules. Chloroplasts and mitochondria cooperate with peroxisomes in certain metabolic functions (TEM).

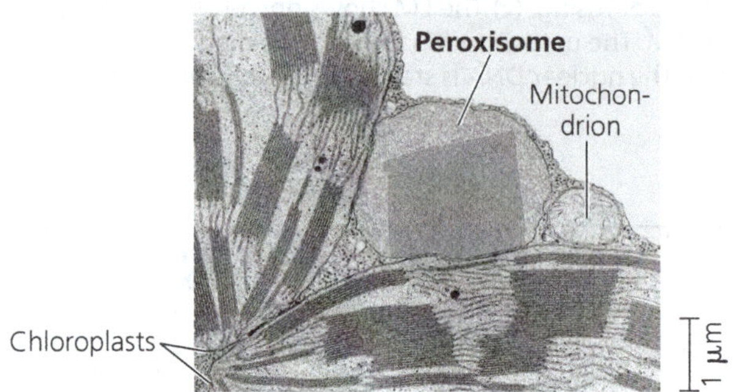

Peroxisomes: Oxidation

The **peroxisome** is a specialised metabolic compartment bounded by a single membrane **(Figure 6.19)**. Peroxisomes contain enzymes that remove hydrogen atoms from various substrates and transfer them to oxygen (O_2), producing hydrogen peroxide (H_2O_2) as a by-product (from which the organelle derives its name). These reactions have many different functions. Some peroxisomes use oxygen to break fatty acids down into smaller molecules that are transported to mitochondria and used as fuel for cellular respiration. Peroxisomes in the liver detoxify alcohol and other harmful compounds by transferring hydrogen from the poisonous compounds to oxygen. The H_2O_2 formed by peroxisomes is itself toxic, but the organelle also contains an enzyme that converts H_2O_2 to water. This is an excellent example of how the cell's compartmental structure is crucial to its functions: The enzymes that produce H_2O_2 and those that dispose of this toxic compound are sequestered away from other cellular components that could be damaged.

Specialised peroxisomes called *glyoxysomes* are found in the fat-storing tissues of plant seeds. These organelles contain enzymes that initiate the conversion of fatty acids to sugar, which the emerging seedling uses as a source of energy and carbon until it can produce its own sugar by photosynthesis.

Peroxisomes grow larger by incorporating proteins made in the cytosol and ER, as well as lipids made in the ER and within the peroxisome itself. But how peroxisomes increase in number and how they arose in evolution—as well as what organelles they are related to—are still open questions.

CONCEPT CHECK 6.5

1. Describe two characteristics shared by chloroplasts and mitochondria. Consider both function and membrane structure.
2. Do plant cells have mitochondria? Explain.
3. **WHAT IF?** A classmate proposes that mitochondria and chloroplasts should be classified in the endomembrane system. Argue against the proposal.

For suggested answers, see Appendix A.

CONCEPT 6.6

The cytoskeleton is a network of fibres that organises structures and activities in the cell

In the early days of electron microscopy, biologists thought that the organelles of a eukaryotic cell floated freely in the cytosol. But improvements in both light microscopy and electron microscopy have revealed the **cytoskeleton**, a network of fibres extending throughout the cytoplasm **(Figure 6.20)**. Bacterial cells also have fibres that form a type of cytoskeleton, constructed of proteins similar to eukaryotic ones, but here we will concentrate on eukaryotes. The eukaryotic cytoskeleton plays a major role in organising the structures and activities of the cell.

Roles of the Cytoskeleton: Support and Motility

The most obvious function of the cytoskeleton is to give mechanical support to the cell and maintain its shape. This is especially important for animal cells, which lack walls. The remarkable strength and resilience of the cytoskeleton as a whole are based on its architecture. Like a dome tent, the cytoskeleton is stabilised by a balance between opposing forces exerted by its elements. And just as the skeleton of an animal helps fix the positions of other body parts, the cytoskeleton provides anchorage for many organelles and even cytosolic enzyme molecules. The cytoskeleton is more dynamic than an animal skeleton, however. It can be quickly dismantled in one part of the cell and reassembled in a new location, changing the shape of the cell.

Some types of cell motility (movement) also involve the cytoskeleton. The term *cell motility* includes both changes in

▼ **Figure 6.20 The cytoskeleton.** As shown in this fluorescence micrograph, the cytoskeleton extends throughout the cell. The cytoskeletal elements have been tagged with different fluorescent molecules: green for microtubules and reddish-orange for microfilaments. A third component of the cytoskeleton, intermediate filaments, is not evident here. (The blue colour tags the DNA in the nucleus.)

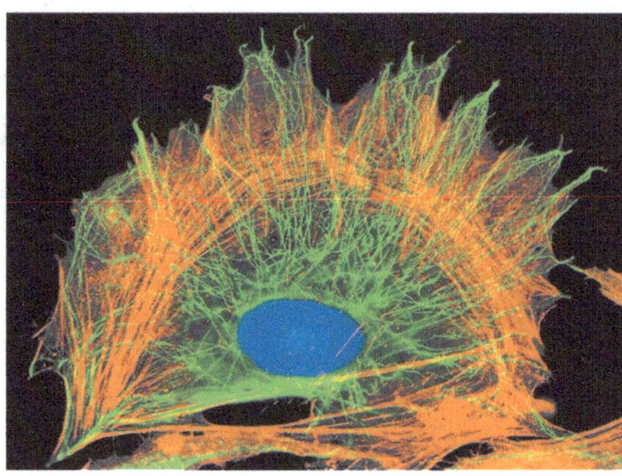

cell location and movements of cell parts. Cell motility generally requires interaction of the cytoskeleton with **motor proteins**. There are many such examples: Cytoskeletal elements and motor proteins work together with plasma membrane molecules to allow whole cells to move along fibres outside the cell. Inside the cell, vesicles and other organelles often use motor protein "feet" to "walk" to their destinations along a track provided by the cytoskeleton. For example, this is how vesicles containing neurotransmitter molecules migrate to the tips of axons, the long extensions of nerve cells that release these molecules as chemical signals to adjacent nerve cells **(Figure 6.21)**. The cytoskeleton also manipulates the plasma membrane, bending it inwards to form food vacuoles or other phagocytic vesicles.

▼ **Figure 6.21 Motor proteins and the cytoskeleton.**

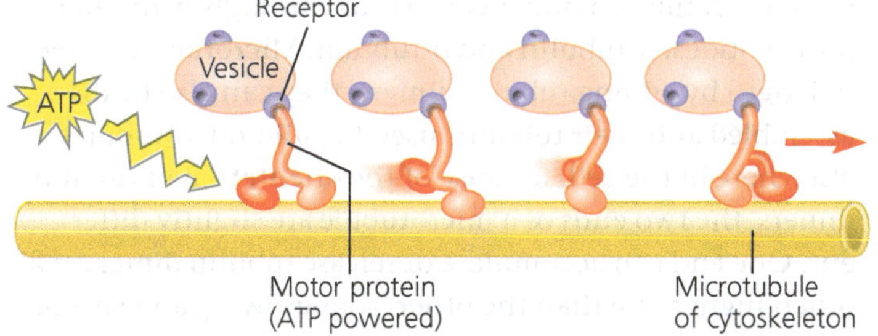

(a) A motor protein that attaches to a receptor on a vesicle can "walk" the vesicle along a microtubule or microfilament.

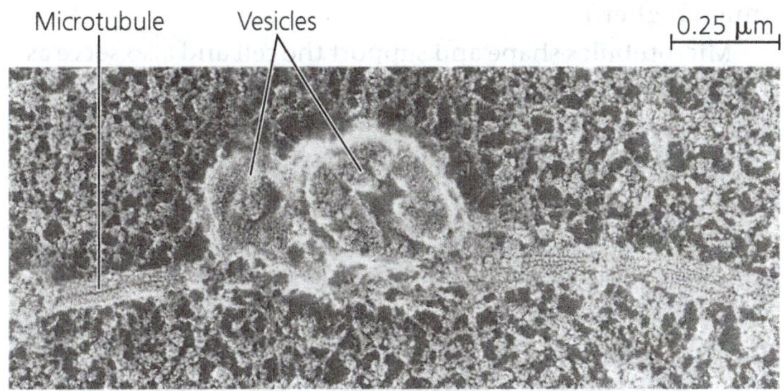

(b) Two vesicles containing neurotransmitters move along a microtubule towards the tip of a nerve cell extension called an axon (SEM).

Components of the Cytoskeleton

Now let's look more closely at the three main types of fibres that make up the cytoskeleton: *Microtubules* are the thickest of the three types; *microfilaments* (also called actin filaments) are the thinnest; and *intermediate filaments* are fibres with diameters in a middle range **(Table 6.1)**.

Microtubules

All eukaryotic cells have **microtubules**, hollow rods constructed from globular proteins called tubulins. Each

Table 6.1 The Structure and Function of the Cytoskeleton

Property	Microtubules (Tubulin Polymers)	Microfilaments (Actin Filaments)	Intermediate Filaments
Structure	Hollow tubes	Two intertwined strands of actin	Fibrous proteins coiled into cables
Diameter	25 nm with 15-nm lumen	7 nm	8–12 nm
Protein subunits	Tubulin, a dimer consisting of an α-tubulin and a β-tubulin	Actin	One of several different proteins (including keratins)
Main functions	Maintenance of cell shape; cell motility; chromosome movements in cell division; organelle movements	Maintenance of cell shape; changes in cell shape; muscle contraction; cytoplasmic streaming (plant cells); cell motility; cell division (animal cells)	Maintenance of cell shape; anchorage of nucleus and certain other organelles; formation of nuclear lamina
Fluorescence micrographs of fibroblasts. Connective tissue cells called fibroblasts are a favourite cell type for cell biology studies because they spread out flat and their internal structures are easy to see. In each fibroblast shown here, the structure of interest has been tagged with fluorescent molecules. In the third micrograph, the DNA in the nucleus has also been tagged (orange).	10 μm; Microtubule; Row of tubulin dimers; 25 nm; α; β; Tubulin dimer	10 μm; Microfilament; Actin subunit; 7 nm	5 μm; Intermediate filament; Keratin proteins; Fibrous subunit (keratins coiled together); 8–12 nm

VISUAL SKILLS *How many tubulin dimers are in the boxed row?*

tubulin protein is a *dimer*, a molecule made up of two components. A tubulin dimer consists of two slightly different polypeptides, α-tubulin and β-tubulin. Microtubules grow in length by adding tubulin dimers; they can also be disassembled and their tubulins used to build microtubules elsewhere in the cell. Because of the orientation of tubulin dimers, the two ends of a microtubule are slightly different. One end can accumulate or release tubulin dimers at a much higher rate than the other, thus growing and shrinking significantly during cellular activities. (This is called the "plus end," not because it can only add tubulin proteins but because it's the end where both "on" and "off" rates are much higher.)

Microtubules shape and support the cell and also serve as tracks along which organelles equipped with motor proteins can move. In addition to the example in Figure 6.21, microtubules guide vesicles from the ER to the Golgi apparatus and from the Golgi to the plasma membrane. Microtubules are also involved in the separation of chromosomes during cell division, as shown in Figure 12.7.

Centrosomes and Centrioles In animal cells, microtubules grow out from a **centrosome**, a region that is often located near the nucleus. These microtubules function as compression-resisting girders of the cytoskeleton. Within the centrosome is a pair of **centrioles**, each composed of nine sets of triplet microtubules arranged in a ring **(Figure 6.22)**. Although centrosomes with centrioles may help organise microtubule assembly in animal cells, many other eukaryotic cells lack centrosomes with centrioles and instead organise microtubules by other means.

▼ **Figure 6.22 Centrosome containing a pair of centrioles.** Most animal cells have a centrosome, a region near the nucleus where the cell's microtubules are initiated. Within the centrosome is a pair of centrioles, each about 250 nm (0.25 μm) in diameter. The two centrioles are at right angles to each other, and each is made up of nine sets of three microtubules. The blue portions of the drawing represent nontubulin proteins that connect the microtubule triplets.

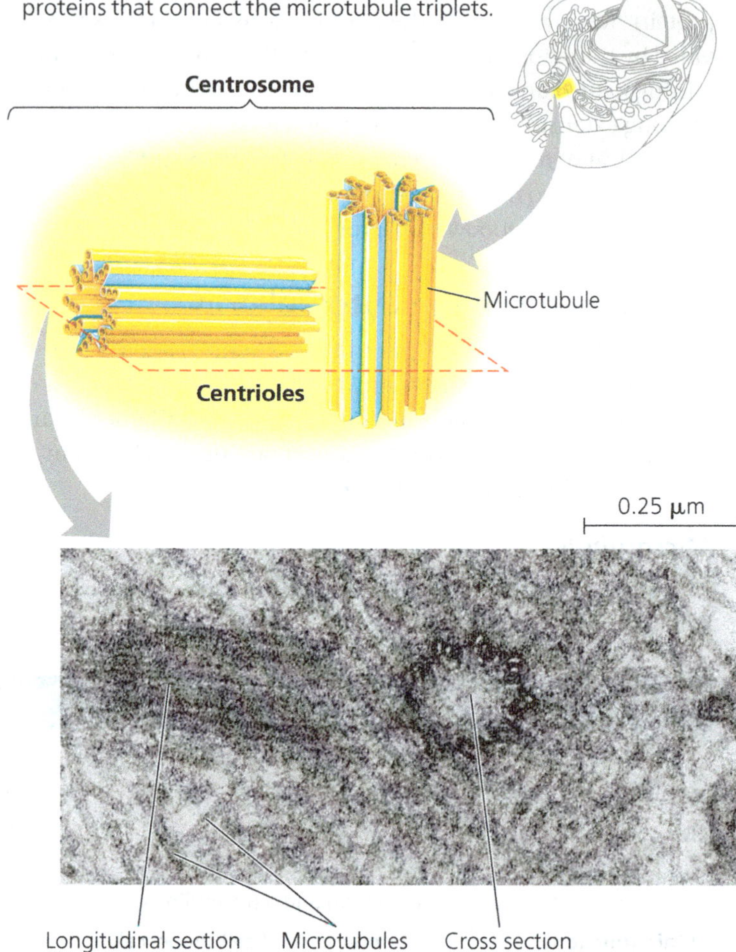

VISUAL SKILLS *How many microtubules are in a centrosome? In the drawing, circle and label one microtubule and describe its structure. Circle and label a triplet.*

Cilia and Flagella Some eukaryotic cells have **flagella** (singular, *flagellum*) and **cilia** (singular, *cilium*), cellular extensions that contain microtubules. A specialised arrangement of the microtubules is responsible for the beating of these structures. (The bacterial flagellum, shown in Figure 6.5, has a completely different structure.) Many unicellular protists are propelled through water by cilia or flagella that act as locomotor appendages, and the sperm of animals, algae, and some plants have flagella. When cilia or flagella extend from cells that are attached tightly together in a sheet that is part of a tissue layer, they can move fluid over the surface of the tissue. For example, the ciliated lining of the trachea (windpipe) sweeps mucus containing trapped debris out of the lungs (see the EMs in Figure 6.3). In a woman's reproductive tract, the cilia lining the oviducts help move an egg towards the uterus.

Motile cilia usually occur in large numbers on the cell surface. Flagella are usually limited to just one or a few per cell, and they are longer than cilia. Flagella and cilia differ in their beating patterns. A flagellum has an undulating motion like the tail of a fish. In contrast, cilia have alternating power and recovery strokes, much like the oars of a racing crew boat **(Figure 6.23)**.

A cilium may also act as a signal-receiving "antenna" for the cell. Cilia that have this function are generally nonmotile, and there is only one per cell. (In fact, in vertebrate animals, it appears that almost all cells have such a cilium, which is called a *primary cilium*.) Membrane proteins on this kind of cilium transmit molecular signals from the cell's environment to its interior, triggering signalling pathways that may lead to changes in the cell's activities. Cilium-based signalling appears to be crucial to brain function and to embryonic development.

▼ **Figure 6.23 A comparison of the beating of flagella and motile cilia.**

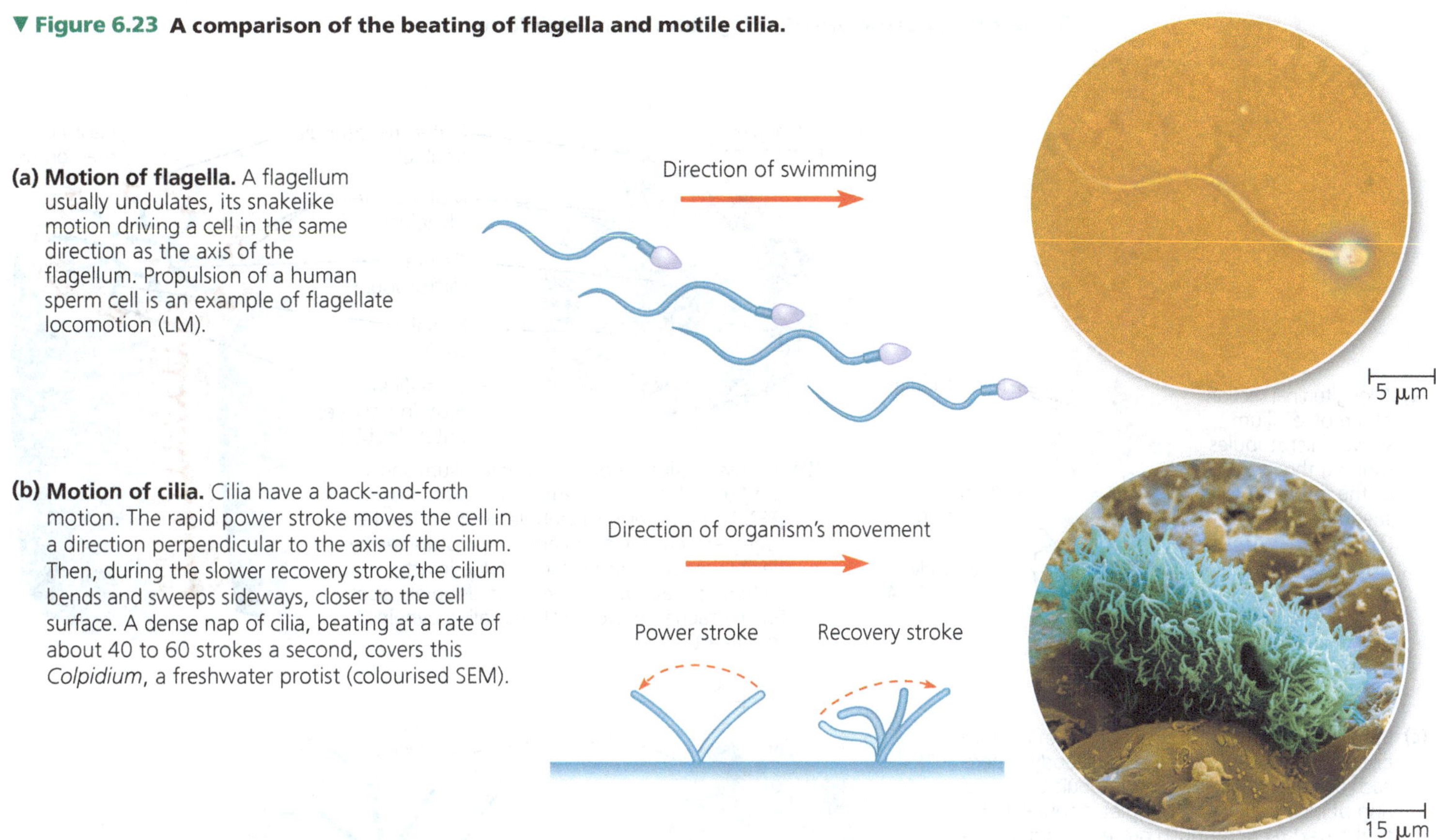

(a) Motion of flagella. A flagellum usually undulates, its snakelike motion driving a cell in the same direction as the axis of the flagellum. Propulsion of a human sperm cell is an example of flagellate locomotion (LM).

(b) Motion of cilia. Cilia have a back-and-forth motion. The rapid power stroke moves the cell in a direction perpendicular to the axis of the cilium. Then, during the slower recovery stroke,the cilium bends and sweeps sideways, closer to the cell surface. A dense nap of cilia, beating at a rate of about 40 to 60 strokes a second, covers this *Colpidium*, a freshwater protist (colourised SEM).

Though different in length, number per cell, and beating pattern, motile cilia and flagella share a common structure. Each motile cilium or flagellum has a group of microtubules sheathed in an extension of the plasma membrane **(Figure 6.24a)**. Nine doublets of microtubules are arranged in a ring with two single microtubules in its centre **(Figure 6.24b)**. This arrangement, referred to as the "9 + 2" pattern, is found in nearly all eukaryotic flagella and motile cilia. (Nonmotile primary cilia have a "9 + 0" pattern, lacking the central pair of microtubules.) The microtubule assembly of a cilium or flagellum is anchored in the cell by a **basal body**, which is structurally very similar to a centriole, with microtubule triplets in a "9 + 0" pattern **(Figure 6.24c)**. In fact, in many animals (including humans), the basal body of the fertilising sperm's flagellum enters the egg and becomes a centriole.

How does the microtubule assembly produce the bending movements of flagella and motile cilia? Bending involves large motor proteins called **dyneins** (red in the diagram in Figure 6.24) that are attached along each outer microtubule doublet. A typical dynein protein has two "feet" that "walk" along the microtubule of the adjacent doublet, using ATP for energy. One foot maintains contact, while the other releases and reattaches one step further along the microtubule (see Figure 6.21). The outer doublets and two central microtubules are held together by flexible cross-linking proteins (blue in the diagram in Figure 6.24), and the walking movement is coordinated so that it happens on one side of the circle at a time. If the doublets were not held in place, the walking action would make them slide past each other. Instead, the movements of the dynein feet cause the microtubules—and the organelle as a whole—to bend.

Microfilaments (Actin Filaments)

Microfilaments are thin solid rods. They are also called actin filaments because they are built from molecules of **actin**, a globular protein. A microfilament is a twisted double chain of actin subunits (see Table 6.1). Besides occurring as linear filaments, microfilaments can form structural networks when certain proteins bind along the side of such a filament and allow a new filament to extend as a branch. Like microtubules, microfilaments seem to be present in all eukaryotic cells.

In contrast to the compression-resisting role of microtubules, the structural role of microfilaments in the cytoskeleton is to bear tension (pulling forces). A three-dimensional

▼ Figure 6.24 Structure of a flagellum or motile cilium.

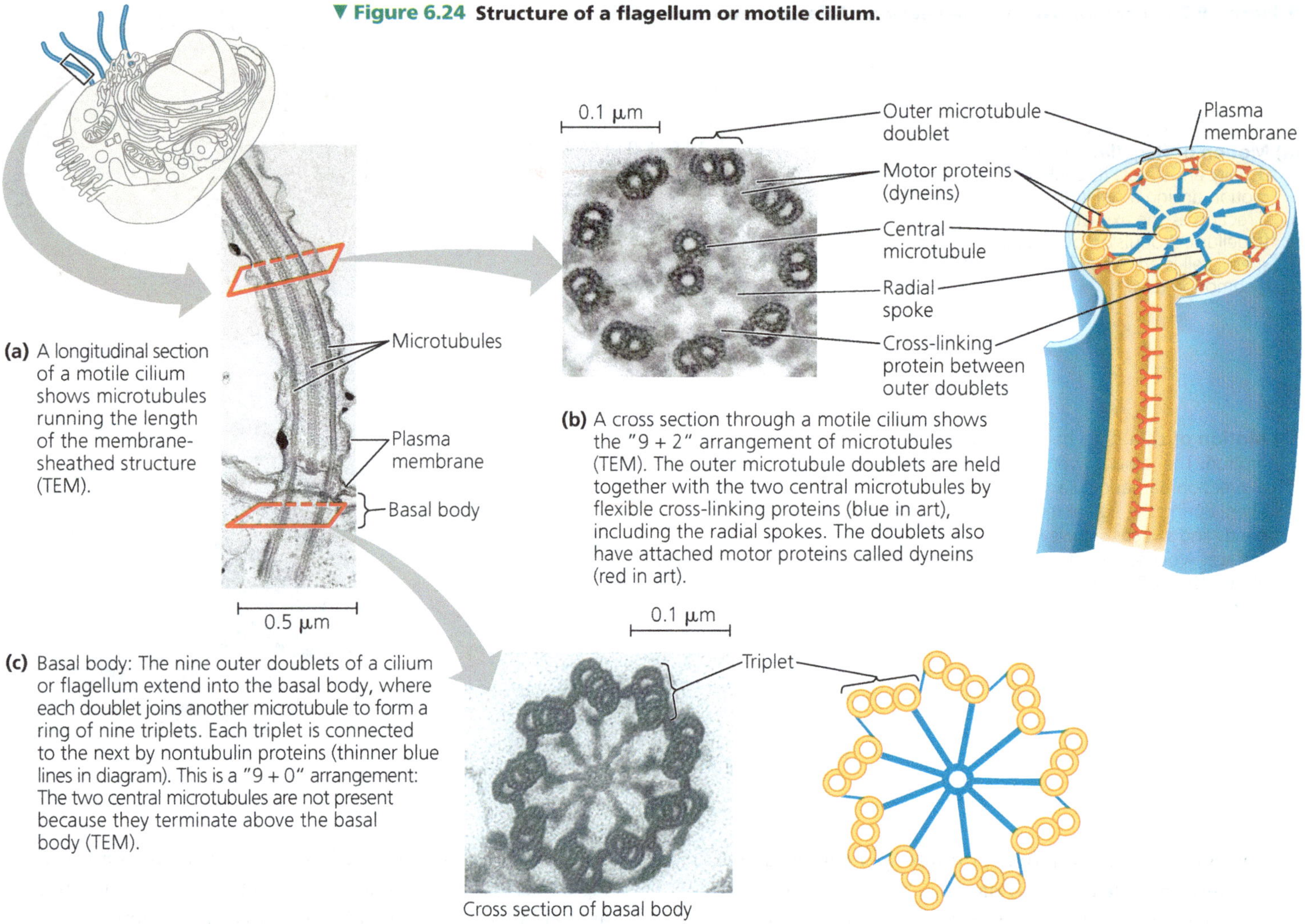

Microtubules

Plasma membrane

Basal body

0.5 μm

(a) A longitudinal section of a motile cilium shows microtubules running the length of the membrane-sheathed structure (TEM).

0.1 μm

Outer microtubule doublet

Motor proteins (dyneins)

Central microtubule

Radial spoke

Cross-linking protein between outer doublets

Plasma membrane

(b) A cross section through a motile cilium shows the "9 + 2" arrangement of microtubules (TEM). The outer microtubule doublets are held together with the two central microtubules by flexible cross-linking proteins (blue in art), including the radial spokes. The doublets also have attached motor proteins called dyneins (red in art).

0.1 μm

Triplet

Cross section of basal body

(c) Basal body: The nine outer doublets of a cilium or flagellum extend into the basal body, where each doublet joins another microtubule to form a ring of nine triplets. Each triplet is connected to the next by nontubulin proteins (thinner blue lines in diagram). This is a "9 + 0" arrangement: The two central microtubules are not present because they terminate above the basal body (TEM).

DRAW IT *In (a) and (b), circle and label the central pair of microtubules. In (a), show where they terminate, and explain why they aren't seen in the cross section of the basal body in (c).*

network formed by microfilaments just to the inside of the plasma membrane (*cortical microfilaments*) helps support the cell's shape (see Figure 6.8). This network gives the outer cytoplasmic layer of a cell, called the **cortex**, the semisolid consistency of a gel, in contrast with the more fluid state of the interior cytoplasm. In some kinds of animal cells, such as nutrient-absorbing intestinal cells, bundles of microfilaments make up the core of microvilli, delicate projections that increase the cell's surface area **(Figure 6.25)**.

Microfilaments are well known for their role in cell motility. Thousands of actin filaments and thicker filaments made of a protein called **myosin** interact to cause contraction of muscle cells **(Figure 6.26a)**; muscle contraction is described in

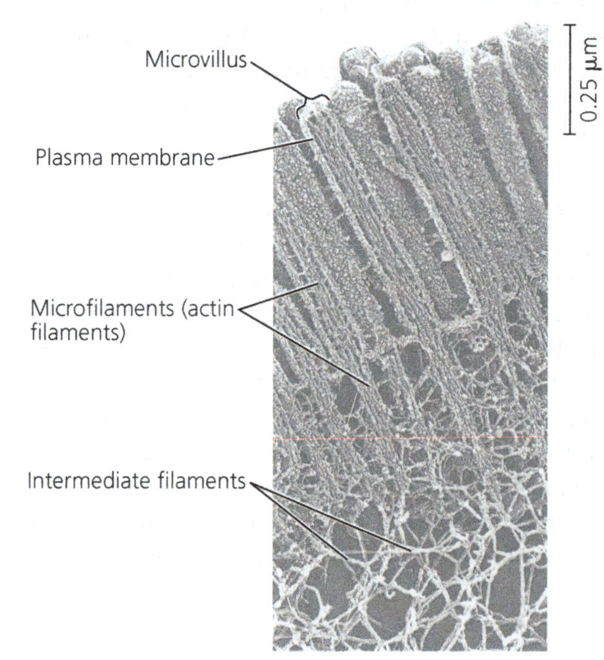

▶ Figure 6.25 A structural role of microfilaments. The surface area of this nutrient-absorbing intestinal cell is increased by its many microvilli (singular, *microvillus*), cellular extensions reinforced by bundles of microfilaments. These actin filaments are anchored to a network of intermediate filaments (TEM).

detail in Concept 50.5. In the unicellular protist *Amoeba* and some of our white blood cells, localised contractions brought about by actin and myosin are involved in the amoeboid (crawling) movement of the cells. The cell crawls along a surface by extending cellular extensions called **pseudopodia** (from the Greek *pseudes*, false, and *pod*, foot) and moving towards them **(Figure 6.26b)**. In plant cells, actin-protein interactions contribute to **cytoplasmic streaming**, a circular flow of cytoplasm within cells **(Figure 6.26c)**. This movement, which is especially common in large plant cells, speeds the movement of organelles and the distribution of materials within the cell.

Intermediate Filaments

Intermediate filaments are named for their diameter, which is larger than the diameter of microfilaments but smaller than that of microtubules (see Table 6.1). While microtubules and microfilaments are found in all eukaryotic cells, intermediate filaments are only found in the cells of some animals, including vertebrates. Specialised for bearing tension (like microfilaments), intermediate filaments are a diverse class of cytoskeletal elements. Each type is constructed from a particular molecular subunit belonging to a family of proteins whose members include the keratins. Microtubules and microfilaments, in contrast, are consistent in diameter and composition in all eukaryotic cells.

Intermediate filaments are more permanent fixtures of cells than are microfilaments and microtubules, which are often disassembled and reassembled in various parts of a cell. Even after cells die, intermediate filament networks often persist; for example, the outer layer of our skin consists of dead skin cells full of keratin filaments. Intermediate filaments are especially sturdy and play an important role in reinforcing the shape of a cell and fixing the position of certain organelles. For instance, the nucleus typically sits within a cage made of intermediate filaments. Other intermediate filaments make up the nuclear lamina, which lines the interior of the nuclear envelope (see Figure 6.9). In general, the various kinds of intermediate filaments seem to function together as the permanent framework of the entire cell.

CONCEPT CHECK 6.6

1. Describe how cilia and flagella bend.
2. **WHAT IF?** Males afflicted with Kartagener's syndrome are sterile because of immotile sperm, and they tend to suffer from lung infections. This disorder has a genetic basis. Suggest what the underlying defect might be.

For suggested answers, see Appendix A.

▼ **Figure 6.26 Microfilaments and motility.** In these three examples, interactions between actin filaments and motor proteins bring about cell movement.

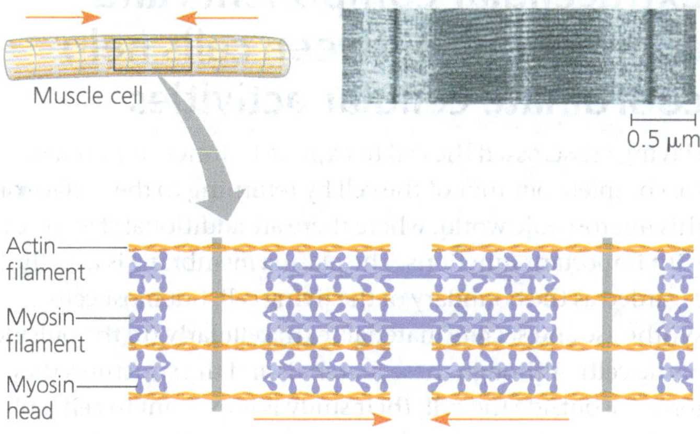

(a) Myosin motors in muscle cell contraction. The "walking" of myosin projections (the so-called heads) drives the parallel myosin and actin filaments past each other so that the actin filaments approach each other in the middle (red arrows). This shortens the muscle cell. Muscle contraction involves the shortening of many muscle cells at the same time (TEM).

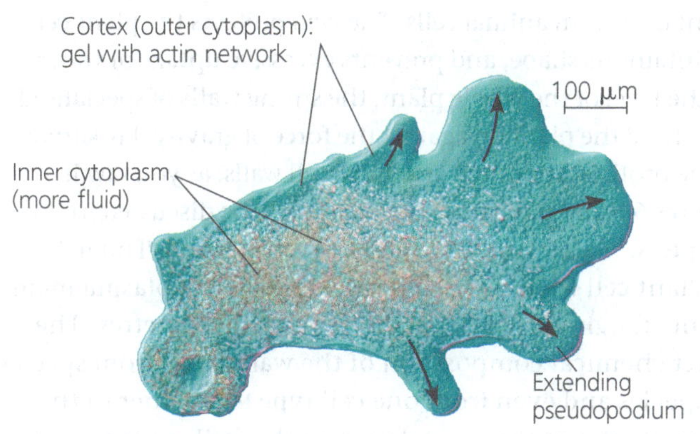

(b) Amoeboid movement. Interaction of actin filaments with myosin causes contraction of the cell, pulling the cell's trailing end (at left) forwards (to the right) (LM).

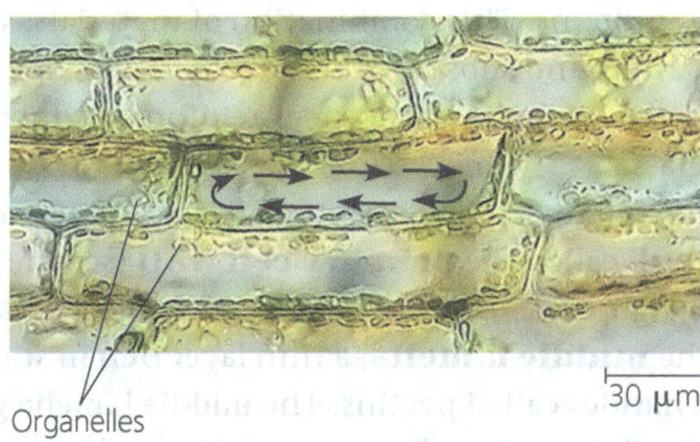

(c) Cytoplasmic streaming in plant cells. A layer of cytoplasm cycles around the cell, moving over tracks of actin filaments. Myosin motors attached to some organelles drive the streaming by interacting with the actin (LM).

CONCEPT 6.7

Extracellular components and connections between cells help coordinate cellular activities

Having crisscrossed the cell to explore its inner components, we complete our tour of the cell by returning to the surface of this microscopic world, where there are additional structures with important functions. The plasma membrane is usually regarded as the boundary of the living cell, but most cells synthesise and secrete materials extracellularly (to the outside of the cell). Although these materials and the structures they form are outside the cell, their study is important to cell biology because they are involved in a great many essential cellular functions.

Cell Walls of Plants

The **cell wall** is an extracellular structure of plant cells **(Figure 6.27)**. This is one of the features that distinguishes plant cells from animal cells. The wall protects the plant cell, maintains its shape, and prevents excessive uptake of water. At the level of the whole plant, the strong walls of specialised cells hold the plant up against the force of gravity. Prokaryotes, some protists, and fungi also have cell walls, as you saw in Figures 6.5 and 6.8. Those organisms will be discussed in the chapters, "Bacteria and Archaea," "Protists," and "Fungi."

Plant cell walls are much thicker than the plasma membrane, ranging from 0.1 μm to several micrometres. The exact chemical composition of the wall varies from species to species and even from one cell type to another in the same plant, but the basic design of the wall is consistent. Microfibrils made of the polysaccharide cellulose (see Figure 5.6) are synthesised by an enzyme called cellulose synthase and secreted to the extracellular space, where they become embedded in a matrix of other polysaccharides and proteins. This combination of materials, strong fibres in a "ground substance" (matrix), is the same basic architectural design found in steel-reinforced concrete and in fibreglass.

A young plant cell first secretes a relatively thin and flexible wall called the **primary cell wall** (see the micrograph in Figure 6.27). Between primary walls of adjacent cells is the **middle lamella**, a thin layer rich in sticky polysaccharides called pectins. The middle lamella glues adjacent cells together. (Pectin is used in cooking as a thickening agent in jams and jellies.) When the cell matures and stops growing, it strengthens its wall. Some plant cells do this simply by secreting hardening substances into the primary wall. Other cells add a **secondary cell wall** between the plasma membrane and the primary wall. The secondary wall, often deposited in several laminated layers, has a strong and durable matrix that affords the cell protection and support. Wood, for example, consists mainly of secondary walls.

▼ **Figure 6.27 Plant cell walls.** The drawing shows the relationship between primary and secondary cell walls in several mature plant cells. (Organelles aren't shown because many cells with secondary walls, such as the water-conducting cells, lack organelles.) The TEM shows the cell walls where two cells come together. The multilayered partition between plant cells consists of adjoining walls individually secreted by the cells. Adjacent cells are glued together by a very thin layer called the middle lamella.

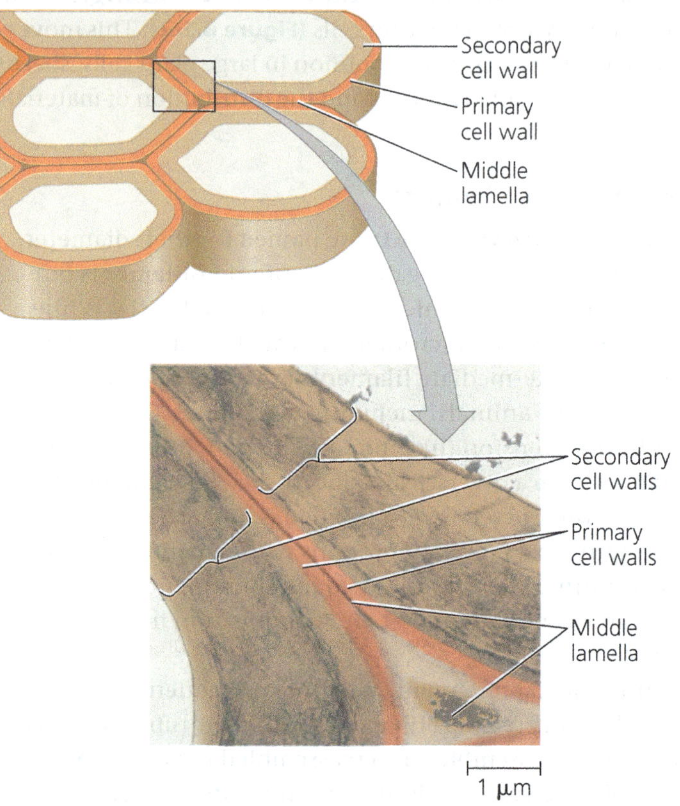

The Extracellular Matrix (ECM) of Animal Cells

Although animal cells lack walls akin to those of plant cells, they do have an elaborate **extracellular matrix (ECM)**. The main ingredients of the ECM are glycoproteins and other carbohydrate-containing molecules secreted by the cells. (Recall that glycoproteins are proteins with covalently bonded carbohydrates.) The most abundant glycoprotein in the ECM of most animal cells is **collagen**, which forms strong fibres outside the cells (see Figure 5.18). In fact, collagen accounts for about 40% of the total protein in the human body. The collagen fibres are embedded in a network woven out of **proteoglycans** secreted by

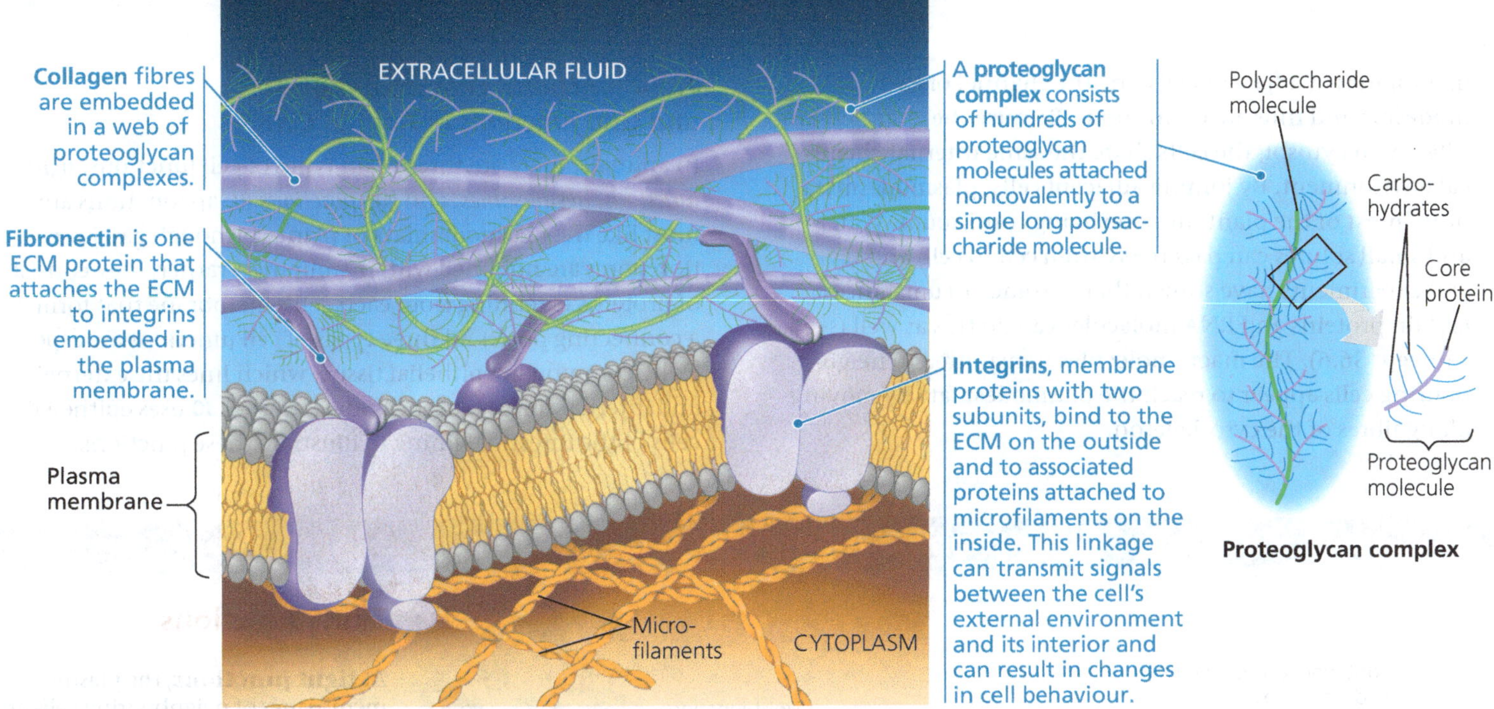

▲ **Figure 6.28 Extracellular matrix (ECM) of an animal cell.** The molecular composition and structure of the ECM vary from one cell type to another. In this example, three different types of ECM molecules are present: collagen, fibronectin, and proteoglycans.

cells **(Figure 6.28)**. A proteoglycan molecule consists of a small core protein with many carbohydrate chains covalently attached, so that it may be up to 95% carbohydrate. Large proteoglycan complexes can form when hundreds of proteoglycan molecules become noncovalently attached to a single long polysaccharide molecule, as shown in Figure 6.28. Some cells are attached to the ECM by ECM glycoproteins such as **fibronectin**. Fibronectin and other ECM proteins bind to cell-surface receptor proteins called **integrins** that are built into the plasma membrane. Integrins span the membrane and bind on their cytoplasmic side to associated proteins attached to microfilaments of the cytoskeleton. The name *integrin* is based on the word *integrate*: Integrins are in a position to transmit signals between the ECM and the cytoskeleton and thus to integrate changes occurring outside and inside the cell.

Current research on fibronectin, other ECM molecules, and integrins reveals the influential role of the ECM in the lives of cells. By communicating with a cell through integrins, the ECM can regulate a cell's behaviour. For example, some cells in a developing embryo migrate along specific pathways by matching the orientation of their microfilaments to the "grain" of fibres in the extracellular matrix. Researchers have also learned that the extracellular matrix around a cell can influence the activity of genes in the nucleus. Information about the ECM probably reaches the nucleus by a combination of mechanical and chemical signalling pathways. Mechanical signalling involves fibronectin, integrins, and microfilaments of the cytoskeleton. Changes in the cytoskeleton may in turn trigger signalling pathways inside the cell, leading to changes in the set of proteins being made by the cell and therefore changes in the cell's function. In this way, the extracellular matrix of a particular tissue may help coordinate the behaviour of all the cells of that tissue. Direct connections between cells also function in this coordination, as we'll explore next.

Cell Junctions

Cells in an animal or plant are organised into tissues, organs, and organ systems. Neighbouring cells often adhere, interact, and communicate via sites of direct physical contact.

Plasmodesmata in Plant Cells

It might seem that the nonliving cell walls of plants would isolate plant cells from one another. But in fact, as shown in **Figure 6.29**, many plant cell walls are perforated with **plasmodesmata** (singular, *plasmodesma*; from the Greek *desma*, bond), channels that connect cells. The plasma

▼ **Figure 6.29 Plasmodesmata between plant cells.** The cytoplasm of one plant cell is continuous with the cytoplasm of its neighbours via plasmodesmata, cytoplasmic channels through the cell walls (TEM).

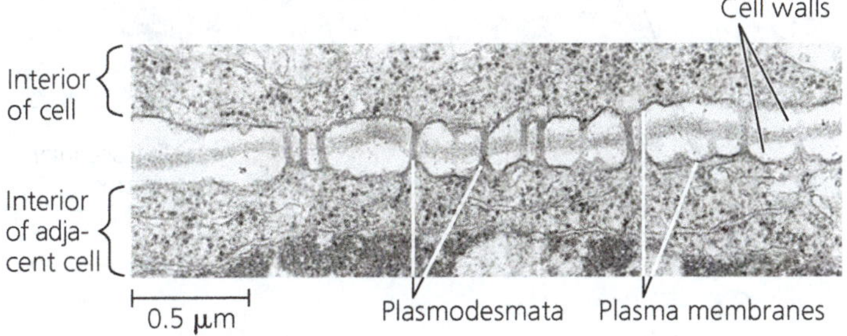

membranes of adjacent cells line the channel of each plasmodesma and thus are continuous. Because the channels are filled with cytosol, the cells share the same internal chemical environment. By joining adjacent cells, plasmodesmata unify most of the plant into one living continuum. Water and small solutes can pass freely from cell to cell, and several experiments have shown that in some circumstances, certain proteins and RNA molecules can do this as well (see Concept 36.6). The macromolecules transported to neighbouring cells appear to reach the plasmodesmata by moving along fibres of the cytoskeleton.

Tight Junctions, Desmosomes, and Gap Junctions in Animal Cells

In animals, there are three main types of cell junctions: *tight junctions, desmosomes*, and *gap junctions*. (Gap junctions are most like the plasmodesmata of plants, although gap junction pores are not lined with membrane—rather, they consist of proteins extending from each cell's membrane that form a connecting pore.) All three types of cell junctions are especially common in epithelial tissue, which lines the external and internal surfaces of the body. **Figure 6.30** uses epithelial cells of the intestinal lining to illustrate these junctions.

▼ Figure 6.30 Exploring Cell Junctions in Animal Tissues

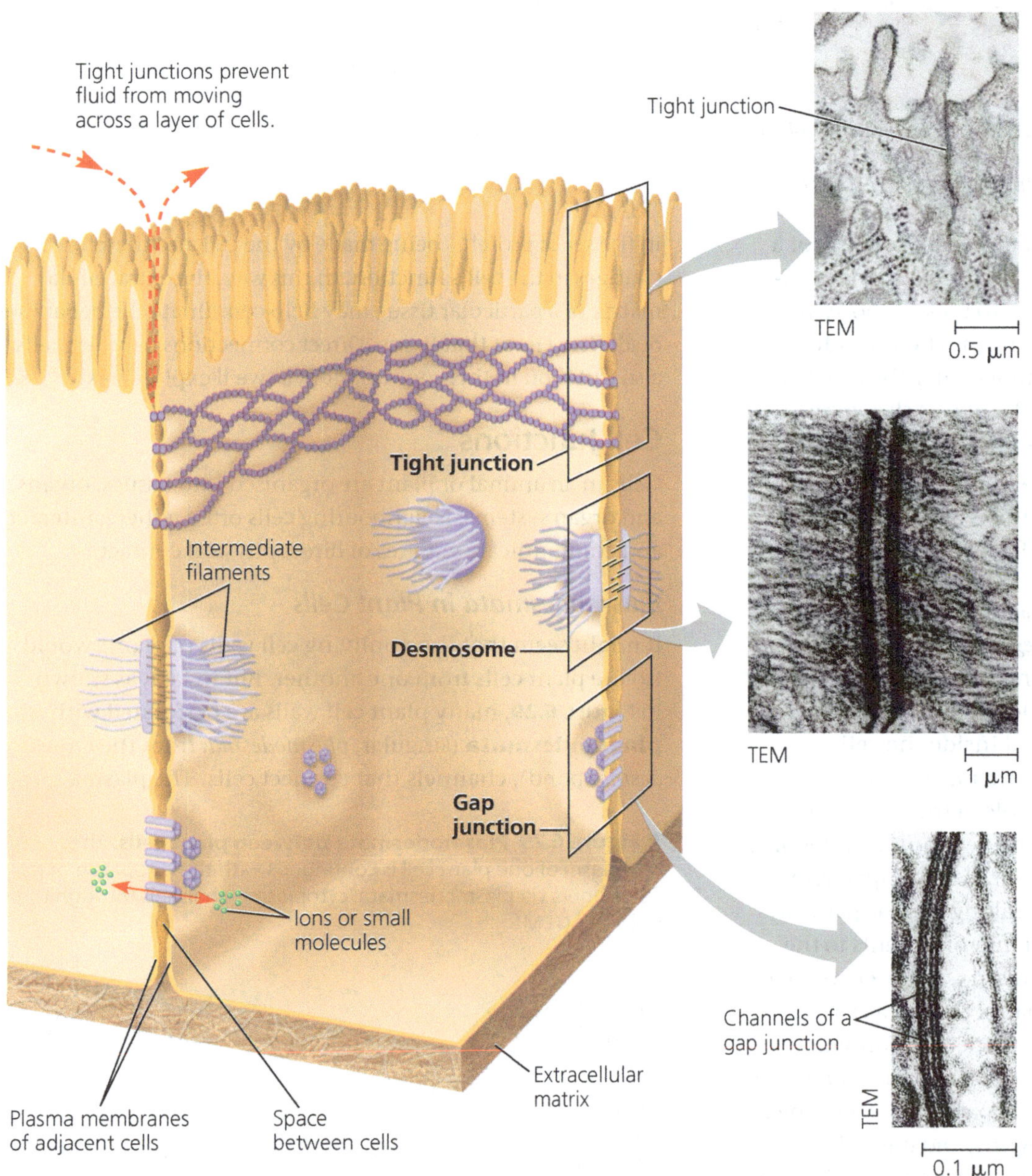

Tight Junctions

At **tight junctions**, the plasma membranes of neighbouring cells are very tightly pressed against each other, bound together by specific proteins. Forming continuous seals around the cells, tight junctions establish a barrier that prevents leakage of extracellular fluid across a layer of epithelial cells (see red dashed arrow). For example, tight junctions between skin cells make us watertight.

Desmosomes

Desmosomes (one type of *anchoring junction*) function like rivets, fastening cells together into strong sheets. Intermediate filaments made of sturdy keratin proteins anchor desmosomes in the cytoplasm. Desmosomes attach muscle cells to each other in a muscle. Some "muscle tears" involve the rupture of desmosomes.

Gap Junctions

Gap junctions (also called *communicating junctions*) provide cytoplasmic channels from one cell to an adjacent cell and in this way are similar in their function to the plasmodesmata in plants. Gap junctions consist of membrane proteins extending from the membranes of the two cells. These proteins create pores through which ions, sugars, amino acids, and other small molecules may pass. Gap junctions are necessary for communication between cells in many types of tissues, such as heart muscle, and in animal embryos.

CONCEPT CHECK 6.7

1. In what way are the cells of plants and animals structurally different from single-celled eukaryotes?
2. **WHAT IF?** If the plant cell wall or the animal extracellular matrix were impermeable, what effect would this have on cell function?
3. **MAKE CONNECTIONS** The polypeptide chain that makes up a tight junction weaves back and forth through the membrane four times, with two extracellular loops and one loop plus short C-terminal and N-terminal tails in the cytoplasm. Looking at Figure 5.14, what would you predict about the amino acid sequence of the tight junction protein?

For suggested answers, see Appendix A.

CONCEPT 6.8

A cell is greater than the sum of its parts

From our panoramic view of the cell's compartmental organisation to our close-up inspection of each organelle's architecture, this tour of the cell has shown the correlation of structure with function. (See Figure 6.8 to review cell structure.)

Remember that none of a cell's components work alone. For example, consider the microscopic scene in **Figure 6.31**. The large cell is a macrophage (see Figure 6.13a). It helps defend the mammalian body against infections by ingesting bacteria (the smaller cells) into phagocytic vesicles. The macrophage crawls along a surface and reaches out to the bacteria with thin pseudopodia (specifically, filopodia). Actin filaments interact with other elements of the cytoskeleton in these movements. After the macrophage engulfs the bacteria, they are destroyed by lysosomes produced by the elaborate endomembrane system. The digestive enzymes of the lysosomes and the proteins of the cytoskeleton are all made by ribosomes. And the synthesis of these proteins is programmed by genetic messages dispatched from the DNA in the nucleus. All these processes require energy, which mitochondria supply in the form of ATP.

Cellular functions arise from cellular order: The cell is a living unit greater than the sum of its parts. The cell in Figure 6.31 is a good example of integration of cellular processes, seen from the outside. But what about the internal organisation of a cell? As you proceed in your study of biology to consider different cellular processes, it will be helpful to try to visualise the architecture and furnishings inside a cell. **Figure 6.32** is designed to introduce you to some important biological molecules and molecules and to help you get a sense of their relative sizes and organisation in the context of cellular structures and organelles. As you study this figure, see if you can shrink yourself down to the size of a protein and contemplate your surroundings.

CONCEPT CHECK 6.8

1. *Colpidium colpoda* is a unicellular protist that lives in freshwater, eats bacteria, and moves by cilia (see Figure 6.23b). Describe how the parts of this cell work together in the functioning of *C. colpoda*, including as many organelles and other cell structures as you can.

For suggested answers, see Appendix A.

▼ **Figure 6.31 Coordination of activities in a cell.** The ability of this macrophage to recognise, apprehend, and destroy bacteria involves coordination among components such as the cytoskeleton, lysosomes, and plasma membrane (colourised SEM).

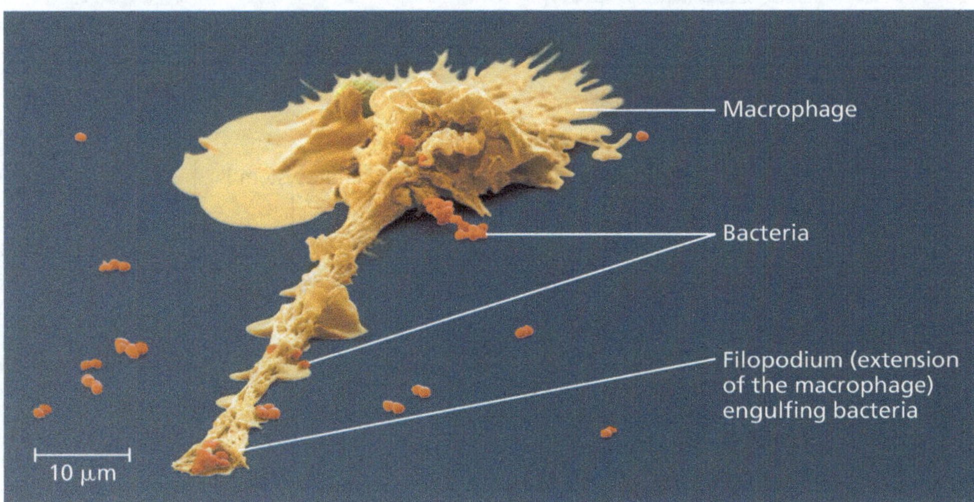

▼ Figure 6.32 Visualising the Scale of the Molecular Machinery in a Cell

A slice of a plant cell's interior is illustrated in the centre panel, with all structures and molecules drawn to scale. Selected molecules and structures are shown above and below, all enlarged by the same factor so you can compare their sizes. All protein and nucleic acid structures are based on data from the Protein Data Bank; regions whose structure has not yet been determined are shown in grey.

This figure introduces a cast of characters that you will learn more about as you study biology. Refer back to this figure as you encounter these molecules in your studies.

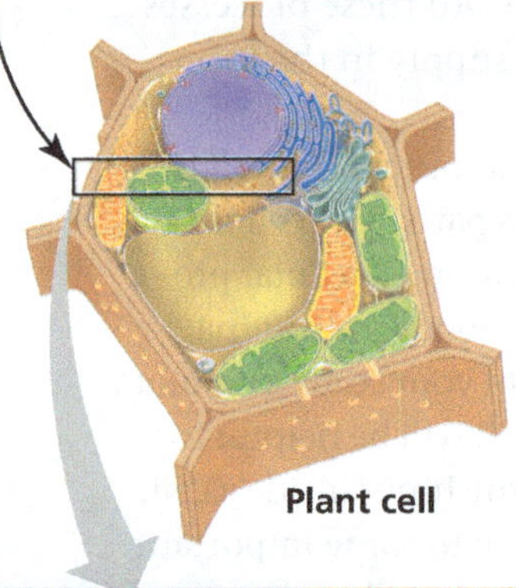

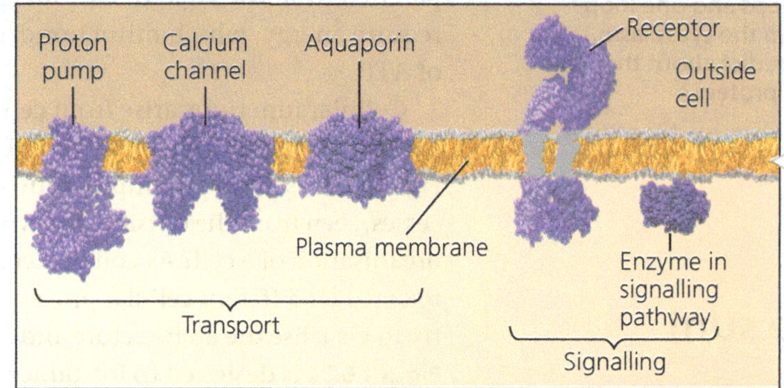

(a) Membrane proteins (Chapter 7) Proteins embedded in cellular membranes help transport substances and conduct signals across membranes. They also participate in other crucial cellular functions. Many proteins are able to move within the membrane.

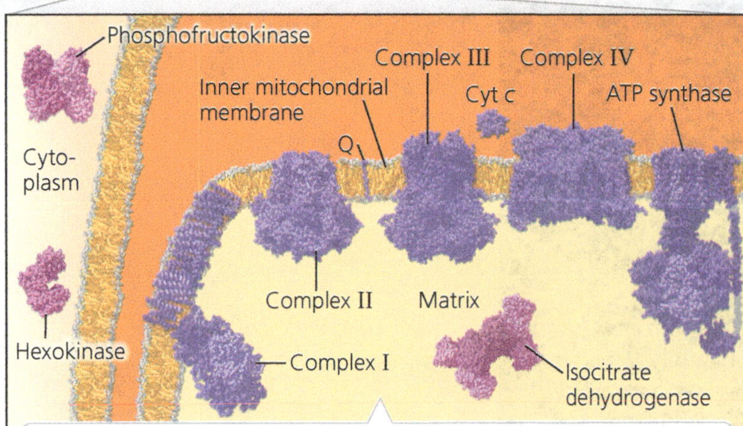

(b) Cellular respiration (Chapter 9) Cellular respiration, a multi-step process, generates ATP from food molecules. The first two stages are carried out by enzymes in the cytoplasm and mitochondrial matrix; a few of these enzymes (pinkish-purple) are shown. The final stage is carried out by proteins (bluish-purple) that form a "chain" in the inner mitochondrial membrane.

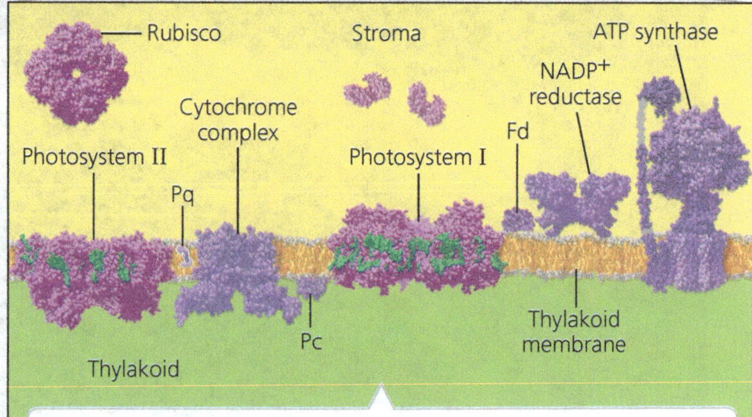

(c) Photosynthesis (Chapter 10) Photosynthesis produces sugars that provide food for all life on the planet. The process begins with large complexes of proteins and chlorophyll (green) embedded in the thylakoid membranes. These complexes trap light energy in molecules that are used by rubisco and other proteins in the stroma to make sugars.

(d) Transcription (Chapter 17) In the nucleus, the information contained in a DNA sequence is transferred to messenger RNA (mRNA) by an enzyme called RNA polymerase. After their synthesis, mRNA molecules leave the nucleus via nuclear pores.

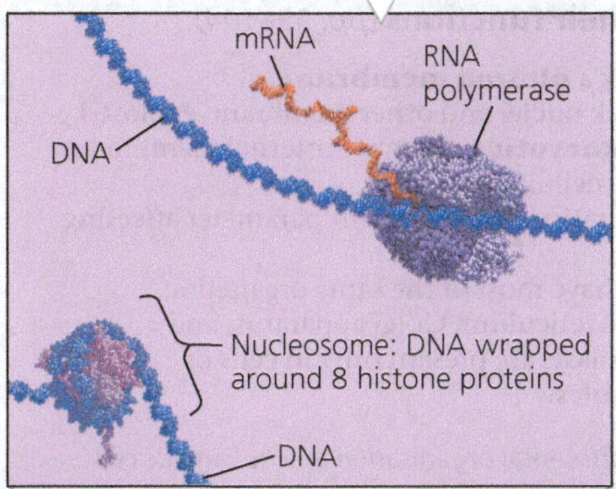

(e) Nuclear pore (Concept 6.3) The nuclear pore complex regulates molecular traffic in and out of the nucleus, which is bounded by a double membrane. Among the largest structures that pass through the pore are the ribosomal subunits, which are built in the nucleus.

(f) Translation (Chapter 17) In the cytoplasm, the information in mRNA is used to assemble a polypeptide with a specific sequence of amino acids. Both transfer RNA (tRNA) molecules and ribosomes play a role. The eukaryotic ribosome, which includes a large subunit and a small subunit, is a colossal complex composed of four large ribosomal RNA (rRNA) molecules and more than 80 proteins. Through transcription and translation, the nucleotide sequence of DNA in a gene determines the amino acid sequence of a polypeptide, via the intermediary mRNA.

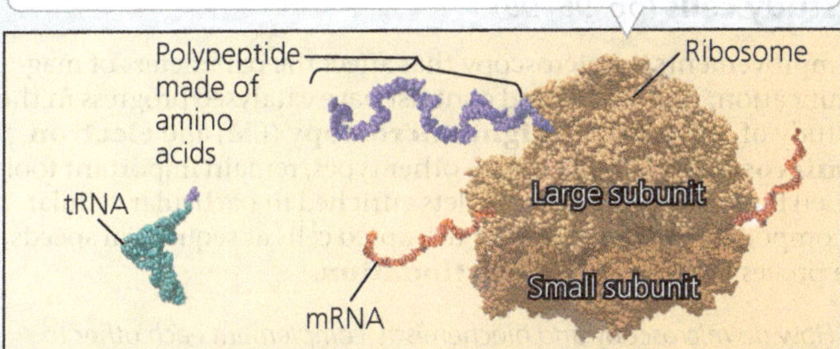

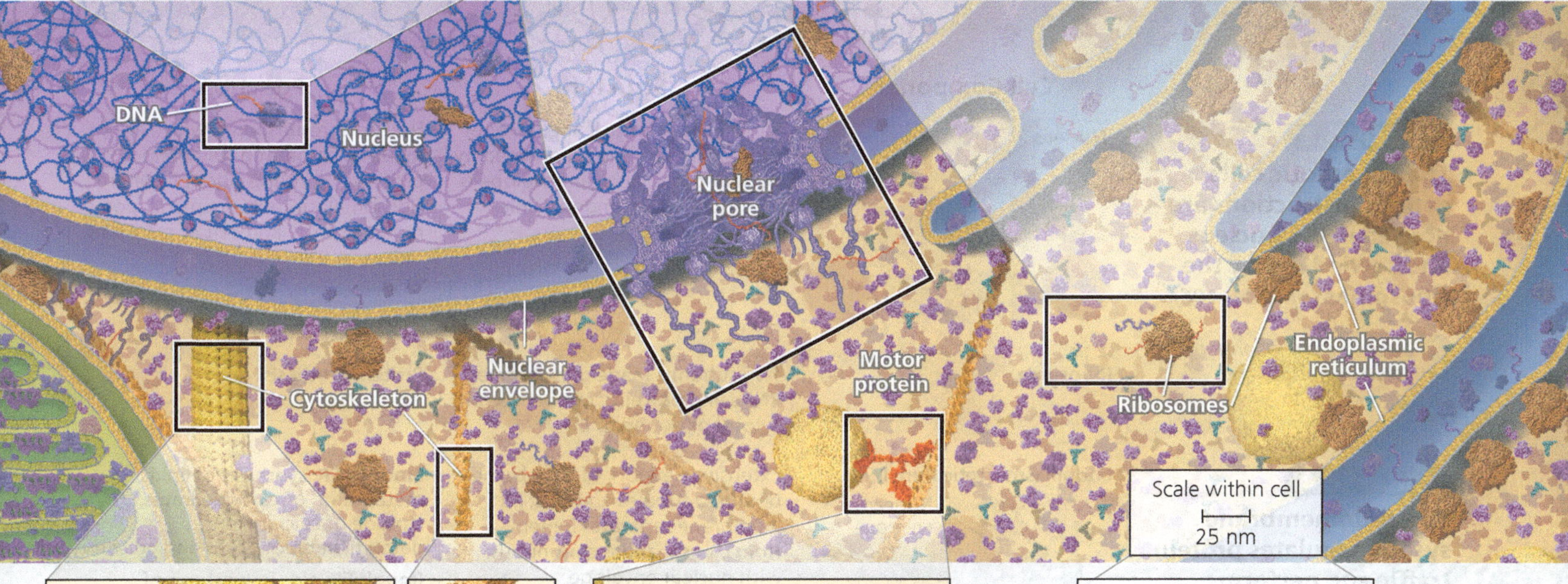

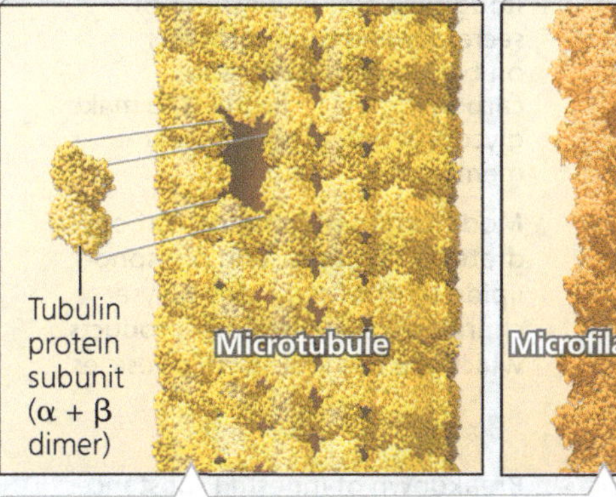

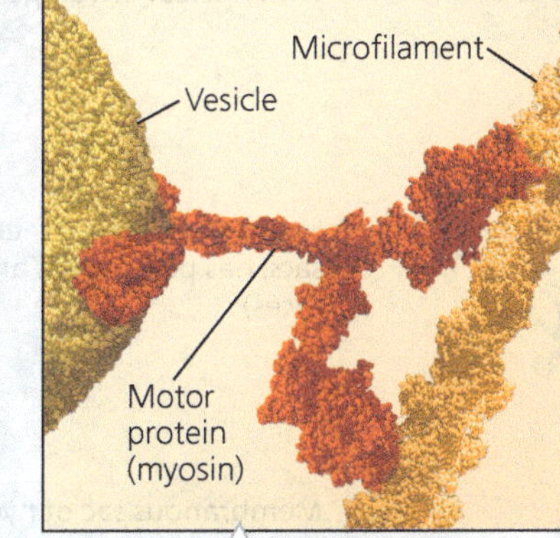

(g) Cytoskeleton (Concept 6.6)
Cytoskeletal structures are polymers of protein subunits. Microtubules are hollow structural rods made of tubulin protein subunits, while microfilaments are cables that have two chains of actin proteins wound around each other.

(h) Motor proteins (Concept 6.6) Motor proteins, such as myosin, are responsible for transport of vesicles and movement of organelles within the cell.

1. ***List the following structures from largest to smallest: proton pump, nuclear pore, cyt c, ribosome.***
2. ***Considering the structures of a nucleosome and of RNA polymerase, speculate about what must happen before RNA polymerase can transcribe the DNA that is wrapped around the histone proteins of a nucleosome.***
3. ***Find another myosin motor protein walking on a microfilament in this figure. What organelle is being moved by that myosin protein?***

6 Chapter Review

SUMMARY OF KEY CONCEPTS

CONCEPT **6.1**

Biologists use microscopes and biochemistry to study cells *(pp. 96–99)*

- Improvements in microscopy that affect the parameters of magnification, resolution, and contrast have catalysed progress in the study of cell structure. **Light microscopy** (LM) and **electron microscopy** (EM), as well as other types, remain important tools.
- Cell biologists can obtain pellets enriched in particular cellular components by centrifuging disrupted cells at sequential speeds, a process known as **cell fractionation**.

? *How do microscopy and biochemistry complement each other to reveal cell structure and function?*

CONCEPT **6.2**

Eukaryotic cells have internal membranes that compartmentalise their functions *(pp. 99–104)*

- All cells are bounded by a **plasma membrane**.
- **Prokaryotic cells** lack nuclei and other membrane-enclosed **organelles**, while **eukaryotic cells** have internal membranes that compartmentalise cellular functions.
- The surface-to-volume ratio is an important parameter affecting cell size and shape.
- Plant and animal cells have most of the same organelles: a nucleus, endoplasmic reticulum, Golgi apparatus, and mitochondria. Chloroplasts are present only in cells of photosynthetic eukaryotes.

? *Explain how the compartmental organisation of a eukaryotic cell contributes to its biochemical functioning.*

	Cell Component	Structure	Function
CONCEPT **6.3** **The eukaryotic cell's genetic instructions are housed in the nucleus and carried out by the ribosomes** *(pp. 104–106)* ? *Describe the relationship between the nucleus and ribosomes.*	**Nucleus** ER	Surrounded by nuclear envelope (double membrane) perforated by nuclear pores; nuclear envelope is continuous with endoplasmic reticulum (ER)	Houses chromosomes, which are made of chromatin (DNA and proteins); contains nucleoli, where ribosomal subunits are made; pores regulate entry and exit of materials
	Ribosome	Two subunits made of ribosomal RNAs and proteins; can be free in cytosol or bound to ER	Protein synthesis
CONCEPT **6.4** **The endomembrane system regulates protein traffic and performs metabolic functions** *(pp. 106–111)* ? *Describe the key role played by transport vesicles in the endomembrane system.*	**Endoplasmic reticulum (ER)** Nuclear envelope	Extensive network of membrane-bounded tubules and sacs; ER membrane separates lumen from cytosol and is continuous with nuclear envelope	**Smooth ER:** synthesis of lipids, metabolism of carbohydrates, storage of calcium ions, detoxification of drugs and poisons **Rough ER:** aids in synthesis of secretory and other proteins on bound ribosomes; adds carbohydrates to proteins to make glycoproteins; produces new membrane
	Golgi apparatus	Stacks of flattened membranous sacs; has polarity (*cis* and *trans* faces)	Modification of proteins, carbohydrates on proteins, and phospholipids; synthesis of many polysaccharides; sorting of Golgi products, which are then released in vesicles
	Lysosome	Membranous sac of hydrolytic enzymes (in animal cells)	Breakdown of ingested substances, cell macromolecules, and damaged organelles for recycling
	Vacuole	Large membrane-bounded vesicle	Digestion, storage, waste disposal, water balance, cell growth, and protection

	Cell Component	Structure	Function
CONCEPT 6.5 **Mitochondria and chloroplasts change energy from one form to another** *(pp. 111–114)* ? *What does the serial endosymbiont theory propose as the origin for mitochondria and chloroplasts? Explain.*	Mitochondrion	Bounded by double membrane; inner membrane has infoldings	Cellular respiration
	Chloroplast	Typically two membranes around fluid stroma, which contains thylakoids stacked into grana	Photosynthesis (chloroplasts are present in cells of photosynthetic eukaryotes, including plants)
	Peroxisome	Specialised metabolic compartment bounded by a single membrane	Contains enzymes that transfer H atoms from substrates to oxygen, producing H_2O_2 (hydrogen peroxide), which is converted to H_2O

CONCEPT 6.6

The cytoskeleton is a network of fibres that organises structures and activities in the cell *(pp. 114–119)*

- The **cytoskeleton** functions in structural support for the cell and in motility and signal transmission.
- **Microtubules** shape the cell, guide organelle movement, and separate chromosomes in dividing cells. **Cilia** and **flagella** are motile appendages containing microtubules. Primary cilia also play sensory and signalling roles. **Microfilaments** are thin rods that function in muscle contraction, amoeboid movement, **cytoplasmic streaming**, and support of microvilli. **Intermediate filaments** support cell shape and fix organelles in place.

? *Describe the role of motor proteins inside the eukaryotic cell and in whole-cell movement.*

CONCEPT 6.7

Extracellular components and connections between cells help coordinate cellular activities *(pp. 120–123)*

- Plant **cell walls** are made of cellulose fibres embedded in other polysaccharides and proteins.
- Animal cells secrete glycoproteins and **proteoglycans** that form the **extracellular matrix (ECM)**, which functions in support, adhesion, movement, and regulation.
- Cell junctions connect neighbouring cells. Plants have **plasmodesmata** that pass through adjoining cell walls. Animal cells have **tight junctions, desmosomes**, and **gap junctions**.

? *Compare the structure and functions of a plant cell wall and the extracellular matrix of an animal cell.*

CONCEPT 6.8

A cell is greater than the sum of its parts *(pp. 123–125)*

Many components work together in a functioning cell.

? *When a cell ingests a bacterium, what role does the nucleus play?*

TEST YOUR UNDERSTANDING

Levels 1-2: Remembering/Understanding

1. Which structure is part of the endomembrane system?
   (A) mitochondrion (C) chloroplast
   (B) Golgi apparatus (D) centrosome
2. Which structure is common to plant *and* animal cells?
   (A) chloroplast (C) mitochondrion
   (B) central vacuole (D) centriole
3. Which of the following is present in a prokaryotic cell?
   (A) mitochondrion (C) nuclear envelope
   (B) ribosome (D) chloroplast

Levels 3-4: Applying/Analysing

4. Cyanide binds to at least one molecule involved in producing ATP. In a cell exposed to cyanide, most of the cyanide will be in
   (A) mitochondria. (C) peroxisomes.
   (B) ribosomes. (D) lysosomes.
5. Which cell would be best for studying lysosomes?
   (A) muscle cell (C) bacterial cell
   (B) nerve cell (D) phagocytic white blood cell
6. **DRAW IT** Draw two eukaryotic cells. Label the structures listed here and show any physical connections between the structures of each cell: nucleus, rough ER, smooth ER, mitochondrion, centrosome, chloroplast, vacuole, lysosome, microtubule, cell wall, ECM, microfilament, Golgi apparatus, intermediate filament, plasma membrane, peroxisome, ribosome, nucleolus, nuclear pore, vesicle, flagellum, microvilli, plasmodesma.

Levels 5-6: Evaluating/Creating

7. **EVOLUTION CONNECTION** (a) What cell structures best reveal evolutionary unity? (b) Give an example of diversity related to specialised cellular modifications.
8. **SCIENTIFIC INQUIRY** Imagine protein X, destined to span the plasma membrane; assume the mRNA carrying the genetic message for protein X has been translated by ribosomes in a cell culture. If you fractionate the cells (see Figure 6.4), in which fraction would you find protein X? Explain by describing its transit.
9. **WRITE ABOUT A THEME: ORGANISATION** Write a short essay (100–150 words) on this topic: Life is an emergent property that appears at the level of the cell. (See Concept 1.1.)
10. **SYNTHESISE YOUR KNOWLEDGE**

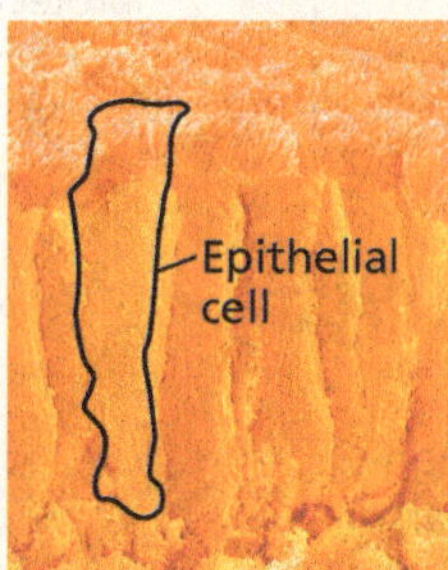

The cells in this SEM are epithelial cells from the small intestine. Discuss how their cellular structure contributes to their specialised functions of nutrient absorption and as a barrier between the intestinal contents and the blood supply on the other side of the sheet of epithelial cells.

For selected answers, see Appendix A.

7 Membrane Structure and Function

KEY CONCEPTS

Study Tip

Make a visual study guide: Draw a plasma membrane (two lines) all the way down a piece of paper (or digitally). Label the cytoplasm and the extracellular fluid. At the top, draw the phospholipid bilayer in detail. As you read the chapter, draw and label membrane proteins that you encounter and diagram the different ways materials can enter or leave a cell.

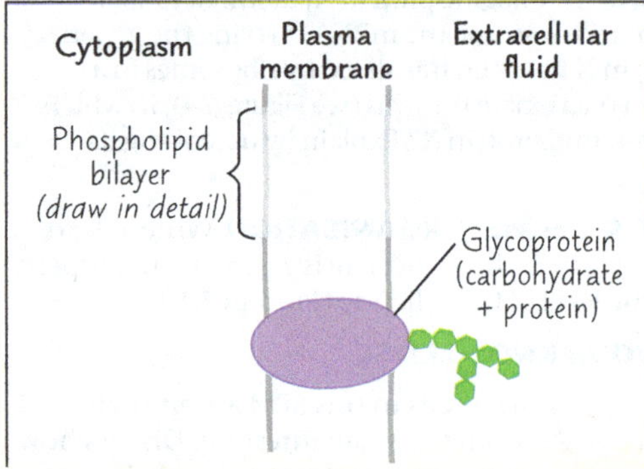

Go to Mastering Biology

to access Dynamic Study Modules for revision, 3D BioFlix® animations and high-quality videos, and your interactive Pearson eText.

Figure 7.1 Successful learning relies on communication between brain cells. Here, the vesicles fusing with the plasma membrane of the top cell release molecules (yellow) that bind to membrane proteins (light green) on the surface of the bottom cell, triggering the proteins to change shape. The plasma membrane that surrounds each cell regulates its exchanges with its environment and surrounding cells.

How does the plasma membrane regulate inbound and outbound traffic?

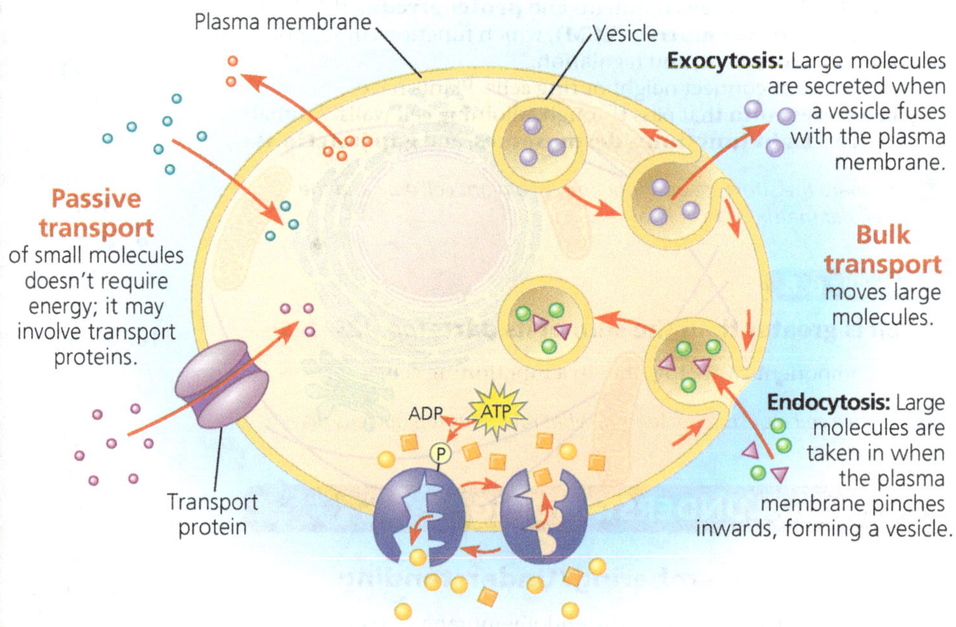

CONCEPT 7.1

Cellular membranes are fluid mosaics of lipids and proteins

Lipids and proteins are the staple ingredients of membranes, although carbohydrates are also important. The most abundant lipids in most membranes are phospholipids. Their ability to form membranes is inherent in their molecular structure. A phospholipid is an **amphipathic** molecule, meaning it has both a hydrophilic ("water-loving") region and a hydrophobic ("water-fearing") region (see Figure 5.11). A phospholipid bilayer can exist as a stable boundary between two aqueous compartments because the molecular arrangement shelters the hydrophobic tails of the phospholipids from water while exposing the hydrophilic heads to water **(Figure 7.2)**.

Like membrane lipids, most membrane proteins are amphipathic. Such proteins can reside in the phospholipid bilayer with their hydrophilic regions protruding. This molecular orientation maximises contact of hydrophilic regions of a protein with water in the cytosol and extracellular fluid, while providing their hydrophobic parts with a nonaqueous environment.

▼ **Figure 7.2 Phospholipid bilayer (cross section).**

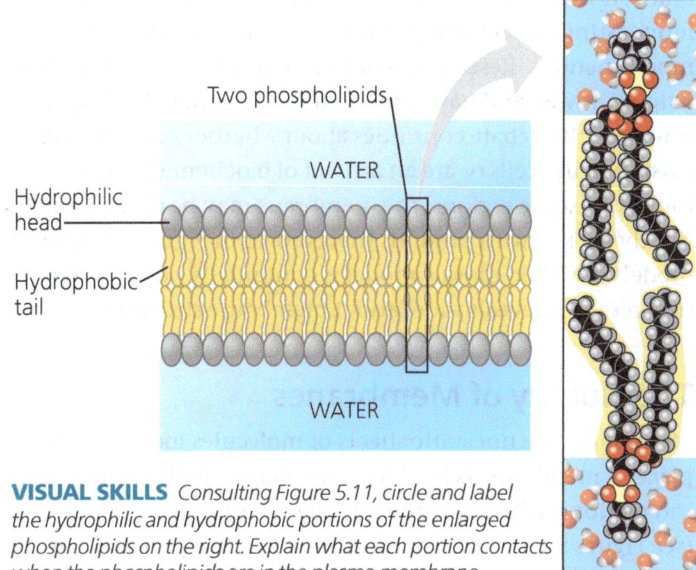

VISUAL SKILLS *Consulting Figure 5.11, circle and label the hydrophilic and hydrophobic portions of the enlarged phospholipids on the right. Explain what each portion contacts when the phospholipids are in the plasma membrane.*

Figure 7.3 shows the currently accepted model of the arrangement of molecules in the plasma membrane. In this **fluid mosaic model**, the membrane is a mosaic of protein molecules bobbing in a fluid bilayer of phospholipids.

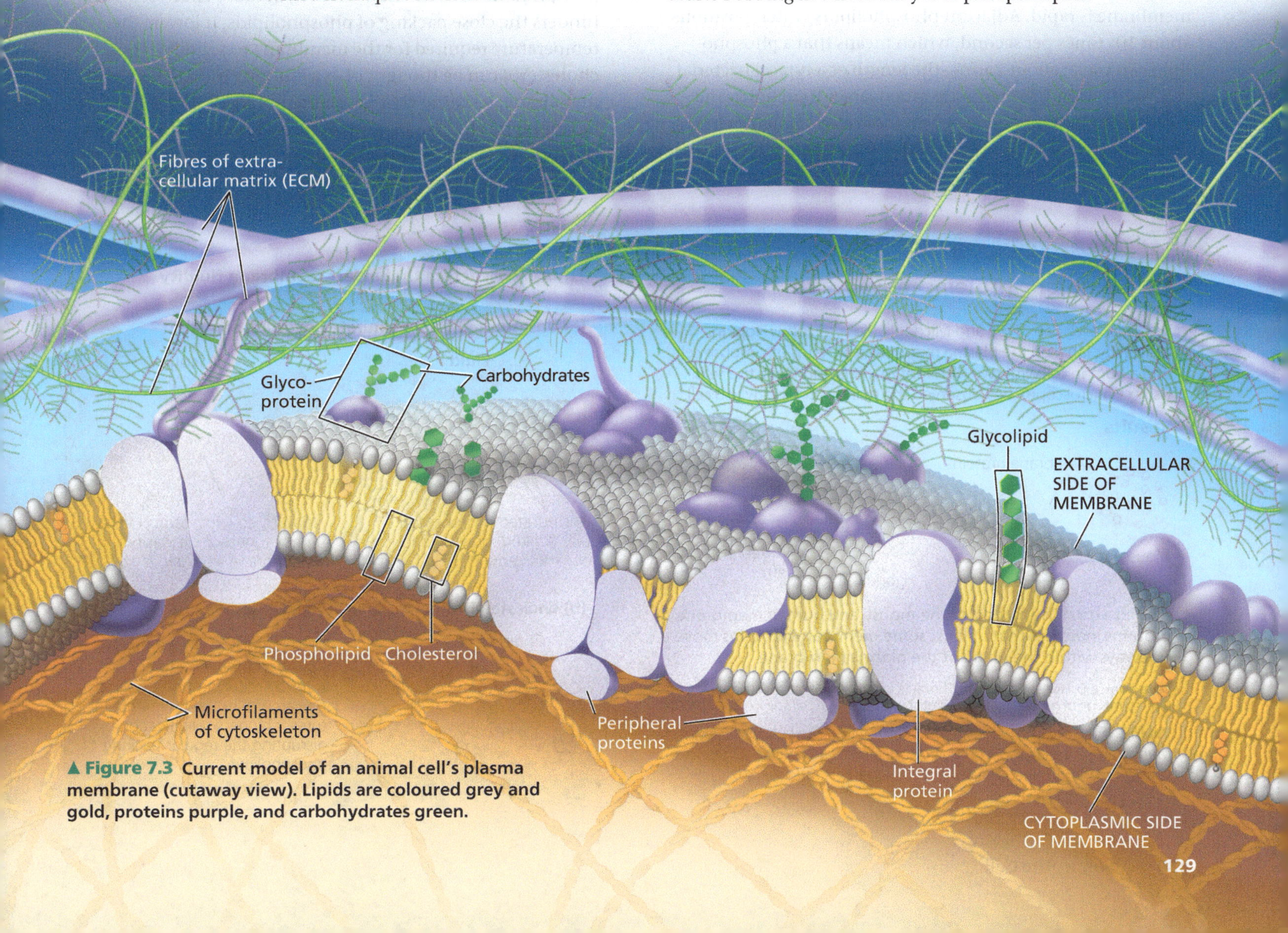

Fibres of extracellular matrix (ECM)
Glycoprotein
Carbohydrates
Glycolipid
EXTRACELLULAR SIDE OF MEMBRANE
Phospholipid
Cholesterol
Microfilaments of cytoskeleton
Peripheral proteins
Integral protein
CYTOPLASMIC SIDE OF MEMBRANE

▲ **Figure 7.3 Current model of an animal cell's plasma membrane (cutaway view). Lipids are coloured grey and gold, proteins purple, and carbohydrates green.**

The proteins are not randomly distributed in the membrane, however. Groups of proteins are often associated in long-lasting, specialised patches, where they carry out common functions. Researchers have found specific lipids in these patches as well and have proposed naming them *lipid rafts*; however, the debate continues about whether such structures exist in living cells or are an artifact of biochemical techniques. In some regions, the membrane may be much more tightly packed with proteins than shown in Figure 7.3. Like all models, the fluid mosaic model is continually being refined as new research reveals more about membrane structure.

The Fluidity of Membranes

Membranes are not static sheets of molecules locked rigidly in place. A membrane is held together mainly by hydrophobic interactions, which are much weaker than covalent bonds (see Figure 5.18). Most of the lipids and some proteins can shift about sideways—that is, in the plane of the membrane, like partygoers elbowing their way through a crowded room. Very rarely, also, a lipid may flip-flop across the membrane, switching from one phospholipid layer to the other.

The sideways movement of phospholipids within the membrane is rapid. Adjacent phospholipids switch positions about 10^7 times per second, which means that a phospholipid can travel about 2 μm—the length of a typical bacterial cell—in 1 second. Proteins are much larger than lipids and move more slowly, when they do move. Many membrane proteins seem to be held immobile by their attachment to the cytoskeleton or to the extracellular matrix (see Figure 7.3). Some membrane proteins seem to move in a highly directed manner, perhaps driven along cytoskeletal fibres by motor proteins. And other proteins simply drift in the membrane, as shown the classic experiment described in **Figure 7.4**.

A membrane remains fluid as temperature decreases until the phospholipids settle into a closely packed arrangement and the membrane solidifies, much as bacon fat forms lard when it cools. The temperature at which a membrane solidifies depends on the types of lipids it is made of. As the temperature decreases, the membrane remains fluid to a lower temperature if it is rich in phospholipids with unsaturated hydrocarbon tails (see Figures 5.10 and 5.11). Because of kinks in the tails where double bonds are located, unsaturated hydrocarbon tails cannot pack together as closely as saturated hydrocarbon tails, making the membrane more fluid **(Figure 7.5a)**.

The steroid cholesterol, which is wedged between phospholipid molecules in the plasma membranes of animal cells, has different effects on membrane fluidity at different temperatures **(Figure 7.5b)**. At relatively high temperatures—at 37°C, the body temperature of humans, for example—cholesterol makes the membrane less fluid by restraining phospholipid movement. However, because cholesterol also hinders the close packing of phospholipids, it lowers the temperature required for the membrane to solidify. Thus, cholesterol can be thought of as a "fluidity buffer" for the membrane, resisting changes in membrane fluidity that can be caused by changes in temperature. Compared to animals, plants have very low levels of cholesterol; rather, related steroid lipids buffer membrane fluidity in plant cells.

▼ Figure 7.4 Inquiry

Do membrane proteins move?

Experiment Larry Frye and Michael Edidin, at Johns Hopkins University, labelled the plasma membrane proteins of a mouse cell and a human cell with two different markers and fused the cells. Using a microscope, they observed the markers on the hybrid cell.

Results

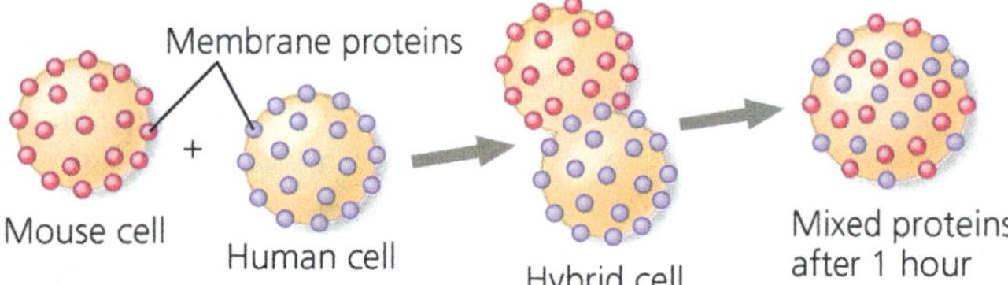

Conclusion The mixing of the mouse and human membrane proteins indicates that at least some membrane proteins move sideways within the plane of the plasma membrane.

Data from L. D. Frye and M. Edidin, The rapid intermixing of cell surface antigens after formation of mouse-human heterokaryons, *Journal of Cell Science* 7:319 (1970).

WHAT IF? *Suppose the proteins did not mix in the hybrid cell, even many hours after fusion. Could you conclude that proteins don't move within the membrane? What other explanation could there be?*

▼ Figure 7.5 Factors that affect membrane fluidity.

(a) Unsaturated versus saturated hydrocarbon tails.

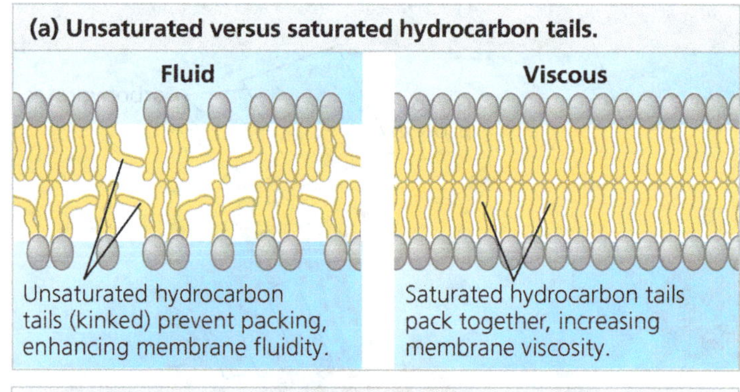

(b) Cholesterol within the animal cell membrane.

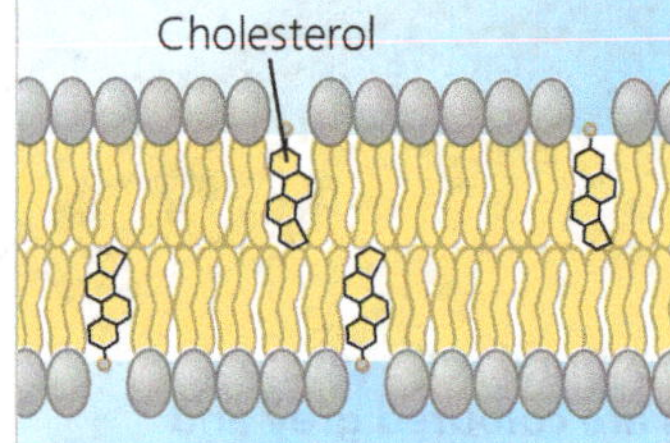

Cholesterol reduces membrane fluidity at moderate temperatures by reducing phospholipid movement, but at low temperatures it hinders solidification by disrupting the regular packing of phospholipids.

Membranes must be fluid to work properly; the fluidity of a membrane affects both its permeability and the ability of membrane proteins to move to where their function is needed. Usually, membranes are about as fluid as olive oil. When a membrane solidifies, its permeability changes, and enzymatic proteins in the membrane may become inactive if their activity requires movement within the membrane. However, membranes that are too fluid cannot support protein function either. Therefore, extreme environments (for example, those with extreme temperatures) pose a challenge for life, resulting in evolutionary adaptations that include differences in membrane lipid composition.

Evolution of Differences in Membrane Lipid Composition

EVOLUTION Variations in the cell membrane lipid compositions of many species appear to be evolutionary adaptations that maintain the appropriate membrane fluidity under specific environmental conditions. For instance, fishes that live in extreme cold have membranes with a high proportion of unsaturated hydrocarbon tails, enabling their membranes to remain fluid in spite of the low temperature (see Figure 7.5a). At the other extreme, some bacteria and archaea thrive at temperatures greater than 90°C in thermal hot springs and geysers. Their membranes include unusual lipids that may prevent excessive fluidity at such high temperatures.

The ability to change the lipid composition of cell membranes in response to changing temperatures has evolved in organisms that live where temperatures vary. In many plants that tolerate extreme cold, such as winter wheat, the percentage of unsaturated phospholipids increases in autumn, an adjustment that keeps the membranes from solidifying during winter. Some bacteria and archaea also exhibit different proportions of unsaturated phospholipids in their cell membranes, depending on the temperature at which they are growing. Overall, natural selection has apparently favoured organisms whose mix of membrane lipids ensures an appropriate level of membrane fluidity for their environment.

Membrane Proteins and Their Functions

Now we come to the *mosaic* aspect of the fluid mosaic model. Somewhat like a tile mosaic (shown here), a membrane is a collage of different proteins, often clustered together in groups, embedded in the fluid matrix of the lipid bilayer (see Figure 7.3). More than 50 kinds of proteins have been found in the plasma membrane of red blood cells, for example. Phospholipids form the main fabric of the membrane, but proteins determine most of the membrane's functions. Different types of cells contain different sets of membrane proteins, and the various membranes within a cell each have a unique collection of proteins.

◀ **Tile mosaic.**

Notice in Figure 7.3 that there are two major populations of membrane proteins: integral proteins and peripheral proteins. **Integral proteins** penetrate the hydrophobic interior of the lipid bilayer. The majority are **transmembrane proteins**, which span the membrane; other integral proteins extend only partway into the hydrophobic interior. The hydrophobic regions of an integral protein consist of one or more stretches of nonpolar amino acids (see Figure 5.14), typically 20–30 amino acids in length, usually coiled into α helices **(Figure 7.6)**. The hydrophilic parts of the molecule are exposed to the aqueous solutions on either side of the membrane. Some proteins also have one or more hydrophilic channels that allow passage through the membrane of hydrophilic substances (even of water itself). **Peripheral proteins** are not embedded in the lipid bilayer at all; they are loosely bound to the surface of the membrane, often to exposed parts of integral proteins (see Figure 7.3).

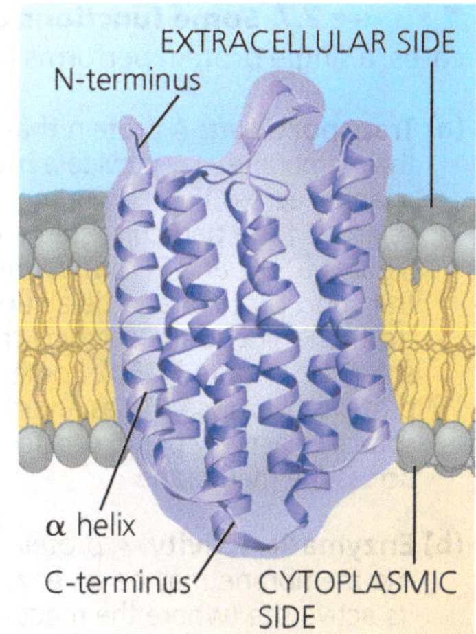

▶ **Figure 7.6 The structure of a transmembrane protein.** Bacteriorhodopsin (a bacterial transport protein) has a distinct orientation in the membrane, with its N-terminus outside the cell and its C-terminus inside. This ribbon model highlights the secondary structure of the hydrophobic parts, including seven transmembrane α helices, which lie mostly within the hydrophobic interior of the membrane. The nonhelical hydrophilic segments are in contact with the aqueous solutions on the extracellular and cytoplasmic sides of the membrane.

On the cytoplasmic side of the plasma membrane, some membrane proteins are held in place by attachment to the cytoskeleton. And on the extracellular side, certain membrane proteins may attach to materials outside the cell. For example, in animal cells, membrane proteins may be attached to fibres of the extracellular matrix (see Figure 6.28; *integrins* are one type of integral, transmembrane protein). These attachments combine to give animal cells a stronger framework than the plasma membrane alone could provide.

Figure 7.7 illustrates six major functions performed by proteins of the plasma membrane. A single cell may have different membrane proteins that carry out various functions, and one protein may itself carry out multiple functions. Thus, the membrane is a functional mosaic as well as a structural one.

Proteins on a cell's surface are important in the medical field. For example, a protein called CD4 on the surface

▼ **Figure 7.7 Some functions of membrane proteins.** In many cases, a single protein performs multiple tasks.

(a) Transport. *Left:* A protein that spans the membrane may provide a hydrophilic channel across the membrane that is selective for a particular solute (see Figures 6.32a and 7.15a). *Right:* Other transport proteins shuttle a substance from one side to the other by changing shape (see Figure 7.15b). Some of these proteins hydrolyse ATP as an energy source to actively pump substances across the membrane.

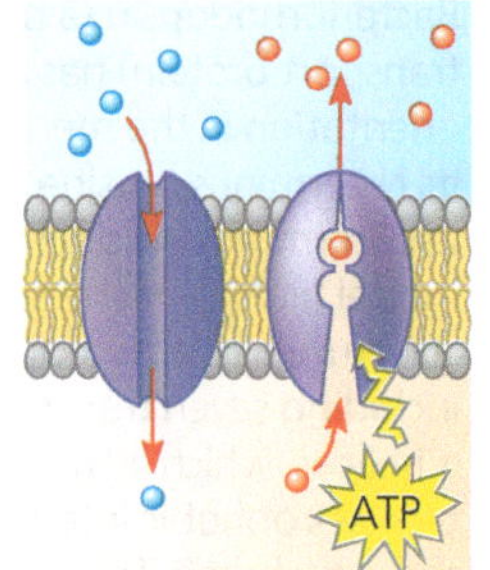

(b) Enzymatic activity. A protein built into the membrane may be an enzyme with its active site (where the reactant binds) exposed to substances in the adjacent solution. In some cases, several enzymes in a membrane are organised as a team that carries out sequential steps of a metabolic pathway.

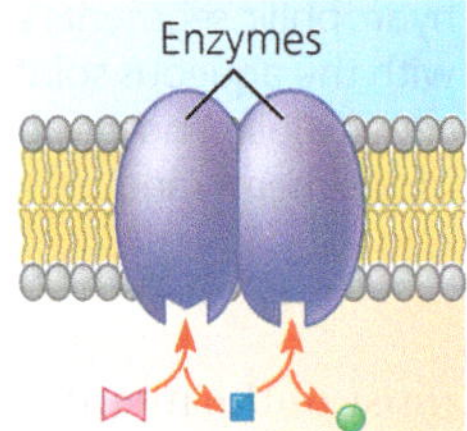

(c) Signal transduction. A membrane protein (receptor) may have a binding site with a specific shape that fits the shape of a chemical messenger, such as a hormone. The external messenger (signalling molecule) may cause the protein to change shape, allowing it to relay the message to the inside of the cell, usually by binding to a cytoplasmic protein (see Figures 6.32a and 11.6).

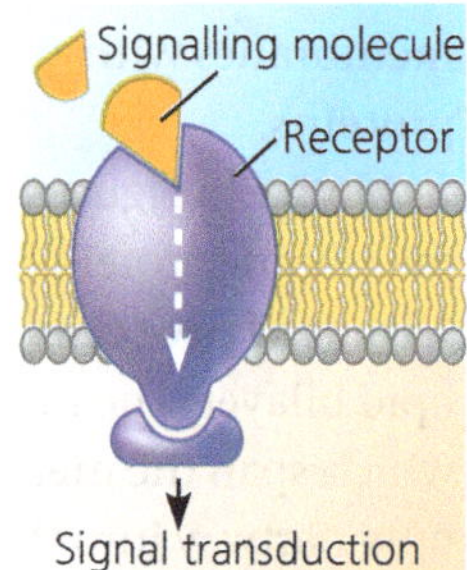

(d) Cell-cell recognition. Some glycoproteins serve as identification tags that are specifically recognised by membrane proteins of other cells. This type of cell-cell binding is usually short-lived compared with that shown in (e).

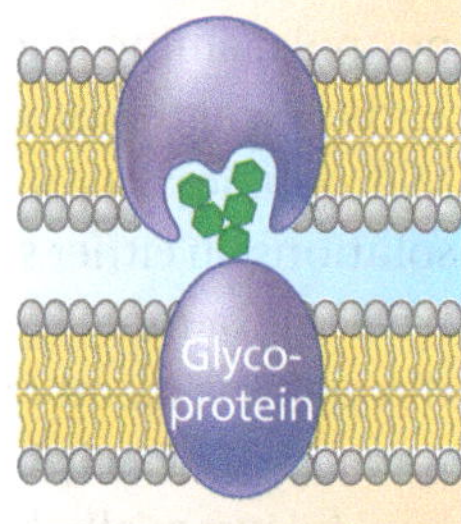

(e) Intercellular joining. Membrane proteins of adjacent cells may hook together in various kinds of junctions, such as gap junctions or tight junctions (see Figure 6.30). This type of binding is more long-lasting than that shown in (d).

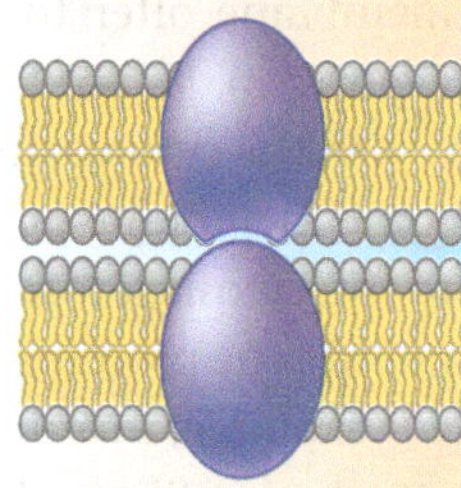

(f) Attachment to the cytoskeleton and extracellular matrix (ECM). Microfilaments or other elements of the cytoskeleton may be noncovalently bound to membrane proteins, a function that helps maintain cell shape and stabilises the location of certain membrane proteins. Proteins that can bind to ECM molecules can coordinate extracellular and intracellular changes (see Figure 6.28).

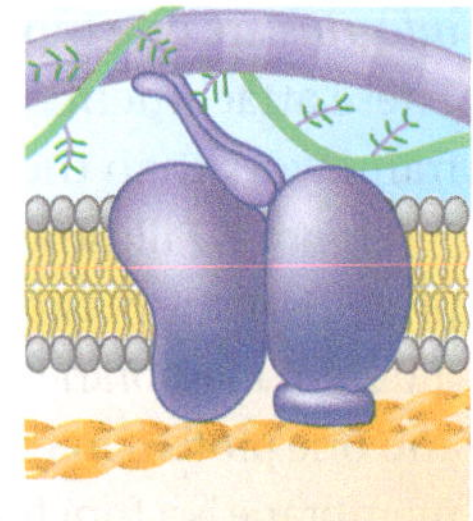

VISUAL SKILLS *Some transmembrane proteins can bind to a particular ECM molecule and, when bound, transmit a signal into the cell. Use the proteins shown in (c) and (f) to explain how this might occur.*

of immune cells helps the human immunodeficiency virus (HIV) infect these cells, leading to acquired immune deficiency syndrome (AIDS). Despite multiple exposures to HIV, however, a small number of people do not develop AIDS and show no evidence of HIV-infected cells. Comparing their genes with the genes of infected individuals, researchers learned that resistant people have an unusual form of a gene that codes for an immune cell-surface protein called CCR5. Further work showed that although CD4 is the main HIV receptor, HIV must also bind to CCR5 as a "co-receptor" to infect most cells **(Figure 7.8a)**. An absence of CCR5 on the cells of resistant individuals, due to the gene alteration, prevents the virus from entering the cells **(Figure 7.8b)**.

This information has been key to developing a treatment for HIV infection. Interfering with CD4 causes dangerous side effects because of its many important functions in cells. Discovery of the CCR5 co-receptor provided a safer target for development of drugs that mask this protein and block HIV entry. One such drug, maraviroc (brand name Celsentri), was approved for treatment of HIV in 2007; ongoing trials to determine its ability to prevent HIV infection in uninfected, at-risk patients have been disappointing.

The Role of Membrane Carbohydrates in Cell-Cell Recognition

Cell-cell recognition, a cell's ability to distinguish one type of neighbouring cell from another, is crucial to the functioning of an organism. It is important, for example, in the sorting of cells into tissues and organs in an animal embryo. It is also the basis for the rejection of foreign cells by the immune system, an important line of defence in vertebrate animals (see Concept 43.1). Cells recognise other cells by binding to molecules, often containing carbohydrates, on the extracellular surface of the plasma membrane (see Figure 7.7d).

▼ **Figure 7.8 The genetic basis for HIV resistance.**

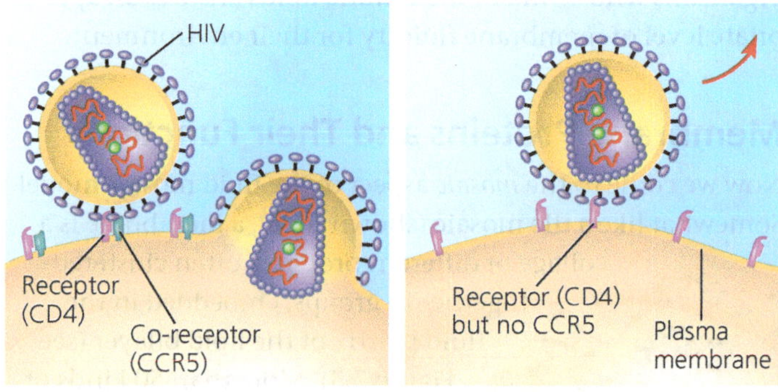

(a) HIV can infect a cell with CCR5 on its surface, as in most people.

(b) HIV cannot infect a cell lacking CCR5 on its surface, as in resistant individuals.

MAKE CONNECTIONS *Study Figures 2.16 and 5.17; each shows pairs of molecules binding to each other. What would you predict about CCR5 that would allow HIV to bind to it? How could a drug molecule interfere with this binding?*

Membrane carbohydrates are usually short, branched chains of fewer than 15 sugar units. Some are covalently bonded to lipids, forming molecules called **glycolipids**. (Recall that *glyco* refers to carbohydrate.) However, most are covalently bonded to proteins, which are thereby **glycoproteins** (see Figure 7.3).

The carbohydrates on the extracellular side of the plasma membrane vary from species to species, among individuals of the same species, and even from one cell type to another in a single individual. The diversity of the molecules and their location on the cell's surface enable membrane carbohydrates to function as markers that distinguish one cell from another. For example, the four human blood types designated A, B, AB, and O reflect variation in the carbohydrate part of glycoproteins on the surface of red blood cells.

Synthesis and Sidedness of Membranes

Membranes have distinct inside and outside faces. The two lipid layers may differ in lipid composition, and each protein has directional orientation in the membrane (see Figure 7.6). **Figure 7.9** shows how membrane sidedness arises: The asymmetrical arrangement of proteins, lipids, and their associated carbohydrates in the plasma membrane is determined as the membrane is being built.

CONCEPT CHECK 7.1

1. **VISUAL SKILLS** Carbohydrates are attached to plasma membrane proteins in the ER (see Figure 7.9). On which side of the vesicle membrane are the carbohydrates during transport to the cell surface?
2. **WHAT IF?** How might the membrane lipid composition of a native grass found in very warm soil around hot springs differ from that of a native grass found in cooler soil? Explain.

For suggested answers, see Appendix A.

CONCEPT 7.2

Membrane structure results in selective permeability

The biological membrane has emergent properties beyond those of the many individual molecules that make it up. The remainder of this chapter focuses on one of those properties: A membrane exhibits **selective permeability**; that is, it allows some substances to cross more easily than others. The ability to regulate transport across cellular boundaries is essential to the cell's existence. We will see once again that form fits function: The fluid mosaic model helps explain how membranes regulate the cell's molecular traffic.

A steady traffic of small molecules and ions moves across the plasma membrane in both directions. Consider the chemical

▼ Figure 7.9 Synthesis of membrane components and their orientation in the membrane.
The cytoplasmic (orange) face of the plasma membrane differs from the extracellular (aqua) face. The latter arises from the inside face of ER, Golgi, and vesicle membranes.

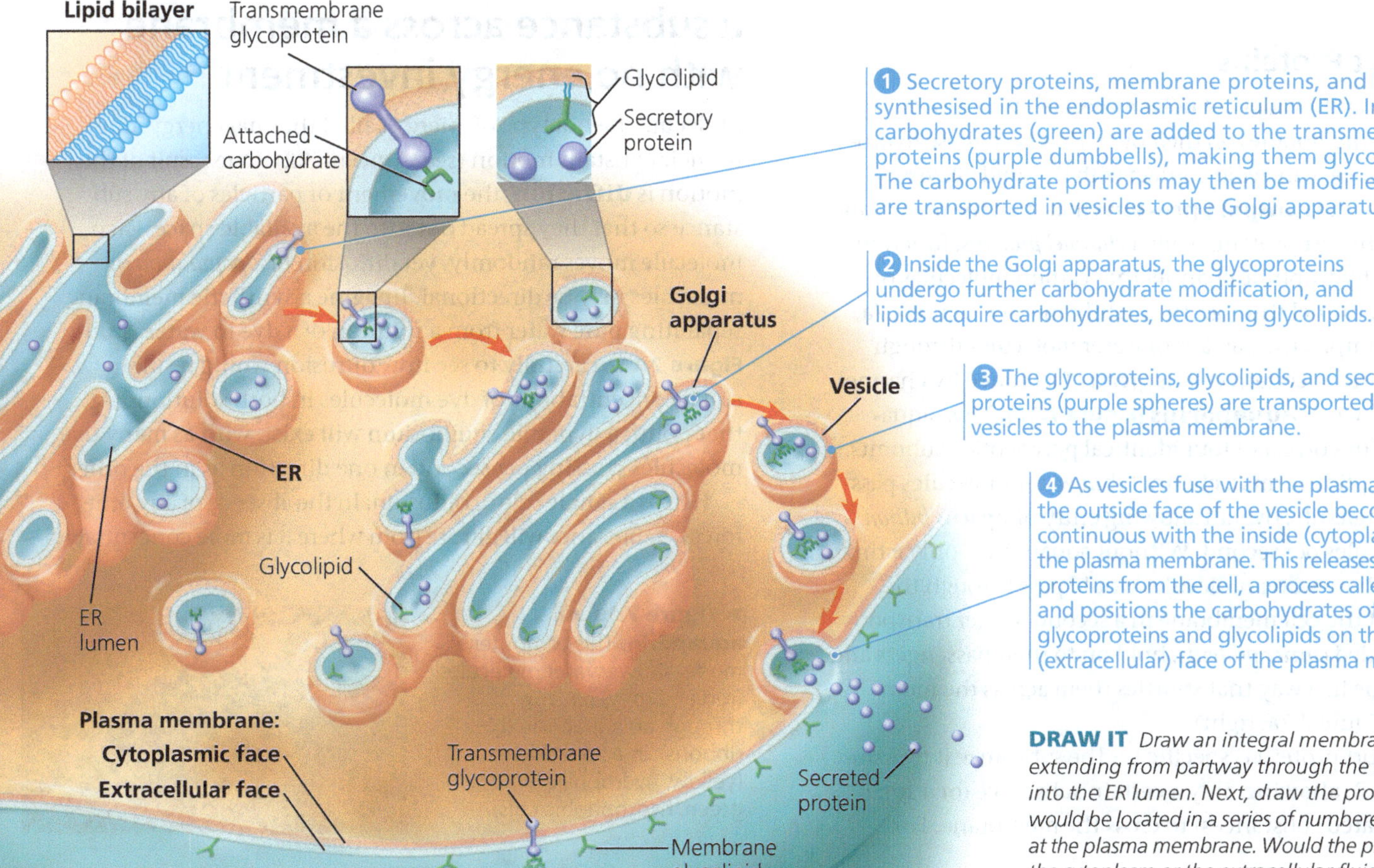

1 Secretory proteins, membrane proteins, and lipids are synthesised in the endoplasmic reticulum (ER). In the ER, carbohydrates (green) are added to the transmembrane proteins (purple dumbbells), making them glycoproteins. The carbohydrate portions may then be modified. Materials are transported in vesicles to the Golgi apparatus.

2 Inside the Golgi apparatus, the glycoproteins undergo further carbohydrate modification, and lipids acquire carbohydrates, becoming glycolipids.

3 The glycoproteins, glycolipids, and secretory proteins (purple spheres) are transported in vesicles to the plasma membrane.

4 As vesicles fuse with the plasma membrane, the outside face of the vesicle becomes continuous with the inside (cytoplasmic) face of the plasma membrane. This releases the secretory proteins from the cell, a process called *exocytosis*, and positions the carbohydrates of membrane glycoproteins and glycolipids on the outside (extracellular) face of the plasma membrane.

DRAW IT *Draw an integral membrane protein extending from partway through the ER membrane into the ER lumen. Next, draw the protein where it would be located in a series of numbered steps ending at the plasma membrane. Would the protein contact the cytoplasm or the extracellular fluid? Explain.*

exchanges between a muscle cell and the extracellular fluid that bathes it. Sugars, amino acids, and other nutrients enter the cell, and metabolic waste products leave it. The cell takes in O_2 for use in cellular respiration and expels CO_2. Also, the cell regulates its concentrations of inorganic ions, such as Na^+, K^+, Ca^{2+}, and Cl^-, by shuttling them one way or the other across the plasma membrane. In spite of heavy traffic through them, cell membranes are selective in their permeability: Substances do not cross the barrier indiscriminately. The cell is able to take up some small molecules and ions and exclude others.

The Permeability of the Lipid Bilayer

Nonpolar molecules, such as hydrocarbons, CO_2, and O_2, are hydrophobic, as are lipids. They can all therefore dissolve in the lipid bilayer of the membrane and cross it easily, without the aid of membrane proteins. However, the hydrophobic interior of the membrane impedes direct passage through the membrane of ions and polar molecules, which are hydrophilic. Polar molecules such as glucose and other sugars pass only slowly through a lipid bilayer, and even water, a very small polar molecule, does not cross rapidly relative to nonpolar molecules. A charged atom or molecule and its surrounding shell of water (see Figure 3.8) are even less likely to penetrate the hydrophobic interior of the membrane. Furthermore, the lipid bilayer is only one aspect of the gatekeeper system responsible for a cell's selective permeability. Proteins built into the membrane play key roles in regulating transport.

Transport Proteins

Specific ions and a variety of polar molecules can't move through cell membranes on their own. However, these hydrophilic substances can avoid contact with the lipid bilayer by passing through **transport proteins** that span the membrane.

Some transport proteins, called *channel proteins*, function by having a hydrophilic channel that certain molecules or ions use as a tunnel through the membrane (see Figure 7.7a, left). For example, the passage of water molecules through the membrane in certain cells is greatly facilitated by channel proteins known as **aquaporins** **(Figure 7.10)**. Most aquaporin proteins consist of four identical polypeptide subunits. Each polypeptide forms a channel that water molecules pass through, single-file, overall allowing entry of up to *3 billion* water molecules per second. Without aquaporins, only a tiny fraction of these water molecules would pass through the same area of the cell membrane in a second. Other transport proteins, called *carrier proteins*, hold on to their passengers and change shape in a way that shuttles them across the membrane (see Figure 7.7a, right).

A transport protein is specific for the substance it translocates (moves), allowing only a certain substance (or a small group of related substances) to cross the membrane. For example, a glucose carrier protein in the plasma membrane of red blood cells transports glucose across the membrane 50,000 times faster than glucose can pass through on its own. This "glucose transporter" is so selective that it even rejects fructose, a structural isomer of glucose (see Figure 5.3). Thus, the selective permeability of a membrane depends on both the discriminating barrier of the lipid bilayer and the specific transport proteins built into the membrane.

What establishes the *direction* of traffic across a membrane? And what mechanisms drive molecules across membranes? We will address these questions next as we explore two modes of membrane traffic: passive transport and active transport.

CONCEPT CHECK 7.2

1. What property allows O_2 and CO_2 to cross a lipid bilayer without the aid of membrane proteins?
2. **VISUAL SKILLS** Examine Figure 7.2. Why is a transport protein needed to move many water molecules rapidly across a membrane?
3. **MAKE CONNECTIONS** Aquaporins exclude passage of hydronium ions (H_3O^+), but some aquaporins allow passage of glycerol, a three-carbon alcohol (see Figure 5.9), as well as H_2O. Since H_3O^+ is closer in size to water than glycerol is, yet cannot pass through, what might be the basis of this selectivity?

For suggested answers, see Appendix A.

CONCEPT 7.3

Passive transport is diffusion of a substance across a membrane with no energy investment

Molecules have a type of energy called thermal energy, due to their constant motion (see Concept 3.2). One result of this motion is **diffusion**, the movement of particles of any substance so that they spread out into the available space. Each molecule moves randomly, yet diffusion of a *population* of molecules may be directional. Imagine a synthetic membrane separating pure water from a solution of a dye in water. Study **Figure 7.11a** carefully to see how diffusion would result in equal concentrations of dye molecules in both solutions. At that point, a dynamic equilibrium will exist, with as many dye molecules crossing per second in one direction as in the other.

Here is a simple rule of diffusion: In the absence of any other forces, a substance will diffuse from where it is more concentrated

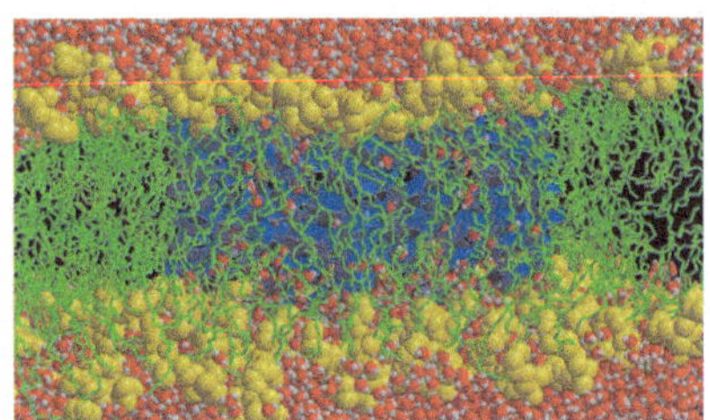

▶ **Figure 7.10 An aquaporin.** This computer model shows water molecules (red and grey) passing through an aquaporin (blue ribbons), in a lipid bilayer (yellow, hydrophilic heads; green, hydrophobic tails).

to where it is less concentrated. Put another way, a substance diffuses down its **concentration gradient**, the region along which the density of a chemical substance increases or decreases (in this case, decreases). Diffusion is a spontaneous process, needing no input of energy. Each substance diffuses down its *own* concentration gradient, unaffected by the concentration gradients of other substances **(Figure 7.11b)**.

Much of the traffic across cell membranes occurs by diffusion. When a substance is more concentrated on one side of a membrane than on the other, there is a tendency for it to diffuse across, down its concentration gradient (assuming that the membrane is permeable to that substance). One important example is the uptake of oxygen by a cell performing cellular respiration. Dissolved oxygen diffuses into the cell across the plasma membrane. As long as cellular respiration consumes the O_2 as it enters, diffusion into the cell will continue because the concentration gradient favours movement in that direction.

The diffusion of a substance across a biological membrane is called **passive transport** because it requires no energy. The concentration gradient itself represents potential energy (see Concept 2.2 and Figure 8.5b) and drives diffusion. Remember, though, that membranes are selectively permeable and therefore have different effects on the rates of diffusion of various molecules. Water can diffuse very rapidly across the membranes of cells with aquaporins, compared with diffusion in the absence of aquaporins. The movement of water across the plasma membrane has important consequences for cells.

▼ Figure 7.11 Diffusion of solutes across a synthetic membrane. Each large arrow shows net diffusion of dye molecules of that colour.

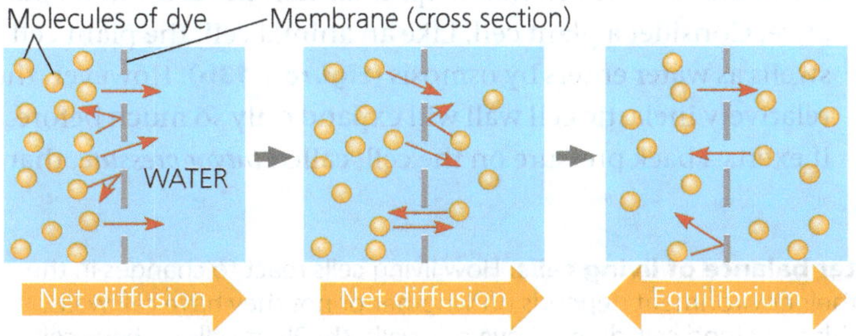

(a) Diffusion of one solute. Molecules of dye can pass through membrane pores. Random movement of dye molecules will cause some to pass through the pores; this happens more often on the side with more dye molecules. The dye diffuses from the more concentrated side to the less concentrated side (called diffusing down a concentration gradient). A dynamic equilibrium results: Solute molecules still cross, but at roughly equal rates in both directions.

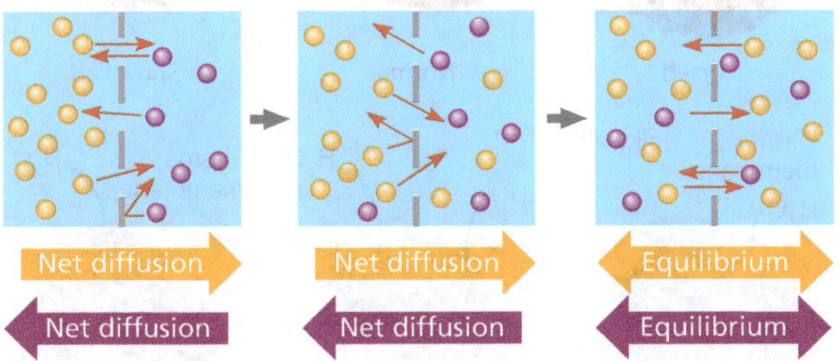

(b) Diffusion of two solutes. Solutions of two different dyes are separated by a membrane that is permeable to both. Each dye diffuses down its own concentration gradient. There will be a net diffusion of the purple dye towards the left, even though the *total* solute concentration was initially greater on the left side.

Effects of Osmosis on Water Balance

To see how two solutions with different solute concentrations interact, picture a U-shaped glass tube with a selectively permeable artificial membrane separating two sugar solutions **(Figure 7.12)**. Pores in this synthetic membrane are too small for sugar molecules to pass through but large enough for water molecules. However, tight clustering of water molecules around the hydrophilic solute molecules makes some

▼ Figure 7.12 Osmosis. Two sugar solutions of different concentrations are separated by a membrane that the solvent (water) can pass through but the solute (sugar) cannot. Water molecules move randomly and may cross in either direction, but overall, water diffuses from the solution with less concentrated solute to that with more concentrated solute. This passive transport of water, or osmosis, makes the sugar concentrations on both sides roughly equal.

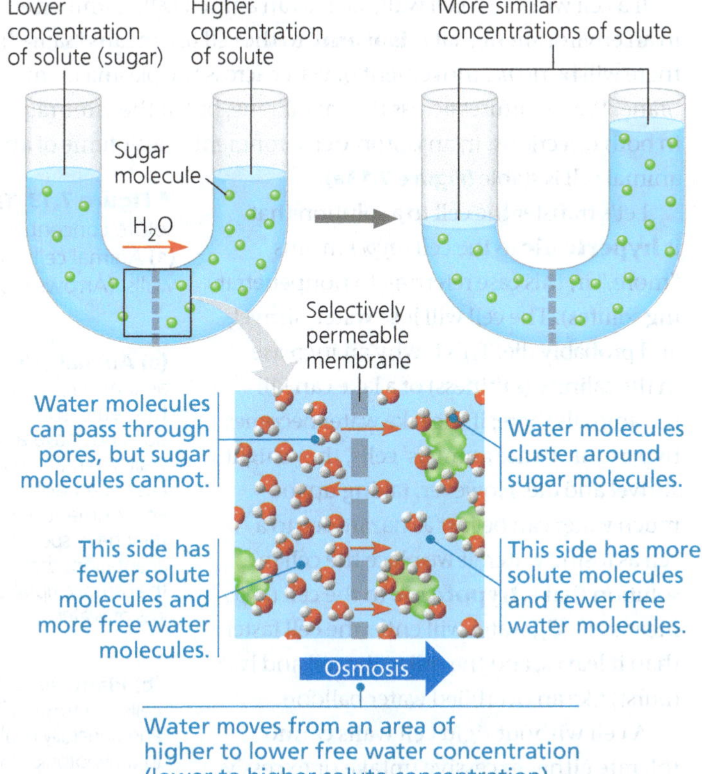

VISUAL SKILLS *If an orange dye capable of passing through the membrane was added to the left side of the tube above, how would it be distributed at the end of the experiment? (See Figure 7.11.) Would the final solution levels in the tube be affected? What cellular component does the membrane represent in this experiment?*

of the water unavailable to cross the membrane. As a result, the solution with a higher solute concentration has a lower *free* water concentration. Water diffuses across the membrane from the region of higher free water concentration (lower solute concentration) to that of lower free water concentration (higher solute concentration) until the solute concentrations on both sides of the membrane are more nearly equal. The diffusion of free water across a selectively permeable membrane, whether artificial or cellular, is called **osmosis**. The movement of water across cell membranes and the balance of water between the cell and its environment are crucial to organisms. Let's now apply what we've learned about osmosis in this system to living cells.

Water Balance of Cells Without Cell Walls

To explain the behaviour of a cell in a solution, we must consider both solute concentration and membrane permeability. Both factors are taken into account in the concept of **tonicity**, the ability of a surrounding solution to cause a cell to gain or lose water. The tonicity of a solution depends in part on its concentration of solutes that cannot cross the membrane (nonpenetrating solutes) relative to that inside the cell. If there is a higher concentration of nonpenetrating solutes in the surrounding solution, water will tend to leave the cell, and vice versa.

If a cell without a cell wall, such as an animal cell, is immersed in an environment that is **isotonic** to the cell (*iso* means "same"), there will be no *net* movement of water across the plasma membrane. Water diffuses across the membrane, but at the same rate in both directions. In an isotonic environment, the volume of an animal cell is stable **(Figure 7.13a)**.

Let's transfer the cell to a solution that is **hypertonic** to the cell (*hyper* means "more," in this case referring to nonpenetrating solutes). The cell will lose water, shrivel, and probably die. This is why an increase in the salinity (saltiness) of a lake can kill the animals there; if the lake water becomes hypertonic to the animals' cells, they might shrivel and die. However, taking up too much water can be just as hazardous to a cell as losing water. If we place the cell in a solution that is **hypotonic** to the cell (*hypo* means "less"), water will enter the cell faster than it leaves, and the cell will swell and lyse (burst) like an overfilled water balloon.

A cell without rigid cell walls cannot tolerate either excessive uptake or excessive loss of water. This problem of water balance is automatically solved if such a cell lives in isotonic surroundings. Seawater is isotonic to many marine invertebrates. The cells of most terrestrial (land-dwelling) animals are bathed in an extracellular fluid that is isotonic to the cells. In hypertonic or hypotonic environments, however, organisms that lack rigid cell walls must have other adaptations for **osmoregulation**, the control of solute concentrations and water balance. For example, the unicellular protist *Paramecium caudatum* lives in pond water, which is hypotonic to the cell. *Paramecium* has a plasma membrane that is much less permeable to water than the membranes of most other cells, but this only slows the uptake of water, which continually enters the cell. The reason the *Paramecium* cell doesn't burst is that it has a contractile vacuole, an organelle that functions as a pump to force water out of the cell as fast as it enters by osmosis **(Figure 7.14)**. In contrast, the bacteria and archaea that live in hypersaline (excessively salty) environments (see Figure 27.1) have cellular mechanisms that balance the internal and external solute concentrations to ensure that water does not move out of the cell. We'll examine other evolutionary adaptations for osmoregulation by animals in Concept 44.1.

Water Balance of Cells with Cell Walls

The cells of plants, prokaryotes, fungi, and some protists are surrounded by cell walls (see Figure 6.27). When such a cell is immersed in a hypotonic solution—bathed in rainwater, for example—the cell wall helps maintain the cell's water balance. Consider a plant cell. Like an animal cell, the plant cell swells as water enters by osmosis **(Figure 7.13b)**. However, the relatively inelastic cell wall will expand only so much before it exerts a back pressure on the cell, called *turgor pressure*, that

▼ **Figure 7.13 The water balance of living cells.** How living cells react to changes in the solute concentration of their environment depends on whether or not they have cell walls. **(a)** Animal cells, such as this red blood cell, do not have cell walls. **(b)** Plant cells do have cell walls. (Arrows indicate net water movement after the cells were first placed in these solutions.)

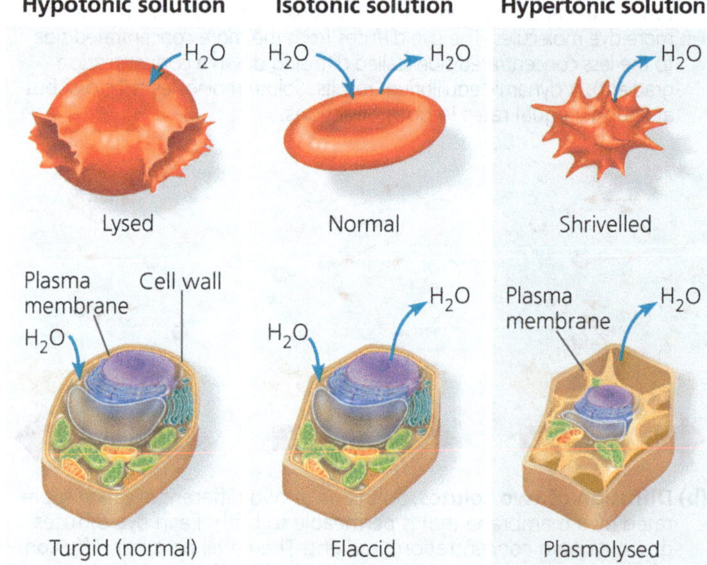

(a) Animal cell. An animal cell, such as this red blood cell, does not have a cell wall. Animal cells fare best in an isotonic environment unless they have special adaptations that offset the osmotic uptake or loss of water.

(b) Plant cell. Plant cells are turgid (firm) and generally healthiest in a hypotonic environment, where the uptake of water is eventually balanced by the wall pushing back on the cell.

? *Why do limp celery stalks become crisp when placed in a glass of water?*

▼ **Figure 7.14 The contractile vacuole of *Paramecium*.** The vacuole collects fluid from canals in the cytoplasm. When full, the vacuole and canals contract, expelling fluid from the cell (LM).

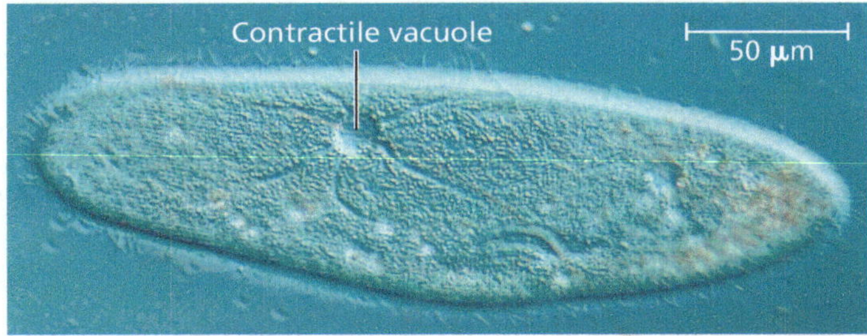

opposes further water uptake. At this point, the cell is **turgid** (very firm), which is the healthy state for most plant cells. Plants that are not woody, such as most houseplants, depend for mechanical support on cells kept turgid by a surrounding hypotonic solution. If a plant's cells and surroundings are isotonic, there is no net tendency for water to enter and the cells become **flaccid** (limp); the plant wilts.

However, a cell wall is of no advantage if the cell is immersed in a hypertonic environment. In this case, a plant cell, like an animal cell, will lose water to its surroundings and shrink. As the plant cell shrivels, its plasma membrane pulls away from the cell wall at multiple places. This phenomenon, called **plasmolysis**, causes the plant to wilt and can lead to plant death. The walled cells of bacteria and fungi also plasmolyse in hypertonic environments.

Facilitated Diffusion: Passive Transport Aided by Proteins

Let's look more closely at how water and certain hydrophilic solutes cross a membrane. As mentioned earlier, many polar molecules and ions blocked by the lipid bilayer of the membrane diffuse passively with the help of transport proteins that span the membrane. This phenomenon is called **facilitated diffusion**. Cell biologists are still trying to learn exactly how various transport proteins facilitate diffusion. Most transport proteins are very specific: They transport some substances but not others.

As mentioned earlier, the two types of transport proteins are channel proteins and carrier proteins. Channel proteins simply provide corridors that allow specific molecules or ions to cross the membrane **(Figure 7.15a)**. The hydrophilic passageways provided by these proteins can allow water molecules or small ions to diffuse very quickly from one side of the membrane to the other. Aquaporins, the water channel proteins, facilitate the massive levels of diffusion of water (osmosis) that occur in plant cells and in animal cells such as red blood cells (see Figure 7.13). Certain kidney cells also have a high number of aquaporins, allowing them to reclaim water from urine before it is excreted. If the kidneys didn't perform this function, you would excrete about 180 L of urine per day—and have to drink an equal volume of water!

▼ **Figure 7.15 Two types of transport proteins that carry out facilitated diffusion.** In both cases, the protein can transport the solute in either direction, but the net movement is down the concentration gradient of the solute.

(a) A channel protein has a channel through which water molecules or a specific solute can pass.

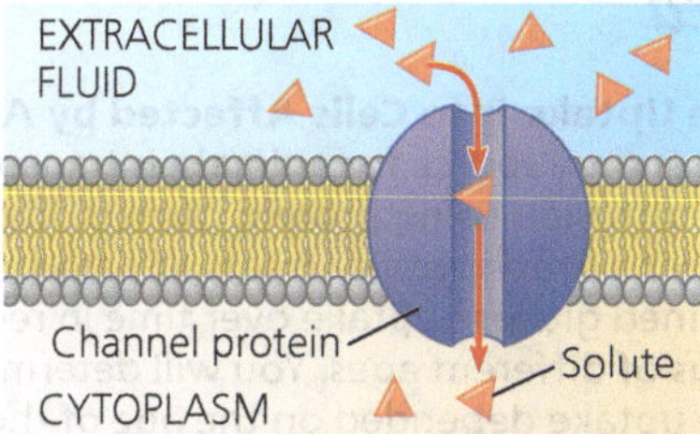

(b) A carrier protein alternates between two shapes, moving a solute across the membrane during the shape change.

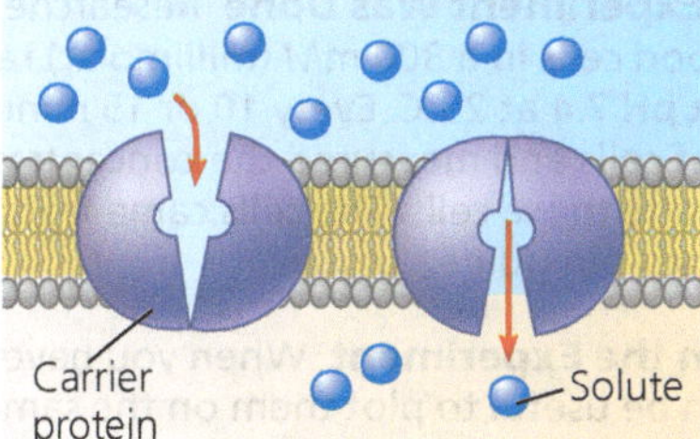

Channel proteins that transport ions are called **ion channels**. Many ion channels function as **gated channels**, which open or close in response to a stimulus (see Figure 11.8). For some gated channels, the stimulus is electrical. In a nerve cell, for example, a potassium ion channel protein (see computer model) opens in response to an electrical stimulus, allowing a stream of potassium ions to leave the cell. This restores the cell's ability to fire again. Other gated channels have a chemical stimulus: They open or close when a specific substance (not the one to be transported) binds to the channel. Ion channels are important in the functioning of the nervous system, as you'll learn in the chapter, "Neurons, Synapses, and Signalling."

▼ **Potassium ion channel protein** (See also the side view of a calcium channel protein in Figure 6.32a.)

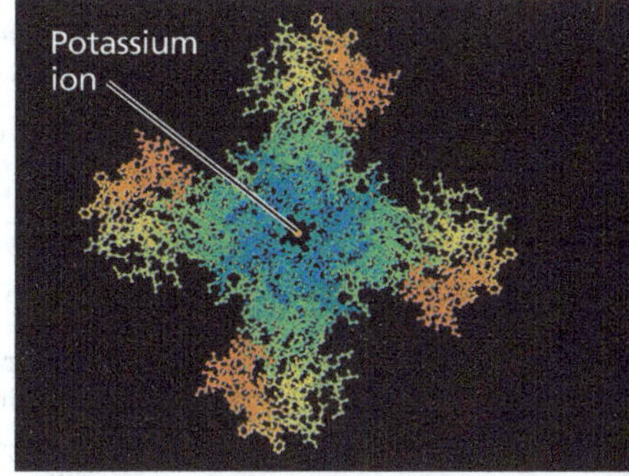

Carrier proteins, such as the glucose transporter mentioned earlier, seem to undergo a subtle change in shape that somehow translocates the solute-binding site across the membrane **(Figure 7.15b)**. Such a change in shape may be triggered by the binding and release of the transported molecule. Like ion channels, carrier proteins involved in facilitated diffusion result in the net movement of a substance down

Scientific Skills Exercise

Interpreting a Scatter Plot with Two Sets of Data

Is Glucose Uptake into Cells Affected by Age? Glucose, an important energy source for animals, is transported into cells by facilitated diffusion using protein carriers. In this exercise, you will interpret a graph with two sets of data from an experiment that examined glucose uptake over time in red blood cells from guinea pigs of different ages. You will determine if the cells' rate of glucose uptake depended on the age of the guinea pig.

How the Experiment Was Done Researchers incubated guinea pig red blood cells in a 300 m*M* (millimolar) radioactive glucose solution at pH 7.4 at 25°C. Every 10 or 15 minutes, they removed a sample of cells and measured the concentration of radioactive glucose inside those cells. The cells came from either a 15-day-old or a 1-month-old guinea pig.

Data from the Experiment When you have multiple sets of data, it can be useful to plot them on the same graph for comparison. In the graph here, each set of dots (of the same colour) forms a *scatter plot*, in which every data point represents two numerical values, one for each variable. For each data set, a curve that best fits the points has been drawn to make it easier to see the trends. (For additional information about graphs, see the Scientific Skills Review in Appendix D.)

▼ **Glucose Uptake over Time in Guinea Pig Red Blood Cells.**

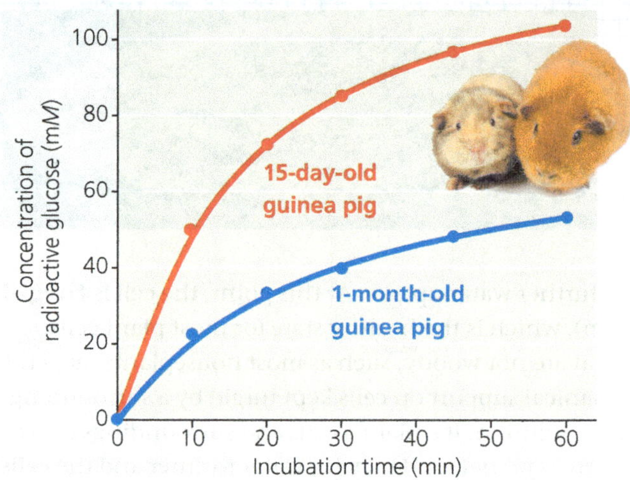

Data from T. Kondo and E. Beutler, Developmental changes in glucose transport of guinea pig erythrocytes, *Journal of Clinical Investigation* 65:1–4 (1980).

INTERPRET THE DATA

1. First make sure you understand the parts of the graph. **(a)** Which variable is the independent variable—the variable controlled by the researchers? **(b)** Which variable is the dependent variable—the variable that depended on the treatment and was measured by the researchers? **(c)** What do the red dots represent? **(d)** The blue dots?
2. From the data points on the graph, construct a table of the data. Put "Incubation Time (min)" in the left column of the table.
3. What does the graph show? Compare and contrast glucose uptake in red blood cells from 15-day-old and 1-month-old guinea pigs.
4. Develop a hypothesis to explain the difference between glucose uptake in red blood cells from 15-day-old and 1-month-old guinea pigs. (Think about how glucose gets into cells.)
5. Design an experiment to test your hypothesis.

its concentration gradient. No energy input is thus required: This is passive transport. The **Scientific Skills Exercise** gives you an opportunity to work with data from an experiment related to glucose transport.

CONCEPT CHECK 7.3

1. Speculate about how a cell performing cellular respiration might rid itself of the resulting CO_2.
2. **WHAT IF?** If a *Paramecium* swims from a hypotonic to an isotonic environment, will its contractile vacuole become more active or less? Why?

For suggested answers, see Appendix A.

CONCEPT 7.4

Active transport uses energy to move solutes against their gradients

Despite the help of transport proteins, facilitated diffusion is considered passive transport because the solute is moving down its concentration gradient, a process that requires no energy. Facilitated diffusion speeds transport of a solute by providing efficient passage through the membrane, but it does not alter the direction of transport. Some other transport proteins, however, can use energy to move solutes *against* their concentration gradients, across the plasma membrane from the side where they are less concentrated (whether inside or outside) to the side where they are more concentrated.

The Need for Energy in Active Transport

To pump a solute across a membrane against its gradient requires work; the cell must expend energy. Therefore, this type of membrane traffic is called **active transport**. The transport proteins that move solutes against their concentration gradients are all carrier proteins rather than channel proteins. This makes sense because when channel proteins are open, they merely allow solutes to diffuse down their concentration gradients rather than picking them up and transporting them against their gradients.

Active transport enables a cell to maintain internal concentrations of small solutes that differ from concentrations in its environment. For example, compared with its surroundings,

▶ **Figure 7.16 The sodium-potassium pump: a specific case of active transport.** This transport system pumps ions against steep concentration gradients. The pump oscillates between two shapes in a cycle that moves Na^+ out of the cell (steps 1–3) and K^+ into the cell (steps 4–6). The two shapes have different binding affinities for Na^+ and K^+. ATP hydrolysis powers the shape change by transferring a phosphate group to the transport protein (phosphorylating the protein).

VISUAL SKILLS *For each ion (Na^+ and K^+), describe its concentration inside the cell relative to outside. How many Na^+ are moved out of the cell and how many K^+ moved in per cycle?*

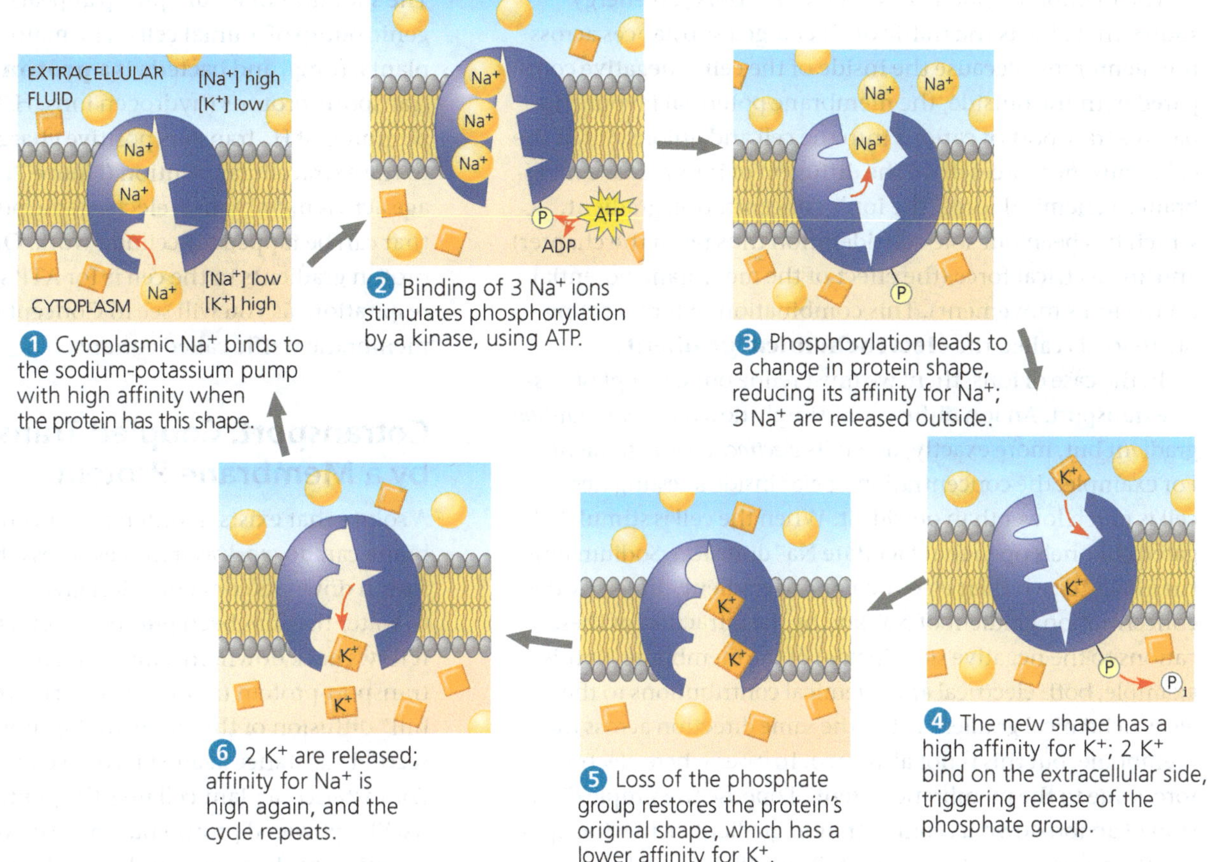

1 Cytoplasmic Na^+ binds to the sodium-potassium pump with high affinity when the protein has this shape.

2 Binding of 3 Na^+ ions stimulates phosphorylation by a kinase, using ATP.

3 Phosphorylation leads to a change in protein shape, reducing its affinity for Na^+; 3 Na^+ are released outside.

4 The new shape has a high affinity for K^+; 2 K^+ bind on the extracellular side, triggering release of the phosphate group.

5 Loss of the phosphate group restores the protein's original shape, which has a lower affinity for K^+.

6 2 K^+ are released; affinity for Na^+ is high again, and the cycle repeats.

an animal cell has a much higher concentration of potassium ions (K^+) and a much lower concentration of sodium ions (Na^+). The plasma membrane helps maintain these steep gradients by pumping Na^+ out of the cell and K^+ into the cell.

As in other types of cellular work, ATP hydrolysis supplies the energy for most active transport. One way ATP can power active transport is when its terminal phosphate group is transferred directly to the transport protein. This can induce the protein to change its shape in a manner that translocates a solute bound to the protein across the membrane. One transport system that works this way is the **sodium-potassium pump**, which exchanges Na^+ for K^+ across the plasma membrane of animal cells (**Figure 7.16**). The distinction between passive transport and active transport is reviewed in **Figure 7.17**.

How Ion Pumps Maintain Membrane Potential

All cells have voltages across their plasma membranes. Voltage is electrical potential energy (see Concept 2.2)—a separation of opposite charges. The cytoplasmic side of the membrane is negative in charge relative to the extracellular side because of an unequal distribution of anions and cations on the two sides. The voltage across a membrane, called a **membrane potential**, ranges from about −50 to −200 millivolts (mV). (The minus sign indicates that the inside of the cell is negative relative to the outside.)

▼ **Figure 7.17 Review: Passive and active transport.**

Passive transport. Substances diffuse spontaneously down their concentration gradients, crossing a membrane with no expenditure of energy by the cell. The rate of diffusion can be greatly increased by transport proteins in the membrane.

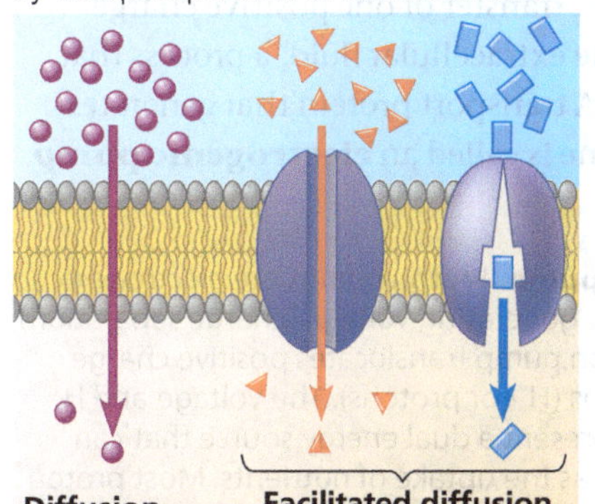

Diffusion. Hydrophobic molecules and (at a slow rate) very small uncharged polar molecules can diffuse through the lipid bilayer.

Facilitated diffusion. Many hydrophilic substances diffuse through membranes with the assistance of transport proteins, either channel proteins (left) or carrier proteins (right).

Active transport. Some transport proteins expend energy and act as pumps, moving substances across a membrane against their concentration (or electrochemical) gradients. Energy is usually supplied by ATP hydrolysis.

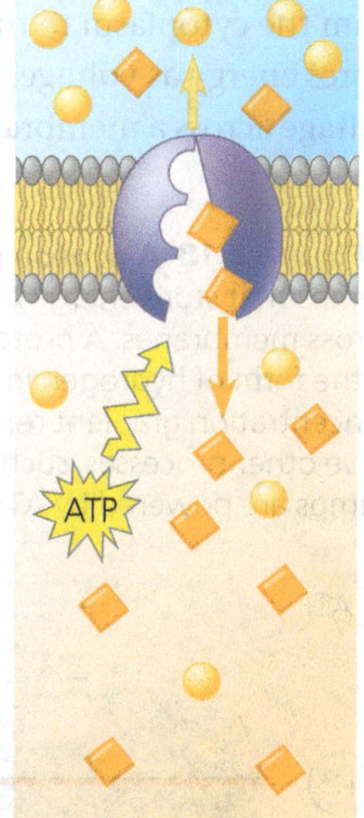

VISUAL SKILLS *For each solute in the right panel, describe its direction of movement, and state whether it is moving with or against its concentration gradient.*

The membrane potential acts like a battery, an energy source that affects the traffic of all charged substances across the membrane. Because the inside of the cell is negative compared with the outside, the membrane potential favours the passive transport of cations into the cell and anions out of the cell. Thus, *two* forces drive the diffusion of ions across a membrane: a chemical force (the ion's concentration gradient, which has been our sole consideration thus far in the chapter) and an electrical force (the effect of the membrane potential on the ion's movement). This combination of forces acting on an ion is called the **electrochemical gradient**.

In the case of ions, then, we must refine our concept of passive transport: An ion diffuses not simply down its *concentration* gradient but, more exactly, down its *electrochemical* gradient. For example, the concentration of Na^+ inside a resting nerve cell is much lower than outside it. When the cell is stimulated, gated channels open that facilitate Na^+ diffusion. Sodium ions then "fall" down their electrochemical gradient, driven by the concentration gradient of Na^+ and by the attraction of these cations to the negative side (inside) of the membrane. In this example, both electrical and chemical contributions to the electrochemical gradient act in the same direction across the membrane, but this is not always so. In cases where electrical forces due to the membrane potential oppose the simple diffusion of an ion down its concentration gradient, active transport may be necessary. In Concepts 48.2 and 48.3, you'll learn about the importance of electrochemical gradients and membrane potentials in the transmission of nerve impulses.

Some membrane proteins that actively transport ions contribute to the membrane potential. Study Figure 7.16 to see if you can see why the sodium-potassium pump is a good example. Notice that the pump does not translocate Na^+ and K^+ one for one, but pumps *three* sodium ions out of the cell for every *two* potassium ions it pumps into the cell. With each "crank" of the pump, there is a net transfer of one positive charge from the cytoplasm to the extracellular fluid, a process that stores energy as voltage. A transport protein that generates voltage across a membrane is called an **electrogenic pump**.

▼ Figure 7.18 A proton pump. Proton pumps are electrogenic pumps that store energy by generating voltage (charge separation) across membranes. A proton pump translocates positive charge in the form of hydrogen ions (H^+, or protons). The voltage and H^+ concentration gradient represent a dual energy source that can drive other processes, such as the uptake of nutrients. Most proton pumps are powered by ATP hydrolysis. (See Figure 6.32a.)

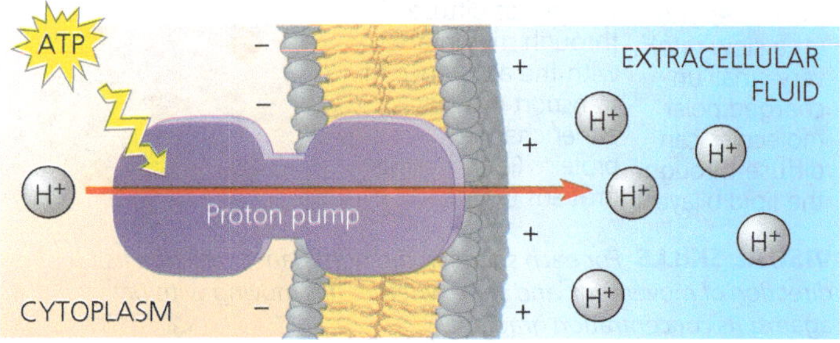

The sodium-potassium pump appears to be the major electrogenic pump of animal cells. The main electrogenic pump of plants, fungi, and bacteria is a **proton pump**, which actively transports protons (hydrogen ions, H^+) out of the cell. The pumping of H^+ transfers positive charge from the cytoplasm to the extracellular solution **(Figure 7.18)**. By generating voltage across membranes, electrogenic pumps help store energy that can be tapped for cellular work. One important use of proton gradients in the cell is for ATP synthesis during cellular respiration, as you will see in Concept 9.4. Another is a type of membrane traffic called cotransport.

Cotransport: Coupled Transport by a Membrane Protein

A solute that exists in different concentrations across a membrane can do work as it moves across that membrane by diffusion down its concentration gradient. This is analogous to water that has been pumped uphill and performs work as it flows back down. In a mechanism called **cotransport**, a transport protein (a cotransporter) can couple the "downhill" diffusion of the solute to the "uphill" transport of a second substance against its own concentration gradient. For instance, a plant cell uses the gradient of H^+ generated by its ATP-powered proton pumps to drive the active transport of amino acids, sugars, and several other nutrients into the cell. In the example shown in **Figure 7.19**, a cotransporter couples the return of H^+ to the transport of sucrose into the cell. This protein can translocate sucrose into the cell against

▼ Figure 7.19 Cotransport: active transport driven by a concentration gradient. A carrier protein, such as this H^+/sucrose cotransporter in a plant cell (top), is able to use the diffusion of H^+ down its electrochemical gradient into the cell to drive the uptake of sucrose. (The cell wall is not shown.) Although not technically part of the cotransport process, an ATP-driven proton pump is shown here (bottom), which concentrates H^+ outside the cell. The resulting H^+ gradient represents potential energy that can be used for active transport—of sucrose, in this case. Thus, ATP hydrolysis indirectly provides the energy necessary for cotransport.

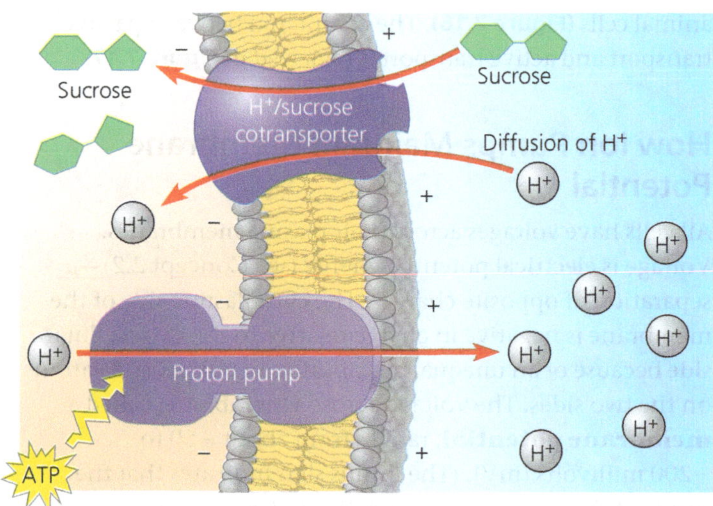

its concentration gradient, but only if the sucrose molecule travels in the company of an H^+. The H^+ uses the transport protein as an avenue to diffuse down its own electrochemical gradient, which is maintained by the proton pump. Plants use H^+/sucrose cotransport to load sucrose produced by photosynthesis into cells in the veins of leaves. The vascular tissue of the plant can then distribute the sugar to roots and other nonphotosynthetic organs that do not make their own sugars.

A similar cotransporter in animals transports Na^+ into intestinal cells together with glucose. The glucose is transported from the gut lumen against its gradient using the potential energy of the inward Na^+ gradient. The Na^+ gradient in turn is maintained by the Na^+/K^+ pump on the opposite side of the cell. Our understanding of the Na^+/glucose cotransporters has helped us find more effective treatments for diarrhoea, a serious problem in developing countries which saw mortality rates of over 30%. Oral rehydration therapies were developed that contain both Na^+ and glucose, which, when mixed with clean drinking water, can be administered easily by non-medical personnel. The cotransport of the Na^+ and glucose across the intestinal cells establishes an osmotic potential that quickly pulls water from the gut lumen, rehydrating the patient and stemming the water and electrolyte losses associated with the diarrhoea. This simple treatment has lowered infant mortality worldwide.

CONCEPT CHECK 7.4

1. Na^+/K^+ pumps help nerve cells establish a voltage across their plasma membranes. Do these pumps use ATP or produce ATP? Explain.
2. **VISUAL SKILLS** Compare the Na^+/K^+ pump in Figure 7.16 with the cotransporter in Figure 7.19. Explain why the Na^+/K^+ pump would not be considered a cotransporter.
3. **MAKE CONNECTIONS** Review the characteristics of the lysosome in Concept 6.4. Given the internal environment of a lysosome, what transport protein might you expect to see in its membrane?

For suggested answers, see Appendix A.

CONCEPT 7.5

Bulk transport across the plasma membrane occurs by exocytosis and endocytosis

Large molecules, such as proteins and polysaccharides, generally don't cross the membrane by diffusion or transport proteins. Instead, they usually enter and leave the cell in bulk, packaged in vesicles.

Exocytosis

The cell secretes certain molecules by the fusion of vesicles with the plasma membrane; this process is called **exocytosis** **(Figure 7.20)**. A transport vesicle that has budded from the Golgi apparatus moves along a microtubule of the cytoskeleton to the plasma membrane. When the vesicle membrane and plasma membrane come into contact, specific proteins in both membranes rearrange the lipid molecules of the two bilayers so that the two membranes fuse. The contents of the vesicle spill out of the cell, and the vesicle membrane becomes part of the plasma membrane.

▼ Figure 7.20 Exocytosis.

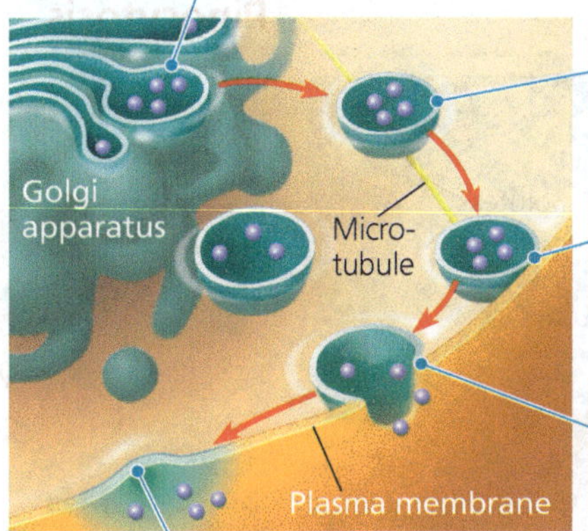

Many secretory cells use exocytosis to export products. For example, cells in the pancreas that make insulin secrete it into the extracellular fluid by exocytosis. In another example, nerve cells use exocytosis to release neurotransmitters that signal other neurons or muscle cells (see Figure 7.1). When plant cells are making cell walls, exocytosis delivers some of the necessary proteins and carbohydrates from Golgi vesicles to the outside of the cell.

Endocytosis

In **endocytosis**, the cell takes in molecules and particulate matter by forming new vesicles from the plasma membrane. Although the proteins involved in the processes are different, the events of endocytosis look like the reverse of exocytosis. First, a small area of the plasma membrane sinks inwards to form a pocket. Then, as the pocket deepens, it pinches in, forming a vesicle containing material that had been outside the cell. Study **Figure 7.21** carefully to understand the three types of endocytosis: phagocytosis ("cellular eating"), pinocytosis ("cellular drinking"), and receptor-mediated endocytosis.

Human cells use receptor-mediated endocytosis to take in cholesterol for membrane synthesis and the synthesis of other steroids. Cholesterol travels in the blood in particles called low-density lipoproteins (LDLs), each a complex of lipids and a protein. LDLs bind to LDL receptors on plasma membranes and then enter the cells by endocytosis. In the inherited disease familial hypercholesterolemia, characterised by a very high level of cholesterol in the blood, LDLs cannot enter cells because the LDL receptor proteins are defective or missing. Cholesterol thus accumulates in the

▼ Figure 7.21 Exploring Endocytosis in Animal Cells

Phagocytosis

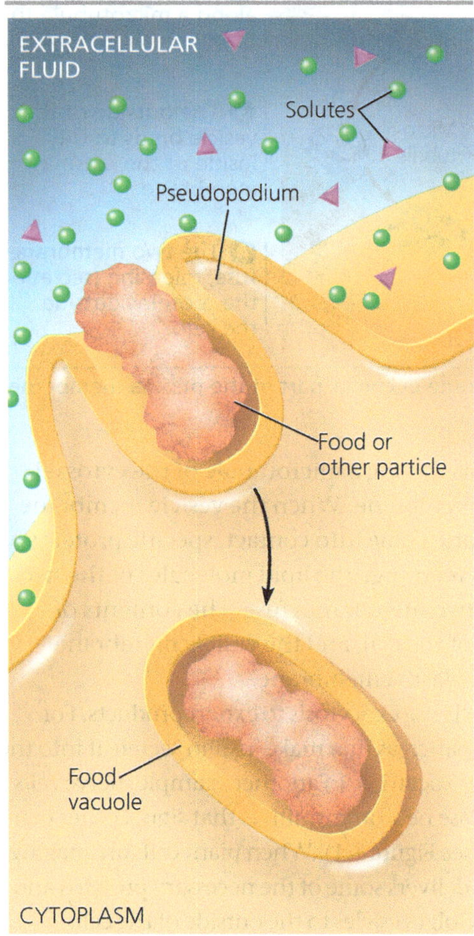

In **phagocytosis**, a cell engulfs a particle by extending pseudopodia (singular, *pseudopodium*) around it and packaging it within a membranous sac called a food vacuole. The particle will be digested after the food vacuole fuses with a lysosome containing hydrolytic enzymes (see Figure 6.13a).

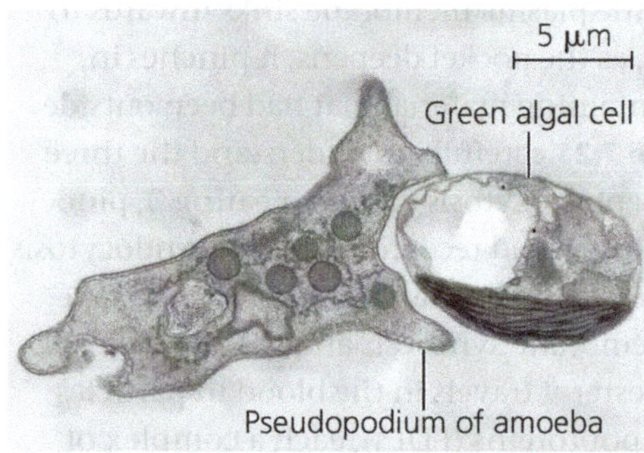

An amoeba engulfing a green algal cell via phagocytosis (TEM).

Pinocytosis

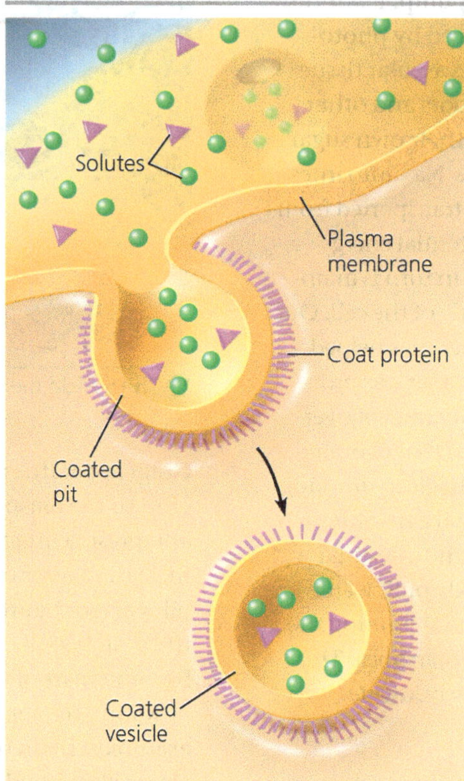

In **pinocytosis**, a cell continually "gulps" droplets of extracellular fluid into tiny vesicles, formed by infoldings of the plasma membrane. In this way, the cell obtains molecules dissolved in the droplets. Because any and all solutes are taken into the cell, pinocytosis as shown here is nonspecific for the substances it transports. In many cases, the parts of the plasma membrane that form vesicles are lined on their cytoplasmic side by a fuzzy layer of coat protein; the "pits" and resulting vesicles are called *coated pits*.

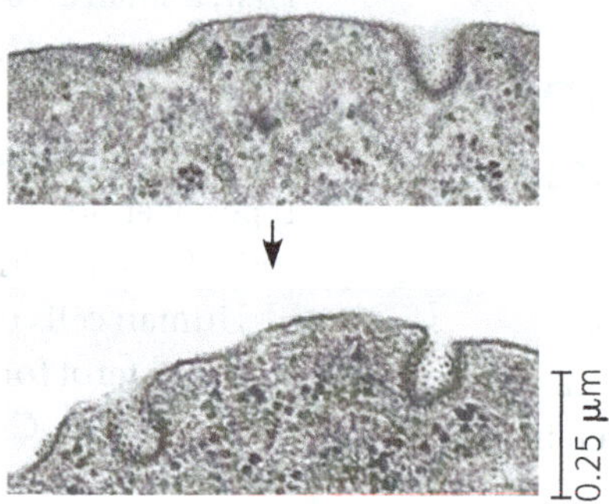

Pinocytotic vesicles forming (TEMs).

Receptor-Mediated Endocytosis

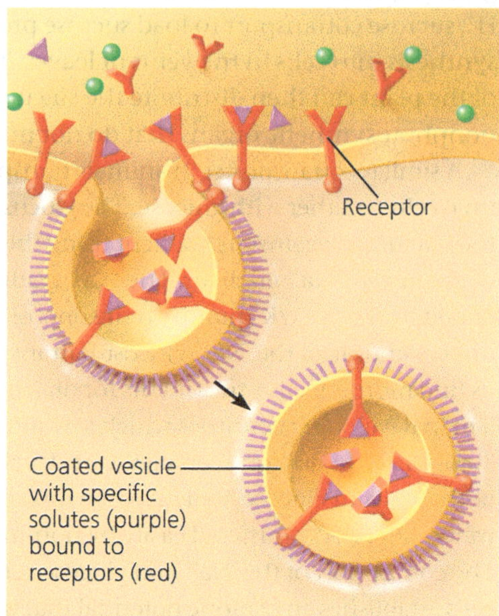

Receptor-mediated endocytosis is a specialised type of pinocytosis that enables the cell to acquire bulk quantities of specific substances, even though those substances may not be very concentrated in the extracellular fluid. Embedded in the plasma membrane are proteins with receptor sites exposed to the extracellular fluid. Specific solutes bind to the receptors. The receptor proteins then cluster in coated pits, and each coated pit forms a vesicle containing the bound molecules. The diagram shows only bound molecules (purple triangles) inside the vesicle, but other molecules from the extracellular fluid are also present. After the ingested material is liberated from the vesicle, the emptied receptors are recycled to the plasma membrane by the same vesicle (not shown).

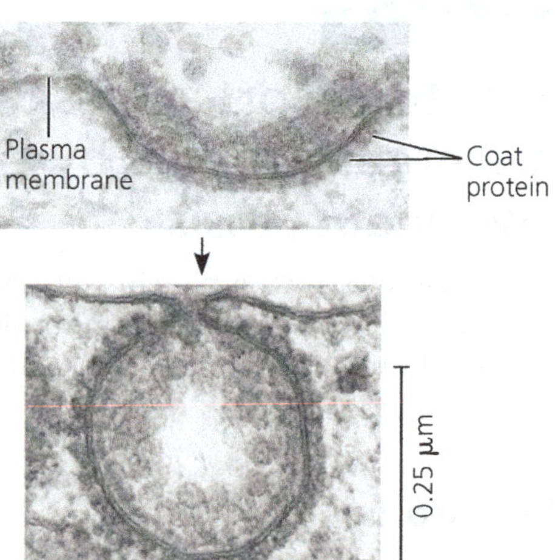

Top: A coated pit. *Bottom*: A coated vesicle forming during receptor-mediated endocytosis (TEMs).

VISUAL SKILLS *Use the scale bars to estimate the diameters of (a) the food vacuole that will form around the algal cell (left micrograph; measure the length, not the width) and (b) the coated vesicle (lower right micrograph). (c) Which is larger, and by what factor?*

blood, contributing to early atherosclerosis, the buildup of lipids within blood vessel walls. This narrows the space in the vessels and impedes blood flow, potentially resulting in heart damage and stroke.

Endocytosis and exocytosis also provide mechanisms for rejuvenating or remodelling the plasma membrane. These processes occur continually in most eukaryotic cells, yet the amount of plasma membrane in a nongrowing cell remains fairly constant. The addition of membrane by one process appears to offset the loss of membrane by the other.

CONCEPT CHECK 7.5

1. As a cell grows, its plasma membrane expands. Does this involve endocytosis or exocytosis? Explain.
2. **DRAW IT** Return to Figure 7.9, and circle a patch of plasma membrane that is coming from a vesicle involved in exocytosis.
3. **MAKE CONNECTIONS** In Concept 6.7, you learned that animal cells make an extracellular matrix (ECM). Describe the cellular pathway of synthesis and deposition of an ECM glycoprotein.

For suggested answers, see Appendix A.

7 Chapter Review

SUMMARY OF KEY CONCEPTS

CONCEPT 7.1

Cellular membranes are fluid mosaics of lipids and proteins *(pp. 129–133)*

- In the **fluid mosaic model, amphipathic** proteins are embedded in the phospholipid bilayer.
- Phospholipids and some proteins move sideways within the membrane. The unsaturated hydrocarbon tails of some phospholipids keep membranes fluid at lower temperatures, while cholesterol helps membranes resist changes in fluidity caused by temperature changes.
- Membrane proteins function in transport, enzymatic activity, signal transduction, cell-cell recognition, intercellular joining, and attachment to the cytoskeleton and extracellular matrix. Short chains of sugars linked to proteins (in **glycoproteins**) and lipids (in **glycolipids**) on the exterior side of the plasma membrane interact with surface molecules of other cells.
- Membrane proteins and lipids are synthesised in the ER and modified in the ER and Golgi apparatus. The inside and outside faces of membranes differ in molecular composition.

? *In what ways are membranes crucial to life?*

CONCEPT 7.2

Membrane structure results in selective permeability *(pp. 133–134)*

- A cell must exchange molecules and ions with its surroundings, a process controlled by the **selective permeability** of the plasma membrane. Hydrophobic substances are soluble in lipids and pass through membranes rapidly, whereas polar molecules and ions generally require specific **transport proteins**.

? *How do aquaporins affect the permeability of a membrane?*

CONCEPT 7.3

Passive transport is diffusion of a substance across a membrane with no energy investment *(pp. 134–138)*

- **Diffusion** is the spontaneous movement of a substance down its **concentration gradient**. Water diffuses out of a cell (**osmosis**) if the solution outside has a higher concentration of nonpenetrating solutes (**hypertonic**); water enters if the solution has a lower solute concentration (**hypotonic**). If the concentrations are equal (**isotonic**), no net osmosis occurs. Cell survival depends on balancing water uptake and loss.
- In **facilitated diffusion**, a transport protein speeds water or solute movement down its concentration gradient across a membrane. **Ion channels** facilitate the diffusion of ions across a membrane. Carrier proteins can undergo changes in shape that translocate bound solutes across the membrane.

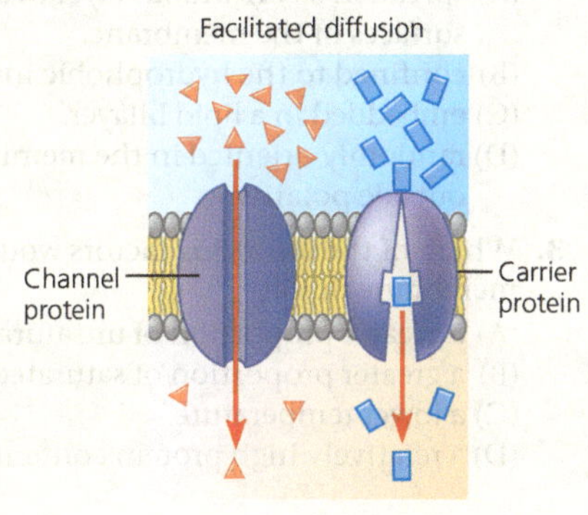

? *What happens to a cell placed in a hypertonic solution? Describe the free water concentration inside and out.*

CONCEPT 7.4

Active transport uses energy to move solutes against their gradients *(pp. 138–141)*

- Specific membrane proteins use energy, usually in the form of ATP, to do the work of **active transport**.
- Ions can have both a concentration (chemical) gradient and an electrical gradient (voltage). These gradients combine in the **electrochemical gradient**, which determines the net direction of ionic diffusion.
- **Cotransport** of two solutes occurs when a membrane protein enables the "downhill" diffusion of one solute to drive the "uphill" transport of the other.

Active transport

ATP

? *ATP is not directly involved in the functioning of a cotransporter. Why, then, is cotransport considered active transport?*

CONCEPT 7.5

Bulk transport across the plasma membrane occurs by exocytosis and endocytosis *(pp. 141–143)*

- In **exocytosis**, transport vesicles migrate to the plasma membrane, fuse with it, and release their contents. In **endocytosis**, molecules enter cells within vesicles that pinch inwards from the plasma membrane. The three types of endocytosis are **phagocytosis, pinocytosis**, and **receptor-mediated endocytosis**.

? *Which type of endocytosis involves the binding of specific substances in the extracellular fluid to membrane proteins? What does this type of transport enable a cell to do?*

TEST YOUR UNDERSTANDING

Levels 1-2: Remembering/Understanding

1. In what way do the membranes of a eukaryotic cell vary?
(A) Phospholipids are found only in certain membranes.
(B) Certain proteins are unique to each membrane.
(C) Only certain membranes of the cell are selectively permeable.
(D) Only certain membranes are constructed from amphipathic molecules.

2. According to the fluid mosaic model of membrane structure, proteins of the membrane are mostly
(A) spread in a continuous layer over the inner and outer surfaces of the membrane.
(B) confined to the hydrophobic interior of the membrane.
(C) embedded in a lipid bilayer.
(D) randomly oriented in the membrane, with no fixed inside-outside polarity.

3. Which of the following factors would tend to increase membrane fluidity?
(A) a greater proportion of unsaturated phospholipids
(B) a greater proportion of saturated phospholipids
(C) a lower temperature
(D) a relatively high protein content in the membrane

Levels 3-4: Applying/Analysing

4. Which of the following processes includes all the others?
(A) osmosis
(B) diffusion of a solute across a membrane
(C) passive transport
(D) transport of an ion down its electrochemical gradient

5. Based on Figure 7.19, which of these experimental treatments would increase the rate of sucrose transport into a plant cell?
(A) decreasing extracellular sucrose concentration
(B) decreasing extracellular pH
(C) decreasing cytoplasmic pH
(D) adding a substance that makes the membrane more permeable to hydrogen ions

6. DRAW IT An artificial "cell" consisting of an aqueous solution enclosed in a selectively permeable membrane is immersed in a beaker containing a different solution, the "environment," as shown in the accompanying diagram. The membrane is permeable to water and to the simple sugars glucose and fructose but impermeable to the disaccharide sucrose.
(a) Draw solid arrows to indicate the net movement of solutes into and/or out of the cell.
(b) Is the solution outside the cell isotonic, hypotonic, or hypertonic?
(c) Draw a dashed arrow to show the net osmosis, if any.
(d) Will the artificial cell become more flaccid, more turgid, or stay the same?
(e) Eventually, will the two solutions have the same or different solute concentrations?

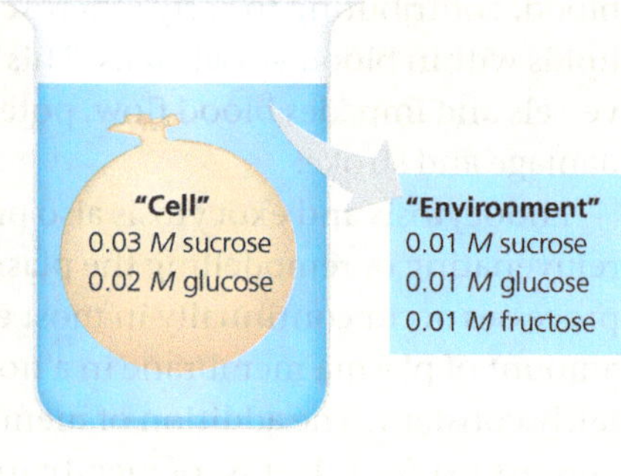

Levels 5-6: Evaluating/Creating

7. EVOLUTION CONNECTION *Paramecium* and other protists that live in hypotonic environments have cell membranes that limit water uptake, while those living in isotonic environments have membranes that are more permeable to water. Describe what water regulation adaptations might have evolved in protists in hypertonic habitats such as the Great Salt Lake and in habitats with changing salt concentration.

8. SCIENTIFIC INQUIRY An experiment is designed to study the mechanism of sucrose uptake by plant cells. Cells are immersed in a sucrose solution, and the pH of the solution is monitored. Samples of the cells are taken at intervals, and their sucrose concentration is measured. The pH is observed to decrease until it reaches a steady, slightly acidic level, and then sucrose uptake begins. (a) Evaluate these results and propose a hypothesis to explain them. (b) Predict what would happen if an inhibitor of ATP regeneration by the cell were added to the beaker once the pH was at a steady level. Explain.

9. SCIENCE, TECHNOLOGY, AND SOCIETY Extensive irrigation in arid regions causes salts to accumulate in the soil. (When water evaporates, salts that were dissolved in the water are left behind in the soil.) Based on what you learned about water balance in plant cells, explain why increased soil salinity (saltiness) might be harmful to crops.

10. WRITE ABOUT A THEME: INTERACTIONS A human pancreatic cell obtains O_2—and necessary molecules such as glucose, amino acids, and cholesterol—from its environment, and it releases CO_2 as a waste product. In response to hormonal signals, the cell secretes digestive enzymes. It also regulates its ion concentrations by exchange with its environment. Based on what you have just learned about the structure and function of cellular membranes, write a short essay (100–150 words) to describe how such a cell accomplishes these interactions with its environment.

11. SYNTHESISE YOUR KNOWLEDGE

In the supermarket, lettuce and other produce are often sprayed with water. Explain why this makes vegetables crisp.

For selected answers, see Appendix A.

8 An Introduction to Metabolism

KEY CONCEPTS

Figure 8.1 The green glowing spots on the outside of this Brazilian termite mound are larvae of the click beetle, *Pyrophorus nyctophanus*. These larvae convert the energy stored in organic molecules to light, a process called bioluminescence, which attracts termites that the larvae eat. Bioluminescence and other metabolic activities in a cell are energy transformations that are subject to physical laws.

Study Tip

Make a table: Fill in the following table for each process you read about in this chapter, such as water spilling over a dam or a chemical reaction.

Process	Starting Materials; Relative Energy Level	Ending Materials; Relative Energy Level	Will this process occur spontaneously (without an input of energy)?
Water spilling over a dam	Water at the top of a dam; higher energy level	Water at the bottom of a dam; lower energy level	Yes
Hydrolysis of ATP (Figure 8.9)			

➔ Go to Mastering Biology

to access Dynamic Study Modules for revision, 3D BioFlix® animations and high-quality videos, and your interactive Pearson eText.

How do the laws of thermodynamics relate to biological processes?

The first law of thermodynamics: *Energy can be transferred and transformed, but it cannot be created or destroyed.* For example:

Light energy from the sun

Transformed by plants to

Chemical energy in organic molecules in plants

Transformed by termites to

Chemical energy in organic molecules in termites

Transformed by click beetle larvae to

Light produced by bioluminescence

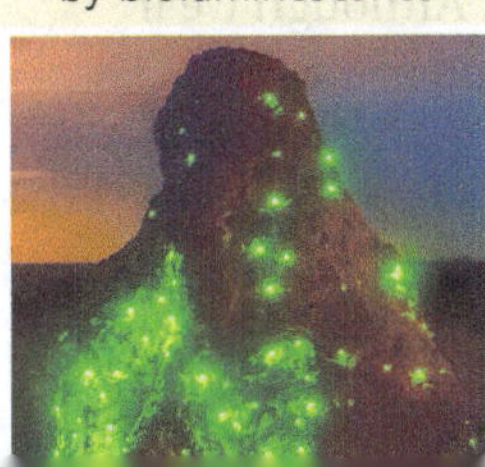

The second law of thermodynamics: *Every energy transfer or transformation increases the entropy (disorder) of the universe.* For example, during every energy transformation, some energy is converted to thermal energy and released as heat:

Plant using light to make organic molecules

Termite digesting plant and making new molecules

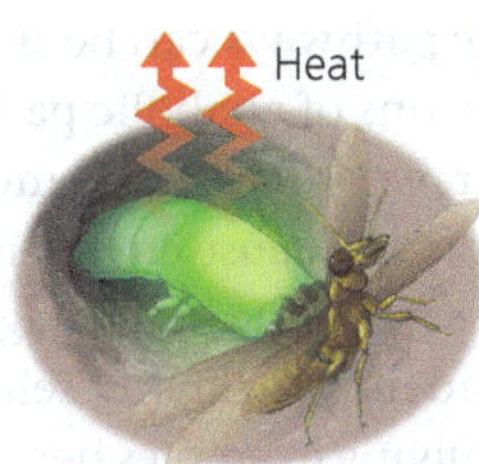

Click beetle larva digesting termite and emitting light

CONCEPT 8.1

An organism's metabolism transforms matter and energy

The totality of an organism's chemical reactions is called **metabolism** (from the Greek *metabole*, change). Metabolism is an emergent property of life that arises from orderly interactions between molecules.

Metabolic Pathways

We can picture a cell's metabolism as an elaborate road map of many chemical reactions, arranged as intersecting metabolic pathways. In a **metabolic pathway**, a specific molecule is altered in a series of defined steps, resulting in a certain product. Each step is catalysed by a specific *enzyme*, a macromolecule that speeds up a chemical reaction:

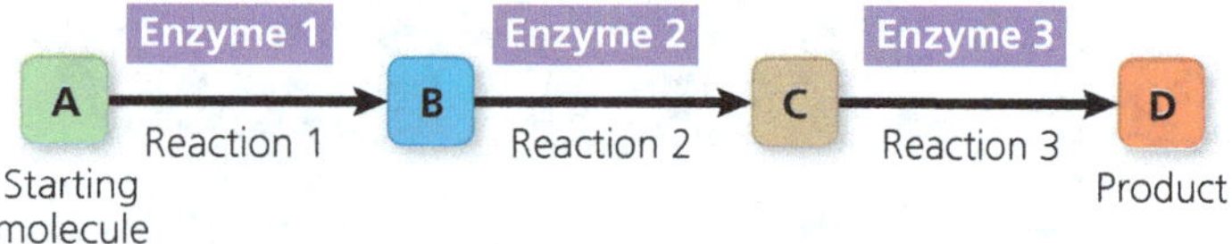

The mechanisms that regulate these enzymes balance metabolic supply and demand, just like traffic lights control the flow of traffic.

Metabolism as a whole manages the material and energy resources of the cell. Some metabolic pathways release energy by breaking down complex molecules to simpler compounds. These degradative processes are called **catabolic pathways**, or breakdown pathways. One major catabolic pathway is cellular respiration, which breaks down glucose and other organic fuels in the presence of oxygen to carbon dioxide and water. (Pathways can have more than one starting molecule and/or product.) Energy stored in the organic molecules becomes available to do cellular work, such as ciliary beating or membrane transport. **Anabolic pathways**, in contrast, consume energy to build complicated molecules from simpler ones; they are sometimes called biosynthetic pathways. Examples of anabolism are synthesis of an amino acid from simpler molecules and synthesis of a protein from amino acids. Catabolic and anabolic pathways are the "downhill" and "uphill" avenues of the metabolic landscape. Energy released from downhill reactions of catabolic pathways can be stored and then used to drive uphill reactions of anabolic pathways.

In this chapter, we will focus on mechanisms common to metabolic pathways. Because energy is fundamental to all metabolic processes, a basic knowledge of energy is necessary to understand how the living cell works. Although we'll look at some nonliving examples here to consider energetic principles, the concepts demonstrated by these examples also apply to **bioenergetics**, the study of how energy flows through living organisms.

Forms of Energy

Energy is the capacity to cause change. In everyday life, energy is important because some forms of energy can be used to do work—that is, to move matter against opposing forces, such as gravity and friction. Put another way, energy is the ability to rearrange a collection of matter. For example, you expend energy to turn the pages of a book, and your cells expend energy in transporting certain substances across membranes. Energy exists in various forms, and the work of life depends on the ability of cells to transform energy from one form to another.

Energy can be associated with the relative motion of objects; this energy is called **kinetic energy**. Moving objects can perform work by imparting motion to other matter: A pool player uses the motion of the cue stick to push the cue ball, which in turn moves the other balls; water gushing through a dam turns turbines; and the contraction of leg muscles pushes bicycle pedals. **Thermal energy** is kinetic energy associated with the random movement of atoms or molecules; thermal energy in transfer from one object to another is called **heat**. Light is also a type of energy that can be harnessed to perform work, such as powering photosynthesis in green plants.

An object not presently moving may still possess energy. Energy that is not kinetic is called **potential energy**; it is energy that matter possesses because of its location or structure. Water behind a dam, for instance, possesses energy because of its altitude above sea level. Molecules possess energy because of the arrangement of electrons in the bonds between their atoms. **Chemical energy** is a term used by biologists to refer to the potential energy available for release in a chemical reaction. Recall that catabolic pathways release energy by breaking down complex molecules. Biologists say that these complex molecules, such as glucose, are high in chemical energy. During a catabolic reaction, some bonds are broken and others are formed, releasing energy and resulting in lower-energy breakdown products. This transformation also occurs in the engine of a car when the hydrocarbons of petrol react explosively with oxygen, releasing the energy that pushes the pistons and producing exhaust. Although less explosive, a similar reaction of food molecules with oxygen provides chemical energy in biological systems, producing carbon dioxide and water as waste products. Biochemical pathways, carried out in the context of cellular structures, enable cells to release chemical energy from food molecules and use the energy to power life processes.

How is energy converted from one form to another? Consider **Figure 8.2**. The woman climbing the ladder to the diving platform is releasing chemical energy from the food

▼ **Figure 8.2 Transformations between potential energy and kinetic energy.**

she ate for lunch and using some of that energy to perform the work of climbing. The kinetic energy of muscle movement is thus being transformed into potential energy due to her increasing height above the water. The man diving is converting his potential energy to kinetic energy, which is then transferred to the water as he enters it, resulting in splashing, noise, and increased movement of water molecules. A small amount of energy is lost as heat due to friction.

Now let's consider the original source of the organic food molecules that provided the necessary chemical energy for these divers to climb the steps. This chemical energy was itself derived from light energy absorbed by plants during photosynthesis. Organisms are energy transformers.

The Laws of Energy Transformation

The study of the energy transformations that occur in a collection of matter is called **thermodynamics**. Scientists use the word *system* to denote the matter under study; they refer to the rest of the universe—everything outside the system—as the *surroundings*. An *isolated system*, such as that approximated by liquid in a thermos bottle, is unable to exchange either energy or matter with its surroundings outside the thermos. In an *open system*, energy and matter can be transferred between the system and its surroundings. Organisms are open systems. They absorb energy—for instance, light energy or chemical energy in the form of organic molecules—and release heat and metabolic waste products, such as carbon dioxide, to the surroundings. Two laws of thermodynamics govern energy transformations in organisms and all other collections of matter.

The First Law of Thermodynamics

According to the **first law of thermodynamics**, the energy of the universe is constant: *Energy can be transferred and transformed, but it cannot be created or destroyed.* The first law is also known as the *principle of conservation of energy*. The electric company does not make energy, but merely converts it to a form that is convenient for us to use. By converting sunlight to chemical energy, a plant acts as an energy transformer, not an energy producer.

The emu in **Figure 8.3a** will convert the chemical energy of the organic molecules in its food to kinetic and other forms of energy as it carries out biological processes. What happens to this energy after it has performed work? The second law of thermodynamics helps to answer this question.

▼ **Figure 8.3 The first two laws of thermodynamics.**

(a) First law of thermodynamics: Energy can be transferred or transformed but neither created nor destroyed. For example, chemical reactions in this emu (*Dromaius novaehollandiae*) will convert the chemical (potential) energy from food (plants) into the kinetic energy of running.

(b) Second law of thermodynamics: Every energy transfer or transformation increases the disorder (entropy) of the universe. For example, as the emu runs, disorder is increased around its body by the release of heat and small molecules that are the by-products of metabolism. An emu can run at speeds up to 56 kilometres per hour—as fast as a racehorse.

The Second Law of Thermodynamics

If energy cannot be destroyed, why can't organisms simply recycle their energy over and over again? It turns out that during every energy transfer or transformation, some energy is converted to thermal energy and released as heat, becoming unavailable to do work. Only a small fraction of the chemical energy from the food in Figure 8.3a is transformed into the motion of the emu shown in **Figure 8.3b**; most is lost as heat, which dissipates rapidly through the surroundings.

A system can put thermal energy to work only when there is a temperature difference that results in thermal energy flowing as heat from a warmer location to a cooler one. If temperature is uniform, as it is in a living cell, then the heat generated during a chemical reaction will simply warm a body of matter, such as the organism. (This can make a room crowded with people uncomfortably warm, as each person is carrying out a multitude of chemical reactions!)

A consequence of the loss of usable energy as heat to the surroundings is that each energy transfer or transformation makes the universe more disordered. We are all familiar with the word "disorder" in the sense of a messy room or a run-down building. The word "disorder" as used by scientists, however, has a specific molecular definition related to how dispersed the energy is in a system and how many different energy levels are present. For simplicity, we use "disorder" in the following discussion because our common understanding (the messy room) is a good analogy for molecular disorder.

Scientists use a quantity called **entropy** as a measure of molecular disorder, or randomness. The more randomly arranged a collection of matter is, the greater its entropy. We can now state the **second law of thermodynamics**: *Every energy transfer or transformation increases the entropy of the universe.* Although order can increase locally, there is an unstoppable trend towards randomisation of the universe as a whole.

The physical disintegration of a system's organised structure is a good analogy for an increase in entropy. For example, you can observe increasing entropy in the gradual decay of an unmaintained building over time. Much of the increasing entropy of the universe is more abstract, however, because it takes the form of increasing amounts of heat and less ordered forms of matter. As the emu in Figure 8.3b converts chemical energy to kinetic energy, it is also increasing the disorder of its surroundings by producing heat and small molecules, such as the CO_2 it exhales, that are the breakdown products of food.

The concept of entropy helps us understand why certain processes are energetically favourable and occur on their own. It turns out that if a given process, by itself, leads to an increase in entropy, that process can proceed without requiring an input of energy. Such a process is called a **spontaneous process**. Note that as we're using it here, the word *spontaneous* does not imply that the process would occur quickly; rather, the word signifies that it is energetically favourable. (In fact, it may be helpful for you to think of the phrase "energetically favourable" when you read the formal term *spontaneous*, the word preferred by chemists.) Some spontaneous processes, such as an explosion, may be virtually instantaneous, while others, such as the rusting of an old car over time, are much slower.

A process that, on its own, leads to a decrease in entropy is said to be nonspontaneous: It will happen only if energy is supplied. We know from experience that certain events occur spontaneously and others do not. For instance, we know that water flows downhill spontaneously but moves uphill only with an input of energy, such as when a machine pumps the water against gravity. Some of that energy is inevitably lost as heat, increasing entropy in the surroundings, so usage of energy during a nonspontaneous process also leads to an increase in the entropy of the universe as a whole.

Biological Order and Disorder

Living systems increase the entropy of their surroundings, as predicted by thermodynamic law. You may wonder, then, how cells create ordered structures from less organised starting materials. For example, simpler molecules are ordered into the more complex structure of an amino acid, and amino acids are ordered into polypeptide chains. At the organismal level as well, complex and beautifully ordered structures result from biological processes that use simpler starting materials **(Figure 8.4)**. How can this happen, if entropy must constantly increase?

▼ **Figure 8.4 Order as a characteristic of life.** Order is evident in the detailed structures of the biscuit starfish and the agave plant shown here. As open systems, organisms can increase their order as long as the order of their surroundings decreases, with an overall increase in entropy in the universe.

This increase in order is balanced by an organism's taking in organised forms of matter and energy from the surroundings and replacing them with less ordered forms. For example, an animal obtains starch, proteins, and other complex molecules from the food it eats. As catabolic pathways break these molecules down, the animal releases CO_2 and H_2O —small molecules that possess less chemical energy than the food did (see Figure 8.3b). The depletion of chemical energy is accounted for by heat generated during metabolism. On a larger scale, energy flows into most ecosystems in the form of light and exits in the form of heat (see Figure 1.9).

During the early history of life, complex organisms evolved from simpler ancestors. For instance, we can trace the ancestry of the plant kingdom from much simpler organisms called green algae to more complex flowering plants. However, this increase in organisation over time in no way violates the second law. The entropy of a particular system, such as an organism, may actually decrease as long as the total entropy of the *universe*—the system plus its surroundings—increases. Thus, organisms are islands of low entropy in an increasingly random universe. The evolution of biological order is perfectly consistent with the laws of thermodynamics.

CONCEPT CHECK 8.1

1. **MAKE CONNECTIONS** How does the second law of thermodynamics help explain the diffusion of a substance across a membrane? (See Figure 7.11.)
2. Describe the forms of energy found in an apple as it grows on a tree, then falls, then is digested by someone who eats it.
3. **WHAT IF?** If you place a teaspoon of sugar in the bottom of a glass of water, it will dissolve completely over time. Left longer, eventually the water will disappear and the sugar crystals will reappear. Explain these observations in terms of entropy.

For suggested answers, see Appendix A.

CONCEPT 8.2

The free-energy change of a reaction tells us whether or not the reaction occurs spontaneously

The laws of thermodynamics that we've just explored apply to the universe as a whole. As biologists, we want to understand the chemical reactions of life—for example, which reactions occur spontaneously and which ones require some input of energy from outside. But how can we know this without assessing the energy and entropy changes in the entire universe for each separate reaction?

Free-Energy Change, ΔG

Recall that the universe is really equivalent to "the system" plus "the surroundings." In 1878, J. Willard Gibbs, a professor at Yale, defined a very useful function called the Gibbs free energy of a system (without considering its surroundings), symbolised by the letter G. We'll refer to the Gibbs free energy simply as free energy. **Free energy** is the portion of a system's energy that can perform work when temperature and pressure are uniform throughout the system, as in a living cell. Let's consider how we determine the free-energy change that occurs when a system changes—for example, during a chemical reaction.

The change in free energy, ΔG, can be calculated for a chemical reaction by applying the following equation:

$$\Delta G = \Delta H - T\Delta S$$

This equation uses only properties of the system (the reaction) itself: ΔH symbolises the change in the system's *enthalpy* (in biological systems, equivalent to total energy); ΔS is the change in the system's entropy; and T is the absolute temperature in Kelvin (K) units (K = °C + 273; see the end of this text).

Using chemical methods, we can measure ΔG for any reaction. (The value will depend on conditions such as pH, temperature, and concentrations of reactants and products.) Once we know the value of ΔG for a process, we can use it to predict whether the process will be spontaneous (that is, whether it is energetically favourable and will occur without an input of energy). More than a century of experiments has shown that only processes with a negative ΔG are spontaneous. For ΔG to be negative, ΔH must be negative (the system gives up enthalpy and H decreases) or $T\Delta S$ must be positive (the system gives up order and S increases), or both: When ΔH and $T\Delta S$ are tallied, ΔG has a negative value ($\Delta G < 0$) for all spontaneous processes. In other words, every spontaneous process decreases the system's free energy, and processes that have a positive or zero ΔG are never spontaneous.

This information is immensely interesting to biologists, for it allows us to predict which kinds of change can happen without an input of energy. Such spontaneous changes can be harnessed by the cell to perform work. This principle is very important in the study of metabolism, where a major goal is to determine which reactions can supply energy for cellular work.

Free Energy, Stability, and Equilibrium

As we saw in the previous section, when a process occurs spontaneously in a system, we can be sure that ΔG is negative. Another way to think of ΔG is to realise that it represents the difference between the free energy of the final state and the free energy of the initial state:

$$\Delta G = G_{\text{final state}} - G_{\text{initial state}}$$

For a reaction to have a negative ΔG, the system must lose free energy during the change from initial state to final state. Because it has less free energy, the system in its final state is less likely to change and is therefore more stable than it was previously.

We can think of free energy as a measure of a system's instability—its tendency to change to a more stable state. Unstable systems (higher G) tend to change in such a way that they become more stable (lower G). For example, a diver on top of a platform is less stable (more likely to fall) than when floating in the water; a drop of concentrated dye is less stable (more likely to disperse) than when the dye is spread randomly through the liquid; and a glucose molecule is less stable (more likely to break down) than the simpler molecules into which it can be split **(Figure 8.5)**. Unless something prevents it, each of these systems will move towards greater stability: The diver falls, the solution becomes uniformly coloured, and the glucose molecule is broken down into smaller molecules.

Another term that describes a state of maximum stability is *equilibrium*, which you learned about in Concept 2.4 in connection with chemical reactions. At equilibrium, the forward and reverse reactions occur at the same rate, and there is no further net change in the relative concentration of products and reactants. For a system at equilibrium, G is at its lowest possible value in that system. Free energy increases when a reaction is somehow pushed away from equilibrium, perhaps by removing some of the products (and thus changing their concentration relative to that of the reactants). We can think of the equilibrium state as a free-energy valley. Any change from the equilibrium position will have a positive ΔG and will not be spontaneous. For this reason, systems never spontaneously move away from equilibrium. Because a system at equilibrium cannot spontaneously change, it can do no work. *A process is spontaneous and can perform work only when it is moving towards equilibrium.*

Free Energy and Metabolism

We can now apply the free-energy concept more specifically to the chemistry of life's processes.

Exergonic and Endergonic Reactions in Metabolism

Based on their free-energy changes, chemical reactions can be classified as either exergonic ("energy outwards") or endergonic ("energy inwards"). An **exergonic reaction** proceeds with a net release of free energy **(Figure 8.6a)**. Because the chemical mixture loses free energy (G decreases), ΔG is negative for an exergonic reaction. Using ΔG as a standard for spontaneity, exergonic reactions are those that occur spontaneously. (Remember, the word *spontaneous* implies that it is energetically favourable, not that it will occur rapidly.) The magnitude of ΔG for an

▼ Figure 8.5 The relationship of free energy to stability, work capacity, and spontaneous change. Unstable systems (top) are rich in free energy, G. They tend to change spontaneously to a more stable state (bottom). This "downhill" change can be harnessed to perform work.

- More free energy (higher G)
- Less stable
- Greater work capacity

In a **spontaneous change**
- The free energy of the system decreases ($\Delta G < 0$).
- The system becomes more stable.
- The released free energy can be harnessed to do work.

- Less free energy (lower G)
- More stable
- Less work capacity

(a) Gravitational motion. Objects move spontaneously from a higher altitude to a lower one.

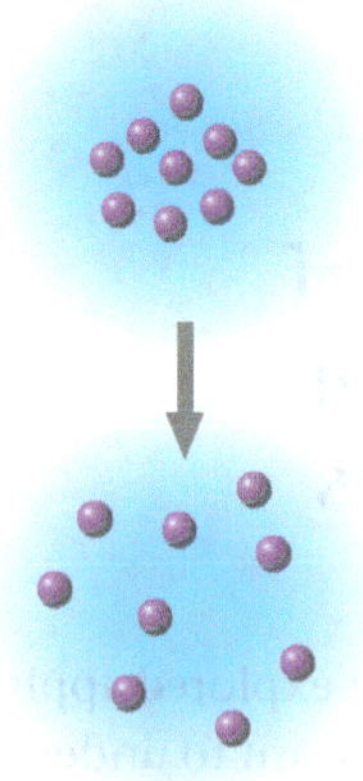

(b) Diffusion. Molecules in a drop of dye diffuse until they are randomly dispersed.

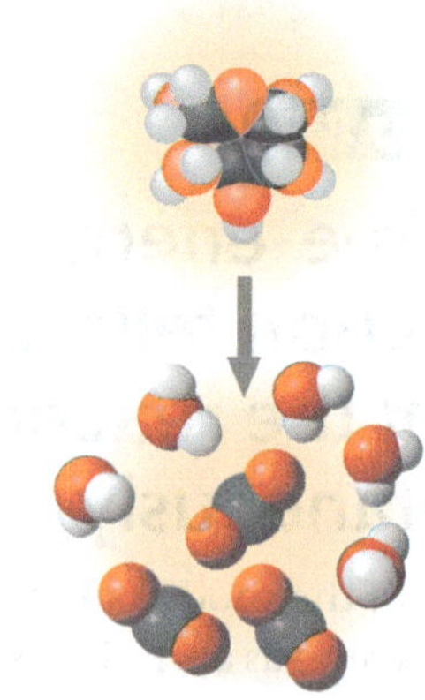

(c) Chemical reaction. In a cell, a glucose molecule is broken down into simpler molecules.

MAKE CONNECTIONS *Compare the redistribution of molecules shown in (b) to the transport of hydrogen ions (H^+) across a membrane by a proton pump, creating a concentration gradient (see Figure 7.18). Which process(es) result(s) in higher free energy? Which system(s) can do work?*

▼ **Figure 8.6 Free energy changes (ΔG) in exergonic and endergonic reactions.**

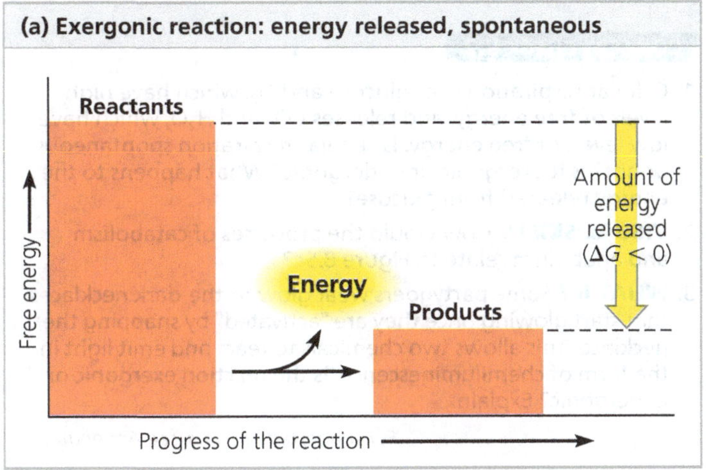

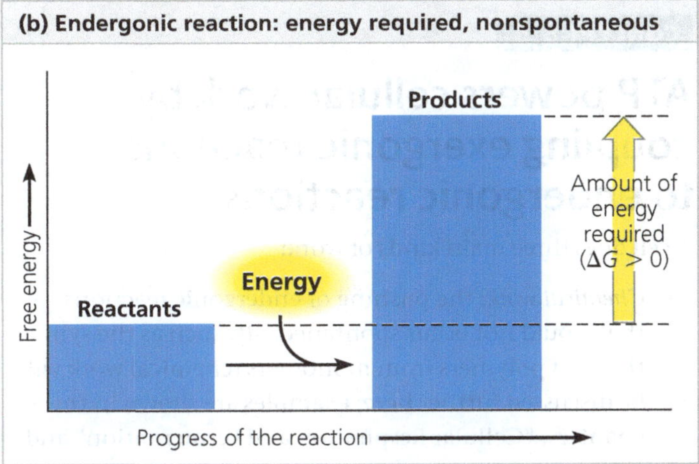

exergonic reaction represents the maximum amount of work the reaction can perform.* The greater the decrease in free energy, the greater the amount of work that can be done.

We can use the overall reaction for cellular respiration as an example:

$$C_6H_{12}O_6 + 6\,O_2 \rightarrow 6\,CO_2 + 6\,H_2O$$
$$\Delta G = -2{,}870 \text{ kJ/mol}$$

For each mole (180 g) of glucose broken down by respiration under what are called "standard conditions" (1 *M* of each reactant and product, 25°C, pH 7), 2,870 kJ of energy is made available for work. Because energy must be conserved, the chemical *products* of respiration store 2,870 kJ less free energy per mole than the *reactants*. The products are, in a sense, the spent exhaust of a process that tapped the free energy stored in the bonds of the sugar molecules.

It is important to realise that the breaking of bonds does not release energy; on the contrary, as you will soon see, it requires energy. The phrase "energy stored in bonds" is shorthand for the potential energy that can be released when new bonds are formed after the original bonds break, as long as the products are of lower free energy than the reactants.

An **endergonic reaction** is one that absorbs free energy from its surroundings **(Figure 8.6b)**. Because this kind of reaction essentially *stores* free energy in molecules (G increases), ΔG is positive. Such reactions are non-spontaneous, and the magnitude of ΔG is the quantity of energy required to drive the reaction. If a chemical process is exergonic (downhill), releasing energy in one direction, then the reverse process must be endergonic (uphill), using energy. A reversible process cannot be downhill in both directions. If $\Delta G = -2{,}870$ kJ/mol for respiration, which converts glucose and oxygen to CO_2 and H_2O, then the reverse process—the conversion of CO_2 and H_2O to glucose and oxygen (O_2)—must be strongly endergonic, with $\Delta G = +2{,}870$ kJ/mol. Such a reaction would never happen by itself.

How, then, do plants make sugar? They get the required energy (2,870 kJ to make a mole of glucose) by capturing light from the sun and converting its energy to chemical energy. Next, in a long series of exergonic steps, they gradually spend that chemical energy to assemble glucose molecules.

Equilibrium and Metabolism

Reactions in an isolated system eventually reach equilibrium and can then do no work, as illustrated by the isolated hydroelectric system in **Figure 8.7**. The chemical reactions of metabolism are reversible, and they, too, would reach equilibrium if they occurred in the isolation of a test tube. Because systems at equilibrium are at a minimum of G and can do no work, a cell that has reached metabolic equilibrium is dead! *The fact that metabolism as a whole is never at equilibrium is one of the defining features of life.*

▼ **Figure 8.7 Equilibrium and work in an isolated hydroelectric system.** Water flowing downhill turns a turbine that drives a generator providing electricity to a lightbulb, but only until the system reaches equilibrium.

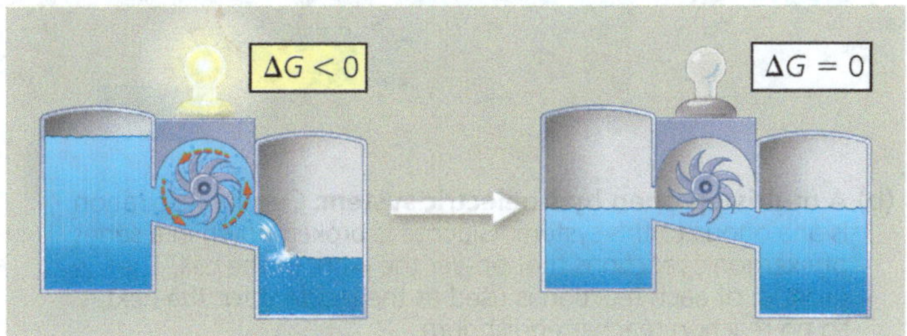

*The word *maximum* qualifies this statement because some of the free energy is released as heat and cannot do work. Therefore, ΔG represents a theoretical upper limit of available energy.

Like most systems, a living cell is not in equilibrium. Materials flow in and out, keeping metabolic pathways from ever reaching equilibrium, and the cell continues to do work throughout its life. This principle is illustrated by the open (and more realistic) hydroelectric system in **Figure 8.8a**. However, unlike this simple single-step system, a catabolic pathway in a cell releases free energy in a series of reactions. An example is cellular respiration, illustrated by analogy in **Figure 8.8b**. Some of the reversible reactions of respiration are constantly "pulled" in one direction—that is, they are kept out of equilibrium. The key to maintaining this lack of equilibrium is that the product of a reaction does not accumulate but instead becomes a reactant in the next step; finally, waste products are expelled from the cell. The overall sequence of reactions is kept going by the huge free-energy difference between glucose and O_2 at the top of the energy "hill" and CO_2 and H_2O at the "downhill" end. As long as our cells have a steady supply of glucose or other fuels and oxygen and are able to expel waste products to the surroundings, their metabolic pathways never reach equilibrium and can continue to do the work of life.

Organisms are open systems. Sunlight provides a daily source of free energy for an ecosystem's plants and other photosynthetic organisms. Animals and other nonphotosynthetic organisms in an ecosystem must have a source of free energy—the organic products of photosynthesis. Now we are ready to see how a cell actually performs the work of life.

▼ **Figure 8.8 Equilibrium and work in open systems.**

(a) An open hydroelectric system. Water flowing through a turbine keeps driving the generator because intake and outflow of water keep the system from reaching equilibrium.

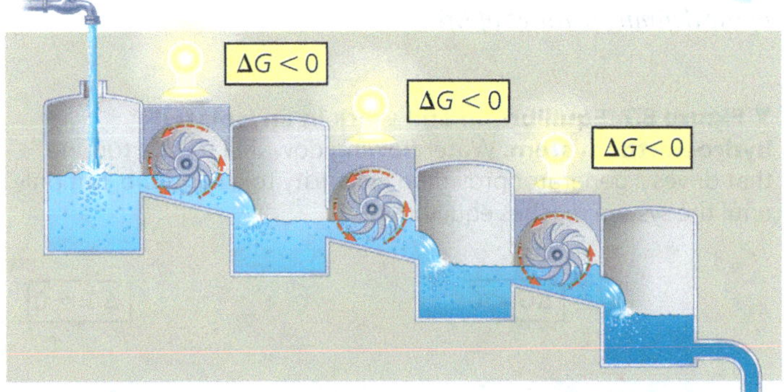

(b) A multistep open hydroelectric system. Cellular respiration is analogous to this system: Glucose is broken down in a series of exergonic reactions that power the work of the cell. The product of each reaction is used as the reactant for the next, so no reaction reaches equilibrium.

CONCEPT CHECK 8.2

1. Cellular respiration uses glucose and O_2, which have high levels of free energy, and releases CO_2 and H_2O, which have low levels of free energy. Is cellular respiration spontaneous or not? Is it exergonic or endergonic? What happens to the energy released from glucose?
2. **VISUAL SKILLS** How would the processes of catabolism and anabolism relate to Figure 8.5c?
3. **WHAT IF?** Some partygoers wear glow-in-the-dark necklaces that start glowing once they are "activated" by snapping the necklace. This allows two chemicals to react and emit light in the form of chemiluminescence. Is the reaction exergonic or endergonic? Explain.

For suggested answers, see Appendix A.

CONCEPT 8.3

ATP powers cellular work by coupling exergonic reactions to endergonic reactions

A cell does three main kinds of work:

- *Chemical work*, the pushing of endergonic reactions that would not occur spontaneously, such as the synthesis of polymers from monomers (chemical work will be discussed further here; examples are shown in the chapters, "Cellular Respiration and Fermentation" and "Photosynthesis")
- *Transport work*, the pumping of substances across membranes against the direction of spontaneous movement (see Concept 7.4)
- *Mechanical work*, such as the beating of cilia (see Concept 6.6), the contraction of muscle cells, and the movement of chromosomes during cellular reproduction

A key feature in the way cells manage their energy resources to do this work is **energy coupling**, the use of an exergonic process to drive an endergonic one. ATP is responsible for mediating most energy coupling in cells, and in most cases it acts as the immediate source of energy that powers cellular work.

The Structure and Hydrolysis of ATP

ATP (adenosine triphosphate; see Concept 4.3) contains the sugar ribose, with the nitrogenous base adenine and a chain of three phosphate groups (the triphosphate group) bonded to it **(Figure 8.9a)**. In addition to its role in energy coupling, ATP is also one of the nucleoside triphosphates used to make RNA (see Figure 5.23).

The bonds between the phosphate groups of ATP can be broken by hydrolysis. When the terminal phosphate bond is broken by addition of a water molecule, a molecule of inorganic phosphate ($HOPO_3^{2-}$, abbreviated as Ⓟ$_i$ throughout this text) leaves the ATP, which becomes adenosine diphosphate, or ADP **(Figure 8.9b)**. The reaction is exergonic and releases 30.5 kJ of energy per mole of ATP hydrolysed:

$$\text{ATP} + H_2O \rightarrow \text{ADP} + Ⓟ_i$$
$$\Delta G = -30.5 \text{ kJ/mol}$$

This is the free-energy change measured under standard conditions. In the cell, conditions do not conform to standard conditions, primarily because reactant and product concentrations differ from 1 *M*. For example, when ATP hydrolysis occurs under typical cellular conditions, the actual ΔG is about −54 kJ/mol, 78% greater than the energy released by ATP hydrolysis under standard conditions.

▼ Figure 8.9 The structure and hydrolysis of adenosine triphosphate (ATP). Throughout this text, the chemical structure of the triphosphate group seen in **(a)** will be represented by the three joined yellow circles shown in **(b)**.

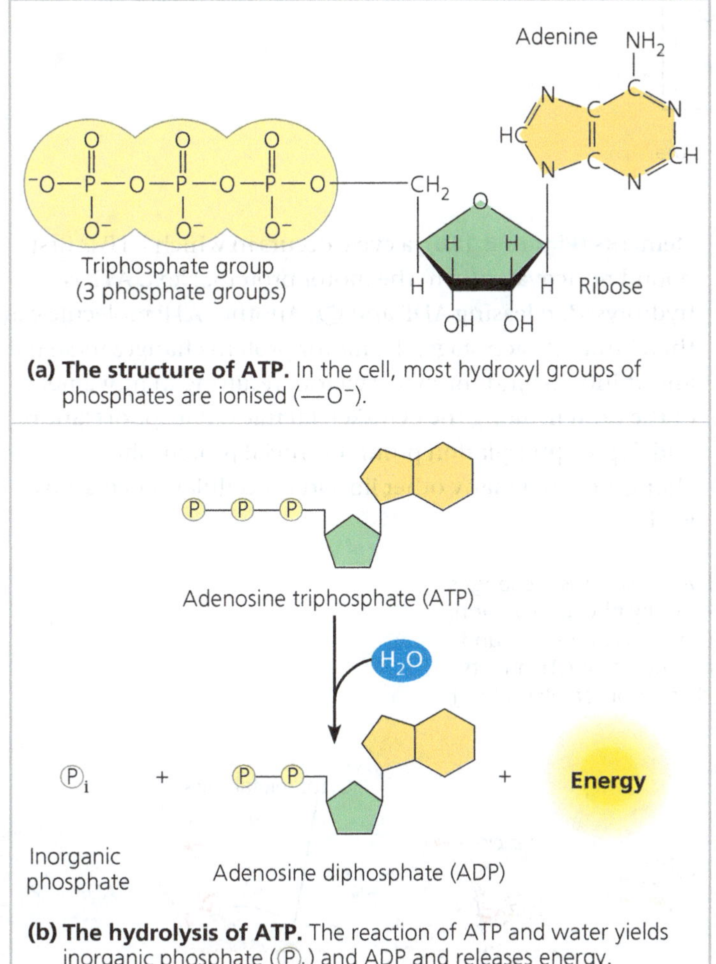

(a) The structure of ATP. In the cell, most hydroxyl groups of phosphates are ionised (—O⁻).

(b) The hydrolysis of ATP. The reaction of ATP and water yields inorganic phosphate (Ⓟ$_i$) and ADP and releases energy.

Because their hydrolysis releases energy, the phosphate bonds of ATP are sometimes referred to as high-energy phosphate bonds, but the term is misleading. The phosphate bonds of ATP are not unusually strong bonds, as "high-energy" may imply; rather, the reactants (ATP and water) themselves have high energy relative to the energy of the products (ADP and Ⓟ$_i$). The release of energy during the hydrolysis of ATP comes from the chemical change of the system to a state of lower free energy, not from the phosphate bonds themselves.

ATP is useful to the cell because the energy it releases on losing a phosphate group is somewhat greater than the energy most other molecules could deliver. But why does this hydrolysis release so much energy? If we reexamine the ATP molecule in Figure 8.9a, we can see that all three phosphate groups are negatively charged. These like charges are crowded together, and their mutual repulsion contributes to the instability of this region of the ATP molecule. The triphosphate tail of ATP is the chemical equivalent of a compressed spring.

How ATP Provides Energy That Performs Work

When ATP is hydrolysed in a test tube, the release of free energy merely heats the surrounding water. In an organism, this same generation of heat can sometimes be beneficial. For instance, the process of shivering uses ATP hydrolysis during muscle contraction to warm the body. In most cases in the cell, however, the generation of heat alone would be an inefficient (and potentially dangerous) use of a valuable energy resource. Instead, the cell's proteins harness the energy released during ATP hydrolysis in several ways to perform the three types of cellular work—chemical, transport, and mechanical.

For example, with the help of specific enzymes, the cell is able to use the high free energy of ATP to drive chemical reactions that, by themselves, are endergonic. If the ΔG of an endergonic reaction is less than the amount of energy released by ATP hydrolysis, then the two reactions can be coupled so that, overall, the coupled reactions are exergonic. This usually involves phosphorylation, the transfer of a phosphate group from ATP to some other molecule, such as the reactant. The recipient molecule with the phosphate group covalently bonded to it is then called a **phosphorylated intermediate**. The key to coupling exergonic and endergonic reactions is the formation of this phosphorylated intermediate, which is more reactive (less stable, with more

▼ **Figure 8.10 How ATP drives chemical work: energy coupling using ATP hydrolysis.** In this example, the exergonic process of ATP hydrolysis drives an endergonic process—synthesis of the amino acid glutamine.

(a) Glutamic acid conversion to glutamine. Glutamine synthesis from glutamic acid (Glu) by itself is endergonic (ΔG is positive), so it is not spontaneous.

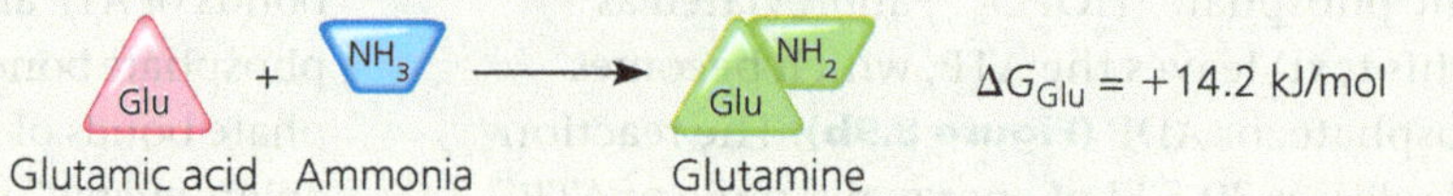

(b) Conversion reaction coupled with ATP hydrolysis. In the cell, glutamine synthesis occurs in two steps, coupled by a phosphorylated intermediate (Glu-Ⓟ). **1** ATP phosphorylates glutamic acid, making it less stable, with more free energy. **2** Ammonia displaces the phosphate group, forming glutamine.

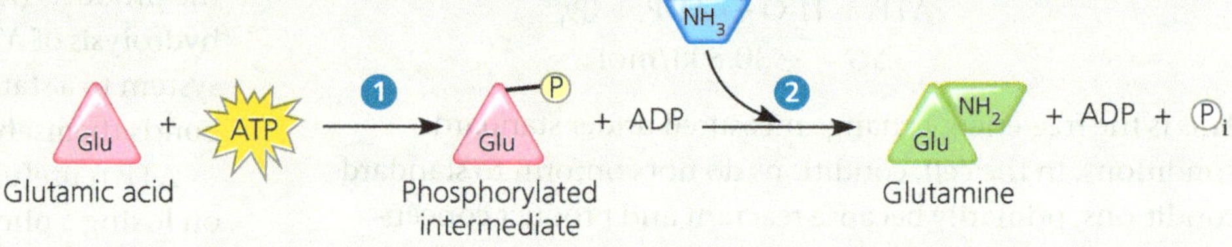

(c) Free-energy change for coupled reaction. ΔG for the glutamic acid conversion to glutamine (+14.2 kJ/mol) plus ΔG for ATP hydrolysis (–30.5 kJ/mol) gives the free-energy change for the overall reaction (–16.3 kJ/mol). Because the overall process is exergonic (net ΔG is negative), it occurs spontaneously.

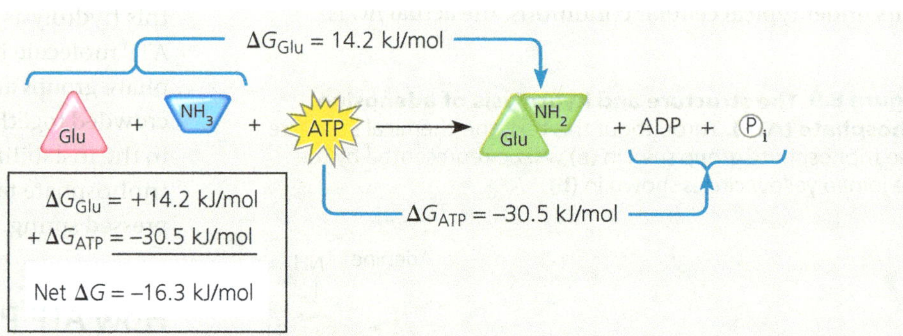

MAKE CONNECTIONS *Referring to Figure 5.14, explain why glutamine (Gln) is drawn in this figure as a glutamic acid (Glu) with an amino group attached.*

free energy) than the original unphosphorylated molecule **(Figure 8.10)**.

Transport and mechanical work in the cell are also nearly always powered by the hydrolysis of ATP. In these cases, ATP hydrolysis leads to a change in a protein's shape and often its ability to bind another molecule. Sometimes this occurs via a phosphorylated intermediate, as seen for the transport protein in **Figure 8.11a**. In most instances of mechanical work involving motor proteins "walking" along cytoskeletal elements **(Figure 8.11b)**, a cycle occurs in which ATP is first bound noncovalently to the motor protein. Next, ATP is hydrolysed, releasing ADP and $Ⓟ_i$. Another ATP molecule can then bind. At each stage, the motor protein changes its shape and ability to bind the cytoskeleton, resulting in movement of the protein along the cytoskeletal track. Phosphorylation and dephosphorylation promote crucial protein shape changes during many other important cellular processes as well.

▼ **Figure 8.11 How ATP drives transport and mechanical work.** ATP hydrolysis causes changes in the shapes and binding affinities of proteins. This can occur either **(a)** directly, by phosphorylation, as shown for a membrane protein carrying out active transport of a solute (see also Figure 7.16 and the proton pump in Figure 6.32, upper left), or **(b)** indirectly, via noncovalent binding of ATP and its hydrolytic products, as is the case for motor proteins that move vesicles (and other organelles) along cytoskeletal "tracks" in the cell (see also Figures 6.21 and 6.32, lower right).

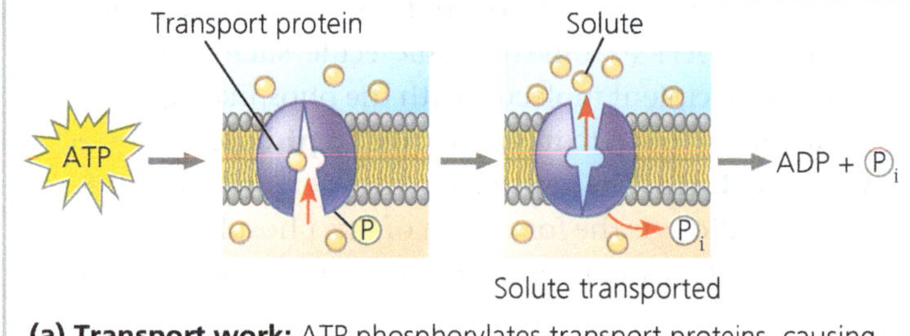

(a) Transport work: ATP phosphorylates transport proteins, causing a shape change that allows transport of solutes.

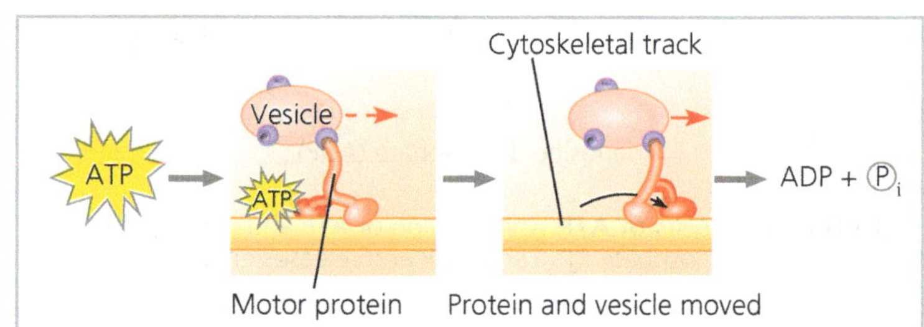

(b) Mechanical work: ATP binds noncovalently to motor proteins and then is hydrolysed, causing a shape change that walks the motor protein forwards.

▼ **Figure 8.12 The ATP cycle.** Energy released by breakdown reactions (catabolism) in the cell is used to phosphorylate ADP, regenerating ATP. Chemical potential energy stored in ATP drives most cellular work.

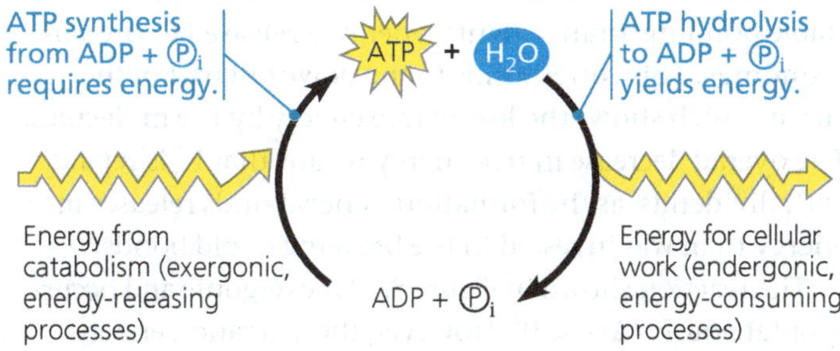

The Regeneration of ATP

An organism at work uses ATP continuously, but ATP is a renewable resource that can be regenerated by the addition of phosphate to ADP **(Figure 8.12)**. The free energy required to phosphorylate ADP comes from exergonic breakdown reactions (catabolism) in the cell. This shuttling of inorganic phosphate and energy is called the ATP cycle, and it couples the cell's energy-yielding (exergonic) processes to the energy-consuming (endergonic) ones. The ATP cycle proceeds at an astonishing pace. For example, a working muscle cell recycles its entire pool of ATP in less than a minute. That turnover represents 10 million molecules of ATP consumed and regenerated per second per cell. If ATP could not be regenerated by the phosphorylation of ADP, humans would use up nearly their body weight in ATP each day.

Because both directions of a reversible process cannot be downhill, the regeneration of ATP from ADP and Ⓟi is necessarily endergonic:

$$\text{ADP} + \text{Ⓟ}_i \rightarrow \text{ATP} + H_2O$$
$$\Delta G = +30.5 \text{ kJ/mol (standard conditions)}$$

Since ATP formation from ADP and Ⓟi is not spontaneous, free energy must be spent to make it occur. Catabolic (exergonic) pathways, especially cellular respiration, provide the energy for the endergonic process of making ATP. Plants also use light energy to produce ATP. Thus, the ATP cycle is a key player in bioenergetics, functioning as a revolving door through which energy passes during its transfer from catabolic to anabolic pathways.

CONCEPT CHECK 8.3

1. How does ATP typically transfer energy from an exergonic to an endergonic reaction in the cell?
2. Which combination has more free energy: glutamic acid + ammonia + ATP or glutamine + ADP + Ⓟi? Explain.
3. **MAKE CONNECTIONS** Does Figure 8.11a show passive or active transport? Explain. (See Concepts 7.3 and 7.4.)

For suggested answers, see Appendix A.

CONCEPT 8.4

Enzymes speed up metabolic reactions by lowering energy barriers

The laws of thermodynamics tell us what will and will not happen under given conditions but say nothing about the rate of these processes. A spontaneous chemical reaction occurs without any requirement for outside energy, but it may occur so slowly that it is imperceptible. For example, even though the hydrolysis of sucrose (table sugar) to glucose and fructose is exergonic, occurring spontaneously with a release of free energy ($\Delta G = -29.3$ kJ/mol), a solution of sucrose dissolved in sterile water will sit for years at room temperature with no appreciable hydrolysis. However, if we add a small amount of the enzyme sucrase to the solution, then all the sucrose may be hydrolysed within seconds, as shown here:

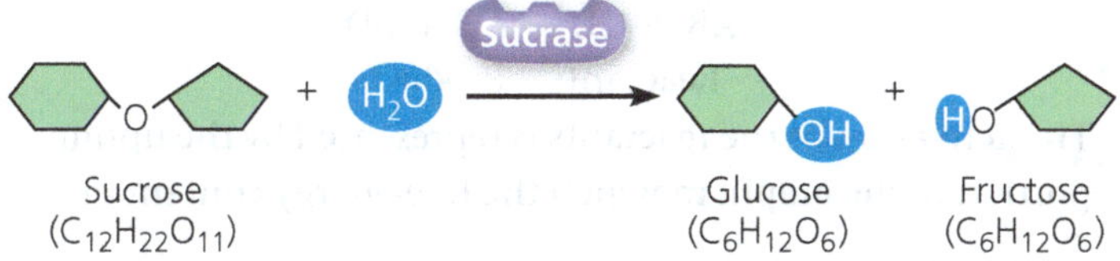

How does the enzyme do this? An **enzyme** is a macromolecule that acts as a **catalyst**, a chemical agent that speeds up a reaction without being consumed by the reaction. In this chapter, we focus on enzymes that are proteins. (Some RNA molecules, called ribozymes, can function as enzymes; these will be discussed in Concepts 17.3 and 25.1.) Without regulation by enzymes, chemical traffic through the pathways of metabolism would become terribly congested because many chemical reactions would take such a long time. In the next two sections, we will see why spontaneous reactions can be slow and how an enzyme changes the situation.

The Activation Energy Barrier

Every chemical reaction between molecules involves both bond breaking and bond forming. For example, the hydrolysis of sucrose involves breaking the bond between glucose and fructose and one of the bonds of a water molecule and then forming two new bonds, as shown above. Changing one molecule into another generally involves contorting the starting molecule into a highly unstable state before the reaction can proceed. This contortion can be compared to the bending of a metal key ring when you pry it open to add a new key. The key ring is highly unstable in its opened form but returns to a stable state once the key is threaded all the way onto the ring. To reach the contorted state where bonds can change, reactant molecules must absorb energy from their surroundings. When the new bonds of the product molecules form, energy is released as heat, and the molecules return to stable shapes with lower energy than the contorted state.

The initial investment of energy for starting a reaction—the energy required to contort the reactant molecules so the bonds can break—is known as the *free energy of activation*, or **activation energy**, abbreviated E_A in this text. We can think of activation energy as the amount of energy needed to push the reactants to the top of an energy barrier, or "uphill," so that the "downhill" part of the reaction can begin. Activation energy is often supplied by heat in the form of thermal energy that the reactant molecules absorb from the surroundings. The absorption of thermal energy accelerates the reactant molecules, so they collide more often and more forcefully. It also agitates the atoms within the molecules, making the breakage of bonds more likely. When the molecules have absorbed enough energy for the bonds to break, the reactants are in an unstable condition known as the *transition state*.

Figure 8.13 graphs the energy changes for a hypothetical exergonic reaction that swaps portions of two reactant molecules:

$$\underset{\text{Reactants}}{AB + CD} \rightarrow \underset{\text{Products}}{AC + BD}$$

The activation of the reactants is represented by the uphill portion of the graph, in which the free-energy content of the reactant molecules is increasing. At the summit, when energy equivalent to E_A has been absorbed, the reactants are in the transition state: They are activated, and their bonds can be broken. As the atoms then settle into their new, more stable bonding arrangements, energy is released to the surroundings. This corresponds to the downhill part of the curve, which shows the loss of free energy by the molecules. The overall decrease in free energy means that E_A is repaid with dividends, as the formation of new bonds releases more energy than was invested in the breaking of old bonds.

The reaction shown in Figure 8.13 is exergonic and occurs spontaneously ($\Delta G < 0$). However, the activation energy provides a barrier that determines the rate of the reaction. The reactants must absorb enough energy to reach the top of the activation energy barrier before the reaction can occur. For some reactions, E_A is modest enough that even at room temperature there is sufficient thermal energy for many of the reactant molecules to reach the transition state in a short time. In most cases, however, E_A is so high and the transition state is reached so rarely that the reaction will hardly proceed at all. In these cases, the reaction will occur at a noticeable rate only if energy is provided, usually by heat. For example, the reaction of petrol and oxygen is exergonic and will occur spontaneously, but energy is required for the molecules to reach the transition state and react. Only when the spark plugs fire in an automobile engine can there be the explosive release of energy that pushes the pistons. Without a spark, a mixture of petrol hydrocarbons and oxygen will not react because the E_A barrier is too high.

▼ Figure 8.13 Energy profile of an exergonic reaction. The "molecules" are hypothetical, with A, B, C, and D representing portions of the molecules. Thermodynamically, this is an exergonic reaction, with a negative ΔG, and the reaction occurs spontaneously. However, the activation energy (E_A) provides a barrier that determines the rate of the reaction.

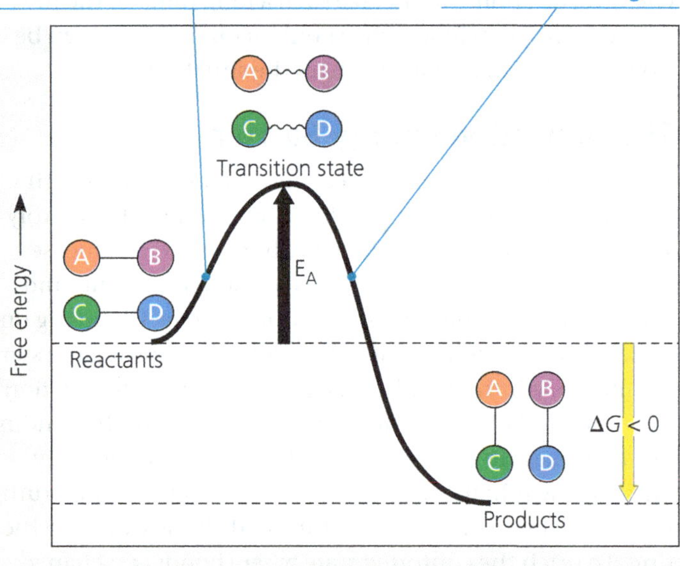

DRAW IT *Graph the progress of an endergonic reaction in which EF and GH form products EG and FH, assuming that the reactants must pass through a transition state.*

How Enzymes Speed Up Reactions

Proteins, DNA, and other complex cellular molecules are rich in free energy and have the potential to decompose spontaneously; that is, the laws of thermodynamics favour their breakdown. These molecules only persist because at temperatures typical for cells, few molecules can make it over the hump of activation energy. The barriers for selected reactions must occasionally be surmounted, however, for cells to carry out the processes needed for life. Heat can increase the rate of a reaction by allowing reactants to attain the transition state more often, but this would not work well in biological systems. First, high temperature denatures proteins and kills cells. Second, heat would speed up *all* reactions, not just those that are needed. Instead of heat, organisms carry out **catalysis**, the process by which a catalyst selectively speeds up a reaction without itself being consumed. (You learned about catalysts earlier in this section.)

An enzyme catalyses a reaction by lowering the E_A barrier **(Figure 8.14)**, enabling the reactant molecules to absorb enough energy to reach the transition state even at moderate temperatures, as we'll see shortly. It is crucial to note that *an enzyme cannot change the ΔG for a reaction; it cannot make an*

▼ Figure 8.14 The effect of an enzyme on activation energy. Without affecting the free-energy change (ΔG) for a reaction, an enzyme speeds the reaction by reducing its activation energy (E_A).

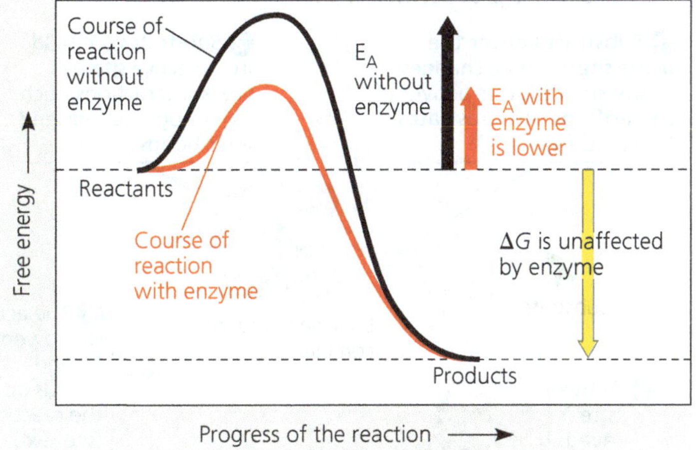

endergonic reaction exergonic. Enzymes can only hasten reactions that would eventually occur anyway, but this enables the cell to have a dynamic metabolism, routing chemicals smoothly through metabolic pathways. Also, enzymes are very specific for the reactions they catalyse, so they determine which chemical processes will be going on in the cell at any given time.

Substrate Specificity of Enzymes

The reactant an enzyme acts on is referred to as the enzyme's **substrate**. The enzyme binds to its substrate (or substrates, when there are two or more reactants), forming an **enzyme-substrate complex**. While enzyme and substrate are joined, the catalytic action of the enzyme converts the substrate to the product (or products) of the reaction. The overall process can be summarised as follows:

Enzyme + Substrate(s)	⇌	Enzyme-substrate complex	⇌	Enzyme + Product(s)

Most enzyme names end in *-ase*. (See if you can find three examples of enzymes in the lower left of Figure 6.32.) For example, the enzyme sucrase catalyses the hydrolysis of the disaccharide sucrose into its two monosaccharides, glucose and fructose (see the diagram at the beginning of Concept 8.4):

Sucrase + Sucrose + H_2O	⇌	Sucrase-sucrose-H_2O complex	⇌	Sucrase + Glucose + Fructose

The reaction catalysed by each enzyme is very specific; an enzyme can recognise its specific substrate even among closely related compounds. For instance, sucrase will act only on sucrose and will not bind to other disaccharides, such as maltose. What accounts for this molecular recognition? Recall that most enzymes are proteins, and that proteins are macromolecules with unique 3-D configurations. The specificity of an enzyme results from its shape, which is a consequence of its amino acid sequence.

Only a restricted region of the enzyme molecule actually binds to the substrate. This region, called the **active site**, is typically a pocket or groove on the surface of the enzyme where catalysis occurs (**Figure 8.15a**; see also Figure 5.16). Usually, the active site is formed by only a few of the enzyme's amino acids, with the rest of the protein molecule providing a framework that determines the shape of the active site. The specificity of an enzyme is attributed to a complementary fit between the shape of its active site and the shape of the substrate.

An enzyme is not a stiff structure locked into a given shape. In fact, recent work by biochemists has shown that enzymes (and other proteins) seem to "dance" between subtly different shapes in a dynamic equilibrium, with slight differences in free energy for each "pose." The shape that best fits the substrate isn't necessarily the one with the lowest energy, but during the very short time the enzyme takes on this shape, its active site can bind to the substrate. The active site itself is also not a rigid receptacle for the substrate. As shown in **Figure 8.15b**, when the substrate enters the active site, the enzyme changes shape slightly due to interactions between the substrate's chemical groups and chemical groups

▼ Figure 8.15 Induced fit between an enzyme and its substrate.

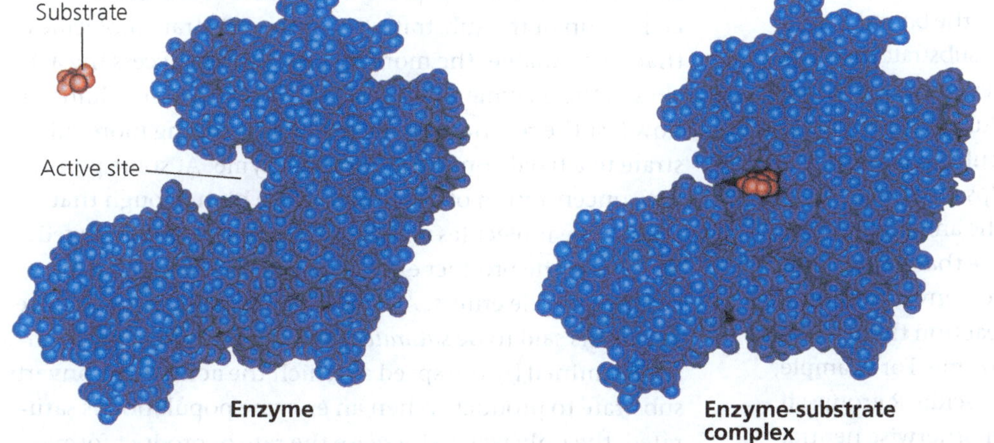

(a) In this space-filling model of the enzyme hexokinase (blue), the active site forms a groove on the surface. The enzyme's substrate is glucose (red).

(b) When the substrate enters the active site, it forms weak bonds with the enzyme, inducing a change in the shape of the protein. This change allows additional weak bonds to form, causing the active site to enfold the substrate and hold it in place.

on the side chains of the amino acids that form the active site. This shape change makes the active site fit even more snugly around the substrate. The tightening of the binding after initial contact—called **induced fit**—is like a clasping handshake. Induced fit brings chemical groups of the active site into positions that enhance their ability to catalyse the chemical reaction.

Catalysis in the Enzyme's Active Site

In most enzymatic reactions, the substrate is held in the active site by so-called weak interactions, such as hydrogen bonds and ionic bonds. The R groups of a few of the amino acids that make up the active site catalyse the conversion of substrate to product, and the product departs from the active site. The enzyme is then free to take another substrate molecule into its active site. The entire cycle happens so fast that a single enzyme molecule typically acts on about 1,000 substrate molecules per second, and some enzymes are even faster. Enzymes, like other catalysts, emerge from the reaction in their original form. Therefore, very small amounts of enzyme can have a huge metabolic impact by functioning over and over again in catalytic cycles. **Figure 8.16** shows a catalytic cycle involving two substrates and two products.

Most metabolic reactions are reversible, and an enzyme can catalyse either the forward or the reverse reaction, depending on which direction has a negative ΔG. This in turn depends mainly on the relative concentrations of reactants and products. The net effect is always in the direction of equilibrium.

Enzymes use a variety of mechanisms that lower activation energy and speed up a reaction (see Figure 8.16, step 3):

- In reactions involving two or more reactants, the active site provides a template on which the substrates can come together in the proper orientation for a reaction to occur between them.
- As the active site of an enzyme clutches the bound substrates, the enzyme may stretch the substrate molecules towards their transition state form, stressing and bending critical chemical bonds to be broken during the reaction. Because E_A is proportional to the difficulty of breaking the bonds, distorting the substrate helps it approach the transition state and thus reduces the amount of free energy that must be absorbed to achieve that state.
- The active site may also provide a microenvironment that is more conducive to a certain reaction than the solution itself would be without the enzyme. For example, if the active site has amino acids with acidic R groups, it may provide a pocket of low pH in an otherwise neutral cell. Here, an acidic amino acid may facilitate H^+ transfer to the substrate as a key step in catalysing the reaction.
- Amino acids in the active site may directly participate in the chemical reaction. Sometimes this process involves brief covalent bonding between the substrate and the side chain of an amino acid of the enzyme. Subsequent steps restore the side chains to their original states so that the active site is the same after the reaction as it was before.

▼ Figure 8.16 The active site and catalytic cycle of an enzyme. An enzyme can convert one or more reactant molecules to one or more product molecules. The enzyme shown here converts two substrate molecules to two product molecules.

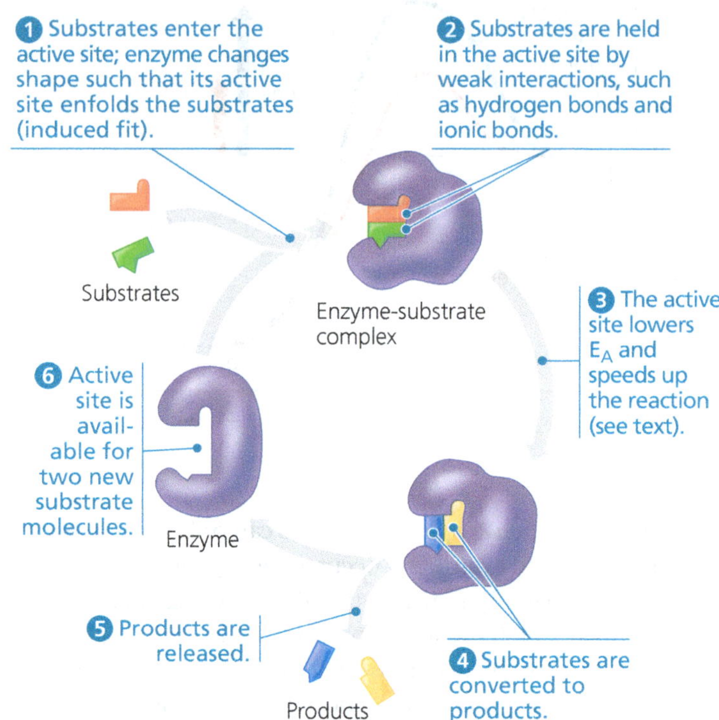

VISUAL SKILLS *The enzyme-substrate complex passes through a transition state (see Figure 8.13). Label the part of the cycle where the transition state occurs.*

The rate at which a particular amount of enzyme converts substrate to product is partly a function of the initial concentration of the substrate: The more substrate molecules that are available, the more frequently they access the active sites of the enzyme molecules. However, there is a limit to how fast the reaction can be pushed by adding more substrate to a fixed concentration of enzyme. At some point, the concentration of substrate will be high enough that all enzyme molecules will have their active sites engaged. As soon as the product exits an active site, another substrate molecule enters. At this substrate concentration, the enzyme is said to be *saturated*, and the rate of the reaction is determined by the speed at which the active site converts substrate to product. When an enzyme population is saturated, the only way to increase the rate of product formation is to add more enzyme. Cells often increase the rate of a reaction by producing more enzyme molecules. You can graph the overall progress of an enzymatic reaction in the **Scientific Skills Exercise**.

Scientific Skills Exercise

Making a Line Graph and Calculating a Slope

Does the Rate of Glucose 6-Phosphatase Activity Change over Time in Isolated Liver Cells? Glucose 6-phosphatase, which is found in mammalian liver cells, is a key enzyme in control of blood glucose level. The enzyme catalyses the breakdown of glucose 6-phosphate into glucose and inorganic phosphate (Ⓟ$_i$). These products are transported out of liver cells into the blood, increasing the blood glucose level. In this exercise, you will graph data from a time-course experiment that measured Ⓟ$_i$ concentration in the buffer outside isolated liver cells, thus indirectly measuring glucose 6-phosphatase activity inside the cells.

How the Experiment Was Done Isolated rat liver cells were placed in a dish with buffer at physiological conditions (pH 7.4, 37°C). Glucose 6-phosphate (the substrate) was added so it could be taken up by cells and broken down by glucose 6-phosphatase. A sample of buffer was removed every 5 minutes and the concentration of Ⓟ$_i$ that had been transported out of the cells was determined.

Data from the Experiment

Time (min)	Concentration of Ⓟ$_i$ (μmol/mL)
0	0
5	10
10	90
15	180
20	270
25	330
30	355
35	355
40	355

Data from S. R. Commerford et al., Diets enriched in sucrose or fat increase gluconeogenesis and G-6-Pase but not basal glucose production in rats, *American Journal of Physiology-Endocrinology and Metabolism* 283:E545–E555 (2002).

INTERPRET THE DATA

1. To see patterns in the data from a time-course experiment like this, it is helpful to graph the data. First, determine which set of data goes on each axis. **(a)** What did the researchers intentionally vary in the experiment? This is the independent variable, which goes on the *x*-axis. **(b)** What are the units (abbreviated) for the independent variable? Explain in words what the abbreviation stands for. **(c)** What was measured by the researchers? This is the dependent variable, which goes on the *y*-axis. **(d)** What does the units abbreviation stand for? Label each axis, including the units.
2. Next, mark off the axes with just enough evenly spaced tick marks to accommodate the full set of data. Determine the span of data values for each axis. **(a)** What is the largest value to go on the *x*-axis? What is a reasonable spacing for the tick marks, and what should be the highest one? **(b)** What is the largest value to go on the *y*-axis? What is a reasonable spacing for the tick marks, and what should be the highest one?
3. Plot the data points on your graph. Match each *x*-value with its partner *y*-value and place a point on the graph at that coordinate. Draw a line that connects the points. (For additional information about graphs, see the Scientific Skills Review in Appendix D.)
4. Examine your graph and look for patterns in the data. **(a)** Does the concentration of Ⓟ$_i$ increase evenly throughout the course of the experiment? To answer this question, describe the pattern you see in the graph. **(b)** What part of the graph shows the highest rate of enzyme activity? Consider that the rate of enzyme activity is related to the slope of the line, $\Delta y/\Delta x$ (the "rise" over the "run"), in μmol/(mL · min), with the steepest slope indicating the highest rate of enzyme activity. Calculate the rate of enzyme activity (slope) where the graph is steepest. **(c)** Can you think of a biological explanation for the pattern you see?
5. If your blood glucose level is low from skipping lunch, what reaction (discussed in this exercise) will occur in your liver cells? Write out the reaction and put the name of the enzyme over the reaction arrow. How will this reaction affect your blood glucose level?

Effects of Local Conditions on Enzyme Activity

The activity of an enzyme—how efficiently the enzyme functions—is affected by general environmental factors, such as temperature and pH. It can also be affected by chemicals that specifically influence that enzyme. In fact, researchers have learned much about enzyme function by employing such chemicals.

Effects of Temperature and pH

The 3-D structures of proteins are sensitive to their environment (see Figure 5.20). As a consequence, each enzyme works better under some conditions than others, because these *optimal conditions* favour the most active shape for the enzyme.

Temperature and pH are environmental factors important in the activity of an enzyme. Up to a point, the rate of an enzymatic reaction increases with increasing temperature, partly because substrates collide with active sites more frequently when the molecules move rapidly. Above that temperature, however, the speed of the enzymatic reaction drops sharply. The thermal agitation of the enzyme molecule disrupts the hydrogen bonds, ionic bonds, and other weak interactions that stabilise the active shape of the enzyme, and the protein molecule eventually denatures. Each enzyme has an optimal temperature at which its reaction rate is greatest. Without denaturing the enzyme, this temperature allows the greatest number of molecular collisions and the fastest conversion of the reactants to product molecules. Most human enzymes have optimal temperatures of about 35–40°C (close to human body temperature). The thermophilic bacteria that live in hot springs contain enzymes with optimal temperatures of 70°C or higher **(Figure 8.17a)**.

▼ Figure 8.17 Environmental factors affecting enzyme activity. Each enzyme has an optimal **(a)** temperature and **(b)** pH that favour the most active shape of the protein molecule.

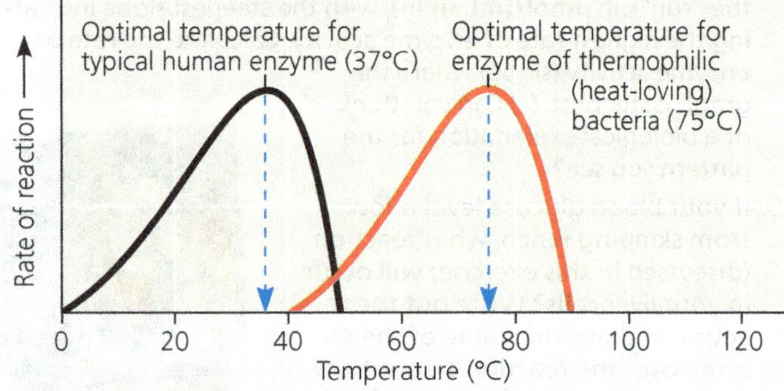

(a) The photo shows thermophilic cyanobacteria (green) thriving in the hot water of a Nevada geyser. The graph compares the optimal temperatures for an enzyme from the thermophilic bacterium *Thermus oshimai* (75°C) and human enzymes (37°C, body temperature).

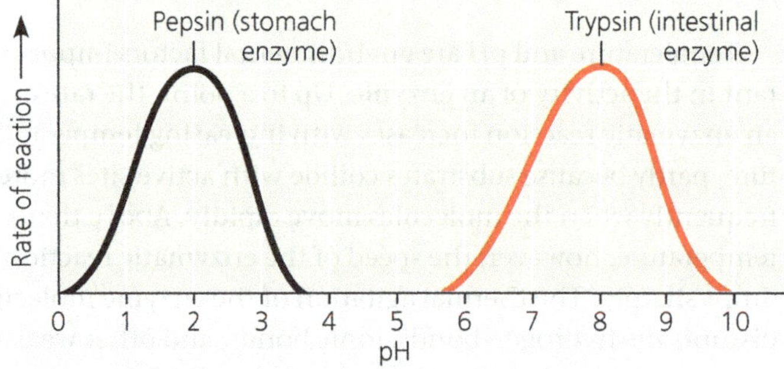

(b) This graph shows the rate of reaction for two digestive enzymes over a range of pH values.

INTERPRET THE DATA *Looking at the graph in (b), what is the optimal pH for pepsin activity? Explain why natural selection might have resulted in the optimal pH for pepsin, a stomach enzyme (see Figure 3.11). What is the optimal pH for trypsin?*

Just as each enzyme has an optimal temperature, it also has a pH at which it is most active. The optimal pH values for most enzymes fall in the range of pH 6–8, but there are exceptions. For example, pepsin, a digestive enzyme in the human stomach, works best at a very low pH. Such an acidic environment denatures most enzymes, but pepsin is well-adapted evolutionarily to maintain its functional 3-D structure in the acidic environment of the stomach. In contrast, trypsin, a digestive enzyme residing in the more alkaline environment of the human intestine would be denatured in the stomach **(Figure 8.17b)**.

Cofactors

Many enzymes require nonprotein helpers for catalytic activity, often for chemical processes like electron transfers that cannot easily be carried out by the amino acids in proteins. These adjuncts, called **cofactors**, may be bound tightly to the enzyme as permanent residents, or they may bind loosely and reversibly along with the substrate. The cofactors of some enzymes are inorganic, such as the metal atoms zinc, iron, and copper in ionic form. If the cofactor is an organic molecule, it is referred to, more specifically, as a **coenzyme**. Most vitamins are important in nutrition because they act as coenzymes or raw materials from which coenzymes are made.

Enzyme Inhibitors

Certain chemicals selectively inhibit the action of specific enzymes. Sometimes the inhibitor attaches to the enzyme by covalent bonds, in which case the inhibition is usually irreversible. Many enzyme inhibitors, however, bind to the enzyme by weak interactions, and when this occurs the inhibition is reversible. Some reversible inhibitors resemble the normal substrate molecule and compete for admission into the active site **(Figure 8.18a** and **b)**. These mimics, called **competitive inhibitors**, reduce the productivity of enzymes by blocking substrates from entering active sites. This kind of inhibition can be overcome by increasing the concentration of substrate so that as active sites become available, more substrate molecules than inhibitor molecules are around to gain entry to the sites.

In contrast, **noncompetitive inhibitors** do not directly compete with the substrate to bind to the enzyme at the active site **(Figure 8.18c)**. Instead, they impede enzymatic reactions by binding to another part of the enzyme. This interaction causes the enzyme molecule to change its shape in such a way that the active site becomes much less effective at catalysing the conversion of substrate to product.

Toxins and poisons are often irreversible enzyme inhibitors. An example is sarin, a nerve gas. In 2017, sarin was used in a chemical attack in Syria, killing about 100 people and injuring hundreds more. This small molecule binds covalently to the R group on the amino acid serine, which is found in the active site of acetylcholinesterase, an enzyme important in the nervous system. Other examples include the pesticides DDT and parathion, inhibitors of key enzymes in the nervous

▼ **Figure 8.18 Inhibition of enzyme activity.**

(a) Normal binding

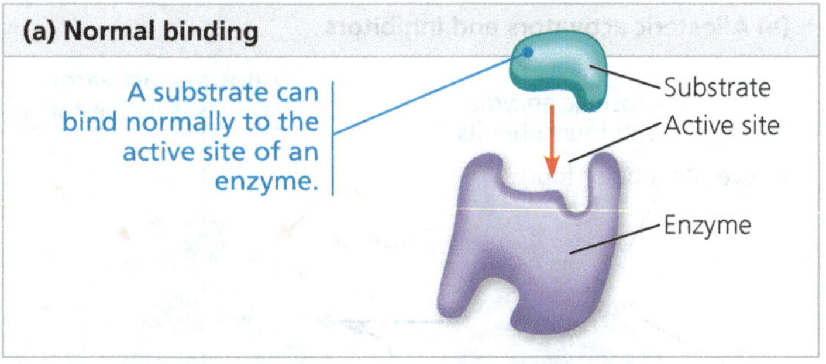

(b) Competitive inhibition

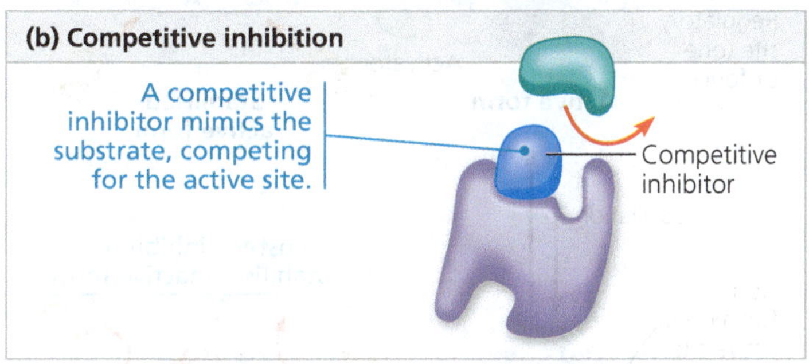

(c) Noncompetitive inhibition

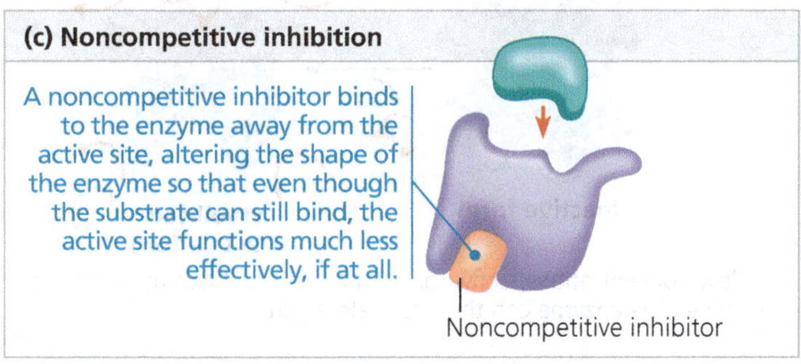

system. Finally, many antibiotics are inhibitors of specific enzymes in bacteria. For instance, penicillin blocks the active site of an enzyme that many bacteria use to make cell walls.

Citing enzyme inhibitors that are metabolic poisons may give the impression that enzyme inhibition is generally abnormal and harmful. In fact, molecules naturally present in the cell often regulate enzyme activity by acting as inhibitors. Such regulation—selective inhibition—is essential to the control of cellular metabolism, as you'll see in Concept 8.5.

The Evolution of Enzymes

EVOLUTION Thus far, biochemists have identified more than 4,000 different enzymes in various species, most likely a very small fraction of all enzymes. How did this grand profusion of enzymes arise? Recall that most enzymes are proteins, and proteins are encoded by genes. A permanent change in a gene, known as a *mutation*, can result in a protein with one or more changed amino acids. In the case of an enzyme, if the changed amino acids are in the active site or some other crucial region, the altered enzyme might have a novel activity or might bind to a different substrate. Under environmental conditions where the new function benefits the organism, natural selection would tend to favour the mutated form of the gene, causing it to persist in the population. This simplified model is generally accepted as the main way in which the multitude of different enzymes arose over the past few billion years of life's history. Data supporting this model have been collected by researchers using a lab procedure that mimics evolution in natural populations **(Figure 8.19)**.

▼ **Figure 8.19 Mimicking evolution of an enzyme with a new function.** Researchers tested whether the function of an enzyme called β-galactosidase, which breaks down the sugar lactose, could change over time in populations of the bacterium *Escherichia coli*. After seven rounds of mutation and selection in the lab, β-galactosidase evolved into an enzyme specialised for breaking down a sugar other than lactose. This ribbon model shows one subunit of the altered enzyme; six amino acids (blue dots) were different.

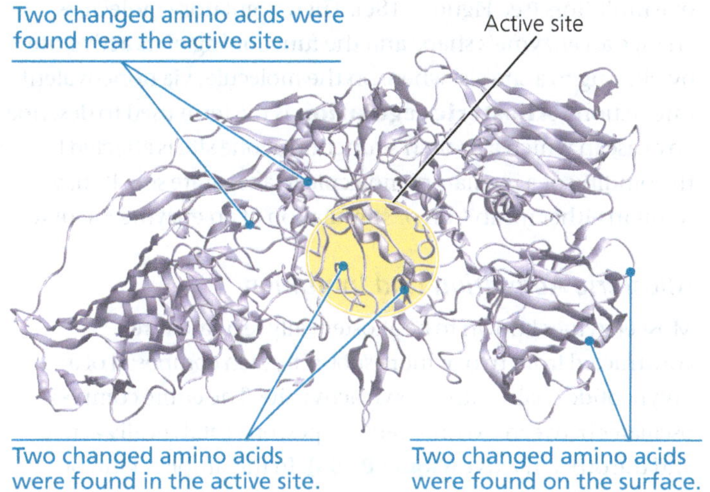

CONCEPT CHECK 8.4

1. Many spontaneous reactions occur very slowly. Why don't all spontaneous reactions occur instantly?
2. Say you are using a Bunsen burner in the lab. Why doesn't the flame creep back into the gas line and set the gas supply on fire?
3. **WHAT IF?** Malonate is an inhibitor of the enzyme succinate dehydrogenase. How would you determine whether malonate is a competitive or noncompetitive inhibitor?
4. **DRAW IT** A mature lysosome has an internal pH of around 4.5. Using Figure 8.17b as a guide, draw a graph showing what you would predict for the rate of reaction for a lysosomal enzyme. Label its optimal pH, assuming its optimal pH matches its environment.

For suggested answers, see Appendix A.

CONCEPT 8.5

Regulation of enzyme activity helps control metabolism

Chemical chaos would result if all of a cell's metabolic pathways were operating simultaneously. Intrinsic to life's processes is a cell's ability to tightly regulate its metabolic pathways by controlling when and where its various enzymes are active. It does this either by switching on and off the genes that encode specific enzymes (which will be discussed in Concepts 18.1 and 18.2) or, as discussed here, by regulating the activity of enzymes once they are made.

Allosteric Regulation of Enzymes

In many cases, the molecules that naturally regulate enzyme activity in a cell behave something like reversible noncompetitive inhibitors (see Figure 8.18c): These regulatory molecules change an enzyme's shape and the functioning of its active site by binding to a site elsewhere on the molecule, via noncovalent interactions. **Allosteric regulation** is the term used to describe any case in which a protein's function at one site is affected by the binding of a regulatory molecule to a separate site. It may result in either inhibition or stimulation of an enzyme's activity.

Allosteric Activation and Inhibition

Most enzymes known to be allosterically regulated are constructed from two or more subunits, each composed of a polypeptide chain with its own active site. The entire complex oscillates between two different shapes, one catalytically active and the other inactive **(Figure 8.20a)**. In the simplest kind of allosteric regulation, an activating or inhibiting regulatory molecule binds to a regulatory site (sometimes called an allosteric site), often located where subunits join. The binding of an *activator* to a regulatory site stabilises the shape that has functional active sites, whereas the binding of an *inhibitor* stabilises the inactive form of the enzyme. The subunits of an allosteric enzyme fit together in such a way that a shape change in one subunit is transmitted to all others. Through this interaction of subunits, a single activator or inhibitor molecule that binds to one regulatory site will affect the active sites of all subunits.

Fluctuating concentrations of regulators can cause a sophisticated pattern of response in the activity of cellular enzymes. The products of ATP hydrolysis (ADP and Ⓟ$_i$), for example, play a complex role in balancing the flow of traffic between anabolic and catabolic pathways by their effects on key enzymes. ATP binds to several catabolic enzymes allosterically, lowering their affinity for substrate and thus inhibiting their activity. ADP, however, functions as an activator of the same enzymes. This is logical because catabolism functions in regenerating ATP. If ATP production lags behind its use, ADP accumulates and activates the enzymes that speed up catabolism, producing more ATP. If the supply of ATP exceeds demand, then catabolism slows down as ATP molecules accumulate and bind to the same enzymes, inhibiting them. (You'll see specific examples of this type of regulation when you learn about cellular respiration in the chapter,"Cellular Respiration and Fermentation"; see, for example, Figure 9.19.) ATP, ADP, and other related molecules also affect key enzymes in anabolic pathways. In this way, allosteric enzymes control the rates of important reactions in both sorts of metabolic pathways.

In another kind of allosteric activation, a *substrate* molecule binding to one active site in a multisubunit enzyme triggers a shape change in all the subunits, thereby increasing catalytic activity at the other active sites **(Figure 8.20b)**. Called **cooperativity**, this mechanism amplifies the response of enzymes to substrates: One substrate molecule primes an enzyme to act on additional substrate molecules more readily. Cooperativity is considered allosteric regulation because even though substrate is binding to an active site, its binding affects catalysis in another active site.

▼ Figure 8.20 Allosteric regulation of enzyme activity.

(a) Allosteric activators and inhibitors

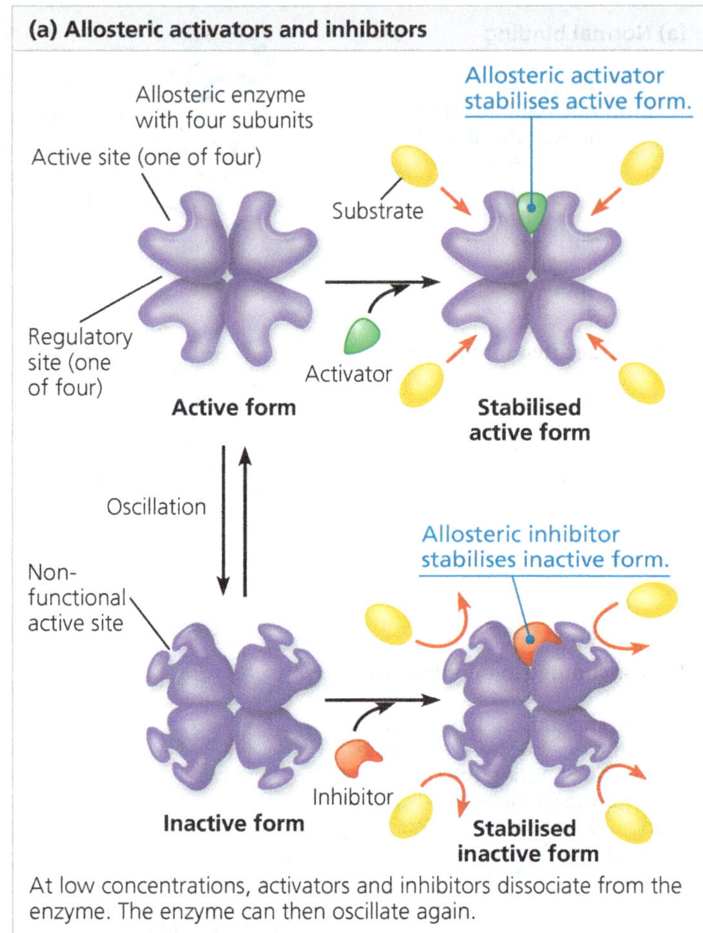

At low concentrations, activators and inhibitors dissociate from the enzyme. The enzyme can then oscillate again.

(b) Cooperativity: another type of allosteric activation

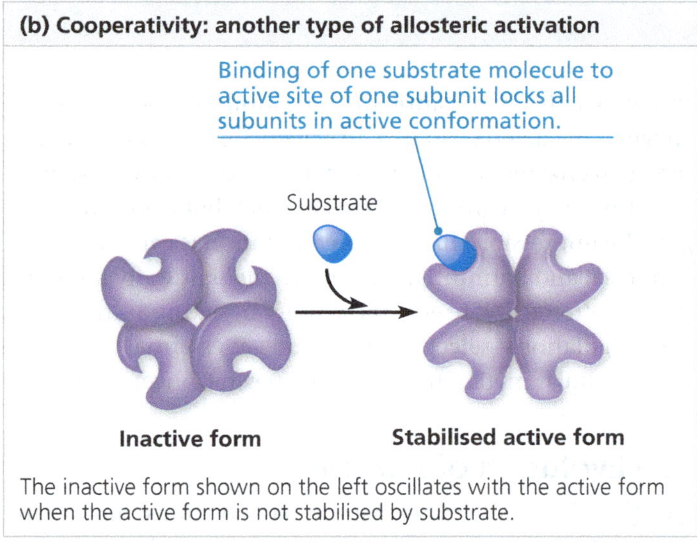

The inactive form shown on the left oscillates with the active form when the active form is not stabilised by substrate.

Although haemoglobin is not an enzyme (it carries O_2 rather than catalysing a reaction), classic studies of haemoglobin have elucidated the principle of cooperativity. Haemoglobin is made up of four subunits, each with an O_2-binding site (see Figure 5.18). The binding of an O_2 to one binding site increases the affinity for O_2 of the remaining binding sites. Thus, where O_2 is at a high level, such as in the lungs or gills, haemoglobin's affinity for O_2 increases as more binding sites are filled. In O_2-deprived tissues, however, the release of each O_2 molecule decreases the O_2 affinity of the other binding sites, resulting in the release of O_2 where it is most needed. Cooperativity works similarly in multisubunit enzymes that have been studied.

Feedback Inhibition

Earlier, we looked at the allosteric inhibition of an enzyme in an ATP-generating pathway by ATP itself. This is a common mode of metabolic control, called **feedback inhibition**, in which a metabolic pathway is halted by the inhibitory binding of its end product to an enzyme that acts early in the pathway. **Figure 8.21** shows an example of feedback inhibition operating on an anabolic pathway. Some cells use this five-step pathway to synthesise the amino acid isoleucine from threonine, another amino acid. As isoleucine accumulates, it slows down its own synthesis by allosterically inhibiting the enzyme for the first step of the pathway. Feedback inhibition thereby prevents the cell from making more isoleucine than is necessary and thus wasting chemical resources.

▼ **Figure 8.21 Feedback inhibition in isoleucine synthesis.**

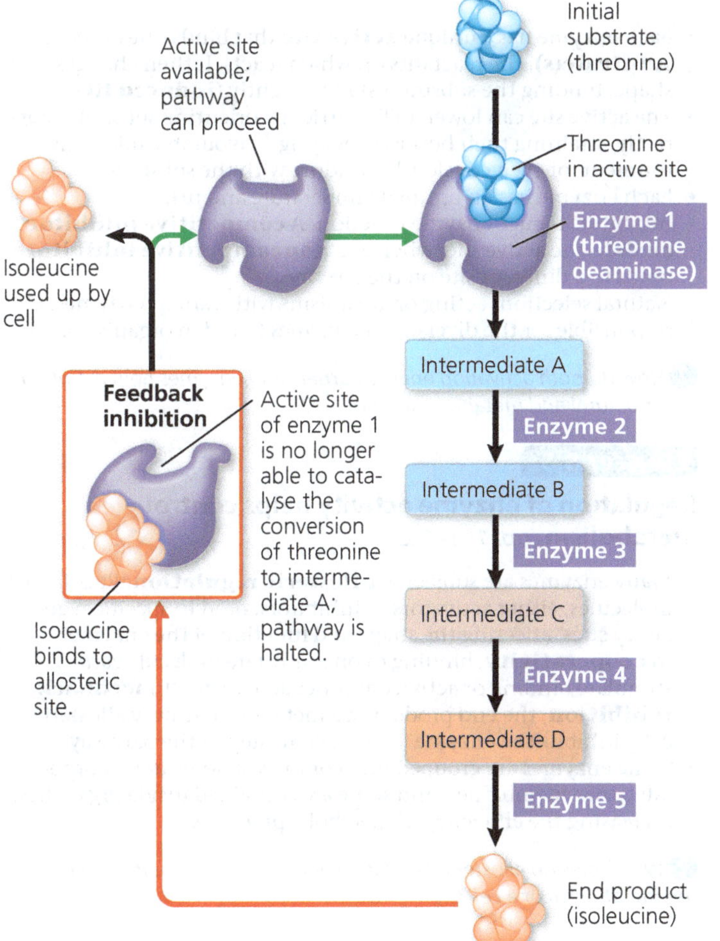

Localisation of Enzymes Within the Cell

The cell is not just a bag of chemicals with thousands of different kinds of enzymes and substrates in a random mix. The cell is compartmentalised, and cellular structures help bring order to metabolic pathways. In some cases, a team of enzymes for several steps of a metabolic pathway are assembled into a multienzyme complex. The arrangement facilitates the sequence of reactions, with the product from the first enzyme becoming the substrate for an adjacent enzyme in the complex, and so on, until the end product is released. Some enzymes and enzyme complexes have fixed locations within the cell and act as structural components of particular membranes. Others are in solution within particular membrane-enclosed eukaryotic organelles, each with its own internal chemical environment. For example, in eukaryotic cells, the enzymes for the second and third stages of cellular respiration reside in specific locations within mitochondria (**Figure 8.22**; see also Figure 6.32b).

▼ **Figure 8.22 Organelles and structural order in metabolism.** Organelles such as the mitochondrion (TEM) contain enzymes that carry out specific functions, in this case the second and third stages of cellular respiration. (See also Figure 6.32b, lower left.)

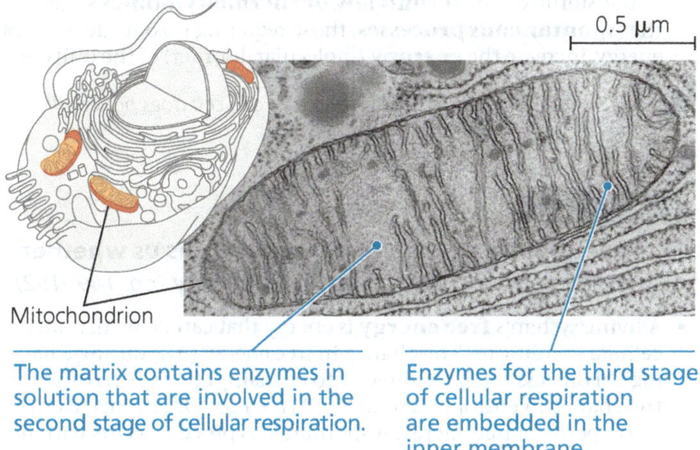

In this chapter, you have learned about the laws of thermodynamics that govern metabolism, the intersecting set of chemical pathways characteristic of life. We have explored the bioenergetics of breaking down and building up biological molecules. To continue the theme of energy flow, and specifically of bioenergetics, we will next examine cellular respiration. This is the major catabolic pathway that breaks down organic molecules and releases energy that can be used for the crucial processes of life.

CONCEPT CHECK 8.5

1. How do an activator and an inhibitor have different effects on an allosterically regulated enzyme?
2. **WHAT IF?** Regulation of isoleucine synthesis is an example of feedback inhibition of an anabolic pathway. With that in mind, explain how ATP might be involved in feedback inhibition of a catabolic pathway.

For suggested answers, see Appendix A.

8 Chapter Review

SUMMARY OF KEY CONCEPTS

CONCEPT 8.1

An organism's metabolism transforms matter and energy *(pp. 146–149)*

- **Metabolism** is the collection of chemical reactions that occur in an organism. Enzymes catalyse reactions in intersecting **metabolic pathways**, which may be **catabolic** (breaking down molecules, releasing energy) or **anabolic** (building molecules, consuming energy). **Bioenergetics** is the study of the flow of energy through living organisms.
- **Energy** is the capacity to cause change; some forms of energy do work by moving matter. **Kinetic energy** is associated with motion and includes **thermal energy** associated with random motion of atoms or molecules. **Heat** is thermal energy in transfer from one object to another. **Potential energy** is related to the location or structure of matter and includes **chemical energy** possessed by a molecule due to its structure.
- The **first law of thermodynamics**, conservation of energy, states that energy cannot be created or destroyed, only transferred or transformed. The **second law of thermodynamics** states that **spontaneous processes**, those requiring no outside input of energy, increase the **entropy** (molecular disorder) of the universe.

? *Explain how the highly ordered structure of a cell does not conflict with the second law of thermodynamics.*

CONCEPT 8.2

The free-energy change of a reaction tells us whether or not the reaction occurs spontaneously *(pp. 149–152)*

- A living system's **free energy** is energy that can do work under cellular conditions. The change in free energy (ΔG) during a biological process is related directly to enthalpy change (ΔH) and to the change in entropy (ΔS): $\Delta G = \Delta H - T\Delta S$. Organisms live at the expense of free energy. A spontaneous process occurs with no energy input; during such a process, free energy decreases and the stability of a system increases. At maximum stability, the system is at equilibrium and can do no work.
- In an **exergonic** (spontaneous) chemical reaction, the products have less free energy than the reactants ($-\Delta G$). **Endergonic** (nonspontaneous) reactions require an input of energy ($+\Delta G$). The addition of starting materials and the removal of end products prevent metabolism from reaching equilibrium.

? *Explain the meaning of each component in the equation for the change in free energy of a spontaneous chemical reaction. Why are spontaneous reactions important in the metabolism of a cell?*

CONCEPT 8.3

ATP powers cellular work by coupling exergonic reactions to endergonic reactions *(pp. 152–155)*

- **ATP** is the cell's energy shuttle. Hydrolysis of its terminal phosphate yields ADP and Ⓟ$_i$ and releases free energy.
- Through **energy coupling**, the exergonic process of ATP hydrolysis drives endergonic reactions by transfer of a phosphate group to specific reactants, forming a **phosphorylated intermediate** that is more reactive. ATP hydrolysis (sometimes with protein phosphorylation) also causes changes in the shape and binding affinities of transport and motor proteins.
- Catabolic pathways drive regeneration of ATP from ADP + Ⓟ$_i$.

? *Describe the ATP cycle: How is ATP used and regenerated in a cell?*

CONCEPT 8.4

Enzymes speed up metabolic reactions by lowering energy barriers *(pp. 155–161)*

- In a chemical reaction, the energy necessary to break the bonds of the reactants is the **activation energy**, E_A.
- **Enzymes** lower the E_A barrier:

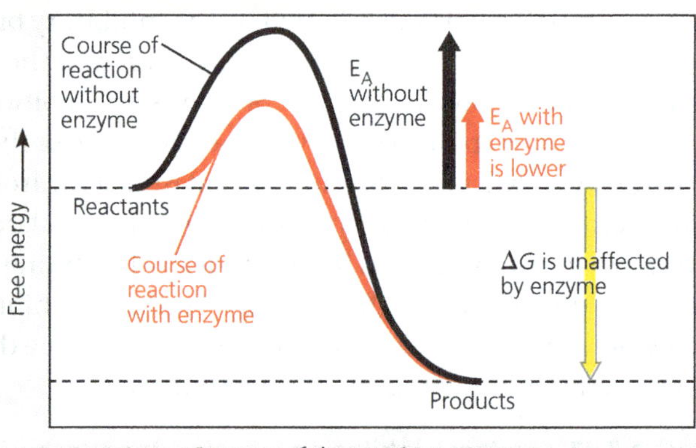

- Each enzyme has a unique **active site** that binds one or more **substrate(s)**, the reactants on which it acts. It then changes shape, binding the substrate(s) more tightly (**induced fit**).
- The active site can lower an E_A barrier by orienting substrates correctly, straining their bonds, providing a favourable microenvironment, or even covalently bonding with the substrate.
- Each enzyme has an optimal temperature and pH.
- Inhibitors reduce enzyme function. A **competitive inhibitor** binds to the active site, whereas a **noncompetitive inhibitor** binds to a different site on the enzyme.
- Natural selection, acting on organisms with variant enzymes, is responsible for the diversity of enzymes found in organisms.

? *How do both activation energy barriers and enzymes help maintain the structural and metabolic order of life?*

CONCEPT 8.5

Regulation of enzyme activity helps control metabolism *(pp. 161–163)*

- Many enzymes are subject to **allosteric regulation**: Regulatory molecules, either activators or inhibitors, bind to specific regulatory sites, affecting the shape and function of the enzyme. In **cooperativity**, binding of one substrate molecule can stimulate binding or activity at other active sites. In **feedback inhibition**, the end product of a metabolic pathway allosterically inhibits the enzyme for a previous step in the pathway.
- Some enzymes are grouped into complexes, some are incorporated into membranes, and some are contained inside organelles, increasing the efficiency of metabolic processes.

? *What roles do allosteric regulation and feedback inhibition play in the metabolism of a cell?*

TEST YOUR UNDERSTANDING

Levels 1-2: Remembering/Understanding

1. Choose the pair of terms that correctly completes this sentence: Catabolism is to anabolism as ___________ is to ___________.
(A) exergonic; spontaneous
(B) exergonic; endergonic
(C) free energy; entropy
(D) work; energy

2. Most cells cannot harness heat to perform work because
(A) heat does not involve a transfer of energy.
(B) cells do not have much thermal energy; they are relatively cool.
(C) temperature is usually uniform throughout a cell.
(D) heat can never be used to do work.

3. Which of the following metabolic processes can occur without a net influx of energy from some other process?
(A) $ADP + Ⓟ_i + \rightarrow ATP + H_2O$
(B) $C_6H_{12}O_6 + 6\,O_2 \rightarrow 6\,CO_2 + 6\,H_2O$
(C) $6\,CO_2 + 6\,H_2O \rightarrow C_6H_{12}O_6 + 6\,O_2$
(D) Amino acids → Protein

4. If an enzyme in solution is saturated with substrate, the most effective way to obtain a faster yield of products is to
(A) add more of the enzyme.
(B) heat the solution to 90°C.
(C) add more substrate.
(D) add a noncompetitive inhibitor.

5. Some bacteria are metabolically active in hot springs because
(A) they are able to maintain a lower internal temperature.
(B) high temperatures make catalysis unnecessary.
(C) their enzymes have high optimal temperatures.
(D) their enzymes are completely insensitive to temperature.

Levels 3-4: Applying/Analysing

6. If an enzyme is added to a solution where its substrate and product are in equilibrium, what will occur?
(A) Additional substrate will be formed.
(B) The reaction will change from endergonic to exergonic.
(C) The free energy of the system will change.
(D) Nothing; the reaction will stay at equilibrium.

Levels 5-6: Evaluating/Creating

7. **DRAW IT** Using a series of arrows, draw the branched metabolic reaction pathway described by the following statements, and then answer the question at the end. Use red arrows and minus signs to indicate inhibition.

L can form either M or N.

M can form O.

O can form either P or R.

P can form Q.

R can form S.

O inhibits the reaction of L to form M.

Q inhibits the reaction of O to form P.

S inhibits the reaction of O to form R.

Which reaction would prevail if both Q and S were present in the cell in high concentrations?
(A) L → M
(B) M → O
(C) L → N
(D) O → P

8. **EVOLUTION CONNECTION** Some people argue that biochemical pathways are too complex to have evolved because all intermediate steps in a given pathway must be present to produce the final product. Critique this argument. How could you use the diversity of metabolic pathways that produce the same or similar products to support your case?

9. **SCIENTIFIC INQUIRY • DRAW IT** A researcher has developed an assay to measure the activity of an important enzyme present in pancreatic cells growing in culture. She adds the enzyme's substrate to a dish of cells and then measures the appearance of reaction products. The results are graphed as the amount of product on the *y*-axis versus time on the *x*-axis. The researcher notes four sections of the graph. For a short period of time, no products appear (section A). Then (section B) the reaction rate is quite high (the slope of the line is steep). Next, the reaction gradually slows down (section C). Finally, the graph line becomes flat (section D). Draw and label the graph, and propose a model to explain the molecular events occurring at each stage of this reaction profile.

10. **WRITE ABOUT A THEME: ENERGY AND MATTER** Life requires energy. In a short essay (100–150 words), describe the basic principles of bioenergetics in an animal cell. How are the flow and transformation of energy different in a photosynthesising cell? Include the role of ATP and enzymes in your discussion.

11. **SYNTHESISE YOUR KNOWLEDGE**

Explain what is happening in this photo in terms of kinetic energy and potential energy. Include the energy conversions that occur when the penguins eat fish and climb back up on the glacier. Describe the role of ATP and enzymes in the underlying molecular processes, including what happens to the free energy of some of the molecules involved.

For selected answers, see Appendix A.

9 Cellular Respiration and Fermentation

KEY CONCEPTS

Study Tip

Make a visual study guide: Draw a cell with a large mitochondrion, labelling the parts of the mitochondrion. As you go through the chapter, add key reactions for each stage of cellular respiration, linking the stages together. Label the carbon molecule(s) with the most energy and the carbon molecule(s) with the least energy. Your cell can be a simple sketch, as shown here.

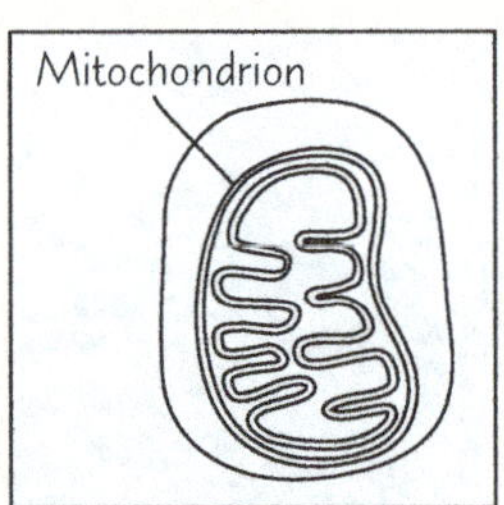

Go to Mastering Biology

to access Dynamic Study Modules for revision, 3D BioFlix® animations and high-quality videos, and your interactive Pearson eText.

Figure 9.1 One of New Zealand's most endangered species, the Canterbury knobbled weevil (*Hadramphus tuberculatus*) obtains energy for its cells when it eats taramea, or golden Spaniard speargrass (*Aciphylla sp*). In the process of cellular respiration, mitochondria in the cells of insects, animals, plants, and other organisms break down organic molecules, generating ATP and waste products: carbon dioxide, water, and heat. Note that energy flows one way, but chemicals are recycled.

How is the chemical energy stored in food used to generate ATP, the molecule that drives most cellular work?

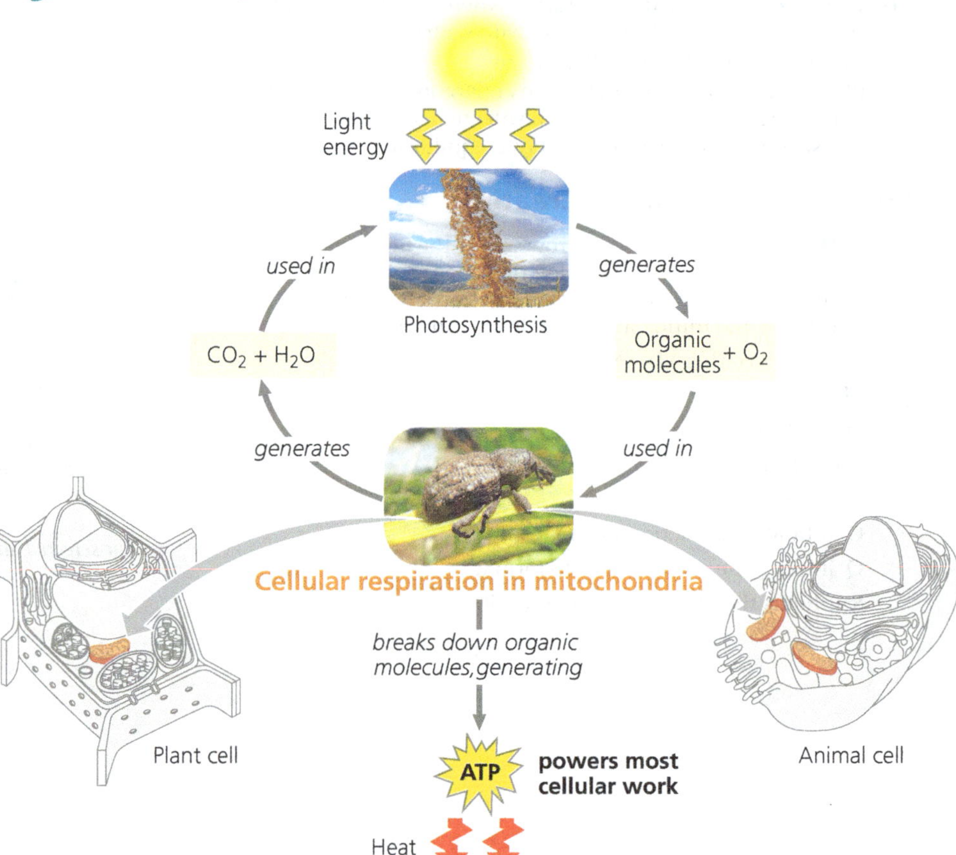

CONCEPT 9.1

Catabolic pathways yield energy by oxidising organic fuels

Living cells require transfusions of energy from outside sources to perform their many tasks—for example, assembling polymers, pumping substances across membranes, moving, and reproducing. The outside source of energy is food, and the energy stored in the organic molecules of food ultimately comes from the sun. As shown in **Figure 9.1**, energy flows into an ecosystem as sunlight and exits as heat; in contrast, the chemical elements essential to life are recycled. Photosynthesis generates oxygen, as well as organic molecules used by the mitochondria of eukaryotes as fuel for cellular respiration. Respiration breaks this fuel down, using oxygen (O_2) and generating ATP. The waste products of this type of respiration, carbon dioxide (CO_2) and water (H_2O), are the raw materials for photosynthesis.

Let's consider how cells harvest the chemical energy stored in organic molecules and use it to generate ATP, the molecule that drives most cellular work. Metabolic pathways that release stored energy by breaking down complex molecules are called catabolic pathways (see Concept 8.1). Transfer of electrons from food molecules (like glucose) to other molecules plays a major role in these pathways. In this section, we consider these processes, which are central to cellular respiration.

Catabolic Pathways and Production of ATP

Organic compounds possess potential energy as a result of the arrangement of electrons in the bonds between their atoms. Compounds that can participate in exergonic reactions can act as fuels. Through the activity of enzymes (see Concept 8.4), a cell systematically degrades complex organic molecules that are rich in potential energy to simpler waste products that have less energy. Some of the energy taken out of chemical storage can be used to do work; the rest is dissipated as heat.

One catabolic process, **fermentation**, is a partial degradation of sugars or other organic fuel that occurs without the use of oxygen. However, the most efficient catabolic pathway is **aerobic respiration**, in which oxygen is consumed as a reactant along with the organic fuel (*aerobic* is from the Greek *aer*, air, and *bios*, life). The cells of most eukaryotic and many prokaryotic organisms can carry out aerobic respiration. Some prokaryotes use substances other than oxygen as reactants in a similar process that harvests chemical energy without oxygen; this process is called *anaerobic respiration* (the prefix *an-* means "without"). Technically, the term **cellular respiration** includes both aerobic and anaerobic processes. However, it originated as a synonym for aerobic respiration because of the relationship of that process to organismal respiration, in which an animal breathes in oxygen. Thus, *cellular respiration* is often used to refer to the aerobic process, a practice we follow in most of this chapter.

Although very different in mechanism, aerobic respiration is in principle similar to the combustion of petrol in a car's engine after oxygen is mixed with the fossil fuel (hydrocarbons). Food provides the fuel for respiration, and the exhaust is carbon dioxide and water. The overall process can be summarised as follows:

$$\begin{matrix}\text{Organic}\\\text{compounds}\end{matrix} + \text{Oxygen} \rightarrow \begin{matrix}\text{Carbon}\\\text{dioxide}\end{matrix} + \text{Water} + \text{Energy}$$

Carbohydrates, fats, and proteins from food can all be processed and consumed as fuel. In animal diets, a major source of carbohydrates is starch, a storage polysaccharide that can be broken down into glucose ($C_6H_{12}O_6$) subunits. Here, we will learn the steps of cellular respiration by tracking the degradation of the sugar glucose:

$$C_6H_{12}O_6 + 6\,O_2 \rightarrow 6\,CO_2 + 6\,H_2O + \text{Energy (ATP + heat)}$$

This breakdown of glucose is exergonic, having a free-energy change of −2,870 kJ per mole of glucose decomposed ($\Delta G = -2{,}870$ kJ/mol). Recall that a negative $\Delta G (\Delta G < 0)$ indicates that the products of the chemical process store less energy than the reactants and that the reaction can happen spontaneously—in other words, without an input of energy (see Concept 8.2).

Catabolic pathways do not directly move flagella, pump solutes across membranes, polymerise monomers, or perform other cellular work. Catabolism is linked to work by a chemical drive shaft—ATP (see Concept 8.3). To keep working, the cell must regenerate its supply of ATP from ADP and Ⓟ$_i$ (see Figure 8.12). To understand how cellular respiration accomplishes this, let's examine the fundamental chemical processes known as oxidation and reduction.

Redox Reactions: Oxidation and Reduction

How do the catabolic pathways that decompose glucose and other organic fuels yield energy? The answer is based on the transfer of electrons during the chemical reactions. The relocation of electrons releases energy stored in organic molecules, and this energy ultimately is used to synthesise ATP.

The Principle of Redox

In many chemical reactions, there is a transfer of one or more electrons (e^-) from one reactant to another. These electron transfers are called oxidation-reduction reactions, or **redox reactions** for short. In a redox reaction, the loss of electrons from one substance is called **oxidation**, and the addition of electrons to another substance is known as **reduction**. (Note that *adding* electrons is called *reduction*; adding negatively charged electrons to an atom *reduces* the amount of positive charge of that atom.)

To take a simple, nonbiological example, consider the reaction between the elements sodium (Na) and chlorine (Cl) that forms table salt:

$$\underbrace{Na}_{\text{becomes oxidised (loses electron)}} + \underbrace{Cl}_{\text{becomes reduced (gains electron)}} \longrightarrow Na^+ + Cl^-$$

We could generalise a redox reaction this way:

$$\underbrace{Xe^-}_{\text{becomes oxidised}} + \underbrace{Y}_{\text{becomes reduced}} \longrightarrow X + Ye^-$$

In the generalised reaction, substance Xe^-, the electron donor, is called the **reducing agent**; it reduces Y, which accepts the donated electron. Substance Y, the electron acceptor, is the **oxidising agent**; it oxidises Xe^- by removing its electron. Because an electron transfer requires both an electron donor and an acceptor, oxidation and reduction always go hand in hand.

Not all redox reactions involve the complete transfer of electrons from one substance to another; some change the degree of electron sharing in covalent bonds. Methane combustion, shown in **Figure 9.2**, is an example. The covalent electrons in methane are shared nearly equally between the bonded atoms because carbon and hydrogen have about the same affinity for valence electrons; they are about equally electronegative (see Concept 2.3). But when methane reacts with O_2, forming CO_2, electrons end up shared less equally between the carbon atom and its new covalent partners, the oxygen atoms, which are very electronegative. In effect, the carbon atom has partially "lost" its shared electrons; thus, methane has been oxidised.

Now let's examine the fate of the reactant O_2. The two atoms of O_2 share their electrons equally. But after the reaction with methane, when each O atom is bonded to two H atoms in H_2O, the electrons of those covalent bonds spend more time near the oxygen (see Figure 9.2). In effect, each O atom has partially "gained" electrons, so the oxygen molecule (O_2) has been reduced. Because the O atom is so electronegative, O_2 is one of the most powerful of all oxidising agents.

Energy must be added to pull an electron away from an atom, just as energy is required to push a ball uphill. The more electronegative the atom (the stronger its pull on electrons), the more energy is required to take an electron away from it. An electron loses potential energy when it shifts from a less electronegative atom towards a more electronegative one, just as a ball loses potential energy when it rolls downhill. A redox

▼ Figure 9.2 Methane combustion as an energy-yielding redox reaction. The reaction releases energy to the surroundings because the electrons lose potential energy when they end up being shared unequally, spending more time near electronegative atoms such as oxygen.

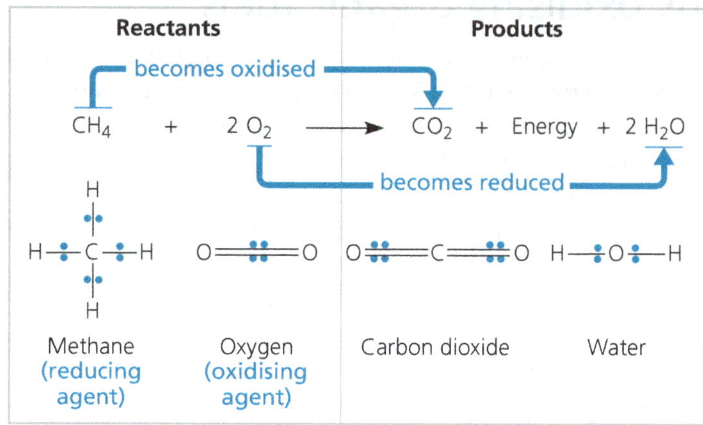

VISUAL SKILLS *Is the carbon atom oxidised or reduced during this reaction? Explain.*

reaction that moves electrons closer to an O atom, such as the burning (oxidation) of methane, therefore releases chemical energy that can be put to work.

Oxidation of Organic Fuel Molecules During Cellular Respiration

The oxidation of methane by O_2 is the main combustion reaction that occurs at the burner of a gas stove. The combustion of petrol in an car's internal combustion engine is also a redox reaction; the energy released pushes the pistons. But the energy-yielding redox process of greatest interest to biologists is respiration: the oxidation of glucose and other molecules in food. Examine again the summary equation for cellular respiration, but this time think of it as a redox process:

$$\underbrace{C_6H_{12}O_6}_{\text{becomes oxidised}} + \underbrace{6\,O_2}_{\text{becomes reduced}} \longrightarrow 6\,CO_2 + 6\,H_2O + \text{Energy}$$

As in the combustion of methane or petrol, the fuel (glucose) is oxidised and O_2 is reduced. The electrons lose potential energy along the way, and energy is released.

In general, organic molecules that have an abundance of hydrogen are excellent fuels because their bonds are a source of "hilltop" electrons, whose energy may be released as these electrons "fall" down an energy gradient during their transfer to oxygen. The summary equation for respiration indicates that hydrogen is transferred from glucose to the O atoms in O_2. But the important point, not visible in the summary equation, is that the energy state of the electron changes as hydrogen (with its electron) is transferred to oxygen. In respiration, the oxidation of glucose transfers electrons to a lower energy state, liberating energy that becomes available for ATP synthesis. So, in general, we see fuels with multiple C—H bonds oxidised into products with multiple C—O bonds.

The main energy-yielding foods—carbohydrates and fats—are reservoirs of electrons associated with hydrogen, often in the form of C—H bonds. Only the barrier of activation energy holds back the flood of electrons to a lower energy state (see Figure 8.13). Without this barrier, a food substance like glucose would combine almost instantaneously with O_2. If we supply the activation energy by igniting glucose, it burns in air, releasing 2,870 kJ of heat per mole of glucose (about 180 g). Body temperature is not high enough to initiate burning, of course. Instead, if you swallow some glucose, enzymes in your cells will lower the barrier of activation energy, allowing the sugar to be oxidised in a series of steps.

Stepwise Energy Harvest via NAD⁺ and the Electron Transport Chain

If energy is released from a fuel all at once, it cannot be harnessed efficiently for constructive work. For example, if a petrol tank explodes, the energy released instantaneously cannot propel a vehicle a car very far. Cellular respiration does not oxidise glucose (or any other organic fuel) in a single explosive step either. Rather, glucose is broken down in a series of steps, each one catalysed by an enzyme. At key steps, electrons are stripped from the glucose. As is often the case in oxidation reactions, each electron travels with a proton—which, you will recall, represents a hydrogen atom. The hydrogen atoms are not transferred directly to O_2. Instead, H atoms connect to an electron carrier, a coenzyme called nicotinamide adenine dinucleotide, a derivative of the vitamin niacin. This coenzyme is well suited as an electron carrier because it can cycle easily between its oxidised form, **NAD⁺**, and its reduced form, **NADH**. As an electron acceptor, NAD^+ functions as an oxidising agent during respiration.

How does NAD^+ trap electrons from glucose and the other organic molecules in food? Enzymes called dehydrogenases remove a pair of hydrogen atoms (2 electrons and 2 protons) from the substrate (glucose, in the preceding example), thereby oxidising it. The enzyme delivers the 2 electrons along with 1 proton to its coenzyme, NAD^+, forming NADH **(Figure 9.3)**. The other proton is released as a hydrogen ion (H^+) into the surrounding solution:

$$\mathrm{H{-}\overset{|}{\underset{|}{C}}{-}OH + NAD^+ \xrightarrow{\text{Dehydrogenase}} \overset{|}{\underset{|}{C}}{=}O + NADH + H^+}$$

By receiving 2 negatively charged electrons but only 1 positively charged proton, the nicotinamide portion of NAD^+ has its charge neutralised when NAD^+ is reduced to NADH. The name NADH shows the hydrogen that has been received in the reaction. NAD^+ is the most versatile electron acceptor in cellular respiration and functions in several of the redox steps during the breakdown of glucose.

Electrons lose very little of their potential energy when they are transferred from glucose to NAD^+. Each NADH molecule formed during respiration represents stored energy that can be tapped to make ATP when the electrons complete their "fall" down an energy gradient from NADH to O_2.

How do electrons that are extracted from glucose and stored as potential energy in NADH finally reach oxygen? It will help to compare the redox chemistry of cellular respiration to a much simpler reaction: the reaction between hydrogen and oxygen to form water **(Figure 9.4a)**. Mix H_2 and O_2, provide a spark for activation energy, and the gases combine explosively. In fact, rockets harness the energy released when cyrogenically cooled liquid H_2 and O_2 mix. The explosion represents a release of energy as the electrons of hydrogen "fall" closer to the electronegative oxygen atoms. Cellular respiration also brings hydrogen and oxygen together to form water, but there are two important differences. First, in cellular respiration, the hydrogen that reacts with oxygen

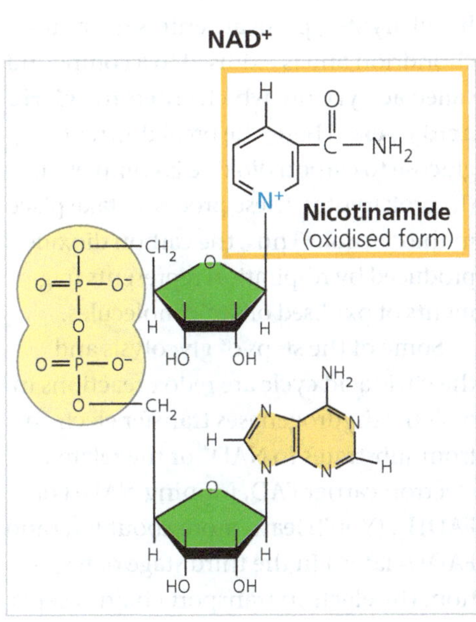

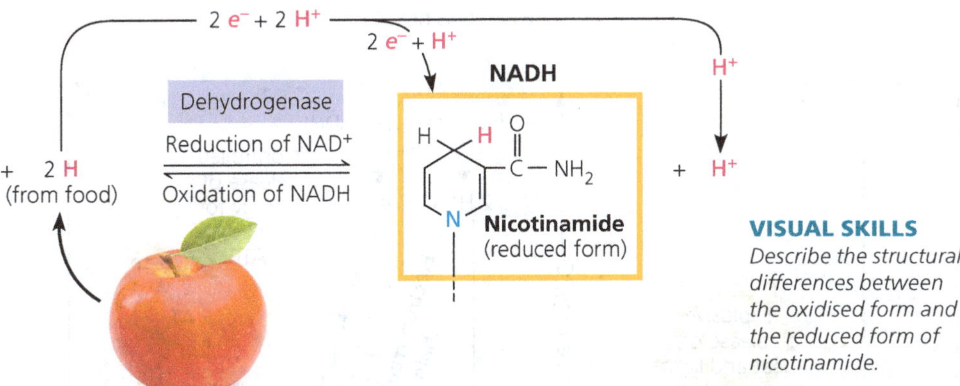

VISUAL SKILLS
Describe the structural differences between the oxidised form and the reduced form of nicotinamide.

▲ **Figure 9.3 NAD⁺ as an electron shuttle.** The full name for NAD^+, nicotinamide adenine dinucleotide, describes its structure—the molecule consists of two nucleotides joined together at their phosphate groups (shown in yellow). (Nicotinamide is a nitrogenous base, although not one that is present in DNA or RNA; see Figure 5.23.) The enzymatic transfer of 2 electrons and 1 proton (H^+) from an organic molecule in food to NAD^+ reduces the NAD^+ to NADH: Most of the electrons removed from food are transferred initially to NAD^+, forming NADH.

is derived from organic molecules rather than H_2. Second, instead of occurring in one explosive reaction, respiration uses an electron transport chain to break the fall of electrons to oxygen into several energy-releasing steps **(Figure 9.4b)**. An **electron transport chain** consists of a number of molecules, mostly proteins, built into the inner membrane of the mitochondria of eukaryotic cells (and the plasma membrane of respiring prokaryotes). Electrons removed from glucose are shuttled by NADH to the "top," higher-energy end of the chain. At the "bottom," lower-energy end, O_2 captures these electrons along with hydrogen nuclei (H^+), forming water. (Anaerobically respiring prokaryotes have an electron acceptor at the end of the chain that is different from O_2.)

Electron transfer from NADH to oxygen is an exergonic reaction with a free-energy change of −222 kJ/mol. Instead of this energy being released and wasted in a single explosive step, electrons cascade down the chain from one carrier molecule to the next in a series of redox reactions, losing a small amount of energy with each step until they finally reach oxygen, the terminal electron acceptor, which has a very great affinity for electrons. Each "downhill" carrier has a greater affinity for electrons than—and is thus capable of accepting electrons from (oxidising)—its "uphill" neighbour, with O_2 at the bottom of the chain. Therefore, the electrons transferred from glucose to NAD^+, reducing it to NADH, fall down an energy gradient in the electron transport chain to a far more stable location in an electronegative oxygen atom from O_2. Put another way, O_2 pulls electrons down the chain in an energy-yielding tumble analogous to gravity pulling objects downhill.

In summary, during cellular respiration, most electrons travel the following "downhill" route: glucose → NADH → electron transport chain → oxygen. Later in this chapter, you will learn more about how the cell uses the energy released from this exergonic electron fall to regenerate its supply of ATP. For now, having covered the basic redox mechanisms of cellular respiration, let's look at the entire process by which energy is harvested from organic fuels.

The Stages of Cellular Respiration: *A Preview*

The harvesting of energy from glucose by cellular respiration is a cumulative function of three metabolic stages. We list them here along with a colour-coding scheme we will use throughout the chapter to help you keep track of the big picture:

1. GLYCOLYSIS (colour-coded blue throughout the chapter)
2. PYRUVATE OXIDATION (light orange) and the CITRIC ACID CYCLE (dark orange)
3. OXIDATIVE PHOSPHORYLATION: Electron transport and chemiosmosis (purple)

Biochemists usually reserve the term *cellular respiration* for stages 2 and 3 together. In this text, however, we include glycolysis as a part of cellular respiration because most respiring cells deriving energy from glucose use glycolysis to produce the starting material for the citric acid cycle.

▼ Figure 9.4 An introduction to electron transport chains.

(a) Uncontrolled reaction. The one-step exergonic reaction of hydrogen with oxygen to form water releases a large amount of energy in the form of heat and light: an explosion.

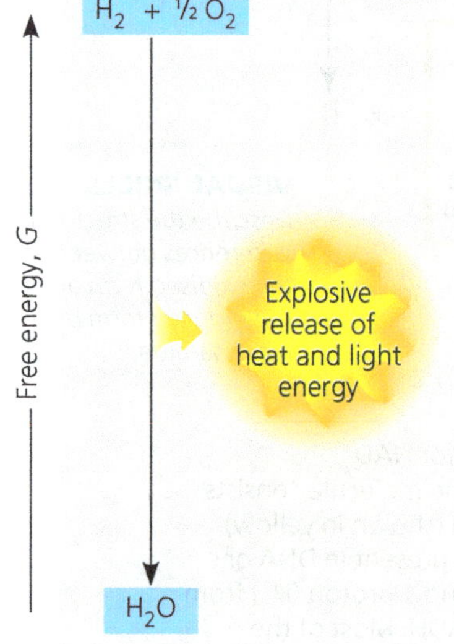

(b) Cellular respiration. In cellular respiration, the same reaction occurs in stages: An electron transport chain breaks the "fall" of electrons in this reaction into a series of smaller steps and stores some of the released energy in a form that can be used to make ATP. (The rest of the energy is released as heat.)

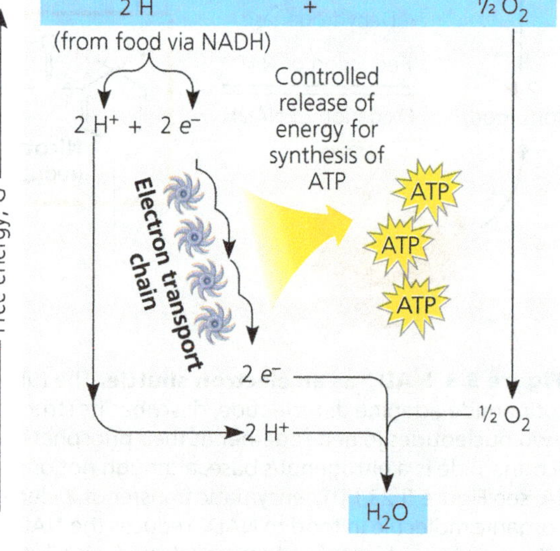

As described in **Figure 9.5**, glycolysis and then pyruvate oxidation and the citric acid cycle are the catabolic pathways that break down glucose and other organic fuels. **Glycolysis**, which occurs in the cytosol, begins the degradation process by breaking glucose into two molecules of a compound called pyruvate. In eukaryotes, pyruvate enters the mitochondrion and is oxidised to a compound called acetyl CoA, which enters the **citric acid cycle**. There, the breakdown of glucose to carbon dioxide is completed. (In prokaryotes, these processes take place in the cytosol.) Thus, the carbon dioxide produced by respiration represents fragments of oxidised organic molecules.

Some of the steps of glycolysis and the citric acid cycle are redox reactions in which dehydrogenases transfer electrons from substrates to NAD^+ or the related electron carrier FAD, forming NADH or $FADH_2$. (You'll learn more about FAD and $FADH_2$ later.) In the third stage of respiration, the electron transport chain accepts

▶ **Figure 9.5 An overview of cellular respiration.** During glycolysis, each glucose molecule is broken down into two molecules of pyruvate. In eukaryotic cells, as shown here, the pyruvate enters the mitochondrion. There it is oxidised to acetyl CoA, which will be further oxidised to CO_2 in the citric acid cycle. The electron carriers NADH and $FADH_2$ transfer electrons derived from glucose to electron transport chains. During oxidative phosphorylation, electron transport chains convert the chemical energy to a form used for ATP synthesis in the process called chemiosmosis. (During earlier steps of cellular respiration, a few molecules of ATP are synthesised in a process called substrate-level phosphorylation.) To visualise these processes in their cellular context, see Figure 6.32b.

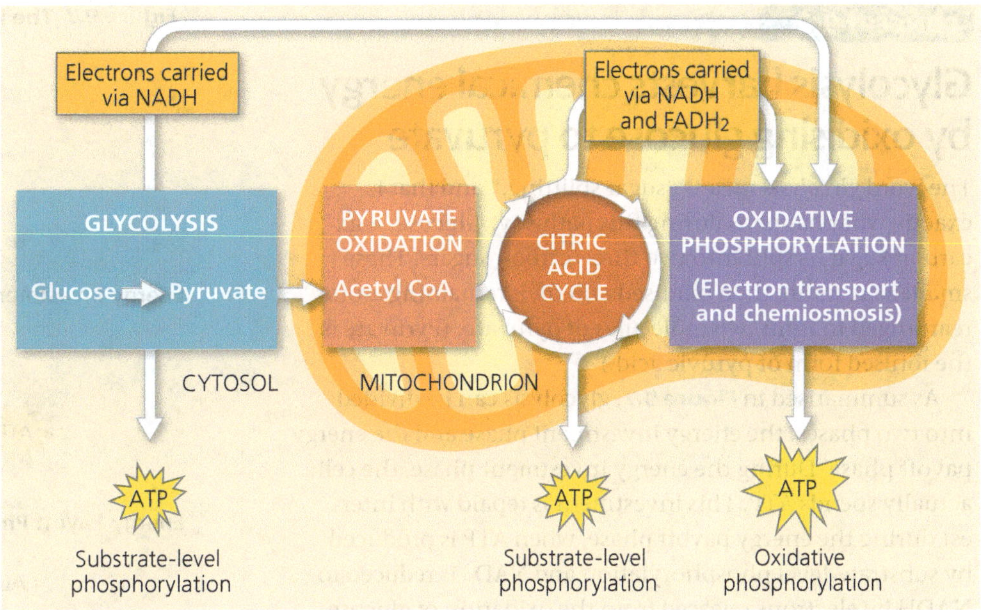

electrons from NADH or $FADH_2$ generated during the first two stages and passes these electrons down the chain. At the end of the chain, the electrons are combined with molecular oxygen (O_2) and hydrogen ions (H^+), forming water (see Figure 9.4b). The energy released at each step of the chain is stored in a form the mitochondrion (or prokaryotic cell) can use to make ATP from ADP. This mode of ATP synthesis is called **oxidative phosphorylation** because it is powered by the redox reactions of the electron transport chain.

In eukaryotic cells, the inner membrane of the mitochondrion is the site of electron transport and another process called *chemiosmosis*, together making up oxidative phosphorylation. (In prokaryotes, these processes take place in the plasma membrane.) Oxidative phosphorylation accounts for almost 90% of the ATP generated by respiration. A smaller amount of ATP is formed directly in a few reactions of glycolysis and the citric acid cycle by a mechanism called **substrate-level phosphorylation** **(Figure 9.6)**. This mode of ATP synthesis occurs when an enzyme transfers a phosphate group from a substrate molecule to ADP, rather than adding an inorganic phosphate to ADP as in oxidative phosphorylation. "Substrate molecule" here refers to an organic molecule generated as an intermediate during the catabolism of glucose. You'll see examples of substrate-level phosphorylation later in the chapter, in both glycolysis and the citric acid cycle.

▼ **Figure 9.6 Substrate-level phosphorylation.** Some ATP is made by direct transfer of a phosphate group from an organic substrate to ADP by an enzyme. (For examples in glycolysis, see Figure 9.8, steps 7 and 10.)

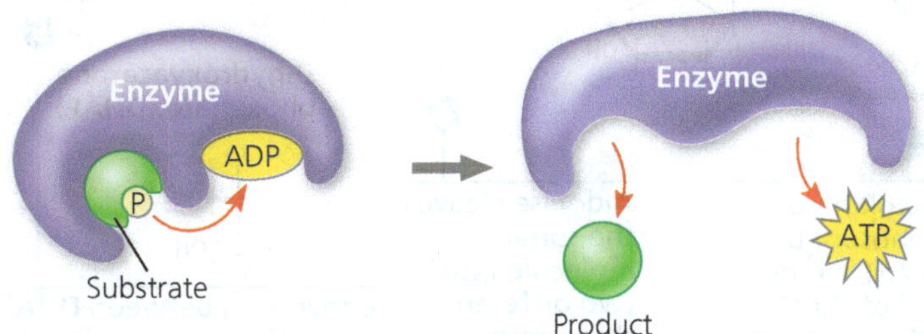

MAKE CONNECTIONS *Review Figure 8.9. In the reaction shown above, is the potential energy higher for the reactants or for the products? Explain.*

You can think of the whole process this way: When you withdraw a relatively large sum of money from an ATM, it is not delivered to you in a single note of a large denomination. Instead, the machine dispenses a number of smaller-denomination notes that you can spend more easily. This is analogous to ATP production during cellular respiration. For each molecule of glucose degraded to CO_2 and H_2O by respiration, the cell makes up to about 32 molecules of ATP, each with 30.5 kJ/mol of free energy. Respiration cashes in the large denomination of energy banked in a single molecule of glucose (2,870 kJ/mol under standard conditions) for the small change of many molecules of ATP, which is more practical for the cell to spend on its work.

This preview has introduced you to how glycolysis, the citric acid cycle, and oxidative phosphorylation fit into the process of cellular respiration so you can keep the big picture in mind as you take a closer look at each of these three stages of respiration. As you read about the chemical reactions, remember that each reaction is catalysed by a specific enzyme, some of which are shown in Figure 6.32b.

CONCEPT CHECK 9.1

1. Compare and contrast aerobic and anaerobic respiration, including the processes involved.
2. **WHAT IF?** If the following redox reaction occurred, which compounds would be oxidised Reduced?

$$C_4H_6O_5 + NAD^+ \rightarrow C_4H_4O_5 + NADH + H^+$$

For suggested answers, see Appendix A.

CONCEPT 9.2

Glycolysis harvests chemical energy by oxidising glucose to pyruvate

The word *glycolysis* means "sugar splitting," and that is exactly what happens during this pathway. Glucose, a six-carbon sugar, is split into two three-carbon sugars. These smaller sugars are then oxidised and their remaining atoms rearranged to form two molecules of pyruvate. (Pyruvate is the ionised form of pyruvic acid.)

As summarised in **Figure 9.7**, glycolysis can be divided into two phases: the energy investment phase and the energy payoff phase. During the energy investment phase, the cell actually spends ATP. This investment is repaid with interest during the energy payoff phase, when ATP is produced by substrate-level phosphorylation and NAD^+ is reduced to NADH by electrons released from the oxidation of glucose. The net energy yield from glycolysis, per glucose molecule, is 2 ATP plus 2 NADH. The ten steps of the glycolytic pathway are shown in **Figure 9.8**.

All of the carbon originally present in glucose is accounted for in the two molecules of pyruvate; no carbon is released as CO_2 during glycolysis. Glycolysis occurs whether or not O_2 is present. However, if O_2 *is* present, the chemical energy stored in pyruvate and NADH can be extracted by pyruvate oxidation, the citric acid cycle, and oxidative phosphorylation.

▼ **Figure 9.7 The inputs and outputs of glycolysis.**

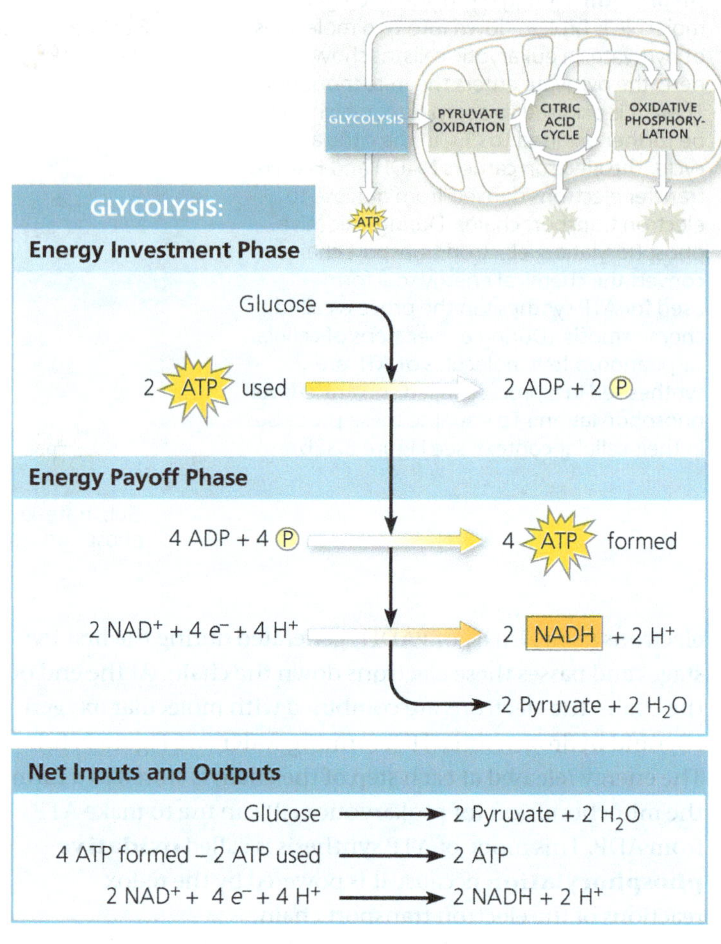

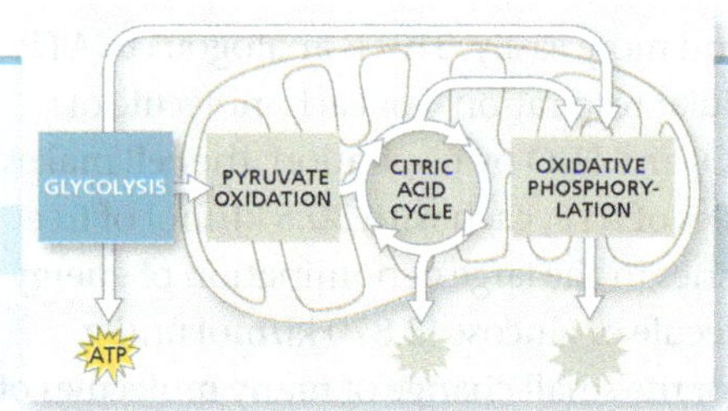

▼ **Figure 9.8 The steps of glycolysis.** Glycolysis, a source of ATP and NADH, takes place in the cytosol. Two of the enzymes (in steps 1 and 3) are shown in Figure 6.32b.

GLYCOLYSIS: Energy Investment Phase

WHAT IF? *What would happen if you removed the dihydroxyacetone phosphate generated in step 4 as fast as it was produced?*

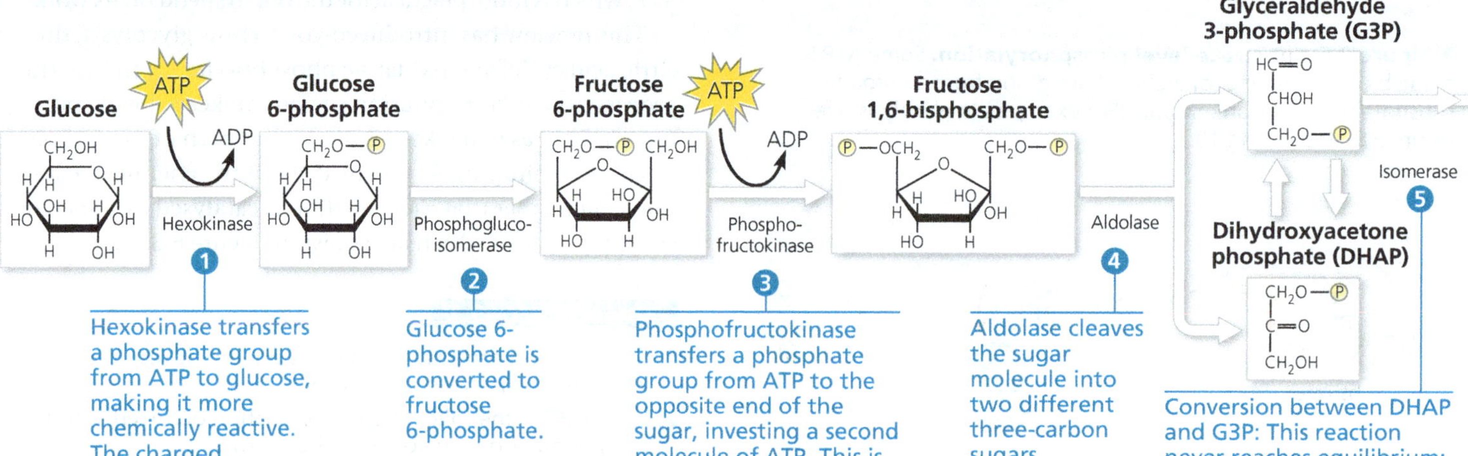

CONCEPT CHECK 9.2

1. **VISUAL SKILLS** During the redox reaction in glycolysis (see step 6 in Figure 9.8), which one of the molecules acts as the oxidising agent? The reducing agent?

For suggested answers, see Appendix A.

CONCEPT 9.3

After pyruvate is oxidised, the citric acid cycle completes the energy-yielding oxidation of organic molecules

Glycolysis releases less than a quarter of the chemical energy in glucose that can be harvested by cells; most of the energy remains stockpiled in the two molecules of pyruvate. When O_2 is present, the pyruvate in eukaryotic cells enters a mitochondrion, where the oxidation of glucose is completed. In aerobically respiring prokaryotic cells, this process occurs in the cytosol. (Later in the chapter, we'll examine other fates for pyruvate—for example, when O_2 is unavailable or in a prokaryote that is unable to use O_2.)

Oxidation of Pyruvate to Acetyl CoA

Upon entering the mitochondrion via active transport, pyruvate is first converted to a compound called acetyl coenzyme A, or **acetyl CoA (Figure 9.9)**. This step, linking glycolysis and

▼ Figure 9.9 Oxidation of pyruvate to acetyl CoA, the step before the citric acid cycle. Pyruvate enters the mitochondrion through a transport protein and is processed by a complex of several enzymes known as pyruvate dehydrogenase (shown in the computer-generated image based on cryo-EMs). This complex catalyses the three numbered steps, which are described in the text. The CO_2 molecule will diffuse out of the cell. The NADH will be used in oxidative phosphorylation. The acetyl group of acetyl CoA will enter the citric acid cycle. (Coenzyme A is abbreviated S-CoA when it is attached to a molecule, emphasising its sulfur atom, S.)

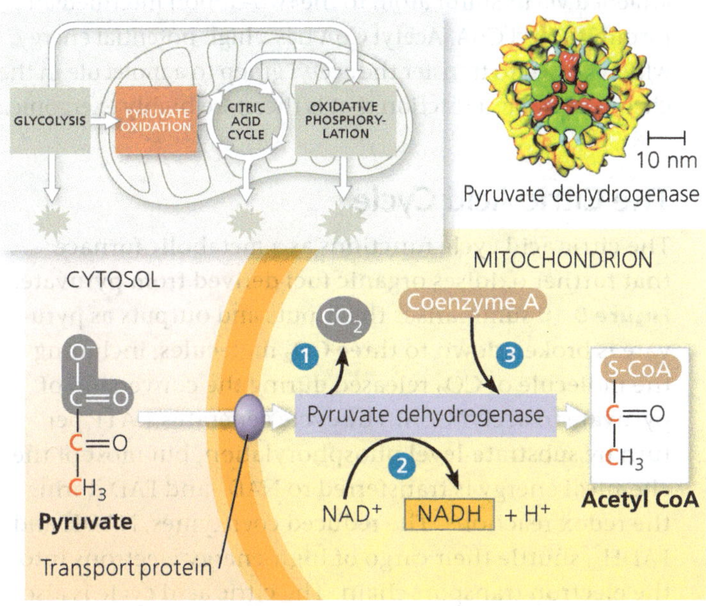

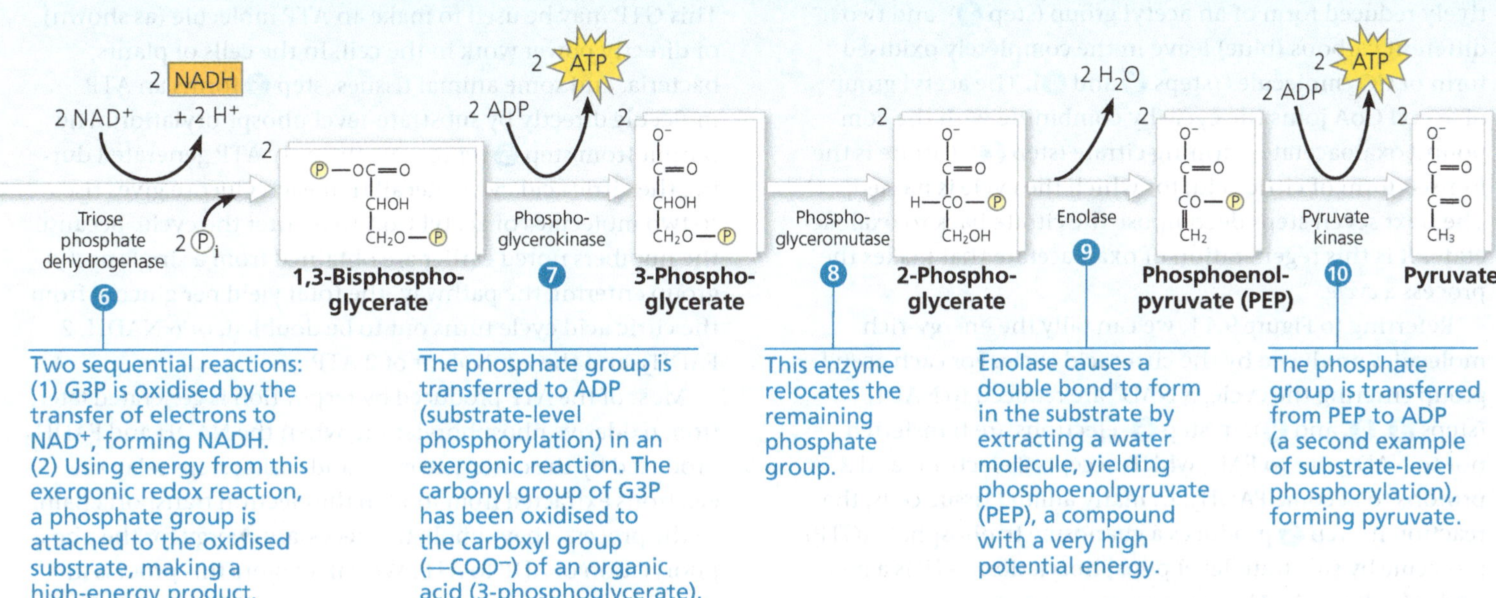

the citric acid cycle, is carried out by a multienzyme complex that catalyses three reactions: 1 Pyruvate's carboxyl group (—COO⁻), already somewhat oxidised and thus carrying little chemical energy, is now fully oxidised and given off as a molecule of CO_2. This is the first step in which CO_2 is released during respiration. 2 Next, the remaining two-carbon fragment is oxidised and the electrons transferred to NAD^+, storing energy in the form of NADH. 3 Finally, coenzyme A (CoA), a sulfur-containing compound derived from a B vitamin, is attached via its sulfur atom to the two-carbon intermediate, forming acetyl CoA. Acetyl CoA has a high potential energy, which is used to transfer the acetyl group to a molecule in the citric acid cycle, a reaction that is therefore highly exergonic.

The Citric Acid Cycle

The citric acid cycle functions as a metabolic furnace that further oxidises organic fuel derived from pyruvate. **Figure 9.10** summarises the inputs and outputs as pyruvate is broken down to three CO_2 molecules, including the molecule of CO_2 released during the conversion of pyruvate to acetyl CoA. The cycle generates 1 ATP per turn by substrate-level phosphorylation, but most of the chemical energy is transferred to NAD^+ and FAD during the redox reactions. The reduced coenzymes, NADH and $FADH_2$, shuttle their cargo of high-energy electrons into the electron transport chain. The citric acid cycle is also called the tricarboxylic acid cycle or the Krebs cycle, the latter honoring Hans Krebs. Krebs was the German-British scientist largely responsible for working out the pathway in the 1930s.

Now let's look at the citric acid cycle in more detail. The cycle has eight steps, each catalysed by a specific enzyme. You can see in **Figure 9.11** that for each turn of the citric acid cycle, two carbons (red) enter in the relatively reduced form of an acetyl group (step 1), and two different carbons (blue) leave in the completely oxidised form of CO_2 molecules (steps 3 and 4). The acetyl group of acetyl CoA joins the cycle by combining with the compound oxaloacetate, forming citrate (step 1). Citrate is the ionised form of citric acid, for which the cycle is named. The next seven steps decompose the citrate back to oxaloacetate. It is this regeneration of oxaloacetate that makes the process a *cycle*.

Referring to Figure 9.11, we can tally the energy-rich molecules produced by the citric acid cycle. For each acetyl group entering the cycle, 3 NAD^+ are reduced to NADH (steps 3, 4, and 8). In step 6, electrons are transferred not to NAD^+, but to FAD, which accepts 2 electrons and 2 protons to become $FADH_2$. In many animal tissue cells, the reaction in step 5 produces a guanosine triphosphate (GTP) molecule by substrate-level phosphorylation. GTP is a molecule similar to ATP in its structure and cellular function.

▼ **Figure 9.10 An overview of pyruvate oxidation and the citric acid cycle.** The inputs and outputs per pyruvate molecule are shown with a focus on the carbon atoms involved. To calculate on a per-glucose basis, multiply by 2 because each glucose molecule is split during glycolysis into two pyruvate molecules.

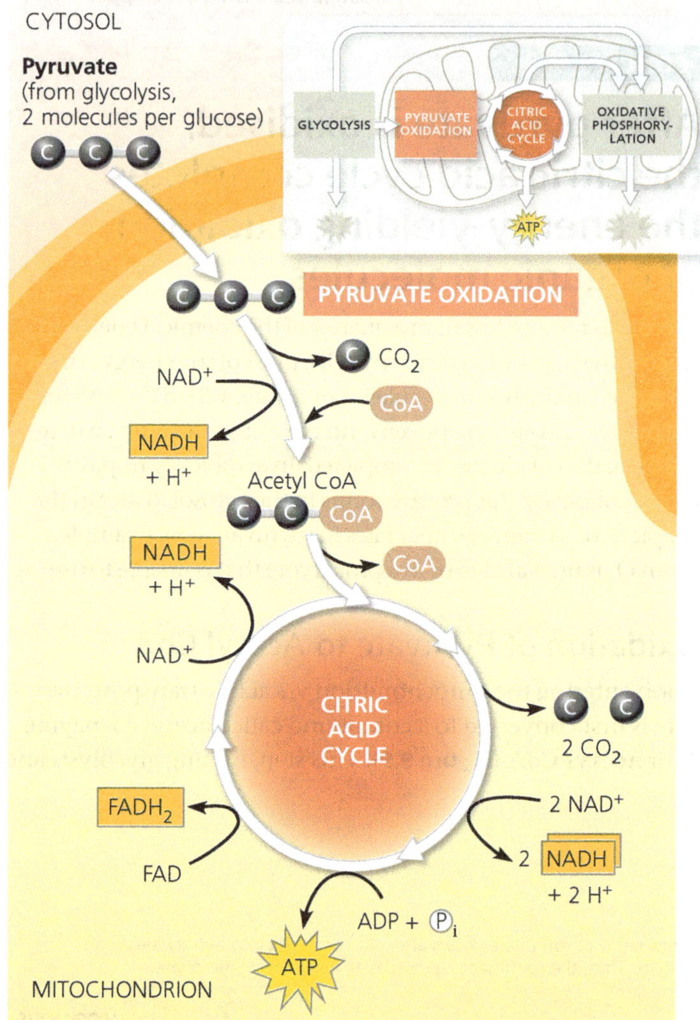

This GTP may be used to make an ATP molecule (as shown) or directly power work in the cell. In the cells of plants, bacteria, and some animal tissues, step 5 forms an ATP molecule directly by substrate-level phosphorylation. The output from step 5 represents the only ATP generated during the citric acid cycle. Recall that each glucose gives rise to two molecules of acetyl CoA that enter the cycle. Because the numbers noted earlier are obtained from a single acetyl group entering the pathway, the total yield per glucose from the citric acid cycle turns out to be doubled, or 6 NADH, 2 $FADH_2$, and the equivalent of 2 ATP.

Most of the ATP produced by respiration is generated later, from oxidative phosphorylation, when the NADH and $FADH_2$ produced by the citric acid cycle and earlier steps relay the electrons extracted from food to the electron transport chain. In the process, they supply the necessary energy for the phosphorylation of ADP to ATP. We will explore this process in the next section.

▼ Figure 9.11 A closer look at the citric acid cycle. In the chemical structures, red type traces the fate of the two carbon atoms that enter the cycle via acetyl CoA (step 1), and blue type indicates the two carbons that exit the cycle as CO_2 in steps 3 and 4. (The red type goes only through step 5 because the succinate molecule is symmetrical; the two ends cannot be distinguished from each other.) Notice that the carbon atoms that enter the cycle from acetyl CoA do not leave the cycle in the same turn. They remain in the cycle, occupying a different location in the molecules on their next turn, after another acetyl group is added. Therefore, the oxaloacetate regenerated at step 8 is made up of different carbon atoms each time around. Carboxylic acids are represented in their ionised forms, as $—COO^-$, because the ionised forms prevail at the pH within the mitochondrion. In eukaryotic cells, all the citric acid cycle enzymes are located in the mitochondrial matrix except for the enzyme that catalyses step 6, which resides in the inner mitochondrial membrane. (The enzyme that catalyses step 3, isocitrate dehydrogenase, is shown in Figure 6.32b.)

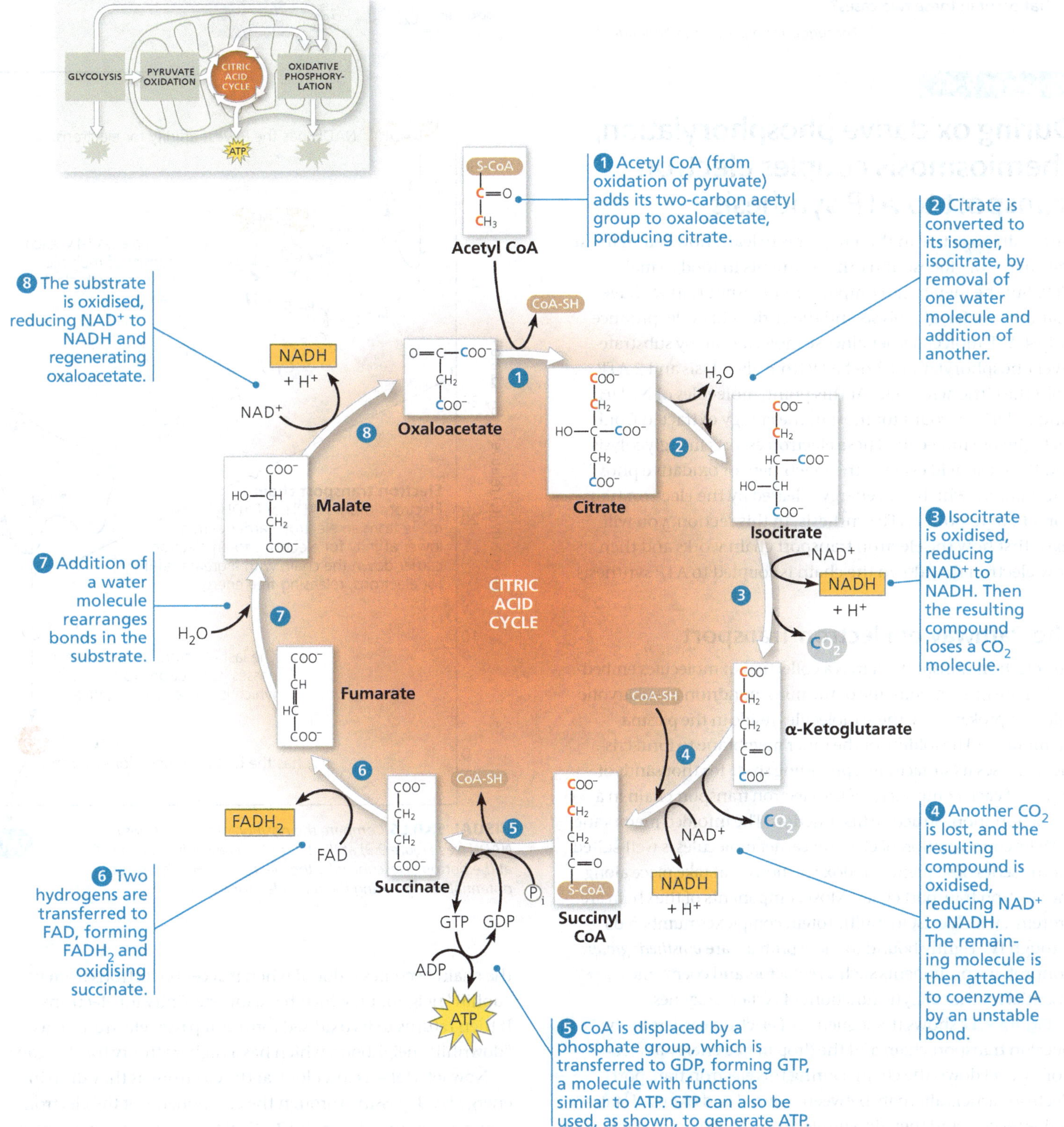

CONCEPT CHECK 9.3

1. **VISUAL SKILLS** In the citric acid cycle shown in Figure 9.11, what molecules capture energy from the redox reactions? How is ATP produced?
2. What processes in your cells produce the CO_2 that you exhale?
3. **VISUAL SKILLS** The conversions shown in Figure 9.9 and step 4 of Figure 9.11 are each catalysed by a large multienzyme complex. What similarities are there in the reactions that occur in these two cases?

For suggested answers, see Appendix A.

CONCEPT 9.4

During oxidative phosphorylation, chemiosmosis couples electron transport to ATP synthesis

Our main objective in this chapter is to learn how cells harvest the energy of glucose and other nutrients in food to make ATP. But the metabolic components of respiration we have examined so far, glycolysis and the citric acid cycle, produce only 4 ATP molecules per glucose molecule, all by substrate-level phosphorylation: 2 net ATP from glycolysis and 2 ATP from the citric acid cycle. At this point, molecules of NADH (and $FADH_2$) account for most of the energy extracted from each glucose molecule. These electron escorts link glycolysis and the citric acid cycle to the machinery of oxidative phosphorylation, which uses energy released by the electron transport chain to power ATP synthesis. In this section, you will learn first how the electron transport chain works and then how electron flow down the chain is coupled to ATP synthesis.

The Pathway of Electron Transport

The electron transport chain is a collection of molecules embedded in the inner membrane of the mitochondrion in eukaryotic cells. (In prokaryotes, these molecules reside in the plasma membrane.) The folding of the inner membrane to form cristae increases its surface area, providing space for thousands of copies of each component of the electron transport chain in a mitochondrion. Structure fits function: The infolded membrane with its concentration of electron carrier molecules is well-suited for the series of sequential redox reactions that take place along the electron transport chain. Most components of the chain are proteins, which exist in multiprotein complexes numbered I through IV. Tightly bound to these proteins are *prosthetic groups*, nonprotein components such as cofactors and coenzymes essential for the catalytic functions of certain enzymes.

Figure 9.12 shows the sequence of electron carriers in the electron transport chain and the drop in free energy as electrons travel down the chain. During this electron transport, electron carriers alternate between reduced and oxidised states as they accept and then donate electrons. Each component of the chain becomes reduced when it accepts electrons from its "uphill" neighbour, which has a lower affinity for electrons. It then returns to its oxidised form as it passes electrons to its "downhill" neighbour, which has a higher affinity for electrons.

▼ **Figure 9.12 Free-energy change during electron transport.** The overall energy drop (ΔG) for electrons travelling from NADH to oxygen is 222 kJ/mol, but this "fall" is broken up into a series of smaller steps by the electron transport chain. (An oxygen atom is represented here as $\frac{1}{2}\,O_2$ to show that O_2 is reduced, not individual oxygen atoms.) To view these proteins in their cellular context, see Figure 6.32b.

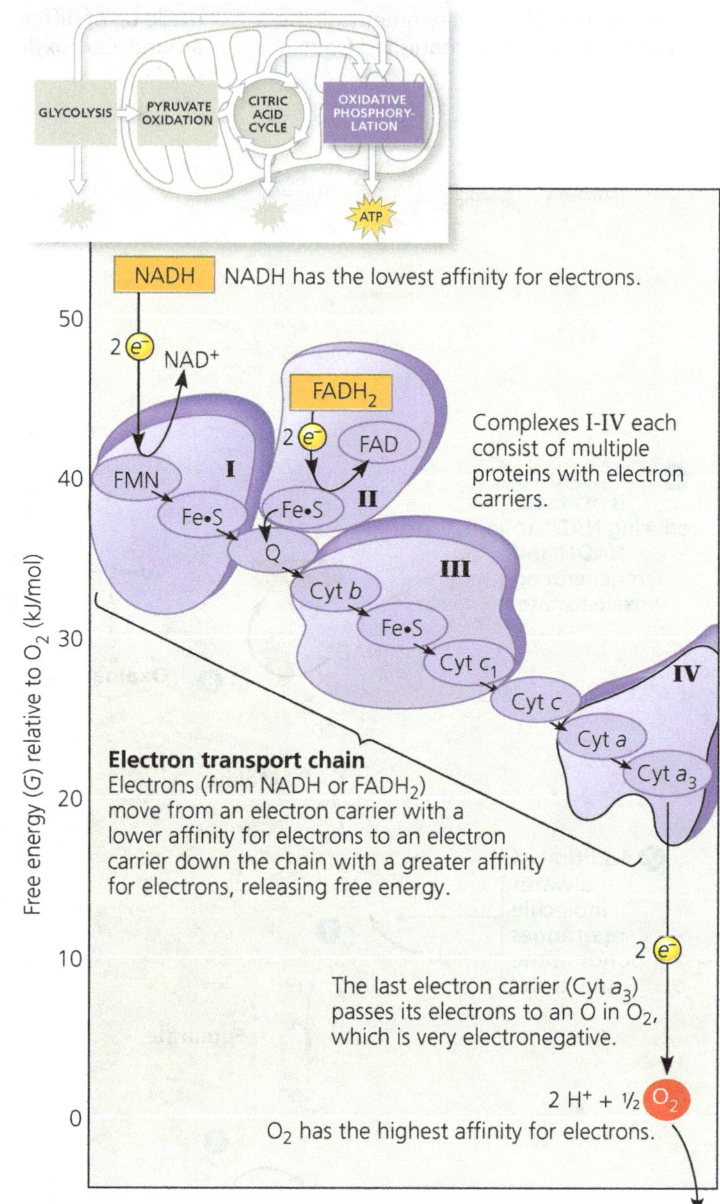

VISUAL SKILLS *Compare the position of the electrons in NADH (see Figure 9.3) at the top of the chain with that in H_2O, at the bottom. Describe why the electrons in H_2O have less potential energy, using the term electronegativity.*

Now let's take a closer look at the electrons as they drop in energy level, passing through the components of the electron transport chain in Figure 9.12. We'll first look at the passage of electrons through complex I in some detail as an illustration of the general principles involved in electron transport. Electrons acquired from glucose by NAD^+ during glycolysis and the citric acid cycle are transferred from NADH to the first molecule of the electron transport chain in complex I. This molecule is a flavoprotein, so named because it has a prosthetic group called flavin mononucleotide (FMN). In the next redox reaction, the flavoprotein returns to its oxidised form as it passes electrons to an iron-sulfur protein (Fe · S in complex I), one of a family of proteins with both iron and sulfur tightly bound. The iron-sulfur protein then passes the electrons to a compound called ubiquinone (Q in Figure 9.12). This electron carrier is a small hydrophobic molecule, the only member of the electron transport chain that is not a protein. Ubiquinone is individually mobile within the membrane rather than residing in a particular complex. (Another name for ubiquinone is coenzyme Q, or CoQ; you may have seen it sold as a nutritional supplement.)

Most of the remaining electron carriers between ubiquinone and oxygen are proteins called **cytochromes**. Their prosthetic group, called a haem group, has an iron atom that accepts and donates electrons. (The haem group in a cytochrome is similar to the haem group in haemoglobin, the protein of red blood cells, except that the iron in haemoglobin carries oxygen, not electrons.) The electron transport chain has several types of cytochromes, each named "cyt" with a letter and number to distinguish it as a different protein with a slightly different electron-carrying haem group. The last cytochrome of the chain, cyt a_3, passes its electrons to oxygen (in O_2), which is *very* electronegative. Each O also picks up a pair of hydrogen ions (protons) from the aqueous solution, neutralising the −2 charge of the added electrons and forming water.

Another source of electrons for the electron transport chain is $FADH_2$, the other reduced product of the citric acid cycle. Notice in Figure 9.12 that $FADH_2$ adds its electrons from within complex II, at a lower energy level than NADH does. Consequently, although NADH and $FADH_2$ each donate an equivalent number of electrons (2) for oxygen reduction, the electron transport chain provides about one-third less energy for ATP synthesis when the electron donor is $FADH_2$ rather than NADH. We'll see why in the next section.

The electron transport chain makes no ATP directly. Instead, it eases the fall of electrons from food to oxygen, breaking a large free-energy drop into a series of smaller steps that release energy in manageable amounts, step by step. How does the mitochondrion (or the plasma membrane in prokaryotes) couple this electron transport and energy release to ATP synthesis? The answer is a mechanism called chemiosmosis.

Chemiosmosis: The Energy-Coupling Mechanism

Populating the inner membrane of the mitochondrion or the prokaryotic plasma membrane are many copies of a protein complex called **ATP synthase**, the enzyme that makes ATP from ADP and inorganic phosphate **(Figure 9.13)**. ATP synthase works like an ion pump running in reverse. Ion pumps usually use ATP as an energy source to transport ions against their gradients. Enzymes can catalyse a reaction in either direction, depending on the ΔG for the reaction, which is affected by the local concentrations of reactants and products (see Concepts 8.2 and 8.3). Under the conditions of cellular respiration, rather than hydrolysing ATP to pump protons against their concentration gradient, ATP synthase uses the energy of an existing ion gradient to power ATP synthesis. The power source for ATP synthase is a difference in the concentration of H^+ (a pH difference) on opposite sides of the inner mitochondrial membrane. This process, in which energy stored in the form of a hydrogen ion gradient across a membrane is used to drive cellular work such as the synthesis of ATP, is called **chemiosmosis** (from the Greek *osmos*, push). The word *osmosis*

▼ **Figure 9.13 ATP synthase, a molecular mill.** Multiple ATP synthases reside in eukaryotic mitochondrial and chloroplast membranes and in prokaryotic plasma membranes. (See Figure 6.32b and c.)

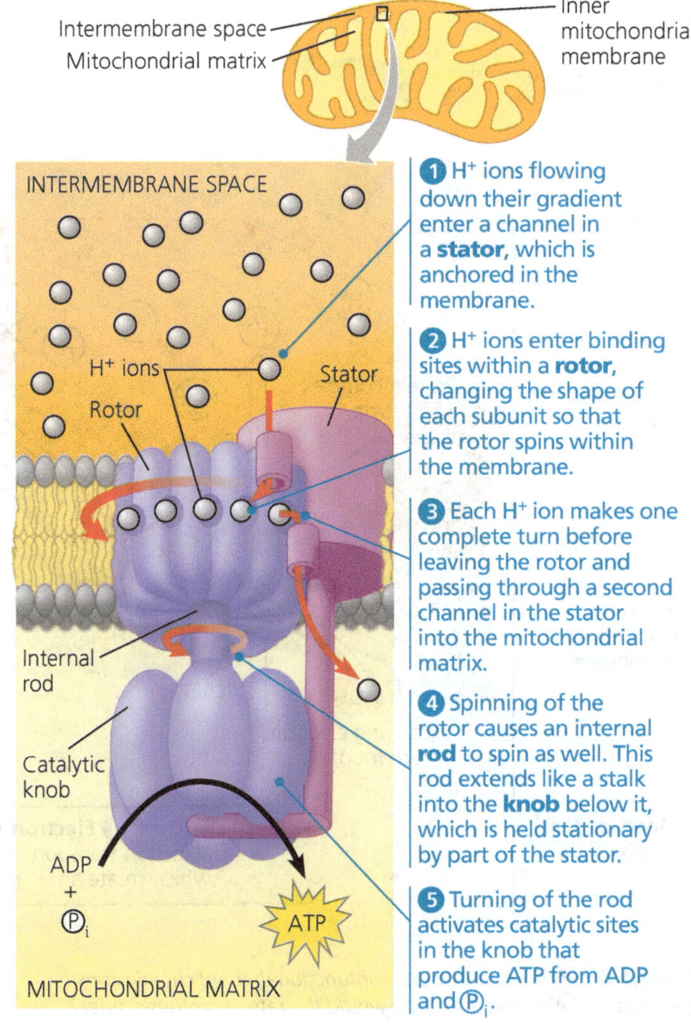

was previously used in discussing water transport, but here it refers to the flow of H^+ across a membrane.

From studying the structure of ATP synthase, scientists have learned how the flow of H^+ through this enzyme powers ATP generation. ATP synthase is a multisubunit complex with four main parts, each made up of multiple polypeptides (see Figure 9.13). Protons move one by one into binding sites on the rotor, causing it to spin in a way that catalyses ATP production from ADP and Ⓟ$_i$. The flow of protons thus behaves somewhat like a rushing stream that turns a waterwheel. ATP synthase is the smallest molecular rotary motor known in nature.

How does the inner mitochondrial membrane (or the prokaryotic plasma membrane) generate and maintain the H^+ gradient that drives ATP synthesis by the ATP synthase protein complex? Establishing the H^+ gradient is a major function of the electron transport chain, which is shown in its mitochondrial location in **Figure 9.14**. The chain is an energy converter that uses the exergonic flow of electrons from NADH and $FADH_2$ to pump H^+ across the membrane, from the mitochondrial matrix into the intermembrane space. The H^+ has a tendency to move back across the membrane, diffusing down its gradient. And the ATP synthases are the only sites that provide a route through the membrane for H^+. As we described previously, the passage

▼ Figure 9.14 Chemiosmosis couples the electron transport chain to ATP synthesis. ① NADH and $FADH_2$ shuttle high-energy electrons extracted from food during glycolysis and the citric acid cycle into an electron transport chain built into the inner mitochondrial membrane. (See Figure 6.32b.) The gold arrows trace the transport of electrons, which are finally passed to a terminal acceptor (O_2, in the case of aerobic respiration) at the "downhill" end of the chain, forming water. Most of the electron carriers of the chain are grouped into four complexes (I–IV). Two mobile carriers, ubiquinone (Q) and cytochrome *c* (Cyt *c*), move rapidly, ferrying electrons between the large complexes. As the complexes shuttle electrons, they pump protons from the mitochondrial matrix into the intermembrane space. $FADH_2$ deposits its electrons via complex II and so results in fewer protons being pumped into the intermembrane space than occurs with NADH. Chemical energy originally harvested from food is transformed into a proton-motive force, a gradient of H^+ across the membrane. ② During chemiosmosis, the protons flow back down their gradient via ATP synthase, which is built into the membrane nearby. The ATP synthase harnesses the proton-motive force to phosphorylate ADP, forming ATP. Together, electron transport and chemiosmosis make up oxidative phosphorylation.

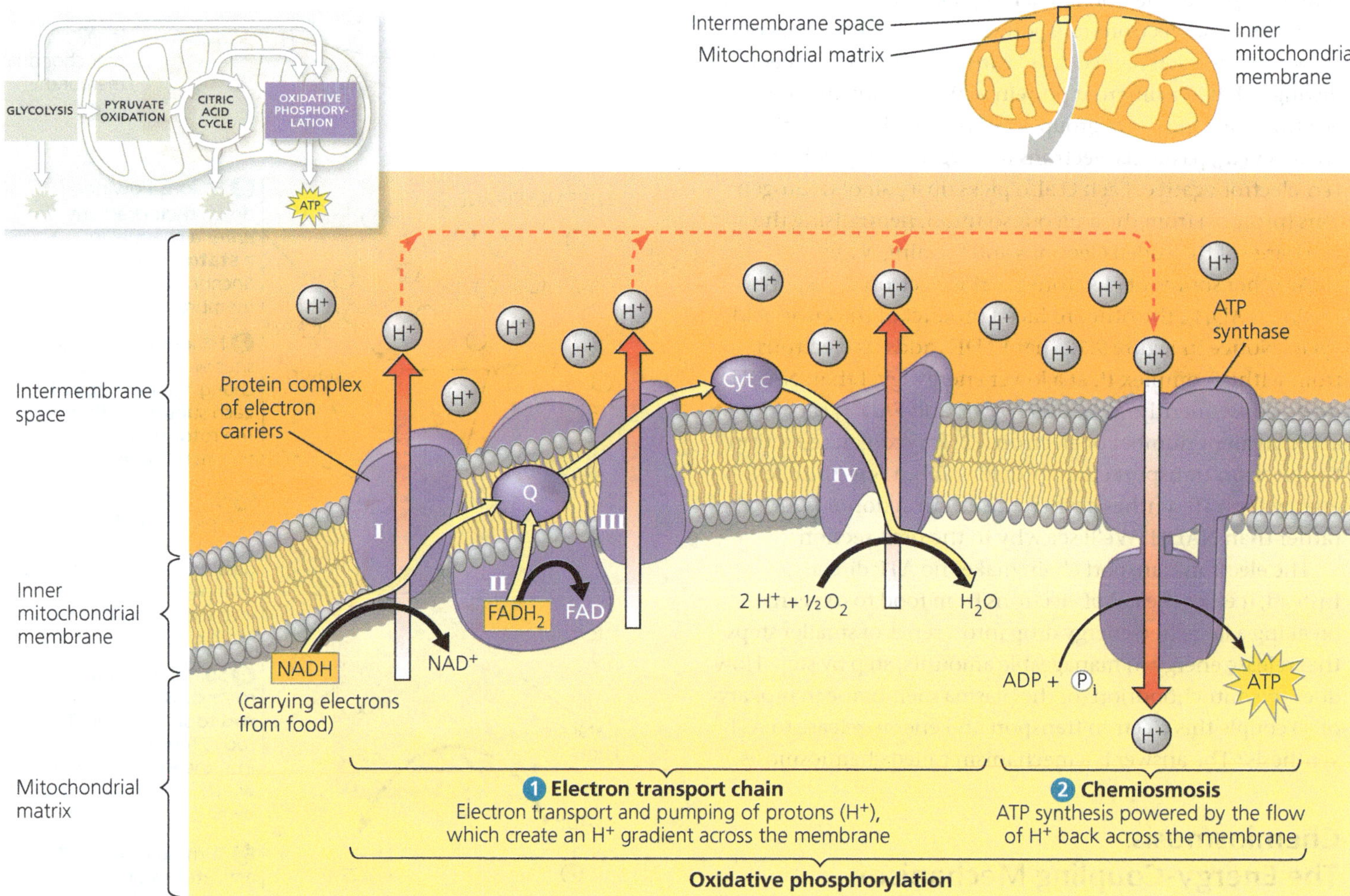

WHAT IF? *If complex IV were nonfunctional, could chemiosmosis produce any ATP, and if so, how would the rate of synthesis differ?*

of H^+ through ATP synthase uses the exergonic flow of H^+ to drive the phosphorylation of ADP. Thus, the energy stored in an H^+ gradient across a membrane couples the redox reactions of the electron transport chain to ATP synthesis.

At this point, you may wonder how the electron transport chain pumps hydrogen ions into the intermembrane space. Researchers have found that certain members of the electron transport chain accept and release protons (H^+) along with electrons. (The aqueous solutions inside and surrounding the cell are a ready source of H^+.) At certain steps along the chain, electron transfers cause H^+ to be taken up and released into the surrounding solution. In eukaryotic cells, the electron carriers are spatially arranged in the inner mitochondrial membrane in such a way that H^+ is accepted from the mitochondrial matrix and deposited in the intermembrane space (see Figure 9.14). The H^+ gradient that results is referred to as a **proton-motive force**, emphasising the capacity of the gradient to perform work. The force drives H^+ back across the membrane through the H^+ channels provided by ATP synthases.

In general terms, *chemiosmosis is an energy-coupling mechanism that uses energy stored in the form of an H^+ gradient across a membrane to drive cellular work.* In mitochondria, the energy for gradient formation comes from exergonic redox reactions along the electron transport chain, and ATP synthesis is the work performed. But chemiosmosis also occurs elsewhere and in other variations. Chloroplasts use chemiosmosis to generate ATP during photosynthesis; in these organelles, light (rather than chemical energy) drives both electron flow down an electron transport chain and the resulting H^+ gradient formation. Prokaryotes, as already mentioned, generate H^+ gradients across their plasma membranes. They then tap the proton-motive force not only to make ATP inside the cell but also to rotate their flagella and to pump nutrients and waste products across the membrane. Because of its central importance to energy conversions in prokaryotes and eukaryotes, chemiosmosis has helped unify the study of bioenergetics. Peter Mitchell was awarded the Nobel Prize in 1978 for originally proposing the chemiosmotic model.

An Accounting of ATP Production by Cellular Respiration

In the last few sections, we have looked rather closely at the key processes of cellular respiration. Now let's take a step back and remind ourselves of its overall function: harvesting the energy of glucose for ATP synthesis.

During respiration, most energy flows in this sequence: glucose → NADH → electron transport chain → proton-motive force → ATP. We can do some bookkeeping to calculate the ATP profit when cellular respiration oxidises a molecule of glucose to six molecules of carbon dioxide. The three main departments of this metabolic enterprise are glycolysis, pyruvate oxidation and the citric acid cycle, and the electron transport chain, which drives oxidative phosphorylation. **Figure 9.15** gives a detailed accounting of the ATP yield for

▼ Figure 9.15 ATP yield per molecule of glucose at each stage of cellular respiration.

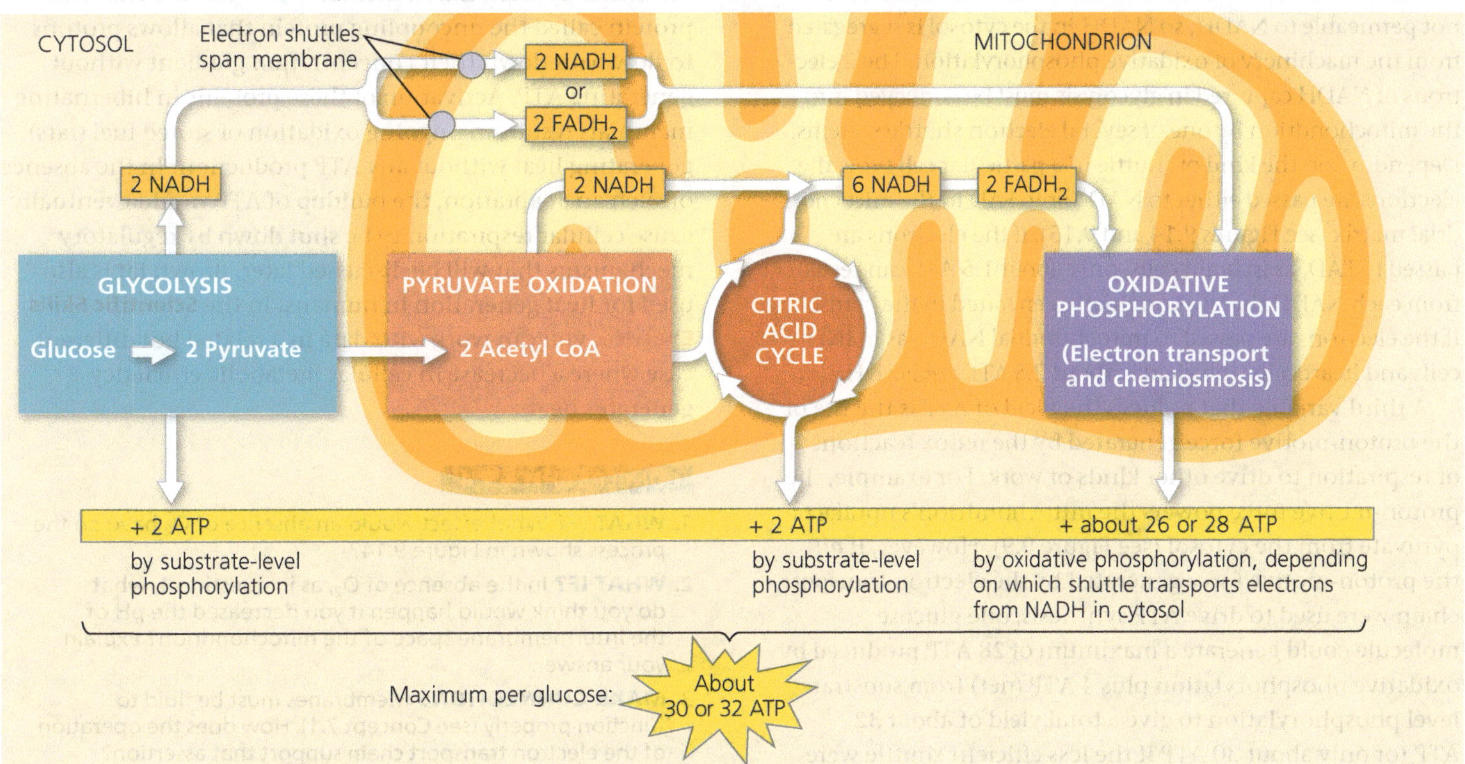

VISUAL SKILLS *After reading the discussion in the text, explain exactly how the total of 26 or 28 ATP from oxidative phosphorylation was calculated (see the yellow bar).*

each glucose molecule that is oxidised. The tally adds the 4 ATP produced directly by substrate-level phosphorylation during glycolysis and the citric acid cycle to the many more molecules of ATP generated by oxidative phosphorylation. Each NADH that transfers a pair of electrons from glucose to the electron transport chain contributes enough to the proton-motive force to generate a maximum of about 3 ATP.

Why are the numbers in Figure 9.15 inexact? There are three reasons we cannot state an exact number of ATP molecules generated by the breakdown of one molecule of glucose. First, phosphorylation and the redox reactions are not directly coupled to each other, so the ratio of the number of NADH molecules to the number of ATP molecules is not a whole number. We know that 1 NADH results in 10 H^+ being transported out across the inner mitochondrial membrane, but the exact number of H^+ that must reenter the mitochondrial matrix via ATP synthase to generate 1 ATP has long been debated. Based on experimental data, however, most biochemists now agree that the most accurate number is 4 H^+. Therefore, a single molecule of NADH generates enough proton-motive force for the synthesis of 2.5 ATP. The citric acid cycle also supplies electrons to the electron transport chain via $FADH_2$, but since its electrons enter later in the chain, each molecule of this electron carrier is responsible for transport of only enough H^+ for the synthesis of 1.5 ATP. These numbers also take into account the slight energetic cost of moving the ATP formed in the mitochondrion out into the cytosol, where it will be used.

Second, the ATP yield varies slightly depending on the type of shuttle used to transport electrons from the cytosol into the mitochondrion. The mitochondrial inner membrane is not permeable to NADH, so NADH in the cytosol is segregated from the machinery of oxidative phosphorylation. The 2 electrons of NADH captured in glycolysis must be conveyed into the mitochondrion by one of several electron shuttle systems. Depending on the kind of shuttle in a particular cell type, the electrons are passed either to NAD^+ or to FAD in the mitochondrial matrix (see Figures 9.14 and 9.15). If the electrons are passed to FAD, as in brain cells, only about 1.5 ATP can result from each NADH that was originally generated in the cytosol. If the electrons are passed to mitochondrial NAD^+, as in liver cells and heart cells, the yield is about 2.5 ATP per NADH.

A third variable that reduces the yield of ATP is the use of the proton-motive force generated by the redox reactions of respiration to drive other kinds of work. For example, the proton-motive force powers the mitochondrion's uptake of pyruvate from the cytosol (see Figure 9.9). However, if *all* the proton-motive force generated by the electron transport chain were used to drive ATP synthesis, one glucose molecule could generate a maximum of 28 ATP produced by oxidative phosphorylation plus 4 ATP (net) from substrate-level phosphorylation to give a total yield of about 32 ATP (or only about 30 ATP if the less efficient shuttle were functioning).

We can now roughly estimate the efficiency of respiration—that is, the percentage of chemical energy in glucose that has been transferred to ATP. Recall that the complete oxidation of a mole of glucose releases 2,870 kJ of energy under standard conditions ($\Delta G = -2{,}870$ kJ/mol). Phosphorylation of ADP to form ATP stores at least 30.5 kJ per mole of ATP. Therefore, the efficiency of respiration is 30.5 kJ per mole of ATP times 32 moles of ATP per mole of glucose divided by 2,870 kJ per mole of glucose, which equals 0.34. Thus, about 34% of the potential chemical energy in glucose has been transferred to ATP; the actual percentage is bound to vary as ΔG varies under different cellular conditions. Cellular respiration is remarkably efficient in its energy conversion. By comparison, even the most efficient car converts only about 25% of the energy stored in petrol to energy that moves the car.

The rest of the energy stored in glucose is lost as heat. We humans use some of this heat to maintain our relatively high body temperature (37°C), and we dissipate the rest through sweating and other cooling mechanisms.

Surprisingly, perhaps, it may be beneficial under certain conditions to reduce the efficiency of cellular respiration. A remarkable adaptation is shown by hibernating mammals, which overwinter in a state of inactivity and lowered metabolism. Although their internal body temperature is lower than normal, it still must be kept significantly higher than the external air temperature. One type of tissue, called brown fat, is made up of cells packed full of mitochondria. The inner mitochondrial membrane contains a channel protein called the uncoupling protein that allows protons to flow back down their concentration gradient without generating ATP. Activation of these proteins in hibernating mammals results in ongoing oxidation of stored fuel (fats), generating heat without any ATP production. In the absence of such an adaptation, the buildup of ATP would eventually cause cellular respiration to be shut down by regulatory mechanisms that will be discussed later. Brown fat is also used for heat generation in humans. In the **Scientific Skills Exercise**, you can work with data in a related but different case where a decrease in cellular metabolic efficiency generates heat.

CONCEPT CHECK 9.4

1. **WHAT IF?** What effect would an absence of O_2 have on the process shown in Figure 9.14?
2. **WHAT IF?** In the absence of O_2, as in question 1, what do you think would happen if you decreased the pH of the intermembrane space of the mitochondrion? Explain your answer.
3. **MAKE CONNECTIONS** Membranes must be fluid to function properly (see Concept 7.1). How does the operation of the electron transport chain support that assertion?

For suggested answers, see Appendix A.

Scientific Skills Exercise

Making a Bar Graph and Evaluating a Hypothesis

Does Thyroid Hormone Level Affect O_2 Consumption in Cells? Some animals, such as mammals and birds, maintain a relatively constant body temperature, above that of their environment, by using heat produced as a by-product of metabolism. When the core temperature of these animals drops below an internal set point, their cells are triggered to reduce the efficiency of ATP production by the electron transport chains in mitochondria. At lower efficiency, extra fuel must be consumed to produce the same number of ATPs, generating additional heat. This response is moderated by the endocrine system, and researchers hypothesised that it might be triggered by thyroid hormone. In this exercise, you will use a bar graph to visualise data from an experiment that compared the metabolic rates (by measuring O_2 consumption) in mitochondria of cells from animals with different levels of thyroid hormone.

How the Experiment Was Done Liver cells were isolated from sibling rats that had low, normal, or elevated thyroid hormone levels. The oxygen consumption rate due to activity of the mitochondrial electron transport chains of each type of cell was measured under controlled conditions.

Data from the Experiment

Thyroid Hormone Level	Oxygen Consumption Rate [nmol O_2/(min · mg cells)]
Low	4.3
Normal	4.8
Elevated	8.7

Data from M. E. Harper and M. D. Brand, The quantitative contributions of mitochondrial proton leak and ATP turnover reactions to the changed respiration rates of hepatocytes from rats of different thyroid status, *Journal of Biological Chemistry* 268:14850–14860 (1993).

INTERPRET THE DATA

1. To visualise any differences in O_2 consumption between cell types, it will be useful to graph the data in a bar graph. First, set up the axes. **(a)** What is the independent variable (intentionally varied by the researchers), which goes on the *x*-axis? List the categories along the *x*-axis; because they are discrete rather than continuous, you can list them in any order. **(b)** What is the dependent variable (measured by the researchers), which goes on the *y*-axis? **(c)** What units (abbreviated) should go on the *y*-axis? Label the *y*-axis, including the units specified in the data table. Determine the range of values of the data that will need to go on the *y*-axis. What is the largest value? Draw evenly spaced tick marks and label them, starting with 0 at the bottom.
2. Graph the data for each sample. Match each *x*-value with its *y*-value and place a mark on the graph at that coordinate, then draw a bar from the *x*-axis up to the correct height for each sample. Why is a bar graph more appropriate than a scatter plot or line graph? (For additional information about graphs, see the Scientific Skills Review in Appendix D.)
3. Examine your graph and look for a pattern in the data. **(a)** Which cell type had the highest rate of O_2 consumption, and which had the lowest? **(b)** Does this support the researchers' hypothesis? Explain. **(c)** Based on what you know about mitochondrial electron transport and heat production, predict which rats had the highest, and which had the lowest, body temperature.

CONCEPT 9.5

Fermentation and anaerobic respiration enable cells to produce ATP without the use of oxygen

Because most of the ATP generated by cellular respiration is due to the work of oxidative phosphorylation, our estimate of ATP yield from aerobic respiration depends on an adequate supply of O_2 to the cell. Without the electronegative oxygen atoms in O_2 to pull electrons down the transport chain, oxidative phosphorylation eventually ceases. However, there are two general mechanisms by which certain cells can oxidise organic fuel and generate ATP *without* the use of O_2: anaerobic respiration and fermentation. The distinction between these two is that an electron transport chain is used in anaerobic respiration but not in fermentation. (The electron transport chain is also called the respiratory chain because of its role in both types of cellular respiration.)

We have already mentioned anaerobic respiration, which takes place in certain prokaryotic organisms that live in environments without O_2. These organisms have an electron transport chain but do not use O_2 as a final electron acceptor at the end of the chain. O_2 performs this function very well because it consists of two extremely electronegative atoms, but other substances can also serve as final electron acceptors. Some "sulfate-reducing" marine bacteria, for instance, use the sulfate ion (SO_4^{2-}) at the end of their respiratory chain. Operation of the chain builds up a proton-motive force used to produce ATP, but H_2S (hydrogen sulfide) is made as a by-product rather than water. The rotten-egg odour you may have smelled while walking through a salt marsh or a mudflat signals the presence of sulfate-reducing bacteria.

Fermentation is a way of harvesting chemical energy without using either O_2 or any electron transport chain—in other words, without cellular respiration. How can food be oxidised without cellular respiration? Remember, oxidation simply refers to the loss of electrons to an electron acceptor, so it does not need to involve O_2. Glycolysis oxidises glucose to two molecules of pyruvate. The oxidising agent of glycolysis is NAD^+, and neither O_2 nor any electron transfer chain is involved. Overall, glycolysis is exergonic, and some of the energy made available is used to produce 2 ATP (net) by substrate-level phosphorylation. If O_2 *is* present, then additional ATP is made by oxidative phosphorylation when NADH passes electrons removed from glucose to the electron transport chain. But glycolysis generates 2 ATP whether oxygen is present or not—that is, whether conditions are aerobic or anaerobic.

As an alternative to respiratory oxidation of organic nutrients, fermentation is an extension of glycolysis that allows continuous generation of ATP by the substrate-level phosphorylation of glycolysis. For this to occur, there must be a sufficient supply of NAD^+ to accept electrons during the oxidation step of glycolysis. Without some mechanism to recycle NAD^+ from NADH, glycolysis would soon deplete the cell's pool of NAD^+ by reducing it all to NADH and would shut itself down for lack of an oxidising agent. Under aerobic conditions, NAD^+ is recycled from NADH by the transfer of electrons to the electron transport chain. An anaerobic alternative is to transfer electrons from NADH to pyruvate, the end product of glycolysis.

Types of Fermentation

Fermentation consists of glycolysis plus reactions that regenerate NAD^+ by transferring electrons from NADH to pyruvate or derivatives of pyruvate. The NAD^+ can then be reused to oxidise sugar by glycolysis, which nets two molecules of ATP by substrate-level phosphorylation. There are many types of fermentation, differing in the end products formed from pyruvate. Two types are alcohol fermentation and lactic acid fermentation, and both are harnessed by humans for food and industrial production.

In **alcohol fermentation** (Figure 9.16a), pyruvate is converted to ethanol (ethyl alcohol) in two steps. The first step releases CO_2 from the pyruvate, which is converted to the two-carbon compound acetaldehyde. In the second step, acetaldehyde is reduced by NADH to ethanol. This regenerates the supply of NAD^+ needed for the continuation of glycolysis. Many bacteria carry out alcohol fermentation under anaerobic conditions. Yeast (a fungus), in addition to aerobic respiration, also carries out alcohol fermentation. For thousands of years, humans have used yeast in brewing, winemaking, and baking. The CO_2 bubbles generated by baker's yeast during alcohol fermentation allow bread to rise.

▼ **Figure 9.16 Fermentation.** In the absence of oxygen, many cells use fermentation to produce ATP by substrate-level phosphorylation. NAD^+ is regenerated for use in glycolysis when pyruvate, the end product of glycolysis, serves as an electron acceptor for oxidising NADH. Two of the common end products formed from fermentation are **(a)** ethanol and **(b)** lactate, the ionised form of lactic acid.

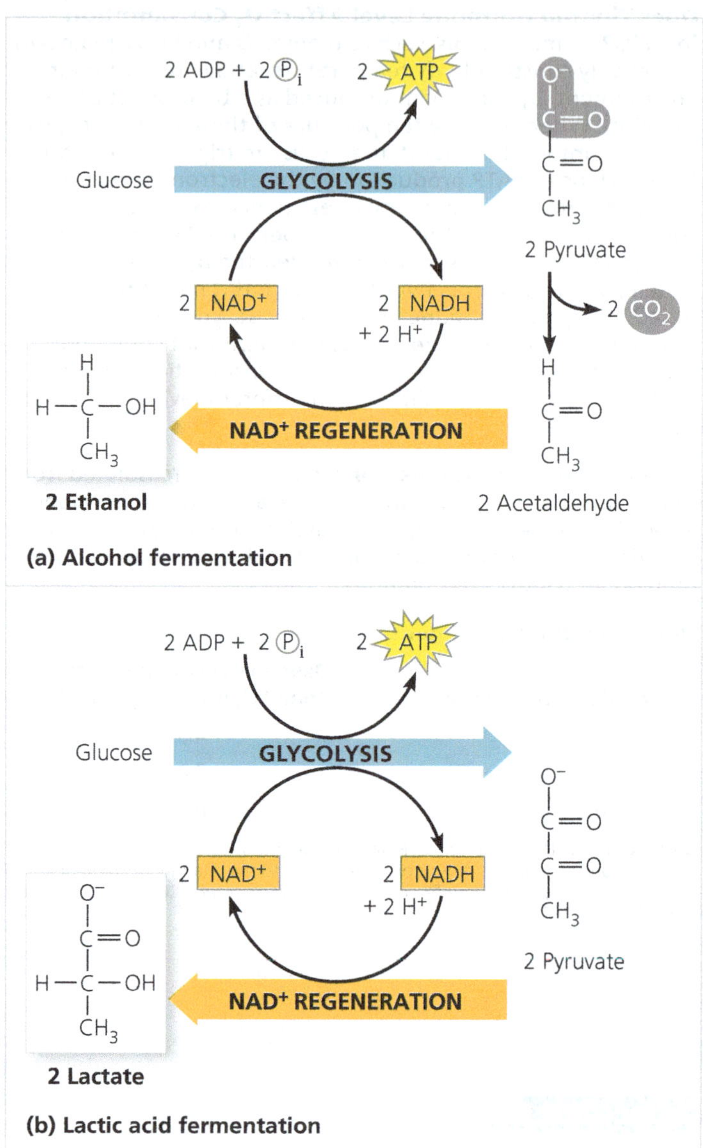

During **lactic acid fermentation** (Figure 9.16b), pyruvate is reduced directly by NADH to form lactate as an end product, regenerating NAD^+ with no release of CO_2. (Lactate is the ionised form of lactic acid.) Lactic acid fermentation by certain fungi and bacteria is used in the dairy industry to make cheese and yoghurt. A complex series of fermentation and aerobic respiration pathways carried out by yeasts and bacteria on cacao beans is responsible for the production of chocolate.

▼ **Cacao beans and fruits**

What about lactate production in humans? Previously, we thought that human muscle cells only produced lactate when O_2 was in short supply, such as during intense exercise. Research done over the last few decades, though, indicates that the lactate story, in mammals at least, is more complicated. There are two types of skeletal muscle fibres. One (red muscle) preferentially oxidises glucose completely to CO_2; the other (white muscle) produces significant amounts of lactate from the pyruvate made during glycolysis, even under aerobic conditions, offering fast but energetically inefficient ATP production. The lactate product is then mostly oxidised by red muscle cells in the vicinity, with the remainder exported to liver or kidney cells for glucose formation. Because this lactate production is not anaerobic, but the result of glycolysis in these cells, exercise physiologists prefer not to use the term *fermentation.*

During strenuous exercise, when carbohydrate catabolism outpaces the supply of O_2 from the blood to the muscle, lactate can't be oxidised to pyruvate. The lactate that accumulates was once thought to cause muscle fatigue during intense exercise and pain a day or so later. However, research suggests that, contrary to popular opinion, lactate production actually improves performance during exercise! Furthermore, within an hour, excess lactate is shuttled to other tissues for oxidation or to the liver and kidneys for production of glucose or its storage molecule, glycogen. (Nextday muscle soreness is more likely caused by trauma to cells in small muscle fibres, which leads to inflammation and pain.)

Comparing Fermentation with Anaerobic and Aerobic Respiration

Fermentation, anaerobic respiration, and aerobic respiration are three alternative cellular pathways for producing ATP by harvesting the chemical energy of food. All three use glycolysis to oxidise glucose and other organic fuels to pyruvate, with a net production of 2 ATP by substrate-level phosphorylation. And in all three pathways, NAD^+ is the oxidising agent that accepts electrons from food during glycolysis.

A key difference is the contrasting mechanisms for oxidising NADH back to NAD^+, which is required to sustain glycolysis. In fermentation, the final electron acceptor is an organic molecule such as pyruvate (lactic acid fermentation) or acetaldehyde (alcohol fermentation). In cellular respiration, by contrast, electrons carried by NADH are transferred to an electron transport chain, which regenerates the NAD^+ required for glycolysis.

Another major difference is the amount of ATP produced. Fermentation yields two molecules of ATP, produced by substrate-level phosphorylation. In the absence of an electron transport chain, the energy stored in pyruvate is unavailable. In cellular respiration, however, pyruvate is completely oxidised in the mitochondrion. Most of the chemical energy from this process is shuttled by NADH and $FADH_2$ in the form of electrons to the electron transport chain. There, the electrons move stepwise down a series of redox reactions to a final electron acceptor. (In aerobic respiration, the final electron acceptor is O_2; in anaerobic respiration, the final acceptor is another molecule with a high affinity for electrons, although less so than O_2.) Stepwise electron transport drives oxidative phosphorylation, yielding ATP. Thus, cellular respiration harvests much more energy from each sugar molecule than fermentation can. In fact, aerobic respiration yields up to 32 molecules of ATP per glucose molecule—up to 16 times as much as does fermentation.

Some organisms, called **obligate anaerobes**, carry out only fermentation or anaerobic respiration. In fact, these organisms cannot survive in the presence of oxygen, some forms of which can actually be toxic if protective systems are not present in the cell. A few cell types, such as cells of the vertebrate brain, can carry out only aerobic oxidation of pyruvate, and need O_2 to survive. Other organisms, including yeasts and many bacteria, can make enough ATP to survive using either fermentation or respiration. Such species are called **facultative anaerobes**. In yeast cells, for example, pyruvate is a fork in the metabolic road that leads to two alternative catabolic routes **(Figure 9.17)**. Under aerobic conditions, pyruvate can be converted to acetyl CoA, and oxidation continues in the citric acid cycle via aerobic respiration. Under anaerobic conditions, lactic acid fermentation occurs. Pyruvate is diverted from the citric acid cycle, serving instead

▼ Figure 9.17 Pyruvate as a key juncture in catabolism. Glycolysis is common to fermentation and cellular respiration. The end product of glycolysis, pyruvate, represents a fork in the catabolic pathways of glucose oxidation. In a facultative anaerobe, capable of both aerobic cellular respiration and fermentation, pyruvate is committed to one of those two pathways, usually depending on whether or not oxygen is present.

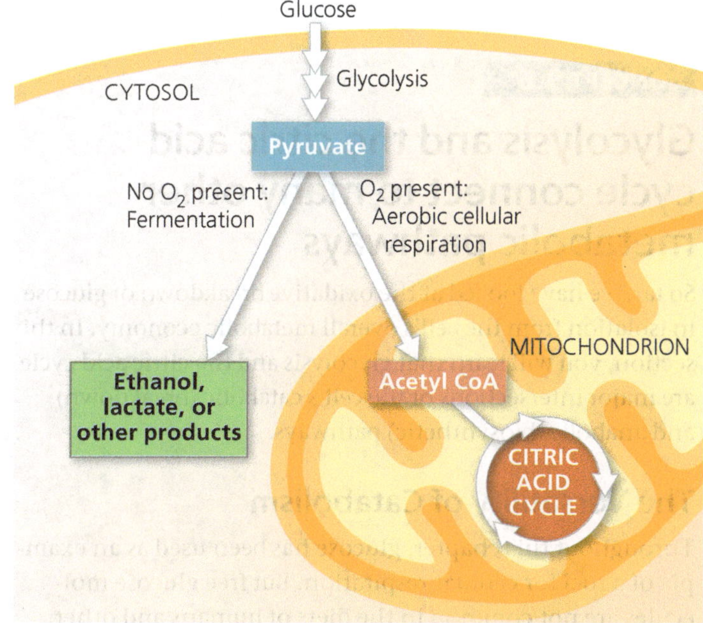

as an electron acceptor to recycle NAD^+. To make the same amount of ATP, a facultative anaerobe has to consume sugar at a much faster rate when fermenting than when respiring.

The Evolutionary Significance of Glycolysis

EVOLUTION The role of glycolysis in both fermentation and respiration has an evolutionary basis. Ancient prokaryotes are thought to have used glycolysis to make ATP long before oxygen was present in Earth's atmosphere. The oldest known fossils of bacteria date back 3.5 billion years, but appreciable quantities of oxygen probably did not begin to accumulate in the atmosphere until about 2.7 billion years ago. Cyanobacteria produced this O_2 as a by-product of photosynthesis. Therefore, early prokaryotes may have generated ATP exclusively from glycolysis. The fact that glycolysis is today the most widespread metabolic pathway among Earth's organisms suggests that it evolved very early in the history of life. The cytosolic location of glycolysis also implies great antiquity; the pathway does not require any of the membrane-enclosed organelles of the eukaryotic cell, which evolved approximately 1 billion years after the first prokaryotic cell. Glycolysis is a metabolic heirloom from early cells that continues to function in fermentation and as the first stage in the breakdown of organic molecules by respiration.

CONCEPT CHECK 9.5

1. Consider the NADH formed during glycolysis. What is the final acceptor for its electrons during fermentation? During aerobic respiration? During anaerobic respiration?
2. **WHAT IF?** A glucose-fed yeast cell is moved from an aerobic environment to an anaerobic one. How would its rate of glucose consumption change if ATP were to be generated at the same rate?

For suggested answers, see Appendix A.

CONCEPT 9.6

Glycolysis and the citric acid cycle connect to many other metabolic pathways

So far, we have looked at the oxidative breakdown of glucose in isolation from the cell's overall metabolic economy. In this section, you will learn that glycolysis and the citric acid cycle are major intersections of the cell's catabolic (breakdown) and anabolic (biosynthetic) pathways.

The Versatility of Catabolism

Throughout this chapter, glucose has been used as an example of a fuel for cellular respiration. But free glucose molecules are not common in the diets of humans and other animals. We obtain most of our calories in the form of fats, proteins, and carbohydrates such as sucrose and other disaccharides, and starch, a polysaccharide. All these organic molecules in food can be used by cellular respiration to make ATP (**Figure 9.18**).

▼ **Figure 9.18 The catabolism of various molecules from food.** Carbohydrates, fats, and proteins can all be used as fuel for cellular respiration. Monomers of these molecules enter glycolysis or the citric acid cycle at various points. Glycolysis and the citric acid cycle are catabolic funnels through which electrons from all kinds of organic molecules flow on their exergonic fall to oxygen.

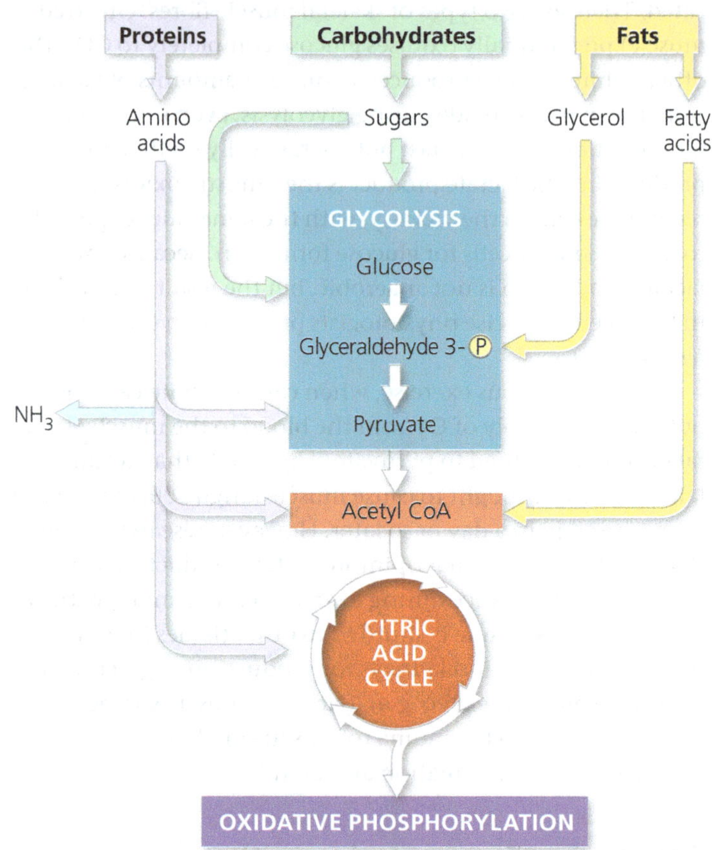

Glycolysis can accept a wide range of carbohydrates for catabolism. In the digestive tract, starch is hydrolysed to glucose, which is broken down in cells by glycolysis and the citric acid cycle. Glycogen, the polysaccharide that humans and many other animals store in their liver and muscle cells, can be hydrolysed to glucose between meals as fuel for respiration. Digestion of disaccharides, including sucrose, provides glucose and other monosaccharides as fuel for respiration.

Proteins can also be used for fuel, but first they must be digested to their constituent amino acids. Many of the amino acids are used by the organism to build new proteins. Amino acids present in excess are converted by enzymes to intermediates of glycolysis and the citric acid cycle. Before amino acids can feed into glycolysis or the citric acid cycle, their amino groups must be removed, a process called *deamination*. The nitrogenous waste is excreted from the animal in the form of ammonia (NH_3), urea, or other waste products.

Catabolism can also harvest energy stored in fats obtained either from food or from fat cells in the body. After fats are digested to glycerol and fatty acids, the glycerol is converted to glyceraldehyde 3-phosphate, an intermediate of glycolysis. Most of the energy of a fat is stored in the fatty acids. A metabolic sequence called **beta oxidation** breaks the fatty acids down to two-carbon fragments, which enter the citric acid cycle as acetyl CoA. NADH and $FADH_2$ are also generated during beta oxidation; they can enter the electron transport chain, leading to further ATP production. Fats make excellent fuels, in large part due to their chemical structure and the high energy level of their electrons (present in many C—H bonds, equally shared between C and H) compared to those of carbohydrates. A gram of fat oxidised by respiration produces more than twice as much ATP as a gram of carbohydrate. Unfortunately, this also means that a person trying to lose weight must work hard to use up fat stored in the body because so many calories are stockpiled in each gram of fat.

Biosynthesis (Anabolic Pathways)

Cells need substance as well as energy. Not all the organic molecules of food are destined to be oxidised as fuel to make ATP. In addition to calories, food must also provide the carbon skeletons that cells require to make their own molecules. Some organic monomers obtained from digestion can be used directly. For example, as previously mentioned, amino acids from the hydrolysis of proteins in food can be incorporated into the organism's own proteins. Often, however, the body needs specific molecules that are not present as such in food. Compounds formed as intermediates of glycolysis and the citric acid cycle can be diverted into anabolic pathways as precursors from which the cell can synthesise the molecules it requires. For example, humans can make about half of the 20 amino acids in proteins by modifying compounds siphoned away from the citric acid cycle; the rest are "essential amino acids" that must be obtained in the diet. Also, glucose can be made from pyruvate, and fatty acids can be synthesised from acetyl CoA. Of course, these anabolic, or biosynthetic, pathways do not generate ATP, but instead consume it.

In addition, glycolysis and the citric acid cycle function as metabolic interchanges that enable our cells to convert some kinds of molecules to others as we need them. For example, an intermediate compound generated during glycolysis, dihydroxyacetone phosphate (see Figure 9.8, step 5), can be converted to one of the major precursors of fats. If we eat more food than we need, we store fat even if our diet is fat-free. Metabolism is remarkably versatile and adaptable.

Regulation of Cellular Respiration via Feedback Mechanisms

Basic principles of supply and demand regulate the metabolic economy. The cell does not waste energy making more of a particular substance than it needs. If there is a surplus of a certain amino acid, for example, the anabolic pathway that synthesises that amino acid from an intermediate of the citric acid cycle is switched off. The most common mechanism for this control is feedback inhibition: The end product of the anabolic pathway inhibits the enzyme that catalyses an early step of the pathway (see Figure 8.21). This prevents the needless diversion of key metabolic intermediates from uses that are more urgent.

The cell also controls its catabolism. If the cell is working hard and its ATP concentration begins to drop, cellular respiration speeds up. When there is plenty of ATP to meet demand, respiration slows down, sparing valuable organic molecules for other functions. Again, control is based mainly on regulating the activity of enzymes at strategic points in the catabolic pathway. As shown in **Figure 9.19**, one important switch is phosphofructokinase, the enzyme that catalyses step 3 of glycolysis (see Figure 9.8). That is the first step that

▼ Figure 9.19 The control of cellular respiration. Allosteric enzymes at certain points in the respiratory pathway respond to inhibitors and activators that help set the pace of glycolysis and the citric acid cycle. Phosphofructokinase, which catalyses an early step in glycolysis (see Figure 9.8, step 3 and Figure 6.32b), is one such enzyme. It is stimulated by AMP (derived from ADP) but is inhibited by ATP and by citrate. This feedback regulation adjusts the rate of respiration as the cell's catabolic and anabolic demands change.

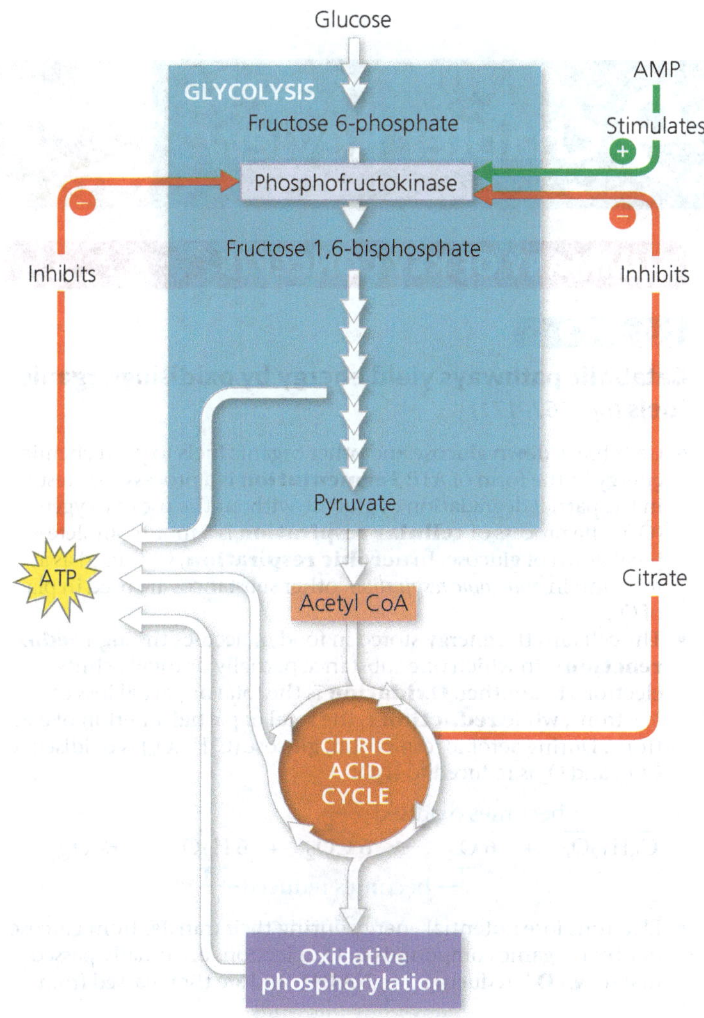

commits the substrate irreversibly to the glycolytic pathway. By controlling the rate of this step, the cell can speed up or slow down the entire catabolic process. Phosphofructokinase can thus be considered the pacemaker of cellular respiration.

Phosphofructokinase is an allosteric enzyme with receptor sites for specific inhibitors and activators. It is inhibited by ATP and stimulated by AMP (adenosine monophosphate), which the cell derives from ADP. As ATP accumulates, inhibition of the enzyme slows down glycolysis. The enzyme becomes active again as cellular work converts ATP to ADP (and AMP) faster than ATP is being regenerated. Phosphofructokinase is also sensitive to citrate, the first product of the citric acid cycle. If citrate accumulates in mitochondria, some of it passes into the cytosol and inhibits phosphofructokinase. This mechanism helps synchronise the rates of glycolysis and the citric acid cycle. As citrate accumulates, glycolysis slows down, and the supply of pyruvate and thus acetyl groups to the citric acid cycle decreases. If citrate consumption increases, either because of a demand for more ATP or because anabolic pathways are draining off intermediates of the citric acid cycle, glycolysis accelerates and meets the demand. Metabolic balance is augmented by the control of enzymes that catalyse other key steps of glycolysis and the citric acid cycle. Cells are thrifty, expedient, and responsive in their metabolism.

Review the first page of this chapter to put cellular respiration into the broader context of energy flow and chemical cycling in ecosystems. The energy that keeps us alive is *released*, not *produced*, by cellular respiration. We are tapping energy that was stored in food by photosynthesis, which captures light and converts it to chemical energy, a process you will learn about next, in the chapter, "Photosynthesis."

CONCEPT CHECK 9.6

1. **MAKE CONNECTIONS** Compare the structures of a carbohydrate and a fat (see Figures 5.3 and 5.9). What features make fat a much better fuel?
2. Under what circumstances might your body synthesise fat molecules?
3. **VISUAL SKILLS** What will happen in a muscle cell that has used up its supply of O_2 and ATP? (Review Figures 9.17 and 9.19.)
4. **VISUAL SKILLS** During intense exercise, can a muscle cell use fat as a concentrated source of chemical energy? Explain. (Review Figures 9.17 and 9.18.)

For suggested answers, see Appendix A.

9 Chapter Review

SUMMARY OF KEY CONCEPTS

CONCEPT 9.1

Catabolic pathways yield energy by oxidising organic fuels *(pp. 167–171)*

- Cells break down glucose and other organic fuels to yield chemical energy in the form of ATP. **Fermentation** is a process that results in the partial degradation of glucose without the use of oxygen (O_2). The process of **cellular respiration** is a more complete breakdown of glucose. In **aerobic respiration**, O_2 is used as a reactant; in *anaerobic respiration*, other substances are used in place of O_2.
- The cell taps the energy stored in food molecules through **redox reactions**, in which one substance partially or totally shifts electrons to another. **Oxidation** is the total or partial loss of electrons, while **reduction** is the total or partial addition of electrons. During aerobic respiration, glucose ($C_6H_{12}O_6$) is oxidised to CO_2, and O_2 is reduced to H_2O:

$$\overbrace{C_6H_{12}O_6 \; + \; 6\,O_2 \longrightarrow 6\,CO_2}^{\text{becomes oxidised}} \; + \; \underbrace{6\,H_2O}_{\text{becomes reduced}} \; + \; \text{Energy}$$

- Electrons lose potential energy during their transfer from glucose or other organic compounds to O_2. Electrons are usually passed first to **NAD**$^+$, reducing it to **NADH**, and are then passed from NADH to an **electron transport chain**, which conducts the electrons to O_2 in energy-releasing steps. The energy that is released is used to make ATP.
- Aerobic respiration occurs in three stages: (1) **glycolysis**, (2) pyruvate oxidation and the **citric acid cycle**, and (3) **oxidative phosphorylation** (electron transport and chemiosmosis).

? *Describe the difference between the two processes in cellular respiration that produce ATP: oxidative phosphorylation and substrate-level phosphorylation.*

CONCEPT 9.2

Glycolysis harvests chemical energy by oxidising glucose to pyruvate *(pp. 172–173)*

- Glycolysis ("splitting of sugar") is a series of reactions that breaks down glucose into two pyruvate molecules, which may go on to enter the citric acid cycle, and nets 2 ATP and 2 NADH per glucose molecule.

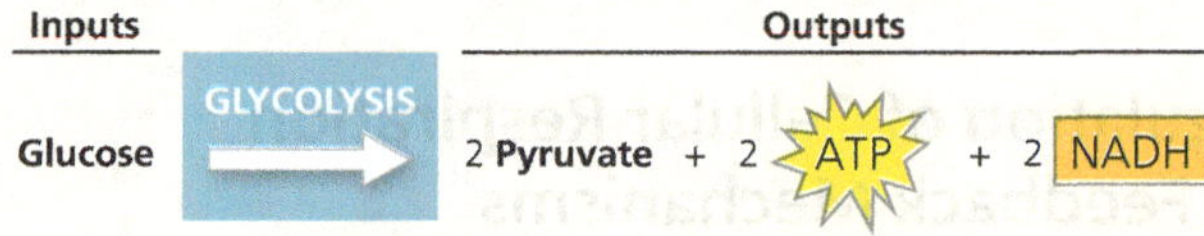

? *Which reactions in glycolysis are the source of energy for the formation of ATP and NADH?*

CONCEPT 9.3

After pyruvate is oxidised, the citric acid cycle completes the energy-yielding oxidation of organic molecules *(pp. 173–176)*

- In eukaryotic cells, pyruvate enters the mitochondrion and is oxidised to **acetyl CoA**, which is further oxidised in the citric acid cycle.

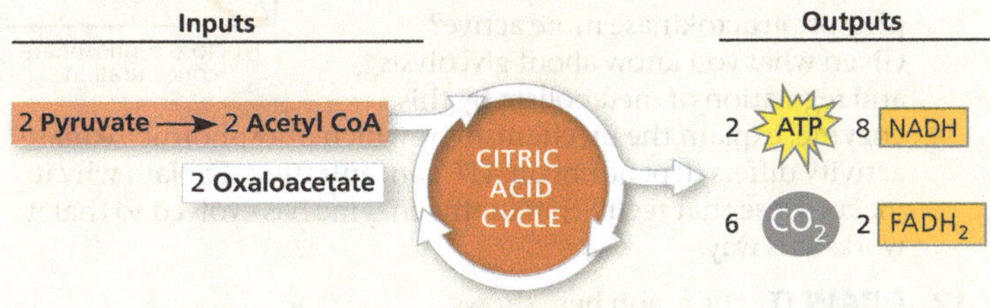

? *What molecular products indicate the complete oxidation of glucose during cellular respiration?*

CONCEPT 9.4

During oxidative phosphorylation, chemiosmosis couples electron transport to ATP synthesis *(pp. 176–181)*

- NADH and $FADH_2$ transfer electrons to the electron transport chain. Electrons move down the chain, losing energy in several energy-releasing steps. Finally, electrons are passed to O_2, reducing it to H_2O.

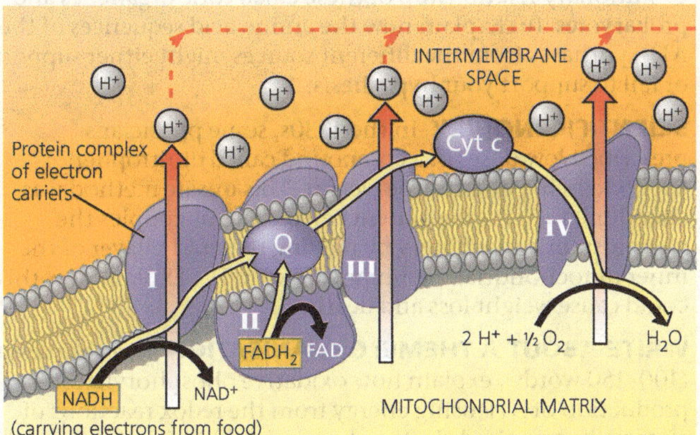

- Along the electron transport chain, electron transfer causes protein complexes to move H^+ from the mitochondrial matrix (in eukaryotes) to the intermembrane space, storing energy as a **proton-motive force** (H^+ gradient). As H^+ diffuses back into the matrix through **ATP synthase**, its passage drives the phosphorylation of ADP to form ATP, called **chemiosmosis**.
- About 34% of the energy stored in a glucose molecule is transferred to ATP during cellular respiration, producing a maximum of about 32 ATP.

? *Briefly explain the mechanism by which ATP synthase produces ATP. List three locations in which ATP synthases are found.*

CONCEPT 9.5

Fermentation and anaerobic respiration enable cells to produce ATP without the use of oxygen *(pp. 181–184)*

- Glycolysis nets 2 ATP by substrate-level phosphorylation, whether O_2 is present or not. Under anaerobic conditions, anaerobic respiration or fermentation can take place. In anaerobic respiration, an electron transport chain is present with a final electron acceptor other than oxygen. In fermentation, the electrons from NADH are passed to pyruvate or a derivative of pyruvate, regenerating the NAD^+ required to oxidise more glucose. Two common types of fermentation are **alcohol fermentation** and **lactic acid fermentation**.
- Fermentation, anaerobic respiration, and aerobic respiration all use glycolysis to oxidise glucose, but they differ in their final electron acceptor and whether an electron transport chain is used (respiration) or not (fermentation). Respiration yields more ATP; aerobic respiration, with O_2 as the final electron acceptor, yields about 16 times as much ATP as does fermentation.
- Glycolysis occurs in nearly all organisms and is thought to have evolved in ancient prokaryotes before there was O_2 in the atmosphere.

? *Which process yields more ATP, fermentation or anaerobic respiration? Explain.*

CONCEPT 9.6

Glycolysis and the citric acid cycle connect to many other metabolic pathways *(pp. 184–186)*

- Catabolic pathways funnel electrons from many kinds of organic molecules into cellular respiration. Many carbohydrates can enter glycolysis, most often after conversion to glucose. Amino acids of proteins must be deaminated before being oxidised. The fatty acids of fats undergo **beta oxidation** to two-carbon fragments and then enter the citric acid cycle as acetyl CoA. Anabolic pathways can use small molecules from food directly or build other substances using intermediates of glycolysis or the citric acid cycle.
- Cellular respiration is controlled by allosteric enzymes at key points in glycolysis and the citric acid cycle.

? *Describe how the catabolic pathways of glycolysis and the citric acid cycle intersect with anabolic pathways in the metabolism of a cell.*

TEST YOUR UNDERSTANDING

Levels 1-2: Remembering/Understanding

1. The *immediate* energy source that drives ATP synthesis by ATP synthase during oxidative phosphorylation is the
   (A) oxidation of glucose and other organic compounds.
   (B) flow of electrons down the electron transport chain.
   (C) H^+ concentration gradient across the membrane holding ATP synthase.
   (D) transfer of phosphate to ADP.
2. Which metabolic pathway is common to both fermentation and cellular respiration of a glucose molecule?
   (A) the citric acid cycle
   (B) the electron transport chain
   (C) glycolysis
   (D) reduction of pyruvate to lactate

3. The final electron acceptor of the electron transport chain that functions in aerobic oxidative phosphorylation is
(A) O_2.
(B) water.
(C) NAD^+.
(D) pyruvate.

4. In mitochondria, exergonic redox reactions
(A) are the source of energy driving prokaryotic ATP synthesis.
(B) provide the energy that establishes the proton gradient.
(C) reduce carbon atoms to carbon dioxide.
(D) are coupled via phosphorylated intermediates to endergonic processes.

Levels 3-4: Applying/Analysing

5. What is the oxidising agent in the following reaction?

$$\text{Pyruvate} + \text{NADH} + H^+ \rightarrow \text{Lactate} + NAD^+$$

(A) oxygen
(B) NADH
(C) lactate
(D) pyruvate

6. When electrons flow along the electron transport chains of mitochondria, which of the following changes occurs?
(A) The pH of the matrix increases.
(B) ATP synthase pumps protons by active transport.
(C) The electrons gain free energy.
(D) NAD^+ is oxidised.

7. Most CO_2 from catabolism is released during
(A) glycolysis.
(B) the citric acid cycle.
(C) lactate fermentation.
(D) electron transport.

8. **MAKE CONNECTIONS** Step 3 in Figure 9.8 is a major point of regulation of glycolysis. The enzyme phosphofructokinase is allosterically regulated by ATP and related molecules (see Concept 8.5). Considering the overall result of glycolysis, would you expect ATP to inhibit or stimulate activity of this enzyme? Explain. (Hint: Make sure you consider the role of ATP as an allosteric regulator, not as a substrate of the enzyme.)

9. **MAKE CONNECTIONS** The proton pump shown in Figures 7.18 and 7.19 is a type of ATP synthase like that in Figure 9.13. Compare the processes shown in the three figures, and say whether they are involved in active or passive transport (see also Concepts 7.3 and 7.4).

10. **VISUAL SKILLS** This computer model shows the four parts of ATP synthase, each part consisting of a number of polypeptide subunits (the solid grey part is still an area of active research). Using Figure 9.13 as a guide, label the rotor, stator, internal rod, and catalytic knob of this molecular motor.

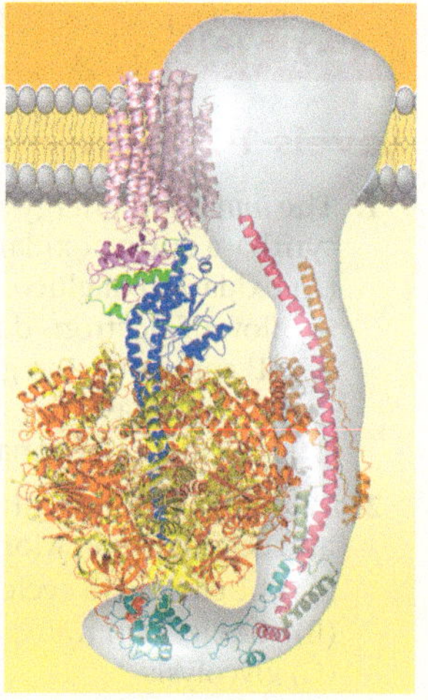

Levels 5-6: Evaluating/Creating

11. **INTERPRET THE DATA** Phosphofructokinase is an enzyme that acts on fructose 6-phosphate at an early step in glucose breakdown. Regulation of this enzyme controls whether the sugar will continue on in the glycolytic pathway. Considering this graph, under which condition is phosphofructokinase more active? Given what you know about glycolysis and regulation of metabolism by this enzyme, explain the mechanism by which phosphofructokinase activity differs depending on ATP concentration. Explain why it makes sense that regulation of this enzyme has evolved so that it works this way.

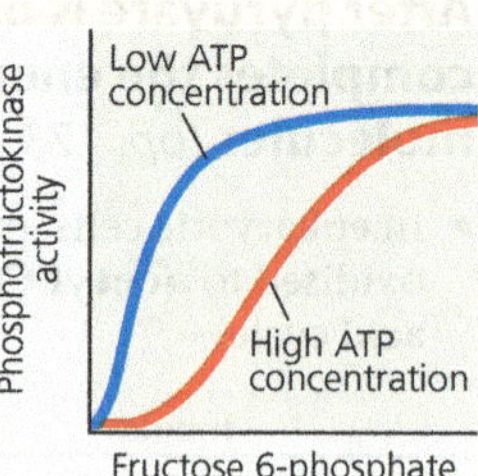

12. **DRAW IT** The graph here shows the pH difference across the inner mitochondrial membrane over time in an actively respiring cell. At the time indicated by the vertical arrow, a metabolic poison is added that specifically and completely inhibits all function of mitochondrial ATP synthase. Draw what you would expect to see for the rest of the graphed line over the next short period of time. Explain.

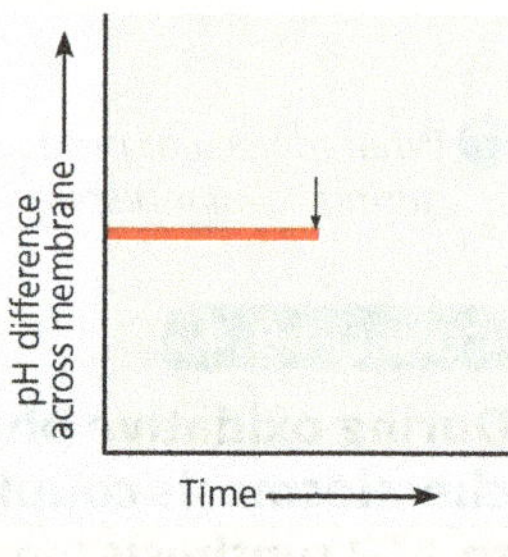

13. **EVOLUTION CONNECTION** ATP synthases are found in the prokaryotic plasma membrane and in mitochondria and chloroplasts. (a) Propose a hypothesis to account for an evolutionary relationship of these eukaryotic organelles and prokaryotes. (b) Explain how the amino acid sequences of the ATP synthases from the different sources might either support or fail to support your hypothesis.

14. **SCIENTIFIC INQUIRY** In the 1930s, some physicians prescribed low doses of a compound called dinitrophenol (DNP) to help patients lose weight. This unsafe method was abandoned after some patients died. DNP uncouples the chemiosmotic machinery by making the lipid bilayer of the inner mitochondrial membrane leaky to H^+. Explain how this could cause weight loss and death.

15. **WRITE ABOUT A THEME: ORGANISATION** In a short essay (100–150 words), explain how oxidative phosphorylation—production of ATP using energy from the redox reactions of a spatially organised electron transport chain followed by chemiosmosis—is an example of how new properties emerge at each level of the biological hierarchy.

16. **SYNTHESISE YOUR KNOWLEDGE**

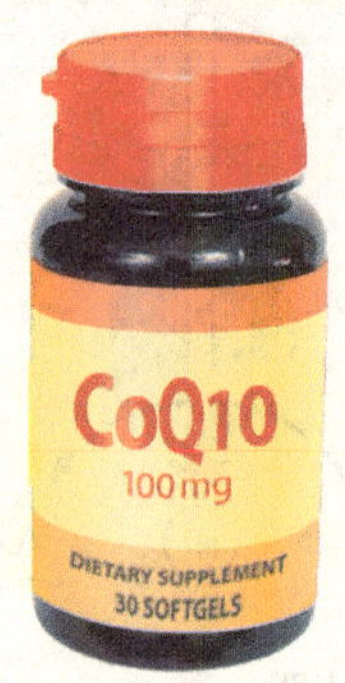

Coenzyme Q (CoQ) is sold as a nutritional supplement. One company uses this marketing slogan for CoQ: "Give your heart the fuel it craves most." Considering the role of coenzyme Q, critique this claim. How do you think this product might function to benefit the heart? Is CoQ used as a "fuel" during cellular respiration?

For selected answers, see Appendix A.

10 Photosynthesis

KEY CONCEPTS

Study Tip

Make a visual study guide: Draw a cell with a large chloroplast, labelling the parts of the chloroplast. As you go through the chapter, add key reactions for each stage of photosynthesis, linking the stages together. Label the carbon molecule(s) with the most energy and the carbon molecule(s) with the least energy.

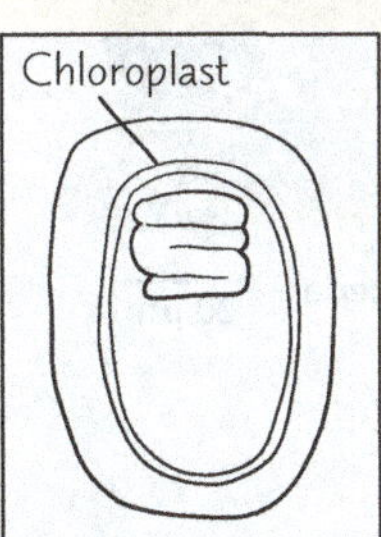

Go to Mastering Biology

to access Dynamic Study Modules for revision, 3D BioFlix® animations and high-quality videos, and your interactive Pearson eText.

Figure 10.1 Each leaf of this tree is harvesting energy from sunlight and using it to convert CO_2 and H_2O into chemical energy stored in sugar and other organic molecules. The tree uses these sugars for energy and to build its own trunk, branches, and leaves. Remarkably, enough sugars are left over to feed other organisms (like moth larvae, shown below) that cannot carry out this extraordinary transformation.

How do photosynthetic cells use light to change carbon dioxide and water into organic molecules and oxygen?

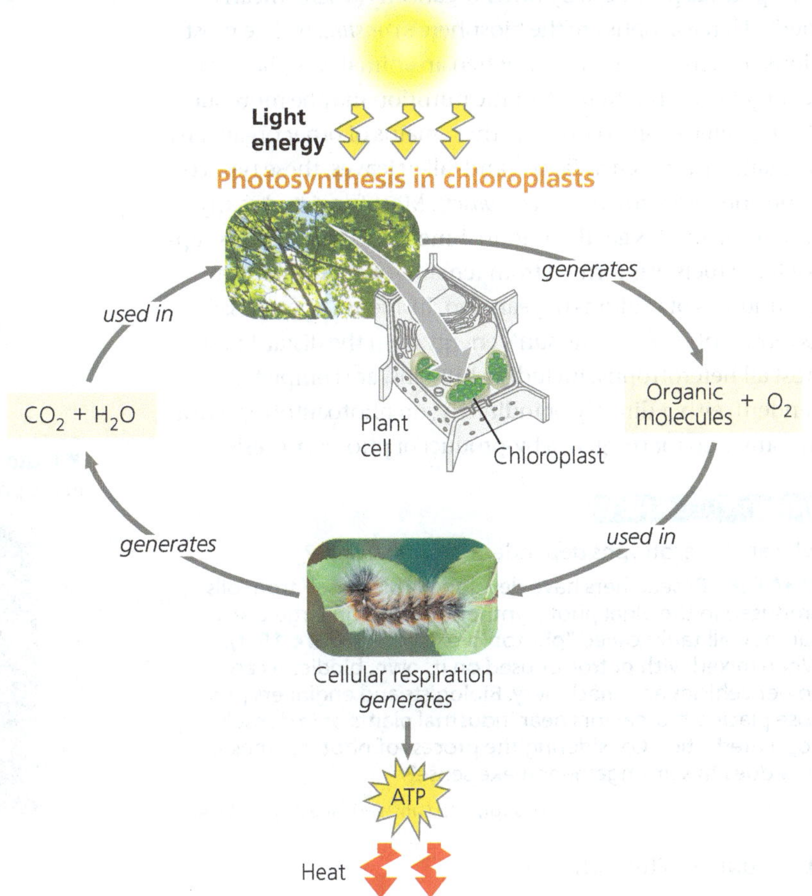

CONCEPT 10.1

Photosynthesis feeds the biosphere

The conversion process that transforms the energy of sunlight into chemical energy stored in sugars and other organic molecules is called **photosynthesis**. Let's begin by placing photosynthesis in its ecological context.

Photosynthesis nourishes almost the entire living world directly or indirectly. An organism acquires the organic compounds it uses for energy and carbon skeletons by one of two major modes: autotrophic nutrition or heterotrophic nutrition. **Autotrophs** are "self-feeders" (*auto-* means "self," and *trophos* means "feeder"); they sustain themselves without eating anything derived from other living beings. Autotrophs produce their organic molecules from CO_2 and other inorganic raw materials obtained from the environment. They are the ultimate sources of organic compounds for all nonautotrophic organisms, and for this reason, biologists refer to autotrophs as the *producers* of the biosphere.

Almost all plants are autotrophs; the only nutrients they require are water and minerals from the soil and CO_2 from the air. Specifically, plants are *photo*autotrophs, organisms that use light as a source of energy to synthesise organic substances **(Figure 10.1)**. Photosynthesis also occurs in algae, certain other protists, and some prokaryotes **(Figure 10.2)**. In this chapter, we will touch on these other groups in passing, but our emphasis will be on plants. Variations in autotrophic nutrition that occur in prokaryotes and algae will be described in Concept 27.3.

Heterotrophs are unable to make their own food; they live on compounds produced by other organisms (*hetero-* means "other"). Heterotrophs are the biosphere's *consumers*. The most obvious "other-feeding" occurs when an animal eats plants or other organisms, but heterotrophic nutrition may be more subtle. Some heterotrophs consume the remains of other organisms and organic litter such as faeces, and fallen leaves; these types of heterotrophs are known as *decomposers*. Most fungi and many types of prokaryotes get their nourishment this way. Earth's supply of fossil fuels was formed from remains of organisms that died hundreds of millions of years ago. In a sense, then, fossil fuels represent stores of the sun's energy from the distant past. Almost all heterotrophs, including humans, are completely dependent, either directly or indirectly, on photoautotrophs for food—and also for oxygen, a by-product of photosynthesis.

CONCEPT CHECK 10.1

1. Why are heterotrophs dependent on autotrophs?
2. **WHAT IF?** Researchers have developed "biodiesel" from oils produced in the algal photosynthetic process, in large transparent wall tanks called "photobioreactors" **(Figure 10.3)**. When mixed with petrol, or used on its own, biodiesel can power vehicles and machinery. Biologists and engineers propose placing bioreactors near industrial plants or in densely populated cities. Considering the process of photosynthesis, how does this arrangement make sense?

For suggested answers, see Appendix A.

▼ **Figure 10.2 Photoautotrophs.** These organisms use light energy to drive the synthesis of organic molecules from carbon dioxide and (in most cases) water. They feed themselves and the entire living world. On land, **(a)** plants are the predominant producers of food. In aquatic environments, photoautotrophs include unicellular and **(b)** multicellular algae, such as this kelp; **(c)** some non-algal unicellular protists, such as *Euglena*; **(d)** the prokaryotes called cyanobacteria; and **(e)** other photosynthetic prokaryotes, such as these purple sulfur bacteria, which produce sulfur (the yellow globules within the cells) (c–e, LMs).

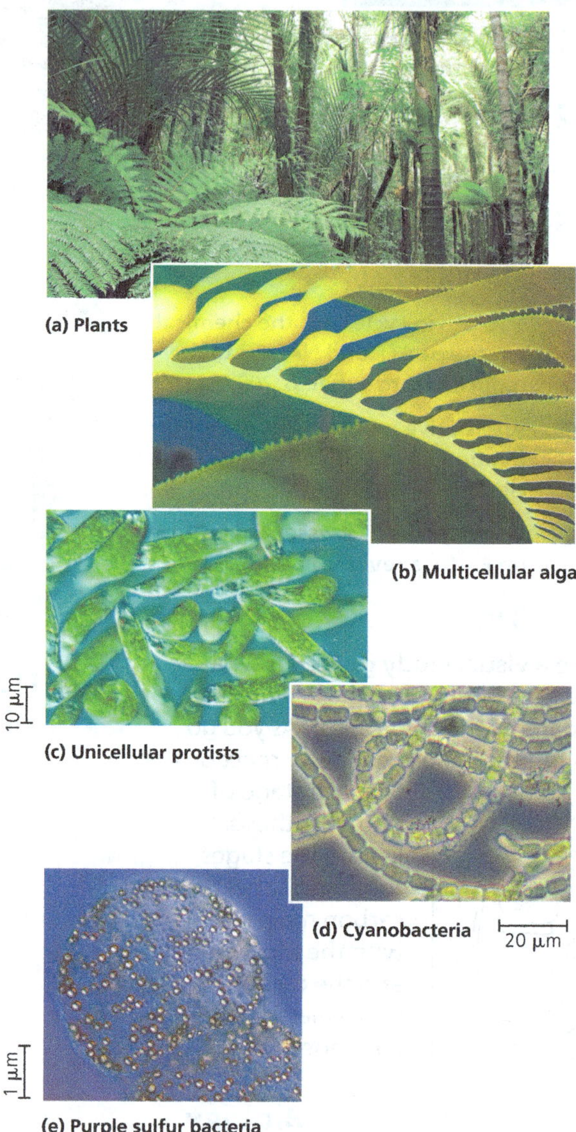

▼ **Figure 10.3 Photobioreactors.** These tanks use the oils produced in the photosynthetic products of algae to produce "biodiesel."

CONCEPT 10.2

Photosynthesis converts light energy to the chemical energy of food

The remarkable ability of an organism to harness light energy and use it to drive the synthesis of organic compounds emerges from structural organisation in the cell: Photosynthetic enzymes and other molecules are grouped together as specialised molecular complexes in a biological membrane, enabling the necessary series of chemical reactions to be carried out efficiently. The process of photosynthesis most likely originated in a group of bacteria that had infolded regions of the plasma membrane containing clusters of such molecules. In existing photosynthetic bacteria, infolded photosynthetic membranes function similarly to the internal membranes of the **chloroplast**, the eukaryotic organelle that absorbs energy from sunlight and uses it to drive the synthesis of organic compounds from carbon dioxide (CO_2) and water (H_2O). According to what has come to be known as the *serial endosymbiont theory*, the original chloroplast was a photosynthetic prokaryote that lived inside an ancestor of eukaryotic cells. (You learned about this theory in Concept 6.5, and it will be described more fully in Concept 25.3.) Chloroplasts are present in a variety of photosynthesising organisms (see some examples in Figure 10.2); in this chapter we focus on chloroplasts in plants, while mentioning prokaryotes from time to time.

Chloroplasts: The Sites of Photosynthesis in Plants

All green parts of a plant, including green stems and unripened fruit, have chloroplasts, but the leaves are the major sites of photosynthesis in most plants **(Figure 10.4)**. There are about half a million chloroplasts in a chunk of leaf with a top surface area of 1 mm^2. Chloroplasts are found mainly in the cells of the **mesophyll**, the tissue in the interior of the leaf. CO_2 enters the leaf, and O_2 exits, by way of microscopic pores called **stomata** (singular, *stoma*; from the Greek, meaning "mouth"). Water absorbed by the roots is delivered to the leaves in veins. Leaves also use veins to export sugar to roots and other nonphotosynthetic parts of the plant.

A typical mesophyll cell has about 30–40 chloroplasts, each measuring about 2–4 μm by 4–7 μm. A chloroplast has two membranes surrounding a dense fluid called the **stroma**. Suspended within the stroma is a third membrane system, made up of sacs called **thylakoids**, which segregates the stroma from the *thylakoid space* inside these sacs. In some places, thylakoid sacs are stacked in columns called *grana* (singular, *granum*). **Chlorophyll**, the green pigment that gives leaves their colour, resides in the thylakoid membranes of the chloroplast. (The internal photosynthetic membranes of some prokaryotes are also called thylakoid membranes; see Figure 27.8b.) It is the light energy absorbed by chlorophyll

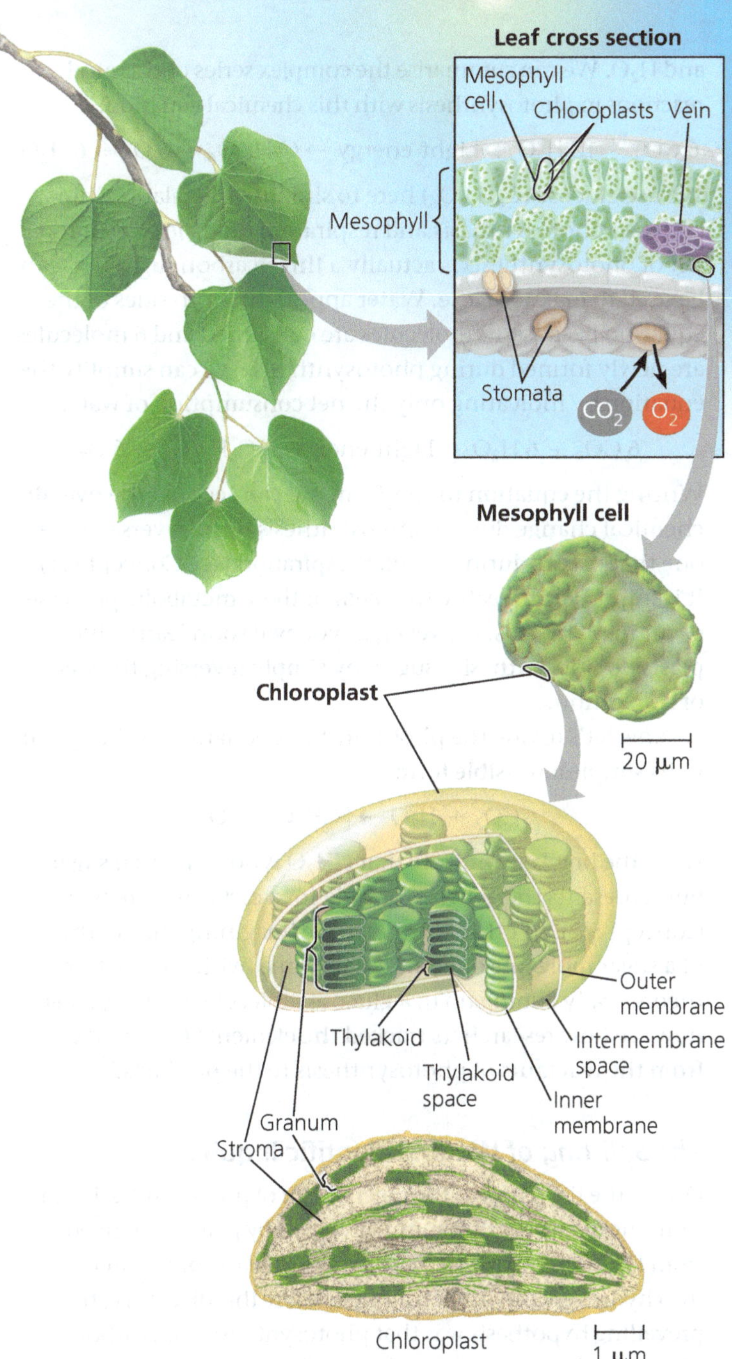

▲ **Figure 10.4 Zooming in on the location of photosynthesis in a plant.** Leaves are the major organs of photosynthesis in plants. These images take you into a leaf, then into a cell, and finally into a chloroplast, the organelle where photosynthesis occurs (middle, LM; bottom, TEM).

that drives the synthesis of organic molecules in the chloroplast. Now that we have looked at the sites of photosynthesis in plants, we are ready to look more closely at the process itself.

Tracking Atoms Through Photosynthesis

Scientists have tried for centuries to piece together the process by which plants make food. Although some of the steps are still not completely understood, the overall photosynthetic equation has been known since the 1800s: In the presence of light, the green parts of plants produce organic compounds and O_2 from CO_2

and H_2O. We can summarise the complex series of chemical reactions in photosynthesis with this chemical equation:

$$6\,CO_2 + 12\,H_2O + \text{Light energy} \rightarrow C_6H_{12}O_6 + 6\,O_2 + 6\,H_2O$$

We use glucose ($C_6H_{12}O_6$) here to simplify the relationship between photosynthesis and respiration, but the direct product of photosynthesis is actually a three-carbon sugar that can be used to make glucose. Water appears on both sides of the equation because 12 molecules are consumed and 6 molecules are newly formed during photosynthesis. We can simplify the equation by indicating only the net consumption of water:

$$6\,CO_2 + 6\,H_2O + \text{Light energy} \rightarrow C_6H_{12}O_6 + 6\,O_2$$

Writing the equation in this form, we can see that the overall chemical change during photosynthesis is the reverse of the one that occurs during cellular respiration (see Concept 9.1). It is important to realise that *both* of these metabolic processes occur in plant cells. However, as you will soon learn, chloroplasts do not synthesise sugars by simply reversing the steps of respiration.

Now let's divide the photosynthetic equation by 6 to put it in its simplest possible form:

$$CO_2 + H_2O \rightarrow [CH_2O] + O_2$$

Here, the brackets indicate that CH_2O is not an actual sugar but represents the general formula for a carbohydrate (see Concept 5.2). In other words, we are imagining the synthesis of a sugar molecule one carbon at a time (with six repetitions theoretically adding up to a glucose molecule: $C_6H_{12}O_6$). Let's now see how researchers tracked the elements C, H, and O from the reactants of photosynthesis to the products.

The Splitting of Water: Scientific Inquiry

One of the first clues to the mechanism of photosynthesis came from the discovery that the O_2 given off by plants is derived from H_2O and not from CO_2. The chloroplast splits water into hydrogen and oxygen atoms. Before this discovery, the prevailing hypothesis was that photosynthesis split carbon dioxide ($CO_2 \rightarrow C + O_2$) and then added water to the carbon ($C + H_2O \rightarrow [CH_2O]$). That hypothesis predicted that the O_2 released during photosynthesis came from CO_2, an idea challenged in the 1930s by C. B. van Niel, of Stanford University. Van Niel was investigating photosynthesis in bacteria that make their carbohydrate from CO_2 but do not release O_2. He concluded that, at least in these bacteria, CO_2 is not split into carbon and oxygen. One group of bacteria used hydrogen sulfide (H_2S) rather than water for photosynthesis, forming yellow globules of sulfur as a waste product (these globules are visible in **Figure 10.2e**). Here is the chemical equation for photosynthesis in these sulfur bacteria:

$$CO_2 + 2\,H_2S \rightarrow [CH_2O] + H_2O + 2\,S$$

Van Niel reasoned that the bacteria split H_2S and used the hydrogen atoms to make sugar. He then generalised that idea, proposing that all photosynthetic organisms require a hydrogen source but that the source varies:

$$\text{Sulfur bacteria: } CO_2 + 2\,H_2S \rightarrow [CH_2O] + H_2O + 2\,S$$
$$\text{Plants: } CO_2 + 2\,H_2O \rightarrow [CH_2O] + H_2O + O_2$$
$$\text{General: } CO_2 + 2\,H_2X \rightarrow [CH_2O] + H_2O + 2\,X$$

Thus, van Niel hypothesised that plants split H_2O as a source of electrons from hydrogen atoms, releasing O_2 as a by-product.

Nearly 20 years later, scientists confirmed van Niel's hypothesis by using oxygen-18 (^{18}O), a heavy isotope, as a tracer to follow the path of oxygen atoms during photosynthesis. The experiments showed that the O_2 produced by plants was labelled with ^{18}O *only* if water was the source of the tracer (experiment 1). If the ^{18}O was introduced to the plant in the form of CO_2, the label did not turn up in the released O_2 (experiment 2). In the following summary, green denotes labelled atoms of oxygen (^{18}O):

$$\text{Experiment 1: } CO_2 + 2\,H_2\mathbf{O} \rightarrow [CH_2O] + H_2O + \mathbf{O_2}$$
$$\text{Experiment 2: } C\mathbf{O_2} + 2\,H_2O \rightarrow [CH_2\mathbf{O}] + H_2\mathbf{O} + O_2$$

A significant result of the shuffling of atoms during photosynthesis is the extraction of hydrogen from water and its incorporation into sugar. The waste product of photosynthesis, O_2, is released to the atmosphere. **Figure 10.5** shows the fates of all atoms in photosynthesis.

Photosynthesis as a Redox Process

Let's briefly compare photosynthesis with cellular respiration. Both processes involve redox reactions. During cellular respiration, energy is released from sugar when electrons associated with hydrogen are transported by carriers to oxygen, forming water as a by-product (see Concept 9.1). The electrons lose potential energy as they "fall" down the electron transport chain towards electronegative oxygen, and the mitochondrion harnesses that energy to synthesise ATP (see Figure 9.14). Photosynthesis reverses the direction of electron flow. Water is split, and its electrons are transferred along with hydrogen ions (H^+) from the water to carbon dioxide, reducing it to sugar.

becomes reduced ($CO_2 \rightarrow C_6H_{12}O_6$)

$$\text{Energy} + 6\,CO_2 + 6\,H_2O \longrightarrow C_6H_{12}O_6 + 6\,O_2$$

becomes oxidised ($H_2O \rightarrow O_2$)

▼ Figure 10.5 Tracking atoms through photosynthesis. The atoms from CO_2 are shown in magenta, and the atoms from H_2O are shown in blue.

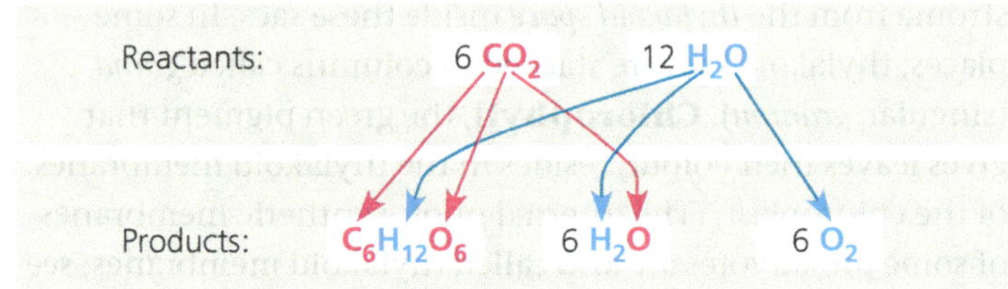

Because the electrons increase in potential energy as they move from water to sugar, this process requires energy—in other words, it is endergonic. This energy boost that occurs during photosynthesis is provided by light.

The Two Stages of Photosynthesis: *A Preview*

The equation for photosynthesis is a deceptively simple summary of a very complex process. Actually, photosynthesis is not a single process, but two processes, each of which has multiple steps. These two stages of photosynthesis are known as the **light reactions** (the *photo* part of photosynthesis) and the **Calvin cycle** (the *synthesis* part) **(Figure 10.6)**.

The light reactions are the steps of photosynthesis that convert solar energy to chemical energy. Water is split, providing a source of electrons and protons (hydrogen ions, H^+) and giving off O_2 as a by-product. Light absorbed by chlorophyll drives a transfer of the electrons and hydrogen ions from water to an acceptor called **$NADP^+$** (nicotinamide adenine dinucleotide phosphate), where they are temporarily stored. (The electron acceptor $NADP^+$ is first cousin to NAD^+, which functions as an electron carrier in cellular respiration; the two molecules differ only by the presence of an extra phosphate group in the $NADP^+$ molecule.) The light reactions use solar energy to reduce $NADP^+$ to **NADPH** by adding a pair of electrons along with an H^+. The light reactions also generate ATP, using chemiosmosis to power the addition of a phosphate group to ADP, a process called **photophosphorylation**. Thus, light energy is initially converted to chemical energy in the form of two compounds: NADPH and ATP. NADPH, a source of electrons, acts as "reducing power" that can be passed along to an electron acceptor, reducing it, while ATP is the versatile energy currency of cells. Notice that the light reactions produce no sugar; that happens in the second stage of photosynthesis, the Calvin cycle.

The Calvin cycle is named for Melvin Calvin, who, along with his colleagues James Bassham and Andrew Benson, began to elucidate its steps in the late 1940s. The cycle begins by incorporating CO_2 from the air into organic molecules already present in the chloroplast. This initial incorporation of carbon into organic compounds is known as **carbon fixation**. The Calvin cycle then reduces the fixed carbon to carbohydrate by the addition of electrons. The reducing power is provided by NADPH, which acquired its cargo of electrons in the light reactions. To convert CO_2 to carbohydrate, the Calvin cycle also requires chemical energy in the form of ATP, which is also generated by the light reactions. Thus, it is the Calvin cycle that makes sugar, but it can do so only with the help of the NADPH and ATP produced by the light reactions. The metabolic steps of the Calvin cycle are sometimes referred to as the dark reactions, or light-independent reactions, because none of the steps requires light *directly*. Nevertheless, the Calvin cycle in most plants occurs during daylight, for only then can the light reactions provide the NADPH and ATP that the Calvin cycle requires. In essence, the chloroplast uses light energy to make sugar by coordinating the two stages of photosynthesis.

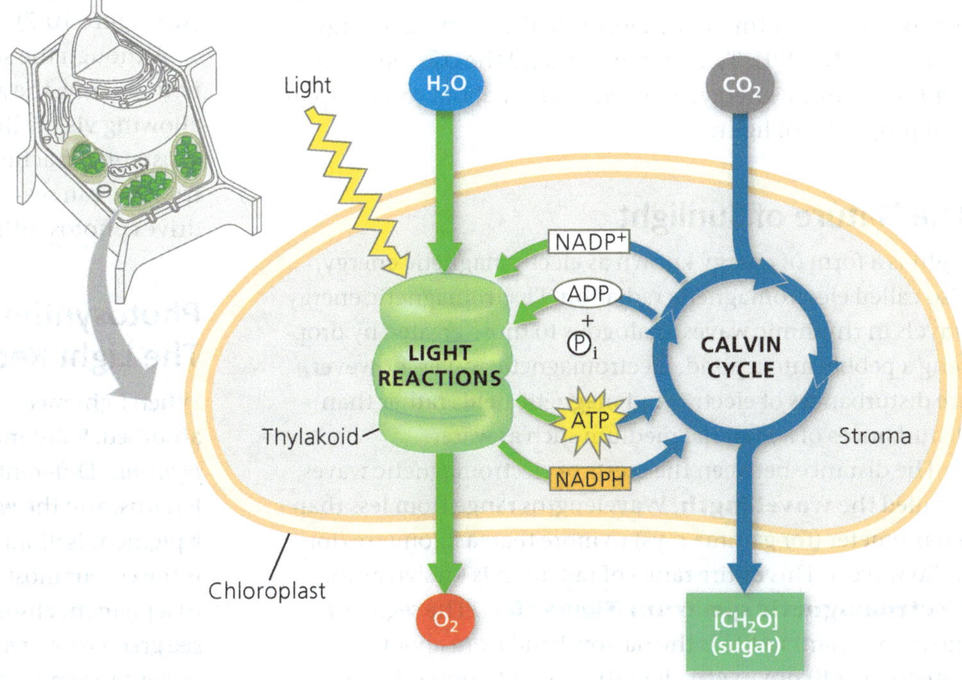

▶ Figure 10.6 An overview of photosynthesis: cooperation of the light reactions and the Calvin cycle. In the chloroplast, the thylakoid membranes (green) are the sites of the light reactions, whereas the Calvin cycle occurs in the stroma (grey). The light reactions use solar energy to make ATP and NADPH, which supply chemical energy and reducing power, respectively, to the Calvin cycle. The Calvin cycle incorporates CO_2 into organic molecules, which are converted to sugar. (Recall that most simple sugars have formulas that are some multiple of CH_2O.) To visualise these processes in their cellular context, see Figure 6.32.

As Figure 10.6 indicates, the thylakoids of the chloroplast are the sites of the light reactions, while the Calvin cycle occurs in the stroma. On the outside of the thylakoids, during the light reactions, molecules of $NADP^+$ and ADP pick up electrons and phosphate, respectively, and NADPH and ATP are then released to the stroma, where they play crucial roles in the Calvin cycle. The two stages of photosynthesis are treated in this figure as metabolic modules that take in ingredients and crank out products. In the next two sections, we'll look more closely at how the two stages work, beginning with the light reactions.

CONCEPT CHECK 10.2

1. **MAKE CONNECTIONS** How do the CO_2 molecules used in photosynthesis reach and enter the chloroplasts inside leaf cells? (See Concept 7.2.)
2. Explain how the use of an oxygen isotope helped elucidate the chemistry of photosynthesis.
3. **WHAT IF?** The Calvin cycle requires ATP and NADPH, products of the light reactions. If a classmate asserted that the light reactions don't depend on the Calvin cycle and, with continual light, could just keep on producing ATP and NADPH, how would you respond?

For suggested answers, see Appendix A.

CONCEPT 10.3

The light reactions convert solar energy to the chemical energy of ATP and NADPH

Chloroplasts are chemical factories powered by the sun. Their thylakoids transform light energy into the chemical energy of ATP and NADPH. To better understand the conversion of light to chemical energy, we need to know about some important properties of light.

The Nature of Sunlight

Light is a form of energy known as electromagnetic energy, also called electromagnetic radiation. Electromagnetic energy travels in rhythmic waves analogous to those created by dropping a pebble into a pond. Electromagnetic waves, however, are disturbances of electric and magnetic fields rather than disturbances of a material medium such as water.

The distance between the crests of electromagnetic waves is called the **wavelength**. Wavelengths range from less than a nanometer (for gamma rays) to more than a kilometre (for radio waves). This entire range of radiation is known as the **electromagnetic spectrum** (Figure 10.7). The segment most important to life is the narrow band from about 380 nm to 740 nm in wavelength. This radiation is known as

▼ **Figure 10.7 The electromagnetic spectrum.** White light is a mixture of all wavelengths of visible light. A prism can sort white light into its component colours by bending light of different wavelengths at different angles. (Droplets of water in the atmosphere can act as prisms, causing a rainbow to form.) Visible light drives photosynthesis.

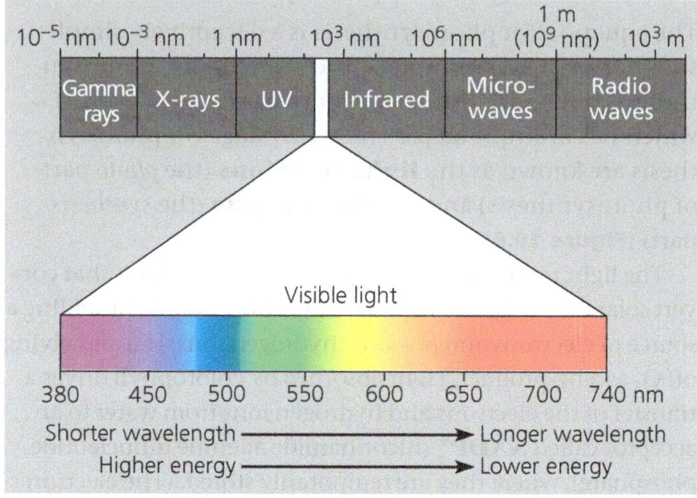

visible light because it can be detected as various colours by the human eye.

The model of light as waves explains many of light's properties, but in certain respects light behaves as though it consists of discrete particles, called **photons**. Photons are not tangible objects, but they act like objects in that each of them has a fixed quantity of energy. The amount of energy is inversely related to the wavelength of the light: The shorter the wavelength, the greater the energy of each photon of that light. Thus, a photon of violet light packs nearly twice as much energy as a photon of red light (see Figure 10.7).

Although the sun radiates the full spectrum of electromagnetic energy, the atmosphere acts like a selective window, allowing visible light to pass through while screening out a substantial fraction of other radiation. The part of the spectrum we can see—visible light—is also the radiation that drives photosynthesis.

Photosynthetic Pigments: The Light Receptors

When light meets matter, it may be reflected, transmitted, or absorbed. Substances that absorb visible light are known as *pigments*. Different pigments absorb light of different wavelengths, and the wavelengths that are absorbed disappear. If a pigment is illuminated with white light, the colour we see is the colour most reflected or transmitted by the pigment. (If a pigment absorbs all wavelengths, it appears black.) We see green when we look at a leaf because chlorophyll absorbs violet-blue and red light while transmitting and reflecting

▼ Figure 10.8 Why leaves are green: interaction of light with chloroplasts. The chlorophyll molecules of chloroplasts absorb violet-blue and red light (the colours most effective in driving photosynthesis) and reflect or transmit green light. This is why leaves appear green.

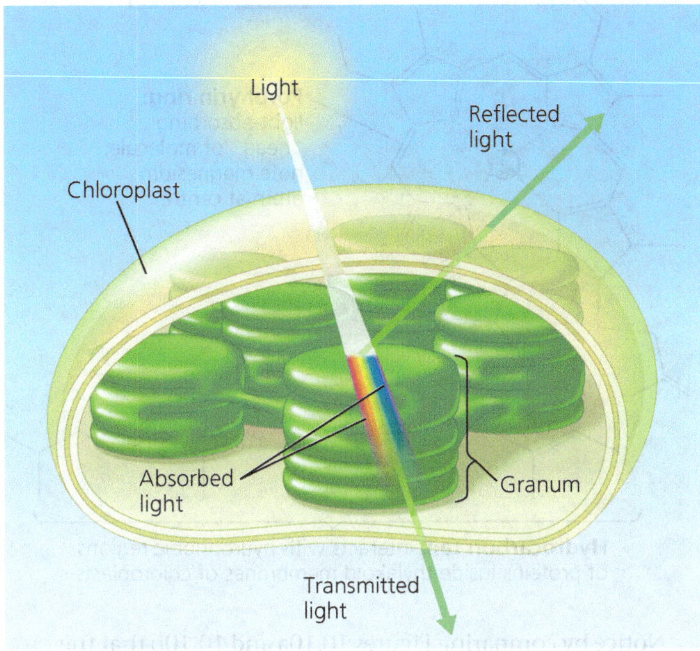

green light **(Figure 10.8)**. The ability of a pigment to absorb various wavelengths of light can be measured with an instrument called a **spectrophotometer**. This machine directs beams of light of different wavelengths through a solution of the pigment and measures the fraction of the light transmitted at each wavelength. A graph plotting a pigment's light absorption versus wavelength is called an **absorption spectrum (Figure 10.9)**.

The absorption spectra of chloroplast pigments provide clues to the relative effectiveness of different wavelengths for driving photosynthesis since light can perform work in chloroplasts only if it is absorbed. **Figure 10.10a** shows the absorption spectra of three types of pigments in chloroplasts: **chlorophyll *a***, the key light-capturing pigment that participates directly in the light reactions; the accessory pigment **chlorophyll *b***; and a separate group of accessory pigments called carotenoids. The spectrum of chlorophyll *a* suggests that violet-blue and red light work best for photosynthesis, since they are absorbed, while green is the least effective colour. This is confirmed by an **action spectrum** for photosynthesis **(Figure 10.10b)**, which profiles the relative effectiveness of different wavelengths of radiation in driving the process. An action spectrum is prepared by illuminating chloroplasts with light of different colours and then plotting wavelength against some measure of photosynthetic rate, such as CO_2 consumption or O_2 release. The action spectrum for photosynthesis was first demonstrated by Theodor W. Engelmann, a German botanist, in 1883.

▼ Figure 10.9 Research Method

Determining an Absorption Spectrum

Application An absorption spectrum is a visual representation of how well a particular pigment absorbs different wavelengths of visible light. Absorption spectra of various chloroplast pigments help scientists decipher the role of each pigment in a plant.

Technique A spectrophotometer measures the relative amounts of light of different wavelengths absorbed and transmitted by a pigment solution.

1. White light is separated into colours (wavelengths) by a prism.
2. One by one, the different colours of light are passed through the sample (chlorophyll in this example). Green light and blue light are shown here.
3. The transmitted light strikes a photoelectric tube, which converts the light energy to electricity.
4. The electric current is measured by a galvanometer. The meter indicates the fraction of light transmitted through the sample, from which we can determine the amount of light absorbed.

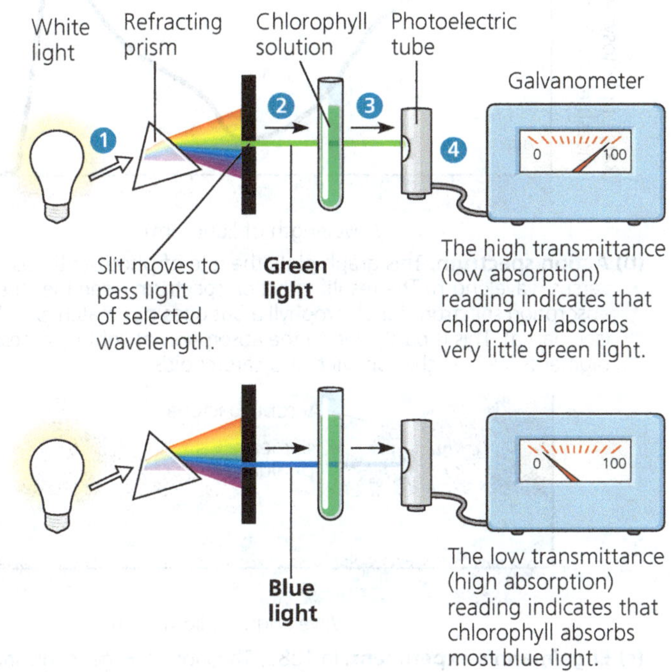

Results See Figure 10.10a for absorption spectra of three types of chloroplast pigments.

Before equipment for measuring O_2 levels had even been invented, Engelmann performed a clever experiment in which he used O_2-requiring bacteria to measure rates of photosynthesis in filamentous algae **(Figure 10.10c)**. His results are a striking match to the modern action spectrum shown in Figure 10.10b.

▼ Figure 10.10 Inquiry

Which wavelengths of light are most effective in driving photosynthesis?

Experiment Absorption and action spectra, along with a classic experiment by Theodor W. Engelmann, reveal which wavelengths of light are photosynthetically important.

Results

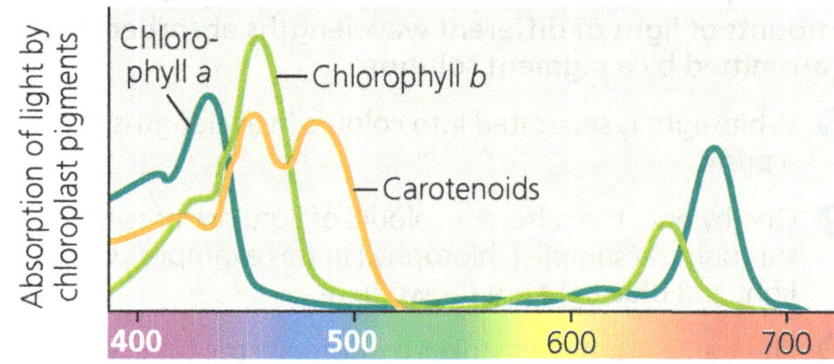

(a) Absorption spectra. The three curves show the wavelengths of light best absorbed by three types of chloroplast pigments.

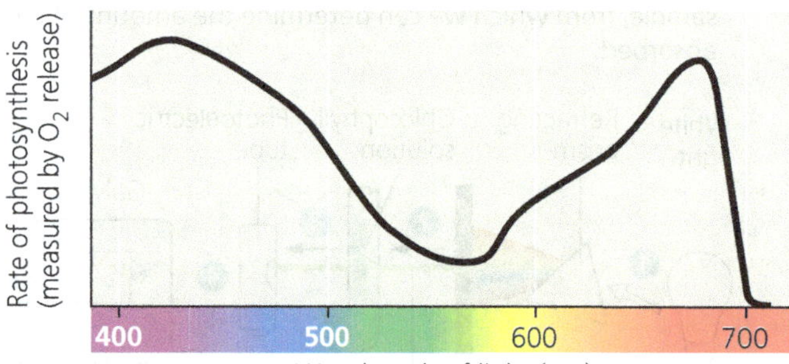

(b) Action spectrum. This graph plots the rate of photosynthesis versus wavelength. The resulting action spectrum resembles the absorption spectrum for chlorophyll *a* but does not match exactly (see part a). This is partly due to the absorption of light by accessory pigments such as chlorophyll *b* and carotenoids.

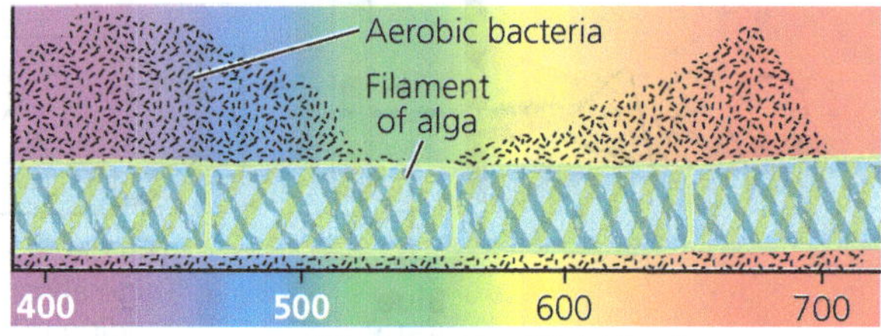

(c) Engelmann's experiment. In 1883, Theodor W. Engelmann shined light through a prism onto an alga, exposing different segments of the alga to different wavelengths. He used aerobic bacteria, which congregate near an oxygen source, to determine which segments of the alga were releasing the most O_2 and thus photosynthesising most. Bacteria congregated in greatest numbers around the parts of the alga illuminated with violet-blue or red light.

Conclusion The action spectra, confirmed by Engelmann's experiment, show which portions of the spectrum are most effective in driving photosynthesis.

Data from T. W. Engelmann, *Bacterium photometricum.* Ein Beitrag zur vergleichenden Physiologie des Licht-und Farbensinnes, *Archiv. für Physiologie* 30:95–124 (1883).

INTERPRET THE DATA *According to the graph, which wavelengths of light drive the highest rates of photosynthesis?*

▼ Figure 10.11 Structure of chlorophyll *a* and *b*.

Chlorophyll *a* and *b* differ in only one functional group:

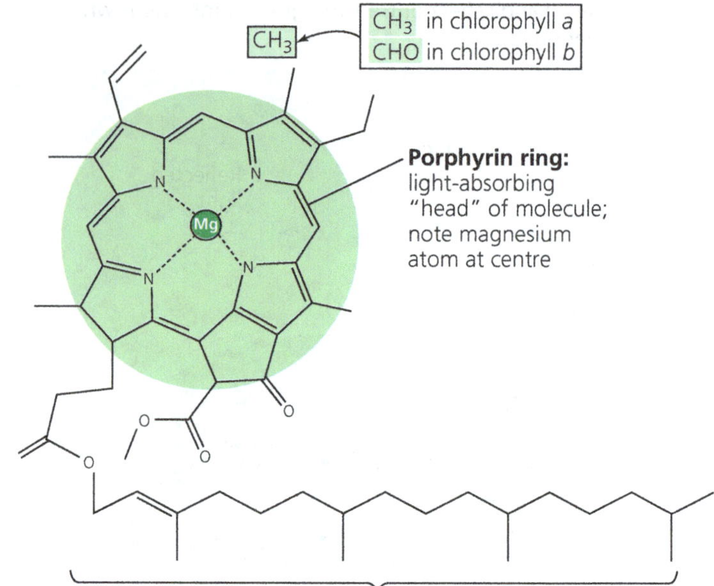

Notice by comparing Figures 10.10a and 10.10b that the action spectrum for photosynthesis is much broader than the absorption spectrum of chlorophyll *a*. The absorption spectrum of chlorophyll *a* alone underestimates the effectiveness of certain wavelengths in driving photosynthesis. This is partly because accessory pigments with different absorption spectra—including chlorophyll *b* and carotenoids—broaden the spectrum of colour that can be used for photosynthesis. **Figure 10.11** shows the structure of chlorophyll *a* compared with that of chlorophyll *b*. A slight structural difference between them is enough to cause the two pigments to absorb at slightly different wavelengths in the red and blue parts of the spectrum (see Figure 10.10a). As a result, chlorophyll *a* appears blue green and chlorophyll *b* olive green under visible light.

In the last decade, researchers have identified two other forms of chlorophyll—chlorophyll *d* and chlorophyll *f*—that absorb higher wavelengths of light. In 2018, researchers grew a species of cyanobacterium called *Chroococcidiopsis thermalis* under only far-red light with a wavelength of 750

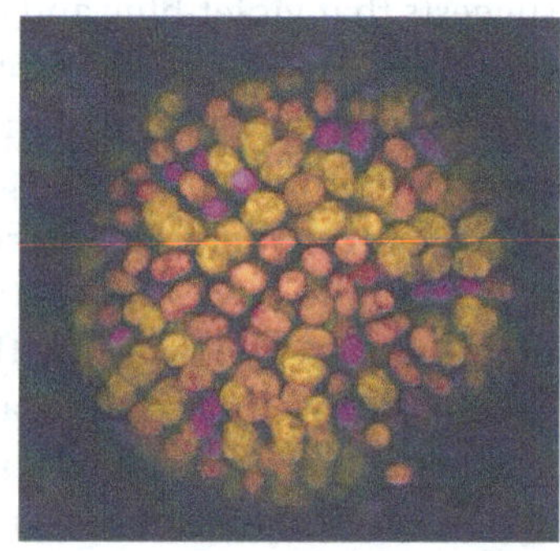

▶ Colony of the cyanobacteria *Chroococcidiopsis thermalis.* In this micrograph of cells becoming acclimated to higher wavelengths of light, cells that haven't yet acclimated and are still using chlorophyll *a* for photosynthesis appear purplish-pink, and cells that have acclimated and are using chlorophyll *f* appear yellow.

nm. They concluded that chlorophyll *f* was functioning in place of chlorophyll *a*, enabling this cyanobacterium to flourish in very shaded conditions. Higher-wavelength light has less energy, so this observation extends the lower limit of energy (the "red limit") needed for photosynthesis to occur.

Other accessory pigments include **carotenoids**, hydrocarbons that are various shades of yellow and orange because they absorb violet and blue-green light (see Figure 10.10a). Carotenoids may broaden the spectrum of colours that can drive photosynthesis. However, a more important function of at least some carotenoids seems to be *photoprotection*: These compounds absorb and dissipate excessive light energy that would otherwise damage chlorophyll or interact with oxygen, forming reactive oxidative molecules that are dangerous to the cell. Interestingly, carotenoids similar to the photoprotective ones in chloroplasts have a photoprotective role in the human eye. (Carrots, known for aiding night vision, are rich in carotenoids.) These and related molecules are, of course, found naturally in many vegetables and fruits. They are also often advertised in health food products as "phytochemicals" (from the Greek *phyton*, plant), some of which have antioxidant properties. Plants can synthesise all the antioxidants they require, but humans and other animals must obtain some of them from their diets.

Excitation of Chlorophyll by Light

What exactly happens when chlorophyll and other pigments absorb light? The colours corresponding to the absorbed wavelengths disappear from the spectrum of the transmitted and reflected light, but energy cannot disappear. When a molecule absorbs a photon of light, one of the molecule's electrons is elevated to an orbital where it has more potential energy (see Figure 2.6b). When the electron is in its normal orbital, the pigment molecule is said to be in its ground state. Absorption of a photon boosts an electron to an orbital of higher energy, and the pigment molecule is then said to be in an excited state. The only photons absorbed are those whose energy is exactly equal to the energy difference between the ground state and an excited state, and this energy difference varies from one kind of molecule to another. Thus, a particular compound absorbs only photons corresponding to specific wavelengths, which is why each pigment has a unique absorption spectrum.

Once absorption of a photon raises an electron to an excited state, the electron cannot stay there long **(Figure 10.12a)**. The excited state, like all high-energy states, is unstable. Generally, when isolated pigment molecules absorb light, their excited electrons drop back down to the ground-state orbital in a billionth of a second, releasing their excess energy as heat. This conversion of light energy to heat is what makes the top of a car so hot on a sunny day. (White cars are coolest because their paint reflects all wavelengths of visible light.) In isolation, some pigments, including chlorophyll, emit light as well as heat after absorbing photons. As excited electrons fall back to the ground state, photons are given off, an afterglow called fluorescence. An illuminated solution of chlorophyll isolated from chloroplasts will fluoresce in the red part of the spectrum and also give off heat **(Figure 10.12b)**. This is best seen by illuminating with ultraviolet light, which chlorophyll can also absorb (see Figures 10.7 and 10.10a). Viewed under visible light, the fluorescence would be difficult to see against the green of the solution.

A Photosystem: A Reaction-Centre Complex Associated with Light-Harvesting Complexes

Chlorophyll molecules excited by the absorption of light energy produce very different results in an intact chloroplast than they do in isolation (see Figure 10.12). In their native environment of the thylakoid membrane, chlorophyll

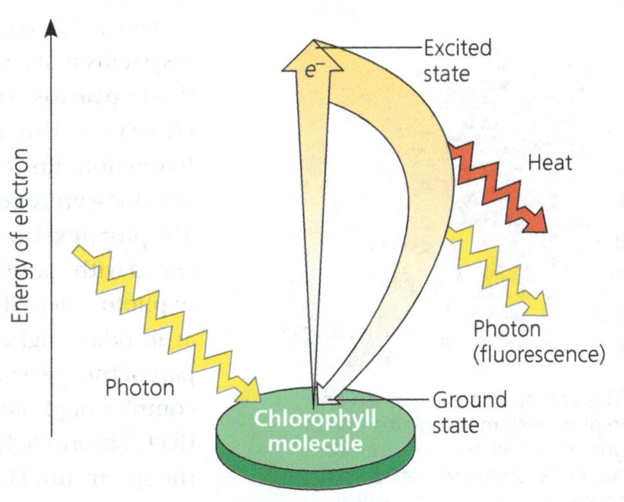

(a) Excitation of isolated chlorophyll molecule

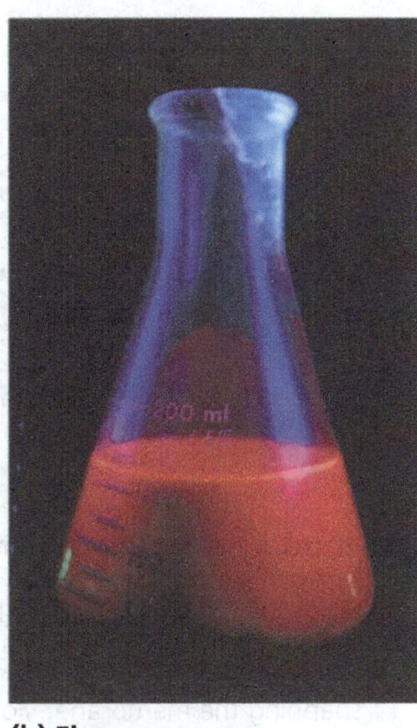

(b) Fluorescence

▶ **Figure 10.12 Excitation of isolated chlorophyll by light. (a)** Absorption of a photon causes a transition of the chlorophyll molecule from its ground state to its excited state. The photon boosts an electron to an orbital where it has more potential energy. If the illuminated molecule exists in isolation, its excited electron immediately drops back down to the ground-state orbital, and its excess energy is given off as heat and fluorescence (light). **(b)** A chlorophyll solution excited with ultraviolet light fluoresces with a red-orange glow.

WHAT IF? *If a leaf containing the same concentration of chlorophyll as in the solution was exposed to the same ultraviolet light, no fluorescence would be seen. Propose an explanation for the difference in fluorescence emission between the solution and the leaf.*

molecules are organised along with other small organic molecules and proteins into complexes called photosystems.

A **photosystem** is composed of a reaction-centre complex surrounded by several light-harvesting complexes **(Figure 10.13)**. The **reaction-centre complex** is an organised association of proteins holding a special pair of chlorophyll *a* molecules and a primary electron acceptor. Each **light-harvesting complex** consists of various pigment molecules (which may include chlorophyll *a*, chlorophyll *b*, and multiple carotenoids) bound to proteins. The number and variety of pigment molecules enable a photosystem to harvest light over a larger surface area and a larger portion of the spectrum than could any single pigment molecule alone. Together, these light-harvesting complexes act as an antenna for the reaction-centre complex. When a pigment molecule absorbs a photon, the energy is transferred from pigment molecule to pigment molecule within a light-harvesting complex, like a human "wave" at a sports arena, until it is passed to the pair of chlorophyll *a* molecules in the reaction-centre complex. This pair of chlorophyll *a* molecules is special because their molecular environment—their location and the other molecules with which they are associated—enables them to use the energy from light not only to boost one of their electrons to a higher energy level, but also to transfer it to a different molecule—the **primary electron acceptor**, which is a molecule capable of accepting electrons and becoming reduced.

▼ Figure 10.13 The structure and function of a photosystem.

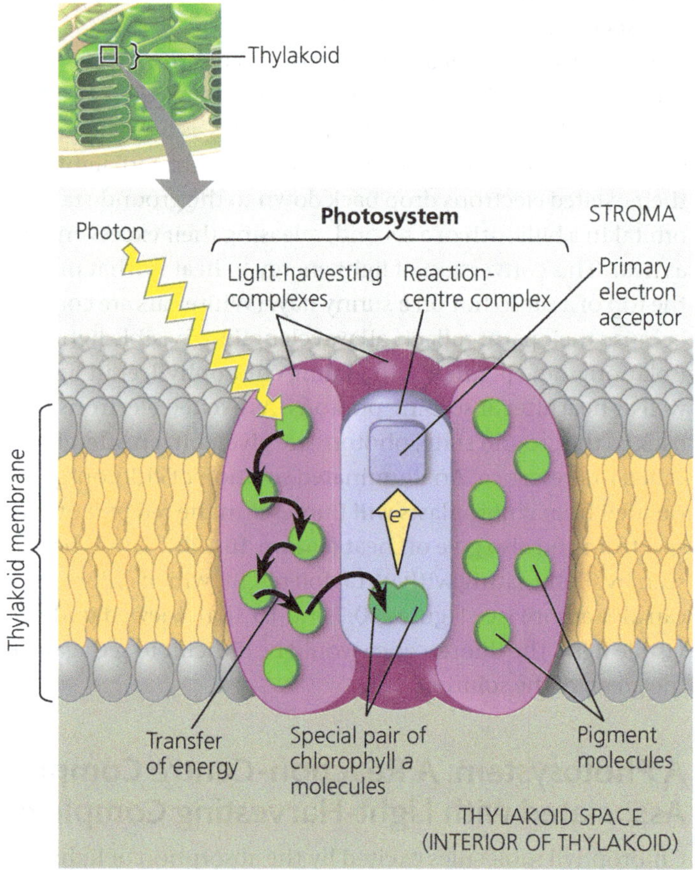

(a) How a photosystem harvests light. When a photon strikes a pigment molecule in a light-harvesting complex, the energy is passed from molecule to molecule until it reaches the reaction-centre complex. Here, an excited electron from the special pair of chlorophyll *a* molecules is transferred to the primary electron acceptor.

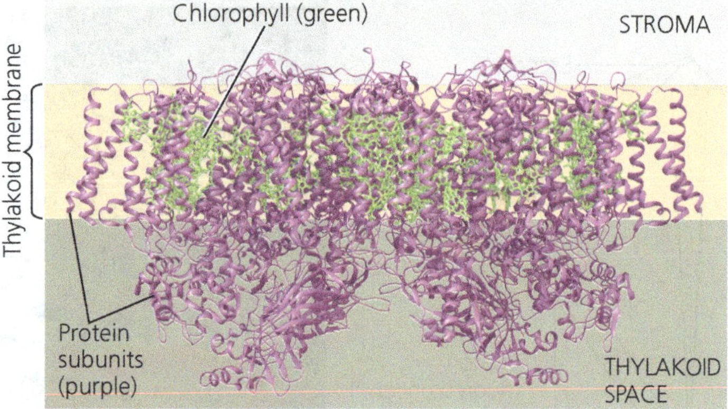

(b) Structure of a photosystem. This computer model, based on X-ray crystallography, shows two photosystem complexes side by side. Chlorophyll molecules (bright green ball-and-stick models within the membrane; the tails are not shown) are interspersed with protein subunits (purple ribbons; notice the many α helices spanning the membrane). For simplicity, a photosystem will be shown as a single complex in the rest of the chapter.

The solar-powered transfer of an electron from the reaction-centre chlorophyll *a* pair to the primary electron acceptor is the first step of the light reactions. As soon as the chlorophyll electron is excited to a higher energy level, the primary electron acceptor captures it; this is a redox reaction. In the flask shown in Figure 10.12b, isolated chlorophyll fluoresces because there is no electron acceptor, so electrons of photoexcited chlorophyll drop right back to the ground state. In the structured environment of a chloroplast, however, an electron acceptor is readily available, and the potential energy represented by the excited electron is not dissipated as light and heat. Thus, each photosystem—a reaction-centre complex surrounded by light-harvesting complexes—functions in the chloroplast as a unit. It converts light energy to chemical energy, which will ultimately be used for the synthesis of sugar.

The thylakoid membrane is populated by two types of photosystems that cooperate in the light reactions of photosynthesis: **photosystem II (PS II)** and **photosystem I (PS I)**. (They were named in order of their discovery, but photosystem II functions first in the light reactions.) Each has a characteristic reaction-centre complex—a particular kind of primary electron acceptor next to a special pair of chlorophyll *a* molecules associated with specific proteins. The reaction-centre chlorophyll *a* of photosystem II is known as P680 because this pigment is best at absorbing light having a wavelength of 680 nm (in the red part of the spectrum). The chlorophyll *a* at the reaction-centre complex of photosystem I is called P700 because it most effectively absorbs light of wavelength 700 nm (in the far-red part of the spectrum). These two pigments, P680 and P700, are nearly identical chlorophyll *a* molecules. However, their association with different proteins in the thylakoid membrane affects the electron distribution in the two pigments and accounts for the slight differences in their light-absorbing properties. Now let's see how the two types of photosystems work together in using light energy to generate ATP and NADPH, the two main products of the light reactions.

Linear Electron Flow

Light drives the synthesis of ATP and NADPH by energising the two types of photosystems embedded in the thylakoid membranes of chloroplasts. The key to this energy transformation is a flow of electrons through the photosystems and other molecular components built into the thylakoid membrane. This is called **linear electron flow**, and it occurs during the light reactions of photosynthesis, as shown in **Figure 10.14**. The numbered steps in the text correspond to the numbered steps in the figure.

1. A photon of light strikes one of the pigment molecules in a light-harvesting complex of PS II, boosting one of its electrons to a higher energy level. As this electron falls back to its ground state, an electron in a nearby pigment molecule is simultaneously raised to an excited state. The process continues, with the energy being relayed to other pigment molecules until it reaches the P680 pair of chlorophyll *a* molecules in the PS II reaction-centre complex. It excites an electron in this pair of chlorophylls to a higher energy state.

2. This electron is transferred from the excited P680 to the primary electron acceptor. We can refer to the resulting form of P680, missing the negative charge of an electron, as P680$^+$.

3. An enzyme catalyses the splitting of a water molecule into two electrons, two hydrogen ions (H^+), and an oxygen atom. The electrons are supplied one by one to the P680$^+$ pair, each electron replacing one transferred to the primary electron acceptor. (P680$^+$ is the strongest biological oxidising agent known; its electron "hole" must be filled. This greatly facilitates the transfer of electrons from the split water molecule.) The H^+ are released into the thylakoid space (interior of the thylakoid). The oxygen atom immediately combines with an oxygen atom generated by the splitting of another water molecule, forming O_2.

▼ **Figure 10.14 How linear electron flow during the light reactions generates ATP and NADPH.** The gold arrows trace the flow of light-driven electrons from water to NADPH. The black arrows trace the transfer of energy from pigment molecule to pigment molecule. To see these proteins in their cellular context, see Figure 6.32b.

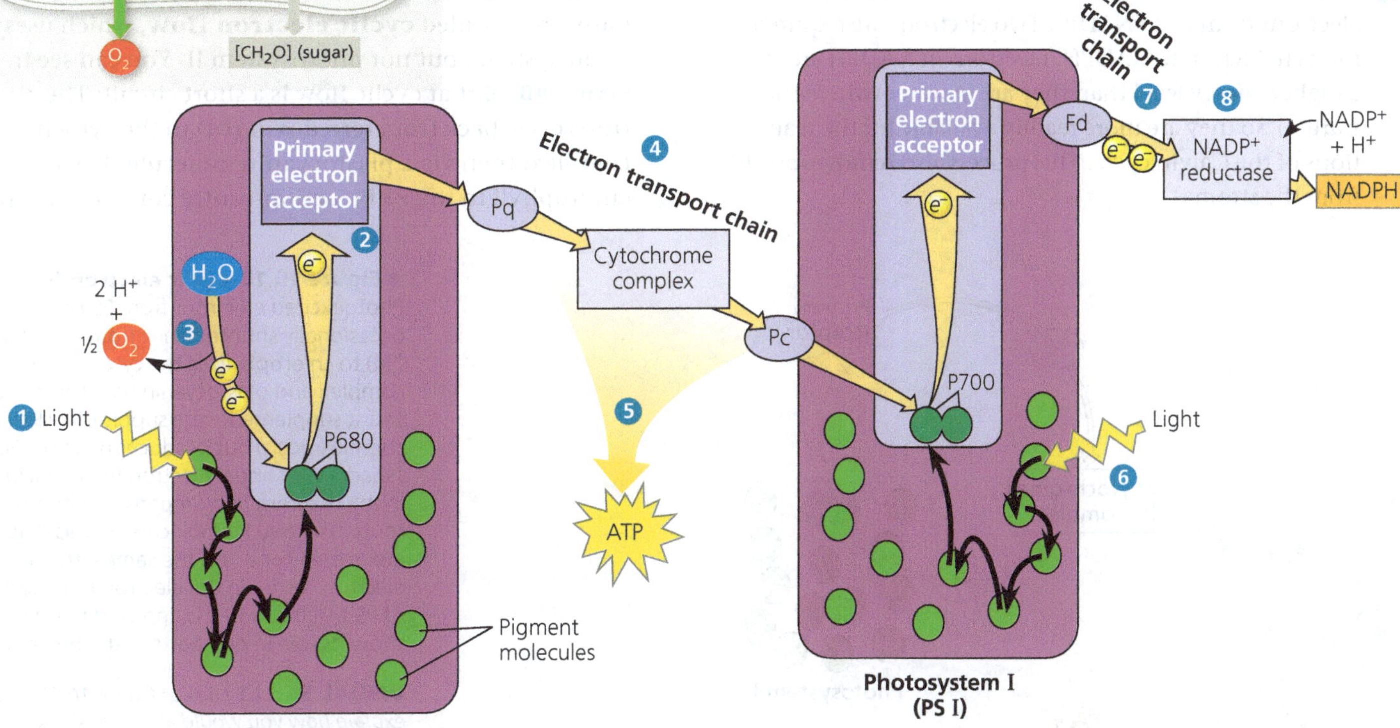

4 Each photoexcited electron passes from the primary electron acceptor of PS II to PS I via an electron transport chain, the components of which are similar to those of the electron transport chain that functions in cellular respiration. The electron transport chain between PS II and PS I is made up of the electron carrier plastoquinone (Pq), a cytochrome complex, and a protein called plastocyanin (Pc). Each component carries out redox reactions as electrons flow down the electron transport chain, releasing free energy that is used to pump protons (H^+) into the thylakoid space, contributing to a proton gradient across the thylakoid membrane.

5 The potential energy stored in the proton gradient is used to make ATP in a process called chemiosmosis, to be discussed shortly.

6 Meanwhile, light energy has been transferred via light-harvesting complex pigments to the PS I reaction-centre complex, exciting an electron of the P700 pair of chlorophyll *a* molecules located there. The photoexcited electron is then transferred to PS I's primary electron acceptor, creating an electron "hole" in the P700—which we now can call $P700^+$. In other words, $P700^+$ can now act as an electron acceptor, accepting an electron that reaches the bottom of the electron transport chain from PS II.

7 Photoexcited electrons are passed in a series of redox reactions from the primary electron acceptor of PS I down a second electron transport chain through the protein ferredoxin (Fd). (This chain does not create a proton gradient and thus does not produce ATP.)

8 The enzyme $NADP^+$ reductase catalyses the transfer of electrons from Fd to $NADP^+$. Two electrons are required for its reduction to NADPH. Electrons in NADPH are at a higher energy level than they are in water (where they started), so they are more readily available for the reactions of the Calvin cycle. This process also removes an H^+ from the stroma.

▼ **Figure 10.15 A mechanical analogy for linear electron flow during the light reactions.**

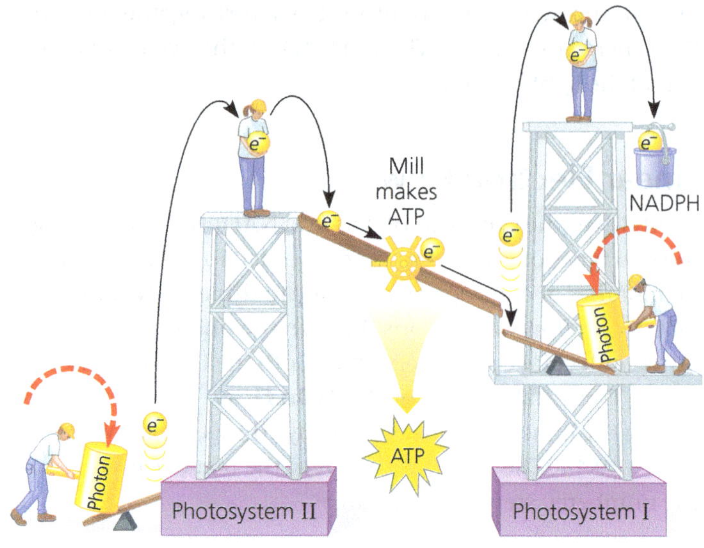

The energy changes of electrons during their linear flow through the light reactions are shown in a mechanical analogy in **Figure 10.15**. Although the scheme shown in Figures 10.14 and 10.15 may seem complicated, do not lose track of the big picture: The light reactions use solar power to generate ATP and NADPH, which provide chemical energy and reducing power, respectively, to the carbohydrate-synthesising reactions of the Calvin cycle.

Cyclic Electron Flow

In certain cases, photoexcited electrons can take an alternative path called **cyclic electron flow**, which uses photosystem I but not photosystem II. You can see in **Figure 10.16** that cyclic flow is a short circuit: The electrons cycle back from ferredoxin (Fd) to the cytochrome complex, then via a plastocyanin molecule (Pc) to a P700 chlorophyll in the PS I reaction-centre complex. There is

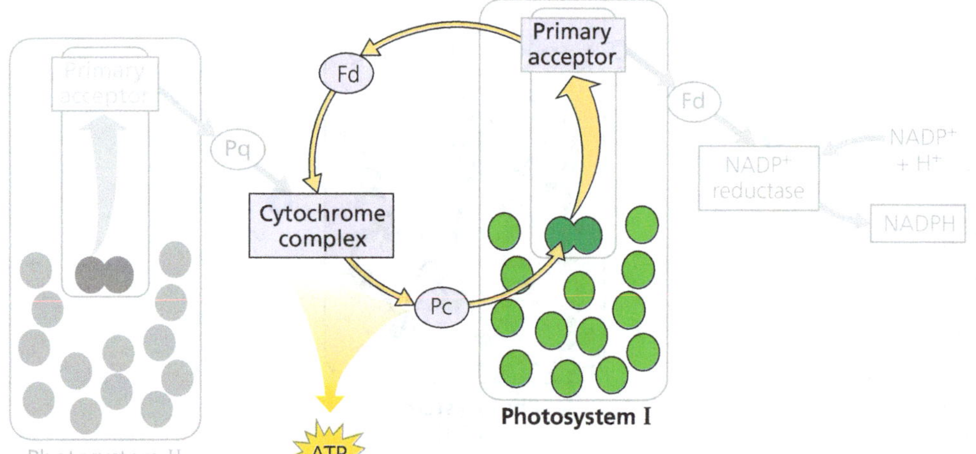

◀ **Figure 10.16 Cyclic electron flow.** Photoexcited electrons from PS I are occasionally shunted back from ferredoxin (Fd) to chlorophyll via the cytochrome complex and plastocyanin (Pc). This electron shunt supplements the supply of ATP (via chemiosmosis) but produces no NADPH. The "shadow" of linear electron flow is included in the diagram for comparison with the cyclic route. The two Fd molecules in this diagram are actually one and the same—the final electron carrier in the electron transport chain of PS I—although it is depicted twice to clearly show its role in two parts of the process.

VISUAL SKILLS *Look at Figure 10.15 and explain how you would alter it to show a mechanical analogy for cyclic electron flow.*

no production of NADPH and no release of oxygen that results from this process. On the other hand, cyclic flow does generate ATP.

Rather than having both PS II and PS I, several of the currently existing groups of photosynthetic bacteria are known to have a single photosystem related to either PS II or PS I. For these species, which include the purple sulfur bacteria (see Figure 10.2e) and the green sulfur bacteria, cyclic electron flow is the one and only means of generating ATP during the process of photosynthesis. Evolutionary biologists hypothesise that these bacterial groups are descendants of ancestral bacteria in which photosynthesis first evolved, in a form similar to cyclic electron flow.

Cyclic electron flow can also occur in photosynthetic species that possess both photosystems; this includes some prokaryotes, such as the cyanobacteria shown in Figure 10.2d, as well as the eukaryotic photosynthetic species that have been tested thus far. Although the process is probably in part an "evolutionary leftover," research suggests it plays at least one beneficial role for these organisms. Plants with mutations that render them unable to carry out cyclic electron flow are capable of growing well in low light, but do not grow well where light is intense. This is evidence for the idea that cyclic electron flow may be photoprotective. Later you'll learn more about cyclic electron flow as it relates to a particular adaptation of photosynthesis (C_4 plants; see Concept 10.5).

Whether ATP synthesis is driven by linear or cyclic electron flow, the actual mechanism is the same. Before we move on to consider the Calvin cycle, let's review chemiosmosis, the process that uses membranes to couple redox reactions to ATP production.

A Comparison of Chemiosmosis in Chloroplasts and Mitochondria

Chloroplasts and mitochondria generate ATP by the same basic mechanism: chemiosmosis (see Figure 9.14). An electron transport chain pumps protons (H^+) across a membrane as electrons are passed through a series of carriers that have progressively more affinity for electrons. Thus, electron transport chains transform redox energy to a proton-motive force, potential energy stored in the form of an H^+ gradient across a membrane. An ATP synthase complex in the same membrane couples the diffusion of hydrogen ions down their gradient to the phosphorylation of ADP, forming ATP.

Some of the electron carriers, including the iron-containing proteins called cytochromes, are very similar in mitochondria and chloroplasts (see Figure 6.32b and c). The ATP synthase complexes of the two organelles are also quite similar. But there are noteworthy differences between photophosphorylation in chloroplasts and oxidative phosphorylation in mitochondria. Both work by way of chemiosmosis, but in chloroplasts, the high-energy electrons dropped down the transport chain come from water, while in mitochondria, they are extracted from organic molecules (which are thus oxidised). Chloroplasts do not need molecules from food to make ATP; their photosystems capture light energy and use it to drive the electrons from water to the top of the transport chain. In other words, mitochondria use chemiosmosis to transfer chemical energy from food molecules to ATP, whereas chloroplasts use it to transform light energy into chemical energy in ATP.

Although the spatial organisation of chemiosmosis differs slightly between chloroplasts and mitochondria, it is easy to

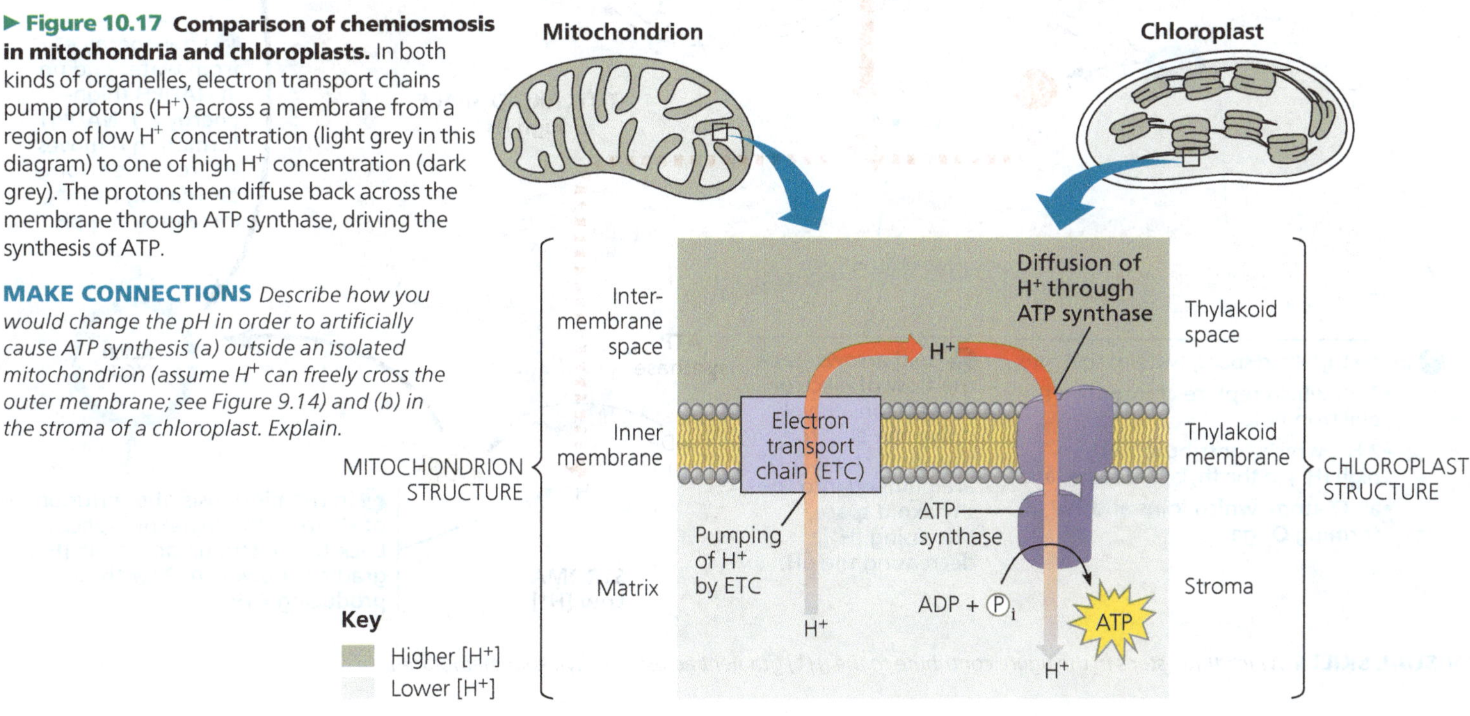

▶ Figure 10.17 Comparison of chemiosmosis in mitochondria and chloroplasts. In both kinds of organelles, electron transport chains pump protons (H^+) across a membrane from a region of low H^+ concentration (light grey in this diagram) to one of high H^+ concentration (dark grey). The protons then diffuse back across the membrane through ATP synthase, driving the synthesis of ATP.

MAKE CONNECTIONS *Describe how you would change the pH in order to artificially cause ATP synthesis (a) outside an isolated mitochondrion (assume H^+ can freely cross the outer membrane; see Figure 9.14) and (b) in the stroma of a chloroplast. Explain.*

see similarities in the two **(Figure 10.17)**. Electron transport chain proteins in the inner membrane of the mitochondrion pump protons from the mitochondrial matrix out to the intermembrane space, which then serves as a reservoir of hydrogen ions. Similarly, electron transport chain proteins in the thylakoid membrane of the chloroplast pump protons from the stroma into the thylakoid space, which functions as the H^+ reservoir. If you imagine the cristae of mitochondria pinching off from the inner membrane, this may help you see how the thylakoid space and the intermembrane space are comparable spaces in the two organelles, while the mitochondrial matrix is analogous to the stroma of the chloroplast.

In the mitochondrion, protons diffuse down their concentration gradient from the intermembrane space through ATP synthase to the matrix, driving ATP synthesis. In the chloroplast, ATP is synthesised as the hydrogen ions diffuse from the thylakoid space back to the stroma through ATP synthase complexes **(Figure 10.18)**, whose catalytic knobs are on the stroma side of the membrane. Thus, ATP forms in the stroma, where it is used to help drive sugar synthesis during the Calvin cycle.

▼ **Figure 10.18 The light reactions: organisation of the thylakoid membrane.**

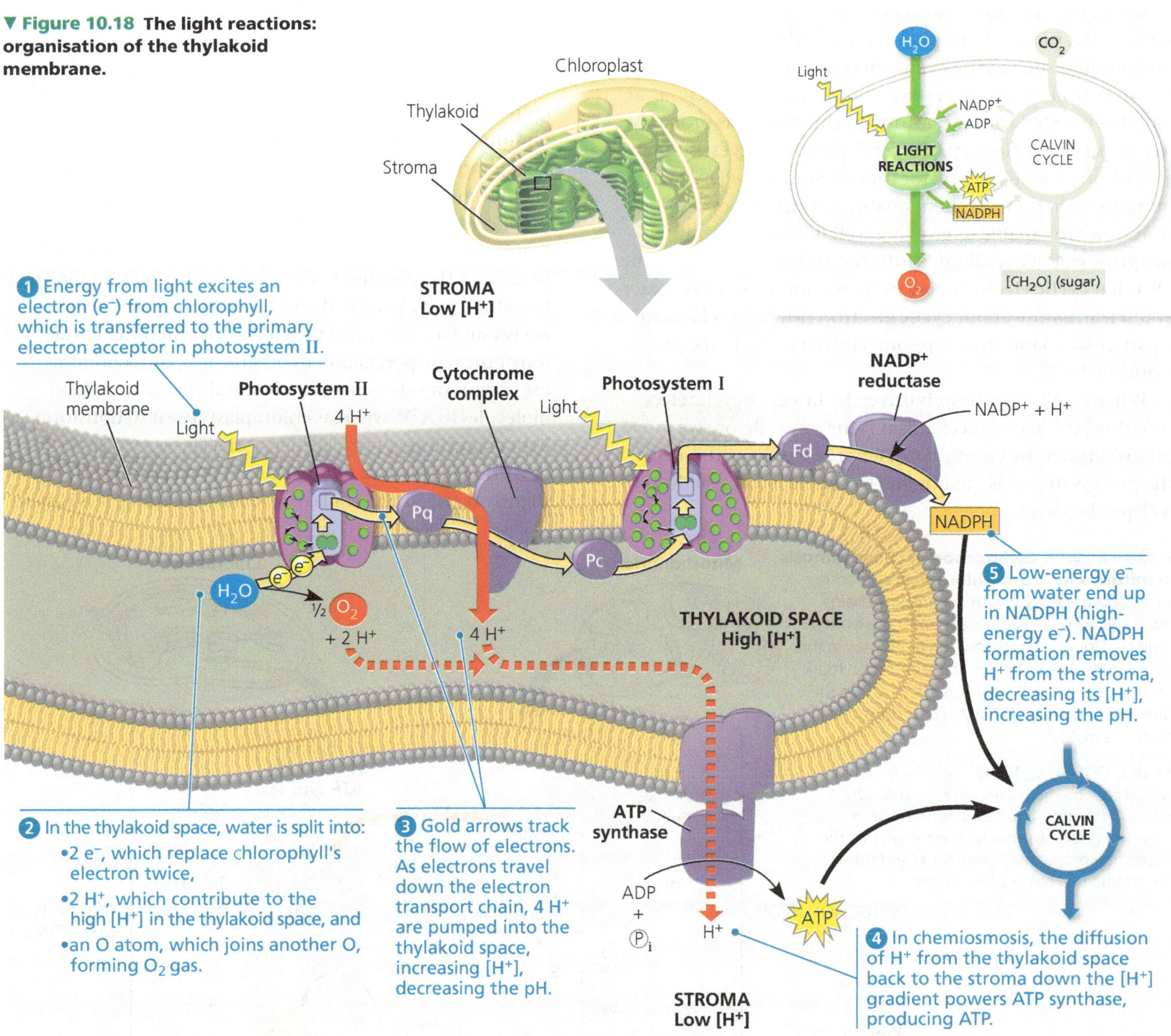

VISUAL SKILLS *Which three steps in the figure contribute to the [H^+] gradient across the thylakoid membrane?*

The proton (H^+) gradient, or pH gradient, across the thylakoid membrane is substantial. When chloroplasts in an experimental setting are illuminated, the pH in the thylakoid space drops to about 5 (the H^+ concentration increases), and the pH in the stroma increases to about 8 (the H^+ concentration decreases). This gradient of three pH units corresponds to a thousandfold difference in H^+ concentration. If the lights are then turned off, the pH gradient is abolished, but it can quickly be restored by turning the lights back on. Experiments such as this provided strong evidence in support of the chemiosmotic model.

The currently accepted model for the organisation of the light-reaction "machinery" within the thylakoid membrane is based on several research studies. Each of the molecules and molecular complexes in Figure 10.18 is present in numerous copies in each thylakoid. Notice that NADPH, like ATP, is produced on the side of the membrane facing the stroma, where the Calvin cycle reactions take place.

Let's step back and see the big picture by summarising the light reactions. Electron flow pushes electrons from water, where they are at a state of low potential energy, ultimately to NADPH, where they are stored at a state of high potential energy. The light-driven electron flow also generates ATP. Thus, the equipment of the thylakoid membrane converts light energy to chemical energy stored in ATP and NADPH. O_2 is produced as a by-product. Let's now see how the enzymes of the Calvin cycle use the products of the light reactions to synthesise sugar from CO_2.

CONCEPT CHECK 10.3

1. What colour of light is *least* effective in driving photosynthesis? Explain.
2. In the light reactions, what is the initial electron donor? Where do the electrons finally end up?
3. **WHAT IF?** In an experiment, isolated chloroplasts placed in an illuminated solution with the appropriate chemicals can carry out ATP synthesis. Predict what would happen to the rate of synthesis if a compound is added to the solution that makes membranes freely permeable to hydrogen ions.

For suggested answers, see Appendix A.

CONCEPT 10.4

The Calvin cycle uses the chemical energy of ATP and NADPH to reduce CO_2 to sugar

The Calvin cycle takes place in the stroma; it is similar to the citric acid cycle in that a starting material is regenerated after some molecules enter and others exit the cycle. However, the citric acid cycle is catabolic, oxidising acetyl CoA and using the energy to synthesise ATP (see Figure 9.11), while the Calvin cycle is anabolic, building carbohydrates from smaller molecules and consuming energy. Carbon enters the Calvin cycle in the form of CO_2 and leaves in the form of sugar. The cycle spends ATP as an energy source and consumes NADPH as reducing power for adding high-energy electrons to make the sugar.

As we mentioned in Concept 10.2, the carbohydrate produced directly from the Calvin cycle is not glucose. It is actually a three-carbon sugar called **glyceraldehyde 3-phosphate (G3P)**. For the net synthesis of *one* molecule of G3P, the cycle must take place three times, fixing *three* molecules of CO_2—one per turn of the cycle. (Recall that the term *carbon fixation* refers to the initial incorporation of CO_2 into organic material.) As we trace the steps of the Calvin cycle, keep in mind that we are following three molecules of CO_2 through the reactions. **Figure 10.19** divides the Calvin cycle into three phases: carbon fixation, reduction, and regeneration of the CO_2 acceptor.

Phase 1: Carbon fixation. The Calvin cycle incorporates each CO_2 molecule, one at a time, by attaching it to a five-carbon sugar named ribulose bisphosphate (which is abbreviated RuBP). The enzyme that catalyses this first step is RuBP carboxylase-oxygenase, or **rubisco** (see Figure 6.32c). (This is the most abundant protein in chloroplasts and is also thought to be the most abundant protein on Earth.) The product of the reaction is a six-carbon intermediate that is short-lived because it is so energetically unstable that it immediately splits in half, forming two molecules of 3-phosphoglycerate (for each CO_2 fixed).

Phase 2: Reduction. Each molecule of 3-phosphoglycerate receives an additional phosphate group from ATP, becoming 1,3-bisphosphoglycerate. Next, a pair of electrons donated from NADPH reduces 1,3-bisphosphoglycerate, which also loses a phosphate group in the process, becoming glyceraldehyde 3-phosphate (G3P). Specifically, the electrons from NADPH reduce a carboxyl group on 1,3-bisphosphoglycerate to the aldehyde group of G3P, which stores more potential energy. G3P is a sugar—the same three-carbon sugar formed in glycolysis by the splitting of glucose (see Figure 9.8). Notice in Figure 10.19 that for every *three* molecules of CO_2 that enter the cycle, there are *six* molecules of G3P formed. But only one molecule of this three-carbon sugar can be counted as a net gain of carbohydrate because the rest are required to complete the cycle. The cycle began with 15 carbons' worth of carbohydrate in the form of three molecules of the five-carbon sugar RuBP. Now there are 18 carbons' worth of carbohydrate in the form of six molecules of G3P. One molecule exits the cycle to be used by the plant cell, but the other five molecules must be recycled to regenerate the three molecules of RuBP.

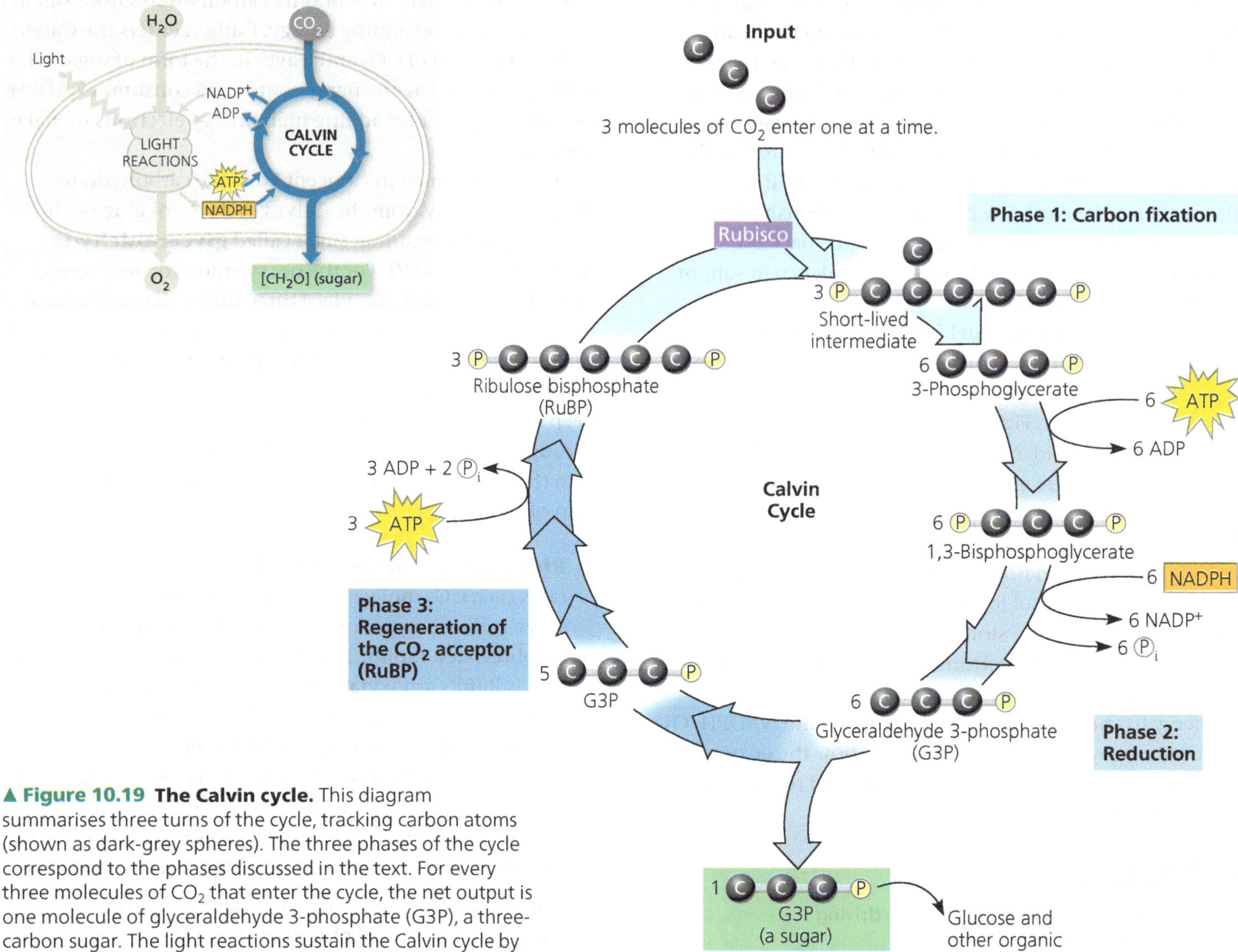

▲ **Figure 10.19 The Calvin cycle.** This diagram summarises three turns of the cycle, tracking carbon atoms (shown as dark-grey spheres). The three phases of the cycle correspond to the phases discussed in the text. For every three molecules of CO_2 that enter the cycle, the net output is one molecule of glyceraldehyde 3-phosphate (G3P), a three-carbon sugar. The light reactions sustain the Calvin cycle by regenerating the required ATP and NADPH.

Phase 3: Regeneration of the CO_2 acceptor (RuBP). In a complex series of reactions, the carbon skeletons of five molecules of G3P are rearranged by the last steps of the Calvin cycle into three molecules of RuBP. To accomplish this, the cycle spends three more molecules of ATP. The RuBP is now prepared to receive CO_2 again, and the cycle continues.

For the net synthesis of one G3P molecule, the Calvin cycle consumes a total of nine molecules of ATP and six molecules of NADPH. The light reactions regenerate the ATP and NADPH. The G3P spun off from the Calvin cycle becomes the starting material for metabolic pathways that synthesise other organic compounds, including glucose (from two molecules of G3P), sucrose, and other carbohydrates. Neither the light reactions nor the Calvin cycle alone can make sugar from CO_2. Photosynthesis is an emergent property of the intact chloroplast, which integrates the two stages of photosynthesis.

CONCEPT CHECK 10.4

1. To synthesise one glucose molecule, the Calvin cycle uses ________ molecules of CO_2, ________ molecules of ATP, and ________ molecules of NADPH.
2. How are the large numbers of ATP and NADPH molecules used during the Calvin cycle consistent with the high value of glucose as an energy source?
3. **WHAT IF?** Consider a poison that inhibits an enzyme of the Calvin cycle. Do you think such a poison will also inhibit the light reactions? Explain.
4. **DRAW IT** Draw a simple version of the Calvin cycle that shows only the number of carbon atoms at each step, using numerals rather than grey spheres (for example, "$3 \times 1C = 3C$" at the beginning of the cycle). Explain how the total number of carbon atoms remains constant for every three turns of the cycle, noting the forms in which the carbon atoms enter and leave the cycle.
5. **MAKE CONNECTIONS** Review Figures 9.8 and 10.19, and then discuss the roles of intermediate and product played by glyceraldehyde 3-phosphate (G3P) in the two processes shown in these figures.

For suggested answers, see Appendix A.

CONCEPT 10.5

Alternative mechanisms of carbon fixation have evolved in hot, arid climates

EVOLUTION Ever since plants first moved onto land about 475 million years ago, they have been adapting to the problems of terrestrial life, particularly the problem of dehydration. In Concept 36.4, we will consider anatomical adaptations that help plants conserve water, while in this chapter we are concerned with metabolic adaptations. The solutions often involve trade-offs. An important example is the balance between photosynthesis and the prevention of excessive water loss from the plant. The CO_2 required for photosynthesis enters a leaf (and the resulting O_2 exits) via stomata, the pores on the leaf surface (see Figure 10.4). However, stomata are also the main avenues of transpiration, the evaporative loss of water from leaves. On a hot, dry day, most plants close their stomata, a response that conserves water but also reduces CO_2 levels. With stomata even partially closed, CO_2 concentrations begin to decrease in the air spaces within the leaf, and the concentration of O_2 released from the light reactions begins to increase. These conditions within the leaf favour an apparently wasteful process called photorespiration.

Photorespiration: An Evolutionary Relic?

In most plants, initial fixation of carbon occurs via rubisco, the Calvin cycle enzyme that adds CO_2 to ribulose bisphosphate. Such plants are called **C_3 plants** because the first organic product of carbon fixation is a three-carbon compound, 3-phosphoglycerate (see Figure 10.19). C_3 plants include important agricultural plants such as rice, wheat, and soybeans. When their stomata close on hot, dry days, C_3 plants produce less sugar because the declining level of CO_2 in the leaf starves the Calvin cycle. In addition, rubisco is capable of binding O_2 in place of CO_2. As CO_2 becomes scarce within the air spaces of the leaf and O_2 builds up, rubisco adds O_2 to the Calvin cycle instead of CO_2. The product splits, and a two-carbon compound leaves the chloroplast. Peroxisomes and mitochondria within the plant cell rearrange and split this compound, releasing CO_2. The process is called **photorespiration** because it occurs in the light (*photo*) and consumes O_2 while producing CO_2 (*respiration*). However, unlike normal cellular respiration, photorespiration uses ATP rather than generating it. And unlike photosynthesis, photorespiration produces no sugar. In fact, photorespiration *decreases* photosynthetic output by siphoning organic material from the Calvin cycle and releasing CO_2 that would otherwise be fixed. This CO_2 can eventually be fixed if it is still in the leaf once the CO_2 concentration builds up to a high enough level. In the meantime, though, the process is energetically costly, much like a hamster running on its wheel.

How can we explain the existence of a metabolic process that seems to be counterproductive for the plant? According to one hypothesis, photorespiration is evolutionary baggage—a metabolic relic from a much earlier time when the atmosphere had less O_2 and more CO_2 than it does today. In the ancient atmosphere that prevailed when rubisco first evolved, the ability of the enzyme's active site to bind O_2 would have made little difference. The hypothesis suggests that modern rubisco retains some of its chance affinity for O_2, which is now so concentrated in the atmosphere that a certain amount of photorespiration is inevitable. There is also some evidence that photorespiration may provide protection against the damaging products of the light reactions, which build up when the Calvin cycle slows due to low CO_2.

In many types of plants—including a significant number of crop plants—photorespiration drains away as much as 50% of the carbon fixed by the Calvin cycle. Indeed, if photorespiration could be reduced in certain plant species without otherwise affecting photosynthetic productivity, crop yields and food supplies might increase.

In some plant species, alternate modes of carbon fixation have evolved that minimise photorespiration and optimise the Calvin cycle—even in hot, arid climates. The two most important of these photosynthetic adaptations are C_4 photosynthesis and crassulacean acid metabolism (CAM).

C_4 Plants

The **C_4 plants** are so named because they preface the Calvin cycle with an alternate mode of carbon fixation that forms a four-carbon compound as its first product. The C_4 pathway is believed to have evolved independently at least 45 separate times and is used by several thousand species in at least 19 plant families. Among the C_4 plants important to agriculture are sugarcane and corn, members of the grass family.

When the weather is hot and dry, a C_4 plant partially closes its stomata, conserving water but reducing the CO_2 concentration in the leaves. However, sugar is still made because C_4 plants use a multistep process that operates even under low CO_2 conditions. Photosynthesis begins in mesophyll cells but is completed in **bundle-sheath cells**, cells that are arranged into tightly packed sheaths around the veins of the leaf **(Figure 10.20)**. In C_4 leaves, the loosely arranged mesophyll cells are located between the bundle-sheath cells and the leaf surface, no more than two to three cells away from the bundle-sheath cells. In mesophyll cells, CO_2 is incorporated into organic compounds that then move into the bundle-sheath cells, where the Calvin cycle takes place. See the numbered steps in Figure 10.20, which are also described here:

1. The first step is carried out by an enzyme present only in mesophyll cells called **PEP carboxylase**, which has a much higher affinity for CO_2 than does rubisco, and no affinity for O_2. This enzyme adds CO_2 to

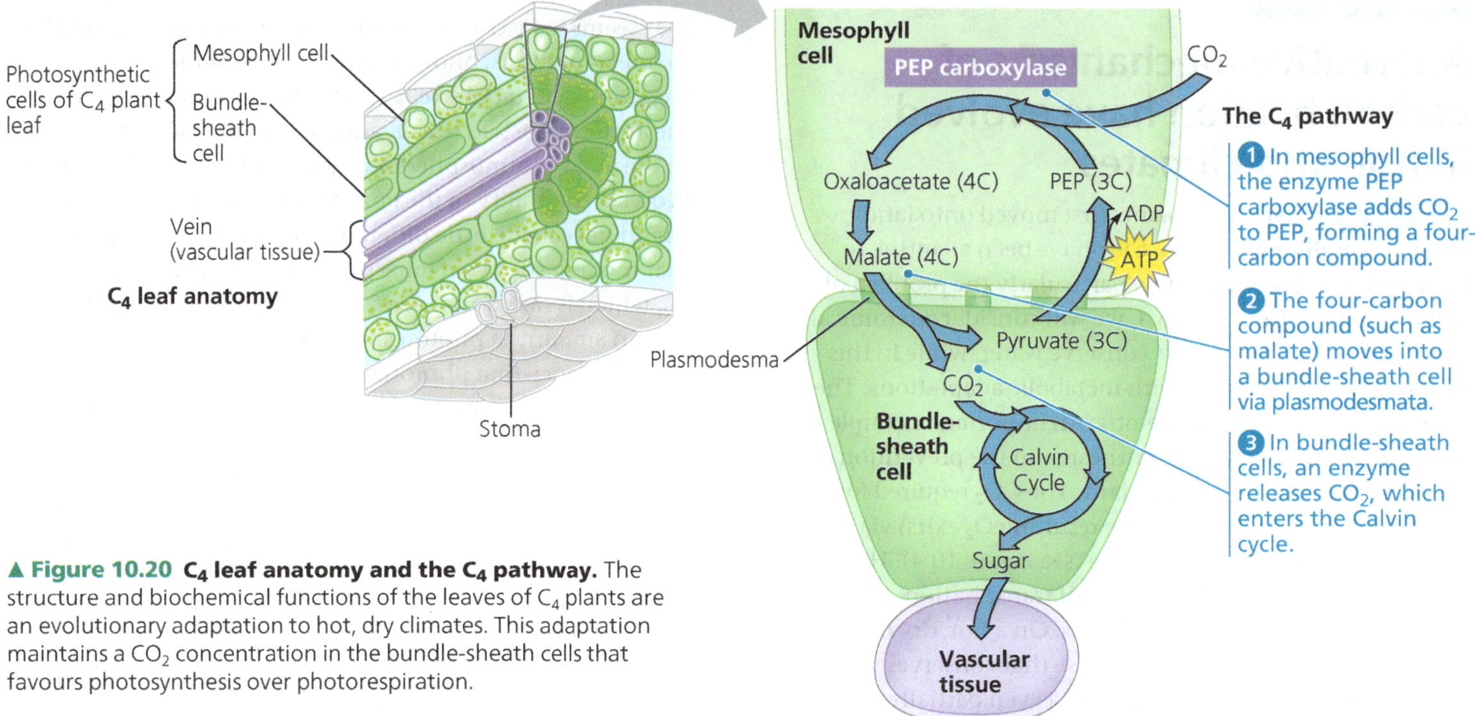

▲ **Figure 10.20 C_4 leaf anatomy and the C_4 pathway.** The structure and biochemical functions of the leaves of C_4 plants are an evolutionary adaptation to hot, dry climates. This adaptation maintains a CO_2 concentration in the bundle-sheath cells that favours photosynthesis over photorespiration.

phosphoenolpyruvate (PEP), forming the four-carbon product oxaloacetate; it does this even under conditions of lower CO_2 concentration and relatively higher O_2 concentration.

2. After the CO_2 is fixed in the mesophyll cells, the four-carbon products (malate in the example shown in Figure 10.20) are exported to bundle-sheath cells through plasmodesmata (see Figure 6.29).

3. Within the bundle-sheath cells, an enzyme releases CO_2 from the four-carbon compounds; the CO_2 is re-fixed into organic material by rubisco and the Calvin cycle. The same reaction regenerates pyruvate, which is transported to mesophyll cells. There, ATP is used to convert pyruvate to PEP, which can then accept addition of another CO_2, allowing the reaction cycle to continue. This ATP can be thought of, in a sense, as the "price" of concentrating CO_2 in the bundle-sheath cells. To generate this extra ATP, bundle-sheath cells carry out cyclic electron flow, the process described earlier in this chapter (see Figure 10.16). In fact, these cells contain PS I but no PS II, so cyclic electron flow is their only photosynthetic mode of generating ATP.

In effect, the mesophyll cells of a C_4 plant pump CO_2 into the bundle-sheath cells, keeping the CO_2 concentration in those cells high enough for rubisco to bind CO_2 rather than O_2. The cyclic series of reactions involving PEP carboxylase and the regeneration of PEP can be thought of as an ATP-powered pump that concentrates CO_2 where it can be fixed. In this way, C_4 photosynthesis spends ATP energy to minimise photorespiration and enhance sugar production. This adaptation is especially advantageous in hot regions with intense sunlight, and it is in such environments that C_4 plants evolved and thrive today.

The concentration of CO_2 in the atmosphere has drastically increased since the Industrial Revolution began in the 1800s, and it continues to rise today due to human activities such as the burning of fossil fuels (see Concept 1.1). The resulting global climate change, including an increase in average temperatures around the planet, may have far-reaching effects on plant species. Scientists are concerned that increasing CO_2 concentration and temperature may affect C_3 and C_4 plants differently, thus changing the relative abundance of these species in a given plant community.

Which type of plant would stand to gain more from increasing CO_2 levels? Recall that in C_3 plants, the binding of O_2 rather than CO_2 by rubisco leads to photorespiration, lowering the efficiency of photosynthesis. C_4 plants overcome this problem by concentrating CO_2 in the bundle-sheath cells at the cost of ATP. Rising CO_2 levels should benefit C_3 plants by lowering the amount of photorespiration that occurs. At the same time, rising temperatures have the opposite effect, increasing photorespiration. (Other factors such as water availability may also come into play.) In contrast, many C_4 plants could be largely unaffected by increasing CO_2 levels or temperature. Researchers have investigated aspects of this question in several studies; you can work with data from one such experiment in the **Scientific Skills Exercise**. In different regions, the particular combination of CO_2 concentration and temperature is likely to alter the balance of C_3 and C_4 plants in varying ways. The effects of such a widespread and variable change in community structure are unpredictable and thus a cause of legitimate concern.

Scientific Skills Exercise

Making Scatter Plots with Regression Lines

Does Atmospheric CO_2 Concentration Affect the Productivity of Agricultural Crops? The atmospheric concentration of CO_2 has been rising globally, and scientists wondered whether this would affect C_3 and C_4 plants differently. In this exercise, you will make a scatter plot to examine the relationship between CO_2 concentration and growth of both corn, a C_4 crop plant, and velvetleaf, a C_3 weed found in cornfields.

How the Experiment Was Done For 45 days, researchers grew corn and velvetleaf plants under controlled conditions, where all plants received the same amounts of water and light. The plants were divided into three groups, and each was exposed to a different concentration of CO_2 in the air: 350, 600, or 1,000 ppm (parts per million).

Data from the Experiment The table shows the dry mass (in grams) of corn and velvetleaf plants grown at the three concentrations of CO_2. The dry mass values are averages calculated from the leaves, stems, and roots of eight plants.

	350 ppm CO_2	600 ppm CO_2	1,000 ppm CO_2
Average dry mass of one corn plant (g)	91	89	80
Average dry mass of one velvetleaf plant (g)	35	48	54

Data from D. T. Patterson and E. P. Flint, Potential effects of global atmospheric CO_2 enrichment on the growth and competitiveness of C_3 and C_4 weed and crop plants, *Weed Science* 28(1):71–75 (1980).

▶ **Corn plant surrounded by invasive velvetleaf plants.**

INTERPRET THE DATA

1. To explore the relationship between the two variables, it is useful to graph the data in a scatter plot, and then draw a regression line. **(a)** First, place labels for the dependent and independent variables on the appropriate axes. Explain your choices. **(b)** Plot the data points for corn and velvetleaf using different symbols for each set of data, and add a key for the two symbols. (For additional information about graphs, see the Scientific Skills Review in Appendix D.)
2. Draw a "best-fit" line for each set of points. A best-fit line does not necessarily pass through all or even most points. Instead, it is a straight line that passes as close as possible to all data points from that set. Draw a best-fit line for each set of data. Because placement of the line is a matter of judgment, two individuals may draw two slightly different lines for a given set of points. The line that actually fits best, a regression line, can be identified by squaring the distances of all points to any candidate line, then selecting the line that minimises the sum of the squares. (See the graph in the Scientific Skills Exercise in the chapter, "Water and Life" for an example of a linear regression line.) Excel or other software programs, including those on a graphing calculator, can plot a regression line once data points are entered. Using either Excel or a graphing calculator, enter the data points for each data set and have the program draw the two regression lines. Compare them to the lines you drew.
3. Describe the trends shown by the regression lines in your scatter plot. **(a)** Compare the relationship between increasing concentration of CO_2 and the dry mass of corn to that for velvetleaf. **(b)** Considering that velvetleaf is a weed invasive to cornfields, predict how increased CO_2 concentration may affect interactions between the two species.
4. Based on the data in the scatter plot, estimate the percentage change in dry mass of corn and velvetleaf plants if atmospheric CO_2 concentration increased from 390 ppm (current levels) to 800 ppm. **(a)** What is the estimated dry mass of corn and velvetleaf plants at 390 ppm? 800 ppm? **(b)** To calculate the percentage change in mass for each plant, subtract the mass at 390 ppm from the mass at 800 ppm (change in mass), divide by the mass at 390 ppm (initial mass), and multiply by 100. What is the estimated percentage change in dry mass for corn? For velvetleaf? **(c)** Do these results support the conclusion from other experiments that C_3 plants grow better than C_4 plants under increased CO_2 concentration? Why or why not?

C_4 photosynthesis is considered more efficient than C_3 photosynthesis because it uses less water and resources. On our planet today, the world population and demand for food are rapidly increasing. At the same time, the amount of land suitable for growing crops is decreasing due to the effects of global climate change, which include an increase in sea level as well as a hotter, drier climate in many regions. To address issues of food supply, scientists in the Philippines have been working on genetically modifying rice—an important food staple that is a C_3 crop—so that it can instead carry out C_4 photosynthesis. Results so far seem promising, and these researchers estimate that the yield of C_4 rice might be 30–50% higher than C_3 rice with the same input of water and resources.

CAM Plants

A second photosynthetic adaptation to arid conditions has evolved in pineapples, many cacti, and other succulent (water-storing) plants, such as aloe and jade plants. These plants open their stomata at night and close them during the day, just the reverse of how other plants behave. Closing stomata during the day helps desert plants conserve water, but it also prevents CO_2 from entering the leaves. During the night, when their stomata are open, these plants take up CO_2 and incorporate it into a variety of organic acids. This mode of carbon fixation is called **crassulacean acid metabolism**, or **CAM**, after the plant family Crassulaceae, the succulents in which the process was first discovered. The mesophyll cells of **CAM plants** store the organic acids they make during the night in their vacuoles until morning, when the stomata close. During the day, when the light reactions can supply ATP and NADPH for the Calvin cycle, CO_2 is released from the organic acids made the night before to become incorporated into sugar in the chloroplasts.

Notice in **Figure 10.21** that the CAM pathway is similar to the C_4 pathway in that CO_2 is first incorporated into organic intermediates before it enters the Calvin cycle. The difference is that in C_4 plants, the initial steps of carbon fixation are separated structurally from the Calvin cycle, whereas in CAM plants, the two steps occur within the same cell but at separate times. (Keep in mind that CAM, C_4, and C_3 plants all eventually use the Calvin cycle to make sugar from carbon dioxide.)

▼ **Figure 10.21 C_4 and CAM photosynthesis compared.** The C_4 and CAM pathways are two evolutionary solutions to the problem of maintaining photosynthesis with stomata partially or completely closed on hot, dry days. Both adaptations are characterised by: ① preliminary incorporation of CO_2 into organic acids, followed by ② transfer of CO_2 to the Calvin cycle.

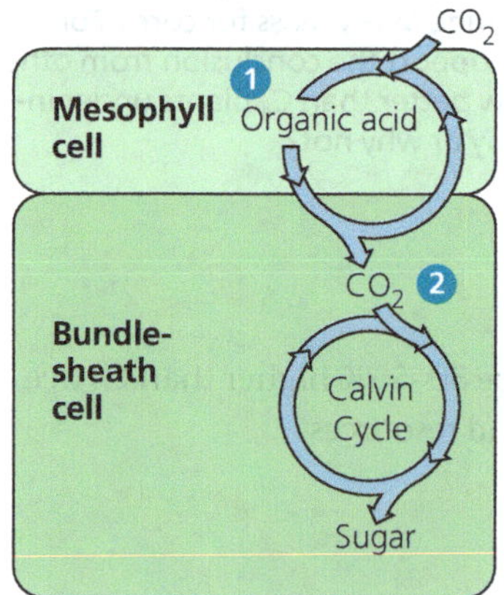

(a) Spatial separation of steps. In C_4 plants, carbon fixation and the Calvin cycle occur in different types of cells.

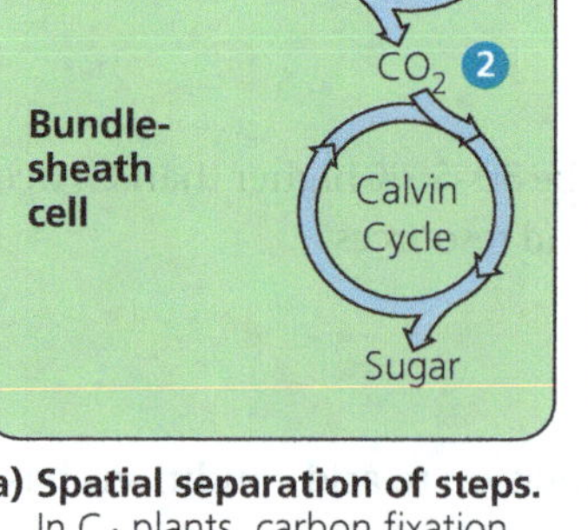

(b) Temporal separation of steps. In CAM plants, carbon fixation and the Calvin cycle occur in the same cell at different times.

CONCEPT CHECK 10.5

1. Describe how photorespiration lowers photosynthetic output for plants.
2. The presence of only PS I, not PS II, in the bundle-sheath cells of C_4 plants has an effect on O_2 concentration. What is that effect, and how might that benefit the plant?
3. **MAKE CONNECTIONS** Refer to the discussion of ocean acidification in Concept 3.3. Ocean acidification and changes in the distribution of C_3 and C_4 plants may seem to be two very different problems, but what do they have in common? Explain.
4. **WHAT IF?** How would you expect the relative abundance of C_3 versus C_4 and CAM species to change in a geographic region whose climate becomes much hotter and drier, with no change in CO_2 concentration?

For suggested answers, see Appendix A.

CONCEPT 10.6

Photosynthesis is essential for life on Earth: a review

In this chapter, we have followed photosynthesis from photons to food. The light reactions capture solar energy and use it to make ATP and transfer electrons from H_2O to $NADP^+$, forming NADPH. The Calvin cycle uses the ATP and NADPH to produce sugar from CO_2. The energy that enters the chloroplasts as sunlight becomes stored as chemical energy in organic compounds. The entire process is reviewed visually in **Figure 10.22**, where photosynthesis is also put in its natural context.

As for the fates of photosynthetic products, enzymes in the chloroplast and cytosol convert the G3P made in the Calvin cycle to many other organic compounds. In fact, the sugar made in the chloroplasts supplies the entire plant with chemical energy and carbon skeletons for the synthesis of all the major organic molecules of plant cells. About 50% of the organic material made by photosynthesis is consumed as fuel for cellular respiration in plant cell mitochondria.

Technically, green cells are the only autotrophic parts of the plant. The rest of the plant depends on organic molecules exported from leaves through veins (see Figure 10.22, top). In most plants, carbohydrate is transported out of the leaves to the rest of the plant in the form of sucrose, a disaccharide. After arriving at nonphotosynthetic cells, the sucrose provides raw material for cellular respiration and a multitude of anabolic pathways that synthesise proteins, lipids, and other products. A considerable amount of sugar in the form of glucose is linked together to make the polysaccharide cellulose (see Figure 5.6c), especially in plant cells that are still growing and maturing. Cellulose, the main ingredient of cell walls, is the most abundant organic molecule in the plant—and probably on the surface of the planet.

Most plants and other photosynthesisers make more organic material per day than they need to use as respiratory fuel and precursors for biosynthesis. For example, flowering plants stockpile the extra sugar in the form of starch, storing some in the chloroplasts and some in cells of roots, tubers, seeds, and fruits. In accounting for the use of food molecules produced by photosynthesis, note that most flowering plants lose leaves, roots, stems, fruits, and often their entire bodies to heterotrophs, including humans.

On a global scale, photosynthesis is the process responsible for the presence of O_2 in our atmosphere. Furthermore, although each chloroplast is minuscule, their collective productivity in terms of food production is enormous: Photosynthesis makes an estimated 150 billion metric tons of carbohydrate per year (a metric ton is 1,000 kg). That's organic matter equivalent in mass to a stack of about 60 trillion biology textbooks—and the stack would reach 17 times the distance from Earth to the sun! No chemical process is more important than photosynthesis to the life on Earth.

In Chapters 5 through 10, you have learned about many activities of cells. **Figure 10.23** integrates these cellular processes into the context of a working plant cell. As you study the figure, reflect on how each process fits into the big picture: As the most basic unit of living organisms, a cell performs all functions characteristic of life.

CONCEPT CHECK 10.6

1. **MAKE CONNECTIONS** How do plants use the sugar they produce during photosynthesis to directly power the work of the cell? Provide some examples of cellular work. (See Figures 8.10, 8.11, and 9.5.)

For suggested answers, see Appendix A.

▼ Figure 10.22 A review of photosynthesis. This diagram shows the main reactants and products of photosynthesis as they move through the tissues of a tree (left) and a chloroplast (right).

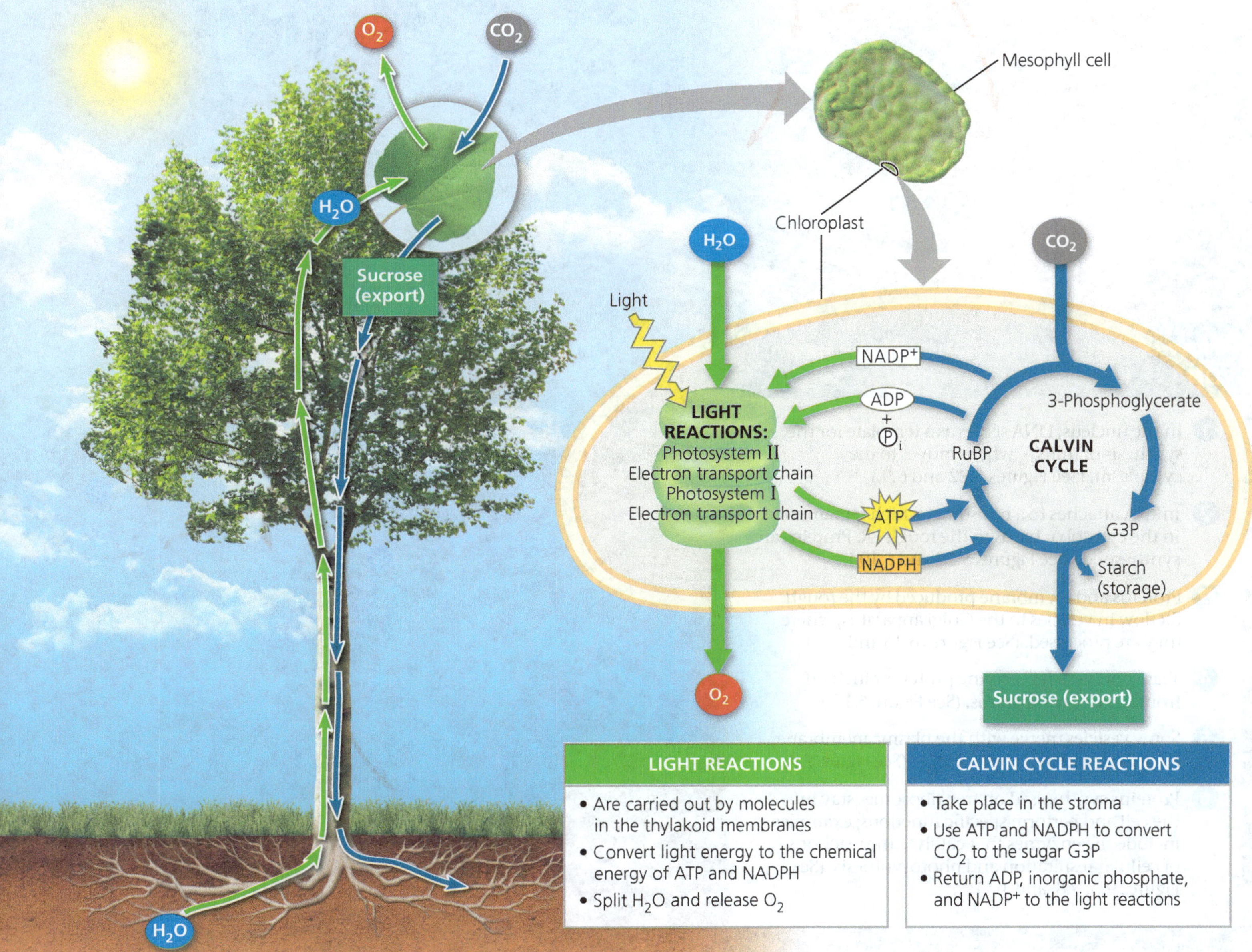

▼ **Figure 10.23**

MAKE CONNECTIONS

The Working Cell

This figure illustrates how a generalised plant cell functions, integrating the cellular activities you learned about in Chapters 5–10. To see some of the enzymes involved, look at Figure 6.32.

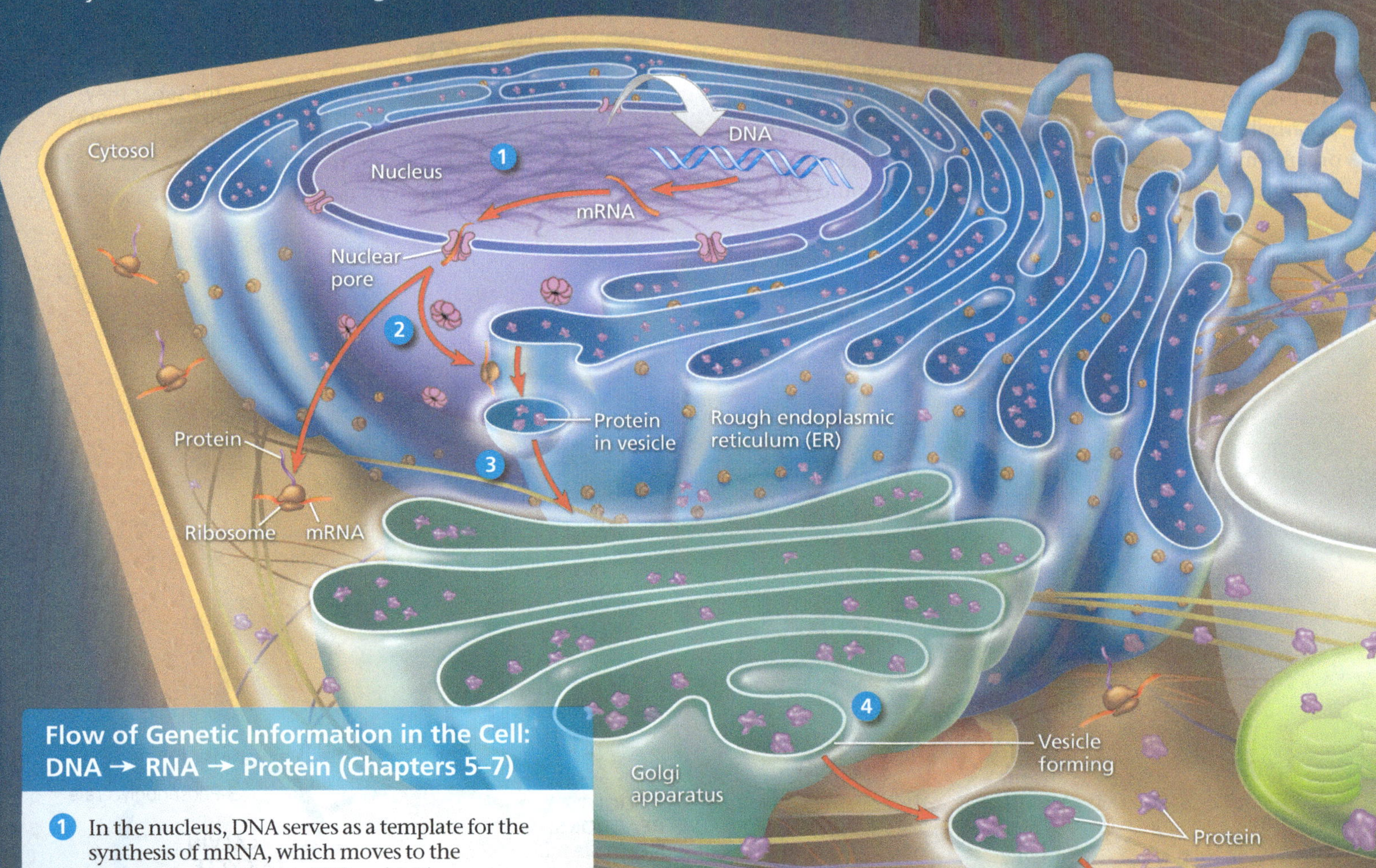

Flow of Genetic Information in the Cell: DNA → RNA → Protein (Chapters 5–7)

1. In the nucleus, DNA serves as a template for the synthesis of mRNA, which moves to the cytoplasm. (See Figures 5.22 and 6.9.)
2. mRNA attaches to a ribosome, which remains free in the cytosol or binds to the rough ER. Proteins are synthesised. (See Figures 5.22 and 6.10.)
3. Proteins and membrane produced by the rough ER flow in vesicles to the Golgi apparatus, where they are processed. (See Figures 6.15 and 7.9.)
4. Transport vesicles carrying proteins pinch off from the Golgi apparatus. (See Figure 6.15.)
5. Some vesicles merge with the plasma membrane, releasing proteins by exocytosis. (See Figure 7.9.)
6. Proteins synthesised on free ribosomes stay in the cell and perform specific functions; examples include the enzymes that catalyse the reactions of cellular respiration and photosynthesis. (See Figures 9.6, 9.8, and 10.19.)

Movement Across Cell Membranes (Chapter 7)

7 Water diffuses into and out of the cell directly through the plasma membrane and by facilitated diffusion through aquaporins. (See Figure 7.10.)

8 By passive transport, the CO_2 used in photosynthesis diffuses into the cell, and the O_2 formed as a by-product of photosynthesis diffuses out of the cell. Both solutes move down their concentration gradients. (See Figures 7.11 and 10.22.)

9 In active transport, energy (usually supplied by ATP) is used to transport a solute against its concentration gradient. (See Figure 7.17.)

Exocytosis (shown in step 5) and endocytosis move larger materials out of and into the cell. (See Figures 7.20 and 7.21.)

Energy Transformations in the Cell: Photosynthesis and Cellular Respiration (Chapters 8–10)

10 In chloroplasts, the process of photosynthesis uses the energy of light to convert CO_2 and H_2O to organic molecules, with O_2 as a by-product. (See Figures 10.1 and 10.22.)

11 In mitochondria, organic molecules are broken down by cellular respiration, capturing energy in molecules of ATP, which are used to power the work of the cell, such as protein synthesis and active transport. CO_2 and H_2O are by-products. (See Figures 8.9, 8.11, 9.1, and 9.15.)

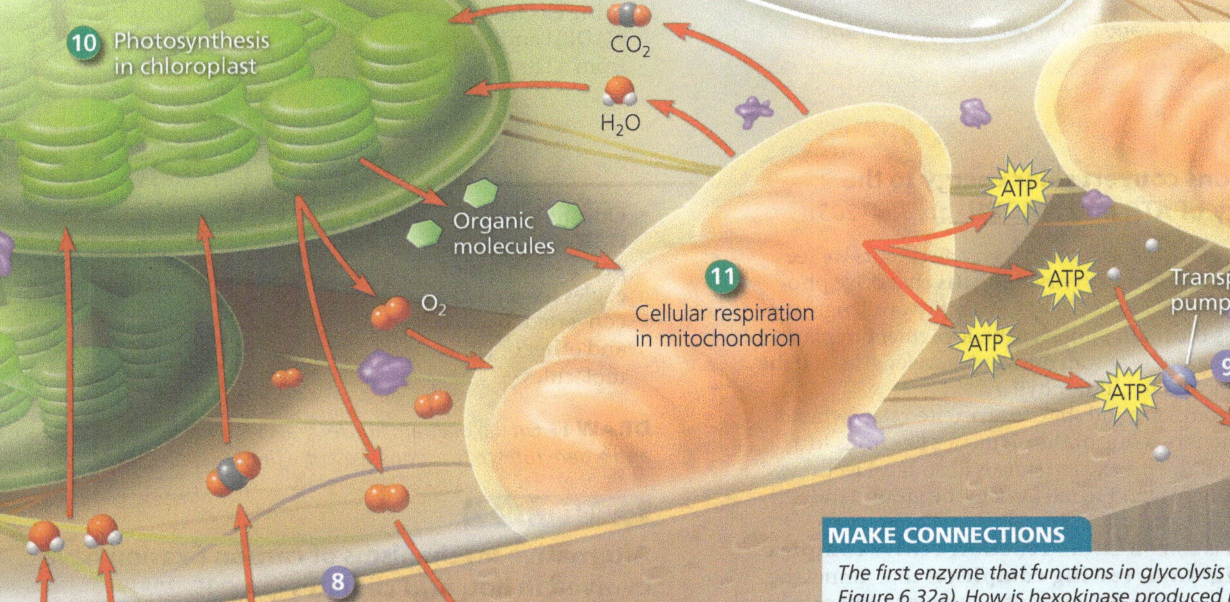

MAKE CONNECTIONS

The first enzyme that functions in glycolysis is hexokinase (see Figure 6.32a). How is hexokinase produced in this plant cell? Specify the locations of each step of the process. Where does hexokinase function? (See Figures 5.18, 5.22, and 9.8.)

10 Chapter Review

SUMMARY OF KEY CONCEPTS

CONCEPT 10.1

Photosynthesis feeds the biosphere *(p. 190)*

- **Photosynthesis** converts the energy of sunlight into chemical energy stored in sugars.
- **Autotrophs** ("producers") produce their organic molecules from CO_2 and other inorganic raw materials; plants are *photoautotrophs* that use light energy to do this. **Heterotrophs** (consumers, including decomposers) are unable to make their own food and live on compounds produced by others.

? *Using the terms producers, consumers, and decomposers, explain how photosynthesis feeds organisms, either directly or indirectly.*

CONCEPT 10.2

Photosynthesis converts light energy to the chemical energy of food *(pp. 191–194)*

- In plants and other eukaryotic autotrophs, photosynthesis occurs in **chloroplasts**, organelles containing **thylakoids**. Stacks of thylakoids form grana. Photosynthesis is summarised as

  $6\,CO_2 + 12\,H_2O + \text{Light energy} \rightarrow C_6H_{12}O_6 + 6\,O_2 + 6\,H_2O.$

  Chloroplasts split water into hydrogen and oxygen, incorporating the electrons of hydrogen into sugar molecules. Photosynthesis is a redox process: H_2O is oxidised, and CO_2 is reduced. The **light reactions** in the thylakoid membranes split water, releasing O_2, producing ATP, and forming **NADPH**. The **Calvin cycle** in the **stroma** forms sugar from CO_2, using ATP for energy and NADPH for reducing power.

? *Compare the roles of CO_2 and H_2O in cellular respiration and photosynthesis.*

CONCEPT 10.3

The light reactions convert solar energy to the chemical energy of ATP and NADPH *(pp. 194–203)*

- Light is a form of electromagnetic energy. The colours we see as **visible light** include those **wavelengths** that drive photosynthesis. A pigment absorbs light of specific wavelengths; **chlorophyll *a*** is the main photosynthetic pigment in plants. Other accessory pigments absorb different wavelengths of light and pass the energy on to chlorophyll *a*.
- A pigment goes from a ground state to an excited state when a **photon** of light boosts one of the pigment's electrons to a higher-energy orbital. This excited state is unstable. Electrons from isolated pigments tend to fall back to the ground state, giving off heat and/or light.
- A **photosystem** is composed of a **reaction-centre complex** surrounded by **light-harvesting complexes** that funnel the energy of photons to the reaction-centre complex. When a special pair of reaction-centre chlorophyll *a* molecules absorbs energy, one of its electrons is boosted to a higher energy level and transferred to the **primary electron acceptor. Photosystem II** contains P680 chlorophyll *a* molecules in the reaction-centre complex; **photosystem I** contains P700 molecules.
- **Linear electron flow** during the light reactions uses both photosystems and produces NADPH, ATP, and oxygen:

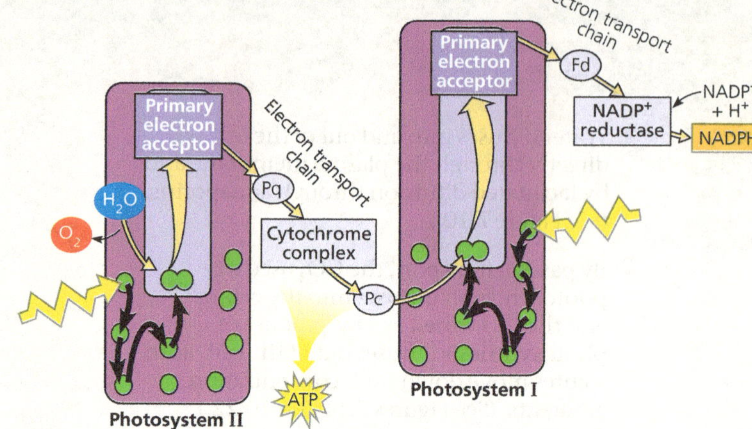

- **Cyclic electron flow** employs only one photosystem, producing ATP but no NADPH or O_2.
- During chemiosmosis in both mitochondria and chloroplasts, electron transport chains generate an H^+ gradient across a membrane. ATP synthase uses this proton-motive force to make ATP.

? *The absorption spectrum of chlorophyll a differs from the action spectrum of photosynthesis. Explain this observation.*

CONCEPT 10.4

The Calvin cycle uses the chemical energy of ATP and NADPH to reduce CO_2 to sugar *(pp. 203–204)*

- The Calvin cycle occurs in the stroma, using electrons from NADPH and energy from ATP.
- One molecule of **G3P** exits the cycle per three CO_2 molecules fixed and is converted to glucose and other organic molecules.

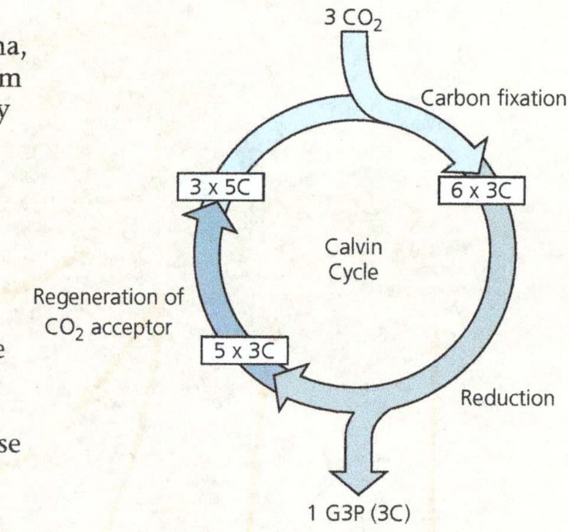

DRAW IT *On the diagram above, draw where ATP and NADPH are used and where rubisco functions. Describe these steps.*

CONCEPT 10.5

Alternative mechanisms of carbon fixation have evolved in hot, arid climates *(pp. 205–208)*

- On dry, hot days, **C_3 plants** close their stomata, conserving water but keeping CO_2 out and O_2 in. Under these conditions, **photorespiration** can occur: **Rubisco** binds O_2 instead of CO_2, consuming ATP and releasing CO_2 without producing ATP or carbohydrate. Photorespiration may be an evolutionary relic, and it may play a photoprotective role.
- C_4 and CAM plants are adopted to hot, dry climates.

- **C_4 plants** minimise the cost of photorespiration by incorporating CO_2 into four-carbon compounds in mesophyll cells. These compounds are exported to **bundle-sheath cells**, where they release carbon dioxide for use in the Calvin cycle.
- **CAM plants** open their stomata at night, incorporating CO_2 into organic acids, which are stored in mesophyll cells. During the day, the stomata close, and the CO_2 is released from the organic acids for use in the Calvin cycle.
- Organic compounds produced by photosynthesis provide the energy and building material for Earth's ecosystems.

? *Why are C_4 and CAM photosynthesis more energetically expensive than C_3 photosynthesis? What climate conditions would favour C_4 and CAM plants?*

CONCEPT **10.6**

Photosynthesis is essential for life on Earth: a review

(pp. 208–211)

- Organic compounds produced by photosynthesis provide the energy and building material for Earth's ecosystems.

? *How do plants use the products of photosynthesis?*

TEST YOUR UNDERSTANDING

Levels 1-2: Remembering/Understanding

1. The light reactions supply the Calvin cycle with
(A) light energy.
(B) CO_2 and ATP.
(C) H_2O and NADPH.
(D) ATP and NADPH.

2. Which of the following sequences correctly represents the flow of electrons during photosynthesis?
(A) NADPH → O_2 → CO_2
(B) H_2O → NADPH → Calvin cycle
(C) H_2O → photosystem I → photosystem II
(D) NADPH → electron transport chain → O_2

3. How is photosynthesis similar in C_4 plants and CAM plants?
(A) In both cases, only photosystem I is used.
(B) Both types of plants make sugar without the Calvin cycle.
(C) In both cases, rubisco is not used to fix carbon initially.
(D) Both types of plants make most of their sugar in the dark.

4. Which of the following statements is a correct distinction between autotrophs and heterotrophs?
(A) Autotrophs, but not heterotrophs, can nourish themselves beginning with CO_2 and other nutrients that are inorganic.
(B) Only heterotrophs require chemical compounds from the environment.
(C) Cellular respiration is unique to heterotrophs.
(D) Only heterotrophs have mitochondria.

5. Which of the following occurs during the Calvin cycle?
(A) carbon fixation
(B) reduction of $NADP^+$
(C) release of oxygen
(D) generation of CO_2

Levels 3-4: Applying/Analysing

6. In mechanism, photophosphorylation is most similar to
(A) substrate-level phosphorylation in glycolysis.
(B) oxidative phosphorylation in cellular respiration.
(C) carbon fixation.
(D) reduction of $NADP^+$.

7. Which process is most directly driven by light energy?
(A) creation of a pH gradient by pumping protons across the thylakoid membrane
(B) reduction of $NADP^+$ molecules
(C) transfer of energy from pigment molecule to pigment molecule
(D) ATP synthesis

Levels 5-6: Evaluating/Creating

8. SCIENCE, TECHNOLOGY, AND SOCIETY Scientific evidence indicates that the CO_2 added to the air by the burning of wood and fossil fuels is contributing to global warming, a rise in global temperature. Tropical rain forests are estimated to be responsible for approximately 20% of global photosynthesis, yet the consumption of large amounts of CO_2 by living trees is thought to make little or no *net* contribution to reduction of global warming. Explain why this might be the case. (*Hint*: What processes in both living and dead trees produce CO_2?)

9. EVOLUTION CONNECTION Photorespiration can decrease soybeans' photosynthetic output by about 50%. Would this figure be higher or lower in wild relatives of soybeans? Why?

10. SCIENTIFIC INQUIRY • DRAW IT The following diagram represents an experiment with isolated thylakoids. The thylakoids were first made acidic by soaking them in a solution at pH 4. After the thylakoid space reached pH 4, the thylakoids were transferred to a basic solution at pH 8. The thylakoids then made ATP in the dark. (See Concept 3.3 to review pH.)

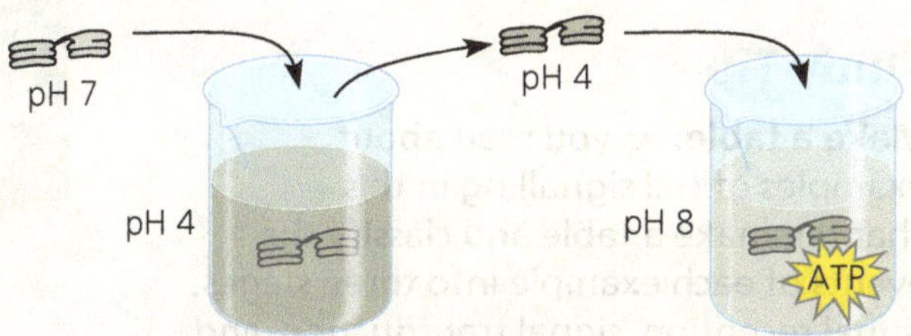

Draw an enlargement of part of the thylakoid membrane in the beaker with the solution at pH 8. Draw ATP synthase. Label the areas of high H^+ concentration and low H^+ concentration. Show the direction protons flow through the enzyme, and show the reaction where ATP is synthesised. Would ATP end up in the thylakoid or outside of it? Explain why the thylakoids in the experiment were able to make ATP in the dark.

11. WRITE ABOUT A THEME: ENERGY AND MATTER In a short essay (100–150 words), describe how photosynthesis transforms solar energy into the chemical energy of sugar molecules.

12. SYNTHESISE YOUR KNOWLEDGE

"Watermelon snow" in Antarctica is caused by a photosynthetic green algae (*Chlamydomonas nivalis*) that thrives in subzero temperatures. The same algae also occur in high-altitude, year-round snowfields. In both locations, UV light levels tend to be high. Propose an explanation for why this alga appears reddish-pink.

For selected answers, see Appendix A.

11 Cell Communication

KEY CONCEPTS

Figure 11.1 This impala is fleeing for its life, racing to escape the predatory cheetah nipping at its heels. The impala is breathing rapidly, its heart pounding and its legs pumping furiously. These physiological functions are all part of the impala's "fight-or-flight" response, driven by hormones released from its adrenal glands at times of stress—in this case, upon sensing the cheetah.

Study Tip

Make a table: As you read about examples of cell signalling in this chapter, make a table and classify the events of each example into three stages: signal reception, signal transduction, and cellular response.

Example of Cell Signalling	Signal Reception	Signal Trans-duction	Cellular Response
Adrenaline	Adrenaline binds cell-surface receptor.	Relay molecules each activate the next molecule.	An enzyme is activated that breaks down glycogen into glucose for energy to fight or flee.

How does cell signalling fuel the desperate flight of an impala?

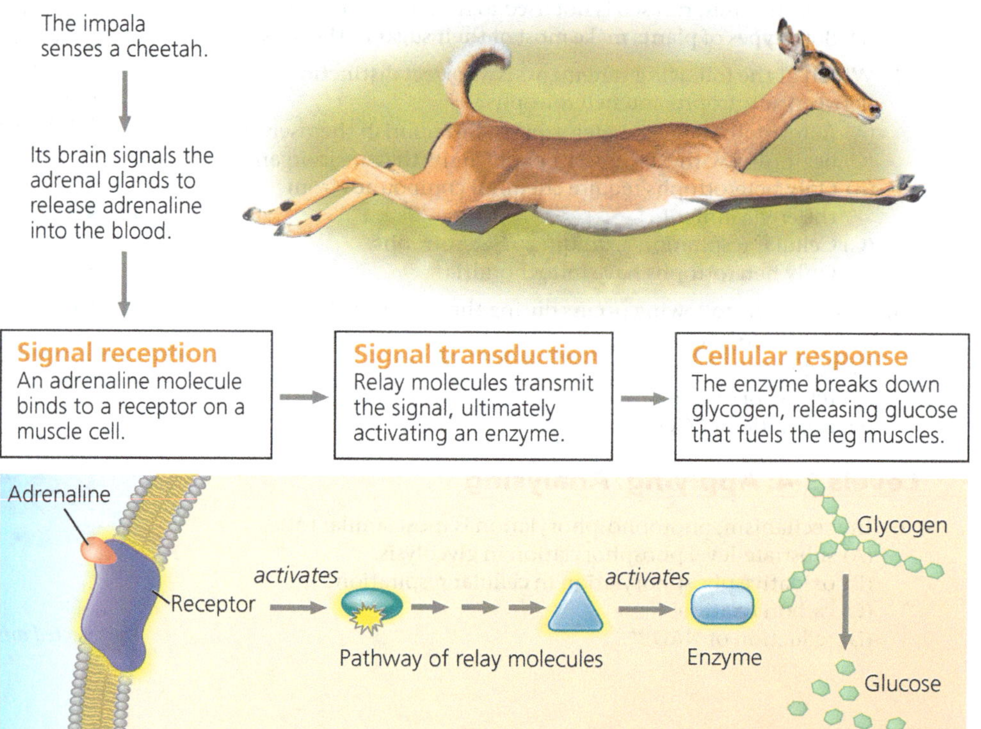

Go to Mastering Biology

to access Dynamic Study Modules for revision, 3D BioFlix® animations and high-quality videos, and your interactive Pearson eText.

CONCEPT 11.1

External signals are converted to responses within the cell

Scientists think that signalling mechanisms first evolved hundreds of millions of years ago in ancient prokaryotes and single-celled eukaryotes and then were adapted for new uses in their multicellular descendants. So let's begin by considering signalling in some examples of single-celled organisms: bacteria and yeasts.

Evolution of Cell Signalling

EVOLUTION Research during the 1970s suggested that bacterial cells—somewhat surprisingly, since they are single-celled organisms—were capable of signalling to each other. Since then, we have come to understand that cell signalling is critical among prokaryotes. Bacterial cells secrete molecules that can be detected by other bacterial cells **(Figure 11.2)**. Sensing the concentration of such signalling molecules allows bacteria to monitor their own local cell density, a phenomenon called *quorum sensing*.

Quorum sensing allows bacterial populations to coordinate the behaviour of all cells in a population in activities that require a given density of cells acting at the same time. One example is formation of a *biofilm*, an aggregation of bacterial cells attached to a surface by molecules secreted by the cells, but only after the cells have reached a certain density. The biofilm protects the cells in it, and they often derive nutrition from the surface they are on. Biofilms are believed to be involved in up to 80% of all human bacterial infections. You have probably encountered biofilms many times, perhaps without realising it. The slimy coating on a fallen log or on leaves lying on a forest path, and even the film on your teeth each morning, are examples of bacterial biofilms. In fact, tooth-brushing and flossing disrupt biofilms that would otherwise cause cavities and gum disease.

Another example of bacterial behaviour coordinated by quorum sensing is the secretion of toxins by infectious bacteria, which has serious medical implications. Sometimes treatment by antibiotics doesn't work with such infections because of antibiotic resistance that has evolved in a particular strain of bacteria. A promising alternative treatment would be to disrupt toxin production by interfering with the signalling pathways used in quorum sensing. In the **Problem-Solving Exercise**, you can participate in the process of scientific thinking involved in this novel approach.

Now let's look at an example of cell signalling in yeasts (single-celled fungi). Cells of the yeast *Saccharomyces cerevisiae*—which are used to make bread, wine, and beer—identify their sexual mates by chemical signalling when they reproduce sexually. There are two sexes, or mating types, called **a** and **α** **(Figure 11.3)**. Each type secretes a specific

▼ **Figure 11.2 Communication among bacteria.** Soil-dwelling bacteria called myxobacteria ("slime bacteria") use chemical signals to share information about nutrient availability. When food is scarce, starving cells secrete a signalling molecule that stimulates neighbouring cells to aggregate. The cells form a structure called a fruiting body that produces spores, thick-walled cells that can survive until the environment improves. The myxobacteria shown here are the species *Myxococcus xanthus* (steps 1–3, SEMs; lower photo, LM).

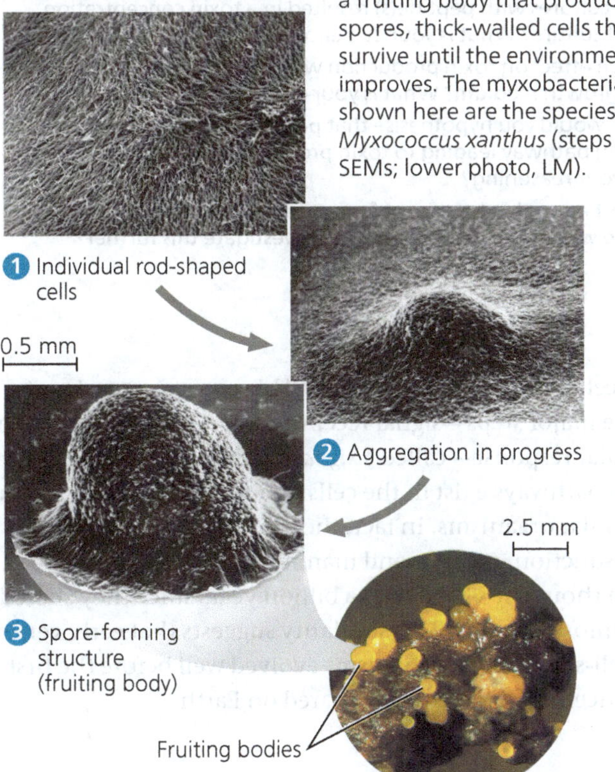

▼ **Figure 11.3 Communication between mating yeast cells.** *Saccharomyces cerevisiae* cells use chemical signalling to identify cells of the opposite mating type and initiate the mating process. The two mating types and their corresponding chemical signalling molecules, or mating factors, are called **a** and **α**.

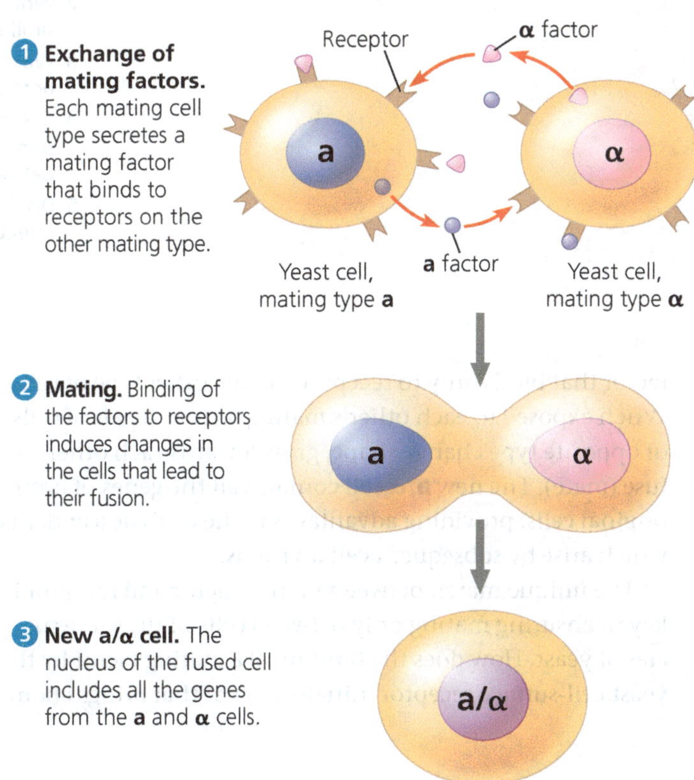

PROBLEM-SOLVING EXERCISE

Can a skin wound turn deadly?

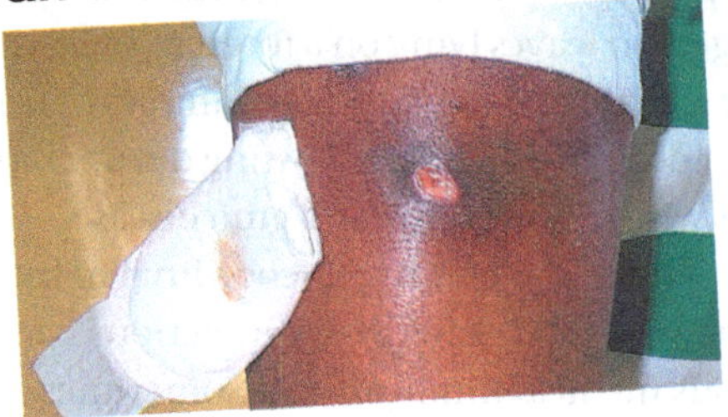

Staphylococcus aureus (S. aureus) is a common bacterial species found on the surface of healthy skin that can turn into a serious pathogen if introduced into tissue through a cut or abrasion. Once *S. aureus* cells enter the body and reach a certain density, they secrete a toxin that kills body cells and contributes significantly to inflammation and damage. Because about one in 100 people carry a strain of *S. aureus* that is resistant to common antibiotics, a minor infection can turn permanently harmful or even deadly.

Cells sense their own population density by *quorum sensing*, and at a certain density they start to secrete toxin. In this exercise, you will analyse whether blocking quorum sensing can stop *S. aureus* from producing toxin.

Your Approach In *S. aureus,* quorum sensing involves two separate signal transduction pathways. Two candidate synthetic peptides (short proteins), called peptides 1 and 2, have been proposed to interfere with *S. aureus* quorum-sensing pathways. Your job is to test each potential inhibitor of quorum sensing to see if it blocks either or both of the pathways that lead to toxin production.

For your experiment, you grow four cultures of *S. aureus* to a standardised high density and measure the concentration of toxin in the culture. The control culture contains no peptide. The other cultures have one or both candidate inhibitory peptides mixed into the growth medium before starting the cultures.

Your Data

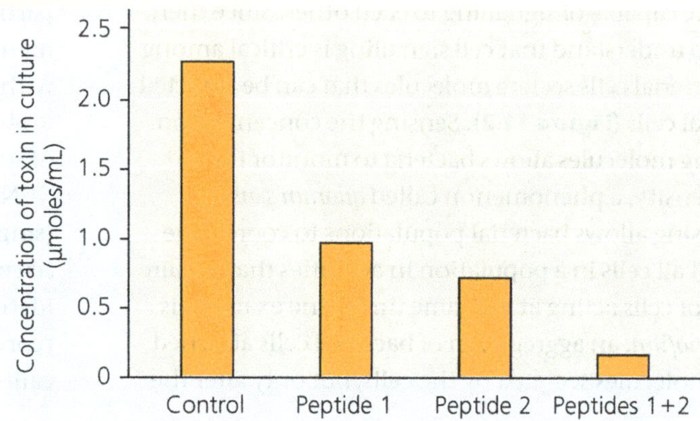

Data from N. Balaban et al., Treatment of *Staphylococcus aureus* biofilm infection by the quorum-sensing inhibitor RIP, *Antimicrobial Agents and Chemotherapy* 51(6):2226–2229 (2007).

Your Analysis

1. Rank the cultures according to toxin production, from most to least.
2. Which, if any, of the cultures with peptide(s) resulted in a toxin concentration similar to the control culture? What is your evidence for this?
3. Was there an additive effect on toxin production when peptides 1 and 2 were both present in the growth medium? What is your evidence for this?
4. Based on these data, would you hypothesise that peptides 1 and 2 act on the same quorum-sensing pathway leading to toxin production or on two different pathways? What is your reasoning?
5. Do these data suggest a possible treatment for antibiotic-resistant *S. aureus* infections? What else would you want to know to investigate this further?

factor that binds only to receptors on the other type of cell. When exposed to each other's mating factors, a pair of cells of opposite type change shape, grow towards each other, and fuse (mate). The new **a**/**α** cell contains all the genes of both original cells, providing advantages to the cell's descendants, which arise by subsequent cell divisions.

The unique match between mating factor and receptor is key to ensuring mating only between cells of the same species of yeast. How does the binding of a mating factor by the yeast cell-surface receptor initiate a signal that brings about the cellular response of mating? This occurs in a series of three major steps—signal reception, signal transduction, and cellular response—called a *signal transduction pathway*. Many such pathways exist in the cells of both unicellular and multicellular organisms. In fact, the molecular details of signal transduction in yeasts and mammals are strikingly similar, even though it's been over a billion years since they shared a common ancestor. This similarity suggests that early versions of cell-signalling mechanisms evolved well before the first multicellular organisms appeared on Earth.

Local and Long-Distance Signalling

Like bacteria or yeast cells, cells in a multicellular organism communicate by way of a signalling molecules targeted for cells that may or may not be immediately adjacent. As we saw in Concepts 6.7 and 7.1, eukaryotic cells may communicate by direct contact, which is one type of local signalling. Many animal and plant cells have cell junctions that directly connect the cytoplasms of adjacent cells **(Figure 11.4a)**. In these cases, signalling substances dissolved in the cytosol can pass between neighbouring cells. Moreover, some animal cells may communicate by direct contact between cell-surface molecules, as shown in **Figure 11.4b**. This type of local signalling is especially important in embryonic development, the immune response, and in maintaining adult stem cell populations.

In many other cases of local signalling, signalling molecules are secreted by the signalling cell. Some molecules travel only short distances; such local regulators influence cells that are nearby. This type of local signalling in animals is called *paracrine signalling* **(Figure 11.5a)**. One class of local regulators in animals, *growth factors*, consists of compounds that stimulate nearby target cells to grow and divide. Numerous cells can simultaneously receive and respond to the growth factors produced by a single cell in their vicinity.

A highly specialised type of local signalling called *synaptic signalling* occurs in the animal nervous system **Figure 11.5b**; see Concept 48.4. An electrical signal along a nerve cell triggers the secretion of neurotransmitter molecules. These molecules act as chemical signals, diffusing across the synapse—the narrow space between the nerve cell and its target cell—triggering a response in the target cell. Drugs to treat depression, anxiety, and post-traumatic stress disorder (PTSD) affect this signalling process.

▼ Figure 11.4 Communication requiring contact between cells.

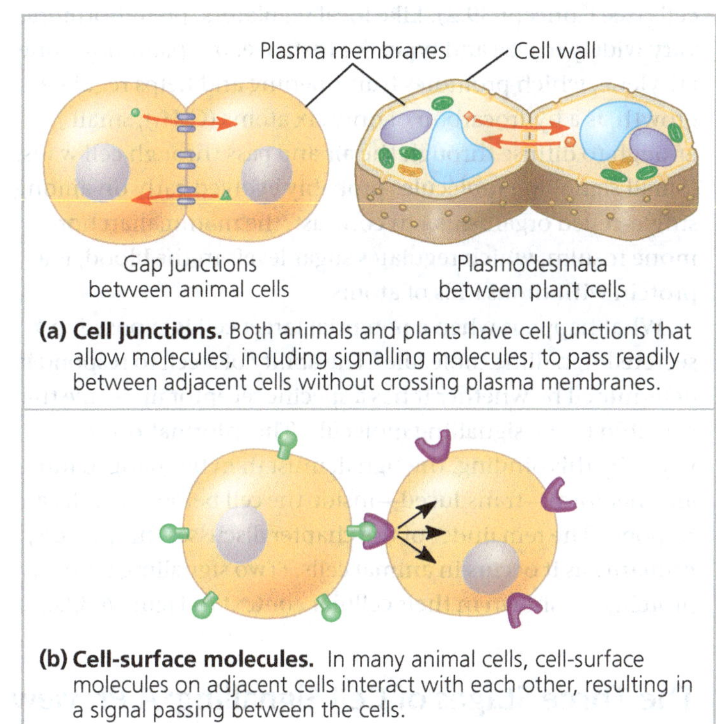

(a) Cell junctions. Both animals and plants have cell junctions that allow molecules, including signalling molecules, to pass readily between adjacent cells without crossing plasma membranes.

(b) Cell-surface molecules. In many animal cells, cell-surface molecules on adjacent cells interact with each other, resulting in a signal passing between the cells.

Both animals and plants use molecules called **hormones** for long-distance signalling. In hormonal signalling in animals, also known as *endocrine signalling*, specialised cells release hormones, which travel through the circulatory system to other parts of the body, where they reach target cells that can recognise and respond to them **(Figure 11.5c)**. Many

▼ Figure 11.5 Local and long-distance cell signalling by secreted molecules in animals. In both local and long-distance signalling, only specific target cells that can recognise a given signalling molecule will respond to it.

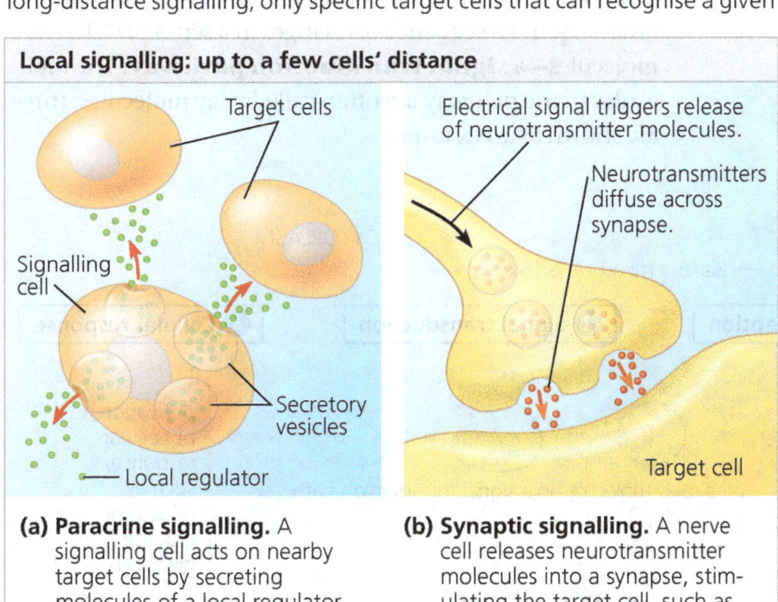

(a) Paracrine signalling. A signalling cell acts on nearby target cells by secreting molecules of a local regulator (a growth factor, for example).

(b) Synaptic signalling. A nerve cell releases neurotransmitter molecules into a synapse, stimulating the target cell, such as a muscle or another nerve cell.

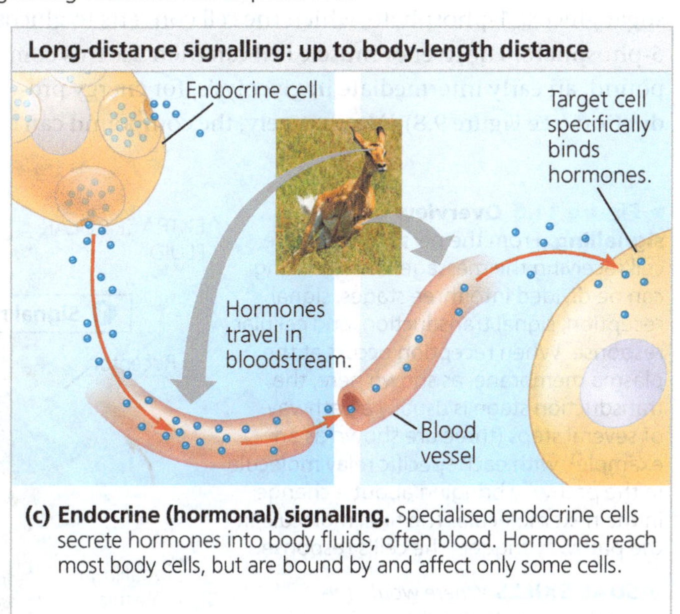

(c) Endocrine (hormonal) signalling. Specialised endocrine cells secrete hormones into body fluids, often blood. Hormones reach most body cells, but are bound by and affect only some cells.

plant hormones reach distant targets by travelling through cells (see Concept 39.2). Like local regulators, plant hormones vary widely in size and type. For instance, the plant hormone ethylene, which promotes fruit ripening and helps regulate growth, is a hydrocarbon of only six atoms (C_2H_4), small enough to diffuse through the air and pass through cell walls. (Small signalling molecules probably evolved early on among single-celled organisms.) In contrast, the mammalian hormone insulin, which regulates sugar levels in the blood, is a protein with thousands of atoms.

What happens when a potential target cell is exposed to a secreted signalling molecule? The ability of a cell to respond is determined by whether it has a specific receptor molecule that can bind to the signalling molecule. The information conveyed by this binding, the signal, must then be changed into another form—transduced—inside the cell before the cell can respond. The remainder of the chapter discusses this process, primarily as it occurs in animal cells. (Two signalling pathway proteins are shown in their cellular context in Figure 6.32a.)

The Three Stages of Cell Signalling: *A Preview*

Our current understanding of how signalling molecules act through signal transduction pathways had its origins in the pioneering work of Earl W. Sutherland, whose research led to a Nobel Prize in 1971. Sutherland and his colleagues at Vanderbilt University were investigating how the animal hormone adrenaline (also called epinephrine) triggers the "fight-or-flight" response in animals like the impala in Figure 11.1. One effect of adrenaline is to mobilise fuel reserves, which can be used by the animal to either defend itself (fight) or try to escape (flight). Adrenaline stimulates the breakdown of the storage polysaccharide glycogen within liver cells and skeletal muscle cells. The breakdown of glycogen releases the sugar glucose 1-phosphate, which the cell converts to glucose 6-phosphate. The liver or muscle cell can then use this compound, an early intermediate in glycolysis, for energy production (see Figure 9.8). Alternatively, the compound can be stripped of phosphate and released from the liver cell into the blood as glucose, which can fuel cells throughout the body.

But how, exactly, does adrenaline mobilise glucose for use? Sutherland's research team discovered that adrenaline outside the cell stimulates glycogen breakdown by somehow activating an enzyme, glycogen phosphorylase, inside the cell. However, when adrenaline was added to a solution containing the enzyme and its substrate, glycogen, no breakdown occurred. Glycogen phosphorylase could be activated by adrenaline only when the hormone was added to intact cells. This result told Sutherland two things. First, adrenaline does not interact directly with the enzyme responsible for glycogen breakdown; an intermediate step or series of steps must be occurring in the cell. Second, an intact, membrane-bound cell must be present for transmission of the signal to take place.

Sutherland's work suggested that the process going on at the receiving end of a cellular communication can be dissected into three stages: signal reception, signal transduction, and cellular response **(Figure 11.6)**:

1. **Signal reception.** Reception is the target cell's detection of a signalling molecule coming from outside the cell. A chemical signal is "detected" when the signalling molecule binds to a receptor protein located at the cell's surface (or inside the cell, to be discussed later).
2. **Signal transduction.** The binding of the signalling molecule changes the receptor protein in some way, initiating the process of transduction. The transduction stage converts the signal to a form that can bring about a specific cellular response. In Sutherland's system, the binding of adrenaline to a receptor protein in a liver cell's plasma membrane leads to activation of glycogen phosphorylase in the cytosol. Transduction sometimes occurs in a single step but more often requires a sequence of changes in a series of different molecules—a **signal transduction pathway**. The molecules in the pathway are often called relay molecules; three are shown as an example.

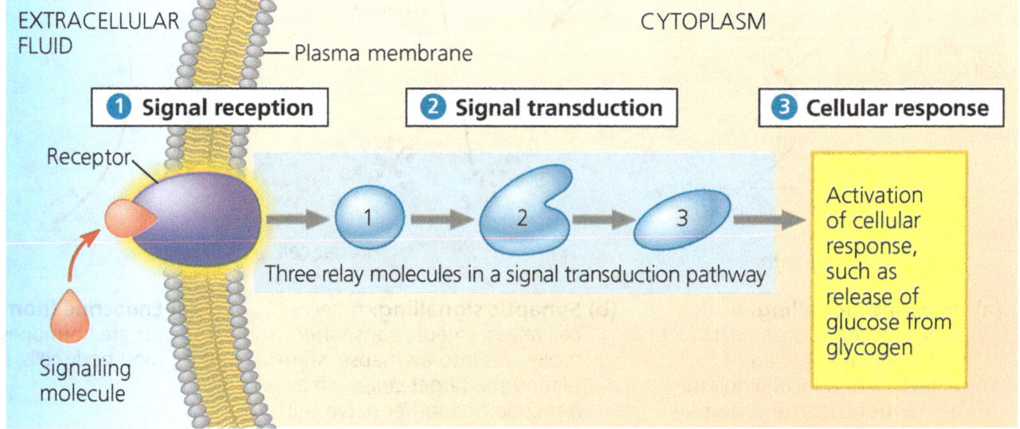

▶ **Figure 11.6 Overview of cell signalling.** From the perspective of the cell receiving the message, cell signalling can be divided into three stages: signal reception, signal transduction, and cellular response. When reception occurs at the plasma membrane, as shown here, the transduction stage is usually a pathway of several steps (three are shown as an example), with each specific relay molecule in the pathway bringing about a change in the next molecule. The final molecule in the pathway triggers the cell's response.

VISUAL SKILLS *Where would the adrenaline in Sutherland's experiment fit into this diagram of cell signalling?*

3 **Cellular response.** The transduced signal finally triggers a specific cellular response. The response may be almost any imaginable cellular activity—such as catalysis by an enzyme (for example, glycogen phosphorylase), rearrangement of the cytoskeleton, or activation of specific genes in the nucleus. The cell-signalling process helps ensure that crucial activities like these occur in the right cells, at the right time, and in proper coordination with the activities of other cells of the organism. We'll now explore the mechanisms of cell signalling in more detail, including a discussion of regulation and termination of the process.

CONCEPT CHECK **11.1**

1. Explain how signalling is involved in ensuring that yeast cells fuse only with cells of the opposite mating type.
2. In liver cells, glycogen phosphorylase acts in which of the three stages of the signalling pathway associated with an adrenaline-initiated signal?
3. **WHAT IF?** If adrenaline were mixed with glycogen phosphorylase and glycogen in a cell-free mixture in a test tube, would glucose 1-phosphate be generated? Why or why not?

For suggested answers, see Appendix A.

CONCEPT **11.2**

Signal reception: A signalling molecule binds to a receptor, causing it to change shape

A wireless router may broadcast its network signal indiscriminately, but only computers with the correct password can connect to it: Reception of the signal depends on the receiver. Similarly, the signals emitted by an **a** mating type yeast cell are "heard" only by its prospective mates, **α** cells. In the case of the adrenaline circulating throughout the bloodstream of the impala in Figure 11.1, the hormone encounters many types of cells, but only certain target cells, those with the corresponding receptor protein, detect and respond to the adrenaline molecule. The signalling molecule is complementary in shape to a specific site on the receptor and attaches there, like a hand in a glove. The signalling molecule acts as a **ligand**, the term for a molecule that specifically binds to another (often larger) molecule. Ligand binding generally causes a receptor protein to undergo a change in shape. For many receptors, this shape change directly activates the receptor, enabling it to interact with other molecules in or on the cell. For other receptors, the immediate effect of ligand binding is to cause the aggregation of two or more receptor proteins, which leads to further molecular events inside the cell. Most signal receptors are plasma membrane proteins, but others are located inside the cell. We'll look at both of these types next.

Receptors in the Plasma Membrane

Cell-surface transmembrane receptors play crucial roles in the biological systems of animals. The largest family of human cell-surface receptors is that of the G protein-coupled receptors (GPCRs). There are more than 800 GPCRs; an example is shown in **Figure 11.7**. Another example is the co-receptor hijacked by HIV to enter immune cells (see Figure 7.8); this GPCR is the target of the drug maraviroc, which has shown some success at treating AIDS.

Most water-soluble signalling molecules bind to specific sites on transmembrane receptor proteins that transmit information from the extracellular environment to the inside of the cell. We can see how cell-surface transmembrane receptors work by looking at three major types: G protein-coupled receptors (GPCRs), receptor tyrosine kinases (RTKs), and ion channel receptors. These receptors are discussed and illustrated in **Figure 11.8**; study this figure before going on.

Given the many important functions of cell-surface receptors, it is not surprising that their malfunctions are associated with many human diseases, including cancer, heart disease, and asthma. To better understand and treat these conditions, a major focus of both university research teams and the pharmaceutical industry has been to analyse the structure of these receptors.

Although cell-surface receptors (half of which are GPCRs) represent 30% of all human proteins, determining their structures by X-ray crystallography (see Figure 5.21) has proved challenging. For one thing, cell-surface receptors tend to be flexible and inherently unstable, thus difficult to crystallise. It took years of persistent efforts for researchers

▼ Figure 11.7 The structure of a G protein-coupled receptor (GPCR). This is a model of the human β_2-adrenergic receptor, which binds adrenaline. The receptor was crystallised (discussed later in this section) in the presence of both a molecule that mimics adrenaline (green in the model) and cholesterol in the membrane (orange). Two receptors (blue) are shown as ribbon models in a side view. Caffeine can also bind to this receptor; see question 10 at the end of the chapter.

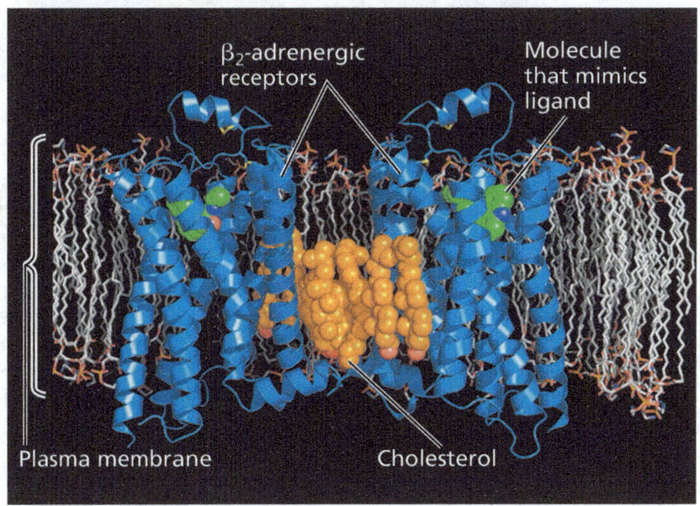

▼ Figure 11.8 Exploring Cell-Surface Transmembrane Receptors

G Protein-Coupled Receptors

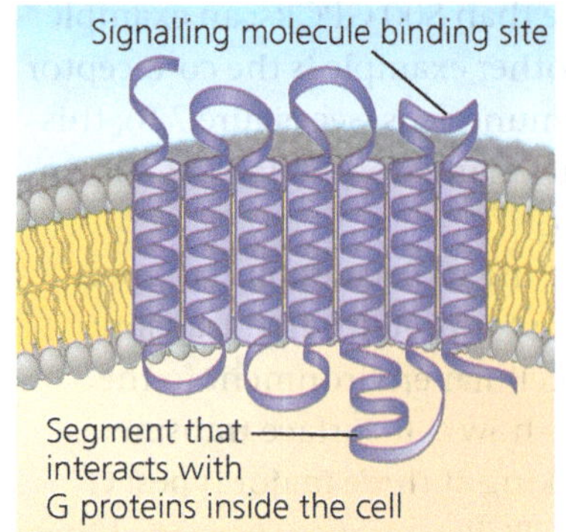

G protein-coupled receptor

A **G protein-coupled receptor** (GPCR) is a cell-surface transmembrane receptor that works with the help of a **G protein**, a protein that binds the energy-rich molecule GTP. Many different signalling molecules—including yeast mating factors, neurotransmitters, and adrenaline (epinephrine) and many other hormones—use GPCRs.

G protein-coupled receptors vary in the binding sites for their ligands and also for different types of G proteins inside the cell. Nevertheless, GPCR proteins are all remarkably similar in structure. In fact, they make up a large family of eukaryotic receptor proteins with a secondary structure in which the single polypeptide, represented here in a ribbon model, has seven transmembrane α helices (outlined with cylinders and depicted in a row for clarity). Specific loops between the helices (here, the loops on the right) form binding sites for signalling molecules (outside the cell) and G proteins (on the cytoplasmic side).

GPCR-based signalling systems are extremely widespread and diverse in their functions, including roles in embryonic development and sensory reception. In humans, for example, vision, smell, and taste depend on GPCRs (see Concept 50.4). Similarities in the structures of G proteins and GPCRs in diverse organisms suggest that G proteins and their associated receptors evolved very early among eukaryotes.

Malfunctions of the associated G proteins themselves are involved in many human diseases, including bacterial infections. The bacteria that cause cholera, pertussis (whooping cough), and botulism, among others, make their victims ill by producing toxins that interfere with G protein function. Up to 60% of all medicines used today exert their effects by influencing G protein pathways.

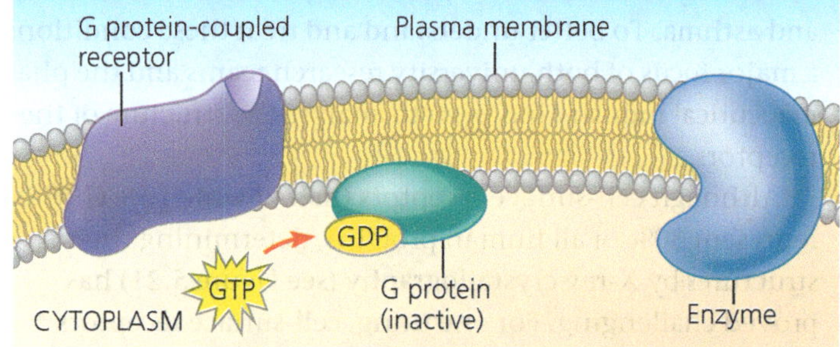

1 Attached but able to move along the cytoplasmic side of the membrane, a G protein functions as a molecular switch that is either on or off, depending on whether GDP or GTP is attached—hence the term *G protein*. (GTP, or guanosine triphosphate, is similar to ATP.) When GDP is bound to the G protein, as shown above, the G protein is inactive. The receptor and G protein work together with another protein, usually an enzyme.

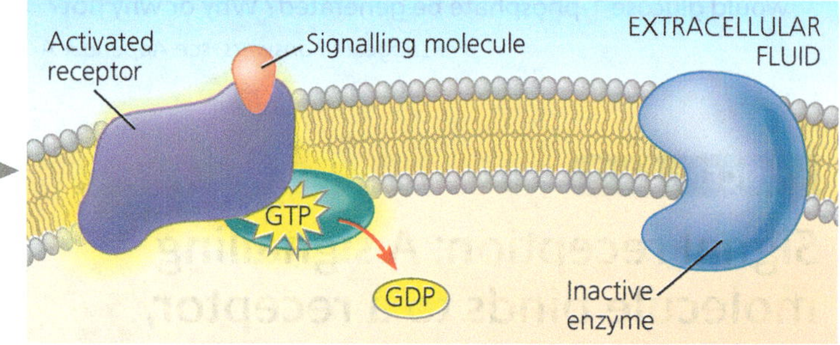

2 When the appropriate signalling molecule binds to the extracellular side of the receptor, the receptor is activated and changes shape. Its cytoplasmic side then binds an inactive G protein, causing a GTP to displace the GDP. This activates the G protein.

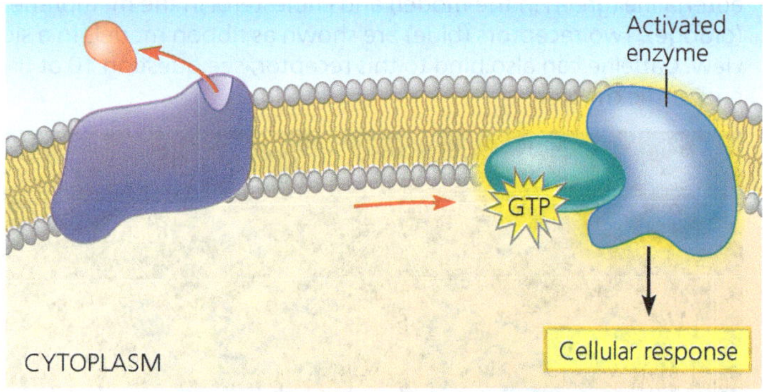

3 The activated G protein dissociates from the receptor, diffuses along the membrane, and then binds to an enzyme, altering the enzyme's shape and activity. Once activated, the enzyme can trigger the next step leading to a cellular response. Binding of signalling molecules is reversible: Like other ligands, they bind and dissociate many times. The ligand concentration outside the cell determines how often a ligand is bound and initiates signalling.

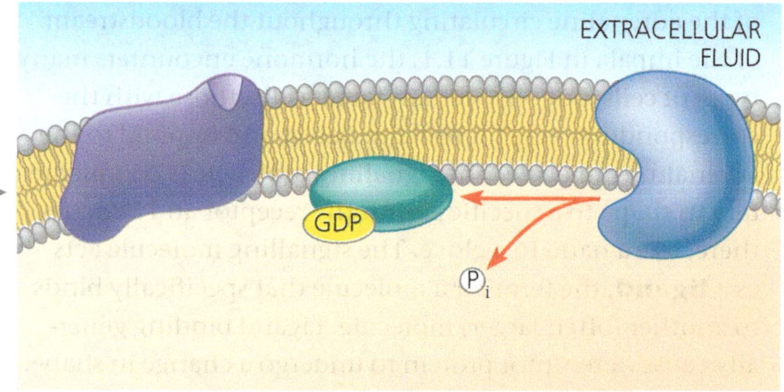

4 The changes in the enzyme and G protein are only temporary: The G protein also acts as a GTPase, hydrolysing its bound GTP to GDP and P_i; as a result, it can no longer activate the enzyme. The G protein leaves the enzyme, which returns to its inactive state. The G protein is now available for reuse. Its GTPase function allows the pathway shut down rapidly when the signalling molecule is no longer present.

Receptor Tyrosine Kinases

Receptor tyrosine kinases (RTKs) belong to a major class of plasma membrane receptors characterised by having enzymatic activity. An RTK is a *protein kinase*—an enzyme that catalyses the transfer of phosphate groups from ATP to another protein. The part of the receptor protein extending into the cytoplasm functions more specifically as a tyrosine kinase, an enzyme that catalyses the transfer of a phosphate group from ATP to the amino acid tyrosine of a substrate protein. Thus, RTKs are membrane receptors that attach phosphates to tyrosines.

Upon binding a ligand such as a growth factor, one RTK may activate ten or more different transduction pathways and cellular responses. Often, more than one signal transduction pathway can be triggered at once, helping the cell regulate and coordinate many aspects of cell growth and cell reproduction. The ability of a single ligand-binding event to trigger so many pathways is a key difference between RTKs and GPCRs; GPCRs generally activate a single transduction pathway. Abnormal RTKs that function even in the absence of signalling molecules are associated with many kinds of cancer.

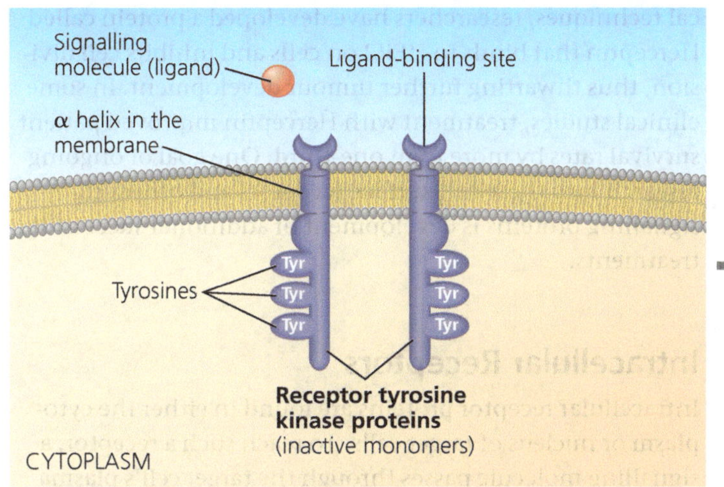

1 Many receptor tyrosine kinases have the structure depicted schematically here. Before the signalling molecule binds, the receptors exist as individual units referred to as monomers. Notice that each monomer has an extracellular ligand-binding site, an α helix spanning the membrane, and an intracellular tail containing multiple tyrosines.

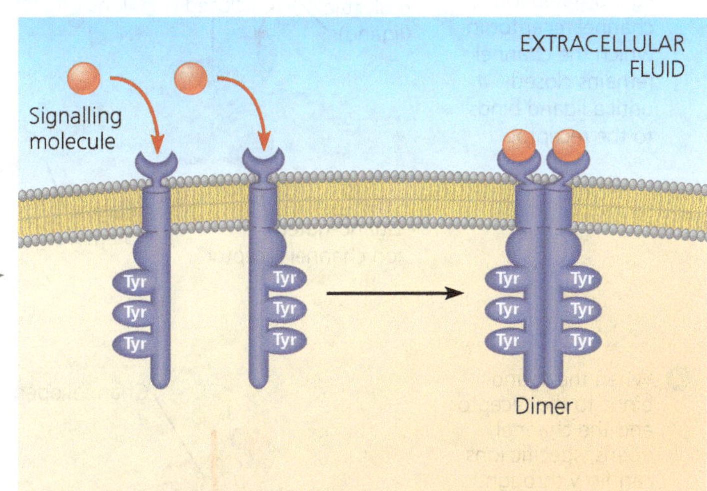

2 The binding of a signalling molecule (such as a growth factor) causes two receptor monomers to associate closely with each other, forming a complex known as a dimer, a process called dimerisation. (In some cases, larger clusters form. The details of monomer association are a focus of current research.)

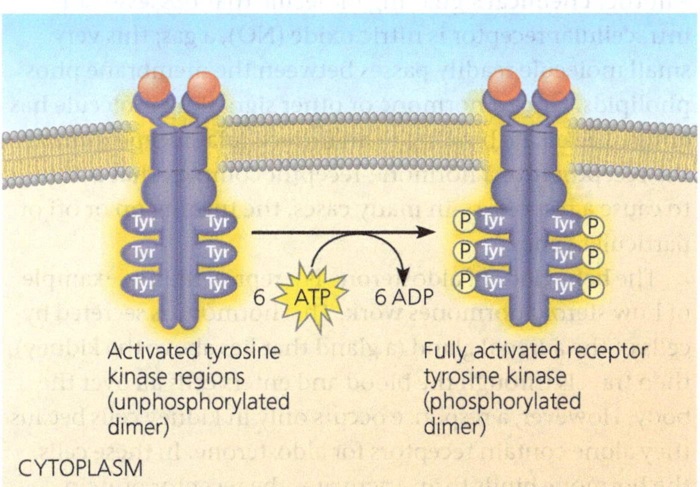

3 Dimerisation activates the tyrosine kinase region of each monomer; each tyrosine kinase adds a phosphate from an ATP molecule to a tyrosine that is part of the tail of the other monomer.

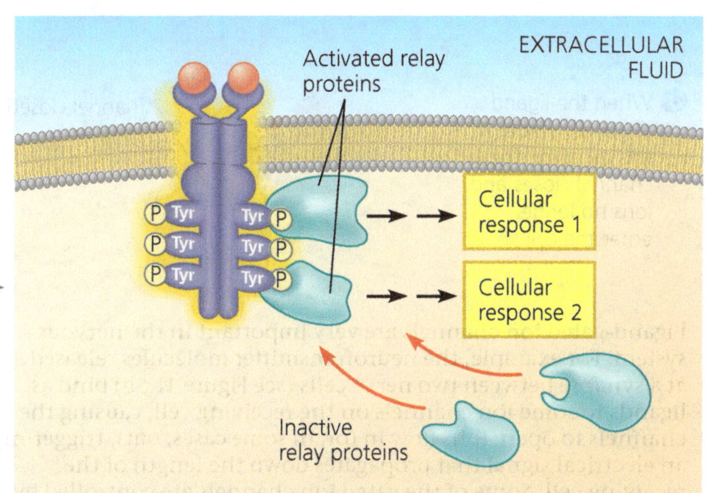

4 Now that the receptor is fully activated, it is recognised by specific relay proteins inside the cell. Each such protein binds to a specific phosphorylated tyrosine, undergoing a resulting structural change that activates the bound relay protein. Each activated protein triggers a transduction pathway, leading to a cellular response.

Continued on next page

▼ Figure 11.8 (continued)

Ion Channel Receptors

A **ligand-gated ion channel** is a type of membrane channel receptor containing a region that can act as a "gate," opening or closing the channel when the receptor changes shape. When a signalling molecule binds as a ligand to the channel receptor, the channel opens or closes, allowing or blocking the flow of specific ions, such as Na^+ or Ca^{2+}. Like the other receptors we have discussed, these proteins bind the ligand at a specific site on their extracellular sides.

1 Here we show a ligand-gated ion channel receptor in which the channel remains closed until a ligand binds to the receptor.

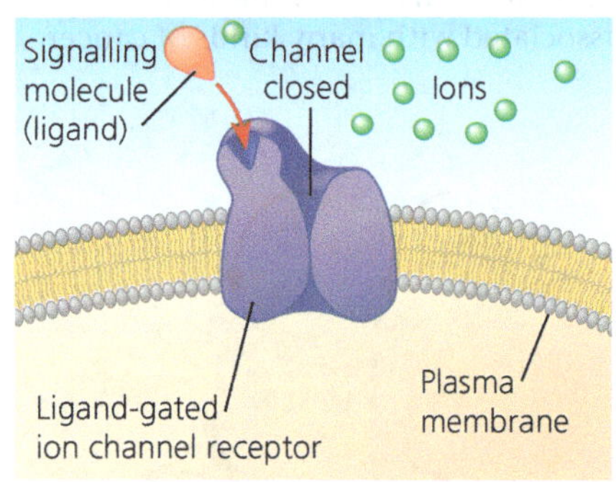

2 When the ligand binds to the receptor and the channel opens, specific ions can flow through the channel and rapidly change the local concentration of that ion inside the cell. This change may directly affect the activity of the cell in some way.

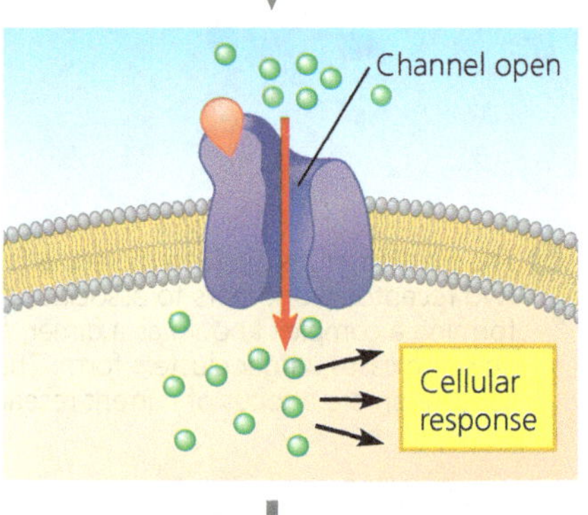

3 When the ligand dissociates from this receptor, the channel closes and ions no longer enter the cell.

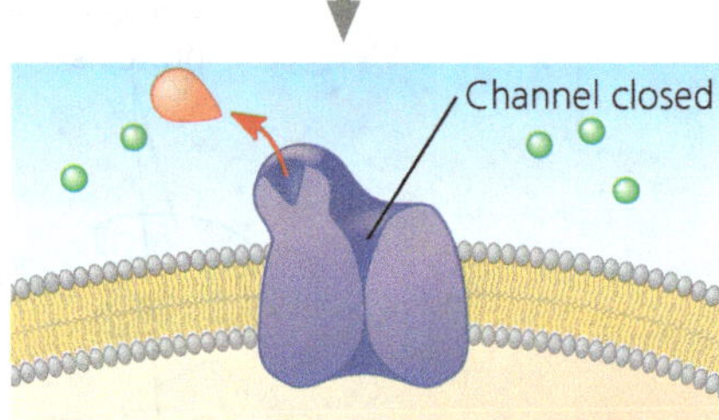

Ligand-gated ion channels are very important in the nervous system. For example, the neurotransmitter molecules released at a synapse between two nerve cells (see Figure 11.5b) bind as ligands to some ion channels on the receiving cell, causing the channels to open. Ions flow in (or, in some cases, out), triggering an electrical signal that propagates down the length of the receiving cell. Some of the gated ion channels are controlled by electrical signals instead of ligands; these voltage-gated ion channels are crucial to the functioning of the nervous system, as you'll see in the chapter, "Neurons, Synapses, and Signalling." Some ion channels are present on membranes of organelles, such as the ER.

MAKE CONNECTIONS *Is the flow of ions through a ligand-gated channel an example of active or passive transport? (Review Concepts 7.3 and 7.4.)*

to determine the first few of these structures, such as the GPCR shown in Figure 11.7. In that case, the β-adrenergic receptor was stable enough to be crystallised only while it was among membrane molecules and in the presence of a molecule mimicking its ligand. Newer techniques that do not require crystallisation, like cryo-EM (see Figure 6.3), have shown promise in determining the structures of cell-surface receptors.

Abnormal functioning of receptor tyrosine kinases (RTKs) is associated with many types of cancers. For example, some breast cancer patients have tumour cells with excessive levels of a receptor tyrosine kinase called HER2 (see Concept 12.3 and Figure 18.27). Using molecular biological techniques, researchers have developed a protein called Herceptin that binds to HER2 on cells and inhibits cell division, thus thwarting further tumour development. In some clinical studies, treatment with Herceptin improved patient survival rates by more than one-third. One goal of ongoing research into these cell-surface receptors and other cell-signalling proteins is development of additional successful treatments.

Intracellular Receptors

Intracellular receptor proteins are found in either the cytoplasm or nucleus of target cells. To reach such a receptor, a signalling molecule passes through the target cell's plasma membrane. A number of important signalling molecules can do this because they are either hydrophobic enough or small enough to cross the hydrophobic interior of the membrane (see Concept 7.1). Hydrophobic signalling molecules include steroid hormones and thyroid hormones of animals. Another chemical signalling molecule that possesses an intracellular receptor is nitric oxide (NO), a gas; this very small molecule readily passes between the membrane phospholipids. Once a hormone or other signalling molecule has entered a cell, its binding to an intracellular receptor changes the receptor into a hormone-receptor complex that is able to cause a response—in many cases, the turning on or off of particular genes.

The behaviour of aldosterone is a representative example of how steroid hormones work. This hormone is secreted by cells of the adrenal gland (a gland that lies above the kidney), then travels through the blood and enters cells all over the body. However, a response occurs only in kidney cells because they alone contain receptors for aldosterone. In these cells, the hormone binds to and activates the receptor protein. With aldosterone attached, the active form of the receptor protein then enters the nucleus and turns on specific genes that control water and sodium flow in kidney cells, ultimately affecting blood volume **(Figure 11.9)**.

How does the activated hormone-receptor complex turn on genes? Recall that the genes in a cell's DNA

▼ **Figure 11.9 Steroid hormone interacting with an intracellular receptor.**

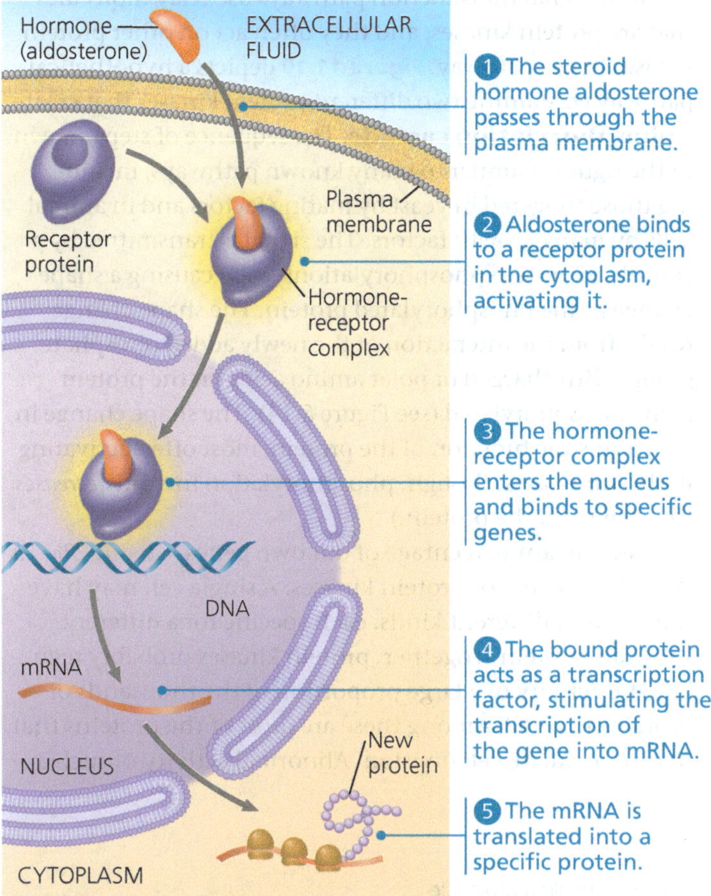

MAKE CONNECTIONS *Why is a cell-surface receptor protein not required for this steroid hormone to enter the cell? (See Concept 7.2.)*

function by being transcribed and processed into messenger RNA (mRNA), which leaves the nucleus and is translated into a specific protein by ribosomes in the cytoplasm (see Figure 5.22). Special proteins called *transcription factors* control which genes are turned on—that is, which genes are transcribed into mRNA—in a particular cell at a particular time. When the aldosterone receptor is activated, it acts as a transcription factor that turns on specific genes. (You'll learn more about transcription factors in the chapters, "Gene Expression: From Gene to Protein" and "Regulation of Gene Expression.")

By acting as a transcription factor, the aldosterone receptor itself carries out two parts of the signalling pathway, as receptor and transducer. Most other intracellular receptors function in the same way, although many of them, such as the thyroid hormone receptor, are already in the nucleus before the signalling molecule reaches them. Interestingly, many of these intracellular receptor proteins are structurally similar, suggesting an evolutionary kinship.

CONCEPT CHECK 11.2

1. Nerve growth factor (NGF) is a water-soluble signalling molecule. Would you expect the receptor for NGF to be intracellular or in the plasma membrane? Explain.
2. **WHAT IF?** What would the effect be if a cell made defective receptor tyrosine kinase proteins that were unable to dimerise?
3. **MAKE CONNECTIONS** How is ligand binding similar to the process of allosteric regulation of enzymes? (See Figure 8.20.)

For suggested answers, see Appendix A.

CONCEPT 11.3

Signal transduction: Cascades of molecular interactions transmit signals from receptors to relay molecules in the cell

When receptors for signalling molecules are plasma membrane proteins, like most of those we have discussed, the transduction stage of cell signalling is usually a multistep pathway involving many molecules. Steps often include activation of proteins by addition or removal of phosphate groups or release of other small molecules or ions that act as signalling molecules. One benefit of using multiple steps is that a signal caused by a small number of signalling molecules can be greatly amplified. If each molecule transmits the signal to numerous molecules at the next step in the series, the result is a geometric increase in the number of activated molecules by the end (see Figure 11.16). A second benefit of using multistep pathways is that they provide more opportunities for coordination and control than do simpler systems. This allows regulation of the response, as we'll see later in the chapter.

Signal Transduction Pathways

The binding of a specific signalling molecule to a receptor in the plasma membrane triggers the first step in the signal transduction pathway—the chain of molecular interactions that leads to a particular response within the cell. Like falling dominoes, the signal-activated receptor activates another molecule, which activates yet another molecule, and so on, until the protein that produces the final cellular response is activated. The molecules that relay a signal from receptor to response, which we call relay molecules in this text, are often proteins. Protein-protein interactions are a major theme of cell signalling—indeed, a unifying theme of all cellular activities.

Keep in mind that the original signalling molecule is not physically passed along a signalling pathway; in most cases, it never even enters the cell. When we say that the signal is relayed along a pathway, we mean that certain information is passed on. At each step, the signal is transduced into a different form, commonly a shape change in the next protein. Very often, the shape change is brought about by phosphorylation.

Protein Phosphorylation and Dephosphorylation

Previous chapters introduced the concept of activating a protein by adding one or more phosphate groups to it (see Figure 8.11a). In Figure 11.8, you have already seen how phosphorylation is involved in the activation of receptor tyrosine kinases. In fact, phosphorylation of proteins and its reverse, dephosphorylation, are commonly used in cells to regulate protein activity. An enzyme that transfers phosphate groups from ATP to a protein is generally known as a **protein kinase**. Recall that a receptor tyrosine kinase (RTK) is a specific kind of protein kinase that phosphorylates tyrosines on the other RTK in a dimer. Most cytoplasmic protein kinases, however, act on proteins different from themselves. Another distinction is that most cytoplasmic protein kinases phosphorylate either of two other amino acids, serine or threonine, rather than tyrosine. Serine/threonine kinases are widely involved in signalling pathways in animals, plants, and fungi.

Many signal transduction pathways use relay molecules that are protein kinases, and they often act on other protein kinases in the pathway. **Figure 11.10** depicts a hypothetical pathway containing two different protein kinases that create a **phosphorylation cascade**. The sequence of steps shown in the figure is similar to many known pathways, including those triggered in yeast by mating factors and in animal cells by many growth factors. The signal is transmitted by a cascade of protein phosphorylations, each causing a shape change in the phosphorylated protein. The shape change results from the interaction of the newly added phosphate groups with charged or polar amino acids on the protein being phosphorylated (see Figure 5.14). The shape change in turn alters the function of the protein, most often activating it. (In some cases, though, phosphorylation instead *decreases* the activity of the protein.)

A significant percentage of our own genes—about 2%—is thought to code for protein kinases. A single cell may have hundreds of different kinds, each specific for a different substrate protein. Together, protein kinases probably regulate the activity of a large proportion of the thousands of proteins in a cell. Among these are most of the proteins that, in turn, regulate cell division. Abnormal activity of such a

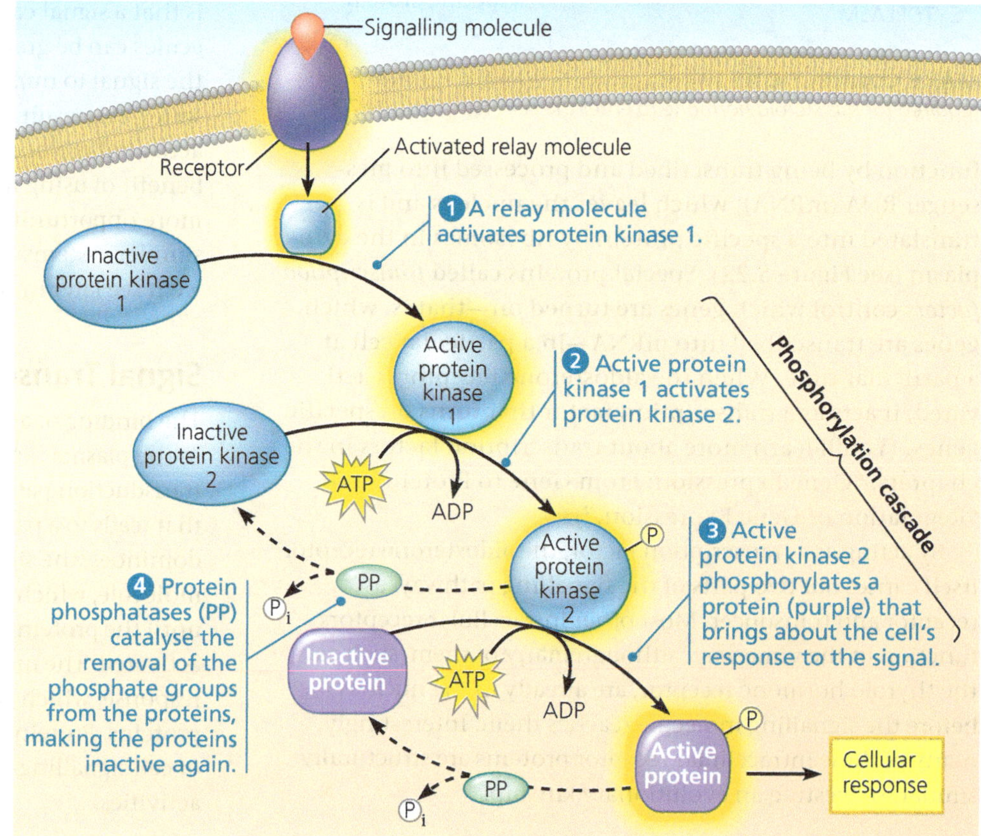

▶ **Figure 11.10 A phosphorylation cascade.** In a phosphorylation cascade, a series of different proteins in a pathway are phosphorylated in turn, each protein adding a phosphate group to the next one in line. Here, phosphorylation activates each protein, and dephosphorylation returns it to its inactive form. The active and inactive forms of each protein are represented by different shapes to remind you that activation is usually associated with a change in molecular shape.

WHAT IF? *What would happen if a mutation in protein kinase 2 made it incapable of being phosphorylated?*

kinase can cause abnormal cell division and contribute to the development of cancer.

Equally important in the phosphorylation cascade are the **protein phosphatases**, enzymes that can rapidly remove phosphate groups from proteins, a process called *dephosphorylation*. By dephosphorylating and thus inactivating protein kinases, phosphatases provide the mechanism for turning off the signal transduction pathway when the initial signal is no longer present. Phosphatases also make the protein kinases available for reuse, enabling the cell to respond again to an extracellular signal. The phosphorylation-dephosphorylation system acts as a molecular switch in the cell, turning activities on or off, or up or down, as required. At any given moment, the activity of a protein regulated by phosphorylation depends on the balance in the cell between active kinase molecules and active phosphatase molecules.

Small Molecules and Ions as Second Messengers

Not all components of signal transduction pathways are proteins. Many signalling pathways also involve small, nonprotein, water-soluble molecules or ions called **second messengers**. (The pathway's "first messenger" is considered to be the extracellular signalling molecule—the ligand—that binds to the membrane receptor.) Because they are small and water-soluble, they can readily spread through regions of the cell by diffusion. For example, as we'll see shortly, a second messenger called cyclic AMP carries the signal initiated by adrenaline from the plasma membrane of a liver or muscle cell into the cell's interior, where the signal eventually brings about glycogen breakdown. Second messengers participate in pathways that are initiated by both G protein-coupled receptors and receptor tyrosine kinases. The two most common second messengers are cyclic AMP and calcium ions, Ca^{2+}. A large variety of relay proteins respond to changes in the cytosolic concentration of one or the other of these second messengers.

Cyclic AMP

Having discovered that adrenaline somehow causes glycogen breakdown within cells, Earl Sutherland next looked for a second messenger that transmits the signal from the plasma membrane to the metabolic machinery in the cytoplasm. He found that binding of adrenaline to the G protein-coupled receptor (GPCR) in the plasma membrane results in a rise in the cytosolic concentration of **cyclic AMP (cAMP**; cyclic adenosine monophosphate), a small molecule produced from ATP. As shown in **Figure 11.11**, an enzyme embedded in the plasma membrane, **adenylyl cyclase** (also known as adenylate cyclase), converts ATP to cAMP in response to an extracellular signal—in this case, provided by adrenaline. When adrenaline outside the cell binds to a GPCR, it activates a G protein that in turn activates adenylyl cyclase. Adenylyl cyclase can then catalyse the synthesis of many molecules of cAMP. In this way, the normal cellular concentration of cAMP can be boosted 20-fold in a matter of seconds. The cAMP broadcasts the signal to the cytoplasm. It does not persist for long in the absence of the hormone because a different enzyme, called phosphodiesterase, converts cAMP to AMP. Another surge of adrenaline is needed to boost the cytosolic concentration of cAMP again.

Subsequent research has revealed that adrenaline is only one of many hormones and many other signalling molecules that lead to activation of adenylyl cyclase by G proteins and formation of cAMP **(Figure 11.12)**. The immediate effect of an elevation in cAMP level is usually the activation of a serine/threonine kinase called *protein kinase A*. The activated protein kinase A then phosphorylates various other proteins, depending on the cell type. (The complete pathway for adrenaline's stimulation of glycogen breakdown is shown later, in Figure 11.16.)

▼ Figure 11.11 Cyclic AMP. The second messenger cyclic AMP (cAMP) is made from ATP by adenylyl cyclase, an enzyme embedded in the plasma membrane. The phosphate group in cAMP is attached to both the 5′ and the 3′ carbons; cAMP's cyclic arrangement is the basis for its name. cAMP is inactivated by phosphodiesterase, an enzyme that converts it to AMP.

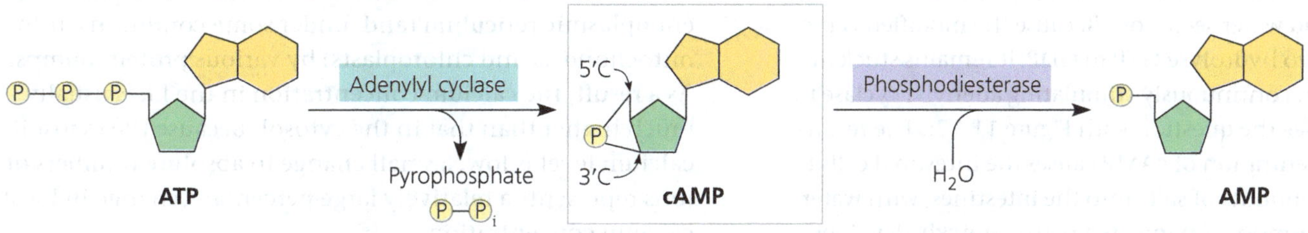

WHAT IF? *What would happen if a molecule that inactivated phosphodiesterase were introduced into the cell?*

▼ Figure 11.12 cAMP as a second messenger in a G protein signalling pathway.

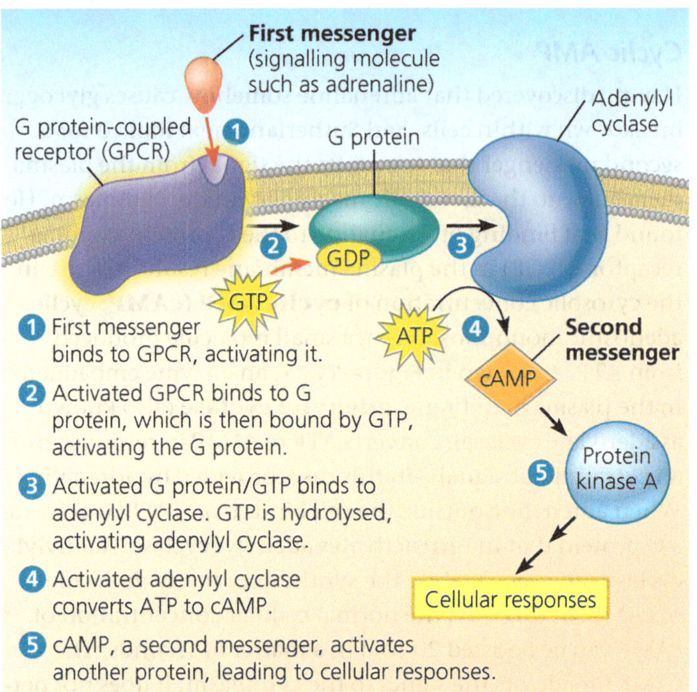

DRAW IT *The bacterium that causes the disease cholera produces a toxin that locks the G protein in its activated state. Review Figure 11.8, then draw this figure as it would be if cholera toxin were present. (You do not need to draw the cholera toxin molecule.)*

Further regulation of cell metabolism is provided by other G protein systems that *inhibit* adenylyl cyclase. In these systems, a different signalling molecule activates a different receptor, which in turn activates an *inhibitory* G protein that blocks activation of adenylyl cyclase. Cell activities can be fine-tuned by the balance between these systems.

Now that we know about the role of cAMP in G protein signalling pathways, we can explain in molecular detail how certain microorganisms cause disease. Consider cholera, a disease that is frequently epidemic in places where the water supply is contaminated with human faeces. People acquire the cholera bacterium, *Vibrio cholerae*, by drinking contaminated water. The bacteria form a biofilm on the lining of the small intestine and produce a toxin. The cholera toxin is an enzyme that chemically modifies a G protein involved in regulating salt and water secretion. Because the modified G protein is unable to hydrolyse GTP to GDP, it remains stuck in its active form, continuously stimulating adenylyl cyclase to make cAMP (see the question with Figure 11.12). The resulting high concentration of cAMP causes the intestinal cells to secrete large amounts of salts into the intestines, with water following by osmosis. An infected person quickly develops profuse diarrhoea and if left untreated can soon die from the loss of water and salts.

Our understanding of signalling pathways involving cyclic AMP or related messengers has allowed us to develop treatments for certain conditions in humans such as erectile dysfunction. In one pathway, the gas nitric oxide (NO) is released by a cell and enters a neighbouring muscle cell, where it causes production of a molecule similar to cAMP called *cyclic GMP (cGMP)*. The cGMP then acts as a second messenger that causes relaxation of muscles, such as those in the walls of arteries. A compound that prolongs the signal (by inhibiting the hydrolysis of cGMP to GMP) was originally prescribed for chest pains because it relaxed blood vessels and increased blood flow to the heart muscle. Under the trade name Viagra, this compound is now widely used as a treatment for erectile dysfunction in human males. Because Viagra leads to dilation of blood vessels, it also allows increased blood flow to the penis, optimising physiological conditions for penile erections.

Calcium Ions and Inositol Trisphosphate (IP_3)

Many of the signalling molecules that function in animals—including neurotransmitters, growth factors, and some hormones—induce responses in their target cells through signal transduction pathways that increase the cytosolic concentration of calcium ions (Ca^{2+}). Calcium is even more widely used than cAMP as a second messenger. Increasing the local cytosolic concentration of Ca^{2+} causes many responses in animal cells, including muscle cell contraction, exocytosis of molecules (secretion), and cell division. In plant cells, a wide range of hormonal and environmental stimuli can cause brief increases in cytosolic Ca^{2+} concentration, triggering various signalling pathways, such as the pathway for greening in response to light (see Figure 39.4). Cells use Ca^{2+} as a second messenger in pathways triggered by both G protein-coupled receptors and receptor tyrosine kinases.

Although cells always contain some Ca^{2+}, this ion can function as a second messenger because its concentration in the cytosol is normally much lower than the concentration outside the cell **(Figure 11.13)**. In fact, the level of Ca^{2+} in the blood and extracellular fluid of an animal is often more than 10,000 times higher than that in the cytosol. Calcium ions are actively transported out of the cell and are actively imported from the cytosol into the endoplasmic reticulum (and, under some conditions, into mitochondria and chloroplasts) by various protein pumps. As a result, the calcium concentration in the ER is usually much higher than that in the cytosol. Because the cytosolic calcium level is low, a small change in absolute numbers of ions represents a relatively large percentage change in local calcium concentration.

In response to a signal relayed by a signal transduction pathway, the cytosolic calcium level may rise, usually by a

▼ **Figure 11.13 The maintenance of calcium ion concentrations in an animal cell.** The Ca^{2+} concentration in the cytosol is usually much lower (light green) than in the extracellular fluid and ER (dark green). Protein pumps in the plasma membrane and the ER membrane, driven by ATP, move Ca^{2+} from the cytosol into the extracellular fluid and into the lumen of the ER. Mitochondrial pumps, driven by chemiosmosis (see Concept 9.4), move Ca^{2+} into mitochondria when the calcium level in the cytosol rises significantly.

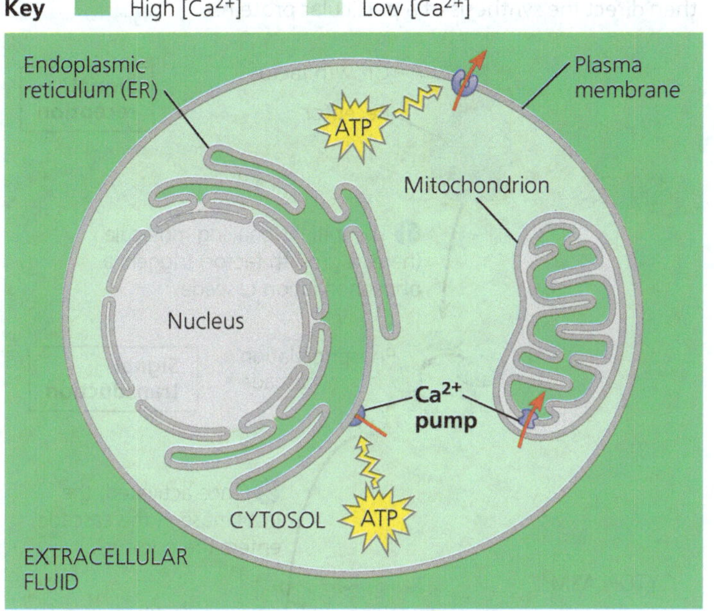

mechanism that releases Ca^{2+} from the cell's ER. The pathways leading to calcium release involve two other second messengers, **inositol trisphosphate (IP_3)** and **diacylglycerol (DAG)**. These two messengers are produced by cleavage of a certain kind of phospholipid in the plasma membrane. **Figure 11.14** shows the complete picture of how a signal causes IP_3 to stimulate the release of calcium from the ER. Because IP_3 acts before calcium in these pathways, calcium could be considered a "*third* messenger." However, scientists use the term *second messenger* for all small, nonprotein components of signal transduction pathways.

CONCEPT CHECK 11.3

1. What is a protein kinase, and what is its role in a signal transduction pathway?
2. When a signal transduction pathway involves a phosphorylation cascade, how does the cell's response get turned off?
3. What is the actual "signal" that is being transduced in any signal transduction pathway, such as those shown in Figures 11.6 and 11.10? In what way is this information being passed from the exterior to the interior of the cell?
4. **WHAT IF?** If you exposed a cell to a ligand that binds to a receptor and activates phospholipase C, predict the effect the IP_3-gated calcium channel would have on Ca^{2+} concentration in the cytosol.

For suggested answers, see Appendix A.

► **Figure 11.14 Calcium and IP_3 in signalling pathways.** Calcium ions (Ca^{2+}) and inositol trisphosphate (IP_3) function as second messengers in many signal transduction pathways. In this figure, the process is initiated by the binding of a signalling molecule to a G protein-coupled receptor. A receptor tyrosine kinase could also initiate this pathway by activating phospholipase C.

MAKE CONNECTIONS *Explain the difference in function (in terms of ion transport) between the Ca^{2+} pump in Figure 11.13 and the Ca^{2+} channel protein shown here. (See Figure 7.17.)*

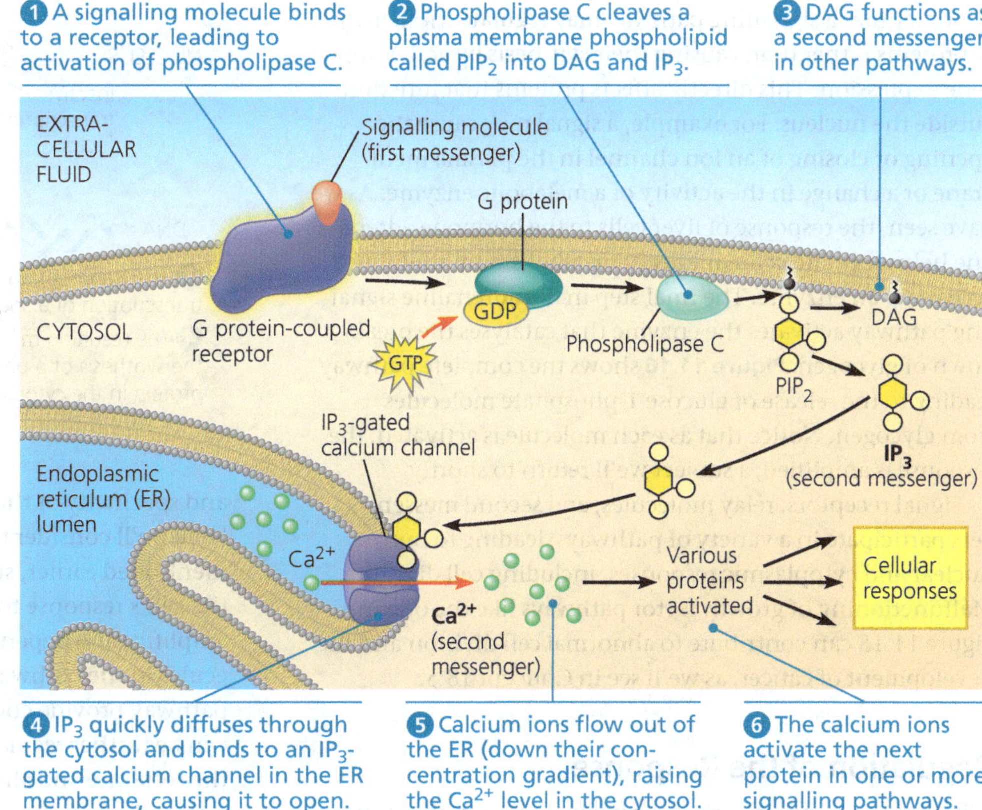

CONCEPT 11.4

Cellular response: Cell signalling leads to regulation of transcription or cytoplasmic activities

We now take a closer look at the cell's subsequent response to an extracellular signal—what some researchers call the "output response." What is the nature of the final step in a signalling pathway?

Nuclear and Cytoplasmic Responses

Ultimately, a signal transduction pathway leads to the regulation of one or more cellular activities. The response may occur in the nucleus or in the cytoplasm of the cell.

Many signalling pathways ultimately regulate protein synthesis, usually by turning specific genes on or off in the nucleus. Like an activated steroid receptor (see Figure 11.9), the final activated molecule in a signalling pathway may function as a transcription factor. **Figure 11.15** shows an example in which a signalling pathway activates a transcription factor that turns a gene on: The response to this growth factor signal is transcription, the synthesis of one or more specific mRNAs, which will be translated in the cytoplasm into specific proteins. In other cases, the transcription factor might regulate a gene by turning it off. Often, a transcription factor regulates several different genes.

Sometimes a signalling pathway may regulate the *activity* of proteins rather than causing their synthesis by activating gene expression. This directly affects proteins that function outside the nucleus. For example, a signal may cause the opening or closing of an ion channel in the plasma membrane or a change in the activity of a metabolic enzyme. As we have seen, the response of liver cells to the hormone adrenaline helps regulate cellular energy metabolism by affecting the activity of an enzyme. The final step in the adrenaline signalling pathway activates the enzyme that catalyses the breakdown of glycogen. **Figure 11.16** shows the complete pathway leading to the release of glucose 1-phosphate molecules from glycogen. Notice that as each molecule is activated, the response is amplified, a subject we'll return to shortly.

Signal receptors, relay molecules, and second messengers participate in a variety of pathways, leading to both nuclear and cytoplasmic responses, including cell division. Malfunctioning of growth factor pathways like the one in Figure 11.15 can contribute to abnormal cell division and the development of cancer, as we'll see in Concept 18.5.

Regulation of the Response

Whether the response occurs in the nucleus or in the cytoplasm, it is not simply turned "on" or "off." Rather, the extent and specificity of the response are regulated in multiple ways. Here we'll consider four aspects of this regulation. First, as mentioned earlier, signalling pathways generally amplify the cell's response to a single signalling event. The degree of amplification depends on the function of the specific molecules in the pathway. Second, the many steps in a multistep pathway provide control points at which the cell's response can be further regulated, contributing to the specificity of the response and allowing coordination with other signalling pathways. Third, the overall efficiency of the response is enhanced by the presence of proteins known as scaffolding

▼ **Figure 11.15 Nuclear responses to a signal: the activation of a specific gene by a growth factor.** This diagram shows a typical signalling pathway that leads to regulation of gene activity in the cell nucleus. The initial signalling molecule, in this case a growth factor, triggers a phosphorylation cascade, as in Figure 11.10. (The ATP molecules and phosphate groups are not shown.) Once phosphorylated, the last kinase in the sequence enters the nucleus and activates a transcription factor, which stimulates transcription of a specific gene (or genes). The resulting mRNAs then direct the synthesis of a particular protein.

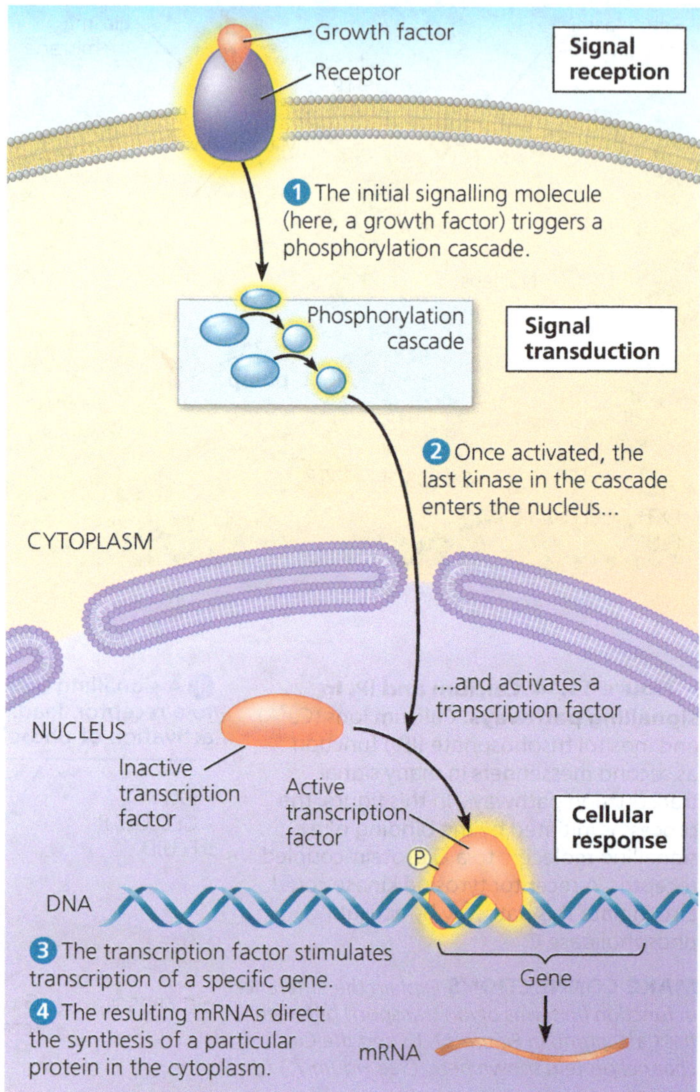

▼ Figure 11.16 Cytoplasmic response to a signal: the stimulation of glycogen breakdown by adrenaline. In this signalling system, the hormone adrenaline acts through a G protein-coupled receptor to activate a succession of relay molecules, including cAMP and two protein kinases (see also Figure 11.12). The final protein activated is the enzyme glycogen phosphorylase, which uses inorganic phosphate to release glucose monomers from glycogen in the form of glucose 1-phosphate molecules. This pathway amplifies the hormonal signal: One receptor protein can activate approximately 100 molecules of G protein, and each enzyme in the pathway, once activated, can act on many molecules of its substrate, the next molecule in the cascade. The number of activated molecules given for each step is approximate.

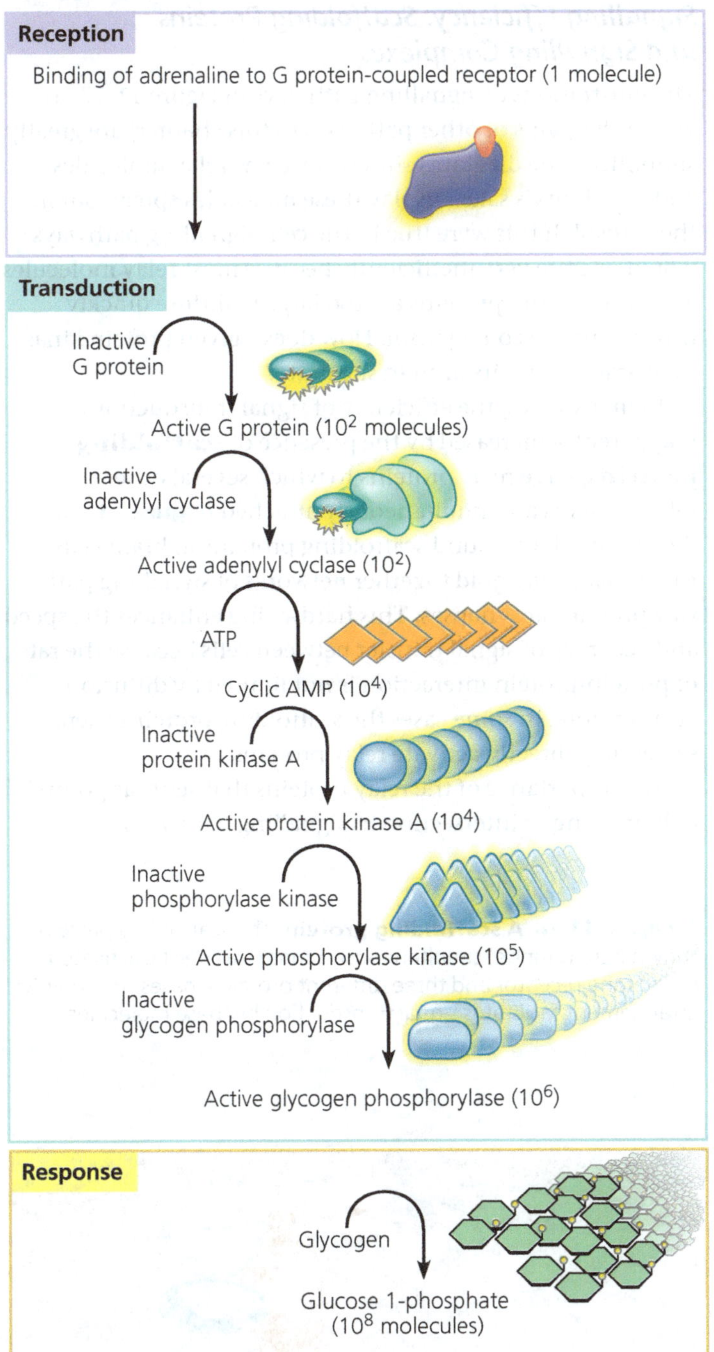

VISUAL SKILLS *In the figure, how many glucose 1-phosphate molecules are released in response to one signalling molecule? Calculate the factor by which the response is amplified in going from each step to the next.*

proteins. Finally, a crucial point in regulating the response is the termination of the signal.

Signal Amplification

Elaborate enzyme cascades amplify the cell's response to a signal. At each catalytic step in the cascade, the number of activated products can be much greater than in the preceding step. For example, in the adrenaline-triggered pathway in Figure 11.16, each adenylyl cyclase molecule catalyses the formation of 100 or so cAMP molecules, each molecule of protein kinase A phosphorylates about 10 molecules of the next kinase in the pathway, and so on. The amplification effect stems from the fact that these proteins persist in their active form long enough to process multiple molecules of substrate before they become inactive again. As a result of the signal's amplification, a small number of adrenaline molecules binding to receptors on the surface of a liver cell or muscle cell can lead to the release of hundreds of millions of glucose molecules from glycogen.

The Specificity of Cell Signalling and Coordination of the Response

Consider two different cells in your body—a liver cell and a heart muscle cell, for example. Both are in contact with your bloodstream and are therefore constantly exposed to many different hormone molecules, as well as to local regulators secreted by nearby cells. Yet the liver cell responds to some signals but ignores others, and the same is true for the heart cell. And some kinds of signals trigger responses in both cells—but different responses. For instance, adrenaline stimulates the liver cell to break down glycogen, but the main response of the heart cell to adrenaline is contraction, leading to a more rapid heartbeat. How do we account for this difference?

The explanation for the specificity exhibited in cellular responses to signals is the same as the basic explanation for virtually all differences between cells: Because different kinds of cells turn on different sets of genes, *different kinds of cells have different collections of proteins*. The response of a cell to a signal depends on its particular collection of signal receptor proteins, relay proteins, and proteins needed to carry out the response. A liver cell, for example, is poised to respond appropriately to adrenaline by having the proteins listed in Figure 11.16 as well as those needed to manufacture glycogen.

Thus, two cells that respond differently to the same signal differ in one or more proteins that respond to the signal. Also, within some cells, there are more complex pathways that branch or converge, as shown in **Figure 11.17**. Notice, though, that these different pathways still have some molecules in common. For example, cells A, B, and C all use the same receptor protein for the red signalling molecule; differences in other proteins account for their differing responses. In cell D, a different receptor protein is used for the same signalling molecule, leading to yet another response. In cell B, a pathway triggered by one signal diverges to produce two responses; such branched pathways often involve receptor tyrosine kinases (which can activate multiple relay proteins) or second messengers (which can regulate numerous proteins). In cell C, two pathways triggered by separate signals converge to modulate a single response. Branching of pathways and "cross-talk" (interaction) between pathways are important in regulating and coordinating a cell's responses to information coming in from different sources in the body. (You'll learn more about this coordination in Concept 11.5.) Moreover, the use of some of the same proteins in more than one pathway allows the cell to economise on the number of different proteins it must make.

▼ Figure 11.17 The specificity of cell signalling. The particular proteins a cell possesses determine what signalling molecules it responds to and the nature of the response. The four cells in these diagrams respond to the same signalling molecule (red) in different ways because each has a different set of proteins (purple and teal). Note, however, that the same kinds of molecules can participate in more than one pathway.

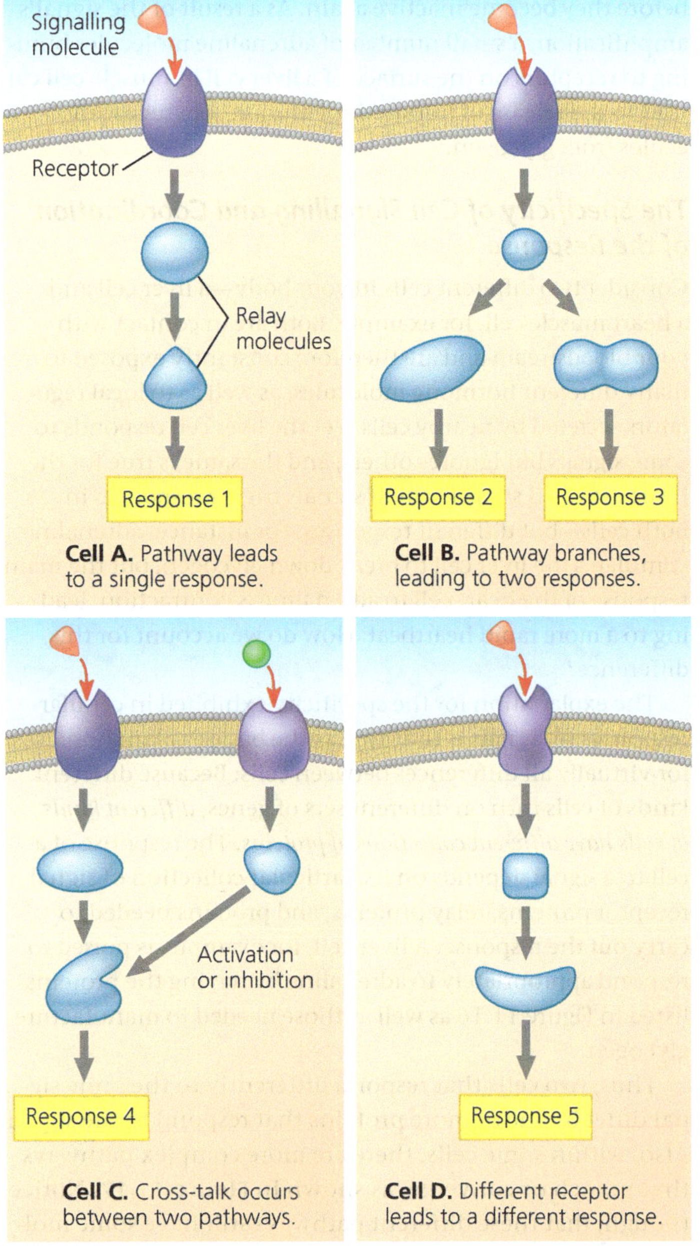

VISUAL SKILLS *Study the signalling pathway shown in Figure 11.14, and explain how the situation pictured for cell B in Figure 11.17 could apply to that pathway.*

Signalling Efficiency: Scaffolding Proteins and Signalling Complexes

The illustrations of signalling pathways in Figure 11.17 (as well as diagrams of other pathways in this chapter) are greatly simplified. The diagrams show only a few relay molecules and, for clarity's sake, display these molecules spread out in the cytosol. If this were true in the cell, signalling pathways would operate very inefficiently because most relay molecules are proteins, and proteins are too large to diffuse quickly through the viscous cytosol. How does a given protein kinase, for instance, find its protein substrate?

In many cases, the efficiency of signal transduction is apparently increased by the presence of **scaffolding proteins**, large relay proteins to which several other relay proteins are simultaneously attached **(Figure 11.18)**. Researchers have found scaffolding proteins in brain cells that *permanently* hold together networks of signalling pathway proteins at synapses. This hardwiring enhances the speed and accuracy of signal transfer between cells because the rate of protein-protein interaction is not limited by diffusion. Furthermore, in some cases the scaffolding proteins themselves may directly activate relay proteins.

The importance of the relay proteins that serve as points of branching or intersection in signalling pathways is

▼ Figure 11.18 A scaffolding protein. The scaffolding protein shown here (orange) simultaneously binds to a specific activated membrane receptor and three different protein kinases. This physical arrangement facilitates signal transduction by these molecules.

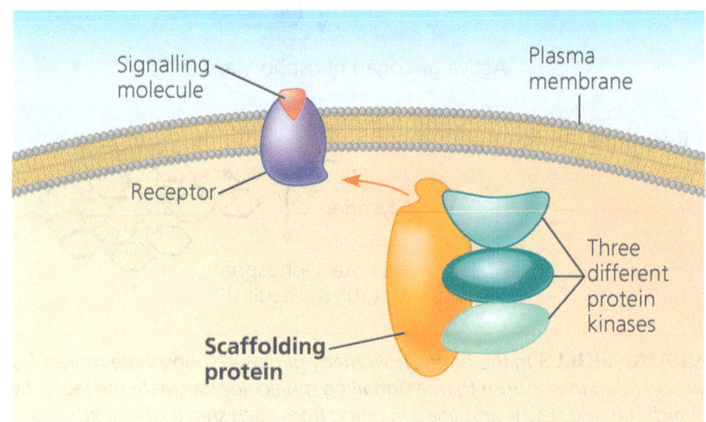

highlighted by the problems arising when these proteins are defective or missing. For instance, in an inherited disorder called Wiskott-Aldrich syndrome (WAS), the absence of a single relay protein leads to such diverse effects as abnormal bleeding, eczema, and a predisposition to infections and leukemia. These symptoms are thought to arise primarily from the absence of the protein in cells of the immune system. By studying normal cells, scientists found that the WAS protein is located just beneath the immune cell surface. The protein interacts both with microfilaments of the cytoskeleton and with several different components of signalling pathways that relay information from the cell surface, including pathways regulating immune cell proliferation. This multifunctional relay protein is thus both a branch point and an important intersection point in a complex signal transduction network that controls immune cell behaviour. When the WAS protein is absent, the cytoskeleton is not properly organised and signalling pathways are disrupted, leading to the WAS symptoms.

Termination of the Signal

In the interest of keeping Figure 11.17 simple, we did not show the *inactivation* mechanisms that are an essential aspect of any cell-signalling pathway. For a cell of a multicellular organism to remain capable of responding to incoming signals, each molecular change in its signalling pathways must last only a short time. As we saw in the cholera example, if a signalling pathway component becomes locked into one state, whether active or inactive, consequences for the organism can be serious.

The ability of a cell to receive new signals depends on reversibility of the changes produced by prior signals. The binding of signalling molecules to receptors is reversible. As the external concentration of signalling molecules falls, fewer receptors are bound at any given moment, and the unbound receptors revert to their inactive form. The cellular response occurs only when the concentration of receptors with bound signalling molecules is above a certain threshold. When the number of active receptors falls below that threshold, the cellular response ceases. Then, by a variety of means, the relay molecules return to their inactive forms: The GTPase activity intrinsic to a G protein hydrolyses its bound GTP; the enzyme phosphodiesterase converts cAMP to AMP; protein phosphatases inactivate phosphorylated kinases and other proteins; and so forth. As a result, the cell is soon ready to respond to a fresh signal.

In this section, we explored the complexity of signalling initiation and termination in a single pathway, and we saw the potential for pathways to intersect with each other. In the next section, we'll consider one especially important network of interacting pathways in the cell.

CONCEPT CHECK 11.4

1. How can a target cell's response to a single hormone molecule result in a response that affects a million other molecules?
2. **WHAT IF?** If two cells have different scaffolding proteins, explain how they might behave differently in response to the same signalling molecule.
3. **WHAT IF?** Some human diseases are associated with malfunctioning protein phosphatases. How would such proteins affect signalling pathways? (Review the discussion of protein phosphatases in Concept 11.3 and see Figure 11.10.)
4. Adrenaline affects heart muscle cells by causing them to mobilise glucose, contract faster, and increase heart rate. The muscle cells in the airways of the lungs, on the other hand, have the opposite response to adrenaline: They relax, allowing more air to be breathed in. What might explain why respiratory (breathing-related) muscle cells can respond so differently from heart muscle cells?

For suggested answers, see Appendix A.

CONCEPT 11.5

Apoptosis requires integration of multiple cell-signalling pathways

When signalling pathways were first discovered, they were thought to be linear, independent pathways. Our understanding of cellular communication has benefited from the realisation that signalling pathway components interact with each other in various ways. For a cell to carry out the appropriate response, cellular proteins often must integrate multiple signals. Let's consider an important cellular process—cellular suicide—as an example.

Cells that are infected, are damaged, or have reached the end of their functional life span often undergo "programmed cell death" (**Figure 11.19**). The best-understood type of this controlled cell suicide is **apoptosis** (from the Greek, meaning "falling off," and used in a classic Greek poem to refer to leaves falling from a tree). During this process, cellular agents chop up

▼ Figure 11.19 Apoptosis of a human white blood cell. On the left is a normal white blood cell, while on the right is a white blood cell undergoing apoptosis. The apoptotic cell is shrinking and forming lobes ("blebs"), which eventually are shed as membrane-bounded cell fragments (colourised SEMs).

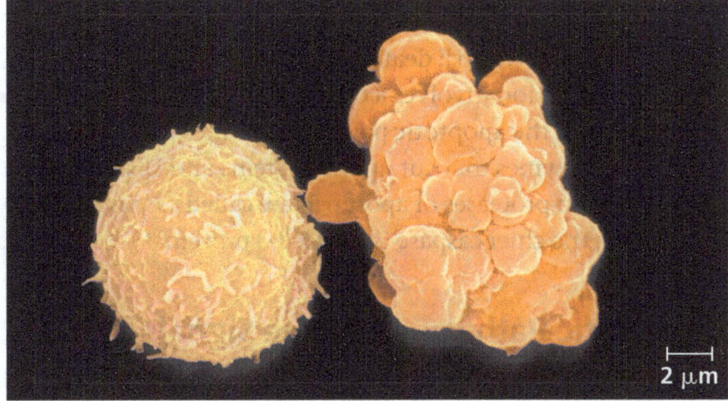

the DNA and fragment the organelles and other cytoplasmic components. The cell shrinks and becomes lobed (a change called "blebbing"), and the cell's parts are packaged up in vesicles that are engulfed and digested by specialised scavenger cells, leaving no trace. Apoptosis protects neighbouring cells from damage that they would otherwise suffer if a dying cell merely leaked out all its contents, including its many digestive enzymes.

The signal that triggers all of the complex events that occur during apoptosis can come from either outside or inside the cell. Outside the cell, signalling molecules released from other cells can initiate a signal transduction pathway that activates the genes and proteins responsible for carrying out cell death. Within a cell whose DNA has been irretrievably damaged, a series of protein-protein interactions can pass along a signal that similarly triggers cell death. Considering some examples of apoptosis can help us to see how signalling pathways are integrated in cells.

Apoptosis in the Soil Worm *Caenorhabditis elegans*

The molecular mechanisms of apoptosis were worked out by researchers studying embryonic development of a small soil worm, a nematode called *Caenorhabditis elegans*. Because the adult worm has only about 1,000 cells, the researchers were able to work out the entire ancestry of each cell. The timely suicide of cells occurs exactly 131 times during normal development of *C. elegans*, at precisely the same points in the cell lineage of each worm. In worms and other species, apoptosis is triggered by signals that activate a cascade of "suicide" proteins in the cells destined to die.

Genetic research on *C. elegans* initially revealed two key apoptosis genes, called *ced-3* and *ced-4* (*ced* stands for "cell death"), which encode proteins essential for apoptosis. The proteins are called Ced-3 and Ced-4, respectively. These and most other proteins involved in apoptosis are continually present in cells, but in inactive form; thus, regulation in this case occurs at the level of protein activity rather than through gene activity and protein synthesis. In *C. elegans*, a protein in the outer mitochondrial membrane, called Ced-9 (the product of the *ced-9* gene), serves as a master regulator of apoptosis, acting as a brake in the absence of a signal promoting apoptosis **(Figure 11.20a)**. When a death signal is received by the cell, signal transduction involves a change in Ced-9 that disables the brake, and the apoptotic pathway activates proteases and nucleases, enzymes that cut up the proteins and DNA of the cell. The main proteases of apoptosis are called *caspases*; in the nematode, the chief caspase is the Ced-3 protein.

▼ Figure 11.20 Molecular basis of apoptosis in *C. elegans*. Three proteins, Ced-3, Ced-4, and Ced-9, are critical to apoptosis and its regulation in the nematode. Apoptosis is more complicated in mammals but involves proteins similar to those in *C. elegans*.

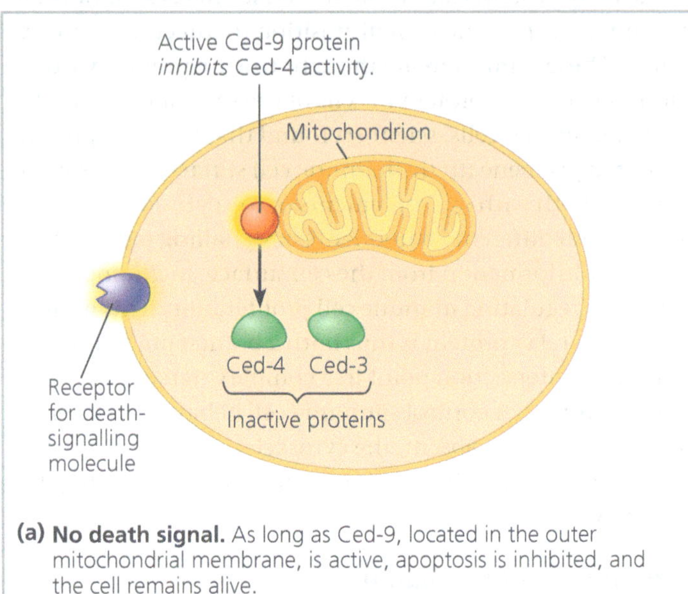

(a) No death signal. As long as Ced-9, located in the outer mitochondrial membrane, is active, apoptosis is inhibited, and the cell remains alive.

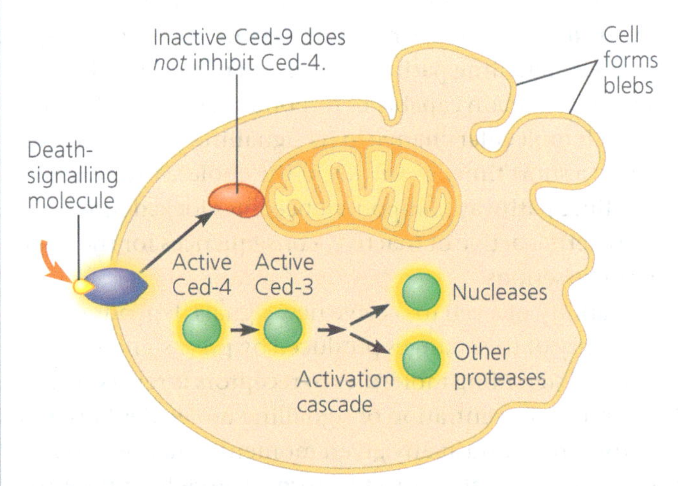

(b) Death signal. When a cell receives a death signal, Ced-9 is inactivated, relieving its inhibition of Ced-4. Active Ced-4 activates Ced-3, a protease, which triggers a cascade of reactions leading to the activation of nucleases and other proteases. The action of these enzymes causes the changes seen in apoptotic cells and eventually leads to cell death.

Apoptotic Pathways and the Signals That Trigger Them

In humans and other mammals, several different pathways, involving about 15 different caspases, can carry out apoptosis. The pathway that is used depends on the type of cell and on the particular signal that initiates apoptosis. One major pathway involves certain mitochondrial proteins that are triggered to form molecular pores in the mitochondrial outer membrane, causing it to leak and release other proteins that promote apoptosis. Perhaps surprisingly, these latter include cytochrome *c*, which functions in mitochondrial electron transport in healthy cells (see Figure 9.14) but acts as a cell death factor when released from mitochondria. The process of mitochondrial apoptosis in mammals uses proteins similar to the nematode proteins Ced-3, Ced-4, and Ced-9. These can be thought of as relay proteins capable of transducing the apoptotic signal.

At key gateways into the apoptotic program, relay proteins integrate signals from several different sources and can send a cell down an apoptotic pathway. Often, the signal originates outside the cell, like the death-signalling molecule depicted in **Figure 11.20b**, which presumably was released by a neighbouring cell. When a death-signalling ligand occupies a cell-surface receptor, this binding leads to activation of caspases and other enzymes that carry out apoptosis, without involving the mitochondrial pathway. This process of signal reception, transduction, and response is similar to what we have discussed throughout this chapter. In a twist on the classic scenario, two other types of alarm signals that can lead to apoptosis originate from *inside* the cell rather than from a cell-surface receptor. One signal comes from the nucleus, generated when the DNA has suffered irreparable damage, and a second comes from the endoplasmic reticulum when excessive protein misfolding occurs. Mammalian cells make life-or-death "decisions" by somehow integrating the death signals and life signals they receive from these external and internal sources.

A built-in cell suicide mechanism is essential to development and maintenance in all animals. The similarities between apoptosis genes in nematodes and those in mammals, as well as the observation that apoptosis occurs in multicellular fungi and even in single-celled yeasts, indicate that the basic mechanism evolved early in the evolution of eukaryotes. In vertebrates, apoptosis is essential for normal development of the nervous system, for normal operation of the immune system, and for normal morphogenesis of hands and feet in humans and paws in other mammals **(Figure 11.21)**. The level of apoptosis between the developing digits is lower in the webbed feet of ducks and other water birds than in the nonwebbed feet of land birds, such as chickens. In the case of humans, the failure of appropriate apoptosis can result in webbed fingers and toes.

Significant evidence points to the involvement of apoptosis in certain degenerative diseases of the nervous system, such as Parkinson's disease and Alzheimer's disease. In Alzheimer's disease, an accumulation of aggregated proteins in neuronal cells activates an enzyme that triggers apoptosis, resulting in the loss of brain function seen in these patients. Furthermore, cancer can result from a failure of cell suicide; some cases of human melanoma, for example, have been linked to faulty forms of the human version of the *C. elegans* Ced-4 protein. It is not surprising, therefore, that the signalling pathways feeding into apoptosis are quite elaborate. After all, the life-or-death question is the most fundamental one imaginable for a cell.

This chapter has introduced you to many of the general mechanisms of cell communication, such as ligand binding, protein-protein interactions and shape changes, cascades of interactions, and protein phosphorylation. Throughout your study of biology, you will encounter numerous examples of cell signalling.

CONCEPT CHECK 11.5

1. Give an example of apoptosis during embryonic development, and explain its function in the developing embryo.
2. **WHAT IF?** If apoptosis occurred when it should not, what types of protein defects might be the cause? What types could result in apoptosis not occurring when it should?

For suggested answers, see Appendix A.

▼ Figure 11.21 Effect of apoptosis during paw development in the mouse. In mice, humans, other mammals, and land birds, the embryonic region that develops into feet or hands initially has a solid, platelike structure. Apoptosis eliminates the cells in the interdigital regions, thus forming the digits. The embryonic mouse paws shown in these fluorescence light micrographs are stained so that cells undergoing apoptosis appear a bright yellowish green. Apoptosis of cells begins at the margin of each interdigital region (left), peaks as the tissue in these regions is reduced (middle), and is no longer visible when the interdigital tissue has been eliminated (right).

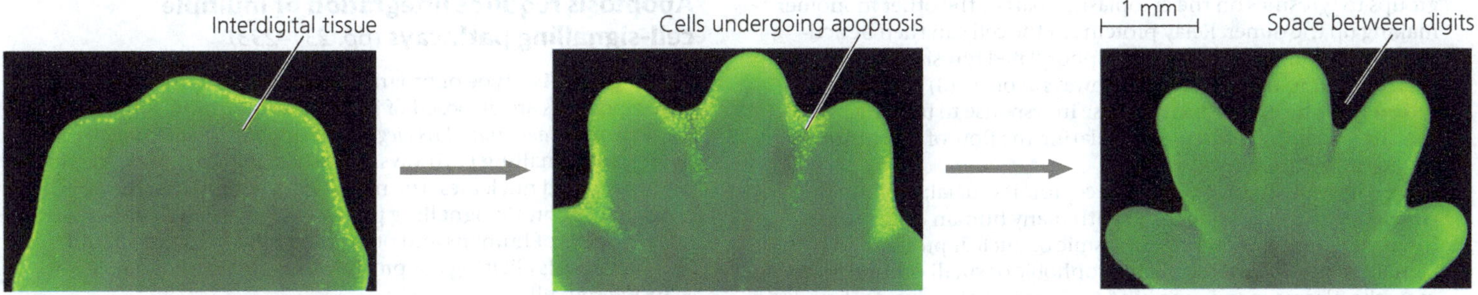

11 Chapter Review

SUMMARY OF KEY CONCEPTS

CONCEPT 11.1

External signals are converted to responses within the cell *(pp. 215–219)*

- **Signal transduction pathways** are crucial for many processes. Bacterial cells can sense the local density of bacterial cells (quorum sensing). Signalling during yeast cell mating has much in common with processes in multicellular organisms, suggesting an early evolutionary origin of signalling mechanisms.
- Local signalling by animal cells involves direct contact or the secretion of local regulators. For long-distance signalling, animal and plant cells use **hormones**; animals also pass signals electrically.
- Like adrenaline, other hormones that bind to membrane receptors trigger a three-stage cell-signalling pathway:

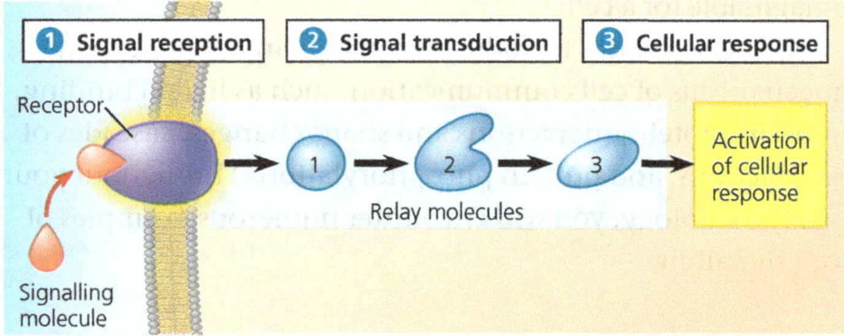

? *What determines whether a cell responds to a hormone such as adrenaline? What determines how a cell responds to such a hormone?*

CONCEPT 11.2

Signal reception: A signalling molecule binds to a receptor, causing it to change shape *(pp. 219–223)*

- The binding between signalling molecule (**ligand**) and receptor is highly specific. A specific shape change in a receptor is often the initial transduction of the signal.
- Three of the major types of cell-surface transmembrane receptors are the following: (1) **G protein-coupled receptors (GPCRs)** work with cytoplasmic **G proteins**. Ligand binding activates the receptor, which then activates a specific G protein, which activates yet another protein, thus propagating the signal. (2) **Receptor tyrosine kinases (RTKs)** react to the binding of signalling molecules by forming dimers and then adding phosphate groups to tyrosines on the cytoplasmic part of the other monomer making up the dimer. Relay proteins in the cell can then be activated by binding to different phosphorylated tyrosines, allowing this receptor to trigger several pathways at once. (3) **Ligand-gated ion channels** open or close in response to binding by specific signalling molecules, regulating the flow of specific ions across the membrane.
- The activity of all three types of receptors is crucial; abnormal GPCRs and RTKs are associated with many human diseases.
- Intracellular receptors are cytoplasmic or nuclear proteins. Signalling molecules that are hydrophobic or small enough to cross the plasma membrane bind to these receptors inside the cell.

? *How are the structures of a GPCR and an RTK similar? How does initiation of signal transduction differ for these two types of receptors?*

CONCEPT 11.3

Signal transduction: Cascades of molecular interactions transmit signals from receptors to relay molecules in the cell *(pp. 223–227)*

- At each step in a signal transduction pathway, the signal is transduced into a different form, which commonly involves a shape change in a protein. Many signal transduction pathways include **phosphorylation cascades**, in which a series of **protein kinases** each add a phosphate group to the next one in line, activating it. Enzymes called **protein phosphatases** remove the phosphate groups. The balance between phosphorylation and dephosphorylation regulates the activity of proteins involved in the sequential steps of a signal transduction pathway.
- **Second messengers**, such as the small molecule **cyclic AMP (cAMP)** and the ion Ca^{2+}, diffuse readily through the cytosol and thus help broadcast signals quickly. Many G proteins activate **adenylyl cyclase**, which makes cAMP from ATP. Cells use Ca^{2+} as a second messenger in GPCR and RTK pathways. The tyrosine kinase pathways can also involve two other second messengers, **diacylglycerol (DAG)** and **inositol trisphosphate** (IP_3). IP_3 can trigger a subsequent increase in Ca^{2+} level.

? *What is the difference between a protein kinase and a second messenger? Can both operate in the same signal transduction pathway?*

CONCEPT 11.4

Cellular response: Cell signalling leads to regulation of transcription or cytoplasmic activities *(pp. 228–231)*

- Some pathways lead to a nuclear response: Specific genes are turned on or off by activated transcription factors. In others, the response involves cytoplasmic regulation.
- Cellular responses are not simply on or off; they are regulated at many steps. Each protein in a signalling pathway amplifies the signal by activating multiple copies of the next component; for long pathways, the total amplification may be over a millionfold. The combination of proteins in a cell confers specificity in the signals it detects and the responses it carries out. **Scaffolding proteins** increase signalling efficiency. Pathway branching further helps the cell coordinate signals and responses. Signal response can be terminated quickly because ligand binding is reversible.

? *What mechanisms in the cell terminate its response to a signal and maintain its ability to respond to new signals?*

CONCEPT 11.5

Apoptosis requires integration of multiple cell-signalling pathways *(pp. 231–233)*

- **Apoptosis** is a type of programmed cell death in which cell components are disposed of in an orderly fashion. Studies of the soil worm *Caenorhabditis elegans* clarified molecular details of the relevant signalling pathways. A death signal leads to activation of caspases and nucleases, the main enzymes involved in apoptosis.
- Several apoptotic signalling pathways with related proteins exist in the cells of humans and other mammals, triggered in different ways. Signals eliciting apoptosis can originate from outside or inside the cell.

? *What is an explanation for the similarities between genes in yeasts, nematodes, and mammals that control apoptosis?*

TEST YOUR UNDERSTANDING

Levels 1-2: Remembering/Understanding

1. Binding of a signalling molecule to which type of receptor leads directly to a change in the distribution of substances on opposite sides of the membrane?
   (A) intracellular receptor
   (B) G protein-coupled receptor
   (C) phosphorylated receptor tyrosine kinase dimer
   (D) ligand-gated ion channel

2. The activation of receptor tyrosine kinases is characterised by
   (A) dimerisation and phosphorylation.
   (B) dimerisation and IP_3 binding.
   (C) a phosphorylation cascade.
   (D) GTP hydrolysis.

3. Lipid-soluble signalling molecules, such as aldosterone, cross the membranes of all cells but affect only target cells because
   (A) only target cells retain the appropriate DNA segments.
   (B) intracellular receptors are present only in target cells.
   (C) only target cells have enzymes that break down aldosterone.
   (D) only in target cells is aldosterone able to initiate the phosphorylation cascade that turns genes on.

4. Consider this pathway: adrenaline → G protein-coupled receptor → G protein → adenylyl cyclase → cAMP. Identify the second messenger.
   (A) cAMP
   (B) G protein
   (C) GTP
   (D) adenylyl cyclase

5. Which of the following occurs during apoptosis?
   (A) lysis of the cell
   (B) direct contact between signalling cells
   (C) fragmentation of the DNA
   (D) release of proteases outside the cell

Levels 3-4: Applying/Analysing

6. Which observation suggested to Sutherland the involvement of a second messenger in adrenaline's effect on liver cells?
   (A) Enzymatic activity was proportional to the amount of calcium added to a cell-free extract.
   (B) Receptor studies indicated that adrenaline was a ligand.
   (C) Glycogen breakdown was observed only when adrenaline was administered to intact cells.
   (D) Glycogen breakdown was observed only when adrenaline and glycogen phosphorylase were mixed.

7. Protein phosphorylation is commonly involved with which of the following?
   (A) ligand binding by receptor tyrosine kinases
   (B) activation of G protein-coupled receptors
   (C) activation of protein kinase molecules
   (D) release of Ca^{2+} from the ER lumen

Levels 5-6: Evaluating/Creating

8. **DRAW IT** Draw the following apoptotic pathway, which operates in human immune cells. A death signal is received when a molecule called Fas binds its cell-surface receptor. The binding of many Fas molecules to receptors causes receptor clustering. The intracellular regions of the receptors, when together, bind proteins called adaptor proteins. These in turn bind to inactive molecules of caspase-8, which become activated and then activate caspase-3. Once activated, caspase-3 initiates apoptosis.

9. **EVOLUTION CONNECTION** Identify the evolutionary mechanisms that might account for the origin and persistence of cell-to-cell signalling systems in prokaryotes.

10. **SCIENTIFIC INQUIRY** Adrenaline initiates a signal transduction pathway that produces cyclic AMP (cAMP) and leads to the breakdown of glycogen to glucose, a major energy source for cells. But glycogen breakdown is only part of the fight-or-flight response that adrenaline brings about; the overall effect on the body includes an increase in heart rate and alertness, as well as a burst of energy. Given that caffeine blocks the activity of cAMP phosphodiesterase, propose a mechanism by which caffeine ingestion leads to heightened alertness and sleeplessness.

11. **SCIENCE, TECHNOLOGY, AND SOCIETY** The ageing process is thought to be initiated at the cellular level. Among the changes that can occur after a certain number of cell divisions is the loss of a cell's ability to respond to growth factors and other signals. Much research into ageing is aimed at understanding such losses, with the ultimate goal of extending the human life span. Not everyone, however, agrees that this is a desirable goal. If life expectancy were greatly increased, discuss what might be the social and ecological consequences.

12. **WRITE ABOUT A THEME: ORGANISATION** The properties of life emerge at the biological level of the cell. The highly regulated process of apoptosis is not simply the destruction of a cell; it is also an emergent property. Write a short essay (about 100–150 words) that briefly explains the role of apoptosis in the development and proper functioning of an animal, and describe how this form of programmed cell death is a process that emerges from the orderly integration of signalling pathways.

13. **SYNTHESISE YOUR KNOWLEDGE**

There are five basic tastes—sour, salty, sweet, bitter, and "umami." Salt is detected when the concentration of salt outside of a taste bud cell is higher than that inside of it, and ion channels allow the passive leakage of Na^+ into the cell. The resulting change in membrane potential (see Concept 7.4) sends the "salty" signal to the brain. Umami is a savoury taste generated by glutamate (glutamic acid, found in monosodium glutamate, or MSG), which is used as a flavour enhancer in foods such as taco-flavoured tortilla chips. The glutamate receptor is a GPCR, which, when bound, initiates a signalling pathway that ends with a cellular response, perceived by you as "taste." If you eat a regular potato chip and then rinse your mouth, you will no longer taste salt. But if you eat a flavoured tortilla chip and then rinse, the taste persists. (Try it!) Propose a possible explanation for this difference.

For selected answers, see Appendix A.

12 The Cell Cycle

KEY CONCEPTS

Study Tip

Make a visual study guide: Figure 12.1 presents the events of the cell cycle as a simplified linear diagram. As you learn more about the cell cycle, draw a detailed linear diagram of the stages of interphase, mitosis, and cytokinesis. Include explanatory labels. Add the circular chart from Figure 12.6 and show how the two diagrams are related.

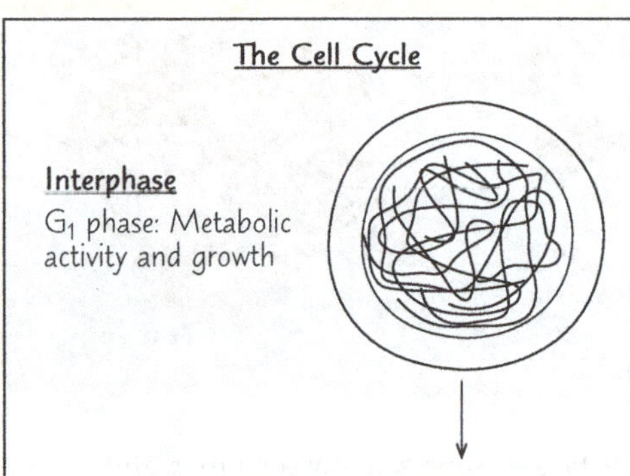

Go to Mastering Biology

to access Dynamic Study Modules for revision, 3D BioFlix® animations and high-quality videos, and your interactive Pearson eText.

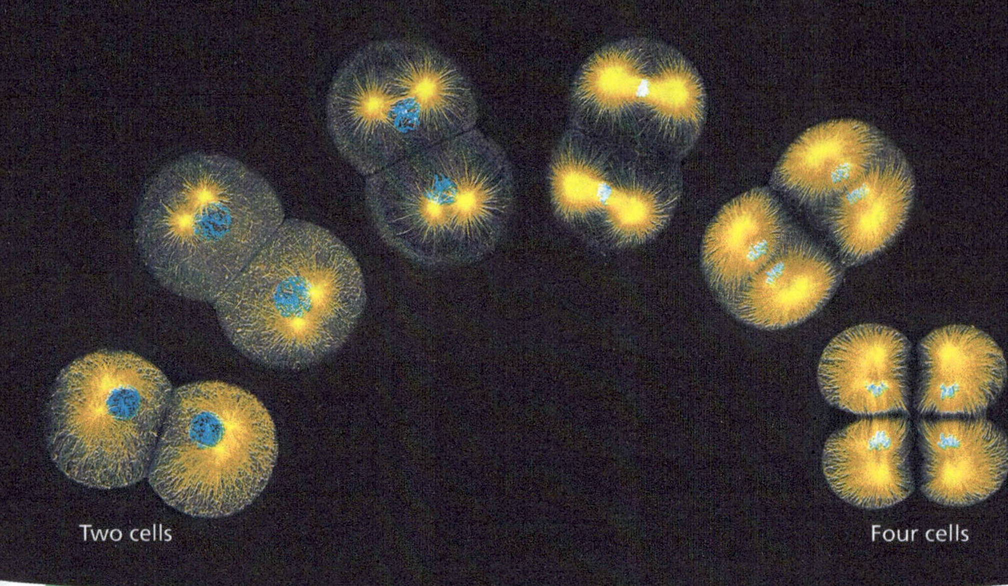

Figure 12.1 A multicellular organism starts out as a single cell that divides into two. Those two cells then divide into four, as shown in these fluorescence micrographs of a marine worm embryo. Cell division continues throughout an organism's life, for growth or to replace worn-out or damaged cells. Each time a cell divides in this way, it is crucial that the daughter cells be genetically identical to the parent cell.

How does one parent cell give rise to two genetically identical daughter cells?

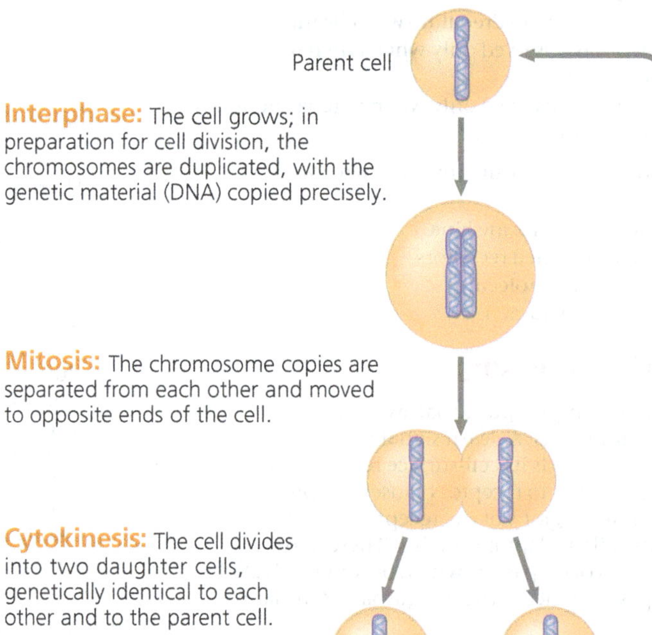

CONCEPT 12.1

Most cell division results in genetically identical daughter cells

The ability of organisms to produce more of their own kind is the one characteristic that best distinguishes living things from nonliving matter. This unique capacity to procreate, like all biological functions, has a cellular basis. The continuity of life is based on the reproduction of cells, or **cell division**.

Key Roles of Cell Division

Cell division plays several important roles in life. When a prokaryotic cell divides, it is actually reproducing because the process gives rise to a new organism (another cell). The same is true of any unicellular eukaryote, such as the amoeba shown in **Figure 12.2a**. As for multicellular eukaryotes, cell division enables each of these organisms to develop from a single cell—the fertilised egg. A two-celled embryo, the first stage in this process, is shown in **Figure 12.2b**. And cell division continues to function in renewal and repair in fully grown multicellular eukaryotes, replacing cells that die from accidents or normal wear and tear. For example, dividing cells in your bone marrow continuously make new blood cells **(Figure 12.2c)**.

The reproduction of a cell, with all of its complexity, cannot occur by a mere pinching in half; a cell is not like a soap bubble that simply enlarges and splits in two. In both prokaryotes and eukaryotes, a crucial function of most cell divisions is the distribution of identical genetic material—DNA—to two daughter cells. (The exception is meiosis, the special type of eukaryotic cell division that can produce sperm and eggs.) What is most remarkable about cell division is the accuracy with which the DNA is passed from one generation of cells to the next. A dividing cell replicates its DNA, distributes the two copies to opposite ends of the cell, and then splits into daughter cells.

▼ Figure 12.2 The functions of cell division.

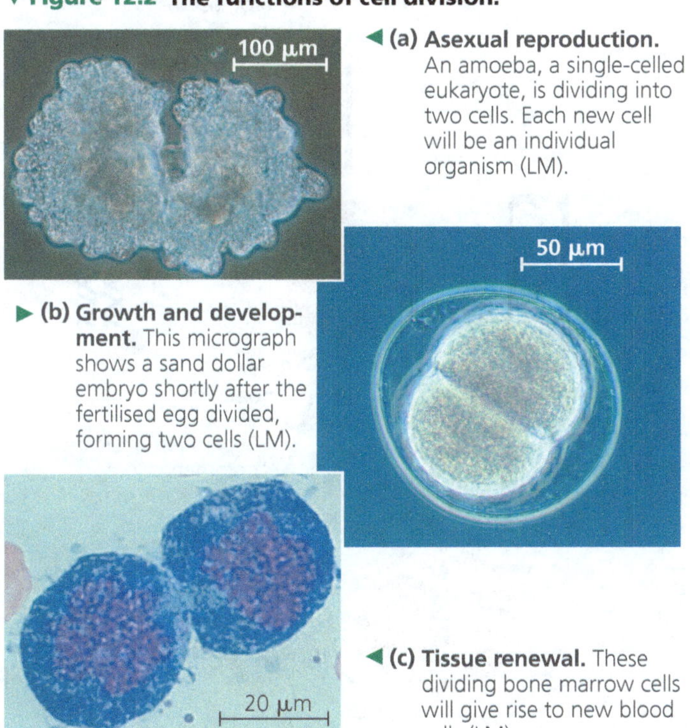

◀ (a) Asexual reproduction. An amoeba, a single-celled eukaryote, is dividing into two cells. Each new cell will be an individual organism (LM).

▶ (b) Growth and development. This micrograph shows a sand dollar embryo shortly after the fertilised egg divided, forming two cells (LM).

◀ (c) Tissue renewal. These dividing bone marrow cells will give rise to new blood cells (LM).

Cellular Organisation of the Genetic Material

A cell's DNA, its genetic information, is called its **genome**. Although a prokaryotic genome is often a single DNA molecule, eukaryotic genomes usually consist of a number of DNA molecules. The overall length of DNA in a eukaryotic cell is enormous. A typical human cell, for example, has about 2 m of DNA—a length about 250,000 times greater than the cell's diameter. Before the cell can divide to form genetically identical daughter cells, all of this DNA must be copied, or replicated, and then the two copies must be separated so that each daughter cell ends up with a complete genome.

The replication and distribution of so much DNA are manageable because the DNA molecules are packaged into structures called **chromosomes** (from the Greek *chroma*, colour, and *soma*, body), so named because they take up certain dyes used in microscopy **(Figure 12.3)**. Each eukaryotic chromosome consists of one very long, linear DNA molecule associated with many proteins (see Figure 6.9). The DNA molecule carries several hundred to a few thousand genes, the units of information that specify an organism's inherited traits. The associated proteins maintain the structure of the chromosome and help control the activity of the genes. Together, the entire complex of DNA and proteins that is the building material of chromosomes is referred to as **chromatin**. As you will soon see, the chromatin of a chromosome varies in its degree of condensation during the process of cell division.

▼ Figure 12.3 Eukaryotic chromosomes. Chromosomes (stained purple) are visible within the nucleus of this cell from an African blood lily. The thinner red threads in the surrounding cytoplasm are the cytoskeleton. The cell is preparing to divide (LM).

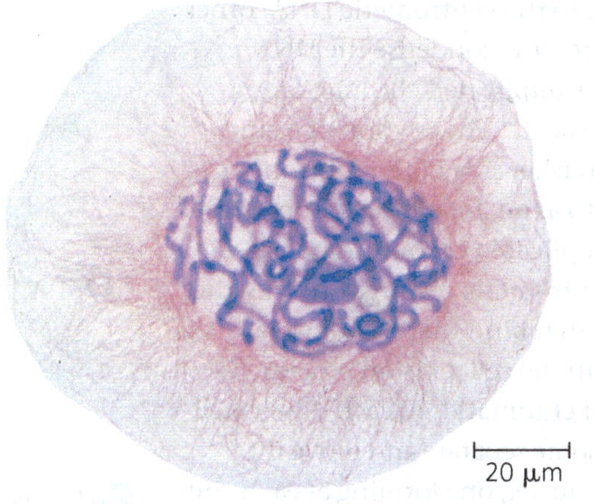

Every eukaryotic species has a characteristic number of chromosomes in each cell's nucleus. For example, the nuclei of human **somatic cells** (all body cells except the reproductive cells) each contain 46 chromosomes, made up of two sets of 23, one set inherited from each parent. Reproductive cells, or **gametes**—such as sperm and eggs—have half as many chromosomes as somatic cells; in our example, human gametes have one set of 23 chromosomes. The number of chromosomes in somatic cells varies widely among species: 18 in cabbage plants, 48 in chimpanzees, 56 in elephants, 52 in platypus and 148 in one species of alga. We'll now consider how these chromosomes behave during cell division.

▼ **Figure 12.4 A highly condensed, duplicated human chromosome. (SEM)**

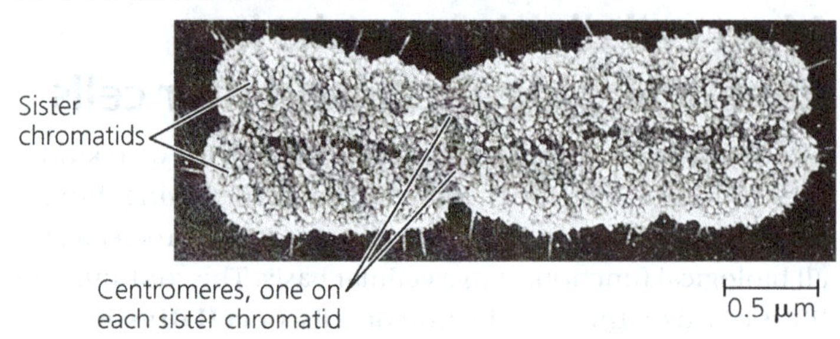

DRAW IT *Circle one sister chromatid of this chromosome.*

Distribution of Chromosomes During Eukaryotic Cell Division

When a cell is not dividing, and even as it replicates its DNA in preparation for cell division, each chromosome is in the form of a long, thin chromatin fibre. After DNA replication, however, the chromosomes condense as a part of cell division: Each chromatin fibre becomes densely coiled and folded, making the chromosomes much shorter and so thick that we can see them with a light microscope.

Each duplicated chromosome consists of two **sister chromatids**, which are joined copies of the original chromosome **(Figure 12.4)**. The two chromatids, each containing an identical DNA molecule, are often attached all along their lengths by protein complexes called *cohesins*; this attachment is known as *sister chromatid cohesion*. Each sister chromatid has a **centromere**, a region made up of repetitive sequences in the chromosomal DNA where the chromatid is attached most closely to its sister chromatid. This attachment is mediated by proteins that recognise and bind to the centromeric DNA; other bound proteins condense the DNA, giving the duplicated chromosome a narrow "waist." The portion of a chromatid to either side of the centromere is referred to as an *arm* of the chromatid. (An unduplicated chromosome has a single centromere, distinguished by the proteins that bind there, and two arms.)

Later in the cell division process, the two sister chromatids of each duplicated chromosome separate and move into two new nuclei, one forming at each end of the cell. Once the sister chromatids separate, they are no longer called sister chromatids but are considered individual chromosomes; this is the step that essentially doubles the number of chromosomes during cell division. Thus, each new nucleus receives a collection of chromosomes identical to that of the parent cell **(Figure 12.5)**. **Mitosis**, the division of the genetic material in the nucleus, is usually followed immediately by **cytokinesis**, the division of the cytoplasm. One cell has become two, each the genetic equivalent of the parent cell.

▼ **Figure 12.5 Chromosome duplication and distribution during cell division.**

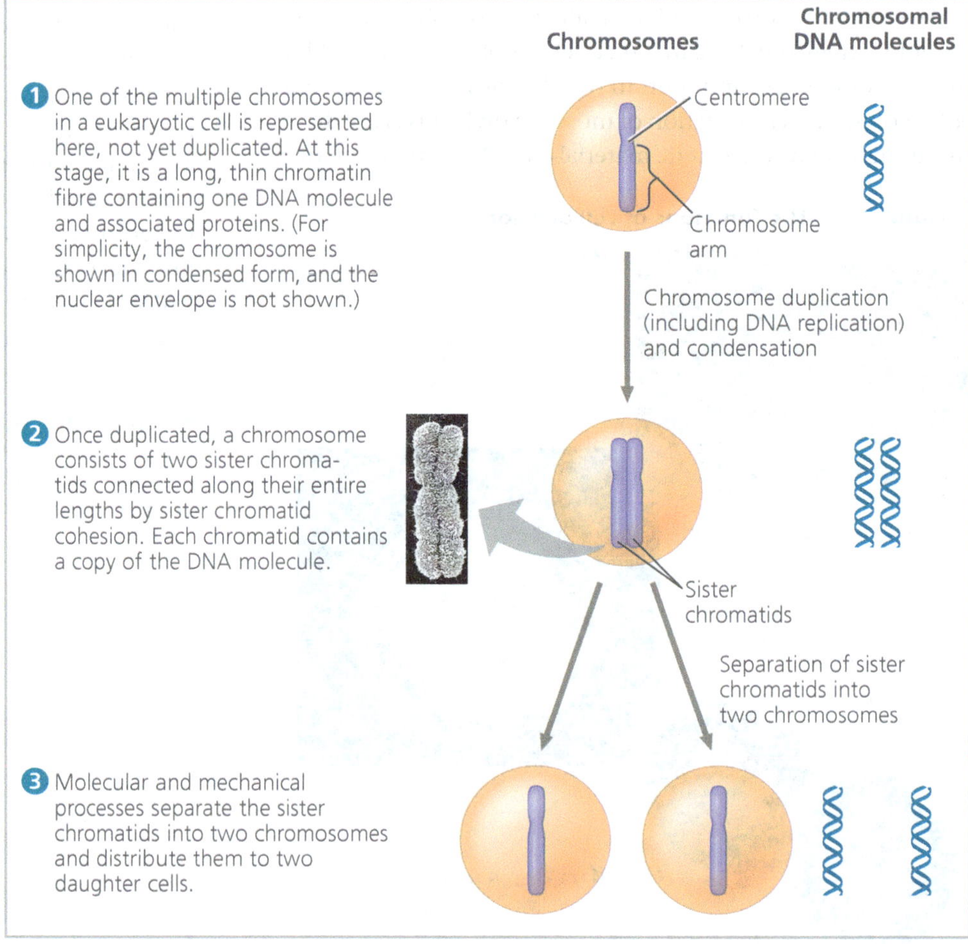

? *How many chromatid arms does the chromosome in* 2 *have? Identify the point in the figure where one chromosome becomes two.*

From a fertilised egg, mitosis and cytokinesis produced the 37 trillion somatic cells that now make up your body, and the same processes continue to generate new cells to replace dead and damaged ones. In contrast, you produce gametes—eggs or sperm—by a variation of cell division called *meiosis*, which yields daughter cells with only one set of chromosomes, half as many chromosomes as the parent cell. Meiosis in humans occurs only in special cells in the ovaries or testes (the gonads). Generating gametes, meiosis reduces the chromosome number from 46 (two sets) to 23 (one set). Fertilisation fuses two gametes together and returns the chromosome number to 46 (two sets). Mitosis then conserves that number in every somatic cell nucleus of the new human individual. In the chapter, "Meiosis and Sexual Life Cycles," we will examine the role of meiosis in reproduction and inheritance in more detail. In the remainder of this chapter, we focus on mitosis and the rest of the cell cycle in eukaryotes.

CONCEPT CHECK 12.1

1. How many chromosomes are drawn in each part of Figure 12.5? (Ignore the micrograph in step 2.)
2. **WHAT IF?** A chicken has 78 chromosomes in its somatic cells. How many chromosomes did the chicken inherit from each parent? How many chromosomes are in each of the chicken's gametes? How many chromosomes will be in each somatic cell of the chicken's offspring?

For suggested answers, see Appendix A.

CONCEPT 12.2

The mitotic phase alternates with interphase in the cell cycle

In 1882, a German anatomist named Walther Flemming developed dyes that allowed him to observe, for the first time, the behaviour of chromosomes during mitosis and cytokinesis. (In fact, Flemming coined the terms *mitosis* and *chromatin*.) During the period between one cell division and the next, it appeared to Flemming that the cell was simply growing larger. But we now know that many critical events occur during this stage in the life of a cell.

Phases of the Cell Cycle

Mitosis is just one part of the **cell cycle**, the life of a cell from the time it is first formed during division of a parent cell until its own division into two daughter cells **(Figure 12.6)**. (Biologists use the words *daughter* or *sister* in relation to cells, but this is not meant to imply gender.) In fact, the **mitotic (M) phase**, which includes both mitosis and cytokinesis, is usually the shortest part of the cell cycle. The mitotic phase alternates with a much longer stage called **interphase**, which often accounts for about 90% of the cycle. Interphase can be divided into three phases: the **G_1 phase** ("first gap"), the **S phase** ("synthesis"), and the **G_2 phase** ("second gap").

▼ **Figure 12.6 The cell cycle.** In a dividing cell, the mitotic (M) phase alternates with interphase, a growth period.

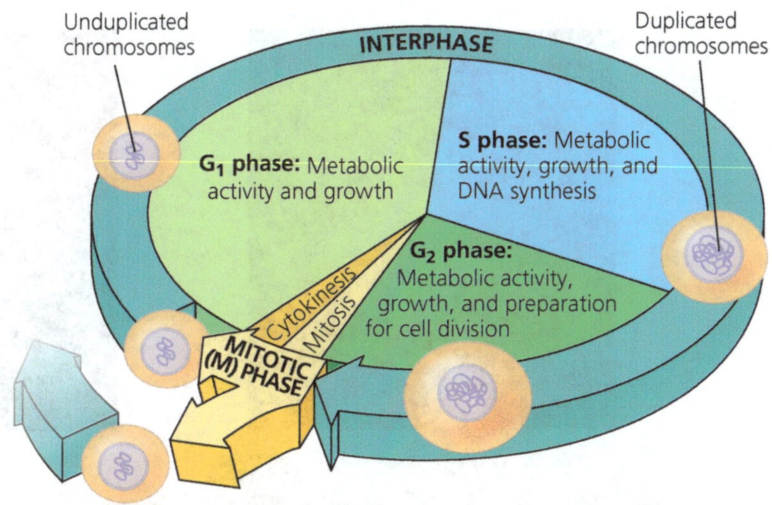

MITOTIC (M) PHASE:
Mitosis: Distribution of chromosomes into two daughter nuclei
Cytokinesis: Division of cytoplasm, producing two daughter cells. Each daughter cell can start a new cell cycle.

The G phases were misnamed as "gaps" when they were first observed because the cells appeared inactive, but we now know that intense metabolic activity and growth occur throughout interphase. During all three phases of interphase, actually, a cell grows by producing proteins and cytoplasmic organelles such as mitochondria and endoplasmic reticulum. Duplication of the chromosomes, crucial for eventual division of the cell, occurs entirely during the S phase. (We'll explore synthesis, or replication, of DNA in Concept 16.2.) Thus, a cell grows (G_1), continues to grow as it copies its chromosomes (S), grows more as it completes preparations for cell division (G_2), and divides (M). The daughter cells may then repeat the cycle.

A particular human cell might undergo one division in 24 hours. Of this time, the M phase would occupy less than 1 hour, while the S phase might occupy 10–12 hours, or about half the cycle. The rest of the time would be apportioned between the G_1 and G_2 phases. The G_2 phase usually takes 4–6 hours; in our example, G_1 would occupy about 5–6 hours. G_1 is the most variable in length in different types of cells. Some cells in a multicellular organism divide very infrequently or not at all. These cells spend their time in G_1 (or a related phase called G_0, to be discussed later in the chapter) doing their job in the organism—a cell of the pancreas secretes digestive enzymes, for example.

Mitosis is conventionally broken down into five stages: **prophase**, **prometaphase**, **metaphase**, **anaphase**, and **telophase**. Overlapping with the latter stages of mitosis, cytokinesis completes the mitotic phase. **Figure 12.7** describes these stages in an animal cell. Study this figure thoroughly before proceeding to the next two sections, which examine mitosis and cytokinesis more closely.

▼ Figure 12.7 Exploring Mitosis in an Animal Cell

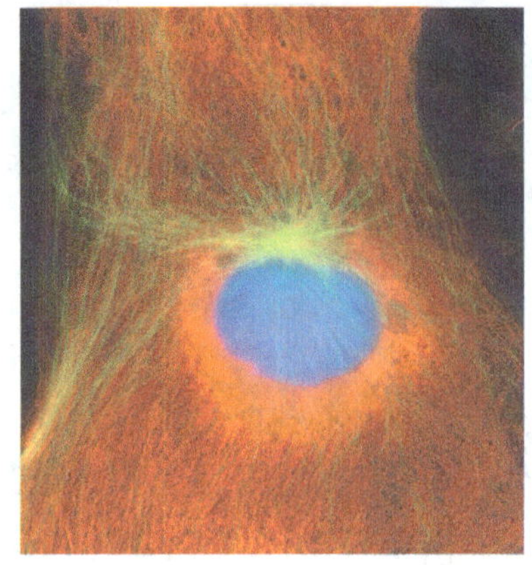

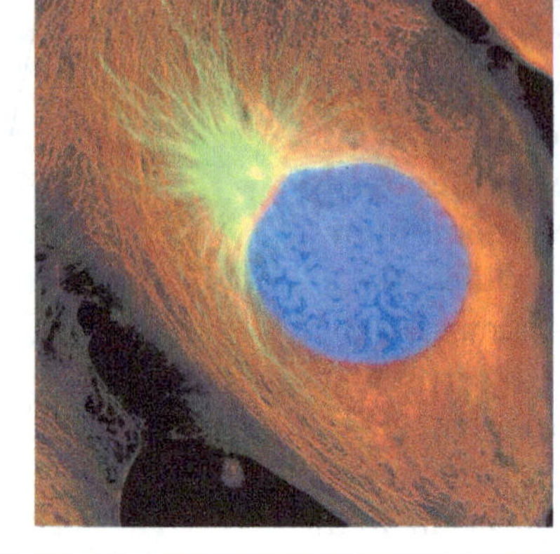

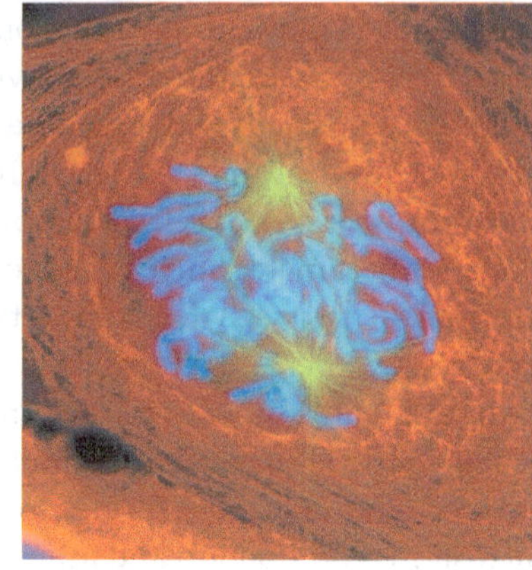

G$_2$ of Interphase

Prophase

Prometaphase

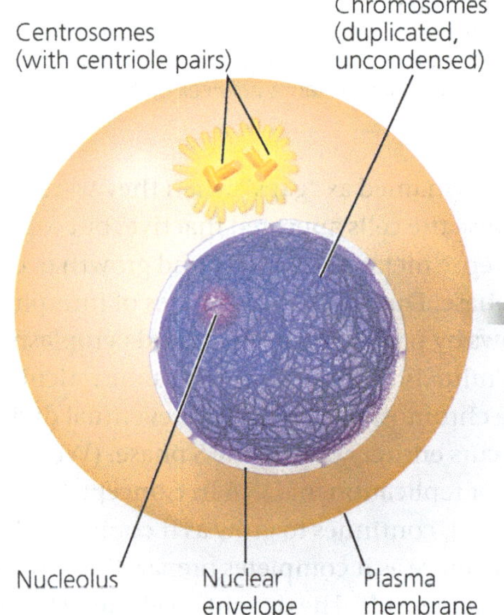

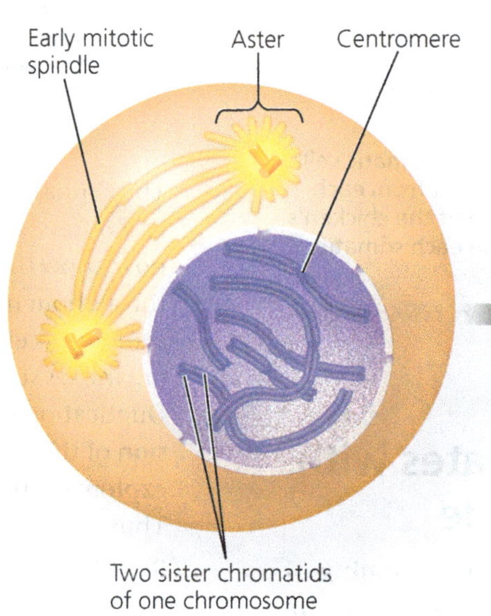

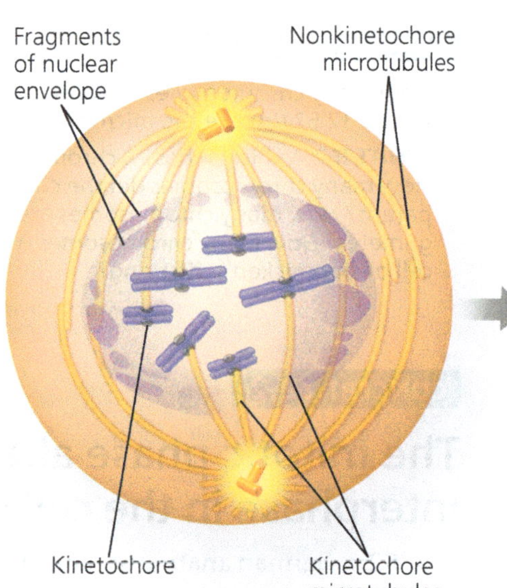

G$_2$ of Interphase

- A nuclear envelope encloses the nucleus.
- The nucleus contains one or more nucleoli (singular, *nucleolus*).
- Two centrosomes have formed by duplication of a single centrosome. Centrosomes are regions in animal cells that organise the microtubules of the spindle. Each centrosome contains two centrioles.
- Chromosomes, duplicated during S phase, cannot be seen individually because they have not yet condensed.

The fluorescence micrographs show dividing lung cells from a newt; this species has 22 chromosomes. Chromosomes appear blue, microtubules green, and intermediate filaments red. For simplicity, the drawings show only 6 chromosomes.

Prophase

- The chromatin fibres become more tightly coiled, condensing into discrete chromosomes observable with a light microscope.
- The nucleoli disappear.
- Each duplicated chromosome appears as two identical sister chromatids joined at their centromeres and, often, all along their arms by cohesins, resulting in sister chromatid cohesion.
- The mitotic spindle (named for its shape) begins to form. It is composed of the centrosomes and the microtubules that extend from them. The radial arrays of shorter microtubules that extend from the centrosomes are called asters ("stars").
- The centrosomes move away from each other, propelled partly by the lengthening microtubules between them.

Prometaphase

- The nuclear envelope fragments.
- The microtubules extending from each centrosome can now invade the nuclear area.
- The chromosomes have become even more condensed.
- A kinetochore, a specialised protein structure, has now formed at the centromere of each chromatid (thus, two per chromosome).
- Some of the microtubules attach to the kinetochores, becoming "kinetochore microtubules," which jerk the chromosomes back and forth.
- Nonkinetochore microtubules interact with those from the opposite pole of the spindle, lengthening the cell.

? *How many molecules of DNA are in the prometaphase drawing? How many molecules per chromosome? How many double helices are there per chromosome? Per chromatid?*

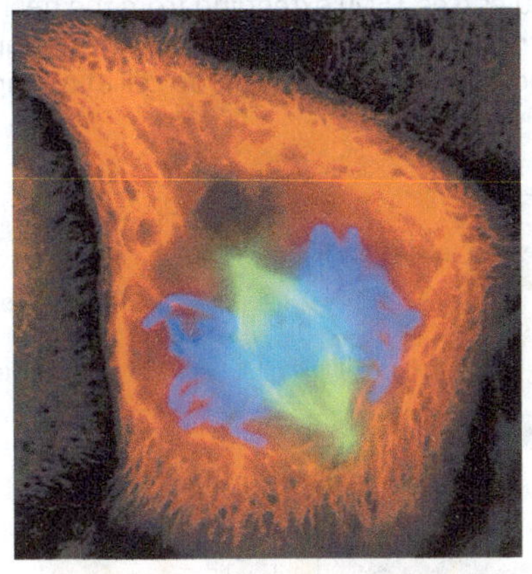

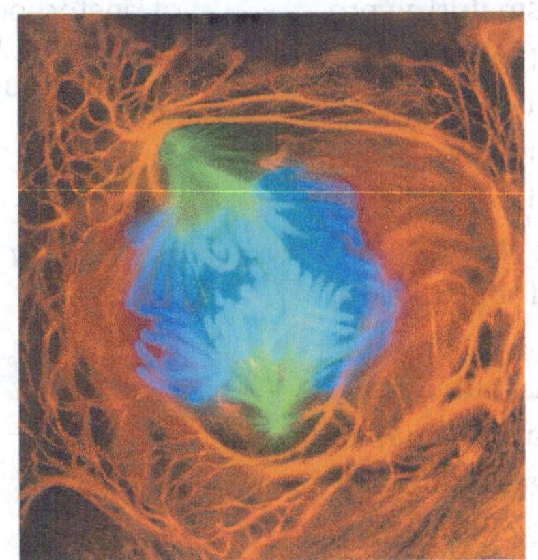

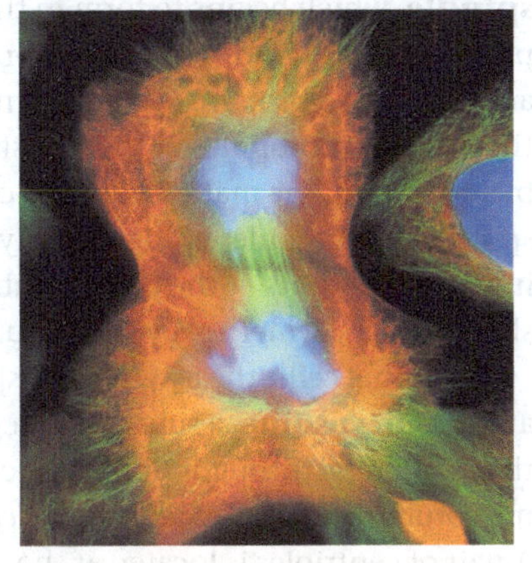

Metaphase

Anaphase

Telophase and Cytokinesis

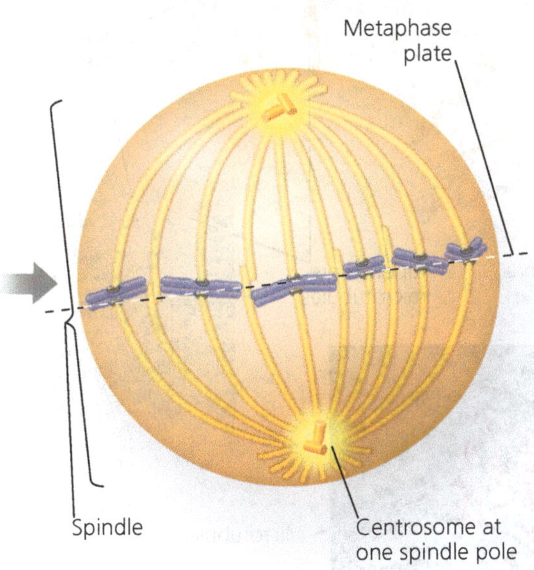

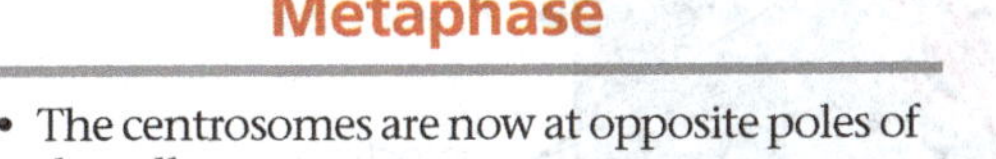

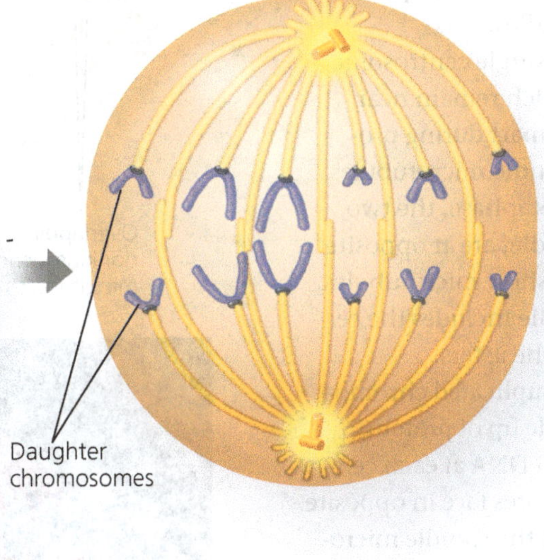

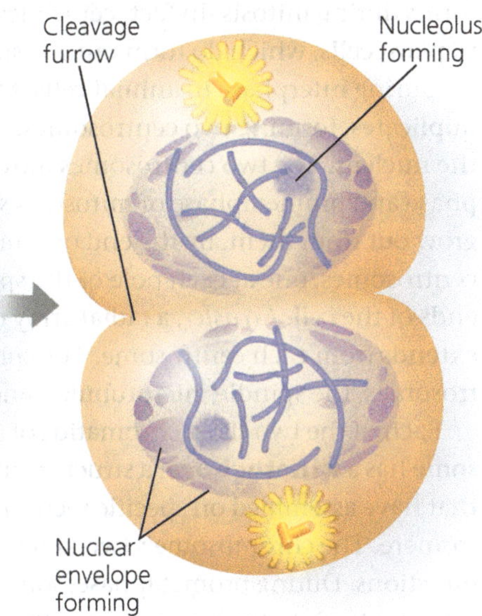

Metaphase

- The centrosomes are now at opposite poles of the cell.
- The chromosomes have all arrived at the *metaphase plate*, a plane that is equidistant between the spindle's two poles. The chromosomes' centromeres lie at the metaphase plate.
- For each chromosome, the kinetochores of the sister chromatids are attached to kinetochore microtubules coming from opposite poles.

Anaphase

- Anaphase is the shortest stage of mitosis, often lasting only a few minutes.
- Anaphase begins when the cohesin proteins are cleaved. This allows the two sister chromatids of each pair to part suddenly. Each chromatid thus becomes an independent chromosome.
- The two new daughter chromosomes begin moving towards opposite ends of the cell as their kinetochore microtubules shorten. Because these microtubules are attached at the centromere region, the centromeres are pulled ahead of the arms, moving at a rate of about 1 μm/min.
- The cell elongates as the nonkinetochore microtubules lengthen.
- By the end of anaphase, the two ends of the cell have identical—and complete—collections of chromosomes.

Telophase

- Two daughter nuclei form in the cell. Nuclear envelopes arise from the fragments of the parent cell's nuclear envelope and other portions of the endomembrane system.
- Nucleoli reappear.
- The chromosomes become less condensed.
- Any remaining spindle microtubules are depolymerised.
- Mitosis, the division of one nucleus into two genetically identical nuclei, is now complete.

Cytokinesis

- The division of the cytoplasm is usually well under way by late telophase, so the two daughter cells appear shortly after the end of mitosis.
- In animal cells, cytokinesis involves the formation of a cleavage furrow, which pinches the cell in two.

The Mitotic Spindle: *A Closer Look*

Many of the events of mitosis depend on the **mitotic spindle**, which begins to form in the cytoplasm during prophase. This structure consists of fibres made of microtubules and associated proteins. While the mitotic spindle assembles, the other microtubules of the cytoskeleton partially disassemble, providing the material used to construct the spindle. The spindle microtubules elongate (polymerise) by incorporating more subunits of the protein tubulin (see Table 6.1) and shorten (depolymerise) by losing subunits.

In animal cells, the assembly of spindle microtubules starts at the **centrosome**, a subcellular region containing material that functions throughout the cell cycle to organise the cell's microtubules. (It is also a type of *microtubule-organising-centre.*) A pair of centrioles is located at the centre of the centrosome, but they are not essential for cell division: If the centrioles are destroyed with a laser microbeam, a spindle nevertheless forms during mitosis. In fact, centrioles are not even present in plant cells, which do form mitotic spindles.

During interphase in animal cells, the single centrosome duplicates, forming two centrosomes, which remain near the nucleus. The two centrosomes move apart during prophase and prometaphase of mitosis as spindle microtubules grow out from them. By the end of prometaphase, the two centrosomes, one at each pole of the spindle, are at opposite ends of the cell. An *aster*, a radial array of short microtubules, extends from each centrosome. The spindle includes the centrosomes, the spindle microtubules, and the asters.

Each of the two sister chromatids of a duplicated chromosome has a **kinetochore**, a structure made up of proteins that have assembled on specific sections of DNA at each centromere. The chromosome's two kinetochores face in opposite directions. During prometaphase, some of the spindle microtubules attach to the kinetochores; these are called *kinetochore microtubules*. (The number of microtubules attached to a kinetochore varies among species, from one microtubule in yeast cells to 40 or so in some mammalian cells.) When one of a chromosome's kinetochores is "captured" by microtubules, the chromosome begins to move towards the pole from which those microtubules extend. However, this movement comes to a halt as soon as microtubules from the opposite pole attach to the kinetochore on the other chromatid. What happens next is like a tug-of-war that ends in a draw. The chromosome moves first in one direction and then in the other, back and forth, finally settling midway between the two ends of the cell. At metaphase, the centromeres of all the duplicated chromosomes are on a plane midway between the spindle's two poles. This plane is called the **metaphase plate**, which is an imaginary plate rather than an actual cellular structure **(Figure 12.8)**. Meanwhile, microtubules that do not attach to kinetochores have been elongating, and by metaphase they overlap and interact with other *nonkinetochore microtubules* from the opposite pole of the spindle. By metaphase, the microtubules

▼ Figure 12.8 The mitotic spindle at metaphase. The kinetochores of each chromosome's two sister chromatids face in opposite directions. Here, each kinetochore is attached to a cluster of kinetochore microtubules (see TEM) extending from the nearest centrosome. Nonkinetochore microtubules overlap at the metaphase plate. The fluorescent micrograph is a rat kangaroo cell at metaphase.

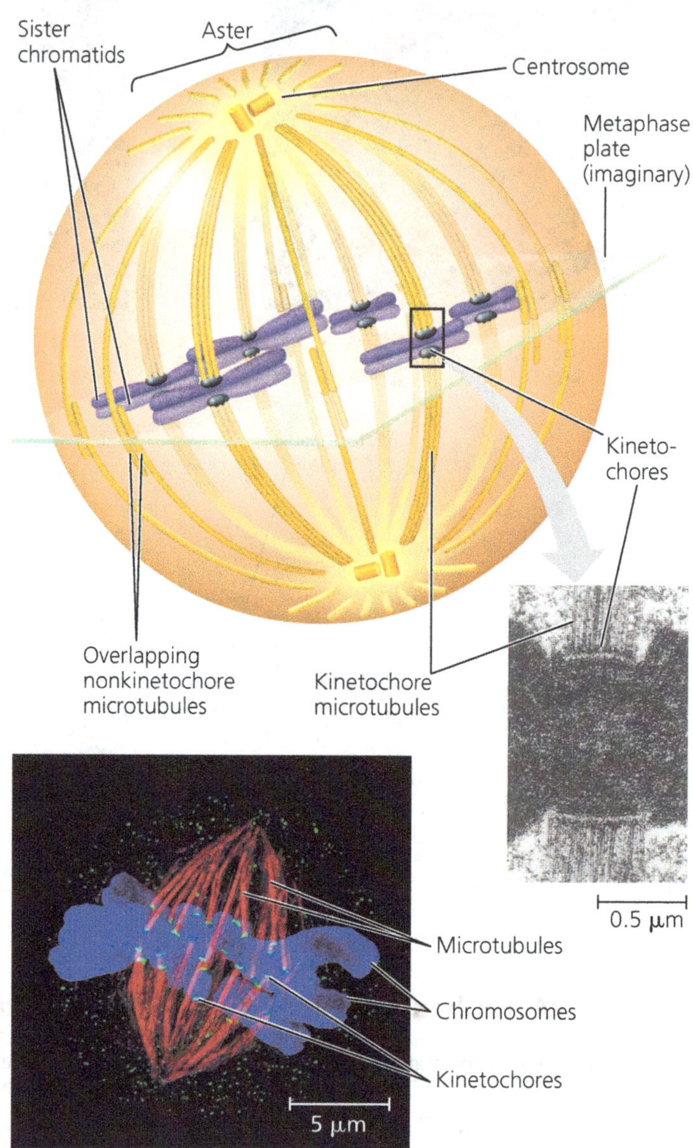

DRAW IT *On the lower micrograph, draw a line indicating the position of the metaphase plate. Draw arrows showing the directions of chromosome movement when anaphase begins.*

of the asters have also grown and are in contact with the plasma membrane. The spindle is now complete.

The structure of the spindle correlates well with its function during anaphase. Anaphase begins suddenly when the cohesins holding together the sister chromatids of each chromosome are cleaved by an enzyme called *separase*. Once separated, the chromatids become individual chromosomes that move towards opposite ends of the cell.

How do the kinetochore microtubules function in this poleward movement of chromosomes? Apparently, two mechanisms are in play, both involving motor proteins. (To review how motor proteins move an object along a microtubule, see Figure 6.21.) Results of a cleverly designed experiment suggested that motor proteins on the kinetochores "walk" the chromosomes along the microtubules, which depolymerise at their kinetochore ends after the motor proteins have passed **(Figure 12.9)**. (This is referred to as the "Pac-man" mechanism because of its resemblance to the arcade game character that moves by eating all the dots in its path.) However, other researchers, working with different cell types or cells from other species, have shown that chromosomes are "reeled in" by motor proteins at the spindle poles and that the microtubules depolymerise after they pass by these motor proteins at the poles. The general consensus now is that both mechanisms are used and that their relative contributions vary among cell types.

In a dividing animal cell, the nonkinetochore microtubules are responsible for elongating the whole cell during anaphase. Nonkinetochore microtubules from opposite poles overlap each other extensively during metaphase (see Figure 12.8). During anaphase, the region of overlap is reduced as motor proteins attached to the microtubules walk them away from one another, using energy from ATP. As the microtubules push apart from each other, their spindle poles are pushed apart, elongating the cell. At the same time, the microtubules lengthen somewhat by the addition of tubulin subunits to their overlapping ends. As a result, the microtubules continue to overlap.

At the end of anaphase, duplicate groups of chromosomes have arrived at opposite ends of the elongated parent cell. Nuclei re-form during telophase. Cytokinesis generally begins during anaphase or telophase, and the spindle eventually disassembles by depolymerisation of microtubules.

Cytokinesis: *A Closer Look*

In animal cells, cytokinesis occurs by a process known as **cleavage**. The first sign of cleavage is the appearance of a **cleavage furrow**, a shallow groove in the cell surface near the old metaphase plate **(Figure 12.10a)**. On the cytoplasmic side of the furrow is a contractile ring of actin microfilaments associated with molecules of the protein myosin. The actin microfilaments interact with the myosin molecules, causing the ring to contract. The contraction of the dividing cell's ring of microfilaments is like the pulling of a drawstring. The cleavage furrow deepens until the parent cell is pinched in two, producing two completely separated cells, each with its own nucleus and its own share of cytosol, organelles, and other subcellular structures.

Cytokinesis in plant cells, which have cell walls, is markedly different. There is no cleavage furrow. Instead, during telophase, vesicles derived from the Golgi apparatus move along microtubules to the middle of the cell, where they coalesce, producing a **cell plate**. Cell wall materials carried in the vesicles collect inside the cell plate as it grows **(Figure 12.10b)**. The cell plate enlarges until its surrounding membrane fuses

▼ Figure 12.9 Inquiry

At which end do kinetochore microtubules shorten during anaphase?

Experiment Gary Borisy and colleagues at the University of Wisconsin wanted to determine whether kinetochore microtubules depolymerise at the kinetochore end or the pole end as chromosomes move towards the poles during mitosis. First they labelled the microtubules of a pig kidney cell in early anaphase with a yellow fluorescent dye. (Nonkinetochore microtubules are not shown.)

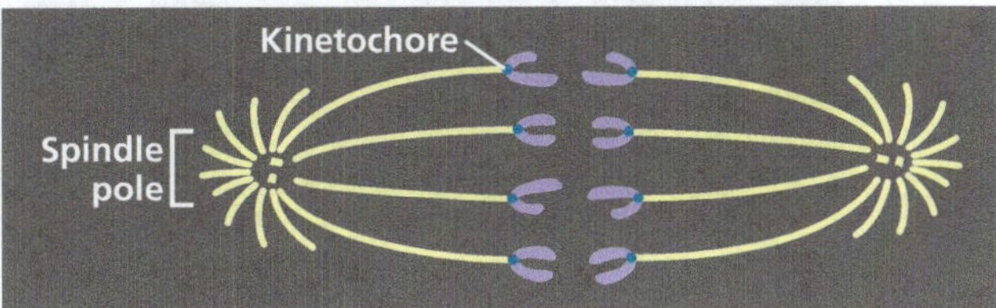

Then they marked a region of the kinetochore microtubules between one spindle pole and the chromosomes by using a laser to eliminate the fluorescence from that region, while leaving the microtubules intact (see below). As anaphase proceeded, they monitored the changes in microtubule length on either side of the non-fluorescent mark.

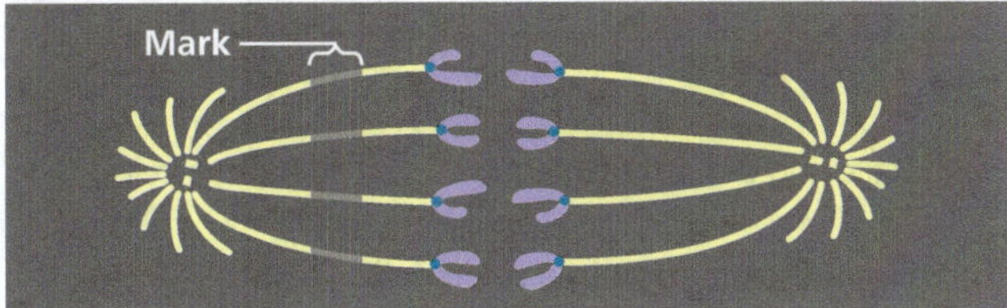

Results As the chromosomes moved polewards, the microtubule segments on the kinetochore side of the mark shortened, while those on the spindle pole side stayed the same length.

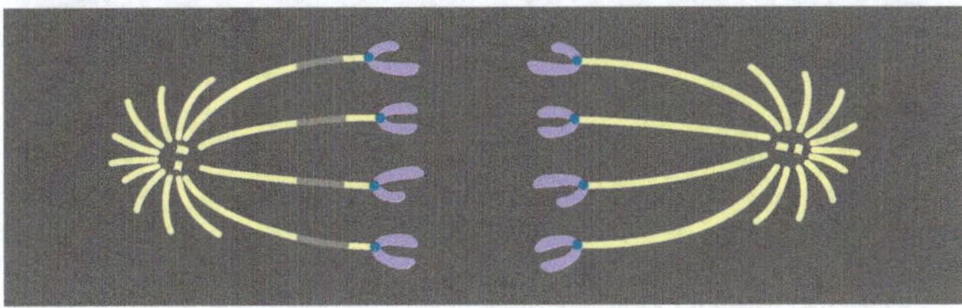

Conclusion During anaphase in this cell type, chromosome movement is correlated with kinetochore microtubules shortening at their kinetochore ends and not at their spindle pole ends. This experiment supports the hypothesis that during anaphase, a chromosome is walked along a microtubule as the microtubule depolymerises at its kinetochore end, releasing tubulin subunits.

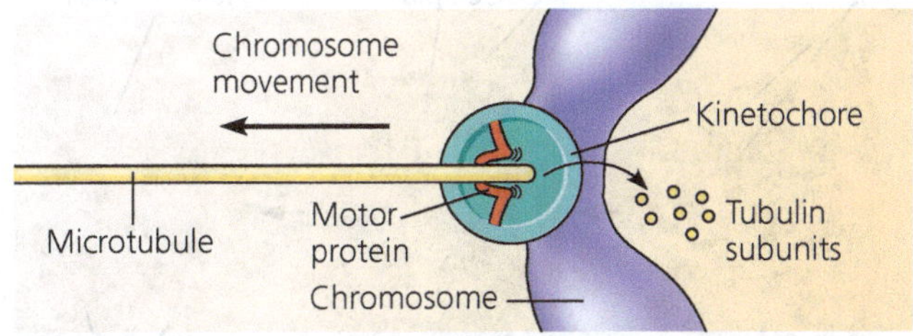

Data from G. J. Gorbsky, P. J. Sammak, and G. G. Borisy, Chromosomes move poleward in anaphase along stationary microtubules that coordinately disassemble from their kinetochore ends, *Journal of Cell Biology* 104:9–18 (1987).

WHAT IF? *If this experiment had been done on a cell type in which "reeling in" at the poles was the main cause of chromosome movement, how would the mark have moved relative to the poles? How would the microtubule portions on either side of the mark have changed?*

▼ Figure 12.10 Cytokinesis in animal and plant cells.

(a) Cleavage of an animal cell (SEM)

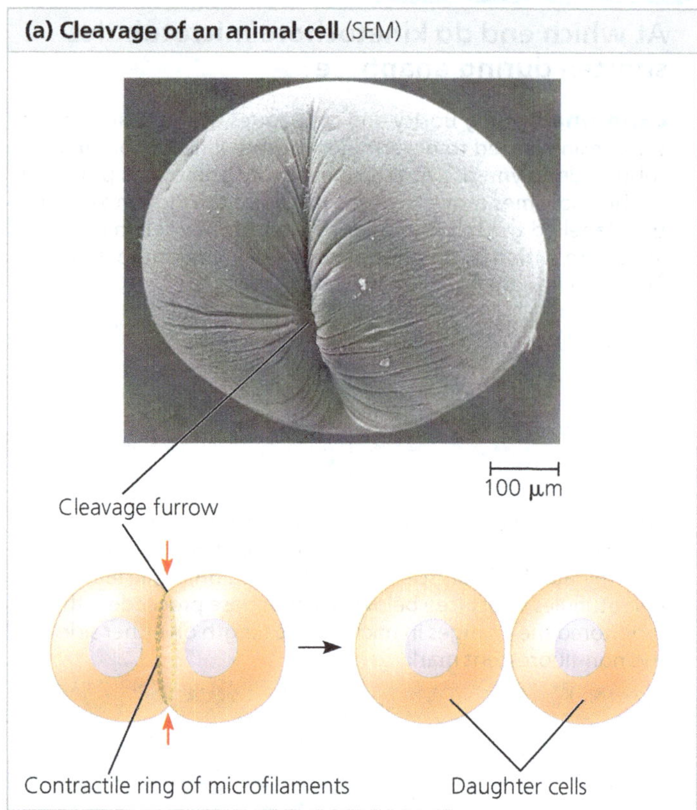

(b) Cell plate formation in a plant cell (TEM)

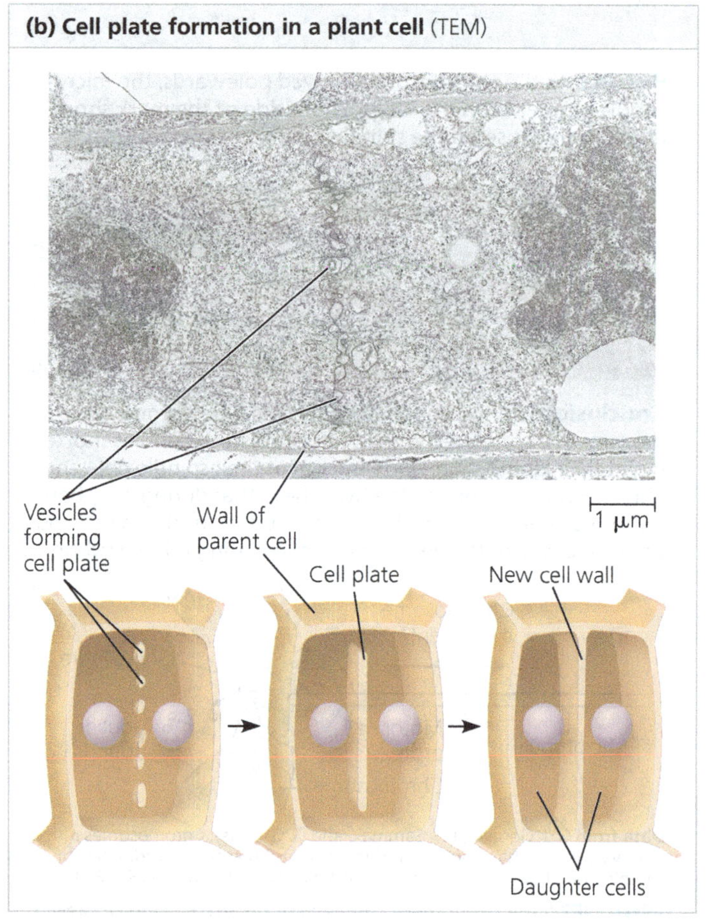

with the plasma membrane along the perimeter of the cell. Two daughter cells result, each with its own plasma membrane. Meanwhile, a new cell wall arising from the contents of the cell plate forms between the daughter cells.

Figure 12.11 is a series of micrographs of a dividing plant cell. Examining this figure will help you review mitosis and cytokinesis.

Binary Fission in Bacteria

Prokaryotes (bacteria and archaea) can undergo a type of reproduction in which the cell grows to roughly double its size and then divides to form two cells. The term **binary fission**, meaning "division in half," refers to this process and to the asexual reproduction of single-celled eukaryotes, such as the amoeba in Figure 12.2a. However, the process in eukaryotes involves mitosis, while that in prokaryotes does not.

In bacteria, most genes are carried on a single bacterial chromosome that consists of a circular DNA molecule and associated proteins. Although bacteria are smaller and simpler than eukaryotic cells, the challenge of replicating their genomes in an orderly fashion and distributing the copies equally to two daughter cells is still formidable. For example, when it is fully stretched out, the chromosome of the bacterium *Escherichia coli* is about 500 times as long as the cell. For such a long chromosome to fit within the cell, it must be highly coiled and folded.

In some bacteria, the process of cell division is initiated when the DNA of the bacterial chromosome begins to replicate at a specific place on the chromosome called the **origin of replication**, producing two origins. As the chromosome continues to replicate, one origin moves rapidly towards the opposite end of the cell **(Figure 12.12)**. While the chromosome is replicating, the cell elongates. When replication is complete and the bacterium has reached about twice its initial size, proteins cause its plasma membrane to pinch inwards, dividing the parent bacterial cell into two daughter cells. In this way, each cell inherits a complete genome.

Using the techniques of modern DNA technology to tag the origins of replication with molecules that glow green in fluorescence microscopy (see Figure 6.3), researchers have directly observed the movement of bacterial chromosomes. This movement is reminiscent of the poleward movements of the centromere regions of eukaryotic chromosomes during anaphase of mitosis, but bacteria don't have visible mitotic spindles or even microtubules. In most bacterial species studied, the two origins of replication end up at opposite ends of the cell or in some other very specific location, possibly anchored there by one or more proteins. How bacterial chromosomes move and how their specific location is established and maintained are active areas of research. Several proteins that play important roles have been identified. Polymerisation of one protein resembling eukaryotic actin apparently functions in bacterial

▼ Figure 12.11 Mitosis in a plant cell. These light micrographs show mitosis in cells of an onion root.

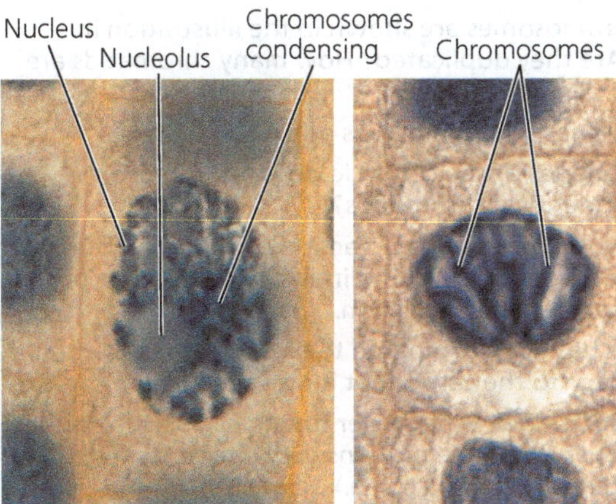

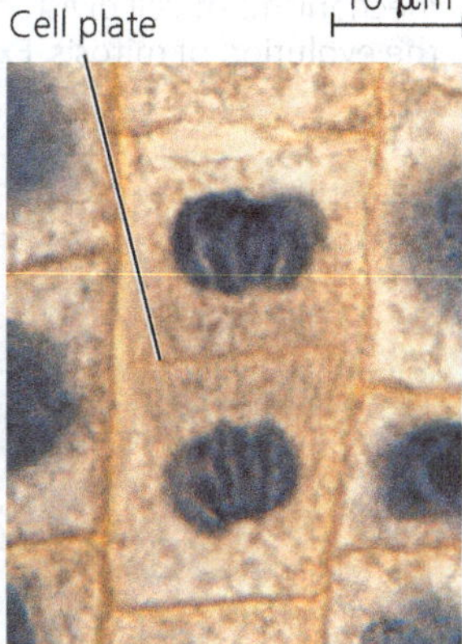

1 Prophase. The chromosomes are condensing and the nucleolus is beginning to disappear. Although not yet visible in the micrograph, the mitotic spindle is starting to form.

2 Prometaphase. Discrete chromosomes are now visible; each consists of two aligned, identical sister chromatids. Later in prometaphase, the nuclear envelope will fragment.

3 Metaphase. The spindle is complete, and the chromosomes, attached to microtubules at their kinetochores, are all at the metaphase plate.

4 Anaphase. Chromatids of each chromosome have separated. The daughter chromosomes are moving to the ends of the cell as their kinetochore microtubules shorten.

5 Telophase. Daughter nuclei are forming. Cytokinesis has started: The cell plate, which will divide the cytoplasm in two, grows towards the perimeter of the parent cell.

▼ Figure 12.12 Bacterial cell division by binary fission. The bacterium *E. coli*, shown here, has a single, circular chromosome.

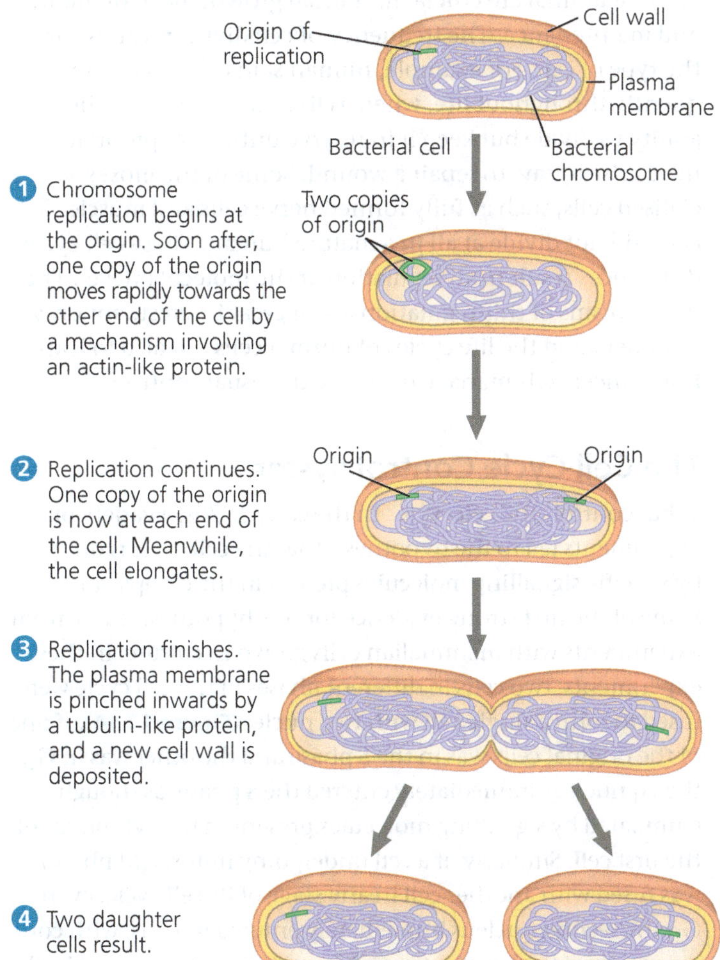

chromosome movement during cell division, and another protein that is related to tubulin helps pinch the plasma membrane inwards, separating the two bacterial daughter cells.

The Evolution of Mitosis

EVOLUTION Given that prokaryotes preceded eukaryotes on Earth by more than a billion years, we might hypothesise that mitosis evolved from simpler prokaryotic mechanisms of cell reproduction. The fact that some of the proteins involved in bacterial binary fission are related to eukaryotic proteins that function in mitosis supports that hypothesis.

As eukaryotes with nuclear envelopes and larger genomes evolved, the ancestral process of binary fission, seen today in bacteria, somehow gave rise to mitosis. Variations on cell division exist in different groups of organisms. These variant processes may be similar to mechanisms used by ancestral species and thus may resemble steps in the evolution of mitosis from a binary fission-like process presumably carried out by very early bacteria. Possible intermediate stages are suggested by two unusual types of nuclear division found today in certain unicellular eukaryotes—dinoflagellates, diatoms, and some yeasts **(Figure 12.13)**. These two modes of nuclear division are thought to be cases where ancestral mechanisms have remained relatively unchanged over evolutionary time. In both types, the nuclear envelope remains intact, in contrast to what happens in most eukaryotic cells. Keep in mind, however, that we can't observe cell division in cells of extinct species. This hypothesis uses only currently existing species as some possible examples of intermediate mechanisms. Other mechanisms may have existed in species that have gone extinct; we simply have no way of knowing.

▼ Figure 12.13 Mechanisms of cell division in several groups of organisms. Some unicellular eukaryotes existing today have mechanisms of cell division that may resemble intermediate steps in the evolution of mitosis. Except for **(a)**, cell walls are not shown.

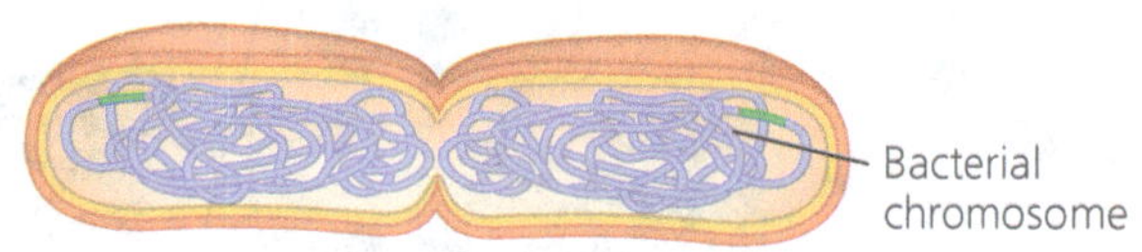

(a) Bacteria. During binary fission in bacteria, the origins of the daughter chromosomes move to opposite ends of the cell. The mechanism involves polymerisation of actin-like molecules and possibly proteins that may anchor the daughter chromosomes to specific sites on the plasma membrane.

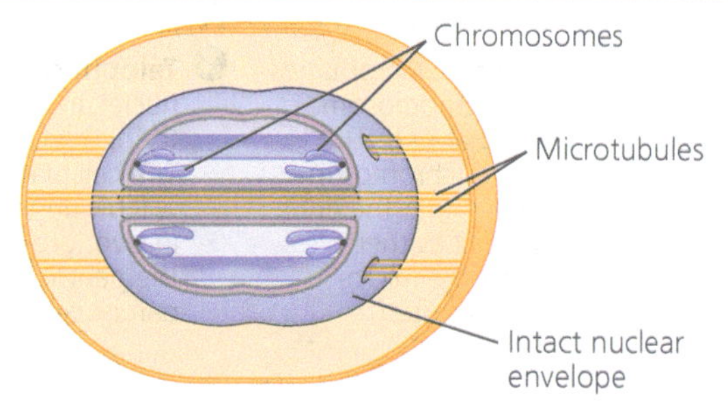

(b) Dinoflagellates. In unicellular eukaryotes called dinoflagellates, the chromosomes attach to the nuclear envelope, which remains intact during cell division. Microtubules pass through the nucleus inside cytoplasmic tunnels, reinforcing the spatial orientation of the nucleus, which then divides in a process reminiscent of bacterial binary fission.

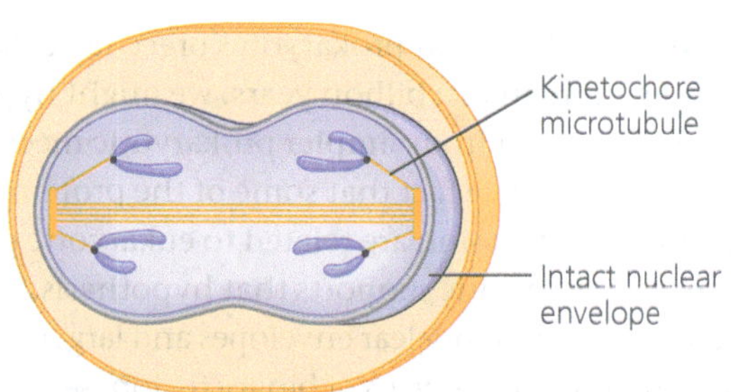

(c) Diatoms and some yeasts. In these two other groups of unicellular eukaryotes, the nuclear envelope also remains intact during cell division. In these organisms, the microtubules form a spindle *within* the nucleus. Microtubules separate the chromosomes, and the nucleus splits into two daughter nuclei.

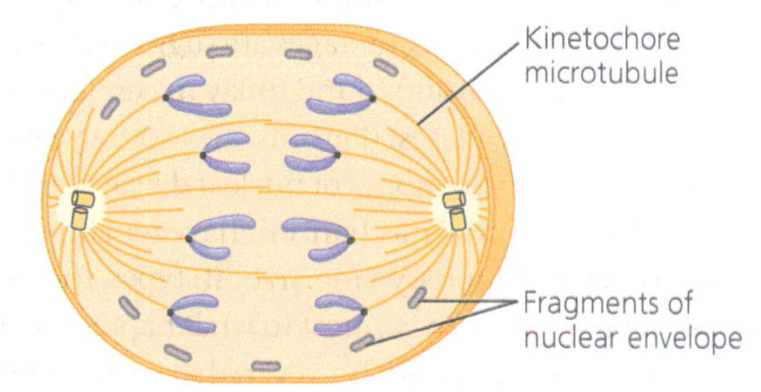

(d) Most eukaryotes. In most other eukaryotes, including plants and animals, the spindle forms outside the nucleus, and the nuclear envelope breaks down during mitosis. Microtubules separate the chromosomes, and two nuclear envelopes then form.

CONCEPT CHECK **12.2**

1. How many chromosomes are shown in the illustration in Figure 12.8? Are they duplicated? How many chromatids are shown?
2. Compare cytokinesis in animal cells and plant cells.
3. During which stages of the cell cycle does a chromosome consist of two identical chromatids?
4. Compare the roles of tubulin and actin during eukaryotic cell division with the roles of tubulin-like and actin-like proteins during bacterial binary fission.
5. A kinetochore has been compared to a coupling device that connects a motor to the cargo that it moves. Explain.
6. **MAKE CONNECTIONS** What other functions do actin and tubulin carry out? Name the proteins they interact with to do so. (Review Figures 6.21a and 6.26a.)

For suggested answers, see Appendix A.

CONCEPT **12.3**

The eukaryotic cell cycle is regulated by a molecular control system

The timing and rate of cell division in different parts of a plant or animal are crucial to normal growth, development, and maintenance. The frequency of cell division varies with the type of cell. For example, human skin cells divide frequently throughout life, whereas liver cells maintain the ability to divide but keep it in reserve until an appropriate need arises—say, to repair a wound. Some of the most specialised cells, such as fully formed nerve cells and muscle cells, do not divide at all in a mature human. These cell cycle differences result from regulation at the molecular level. The mechanisms of this regulation are of great interest, not only to understand the life cycles of normal cells but also to learn how cancer cells manage to escape the usual controls.

The Cell Cycle Control System

What controls the cell cycle? In the early 1970s, a variety of experiments led to the hypothesis that the cell cycle is driven by specific signalling molecules present in the cytoplasm. Some of the first strong evidence for this hypothesis came from experiments with mammalian cells grown in culture. In these experiments, two cells in different phases of the cell cycle were fused to form a single cell with two nuclei **(Figure 12.14)**. If one of the original cells was in the S phase and the other was in G_1, the G_1 nucleus immediately entered the S phase, as though stimulated by signalling molecules present in the cytoplasm of the first cell. Similarly, if a cell undergoing mitosis (M phase) was fused with another cell in any stage of its cell cycle, even G_1, the second nucleus immediately entered mitosis, with condensation of the chromatin and formation of a mitotic spindle.

▼ Figure 12.14 Inquiry

Do molecular signals in the cytoplasm regulate the cell cycle?

Experiment Researchers at the University of Colorado wondered whether a cell's progression through the cell cycle is controlled by cytoplasmic molecules. They induced cultured mammalian cells at different phases of the cell cycle to fuse. Two experiments are shown.

Results

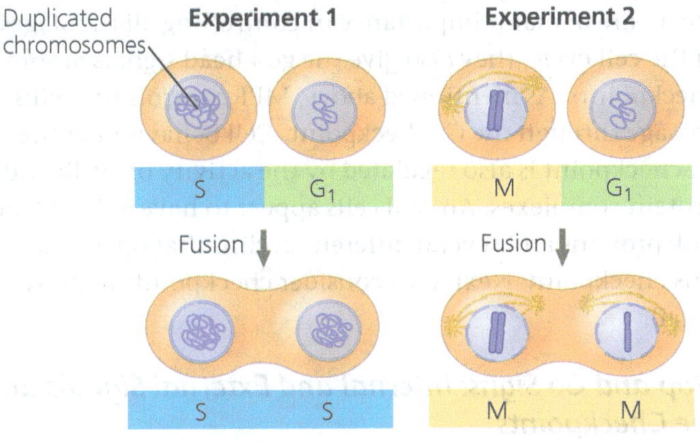

When a cell in S phase was fused with a cell in G_1, the G_1 nucleus immediately entered S phase—DNA was synthesised.

When a cell in M phase was fused with a cell in G_1, the G_1 nucleus immediately began mitosis—a spindle formed and the chromosomes condensed, even though the chromosomes had not been duplicated.

Conclusion The results of fusing a G_1 cell with a cell in the S or M phase of the cell cycle suggest that molecules present in the cytoplasm during the S or M phase control the progression to those phases.

Data from R. T. Johnson and P. N. Rao, Mammalian cell fusion: Induction of premature chromosome condensation in interphase nuclei, *Nature* 226:717–722 (1970).

WHAT IF? *If the progression of phases did not depend on cytoplasmic molecules and each phase began when the previous one was complete, how would the results have differed?*

The experiment shown in Figure 12.14 and other experiments on animal cells and yeasts demonstrated that the sequential events of the cell cycle are directed by a distinct **cell cycle control system**, a cyclically operating set of molecules in the cell that both triggers and coordinates key events in the cell cycle **(Figure 12.15)**. The cell cycle control system has been compared to the control device of a washing machine. Like the washer's timing device, the cell cycle control system proceeds on its own, according to a built-in clock. However, just as a washer's cycle is subject to both internal control (such as the sensor that detects when the tub is filled with water) and external adjustment (such as starting or stopping the machine), the cell cycle is regulated at certain checkpoints by both internal and external signals. A **checkpoint** in the cell cycle is a control point where stop and go-ahead signals can regulate the cycle. Three important checkpoints are found in the G_1, G_2, and M phases (red gates in Figure 12.15), which will be discussed shortly.

▼ Figure 12.15 Mechanical analogy for the cell cycle control system. In this diagram, the flat "stepping stones" around the perimeter represent sequential events. Like the control device of a washing machine, the cell cycle control system proceeds on its own, driven by a built-in clock. However, the system is subject to internal and external regulation at various checkpoints; three important checkpoints are shown (red).

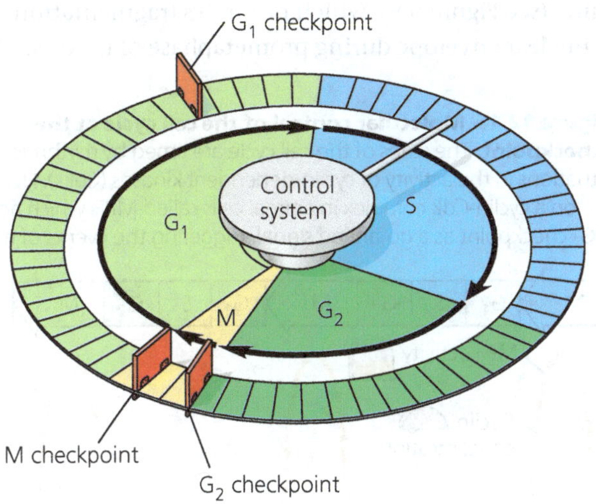

To understand how cell cycle checkpoints work, we'll first identify some of the molecules that make up the cell cycle control system (the molecular basis for the cell cycle clock) and describe how a cell progresses through the cycle. We'll then consider the internal and external checkpoint signals that can make the clock either pause or continue.

The Cell Cycle Clock: Cyclins and Cyclin-Dependent Kinases

Rhythmic fluctuations in the abundance and activity of cell cycle control molecules pace the sequential events of the cell cycle. These regulatory molecules are mainly proteins of two types: protein kinases and cyclins. As you learned in Concept 11.3, protein kinases are enzymes that activate or inactivate other proteins by phosphorylating them.

Many of the kinases that drive the cell cycle are actually present at a constant concentration in the growing cell, but much of the time they are in an inactive form. To be active, such a kinase must be attached to a **cyclin**, a protein that gets its name from its cyclically fluctuating concentration in the cell. Because of this requirement, these kinases are called **cyclin-dependent kinases**, or **Cdks**. The activity of a Cdk rises and falls with changes in the concentration of its cyclin partner. **Figure 12.16a** shows the fluctuating activity of **MPF**, the cyclin-Cdk complex that was discovered first (in frog eggs). Note that the peaks of MPF activity correspond to the peaks of cyclin concentration. The cyclin level rises during the S and G_2 phases and then falls abruptly during M phase.

The initials MPF stand for "maturation-promoting factor," but we can think of MPF as "M-phase-promoting factor" because it triggers the cell's passage into the M phase, past the G_2 checkpoint. When cyclins that accumulate during G_2 associate with Cdk molecules, the resulting MPF complex is active—it phosphorylates a variety of proteins, initiating mitosis **(Figure 12.16b)**. MPF acts both directly as a kinase and indirectly by activating other kinases. For example, MPF causes phosphorylation of various proteins of the nuclear lamina (see Figure 6.9), which promotes fragmentation of the nuclear envelope during prometaphase of mitosis. There is also evidence that MPF contributes to molecular events required for chromosome condensation and spindle formation during prophase.

▼ Figure 12.16 Molecular control of the cell cycle at the G_2 checkpoint. The steps of the cell cycle are timed by rhythmic fluctuations in the activity of cyclin-dependent kinases (Cdks). Here we focus on a cyclin-Cdk complex in animal cells called MPF, which acts at the G_2 checkpoint as a go-ahead signal, triggering the events of mitosis.

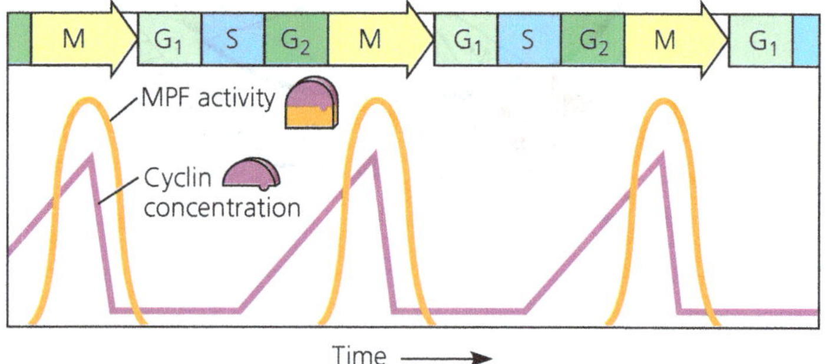

(a) Fluctuation of MPF activity and cyclin concentration during the cell cycle

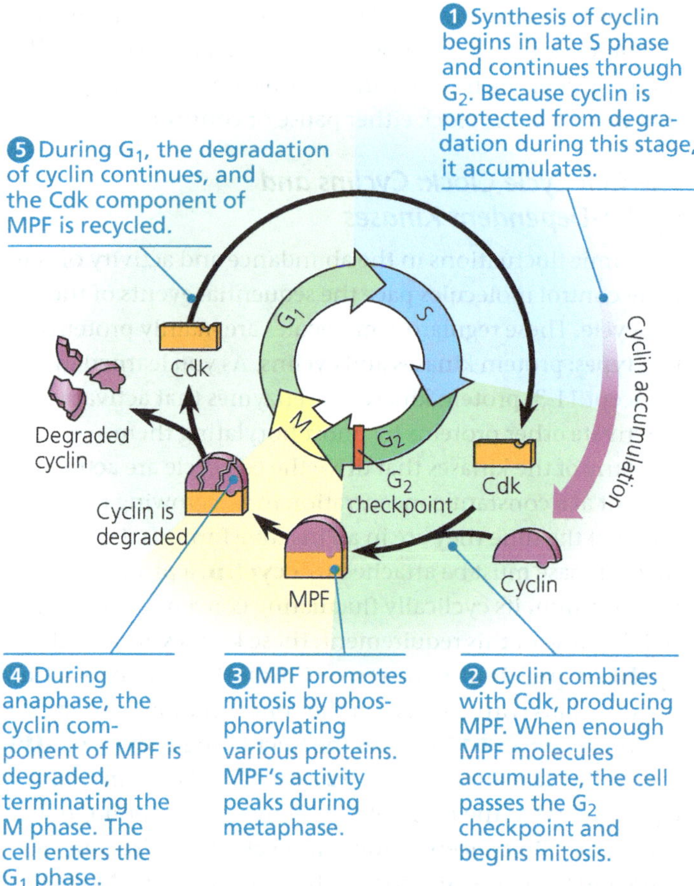

(b) Molecular mechanisms that help regulate the cell cycle

VISUAL SKILLS *Explain how the events in the diagram in (b) are related to the "Time" axis of the graph in (a), beginning at the left.*

During anaphase, MPF helps switch itself off by initiating a process that leads to the destruction of its own cyclin. The noncyclin part of MPF, the Cdk, persists in the cell, inactive until it becomes part of MPF again by associating with new cyclin molecules synthesised during the S and G_2 phases of the next round of the cycle.

The fluctuating activities of different cyclin-Cdk complexes are of major importance in controlling all the stages of the cell cycle; they also give the go-ahead signals at some checkpoints. As mentioned above, MPF controls the cell's passage through the G_2 checkpoint. Cell behaviour at the G_1 checkpoint is also regulated by the activity of cyclin-Cdk protein complexes. Animal cells appear to have at least three Cdk proteins and several different cyclins that operate at this checkpoint. Next, let's consider checkpoints in more detail.

Stop and Go Signs: Internal and External Signals at the Checkpoints

Animal cells generally have built-in stop signals that halt the cell cycle at checkpoints until overridden by go-ahead signals. Many signals registered at checkpoints come from cellular surveillance mechanisms inside the cell. These signals report whether crucial cellular processes that should have occurred by that point have in fact been completed correctly and thus whether or not the cell cycle should proceed. Checkpoints also register signals from outside the cell. The signals are transmitted within the cell by signal transduction pathways (see Figure 11.6). Three important checkpoints are those in the G_1, G_2, and M phases (see Figure 12.15).

For many cells, the G_1 checkpoint seems to be the most important. If a cell receives a go-ahead signal at the G_1 checkpoint, it will usually complete the G_1, S, G_2, and M phases and divide. If it does not receive a go-ahead signal at that point, it may exit the cycle, switching into a nondividing state called the **G_0 phase** **(Figure 12.17a)**. Most cells of the human body are actually in the G_0 phase. As mentioned earlier, mature nerve cells and muscle cells never divide. Other cells, such as liver cells, can be "called back" from the G_0 phase to the cell cycle by external cues, such as growth factors released during injury.

Biologists are currently working out the pathways that link signals originating inside and outside the cell with the responses by cyclin-dependent kinases and other proteins. An example of an internal signal occurs at the third important checkpoint, the M checkpoint **(Figure 12.17b)**. Anaphase, the separation of sister chromatids, does not begin until all the chromosomes are properly attached to the spindle at the metaphase plate. Researchers have learned that as long as some kinetochores are unattached to spindle microtubules, the sister chromatids remain together, delaying anaphase. Only when the kinetochores of all the chromosomes are properly attached to the spindle does the appropriate regulatory protein complex become activated. (In this case, the regulatory molecule is not a cyclin-Cdk complex but, instead, a different complex made up of several proteins.) Once activated, the complex sets off a chain of molecular events that activates the enzyme separase, which cleaves the cohesins, allowing the sister chromatids to separate. This mechanism ensures that daughter cells do not end up with missing or extra chromosomes.

There are checkpoints in addition to those in G_1, G_2, and M. For instance, a checkpoint in S phase stops cells with DNA damage from proceeding in the cell cycle. And, in 2014, researchers presented evidence for another checkpoint between anaphase and telophase that ensures anaphase is completed and the chromosomes are well separated before cytokinesis can begin, thus avoiding chromosomal damage.

What about the stop and go-ahead signals themselves—what are the signalling molecules? Studies using animal cells in culture have led to the identification of many external factors, both chemical and physical, that can influence cell division. For example, cells fail to divide if an essential nutrient is lacking in the culture medium. (This is analogous to trying to run a washing machine without the water supply hooked up; an internal sensor won't allow the machine to continue past the point where water is needed.) And even if all other conditions are favourable, most types of mammalian cells divide in culture only if the growth medium includes specific growth factors. As mentioned in Concept 11.1, a **growth factor** is a protein released by certain cells that stimulates other cells to divide. Different cell types respond specifically to different growth factors or combinations of growth factors.

Consider, for example, *platelet-derived growth factor (PDGF)*, which is made by blood cell fragments called platelets. When an injury occurs, platelets release PDGF in the vicinity. The experiment illustrated in **Figure 12.18** demonstrates that PDGF is required for the division of cultured fibroblasts, a type of connective tissue cell. Fibroblasts have PDGF receptors on their plasma membranes. The binding of PDGF molecules to these receptors (which are receptor tyrosine kinases; see Figure 11.8) triggers a signal transduction pathway that allows the cells to pass the G_1 checkpoint and divide. PDGF stimulates fibroblast division not only in the artificial conditions of cell culture but also in an animal's body. Thus, injury results in a proliferation of fibroblasts that helps heal the wound.

The effect of an external physical factor on cell division is clearly seen in **density-dependent inhibition**, a phenomenon in which crowded cells stop dividing **(Figure 12.19a)**. Studies done many years ago showed that cultured cells normally divide until they form a single layer of cells on the inner surface of a culture flask, at which point the cells stop dividing. If some cells are removed, those bordering the open space begin dividing again and continue until the vacancy is filled. Follow-up studies revealed that the binding of a cell-surface protein to its counterpart on an adjoining cell sends a signal to both cells that inhibits cell division, preventing them from moving forwards in the cell cycle, even in the presence of growth factors.

Most animal cells also exhibit **anchorage dependence** (see Figure 12.19a). To divide, they must be attached to something, such as the inside of a culture flask or the extracellular matrix of a tissue. Experiments suggest that, like cell density, anchorage is signalled to the cell cycle control system via pathways involving plasma membrane proteins and elements of the cytoskeleton linked to them.

Density-dependent inhibition and anchorage dependence appear to function not only in cell culture but also in the

▼ Figure 12.17 Two important checkpoints. At certain checkpoints in the cell cycle (red gates), cells do different things depending on the signals they receive. Events of the **(a)** G_1 and **(b)** M checkpoints are shown. In **(b)**, the G_2 checkpoint has already been passed by the cell.

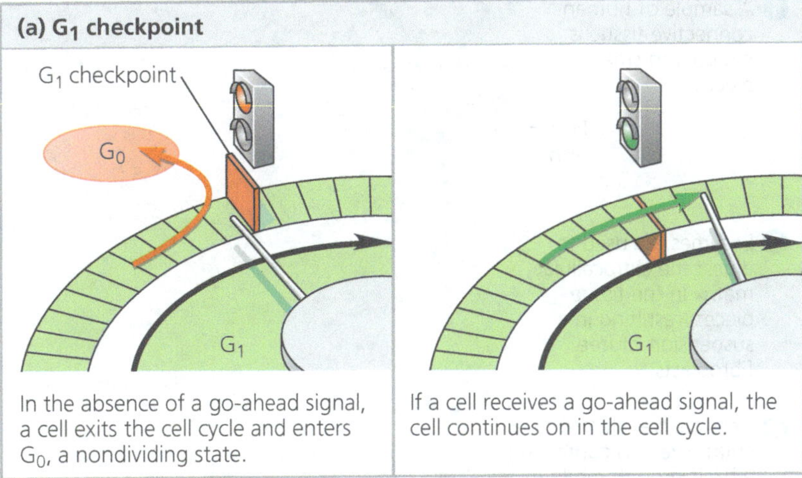

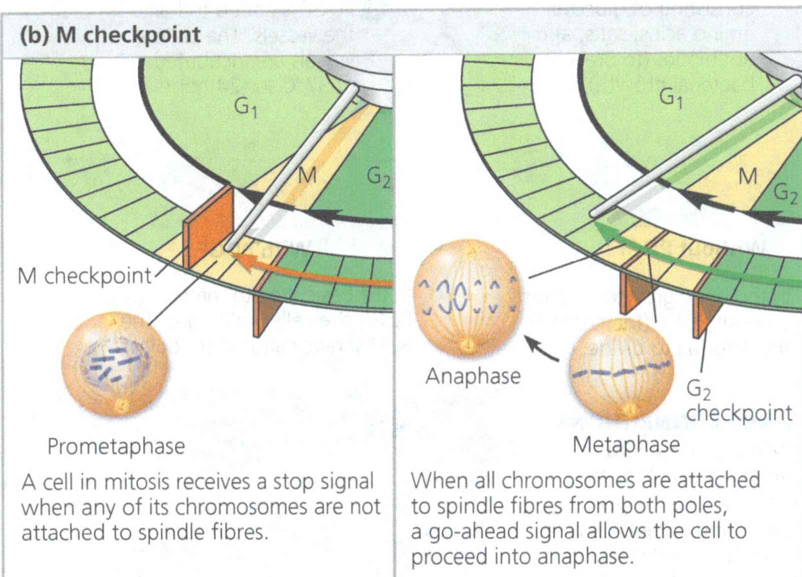

WHAT IF? *What might be the result if the cell ignored either checkpoint and progressed through the cell cycle?*

▼ Figure 12.18 The effect of platelet-derived growth factor (PDGF) on cell division.

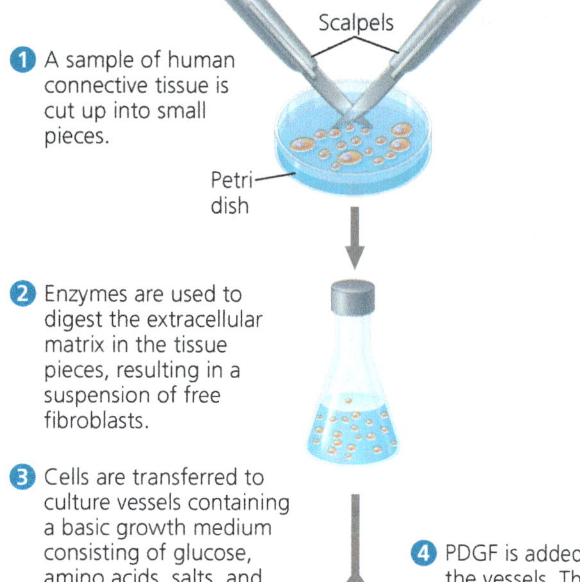

3 Cells are transferred to culture vessels containing a basic growth medium consisting of glucose, amino acids, salts, and antibiotics (to prevent bacterial growth).

4 PDGF is added to half the vessels. The culture vessels are incubated at 37°C for 24 hours.

Without PDGF

In the basic growth medium without PDGF (the control), the cells fail to divide.

With PDGF

In the basic growth medium plus PDGF, the cells proliferate. The SEM shows cultured fibroblasts.

MAKE CONNECTIONS *PDGF signals cells by binding to a cell-surface receptor tyrosine kinase. If you added a chemical that blocked phosphorylation, how would the results differ? (See Figure 11.8.)*

▼ Figure 12.19 Density-dependent inhibition and anchorage dependence of cell division. Individual cells are shown disproportionately large in the drawings.

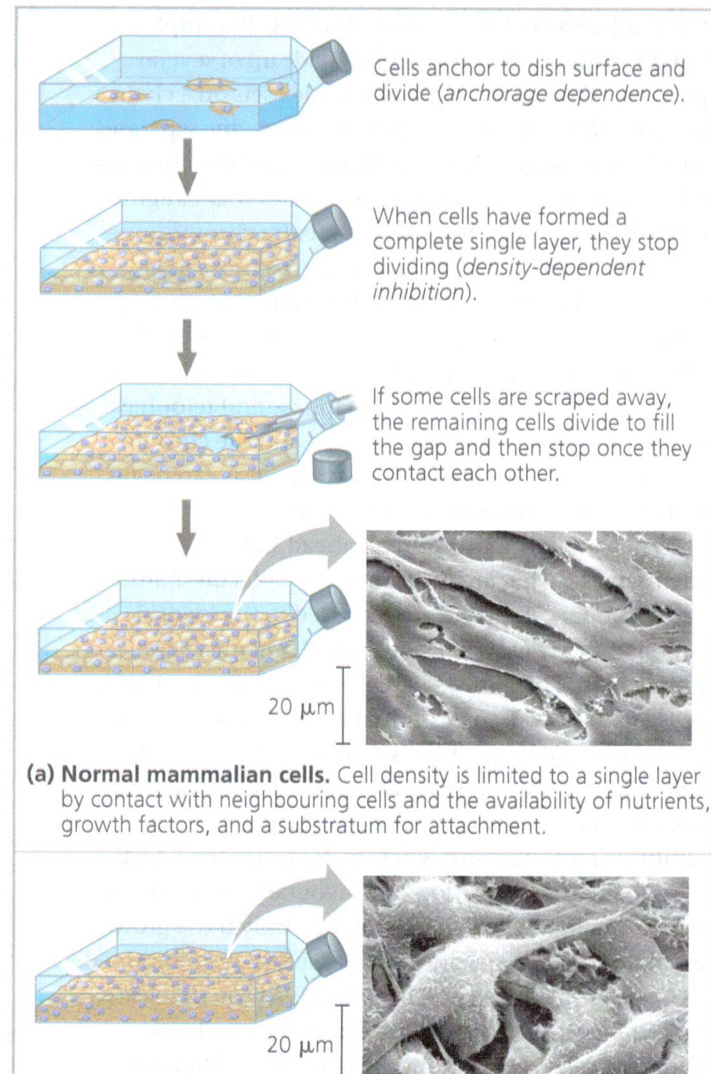

(a) Normal mammalian cells. Cell density is limited to a single layer by contact with neighbouring cells and the availability of nutrients, growth factors, and a substratum for attachment.

20 μm

(b) Cancer cells. Cancer cells usually continue to divide well beyond a single layer, forming a clump of overlapping cells. They do not exhibit anchorage dependence or density-dependent inhibition.

body's tissues, checking the growth of cells at some optimal density and location during embryonic development and throughout an organism's life. Cancer cells, which we examine next, exhibit neither density-dependent inhibition nor anchorage dependence (**Figure 12.19b**).

Loss of Cell Cycle Controls in Cancer Cells

Cancer cells do not heed the normal signals that regulate the cell cycle. In culture, they do not stop dividing when growth factors are depleted. A logical hypothesis is that cancer cells do not need growth factors in their culture medium to grow and divide. They may make a required growth factor themselves, or they may have an abnormality in the signalling pathway that conveys the growth factor's signal to the cell cycle control system even in the absence of that factor. Another possibility is an abnormal cell cycle control system.

In these scenarios, the underlying basis of the abnormality is almost always a change in one or more genes (for example, a mutation) that alters the function of their protein products, resulting in faulty cell cycle control. We'll explore the molecular basis for such changes in Concept 18.5.

There are other important differences between normal cells and cancer cells that reflect malfunctions of the cell cycle. If and when they stop dividing, cancer cells do so at random points in the cycle, rather than at the normal checkpoints. Moreover, cancer cells can go on dividing indefinitely in culture if they are given a continual supply of nutrients; in essence, they are "immortal." A striking example is a cell line that has been reproducing in culture since 1951. Cells of this line are called HeLa cells because their original source was a tumour removed from a woman named Henrietta Lacks. (Neither Ms. Lacks nor her family gave permission or even knew about the propagation and use of her cells, which have helped biologists make countless significant discoveries over the years.) Cells in culture that acquire the ability to divide indefinitely are said to have undergone a process called **transformation**, causing them to behave (in cell division, at least) like cancer cells. By contrast, nearly all normal, non-transformed mammalian cells growing in culture divide only about 20 to 50 times before they stop dividing, age, and die. Finally, cancer cells evade the normal controls that trigger a cell to undergo apoptosis when something is wrong—for example, when an irreparable mistake has occurred during DNA replication preceding mitosis.

Abnormal cell behaviour in the body can be catastrophic. The problem begins when a single cell in a tissue undergoes the first of many steps that convert a normal cell to a cancer cell. Such a cell often has altered proteins on its surface, and the body's immune system normally recognises the cell as "non-self"—an insurgent—and destroys it. However, if the cell evades destruction, it may proliferate and form a tumour, a mass of abnormal cells within otherwise normal tissue. The abnormal cells may remain at the original site if their genetic and cellular changes don't allow them to move to or survive at another site. In that case, the tumour is called a **benign tumour**. Most benign tumours do not cause serious problems (depending on their location) and can be removed by surgery. In contrast, a **malignant tumour** includes cells whose genetic and cellular changes enable them to spread to new tissues and impair the functions of one or more organs; these cells are also sometimes called *transformed* cells (although usage of this term is generally restricted to cells in culture). An individual with a malignant tumour is said to have cancer **(Figure 12.20)**.

The changes that have occurred in cells of malignant tumours show up in many ways besides excessive proliferation. These cells may have unusual numbers of chromosomes, though whether this is a cause or an effect of tumour-related changes is an ongoing debate. Their metabolism may be altered, and they may cease to function in any constructive way. Abnormal changes on the cell surface cause cancer cells to lose attachments to neighbouring cells and the extracellular matrix, allowing them to spread into nearby tissues. Cancer cells may also secrete signalling molecules that cause blood vessels to grow towards the tumour. A few tumour cells may separate from the original tumour, enter blood vessels and lymph vessels, and travel to other parts of the body. There, they may proliferate and form a new tumour. This spread of cancer cells to locations distant from their original site is called **metastasis** (see Figure 12.20).

A tumour that appears to be localised may be treated with high-energy radiation, which damages DNA in cancer cells much more than DNA in normal cells, apparently because the majority of cancer cells have lost the ability to repair such damage. To treat known or suspected metastatic tumours, chemotherapy is used, in which drugs that are toxic to actively dividing cells are administered through the circulatory system. As you might expect, chemotherapeutic drugs interfere with specific steps in the cell cycle. For example, the drug Taxol freezes the mitotic spindle by preventing microtubule depolymerisation, which stops actively dividing cells from proceeding past metaphase and leads to their destruction. The side effects of chemotherapy are due to the effects of the drugs on normal cells that divide frequently, due to the function of that cell type in the organism. For example, nausea results from chemotherapy's effects on intestinal cells, hair loss from effects on hair follicle cells, and susceptibility to infection from effects on immune system cells. You can work with data from an experiment involving a potential chemotherapeutic agent in the **Scientific Skills Exercise**.

Over the past several decades, researchers have produced a flood of valuable information about cell-signalling pathways and how their malfunction contributes to the development of cancer through effects on the cell cycle. Coupled with new

▼ Figure 12.20 The growth and metastasis of a malignant breast tumour. A series of genetic and cellular changes contribute to a tumour becoming malignant (cancerous). The cells of malignant tumours grow in an uncontrolled way and can spread to neighbouring tissues and, via lymph and blood vessels, to other parts of the body (metastasis).

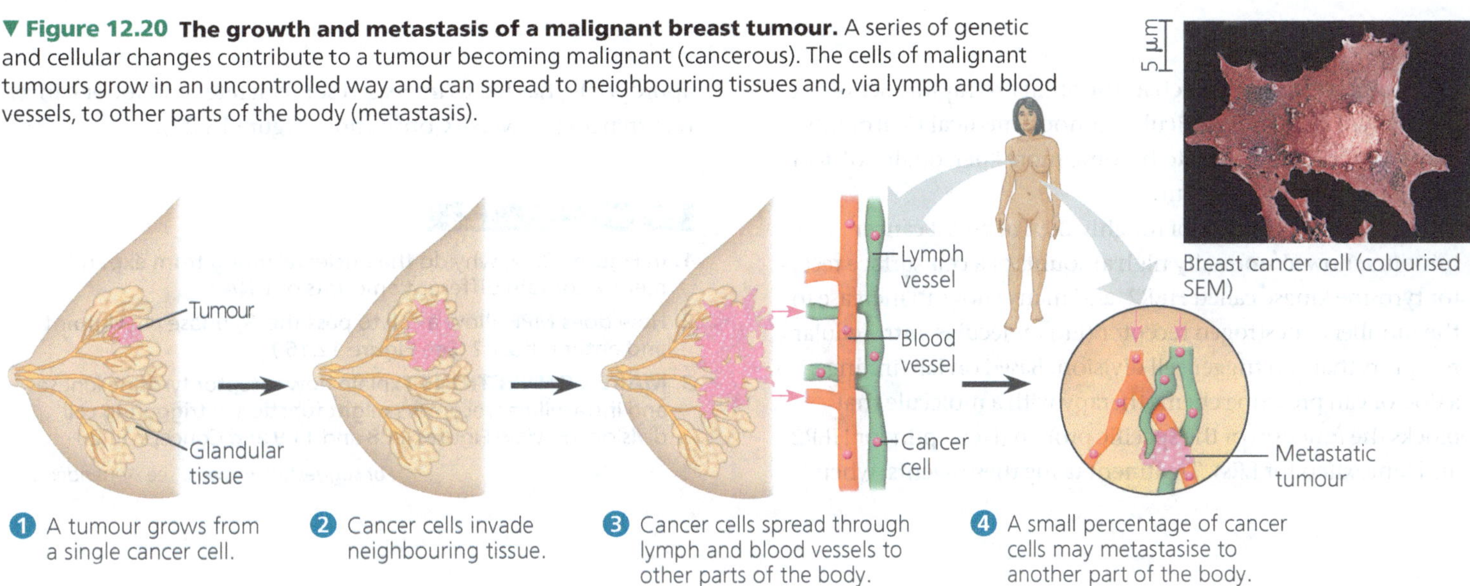

Scientific Skills Exercise

Interpreting Histograms

At What Phase Is the Cell Cycle Arrested by an Inhibitor? Many medical treatments are aimed at stopping cancer cell proliferation by blocking the cell cycle of cancerous tumour cells. One potential treatment is a cell cycle inhibitor derived from human umbilical cord stem cells. In this exercise, you will compare two histograms to determine where in the cell cycle the inhibitor blocks the division of cancer cells.

How the Experiment Was Done In the treated sample, human glioblastoma (brain cancer) cells were grown in tissue culture in the presence of the inhibitor, while a control sample of glioblastoma cells was grown in its absence. After 72 hours of growth, the two cell samples were harvested. To get a "snapshot" of the phase of the cell cycle each cell was in at that time, the samples were treated with a fluorescent chemical that binds to DNA and then run through a flow cytometer, an instrument that records the fluorescence level of each cell. Computer software then graphed the number of cells in each sample with a particular fluorescence level, as shown below.

Data from the Experiment

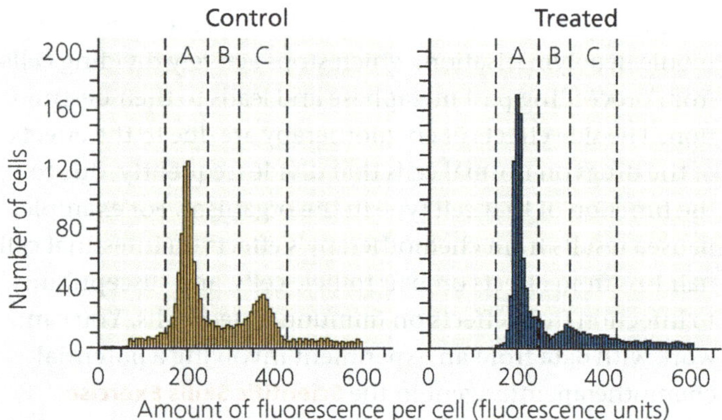

The data are plotted in a type of graph called a histogram (above), which groups the values for a numeric variable on the *x*-axis into intervals. A histogram allows you to see how all the experimental subjects (cells, in this case) are distributed along a continuous variable (amount of fluorescence). In these histograms, the bars are so narrow that the data appear to follow a curve for which you can detect peaks and dips. Each narrow bar represents the number of cells observed to have a level of fluorescence in the range of that interval. This in turn indicates the relative amount of DNA in those cells. Overall, comparing the two histograms allows you to see how the DNA content of this cell population is altered by the treatment.

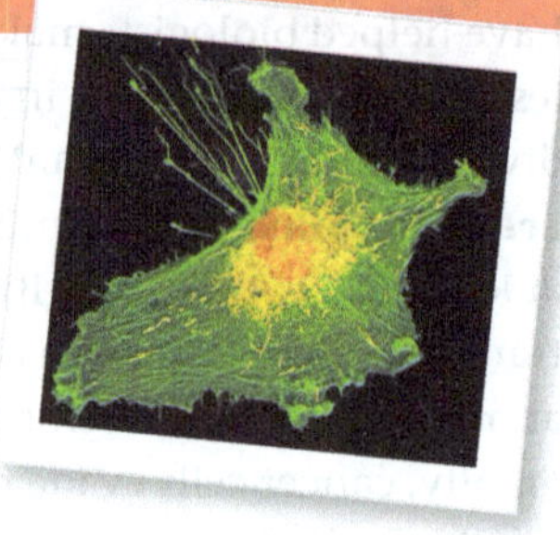

▶ **Human brain cancer cell**

INTERPRET THE DATA

1. Study the data in the histograms. **(a)** Which axis indirectly shows the relative amount of DNA per cell? Explain your answer. **(b)** In the control sample, compare the first peak in the histogram (in region A) to the second peak (in region C). Which peak shows the population of cells with the higher amount of DNA per cell? Explain. (For additional information about histograms, see the Scientific Skills Review in Appendix D.)
2. **(a)** In the control sample histogram, identify the phase of the cell cycle (G_1, S, or G_2) of the population of cells in each region delineated by vertical lines. Label the histogram with these phases and explain your answer. **(b)** Does the S phase population of cells show a distinct peak in the histogram? Why or why not?
3. The histogram representing the treated sample shows the effect of growing the cancer cells alongside human umbilical cord stem cells that produce the potential inhibitor. **(a)** Label the histogram with the cell cycle phases. Which phase of the cell cycle has the greatest number of cells in the treated sample? Explain. **(b)** Compare the distribution of cells among G_1, S, and G_2 phases in the control and treated samples. What does this tell you about the cells in the treated sample? **(c)** Based on what you learned in Concept 12.3, propose a mechanism by which the stem cell–derived inhibitor might arrest the cancer cell cycle at this stage. (More than one answer is possible.)

Data from K. K. Velpula et al., Regulation of glioblastoma progression by cord blood stem cells is mediated by downregulation of cyclin D1, *PLoS ONE* 6(3):e18017 (2011).

molecular techniques, such as the ability to rapidly sequence the DNA of cells in a particular tumour, medical treatments for cancer are beginning to become more "personalised" to a particular patient's tumour.

For example, the cells of roughly 20% of breast cancer tumours show abnormally high amounts of a cell-surface receptor tyrosine kinase called HER2, and many show an increase in the number of oestrogen receptor (ER) molecules, intracellular receptors that can trigger cell division. Based on lab findings, a doctor can prescribe chemotherapy with a molecule that blocks the function of the specific protein (Herceptin for HER2 and tamoxifen for ERs). Treatment using these agents, when appropriate, has led to increased survival rates and fewer cancer recurrences (see Make Connections Figure 18.27).

CONCEPT CHECK 12.3

1. In Figure 12.14, why do the nuclei resulting from experiment 2 contain different amounts of DNA?
2. How does MPF allow a cell to pass the G_2 phase checkpoint and enter mitosis? (See Figure 12.16.)
3. **MAKE CONNECTIONS** Explain how receptor tyrosine kinases and intracellular receptors might function in triggering cell division. (Review Figures 11.8 and 11.9 and Concept 11.2.)

For suggested answers, see Appendix A.

12 Chapter Review

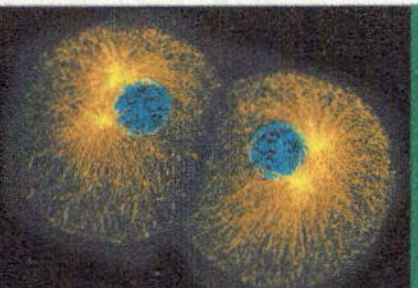

SUMMARY OF KEY CONCEPTS

CONCEPT 12.1

Most cell division results in genetically identical daughter cells *(pp. 237–239)*

- Unicellular organisms reproduce by **cell division**; multicellular organisms depend on cell division for their development from a fertilised egg and for growth and repair.
- The genetic material (DNA) of a cell—its **genome**—is partitioned among **chromosomes**. Each eukaryotic chromosome consists of one DNA molecule associated with many proteins. Together, the complex of DNA and associated proteins is called **chromatin**. The chromatin of a chromosome exists in different states of condensation at different times. In animals, **gametes** have one set of chromosomes and **somatic cells** have two sets.
- Cells replicate their genetic material before they divide, each daughter cell receiving a copy of the DNA. Prior to cell division, chromosomes are duplicated. Each one then consists of two identical **sister chromatids** joined along their lengths by sister chromatid cohesion and held most tightly together at a constricted region at the **centromeres**. When this cohesion is broken, the chromatids separate during cell division, becoming the chromosomes of the daughter cells. Eukaryotic cell division consists of **mitosis** (division of the nucleus) and **cytokinesis** (division of the cytoplasm).

? *Differentiate between these terms: chromosome, chromatin, and chromatid.*

CONCEPT 12.2

The mitotic phase alternates with interphase in the cell cycle *(pp. 239–246)*

- Cell division is part of the **cell cycle**, an ordered sequence of events in the life of a cell.
- Between divisions, a cell is in **interphase**: the $\mathbf{G_1}$, **S**, and $\mathbf{G_2}$ phases. The cell grows throughout interphase, with DNA being replicated during the synthesis (S) phase. Mitosis and cytokinesis make up the **mitotic (M) phase** of the cell cycle.

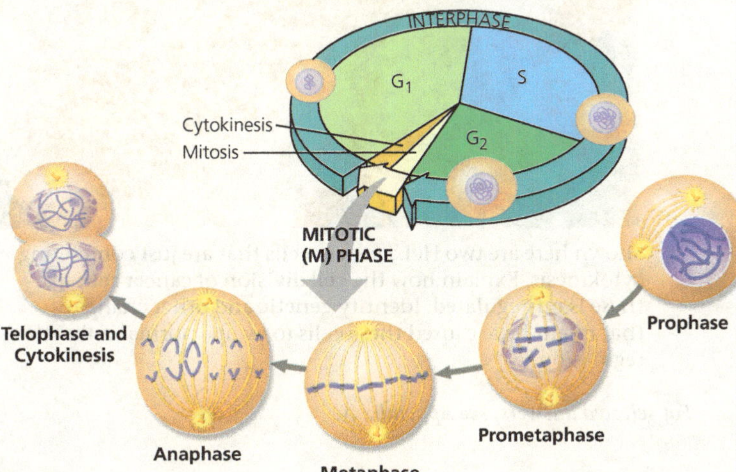

- The **mitotic spindle**, made up of microtubules, controls chromosome movement during mitosis. In animal cells, it arises from the **centrosomes** and includes spindle microtubules and asters. Some spindle microtubules attach to the **kinetochores** of chromosomes and move the chromosomes to the **metaphase plate**. After sister chromatids separate, motor proteins move them along kinetochore microtubules towards opposite ends of the cell. The cell elongates when motor proteins push nonkinetochore microtubules from opposite poles away from each other.
- Mitosis is usually followed by cytokinesis. Animal cells carry out cytokinesis by **cleavage**, and plant cells form a **cell plate**.
- During **binary fission** in bacteria, the chromosome replicates and the daughter chromosomes actively move apart. Some of the proteins involved in bacterial binary fission are related to eukaryotic actin and tubulin.
- Since prokaryotes preceded eukaryotes by more than a billion years, it is likely that mitosis evolved from prokaryotic cell division. Certain unicellular eukaryotes exhibit mechanisms of cell division that may be similar to those of ancestors of existing eukaryotes. Such mechanisms might represent intermediate steps in the evolution of mitosis.

? *In which of the three phases of interphase and the stages of mitosis do chromosomes exist as single DNA molecules?*

CONCEPT 12.3

The eukaryotic cell cycle is regulated by a molecular control system *(pp. 246–252)*

- Signalling molecules present in the cytoplasm regulate progress through the cell cycle.
- In the **cell cycle control system**, cyclic changes in regulatory proteins—including **cyclins** and **cyclin-dependent kinases (Cdks)**—work as a cell cycle clock. The cell cycle stops at specific **checkpoints** until a go-ahead signal is received; important checkpoints occur in G_1, G_2, and M phases. Internal and external signals control the cell cycle checkpoints via signal transduction pathways. Most cells exhibit **density-dependent inhibition** of cell division as well as **anchorage dependence**.
- Cancer cells elude normal cell cycle regulation and divide unchecked, forming tumour. **Malignant tumours** invade nearby tissues and can undergo **metastasis**, exporting cancer cells to other sites, where they may form secondary tumours.

? *Explain the significance of the G_1, G_2, and M checkpoints and the go-ahead signals involved in the cell cycle control system.*

TEST YOUR UNDERSTANDING

Levels 1-2: Remembering/Understanding

1. Through a microscope, you can see a cell plate beginning to develop across the middle of a cell and nuclei forming on either side of the cell plate. This cell is most likely
   (A) an animal cell in the process of cytokinesis.
   (B) a plant cell in the process of cytokinesis.
   (C) a bacterial cell dividing.
   (D) a plant cell in metaphase.

2. Vinblastine is a standard chemotherapeutic drug used to treat cancer. Because it interferes with the assembly of microtubules, its effectiveness must be related to
   (A) disruption of mitotic spindle formation.
   (B) suppression of cyclin production.
   (C) myosin denaturation and inhibition of cleavage furrow formation.
   (D) inhibition of DNA synthesis.

3. One difference between cancer cells and normal cells is that cancer cells
   (A) are unable to synthesise DNA.
   (B) are arrested at the S phase of the cell cycle.
   (C) continue to divide even when they are tightly packed together.
   (D) cannot function properly because they are affected by density-dependent inhibition.

4. The decline of MPF activity at the end of mitosis is due to
   (A) the destruction of the protein kinase Cdk.
   (B) decreased synthesis of Cdk.
   (C) the degradation of cyclin.
   (D) the accumulation of cyclin.

5. In the cells of some organisms, mitosis occurs without cytokinesis. This will result in
   (A) cells with more than one nucleus.
   (B) cells that are unusually small.
   (C) cells lacking nuclei.
   (D) cell cycles lacking an S phase.

6. Which of the following occurs during S phase?
   (A) condensation of the chromosomes
   (B) replication of the DNA
   (C) separation of sister chromatids
   (D) spindle formation

Levels 3-4: Applying/Analysing

7. Cell A has half as much DNA as cells B, C, and D in a mitotically active tissue. Cell A is most likely in
   (A) G_1.
   (B) G_2.
   (C) prophase.
   (D) metaphase.

8. The drug cytochalasin B blocks the function of actin. Which of the following aspects of the animal cell cycle would be most disrupted by cytochalasin B?
   (A) spindle formation
   (B) spindle attachment to kinetochores
   (C) cell elongation during anaphase
   (D) cleavage furrow formation and cytokinesis

9. **VISUAL SKILLS** The light micrograph shows dividing cells near the tip of an onion root. Identify a cell in each of the following stages: prophase, prometaphase, metaphase, anaphase, and telophase. Describe the major events occurring at each stage.

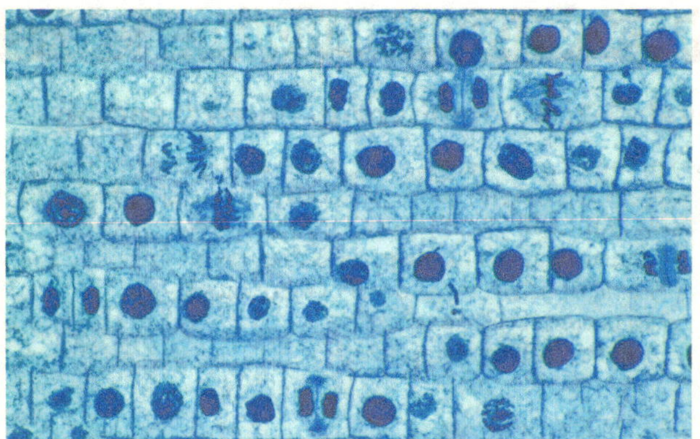

10. **DRAW IT** Draw one eukaryotic chromosome as it would appear during interphase, during each of the stages of mitosis, and during cytokinesis. Also draw and label the nuclear envelope and any microtubules attached to the chromosome(s).

Levels 5-6: Evaluating/Creating

11. **EVOLUTION CONNECTION** The result of mitosis is that the daughter cells end up with the same number of chromosomes that the parent cell had. Another potential way to maintain the number of chromosomes would be to carry out cell division first and then duplicate the chromosomes in each daughter cell. Assess whether this would be an equally good way of organising the cell cycle. Explain why evolution has not led to this alternative.

12. **SCIENTIFIC INQUIRY** Although both ends of a microtubule can gain or lose subunits, one end (called the plus end) polymerises and depolymerises at a higher rate than the other end (the minus end). For spindle microtubules, the plus ends are in the centre of the spindle, and the minus ends are at the poles. Motor proteins that move along microtubules specialise in walking either towards the plus end or towards the minus end; the two types are called plus end–directed and minus end–directed motor proteins, respectively. Given what you know about chromosome movement and spindle changes during anaphase, predict which type of motor proteins would be present on (a) kinetochore microtubules and (b) nonkinetochore microtubules.

13. **WRITE ABOUT A THEME: INFORMATION** The continuity of life is based on heritable information in the form of DNA. In a short essay (100–150 words), explain how the process of mitosis faithfully parcels out exact copies of this heritable information in the production of genetically identical daughter cells.

14. **SYNTHESISE YOUR KNOWLEDGE**

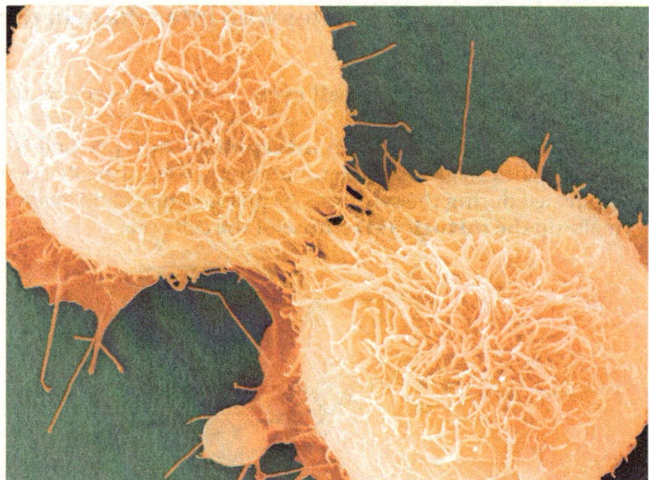

Shown here are two HeLa cancer cells that are just completing cytokinesis. Explain how the cell division of cancer cells like these is misregulated. Identify genetic and other changes that might have caused these cells to escape normal cell cycle regulation.

For selected answers, see Appendix A.

Unit 3 GENETICS

Dr Francisco Mojica is a Professor of Microbiology at the Universidad de Alicante in Spain, where he studies CRISPR sequences. His parents fostered in him a love of nature, and he realised in high school that he was most interested in biology. In college, when he first looked through a microscope and saw small, single-celled "living beings" moving around, he was captivated, and he has spent his career focusing on prokaryotes. While in graduate school in the 1990s, he found the prokaryotic DNA repeats that he later named CRISPR repeats, and he spent the next decade establishing their role as a prokaryotic immune system. Dr Mojica's lab currently studies the mysteries that still surround CRISPR, trying to understand every single aspect of its mechanism of action and regulation.

AN INTERVIEW WITH

Francisco Mojica

When did you first decide to be a biologist?

In high school, my parents asked me, "What are you thinking you'll do with your life—become a doctor or a lawyer?" I said, "Not at all. I really like biology." In spite of his concern about my future career, my father told me I must study what I loved, so that I could be happy in my life. I really appreciated his words! So I went to university and studied biology. When you like biology, initially you get attracted because of the things you see—animals and plants. But when you see microscopic living beings for the first time, and they're moving around, that's really amazing! So, I decided to focus on microbiology.

▲ Dr Mojica cultured and analysed prokaryotes from this salt lake in Spain, leading to his discovery of CRISPR.

What did you study as a graduate student?

I studied prokaryotes that live in salt ponds. These weird organisms can tolerate very high salt concentrations. I was interested in doing molecular biology, which had not been done in that university before—I knew that would be a challenge, and I love challenges. There weren't many genome sequences available at that time, in the early 1990s, but we had cloned a few sequences, and we had indirect evidence they might be related to adaptation to salinity. One of the DNA sequences had these regularly spaced repeats that were partially palindromic (reading the same forward and backward), so they could form secondary structures. We thought maybe these repeats were affected by salinity and would in turn affect gene expression, but we were absolutely wrong! At the end of my PhD thesis, all I could say was, "We have found regularly interspaced repeats that are expressed by the cell, and they could be very important."

"Studying biology is basically just saying 'Wow!' all the time."

What made you continue to study these repeats?

When you find something in your thesis, you believe that it belongs to you—you have to try to answer that question. No one could figure out the role that these repeats played. However, others had found similar repeats in bacteria like *E. coli* and another very distantly-related bacterial species. That was an important finding that pushed me to keep working on the repeats, because it meant the repeats must be doing something important for them to stay in the genomes of various prokaryotes over such long evolutionary time. Evolution is so fundamental to biology, to understanding everything.

How did you figure out what the repeats were doing?

I started wondering about the sequences *between* the repeats, the so-called "spacers." Where did they come from? I compared the spacer sequences to genome database sequences. By 2000 or 2001, many more genome sequences were available. Finally, in 2003, I found a spacer in *E. coli* whose sequence was identical to the sequence of a virus that infects *E. coli*. Even more interesting, the strain of *E. coli* with that spacer could not be infected by that virus. I checked more and more spacer sequences, and when a spacer sequence matched a sequence in the database, the match was always to the sequence of a virus that could no longer infect that species. So I realised these bacteria keep a "memory" of the viruses that infect them, and that the repeat regions were part of an immune system—an unexpected and very surprising finding. I proposed naming the repeats Clustered Regularly Interspaced Short Palindromic Repeats, or CRISPR, and to my surprise, the name took hold.

What is your advice to a student considering a career in biology?

Studying biology is basically just saying "Wow!" all the time. In science, the truth changes very fast. The most important thing is to be open. Be ready to hear something that destroys what you believe—because biology is a fast-moving field, a field that evolves a lot. Knowledge will change.

13 Meiosis and Sexual Life Cycles

KEY CONCEPTS

Study Tip

Make a visual study guide: Figure 13.1 presents the events of meiosis as a simplified diagram. As you learn more about meiosis, draw a detailed diagram of the stages of meiosis, starting with the sketch below. Include explanatory labels. Compare your visual study guides of mitosis (from the chapter, "The Cell Cycle") and meiosis, listing similarities and differences.

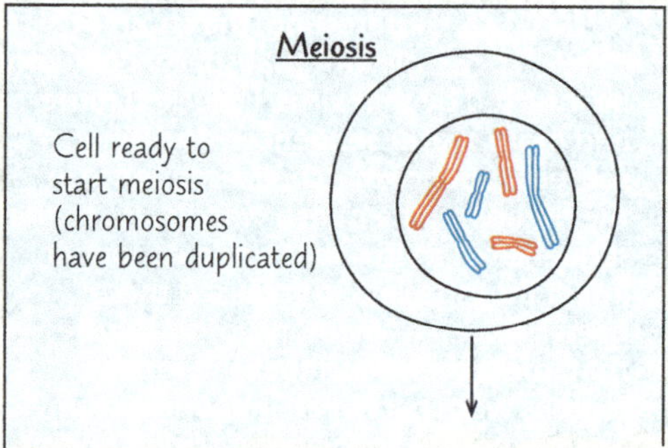

Go to Mastering Biology

to access Dynamic Study Modules for revision, 3D BioFlix® animations and high-quality videos, and your interactive Pearson eText.

Figure 13.1 The family members shown in this photo have some similar features. Offspring resemble their parents more than they do unrelated individuals.

What biological mechanisms account for the resemblance between offspring and their parents?

Meiosis is a special type of cell division that produces cells with half the chromosomes of the parent cell. It occurs only in specialised cells, such as the cells of testes and ovaries in humans.

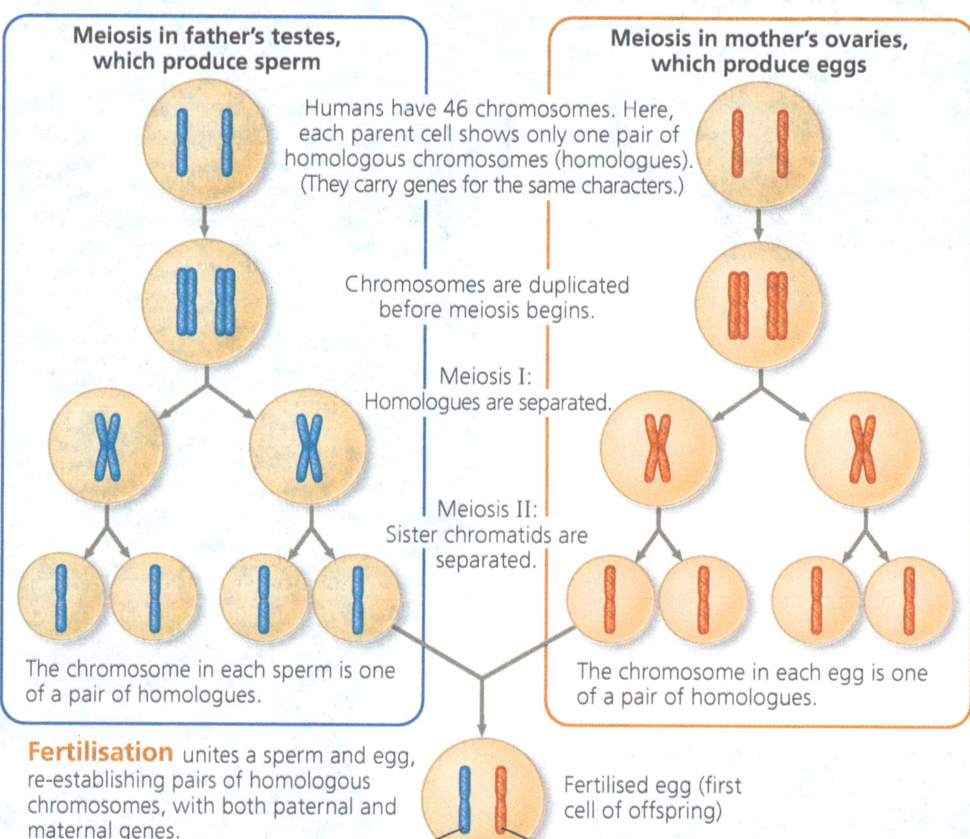

CONCEPT 13.1

Offspring acquire genes from parents by inheriting chromosomes

Friends may tell you that you have your mother's nose or your father's eyes—although they don't mean that literally, of course. The transmission of traits from one generation to the next is called inheritance, or **heredity** (from the Latin *heres*, heir). At the same time, sons and daughters are not identical copies of either parent or of their siblings. Along with inherited similarity, there is also **variation**. The study of both heredity and inherited variation is called **genetics**.

Inheritance of Genes

Parents endow their offspring with coded information in the form of hereditary units called **genes**. The genes we inherit from our mothers and fathers are our genetic link to our parents, and they account for family resemblances such as shared eye colour or freckles. Our genes program specific traits that emerge as we develop from fertilised eggs into adults.

The genetic program is written in the language of DNA, the polymer of four different nucleotides you learned about in Concepts 1.1 and 5.5. Inherited information is passed on in the form of each gene's specific sequence of DNA nucleotides, much as printed information is communicated in the form of meaningful sequences of letters. In both cases, the language is symbolic. Just as your brain translates the word *apple* into a mental image of the fruit, cells translate genes into freckles and other features. Most genes program cells to synthesise specific enzymes and other proteins, whose cumulative action produces an organism's inherited traits. The programming of these traits in the form of DNA is one of the unifying themes of biology.

The transmission of hereditary traits has its molecular basis in the replication of DNA, which produces copies of genes that can be passed from parents to offspring. In animals and plants, reproductive cells called **gametes** are the vehicles that transmit genes from one generation to the next. During fertilisation, male and female gametes (sperm and eggs) unite, passing on genes of both parents to their offspring.

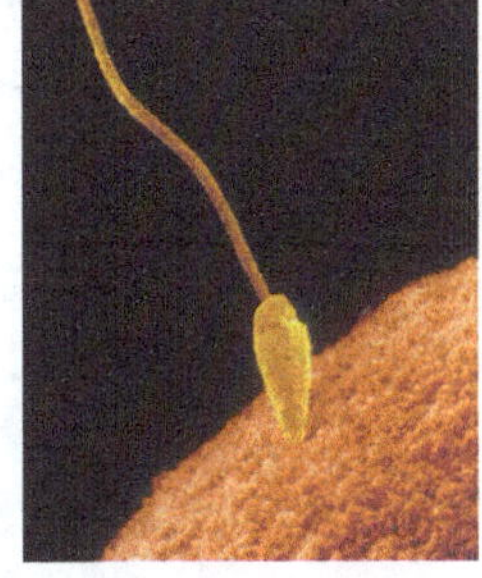

▲ Sperm and egg

Except for small amounts of DNA in mitochondria and chloroplasts, the DNA of a eukaryotic cell is packaged into chromosomes within the nucleus. Every species has a characteristic number of chromosomes. For example, humans have 46 chromosomes in their **somatic cells**—all cells of the body except the gametes and their precursors. Each chromosome consists of a single long DNA molecule, elaborately coiled in association with various proteins. One chromosome includes several hundred to a few thousand genes, each gene being a precise sequence of nucleotides along the DNA molecule. A gene's specific location along the length of a chromosome is called the gene's **locus** (plural, *loci*; from the Latin, meaning "place"). Our genetic endowment (our genome) consists of the genes and other DNA that make up the chromosomes we inherited from our parents.

Comparison of Asexual and Sexual Reproduction

Only organisms that reproduce asexually have offspring that are exact genetic copies of themselves. In **asexual reproduction**, a single individual (like a yeast cell or an amoeba; see Figure 12.2a) is the sole parent and passes copies of all its genes to its offspring without the fusion of gametes. For example, single-celled eukaryotic organisms can reproduce asexually by mitotic cell division, in which DNA is copied and allocated equally to two daughter cells. The genomes of the offspring are virtually exact copies of the parent's genome. Some multicellular organisms are also capable of reproducing asexually **(Figure 13.2)**. Because the cells of the offspring arise via mitosis in the parent, the offspring is usually genetically identical to its parent. An individual that reproduces asexually gives rise to a **clone**, an individual or group of individuals that are genetically identical to the parent. Genetic differences occasionally arise in asexually reproducing organisms as a result of changes in the DNA called mutations, which we will discuss in Concept 17.5.

In **sexual reproduction**, two parents give rise to offspring that have unique combinations of genes inherited from the two parents. In contrast to a clone, offspring of sexual reproduction vary genetically from their siblings and both parents: They are variations on a common theme of family resemblance, not exact replicas. Genetic variation like that shown in Figure 13.1 is an important consequence of sexual reproduction. What mechanisms generate this genetic variation? The key is the behaviour of chromosomes during the sexual life cycle.

▼ **Figure 13.2 Asexual reproduction in two multicellular organisms. (a)** This relatively simple animal, a hydra, reproduces by budding. The bud, a localised mass of mitotically dividing cells, develops into a small hydra, which detaches from the parent (LM). **(b)** All the trees in this circle of black beech arose asexually from a single parent tree.

(a) Hydra

(b) New Zealand black beech

CONCEPT CHECK 13.1

1. **MAKE CONNECTIONS** Using your knowledge of gene expression in a cell, explain what causes the traits of parents (such as hair colour) to show up in their offspring. (See Concept 5.5.)
2. Describe how an asexually reproducing eukaryotic organism produces offspring that are genetically identical to each other and to their parents.
3. **WHAT IF?** A horticulturalist breeds orchids, trying to obtain a plant with a unique combination of desirable traits. After many years, she finally succeeds, and wants to produce more plants like this one. Discuss whether she should crossbreed it with another plant or cause it to undergo asexual reproduction (forming a clone), and why.

For suggested answers, see Appendix A.

CONCEPT 13.2

Fertilisation and meiosis alternate in sexual life cycles

A **life cycle** is the generation-to-generation sequence of stages in the reproductive history of an organism, from conception to production of its own offspring. In this section, we use humans as an example to track the behaviour of chromosomes through the sexual life cycle. We begin by considering the chromosome count in human somatic cells and gametes. We will then explore how the behaviour of chromosomes relates to the human life cycle and other types of sexual life cycles.

Sets of Chromosomes in Human Cells

In humans, each somatic cell has 46 chromosomes. Before mitosis begins, the chromosomes are duplicated. During mitosis, the chromosomes become condensed enough to be visible under a light microscope. At this point, they can be distinguished from one another by their size, the position of their centromeres, and the pattern of coloured bands produced by certain chromatin-binding stains.

Careful examination of a micrograph of the 46 human chromosomes from a single cell in mitosis reveals that there are two chromosomes of each of 23 types. This becomes clear when images of the chromosomes are arranged in pairs, starting with the longest chromosomes. The resulting ordered display is called a **karyotype** (**Figure 13.3**). The two chromosomes of a pair have the same length, centromere position, and staining pattern: These are called **homologous chromosomes** (or **homologues**). Both chromosomes of each pair carry genes controlling the same inherited characters. For example, if a gene for eye colour is situated at a particular locus on a certain chromosome, then its homologous chromosome (its homologue) will also have a version of the eye-colour gene at the equivalent locus.

The two chromosomes referred to as X and Y are an important exception to the general pattern of homologous chromosomes in human somatic cells. Typically, human females have a homologous pair of X chromosomes (XX), while males have one X and one Y chromosome (XY; see Figure 13.3). Only small

▼ Figure 13.3 Research Method

Preparing a Karyotype

Application A karyotype is a display of condensed chromosomes arranged in pairs. Karyotyping can be used to screen for defective chromosomes or abnormal numbers of chromosomes associated with certain congenital disorders, such as Down syndrome.

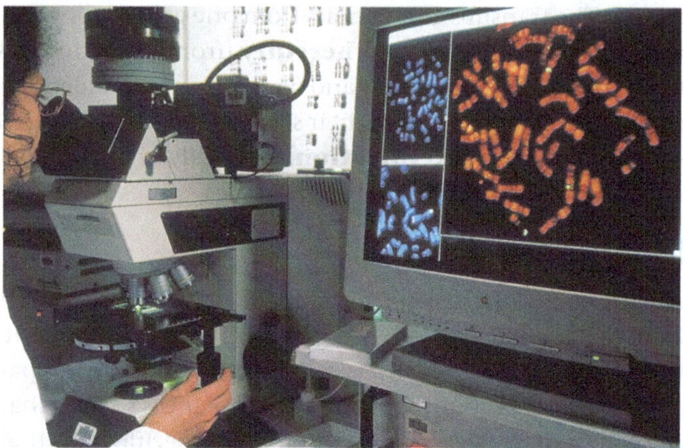

Technique Karyotypes are prepared from isolated somatic cells, which are treated with a drug to stimulate mitosis and then grown in culture for several days. Cells arrested when the chromosomes are most highly condensed—at metaphase—are stained and then viewed with a microscope equipped with a digital camera. An image of the chromosomes is displayed on a computer monitor, and digital software is used to arrange them in pairs according to their appearance.

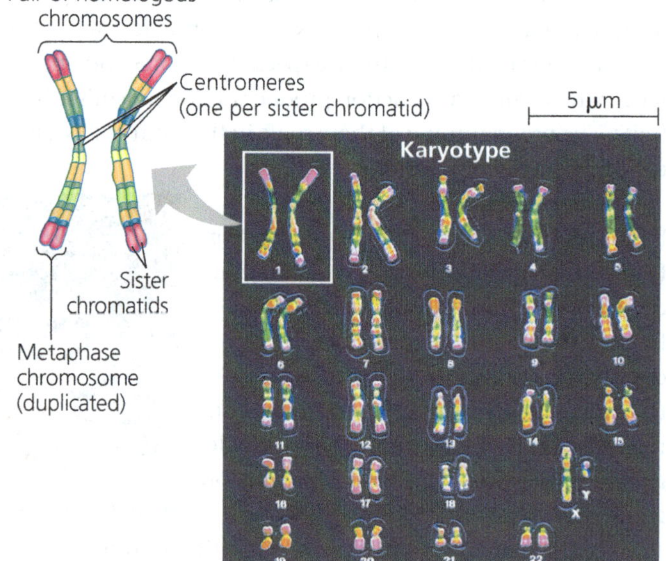

Results This karyotype shows the chromosomes from a human male (as seen by the presence of the XY chromosome pair), coloured to emphasise their chromosome banding patterns. The size of the chromosome, position of the centromere, and pattern of stained bands help identify specific chromosomes. Although difficult to discern in the karyotype, each metaphase chromosome consists of two closely attached sister chromatids (see the diagram of the first pair of homologous duplicated chromosomes).

MAKE CONNECTIONS *Explain why each chromosome in the karyotype consists of two sister chromatids. (See Figures 12.6 and 12.7.)*

parts of the X and Y are homologous. Most of the genes carried on the X chromosome do not have counterparts on the tiny Y, and the Y chromosome has genes lacking on the X. Due to their role in sex determination, the X and Y chromosomes are called **sex chromosomes**. The other chromosomes are called **autosomes**.

The occurrence of pairs of homologous chromosomes in each human somatic cell is a consequence of our sexual origins. We inherit one chromosome of a pair from each parent. Thus, the 46 chromosomes in our somatic cells are actually two sets of 23 chromosomes—a maternal set (from our mother) and a paternal set (from our father). The number of chromosomes in a single set is represented by n. Any cell with two chromosome sets is called a **diploid cell** and has a diploid number of chromosomes, abbreviated $2n$. For humans, the diploid number is 46 ($2n = 46$), the number of chromosomes in our somatic cells. In a cell in which DNA synthesis has occurred, all the chromosomes are duplicated, and therefore each consists of two identical sister chromatids, associated closely at the centromere and along the arms. (Even though the chromosomes are duplicated, we still say the cell is diploid, or $2n$. This is because it has only two sets of information regardless of the number of chromatids, which are merely copies of the information in one set.) **Figure 13.4** helps clarify the various terms that we use to describe duplicated chromosomes in a diploid cell.

Unlike somatic cells, gametes contain a single set of chromosomes. Such cells are called **haploid cells**, and each has a haploid number of chromosomes (n). For humans, the haploid number is 23 ($n = 23$). The set of 23 consists of the 22 autosomes plus a single sex chromosome. An unfertilised egg contains an X chromosome; a sperm contains either an X or a Y chromosome.

Each sexually reproducing species has a characteristic diploid and haploid number. For example, the fruit fly *Drosophila melanogaster* has a diploid number ($2n$) of 8 and a haploid number (n) of 4, while for dogs, $2n$ is 78 and n is 39. The chromosome number generally does not correlate with the size or complexity of a species' genome; it simply reflects how many linear pieces of DNA make up the genome, which is a function of the evolutionary history of that species (see Concept 21.5). Now let's consider chromosome behaviour during sexual life cycles. We'll use the human life cycle as an example.

▼ **Figure 13.4 Describing chromosomes.** A cell from an organism with a diploid number of 6 ($2n = 6$) is depicted here following chromosome duplication and condensation. Each of the six duplicated chromosomes consists of two sister chromatids associated closely along their lengths. Each homologous pair is composed of one chromosome from the maternal set (red) and one from the paternal set (blue). Each set is made up of three chromosomes in this example (long, medium, and short). Together, one maternal and one paternal chromatid in a pair of homologous chromosomes are called nonsister chromatids.

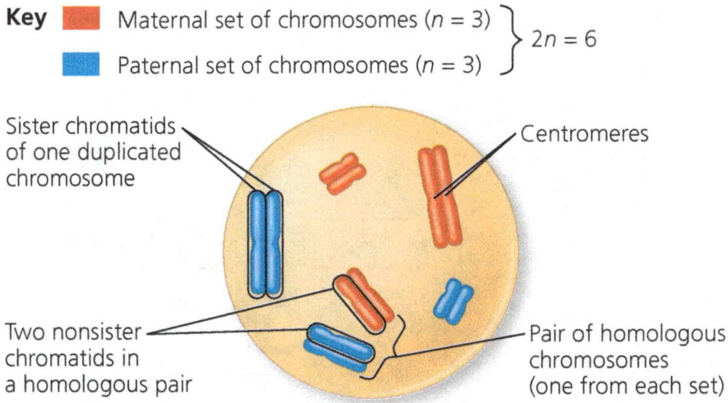

VISUAL SKILLS *How many sets of chromosomes are present in this diagram? How many pairs of homologous chromosomes are present? If the phases of mitosis in Figure 12.7 were redrawn using the colour scheme shown in this figure, describe how the six chromosomes would be coloured.*

Behaviour of Chromosome Sets in the Human Life Cycle

The human life cycle begins when a haploid sperm from the father fuses with a haploid egg from the mother **(Figure 13.5)**. This union of gametes, culminating in fusion of their nuclei, is

▼ **Figure 13.5 The human life cycle.** In each generation, the number of chromosome sets per cell is halved during meiosis but doubles at fertilisation. For humans, the number of chromosomes in a haploid cell is 23, consisting of one set ($n = 23$); the number of chromosomes in the diploid zygote and all somatic cells arising from it is 46, consisting of two sets ($2n = 46$).

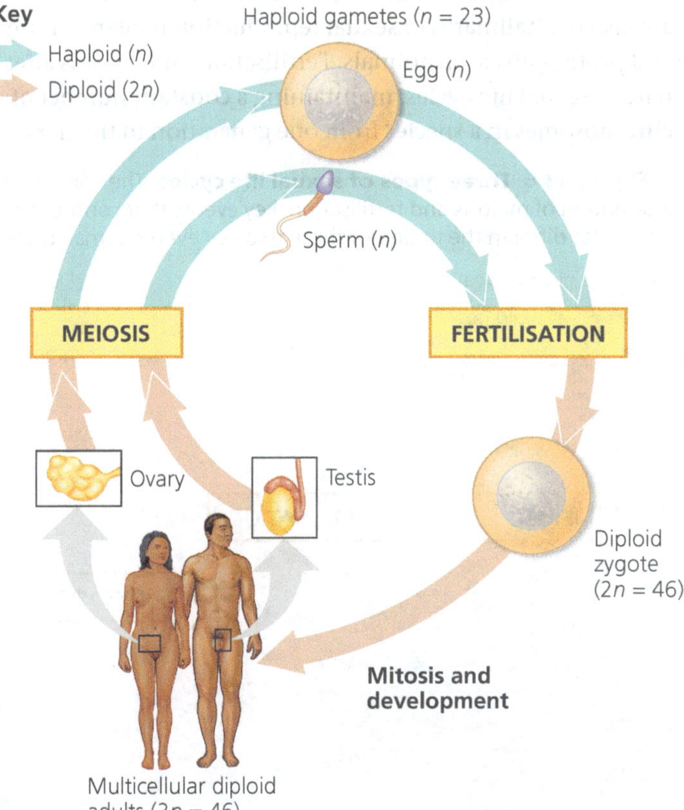

This figure introduces a colour code that will be used for other life cycles later in this text. The aqua arrows identify haploid stages of a life cycle, and the tan arrows identify diploid stages.

called **fertilisation**. The resulting fertilised egg, or **zygote**, is diploid because it contains two haploid sets of chromosomes bearing genes representing the maternal and paternal family lines. As a human develops into a sexually mature adult, mitosis of the zygote and its descendant cells generates all the somatic cells of the body. Both chromosome sets in the zygote and all the genes they carry are passed with precision to the somatic cells.

The only cells of the human body not produced by mitosis are the gametes, which develop from specialised cells called *germ cells* in the gonads—ovaries in females and testes in males (see Figure 13.5). Imagine what would happen if human gametes were made by mitosis: They would be diploid like the somatic cells. At the next round of fertilisation, when two gametes fused, the normal chromosome number of 46 would double to 92, and each subsequent generation would double the number of chromosomes yet again. This does not happen, however, because in sexually reproducing organisms, gamete formation involves a type of cell division called **meiosis**. This type of cell division reduces the number of sets of chromosomes from two in the parent cell to one in each gamete, counterbalancing the doubling that occurs at fertilisation. As a result of meiosis, each human sperm and egg is haploid ($n = 23$). Fertilisation restores the diploid condition by combining two sets of chromosomes, and the human life cycle is repeated, generation after generation (see Figure 13.5).

The steps of the human life cycle are typical of many sexually reproducing animals. Indeed, fertilisation and meiosis are also the hallmarks of sexual reproduction in plants, fungi, and protists just as in animals. Fertilisation and meiosis alternate in sexual life cycles, maintaining a constant number of chromosomes in a species from one generation to the next.

The Variety of Sexual Life Cycles

Although the alternation of meiosis and fertilisation is common to all organisms that reproduce sexually, the timing of these two events in the life cycle varies, depending on the species. These variations can be grouped into three main types of life cycles. In the type that occurs in humans and most other animals, gametes are the only haploid cells **(Figure 13.6a)**. Meiosis occurs in germ cells during the production of gametes, which undergo no further cell division prior to fertilisation. After fertilisation, the diploid zygote divides by mitosis, producing a multicellular organism that is diploid.

Plants and some species of algae exhibit a second type of life cycle called **alternation of generations (Figure 13.6b)**. This type includes both diploid and haploid stages that are multicellular. The multicellular diploid stage is called the *sporophyte*. Meiosis in the sporophyte produces haploid cells called *spores*. Unlike a gamete, a haploid spore doesn't fuse with another cell but divides mitotically, generating a multicellular haploid stage called the *gametophyte*. Cells of the gametophyte give rise to gametes by mitosis. Fusion of two haploid gametes at fertilisation results in a diploid zygote, which develops into the next sporophyte generation. Therefore, in this type of life cycle, the sporophyte generation produces a gametophyte as its offspring, and the gametophyte generation produces a sporophyte. The name *alternation of generations* fits well for this type of life cycle.

A third type of life cycle occurs in most fungi and some protists, including some algae **(Figure 13.6c)**. After gametes fuse and form a diploid zygote, meiosis occurs without a multicellular diploid offspring developing. Meiosis produces not gametes but

▼ Figure 13.6 Three types of sexual life cycles. The common feature of all three cycles is the alternation of meiosis and fertilisation, key events that contribute to genetic variation among offspring. The cycles differ in the timing of these two key events. (Small circles are cells; large circles are organisms.)

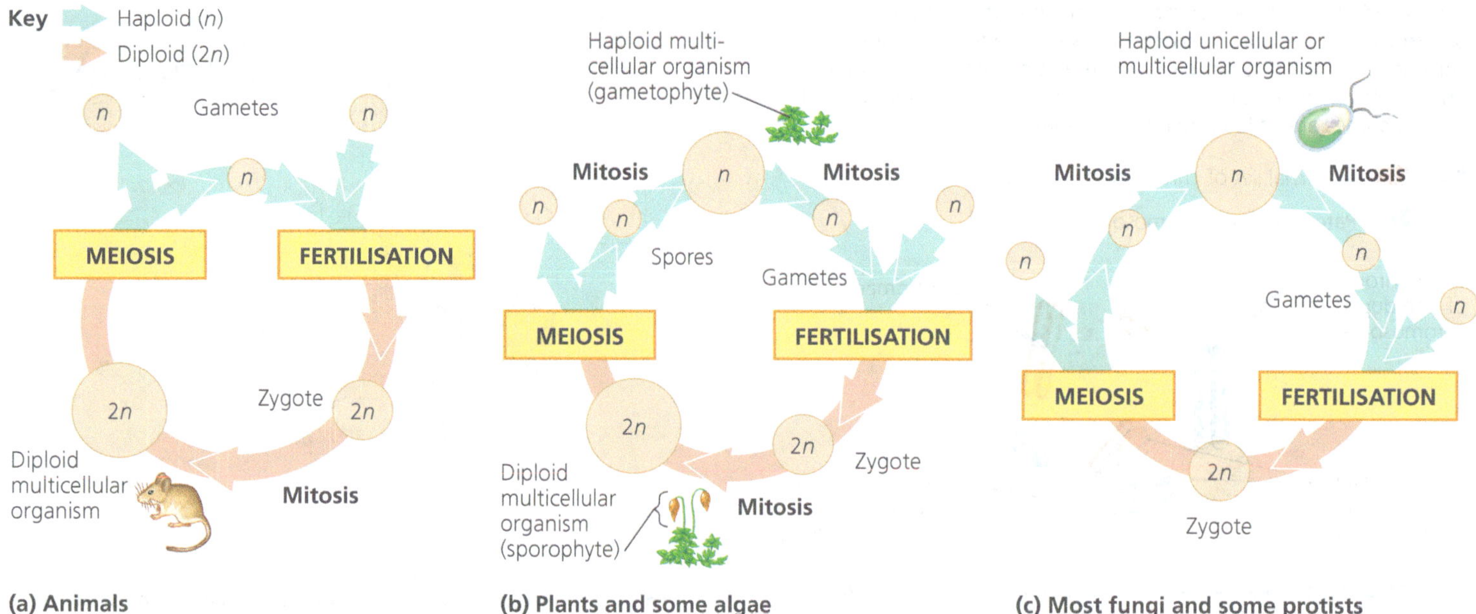

VISUAL SKILLS *For each type of life cycle, indicate whether haploid cells undergo mitosis, and if they do, describe the cells that are formed.*

haploid cells that then divide by mitosis and give rise to either unicellular descendants or a haploid multicellular adult organism. Subsequently, the haploid organism carries out further mitoses, producing the cells that develop into gametes. The only diploid stage found in these species is the single-celled zygote.

Note that *either* haploid or diploid cells can divide by mitosis, depending on the type of life cycle. Only diploid cells, however, can undergo meiosis: Haploid cells can't because they already have a single set of chromosomes that cannot be further reduced. Though the three types of sexual life cycles differ in the timing of meiosis and fertilisation, they share a fundamental result: genetic variation among offspring.

CONCEPT CHECK 13.2

1. **MAKE CONNECTIONS** In Figure 13.4, how many DNA molecules (double helices) are present (see Figure 12.5)? What is the haploid number of this cell? Is a set of chromosomes haploid or diploid?
2. **VISUAL SKILLS** In the karyotype in Figure 13.3, how many pairs of chromosomes are present? How many sets?
3. Using shoes as an analogy for chromosomes, how would you describe the collection of "shoes" in human diploid and haploid cells?
4. **WHAT IF?** A certain eukaryote lives as a unicellular organism, but during environmental stress, it produces gametes. The gametes fuse, and the resulting zygote undergoes meiosis, generating new single cells. What type of organism could this be?

For suggested answers, see Appendix A.

CONCEPT 13.3

Meiosis reduces the number of chromosome sets from diploid to haploid

Several steps of meiosis closely resemble corresponding steps in mitosis. Meiosis, like mitosis, is preceded by interphase, which includes S phase (the duplication of chromosomes). However, this is followed by not one but two consecutive cell divisions, called **meiosis I** and **meiosis II**. These two divisions result in four daughter cells (rather than the two daughter cells of mitosis), each with only half as many chromosomes as the parent cell—one set, rather than two.

The Stages of Meiosis

The overview of meiosis in **Figure 13.7** shows, for a single pair of homologous chromosomes in a diploid cell, that both members of the pair are duplicated and the copies sorted into four haploid daughter cells. Recall that sister chromatids are two copies of *one* chromosome, closely associated all along their lengths; this association is called *sister chromatid cohesion*. Together, the sister chromatids make up one duplicated chromosome (see Figure 13.4). In contrast, the two chromosomes of a homologous pair are individual chromosomes that were inherited from each parent. Homologues appear alike in the microscope, but they may have different versions of genes at corresponding loci; each version is called an *allele* of that gene (see Figure 14.4). For example, one chromosome might have an allele for freckles, but the homologous chromosome may have an allele for the absence of freckles at the same locus. Homologues are not associated with each other in any obvious way except during meiosis.

▼ Figure 13.7 Overview of meiosis: how meiosis reduces chromosome number. After the chromosomes duplicate in interphase, the diploid cell divides *twice*, yielding four haploid daughter cells. This overview tracks just one pair of homologous chromosomes, which for the sake of simplicity are drawn in the condensed state throughout. (They would not normally be condensed during interphase.)

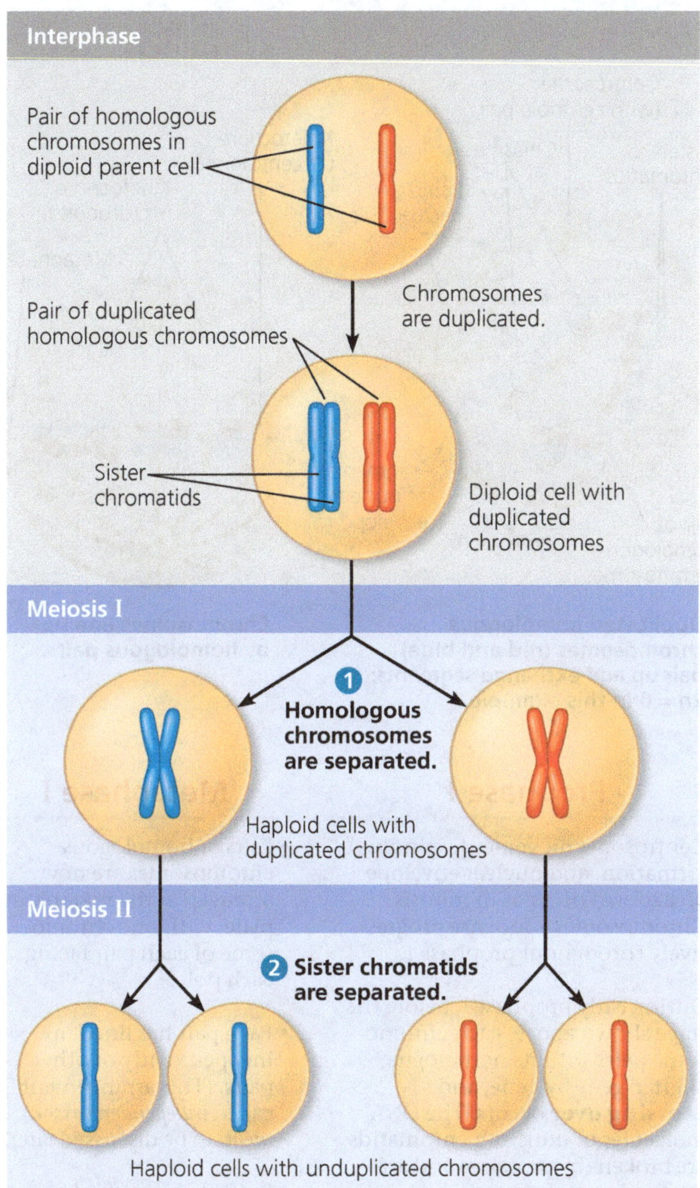

DRAW IT *Redraw the cells in this figure using a simple double helix to represent each DNA molecule.*

Figure 13.8 describes in detail the stages of the two divisions of meiosis for an animal cell whose diploid number is 6. Study this figure thoroughly before going on.

▼ Figure 13.8 Exploring Meiosis in an Animal Cell

MEIOSIS I: Separates homologous chromosomes

Prophase I	Metaphase I	Anaphase I	Telophase I and Cytokinesis

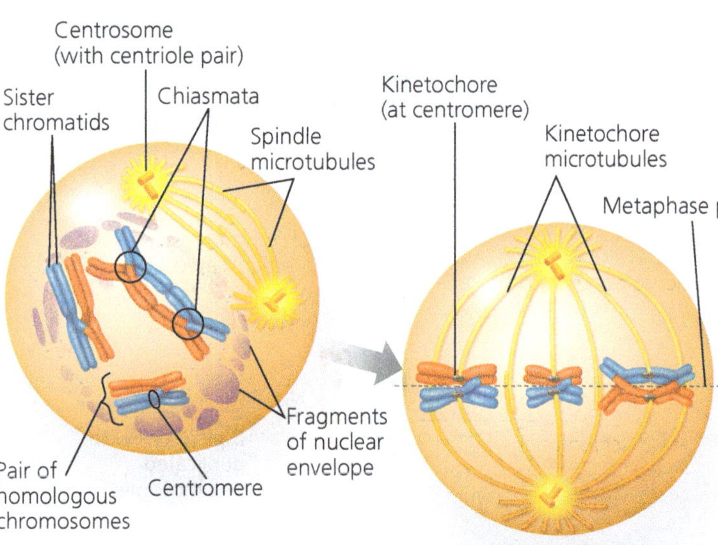

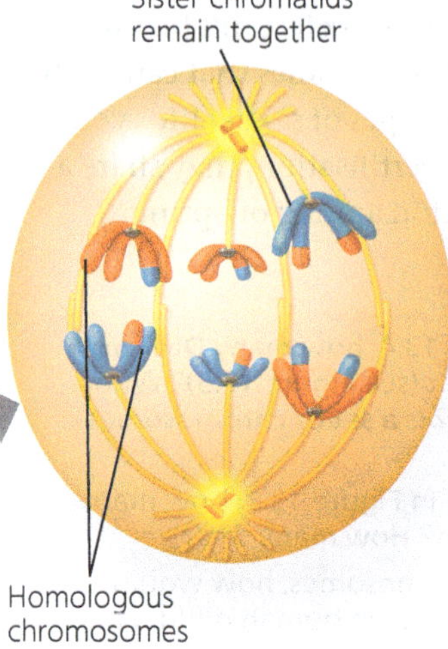

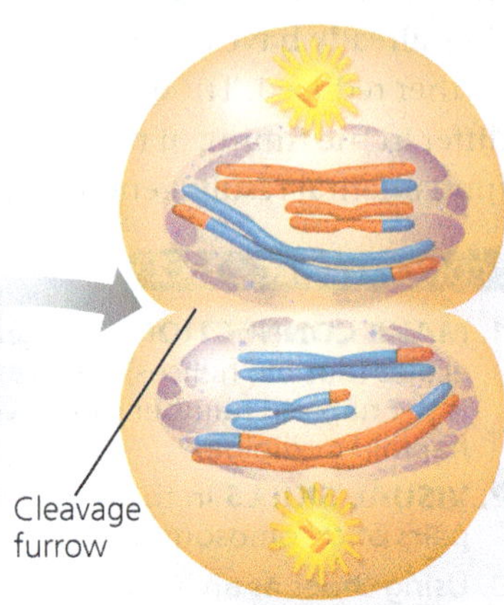

Duplicated homologous chromosomes (red and blue) pair up and exchange segments; 2*n* = 6 in this example.

Chromosomes line up by homologous pairs.

The two homologous chromosomes of each pair separate.

Two haploid cells form; each chromosome still consists of two sister chromatids.

Prophase I

- Centrosome movement, spindle formation, and nuclear envelope breakdown occur as in mitosis. Chromosomes condense progressively throughout prophase I.
- During early prophase I, before the stage shown above, each chromosome pairs with its homologue, aligned gene by gene, and **crossing over** occurs: The DNA molecules of nonsister chromatids are broken (by proteins) and are rejoined to each other.
- At the stage shown above, each homologous pair has one or more X-shaped regions called **chiasmata** (singular, *chiasma*), where crossovers have occurred.
- Later in prophase I, microtubules from one pole or the other attach to the kinetochores, one at the centromere of each homologue. (The two kinetochores on the sister chromatids of a homologue are linked together by proteins and act as a single kinetochore.) Microtubules move the homologous pairs towards the metaphase plate (see the metaphase I diagram).

Metaphase I

- Pairs of homologous chromosomes are now arranged at the metaphase plate, with one chromosome of each pair facing each pole.
- Each pair has lined up independently of other pairs. (This arrangement is called *independent assortment*, to be discussed later.)
- Both chromatids of one homologue are attached to kinetochore microtubules from one pole; the chromatids of the other homologue are attached to microtubules from the opposite pole.

Anaphase I

- Breakdown of proteins that are responsible for sister chromatid cohesion along chromatid arms allows homologues to separate.
- The homologues move towards opposite poles, guided by the spindle apparatus.
- Sister chromatid cohesion persists at the centromere, causing the two chromatids of each chromosome to move as a unit towards the same pole.

Telophase I and Cytokinesis

- When telophase I begins, each half of the cell has a complete haploid set of duplicated chromosomes. Each chromosome is composed of two sister chromatids; one or both chromatids include regions of nonsister chromatid DNA.
- Cytokinesis (division of the cytoplasm) usually occurs simultaneously with telophase I, forming two haploid daughter cells.
- In animal cells like these, a cleavage furrow forms. (In plant cells, a cell plate forms.)
- In some species, chromosomes decondense and nuclear envelopes form.
- No chromosome duplication occurs between meiosis I and meiosis II.

MEIOSIS II: Separates sister chromatids

Prophase II | Metaphase II | Anaphase II | Telophase II and Cytokinesis

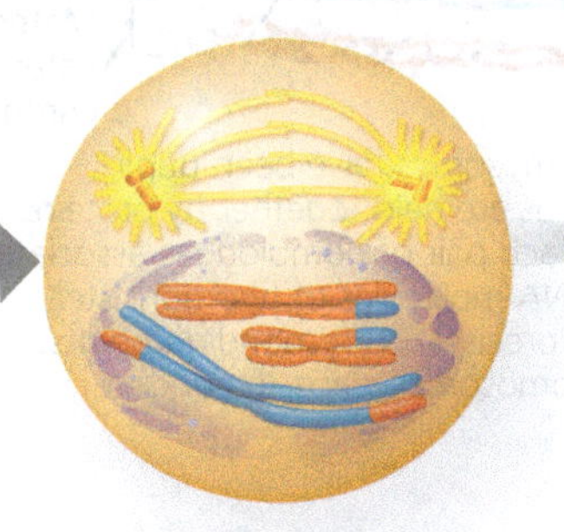
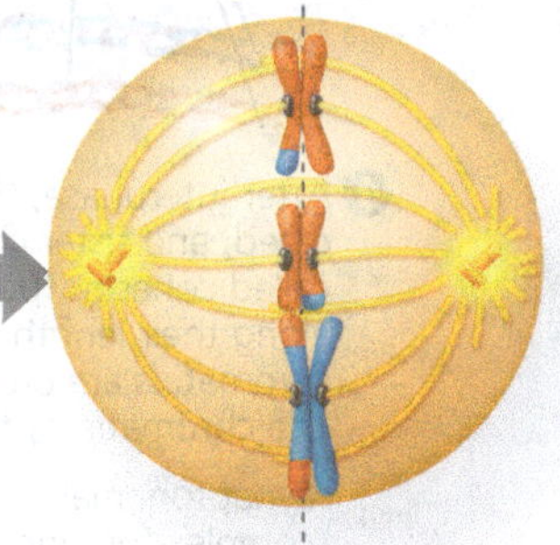
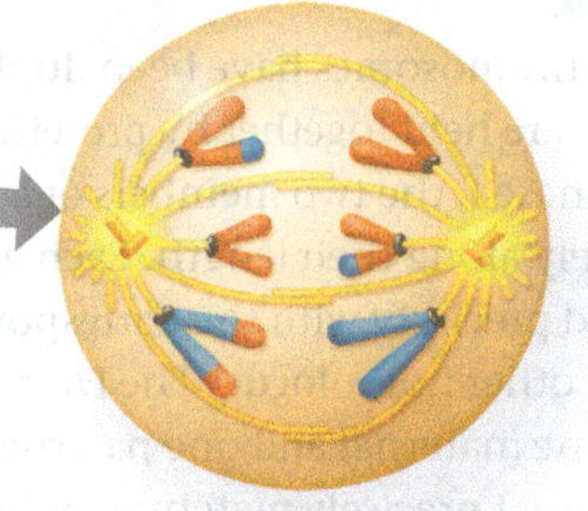
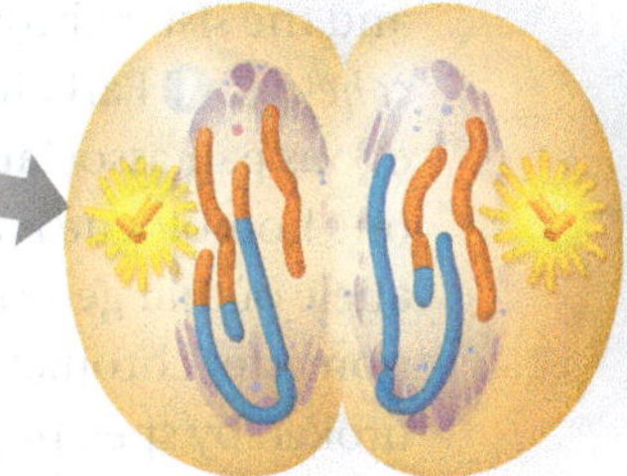

During another round of cell division, the sister chromatids finally separate; four haploid daughter cells result, containing unduplicated chromosomes.

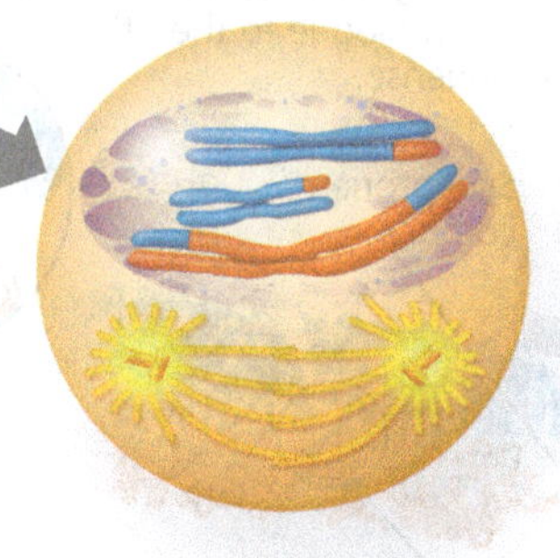
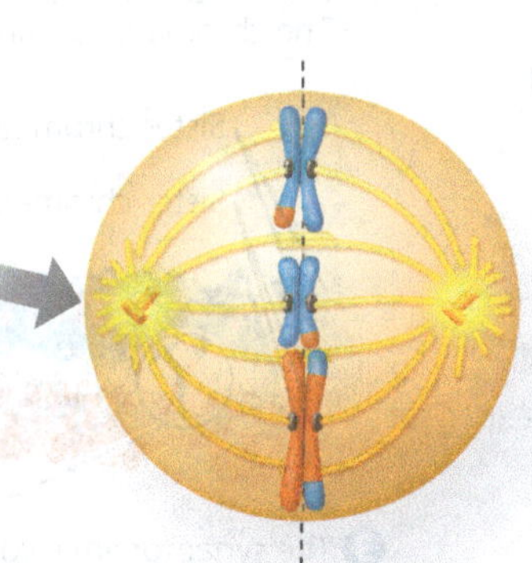
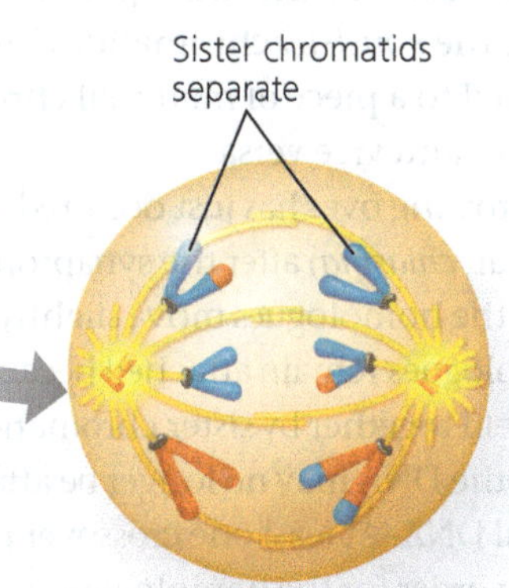

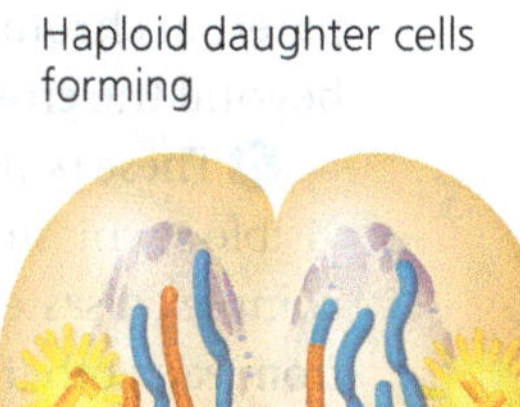

Prophase II

- A spindle apparatus forms.
- In late prophase II (not shown here), chromosomes, each still composed of two chromatids associated at the centromere, are moved by microtubules towards the metaphase II plate.

Metaphase II

- The chromosomes are positioned at the metaphase plate as in mitosis.
- Because of crossing over in meiosis I, the two sister chromatids of each chromosome are not genetically identical.
- The kinetochores of sister chromatids are attached to microtubules extending from opposite poles.

Anaphase II

- Breakdown of proteins holding the sister chromatids together at the centromere allows the chromatids to separate and move towards opposite poles. Each chromatid has now become an individual chromosome.

Telophase II and Cytokinesis

- Nuclei form, the chromosomes begin decondensing, and cytokinesis occurs.
- The meiotic division of one parent cell produces four daughter cells, each with a haploid set of (unduplicated) chromosomes.
- The four daughter cells are genetically distinct from one another and from the parent cell.

MAKE CONNECTIONS *Look at Figure 12.7 and imagine the two daughter cells undergoing another round of mitosis, yielding four cells. Compare the number of chromosomes in each of those four cells, after mitosis, with the number in each cell in Figure 13.8, after meiosis. Explain how the process of meiosis results in this difference, even though meiosis also includes two cell divisions.*

Crossing Over and Synapsis During Prophase I

Prophase I of meiosis is a very busy time. The prophase I cell shown in Figure 13.8 is at a point fairly late in prophase I, when pairing of homologous chromosomes, crossing over, and chromosome condensation have already taken place. The sequence of events leading up to that point is shown in more detail in **Figure 13.9**.

After interphase, the chromosomes have been duplicated and the sister chromatids are held together by proteins called *cohesins*. ❶ Early in prophase I, the two members of a homologous pair associate loosely along their length. Each gene on one homologue is aligned precisely with the corresponding allele of that gene on the other homologue. The DNA of two nonsister chromatids—one maternal and one paternal—is broken by specific proteins at precisely matching points. ❷ Next, the formation of a zipper-like structure called the **synaptonemal complex** holds one homologue tightly to the other. ❸ During this association, called **synapsis**, the DNA breaks are closed up so that each broken end is joined to the corresponding segment of the *nonsister* chromatid. Thus, a paternal chromatid is joined to a piece of maternal chromatid beyond the crossover point, and vice versa.

❹ These points where crossing over has just occurred become visible as chiasmata (singular, *chiasma*) after the synaptonemal complex disassembles and the homologues move slightly apart from each other. The homologues remain attached because sister chromatids are still held together by sister chromatid cohesion, even though some of the DNA may no longer be attached to its original chromosomal DNA. At least one crossover per chromosome must occur in order for the homologous pair to stay together as it moves to the metaphase I plate, for reasons that will be explained shortly.

▼ Figure 13.9 Crossing over and synapsis in prophase I: a closer look.

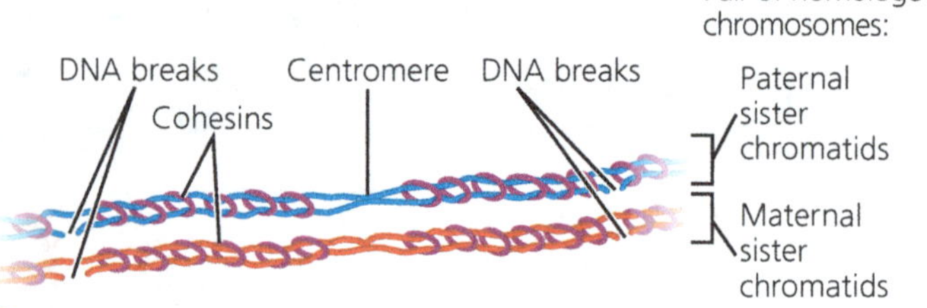

❶ After interphase, the chromosomes have been duplicated, and sister chromatids are held together by proteins called cohesins (purple). Each pair of homologues associate along their length. The DNA molecules of two nonsister chromatids are broken at precisely corresponding points. The chromatin of the chromosomes starts to condense.

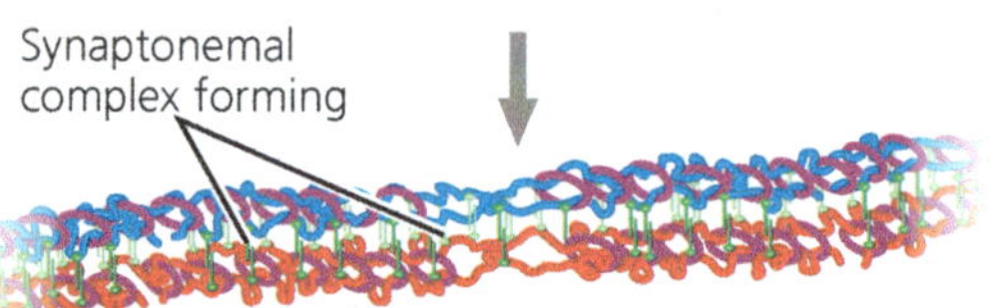

❷ A zipper-like protein complex, the synaptonemal complex (green), begins to form, attaching one homologue to the other. The chromatin continues to condense.

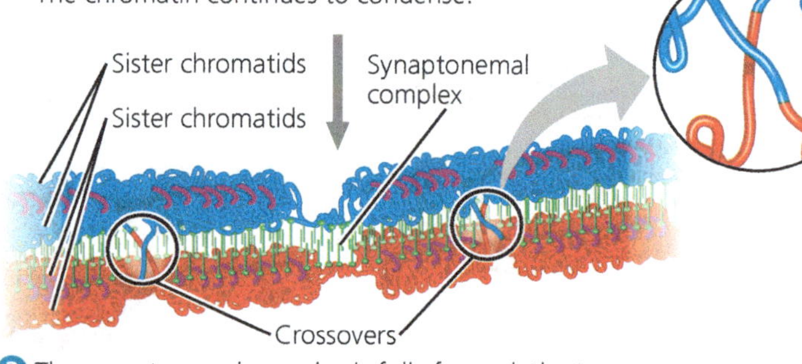

❸ The synaptonemal complex is fully formed; the two homologues are said to be in synapsis. During synapsis, the DNA breaks are closed up when each broken end is joined to the corresponding segment of the nonsister chromatid, producing crossovers.

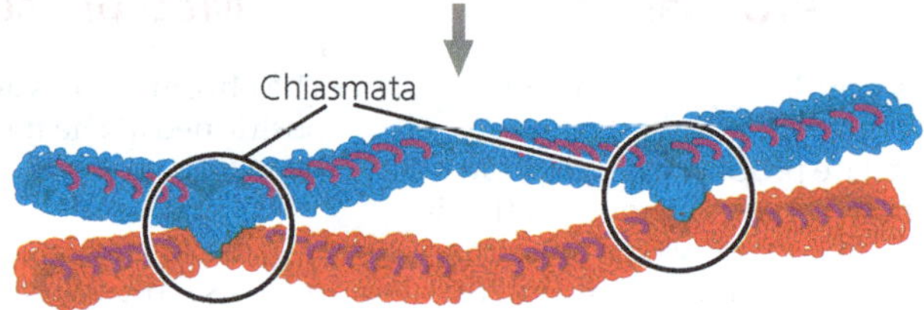

❹ After the synaptonemal complex disassembles, the homologues move slightly apart from each other but remain attached because of sister chromatid cohesion, even though some of the DNA may no longer be attached to its original chromosome. The points of attachment where crossovers have occurred show up as chiasmata. The chromosomes continue to condense as they move towards the metaphase plate.

VISUAL SKILLS *Label the four DNA breaks in step 2.*

A Comparison of Mitosis and Meiosis

Figure 13.10 summarises the key differences between meiosis and mitosis in diploid cells. Meiosis produces four cells and reduces the number of chromosome sets from two (diploid) to one (haploid), whereas mitosis produces two cells and conserves the number of chromosome sets. Meiosis produces cells that differ genetically from the parent cell and from each other, whereas mitosis produces daughter cells that are genetically identical to their parent cell and to each other.

Three events unique to meiosis occur during meiosis I:

1. **Synapsis and crossing over.** During prophase I, duplicated homologues pair up and crossing over occurs as described previously and in Figure 13.9. Synapsis and crossing over do not occur during prophase of mitosis.
2. **Alignment of homologous pairs at the metaphase plate.** At metaphase I of meiosis, pairs of homologues are positioned at the metaphase plate, rather than individual chromosomes, as in metaphase of mitosis.
3. **Separation of homologues.** At anaphase I of meiosis, the duplicated chromosomes of each homologous pair move towards opposite poles, but the sister chromatids of each duplicated chromosome remain attached. In anaphase of mitosis, by contrast, sister chromatids separate.

▼ Figure 13.10 A comparison of mitosis and meiosis.

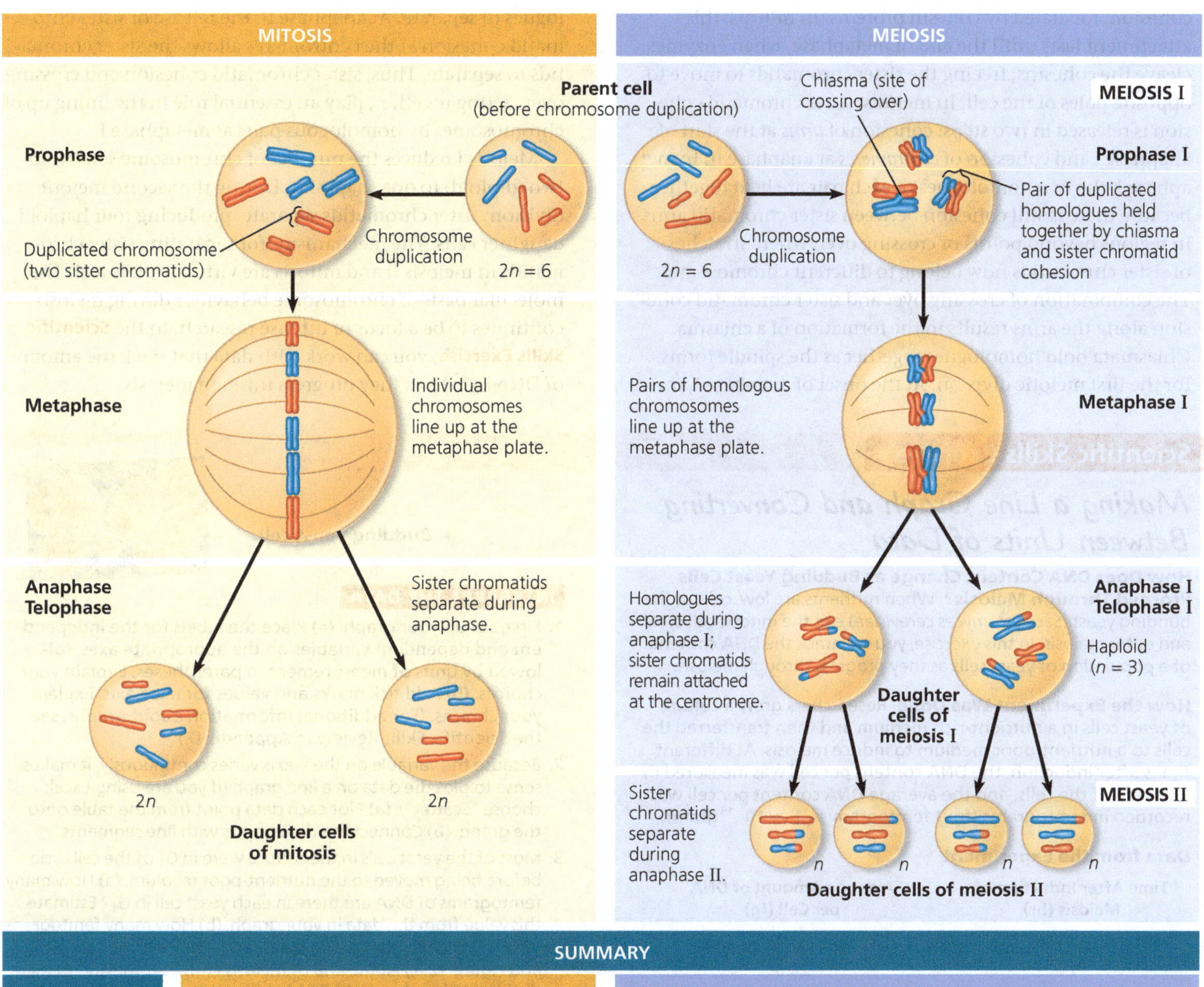

SUMMARY		
Property	**Mitosis** (occurs in both diploid and haploid cells)	**Meiosis** (can only occur in diploid cells)
DNA replication	Occurs during interphase, before mitosis begins	Occurs during interphase before meiosis I but not before meiosis II
Number of divisions	One, including prophase, prometaphase, metaphase, anaphase, and telophase	Two, each including prophase, metaphase, anaphase, and telophase
Synapsis of homologous chromosomes	Does not occur	Occurs during prophase I along with crossing over between nonsister chromatids; resulting chiasmata hold pairs together due to sister chromatid cohesion
Number of daughter cells and genetic composition	Two, each genetically identical to the parent cell, with the same number of chromosomes	Four, each haploid (*n*); genetically different from the parent cell and from each other
Role in animals, fungi, and plants	Enables multicellular animal, fungus, or plant (gametophyte or sporophyte) to arise from a single cell; produces cells for growth, repair, and, in some species, asexual reproduction; produces gametes in the plant gametophyte	Produces gametes (in animals) or spores (in fungi and in plant sporophytes); reduces number of chromosome sets by half and introduces genetic variability among the gametes or spores

DRAW IT *Could any other combinations of chromosomes be generated during meiosis II from the specific cells shown in telophase I? Explain. (Hint: Draw the cells as they would appear in metaphase II.)*

Sister chromatids stay together due to sister chromatid cohesion, mediated by cohesin proteins. In mitosis, this attachment lasts until the end of metaphase, when enzymes cleave the cohesins, freeing the sister chromatids to move to opposite poles of the cell. In meiosis, sister chromatid cohesion is released in two steps: cohesion of *arms* at the start of anaphase I and cohesion of *centromeres* at anaphase II. In metaphase I, the two homologues of each pair are held together because there is still cohesion between sister chromatid arms in regions beyond points of crossing over, where stretches of sister chromatids now belong to different chromosomes. The combination of crossing over and sister chromatid cohesion along the arms results in the formation of a chiasma. Chiasmata hold homologues together as the spindle forms for the first meiotic division. At the onset of anaphase I, the release of cohesion along sister chromatid arms allows homologues to separate. At anaphase II, the release of sister chromatid cohesion at the centromeres allows the sister chromatids to separate. Thus, sister chromatid cohesion and crossing over, acting together, play an essential role in the lining up of chromosomes by homologous pairs at metaphase I.

Meiosis I reduces the number of chromosome sets: from two (diploid) to one (haploid). During the second meiotic division, sister chromatids separate, producing four haploid daughter cells. The mechanisms for separating sister chromatids in meiosis II and mitosis are virtually identical. The molecular basis of chromosome behaviour during meiosis continues to be a focus of intense research. In the **Scientific Skills Exercise**, you can work with data that track the amount of DNA in cells as they progress through meiosis.

Scientific Skills Exercise

Making a Line Graph and Converting Between Units of Data

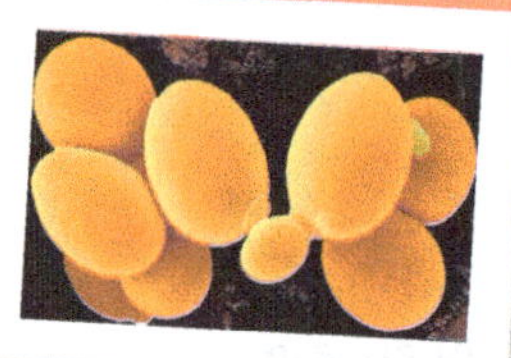

▶ **Budding yeast cells**

How Does DNA Content Change as Budding Yeast Cells Proceed Through Meiosis? When nutrients are low, cells of the budding yeast (*Saccharomyces cerevisiae*) exit the mitotic cell cycle and enter meiosis. In this exercise, you will track the DNA content of a population of yeast cells as they progress through meiosis.

How the Experiment Was Done Researchers grew a culture of yeast cells in a nutrient-rich medium and then transferred the cells to a nutrient-poor medium to induce meiosis. At different times after induction, the DNA content per cell was measured in a sample of the cells, and the average DNA content per cell was recorded in femtograms (fg; 1 femtogram = 1×10^{-15} gram).

Data from the Experiment

Time After Induction of Meiosis (hr)	Average Amount of DNA per Cell (fg)
0.0	24.0
1.0	24.0
2.0	40.0
3.0	47.0
4.0	47.5
5.0	48.0
6.0	48.0
7.0	47.5
7.5	25.0
8.0	24.0
9.0	23.5
9.5	14.0
10.0	13.0
11.0	12.5
12.0	12.0
13.0	12.5
14.0	12.0

INTERPRET THE DATA

1. First, set up your graph. **(a)** Place the labels for the independent and dependent variables on the appropriate axes, followed by units of measurement in parentheses. Explain your choices. **(b)** Add tick marks and values for each axis. Explain your choices. (For additional information about graphs, see the Scientific Skills Review in Appendix D.)
2. Because the variable on the *x*-axis varies continuously, it makes sense to plot the data on a line graph (if you are using Excel, choose "Scatter"). **(a)** Plot each data point from the table onto the graph. **(b)** Connect the data points with line segments.
3. Most of the yeast cells in the culture were in G_1 of the cell cycle before being moved to the nutrient-poor medium. **(a)** How many femtograms of DNA are there in each yeast cell in G_1? Estimate this value from the data in your graph. **(b)** How many femtograms of DNA should be present in each cell in G_2? (See Concept 12.2 and Figure 12.6.) At the end of meiosis I (MI)? At the end of meiosis II (MII)? (See Figure 13.7.) **(c)** Using these values as a guideline, distinguish the different phases by inserting vertical dashed lines in the graph between phases and label each phase (G_1, S, G_2, MI, MII). You can figure out where to put the dividing lines based on what you know about the DNA content of each phase (see Figure 13.7). **(d)** Think carefully about the point where the line at the highest value begins to slope downwards. What specific point of meiosis does this "corner" represent? What stage(s) correspond(s) to the downward-sloping line?
4. Given the fact that 1 fg of DNA = 9.78×10^5 base pairs (on average), you can convert the amount of DNA per cell to the length of DNA in numbers of base pairs. **(a)** Calculate the number of base pairs of DNA in the haploid yeast genome. Express your answer in millions of base pairs (Mb), a standard unit for expressing genome size. Show your work. **(b)** How many base pairs per minute were synthesised during the S phase of these yeast cells?

Further Reading G. Simchen, Commitment to meiosis: What determines the mode of division in budding yeast? *BioEssays* 31:169–177 (2009).

CONCEPT CHECK 13.3

1. **MAKE CONNECTIONS** Compare the chromosomes in a cell at metaphase of mitosis with those in a cell at metaphase II. (See Figures 12.7 and 13.8.)
2. **WHAT IF?** After the synaptonemal complex disappears, how would any pair of homologous chromosomes be associated if crossing over did not occur? What effect might this have on gamete formation?

For suggested answers, see Appendix A.

CONCEPT 13.4

Genetic variation produced in sexual life cycles contributes to evolution

How do we account for the genetic variation of the family members in Figure 13.1? As you will learn more about in later chapters, mutations are the original source of genetic diversity. These changes in an organism's DNA create the different versions of genes, known as alleles. Once these differences arise, reshuffling of the alleles during sexual reproduction produces the variation that results in each member of a sexually reproducing population having a unique combination of traits.

Origins of Genetic Variation Among Offspring

In species that reproduce sexually, the behaviour of chromosomes during meiosis and fertilisation is responsible for most of the variation that arises in each generation. Three mechanisms contribute to the genetic variation arising from sexual reproduction: independent assortment of chromosomes, crossing over, and random fertilisation.

Independent Assortment of Chromosomes

One aspect of sexual reproduction that generates genetic variation is the random orientation of pairs of homologous chromosomes at metaphase of meiosis I. At metaphase I, the homologous pairs, each consisting of one maternal and one paternal chromosome, are situated at the metaphase plate. (Note that the terms *maternal* and *paternal* refer, respectively, to whether the chromosome in question was contributed by the mother or the father of the individual whose cells are undergoing meiosis.) Each pair may orient with either its maternal or paternal homologue closer to a given pole—its orientation is as random as the flip of a coin. Thus, there is a 50% chance that a particular daughter cell of meiosis I will get the maternal chromosome of a certain homologous pair and a 50% chance that it will get the paternal chromosome.

Because each pair of homologous chromosomes is positioned independently of the other pairs at metaphase I, the first meiotic division results in each pair sorting its maternal and paternal homologues into daughter cells independently of every other pair. This is called *independent assortment*. Each daughter cell represents one outcome of all possible combinations of maternal and paternal chromosomes. As shown in **Figure 13.11**, the number of combinations possible for daughter cells formed by meiosis of a diploid cell with two pairs of homologous chromosomes ($n = 2$) is four: two possible arrangements for the first pair times two possible arrangements for the second pair. Note that only two of the four combinations of daughter cells shown in the figure would result from meiosis of a *single* diploid cell because a single parent cell would have one or the other possible chromosomal arrangement at metaphase I, but not both. However, the population of daughter cells resulting from meiosis of a large number of diploid cells contains all four types in approximately equal numbers. In the case of $n = 3$, eight combinations ($2 \times 2 \times 2 = 2^3$) of chromosomes are possible for daughter cells. More generally, the number of possible combinations when chromosomes sort independently during meiosis is 2^n, where n is the haploid number of the species.

In the case of humans ($n = 23$), the number of possible combinations of maternal and paternal chromosomes in the resulting gametes is 2^{23}, or about 8.4 million. Each gamete that you produce in your lifetime contains one of roughly 8.4 million possible combinations of chromosomes. This is an underestimate, because it doesn't take into account crossing over, which we'll consider next.

Crossing Over

As a consequence of the independent assortment of chromosomes during meiosis, each of us produces a collection

▼ Figure 13.11 The independent assortment of homologous chromosomes in meiosis.

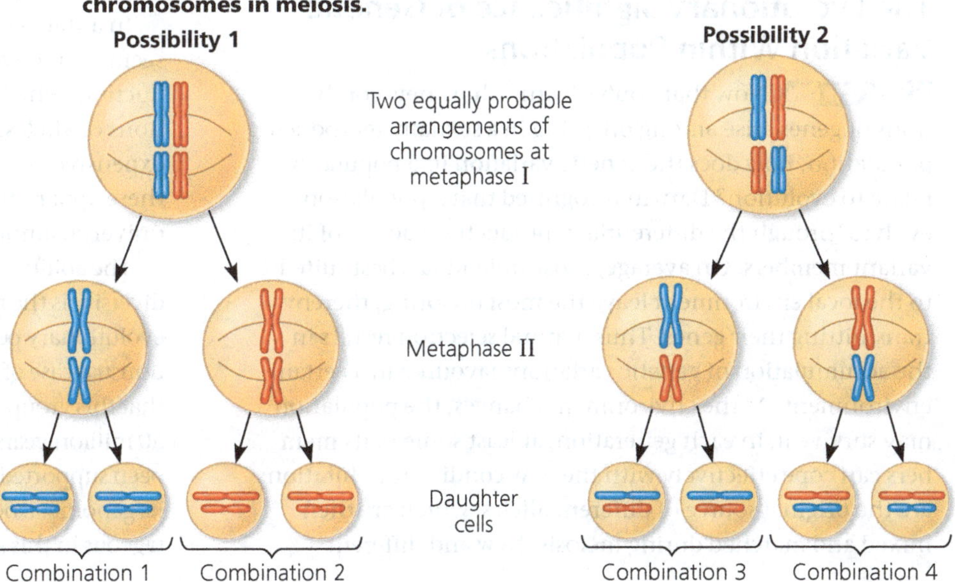

of gametes differing greatly in their combinations of the chromosomes we inherited from our two parents. Figure 13.11 suggests that each chromosome in a gamete is exclusively maternal or paternal in origin. In fact, this is *not* the case, because crossing over produces **recombinant chromosomes**, individual chromosomes that carry genes (DNA) from two different parents **(Figure 13.12)**. In meiosis in humans, an average of one to three crossover events occurs per chromosome pair, depending on the size of the chromosomes and the position of their centromeres.

As you learned in Figure 13.9, crossing over produces chromosomes with new combinations of maternal and paternal alleles. At metaphase II, chromosomes that contain one or more recombinant chromatids can be oriented in two alternative, nonequivalent ways with respect to other chromosomes because their sister chromatids are no longer identical (see Figure 13.12). The different possible arrangements of nonidentical sister chromatids during meiosis II further increase the number of genetic types of daughter cells that can result from meiosis.

You'll learn more about crossing over in Concept 15.3. The important point for now is that crossing over, by combining DNA inherited from two parents into a single chromosome, is an important source of genetic variation in sexual life cycles.

Random Fertilisation

The random nature of fertilisation adds to the genetic variation arising from meiosis. In humans, each male and female gamete represents one of about 8.4 million (2^{23}) possible chromosome combinations due to independent assortment. The fusion of a male gamete with a female gamete during fertilisation will produce a zygote with any of about 70 trillion ($2^{23} \times 2^{23}$) diploid combinations. If we factor in the variation brought about by crossing over, the number of possibilities is truly astronomical. You really *are* unique.

▼ Figure 13.12 The results of crossing over during meiosis.

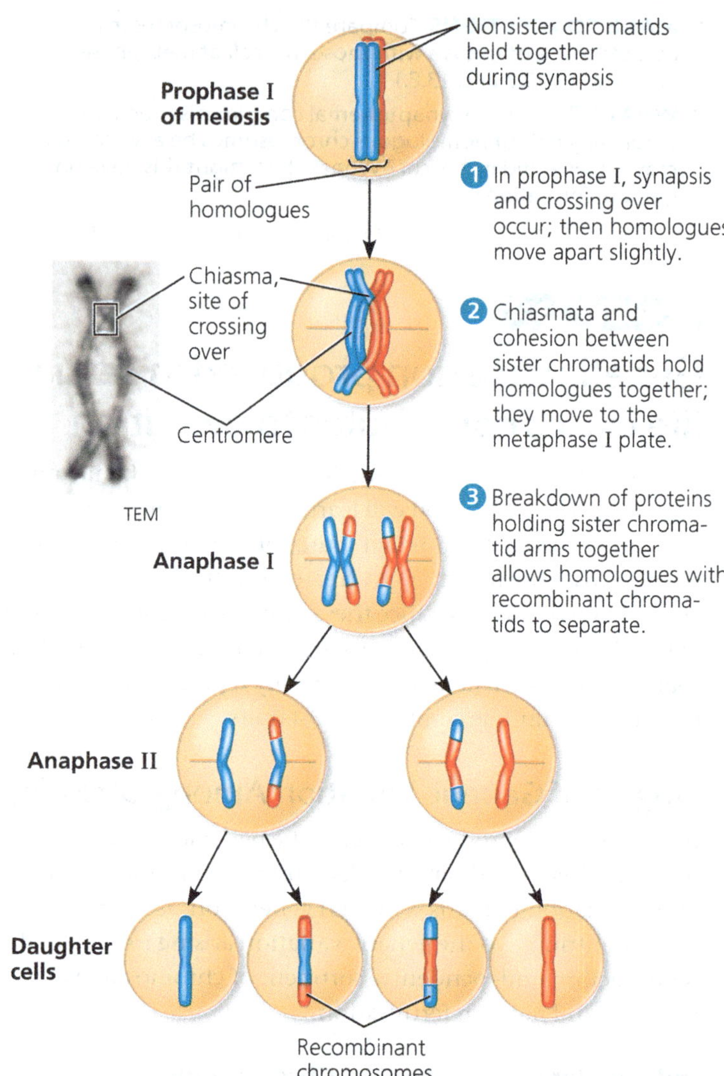

The Evolutionary Significance of Genetic Variation Within Populations

EVOLUTION Now that you've learned how new combinations of genes arise among offspring in a sexually reproducing population, how does the genetic variation in a population relate to evolution? Darwin recognised that a population evolves through the differential reproductive success of its variant members. On average, those individuals best suited to the local environment leave the most offspring, thereby transmitting their genes. Thus, natural selection results in the accumulation of genetic variations favoured in a certain environment. As the environment changes, the population may survive if, in each generation, at least some of its members can cope effectively with the new conditions. Mutations are the original source of different alleles, which are then mixed and matched during meiosis. New and different combinations of alleles may work better than those that previously prevailed.

In a stable environment, though, sexual reproduction seems as if it would be less advantageous than asexual reproduction, which ensures perpetuation of successful combinations of alleles. Furthermore, sexual reproduction is more expensive energetically than asexual reproduction. In spite of these apparent disadvantages, sexual reproduction is almost universal among animals. Why is this?

The ability of sexual reproduction to generate genetic diversity is the most commonly proposed explanation for the evolutionary persistence of this process. However, consider the unusual case of the bdelloid rotifer **(Figure 13.13)**. It appears that this group may not have reproduced sexually for more than 50 million years of their evolutionary history, a model that has been supported by recent analysis of the genetic sequences in its genome. Does this mean that genetic diversity is not advantageous in this species? It turns out that bdelloid rotifers are an exception to the "rule" that sex alone generates genetic diversity: Bdelloids have mechanisms other than sexual reproduction for generating genetic diversity. For example, they live in environments that can dry up for long periods of time, during which they can enter a state of suspended animation. In this state, their cell membranes may crack in places, allowing entry of DNA from other rotifer species and even from more distantly related species. Evidence suggests that this foreign DNA can become incorporated into the genome of the bdelloid, leading to increased genetic diversity. In fact, the genomic analysis shows that bdelloid rotifers pick up non-bdelloid DNA at a much higher rate than most other species pick up foreign DNA. The conclusion that bdelloid rotifers have developed other ways of achieving genetic diversity supports the idea that genetic diversity is advantageous, but points out as well that sexual reproduction is not the only way of generating such diversity.

▼ Figure 13.13 A bdelloid rotifer, an animal that reproduces only asexually.

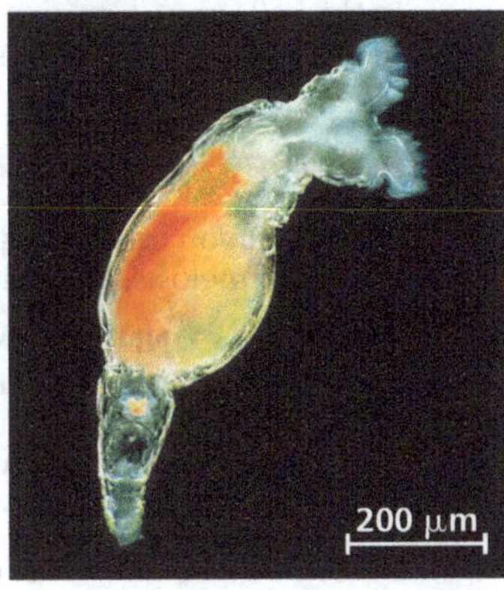

In this chapter, we have seen how sexual reproduction greatly increases the genetic variation present in a population. Although Darwin realised that heritable variation is what makes evolution possible, he could not explain why offspring resemble—but are not identical to—their parents. Ironically, Gregor Mendel, a contemporary of Darwin, published a theory of inheritance that helps explain genetic variation, but his discoveries had no impact on biologists until 1900, more than 15 years after Darwin (1809–1882) and Mendel (1822–1884) had died. In the chapter, "Mendel and the Gene Idea" you'll learn how Mendel discovered the basic rules governing the inheritance of specific traits.

CONCEPT CHECK 13.4

1. What is the original source of variation among the different alleles of a gene?
2. The diploid number for fruit flies is 8, and the diploid number for grasshoppers is 46. If no crossing over took place, would the genetic variation among offspring from a given pair of parents be greater in fruit flies or grasshoppers? Explain.
3. **WHAT IF?** If maternal and paternal chromatids have the same two alleles for every gene, will crossing over lead to genetic variation?

For suggested answers, see Appendix A.

13 Chapter Review

SUMMARY OF KEY CONCEPTS

CONCEPT 13.1

Offspring acquire genes from parents by inheriting chromosomes *(pp. 257–258)*

- Each **gene** in an organism's DNA exists at a specific **locus** on a certain chromosome.
- In **asexual reproduction**, a single parent produces genetically identical offspring by mitosis. **Sexual reproduction** combines genes from two parents, leading to genetically diverse offspring.

? *Explain why human offspring resemble their parents but are not identical to them.*

CONCEPT 13.2

Fertilisation and meiosis alternate in sexual life cycles *(pp. 258–261)*

- Normal human **somatic cells** are **diploid**. They have 46 chromosomes made up of two sets of 23 chromosomes, one set from each parent. Human diploid cells have 22 pairs of **homologues** that are **autosomes**, and one pair of **sex chromosomes**; the latter typically determines whether the person is female (XX) or male (XY).
- In humans, ovaries and testes produce **haploid gametes** by **meiosis**, each gamete containing a single set of 23 chromosomes ($n = 23$). During **fertilisation**, an egg and sperm unite, forming a diploid ($2n = 46$) single-celled **zygote**, which develops into a multicellular organism by mitosis.
- Sexual life cycles differ in the timing of meiosis relative to fertilisation and in the point(s) of the cycle at which a multicellular organism is produced by mitosis.

? *Compare and contrast the life cycles of animals and plants.*

CONCEPT 13.3

Meiosis reduces the number of chromosome sets from diploid to haploid *(pp. 261–267)*

- The two cell divisions of meiosis, **meiosis I** and **meiosis II**, produce four haploid daughter cells. The number of chromosome sets is reduced from two (diploid) to one (haploid) during meiosis I.
- Meiosis is distinguished from mitosis by three events of meiosis I:

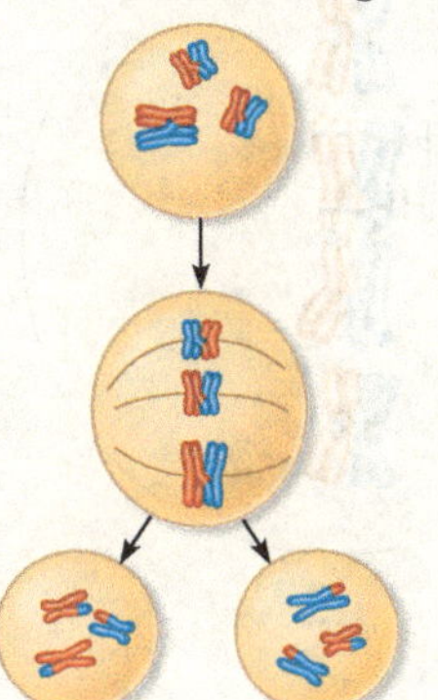

Prophase I: Each pair of homologous chromosomes undergoes **synapsis** and **crossing over** between nonsister chromatids with the subsequent appearance of **chiasmata**.

Metaphase I: Chromosomes line up as homologous pairs on the metaphase plate.

Anaphase I: Homologues separate from each other; sister chromatids remain joined at the centromere.

- Meiosis II separates the sister chromatids.
- Sister chromatid cohesion and crossing over allow chiasmata to hold homologues together until anaphase I. Cohesins are cleaved along the arms at anaphase I, allowing homologues to separate, and at the centromeres in anaphase II, releasing sister chromatids from each other.

? *In prophase I, homologous chromosomes pair up and undergo synapsis and crossing over. Can this also occur during prophase II? Explain.*

CONCEPT **13.4**

Genetic variation produced in sexual life cycles contributes to evolution *(pp. 267–269)*

- Three events in sexual reproduction contribute to genetic variation in a population: independent assortment of chromosomes during meiosis I, crossing over during meiosis I, and random fertilisation of egg cells by sperm. During crossing over, DNA of nonsister chromatids in a homologous pair is broken and rejoined.
- Genetic variation is the raw material for evolution by natural selection. Mutations are the original source of this variation; recombination of variant genes generates additional genetic diversity.

? *Explain how three processes unique to sexual reproduction generate a great deal of genetic variation.*

TEST YOUR UNDERSTANDING

Levels 1-2: Remembering/Understanding

1. A human cell containing 22 autosomes and a Y chromosome is
(A) a sperm.
(B) an egg.
(C) a zygote.
(D) a somatic cell of a male.

2. The two homologues of a pair move towards opposite poles of dividing cell during
(A) mitosis.
(B) meiosis I.
(C) meiosis II.
(D) fertilisation.

Levels 3-4: Applying/Analysing

3. Meiosis II is similar to mitosis in that
(A) sister chromatids separate during anaphase.
(B) DNA replicates before the division.
(C) the daughter cells are diploid.
(D) homologous chromosomes synapse.

4. If the DNA content of a diploid cell in the G_1 phase of the cell cycle is x, then the DNA content of the same cell at metaphase of meiosis I will be
(A) $0.25x$. (B) $0.5x$. (C) x. (D) $2x$.

5. If we continue to follow the cell lineage from question 4, then the DNA content of a single cell at metaphase of meiosis II will be
(A) $0.25x$. (B) $0.5x$. (C) x. (D) $2x$.

6. **DRAW IT** The diagram shows a cell in meiosis.

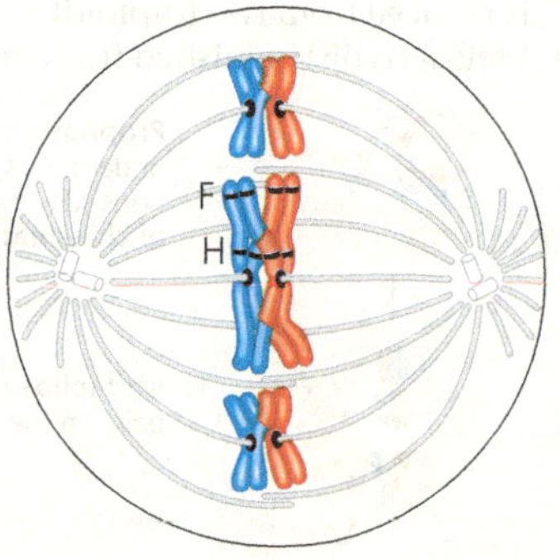

(a) Label the appropriate structures with these terms: chromosome (label as duplicated or unduplicated), centromere, kinetochore, sister chromatids, nonsister chromatids, homologous pair (use a bracket when labelling), homologue (label each one), chiasma, sister chromatid cohesion, and gene loci. Circle and label the alleles of the F gene.
(b) Precisely describe the makeup of a haploid set and a diploid set in this cell.
(c) Identify the stage of meiosis shown.

Levels 5-6: Evaluating/Creating

7. Explain how you can tell that the cell in question 6 is undergoing meiosis, not mitosis.

8. **EVOLUTION CONNECTION** Many species can reproduce either asexually or sexually. Explain what you think might be the evolutionary significance of the switch from asexual to sexual reproduction that occurs in some species when the environment becomes unfavourable.

9. **SCIENTIFIC INQUIRY** The diagram in question 6 represents just a few of the chromosomes of a meiotic cell in a certain person. Assume that the freckles gene is located at the locus marked F, and the hair-colour gene is located at the locus marked H, both on the long chromosome. The individual from whom this cell was taken has inherited different alleles for each gene ("freckles" and "black hair" from one parent, and "no freckles" and "blond hair" from the other). Predict allele combinations in the gametes resulting from this meiotic event. (It will help if you draw out the rest of meiosis and label the alleles by name.) List other possible combinations of these alleles in this individual's gametes.

10. **WRITE ABOUT A THEME: INFORMATION** The continuity of life is based on heritable information in the form of DNA. In a short essay (100–150 words), explain how chromosome behaviour during sexual reproduction in animals ensures perpetuation of parental traits in offspring and, at the same time, genetic variation among offspring.

11. **SYNTHESISE YOUR KNOWLEDGE**

The Cavendish banana is currently threatened by extinction due to a fungus. This banana variety is "triploid" ($3n$, with three sets of chromosomes) and can only reproduce through cloning by cultivators. Given what you know about meiosis, explain how the banana's triploid number accounts for its inability to form normal gametes. Considering genetic diversity, discuss how the absence of sexual reproduction might make this domesticated species vulnerable to infectious agents.

For selected answers, see Appendix A.

14 Mendel and the Gene Idea

KEY CONCEPTS

Study Tip

Solving genetics problems: In Figure 14.5, you'll learn how to predict the offspring of a genetic cross with a Punnett square, like the example shown below. For crosses shown in other figures, cover up the Punnett square and draw your own. Also, check out the Tips for Genetics Problems at the end of the chapter.

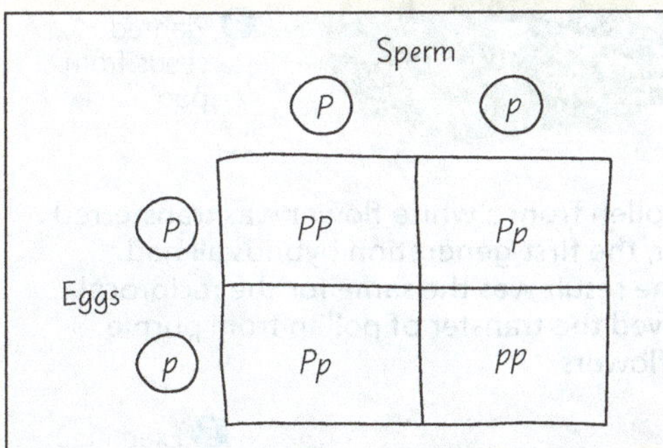

Go to Mastering Biology

to access Dynamic Study Modules for revision, 3D BioFlix® animations and high-quality videos, and your interactive Pearson eText.

Figure 14.1 Flower colour was one of the many characteristics of pea plants studied by Gregor Mendel. By breeding pea plants over many generations and carefully counting the types of offspring, Mendel developed a theory of inheritance that is the basis of modern genetics.

How are traits, such as the purple or white colour of flowers, transmitted from parents to offspring?

Alternative versions of a gene (alleles) account for different traits.

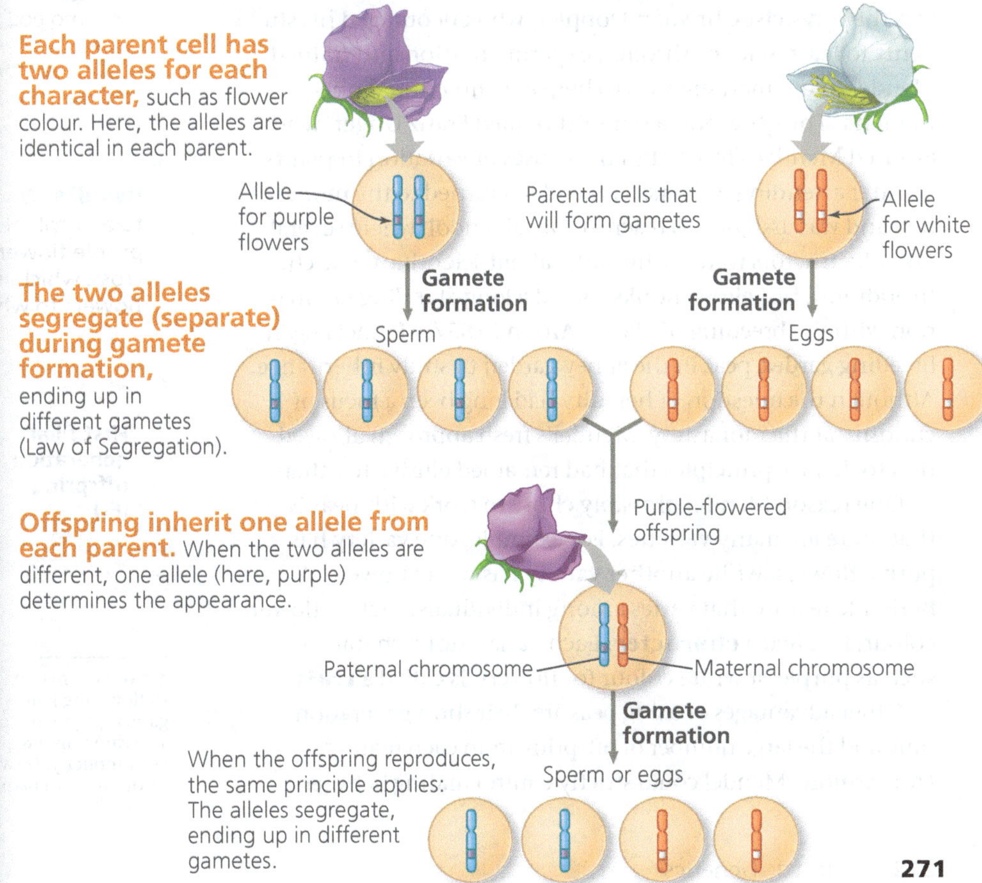

CONCEPT 14.1

Mendel used the scientific approach to identify two laws of inheritance

Modern genetics began during the mid-1800s with a monk named Gregor Mendel. He discovered the basic principles of heredity by breeding garden peas in carefully planned experiments. As we retrace his work, you will recognise the key elements of the scientific process that were introduced in Concept 1.3.

Mendel's Experimental, Quantitative Approach

Mendel grew up on his parents' small farm in a region of Austria that is now part of the Czech Republic. In this agricultural area, Mendel received agricultural training in school along with a basic education. As an adolescent, Mendel overcame financial hardship and illness to excel in high school and, later, at the Olmutz Philosophical Institute.

In 1843, at the age of 21, Mendel entered an Augustinian monastery, a reasonable choice at that time for someone who valued the life of the mind. He considered becoming a teacher but failed the necessary examination. In 1851, he left the monastery to pursue two years of study in physics and chemistry at the University of Vienna. These were very important years for Mendel's development as a scientist, in large part due to the strong influence of two professors. One was the physicist Christian Doppler, who encouraged his students to learn science through experimentation and trained Mendel to use mathematics to help explain natural phenomena. The other was a botanist named Franz Unger, who aroused Mendel's interest in the causes of variation in plants.

After attending university, Mendel returned to the monastery and was assigned to teach at a local school, where several other instructors were enthusiastic about scientific research. In addition, his fellow monks shared a long-standing fascination with the breeding of plants. Around 1857, Mendel began breeding garden peas in the abbey garden to study inheritance. Although the question of heredity had long been a focus of curiosity at the monastery, Mendel's fresh approach allowed him to deduce principles that had remained elusive to others.

One reason Mendel probably chose to work with peas is that there are many varieties. For example, one variety has purple flowers, while another variety has white flowers. A heritable feature that varies among individuals, such as flower colour, is called a **character**. Each variant for a character, such as purple or white colour for flowers, is called a **trait**.

Other advantages of using peas are their short generation time and the large number of offspring from each mating. Furthermore, Mendel could strictly control mating between plants (**Figure 14.2**). Each pea flower has both pollen-producing organs (stamens) and an egg-bearing organ (carpel). In nature, pea plants usually self-fertilise: Pollen grains from the stamens land on the carpel of the same flower, and sperm released from the pollen grains fertilise eggs present in the carpel.* To achieve

▼ **Figure 14.2** Research Method

Crossing Pea Plants

Application By crossing, or mating, two true-breeding varieties of an organism, scientists can study patterns of inheritance. In this example, Mendel crossed pea plants that varied in flower colour.

Technique

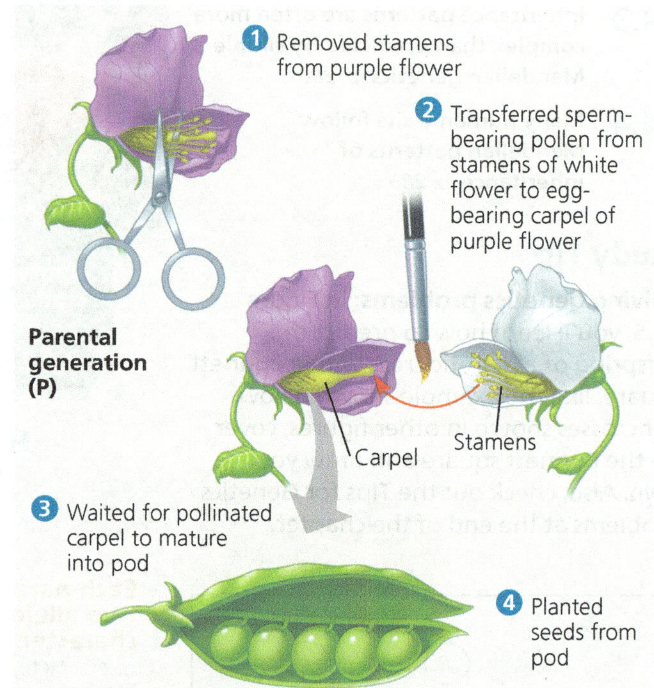

Results When pollen from a white flower was transferred to a purple flower, the first-generation hybrids all had purple flowers. The result was the same for the reciprocal cross, which involved the transfer of pollen from purple flowers to white flowers.

*As you learned in Figure 13.6b, meiosis in plants produces spores, not gametes. In flowering plants like the pea, each spore develops into a microscopic haploid gametophyte that contains only a few cells and is located on the parent plant. The gametophyte produces sperm, in pollen grains, and eggs, in the carpel. For simplicity, we will not include the gametophyte stage in our discussion of fertilisation in plants.

cross-pollination of two plants, Mendel removed the immature stamens of a plant before they produced pollen and then dusted pollen from another plant onto the altered flowers (see Figure 14.2). Each resulting zygote then developed into a plant embryo encased in a seed (a pea). His method allowed Mendel to always be sure of the parentage of new seeds.

Mendel chose to track only those characters that occurred in two distinct, alternative forms, such as purple or white flower colour. He also made sure that he started his experiments with varieties that were **true-breeding**—that is, over many generations of self-pollination, these plants had produced only the same variety as the parent plant. For example, a plant with purple flowers is true-breeding if the seeds produced by self-pollination in successive generations all give rise to plants that also have purple flowers.

In a typical breeding experiment, Mendel cross-pollinated two contrasting, true-breeding pea varieties—for example, purple-flowered plants and white-flowered plants (see Figure 14.2). This mating, or *crossing*, of two true-breeding varieties is called **hybridisation**. The true-breeding parents are referred to as the **P generation** (parental generation), and their hybrid offspring are the **F_1 generation** (first filial generation, the word *filial* from the Latin word for "son"). Allowing these F_1 hybrids to self-pollinate (or to cross-pollinate with other F_1 hybrids) produces an **F_2 generation** (second filial generation). Mendel usually followed traits for at least the P, F_1, and F_2 generations. Had he stopped his experiments with the F_1 generation, the basic patterns of inheritance would have eluded him. Mendel's quantitative analysis of the F_2 plants from thousands of genetic crosses like these allowed him to deduce two fundamental principles of heredity, now called the law of segregation and the law of independent assortment.

The Law of Segregation

The explanation of heredity most widely in favour during the 1800s was the "blending" hypothesis, the idea that genetic material contributed by the two parents mixes just as blue and yellow paints blend to make green. This hypothesis predicts that over many generations, a freely mating population will give rise to a uniform population of individuals, something we don't see.

If the blending model were correct, the F_1 hybrids from Mendel's cross between purple-flowered and white-flowered pea plants would have been pale purple flowers, a trait intermediate between those of the P generation. Notice in Figure 14.2 that the experiment produced a very different result: All the F_1 offspring had flowers of the same colour as the purple-flowered parents. What happened to the white-flowered plants' genetic contribution to the hybrids? If it were lost, then the F_1 plants could produce only purple-flowered offspring in the F_2 generation. But when Mendel allowed the F_1 plants to self- or cross-pollinate and planted their seeds, the white-flower trait reappeared in the F_2 generation. The blending hypothesis is also inconsistent with this reappearance of traits after they've skipped a generation.

Mendel used very large sample sizes and kept accurate records of his results: 705 of the F_2 plants had purple flowers, and 224 had white flowers. These data fit a ratio of approximately three purple to one white (**Figure 14.3**). Mendel reasoned that the heritable factor for white flowers did not disappear in the F_1 plants but was somehow hidden, or masked,

▼ **Figure 14.3 Inquiry**

When F_1 hybrid pea plants self- or cross-pollinate, which traits appear in the F_2 generation?

Experiment Mendel crossed true-breeding purple-flowered plants and white-flowered plants (crosses are symbolised by ×). The resulting F_1 hybrids were allowed to self-pollinate or were cross-pollinated with other F_1 hybrids. The F_2 generation plants were then observed for flower colour.

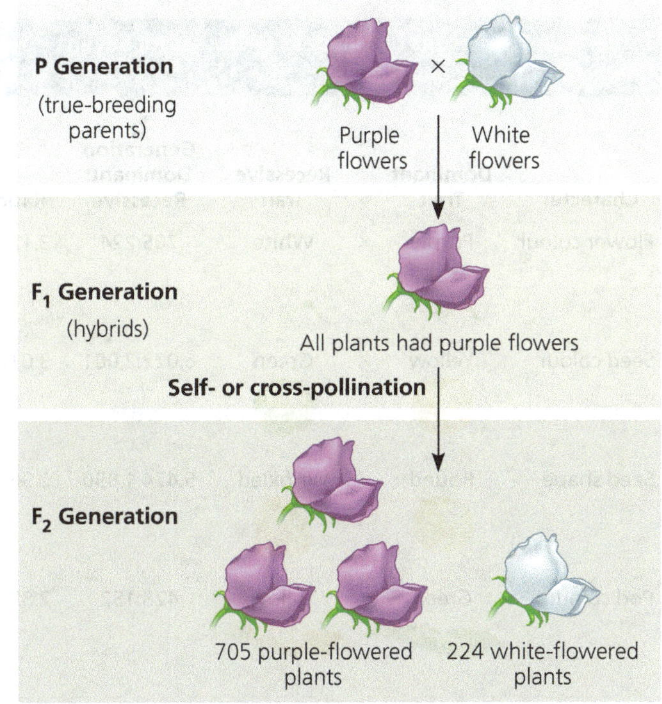

Results Both purple-flowered and white-flowered plants appeared in the F_2 generation, in a ratio of approximately 3:1.

Conclusion The "heritable factor" for the recessive trait (white flowers) had not been destroyed, deleted, or "blended" in the F_1 generation but was merely masked by the presence of the factor for purple flowers, which is the dominant trait.

Data from G. Mendel, Experiments in plant hybridization, *Proceedings of the Natural History Society of Brünn* 4:3–47 (1866).

WHAT IF? *If you mated two purple-flowered plants from the P generation, what ratio of traits would you expect to observe in the offspring? Explain. What might Mendel have concluded if he stopped his experiment after the F_1 generation?*

when the purple-flower factor was present. In Mendel's terminology, purple flower colour is a *dominant* trait, and white flower colour is a *recessive* trait. The reappearance of white-flowered plants in the F_2 generation was evidence that the heritable factor causing white flowers had not been diluted or destroyed by coexisting with the purple-flower factor in the F_1 hybrids.

Mendel observed the same pattern of inheritance in six other characters, each represented by two distinctly different traits **(Table 14.1)**. For example, when Mendel crossed a true-breeding variety that produced smooth, round pea seeds with one that produced wrinkled seeds, all the F_1 hybrids produced round seeds; this is the dominant trait for seed shape. In the F_2 generation, approximately 75% of the seeds were round and 25% were wrinkled—a 3:1 ratio, as in Figure 14.3.

Now let's see how Mendel deduced the law of segregation from his experimental results. In the discussion that follows, we will use modern terms instead of some of the terms used by Mendel. (For example, we'll use "gene" instead of Mendel's "heritable factor.")

Table 14.1 The Results of Mendel's F_1 Crosses for Seven Characters in Pea Plants

Character	Dominant Trait	×	Recessive Trait	F_2 Generation Dominant: Recessive	Ratio
Flower colour	Purple	×	White	705:224	3.15:1
Seed colour	Yellow	×	Green	6,022:2,001	3.01:1
Seed shape	Round	×	Wrinkled	5,474:1,850	2.96:1
Pod colour	Green	×	Yellow	428:152	2.82:1
Pod shape	Inflated	×	Constricted	882:299	2.95:1
Flower position	Axial	×	Terminal	651:207	3.14:1
Stem length	Tall	×	Dwarf	787:277	2.84:1

Mendel's Model

Mendel developed a model to explain the 3:1 inheritance pattern that he consistently observed among the F_2 offspring in his pea experiments. We'll now explore four related concepts making up Mendel's model, the fourth of which is the law of segregation.

First, *alternative versions of genes account for variations in inherited characters.* The gene for flower colour in pea plants, for example, exists in two versions, one for purple flowers and the other for white flowers. These alternative versions of a gene are called **alleles**. Today, we can relate this concept to chromosomes and DNA. As shown in **Figure 14.4**, each gene is a sequence of nucleotides at a specific place, or locus, along a particular chromosome. The DNA at that locus, however, can vary slightly in its nucleotide sequence. This variation in information content can affect the function of the encoded protein and thus an inherited character of the organism. The purple-flower allele and the white-flower allele are two DNA sequence variations possible at the flower-colour locus on a pea plant's chromosomes. The purple-flower allele sequence allows synthesis of purple pigment, and the white-flower allele sequence does not.

Second, *for each character, an organism inherits two versions (that is, two alleles) of a gene, one from each parent.* Remarkably, Mendel made this deduction without knowing about the role, or even the existence, of chromosomes. Each somatic cell in a diploid organism has two sets of chromosomes, one set inherited from each parent (see Figure 13.4). Thus, a genetic locus is actually represented twice in a diploid cell, once on each homologue of a specific pair of chromosomes. The two alleles at a particular locus may be identical, as in the true-breeding plants of Mendel's P generation. Or the alleles may differ, as in the F_1 hybrids (see Figure 14.4).

Third, *if the two alleles at a locus differ, then one, the* ***dominant allele****, determines the organism's appearance; the other, the* ***recessive allele****, has no noticeable effect on the organism's appearance.* Accordingly, Mendel's F_1 plants had purple flowers because the allele for that trait is dominant and the allele for white flowers is recessive.

The fourth and final part of Mendel's model, the **law of segregation**, states that *the two alleles for a heritable character segregate (in other words, separate from each other) during gamete formation and end up in different gametes.* Thus, an egg or a sperm gets only one of the two alleles that are present in the diploid cells of the organism making the gamete. In terms of chromosomes, this segregation corresponds to the distribution of copies of the two members of a pair of homologous chromosomes to different gametes in meiosis (see Figure 13.7). Note that if an organism has identical alleles for a particular character, then that allele is present in all gametes. Because it is the only allele that can be passed on to offspring when the plant self-pollinates, the offspring always look the same as their parents in regard to that characteristic; this explains why these plants are true-breeding. But if different alleles are

▶ **Figure 14.4 Alleles, alternative versions of a gene.** This diagram shows a pair of homologous chromosomes in an F_1 hybrid pea plant, with the actual DNA sequence from the flower-colour allele of each chromosome. The paternally inherited chromosome (blue) has an allele for purple flowers, which codes for a protein that indirectly controls synthesis of purple pigment. The maternally inherited chromosome (red) has an allele for white flowers, which results in no functional protein being made.

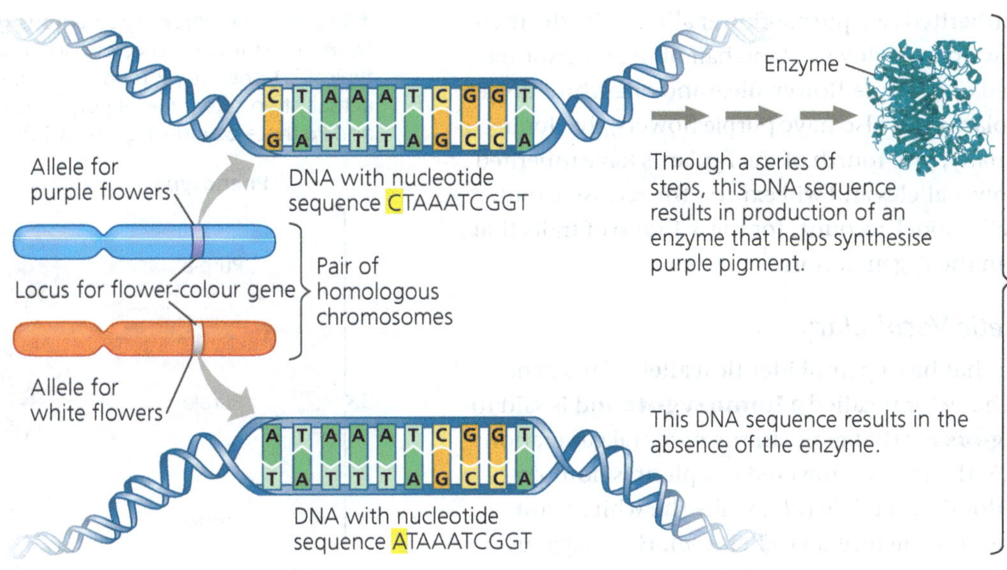

present, as in the F_1 hybrids, then 50% of the gametes receive the dominant allele and 50% receive the recessive allele.

Does Mendel's segregation model account for the 3:1 ratio he observed in the F_2 generation of his numerous crosses? For the flower-colour character, the model predicts that the two different alleles present in an F_1 individual will segregate into gametes such that half the gametes will have the purple-flower allele and half will have the white-flower allele. During self-pollination, gametes of each class unite randomly. An egg with a purple-flower allele has an equal chance of being fertilised by a sperm with a purple-flower allele or by a sperm with a white-flower allele. Since the same is true for an egg with a white-flower allele, there are four equally likely combinations of sperm and egg. **Figure 14.5** illustrates these combinations using a **Punnett square**, a handy diagrammatic device for predicting the allele composition of offspring from a cross between individuals of known genetic makeup. Notice that we use a capital letter to symbolise a dominant allele and a lowercase letter for a recessive allele. In our example, P is the purple-flower allele, and p is the white-flower allele; it is often useful as well to be able to refer to the gene itself as the P/p gene.

In the F_2 offspring, what colour will the flowers be? One-fourth of the

▼ **Figure 14.5 Mendel's law of segregation.** This diagram shows the genetic makeup of the generations in Figure 14.3. It illustrates Mendel's model for inheritance of the alleles of a single gene. Each plant has two alleles for the gene controlling flower colour, one allele inherited from each of the plant's parents. To construct a Punnett square that predicts the F_2 generation offspring, we list all the possible gametes from one parent (here, the F_1 female) along the left side of the square and all the possible gametes from the other parent (here, the F_1 male) along the top. The boxes represent the offspring resulting from all the possible unions of male and female gametes.

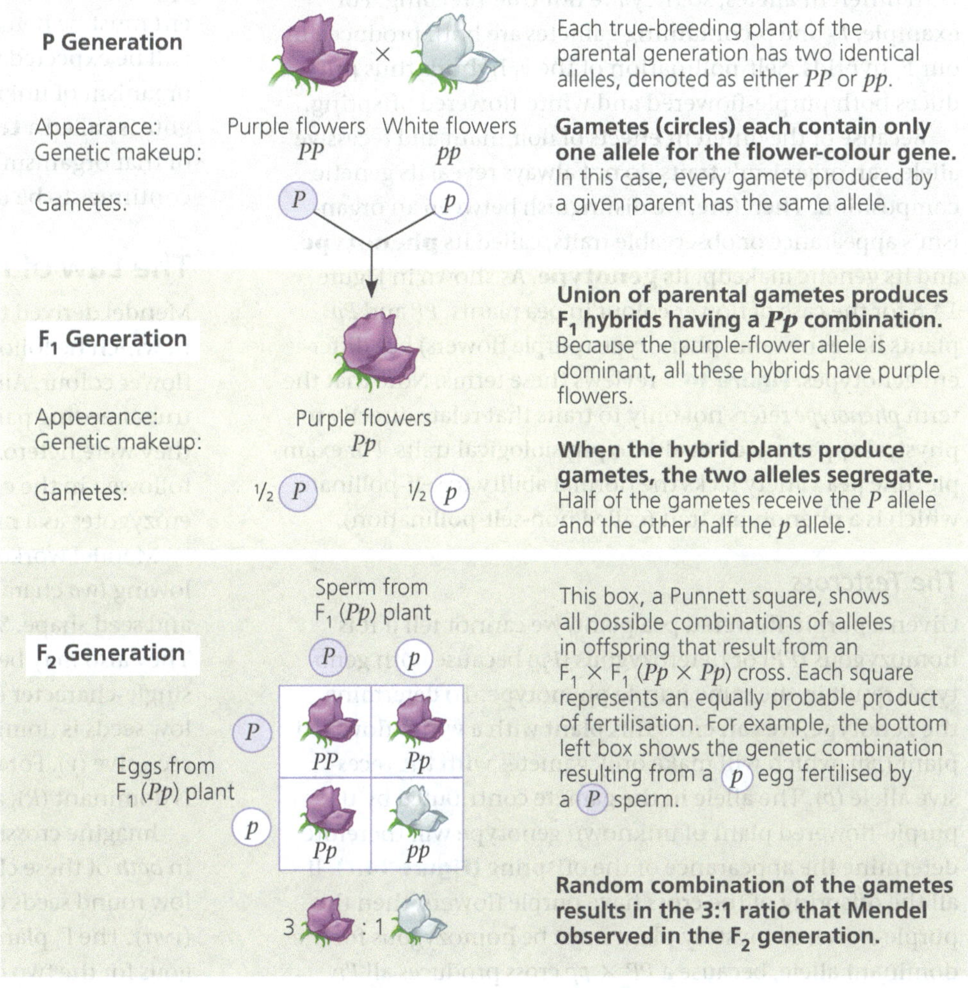

plants have inherited two purple-flower alleles; clearly, these plants will have purple flowers. One-half of the F_2 offspring have inherited one purple-flower allele and one white-flower allele; these plants will also have purple flowers, the dominant trait. Finally, one-fourth of the F_2 plants have inherited two white-flower alleles and will express the recessive trait. Thus, Mendel's model accounts for the 3:1 ratio of traits that he observed in the F_2 generation.

Useful Genetic Vocabulary

An organism that has a pair of identical alleles for a gene encoding a character is called a **homozygote** and is said to be **homozygous** for that gene. In the parental generation in Figure 14.5, the purple-flowered pea plant is homozygous for the dominant allele (*PP*), while the white plant is homozygous for the recessive allele (*pp*). Homozygous plants "breed true" because all of their gametes contain the same allele—either *P* or *p* in this example. If we cross dominant homozygotes with recessive homozygotes, every offspring will have two different alleles—*Pp* in the case of the F_1 hybrids of our flower-colour experiment (see Figure 14.5). An organism that has two different alleles for a gene is called a **heterozygote** and is said to be **heterozygous** for that gene. Unlike homozygotes, heterozygotes produce gametes with different alleles, so they are not true-breeding. For example, *P*- and *p*-containing gametes are both produced by our F_1 hybrids. Self-pollination of the F_1 hybrids thus produces both purple-flowered and white-flowered offspring.

Because of the different effects of dominant and recessive alleles, an organism's traits do not always reveal its genetic composition. Therefore, we distinguish between an organism's appearance or observable traits, called its **phenotype**, and its genetic makeup, its **genotype**. As shown in Figure 14.5 for the case of flower colour in pea plants, *PP* and *Pp* plants have the same phenotype (purple flowers) but different genotypes. **Figure 14.6** reviews these terms. Note that the term *phenotype* refers not only to traits that relate directly to physical appearance but also to physiological traits. For example, one pea variety lacks the normal ability to self-pollinate, which is a phenotypic trait (called non-self-pollination).

The Testcross

Given a purple-flowered pea plant, we cannot tell if it is homozygous (*PP*) or heterozygous (*Pp*) because both genotypes result in the same purple phenotype. To determine the genotype, we can cross this plant with a white-flowered plant (*pp*), which will make only gametes with the recessive allele (*p*). The allele in the gamete contributed by the purple-flowered plant of unknown genotype will therefore determine the appearance of the offspring **(Figure 14.7)**. If all the offspring of the cross have purple flowers, then the purple-flowered mystery plant must be homozygous for the dominant allele, because a *PP* × *pp* cross produces all *Pp* offspring. But if both the purple and the white phenotypes

▼ **Figure 14.6 Phenotype versus genotype.** Grouping F_2 offspring from a cross for flower colour according to phenotype results in the typical 3:1 phenotypic ratio. In terms of genotype, however, there are actually two categories of purple-flowered plants, *PP* (homozygous) and *Pp* (heterozygous), giving a 1:2:1 genotypic ratio.

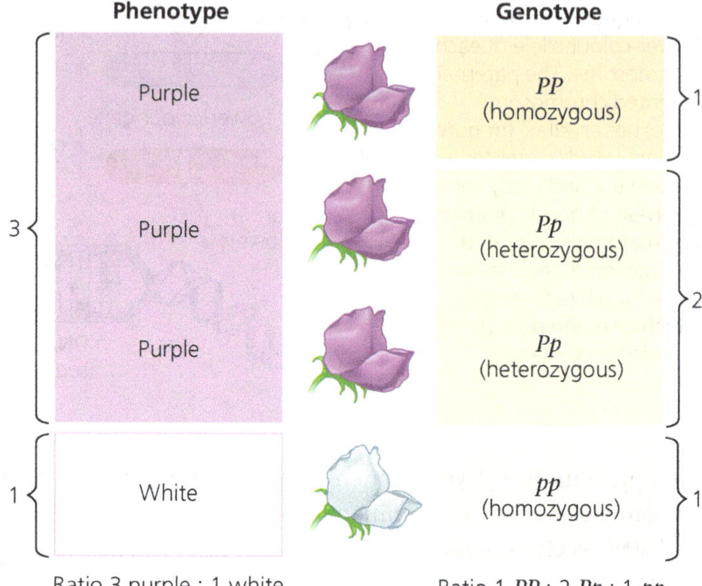

appear among the offspring, then the purple-flowered parent must be heterozygous. The offspring of a *Pp* × *pp* cross will be expected to have a 1:1 phenotypic ratio. Breeding an organism of unknown genotype with a recessive homozygote is called a **testcross** because it can reveal the genotype of that organism. The testcross was devised by Mendel and continues to be used by geneticists.

The Law of Independent Assortment

Mendel derived the law of segregation from experiments in which he followed only a *single* character, such as flower colour. All the F_1 progeny produced in his crosses of true-breeding parents were **monohybrids**, meaning that they were heterozygous for the one particular character being followed in the cross. We refer to a cross between such heterozygotes as a **monohybrid cross**.

Mendel worked out the second law of inheritance by following *two* characters at the same time, such as seed colour and seed shape. Seeds (peas) may be either yellow or green. They also may be either round (smooth) or wrinkled. From single-character crosses, Mendel knew that the allele for yellow seeds is dominant (*Y*), and the allele for green seeds is recessive (*y*). For the seed-shape character, the allele for round is dominant (*R*), and the allele for wrinkled is recessive (*r*).

Imagine crossing two true-breeding pea varieties that differ in *both* of these characters—a cross between a plant with yellow round seeds (*YYRR*) and a plant with green wrinkled seeds (*yyrr*). The F_1 plants will be **dihybrids**, individuals heterozygous for the two characters being followed in the cross (*YyRr*). But are these two characters transmitted from parents to

▼ Figure 14.7 Research Method

The Testcross

Application An organism that shows a dominant trait in its phenotype, such as purple flowers in pea plants, can be either homozygous for the dominant allele or heterozygous. To determine the organism's genotype, geneticists can perform a testcross.

Technique In a testcross, the individual with the unknown genotype is crossed with a homozygous individual expressing the recessive trait (white flowers in this example), and Punnett squares are used to predict the possible outcomes.

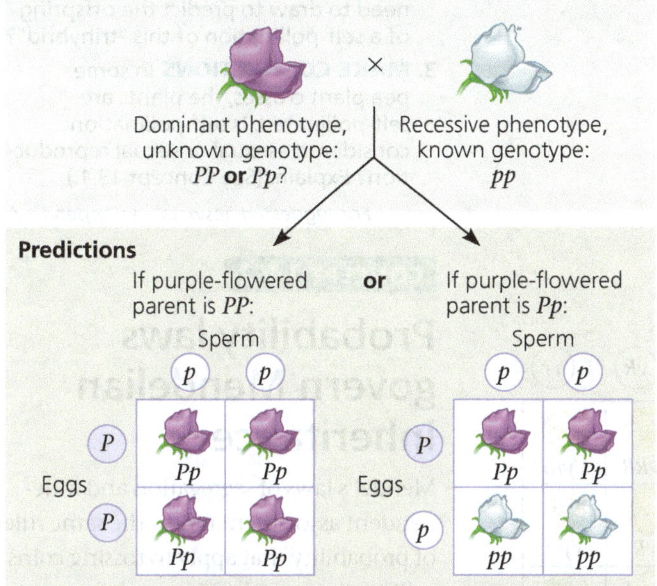

Results Matching the results to either prediction identifies the unknown parental genotype (either *PP* or *Pp* in this example). In this testcross, we transferred pollen from a white-flowered plant to the carpels of a purple-flowered plant; the opposite (reciprocal) cross would have led to the same results.

offspring as a package? That is, will the *Y* and *R* alleles always stay together, generation after generation? Or are seed colour and seed shape inherited independently? **Figure 14.8** shows how a **dihybrid cross**, a cross between F_1 dihybrids, can determine which of these two hypotheses is correct.

The F_1 plants, of genotype *YyRr*, exhibit both dominant phenotypes, yellow seeds with round shapes, no matter which hypothesis is correct. The key step in the experiment is to see what happens when F_1 plants self-pollinate and produce F_2 offspring. If the hybrids must transmit their alleles in the same combinations in which the alleles were inherited from the P generation, then the F_1 hybrids will produce only two classes of gametes: *YR* and *yr*. As shown on the left side of Figure 14.8, this "dependent assortment" hypothesis predicts that the phenotypic ratio of the F_2 generation will be 3:1, just as in a monohybrid cross:

The alternative hypothesis is that the two pairs of alleles segregate independently of each other. In other words, genes are packaged into gametes in all possible allelic combinations, as long as each gamete has one allele for each gene (see Figure 13.11). In our example, an F_1 plant will produce four classes of gametes in equal quantities: *YR*, *Yr*, *yR*, and *yr*. If sperm of the four classes fertilise eggs of the four classes, there will be 16 (4×4) equally probable ways in which the alleles can combine in the F_2 generation, as shown on the right side of Figure 14.8. These combinations result in four phenotypic categories with a ratio of 9:3:3:1 (9 yellow round to 3 green round to 3 yellow wrinkled to 1 green wrinkled):

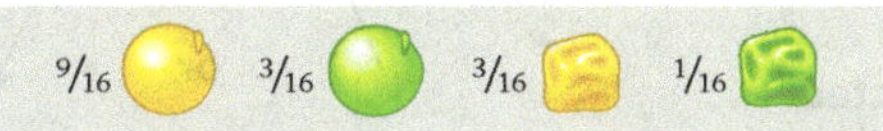

When Mendel did the experiment and classified the F_2 offspring, his results were close to the predicted 9:3:3:1 phenotypic ratio, supporting the hypothesis that the alleles for one gene—the gene for seed colour, for example—segregate into gametes independently of the alleles of any other gene, such as the gene for seed shape.

Mendel tested his seven pea characters in various dihybrid combinations and always observed a 9:3:3:1 phenotypic ratio in the F_2 generation. Is this consistent with the 3:1 phenotypic ratio seen for the monohybrid cross shown in Figure 14.5? To answer this question, count the number of yellow and green peas, ignoring shape, and calculate the ratio. The results of Mendel's dihybrid experiments are the basis for what we now call the **law of independent assortment**, which states that *two or more genes assort independently—that is, each pair of alleles segregates independently of any other pair of alleles—during gamete formation.*

This law applies only to genes (allele pairs) located on different chromosomes (that is, on chromosomes that are not homologous) or, alternatively, to genes that are very far apart on the same chromosome. (This will be explained in Concept 15.3, along with the more complex inheritance patterns of genes located near each other, alleles of which tend to be inherited together.) All the pea characters Mendel chose for analysis were controlled by genes on different chromosomes or were far apart on the same chromosome; this situation greatly simplified interpretation of his multicharacter pea crosses. All the examples we consider in the rest of this chapter involve genes located on different chromosomes.

▼ Figure 14.8 Inquiry

Do the alleles for one character segregate into gametes dependently or independently of the alleles for a different character?

Experiment To follow the characters of seed colour and seed shape through the F_2 generation, Mendel crossed a true-breeding plant with yellow round seeds with a true-breeding plant with green wrinkled seeds, producing dihybrid F_1 plants. Self-pollination of the F_1 dihybrids produced the F_2 generation. The two hypotheses (dependent and independent "assortment" of the two genes) predict different phenotypic ratios.

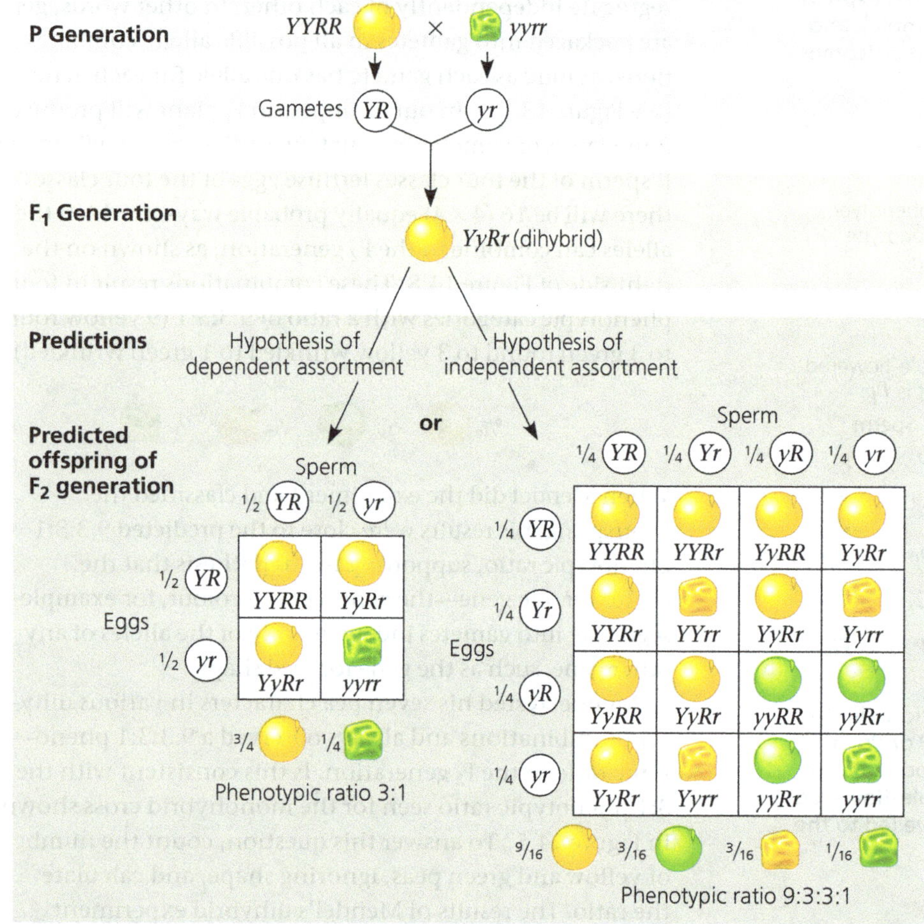

Results

Conclusion The results support the hypothesis of independent assortment, the only one that predicts two newly observed phenotypes: green round seeds and yellow wrinkled seeds (see the right-hand Punnett square). The alleles for each gene segregate independently of those of the other, and the two genes are said to assort independently.

Data from G. Mendel, Experiments in plant hybridization, *Proceedings of the Natural History Society of Brünn* 4:3–47 (1866).

WHAT IF? *Suppose Mendel had transferred pollen from an F_1 plant to the carpel of a plant that was homozygous recessive for both genes. Set up the cross and draw Punnett squares that predict the offspring for both hypotheses. Would this cross have supported the hypothesis of independent assortment equally well?*

CONCEPT CHECK 14.1

1. **DRAW IT** Pea plants heterozygous for flower position and stem length (*AaTt*) are allowed to self-pollinate, and 400 of the resulting seeds are planted. Draw a Punnett square for this cross. How many offspring would be predicted to have terminal flowers and be dwarf? (See Table 14.1.)
2. **WHAT IF?** List all gametes that could be made by a pea plant heterozygous for seed colour, seed shape, and pod shape (*YyRrIi*; see Table 14.1). How large a Punnett square would you need to draw to predict the offspring of a self-pollination of this "trihybrid"?
3. **MAKE CONNECTIONS** In some pea plant crosses, the plants are self-pollinated. Is self-pollination considered asexual or sexual reproduction? Explain. (See Concept 13.1.)

For suggested answers, see Appendix A.

CONCEPT **14.2**

Probability laws govern Mendelian inheritance

Mendel's laws of segregation and independent assortment reflect the same rules of probability that apply to tossing coins, rolling dice, and drawing cards from a deck. The probability scale ranges from 0 to 1. An event that is certain to occur has a probability of 1, while an event that is certain *not* to occur has a probability of 0. With a coin that has heads on both sides, the probability of tossing heads is 1, and the probability of tossing tails is 0. With a normal coin, the chance of tossing heads is $\frac{1}{2}$, and the chance of tossing tails is $\frac{1}{2}$. The probability of drawing the ace of spades from a 52-card deck is $\frac{1}{52}$. The probabilities of all possible outcomes for an event must add up to 1. With a deck of cards, the chance of picking a card other than the ace of spades is $\frac{51}{52}$.

Tossing a coin illustrates an important lesson about probability. For every toss, the probability of heads is $\frac{1}{2}$. The outcome of any particular toss is unaffected by what has happened on previous trials. We refer to phenomena such as coin tosses as independent events. Each toss of a coin, whether done sequentially with one coin or simultaneously with many, is independent of every other toss. And like two separate coin tosses, the alleles of one gene segregate into gametes independently of another gene's alleles (the law of independent assortment). We'll now look at two basic rules of probability that help us predict the outcome of the fusion of such gametes in simple monohybrid crosses and more complicated crosses as well.

The Multiplication and Addition Rules Applied to Monohybrid Crosses

How do we determine the probability that two or more independent events will occur together in some specific combination? For example, what is the chance that two coins tossed simultaneously will both land heads up? The **multiplication rule** states that to determine the probability of one event *and* the other occurring, we multiply the probability of one event (one coin coming up heads) by the probability of the other event (the other coin coming up heads). By the multiplication rule, then, the probability that both coins will land heads up is $\frac{1}{2} \times \frac{1}{2} = \frac{1}{4}$.

We can apply the same reasoning to an F_1 monohybrid cross. With seed shape in pea plants as the heritable character, the genotype of F_1 plants is *Rr*. Segregation in a heterozygous plant is like flipping a coin in terms of calculating the probability of each outcome: Each egg produced has a $\frac{1}{2}$ chance of carrying the dominant allele (*R*) and a $\frac{1}{2}$ chance of carrying the recessive allele (*r*). The same odds apply to each sperm cell produced. For a particular F_2 plant to have wrinkled seeds, the recessive trait, both the egg and the sperm that come together must carry the *r* allele. The probability that an *r* allele will be present in the egg *and* in the sperm at fertilisation is found by multiplying $\frac{1}{2}$ (the probability that the egg will have an *r*) $\times$ $\frac{1}{2}$ (the probability that the sperm will have an *r*). Thus, the multiplication rule tells us that the probability of an F_2 plant having wrinkled seeds (*rr*) is $\frac{1}{4}$ **(Figure 14.9)**. Likewise, the probability of an F_2 plant carrying both dominant alleles for seed shape (*RR*) is $\frac{1}{4}$.

To figure out the probability that an F_2 plant from a monohybrid cross will be heterozygous rather than homozygous, we need to invoke a second rule. Notice in Figure 14.9 that the dominant allele can come from the egg and the recessive allele from the sperm, or vice versa. That is, F_1 gametes can combine to produce *Rr* offspring in two *mutually exclusive* ways: For any particular heterozygous F_2 plant, the dominant allele can come from the egg *or* the sperm, but not from both. According to the **addition rule**, the probability that any one of two or more mutually exclusive events (one event *or* the other) will occur is calculated by adding their individual probabilities. As we have just seen, the multiplication rule gives us the individual probabilities that we will now add together. The probability for one possible way of obtaining an F_2 heterozygote—the dominant allele from the egg and the recessive allele from the sperm—is $\frac{1}{4}$. The probability for the other possible way—the recessive allele from the egg and the dominant allele from the sperm—is also $\frac{1}{4}$ (see Figure 14.9). Using the rule of addition, then, we can calculate the probability of an F_2 heterozygote as $\frac{1}{4} + \frac{1}{4} = \frac{1}{2}$. (Note that use of the word *and* in the statement is a clue that you should use the multiplication rule, while *or* suggests use of the addition rule.)

▼ **Figure 14.9 Segregation of alleles and fertilisation as chance events.** When a heterozygote (*Rr*) forms gametes, whether a particular gamete ends up with an *R* or an *r* is like the toss of a coin. We can determine the probability for any genotype among the offspring of two heterozygotes by multiplying together the individual probabilities of an egg and sperm having a particular allele (*R* or *r* in this example).

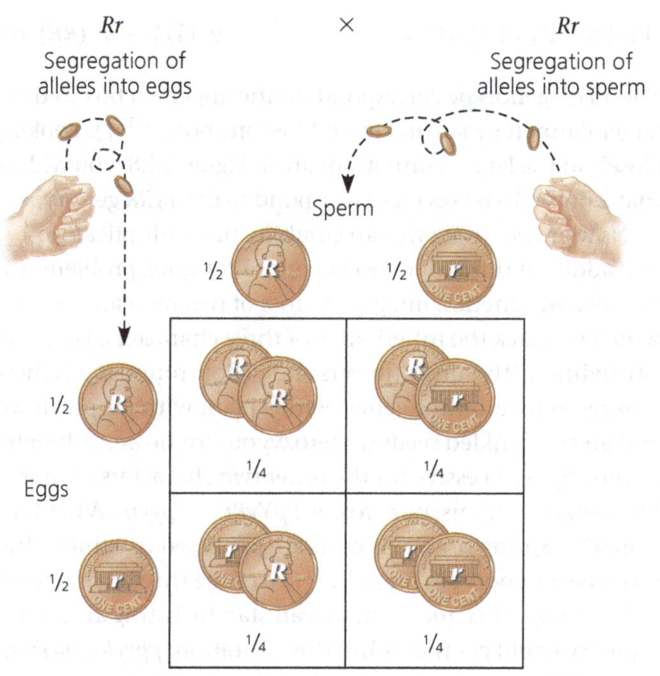

Solving Complex Genetics Problems with the Rules of Probability

We can also apply the rules of probability to predict the outcome of crosses involving multiple characters. Recall that each allelic pair (gene) segregates independently during gamete formation (the law of independent assortment). Thus, a dihybrid or other multicharacter cross is equivalent to two or more independent monohybrid crosses occurring simultaneously. By applying what we have learned about monohybrid crosses, we can determine the probability of specific genotypes occurring in the F_2 generation without having to construct unwieldy Punnett squares.

Consider the dihybrid cross between *YyRr* heterozygotes shown in Figure 14.8. We will focus first on the seed-colour character. For a monohybrid cross of *Yy* plants, we can use a simple Punnett square to determine that the probabilities of the offspring genotypes are $\frac{1}{4}$ for *YY*, $\frac{1}{2}$ for *Yy*, and $\frac{1}{4}$ for *yy*. We can draw a second Punnett square to determine that the same probabilities apply to the offspring genotypes for seed shape: $\frac{1}{4}$ *RR*, $\frac{1}{2}$ *Rr*, and $\frac{1}{4}$ *rr*. Knowing these probabilities, we can simply use the multiplication rule to determine the probability of each of the genotypes in the F_2 generation. To give two examples, the calculations for finding the probabilities of two of the possible F_2 genotypes (*YYRR* and *YyRR*) are shown below:

Probability of *YYRR* = ¼ (probability of *YY*) × ¼ (*RR*) = 1/16

Probability of *YyRR* = ½ (*Yy*) × ¼ (*RR*) = ⅛

The *YYRR* genotype corresponds to the upper left box in the larger Punnett square in Figure 14.8 (one box = 1/16). Looking closely at the larger Punnett square in Figure 14.8, you will see that 2 of the 16 boxes (⅛) correspond to the *YyRR* genotype.

Now let's see how we can combine the multiplication and addition rules to solve even more complex problems in Mendelian genetics. Imagine a cross of two pea varieties in which we track the inheritance of three characters. Let's cross a trihybrid with purple flowers and yellow round seeds (heterozygous for all three genes) with a plant with purple flowers and green wrinkled seeds (heterozygous for flower colour but homozygous recessive for the other two characters). Using Mendelian symbols, our cross is *PpYyRr* × *Ppyyrr*. What fraction of offspring from this cross are predicted to exhibit the recessive phenotypes for *at least two* of the three characters?

To answer this question, we can start by listing all genotypes we could get that fulfill this condition: *ppyyRr*, *ppYyrr*, *Ppyyrr*, *PPyyrr*, and *ppyyrr*. (Because the condition is *at least two* recessive traits, it includes the last genotype, which shows all three recessive traits.) Next, we calculate the probability for each of these genotypes resulting from our *PpYyRr* × *Ppyyrr* cross by multiplying together the individual probabilities for the allele pairs, just as we did in our dihybrid example. Note that in a cross involving heterozygous and homozygous allele pairs (for example, *Yy* × *yy*), the probability of heterozygous (*Yy*) offspring is ½ and the probability of homozygous (in this case, *yy*) offspring is ½. Finally, we use the addition rule to add the probabilities for all the different genotypes that fulfill the condition of at least two recessive traits resulting from our *PpYyRr* × *Ppyyrr* cross, as shown below:

ppyyRr	¼ (probability of *pp*) × ½ (*yy*) × ½ (*Rr*) = 1/16
ppYyrr	¼ (*pp*) × ½ (*Yy*) × ½ (*rr*) = 1/16
Ppyyrr	½ (*Pp*) × ½ (*yy*) × ½ (*rr*) = 2/16
PPyyrr	¼ (*PP*) × ½ (*yy*) × ½ (*rr*) = 1/16
ppyyrr	¼ (*pp*) × ½ (*yy*) × ½ (*rr*) = 1/16
Chance of *at least two* recessive traits	= 6/16 = ⅜

In time, you'll be able to solve genetics problems faster by using the rules of probability than by filling in Punnett squares.

We cannot predict with certainty the exact numbers of progeny of different genotypes resulting from a genetic cross. But the rules of probability give us the *likelihood* of various outcomes. Usually, the larger the sample size, the closer the results will conform to our predictions. Mendel understood this statistical feature of inheritance and had a keen sense of the rules of chance. It was for this reason that he set up his experiments so as to generate, and then count, large numbers of offspring from his crosses.

CONCEPT CHECK 14.2

1. For any gene with a dominant allele *A* and recessive allele *a*, what proportions of the offspring from an *AA* × *Aa* cross are expected to be homozygous dominant, homozygous recessive, and heterozygous?
2. Two organisms, with genotypes *BbDD* and *BBDd*, are mated. Assuming independent assortment of the *B*/*b* and *D*/*d* genes, write the genotypes of all possible offspring from this cross and use the rules of probability to calculate the chance of each genotype occurring.
3. **WHAT IF?** Three characters (flower colour, seed colour, and pod shape) are considered in a cross between two pea plants: *PpYyIi* × *ppYyIi*. Using the rules of probability, determine the fraction of offspring predicted to be homozygous recessive for at least two of the three characters.

For suggested answers, see Appendix A.

CONCEPT 14.3

Inheritance patterns are often more complex than predicted by simple Mendelian genetics

In the 1900s, geneticists extended Mendelian principles not only to diverse organisms, but also to patterns of inheritance more complex than those described by Mendel. For the work that led to his two laws of inheritance, Mendel chose pea plant characters that turn out to have a relatively simple genetic basis: Each character is determined by one gene, for which there are only two alleles, one completely dominant and the other completely recessive. (There is one exception: Mendel's pod shape character is actually determined by two genes.) Not all heritable characters are determined so simply, and the relationship between genotype and phenotype is rarely so straightforward. Mendel himself realised that he could not explain the more complicated patterns he observed in crosses involving other pea characters or other plant species. This does not diminish the utility of Mendelian genetics, however, because the basic principles of segregation and independent assortment apply even to more complex patterns of inheritance. In this section, we will extend Mendelian genetics to hereditary patterns that were not reported by Mendel.

Extending Mendelian Genetics for a Single Gene

The inheritance of characters determined by a single gene deviates from simple Mendelian patterns when alleles are not completely dominant or recessive, when a particular gene has more than two alleles, or when a single gene produces multiple phenotypes. We will describe examples of each of these situations in this section.

Degrees of Dominance

Alleles can show different degrees of dominance and recessiveness in relation to each other. In Mendel's classic pea crosses, the F_1 offspring always looked like one of the two parental varieties because one allele in a pair showed **complete dominance** over the other. In such situations, the phenotypes of the heterozygote and the dominant homozygote are indistinguishable (see Figure 14.6).

For some genes, however, neither allele is completely dominant, and the F_1 hybrids have a phenotype somewhere between those of the two parental varieties. This phenomenon, called **incomplete dominance**, is seen when red snapdragons are crossed with white snapdragons: All the F_1 hybrids have pink flowers **(Figure 14.10)**. This third, intermediate phenotype results from flowers of the heterozygotes having less red pigment than the red homozygotes. (This is unlike the case of Mendel's pea plants, where the *Pp* heterozygotes make enough pigment for the flowers to be purple, indistinguishable from those of *PP* plants.)

At first glance, incomplete dominance of either allele seems to provide evidence for the blending hypothesis of inheritance, which would predict that the red or white trait could never reappear among offspring of the pink hybrids. In fact, interbreeding F_1 hybrids produces F_2 offspring with a phenotypic ratio of one red to two pink to one white. (Because heterozygotes have a separate phenotype from homozygotes, the genotypic and phenotypic ratios for the F_2 generation are the same, 1:2:1.) The segregation of the red-flower and white-flower alleles in the gametes produced by the pink-flowered plants confirms that the alleles for flower colour are heritable factors that maintain their identity in the hybrids; that is, that the factors are discrete rather than "blendable."

Another variation on dominance relationships between alleles is called **codominance**; in this variation, the two alleles each affect the phenotype in separate, distinguishable ways. For example, the human MN blood group is determined by codominant alleles for two specific molecules located on the surface of red blood cells, the M and N molecules. A single gene (L), for which two allelic variations are possible (L^M or L^N), determines the phenotype of this blood group. Individuals homozygous for the L^M allele (L^ML^M) have red blood cells with only M molecules; individuals homozygous for the L^N allele (L^NL^N) have red blood cells with only N molecules. But *both* M and N molecules are present on the red blood cells of individuals heterozygous for the *M* and *N* alleles (L^ML^N). Note that the MN phenotype is *not* intermediate between the M and N phenotypes, which distinguishes codominance from incomplete dominance. Rather, *both* M and N phenotypes are exhibited by heterozygotes, since both molecules are present.

▼ Figure 14.10 Incomplete dominance in snapdragon colour. When red snapdragons are crossed with white ones, the F_1 hybrids have pink flowers. Segregation of alleles into gametes of the F_1 plants results in an F_2 generation with a 1:2:1 ratio for both genotype and phenotype. Neither allele is dominant, so rather than using upper- and lowercase letters, we use the letter C with a superscript to indicate an allele for flower colour: C^R for red and C^W for white.

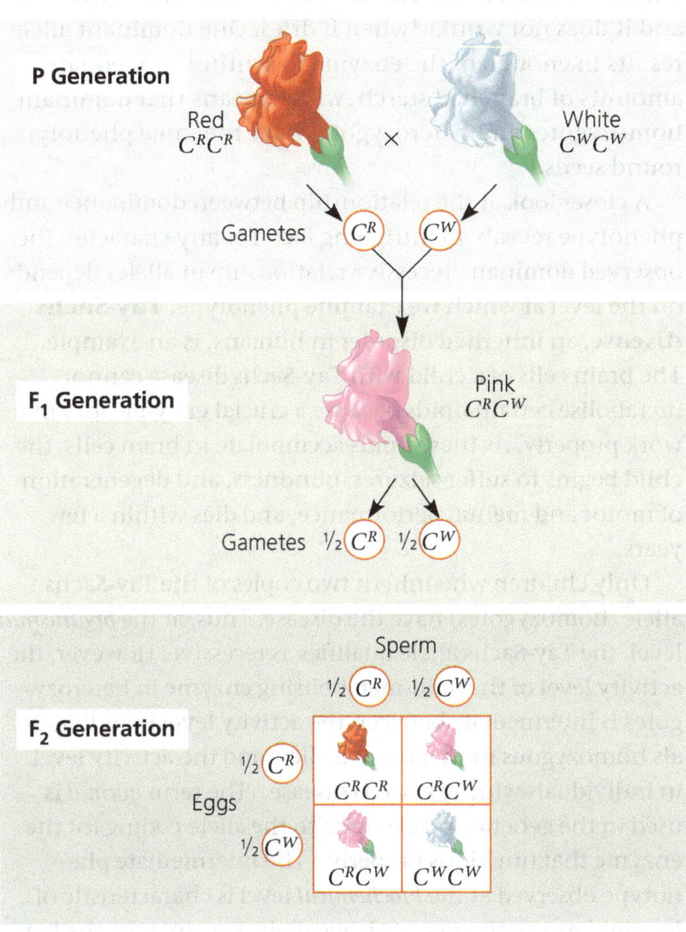

? *Suppose a classmate argues that this figure supports the blending hypothesis for inheritance. What might your classmate say, and how would you respond?*

The Relationship Between Dominance and Phenotype We've now seen that the relative effects of two alleles range from complete dominance of one allele to incomplete dominance of either allele to codominance of both alleles. It is important to understand that an allele is called *dominant* because it is seen in the phenotype, not because it somehow subdues a recessive allele. Alleles are simply variations in a gene's nucleotide sequence (see Figure 14.4). When a dominant allele coexists with a recessive allele in a heterozygote, they do not actually interact at all. It is in the pathway from genotype to phenotype that dominance and recessiveness come into play.

To illustrate the relationship between dominance and phenotype, we can use one of the characters Mendel studied—round versus wrinkled pea seed shape. The dominant allele (round) codes for an enzyme that helps convert an unbranched form of starch to a branched form in the seed. The recessive allele (wrinkled) codes for a defective form of this enzyme, leading to an accumulation of unbranched starch, which causes excess water to enter the seed by osmosis. Later, when the seed dries, it wrinkles. If a dominant allele is present, no excess water enters the seed and it does not wrinkle when it dries. One dominant allele results in enough of the enzyme to synthesise adequate amounts of branched starch, which means that dominant homozygotes and heterozygotes have the same phenotype: round seeds.

A closer look at the relationship between dominance and phenotype reveals an intriguing fact: For any character, the observed dominant/recessive relationship of alleles depends on the level at which we examine phenotype. **Tay-Sachs disease**, an inherited disorder in humans, is an example. The brain cells of a child with Tay-Sachs disease cannot metabolise certain lipids because a crucial enzyme does not work properly. As these lipids accumulate in brain cells, the child begins to suffer seizures, blindness, and degeneration of motor and mental performance, and dies within a few years.

Only children who inherit two copies of the Tay-Sachs allele (homozygotes) have the disease. Thus, at the *organismal* level, the Tay-Sachs allele qualifies as recessive. However, the activity level of the lipid-metabolising enzyme in heterozygotes is intermediate between the activity level in individuals homozygous for the normal allele and the activity level in individuals with Tay-Sachs disease. (The term *normal* is used in the genetic sense to refer to the allele coding for the enzyme that functions properly.) The intermediate phenotype observed at the *biochemical* level is characteristic of incomplete dominance of either allele. Fortunately, the heterozygote condition does not lead to disease symptoms, apparently because half the normal enzyme activity is sufficient to prevent lipid accumulation in the brain. Extending our analysis to yet another level, we find that heterozygous individuals produce equal numbers of normal and dysfunctional enzyme molecules. Thus, at the *molecular* level, the normal allele and the Tay-Sachs allele are codominant. As you can see, whether alleles appear to be completely dominant, incompletely dominant, or codominant depends on the level at which the phenotype is analysed.

Frequency of Dominant Alleles While you might assume that the dominant allele for a particular character would be more common than the recessive allele, this is not always the case. For an example of a rare dominant allele, about one baby in 1,000 in New Zealand or Australia is born with extra fingers or toes, a condition known as polydactyly. Some cases are caused by the presence of a dominant allele. The low frequency of these types of polydactyly indicates that the recessive allele, which results in five digits per appendage, is far more prevalent than the dominant allele in the population.

Multiple Alleles

Only two alleles exist for the pea characters that Mendel studied, but most genes have more than two alleles. The ABO blood groups in humans, for instance, are determined by the two alleles a person has of the blood group gene; the three possible alleles are I^A, I^B, and i. A person's blood group may be one of four types: A, B, AB, or O. These letters refer to two carbohydrates—A and B—that may be found attached to specific cell-surface molecules on red blood cells. An individual's blood cells may have carbohydrate A (type A blood), carbohydrate B (type B), both (type AB), or neither (type O), as shown in **Figure 14.11**, along with the relevant genotypes. Matching compatible blood groups is critical for safe blood transfusions (see Concept 43.3).

▼ **Figure 14.11 Multiple alleles for the ABO blood groups.** The four blood groups result from different combinations of three alleles.

(a) The three alleles for the ABO blood groups and their carbohydrates. Each allele codes for an enzyme that may add a specific carbohydrate (designated by the superscript on the allele and shown as a triangle or circle) to red blood cells.

Allele	I^A	I^B	i
Carbohydrate	A △	B ○	none

(b) Blood group genotypes and phenotypes. There are six possible genotypes, resulting in four different phenotypes.

Genotype	I^AI^A or I^Ai	I^BI^B or I^Bi	I^AI^B	ii
Red blood cell with surface carbohydrates		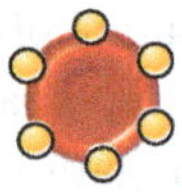		
Phenotype (blood group)	A	B	AB	O

VISUAL SKILLS *Based on the surface carbohydrate phenotype in part (b), what are the dominance relationships among the alleles?*

Pleiotropy

So far, we have treated Mendelian inheritance as though each gene affects only one phenotypic character. Most genes, however, have multiple phenotypic effects, a property called **pleiotropy** (from the Greek *pleion*, more). In humans, for example, pleiotropic alleles are responsible for the multiple symptoms associated with certain hereditary diseases, such as cystic fibrosis and sickle-cell disease, discussed later in this chapter. In the garden pea, the gene that determines flower colour also affects the colour of the coating on the outer surface of the seed, which can be grey or white. Given the intricate molecular and cellular interactions responsible for an organism's development and physiology, it isn't surprising that a single gene can affect a number of characters.

Extending Mendelian Genetics for Two or More Genes

Dominance relationships, multiple alleles, and pleiotropy all have to do with the effects of the alleles of a single gene. We now consider two situations in which two or more genes are involved in determining a particular phenotype. In the first case, called epistasis, one gene affects the phenotype of another because the two gene products interact; in the second case, called polygenic inheritance, multiple genes independently affect a single trait.

Epistasis

In **epistasis** (from the Greek for "standing upon"), the phenotypic expression of a gene at one locus alters that of a gene at a second locus. An example will help clarify this concept. In Labrador retrievers (commonly called "Labs"), black coat colour is dominant to brown. Let's designate *B* and *b* as the two alleles for this character. For a Lab to have brown fur, its genotype must be *bb*; these dogs are called chocolate Labs. But there is more to the story. A second gene determines whether or not pigment will be deposited in the hair. The dominant allele, symbolised by *E*, results in the deposition of either black or brown pigment, depending on the genotype at the first locus. But if the Lab is homozygous recessive for the second locus (*ee*), then the coat is yellow, regardless of the genotype at the black/brown locus (yellow Labs). In this case, the gene for pigment deposition (*E*/*e*) is said to be *epistatic to* the gene that codes for black or brown pigment (*B*/*b*).

What happens if we mate black Labs that are heterozygous for both genes (*BbEe*)? Although the two genes affect the same phenotypic character (coat colour), they follow the law of independent assortment. Thus, our breeding experiment represents an F_1 dihybrid cross, like those that produced a 9:3:3:1 ratio in Mendel's experiments. We can use a Punnett square to represent the genotypes of the F_2 offspring **(Figure 14.12)**. As a result of epistasis, the phenotypic ratio among the F_2 offspring is 9 black to 3 chocolate to 4 yellow Labs. Other types of epistatic interactions produce different ratios, but all are modified versions of 9:3:3:1.

▼ Figure 14.12 An example of epistasis. This Punnett square illustrates the genotypes and phenotypes predicted for offspring of matings between two black Labrador retrievers of genotype *BbEe*. The *E*/*e* gene, which is epistatic to the *B*/*b* gene coding for hair pigment, controls whether or not pigment of any colour will be deposited in the hair.

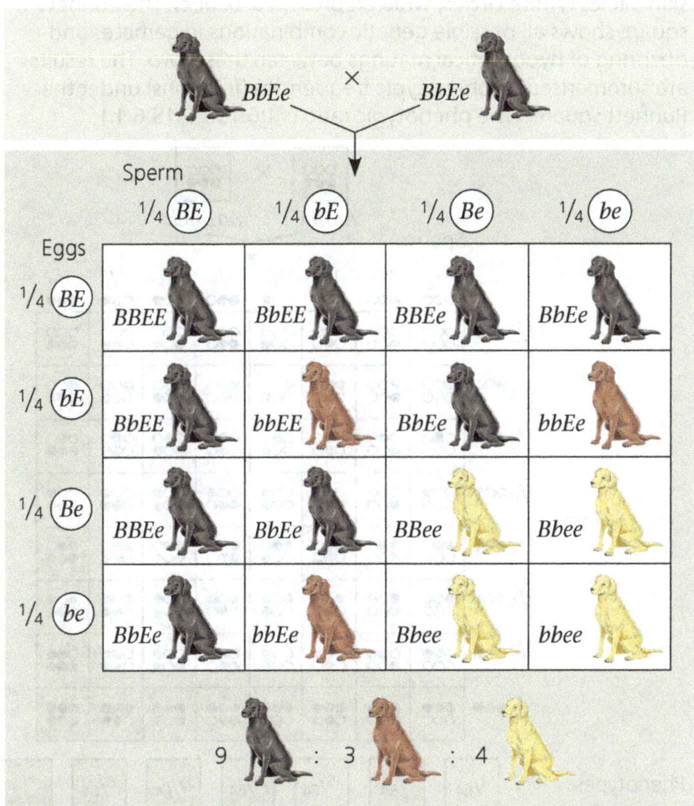

VISUAL SKILLS *Compare the four squares in the lower right of this Punnett square with those in Figure 14.8. Explain the genetic basis for the difference between the ratio (9:3:4) of phenotypes seen in this cross and the 9:3:3:1 ratio seen in Figure 14.8.*

Polygenic Inheritance

Mendel studied characters that could be classified on an either-or basis, such as purple versus white flower colour. But many characters, such as human skin colour and height, are not one of two discrete characters, but instead vary in the population in gradations along a continuum. These are called **quantitative characters**. Quantitative variation usually indicates **polygenic inheritance**, an additive effect of two or more genes on a single phenotypic character. (In a way, this is the converse of pleiotropy, where a single gene affects several phenotypic characters.) Height is a good example of polygenic inheritance: In 2014, a genomic study of over 250,000 people found almost 700 genetic variations in over 180 genes that affect height. Many variations were in or near genes involved in biochemical pathways affecting growth of the skeleton, but others were associated with genes not obviously related to growth. Another study in 2018, of 300,000 individuals, identified 124 genes that affect hair colour. Although simplifying can sometimes be valuable in understanding genetic principles, as we'll see shortly, the classic idea of a single gene determining eye colour or hair colour or other such traits is an oversimplification. (Earlobe attachment, an example used until a decade ago in this text, was shown in 2017 to be affected by nearly 50 genes!)

Skin pigmentation in humans is also controlled by many separately inherited genes—378 at latest count, many of which are involved in the production of melanin skin pigments. Here, we'll simplify the story in order to understand the concept of polygenic inheritance. Let's consider three genes, with a dark-skin allele for each gene (*A*, *B*, or *C*) contributing one "unit" of darkness (also a simplification) to the phenotype and being incompletely dominant to the other, light-skin allele (*a*, *b*, or *c*). In our model, an *AABBCC* person

▼ Figure 14.13 A simplified model for polygenic inheritance of skin colour. In this model, three separately inherited genes affect skin colour. (The reported number is actually 378 genes.) The heterozygous individuals (*AaBbCc*) represented by the two rectangles at the top of this figure each carry three dark-skin alleles (black circles, which represent *A*, *B*, or *C*) and three light-skin alleles (white circles, which represent *a*, *b*, or *c*). The Punnett square shows all possible genetic combinations in gametes and offspring of hypothetical matings between these two. The results are summarised by phenotypic frequencies (fractions) under the Punnett square. (The phenotypic ratio is 1:6:15:20:15:6:1.)

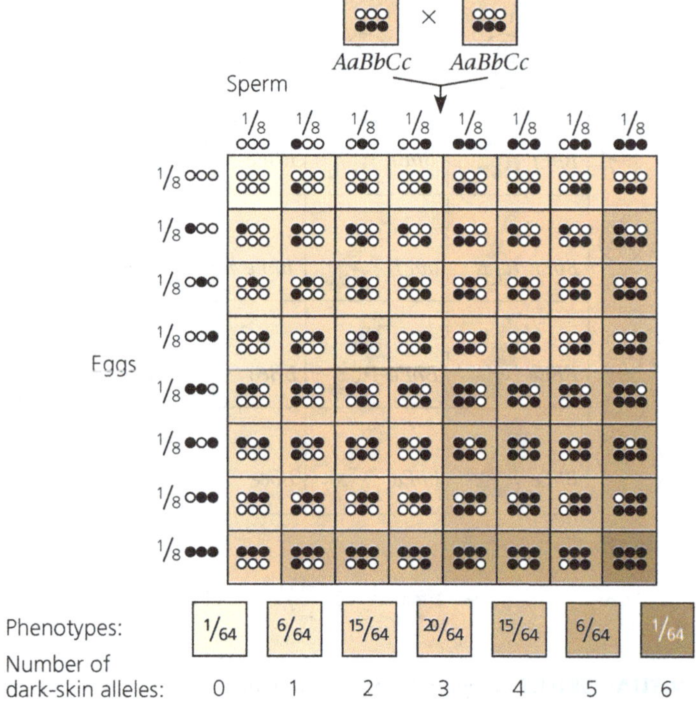

would be very dark, whereas an *aabbcc* individual would be very light. An *AaBbCc* person would have skin of an intermediate shade. Because the alleles have a cumulative effect, the genotypes *AaBbCc* and *AABbcc* would make the same genetic contribution (three units) to skin darkness. There are seven skin colour phenotypes that could result from a mating between *AaBbCc* heterozygotes, as shown in **Figure 14.13**. In a large number of such matings, the majority of offspring would be expected to have intermediate phenotypes (skin colour in the middle range). You can graph the predictions from the Punnett square in the **Scientific Skills Exercise**. Environmental factors, such as exposure to the sun, also affect the skin colour phenotype.

Nature and Nurture: The Environmental Impact on Phenotype

Another departure from simple Mendelian genetics arises when the phenotype for a character depends on environment as well as genotype. A single tree, locked into its inherited genotype, has leaves that vary in size, shape, and greenness, depending on their exposure to wind and sun. For humans, nutrition influences height, exercise alters build, sun-tanning darkens the skin, and experience improves performance on intelligence tests. Even identical twins, who are genetic equals, accumulate phenotypic differences as a result of their unique experiences.

Whether human characters are more influenced by genes or the environment—in everyday terms, nature versus nurture—is a debate that we will not attempt to settle here. We can say, however, that a genotype generally is not associated with a rigidly defined phenotype, but rather with a range of phenotypic possibilities due to environmental influences **(Figure 14.14)**. For some characters, such as the ABO blood group system, the phenotypic range is extremely narrow; that is, a given genotype mandates a very specific phenotype. Other characters, such as a person's blood count of red and white cells, vary quite a bit, depending on such factors as the altitude, the customary level of physical activity, and the presence of infectious agents.

▼ Figure 14.14 The effect of environment on phenotype. The outcome of a genotype lies within a phenotypic range that depends on the environment in which the genotype is expressed. For example, the acidity and free aluminum content of the soil affect the colour of hydrangea flowers, which range from pink (basic soil) to blue-violet (acidic soil). Free aluminum is necessary for bluer colours.

(a) Hydrangeas grown in basic soil

(b) Hydrangeas of the same genetic variety grown in acidic soil with free aluminum

Generally, the phenotypic range is broadest for polygenic characters. Environment contributes to the quantitative nature of these characters, as we have seen in the continuous variation of skin colour. Geneticists refer to such characters as **multifactorial**, meaning that many factors, both genetic and environmental, collectively influence phenotype.

A Mendelian View of Heredity and Variation

We have now broadened our view of Mendelian inheritance by exploring degrees of dominance as well as multiple alleles, pleiotropy, epistasis, polygenic inheritance, and the phenotypic impact of the environment. How can we integrate these refinements into a comprehensive theory of Mendelian genetics? The key is to make the transition from the reductionist emphasis on single genes and phenotypic characters to the emergent properties of the organism as a whole, one of the themes of this text.

The term *phenotype* can refer not only to specific characters, such as flower colour and blood group, but also

Scientific Skills Exercise

Making a Histogram and Analysing a Distribution Pattern

What Is the Distribution of Phenotypes Among Offspring of Two Parents Who Are Both Heterozygous for Three Additive Genes? Human skin colour is a polygenic trait that is determined by the additive effects of many different genes. In this exercise, you will work with a simplified model of skin colour genetics where only three genes are assumed to affect the darkness of skin colour and where each gene has two alleles—dark or light (see Figure 14.13). In this model, each dark allele contributes equally to the darkness of skin colour, and each pair of alleles segregates independently of any other pair. Using a type of graph called a histogram, you will determine the distribution of phenotypes of offspring with different numbers of dark-skin alleles. (For additional information about graphs, see the Scientific Skills Review in Appendix D.)

How This Model Is Analysed To predict the phenotypes of the offspring of parents heterozygous for the three genes in our simplified model, we can use the Punnett square in Figure 14.13. The heterozygous individuals (*AaBbCc*) represented by the two rectangles at the top of that figure each carry three dark-skin alleles (black circles, which represent *A, B*, or *C*) and three light-skin alleles (white circles, which represent *a, b*, or *c*). The Punnett square shows all the possible genetic combinations in gametes and in offspring of a large number of hypothetical matings between these heterozygotes.

Predictions from the Punnett Square If we assume that each square in the Punnett square represents one offspring of the heterozygous *AaBbCc* parents, then the squares below show the possible skin colour phenotypes and their predicted frequencies. Below the squares is the number of dark-skin alleles for each phenotype.

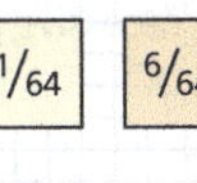

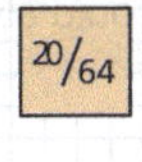
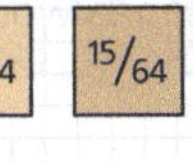
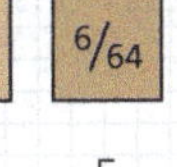
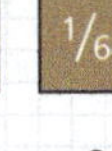

Phenotypes:	1/64	6/64	15/64	20/64	15/64	6/64	1/64
Number of dark-skin alleles:	0	1	2	3	4	5	6

INTERPRET THE DATA

1. A histogram is a bar graph that shows the distribution of numeric data (here, the number of dark-skin alleles). To make a histogram of the allele distribution, put skin colour (as the number of dark-skin alleles) along the *x*-axis and predicted number of offspring (out of 64) with each phenotype on the *y*-axis. There are no gaps in these allele data, so draw the bars next to each other with no space in between.
2. You can see that the skin-colour phenotypes are not distributed uniformly. **(a)** Which phenotype has the highest frequency? Draw a vertical dashed line through that bar. **(b)** Distributions of values like this one tend to show one of several common patterns. Sketch a rough curve that approximates the values and look at its shape. Is it symmetrically distributed around a central peak value (a "normal distribution," sometimes called a bell curve); is it skewed to one end of the *x*-axis or the other (a "skewed distribution"); or does it show two apparent groups of frequencies (a "bimodal distribution")? Explain the reason for the curve's shape. (It will help to read the text description that supports Figure 14.13.)

Further Reading R. A. Sturm, A golden age of human pigmentation genetics, *Trends in Genetics* 22:464–468 (2006).

to an organism in its entirety—*all* aspects of its physical appearance, internal anatomy, physiology, and behaviour. Similarly, the term *genotype* can refer to an organism's entire genetic makeup, not just its alleles for a single genetic locus. In most cases, a gene's impact on phenotype is affected by other genes and by the environment. In this integrated view of heredity and variation, an organism's phenotype reflects its overall genotype and unique environmental history.

Considering all that can occur in the pathway from genotype to phenotype, it is indeed impressive that Mendel could uncover the fundamental principles governing the transmission of individual genes from parents to offspring. Mendel's laws of segregation and of independent assortment explain heritable variations in terms of alternative forms of genes (hereditary "particles," now known as the alleles of genes) that are passed along, generation after generation, according to simple rules of probability. This theory of inheritance is equally valid for peas, flies, fishes, birds, and human beings—indeed, for any organism with a sexual life cycle. Furthermore, by extending the principles of segregation and independent assortment to help explain such hereditary patterns as epistasis and quantitative characters, we begin to see how broadly Mendelian genetics applies. From Mendel's abbey garden came a theory of inheritance that anchors modern genetics. In the last section of this chapter, we will apply Mendelian genetics to human inheritance, with emphasis on the transmission of hereditary diseases.

CONCEPT CHECK 14.3

1. *Incomplete dominance* and *epistasis* are both terms that define genetic relationships. What is the most basic distinction between these terms?
2. If a man with type AB blood marries a woman with type O, what blood types would you expect in their children? What fraction would you expect of each type?
3. **WHAT IF?** A rooster with grey feathers and a hen of the same phenotype produce 15 grey, 6 black, and 8 white chicks. What is the simplest explanation for the inheritance of these colours in chickens? What phenotypes would you expect in the offspring of a cross between a grey rooster and a black hen?

For suggested answers, see Appendix A.

CONCEPT 14.4

Many human traits follow Mendelian patterns of inheritance

The study of human genetics is fueled by our desire to understand our own inheritance and to develop treatments for genetically based diseases. While peas are convenient subjects for genetic research, humans are not. The human generation span is long (about 20 years), human parents produce many fewer offspring than peas, and breeding experiments would be unethical!

Pedigree Analysis

In place of breeding experiments, geneticists analyse the results of human matings that have already occurred. They collect information about a family's history for a particular trait and assemble this information into a family tree describing the trait across the generations—a family **pedigree**.

Figure 14.15a shows a three-generation pedigree that traces the occurrence of a pointed contour of the hairline on the forehead. We simplify the genetics of this trait, called a widow's peak, by assuming it is due to a dominant allele, *W*. (In reality, other genes are probably also involved.) In our model, because the widow's peak allele is dominant, all individuals who lack a widow's peak must be homozygous recessive (*ww*). The two grandparents with widow's peaks must have the *Ww* genotype since some of their offspring are homozygous recessive. The offspring in the second generation who *do* have widow's peaks must also be heterozygous because they are the products of *Ww* × *ww* matings. The third generation in this pedigree consists of two sisters. The one who has a widow's peak could be either homozygous (*WW*) or heterozygous (*Ww*), given what we know about the genotypes of her parents (both *Ww*).

Figure 14.15b is a pedigree of the same family, but this time we focus on a recessive trait, the inability of individuals to taste a chemical called PTC (phenylthiocarbamide). Compounds similar to PTC are found in broccoli, brussels sprouts, and related vegetables and account for the bitter taste some people report when eating these foods. We'll use *t* for the recessive allele and *T* for the dominant allele, which results in the ability to taste PTC. As you work your way through the pedigree, notice once again that you can apply what you have learned about Mendelian inheritance to understand the genotypes shown for the family members.

An important application of a pedigree is to help us calculate the probability that a future child will have a particular genotype and phenotype. Suppose that the couple represented in the second generation of Figure 14.15 decides to have one more child. What is the probability that the child will have a widow's peak? This is equivalent to a Mendelian F_1 monohybrid cross (*Ww* × *Ww*), and therefore the probability that a child will inherit a dominant allele and have a widow's peak is $\frac{3}{4}$ ($\frac{1}{4}$ *WW* + $\frac{1}{2}$ *Ww*). What is the probability that the

▼ **Figure 14.15 Pedigree analysis.** Each pedigree traces a trait through three generations of the same family; the two traits have different inheritance patterns. For the sake of understanding genetic principles, each trait is shown as being determined by two alleles of a single gene—a simplification, since other genes likely also affect these two characters.

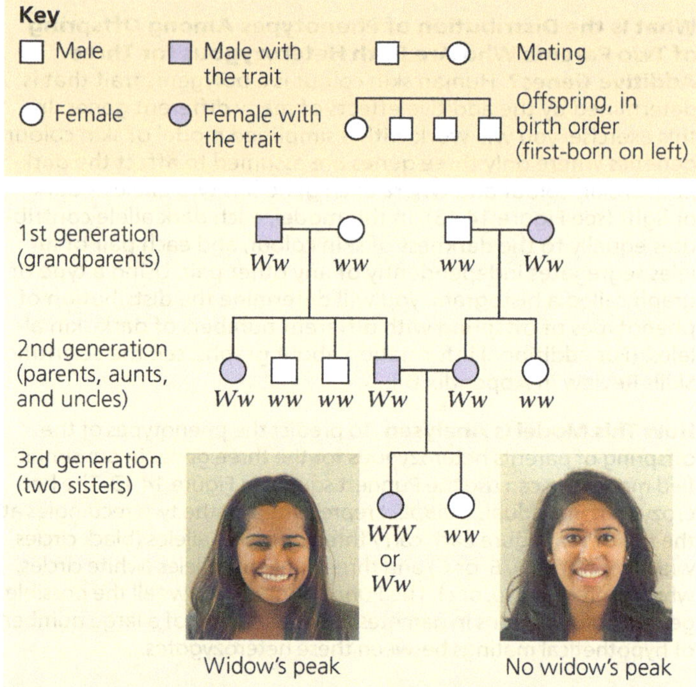

(a) Is a widow's peak a dominant or recessive trait?

Tips for pedigree analysis: Notice in the third generation that the second-born daughter lacks a widow's peak, although both of her parents had the trait. Such a pattern indicates that the trait is due to a dominant allele. If it were due to a *recessive* allele, and both parents had the recessive phenotype (straight hairline), *all* of their offspring would also have the recessive phenotype.

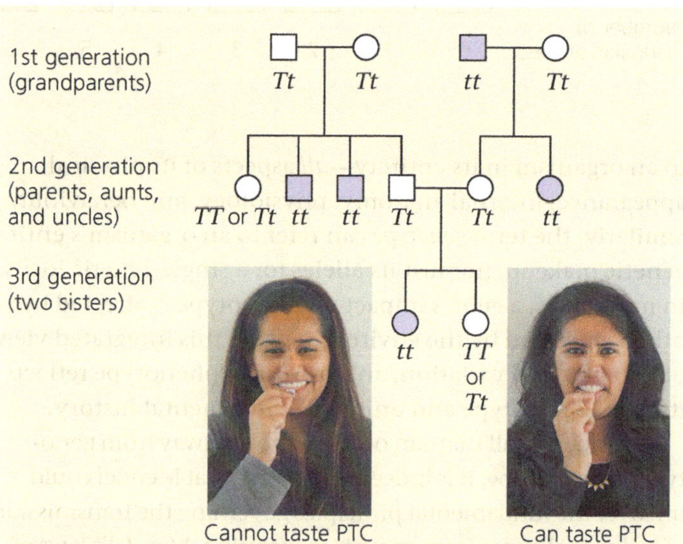

(b) Is the inability to taste a chemical called PTC a dominant or recessive trait?

Tips for pedigree analysis: Notice that the first-born daughter in the third generation has the trait (is unable to taste PTC), although both parents lack that trait (they *can* taste PTC). Such a pattern is explained if the non-taster phenotype is due to a recessive allele. If it were due to a *dominant* allele, then at least one parent would also have had the trait.

child will be unable to taste PTC? We can also treat this as a monohybrid cross ($Tt \times Tt$), but this time we want to know the chance that the offspring will be homozygous recessive (tt). That probability is $\frac{1}{4}$. Finally, what is the chance that the child will have a widow's peak *and* be unable to taste PTC? Assuming that the genes for these two characters are on different chromosomes, the two pairs of alleles will assort independently in this dihybrid cross ($WwTt \times WwTt$). Therefore, we can use the multiplication rule: $\frac{3}{4}$ (chance of widow's peak) $\times$ $\frac{1}{4}$ (chance of inability to taste PTC) $= \frac{3}{16}$ (chance of widow's peak and inability to taste PTC).

Pedigrees are a more serious matter when the alleles in question cause disabling or deadly diseases instead of innocuous human variations such as hairline or inability to taste a harmless chemical. However, for disorders inherited as simple Mendelian traits, the same techniques of pedigree analysis apply.

Recessively Inherited Disorders

Thousands of genetic disorders are known to be inherited as simple recessive traits. These disorders range in severity from relatively mild, such as albinism (lack of pigmentation, which results in susceptibility to skin cancers and vision problems), to life-threatening, such as cystic fibrosis.

The Behaviour of Recessive Alleles

How can we account for the behaviour of alleles that cause recessively inherited disorders? Recall that genes code for proteins of specific function. An allele that causes a genetic disorder (let's call it allele *a*) codes for either a malfunctioning protein or no protein at all. In the case of disorders classified as recessive, heterozygotes (*Aa*) typically have the normal phenotype because one copy of the normal allele (*A*) produces a sufficient amount of the specific protein. Thus, a recessively inherited disorder shows up only in the homozygous individuals (*aa*) who inherit a recessive allele from each parent. Although phenotypically normal with regard to the disorder, heterozygotes may transmit the recessive allele to their offspring and thus are called **carriers**. **Figure 14.16** illustrates these ideas using albinism as an example.

▼ **Figure 14.16 Albinism: a recessive trait.** One of the two sisters shown here does not have albinism; the other does. Most recessive homozygotes are born to parents who are carriers of the disorder but themselves have a normal phenotype, the case shown in the Punnett square.

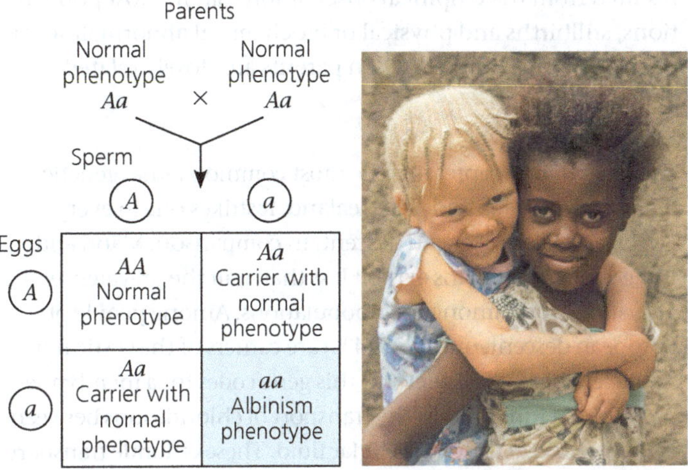

? *What is the probability that the sister without albinism is a carrier of the albinism allele?*

Most people who have recessive disorders are born to parents who are carriers of the disorder but have a normal phenotype, as is the case shown in the Punnett square in Figure 14.16. A mating between two carriers corresponds to a Mendelian F_1 monohybrid cross, so the predicted genotypic ratio for the offspring is 1 *AA* : 2 *Aa* : 1 *aa*. Thus, each child has a $\frac{1}{4}$ chance of inheriting a double dose of the recessive allele; in the case of albinism, such a child will have albinism. From the genotypic ratio, we also can see that out of three offspring with the *normal* phenotype (one *AA* plus two *Aa*), two are predicted to be heterozygous carriers, $\frac{2}{3}$ chance. Recessive homozygotes could also result from $Aa \times aa$ and $aa \times aa$ matings, but if the disorder is lethal before reproductive age or results in sterility (neither of which is true for albinism), no *aa* individuals will reproduce. Even if recessive homozygotes are able to reproduce, this will occur relatively rarely because such individuals are much less common in the population than heterozygous carriers (for reasons we'll examine in Concept 23.2).

In general, genetic disorders are not evenly distributed among all groups of people. For example, the incidence of Tay-Sachs disease, which we described earlier in this chapter, is disproportionately high among Ashkenazic Jews, Jewish people whose ancestors lived in central Europe. In that population, Tay-Sachs disease occurs in one out of 3,600 births, an incidence about 100 times greater than that among non-Jews or Mediterranean (Sephardic) Jews. This uneven distribution results from the different genetic histories of the world's peoples during less technological times, when populations were more geographically (and hence genetically) isolated.

When a disease-causing recessive allele is rare, it is relatively unlikely that two carriers of the same harmful allele will meet and mate. The probability of passing on recessive traits increases greatly, however, if the man and woman are close relatives (for example, siblings or first cousins). This is because people with recent common ancestors are more likely to carry the same recessive alleles than are unrelated people. Thus, these consanguineous ("same blood") matings, indicated in pedigrees by double lines, are more likely to produce offspring homozygous for recessive traits—including harmful ones. Such effects can be observed in many types of domesticated and zoo animals that have become inbred.

Although geneticists generally agree that inbreeding causes an increase in autosomal recessive conditions compared to those resulting from matings between unrelated parents, they debate exactly how much human consanguinity increases the risk of inherited diseases. For one thing, many harmful alleles have such severe effects that a homozygous embryo spontaneously aborts long before birth. Most societies and cultures have laws or taboos forbidding marriages between close relatives, some for social or economic reasons. These rules may have also resulted from the empirical observation that in most populations, stillbirths and physical or biochemical abnormalities at birth are more common when parents are closely related.

Cystic Fibrosis

Cystic fibrosis represents the most common lethal genetic disease in Australia and New Zealand. It strikes one in every 3,700 people of European descent. In comparison, Māori and Indigenous Australians exhibit less than half the prevalence of cystic fibrosis among their populations. Among people of European descent, one in 25 (4%) are carriers of the cystic fibrosis allele. The normal allele for this gene codes for a membrane protein that functions in the transport of chloride ions between certain cells and the extracellular fluid. These chloride transport channels are defective or absent in the plasma membranes of children who inherit two recessive alleles for cystic fibrosis. The result is an abnormally high concentration of intracellular chloride, which causes an uptake of water due to osmosis. This in turn causes the mucus that coats certain cells to become thicker and stickier than normal and to build up in the pancreas, lungs, digestive tract, and other organs. Multiple (pleiotropic) effects result, including poor absorption of nutrients from the intestines, chronic bronchitis, and recurrent bacterial infections.

Untreated, cystic fibrosis can cause death by the age of 5. Daily doses of antibiotics to stop infection, gentle pounding on the chest to clear mucus from clogged airways, supplemented diets, and other therapies can prolong life. Australians and New Zealanders with cystic fibrosis now survive into their 40s.

Sickle-Cell Disease: A Genetic Disorder with Evolutionary Implications

EVOLUTION The most common inherited disorder among people of African descent is **sickle-cell disease**, which affects one in 400 African-Americans. Sickle-cell disease is caused by the substitution of a single amino acid in the haemoglobin protein of red blood cells; in homozygous individuals, all haemoglobin is of the sickle-cell (abnormal) variety. When the oxygen content of an affected individual's blood is low (at high altitudes or under physical stress, for instance), the sickle-cell haemoglobin proteins aggregate into long fibres that deform the red cells into a sickle shape (see Figure 5.19). Sickled cells may clump and clog small blood vessels, often leading to other symptoms throughout the body, including physical weakness, pain, organ damage, and even stroke and paralysis. Regular blood transfusions can ward off brain damage in children with sickle-cell disease, and new drugs can help prevent or treat other problems. There is currently no widely available cure, but the disease is the target of ongoing gene therapy research.

Although two sickle-cell alleles are necessary for an individual to have sickle-cell disease and thus the condition is considered recessive, the presence of one sickle-cell allele can affect the phenotype. Thus, at the organismal level, the normal allele is incompletely dominant to the sickle-cell allele **(Figure 14.17)**. At the molecular level, the two alleles are codominant; both normal and abnormal (sickle-cell) haemoglobins are made in heterozygotes (carriers), who are said to have *sickle-cell trait*. (Here the word "trait" is used to distinguish this condition from sickle-cell disease, not as it was defined earlier—as any variant of a character.) Heterozygotes are usually healthy but may suffer some symptoms during long periods of reduced blood oxygen.

About one in ten African-Americans have sickle-cell trait, an unusually high frequency of heterozygotes for an allele with severe detrimental effects in homozygotes. Why haven't evolutionary processes resulted in the disappearance of the allele among this population? One explanation is that having a single copy of the sickle-cell allele reduces the frequency and severity of malaria attacks, especially among young children. The malaria parasite spends part of its life cycle in red blood cells (see Figure 28.18), and the presence of even heterozygous amounts of sickle-cell haemoglobin results in lower parasite densities and hence reduced malaria symptoms. Thus, in tropical Africa, where infection with the malaria parasite is common, the sickle-cell allele confers an advantage to heterozygotes even though it is harmful in the homozygous state. (The balance between these two effects will be discussed in Concept 23.4; see Make Connections Figure 23.18.) The relatively high frequency of African-Americans with sickle-cell trait is a vestige of their tropical African ancestry.

▼ **Figure 14.17 Sickle-cell disease and sickle-cell trait.**

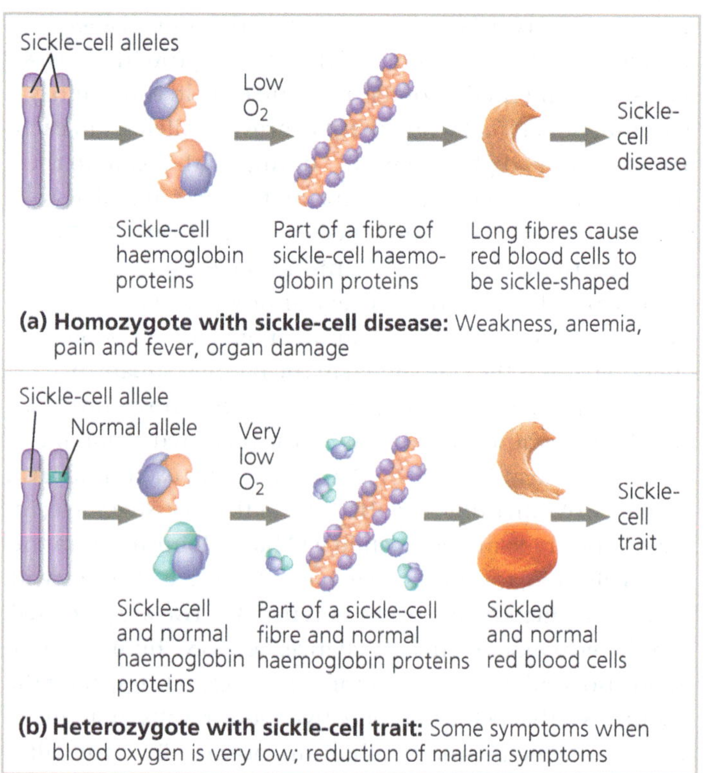

(a) Homozygote with sickle-cell disease: Weakness, anemia, pain and fever, organ damage

(b) Heterozygote with sickle-cell trait: Some symptoms when blood oxygen is very low; reduction of malaria symptoms

Dominantly Inherited Disorders

Although many harmful alleles are recessive, some disorders are due to dominant alleles. One example is *achondroplasia*, a form of dwarfism that occurs in one in every 25,000 people. Heterozygous individuals have the dwarf phenotype **(Figure 14.18)**. Therefore, all people who do not have achondroplasia—99.99% of the population—are homozygous recessive. Like the presence of extra fingers or toes mentioned earlier, achondroplasia is a trait for which the recessive allele is much more prevalent than the corresponding dominant allele.

Unlike achondroplasia, which is relatively harmless, some dominant alleles cause lethal diseases. Those that do are much less common than recessive alleles that have lethal effects. A lethal recessive allele is only lethal when homozygous; it can be passed from one generation to the next by heterozygous carriers because the carriers themselves have normal phenotypes. A lethal dominant allele, however, often causes the death of afflicted individuals before they can mature and reproduce, and in this case the allele is not passed on to future generations.

A lethal dominant allele may be passed on, though, if the lethal disease symptoms first appear after reproductive age. In these cases, the individual may already have transmitted the allele to his or her children. For example, a degenerative disease of the nervous system, called **Huntington's disease**, is caused by a lethal dominant allele that has no obvious phenotypic effect until the individual is about 35 to 45 years old. Once the deterioration of the nervous system begins, it is irreversible and inevitably fatal. As with other dominant traits, a child born to a parent with the Huntington's disease allele has a 50% chance of inheriting the allele and the disorder (see the Punnett square in Figure 14.18). In Australia, this disease afflicts about seven in 10,000 people. We could source no comparable data for New Zealand.

▼ Figure 14.18 Achondroplasia: a dominant trait. Dr. Michael C. Ain has achondroplasia, a form of dwarfism caused by a dominant allele. This has inspired his work: He is a specialist in the repair of bone defects caused by achondroplasia and other disorders. The dominant allele (*D*) might have arisen as a mutation in the egg or sperm of a parent or could have been inherited from an affected parent, as shown for an affected father in the Punnett square.

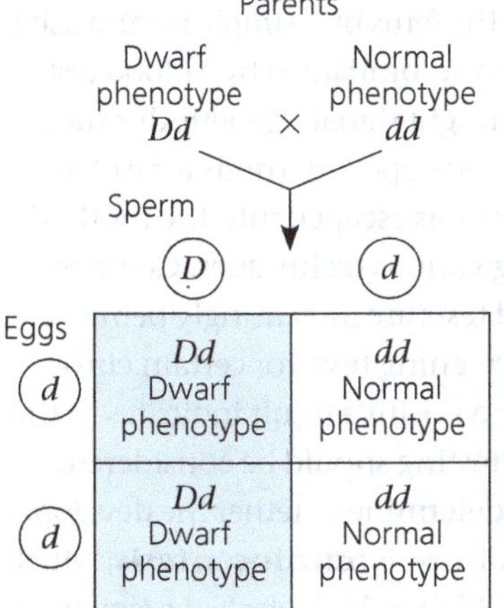

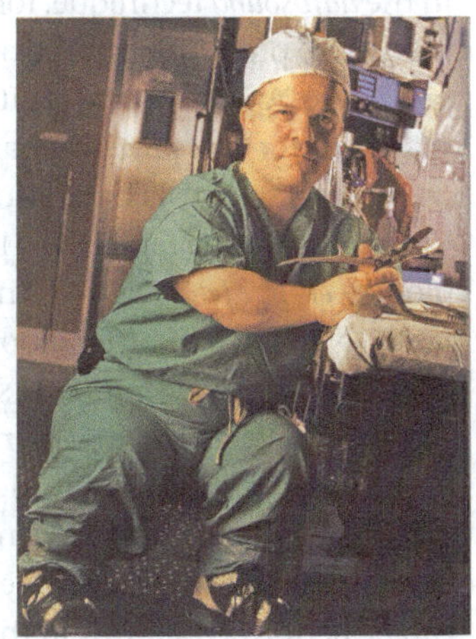

At one time, the onset of symptoms was the only way to know if a person had inherited the Huntington's allele, but this is no longer the case. By analysing DNA samples from a large family with a high incidence of the disorder, geneticists tracked the Huntington's allele to a locus near the tip of chromosome 4, and the gene was sequenced in 1993. This information led to the development of a test that could detect the presence of the Huntington's allele in an individual's genome. (The methods that make such tests possible are discussed in Concepts 20.1 and 20.4.) The availability of this test poses an agonising dilemma for those with a family history of Huntington's disease. Some individuals may want to be tested for this disease, whereas others may decide it would be too stressful to find out.

Multifactorial Disorders

The hereditary diseases we have discussed so far are sometimes described as simple Mendelian disorders because they result from abnormality of one or both alleles at a single genetic locus. Many more people are susceptible to diseases that have a multifactorial basis—a genetic component plus a significant environmental influence. Heart disease, diabetes, cancer, alcoholism, certain mental illnesses such as schizophrenia and bipolar disorder, and many other diseases are multifactorial. In these cases, the hereditary component is polygenic. For example, many genes affect cardiovascular health, making some of us more prone than others to heart attacks and strokes. No matter what our genotype, however, our lifestyle has a tremendous effect on phenotype for cardiovascular health and other multifactorial characters. Exercise, a healthful diet, abstinence from smoking, and an ability to handle stressful situations all reduce our risk of heart disease and some types of cancer.

Genetic Testing and Counselling

Avoiding simple Mendelian disorders is possible when the risk of a particular genetic disorder can be assessed before a child is conceived or during the early stages of the pregnancy. Many hospitals have genetic counsellors who can provide information to prospective parents concerned about a family history for a specific disease. Fetal and newborn testing can also reveal genetic disorders.

Counselling Based on Mendelian Genetics and Probability Rules

Consider the case of a hypothetical couple, Tyler and Lily. Each had a brother who died from the same recessively inherited lethal disease. Before conceiving their first child, Tyler and Lily seek genetic counselling to determine the risk of having a child with the disease. From the information about their brothers, we know that both parents of Tyler and both parents of Lily must have been carriers of the recessive allele. Thus, Tyler and Lily are both products of $Aa \times Aa$ crosses, where *a* symbolises the allele that causes this particular disease. We also know that Tyler and Lily are not homozygous recessive (*aa*) because they do not have the disease. Therefore, their genotypes are either *AA* or *Aa*.

Given a genotypic ratio of 1 *AA*:2 *Aa*:1 *aa* for offspring of an $Aa \times Aa$ cross, Tyler and Lily each have a $\frac{2}{3}$ chance of being carriers (*Aa*). According to the rule of multiplication, the overall probability of their firstborn having the disorder is $\frac{2}{3}$ (the chance that Tyler is a carrier) times $\frac{2}{3}$ (the chance that Lily is a carrier) times $\frac{1}{4}$ (the chance of two carriers having a child with the disease), which equals $\frac{1}{9}$. Suppose that Lily and Tyler decide to have a child—after all, there is an $\frac{8}{9}$ chance that their baby will not have the disorder. If, despite these odds, their child is born with the disease, then we would know that *both* Tyler and Lily are, in fact, carriers (*Aa* genotype). If both Tyler and Lily are carriers, there is a $\frac{1}{4}$ chance that any subsequent child this couple has will have the disease. The probability is higher for subsequent children because the diagnosis of the disease in the first child established that both parents are carriers, not because the genotype of the first child affects in any way that of future children.

When we use Mendel's laws to predict possible outcomes of matings, it is important to remember that each child represents an independent event in the sense that its genotype is unaffected by the genotypes of older siblings. Suppose that Tyler and Lily have three more children, and *all three* have the hypothetical hereditary disease. There is only one chance in 64 ($\frac{1}{4} \times \frac{1}{4} \times \frac{1}{4}$) that such an outcome will occur. Despite these circumstances, the chance that a fourth child of this couple will have the disease remains $\frac{1}{4}$.

Tests for Identifying Carriers

Most children with recessive disorders are born to parents with normal phenotypes. The key to accurately assessing the genetic risk for a particular disease is therefore to find out whether the prospective parents are heterozygous carriers of the recessive allele. Tests are available that can distinguish individuals of normal phenotype who are dominant homozygotes from those who are heterozygous carriers, using blood or cells from the inside of the cheek. Hundreds of different recessive alleles can be detected, including those for Tay-Sachs disease, sickle-cell disease, and cystic fibrosis.

These tests for identifying carriers enable people with family histories of genetic disorders to make informed decisions about having children, including whether to do genetic testing of the fetus, should they decide to become pregnant. The results can also lead carriers to alert family members that they may also be carriers and might choose to be tested. The tests raise other issues as well: Could carriers be denied health or life insurance or lose the jobs providing those benefits, even though they themselves are healthy? In 2008, the United States passed into law the *Genetic Information Nondiscrimination Act*. The Act prohibits discrimination in employment or health insurance coverage based on genetic test results. Unfortunately, neither New Zealanders nor Australians enjoy these rights. New Zealand's *Privacy Act 1993* and the *Health Information Privacy Code 1994* protect an individual from the need to disclose health or other details to third parties. However, some insurance companies mange to circumvent these legislative obligations, and insurance-seekers may have to sign away their rights to genetic privacy as part of the application process. Discrimination could take the form of refusal to offer insurance, or the imposition of significantly higher premiums. Similar outcomes occur in Australia. Federal and state legislation protects an individual's privacy, as well as prohibiting discrimination (except in actuarial decisions). Jane Tiller at Monash University reports cases of legal (but morally questionable) and illegal discrimination against people based on their genetic tests. Despite a 2018 Parliamentary enquiry into genetic privacy that recommended US-style protections, Australian citizens enjoy only the loose benefits associated with an industry-based, self-regulatory protection system.

Genetic counsellors, trained in a growing number of professional programs, can help individuals understand their genetic test results and their rights and obligations under the law. Once test results are clearly understood, those with certain results may face difficult decisions; preventative surgery, for example. People may choose preventative surgery to remove the risks of a disease that their genome makes them more likely to contract. While advances in biotechnology offer enriched diagnostic and preventative decisions years ahead of the first symptom, the technologies carry ethical and legislative challenges. Australia and New Zealand continue to lag behind other nations, such as Colombia and the United States, in the introduction and enforcement of legislation to address this issue.

Fetal Testing

Two types of testing can be done to look for genetic disorders in the fetus: screening tests and diagnostic tests. Screening tests are generally noninvasive and can be followed by diagnostic tests if screening results suggest that a genetic disorder might be present or if there are other risks, as when both parents are carriers.

Screening tests include imaging and blood tests. Imaging techniques allow a physician to examine a fetus directly for major anatomical abnormalities that might not show up in genetic tests. In the *ultrasound* technique, for example, reflected sound waves are used to produce an image of the fetus by a simple noninvasive procedure. An ultrasound is often accompanied by a blood test that looks for fetal proteins that might signal a genetic disorder.

Medical scientists have also developed methods for isolating and analysing fetal DNA that has escaped into the mother's blood (*cell-free fetal DNA*), using various techniques. Cell-free fetal DNA tests and other blood tests are increasingly being used as noninvasive prenatal screening tests for certain chromosomal defects and disorders; a positive result indicates to the parents that further diagnostic testing should be considered.

One diagnostic test that can determine whether the developing fetus has a serious recessive disease is **amniocentesis**, which can be performed starting at the 15th or 16th week of pregnancy **(Figure 14.19a)**. In this procedure, a physician inserts a needle

▼ **Figure 14.19 Diagnostic fetal testing for genetic disorders.** Most testing for genetic abnormalities is now DNA-based. In some cases biochemical tests are used to detect particular disorders, and karyotyping may be done to see whether the chromosomes of the fetus are normal in number and appearance.

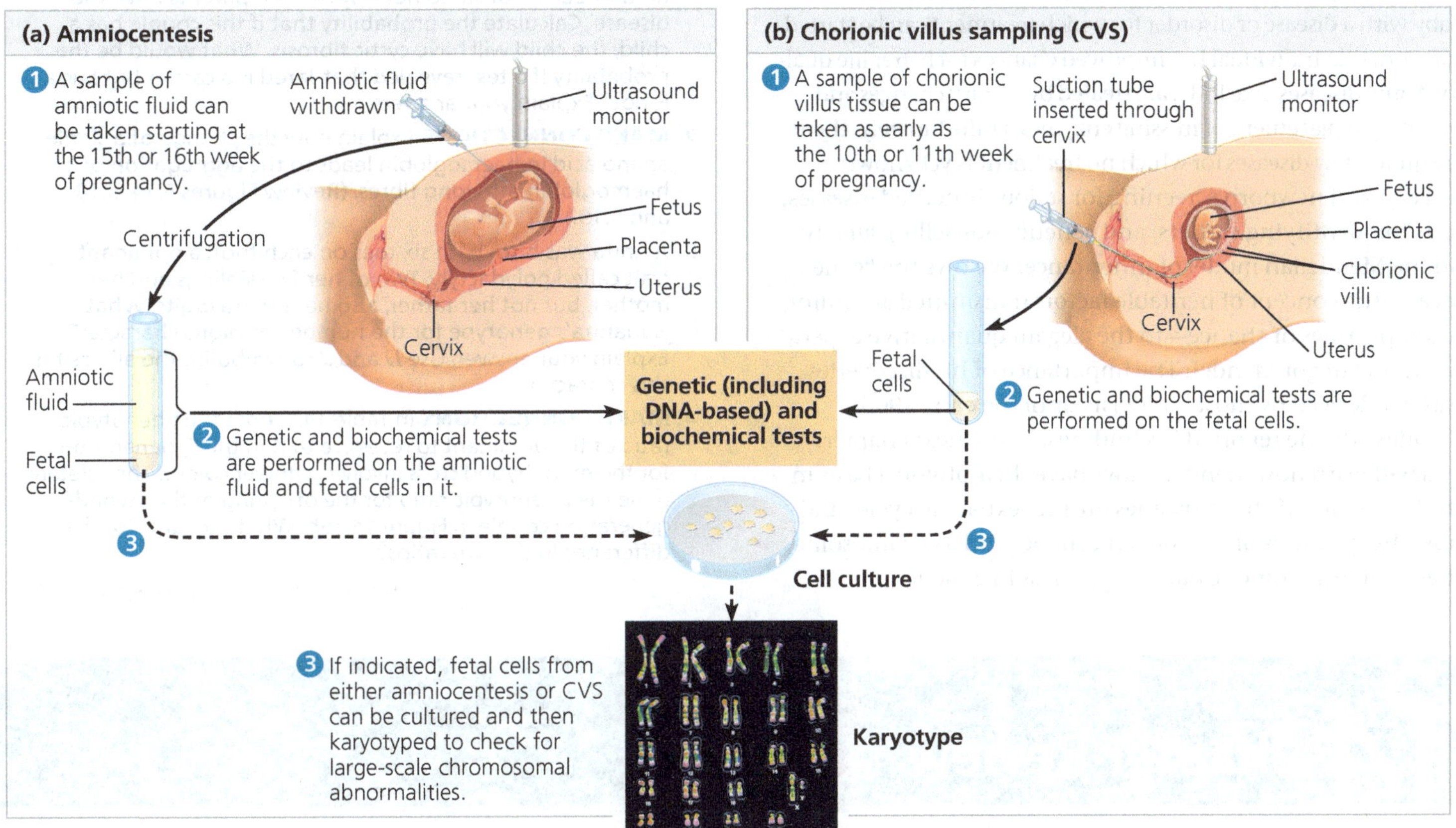

into the uterus and extracts about 10 mL of amniotic fluid, the liquid that bathes the fetus. Some genetic disorders can be detected from the presence of certain proteins or hormones in the amniotic fluid itself. Tests for thousands of other disorders (such as cystic fibrosis) are performed on DNA from fetal cells sloughed off into the amniotic fluid using targeted DNA sequencing or a technique called a *DNA micro-array* (see Figure 20.13) that tests for many alleles at the same time. A karyotype of these cultured cells can also confirm certain chromosomal defects that may not have been suggested by cell-free fetal DNA tests.

In an alternative diagnostic technique that is called **chorionic villus sampling (CVS)**, a physician inserts a narrow tube through the cervix into the uterus and suctions out a tiny sample of tissue from the placenta, the organ that transmits nutrients and fetal wastes between the fetus and the mother **(Figure 14.19b)**. The cells of the chorionic villi of the placenta—the portion sampled—are derived from the fetus and have the same genotype and DNA analysed as the new individual. DNA from these cells can be analysed for the presence of many genetic disorders. The main advantage of CVS is that it can be performed as early as the 10th or 11th week of pregnancy.

Ultrasound and isolation of fetal cells or DNA from maternal blood pose no known risk to either mother or fetus, while amniocentesis and CVS can cause complications in a small percentage of cases (for both, between 1 in 1,000 and 1 in 500 cases). Amniocentesis or CVS for diagnostic testing is generally offered to women over age 35, due to their increased risk of bearing a child with Down syndrome or other chromosomal disorders, and may also be offered to younger women if there are risk factors or conditions suggested by family history. If the fetal tests reveal a serious disorder, the parents face the difficult choice of either terminating the pregnancy or preparing to care for a child with a genetic disorder, one that might even be fatal. Parental and fetal screening for Tay-Sachs alleles done since 1980 has reduced the number of children born with this incurable disease by 90%.

Newborn Screening

Some genetic diseases and disorders can be detected at birth by simple biochemical tests that are routinely performed on blood from the baby's heel in most hospitals in New Zealand and Australia. One common screening program is for phenylketonuria (PKU), a recessively inherited disorder that occurs in about one in every 6,000 births in Australia and New Zealand. Children with this disease cannot properly metabolise the amino acid phenylalanine. This compound and its by-product, phenylpyruvate, can accumulate to toxic levels in the blood, causing severe intellectual disability. However, if PKU is detected in the newborn, a special diet low in phenylalanine will usually allow for typical development. (Among many other substances, this diet excludes the artificial sweetener aspartame, which contains phenylalanine.)

The number of newborn genetic tests, many for metabolic disorders or enzyme deficiencies, is over 100 and ever-increasing; each state regulates which tests are used. If screening identifies a baby with a disease or disorder for which treatment can be started early on, that individual has improved chances for better life quality. Some diseases, like PKU, are treated by dietary changes and others by replacement of missing enzymes. Unfortunately, there are quite a few diseases for which no treatment is yet known.

Fetal and newborn screening for serious inherited diseases, tests for identifying carriers, and genetic counselling all rely on the Mendelian model of inheritance. We owe the "gene idea"—the concept of heritable factors transmitted according to simple rules of chance—to the elegant quantitative experiments of Gregor Mendel. The importance of his discoveries was overlooked by most biologists until the early 1900s, decades after he reported his findings. In the next chapter, you will learn how Mendel's laws have their physical basis in the behaviour of chromosomes during sexual life cycles and how the synthesis of Mendelian genetics and a chromosome theory of inheritance catalysed progress in genetics.

CONCEPT CHECK 14.4

1. Lucia and Jared each have a sibling with cystic fibrosis, but neither Lucia nor Jared nor any of their parents have the disease. Calculate the probability that if this couple has a child, the child will have cystic fibrosis. What would be the probability if a test revealed that Jared is a carrier but Lucia is not? Explain your answers.
2. **MAKE CONNECTIONS** Explain how the change of a single amino acid in haemoglobin leads to the aggregation of haemoglobin into long fibres. (Review Figures 5.14, 5.18, and 5.19.)
3. Juanita was born with six toes on each foot, a dominant trait called polydactyly. Two of her five siblings and her mother, but not her father, also have extra digits. What is Juanita's genotype for the number-of-digits character? Explain your answer. Use D and d to symbolise the alleles for this character.
4. **MAKE CONNECTIONS** In Table 14.1, note the phenotypic ratio of the dominant to recessive trait in the F_2 generation for the monohybrid cross involving flower colour. Then determine the phenotypic ratio for the offspring of the second-generation couple in Figure 14.15b. What accounts for the difference in the two ratios?

For suggested answers, see Appendix A.

14 Chapter Review

SUMMARY OF KEY CONCEPTS

CONCEPT 14.1

Mendel used the scientific approach to identify two laws of inheritance *(pp. 272–278)*

- Gregor Mendel formulated a theory of inheritance based on experiments with garden peas, proposing that parents pass on to their offspring discrete genes that retain their identity through generations. This theory includes two "laws."
- The **law of segregation** states that genes have alternative forms, or **alleles**. In a diploid organism, the two alleles of a gene segregate (separate) during meiosis and gamete formation; each sperm or egg carries only one allele of each pair. This law explains the 3:1 ratio of F_2 **phenotypes** observed when **monohybrids** cross-or self-pollinate. Each organism inherits one allele for each gene from each parent. In **heterozygotes**, the two alleles are different: Expression of the **dominant allele** masks the phenotypic effect of the **recessive allele. Homozygotes** have identical alleles of a given gene and are therefore **true-breeding**.
- The **law of independent assortment** states that the pair of alleles for a given gene segregates into gametes independently of the pair of alleles for any other gene. In a cross between **dihybrids** (individuals heterozygous for two genes), the offspring have four phenotypes in a 9:3:3:1 ratio.

? *When Mendel did crosses of true-breeding purple- and white-flowered pea plants, the white-flowered trait disappeared from the F_1 generation but reappeared in the F_2 generation. Use genetic terms to explain why that happened.*

CONCEPT 14.2

Probability laws govern Mendelian inheritance *(pp. 278–280)*

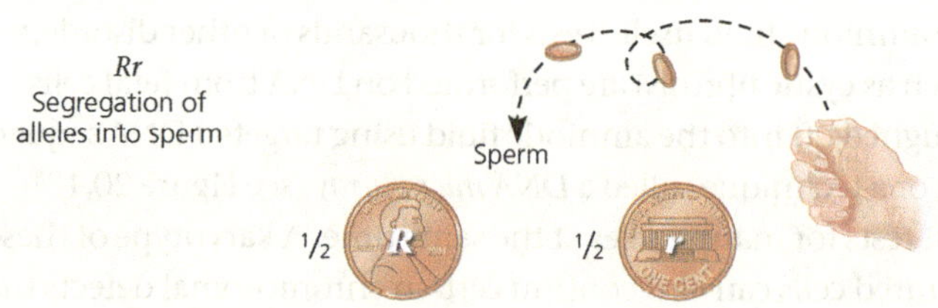

- The **multiplication rule** states that the probability of two or more events occurring together (a statement using *and*) is equal to the product of the individual probabilities of the independent single events. The **addition rule** states that the probability of an event that can occur in two or more independent, mutually exclusive ways (using *or*) is the sum of the individual probabilities.
- The rules of probability can be used to solve complex genetics problems. A dihybrid or other multicharacter cross is equivalent to two or more independent monohybrid crosses occurring simultaneously. In calculating the chances of the various offspring genotypes from such crosses, each character is first considered separately and then the individual probabilities are multiplied.

DRAW IT *Redraw the Punnett square on the right side of Figure 14.8 as two smaller monohybrid Punnett squares, one for each gene. Below each square, list the fractions of each phenotype produced. Use the rule of multiplication to compute the overall fraction of each of the possible dihybrid phenotypes. What is the phenotypic ratio?*

CONCEPT 14.3

Inheritance patterns are often more complex than predicted by simple Mendelian genetics *(pp. 280–285)*

- Extensions of Mendelian genetics for a single gene:

Relationship among alleles of a single gene	Description	Example
Complete dominance of one allele	Heterozygous phenotype same as that of homozygous dominant	*PP* *Pp*
Incomplete dominance of either allele	Heterozygous phenotype intermediate between the two homozygous phenotypes	C^RC^R C^RC^W C^WC^W
Codominance	Both phenotypes expressed in heterozygotes	I^AI^B
Multiple alleles	In the population, some genes have more than two alleles	ABO blood group alleles I^A, I^B, i
Pleiotropy	One gene affects multiple phenotypic characters	Sickle-cell disease

- Extensions of Mendelian genetics for two or more genes:

Relationship among two or more genes	Description	Example
Epistasis	The phenotypic expression of one gene affects the expression of another gene	*BbEe* × *BbEe*; *BE* *bE* *Be* *be*; 9 : 3 : 4
Polygenic inheritance	A single phenotypic character is affected by two or more genes	*AaBbCc* × *AaBbCc*

- The expression of a genotype can be affected by environmental influences, resulting in a range of phenotypes. Polygenic characters that are also influenced by the environment are called **multifactorial** characters.
- An organism's overall phenotype reflects its overall genotype and unique environmental history. Even in more complex inheritance patterns, Mendel's fundamental laws still apply.

? *Which genetic relationships listed in the first column of the two tables above are demonstrated by the inheritance pattern of the ABO blood group alleles? For each genetic relationship, explain why this inheritance pattern is or is not an example.*

CONCEPT 14.4

Many human traits follow Mendelian patterns of inheritance *(pp. 286–292)*

- Analysis of family **pedigrees** can be used to deduce the possible genotypes of individuals and make predictions about future offspring. Such predictions are statistical probabilities rather than certainties.

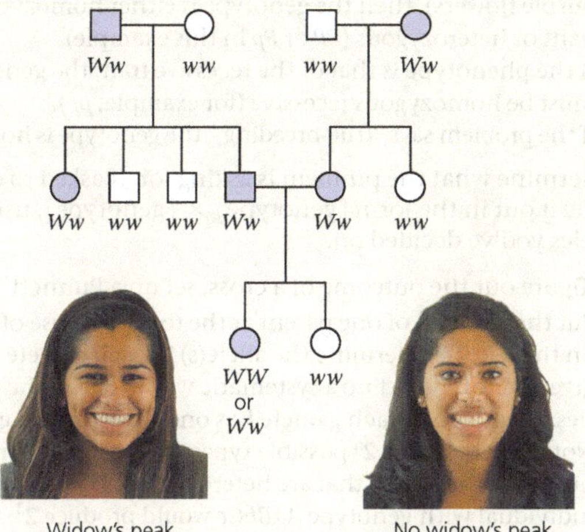

- Many genetic disorders are inherited as simple recessive traits. Most affected (homozygous recessive) individuals are children of phenotypically normal, heterozygous **carriers**.
- The sickle-cell allele has probably persisted for evolutionary reasons: Homozygotes have **sickle-cell disease**, but heterozygotes have an advantage because one copy of the sickle-cell allele reduces both the frequency and severity of malaria attacks.

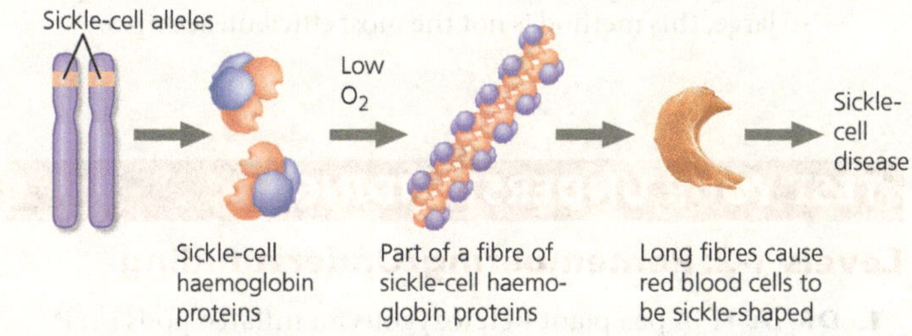

- Lethal dominant alleles are eliminated from the population if affected people die before reproducing. Nonlethal dominant alleles and lethal ones that strike relatively late in life can be inherited in a Mendelian way.
- Many human diseases are multifactorial—that is, they have both genetic and environmental components and do not follow simple Mendelian inheritance patterns.
- Using family histories, genetic counsellors help couples determine the probability that their children will have genetic disorders. Genetic testing of prospective parents can reveal whether they are carriers of recessive alleles associated with specific disorders. Ultrasounds and blood tests can screen for certain disorders in a fetus. **Amniocentesis** and **chorionic villus sampling** can diagnose whether a suspected genetic disorder is present in a fetus. Other genetic tests can be performed after birth.

? *Both members of a couple know that they are carriers of the cystic fibrosis allele. None of their three children have cystic fibrosis, but any one of them might be a carrier. The couple would like to have a fourth child but are worried that he or she would very likely have the disease, since the first three do not. What would you tell the couple? Would it remove some uncertainty from their prediction if they could find out whether the three children are carriers?*

TIPS FOR GENETICS PROBLEMS

1. Write down symbols for the alleles. (These may be given in the problem.) When represented by single letters, the dominant allele is uppercase and the recessive is lowercase.
2. Write down the possible genotypes, as determined by the phenotype.
   a. If the phenotype is that of the dominant trait (for example, purple flowers), then the genotype is either homozygous dominant or heterozygous (*PP* or *Pp* in this example).
   b. If the phenotype is that of the recessive trait, the genotype must be homozygous recessive (for example, *pp*).
   c. If the problem says "true-breeding," the genotype is homozygous.
3. Determine what the problem is asking for. If asked to do a cross, write it out in the form [genotype] × [genotype], using the alleles you've decided on.
4. To figure out the outcome of a cross, set up a Punnett square.
   a. Put the gametes of one parent at the top and those of the other on the left. To determine the allele(s) in each gamete for a given genotype, set up a systematic way to list all the possibilities. (Remember, each gamete has one allele of each gene.) Note that there are 2^n possible types of gametes, where n is the number of gene loci that are heterozygous. For example, an individual with genotype *AaBbCc* would produce $2^3 = 8$ types of gametes. Write the genotypes of the gametes in circles above the columns and to the left of the rows.
   b. Fill in the Punnett square as if each possible sperm were fertilising each possible egg, making all of the possible offspring. In a cross of *AaBbCc* × *AaBbCc*, for example, the Punnett square would have 8 columns and 8 rows, so there are 64 different offspring; you would know the genotype of each and thus the phenotype. Count genotypes and phenotypes to obtain the genotypic and phenotypic ratios. Because the Punnett square is so large, this method is not the most efficient. See tip 5.
5. You can use the rules of probability if a Punnett square would be too big. (For example, see the question at the end of the summary for Concept 14.2 and question 7 below.) You can consider each gene separately (see the section Solving Complex Genetics Problems with the Rules of Probability in Concept 14.2).
6. If the problem gives you the phenotypic ratios of offspring but not the genotypes of the parents in a given cross, the phenotypes can help you deduce the parents' unknown genotypes.
   a. For example, if ½ of the offspring have the recessive phenotype and ½ the dominant, you know that the cross was between a heterozygote and a homozygous recessive.
   b. If the ratio is 3:1, the cross was between two heterozygotes.
   c. If two genes are involved and you see a 9:3:3:1 ratio in the offspring, you know that each parent is heterozygous for both genes. Caution: Don't assume that the reported numbers will exactly equal the predicted ratios. For example, if there are 13 offspring with the dominant trait and 11 with the recessive, assume that the ratio is one dominant to one recessive.
7. For pedigree problems, use the tips in Figure 14.15 and below to determine what kind of trait is involved.
   a. If parents without the trait have offspring with the trait, the trait must be recessive and the parents both carriers.
   b. If the trait is seen in every generation, it is most likely dominant (see the next possibility, though).
   c. If both parents have the trait, then in order for it to be recessive, all offspring must show the trait.
   d. To determine the likely genotype of a certain individual in a pedigree, first label the genotypes of all the family members you can. Even if some of the genotypes are incomplete, label what you do know. For example, if an individual has the dominant phenotype, the genotype must be *AA* or *Aa*; you can write this as *A*–. Try different possibilities to see which fits the results. Use the rules of probability to calculate the probability of each possible genotype being the correct one.

TEST YOUR UNDERSTANDING

Levels 1-2: Remembering/Understanding

1. **DRAW IT** A pea plant heterozygous for inflated pods (*Ii*) is crossed with a plant homozygous for constricted pods (*ii*). Draw a Punnett square for this cross to predict genotypic and phenotypic ratios. Assume that pollen comes from the *ii* plant.
2. A man with type A blood marries a woman with type B blood. Their child has type O blood. What are the genotypes of these three individuals? What genotypes, and in what frequencies, would you expect in future offspring from this marriage?
3. A man has six fingers on each hand and six toes on each foot. His wife and their daughter have the typical number of digits. Remember that extra digits is a dominant trait. What fraction of this couple's children would be expected to have extra digits?
4. **DRAW IT** Two pea plants heterozygous for the characters of pod colour and pod shape are crossed. (See Table 14.1.) Draw a Punnett square to determine the phenotypic ratios of the offspring.

Levels 3-4: Applying/Analysing

5. Flower position, stem length, and seed shape are three characters that Mendel studied. Each is controlled by an independently assorting gene and has dominant and recessive expression as indicated in Table 14.1. If a plant that is heterozygous for all three characters is allowed to self-fertilise, what proportion of the offspring would you expect to be each of the following? (Note: Use the rules of probability instead of a huge Punnett square.)
   (a) homozygous for the three dominant traits
   (b) homozygous for the three recessive traits
   (c) heterozygous for all three characters
   (d) homozygous for axial and tall, heterozygous for seed shape
6. Haemochromatosis is an inherited disease caused by a recessive allele. If a woman and her husband, who are both carriers, have three children, what is the probability of each of the following?
   (a) All three children are of normal phenotype.
   (b) One or more of the three children have the disease.
   (c) All three children have the disease.
   (d) At least one child is phenotypically normal.
   (*Note*: It will help to remember that the probabilities of all possible outcomes always add up to 1.)
7. The genotype of F_1 individuals in a tetrahybrid cross is *AaBbCcDd*. Assuming independent assortment of these four genes, what are the probabilities that F_2 offspring will have the following genotypes?
   (a) *aabbccdd*
   (b) *AaBbCcDd*
   (c) *AABBCCDD*
   (d) *AaBBccDd*
   (e) *AaBBCCdd*

8. What is the probability that each of the following pairs of parents will produce the indicated offspring? (Assume independent assortment of all gene pairs.)
   (a) $AABBCC \times aabbcc \rightarrow AaBbCc$
   (b) $AABbCc \times AaBbCc \rightarrow AAbbCC$
   (c) $AaBbCc \times AaBbCc \rightarrow AaBbCc$
   (d) $aaBbCC \times AABbcc \rightarrow AaBbCc$

9. Keisha and Jerome each have a sibling with sickle-cell disease. Neither Keisha nor Jerome nor any of their parents have the disease, and none of them have ever been tested to see if they carry the sickle-cell allele. Based on this incomplete information, calculate the probability that if this couple has a child, the child will have sickle-cell disease.

10. In 1981, a stray black cat with unusual rounded, curled-back ears was adopted by a family in California. Thousands of descendants of the cat have since been born, and cat fanciers have developed the curl cat into a show breed. Suppose you owned the first curl cat and wanted to develop a true-breeding variety. How would you determine whether the curl allele is dominant or recessive? How would you obtain true-breeding curl cats? How could you be sure they are true-breeding?

11. In tigers, a recessive allele of a particular gene causes both an absence of fur pigmentation (a white tiger) and a cross-eyed condition. If two phenotypically normal tigers that are heterozygous at this locus are mated, what percentage of their offspring will be cross-eyed? What percentage of tigers with crossed eyes will also be white?

12. In corn plants, a dominant allele *I* inhibits kernel colour, while the recessive allele *i* permits colour when homozygous. At a different locus, the dominant allele *P* causes purple kernel colour, while the homozygous recessive genotype *pp* causes red kernels. If plants heterozygous at both loci are crossed, what will be the phenotypic ratio of the offspring?

13. The pedigree below traces the inheritance of alkaptonuria, a biochemical disorder. Affected individuals, indicated here by the coloured circles and squares, are unable to metabolise a substance called alkapton, which colours the urine and stains body tissues. Does alkaptonuria appear to be caused by a dominant allele or by a recessive allele? Fill in the genotypes of the individuals whose genotypes can be deduced. What genotypes are possible for each of the other individuals?

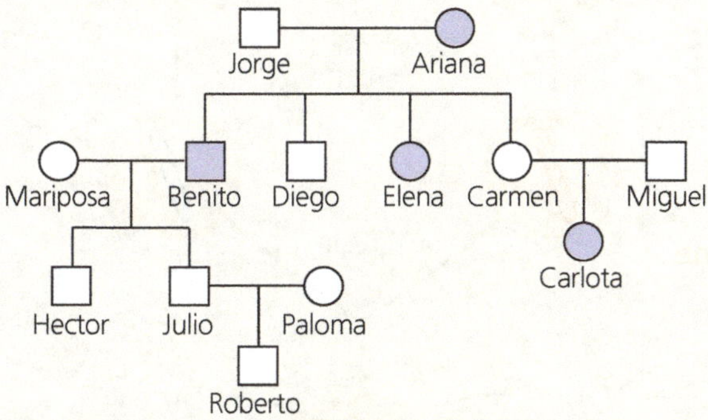

14. Imagine that you are a genetic counsellor, and a couple planning to start a family comes to you for information. Charles was married once before, and he and his first wife had a child with cystic fibrosis. The brother of his current wife, Elaine, died of cystic fibrosis. What is the probability that Charles and Elaine will have a baby with cystic fibrosis? (Neither Charles, Elaine, nor their parents have cystic fibrosis.)

Levels 5-6: Evaluating/Creating

15. **EVOLUTION CONNECTION** Over the past half century, there has been a trend in New Zealand, Australia, and other developed countries for people to marry and start families later in life than did their parents and grandparents. What effects might this trend have on the incidence (frequency) of late-acting dominant lethal alleles in the population?

16. **SCIENTIFIC INQUIRY** You are handed a mystery pea plant with tall stems and axial flowers and asked to determine its genotype as quickly as possible. You know that the allele for tall stems (*T*) is dominant to that for dwarf stems (*t*) and that the allele for axial flowers (*A*) is dominant to that for terminal flowers (*a*).
   (a) Identify all the possible genotypes for your mystery plant.
   (b) Describe the one cross you would do in your garden to determine the exact genotype of your mystery plant.
   (c) While waiting for the results of your cross, you predict the results for each possible genotype listed in part a. Explain how you do this and why this is not called "performing a cross."
   (d) Explain how the results of your cross and your predictions will help you learn the genotype of your mystery plant.

17. **WRITE ABOUT A THEME: INFORMATION** The continuity of life is based on heritable information in the form of DNA. In a short essay (100–150 words), explain how the passage of genes from parents to offspring, in the form of particular alleles, ensures perpetuation of parental traits in offspring and, at the same time, genetic variation among offspring. Use genetic terms in your explanation.

18. **SYNTHESISE YOUR KNOWLEDGE**

Just for fun, imagine that "shirt-striping" is a phenotypic character caused by a single gene. Construct a genetic explanation for the appearance of the family in the above photograph, consistent with their "shirt phenotypes." Include in your answer the presumed allele combinations for "shirt-striping" in each family member. Identify the inheritance pattern shown by the child.

For selected answers, see Appendix A.

15 The Chromosomal Basis of Inheritance

KEY CONCEPTS

15.1 **Mendelian inheritance has its physical basis in the behaviour of chromosomes** *p. 297*

15.2 **Sex-linked genes exhibit unique patterns of inheritance** *p. 300*

15.3 **Linked genes tend to be inherited together because they are located near each other on the same chromosome** *p. 303*

15.4 **Alterations of chromosome number or structure cause some genetic disorders** *p. 308*

15.5 **Some inheritance patterns are exceptions to standard Mendelian inheritance** *p. 312*

Study Tip

Draw chromosomes: As you work on the figure legend questions in this chapter, draw the chromosomes for the crosses described in the questions. Below is a starting sketch for the What If? question that follows Figure 15.3. (Here you are including sex chromosomes to keep track of the offspring.)

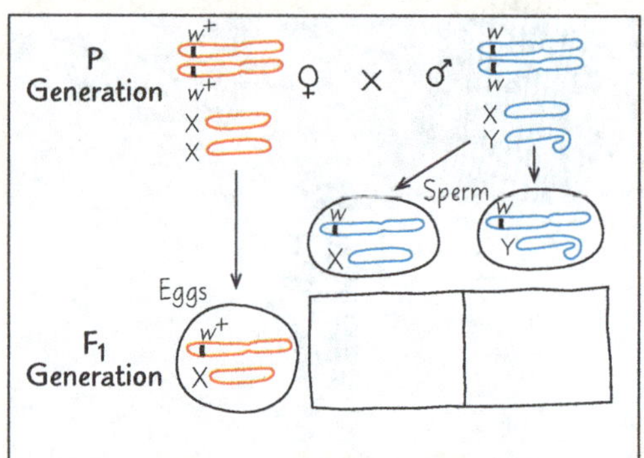

Go to Mastering Biology to access Dynamic Study Modules for revision, 3D BioFlix® animations and high-quality videos, and your interactive Pearson eText.

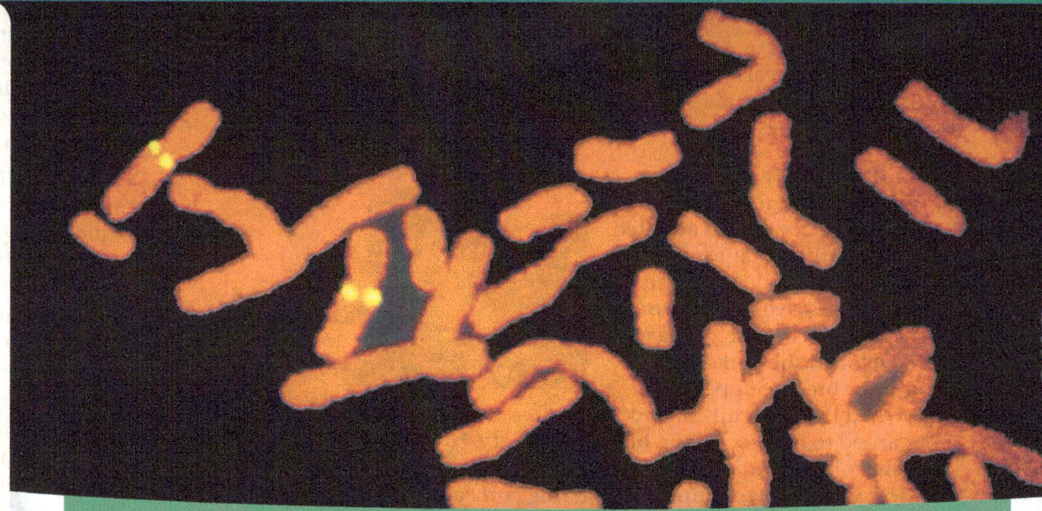

Figure 15.1 The four yellow dots mark the locations of a specific gene, tagged with a fluorescent yellow dye, on a pair of homologous chromosomes. The chromosomes have duplicated, so each chromosome has two sister chromatids, each with a copy of the gene. This provides a visual demonstration that genes—Mendel's "factors"—are segments of DNA located along chromosomes.

What is the relationship between genes and chromosomes?

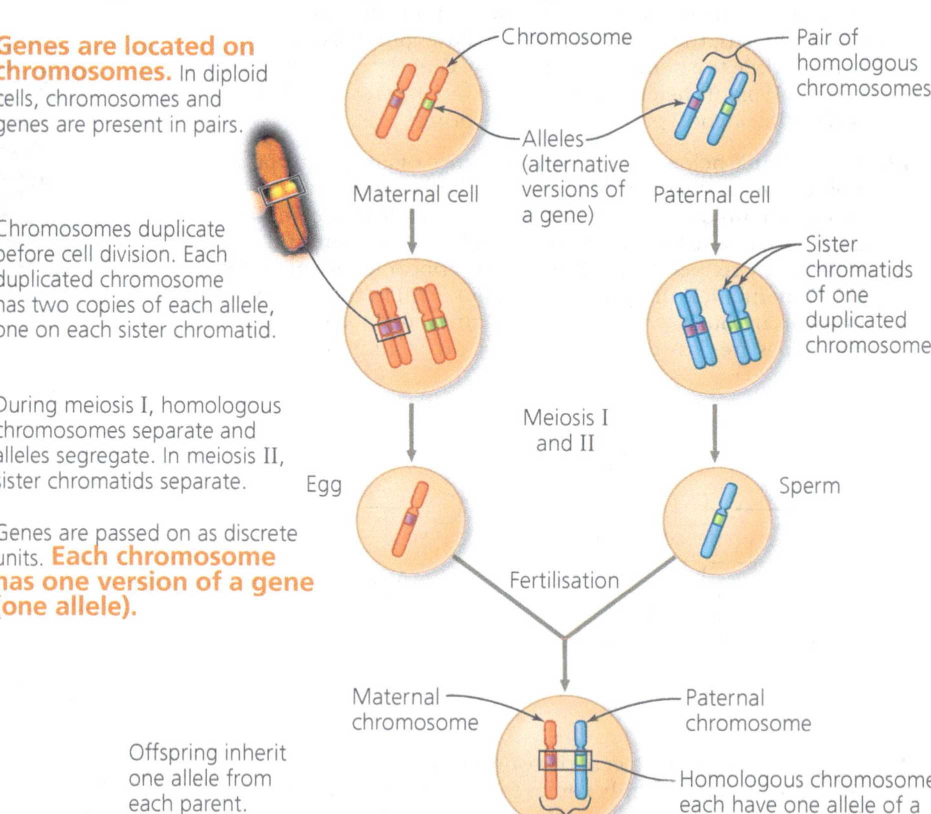

CONCEPT 15.1

Mendelian inheritance has its physical basis in the behaviour of chromosomes

When Gregor Mendel proposed the existence of "hereditary factors" in 1860, no cellular structures had been identified that could house these imaginary units, and most biologists were skeptical. When the processes of mitosis and meiosis were worked out later that century, biologists saw parallels between the behaviour of Mendel's proposed factors (genes) during sexual life cycles and the behaviour of chromosomes: As shown in **Figure 15.1**, chromosomes and genes are both present in pairs in diploid cells, and homologous chromosomes separate and alleles segregate during the process of meiosis. After meiosis, fertilisation restores the paired condition for both chromosomes and genes. Around 1902, Walter S. Sutton, Theodor Boveri, and others independently noted these parallels and began to develop the **chromosome theory of inheritance**. According to this theory, Mendelian genes have specific loci (sites) along chromosomes, and it is the chromosomes that undergo segregation and independent assortment.

The first solid evidence associating a specific gene with a specific chromosome came early in the 1900s from the work of Thomas Hunt Morgan, an experimental embryologist at Columbia University. Although Morgan was initially doubtful about both Mendelian genetics and the chromosome theory, his early experiments provided convincing evidence that chromosomes are indeed the location of Mendel's heritable factors.

Morgan's Choice of Experimental Organism

Many times in the history of biology, important discoveries have come to those insightful or lucky enough to choose an experimental organism suitable for the research problem being tackled. Mendel chose the garden pea because a number of distinct varieties were available. For his work, Morgan selected a species of fruit fly, *Drosophila melanogaster*, a common insect that feeds on the fungi growing on fruit. Fruit flies are prolific breeders; a single mating will produce hundreds of offspring, and a new generation can be bred every two weeks. Morgan's laboratory began using this convenient organism for genetic studies in 1907 and soon became known as "the fly room."

Another advantage of the fruit fly is that it has only four pairs of chromosomes, which are easily distinguishable with a light microscope. There are three pairs of autosomes and one pair of sex chromosomes. Female fruit flies have a pair of homologous X chromosomes, and males have one X chromosome and one Y chromosome.

While Mendel could readily obtain different pea varieties from seed suppliers, Morgan was probably the first person to want different varieties of the fruit fly. He faced the tedious task of carrying out many matings and then microscopically inspecting large numbers of offspring in search of naturally occurring variant individuals. After many months of this, he complained, "Two years' work wasted. I have been breeding those flies for all that time and I've got nothing out of it." Morgan persisted, however, and was finally rewarded with the discovery of a single male fly with white eyes instead of the usual red. The phenotype for a character most commonly observed in natural populations, such as red eyes in *Drosophila*, is called the **wild type** (**Figure 15.2**). Traits that are alternatives to the wild type, such as white eyes in *Drosophila*, are called *mutant phenotypes* because they are due to alleles assumed to have originated as changes, or mutations, in the wild-type allele.

▼ **Figure 15.2 Morgan's first mutant.** Wild-type *Drosophila* flies have red eyes (left). Among his flies, Morgan discovered a mutant male with white eyes (right). This variation made it possible for Morgan to trace a gene for eye colour to a specific chromosome.

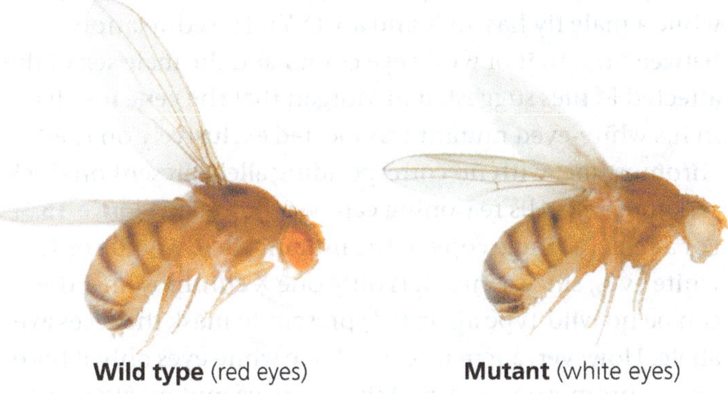

Morgan and his students invented a notation for symbolising alleles in *Drosophila* that is still widely used for fruit flies. For a given character in flies, the gene takes its symbol from the first mutant (non–wild type) discovered. Thus, the allele for white eyes in *Drosophila* is symbolised by *w*. A superscript + identifies the allele for the wild-type trait: w^+ for the allele for red eyes, for example. Over the years, a variety of gene notation systems have been developed for different organisms. For example, human genes are usually written in all capitals, such as *HTT* for the gene involved in Huntington's disease. (More than one letter may be used for an allele, as in *HTT*.)

Correlating Behaviour of a Gene's Alleles with Behaviour of a Chromosome Pair: *Scientific Inquiry*

Morgan mated his white-eyed male fly with a red-eyed female. All the F_1 offspring had red eyes, suggesting that the wild-type allele is dominant. When Morgan bred the F_1 flies to each other, he observed the classical 3:1 phenotypic ratio among the F_2 offspring. However, there was a surprising additional result: The white-eye trait showed up only in males. All the F_2 females had red eyes, while half the males had red eyes and half had white eyes. Therefore, Morgan concluded that somehow a fly's eye colour was linked to its sex. (If the eye colour gene were unrelated to sex, half of the white-eyed flies would have been female.)

Recall that a female fly has two X chromosomes (XX), while a male fly has an X and a Y (XY). The correlation between the trait of white eye colour and the male sex of the affected F_2 flies suggested to Morgan that the gene involved in his white-eyed mutant was located exclusively on the X chromosome, with no corresponding allele present on the Y chromosome. His reasoning can be followed in **Figure 15.3**. For a male, a single copy of the mutant allele would confer white eyes; since a male has only one X chromosome, there can be no wild-type allele (w^+) present to mask the recessive allele. However, a female could have white eyes only if both her X chromosomes carried the recessive mutant allele (w). This was impossible for the F_2 females in Morgan's experiment because all the F_1 fathers had red eyes, so each F_2 female received a w^+ allele on the X chromosome inherited from her father.

Morgan's finding of the correlation between a particular trait and an individual's sex provided support for the chromosome theory of inheritance: namely, that a specific gene is carried on a specific chromosome. In this case, an eye colour gene is carried on the X chromosome.

Figure 15.4 illustrates the relationship between the chromosome theory of inheritance and Mendel's laws. The separation of homologues during anaphase I accounts for the segregation of the two alleles of a gene into separate gametes, and the random arrangement of chromosome pairs at metaphase I accounts for independent assortment of the alleles for two or more genes located on different homologue pairs. This figure traces the same dihybrid pea cross you learned about in Figure 14.8. By carefully studying Figure 15.4, you can see how the behaviour of chromosomes during meiosis in the F_1 generation and subsequent random fertilisation give rise to the F_2 phenotypic ratio observed by Mendel.

In addition to showing that a specific gene is carried on a specific chromosome, Morgan's work indicated that genes located on a sex chromosome exhibit unique inheritance patterns, which we will discuss in the next section. Recognising the importance of Morgan's early work, many bright students were attracted to his fly room.

CONCEPT CHECK 15.1

1. **WHAT IF?** Propose a possible reason that the first naturally occurring mutant fruit fly Morgan saw involved a gene on a sex chromosome and was found in a male.
2. Which one of Mendel's laws describes the inheritance of alleles for a single character? Which law relates to the inheritance of alleles for two characters in a dihybrid cross?
3. **MAKE CONNECTIONS** Review the description of meiosis (see Figure 13.8) and Mendel's laws of segregation and independent assortment (see Concept 14.1). What is the physical basis for each of Mendel's laws?

For suggested answers, see Appendix A.

▼ Figure 15.3 Inquiry

In a cross between a wild-type female fruit fly and a mutant white-eyed male, what colour eyes will the F_1 and F_2 offspring have?

Experiment Thomas Hunt Morgan wanted to analyse the behaviour of two alleles of a fruit fly eye colour gene. In crosses similar to those done by Mendel with pea plants, Morgan and his colleagues mated a wild-type (red-eyed) female with a mutant white-eyed male.

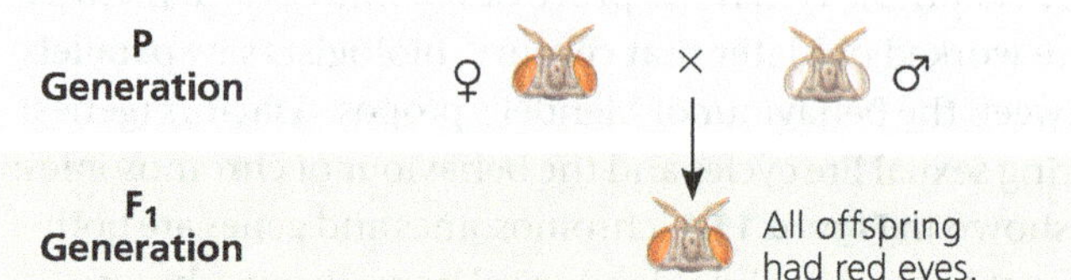

Morgan then bred an F_1 red-eyed female to an F_1 red-eyed male to produce the F_2 generation.

Results The F_2 generation showed a typical Mendelian ratio of 3 red-eyed flies: 1 white-eyed fly. However, all white-eyed flies were males; no females displayed the white-eye trait.

Conclusion All F_1 offspring had red eyes, so the mutant white-eye trait (w) must be recessive to the wild-type red-eye trait (w^+). Since the recessive trait—white eyes—was expressed only in males in the F_2 generation, Morgan deduced that this eye colour gene is located on the X chromosome and that there is no corresponding locus on the Y chromosome.

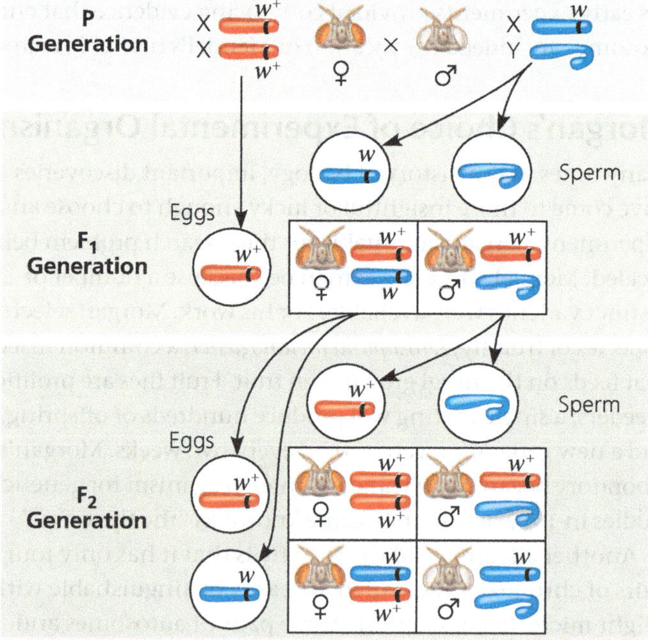

Data from T. H. Morgan, Sex-limited inheritance in *Drosophila*, *Science* 32: 120–122 (1910).

WHAT IF? *Suppose this eye colour gene were located on an autosome rather than a sex chromosome. Predict the phenotypes (including sex) of the F_2 flies in this hypothetical cross. (Hint: Draw Punnett squares, including chromosomes, for the F_1 and F_2 generations; see the Study Tip.)*

▼ Figure 15.4 The chromosomal basis of Mendel's laws. Here we correlate the results of one of Mendel's dihybrid crosses (see Figure 14.8) with the behaviour of chromosomes during meiosis (see Figure 13.8). The arrangement of chromosomes at metaphase I of meiosis and their movement during anaphase I account for, respectively, the independent assortment and segregation of the alleles for seed colour and shape. Each cell that undergoes meiosis in an F_1 plant produces two kinds of gametes. If we count the results for all cells, however, each F_1 plant produces equal numbers of all four kinds of gametes because the alternative chromosome arrangements at metaphase I are equally likely.

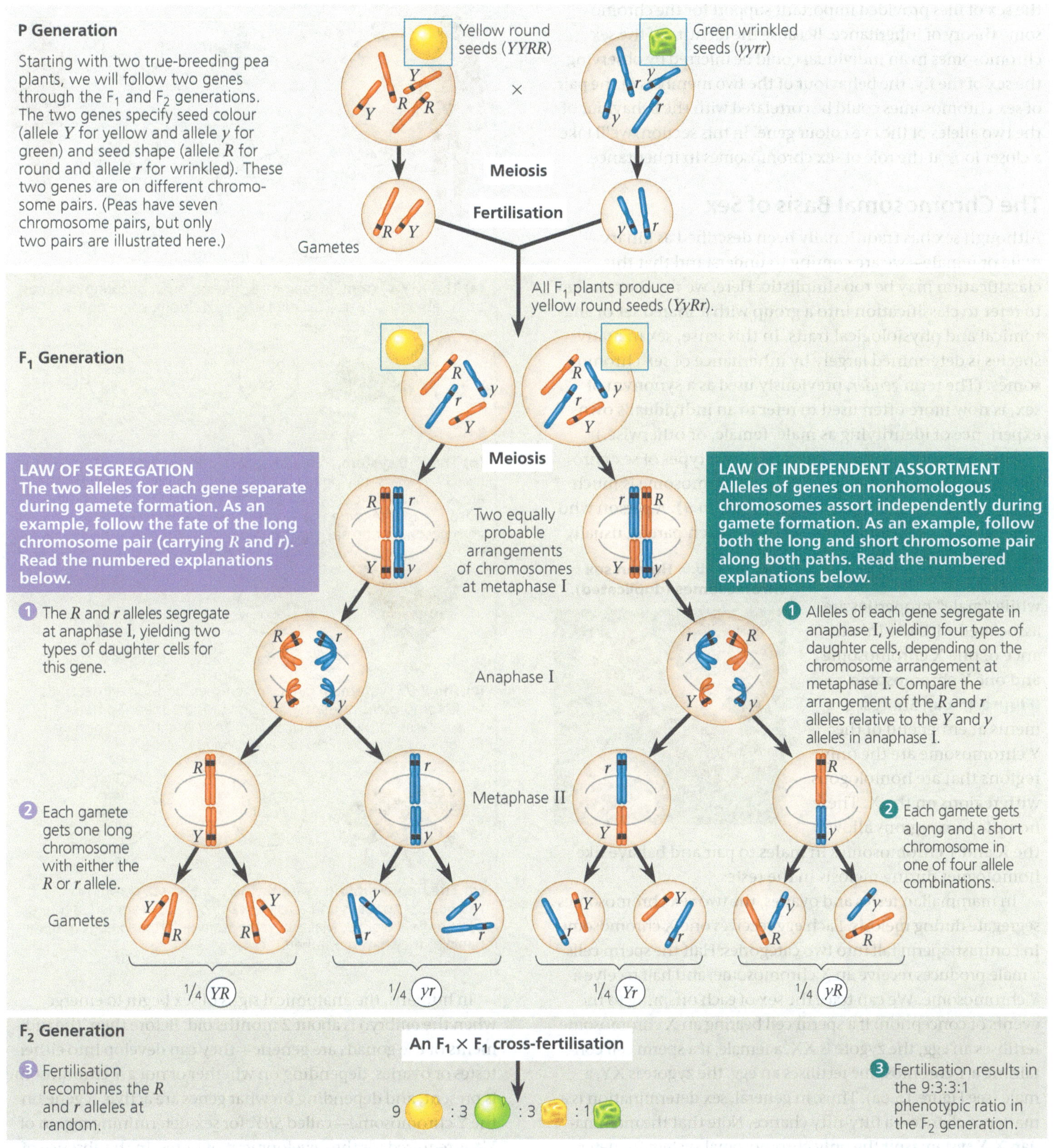

? *If you crossed an F_1 plant above with a plant that was homozygous recessive for both genes (yyrr), how would the phenotypic ratio of the offspring compare with the 9:3:3:1 ratio seen here?*

CONCEPT 15.2

Sex-linked genes exhibit unique patterns of inheritance

Morgan's discovery of a trait (white eyes) that correlated with the sex of flies provided important support for the chromosome theory of inheritance. Because the identity of the sex chromosomes in an individual could be inferred by observing the sex of the fly, the behaviour of the two members of the pair of sex chromosomes could be correlated with the behaviour of the two alleles of the eye colour gene. In this section, we'll take a closer look at the role of sex chromosomes in inheritance.

The Chromosomal Basis of Sex

Although sex has traditionally been described as binary—male or female—we are coming to understand that this classification may be too simplistic. Here, we use the term *sex* to refer to classification into a group with a shared set of anatomical and physiological traits. In this sense, sex in many species is determined largely by inheritance of sex chromosomes. (The term *gender*, previously used as a synonym of sex, is now more often used to refer to an individual's own experience of identifying as male, female, or otherwise.)

Humans and other mammals have two types of sex chromosomes, designated X and Y. The Y chromosome is much smaller than the X chromosome **(Figure 15.5)**. A person who inherits two X chromosomes, one from each parent, usually develops anatomy we associate with the "female" sex, while "male" properties are associated with the inheritance of one X chromosome and one Y chromosome **(Figure 15.6a)**. Short segments at either end of the Y chromosome are the only regions that are homologous with regions on the X. These homologous regions allow the X and Y chromosomes in males to pair and behave like homologues during meiosis in the testes.

▼ **Figure 15.5 Human sex chromosomes (duplicated).**

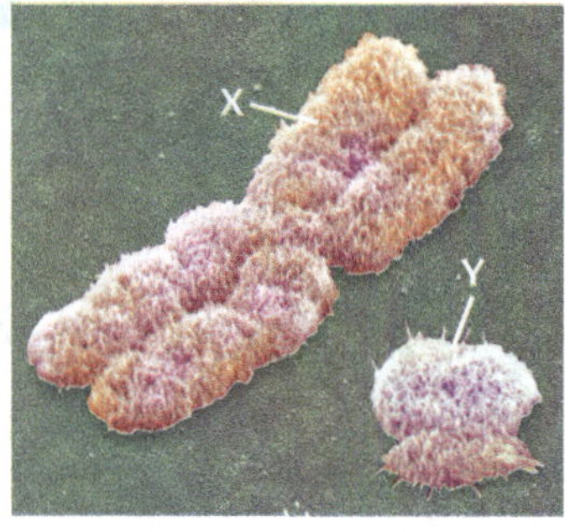

In mammalian testes and ovaries, the two sex chromosomes segregate during meiosis. Each egg receives one X chromosome. In contrast, sperm fall into two categories: Half the sperm cells a male produces receive an X chromosome, and half receive a Y chromosome. We can trace the sex of each offspring to the events of conception: If a sperm cell bearing an X chromosome fertilises an egg, the zygote is XX, a female; if a sperm cell containing a Y chromosome fertilises an egg, the zygote is XY, a male (see Figure 15.6a). Thus, in general, sex determination is a matter of chance—a fifty-fifty chance. Note that the mammalian X-Y system isn't the only chromosomal system for determining sex. **Figure 15.6b–d** illustrates three other systems.

▼ **Figure 15.6 Some chromosomal systems of sex determination.** Numerals indicate the number of autosomes in the species pictured. In *Drosophila*, males are XY, but sex depends on the ratio of the number of X chromosomes to the number of autosome sets, not simply on the presence of a Y chromosome. In some species (not shown here), sex is determined not by chromosomes but by environmental factors such as temperature.

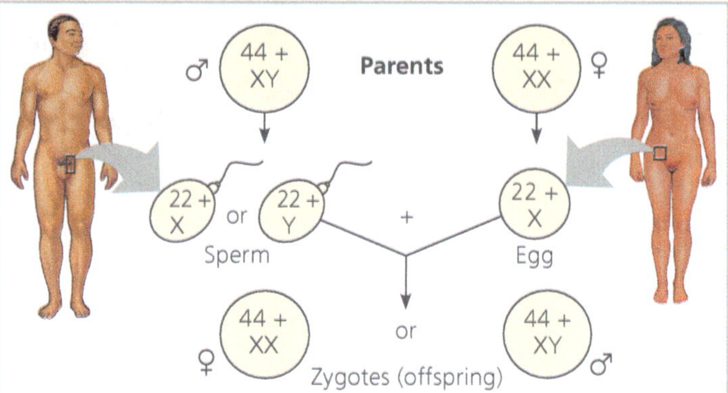

(a) The X-Y system. In mammals, the sex of an offspring depends on whether the sperm cell contains an X chromosome or a Y.

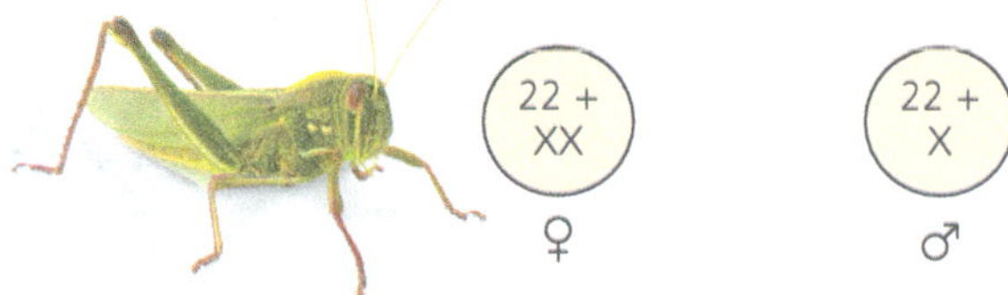

(b) The X-0 system. In grasshoppers, cockroaches, and some other insects, there is only one type of sex chromosome, the X. Females are XX; males have only one sex chromosome (X0). Sex of the offspring is determined by whether the sperm cell contains an X chromosome or no sex chromosome.

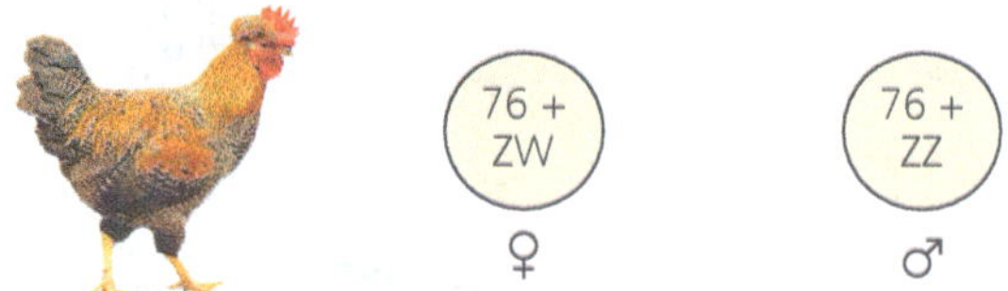

(c) The Z-W system. In birds, some fishes, and some insects, the sex chromosomes present in the egg (not the sperm) differ, and thus determine the sex of offspring. The sex chromosomes are designated Z and W. Females are ZW and males are ZZ.

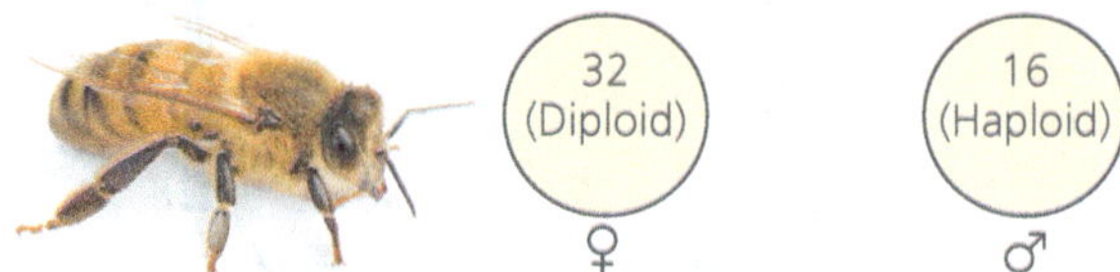

(d) The haplo-diploid system. There are no sex chromosomes in most species of bees and ants. Females develop from fertilised eggs and are thus diploid. Males develop from unfertilised eggs and are haploid; they have no fathers.

In humans, the anatomical signs of sex begin to emerge when the embryo is about 2 months old. Before then, the rudiments of the gonads are generic—they can develop into either testes or ovaries, depending on whether or not a Y chromosome is present, and depending on what genes are active. A gene on the Y chromosome—called *SRY*, for sex-determining region of Y—is required for the development of testes. In the absence of *SRY*, the gonads develop into ovaries, even in an XY embryo.

A gene located on either sex chromosome is called a **sex-linked gene**. The human X chromosome contains approximately 1,100 genes, which are called **X-linked genes**, while genes located on the Y chromosome are called *Y-linked genes*. On the human Y chromosome, researchers have identified 78 genes that code for about 25 proteins (some genes are duplicates). About half of these genes are expressed only in the testis, and some are required for normal testicular functioning and the production of normal sperm. The Y chromosome is passed along virtually intact from a father to all his sons. Because there are so few Y-linked genes, very few disorders are transferred from father to son on the Y chromosome.

The development of female gonads in humans requires a gene called *WNT4* (on chromosome 1, an autosome), which encodes a protein that promotes ovary development. An embryo that is XY but has extra copies of the *WNT4* gene can develop rudimentary female gonads. Overall, sex is determined by the interactions of a network of gene products like these.

The biochemical, physiological, and anatomical features associated with "males" and "females" are turning out to be more complicated than previously thought, with many genes involved in their development. Because of the complexity of this process, many variations exist: Some individuals vary in the number of sex chromosomes in their cells (see Concept 15.4), and others are born with intermediate sexual (*intersex*) characteristics, or with anatomical features that do not match an individual's sense of their own gender (*transgender* individuals). Sex determination is an active area of research that will likely yield a more sophisticated understanding in years to come.

Inheritance of X-Linked Genes

The fact that males and females inherit a different number of X chromosomes leads to a pattern of inheritance different from that produced by genes located on autosomes. While there are very few Y-linked genes, many of which help determine sex, the X chromosomes have numerous genes for characters unrelated to sex. X-linked genes in humans follow the same pattern of inheritance that Morgan observed for the eye colour locus he studied in *Drosophila* (see Figure 15.3). Fathers pass X-linked alleles to all of their daughters but to none of their sons. In contrast, mothers can pass X-linked alleles to both sons and daughters, as shown in **Figure 15.7** for the inheritance of a mild X-linked disorder, red-green colour blindness.

If an X-linked trait is due to a recessive allele, a female will express the phenotype only if she is homozygous for that allele. Because males have only one locus, the terms *homozygous* and *heterozygous* lack meaning for describing their X-linked genes; the term *hemizygous* is used in such cases. Any male receiving the recessive allele from his mother will express the trait. For this reason, far more males than females have X-linked recessive disorders. However, even though the chance of a female inheriting a double dose of the mutant allele is much less than the probability of a male inheriting a single dose, there *are* females with X-linked disorders. For instance, colour blindness is almost always inherited as an X-linked trait. A colour-blind daughter may be born to a colour-blind father whose mate is a carrier (see Figure 15.7c). Because the X-linked allele for colour blindness is relatively rare, though, the probability that such a man and woman will mate is low.

A number of human X-linked disorders are much more serious than colour blindness, such as **Duchenne's muscular dystrophy**, which affects about one in 3,500 males and one in 50,000 females born in New Zealand and Australia. The disease is characterised by a progressive weakening of the muscles and loss of coordination. Affected individuals' life expectancy is into the mid-20s. Researchers have traced the disorder to the absence of a key muscle protein called dystrophin and have mapped the gene for this protein to a specific locus on the X chromosome. Since the gene is known, gene therapy is being explored.

▼ Figure 15.7 The transmission of X-linked recessive traits. In this diagram, red-green colour blindness is used as an example. The superscript *N* represents the dominant allele for normal colour vision carried on the X chromosome, while *n* represents the recessive allele, which has a mutation causing colour blindness. White boxes indicate unaffected individuals, light orange boxes indicate carriers, and dark orange boxes indicate colour-blind individuals.

? *If a colour-blind woman married a man who had normal colour vision, what would be the probable phenotypes of their children?*

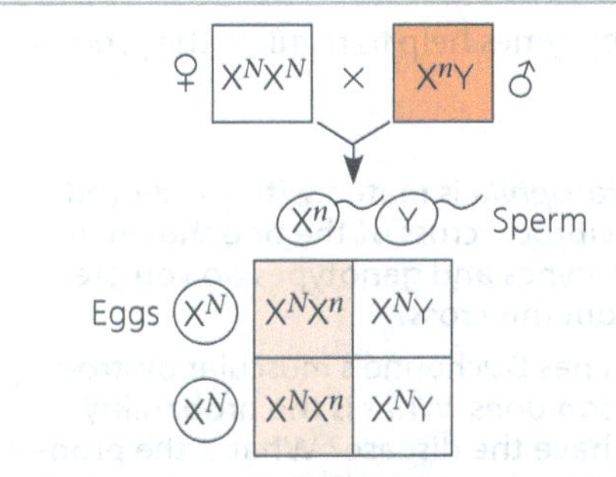

(a) A father with colour-blindness will transmit the mutant allele to all daughters but to no sons. When the mother is a dominant homozygote, the daughters will have the normal phenotype but will be carriers of the mutation.

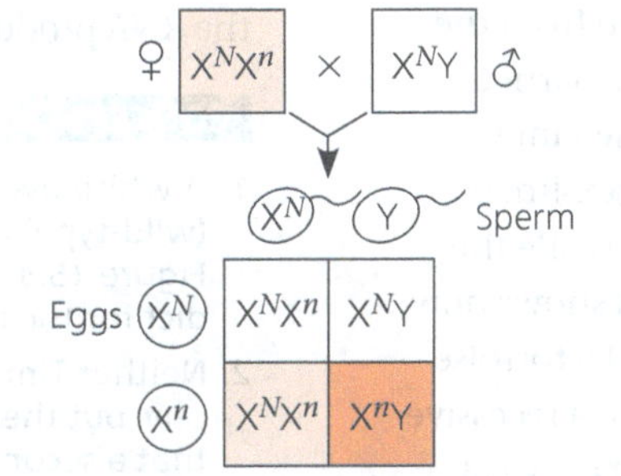

(b) If a carrier mates with a male who has normal colour vision, there is a 50% chance that each daughter will be a carrier like her mother and a 50% chance that each son will have the disorder.

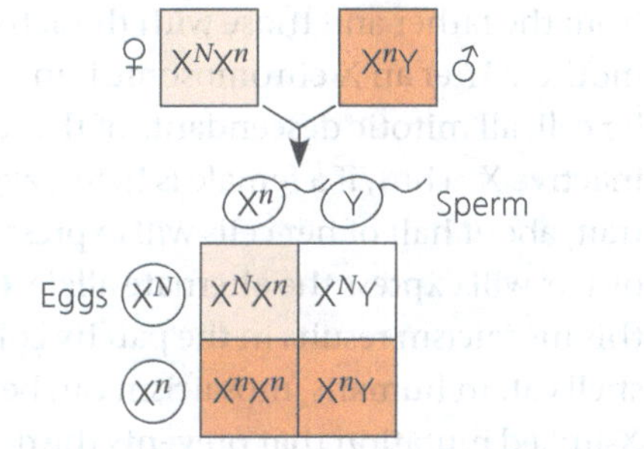

(c) If a carrier mates with a colour-blind male, there is a 50% chance that each child born to them will have the disorder, regardless of sex. Daughters who have normal colour vision will be carriers, whereas males who have normal colour vision will be free of the recessive allele.

Haemophilia is an X-linked recessive disorder defined by the absence of one or more of the proteins required for blood clotting. When a person with haemophilia is injured, bleeding is prolonged because a firm clot is slow to form. Small cuts in the skin are usually not a problem, but bleeding in the muscles or joints can be painful and can lead to serious damage. In the 1800s, haemophilia was widespread among the royal families of Europe. Queen Victoria of England is known to have passed the allele to several of her descendants. Subsequent intermarriage with royal family members of other nations, such as Spain and Russia, further spread this X-linked trait, and its incidence is well documented in royal pedigrees. A few years ago, new genomic techniques allowed sequencing of DNA from tiny amounts isolated from the buried remains of royal family members. The genetic basis of the mutation, and how it resulted in a nonfunctional blood-clotting factor, is now understood. Today, people with haemophilia are treated as needed with intravenous injections of the protein that is missing.

X Inactivation in Female Mammals

Female mammals, including human females, inherit two X chromosomes—twice the number inherited by males—so you may wonder whether females make twice as many of the proteins encoded by X-linked genes as males. In fact, almost all of one X chromosome in each cell in female mammals becomes inactivated during early embryonic development. As a result, the cells of females and males have the same effective dose (one active copy) of most X-linked genes. The inactive X in each cell of a female condenses into a compact object called a **Barr body** (discovered by Canadian anatomist Murray Barr), which lies along the inside of the nuclear envelope. Most of the genes of the X chromosome that forms the Barr body are not expressed. In the ovaries, however, Barr body chromosomes are reactivated in the cells that give rise to eggs, resulting in every female gamete (egg) having an active X after meiosis.

British geneticist Mary Lyon demonstrated that the selection of which X chromosome will form the Barr body occurs randomly and independently in each embryonic cell present at the time of X inactivation. As a consequence, females consist of a *mosaic* of two types of cells: those with the active X derived from the father and those with the active X derived from the mother. After an X chromosome is inactivated in a particular cell, all mitotic descendants of that cell have the same inactive X. Thus, if a female is heterozygous for a sex-linked trait, about half of her cells will express one allele, while the others will express the alternate allele. **Figure 15.8** shows how this mosaicism results in the patchy colouration of a tortoiseshell cat. In humans, mosaicism can be observed in a recessive X-linked mutation that prevents the development of sweat glands. A woman who is heterozygous for this trait has patches of normal skin and patches of skin lacking sweat glands.

Inactivation of an X chromosome involves modification of the DNA and proteins bound to it called histones, including attachment of methyl groups ($-CH_3$) to DNA nucleotides.

▼ Figure 15.8 X inactivation and the tortoiseshell cat. The tortoiseshell gene is on the X chromosome, and the tortoiseshell phenotype requires the presence of two different alleles, one for orange fur and one for black fur. Normally, only females can have both alleles because only they have two X chromosomes. If a female cat is heterozygous for the tortoiseshell gene, she is tortoiseshell. Orange patches are formed by populations of cells in which the X chromosome with the orange allele is active; black patches have cells in which the X chromosome with the black allele is active. ("Calico" cats also have white areas, which are determined by another gene.)

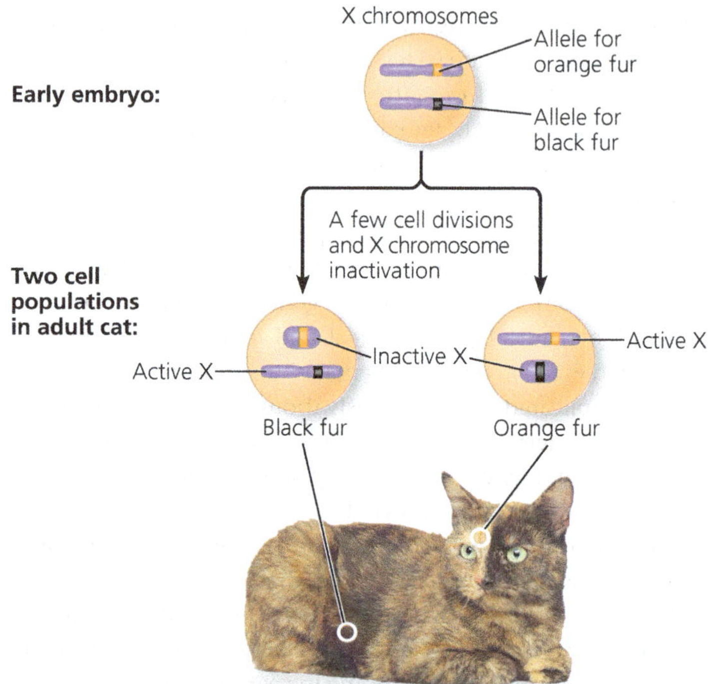

(The regulatory role of DNA methylation is discussed in Concept 18.2.) A particular region of each X chromosome contains several genes involved in the inactivation process. The two regions, one on each X chromosome, associate briefly with each other in each cell at an early stage of embryonic development. Then one of the genes, called *XIST* (for X-inactive specific transcript), becomes active *only* on the chromosome that will become the Barr body (the inactive X). Multiple copies of the RNA product of this gene apparently attach to the X chromosome on which they are made, eventually almost covering it. Interaction of this RNA with the chromosome initiates X inactivation, and the RNA products of nearby genes help to regulate the process.

CONCEPT CHECK 15.2

1. A white-eyed female *Drosophila* is mated with a red-eyed (wild-type) male, the reciprocal cross of the one shown in Figure 15.3. What phenotypes and genotypes do you predict for the offspring from this cross?
2. Neither Tim nor Shonda has Duchenne's muscular dystrophy, but their firstborn son does. What is the probability that a second child will have the disease? What is the probability if the second child is a boy? A girl?
3. **MAKE CONNECTIONS** Consider what you learned about dominant and recessive alleles in Concept 14.1. If a disorder were caused by a dominant X-linked allele, how would the inheritance pattern differ from what we see for recessive X-linked disorders?

For suggested answers, see Appendix A.

CONCEPT 15.3

Linked genes tend to be inherited together because they are located near each other on the same chromosome

The number of genes in a cell is far greater than the number of chromosomes; in fact, each chromosome (except the Y) has hundreds or thousands of genes. Genes located near each other on the same chromosome tend to be inherited together in genetic crosses; such genes are said to be genetically linked and are called **linked genes**. When geneticists follow linked genes in breeding experiments, the results deviate from those expected from Mendel's law of independent assortment.

How Linkage Affects Inheritance

To see how linkage between genes affects the inheritance of two different characters, let's examine another of Morgan's *Drosophila* experiments. In this case, the characters are body colour and wing size, each with two different phenotypes. Wild-type flies have grey bodies and normal-sized wings. In addition to these flies, Morgan had managed to obtain, through breeding, doubly mutant flies with black bodies and wings much smaller than normal, called vestigial wings. The mutant alleles are recessive to the wild-type alleles, and neither gene is on a sex chromosome. In his investigation of these two genes, Morgan carried out the crosses shown in **Figure 15.9**. The first was a P generation cross to generate F_1 dihybrid flies, and the second was essentially a testcross.

▼ Figure 15.9 Inquiry

How does linkage between two genes affect inheritance of characters?

Experiment Morgan wanted to know whether the genes for body colour and wing size are genetically linked and, if so, how this affects their inheritance. The alleles for body colour are b^+ (grey) and b (black), and those for wing size are vg^+ (normal) and vg (vestigial).

Morgan mated true-breeding P (parental) generation flies—wild-type flies with black, vestigial-winged flies—to produce heterozygous F_1 dihybrids ($b^+\ b\ \ vg^+\ vg$), all of which are wild-type in appearance.

He then mated wild-type F_1 dihybrid females with homozygous recessive males. This testcross will reveal the genotype of the eggs made by the dihybrid female.

The male's sperm contributes only recessive alleles, so the phenotype of the offspring reflects the genotype of the female's eggs.

Note: Although only females (with pointed abdomens) are shown, half the offspring in each class would be males (with rounded abdomens).

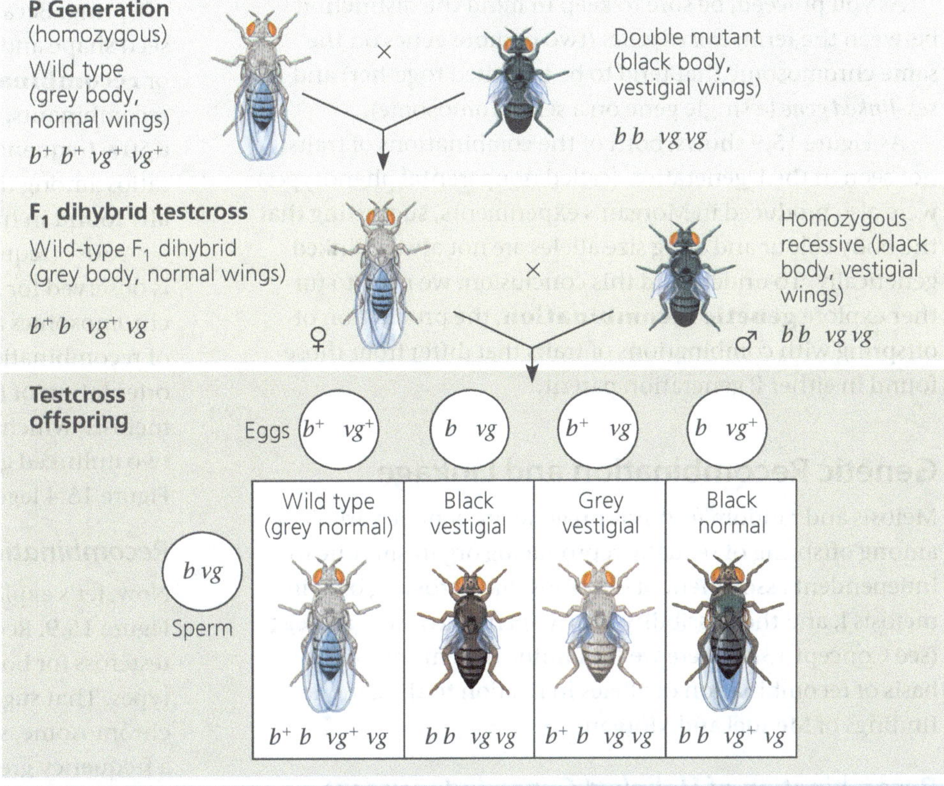

		Wild type		Black vestigial		Grey vestigial		Black normal
Predicted ratios of testcross offspring	if genes are located on different chromosomes:	1	:	1	:	1	:	1
	if genes are located on the same chromosome *and* parental alleles are always inherited together:	1	:	1	:	0	:	0
Results	Data from Morgan's experiment:	965	:	944	:	206	:	185

Conclusion Since most offspring had a parental (P generation) phenotype, Morgan concluded that the genes for body colour and wing size are genetically linked on the same chromosome. However, the production of a relatively small number of offspring with nonparental phenotypes indicated that some mechanism occasionally breaks the linkage between specific alleles of genes on the same chromosome.

Data from T. H. Morgan and C. J. Lynch, The linkage of two factors in *Drosophila* that are not sex-linked, *Biological Bulletin* 23:174–182 (1912).

WHAT IF? *If the parental (P generation) flies had been true-breeding for grey body with vestigial wings and black body with normal wings, which phenotypic class(es) would be largest among the testcross offspring?*

The resulting flies had a much higher proportion of the combinations of traits seen in the P generation flies (called parental phenotypes) than would be expected if the two genes assorted independently. Morgan thus concluded that body colour and wing size are usually inherited together in specific (parental) combinations because the genes are linked; they are near each other on the same chromosome:

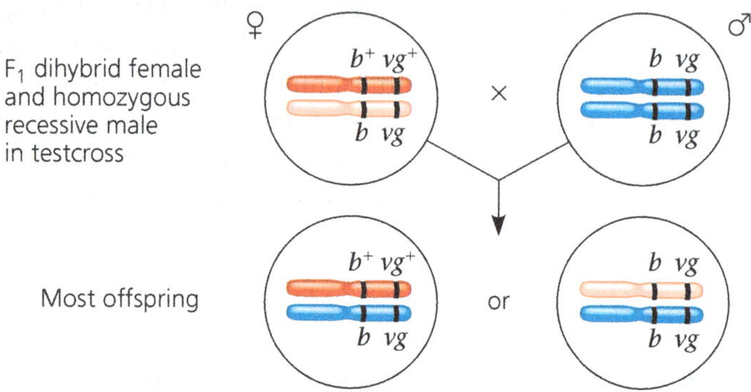

As you proceed, be sure to keep in mind the distinction between the terms *linked genes* (two or more genes on the same chromosome that tend to be inherited together) and *sex-linked gene* (a single gene on a sex chromosome).

As Figure 15.9 shows, both of the combinations of traits not seen in the P generation (called nonparental phenotypes) were also produced in Morgan's experiments, suggesting that the body colour and wing size alleles are not always linked genetically. To understand this conclusion, we need to further explore **genetic recombination**, the production of offspring with combinations of traits that differ from those found in either P generation parent.

Genetic Recombination and Linkage

Meiosis and random fertilisation generate genetic variation among offspring of sexually reproducing organisms due to independent assortment of chromosomes, crossing over in meiosis I, and the possibility of any sperm fertilising any egg (see Concept 13.4). Here we'll examine the chromosomal basis of recombination of alleles in relation to the genetic findings of Mendel and Morgan.

Recombination of Unlinked Genes: Independent Assortment of Chromosomes

Mendel learned from crosses in which he followed two characters that some offspring have combinations of traits that do not match those of either parent. For example, consider a cross of a dihybrid pea plant with yellow round seeds, heterozygous for both seed colour and seed shape (*YyRr*), with a plant homozygous for both recessive alleles (with green wrinkled seeds, *yyrr*). (This cross acts as a testcross because the results will reveal the genotype not only of the dihybrid *YyRr* plant, which we know, but of the gametes made in that plant.) Let's represent the cross by the following Punnett square:

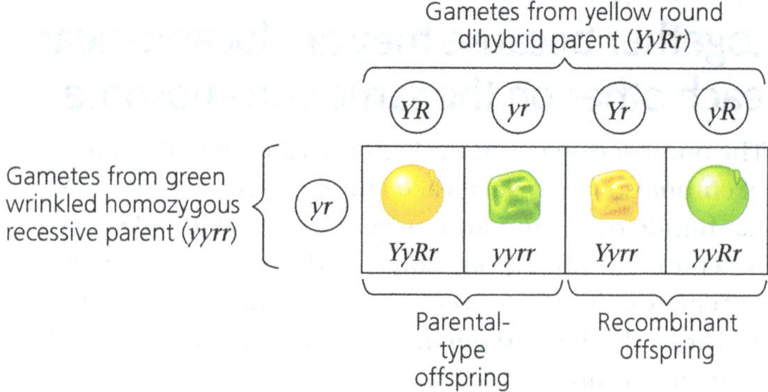

Notice in this Punnett square that one-half of the offspring are expected to inherit a phenotype that matches either of the phenotypes of the P (parental) generation originally crossed to produce the F$_1$ dihybrid (see Figure 15.2). These matching offspring are called **parental types** (short for phenotypes). But two nonparental phenotypes are also found among the offspring. Because these offspring have new combinations of seed shape and colour, they are called **recombinant types**, or **recombinants** for short. When 50% of all offspring are recombinants, as in this example, geneticists say that there is a 50% frequency of recombination. The predicted phenotypic ratios among the offspring are similar to what Mendel actually found in his *YyRr* × *yyrr* crosses.

A 50% frequency of recombination in such testcrosses is observed for any two genes that are located on different chromosomes and thus cannot be linked. The physical basis of recombination between unlinked genes is the random orientation of homologous chromosomes at metaphase I of meiosis, which leads to the independent assortment of the two unlinked genes (see Figure 13.11 and the question in the Figure 15.4 legend).

Recombination of Linked Genes: Crossing Over

Now, let's explain the results of the *Drosophila* testcross in Figure 15.9. Recall that most (83%) of the offspring from the testcross for body colour and wing size had parental phenotypes. That suggested that the two genes were on the same chromosome, since the occurrence of parental types with a frequency greater than 50% indicates that the genes are linked. About 17% of offspring, however, were recombinants.

Seeing these results, Morgan proposed that some process must occasionally break the physical connection between specific alleles of genes on the same chromosome. Later experiments showed that this process, now called **crossing over**, accounts for the recombination of linked genes. In crossing over, which occurs while replicated homologous chromosomes are paired during prophase of meiosis I, a set of proteins breaks the DNA molecules of one maternal and one paternal chromatid and rejoins each to the other

▶ Figure 15.10 Chromosomal basis for recombination of linked genes. In these diagrams re-creating the testcross in Figure 15.9, we track chromosomes as well as genes. The maternal chromosomes (those present in the wild-type F_1 dihybrid) are colour-coded red and pink to distinguish one homologue from the other before any meiotic crossing over has occurred. Because crossing over between the b^+/b and vg^+/vg loci occurs in some, but not most, egg-producing cells, more eggs with parental-type chromosomes than with recombinant ones are produced in the mating females. Fertilisation of the eggs by sperm of genotype *b vg* gives rise to some recombinant offspring. The recombination frequency is the percentage of recombinant flies in the total pool of offspring.

DRAW IT *Suppose, as in the question at the bottom of Figure 15.9, the parental (P generation) flies were true-breeding for grey body with vestigial wings and black body with normal wings. Draw the chromosomes in each of the four possible kinds of eggs from an F_1 female, and label each chromosome as "parental" or "recombinant."*

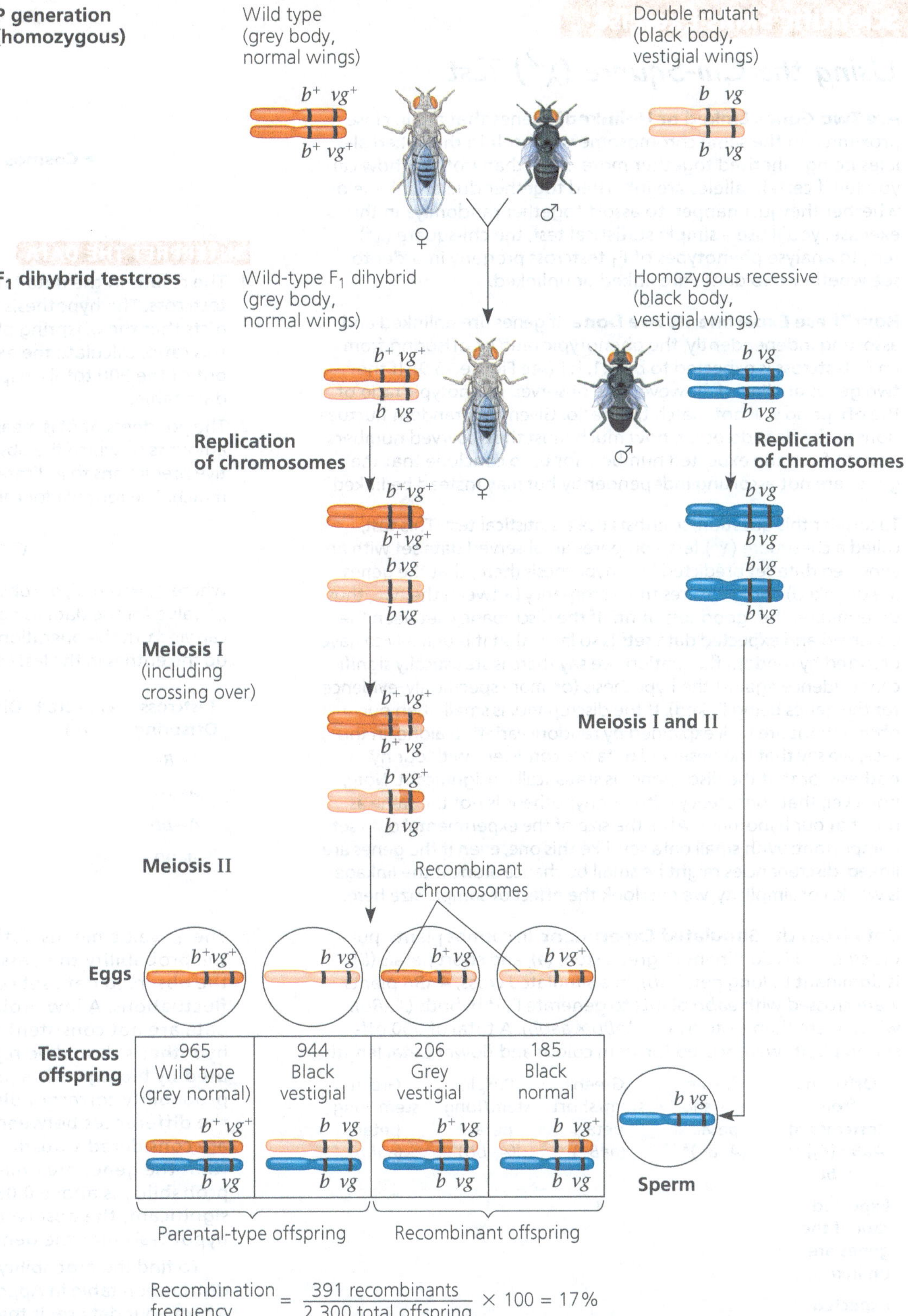

$$\text{Recombination frequency} = \frac{391 \text{ recombinants}}{2{,}300 \text{ total offspring}} \times 100 = 17\%$$

(see Figure 13.9). In effect, when a single crossover occurs, end portions of two nonsister chromatids trade places.

Figure 15.10 shows how crossing over in a dihybrid female fly resulted in recombinant eggs and ultimately recombinant offspring in Morgan's testcross. Most eggs had a chromosome with either the $b^+ vg^+$ or $b\ vg$ parental genotype, but some eggs had a recombinant chromosome ($b^+ vg$ or $b\ vg^+$). Fertilisation of all classes of eggs by homozygous recessive sperm (*b vg*) produced an offspring population in which 17% exhibited a nonparental, recombinant phenotype, reflecting combinations of alleles not seen before in either P generation parent. In the **Scientific Skills Exercise**, you can use a statistical test to analyse the results from an F_1 dihybrid testcross and see whether the two genes assort independently or are linked.

Scientific Skills Exercise

Using the Chi-Square (χ^2) Test

Are Two Genes Linked or Unlinked? Genes that are in close proximity on the same chromosome will result in the linked alleles being inherited together more often than not. But how can you tell if certain alleles are inherited together due to linkage or whether they just happen to assort together randomly? In this exercise, you'll use a simple statistical test, the chi-square (χ^2) test, to analyse phenotypes of F_1 testcross progeny in order to see whether two genes are linked or unlinked.

How These Experiments Are Done If genes are unlinked and assorting independently, the phenotypic ratio of offspring from an F_1 testcross is expected to be 1:1:1:1 (see Figure 15.9). If the two genes are linked, however, the observed phenotypic ratio of the offspring will not match that ratio. Given that random fluctuations in the data do occur, how much must the observed numbers deviate from the expected numbers for us to conclude that the genes are not assorting independently but may instead be linked?

To answer this question, scientists use a statistical test. This test, called a chi-square (χ^2) test, compares an observed data set with an expected data set predicted by a hypothesis (here, that the genes are unlinked) and measures the discrepancy between the two, thus determining the "goodness of fit." If the discrepancy between the observed and expected data sets is so large that it is unlikely to have occurred by random fluctuation, we say there is statistically significant evidence against the hypothesis (or, more specifically, evidence for the genes being linked). If the discrepancy is small, then our observations are well explained by random variation alone. In this case, we say that the observed data are consistent with our hypothesis, or that the discrepancy is statistically insignificant. Note, however, that consistency with our hypothesis is not the same as proof of our hypothesis. Also, the size of the experimental data set is important: With small data sets like this one, even if the genes are linked, discrepancies might be small by chance alone if the linkage is weak. For simplicity, we overlook the effect of sample size here.

Data from the Simulated Experiment In cosmos plants, purple stem (*A*) is dominant to green stem (*a*), and short petals (*B*) is dominant to long petals (*b*). In a simulated cross, *AABB* plants were crossed with *aabb* plants to generate F_1 dihybrids (*AaBb*), which were then testcrossed (*AaBb* × *aabb*). A total of 900 offspring plants were scored for stem colour and flower petal length.

Offspring from testcross of *AaBb* (F_1) × *aabb*	Purple stem/short petals (*A–B–*)*	Green stem/short petals (*aaB–*)	Purple stem/long petals (*A–bb*)	Green stem/long petals (*aabb*)
Expected ratio if the genes are unlinked	1	1	1	1
Expected number of offspring (of 900)				
Observed number of offspring (of 900)	220	210	231	239

*If the phenotype is dominant, a dash is used for the second allele; it could be either the dominant or recessive allele.

▶ **Cosmos plants**

INTERPRET THE DATA

1. The results in the data table are from a simulated F_1 dihybrid testcross. The hypothesis that the two genes are unlinked predicts that the offspring phenotypic ratio will be 1:1:1:1. Using this ratio, calculate the expected number of each phenotype out of the 900 total offspring, and enter the values in that data table.
2. The goodness of fit is measured by χ^2. This statistic measures the amounts by which the observed values differ from their respective predictions to indicate how closely the two sets of values match. The formula for calculating this value is

$$\chi^2 = \sum \frac{(o - e)^2}{e}$$

where $\sum$ = sum of, o = observed and e = expected. Calculate the χ^2 value for the data using the table below. Fill out that table, carrying out the operations indicated in the top row. Then add up the entries in the last column to find the χ^2 value.

Testcross Offspring	Expected (e)	Observed (o)	Deviation ($o - e$)	$(o - e)^2$	$(o - e)^2/e$
A–B–		220			
aaB–		210			
A–bb		231			
aabb		239			
				χ^2 = Sum	

3. The χ^2 value means nothing on its own—it is used to find the probability that, assuming the hypothesis is true, the observed data set could have resulted from random fluctuations. A low probability suggests that the observed data are not consistent with the hypothesis and thus the hypothesis should be rejected. A standard cutoff point used by biologists is a probability of 0.05 (5%). If the probability corresponding to the χ^2 value is 0.05 or less, the differences between observed and expected values are considered statistically significant and the hypothesis (that the genes are unlinked) should be rejected. If the probability is above 0.05, the results are not statistically significant; the observed data are consistent with the hypothesis that the genes are unlinked.

   To find the probability, locate your χ^2 value in the χ^2 distribution table in Appendix D. The "degrees of freedom" (df) of your data set is the number of categories (here, 4 phenotypes) minus 1, so df = 3. **(a)** Determine which values on the df = 3 line of the table your calculated χ^2 value lies between. **(b)** The column headings for these values show the probability range for your χ^2 number. Based on whether there are nonsignificant ($p > 0.05$) or significant ($p \leq 0.05$) differences between the observed and expected values, are the data consistent with the hypothesis that the two genes are unlinked and assorting independently, or is there enough evidence to reject this hypothesis?

New Combinations of Alleles: Variation for Natural Selection

EVOLUTION The physical behaviour of chromosomes during meiosis contributes to the generation of variation in offspring (see Concept 13.4). Each pair of homologous chromosomes lines up independently of other pairs during metaphase I, and crossing over prior to that, during prophase I, can mix and match parts of maternal and paternal homologues. Mendel's elegant experiments show that the behaviour of the abstract entities known as genes—or, more concretely, alleles of genes—also leads to variation in offspring (see Concept 14.1). Now, putting these different ideas together, you can see that the recombinant chromosomes resulting from crossing over may bring alleles together in new combinations, and the subsequent events of meiosis distribute to gametes the recombinant chromosomes in a multitude of combinations, such as the new variants seen in Figures 15.9 and 15.10. Random fertilisation then increases even further the number of variant allele combinations that can be created.

This abundance of genetic variation provides the raw material on which natural selection works. If the traits conferred by particular combinations of alleles are better suited for a given environment, organisms possessing those genotypes will be expected to thrive and leave more offspring, ensuring the continuation of their genetic complement. In the next generation, of course, the alleles will be shuffled anew. Ultimately, the interplay between environment and phenotype (and thus genotype) will determine which genetic combinations persist over time.

Mapping the Distance Between Genes Using Recombination Data: *Scientific Inquiry*

The discovery of linked genes and recombination due to crossing over motivated one of Morgan's students, Alfred H. Sturtevant, to work out a method for constructing a **genetic map**, an ordered list of the genetic loci along a particular chromosome.

Sturtevant hypothesised that the percentage of recombinant offspring, the *recombination frequency*, calculated from experiments like the one in Figures 15.9 and 15.10, depends on the distance between genes on a chromosome. He assumed that crossing over is a random event, with the chance of crossing over approximately equal at all points along a chromosome. Based on these assumptions, Sturtevant predicted that *the farther apart two genes are, the higher the probability that a crossover will occur between them and therefore the higher the recombination frequency*. His reasoning was simple: The greater the distance between two genes, the more points there are between them where crossing over can occur. Using recombination data from various fruit fly crosses, Sturtevant proceeded to assign relative positions to genes on the same chromosomes—that is, to *map* genes.

A genetic map based on recombination frequencies is called a **linkage map**. **Figure 15.11** shows Sturtevant's linkage map of three genes: the body colour (*b*) and wing size (*vg*) genes depicted in Figure 15.10 and a third gene, called cinnabar (*cn*). Cinnabar is one of many *Drosophila* genes affecting eye colour. Cinnabar eyes, a mutant phenotype, are a brighter red than the wild-type colour. The recombination frequency between *cn* and *b* is 9%; that between *cn* and *vg*, 9.5%; and that between *b* and *vg*, 17%. In other words, crossovers between *cn* and *b* and between *cn* and *vg* are about half as frequent as crossovers between *b* and *vg*. Only a map that locates *cn* about midway between *b* and *vg* is consistent with these data, as you can prove to yourself by drawing alternative maps. Sturtevant expressed the distances between genes in **map units**, defining one map unit as equivalent to a 1% recombination frequency.

In practice, the interpretation of recombination data is more complicated than this example suggests. Some genes

▼ Figure 15.11 Research Method

Constructing a Linkage Map

Application A linkage map shows the relative locations of genes along a chromosome.

Technique A linkage map is based on the assumption that the probability of a crossover between two genetic loci is proportional to the distance separating the loci. The recombination frequencies used to construct a linkage map for a particular chromosome are obtained from experimental crosses, such as the cross depicted in Figures 15.9 and 15.10. The distances between genes are expressed as map units, with one map unit equivalent to a 1% recombination frequency. Genes are arranged on the chromosome in the order that best fits the data.

Results In this example, the observed recombination frequencies between three *Drosophila* gene pairs (*b* and *cn*, 9%; *cn* and *vg*, 9.5%; and *b* and *vg*, 17%) best fit a linear order in which *cn* is positioned about halfway between the other two genes:

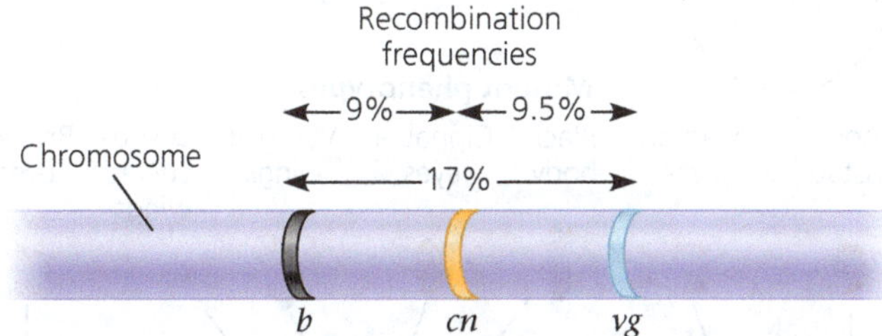

The *b*-*vg* recombination frequency (17%) is slightly less than the sum of the *b*-*cn* and *cn*-*vg* frequencies (9 + 9.5 = 18.5%) because of the few times that one crossover occurs between *b* and *cn* and another crossover occurs between *cn* and *vg*. The second crossover would "cancel out" the first, reducing the observed *b*-*vg* recombination frequency while contributing to the frequency between each of the closer pairs of genes. The value of 18.5% (18.5 map units) is closer to the actual distance between the genes. In practice, a geneticist would add the smaller distances in constructing a map.

on a chromosome are so far from each other that a crossover between them is virtually certain. The observed frequency of recombination in crosses involving two such genes can have a maximum value of 50%, a result indistinguishable from that for genes on different chromosomes. In this case, the physical connection between genes on the same chromosome is not reflected in the results of genetic crosses. Despite being on the same chromosome and thus being *physically connected*, the genes are *genetically unlinked*; alleles of such genes assort independently, as if they were on different chromosomes. In fact, at least two of the genes for pea characters that Mendel studied are now known to be on the same chromosome, but the distance between them is so great that linkage is not observed in genetic crosses. Consequently, the two genes behaved as if they were on different chromosomes in Mendel's experiments. Genes located far apart on a chromosome are mapped by adding the recombination frequencies from crosses involving closer pairs of genes lying between the two distant genes.

Using recombination data, Sturtevant and his colleagues were able to map numerous *Drosophila* genes in linear arrays. They found that the genes clustered into four groups of linked genes (*linkage groups*). Light microscopy had revealed four pairs of chromosomes in *Drosophila*, so the linkage map provided additional evidence that genes are located on chromosomes. Each chromosome has a linear array of specific genes, each gene with its own locus **(Figure 15.12)**.

Because a linkage map is based strictly on recombination frequencies, it gives only an approximate picture of a chromosome. The frequency of crossing over is not actually uniform

▼ Figure 15.12 A partial genetic (linkage) map of a *Drosophila* chromosome. This partial map shows just seven mapped genes on *Drosophila* chromosome II. (DNA sequencing has revealed over 9,000 genes on that chromosome.) The number at each gene locus is the number of map units from the arista length gene (far left).

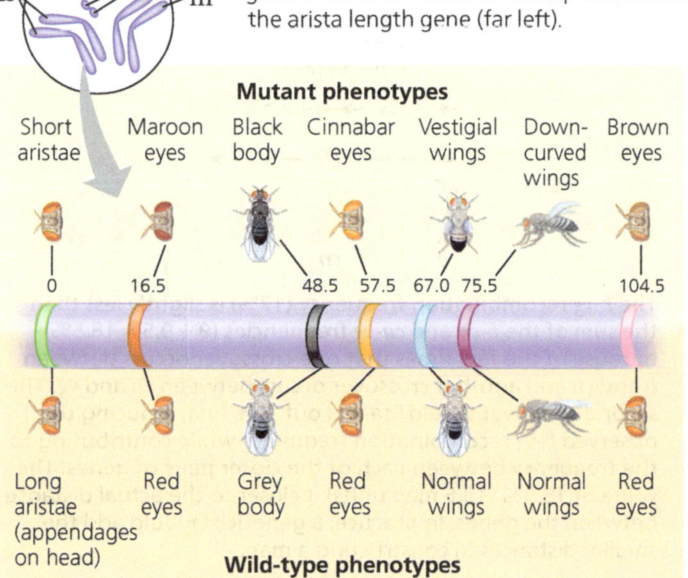

over the length of a chromosome, as Sturtevant assumed, and therefore map units do not correspond to actual physical distances (in nanometres, for instance). A linkage map does portray the order of genes along a chromosome, but it does not accurately portray the precise locations of those genes. Other methods enable geneticists to construct *cytogenetic maps* of chromosomes, which locate genes with respect to chromosomal features, such as stained bands, that can be seen in the microscope. Technical advances over recent decades have enormously increased the rate and affordability of DNA sequencing. Therefore, today, most researchers sequence whole genomes to map the locations of genes of a given species. The entire nucleotide sequence is the ultimate physical map of a chromosome, revealing the physical distances in DNA nucleotides between gene loci (see Concept 21.1). Comparing a linkage map with such a physical map or with a cytogenetic map of the same chromosome, we find that the linear order of genes is identical in all the maps, but the spacing between genes is not.

CONCEPT CHECK 15.3

1. When two genes are located on the same chromosome, what is the physical basis for the production of recombinant offspring in a testcross between a dihybrid parent and a double-mutant (recessive) parent?
2. **VISUAL SKILLS** For each type of offspring of the testcross in Figure 15.9, explain the relationship between its phenotype and the alleles contributed by the female parent. (It will be useful to draw out the chromosomes of each fly and follow the alleles throughout the cross.)
3. **WHAT IF?** Genes *A*, *B*, and *C* are located on the same chromosome. Testcrosses show that the recombination frequency between *A* and *B* is 28% and that between *A* and *C* is 12%. Can you determine the linear order of these genes? Explain.

For suggested answers, see Appendix A.

CONCEPT 15.4

Alterations of chromosome number or structure cause some genetic disorders

As you have learned so far in this chapter, the phenotype of an organism can be affected by small-scale changes involving individual genes. Random mutations are the source of all new alleles, which can lead to new phenotypic traits.

Large-scale chromosomal changes can also affect an organism's phenotype. Physical and chemical disturbances, as well as errors during meiosis, can damage chromosomes in major ways or alter their number in a cell. Large-scale chromosomal alterations in humans and other mammals often lead to spontaneous abortion (miscarriage) of a fetus, and individuals born with these types of genetic defects commonly exhibit various developmental disorders. Plants appear to tolerate such genetic defects better than animals do.

Abnormal Chromosome Number

Ideally, the meiotic spindle distributes chromosomes to daughter cells without error. But there is an occasional mishap, called a **nondisjunction**, in which the members of a pair of homologous chromosomes do not move apart properly during meiosis I or sister chromatids fail to separate during meiosis II **(Figure 15.13)**. In nondisjunction, one gamete receives two of the same type of chromosome and another gamete receives no copy. The other chromosomes are usually distributed normally.

If either of the aberrant gametes unites with a normal one at fertilisation, the zygote will also have an abnormal number of a particular chromosome, a condition known as **aneuploidy**. Fertilisation involving a gamete that has no copy of a particular chromosome will lead to a missing chromosome in the zygote (so that the cell has $2n-1$ chromosomes); the aneuploid zygote is said to be **monosomic** for that chromosome. If a chromosome is present in triplicate in the zygote (so that the cell has $2n+1$ chromosomes), the aneuploid cell is **trisomic** for that chromosome. Mitosis will subsequently transmit the anomaly to all embryonic cells. Monosomy and trisomy are estimated to occur in 10–25% of human conceptions and are the main reason for pregnancy loss. If the organism survives, it usually has a set of traits caused by the abnormal dose of the genes associated with the extra or missing chromosome. Down syndrome is an example of trisomy in humans that will be discussed later.

▼ **Figure 15.13 Meiotic nondisjunction.** Gametes with an abnormal chromosome number can arise by nondisjunction in either meiosis I or meiosis II. For simplicity, the figure does not show the spores formed by meiosis in plants. Ultimately, spores form gametes that have the defects shown. (See Figure 13.6.)

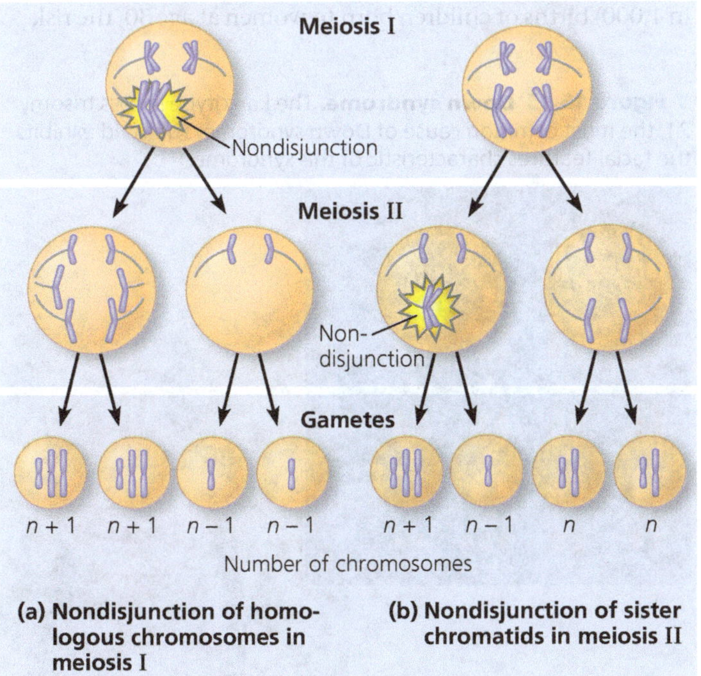

Nondisjunction can also occur during mitosis. If such an error takes place early in embryonic development, then the aneuploid condition is passed along by mitosis to a large number of cells and is likely to have a substantial effect on the organism.

Some organisms have more than two complete chromosome sets in all somatic cells. The general term for this chromosomal alteration is **polyploidy**; the specific terms *triploidy* ($3n$) and *tetraploidy* ($4n$) indicate three and four chromosomal sets, respectively. One way a triploid cell may arise is by the fertilisation of an abnormal diploid egg produced by nondisjunction of all its chromosomes. Tetraploidy could result from the failure of a $2n$ zygote to divide after replicating its chromosomes. Subsequent normal mitotic divisions would then produce a $4n$ embryo.

Polyploidy is fairly common in the plant kingdom. The spontaneous origin of polyploid individuals plays an important role in plant evolution (see Concept 24.2). Many species we eat are polyploid: Bananas are triploid, wheat hexaploid ($6n$), and strawberries octoploid ($8n$). Polyploid animal species are much less common, but there are a few fishes and amphibians known to be polyploid. In general, polyploids are more nearly normal in appearance than aneuploids. One extra (or missing) chromosome apparently disrupts genetic balance more than does an entire extra set of chromosomes.

Alterations of Chromosome Structure

Errors in meiosis or damaging agents such as radiation can cause breakage of a chromosome, which can lead to four types of changes in chromosome structure **(Figure 15.14)**. A **deletion** occurs when a chromosomal fragment is lost. The affected chromosome is then missing certain genes. A broken fragment may become reattached as an extra segment to a sister or nonsister chromatid, producing a **duplication** of a portion of that chromosome. A chromosomal fragment may also reattach to the original chromosome but in the reverse orientation, producing an **inversion**. A fourth possible result of chromosomal breakage is for the fragment to join a nonhomologous chromosome, a rearrangement called a **translocation**.

Deletions and duplications are especially likely to occur during meiosis. In crossing over, nonsister chromatids sometimes exchange unequal-sized segments of DNA, so that one partner gives up more genes than it receives (see Figure 21.13). The products of such an unequal crossover are one chromosome with a deletion and one chromosome with a duplication.

A diploid embryo that is homozygous for a large deletion (both homologues have the deletion) or, if a male, has a single X chromosome with a large deletion is usually missing a number of essential genes. This condition is typically lethal.

Duplications and translocations also tend to be harmful. In reciprocal translocations, in which segments are exchanged between nonhomologous chromosomes, and in

▼ Figure 15.14 Alterations of chromosome structure. Red arrows indicate breakage points. Dark purple highlights the chromosomal parts affected by the rearrangements.

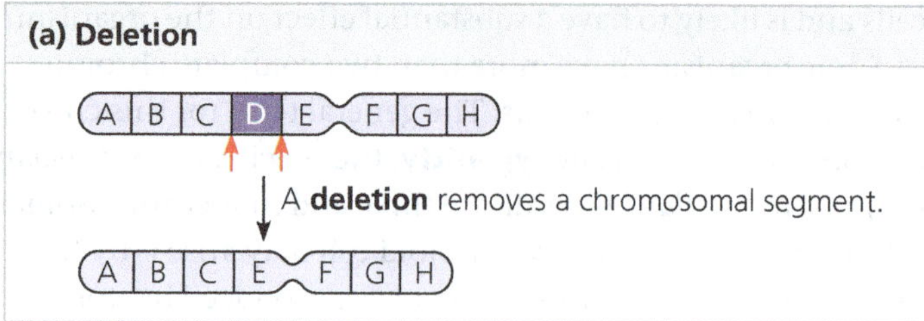

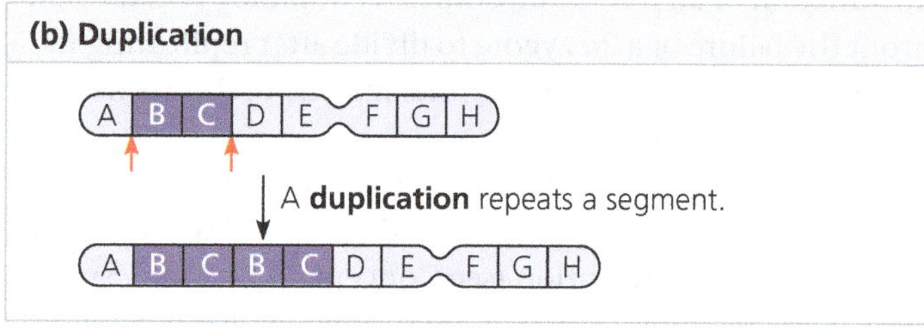

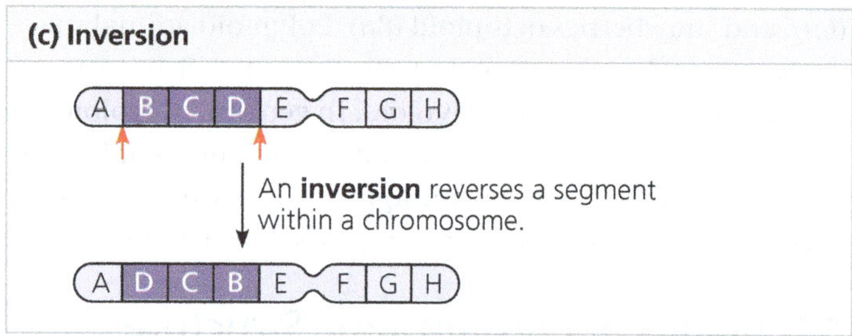

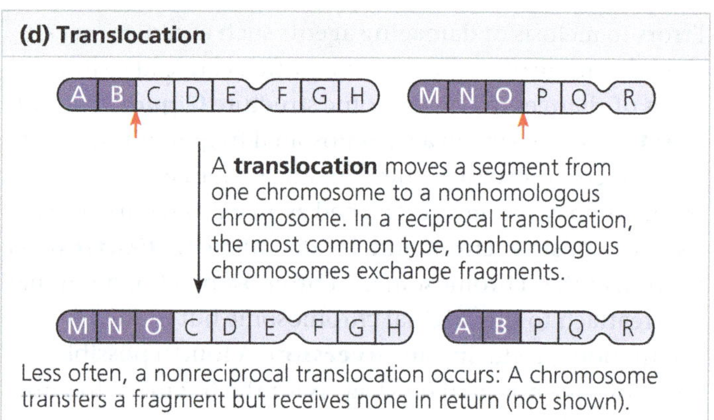

inversions, the balance of genes is not abnormal—all genes are present in their normal doses. Nevertheless, translocations and inversions can alter phenotype because a gene's expression can be influenced by its location among neighbouring genes, which can have devastating effects.

Human Conditions Due to Chromosomal Alterations

Alterations of chromosome number and structure are associated with a number of human conditions, which vary in severity. As described earlier, nondisjunction in meiosis results in aneuploidy in gametes and any resulting zygotes. Although the frequency of aneuploid zygotes may be quite high in humans, most of these chromosomal alterations are so disastrous to development that the affected embryos are spontaneously aborted long before birth. However, some types of aneuploidy appear to upset the genetic balance less than others, so individuals with certain aneuploid conditions can survive to birth and beyond. These individuals have a set of traits—a *syndrome*—characteristic of the type of aneuploidy. Genetic disorders caused by aneuploidy can be diagnosed before birth by fetal testing (see Figure 14.19).

Down Syndrome (Trisomy 21)

One aneuploid condition, **Down syndrome**, affects approximately one in every 1,130 children born in Australia and New Zealand **(Figure 15.15)**. Down syndrome is usually the result of an extra chromosome 21, so that each body cell has a total of 47 chromosomes. Because the cells are trisomic for chromosome 21, Down syndrome is often called *trisomy 21*. Down syndrome includes characteristic facial features, short stature, correctable heart defects, and developmental delays. Individuals with Down syndrome have an increased chance of developing leukaemia and Alzheimer's disease but have a lower rate of high blood pressure, atherosclerosis (hardening of the arteries), stroke, and many types of solid tumours. People with Down syndrome, on average, have a life span shorter than is typical; however, most live to middle age and beyond with proper medical care. Many live independently or at home with their families, are employed, and are valuable contributors to their communities. Almost all males and about half of females with Down syndrome are sexually underdeveloped and sterile.

The frequency of Down syndrome increases with the age of the mother. While the syndrome occurs in just 0.04% (one in 1,000) births of children born to women at age 30, the risk

▼ Figure 15.15 Down syndrome. The karyotype shows trisomy 21, the most common cause of Down syndrome. The child exhibits the facial features characteristic of this syndrome.

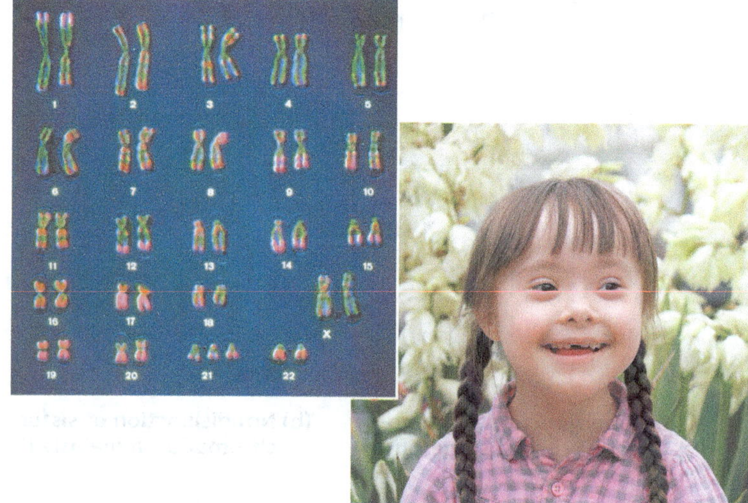

climbs to 1% (one in 100) for mothers at age 40 and is even higher for older mothers. The correlation of Down syndrome with maternal age has not yet been explained. Most cases result from nondisjunction during meiosis I, and some research points to an age-dependent abnormality in meiosis. Trisomies of some other chromosomes (13 and 18) also increase in incidence with maternal age, although infants with other autosomal trisomies rarely survive for long. An increasing number of New Zealand and Australian first-time mothers give birth in their early to mid-30s. Consequently, medical experts recommend all pregnant women receive prenatal screening for trisomies in the embryo, because the test carries low risk and provides useful results. As a minimum, any prenatal or postnatal diagnosis allows medical practitioners to connect parents with appropriate support services.

Aneuploidy of Sex Chromosomes

Aneuploid conditions involving sex chromosomes appear to upset the genetic balance less than those involving autosomes. This may be because the Y chromosome carries relatively few genes. Also, extra copies of the X chromosome simply become inactivated as Barr bodies.

An extra X chromosome in a male, producing XXY, occurs approximately once in every 650 live male births. People with this condition, called *Klinefelter syndrome*, have male sex organs, but their testes are small and produce little or no sperm. Signs and symptoms vary widely, but may include taller than average height, less muscle mass, and enlarged breast tissue. Affected individuals may have learning disabilities. Testosterone replacement therapy helps stimulate changes that typically occur at puberty in males. About one in 1,000 males is born with an extra Y chromosome (XYY). These males undergo typical sexual development and do not exhibit any well-defined syndrome, but tend to be taller than average.

Females with trisomy X (XXX), which occurs once in approximately 1,000 live female births, are generally healthy and have no unusual physical features other than being slightly taller than average. Females with trisomy X are at risk for learning disabilities. Monosomy X, which is called *Turner syndrome*, occurs about once in every 2,500 female births and is the only known viable monosomy in humans. Although these X0 individuals are phenotypically female, they are usually sterile because their sex organs do not mature. When provided with estrogen replacement therapy, girls with Turner syndrome do develop secondary sex characteristics. Most have typical intelligence.

Disorders Caused by Structurally Altered Chromosomes

Many deletions in human chromosomes, even in a heterozygous state, cause severe problems. One such syndrome, known as *cri du chat* ("cry of the cat"), results from a specific deletion in chromosome 5. A child born with this deletion has a small head with unusual facial features, severe intellectual disabilities, and a cry that sounds like the mewing of a distressed cat. Such individuals usually die in infancy or early childhood.

Chromosomal translocations can also occur during mitosis; some have been implicated in certain cancers, including *chronic myelogenous leukaemia* (*CML*). This disease occurs when a reciprocal translocation happens during mitosis of cells that are precursors of white blood cells. The exchange of a large portion of chromosome 22 with a small fragment from a tip of chromosome 9 produces a much shortened, easily recognised chromosome 22, called the *Philadelphia chromosome* **(Figure 15.16)**. Such an exchange causes cancer by creating a new "fused" gene that leads to uncontrolled cell cycle progression. (The mechanism of gene activation will be discussed in Concept 18.5.)

▼ Figure 15.16 Translocation associated with chronic myelogenous leukaemia (CML). The cancerous cells in nearly all CML patients contain an abnormally short chromosome 22, the so-called Philadelphia chromosome, and an abnormally long chromosome 9. These altered chromosomes result from the reciprocal translocation shown here, which presumably occurred in a single white blood cell precursor undergoing mitosis and was then passed along to all descendant cells.

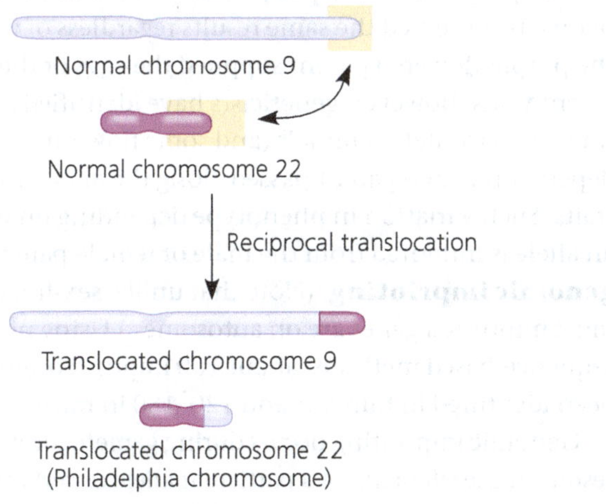

CONCEPT CHECK 15.4

1. About 5% of individuals with Down syndrome have a chromosomal translocation in which a third copy of chromosome 21 is attached to chromosome 14. If this translocation occurred in a parent's gonad, how could it lead to Down syndrome in a child?
2. **MAKE CONNECTIONS** The ABO blood type locus has been mapped on chromosome 9. A father who has type AB blood and a mother who has type O blood have a child with trisomy 9 and type A blood. Using this information, can you tell in which parent the nondisjunction occurred? Explain your answer. (See Figures 14.11 and 15.13.)
3. **MAKE CONNECTIONS** The gene that is activated on the Philadelphia chromosome codes for an intracellular tyrosine kinase. Review the discussion of cell cycle control in Concept 12.3, and explain how the activation of this gene could contribute to the development of cancer.

For suggested answers, see Appendix A.

CONCEPT 15.5

Some inheritance patterns are exceptions to standard Mendelian inheritance

In the previous section, you learned about deviations from the usual patterns of chromosomal inheritance due to abnormal events in meiosis and mitosis. We conclude this chapter by describing two normally occurring exceptions to Mendelian genetics, one involving genes located in the nucleus and the other involving genes located outside the nucleus. In both cases, the sex of the parent contributing an allele is a factor in the pattern of inheritance.

Genomic Imprinting

Throughout our discussions of Mendelian genetics and the chromosomal basis of inheritance, we have assumed that a given allele will have the same effect whether it was inherited from the mother or the father. This is probably a safe assumption most of the time. For example, when Mendel crossed purple-flowered pea plants with white-flowered pea plants, he observed the same results regardless of whether the purple-flowered parent supplied the eggs or the sperm. In recent years, however, geneticists have identified a number of traits in placental mammals (and some flowering plants) that depend on which parent passed along the alleles for those traits. Such variation in phenotype depending on whether an allele is inherited from the male or female parent is called **genomic imprinting**. (Note that unlike sex-linked genes, most imprinted genes are on autosomes.) Using newer DNA sequence-based methods, about 100 imprinted genes have been identified in humans and 120–180 in mice.

Genomic imprinting occurs during gamete formation and results in the silencing of a particular allele of certain genes. Because these genes are imprinted differently in sperm and eggs, the offspring expresses only one allele of an imprinted gene, the one that has been inherited from a specific parent—either the female parent or the male parent, depending on the particular gene. The imprints are then transmitted to all body cells during early development. In each generation, the old imprints are "erased" in gamete-producing cells, and the chromosomes of the developing gametes are newly imprinted according to the sex of the individual forming the gametes. In a given species, the imprinted genes are always imprinted in the same way. For instance, a gene imprinted for maternal allele expression is always imprinted this way, generation after generation.

Consider, for example, the mouse gene for insulin-like growth factor 2 (*Igf2*), one of the first imprinted genes to be identified. Although this growth factor is required for normal prenatal growth, only the paternal allele is expressed **(Figure 15.17a)**. Evidence that the *Igf2* gene is imprinted came initially from

▼ **Figure 15.17 Genomic imprinting of the mouse *Igf2* gene.**

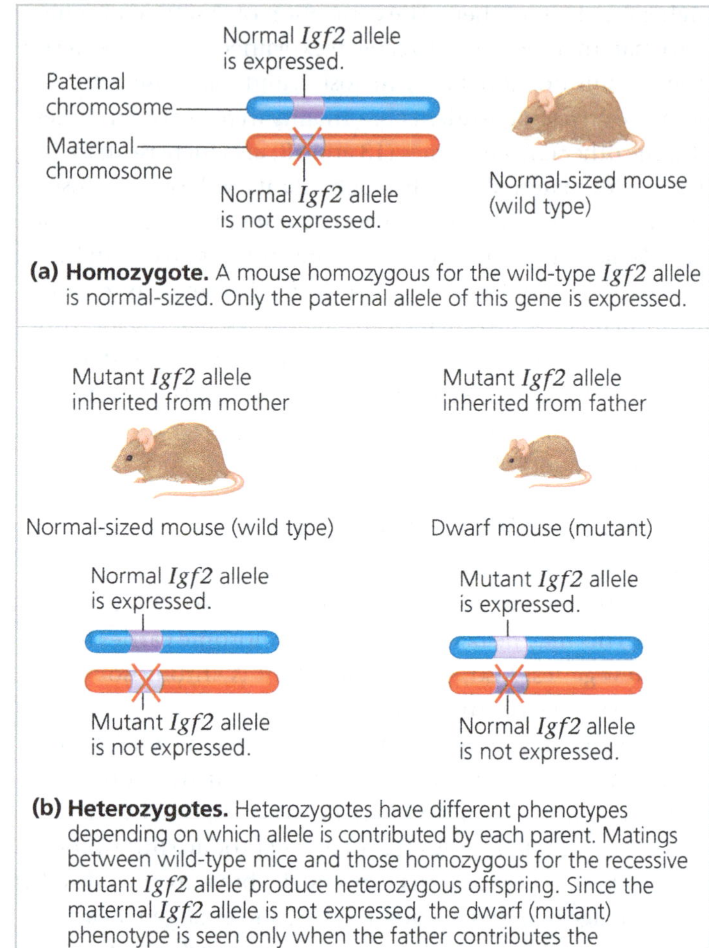

(a) Homozygote. A mouse homozygous for the wild-type *Igf2* allele is normal-sized. Only the paternal allele of this gene is expressed.

(b) Heterozygotes. Heterozygotes have different phenotypes depending on which allele is contributed by each parent. Matings between wild-type mice and those homozygous for the recessive mutant *Igf2* allele produce heterozygous offspring. Since the maternal *Igf2* allele is not expressed, the dwarf (mutant) phenotype is seen only when the father contributes the mutant allele.

crosses between normal-sized (wild-type) mice and dwarf (mutant) mice homozygous for a recessive mutation in the *Igf2* gene. The phenotypes of heterozygous offspring (with one normal allele and one mutant) differed depending on whether the mutant allele came from the mother or the father **(Figure 15.17b)**.

What exactly is a genomic imprint? It turns out that imprinting can involve either silencing an allele in one type of gamete (egg or sperm) or activating it in the other. In many cases, the imprint seems to consist of methyl ($—CH_3$) groups that are added to cytosine nucleotides of one of the alleles. Such methylation may silence the allele, an effect consistent with evidence that heavily methylated genes are usually inactive (see Concept 18.2). However, for a few genes, methylation has been shown to *activate* expression of the allele. This is the case for the *Igf2* gene: Methylation of certain cytosines on the paternal chromosome leads to expression of the paternal *Igf2* allele, by an indirect mechanism involving chromatin structure and protein-DNA interactions.

Genomic imprinting may affect only a small fraction of the genes in mammalian genomes, but most of the known imprinted genes are critical for embryonic development. Recent research has revealed that imprinted genes are involved in regulation of body temperature, sleep, and other metabolic functions. Some evolutionary biologists have proposed that genomic imprinting represents a kind of evolved "competition" between males (who benefit from larger offspring more likely to pass on the father's genes) and female mammals (who benefit from smaller offspring during birth). In experiments with mice, embryos engineered to inherit both copies of certain chromosomes from the same parent usually die before birth, whether that parent is male or female. Normal development seems to require that embryonic cells have exactly one active copy—not zero, not two—of certain genes. The association of improper imprinting with abnormal development and certain cancers has stimulated ongoing studies of how different genes are imprinted.

Inheritance of Organelle Genes

Although our focus in this chapter has been on the chromosomal basis of inheritance, we end with an important amendment: Not all of a eukaryotic cell's genes are located on nuclear chromosomes, or even in the nucleus; some genes are located in organelles in the cytoplasm. Because they are outside the nucleus, these genes are sometimes called *extranuclear genes* or *cytoplasmic genes*. Mitochondria, as well as chloroplasts and other plastids in plants, contain small circular DNA molecules that carry a number of genes. These organelles reproduce themselves and transmit their genes to daughter organelles. Genes on organelle DNA are not distributed to offspring according to the same rules that direct the distribution of nuclear chromosomes during meiosis, so they do not display Mendelian inheritance.

The first hint that extranuclear genes exist came from studies by the German scientist Carl Correns on the inheritance of yellow or white patches on the leaves of an otherwise green plant. In 1909, he observed that the colouration of the offspring was determined only by the maternal parent (the source of eggs) and not by the paternal parent (the source of sperm). Subsequent research showed that such colouration patterns, or variegation **(Figure 15.18)**, are due to mutations in genes controlling pigmentation that are located in the DNA circles in the organelles called plastids (see Concept 6.5). In most plants, a zygote receives all its plastids from the cytoplasm of the egg and none from the sperm, which contributes little more than a haploid set of chromosomes. An egg may contain plastids with different alleles for a pigmentation gene. As the zygote develops, plastids containing wild-type or mutant pigmentation genes are distributed randomly to daughter cells. The pattern of leaf colouration exhibited by a plant depends on the ratio of wild-type to mutant plastids in its various tissues.

Similar maternal inheritance is also the rule for mitochondrial genes in most animals and plants, because the mitochondria passed on to a zygote come from the cytoplasm of the egg. (The few mitochondria contributed by the sperm appear to be destroyed in the egg.) The products of most mitochondrial genes help make up some of the protein complexes of the electron transport chain and ATP synthase (see Figure 9.14). Defects in one or more of these proteins, therefore, reduce the amount of ATP the cell can make and have been shown to cause a number of human disorders in as many as one in every 5,000 births. Because the parts of the body most susceptible to energy deprivation are the nervous system and the muscles, most mitochondrial diseases primarily affect these systems. For example, *mitochondrial myopathy* causes weakness, intolerance of exercise, and muscle deterioration. Another mitochondrial disorder is *Leber's hereditary optic neuropathy*, which can produce sudden blindness in people as young as 20–30 years of age. The four mutations found thus far to cause this disorder affect oxidative phosphorylation during cellular respiration, a crucial function for the cell (see Concept 9.4).

The fact that mitochondrial disorders are inherited only from the mother suggests a way to avoid passing along these disorders. The chromosomes from the egg of an affected mother could be transferred to an egg of a healthy donor that has had its own chromosomes removed. This "two-mother" egg could then be fertilised by a sperm from the prospective father and transplanted into the womb of the prospective mother, becoming an embryo with three parents. After optimising conditions for this approach in monkeys, researchers reported in 2013 that they have successfully carried out this procedure on human eggs. In 2016, clinics in Mexico and China reported the births of such "three-parent" babies. In 2017, the United Kingdom approved this procedure. Meanwhile, New Zealand and Australian legal and medical ethicists are considering the practical, moral, and legislative frameworks necessary for approval of such techniques.

In addition to the rarer diseases clearly caused by defects in mitochondrial DNA, mitochondrial mutations inherited from a person's mother may contribute to at least some types of diabetes and heart disease, as well as to other disorders that commonly debilitate the elderly, such as Alzheimer's disease. In the course of a lifetime, new mutations gradually accumulate in our mitochondrial DNA, and some researchers think that these mutations play a role in the normal aging process.

◀ **Figure 15.18 A painted nettle coleus plant.** The variegated (patterned) leaves on this coleus plant (*Plectranthus scutellarioides*) result from mutations that affect expression of pigment genes located in plastids, which generally are inherited from the maternal parent.

CONCEPT CHECK 15.5

1. Gene dosage—the number of copies of a gene that are actively being expressed—is important to proper development. Identify and describe two processes that establish the proper dosage of certain genes.
2. Reciprocal crosses between two primrose varieties, A and B, produced the following results: A female × B male → offspring with all green (nonvariegated) leaves; B female × A male → offspring with patterned (variegated) leaves. Explain these results.
3. **WHAT IF?** Mitochondrial genes are critical to the energy metabolism of cells, but mitochondrial disorders caused by mutations in these genes are generally not lethal. Why not?

For suggested answers, see Appendix A.

15 Chapter Review

SUMMARY OF KEY CONCEPTS

CONCEPT 15.1

Morgan showed that Mendelian inheritance has its physical basis in the behaviour of chromosomes *(pp. 297–299)*

- Morgan's work with an eye colour gene in *Drosophila* led to the **chromosome theory of inheritance**, which states that genes are located on chromosomes and that the behaviour of chromosomes during meiosis accounts for Mendel's laws.

? *What characteristic of the sex chromosomes allowed Morgan to correlate their behaviour with that of the alleles of the eye colour gene?*

CONCEPT 15.2

Sex-linked genes exhibit unique patterns of inheritance *(pp. 300–302)*

- Sex is often chromosomally based. Humans and other mammals have an X-Y system in which sex is largely determined by whether a Y chromosome is present. Other systems are found in birds, fishes, and insects.
- The sex chromosomes carry **sex-linked genes**, virtually all of which are on the X chromosome (**X-linked**). Any male who inherits a recessive X-linked allele (from his mother) will express the trait, such as colour blindness.
- In mammalian females, one of the two X chromosomes in each cell is randomly inactivated during early embryonic development, becoming highly condensed into a **Barr body**.

? *Why are males affected by X-linked disorders much more often than females?*

CONCEPT 15.3

Linked genes tend to be inherited together because they are located near each other on the same chromosome *(pp. 303–308)*

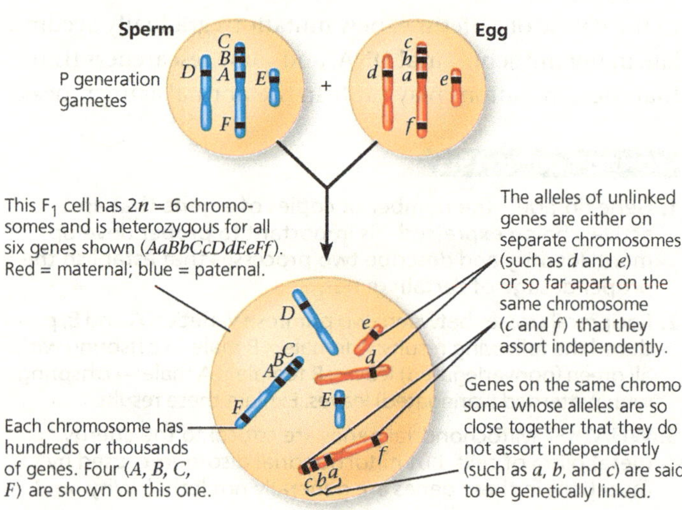

- An F_1 dihybrid testcross yields **parental types** with the same combination of traits as those in the P generation parents and **recombinant types** (**recombinants**) with new combinations of traits not seen in either P generation parent. Because of the independent assortment of chromosomes, unlinked genes exhibit a 50% frequency of recombination in the gametes. For genetically **linked genes, crossing over** between nonsister chromatids during meiosis I accounts for the observed recombinants, always less than 50%.
- The order of genes on a chromosome and the relative distances between them can be deduced from recombination frequencies observed in genetic crosses. These data allow construction of a **linkage map** (a type of **genetic map**). The further apart genes are, the more likely their allele combinations will be recombined during crossing over.

? *Why are specific alleles of two distant genes more likely to show recombination than those of two closer genes?*

CONCEPT 15.4

Alterations of chromosome number or structure cause some genetic disorders *(pp. 308–311)*

- **Aneuploidy**, an abnormal chromosome number, can result from **nondisjunction** during meiosis. When a normal gamete unites with one containing two copies or no copies of a particular chromosome, the resulting zygote and its descendant cells either have one extra copy of that chromosome (**trisomy**, $2n+1$) or are missing a copy (**monosomy**, $2n-1$). **Polyploidy** (extra sets of chromosomes) can result from nondisjunction of all chromosomes.
- Chromosome breakage can result in alterations of chromosome structure: **deletions, duplications, inversions**, and **translocations**. Translocations can be reciprocal or nonreciprocal.
- Changes in the number of chromosomes per cell or in the structure of individual chromosomes can affect the phenotype and, in some cases, lead to disorders. Such alterations cause **Down syndrome** (usually due to trisomy of chromosome 21), certain cancers associated with chromosomal translocations that occur during mitosis, and various other human disorders.

? *Why are inversions and reciprocal translocations less likely to be lethal than are aneuploidy, duplications, deletions, and nonreciprocal translocations?*

CONCEPT 15.5

Some inheritance patterns are exceptions to standard Mendelian inheritance *(pp. 312–313)*

- In mammals, the phenotypic effects of a small number of particular genes depend on which allele is inherited from each parent, a phenomenon called **genomic imprinting**. Imprints are formed during gamete production, with the result that one allele (either maternal or paternal) is not expressed in offspring.
- The inheritance of traits controlled by the genes present in mitochondria and plastids depends solely on the maternal parent because the zygote's cytoplasm containing these organelles comes from the egg. Some diseases affecting the nervous and muscular systems are caused by defects in mitochondrial genes that prevent cells from making enough ATP.

? *Explain how genomic imprinting and inheritance of mitochondrial and chloroplast DNA are exceptions to standard Mendelian inheritance.*

TEST YOUR UNDERSTANDING

Levels 1-2: Remembering/Understanding

1. A man with haemophilia (a recessive, sex-linked condition) has a daughter without haemophilia who marries a man without haemophilia. What is the probability of their daughter having haemophilia? Their son? If they have four sons, that all will be affected?

2. Pseudohypertrophic muscular dystrophy is an inherited disorder that causes gradual deterioration of the muscles. It is seen almost exclusively in boys born to unaffected parents and usually results in death in the early teens. Is this disorder caused by a dominant or a recessive allele? Is its inheritance sex-linked or autosomal? How do you know? Explain why this disorder is almost never seen in girls.

3. A wild-type fruit fly (heterozygous for grey body colour and normal wings) is mated with a black fly with vestigial wings. The offspring have the following phenotypic distribution: wild-type, 778; black vestigial, 785; black normal, 158; grey vestigial, 162. What is the recombination frequency between these genes for body colour and wing size? Is this consistent with the results of the experiment in Figure 15.9? Draw the chromosomes in the wild-type and black parents.

4. A planet is inhabited by creatures that reproduce with the same hereditary patterns seen in humans. Three phenotypic characters are height (T = tall, t = dwarf), head appendages (A = antennae, a = no antennae), and nose shape (S = upturned snout, s = downturned snout). Since the creatures are not "intelligent," Earth scientists are able to do some controlled breeding experiments using various heterozygotes in testcrosses. For tall heterozygotes with antennae, the offspring are tall/antennae, 46; dwarf/antennae, 7; dwarf/no antennae, 42; tall/no antennae, 5. For heterozygotes with antennae and an upturned snout, the offspring are antennae/upturned snout, 47; antennae/downturned snout, 2; no antennae/downturned snout, 48; no antennae/upturned snout, 3. Calculate the recombination frequencies for both experiments.

Levels 3-4: Applying/Analysing

5. Scientists do a testcross of the creatures from problem 4 using heterozygotes for height and nose shape. The offspring are tall/upturned snout, 41; dwarf/upturned snout, 8; dwarf/downturned snout, 43; tall/downturned snout, 8. Calculate the recombination frequency from these data, and then use your answer from problem 4 to determine the correct order of the three linked genes.

6. A wild-type fruit fly (heterozygous for grey body colour and red eyes) is mated with a black fruit fly with purple eyes. The offspring are wild-type, 721; black/purple, 751; grey/purple, 49; black/red, 45. What is the recombination frequency between these genes for body colour and eye colour? Using information from problem 3, what fruit flies (genotypes and phenotypes) would you mate to determine the order of the body colour, wing size, and eye colour genes on the chromosome?

7. Assume that genes A and B are 50 map units apart on the same chromosome. An animal heterozygous at both loci is crossed with one that is homozygous recessive at both loci. What percentage of the offspring will show recombinant phenotypes resulting from crossovers? Without knowing these genes are on the same chromosome, how would you interpret the results of this cross?

8. Two genes of a flower, one controlling blue (B) versus white (b) petals and the other controlling round (R) versus oval (r) stamens, are linked and are 10 map units apart. You cross a homozygous blue/oval plant with a homozygous white/round plant. The resulting F_1 progeny are crossed with homozygous white/oval plants, and 1,000 offspring plants are obtained. How many F_2 plants of each of the four phenotypes do you expect?

9. You design *Drosophila* crosses to provide recombination data for gene a, which is located on the chromosome shown in Figure 15.12. Gene a has recombination frequencies of 14% with the vestigial wing locus and 26% with the brown eye locus. Approximately where is a located along the chromosome relative to these two loci?

Levels 5-6: Evaluating/Creating

10. Banana plants, which are triploid, are seedless and therefore sterile. Thinking about meiosis, propose a possible explanation.

11. **EVOLUTION CONNECTION** Crossing over is thought to be evolutionarily advantageous because it continually shuffles genetic alleles into novel combinations. It was thought that Y-linked genes might degenerate because they lack homologous genes on the X chromosome with which to pair up prior to crossing over. However, when the Y chromosome was sequenced, scientists discovered eight large regions that were internally homologous to each other, and several genes are duplicates. (Y chromosome researcher David Page has called it a "hall of mirrors.") How might this be beneficial?

12. **SCIENTIFIC INQUIRY DRAW IT** Assume you are mapping genes A, B, C, and D in *Drosophila*. You know that these genes are linked on the same chromosome, and you determine the recombination frequencies between each pair of genes to be as follows: A and B, 8%; A and C, 28%; A and D, 25%; B and C, 20%; B and D, 33%.

(a) Describe how you determined the recombination frequency for each pair of genes.

(b) Draw a chromosome map based on your data.

13. **WRITE ABOUT A THEME: INFORMATION** The continuity of life is based on heritable information in the form of DNA. In a short essay (100–150 words), relate the structure and behaviour of chromosomes to inheritance in both asexually and sexually reproducing species.

14. **SYNTHESISE YOUR KNOWLEDGE**

Butterflies have an X-Y sex determination system that is different from that of flies or humans. Female butterflies may be either XY or X0, while butterflies with two or more X chromosomes are males. This photograph shows a tiger swallowtail *gynandromorph*, which is half male (left side) and half female (right side). Given that the first division of the zygote divides the embryo into the future right and left halves of the butterfly, propose a hypothesis that explains how nondisjunction during the first mitosis might have produced this unusual-looking butterfly.

For selected answers, see Appendix A.

16 The Molecular Basis of Inheritance

KEY CONCEPTS

Study Tip

Draw DNA replication: Make up a single-stranded DNA sequence that is 40 nucleotides in length. As you go through Concepts 16.1 and 16.2, use your sequence to draw a double-stranded DNA molecule, labelling the ends. Then diagram the process of replication of your DNA molecule.

DNA Structure and Replication

T C A G C G T A A G G T G A T G T A T A...

to access Dynamic Study Modules for revision, 3D BioFlix® animations and high-quality videos, and your interactive Pearson eText.

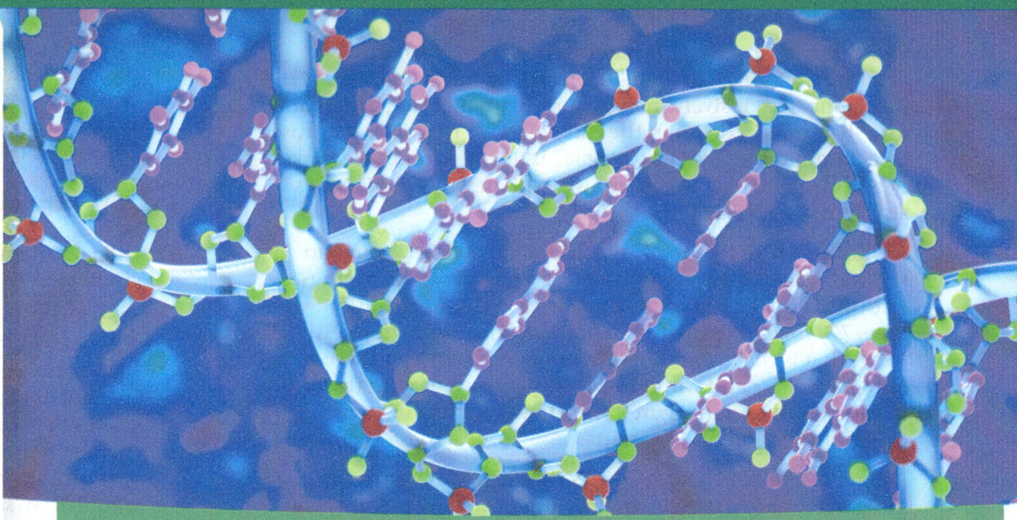

Figure 16.1 The elegant double-helical structure of deoxyribonucleic acid (DNA) shook the scientific world when it was proposed in April 1953 by James Watson and Francis Crick. The DNA you inherited from your parents contains all your genes—your genetic information.

How does DNA replication transmit genetic information?

DNA replication allows genetic information to be inherited from a parent cell to daughter cells (by mitosis) and from generation to generation (starting with meiosis).

Unduplicated chromosome (one DNA molecule and proteins)

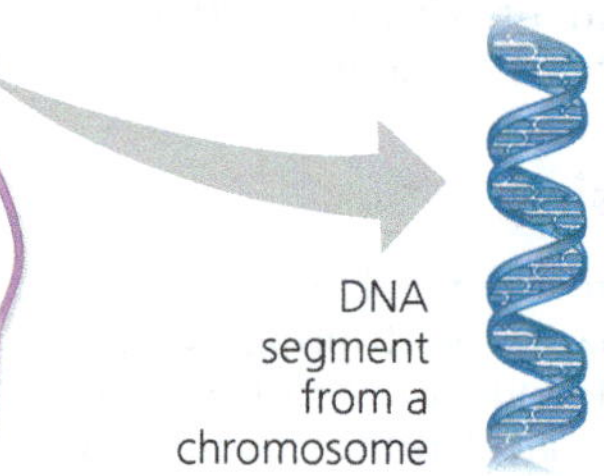

Each gene is a unit of hereditary information consisting of a specific DNA sequence.

DNA replication begins.

Replication begins at multiple sites (origins), each forming a replication bubble with a fork at each end.

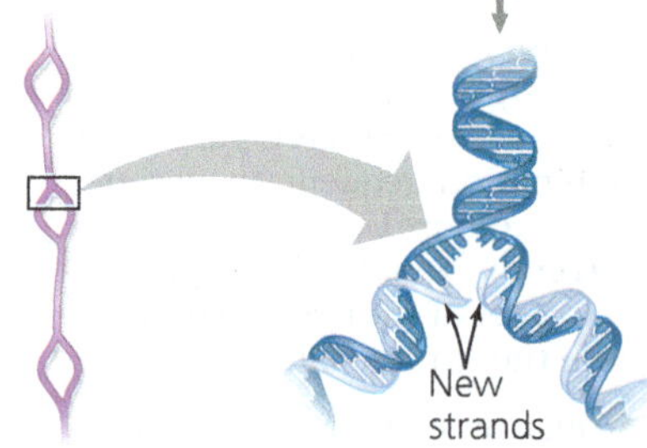

DNA replication and chromosome condensation is completed.

Duplicated and condensed chromosome (two DNA molecules and proteins)

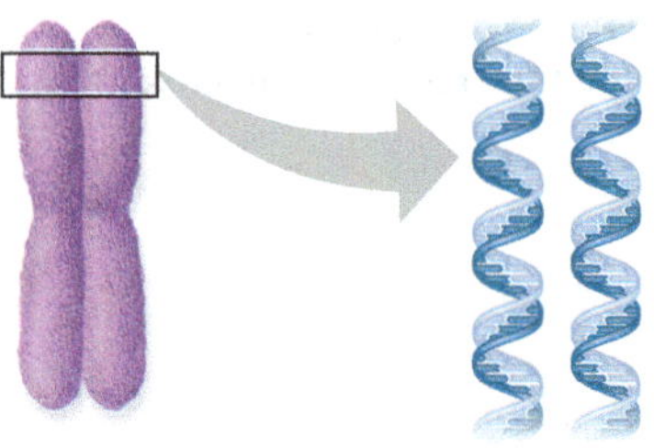

Two DNA molecules, which are distributed to daughter cells.

CONCEPT 16.1

DNA is the genetic material

Today, even schoolchildren have heard of DNA, and scientists routinely manipulate DNA in the laboratory. Early in the 20th century, however, identifying the molecules of inheritance posed a major challenge to biologists.

The Search for the Genetic Material: *Scientific Inquiry*

Once T. H. Morgan's group showed that genes exist as parts of chromosomes (described in Concept 15.1), the two chemical components of chromosomes—DNA and protein—emerged as the leading candidates for the genetic material. Until the 1940s, the case for proteins seemed stronger: Biochemists had identified proteins as a class of macromolecules with great heterogeneity and specificity of function, essential requirements for the hereditary material. Moreover, little was known about nucleic acids, whose physical and chemical properties seemed far too uniform to account for the multitude of specific inherited traits exhibited by every organism. This view gradually changed as the role of DNA in heredity was worked out in studies of bacteria and the viruses that infect them, systems far simpler than fruit flies or humans. Let's trace the search for the genetic material as a case study in scientific inquiry.

Evidence That DNA Can Transform Bacteria

In 1928, a British medical officer named Frederick Griffith was trying to develop a vaccine against pneumonia. He was studying *Streptococcus pneumoniae*, a bacterium that causes pneumonia in mammals. Griffith had two strains (varieties) of the bacterium, one pathogenic (disease-causing) and one nonpathogenic (harmless). He was surprised to find that when he killed the pathogenic bacteria with heat and then mixed the cell remains with living bacteria of the nonpathogenic strain, some of the living cells became pathogenic **(Figure 16.2)**. Furthermore, this newly acquired trait of pathogenicity was inherited by all the descendants of the transformed bacteria. Apparently, some chemical component of the dead pathogenic cells caused this heritable change, although the identity of the substance was not known. Griffith called the phenomenon **transformation**, now defined as a change in genotype and phenotype due to the assimilation of external DNA by a cell. Later work by Oswald Avery, Maclyn McCarty, and Colin MacLeod identified the transforming substance as DNA.

Scientists remained skeptical, however, since many still viewed proteins as better candidates for the genetic material. Also, many biologists were not convinced that bacterial genes would be similar in composition and function to those of more complex organisms. But the major reason for the continued doubt was that so little was known about DNA.

▼ **Figure 16.2 Inquiry**

Can a genetic trait be transferred between different bacterial strains?

Experiment Frederick Griffith studied two strains of the bacterium *Streptococcus pneumoniae*. The S (smooth) strain can cause pneumonia in mice; it is pathogenic because the cells have an outer capsule that protects them from an animal's immune system. Cells of the R (rough) strain lack a capsule and are nonpathogenic. To test for the trait of pathogenicity, Griffith injected mice with the two strains:

Living S cells (pathogenic control)
Living R cells (nonpathogenic control)
Heat-killed S cells (nonpathogenic control)
Mixture of heat-killed S cells and living R cells

Results

Mouse dies | Mouse healthy | Mouse healthy | Mouse dies

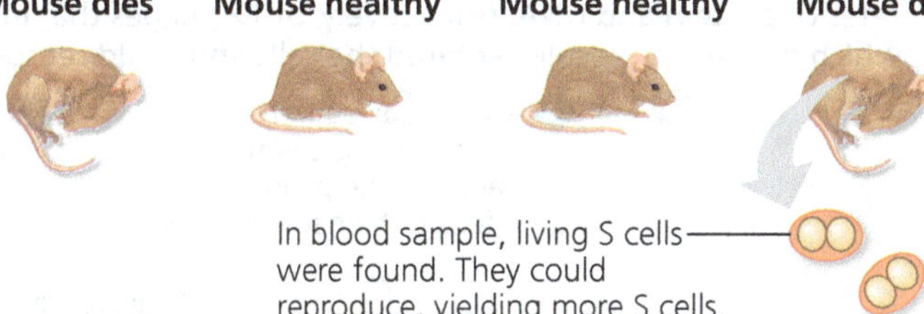

Conclusion The living R bacteria had been transformed into pathogenic S bacteria by an unknown, heritable substance from the dead S cells that enabled the R cells to make capsules.

Data from F. Griffith, The significance of pneumococcal types, *Journal of Hygiene* 27:113–159 (1928).

WHAT IF? *How did this experiment rule out the possibility that the R cells simply used the dead S cells' capsules to become pathogenic?*

Evidence That Viral DNA Can Program Cells

Additional evidence that DNA was the genetic material came from studies of viruses that infect bacteria **(Figure 16.3)**. These viruses are called **bacteriophages** (meaning "bacteria-eaters"), or **phages** for short. Viruses are much simpler than cells. A **virus** is little more than DNA (or sometimes RNA) enclosed by a protective coat, which is often simply protein. To produce more viruses, a virus must infect a cell and take over the cell's metabolic machinery.

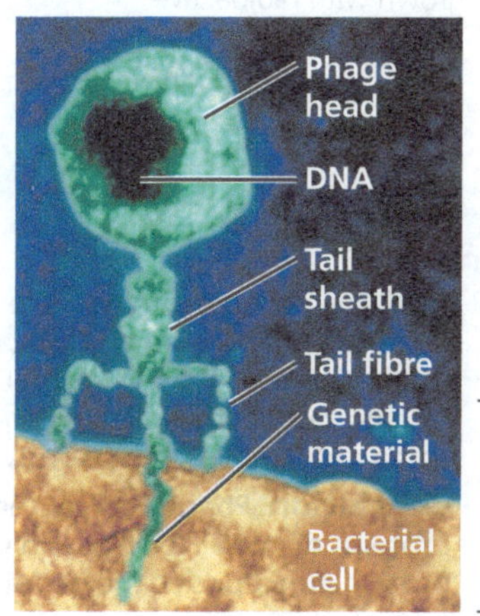

► **Figure 16.3 A virus infecting a bacterial cell.** A phage called T2 attaches to a host cell and injects its genetic material through the plasma membrane, while the head and tail parts remain on the outer bacterial surface (colourised TEM).

Phages have been widely used as tools by researchers in molecular genetics. In 1952, Alfred Hershey and Martha Chase performed experiments showing that DNA is the genetic material of a phage known as T2. This is one of many phages that infect *Escherichia coli* (*E. coli*), a bacterium that normally lives in the intestines of mammals and is a model organism for molecular biologists. At that time, biologists already knew that T2, like many other phages, was composed almost entirely of DNA and protein. They also knew that the T2 phage could quickly turn an *E. coli* cell into a T2-producing factory that released many copies of new phages when the cell ruptured. Somehow, T2 could reprogram its host cell to produce viruses. But which viral component—protein or DNA—was responsible?

Hershey and Chase answered this question by devising an experiment showing that only one of the two components of T2 actually enters the *E. coli* cell during infection **(Figure 16.4)**. In their experiment, they used a radioactive isotope of sulfur to tag protein in one batch of T2 and a radioactive isotope of phosphorus to tag DNA in a second batch. Because protein, but not DNA, contains sulfur, radioactive sulfur atoms were incorporated only into the protein of the phage. In a similar way, the atoms of radioactive phosphorus labelled only the DNA, not the protein, because nearly all the phage's phosphorus is in its DNA. In the experiment, separate samples of nonradioactive *E. coli* cells were infected separately with the protein-labelled and DNA-labelled batches of T2. The researchers then tested the two samples shortly

▼ Figure 16.4 Inquiry

Is protein or DNA the genetic material of phage T2?

Experiment Alfred Hershey and Martha Chase used radioactive sulfur and phosphorus to trace the fates of protein and DNA, respectively, of T2 phages that infected bacterial cells. They wanted to see which of these molecules entered the cells and could reprogram them to make more phages.

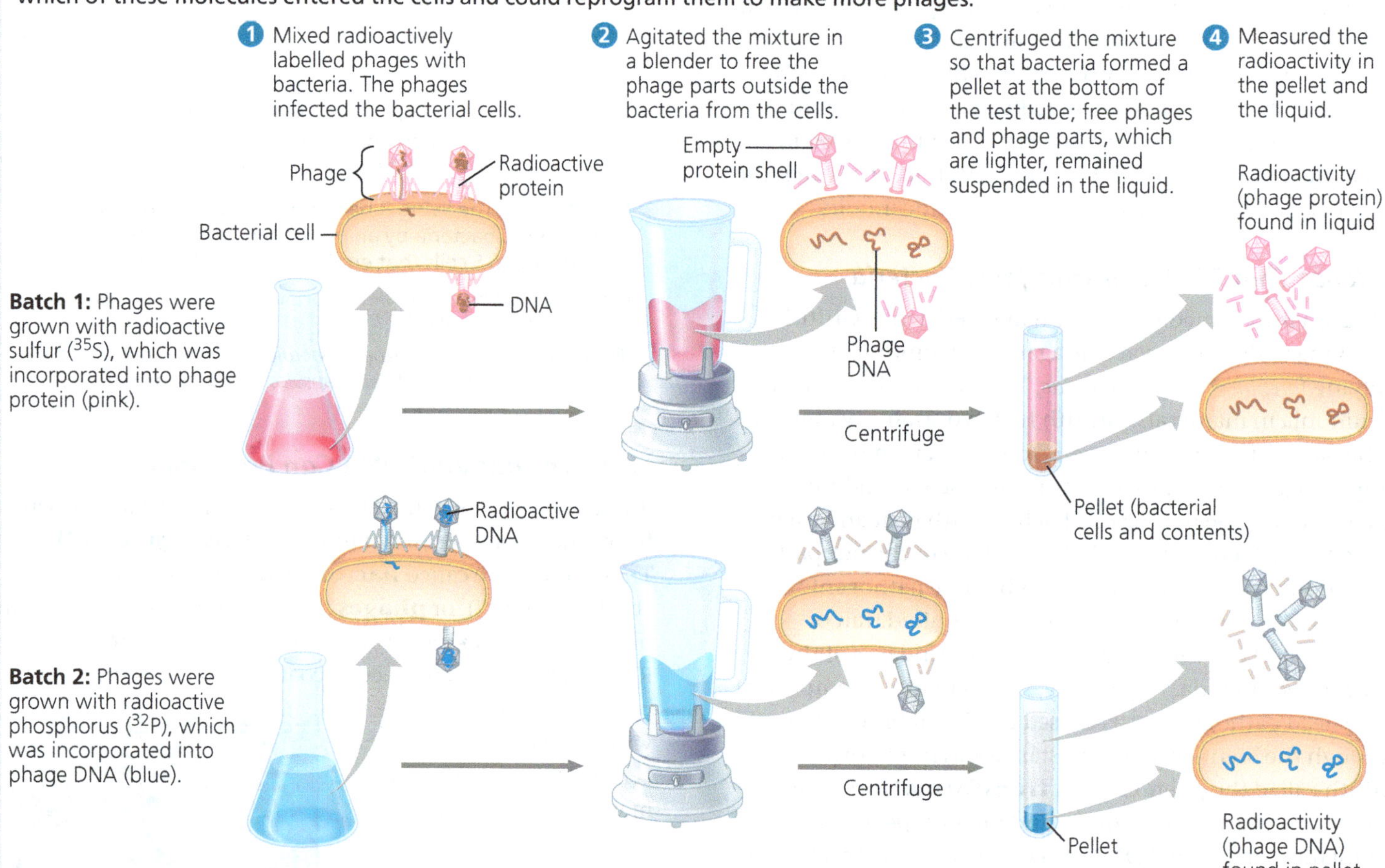

Results When proteins were labelled (batch 1), radioactivity remained outside the cells, but when DNA was labelled (batch 2), radioactivity was found inside the cells. Cells containing radioactive phage DNA released new phages with some radioactive phosphorus.

Conclusion Phage DNA entered bacterial cells, but phage proteins did not. Hershey and Chase concluded that DNA, not protein, functions as the genetic material of phage T2.

Data from A. D. Hershey and M. Chase, Independent functions of viral protein and nucleic acid in growth of bacteriophage, *Journal of General Physiology* 36:39–56 (1952).

WHAT IF? *How would the results have differed if proteins carried the genetic information?*

after the onset of infection to see which type of radioactively labelled molecule—protein or DNA—had entered the bacterial cells and would therefore be capable of reprogramming them.

Hershey and Chase found that the phage DNA entered the host cells, but the phage protein did not. Moreover, when these bacteria were returned to a culture medium and the infection ran its course, the *E. coli* released phages that contained some radioactive phosphorus. This result further showed that the DNA inside the cell played an ongoing role during the infection process. They concluded that the DNA injected by the phage must be the molecule carrying the genetic information that makes the cells produce new viral DNA and proteins. The Hershey-Chase experiment was a landmark study because it provided powerful evidence that nucleic acids, rather than proteins, are the hereditary material, at least for certain viruses.

Additional Evidence That DNA Is the Genetic Material

Further evidence that DNA is the genetic material came from the laboratory of biochemist Erwin Chargaff. DNA was known to be a polymer of nucleotides, each having three components: a nitrogenous (nitrogen-containing) base, a pentose sugar called deoxyribose, and a phosphate group **(Figure 16.5)**. The base can be adenine (A), thymine (T), guanine (G), or cytosine (C). Chargaff analysed the base composition of DNA from a number of different organisms. In 1950, he reported that the base composition of DNA varies from one species to another. For example, he found that 32.8% of sea urchin DNA nucleotides have the base A, whereas only 24.7% of those from the bacterium *E. coli* have an A. This evidence of molecular diversity among species, which most scientists had presumed to be absent from DNA, made DNA a more credible candidate for the genetic material.

Chargaff also noticed a peculiar regularity in the ratios of nucleotide bases. In the DNA of each species he studied, the number of adenines approximately equalled the number of thymines, and the number of guanines approximately equaled the number of cytosines. In sea urchin DNA, for example, the four bases are present in these percentages: A = 32.8% and T = 32.1%; G = 17.7% and C = 17.3%. (The percentages are not exactly the same because of limitations in Chargaff's techniques.)

These two findings became known as *Chargaff's rules*: (1) DNA base composition varies between species, and (2) for each species, the percentages of A and T bases are roughly equal, as are those of G and C bases. In the **Scientific Skills Exercise**, you can use Chargaff's rules to predict percentages of nucleotide bases. The basis for these rules remained unexplained until the discovery of the double helix.

▼ **Figure 16.5 The structure of a DNA strand.** Each DNA nucleotide monomer consists of the sugar deoxyribose (blue) attached to both a nitrogenous base (A, T, G, or C), and a phosphate group (yellow). The phosphate group of one nucleotide is attached to the sugar of the next by a covalent bond, forming a "backbone" of alternating phosphates and sugars from which the bases project. A polynucleotide strand has directionality, from the 5′ end (with the phosphate group) to the 3′ end (with the —OH group of the sugar). 5′and 3′ refer to the numbers assigned to the carbons in the sugar ring (see magenta numbers).

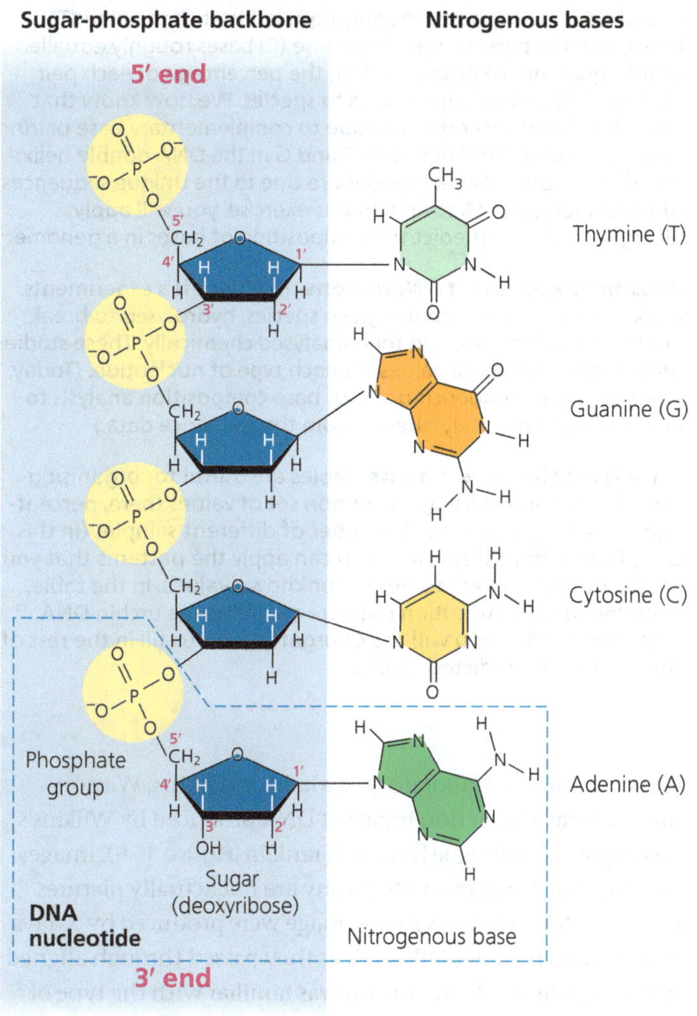

Building a Structural Model of DNA

Once most biologists were convinced that DNA was the genetic material, the challenge was to determine how the structure of DNA could account for its role in inheritance. By the early 1950s, the arrangement of covalent bonds in a nucleic acid polymer was well established (see Figure 16.5), and researchers focused on discovering the three-dimensional structure of DNA. Among the scientists working on the problem were Linus Pauling, at the California Institute of Technology, and Maurice Wilkins and Rosalind Franklin, at King's College in London. First to come up with the complete answer, however, were two scientists who were relatively unknown at the time—the American James Watson and the Englishman Francis Crick.

The collaborative work that solved the puzzle of DNA structure began soon after Watson went to Cambridge University, where Crick was studying protein structure with a technique called X-ray crystallography (see Figure 5.21).

Scientific Skills Exercise

Working with Data in a Table

Given the Percentage Composition of One Nucleotide in a Genome, Can We Predict the Percentages of the Other Three Nucleotides? Even before the structure of DNA was elucidated, Erwin Chargaff and his co-workers noticed a pattern in the base composition of nucleotides from different species: The percentage of adenine (A) bases roughly equaled that of thymine (T) bases, and the percentage of cytosine (C) bases roughly equalled that of guanine (G) bases. Further, the percentage of each pair (A–T or C–G) varied from species to species. We now know that the 1:1 A/T and C/G ratios are due to complementary base pairing between A and T and between C and G in the DNA double helix, and differences between species are due to the unique sequences of bases along a DNA strand. In this exercise, you will apply Chargaff's rules to predict the composition of bases in a genome.

How the Experiments Were Done In Chargaff's experiments, DNA was extracted from the given species, hydrolysed to break apart the nucleotides, and then analysed chemically. These studies provided approximate values for each type of nucleotide. (Today, whole-genome sequencing allows base composition analysis to be done more precisely directly from the sequence data.)

Data from the Experiments Tables are useful for organising sets of data representing a common set of values (here, percentages of A, G, C, and T) for a number of different samples (in this case, from different species). You can apply the patterns that you see in the known data to predict unknown values. In the table, complete base distribution data are given for sea urchin DNA and salmon DNA; you will use Chargaff's rules to fill in the rest of the table with predicted values.

	Base Percentage			
Source of DNA	**Adenine**	**Guanine**	**Cytosine**	**Thymine**
Sea urchin	32.8	17.7	17.3	32.1
Salmon	29.7	20.8	20.4	29.1
Wheat	28.1	21.8	22.7	
E. coli	24.7	26.0		
Human	30.4			30.1
Ox	29.0			
Average %				

Data from several papers by Chargaff: for example, E. Chargaff et al., Composition of the desoxypentose nucleic acids of four genera of sea-urchin, *Journal of Biological Chemistry* 195: 155–160 (1952).

▶ **Sea urchin**

INTERPRET THE DATA

1. Explain how the sea urchin and salmon data demonstrate both of Chargaff's rules.
2. Using Chargaff's rules, fill in the table with your predictions of the missing percentages of bases, starting with the wheat genome and proceeding through *E. coli*, human, and ox. Show how you arrived at your answers.
3. If Chargaff's rule—that the amount of A equals the amount of T and the amount of C equals the amount of G—is valid, then hypothetically we could extrapolate this to the combined DNA of all species on Earth (like one huge Earth genome). To see whether the data in the table support this hypothesis, calculate the average percentage for each base in your completed table by averaging the values in each column. Does Chargaff's equivalence rule still hold true?

While visiting the laboratory of Maurice Wilkins, Watson saw an X-ray diffraction image of DNA produced by Wilkins's accomplished colleague Rosalind Franklin **(Figure 16.6)**. Images produced by X-ray crystallography are not actually pictures of molecules. The spots in the image were produced by X-rays that were diffracted (deflected) as they passed through aligned fibres of purified DNA. Watson was familiar with the type of X-ray diffraction pattern of helical molecules, and seeing the photo that Wilkins showed him confirmed that DNA was helical in shape. The photo also added to earlier data obtained by Franklin and others suggesting the width of the helix and the spacing of the nitrogenous bases along it. The pattern in this photo implied that the helix was made up of two strands, contrary to a three-stranded model proposed by Linus Pauling a short time earlier. The presence of two strands accounts for the now-familiar term **double helix**. DNA is shown in some of its many different representations in **Figure 16.7**.

▼ **Figure 16.6 Rosalind Franklin and her X-ray diffraction photo of DNA.**

(a) Rosalind Franklin

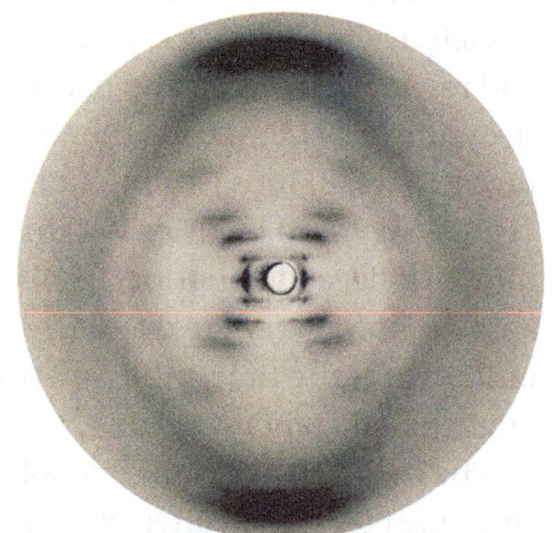

(b) Franklin's X-ray diffraction photograph of DNA

Watson and Crick began building models of a double helix that would conform to the X-ray measurements and what was then known about the chemistry of DNA, including Chargaff's rule of base equivalences. Having also read an unpublished annual report summarising Franklin's work, they knew she had concluded that the sugar-phosphate backbones were on the outside of the DNA molecule, contrary to their working model. Franklin's arrangement was appealing because it put the negatively charged phosphate groups facing the aqueous surroundings, while the relatively hydrophobic nitrogenous bases were hidden in the interior. Watson constructed such a model, in which the two sugar-phosphate backbones are **antiparallel**—that is, their subunits run in opposite directions (see Figure 16.7). You can imagine the overall arrangement as a rope ladder with rigid rungs. The side ropes represent

▼ Figure 16.7 VISUALISING DNA

DNA can be illustrated in many ways, but all diagrams represent the same basic structure. The level of detail shown depends on the process or the type of information being conveyed.

Structural Images

The space-filling model on the left shows the 3-D shape of the DNA double helix; the diagram on the right shows chemical details of DNA's structure. In both images, phosphate groups are yellow, deoxyribose sugars are blue, and nitrogenous bases are shades of green and orange.

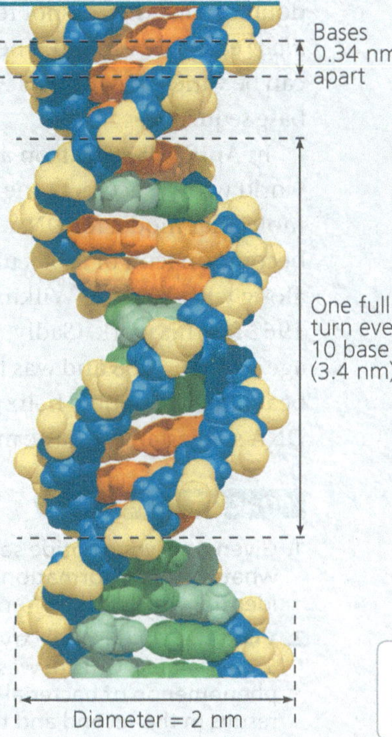

The DNA double helix is "right-handed." Use your right hand as shown to follow the sugar-phosphate backbone up the helix and around to the back. (It won't work with your left hand.)

Here, the two DNA strands are shown untwisted so it's easier to see the chemical details. Note that the strands are **antiparallel**—they are oriented in opposite directions, like the lanes of a divided street.

5′ end | 3′ end

Phosphate group attached to 5′ carbon

Nitrogenous base attached to 1′ carbon

DNA nucleotide

Sugar-phosphate backbone

Sugar

Covalent sugar-phosphate bonds link nucleotides.

Hydrogen bonds between bases hold strands together.

Van der Waals interactions between stacked base pairs help hold the molecule together.

–OH attached to 3′ carbon

3′ end | 5′ end

1. ***Describe the bonds that hold together the nucleotides in one DNA strand. Then compare them with the bonds that hold the two DNA strands together.***

2. ***How do the two ends of a DNA strand differ in structure?***

Simplified Images

When molecular detail is not necessary, DNA is portrayed in a range of simplified diagrams, depending on the focus of the figure.

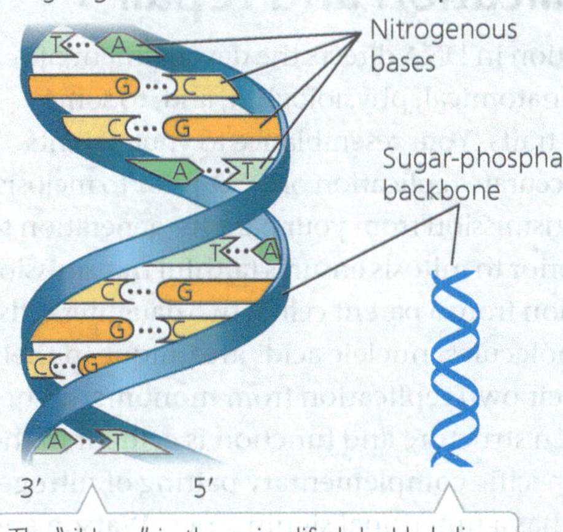

The "ribbons" in these simplified double helix diagrams represent the sugar-phosphate backbones; these models emphasise the 3-D nature of DNA.

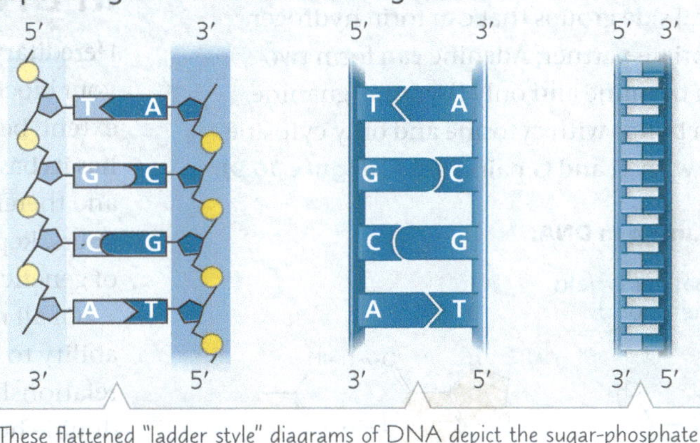

These flattened "ladder style" diagrams of DNA depict the sugar-phosphate backbones like the side rails of a ladder, with the base pairs as rungs. Light blue is used to indicate the more recently synthesised strand.

5′ 3′

3′ 5′

Sometimes the double-stranded DNA molecule is shown simply as two straight lines.

3. ***Compare the information conveyed in the three ladder diagrams.***

DNA Sequences

Genetic information is carried in DNA as a linear sequence of nucleotides that may be transcribed into mRNA and translated into a polypeptide. When focusing on the DNA sequence, each nucleotide can be represented simply as the letter of its base: A, T, C, or G.

3′- A C G T A A G C G G T T A A T -5′
5′- T G C A T T C G C C A A T T A -3′

the sugar-phosphate backbones, and the rungs represent pairs of nitrogenous bases. Now imagine twisting the ladder to form a helix. Franklin's X-ray data indicated that the helix makes one full turn every 3.4 nm along its length. With the bases stacked just 0.34 nm apart, there are ten layers of base pairs, or rungs of the ladder, in each full turn of the helix.

The nitrogenous bases of the double helix are paired in specific combinations: adenine (A) with thymine (T), and guanine (G) with cytosine (C). It was mainly by trial and error that Watson and Crick arrived at this key feature of DNA. At first, Watson imagined that the bases paired like with like—for example, A with A and C with C. But this model did not fit the X-ray data, which suggested that the double helix had a uniform diameter. Why is this requirement inconsistent with like-with-like pairing of bases? Adenine and guanine are purines, nitrogenous bases with two organic rings, while cytosine and thymine are nitrogenous bases called pyrimidines, which have a single ring. Pairing a purine with a pyrimidine is the only combination that results in a uniform diameter for the double helix (**Figure 16.8**).

▼ **Figure 16.8 Possible base pairings in the DNA double helix.**

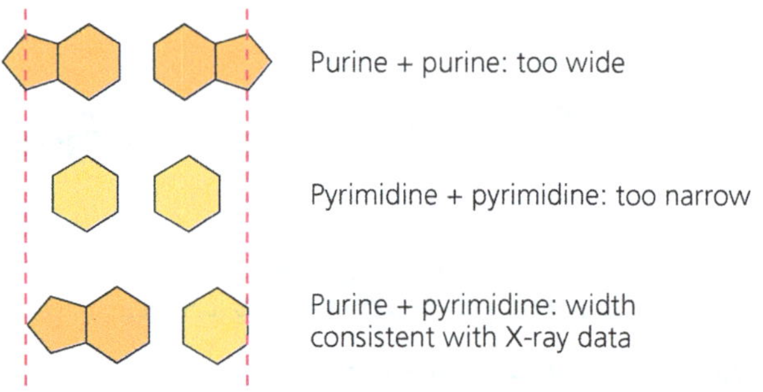

Watson and Crick reasoned that there must be additional specificity of pairing dictated by the structure of the bases. Each base has chemical side groups that can form hydrogen bonds with its appropriate partner: Adenine can form two hydrogen bonds with thymine and only thymine; guanine forms three hydrogen bonds with cytosine and only cytosine. In shorthand, A pairs with T, and G pairs with C (**Figure 16.9**).

▼ **Figure 16.9 Base pairing in DNA.**

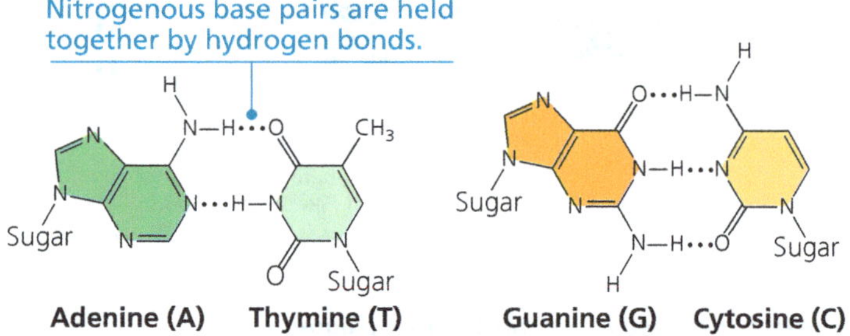

The Watson-Crick model took into account Chargaff's ratios and ultimately explained them. Wherever one strand of a DNA molecule has an A, the partner strand has a T. Similarly, a G in one strand is always paired with a C in the complementary strand. Therefore, in the DNA of any organism, the amount of adenine equals the amount of thymine, and the amount of guanine equals the amount of cytosine. (Modern DNA-sequencing techniques have confirmed that the amounts are exactly equal.) Although the base-pairing rules dictate the combinations of nitrogenous bases that form the "rungs" of the double helix, they do not restrict the sequence of nucleotides *along* each DNA strand. The linear sequence of the four bases can be varied in countless ways, and each gene has a unique base sequence.

In April 1953, Watson and Crick surprised the scientific world with a succinct, one-page paper that reported their molecular model for DNA: the double helix, which has since become an icon of molecular biology. Watson and Crick, along with Maurice Wilkins, were awarded the Nobel Prize in 1962 for this work. (Sadly, Rosalind Franklin had died at the age of 37 in 1958 and was thus ineligible for the prize.) The beauty of the double helix model was that the structure of DNA suggested the basic mechanism of its replication.

CONCEPT CHECK 16.1

1. Given a polynucleotide sequence such as GAATTC, explain what further information you would need in order to identify which is the 5′ end. (See Figure 16.5.)
2. **VISUAL SKILLS** While trying to develop a vaccine for *S. pneumonia*, Griffith was surprised to discover the phenomenon of bacterial transformation. Based on the results in the second and third panels of Figure 16.2, what result was he expecting in the fourth panel? Explain.

For suggested answers, see Appendix A.

CONCEPT 16.2

Many proteins work together in DNA replication and repair

Hereditary information in DNA directs the development of your biochemical, anatomical, physiological, and, to some extent, behavioural traits. Your resemblance to your parents has its basis in the accurate replication of DNA prior to meiosis and therefore its transmission from your parents' generation to yours. Replication prior to mitosis ensures faithful transmission of genetic information from a parent cell to two daughter cells.

Of all nature's molecules, nucleic acids are unique in their ability to dictate their own replication from monomers. The relationship between structure and function is evident in the double helix: The specific complementary pairing of nitrogenous bases in DNA has a functional significance. Watson and Crick ended their classic paper with this statement: "It has not escaped our notice that the specific pairing we have postulated immediately suggests a possible copying mechanism for the genetic material."[1] In this section, you will learn about the basic principle of **DNA replication**, the copying of DNA, as well as some important details of the process.

[1]J. D. Watson and F. H. C. Crick, Molecular structure of nucleic acids: a structure for deoxyribose nucleic acids, *Nature* 171:737–738 (1953).

▼ Figure 16.10 A model for DNA replication: the basic concept. In this simplified illustration, a short segment of DNA has been untwisted. Simple shapes symbolise the four kinds of bases. Dark blue represents DNA strands present in the parental molecule; light blue represents newly synthesised DNA.

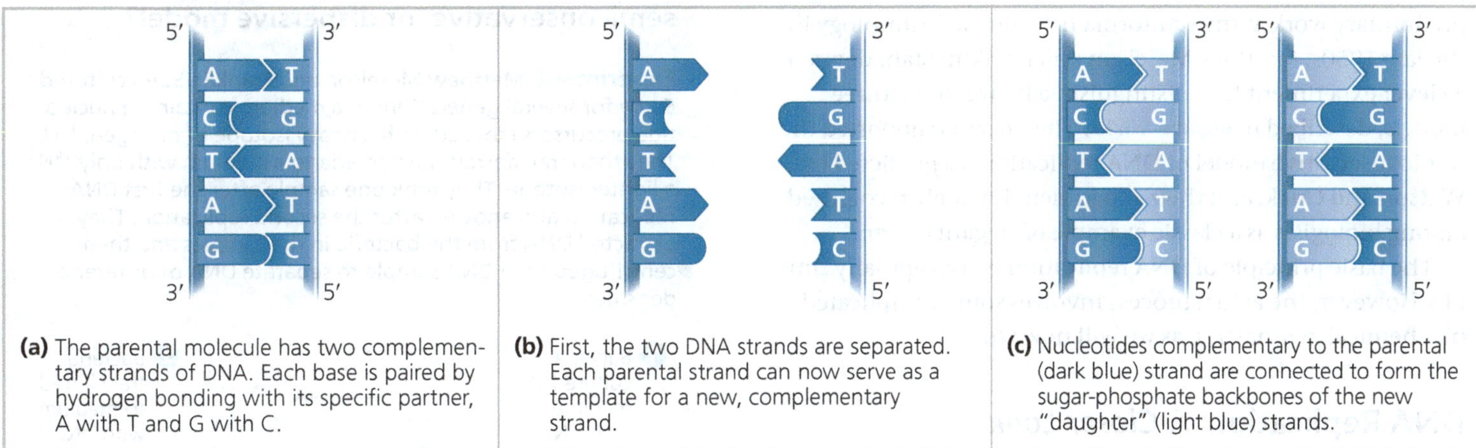

(a) The parental molecule has two complementary strands of DNA. Each base is paired by hydrogen bonding with its specific partner, A with T and G with C.

(b) First, the two DNA strands are separated. Each parental strand can now serve as a template for a new, complementary strand.

(c) Nucleotides complementary to the parental (dark blue) strand are connected to form the sugar-phosphate backbones of the new "daughter" (light blue) strands.

The Basic Principle: Base Pairing to a Template Strand

In a second paper, Watson and Crick stated their hypothesis for how DNA replicates: "Now our model for deoxyribonucleic acid is, in effect, a pair of templates, each of which is complementary to the other. We imagine that prior to duplication the hydrogen bonds are broken, and the two chains unwind and separate. Each chain then acts as a template for the formation on to itself of a new companion chain, so that eventually we shall have two pairs of chains, where we only had one before. Moreover, the sequence of the pairs of bases will have been duplicated exactly."[2]

Figure 16.10 illustrates the basic idea. If you cover a DNA strand in Figure 16.10a, its linear sequence of nucleotides is revealed by applying the base-pairing rules to the uncovered strand. The two strands are complementary; each stores the information necessary to reconstruct the other. When a cell copies a DNA molecule, each strand serves as a template for ordering nucleotides into a new, complementary strand. Nucleotides line up along the template strand according to the base-pairing rules and are linked to form the new strands. One double-stranded DNA molecule becomes two, each an exact replica of the "parental" molecule.

This model of DNA replication remained untested for several years following publication of the DNA structure. The necessary experiments were simple in concept but difficult to perform. Watson and Crick's model predicts that when a double helix replicates, each of the two daughter molecules will have one old strand, from the parental molecule, and one newly made strand. This **semiconservative model** can be distinguished from a *conservative model* of replication, in which the two parental strands somehow come back together after the process (that is, the parental molecule is conserved). In yet a third model, called the *dispersive model*, all four strands of DNA following replication have a mixture of old and new DNA **(Figure 16.11)**.

[2]J. D. Watson and F. H. C. Crick, Genetical implications of the structure of deoxyribonucleic acid, *Nature* 171:964–967 (1953).

▼ Figure 16.11 DNA replication: three alternative models. Each short segment of double helix symbolises the DNA within a cell. Beginning with a parent cell, we follow the DNA for two more generations of cells—two rounds of DNA replication. Parental DNA is dark blue; newly made DNA is light blue.

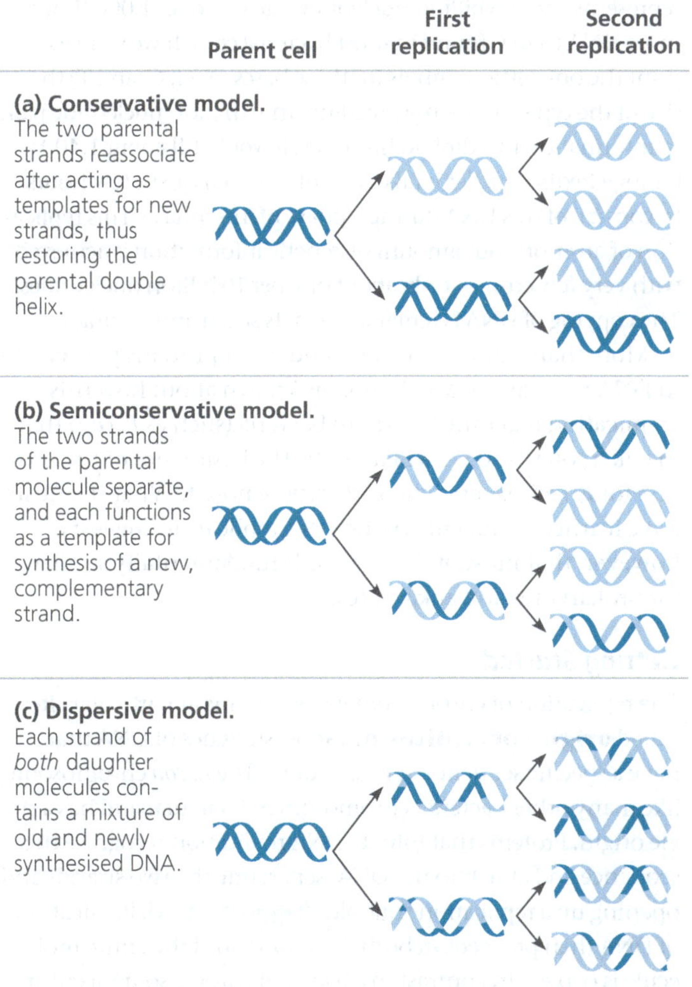

	Parent cell	First replication	Second replication

(a) Conservative model. The two parental strands reassociate after acting as templates for new strands, thus restoring the parental double helix.

(b) Semiconservative model. The two strands of the parental molecule separate, and each functions as a template for synthesis of a new, complementary strand.

(c) Dispersive model. Each strand of *both* daughter molecules contains a mixture of old and newly synthesised DNA.

Although mechanisms for conservative or dispersive DNA replication are not easy to devise, these models remained possibilities until they could be ruled out. After two years of preliminary work at the California Institute of Technology in the late 1950s, Matthew Meselson and Franklin Stahl devised a clever experiment that distinguished between the three models, described in **Figure 16.12**. Their results supported the semiconservative model of DNA replication, as predicted by Watson and Crick, and their experiment is widely recognised among biologists as a classic example of elegant design.

The basic principle of DNA replication is conceptually simple. However, the actual process involves some complicated biochemical gymnastics, as we will now see.

DNA Replication: *A Closer Look*

The bacterium *E. coli* has a single chromosome of about 4.6 million nucleotide pairs. In a favourable environment, an *E. coli* cell can copy all of this DNA and divide to form two genetically identical daughter cells in considerably less than an hour. Each of *your* somatic cells has 46 DNA molecules in its nucleus, one long double-helical molecule per chromosome. In all, that represents about 6 billion nucleotide pairs, or over 1,000 times more DNA than is found in most bacterial cells. If we were to print the one-letter symbols for these bases (A, G, C, and T) the size of the type you are now reading, the 6 billion nucleotide pairs of information in a diploid human cell would fill about 1,400 biology textbooks. Yet it takes one of your cells just a few hours to copy all of this DNA during S phase of interphase. This replication of an enormous amount of genetic information is achieved with very few errors—only about one per 10 billion nucleotides. The copying of DNA is remarkable in its speed and accuracy.

More than a dozen enzymes and other proteins participate in DNA replication. Much more is known about how this "replication machine" works in bacteria (such as *E. coli*) than in eukaryotes, and we will describe the basic steps of the process for *E. coli*, except where otherwise noted. What scientists have learned about eukaryotic DNA replication suggests, however, that most of the process is fundamentally similar for prokaryotes and eukaryotes.

Getting Started

The replication of chromosomal DNA begins at particular sites called **origins of replication**, short stretches of DNA that have a specific sequence of nucleotides. The *E. coli* chromosome, like many other bacterial chromosomes, is circular and has a single origin. Proteins that initiate DNA replication recognise this sequence and attach to the DNA, separating the two strands and opening up a replication "bubble" **(Figure 16.13a)**. Replication of DNA then proceeds in both directions until the entire molecule is copied. In contrast to a bacterial chromosome, a eukaryotic chromosome may have hundreds or even a few thousand replication origins. Multiple replication bubbles form and eventually fuse, thus speeding up the copying of the very long DNA

▼ Figure 16.12 Inquiry

Does DNA replication follow the conservative, semiconservative, or dispersive model?

Experiment Matthew Meselson and Franklin Stahl cultured *E. coli* for several generations in a medium containing nucleotide precursors labelled with a heavy isotope of nitrogen, ^{15}N. They then transferred the bacteria to a medium with only ^{14}N, a lighter isotope. They took one sample after the first DNA replication and another after the second replication. They extracted DNA from the bacteria in the samples and then centrifuged each DNA sample to separate DNA of different densities.

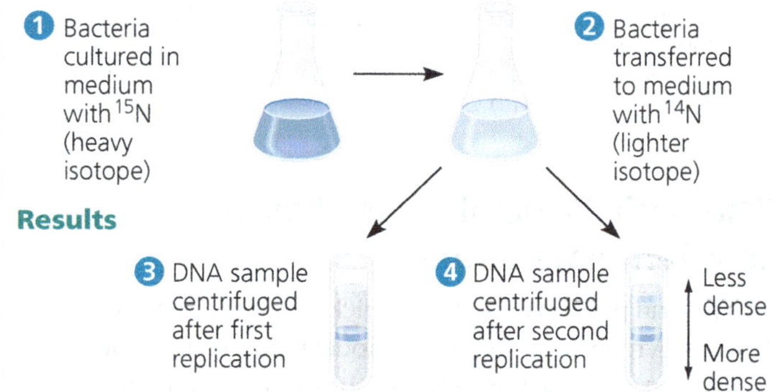

Conclusion Meselson and Stahl compared their results to those predicted by each of the three models in Figure 16.11, as shown below. The first replication in the ^{14}N medium produced a band of many molecules of hybrid (^{15}N-^{14}N) DNA. This result eliminated the conservative model. The second replication produced both light and hybrid DNA, a result that refuted the dispersive model and supported the semiconservative model. They therefore concluded that DNA replication is semiconservative.

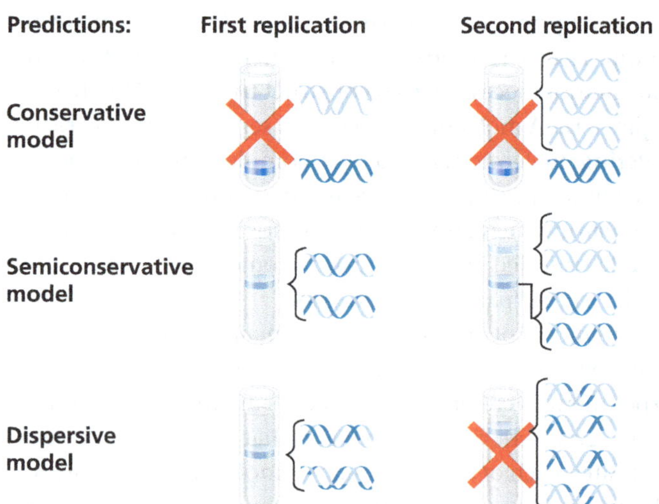

Data from M. Meselson and F. W. Stahl, The replication of DNA in *Escherichia coli*, *Proceedings of the National Academy of Sciences USA* 44:671–682 (1958).

INQUIRY IN ACTION Read and analyse the original paper in *Inquiry in Action: Interpreting Scientific Papers.*

WHAT IF? *If Meselson and Stahl had first grown the cells in ^{14}N-containing medium and then moved them into ^{15}N-containing medium before taking samples, what would have been the result after each of the two replications?*

▼ Figure 16.13 Origins of replication in *E. coli* and eukaryotes. The red arrows indicate the movement of the replication forks and thus the overall directions of DNA replication within each bubble.

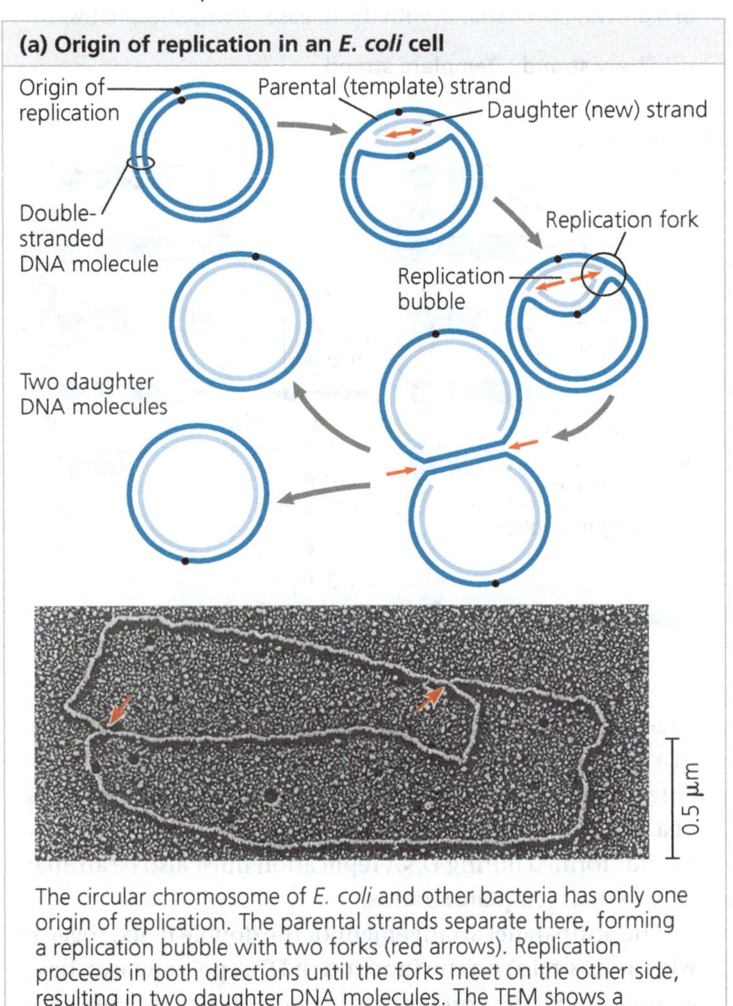

The circular chromosome of *E. coli* and other bacteria has only one origin of replication. The parental strands separate there, forming a replication bubble with two forks (red arrows). Replication proceeds in both directions until the forks meet on the other side, resulting in two daughter DNA molecules. The TEM shows a bacterial chromosome with a replication bubble.

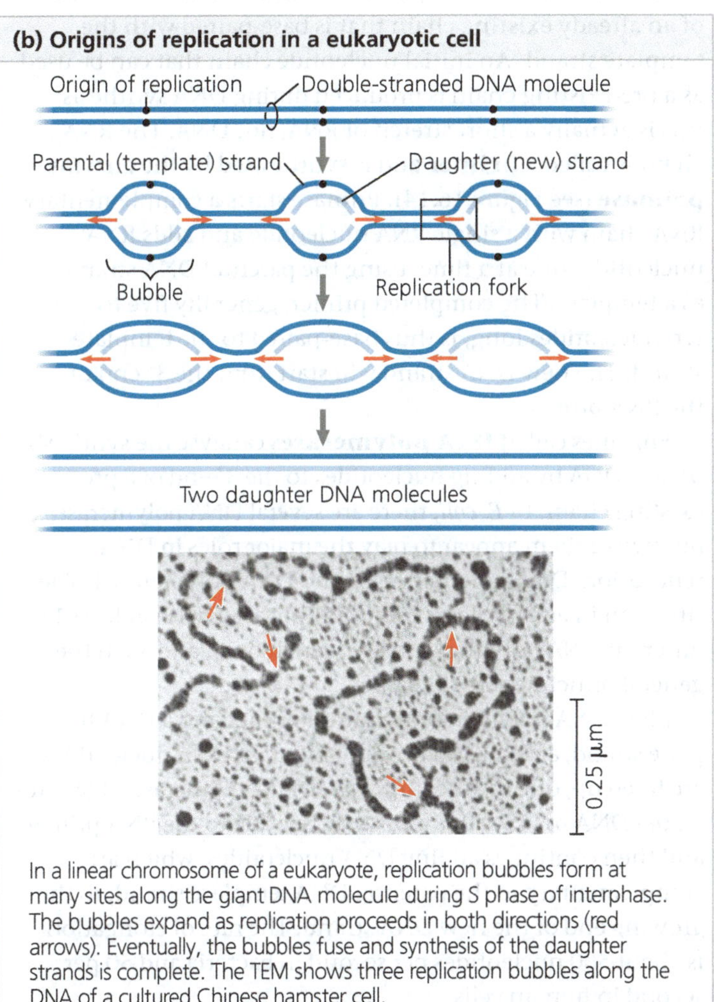

In a linear chromosome of a eukaryote, replication bubbles form at many sites along the giant DNA molecule during S phase of interphase. The bubbles expand as replication proceeds in both directions (red arrows). Eventually, the bubbles fuse and synthesis of the daughter strands is complete. The TEM shows three replication bubbles along the DNA of a cultured Chinese hamster cell.

DRAW IT *In the TEM, add arrows in the forks of the third bubble.*

molecules **(Figure 16.13b)**. As in bacteria, eukaryotic DNA replication proceeds in both directions from each origin.

At each end of a replication bubble is a **replication fork**, a Y-shaped region where the parental strands of DNA are being unwound. Several kinds of proteins participate in the unwinding **(Figure 16.14)**. **Helicases** are enzymes that untwist the double helix at the replication forks, separating the two parental strands and making them available as template strands. After the parental strands separate, **single-strand binding proteins** bind to the unpaired DNA strands, keeping them from re-pairing. The untwisting of the double helix causes tighter twisting and strain ahead of the replication fork. **Topoisomerase** is an enzyme that helps relieve this strain by breaking, swivelling, and rejoining DNA strands.

Synthesising a New DNA Strand

The unwound sections of parental DNA strands are now available to serve as templates for the synthesis of new complementary DNA strands. However, the enzymes that

▼ Figure 16.14 Some of the proteins involved in the initiation of DNA replication. The same proteins function at both replication forks in a replication bubble. For simplicity, only the left-hand fork is shown, and the DNA bases are drawn much larger in relation to the proteins than they are in reality.

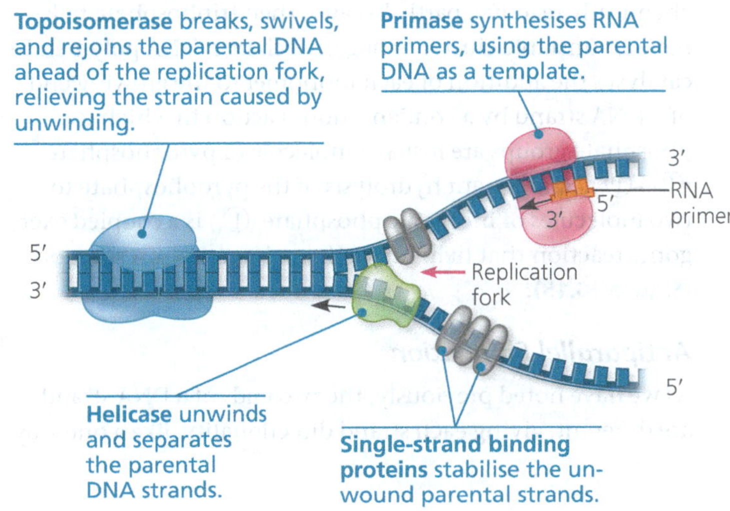

synthesise DNA cannot *initiate* the synthesis of a polynucleotide; they can only add DNA nucleotides to the end of an already existing chain that is base-paired with the template strand. An initial nucleotide chain that can be used as a pre-existing chain is produced during DNA synthesis; this is actually a short stretch of RNA, not DNA. The RNA chain is called a **primer** and is synthesised by the enzyme **primase** (see Figure 16.14). Primase starts a complementary RNA chain with a single RNA nucleotide and adds RNA nucleotides one at a time, using the parental DNA strand as a template. The completed primer, generally five to ten nucleotides long, is thus base-paired to the template strand. The new DNA strand will start from the 3′ end of the RNA primer.

Enzymes called **DNA polymerases** catalyse the synthesis of new DNA by adding nucleotides to the 3′ end of a pre-existing chain. In *E. coli*, there are several DNA polymerases, but two of them appear to play the major roles in DNA replication: DNA polymerase III and DNA polymerase I. The situation in eukaryotes is more complicated, with at least 11 different DNA polymerases discovered so far, although the general principles are the same.

Most DNA polymerases require a primer and a DNA template strand, along which complementary DNA nucleotides are lined up, one by one. In *E. coli*, DNA polymerase III (abbreviated DNA pol III) adds a DNA nucleotide to the RNA primer and then continues adding DNA nucleotides, which are complementary to the parental DNA template strand, to the growing end of the new DNA strand. The rate of elongation is about 500 nucleotides per second in bacteria and 50 per second in human cells.

Each nucleotide to be added to a growing DNA strand consists of a sugar attached to a base and to three phosphate groups. You have already encountered such a molecule—ATP (adenosine triphosphate; see Figure 8.9). The only difference between the ATP of energy metabolism and dATP, the adenine nucleotide used to make DNA, is the sugar component, which is deoxyribose in the building block of DNA but ribose in ATP. Like ATP, the nucleotides used for DNA synthesis are chemically reactive, partly because their triphosphate tails have an unstable cluster of negative charge. DNA polymerase catalyses the addition of each monomer to the growing end of a DNA strand by a condensation reaction in which two phosphate groups are lost as a molecule of pyrophosphate (Ⓟ—$Ⓟ_i$). Subsequent hydrolysis of the pyrophosphate to two molecules of inorganic phosphate ($Ⓟ_i$) is a coupled exergonic reaction that helps drive the polymerisation reaction **(Figure 16.15)**.

▼ Figure 16.15 Addition of a nucleotide to a DNA strand. DNA polymerase catalyses addition of a nucleotide to the 3′ end of a growing DNA strand, with the release of two phosphates.

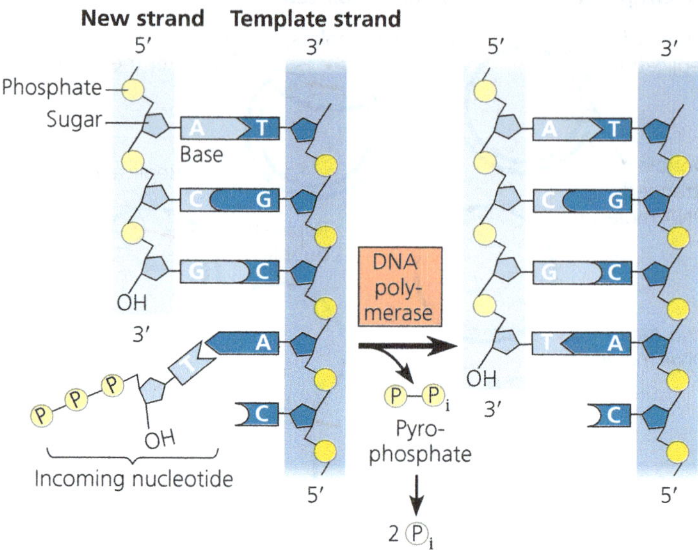

DRAW IT *Circle the area where the new bond was made.*

Antiparallel Elongation

As we have noted previously, the two ends of a DNA strand are different, giving each strand directionality, like a one-way street (see Figure 16.5). In addition, the two strands of DNA in a double helix are antiparallel, meaning that they are oriented in opposite directions to each other, like the two sides of a divided street (see Figure 16.15). Therefore, the two new strands formed during DNA replication must also be antiparallel to their template strands.

The antiparallel arrangement of the double helix, together with a constraint on the function of DNA polymerases, has an important effect on how replication occurs. Because of their structure, DNA polymerases can add nucleotides only to the free 3′ end of a primer or growing DNA strand, never to the 5′ end (see Figure 16.15). Thus, a new DNA strand can elongate only in the 5′ → 3′ direction. With this in mind, let's examine one of the two replication forks in a bubble **(Figure 16.16)**. Along one template strand, DNA polymerase III can synthesise a complementary strand continuously by elongating the new DNA in the mandatory 5′ → 3′ direction. DNA pol III remains in the replication fork on that template strand and continuously adds nucleotides to the new complementary strand as the fork progresses. The DNA strand made by this mechanism is called the **leading strand**. Only one primer is required for DNA pol III to synthesise the entire leading strand (see Figure 16.16).

To elongate the other new strand of DNA in the mandatory 5′ → 3′ direction, DNA pol III must work along the other template strand in the direction *away from* the replication fork. The DNA strand elongating in this direction is called the **lagging strand**. In contrast to the leading strand, which elongates continuously, the lagging strand is synthesised discontinuously, as a series of segments. These segments of the

▼ **Figure 16.16 Synthesis of the leading strand during DNA replication.** This diagram focuses on the left replication fork shown in the overview box. DNA polymerase III (DNA pol III), shaped like a cupped hand, is shown closely associated with a protein called the "sliding clamp" that encircles the newly synthesised double helix like a doughnut. The sliding clamp moves DNA pol III along the DNA template strand.

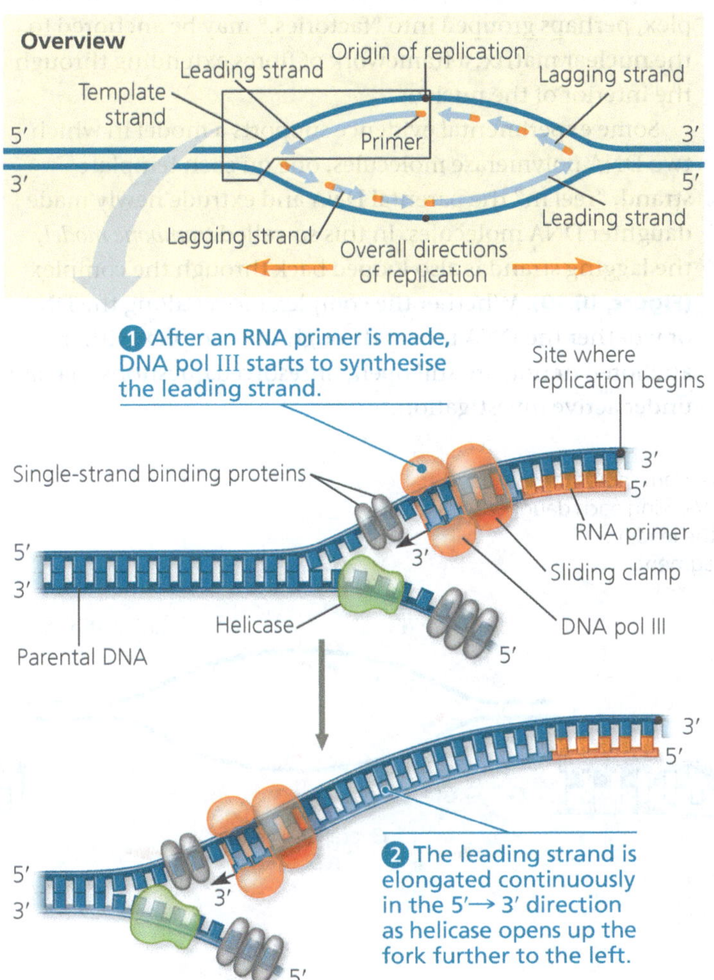

lagging strand are called **Okazaki fragments**, after Reiji Okazaki, the Japanese scientist who discovered them. The fragments are about 1,000–2,000 nucleotides long in *E. coli* and 100–200 nucleotides long in eukaryotes.

Figure 16.17 illustrates the steps in the synthesis of the lagging strand at one fork. Whereas only one primer is required on the leading strand, each Okazaki fragment on the lagging strand must be primed separately (steps 1 and 4). After DNA pol III forms an Okazaki fragment (steps 2 to 4), another DNA polymerase, DNA pol I, replaces the RNA nucleotides of the adjacent primer with DNA nucleotides one at a time (step 5). But DNA pol I cannot join the final nucleotide of this replacement DNA segment to the first DNA nucleotide of the adjacent Okazaki fragment. Another enzyme, **DNA ligase**, accomplishes this task, joining the sugar-phosphate backbones of all the Okazaki fragments into a continuous DNA strand (step 6).

▼ **Figure 16.17 Synthesis of the lagging strand.**

Overview
Origin of replication
Leading strand
Lagging strand
Template strand
Leading strand
Overall directions of replication

1 Primase joins RNA nucleotides into the first primer for the lagging strand.

Site where replication begins

Primer for leading strand

Template strand

2 DNA pol III adds DNA nucleotides to the primer, forming Okazaki fragment 1.

RNA primer for fragment 1

3 After reaching the next RNA primer to the right, DNA pol III detaches.

Okazaki fragment 1

RNA primer for fragment 2

Okazaki fragment 2

4 Fragment 2 is primed. Then DNA pol III adds DNA nucleotides, detaching when it reaches the fragment 1 primer.

5 DNA pol I replaces the RNA with DNA, adding nucleotides to the 3′ end of fragment 1 (and, later, of fragment 2).

6 DNA ligase forms a bond between the newest DNA and the DNA of fragment 1.

7 The lagging strand in this region is now complete.

Overall direction of replication

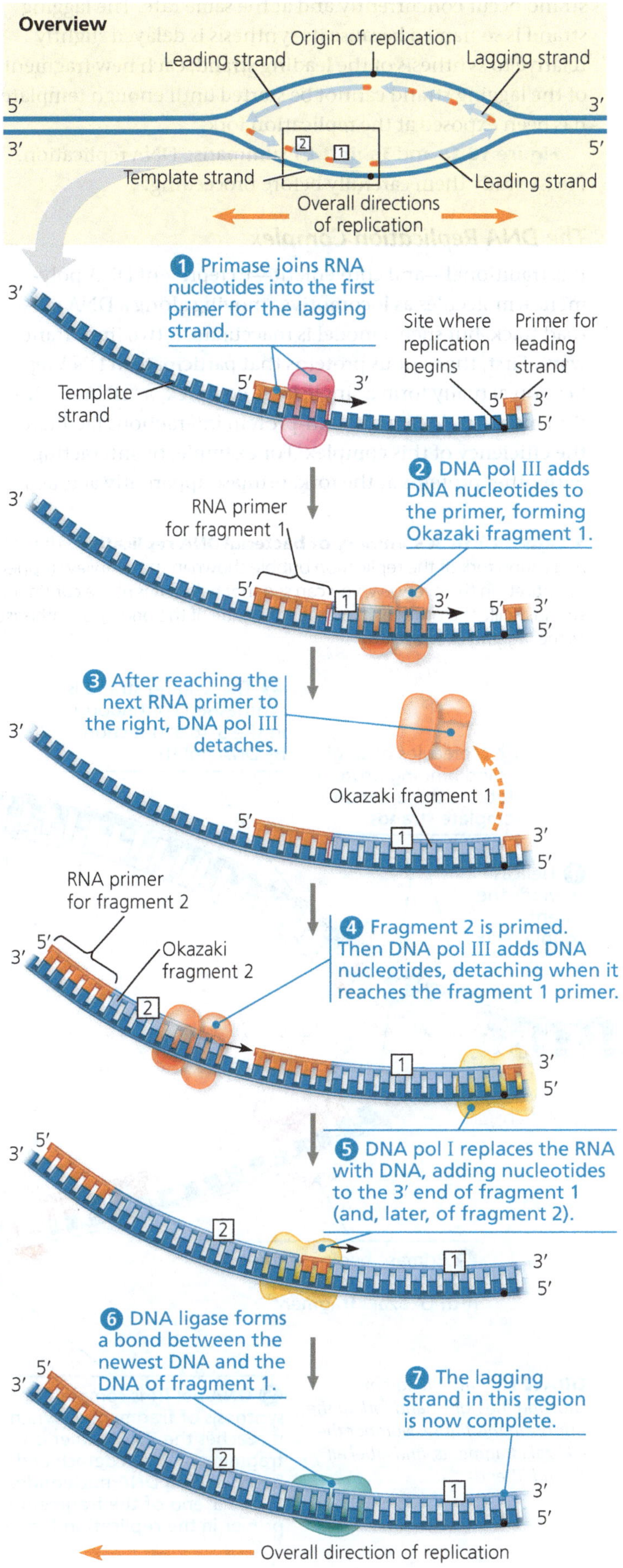

Synthesis of the leading strand and synthesis of the lagging strand occur concurrently and at the same rate. The lagging strand is so named because its synthesis is delayed slightly relative to synthesis of the leading strand; each new fragment of the lagging strand cannot be started until enough template has been exposed at the replication fork.

Figure 16.18 and **Table 16.1** summarise DNA replication. Please study them carefully before proceeding.

The DNA Replication Complex

It is traditional—and convenient—to represent DNA polymerase molecules as locomotives moving along a DNA railroad track, but such a model is inaccurate in two important ways. First, the various proteins that participate in DNA replication actually form a single large complex, a "DNA replication machine." Many protein-protein interactions facilitate the efficiency of this complex. For example, by interacting with other proteins at the fork, primase apparently acts as a molecular brake, slowing progress of the replication fork and coordinating the placement of primers and the rates of replication on the leading and lagging strands. Second, the DNA replication complex may not move along the DNA; rather, the DNA may move through the complex during the replication process. In eukaryotic cells, multiple copies of the complex, perhaps grouped into "factories," may be anchored to the nuclear matrix, a framework of fibres extending through the interior of the nucleus.

Some experimental evidence supports a model in which two DNA polymerase molecules, one on each template strand, "reel in" the parental DNA and extrude newly made daughter DNA molecules. In this so-called *trombone model*, the lagging strand is also looped back through the complex **(Figure 16.19)**. Whether the complex moves along the DNA or whether the DNA moves through the complex, either anchored or not, are still open, unresolved questions that are under active investigation.

▼ **Figure 16.18 A summary of bacterial DNA replication.** The detailed diagram shows the left-hand replication fork of the replication bubble shown in the overview (upper right). Viewing each daughter strand in its entirety in the overview, you can see that half of it is made continuously as the leading strand, while the other half (on the other side of the origin) is synthesised in fragments as the lagging strand.

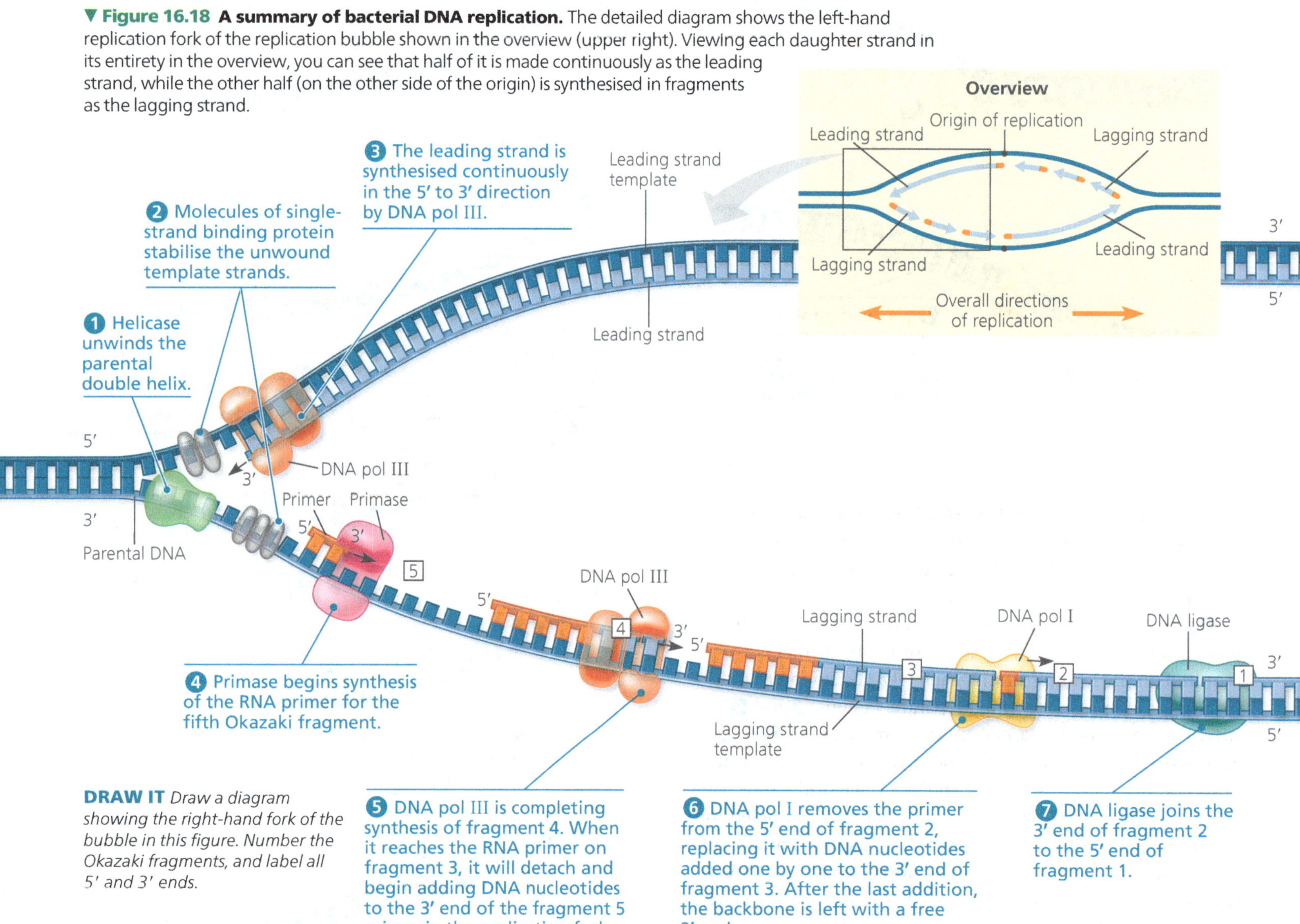

DRAW IT *Draw a diagram showing the right-hand fork of the bubble in this figure. Number the Okazaki fragments, and label all 5′ and 3′ ends.*

Table 16.1 Bacterial DNA Replication Proteins and Their Functions

Protein	Function
Helicase	Unwinds parental double helix at replication forks
Single-strand binding protein	Binds to and stabilises single-stranded DNA until it is used as a template
Topoisomerase	Relieves overwinding strain ahead of replication forks by breaking, swivelling, and rejoining DNA strands
Primase	Synthesises an RNA primer at 5′ end of leading strand and at 5′ end of each Okazaki fragment of lagging strand
DNA pol III	Using parental DNA as a template, synthesises new DNA strand by adding nucleotides to an RNA primer or a pre-existing DNA strand
DNA pol I	Removes RNA nucleotides of primer from 5′ end and replaces them with DNA nucleotides added to 3′ end of adjacent fragment
DNA ligase	Joins Okazaki fragments of lagging strand; on leading strand, joins 3′ end of DNA that replaces primer to rest of leading strand DNA

▼ **Figure 16.19 The "trombone" model of the DNA replication complex.** In this proposed model, two molecules of DNA polymerase III work together in a complex, one on each strand, with helicase and other proteins. The lagging strand template DNA loops through the complex, resembling the slide of a trombone.

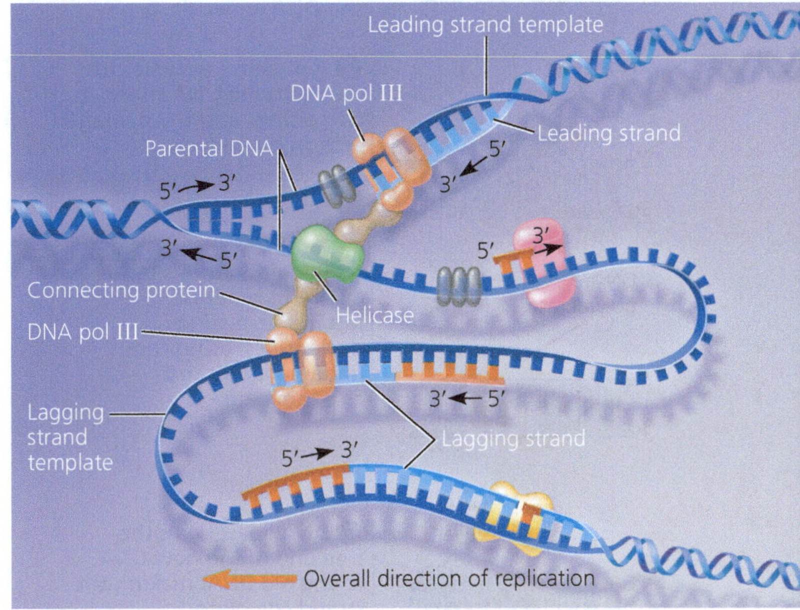

DRAW IT *Draw a line tracing the lagging strand template in this figure.*

Proofreading and Repairing DNA

We cannot attribute the accuracy of DNA replication solely to the specificity of base pairing. Initial pairing errors between incoming nucleotides and those in the template strand occur at a rate of one in 10^5 nucleotides. However, errors in the completed DNA molecule amount to only one in 10^{10} (10 billion) nucleotides, an error rate that is 100,000 times lower. This is because during DNA replication, many DNA polymerases proofread each nucleotide against its template as soon as it is covalently bonded to the growing strand. Upon finding an incorrectly paired nucleotide, the polymerase removes the nucleotide and then resumes synthesis. (This action is similar to fixing a texting error by deleting the wrong letter and then entering the correct one.)

Mismatched nucleotides sometimes evade proofreading by a DNA polymerase. In **mismatch repair**, other enzymes remove and replace incorrectly paired nucleotides that have resulted from replication errors. Researchers highlighted the importance of such repair enzymes when they found that a hereditary defect in one of them is associated with a form of colon cancer. Apparently, this defect allows cancer-causing errors to accumulate in the DNA faster than normal.

Incorrectly paired or altered nucleotides can also arise after replication. In fact, maintenance of the genetic information encoded in DNA requires frequent repair of various kinds of damage to existing DNA. DNA molecules are constantly subjected to potentially harmful chemical and physical agents, such as X-rays, as we'll discuss in Concept 17.5. In addition, DNA bases may undergo spontaneous chemical changes under normal cellular conditions. However, these changes in DNA are usually corrected before they become permanent changes—*mutations*—perpetuated through successive replications. Each cell continuously monitors and repairs its genetic material. Because repair of damaged DNA is so important to the survival of an organism, it is no surprise that many different DNA repair enzymes have evolved. Almost 100 are known in *E. coli*, and about 170 have been identified so far in humans.

Most cellular systems for repairing incorrectly paired nucleotides, whether they are due to DNA damage or to replication errors, use a mechanism that takes advantage of the base-paired structure of DNA. In many cases, a segment of the strand containing the damage is cut out (excised) by a DNA-cutting enzyme—a **nuclease**—and the resulting gap is then filled in with nucleotides, using the undamaged strand as a template. The enzymes involved in filling the gap are a DNA polymerase and DNA ligase. There are several such DNA repair systems; one is called **nucleotide excision repair**.

▼ **Figure 16.20 Nucleotide excision repair of DNA damage.**

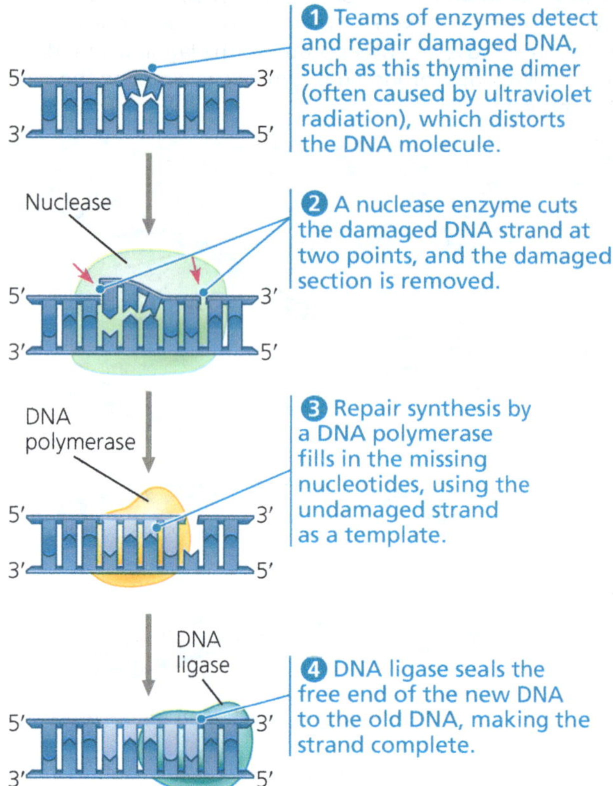

An example is shown in **Figure 16.20**. An important function of the DNA repair enzymes in our skin cells is to repair genetic damage caused by the UV rays of sunlight: For example, adjacent thymine bases on a DNA strand can become covalently linked into *thymine dimers*, causing the DNA to buckle and interfere with DNA replication. The importance of repairing this kind of damage is underscored by a disorder called xeroderma pigmentosum (XP), which in most cases is caused by an inherited defect in a nucleotide excision repair enzyme. Individuals with XP are hypersensitive to sunlight; mutations in their skin cells caused by ultraviolet light are left uncorrected, often resulting in skin cancer. The effects are extreme: Without sun protection, children who have XP can develop skin cancer by age 10.

Evolutionary Significance of Altered DNA Nucleotides

EVOLUTION Faithful replication of the genome and repair of DNA damage are important for the functioning of the organism and for passing on a complete, accurate genome to the next generation. The error rate after proofreading and repair is extremely low, but rare mistakes do slip through. Once a mismatched nucleotide pair is replicated, the sequence change is permanent in the daughter molecule that has the incorrect nucleotide as well as in any subsequent copies. As we mentioned earlier, a permanent change in the DNA sequence is called a mutation.

Mutations can change the phenotype of an organism (see Concept 17.5). And if they occur in germ cells, which give rise to gametes, mutations can be passed on from generation to generation. The vast majority of such changes either have no effect or are harmful, but a very small percentage can be beneficial. In either case, mutations are the original source of the variation on which natural selection operates during evolution and are ultimately responsible for the appearance of new species. (You'll learn more about this process in Unit 4.) The balance between complete fidelity of DNA replication or repair and a low mutation rate has resulted in new proteins that contribute to different phenotypes. Ultimately, over long periods of time, this process leads to new species and thus to the rich diversity of life on Earth today.

Replicating the Ends of DNA Molecules

For linear DNA, such as the DNA of eukaryotic chromosomes, the usual replication machinery cannot complete the 5′ ends of daughter DNA strands because there is no 3′ end of a preexisting polynucleotide for DNA polymerase to add onto. This is another consequence of the enzyme's requirements. Even if an Okazaki fragment can be started with an RNA primer hydrogen-bonded to the very end of the template strand, once that primer is removed, it cannot be replaced with DNA because there is no 3′ end available for nucleotide addition **(Figure 16.21)**. As a result, repeated rounds of replication produce shorter and shorter DNA molecules with uneven ("staggered") ends.

Most prokaryotes have a circular chromosome, with no ends, so the shortening of DNA does not occur. But what protects the genes of linear eukaryotic chromosomes from being eroded away during successive rounds of DNA replication? Eukaryotic chromosomal DNA molecules have special nucleotide sequences called **telomeres** at their ends **(Figure 16.22)**. Telomeres do not contain genes; instead, the DNA typically consists of multiple repetitions of one short nucleotide sequence. In each human telomere, for example, the six-nucleotide sequence TTAGGG is repeated between 100 and 1,000 times.

Telomeres have two protective functions. First, specific proteins associated with telomeric DNA prevent the staggered ends of the daughter molecule from activating the cell's systems for monitoring DNA damage. (Staggered ends of a DNA molecule, which often result from double-strand breaks, can trigger signal transduction pathways leading to cell cycle arrest or cell death.) Second, telomeric DNA acts as a kind of buffer zone that provides some protection against the organism's genes shortening, somewhat like how the plastic-wrapped ends of a shoelace slow down its unravelling. Telomeres do not prevent the erosion of genes near the ends of chromosomes; they merely postpone it.

▼ Figure 16.21 Shortening of the ends of linear DNA molecules. Here we follow the left end of one DNA molecule through two rounds of replication. After the first round, the new lagging strand is shorter than its template. After a second round, both the leading and lagging strands have become shorter than the original parental DNA. Although not shown here, the other ends of these chromosomal DNA molecules (not shown) also become shorter.

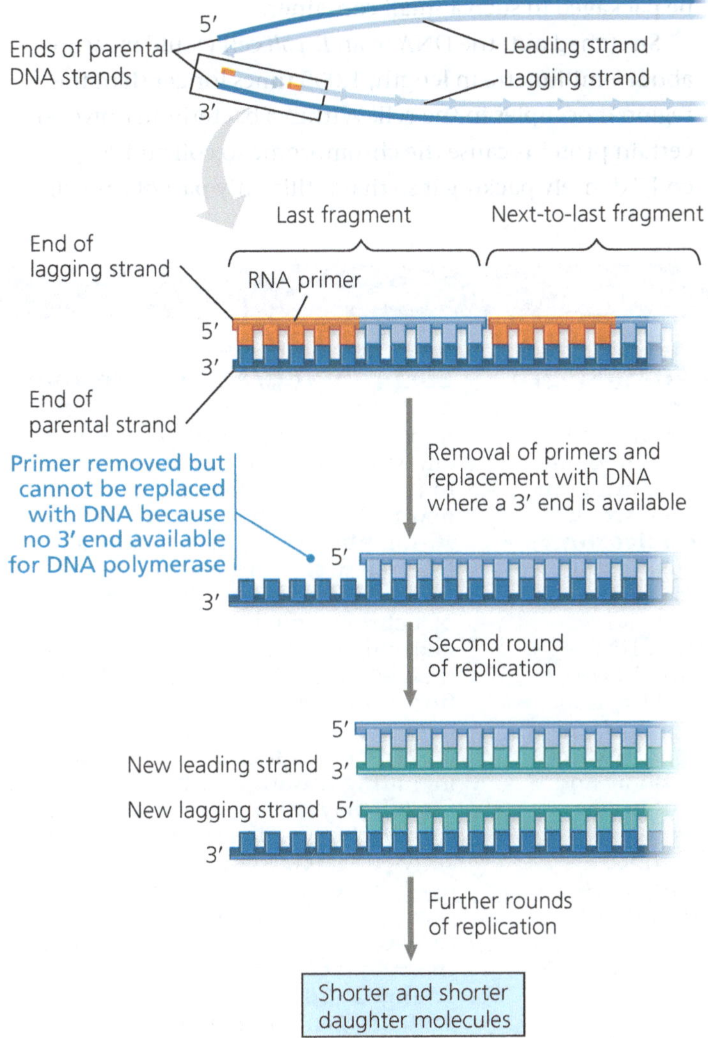

▼ Figure 16.22 Telomeres. Eukaryotes have repetitive, noncoding sequences called telomeres at the ends of their DNA. Telomeres are stained orange in these mouse chromosomes (LM).

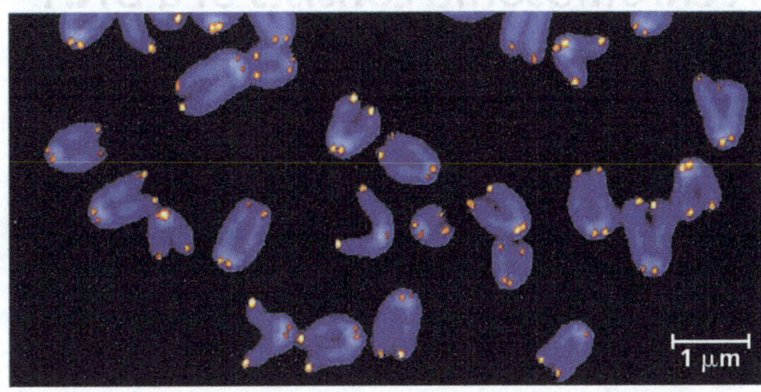

As shown in Figure 16.21, telomeres become shorter during every round of replication. Thus, as expected, telomeric DNA tends to be shorter in dividing somatic cells of older individuals and in cultured cells that have divided many times. It has been proposed that shortening of telomeres is somehow connected to the aging process of certain tissues and even to aging of the organism as a whole.

But what about germ cells, whose genome must persist virtually unchanged from an organism to its offspring over many generations? If the chromosomes of germ cells became shorter in every cell cycle, essential genes would eventually be missing from the gametes they produce. However, this does not occur: An enzyme called *telomerase* catalyses the lengthening of telomeres in eukaryotic germ cells, thus restoring their original length and compensating for the shortening that occurs during DNA replication. This enzyme contains its own RNA molecule that it uses as a template to artificially "extend" the leading strand, allowing the lagging strand to maintain a given length. Telomerase is not active in most human somatic cells, but its activity varies from tissue to tissue. The activity of telomerase in germ cells results in telomeres of maximum length in the zygote.

Normal shortening of telomeres may protect organisms from cancer by limiting the number of divisions that somatic cells can undergo. Cells from large tumours often have unusually short telomeres, as we would expect for cells that have undergone many cell divisions. Further shortening would presumably lead to self-destruction of the tumour cells. Telomerase activity is abnormally high in cancerous somatic cells, suggesting that its ability to stabilise telomere length may allow these cancer cells to persist. Many cancer cells do seem capable of unlimited cell division, as do immortal strains of cultured cells (see Concept 12.3). For several years, researchers have studied inhibition of telomerase as a possible cancer therapy. While studies that inhibited telomerase in mice with tumours have led to the death of cancer cells, eventually the cells have restored the length of their telomeres by an alternative pathway. This is an area of ongoing research that may eventually yield useful cancer treatments.

CONCEPT CHECK 16.2

1. What role does complementary base pairing play in the replication of DNA?
2. Identify two major functions of DNA pol III in DNA replication.
3. **MAKE CONNECTIONS** What is the relationship between DNA replication and the S phase of the cell cycle? See Figure 12.6.
4. **VISUAL SKILLS** If the DNA pol I in a given cell were nonfunctional, how would that affect the synthesis of a *leading* strand? In the overview box in Figure 16.18, point out where DNA pol I would normally function on the top leading strand.

For suggested answers, see Appendix A.

CONCEPT 16.3

A chromosome consists of a DNA molecule packed together with proteins

Now let's examine how DNA is packaged into chromosomes, the structures that carry genetic information. The main component of the genome in most bacteria is a double-stranded, circular DNA molecule that is associated with specific proteins. A bacterial chromosome differs from a eukaryotic chromosome in that a eukaryotic chromosome consists of a single linear DNA molecule associated with a large number of proteins. In *E. coli*, the chromosomal DNA consists of about 4.6 million nucleotide pairs, representing about 4,400 genes. This is 100 times more DNA than is found in a typical virus, but only about one-thousandth as much DNA as in a human somatic cell. Even so, that is a tremendous amount of DNA to be packaged in such a small container.

Stretched out, the DNA of an *E. coli* cell would measure about a millimetre in length, 1,000 times longer than the region it occupies in the cell. Within a bacterium, however, certain proteins cause the chromosome to coil and "supercoil," densely packing it so that it fills only part of the cell.

▼ Figure 16.23 Exploring Chromatin Packing in a Eukaryotic Chromosome

DNA, the Double Helix

Shown below is a ribbon model of DNA, with each ribbon representing one of the polynucleotide strands. The TEM shows a molecule of naked (protein-free) DNA; the double helix is 2 nm across.

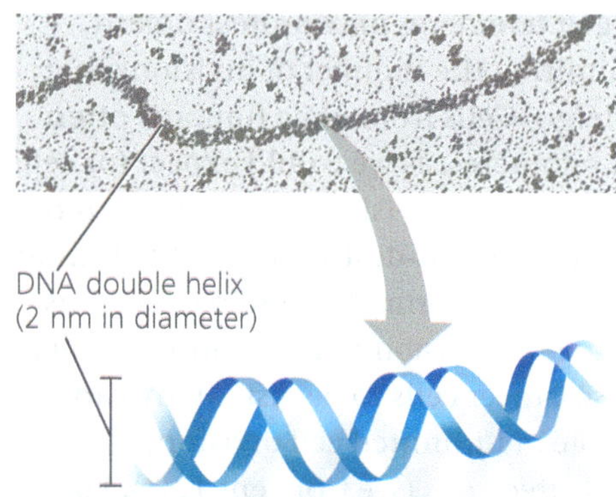

Histones

Proteins called **histones** are responsible for the main level of DNA packing in interphase chromatin. More than a fifth of a histone's amino acids are positively charged (Lys or Arg) and therefore bind tightly to the negatively charged DNA.

Four types of histones are most common in chromatin. The histones are very similar among eukaryotes, probably reflecting their important role in chromatin structure.

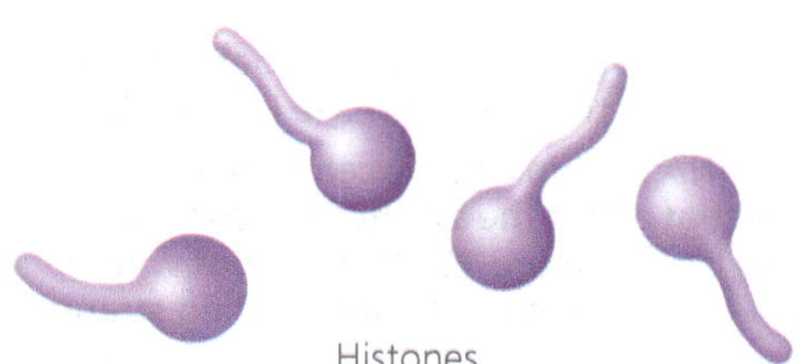
Histones

Nucleosomes in a 10-nm Fibre

In electron micrographs, unfolded chromatin is roughly 10 nm in diameter (the *10-nm fibre*). Such chromatin resembles beads on a string (see the TEM). Each "bead" is a **nucleosome**, the basic unit of DNA packing; the "string" between beads is called *linker DNA*.

A nucleosome consists of DNA wound twice around a protein core of eight histones, two each of the four main histone types. The amino end of each histone (the *histone tail*) extends outwards from the nucleosome and is involved in regulation of gene expression.

In the cell cycle, the histones leave the DNA only briefly during DNA replication. Generally, they do the same during the process of transcription, which requires access to the DNA by transcription proteins.

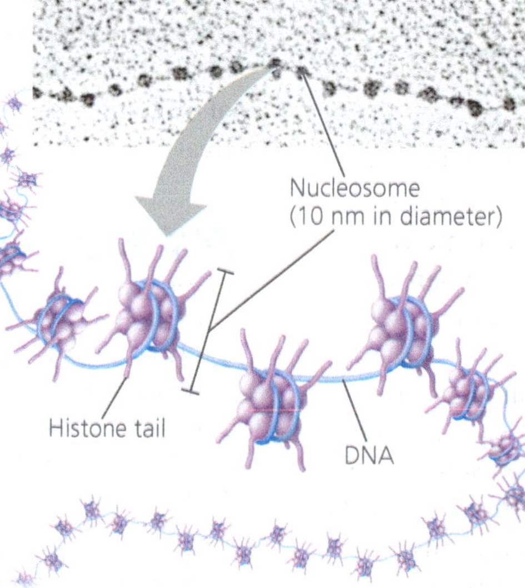

Euchromatin (loosely arranged 10-nm fibre)

Euchromatin/Heterochromatin

A recent technique called ChromEMT allows scientists to visualise chromatin in intact cells. ChromEMT and other new techniques have shown that the 10-nm fibre is the basic constituent of interphase chromatin.

During interphase, different regions of a chromosome may exist as euchromatin (below left) or heterochromatin (below). In euchromatin, the 10-nm fibre is loosely arranged in a more open configuration than heterochromatin; higher degrees of organisation, including the 30-nm fibre in one older model, may exist in specific cells or at certain times. (This is an area of very active research.) The DNA in euchromatin is accessible to the proteins that carry out transcription, and its genes can be expressed. In heterochromatin, the 10-nm fibre is more densely arranged and less accessible to these proteins; genes in heterochromatin are generally not expressed.

Euchromatin and heterochromatin are organised into regions by other proteins not shown here; this organisation is dynamic but disappears once mitosis begins.

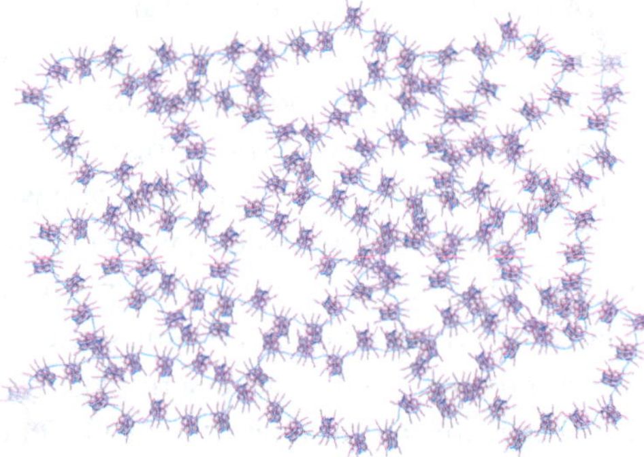
Heterochromatin (densely arranged 10-nm fibre)

Unlike the nucleus of a eukaryotic cell, this dense region of DNA in a bacterium, called the nucleoid, is not bounded by membrane (see Figure 6.5).

Each eukaryotic chromosome contains a single linear DNA double helix that, in humans, averages about 1.5×10^8 nucleotide pairs. This is an enormous amount of DNA relative to a chromosome's condensed length. If completely stretched out, such a DNA molecule would be about 4 cm long, thousands of times the diameter of a cell nucleus—and that's not even considering the DNA of the other 45 human chromosomes!

In the cell, eukaryotic DNA is precisely combined with a large amount of protein. The exact way in which this complex of DNA and protein, called **chromatin**, fits into the nucleus has long been debated. **Figure 16.23** outlines the current view of the organisation of chromatin during interphase and how it is condensed during mitosis into the metaphase chromosome. Study this figure carefully before reading further.

Chromatin undergoes striking changes in how densely it is packed during the course of the cell cycle (see Figure 12.7).

Prophase

When mitosis begins, DNA replication has already occurred, so each chromosome consists of two sister chromatids. During prophase, the chromatin of each sister chromatid begins to condense. Two related proteins called condensin II and condensin I play important roles. First, condensin II proteins (red) bind to the 10-nm fibre of DNA and form DNA loops that get larger and larger. The condensin II proteins form a central scaffold from which the loops extend. As the loops grow, the chromosome gets wider and shorter. By the end of prophase, the chromosome is half as long.

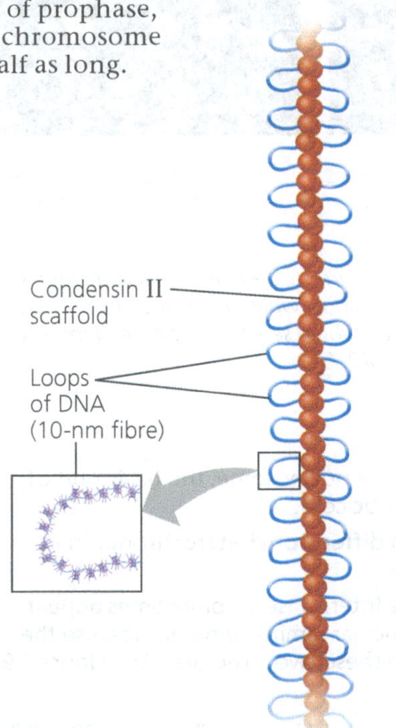

Prometaphase

When prometaphase begins, condensin I proteins (green) bind to DNA outside the central scaffold, making smaller loops (not shown) out of the larger loops generated by condensin II. The process continues, with more and more loops extending outwards, and the chromosome getting denser, shorter, and wider. The scaffold itself also begins to twist into a helix (suggested by the curved grey arrows), allowing even more loops per given length of chromosome.

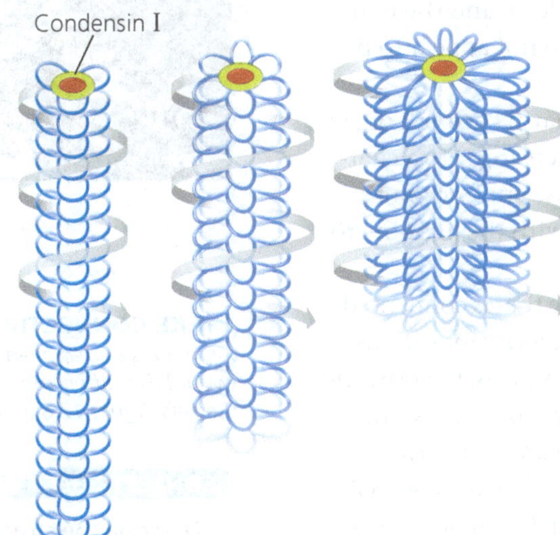

Metaphase

At metaphase, the chromosome is at its most dense, with the most loops per turn, and therefore is at its shortest length. The two sister chromatids are fully condensed.

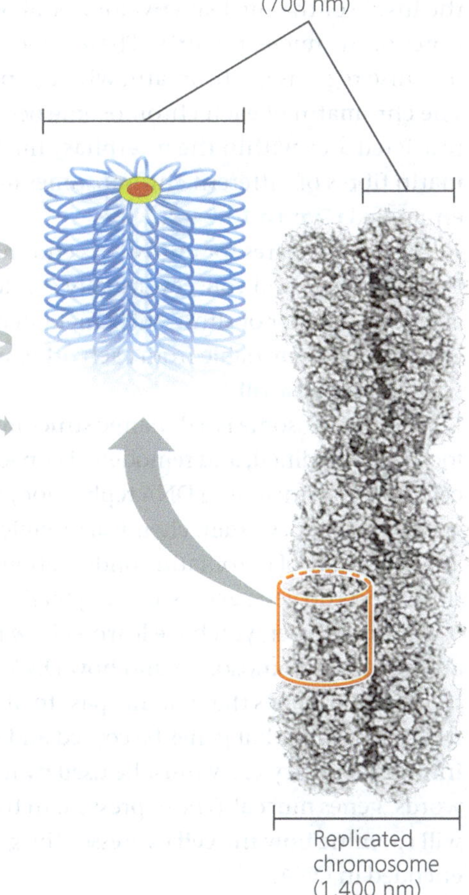

In interphase cells stained for light microscopy, the chromatin usually appears as a diffuse mass within the nucleus, with some denser clumps, including in regions of centromeres and telomeres. The less compacted, more dispersed interphase chromatin is called **euchromatin** ("true chromatin") to distinguish it from the more compacted, denser-appearing **heterochromatin** (see Figure 16.23). Recent research has clarified our understanding of chromatin structure: For both types of chromatin, the basic organising unit is the 10-nm fibre—nucleosomes joined by linker DNA. In heterochromatin, this 10-nm fibre is folded and bent back on itself to a much greater degree than in euchromatin, accounting for its denser appearance. Because heterochromatin is so compacted, it is largely inaccessible to the proteins responsible for transcribing the genetic information, a crucial early step in gene expression. In contrast, the looser packing of euchromatin makes its DNA accessible to those proteins, and the genes present in euchromatin are available for transcription.

Early on, biologists assumed that interphase chromatin was a tangled mass in the nucleus, like a bowl of spaghetti, but this is far from the case. Although an interphase chromosome lacks an obvious scaffold, there are proteins that further organise the 10-nm fibre into larger compartments and smaller looped domains. Some of the looped domains appear to be attached to the nuclear lamina, on the inside of the nuclear envelope, and perhaps also to fibres of the nuclear matrix. These attachments may help organise regions of chromatin where genes are active. The chromatin of each chromosome occupies a specific restricted area within the interphase nucleus, and the chromatin fibres of different chromosomes do not appear to be entangled **(Figure 16.24a)**.

As a cell prepares for mitosis, its chromatin becomes organised into loops and coils, eventually condensing into a characteristic number of short, thick metaphase chromosomes that are distinguishable from each other with the light microscope **(Figure 16.24b)**.

The chromosome is a dynamic structure that is condensed, loosened, modified, and remodelled as necessary for various cell processes, including DNA replication, mitosis, meiosis, and gene expression. Certain chemical modifications of histones affect the state of chromatin condensation and also have multiple effects on gene expression, as you'll see in Concept 18.2.

In this chapter, you have learned how DNA molecules are arranged in chromosomes and how DNA replication provides the copies of genes that parents pass to offspring. However, it is not enough that genes be copied and transmitted; the information they carry must be used by the cell. In other words, genes must also be expressed. In the next chapter, we will examine how the cell expresses the genetic information encoded in DNA.

▼ **Figure 16.24 "Painting" chromosomes.** Researchers can treat ("paint") human chromosomes with molecular tags that cause each chromosome pair to appear a different colour.

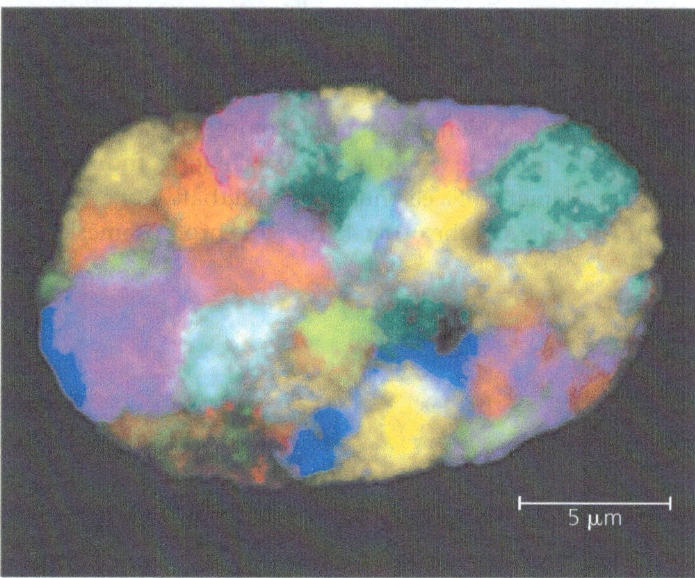

(a) The ability to visually distinguish among chromosomes makes it possible to see how the chromosomes are arranged in the interphase nucleus. Each chromosome appears to occupy a specific territory during interphase. In general, the two homologues of a pair are not located together.

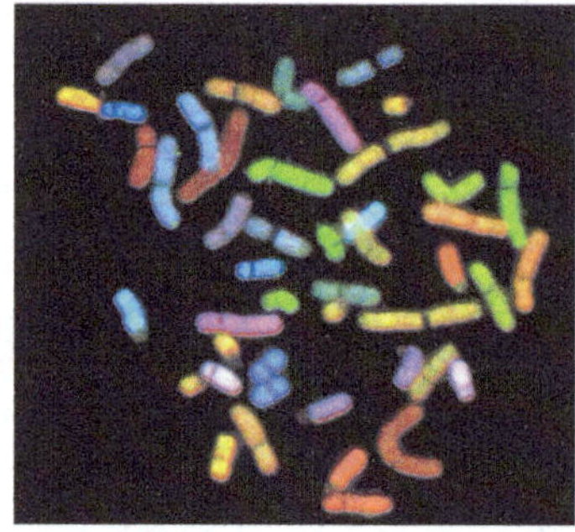

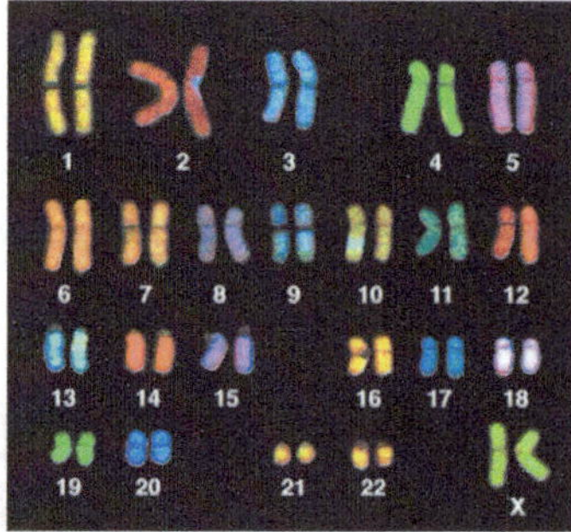

(b) These metaphase chromosomes have been "painted" so that the two homologues of a pair are the same colour. On the left is a spread of treated chromosomes; on the right, they have been organised into a karyotype.

MAKE CONNECTIONS *If you arrested a human cell in metaphase I of meiosis and applied this technique, what would you observe? How would this differ from what you would see in metaphase of mitosis? Review Figure 13.8 and Figure 12.7.*

CONCEPT CHECK 16.3

1. Describe the structure of a nucleosome, the basic unit of DNA packing in eukaryotic cells.
2. How does euchromatin differ from heterochromatin in structure and function?
3. **MAKE CONNECTIONS** Interphase chromosomes appear to be attached to the nuclear lamina and perhaps also the nuclear matrix. Describe these two structures. See Figure 6.9 and the associated text.

For suggested answers, see Appendix A.

16 Chapter Review

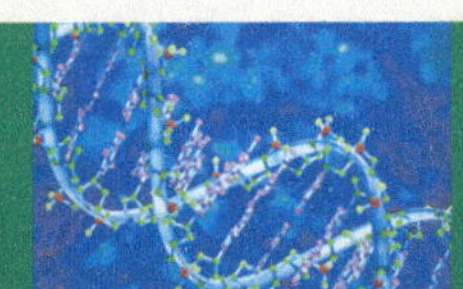

SUMMARY OF KEY CONCEPTS

CONCEPT 16.1

DNA is the genetic material *(pp. 317–322)*

- Experiments with bacteria and with **phages** provided the first strong evidence that the genetic material is DNA.
- Watson and Crick deduced that DNA is a **double helix** and built a structural model. Two **antiparallel** sugar-phosphate chains wind around the outside of the molecule; the nitrogenous bases project into the interior, where they hydrogen-bond in specific pairs, A with T, G with C.

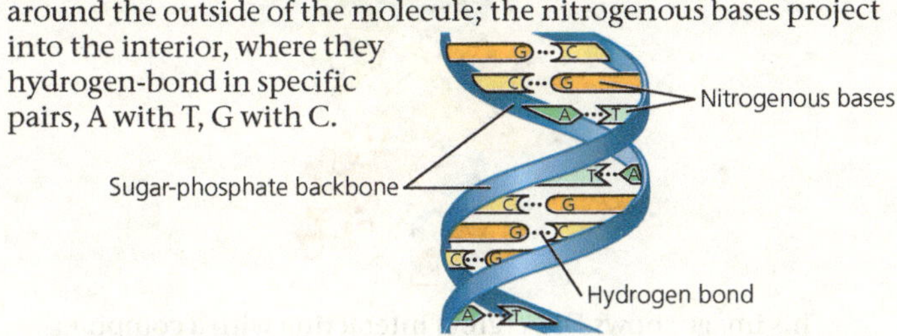

? *What does it mean when we say that the two DNA strands in the double helix are antiparallel? What would an end of the double helix look like if the strands were parallel?*

CONCEPT 16.2

Many proteins work together in DNA replication and repair *(pp. 322–331)*

- The Meselson-Stahl experiment showed that **DNA replication** is **semiconservative**: The parental molecule unwinds, and each strand then serves as a template for the synthesis of a new strand according to base-pairing rules.
- DNA replication at one **replication fork** is summarised here:

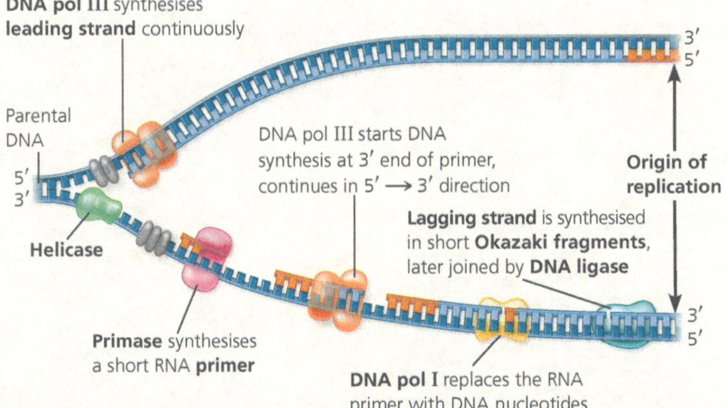

- **DNA polymerases** proofread new DNA, replacing incorrect nucleotides. In **mismatch repair**, enzymes correct errors that persist. **Nucleotide excision repair** is a process by which **nucleases** cut out and other enzymes replace damaged stretches of DNA.
- The ends of eukaryotic chromosomal DNA get shorter with each round of replication. The presence of **telomeres**, repetitive sequences at the ends of linear DNA molecules, postpones the erosion of genes. Telomerase catalyses the lengthening of telomeres in germ cells.

? *Compare DNA replication on the leading and lagging strands, including similarities and differences.*

CONCEPT 16.3

A chromosome consists of a DNA molecule packed together with proteins *(pp. 332–334)*

- The chromosome of most bacterial species is a circular DNA molecule with some associated proteins, making up the nucleoid.
- The **chromatin** making up a eukaryotic chromosome is composed of DNA, **histones**, and other proteins. The histones bind to each other and to the DNA to form **nucleosomes**, the most basic units of DNA packing, which exist as part of the 10-nm fibre. Histone tails extend outwards from each bead-like nucleosome core. Additional coiling and folding lead ultimately to the highly condensed chromatin of the metaphase chromosome.
- Chromosomes occupy restricted areas in the interphase nucleus. In interphase cells, most 10-nm fibre chromatin is loosely arranged (**euchromatin**), but some is more densely arranged (**heterochromatin**). Euchromatin, but not heterochromatin, is generally accessible for gene transcription.

? *Describe the levels of chromatin packing you'd expect to see in an interphase nucleus.*

TEST YOUR UNDERSTANDING

Levels 1-2: Remembering/Understanding

1. In his work with pneumonia-causing bacteria and mice, Griffith found that
   (A) the protein coat from pathogenic cells was able to transform nonpathogenic cells.
   (B) heat-killed pathogenic cells caused pneumonia.
   (C) some substance from pathogenic cells was transferred to nonpathogenic cells, making them pathogenic.
   (D) the polysaccharide coat of bacteria caused pneumonia.
2. What is the basis for the difference in how the leading and lagging strands of DNA molecules are synthesised?
   (A) The origins of replication occur only at the 5′ end.
   (B) Helicases and single-strand binding proteins work at the 5′ end.
   (C) DNA polymerase can join new nucleotides only to the 3′ end of a pre-existing strand, and the strands are antiparallel.
   (D) DNA ligase works only in the 3′ → 5′direction.
3. In analysing the number of different bases in a DNA sample, which result would be consistent with the base-pairing rules?
   (A) A = G
   (B) A + G = C + T
   (C) A + T = G + C
   (D) A = C
4. The elongation of the leading strand during DNA synthesis
   (A) progresses away from the replication fork.
   (B) occurs in the 3′ → 5′ direction.
   (C) produces Okazaki fragments.
   (D) depends on the action of DNA polymerase.
5. In a nucleosome, the DNA is wrapped around
   (A) histones.
   (B) ribosomes.
   (C) polymerase molecules.
   (D) a thymine dimer.

Levels 3-4: Applying/Analysing

6. *E. coli* cells grown on ^{15}N medium are transferred to ^{14}N medium and allowed to grow for two more generations (two rounds of DNA replication). DNA extracted from these cells is centrifuged. What density distribution of DNA would you expect in this experiment?
   (A) one high-density and one low-density band
   (B) one intermediate-density band
   (C) one high-density and one intermediate-density band
   (D) one low-density and one intermediate-density band

7. A student isolates, purifies, and combines in a test tube a variety of molecules needed for DNA replication. When she adds some DNA to the mixture, replication occurs, but each DNA molecule consists of a normal strand paired with numerous segments of DNA a few hundred nucleotides long. What has she probably left out of the mixture?
   (A) DNA polymerase
   (B) DNA ligase
   (C) Okazaki fragments
   (D) primase

8. The spontaneous loss of amino groups from adenine in DNA results in hypoxanthine, an uncommon base, opposite thymine. What combination of proteins could repair such damage?
   (A) nuclease, DNA polymerase, DNA ligase
   (B) telomerase, primase, DNA polymerase
   (C) telomerase, helicase, single-strand binding protein
   (D) DNA ligase, replication fork proteins, adenylyl cyclase

9. **MAKE CONNECTIONS** Although the proteins that cause the *E. coli* chromosome to coil are not histones, what property would you expect them to share with histones, given their ability to bind to DNA (see Figure 5.14)?

Levels 5-6: Evaluating/Creating

10. **EVOLUTION CONNECTION** Some bacteria may be able to respond to environmental stress by increasing the rate at which mutations occur during cell division. How might this be accomplished? Might there be an evolutionary advantage to this ability? Explain.

11. **SCIENTIFIC INQUIRY**

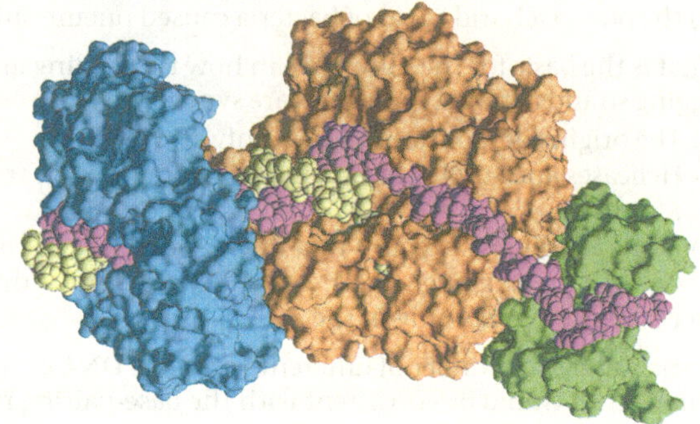

DRAW IT Model building can be an important part of the scientific process. The illustration shown above is a computer-generated model of a DNA replication complex. The parental and newly synthesised DNA strands are colour-coded differently, as are each of the following three proteins: DNA pol III, the sliding clamp, and single-strand binding protein.
   (a) Using what you've learned in this chapter to clarify this model, label each DNA strand and each protein.
   (b) Draw an arrow to indicate the overall direction of DNA replication.

12. **WRITE ABOUT A THEME: ORGANISATION** The continuity of life is based on heritable information in the form of DNA, and structure and function are correlated at all levels of biological organisation. In a short essay (100–150 words), describe how the structure of DNA is correlated with its role as the molecular basis of inheritance.

13. **SYNTHESISE YOUR KNOWLEDGE**

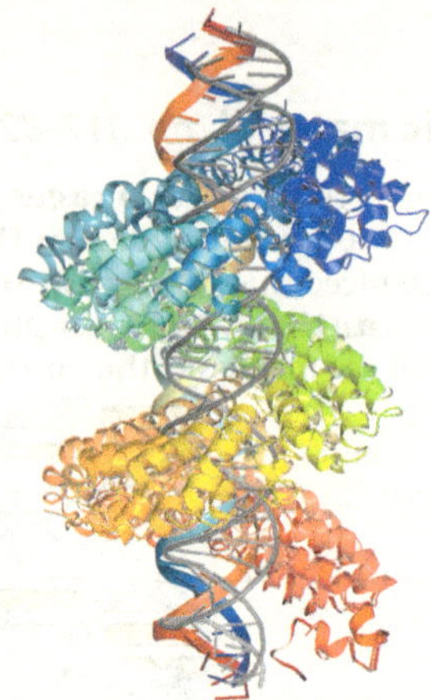

This image shows DNA (grey) interacting with a computer-generated model of a TAL protein (multicoloured), one of a family of proteins found only in a species of the bacterium *Xanthomonas*. The bacterium uses proteins like this one to find specific gene sequences in cells of the organisms it infects, such as tomatoes, rice, and citrus fruits. Given what you know about DNA structure and considering the image above, discuss how the TAL protein's structure suggests that it functions.

For selected answers, see Appendix A.

17 Gene Expression: From Gene to Protein

KEY CONCEPTS

Study Tip

Make a visual study guide: Sketch the process shown below, and add labels and details as you read the chapter. (In this exercise, assume all processes take place in a eukaryotic cell.)

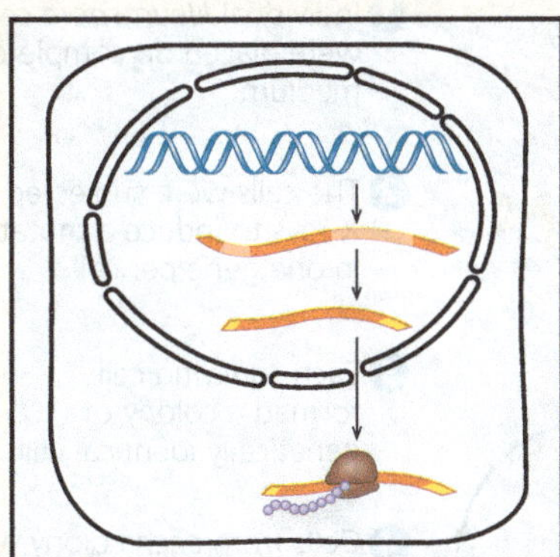

Go to Mastering Biology

to access Dynamic Study Modules for revision, 3D BioFlix® animations and high-quality videos, and your interactive Pearson eText.

Figure 17.1 An albino joey (the name for a young koala) with its regularly coloured mother. The joey's albino colouration arose because a recessive mutation occurred in the DNA of an ancestor koala and passed through several generations until it manifested in this mother. The recessive mutation disables pigment synthesis. When a male with the same heterozygous gene mates with a similarly genetically endowed mother, there remains a small chance that their offspring will become a homozygous albino koala.

How can one change in DNA result in such a dramatic change in appearance?

Proteins are the link between genotype and phenotype. Gene expression is the process by which DNA directs the synthesis of proteins:

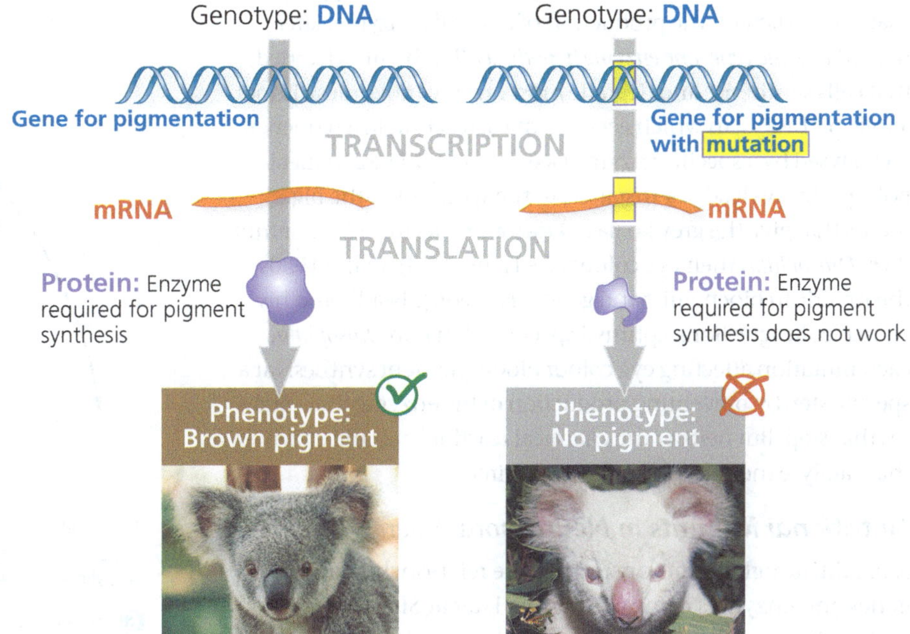

CONCEPT 17.1

Genes specify proteins via transcription and translation

Inherited traits like albinism are determined by genes, which have information content in the form of specific sequences of nucleotides along stretches of DNA, the genetic material. The DNA inherited by an organism leads to specific traits by dictating the synthesis of proteins and of RNA molecules involved in protein synthesis. In the case of coat and skin colour, the gene containing the information to synthesise an enzyme that makes pigment allows normal colouring, and a faulty version of that gene leads to lack of colour, or the albino phenotype. Proteins are the link between genotype and phenotype. **Gene expression** is the process by which DNA directs the synthesis of proteins (or, in some cases, just RNAs). The expression of genes that code for proteins includes two stages: transcription and translation.

Before going into the details of how genes direct protein synthesis, let's step back and examine how the fundamental relationship between genes and proteins was discovered.

Evidence from Studying Metabolic Defects

In 1902, British physician Archibald Garrod was the first to suggest that genes dictate phenotypes through enzymes, proteins that catalyse specific chemical reactions in the cell. He postulated that the symptoms of an inherited disease reflect an inability to make a particular enzyme. He later referred to such diseases as "inborn errors of metabolism." For example, people with a disease called alkaptonuria have black urine because it contains a chemical called alkapton, which darkens upon exposure to air. Garrod reasoned that these people cannot make an enzyme that breaks down alkapton, so alkapton is expelled in their urine.

Several decades later, research supported Garrod's hypothesis that a gene dictates the production of a specific enzyme, later named the *one gene–one enzyme hypothesis*. Biochemists learned that cells synthesise and degrade most organic molecules via metabolic pathways, in which each chemical reaction in a sequence is catalysed by a specific enzyme (see Concept 8.1). Such metabolic pathways lead, for instance, to the synthesis of the pigments that give the grey koala in **Figure 17.1** its fur colour or fruit flies (*Drosophila*) their eye colour (see Figure 15.3). In the 1930s, the American biochemist and geneticist George Beadle and his French colleague Boris Ephrussi speculated that in *Drosophila*, each mutation affecting eye colour blocks pigment synthesis at a specific step by preventing production of the enzyme that catalyses that step. But neither the chemical reactions nor the enzymes that catalyse them were known at the time.

Nutritional Mutants in Neurospora: Scientific Inquiry

A breakthrough in demonstrating the relationship between genes and enzymes came a few years later at Stanford University, where Beadle and Edward Tatum began working with a bread mould, *Neurospora crassa*. Unlike the diploid organisms studied by Mendel (peas) and T. H. Morgan (fruit flies), *Neurospora* is a haploid species. Because its genome contains only one copy of each gene, that single copy determines the individual's expressed phenotype. Therefore, when Beadle and Tatum wanted to discover the function of any gene, they only needed to mutate and disable that one allele. They chose to study the protein-coding genes required for a specific nutritional activity. They caused mutations in genes by bombarding *Neurospora* with X-rays and looked among the survivors for mutants that differed in their nutritional needs from the wild-type bread mould.

Wild-type *Neurospora* has modest food requirements. It can grow in the laboratory on *minimal medium*—a simple solution containing minimal nutrients for growth—inorganic salts, glucose, and the vitamin biotin—incorporated into agar, a support medium. From these basics, wild-type mould cells use their metabolic pathways to produce all the other nutrient molecules (like amino acids) they need for growth, dividing repeatedly and forming visible colonies of genetically identical cells. As shown in **Figure 17.2**, Beadle and Tatum generated different "nutritional mutants" of *Neurospora* cells, each of which had a mutation in one gene and was unable

▼ Figure 17.2 Beadle and Tatum's experimental approach. To obtain nutritional mutants, Beadle and Tatum exposed *Neurospora* cells to X-rays, inducing mutations, then screened for (tested for) mutants that had new nutritional requirements, such as arginine, as shown here.

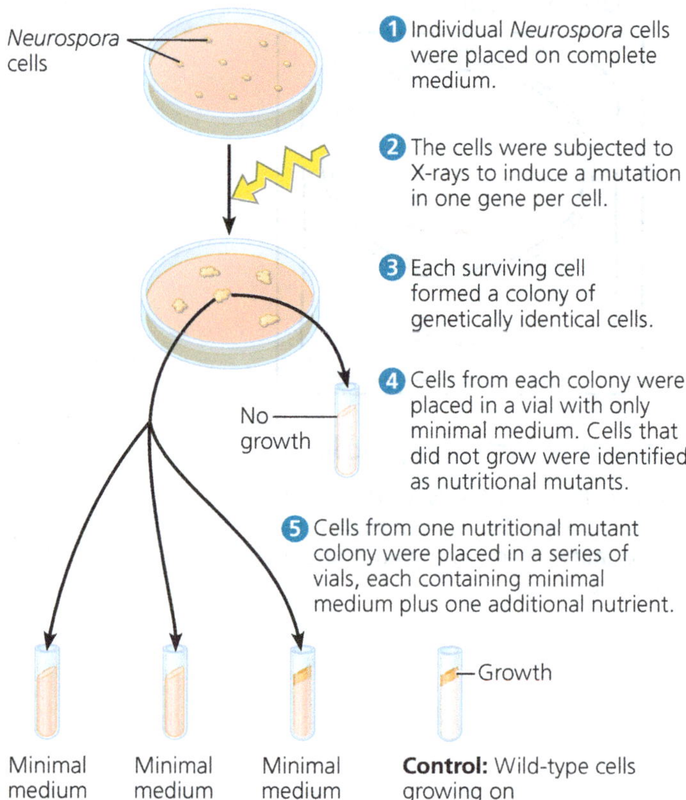

to synthesise a particular essential nutrient. Such cells could not grow on minimal medium but could grow on *complete medium*, which contains all nutrients needed for growth. For *Neurospora*, the complete medium consists of the minimal medium supplemented with all 20 amino acids and a few other nutrients. Beadle and Tatum hypothesised that in each nutritional mutant, the gene for the enzyme that synthesises a particular nutrient had been disabled by the mutation. They were able to determine experimentally which nutrient each mutant strain (original cell and its descendants) was unable to synthesise.

This approach resulted in a valuable collection of mutant strains of *Neurospora*, each catalogued by its defect in a particular pathway. Two colleagues, Adrian Srb and Norman Horowitz, used a collection of arginine-requiring mutants to investigate the biochemical pathway for arginine synthesis in *Neurospora* **(Figure 17.3)**. Srb and Horowitz pinned down each mutant's defect more specifically, using additional tests to distinguish among three classes of arginine-requiring mutants. Mutants in each class required a different set of compounds along the arginine-synthesising pathway, which has three steps. These results, and those of many similar experiments done by Beadle and Tatum, suggested that each class was blocked at a different step in this pathway because mutants in that class lacked the enzyme that catalyses the blocked step.

Because Beadle and Tatum set up their experimental conditions so that each mutant was defective in a single gene, the collected results, taken together, provided strong support for the *one gene–one enzyme hypothesis*, as they dubbed it: that the function of a gene is to dictate the production of a specific enzyme. Further support for this hypothesis came from experiments that identified the specific enzymes lacking in the mutants. Beadle and Tatum shared a Nobel Prize in 1958 for "their discovery that genes act by regulating definite chemical events" (in the words of the Nobel committee).

Today, we know of countless examples in which a mutation in a gene causes a faulty enzyme that in turn leads to an identifiable condition. The albino koala in Figure 17.1 lacks a key enzyme called tyrosinase in the metabolic pathway that produces melanin, a dark pigment. The genetic expression of the albino appearance results from a lack of the same enzymes. Its eyes are pink because no melanin is present to mask the reddish colour of the blood vessels that run through those structures.

The Products of Gene Expression: A Developing Story

As researchers learned more about proteins, they made revisions to the one gene–one enzyme hypothesis. First of all, not all proteins are enzymes. Keratin, the structural protein of animal hair, and the hormone insulin are two examples of nonenzyme proteins. Because proteins that are not enzymes are nevertheless gene products, molecular biologists began to think in terms of one gene–one protein. However, many proteins are constructed from two or more different polypeptide chains, and each polypeptide is specified by its own gene. For example, haemoglobin—the oxygen-transporting protein of vertebrate red blood cells—contains two kinds of polypeptides, and thus two genes code for this protein, one for each type of polypeptide (see Figure 5.18). Beadle and Tatum's idea was therefore restated as the *one gene–one polypeptide hypothesis*. Even this description is not entirely accurate, though. First, in many cases, a eukaryotic gene can code for a set of closely related polypeptides via a process called alternative splicing, which you will learn about later in this chapter. Second, quite a few genes code for RNA molecules that have important functions in cells even though they are never translated into protein. For now, we will focus on genes that do code for polypeptides. (Note that it is common to refer to these gene products as proteins—a practice you will encounter in this text—rather than more precisely as polypeptides.)

Basic Principles of Transcription and Translation

Genes provide the instructions for making specific proteins, but a gene does not build a protein directly. The bridge between DNA and protein synthesis is the nucleic acid RNA. RNA is chemically similar to DNA except that it contains ribose instead of deoxyribose as its sugar and has the nitrogenous base uracil rather than thymine (see Figure 5.23). Thus, each nucleotide along a DNA strand has A, G, C, or T as its base, and each nucleotide along an RNA strand has A, G, C, or U as its base. An RNA molecule usually consists of a single strand.

We often describe the flow of information from gene to protein in terms of languages. Just as specific sequences of letters communicate information in a language such as English, specific sequences of monomers convey information in polymers like nucleic acids and proteins. In DNA or RNA, the monomers are the four types of nucleotides listed previously, which differ in their nitrogenous bases. Genes are typically hundreds or thousands of nucleotides long, each gene having a specific sequence of nucleotides. Each polypeptide of a protein also has monomers arranged in a particular linear order (the protein's primary structure; see Figure 5.18), but its monomers are amino acids. Thus, nucleic acids and proteins contain information written in two different chemical languages. Getting from DNA to protein involves two major stages: transcription and translation.

Transcription is the synthesis (production) of RNA using information in the DNA. The two nucleic acids are written in different forms of the same language, and the information is simply transcribed, or "rewritten," from DNA to RNA. Just as a DNA strand provides a template for making a new complementary strand during DNA replication (see Concept 16.2), it also can serve as a template for assembling a complementary sequence of RNA nucleotides. For a protein-coding gene, the resulting RNA molecule is a faithful transcript of the gene's protein-building instructions. This type of RNA molecule

▼ **Figure 17.3** Inquiry

Do individual genes specify the enzymes that function in a biochemical pathway?

Experiment Working with the mould *Neurospora crassa*, Adrian Srb and Norman Horowitz, then at Stanford University, used Beadle and Tatum's experimental approach (see Figure 17.2) to isolate mutants that required arginine in their growth medium. The researchers showed that these mutants fell into three classes, each defective in a different gene. From studies by others on mammalian liver cells, they suspected that the metabolic pathway of arginine biosynthesis involved a precursor nutrient and the intermediate molecules ornithine and citrulline, as shown in the diagram on the right.

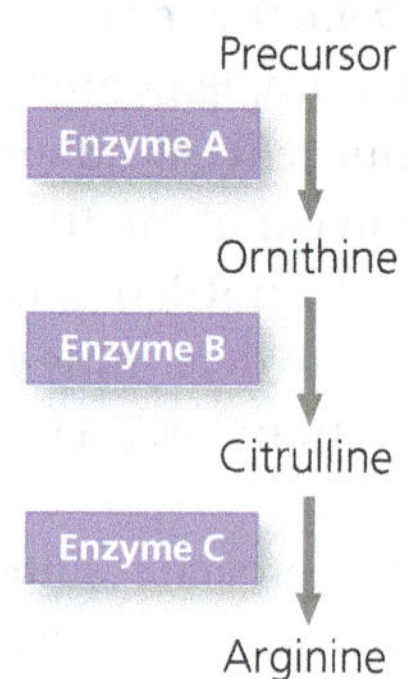

Their most famous experiment, shown here, tested both the *one gene–one enzyme hypothesis* and their postulated arginine-synthesising pathway. In this experiment, they grew their three classes of mutants under the four different conditions shown in the results table below. They included minimal medium (MM) as a control, knowing that wild-type cells could grow on MM but mutant cells could not. (See test tubes below.)

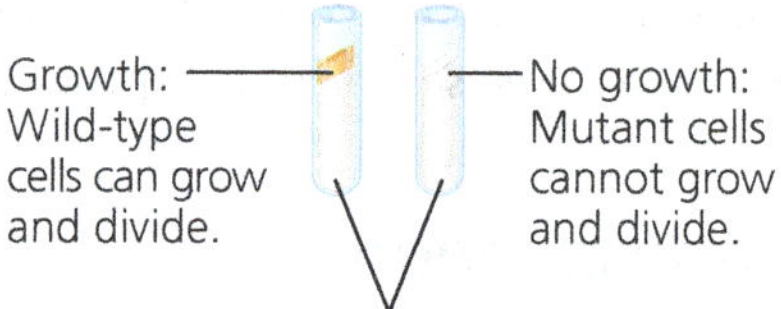

Control: Minimal medium

Results As shown in the table on the right, the wild-type strain was capable of growth under all experimental conditions, requiring only the minimal medium. The three classes of mutants each had a specific set of growth requirements. For example, class II mutants could not grow when ornithine alone was added but could grow when either citrulline or arginine was added.

Results Table

Condition	Classes of *Neurospora crassa*: Wild type	Class I mutants	Class II mutants	Class III mutants
Minimal medium (MM) (control)				
MM + ornithine				
MM + citrulline				
MM + arginine (control)				
Summary of results	Can grow with or without any supplements	Can grow on ornithine, citrulline, or arginine	Can grow only on citrulline or arginine	Require arginine to grow

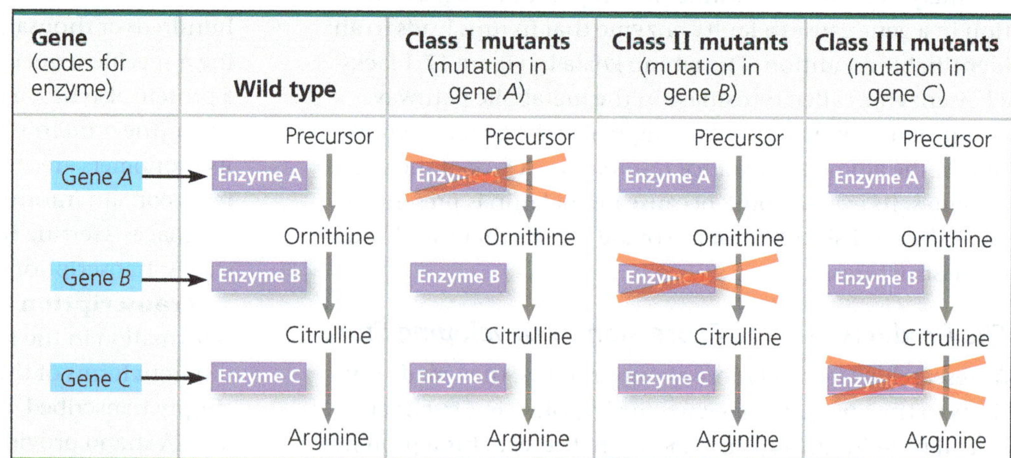

Conclusion From the growth requirements of the mutants, Srb and Horowitz deduced that each class of mutant was unable to carry out one step in the pathway for synthesising arginine, presumably because it lacked the necessary enzyme, as shown in the table on the right. Because each of their mutants was mutated in a single gene, they concluded that each mutated gene must normally dictate the production of one enzyme. Their results supported the one gene–one enzyme hypothesis, proposed by Beadle and Tatum, and also confirmed that the arginine pathway described in the mammalian liver also operates in *Neurospora*. (Notice in the results table that a mutant can grow only if supplied with a compound made *after* the defective step because this bypasses the defect.)

Data from A. M. Srb and N. H. Horowitz, The ornithine cycle in *Neurospora* and its genetic control, *Journal of Biological Chemistry* 154:129–139 (1944).

WHAT IF? *Suppose the experiment had shown that class I mutants could grow only in MM supplemented by ornithine or arginine and that class II mutants could grow in MM supplemented by citrulline, ornithine, or arginine. What conclusions would the researchers have drawn from those results regarding the biochemical pathway and the defect in class I and class II mutants?*

is called **messenger RNA (mRNA)** because it carries a genetic message from the DNA to the protein-synthesising machinery of the cell. (Transcription is the general term for the synthesis of *any* kind of RNA on a DNA template. Later, you will learn about some other types of RNA produced by transcription.)

Translation is the synthesis of a polypeptide using the information in the mRNA. During this stage, there is a change in language: The cell must translate the nucleotide sequence of an mRNA molecule into the amino acid sequence of a polypeptide. The sites of translation are **ribosomes**, molecular complexes that facilitate the orderly linking of amino acids into polypeptide chains.

Transcription and translation occur in all organisms. Because most studies have involved bacteria and eukaryotic cells, they are our main focus in this chapter. Our understanding of transcription and translation in archaea lags behind, but we do know that archaeal cells share some features of gene expression with bacteria and others with eukaryotes.

The basic mechanics of transcription and translation are similar for bacteria and eukaryotes, but there is an important difference in the flow of genetic information within the cells. Bacteria do not have nuclei. Therefore, nuclear membranes do not separate bacterial DNA and mRNA from ribosomes and the other protein-synthesising equipment **(Figure 17.4a)**. This lack of compartmentalisation allows translation of an mRNA to begin while its transcription is still in progress, as you'll see later. By contrast, eukaryotic cells have nuclei. The presence of a nuclear envelope separates transcription from translation in space and time **(Figure 17.4b)**. Transcription occurs in the nucleus, but the mRNA must be transported to the cytoplasm for translation. In eukaryotes, before RNA transcripts from protein-coding genes can leave the nucleus, they are modified in various ways to produce the final, functional mRNA. The transcription of a protein-coding eukaryotic gene results in *pre-mRNA*, and further RNA processing yields the finished mRNA. The initial RNA transcript from any gene, including those specifying RNA that is not translated into protein (such as tRNA and rRNA, which will be described in Concept 17.4), is more generally called a **primary transcript**.

To summarise: Genes program protein synthesis via genetic messages in the form of messenger RNA. Put another way, cells are governed by a molecular chain of command with a directional flow of genetic information:

This idea, that the flow of information went only one way, was named the *central dogma* by Francis Crick in 1956. In the 1970s, however, scientists were surprised to discover some enzymes that use RNA molecules as templates for DNA synthesis, an example of information flow from RNA to DNA (see Concept 19.2). Still, these exceptions do not discredit the idea that, in general, genetic information flows from DNA to RNA to protein. In the next section, we discuss how the instructions for assembling amino acids into a specific order are encoded in nucleic acids.

▼ Figure 17.4 Overview: the roles of transcription and translation in the flow of genetic information. In a cell, inherited information flows from DNA to RNA to protein. The two main stages of information flow are transcription and translation. A miniature version of part (a) or (b) accompanies several figures later in the chapter as an orientation diagram to help you see where a particular figure fits into the overall scheme of gene expression.

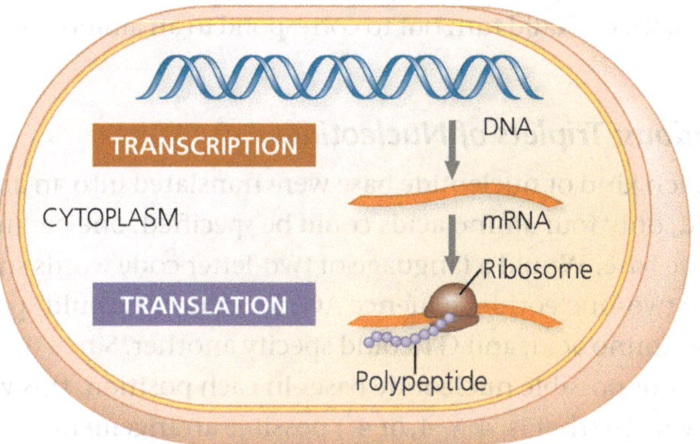

(a) Bacterial cell. In a bacterial cell, which lacks a nucleus, mRNA produced by transcription is immediately translated without additional processing.

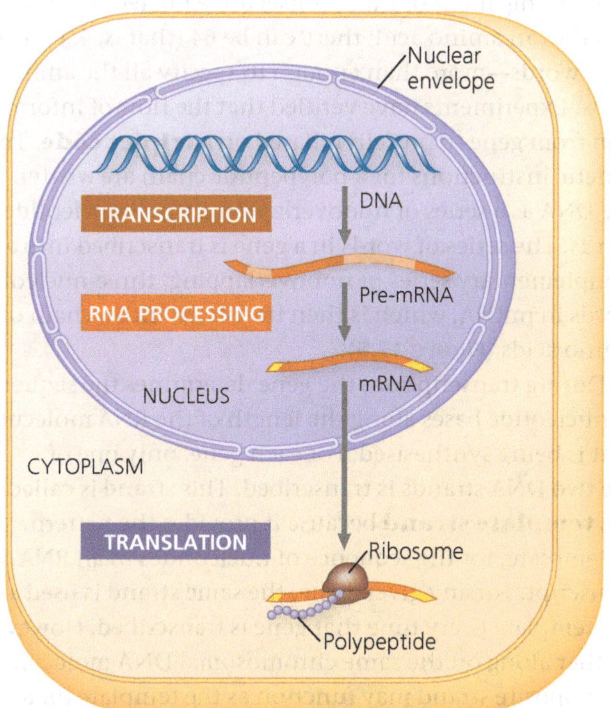

(b) Eukaryotic cell. The nucleus provides a separate compartment for transcription. The original RNA transcript, called pre-mRNA, is processed in various ways before leaving the nucleus as mRNA.

The Genetic Code

When biologists began to suspect that the instructions for protein synthesis were encoded in DNA, they recognised a problem: There are only four nucleotide bases to specify 20 amino acids. Thus, the genetic code cannot be a language like Chinese, where each written symbol corresponds to a word. How many nucleotides, then, would turn out to correspond to an amino acid?

Codons: Triplets of Nucleotides

If each kind of nucleotide base were translated into an amino acid, only four amino acids could be specified, one per nucleotide base. Would a language of two-letter code words suffice? The two-nucleotide sequence AG, for example, could specify one amino acid, and GT could specify another. Since there are four possible nucleotide bases in each position, this would give us 16 (that is, 4×4, or 4^2) possible arrangements—still not enough to code for all 20 amino acids.

Triplets of nucleotide bases are the smallest units of uniform length that can code for all the amino acids. If each arrangement of three consecutive nucleotide bases specifies an amino acid, there can be 64 (that is, 4^3) possible code words—more than enough to specify all the amino acids. Experiments have verified that the flow of information from gene to protein is based on a **triplet code**: The genetic instructions for a polypeptide chain are written in the DNA as a series of nonoverlapping, three-nucleotide words. The series of words in a gene is transcribed into a complementary series of nonoverlapping, three-nucleotide words in mRNA, which is then translated into a chain of amino acids **(Figure 17.5)**.

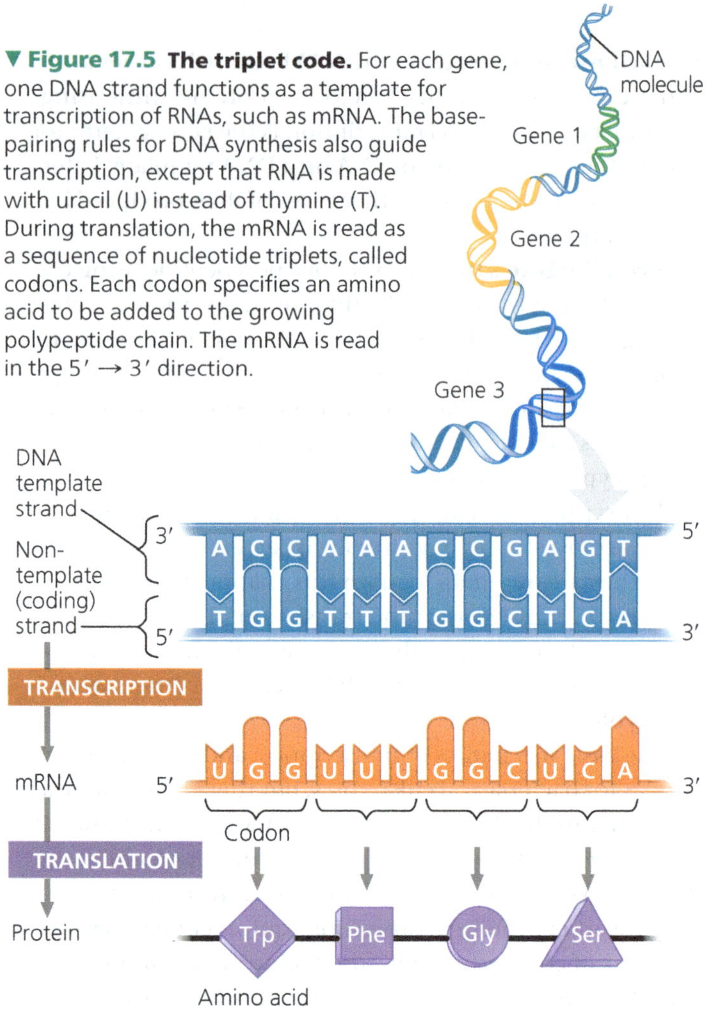

During transcription, the gene determines the sequence of nucleotide bases along the length of the RNA molecule that is being synthesised. For each gene, only one of the two DNA strands is transcribed. This strand is called the **template strand** because it provides the pattern, or template, for the sequence of nucleotides in an RNA transcript. For any given gene, the same strand is used as the template every time that gene is transcribed. However, further along on the same chromosomal DNA molecule, the opposite strand may function as the template for a different gene. Particular DNA sequences associated with a gene determine how the enzyme that transcribes genes is oriented when it binds, and this establishes which strand will be used as the template.

An mRNA molecule is complementary rather than identical to its DNA template because RNA nucleotides are assembled on the template according to base-pairing rules (see Figure 17.5). The pairs are similar to those that form during DNA replication, except that U (the RNA substitute for T) pairs with A and the mRNA nucleotides contain ribose instead of deoxyribose. Like a new strand of DNA, the RNA molecule is synthesised in an antiparallel direction to

▼ Figure 17.5 The triplet code. For each gene, one DNA strand functions as a template for transcription of RNAs, such as mRNA. The base-pairing rules for DNA synthesis also guide transcription, except that RNA is made with uracil (U) instead of thymine (T). During translation, the mRNA is read as a sequence of nucleotide triplets, called codons. Each codon specifies an amino acid to be added to the growing polypeptide chain. The mRNA is read in the $5' \rightarrow 3'$ direction.

VISUAL SKILLS *By convention, the nontemplate strand, also called the coding strand, is used to represent a DNA sequence. Write the sequence of the mRNA strand and the nontemplate strand—in both cases reading from 5′ to 3′—and compare them. Why do you think this convention was adopted? (Hint: Why is this called the coding strand?)*

the template strand of DNA. (To review what is meant by "antiparallel" and the 5′ and 3′ ends of a nucleic acid chain, see Figure 16.7.) In the example in Figure 17.5, the nucleotide triplet ACC along the DNA template strand (written as 3′-ACC-5′) provides a template for 5′-UGG-3′ in the mRNA molecule. The mRNA nucleotide triplets are called **codons**, and they are customarily written in the $5' \rightarrow 3'$ direction. In our example, UGG is the codon for the amino acid tryptophan (abbreviated Trp, or W). The term *codon* is also used for the DNA nucleotide triplets along the *nontemplate* strand. These codons are complementary to the template strand and thus identical in sequence to the mRNA (except for T's instead of U's). For this reason, the nontemplate DNA strand is often called the **coding strand**; by convention, the sequence of the coding strand is used when a gene's sequence is reported.

During translation, the sequence of codons along an mRNA molecule is decoded, or translated, into a sequence of amino acids making up a polypeptide chain. The codons are read by the translation machinery in the $5' \rightarrow 3'$

direction along the mRNA. Each codon specifies which one of the 20 amino acids will be incorporated at the corresponding position along a polypeptide. Because codons are nucleotide triplets, the number of nucleotides making up a genetic message must be three times the number of amino acids in the protein product. For example, 300 nucleotides along an mRNA strand code for 100 amino acids in the polypeptide it encodes.

Cracking the Code

Molecular biologists cracked the genetic code of life in the early 1960s when a series of elegant experiments disclosed the amino acid translations of each of the mRNA codons. The first codon was deciphered in 1961 by Marshall Nirenberg, of the National Institutes of Health, and colleagues. They synthesised an artificial mRNA made up of only uracil-bearing nucleotides, thus containing only one codon (UUU) over and over. When added to a test-tube mixture that contained all 20 amino acids, ribosomes, and the other components required for protein synthesis, the "poly-U" mRNA was translated into a polypeptide containing many units of the phenylalanine (Phe, or F) amino acids, strung together as a polyphenylalanine chain. Thus, Nirenberg determined that the mRNA codon UUU specifies the amino acid phenylalanine. Soon, the amino acids specified by the codons AAA, GGG, and CCC were similarly identified.

Although more elaborate techniques were required to decode mixed triplets such as AUA and CGA, all 64 codons were deciphered by the mid-1960s. As **Figure 17.6** shows, 61 of the 64 triplets code for amino acids. The other three codons act as "stop" signals, or termination codons, marking the end of translation. Notice that the codon AUG has a dual function: It codes for the amino acid methionine (Met, or M) and also functions as a "start" signal, or initiation codon. Genetic messages usually begin with the mRNA codon AUG, which signals the protein-synthesising machinery to begin translating the mRNA at that location. (Because AUG also stands for methionine, polypeptide chains all begin with methionine when they are synthesised. However, an enzyme may subsequently remove this starter amino acid from the chain.)

Notice in Figure 17.6 that there is redundancy in the genetic code, but no ambiguity. For example, although codons GAA and GAG both specify glutamic acid (redundancy), neither of them ever specifies any other amino acid (no ambiguity). The redundancy in the code is not altogether random. In many cases, codons that are synonyms for a particular amino acid differ only in the third nucleotide base of the triplet. We will see why this is later in the chapter.

Our ability to extract the intended message from a written language depends on reading the symbols in the correct groupings—that is, in the correct **reading frame**. Consider this statement: "The red dog ate the bug." Group the letters

▼ Figure 17.6 The codon table for mRNA. The three nucleotide bases of an mRNA codon are designated here as the first, second, and third bases, reading in the 5′ → 3′ direction along the mRNA. The codon AUG not only stands for the amino acid methionine (Met, or M) but also functions as a "start" signal for ribosomes to begin translating the mRNA at that point. Three of the 64 codons function as "stop" signals, marking where ribosomes end translation. Both one- and three-letter codes are shown for the amino acids; see Figure 5.14 for their full names.

First mRNA base (5′ end of codon)	Second mRNA base: U	C	A	G	Third mRNA base (3′ end of codon)
U	UUU Phe (F)	UCU Ser (S)	UAU Tyr (Y)	UGU Cys (C)	U
	UUC	UCC	UAC	UGC	C
	UUA Leu (L)	UCA	UAA Stop	UGA Stop	A
	UUG	UCG	UAG Stop	UGG Trp (W)	G
C	CUU Leu (L)	CCU Pro (P)	CAU His (H)	CGU Arg (R)	U
	CUC	CCC	CAC	CGC	C
	CUA	CCA	CAA Gln (Q)	CGA	A
	CUG	CCG	CAG	CGG	G
A	AUU Ile (I)	ACU Thr (T)	AAU Asn (N)	AGU Ser (S)	U
	AUC	ACC	AAC	AGC	C
	AUA	ACA	AAA Lys (K)	AGA Arg (R)	A
	AUG Met (M) or start	ACG	AAG	AGG	G
G	GUU Val (V)	GCU Ala (A)	GAU Asp (D)	GGU Gly (G)	U
	GUC	GCC	GAC	GGC	C
	GUA	GCA	GAA Glu (E)	GGA	A
	GUG	GCG	GAG	GGG	G

VISUAL SKILLS *A segment in the middle of an mRNA has the sequence 5′-AGAGAACCGCGA-3′. Using the codon table, translate this sequence, assuming the first three nucleotides are a codon.*

incorrectly by starting at the wrong point, and the result will probably be gibberish: for example, "her edd oga tet heb ug." The reading frame is also important in the molecular language of cells. The short stretch of polypeptide shown in Figure 17.5, for instance, will be made correctly only if the mRNA nucleotides are read from left to right (5′ → 3′) in the groups of three shown in the figure: UGG UUU GGC UCA. Although a genetic message is written with no spaces between the codons, the cell's protein-synthesising machinery reads the message as a series of nonoverlapping three-letter words. The message is *not* read as a series of overlapping words—UGGUUU, and so on—which would convey a very different message.

Evolution of the Genetic Code

EVOLUTION The genetic code is nearly universal, shared by organisms from the simplest bacteria to the most complex plants and animals. The mRNA codon CCG, for instance, is translated as the amino acid proline in all organisms whose

▼ Figure 17.7 Evidence for evolution: expression of genes from different species. Because diverse forms of life share a common genetic code due to their shared ancestry, one species can be programmed to produce proteins characteristic of a second species by introducing DNA from the second species into the first.

(a) Tobacco plant expressing a firefly gene. The yellow glow is produced by a chemical reaction catalysed by the protein product of the firefly gene.

(b) Mosquito larva expressing a jellyfish gene. The gene codes for green fluorescent protein (GFP) and is inserted into organisms as a *reporter gene* so researchers can tell if a gene of interest is being expressed.

genetic code has been examined. In laboratory experiments, genes can be transcribed and translated after being transplanted from one species to another, sometimes with quite striking results, as shown in **Figure 17.7**. Bacteria can be programmed by the insertion of human genes to synthesise certain human proteins for medical use, such as insulin. Such applications have produced many exciting developments in the area of biotechnology (see Concept 20.4).

The evolutionary significance of the code's near universality is clear. A language shared by all living things must have been operating very early in the history of life—early enough to be present in the common ancestor of all present-day organisms. A shared genetic vocabulary is a reminder of the kinship of all life.

CONCEPT CHECK 17.1

1. **MAKE CONNECTIONS** In a research article about alkaptonuria published in 1902, Garrod suggested that humans inherit two "characters" (alleles) for a particular enzyme and that both parents must contribute a faulty version for the offspring to have alkaptonuria. Today, would this disorder be called dominant or recessive? (See Concept 14.4.)
2. Describe the polypeptide product you would expect from a poly-G mRNA that is 30 nucleotides long.
3. **DRAW IT** The template strand of a gene contains the sequence 3′-TTCAGTCGT-5′. Suppose that the nontemplate sequence was transcribed instead of the template sequence. Draw the nontemplate sequence in 3′ to 5′ order. Then draw the mRNA sequence and translate it using Figure 17.6. Predict how well the protein synthesised from the nontemplate strand would function, if at all.

For suggested answers, see Appendix A.

CONCEPT 17.2

Transcription is the DNA-directed synthesis of RNA: *A Closer Look*

Now that we have considered the linguistic logic and evolutionary significance of the genetic code, we are ready to reexamine transcription, the first stage of gene expression, in greater detail.

Molecular Components of Transcription

Messenger RNA, the carrier of information from DNA to the cell's protein-synthesising machinery, is transcribed from the template strand of a gene. An enzyme called an **RNA polymerase** pries the two strands of DNA apart and joins together RNA nucleotides complementary to the DNA template strand, thus elongating the RNA polynucleotide **(Figure 17.8)**. Like the DNA polymerases that function in DNA replication, RNA polymerases can assemble a polynucleotide only in its 5′ → 3′ direction, adding onto its 3′ end. Unlike DNA polymerases, however, RNA polymerases are able to start a chain from scratch; they don't need to add the first nucleotide onto a pre-existing primer.

Specific sequences of nucleotides along the DNA mark where transcription of a gene begins and ends. The DNA sequence where RNA polymerase attaches and initiates transcription is known as the **promoter**; in bacteria, the sequence that signals the end of transcription is called the **terminator**. (The termination mechanism in eukaryotes will be described later.) Molecular biologists refer to the direction of transcription as "downstream" and the other direction as "upstream." These terms are also used to describe the positions of nucleotide sequences within the DNA or RNA. Thus, the promoter sequence in DNA is said to be upstream from the terminator. The stretch of DNA downstream from the promoter that is transcribed into an RNA molecule is called a **transcription unit**.

Bacteria have a single type of RNA polymerase that synthesises not only mRNA but also other types of RNA that function in gene expression, such as ribosomal RNA. In contrast, eukaryotes have at least three types of RNA polymerase in their nuclei; the one used for pre-mRNA synthesis is called RNA polymerase II. The other RNA polymerases transcribe RNA molecules that are not translated into protein. In the discussion that follows, we start with the features of mRNA synthesis common to both bacteria and eukaryotes and then describe some key differences.

Synthesis of an RNA Transcript

The three stages of transcription, as shown in Figure 17.8 and described next, are initiation, elongation, and termination of the RNA chain. Study Figure 17.8 to familiarise yourself with the stages and the terms used to describe them.

RNA Polymerase Binding and Initiation of Transcription

The promoter sequence of a gene includes within it the transcription **start point**—the nucleotide where RNA polymerase actually begins synthesising the mRNA—and typically extends several dozen or so nucleotide pairs upstream from the start point **(Figure 17.9)**. Based on interactions with proteins (transcription factors), RNA polymerase binds in a precise location and orientation on the promoter. This binding determines where transcription starts and the direction it will travel, thus which strand of DNA is used as the template.

▼ **Figure 17.8 The stages of transcription: initiation, elongation, and termination.** This general depiction of transcription applies to both bacteria and eukaryotes, but the details of termination differ, as described in the text. Also, in a bacterium, the RNA transcript is immediately usable as mRNA; in a eukaryote, the RNA transcript must first undergo processing.

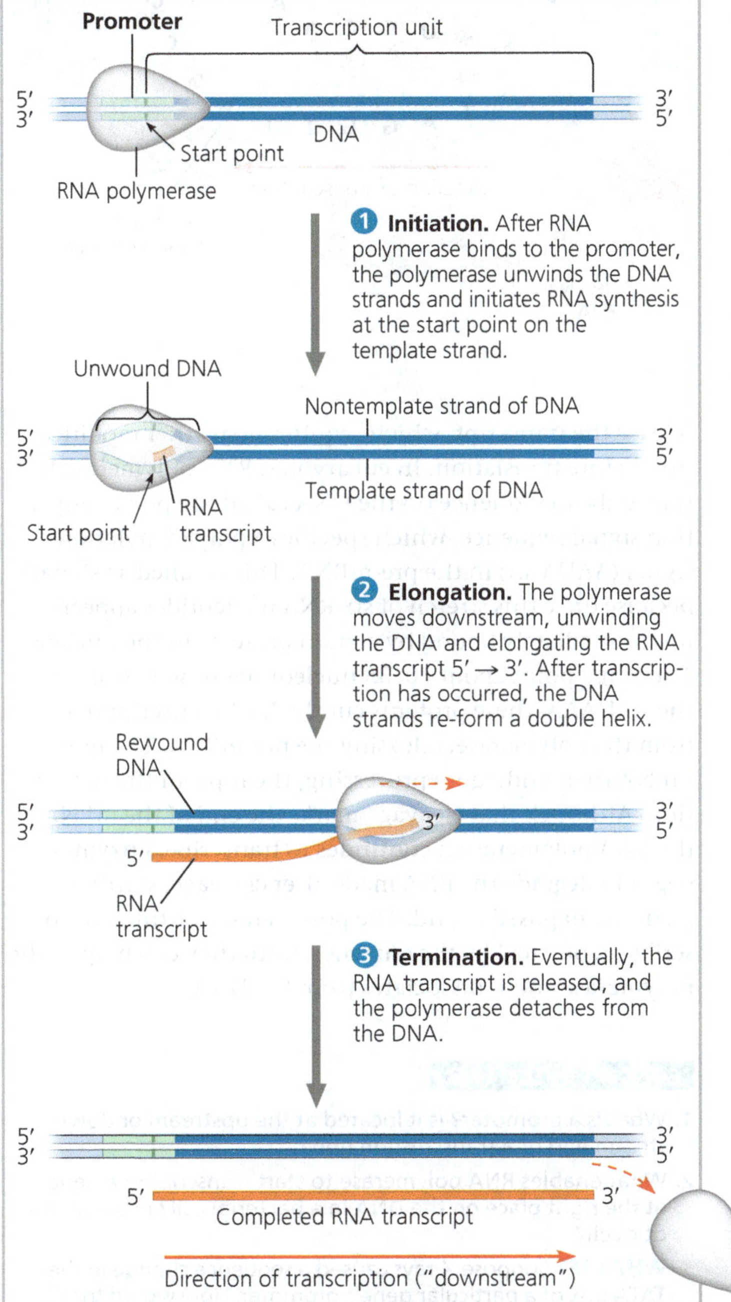

MAKE CONNECTIONS *Compare the use of a template strand during transcription and replication. See Figure 16.18.*

▼ **Figure 17.9 The initiation of transcription at a eukaryotic promoter.** In eukaryotic cells, proteins called transcription factors mediate the initiation of transcription by RNA polymerase II.

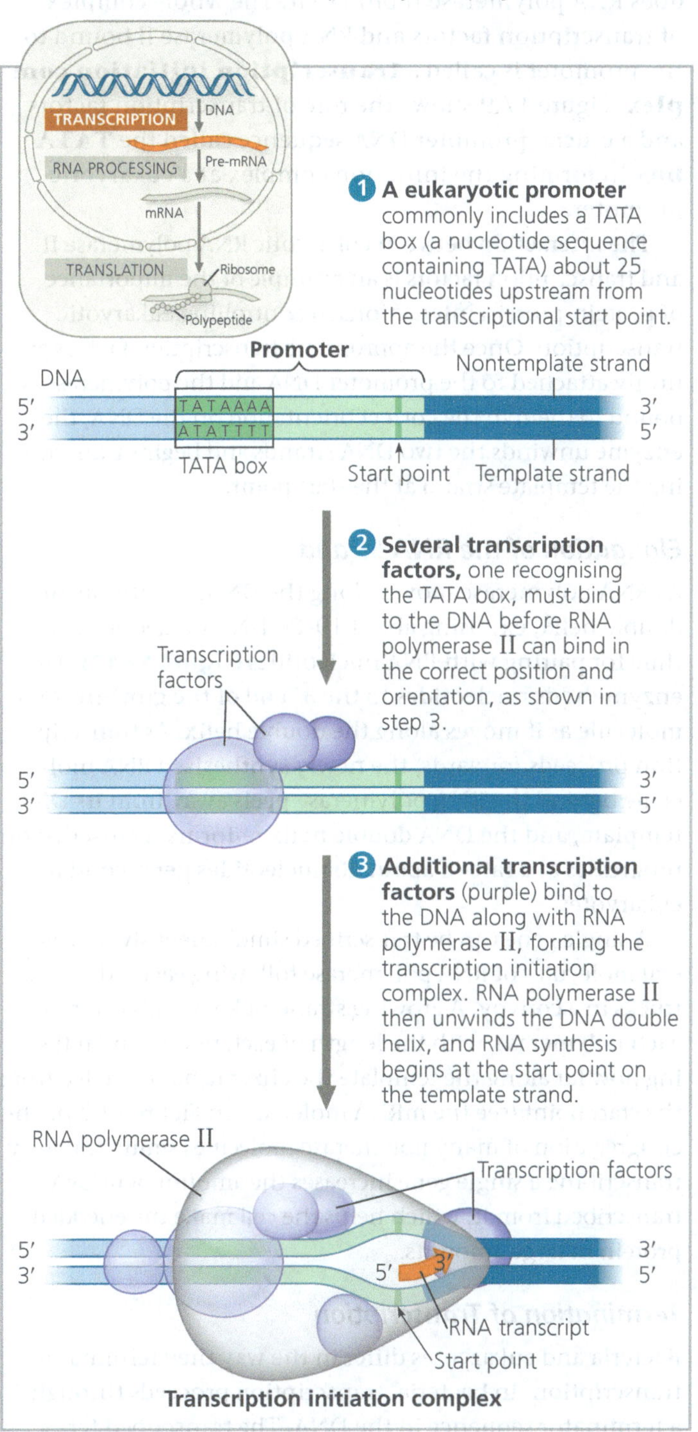

? *Explain how the interaction of RNA polymerase with the promoter would differ if the figure showed transcription initiation for bacteria.*

Certain sections of a promoter sequence are especially important for binding RNA polymerase in a way that ensures that transcription will begin at the correct start point. In bacteria, part of the RNA polymerase itself specifically recognises and binds to the promoter. In eukaryotes, a collection of proteins called **transcription factors** (purple proteins in Figure 17.9), help guide the binding of RNA polymerase and the initiation of transcription. Only after transcription factors are attached to the promoter does RNA polymerase II bind to it. The whole complex of transcription factors and RNA polymerase II bound to the promoter is called a **transcription initiation complex**. Figure 17.9 shows the role of transcription factors and a crucial promoter DNA sequence called the **TATA box** in forming the initiation complex at a eukaryotic promoter.

The interaction between eukaryotic RNA polymerase II and transcription factors is an example of the importance of protein-protein interactions in controlling eukaryotic transcription. Once the appropriate transcription factors are firmly attached to the promoter DNA and the polymerase is bound to them in the correct orientation on the DNA, the enzyme unwinds the two DNA strands and begins transcribing the template strand at the start point.

Elongation of the RNA Strand

As RNA polymerase moves along the DNA, it untwists the double helix, exposing about 10–20 DNA nucleotides at a time for pairing with RNA nucleotides **(Figure 17.10)**. The enzyme adds nucleotides to the 3′ end of the growing RNA molecule as it moves along the double helix. As transcription proceeds forwards, the newly synthesised RNA molecule behind the RNA polymerase peels away from its DNA template, and the DNA double helix re-forms. Transcription progresses at a rate of about 40 nucleotides per second in eukaryotes.

A single gene can be transcribed simultaneously by several molecules of RNA polymerase following each other like trucks in a convoy. A growing strand of RNA trails off from each polymerase, with the length of each new strand reflecting how far along the template the enzyme has travelled from the start point (see the mRNA molecules in Figure 17.23). The congregation of many polymerase molecules simultaneously transcribing a single gene increases the amount of mRNA transcribed from it, which helps the cell make the encoded protein in large amounts.

Termination of Transcription

Bacteria and eukaryotes differ in the way they terminate transcription. In bacteria, transcription proceeds through a terminator sequence in the DNA. The transcribed terminator (an RNA sequence) functions as the termination signal, causing the polymerase to detach from the DNA and

▼ **Figure 17.10 Transcription elongation.** RNA polymerase moves along the DNA template strand, joining complementary RNA nucleotides to the 3′ end of the growing RNA transcript. Behind the polymerase, the new RNA peels away from the template strand, which re-forms a double helix with the nontemplate strand.

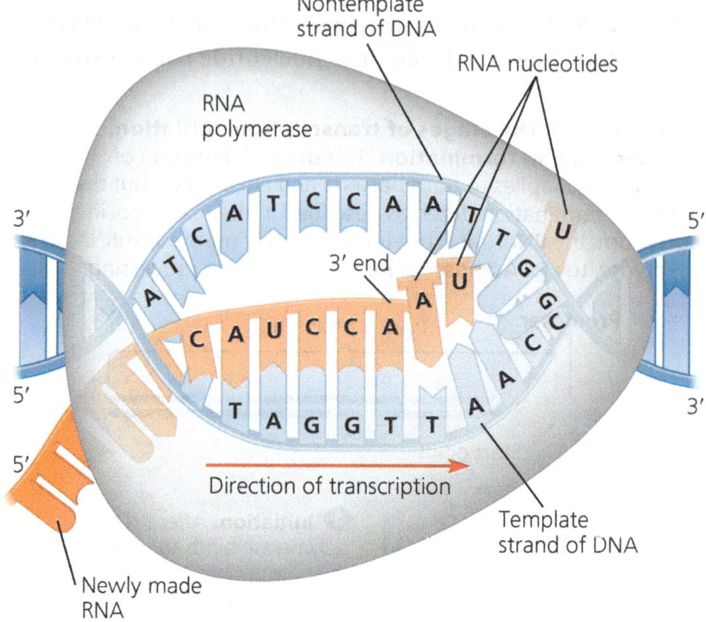

release the transcript, which requires no further modification before translation. In eukaryotes, RNA polymerase II transcribes a sequence on the DNA called the polyadenylation signal sequence, which specifies a polyadenylation signal (AAUAAA) in the pre-mRNA. This is called a "signal" because once this stretch of six RNA nucleotides appears, it is immediately bound by certain proteins in the nucleus. Then, at a point about 10–35 nucleotides downstream from the AAUAAA, these proteins cut the RNA transcript free from the polymerase, releasing the pre-mRNA. The pre-mRNA then undergoes processing, the topic of the next section. Although that cleavage marks the end of the mRNA, the RNA polymerase II continues to transcribe. Enzymes begin to degrade the RNA made after cleavage, starting at its newly exposed 5′ end. The polymerase continues transcribing, pursued by the enzymes, until they catch up to the polymerase and it dissociates from the DNA.

CONCEPT CHECK 17.2

1. What is a promoter? Is it located at the upstream or downstream end of a transcription unit?
2. What enables RNA polymerase to start transcribing a gene at the right place on the DNA in a bacterial cell? In a eukaryotic cell?
3. **WHAT IF?** Suppose X-rays caused a sequence change in the TATA box of a particular gene's promoter. How would that affect transcription of the gene? (See Figure 17.9.)

For suggested answers, see Appendix A.

CONCEPT 17.3

Eukaryotic cells modify RNA after transcription

Enzymes in the eukaryotic nucleus modify pre-mRNA in specific ways before the genetic message is dispatched to the cytoplasm. During this **RNA processing**, both ends of the primary transcript are altered. Also, in most cases, certain interior sections of the RNA molecule are cut out and the remaining parts spliced together. These modifications produce an mRNA molecule ready for translation.

Alteration of mRNA Ends

Each end of a pre-mRNA molecule is modified in a particular way **(Figure 17.11)**. The 3′ end, which is synthesised first, receives a **5′ cap**, a modified form of a guanine (G) nucleotide added onto the 5′ end after transcription of the first 20–40 nucleotides have been transcribed. The 3′ end of the pre-mRNA molecule is also modified before the mRNA exits the nucleus. Recall that the pre-mRNA is cut and released soon after the polyadenylation signal, AAUAAA, is transcribed. At the 3′ end, an enzyme then adds 50–250 more adenine (A) nucleotides, forming a **poly-A tail**. The 5′ cap and poly-A tail share several important functions. First, they seem to facilitate the export of the mature mRNA from the nucleus. Second, they help protect the mRNA from degradation by hydrolytic enzymes. And third, they help ribosomes attach to the 5′ end of the mRNA once it reaches the cytoplasm. In addition to the cap and tail on a eukaryotic mRNA molecule, Figure 17.11 shows the untranslated regions (UTRs) at the 5′ and 3′ ends of the mRNA (referred to as the 5′ UTR and 3′ UTR). The UTRs are parts of the mRNA that will not be translated into protein, but they have other functions, such as ribosome binding.

Split Genes and RNA Splicing

A remarkable stage of RNA processing in the eukaryotic nucleus is called **RNA splicing (Figure 17.12)**, in which large portions of the RNA primary transcript molecules are

▼ Figure 17.11 RNA processing: addition of the 5′ cap and poly-A tail. Enzymes modify the two ends of a eukaryotic pre-mRNA molecule. The modified ends may promote the export of mRNA from the nucleus, and they help protect the mRNA from degradation. When the mRNA reaches the cytoplasm, the modified ends, in conjunction with certain cytoplasmic proteins, help with ribosome attachment. The 5′ cap and poly-A tail are not translated into protein, nor are the regions called the 5′ untranslated region (5′ UTR) and 3′ untranslated region (3′ UTR). The pink segments are introns, which will be described shortly (see Figure 17.12).

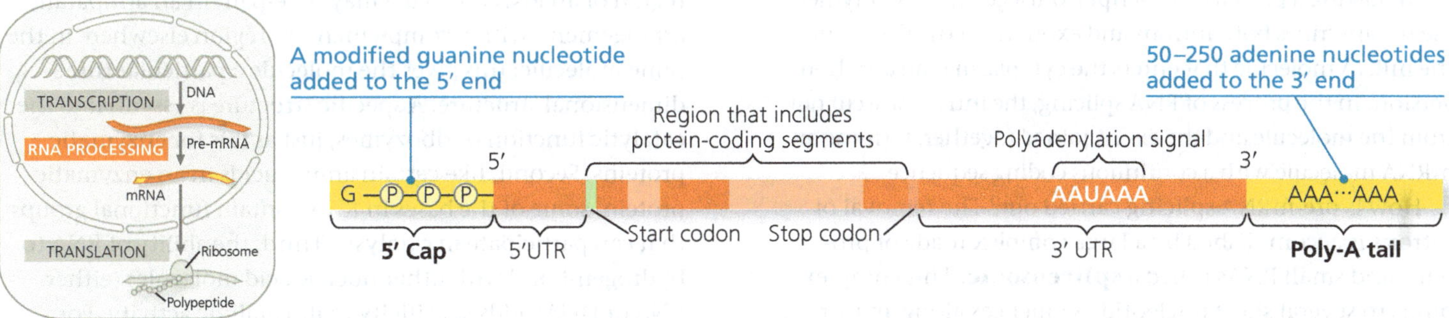

▼ Figure 17.12 RNA processing: RNA splicing. The RNA molecule shown here codes for β-globin, one of the polypeptides of haemoglobin. The numbers under the RNA refer to codons; β-globin is 146 amino acids long. The β-globin gene and its pre-mRNA transcript have three exons, corresponding to sequences that will exit the nucleus as mRNA. (The 5′ UTR and 3′ UTR do not code for proteins, but they are parts of exons because they remain in the mRNA.) During RNA processing, the introns are cut out and the exons spliced together. In many genes, the introns are much longer than the exons.

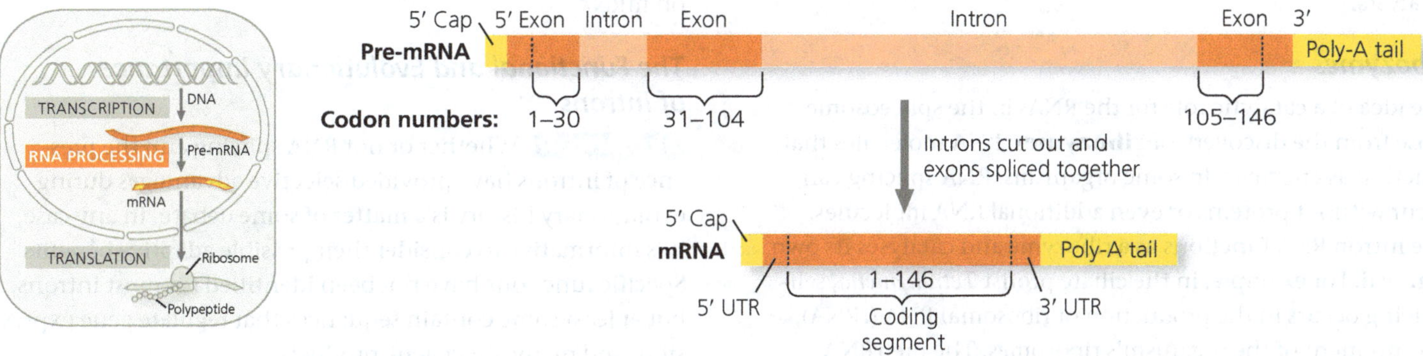

DRAW IT *On the mRNA, indicate the sites of the start and stop codons.*

removed and the remaining portions are reconnected. This cut-and-paste job is similar to editing a movie. The average length of a transcription unit along a human DNA molecule is about 27,000 nucleotide pairs, so the primary RNA transcript is also that long. However, the average-sized protein of 400 amino acids requires only 1,200 nucleotides in RNA to code for it. (Remember, each amino acid is encoded by a *triplet* of nucleotides.) This is because most eukaryotic genes and their RNA transcripts have long noncoding stretches of nucleotides, regions that are not translated. Even more surprising is that most of these noncoding sequences are interspersed between coding segments of the gene and thus between coding segments of the pre-mRNA. In other words, the sequence of DNA nucleotides that codes for a eukaryotic polypeptide is usually not continuous; it is split into segments. The noncoding segments of nucleic acid that lie between coding regions are called intervening sequences, or **introns**. The other regions are called **exons**, because they are eventually expressed, usually by being translated into amino acid sequences. (Exceptions include the UTRs of the exons at the ends of the RNA, which make up part of the mRNA but are not translated into protein. Because of these exceptions, you may prefer to think of exons as sequences of RNA that exit the nucleus.) The terms *intron* and *exon* are used for both RNA sequences and the DNA sequences that specify them.

In making a primary transcript from a gene, RNA polymerase II transcribes both introns and exons from the DNA, but the mRNA molecule that enters the cytoplasm is an abridged version. In the process of RNA splicing, the introns are cut out from the molecule and the exons joined together, forming an mRNA molecule with a continuous coding sequence.

How is pre-mRNA splicing carried out? The removal of introns is accomplished by a large complex made of proteins and small RNAs called a **spliceosome**. This complex binds to several short nucleotide sequences along an intron, including key sequences at each end **(Figure 17.13)**. The intron is then released (and rapidly degraded), and the spliceosome joins together the two exons that flanked the intron. The small RNAs in the spliceosome not only participate in spliceosome assembly and splice site recognition, but also catalyse the splicing reaction; like proteins, RNAs can act as catalysts.

Ribozymes

The idea of a catalytic role for the RNAs in the spliceosome arose from the discovery of **ribozymes**, RNA molecules that function as enzymes. In some organisms, RNA splicing can occur without proteins or even additional RNA molecules: The intron RNA functions as a ribozyme and catalyses its own removal. For example, in the ciliate protist *Tetrahymena*, self-splicing occurs in the production of ribosomal RNA (rRNA), a component of the organism's ribosomes. The pre-rRNA

▼ **Figure 17.13 A spliceosome splicing a pre-mRNA.** The diagram shows a portion of a pre-mRNA transcript, with an intron (pink) flanked by two exons (red). Small RNAs within the spliceosome base-pair with nucleotides at specific sites along the intron. Next, small spliceosome RNAs catalyse cutting of the pre-mRNA and the splicing together of the exons, releasing the intron for rapid degradation.

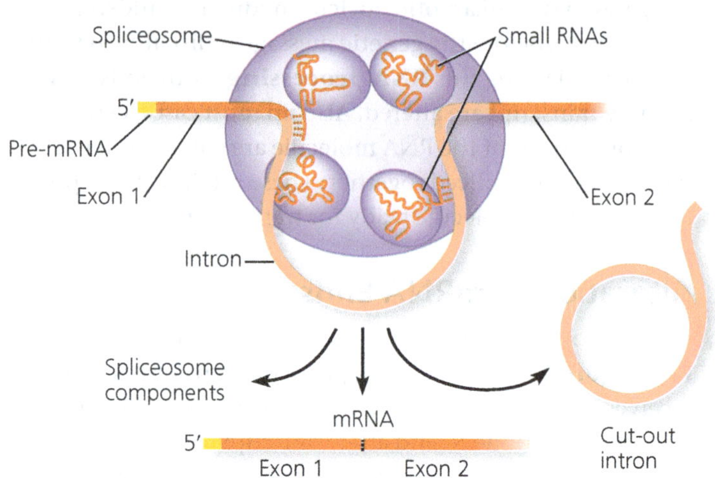

actually removes its own introns! The discovery of ribozymes invalidated the idea that all biological catalysts are proteins.

Three properties of RNA enable some RNA molecules to function as enzymes. First, because RNA is single-stranded, a region of an RNA molecule may base-pair, in an antiparallel arrangement, with a complementary region elsewhere in the same molecule; this gives the molecule a particular three-dimensional structure. A specific structure is essential to the catalytic function of ribozymes, just as it is for enzymatic proteins. Second, like certain amino acids in an enzymatic protein, some of the bases in RNA contain functional groups that can participate in catalysis. Third, the ability of RNA to hydrogen-bond with other nucleic acid molecules (either RNA or DNA) adds specificity to its catalytic activity. For example, complementary base pairing between the RNA of the spliceosome and the RNA of a primary RNA transcript precisely locates the region where the ribozyme catalyses splicing. Later in this chapter, you will see how these properties of RNA also allow it to perform important noncatalytic roles in the cell, such as recognition of the three-nucleotide codons on mRNA.

The Functional and Evolutionary Importance of Introns

EVOLUTION Whether or not RNA splicing and the presence of introns have provided selective advantages during evolutionary history is a matter of some debate. In any case, it is informative to consider their possible adaptive benefits. Specific functions have not been identified for most introns, but at least some contain sequences that regulate gene expression, and many affect gene products.

One important consequence of the presence of introns in genes is that a single gene can encode more than one kind of polypeptide. Many genes are known to give rise to two or more different polypeptides, depending on which segments are treated as exons during RNA processing; this is called **alternative RNA splicing** (see Figure 18.14). Results from the Human Genome Project (discussed in Concept 21.1) suggest that alternative RNA splicing is one reason humans can get along with about the same number of genes as a nematode (roundworm). Because of alternative splicing, the number of different protein products an organism produces can be much greater than its number of genes.

Proteins often have a modular architecture consisting of discrete structural and functional regions called **domains**. One domain of an enzyme, for example, might include the active site, while another might allow the enzyme to bind to a cellular membrane. In quite a few cases, different exons code for the different domains of a protein **(Figure 17.14)**.

The presence of introns in a gene may facilitate the evolution of new and potentially beneficial proteins as a result of a process known as *exon shuffling* (see Figure 21.16). Introns increase the probability of crossing over between the exons of alleles of a gene—simply by providing more terrain for crossovers without interrupting coding sequences. This might result in new combinations of exons and proteins with altered structure and function. We can also imagine the occasional mixing and matching of exons between completely different (nonallelic) genes. Exon shuffling of either sort could lead to new proteins with novel combinations of functions. While most of the shuffling would result in nonbeneficial changes, occasionally a beneficial variant might arise.

▼ Figure 17.14 Correspondence between exons and protein domains.

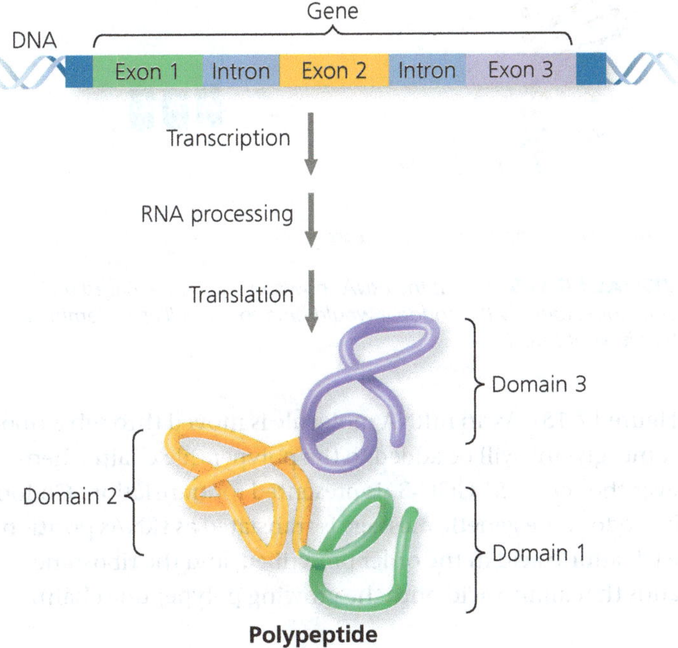

CONCEPT CHECK 17.3

1. Given that there are about 20,000 human protein-coding genes, how can human cells make 75,000–100,000 different proteins?
2. Compare RNA splicing to how you would watch a prerecorded television show. What would introns correspond to in this analogy?
3. **WHAT IF?** What would be the effect of treating cells with an agent that removed the 5′ cap from mRNAs?

For suggested answers, see Appendix A.

CONCEPT 17.4

Translation is the RNA-directed synthesis of a polypeptide: *A Closer Look*

We will now examine how genetic information flows from mRNA to protein—the process of translation **(Figure 17.15)**. We'll focus on the basic steps of translation that occur in both bacteria and eukaryotes, while pointing out key differences.

▼ Figure 17.15 Translation: the basic concept. As a molecule of mRNA is moved through a ribosome, codons are translated into amino acids, one by one. The translators, or interpreters, are tRNA molecules, each type with a specific anticodon at one end and a corresponding amino acid at the other end. A tRNA adds its amino acid cargo to a growing polypeptide chain when the anticodon hydrogen-bonds to the complementary codon on the mRNA.

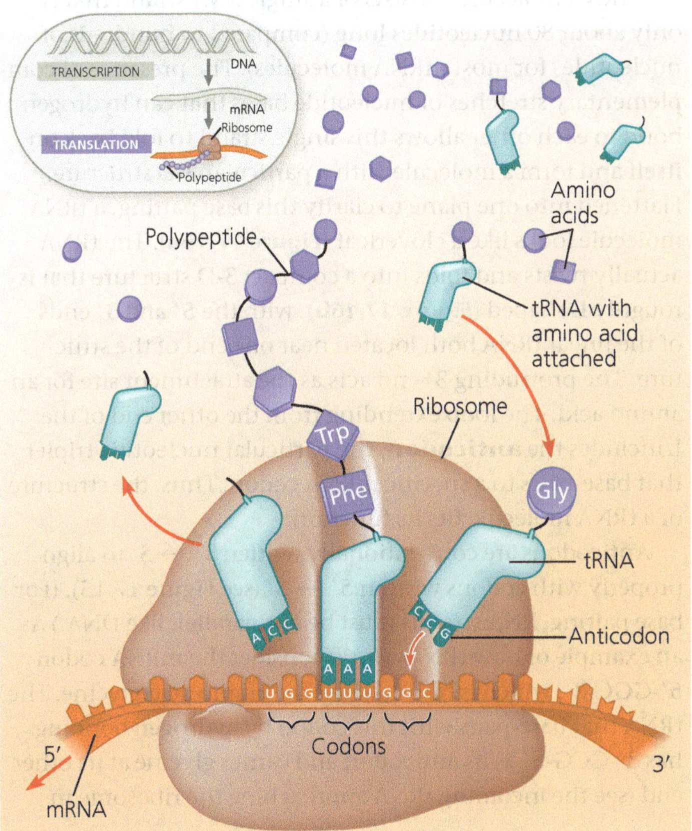

Molecular Components of Translation

In the process of translation, a cell "reads" a genetic message and builds a polypeptide accordingly. The message is a series of codons along an mRNA molecule, and the translator is called a **transfer RNA (tRNA)**. The function of a tRNA is to transfer an amino acid from the cytoplasmic pool of amino acids to a growing polypeptide in a ribosome. A cell keeps its cytoplasm stocked with all 20 amino acids, either by synthesising them from other compounds or by taking them up from the surrounding solution. The ribosome, a structure made of proteins and RNAs, adds each amino acid brought to it by a tRNA to the growing end of a polypeptide chain (see Figure 17.15).

Translation is simple in principle but complex in its biochemistry and mechanics, especially in the eukaryotic cell. In dissecting translation, we'll focus on the slightly less complicated version of the process that occurs in bacteria. We'll first look at the major players in this process.

The Structure and Function of Transfer RNA

The key to translating a genetic message into a specific amino acid sequence is the fact that each tRNA molecule enables translation of a given mRNA codon into a certain amino acid. This is possible because a tRNA bears a specific amino acid at one end of its three-dimensional structure, while at the other end is a nucleotide triplet that can base-pair with the complementary codon on mRNA.

A tRNA molecule consists of a single RNA strand that is only about 80 nucleotides long (compared to hundreds of nucleotides for most mRNA molecules). The presence of complementary stretches of nucleotide bases that can hydrogen-bond to each other allows this single strand to fold back on itself and form a molecule with a particular 3-D structure. Flattened into one plane to clarify this base pairing, a tRNA molecule looks like a cloverleaf **(Figure 17.16a)**. The tRNA actually twists and folds into a compact 3-D structure that is roughly L-shaped **(Figure 17.16b)**, with the 5′ and 3′ ends of the linear tRNA both located near one end of the structure. The protruding 3′ end acts as the attachment site for an amino acid. The loop extending from the other end of the L includes the **anticodon**, the particular nucleotide triplet that base-pairs to a specific mRNA codon. Thus, the structure of a tRNA molecule fits its function.

Anticodons are conventionally written 3′ → 5′ to align properly with codons written 5′ → 3′ (see Figure 17.15). (For base pairing, RNA strands must be antiparallel, like DNA.) As an example of how tRNAs work, consider the mRNA codon 5′-GGC-3′, which is translated as the amino acid glycine. The tRNA that base-pairs with this codon by hydrogen bonding has 3′-CCG-5′ as its anticodon and carries glycine at its other end (see the incoming tRNA approaching the ribosome in Figure 17.15). As an mRNA molecule is moved through a ribosome, glycine will be added to the polypeptide chain whenever the codon 5′-GGC-3′ is presented for translation. Codon by codon, the genetic message is translated as tRNAs position each amino acid in the order prescribed, and the ribosome adds that amino acid onto the growing polypeptide chain.

▼ Figure 17.16 The structure of transfer RNA (tRNA).

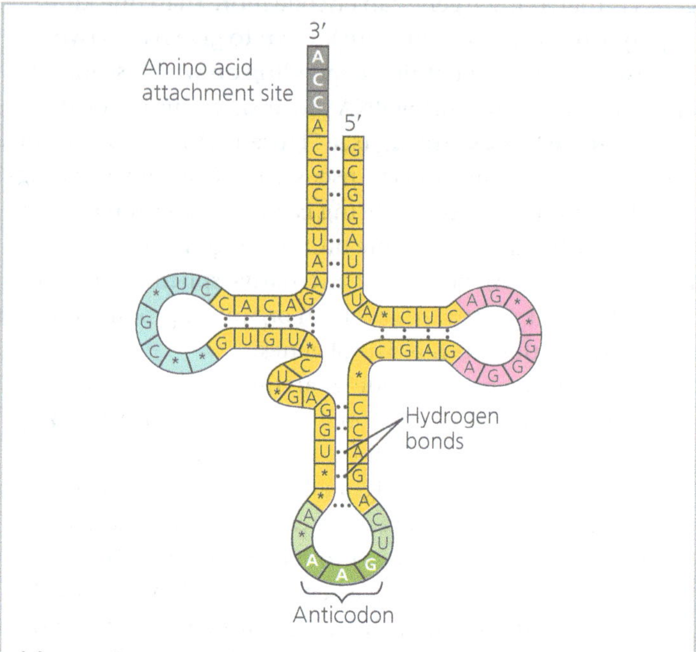

(a) Two-dimensional structure. The four base-paired regions and three loops are characteristic of all tRNAs, as is the base sequence of the amino acid attachment site at the 3′ end. The anticodon triplet is unique to each tRNA type, as are some sequences in the other two loops. (The asterisks mark bases that have been chemically modified, a characteristic of tRNA. The modified bases contribute to tRNA function in a way that is not yet understood.)

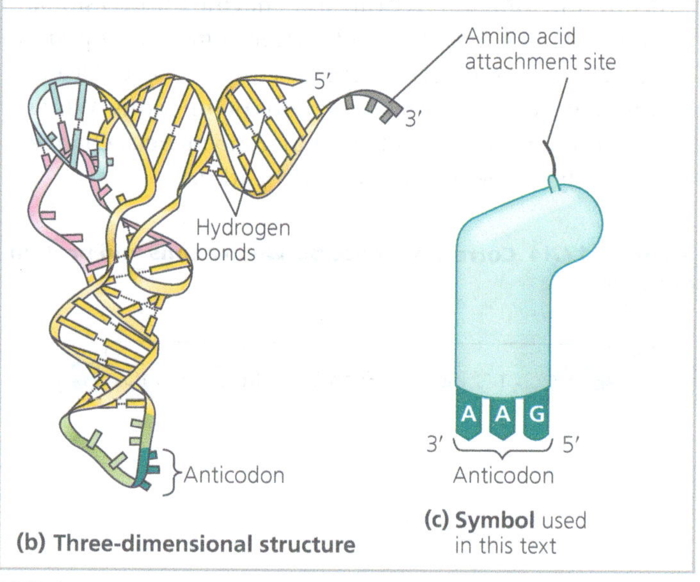

(b) Three-dimensional structure

(c) Symbol used in this text

VISUAL SKILLS *Look at the tRNA shown in this figure. Based on its anticodon, identify the codon it would bind to, as well as the amino acid that it would carry.*

The tRNA molecule is a translator in the sense that, in the context of the ribosome, it can read a nucleic acid word (the mRNA codon) and interpret it as a protein word (the amino acid).

Like mRNA and other types of cellular RNA, transfer RNA molecules are transcribed from DNA templates. In a eukaryotic cell, tRNA, like mRNA, is made in the nucleus and then travels to the cytoplasm, where it will participate in the process of translation. In both bacterial and eukaryotic cells, each tRNA molecule is used repeatedly, picking up its designated amino acid in the cytosol, depositing this cargo onto a polypeptide chain at the ribosome, and then leaving the ribosome, ready to pick up another of the same amino acid.

The accurate translation of a genetic message requires two instances of molecular recognition. First, a tRNA that binds to an mRNA codon specifying a particular amino acid must carry that amino acid, and no other, to the ribosome. The correct matching up of tRNA and amino acid is carried out by a family of related enzymes that are aptly named **aminoacyl-tRNA synthetases** (**Figure 17.17**). The active site of each type of aminoacyl-tRNA synthetase fits only a specific combination of amino acid and tRNA. (Regions of both the amino acid attachment end and the anticodon end of the tRNA ensure the specific fit.) There are 20 different synthetases, one for each amino acid. A synthetase joins a given amino acid to an appropriate tRNA; one synthetase is able to bind to all the different tRNAs for its particular amino acid. The synthetase catalyses the covalent attachment of the amino acid to its tRNA in a process driven by the hydrolysis of ATP. The resulting aminoacyl tRNA, also called a *charged tRNA*, is released from the enzyme and is then available to deliver its amino acid to a growing polypeptide chain on a ribosome.

The second instance of molecular recognition is the pairing of the tRNA anticodon with the appropriate mRNA codon. If one tRNA type existed for each mRNA codon specifying an amino acid, there would be 61 tRNAs (see Figure 17.6). In bacteria, however, there are only about 45 tRNAs, signifying that some tRNAs must be able to bind to more than one codon. This is possible because the rules for base pairing between the third nucleotide base of a codon and the corresponding base of a tRNA anticodon are relaxed compared to those at other codon positions. For example, the nucleotide base U at the 5′ end of a tRNA anticodon can pair with either A or G in the third position (at the 3′ end) of an mRNA codon. The flexible base pairing at this codon position is called **wobble**. Wobble explains why the synonymous codons for a given amino acid most often differ in their third nucleotide base. Accordingly, a tRNA with the anticodon 3′-UCU-5′ can base-pair with either the mRNA codon 5′-AGA-3′ or 5′-AGG-3′, both of which code for arginine (see Figure 17.6).

▼ **Figure 17.17 Aminoacyl-tRNA synthetases provide specificity in joining amino acids to their tRNAs.** Linkage of a tRNA to its amino acid is an endergonic process that occurs at the expense of ATP, which loses two phosphate groups, becoming AMP.

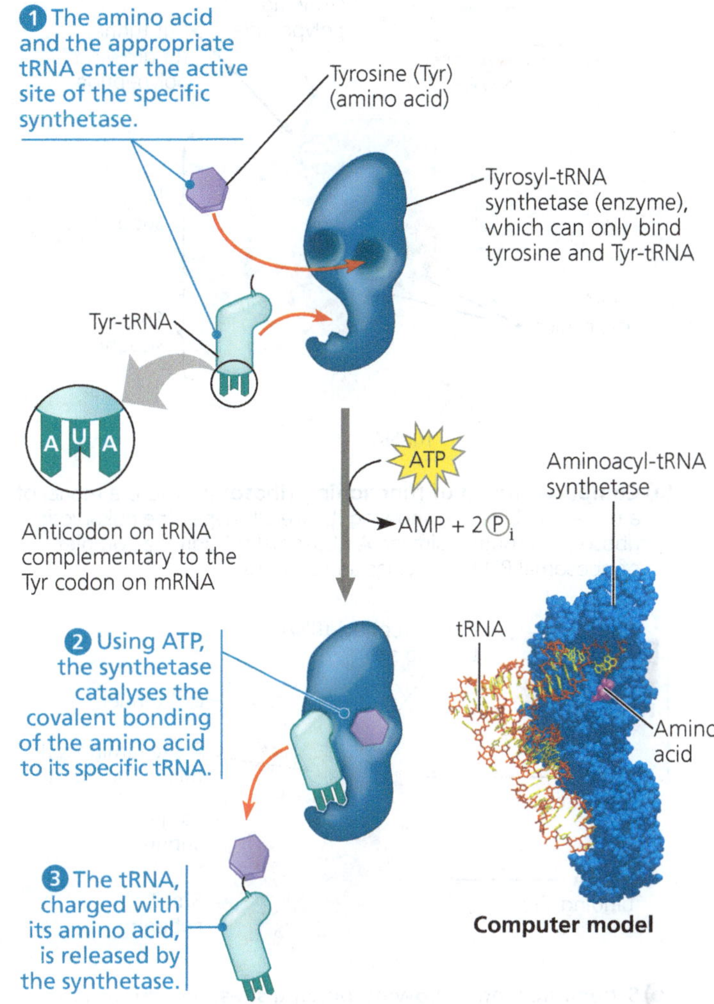

The Structure and Function of Ribosomes

Although the ribosomes of bacteria and eukaryotes are very similar in structure and function, eukaryotic ribosomes are slightly larger and differ somewhat from bacterial ribosomes in their molecular composition. The differences are medically significant. Certain antibiotic drugs can inactivate bacterial ribosomes without affecting the ability of eukaryotic ribosomes to make proteins. These drugs, including tetracycline and streptomycin, are used to combat bacterial infections.

Ribosomes facilitate the specific coupling of tRNA anticodons with mRNA codons during protein synthesis. A ribosome consists of a large subunit and a small subunit, each made up of proteins and one or more **ribosomal RNAs**, or **rRNAs** (**Figure 17.18**). In eukaryotes, the subunits are made in the nucleolus. Ribosomal RNA genes are transcribed, and the RNA is processed and assembled with proteins imported from the cytoplasm. Completed ribosomal subunits are then exported

▼ **Figure 17.18 The anatomy of a functioning ribosome.**

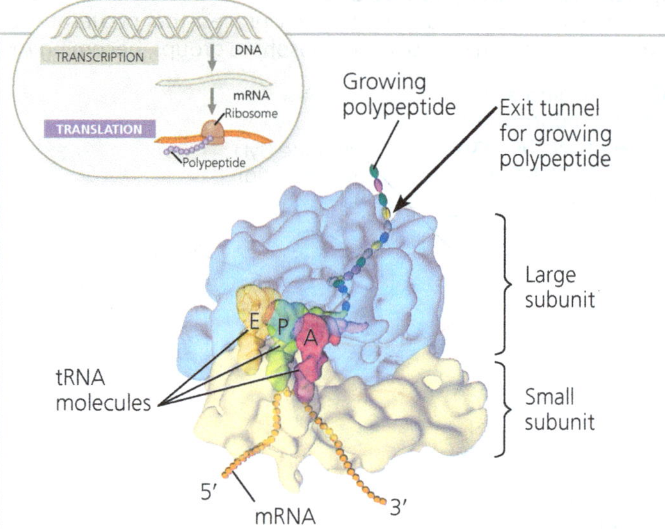

(a) Computer model of functioning ribosome. This is a model of a bacterial ribosome, showing its overall shape. The eukaryotic ribosome is roughly similar. A ribosomal subunit is a complex of ribosomal RNA molecules and proteins.

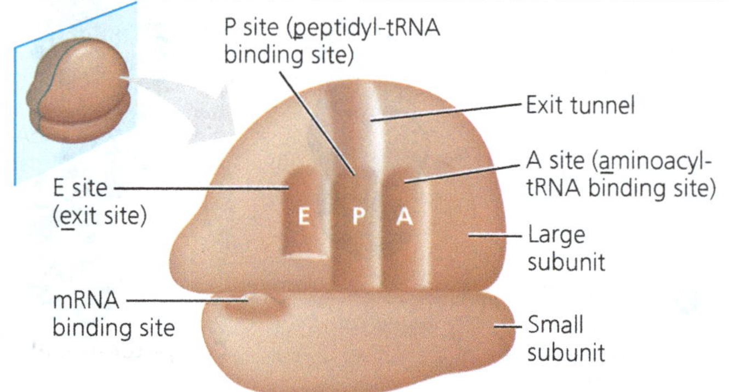

(b) Schematic model showing binding sites. A ribosome has an mRNA binding site and three tRNA binding sites, known as the A, P, and E sites. This schematic ribosome will appear in later diagrams.

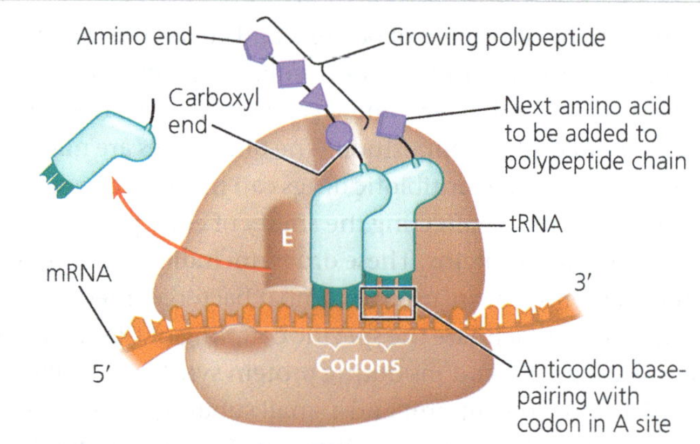

(c) Schematic model with mRNA and tRNA. A tRNA fits into the A site when its anticodon base-pairs with an mRNA codon. The P site holds the tRNA attached to the growing polypeptide. The A site holds the tRNA carrying the next amino acid to be added to the polypeptide chain. Discharged tRNAs leave from the E site. The polypeptide grows at its carboxyl end.

via nuclear pores to the cytoplasm. In both bacteria and eukaryotes, a large and a small subunit join to form a functional ribosome only when attached to an mRNA molecule. About one-third of the mass of a ribosome is made up of proteins; the rest consists of three rRNA molecules (in bacteria) or four (in eukaryotes). Because most cells contain thousands of ribosomes, rRNA is the most abundant type of cellular RNA.

The structure of a ribosome reflects its function of bringing an mRNA molecule together with tRNAs carrying amino acids. The mRNA itself has a binding site for the ribosome. (In the **Scientific Skills Exercise**, you can work with DNA sequences representing this binding site in a group of *Escherichia coli* genes.) The ribosome, in turn, has a binding site for mRNA, as well as three binding sites for tRNA (see Figure 17.18). The **P site** (peptidyl-tRNA binding site) holds the tRNA carrying the growing polypeptide chain, while the **A site** (aminoacyl-tRNA binding site) holds the tRNA carrying the next amino acid to be added to the chain. Discharged tRNAs leave the ribosome from the **E site** (exit site). The ribosome holds the tRNA and mRNA in close proximity and positions the new amino acid so that it can be added to the carboxyl end of the growing polypeptide. It then catalyses the formation of the peptide bond. As the polypeptide becomes longer, it passes through an *exit tunnel* in the ribosome's large subunit. When the polypeptide is complete, it is released through the exit tunnel.

The widely accepted model is that rRNAs, rather than ribosomal proteins, are primarily responsible for both the structure and the function of the ribosome. The proteins, which are largely on the exterior, support the shape changes of the rRNA molecules as they carry out catalysis during translation. Ribosomal RNA is the main constituent of the A and P sites and of the interface between the two subunits; it also acts as the catalyst of peptide bond formation. Thus, a ribosome could actually be considered one colossal ribozyme!

Building a Polypeptide

We can divide translation, the synthesis of a polypeptide, into three stages: initiation, elongation, and termination. All three require protein "factors" that aid in the translation process. Some steps of initiation and elongation also require energy, provided by the hydrolysis of guanosine triphosphate (GTP).

Ribosome Association and Initiation of Translation

The initiation stage of translation brings together an mRNA, a tRNA bearing the first amino acid of the polypeptide, and the two subunits of a ribosome. First, a small ribosomal subunit binds to both the mRNA and a specific initiator tRNA, which carries the amino acid methionine. In bacteria, the small subunit can bind the two in either order; it binds the mRNA at a specific RNA sequence, just upstream of the AUG start codon. In eukaryotes, the small subunit, with the initiator tRNA already bound, binds to the 5′ cap of the mRNA and then moves, or *scans*,

Scientific Skills Exercise

Interpreting a Sequence Logo

How Can a Sequence Logo Be Used to Identify Ribosome Binding Sites on Bacterial mRNAs? When initiating translation, ribosomes bind to an mRNA at a ribosome binding site upstream of the AUG start codon. Because mRNAs from different genes all bind to a ribosome, the genes encoding these mRNAs are likely to have a similar base sequence where the ribosomes bind. Therefore, candidate ribosome binding sites on mRNA can be identified by comparing DNA sequences (and thus the mRNA sequences) of multiple genes in a species, searching the region upstream of the start codon for shared (conserved) stretches of bases. In this exercise, you will analyse DNA sequences from multiple such genes, represented by a visual graphic called a sequence logo.

How the Experiment Was Done The DNA sequences of 149 genes from the *E. coli* genome were aligned using computer software. The aim was to identify similar base sequences—at the appropriate location in each gene—as potential ribosome binding sites. Rather than presenting the data as a series of 149 sequences aligned in a column (a sequence alignment), the researchers used a sequence logo.

Data from the Experiment To show how sequence logos are made, the potential ribosome binding regions from ten *E. coli* genes are shown below in a sequence alignment, followed by the sequence logo derived from the aligned sequences. Note that the DNA shown is the nontemplate (coding) strand, which is how DNA sequences are typically presented. (All data from Thomas D. Schneider.)

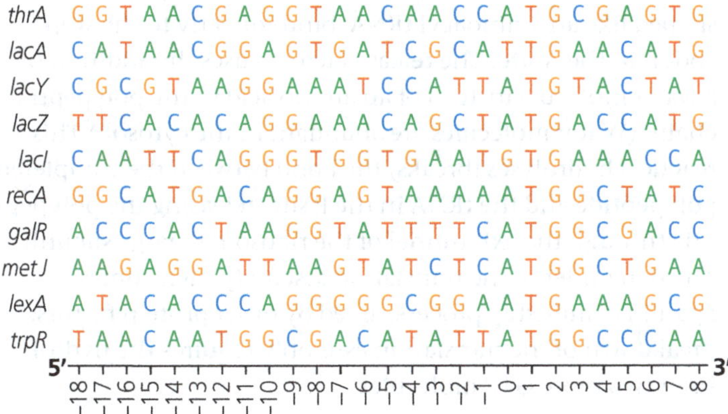

▲ **Sequence alignment**

▼ **Sequence logo**

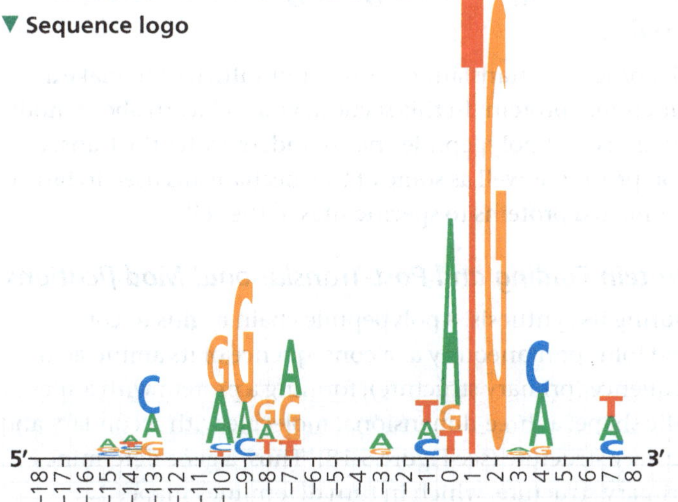

Further Reading T. D. Schneider and R. M. Stephens, Sequence logos: A new way to display consensus sequences, *Nucleic Acids Research* 18:6097–6100 (1990).

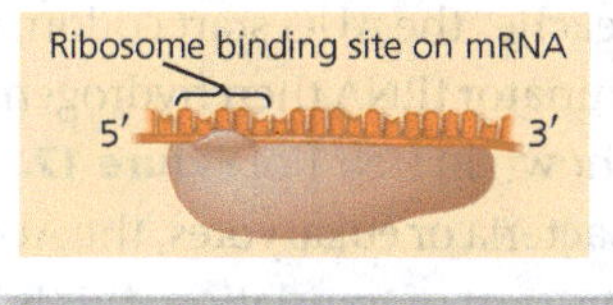

INTERPRET THE DATA

1. In the sequence logo, the horizontal axis shows the primary sequence of the DNA by nucleotide position. Letters for each base are stacked on top of each other according to their relative frequency at that position among the aligned sequences, with the most common base as the largest letter at the top of the stack. The height of each letter represents the relative frequency of that base *at that position*. **(a)** In the sequence alignment, count the number of each base at position –9 and order them from most to least frequent. Compare this to the size and placement of each base at –9 in the logo. **(b)** Do the same for positions 0 and 1.
2. The height of a stack of letters in a logo indicates the predictive power of that stack (determined statistically). If the stack is tall, we can be more confident in predicting what base will be in that position if a new sequence is added to the logo. For example, at position 2 in the sequence alignment, all ten sequences have a G; the probability of finding a G there in a new sequence is very high, as is the stack in the sequence logo. For short stacks, the bases all have about the same frequency, so it's hard to predict a base at those positions. **(a)** Looking at the sequence logo, which two positions have the most predictable bases? What bases do you predict would be at those positions in a newly sequenced gene? **(b)** Which 12 positions have the least predictable bases? How do you know? How does this reflect the relative frequencies of the bases shown at these positions in the sequence alignment? Use the two leftmost positions of the 12 as examples in your answer.
3. In the actual experiment, the researchers used 149 sequences to build their sequence logo, which is shown below. There is a stack at each position, even if short, because the sequence logo includes more data. **(a)** Which three positions in this sequence logo have the most predictable bases? Name the most frequent base at each. **(b)** Which four positions have the least predictable bases? How can you tell?

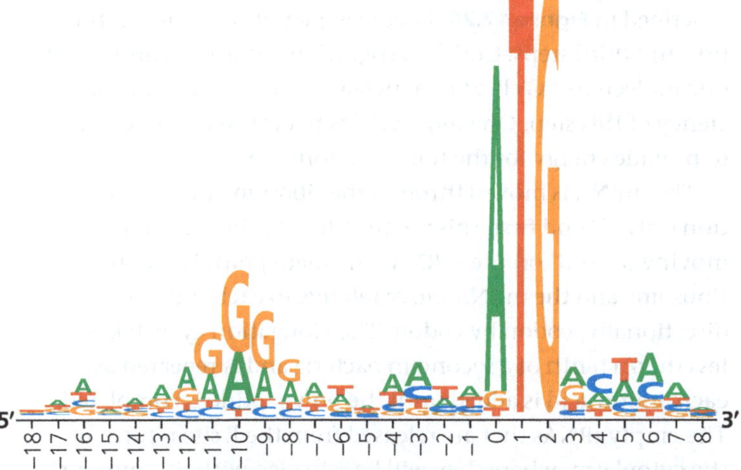

4. A consensus sequence identifies the base occurring most often at each position in the set of sequences. **(a)** Write out the consensus sequence of this (the nontemplate) strand. In any position where the base can't be determined, put a dash. **(b)** Which provides more information—the consensus sequence or the sequence logo? What is lost in the less informative method?
5. **(a)** Based on the logo, what five adjacent base positions in the 5′ UTR region are most likely to be involved in ribosome binding? Explain. **(b)** What is represented by the bases in positions 0–2?

downstream along the mRNA until it reaches the AUG start codon, where the initiator tRNA then hydrogen-bonds, as shown in step 1 of **Figure 17.19**. In either bacteria or eukaryotes, the AUG signals the start of translation; this is important because it establishes the codon reading frame for the mRNA.

The union of mRNA, initiator tRNA, and the small ribosomal subunit is followed by the attachment of a large ribosomal subunit, completing the *translation initiation complex* (see Figure 17.19). Proteins called *initiation factors* are required to bring all these components together. The cell also expends energy obtained by hydrolysis of a GTP molecule to form the initiation complex. At the completion of the initiation process, the initiator tRNA sits in the P site of the ribosome, and the vacant A site is ready for the next aminoacyl tRNA. Note that a polypeptide is always synthesised in one direction, from the initial methionine at the amino end, also called the N-terminus, towards the final amino acid at the carboxyl end, also called the C-terminus (see Figure 5.15).

▼ Figure 17.19 The initiation of translation.

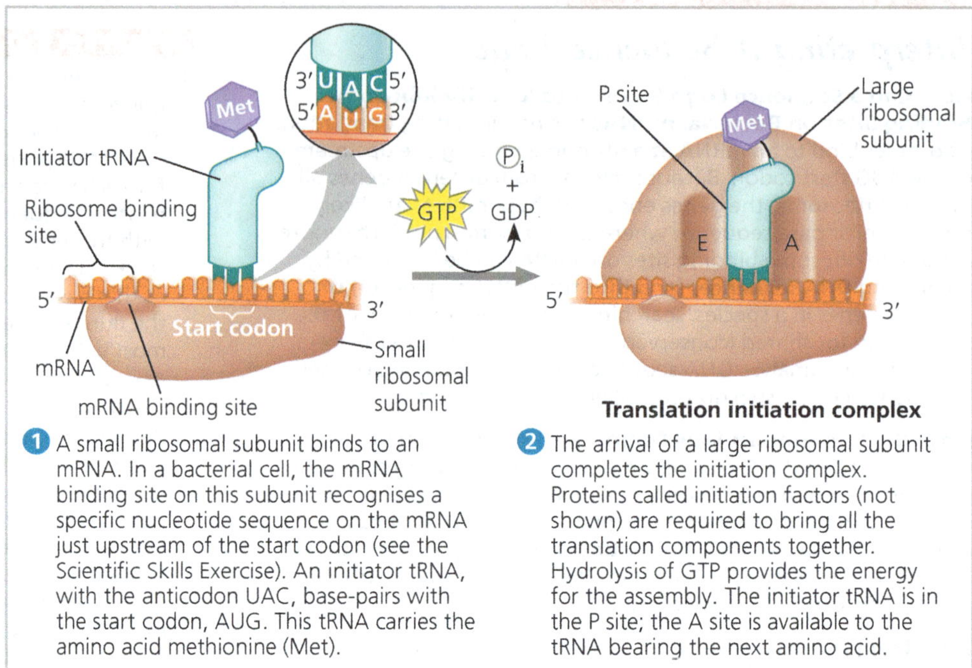

1 A small ribosomal subunit binds to an mRNA. In a bacterial cell, the mRNA binding site on this subunit recognises a specific nucleotide sequence on the mRNA just upstream of the start codon (see the Scientific Skills Exercise). An initiator tRNA, with the anticodon UAC, base-pairs with the start codon, AUG. This tRNA carries the amino acid methionine (Met).

2 The arrival of a large ribosomal subunit completes the initiation complex. Proteins called initiation factors (not shown) are required to bring all the translation components together. Hydrolysis of GTP provides the energy for the assembly. The initiator tRNA is in the P site; the A site is available to the tRNA bearing the next amino acid.

Elongation of the Polypeptide Chain

In the elongation stage of translation, amino acids are added one by one to the previous amino acid at the C-terminus of the growing chain. Each addition involves several proteins called *elongation factors* and occurs in a three-step cycle described in **Figure 17.20**. Energy expenditure occurs in the first and third steps. Codon recognition requires hydrolysis of one molecule of GTP, which increases the accuracy and efficiency of this step. One more GTP is hydrolysed (broken up) to provide energy for the translocation step.

The mRNA is moved through the ribosome in one direction only, 5′ end first; this is equivalent to the ribosome moving 5′ → 3′ on the mRNA. The main point is that the ribosome and the mRNA move relative to each other, unidirectionally, codon by codon. The elongation cycle takes less than a tenth of a second in bacteria and is repeated as each amino acid is added until the polypeptide is completed. The empty tRNAs that are released from the E site return to the cytoplasm, where they will be reloaded with the appropriate amino acid (see Figure 17.17).

Termination of Translation

The final stage of translation is termination **(Figure 17.21)**. Elongation continues until a stop codon in the mRNA reaches the A site. The nucleotide base triplets UAG, UAA, and UGA (all written 5′ → 3′) do not code for amino acids but instead act as signals to stop translation. A *release factor*, a protein shaped like an aminoacyl tRNA, binds directly to the stop codon in the A site. The release factor causes the addition of a water molecule instead of an amino acid to the polypeptide chain. (Water molecules are abundant in the cytosol.) This reaction hydrolyses (breaks) the bond between the completed polypeptide and the tRNA in the P site, releasing the polypeptide through the exit tunnel of the ribosome's large subunit. The remainder of the translation assembly then comes apart in a multistep process, aided by other protein factors. Breakdown of the translation assembly requires the hydrolysis of two more GTP molecules.

Completing and Targeting the Functional Protein

The process of translation is often not sufficient to make a functional protein. In this section, you will learn about modifications that polypeptide chains undergo after the translation process as well as some of the mechanisms used to target completed proteins to specific sites in the cell.

Protein Folding and Post-Translational Modifications

During its synthesis, a polypeptide chain begins to coil and fold spontaneously as a consequence of its amino acid sequence (primary structure), forming a protein with a specific shape: a three-dimensional molecule with secondary and tertiary structure (see Figure 5.18). Thus, a gene determines primary structure, which in turn determines shape.

▼ Figure 17.20 The elongation cycle of translation. The hydrolysis of GTP plays an important role in the elongation process; elongation factors are not shown.

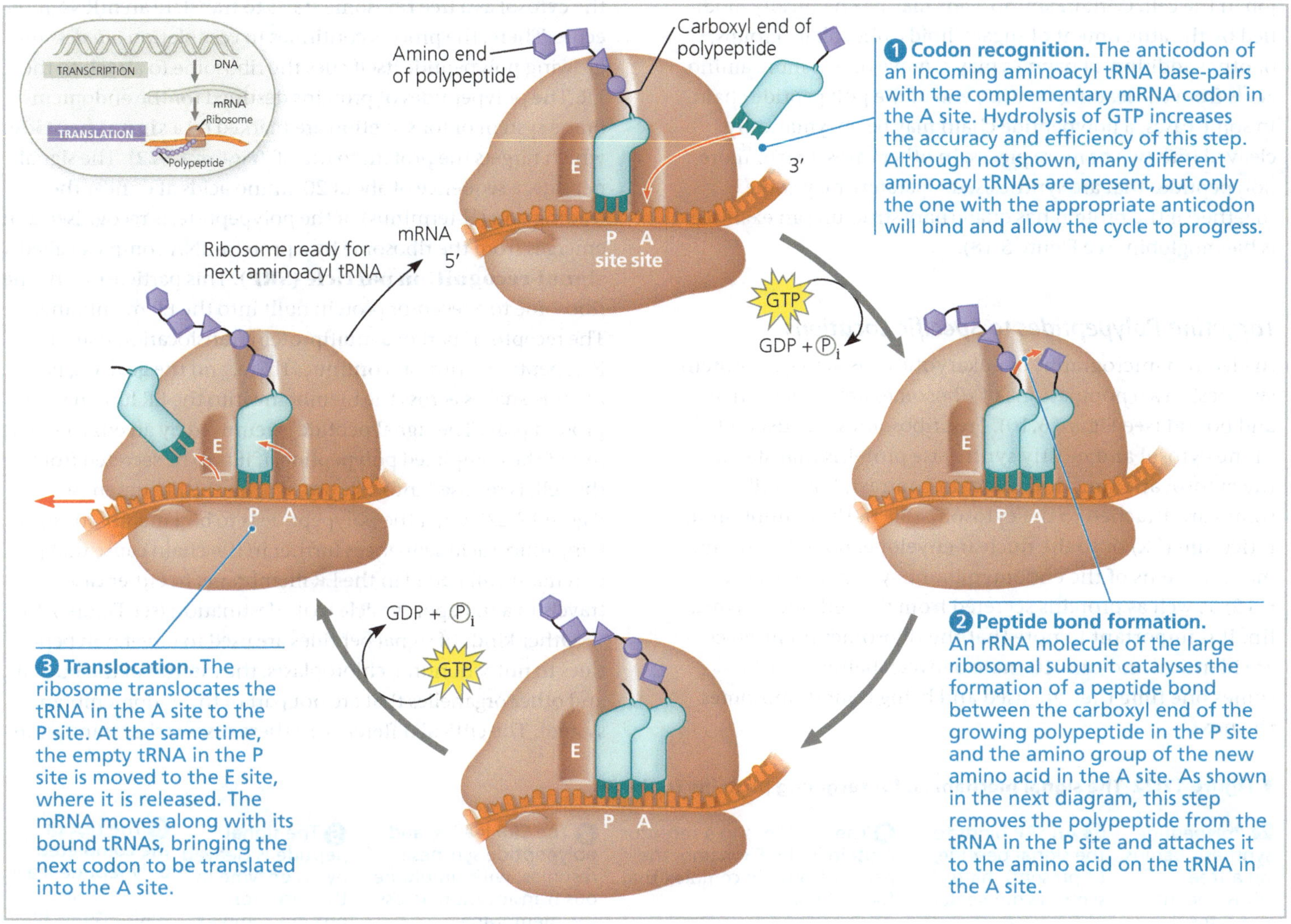

TRANSCRIPTION

DNA

mRNA

Ribosome

TRANSLATION

Polypeptide

Amino end of polypeptide

Carboxyl end of polypeptide

E

P site

A site

mRNA

5′

3′

1 Codon recognition. The anticodon of an incoming aminoacyl tRNA base-pairs with the complementary mRNA codon in the A site. Hydrolysis of GTP increases the accuracy and efficiency of this step. Although not shown, many different aminoacyl tRNAs are present, but only the one with the appropriate anticodon will bind and allow the cycle to progress.

GTP

GDP + Ⓟ$_i$

2 Peptide bond formation. An rRNA molecule of the large ribosomal subunit catalyses the formation of a peptide bond between the carboxyl end of the growing polypeptide in the P site and the amino group of the new amino acid in the A site. As shown in the next diagram, this step removes the polypeptide from the tRNA in the P site and attaches it to the amino acid on the tRNA in the A site.

GTP

GDP + Ⓟ$_i$

3 Translocation. The ribosome translocates the tRNA in the A site to the P site. At the same time, the empty tRNA in the P site is moved to the E site, where it is released. The mRNA moves along with its bound tRNAs, bringing the next codon to be translated into the A site.

Ribosome ready for next aminoacyl tRNA

▼ Figure 17.21 The termination of translation. Like elongation, termination requires GTP hydrolysis as well as additional protein factors, which are not shown here.

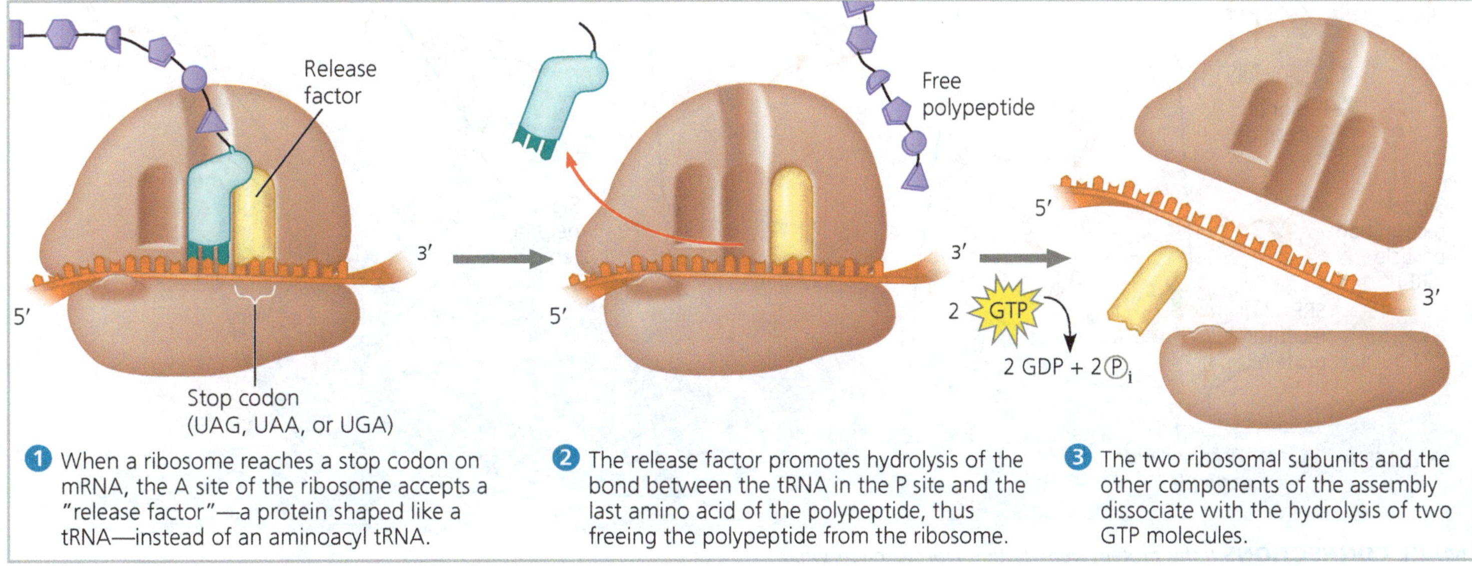

Additional steps—*post-translational modifications*—may be required before the protein can begin doing its particular job in the cell. Certain amino acids may be chemically modified by the attachment of sugars, lipids, phosphate groups, or other additions. Enzymes may remove one or more amino acids from the leading (amino) end of the polypeptide chain. In some cases, a polypeptide chain may be enzymatically cleaved into two or more pieces. In other cases, two or more polypeptides that are synthesised separately may come together, if the protein has quaternary structure; an example is haemoglobin (see Figure 5.18).

Targeting Polypeptides to Specific Locations

In electron micrographs of eukaryotic cells active in protein synthesis, two populations of ribosomes are evident: free and bound (see Figure 6.10). Free ribosomes are suspended in the cytosol and mostly synthesise proteins that stay in the cytosol and function there. In contrast, bound ribosomes are attached to the cytosolic side of the endoplasmic reticulum (ER) or to the nuclear envelope. Bound ribosomes make proteins of the endomembrane system (see Figure 6.15) as well as proteins secreted from the cell, such as insulin. It is important to note that the ribosomes themselves are identical and can alternate between being free ribosomes one time they are used and being bound ribosomes the next.

What determines whether a ribosome is free in the cytosol or bound to rough ER? Polypeptide synthesis always begins in the cytosol as a free ribosome starts to translate an mRNA molecule. There, the process continues to completion—*unless* the growing polypeptide itself cues the ribosome to attach to the ER. The polypeptides of proteins destined for the endomembrane system or for secretion are marked by a **signal peptide**, which targets the protein to the ER **(Figure 17.22)**. The signal peptide, a sequence of about 20 amino acids at or near the leading end (N-terminus) of the polypeptide, is recognised as it emerges from the ribosome by a protein-RNA complex called a **signal-recognition particle (SRP)**. This particle escorts the ribosome to a receptor protein built into the ER membrane. The receptor is part of a multiprotein translocation complex. Polypeptide synthesis continues there, and the growing polypeptide snakes across the membrane into the ER lumen via a protein pore. The signal peptide is removed by an enzyme. The rest of the completed polypeptide, if it is to be secreted from the cell, is released into solution within the ER lumen (see Figure 17.22). Or, if the polypeptide is to be a membrane protein, amino acid sequences further in the chain cause that part to remain embedded in the ER membrane. In either case, it travels in a transport vesicle to its destination (see Figure 7.9).

Other kinds of signal peptides are used to target polypeptides to mitochondria, chloroplasts, the interior of the nucleus, and other organelles that are not part of the endomembrane system. The critical difference in these cases is that translation

▼ **Figure 17.22 The signal mechanism for targeting proteins to the ER.**

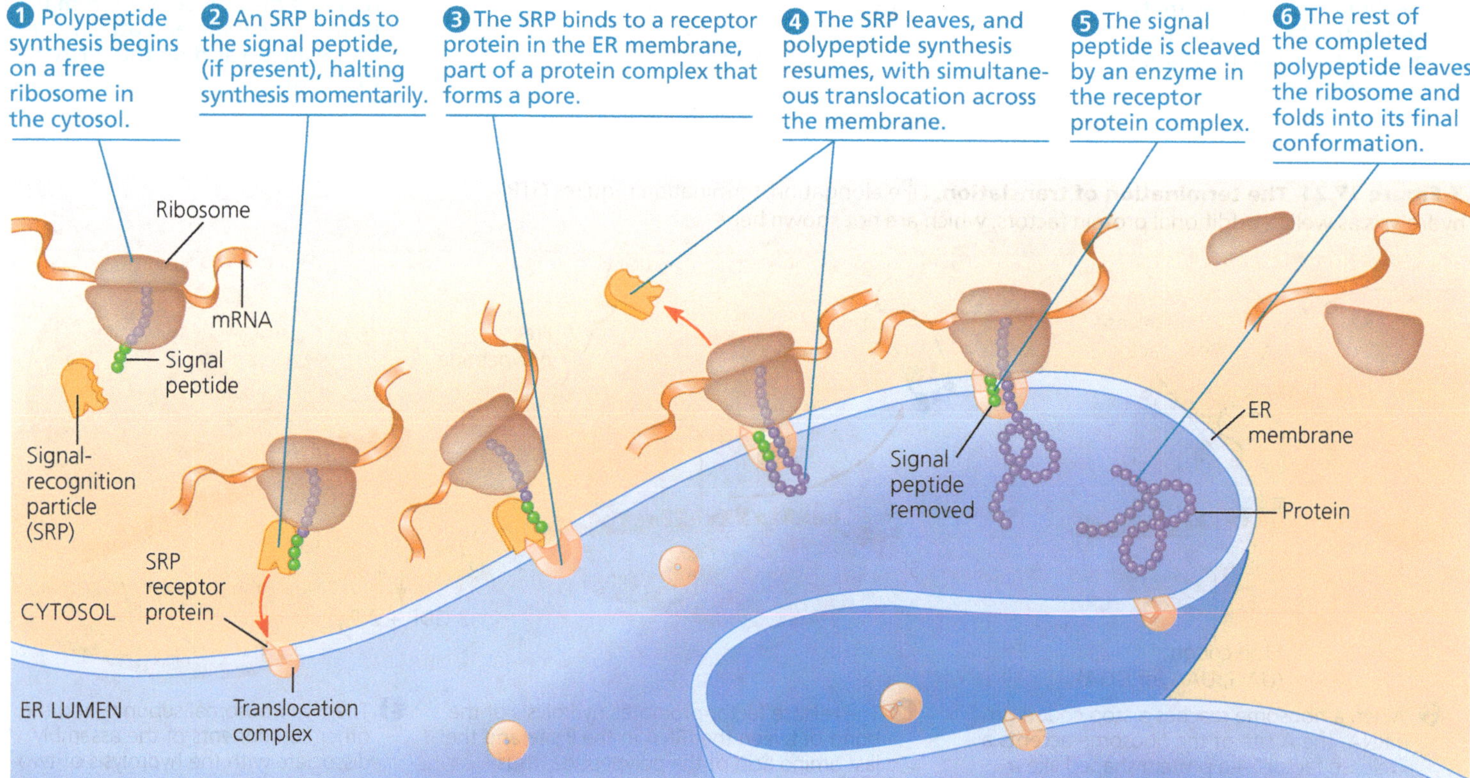

MAKE CONNECTIONS *If this protein were destined for secretion, what would happen to it after its synthesis was completed? See Figure 7.9.*

is completed in the cytosol before the polypeptide is imported into the organelle. Translocation mechanisms also vary, but in all cases studied to date, the "post codes" that address proteins for secretion or to cellular locations are signal peptides of some sort. Bacteria also employ signal peptides to target proteins to the plasma membrane or for secretion.

Making Multiple Polypeptides in Bacteria and Eukaryotes

In previous sections, you learned how a single polypeptide is synthesised using the information encoded in an mRNA molecule. When a polypeptide is required in a cell, though, the need is for many copies, not just one.

In both bacteria and eukaryotes, multiple ribosomes translate an mRNA at the same time **(Figure 17.23)**; that is, a single mRNA is used to make many copies of a polypeptide simultaneously. Once a ribosome is far enough past the start codon, a second ribosome can attach to the mRNA, eventually resulting in a number of ribosomes trailing along the mRNA. Such strings of ribosomes, called **polyribosomes** (or **polysomes**), can be seen with an electron microscope; they can be either free or bound. They enable a cell to rapidly make many copies of a polypeptide.

Another way both bacteria and eukaryotes increase the number of copies of a polypeptide is by transcribing multiple mRNAs from the same gene. However, the coordination of the two processes—transcription and translation—differs in the two groups. The most important differences between bacteria

▼ **Figure 17.23 Polyribosomes.**

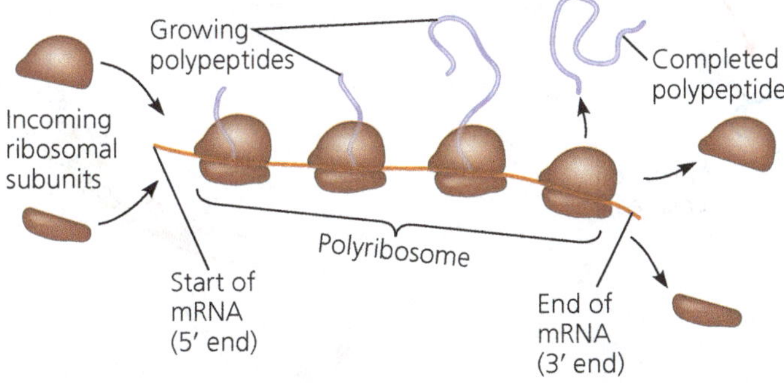

(a) An mRNA molecule is generally translated simultaneously by several ribosomes in clusters called polyribosomes.

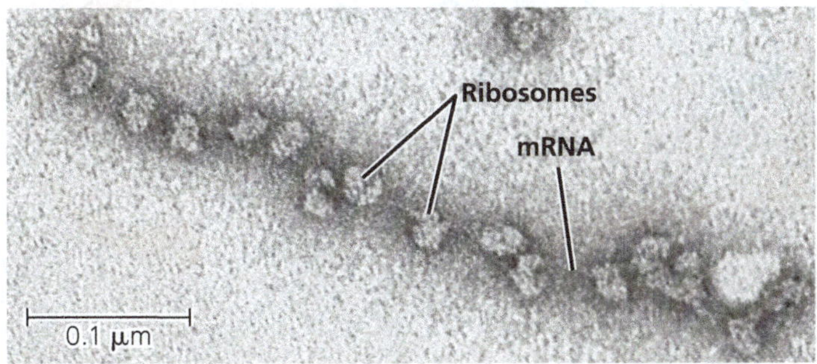

(b) This micrograph shows a large polyribosome in a bacterial cell. Growing polypeptides are not visible here (TEM).

▼ **Figure 17.24 Coupled transcription and translation in bacteria.** In bacterial cells, the translation of mRNA can begin as soon as the leading (5′) end of the mRNA molecule peels away from the DNA template. The micrograph (TEM) shows a strand of *E. coli* DNA being transcribed by RNA polymerase molecules. Attached to each RNA polymerase molecule is a growing strand of mRNA, which is already being translated by ribosomes. The newly synthesised polypeptides are not visible in the micrograph but are shown in the diagram.

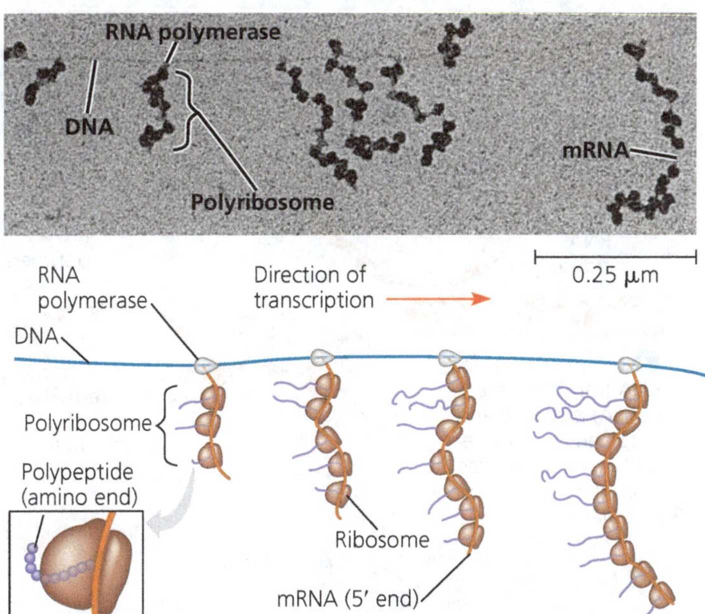

VISUAL SKILLS *Which one of the mRNA molecules started being transcribed first? On that mRNA, which ribosome started translating the mRNA first?*

and eukaryotes arise from the bacterial cell's lack of compartmental organisation. Like a one-room workshop, a bacterial cell ensures a streamlined operation by coupling the two processes. With no nuclear envelope, it can simultaneously transcribe and translate the same gene **(Figure 17.24)**, and the newly made protein can quickly diffuse to its site of function.

In contrast, the eukaryotic cell's nuclear envelope segregates transcription from translation and provides a compartment for extensive RNA processing. This processing stage includes additional steps, discussed earlier, the regulation of which can help coordinate the eukaryotic cell's elaborate activities. **Figure 17.25** summarises the path from gene to polypeptide in a eukaryotic cell.

CONCEPT CHECK 17.4

1. What two processes ensure that the correct amino acid is added to a growing polypeptide chain?
2. Describe how a polypeptide to be secreted reaches the endomembrane system.
3. **DRAW IT** Draw a tRNA with the anticodon 3′-CGU-5′. Given wobble, what two different codons could it bind to? Draw each codon on an mRNA, labelling all 5′ and 3′ ends, the tRNA, and the amino acid it carries.
4. **WHAT IF?** In eukaryotic cells, mRNAs have been found to have a circular arrangement in which proteins hold the poly-A tail near the 5′ cap. How might this increase translation efficiency?

For suggested answers, see Appendix A.

▼ Figure 17.25 A summary of transcription and translation in a eukaryotic cell. This diagram shows the path from one gene to one polypeptide. Each gene in the DNA can be transcribed repeatedly into many identical RNA molecules, and each mRNA can be translated repeatedly to yield many identical polypeptide molecules. (Remember that the final products of some genes are not polypeptides but RNA molecules that don't get translated, including tRNAs and rRNAs.) In general, the steps of transcription and translation are similar in bacterial, archaeal, and eukaryotic cells. The major difference is the occurrence of RNA processing in the eukaryotic nucleus. Other significant differences are found in the initiation stages of both transcription and translation and in the termination of transcription. To visualise these processes in their cellular context, see Figure 6.32d-f.

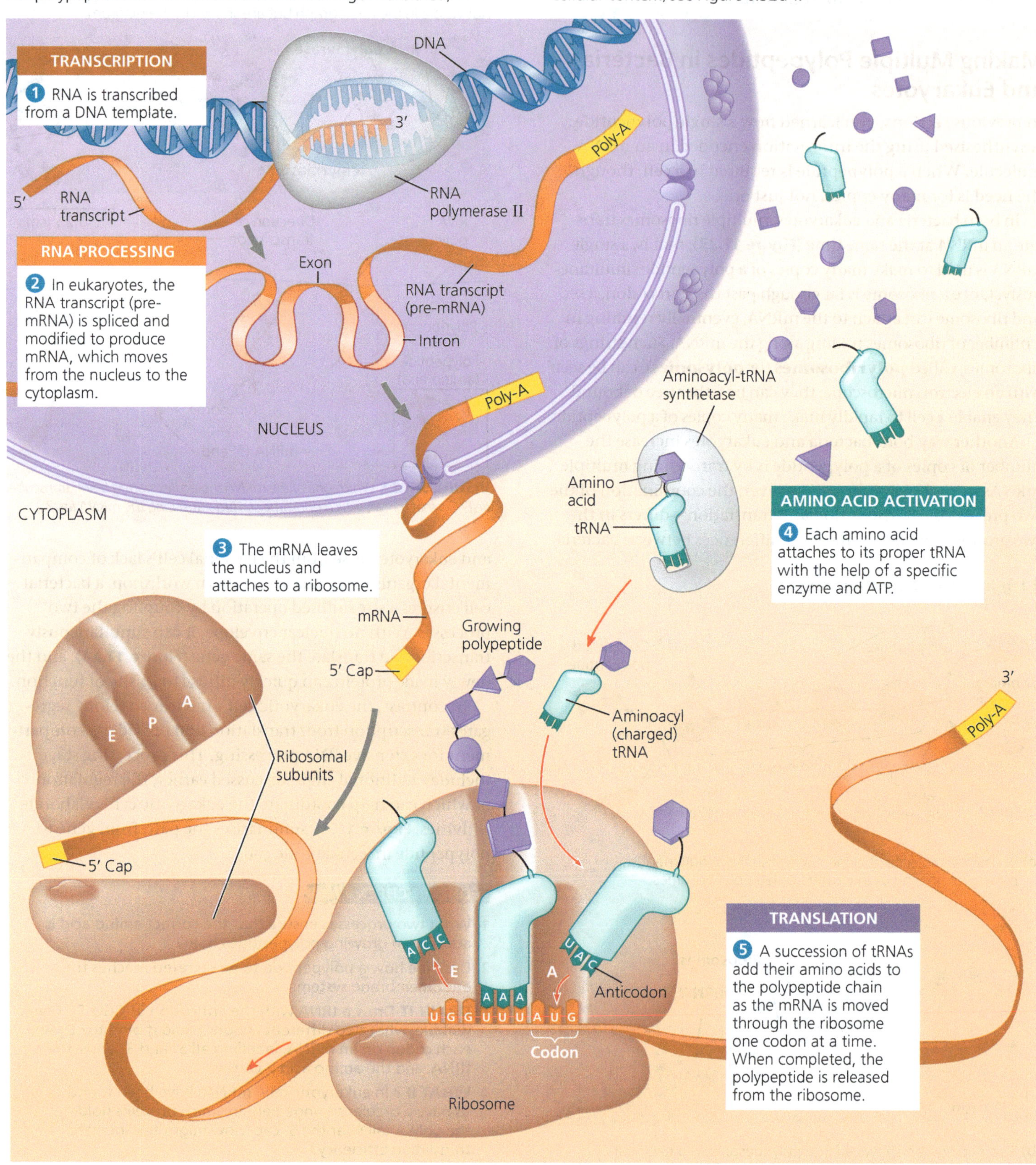

CONCEPT 17.5

Mutations of one or a few nucleotides can affect protein structure and function

Now that you have explored the process of gene expression, you can consider the effects of changes to the genetic information of a cell. Because these changes, called **mutations**, are the ultimate source of new genes, they are responsible for the huge diversity of genes found among organisms. Earlier, we considered chromosomal rearrangements that affect long segments of DNA (see Figure 15.14); these are considered large-scale mutations. Here we examine small-scale mutations of one or a few nucleotide pairs, including **point mutations**, changes in a single nucleotide pair of a gene.

If a point mutation occurs in a gamete or in a cell that gives rise to gametes, it may be transmitted to offspring and to future generations. If the mutation has an adverse effect on the phenotype of a person, the mutant condition is referred to as a genetic disorder or hereditary disease. For example, we can trace the genetic basis of sickle-cell disease to the mutation of a single nucleotide pair in the gene that encodes the β-globin polypeptide of haemoglobin. The change of a single nucleotide in the DNA's template strand leads to an altered mRNA and the production of an abnormal protein (**Figure 17.26**; also see Figure 5.19). In individuals who are homozygous for the mutant allele, the sickling of red blood cells caused by the altered haemoglobin produces the multiple symptoms associated with sickle-cell disease (see Concept 14.4 and Figure 23.19). Another disorder caused by a point mutation is a heart condition called familial cardiomyopathy, which is responsible for some of the tragic incidents of sudden death in young athletes. Point mutations in several genes encoding muscle proteins have been identified, any of which can lead to this disorder.

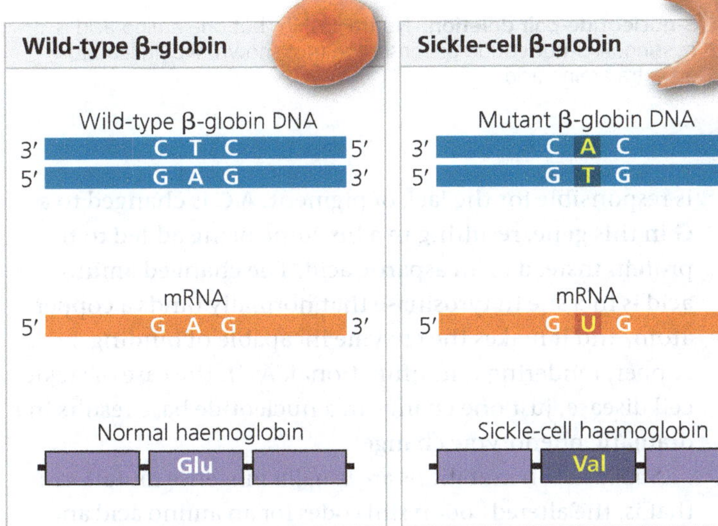

▼ **Figure 17.26 The molecular basis of sickle-cell disease: a point mutation.** The allele that causes sickle-cell disease differs from the wild-type (normal) allele by a single DNA nucleotide pair. The micrographs are SEMs of a normal red blood cell (on the left) and a sickled red blood cell (right) from individuals homozygous for wild-type and mutant alleles, respectively.

Types of Small-Scale Mutations

Many mutations occur in regions outside of protein-coding genes, and any potential effect they have on the phenotype of the organism may be subtle and hard to detect. For this reason, here we'll concentrate on mutations within protein-coding genes. Small-scale mutations within a gene can be divided into two general categories: (1) single nucleotide-pair substitutions and (2) nucleotide-pair insertions or deletions. Insertions and deletions can involve one or more nucleotide pairs.

Substitutions

A **nucleotide-pair substitution** is the replacement of one nucleotide and its partner with another pair of nucleotides (**Figure 17.27a**). Some substitutions have no effect on the encoded protein, owing to the redundancy of the genetic code. For example, if 3′-CCG-5′ on the template strand mutated to 3′-CCA-5′, the mRNA codon that used to be GGC would become GGU, but a glycine would still be inserted at the proper location in the protein (see Figure 17.6). In other words, a change in a nucleotide pair may transform one codon into another that is translated into the same amino acid. Such a change is an example of a **silent mutation**, which has no observable effect on the phenotype. (Silent mutations can occur outside genes as well.) Interestingly, there is evidence that some silent mutations may indirectly affect where or at what level the gene gets expressed, even though the actual protein is the same.

Substitutions that change one amino acid to another one are called **missense mutations**. Such a mutation may have little effect on the protein: The new amino acid may have properties similar to those of the amino acid it replaces, or it may be in a region of the protein where the exact sequence of amino acids is not essential to the protein's function.

However, the nucleotide-pair substitutions of greatest interest are those that cause a major change in a protein. The alteration of a single amino acid in a crucial area of a protein—such as in the part of the

▼ Figure 17.27 Types of small-scale mutations that affect mRNA sequence. All but one of the types shown here also affect the amino acid sequence of the encoded polypeptide.

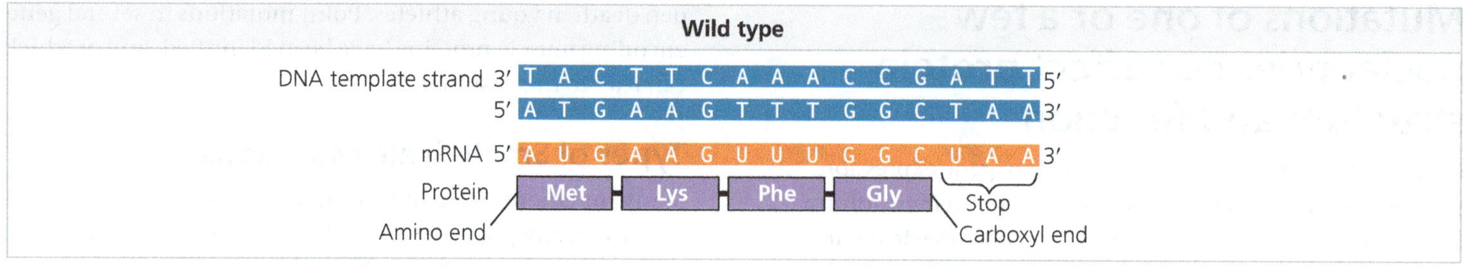

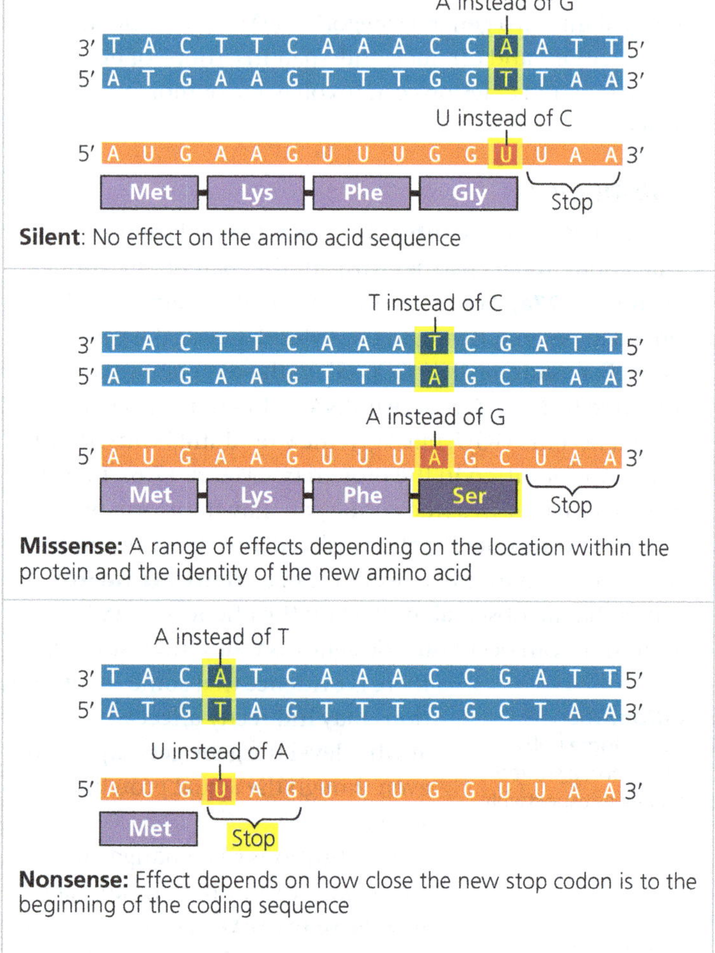

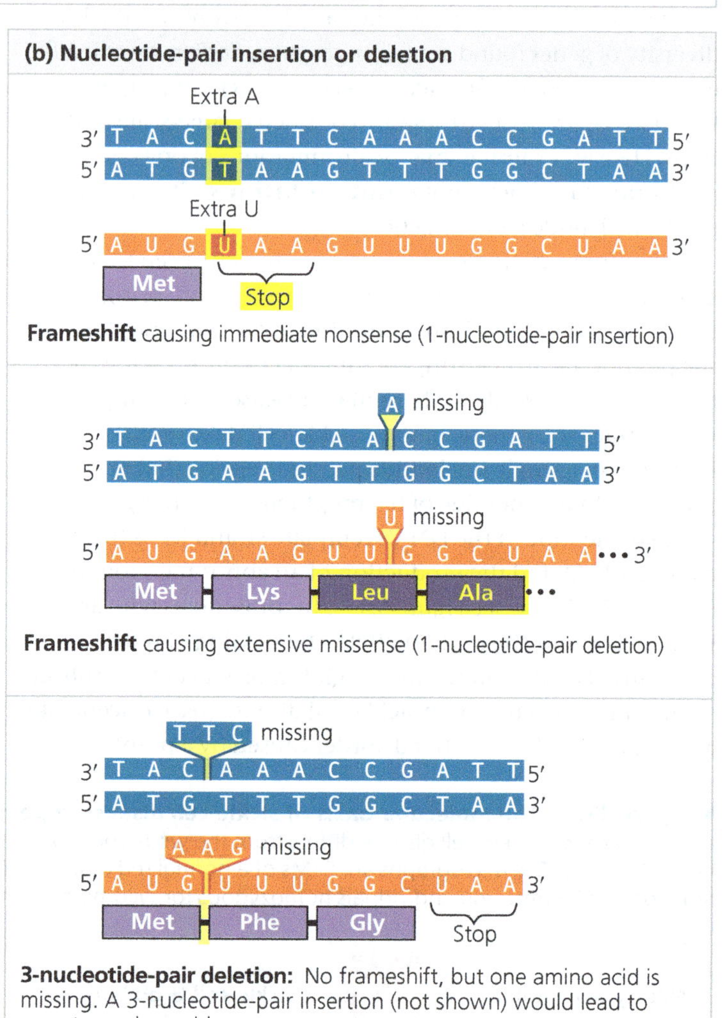

β-globin subunit of haemoglobin shown in Figure 17.26 or in the active site of an enzyme as shown in Figure 8.19—can significantly alter protein activity. Occasionally, such a mutation leads to an improved protein or one with novel capabilities, but much more often such mutations are neutral or detrimental, leading to a useless or less active protein that impairs cellular function. A second example of a missense mutation is a mutation in the tyrosinase gene that causes the albino phenotype in a group of well-researched donkeys on the island of Asinara. Recently, Italian researchers sequenced the tyrosinase gene in donkeys from this wild population and showed that a recessive mutation is responsible for the lack of pigment. A C is changed to a G in this gene, resulting in a histidine being added to the protein instead of an aspartic acid. The changed amino acid is in a site in tyrosinase that normally binds a copper atom, and it makes the enzyme incapable of binding copper, rendering it nonfunctional. As in the case of sickle-cell disease, just one change in a nucleotide base results in a dramatic phenotypic change.

Substitution mutations are usually missense mutations; that is, the altered codon still codes for an amino acid and thus makes sense, although not necessarily the *right* sense. But a point mutation can also change a codon for an amino

PROBLEM-SOLVING EXERCISE

Are insulin mutations the cause of three infants' neonatal diabetes?

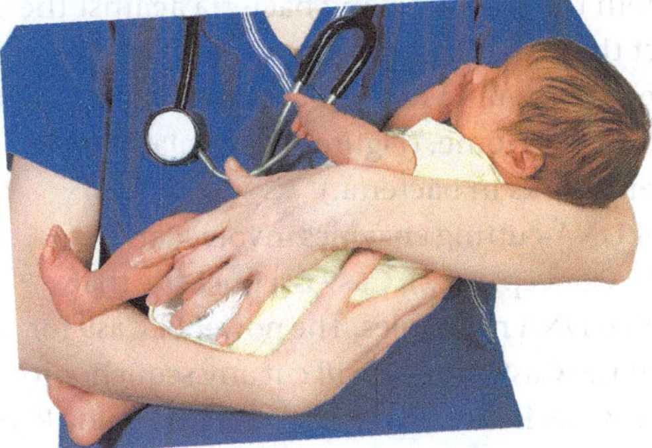

Insulin is a hormone that acts as a key regulator of blood glucose level, and insulin deficiency can lead to diabetes. In some cases of neonatal (newborn) diabetes, the gene coding for the insulin protein has a nucleotide-pair substitution mutation that alters the protein structure enough to cause it to malfunction. Doctors can use DNA sequence information to diagnose diabetes and other diseases. For example, the insulin gene sequence of a patient with neonatal diabetes can be analysed to determine if it has a mutation and, if so, its effect.

In this exercise, you will determine the effect of mutations within insulin gene sequences in infant diabetes patients.

Your Approach Suppose you are a medical geneticist presented with three infant patients, all of whom have a nucleotide-pair substitution in their insulin gene. It is your job to analyse each mutation to figure out its effect on the amino acid sequence of the insulin protein. To identify the mutation in each patient, you will compare his or her individual insulin complementary DNA (cDNA) sequence to that of the wild-type cDNA. (cDNA is a double-stranded DNA molecule that is based on the mRNA sequence and thus contains only the portion of a gene that is translated—introns are not included. cDNA sequences are commonly used to compare the coding regions of genes.) Identifying the codons that have been changed will tell you which, if any, amino acids are altered in the patient's insulin protein.

Your Data You will analyse the cDNA codons for amino acids 35–54 (of the 110 amino acids) of each patient's insulin protein, so the start codon (AUG) is not present. The sequences of the wild-type cDNA and the patients' cDNA are shown below, arranged in codons.

Wild-type cDNA	5′-CTG GTG GAA GCT CTC TAC CTA GTG TGC GGG GAA CGA GGC TTC TTC TAC ACA CCC AAG ACC-3′
Patient 1 cDNA	5′-CTG GTG GAA GCT CTC TAC CTA GTG TGC GGG GAA CGA GGC TGC TTC TAC ACA CCC AAG ACC-3′
Patient 2 cDNA	5′-CTG GTG GAA GCT CTC TAC CTA GTG TGC GGG GAA CGA GGC TCC TTC TAC ACA CCC AAG ACC-3′
Patient 3 cDNA	5′-CTG GTG GAA GCT CTC TAC CTA GTG TGC GGG GAA CGA GGC TTC TTG TAC ACA CCC AAG ACC-3′

Data from N. Nishi and K. Nanjo, Insulin gene mutations and diabetes, *Journal of Diabetes Investigation* 2:92–100 (2011).

Your Analysis

1. Compare each patient's cDNA sequence to the wild-type cDNA sequence. Circle the codons where a nucleotide-pair substitution mutation has occurred.
2. Use a codon table (see Figure 17.6) to identify the amino acid encoded by the codon with the mutation in each patient's insulin sequence, and compare it to the amino acid encoded by the codon in the corresponding wild-type sequence. As is standard practice with DNA sequences, the *coding* (nontemplate) strand of the cDNA has been provided, so to convert it to mRNA for use with the codon table, you only need to change T to U. Classify each patient's nucleotide-pair substitution mutation: Is it a silent, missense, or nonsense mutation? Explain, for each answer.
3. Compare the structure of the amino acid you identified in each patient's insulin sequence to that of the corresponding amino acid in the wild-type sequence (see Figure 5.14). Given that each patient has neonatal diabetes, discuss how the change of amino acid in each might have affected the insulin protein and thus resulted in the disease. (Consider the chemical nature of the side chains.)

acid into a stop codon. This is called a **nonsense mutation**, and it causes translation to be terminated prematurely; the resulting polypeptide will be shorter than the polypeptide encoded by the normal gene. Most nonsense mutations lead to nonfunctional proteins.

In the **Problem-Solving Exercise**, you'll work with a few common single nucleotide-pair substitution mutations in the gene encoding insulin, some or all of which may lead to diabetes. You will classify these mutations into one of the types we just described and characterise the change in the amino acid sequence.

Insertions and Deletions

Insertions and **deletions** are additions or losses of nucleotide pairs in a gene **(Figure 17.27b)**. These mutations have a disastrous effect on the resulting protein more often than substitutions do. Insertion or deletion of nucleotides may alter the reading frame of the genetic message, the triplet grouping of nucleotides on the mRNA that is read during translation. Such a mutation, called a **frameshift mutation**, occurs whenever the number of nucleotides inserted or deleted is not a multiple of three. All nucleotides downstream of the deletion or insertion will be improperly grouped into codons; the result will be extensive missense mutations, usually ending sooner or later in a nonsense mutation that leads to premature termination. Unless the frameshift is very near the end of the gene, the protein is almost certain to be nonfunctional. Insertions and deletions also occur outside of coding regions; these are not called frameshift mutations, but can have effects on the phenotype—for instance, they can affect how a gene is expressed.

New Mutations and Mutagens

Mutations can arise in a number of ways. Errors during DNA replication or recombination can lead to nucleotide-pair substitutions, insertions, or deletions, as well as to mutations affecting longer stretches of DNA. If an incorrect nucleotide is added to a growing chain during replication, for example, the base on that nucleotide will then be mismatched with the nucleotide base on the other strand. In many cases, the error will be corrected by DNA proofreading and repair systems (see Concept 16.2). Otherwise, the incorrect base will be used as a template in the next round of replication, resulting in a mutation. Such mutations are called *spontaneous mutations*. It is difficult to calculate the rate at which such mutations occur. Rough estimates have been made of the rate of mutation during DNA replication for both *E. coli* and eukaryotes, and the numbers are similar: About one nucleotide in every 10^{10} is altered, and the change is passed on to the next generation of cells.

A number of physical and chemical agents, called **mutagens**, interact with DNA in ways that cause mutations. In the 1920s, Hermann Muller discovered that X-rays caused genetic changes in fruit flies, and he used X-rays to make *Drosophila* mutants for his genetic studies (just as Beadle and Tatum did with their *Neurospora* cells). But Muller also recognised an alarming implication of his discovery: X-rays and other forms of high-energy radiation pose hazards to the genetic material of people as well as laboratory organisms. Mutagenic radiation, a physical mutagen, includes ultraviolet (UV) light, which can cause disruptive thymine dimers in DNA (see Figure 16.20).

Chemical mutagens fall into several categories. Nucleotide analogs are chemicals similar to normal DNA nucleotides but that pair incorrectly during DNA replication. Other chemical mutagens interfere with correct DNA replication by inserting themselves into the DNA and distorting the double helix. Still other mutagens cause chemical changes in bases that change their pairing properties.

Researchers have developed a variety of methods to test the mutagenic activity of chemicals. A major application of these tests is the preliminary screening of chemicals to identify those that may cause cancer. This approach makes sense because most carcinogens (cancer-causing chemicals) are mutagenic, and conversely, most mutagens are carcinogenic.

Using CRISPR to Edit Genes and Correct Disease-Causing Mutations

Ever since biologists have understood how disease-causing proteins were the result of mutations in genes, they have sought techniques for **gene editing**—altering genes in a specific, predictable way. Their aim has been to use such a technique to change specific genes in living cells, in part to try to correct genes that cause disease. Over the past 15 years, biologists have developed a powerful new technique for gene editing, called the **CRISPR-Cas9 system**, that is transforming the field of *genetic engineering*, the direct manipulation of genes for practical purposes. Cas9 is a bacterial protein that helps defend bacteria against the viruses that infect them (bacteriophages). In bacterial cells, Cas9 acts together with "guide RNA" made from the CRISPR region of the bacterial genome. (Figure 19.9 explains how this defence system works in bacteria.)

Similar to other DNA-cutting enzymes involved in DNA repair (described in Concept 16.2), Cas9 is a nuclease that cuts double-stranded DNA molecules. The power of Cas9 for gene editing is that the Cas9 protein will cut any sequence to which it is directed. Cas9 is directed to its target by a guide RNA molecule that it binds and uses as a homing device, cutting both strands of any DNA sequence that is complementary to the guide RNA. Scientists have been able to exploit the function of Cas9 by introducing a Cas9–guide RNA complex into a cell they wish to alter **(Figure 17.28)**. The guide RNA in the complex is engineered to be complementary to the "target" gene. Cas9 cuts both strands of the target DNA, and the resulting broken ends of DNA trigger a DNA repair system (similar to that shown in Figure 16.20). When there is no undamaged DNA for the enzymes of the repair system to use as a template, as shown at the bottom left of Figure 17.28, the repair enzymes introduce or remove random nucleotides while rejoining the ends. Generally, this process alters the DNA sequence so that the gene no longer works properly. This technique is a highly successful way for researchers to "knock out" (disable) a given gene to study what that gene does in an organism.

But what about using this system to help treat genetic diseases? Researchers have modified the technique so that the CRISPR-Cas9 system can be used to repair a gene that has a harmful mutation. They introduce a segment from the normal (functional) gene along with the CRISPR-Cas9 system. After Cas9 cuts the target DNA, repair enzymes can use the normal DNA as a template to repair the target DNA at the break point. In this way, the CRISPR-Cas9 system edits the defective gene so that it is corrected (see the bottom right of Figure 17.28).

In 2018, researchers reported promising results using the CRISPR-Cas 9 system in an attempt to correct the genetic defect that causes sickle-cell disease. They edited human cells from patients with sickle-cell disease and injected them into bone marrow in mice. After 19 weeks, the gene remained corrected in 20–40% of the injected cells. Researchers, physicians, and patients alike are excited about the potential of CRISPR technology to treat or even cure human diseases that have a genetic basis, such as sickle-cell, Alzheimer's, and Parkinson's diseases, as well as some types of cancer. However, there are still concerns about using CRISPR-treated

▼ Figure 17.28 Gene editing using the CRISPR-Cas9 system.

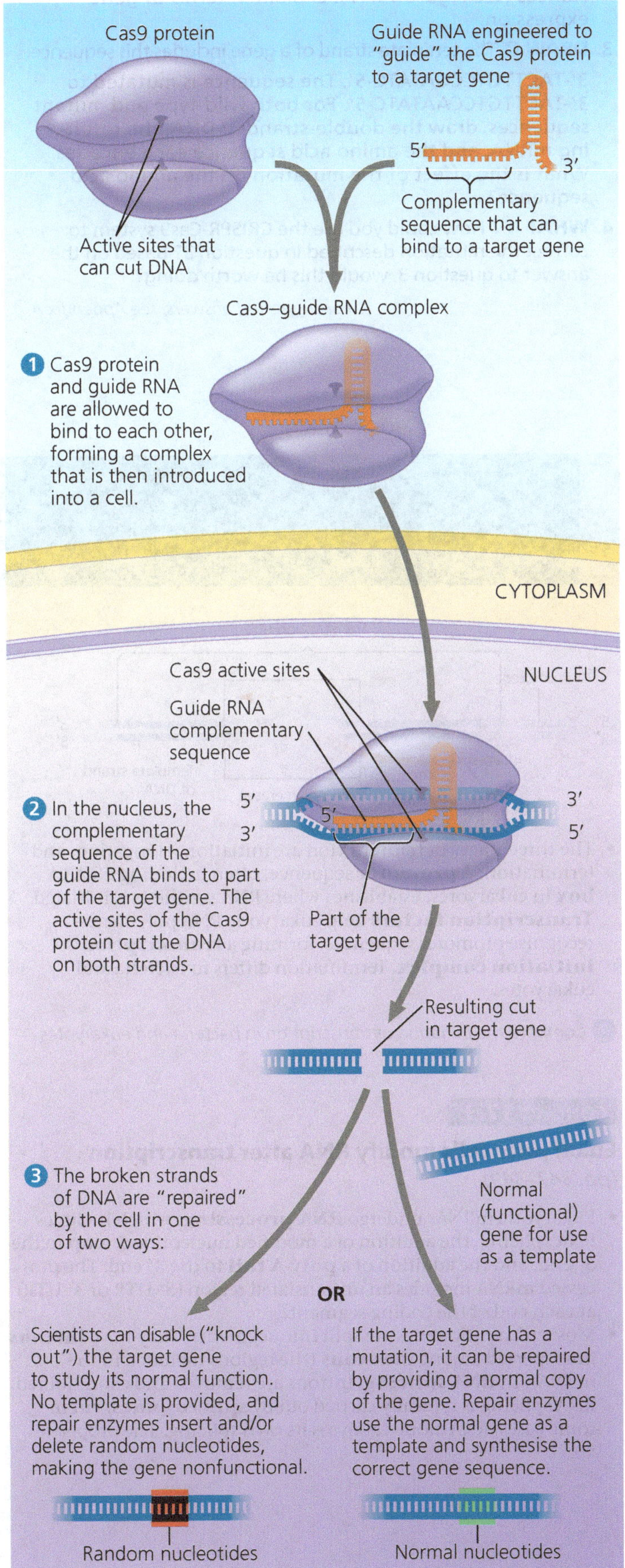

cells in humans because of possible effects on genes that are not being targeted; this is an active area of research.

Jennifer Doudna, a co-discoverer of CRISPR-Cas9, recognised not only its incredible potential but also the danger of its misapplication. After having a dream in which Adolf Hitler asked her what could be done with this gene-editing tool, Doudna realised it was imperative to spend some time reflecting on ethical considerations. In 2015, she convened a conference of biologists who agreed at the end of the meeting to urge extreme caution as the field moves forwards—and the discussion continues.

What Is a Gene? *Revisiting the Question*

Our definition of a gene has evolved over the past few chapters, as it has through the history of genetics. We began with the Mendelian concept of a gene as a discrete unit of inheritance that affects a phenotypic character ("Mendel and the Gene Idea"). We saw that Morgan and his colleagues assigned such genes to specific loci on chromosomes ("The Chromosomal Basis of Inheritance"). We went on to view a gene as a region of specific nucleotide sequence along the length of the DNA molecule of a chromosome ("The Molecular Basis of Inheritance"). Finally, in this chapter, we have considered a functional definition of a gene as a DNA sequence that codes for a specific polypeptide chain or a functional RNA molecule, such as a tRNA. All these definitions are useful, depending on the context in which genes are being studied.

We have noted that merely saying a gene codes for a polypeptide is an oversimplification. Most eukaryotic genes contain noncoding segments (such as introns), so large portions of these genes have no corresponding segments in polypeptides. Molecular biologists also often include promoters and certain other regulatory regions of DNA within the boundaries of a gene. These DNA sequences are not transcribed, but they can be considered part of the functional gene because they must be present for transcription to occur. Our definition of a gene must also be broad enough to include the DNA that is transcribed into rRNA, tRNA, and other RNAs that are not translated. These genes have no polypeptide products but play crucial roles in the cell. Thus, we arrive at the following definition: *A gene is a region of DNA that can be expressed to produce a final functional product that is either a polypeptide or an RNA molecule.*

When considering phenotypes, however, it is often useful to start by focusing on genes that code for polypeptides. In this chapter, you have learned in molecular terms how a typical gene is expressed—by transcription into RNA and then translation into a polypeptide that forms a protein of specific structure and function. Proteins, in turn, bring about an organism's observable phenotype.

A given type of cell expresses only a subset of its genes. This is an essential feature in multicellular organisms: You'd be in trouble if the lens cells in your eyes started expressing the genes for hair proteins, which are normally expressed only in hair follicle cells! Gene expression is precisely regulated, which we'll explore in the next chapter, beginning with the simpler case of bacteria and continuing with eukaryotes.

CONCEPT CHECK 17.5

1. What happens when one nucleotide pair is lost from the middle of the coding sequence of a gene?
2. **MAKE CONNECTIONS** Individuals heterozygous for the sickle-cell allele are generally healthy but show phenotypic effects of the allele under some circumstances (see Figure 14.17). Explain in terms of gene expression.
3. **DRAW IT** The template strand of a gene includes this sequence: 3′-TACTTGTCCGATATC-5′. The sequence is mutated to 3′-TACTTGTCCAATATC-5′. For both wild-type and mutant sequences, draw the double-stranded DNA, the resulting mRNA, and the amino acid sequence each encodes. What is the effect of the mutation on the amino acid sequence?
4. **WHAT IF?** How could you use the CRISPR-Cas9 system to correct the mutation described in question 3? Based on the answer to question 3, would this be worth doing?

For suggested answers, see Appendix A.

17 Chapter Review

SUMMARY OF KEY CONCEPTS

CONCEPT 17.1

Genes specify proteins via transcription and translation *(pp. 338–344)*

- Beadle and Tatum's studies of mutant strains of *Neurospora* led to the one gene–one polypeptide hypothesis. During **gene expression**, the information encoded in genes is used to make specific polypeptide chains (enzymes and other proteins) or RNA molecules.
- **Transcription** is the synthesis of RNA complementary to a **template strand** of DNA. **Translation** is the synthesis of a polypeptide whose amino acid sequence is specified by the nucleotide sequence in **messenger RNA (mRNA)**.
- Genetic information is encoded as a sequence of nonoverlapping nucleotide triplets, or **codons**. A codon in mRNA either is translated into an amino acid (61 of the 64 codons) or serves as a stop signal (3 codons). Codons must be read in the correct **reading frame**.

? *Describe the process of gene expression, by which a gene affects the phenotype of an organism.*

CONCEPT 17.2

Transcription is the DNA-directed synthesis of RNA: *A Closer Look* *(pp. 344–346)*

- RNA synthesis is catalysed by **RNA polymerase**, which links together RNA nucleotides complementary to a DNA template strand. Transcription follows the same base-pairing rules as DNA replication, except that in RNA, uracil (U) substitutes for thymine (T).

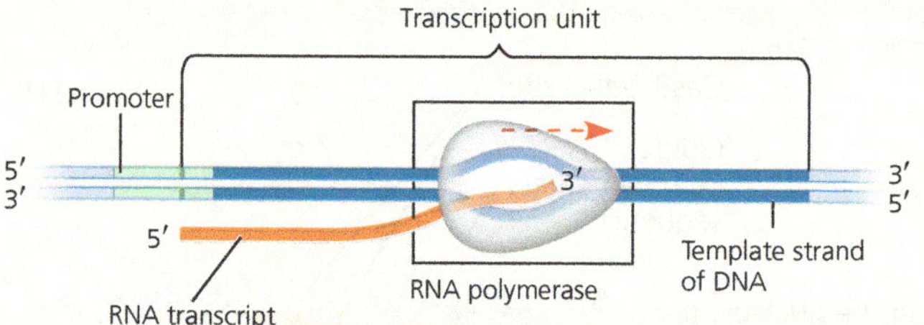

- The three stages of transcription are initiation, elongation, and termination. A **promoter** sequence, often including a **TATA box** in eukaryotes, establishes where RNA synthesis is initiated. **Transcription factors** help eukaryotic RNA polymerase recognise promoter sequences, forming a **transcription initiation complex**. Termination differs in bacteria and eukaryotes.

? *Compare the initiation of transcription in bacteria and eukaryotes.*

CONCEPT 17.3

Eukaryotic cells modify RNA after transcription *(pp. 347–349)*

- Eukaryotic mRNAs undergo **RNA processing**, which includes RNA splicing, the addition of a modified nucleotide **5′ cap** to the 5′ end, and the addition of a **poly-A tail** to the 3′ end. The processed mRNA includes an untranslated region (5′ UTR or 3′ UTR) at each end of the coding segment.
- Most eukaryotic genes are split into segments: They have **introns** interspersed among the **exons** (the regions included in the mRNA). In **RNA splicing**, introns are removed and exons joined. RNA splicing is typically carried out by **spliceosomes**, but in some cases, RNA alone catalyses its own splicing. The properties

of RNA allow some RNAs (called **ribozymes**) to act as catalysts. The presence of introns allows for **alternative RNA splicing**.

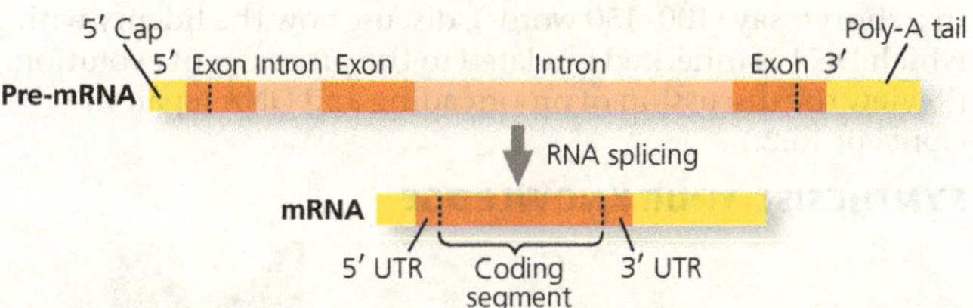

? *What function do the 5′ cap and the poly-A tail serve on a eukaryotic mRNA?*

CONCEPT 17.4

Translation is the RNA-directed synthesis of a polypeptide: *A Closer Look* (pp. 349–358)

- A cell translates an mRNA message into protein using **transfer RNAs (tRNAs)**. After being bound to a specific amino acid by an **aminoacyl-tRNA synthetase**, a tRNA lines up via its **anticodon** at the complementary codon on mRNA. A **ribosome**, made up of **ribosomal RNAs (rRNAs)** and proteins, facilitates this coupling with binding sites for mRNA and tRNA.
- Ribosomes coordinate the three stages of translation: initiation, elongation, and termination. The formation of peptide bonds between amino acids is catalysed by rRNAs as tRNAs move through the **A** and **P sites** and exit through the **E site**.

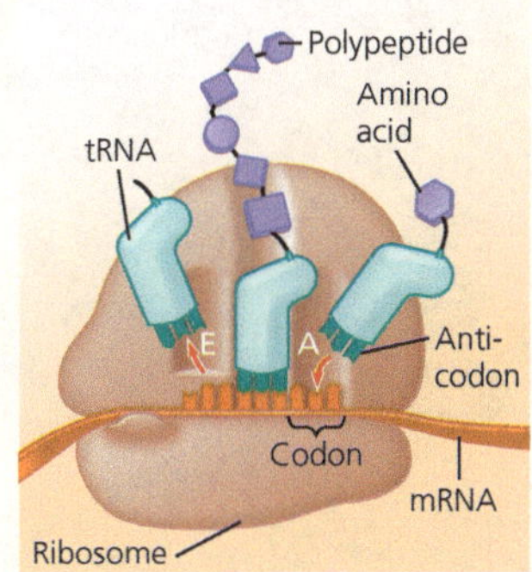

- After translation, during protein processing, proteins may be modified by cleavage or by attachment of sugars, lipids, phosphates, or other chemical groups.
- Free ribosomes in the cytosol initiate synthesis of all proteins, but proteins with a **signal peptide** are synthesised on the ER.
- A gene can be transcribed by multiple RNA polymerases simultaneously. Also, a single mRNA molecule can be translated simultaneously by a number of ribosomes, forming a **polyribosome**. In bacteria, these processes are coupled, but in eukaryotes they are separated in space and time by the nuclear membrane.

? *How do tRNAs function in the context of the ribosome during translation?*

CONCEPT 17.5

Mutations of one or a few nucleotides can affect protein structure and function (pp. 359–364)

- Small-scale **mutations** include **point mutations**, changes in one DNA nucleotide pair, which may lead to production of nonfunctional proteins. **Nucleotide-pair substitutions** can cause **missense** or **nonsense mutations**. Nucleotide-pair **insertions** or **deletions** may produce **frameshift mutations**.
- Spontaneous mutations can occur during DNA replication, recombination, or repair. Chemical and physical **mutagens** cause DNA damage that can alter genes.
- The **CRISPR-Cas9 system** is a powerful new **gene-editing** technique with the potential to correct genetic mutations that cause disease. Significant technical and ethical questions have been raised about how this could or should be used.

? *What will be the results of chemically modifying one nucleotide base of a gene? What role is played by DNA repair systems in the cell?*

TEST YOUR UNDERSTANDING

Levels 1-2: Remembering/Understanding

1. In eukaryotic cells, transcription cannot begin until
(A) the two DNA strands have completely separated and exposed the promoter.
(B) several transcription factors have bound to the promoter.
(C) the 5′ caps are removed from the mRNA.
(D) the DNA introns are removed from the template.

2. Which of the following is true of a codon?
(A) It never codes for the same amino acid as another codon.
(B) It can code for more than one amino acid.
(C) It can be either in DNA or in RNA
(D) It is the basic unit of protein structure.

3. The anticodon of a particular tRNA molecule is
(A) complementary to the corresponding mRNA codon.
(B) complementary to the corresponding triplet in rRNA.
(C) the part of tRNA that bonds to a specific amino acid.
(D) catalytic, making the tRNA a ribozyme.

4. Which of the following is true of RNA processing?
(A) Exons are cut out before mRNA leaves the nucleus.
(B) Nucleotides are added at both ends of the RNA.
(C) Ribozymes may function in the addition of a 5′ cap.
(D) RNA splicing adds a poly-A tail to the mRNA.

5. Which component is directly involved in translation?
(A) RNA polymerase
(B) ribosome
(C) spliceosome
(D) DNA

Levels 3-4: Applying/Analysing

6. Using Figure 17.6, identify a 5′ → 3′ sequence of nucleotides in the DNA template strand for an mRNA coding for the polypeptide sequence Phe-Pro-Lys.
(A) 5′-UUUCCCAAA-3′
(B) 5′-GAACCCCTT-3′
(C) 5′-CTTCGGGAA-3′
(D) 5′-AAACCCUUU-3′

7. Which of the following mutations would be *most* likely to have a harmful effect on an organism?
(A) a deletion of three nucleotides near the middle of a gene
(B) a single nucleotide deletion in the middle of an intron
(C) a single nucleotide deletion near the end of the coding sequence
(D) a single nucleotide insertion downstream of, and close to, the start of the coding sequence

8. Would the coupling of the processes shown in Figure 17.24 be found in a eukaryotic cell? Explain why or why not.

9. Complete the following table:

Type of RNA	Functions
Messenger RNA (mRNA)	
Transfer RNA (tRNA)	
	In a ribosome, plays a structural role; as a ribozyme, plays a catalytic role (catalyses peptide bond formation)
Primary transcript	
Small RNAs in the spliceosome	

Levels 5-6: Evaluating/Creating

10. **EVOLUTION CONNECTION** Most amino acids are coded for by a set of similar codons (see Figure 17.6). Propose at least one evolutionary explanation to account for this pattern.

11. **SCIENTIFIC INQUIRY** Knowing that the genetic code is almost universal, a scientist uses molecular biological methods to insert the human β-globin gene (shown in Figure 17.12) into bacterial cells, hoping the cells will express it and synthesise functional β-globin protein. Instead, the protein produced is nonfunctional and is found to contain many fewer amino acids than does β-globin made by a eukaryotic cell. Explain why.

12. **WRITE ABOUT A THEME: INFORMATION** Evolution accounts for the unity and diversity of life, and the continuity of life is based on heritable information in the form of DNA. In a short essay (100–150 words), discuss how the fidelity with which DNA is inherited is related to the processes of evolution. (Review the discussion of proofreading and DNA repair in Concept 16.2.)

13. **SYNTHESISE YOUR KNOWLEDGE**

Some mutations result in proteins that function well at one temperature but are nonfunctional at a different (usually higher) temperature. Siamese cats have such a "temperature-sensitive" mutation in a gene encoding an enzyme that makes dark pigment in the fur. The mutation results in the breed's distinctive point markings and lighter body colour (see the photo). Using this information and what you learned in the chapter, explain the pattern of the cat's fur pigmentation.

For selected answers, see Appendix A.

This page is intentionally blank.

19 Viruses

KEY CONCEPTS

Study Tip

Make flowcharts: Make a simple flowchart for each viral replicative cycle in this chapter. Below is the beginning of an example for the lysogenic cycle. Note the similarities and differences between the replicative cycles.

Lysogenic cycle

Virus attaches to bacterial cell and injects genome.

↓

Viral genome is integrated into the bacterial chromosome.

↓

Go to Mastering Biology

to access Dynamic Study Modules for revision, 3D BioFlix® animations and high-quality videos, and your interactive Pearson eText.

Figure 19.1 A human cell (deep sea-green colour) infected with severe acute respiratory syndrome coronavirus 2 (SARS-CoV-2). The virus that causes the COVID-19 disease takes over a cell and produces more SARS-CoV-2 viruses (magenta dots) that infect other cells. Cumulatively, the infection produces a severe immune response that leads to progressive failures of individual organs and organ systems. Around 2–3% of infected individuals die. Many survivors continue to experience debilitating illnesses.

How does a virus make more viruses?

A virus consists only of nucleic acid, proteins, and sometimes a membranous envelope. After infecting a host cell, it uses the host cell's molecules to make new viruses, as shown in this simplified diagram.

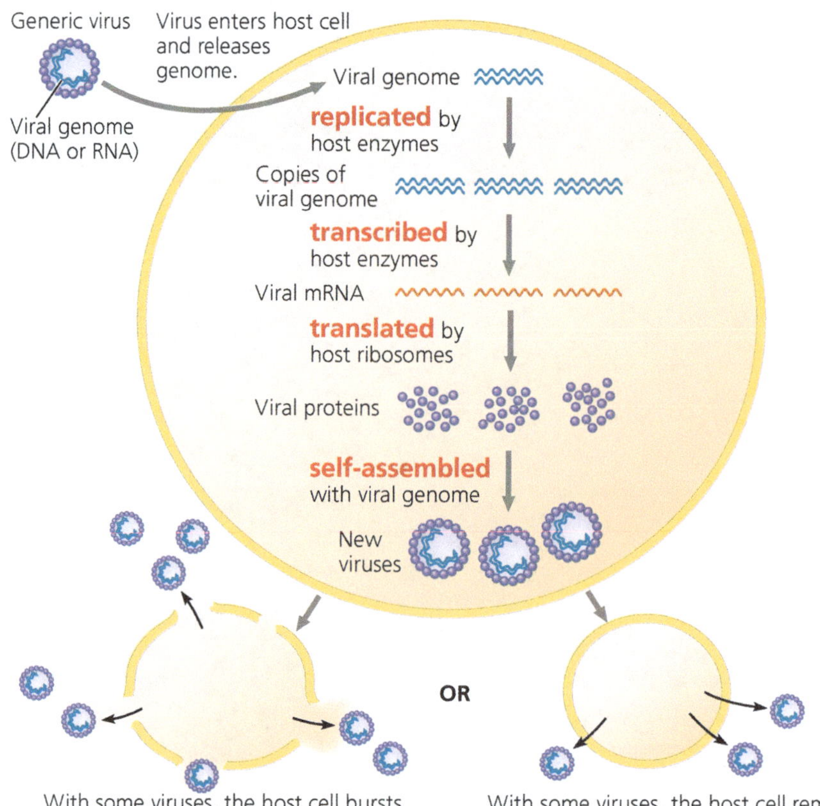

CONCEPT 19.1

A virus consists of a nucleic acid surrounded by a protein coat

Compared with eukaryotic and even prokaryotic cells, most viruses are much smaller and simpler in structure. Lacking the structures and metabolic machinery found in a cell, a **virus** is an infectious particle consisting of little more than genes packaged in a protein coat.

Are viruses living or nonliving? Early on, they were considered biological chemicals; the Latin root for *virus* means "poison." Viruses can cause a wide variety of diseases, so researchers in the late 1800s saw a parallel with bacteria and proposed that viruses were the simplest of living forms. However, viruses cannot reproduce or carry out metabolic activities outside of a host cell. Most biologists would probably agree that viruses are not alive, but instead exist in a shady area between life-forms and chemicals. The simple phrase used by two researchers describes them aptly: Viruses lead "a kind of borrowed life."

If we apply the informal meaning of the term, "going viral," we can describe what a borrowed life can do. "Emergent viruses" seem to appear suddenly, as if from nowhere, and their novelty may render susceptible to disease very large numbers of organisms. A classic, human-centred example of an emergent virus occurred in March 1918, when American biologists first observed the H1N1 virus. Within a year, the virus had infected more than a third of Earth's human population. Without either a vaccine for the virus or anti-bacterials (the discovery of penicillin waited for another 11 years) to treat secondary symptoms associated with the disease, the pandemic killed more than 50 million people.

Over a century after the 1918–1919 pandemic, the severe acute respiratory syndrome coronavirus 2 (SARS-CoV-2) produced the coronavirus disease 2019 (COVID-19). First recognised in Wuhan, China in December 2019, it seemingly appeared out of nowhere. In early March 2020, the World Health Organization (WHO) declared COVID-19 to be a pandemic that reached into every country in the world (**Figure 19.2**). Without the early implementation of public health control measures in many countries, the world would have experienced much higher rates of infection. Arguably, nowhere was this truer than in New Zealand. In both 1918 and during the 2019–2021 pandemic, the disease spread slowly in those nations that required people to wear masks and quarantined infected people. In these countries, the populations became used to the establishment and maintenance of social distance, enhanced personal hygiene, and a ban on large public gatherings.

Much about the COVID-19 disease remains unknown, although, given the worldwide research effort, our knowledge quickly expanded to the point where the rapid development of vaccines occurred. Biologists' pursuit of deeper understandings of the virus have produced swift advances. To research the latest understandings about the virus, cases, its causes, information about the disease's progression, and other information relevant to your studies, please consult the reputable websites listed in **Figure 19.3**.

▼ **Figure 19.2 WHO Coronavirus Disease (COVID-19) Dashboard.** The map visualises the number of COVID-19 cases reported to the World Health Organization (WHO) from nations all over the world.

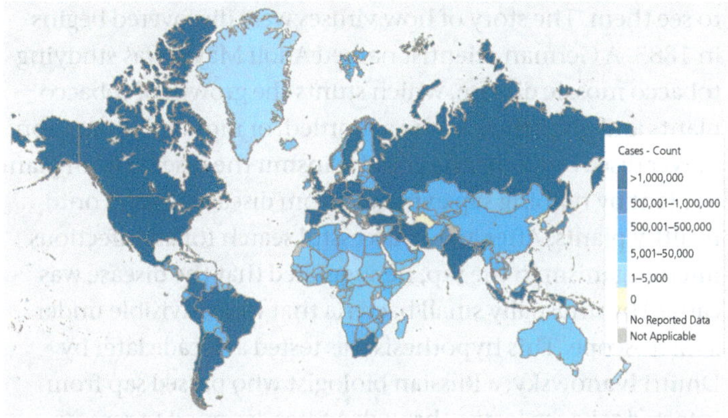

▼ **Figure 19.3 Australian and New Zealand Government health websites.** These provide nation-specific details of the current progression of the COVID-19 pandemic. Each site contains information regarding the current status of the disease, and provides data for analysis and up-to-date information about the progression of the disease in our respective nations.

(a) Australian Department of Health: www.health.gov.au. Go to: Health topics > Coronavirus (COVID-19).

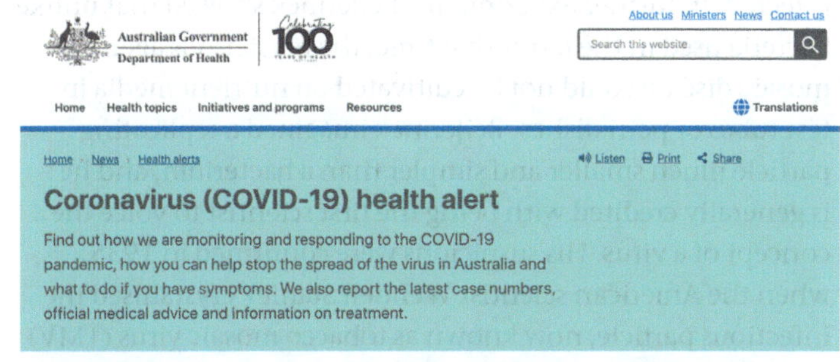

(b) New Zealand Ministry of Health: www.health.govt.nz. Go to: Our work > COVID-19 (novel coronavirus).

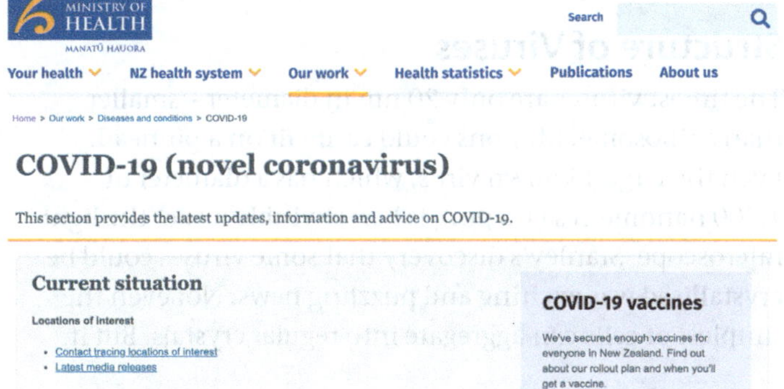

The WHO also provides detailed reports on current and past actions, with up-to-date information about the international scope of the virus, rates of infections, fatalities, discussions, and advice to the public. The data is available for download and personal analysis at: https://covid19.who.int.

In this chapter, we focus on the biology of viruses, so that you can gain knowledge of this important group of "borrowed lives," of which SARS-CoV-2 represents only one.

The Discovery of Viruses: *Scientific Inquiry*

Scientists detected viruses indirectly long before they were able to see them. The story of how viruses were discovered begins in 1883. A German scientist named Adolf Mayer was studying tobacco mosaic disease, which stunts the growth of tobacco plants and gives their leaves a mottled, or mosaic, colouration. Mayer discovered that he could transmit the disease from plant to plant by rubbing sap extracted from diseased leaves onto healthy plants. After an unsuccessful search for an infectious microorganism in the sap, he suggested that the disease was caused by unusually small bacteria that were invisible under a microscope. This hypothesis was tested a decade later by Dmitri Ivanowsky, a Russian biologist who passed sap from infected tobacco leaves through a filter designed to remove bacteria. After filtration, the sap still produced mosaic disease.

But Ivanowsky reasoned that perhaps the bacteria were small enough to pass through the filter or made a toxin that could do so. The second possibility was ruled out when the Dutch botanist Martinus Beijerinck carried out a classic series of experiments that showed that the infectious agent in the filtered sap could replicate **(Figure 19.4)**.

▼ Figure 19.4 Inquiry

What causes tobacco mosaic disease?

Experiment In the late 1800s, Martinus Beijerinck, of the Technical School in Delft, the Netherlands, investigated the properties of the agent that causes tobacco mosaic disease (then called spot disease).

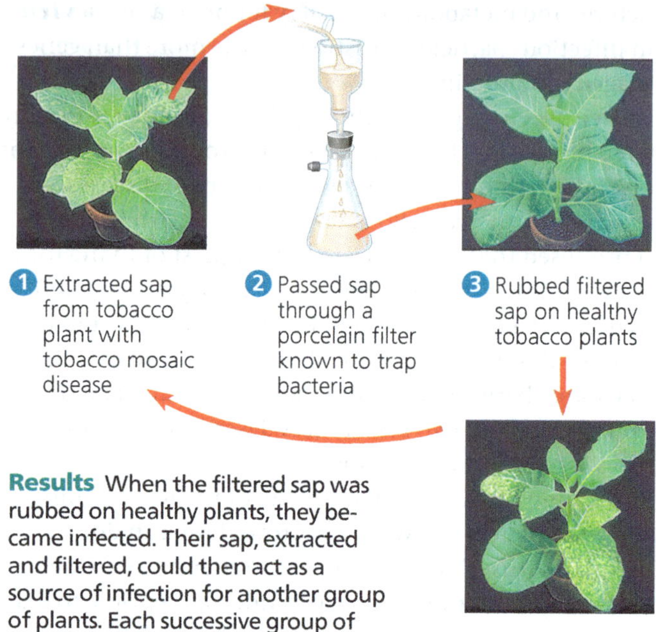

Results When the filtered sap was rubbed on healthy plants, they became infected. Their sap, extracted and filtered, could then act as a source of infection for another group of plants. Each successive group of plants developed the disease to the same extent as earlier groups.

Conclusion The infectious agent was apparently not a bacterium because it could pass through a bacterium-trapping filter. The pathogen must have been replicating in the plants because its ability to cause disease was undiluted after several transfers from plant to plant.

Data from M. J. Beijerinck, Concerning a *contagium vivum fluidum* as cause of the spot disease of tobacco leaves, *Verhandelingen der Koninkyke akademie Wettenschappen te Amsterdam* 65:3–21 (1898). Translation published in English as Phytopathological Classics Number 7 (1942), American Phytopathological Society Press, St. Paul, MN.

WHAT IF? *If Beijerinck had observed that the infection of each group was weaker than that of the previous group and that ultimately the sap could no longer cause disease, what might he have concluded?*

In fact, the pathogen replicated only within the host it infected. In further experiments, Beijerinck showed that unlike bacteria used in the lab at that time, the mysterious agent of mosaic disease could not be cultivated on nutrient media in test tubes or petri dishes. Beijerinck imagined a replicating particle much smaller and simpler than a bacterium, and he is generally credited with being the first scientist to voice the concept of a virus. His suspicions were confirmed in 1935 when the American scientist Wendell Stanley crystallised the infectious particle, now known as tobacco mosaic virus (TMV). Subsequently, TMV and many other viruses were actually seen with the help of the electron microscope.

Structure of Viruses

The tiniest viruses are only 20 nm in diameter—smaller than a ribosome. Millions could easily fit on a pinhead. Even the largest known virus, which has a diameter of 1,500 nanometres (1.5 μm), is barely visible under the light microscope. Stanley's discovery that some viruses could be crystallised was exciting and puzzling news. Not even the simplest of cells can aggregate into regular crystals. But if viruses are not cells, then what are they? Examining the structure of a virus more closely reveals that it is an infectious particle consisting of one or more molecules of a nucleic acid enclosed in a protein coat and, for some viruses, surrounded by a membranous envelope.

The simple structure of viruses make them a useful biological system: To a large extent, molecular biology was born in the laboratories of biologists studying viruses that infect bacteria. Experiments using these viruses provided evidence that genes are made of nucleic acids, and they were critical in working out the molecular mechanisms of the fundamental processes of DNA replication, transcription, and translation.

Viral Genomes

We usually think of genes as being made of double-stranded DNA, but many viruses defy this convention. Their genomes may consist of double-stranded DNA, single-stranded DNA, double-stranded RNA, or single-stranded RNA, depending on the type of virus. A virus is called a DNA virus or an RNA virus based on the kind of nucleic acid that makes up its genome. In either case, the genome is usually organised as a single linear or circular molecule of nucleic acid, although the genomes of some viruses consist of multiple molecules of nucleic acid. The smallest viruses known have only three genes in their genome, while the largest have several hundred to 2,000. For comparison, bacterial genomes contain about 200 to a few thousand genes.

Capsids and Envelopes

The protein shell enclosing the viral genome is called a **capsid**. Depending on the type of virus, the capsid may be rod-shaped, polyhedral, or more complex in shape. Capsids are built from a large number of protein subunits called *capsomeres*, but the number of different *kinds* of proteins in a capsid is usually small. Tobacco mosaic virus has a rigid, rod-shaped capsid made from over 1,000 molecules of a single type of protein arranged in a helix; rod-shaped viruses are commonly called *helical viruses* for this reason **(Figure 19.5a)**. Adenoviruses, which infect the respiratory tracts of animals, have 252 identical protein molecules arranged in a polyhedral capsid with 20 triangular facets—an icosahedron; thus, these and other similarly shaped viruses are referred to as *icosahedral viruses* **(Figure 19.5b)**.

Some viruses have accessory structures that help them infect their hosts. For instance, a membranous envelope surrounds the capsids of influenza viruses and many other viruses found in animals **(Figure 19.5c)**. These **viral envelopes**, which are derived from the membranes of the host cell, contain host cell phospholipids and membrane proteins. They also contain proteins and glycoproteins of viral origin. (Glycoproteins are proteins with carbohydrates covalently attached.) Some viruses carry a few viral enzymes, such as viral polymerase, within their capsids.

▼ **Figure 19.5 Viral structure.** Viruses are made up of nucleic acid (DNA or RNA) enclosed in a protein coat (the capsid) and sometimes further wrapped in a membranous envelope. The individual protein subunits making up the capsid are called capsomeres. Although diverse in size and shape, viruses have many common structural features. (All micrographs are colourised TEMs.)

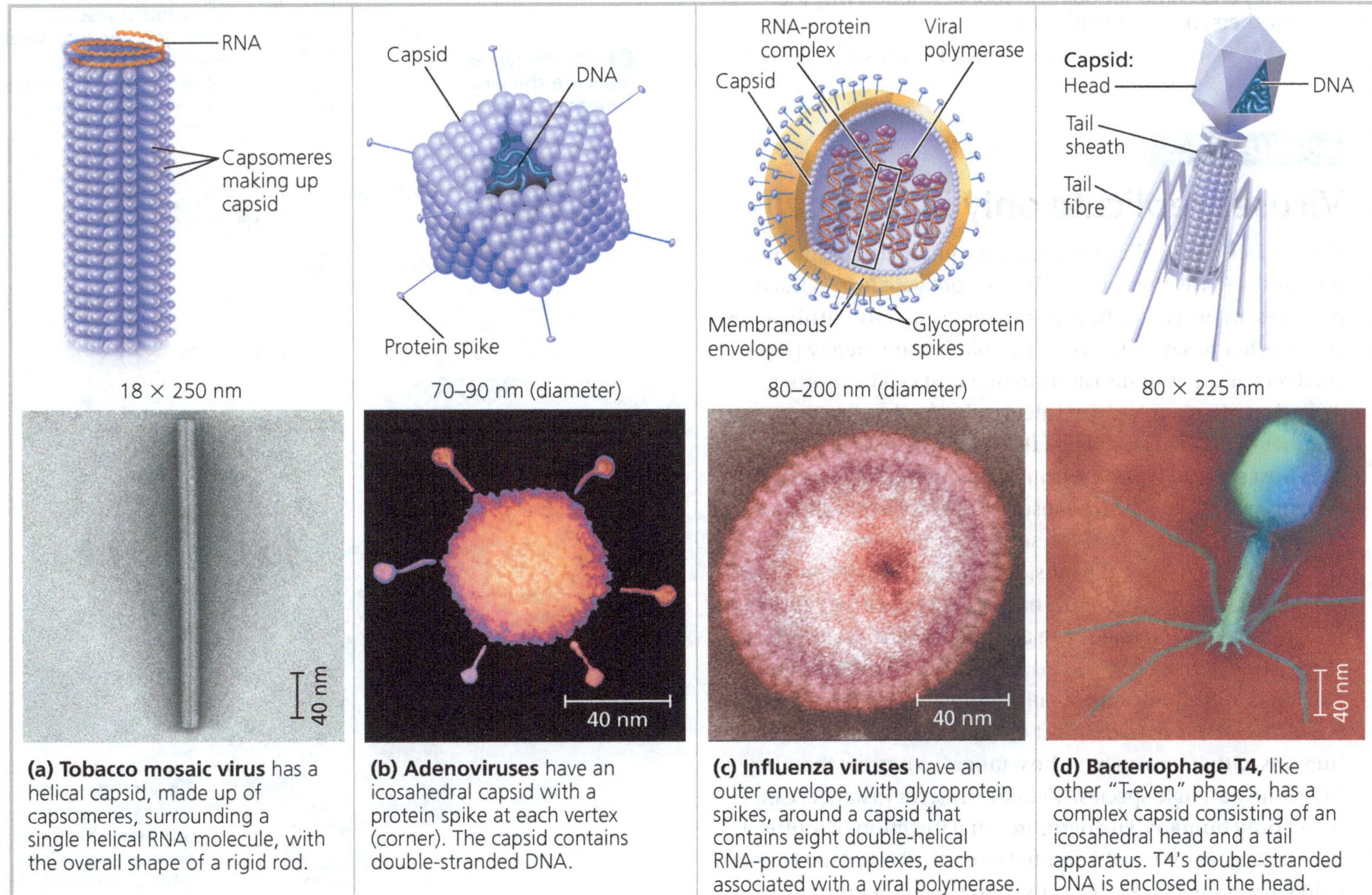

(a) Tobacco mosaic virus has a helical capsid, made up of capsomeres, surrounding a single helical RNA molecule, with the overall shape of a rigid rod.

(b) Adenoviruses have an icosahedral capsid with a protein spike at each vertex (corner). The capsid contains double-stranded DNA.

(c) Influenza viruses have an outer envelope, with glycoprotein spikes, around a capsid that contains eight double-helical RNA-protein complexes, each associated with a viral polymerase.

(d) Bacteriophage T4, like other "T-even" phages, has a complex capsid consisting of an icosahedral head and a tail apparatus. T4's double-stranded DNA is enclosed in the head.

Many of the most complex capsids are found among the viruses that infect bacteria, called **bacteriophages**, or simply **phages**. The first phages studied included seven that infect *Escherichia coli* (*E. coli*). These seven phages were named type 1 (T1), type 2 (T2), and so forth, in the order of their discovery. (The T2 phage was used in the experiment that established DNA as the genetic material; see Figure 16.4.) The three "T-even" phages (T2, T4, and T6) turned out to be very similar in structure. Their capsids have elongated icosahedral heads enclosing their DNA. Attached to the head is a protein tail piece with fibres by which the phages attach to a bacterial cell **(Figure 19.5d)**. In the next section, we'll examine how these few viral parts function together with cellular components to produce large numbers of viral progeny.

CONCEPT CHECK 19.1

1. **VISUAL SKILLS** Describe one similarity and two differences between the structures of tobacco mosaic virus (TMV) and influenza virus (see Figure 19.5).
2. **MAKE CONNECTIONS** Bacteriophages were used to provide evidence that DNA carries genetic information (see Figure 16.4). Briefly describe the experiment carried out by Hershey and Chase, including in your description why the researchers chose to use phages.

For suggested answers, see Appendix A.

CONCEPT 19.2

Viruses replicate only in host cells

Viruses lack metabolic enzymes and equipment for making proteins, such as ribosomes. They are obligate intracellular parasites; in other words, they can replicate only within a host cell. It is fair to say that viruses in isolation are merely packaged sets of genes in transit from one host cell to another.

Each particular virus can infect cells of only a limited number of host species, called the **host range** of the virus. This host specificity results from the evolution of recognition systems by the virus. Viruses usually identify host cells by a "handshake" fit between viral surface proteins and specific receptor molecules on the outside of cells. Such receptor molecules are most often proteins that carry out necessary functions for the host cell and have been adopted by viruses as portals of entry. Some viruses have broad host ranges. For example, West Nile virus and equine encephalitis virus are distinctly different viruses that can each infect mosquitoes, birds, horses, and humans. Other viruses have host ranges so narrow that they infect only a single species. Measles virus, for instance, can infect only humans. Furthermore, viral infection of multicellular eukaryotes is usually limited to particular tissues. Human cold viruses infect only the cells lining the upper respiratory tract, and the human immunodeficiency virus (HIV) binds to receptors present only on certain types of immune cells.

General Features of Viral Replicative Cycles

A viral infection begins when a virus binds to a host cell and the viral genome makes its way inside **(Figure 19.6)**. The mechanism of genome entry depends on the type of virus and the type of host cell. For example, T-even phages use their elaborate tail apparatus to inject DNA into a bacterium (see Figure 19.5d). Other viruses are taken up by endocytosis or, in the case of enveloped viruses, by fusion of the viral envelope with the host's plasma membrane. Once the viral genome is inside, the proteins it encodes can commandeer the host, reprogramming the cell to copy the viral genome and manufacture viral proteins. The host provides the nucleotides for making viral nucleic acids, as well as enzymes, ribosomes, tRNAs, amino acids, ATP, and other components needed for making the viral proteins.

▼ **Figure 19.6 A simplified viral replicative cycle.** A virus is an intracellular parasite that uses the equipment and small molecules of its host cell to replicate. In this simplest of viral cycles, the parasite is a DNA virus with a capsid consisting of a single type of protein.

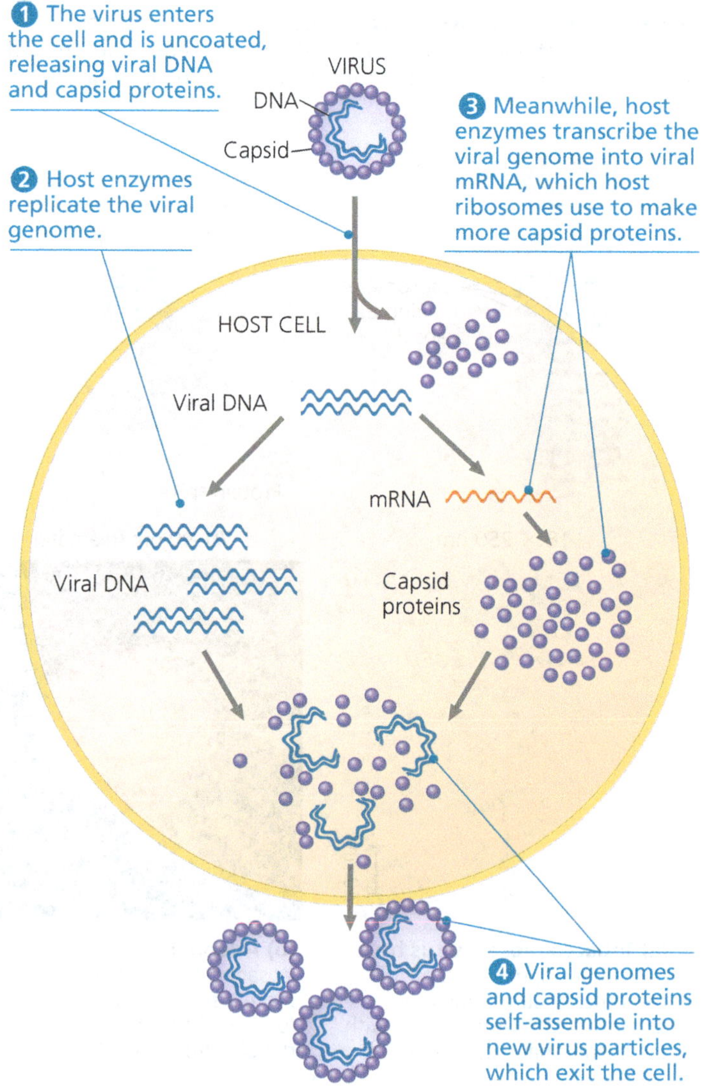

MAKE CONNECTIONS *Label each of the straight grey arrows to identify the process that is occurring. Review Figure 17.25.*

Many DNA viruses use the DNA polymerases of the host cell to synthesise new genomes along the templates provided by the viral DNA. In contrast, to replicate their genomes, RNA viruses use virally encoded RNA polymerases that can use RNA as a template. (Uninfected cells generally make no enzymes for carrying out this process.)

After the viral nucleic acid molecules and capsomeres are produced, they spontaneously self-assemble into new viruses. In fact, researchers can separate the RNA and capsomeres of TMV and then reassemble complete viruses simply by mixing the components together under the right conditions. The simplest type of viral replicative cycle ends with the exit of hundreds or thousands of viruses from the infected host cell, a process that often damages or destroys the cell. Such cellular damage and death, as well as the body's responses to this destruction, cause many of the symptoms associated with viral infections. The viral progeny that exit a cell have the potential to infect additional cells, spreading the viral infection.

There are many variations on the simplified viral replicative cycle we have just described. We will now take a look at some of these variations in bacterial viruses (phages) and animal viruses; later in the chapter, we will consider plant viruses.

Replicative Cycles of Phages

Phages are the best understood of all viruses, although some of them are also among the most complex. Research on phages led to the discovery that some double-stranded DNA viruses can replicate by two alternative mechanisms: the lytic cycle and the lysogenic cycle.

The Lytic Cycle

A phage replicative cycle that culminates in death of the host cell is known as a **lytic cycle**. The term *lytic* refers to the last stage of infection, during which the bacterium lyses (breaks open) and releases the phages that were produced within the cell. Each of these phages can then infect a healthy cell, and a few successive lytic cycles can destroy an entire bacterial population in just a few hours. A phage that replicates only by a lytic cycle is a **virulent phage**. Figure 19.7 illustrates the major steps in the lytic cycle of T4, a typical virulent phage.

The Lysogenic Cycle

Instead of lysing their host cells, many phages coexist with them in a state called lysogeny. In contrast to the lytic cycle, which kills the host cell, the **lysogenic cycle** allows

▶ **Figure 19.7 The lytic cycle of phage T4, a virulent phage.** Phage T4 has almost 300 genes, which are transcribed and translated using the host cell's machinery. One of the first phage genes translated after the viral DNA enters the host cell codes for an enzyme that degrades the host cell's DNA (step 2); the phage DNA is protected from breakdown because it contains a modified form of cytosine that is not recognised by the phage enzyme. The entire lytic cycle, from the phage's first contact with the cell surface to cell lysis, takes only 20–30 minutes at 37°C.

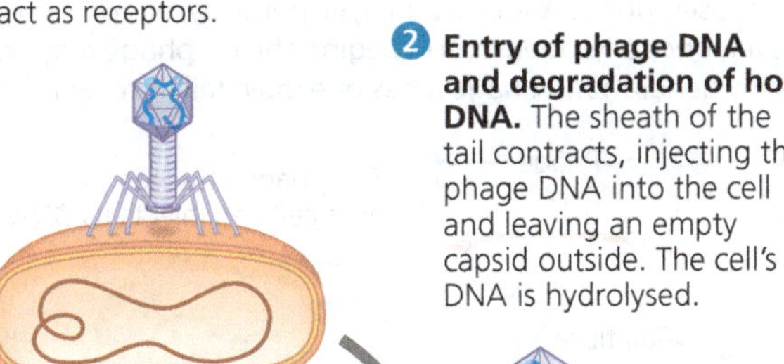

1 **Attachment.** The T4 phage uses its tail fibres to bind to specific surface proteins on an *E. coli* cell that act as receptors.

2 **Entry of phage DNA and degradation of host DNA.** The sheath of the tail contracts, injecting the phage DNA into the cell and leaving an empty capsid outside. The cell's DNA is hydrolysed.

3 **Synthesis of viral genomes and proteins.** The phage DNA directs production of phage proteins and copies of the phage genome by host and viral enzymes, using components within the cell.

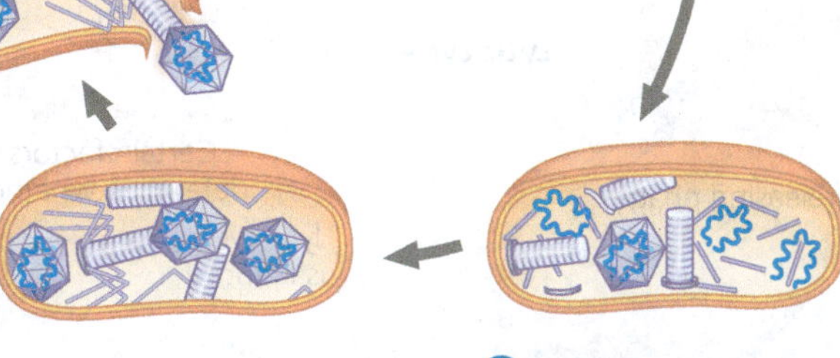

4 **Self-assembly.** Three separate sets of proteins self-assemble to form phage heads, tails, and tail fibres. The phage genome is packaged inside the capsid as the head forms.

5 **Release.** The phage directs production of an enzyme that damages the bacterial cell wall, allowing fluid to enter. The cell swells and finally bursts, releasing 100 to 200 phage particles.

replication of the phage genome without destroying the host. Phages capable of using both modes of replicating within a bacterium are called **temperate phages**. A temperate phage called lambda, written with the Greek letter λ, has been widely used in biological research. Phage λ resembles T4, but its tail has only one short tail fibre.

Infection of an *E. coli* cell by phage λ begins when the phage binds to the surface of the cell and injects its linear DNA genome **(Figure 19.8)**. Within the host, the λ DNA molecule forms a circle. What happens next depends on the replicative mode: lytic cycle or lysogenic cycle. During a lytic cycle, the viral genes immediately turn the host cell into a λ-producing factory, and the cell soon lyses and releases its virus progeny. During a lysogenic cycle, however, the λ DNA molecule is incorporated into a specific site on the *E. coli* chromosome by viral proteins that break both circular DNA molecules and join them to each other. When integrated into the bacterial chromosome in this way, the viral DNA is known as a **prophage**. One prophage gene codes for a protein that prevents transcription of most of the other prophage genes. Thus, the phage genome is mostly silent within the bacterium. Every time the *E. coli* cell prepares to divide, it replicates the phage DNA along with its own chromosome such that each daughter cell inherits a prophage. A single infected cell can quickly give rise to a large population of bacteria carrying the virus in prophage form. This mechanism enables viruses to propagate without killing the host cells on which they depend.

The term *lysogenic* signifies that prophages are capable of generating active phages that lyse their host cells. This occurs when the λ genome (or that of another temperate phage) is induced to exit the bacterial chromosome and initiate a lytic cycle. An environmental signal, such as a certain chemical or high-energy radiation, usually triggers the switchover from the lysogenic to the lytic mode.

In addition to the gene for the viral protein that prevents transcription, a few other prophage genes may be expressed during lysogeny. Expression of these genes may alter the host's phenotype, a phenomenon that can have important medical significance. For example, the three species of bacteria that cause the human diseases diphtheria, botulism, and scarlet fever would not be so harmful to humans without certain prophage genes that cause the host bacteria to make toxins. And the difference between the *E. coli* strain in our intestines and the O157:H7 strain that has caused several deaths by food poisoning appears to be the presence of toxin genes of prophages in the O157:H7 strain.

▼ **Figure 19.8 The lytic and lysogenic cycles of phage λ, a temperate phage.** After entering the bacterial cell and circularising, the λ DNA can immediately initiate the production of a large number of progeny phages (lytic cycle) or integrate into the bacterial chromosome (lysogenic cycle). In most cases, phage λ follows the lytic pathway, which is similar to that detailed in Figure 19.7. However, once a lysogenic cycle begins, the prophage may be carried in the host cell's chromosome for many generations. Phage λ has one main tail fibre, which is short.

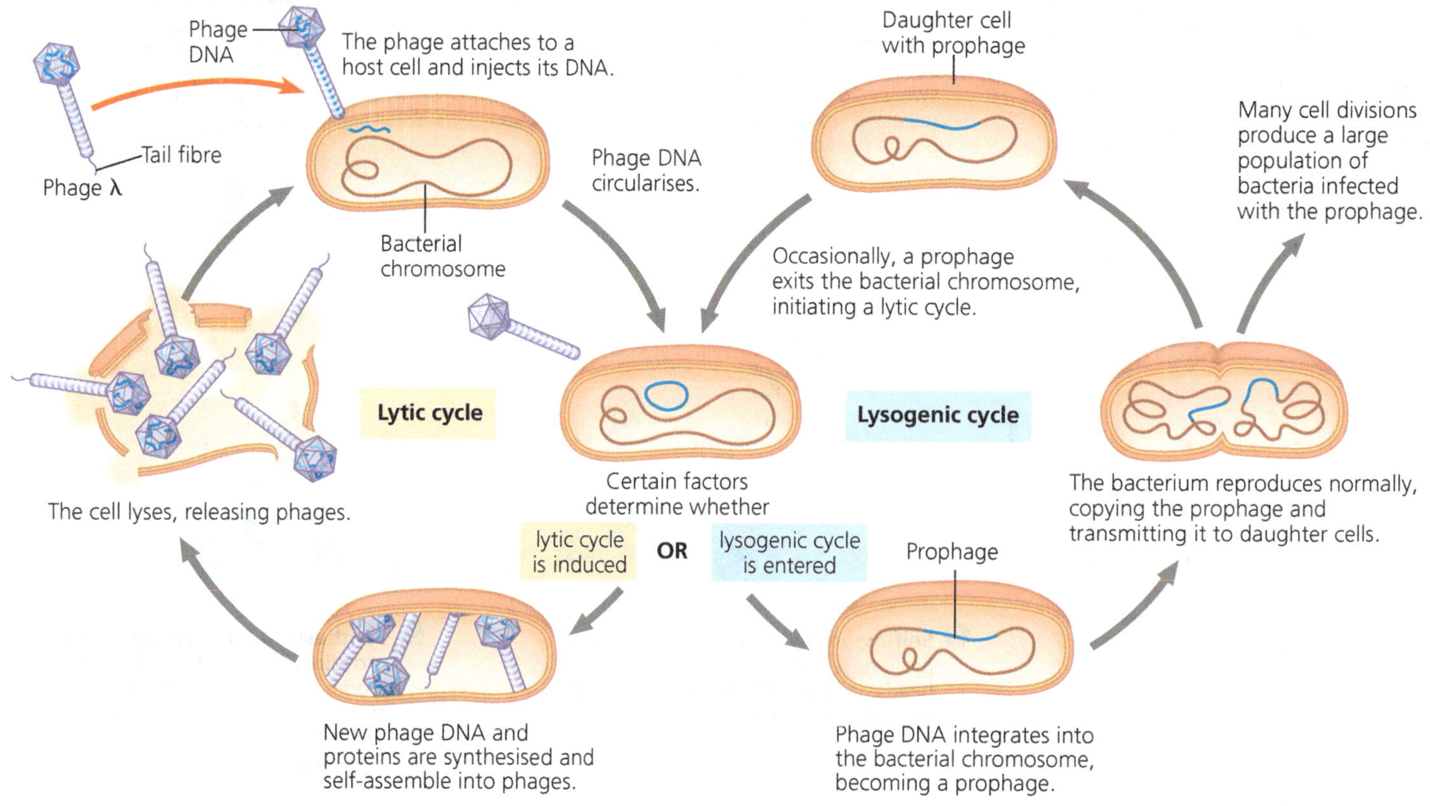

Bacterial Defences Against Phages

After reading about the lytic cycle, you may have wondered why phages haven't exterminated all bacteria. Lysogeny is one major reason why bacteria have been spared from extinction caused by phages. Bacteria also have their own defences against phages. First, natural selection favours bacterial mutants with surface proteins that are no longer recognised as receptors by a particular type of phage. Second, when phage DNA does enter a bacterium, the DNA is often identified as foreign and cut up by cellular enzymes called **restriction enzymes**, which are so named because they *restrict* a phage's ability to replicate within the bacterium. (Restriction enzymes are used in molecular biology and DNA cloning techniques; see Concept 20.1.) The bacterium's own DNA is methylated in a way that prevents attack by its own restriction enzymes. A third defence is a system present in both bacteria and archaea called the *CRISPR-Cas system*, which you learned about in Concept 17.5.

The CRISPR-Cas system was discovered during a study of repetitive DNA sequences present in the genomes of many prokaryotes. These sequences, which puzzled scientists, were named clustered regularly interspaced short palindromic repeats (CRISPRs) because each sequence reads the same forwards and backwards (a palindrome), with different stretches of "spacer DNA" in between the repeats. At first, scientists assumed the spacer DNA sequences were random and meaningless, but analysis by several research groups showed that each spacer sequence corresponded to DNA from a particular phage that had infected the cell. Further studies revealed that particular nuclease proteins interact with the CRISPR region. These nucleases, called Cas (CRISPR-associated) proteins, can identify and cut phage DNA, thereby defending the bacterium against phage infection.

When a phage infects a bacterial cell that has the CRISPR-Cas system, the DNA of the invading phage is stored, integrated into the genome between two repeat sequences (**Figure 19.9**). If the cell survives the infection, any further attempt by the same type of phage to infect this cell (or its offspring) triggers transcription of the CRISPR region into RNA molecules. These RNAs are cut into pieces and then bound by Cas proteins, such as the Cas9 protein (see Figure 17.28). The Cas protein uses a portion of the phage-related RNA as a homing device to identify the invading phage DNA and cut it, leading to its destruction.

Just as natural selection favours bacteria that have receptors altered by mutation or that have enzymes that cut phage DNA, it also favours phage mutants that can bind to altered receptors or that are resistant to enzymes. Thus, the bacterium-phage relationship is in constant evolutionary flux.

Replicative Cycles of Animal Viruses

Everyone has suffered from viral infections, whether cold sores, influenza, or the common cold. Like all viruses, those that cause

▼ **Figure 19.9 The CRISPR-Cas system: a type of bacterial immune system.**

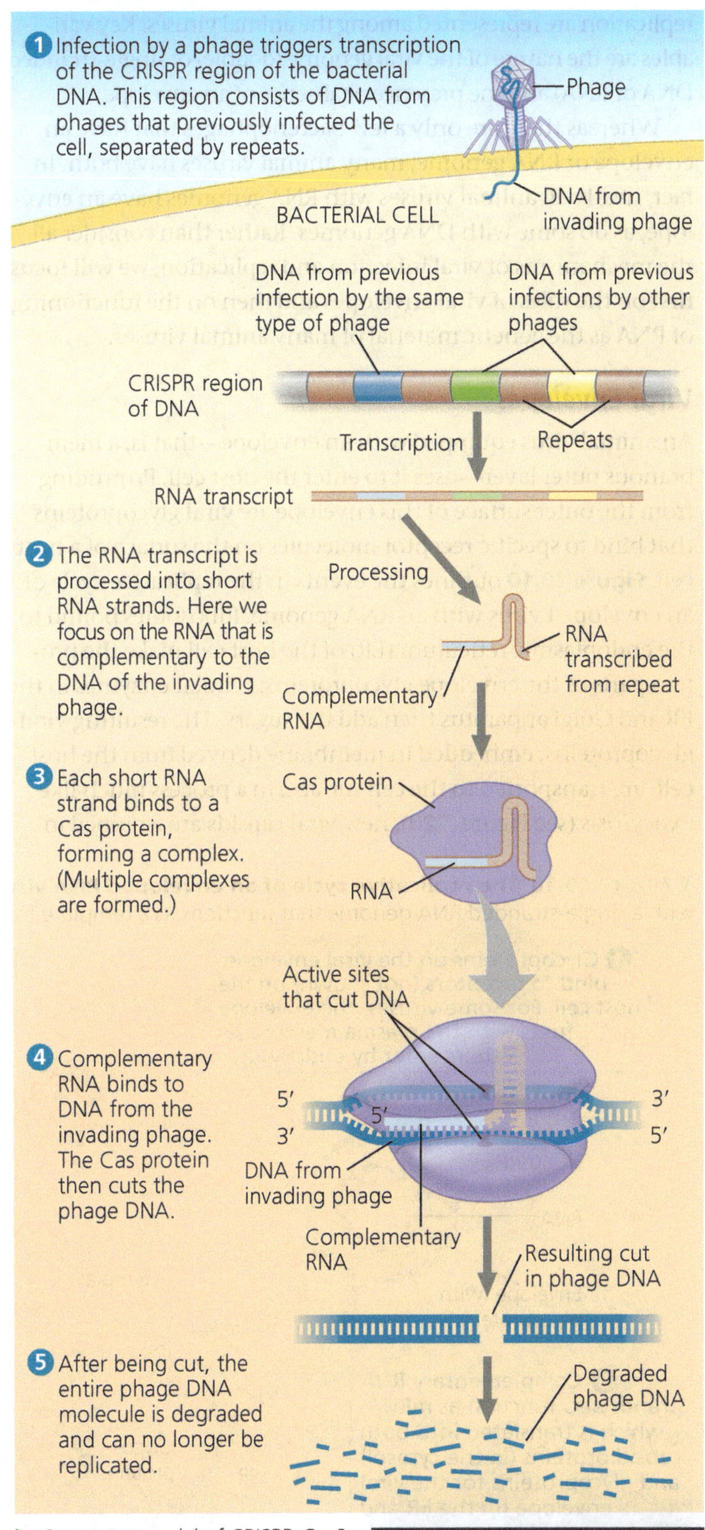

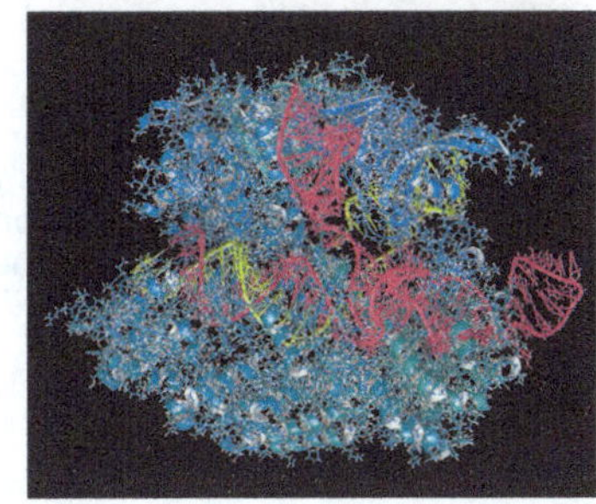

▶ Computer model of CRISPR-Cas9 gene editing complex from *Streptococcus pyogenes*

illness in humans and other animals can replicate only inside host cells. Many variations on the basic scheme of viral infection and replication are represented among the animal viruses. Key variables are the nature of the viral genome (double- or single-stranded DNA or RNA) and the presence or absence of an envelope.

Whereas there are only a few bacteriophages that have an envelope or RNA genome, many animal viruses have both. In fact, nearly all animal viruses with RNA genomes have an envelope, as do some with DNA genomes. Rather than consider all the mechanisms of viral infection and replication, we will focus first on the roles of viral envelopes and then on the functioning of RNA as the genetic material of many animal viruses.

Viral Envelopes

An animal virus equipped with an envelope—that is, a membranous outer layer—uses it to enter the host cell. Protruding from the outer surface of this envelope are viral glycoproteins that bind to specific receptor molecules on the surface of a host cell. **Figure 19.10** outlines the events in the replicative cycle of an enveloped virus with an RNA genome. Ribosomes bound to the endoplasmic reticulum (ER) of the host cell make the protein parts of the envelope glycoproteins; cellular enzymes in the ER and Golgi apparatus then add the sugars. The resulting viral glycoproteins, embedded in membrane derived from the host cell, are transported to the cell surface. In a process much like exocytosis (see Figure 7.20), new viral capsids are wrapped in membrane as they bud from the cell. In other words, the viral envelope is usually derived from the host cell's plasma membrane, although all or most of the molecules of this membrane are specified by viral genes. The enveloped viruses are now free to infect other cells. This replicative cycle does not necessarily kill the host cell, in contrast to the lytic cycles of phages.

Some viruses have envelopes that are not derived from plasma membrane. Herpesviruses, for example, are temporarily cloaked in membrane derived from the nuclear envelope of the host; they then shed this membrane in the cytoplasm and acquire a new envelope made from membrane of the Golgi apparatus. These viruses have a double-stranded DNA genome and replicate within the host cell nucleus, using a combination of viral and cellular enzymes to replicate and transcribe their DNA. In the case of herpesviruses, copies of the viral DNA can remain behind as mini-chromosomes in the nuclei of certain nerve cells. There they remain latent until some sort of physical or emotional stress triggers a new round of active virus production. The infection of other cells by these new viruses causes the blisters characteristic of herpes, such as cold sores or genital sores. Once someone acquires a herpesvirus infection, flare-ups may recur throughout the person's life.

Viral Genetic Material

Table 19.1 shows the common classification system for animal viruses, which is based on their genetic material: double- or

▼ **Figure 19.10 The replicative cycle of an enveloped RNA virus.** Shown here is a virus with a single-stranded RNA genome that functions as a template for synthesis of mRNA.

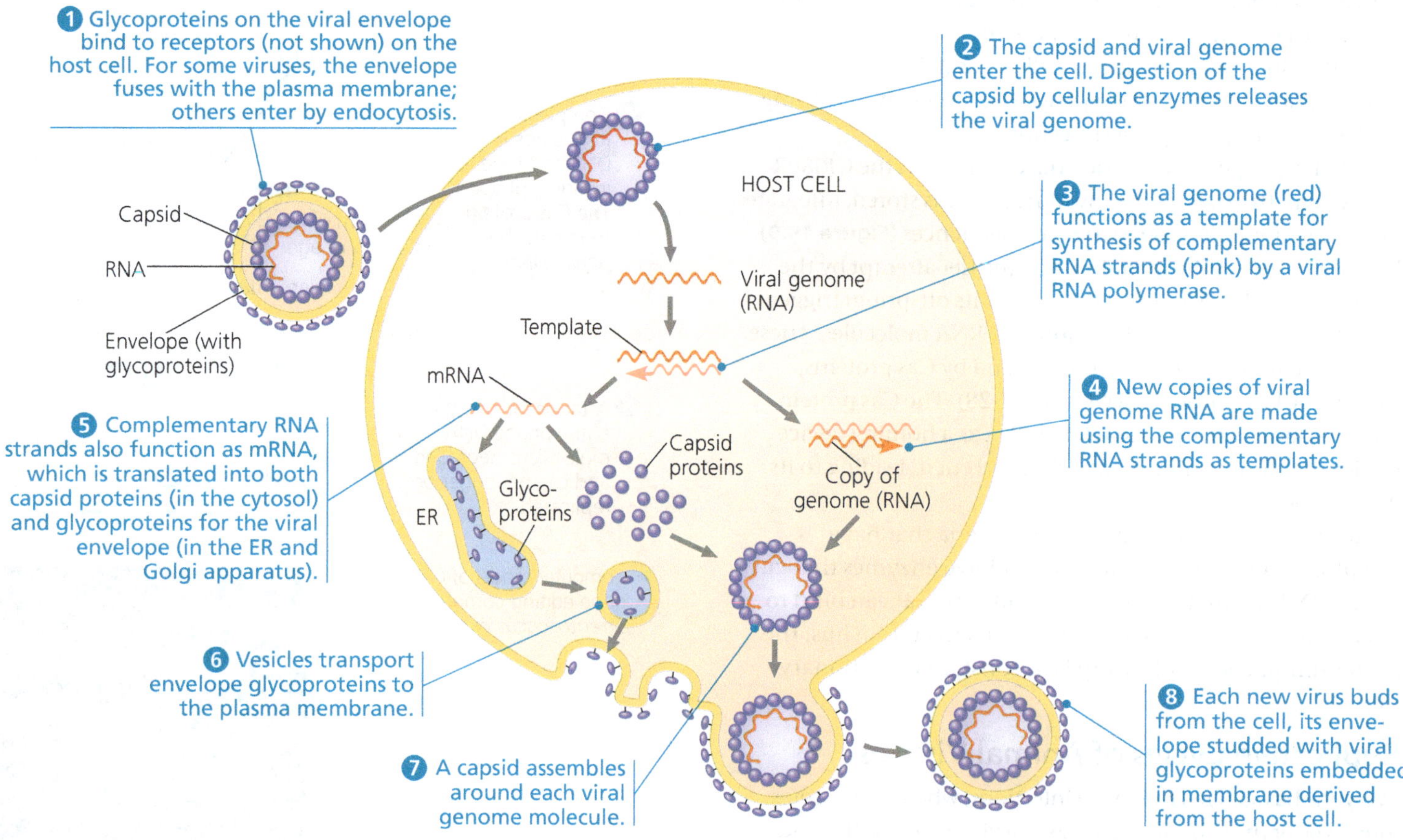

single-stranded DNA, or double- or single-stranded RNA, with further classification of RNA viruses. Although some phages and most plant viruses are RNA viruses, the broadest variety of RNA genomes is found among the viruses that infect animals. There are three types of single-stranded RNA genomes found in animal viruses (classes IV–VI in Table 19.1). The genome of class IV viruses can directly serve as mRNA and thus can be translated into viral protein immediately after infection. Figure 19.10 shows a virus of class V, in which the RNA genome serves instead as a *template* for mRNA synthesis. The RNA genome is transcribed into complementary RNA strands, which function both as mRNA and as templates for the synthesis of additional copies of genomic RNA. All viruses that use an RNA genome as a template for mRNA transcription require RNA → RNA synthesis. These viruses use a viral enzyme capable of carrying out this process; there are no such enzymes in most cells. The enzyme used in this process is encoded by the viral genome. After the protein is synthesised, it is packaged during viral self-assembly with the genome inside the viral capsid.

Table 19.1 Classes of Animal Viruses

Class/Family	Envelope?	Examples That Cause Human Diseases
I. Double-Stranded DNA (dsDNA)		
Adenovirus (see Figure 19.5b)	No	Respiratory viruses
Papillomavirus	No	Warts, cervical cancer
Polyomavirus	No	Tumours
Herpesvirus	Yes	Herpes simplex I and II (cold sores, genital sores); varicella zoster (shingles, chicken pox); Epstein-Barr virus (mononucleosis, Burkitt's lymphoma)
Poxvirus	Yes	Smallpox virus; cowpox virus
II. Single-Stranded DNA (ssDNA)		
Parvovirus	No	B19 parvovirus (mild rash)
III. Double-Stranded RNA (dsRNA)		
Reovirus	No	Rotavirus (diarrhoea); the tentatively named Parry's Lagoon virus
IV. Single-Stranded RNA (ssRNA); Serves as mRNA		
Picornavirus	No	Rhinovirus (common cold); poliovirus; hepatitis A virus; other intestinal viruses
Coronavirus	Yes	Severe acute respiratory syndrome (SARS); Middle East respiratory syndrome (MERS); severe acute respiratory syndrome coronavirus 2 (SARS-CoV-2) (see Figure 19.10c)
Flavivirus	Yes	Zika virus; yellow fever virus; dengue virus; West Nile virus; hepatitis C virus
Togavirus	Yes	Chikungunya virus (see Figure 19.12b); rubella virus; equine encephalitis viruses
V. ssRNA; Serves as Template for mRNA Synthesis		
Filovirus	Yes	Ebola virus (haemorrhagic fever; see Figure 19.12a)
Orthomyxovirus	Yes	Influenza virus (see Figure 19.5c)
Paramyxovirus	Yes	Measles virus; mumps virus
Rhabdovirus	Yes	Rabies virus
VI. ssRNA; Serves as Template for DNA Synthesis		
Retrovirus	Yes	Human immunodeficiency virus (HIV/AIDS; see Figure 19.11); human T-lymphotropic virus type 1 (HTLV-1) (leukaemia)

The RNA animal viruses with the most complicated replicative cycles are the **retroviruses** (class VI). These viruses have an enzyme called **reverse transcriptase** that transcribes an RNA template into a DNA copy, an RNA → DNA information flow that is the opposite of the usual direction. This unusual phenomenon is the source of the name retroviruses (*retro* means "backwards"). Of particular medical importance is **HIV (human immunodeficiency virus)**, the retrovirus that causes **AIDS (acquired immunodeficiency syndrome)**. HIV and other retroviruses are enveloped viruses that contain two identical molecules of single-stranded RNA and two molecules of reverse transcriptase.

The HIV replicative cycle (illustrated in **Figure 19.11**) is typical of a retrovirus. After HIV enters a host cell, its reverse transcriptase molecules are released into the cytoplasm, where they catalyse synthesis of viral DNA. The newly made viral DNA then enters the cell's nucleus and integrates into the DNA of a chromosome. The integrated viral DNA, called a **provirus**, never leaves the host's genome, remaining a permanent resident of the cell. (Recall that a prophage, in contrast, leaves the host's genome at the start of a lytic cycle.) The RNA polymerase of the host transcribes the proviral DNA into RNA molecules, which can function both as mRNA for the synthesis of viral proteins and as genomes for the new viruses that will be assembled and released from the cell. In Concept 43.4, we describe how HIV causes the deterioration of the immune system that occurs in AIDS.

Evolution of Viruses

EVOLUTION We began this chapter by asking whether or not viruses are alive. Viruses do not really fit our definition of living organisms. An isolated virus is biologically inert, unable to replicate its genes or regenerate its own ATP. Yet it has a genetic program written in the universal language of life. Do we think of viruses as nature's most complex associations of molecules or as the simplest forms of life? Either way, we must bend our usual definitions. Although viruses cannot replicate or carry out metabolic activities independently, their use of the genetic code makes it hard to deny their evolutionary connection to the living world.

How did viruses originate? Viruses have been found that infect every form of life—not only bacteria, animals, and plants,

▼ Figure 19.11 The replicative cycle of HIV, the retrovirus that causes AIDS. The photos on the left (artificially coloured TEMs) show HIV entering and leaving a human white blood cell. See Figure 7.8 for the cell-surface proteins that act as receptors for HIV. Note in step 6 that DNA synthesised from the viral RNA genome is integrated as a provirus into the host cell chromosomal DNA, a characteristic unique to retroviruses.

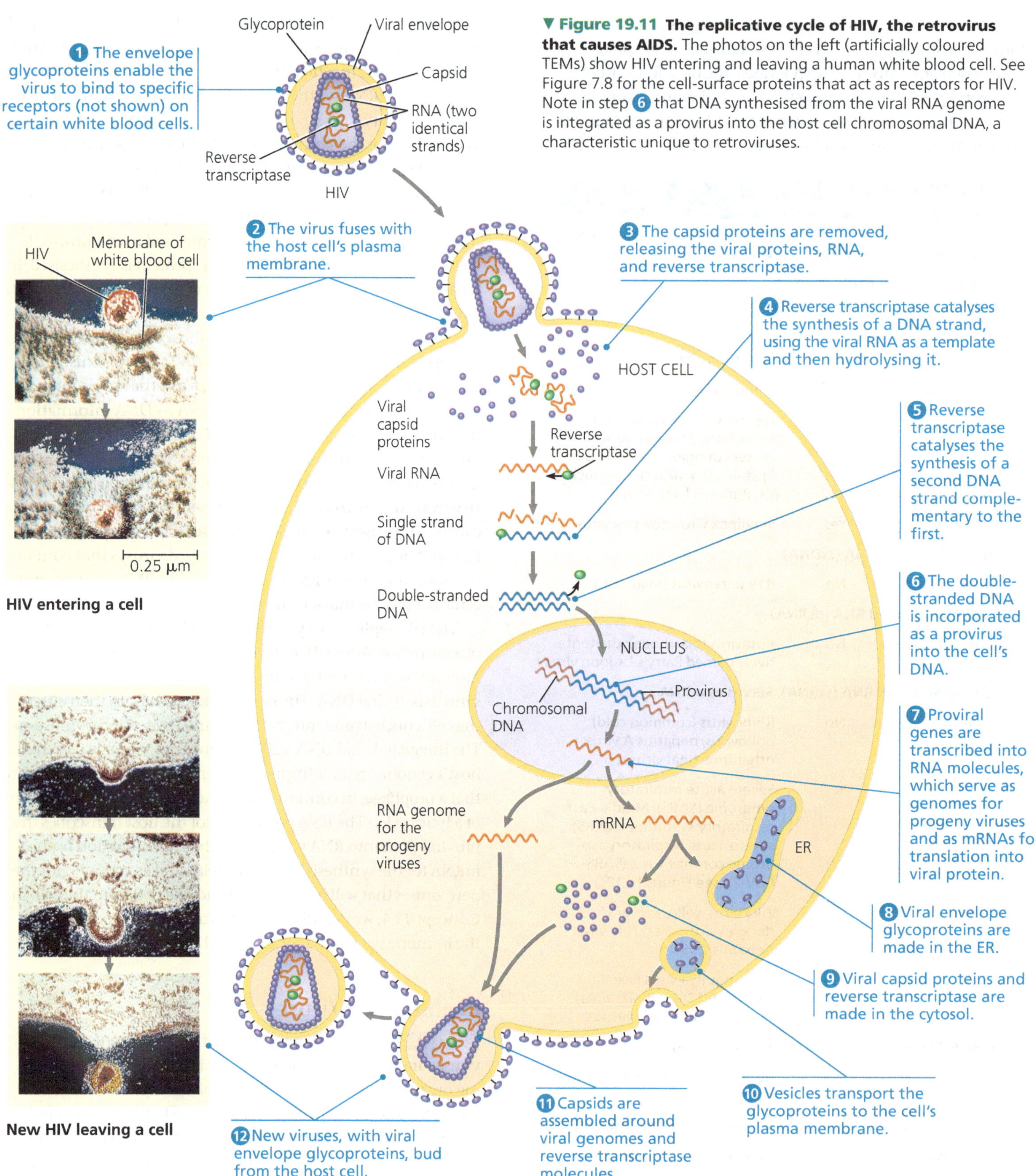

MAKE CONECTIONS *Describe what is known about binding of HIV to immune system cells. (See Figure 7.8.) How was this discovered?*

but also archaea, fungi, and algae and other protists. Researchers estimate that each millilitre (one-fifth of a teaspoon) of seawater contains between one and 100 million viruses, ten times the number of microorganisms! Because viruses depend on cells for their own propagation, it seems likely that viruses are not the descendants of precellular forms of life but evolved—possibly multiple times—*after* the first cells appeared. Most molecular biologists favour the hypothesis that viruses originated from naked bits of cellular nucleic acids that moved from one cell to another, perhaps via injured cell surfaces. The evolution of genes coding for capsid proteins may have allowed viruses to bind cell membranes, thus facilitating the infection of uninjured cells.

Candidates for the original sources of viral genomes include plasmids and transposons. *Plasmids* are small, circular DNA molecules found in bacteria and in the unicellular fungi called yeasts. Plasmids exist apart from the genome, can replicate independently of the genome, and are occasionally transferred between cells. *Transposons* are DNA segments that can move from one location to another within a cell's genome. Thus, plasmids, transposons, and viruses all share an important feature: They are *mobile genetic elements*. (We'll discuss plasmids in more detail in Concepts 20.1 and 27.2 and transposons in Concept 21.4.)

Consistent with this notion of pieces of DNA shuttling from cell to cell is the observation that a viral genome can have more in common with the genome of its host than with the genomes of viruses that infect other hosts. Indeed, some viral genes are essentially identical to genes of the host.

Debate about the origin of viruses was reinvigorated about 20 years ago when an extremely large virus was discovered: Mimivirus is a double-stranded DNA (dsDNA) virus with an icosahedral capsid that is 400 nm in diameter, the size of a small bacterium. Its genome contains 1.2 million bases (Mb)—about 100 times as many as the influenza virus genome—and an estimated 1,000 genes. Perhaps the most surprising aspect of mimivirus, however, was that its genome included genes previously found only in cellular genomes. Some of these genes code for proteins involved in translation, DNA repair, protein folding, and polysaccharide synthesis. Whether mimivirus evolved *before* the first cells and then developed an exploitative relationship with them or evolved more recently and simply scavenged genes from its hosts is not yet settled.

In the past decade, several even larger viruses have been discovered that cannot be classified with any existing known virus. One such virus is 1 μm (1,000 nm) in diameter, with a dsDNA genome of around 2–2.5 Mb, larger than that of some small eukaryotes. What's more, over 90% of its 2,000 or so genes are unrelated to cellular genes, inspiring the name it was given, pandoravirus. The number of genes in pandoraviruses varies from 1,500 to 2,500 genes. A second virus, called *Pithovirus sibericum*, with a diameter of 1.5 μm and 500 genes, was discovered in permanently frozen soil in Siberia. This virus, once thawed, was able to infect an amoeba after being frozen for 30,000 years! How these and all other viruses fit in the tree of life is an intriguing, unresolved question.

The ongoing evolutionary relationship between viruses and the genomes of their host cells is an association that continues to make viruses very useful experimental systems in molecular biology. Knowledge about viruses also allows many practical applications, since viruses have a tremendous impact on all organisms through their ability to cause disease.

CONCEPT CHECK 19.2

1. Compare the effect on the host cell of a lytic (virulent) phage and a lysogenic (temperate) phage.
2. **MAKE CONNECTIONS** Compare the CRISPR-Cas system with the miRNA system discussed in Concept 18.3, including their mechanisms and their functions.
3. **MAKE CONNECTIONS** The RNA virus in Figure 19.10 has a viral RNA polymerase that functions in step 3 of the virus's replicative cycle. Compare this with a cellular RNA polymerase in terms of template and overall function (see Figure 17.10).
4. Why is HIV called a retrovirus?
5. **VISUAL SKILLS** Looking at Figure 19.11, imagine you are a researcher trying to combat HIV infection. What molecular processes could you attempt to block?

For suggested answers, see Appendix A.

CONCEPT 19.3

Viruses and prions are formidable pathogens in animals and plants

Diseases caused by viral infections afflict humans, agricultural crops, and livestock worldwide. Other smaller, less complex entities known as prions also cause disease in animals. We'll first consider animal viruses.

Viral Diseases in Animals

A viral infection can produce symptoms by a number of different mechanisms. Viruses may damage or kill cells by causing the release of hydrolytic enzymes from lysosomes. Some viruses cause infected cells to produce toxins that lead to disease symptoms, and some have molecular components that are toxic, such as envelope proteins. How much damage a virus causes depends partly on the ability of the infected tissue to regenerate by cell division. People usually recover completely from colds because the epithelium of the respiratory tract, which the viruses infect, can efficiently repair itself. In contrast, damage inflicted by poliovirus to mature nerve cells is permanent because these cells do not divide and usually cannot be replaced. Many of the temporary symptoms associated with viral infections, such as fever and body aches, actually result from the body's own efforts to defend itself against infection rather than from cell death caused by the virus.

The immune system is a critical part of the body's natural defences (see the chapter, "The Immune System"). It is also the basis for the major medical tool used to prevent viral infections—vaccines. A **vaccine** is a harmless derivative of a pathogen that stimulates the immune system to mount defences against the harmful pathogen. Smallpox, a viral disease that was once a devastating scourge in many parts of the world, was eradicated by 1980 due to a vaccination program carried out by the World Health Organization (WHO). The very narrow host range of the smallpox virus—it infects only humans—was a critical factor in the success of this program. Similar worldwide vaccination campaigns are under way to eradicate polio, the incidence of which has dropped by 99%, and measles. Although an effective vaccine exists for measles, a large outbreak occurred in the Pacific Northwest region of North America in 2019, correlated with lower vaccination rates in that region. Effective vaccines are also available to protect against rubella, mumps, hepatitis B, and a number of other viral diseases.

Although vaccines can prevent some viral illnesses, medical care can do little, at present, to cure most viral infections once they occur. The antibiotics that help us recover from bacterial infections are powerless against viruses. Antibiotics kill bacteria by inhibiting enzymes specific to bacteria but have little or no effect on eukaryotic or virally encoded enzymes. However, the few enzymes that are encoded only by viruses have provided targets for other drugs. Most antiviral drugs resemble nucleosides and thus interfere with viral nucleic acid synthesis. One such drug is acyclovir, which impedes herpesvirus replication by inhibiting the viral polymerase that with viral nucleic acid synthesises viral DNA but not the eukaryotic one. Similarly, azidothymidine (AZT) curbs HIV replication by interfering with the synthesis of DNA by reverse transcriptase. In the past 30 years, much effort has gone into developing drugs to treat HIV. Currently, multidrug treatments, sometimes called "cocktails," are considered to be most effective. Such treatments commonly include a combination of two nucleoside mimics and a protease inhibitor, which interferes with an enzyme required for assembly of the viruses. Multidrug treatments originally involved taking up to 20 pills multiple times per day but now usually consist of a single daily tablet. Another effective treatment involves a drug called maraviroc, which blocks a protein on the surface of human immune cells that helps bind the HIV virus (see Figure 7.8). This drug has also been used successfully to prevent infection in individuals who either have been exposed to, or are at risk of exposure to, HIV.

Emerging Viral Diseases

Viruses that suddenly become apparent are often referred to as *emerging viruses*. HIV, the AIDS virus, is a classic example: It appeared in San Francisco in the early 1980s, seemingly out of nowhere, although later studies uncovered a case in the Belgian Congo in 1959. Some other dangerous emerging viruses cause encephalitis, inflammation of the brain. One example is the West Nile virus, which appeared in North America in 1999 and has spread to 49 US states, by 2019 resulting in over 50,000 cases and about 2,000 deaths.

The deadly Ebola virus **(Figure 19.12a)**, recognised initially in 1976 in central Africa, is one of several emerging viruses that cause *haemorrhagic fever*, an often fatal illness characterised by fever, vomiting, massive bleeding, and circulatory system collapse. In 2014, a widespread outbreak, or **epidemic**, of Ebola virus occurred. By 2016, it had resulted in over 11,000 deaths. In 2017, 2018, and 2019, smaller outbreaks occurred in the Democratic Republic of the Congo.

The mosquito-borne chikungunya virus **(Figure 19.12b)** causes an acute illness with fever, rashes, and persistent joint pain. Chikungunya has long been considered a tropical virus, but it has now appeared in Italy, France, and Spain. The Zika virus **(Figure 19.12c)** was first observed in Uganda in 1947, but for decades only a few cases occurred per year. In the spring of 2015, however, it became an emerging virus when it caused a large outbreak in Brazil. Although symptoms of Zika are often mild, the infection of pregnant women was correlated with a striking increase in the number of babies born with abnormally small brains, a condition called microcephaly. Zika is a mosquito-borne flavivirus (like West Nile virus) that infects neural cells, posing a particular danger to fetal brain development.

Where do these new strains of virus come from? One cause of rapidly emerging viral diseases in humans is mutation of existing viruses into new viruses that can spread more easily. RNA viruses have a high rate of mutation because viral RNA polymerases do not proofread and correct errors in replicating their RNA genomes. Some mutations change existing viruses into new viral strains that can cause disease, even in people immune to the original virus. A well-known related example

▼ **Figure 19.12 Emerging viruses.**

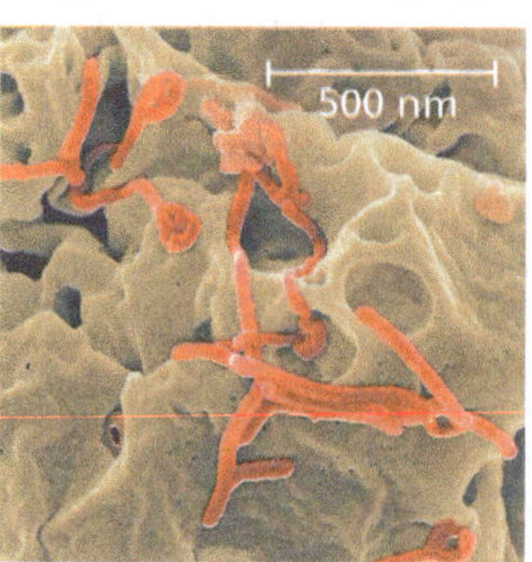

(a) Ebola viruses budding from a monkey cell (colourised SEM).

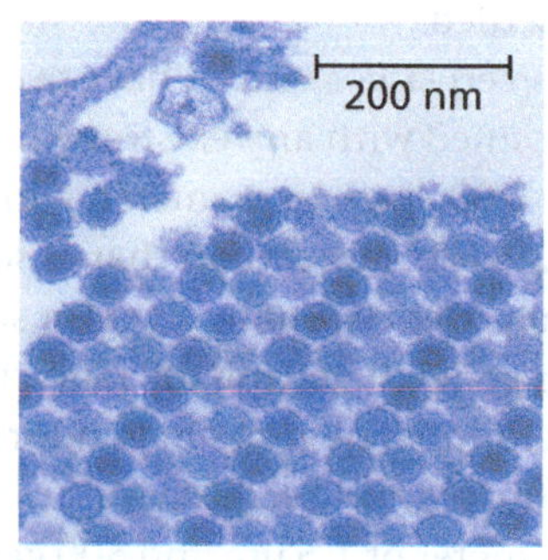

(b) Chikungunya viruses emerging from a cell in the upper left and packing together (colourised TEM).

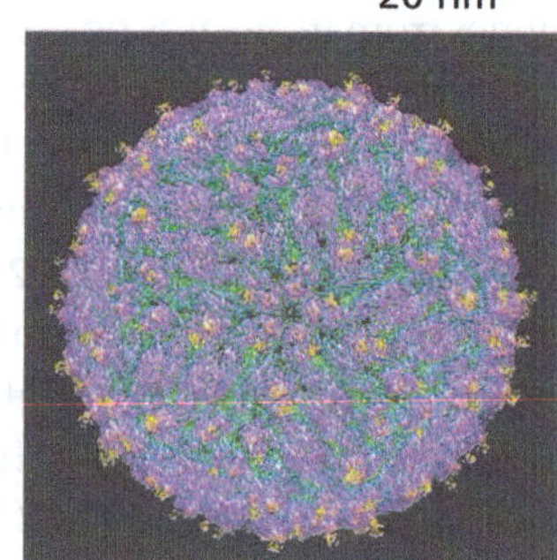

(c) Computer-generated image of a **Zika virus**, based on a technique called cryo-electron microscopy.

is how three or four mutations causing changes in a surface protein of a cat virus (feline panleukopenia virus) resulted in the emergence in 1978 of canine parvovirus, a very contagious deadly virus infecting dogs.

A second cause of the emergence of viral diseases is the spread of a viral disease from a small, isolated human population. HIV, the virus that causes AIDS, went unnamed and virtually unnoticed for decades before spreading around the world. In this case, technological and social factors, including affordable international travel, blood transfusions, unprotected sexual intercourse, and reuse of needles to inject intravenous drugs, allowed a previously rare human disease to become a global scourge.

A third cause of new viral diseases in humans is the spread of existing viruses from other animals. Scientists estimate that about three-quarters of new human diseases originate in this way. Animals infected with a virus that can be transmitted to humans are said to be a natural reservoir for that virus. HIV is an example, as scientists believe that it originated from a version of the virus found in chimpanzees in central Africa, after people ate chimpanzee meat for food and were infected by exposure to chimpanzee blood.

In general, flu epidemics provide an instructive example of these three causes of emerging viruses. There are three types of influenza virus: types B and C, which infect only humans and have never caused an epidemic, and type A, which infects a wide range of animals, including birds, pigs, horses, and humans. The influenza type A viruses present in pigs and wild and domestic birds are potential emerging viruses that represent a long-term threat to human health.

A case in point is the H5N1 strain of avian influenza virus, which is highly contagious and deadly in birds. The first transmission from birds to humans was documented in Hong Kong in 1997. Since then, about 700 people have been infected, with an alarming mortality rate of greater than 50%. The high mortality rate is partly because H5N1 is very different from strains of influenza that have circulated among humans for a long time. Individuals are therefore not able to mount a strong immune response against viruses like H5N1.

Different strains of influenza A are given standardised names; for example, the name H5N1 identifies which forms of two viral surface proteins are present—haemagglutinin and neuraminidase (HA and NA, respectively; see the glycoprotein spikes in Figure 19.5c). These two proteins together help determine the host range and severity of disease caused by each virus. As of 2017, 18 types of haemagglutinin, a protein that helps the flu virus attach to host cells, and 11 types of neuraminidase, an enzyme that helps release new virus particles from infected cells, have been identified. All possible combinations of HA and NA have been found in some waterbirds. In 2018, researchers using advanced microscopy techniques reported that the HA protein is quite dynamic, stretching towards the target host cell, retracting, then refolding to a new shape and re-approaching the target cell.

Although deadly, the H5N1 strain has not yet caused an epidemic because nearly all cases have been the result of transmission from birds to people, rather than person-to-person. Epidemics occur when genetic changes allow a new viral strain to be easily transmitted between humans. An event like this occurred in 2009, when a strain of influenza virus (H1N1) appeared that was very different from the virus that causes the seasonal flu. The H1N1 influenza virus spread rapidly, prompting WHO to declare a global epidemic, or **pandemic**. Within half a year, the disease had reached 207 countries, infecting over 600,000 people and killing almost 8,000. In addition to the 2009 H1N1 pandemic, influenza A strains have caused three other major flu epidemics among humans in the last 100 years. The most notable of these was the "Spanish flu" pandemic of 1918–1919, which killed more than 50 million people worldwide. In the **Scientific Skills Exercise**, you'll analyse genetic changes in variants of the 2009 H1N1 influenza virus and correlate them with spread of the disease.

The disease caused by H1N1 was originally called "swine flu" because parts of the viral genome were very similar to strains of influenza in pigs. However, studies revealed that the virus was not transmitted from pigs to humans. Instead, the story was more complex: H1N1 was a unique combination of swine, avian, and human influenza genes that allowed it to spread among humans.

Influenza viruses can change quickly because they have a genome made up of nine segments of RNA rather than a single RNA molecule. When an animal like a pig or a bird is infected with multiple strains of influenza virus, the RNA molecules making up the viral genomes can mix and match (*reassort*) during viral assembly, resulting in new genetic combinations. If a flu virus from pigs recombines with viruses that circulate widely among humans, it may acquire the ability to spread easily from person to person, dramatically increasing the potential for a major human outbreak. Pigs are believed to have been the main hosts for recombination that led to the 2009 H1N1 flu virus.

Influenza viruses also have a high rate of mutation, for reasons mentioned earlier. Coupled with reassortments, these mutations can lead to the emergence of a viral strain from animals that can infect human cells. Scientists are worried that the H5N1 strain will evolve in a way that enables it to spread as easily as the H1N1 strain. This would represent a major global health threat like that of the 1918 pandemic.

How easily could this capability evolve in the H5N1 strain? In 2011, scientists working with ferrets, small mammals used as model organisms for the human flu, found that only a few mutations of the avian flu virus were sufficient to allow infection of cells in the human nasal cavity and windpipe. Furthermore, when the scientists transferred nasal swabs serially from ferret to ferret, the virus became transmissible through the air. Reports of this startling discovery ignited a firestorm

Scientific Skills Exercise

Analysing a Sequence-Based Phylogenetic Tree to Understand Viral Evolution

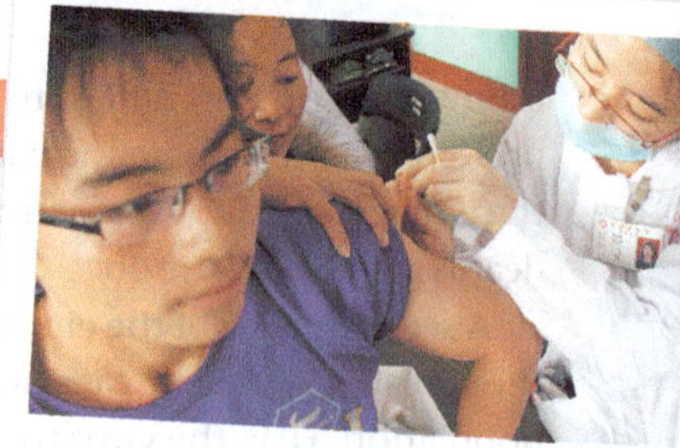

▶ **H1N1 flu vaccination**

How Can Sequence Data Be Used to Track Flu Virus Evolution? In 2009, an influenza A H1N1 virus caused a pandemic, and the virus has continued to resurface in outbreaks across the world. Researchers in Taiwan were curious about why the virus kept appearing despite widespread flu vaccine initiatives. They hypothesised that newly evolved variant strains of the H1N1 virus were able to evade human immune system defences. To test this hypothesis, they needed to determine if each wave of flu infection was caused by a different H1N1 variant strain.

How the Experiment Was Done Scientists obtained the genome sequences for 4,703 virus isolates collected from patients with H1N1 flu in Taiwan, each named by type/location/identifying number/year. They compared the sequences in different strains for the viral haemagglutinin (HA) gene, and based on mutations that had occurred, arranged the isolates into a phylogenetic tree (see Figure 26.5 for information on how to read phylogenetic trees).

Data from the Experiment In the phylogenetic tree, each branch tip is one variant strain of the H1N1 virus with a unique HA gene sequence. The tree is a way to visualise a working hypothesis about the evolutionary relationships between H1N1 variants.

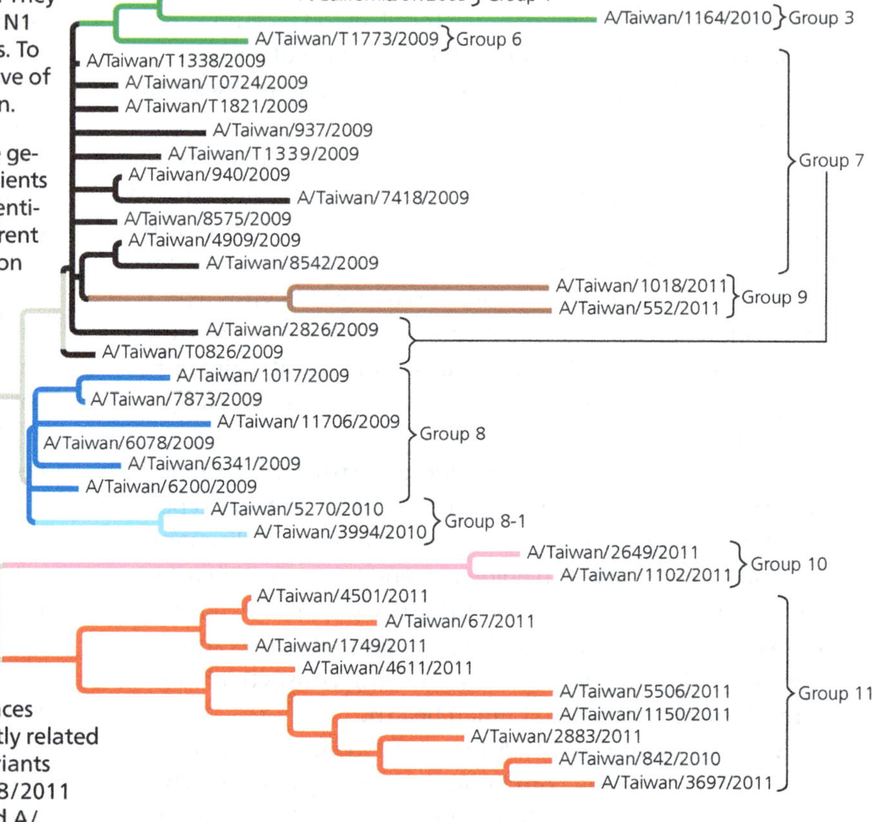

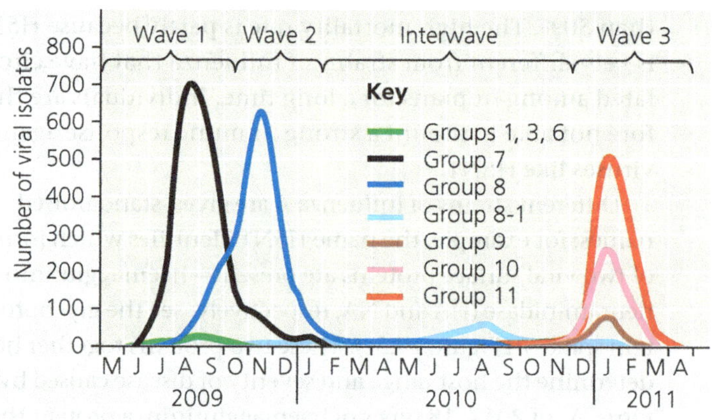

▲ Scientists graphed the number of isolates by the month and year of isolate collection to show the period in which each viral variant was actively causing illness in people.

Data from J.-R. Yang et al., New variants and age shift to high fatality groups contribute to severe successive waves in the 2009 influenza pandemic in Taiwan, *PLoS ONE* 6(11): e28288 (2011).

INTERPRET THE DATA

1. The more closely connected two variants are in the tree, the more alike they are in terms of HA gene sequence. Each fork in a branch, called a node, shows where two lineages separate due to different accumulated mutations. The length of the branches is a measure of how many sequence differences there are between the variants, indicating how distantly related they are. Referring to the phylogenetic tree, which variants are more closely related to each other: A/Taiwan/1018/2011 and A/Taiwan/552/2011 or A/Taiwan/1018/2011 and A/Taiwan/8542/2009? Explain your answer.
2. The scientists arranged the branches into groups made up of one ancestral variant and all of its descendant, mutated variants. They are colour-coded in the figure. Using group 11 as an example, trace the lineage of its variants. **(a)** Do all of the nodes have the same number of branches or branch tips? **(b)** Are all of the branches in the group the same length? **(c)** What do these results indicate?
3. The graph shows the number of isolates collected (each from an ill patient) on the *y*-axis and the month and year that the isolates were collected on the *x*-axis. Each group of variants is plotted separately with a line colour that matches the tree diagram. **(a)** Which group of variants was the earliest to cause the first wave of H1N1 flu in over 100 patients in Taiwan? **(b)** After a group of variants had a peak number of infections, did members of that same group cause another (later) wave of infection? **(c)** One variant in group 1 (green, uppermost branch) was used to make a vaccine that was distributed very early in the pandemic. Based on the graphed data, does it look like the vaccine was effective?
4. Groups 9, 10, and 11 all had H1N1 variants that caused a large number of infections at the same time in Taiwan. Does this mean that the scientists' hypothesis, that new variants cause new waves of infection, was incorrect? Explain your answer.

of debate about whether to publish the results. Ultimately, the scientific community decided the benefits of preventing pandemics outweighed the risks of the information being used for harmful purposes, and the work was published in 2012.

Normal seasonal flu viruses (including influenza types A and B) are not considered emerging viruses because variations of seasonal flu viruses have been circulating among humans for long enough that most components are recognised by the immune system. However, these viruses still undergo mutation and reassortment of genome segments, and variations of the HA protein are used each year to generate vaccines against the strains predicted most likely to occur the following year.

As we have seen, emerging viruses are usually existing viruses that mutate, spread more widely in the current host species, or spread to new host species. Changes in host behaviour or environmental changes can increase the viral traffic responsible for emerging diseases. For instance, new roads built through remote areas can allow viruses to spread between previously isolated human populations. Also, the destruction of forests to expand cropland can bring humans into contact with animals that host infectious viruses. Finally, genetic mutations and changes in host ranges can allow viruses to jump between species. Many viruses are transmitted by mosquitoes. A dramatic expansion of the disease caused by the chikungunya virus occurred in the mid-2000s when a mutation allowed it to infect not only the mosquito species *Aedes aegypti* but also the related *Aedes albopictus*. Insecticides and mosquito netting over beds are crucial tools in public health attempts to prevent diseases carried by mosquitoes **(Figure 19.13)**.

▼ Figure 19.13 Netting as protection against virus-carrying mosquitoes.

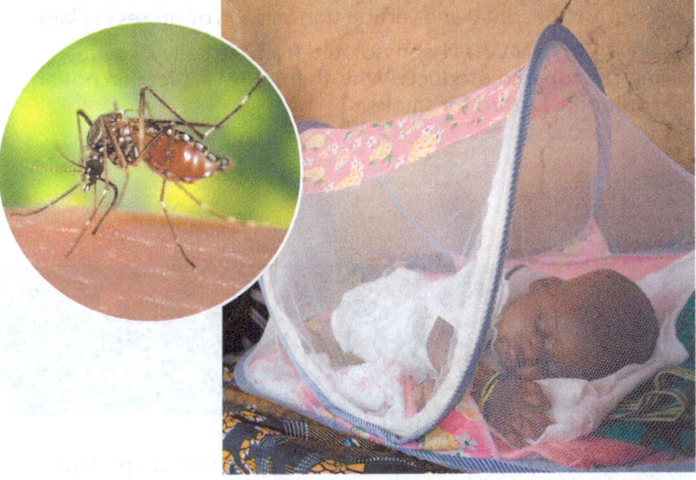

Recently, scientists have become concerned about the possible effects of climate change on worldwide viral transmission. Dengue fever, also mosquito-borne, has appeared in Florida and Portugal, regions where it had not been seen before. The possibility that global climate change has allowed mosquito species carrying these viruses to expand their ranges and interact more is troubling because of the increased chance of a mutation allowing a virus to jump to a new host. This is an area of active research by scientists applying climate change models to what is known about the habitat requirements of mosquito species.

Viral Diseases in Plants

More than 2,000 types of viral diseases of plants are known, accounting for an annual loss of over $30 billion worldwide due to destruction of crops. Common signs of viral infection include bleached or brown spots on leaves and fruits **(Figure 19.14)**, stunted growth, and damaged flowers or roots, all of which can diminish the yield and quality of crops.

► Figure 19.14 Immature tomato infected by a virus.

Plant viruses have the same basic structure and mode of replication as animal viruses. Most known plant viruses, including tobacco mosaic virus (TMV), have an RNA genome. Many have a helical capsid, like TMV, while others have an icosahedral capsid (see Figure 19.5b).

Viral diseases of plants spread by two major routes. In the first route, *horizontal transmission*, an external source infects the plant. Because the invading virus must get past the plant's outer protective layer of cells (the epidermis), a plant becomes more susceptible to viral infections if it has been damaged by wind, injury, or herbivores. Herbivores, especially insects, pose a double threat because they can also carry viruses, transmitting disease from plant to plant. Moreover, gardeners may transmit plant viruses inadvertently on pruning shears and other tools. The other route of viral infection is *vertical transmission*, in which a plant inherits a viral infection from a parent. Vertical transmission can occur in asexual propagation (for example, through cuttings) or in sexual reproduction via infected seeds.

Once a virus enters a plant cell and begins replicating, viral genomes and associated proteins can spread throughout the plant through plasmodesmata, the cytoplasmic connections that penetrate the walls between adjacent plant cells (see Figure 36.19). The passage of viral macromolecules from cell to cell is facilitated by virally encoded proteins that cause enlargement of plasmodesmata. Scientists have not yet devised cures for most viral plant diseases, so research efforts are focused largely on reducing disease transmission and on breeding resistant varieties of crop plants.

Prions: Proteins as Infectious Agents

The viruses discussed in this chapter are infectious agents that spread diseases, and their genetic material is composed of nucleic acids, whose ability to be replicated is well known. Surprisingly, there are also *proteins* that are infectious. Proteins called **prions** appear to cause degenerative brain diseases in various animal species. These diseases include scrapie in sheep; mad cow disease, which plagued the European beef industry about 20 years ago;

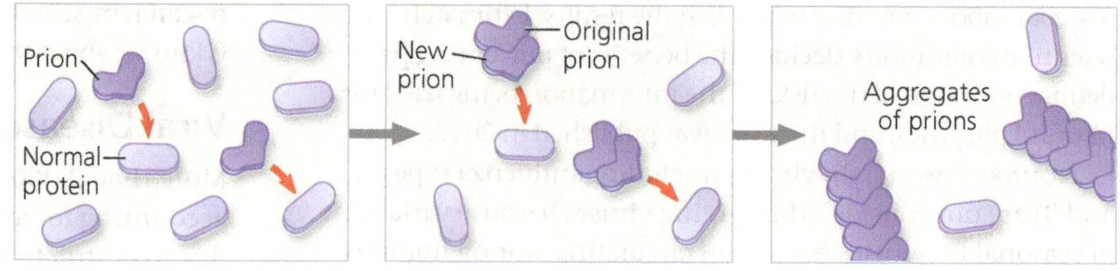

▶ **Figure 19.15 Model for how prions propagate.** Prions are misfolded versions of normal brain proteins. When a prion contacts a normally folded version of the same protein, it may induce the normal protein to assume the abnormal shape. The resulting chain reaction may continue until high levels of prion aggregation cause cellular malfunction and eventual degeneration of the brain.

and Creutzfeldt-Jakob disease in humans, which has caused the death of some 175 people in the United Kingdom since 1996. Prions can be transmitted in food, as may occur when people eat beef from cattle with mad cow disease. Kuru, another human disease caused by prions, was identified in the early 1900s among the South Fore indigenous people of New Guinea. When a kuru epidemic peaked there in the 1960s, scientists at first thought the disease had a genetic basis because family members also often contracted the disease. Eventually, however, investigations revealed a different story: After a death, family members practiced ritual cannibalism, eating organs of the deceased, and prions were transmitted primarily in brain tissue. Women got kuru more often than men because men ate the more "prestigious" organs, like the heart, while women and children ate the brains.

Two characteristics of prions are especially alarming. First, prions act very slowly, with an incubation period of at least ten years before symptoms develop. The lengthy incubation period prevents sources of infection from being identified until long after the first cases appear, allowing many more infections to occur. Second, prions are not destroyed or deactivated by heating to normal cooking temperatures. To date, there is no known cure for prion diseases, and the only hope for developing effective treatments lies in understanding the process of infection.

How can a protein, which cannot replicate itself, be a transmissible pathogen? According to the leading model, a prion is a misfolded form of a protein normally present in brain cells. When the prion gets into a cell containing the normal form of the protein, the prion somehow converts normal protein molecules to the misfolded prion versions. Several prions then aggregate into a complex that can convert other normal proteins to prions, which join the chain **(Figure 19.15)**. Prion aggregation interferes with normal cellular functions and causes disease symptoms. This model was greeted with much skepticism when it was first proposed by Stanley Prusiner in the early 1980s, but it is now widely accepted. Prusiner was awarded the Nobel Prize in 1997 for his work on prions. He has also proposed that prions are involved in neurodegenerative diseases such as Alzheimer's and Parkinson's diseases. There are many outstanding questions about these small infectious agents.

CONCEPT CHECK 19.3

1. Describe two ways in which a preexisting virus can become an emerging virus.
2. Contrast horizontal and vertical transmission of viruses in plants.
3. **WHAT IF?** TMV has been isolated from virtually all commercial tobacco products. Why, then, is TMV infection not an additional hazard for smokers?

For suggested answers, see Appendix A.

19 Chapter Review

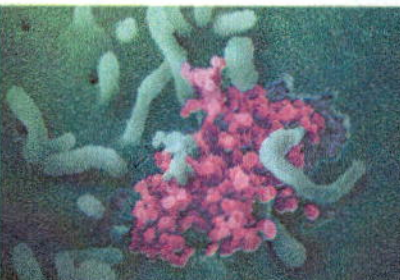

SUMMARY OF KEY CONCEPTS

CONCEPT 19.1

A virus consists of a nucleic acid surrounded by a protein coat *(pp. 401–404)*

- Researchers discovered viruses in the late 1800s by studying a plant disease, tobacco mosaic disease.
- A **virus** is a small nucleic acid genome enclosed in a protein **capsid** and sometimes a membranous **viral envelope**. The genome may be single- or double-stranded DNA or RNA.

? *Are viruses generally considered living or nonliving? Explain.*

CONCEPT 19.2

Viruses replicate only in host cells *(pp. 404–411)*

- Viruses use enzymes, ribosomes, and small molecules of host cells to synthesise progeny viruses during replication.

- Each type of virus has a characteristic **host range**, affected by whether cell-surface proteins are present that viral surface proteins can bind to.

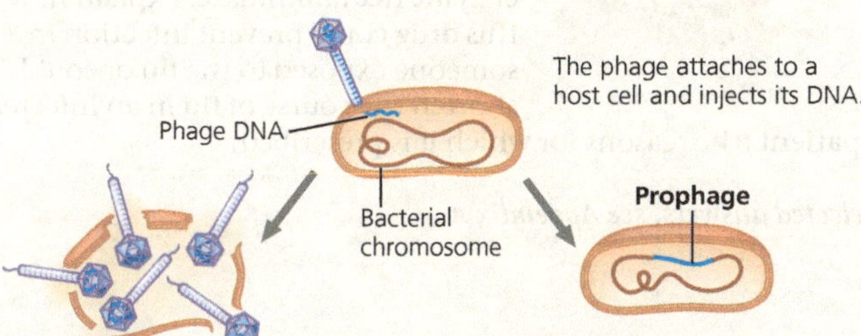

- **Phages** (viruses that infect bacteria) can replicate by two alternative mechanisms: the **lytic cycle** and the **lysogenic cycle**.
- Bacteria have various ways of defending themselves against phage infections, including the CRISPR-Cas system.
- Many animal viruses have an envelope. **Retroviruses** (such as **HIV**) use the enzyme **reverse transcriptase** to copy their RNA genome into DNA, which can be integrated into the host genome as a **provirus**.
- Since viruses can replicate only within cells, they probably evolved after the first cells appeared, perhaps as packaged fragments of cellular nucleic acid.

? *Describe enzymes that are not found in most cells but are necessary for the replication of certain types of viruses.*

CONCEPT 19.3

Viruses and prions are formidable pathogens in animals and plants *(pp. 411–416)*

- Symptoms of viral diseases may be caused by direct viral harm to cells or by the body's immune response. **Vaccines** stimulate the immune system to defend the host against specific viruses.
- An **epidemic**, a widespread outbreak of a disease, can become a **pandemic**, a global epidemic.
- Outbreaks of emerging viral diseases in humans are usually not new, but rather are caused by existing viruses that expand their host territory. The H1N1 2009 flu virus was a new combination of pig, human, and avian viral genes that caused a pandemic. The H5N1 avian flu virus has the potential to cause a high-mortality flu pandemic.
- Viruses enter plant cells through damaged cell walls (horizontal transmission) or are inherited from a parent (vertical transmission).
- **Prions** are slow-acting, virtually indestructible infectious proteins that cause brain diseases in mammals.

? *What aspect of an RNA virus makes it more likely than a DNA virus to become an emerging virus?*

TEST YOUR UNDERSTANDING

Levels 1-2: Remembering/Understanding

1. Which of the following characteristics, structures, or processes is common to both bacteria and viruses?
(A) metabolism
(B) ribosomes
(C) genetic material composed of nucleic acid
(D) cell division

2. Emerging viruses arise by
(A) mutation of existing viruses.
(B) the spread of existing viruses to new host species.
(C) the spread of existing viruses more widely within their host species.
(D) all of the above.

3. To cause a human pandemic, the H5N1 avian flu virus would have to
(A) spread to primates such as chimpanzees.
(B) develop into a virus with a different host range.
(C) become capable of human-to-human transmission.
(D) become much more pathogenic.

Levels 3-4: Applying/Analysing

4. A bacterium is infected with an experimentally constructed bacteriophage composed of the T2 phage protein coat and T4 phage DNA. The new phages produced would have
(A) T2 protein and T4 DNA. (C) T2 protein and T2 DNA.
(B) T4 protein and T2 DNA. (D) T4 protein and T4 DNA.

5. RNA viruses require their own supply of certain enzymes because
(A) host cells rapidly destroy the viruses.
(B) host cells lack enzymes that can replicate the viral genome.
(C) these enzymes translate viral mRNA into proteins.
(D) these enzymes penetrate host cell membranes.

6. **DRAW IT** Redraw Figure 19.10 to show the replicative cycle of a virus with a single-stranded genome that can function as mRNA (a class IV virus).

Levels 5-6: Evaluating/Creating

7. **EVOLUTION CONNECTION** The success of some viruses lies in their ability to evolve rapidly within the host. Such viruses evade the host's defences by mutating and producing many altered progeny viruses before the body can mount an attack. Thus, the viruses present late in infection differ from those that initially infected the body. Discuss this as an example of evolution in microcosm. Which viral lineages tend to predominate?

8. **SCIENTIFIC INQUIRY** When bacteria infect an animal, the number of bacteria in the body increases in an exponential fashion (graph A). After infection by a virulent animal virus with a lytic replicative cycle, there is no evidence of infection for a while. Later, however, the number of viruses rises suddenly and subsequently increases in a series of steps (graph B). Explain the difference in the curves.

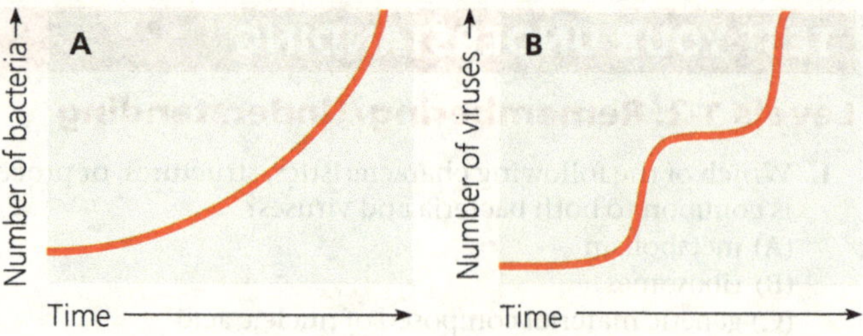

9. **WRITE ABOUT A THEME: ORGANISATION** While viruses are considered by most scientists to be nonliving, they do show some characteristics of life, including the correlation of structure and function. In a short essay (100–150 words), discuss how the structure of a virus correlates with its function.

10. **SYNTHESISE YOUR KNOWLEDGE**

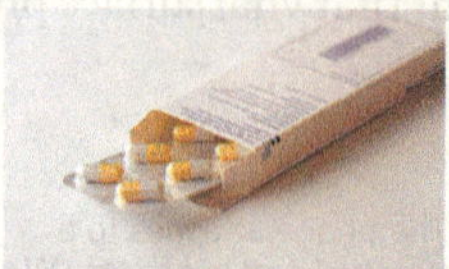

Oseltamivir (Tamiflu), an antiviral drug prescribed for the flu, inhibits the enzyme neuraminidase. Explain how this drug could prevent infection in someone exposed to the flu or could shorten the course of flu in an infected patient (the reasons for which it is prescribed).

For selected answers, see Appendix A.

20 DNA Tools and Biotechnology

KEY CONCEPTS

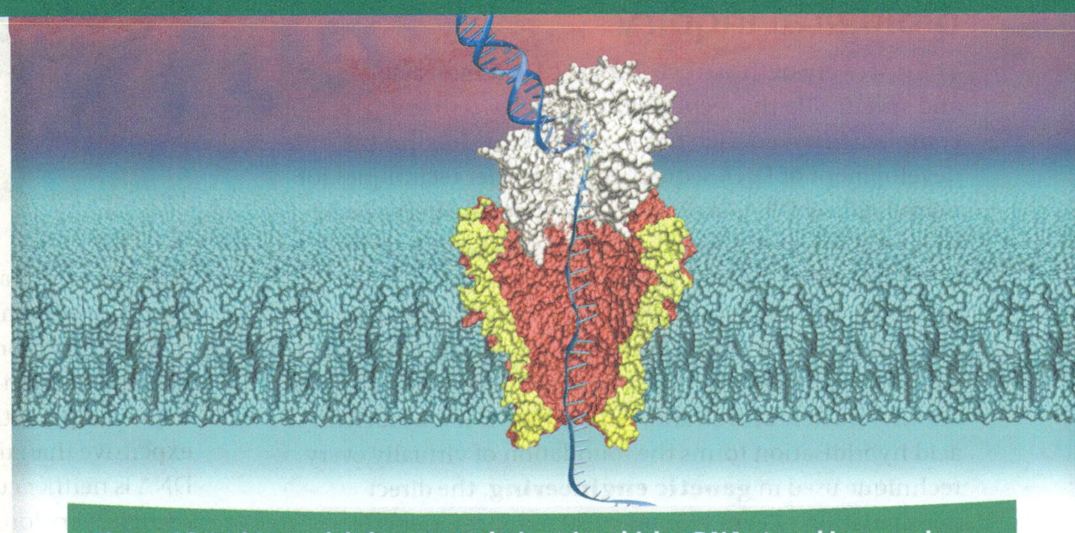

Figure 20.1 This model shows a technique in which a DNA strand is passed through a small pore in a membrane. The resulting changes in an electrical current are used to determine the nucleotide sequence. The first human genome sequence, completed in 2003, took 13 years and cost $1 billion; the time and cost of genome sequencing have been greatly reduced by improved methods like the one shown here.

Study Tip

Apply what you've learned: Write down a gene-related question you've wondered about. Then make a table listing techniques you could use to investigate your question and how you would apply them. Below is an example.

Is autism caused by genes, the environment, or both?	
Technique	Application to question
DNA sequencing	-Compare DNA sequences from people with autism with sequences from nonaffected people.

What are the main techniques and applications of biotechnology?

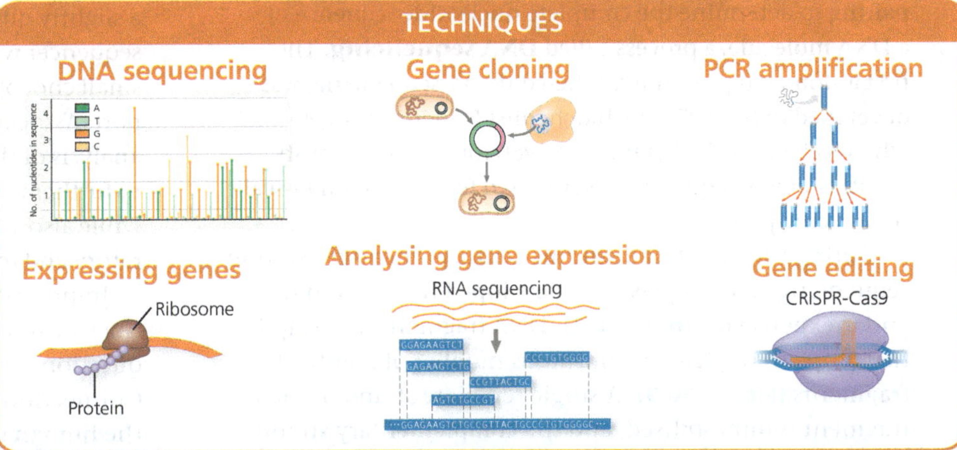

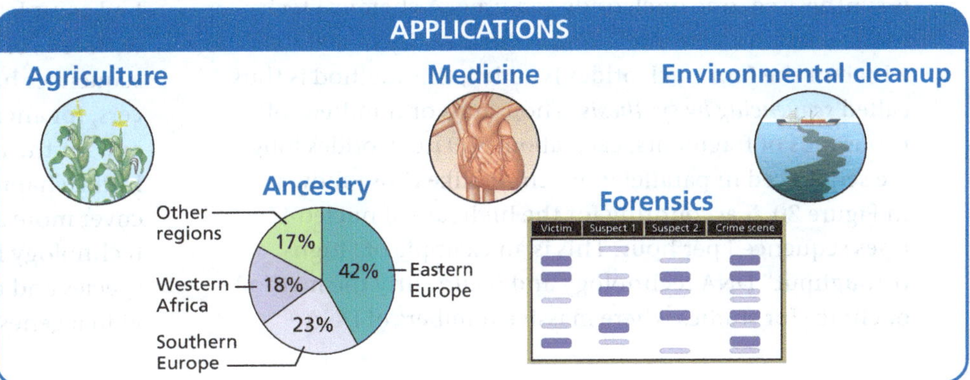

➔ Go to Mastering Biology to access Dynamic Study Modules for revision, 3D BioFlix® animations and high-quality videos, and your interactive Pearson eText.

CONCEPT 20.1

DNA sequencing and DNA cloning are valuable tools for genetic engineering and biological inquiry

The discovery of the structure of the DNA molecule, and specifically the recognition that its two strands are complementary to each other, opened the door for the development of DNA sequencing and other techniques for manipulating DNA—known as **DNA technology**—used today in biological research and beneficial applications that range from the production of lifesaving drugs to environmental bioremediation. Key to these techniques is **nucleic acid hybridisation**, the base pairing of one strand of a nucleic acid to a complementary sequence from another nucleic acid strand, either DNA or RNA. Nucleic acid hybridisation forms the foundation of virtually every technique used in **genetic engineering**, the direct manipulation of genes for practical purposes. Genetic engineering has launched a revolution in fields as varied as criminal law, medicine, and basic biological research. In this section, we'll explore several important techniques and their uses.

DNA Sequencing

Researchers can exploit the principle of complementary base pairing to determine the complete nucleotide sequence of a DNA molecule, a process called **DNA sequencing**. The first automated procedure, called dideoxy sequencing, was developed in the 1970s by biochemist Frederick Sanger, who received the Nobel Prize in 1980 for this accomplishment. Dideoxy sequencing is still used for routine small-scale sequencing jobs.

During the first decade of this century, "next-generation sequencing" techniques were developed that are rapid and inexpensive **(Figure 20.2)**. DNA fragments are amplified (copied) to yield an enormous number of identical fragments **(Figure 20.3)**. A single template strand of each fragment is immobilised, and the complementary strand is synthesised, one nucleotide at a time. A chemical technique enables electronic monitors to identify in real time which of the four nucleotides is added; this method is thus called *sequencing by synthesis*. Thousands or hundreds of thousands of fragments, each about 300 nucleotides long, are sequenced in parallel in machines like those shown in Figure 20.2, accounting for the high rate of nucleotides sequenced per hour. This is an example of "high-throughput" DNA technology and is currently the method of choice for studies where massive numbers of DNA

▼ **Figure 20.2 Next-generation DNA sequencing machines.**

samples—even a set of numerous fragments representing an entire genome—are being sequenced.

More and more often, next-generation sequencing is complemented (or in some cases replaced) by "third-generation sequencing," with each new technique being faster and less expensive than the previous one. In some new methods, the DNA is neither cut into fragments nor amplified. Instead, a single, very long DNA molecule is sequenced on its own. Several groups have developed techniques that move a single strand of a DNA molecule through a very small pore (a *nanopore*) in a membrane, identifying the bases one by one by the distinct way each interrupts an electrical current. One model of this concept is shown in Figure 20.1, in which the pore is a protein channel embedded in a lipid membrane. (Other researchers are using artificial membranes and nanopores.) Each type of base interrupts the electrical current for a slightly different length of time. In 2015, the first nanopore sequencer went on the market; this device is the size of a small chocolate bar and connects to a computer via a USB-A port. Associated software allows immediate identification and analysis of the sequence. This is only one of many approaches to further increase the rate and cut the cost of sequencing, while also allowing the methodology to move out of the laboratory and into the field.

Improved DNA-sequencing techniques have transformed the way in which we can explore fundamental biological questions about evolution and how life works (see Make Connections Figure 5.26). Little more than 15 years after the human genome sequence was announced, researchers had completed the sequencing of thousands of genomes, with tens of thousands in progress. Complete genome sequences have been determined for cells from several cancers, for ancient humans, for extinct woolly mammoths, and for the many bacteria that live in the human intestine. In the chapter, "Genomes and Their Evolution," you'll discover more about how this rapid acceleration of sequencing technology has revolutionised our study of the evolution of species and the genome itself. Now, let's consider how individual genes are studied.

▼ Figure 20.3 Research Method

Sequencing by Synthesis: Next-Generation Sequencing

Application In current next-generation sequencing techniques, each fragment is about 300 nucleotides long; by sequencing the fragments in parallel, about 2 billion nucleotides can be sequenced in 24 hours.

Technique See numbered steps and diagrams.

Results Each of the 2 million wells in the multiwell plate, which holds a different fragment, yields a different sequence. The results for one fragment are shown below as a "flow-gram." The sequences of the entire set of fragments are analysed using computer software, which "stitches" them together into a whole sequence—here, an entire genome.

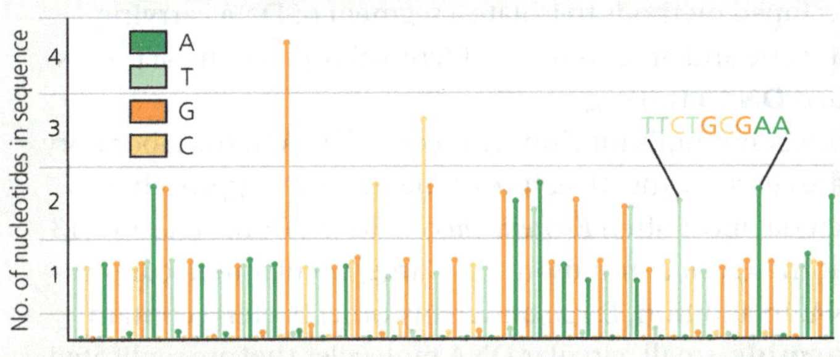

INTERPRET THE DATA *If the template strand has two or more identical nucleotides in a row, their complementary nucleotides will be added one after the other in the same flow step. How are two or more of the same nucleotide (in a row) detected in the flow-gram? (See sample on the right.) Write out the sequence of the first 25 nucleotides in the flow-gram above, starting from the left. (Ignore the very short lines.)*

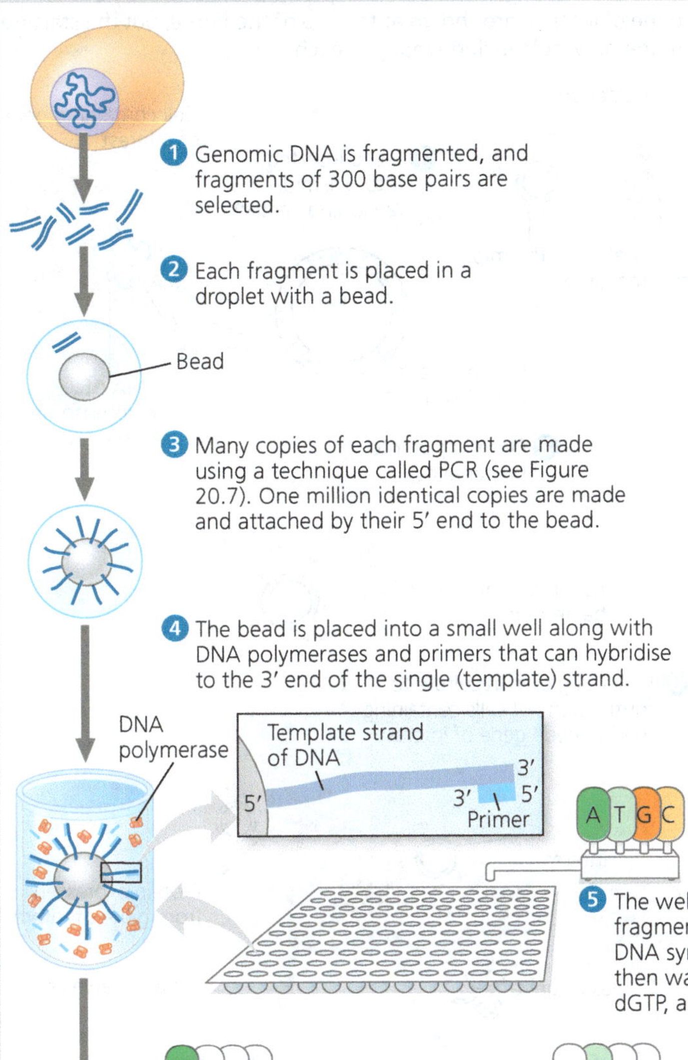

1 Genomic DNA is fragmented, and fragments of 300 base pairs are selected.

2 Each fragment is placed in a droplet with a bead.

3 Many copies of each fragment are made using a technique called PCR (see Figure 20.7). One million identical copies are made and attached by their 5′ end to the bead.

4 The bead is placed into a small well along with DNA polymerases and primers that can hybridise to the 3′ end of the single (template) strand.

5 The well is one of 2 million on a multiwell plate, each containing a different DNA fragment to be sequenced. A solution of one of the four nucleotides required for DNA synthesis (deoxynucleoside triphosphates, or dNTPs) is added to all wells and then washed off. This is done sequentially for all four nucleotides: dATP, dTTP, dGTP, and then dCTP. The entire process is then repeated over and over again.

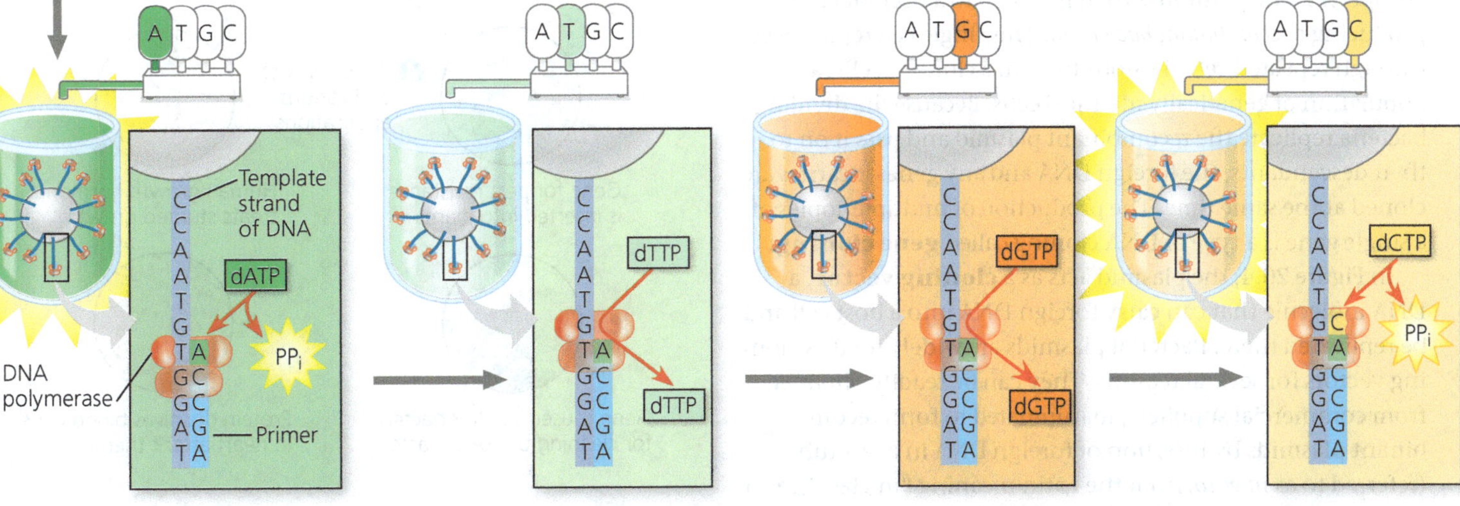

6 In each well, if the next base on the template strand (T in this example) is complementary to the added nucleotide (A, here), the nucleotide is joined to the growing strand, releasing PP$_i$, which causes a flash of light that is recorded.

7 The nucleotide is washed off and a different nucleotide (dTTP, here) is added. If the nucleotide is not complementary to the next template base (G, here), it is not joined to the strand, no reaction occurs, and there is no flash.

8 The process of adding and washing off the four nucleotides is repeated until every fragment has a complete complementary strand. The pattern of flashes reveals the sequence of the original fragment in each well.

Making Multiple Copies of a Gene or Other DNA Segment

A molecular biologist studying a particular gene or group of genes faces a challenge. Naturally occurring DNA molecules are very long, and a single molecule usually carries hundreds or even thousands of genes. Moreover, in many eukaryotic genomes, protein-coding genes occupy only a small proportion of the chromosomal DNA, the rest being noncoding nucleotide sequences. A single human gene, for example, might constitute only 1/100,000 of a chromosomal DNA molecule. As a further complication, it's not easy to distinguish a gene from the surrounding DNA because they differ only in nucleotide sequence. To study a specific gene, scientists have developed methods to isolate a segment of DNA carrying that gene and make multiple identical copies of it—a process called **DNA cloning**.

Most methods for cloning pieces of DNA in the laboratory share certain general features. One common approach uses bacteria, most often *Escherichia coli*. Recall from Figure 16.13 that the *E. coli* chromosome is a large, circular molecule of DNA. In addition, *E. coli* and many other bacteria contain **plasmids**, small, circular DNA molecules that are replicated separately. A plasmid has only a small number of genes; these genes may be useful when the bacterium is in a particular environment but may not be required for survival or reproduction under most conditions.

To clone pieces of DNA using bacteria, scientists have isolated plasmids from bacterial cells and altered them by genetic engineering. Researchers insert DNA they want to study ("foreign" DNA) into the plasmid **(Figure 20.4)**. The resulting plasmid is now a **recombinant DNA molecule**, a molecule containing DNA from two different sources, very often different species. The plasmid is then returned to a bacterial cell, producing a *recombinant bacterium*. This single cell reproduces through repeated cell divisions to form a clone of cells, a population of genetically identical cells. Because the dividing bacteria replicate the recombinant plasmid and pass it on to their descendants, the foreign DNA and any genes it carries are cloned at the same time. The production of multiple copies of a single gene is a type of DNA cloning called **gene cloning**.

In Figure 20.4, the plasmid acts as a **cloning vector**, a DNA molecule that can carry foreign DNA into a host cell and be replicated there. Bacterial plasmids are widely used as cloning vectors for several reasons: They can be readily obtained from commercial suppliers, manipulated to form recombinant plasmids by insertion of foreign DNA in a test tube (referred to as *in vitro*, from the Latin meaning "in glass"), and then easily introduced into bacterial cells. The foreign DNA in Figure 20.4 is a gene from a eukaryotic cell; we will describe in more detail how the foreign DNA segment was obtained later in this section.

▼ **Figure 20.4 Gene cloning and some uses of cloned genes.** In this simplified diagram of gene cloning, we start with a plasmid (originally isolated from a bacterial cell) and a gene of interest from another organism. Only one plasmid and one copy of the gene of interest are shown at the top of the figure, but the starting materials would include many of each.

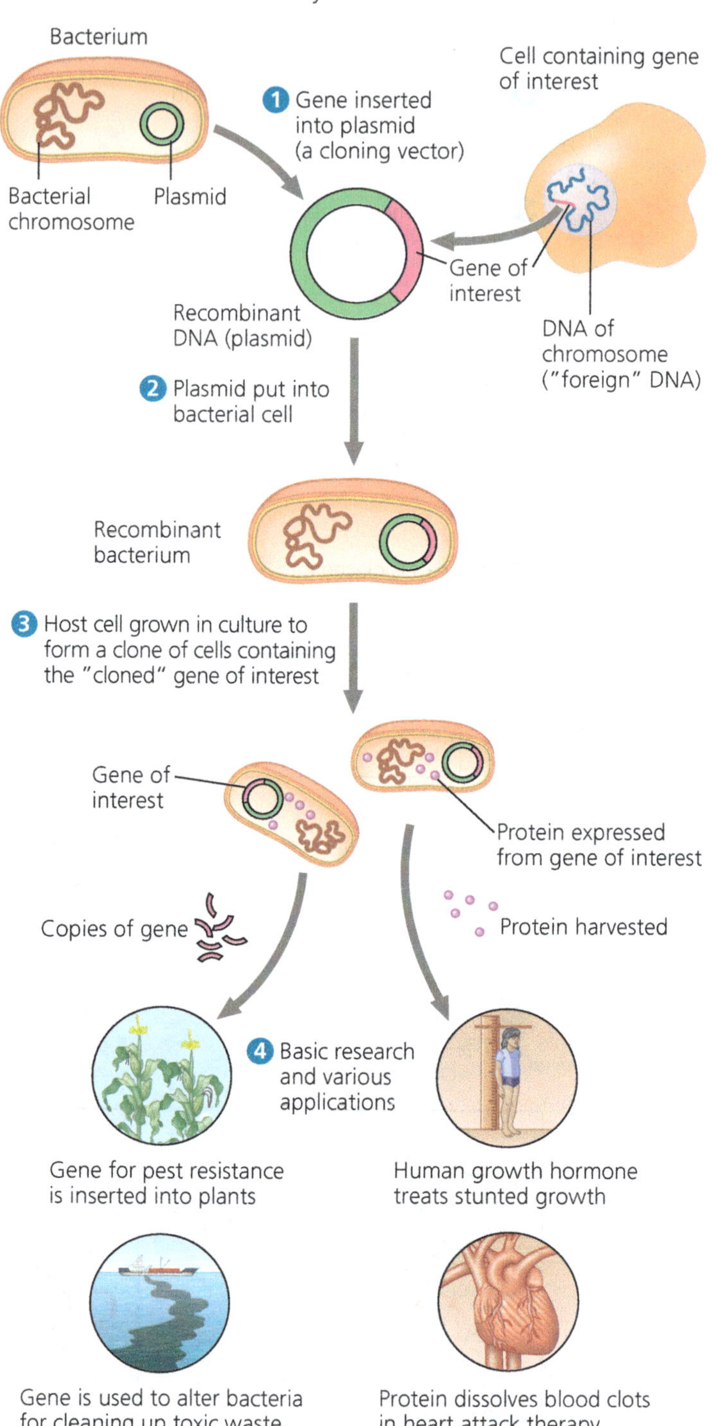

Gene cloning is useful for two basic purposes: to make many copies of, or *amplify*, a particular gene and to produce a protein product from it (see Figure 20.4). Researchers can isolate copies of a cloned gene from bacteria for use in basic research or to endow another organism with a new metabolic capability, such as pest resistance. For example, a resistance gene present in one crop species might be cloned and transferred into plants of another species. (Such organisms are called *genetically modified organisms*, or *GMOs* for short; they will be discussed later in the chapter.) Alternatively, a protein with medical uses, such as human growth hormone, can be harvested in large quantities from cultures of bacteria carrying a cloned gene for the protein. (We'll look at the techniques for expressing cloned genes later.) Since one gene is only a very small part of the total DNA in a cell, the ability to amplify such a DNA fragment is crucial for any application involving a single gene.

Using Restriction Enzymes to Make a Recombinant DNA Plasmid

Gene cloning and genetic engineering generally rely on the use of enzymes that cut DNA molecules at a limited number of specific locations. These enzymes, called restriction endonucleases, or **restriction enzymes**, were discovered in the late 1960s by biologists doing basic research on bacteria. Restriction enzymes protect the bacterial cell by cutting up foreign DNA from other organisms or phages (see Concept 19.2).

Hundreds of different restriction enzymes have been identified and isolated. Each restriction enzyme is very specific, recognising a particular short DNA sequence, or **restriction site**, and cutting both DNA strands at precise points within this restriction site. The DNA of a bacterial cell is protected from the cell's own restriction enzymes by the addition of methyl groups ($—CH_3$) to adenines or cytosines within the sequences recognised by the enzymes.

Figure 20.5 shows how restriction enzymes are used to clone a foreign DNA fragment into a bacterial plasmid. At the top of the figure is a bacterial plasmid (like the one shown in Figure 20.4) that has a single restriction site recognised by a particular restriction enzyme. As shown in this example, most restriction sites are symmetrical. In other words, the sequence of nucleotides is the same on both strands when read in the $5' \rightarrow 3'$ direction. The most commonly used restriction enzymes recognise sequences containing four to eight nucleotide pairs. Because any sequence that is this short usually occurs (by chance) many times in a long DNA molecule, a restriction enzyme will make many cuts in such a DNA molecule, yielding a set of **restriction fragments**. Since restriction enzymes always cut at the same exact DNA sequence, copies of any given DNA molecule exposed to the same restriction enzyme always yield the same set of restriction fragments.

▼ Figure 20.5 Using a restriction enzyme and DNA ligase to make a recombinant DNA plasmid. The restriction enzyme in this example (called *Eco*RI) recognises a single six-base-pair restriction site present in this plasmid. It makes staggered cuts in the sugar-phosphate backbones, producing fragments with "sticky ends." Foreign DNA fragments with complementary sticky ends can base-pair with the plasmid ends; the ligated product is a recombinant plasmid. (If the two plasmid sticky ends base-pair, the original nonrecombinant plasmid is reformed.)

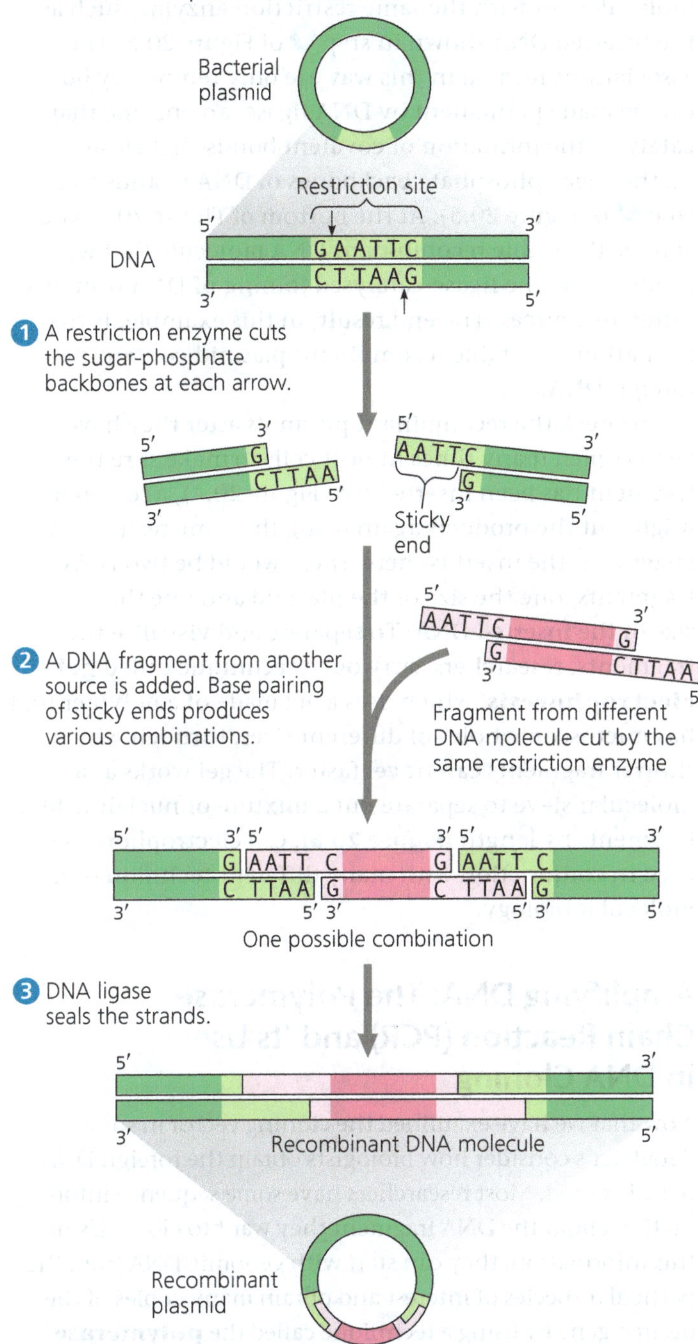

DRAW IT *The restriction enzyme HindIII recognises the sequence 5′-AAGCTT-3′, cutting between the two A's. Draw the double-stranded sequence before and after the enzyme cuts it.*

The most useful restriction enzymes cleave the sugar-phosphate backbones in the two DNA strands in a staggered manner, as shown in step 1 of Figure 20.5. The resulting double-stranded restriction fragments have at least one single-stranded end, called a **sticky end**. These short extensions can form hydrogen-bonded base pairs with complementary sticky ends on any other DNA molecules cut with the same restriction enzyme, such as the inserted DNA shown in step 2 of Figure 20.5. The associations formed in this way are only temporary but can be made permanent by DNA ligase, an enzyme that catalyses the formation of covalent bonds that close up the sugar-phosphate backbones of DNA strands (see step 3 of Figure 20.5). At the bottom of Figure 20.5, you can see the stable recombinant DNA molecule that was produced by the ligase-catalysed joining of DNA from two different sources. The end result, in this example, is the formation of a stable recombinant plasmid containing foreign DNA.

To check the recombinant plasmids after they have been copied many times in host cells to make sure the fragment has been inserted (see Figure 20.4), a researcher might cut the products again using the same restriction enzyme. If the insert is there, there would be two DNA fragments, one the size of the plasmid and one the size of the inserted DNA. To separate and visualise the fragments, researchers carry out a technique called **gel electrophoresis**, which uses a gel made of a polymer that has microscopic holes of different sizes, through which shorter fragments can travel faster. The gel works as a molecular sieve to separate out a mixture of nucleic acid fragments by length **(Figure 20.6)**. Gel electrophoresis is used in conjunction with many different techniques in molecular biology.

Amplifying DNA: The Polymerase Chain Reaction (PCR) and Its Use in DNA Cloning

Now that we have examined the cloning vector in some detail, let's consider how biologists obtain the foreign DNA to be inserted. Most researchers have some sequence information about the DNA fragment they want to clone. Using this information, they can start with genomic DNA from the particular species of interest and obtain many copies of the desired gene by using a technique called the **polymerase chain reaction**, or **PCR**. **Figure 20.7** illustrates the steps in PCR. Within a few hours, this technique can make billions of copies of a specific target DNA segment in a sample, even if that segment makes up less than 0.001% of the total DNA in the sample.

In the PCR procedure, a three-step cycle causes a chain reaction that produces an exponentially growing population of identical DNA molecules. During each cycle, the reaction mixture is 1 heated to high temperatures to denature (separate) the strands of the double-stranded DNA and then 2 cooled to allow annealing (hydrogen bonding) of short, single-stranded DNA primers complementary to sequences on opposite strands at each end of the target sequence; finally, 3 a special DNA polymerase extends the primers in the $5' \rightarrow 3'$ direction. This cycle is then repeated 30–40 times. If a standard DNA polymerase were used, this enzyme would be denatured along with the DNA during the first heating step and would have to be replaced after each cycle.

▼ **Figure 20.6 Gel electrophoresis.** A gel made of a polymer acts as a molecular sieve to separate nucleic acids or proteins differing in size, electrical charge, or other physical properties as they move in an electric field. In the example shown here, DNA molecules are separated by length in a gel made of a polysaccharide called agarose.

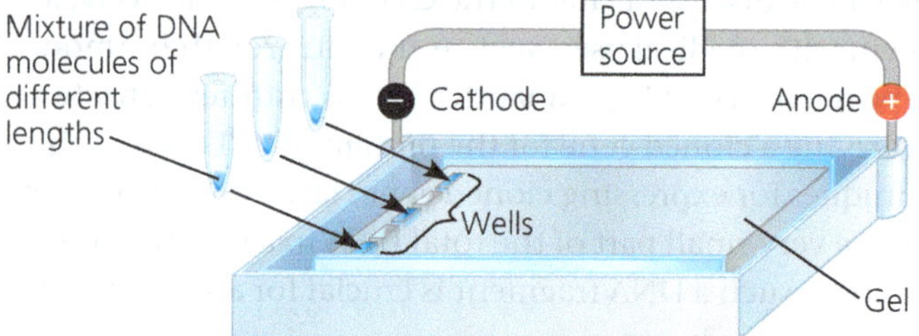

(a) Each sample, a mixture of different DNA molecules, is placed in a separate well near one end of a thin slab of agarose gel. The gel is set into a small plastic support and immersed in an aqueous, buffered solution in a tray with electrodes at each end. The current is then turned on, causing the negatively charged DNA molecules to move towards the positive electrode.

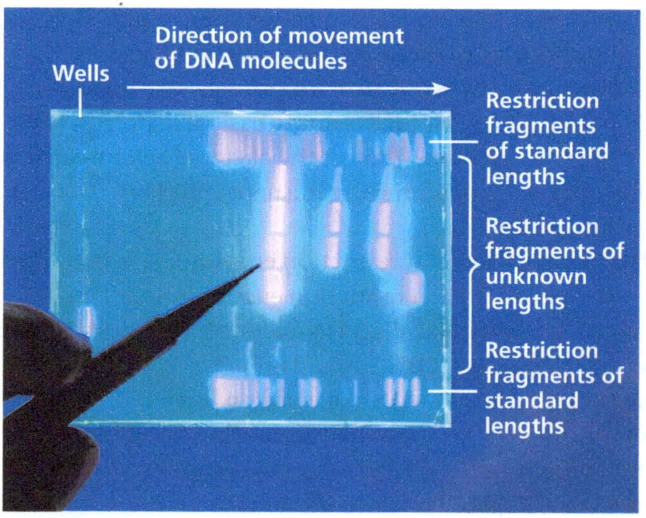

(b) Shorter molecules are slowed down less than longer molecules, so shorter molecules move faster through the gel. After the current is turned off, a DNA-binding dye is added that fluoresces pink in ultraviolet (UV) light. Each pink band corresponds to many thousands of DNA molecules of the same length. The bands at the upper and lower edges of the gel are restriction fragments of standard lengths for comparison with samples of unknown length.

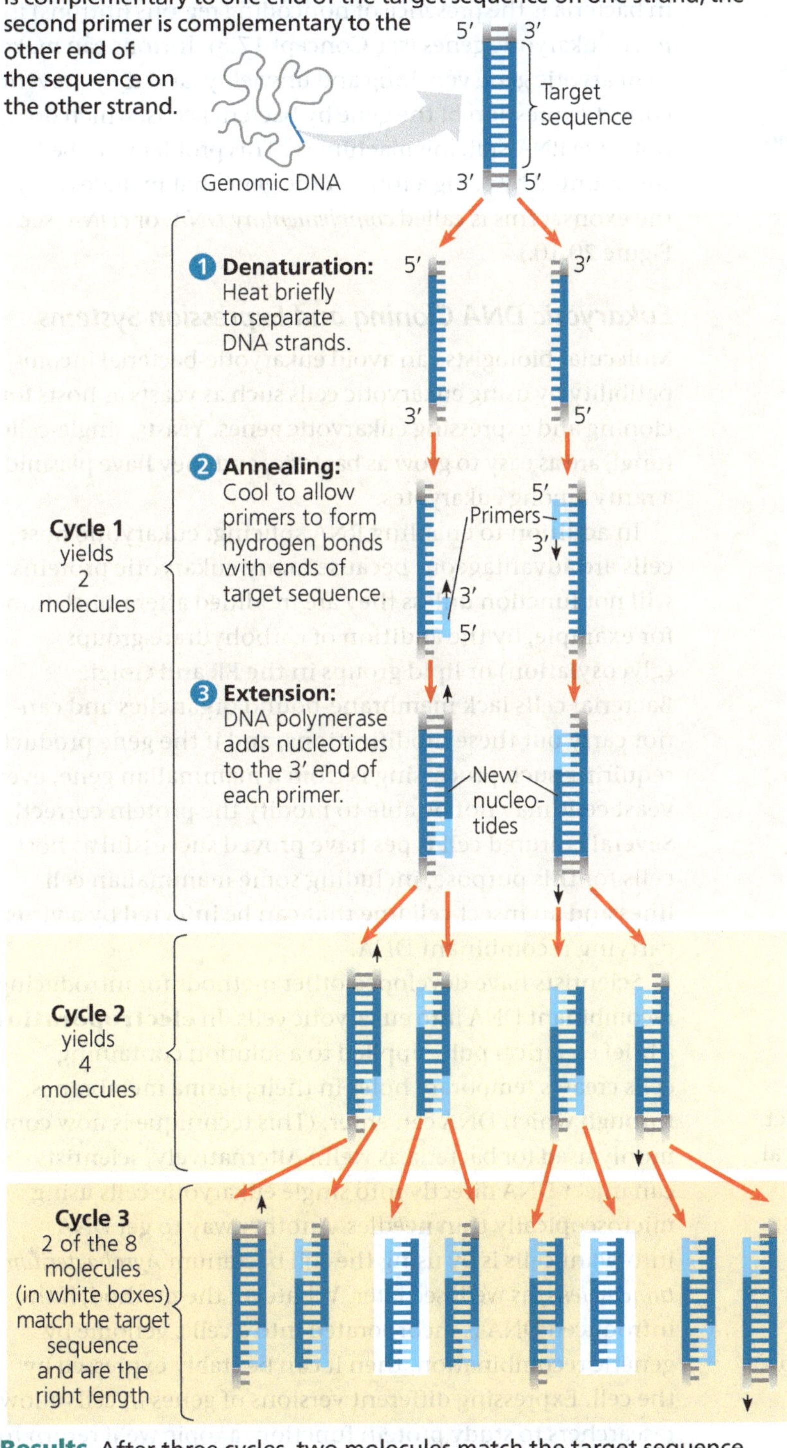

The key to automating PCR was the discovery of an unusual heat-stable DNA polymerase enzyme called *Taq* polymerase, named after the bacterial species from which it was first isolated. This bacterial species, *Thermus aquaticus*, lives in hot springs, and the stability of its DNA polymerase at high temperatures is an evolutionary adaptation that enables the enzyme to function at temperatures up to 95°C. Today, researchers also use a DNA polymerase from the archaean species *Pyrococcus furiosus*. This enzyme, called *Pfu* polymerase, is more accurate and stable but more expensive than *Taq* polymerase.

PCR is speedy and very specific. Only a minuscule amount of DNA need be present in the starting material, and this DNA can be partially degraded, as long as there are a few copies of the complete target sequence. The key to the high specificity is the pair of primers used for each PCR amplification. The primer sequences are chosen so that they hybridise *only* to sequences at opposite ends of the target segment, one on the 3′ end of each strand. (For high specificity, the primers must be at least 15 nucleotides long.) With each successive cycle, the number of target segment molecules of the correct length doubles, so the number of molecules equals 2^n, where n is the number of cycles. After 30 or so cycles, about a billion copies of the target sequence are present!

Despite its speed and specificity, PCR amplification cannot substitute for gene cloning in cells to make large amounts of a gene. This is because the polymerases that are used have no proofreading function, and occasional errors during PCR replication limit the number of good copies and the length of DNA fragments that can be copied. Instead, PCR is used to provide the specific DNA fragment for cloning. PCR primers are synthesised to include a restriction site at each end of the DNA fragment that matches the site in the cloning vector, and the fragment and vector are cut and ligated together **(Figure 20.8)**. The resulting plasmids are sequenced so that those with error-free inserts can be selected.

Devised in 1985, PCR has had a major impact on biological research and genetic engineering. PCR has been used to amplify DNA from a wide variety of sources: a 40,000-year-old frozen woolly mammoth; fingerprints or tiny amounts of blood, tissue, or semen found at crime scenes; single embryonic cells for rapid prenatal diagnosis of genetic disorders (see Figure 14.19); and cells infected with viruses that are difficult to detect,

▼ **Figure 20.8 Use of a restriction enzyme and PCR in gene cloning.** In a closer look at the process shown at the top of Figure 20.4, PCR is used to produce the DNA fragment or gene of interest that will be ligated into a cloning vector, in this case a bacterial plasmid.

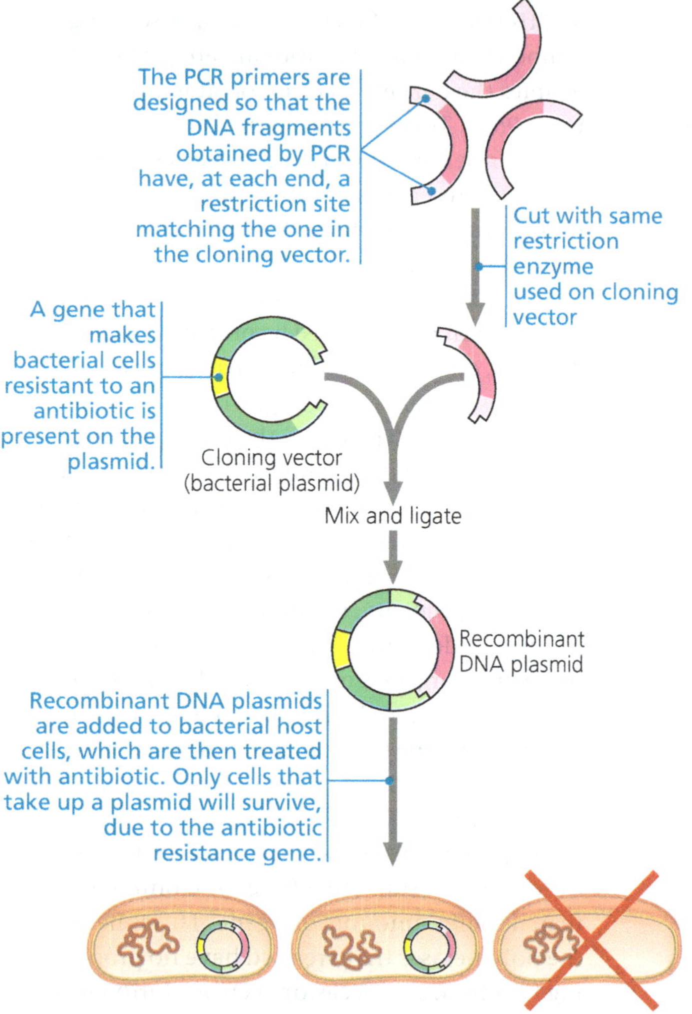

such as HIV. (To test for HIV, viral genes are amplified.) We'll return to applications of PCR later.

Expressing Cloned Eukaryotic Genes

Once a gene has been cloned in host cells, its protein product can be expressed in large amounts for research or for practical applications, which we'll explore in Concept 20.4. Cloned genes can be expressed in either bacterial or eukaryotic cells; each option has advantages and disadvantages.

Bacterial Expression Systems

Getting a cloned eukaryotic gene to function in bacterial host cells can be difficult because certain aspects of gene expression are different in eukaryotes and bacteria. To overcome differences in promoters and other DNA control sequences (see Concept 17.2), scientists usually employ an **expression vector**, a cloning vector that contains a highly active bacterial promoter just upstream of a restriction site where the eukaryotic gene can be inserted in the correct reading frame. The bacterial host cell will recognise the promoter and proceed to express the foreign gene now linked to that promoter. Such expression vectors allow the synthesis of many eukaryotic proteins in bacterial cells.

Another problem with expressing cloned eukaryotic genes in bacteria is the presence of noncoding regions (introns) in most eukaryotic genes (see Concept 17.3). Introns can make a eukaryotic gene very long and unwieldy, and they prevent correct expression of the gene by bacterial cells, which do not have RNA-splicing machinery. This problem can be surmounted by using a form of the gene that includes only the exons. (This is called *complementary DNA*, or *cDNA*; see Figure 20.10.)

Eukaryotic DNA Cloning and Expression Systems

Molecular biologists can avoid eukaryotic-bacterial incompatibility by using eukaryotic cells such as yeasts as hosts for cloning and expressing eukaryotic genes. Yeasts, single-celled fungi, are as easy to grow as bacteria, and they have plasmids, a rarity among eukaryotes.

In addition to enabling RNA splicing, eukaryotic host cells are advantageous because many eukaryotic proteins will not function unless they are modified after translation—for example, by the addition of carbohydrate groups (glycosylation) or lipid groups in the ER and Golgi. Bacterial cells lack membrane-bound organelles and cannot carry out these modifications, and if the gene product requiring such processing is from a mammalian gene, even yeast cells may not be able to modify the protein correctly. Several cultured cell types have proved successful as host cells for this purpose, including some mammalian cell lines and an insect cell line that can be infected by a virus carrying recombinant DNA.

Scientists have developed other methods for introducing recombinant DNA into eukaryotic cells. In **electroporation**, a brief electrical pulse applied to a solution containing cells creates temporary holes in their plasma membranes, through which DNA can enter. (This technique is now commonly used for bacteria as well.) Alternatively, scientists can inject DNA directly into single eukaryotic cells using microscopically thin needles. Another way to get DNA into plant cells is by using the soil bacterium *Agrobacterium tumefaciens*, as we'll see later. Whatever the method, if the introduced DNA is incorporated into a cell's genome by genetic recombination, then it can be stably expressed by the cell. Expressing different versions of genes in cells allows researchers to study protein function, a topic we'll return to in Concept 20.2.

Cross-Species Gene Expression and Evolutionary Ancestry

EVOLUTION The ability to express eukaryotic proteins in bacteria (even if the proteins can't be modified properly) is quite remarkable when we consider how different eukaryotic and bacterial cells are. In fact, examples abound of genes that are taken from one species and function perfectly well when transferred into another very different species, such as a firefly gene in a tobacco plant and a jellyfish gene in a pig (see Figure 17.7). These observations underscore the shared evolutionary ancestry of species living today.

One example involves a gene called *Pax-6*, which has been found in animals as diverse as vertebrates and fruit flies. The vertebrate *Pax-6* gene product (the PAX-6 protein) triggers a complex program of gene expression resulting in formation of the vertebrate eye, which has a single lens. Expression of the fly *Pax-6* gene leads to formation of the compound fly eye, which is quite different from the vertebrate eye. When the mouse *Pax-6* gene was cloned and introduced into a fly embryo so that it replaced the fly's own *Pax-6* gene, researchers were surprised to see that the mouse version of the gene led to formation of a compound fly eye (see Figure 50.16). Conversely, when the fly *Pax-6* gene was transferred into a vertebrate embryo—a frog, in this case—a frog eye formed. Although the genetic programs triggered in vertebrates and flies generate very different eyes, the two versions of the *Pax-6* gene can substitute for each other to trigger lens development, evidence of their evolution from a gene in a very ancient common ancestor. Because of their ancient evolutionary roots, all living organisms share the same basic mechanisms of gene expression. This commonality is the basis of many recombinant DNA techniques described in this chapter.

CONCEPT CHECK 20.1

1. **MAKE CONNECTIONS** The restriction site for an enzyme called *Pvu*I is the following sequence:

   5′-CGATCG-3′
   3′-GCTAGC-5′

   Staggered cuts are made between the T and C on each strand. What type of bonds are being cleaved? (See Concept 5.5.)
2. **DRAW IT** One strand of a DNA molecule has the following sequence:

   5′-CTTGACGATCGTTACCG-3′

   Draw the other strand. Will *Pvu*I (see question 1) cut this molecule? If so, draw the products.
3. What are some potential difficulties in using plasmid vectors and bacterial host cells to produce large quantities of proteins from cloned eukaryotic genes?
4. **VISUAL SKILLS** Compare Figure 20.7 with Figure 16.21. How does replication of DNA ends during PCR proceed without shortening the fragments each time?

For suggested answers, see Appendix A.

CONCEPT 20.2

Biologists use DNA technology to study gene expression and function

To see how a biological system works, scientists seek to understand the functioning of the system's component parts. Analysis of when and where a gene or group of genes is expressed can provide important clues about their function and how they contribute to the organism as a whole.

Analysing Gene Expression

Biologists driven to understand the assorted cell types of a multicellular organism, cancer cells, or the developing tissues of an embryo first try to discover which genes are expressed by the cells of interest. The most straightforward way to do this is usually to identify the mRNAs being made. We'll first examine techniques that look for patterns of expression of specific individual genes. Next, we'll explore ways to characterise groups of genes being expressed by cells or tissues of interest. As you will see, all of these procedures depend in some way on base pairing between complementary nucleotide sequences.

Studying the Expression of Single Genes

Suppose we have cloned a gene that we suspect plays an important role in the embryonic development of *Drosophila melanogaster* (the fruit fly). The first thing we might wonder is which embryonic cells express the gene—in other words, where in the embryo is the corresponding mRNA found? We can detect the mRNA in an embryonic sample using nucleic acid hybridisation with molecules of complementary sequence to the mRNA we want to follow. Using our cloned gene as a template, we can synthesise a short, single-stranded nucleic acid (either RNA or DNA) complementary to the mRNA of interest; this is called a **nucleic acid probe**. For example, if part of the sequence on the mRNA were

5′ ···C U C A U C A C C G G C··· 3′

then we would synthesise this single-stranded DNA probe:

3′ G A G T A G T G G C C G 5′

Each probe molecule is labelled during synthesis with a fluorescent tag so we can follow it. A solution containing probe molecules is applied to *Drosophila* embryos, allowing the probe to hybridise specifically with any complementary sequences on the many mRNAs in embryonic cells in which the gene is being transcribed. Because this technique allows us to see the mRNA in place (or *in situ*, in Latin) in the intact organism, this technique is called ***in situ* hybridisation**.

▼ **Figure 20.9 Determining where single genes are expressed by *in situ* hybridisation analysis.** This *Drosophila* embryo was incubated in a solution containing DNA probes for five different mRNAs, each probe labelled with a different fluorescently coloured tag. The embryo was then viewed using fluorescence microscopy. Each colour marks where a specific gene is expressed as mRNA. The arrows from the groups of yellow and blue cells above the micrograph show a magnified view of nucleic acid hybridisation of the appropriately coloured probe to the mRNA. The thorax (trunk) and abdomen are made up of repeating segments. Yellow cells (expressing the *wg* gene) interact with blue cells (expressing the *en* gene); their interaction helps establish the pattern in a body segment.

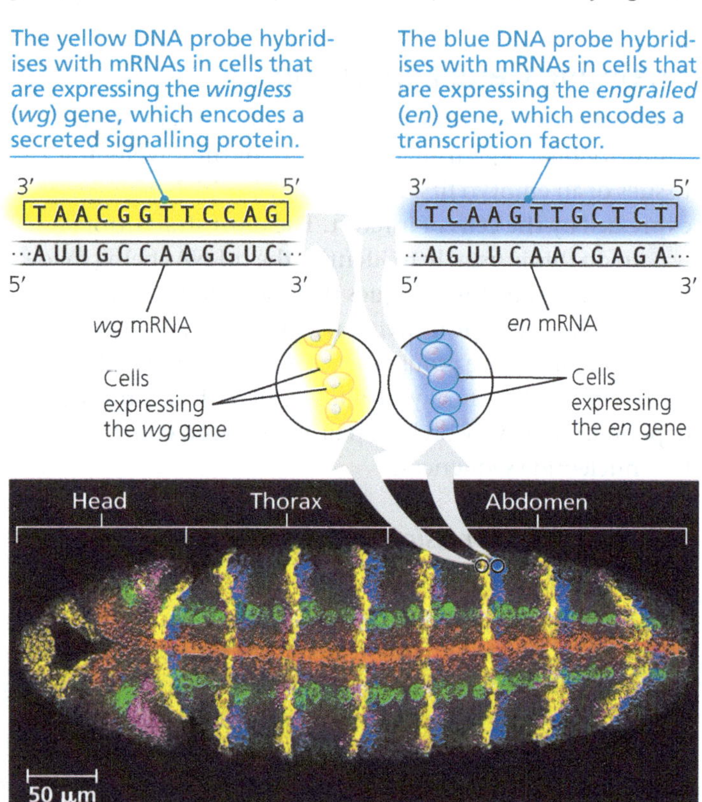

▼ **Figure 20.10 Making complementary DNA (cDNA) from eukaryotic genes.** Complementary DNA is made in a test tube using mRNA as a template for the first strand. Only one mRNA is shown after step 1, but the final collection of cDNAs would reflect all the mRNAs present in the cell.

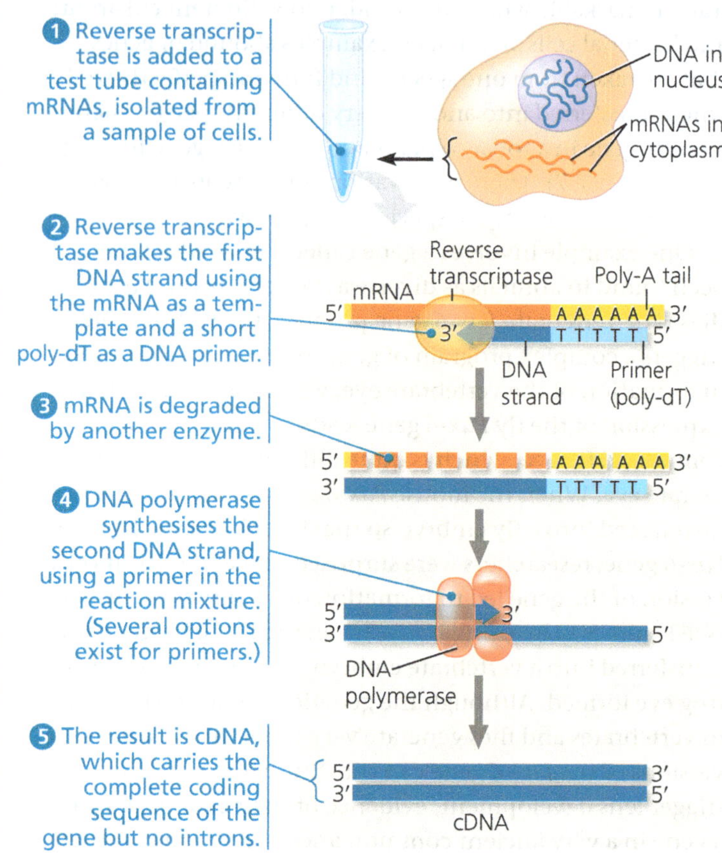

Different probes can be labelled with different fluorescent dyes, sometimes with strikingly beautiful results **(Figure 20.9)**.

Other mRNA detection techniques may be preferable for comparing the amounts of a specific mRNA in several samples at the same time—for example, in different cell types or in embryos of different stages. One method that is widely used is called the **reverse transcriptase polymerase chain reaction**, or **RT-PCR**.

RT-PCR begins by turning sample sets of mRNAs into double-stranded DNAs with the corresponding sequences. First, the enzyme reverse transcriptase (from a retrovirus; see Figure 19.11) is used to synthesise a complementary DNA copy (a *reverse transcript*) of each mRNA in the sample **(Figure 20.10)**. The mRNA is then degraded by addition of a specific enzyme, and a second DNA strand, complementary to the first, is synthesised by DNA polymerase. The resulting double-stranded DNA is called **complementary DNA (cDNA)**. (Made from mRNA, cDNA lacks introns and can be used for protein expression in bacteria, as mentioned earlier.) To analyse the timing of expression of the *Drosophila* gene of interest, for example, we would first isolate all the mRNAs from different stages of *Drosophila* embryos and make cDNA from each stage **(Figure 20.11)**.

Next in RT-PCR is the PCR step (see Figure 20.7). As you will recall, PCR is a way of rapidly making many copies of one specific stretch of double-stranded DNA, using primers that hybridise to the opposite ends of the segment that we are interested in studying. In our case, we would add primers corresponding to a segment of our *Drosophila* gene, using the cDNA from each embryonic stage as a template for PCR amplification in separate samples. When the products are analysed on a gel, copies of the amplified region will be observed as bands only in samples that originally contained mRNA from the gene being studied. This method can tell researchers whether an mRNA is present. However, accurate measurement of mRNA levels requires newer, more quantitative PCR machines that use a dye that fluoresces only when bound to a double-stranded PCR product. This technique, called *quantitative RT-PCR* (*qRT-PCR*), can detect the light and measure the PCR product, thus avoiding the need for electrophoresis while providing quantitative data, a distinct advantage. RT-PCR or qRT-PCR can also be carried out with mRNAs collected from different tissues at one time to discover which tissue is producing a specific mRNA.

Studying the Expression of Interacting Groups of Genes

A major goal of biologists is to learn how genes act together to produce and maintain a functioning organism. Now that the genomes of a number of species have been sequenced, it is possible to study the expression of large groups of genes—an approach called the *systems approach*. Researchers use what is known about the whole genome to investigate which genes are transcribed in different tissues or at different stages of development. One aim is to identify networks of gene expression across an entire genome.

To accomplish such genome-wide expression studies, today's rapid, inexpensive DNA-sequencing methods allow researchers to discover which genes are expressed by simply sequencing the cDNA samples from different tissues or different embryonic stages. This straightforward method is called **RNA sequencing**, or **RNA-seq** (pronounced "RNA-seek"), even though it is the cDNA that is actually sequenced. In RNA-seq, the mRNA (or other RNA) samples are isolated, cut into shorter, similar-sized fragments, and converted into cDNAs **(Figure 20.12)**. These short cDNA stretches are sequenced, and a computer program reassembles them, either mapping them onto the genome of the species in question (when available) or simply putting the fragments in order from scratch based on overlapping sequences of multiple RNAs.

▼ Figure 20.11 Research Method

RT-PCR Analysis of the Expression of Single Genes

Application RT-PCR uses the enzyme reverse transcriptase (RT) in combination with PCR and gel electrophoresis. RT-PCR can be used to compare gene expression between samples—for instance, in different embryonic stages, in different tissues, or in the same type of cell under different conditions.

Technique In this example, samples containing mRNAs from six embryonic stages of *Drosophila* were analysed for a specific mRNA as shown below. (In steps 1 and 2, the mRNA from only one stage is shown.)

1. **cDNA synthesis** is carried out by incubating the mRNAs from each stage with reverse transcriptase and other necessary components.

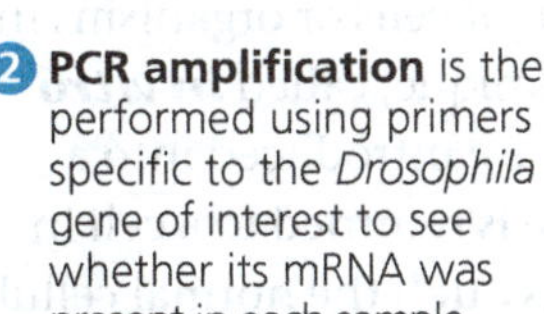

2. **PCR amplification** is then performed using primers specific to the *Drosophila* gene of interest to see whether its mRNA was present in each sample.

3. **Gel electrophoresis** will reveal amplified DNA products only in samples that contained mRNA transcribed from the specific *Drosophila* gene.

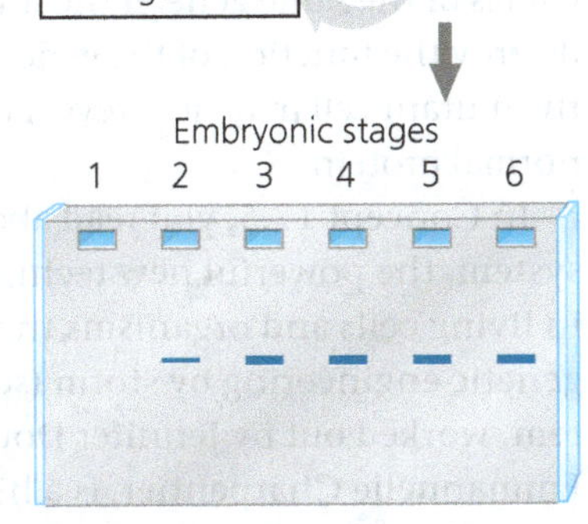

Results The mRNA for this gene is first expressed from stage 2 to stage 6. The size of the amplified fragment (shown by its position on the gel) depends on the distance between the primers that were used (not on the size of the mRNA).

▼ Figure 20.12 Use of RNA sequencing (RNA-seq) to analyse expression of many genes. RNA-seq yields a wide range of information about expression of genes, including their level of expression.

1. mRNAs are isolated from the tissue being studied.
2. mRNAs are cut into similar-sized, small fragments.
3. mRNAs are reverse-transcribed into cDNAs of the same size.
4. cDNAs are sequenced.
5. The short sequences are mapped by computer onto the genome sequence. The resulting data, including the number of times a sequence is present, indicate which genes are expressed in a given tissue and at what level.

GGAGAAGTCT
AGTCTGCCGT
CCCTGTGGGG
CCGTTACTGC
GAGAAGTCTG

···GGAGAAGTCTGCCGTTACTGCCCTGTGGGGC···
Genome sequence

Several characteristics make RNA-seq a powerful technique. First, it is not based on hybridisation with a labelled probe, so it doesn't depend on knowing genomic sequences. Second, it can measure levels of expression over a very wide range. Third, a careful analysis provides a wealth of information about expression of a particular gene, such as relative levels of alternatively spliced mRNAs. In most cases, however, expression of individual genes still needs to be confirmed by RT-PCR.

An older method of genome-wide expression studies, called **DNA micro-array assays**, is less powerful than RNA-seq but is still used for some applications, such as fetal testing (see Figure 14.19). A DNA split micro-array consists of tiny amounts of a large number of single-stranded DNA fragments representing different genes fixed to a glass slide in a tightly spaced array, or grid, of dots. (The micro-array is also called a *DNA chip* by analogy to a computer chip.) Ideally, these fragments represent all the genes of an organism. The mRNAs from the cells being studied are reverse-transcribed into cDNAs (see Figure 20.10), and a fluorescent label is added so the cDNAs can be used as probes on the micro-array. Different fluorescent labels are used for different cell samples so that multiple samples can be tested in the same experiment. The resulting pattern of coloured dots, shown in an actual-size micro-array in **Figure 20.13**, reveals the dots to which each probe was bound and thus the genes that are expressed in the cell samples being tested.

Scientists can now measure the expression of thousands of genes at one time. DNA technology makes such studies possible; with automation, they are easily performed on a large scale. By uncovering gene interactions and providing clues to gene function, DNA micro-array assays and RNA-seq may contribute to a better understanding of diseases and suggest new diagnostic techniques or therapies. For instance, comparing patterns of gene expression in breast cancer tumours and noncancerous breast tissue has already resulted in more informed and effective treatment protocols (see Figure 18.27). Ultimately, information from these methods should provide a grander view of how ensembles of genes interact to form an organism and maintain its vital systems.

▼ **Figure 20.13 Use of split-micro-arrays to analyse expression of many genes.** Researchers extracted mRNAs from two different human tissues and synthesised two sets of cDNAs, fluorescent micro-arrays red (tissue 1) or green (tissue 2). Labelled cDNAs were hybridised with a split-micro-array containing 5,760 human genes (about 25% of human genes), part of which is shown in the enlargement. Red indicates that the gene in that well was expressed in tissue 1, green in tissue 2, yellow in both, and black in neither. The fluorescence intensity at each spot indicates the relative expression of the gene.

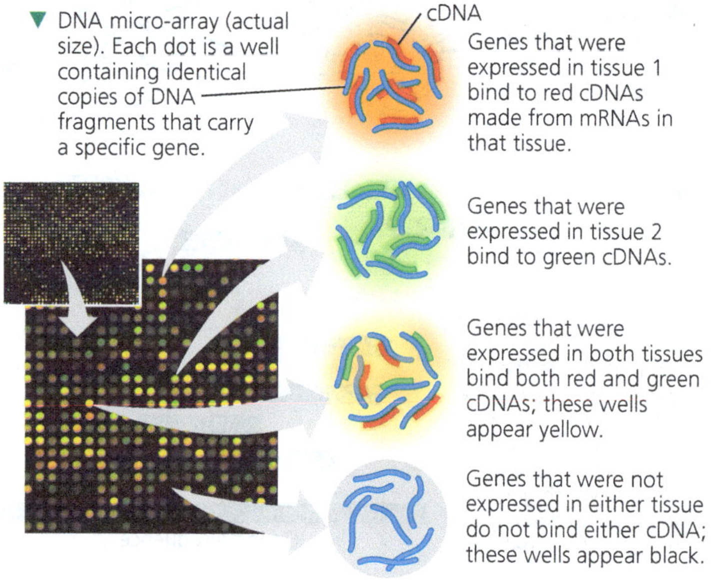

▼ DNA micro-array (actual size). Each dot is a well containing identical copies of DNA fragments that carry a specific gene.

cDNA

Genes that were expressed in tissue 1 bind to red cDNAs made from mRNAs in that tissue.

Genes that were expressed in tissue 2 bind to green cDNAs.

Genes that were expressed in both tissues bind both red and green cDNAs; these wells appear yellow.

Genes that were not expressed in either tissue do not bind either cDNA; these wells appear black.

Determining Gene Function

Once they identify a gene of interest, how do scientists determine its function? A gene's sequence can be compared with sequences in other species. If the function of a similar gene in another species is known, one might suspect that the gene product in question performs a comparable task. Data about the location and timing of gene expression may reinforce the suggested function. To obtain stronger evidence, one approach is to disable the gene and then observe the consequences in the cell or organism.

Editing Genes and Genomes

Molecular biologists have long sought techniques for altering, or editing, the genetic material of cells or organisms in a predictable way. In one such technique, called ***in vitro* mutagenesis**, specific mutations are introduced into a cloned gene, and the mutated gene is returned to a cell in such a way that it disables ("knocks out") the normal cellular copies of the same gene. If the introduced mutations alter or destroy the function of the gene product, the phenotype of the mutant cell may help reveal the function of the missing normal protein.

In Concept 17.5, you read about the CRISPR-Cas9 system, the powerful new technique for gene editing in living cells and organisms that is taking the field of genetic engineering by storm (see Figure 17.28). This system, worked out by Jennifer Doudna **(Figure 20.14)** and Emmanuelle Charpentier, is a highly effective way for researchers to knock out a given gene in order to study what that gene does. It has already been used in many organisms, including bacteria, fish, mice, insects, human cells, and various crop plants. Modifications of the technique allow researchers to repair a gene that has a mutation. This

▼ **Figure 20.14 Jennifer Doudna holding a model of CRISPR-Cas9.**

approach is used for gene therapy, which will be discussed later in the chapter.

In another application of the CRISPR-Cas9 system, scientists are attempting to address the global problem of insect-borne diseases by altering genes in insects so that, for example, they cannot transmit disease. An extra twist to this approach is to engineer the new allele so that it is much more highly favoured for inheritance than the wild-type allele. This technology is called a **gene drive** because the biased inheritance of the engineered gene during reproduction rapidly "drives" the new allele through the population.

Other Methods for Studying Gene Function

Another method for silencing expression of selected genes doesn't alter the genome; instead, it exploits the phenomenon of **RNA interference (RNAi)**, described in Concept 18.3. This experimental approach uses synthetic double-stranded RNA molecules matching the sequence of a particular gene to trigger breakdown of the gene's messenger RNA or to block its translation. In organisms such as the nematode and the fruit fly, RNAi has already proved valuable for analysing the functions of genes on a large scale. This method is quicker than using the CRISPR-Cas9 system, but it leads to only a temporary reduction of gene expression rather than a permanent gene knockout or alteration.

In humans, ethical considerations prohibit knocking out genes to determine their functions. An alternative approach is to analyse the genomes of large numbers of people with a certain phenotypic condition or disease, such as heart disease or diabetes, to try to find differences they all share compared with people without that condition. The logic is that these differences may be associated with one or more malfunctioning genes and thus in a sense are naturally occurring gene knockouts. In these large-scale analyses, called **genome-wide association studies**, researchers look for *genetic markers*, DNA sequences that vary in the population. In a gene, such sequence variation is the basis of different alleles, as we have seen for sickle-cell disease (see Figure 17.26). And just like the coding sequences of genes, noncoding DNA at a specific locus on a chromosome may exhibit small nucleotide differences among individuals. Variations in coding or noncoding DNA sequences among a population are called polymorphisms (from the Greek for "many forms").

Among the most useful genetic markers in tracking down genes that contribute to diseases and disorders are single base-pair variations in the genomes of the human population. For more than 99% of the nucleotides in the human genome, virtually all people have the same nucleotide in each position. However, once in every 100–300 base pairs of both coding and noncoding DNA sequences are positions where the sequence varies between individuals. A single base-pair site where variation is found in at least 1% of the population is called a **single nucleotide polymorphism** (**SNP**, pronounced "snip"); a few million SNPs occur in the human genome. To find SNPs in large numbers of people, it isn't necessary to sequence their DNA; SNPs can be detected by very sensitive micro-array assays, RNA-seq, or PCR.

Once a SNP is identified that is found in all people affected by the disease being studied, researchers focus on that region and sequence it. In nearly all cases, the SNP itself does not contribute directly to the disease in question by altering the encoded protein; in fact, most SNPs are in noncoding regions. Instead, having a particular SNP associated with a disease suggests that the gene whose mutation causes the disease is located very close to that SNP on that chromosome. This closeness (*genetic linkage*; see Concept 15.3) means that crossing over between the SNP and the gene is very unlikely during gamete formation. Therefore, the SNP and gene are almost always inherited together, so the SNP acts as a genetic marker for the disease-causing allele **(Figure 20.15)**. SNPs have been

▼ **Figure 20.15 Single nucleotide polymorphisms (SNPs) as genetic markers for disease-associated alleles.** This diagram depicts the same region of the genome from two groups of individuals, one group having a particular disease or condition with a genetic basis. Unaffected people have an AT pair at a given SNP locus, while affected people have a CG pair there. Once the allele is confirmed as being associated with the disease in question, the SNP that varies in this way can be used as a marker for the disease-associated allele.

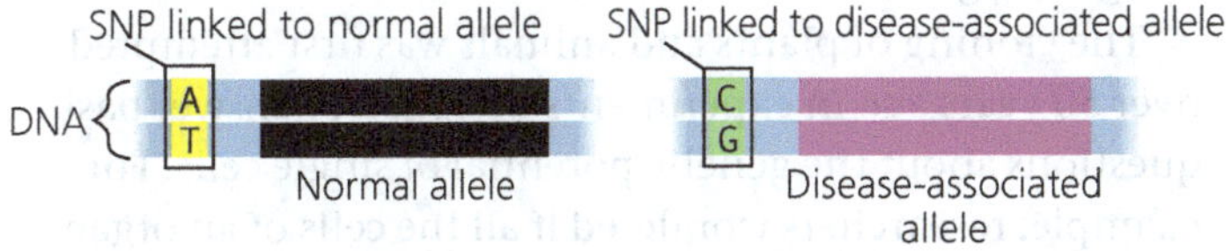

found that correlate with diabetes, heart disease, and several types of cancer, and the search is on for genes that might be involved in these and other conditions that are based on genetics.

The experimental approaches you have learned about thus far focus on working with molecules, mainly DNA and proteins. In a parallel line of inquiry, biologists have been developing powerful techniques for cloning whole multicellular organisms. One aim of this work is to obtain special types of cells, called stem cells, that can give rise to all types of tissues. Being able to manipulate stem cells would allow scientists to use the DNA-based methods previously discussed to alter stem cells for the treatment of diseases. Methods involving the cloning of organisms and production of stem cells are the subject of the next section.

CONCEPT CHECK 20.2

1. Describe the role of complementary base pairing during RT-PCR, RNA sequencing, and DNA micro-array analysis.
2. **VISUAL SKILLS** Consider the micro-array in Figure 20.13. If a sample from normal tissue is labelled with a green fluorescent dye and a sample from cancerous tissue is labelled red, what colour spots would represent genes you would be interested in if you were studying cancer? Explain.

For suggested answers, see Appendix A.

CONCEPT 20.3

Cloned organisms and stem cells are useful for basic research and other applications

Along with advances in DNA technology, scientists have been developing and refining methods for cloning whole multicellular organisms from single cells. In this context, cloning produces one or more organisms that are genetically identical to the "parent" that donated the single cell. This is often called *organismal cloning* to differentiate it from gene cloning and, more significantly, from cell cloning—the division of an asexually reproducing cell such as a bacterium into a group of genetically identical cells. (The common theme is that the product is genetically identical to the parent.) The current interest in organismal cloning is primarily due to its ability to generate stem cells. A **stem cell** is a relatively unspecialised cell that can both reproduce itself indefinitely and, under appropriate conditions, differentiate into specialised cells of one or more types. Stem cells have great potential for regenerating damaged tissues.

The cloning of plants and animals was first attempted over 50 years ago in experiments designed to answer basic questions about the genetic potential of single cells. For example, researchers wondered if all the cells of an organism have the same genes or whether cells lose genes during the process of differentiation (see Concept 18.4). One way to answer this question is to see whether a differentiated cell can generate a whole organism—in other words, whether cloning an organism is possible. Let's discuss these early experiments before we consider more recent progress in organismal cloning and procedures for producing stem cells.

Cloning Plants: Single-Cell Cultures

The successful cloning of whole plants from single differentiated cells was accomplished at Cornell University during the 1950s by F. C. Steward and his students, who worked with carrot plants. They found that differentiated cells taken from the root (the carrot) and incubated in culture medium could grow into normal adult plants, each genetically identical to the parent plant. These results showed that differentiation does not necessarily involve irreversible changes in the DNA. In plants, mature cells can "dedifferentiate" and then give rise to all the specialised cell types of the organism; any cell with this potential is said to be **totipotent**.

Plant cloning is used extensively in agriculture. For plants such as orchids, cloning is the only commercially practical means of producing new plants. In other cases, cloning has been used to reproduce a plant with valuable characteristics, such as resistance to plant pathogens. In fact, you may be come a professional a plant cloner: As it happens, if you have ever grown a new plant from a cutting, you have already practised cloning!

Cloning Animals: Nuclear Transplantation

Differentiated cells from animals generally do not divide in culture, much less develop into the multiple cell types of a new organism. Therefore, early researchers had to use a different approach to answer the question of whether differentiated animal cells are totipotent. Their approach was to remove the nucleus of an egg (creating an *enucleated* egg) and replace it with the nucleus of a differentiated cell, a procedure called *nuclear transplantation*, now more commonly called *somatic cell nuclear transfer*. If the nucleus from the differentiated donor cell retains its full genetic potential, then it should be able to direct development of the recipient cell into all the tissues and organs of an organism. Such experiments were conducted on one species of frog (*Rana pipiens*) by Robert Briggs and Thomas King in the 1950s and on another frog species (*Xenopus laevis*) by John Gurdon in the 1970s **(Figure 20.16)**. These researchers transplanted a nucleus from an embryonic or tadpole cell into an enucleated egg of the same species. In Gurdon's experiments, the transplanted nucleus was often able to support normal development of the egg into a tadpole. However, he found that the potential of a transplanted nucleus to direct normal development was inversely related to the age of the donor: The older the donor nucleus, the lower the percentage of normal tadpoles.

From these results, Gurdon concluded that something in the nucleus *does* change as animal cells differentiate. In frogs and most other animals, nuclear potential tends to be restricted more and more as embryonic development and cell differentiation progress. These were foundational experiments that ultimately led to stem cell technology, and Gurdon received the 2012 Nobel Prize in Medicine for this work.

▼ **Figure 20.16 Inquiry**

Can the nucleus from a differentiated animal cell direct development of an organism?

Experiment John Gurdon and colleagues at Oxford University, in England, destroyed the nuclei of frog (*Xenopus laevis*) eggs by exposing the eggs to ultraviolet light. They then transplanted nuclei from cells of frog embryos and tadpoles into the enucleated eggs.

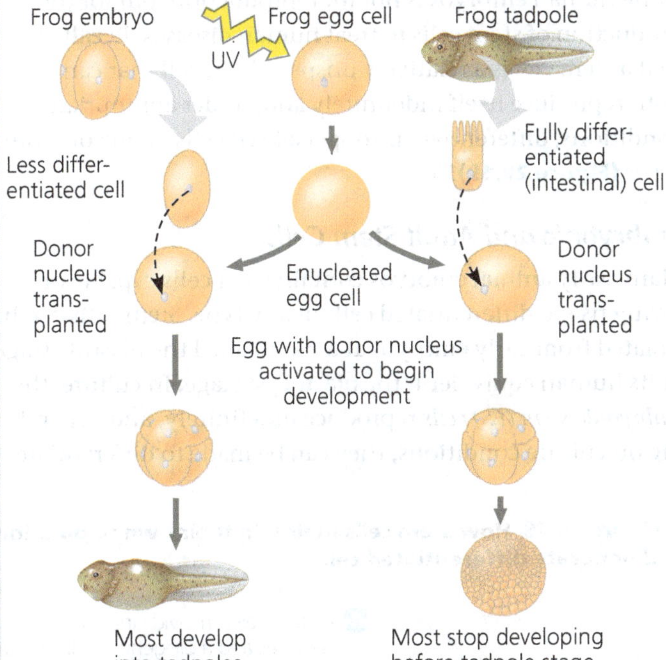

Results When the transplanted nuclei came from an early embryo, the cells of which are relatively undifferentiated, most of the recipient eggs developed into tadpoles. But when the nuclei came from the fully differentiated intestinal cells of a tadpole, fewer than 2% of the eggs developed into normal tadpoles, and most of the embryos stopped developing at a much earlier stage.

Conclusion The nucleus from a differentiated frog cell can direct development of a tadpole. However, its ability to do so decreases as the donor cell becomes more differentiated, presumably because of changes in the nucleus.

Data from J. B. Gurdon et al., The developmental capacity of nuclei transplanted from keratinized cells of adult frogs, *Journal of Embryology and Experimental Morphology* 34:93–112 (1975).

WHAT IF? *If each cell in a four-cell embryo was already so specialised that it was not totipotent, what results would you predict for the experiment on the left side of the figure?*

Reproductive Cloning of Mammals

In addition to cloning frogs, researchers were able to clone mammals using early embryonic cells as a source of donor nuclei. Until about 25 years ago, though, it was not known whether a nucleus from a fully differentiated cell could be reprogrammed successfully to act as a donor nucleus. In 1997, researchers in Scotland announced the birth of Dolly, a lamb cloned from an adult sheep by nuclear transfer from a differentiated mammary gland cell **(Figure 20.17)**. Using a technique related to that in Figure 20.16, the researchers implanted early embryos into surrogate mothers. Out of several hundred embryos, one successfully completed normal development, and Dolly was born, a genetic clone of the nucleus donor. By the age of 6, Dolly had developed premature arthritis, and complications from a lung infection led to her euthanisation. This led to speculation that this sheep's cells were in some way not quite as healthy as those of a normal sheep, possibly reflecting incomplete reprogramming of the original transplanted nucleus. Reprogramming involves epigenetic changes that lead to changes in chromatin structure (see Concept 18.2), to be discussed shortly.

► **Figure 20.17 Reproductive cloning of a mammal by nuclear transfer.** Dolly, shown here as a lamb, has a very different appearance from her surrogate mother, standing beside her.

Since that time, researchers have cloned many other mammals, including mice, cats, cows, horses, pigs, and dogs. In 2018, Chinese biologists reported the first cloning of a primate, the long-tailed macaque. In most cases, the research goal has been the production of new individuals, known as *reproductive cloning*. We have learned from such experiments that cloned animals of the same species are not always identical. For example, in 2016, scientists examined four 7- to 9-year-old clones genetically identical to Dolly and concluded that, unlike Dolly, they were healthy and aging normally. Another example of nonidentity in

▼ **Figure 20.18 CC ("Carbon Copy"), the first cloned cat (right), and her single parent.** Rainbow (left) donated the nucleus in a cloning procedure that resulted in CC. However, the two cats are not identical: Rainbow is a classic calico cat with ginger patches on her fur and has a "reserved personality," while CC has a grey and white coat and is more playful.

clones is the first cloned cat, named CC for Carbon Copy (**Figure 20.18**). The colour and pattern of her calico coat differed from that of her single female parent because of random X chromosome inactivation, which is a normal occurrence during embryonic development (see Figure 15.8). And identical human twins, which are naturally occurring "clones," are always slightly different. Clearly, environmental influences and random phenomena play a significant role during development.

Epigenetic Differences in Cloned Animals

In most nuclear transplantation studies thus far, only a small percentage of cloned embryos develop normally to birth. And like Dolly, many cloned animals exhibit defects. Cloned mice, for instance, are prone to obesity, pneumonia, liver failure, and premature death. Scientists assert that even cloned animals that appear normal are likely to have subtle defects.

Researchers have uncovered some reasons for the low efficiency of cloning and the high incidence of abnormalities. In the nuclei of fully differentiated cells, a small subset of genes is turned on and expression of the rest of the genes is repressed. This regulation often results from epigenetic changes in chromatin, such as acetylation of histones or methylation of DNA (see Figure 18.7). During the nuclear transfer procedure, many of these changes must be reversed in the later-stage nucleus from a donor animal for genes to be expressed or repressed appropriately in earlier stages of development. Researchers have found that the DNA in cells from cloned embryos, like that of differentiated cells, often has abnormally high numbers of methyl groups. This finding suggests that the reprogramming of donor nuclei requires more accurate and complete chromatin restructuring than occurs during cloning procedures. Because DNA methylation helps regulate gene expression, misplaced or extra methyl groups in the DNA of donor nuclei may interfere with the pattern of gene expression necessary for normal embryonic development. In fact, the success of a cloning attempt may depend in large part on whether or not the chromatin in the donor nucleus can be artificially modified to resemble that of a newly fertilised egg.

Stem Cells of Animals

Progress in cloning mammalian embryos, including primates, has heightened speculation about the cloning of humans, which has not yet been achieved past very early embryonic stages. The main reason researchers have been trying to clone human embryos is not for reproduction, but for the production of stem cells to treat human diseases. Recall that a stem cell is a relatively unspecialised cell that can both reproduce itself indefinitely and, under appropriate conditions, differentiate into specialised cells of one or more types (**Figure 20.19**).

Embryonic and Adult Stem Cells

Many early animal embryos contain stem cells capable of giving rise to differentiated cells of any type. Stem cells can be isolated from early embryos at a stage called the blastula stage or its human equivalent, the blastocyst stage. In culture, these *embryonic stem (ES) cells* reproduce indefinitely, and depending on culture conditions, they can be made to differentiate

▼ **Figure 20.19 How stem cells maintain their own population and generate differentiated cells.**

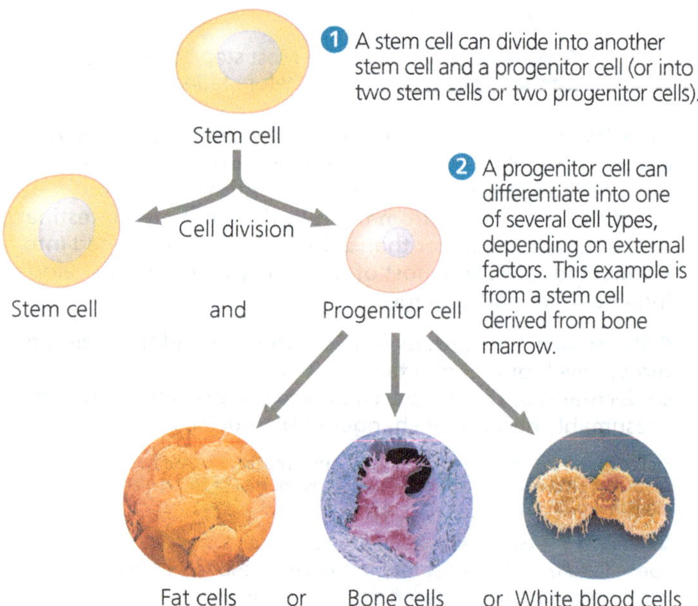

into a wide variety of specialised cells **(Figure 20.20)**, including even eggs and sperm.

The adult body also has stem cells, which serve to replace nonreproducing specialised cells as needed. In contrast to ES cells, *adult stem cells* are not able to give rise to all cell types in the organism, though they can generate several defined types. For example, one of the several types of stem cells in bone marrow can generate all the different kinds of blood cells (see Figure 20.20), and another type of bone marrow stem cell can differentiate into bone, cartilage, fat, muscle, and the linings of blood vessels. To the surprise of many, the adult brain has been found to contain stem cells that continue to produce certain kinds of nerve cells there. Researchers have also reported finding stem cells in skin, hair, eyes, and dental pulp. Although adult animals have only tiny numbers of stem cells, scientists are learning to identify and isolate these cells from various tissues and, in some cases, to grow them in culture. With the right culture conditions (for instance, the addition of specific growth factors), cultured stem cells from adult animals have been made to differentiate into various defined types of specialised cells, although none are as versatile as ES cells.

▼ Figure 20.20 Working with stem cells. Animal stem cells, which can be isolated from early embryos or adult tissues and grown in culture, are self-perpetuating, relatively undifferentiated cells. Embryonic stem cells are easier to grow than adult stem cells and can theoretically give rise to *all* types of cells in an organism. The range of cell types that can arise from adult stem cells is not yet fully understood.

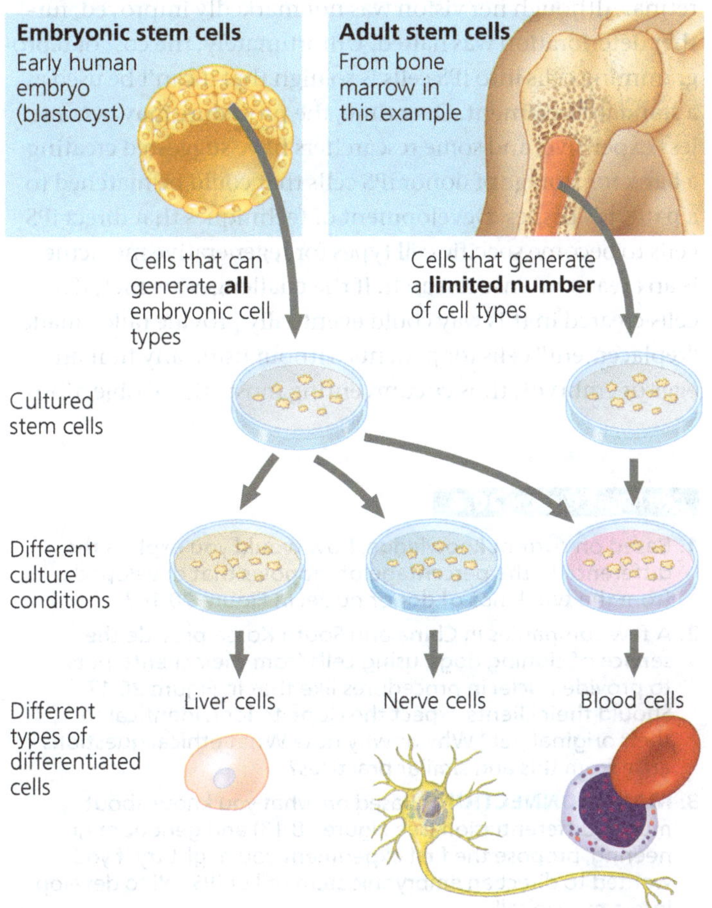

Research with embryonic or adult stem cells is a source of valuable data about differentiation and has enormous potential for medical applications. The ultimate aim is to supply cells for the repair of damaged or diseased organs: for example, insulin-producing pancreatic cells for people with type 1 diabetes or certain kinds of brain cells for people with Parkinson's disease or Huntington's disease. Adult stem cells from bone marrow have long been used in bone marrow transplants as a source of immune system cells in patients whose own immune systems are nonfunctional because of genetic disorders or radiation treatments for cancer.

The developmental potential of adult stem cells is limited to certain tissues. ES cells hold more promise than adult stem cells for most medical applications because ES cells are **pluripotent**, capable of differentiating into many different cell types. One source of ES cells, first reported in 2013, is cloned human blastocysts produced by transferring a nucleus from a differentiated cell into an enucleated egg (similar to cloning Dolly the sheep). Prior to that report, cells were obtained only from embryos donated by patients undergoing infertility treatments or from long-term cell cultures originally established with cells isolated from donated embryos, an issue that prompts ethical and political discussions. Although the techniques for cloning early human embryos are still being optimised and the ethics discussed by researchers, they represent a potential new source for ES cells that may be less controversial. Furthermore, with a donor nucleus from a person with a particular disease, researchers should be able to produce ES cells that match the patient and are thus not rejected by his or her immune system when used for treatment. When the main aim of cloning is to produce ES cells to treat disease, the process is called *therapeutic cloning*. Although most people believe that reproductive cloning of humans is unethical, opinions vary about the morality of therapeutic cloning.

Induced Pluripotent Stem (iPS) Cells

In another approach to make stem cells for research and therapy, researchers succeeded in 2007 in learning how to turn back the clock in fully differentiated cells, reprogramming them to act like ES cells. Differentiated cells can be transformed into a type of ES cell by using a modified retrovirus to introduce extra, cloned copies of four "stem cell" master regulatory genes. The "deprogrammed" cells are known as *induced pluripotent stem* (*iPS*) cells because, in using this fairly simple laboratory technique to return them to their undifferentiated state, pluripotency has been restored. The experiments that first transformed human differentiated cells into iPS cells are described in **Figure 20.21**. Shinya Yamanaka received the 2012 Nobel Prize in Medicine for this work, shared with John Gurdon, whose work you read about in Figure 20.16.

By many criteria, iPS cells can perform most of the functions of ES cells, but there are some differences in gene expression and other cellular functions, such as cell division. At least until these differences are fully understood, the study of ES cells will continue to make important contributions to the development of stem cell therapies. (In fact, it is likely that ES cells will always be a focus of basic research as well.) In the meantime, work is proceeding using the iPS cells that have been experimentally produced.

There are two major potential uses for human iPS cells. First, cells from patients with diseases have been reprogrammed to become iPS cells, which act as model cells for studying the disease and potential treatments. Human iPS cell lines have already been developed from individuals with type 1 diabetes, Parkinson's disease, Huntington's disease, Down syndrome, and many other diseases. Second, in the field of regenerative medicine, a patient's own cells could be reprogrammed into iPS cells and then used to replace nonfunctional tissues, such as cells of the retina of the eye that have been damaged by a condition called age-related macular degeneration (AMD). In fact, in 2014, Japanese researchers made iPS cells out of an AMD patient's skin cells, caused them to differentiate into retinal cells, and implanted them into her retina. Although her vision was not markedly improved, further deterioration was halted. Unfortunately, the cost of reprogramming cells into iPS cells is so high that it can't be used as a standard treatment. Over time, the procedure may become less expensive, and some researchers have suggested creating a bank for storage of donor iPS cells that could be matched to a patient's tissues. Development of techniques that direct iPS cells to become specific cell types for regenerative medicine is an area of intense research. If the challenges are met, iPS cells created in this way could eventually provide tailor-made "replacement" cells for patients without using any human eggs or embryos, thus circumventing most ethical objections.

▼ Figure 20.21 Inquiry

Can a fully differentiated human cell be "deprogrammed" to become a stem cell?

Experiment Shinya Yamanaka and colleagues at Kyoto University, in Japan, used a retroviral vector to introduce four genes into fully differentiated human skin fibroblast cells. The cells were then cultured in a medium that would support growth of stem cells.

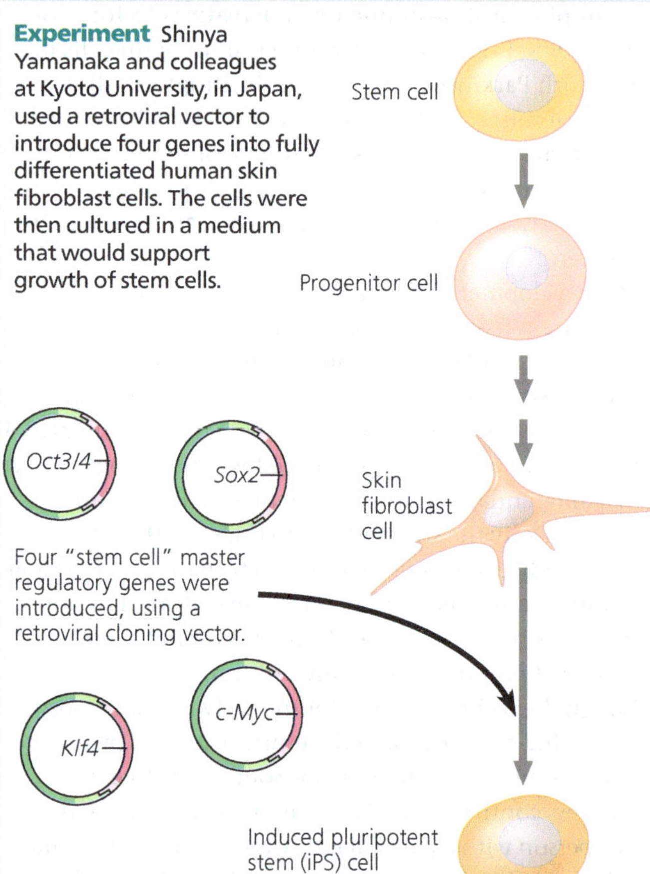

Results Two weeks later, the cells resembled embryonic stem cells in appearance and were actively dividing. Their gene expression patterns, gene methylation patterns, and other characteristics were also consistent with those of embryonic stem cells. The iPS cells were able to differentiate into heart muscle cells, as well as other cell types.

Conclusion The four genes induced differentiated skin cells to become pluripotent stem cells, with characteristics of embryonic stem cells.

Data from K. Takahashi et al., Induction of pluripotent stem cells from adult human fibroblasts by defined factors, *Cell* 131:861–872 (2007).

WHAT IF? *Patients with diseases such as heart disease or Alzheimer's could have their own skin cells reprogrammed to become iPS cells. Once affordable procedures have been developed for this and for converting iPS cells into heart or nervous system cells, the patients' own iPS cells might be used to treat their disease. When organs are transplanted from a donor to a recipient, the recipient's immune system may reject the transplant, a dangerous condition. Would using iPS cells be expected to carry the same risk? Why or why not? Given that these cells are actively dividing, undifferentiated cells, what risks might this procedure carry?*

CONCEPT CHECK 20.3

1. Based on current knowledge, how would you explain the difference in the percentage of tadpoles that developed from the two kinds of donor nuclei in Figure 20.16?
2. A few companies in China and South Korea provide the service of cloning dogs, using cells from their clients' pets to provide nuclei in procedures like that in Figure 20.17. Should their clients expect the clone to look identical to their original pet? Why or why not? What ethical questions arise from this and similar practices?
3. **MAKE CONNECTIONS** Based on what you know about muscle differentiation (see Figure 18.18) and genetic engineering, propose the first experiment you might try if you wanted to direct an embryonic stem cell or iPS cell to develop into a muscle cell.

For suggested answers, see Appendix A.

CONCEPT 20.4

The practical applications of DNA-based biotechnology affect our lives in many ways

In the last section, we'll survey the practical applications of DNA-based **biotechnology**, the manipulation of organisms or their components to make useful products. Today, major applications of DNA technology and genetic engineering include medicine, forensic evidence and genetic profiles, environmental cleanup, and agriculture.

Medical Applications

One important use of DNA technology is the identification of human genes whose mutation plays a role in genetic diseases. These discoveries may lead to ways of diagnosing, treating, and even preventing such conditions. DNA technology has also identified genes that play a role in a number of "nongenetic" diseases, from arthritis to AIDS, by influencing susceptibility to these diseases. Furthermore, a wide variety of diseases involve changes in gene expression within the affected cells and often within the patient's immune system. By using RNA-seq and DNA micro-array assays (see Figures 20.12 and 20.13) or other techniques to compare gene expression in healthy and diseased tissues, researchers are finding genes that are turned on or off in particular diseases. These genes and their products are potential targets for prevention or therapy.

Diagnosis and Treatment of Diseases

A new chapter in the diagnosis of infectious diseases has been opened by DNA technology, in particular the use of PCR and labelled nucleic acid probes to track down pathogens. For example, because the sequence of the RNA genome of HIV is known, RT-PCR can be used to amplify and therefore detect and quantify even a small amount of HIV RNA in blood or tissue samples (see Figure 20.11). RT-PCR is often the best way to detect an otherwise elusive infectious agent.

Medical scientists can now diagnose hundreds of human genetic disorders by using PCR with primers that target the genes associated with these disorders. The amplified DNA product is then sequenced to reveal the presence or absence of the disease-causing mutation. Among the genes for human diseases targeted in this way are those for sickle-cell disease, haemophilia, cystic fibrosis, Huntington's disease, and Duchenne's muscular dystrophy. Individuals with such diseases can often be identified before the onset of symptoms, even before birth (see Figure 14.19). PCR can also be used to identify symptomless carriers of potentially harmful recessive alleles as part of genetic counselling (see Concept 14.4).

Personal Genome Analysis

As you learned earlier, genome-wide association studies have pinpointed SNPs (single nucleotide polymorphisms) that are linked to disease-associated alleles (see Figure 20.15). Individuals can be tested by PCR and sequencing for a SNP that is correlated with the abnormal allele. The presence of particular SNPs is correlated with increased risk for particular adverse health conditions such as heart disease, Alzheimer's disease, and some types of cancer.

Direct-to-consumer genome analysis companies offer kits allowing individuals to send in a swab containing cheek cells that the company will analyse genetically. Genetic testing for risk factors like heart disease, Alzheimer's disease, and some types of cancer is carried out by looking for linked SNPs that have been previously identified (see Figure 20.15). It may be helpful for individuals to learn about their health risks, with the understanding that such genetic tests merely reflect correlations and do not make predictions.

In addition to health-related genetic information, these companies compare a person's DNA segments with those from reference populations around the world, established from thousands of individuals of known ancestry. Based on how closely the segments match up, the report can tell an individual about their likely ancestral breakdown. Females can also learn about their maternal lines of descent, based on comparisons of mitochondrial DNA, which is contributed to the egg only by the mother. For males, an analysis of the sequence of the Y chromosome can trace their paternal lines of descent. As the size of the database used by these companies increases, the results become more refined and more accurate.

Personalised Medicine

The techniques described in this chapter have also prompted improvements in disease treatments—for example, by knowing which specific cancer-related genes have been mutated in a particular patient's tumour. By analysing the expression of many genes in large numbers of breast cancer patients, researchers can refine their understanding of the different subtypes of breast cancer (see Figure 18.27). Knowing the expression levels of particular genes in a given individual can help doctors determine the likelihood that the cancer will recur, thus helping them design an appropriate treatment. Given that some low-risk patients have a 96% survival rate over a ten-year period with no treatment, gene expression analysis allows doctors and patients access to valuable information when they are considering treatment options.

Many researchers envision a future of **personalised medicine**, a type of medical care in which each person's specific genetic profile can provide information about diseases or conditions for which the person is especially at risk and help make health-care decisions. As we will see later in the chapter, a *genetic profile* is currently taken to mean a set of genetic markers such as SNPs. Ultimately, however, it will probably require the complete DNA sequence of an individual—after sequencing becomes inexpensive enough. In fact, a 2016 study suggested that for difficult-to-diagnose patients it might be faster and more economical to sequence just the expressed regions (the exons, together called the *exome*) or even all of the genome right away, rather than running through a standard series of diagnostic tests. The ability to sequence a person's genome quickly and inexpensively is advancing rapidly, perhaps faster than development of appropriate treatments for the conditions. Still, the identification of genes involved in these conditions provides targets for therapeutic interventions.

An individual's genomic information can be used to predict the benefits and risks of particular medications, an approach called *pharmacogenetics.* The United States Food and Drug Administration (FDA) recommends genetic testing for patients prior to their use. Australian and New Zealand medical experts call on the governments of our nations to create legislation based on the standards set out by the International Clinical Pharmacogenetics Implementation Consortium.

Human Gene Therapy and Gene Editing

Gene therapy—the introduction of genes into an afflicted individual for therapeutic purposes—holds great potential for treating the relatively small number of disorders traceable to a single defective gene. The aim of this approach is to insert a normal allele of the defective gene into the somatic cells of the tissue affected by the disorder.

For gene therapy of somatic cells to be permanent, the cells that receive the normal allele must be cells that multiply throughout the patient's life. Bone marrow cells, which include the stem cells that give rise to all the cells of the blood and immune system, are prime candidates. **Figure 20.22** outlines one procedure for gene therapy in an individual whose bone marrow cells do not produce a vital enzyme because of a single defective gene. One type of severe combined immunodeficiency (SCID) is caused by this kind of defect. If the treatment is successful, the patient's bone marrow cells will begin producing the missing protein, and the patient may be cured.

The procedure shown in Figure 20.22 was used in gene therapy trials for SCID in France in 2000, resulting in the first indisputable success of gene therapy. However, three patients subsequently developed leukaemia, a type of blood cell cancer, and one of them died. Researchers have concluded it is likely that the insertion of the retroviral vector occurred near a gene that triggers the proliferation of blood cells. Using a viral vector that does not come from a retrovirus, clinical researchers have treated several other genetic

▼ **Figure 20.22 Gene therapy using a retroviral vector.** A retrovirus that has been rendered harmless is used as a vector in this procedure, which exploits the ability of a retrovirus to insert a DNA transcript of its RNA genome into the chromosomal DNA of its host cell (see Figure 19.9). If the foreign gene carried by the retroviral vector is expressed, the cell and its descendants will possess the gene product. Cells that reproduce throughout life, such as bone marrow cells, are ideal candidates for gene therapy.

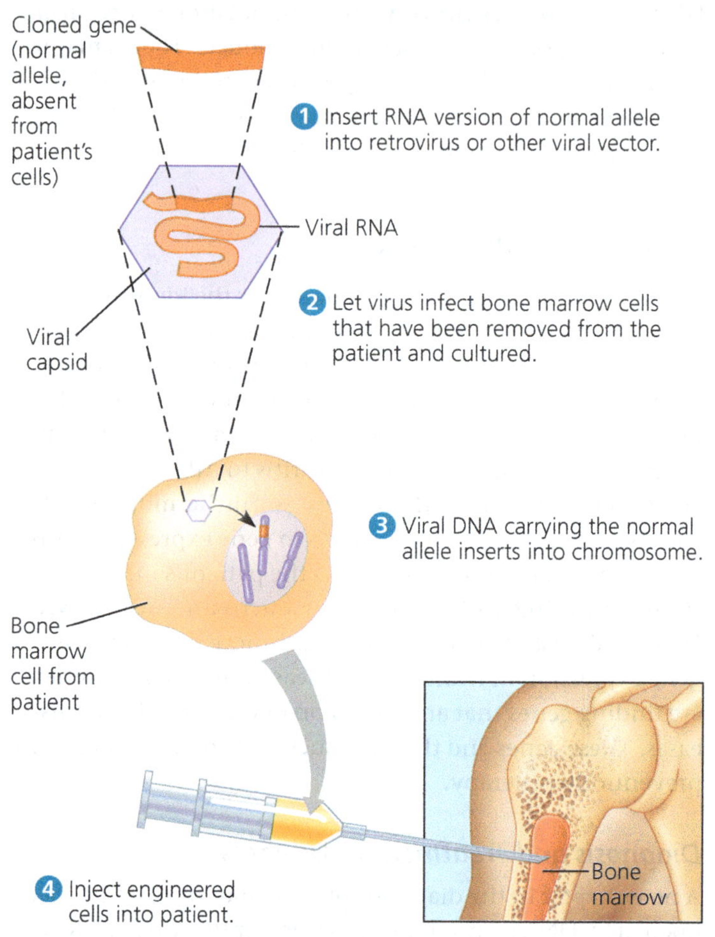

diseases somewhat successfully with gene therapy, including a type of progressive blindness (see Concept 50.3), a degenerative disease of the nervous system, and a blood disorder involving the β-globin gene.

Gene therapy raises many technical issues. For example, how can the activity of the transferred gene be controlled so that cells make appropriate amounts of the gene product at the right time and in the right place? How can we be sure that the insertion of the therapeutic gene does not harm some other necessary cell function? As more is learned about DNA control elements and gene interactions, researchers may be able to answer such questions.

A more direct approach that avoids the complications of using a viral vector in gene therapy is made possible by gene editing, especially using the CRISPR-Cas9 system (see Figure 17.28). In this approach, the existing defective gene is edited to correct the mutation.

In 2018, researchers reported promising results using the CRISPR-Cas9 system in an attempt to correct the genetic defect that causes sickle-cell disease. They edited cells from patients with sickle-cell disease and injected them into bone marrow in mice. After 19 weeks, the gene remained corrected in 20–40% of the injected cells. Researchers, doctors, and patients alike are excited about the potential of CRISPR technology to treat or even cure human diseases that have a genetic basis, such as sickle-cell, Alzheimer's, and Parkinson's diseases, as well as some types of cancer. However, there are still concerns about using CRISPR-treated cells in humans because of possible effects on genes that are not being targeted; this is an active area of research.

In addition to technical challenges, gene therapy and gene editing provoke ethical questions. Some critics believe that tampering with human genes in any way is unethical. Other observers see no fundamental difference between transplanting genes into somatic cells and transplanting organs from one person to another.

An even more pressing question is the issue of engineering human germ-line cells to try to correct a defect in future generations. Such genetic engineering is now routinely done in laboratory mice, and some methods for genetic engineering of human embryos have been developed.

The development of the CRISPR-Cas9 system has engendered much debate about the ethics of gene editing. Jennifer Doudna (Figure 20.14), a co-discoverer of CRISPR-Cas9, and other biologists have agreed to exercise extreme caution as the field moves forwards. Together, they called for the research community to "strongly discourage" any experimental work on human eggs or embryos.

In spite of this general consensus among biologists, a Chinese researcher reported in 2018 that he had used the CRISPR-Cas9 system to edit genes in embryos that completed fetal development and were born as twins and a third individual. He claimed to have edited the *CCR5* gene, which codes for a co-receptor for HIV (see Figure 7.8), so that HIV would be unable to bind and infect the cells. The twins' father was HIV-positive, which was the rationale the scientist used for carrying out this genetic engineering. However, this act was roundly condemned as highly unethical by the biological community; the researcher lost his job and may face a criminal investigation. In 2019, an advisory panel to the World Health Organization proposed establishing a registry to monitor any gene-editing research in humans.

There are many potential technical problems with the CRISPR-Cas9 technique, like the off-target effects mentioned previously. Beyond those technical concerns, however, are underlying ethical considerations: Under what circumstances, if any, should genomes of human germ lines be altered? Would alterations lead to the practice of eugenics, a deliberate effort to control the genetic makeup of human populations? It is imperative to consider these questions.

Pharmaceutical Products

The pharmaceutical industry derives significant benefit from advances in DNA technology and genetic research, applying them to the development of useful drugs to treat diseases. Pharmaceutical products are synthesised using methods of either organic chemistry or biotechnology, depending on the nature of the product.

Synthesis of Small Molecules for Use as Drugs

Determining the sequence and structure of proteins crucial for tumour cell survival has led to the identification of small molecules that combat certain cancers by blocking the function of these proteins. One drug, imatinib (known as Glievec in Australia and New Zealand, and commonly called Gleevec in other parts of the world, including the United States), is a small molecule that inhibits one tyrosine kinase (see Figure 11.8). The overexpression of this kinase, resulting from a chromosomal translocation, is instrumental in causing chronic myelogenous leukaemia (CML; see Figure 15.16). Patients in the early stages of CML who are treated with imatinib have exhibited nearly complete, sustained remission from the cancer. Drugs that work in a similar way have also been used with success to treat a few types of lung and breast cancers. This approach is feasible only for cancers whose molecular basis is fairly well understood.

In many cases of such drug-treated tumours, though, cells later arise that are resistant to the new drug. In one study, the whole genome of the tumour cells was sequenced both before and after the appearance of drug resistance. Comparison of the sequences showed genetic changes that allowed the tumour cells to "get around" the drug-inhibited protein. Here, we can see that cancer cells demonstrate the principles of evolution: Certain tumour cells have a random mutation that allows them to survive in the presence of a particular drug, and as a consequence of natural selection in the presence of the drug, these are the cells that survive and reproduce.

Protein Production in Cell Cultures Pharmaceutical products that are proteins are commonly synthesised on a large scale using cell cultures. You learned earlier in the chapter about DNA cloning and gene expression systems for producing large quantities of a chosen protein that is present naturally in only minute amounts. The host cells used in such expression systems can even be engineered to secrete a protein as it is made, thereby simplifying the task of purifying it by traditional biochemical methods.

Among the first pharmaceutical products manufactured in this way were human insulin and human growth hormone (HGH). Some 2 million people with diabetes in the United States depend on insulin treatment to control their disease. Human growth hormone has been a boon to children born with a form of dwarfism caused by inadequate amounts of HGH, as well as helping patients with AIDS gain weight. Another important pharmaceutical product produced by genetic engineering is tissue plasminogen activator (TPA). If administered shortly after a heart attack, TPA helps dissolve blood clots and reduces the risk of subsequent heart attacks.

Forensic Evidence and Genetic Profiles

In violent crimes, body fluids or small pieces of the perpetrator's tissue may be left at the scene or on the clothes or other possessions of the victim or assailant. If enough blood, semen, or tissue is available, forensic laboratories can determine the blood type or tissue type of the perpetrator using antibodies to detect specific cell-surface proteins. However, such tests require fresh samples in relatively large amounts. Also, because many people have the same blood or tissue type, this approach may only exclude a suspect; it cannot provide strong evidence of guilt.

DNA testing, on the other hand, can identify the guilty individual with a high degree of certainty because the DNA sequence of every person is unique (except for identical twins). Genetic markers that vary in the population can be analysed for a given person to determine that individual's unique set of genetic markers, or **genetic profile**. (Forensic scientists prefer this term to emphasise the heritability of these markers rather than the fact that the genetic material produces a pattern on a gel that, like a fingerprint, is visually recognisable.) The FBI started applying DNA technology to forensics in 1988, using a method involving gel electrophoresis and nucleic acid hybridisation to identify individuals through the detection of similarities and differences in DNA samples. DNA testing required smaller samples of blood or tissue than earlier approaches—only about 1,000 cells.

Today, forensic scientists use an array of even more sensitive approaches that capitalise on improved understandings of DNA. For example, one technique uses variations in the length of genetic markers called **short tandem repeats (STRs)**. STRs comprise tandemly repeated units of two- to five-nucleotide sequences in specific regions of the genome. The number of repeats present in these regions is highly variable from person to person (polymorphic); even for a single individual, the two alleles of an STR may differ from each other. For example, one individual may have the sequence ACAT repeated 30 times at one genome locus and 15 times at the same locus on the other homologue, while another individual may have 18 repeats at this locus on each homologue. (These two genotypes can be expressed by the two repeat numbers: 30,15 and 18,18.) PCR primers that are labelled with coloured fluorescent tags can amplify particular STRs to identify the length of the region, and thus the number of repeats. Forensic scientists then use electrophoresis to discern and quantify STRs. The PCR step allows use of this method even when the DNA is in poor condition or available only in minute quantities: A tissue sample containing as few as 20 cells can be sufficient.

In a murder case, for example, this method can be used to compare DNA samples from the suspect, the victim, and a small amount of blood found at a crime scene. The forensic scientist tests only a few selected portions of the DNA. However, even this small set of markers can provide a forensically useful genetic profile because the probability that two people (who are not identical twins) possess identical genetic profiles remains vanishingly small. The Innocence Project (a not-for-profit organisation founded in the United States in 1992 that now operates in many countries, including New Zealand and Australia) uses genetic evidence to overturn wrongful convictions. The project applies DNA technologies to examine archived samples from crime scenes of old cases. As of 2019, 362 innocent people had been released from prison as a result of DNA-based forensic and legal work. **Figure 20.23** provides one detailed example of the process and evidence used to win these cases.

▼ Figure 20.23 Frank Button, who was released from prison after serving ten months of a six-year sentence for a crime he did not commit. Like the case of Earl Washington (Figure 20.24), Mr Button's DNA was used to conclusively demonstrate he did not commit the crime for which he was imprisoned.

While these cases showcase the strengths of genetic technologies, we offer a counter-example. DNA technologies are almost infallible, as we have noted above—statistically, operationally, and reputationally. That is, unless something goes wrong. Even then, belief in the reliability and infallibility of genetic technologies carries undue weight in legal and other matters. Take the 2009 case of a woman (who currently cannot be named due to changes in Victorian laws) who was given a drink laced with date-rape drugs. Understandably, the victim's recollection of events that night remains hazy. Semen taken from the semi-conscious victim later was shown to positively and unequivocally match the genetic profile of a 22-year-old man, Farrah Jama. The victim recalled no-one at the nightclub matching Mr Jama's appearance, and Mr Jama claimed he was not at the nightclub, had never been there, had not met the victim, and was at home with his family at the time. The courts disagreed. The jury gave conclusive weight to the statistical validity and the sophistication of the genetic technologies in their deliberations. The jury convicted the man, who spent 15 months in prison—all the while protesting his innocence.

The courts subsequently quashed Mr Jama's conviction. The prosecution acknowledged that the forensic technician who prepared the DNA tests on semen from the rape victim had, on the previous day, taken a DNA swab from the mouth of Mr Jama for a completely unrelated and routine matter. Mr Jama's sample and that of the crime scene were processed by the same technician on the same day. The routine of sample preparation had somehow contaminated the crime scene DNA with Mr Jama's sample. Apparently, the likelihood of laboratory handling errors, and the limitations of DNA technologies, were never entered into the record of Mr Jama's original trial. This case, like others before it, produced calls for improved handling of genetic evidence, combined with educative campaigns for the legal fraternity and general public so that everyone can understand the power of genetic techniques, combined with deeper understandings of the process from sample collection to findings.

Genetic profiles can also be useful for other purposes, such as identifying victims of mass casualties. The largest such effort occurred after the attack on the World Trade Center in 2001. Closer to home, the 2002 Bali bombings killed 200 people, 88 of whom were Australian nationals. Where identification of victims' remains proved beyond the scope of regular approaches, the DNA from victims' remains were compared with DNA samples from personal items, such as toothbrushes, provided by families. Ultimately, forensic scientists succeeded in identifying 43 victims of the Bali bombings using these methods.

▼ **Figure 20.24 STR evidence used to release an innocent man from prison, and to incriminate the perpetrator.** In 1999 and early 2000, the Innocence Project used STR analysis to re-examine genetic evidence from a 1982 rape and murder case. Earl Washington had been convicted and sentenced to death for the crime. After serving 17 years in prison, Mr Washington's lawyers and the Innocence Project staff presented new evidence to the Appeals Court, including the genetic evidence that appears in the table below.

Source of sample	STR marker 1	STR marker 2	STR marker 3
Semen on victim	17,19	13,16	12,12
Earl Washington	16,18	14,15	11,12
Kenneth Tinsley	17,19	13,16	12,12

Innocence Project workers compiled evidence from the STR genetic profiles of three samples, taken from semen collected from the body of the victim (Rebecca Williams), from Mr Washington, and from another prisoner (Kenneth Tinsley), whose genetic profile had been collected as part of his conviction for an unrelated crime.

At each of three loci (singular, *locus*), two alleles occurred. Comparison between Mr Washington's STR genetic profile and the semen collected at the crime scene revealed that five of six alleles differed between the two samples. The genetic evidence demonstrated he could not be the perpertrator. The genetics corroborated Mr Washington's long-held claim of innocence. Soon thereafter, he was released from prison.

However, Kenneth Tinsley's STR markers matched those collected at the murder scene. Eventually, the genetic evidence obliged Kenneth Tinsley to confess and plead guilty to rape and murder.

Just how reliable is a genetic profile? The greater the number of markers examined in a DNA sample, the more likely it is that the profile is unique to one individual. In forensic cases using STR analysis with 13 markers, the probability of two people having identical DNA profiles is somewhere between one in 10 billion and one in several trillion. (For comparison, the world's population approaches 8 billion.) The exact probability depends on the frequency of those markers in the general population. Information on how common various markers are in different ethnic groups is critical because these marker frequencies may vary considerably among ethnic groups and between a particular ethnic group and the population as a whole. With the increasing availability of frequency data, forensic scientists can make extremely accurate statistical calculations. Thus, despite problems that can arise from insufficient data, human error, or flawed evidence, improvements in each of these and other aspects of genetic profiles have produced a compelling evidentiary basis among legal experts and scientists alike.

Environmental Cleanup

Increasingly, the diverse abilities of certain microorganisms to transform chemicals are being exploited for environmental cleanup. If the growth needs of such microorganisms make them unsuitable for direct use, scientists can now transfer the genes for their valuable metabolic capabilities into other microorganisms, which can then be used to treat environmental problems. For example, many bacteria can extract heavy metals, such as copper, lead, and nickel, from their environments and incorporate the metals into compounds such as copper sulfate or lead sulfate, which are readily recoverable. Genetically engineered microorganisms may become important in both mining (especially as ore reserves are depleted) and cleaning up highly toxic mining wastes. Biotechnologists are also trying to engineer microorganisms that can degrade chlorinated hydrocarbons and other harmful compounds. These microorganisms could be used in wastewater treatment plants or by manufacturers before the compounds are ever released into the environment.

Agricultural Applications

Scientists are working to learn more about the genomes of agriculturally important plants and animals. For a number of years, they have been using DNA technology in an effort to improve agricultural productivity. The selective breeding of both livestock (animal husbandry) and crops has exploited naturally occurring mutations and genetic recombination for thousands of years.

As we described earlier, DNA technology enables scientists to produce transgenic animals, which speeds up the selective breeding process. The goals of creating a transgenic animal are often the same as the goals of traditional breeding—for instance, to make a sheep with better quality wool, a pig with leaner meat, or a cow that will mature in a shorter time. Scientists might, for example, identify and clone a gene that causes the development of larger muscles (muscles make up most of the meat we eat) in one breed of cattle and transfer it to other cattle or even to sheep. However, health problems are not uncommon among farm animals carrying genes from other species, and modification of the animal's own genes using the CRISPR-Cas9 system will likely emerge as a more useful technique. Animal health and welfare are important issues to consider when genetically altering animals.

Agricultural scientists have already endowed a number of crop plants with genes for desirable traits, such as delayed ripening and resistance to spoilage, disease, and drought. Modifications can also add value to food crops, giving them a longer shelf life or improved flavour or nutritional value. For many plant species, a single tissue cell grown in culture can give rise to an adult plant. Thus, genetic manipulations can be performed on an ordinary somatic cell and the cell then can be used to generate a plant with new traits.

Genetic engineering is rapidly replacing traditional plant-breeding programs, especially for useful traits, such as herbicide or pest resistance, determined by one or a few genes. Crops engineered with a bacterial gene making the plants resistant to an herbicide can grow while weeds are destroyed, and genetically engineered crops that can resist destructive insects reduce the need for chemical insecticides.

The Food and Agriculture Organization of the United Nations has predicted that we will need 70% more food by the year 2050 than the planet is currently producing. One novel approach is being taken by researchers working on the international C_4 Rice Project. The common form of rice, a global food staple, uses the C_3 form of photosynthesis (see Concept 10.5). The aim of researchers engaged in this project is to genetically engineer a strain of rice that can use C_4 photosynthesis, which is more efficient. In 2017, the first step was accomplished when they were able to genetically engineer a corn gene into rice plants.

Safety and Ethical Questions Raised by DNA Technology

Early concerns about potential dangers associated with recombinant DNA technology and genetic engineering focused on the possibility that hazardous new pathogens might be created. What might happen, for instance, if in a research study cancer cell genes were transferred into bacteria or viruses? To guard against such rogue microorganisms, scientists developed a set of guidelines that were adopted as formal government regulations in the United States and some other countries. One safety measure is a set of strict laboratory procedures designed to prevent engineered microorganisms from either infecting researchers or accidentally leaving the laboratory. In addition, strains of microorganisms to be used in recombinant DNA experiments are genetically crippled to ensure that they cannot survive outside the laboratory. Finally, certain obviously dangerous experiments have been banned.

Today, most public concern about possible hazards centres not on recombinant microorganisms but on **genetically modified organisms (GMOs)** used as food. A GMO is a transgenic organism that has acquired one or more genes from another species or from another variety of the same species. Some salmon, for example, have been genetically modified by addition of a more active salmon growth hormone gene. However, the majority of the GMOs that contribute to our food supply are not animals, but crop plants.

GM crops are widespread in the United States, Argentina, and Brazil; together, these countries account for over 80% of the world's acreage devoted to such crops. In the United States, most corn, soybean, and canola crops are genetically modified, and a recent law requires labelling of GM products. The same foods are an ongoing subject of controversy in Europe, where the GM revolution has been met with strong opposition.

In 2020, South Australia lifted most bans on GM crops, and in 2021 New South Wales followed suit. Tasmania remains the only Australian state that prohibits growing them. New Zealand has gone considerably further than this position, and no GM crops are currently grown in any part of the country. Labelling requirements ensure consumers can make informed decisions about imported products that contain GMO-derived materials. New Zealand's approach followed that of the European Union and the legislative frameworks established in 2015. The high degree of consumer distrust in Europe makes the future of GM crops there uncertain.

Advocates of a cautious approach towards GM crops fear that transgenic plants might pass their new genes to close relatives in nearby wild areas. We know that lawn and crop grasses, for example, commonly exchange genes with wild relatives via pollen transfer. If crop plants carrying genes for resistance to herbicides, diseases, or insect pests pollinated wild ones, the offspring might become "super weeds" that are very difficult to control. Another worry involves possible risks to human health from GM foods. Some people fear that the protein products of transgenes might lead to allergic reactions. Although there is some evidence that this could happen, advocates claim that these proteins could be tested in advance to avoid producing ones that cause allergic reactions. (For further discussion of plant biotechnology and GM crops, see Concept 38.3.)

Today, governments and regulatory agencies throughout the world are grappling with how to facilitate the use of biotechnology in agriculture, industry, and medicine while ensuring that new products and procedures are safe. In Australia, such applications of biotechnology must be evaluated for potential risks by various regulatory agencies, including the the Office of the Gene Technology Regulator (OGTR). Food Standards Australia and New Zealand (FSANZ) and other federal and state government departments also hold various jurisdiction over the planting and use of GM organisms. Meanwhile, these same agencies and the public must consider the ethical implications of biotechnology.

Advances in biotechnology have allowed us to obtain complete genome sequences for humans and many other species, providing a vast treasure trove of information about genes. We can ask how certain genes differ from species to species, as well as how genes and, ultimately, entire genomes have evolved. (These are the subjects of the chapter, "Genomes and Their Evolution." At the same time, the increasing speed and falling cost of sequencing the genomes of individuals are raising significant ethical questions. Who should have the right to examine someone else's genetic information? How should that information be used? Should a person's genome be a factor in determining eligibility for a job or insurance? Ethical considerations, as well as concerns about potential environmental and health hazards, will likely slow some applications of biotechnology. There is always a danger that too much regulation will stifle basic research and its potential benefits. On the other hand, genetic engineering—especially gene editing with the CRISPR-Cas9 system—enables us to profoundly and rapidly alter species that have been evolving for millennia. A good example is the potential use of a gene drive that would eliminate the ability of mosquito species to carry diseases or even eradicate certain mosquito species. There would probably be health benefits to this approach, at least initially, but unforeseen problems could easily arise. Given the tremendous power of DNA technology, we must proceed with humility and caution.

CONCEPT CHECK 20.4

1. What is the advantage of using stem cells for gene therapy or gene editing?
2. List at least three different properties that have been acquired by crop plants via genetic engineering.
3. **WHAT IF?** As a physician, you have a patient with symptoms that suggest a hepatitis A infection, but you have not been able to detect viral proteins in the blood. Knowing that hepatitis A is an RNA virus, what lab tests could you perform to support your diagnosis? Explain the results that would support your hypothesis.

For suggested answers, see Appendix A.

20 Chapter Review

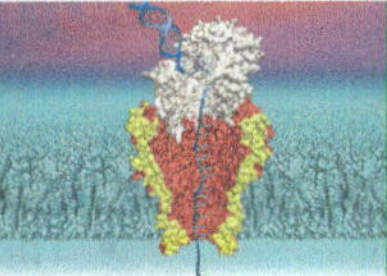

SUMMARY OF KEY CONCEPTS

CONCEPT 20.1

DNA sequencing and DNA cloning are valuable tools for genetic engineering and biological inquiry *(pp. 420–427)*

- **Nucleic acid hybridisation**, the base pairing of one strand of a nucleic acid to the complementary sequence on a strand from another nucleic acid molecule, is widely used in **DNA technology**.
- **DNA sequencing** can be carried out using the dideoxy sequencing method in automated sequencing machines.
- Fast and inexpensive next-generation (high-throughput) techniques for sequencing DNA are based on sequencing by synthesis: DNA polymerase is used to synthesise a stretch of DNA using a single-stranded template, and the order in which nucleotides are added reveals the sequence. Third-generation sequencing methods, including nanopore technology, sequence long DNA molecules one at a time.
- **Gene cloning** (or **DNA cloning**) produces multiple copies of a gene (or DNA segment) that can be used to manipulate and analyse DNA and to produce useful new products or organisms with beneficial traits.
- In **genetic engineering**, bacterial **restriction enzymes** are used to cut DNA molecules within short, specific nucleotide sequences (**restriction sites**), yielding a set of double-stranded **restriction fragments** with single-stranded **sticky ends**:

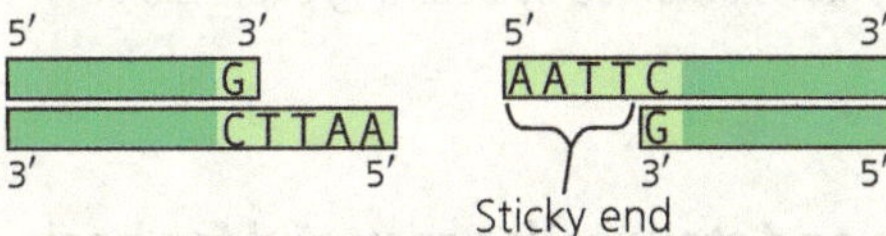

- The sticky ends on restriction fragments from one DNA source can base-pair with complementary sticky ends on fragments from other DNA molecules. Sealing the base-paired fragments with DNA ligase produces **recombinant DNA molecules**.
- DNA restriction fragments of different lengths can be separated by **gel electrophoresis**.
- The **polymerase chain reaction (PCR)** can amplify (produce many copies of) a specific target segment of DNA, using primers that bracket the desired sequence and a heat-resistant DNA polymerase.

- To clone a eukaryotic gene:

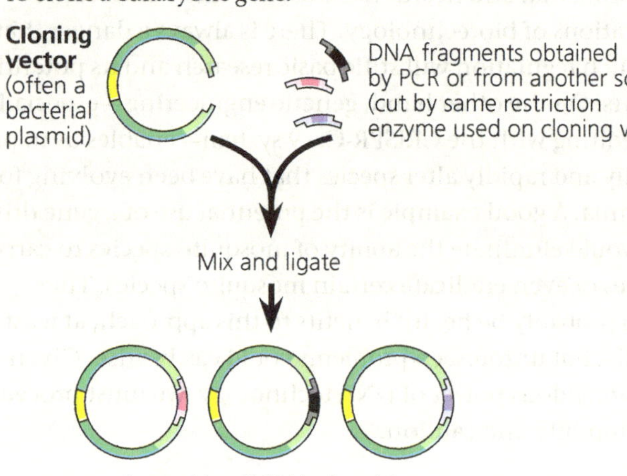

Recombinant **plasmids** are returned to host cells, each of which divides to form a clone of cells.
- Expressing cloned eukaryotic genes in bacterial host cells poses several technical difficulties. The use of cultured eukaryotic cells as host cells, coupled with appropriate **expression vectors**, helps avoid these problems.

? *Describe how the process of gene cloning results in a cell clone containing a recombinant plasmid.*

CONCEPT **20.2**

Biologists use DNA technology to study gene expression and function *(pp. 427–432)*

- Several techniques use hybridisation of a **nucleic acid probe** to detect the presence of specific mRNAs.
- ***In situ* hybridisation** and **RT-PCR** can detect the presence of a given mRNA in a tissue or an RNA sample, respectively.
- Sets of genes co-expressed by a group of cells can be detected by **RNA sequencing (RNA-seq)**—sequencing the **cDNAs** corresponding to mRNAs from the cells. **DNA micro-arrays** are also used for this purpose.
- For a gene of unknown function, experimental inactivation of the gene (a gene knockout) and observation of the resulting phenotypic effects can provide clues to its function. The CRISPR-Cas9 system allows researchers to edit genes in living cells in a specific, desired way. The new alleles can be altered so that they are inherited in a biased way through a population (**gene drive**).
- In humans, **genome-wide association studies** identify and use **single nucleotide polymorphisms (SNPs)** as genetic markers for alleles that are associated with particular conditions.

? *What useful information is obtained by detecting expression of specific genes?*

CONCEPT **20.3**

Cloned organisms and stem cells are useful for basic research and other applications *(pp. 432–436)*

- The question of whether all the cells in an organism have the same genome prompted the first attempts at organismal cloning.
- Single differentiated cells from plants are often **totipotent**: capable of generating all the tissues of a complete new plant.
- Transplantation of the nucleus from a differentiated animal cell into an enucleated egg can sometimes give rise to a new animal.
- Certain embryonic **stem cells** (ES cells) from animal embryos and particular adult stem cells from adult tissues can reproduce and differentiate both in the lab and in the organism, offering the potential for medical use. ES cells are **pluripotent** but difficult to acquire. Induced pluripotent stem (iPS) cells resemble ES cells in their capacity to differentiate; they can be generated by reprogramming differentiated cells. iPS cells hold promise for medical research and regenerative medicine.

? *Describe how, using mice, a researcher could carry out (1) organismal cloning, (2) production of ES cells, and (3) generation of iPS cells, focusing on how the cells are reprogrammed. (The procedures are basically the same in humans and mice.)*

CONCEPT **20.4**

The practical applications of DNA-based biotechnology affect our lives in many ways *(pp. 437–443)*

- DNA technology, including the analysis of genetic markers such as SNPs, is increasingly being used in the diagnosis of genetic and other diseases and in personal genome analysis. Personalised medicine offers potential for an individual minimising their known risk for a disease, as well as better treatment of genetic disorders or cancers. **Gene therapy** or gene editing with the CRISPR-Cas9 system may also lead to permanent cures. DNA technology is used with cell cultures in the large-scale production of protein hormones and other proteins with therapeutic uses. Some therapeutic proteins are being produced in **transgenic** "pharm" animals.
- Analysis of genetic markers such as **short tandem repeats (STRs)** in DNA isolated from tissue or body fluids found at crime scenes leads to a **genetic profile**. Use of genetic profiles can provide definitive evidence that a suspect is innocent or strong evidence of guilt. Such analysis is also useful in parenthood disputes and in identifying the remains of crime victims.
- Genetically engineered microorganisms can be used to extract minerals from the environment or degrade various types of toxic waste materials.
- The aims of developing transgenic plants and animals are to improve agricultural productivity and food quality.
- The potential benefits of genetic engineering must be carefully weighed against the potential for harm to humans or the environment.

? *What factors affect whether a given genetic disease would be a good candidate for successful gene therapy?*

TEST YOUR UNDERSTANDING

Levels 1-2: Remembering/Understanding

1. In DNA technology, the term *vector* can refer to
   (A) the enzyme that cuts DNA into restriction fragments.
   (B) the sticky end of a DNA fragment.
   (C) a SNP marker.
   (D) a plasmid used to transfer DNA into a living cell.
2. Which of the following tools of DNA technology is incorrectly paired with its use?
   (A) electrophoresis—separation of DNA fragments
   (B) DNA ligase—cutting DNA, creating sticky ends of restriction fragments
   (C) DNA polymerase—polymerase chain reaction to amplify sections of DNA
   (D) reverse transcriptase—production of cDNA from mRNA

3. Plants are more readily manipulated by genetic engineering than are animals because
   (A) plant genes do not contain introns.
   (B) more vectors are available for transferring recombinant DNA into plant cells.
   (C) a somatic plant cell can often give rise to a complete plant.
   (D) plant cells have larger nuclei.

4. A paleontologist has recovered a bit of tissue from the 400-year-old preserved skin of an extinct dodo (a bird). To compare a specific region of the DNA from a sample with DNA from living birds, which of the following would be most useful for increasing the amount of dodo DNA available for testing?
   (A) SNP analysis
   (B) polymerase chain reaction (PCR)
   (C) electroporation
   (D) gel electrophoresis

Levels 3-4: Applying/Analysing

5. Which of the following is true of cDNA produced using human brain tissue as the starting material?
   (A) The procedure to make it requires amplification by the polymerase chain reaction.
   (B) It is produced from pre-mRNA using reverse transcriptase.
   (C) It can be labelled and used as a probe to detect genes expressed in the brain.
   (D) It includes the introns of the pre-mRNA.

6. Expression of a cloned eukaryotic gene in a bacterial cell involves many challenges. The use of mRNA and reverse transcriptase is part of a strategy to solve the problem of
   (A) post-transcriptional processing.
   (B) post-translational processing.
   (C) nucleic acid hybridisation.
   (D) restriction fragment ligation.

7. Which of the following sequences in double-stranded DNA is most likely to be recognised as a cutting site for a restriction enzyme?
   (A) AAGG
   TTCC
   (B) GGCC
   CCGG
   (C) ACCA
   TGGT
   (D) AAAA
   TTTT

Levels 5-6: Evaluating/Creating

8. **MAKE CONNECTIONS** Imagine you want to study one of the human crystallins, proteins present in the lens of the eye (see Figure 1.8). To obtain a sufficient amount of the protein of interest, you decide to clone the gene that codes for it. Assume you know the sequence of this gene. Explain how you would go about this.

9. **MAKE CONNECTIONS** Looking at Figure 20.15, what does it mean for a SNP to be "linked" to a disease-associated allele? How does this allow the SNP to be used as a genetic marker? (See Concept 15.3.)

10. **DRAW IT** You are cloning an aardvark gene, using a bacterial plasmid as a vector. The green diagram shows the plasmid, which contains the restriction site for the enzyme used in Figure 20.5. Above the plasmid is a segment of linear aardvark DNA that was synthesised using PCR. Diagram your cloning procedure, and show what would happen to these two molecules during each step. Use one colour for the aardvark DNA and its bases and another colour for those of the plasmid. Label each step and all 5′ and 3′ ends.

5′ GAATTCTAAAGCGCTTATGAATTC 3′
3′ CTTAAGATTTCGCGAATACTTAAG 5′

Aardvark DNA

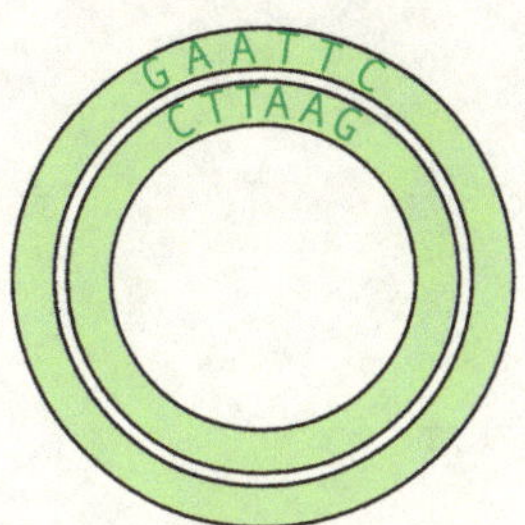

Plasmid

11. **EVOLUTION CONNECTION** Ethical considerations aside, if DNA-based technologies became widely used, discuss how they might change the way evolution proceeds, as compared with the natural evolutionary mechanisms that have operated for the past 4 billion years.

12. **SCIENTIFIC INQUIRY** You hope to study a gene that codes for a neurotransmitter protein produced in human brain cells. You know the amino acid sequence of the protein. Explain how you might (a) identify what genes are expressed in a specific type of brain cell, (b) identify (and isolate) the neurotransmitter gene, (c) produce multiple copies of the gene for study, and (d) produce large quantities of the neurotransmitter for evaluation as a potential medication.

13. **WRITE ABOUT A THEME: INFORMATION** In a short essay (100–150 words), discuss how the genetic basis of life plays a central role in biotechnology.

14. **SYNTHESISE YOUR KNOWLEDGE**

The water in the Yellowstone National Park hot springs shown here is around 160°F (70°C). Biologists assumed that no species of organisms could live in water above about 130°F (55°C), so they were surprised to find several species of bacteria there, now called *thermophiles* ("heat-lovers"). You've learned in this chapter how an enzyme from one species, *Thermus aquaticus*, made feasible one of the most important DNA-based techniques used in labs today. Identify the enzyme, and indicate the value of its being isolated from a thermophile. Suggest other reasons why enzymes from this bacterium (or other thermophiles) might also be valuable.

For selected answers, see Appendix A.

This page is intentionally blank.

Unit 4 MECHANISMS OF EVOLUTION

Professor Ary Hoffman is Chair, Ecological Genetics, in the School of BioSciences and Bio21, at the University of Melbourne. Ary was born in the Netherlands and grew up in New Zealand before moving to Australia. He received his PhD in evolutionary genetics from La Trobe University, then moved to the University of California in Davis, a hot spot in evolutionary biology, where he carried out selection experiments on flies. It was there that he made his seminal discovery of Wolbachia infections in *Drosophila*, which led to the development of a new tool that is now reducing dengue transmission by mosquitoes. Ary then moved back to Australia and has focused on how rapid evolution is important for understanding climate change. He continues to work on Wolbachia and how natural populations evolve in response to environmental stress, mainly using insects as model systems. He also applies evolutionary thinking to develop new ways of controlling agricultural pest species and suppressing diseases. Ary was elected a member of the Australian Academy of Science in 2004.

AN INTERVIEW WITH

Ary Hoffmann

What got you interested in biology?

My father was always interested in nature and he took me for walks in the woods when I was a very young child. I used to collect mushrooms as a hobby. It was their beauty that attracted me; I liked observing their different shapes and colours, and trying to figure out what type they were. My parents bought and ran a market garden in New Zealand and I spent much of my migrant youth working on this farm. I think this helped develop my passion for biology; I particularly loved to watch the insects even though they were eating our crops! I also worked on a sheep farm in New Zealand when I was a bit older and that further expanded my interest in larger organisms. I had a high-school teacher who, unusually for that time, held a master's degree and was interested in research. He introduced me to the idea that organisms evolved to be matched to their environment and that biological diversity could be quantified. This teacher was instrumental in getting me interested in the scientific process, and the question of why certain animals and plants occur in different places. I think high-school teachers are tremendously important and very undervalued.

How do you teach evolution to students?

When I started teaching, we used to run selection experiments in the lab with undergraduate students. We would expose a population of flies to a particular stress, get their offspring, and then breed these to show that after a generation or two they would become more resistant. That was a very powerful way of teaching students about rapid evolution, as they were able to see this process with their own eyes. Evolution is often a random process, and I like to teach students about the way it can lead to different outcomes that contribute to biodiversity. I like to make connections between evolution and everyday life, such as antibiotic resistance, invasive-pest control, and the importance of adaptation in restoring our environments.

Would you call yourself a generalist or a specialist?

I suspect I am a generalist. Evolution used to be regarded as a long-term process and didn't really interact much with ecology. Ecologists tended to focus on populations and species, while evolutionary biologists tended to focus on slow changes in phenotypes that occurred over thousands or millions of years. Yet there were a few examples of rapid evolution, such as the way insects evolved quickly to become resistant to chemical pesticides. When I started my undergraduate studies, populations were thought to be genetically uniform, rather than having the diversity we know they have today. When I was a postdoc [postdoctoral fellow], this view was starting to change and I wanted to see if this could lead to rapid adaptation to different ecological contexts. So I started experiments where I took patchiness in the environment to see if the flies I was studying were genetically adapted to different types of patches. For instance, I started collecting flies from different fruit types and asked questions like, "Are flies adapted to growing on citrus versus apples?" I discovered that they did show genetic differences between patches, which in turn affected their physiology and behaviour. I then recreated this patchiness in the lab and I was able to select for genetic differences very quickly. That seems to make me a generalist!

"... getting evolutionary thinking into the climate change adaptation discussion is an important accomplishment."

Why have you focused much of your initial work on the fruit fly?

My interest in rapid evolution and in genetics made *Drosophila* fruit flies an obvious choice. Their short generation time meant that I could track evolution in action. I could directly select populations of flies to change their responses to different chemicals, levels of aggression, ability to deal with hot and cold conditions, and so on. I could then also test whether these changes had a simple or complicated genetic basis, using the many tools available for manipulating *Drosophila* genomes. And I could test the impact of rapid evolution in nature because fly species occupy so many different environments around the world.

What accomplishment do you feel most proud of in your career?

I think that getting evolutionary thinking into the climate change adaptation discussion is an important accomplishment, although much remains to be done on the ground. I'd also rate the work on Wolbachia bacteria highly, given the substantial impact of this research. And I'm also proud of the many students I've trained who now have careers in science and in the wider community.

This page is intentionally blank.

25 The History of Life on Earth

KEY CONCEPTS

Study Tip

Draw a time line: Draw a time line like the example shown here, and mark when the following key evolutionary events occurred: "oxygen revolution," origin of eukaryotes, first multicellular organisms, colonisation of land by large eukaryotes, origin of tetrapods, and the Permian mass extinction. Briefly describe the significance of each event. You can also add more events to your time line as you study.

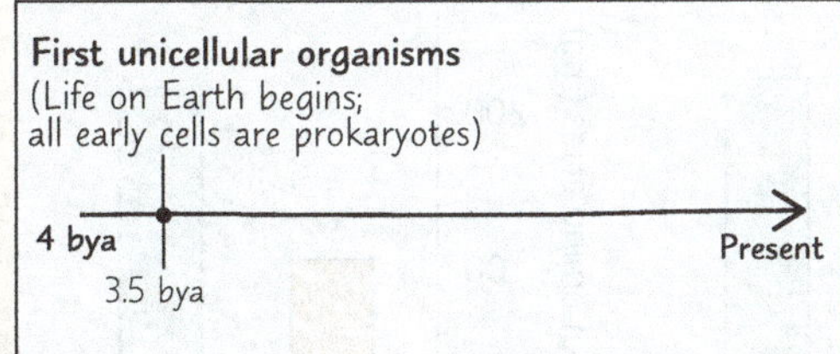

Go to Mastering Biology

to access Dynamic Study Modules for revision, 3D BioFlix® animations and high-quality videos, and your interactive Pearson eText.

Figure 25.1 In the 1870s, researchers working in the searing heat of the Sahara Desert were startled to discover fossils of ancient whales. These fossils were spectacular not only for where they were found, but also for documenting the transition from life on land to life in the sea—just one example of the sweeping changes seen in the history of life on Earth.

How has life on Earth changed over time?

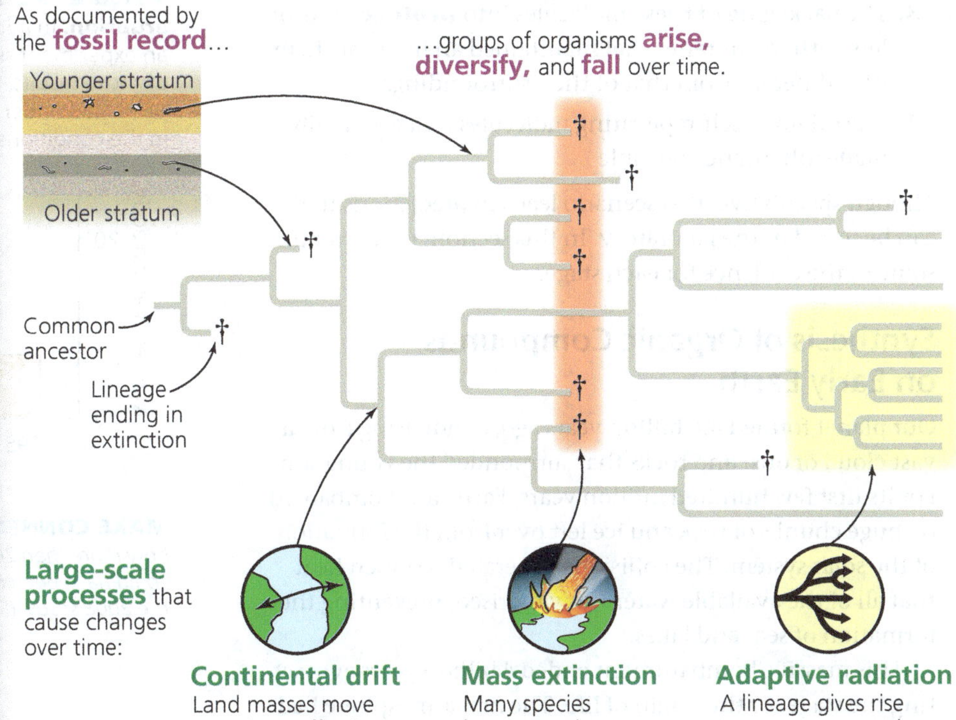

The whale fossils discovered in the Sahara Desert (see Figure 25.1) are one example of evidence that past organisms were very different from those presently living. The sweeping changes in life on Earth as revealed by such fossils illustrate **macroevolution**, the broad pattern of evolution above the species level. Examples of macroevolutionary change include the emergence of terrestrial vertebrates through a series of speciation events, the impact of mass extinctions on biodiversity, and the origin of key adaptations such as flight.

Taken together, such changes provide a grand view of the evolutionary history of life. We'll examine that history in this chapter, beginning with a discussion of hypotheses regarding the origin of life. This is the most speculative topic of the entire unit, because no fossil evidence of that seminal episode exists.

CONCEPT 25.1

Conditions on early Earth made the origin of life possible

Direct evidence of life on early Earth comes from fossils of microorganisms that lived 3.5 billion years ago. But how did the first living cells appear? Observations and experiments in chemistry, geology, and physics have led scientists to propose one scenario that we'll examine here. They hypothesise that chemical and physical processes could have produced simple cells through a sequence of four main stages:

1. The abiotic (nonliving) synthesis of small organic molecules, such as amino acids and nitrogenous bases
2. The joining of these small molecules into macromolecules, such as proteins and nucleic acids
3. The packaging of these molecules into **protocells**, droplets with membranes that maintained an internal chemistry different from that of their surroundings
4. The origin of self-replicating molecules that eventually made inheritance possible

Though speculative, this scenario leads to predictions that can be tested in the laboratory. In this section, we'll examine some of the evidence for each stage.

Synthesis of Organic Compounds on Early Earth

Our planet formed 4.6 billion years ago, condensing from a vast cloud of dust and rocks that surrounded the young sun. For its first few hundred million years, Earth was bombarded by huge chunks of rock and ice left over from the formation of the solar system. The collisions generated so much heat that all of the available water was vaporised, preventing the formation of seas and lakes.

This massive bombardment ended 4 billion years ago, setting the stage for the origin of life. The first atmosphere had little oxygen and was likely thick with water vapour, along with compounds released by volcanic eruptions, such as nitrogen and its oxides, carbon dioxide, methane, ammonia, and hydrogen. As Earth cooled, the water vapour condensed into oceans, and much of the hydrogen escaped into space.

During the 1920s, Russian chemist A. I. Oparin and British scientist J. B. S. Haldane independently hypothesised that Earth's early atmosphere was a reducing (electron-adding) environment, in which organic compounds could have formed from simpler molecules. The energy for this synthesis could have come from lightning and UV radiation. Haldane suggested that the early oceans were a solution of organic molecules, a "primitive soup" from which life arose.

In 1953, Stanley Miller, working with Harold Urey at the University of Chicago, tested the Oparin-Haldane hypothesis by creating laboratory conditions comparable to those that scientists at the time thought existed on early Earth (see Figure 4.2). Miller's apparatus yielded a variety of amino acids found in organisms today, along with other organic compounds. Many laboratories have since repeated Miller's classic experiment using different recipes for the atmosphere, some of which also produced organic compounds.

However, some evidence suggests that the early atmosphere was made up primarily of nitrogen and carbon dioxide and was neither reducing nor oxidising (electron-removing). Recent Miller/Urey-type experiments using such "neutral" atmospheres have also produced organic molecules. In addition, small pockets of the early atmosphere, such as those near the openings of volcanoes, may have been reducing. Perhaps the first organic compounds formed near volcanoes. In a test of this hypothesis **(Figure 25.2)**, researchers used modern equipment

▼ **Figure 25.2 Amino acid synthesis in a simulated volcanic eruption.** In addition to his classic 1953 study, Miller conducted an experiment simulating a volcanic eruption. In a 2008 reanalysis of those results, researchers found that far more amino acids were produced under simulated volcanic conditions than were produced in the conditions of the classic 1953 study.

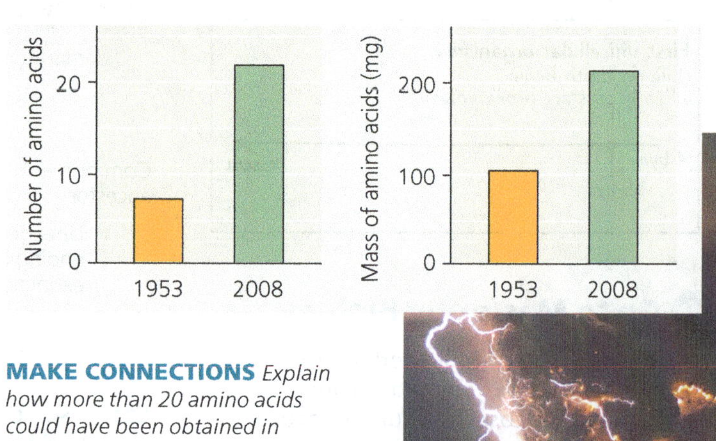

MAKE CONNECTIONS *Explain how more than 20 amino acids could have been obtained in the 2008 study. (See Concept 5.4.)*

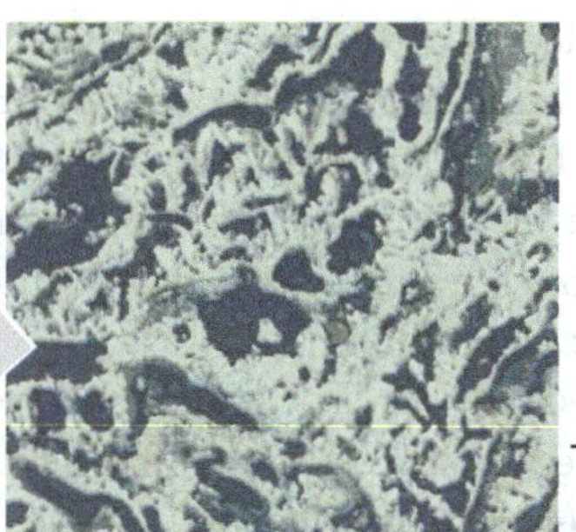

◀ **Figure 25.3 Did life originate in deep-sea alkaline vents?** The first organic compounds may have arisen in warm alkaline vents similar to this one from the 40,000-year-old "Lost City" vent field in the mid-Atlantic Ocean. These vents contain hydrocarbons and are full of tiny pores (inset) lined with iron and other catalytic minerals. Early oceans were acidic, so a pH gradient would have formed between the interior of the vents and the surrounding ocean water. Energy for the synthesis of organic compounds could have been harnessed from this pH gradient.

to reanalyse molecules that Miller had saved from one of his experiments. They found that numerous amino acids had formed under conditions that simulated a volcanic eruption.

Another hypothesis is that organic compounds were first produced in deep-sea **hydrothermal vents**, areas on the seafloor where heated water and minerals gush from Earth's interior into the ocean. Some of these vents, known as "black smokers," release water so hot (300–400°C) that organic compounds formed there may have been unstable. But other deep-sea vents, called **alkaline vents**, release water that has a high pH (9–11) and is warm (40–90°C) rather than hot, an environment that may have been more suitable for the origin of life **(Figure 25.3)**.

Studies related to the volcanic atmosphere and alkaline vent hypotheses show that the abiotic synthesis of organic molecules is possible under various conditions. Another source of organic molecules may have been meteorites. For example, fragments of the Murchison meteorite, a 4.5-billion-year-old rock that landed in Australia in 1969, contain more than 80 amino acids, some in large amounts. These amino acids cannot be contaminants from Earth because they consist of an equal mix of D and L isomers (see Figure 4.7). Organisms make and use only L isomers, with a few rare exceptions. Studies have shown that the Murchison meteorite also contained other key organic molecules, including lipids, simple sugars, and nitrogenous bases such as uracil.

Abiotic Synthesis of Macromolecules

The presence of small organic molecules, such as amino acids and nitrogenous bases, is not sufficient for the emergence of life as we know it. Every cell has many types of macromolecules, including enzymes and other proteins and the nucleic acids needed for self-replication. Could such macromolecules have formed on early Earth? A 2016 study demonstrated that one key step, the abiotic synthesis of RNA's two purine bases, adenine (A) and guanine (G), can occur spontaneously from simple precursor molecules; the abiotic synthesis of the smaller cytosine (C) and uracil (U) bases was accomplished in 2009. In addition, by dripping solutions of amino acids or RNA nucleotides onto hot sand, clay, or rock, researchers have produced polymers of these molecules. The polymers formed spontaneously, without the help of enzymes or ribosomes. Unlike proteins, the amino acid polymers are a complex mix of linked and cross-linked amino acids. Still, it is possible that such polymers acted as weak catalysts for a variety of chemical reactions on early Earth.

Protocells

All organisms must be able to carry out both reproduction and energy processing (metabolism). DNA molecules carry genetic information, including the instructions needed to replicate themselves accurately during reproduction. But DNA replication requires elaborate enzymatic machinery, along with an abundant supply of nucleotide building blocks provided by the cell's metabolism. This suggests that self-replicating molecules and a metabolic source of building blocks may have appeared together in early protocells. The necessary conditions may have been met in *vesicles*, fluid-filled compartments enclosed by a membrane-like structure. Recent experiments show that abiotically produced vesicles can exhibit certain properties of life, including simple reproduction and metabolism, as well as the maintenance of an internal chemical environment different from that of their surroundings **(Figure 25.4)**.

For example, vesicles can form spontaneously when lipids or other organic molecules are added to water. When this occurs, molecules that have both a hydrophobic region and a hydrophilic region can organise into a bilayer similar to the lipid bilayer of a plasma membrane. Adding substances such as *montmorillonite*, a soft mineral clay produced by the weathering of volcanic ash, greatly increases the rate of vesicle self-assembly (see Figure 25.4a). This clay, which is thought to have been common on early Earth, provides surfaces on which organic molecules become concentrated, increasing the likelihood that the molecules will react with each other and form vesicles. Abiotically produced vesicles can "reproduce" on their own (see Figure 25.4b), and they can increase in size ("grow") without dilution of their contents. Vesicles also can absorb montmorillonite particles, including those on which RNA and other organic molecules have become attached (see Figure 25.4c). Finally, experiments have shown that some vesicles have a selectively permeable bilayer and can perform metabolic reactions using an external source of reagents—another important prerequisite for life.

▼ Figure 25.4 Features of abiotically produced vesicles.

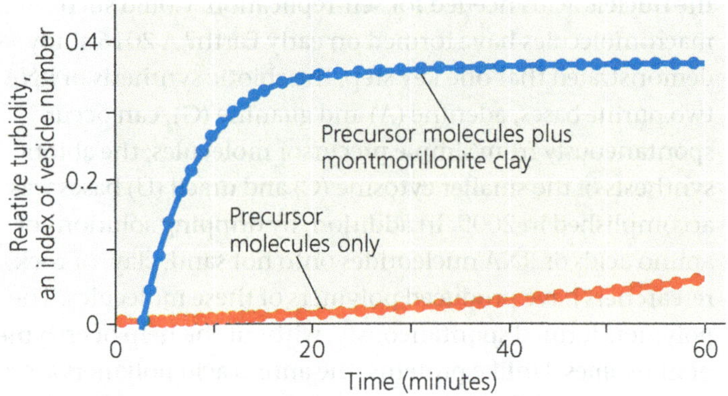

(a) Self-assembly. The presence of montmorillonite clay greatly increases the rate of vesicle self-assembly.

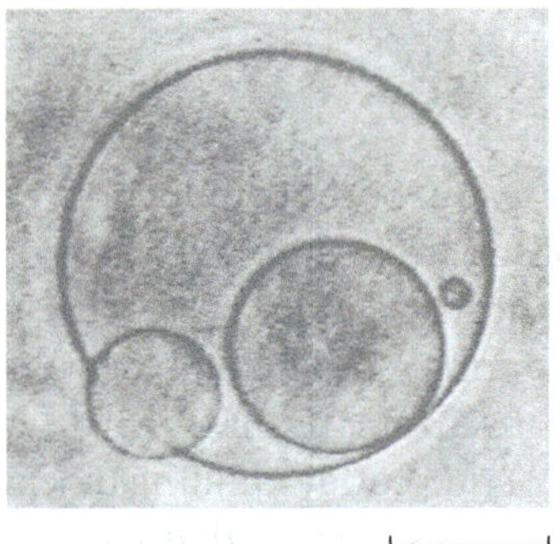

(b) Reproduction. Vesicles can divide on their own, as in this vesicle "giving birth" to smaller vesicles (LM).

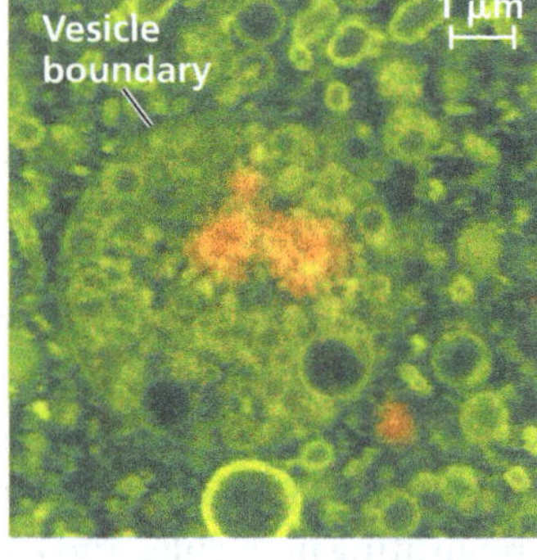

(c) Absorption of RNA. This vesicle has incorporated montmorillonite clay particles coated with RNA (orange).

MAKE CONNECTIONS *Explain how molecules with both a hydrophobic region and a hydrophilic region can self-assemble into a bilayer when in water. (See Concept 5.3.)*

Self-Replicating RNA

The first genetic material was most likely RNA, not DNA. RNA plays a central role in protein synthesis, but it can also function as an enzyme-like catalyst (see Concept 17.3). Such RNA catalysts are called **ribozymes**. Some ribozymes can make complementary copies of short pieces of RNA, provided that they are supplied with nucleotide building blocks.

Natural selection on the molecular level has produced ribozymes capable of self-replication in the laboratory. How does this occur? Unlike double-stranded DNA, which takes the form of a uniform helix, single-stranded RNA molecules assume a variety of specific three-dimensional shapes mandated by their nucleotide sequences. In a given environment, RNA molecules with certain nucleotide sequences may have shapes that enable them to replicate faster and with fewer errors than other sequences. The RNA molecule with the greatest ability to replicate itself will leave the most descendant molecules. Occasionally, a copying error will result in a molecule with a shape that is even more adept at self-replication. Similar selection events may have occurred on early Earth. Thus, life as we know it may have been preceded by an "RNA world," in which small RNA molecules were able to replicate and to store genetic information about the vesicles that carried them.

Recently, Dr Jack Szostak and colleagues at Harvard University succeeded in building a vesicle in which copying of a template strand of RNA could occur—a key step towards constructing a vesicle with self-replicating RNA. On early Earth, a vesicle with such self-replicating, catalytic RNA would differ from its many neighbours that lacked such molecules. If that vesicle could grow, split, and pass its RNA molecules to its "daughters," the daughters would be protocells. Although the first such protocells probably carried only limited amounts of genetic information, specifying only a few properties, their inherited characteristics could have been acted on by natural selection. The most successful of the early protocells would have increased in number because they could exploit their resources effectively and pass their abilities on to subsequent generations.

Once RNA sequences that carried genetic information appeared in protocells, many additional changes would have been possible. For example, RNA could have provided the template on which DNA nucleotides were assembled. Double-stranded DNA is a more chemically stable repository for genetic information than is the more fragile RNA. DNA also can be replicated more accurately. Accurate replication was advantageous as genomes grew larger through gene duplication and other processes and as more properties of the protocells became coded in genetic information. Once DNA appeared, the stage was set for a blossoming of new forms of life—a change we see documented in the fossil record.

CONCEPT CHECK 25.1

1. What hypothesis did Miller test in his classic experiment?
2. How would the appearance of protocells have represented a key step in the origin of life?
3. **MAKE CONNECTIONS** In changing from an "RNA world" to today's "DNA world," genetic information must have flowed from RNA to DNA. After reviewing Figures 17.4 and 19.9, suggest how this could have occurred. Does such a flow occur today?

For suggested answers, see Appendix A.

CONCEPT 25.2

The fossil record documents the history of life

Starting with the earliest traces of life, the fossil record opens a window into the world of long ago and provides glimpses of the evolution of life over billions of years. In this section,

▶ **Figure 25.5 A gallery of fossil types. (a) Sedimentary rock.** Most fossils are found in sedimentary rocks, such as this plant fossil. **(b) Mineralised organic matter.** Some fossils, such as this petrified tree trunk, form as minerals seep into and replace organic matter. **(c) Trace fossils.** Footprints, burrows, or other traces of an organism's activities can be preserved in the fossil record. **(d) Amber.** Entire organisms are sometimes found preserved in hardened resin from a tree. **(e) Frozen soil, ice, and acid bogs.** Rarely, frozen soil, ice, or an acid bog can preserve the body of larger organisms, such as this wolf pup.

(a) Wollemi tree from New South Wales, Australia, from around 90 million years ago (mya)

(b) 180-million-year-old petrified tree at Curio Bay in New Zealand's South Island

(c) 90-million-year-old dinosaur tracks in Queensland, Australia

(d) Insect in amber, about 40 million years old

(e) A 50,000-year-old wolf pup found in frozen soil in Yukon, Canada

we'll examine fossils as a form of scientific evidence: how fossils form, how scientists date and interpret them, and what they can and cannot tell us about changes in the history of life.

The Fossil Record

Sedimentary rocks are the richest source of fossils. As a result, the fossil record is based primarily on the sequence in which fossils have accumulated in sedimentary rock layers, called *strata* (see Figure 22.3). Useful information is also provided by other types of fossils, such as insects preserved in amber (fossilised tree sap) and mammals frozen in soil **(Figure 25.5)**.

The fossil record shows that there have been great changes in the kinds of organisms on Earth at different points in time. Many past organisms were unlike organisms living today, and many organisms that once were common are now extinct. As we'll see later in this section, fossils also document how new groups of organisms arose from previously existing ones.

As substantial and significant as the fossil record is, keep in mind that it is an incomplete chronicle of evolution. Many of Earth's organisms did not die in the right place and time to be preserved as fossils. Of the fossils that were formed, many were destroyed by later geological processes, and only a fraction of the others have been discovered. As a result, the known fossil record is biased in favour of species that existed for a long time, were abundant and widespread in certain kinds of environments, and had hard shells, skeletons, or other parts that facilitated their fossilisation. Even with its limitations, however, the fossil record is a remarkably detailed account of biological change over the vast scale of geological time. Furthermore, as shown by recently unearthed fossils of whale ancestors with hind limbs (see Figure 22.21 and Figure 25.1), gaps in the fossil record continue to be filled by new discoveries.

How Rocks and Fossils Are Dated

Fossils provide valuable data for reconstructing the history of life, but only if we can determine where they fit in that story. While the order of fossils in rock strata tells us the sequence in which the fossils were laid down—their relative ages—it does not tell us their actual ages. Examining

▼ Figure 25.6 Radiometric dating. In this diagram, each unit of time represents one half-life of a radioactive isotope.

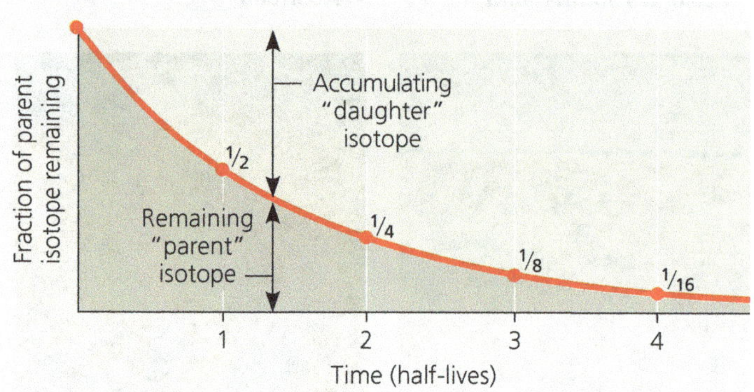

DRAW IT *Relabel the x-axis of this graph in years to illustrate the radioactive decay of uranium-238 (half-life = 4.5 billion years).*

the relative positions of fossils is like peeling off layers of wallpaper in an old house. You can infer the sequence in which the layers were applied, but not the year each layer was added.

How can we determine the age of a fossil? One of the most common techniques is **radiometric dating**, which is based on the decay of radioactive isotopes (see Concept 2.2). In this process, a radioactive "parent" isotope decays to a "daughter" isotope at a characteristic rate. The rate of decay is expressed by the **half-life**, the time required for 50% of the parent isotope to decay **(Figure 25.6)**. Each type of radioactive isotope has a characteristic half-life, which is not affected by temperature, pressure, or other environmental variables. For example, carbon-14 decays relatively quickly; its half-life is 5,730 years. Uranium-238 decays slowly; its half-life is 4.5 billion years.

Fossils contain isotopes of elements that accumulated in the organisms when they were alive. For example, a living organism contains the most common carbon isotope, carbon-12, as well as a radioactive isotope, carbon-14. When the organism dies, it stops accumulating carbon, and the amount of carbon-12 in its tissues does not change over time. However, the carbon-14 that it contains at the time of death slowly decays into another element, nitrogen-14. Thus, by measuring the ratio of carbon-14 to carbon-12 in a fossil, we can determine the fossil's age. This method works for fossils up to about 75,000 years old; fossils older than that contain too little carbon-14 to be detected with current techniques. Radioactive isotopes with longer half-lives are used to date older fossils.

Determining the age of these older fossils in sedimentary rocks can be challenging. Organisms do not use radioisotopes with long half-lives, such as uranium-238, to build their bones or shells. In addition, sedimentary rocks are often composed of sediments of differing ages. Although we cannot date these older fossils directly, an indirect method can be used to infer the age of fossils that are sandwiched between two layers of volcanic rock. As lava cools into volcanic rock, radioisotopes from the surrounding environment become trapped in the newly formed rock. Some of the trapped radioisotopes have long half-lives, allowing geologists to estimate the ages of ancient volcanic rocks. If two volcanic layers surrounding fossils are found to be 525 million and 535 million years old, for example, then the fossils are roughly 530 million years old.

The study of fossils has helped geologists establish a geological record: a standard time scale that divides Earth's history into four eons and further subdivisions **(Table 25.1)**. We'll describe key events that occurred during these different time periods in Concept 25.3.

The Origin of New Groups of Organisms

Some fossils document the origin of new groups of organisms. Such fossils are central to our understanding of evolution; they illustrate how new features arise and how long it takes for such changes to occur. We'll examine one such case here and in **Figure 25.7**: the origin of mammals.

Along with amphibians and reptiles, mammals belong to the group of animals called *tetrapods* (from the Greek *tetra*, four, and *pod*, foot), named for having four limbs. Mammals have a number of unique anatomical features that fossilise readily, allowing scientists to trace their origin. For example, the lower jaw is composed of one bone (the dentary) in mammals but several bones in other tetrapods. In addition, the lower and upper jaws in mammals hinge between a different set of bones than in other tetrapods. Mammals also have a unique set of three bones that transmit sound in the middle ear, the hammer, anvil, and stirrup, whereas other tetrapods have only one such bone, the stirrup (see Concept 34.6). Finally, the teeth of mammals are differentiated into incisors (for tearing), canines (for piercing), and the multi-pointed premolars and molars (for crushing and grinding). In contrast, the teeth of other tetrapods usually consist of a row of undifferentiated, single-pointed teeth.

As detailed in Figure 25.7, the fossil record shows that the unique features of mammalian jaws and teeth evolved gradually, in a series of steps. Keep in mind that the figure includes just a few examples of the fossil skulls that document the origin of mammals. If all the known fossils in the sequence were arranged by shape and placed side by side, their features would blend smoothly from one group to the next. Some of these fossils would reflect how the features of a group that dominates life today, the mammals, gradually arose in a previously existing group, the cynodonts. Others would reveal side branches on the

Table 25.1 The Geological Record

Eons (duration not to scale)	Era	Period	Epoch	Age (Millions of Years Ago)	Some Important Events in the History of Life
Phanerozoic	Cainozoic	Quaternary	Holocene	0.01	Historical time
			Pleistocene	2.6	Ice ages; origin of genus *Homo*
		Neogene	Pliocene	5.3	Appearance of bipedal human ancestors
			Miocene	23	Continued radiation of mammals and angiosperms; earliest direct human ancestors
		Paleogene	Oligocene	34	Origins of many primate groups
			Eocene	56	Angiosperm dominance increases; continued radiation of most present-day mammalian orders
			Paleocene	66	Major radiation of mammals, birds, and pollinating insects
	Mesozoic	Cretaceous		145	Flowering plants (angiosperms) appear and diversify; many groups of organisms, including most dinosaurs, become extinct at end of period
		Jurassic		201	Gymnosperms continue as dominant plants; dinosaurs abundant and diverse
		Triassic		252	Cone-bearing plants (gymnosperms) dominate landscape; dinosaurs evolve and radiate; origin of mammals
	Paleozoic	Permian		299	Radiation of reptiles; origin of most present-day groups of insects; extinction of many marine and terrestrial organisms at end of period
		Carboniferous		359	Extensive forests of vascular plants form; first seed plants appear; origin of reptiles; amphibians dominant
		Devonian		419	Diversification of bony fishes; first tetrapods and insects appear
		Silurian		444	Diversification of early vascular plants
		Ordovician		485	Marine algae abundant; colonisation of land by diverse fungi, plants, and animals
		Cambrian		541	Sudden increase in diversity of many animal phyla (Cambrian explosion)
Proterozoic	Neoproterozoic	Ediacaran		635	Diverse algae and soft-bodied invertebrate animals appear
				1,000	
				1,800	Oldest fossils of eukaryotic cells
				2,500	
Archaean				2,700	Concentration of atmospheric oxygen begins to increase
				3,500	Oldest fossils of cells (prokaryotes)
				4,000	Oldest known rocks on Earth's surface
Hadean				Approx. 4,600	Origin of Earth

tree of life—groups of organisms that thrived for millions of years but ultimately left no descendants that survive today.

CONCEPT CHECK 25.2

1. Describe an example from the fossil record that shows how life has changed over time.
2. **WHAT IF?** A fossilised skull you unearthed has a carbon-14/carbon-12 ratio about $\frac{1}{16}$ that of the skulls of present-day animals. Approximately how many years old is the fossilised skull?

For suggested answers, see Appendix A.

CONCEPT 25.3

Key events in life's history include the origins of unicellular and multicellular organisms and the colonisation of land

The first three eons in the geological record—the Hadean, the Archaean, and the Proterozoic—together lasted about 4 billion years. The Phanerozoic eon, roughly the last half

▼ Figure 25.7 Exploring the Origin of Mammals

Over the course of 120 million years, mammals originated gradually from a group of tetrapods called synapsids. Shown here are a few of the many fossil organisms whose morphological features represent intermediate steps between living mammals and their early synapsid ancestors. The evolutionary context of the origin of mammals is shown in the tree diagram at right (the dagger symbol † indicates extinct lineages).

Key to skull bones

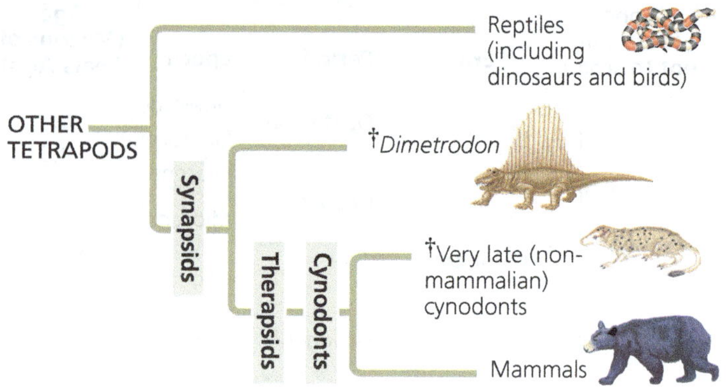

Synapsid (300 million years ago (mya))

Early synapsids had multiple bones in the lower jaw and single-pointed teeth. The jaw hinge was formed by the articular and quadrate bones. Early synapsids also had an opening called the *temporal fenestra* behind the eye socket. Powerful cheek muscles for closing the jaws probably passed through the temporal fenestra. Over time, this opening enlarged and moved in front of the hinge between the lower and upper jaws, thereby increasing the power and precision with which the jaws could be closed (much as moving a doorknob away from the hinge makes a door easier to close).

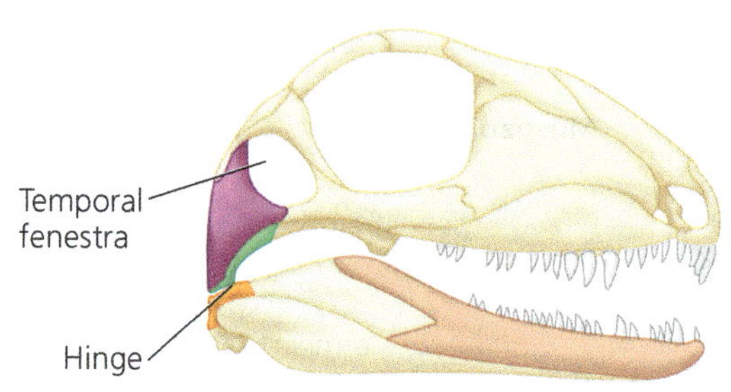

Therapsid (280 million years ago (mya))

Later, a group of synapsids called therapsids appeared. Therapsids had large dentary bones, long faces, and the first examples of specialised teeth, large canines. These trends continued in a group of therapsids called cynodonts.

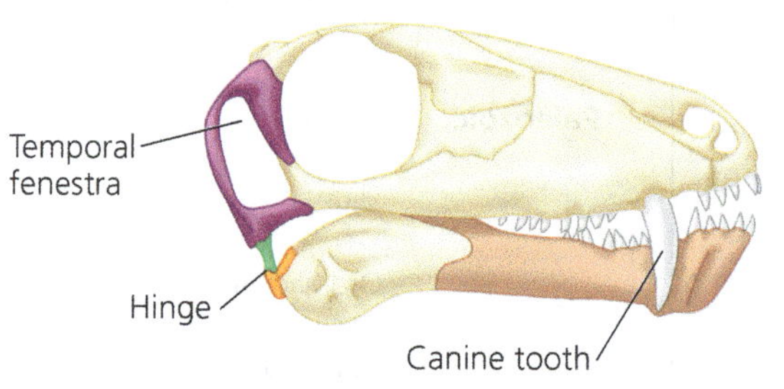

Early cynodont (260 million years ago (mya))

In early cynodont therapsids, the dentary was the largest bone in the lower jaw, the temporal fenestra was large and positioned forwards of the jaw hinge, and teeth with several cusps first appeared (not visible in the diagram). As in earlier synapsids, the jaw had an articular-quadrate hinge.

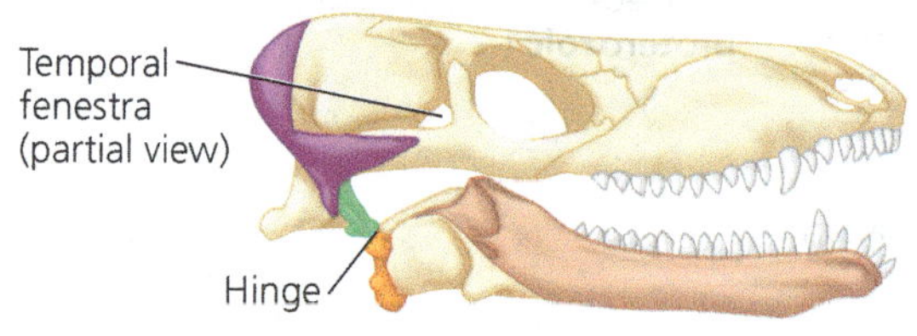

Later cynodont (220 million years ago (mya))

Later cynodonts had teeth with complex cusp patterns, and their lower and upper jaws hinged in two locations: They retained the original articular-quadrate hinge and formed a new, second hinge between the dentary and squamosal bones. (The temporal fenestra is not visible in this or the below cynodont skull at the angles shown.)

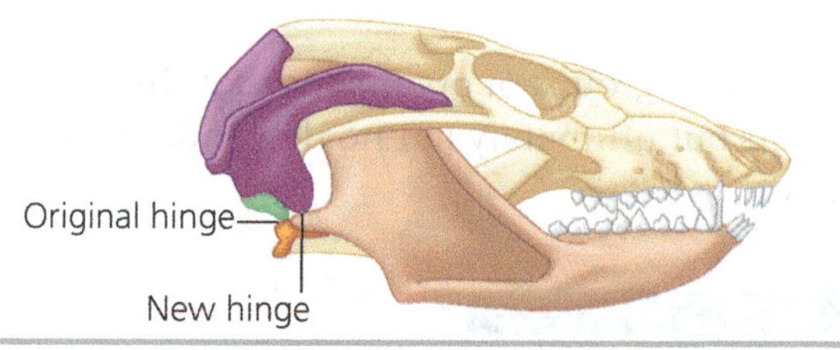

Very late cynodont (195 million years ago (mya))

In some very late (nonmammalian) cynodonts and early mammals, the original articular-quadrate hinge was lost, leaving the dentary-squamosal hinge as the only hinge between the lower and upper jaws, as in living mammals. The articular and quadrate bones migrated into the ear region (not shown), where they functioned in transmitting sound. In the mammal lineage, these two bones later evolved into the familiar hammer (malleus) and anvil (incus) bones of the ear.

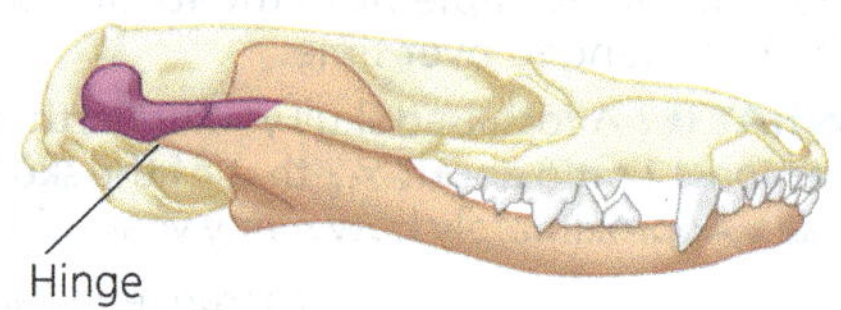

billion years, encompasses most of the time that animals have existed on Earth. It is divided into three eras: the Paleozoic, Mesozoic, and Cenozoic. Each era represents a distinct age in the history of Earth and its life. For example, the Mesozoic era is sometimes called the "age of reptiles" because of its abundance of reptilian fossils, including those of dinosaurs. The boundaries between the eras correspond to major extinction events, when many forms of life disappeared and were replaced by forms that evolved from the survivors.

As we've seen, the fossil record provides a sweeping overview of the history of life over geological time. Here we will focus on a few major events in that history, returning to study the details in Unit 5. **Figure 25.8** will help you visualise how long ago these key events occurred against the vast backdrop of geological time.

The First Single-Celled Organisms

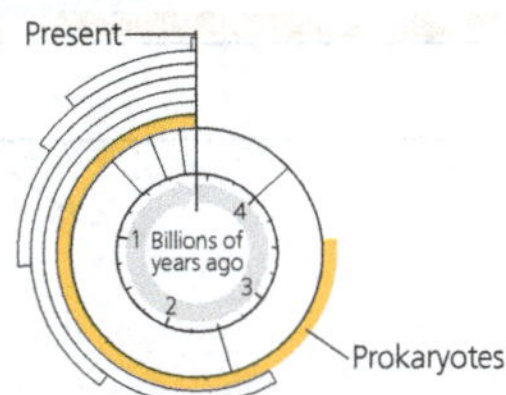

Earth's first organisms were single-celled prokaryotes that lived in the ocean. The earliest direct evidence of these organisms, dating from 3.5 billion years ago, comes from fossilised stromatolites. **Stromatolites** are layered rocks that form when certain prokaryotes bind thin films of sediment together. Stromatolites and other early prokaryotes were Earth's sole inhabitants for about 1.5 billion years. As we will see, these prokaryotes transformed life on our planet.

Photosynthesis and the Oxygen Revolution

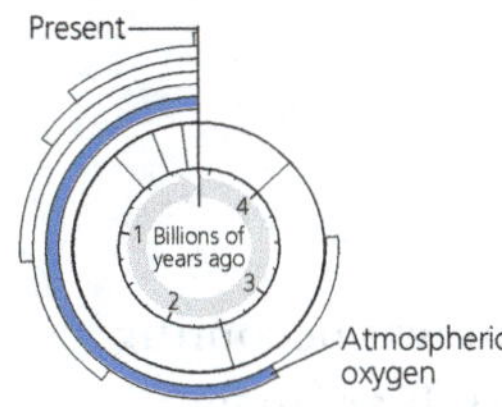

Most atmospheric oxygen gas (O_2) is of biological origin, produced during the water-splitting step of photosynthesis. When oxygenic photosynthesis first evolved—in photosynthetic prokaryotes similar to today's cyanobacteria—the free O_2 it produced probably dissolved in the surrounding water until it reached a high enough concentration to react with elements dissolved in water, including iron. This would have caused the iron to precipitate as iron oxide, which accumulated as sediments. These sediments were compressed into banded iron formations, red layers of rock containing iron oxide that are a source of iron ore today. Once all of the dissolved iron had precipitated, additional O_2 dissolved in the water until the seas and lakes became saturated with O_2. After this occurred, the O_2 finally began to "gas out" of the water and enter the atmosphere. This change left its mark in the rusting of iron-rich terrestrial rocks, a process that began about 2.7 billion years ago.

As shown in **Figure 25.9**, the amount of atmospheric O_2 increased gradually from about 2.7 to 2.4 billion years ago, but then shot up relatively rapidly to between 1% and 10% of its present level. This "oxygen revolution" had an enormous impact on life. In some of its chemical forms, oxygen attacks chemical bonds and can inhibit enzymes and damage cells. As a result, the rising concentration of atmospheric O_2 probably doomed many prokaryotic groups. Some species survived in habitats that remained anaerobic, where we find their descendants living today (see Concept 27.4). Among other survivors, diverse adaptations to the changing atmosphere evolved, including cellular respiration, which uses O_2 in the process of harvesting the energy stored in organic molecules.

The rise in atmospheric O_2 levels left a huge imprint on the history of life. A few hundred million years later, another fundamental change occurred: the origin of the eukaryotic cell.

The First Eukaryotes

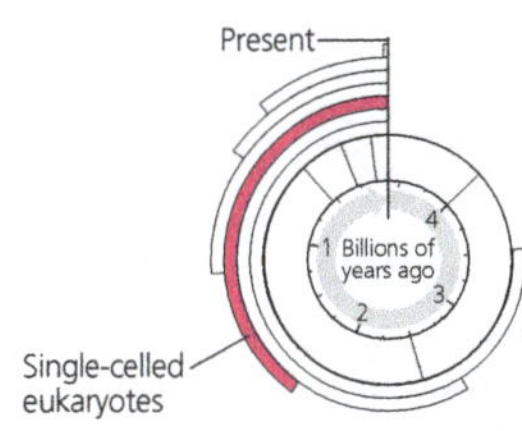

The oldest widely accepted fossils of eukaryotes are of single-celled organisms that lived 1.8 billion years ago. Recall that eukaryotic cells have more complex organisation than prokaryotic cells: Eukaryotic cells have a nuclear envelope, mitochondria, endoplasmic reticulum, and other internal structures that prokaryotes lack. Also, unlike prokaryotic cells, eukaryotic cells have a well-developed cytoskeleton, a feature that enables eukaryotic cells to change their shape and thereby surround and engulf other cells.

How did the eukaryotes evolve from their prokaryotic ancestors? Current evidence indicates that the eukaryotes originated by **endosymbiosis** when a prokaryotic cell engulfed a small cell that would evolve into an organelle found in all eukaryotes, the mitochondrion. The small, engulfed cell is an example of an *endosymbiont*, a cell that lives within another cell, called the *host cell*. The prokaryotic ancestor of the mitochondrion probably entered the host cell as undigested prey or an internal parasite. Though such a process may seem unlikely, scientists have directly observed cases in which endosymbionts that began as prey or parasites developed a mutually beneficial relationship with a host in as little as five years.

By whatever means the relationship began, we can hypothesise how the symbiosis could have become

▼ Figure 25.8 VISUALISING THE SCALE OF GEOLOGICAL TIME

Geological time is so vast that it can be difficult to visualise when key events in the history of life on Earth occurred. In this time line, time runs from left to right, from 4.6 billion years ago (bya) to the present. The fossils along the top illustrate representative organisms from different points in time.

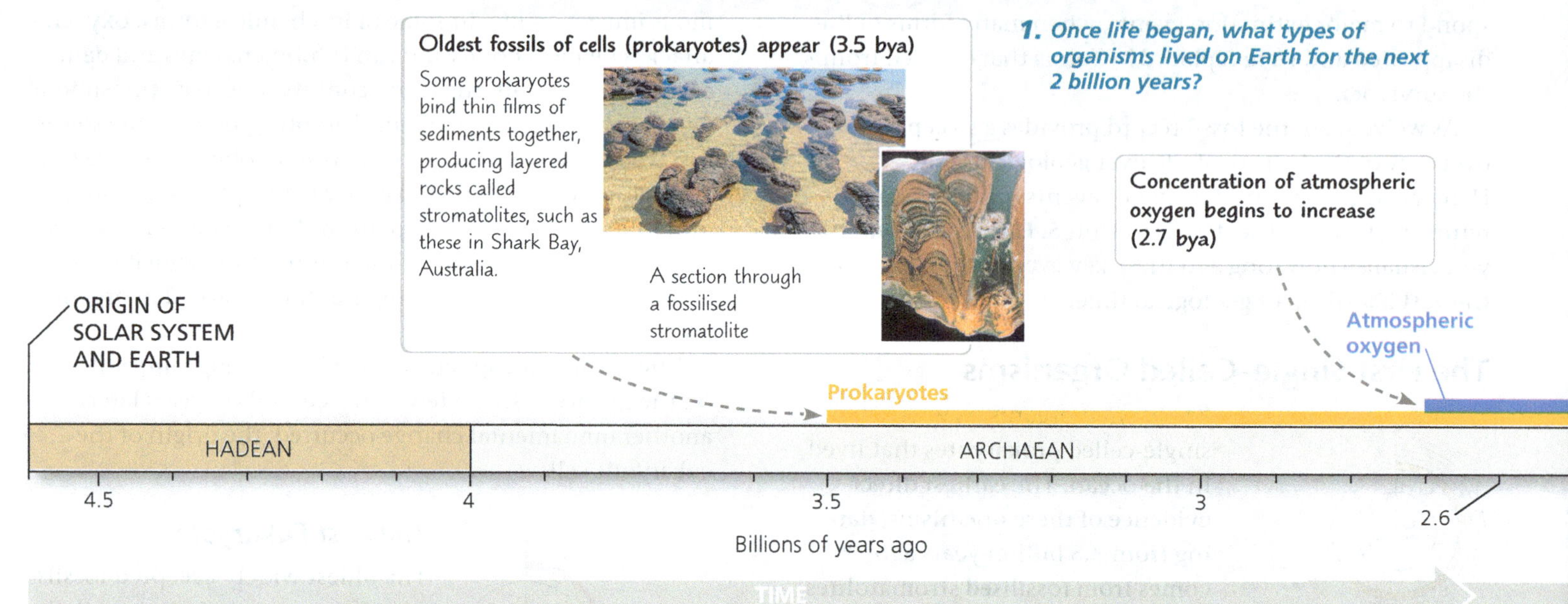

The large time line in this figure shows all of geological time to scale on an unbroken time line, but sometimes it is helpful to "break" a time line to limit a figure's size and highlight key details. As shown at right, hatch marks are often used to represent such an interruption.

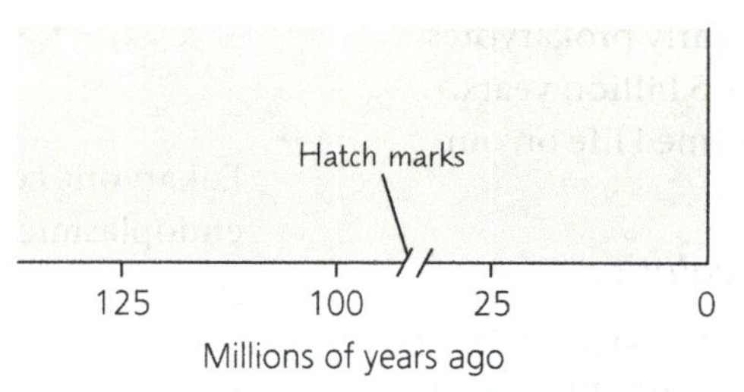

2. Look for a figure in this chapter that uses hatch marks on one of its axes. Explain what the hatch marks represent in that figure.

▼ Figure 25.9 The rise of atmospheric oxygen. Chemical analyses of ancient rocks have enabled this reconstruction of atmospheric oxygen levels during Earth's history.

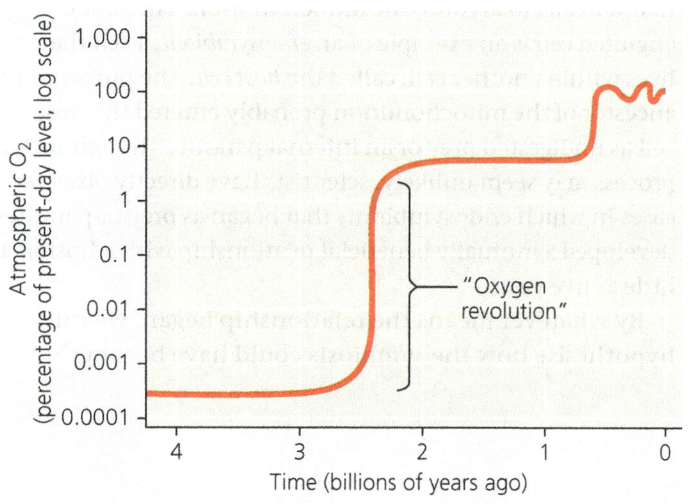

beneficial. For example, in a world that was becoming increasingly aerobic, a host that was itself an anaerobe would have benefited from endosymbionts that could make use of the oxygen. Over time, the host and endosymbionts would have become a single organism, its parts inseparable. Although all eukaryotes have mitochondria or remnants of these organelles, they do not all have plastids (a general term for chloroplasts and related organelles). Thus, the **serial endosymbiosis** hypothesis supposes that mitochondria evolved before plastids through a sequence of endosymbiotic events. As shown in **Figure 25.10**, both mitochondria and plastids are thought to have descended from bacterial cells. The original host—the cell that engulfed the bacterium whose descendants gave rise to the mitochondrion—is thought to have been an archaean or a close relative of the archaea.

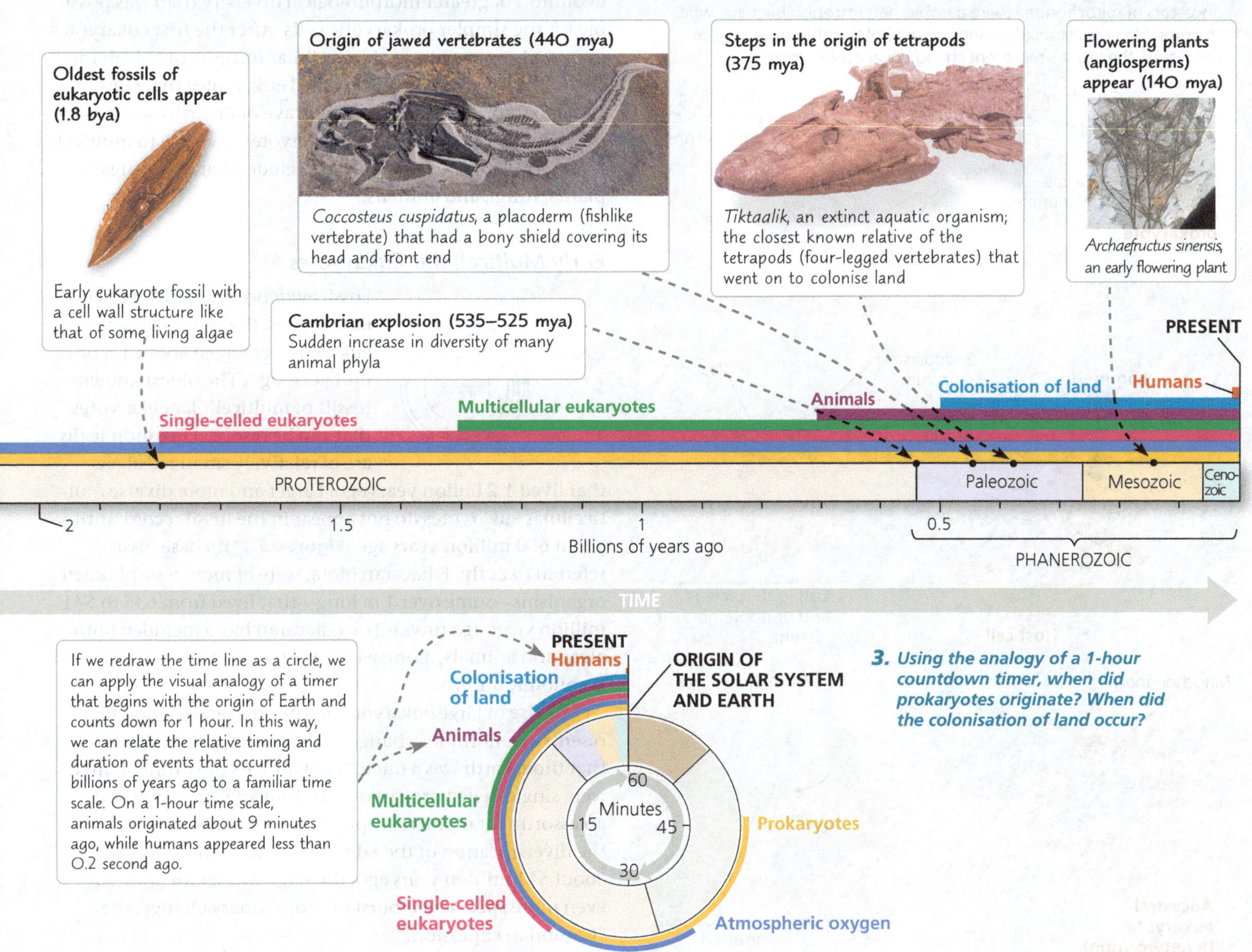

A great deal of evidence supports the endosymbiotic origin of mitochondria and plastids:

- The inner membranes of both organelles have enzymes and transport systems that are homologous to those found in the plasma membranes of living bacteria.
- Mitochondria and plastids replicate by a splitting process that is similar to that of certain bacteria. In addition, each of these organelles contains circular DNA molecules that, like the chromosomes of bacteria, are not associated with histones or large amounts of other proteins.
- As might be expected of organelles descended from free-living organisms, mitochondria and plastids also have the cellular machinery (including ribosomes) needed to transcribe and translate their DNA into proteins.
- Finally, in terms of size, RNA sequences, and sensitivity to certain antibiotics, the ribosomes of mitochondria and plastids are more similar to bacterial ribosomes than they are to the cytoplasmic ribosomes of eukaryotic cells.

In Concept 28.1, we'll return to the origin of eukaryotes, focusing on what genomic data have revealed about the prokaryotic lineages that gave rise to the host and endosymbiont cells.

▼ **Figure 25.10 A hypothesis for the origin of mitochondria and plastids through serial endosymbiosis.** The proposed host cell was an archaean or a close relative of the archaea. The proposed ancestors of mitochondria were aerobic, heterotrophic bacteria, while the proposed ancestors of plastids were photosynthetic bacteria. In this figure, the arrows represent change over evolutionary time.

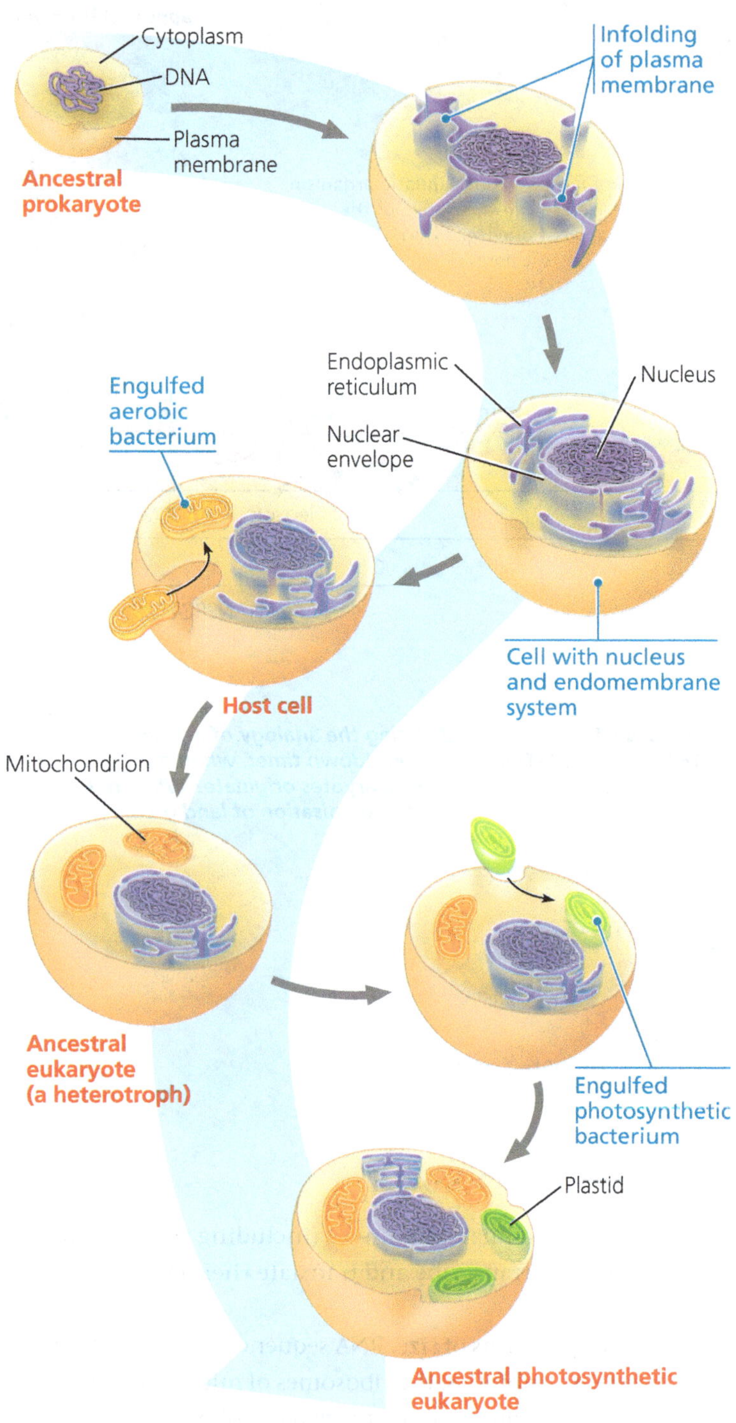

The Origin of Multicellularity

An orchestra can play a greater variety of musical compositions than a violin soloist can; the increased complexity of the orchestra makes more variations possible. Likewise, the origin of structurally complex eukaryotic cells sparked the evolution of greater morphological diversity than was possible for the simpler prokaryotic cells. After the first eukaryotes appeared, a great range of unicellular forms evolved, giving rise to the diversity of single-celled eukaryotes that continue to flourish today. Another wave of diversification also occurred: Some single-celled eukaryotes gave rise to multicellular forms, whose descendants include a variety of algae, plants, fungi, and animals.

Early Multicellular Eukaryotes

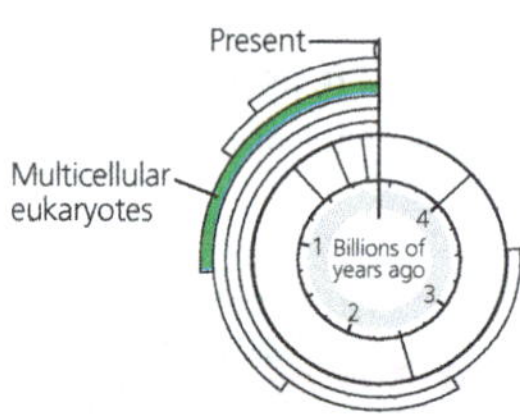

Fossil evidence and DNA sequence data suggest that multicellular eukaryotes emerged about 1.3 billion years ago. The oldest known fossils of multicellular eukaryotes that can be resolved taxonomically are of relatively small red algae that lived 1.2 billion years ago. Larger and more diverse multicellular eukaryotes do not appear in the fossil record until about 600 million years ago **(Figure 25.11)**. These fossils, referred to as the Ediacaran biota, were of mostly soft-bodied organisms—some over 1 m long—that lived from 635 to 541 million years ago (mya). The Ediacaran biota included both algae and animals, along with various organisms of unknown taxonomic affinity.

The rise of large eukaryotes in the Ediacaran period represents an enormous change in the history of life. Before that time, Earth was a microbial world: Its only inhabitants were single-celled prokaryotes and eukaryotes, along with an assortment of microscopic, multicellular eukaryotes. As the diversification of the Ediacaran biota came to a close about 541 million years ago, the stage was set for another, even more spectacular burst of evolutionary change: the "Cambrian explosion."

▼ **Figure 25.11 Life becomes large.** Fossils from the Ediacaran period document the oldest known macroscopic eukaryotes, such as **(a)** *Doushantuophyton*, an alga that lived 600 mya, and **(b)** *Kimberella*, a mollusc (or close relative) that lived 560 mya.

The Cambrian Explosion

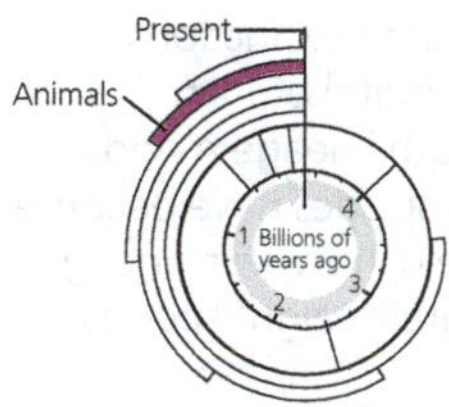

Many present-day animal phyla appear suddenly in fossils formed 535–525 million years ago, early in the Cambrian period. This phenomenon is referred to as the **Cambrian explosion**. Fossils of several animal groups—sponges, cnidarians (sea anemones and their relatives), and molluscs (snails, clams, and their relatives)—appear in even older rocks dating from the late Proterozoic **(Figure 25.12)**.

Prior to the Cambrian explosion, all large animals were soft-bodied. The fossils of large pre-Cambrian animals reveal little evidence of predation. Instead, these animals appear to have been grazers (feeding on algae), filter feeders, or scavengers, not hunters. The Cambrian explosion changed all of that. In a relatively short period of time (10 million years), predators over 1 m in length emerged that had claws and other features for capturing prey; simultaneously, new defensive adaptations, such as sharp spines and heavy body armour, appeared in their prey.

Although the Cambrian explosion had an enormous impact on life on Earth, it appears that many animal phyla originated long before that time. Recent DNA analyses suggest that sponges had evolved by 700 million years ago; such analyses also indicate that the common ancestor of arthropods, chordates, and other animal phyla that radiated during the Cambrian explosion lived 670 million years ago. Researchers have unearthed 710-million-year-old sediments containing steroids indicative of a particular group of sponges—a finding that supports the molecular data. In contrast, the oldest macroscopic fossils of animals date from 560 million years ago and include *Kimberella*, shown in Figure 25.11. Overall, fossils and DNA analyses suggest that animals originated about 700 million years ago and then remained small for over 100 million years—until they diversified explosively during the Cambrian and beyond.

▼ **Figure 25.12 Appearance of selected animal groups.** The white bars indicate the earliest fossil record of these animal groups.

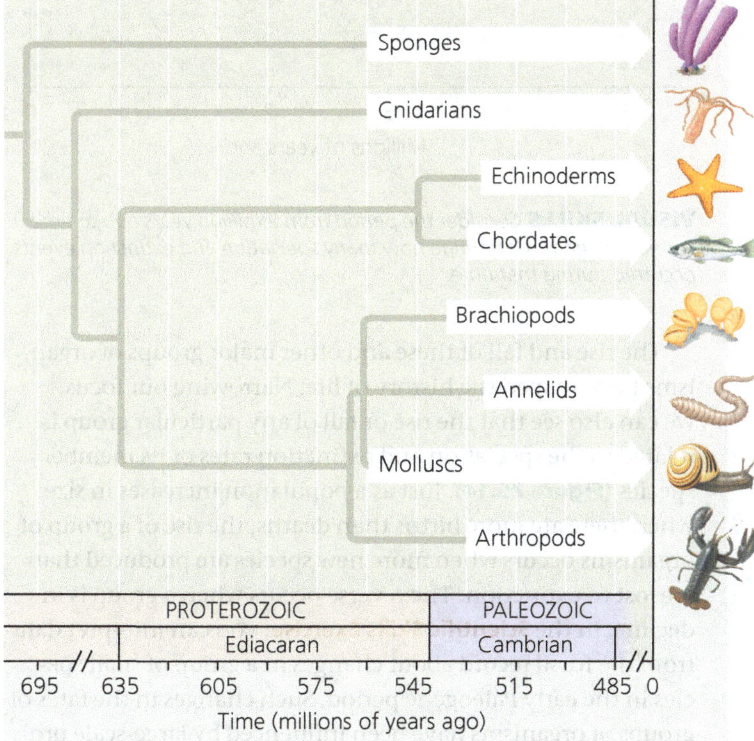

VISUAL SKILLS *Circle the branch point that represents the most recent common ancestor of chordates and annelids. What is a minimum estimate of that ancestor's age?*

The Colonisation of Land

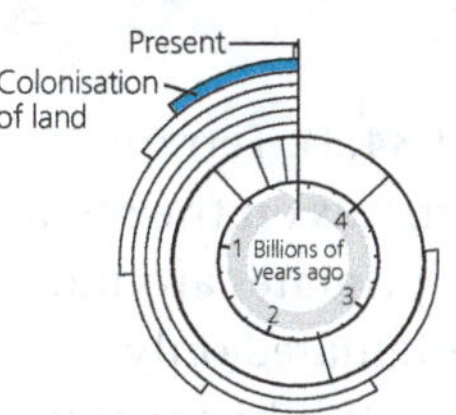

The colonisation of land was another milestone in the history of life. There is fossil evidence that some prokaryotes lived on terrestrial surfaces as early as 3.2 billion years ago. However, larger forms of life, such as fungi, plants, and animals, did not begin to colonise land until about 500 million years ago. This gradual evolutionary venture out of aquatic environments was associated with adaptations that made it possible to reproduce on land and that helped prevent dehydration. For example, many plants today have a vascular system for transporting materials internally and a waterproof coating of wax on their leaves that slows the loss of water to the air. Early signs of these adaptations were present 420 million years ago, at which time small plants (about 10 cm high) existed that had a vascular system but lacked true roots or leaves. By 40 million years later, plants had diversified greatly and included trees and many other plants with true roots and leaves.

Plants appear to have colonised land in the company of fungi. Even today, the roots of most plants are associated with fungi that aid in the absorption of water and minerals from the soil (see Concept 31.1). These root fungi (or *mycorrhizae*), in turn, obtain their organic nutrients from the plants. Such mutually beneficial associations of plants and fungi are evident in some of the oldest fossilised plants, dating this relationship back to the early spread of life onto land **(Figure 25.13)**.

Although many animal groups are now represented in terrestrial environments, the most widespread and diverse land animals are arthropods (particularly insects and spiders) and tetrapods. Arthropods were among the first animals to colonise land, roughly 450 million years ago. The earliest tetrapods found in the fossil record lived about 365 million years ago and appear to have evolved from a

▼ Figure 25.13 An ancient symbiosis. This 405-million-year-old fossil stem (cross section) documents mycorrhizae in the early plant *Aglaophyton major*. The inset shows an enlarged view of a cell containing a branched fungal structure called an arbuscule; the fossil arbuscule resembles those seen in plant cells today.

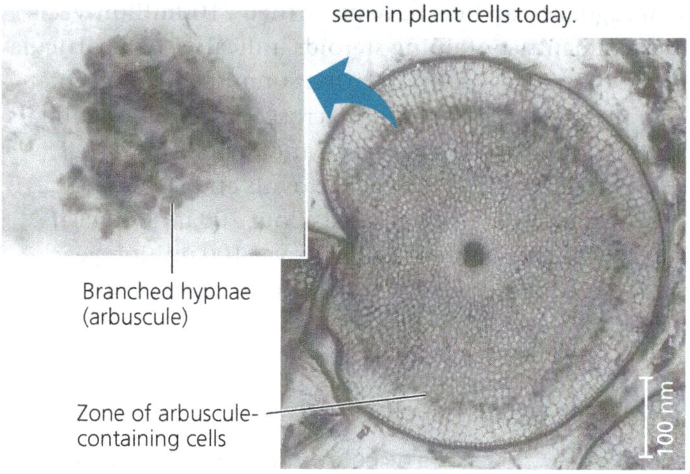

Branched hyphae (arbuscule)

Zone of arbuscule-containing cells

100 nm

group of lobe-finned fishes (see Concept 34.3). Tetrapods include humans, although we are late arrivals on the scene. The human lineage diverged from other primates around 6–7 million years ago, and our species originated only about 195,000 years ago. If the clock of Earth's history were rescaled to represent an hour, humans appeared less than 0.2 seconds ago.

CONCEPT CHECK 25.3

1. The first appearance of free oxygen in the atmosphere triggered a massive wave of extinctions among the prokaryotes of the time. Why?
2. What evidence supports the hypothesis that mitochondria preceded plastids in the evolution of eukaryotic cells?
3. **WHAT IF?** What would a fossil record of life today look like?

For suggested answers, see Appendix A.

CONCEPT 25.4

The rise and fall of groups of organisms reflect differences in speciation and extinction rates

From its beginnings, life on Earth has been marked by the rise and fall of groups of organisms. Anaerobic prokaryotes originated, flourished, and then declined as the oxygen content of the atmosphere rose. Billions of years later, the first tetrapods emerged from the sea, giving rise to several major new groups of organisms. One of these, the amphibians, went on to dominate life on land for 100 million years, until other tetrapods (including dinosaurs and, later, mammals) replaced them as the dominant terrestrial vertebrates.

▼ Figure 25.14 How speciation and extinction affect diversity. The species diversity of an evolutionary lineage will increase when more new member species originate than are lost to extinction. In this hypothetical phylogenetic tree, each horizontal branch represents one species. By 2 million years ago both lineage A and lineage B have given rise to four species, and no species have become extinct (denoted by a dagger symbol, †). By time 0, however, lineage A contains only one species, while lineage B contains eight species.

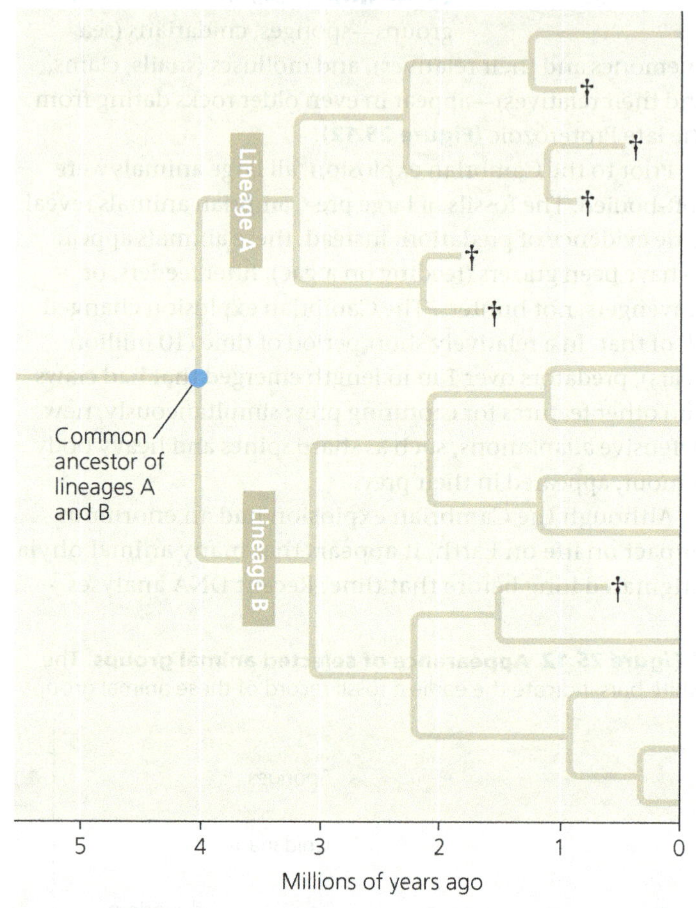

VISUAL SKILLS *Consider the period from 2 million years ago to time 0. For each lineage, determine how many speciation and extinction events occurred during that time.*

The rise and fall of these and other major groups of organisms have shaped the history of life. Narrowing our focus, we can also see that the rise or fall of any particular group is related to the speciation and extinction rates of its member species **(Figure 25.14)**. Just as a population increases in size when there are more births than deaths, the rise of a group of organisms occurs when more new species are produced than are lost to extinction. The reverse occurs when a group is in decline. In the **Scientific Skills Exercise**, you can interpret data from the fossil record about changes in a group of snail species in the early Paleogene period. Such changes in the fates of groups of organisms have been influenced by large-scale processes such as plate tectonics, mass extinctions, and adaptive radiations.

Scientific Skills Exercise

Estimating Quantitative Data from a Graph and Developing Hypotheses

Do Ecological Factors Affect Evolutionary Rates? Researchers studied the fossil record to investigate whether differing modes of larval dispersal might explain species longevity within one taxon of marine snails, the family Volutidae. Some of the snail species had nonplanktonic larvae: They developed directly into adults without a swimming stage. Other species had planktonic larvae: They had a swimming stage and could disperse very long distances. The adults of these planktonic species tended to have broad geographic distributions, whereas nonplanktonic species tended to be more isolated.

How the Research Was Done The researchers studied the stratigraphic distribution of volutes in outcrops of sedimentary rocks located along North America's Gulf coast. These rocks, which formed from 66 to 37 million years ago, early in the Paleogene period, are an excellent source of well-preserved snail fossils. The researchers were able to classify each fossil species of volute snail as having planktonic or nonplanktonic larvae based on features of the earliest formed whorls of the snail's shell. Each bar in the graph shows how long one species of snail persisted in the fossil record.

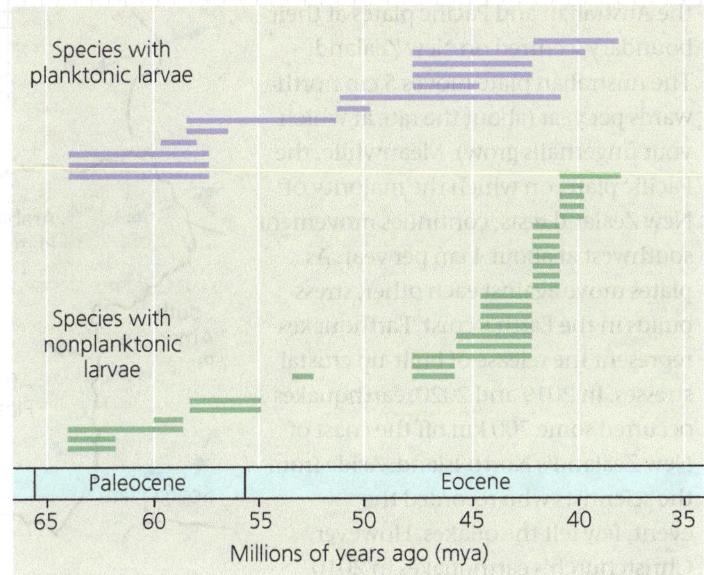

INTERPRET THE DATA

1. You can estimate quantitative data (fairly precisely) from a graph. The first step is to obtain a conversion factor by measuring along an axis that has a scale. In this case, 25 million years (my; from 60 to 35 million years ago [mya] on the *x*-axis) is represented by a distance of 7.0 cm. This yields a conversion factor (a ratio) of 25 my/7.0 cm = 3.6 my/cm. To estimate the time period represented by a horizontal bar on this graph, measure the length of that bar in centimetres and multiply that measurement by the conversion factor, 3.6 my/cm. For example, a bar that measures 1.1 cm on the graph represents a persistence time of 1.1 cm × 3.6 my/cm = 4 million years.
2. Calculate the mean (average) persistence times for species with planktonic larvae and species with nonplanktonic larvae.
3. Count the number of new species that form in each group beginning at 60 mya (the first three species in each group were present around 64 mya, the first time period sampled, so we don't know when those species first appear in the fossil record).
4. Propose a hypothesis to explain the differences in longevity of snail species with planktonic and nonplanktonic larvae.

Data from T. A. Hansen, Larval dispersal and species longevity in Lower Tertiary gastropods, *Science* 199:885–887 (1978). Reprinted with permission from AAAS.

Plate Tectonics

If photographs of Earth were taken from space every 10,000 years and spliced together to make a movie, it would show something many of us find hard to imagine: The seemingly "rock solid" continents we live on move over time. Over the past billion years, there have been three occasions (1 billion, 600 million, and 250 million years ago) when most of the landmasses of Earth came together to form a supercontinent, then later broke apart. Each time, this breakup yielded a different configuration of continents. Based on the directions in which the continents are moving today, some geologists have estimated that a new supercontinent will form roughly 250 million years from now.

According to the theory of **plate tectonics**, the continents are part of great plates of Earth's crust that essentially float on the hot, underlying portion of the mantle **(Figure 25.15)**. Movements in the mantle cause the plates to move over time in a process called *continental drift*. Geologists can measure the rate at which the plates are moving now, usually only a few centimetres per year. They can also infer the past locations of the continents using the magnetic signal recorded in rocks at the time of their formation. This method works because as a continent shifts its position over time, the direction of magnetic north recorded in its newly formed rocks also changes.

▼ **Figure 25.15 Cutaway view of Earth.** The thickness of the crust is exaggerated here.

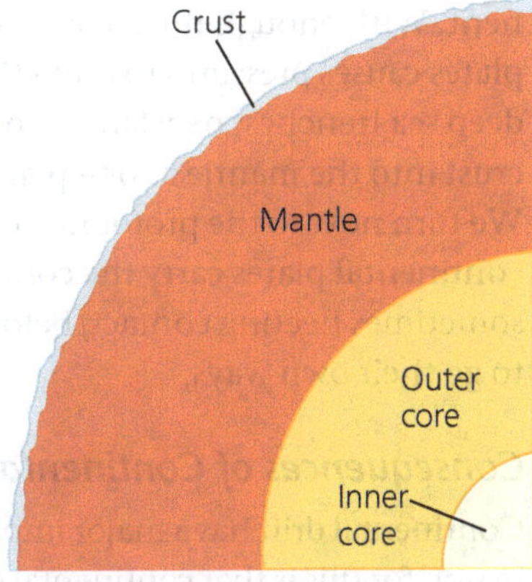

Earth's major tectonic plates are shown in **Figure 25.16**. Many important geological processes, including the formation of mountains and islands, occur at plate boundaries. In some cases, two plates separate from each other. We'll consider the separation of the Australian and Pacific plates at their boundary, centred on New Zealand. The Australian plate moves 5 cm northwards per year (about the rate at which your fingernails grow). Meanwhile, the Pacific plate, on which the majority of New Zealand rests, continues movement southwest at about 4 cm per year. As plates move against each other, stress builds in the Earth's crust. Earthquakes represent the release of built-up crustal stresses. In 2019 and 2020, earthquakes occurred some 700 km off the coast of New Zealand's North Island. Aside from the scientists who recorded the event, few felt the quakes. However, Christchurch's earthquakes in 2010, 2011, and 2016 caused great damage to the social fabric of the region, combined with environmental and economic costs that New Zealanders continue to experience. In still other cases, two plates collide, producing violent upheavals and forming new mountains along the plate boundaries. One spectacular example of this occurred 45 million years ago, when the Indian plate crashed into the Eurasian plate, starting the formation of the Himalayan mountains.

▼ **Figure 25.16 Earth's major tectonic plates.** The blue arrows indicate direction of movement. The reddish orange dots represent zones of violent tectonic activity.

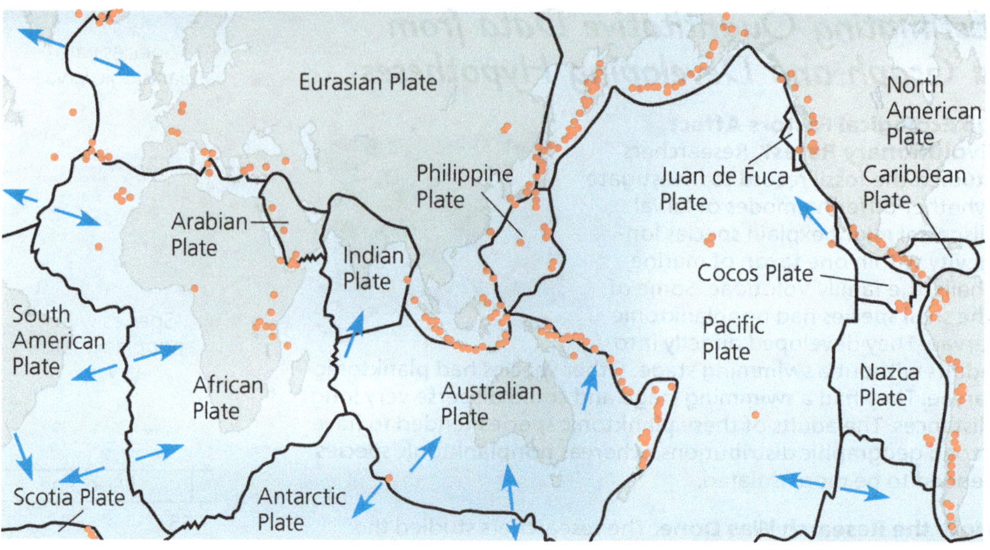

Until 2017, scientists recognised seven distinct continents that floated on or across tectonic plates. Then Nick Mortimer and fellow New Zealand scientists identified a major new continent, Te Riu-a-Māui/Zealandia **(Figure 25.17)**. The continent comprises an area roughly two-thirds the land area of Australia, although 94% of it remains submerged. The 6% above water comprises two land masses: New Zealand and New Caledonia. As discussed earlier, an active plate boundary bisects the continent. With enough time, contact between slowly moving plates causes pressures to build that create mountains, deep sea trenches, new lands, and the recycling of old crust into the mantle as one plate slides beneath another. We turn next to the profound consequences for life as continental plates carry the continents into long and sometimes fleeting contact, before separating once more to go their own ways.

▼ **Figure 25.17 The spatial limits of Zealandia.** (NC, New Caledonia; WTP, West Torres Plateau; CT, Cato Trough; Cf, Chesterfield Islands; L, Lord Howe Island; N, Norfolk Island; K, Kermadec Islands; Ch, Chatham Islands; B, Bounty Islands; An, Antipodes Islands; Au, Auckland Islands; Ca, Campbell Island.)

Consequences of Continental Drift

Continental drift has a major impact on life on Earth. One reason for this is that continental drift alters the habitats in which organisms live. Consider the changes shown in **Figure 25.18**. About 250 million years ago, plate movements brought previously separated landmasses together into a supercontinent named **Pangaea**. Ocean basins

▼ **Figure 25.18 The history of continental drift during the Phanerozoic eon.**

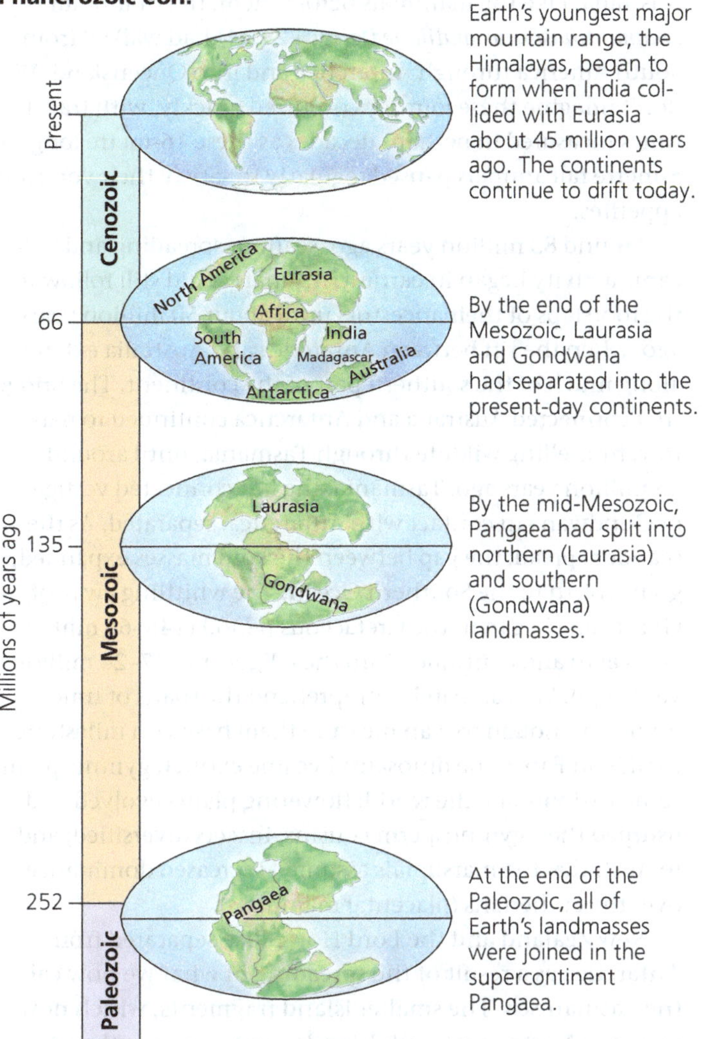

VISUAL SKILLS *Is the Australian plate's current direction of movement (see Figure 25.16) similar to the direction it travelled over the past 66 million years?*

became deeper, which drained shallow coastal seas. At that time, as now, most marine species inhabited shallow waters, and the formation of Pangaea destroyed much of that habitat. Pangaea's interior was cold and dry, probably an even more severe environment than that of central Asia today. Overall, the formation of Pangaea greatly altered the physical environment and climate, which drove some species to extinction and provided new opportunities for groups of organisms that survived the crisis.

Organisms are also affected by the climate change that results when a continent shifts its location. For example, the subtropical rain forests now found in southeast Queensland once occurred in the current location of the Victoria–New South Wales border. As the continent moved progressively towards the tropical region of our planet, the landscapes experienced progressive shifts in environmental conditions.

Continental drift also promotes allopatric speciation on a grand scale. When supercontinents break apart, regions that once were connected become isolated. As the continents drifted apart over the last 200 million years, each became a separate evolutionary arena, with lineages of plants and animals that diverged from those on other continents. For example, marsupial mammals fill ecological roles in Australia analogous to those filled by eutherians (placental mammals) on other continents (see Figure 22.18). Fossil evidence shows that marsupial mammals originated in what is now Asia and reached Australia via South America and Antarctica while these continents were still joined as a part of Gondwana. The subsequent breakup of the southern continents set Australia "afloat," like a giant raft that carried marsupials. In Australia, marsupials diversified, and the few eutherians that lived there became extinct; on other continents, most marsupials became extinct while the eutherians diversified.

Finally, continental drift can help explain puzzling geographic distributions of extinct organisms, such as fossils of the same species of Permian freshwater reptiles found in both Brazil and the West African nation of Ghana. These two parts of the world, now separated by 3,000 km of ocean, were joined together when these reptiles were living. Continental drift also explains the long-term isolation of Australia and New Zealand that produced fauna and flora so different from that in much of the rest of the world.

Continental drift and the associated shifts in climate shaped relatively few of Australia and New Zealand's ancestral families into the biological diversity we see today. While Australia and New Zealand have witnessed the evolution of an astonishing variety of plants and animals, some species have remained relatively unchanged over hundreds of millions of years. We will begin our examination at the start of the Jurassic period (205 to 141 million years ago). Gondwana is named after the Gond region in India, where important fossils of the ancient seed fern *Glossopteris* spp. occur in Carboniferous–Permian coal measures. **Figure 25.19** illustrates the likely appearance of

▼ **Figure 25.19 Permian *Glossopteris* forest.** This Permian forest featured *Glossopteris* as the tallest trees in the often perpetually wet soils. *Glossopteris* saplings grew in the understorey, as did the horsetails, which featured canopies that emerged at uniform lengths up the trunk. Everywhere, you would have seen ferns as the dominant understorey and mid-storey species.

those ancient forests. We owe coal and a range of other fossil fuels to the *Glossopteris* forests, which contained algae, mosses, fungi, ferns, lycopods, horsetails, cycads, and other primitive gymnosperms. These important fossils provide some of the evidence scientists use to reconstruct the form of the supercontinent. Travelling to Australia today, you can see *Glossopteris* fossils in shale and coal beds in the Sydney basin, and in Antarctica, that are similar to those in India's Gond region.

Breakup of Gondwana

In the Jurassic period, Gondwana's landmass encompassed several areas we today recognise as continents and large islands, including Australia, Antarctica, Africa, Arabia (including parts of Turkey and Iran), India, Madagascar, South America, New Zealand, New Caledonia and New Guinea, and fragments of South-East Asia (including most of Indonesia) **(Figure 25.20)**.

With a continuous landmass on which to live, plants dispersed, while animals travelled extensively. Some of the most consequential journeys in the history of life on Earth occurred across Gondwana. One hundred million years ago, monotremes had become established in what became Australia. An opalised jaw of a platypus-like mammal may represent earliest evidence of inhabitation of this group in Australia. Like the mammals before them, the titanosaur *Diamantinasaurus matildae* **(Figure 25.21)** had walked from South America through Antarctica and into Queensland. We don't imagine these journeys occurred quickly, with travel times measured in perhaps decades as these 16-metre long, 5-metre tall animals paused regularly to satisfy their voracious appetites.

Around 83 million years ago, seafloor spreading and volcanic activity began in earnest. Animals could still follow in the footsteps of their ancestors until about 50 million years ago, when the rift between Antarctica and Australia extended along much of the southern part of the continent. The bridge that connected Australia and Antarctica continued to narrow, funnelling wildlife through Tasmania, until around 35 million years ago. Tasmania, the last connected vestige of Gondwana in contact with Antarctica, separated. As the seafloor spread, the gap between the land masses expanded, giving birth to the Southern Ocean. The whittling away of Gondwana began in the Cretaceous period (145–66 million years ago) and continued into the Oligocene (37–24 million years ago). We can barely comprehend the spans of time involved, though we can measure them based on milestones for life on Earth: the dinosaurs became extinct; gymnosperms came to dominate the world; flowering plants evolved and usurped their gymnosperm cousins; insects diversified; and, in Australia, the marsupials assumed increased dominance over the eutherians (placental mammals).

New Zealand and the Lord Howe Rise separated from Antarctica as a result of the spreading of what we now call the Tasman Sea. The smaller island fragments, which now form the North and South Islands, journeyed northeastwards for more than 80 million years. Historically, the islands that we now know as New Zealand varied in shape and size before attaining their current configurations. **Figure 25.22** reveals the shifting shape of New Zealand

▼ **Figure 25.20 The Jurassic Earth.** Two hundred million years ago, in the Jurassic period, a continuous landmass stretched over much of what would become the Southern Hemisphere. In the Tertiary period, and the Oligocene epoch (35 million years ago), continental drift had separated Gondwana into landmasses recognisable to modern geographers.

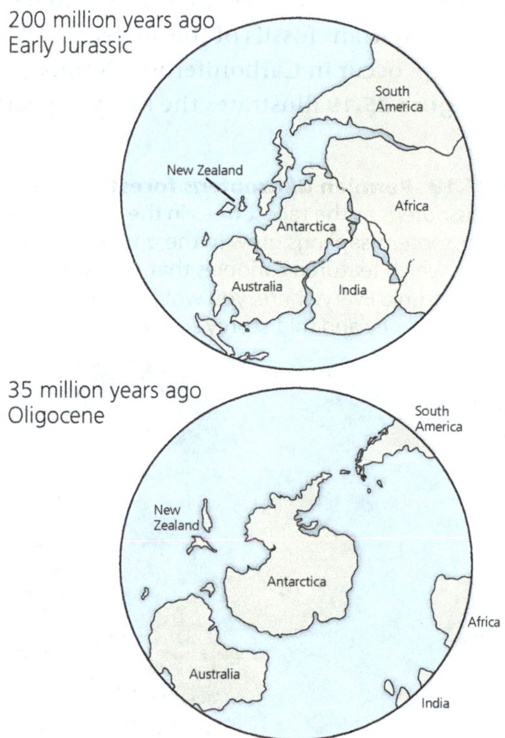

▼ **Figure 25.21 The titanosaur *Diamantinasaurus matildae*.** One of the first of Australia's titanosaurs walked here from South America.

▼ **Figure 25.22 New Zealand during the later Cainozoic Era.** New Zealand's coastlines have shifted repeatedly in response to changes in sea levels over the last 45 million years. New Zealand's current coastline appears in white. The green areas indicate the land area over time.

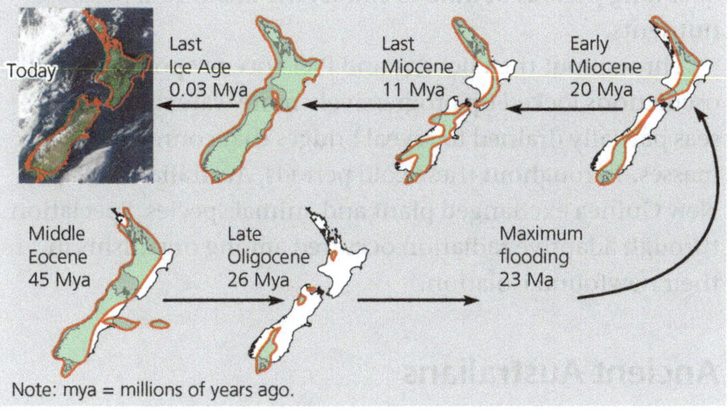

over the last 45 million years. The rise and fall of sea levels, combined with geological activities, have submerged and revealed aspects of New Zealand's landscape. The areas shaded green in the figure represent the exposed landscape. Geologists and biologists estimate the almost total submergence of New Zealand to have occurred around 23 million years ago. Hamish Campbell's interview at the beginning of this unit provides insights into the recolonisation of New Zealand, largely through migration and the establishment of Australian (and perhaps New Caledonian) flora and fauna since the time of maximum flooding.

In particular, the South Island is an amalgam of fused fragments. New Zealand has been described as a changing archipelago that first began to resemble its current form approximately 15 million years ago. These events occurred during the Miocene period (23–5 million years ago) after maximum flooding, with land rising and sediments filling in small basins. Extensive volcanic activity and contact between transverse continental plates over the next 5 million years created high mountain ranges from the low-lying Miocene plains. The longer isolation, lack of recent linkages, and closer affinities between New Zealand, New Caledonia, and Lord Howe Island earn separate recognition from biogeographers, who refer to this group of landmasses as Tasmantis.

Australia's Changing Climate

Today, the magnetic South Pole features a pole painted with red-coloured helices (like those on a barber's pole) that rise to a golden globe. As far as the eye can see, and out to the edge of the continent, ice and snow cover most of the landscape. If you could transport that same pole, where it stands, back in time to when Gondwana still existed, you would immediately feel the warm rain; the climate featured year-round rainfall. A luxuriant rain forest would surround you.

Then, as now, nights at the South Pole lasted for months at a time. With the final separation of Australia from Gondwana, the circulation patterns of the oceans shifted. Polar waters cooled, freed from the mitigating warmth of tropical waters, and global temperatures declined. Life on Earth experienced profound change. As the Earth cooled, ice sheets expanded from the highest peaks. Until recently, biologists believed that the ice sheets swept the forests and attendant species to extinction. However, as recently as 2–3 million years ago, some 400 kilometres from the South Pole, forests with trees perhaps no more 2 metres tall still survived, clinging to rocky slopes. These forests may have closely resembled modern Tasmania's deciduous beech forests (*Nothofagus gunnii*). Perhaps, like their modern Tasmanian cousins, the species of the time, such as *N. beardmorensis*, may have possessed leaves that turned the colour of burnished gold before being shed ahead of freezing winter temperatures and months of darkness.

During the early phases of Australia's separation and subsequent northwards journey, global climates shifted from warm and wet to cold and dry. Three additional distinct warm and moist climatic optima occurred; two in the early Miocene and a third in the middle Miocene, 16–15 million years ago. For the majority of the past 80 million years, however, uniformly cool global temperatures prevailed, due in part to ocean circulation patterns and newly formed mountain ranges. Cooler air cascading down from the newly upthrust Himalayan plateau would have contributed to the cooler climate, as did the development of the Southern Ocean. Local uplifts of the mountains in highland Papua New Guinea, capped by Puncak Jaya/Carstensz Pyramid (4,884 m), developed as the Australian plate pushed against the Pacific plate to the north. **Figure 25.23** illustrates the comparative geography of Australia, New Guinea, and parts of Indonesia. The light blue colouration around Australia reveals the comparatively shallow water covering the continental shelf. Lowered sea levels revealed these areas of continental shelf, which plants colonised and animals used to traverse between formerly isolated land masses. Exposure of large areas of the continental shelf may have played a part in the extinction of the Tasmanian megafauna—as we will learn shortly.

Figure 25.23 also provides a graphic illustration of the comparative height and extent of Papua New Guinea's mountains, which formed a rain shadow across northern Australia. Many scientists attribute the increased aridity and contraction of Australia's rain forests to the uplift of these mountains. The rain forests were replaced with newly evolved vegetation that coped better with aridity and the increased frequency of fire **(Figure 25.24)**.

▼ Figure 25.23 Australia, Papua New Guinea, and the Indonesian islands illustrate the outcomes of mountain building at the boundaries of continental plates. Puncak Jaya/Carstensz Pyramid (4,884 m) appears on the left side of Papua New Guinea, in Indonesian Papua. For millions of years, glaciers covered much of Puncak Jaya. Recent global warming has caused the melting and diminishment of these glaciers to only 6% of their former extent.

▼ Figure 25.24 Australian sclerophyll forests. Australia's drying climate favoured the evolution of plants that were tolerant of less fertile soils, seasonal rains, and much reduced water availability.

Australia's arid-adapted plants developed sclerophylly (tough, leathery leaves), which helped reduce water loss from the plant. As Australia became increasingly parched, rain forests gave way to open forests, then woodlands, savanna, grasslands, and finally, in the later Neogene, deserts.

Changed environments produced the extinction of many animals, while adaptations to the new circumstances evolved in others. For example, the early evolution of sclerophyllous plants occurred in environments free from the activities of herbivores. Initially, few animals could digest and gain sufficient nourishment from the leaves of these new plants. However, several herbivorous lineages evolved that became sclerophyll-digesting specialists. One strategy was to increase in size: Being larger meant that the herbivores could take in more food. Progressive enlargement in the length of the digestive system and a slower metabolism enabled these animals to process low-quality food (such as sclerophyll leaves) over long periods of time to extract the maximum possible nutrients.

Throughout the Pliocene and Pleistocene epochs, climatic oscillations locked up progressively more water at the poles; seas partially drained to reveal bridges that connected land masses. Throughout these cold periods, Australia and Papua New Guinea exchanged plant and animal species. Speciation through adaptive radiation occurred among organisms in their newfound isolation.

Ancient Australians

Looking around the world during the Triassic period (251–199 million years ago), in what would become Australia and New Zealand, we would have seen biological wonders as well as some species remarkably unchanged from those around us today. For example, in rock pools, outcrops, and forest patches, you might have been lucky to observe Australia's freshwater shrimp (*Anaspides tasmaniae*) **(Figure 25.25)**, and also New Zealand's tuatara species. Both species survived the breakup of Pangaea, then Laurasia, and subsequently Gondwana. They survive today.

Marsupial mammals fill ecological roles in Australia analogous to those filled by eutherians on other continents (see Figure 22.18). Fossil evidence suggests that marsupials originated in what is now Asia and reached Australia via South America and Antarctica while the continents were still joined.

▼ Figure 25.25 Tasmania's mountain shrimp (*Anaspides* spp.). Like New Zealand's tuatara, the mountain shrimp has survived largely unchanged for over 200 million years. Both are living fossils, and both are endemic to their island homes. The mountain shrimp occurs in highland creeks, tarns, and rockpools. These animals survive by eating pretty much anything in front of them, from mosses to others of their own species. At adulthood, they grow more than 6 cm in length—largely unaltered over the last 250 million years.

The subsequent breakup of the southern continents set Australia "afloat," like a giant raft of marsupials. In Australia, marsupials diversified, and the few eutherians that lived there became extinct; on other continents, most marsupials became extinct, and the eutherians diversified.

However, mammals represented only a small component of the mixed cargo inherited from Gondwana. Other passengers aboard these ancient landscapes include lineages that survive today—for example, ratites (flightless birds) such as ostriches (Africa), emus (Australia), cassowaries (Australia and New Guinea), moas and kiwis (New Zealand), and rheas (South America). Lungfish also survive in countries derived from Gondwana. They are represented by a single genus in each of the following countries: Australia (*Neoceratodus*), South America (*Lepidosiren*), and Africa (*Proptopterus*). Yet other passengers were the 2-metre long giant meiolaniid turtles with horns and a spiky tail, which died out in South America in the Eocene (about 55–33 million years ago).

Australia also provided a refuge for these creatures, and other animals of unusual size, that we collectively call the **megafauna**.

Australia's Megafauna

An ice age ushered in the conditions conducive to the dominance of the megafauna across the world, in Australia particularly, and to a lesser extent in New Zealand. The "mega" epithet evokes well-deserved reputations in the public's imagination of colossal animals, with commensurate attitudes and influence on the Australian environment.

By about 11,000 years ago, Australia's megafauna were extinct, while the somewhat more isolated giant birds of New Zealand lasted into the 15th century. However, you may have seen some of the survivors of that extinction event in Africa (hippopotamuses, elephants, giraffes, lions, tigers), Northern America (bison, bears, and various mountain lions), Indonesia (Asian elephants, white rhinoceroses, and the Komodo dragon), Australia (estuarine crocodiles, red kangaroos), sub-Antarctic islands (various large seals), and transient whales (right, baleen, blue).

Any trip to a museum that features exhibits of skeletonic representations of the megafauna elicits three, nearly always the same, breathlessly whispered questions among groups of people who range in age from four to 100 years young: How did the megafauna evolve, what were they like, and why did they become extinct? Let's answer those questions in turn.

Evolution of the Megafauna

Somewhere between 217 and 161 million years ago, a remarkable divergence began to occur within an apparently unremarkable group of animals. These nocturnal animals hid during the day in burrows, tree hollows, and other out of the way places, with perhaps some conception of hope that they could avoid the attentions of the most conspicuous animals of the day—dinosaurs that ranged in size from 50 grams to many tonnes. At night, these egg-laying, rodent-like ancestors of today's platypus and echidna would emerge to feed on insects, fungi, decaying plant material, and each other.

According to Professor Guojie Zhang and his team at the University of Copenhagen, the genomes of the iconic platypus and echidna, the monotremes (egg-laying mammals), represented the ancestral group from which evolved all marsupial and placental mammals. (You will learn about those animals in greater detail later in this text.) About 125 million years ago, in North America, the divergence began between the marsupials (animals that gave birth to live, although little developed, young that they carried in pouches, much as modern koalas or Tasmanian devils do) and the placental mammals that gave birth to well-developed young (humans, dingos). Each group shared a noteworthy trait: In their early development, offspring fed on milk their mothers produced in specialised glands. We'll examine more fully the ways these animals reproduced in the chapter, "Animal Reproduction," although we will look specifically at aspects of reproduction among members of Australia's megafauna shortly.

Around 66 million years ago, marsupials followed the trail of the dinosaurs from North America, through South America, across Antarctica, and into Australia. With only a million or so years to settle into their new home, the world might have continued much as it had for a long time. Except that, 65 million years ago, a meteorite struck Earth. For years afterwards, dust darkened the skies, whole ecosystems collapsed, and serial extinctions began; by the time the clouds cleared, the largest land-based animals had become consigned to fossils and, much later, the public's imagination. Animals smaller in mass than a mid-sized dog survived, as did a few dinosaurs. Today, the descendants of dinosaurs carry a single, collective name: birds.

With the passing of the dinosaurs, opportunities arose for the meteorite survivors and their offspring to flourish in the recently vacated and now unexploited parts of the landscape. Nature abhors a vacuum and, along with the birds, monotremes, marsupials, placental mammals, and other animals diversified. Over time, each became specialised to exploit specific resources in the environment—a phenomenon known as an adaptive radiation. (You have already learned about this process through your study of Darwin's finches in the Galápagos Islands in other sections of this text; see for example, the chapter, "Descent with Modification.") While some animals remained small, others increased exponentially in size until about 40 million years ago. Biologists suspect that, at that point, the largest animals may have reached the physiological and/or ecological maxima beyond which available resources could not support them. From these large animals, the foundations for the megafauna had become established in a landscape unlike any other. Around the world, placental mammals succeeded, while the marsupials and monotremes became living relics, curiosities and minor parts of the fauna. In Australia, the opposite happened: The marsupials (with a small number of monotremes) came to own Australia. Examination of the fossil record shows us something of this process: A 55-million-year-old fossil *Djarthia murgonensis* (from Murgon in Queensland) provides a glimpse of the possible ancestors of all subsequent insect-eating Australian marsupials.

Two sites in Australia earned their place on the World Heritage Area list because of their rich animal fossil deposits. The Riversleigh World Heritage area fossil deposits provide a snapshot of Australia from 25 to 15 million years ago **(Figure 25.26a)**. We owe our glimpse of this epoch to the work of Professor Michael Archer, colleagues from the University of New South Wales, and international collaborators. They showed us an almost unimaginable diversity of animals.

▼ Figure 25.26 The Riversleigh fossil fields in Boodjamulla National Park, Western Queensland. The Riversleigh World Heritage site provides access to deposits that encase fossils from around 25–15 million years ago.

(a) This Riversleigh outcrop features rocks that bristle with fossil bones.

(b) Thirty-five years of excavations, painstaking research, and expert reconstruction from fossils all inform this artist's conception of the forests and animals present at the Riversleigh site some 15 million years ago.

Figure 25.26b illustrates a 15-million-year-old tropical forest and its mostly marsupial inhabitants. These included kangaroos that ate meat, others that inhabited trees, and still others that thrived on a diet of slugs; a variety of quolls and small marsupials; reptiles; birds that pollinated plants; and large birds with affinities to the present-day critically endangered cassowary.

Let's journey to the Riversleigh forest together, in our imaginations. Pick a moderately sized tree, any one will do. Kick it hard with the sole of your foot. The immediate effect would be an avalanche of possums, some of which appear in Figure 25.26. At dusk on your first day in the area, make sure to keep your head down: One of the more than 40 bat species that lived here may collide with you. As you set up camp, make sure you have a fire near you, and that you have allocated watch duty among your team. We would not wish a stealth predator to sneak up on any of you in the dark, to kill and eat you with the passage of night.

Our window into the past slams shut with the last of the Riversleigh deposits—at 15 million years ago. However, another opens via the Naracoorte caves site in South Australia, sites on the Darling Downs in Queensland, and parts of Western Australia and the Northern Territory. These allow us to glimpse the world of the Pleistocene epoch (2.6 million—11,700 years ago). In that vast span of time, the megafauna lived, thrived, and largely died. The Pleistocene witnessed environmental fluctuations that varied between glacial and interglacial periods. Glacials featured global temperatures between 5 and 15°C colder than today. Much of Earth's water froze into ice sheets that covered much of the Northern Hemisphere and some of the Southern Hemisphere, including parts of Australia, New Zealand, and most of Antarctica. Sea levels declined between 85 and 120 m below current levels. Interglacials saw much of the ice melt and sea levels approach their modern-day equivalents, as temperatures increased to around 2–5°C colder than today.

The summer of the first glacial in the Pleistocene featured cold days throughout the world, with even colder nights. In the seasons and years that followed, temperatures continued to decline, until much of the world featured ice-coated landscapes throughout the year. Biologists recognise that the cold conditions favoured animals with a large volume to comparatively small surface-area ratio. With less surface area to radiate heat, the animals enjoyed a metabolic and physical advantage that conferred the ability to tolerate, and subsequently thrive in, the cold. While small animals persisted—perhaps in shelters, or burrows, or with activity in the comparatively "warmer" part of the day—glacial periods served as an evolutionary sieve that favoured some adaptations over others. The ancestors of the megafauna shared traits that included greater metabolic efficiency, more extensive fat reserves, and longer lives. The modern poster child for these traits is one of the megafaunal extinction survivors: North American bears. Bears build up extensive fat reserves over the summer. They can survive on those stored reserves through the winter, when they hibernate—a process that involves slower metabolism, lowered body temperatures, little or no food intake, and a kind of long-duration sleep that lasts throughout the inhospitable months of the winter. Other megafauna survivors like the North American bison migrated to warmer, more hospitable climates for at least part of the year.

What Were the Megafauna Like?

Any student of the megafauna needs learn from the bones, and consider the animal and the environment in which it lived. However, that remains insufficient. The megafauna were more than fossils. They lived, breathed, reproduced, and apparently engaged in complex social and other interactions with each other and the environment. For at least the last 60,000 years or so, megafauna would have interacted with humans. In this section, we share examplars of the behaviours, structures, and strategies of the megafaunal species to address the question most asked: What were they like?

Without the extinction of the largest dinosaurs, the monotremes and marsupials would not have evolved, diversified, and become Australia's megafauna from 2.6 million to around 11,000 years ago. Perhaps we could choose no better member of the megafauna with which to begin our discussion than a descendent of the dinosaurs; in this case, a bird—a thunder bird—as it became known. *Genyornis newtoni* **(Figure 25.27)** shared traits with other flightless birds, such as the more recently extinct moas of New Zealand. When *G. newtoni* stood erect, it could exceed 2 m in height—although, like emus, ostriches, and, we suspect, moas, it spent most of its time with its head down, picking up grasses, seeds, nuts, and succulent leaves on trees.

Like dinosaurs, birds possess various skeletal features that include largely hollow bones (reinforced internally with bony struts). The thickly boned and reinforced beak of the thunder bird makes it both different from other birds, and all the more noteworthy. Strong reinforcement of the beak and skull added mass and supports for the musculature required for the birds to break apart some of the toughest nuts. Birds generally lack teeth, and so it was with this animal. Like other herbivorous birds, *G. newtoni* swallowed stones that would remain in its stomach (thereafter called gastroliths)—like modern herbivorous birds. With their muscular stomach walls, birds grind their gastroliths against food items until the process produces a thin, digestible paste. *G. newtoni* would often swallow avocado-sized stones to add to its gastrolith collection. Some seeds (although not many) would pass through the grinder of a thunder bird's stomach, to be deposited far from their parent tree or plant and in a nutrient-rich pile of manure that, along with the scars on the outer seed coat, would enhance seed germination. We observe the same process today among the cassowaries that eat, disperse, and deposit seeds in the tropical rain forests of North Queensland.

▼ **Figure 25.27** ***Genyornis newtoni.*** This giant bird became extinct around 45,000 years ago, after a 100,000-year uninterrupted habitation of parts of South Australia, Victoria, and New South Wales.

Like many birds, stranded in a land of comparatively low-nutritional-value food items, the ancestors of the thunder birds evolved the physiological and behavioural capacities to process higher calorie foods. Adaptations of greater size, larger gastroliths, and enhanced body mass diminished the capacity of these birds for flight. Finally, the evolutionary trade-off between diet and flight passed a boundary: *G. newtoni*'s ancestors could no longer fly. The wings of flightless birds shrank, or atrophied, into limited usefulness. While *G. newtoni* possessed scant wings, its pelvis, musculature, legs and feet attained outsized proportions. Although the animal could weigh up to 250 kg it could run fast, perhaps up to the 50 km per hour top speed of emus.

We know *G. newtoni* lived in flocks, for at least part of the time. In South Australia, vast fossilised nesting grounds exist.

Some evidence suggests *Genyornis* built mounds in which they laid and deposited eggs. Like modern mound-builders, these birds used decaying vegetation as a heat source to maintain a constant temperature that incubates eggs. Modern birds that engage this strategy leave the nest occasionally to search for food. Perhaps when *Genyornis* left the nest, egg predators made the most of those times to sneak in, dig out an egg, and remove it from the nest, or to eat it in place. We see large numbers of eggshell fragments bearing the teeth marks of Tasmanian devils (which once roamed across all of the Australian landmass) and quokkas. However, humans probably represented the most successful of egg predators. A single *Genyornis* egg could weigh up to 1.6 kg—the equivalent of 25 or more chicken eggs. The calorific attraction to hunters of such a feast is obvious.

Although emus and *G. newtoni* had lived contemporaneously in the Pleistocene, only the iconic emu survived into the present. Interestingly, two survivors of the times of the megafauna appear on Australia's coat of arms: the emu and (one of the genuine survivors of the megafauna) the red kangaroo.

The remarkable diversity among the megafauna was built on diversification of a few body types. Let's explore three variations of the kangaroo form: the giant *Procoptodon goliah*, which reached into the branches of the tallest trees to eat leaves and succulent branches; the carnivorous or omnivorous kangaroo (*Propleopus* spp.); and essentially modern kangaroos, which ate grass, herbs, and other plants that grow close to the ground. We can recognise a pattern in these three distinct animals: One body form could adapt to maximise the use of a very specific part of the environment.

The red kangaroo (*Osphranter rufus*, formerly in the genus *Macropus*) that we see today in Australia's dry interior lived contemporaneously with the megafauna. Red kangaroos remain the largest marsupials alive; the heaviest of these animals weigh between 80 and 100 kg. Perhaps these modern giants survived because they possessed adaptations that allowed them to digest grasses that became more prevalent as the landscape dried out—or else they were lucky. Red kangaroos (as you will learn in the chapter, "Basic Principles of Animal Form and Function") possess locomotion that increases in metabolic efficiency as they translate from movement with their five limbs (two legs, two forelimbs, and a tail) to their iconic hop. They use their keen sense of smell to detect distant food and water, and their efficient locomotion to travel often vast distances to access those resources. Red kangaroos represent one of four present-day kangaroo species that all exploit grass as their primary diet.

While *O. rufus* and its cousins utilised a diet of grasses and herbs, another group of contemporaneous kangaroos evolved the structures necessary to take advantage of an omnivorous diet that included meat. Three meat-eating kangaroo species, all in the genus *Propleopus* spp., possessed similar dimensions and mass to a border collie dog. To imagine these animals, think no farther than a kangaroo chasing after you at speeds of up to 40 km per hour, with a hungry look in its eyes. While the chase would have scared most of us, the method used to incapacitate and kill prey remains the stuff of nightmares. Any prey unlucky enough to come within reach would have felt the vice-like grip of strong muscular forelimbs, while the dagger-like, razor-sharp teeth that projected from the front of *Propleopus* spp.'s lower jaw would plunge repeatedly into flesh. The form and wear patterns of these animals' teeth attest to its capacity to rip chunks of meat from live prey or carrion—whatever was easiest to procure for that day's menu.

We now turn to one of the gentle, extinct giants of the megafauna. *Procoptodon goliah* **(Figure 25.28)** represented perhaps the largest of nine megafaunal kangaroos that preferred a diet of leaves. This animal weighed around 250 kg—about three times more than the red kangaroo. At nearly 3 m tall while standing on the toes of its hind legs, you would need to climb a ladder to look at the animal's flattened face. Your observation would immediately reveal the forward orientation of the eyes; all the better to see and focus on objects immediately in front of it. You might recognise similarities with a human's forward-facing eyes, as well as a close resemblance to the face of a koala.

While the red kangaroo earned the reputation as an expert hopper, *P. goliah* could make no such claim. It walked in the same way humans do, with a bipedal gait. It also used its tail as an additional point to aid balance on uneven ground, like some humans use walking sticks. *P. goliah*'s pelvis and rear legs provided the solid foundations it needed to stand on the tips of its hind legs and reach leaves high in the canopy of trees. A recent study concluded that *P. goliah* relied on trees in the Chenopod family (plants that live in arid, dry, and often salty conditions).

P. goliah proved an early exponent of a bush tucker diet, consuming plants like the old man saltbush (*Atriplex nummularia*). The protein- and mineral-rich leaves and shoots, combined with antioxidants and a pleasantly salty flavour, made this plant (and its relatives) a favourite for *P. goliah*. A diet oversaturated in salt would have required several physiological adaptations to concentrate and excrete the salt to avoid dehydration. Additional adaptations allowed *P. goliah* to reach for the highest leaves: The two middle digits of their forelimbs carried long, hook-shaped nails. Grasping objects with its forelimbs and nails allowed the animal to pull tall branches down to eye level for inspection, and perhaps consumption. Without the ability to hop, or run at any speed, combined with its specialisation as a forest dweller, Indigenous Australians would have found this animal

▼ **Figure 25.28** ***Procoptodon goliah.*** The largest of the megafauna's kangaroos preferred a diet of leaves, which its many adaptations allowed it to access and consume in large quantities.

comparatively easy to catch. That is, once the hunters had overcome the animal's strength, long reach, and ability to fight and kick hard enough to fracture bones easily with one powerful kick.

While *P. goliah* and its megafauna cousins fell to an extinction-level event, a close relative survives today: the banded hare wallaby (*Lagostrophus fasciatus*). At only 1.7 kg, this wallaby shares little in common with its giant relative. The banded hare wallaby survives in only two isolated populations in Western Australia, and resides on the list of species threatened with extinction, which overcame its larger cousin.

Given the harsh, often barren, and seasonally water-scarce environments in which *Propleopus* spp. and *P. goliah* coexisted, we should remain unsurprised if these animals used embryonic diapause like their kangaroo counterparts. In times of reduced water availability, female kangaroos excrete a cocktail of hormones that suspend the development of an embryo in the womb. If bad times persist, the embryo will be aborted, and any offspring in the kangaroo's pouch is abandoned. But in the good times, a modern female kangaroo can respond rapidly to the sudden availability of water and food by having a developing embryo, a joey in the pouch, and another joey "at foot"—recently kicked out of the pouch to make way for the pouch young, yet still insufficiently capable of making it in the mob on its own.

We know about kangaroos and other megafauna through eyewitness accounts that supplement the fossil record. For example, in the artistic equivalent of today's YouTube and 3D digital sculptors, content creators among the Indigenous peoples of Arnhem land, in Australia's Northern Territory, depicted possibly the largest monotreme in history **(Figure 25.29)**. Hackett's giant echidna (*Zaglossus hacketti*) lived during the Pleistocene epoch. It built up a body mass of nearly 30 kg and attained sizes similar to merino sheep today. We know that another echidna, *Z. bruijnii*, survived in Australia until at least the 20th century. A collector captured, killed, and preserved one of the animals in 1901 near Mount Anderson in the Kimberley region of Western Australia. The traditional owners of that land, the Mirrwong Gadjerong, give us hope the animal still survives in that remote, incompletely documented region. They recognise two species of echidnas: the short-beaked echidna (*Tachyglossus aculeatus*, which is common throughout much of Australia) and the "other one" (*Z. bruijnii*, referred to commonly as the long-beaked echidna). We know *Z. bruijnii* survives along with two other members of the same genus in the Indonesian Province of Papua, on the island of New Guinea, although three species remain critically endangered.

▼ **Figure 25.29 Hackett's giant echidna, one of the extinct megafauna, depicted from human observation.**

Echidnas, such as Hackett's giant echidna and other survivors, feature stiffened hair follicles that we call quills. These often sharp, curved spikes lie flattened on an echidna's back. When the animal feels threatened, it will erect its quills in a fair impression of a cushion of sharp, outward-facing pins. *Z. bruijnii* in Papua, like its Australian counterparts, digs into ground where it stands and uses its quills and sharp claws to wedge itself to the spot. When dug in like this, determined echidnas remain remarkably difficult to dislodge.

When not frightened, the animals emerge from their burrows in the evening, while temperatures are comparatively low, and return to the burrow before dawn. Echidnas lack sweat glands, denying them the opportunity to cool themselves evaporatively as humans do when we sweat. Instead, their nocturnal habit allows them to dump heat through radiative cooling—the transfer of heat from their bodies directly into the air.

Echidnas exhibit dietary preferences for earthworms and other soil invertebrates that they locate with their exceptionally refined sense of smell. Once in close proximity to their prey, echidnas rely on the sensitivity of electroreceptive nerve endings in the mucus glands of their beaks. The minuscule electrical currents generated as their prey contract and relax their muscles provide all the signal echidnas need to locate precisely their prey—and their meal begins.

From commonalities in the behaviour and reproduction of survivors of the echidna line, we can infer aspects of the behaviour of Hackett's giant echidna and others. These are worthy of detailing because the monotremes are the only survivors of a once more extensive group of egg-laying mammals. Perhaps 200 million years of evolutionary history warrants recounting, as many members of that monotreme group went extinct in the last 40,000 to 11,000 years. Moreover, the monotremes represent part of Australia's now largely extinct megafauna. Monotremes engage in a process known as torpor to conserve energy. The telltale indicators of torpor include inactivity for extended periods and lowered body temperature, respiration rate, heart rates, and metabolism as mechanisms to conserve energy. Interestingly, echidnas exhibit torpor throughout their range in the summer months—yes, you read that correctly. They usually build up fat reserves prior to torpor, and male echidnas may produce most, if not all, their sperm in periods when prey items remain plentiful.

Echidna courtship sometimes begins during, although usually after, a period of torpor. Males often awaken first and go in search of the burrows of females. Males may mate with the females as they remain in torpor. When echidnas live in areas of slightly greater density, the keen observer may see the formation of an "echidna train." A female often attracts a conga line of ten or more males that follow her one after another, within a quill's length of each other. The female—whether through exasperation, resignation, or desire—digs herself a rutting trench with her sharp claws, lays down her quills against her body, and relaxes. Often, the most dominant male, or the most persistent, will mate with the female.

The monotremes differ from other mammals because they possess a cloaca: a single opening through which excretion, defecation, sex, and egg-laying occur. Other mammals, which include the marsupials, feature separated reproductory and excretory apparatus. Female and male echidnas mate cloaca to cloaca. The male penis is used only to transfer sperm, and when not in use resides in an infold of the cloaca. However, in the heat of the mating moment, the male's four-headed penis extends through his cloaca and into the female's. Mating represents a drawn out affair, often lasting over an hour. Two of the male's four penis heads remain inactive, in reserve against the chance of a subsequent mating. The other two penis heads deposit sperm directly into the female's ovary ducts (see Figure 46.7). Echidna and monotreme sperm possess the unusual adaptation that they conglomerate and use their collective flagella to swim more quickly towards the ovule. With the completion of mating, the male remains in proximity to the female for an extended time. Biologists believe such possessive behaviour relates to minimisation of opportunities for another male to mate with the female and create opportunities for competition with the first male's sperm. Should fertilisation of the ovule fail, a female can ovulate up to three times in a mating season, although she produces only one egg per year.

For a time, female echidnas carry their young in pouches—although, for most of the year, they have no pouch. Soon after conception, several physiological and physical changes begin to form a temporary pouch. The outer part of the pouch forms with the swelling of the milk glands on the echidna's stomach. These swellings soon join tightly together to form the outer, protective pouch wall. As the pregnancy progresses, the female echidna begins to clench her stomach muscles to create a cavity beneath the swollen milk glands. Like the wombat, the pouch faces rearwards, all the better to prevent filling with dirt as the echidna excavates her burrow, or forages for soil invertebrates. Prior to birth, female echidnas begin to produce highly effective antimicrobial secretions within the pouch in preparation for receiving the egg.

The gestation of an echidna egg takes around 22 days from conception to birth. Echidna birth involves bodily contortions that would make any human world-famous. Prior to birth, the female rolls herself into a ball, so that her cloaca comes close to the rear entrance of her pouch. The birth involves extension of the cloaca into the pouch, where she deposits the leathery-shelled, grape-sized egg. For ten days, the echidna inside the egg continues to develop. With the depletion of the in-egg reserves, the puggle (name for a baby echidna) uses a "milk tooth" to break the egg's shell. The newly hatched puggle already has well developed olfactory nerves, and follows its sense of smell towards the mother's lactation areole (milk pore) using a kind of commando crawl. Female echidnas provide milk for their young, although they do so differently from both marsupials and eutherian mammals. While the female does not possess teats, the puggle attaches its mouth to the milk pore where, after only a few moments, it will begin to suckle. Antimicrobial secretions maintain a largely sterile environment in the pouch as the puggle develops, until it reaches about six to seven weeks after emergence from the egg. With some justification, the mother ejects the by now sharp-quilled puggle from the pouch into a 1–2 metre-long burrow. While still suckling milk from its mother, the puggle will explore the area outside its burrow to forage and supplement its diet with soil invertebrates. Mothers wean young echidnas at about seven months of age and soon after eject them into the world, where they begin to make their own way. A young echidna can breed in its second year, and may live for up to 16 years, though ten years is more common.

While the megafauna formed a continent that brimmed with curious and unusual species, often the top-order predators attract the most attention—just as other survivors of the megafauna extinction, the lions and tigers of Africa, fascinate us today. Often the marsupial lion, or pouch lion (*Thylacoleo carnifex*), vies for the title of most ferocious marsupial predator among Australia's megafauna. This lion measured 1.5 m from head to tail. It stood 75 cm tall at its shoulders. **Figure 25.30** provides a reconstruction of the animal's appearance. A mature adult would have weighed up to 160 kg, although a female that carried young in her pouch might have weighed more. Three *T. carnifex* skeletons found at Moree in New South Wales evidence the composition of a lion family. Like red kangaroos (*Osphranter rufus*), females might have carried a pouch young as well as another offspring that followed in proximity (called "at foot").

Let's examine the animal's features in Figure 25.30. First, note the elongated sharp, canine teeth—all the better to grasp and hold its prey. Behind those incisors, you can see the outline of the cheek teeth, which acted like carving knives to slice into the flesh of prey. Wear patterns on the animal's teeth demonstrate the consumption of meat and bones, much like the Tasmanian devil. Biomechanical studies of the animal's skull reveal it possessed the most powerful bite of any carnivore, living or dead, in at least the past 30 million years. The strong forearms with their sharp claws, particularly the inner claw on the front limbs, could rapidly disembowel prey. Some biologists recognise the highly dextrous inner claw as having the capacity to act in opposition to the rest of the paw (much like the human thumb), which would have provided the animal with the capacity to manipulate objects. Likewise, the inward-facing and sharp claws allowed the animal to climb into the upper branches of trees in open woodlands. On heavily wooded branches, this stealth predator would wait for prey to pass beneath it.

▼ Figure 25.30 The marsupial lion, *Thylacoleo carnifex*. This ferocious animal represents the megafauna's largest top-order marsupial predator.

The marsupial lion lived in Australia concurrently with Indigenous Australians for perhaps 20,000 years. Several cases in South America, New Zealand, and Australia illustrate the capacity of native languages to share details of historical events and animals over long periods of time as myths. Conceivably, Australia's urban myth of the "drop bear," often shared with unsuspecting international students, came from observations of marsupial lions that dropped onto unsuspecting prey beneath a tree.

When on the ground with a recent kill, the marsupial lion needed to move quickly with its meal to somewhere it could dine in peace. A small den dug into a steep embankment would protect the lion while it ate. Or perhaps it avoided other animals that might attempt to take its meal as it dragged its prey, or parts of it, into the canopies of tall trees, where it could eat and keep watchful vigilance. The reason for the hasty relocation of its meal was that, logically, *T. carnifex* would have relinquished a kill to one of the most fearsome of Australia's megafauna in the Pleistocene: *Quinkana fortirostrum*. These land-dwelling crocodiles ranged in length from between 3 and 6 metres. They possessed adaptations that allowed them to run quickly in pursuit of prey, and could change direction quickly, to counter any prey's evasive manoeuvres. Worse still, *Q. fortirostrum* possessed an endurance that whittled away at their prey's strength and reserves. Their bite quickly dispatched any animal they caught. Razor-sharp, serrated teeth that easily ripped flesh into bite-sized morsels added to the animal's charm. The size of the animals alone would have intimidated even the most determined marsupial predator into relinquishing a recent kill. Large, top-order predators like *Q. fortirostrum* would have avoided a long chase if they could. We see this pattern of behaviour today, in the North American black bear. These megafaunal survivors will similarly scare away bobcats and mountain lions from a recent kill.

Even in a tree, *T. carnifex* could take little comfort in its high position. Perhaps the largest predator of the time was *Varanus prisca*, a 2000-kg goanna that climbed trees with the same dexterity as its modern-day cousins. Although the presence of these true top-order reptilian predators would have potentially cramped the style of *T. carnifex* where they co-occurred, tooth marks on the bones of sub-adult diprotodons illustrate the pure killing power and strength of the marsupial lion.

▼ **Figure 25.31** ***Diprotodon optatum*, the largest marsupial ever.** This hippopotamus-sized wombat lookalike browsed the grasslands and into open forests and woodlands until around 25,000 years ago. According to traditional knowledge, Indigenous Australians referred to these animals as "bunyips"—a name that has long echoed around campfires, as children and adults tell tall tales in a good-natured game to scare each other before bed.

The largest marsupial that ever lived, *Diprotodon optatum*, measured nearly 3 m from head to tail and stood more than 2 m tall; adults weighed nearly six tonnes **(Figure 25.31)**. Female diprotodons carried young in their pouches. Like many marsupials, biologists suspect these animals carried a pouch young and a young "at foot." Perhaps one of the best-known members of Australia's megafauna, these massive herbivores wandered eastern Australia's semi-arid savanna, open grasslands, and some woodlands from about 1.6 million to 25,000 years ago (a detailed discussion of these "biomes" is given in the chapter, "An Introduction to Ecology and the Biosphere"). Diprotodons' inward-facing front feet produced a somewhat ungainly stride, similar to the modern wombat. Despite their size and strange gait, some biologists suggest diprotodons travelled large distances in herds—a useful trait when animals needed to migrate seasonally from colder climates to warmer parts of the country where more plentiful food sources existed. Estimates of the food intake required to sustain an adult diprotodon vary from 60 to 110 kg of plant material per day. A herd of these animals would have ranged over large areas to secure their daily dietary needs. To obtain food, diprotodons consumed grass, seeds, and fruit, while mature adults could reach 4 m into the canopies of woodland trees. To accomplish this feat, they balanced their weight on their hind legs, leaned their front legs against the tree trunk, and tore succulent leaves and branches from the tree with their mouths. A diprotodon herd would create the appearance of an almost manicured forest: bare trunks to 4 m, with luxuriant canopies above that level.

What Happened to the Megafauna?

What paleontologists know is that 85% of animals with masses greater than 44 kg became extinct in the Pleistocene epoch. The reasons for the extinction of the megafauna in Australia and throughout the world create heated debate. The explanations include disease, human hunting, aspects of climate change, or cosmic collisions. While consensus among researchers of the causes of megafaunal extinctions remains in the future, nearly all researchers contend that the extinctions share one or more causes that spanned the Earth. New Zealand represents an interesting exception.

What we don't know, yet, is the root cause of the extinctions. Explanations remain polarised: At one end of the spectrum, some scientists argue that human colonisation and over-hunting caused the extinction. Scientists at the other end of the spectrum cite climate change as the cause. Attributing the different perspectives on the extinction to particular individuals may assist the reader with further research, we have arbitrarily identified scientists associated with each contention; however, others might have easily served the same role.

Tim Flannery and other scientists have recognised a remarkable correspondence between Aboriginal colonisation of Australia and the rapid extinction of the megafauna. Scott Hucknell and others cite shifts in climate (from cool and dry to hot and arid) as the predominant cause of the megafaunal extinctions. Fossil evidence drawn from multiple sites in Australia paints a portrait of a drying continent, where eucalyptus-dominated forests gave way to semi-arid woodlands and grasslands, as forest specialists became rarer while grassland specialists came to dominate more recent fossil deposits.

As there remains no consensus about the root cause of the extinctions, a conceptual framework to contextualise the issues may prove helpful. We will consider the proximate and ultimate causes of the events. For the sake of definition, *proximate causes* occur in the lifespan of individuals, while *ultimate causes* tend to occur across multiple generations of a species that carry a cumulative effect on each successive generation.

Let's consider more recent extinctions to practise the allocation of proximate and ultimate causes. Between 1788 and 2019, 34 Australian mammals became extinct. Seven more will become extinct in the next seven years without urgent management. An additional 134 mammals remain threatened with extinction, according to guidelines by the Australian states and territories and the International Union for the Conservation of Nature (IUCN). The absence of detailed surveys and knowledge of the Australian fauna during the first century since European colonisation means these numbers probably undercount the true number of extinctions (we discuss this issue in detail later in this text). However, for our purposes, we recognise several factors that contribute to these extinctions and the percentages of extinct species to which these factors apply: invasive species (82% of species), ecosystem modification (74%), agriculture (57%), human disturbance (38%), and climate change (35%). Each of these factors represents a proximate cause of the extinctions, or threats to the extinction of species during the past 230 or more years. What do all these factors share? The ultimate cause comes directly from human colonisation and land uses before and since 1788.

Let's use another concrete hypothetical example to illustrate the point. Modern northern hairy-nosed wombats live in several isolated communities. Their numbers remain low. In a climate-change impacted landscape, where fires might sweep through the only remaining areas in which these animals live, the entire wild population may become extinct. The proximate cause of the hypothetical extinction: fire.

The ultimate cause, however, is different. Northern hairy-nosed wombats once inhabited the grasslands that sprawled down much of the east coast of Australia. The contributing factors to the extinction—the introduction of predators and competitors for the wombat's favoured food resources, habitat modification and fragmentation through land clearance, and hunting—all contributed to a decline in the numbers of this species, and its spatial retreat into several isolated populations. The ultimate cause of the extinction: Indigenous and European settlement and management of the landscape.

While we can debate the relative merits of the proximate and ultimate causes noted for this hypothetical example, the megafauna provide a grand stage from which to consider and apply conceptions of the proximate and ultimate causes of the extinction. Let's now take our learning on the road, and apply it to one Australian megafaunal extinction that occurred in Tasmania 43,000 years ago, as the world experienced a glacial maximum.

With much of the Earth's water frozen at the poles and in glaciers, sea levels were 140 m lower than today. Surf beaches and coastlines occurred far out onto the continental shelf, exposing large areas of formerly shallow seafloor. The Sahul Shelf **(Figure 25.32)** provided an ephemeral land bridge for the passage of plants and animals to land masses around Australia. The bridge between Australia and Tasmania concerns us here, particularly for allowing humans to traverse and settle onto the formerly isolated island. Some biologists note that the Tasmanian extinction of the megafauna corresponded with the arrival of humans on the island. Temperatures 10–15°C lower than today produced glaciers and permanent snow over 18% of the island that in turn would have concentrated the megafauna into smaller areas, with commensurate increases in competition for habitat, food, and other resources. When humans arrived, perhaps the megafauna had already experienced challenges that lowered their collective resilience. Could well-organised groups of recent Aboriginal settlers to Tasmania have hunted the megafauna to extinction? The Māori in New Zealand had achieved that result with members of the megafauna in less than 150 years. Could the few newly established, Indigenous human Tasmanians have hunted megafauna to extinction, in a land area only 25% that of New Zealand? Certainly, but did they? In New Zealand, the climate had remained stable for nearly 4,000 years before the Māori became established,

▼ **Figure 25.32 The Sunda and Sahul landmasses.** At the height of a glacial maximum (43,000 years ago), sea levels declined to reveal the Sahul Shelf, which provided both habitat and natural land-based corridors for plants and animals between formerly isolated landmasses.

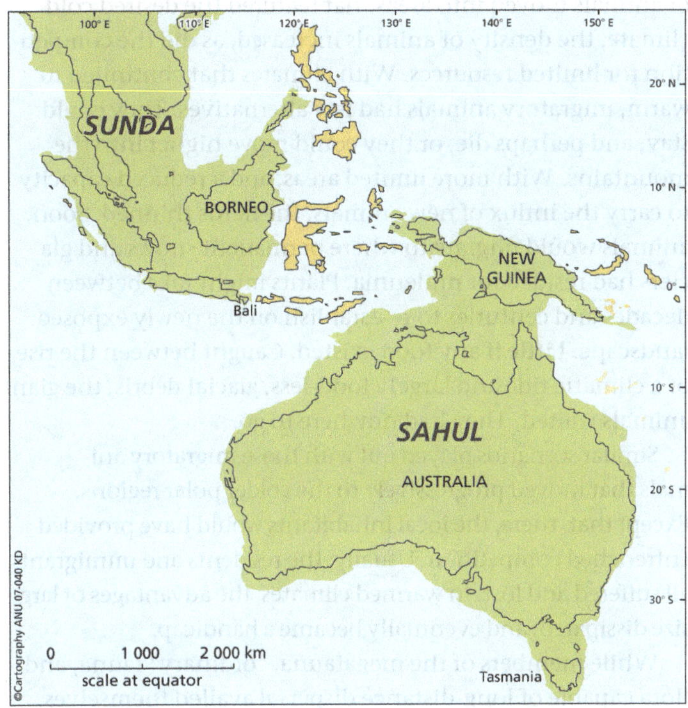

while Aboriginal establishment in Tasmania occurred concurrently with a rapid climatic shift. The other difference is that the Māori had accumulated at least 420 centuries more technology, knowhow, and capability than the first Tasmanian Aboriginal hunter-gathers. Disentanglement of the proximate and ultimate causes of the Tasmanian megafaunal extinction requires further research.

Hypotheses designed to synthesise the evidence of megafaunal extinctions remain rare, although some show promise. Daniel Mann and his colleagues at the University of Alaska seek to synthesise the various lines of evidence into a holistic hypothesis to explain megafaunal extinctions, in terms of the interactions between adaptations, climatic variation, and stability. To begin, we need to recognise several broad generalisations about the megafauna. They exhibited large body size, with high volume to surface area; metabolic efficiency; extensive reserves of body fat that allowed them to go for periods with little food; long lifespans; the capacity for migration in herds or individually; and, sometimes, an enhanced sensory acuity to detect and catch prey. Each trait suited these animals to cold, glacial climates.

What we know is that short interglacial periods punctuated the glacial climates, during which these animals exhibited remarkable adaptations. The interglacials would feature temperatures that rose comparatively quickly. The areas that retained suitable cold climates for the majority of the plants and megafauna occurred in two areas: at higher altitudes, and towards the poles. Those animals that could, did migrate. Several challenges faced them. Higher altitude sites generally comprised smaller areas than the lowlands. As large numbers of animals moved into areas that featured the desired cold climate, the density of animals increased, as did the competition for limited resources. With climates that continued to warm, migratory animals had few alternatives: They could stay, and perhaps die, or they could move higher into the mountains. With more limited areas, and a reduced capacity to carry the influx of new animals, the herds thinned. Soon, animals would migrate to where permanent snows and glaciers had resided for millennia. Plants might take between decades and centuries to re-establish on the newly exposed landscape. Little if any food existed. Caught between the rise of a climatic tide and largely food-less, glacial debris, the giant animals waited. They had nowhere to go.

Similar scenarios played out with those migratory animals that moved progressively to the colder polar regions. Except that, there, the local inhabitants would have provided entrenched competition. Usually, the residents and immigrants all suffered and lost. In warmed climates, the advantages of large size dissipated, and eventually became a handicap.

While members of the megafauna, "ordinary" fauna, and flora capable of long-distance dispersal availed themselves of travel, many animals and plants lacked that capacity. For those animals unable to move, and those plants rooted to the ground, no choice existed: They tolerated and survived the new conditions, or they didn't. Plants, in particular, experienced great challenges, because the climatic conditions where they could flourish changed rapidly, and the capacity for seed dispersal and establishment into suitable and often distant habitats never quite caught up with the contemporary climates. Mann and colleagues recognise that transient ecosystems formed. The inhabitants of these transient systems never quite adapted to the changed conditions.

As an interglacial period wound down, and a glacial approached, those few survivors among the migratory megafauna may have travelled away from the increasingly cold, ice-choked polar and high-altitude sites. Those species that could moved towards more suitable locales in the lowlands, where they would have encountered a different landscape than the one their ancestors left.

Throughout the later stages of the Pleistocene, in particular, multiple and often unpredictable switches occurred between glacial and interglacial climates. Animals and plants that had tolerated one extreme, or had become adapted to it, survived into the switch to the next extreme. Sometimes, adaptations remained unsuited to the new conditions. Ecosystems remained in disequilibrium; some members adapted while others tolerated the new conditions, though few were well-adapted to the current circumstances. Transient ecosystems formed throughout the world from an amalgam of glacially and interglacially adapted species.

Repetitions of these rapid switches between glacial and interglacial climates lowered the resilience threshold and tolerance of new conditions among members of the megafauna. When the ice age ended, the advantage of being large eroded, and thresholds to extinctions lowered too; even small perturbations might have pushed the systems into a spiral of decay.

So, the susceptibility of species to proximal threats increased. For example, human over-exploitation of plants and animals may have provided all the disturbance required for a cascade of extinctions to sweep through the megafauna. However, the ultimate cause for the extinction, according to Mann and colleagues, relates to the rapid fluctuations between climatic extremes that over time cumulatively lowered the megafauna's threshold for extinction until, when on that precipice, a final proximal nudge sent them to oblivion.

New Zealand's Fauna–Human Impacts and Evolution in Isolation?

To begin, we look at New Zealand's iconic tuatara (*Sphenodon punctatus*, **Figure 25.33**). Tuataras survived the disintegration of Pangaea, then Laurasia's breakup, and subsequently Gondwana's disassembly. From the middle of the Jurassic period (199–145 million years ago), tuataras have changed remarkably little: Fossils appear almost indistinguishable from living animals. Although the appearance of these animals gives the impression of close affinities with lizards, various skeletal features ally them with the sphenodontids, an ancient sister group to the squamates (lizards and snakes) that disappeared when Gondwana was still intact. They survived the separation from Gondwana of the island arcs that became New Zealand, the periodic reduction in land areas to chains of remarkably small islands, and changes in Earth's climate associated with continental drift, mountain building, and factors external to the planet.

▼ **Figure 25.33 New Zealand's tuatara.** Tuataras have changed little over the last 200 million years. Today, these endemic New Zealanders survive on 32 offshore islands, and have been recently released into the Zealandia Wildlife Sanctuary on New Zealand's North Island.

Recognisable members of *S. punctatus* have lived on Earth for 1,000 times longer than humans (*Homo sapiens*). Yet despite our youth, humans set in motion the events that would cause the extinction of tuataras on New Zealand's North and South Islands. Around 700 years ago, Polynesian mariners colonised New Zealand, and brought with them stowaways like one of the most widespread rat species, the Pacific rat, which the Māori call the kiore (*Rattus exulans*). When Europeans settled in the 1840s, kiore predation brought tuataras to the brink of extinction on mainland New Zealand. Some islands close to shore provided temporary refuges until rats and other introduced species gained a foothold, with predictable outcomes for tuataras and other species that were extinct on the mainland. Some 30 offshore islands and the oceans that acted as moats around them were all that stood between this species and extinction. Human interventions turned around the decline. New Zealand's Department of Conservation undertook active management that produced a resurgence of the species—so much so that tuataras have been reintroduced to predator-free areas of the North and South Islands. In what has become a theme throughout the world, the loss of species that evolved over millions of years has erased nature's handiwork in a short span of time. Active management, which you will read about later in this text, seeks to reverse this trend.

New Zealand's iconic moas and kiwis (ratites) **(Figure 25.34)** share a Gondwanan origin. Taxonomists recognise nine moa species from their plentiful remains. The largest birds weighed around 250 kg and stood 3.6 m tall. Along with two sister species, collectively termed the dinornithids, these enormous birds browsed small trees and large shrubs. They had long and stout necks. Like Australia's cassowaries, moas occupied the ecological niche that goats and deer occupy in other regions of the world. The other moa group, the emeids, weighed between 20 and 200 kg. They lived by browsing, much like various kangaroos and wallabies of Australia.

Moas are survived by five species of small ratites known as kiwis, which weigh between 1 and 3 kg. It is likely that kiwis evolved from a much larger ancestor, because they produce the largest eggs in proportion to body size of any living bird. The egg is five to ten times larger than that of any similar-sized bird. Not surprisingly, the females are about 20% larger than the males. Kiwis have long, rather cylindrical bills, not unlike the snout of the long-beaked echidna (*Zaglossus* spp.). Like the latter, kiwis prod through the litter and soil to find earthworms, although they also eat small invertebrates, as well as seeds and fallen fruit.

During the time that the moas were sliding towards extinction (about 700–650 years ago), their largest likely

▼ Figure 25.34 Size relationships among some extant and extinct New Zealand ratites and their predator, the Haast eagle (*Harpagornis moorei*). From the left: The little spotted kiwi (*Apteryx owenii*), brown kiwi (*Apteryx mantelli*), kakapo (*Strigops habroptilus*), little bush moa (*Anomalopteryx didiformis*), Eastern moa (*Emeus crassus*), and New Zealand's largest moa (*Dinornis robustus*), although North Islanders may argue their own species to be the largest (*D. novaezealandiae*).

predator, the enormous Haast's eagle (*Harpagornis moorei*) also became extinct. Weighing nearly 17 kg and with a wingspan of up to 2.6 m, this raptor outsized any living eagle. No doubt the eagle's massive talons, incredible grip strength, muscular legs, and enormous lift and manoeuvrability conveyed from its wings helped it to successfully prey on the much larger moas.

Another large bird, the adzebill (*Aptornis otidiformis*)—a flightless rail—might also have been a carnivore. Given that the adzebill was about the size of the smallest moas, it could possibly have preyed on moas, although it seems more likely it targeted smaller prey such as tuataras, lizards, and seabirds. It possessed an adze-like beak similar to that of other predators, although its talons bear no resemblance to those of comparable predators. The feeding mode and ecological niche of this animal remain uncertain.

Many birds that developed unusual specialisations (in comparison with their relatives elsewhere) probably evolved from early immigrants to New Zealand. Native wrens, thrushes, and wattlebirds must have been among the earliest colonisers. Immigration has no doubt been continuous over time, with some visitors failing to become established. The most recent known immigrants from Australia are silvereyes (1856), spur-winged plovers (1932), white-faced herons (1941), and welcome swallows (1958). Freshwater fish, frogs, and lizards are additional remnants of Gondwana that inhabit today's New Zealand. There are 29 species of native freshwater fish in New Zealand, 90% of which are endemic. All New Zealand frogs belong to the endemic genus *Leiopelma*. This genus is unrelated to surviving Australian frogs and is a Gondwanan remnant. Native frogs are extinct on the South Island and have disappeared from the southern half of the North Island.

Two factors laid the foundation for these and other unique New Zealand animals after separation from the bulk of Gondwana around 86–82 million years ago (during the Cretaceous period). First, New Zealand separated on a plate that included New Caledonia and Lord Howe, Norfolk, Kermadec and Chatham Islands. Collectively, the components of this plate are known as Tasmantis. At the time of separation of Tasmantis, mammals appeared rarely, if at all, in that part of Gondwana. If mammals were taken into isolation aboard the New Zealand landmass as it drifted northwards, we find limited evidence of their influence on the evolution of the flora and fauna. New Zealand's current tally of native mammals comprises two survivors: the widely distributed long-tailed bat (*Chalinolobus tuberculatus*) and the ancient lineage of the now endangered lesser short-tailed bat (*Mystacina tuberculata*). Arguably, in addition to hunting the tuataras to near extinction, the kiore caused the extinction of the greater long-tailed bat (*M. robusta*) **(Figure 25.35)**.

Other Gondwanan elements, such as Pleurodira turtles, are also lacking in New Zealand. However, owing to the lack of a fossil record, it is uncertain what Gondwanan species occurred on the landmasses that became New Zealand. We may struggle to find evidence of those early colonists because they left few traces. Scientists continue to debate whether some living groups, especially lizards, have Gondwanan origins. In its slow journey northwards, New Zealand's climate remained comparatively cool. In the Quaternary (2.5 million years ago to the present), New Zealand was in the glaciation zone, while Australia resided north of that line. During the last glacial period, much of New Zealand was covered with ice sheets, and glaciers remain in the alpine areas of the South Island. New Zealand's Southern Alps, presided over by Aorangi/Mount Cook (3,760 m), formed only 5 million years ago.

New Zealand's landmass is also comparatively small—only 265,000 km^2, which is about 3% of Australia's land area, and less than 0.2% of the total world land area. At times, the area had higher sea levels. Extensive marine transgressions during the Miocene (about 23 million years ago) reduced New Zealand to a few relatively small islands, or the landmass may have drowned completely. Today, New Zealand consists of the North Island and the South Island along with about 700

▼ Figure 25.35 Two surviving New Zealand bats.

(a) The long-tailed bat (*Chalinolobus tuberculatus*).

(b) The endangered lesser short-tailed bat (*Mystacina tuberculata*).

much smaller islands. Only 228 islands are larger than 50,000 m^2. New Zealand's current landscapes developed thanks to the Kaikoura orogeny (an area where tectonic plates interact, deforming Earth's crust and mantle), which occurred in response to uplift along the Alpine Fault and lasted from about 12 to 5 million years ago.

Isolation also shaped the New Zealand fauna. Terrestrial snakes, for example, never made it to New Zealand, although sea snakes (hydropids) are occasionally found. Birds provide one example of mobile species able to establish themselves in New Zealand with apparent regularity: In New Zealand, many niches that elsewhere in the world are filled by mammals are filled by birds, especially flightless forms that may have evolved from formerly flighted ancestors.

Mass Extinctions

The fossil record shows that the overwhelming majority of species that ever lived are now extinct. A species may become extinct for many reasons. Its habitat may have been destroyed, or its environment may have changed in a manner unfavourable to the species. For example, if ocean temperatures fall by even a few degrees, species that are otherwise well adapted may perish. Even if physical factors in the environment remain stable, biological factors may change—the origin of one species can spell doom for another.

Although extinction occurs regularly, at certain times disruptive changes to the global environment have caused the rate of extinction to increase dramatically. The result is a **mass extinction**, in which large numbers of species become extinct worldwide.

The "Big Five" Mass Extinction Events

Five mass extinctions are documented in the fossil record over the past 500 million years **(Figure 25.36)**. These events are particularly well documented for the decimation of hard-bodied animals that lived in shallow seas, the organisms for which the fossil record is most complete. In each mass extinction, 50% or more of marine species became extinct.

Two mass extinctions—the Permian and the Cretaceous—have received the most attention. The Permian mass extinction, which defines the boundary between the Paleozoic and Mesozoic eras (252 million years ago), claimed about 96% of marine animal species and drastically altered life in the ocean. Terrestrial life was also affected. For example, eight out of 27 known orders of insects were wiped out. This mass extinction occurred in less than 500,000 years, possibly in just a few thousand years—an instant in the context of geological time.

The Permian mass extinction occurred during the most extreme episode of volcanism in the past 500 million years. Geological data indicate that 1.6 million km^2 (roughly half the size of Western Europe) in Siberia was covered with lava hundreds of metres thick. The eruptions are thought to have produced enough carbon dioxide to warm the global climate by an estimated 6°C, harming many temperature-sensitive species. The rise in atmospheric CO_2 levels would also have led to ocean acidification, thereby reducing the availability of calcium carbonate, which is required by reef-building corals and many shell-building species (see Figure 3.13). The eruptions would also have added nutrients such as phosphorus to marine ecosystems, stimulating the growth of microorganisms. Upon their deaths, these microorganisms would have provided food for bacterial decomposers. Bacteria use oxygen as they decompose the bodies of dead organisms, thus causing oxygen concentrations to drop. This would have harmed oxygen-breathers and promoted the growth of anaerobic bacteria that emit a poisonous metabolic by-product, hydrogen sulfide (H_2S) gas. Overall, the volcanic eruptions may have triggered a series of catastrophic events that together resulted in the Permian mass extinction.

▼ Figure 25.36 Mass extinction and the diversity of life. The five generally recognised mass extinction events, indicated by red arrows, represent peaks in the extinction rate of marine animal families (red line and left vertical axis). These mass extinctions interrupted the overall increase, over time, in the number of extant families of marine animals (blue line and right vertical axis).

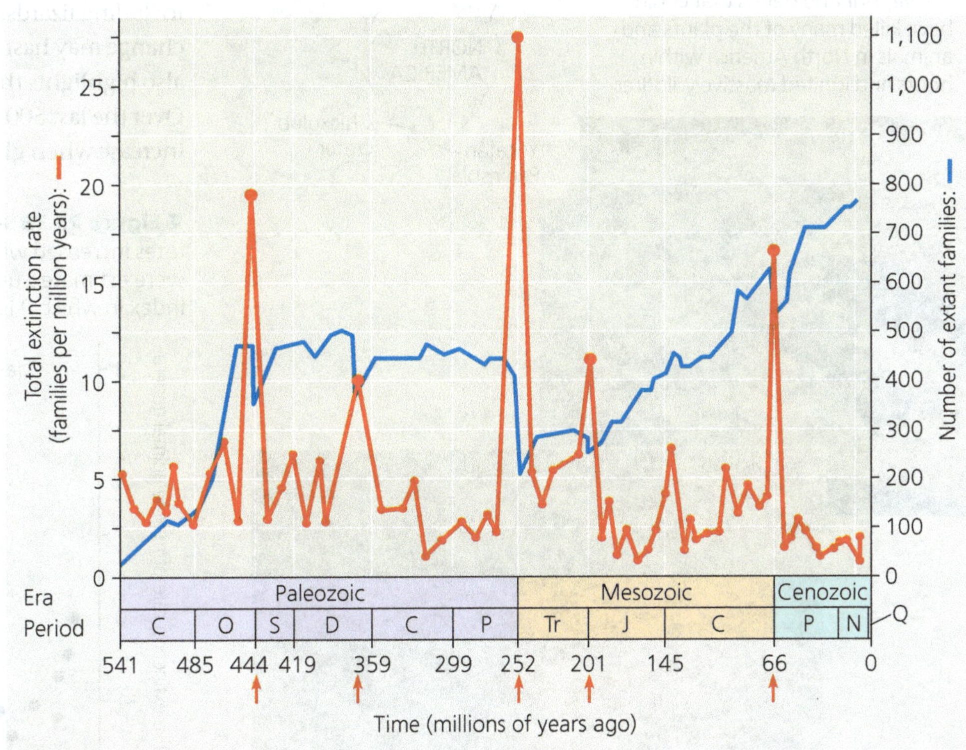

INTERPRET THE DATA *As mentioned in the text, 96% of marine animal species became extinct in the Permian mass extinction (252 million years ago). Explain why the blue curve shows only a 50% drop at that time.*

The Cretaceous mass extinction occurred 66 million years ago. This event extinguished more than half of all marine species and eliminated many families of terrestrial plants and animals, including all dinosaurs (except birds, which are members of the same group; see Figure 34.24). One clue to a possible cause of the Cretaceous mass extinction is a thin layer of clay enriched in iridium that dates to the time of the mass extinction. Iridium is an element that is very rare on Earth but common in many of the meteorites and other extraterrestrial objects that occasionally fall to Earth. As a result, researchers proposed that this clay is fallout from a huge cloud of debris that billowed into the atmosphere when an asteroid collided with Earth. This cloud would have blocked sunlight and caused a sudden drop in global temperatures lasting for several months to years.

Is there evidence of such an asteroid? Research has focused on the Chicxulub crater, a 66-million-year-old scar beneath sediments off the coast of Mexico **(Figure 25.37)**. The crater is the right size to have been caused by an object with a diameter of 10 km. When the impact occurred, many species may have been particularly vulnerable to extinction because recent large-scale volcanic eruptions had already caused their populations to decline. Moreover, a 2018 study estimated that bushfires resulting from the impact contributed to a sharp rise in atmospheric CO_2 levels, causing a period of global warming that lasted 100,000 years—yet another stress for populations already in decline. Critical evaluation of these and other hypotheses for mass extinctions continues.

▼ **Figure 25.37 A trauma for Cretaceous life.** Beneath the Caribbean Sea, the 66-million-year-old Chicxulub crater measures 180 km across. The horseshoe shape of the crater and the pattern of debris in sedimentary rocks indicate that an asteroid struck at a low angle. This drawing represents the impact and its immediate effect: a cloud of hot vapour and debris that could have killed many of the plants and animals in North America within hours and ignited massive wildfires.

Is a Sixth Mass Extinction Under Way?

As you will read further in Concept 56.1, human actions, such as habitat destruction, are modifying the global environment to such an extent that many species are threatened with extinction. More than 1,000 species have become extinct in the last 400 years. Scientists estimate that this rate is 100 to 1,000 times the typical background rate seen in the fossil record. Is a sixth mass extinction now in progress?

This question is difficult to answer, in part because it is hard to document the total number of extinctions occurring today. Tropical rain forests, for example, harbour many undiscovered species. As a result, destroying tropical forest may drive species to extinction before we even learn of them.

Such uncertainties make it hard to assess the full extent of the current extinction crisis. Even so, it is clear that losses to date have not reached those of the "big five" mass extinctions, in which large percentages of Earth's species became extinct. This does not in any way discount the seriousness of today's situation. Monitoring programs show that many species are declining at an alarming rate due to habitat loss, introduced species, overharvesting, and other factors. Recent studies on a variety of organisms, including lizards, pine trees, and polar bears, suggest that climate change may hasten some of these declines. The fossil record also highlights the potential importance of climate change: Over the last 500 million years, extinction rates have tended to increase when global temperatures were high **(Figure 25.38)**.

▼ **Figure 25.38 Fossil extinctions and temperature.** Extinction rates increased when global temperatures were high. Temperatures were estimated using ratios of oxygen isotopes and converted to an index in which 0 is the overall average temperature.

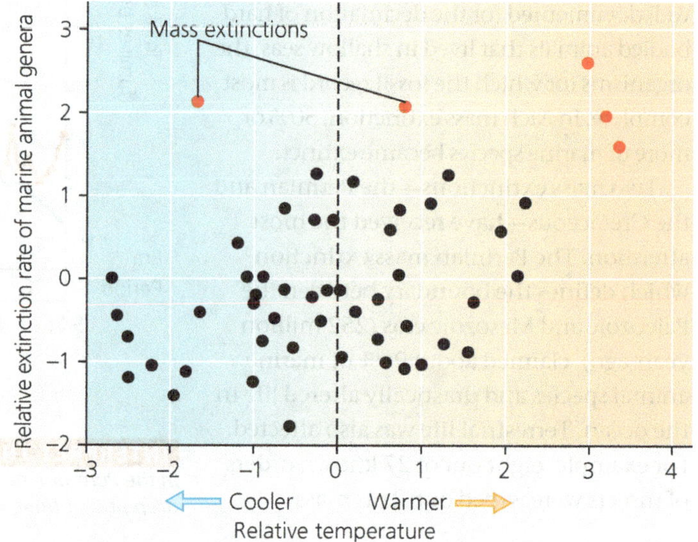

Overall, the evidence suggests that unless dramatic actions are taken, a sixth, human-caused mass extinction is likely to occur within the next few centuries.

Consequences of Mass Extinctions

Mass extinctions have significant and long-term effects. By eliminating large numbers of species, a mass extinction can reduce a thriving and complex ecological community to a pale shadow of its former self. And once an evolutionary lineage disappears, it cannot reappear. The course of evolution is changed forever. Consider what would have happened if the early primates living 66 million years ago had died out in the Cretaceous mass extinction. Humans would not exist, and life on Earth would differ greatly from what it is today.

The fossil record shows that it typically takes 5–10 million years for the diversity of life to recover to previous levels after a mass extinction. In some cases, it has taken much longer than that: It took about 100 million years for the number of marine families to recover after the Permian mass extinction (see Figure 25.36). These data have sobering implications. If current trends continue and a sixth mass extinction occurs, it will take millions of years for life on Earth to recover.

Mass extinctions can also alter ecological communities by changing the types of organisms residing there. For example, after the Permian and Cretaceous mass extinctions, the percentage of marine organisms that were predators increased substantially **(Figure 25.39)**. A rise in the number of predators can increase both the risks faced by prey and the competition among predators for food. In addition, mass extinctions can curtail lineages with novel and advantageous features. For example, in the late Triassic period, a group of gastropods (snails and their relatives) arose that could drill through the shells of bivalves (such as clams) and feed on the animals inside. Although shell drilling provided access to a new and abundant source of food, this newly formed group was wiped out during the mass extinction at the end of the Triassic (about 200 million years ago). Another 120 million years passed before another group of gastropods (the oyster drills) exhibited the ability to drill through shells. As their predecessors might have done if they had not originated at an unfortunate time, oyster drills have since diversified into many new species. Finally, by eliminating so many species, mass extinctions can pave the way for adaptive radiations, in which new groups of organisms proliferate.

▼ Figure 25.39 Mass extinctions and ecology. The Permian and Cretaceous mass extinctions (indicated by red arrows) altered the ecology of the oceans by increasing the percentage of marine genera that were predators.

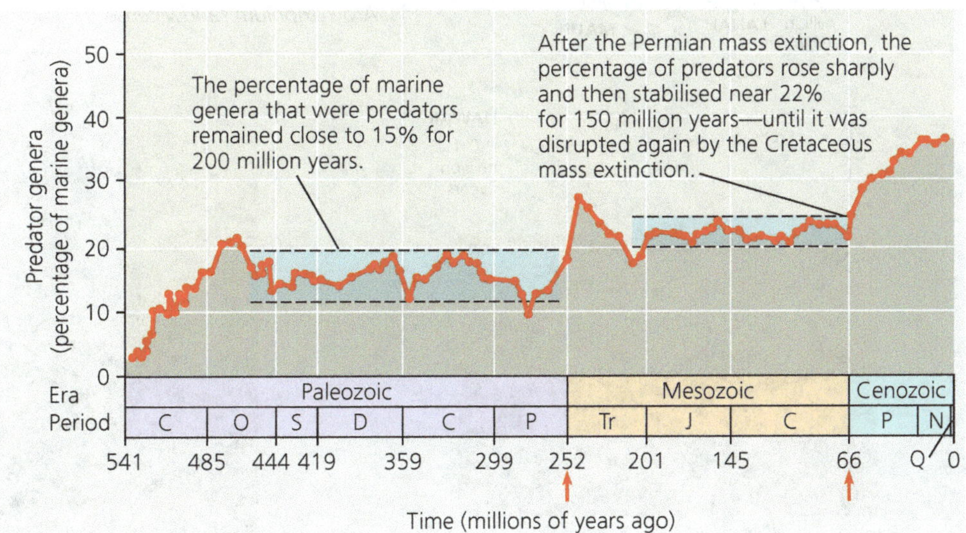

Adaptive Radiations

The fossil record shows that the diversity of life has increased over the past 250 million years (see blue line in Figure 25.36). This increase has been fueled by **adaptive radiations**, periods of evolutionary change in which groups of organisms form many new species whose adaptations allow them to fill different ecological roles, or niches, in their communities. Large-scale adaptive radiations occurred after each of the big five mass extinctions, when survivors became adapted to the many vacant ecological niches. Adaptive radiations have also occurred in groups of organisms that possessed major evolutionary innovations, such as seeds or armoured body coverings, or that colonised regions in which they faced little competition from other species.

Worldwide Adaptive Radiations

Fossil evidence indicates that mammals underwent a dramatic adaptive radiation after the extinction of terrestrial dinosaurs 66 million years ago **(Figure 25.40)**. Although mammals originated about 180 million years ago, the mammal fossils older than 66 million years are mostly small and show less morphological diversity than found today. Many species appear to have been nocturnal based on their large eye sockets, similar to those in living nocturnal mammals. A few early mammals were intermediate in size, such as *Repenomamus giganticus*, a 1-m-long predator that lived 130 million years ago—but none approached the size of many dinosaurs. Early mammals may have been restricted in size and diversity because they were eaten or outcompeted by the larger and more diverse dinosaurs. With the disappearance of the dinosaurs (except for birds), mammals expanded greatly in both diversity and size, filling the ecological roles once occupied by terrestrial dinosaurs.

The history of life has also been greatly altered by radiations in which groups of organisms increased in diversity as they came to play entirely new ecological roles in their communities. As we'll explore in later chapters, examples include the rise of photosynthetic prokaryotes, the evolution

▼ Figure 25.40 Adaptive radiation of mammals.

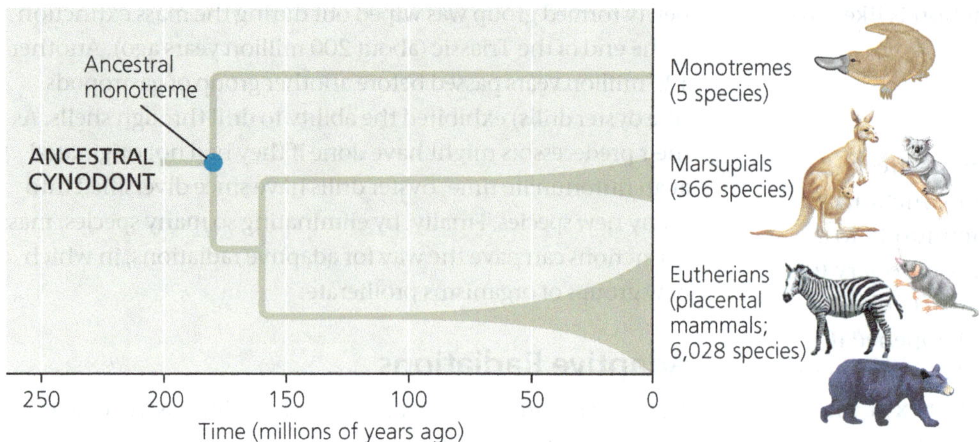

of large predators in the Cambrian explosion, and the radiations following the colonisation of land by plants, insects, and tetrapods. Each of these last three radiations was associated with major evolutionary innovations that facilitated life on land. The radiation of plants, for example, was associated with key adaptations, such as stems that support plants against gravity and a waxy coat that protects leaves from water loss. Finally, organisms that arise in an adaptive radiation can serve as a new source of food for still other organisms. In fact, the diversification of plants stimulated a series of adaptive radiations in insects that ate or pollinated plants, one reason that insects are the most diverse group of animals on Earth today.

Regional Adaptive Radiations

Striking adaptive radiations have also occurred over more limited geographic areas. Such radiations can be initiated when a few organisms make their way to a new, often distant location in which they face relatively little competition from other organisms. The Hawaiian archipelago is one of the world's great showcases of this type of adaptive radiation **(Figure 25.41)**.

▼ Figure 25.41 Adaptive radiation on the Hawaiian Islands. Molecular analysis indicates that these remarkably varied Hawaiian plants, known collectively as the "silversword alliance," are all descended from an ancestral tarweed that arrived on the islands from North America about 5 million years ago. Silverswords have since spread into different habitats and formed new species with strikingly different adaptations.

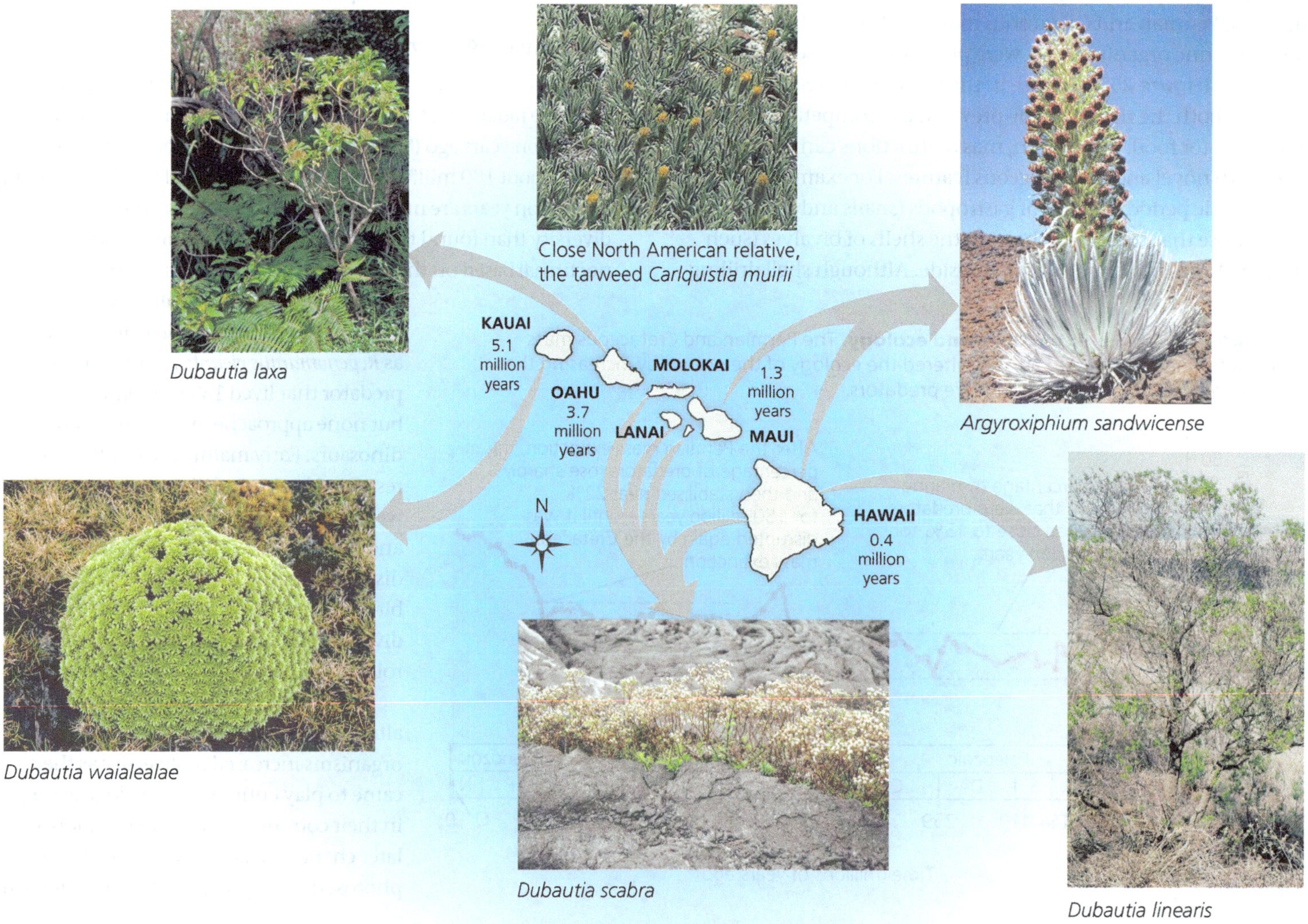

Close North American relative, the tarweed *Carlquistia muirii*

KAUAI 5.1 million years

OAHU 3.7 million years

MOLOKAI 1.3 million years

LANAI

MAUI

HAWAII 0.4 million years

N

Dubautia laxa

Argyroxiphium sandwicense

Dubautia waialealae

Dubautia scabra

Dubautia linearis

Located about 3,500 km from the nearest continent, the volcanic islands are progressively older as one follows the chain towards the northwest; the youngest island, Hawaii, is less than a million years old and still has active volcanoes. Each island was born "naked" and was gradually populated by stray organisms that rode the ocean currents and winds either from far-distant land areas or from older islands of the archipelago itself. The physical diversity of each island, including immense variation in soil conditions, elevation, and rainfall, provides many opportunities for evolutionary divergence by natural selection. Multiple invasions followed by speciation events have ignited an explosion of adaptive radiation in Hawaii. As a result, thousands of species that inhabit the islands are found nowhere else on Earth. Among plants, for example, about 1,100 species are unique to the Hawaiian Islands. Unfortunately, many of these species are now facing an elevated risk of extinction due to human actions such as habitat destruction and the introduction of non-native plant species.

CONCEPT CHECK 25.4

1. Explain the consequences of plate tectonics for life on Earth.
2. What factors promote adaptive radiations?
3. **WHAT IF?** Based on evidence from previous mass extinctions, describe ecological and evolutionary consequences that could result if a sixth human-caused mass extinction occurs.

For suggested answers, see Appendix A.

CONCEPT 25.5

Major changes in body form can result from changes in the sequences and regulation of developmental genes

The fossil record tells us what the great changes in the history of life have been and when they occurred. Moreover, an understanding of plate tectonics, mass extinction, and adaptive radiation provides a picture of how those changes came about. But we can also seek to understand the intrinsic biological mechanisms that underlie changes seen in the fossil record. For this, we turn to genetic mechanisms of change, paying particular attention to genes that influence development.

Effects of Developmental Genes

As you read in Concept 21.6, "evo-devo"—research at the interface between evolutionary biology and developmental biology—is illuminating how slight genetic differences can produce major morphological differences between species. In particular, large morphological differences can result from genes that alter the rate, timing, and spatial pattern of change in an organism's form as it develops from a zygote into an adult.

Changes in Rate and Timing

Many striking evolutionary transformations are the result of **heterochrony** (from the Greek *hetero*, different, and *chronos*, time), an evolutionary change in the rate or timing of developmental events. For example, an organism's shape depends in part on the relative growth rates of different body parts during development. Changes to these rates can alter the adult form substantially, as seen in the contrasting shapes of human and chimpanzee skulls **(Figure 25.42)**. Other examples of the dramatic evolutionary effects of heterochrony include how increased growth rates of finger bones yielded the skeletal structure of wings in bats (see Figure 22.16) and how slowed growth of leg and pelvic bones led to the reduction and eventual loss of hind limbs in whales (see Figure 22.21).

Heterochrony can also alter the timing of reproductive development relative to the development of nonreproductive organs. If the development of reproductive organs accelerates compared to that of other organs, the sexually mature stage of a species may retain body features that

▼ Figure 25.42 Relative skull growth rates. In the human evolutionary lineage, mutations slowed the growth of the jaw relative to other parts of the skull. As a result, in humans the skull of an adult is more similar to the skull of an infant than is the case for chimpanzees.

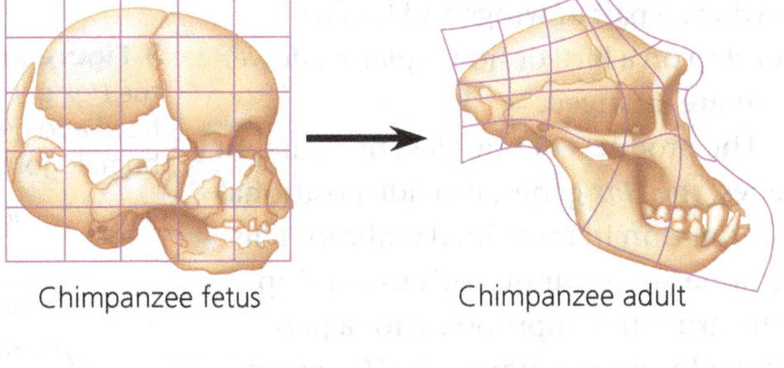

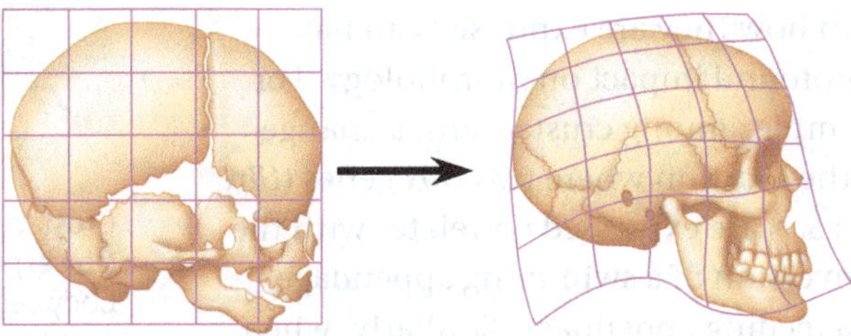

▲ **Figure 25.43 Paedomorphosis.** The adults of some species retain features that were juvenile in ancestors. This salamander is an axolotl, an aquatic species that becomes a sexually mature adult while retaining certain larval (tadpole) characteristics, including gills.

were juvenile structures in an ancestral species, a condition called **paedomorphosis** (from the Greek *paedos*, of a child, and *morphosis*, formation). For example, most salamander species have aquatic larvae that undergo metamorphosis in becoming adults. But some species grow to adult size and become sexually mature while retaining gills and other larval features **(Figure 25.43)**. Such an evolutionary alteration of developmental timing can produce animals that appear very different from their ancestors, even though the overall genetic change may be small. Indeed, recent evidence indicates that a change at a single locus was probably sufficient to bring about paedomorphosis in the axolotl salamander, although other genes may have contributed as well.

Changes in Spatial Pattern

Substantial evolutionary changes can also result from alterations in genes that control the spatial organisation of body parts. For example, master regulatory genes called **homeotic genes** (see Concept 18.4) determine such basic features as where a pair of wings and legs will develop on a bird or how a plant's flower parts are arranged.

The products of one class of homeotic genes, the *Hox* genes, provide positional information in an animal embryo. This information prompts cells to develop into structures appropriate for a particular location. Changes in *Hox* genes or in how they are expressed can have a profound impact on morphology. For example, among crustaceans, a change in the location where two *Hox* genes (*Ubx* and *Scr*) are expressed correlates with the conversion of a swimming appendage to a feeding appendage. Similarly, when comparing plant species, changes to the expression of homeotic genes known as *MADS-box* genes can produce flowers that differ dramatically in form (see Concept 35.5).

The Evolution of Development

The 560-million-year-old fossils of Ediacaran animals (see Figure 25.11) suggest that a set of genes sufficient to produce complex animals existed at least 25 million years *before* the Cambrian explosion. If such genes have existed for so long, how can we explain the astonishing increases in diversity seen during and since the Cambrian explosion?

Adaptive evolution by natural selection provides one answer to this question. As we've seen throughout this unit, by sorting among differences in the sequences of protein-encoding genes, selection can improve adaptations rapidly. In addition, new genes (created by gene duplication events) can take on new metabolic and structural functions, as can existing genes that are regulated in new ways.

Examples in the previous section suggest that developmental genes may have been particularly important. Thus, we'll turn next to how new morphological forms can arise from changes in the nucleotide sequences or regulation of developmental genes.

Changes in Gene Sequence

New developmental genes arising after gene duplication events probably facilitated the origin of novel morphological forms. But since other genetic changes also may have occurred at such times, it can be difficult to establish causal links between genetic and morphological changes that occurred in the past.

This difficulty was sidestepped in a study of developmental changes associated with the divergence of six-legged insects from crustacean ancestors that had more than six legs. (As discussed in Concept 33.4, insects arose from within a subgroup of the crustaceans, the traditional name for organisms such as shrimp, crabs, and lobsters.) Crustaceans

▼ **Figure 25.44 Effects of the *Hox* gene *Ubx* on the insect body plan.** In crustaceans, the *Hox* gene *Ubx* is expressed in the region shaded green, the body segments between the head and genital segments. In insects, *Ubx* is expressed in only a subset (shaded pink) of the homologous body segments, where it suppresses leg formation.

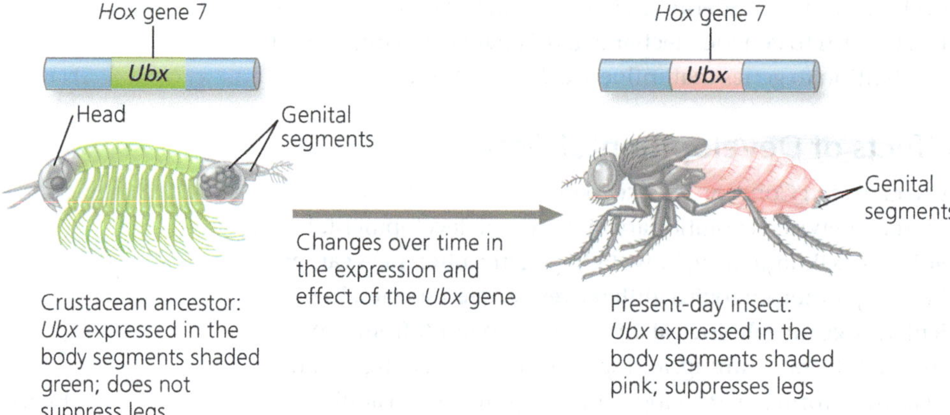

and insects differ in the pattern of expression and the effects of the *Hox* gene *Ubx*: In particular, in insects, *Ubx* suppresses leg formation where it is expressed **(Figure 25.44)**.

To examine the workings of this gene, researchers cloned the *Ubx* gene from an insect, the fruit fly *Drosophila*, and from a crustacean, the brine shrimp *Artemia*. Next, they genetically engineered fruit fly embryos to express either the *Drosophila Ubx* gene or the *Artemia Ubx* gene throughout their bodies. The *Drosophila* gene suppressed 100% of the limbs in the embryos, as expected, whereas the *Artemia* gene suppressed only 15%.

The researchers then sought to uncover key steps involved in the evolutionary transition from an ancestral *Ubx* gene to an insect *Ubx* gene. Their approach was to identify mutations that would cause the *Artemia Ubx* gene to suppress leg formation, thus making its gene act more like an insect *Ubx* gene. To do this, they constructed a series of "hybrid" *Ubx* genes, each of which contained known segments of the *Drosophila Ubx* gene and known segments of the *Artemia Ubx* gene. By inserting these hybrid genes into fruit fly embryos (one hybrid gene per embryo) and observing their effects on leg development, the researchers were able to pinpoint the exact amino acid changes responsible for the suppression of additional limbs in insects. In so doing, this study provided evidence that particular changes in the nucleotide sequence of a developmental gene contributed to a major evolutionary change: the origin of the six-legged insect body plan.

Changes in Gene Regulation

A change in the nucleotide sequence of a gene may affect its function wherever the gene is expressed, while changes in the regulation of gene expression can be limited to one cell type (see Concept 18.4). Thus, a change in the regulation of a developmental gene may have fewer harmful side effects than a change to the sequence of the gene. This reasoning has prompted researchers to suggest that changes in the form of

▼ Figure 25.45 Inquiry

What causes the loss of spines in lake stickleback fish?

Experiment Marine populations of the threespine stickleback fish (*Gasterosteus aculeatus*) have a set of protective spines on their lower (ventral) surface; however, these spines have been lost or reduced in some lake populations of this fish. Researchers performed genetic crosses and found that most of the reduction in spine size resulted from the effects of a single developmental gene, *Pitx1*. The researchers then tested two hypotheses about how *Pitx1* causes this morphological change.

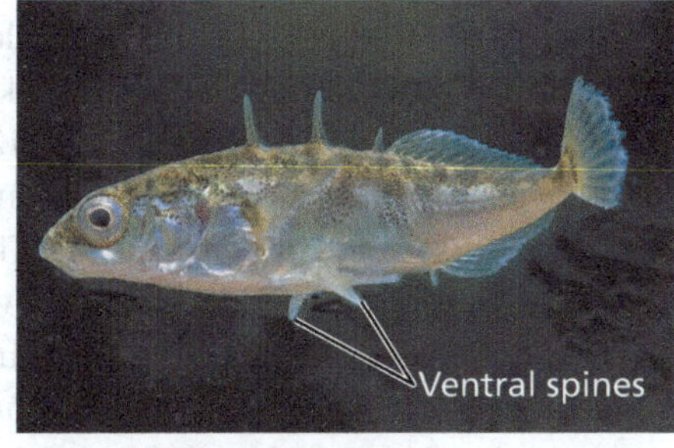

▲ Threespine stickleback (*Gasterosteus aculeatus*)

Hypothesis A: A change in the DNA sequence of *Pitx1* had caused spine reduction in lake populations. To test this idea, the team used DNA sequencing to compare the coding sequence of the *Pitx1* gene between marine and lake stickleback populations.

Hypothesis B: A change in the regulation of the expression of *Pitx1* had caused spine reduction. To test this idea, the researchers monitored where in the developing embryo the *Pitx1* gene was expressed. They conducted whole-body *in situ* hybridisation experiments (see Concept 20.2) using *Pitx1* DNA as a probe to detect *Pitx1* mRNA in the fish.

Results

Test of Hypothesis A:	Are there differences in the coding sequence of the *Pitx1* gene in marine and lake stickleback fish?	**Result: No** →	The 283 amino acids of the *Pitx1* protein are identical in marine and lake stickleback populations.
Test of Hypothesis B:	Are there any differences in the regulation of expression of *Pitx1*?	**Result: Yes** →	Red arrows (→) indicate regions of *Pitx1* gene expression in the photographs below. *Pitx1* is expressed in the ventral spine and mouth regions of developing marine stickleback fish but only in the mouth region of developing lake stickleback fish.

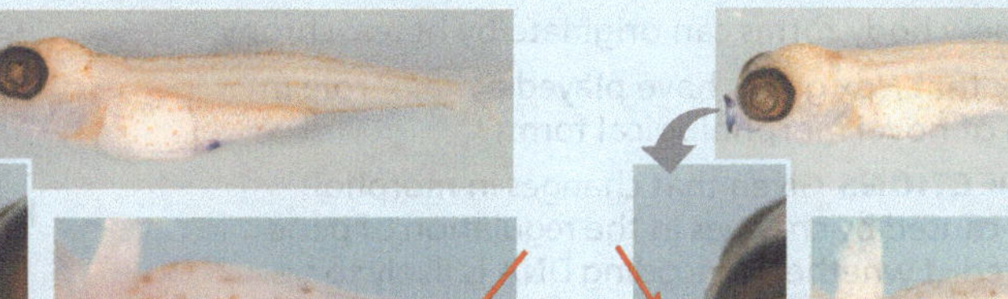

Conclusion The loss or reduction of ventral spines in lake populations of threespine stickleback fish appears to have resulted primarily from a change in the regulation of *Pitx1* gene expression, not from a change in the gene's sequence.

Data from M. D. Shapiro et al., Genetic and developmental basis of evolutionary pelvic reduction in threespine sticklebacks, *Nature* 428:717–723 (2004).

WHAT IF? *Describe the set of results that would have led researchers to the conclusion that a change in the coding sequence of the* Pitx1 *gene was more important than a change in regulation of gene expression.*

organisms may often be caused by mutations that affect the regulation of developmental genes—not their sequences.

This idea is supported by studies of a variety of species, including threespine stickleback fish **(Figure 25.45)**. These fish live in the open ocean and in shallow, coastal waters. In western Canada, they also live in lakes formed when the coastline receded during the past 12,000 years. Marine stickleback fish have a pair of spines on their ventral (lower) surface, which deter some predators. These spines are often reduced or absent in stickleback fish living in lakes that lack predatory fishes and that are also low in calcium. Spines may have been lost in such lakes because they are not advantageous in the absence of predators, and the limited calcium is needed for purposes other than constructing spines.

At the genetic level, the developmental gene *Pitx1* was known to influence whether stickleback fish have ventral spines. Was the reduction of spines in some lake populations due to changes in the *Pitx1* gene or to changes in how the gene is expressed (Figure 25.45)? The researchers' results indicate that the regulation of gene expression has changed, not the DNA sequence. Moreover, lake stickleback fish do express the *Pitx1* gene in tissues not related to the production of spines (for example, the mouth), illustrating how morphological change can be caused by altering the expression of a developmental gene in some parts of the body but not others. In a follow-up study, researchers showed that changes to the *Pel* enhancer, a noncoding DNA region that affects expression of the *Pitx1* gene, resulted in the reduction of ventral spines in lake sticklebacks. Overall, results from studies on stickleback fish provide a clear and detailed example of how changes in gene regulation can alter the form of individual organisms and ultimately lead to evolutionary change in populations.

CONCEPT CHECK 25.5

1. Explain how new body forms can originate by heterochrony.
2. Why is it likely that *Hox* genes have played a major role in the evolution of novel morphological forms?
3. **MAKE CONNECTIONS** Given that changes in morphology are often caused by changes in the regulation of gene expression, predict whether noncoding DNA is likely to be affected by natural selection. (Review Concept 18.3.)

For suggested answers, see Appendix A.

CONCEPT 25.6

Evolution is not goal oriented

What does our study of macroevolution tell us about how evolution works? One lesson is that throughout the history of life, the origin of new species has been affected by both the small-scale factors described in Concept 23.3 (such as natural selection operating in populations) and the large-scale factors described in this chapter (such as continental drift promoting bursts of speciation throughout the globe). Moreover, to paraphrase the Nobel Prize–winning geneticist François Jacob, evolution is like tinkering—a process in which new forms arise by the modification of existing structures or existing developmental genes. Over time, such tinkering has led to the three key features of the natural world described on the opening page of the chapter, "Descent with Modification: A Darwinian View of Life": the striking ways in which organisms are suited for life in their environments, the many shared characteristics of life, and the rich diversity of life.

Evolutionary Novelties

François Jacob's view of evolution harkens back to Darwin's concept of descent with modification. As new species form, novel and complex structures can arise as gradual modifications of ancestral structures. In many cases, complex structures have evolved in increments from simpler versions that performed the same basic function. For example, consider the human eye, an intricate organ constructed from numerous parts that work together in forming an image and transmitting it to the brain. How could the human eye have evolved in gradual increments? Some argue that if the eye needs all of its components to function, a partial eye could not have been of use to our ancestors.

The flaw in this argument, as Darwin himself noted, lies in the assumption that only complicated eyes are useful. In fact, many animals depend on eyes that are far less complex than our own. The simplest eyes that we know of are patches of light-sensitive photoreceptor cells. These simple eyes appear to have had a single evolutionary origin and are now found in a variety of animals, including small molluscs called limpets. Such eyes have no equipment for focusing images, but they do enable the animal to distinguish light from dark. Limpets cling more tightly to their rock when a shadow falls on them, a behavioural adaptation that reduces the risk of being eaten **(Figure 25.46)**. Limpets have had a long evolutionary history, demonstrating that their "simple" eyes are quite adequate to support their survival and reproduction.

In the animal kingdom, complex eyes have evolved independently from such basic structures many times. Some

▼ Figure 25.46 Limpets (*Patella vulgata*), molluscs that can sense light and dark with a simple patch of photoreceptor cells.

▼ Figure 25.47 A range of eye complexity among molluscs.

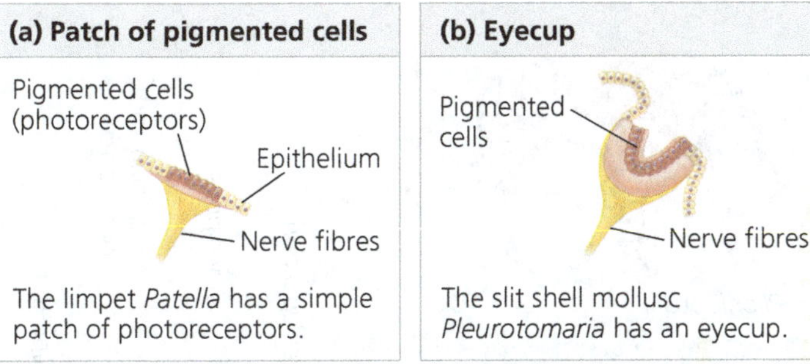

(a) The limpet *Patella* has a simple patch of photoreceptors.

(b) The slit shell mollusc *Pleurotomaria* has an eyecup.

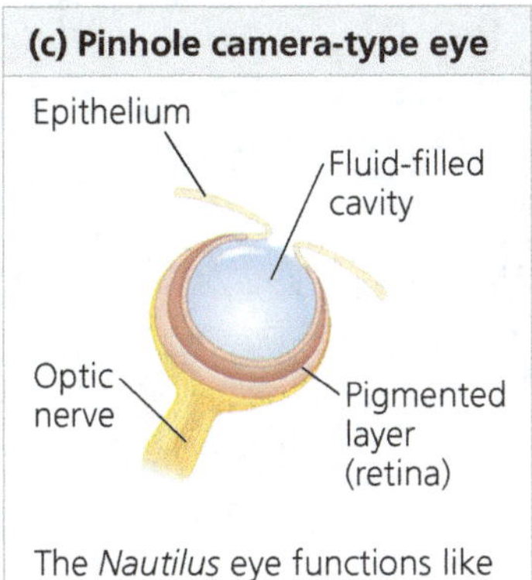

(c) The *Nautilus* eye functions like a pinhole camera (an early type of camera lacking a lens).

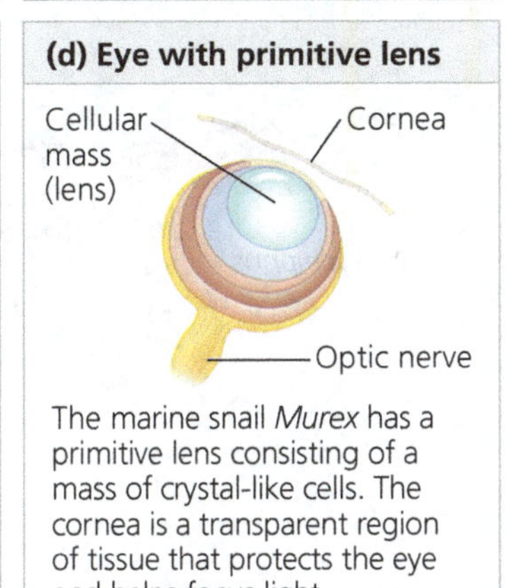

(d) The marine snail *Murex* has a primitive lens consisting of a mass of crystal-like cells. The cornea is a transparent region of tissue that protects the eye and helps focus light.

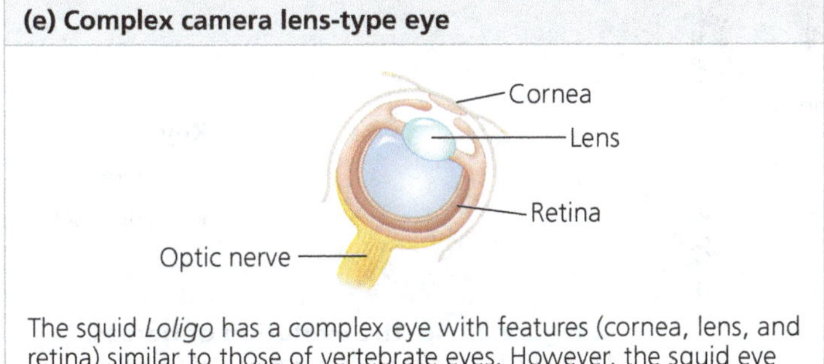

(e) The squid *Loligo* has a complex eye with features (cornea, lens, and retina) similar to those of vertebrate eyes. However, the squid eye evolved independently from vertebrate eyes.

molluscs, such as squids and octopuses, have eyes as complex as those of humans and other vertebrates **(Figure 25.47)**. Although complex mollusc eyes evolved independently of vertebrate eyes, both evolved from a simple cluster of photoreceptor cells present in a common ancestor. In each case, the complex eye evolved through a series of steps that benefited the eyes' owners at every stage. Evidence of their independent evolution can be found in their structure: Vertebrate eyes detect light at the back layer of the retina and conduct nerve impulses towards the front, while complex mollusc eyes do the reverse.

Throughout their evolutionary history, eyes retained their basic function of vision. But evolutionary novelties can also arise when structures that originally played one role gradually acquire a different one. For example, as cynodonts gave rise to early mammals, bones that formerly comprised the jaw hinge (the articular and quadrate; see Figure 25.7) were incorporated into the ear region of mammals, where they eventually took on a new function: the transmission of sound (see Concept 34.6). Structures that evolve in one context but become co-opted for another function are sometimes called *exaptations* to distinguish them from the adaptive origin of the original structure. Note that the concept of exaptation does not imply that a structure somehow evolves in anticipation of future use. Natural selection cannot predict the future; it can only improve a structure in the context of its *current* utility. Novel features, such as the new jaw hinge and ear bones of early mammals, can arise gradually via a series of intermediate stages, each of which has some function in the organism's current context.

Evolutionary Trends

What else can we learn from patterns of macroevolution? Consider evolutionary "trends" observed in the fossil record. For instance, some evolutionary lineages exhibit a trend towards larger or smaller body size. An example is the evolution of the present-day horse (genus *Equus*), a descendant of the 55-million-year-old *Hyracotherium* **(Figure 25.48)**. About the size of a large dog, *Hyracotherium* had four toes on its front feet, three toes on its hind feet, and teeth adapted for browsing on bushes and trees. In comparison, present-day horses are larger, have only one toe on each foot, and possess teeth modified for grazing on grasses.

Extracting a single evolutionary progression from the fossil record can be misleading, however; it is like describing a bush as growing towards a single point by tracing only the branches that lead to that twig. For example, by selecting certain species from the available fossils, it is possible to arrange a succession of animals intermediate between *Hyracotherium* and living horses that shows a trend towards large, single-toed species (follow the yellow highlighting in Figure 25.48). However, if we consider *all* fossil horses known today, this apparent trend vanishes. The genus *Equus* did not evolve in a straight line; it is the only surviving twig of an evolutionary tree that is so branched that it is more like a bush. *Equus* actually descended through a series of speciation episodes that included several adaptive radiations, not all of which led to large, one-toed, grazing horses. In fact, phylogenetic analyses suggest that all lineages that include grazers are closely related to *Parahippus*; the many other horse lineages, all of which are now extinct, remained multi-toed browsers for 35 million years.

Branching evolution *can* result in a real evolutionary trend even if some species counter the trend. One model of long-term trends views species as analogous to individuals: Speciation is their birth, extinction is their death, and new species that diverge from them are their offspring. In this model, just as populations of individual organisms undergo natural selection, species undergo *species selection*. The species that endure the longest and generate the most new offspring species determine the direction of major evolutionary trends. The species selection model suggests that "differential speciation success" plays a role in macroevolution similar to the role of differential reproductive success

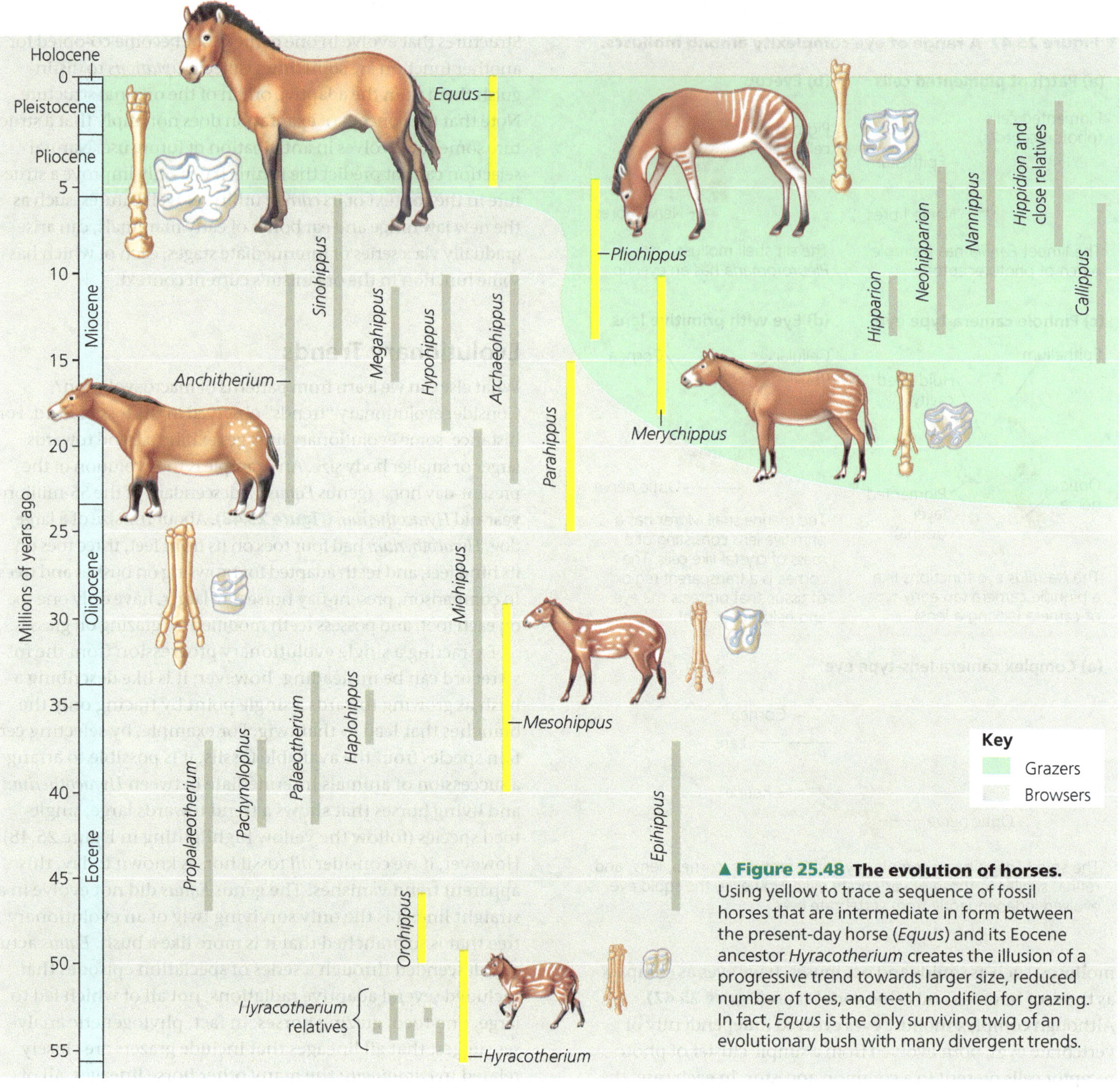

▲ Figure 25.48 The evolution of horses. Using yellow to trace a sequence of fossil horses that are intermediate in form between the present-day horse (*Equus*) and its Eocene ancestor *Hyracotherium* creates the illusion of a progressive trend towards larger size, reduced number of toes, and teeth modified for grazing. In fact, *Equus* is the only surviving twig of an evolutionary bush with many divergent trends.

in microevolution. Evolutionary trends can also result directly from natural selection. For example, when horse ancestors invaded the grasslands that spread during the mid-Cenozoic, there was strong selection for grazers that could escape predators by running faster. This trend would not have occurred without open grasslands.

Whatever its cause, an evolutionary trend does not imply that there is some intrinsic drive towards a particular phenotype. Evolution is the result of the interactions between organisms and their current environments; if environmental conditions change, an evolutionary trend may cease or even reverse itself. The cumulative effect of these ongoing interactions between organisms and their environments is enormous: It is through them that the staggering diversity of life—Darwin's "endless forms most beautiful"—has arisen.

CONCEPT CHECK 25.6

1. How can the Darwinian concept of descent with modification explain the evolution of such complex structures as the vertebrate eye?
2. **WHAT IF?** The myxoma virus kills up to 99.8% of infected European rabbits in populations with no previous exposure to the virus. The virus is transmitted between living rabbits by mosquitoes. Describe an evolutionary trend (in either the rabbit or virus) that might occur after a rabbit population first encounters the virus.

For suggested answers, see Appendix A.

25 Chapter Review

SUMMARY OF KEY CONCEPTS

CONCEPT **25.1**

Conditions on early Earth made the origin of life possible *(pp. 534–536)*

- Experiments simulating possible early atmospheres have produced organic molecules from inorganic precursors. Amino acids, lipids, sugars, and nitrogenous bases have also been found in meteorites.
- Amino acids and RNA nucleotides polymerise when dripped onto hot sand, clay, or rock. Organic compounds can spontaneously assemble into **protocells**, membrane-bounded droplets that have some properties of cells.
- The first genetic material may have self-replicating, catalytic RNA. Early protocells containing such RNA would have increased through natural selection.

? *Describe the roles that montmorillonite clay and vesicles may have played in the origin of life.*

CONCEPT **25.2**

The fossil record documents the history of life *(pp. 536–539)*

- The fossil record, based largely on fossils found in sedimentary rocks, documents the rise and fall of different groups of organisms over time.
- Sedimentary strata reveal the relative ages of fossils. The ages of fossils can be estimated by **radiometric dating** and other methods.
- The fossil record shows how new groups of organisms can arise via the gradual modification of pre-existing organisms.

? *What are the challenges of estimating the ages of old fossils? Explain how these challenges may be overcome in some circumstances.*

CONCEPT **25.3**

Key events in life's history include the origins of unicellular and multicellular organisms and the colonisation of land *(pp. 539–546)*

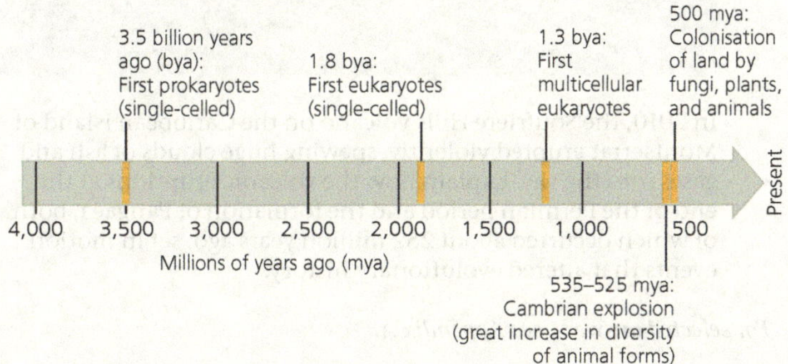

? *What is the "Cambrian explosion," and why is it significant?*

CONCEPT **25.4**

The rise and fall of groups of organisms reflect differences in speciation and extinction rates *(pp. 546–569)*

- In **plate tectonics**, continental plates move gradually over time, altering the physical geography and climate of Earth, leading to extinctions in some groups and speciation in others.
- Evolutionary history has been punctuated by five **mass extinctions** that radically altered life's history. Possible causes for these extinctions include continental drift, volcanic activity, and impacts from asteroids.
- Large increases in the diversity of life have resulted from **adaptive radiations** that followed mass extinctions. Adaptive radiations have also occurred in groups of organisms that possessed major evolutionary innovations or that colonised new regions in which there was little competition from other organisms.

? *Explain how the broad evolutionary changes seen in the fossil record are the cumulative result of speciation and extinction events.*

CONCEPT **25.5**

Major changes in body form can result from changes in the sequences and regulation of developmental genes *(pp. 569–572)*

- Developmental genes affect morphological differences between species by influencing the rate, timing, and spatial patterns of change in an organism's form as it develops into an adult.
- The evolution of new forms can be caused by changes in the nucleotide sequences or regulation of developmental genes.

? *How could changes in a single gene or DNA region ultimately lead to the origin of a new group of organisms?*

CONCEPT **25.6**

Evolution is not goal oriented *(pp. 572–574)*

- Novel and complex biological structures can evolve through a series of incremental modifications, each of which benefits the organism that possesses it.
- Evolutionary trends can be caused by natural selection in a changing environment or species selection, resulting from interactions between organisms and their current environments.

? *Explain the reasoning behind the statement "Evolution is not goal oriented."*

TEST YOUR UNDERSTANDING

Levels 1-2: Remembering/Understanding

1. Fossilised stromatolites
   (A) formed around deep-sea vents.
   (B) resemble structures formed by bacterial communities that are found today in some shallow marine bays.
   (C) provide evidence that plants moved onto land in the company of fungi around 500 million years ago.
   (D) contain the first undisputed fossils of eukaryotes.

2. The oxygen revolution changed Earth's environment dramatically. Which of the following took advantage of the presence of free oxygen in the oceans and atmosphere?
   (A) the evolution of cellular respiration, which used oxygen to help harvest energy from organic molecules
   (B) the persistence of some animal groups in anaerobic habitats
   (C) the evolution of photosynthetic pigments that protected early algae from the corrosive effects of oxygen
   (D) the evolution of chloroplasts after early protists incorporated photosynthetic cyanobacteria

3. Which factor most likely caused animals and plants in India to differ greatly from species in nearby Southeast Asia?
   (A) The species became separated by convergent evolution.
   (B) The climates of the two regions are similar.
   (C) India is in the process of separating from the rest of Asia.
   (D) India was a separate continent until 45 million years ago.

4. Large-scale, worldwide adaptive radiations have occurred in which of the following situations?
   (A) when there are no available ecological niches
   (B) after each of the big five mass extinctions
   (C) after colonisation of an isolated island that contains suitable habitat and few competitor species
   (D) whenever an evolutionary innovation was needed for organisms to thrive

5. Scientists studying the origin of life have accomplished which of the following steps?
   (A) abiotic synthesis of protocells with self-replicating, catalytic RNA
   (B) formation of vesicles that use RNA as a template for DNA synthesis
   (C) formation of protocells that use DNA to direct the polymerisation of amino acids
   (D) abiotic synthesis of RNA's bases (A, C, G, U)

Levels 3-4: Applying/Analysing

6. A genetic change that caused a certain *Hox* gene to be expressed along the tip of a vertebrate limb bud instead of farther back helped make possible the evolution of the tetrapod limb. This type of change is illustrative of
   (A) the influence of environment on development.
   (B) paedomorphosis.
   (C) a change in a developmental gene or in its regulation that altered the spatial organisation of body parts.
   (D) heterochrony.

7. A swim bladder is a gas-filled sac that helps fish maintain buoyancy. The evolution of the swim bladder from the air-breathing organ (a simple lung) of an ancestral fish is an example of
   (A) exaptation.
   (B) changes in *Hox* gene expression.
   (C) paedomorphosis.
   (D) adaptive radiation.

Levels 5-6: Evaluating/Creating

8. **EVOLUTION CONNECTION** Describe how gene flow, genetic drift, and natural selection all can influence macroevolution.

9. **SCIENTIFIC INQUIRY** Herbivory (plant eating) has evolved repeatedly in insects, typically from meat-eating or detritus-feeding ancestors (detritus is dead organic matter). Moths and butterflies, for example, eat plants, whereas their "sister group" (the insect group to which they are most closely related), the caddisflies, feed on animals, fungi, or detritus. As illustrated in the following phylogenetic tree, the combined moth/butterfly and caddisfly group shares a common ancestor with flies and fleas. Like caddisflies, flies and fleas are thought to have evolved from ancestors that did not eat plants.

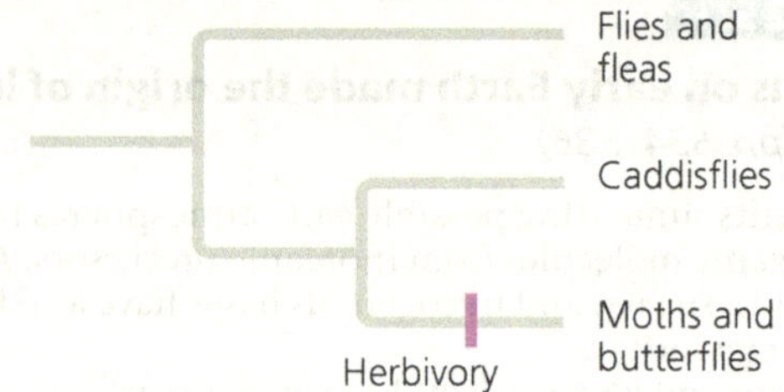

   There are 140,000 species of moths and butterflies and 7,000 species of caddisflies. State a hypothesis about the impact of herbivory on adaptive radiations in insects. How could this hypothesis be tested?

10. **WRITE ABOUT A THEME: ORGANISATION** You have seen many examples of how form fits function at all levels of the biological hierarchy. However, we can imagine forms that would function better than some forms actually found in nature. For example, if the wings of a bird were not formed from its forelimbs, such a hypothetical bird could fly yet also hold objects with its forelimbs. In a short essay (100–150 words), use the concept of "evolution as tinkering" to explain why there are limits to the functionality of forms in nature.

11. **SYNTHESISE YOUR KNOWLEDGE**

   In 2010, the Soufriere Hills volcano on the Caribbean island of Montserrat erupted violently, spewing huge clouds of ash and gases into the sky. Explain how the volcanic eruptions at the end of the Permian period and the formation of Pangaea, both of which occurred about 252 million years ago, set in motion events that altered evolutionary history.

For selected answers, see Appendix A.

Unit 5 THE EVOLUTIONARY HISTORY OF BIOLOGICAL DIVERSITY

Dr Hamish Campbell is a palaeontologist and geologist based in Wellington, New Zealand. He is an Emeritus Scientist with GNS Science. He was educated at Otago (NZ), Auckland (NZ) and Cambridge (UK) universities. He was employed as a professional palaeontologist for 20 years (1978–1998) and then became a science communicator as "the geologist" at Te Papa Tongarewa, the National Museum of New Zealand (1998–2019). Hamish is best known as a co-presenter on the television series, *Coast New Zealand*, and as an author of popular books. He is especially interested in the origin (provenance) of the older (Palaeozoic and Mesozoic) sedimentary rocks of New Zealand and New Caledonia. His research has contributed to our understanding of the geological history of the Zealandia continent, and the origin of the modern endemic biota of Zealandia.

AN INTERVIEW WITH

Hamish Campbell

How did you come to study New Zealand's geological and paleo history?

During my MSc studies, I slept in a tent under a big tree in the middle of a small village in New Caledonia. During the day, I studied the island's geology and fossils. I wanted to know whether New Caledonian and New Zealand fossil assemblages were the same—and they were. It took decades, and the discovery of Zealandia [see the chapter, "The History of Life on Earth"], to help me understand why those similarities existed.

How does the discovery of Zealandia help us understand the unique flora and fauna of New Zealand?

New Zealand and New Caledonia are emergent remnants of a sunken continent: Zealandia. Once, the continent was just another part of the land on Gondwana's southern coastline. After separation from Gondwana, ongoing tectonism stretched and thinned the crust, and reduced crustal buoyancy caused most of Zealandia to submerge. This rifting and sinking prevailed for about 80 million years, from 105 million to 25 million years ago. We wondered whether land had always existed where New Zealand now sits. Could it be that all of Zealandia was submerged about 25 million years ago?

My colleagues and I used the Chatham Islands off New Zealand's eastern coast to investigate the question at a smaller scale. After several years of geological and biological research, we were surprised to learn of the island's youth. The land surface was less than 4 million years old. Likewise, terrestrial animals (insects) and plants with low dispersal potential had arrived and evolved into the species we see today, in less than 4 million years. Molecular biology confirmed that many of these species evolved in the last 1 million years. Their nearest relatives live on nearby landmasses, such as mainland New Zealand.

Encouraged by these results, we focused on New Zealand, using the same techniques. Knowing that the peak ocean transgression occurred between 20 million and 30 million years ago, we examined all the sedimentary rocks from this time. They were almost all of marine origin. For maximum flooding by the sea, we suggested a geological moment in time—23 million years ago—that implied most, if not all, of New Zealand's land surface was less than 23 million years old. But, for terrestrial life, to drown is to die. Our geological interpretation suggests the ancestors of all native New Zealand terrestrial biota somehow arrived, established, and evolved within the past 23 million years.

How might New Zealand's flora and fauna have established after maximum flooding?

Perhaps some short-lived islands existed, although there is no surviving evidence of them. Maybe, during maximum submergence, species hopped between ephemeral islands—provided they could disperse over long distances. Transport of organisms between islands, or across the Tasman Sea from Australia, involves two variations on a theme that we see today: either floating/rafting on the sea surface, or flying (being flown or windblown). Birds can carry all sorts of materials—seeds, eggs, and small passengers such as invertebrates and crustaceans. The latest interpretations suggest the ancestors of the moa and kiwi were smaller and could fly. Once in the relatively predator-free New Zealand environment, they quickly adapted and evolved into flightless birds. Birds are probably also responsible for the transport and introduction of many small invertebrates (molluscs and crustaceans), and maybe vertebrates too (frogs). Floating woody debris or logs have the potential to transport animals; for example, reptiles can survive with little food for long periods. The ancestor of the tuatara may have arrived in New Zealand by log. Other high-energy methods of dispersal may also have played a role: strong winds, tsunamis, volcanic eruptions, and cyclones. And if organisms can get somewhere once, they can surely get there repeatedly. There could therefore have been multiple "arrivals" of the same species.

From our work on the Chatham Islands, we knew species could arrive and evolve in relatively short periods of time. One fossil lake site in the South Island, dated at 19–16 million years, postdates maximum submergence. The dominant fossils include fish and waterfowl, numerous other birds, turtles, crocodiles, reptiles, frogs, and bats. Plant fossils and fossil pollen and spores are also present. Importantly, fossilised remains of moa, kiwi, and tuatara have been recovered. This site shows that most of the major elements of the present-day New Zealand fauna were established in New Zealand during the Miocene period (25–5 million years ago). What's more, molecular biologists have found that more than 95% of all New Zealand native biota descends from *Australian*—not Zealandian or Gondwanan—ancestors. This result supports the idea of substantial dispersal and colonisation since maximum drowning.

If life disperses so easily, why aren't more Australian animals (reptiles, frogs, insects, birds, mammals) and plants in New Zealand?

We need to ask that and also another question: Why does New Zealand have such strong endemism if dispersal is easy? Perhaps establishment is easier initially when there is little competition, but as niches fill up it gets harder to establish new species.

27 Bacteria and Archaea

KEY CONCEPTS

Study Tip

Make a table: As you read the chapter, identify examples of challenging environments in which prokaryotes thrive. List the structural, metabolic, or other features of prokaryotes that contribute to their success in each of these environments.

Environmental challenge	Prokaryotic feature enabling success
Lack of water	Capsule or slime layer
Exposure to antibiotics	

Go to Mastering Biology

to access Dynamic Study Modules for revision, 3D BioFlix® animations and high-quality videos, and your interactive Pearson eText.

Figure 27.1 Lake Bumbunga, South Australia. The pink colour of this lake is a sign that its waters are much saltier than seawater. Its colour is caused not by minerals or other nonliving sources, but by trillions of prokaryotes that thrive in salinities that kill other cells. Other prokaryotes live in environments that are far too cold or hot or acidic for most other organisms.

What characteristics enable prokaryotes to reach huge population sizes and thrive in diverse environments?

Due to the organisms' **small size** and **rapid reproduction**, prokaryotic populations can reach enormous numbers.

Population size (*N*)

Time →

Because prokaryotic populations are so large, **mutations** produce high genetic diversity, enabling rapid evolution.

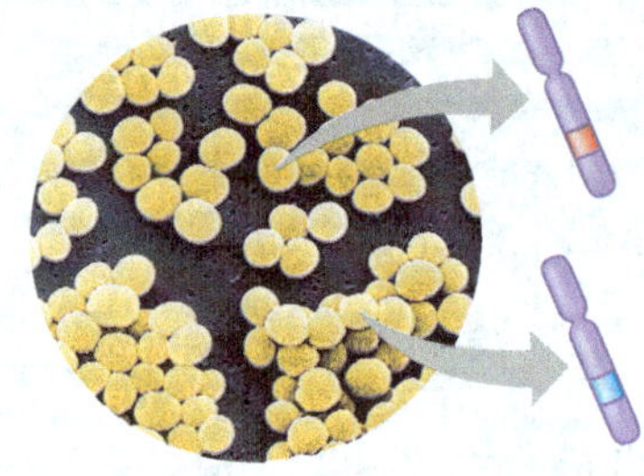

Their **diverse adaptations** enable prokaryotes to live in a wide range of environments.

Rapid evolution in prokaryotic populations results in diverse metabolic and structural adaptations.

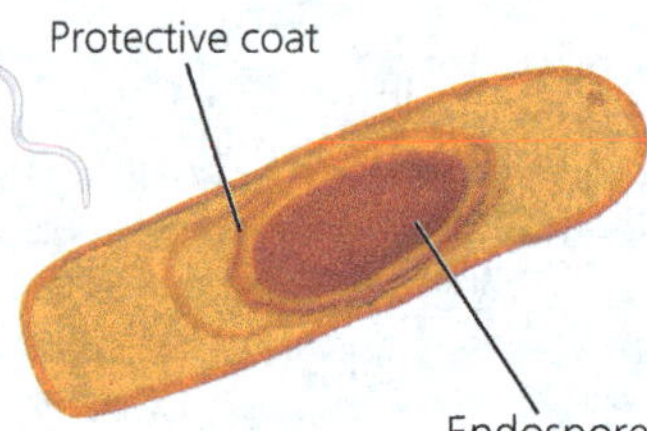

Prokaryotes, the single-celled organisms that make up domains Bacteria and Archaea, can thrive in a wide range of extreme environments. Prokaryotic species are also very well adapted to more "normal" habitats—the lands and waters in which most other species are found. Their ability to adapt to a broad range of habitats helps explain why prokaryotes are the most abundant organisms on Earth. Indeed, the number of prokaryotes in a handful of fertile soil is greater than the number of people who have ever lived. In this chapter, we'll examine the adaptations, diversity, and enormous ecological impact of these remarkable organisms.

CONCEPT 27.1

Structural and functional adaptations contribute to prokaryotic success

The first organisms to inhabit Earth were prokaryotes that lived 3.5 billion years ago (see Concept 25.3). Throughout their long evolutionary history, prokaryotic populations have been (and continue to be) subjected to natural selection in all kinds of environments, resulting in their enormous diversity today.

We'll begin by describing prokaryotes. Most prokaryotes are unicellular, although the cells of some species remain attached to each other after cell division. Prokaryotic cells typically have diameters of 0.5–5 µm, much smaller than the 10- to 100-µm diameter of many eukaryotic cells. (One notable exception, *Thiomargarita namibiensis*, can be as large as 750 µm in diameter—bigger than a poppy seed.) Prokaryotic cells have a variety of shapes **(Figure 27.2)**.

▼ Figure 27.2 The most common shapes of prokaryotes.
(a) Cocci (singular, *coccus*) are spherical prokaryotes. They can occur singly, in chains of two or more cells, and in clusters resembling bunches of grapes. **(b)** Bacilli (singular, *bacillus*) are rod-shaped prokaryotes. They are usually solitary, but in some forms the rods are arranged in chains. **(c)** Spiral prokaryotes include spirochetes (shown here), which are corkscrew-shaped; other spiral prokaryotes resemble commas or loose coils (colourised SEMs).

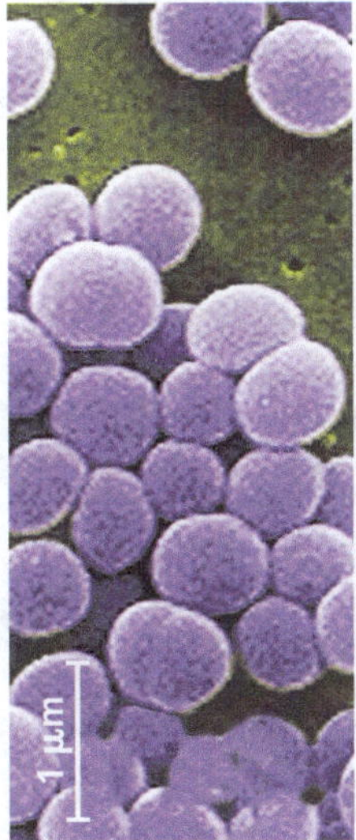

(a) Spherical

1 µm

(b) Rod-shaped

(c) Spiral

Finally, although they are unicellular and small, prokaryotes are well organised, achieving all of an organism's life functions within a single cell.

Cell-Surface Structures

A key feature of nearly all prokaryotic cells is the cell wall, which maintains cell shape, protects the cell, and prevents it from bursting in a hypotonic environment (see Figure 7.13). In a hypertonic environment, most prokaryotes lose water and shrink away from their wall. Such water losses can inhibit cell reproduction. Thus, salt can be used to preserve foods because it causes food-spoiling prokaryotes to lose water, preventing them from reproducing rapidly.

The cell walls of prokaryotes differ in structure from those of eukaryotes. In eukaryotes that have cell walls, such as plants and fungi, the walls are usually made of cellulose or chitin (see Concept 5.2). In contrast, most bacterial cell walls contain **peptidoglycan**, a polymer composed of modified sugars cross-linked by short polypeptides. This molecular fabric encloses the entire bacterium and anchors other molecules that extend from its surface. Archaeal cell walls contain a variety of polysaccharides and proteins but lack peptidoglycan.

Using a technique called the **Gram stain**, developed by the 19th-century Danish physician Hans Christian Gram, scientists can categorise many bacterial species according to differences in cell wall composition. To do this, samples are first stained with crystal violet dye and iodine, then rinsed in alcohol, and finally stained with the red dye safranin, which enters the cell and binds to the cell's DNA. The structure of a bacterium's cell wall determines the staining response **(Figure 27.3)**. **Gram-positive** bacteria have relatively simple walls composed of a thick layer of peptidoglycan. The walls of **gram-negative** bacteria have less peptidoglycan and are structurally more complex, with an outer membrane that contains lipopolysaccharides (carbohydrates bonded to lipids).

Gram staining is a valuable tool in medicine for quickly determining if a patient's infection is due to gram-negative or to gram-positive bacteria. The outer membrane of gram-negative bacteria is essential for protection against the body's immune defence system. Knowledge about gram staining has treatment implications. Gram-negative bacteria tend to be more resistant than gram-positive species to antibiotics because the outer membrane impedes the entry of some drugs. However, some gram-positive species have virulent strains that are resistant to one or more antibiotics. (Figure 22.15 discusses one example: methicillin-resistant *Staphylococcus aureus*, or MRSA, which can cause lethal skin infections.) The lipid portions of the lipopolysaccharides in the walls of many gram-negative bacteria are toxic and can cause both fever and shock, which is why it important to treat quickly with an appropriate antibiotic.

▼ **Figure 27.3 Gram staining.**

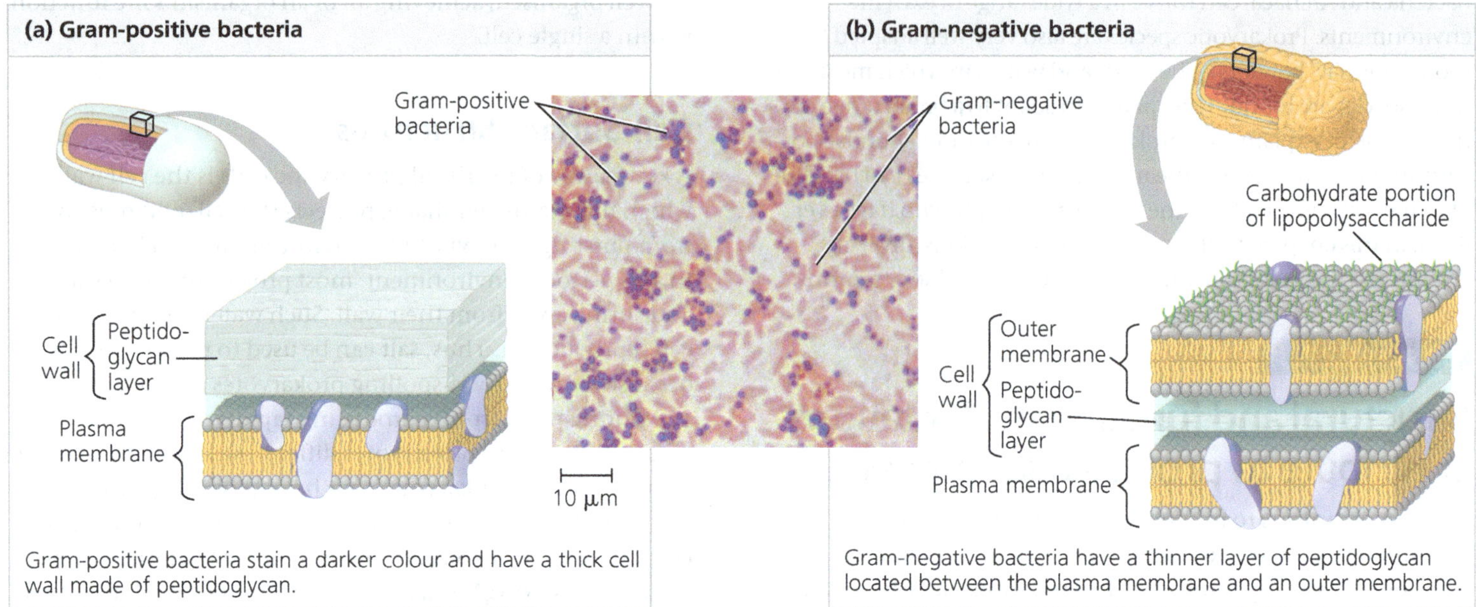

Gram-positive bacteria stain a darker colour and have a thick cell wall made of peptidoglycan.

Gram-negative bacteria have a thinner layer of peptidoglycan located between the plasma membrane and an outer membrane.

The effectiveness of certain antibiotics, such as penicillin, derives from their inhibition of peptidoglycan cross-linking. The resulting cell wall may not be functional, particularly in gram-positive bacteria. Such drugs destroy many species of pathogenic bacteria without adversely affecting human cells, which do not have peptidoglycan.

The cell wall of many prokaryotes is surrounded by a sticky layer of polysaccharide or protein. This layer is called a **capsule** if it is dense and well-defined **(Figure 27.4)** or a *slime layer* if it is not as well organised. Both kinds of sticky outer layers enable prokaryotes to adhere to their substrate or to other individuals in a colony. Capsules and slime layers also protect against dehydration, and some capsules shield pathogenic prokaryotes from attacks by their host's immune system.

In another way of withstanding harsh conditions, certain bacteria develop resistant cells called **endospores** when they lack water or essential nutrients **(Figure 27.5)**. The original cell produces a copy of its chromosome and surrounds that copy with a multilayered structure, forming the endospore. Water is removed from the endospore, and its metabolism halts. The original cell then lyses, releasing the endospore. Most endospores are so durable that they can survive in boiling water; killing them requires heating lab equipment to 121°C under high pressure. In less hostile environments, endospores can remain dormant but viable for centuries, able to rehydrate and resume metabolism when their environment improves.

Finally, some prokaryotes stick to their substrate or to one another by means of hairlike appendages called **fimbriae**

▼ **Figure 27.4 Capsule.** The polysaccharide capsule around this *Streptococcus* bacterium enables the prokaryote to attach to cells in the respiratory tract—in this colourised TEM, a tonsil cell.

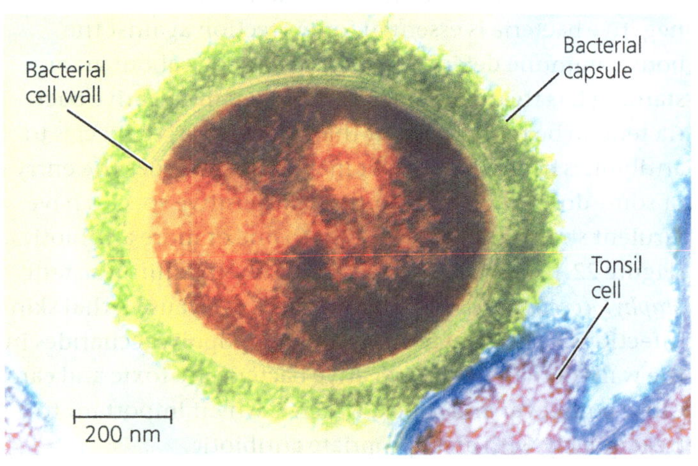

▼ **Figure 27.5 An endospore.** *Bacillus anthracis*, the bacterium that causes the disease anthrax, produces endospores (TEM). An endospore's protective, multilayered coat helps it survive in the soil for years.

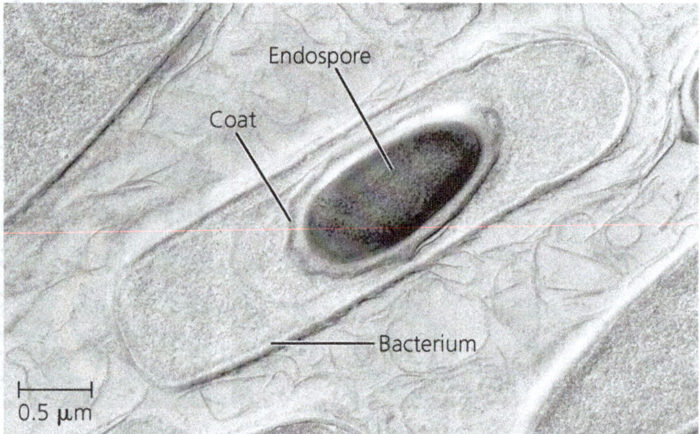

▼ **Figure 27.6 Fimbriae.** These numerous protein-containing appendages enable some prokaryotes to attach to surfaces or to other cells (colourised TEM).

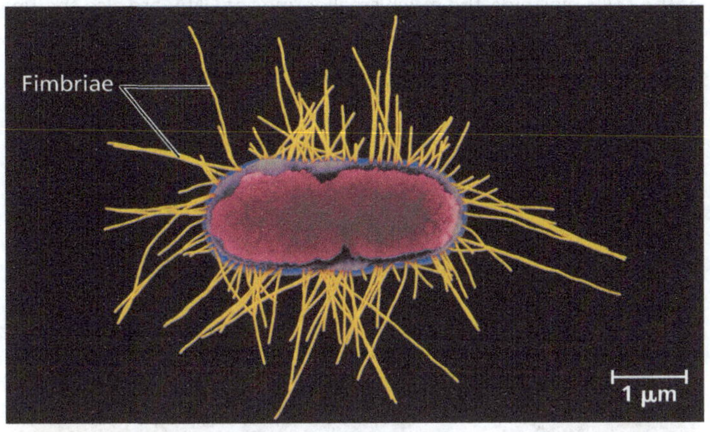

(singular, *fimbria*) **(Figure 27.6)**. For example, the bacterium that causes gonorrhea, *Neisseria gonorrhoeae*, uses fimbriae to fasten itself to the mucous membranes of its host. Fimbriae are usually shorter and more numerous than **pili** (singular, *pilus*), appendages that pull two cells together prior to DNA transfer from one cell to the other (see Figure 27.12); pili are sometimes referred to as *sex pili*.

Motility

About half of all prokaryotes are capable of **taxis**, a directed movement towards or away from a stimulus (from the Greek *taxis*, to arrange). For example, prokaryotes that exhibit *chemotaxis* change their movement pattern in response to chemicals. They may move *towards* nutrients or oxygen (positive chemotaxis) or *away from* a toxic substance (negative chemotaxis). Some species can move at velocities exceeding 50 μm/sec—up to 50 times their body length per second. For perspective, consider that a person 1.7 m tall moving that fast would be running 306 km per hour!

Of the various structures that enable prokaryotes to move, the most common are flagella **(Figure 27.7)**. Flagella (singular, *flagellum*) may be scattered over the entire surface of the cell or concentrated at one or both ends. Prokaryotic flagella differ greatly from eukaryotic flagella: They are one-tenth the width and typically are not covered by an extension of the plasma membrane (see Figure 6.24).

The flagella of prokaryotes and eukaryotes also differ in their molecular composition and their mechanism of propulsion. Among prokaryotes, bacterial and archaeal flagella are similar in size and rotational mechanism, but they are composed of entirely different and unrelated proteins. Overall, these structural and molecular comparisons indicate that the flagella of bacteria, archaea, and eukaryotes arose independently. Since current evidence shows that the flagella of organisms in the three domains perform similar functions but are not related by common descent, they are described as analogous, not homologous, structures (see Concept 22.3).

Evolutionary Origins of Bacterial Flagella

The bacterial flagellum shown in Figure 27.7 has three main parts (the motor, hook, and filament) that are themselves composed of 42 different kinds of proteins. How could such a complex structure evolve? In fact, much evidence indicates that bacterial flagella originated as simpler structures that were modified in a stepwise fashion over time. As in the case of the human eye (see Concept 25.6), biologists asked whether a less complex version of the flagellum could still benefit its owner. Analyses of hundreds of bacterial genomes indicate that only half of the flagellum's protein components appear to be necessary for it to function; the others are inessential or not encoded

▼ **Figure 27.7 A prokaryotic flagellum.** The motor of a prokaryotic flagellum consists of a system of protein rings embedded in the cell wall and plasma membrane (TEM). The electron transport chain pumps protons out of the cell. The diffusion of protons back into the cell provides the force that turns a curved hook and thereby causes the attached filament to rotate and propel the cell. (This diagram shows flagellar structures characteristic of gram-negative bacteria.)

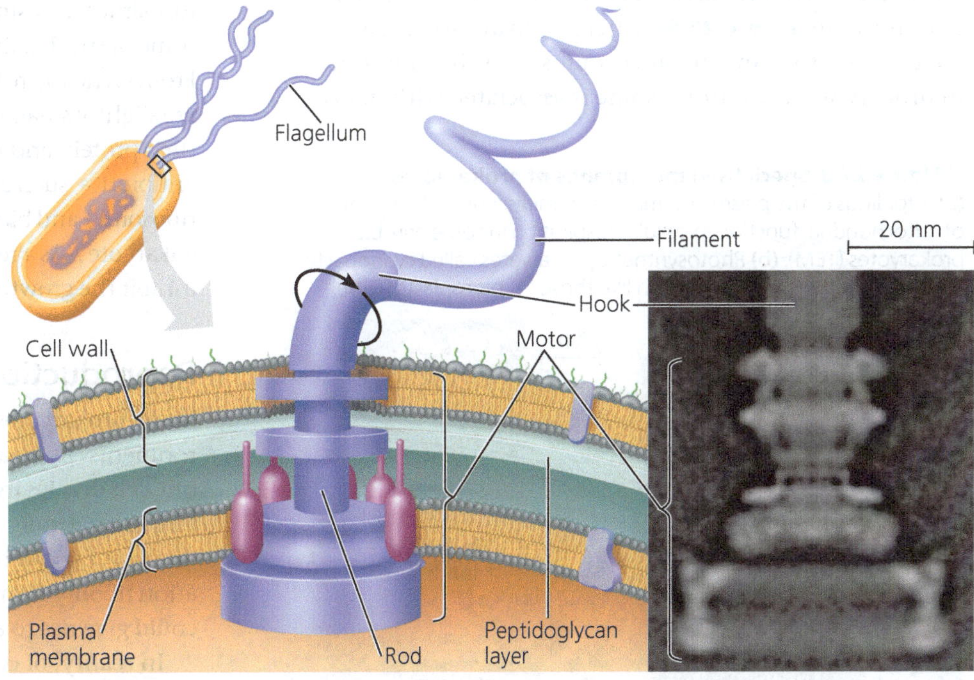

VISUAL SKILLS *Predict which of the four protein rings shown in this diagram are likely hydrophobic. Explain your answer.*

in the genomes of some species. Of the 21 proteins required by all species studied to date, 19 are modified versions of proteins that perform other tasks in bacteria. For example, a set of 10 proteins in the motor is homologous to 10 similar proteins in a secretory system found in bacteria. (A secretory system is a protein complex that enables a cell to produce and release certain macromolecules.) Two other proteins in the motor are homologous to proteins that function in ion transport. The proteins that comprise the rod, hook, and filament are all related to each other and are descended from an ancestral protein that formed a pilus-like tube. These findings suggest that the bacterial flagellum evolved as other proteins were added to an ancestral secretory system. This is an example of *exaptation*, the process in which structures originally adapted for one function take on new functions through descent with modification.

Internal Organisation and DNA

The cells of prokaryotes are simpler than those of eukaryotes in both their internal structure and the physical arrangement of their DNA (see Figure 6.5). Prokaryotic cells lack the complex compartmentalisation associated with the membrane-enclosed organelles found in eukaryotic cells. However, some prokaryotic cells do have specialised membranes that perform metabolic functions **(Figure 27.8)**. These membranes are usually infoldings of the plasma membrane. Recent discoveries also indicate that some prokaryotes can store metabolic by-products in simple compartments that are made out of proteins; these compartments do not have a membrane.

The genome of a prokaryote is structurally different from a eukaryotic genome and in most cases has considerably less DNA. Prokaryotes typically have one circular chromosome **(Figure 27.9)**, whereas eukaryotes usually have several to many linear chromosomes. In addition, in prokaryotes the chromosome is associated with many fewer proteins than are the chromosomes of eukaryotes. Also unlike eukaryotes, prokaryotes lack a nucleus; their chromosome is located in the **nucleoid**, a region of cytoplasm that is not enclosed by a membrane. In addition to its single chromosome, a typical prokaryotic cell may also have much smaller rings of independently replicating DNA molecules called **plasmids** (see Figure 27.9), most carrying only a few genes.

Although DNA replication, transcription, and translation are fundamentally similar processes in prokaryotes and eukaryotes, some of the details differ (see the chapter, "Gene Expression: From Gene to Protein"). For example, prokaryotic ribosomes are slightly smaller than eukaryotic ribosomes and differ in their protein and RNA content. These differences allow certain antibiotics, such as erythromycin and tetracycline, to bind to ribosomes and block protein synthesis in prokaryotes but not in eukaryotes. As a result, people can use these antibiotics to kill or inhibit the growth of bacteria without harming themselves.

▼ Figure 27.8 Specialised membranes of prokaryotes. **(a)** Infoldings of the plasma membrane, reminiscent of the cristae of mitochondria, function in cellular respiration in some aerobic prokaryotes (TEM). **(b)** Photosynthetic prokaryotes called cyanobacteria have thylakoid membranes, much like those in chloroplasts (TEM).

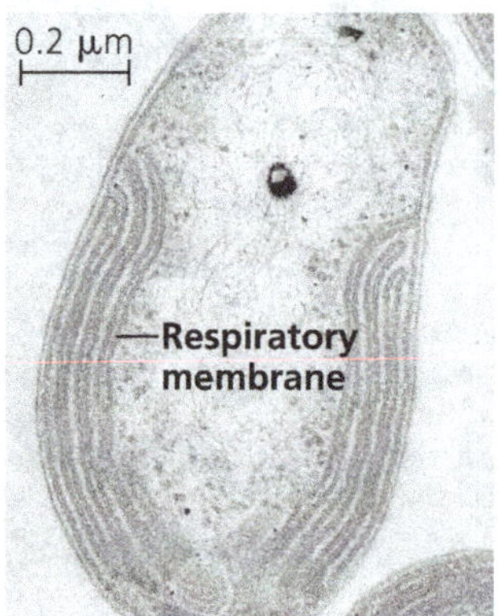

(a) Aerobic prokaryote

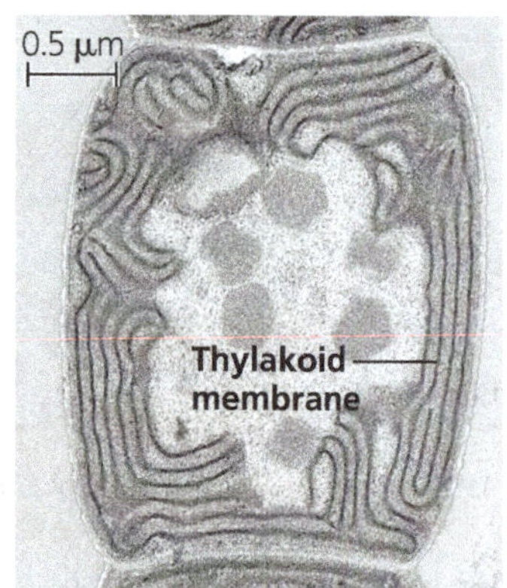

(b) Photosynthetic prokaryote

▼ Figure 27.9 A prokaryotic chromosome and plasmids. The thin, tangled loops surrounding this ruptured *Escherichia coli* cell are parts of the cell's large, circular chromosome (colourised TEM). Three of the cell's plasmids, the much smaller rings of DNA, are also shown.

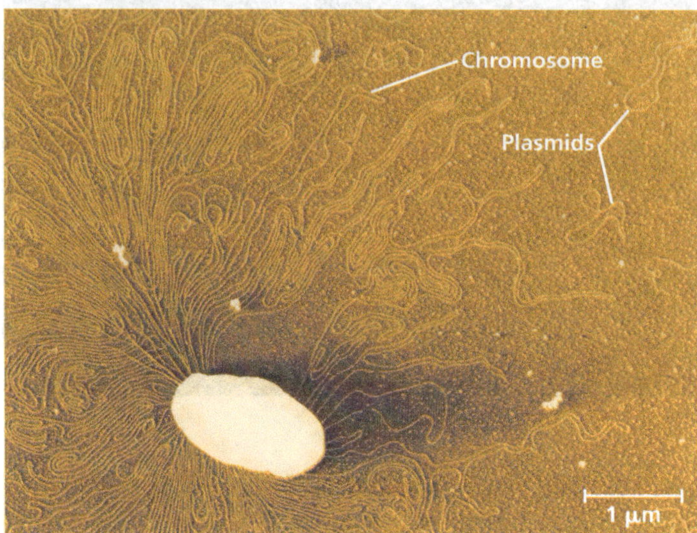

VISUAL SKILLS *Identify and label the third plasmid visible in this micrograph.*

Reproduction

Many prokaryotes can reproduce quickly in favourable environments. By *binary fission* (see Figure 12.12), a single prokaryotic cell divides into 2 cells, which then divide into 4, 8, 16, and so on. Under optimal conditions, many prokaryotes can divide every 1–3 hours; some species can produce a new generation in only 20 minutes. At this rate, a single prokaryotic cell could give rise to a colony outweighing Earth in only two days!

In reality, of course, this does not occur. The cells eventually exhaust their nutrient supply, poison themselves with metabolic wastes, face competition from other microorganisms, or are consumed by other organisms. Still, the potential of many prokaryotic species for rapid population growth emphasises three key features of their biology: *They are small, they reproduce by binary fission, and they often have short generation times*. As a result, prokaryotic populations can consist of many trillions of individuals—far more than populations of multicellular eukaryotes, such as plants or animals.

CONCEPT CHECK 27.1

1. Describe two adaptations that enable prokaryotes to survive in environments too harsh for other organisms.
2. Contrast the cellular and DNA structures of prokaryotes and eukaryotes.
3. **MAKE CONNECTIONS** Suggest a hypothesis that explains why the thylakoid membranes of chloroplasts resemble those of cyanobacteria. Refer to Figures 6.18 and 26.21.

For suggested answers, see Appendix A.

CONCEPT 27.2

Rapid reproduction, mutation, and genetic recombination promote genetic diversity in prokaryotes

As we discussed in Unit 4, evolution cannot occur without genetic variation. The diverse adaptations exhibited by prokaryotes suggest that their populations must have considerable genetic variation—and they do. In this section, we'll examine three factors that give rise to high levels of genetic diversity in prokaryotes: rapid reproduction, mutation, and genetic recombination.

Rapid Reproduction and Mutation

Most of the genetic variation in sexually reproducing species results from the way existing alleles are arranged in new combinations during meiosis and fertilisation (see Concept 13.4). Prokaryotes do not reproduce sexually, so at first glance their extensive genetic variation may seem puzzling. But in many species, this variation can result from a combination of rapid reproduction and mutation and other processes with similar effects.

Consider the bacterium *Escherichia coli* as it reproduces by binary fission in a human intestine, one of its natural environments. After repeated rounds of division, most of the offspring cells are genetically identical to the original parent cell. However, if errors occur during DNA replication, some of the offspring cells may differ genetically. The probability of such a mutation occurring in a given *E. coli* gene is about one in 10 million (1×10^{-7}) per cell division. But among the 2×10^{10} new *E. coli* cells that arise each day in a person's intestine, there will be approximately $(2 \times 10^{10}) \times (1 \times 10^{-7}) = 2{,}000$ bacteria that have a mutation in that gene. The total number of new mutations when all 4,300 *E. coli* genes are considered is about $4{,}300 \times 2{,}000$—more than 8 million per day per human host.

The key point is that new mutations, though rare on a per gene basis, can increase genetic diversity quickly in species with short generation times and large populations. This diversity, in turn, can lead to rapid evolution **(Figure**

▼ Figure 27.10 Inquiry

Can prokaryotes evolve rapidly in response to environmental change?

Experiment Researchers tested the ability of *E. coli* populations to adapt to a new environment. They established 12 populations, each founded by a single cell from an *E. coli* strain, and followed these populations for 20,000 generations (3,000 days).

Daily serial transfer

0.1 mL (population sample)

Old tube (discarded after transfer)

New tube (9.9 mL growth medium)

To maintain a continual supply of resources, each day the researchers performed a *serial transfer*: They transferred 0.1 mL of each population to a new tube containing 9.9 mL of fresh growth medium. The growth medium used throughout the experiment provided a challenging environment that contained only low levels of glucose and other resources needed for growth.

Samples were periodically removed from the 12 populations and grown in competition with the common ancestral strain in the experimental (low-glucose) environment.

Results The fitness of the experimental populations, as measured by the growth rate of each population, increased rapidly for the first 5,000 generations (two years) and more slowly for the next 15,000 generations. The graph shows the averages for the 12 populations.

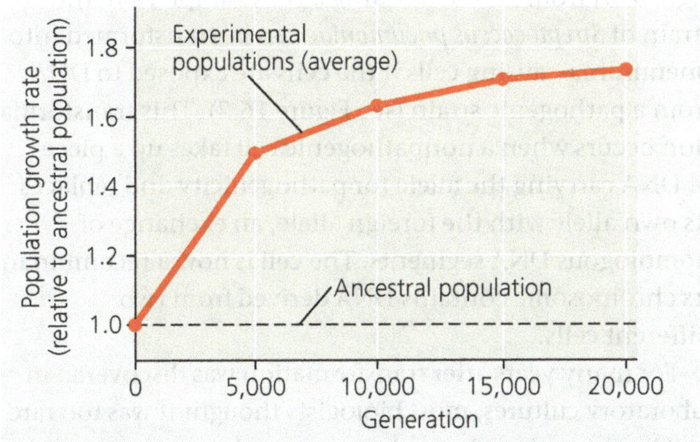

Conclusion Populations of *E. coli* continued to accumulate beneficial mutations for 20,000 generations, allowing rapid evolution of increased population growth rates in their new environment.

Data from V. S. Cooper and R. E. Lenski, The population genetics of ecological specialization in evolving *Escherichia coli* populations, *Nature* 407:736–739 (2000).

WHAT IF? *Suggest possible functions of the genes whose sequence or expression was altered as the experimental populations evolved in the low-glucose environment.*

27.10): Individuals that are genetically better equipped for their environment tend to survive and reproduce at higher rates than other individuals. The ability of prokaryotes to adapt rapidly to new conditions highlights the point that although the structure of their cells is simpler than that of eukaryotic cells, prokaryotes are not "primitive" or "inferior" in an evolutionary sense. They are, in fact, highly evolved: For 3.5 billion years, prokaryotic populations have responded successfully to many types of environmental challenges.

Genetic Recombination

Although new mutations are a major source of variation in prokaryotic populations, additional diversity arises from *genetic recombination*, the combining of DNA from two sources. In eukaryotes, the sexual processes of meiosis and fertilisation combine DNA from two individuals in a single zygote. But meiosis and fertilisation do not occur in prokaryotes. Instead, three other mechanisms—transformation, transduction, and conjugation—can bring together prokaryotic DNA from different individuals (that is, different cells). When the individuals are members of different species, this movement of genes from one organism to another is called *horizontal gene transfer*. Scientists have found evidence that each of these mechanisms can transfer DNA within and between species in both domain Bacteria and domain Archaea. To date, however, most of our knowledge comes from research on bacteria.

Transformation and Transduction

In **transformation**, the genotype and possibly phenotype of a prokaryotic cell are altered by the uptake of foreign DNA from its surroundings. For example, a harmless strain of *Streptococcus pneumoniae* can be transformed into pneumonia-causing cells if the cells are exposed to DNA from a pathogenic strain (see Figure 16.2). This transformation occurs when a nonpathogenic cell takes up a piece of DNA carrying the allele for pathogenicity and replaces its own allele with the foreign allele, an exchange of homologous DNA segments. The cell is now a recombinant: Its chromosome contains DNA derived from two different cells.

For many years after transformation was discovered in laboratory cultures, most biologists thought it was too rare and haphazard to play an important role in natural bacterial populations. But researchers have since learned that many bacteria have cell-surface proteins that recognise DNA from closely related species and transport it into the cell. Once inside the cell, the foreign DNA can be incorporated into the genome by homologous DNA exchange.

In **transduction**, phages (short for "bacteriophages," the viruses that infect bacteria) carry prokaryotic genes from one host cell to another. In most cases, transduction

▼ **Figure 27.11 Transduction.** Phages may carry pieces of a bacterial chromosome from one cell (the donor) to another (the recipient). If crossing over occurs after the transfer, genes from the donor may be incorporated into the recipient's genome.

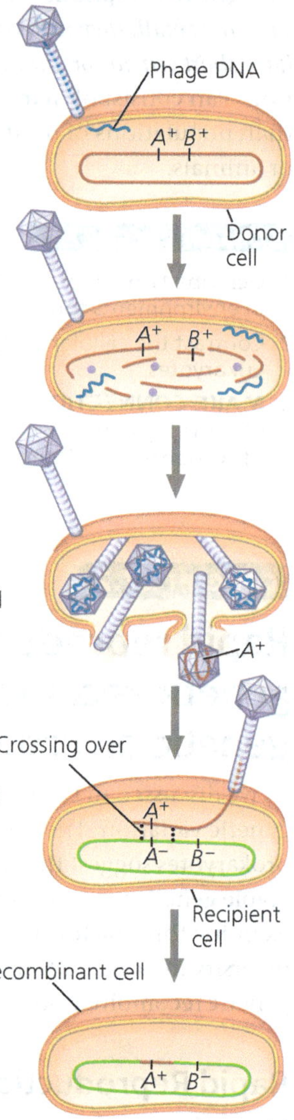

1. A phage infects a bacterial cell that carries the A^+ and B^+ alleles on its chromosome (brown). This bacterium will be the "donor" cell.
2. The phage DNA is replicated, and the cell makes many copies of phage proteins (represented as purple dots). Certain phage proteins halt the synthesis of proteins encoded by the host cell's DNA, and the host cell's DNA may be fragmented, as shown here.
3. As new phage particles assemble, a fragment of bacterial DNA carrying the A^+ allele happens to be packaged in a phage capsid.
4. The phage carrying the A^+ allele from the donor cell infects a recipient cell with alleles A^- and B^-. Crossing over at two sites (dotted lines) allows donor DNA (brown) to be incorporated into recipient DNA (green).
5. The genotype of the resulting recombinant cell (A^+B^-) differs from the genotypes of both the donor (A^+B^+) and the recipient (A^-B^-).

Phage DNA
Donor cell
Crossing over
Recipient cell
Recombinant cell

VISUAL SKILLS *Based on this diagram, describe the circumstances in which a transduction event would result in horizontal gene transfer.*

results from accidents that occur during the phage replicative cycle **(Figure 27.11)**. A virus that carries prokaryotic DNA may not be able to replicate because it lacks some or all of its own genetic material. However, the virus can attach to another prokaryotic cell (a recipient) and inject prokaryotic DNA acquired from the first cell (the donor). If some of this DNA is then incorporated into the recipient cell's chromosome by crossing over, a recombinant cell is formed.

Conjugation and Plasmids

In a process called **conjugation**, DNA is transferred between two prokaryotic cells (usually of the same species) that are temporarily joined. In bacteria, the DNA transfer is always one-way: One cell donates the DNA, and the other

▼ **Figure 27.12 Bacterial conjugation.** The *E. coli* donor cell (left) extends a pilus that attaches to a recipient cell, a key first step in the transfer of DNA. The pilus is a flexible tube of protein subunits (TEM).

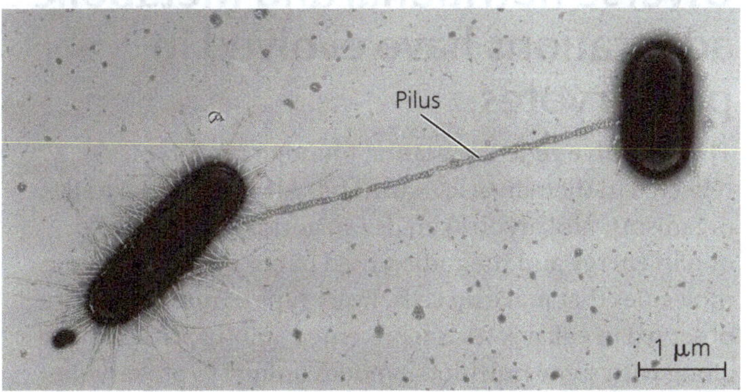

receives it. We'll focus here on the mechanism used by *E. coli*.

First, a pilus of the donor cell attaches to the recipient **(Figure 27.12)**. The pilus then retracts, pulling the two cells together, like a grappling hook. The next step is thought to be the formation of a temporary structure between the two cells, a "mating bridge" through which the donor may transfer DNA to the recipient. However, the mechanism by which DNA transfer occurs is unclear; indeed, recent evidence indicates that DNA may pass directly through the hollow pilus.

The ability to form pili and donate DNA during conjugation results from the presence of a particular piece of DNA called the **F factor** (F for fertility). The F factor

▼ **Figure 27.13 Conjugation and recombination in *E. coli*.** The DNA replication that accompanies transfer of an F plasmid or part of an Hfr bacterial chromosome is called *rolling circle replication*. In effect, the intact circular parental DNA strand "rolls" as its other strand peels off and a new complementary strand is synthesised.

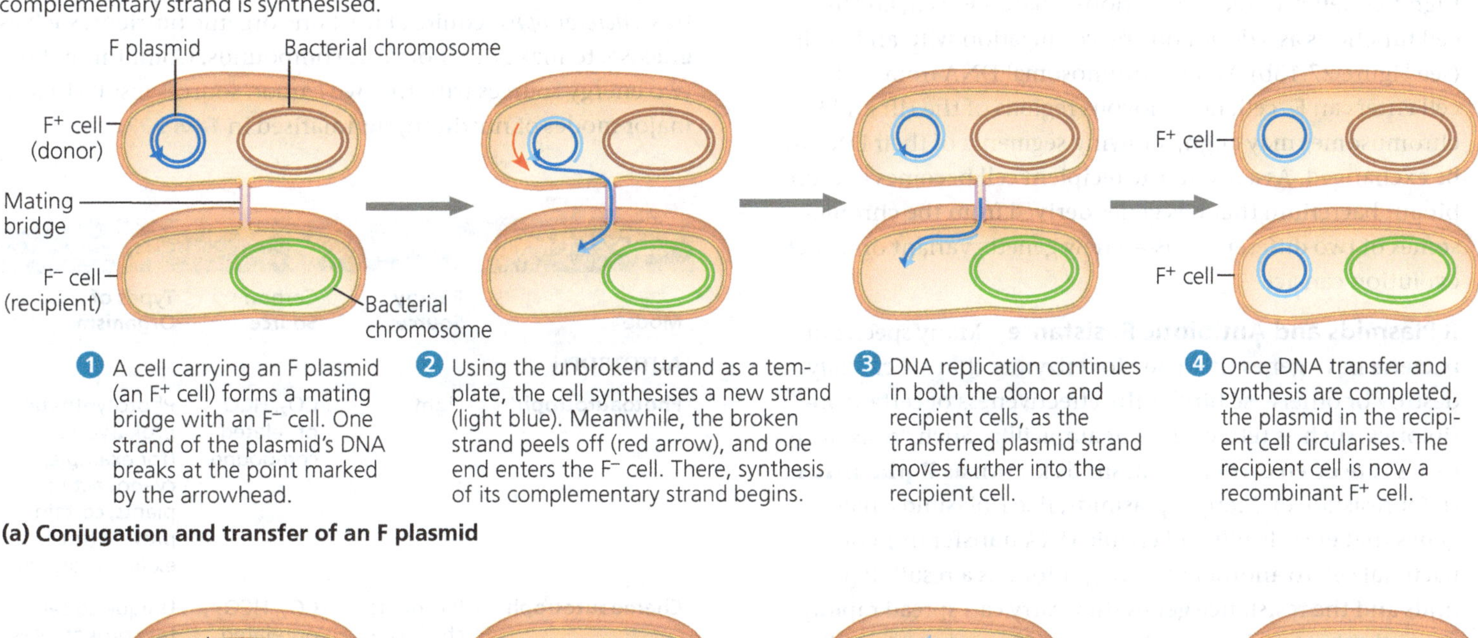

1. A cell carrying an F plasmid (an F^+ cell) forms a mating bridge with an F^- cell. One strand of the plasmid's DNA breaks at the point marked by the arrowhead.
2. Using the unbroken strand as a template, the cell synthesises a new strand (light blue). Meanwhile, the broken strand peels off (red arrow), and one end enters the F^- cell. There, synthesis of its complementary strand begins.
3. DNA replication continues in both the donor and recipient cells, as the transferred plasmid strand moves further into the recipient cell.
4. Once DNA transfer and synthesis are completed, the plasmid in the recipient cell circularises. The recipient cell is now a recombinant F^+ cell.

(a) Conjugation and transfer of an F plasmid

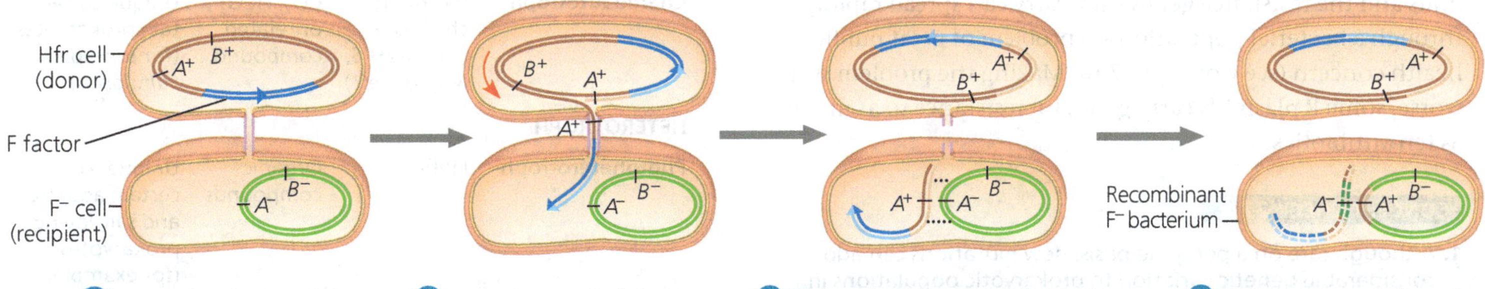

1. In an Hfr cell, the F factor (dark blue) is integrated into the bacterial chromosome. Since an Hfr cell has all of the F factor genes, it can form a mating bridge with an F^- cell and transfer DNA.
2. A single strand of the F factor breaks and begins to move through the bridge. DNA replication occurs in both donor and recipient cells, resulting in double-stranded DNA (daughter strands shown in lighter colour).
3. The mating bridge usually breaks before the entire chromosome is transferred. Crossing over at two sites (dotted lines) can result in the exchange of homologous genes (here, A^+ and A^-) between the transferred DNA (brown) and the recipient's chromosome (green).
4. Cellular enzymes degrade any linear DNA not incorporated into the chromosome. The recipient cell, with a new combination of genes but no F factor, is now a recombinant F^- cell.

(b) Conjugation and transfer of part of an Hfr bacterial chromosome, resulting in recombination. A^+/A^- and B^+/B^- indicate alleles for gene *A* and gene *B*, respectively.

of *E. coli* consists of about 25 genes, most required for the production of pili. As shown in **Figure 27.13**, the F factor can exist either as a plasmid or as a segment of DNA within the bacterial chromosome.

The F Factor as a Plasmid The F factor in its plasmid form is called the **F plasmid**. Cells containing the F plasmid, designated F^+ cells, function as DNA donors during conjugation (see Figure 27.13a). Cells lacking the F factor, designated F^-, function as DNA recipients during conjugation. The F^+ condition is transferable in the sense that an F^+ cell converts an F^- cell to F^+ if a copy of the entire F plasmid is transferred. Even if this does not occur, as long as some of the F plasmid's DNA is transferred successfully to the recipient cell, that cell is now a recombinant cell.

The F Factor in the Chromosome Chromosomal genes can be transferred during conjugation when the donor cell's F factor is integrated into the chromosome. A cell with the F factor built into its chromosome is called an *Hfr cell* (for high frequency of recombination). Like an F^+ cell, an Hfr cell functions as a donor during conjugation with an F^- cell (see Figure 27.13b). When chromosomal DNA from an Hfr cell enters an F^- cell, homologous regions of the Hfr and F^- chromosomes may align, allowing segments of their DNA to be exchanged. As a result, the recipient cell becomes a recombinant bacterium that has genes derived from the chromosomes of two different cells—a new genetic variant on which evolution can act.

R Plasmids and Antibiotic Resistance Many species of bacteria have genes that code for enzymes that specifically destroy or otherwise hinder the effectiveness of certain antibiotics, such as tetracycline or ampicillin. Such "resistance genes" are often carried by plasmids known as **R plasmids** (R for resistance). Many R plasmids, like F plasmids, have genes that encode pili and enable DNA transfer from one bacterial cell to another by conjugation. As a result, R plasmids and the resistance genes they carry can spread rapidly through a bacterial population—a problem of great public health concern (see Concept 27.6). Making the problem still worse, some R plasmids carry genes for resistance to as many as ten antibiotics.

CONCEPT CHECK 27.2

1. Although rare on a per gene basis, new mutations can add considerable genetic variation to prokaryotic populations in each generation. Explain how this occurs.
2. Distinguish between the three mechanisms by which bacteria can transfer DNA from one bacterial cell to another.
3. **WHAT IF?** If a nonpathogenic bacterium were to acquire resistance to antibiotics, could this strain pose a health risk to people? In general, how does DNA transfer among bacteria affect the spread of resistance genes?

For suggested answers, see Appendix A.

CONCEPT 27.3

Diverse nutritional and metabolic adaptations have evolved in prokaryotes

The extensive genetic variation found in prokaryotes is reflected in their diverse nutritional adaptations. Like all organisms, prokaryotes can be categorised by how they obtain energy and the carbon used in building the organic molecules that make up cells. Every type of nutrition observed in eukaryotes is represented among prokaryotes, along with some nutritional modes unique to prokaryotes. In fact, prokaryotes have an astounding range of metabolic adaptations, much broader than that found in eukaryotes.

Organisms that obtain energy from light are called *phototrophs*, and those that obtain energy from chemicals are called *chemotrophs*. Organisms that need only CO_2 or related compounds as a carbon source are called *autotrophs*. In contrast, *heterotrophs* require at least one organic nutrient, such as glucose, to make other organic compounds. Combining the two energy sources with the two carbon sources results in four major modes of nutrition, summarised in **Table 27.1**.

Table 27.1 Major Nutritional Modes

Mode	Energy Source	Carbon Source	Types of Organisms
AUTOTROPH			
Photoautotroph	Light	CO_2, HCO_3^-, or related compound	Photosynthetic prokaryotes (for example, cyanobacteria); plants; certain protists (for example, algae)
Chemoautotroph	Inorganic chemicals (such as H_2S, NH_3, or Fe^{2+})	CO_2, HCO_3^-, or related compound	Unique to certain prokaryotes (for example, *Sulfolobus*)
HETEROTROPH			
Photoheterotroph	Light	Organic compounds	Unique to certain aquatic and salt-loving prokaryotes (for example, *Rhodobacter, Chloroflexus*)
Chemoheterotroph	Organic compounds	Organic compounds	Many prokaryotes (for example, *Clostridium*) and protists; fungi; animals; some plants

The Role of Oxygen in Metabolism

Prokaryotic metabolism also varies with respect to oxygen (O_2). **Obligate aerobes** must use O_2 for cellular respiration and cannot grow without it. **Obligate anaerobes**, on the other hand, are poisoned by O_2. Some obligate anaerobes live exclusively by fermentation; others extract chemical energy by **anaerobic respiration**, in which substances other than O_2, such as nitrate ions (NO_3^-) or sulfate ions (SO_4^{2-}), accept electrons at the "downhill" end of electron transport chains. **Facultative anaerobes** use O_2 if it is present but can also carry out fermentation or anaerobic respiration in an anaerobic environment.

Nitrogen Metabolism

Nitrogen is essential for the production of amino acids and nucleic acids in all organisms. Whereas eukaryotes can obtain nitrogen only from a limited group of nitrogen compounds, prokaryotes can metabolise nitrogen in many forms. For example, some cyanobacteria and some methanogens (various archaea) convert atmospheric nitrogen (N_2) to ammonia (NH_3), a process called **nitrogen fixation**. The cells can then incorporate this "fixed" nitrogen into amino acids and other organic molecules. In terms of nutrition, nitrogen-fixing cyanobacteria are some of the most self-sufficient organisms since they need only light, CO_2, N_2, water, and some minerals to grow.

Nitrogen fixation has a large impact on other organisms. For example, nitrogen-fixing prokaryotes can increase the nitrogen available to plants, which cannot use atmospheric nitrogen but can use the nitrogen compounds that the prokaryotes produce from ammonia. Concept 55.4 discusses this and other essential roles that prokaryotes play in the nitrogen cycles of ecosystems.

Metabolic Cooperation

Cooperation between prokaryotic cells allows them to use environmental resources they could not use as individual cells. In some cases, this cooperation takes place between specialised cells of a filament. For instance, the cyanobacterium *Anabaena* has genes that encode proteins for photosynthesis and for nitrogen fixation. However, a single cell cannot carry out both processes at the same time because photosynthesis produces O_2, which inactivates the enzymes involved in nitrogen fixation. Instead of living as isolated cells, *Anabaena* forms filamentous chains **(Figure 27.14)**. Most cells in a filament carry out only photosynthesis, while a few specialised cells called **heterocysts** (sometimes called *heterocytes*) carry out only nitrogen fixation. Each heterocyst is surrounded by a thickened cell wall that restricts entry of O_2 produced by neigbouring photosynthetic cells. Intercellular connections allow heterocysts to transport fixed nitrogen to neighbouring cells and to receive carbohydrates.

Metabolic cooperation among the cells of one or more prokaryotic species often occurs in surface-coating colonies known as **biofilms**. Cells in a biofilm secrete signalling molecules that recruit nearby cells, causing the colonies to grow.

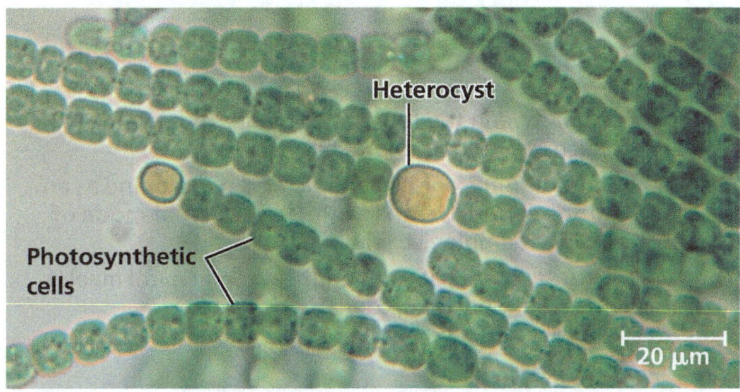

▲ **Figure 27.14 Metabolic cooperation in a prokaryote.** In the filamentous freshwater cyanobacterium *Anabaena*, heterocysts fix nitrogen, while the other cells carry out photosynthesis (LM).

The cells also produce polysaccharides and proteins that stick the cells to the substrate and to one another; these polysaccharides and proteins form the capsule, or slime layer, mentioned earlier in the chapter. Channels in the biofilm allow nutrients to reach cells in the interior and wastes to be expelled.

Biofilms are common in nature, but they can cause a wide range of problems. For example, biofilms growing in pipes slow the flow of liquids such as water or oil and degrade the pipes themselves. Biofilms also corrode boat hulls, oil platforms, and many other structures and industrial products. In medical settings, biofilms can contaminate contact lenses, catheters, pacemakers, heart valves, artificial joints, and many other devices; they have recently been found to adhere to marine microplastics, causing them to sink and accumulate in the sediment on the seafloor. They also contribute to tooth decay and a wide range of chronic infections **(Figure 27.15)**, some of which can be lethal. Many biofilms can evade host immune responses and are extremely resistant to antibiotics, making them hard to treat. Altogether, each year biofilms cause billions of dollars in damage and inflict tens of millions of people with chronic infections.

▼ **Figure 27.15 *Pseudomonas aeruginosa* forming a biofilm (colourised SEM).** This widespread bacterial species can be found in soil and water and as part of the normal community in the human intestines. *P. aeruginosa* can cause serious infections of other human organs, including the skin, urinary tract, and lungs. These infections can be difficult to treat, in part because the bacteria can grow as biofilms that have increased resistance to antibiotics.

CONCEPT CHECK 27.3

1. Distinguish between the four major modes of nutrition, noting which are unique to prokaryotes.
2. A bacterium requires only the amino acid methionine as an organic nutrient and lives in lightless caves. What mode of nutrition does it employ? Explain.
3. **WHAT IF?** Describe what you might eat for a typical meal if humans, like cyanobacteria, could fix nitrogen.

For suggested answers, see Appendix A.

CONCEPT 27.4

Prokaryotes have radiated into a diverse set of lineages

Since their origin 3.5 billion years ago, prokaryotic populations have radiated extensively as a wide range of structural and metabolic adaptations have evolved in them. Collectively, these adaptations have enabled prokaryotes to inhabit every environment known to support life—if there are organisms in a particular place, some of those organisms are prokaryotes. Yet despite their obvious success, it is only in recent decades that advances in genomics have begun to reveal the full extent of prokaryotic diversity.

An Overview of Prokaryotic Diversity

In the 1970s, microbiologists began using small-subunit ribosomal RNA as a marker for evolutionary relationships. Their results indicated that many prokaryotes once classified as bacteria are actually more closely related to eukaryotes and belong in a domain of their own: Archaea. Microbiologists have since analysed larger amounts of genetic data—including more than 1,700 entire genomes—and have concluded that a few traditional groups, such as cyanobacteria, are monophyletic. However, other traditional groups, such as gram-negative bacteria, are scattered throughout several lineages. **Figure 27.16** shows one phylogenetic hypothesis for some of the major taxa of prokaryotes based on molecular systematics.

▼ Figure 27.16 A simplified phylogeny of prokaryotes. This tree shows relationships among major prokaryotic groups; some of these relationships are shown as polytomies to reflect their uncertain order of divergence. Recent studies indicate that within Archaea, the thaumarchaeotes, aigarchaeotes, crenarchaeotes, and korarchaeotes are closely related; systematists have placed them in a supergroup called "TACK" in reference to the first letters of their names.

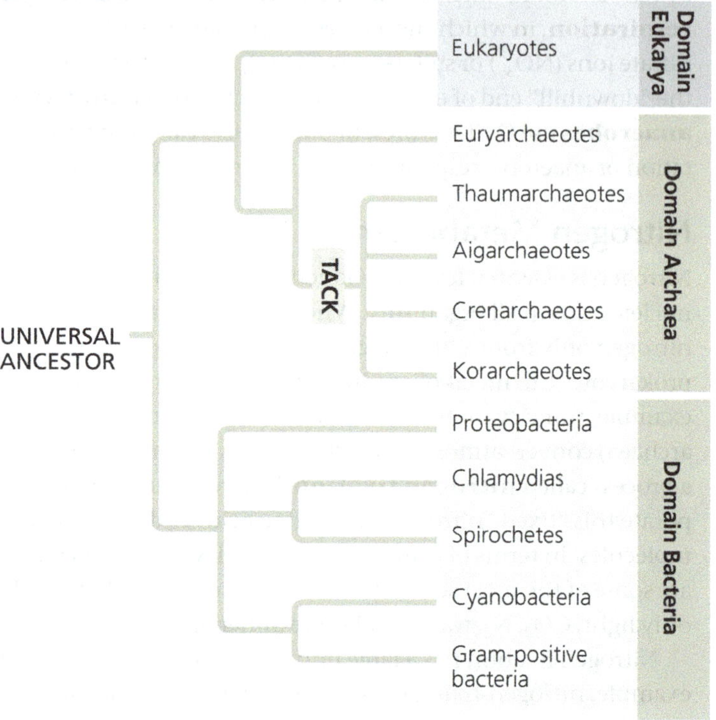

VISUAL SKILLS *Based on this phylogenetic tree diagram, which domain is the sister group of Archaea?*

One lesson from studying prokaryotic phylogeny is that the genetic diversity of prokaryotes is immense. When researchers began to sequence the genes of prokaryotes, they could investigate only the small fraction of species that could be cultured in the laboratory. In the 1980s, researchers began using the polymerase chain reaction (PCR; see Figure 20.8) to analyse the genes of prokaryotes collected from the environment (such as from soil or water samples). Such "genetic prospecting" is now widely used; in fact, today entire prokaryotic genomes can be obtained from environmental samples using *metagenomics* (see Concept 21.1). Each year these techniques add new branches to the tree of life. While only about 16,000 prokaryotic species worldwide have been assigned scientific names, a single handful of soil could contain 10,000 prokaryotic species by some estimates. Taking full stock of this diversity will require many years of research.

Another important lesson from molecular systematics is that horizontal gene transfer has played a key role in the evolution of prokaryotes. Over hundreds of millions of years, prokaryotes have acquired genes from even distantly related species, and they continue to do so today. As a result, significant portions of the genomes of many prokaryotes are actually mosaics of genes imported from other species. For example, a study of 329 sequenced bacterial genomes found that an average of 75% of the genes in each genome had been transferred horizontally at some point in their evolutionary history. As we saw in Concept 26.6, such gene transfers can make it difficult to determine phylogenetic relationships. Still, it is clear that for billions of years, the prokaryotes have evolved in two separate lineages, the bacteria and the archaea (see Figure 27.16).

Bacteria

As surveyed in **Figure 27.17**, bacteria include the vast majority of prokaryotic species familiar to most people, from the pathogenic species that cause strep throat and tuberculosis to the beneficial species used to make Swiss cheese and yoghurt.

▼ Figure 27.17 Exploring Bacterial Diversity

This figure highlights a few key groups of bacteria, but their actual diversity is far greater than shown here. Likewise, gene sequence data indicate that the 16,000 known species of bacteria represent a small fraction of the actual number, estimated at 700,000–1.4 million species.

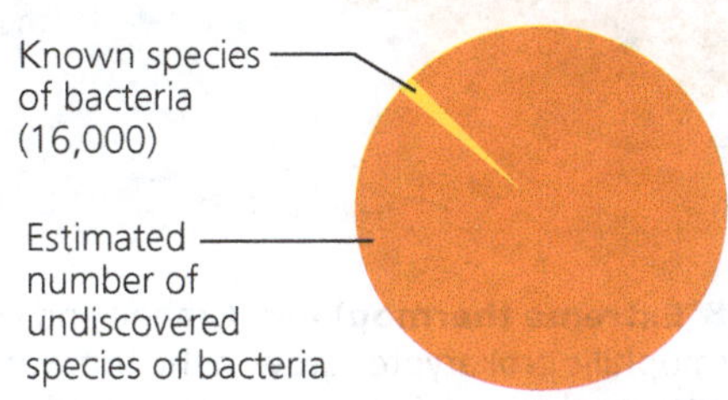

Proteobacteria

This large and diverse clade of gram-negative bacteria includes photoautotrophs, chemoautotrophs, and heterotrophs. One autotroph, the sulfur bacterium *Thiomargarita namibiensis*, obtains energy by oxidising H_2S, producing sulfur as a waste product (the small globules in the photograph below). Heterotrophs include pathogens such as *Neisseria gonorrhoeae*, which causes gonorrhea; *Vibrio cholerae*, which causes cholera; and *Helicobacter pylori*, which causes stomach ulcers. Current evidence indicates that mitochondria evolved by endosymbiosis from a heterotroph in the subgroup alpha proteobacteria.

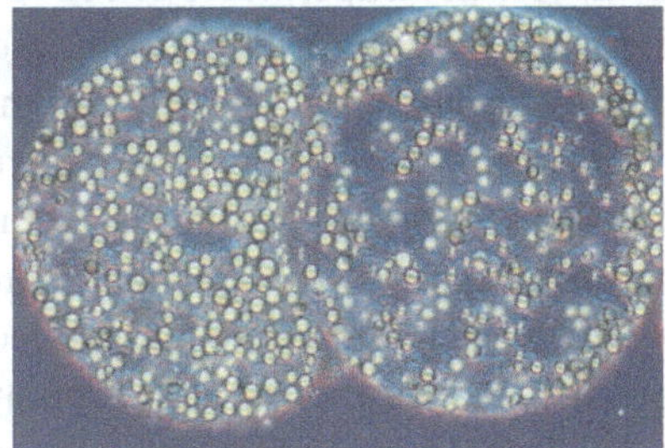

Thiomargarita namibiensis containing sulfur wastes (LM)

Chlamydias

These parasites can survive only within animal cells, depending on their hosts for resources as basic as ATP. The gram-negative walls of chlamydias are unusual in that they lack peptidoglycan. One species, *Chlamydia trachomatis*, is the most common cause of blindness in the world and also causes nongonococcal urethritis, one of the most common sexually transmitted infections in Australia and New Zealand.

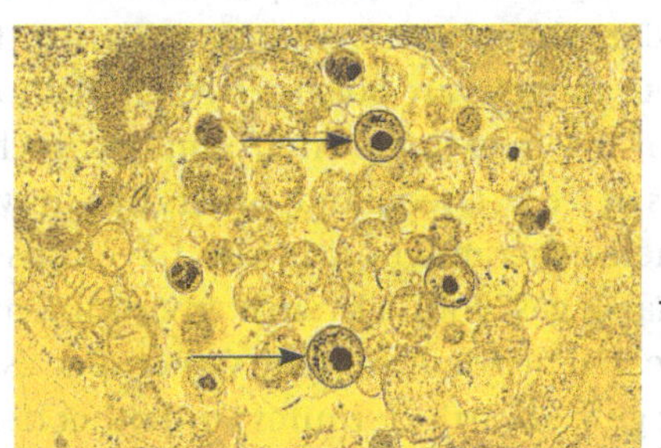

Chlamydia (arrows inside an animal cell (colourised TEM)

Spirochetes

These helical gram-negative heterotrophs spiral through their environment by means of rotating, internal, flagellum-like filaments. Many spirochetes are free-living, but others are notorious pathogenic parasites: *Treponema pallidum* causes syphilis, and *Borrelia burgdorferi* causes Lyme disease.

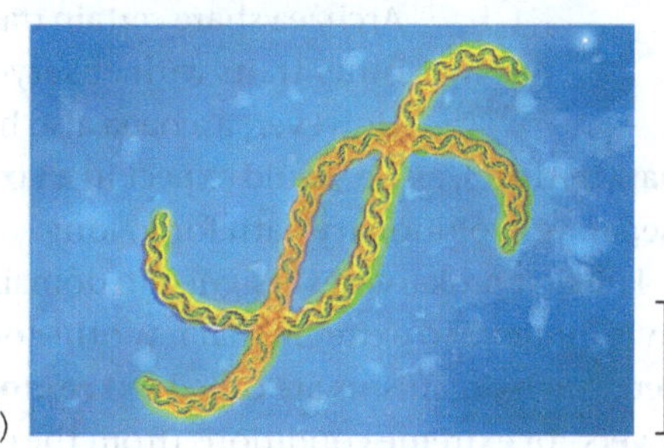

Leptospira, a spirochete (colourised TEM)

Cyanobacteria

These gram-negative photoautotrophs are the only prokaryotes with plantlike, oxygen-generating photosynthesis. (In fact, chloroplasts are thought to have evolved from an endosymbiotic cyanobacterium.) Both solitary and filamentous cyanobacteria are abundant components of freshwater and marine *phytoplankton*, the collection of small photosynthetic organisms that drift near the water's surface.

Cylindrospermum, a filamentous cyanobacterium

Gram-Positive Bacteria

Gram-positive bacteria rival the proteobacteria in diversity. Species in one subgroup, the actinomycetes, form colonies containing branched chains of cells; two of these species cause tuberculosis and leprosy, but most are decomposers living in soil. Soil-dwelling species in the genus *Streptomyces* are cultured as a source of antibiotics, including tetracycline and erythromycin. Other subgroups of gram-positive bacteria include pathogens such as *Staphylococcus aureus* (see Figure 22.15), *Bacillus anthracis*, which causes anthrax, and *Clostridium botulinum*, which causes botulism.

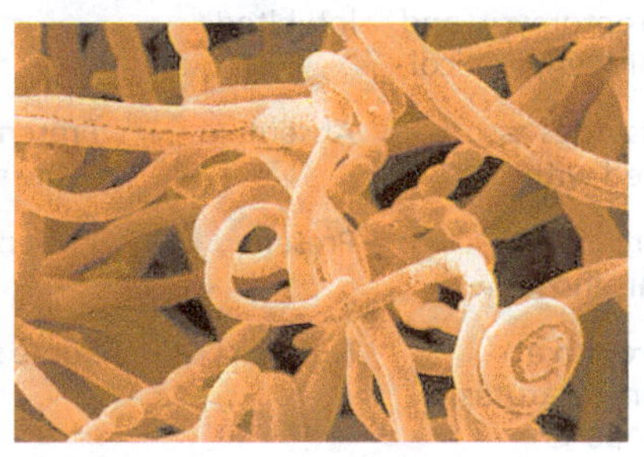

Streptomyces, the source of many antibiotics (SEM)

Every major mode of nutrition and metabolism is represented among the 80 or so phyla of bacteria, and even a small group of bacteria may contain species exhibiting many different nutritional modes. As we'll see, the diverse nutritional and metabolic capabilities of bacteria—and archaea—are behind the great impact these organisms have on Earth and its life.

Archaea

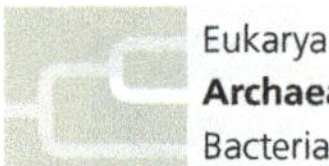

Archaea share certain traits with bacteria and other traits with eukaryotes **(Table 27.2)**. However, archaea also have many unique characteristics, as we would expect in a taxon that has followed a separate evolutionary path for so long.

The first prokaryotes assigned to domain Archaea live in environments so extreme that few other organisms can survive there. Such organisms are called **extremophiles**, meaning "lovers" of extreme conditions (from the Greek *philos*, lover), and include extreme halophiles and extreme thermophiles.

Extreme halophiles (from the Greek *halo*, salt) live in highly saline environments, such as the Great Salt Lake in Utah, the Dead Sea in Israel, and Lake Bumbunga in South Australia (shown in Figure 27.1). Some species merely tolerate salinity, while others require an environment that is several times saltier than seawater (which has a salinity of 3.5%). For example, the proteins and cell wall of *Halobacterium* have unusual features that improve function in extremely salty environments but render these organisms incapable of survival if the salinity drops below 9%.

Table 27.2 A Comparison of the Three Domains of Life

	DOMAIN		
CHARACTERISTIC	**Bacteria**	**Archaea**	**Eukarya**
Nuclear envelope	Absent	Absent	Present
Membrane-enclosed organelles	Absent	Absent	Present
Peptidoglycan in cell wall	Present	Absent	Absent
Membrane lipids	Unbranched hydrocarbons	Some branched hydrocarbons	Unbranched hydrocarbons
RNA polymerase	One kind	Several kinds	Several kinds
Initiator amino acid for protein synthesis	Formyl-methionine	Methionine	Methionine
Introns in genes	Very rare	Present in some genes	Present in many genes
Response to the antibiotics streptomycin and chloramphenicol	Growth usually inhibited	Growth not inhibited	Growth not inhibited
Histones associated with DNA	Absent	Present in some species	Present
Circular chromosome	Present	Present	Absent
Growth at temperatures > 100°C	No	Some species	No

▲ **Figure 27.18 Extreme thermophiles.** Orange and yellow colonies of thermophilic prokaryotes grow in the hot water of Wai-O-Tapu geothermal area in Rotorua, New Zealand.

MAKE CONNECTIONS *How might the enzymes of thermophiles differ from those of other organisms? (Review enzymes in Concept 8.4.)*

Extreme thermophiles (from the Greek *thermos*, hot) thrive in very hot environments. For example, the archaean *Pyrococcus furiosus* lives in geothermally heated marine sediments that reach temperatures of 100°C. At temperatures this high, the cells of most organisms die because their DNA does not remain in a double helix and many of their proteins denature. *P. furiosus* and other extreme thermophiles avoid this fate because they have structural and biochemical adaptations that make their DNA and proteins stable at high temperatures—a feature that has enabled *P. furiosus* to be used in biotechnology as a source of DNA polymerase for the PCR technique (see Figure 20.8). Other examples of extreme thermophiles include archaea in the genus *Sulfolobus*, which live in sulfur-rich volcanic springs **(Figure 27.18)**. One extreme thermophile that lives near deep-sea hot springs called *hydrothermal vents* is informally known as "strain 121" since it can reproduce even at 121°C.

Many other archaea live in more moderate environments. Consider **methanogens**, archaea that release methane as a by-product of how they obtain energy. Many methanogens use CO_2 to oxidise H_2, a process that produces both energy and methane waste. Among the strictest of anaerobes, methanogens are poisoned by O_2. Although some methanogens live in extreme environments, such as under kilometres of ice in Greenland, others live in swamps and marshes where other microorganisms have consumed all the O_2. The "marsh gas" found in such environments is the methane released by these archaea. Other species inhabit the anaerobic guts of cattle, termites, and other herbivores, playing an essential role in the nutrition of these animals. Methanogens are also useful to humans as decomposers in sewage treatment facilities.

Many extreme halophiles and most known methanogens are archaea in the clade Euryarchaeota (from the Greek *eurys*, broad, a reference to their wide habitat range). The euryarchaeotes also include some extreme thermophiles, though most thermophilic species belong to a second clade, Crenarchaeota (*cren* means "spring," such as a hydrothermal spring). Metagenomic studies have identified many species of euryarchaeotes and crenarchaeotes that are not extremophiles. These archaea exist in habitats ranging from farm soils to lake sediments to the surface of the open ocean.

New findings continue to inform our understanding of archaeal phylogeny. For example, recent metagenomic studies have uncovered the genomes of many species that are not members of Euryarchaeota or Crenarchaeota. Moreover, phylogenomic analyses show that three of these newly discovered groups—the Thaumarchaeota, Aigarchaeota, and Korarchaeota—are more closely related to the Crenarchaeota than they are to the Euryarchaeota. These findings have led to the identification of a "supergroup" that contains the Thaumarchaeota, Aigarchaeota, Crenarchaeota, and Korarchaeota (see Figure 27.16). This supergroup is referred to as "TACK" based on the names of the groups it includes.

The importance of the TACK supergroup was highlighted by the recent discovery of the lokiarchaeotes, a group that is closely related to TACK archaea and that may represent the long-sought-after sister group of the eukaryotes. As such, the characteristics of lokiarchaeotes may shed light on one of the major puzzles of biology today—how eukaryotes arose from their prokaryotic ancestors. The pace of these and other recent discoveries suggests that as metagenomic prospecting continues, the tree in Figure 27.16 will likely undergo further changes.

CONCEPT CHECK 27.4

1. Explain how molecular systematics and metagenomics have contributed to our understanding of the phylogeny and evolution of prokaryotes.
2. **DRAW IT** Redraw Figure 27.16 assuming that eukaryotes are more closely related to TACK archaea than to euryarchaeotes. Under this assumption, would the three domains shown in Figure 27.16 be valid? Explain.

For suggested answers, see Appendix A.

CONCEPT 27.5

Prokaryotes play crucial roles in the biosphere

If people were to disappear from the planet tomorrow, life on Earth would change for many species, but few would be driven to extinction. In contrast, prokaryotes are so important to the biosphere that if they were to disappear, the prospects of survival for many other species would be dim.

Chemical Recycling

The atoms that make up the organic molecules in all living things were at one time part of inorganic substances in the soil, air, and water. Sooner or later, those atoms will return to the nonliving environment. Ecosystems depend on the continual recycling of chemical elements between the living and nonliving components of the environment, and prokaryotes play a major role in this process. For example, some chemoheterotrophic prokaryotes function as **decomposers**, breaking down dead organisms as well as waste products and thereby unlocking supplies of carbon, nitrogen, and other elements. Without the actions of prokaryotes and other decomposers such as fungi, life as we know it would cease. (See Concept 55.4 for a detailed discussion of chemical cycles.)

Prokaryotes also convert some molecules to forms that can be taken up by other organisms. Cyanobacteria and other autotrophic prokaryotes use CO_2 to make organic compounds such as sugars, which are then passed up through food chains. Cyanobacteria also produce atmospheric O_2, and a variety of prokaryotes fix atmospheric nitrogen (N_2) into forms that other organisms can use to make the building blocks of proteins and nucleic acids. Under some conditions, prokaryotes can increase the availability of nutrients that plants require for growth, such as nitrogen, phosphorus, and potassium **(Figure 27.19)**. Prokaryotes can also *decrease* the availability of key plant

▼ Figure 27.19 Impact of bacteria on soil nutrient availability. Pine seedlings grown in sterile soils to which one of three strains of the bacterium *Burkholderia glathei* had been added absorbed more potassium (K^+) than did seedlings grown in soil without any bacteria. Other results (not shown) demonstrated that strain 3 increased the amount of K^+ released from mineral crystals to the soil.

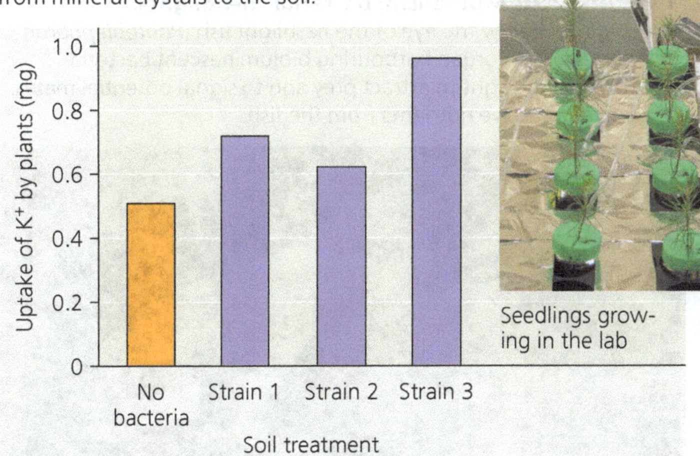

WHAT IF? *Estimate the average uptake of K^+ for seedlings in soils with bacteria. What would you expect this average to be if bacteria had no effect on nutrient availability?*

nutrients; this occurs when prokaryotes "immobilise" nutrients by using them to synthesise molecules that remain within their cells. Thus, prokaryotes can have complex effects on soil nutrient concentrations. In marine environments, an archaean from the clade Crenarchaeota can perform nitrification, a key step in the nitrogen cycle (see Figure 55.13). Crenarchaeotes dominate the oceans by numbers, comprising an estimated 10^{28} cells. The sheer abundance of these organisms suggests that they may have a large impact on the global nitrogen cycle.

Ecological Interactions

Prokaryotes play a central role in many ecological interactions. Consider **symbiosis** (from a Greek word meaning "living together"), an ecological relationship in which two species live in close contact with each other. Prokaryotes often form symbiotic associations with much larger organisms. In general, the larger organism in a symbiotic relationship is known as the **host**, and the smaller is known as the **symbiont**. There are many cases in which a prokaryote and its host participate in **mutualism**, an ecological interaction between two species in which both benefit **(Figure 27.20)**. Other interactions take the form of **commensalism**, an ecological relationship in which one species benefits while the other is not harmed or helped in any significant way. For example, more than 150 bacterial species live on the outer surface of your body, covering portions of your skin with up to 10 million cells per square centimetre. Some of these species are commensalists: You provide them with food, such as the oils that exude from your pores, and a place to live, while they neither harm nor benefit you. Finally, some symbiotic prokaryotes engage in **parasitism**, an ecological relationship in which a **parasite** feeds on the cell contents, tissues, or body fluids of its host.

▼ **Figure 27.20 Mutualism: bacterial "headlights."** The glowing oval below the eye of the flashlight fish (*Photoblepharon palpebratus*) is an organ harbouring bioluminescent bacteria. The fish uses the light to attract prey and to signal potential mates. The bacteria receive nutrients from the fish.

As a group, parasites harm but usually do not kill their host, at least not immediately (unlike a predator). Parasites that cause disease are known as **pathogens**, many of which are prokaryotic. (We'll discuss mutualism, commensalism, and parasitism in greater detail in Concept 54.1.)

The very existence of an ecosystem can depend on prokaryotes. For example, consider the diverse ecological communities found at hydrothermal vents. These communities are densely populated by many different kinds of animals, including worms, clams, crabs, and fish. But since sunlight does not penetrate to the deep ocean floor, the community does not include photosynthetic organisms. Instead, the energy that supports the community is derived from the metabolic activities of chemoautotrophic bacteria. These bacteria harvest chemical energy from compounds such as hydrogen sulfide (H_2S) that are released from the vent. An active hydrothermal vent may support hundreds of eukaryotic species, but when the vent stops releasing chemicals, the chemoautotrophic bacteria cannot survive. As a result, the entire vent community collapses.

CONCEPT CHECK 27.5

1. Explain how prokaryotes, though small, can be considered giants in their collective impact on Earth and its life.
2. **MAKE CONNECTIONS** Review Figure 10.6. Then summarise the main steps by which cyanobacteria produce O_2 and use CO_2 to make organic compounds.

For suggested answers, see Appendix A.

CONCEPT 27.6

Prokaryotes have both beneficial and harmful impacts on humans

Although the best-known prokaryotes tend to be the bacteria that cause human illness, these pathogens represent only a small fraction of prokaryotic species. Many other prokaryotes have positive interactions with people, and some play essential roles in agriculture and industry.

Mutualistic Bacteria

As is true for many other eukaryotes, human well-being can depend on mutualistic prokaryotes. For example, our intestines are home to an estimated 500–1,000 species of bacteria; collectively, their cells outnumber all human cells in the body by a factor of ten. Different species live in different portions of the intestines, and they vary in their ability to process different foods. Many of these species are mutualists, digesting food that our own intestines cannot break down. The genome of one of these gut mutualists, *Bacteroides thetaiotaomicron*, includes a large array of genes involved in synthesising carbohydrates, vitamins, and other nutrients needed by

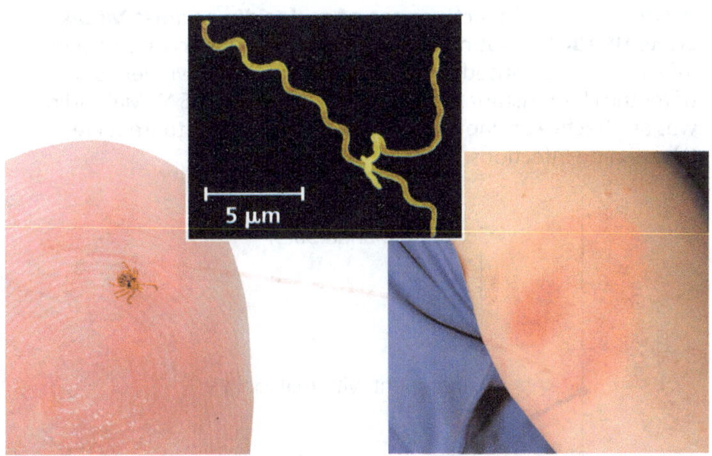

▲ **Figure 27.21 Lyme disease.** Ticks in the genus *Ixodes* spread the disease by transmitting the spirochete *Borrelia burgdorferi* (colourised SEM). A rash may develop at the site of the tick's bite; the rash may be large and ring-shaped (as shown) or much less distinctive.

humans. Signals from the bacterium activate human genes that build the network of intestinal blood vessels necessary to absorb nutrient molecules. Other signals induce human cells to produce antimicrobial compounds to which *B. thetaiotaomicron* is not susceptible. This action may reduce the population sizes of other, competing species, thus potentially benefiting both *B. thetaiotaomicron* and its human host.

Pathogenic Bacteria

All the pathogenic prokaryotes known to date are bacteria, and they deserve their negative reputation. Bacteria cause about half of all human diseases. For example, more than 1.5 million people die each year of the lung disease tuberculosis, caused by *Mycobacterium tuberculosis*. And another 2 million people die each year from diarrheal diseases caused by various bacteria.

Some bacterial diseases are transmitted by other species, such as fleas or ticks. In the United States, the most widespread pest-carried disease is Lyme disease, which infects about 300,000 people each year according to a recent CDC estimate **(Figure 27.21)**. Caused by a bacterium carried by ticks, Lyme disease can result in debilitating arthritis, heart disease, nervous disorders, and death if untreated.

Pathogenic prokaryotes usually cause illness by producing poisons, which are classified as exotoxins or endotoxins. **Exotoxins** are proteins secreted by certain bacteria and other organisms. Cholera, a dangerous diarrhoeal disease, is caused by an exotoxin secreted by the proteobacterium *Vibrio cholerae*. The exotoxin stimulates intestinal cells to release chloride ions into the gut, and water follows by osmosis. In another example, the potentially fatal disease botulism is caused by botulinum toxin, an exotoxin secreted by the gram-positive bacterium *Clostridium botulinum* as it ferments various foods, including improperly canned meat, seafood, and vegetables. Like other exotoxins, the botulinum toxin can produce disease even if the bacteria that manufacture it are no longer present when the food is eaten.

Endotoxins are lipopolysaccharide components of the outer membrane of gram-negative bacteria. In contrast to exotoxins, endotoxins are released only when the bacteria die and their cell walls break down. Endotoxin-producing bacteria include species in the genus *Salmonella*, such as *Salmonella typhi*, which causes typhoid fever. You might have heard of food poisoning caused by other *Salmonella* species that can be found in poultry and some fruits and vegetables.

Finally, horizontal gene transfer can spread genes associated with virulence, turning normally harmless bacteria into potent pathogens. *E. coli*, for instance, is ordinarily a harmless symbiont in the human intestines, but pathogenic strains that cause bloody diarrhoea have emerged. One of the most dangerous strains, O157:H7, is a global threat; in the United States alone, there are around 75,000 cases of O157:H7 infection per year, often from contaminated beef or produce. Scientists have discovered over 1,000 genes in O157:H7 that have no counterpart in harmless *E. coli* strains. Some of the genes found only in O157:H7 are associated with virulence and were likely transferred by phage-mediated horizontal gene transfer (transduction) from pathogenic bacterial species.

Antibiotic Resistance

Since their initial use in the 1940s, antibiotics have saved many lives and reduced the incidence of disease caused by pathogenic bacteria. Unfortunately, as **Figure 27.22** shows,

▼ **Figure 27.22 The rise of antibiotic resistance.** As illustrated by the examples shown here, bacteria have developed resistance to every antibiotic currently in clinical use, often within a few years. It took more than 50 years for resistance to develop to colistin, most likely because this antibiotic has toxic side effects in humans and hence was used rarely (as a treatment of last resort).

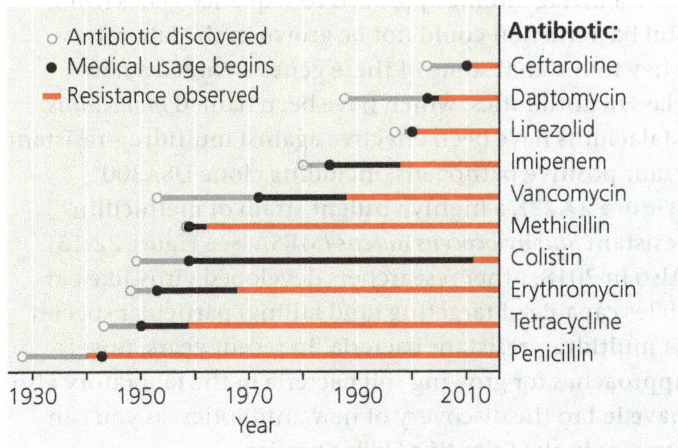

VISUAL SKILLS *Based on this diagram, identify the antibiotic for which resistance appeared before the antibiotic began to be used medicinally.*

resistance to antibiotics has evolved in many bacteria, often rapidly. Compounding this problem, the discovery of new antibiotics has not kept pace with the rate at which resistance has evolved in bacteria. We now face a major public health concern: For every antibiotic now in use, at least one species of bacteria has developed resistance to it.

The rise of antibiotic resistance is driven by several factors. The rapid reproduction of bacteria enables cells carrying resistance genes to quickly produce large numbers of resistant offspring. Thus, when humans use antibiotics in medical or agricultural settings, natural selection can cause the fraction of a bacterial population carrying genes for antibiotic resistance to increase rapidly. Resistance genes also can spread within and among bacterial species by horizontal gene transfer. As a result, bacterial strains that are resistant to multiple antibiotics are becoming more common, making the treatment of certain bacterial infections very difficult.

Consider tuberculosis, the leading cause of death by infectious disease worldwide. The recent rise of drug-resistant strains of the tuberculosis bacterium (*Mycobacterium tuberculosis*) threatens decades of progress in combating tuberculosis. "Superbug" tuberculosis strains came to public notice in 2006, when an outbreak in a rural area of South Africa killed 52 of the 53 patients infected with what is now called extensively drug-resistant tuberculosis (XDR-TB). As of 2017, more than 100 countries had confirmed cases of XDR-TB. There are few treatments for XDR-TB, and those we have are difficult to administer and more toxic than are treatments for standard strains of the tuberculosis bacterium.

M. tuberculosis is not the only bacterium in which resistance to multiple antibiotics has evolved. Highly resistant strains have also been found in species that cause potentially lethal infections of the skin, lungs, digestive tract, and other human organs. Even so, recent discoveries give cause for hope. In 2018, for example, researchers used a metagenomic approach to sequence genes from soil bacteria that could not be grown in the laboratory. They found that some of these genes encoded a new class of antibiotics, which have been named *malacidins*. Malacidins have been effective against multidrug-resistant gram-positive pathogens, including clone USA300 **(Figure 27.23)**, a highly virulent strain of methicillin-resistant *Staphylococcus aureus* (MRSA; see Figure 22.15). Also in 2018, other researchers developed virus-like particles capable of targeting (and killing) particular species of multidrug-resistant bacteria. In recent years, novel approaches for growing soil bacteria in the laboratory also have led to the discovery of new antibiotics, as you can explore in the **Scientific Skills Exercise**.

▼ **Figure 27.23 Effectiveness of malacidin against MRSA clone USA300.** Treatment with malacidin rapidly cleared a skin infection in rats caused by clone USA300, a highly virulent strain of methicillin-resistant *Staphylococcus aureus* (MRSA). Malacidin was as effective as daptomycin, an antibiotic used to treat life-threatening infections.

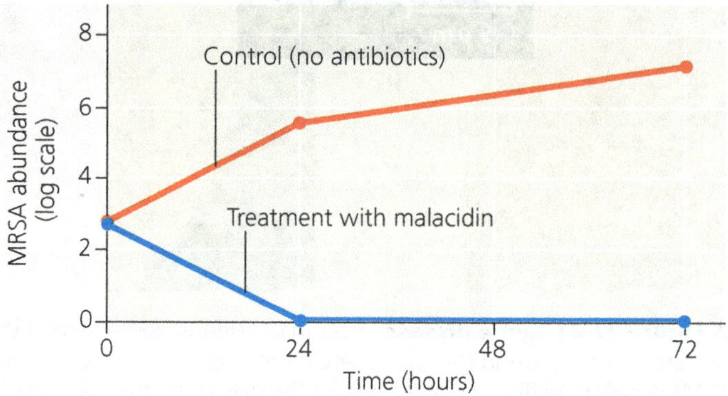

Prokaryotes in Research and Technology

On a positive note, we reap many benefits from the metabolic capabilities of both bacteria and archaea. For example, people have long used bacteria to convert milk to cheese and yoghurt. Bacteria are also used in the production of beer and wine, pepperoni, fermented cabbage (sauerkraut), and soy sauce. In recent decades, our greater understanding of prokaryotes has led to an explosion of new applications in biotechnology. Examples include the use of *E. coli* in gene cloning (see Figure 20.4) and the use of DNA polymerase from *Pyrococcus furiosus* in the PCR technique (see Figure 20.8). Through genetic engineering, we can modify bacteria to produce vitamins, antibiotics, hormones, and other products (see Concept 20.1).

Recently, the prokaryotic CRISPR-Cas system, which helps bacteria and archaea defend against attack by viruses (see Figure 19.9), has been developed into a powerful new tool for altering genes in virtually any organism. The genomes of many prokaryotes contain short DNA repeats, called CRISPRs, that interact with proteins known as the Cas (CRISPR-associated) proteins. Cas proteins, acting together with "guide RNA" made from the CRISPR region, can cut any DNA sequence to which they are directed. Scientists have been able to exploit this system by introducing a Cas protein (Cas9) to guide RNA into cells whose DNA they want to alter (see Figure 17.28). Among other applications, this **CRISPR-Cas9 system** has already opened new lines of research on HIV, the virus that causes AIDS **(Figure 27.24)**. While the CRISPR-Cas9 system can potentially be used in many different ways, care must be taken to guard against the unintended consequences that could arise when applying such a new and powerful technology.

Another valuable application of bacteria is to reduce our use of petroleum. Consider the plastics industry. Globally,

Scientific Skills Exercise

Calculating and Interpreting Means and Standard Errors

Can Antibiotics Obtained from Soil Bacteria Help Fight Drug-Resistant Bacteria? Soil bacteria synthesise antibiotics, which they use against species that attack or compete with them. To date, these species have been inaccessible as sources for new medicines because 99% of soil bacteria cannot be grown using standard laboratory techniques. To address this problem, researchers developed a method in which soil bacteria grow in a simulated version of their natural environment; this led to the discovery of a new antibiotic, teixobactin. In this exercise, you'll calculate means and standard errors from an experiment that tested teixobactin's effectiveness against MRSA (methicillin-resistant *Staphylococcus aureus*; see Figure 22.15).

How the Experiment Was Done Researchers drilled tiny holes into a small plastic chip and filled the holes with a dilute aqueous solution containing soil bacteria and agar. The dilution had been calibrated so that only one bacterium was likely to grow in each hole. After the agar solidified, the chip was placed in a container containing the original soil; nutrients and other essential materials from the soil diffused into the agar, allowing the bacteria to grow.

After isolating teixobactin from a soil bacterium, researchers performed the following experiment: Mice infected with MRSA were given low (1 mg/kg) or high (5 mg/kg) doses of teixobactin or vancomycin, an existing antibiotic; in the control, mice infected with MRSA were not given an antibiotic. After 26 hours, researchers sampled infected mice and estimated the number of *S. aureus* colonies in each sample. Results were reported on a log scale; note that a decrease of 1.0 on this scale reflects a tenfold decrease in MRSA abundance.

Data from the Experiment

Treatment	Dose (mg/kg)	Log of Number of Colonies	Mean ($\bar{x}$)
Control	—	9.0, 9.5, 9.0, 8.9	
Vancomycin	1.0	8.5, 8.4, 8.2	
	5.0	5.3, 5.9, 4.7	
Teixobactin	1.0	8.5, 6.0, 8.4, 6.0	
	5.0	3.8, 4.9, 5.2, 4.9	

▶ **Plastic chip used to grow soil bacteria**

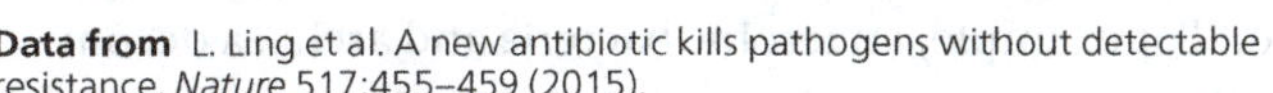
Data from L. Ling et al. A new antibiotic kills pathogens without detectable resistance, *Nature* 517:455–459 (2015).

INTERPRET THE DATA

1. The mean ($\bar{x}$) of a variable is the sum of the data values divided by the number of observations (n):
$$\bar{x} = \frac{\sum x_i}{n}$$
In this formula, x_i is the value of the ith observation of the variable (observation number "i" out of a total of n such observations); the $\sum$ symbol indicates that the n values of x are to be added together. Calculate the mean for each treatment.
2. Use your results from question 1 to evaluate the effectiveness of vancomycin and teixobactin.
3. The variation found in a set of data can be estimated by the standard deviation, s:
$$s = \sqrt{\frac{\sum (x_i - \bar{x})^2}{n - 1}}$$
Calculate the standard deviation for each treatment.
4. The standard error (SE), which indicates how greatly the mean would likely vary if the experiment was repeated, is calculated as:
$$SE = \frac{s}{\sqrt{n}}$$
As a rough rule of thumb, if an experiment were to be repeated, the new mean typically would lie within two standard errors of the original mean (that is, within the range $\bar{x} \pm 2SE$). Calculate $\bar{x} \pm 2SE$ for each treatment, determine whether these ranges overlap, and interpret your results.

▼ **Figure 27.24 CRISPR: opening new avenues of research for treating HIV infection. (a)** In laboratory experiments, untreated (control) human cells were susceptible to infection by HIV, the virus that causes AIDS. **(b)** In contrast, cells treated with a CRISPR-Cas9 system that targets HIV were resistant to viral infection. The CRISPR-Cas9 system was also able to remove HIV proviruses (see Figure 19.10) that had become incorporated into the DNA of human cells.

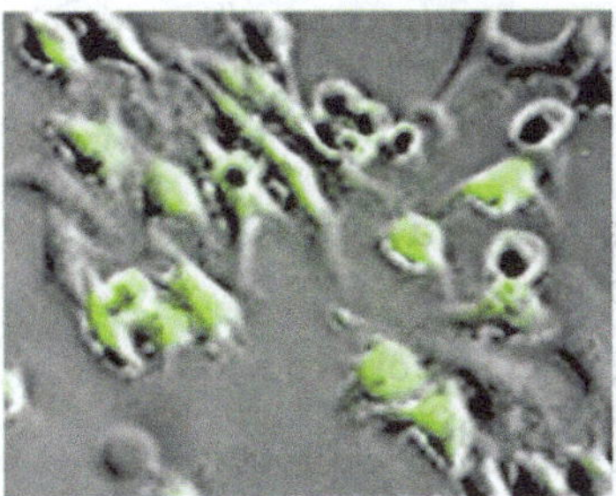

(a) Control cells. The green colour indicates infection by HIV.

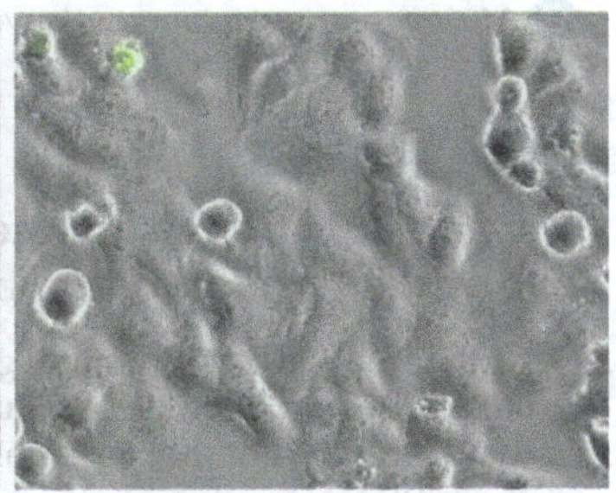

(b) Experimental cells. These cells were treated with a CRISPR-Cas9 system that targets HIV.

each year more than 340 billion kilograms of plastic are produced from petroleum and used to make toys, storage containers, soft drink bottles, and many other items. Many of these products end up as plastic waste that contaminates natural habitats and degrades slowly, creating environmental problems (see Concept 56.4). One promising avenue for addressing these problems is to use bacteria that produce natural plastics **(Figure 27.25)**. For example, some bacteria synthesise a type of polymer known as PHA (polyhydroxyalkanoate), which they use to store chemical energy. The PHA can be extracted, formed into pellets, and used to make durable, yet biodegradable, plastics. Researchers are also seeking to reduce the use of petroleum and other fossil fuels by engineering bacteria that can produce ethanol from various forms of biomass, including agricultural waste, switchgrass, and corn.

Another way to harness prokaryotes is **bioremediation**, the use of organisms to remove pollutants from soil, air, or

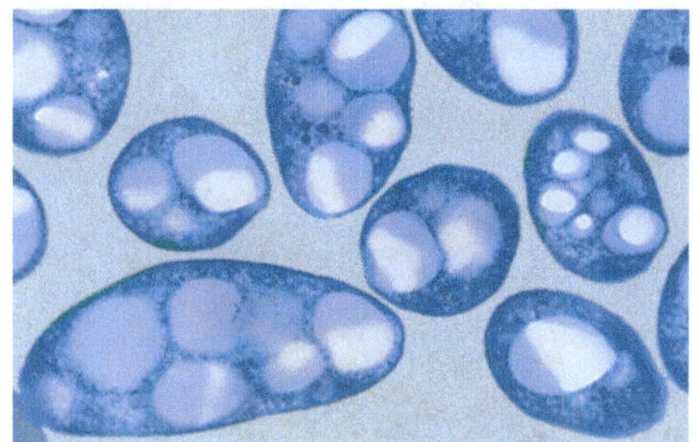

◀ **Figure 27.25 Bacteria synthesising and storing PHA, a component of biodegradable plastics.**

water. For example, anaerobic bacteria and archaea decompose the organic matter in sewage, converting it to material that can be used as landfill or fertiliser after chemical sterilisation. Other bioremediation applications include cleaning up oil spills **(Figure 27.26)** and precipitating radioactive material (such as uranium) out of groundwater.

The usefulness of prokaryotes largely derives from their diverse forms of nutrition and metabolism. All this metabolic versatility evolved prior to the appearance of the structural novelties that heralded the evolution of eukaryotic organisms, discussed in the rest of this unit.

▶ **Figure 27.26 Bioremediation of an oil spill.** Spraying fertiliser stimulates the growth of native bacteria that metabolise oil, increasing the speed of the breakdown process up to fivefold.

CONCEPT CHECK 27.6

1. Identify at least two ways that prokaryotes have affected you positively today.
2. A pathogenic bacterium's toxin causes symptoms that increase the bacterium's chance of spreading from host to host. Does this information indicate whether the poison is an exotoxin or endotoxin? Explain.
3. **WHAT IF?** How might a sudden and dramatic change in your diet affect the diversity of prokaryotic species that live in your digestive tract?

For suggested answers, see Appendix A.

27 Chapter Review

SUMMARY OF KEY CONCEPTS

CONCEPT 27.1

Structural and functional adaptations contribute to prokaryotic success *(pp. 599–603)*

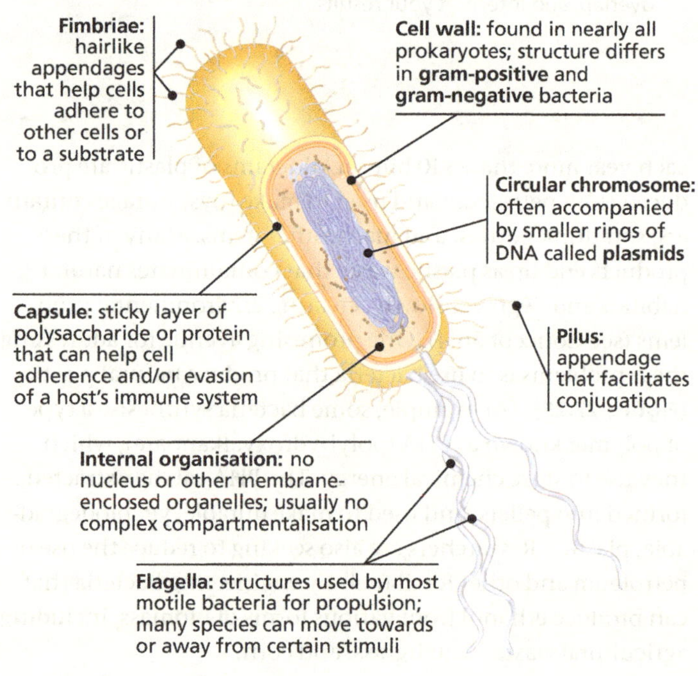

- Many **prokaryotic** species can reproduce quickly by binary fission, leading to the formation of extremely large populations.

? *Describe features of prokaryotes that enable them to thrive in a wide range of different environments.*

CONCEPT 27.2

Rapid reproduction, mutation, and genetic recombination promote genetic diversity in prokaryotes *(pp. 603–606)*

- Because prokaryotes can often proliferate rapidly, mutations can quickly increase a population's genetic variation. As a result, prokaryotic populations often can evolve in short periods of time in response to changing conditions.
- Genetic diversity in prokaryotes also can arise by recombination of the DNA from two different cells (via **transformation, transduction**, or **conjugation**). By transferring advantageous alleles, such as ones for antibiotic resistance, recombination can promote adaptive evolution in prokaryotic populations.

? *Mutations are rare and prokaryotes reproduce asexually, yet their populations can have high genetic diversity. Explain how this can occur.*

CONCEPT 27.3

Diverse nutritional and metabolic adaptations have evolved in prokaryotes *(pp. 606–608)*

- Nutritional diversity is much greater in prokaryotes than in eukaryotes and includes all four modes of nutrition: photoautotrophy, chemoautotrophy, photoheterotrophy, and chemoheterotrophy.
- Among prokaryotes, **obligate aerobes** require O_2, **obligate anaerobes** are poisoned by O_2, and **facultative anaerobes** can survive with or without O_2.

- Unlike eukaryotes, prokaryotes can metabolise nitrogen in many different forms. Some can convert atmospheric nitrogen to ammonia, a process called **nitrogen fixation**.
- Prokaryotic cells and even species may cooperate metabolically, including in surface-coating **biofilms**.

? *Describe the range of prokaryotic metabolic adaptations.*

CONCEPT **27.4**

Prokaryotes have radiated into a diverse set of lineages *(pp. 608–611)*

- Molecular systematics is helping biologists classify prokaryotes and identify new clades.
- Diverse nutritional types are scattered among the major groups of bacteria. The two largest groups are proteobacteria and gram-positive bacteria.
- Some archaea, such as **extreme thermophiles** and **extreme halophiles**, live in extreme environments. Other archaea live in moderate environments such as soils and lakes.

? *How have molecular data informed prokaryotic phylogeny?*

CONCEPT **27.5**

Prokaryotes play crucial roles in the biosphere *(pp. 611–612)*

- Decomposition by heterotrophic prokaryotes and the synthetic activities of autotrophic and nitrogen-fixing prokaryotes contribute to the recycling of elements in ecosystems.
- Many prokaryotes have a **symbiotic** relationship with a **host**; the relationships between prokaryotes and their hosts range from **mutualism** to **commensalism** to **parasitism**.

? *In what ways are prokaryotes key to the survival of many species?*

CONCEPT **27.6**

Prokaryotes have both beneficial and harmful impacts on humans *(pp. 612–616)*

- People depend on mutualistic prokaryotes, including hundreds of species that live in our intestines and help digest food.
- Pathogenic bacteria typically cause disease by releasing **exotoxins** or **endotoxins**. Horizontal gene transfer can spread genes associated with virulence to harmless species or strains.
- Resistance to multiple antibiotics has evolved in many pathogenic bacteria, a problem of great concern for public health.
- Prokaryotes can be used in **bioremediation** and production of plastics, vitamins, antibiotics, and other products.

? *Describe beneficial and harmful impacts of prokaryotes on humans.*

TEST YOUR UNDERSTANDING

Levels 1-2: Remembering/Understanding

1. A process that cannot produce genetic variation in bacterial populations is
(A) transduction. (C) mutation.
(B) conjugation. (D) meiosis.

2. Photoautotrophs use
(A) light as an energy source and CO_2 as a carbon source.
(B) light as an energy source and methane as a carbon source.
(C) N_2 as an energy source and CO_2 as a carbon source.
(D) CO_2 as both an energy source and a carbon source.

3. Which of the following statements is true?
(A) Archaea and bacteria have identical membrane lipids.
(B) The cell walls of archaea lack peptidoglycan.
(C) Prokaryotes have low levels of genetic diversity.
(D) No archaea are capable of using CO_2 to oxidise H_2, releasing methane.

4. Which of the following involves metabolic cooperation among prokaryotic cells?
(A) binary fission (C) biofilms
(B) endospore formation (D) photoautotrophy

5. Which of the following describes a bacterium that lives in the human intestine and causes disease?
(A) commensalist (C) gut mutualist
(B) decomposer (D) symbiotic pathogen

6. Photosynthesis that releases O_2 occurs in
(A) cyanobacteria. (C) gram-positive bacteria.
(B) archaea. (D) chemoautotrophic bacteria.

Levels 3-4: Applying/Analysing

7. EVOLUTION CONNECTION In patients with nonresistant strains of the tuberculosis bacterium, antibiotics can relieve symptoms in a few weeks. However, it takes much longer to halt the infection, and patients may discontinue treatment while bacteria are still present. Explain how this could result in the evolution of drug-resistant pathogens.

Levels 5-6: Evaluating/Creating

8. SCIENTIFIC INQUIRY • INTERPRET THE DATA The nitrogen-fixing bacterium *Rhizobium* infects the roots of some plant species, forming a mutualism in which the bacterium provides nitrogen, and the plant provides carbohydrates. Scientists measured the 12-week growth of one such plant species (*Acacia irrorata*) when infected by six different *Rhizobium* strains. (a) Graph the data. (b) Interpret the graph.

Rhizobium strain	1	2	3	4	5	6
Plant mass (g)	0.91	0.06	1.56	1.72	0.14	1.03

Data from J. J. Burdon et al., Variation in the effectiveness of symbiotic associations between native rhizobia and temperate Australian *Acacia*: within species interactions, *Journal of Applied Ecology* 36:398–408 (1999).

Note: Without *Rhizobium*, after 12 weeks, *Acacia* plants have a mass of about 0.1 g.

9. WRITE ABOUT A THEME: ENERGY In a short essay (about 100–150 words), discuss how prokaryotes and other members of hydrothermal vent communities transfer and transform energy.

10. SYNTHESISE YOUR KNOWLEDGE

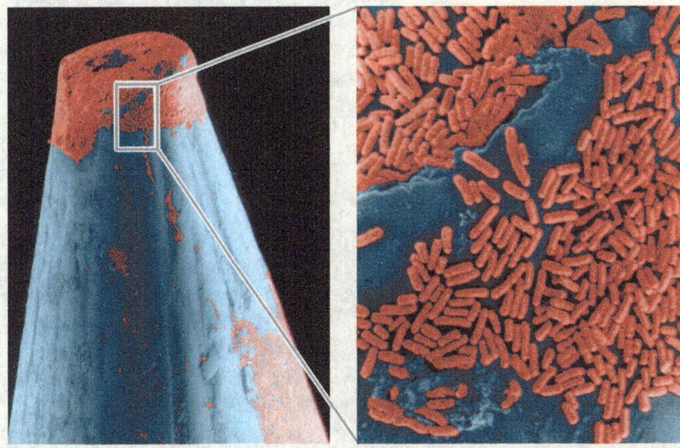

Explain how the small size and rapid reproduction rate of bacteria (such as the population shown here on the tip of a pin) contribute to their large population sizes and high genetic variation.

For selected answers, see Appendix A.

Unit 6 PLANT FORM AND FUNCTION

Meet the man who saved the papaya industry in the US state of Hawaii, Dennis Gonsalves. Dr Gonsalves received a BS in Horticulture and a MS in Plant Pathology from the University of Hawaii, Manoa, and a PhD in Plant Pathology from the University of California, Davis. He became an Associate Professor of plant pathology at the University of Florida and later a Liberty Hyde Bailey Professor at Cornell University. He and his colleagues created virus-resistant papaya plants that rescued the papaya-growing industry in Hawaii from the devastating effects of the ringspot virus. He then served as the Director of the Agricultural Research Center in Hilo, Hawaii, until he retired in 2012. He received the Humboldt Prize for renowned scientific research in 1992 and the Lee Hutchison Award for accomplishments in research, mentoring, and outreach to developing countries.

AN INTERVIEW WITH

Dennis Gonsalves

How did you get interested in plant biology?

I was born on a sugar plantation in Kohala, Hawaii, so I grew up in an agricultural setting. My dad was a labourer on the Kohala Sugar Plantation. I attended the University of Hawaii to study agriculture engineering because my goal was to get a degree and then work on the sugar plantations as an agriculture engineer. But when I was a sophomore [second-year student], the sugar plantations were closing in Hawaii, and the program was disbanded. I took up horticulture as an alternative. After graduation, I moved to Kauai, where there was a University of Hawaii experiment station where I worked as a technician in a plant pathology lab. There, I was a given a small project of my own to work on, and I fell in love with research.

◀ Top: A papaya tree infected with ringspot virus. Inset: An infected papaya fruit. Bottom: A "Rainbow" papaya tree that is resistant to ringspot virus.

Tell us about your experiences in graduate school.

I went to UC Davis for my PhD. I understood the material. I could do the research. But when it came to giving lectures, I was hampered by a stuttering problem. A few months before I got my PhD, I applied for a job at the University of Florida and was invited for an interview. And then I thought, "I have to interview. I am a poor speaker. I don't stand a chance if I do a poor job." I was sick with stress. I practiced and practiced, and then it dawned on me, "I know more about this topic than my professors do. I'm the expert." So, I flew to Florida, and I will never forget that day: I got up and gave my talk, and I have not stuttered since!

What is the ringspot virus and how did you become interested in it?

The virus produces a ring spot on the fruit, but more importantly, the leaves turn yellow, the growth gets stunted, and often the plant won't produce fruit. It is rapidly transmitted by aphids, small sap-sucking insects. An aphid can land on an infected papaya, feed for less than a minute, then jump to a healthy papaya plant and infect it. In 1978, I returned to Hawaii for a short vacation and I met a colleague who told me, "Dennis, there is this virus disease on the Big Island where they grow 95% of the papaya, and it's close to the papaya-growing areas. People are trying to prevent it from spreading." So, when I returned to Cornell, I tried to figure out how to control the papaya ringspot virus.

What was your strategy in fighting the disease?

In our initial attempt, we used artificial selection to isolate a mild strain of papaya ringspot virus to protect plants against infection, but it didn't grow well under field conditions. I have learned, however, as a scientist, sometimes you have to change directions. I was a classical virologist, but I decided to retrain myself as a molecular biologist. Research by others with tobacco had established that certain viral genes, if introduced into the host plant genome, induce resistance. Other colleagues at Cornell had invented a "gene gun" that uses fine, DNA-coated tungsten beads to literally shoot DNA into cells. We shot tungsten beads coated with DNA of the coat protein gene of the ringspot virus into papaya embryos growing in culture. Some of these genetically modified embryos could be grown into plants. When I inoculated them with the virus, one line was completely resistant; it became a parent of the resistant "Rainbow" variety. "Rainbow" was developed just in time because the virus had spread into the main papaya-growing areas and was decimating the papaya crop.

"As a scientist, sometimes you have to change directions."

Have foes of genetically modified foods opposed your research?

In 1994, we teamed up with Thai researchers to repeat our success using a large-fruited papaya variety popular in Thailand. Ten years later, we were successful, but opponents of genetic engineering stormed the fences surrounding our field trials and destroyed many of the fruit. That transgenic papaya has never seen the light of day. Although I've had a pretty good career, that was a heartbreaking thing for me.

35 Vascular Plant Structure, Growth, and Development

KEY CONCEPTS

Study Tip

Make a table: To help keep track of what different plant cells do, make the following table:

Type of plant cell	What it does	How structure fits function

Go to Mastering Biology

to access Dynamic Study Modules for revision, 3D BioFlix® animations and high-quality videos, and your interactive Pearson eText.

Figure 35.1 There is beauty to behold at every level of plant organisation: Every cell, every tissue, and every organ has a function, and the structure of each has been moulded by natural selection.

How does structure fit function in vascular plants?

At the organ level

Leaves provide surface area for absorbing sunlight and exchanging gases.

Stems support and elevate leaves, maximising photosynthesis.

Roots anchor the plant and absorb water and minerals.

At the tissue level

Dermal tissue protects organs.

Vascular tissue provides support and transports resources.

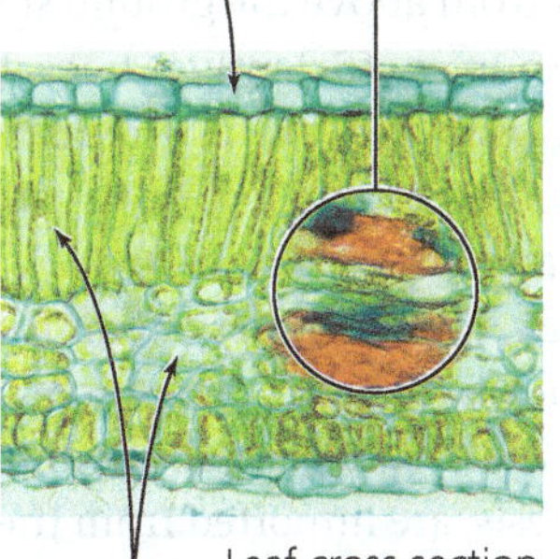

Ground tissue includes cells that carry out photosynthesis and store sugars.

At the cellular level

Photosynthetic cells are packed with chloroplasts that convert sunlight into chemical energy.

Tube-shaped cells transport resources. The cell shown here carries water and minerals. Others conduct sugars.

Cells with root hairs near the tips of roots increase the surface area for absorbing water and minerals.

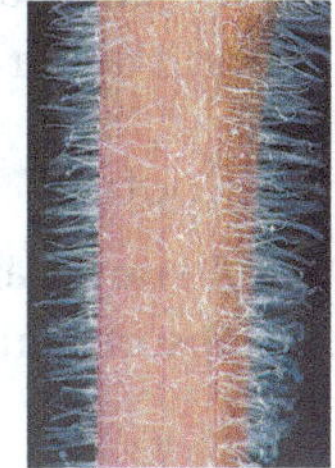

The chapters, "Plant Diversity I: How Plants Colonised Land" and "Plant Diversity II: The Evolution of Seed Plants" provided an overview of plant diversity, including both nonvascular and vascular plants. In this chapter and throughout Unit 6, we'll focus on vascular plants, especially angiosperms because flowering plants are the primary producers in many terrestrial ecosystems and are of great agricultural importance. This chapter mainly explores nonreproductive growth—roots, stems, and leaves—and focuses primarily on the two main groups of angiosperms: eudicots and monocots (see Figure 30.18). Later, in the chapter, "Angiosperm Reproduction and Biotechnology," we'll examine angiosperm reproductive growth: flowers, seeds, and fruits.

CONCEPT 35.1

Plants have a hierarchical organisation consisting of organs, tissues, and cells

Plants, like most animals, are composed of cells, tissues, and organs. A **cell** is the fundamental unit of life. A **tissue** is a group of cells consisting of one or more cell types that together perform a specialised function. An **organ** consists of several types of tissues that together carry out particular functions. As you learn about plant structure, keep in mind how natural selection has produced plant forms that fit plant function at all levels of structure. We begin by discussing plant organs because their structures are most familiar.

Vascular Plant Organs: Roots, Stems, and Leaves

EVOLUTION The basic morphology, or shape, of vascular plants reflects their evolutionary history as terrestrial organisms that inhabit and draw resources from two very different environments—below the ground and above the ground. They must absorb water and minerals from below the ground surface and CO_2 and light from above the ground surface. The ability to acquire these resources efficiently is traceable to the evolution of roots, stems, and leaves as the three basic organs. These organs form a **root system** and a **shoot system**, the latter consisting of stems and leaves **(Figure 35.2)**. Vascular plants, with few exceptions, rely on both systems for survival. Roots are almost never photosynthetic; they starve unless *photosynthates*, the sugars and the other carbohydrates produced during photosynthesis, are imported from the shoot system. Conversely, the shoot system depends on the water and minerals that roots absorb from the soil.

Roots

A **root** is an organ that anchors a vascular plant in the soil, absorbs minerals and water, and often stores carbohydrates and other reserves. The *primary root*, originating in the seed embryo, is the first root (and the first organ) to emerge from a germinating seed. It soon branches to form **lateral roots** (see Figure 35.2) that can also branch, greatly enhancing the ability of the root system to anchor the plant and to acquire resources such as water and minerals from the soil.

▼ **Figure 35.2 An overview of a flowering plant.** The plant body is divided into a root system and a shoot system, connected by vascular tissue (purple strands in this diagram) that is continuous throughout the plant. The plant shown is an idealised eudicot.

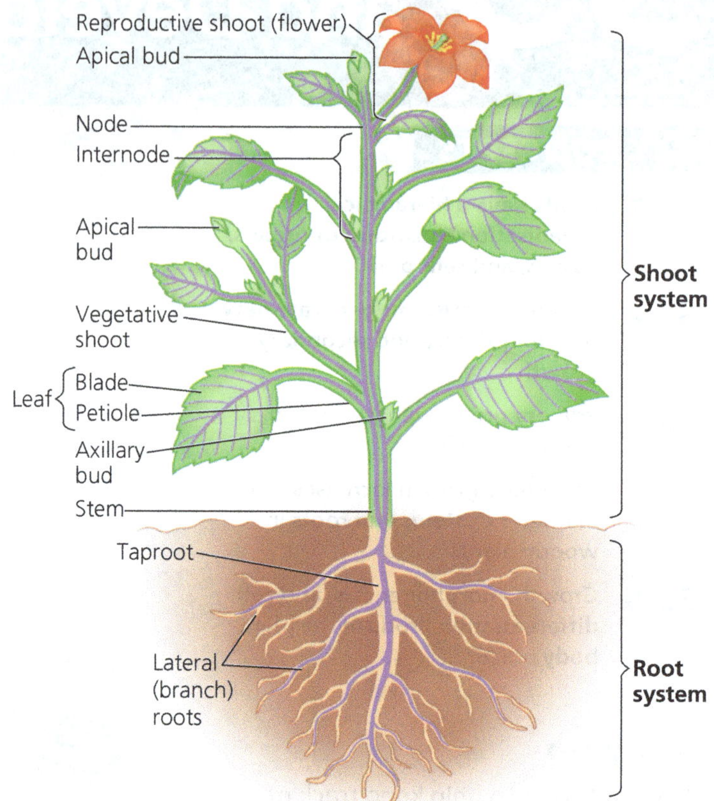

Tall, erect plants with large shoot masses generally have a *taproot system*, consisting of one main vertical root, the **taproot**, which usually develops from the primary root. In taproot systems, the role of absorption is restricted largely to the tips of lateral roots. A taproot, although energetically expensive to make, facilitates the anchorage of the plant in the soil. By preventing toppling, the taproot enables the plant to grow taller, thereby giving it access to more favourable light conditions and, in some cases, providing an advantage for pollen and seed dispersal. Taproots can also be specialised for food storage.

Small vascular plants or those that have a trailing growth habit are particularly susceptible to grazing animals that can uproot the plant and kill it. Such plants are most efficiently anchored by a *fibrous root system*, a thick mat of slender roots spreading out below the soil surface (see Figure 30.18). In plants that have fibrous root systems, including most monocots, the primary root dies early on and does not

▶ **Figure 35.3 Root hairs of a radish seedling.** Root hairs grow by the thousands just behind the tip of each root. By increasing the root's surface area, they greatly enhance the absorption of water and minerals from the soil.

form a taproot. Instead, many small roots emerge from the stem. Such roots are said to be *adventitious*, a term describing a plant organ that grows in an unusual location, such as roots arising from stems or leaves. Each root forms its own lateral roots, which in turn form their own lateral roots. Because this mat of roots holds the topsoil in place, plants such as grasses that have dense fibrous root systems are especially good at preventing soil erosion.

In most plants, the absorption of water and minerals occurs primarily near the tips of elongating roots, where vast numbers of **root hairs**, thin, finger-like extensions of root epidermal cells, emerge and increase the surface area of the root enormously **(Figure 35.3)**. Most root systems also form *mycorrhizal associations*, symbiotic interactions with soil fungi that increase a plant's ability to absorb minerals (see Figure 37.22). The roots of many plants are adapted for specialised functions **(Figure 35.4)**.

▼ **Figure 35.4 Evolutionary adaptations of roots.**

▲ **Prop roots.** The aerial, adventitious roots of corn are prop roots, so named because they support tall, top-heavy plants. All roots of a mature corn plant are adventitious whether they emerge above or below ground.

▲ **Storage roots.** Many plants, such as the common beetroot, store food and water in their roots.

▲ **Pneumatophores.** Also known as air roots, pneumatophores are produced by trees such as mangroves that inhabit tidal swamps. By projecting above the water's surface at low tide, they enable the root system to obtain oxygen, which is lacking in the thick, waterlogged mud.

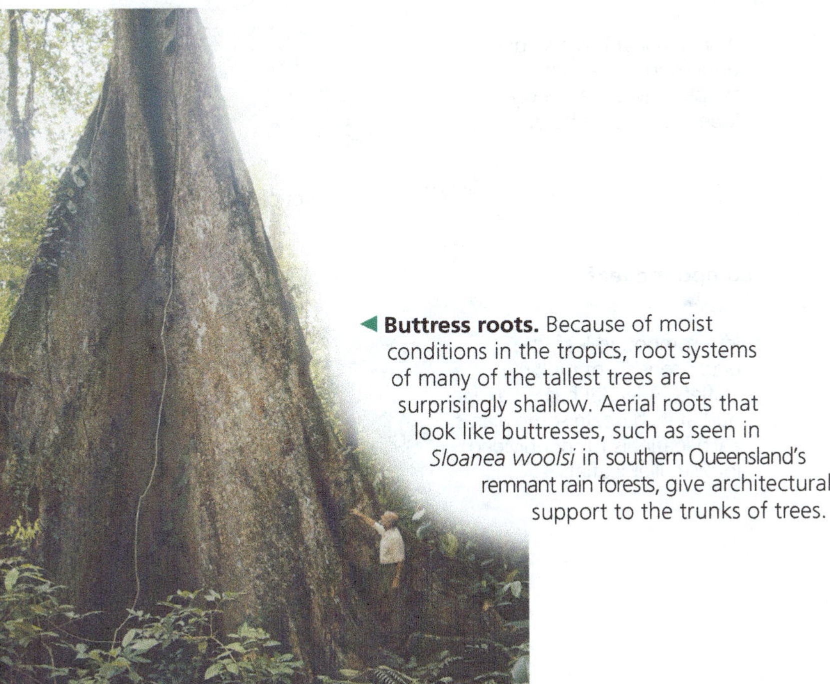

◀ **Buttress roots.** Because of moist conditions in the tropics, root systems of many of the tallest trees are surprisingly shallow. Aerial roots that look like buttresses, such as seen in *Sloanea woolsi* in southern Queensland's remnant rain forests, give architectural support to the trunks of trees.

▶ **"Strangling" aerial roots.** Strangler fig seeds germinate in the crevices of tall trees. Aerial roots grow to the ground, wrapping around the host tree and objects such as this Cambodian temple. Shoots grow upwards and shade out the host tree, killing it.

Stems

A **stem** is a plant organ bearing leaves and buds. Its chief function is to elongate and orient the shoot in a way that maximises photosynthesis by the leaves. Another function of stems is to elevate reproductive structures, thereby facilitating the dispersal of pollen and fruit. Green stems may also perform a limited amount of photosynthesis. Each stem consists of an alternating system of **nodes**, the points at which leaves are attached, and **internodes**, the stem segments between nodes (see Figure 35.2). Most of the growth of a young shoot is concentrated near the growing shoot tip, or **apical bud**. Apical buds are not the only types of buds found in shoots. In the upper angle (axil) formed by each leaf and the stem is an **axillary bud**, which can potentially form a lateral branch or, in some cases, a thorn or flower.

Some plants have stems with alternative functions, such as food storage or asexual reproduction. Many of these modified stems, including rhizomes, stolons, and tubers, are often mistaken for roots **(Figure 35.5)**.

▼ Figure 35.5 Evolutionary adaptations of stems.

Rhizome

Root

◀ **Rhizomes.** The base of this iris plant is an example of a rhizome, a horizontal shoot that grows just below the surface. Vertical shoots emerge from axillary buds on the rhizome.

▶ **Stolons.** Shown here on a strawberry plant, stolons are horizontal shoots that grow along the surface. These "runners" enable a plant to reproduce asexually, as plantlets grow from axillary buds along each runner.

◀ **Tubers.** Tubers, such as these potatoes, are enlarged ends of rhizomes or stolons specialised for storing food. The "eyes" of a potato are clusters of axillary buds.

? *Which of these three examples has nodes?*

Leaves

In most vascular plants, the **leaf** is the main photosynthetic organ. In addition to intercepting light, leaves exchange gases with the atmosphere, dissipate heat, and defend themselves from herbivores and pathogens. These functions may have conflicting anatomical and physiological requirements. For example, a dense covering of hairs may help repel herbivorous insects but may also trap air near the leaf surface, thereby reducing gas exchange and, consequently, photosynthesis. Because of these conflicting demands and trade-offs, leaves vary extensively in form. In general, however, a leaf consists of a flattened **blade** and a stalk, the **petiole**, which joins the leaf to the stem at a node (see Figure 35.2). Grasses and many other monocots lack petioles; instead, the base of the leaf forms a sheath that envelops the stem.

Monocots and eudicots differ in the arrangement of **veins**, the vascular tissue of leaves. Most monocots have parallel major veins of equal diameter that run the length of the blade. Eudicots generally have a branched network of veins arising from a major vein (the *midrib*) that runs down the centre of the blade (see Figure 30.18).

In identifying angiosperms according to structure, taxonomists rely mainly on floral morphology, but they also use variations in leaf morphology, such as leaf shape, the branching pattern of veins, and the spatial arrangement of leaves. **Figure 35.6** illustrates a difference in leaf shape: simple versus compound. Unlike leaves, the leaflets of compound leaves are not associated with axillary buds. Compound leaves may help confine invading pathogens to a single leaflet, rather than allowing them to spread to the entire leaf.

▼ Figure 35.6 Simple versus compound leaves.

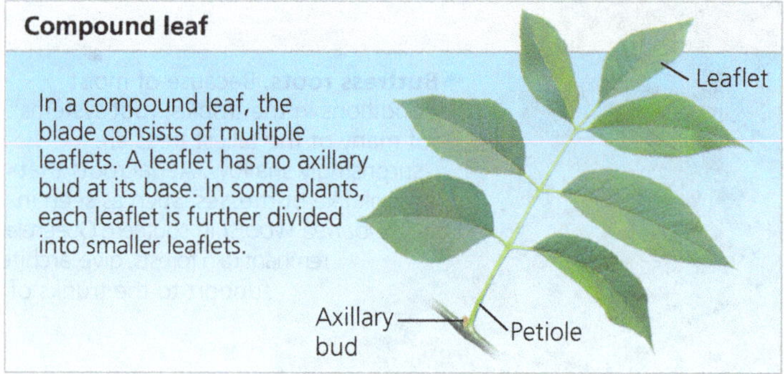

▼ Figure 35.7 Evolutionary adaptations of leaves.

► **Tendrils.** The tendrils by which this pea plant clings to a support are modified leaves. After it has "lassoed" a support, a tendril forms a coil that brings the plant closer to the support. Tendrils are typically modified leaves, but some tendrils are modified stems, as in grapevines.

◄ **Spines.** The spines of cacti, such as this prickly pear, are actually leaves; photosynthesis is carried out by the fleshy green stems.

◄ **Storage leaves.** Bulbs, such as this cut onion, have a short underground stem and modified leaves that store food.

Plantlet

Storage leaves

Stem

◄ **Carnivorous leaves.** In areas of extremely low nutrient availability, members of the sundew genus (*Drosera* spp.) thrive. The leaves exude compounds that attract and ensnare insects. The exudate contains enzymes that dissolve the insect. The ensuing nutrients are absorbed directly through the leaf surface.

The shapes of leaves are often products of genetic programs that are tweaked by environmental influences. Interpret the data in the **Scientific Skills Exercise** to explore the roles of genetics and the environment in determining leaf morphology in red maple trees.

Almost all leaves are specialised for photosynthesis. However, in some species evolution has resulted in additional functions, such as support, protection, storage, or absorption of nutrients **(Figure 35.7)**. Some are sporophylls, leaves highly specialised for sexual reproduction, such as carpels and stamens in flowers (see Figure 30.13).

Dermal, Vascular, and Ground Tissues

All three basic vascular plant organs—roots, stems, and leaves—are composed of three fundamental tissue types: dermal, vascular, and ground tissues. Each of these general types forms a **tissue system** that is continuous throughout the plant, connecting all the organs. However, specific characteristics of the tissues and the spatial relationships of tissues to one another vary in different organs **(Figure 35.8)**.

Dermal tissue serves as the outer protective covering of the plant. Like our skin, it forms the first line of defence against physical damage and pathogens. In nonwoody plants,

Scientific Skills Exercise

Using Bar Graphs to Interpret Data

Nature Versus Nurture: Why Are Leaves from Northern Red Maples "Toothier" Than Leaves from Southern Red Maples? Not all leaves of the red maple (*Acer rubrum*) are the same. The "teeth" along the margins of leaves growing in northern locations differ in size and number from those of their southern counterparts. (The leaf seen here has an intermediate appearance.) Are these differences due to genetic differences between northern and southern *Acer rubrum* populations, or do they arise from environmental differences between northern and southern locations, such as average temperature, that affect gene expression?

How the Experiment Was Done Seeds of *Acer rubrum* were collected from four latitudinally distinct sites: Ontario (Canada), Pennsylvania, South Carolina, and Florida. The seeds from the four sites were then grown in a northern location (Rhode Island) and a southern location (Florida). After a few years of growth, leaves were harvested from the four sets of plants growing in the two locations. The average area of single teeth and the average number of teeth per leaf area were determined.

Data from the Experiment

Seed Collection Site	Average Area of a Single Tooth (cm^2)		Number of Teeth per cm^2 of Leaf Area	
	Grown in Rhode Island	Grown in Florida	Grown in Rhode Island	Grown in Florida
Ontario (43.32°N)	0.017	0.017	3.9	3.2
Pennsylvania (42.12°N)	0.020	0.014	3.0	3.5
South Carolina (33.45°N)	0.024	0.028	2.3	1.9
Florida (30.65°N)	0.027	0.047	2.1	0.9

Data from D. L. Royer et al., Phenotypic plasticity of leaf shape along a temperature gradient in *Acer rubrum, PLoS ONE* 4(10):e7653 (2009).

INTERPRET THE DATA

1. Make a bar graph for tooth size and a bar graph for number of teeth. (For information on bar graphs, see the Scientific Skills Review in Appendix D.) From north to south, what is the general trend in tooth size and number of teeth in leaves of *Acer rubrum?*
2. Based on the data, would you conclude that leaf tooth traits in the red maple are largely determined by genetic heritage (genotype), by the capacity for responding to environmental change within a single genotype (phenotypic plasticity), or by both? Make specific reference to the data in answering the question.
3. The "toothiness" of leaf fossils of known age has been used by paleoclimatologists to estimate past temperatures in a region. If a 10,000-year-old fossilised red maple leaf from South Carolina had an average of 4.2 teeth per square centimetre of leaf area, what could you infer about the temperature of South Carolina 10,000 years ago compared with the temperature today? Explain your reasoning.

▼ Figure 35.10 Exploring Examples of Differentiated Plant Cells

Parenchyma Cells

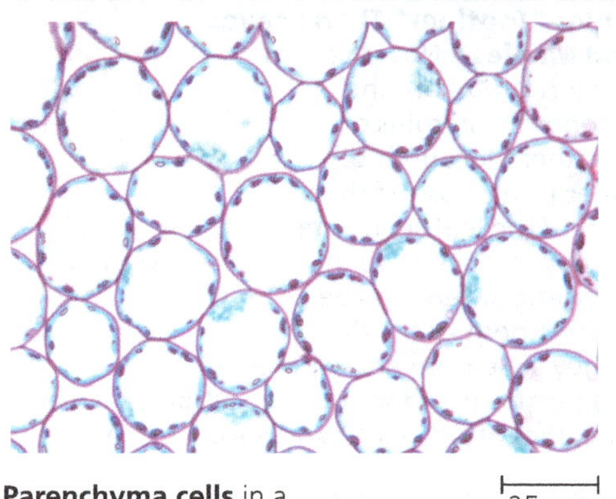

Parenchyma cells in a privet (*Ligustrum*) leaf (LM)

Mature **parenchyma cells** have primary walls that are relatively thin and flexible, and most lack secondary walls. (See Figure 6.27 to review primary and secondary cell walls.) When mature, parenchyma cells generally have a large central vacuole. Parenchyma cells perform most of the metabolic functions of the plant, synthesising and storing various organic products. For example, photosynthesis occurs within the chloroplasts of parenchyma cells in the leaf. Some parenchyma cells in stems and roots have colourless plastids called amyloplasts that store starch. The fleshy tissue of many fruits is composed mainly of parenchyma cells. Most parenchyma cells retain the ability to divide and differentiate into other types of plant cells under particular conditions—during wound repair, for example. It is even possible to grow an entire plant from a single parenchyma cell.

Collenchyma Cells

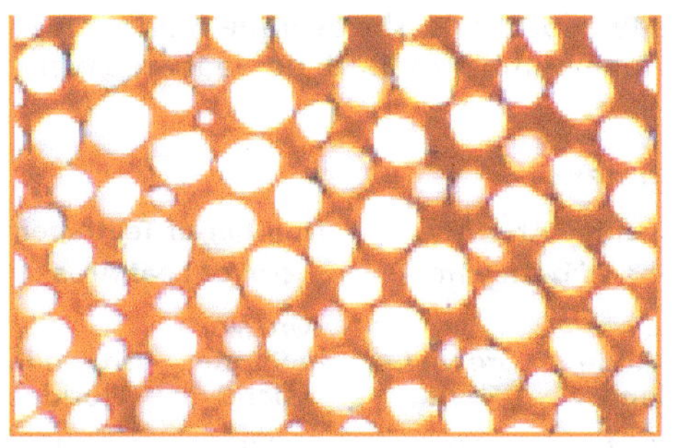

Collenchyma cells in a common nettle (*Urtica dioica*) stem (LM)

Grouped in strands, **collenchyma cells** (seen here in cross section) help support young parts of the plant shoot. Collenchyma cells are generally elongated cells that have thicker primary walls than parenchyma cells, though the walls are unevenly thickened. Young stems and petioles often have strands of collenchyma cells just below their epidermis. Collenchyma cells provide flexible support without restraining growth. At maturity, these cells are living and flexible, elongating with the stems and leaves they support—unlike sclerenchyma cells, which we discuss next.

Sclerenchyma Cells

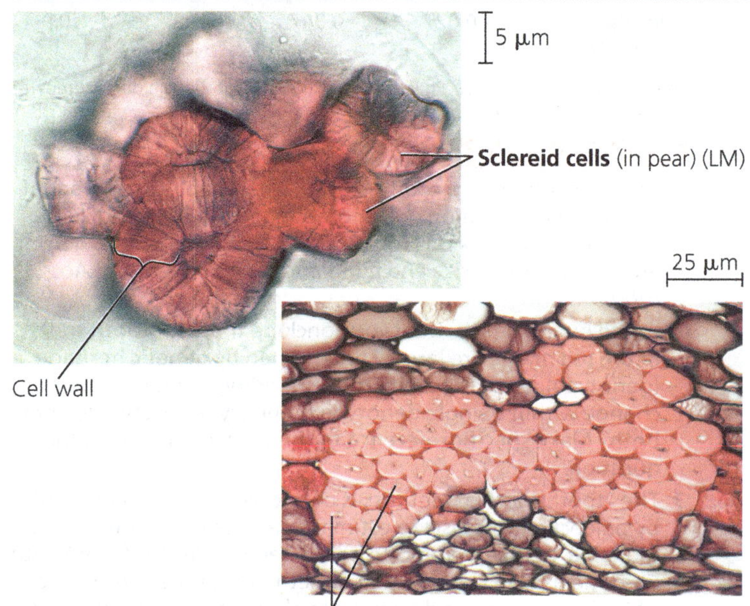

Fibre cells (cross section from ash tree) (LM)

Sclerenchyma cells also function as supporting elements in the plant but are much more rigid than collenchyma cells. In sclerenchyma cells, the secondary cell wall, produced after cell elongation has ceased, is thick and contains large amounts of **lignin**, a relatively indigestible strengthening polymer that accounts for more than a quarter of the dry mass of wood. Lignin is present in all vascular plants but not in bryophytes. Mature sclerenchyma cells cannot elongate, and they occur in regions of the plant that have stopped growing in length. Sclerenchyma cells are so specialised for support that many are dead at functional maturity, but they produce secondary walls before the protoplast (the living part of the cell) dies. The rigid walls remain as a "skeleton" that supports the plant, in some cases for hundreds of years.

Two types of sclerenchyma cells, known as *sclereids* and *fibres*, are specialised entirely for support and strengthening. Sclereids, which are boxier than fibres and irregular in shape, have very thick, lignified secondary walls. Sclereids impart the hardness to nutshells and seed coats and the gritty texture to pear fruits. Fibres, which are usually grouped in strands, are long, slender, and tapered. Some are used commercially, such as hemp fibres for making rope and flax fibres for weaving into linen.

Water-Conducting Cells of the Xylem

The two types of water-conducting cells, **tracheids** and **vessel elements**, are tubular, elongated cells that are dead and lignified at functional maturity. Tracheids occur in the xylem of all vascular plants. In addition to tracheids, most angiosperms, as well as a few gymnosperms and a few seedless vascular plants, have vessel elements. When the living cellular contents of a tracheid or vessel element disintegrate, the cell's thickened walls remain behind, forming a nonliving conduit through which water can flow. The secondary walls of tracheids and vessel elements are often interrupted by pits, thinner regions where only primary walls are present (see Figure 6.27 to review primary and secondary walls). Water can migrate laterally between neighbouring cells through pits.

Tracheids are long, thin cells with tapered ends. Water moves from cell to cell mainly through the pits, where it does not have to cross thick secondary walls.

Vessel elements are generally wider, shorter, thinner walled, and less tapered than the tracheids. They are aligned end to end, forming long pipes known as **vessels** that in some cases are visible with the naked eye. The end walls of vessel elements have perforation plates that enable water to flow freely through the vessels.

The secondary walls of tracheids and vessel elements are hardened with lignin. This hardening provides support and prevents collapse under the tension of water transport.

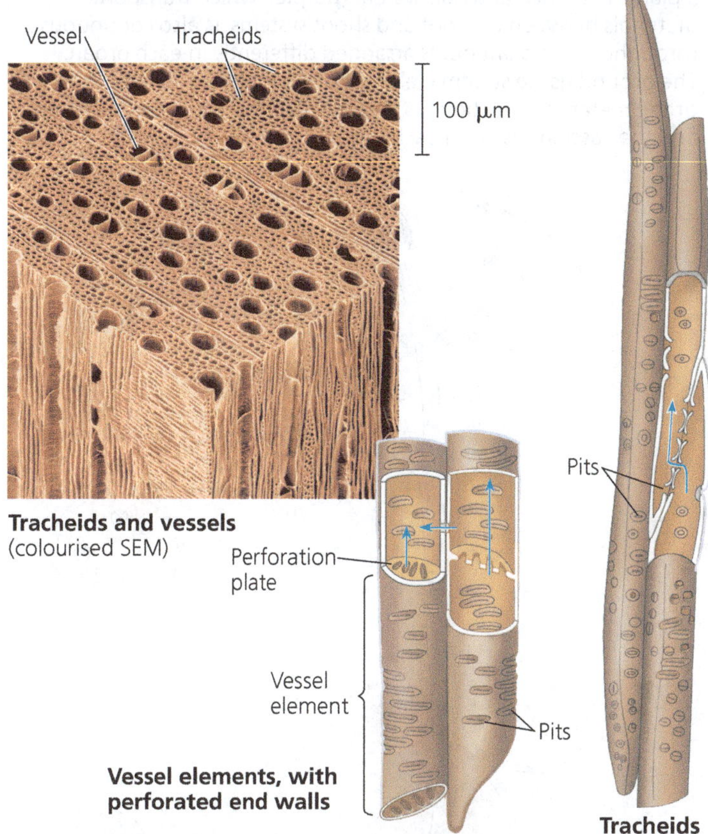

Sugar-Conducting Cells of the Phloem

Unlike the water-conducting cells of the xylem, the sugar-conducting cells of the phloem are alive at functional maturity. In seedless vascular plants and gymnosperms, sugars and other organic nutrients are transported through long, narrow cells called sieve cells. In the phloem of angiosperms, these nutrients are transported through sieve tubes, which consist of chains of cells that are called **sieve-tube elements**, or sieve-tube members.

Though alive, sieve-tube elements lack a nucleus, ribosomes, a distinct vacuole, and cytoskeletal elements. This reduction in cell contents enables nutrients to pass more easily through the cell. The end walls between sieve-tube elements, called **sieve plates**, have pores that facilitate the flow of fluid from cell to cell along the sieve tube. Alongside each sieve-tube element is a nonconducting cell called a **companion cell**, which is connected to the sieve-tube element by numerous plasmodesmata (see Figure 6.27). The nucleus and ribosomes of the companion cell serve not only that cell itself but also the adjacent sieve-tube element. In some plants, the companion cells in leaves also help load sugars into the sieve-tube elements, which then transport the sugars to other parts of the plant.

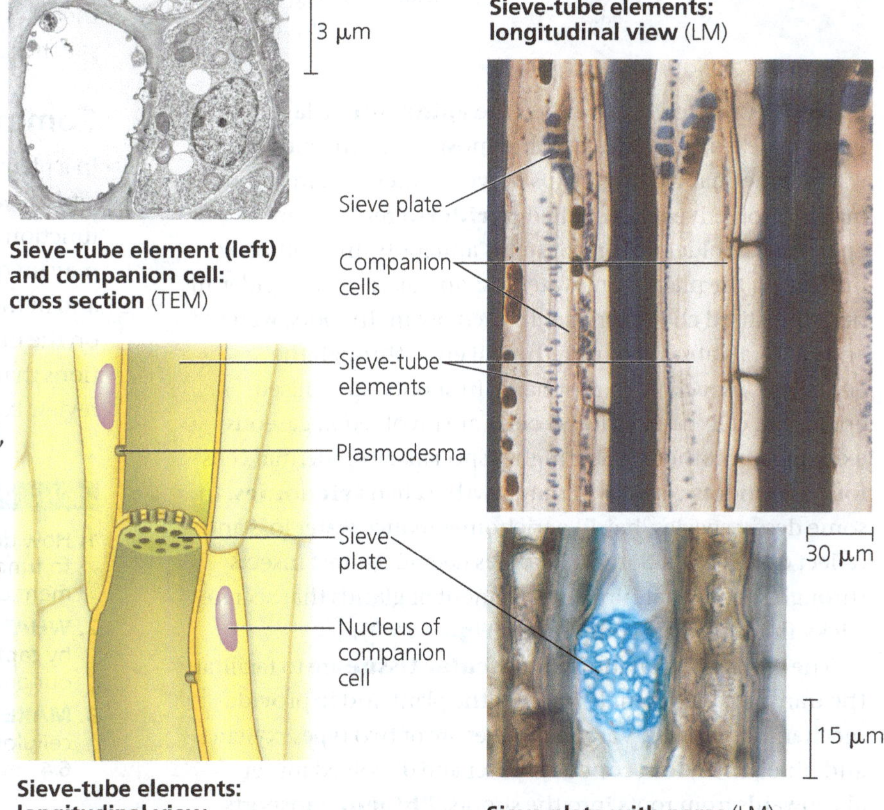

▼ Figure 35.8 The three tissue systems. The dermal tissue system (blue) provides a protective cover for the entire body of a plant. The vascular tissue system (purple), which transports materials between the root and shoot systems, is also continuous throughout the plant but is arranged differently in each organ. The ground tissue system (yellow), which is responsible for most of the metabolic functions, is located between the dermal tissue and the vascular tissue in each organ.

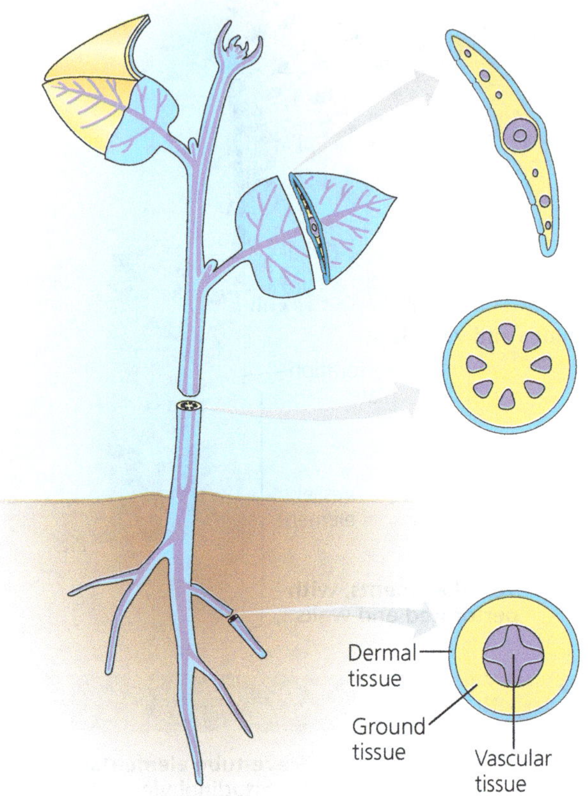

▼ Figure 35.9 Trichome diversity on the surface of a leaf. Three types of trichomes are found on the surface of marjoram (*Origanum majorana*). Spear-like trichomes help hinder the movement of crawling insects, while the other two types of trichomes secrete oils and other chemicals involved in defence (colourised SEM).

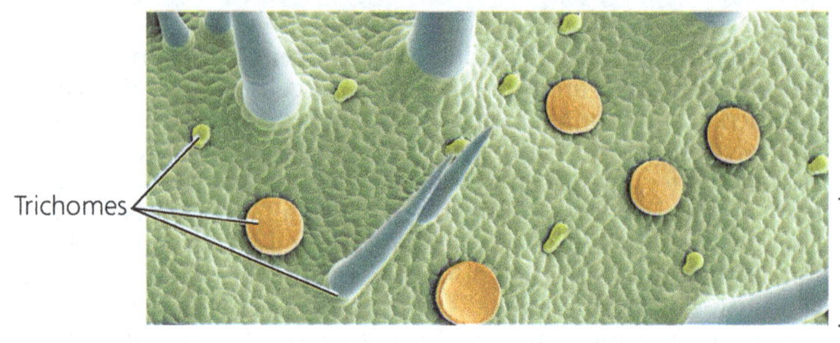

it is usually a single tissue called the **epidermis**, a layer of tightly packed cells. In leaves and most stems, the **cuticle**, a waxy epidermal coating, helps prevent water loss. In woody plants, protective tissues called **periderm** replace the epidermis in older regions of stems and roots. In addition to protecting the plant from water loss and disease, the epidermis has specialised characteristics in each organ. In roots, water and minerals absorbed from the soil enter through the epidermis, especially in root hairs. In shoots, specialised epidermal cells called **guard cells** are involved in gaseous exchange. Another class of highly specialised epidermal cells found in shoots consists of outgrowths called **trichomes**. In some desert species, hairlike trichomes reduce water loss and reflect excess light. Some trichomes defend against insects through shapes that hinder movement or glands that secrete sticky fluids or toxic compounds **(Figure 35.9)**.

The two major functions of **vascular tissue** are to facilitate the transport of materials through the plant and to provide mechanical support. Vascular tissues are of two types: xylem and phloem. **Xylem** conducts water and dissolved minerals upwards from roots into the shoots. **Phloem** transports sugars, the products of photosynthesis, from where they are made (usually the leaves) to where they are needed or stored—usually roots and sites of growth, such as developing leaves and fruits. The vascular tissue of a root or stem is collectively called the **stele** (the Greek word for "pillar"). The arrangement of the stele varies, depending on the species and organ. In angiosperms, for example, the root stele is a solid central *vascular cylinder* of xylem and phloem, whereas the stele of stems and leaves consists of *vascular bundles*, separate strands containing xylem and phloem (see Figure 35.8). Both xylem and phloem are composed of a variety of cell types, including cells that are highly specialised for transport or support.

Tissue that is neither dermal nor vascular is **ground tissue**. Ground tissue that is internal to the vascular tissue is known as **pith**, and ground tissue that is external to the vascular tissue is called **cortex**. Ground tissue is not just filler: It includes cells specialised for functions such as storage, photosynthesis, support, and short-distance transport.

Common Types of Plant Cells

In a plant, as in any multicellular organism, cells undergo cell *differentiation*; that is, they become specialised in structure and function during the course of development. Cell differentiation may involve changes both in the cytoplasm and its organelles and in the cell wall. **Figure 35.10**, on the next two pages, focuses on the major types of plant cells. Notice the structural adaptations that make specific functions possible. You may also wish to review basic plant cell structure (see Figures 6.8 and 6.27).

CONCEPT CHECK 35.1

1. How does the vascular tissue system enable leaves and roots to function together in supporting growth and development of the whole plant?
2. **WHAT IF?** If humans were photoautotrophs, making food by capturing light energy for photosynthesis, how might our anatomy be different?
3. **MAKE CONNECTIONS** Explain how central vacuoles and cellulose cell walls contribute to plant growth (see Concepts 6.4 and 6.7).

For suggested answers, see Appendix A.

CONCEPT 35.2

Different meristems generate new cells for primary and secondary growth

A major difference between plants and most animals is that plant growth is not limited to an embryonic or juvenile period. Instead, growth occurs throughout the plant's life, a process called **indeterminate growth**. Plants can keep growing because they have undifferentiated tissues called **meristems** containing cells that can divide, leading to new cells that elongate and become differentiated **(Figure 35.11)**. Except for dormant periods, most plants grow continuously. In contrast, most animals and some plant organs—such as leaves, thorns, and flowers—undergo **determinate growth**; they stop growing after reaching a certain size.

There are two main types of meristems: apical meristems and lateral meristems. **Apical meristems**, located at root and shoot tips, provide cells that enable **primary growth**, growth in length. Primary growth allows roots to extend throughout the soil and shoots to increase exposure to light. In herbaceous (non-woody) plants, it produces all, or almost all, of the plant body. Woody plants, however, also grow in circumference in the parts of stems and roots that no longer grow in length. This growth in thickness, known as **secondary growth**, is made possible by **lateral meristems**: the vascular cambium and cork cambium. These cylinders of dividing cells extend along the length of roots and stems. The **vascular cambium** adds vascular tissue called secondary xylem (wood) and secondary phloem. Most of the thickening is from secondary xylem. The **cork cambium** replaces the epidermis with the thicker, tougher periderm.

Cells in apical and lateral meristems divide frequently during the growing season, generating additional cells. Some new cells remain in the meristem and produce more cells, while others differentiate and are incorporated into tissues and organs. Cells that remain as sources of new cells have traditionally been called *initials* but are increasingly being called *stem cells* to correspond to animal stem cells that also divide and remain functionally undifferentiated.

Cells displaced from the meristem may divide several more times as they differentiate into mature cells. During primary growth, these cells give rise to three tissues called **primary meristems**—the *protoderm, ground meristem*, and *procambium*—that will produce, respectively, the three mature tissues of a root or shoot: the dermal, ground, and vascular tissues. The lateral meristems in woody plants also have stem cells, which give rise to all secondary growth.

The relationship between primary and secondary growth is seen in the winter twig of a deciduous tree. At the shoot tip is the dormant apical bud, enclosed by scales that protect its apical meristem **(Figure 35.12)**. In spring, the bud sheds its scales and begins a new spurt of primary growth, producing a series of nodes and internodes. On each growth segment, nodes are marked by scars left when leaves fell. Leaf scars are prominent in many twigs. Above each scar is an axillary bud or a branch formed by an axillary bud. Further down are bud scars from whorls of scales that enclosed the apical bud during the previous winter. In each growing season, primary growth extends shoots, and secondary growth increases the diameter of parts formed in previous years.

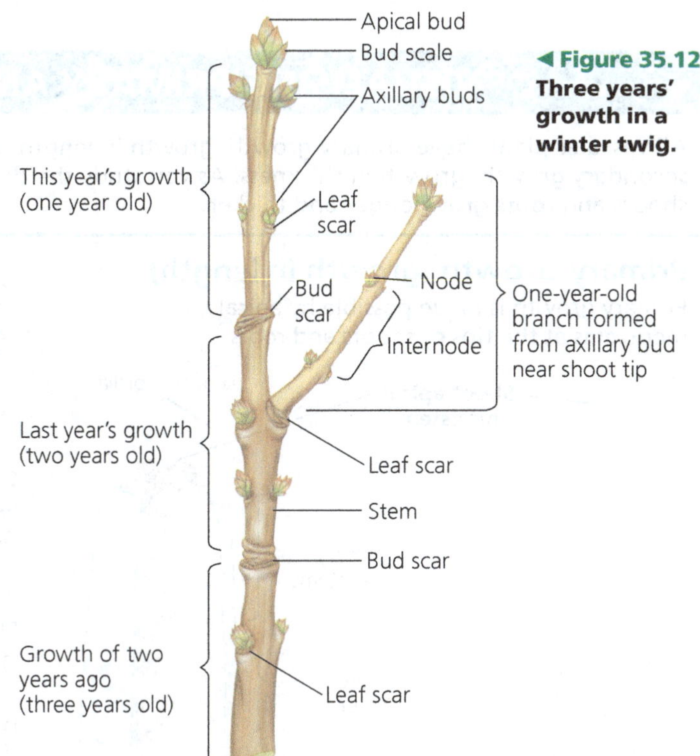

◀ **Figure 35.12 Three years' growth in a winter twig.**

Although meristems enable plants to grow throughout their lives, plants do die, of course. Based on the length of their life cycle, flowering plants can be categorised as annuals, biennials, or perennials. *Annuals* complete their life cycle—from germination to flowering to seed production to death—in a single year or less. Many wildflowers are annuals, as are most staple food crops, including legumes and cereal grains such as wheat and rice. Dying after producing seeds and fruits enables plants to transfer the maximum amount of energy to reproduction. *Biennials*, such as turnips, generally require two growing seasons to complete their life cycle, flowering and fruiting only in their second year. *Perennials* live many years and include trees, shrubs, and some grasses. Some buffalo grass of the North American plains is thought to have been growing for 10,000 years from seeds that sprouted at the close of the last ice age.

CONCEPT CHECK 35.2

1. Would primary and secondary growth ever occur simultaneously in the same plant?
2. Roots and stems grow indeterminately, but leaves do not. How might this benefit the plant?
3. **WHAT IF?** After growing carrots for one season, a gardener decides that the carrots are too small. Since carrots are biennials, the gardener leaves the crop in the ground for a second year, thinking the carrot roots will grow larger. Is this a good idea? Explain.

For suggested answers, see Appendix A.

▼ Figure 35.11 VISUALISING PRIMARY AND SECONDARY GROWTH

All vascular plants have primary growth: growth in length. Woody plants also have secondary growth: growth in thickness. As you study the diagrams, visualise how shoots and roots grow longer and thicker.

Primary Growth (growth in length)

Primary growth is made possible by apical meristems at the tips of shoots and roots.

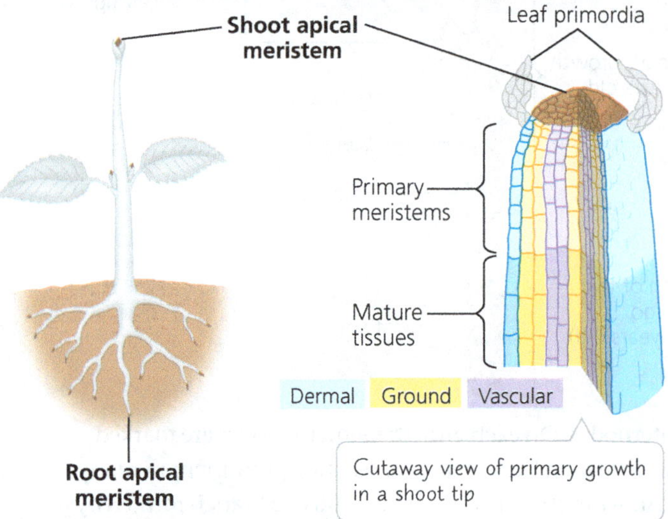

Cutaway view of primary growth in a shoot tip

Apical meristem cells are undifferentiated. When they divide, some daughter cells remain in the apical meristem, ensuring a continuing population of undifferentiated cells. Other daughter cells become partly differentiated as primary meristem cells. After dividing and growing in length, they become fully differentiated cells in the mature tissues.

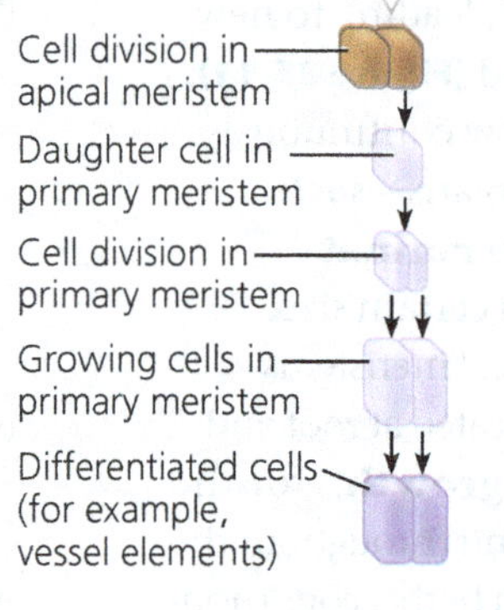

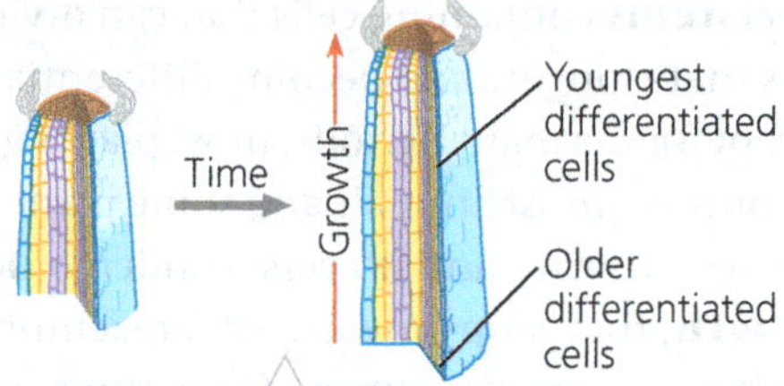

The addition of elongated, differentiated cells lengthens a stem (as shown here) or root.

1. A thimble-like root cap protects each root apical meristem. Draw and label a simple outline of a root divided into four sections: root cap (bottom), root apical meristem, primary meristems, and mature tissues.

Secondary Growth (growth in thickness)

Secondary growth is made possible by two lateral meristems extending along the length of a shoot or root where primary growth has ceased.

The **lateral meristems**, called the vascular cambium and cork cambium, are cylinders of dividing cells that are one cell thick.

Addition of secondary xylem and phloem cells: When a vascular cambium cell divides, sometimes one daughter cell becomes a secondary xylem cell (X) to the inside of the cambium or a secondary phloem cell (P) to the outside. Although xylem and phloem cells are shown being added equally here, usually many more xylem cells are produced.

Increased circumference: When a cambium cell divides, sometimes both daughter cells remain in the cambium and grow, increasing the cambium circumference.

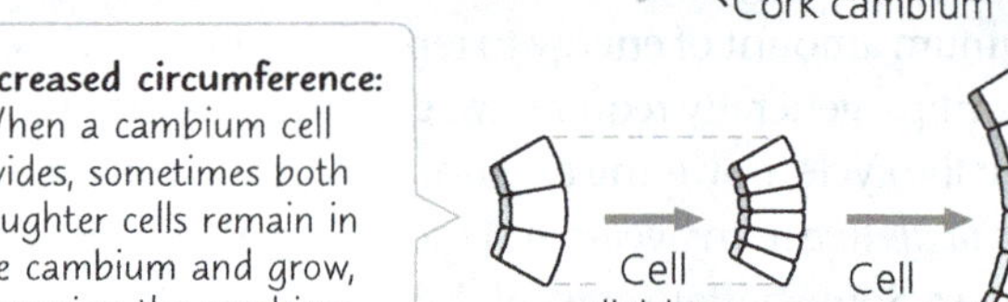

Addition of cork cells: When a cork cambium cell divides, sometimes one daughter cell becomes a cork cell (C) to the outside of the cambium.

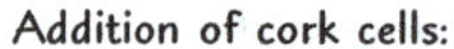

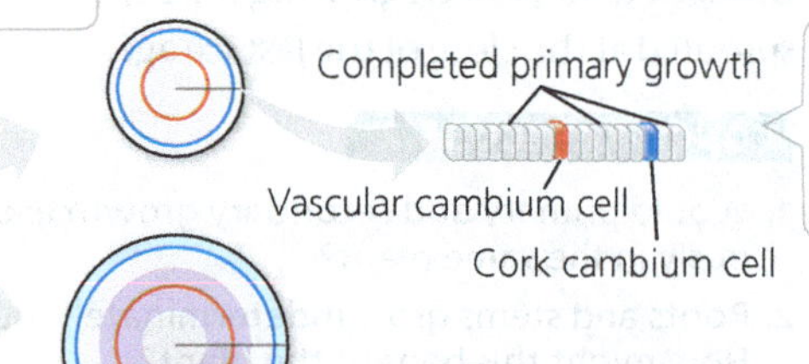

When the vascular cambium and cork cambium become active in a stem (or root), primary growth has ceased in that area.

A stem (or root) thickens as secondary xylem, secondary phloem, and cork cells are added. Most of the cells are secondary xylem (wood).

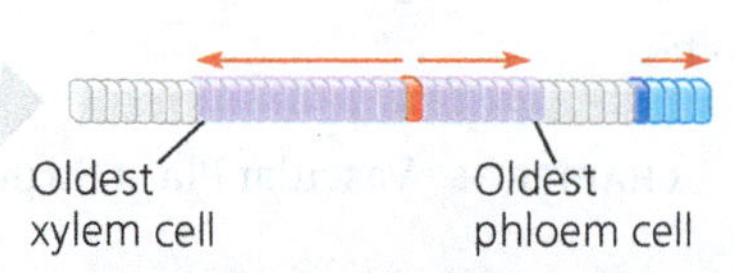

2. Draw the row of cells from the boxed area below and label the vascular cambium cell (V), 5 xylem cells from oldest (X1) to youngest (X5), and 3 phloem cells (P1 to P3). Show what happens after growth continues by drawing and labelling a row with twice as many xylem and phloem cells. How does the vascular cambium's location change?

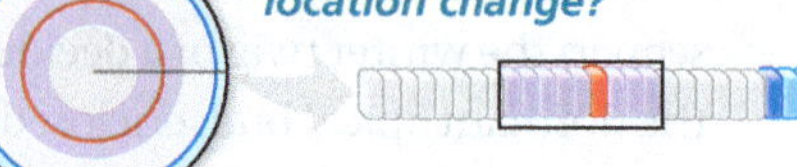

CONCEPT 35.3

Primary growth lengthens roots and shoots

Primary growth arises directly from cells produced by apical meristems. In herbaceous plants, almost the entire plant is created through primary growth, whereas in woody plants only the nonwoody, more recently formed parts of the plant represent primary growth. Although both roots and shoots lengthen as a result of cells derived from apical meristems, the details of their primary growth differ in many ways.

Primary Growth of Roots

The entire biomass of a primary root is derived from the root apical meristem. The root apical meristem also makes a thimble-like **root cap**, which protects the delicate apical meristem as the root pushes through the abrasive soil. The root cap secretes a polysaccharide slime that lubricates the soil around the tip of the root. Growth occurs just behind the tip in three overlapping zones of cells at successive stages of primary growth. These are the zones of cell division, elongation, and differentiation **(Figure 35.13)**.

The *zone of cell division* includes the stem cells of the root apical meristem and their immediate products. New root cells are produced in this region, including cells of the root cap.

▼ Figure 35.13 Primary growth of a eudicot root. In the micrograph, mitotic cells in the apical meristem are revealed by staining for cyclin, a protein involved in cell division (LM).

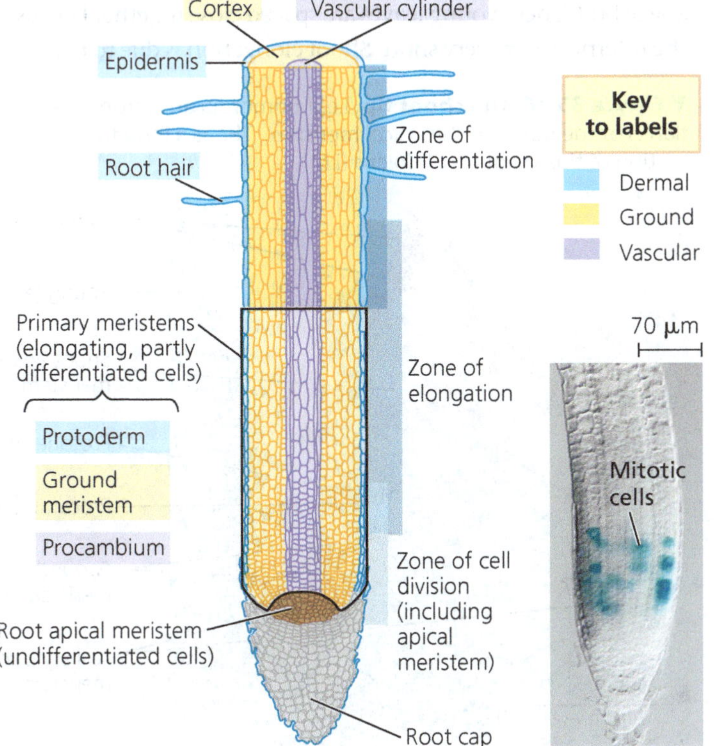

Typically, a few millimetres behind the tip of the root is the *zone of elongation*, where most of the growth occurs as root cells elongate—sometimes to more than ten times their original length. Cell elongation in this zone pushes the tip further into the soil. Meanwhile, the root apical meristem keeps adding cells to the younger end of the zone of elongation. Even before the root cells finish lengthening, many begin specialising in structure and function. As this occurs, the three primary meristems—the protoderm, ground meristem, and procambium—become evident. In the *zone of differentiation*, or zone of maturation, cells complete their differentiation and become distinct cell types.

The protoderm, the outermost primary meristem, gives rise to the epidermis, a single layer of cuticle-free cells covering the root. Root hairs are the most prominent feature of the root epidermis. These modified epidermal cells function in the absorption of water and minerals. Root hairs typically only live a few weeks but together make up 70–90% of the total root surface area. It has been estimated that a four-month-old rye plant has about 14 billion root hairs. Laid end to end, the root hairs of a single rye plant would cover 10,000 km, one-quarter the length of the equator.

Sandwiched between the protoderm and the procambium is the ground meristem, which gives rise to mature ground tissue. The ground tissue of roots, consisting mostly of parenchyma cells, is found in the cortex, the region between the vascular tissue and epidermis. In addition to storing carbohydrates, cells in the cortex transport water and salts from the root hairs to the centre of the root. The cortex also allows for *extracellular* diffusion of water, minerals, and oxygen from the root hairs inwards because there are large spaces between cells. The innermost layer of the cortex is called the **endodermis**, a cylinder one cell thick that forms the boundary with the vascular cylinder. The endodermis is a selective barrier that regulates passage of substances from the soil into the vascular cylinder (see Figure 36.9).

The procambium gives rise to the vascular cylinder, which consists of a solid core of xylem and phloem tissues surrounded by a cell layer called the **pericycle**. In most eudicot roots, the xylem has a starlike appearance in cross section, and the phloem occupies the indentations between the arms of the xylem "star" **(Figure 35.14a)**. In many monocot roots, the vascular tissue consists of a core of undifferentiated parenchyma cells surrounded by a ring of alternating xylem and phloem tissues **(Figure 35.14b)**.

By increasing the length of roots, primary growth facilitates their penetration and exploration of the soil. If a resource-rich pocket is located in the soil, the branching of roots may be stimulated. Branching, too, is a form of primary growth. Lateral (branch) roots arise from meristematically active regions of the pericycle, the outermost cell layer in the vascular cylinder, which is adjacent to and just inside the endodermis (see Figure 35.14). The emerging lateral roots

▼ Figure 35.14 Organisation of primary tissues in young roots. Parts (a) and (b) show cross sections of the roots of a *Ranunculus* (buttercup) species and *Zea mays* (corn), respectively. These represent two basic patterns of root organisation, of which there are many variations, depending on the plant species (all LMs).

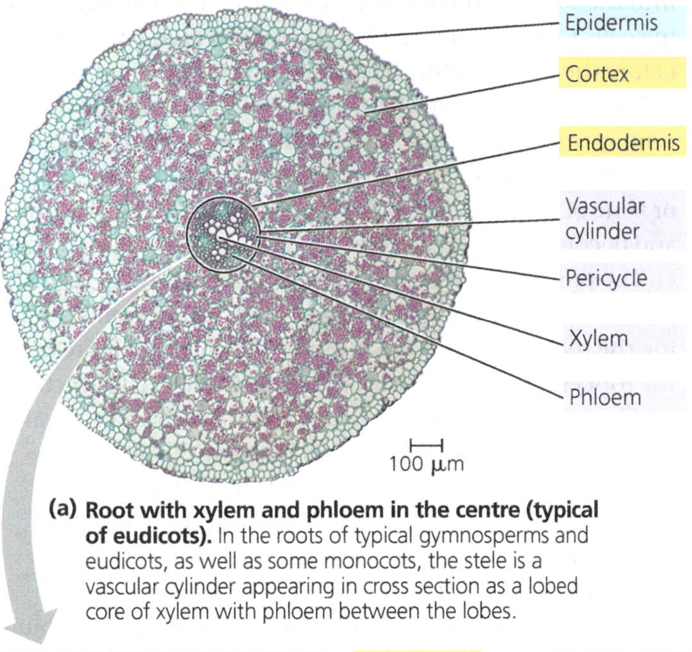

(a) Root with xylem and phloem in the centre (typical of eudicots). In the roots of typical gymnosperms and eudicots, as well as some monocots, the stele is a vascular cylinder appearing in cross section as a lobed core of xylem with phloem between the lobes.

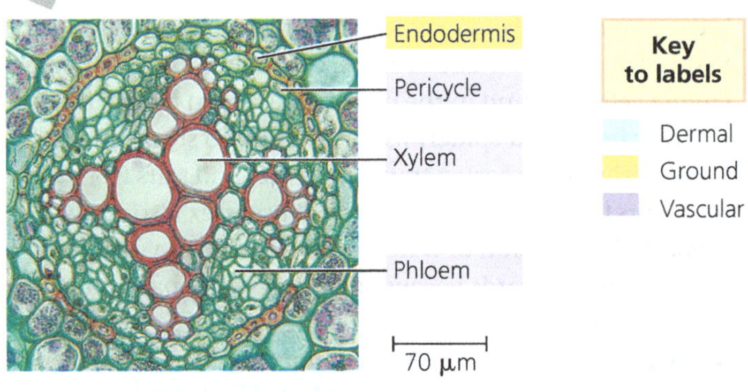

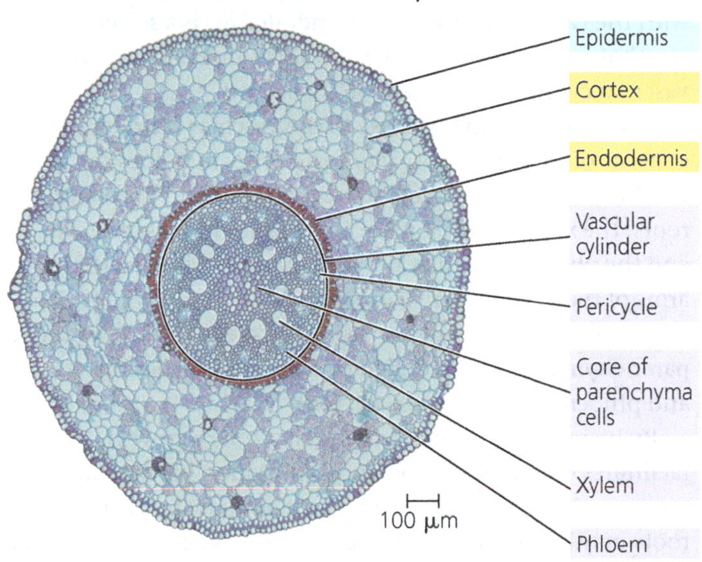

(b) Root with parenchyma in the centre (typical of monocots). The stele of many monocot roots is a vascular cylinder with a core of parenchyma surrounded by a ring of xylem and a ring of phloem.

▼ Figure 35.15 The formation of a lateral root. A lateral root originates in the pericycle, the outermost layer of the vascular cylinder of a root, and destructively pushes through the outer tissues before emerging. In this light micrograph, the view of the original root is a cross section, but the view of the lateral root is a longitudinal section (a view along the length of the lateral root).

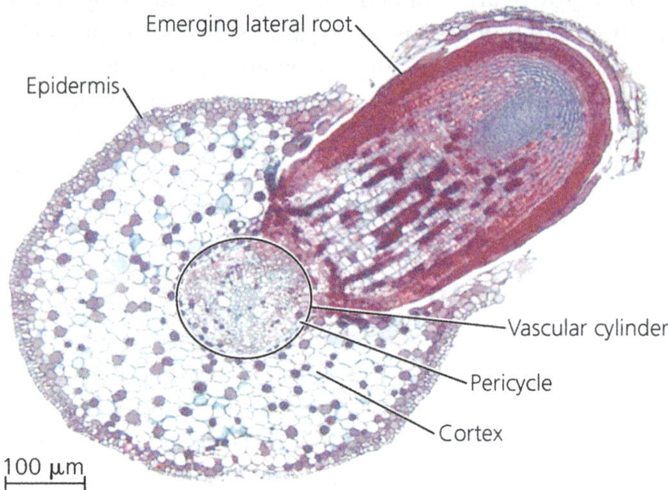

DRAW IT *Draw what the original root and lateral root would look like when viewed from the side, labelling both roots.*

disruptively push through the outer tissues until they emerge from the established root **(Figure 35.15)**.

Primary Growth of Shoots

The entire biomass of a primary shoot—all its leaves and stems—derives from its shoot apical meristem, a dome-shaped mass of dividing cells at the shoot tip **(Figure 35.16)**. The shoot apical meristem is a delicate structure protected by the leaves of the apical bud. These young leaves are spaced close together because the internodes are very short. Shoot elongation is due to the

▼ Figure 35.16 The shoot tip. Leaf primordia arise from the flanks of the dome of the apical meristem. This is a longitudinal section of the shoot tip of *Coleus* (LM).

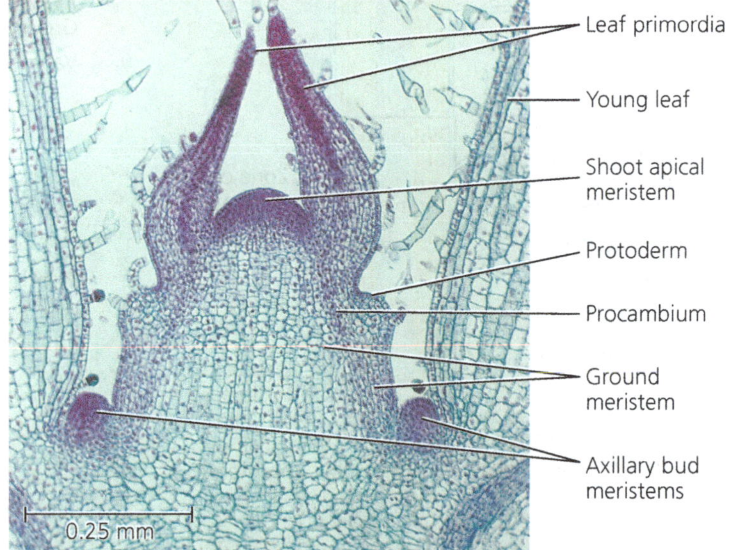

lengthening of internode cells below the shoot tip. As with the root apical meristem, the shoot apical meristem gives rise to three types of primary meristems in the shoot—the protoderm, ground meristem, and procambium. These three primary meristems in turn give rise to the mature primary tissues of the shoot.

The branching of shoots, which is also part of primary growth, arises from the activation of axillary buds, each of which has its own shoot apical meristem. Because of chemical communication by plant hormones, the closer an axillary bud is to an active apical bud, the more inhibited it is, a phenomenon called **apical dominance**. (The specific hormonal changes underlying apical dominance are discussed in Concept 39.2.) If an animal eats the end of the shoot or if shading results in the light being more intense on the side of the shoot, the chemical communication underlying apical dominance is disrupted. As a result, the axillary buds break dormancy and start to grow. Released from dormancy, an axillary bud eventually gives rise to a lateral shoot, complete with its own apical bud, leaves, and axillary buds. When gardeners prune shrubs and pinch back houseplants, they are reducing the number of apical buds a plant has, thereby allowing branches to develop and giving the plants a fuller, bushier appearance.

Stem Growth and Anatomy

The stem is covered by an epidermis that is usually one cell thick and covered with a waxy cuticle that prevents water loss. Some examples of specialised epidermal cells in the stem include guard cells and trichomes.

The ground tissue of stems consists mostly of parenchyma cells. However, collenchyma cells just beneath the epidermis strengthen many stems during primary growth. Sclerenchyma cells, especially fibre cells, also provide support in those parts of the stems that are no longer elongating.

Vascular tissue runs the length of a stem in vascular bundles. Unlike lateral roots, which arise from vascular tissue deep within a root and disrupt the vascular cylinder, cortex, and epidermis as they emerge (see Figure 35.15), lateral shoots develop from axillary bud meristems on the stem's surface and do not disrupt other tissues (see Figure 35.16). Near the soil surface, in the transition zone between shoot and root, the bundled vascular arrangement of the stem converges with the solid vascular cylinder of the root.

The vascular tissue of stems in most eudicot species consists of vascular bundles arranged in a ring **(Figure 35.17a)**. The xylem in each bundle faces the pith, and the phloem in each bundle faces the cortex. In most monocot stems, the vascular bundles do not form a ring but have a more scattered arrangement in the ground tissue **(Figure 35.17b)**.

Leaf Growth and Anatomy

Figure 35.18 provides an overview of leaf anatomy. Leaves develop from **leaf primordia** (singular, *primordium*), projections shaped like a cat's ear that emerge along the sides of the

▼ **Figure 35.17 Organisation of primary tissues in young stems.**

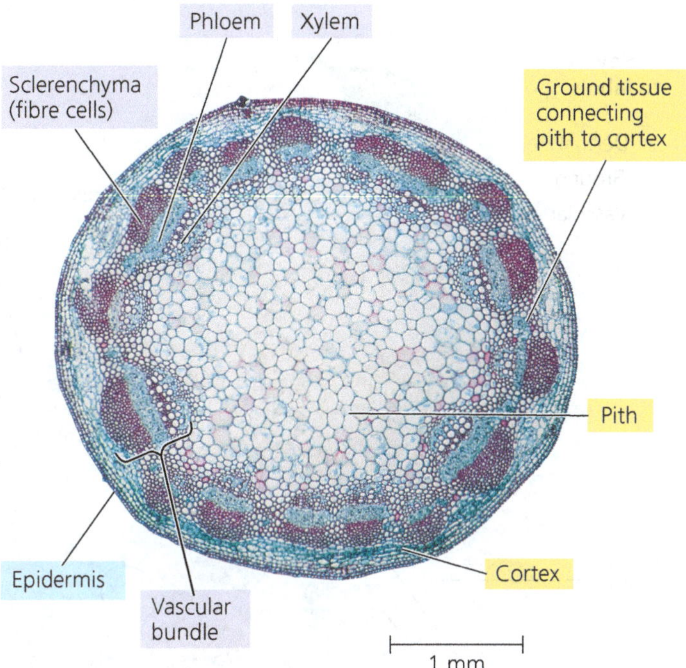

(a) Cross section of stem with vascular bundles forming a ring (typical of eudicots). Ground tissue towards the inside is called pith, and ground tissue towards the outside is called cortex (LM).

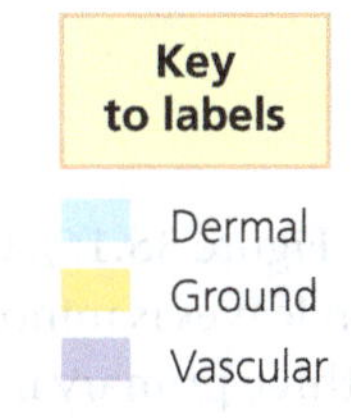

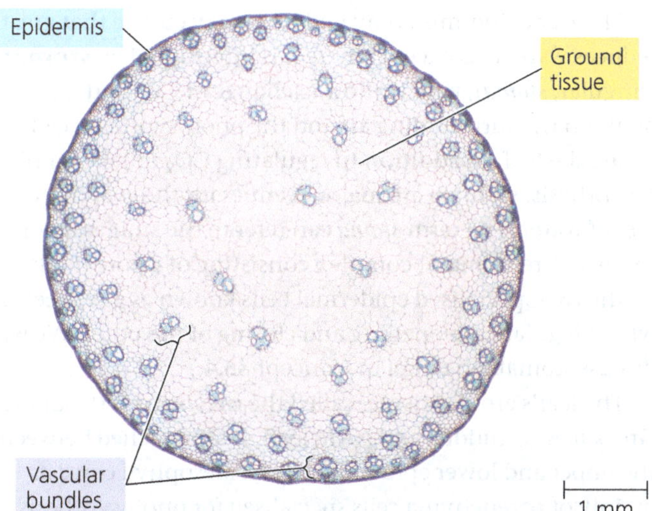

(b) Cross section of stem with scattered vascular bundles (typical of monocots). In such an arrangement, ground tissue is not partitioned into pith and cortex (LM).

VISUAL SKILLS *Compare the locations of the vascular bundles in eudicot and monocot stems. Then explain why the terms pith and cortex are not used in describing the ground tissue of monocot stems.*

▼ **Figure 35.18 Leaf anatomy.**

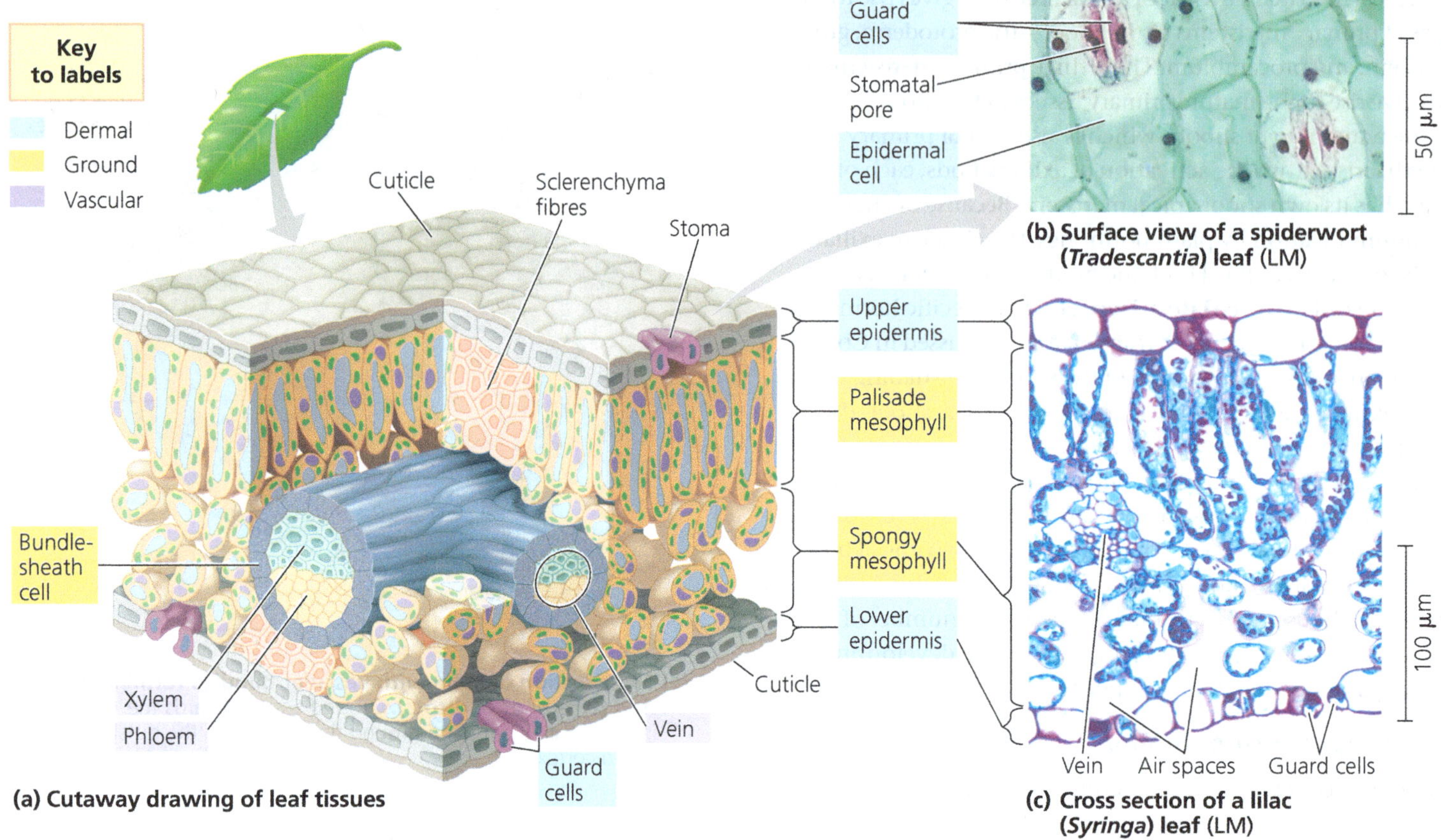

(a) Cutaway drawing of leaf tissues

(b) Surface view of a spiderwort (*Tradescantia*) leaf (LM)

(c) Cross section of a lilac (*Syringa*) leaf (LM)

shoot apical meristem (see Figure 35.16). Unlike roots and stems, secondary growth in leaves is minor or nonexistent. As with roots and stems, the three primary meristems give rise to the tissues of the mature organ.

The leaf epidermis is covered by a waxy cuticle that greatly reduces water loss except where it is interrupted by **stomata** (singular, *stoma*), which allow exchange of CO_2 and O_2 between the surrounding air and the photosynthetic cells inside the leaf. In addition to regulating CO_2 uptake for photosynthesis, stomata are major avenues for the evaporative loss of water. The term *stoma* can refer to the stomatal pore or to the entire stomatal complex consisting of a pore flanked by the two specialised epidermal cells known as guard cells, which regulate the opening and closing of the pore. (We will discuss stomata in detail in Concept 36.4.)

The leaf's ground tissue, called the **mesophyll** (from the Greek *mesos*, middle, and *phyll*, leaf), is sandwiched between the upper and lower epidermal layers. Mesophyll consists mainly of parenchyma cells specialised for photosynthesis. The mesophyll in many eudicot leaves has two distinct layers: palisade and spongy. *Palisade mesophyll*, located beneath the upper epidermis, consists of one or more layers of elongated, chloroplast-rich cells that are specialised for light capture. *Spongy mesophyll*, located inwards from the lower epidermis, consists of irregularly shaped cells that have fewer chloroplasts. These cells form a labyrinth of air spaces through which CO_2 and O_2 circulate to and from the palisade layer. The air spaces are particularly large in the vicinity of stomata, where CO_2 is taken up from the outside air and O_2 is released.

The vascular tissue of each leaf is continuous with the vascular tissue of the stem. Veins subdivide repeatedly and branch throughout the mesophyll. This network brings xylem and phloem into close contact with the photosynthetic tissue, which obtains water and minerals from the xylem and loads its sugars and other organic products into the phloem for transport to other parts of the plant. The vascular structure also functions as a framework that reinforces the shape of the leaf. Each vein is enclosed by a protective *bundle sheath*, a layer of cells that regulates the movement of substances between the vascular tissue and the mesophyll. Bundle-sheath cells are very prominent in leaves of species that carry out C_4 photosynthesis (see Concept 10.5).

CONCEPT CHECK 35.3

1. Contrast primary growth in roots and shoots.
2. **WHAT IF?** A fossil leaf from a region that in the geological past was intermittently very dry and very swampy has stomata only on its upper epidermis. Was the leaf from a desert plant or from a floating aquatic plant? Explain.
3. **MAKE CONNECTIONS** How are root hairs and microvilli analogous structures? (See Figure 6.8 and the discussion of analogy in Concept 26.2.)

For suggested answers, see Appendix A.

CONCEPT 35.4

Secondary growth increases the diameter of stems and roots in woody plants

Many land plants display secondary growth, the growth in thickness produced by lateral meristems. The advent of secondary growth during plant evolution allowed the production of novel plant forms ranging from massive forest trees to woody vines. All gymnosperm species and many eudicot species undergo secondary growth, but it is unusual in monocots. It occurs in stems and roots of woody plants, but rarely in leaves.

Secondary growth consists of the tissues produced by the vascular cambium and cork cambium. The vascular cambium adds secondary xylem (wood) and secondary phloem, thereby increasing vascular flow and support for the shoots. The cork cambium produces a tough, thick covering of waxy cells that protect the stem from water loss and from invasion by insects, bacteria, and fungi.

In woody plants, primary growth and secondary growth occur simultaneously. As primary growth adds leaves and lengthens stems and roots in the younger regions of a plant, secondary growth increases the diameter of stems and roots in older regions where primary growth has ceased. The process is similar in shoots and roots. **Figure 35.19** provides an overview of growth in a woody stem.

▼ Figure 35.19 Secondary growth of a woody stem.

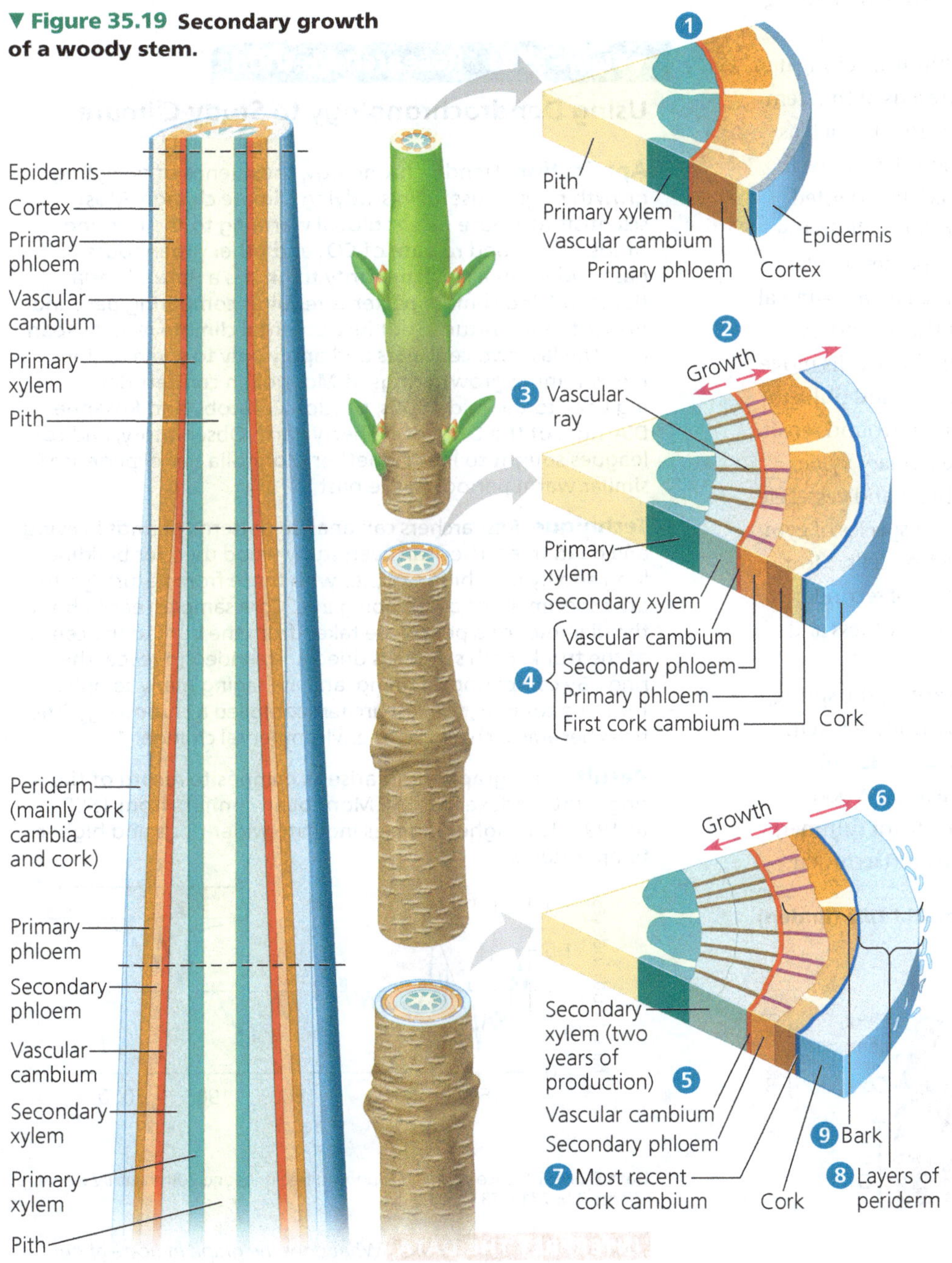

1 Primary growth from the activity of the apical meristem is complete here. The vascular cambium has formed, and its cell divisions will give rise to the bulk of secondary growth.

2 Although primary growth continues in the apical bud, only secondary growth occurs in this region. The stem thickens as the vascular cambium forms secondary xylem to the inside and secondary phloem to the outside.

3 Some stem cells of the vascular cambium give rise to vascular rays.

4 As the vascular cambium's diameter increases, the secondary phloem and other tissues external to the cambium can't keep pace because their cells no longer divide. As a result, these tissues, including the epidermis, will eventually rupture. A second lateral meristem, the cork cambium, develops from parenchyma cells in the cortex. The cork cambium produces cork cells, which replace the epidermis.

5 In year 2 of secondary growth, the vascular cambium produces more secondary xylem and phloem. Most of the thickening is from secondary xylem. Meanwhile, the cork cambium produces more cork.

6 As the stem's diameter increases, the outermost tissues exterior to the cork cambium rupture and are sloughed off.

7 In many cases, the cork cambium re-forms deeper in the cortex. When none of the cortex is left, the cambium develops from phloem parenchyma cells.

8 Each cork cambium and the tissues it produces form a layer of periderm.

9 Bark consists of all tissues exterior to the vascular cambium.

VISUAL SKILLS *Based on the diagram, explain how the vascular cambium causes some tissues to rupture.*

The Vascular Cambium and Secondary Vascular Tissue

The vascular cambium, a cylinder of meristematic cells only one cell thick, is wholly responsible for the production of secondary vascular tissue. In a typical woody stem, the vascular cambium is located outside the pith and primary xylem and to the inside of the primary phloem and the cortex. In a typical woody root, the vascular cambium forms exterior to the primary xylem and interior to the primary phloem and pericycle.

In cross section, the vascular cambium appears as a ring of meristematic cells (see step 4 of Figure 35.19). As these cells divide, they increase the cambium's circumference and add secondary xylem to the inside and secondary phloem to the outside. Each ring is larger than the previous ring, increasing the diameter of roots and stems.

Some of the stem cells in the vascular cambium are elongated and oriented with their long axis parallel to the axis of the stem or root. The cells they produce give rise to mature cells such as the tracheids, vessel elements, and fibres of the xylem, as well as the sieve-tube elements, companion cells, axially oriented parenchyma, and fibres of the phloem. Other stem cells in the vascular cambium are shorter and are oriented perpendicular to the axis of the stem or root: they give rise to *vascular rays*—radial files of mostly parenchyma cells that connect the secondary xylem and phloem (see step 3 of Figure 35.19). These cells move water and nutrients between the secondary xylem and phloem, store carbohydrates and other reserves, and aid in wound repair.

As secondary growth continues, layers of secondary xylem (wood) accumulate, consisting mainly of tracheids and vessel elements (see Figure 35.10), as well as fibres. In most species of gymnosperms, tracheids are the only water-conducting cells. Most angiosperms also have vessel elements. The walls of secondary xylem cells are heavily lignified, giving wood its hardness and strength.

In temperate regions, wood that develops early in the spring, known as early (or spring) wood, usually has secondary xylem cells with large diameters and thin cell walls **(Figure 35.20)**. This structure maximises delivery of water to leaves. Wood produced later in the growing season is called late (or summer) wood. It has thick-walled cells that do not transport as much water but provide more support. Because there is a marked contrast between the large cells of the new early wood and the smaller cells of the late wood of the previous growing season, a year's growth appears as a distinct *growth ring* in cross sections of most tree trunks and roots. Therefore, researchers can estimate a tree's age by counting growth rings. *Dendrochronology* is the science of analysing tree growth ring patterns. Growth rings vary in thickness, depending on seasonal growth. Trees grow well in wet and warm years but may grow hardly at all in cold or dry years. Since a thick ring indicates a warm year and a thin ring indicates a cold or dry one, scientists use ring patterns to study climate changes **(Figure 35.21)**.

As a tree or woody shrub ages, older layers of secondary xylem no longer transport water and minerals (a solution

▼ Figure 35.20 Cross section of a three-year-old *Tilia* (linden) stem. (LM)

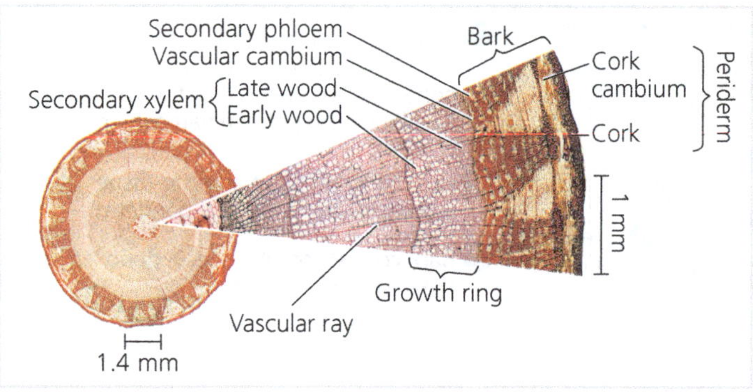

▼ Figure 35.21 Research Method

Using Dendrochronology to Study Climate

Application Dendrochronology, the science of analysing growth rings, is useful in studying climate change. Most scientists attribute recent global warming to the burning of fossil fuels and release of CO_2 and other greenhouse gases, whereas a small minority think it is a natural variation. Studying climate patterns requires comparing past and present temperatures, but instrumental climate records span only the last two centuries and apply only to some regions. By examining growth rings of Mongolian conifers dating back to the mid-1500s, Gordon C. Jacoby and Rosanne D'Arrigo, of the Lamont-Doherty Earth Observatory, and colleagues sought to learn whether Mongolia has experienced similar warm periods in the past.

Technique Researchers can analyse patterns of rings in living and dead trees. They can even study wood used for building long ago by matching samples with those from naturally situated specimens of overlapping age. Core samples, each about the diameter of a pencil, are taken from the bark to the centre of the trunk. Each sample is dried and sanded to reveal the rings. By comparing, aligning, and averaging many samples from the conifers, the researchers compiled a chronology. The trees became a chronicle of environmental change.

Results This graph summarises a composite record of the ring-width indexes for the Mongolian conifers from 1550 to 1993. The higher indexes indicate wider rings and higher temperatures.

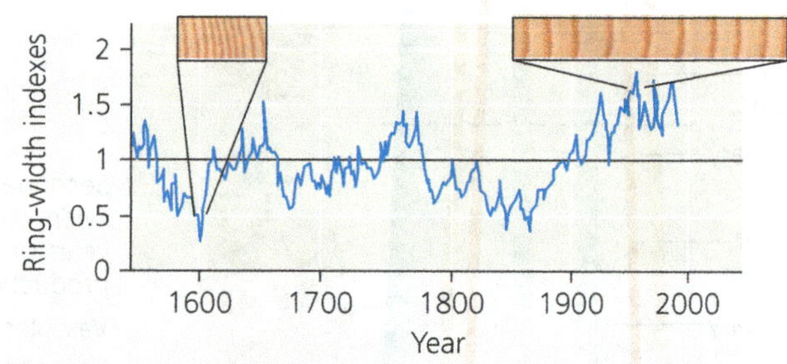

Data from G. C. Jacoby et al., Mongolian tree rings and 20th-century warming, *Science* 273:771–773 (1996).

INTERPRET THE DATA *What does the graph indicate about environmental change during the period 1550–1993?*

called xylem sap). These layers are called *heartwood* because they are closer to the centre of a stem or root **(Figure 35.22)**. The newest, outer layers of secondary xylem still transport xylem sap and are therefore known as *sapwood*. Sapwood allows a large tree to survive even if the centre of its trunk is hollow **(Figure 35.23)**. Because each new layer of secondary xylem has a larger circumference, secondary growth enables the xylem to transport more sap each year, supplying an increasing number of leaves. Heartwood is generally darker than sapwood because of resins and other compounds that permeate the cell cavities and help protect the core of the tree from fungi and wood-boring insects.

▼ **Figure 35.22 Anatomy of a tree trunk.**

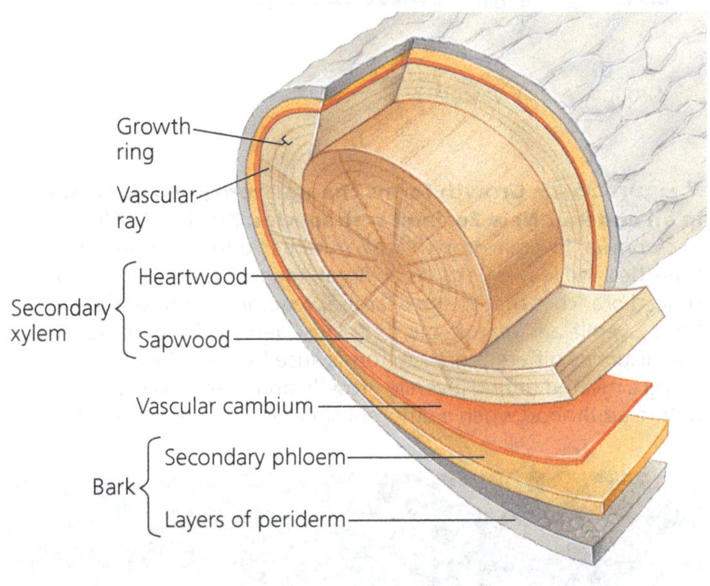

◀ **Figure 35.23 Is this tree living or dead?** The Wawona Sequoia tunnel in Yosemite National Park in California was cut in 1881 as a tourist attraction. This giant sequoia (*Sequoiadendron giganteum*) lived for another 88 years before falling during a severe winter. It was 71.3 m tall and estimated to be 2,100 years old. Though conservation policies today would forbid the mutilation of such an important specimen, the Wawona Sequoia did teach a valuable botanical lesson: Trees can endure the excision of large portions of their heartwood for decades.

VISUAL SKILLS *Name in sequence the tissues that were destroyed as the loggers excavated through the base of the tree to its centre. Refer also to Figure 35.19.*

Only the youngest secondary phloem, closest to the vascular cambium, functions in sugar transport. As a stem or root increases in circumference, the older secondary phloem is sloughed off, which is one reason secondary phloem does not accumulate as extensively as secondary xylem.

The Cork Cambium and the Production of Periderm

During the early stages of secondary growth, the epidermis is pushed outwards, causing it to split, dry, and fall off the stem or root. It is replaced by tissues produced by the first cork cambium, a cylinder of dividing cells that arises in the outer cortex of stems (see Figure 35.19) and in the pericycle in roots. The cork cambium gives rise to cork cells that accumulate to the outside of the cork cambium. As cork cells mature, they deposit a waxy, hydrophobic material called suberin in their walls before dying. Because cork cells have suberin and are usually compacted together, most of the periderm is impermeable to water and gases, unlike the epidermis. Cork thus functions as a barrier that helps protect the stem or root from water loss, physical damage, and pathogens. It should be noted that "cork" is commonly and incorrectly referred to as "bark." In plant biology, **bark** includes all tissues external to the vascular cambium. Its main components are the secondary phloem (produced by the vascular cambium) and, external to that, the most recent periderm and all the older layers of periderm (see Figure 35.22). As this process continues, older layers of periderm are sloughed off, as evident in the cracked, peeling exteriors of many tree trunks.

How can living cells in the interior tissues of woody organs absorb oxygen and respire if they are surrounded by a waxy periderm? Dotting the periderm are small, raised areas called **lenticels**, in which there is more space between cork cells, enabling living cells within a woody stem or root to exchange gases with the outside air. Lenticels often appear as horizontal slits, as shown on the stem in Figure 35.19.

Figure 35.24 summarises the relationships between the primary and secondary tissues of a woody shoot.

Evolution of Secondary Growth

EVOLUTION Surprisingly, some insights into the evolution of secondary growth have been achieved by studying the herbaceous plant *Arabidopsis thaliana*. Researchers have found that they can stimulate some secondary growth in *Arabidopsis* stems by adding weights to the plant. These findings suggest that weight carried by the stem activates a developmental program leading to wood formation. Moreover, several developmental genes that regulate shoot apical meristems in *Arabidopsis* have been found to regulate vascular cambium activity in poplar (*Populus*) trees. This suggests that the processes of primary and secondary growth are evolutionarily more closely related than was previously thought.

▼ **Figure 35.24 A summary of primary and secondary growth in a woody shoot.** Woody roots have the same meristems and tissues. However, the ground tissue of a root is not divided into pith and cortex, and the cork cambium arises instead from the pericycle, the outermost layer of the vascular cylinder.

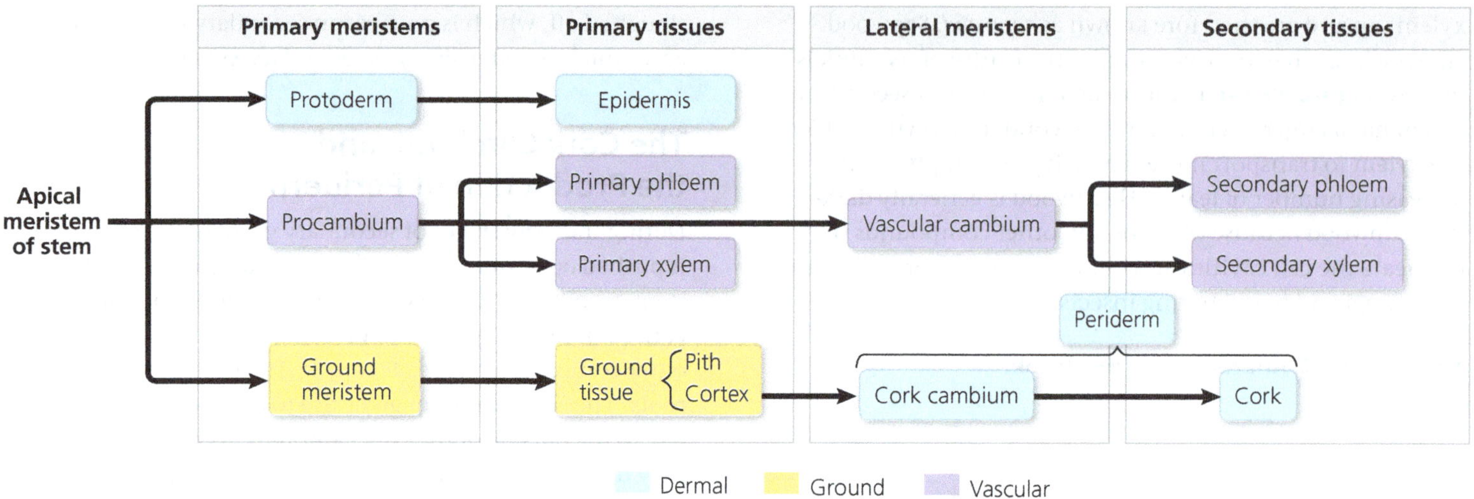

CONCEPT CHECK 35.4

1. A sign is hammered into a tree 2 m from the tree's base. If the tree is 10 m tall and elongates 1 m each year, how high will the sign be after ten years?
2. Stomata and lenticels are both involved in exchange of CO_2 and O_2. Why do stomata need to be able to close, but lenticels do not?
3. Would you expect a tropical tree to have distinct growth rings? Why or why not?
4. **WHAT IF?** If a complete ring of bark is removed from around a tree trunk (a technique called girdling), would the tree die slowly (in weeks) or quickly (in days)? Explain why.

For suggested answers, see Appendix A.

CONCEPT 35.5

Growth, morphogenesis, and cell differentiation produce the plant body

The specific series of changes by which cells form tissues, organs, and organisms is called **development**. Development unfolds according to the genetic information that an organism inherits from its parents but is also influenced by the external environment. A single genotype can produce different phenotypes when exposed to different environments. For example, juveniles of New Zealand's kōwhai (*Sophora microphylla*) grow into thorny, dense, tightly branched canopies close to the ground, while adults develop open, spreading canopies far above the ground **(Figure 35.25)**. Perhaps kōwhai's juvenile form evolved in a "biological arms race" with the moa, which is now extinct. The plant's thorny, dense canopy may have protected leaves from all but the most determined of these birds. Over time, the young kōwhai produced shoots beyond the reach of the long-necked moa.

▼ **Figure 35.25 Growth forms change from juvenile to adult in an endemic New Zealand tree species.** Kōwhai (*Sophora microphylla*) juvenile growth form exhibits tightly interwoven branches and leaves (left). As an adult, the tree produces widely spaced branches that create a large spreading canopy (right). Yet both juvenile and mature plants possess genetically identical cells. The distinctly different growth forms arise from a suite of genes being activated during juvenile growth, and then deactivated before adulthood when a different suite of genes become active.

In that new environment, pressures on survival shifted from avoiding predation to maximising light interception to compete more effectively with other plants. A large, spreading canopy maximises photosynthesis and shades out potential competitors in the understorey. Changes in the kōwhai's growth form from juvenile to adult may represent the outcomes of battles fought with the moa in the near past. The evolution of dramatically different growth forms from juvenile to adult occurs more commonly in plants than in animals. These, and similar differences in the evolution of plant growth forms, may help compensate for the inability of plants to escape adverse conditions by moving.

The three overlapping processes involved in the development of a multicellular organism are growth, morphogenesis, and cell differentiation. *Growth* is an irreversible increase in size. *Morphogenesis* (from the Greek *morphê*, shape, and *genesis*, creation) is the process that gives a tissue, organ, or organism its shape and determines the positions of cell types. *Cell differentiation* is the process by which cells with the same genes become different from one another. We'll examine these three processes in turn, but first we'll discuss how applying techniques of modern molecular biology to model organisms, particularly *Arabidopsis thaliana*, has revolutionised the study of plant development.

Model Organisms: Revolutionising the Study of Plants

As in other branches of biology, techniques of molecular biology and a focus on model organisms such as *Arabidopsis thaliana* have catalysed a research explosion in the last few decades. *Arabidopsis*, a tiny weed in the mustard family, has no inherent agricultural value but is a favoured model organism of plant geneticists and molecular biologists for many reasons. It is so small that thousands of plants can be cultivated in a few square metres of lab space. It also has a short generation time, taking about six weeks for a seed to grow into a mature plant that produces more seeds. This rapid maturation enables biologists to conduct genetic cross experiments in a relatively short time. One plant can produce over 5,000 seeds, another property that makes *Arabidopsis* useful for genetic analysis.

Beyond these basic traits, the plant's genome makes it particularly well suited for analysis by molecular genetic methods. The *Arabidopsis* genome, which includes about 27,000 protein-encoding genes, is among the smallest known in plants. Furthermore, the plant has only five pairs of chromosomes, making it easier for geneticists to locate specific genes. Because *Arabidopsis* has such a small genome, it was the first plant to have its entire genome sequenced.

The natural range of *Arabidopsis* includes varied climates and elevations, from the high mountains of Central Asia to the European Atlantic coast, and from North Africa to the Arctic Circle. These local varieties can differ markedly in outward appearance **(Figure 35.26)**. Genome-sequencing efforts are being expanded to include hundreds of populations of Arabidopsis from throughout its natural range in Eurasia. Contained in the genomes of these populations is information about evolutionary adaptations that enabled Arabidopsis to expand its range into new environments following the retreat of the last ice age. This information may provide plant breeders with new insights and strategies for crop improvement.

Another property that makes *Arabidopsis* attractive to molecular biologists is that its cells can be easily transformed with *transgenes*, genes from a different organism that are stably introduced into the genome of another. CRISPR technology (see Figure 20.14), which is rapidly becoming the technique of choice for creating plants with specific mutations, has been used successfully in *Arabidopsis*. By disrupting or "knocking out" a specific gene, scientists can garner important information about the gene's normal function.

▼ Figure 35.26 Variations in leaf arrangement, leaf shape, and shoot growth between different populations of *Arabidopsis thaliana*. Information in the genomes of these populations may provide insights into strategies for expanding crop production into new environments.

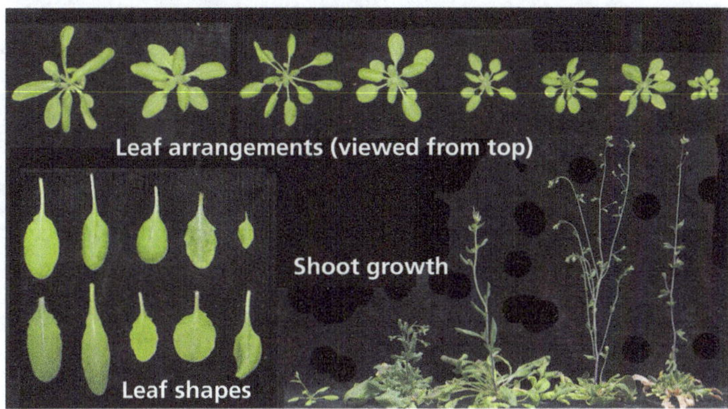

Large-scale projects are under way to determine the function of every gene in *Arabidopsis*. By identifying each gene's function and tracking every biochemical pathway, researchers aim to determine the blueprints for plant development, a major goal of systems biology. It may one day be possible to make a computer-generated "virtual plant" that enables researchers to visualise which genes are activated in different parts of the plant as the plant develops.

Basic research on model organisms such as *Arabidopsis* has accelerated the pace of discovery in the plant sciences, including the identification of the complex genetic pathways underlying plant structure. As you read more about this, you'll be able to appreciate not only the power of studying model organisms but also the history of investigation that underpins all modern plant research.

Growth: Cell Division and Cell Expansion

Cell division enhances the potential for growth by increasing the number of cells, but plant growth itself is brought about by cell enlargement. The process of plant cell division is described more fully in the chapter, "The Cell Cycle" (see Figure 12.10), and the chapter, "Plant Responses to Internal and External Signals" discusses the process of cell elongation (see Figure 39.7). Here we are concerned with how cell division and enlargement contribute to plant form.

Cell Division

The new cell walls that bisect plant cells during cytokinesis develop from the cell plate (see Figure 12.10). The precise plane of cell division, determined during late interphase, usually corresponds to the shortest path that halves the cytoplasm of the parent cell. However, during certain points in development the cytoplasm may not be divided equally, resulting in one daughter cell being larger than the other, even though they have the same number of chromosomes. Such cases of *asymmetrical cell division* usually signal a key event in development. For example, the formation of guard cells involves an asymmetrical cell division. An epidermal cell divides asymmetrically, forming a large cell that remains an undifferentiated epidermal cell and a small cell that becomes the guard cell "mother cell." Guard cells form when this small mother cell divides in a plane perpendicular to the first cell division **(Figure 35.27)**. Thus, asymmetrical cell division generates cells with different fates—that is, cells that mature into different types.

Cell Expansion

Cell divisions do not constitute growth because there is no increase of mass involved. Rather, it is cell expansion that is responsible for plant growth. Before discussing how cell expansion contributes to plant growth and form, it is useful to consider the difference in cell expansion between plants and animals. Animal cells grow mainly by synthesising protein-rich cytoplasm, a metabolically expensive process. Growing plant cells also produce additional protein-rich material in their cytoplasm, but water uptake typically accounts for about 90% of expansion. Most of this water is stored in the large central vacuole. The vacuolar solution, or *vacuolar sap*, is very dilute and nearly devoid of the energetically expensive macromolecules that are found in great abundance in the rest of the cytoplasm. Large vacuoles are therefore a "cheap" way of filling space, enabling a plant to grow rapidly and economically. Bamboo shoots, for instance, can elongate more than 2 m per week. Rapid and efficient extensibility of shoots and roots was an important evolutionary adaptation that increased their exposure to light and soil.

Plant cells rarely expand equally in all directions. Their greatest expansion is usually oriented along the plant's main axis. For example, cells near the tip of the root may elongate 20 times or more their original length, with relatively little increase in width. The orientation of cellulose microfibrils in the innermost layers of the cell wall causes this differential growth. The microfibrils do not stretch, so the cell expands mainly perpendicular to the main orientation of the microfibrils, as shown in **Figure 35.28**. A leading hypothesis proposes that microtubules positioned just beneath the plasma membrane organise the cellulose-synthesising enzyme complexes and guide their movement through the plasma membrane as they create the microfibrils that form much of the cell wall.

▼ Figure 35.27 Asymmetrical cell division and stomatal development. An asymmetrical cell division precedes the development of epidermal guard cells, the cells that border stomata (see Figure 35.18).

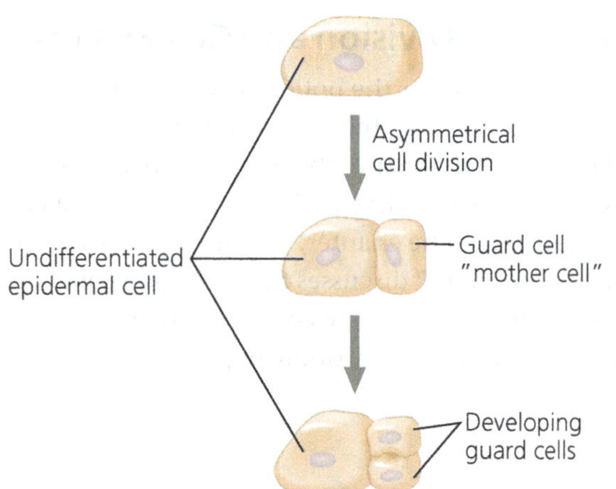

Morphogenesis and Pattern Formation

A plant's body is more than a collection of dividing and expanding cells. During morphogenesis, cells acquire different identities in an ordered spatial arrangement. For example, dermal tissue forms on the exterior and vascular tissue in the interior—never the other way around. The development of specific structures in specific locations is called **pattern formation**.

▼ Figure 35.28 The orientation of plant cell expansion. Growing plant cells expand mainly through water uptake. In a growing cell, enzymes weaken cross-links in the cell wall, allowing it to expand as water diffuses into the vacuole by osmosis; at the same time, more microfibrils are made. The orientation of the cell expansion is mainly perpendicular to the orientation of cellulose microfibrils in the wall. The orientation of microtubules in the cell's outermost cytoplasm determines the orientation of cellulose microfibrils (fluorescent LM). The microfibrils are embedded in a matrix of other (noncellulose) polysaccharides, some of which form the cross-links visible in the TEM.

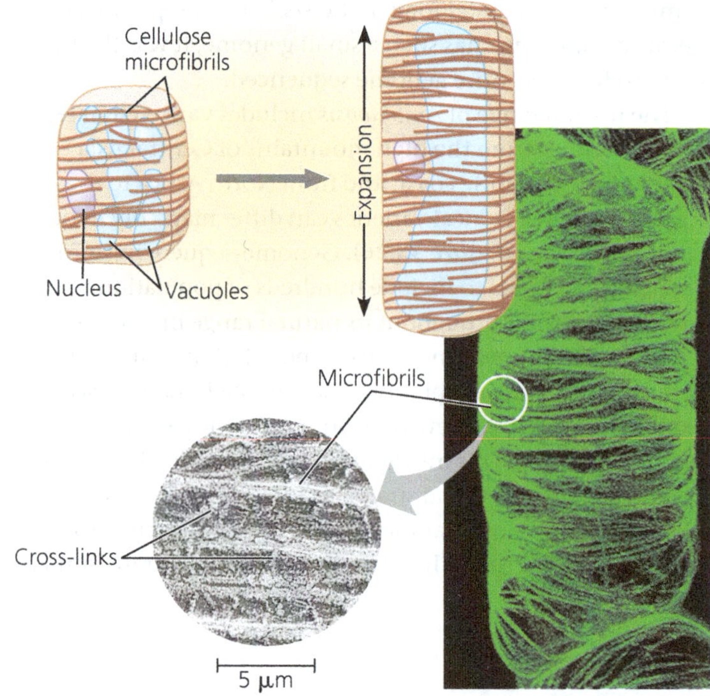

Two types of hypotheses have been put forward to explain how the fate of plant cells is determined during pattern formation. Hypotheses based on *lineage-based mechanisms* propose that cell fate is determined early in development and that cells pass on this destiny to their progeny. In this view, the basic pattern of cell differentiation is mapped out according to the directions in which meristematic cells divide and expand. On the other hand, hypotheses based on *position-based mechanisms* propose that the cell's final position in an emerging organ determines what kind of cell it will become. In support of this view, experiments in which neighbouring cells have been destroyed with lasers have demonstrated that a plant cell's fate is established late in the cell's development and largely depends on signalling from its neighbours.

In contrast, cell fate in animals is largely determined by lineage-dependent mechanisms involving transcription factors. The homeotic (*Hox*) genes that encode such transcription factors are critical for the proper number and placement of embryonic structures, such as legs and antennae, in the fruit fly *Drosophila* (see Figure 18.19). Interestingly, corn has a homologue of *Hox* genes called *KNOTTED-1*, but unlike its counterparts in the animal world, *KNOTTED-1* does not affect the number or placement of plant organs. As you will see, an unrelated class of transcription factors called *MADS-box* proteins plays that role in plants. *KNOTTED-1* is, however, important in the development of leaf shape, including the production of compound leaves. If the *KNOTTED-1* gene is expressed in greater quantity than normal in the genome of tomato plants, the normally compound leaves will then become "super-compound" **(Figure 35.29)**.

▼ Figure 35.29 Overexpression of a *Hox*-like gene in leaf formation. *KNOTTED-1* is a gene that is involved in leaf and leaflet formation. An increase in its expression in tomato plants results in leaves that are "super-compound" (right) compared with normal leaves (left).

Gene Expression and the Control of Cell Differentiation

The cells of a developing organism can synthesise different proteins and diverge in structure and function even though they share a common genome. If a mature cell removed from a root or leaf can dedifferentiate in tissue culture and give rise to the diverse cell types of a plant, then it must possess all the genes necessary to make any kind of cell in the plant. Therefore, cell differentiation depends, to a large degree, on the control of gene expression—the regulation of transcription and translation, resulting in the production of specific proteins.

Evidence suggests that the activation or inactivation of specific genes involved in cell differentiation results largely from cell-to-cell communication. Cells receive information about how they should specialise from neighbouring cells. For example, two cell types arise in the root epidermis of *Arabidopsis*: root hair cells and hairless epidermal cells. Cell fate is associated with the position of the epidermal cells relative to other plant cells. The immature epidermal cells that are in contact with two underlying cells of the root cortex differentiate into root hair cells, whereas the immature epidermal cells in contact with only one cell in the cortex differentiate into mature hairless cells. The differential expression of a homeotic gene called *GLABRA-2* (from the Latin *glaber*, bald) is needed for proper distribution of root hairs **(Figure 35.30)**. Researchers have demonstrated this requirement by coupling the *GLABRA-2* gene to a "reporter gene" that causes every cell expressing *GLABRA-2* in the root to turn pale blue following a certain protocol. The *GLABRA-2* gene is normally expressed only in epidermal cells that will not develop root hairs.

▼ Figure 35.30 Control of root hair differentiation by a homeotic gene. (LM)

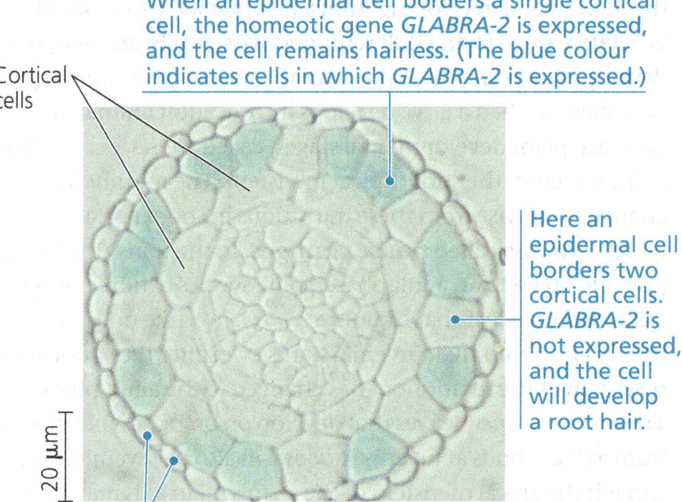

WHAT IF? *What would the roots look like if* GLABRA-2 *were rendered dysfunctional by a mutation?*

Shifts in Development: Phase Changes

Multicellular organisms generally pass through developmental stages. In humans, these are infancy, childhood, adolescence, and adulthood, with puberty as the dividing line between

▼ Figure 35.31 Phase change in *Muehlenbeckia australis*. This native climbing vine of New Zealand has two types of leaves: fiddle-shaped juvenile leaves (top) and oval, smooth-margined mature leaves. This dual foliage reflects a phase change in the development of the apical meristem of each shoot. Once a node forms, the developmental phase—juvenile or adult—is fixed; fiddle-shaped leaves do not mature into oval-shaped leaves.

Juvenile leaf

Mature leaf

the nonreproductive and reproductive stages. Plants also pass through stages, developing from a juvenile stage to an adult vegetative stage to an adult reproductive stage. In animals, the developmental changes take place throughout the entire organism, such as when a larva develops into an adult animal. In contrast, plant developmental stages, called *phases*, occur within a single region, the shoot apical meristem. The morphological changes that arise from these transitions in shoot apical meristem activity are called **phase changes**. In the transition from a juvenile phase to an adult phase, some species exhibit conspicuous changes in leaf shape **(Figure 35.31)**. Juvenile nodes and internodes retain their juvenile status even after the shoot apical meristem of the main shoot has changed to the adult phase. Therefore, any *new* leaves that develop on branches that emerge from axillary buds at juvenile nodes will also be juvenile, even though the apical meristem of the stem's main axis may have been producing mature nodes for years.

If environmental conditions permit, an adult plant is induced to flower. Biologists have made great progress in explaining the genetic control of floral development—the topic of the next section.

Genetic Control of Flowering

Flower formation involves a phase change from vegetative growth to reproductive growth. This transition is triggered by a combination of environmental cues, such as day length, and internal signals, such as hormones. (You will learn more about the roles of these signals in flowering in Concept 39.3.) Unlike vegetative growth, which is indeterminate, floral growth is usually determinate: The production of a flower by a shoot apical meristem generally stops the primary growth of that shoot. The transition from vegetative growth to flowering is associated with the switching on of flower-inducing genes. The protein products of these genes are transcription factors that regulate the genes required for the conversion of the indeterminate vegetative meristems to determinate floral meristems.

When a shoot apical meristem is induced to flower, the order of each primordium's emergence determines its development into a specific type of floral organ—a sepal, petal, stamen, or carpel (see Figure 30.9 to review basic flower structure). These floral organs form four whorls that can be described roughly as concentric "circles" when viewed from above. Sepals form the first (outermost) whorl; petals form the second; stamens form the third; and carpels form the fourth (innermost) whorl. Plant biologists have identified several genes that encode transcription factors that regulate the development of this characteristic floral pattern. Positional information determines which genes are expressed in a particular floral organ primordium. The result is the development of an emerging floral primordium into a specific floral organ. A mutation in a flower-inducing gene can cause abnormal floral development, such as petals growing in place of stamens **(Figure 35.32)**. Some homeotic mutants with increased petal numbers produce showier flowers that are prized by gardeners.

▼ Figure 35.32 Genes and pattern formation in flower development.

▲ Normal *Arabidopsis* flower. *Arabidopsis* normally has four whorls of flower parts: sepals (Se), petals (Pe), stamens (St), and carpels (Ca).

▶ Abnormal *Arabidopsis* flower. Researchers have identified several mutations that cause abnormal flowers to develop. This flower has an extra set of petals in place of stamens and an internal flower where normal plants have carpels.

MAKE CONNECTIONS *Provide another example of a homeotic gene mutation that leads to organs being produced in the wrong place (see Concept 18.4).*

By studying mutants with abnormal flowers, researchers have identified and cloned three classes of genes that determine floral organ identity. **Figure 35.33a** shows a simplified version of the **ABC hypothesis** of flower formation, which proposes that three classes of genes direct the formation of the four types of floral organs. According to the ABC hypothesis, each class of genes is switched on in two specific whorls of the floral meristem. Normally, *A* genes are switched on in the two outer whorls (sepals and petals); *B* genes are switched on in the two middle whorls (petals and stamens); and *C* genes are switched on in the two inner whorls (stamens and carpels). Sepals arise from those parts of floral meristems in which only *A* genes are active; petals arise where *A* and *B* genes are active; stamens where *B* and *C* genes are active; and carpels where only *C* genes are active. The ABC hypothesis can account for the phenotypes of mutants lacking *A*, *B*, or *C* gene activity, with one addition: Where *A* gene activity is present, it inhibits *C*, and vice versa. If either the *A* gene or *C* gene is suppressed, the other gene is expressed. **Figure 35.33b** shows the floral patterns of mutants lacking each of the three classes of genes and depicts how the hypothesis accounts for the floral phenotypes. By constructing such hypotheses and designing experiments to test them, researchers are tracing the genetic basis of plant development.

In dissecting the plant to examine its parts, as we have done in this chapter, we must remember that the whole plant functions as an integrated organism. Plant structures largely reflect evolutionary adaptations to the challenges of a photo-autotrophic existence on land.

CONCEPT CHECK 35.5

1. How can two cells in a plant have vastly different structures even though they have the same genome?
2. What are three differences between animal development and plant development?
3. **WHAT IF?** In some species (for example, Western Australia's critically endangered small-flowered snottygobble *(Persoonia micranthera)*), sepals look like petals, and both are collectively called "tepals." Suggest an extension to the ABC hypothesis that could account for tepals.

For suggested answers, see Appendix A.

▼ Figure 35.33 The ABC hypothesis for the functioning of genes in flower development.

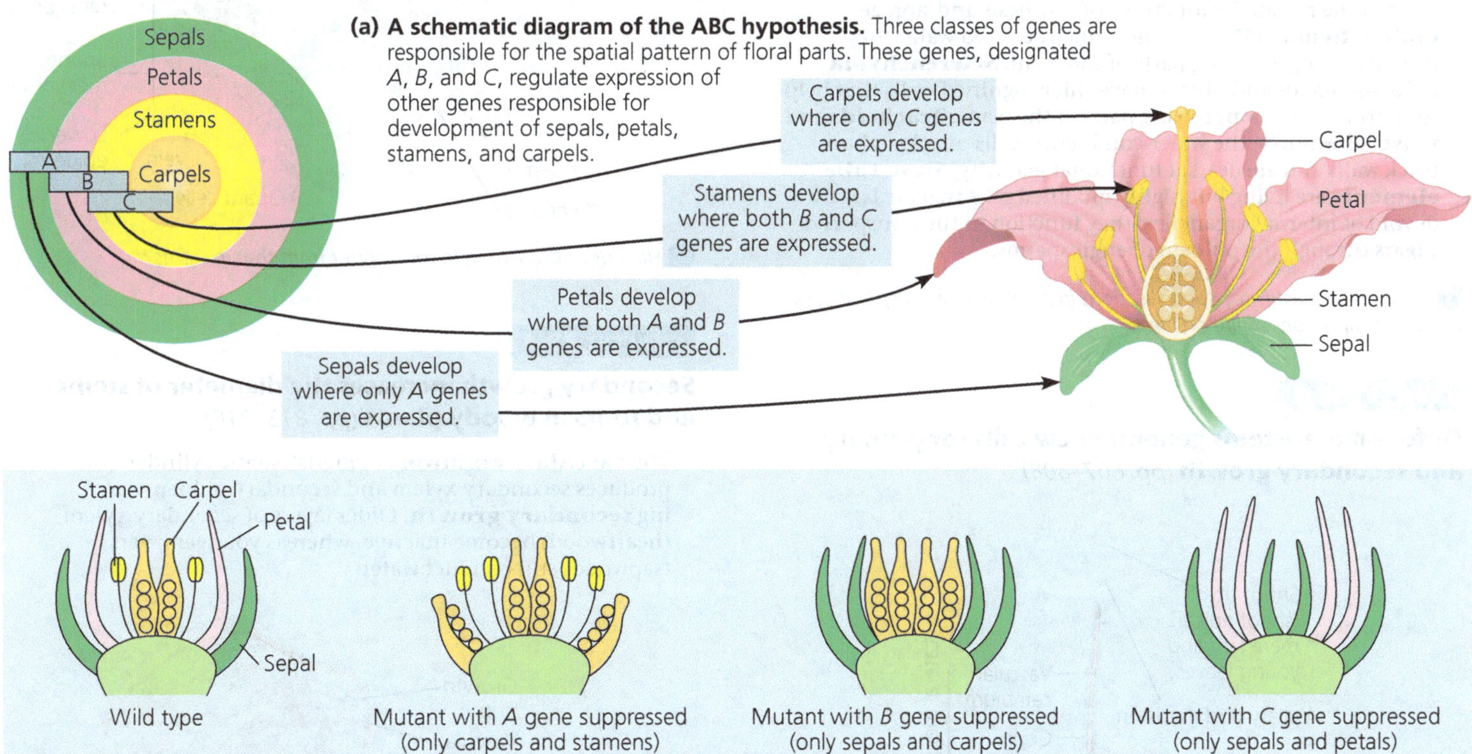

(a) A schematic diagram of the ABC hypothesis. Three classes of genes are responsible for the spatial pattern of floral parts. These genes, designated *A*, *B*, and *C*, regulate expression of other genes responsible for development of sepals, petals, stamens, and carpels.

(b) Side view of wild-type flower and flowers with organ identity mutations. The phenotype of mutants lacking a functional *A*, *B*, or *C* gene can be explained by the model in part (a) and the observation that if the *A* gene or *C* gene is suppressed, then the other gene is expressed in that whorl. For example, if the *A* gene is suppressed in a mutant, then the *C* gene is expressed where the *A* gene would normally be expressed. Therefore, carpels (*C* gene expressed) develop in the outermost whorl, and stamens (*B* and *C* genes expressed) develop in the next whorl.

DRAW IT *(a) For each mutant, draw a "bull's-eye" diagram like the one in part (a), labelling the type of organ and gene(s) expressed in each whorl. (b) Draw and label a "bull's-eye" diagram for a mutant flower in which the A and B genes were suppressed.*

35 Chapter Review

SUMMARY OF KEY CONCEPTS

CONCEPT 35.1

Plants have a hierarchical organisation consisting of organs, tissues, and cells *(pp. 800–806)*

- Vascular plants have shoots consisting of **stems**, **leaves**, and, in angiosperms, flowers. **Roots** anchor the plant, absorb and conduct water and minerals, and store food. Leaves are attached to stem **nodes** and are the main **organs** of photosynthesis. The **axillary buds**, in axils of leaves and stems, give rise to branches. Plant organs may be adapted for specialised functions.
- Vascular plants have three **tissue systems**—dermal, vascular, and ground—which are continuous throughout the plant. The **dermal tissue** is a continuous layer of cells that covers the plant exterior. **Vascular tissues (xylem** and **phloem)** facilitate the long-distance transport of substances. **Ground tissues** function in storage, metabolism, and regeneration.
- **Parenchyma cells** are relatively undifferentiated and thin-walled cells that retain the ability to divide; they perform most of the metabolic functions of synthesis and storage. **Collenchyma cells** have unevenly thickened walls; they support young, growing parts of the plant. **Sclerenchyma cells**—sclereids and fibres—have thick, lignified walls that help support mature, nongrowing parts of the plant. **Tracheids** and **vessel elements**, the water-conducting cells of xylem, have thick walls and are dead at functional maturity. **Sieve-tube elements** are living but highly modified cells that are largely devoid of internal organelles; they function in the transport of sugars through the phloem of angiosperms.

? *Describe at least three specialisations in plant organs and plant cells that are adaptations to life on land.*

CONCEPT 35.2

Different meristems generate new cells for primary and secondary growth *(pp. 807–808)*

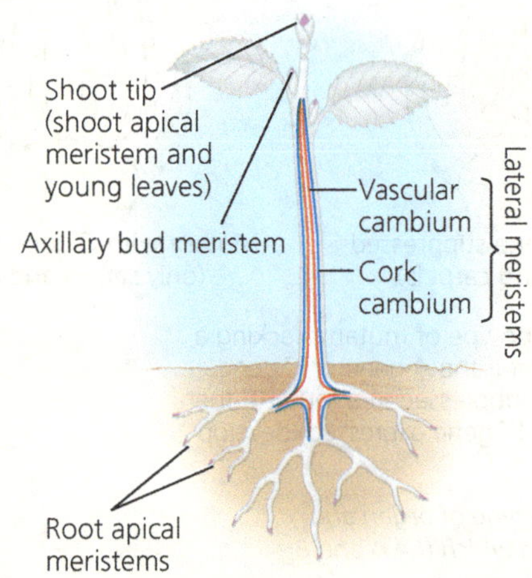

? *What is the difference between primary and secondary growth?*

CONCEPT 35.3

Primary growth lengthens roots and shoots *(pp. 809–812)*

- The root **apical meristem** is located near the tip of the root, where it generates cells for the growing root axis and the **root cap**.
- The apical meristem of a shoot is located in the **apical bud**, where it gives rise to alternating **internodes** and leaf-bearing nodes.
- Eudicot stems have vascular bundles in a ring, whereas monocot stems have scattered vascular bundles.
- **Mesophyll** cells are adapted for photosynthesis. **Stomata**, epidermal pores formed by pairs of **guard cells**, allow for gaseous exchange and are major avenues for water loss.

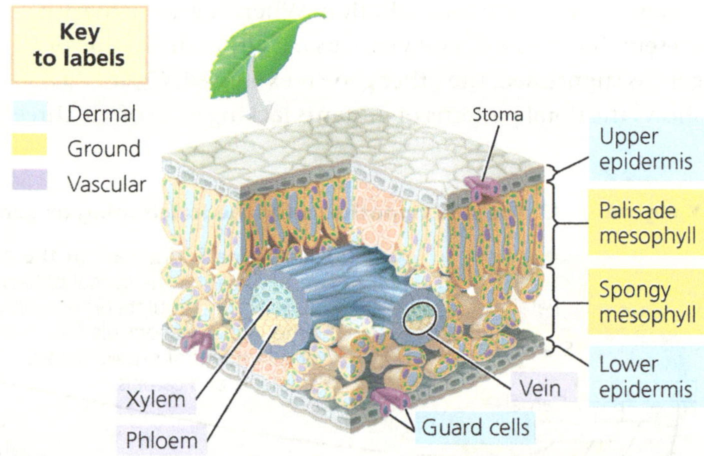

? *How does branching in roots differ from that in stems?*

CONCEPT 35.4

Secondary growth increases the diameter of stems and roots in woody plants *(pp. 813–816)*

- The **vascular cambium** is a meristematic cylinder that produces secondary xylem and secondary phloem during **secondary growth**. Older layers of secondary xylem (heartwood) become inactive, whereas younger layers (sapwood) still conduct water.

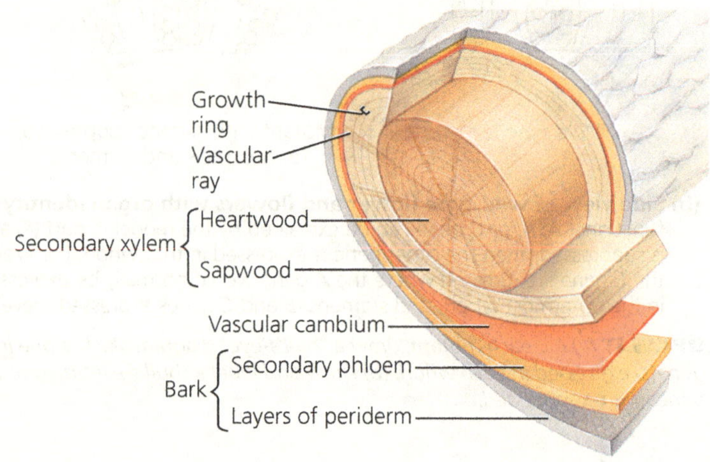

- The **cork cambium** gives rise to a thick protective covering called the periderm, which consists of the cork cambium plus the layers of cork cells it produces.

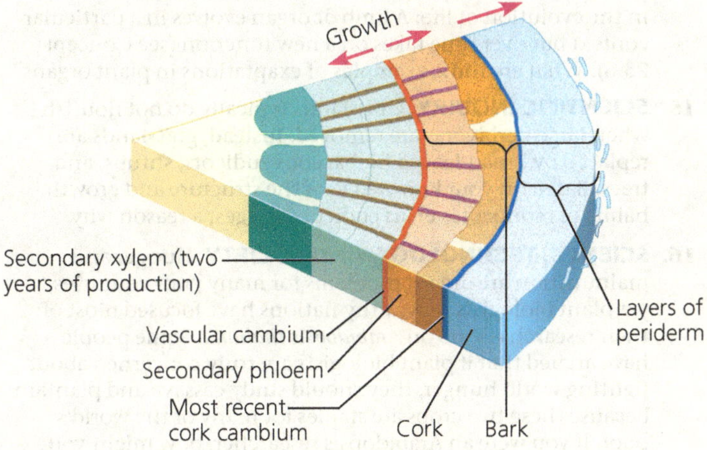

? *What advantages did plants gain from the evolution of secondary growth?*

CONCEPT 35.5

Growth, morphogenesis, and cell differentiation produce the plant body *(pp. 816–821)*

- Cell division and cell expansion are the primary determinants of growth.
- Morphogenesis, the development of body shape and organisation, depends on cells responding to positional information from their neighbours.
- Cell differentiation, arising from differential gene activation, enables cells within the plant to assume different functions despite having identical genomes. The way in which a plant cell differentiates is determined largely by the cell's position in the developing plant.
- Internal or environmental cues may cause a plant to switch from one developmental stage to another—for example, from developing juvenile leaves to developing mature leaves. Such changes in shape are called **phase changes**.
- Studies of floral development have provided a model system for studying **pattern formation**. The **ABC hypothesis** identifies how three classes of genes control formation of sepals, petals, stamens, and carpels.

? *By what mechanism do plant cells tend to elongate along one axis instead of expanding in all directions?*

TEST YOUR UNDERSTANDING

Levels 1-2: Remembering/Understanding

1. Most of the growth of a plant body is the result of
(A) cell differentiation.
(B) morphogenesis.
(C) cell division.
(D) cell elongation.

2. The innermost layer of the root cortex is the
(A) core. (C) endodermis.
(B) pericycle. (D) pith.

3. Heartwood and sapwood consist of
(A) bark. (C) secondary xylem.
(B) periderm. (D) secondary phloem.

4. The phase change of an apical meristem from the juvenile to the mature vegetative phase is often revealed by
(A) a change in the shape of the leaves produced.
(B) the initiation of secondary growth.
(C) the formation of lateral roots.
(D) the activation of flower-inducing genes.

5. The vascular cambium gives rise to
(A) all xylem.
(B) all phloem.
(C) primary xylem and phloem.
(D) secondary xylem and phloem.

6. The root pericycle is the site where
(A) secondary growth originates.
(B) root hairs originate.
(C) lateral roots originate.
(D) the endodermis originates.

7. Root apical meristems are found
(A) only in taproots.
(B) only in lateral roots.
(C) only in adventitious roots.
(D) in all roots.

Levels 3-4: Applying/Analysing

8. Suppose a flower had normal expression of genes A and C and expression of gene B in all four whorls. Based on the ABC hypothesis, what would be the structure of that flower, starting at the outermost whorl?
(A) carpel-petal-petal-carpel
(B) petal-petal-stamen-stamen
(C) sepal-carpel-carpel-sepal
(D) sepal-sepal-carpel-carpel

9. Which of the following arise(s), directly or indirectly, from meristematic activity?
(A) secondary xylem
(B) leaves
(C) dermal tissue
(D) all of the above

10. A strawberry plant mutant that fails to make stolons would suffer from
(A) too little mineral absorption.
(B) a tendency to topple over.
(C) too little water absorption.
(D) a reduction in asexual reproduction.

11. **DRAW IT** On this cross section from a woody eudicot, label a growth ring, late wood, early wood, and a vessel element. Then draw an arrow in the pith-to-cork direction.

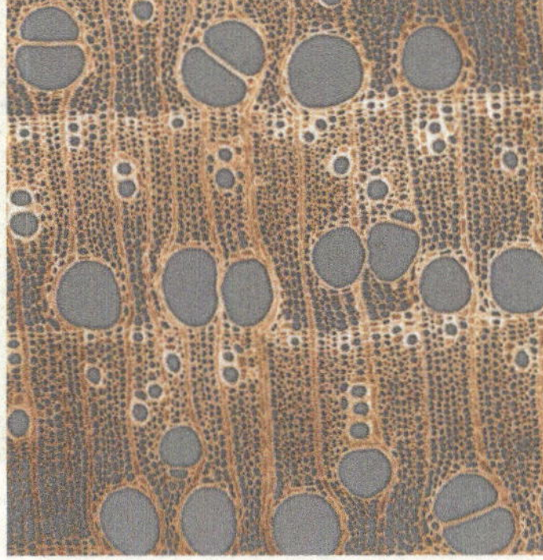

12. **VISUAL SKILLS** How does the internal anatomy of a tea leaf differ from that of an iris leaf? (Review Figure 35.18.) How might this difference relate to the orientations of the leaves? Explain how the difference shows that structure fits function.

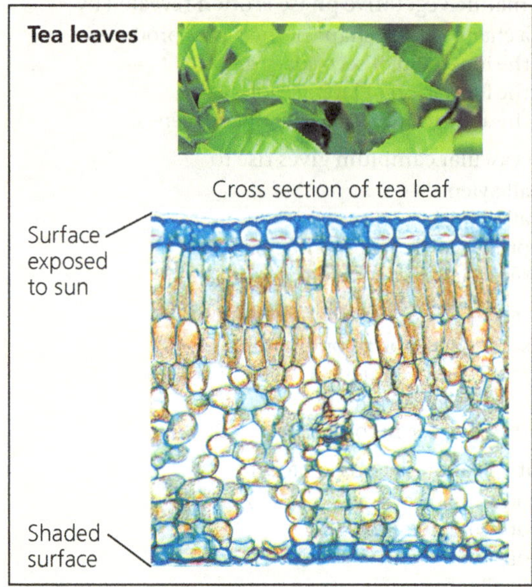

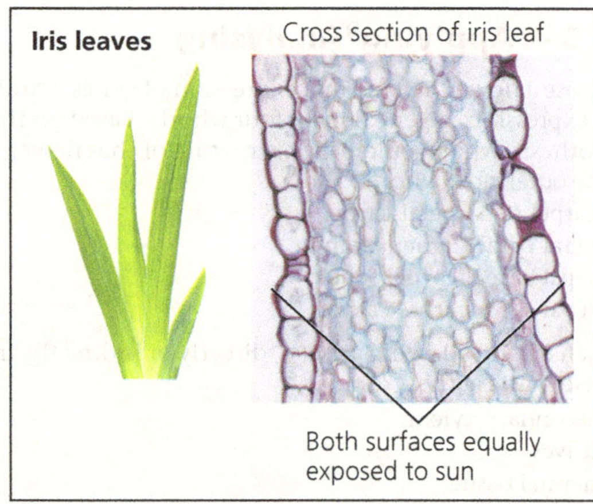

13. **VISUAL SKILLS** Visitors to the state of Washington can marvel at an unusual sight: a bicycle embedded 4 m above the ground in a living tree. Use your knowledge of primary and secondary growth to visualise and explain how this happened.

Levels 5-6: Evaluating/Creating

14. **EVOLUTION CONNECTION** Evolutionary biologists have coined the term *exaptation* to describe a common occurrence in the evolution of life: A limb or organ evolves in a particular context but over time takes on a new function (see Concept 25.6). What are some examples of exaptations in plant organs?

15. **SCIENTIFIC INQUIRY** Grasslands typically do not flourish when large herbivores are removed. Instead, grasslands are replaced by broad-leaved herbaceous eudicots, shrubs, and trees. Based on your knowledge of the structure and growth habits of monocots versus eudicots, suggest a reason why.

16. **SCIENCE, TECHNOLOGY, AND SOCIETY** Hunger and malnutrition are urgent problems for many poor countries, yet plant biologists in wealthy nations have focused most of their research efforts on *Arabidopsis thaliana*. Some people have argued that if plant biologists are truly concerned about fighting world hunger, they should study cassava and plantain because these two crops are staples for many of the world's poor. If you were an Arabidopsis researcher, how might you respond to this argument?

17. **WRITE ABOUT A THEME: ORGANISATION** In a short essay (100–150 words), explain how the evolution of lignin affected vascular plant structure and function.

18. **SYNTHESISE YOUR KNOWLEDGE**

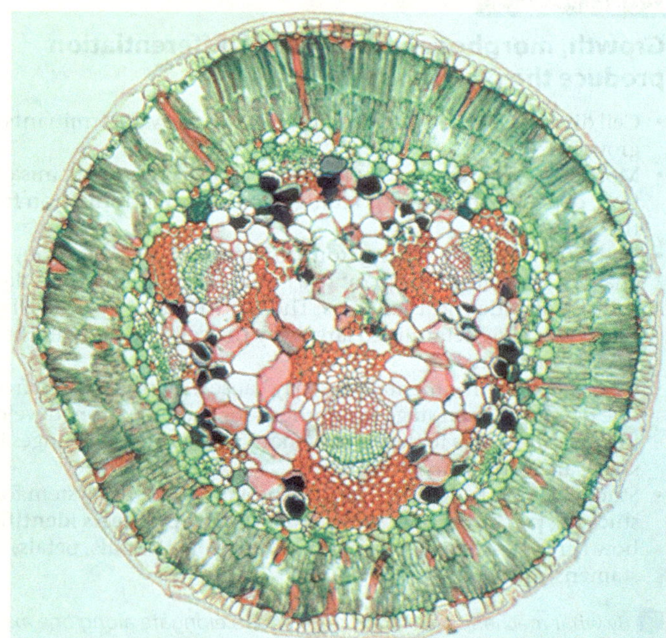

This stained light micrograph shows a cross section through a plant organ from Hakea purpurea, a shrub native to some arid regions of Australia. (a) Review Figures 35.14, 35.17, and 35.18 to identify whether this is a root, stem, or leaf. Explain your reasoning. (b) How might this organ be an adaptation for dry conditions?

For selected answers, see Appendix A.

36 Resource Acquisition and Transport in Vascular Plants

Figure 36.1 This English ivy has covered every square centimetre of a wall with foliage, an effective way to acquire light energy for photosynthesis. The sugars produced by the leaves and the water and minerals absorbed by the roots have to be transported throughout the plant.

KEY CONCEPTS

Study Tip

Make a flowchart: Make a simple flowchart of the movement of water molecules from the soil through the tissues of a plant to the outside air.

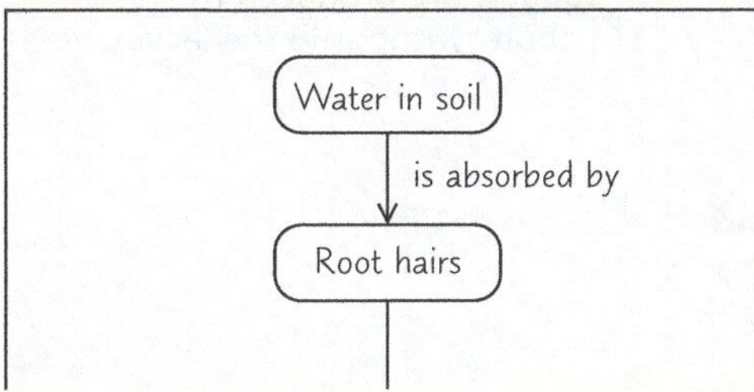

Go to Mastering Biology

to access Dynamic Study Modules for revision, 3D BioFlix® animations and high-quality videos, and your interactive Pearson eText.

What causes the movement of water, minerals, and sugars in most vascular plants?

Water and minerals are pulled up from the roots by **negative pressure** (tension) generated by evaporation from the leaves.

Water molecules are joined in chains as they move upwards

Sugars are pushed by **positive pressure** from where they are produced or stored to where they are needed. They can move both ways between leaves and roots.

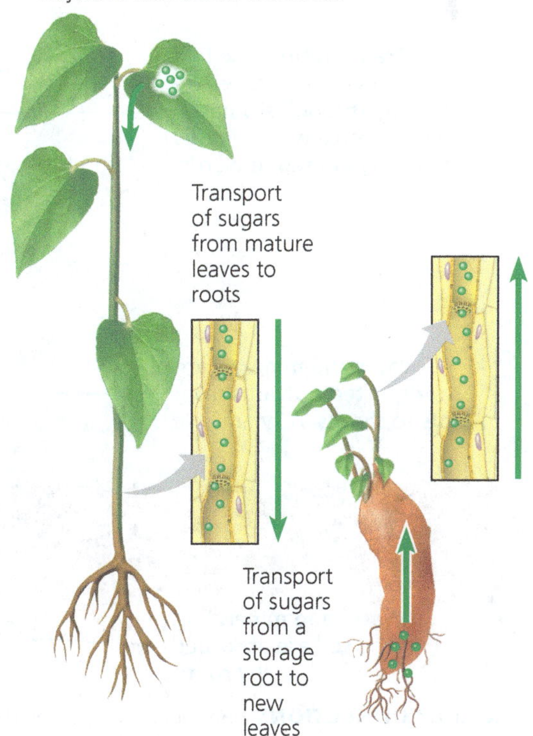

CONCEPT 36.1

Adaptations for acquiring resources were key steps in the evolution of vascular plants

EVOLUTION Most plants grow in soil and therefore inhabit two worlds—above ground, where shoots acquire sunlight and CO_2, and below ground, where roots acquire water and minerals. The successful colonisation of the land by plants depended on adaptations that allowed early plants to acquire resources from these two different settings.

The algal ancestors of plants absorbed water, minerals, and CO_2 directly from the water in which they lived. Transport in these algae was relatively simple because every cell was either close to or in intimate contact with the water that contained these substances. The earliest plants were nonvascular and produced photosynthetic shoots above the shallow fresh water in which they lived. These leafless shoots typically had waxy cuticles and few stomata, which allowed them to avoid excessive water loss while still permitting some exchange of CO_2 and O_2 for photosynthesis. The anchoring and absorbing functions of early plants were assumed by the base of the stem or by threadlike rhizoids (see Figure 29.11).

As plants evolved and increased in number, competition for light, water, and nutrients intensified. Taller plants with broad, flat appendages had an advantage in absorbing light. This increase in surface area, however, resulted in more evaporation and therefore a greater need for water. Larger shoots also required stronger anchorage. These needs favoured the production of multicellular, branching roots. Meanwhile, as greater shoot heights further separated the top of the photosynthetic shoot from the nonphotosynthetic parts below ground, natural selection favoured plants capable of efficient long-distance transport of water, minerals, and products of photosynthesis.

The evolution of vascular tissue consisting of xylem and phloem made possible the development of extensive root and shoot systems that carry out long-distance transport (see Figure 35.10). The **xylem** transports water and minerals from roots to shoots. The **phloem** transports products of photosynthesis from where they are made or stored to where they are needed. **Figure 36.2** provides an overview of resource acquisition and transport in an actively photosynthesising plant.

Shoot Architecture and Light Capture

Because most plants are photoautotrophs, their success depends ultimately on their ability to photosynthesise. Over the course of evolution, plants have developed a wide variety

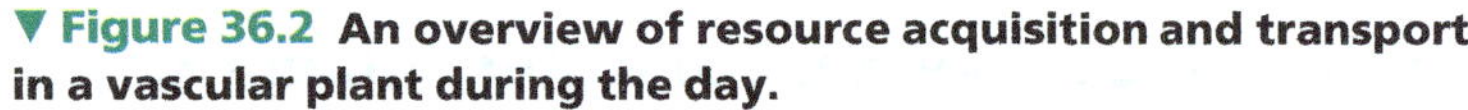

▼ **Figure 36.2 An overview of resource acquisition and transport in a vascular plant during the day.**

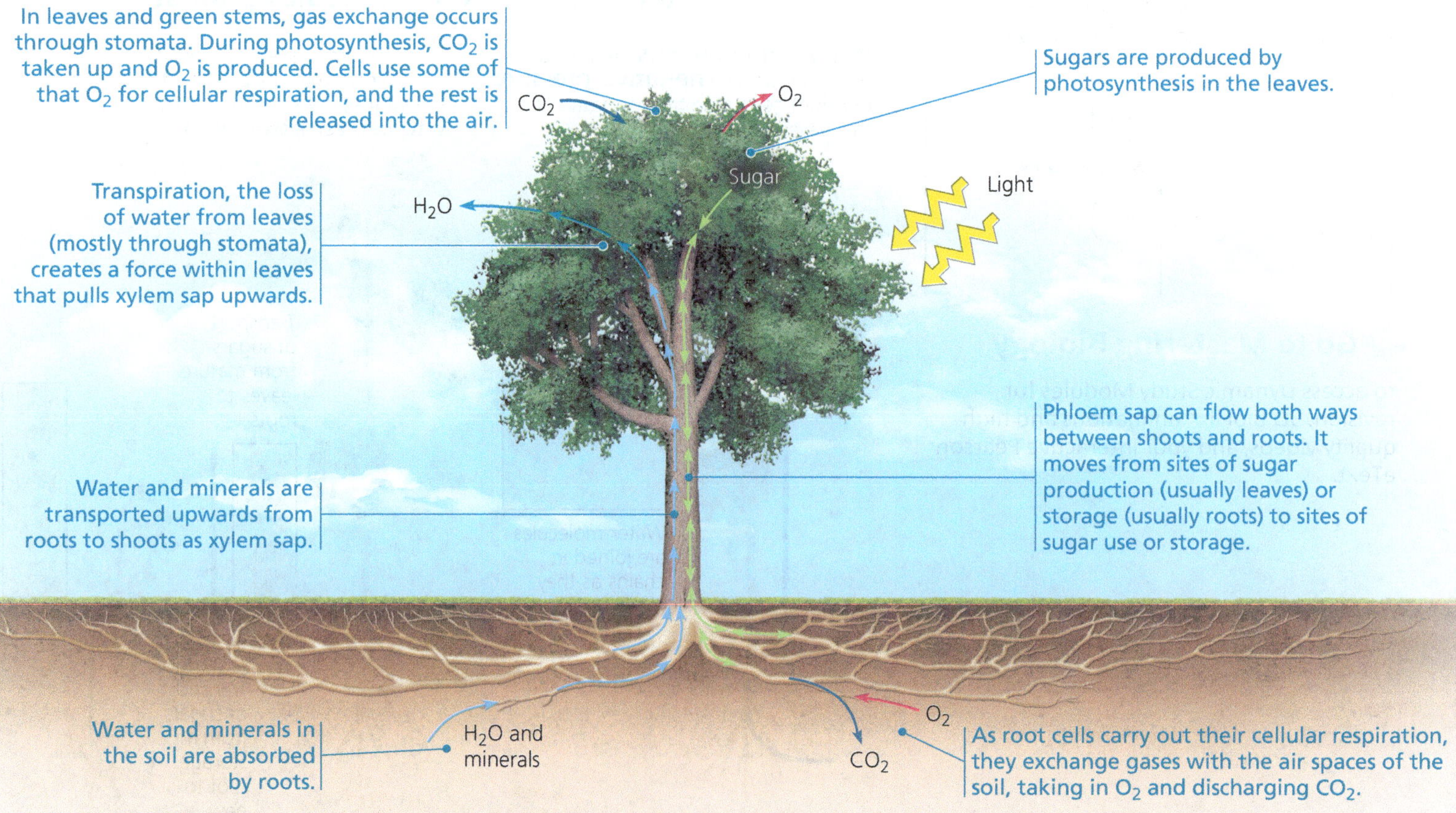

MAKE CONNECTIONS *How would you change the figure in order to show gas exchanges at night? (See Figure 10.23.)*

of shoot architectures that enable each species to compete successfully for light absorption in the ecological niche it occupies. For example, the lengths and widths of stems, as well as the branching pattern of shoots, are all architectural features affecting light capture.

Stems serve as supporting structures for leaves and as conduits for the transport of water and nutrients. Plants that grow tall avoid shading from neighbouring plants. Most tall plants require thick stems, which enable greater vascular flow to and from the leaves and stronger mechanical support for them. Vines are an exception, relying on other objects (usually other plants) to support their stems. In woody plants, stems become thicker through secondary growth (see Figure 35.11). Branching generally enables plants to harvest sunlight for photosynthesis more effectively. However, some species, such as the coconut palm, do not branch at all. Why is there so much variation in branching patterns? Plants have only a finite amount of energy to devote to shoot growth. If most of that energy goes into branching, there is less available for growing tall, and the risk of being shaded by taller plants increases. Conversely, if most of the energy goes into growing tall, the plants are not optimally harvesting sunlight.

Leaf size and structure account for much of the outward diversity in plant form. Leaves range in length from 1 mm in the perpetually damp rain forests of Australia and New Zealand to 20 m in the palm *Raphia regalis*, a native of African rain forests. These species represent extreme examples of a general correlation observed between water availability and leaf size. The largest leaves are typically found in species from tropical rain forests, whereas the smallest are usually found in species from dry or very cold environments, where liquid water is scarce and evaporative loss is more problematic.

The arrangement of leaves on a stem known as *phyllotaxy*, is an architectural feature important in light capture. Phyllotaxy is determined by the shoot apical meristem (see Figure 35.16) and is specific to each species **(Figure 36.3)**. A species may have one leaf per node (alternate, or spiral, phyllotaxy), two leaves per node (opposite phyllotaxy), or more (whorled phyllotaxy). Most angiosperms have alternate phyllotaxy, with leaves arranged in an ascending spiral around the stem, each successive leaf emerging about 137.5° from the site of the previous one. Why 137.5°? One hypothesis is that this angle minimises shading of the lower leaves by those above. In environments where intense sunlight can harm leaves, the greater shading provided by oppositely arranged leaves may be advantageous.

The total area of the leafy portions of all the plants in a community, from the top layer of vegetation to the bottom layer, affects the productivity of each plant. When there are many layers of vegetation, the shading of the lower leaves is so great that they photosynthesise less than they respire. When this happens, the nonproductive leaves or branches undergo programmed cell death and are eventually shed, a process called *self-pruning*.

▼ **Figure 36.3 Emerging phyllotaxy of Norway spruce.** This SEM, taken from above a shoot tip, shows the pattern of emergence of leaves. The leaves are numbered, with 1 being the youngest. (Some numbered leaves are not visible in the close-up.)

VISUAL SKILLS *With your finger, trace the progression of leaf emergence, moving from leaf number 29 to 28 and so on. What is the pattern? Based on this pattern of phyllotaxy, predict between which two developing leaf primordia the next primordium will emerge.*

Plant features that reduce self-shading increase light capture. A useful measurement in this regard is the *leaf area index*, the ratio of the total upper leaf surface of a single plant or an entire crop divided by the surface area of the land on which the plant or crop grows **(Figure 36.4)**. Leaf area index values of up to 7 are common for many mature crops, and there is little agricultural benefit to leaf area indexes higher than this value. Adding more leaves increases shading of lower leaves to the point that self-pruning occurs.

▼ **Figure 36.4 Leaf area index.** The leaf area index of a single plant is the ratio of the total area of the top surfaces of the leaves to the area of ground covered by the plant, as shown in this illustration of two plants viewed from the top. With many layers of leaves, a leaf area index value can easily exceed 1.

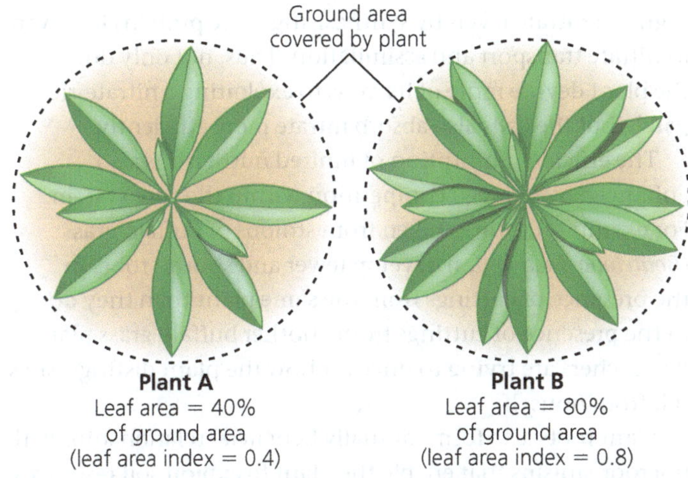

? *Would a higher leaf area index always increase the amount of photosynthesis? Explain.*

Another factor affecting light capture is leaf orientation. Some plants have horizontally oriented leaves; others, such as grasses, have leaves that are vertically oriented. In low-light conditions, horizontal leaves capture sunlight much more effectively than vertical leaves. In grasslands or other sunny regions, however, horizontal orientation may expose upper leaves to overly intense light, injuring leaves and reducing photosynthesis. But if a plant's leaves are nearly vertical, light rays are essentially parallel to the leaf surfaces, so no leaf receives too much light, and light penetrates more deeply to the lower leaves.

The Photosynthesis-Water Loss Compromise

The broad surface of most leaves favours light capture, while open stomatal pores allow for the diffusion of CO_2 into the photosynthetic tissues. Open stomatal pores, however, also promote evaporation of water from the plant. Over 90% of the water lost by plants is by evaporation from stomatal pores. Consequently, shoot adaptations represent compromises between enhancing photosynthesis and minimising water loss, particularly in environments where water is scarce. Later in the chapter, we'll discuss the mechanisms by which plants enhance CO_2 uptake and minimise water loss by regulating the opening of stomatal pores.

Root Architecture and Acquisition of Water and Minerals

Just as carbon dioxide and sunlight are resources exploited by the shoot system, soil contains resources mined by the root system. Plants rapidly adjust the architecture and physiology of their roots to exploit patches of available nutrients in the soil. The roots of many plants, for example, respond to pockets of low nitrate availability in soils by extending straight through the pockets instead of branching within them. Conversely, when encountering a pocket rich in nitrate, a root will often branch extensively there. Root cells also respond to high soil nitrate levels by synthesising more proteins involved in nitrate transport and assimilation. Thus, not only does the plant devote more of its mass to exploiting a nitrate-rich patch, but the cells also absorb nitrate more efficiently.

The efficient absorption of limited nutrients is also enhanced by reduced competition within the root system. For example, cuttings taken from stolons of buffalo grass (*Bouteloua dactyloides*) develop fewer and shorter roots in the presence of cuttings from the same plant than they do in the presence of cuttings from another buffalo grass plant. Researchers are trying to uncover how the plant distinguishes self from nonself.

Plant roots also form mutually beneficial relationships with microorganisms that enable the plant to exploit soil resources more efficiently. For example, the evolution of mutualistic associations between roots and fungi called **mycorrhizae** was a critical step in the successful colonisation of land by plants. Mycorrhizae were particularly important to the first plants to colonise the land. Ancient soils contained few of the life-sustaining bacteria, microbes, nutrients, and other essentials that modern plants rely on for survival. Then, like now, mycorrhizal hyphae indirectly endow the root systems of many plants with an enormous surface area for absorbing water and minerals, particularly phosphate. We will examine the role of mycorrhizal associations with plant nutrition in the chapter, "Soil and Plant Nutrition."

Once acquired, resources must be transported to other parts of the plant that need them. In the next section, we examine the processes and pathways that enable resources such as water, minerals, and sugars to be transported throughout the plant.

CONCEPT CHECK 36.1

1. Why is long-distance transport important for vascular plants?
2. Some plants can detect increased levels of light reflected from leaves of encroaching neighbours. This detection elicits stem elongation, production of erect leaves, and reduced lateral branching. How do these responses help the plant compete?
3. **WHAT IF?** If you prune a plant's shoot tips, what will be the short-term effect on the plant's branching and leaf area index?

For suggested answers, see Appendix A.

CONCEPT 36.2

Different mechanisms transport substances over short or long distances

Given the diversity of substances that move through plants and the great range of distances and barriers over which such substances must be transported, it is not surprising that plants employ a variety of transport processes. Before examining these processes, however, we'll look at the two major pathways of transport: the apoplast and the symplast.

The Apoplast and Symplast: Transport Continuums

Plant tissues have two major compartments—the apoplast and the symplast. The **apoplast** consists of everything external to the plasma membranes of living cells and includes cell walls, extracellular spaces, and the interior of dead cells such as vessel elements and tracheids (see Figure 35.10). The **symplast** consists of the entire mass of cytosol of all the living cells in a plant, as well as the plasmodesmata, the cytoplasmic channels that interconnect them.

▼ Figure 36.5 Cell compartments and routes for short-distance transport. Some substances may use more than one transport route.

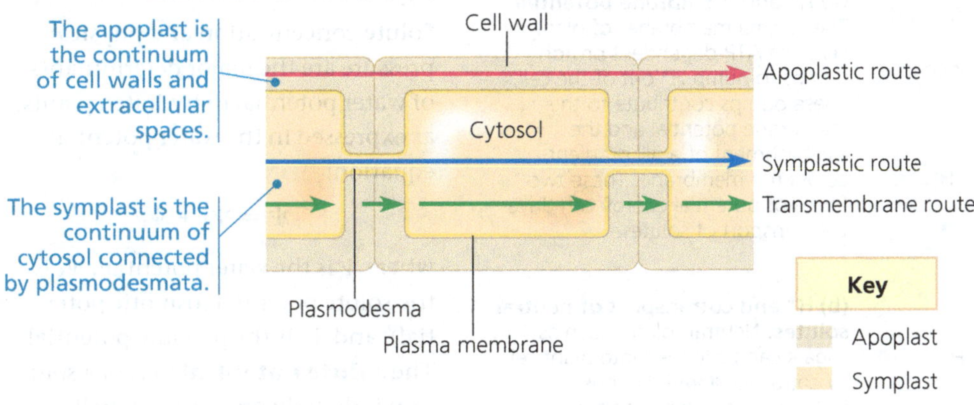

The compartmental structure of plants provides three routes for transport within a plant tissue or organ: the apoplastic, symplastic, and transmembrane routes **(Figure 36.5)**. In the *apoplastic route*, water and solutes (dissolved chemicals) move along the continuum of cell walls and extracellular spaces. In the *symplastic route*, water and solutes move along the continuum of cytosol. This route requires substances to cross a plasma membrane once, when they first enter the plant. After entering one cell, substances can move from cell to cell via plasmodesmata. In the *transmembrane route*, water and solutes move out of one cell, across the cell wall, and into the neighbouring cell, which may pass them to the next cell in the same way. The transmembrane route requires repeated crossings of plasma membranes as substances exit one cell and enter the next. These three routes are not mutually exclusive, and some substances may use more than one route to varying degrees.

Short-Distance Transport of Solutes Across Plasma Membranes

In plants, as in any organism, the selective permeability of the plasma membrane controls the short-distance movement of substances into and out of cells (see Concept 7.2). Both active and passive transport mechanisms occur in plants, and plant cell membranes are equipped with the same *general* types of pumps and transport proteins (channel proteins, carrier proteins, and cotransporters) that function in other cells. There are, however, *specific* differences between the membrane transport processes of plant and animal cells. In this section, we'll focus on some of those differences.

Unlike in animal cells, hydrogen ions (H^+) rather than sodium ions (Na^+) play the primary role in basic transport processes in plant cells. For example, in plant cells the membrane potential (the voltage across the membrane) is established mainly through the pumping of H^+ by proton pumps **(Figure 36.6a)** rather than the pumping of Na^+ by sodium-potassium pumps. Also, H^+ is most often cotransported in plants, whereas Na^+ is typically cotransported in animals. During cotransport, plant cells use the energy in the H^+ gradient and membrane potential to drive the active transport of many different solutes. For instance, cotransport with H^+ is responsible for absorption of neutral solutes, such as the sugar sucrose, by phloem cells and other plant cells. An H^+/sucrose cotransporter couples movement of sucrose against its concentration gradient with movement of H^+ down its electrochemical gradient **(Figure 36.6b)**. Cotransport with H^+ also facilitates movement of ions, as in the uptake of nitrate (NO_3^-) by root cells **(Figure 36.6c)**.

The membranes of plant cells also have ion channels that allow only certain ions to pass **(Figure 36.6d)**. As in animal cells, most channels are gated, opening or closing in response to stimuli such as chemicals, pressure, or voltage. Later in this chapter, we'll discuss how potassium ion channels in guard cells function in opening and closing stomata. Ion channels are also involved in producing electrical signals analogous to the action potentials of animals (see Concept 48.2). However, these signals are 1,000 times slower and employ Ca^{2+}-activated anion channels rather than the sodium ion channels used by animal cells.

Short-Distance Transport of Water Across Plasma Membranes

The absorption or loss of water by a cell occurs by **osmosis**, the diffusion of free water—water that is not bound to solutes or surfaces—across a membrane (see Figure 7.12). The physical property that predicts the direction in which water will flow is called **water potential**, a quantity that includes the effects of solute concentration and physical pressure. Free water moves from regions of higher water potential to regions of lower water potential if there is no barrier to its flow. The word *potential* in the term *water potential* refers to water's potential energy—water's capacity to perform work when it moves from a region of higher water potential to a region of lower water potential. For example, if a plant cell or seed is immersed in a solution that has a higher water potential, water will move into the cell or seed, causing it to expand. The expansion of plant cells and seeds can be a powerful force: The expansion of cells in tree roots can break concrete sidewalks, and the swelling of wet grain seeds within the holds of damaged ships can produce catastrophic hull failure and sink the ships. Given the strong forces generated by swelling seeds, it is interesting to consider whether water uptake by seeds is an active process. This question is

▼ Figure 36.6 Solute transport across plant cell plasma membranes.

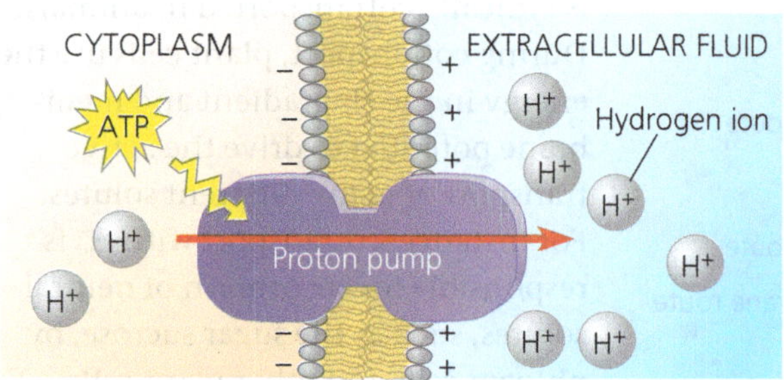

(a) H^+ and membrane potential. The plasma membranes of plant cells use ATP-dependent proton pumps to pump H^+ out of the cell. These pumps contribute to the membrane potential and the establishment of a pH gradient across the membrane. These two forms of potential energy can drive the transport of solutes.

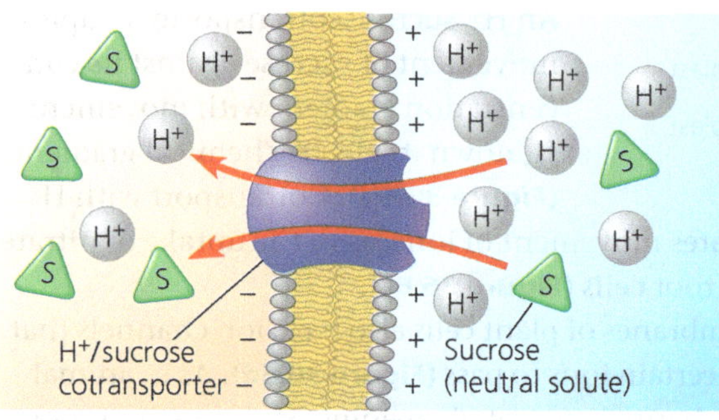

(b) H^+ and cotransport of neutral solutes. Neutral solutes such as sugars can be loaded into plant cells by cotransport with H^+ ions. H^+/sucrose cotransporters, for example, play a key role in loading sugar into the phloem prior to sugar transport throughout the plant.

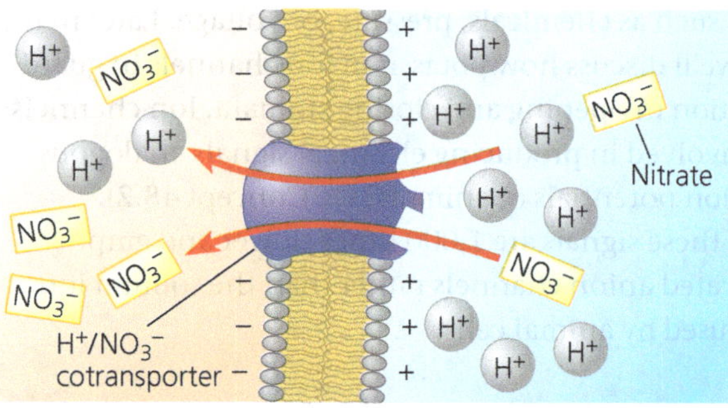

(c) H^+ and cotransport of ions. Cotransport mechanisms involving H^+ also participate in regulating ion fluxes into and out of cells. For example, H^+/NO_3^- cotransporters in the plasma membranes of root cells are important for the uptake of NO_3^- by plant roots.

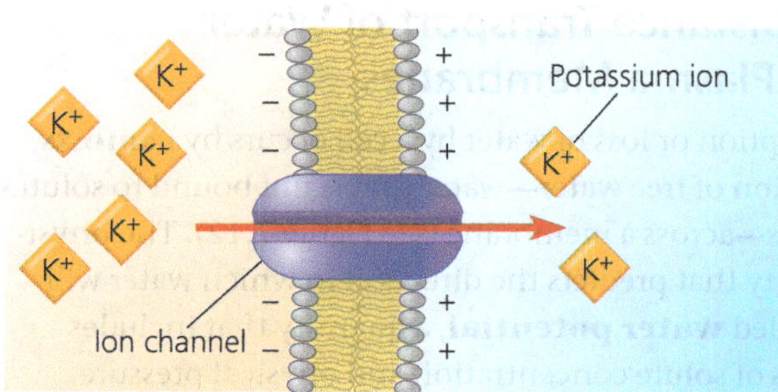

(d) Ion channels. Plant ion channels open and close in response to voltage, stretching of the membrane, and chemical factors. When open, ion channels allow specific ions to diffuse across membranes. For example, a K^+ ion channel is involved in the release of K^+ from guard cells when stomata close.

? *Assume that a plant cell has all four of the plasma membrane transport proteins shown above and that you have a specific inhibitor for each protein. Predict the effect of each inhibitor on the cell's membrane potential.*

examined in the **Scientific Skills Exercise**, which explores the effect of temperature on this process.

Water potential is abbreviated by the Greek letter Ψ (psi, pronounced "sigh"). Plant biologists measure Ψ in a unit of pressure called a **megapascal** (abbreviated MPa). By definition, the Ψ of pure water in a container open to the atmosphere under standard conditions (at sea level and at room temperature) is 0 MPa. One MPa is equal to about 10 times atmospheric pressure at sea level. The internal pressure of a living plant cell due to the osmotic uptake of water is approximately 0.5 MPa, about twice the air pressure inside an inflated car tyre.

How Solutes and Pressure Affect Water Potential

Solute concentration and physical pressure are the major determinants of water potential in hydrated plants, as expressed in the water potential equation:

$$\Psi = \Psi_S + \Psi_P$$

where Ψ is the water potential, Ψ_S is the solute potential (osmotic potential), and Ψ_P is the pressure potential. The **solute potential** (Ψ_S) of a solution is directly proportional to its molarity. Solute potential is also called *osmotic potential* because solutes affect the direction of osmosis. The solutes in plants are typically mineral ions and sugars. By definition, the Ψ_S of pure water is 0. When solutes are added, they bind water molecules. As a result, there are fewer free water molecules, reducing the capacity of the water to move and do work. In this way, an increase in solute concentration has a negative effect on water potential, which is why the Ψ_S of a solution is always expressed as a negative number. For example, a 0.1 *M* solution of a sugar has a Ψ_S of -0.23 MPa. As the solute concentration increases, Ψ_S will become more negative.

Pressure potential (Ψ_P) is the physical pressure on a solution. Unlike Ψ_S, Ψ_P can be positive or negative relative to atmospheric pressure. For example, when a solution is being withdrawn by a syringe, it is under negative pressure; when it is being expelled from a syringe, it is under positive pressure. The water in living cells is usually under positive pressure due to the osmotic uptake of water. Specifically, the **protoplast** (the living part of the cell, which also includes the plasma membrane) presses against the cell wall, creating what is known as **turgor pressure**. This pushing effect of internal pressure, much like the air in an inflated tire, is critical for plant function because it helps maintain the stiffness of plant tissues and also serves as the driving force for cell elongation. Conversely, the water in the hollow nonliving xylem cells (tracheids and vessel elements) of a plant is often under a negative pressure potential (tension) of less than -2 MPa.

Scientific Skills Exercise

Calculating and Interpreting Temperature Coefficients

Does the Initial Uptake of Water by Seeds Depend on Temperature? One way to answer this question is to soak seeds in water at different temperatures and measure the rate of water uptake at each temperature. The data can be used to calculate the temperature coefficient, Q_{10}, the factor by which a physiological reaction (or process) rate increases when the temperature is raised by 10°C:

$$Q_{10} = \left(\frac{k_2}{k_1}\right)^{\frac{10}{t_2 - t_1}}$$

where t_2 is the higher temperature (°C), t_1 is the lower temperature, k_2 is the reaction (or process) rate at t_2, and k_1 is the reaction (or process) rate at t_1. (If $t_2 - t_1 = 10$, as here, the maths is simplified.)

Q_{10} values may be used to make inferences about the physiological process under investigation. Chemical (metabolic) processes involving large-scale protein shape changes are highly dependent on temperature and have higher Q_{10} values, closer to 2 or 3.
In contrast, many, but not all, physical parameters are relatively independent of temperature and have Q_{10} values closer to 1. For example, the Q_{10} of the change in the viscosity of water is 1.2–1.3. In this exercise, you will calculate Q_{10} using data from radish seeds (*Raphanus sativum*) to assess whether the initial uptake of water by seeds is more likely to be a physical or a chemical process.

How the Experiment Was Done Samples of radish seeds were weighed and placed in water at four different temperatures. After 30 minutes, the seeds were removed, blotted dry, and reweighed. The researchers then calculated the percent increase in mass due to water uptake for each sample.

Data from the Experiment

Temperature	% Increase in Mass Due to Water Uptake After 30 Minutes
5°C	18.5
15°C	26.0
25°C	31.0
35°C	36.2

Data from J. D. Murphy and D. L. Noland, Temperature effects on seed imbibition and leakage mediated by viscosity and membranes, *Plant Physiology* 69:428–431 (1982).

INTERPRET THE DATA

1. Based on the data, does the initial uptake of water by radish seeds vary with temperature? What is the relationship between temperature and water uptake?
2. **(a)** Using the data for 35°C and 25°C, calculate Q_{10} for water uptake by radish seeds. Repeat the calculation using the data for 25°C and 15°C and the data for 15°C and 5°C. **(b)** What is the average Q_{10}? **(c)** Do your results imply that the uptake of water by radish seeds is mainly a physical process or a chemical (metabolic) process? **(d)** Given that the Q_{10} for the change in the viscosity of water is 1.2–1.3, could the slight temperature dependence of water uptake by seeds be a reflection of the slight temperature dependence of the viscosity of water?
3. Besides temperature, what other independent variables could you alter to test whether radish seed swelling is essentially a physical process or a chemical process?
4. Would you expect plant growth to have a Q_{10} closer to 1 or 3? Why?

As you learn to apply the water potential equation, keep in mind the key point: *Water moves from regions of higher water potential to regions of lower water potential.*

Water Movement Across Plant Cell Membranes

Now let's consider how water potential affects absorption and loss of water by a living plant cell. First, imagine a cell that is **flaccid** (limp) as a result of losing water. The cell has a Ψ_P of 0 MPa. Suppose this flaccid cell is bathed in a solution of higher solute concentration (more negative solute potential) than the cell itself **(Figure 36.7a)**. Since the external solution has the lower (more negative) water potential, water diffuses out of the cell. The cell's protoplast undergoes **plasmolysis**—that is, it shrinks and pulls away from the cell wall. If we place the same flaccid cell in pure water ($\Psi = 0\,\text{MPa}$) **(Figure 36.7b)**, the cell, because it contains solutes, has a lower water potential than the water, and water enters the cell by osmosis. The contents of the cell begin to swell and press the plasma membrane against the cell wall. The partially elastic wall, exerting turgor pressure, confines the pressurised protoplast. When this pressure is enough to offset the tendency for water to enter because of the solutes in the cell, then Ψ_P and Ψ_S are equal, and $\Psi = 0$. This matches the water potential of the extracellular environment—in this example, 0 MPa. A dynamic equilibrium has been reached, and there is no further *net* movement of water.

In contrast to a flaccid cell, a walled cell with a greater solute concentration than its surroundings is **turgid**, or very firm. When turgid cells in a nonwoody tissue push against each other, the tissue is stiffened. The effects of turgor loss are seen during **wilting**, when leaves and stems droop as a result of cells losing water.

Aquaporins: Facilitating Diffusion of Water

A difference in water potential determines the *direction* of water movement across membranes, but how do water molecules actually cross the membranes? Water molecules are small enough to diffuse across the phospholipid bilayer, even though the bilayer's interior is hydrophobic. However, their movement across biological membranes is too rapid to be explained by unaided diffusion. Transport proteins called **aquaporins** (see Figure 7.10 and Concept 7.2) facilitate the

▼ **Figure 36.7 Water relations in plant cells.** In these experiments, flaccid cells (cells in which the protoplast contacts the cell wall but lacks turgor pressure) are placed in two environments. The blue arrows indicate initial net water movement.

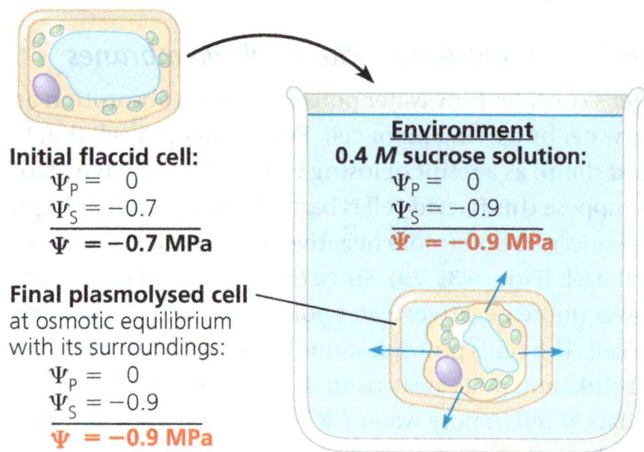

(a) Initial conditions: cellular $\Psi >$ environmental Ψ. The protoplast loses water, and the cell plasmolyses. After plasmolysis is complete, the water potentials of the cell and its surroundings are the same.

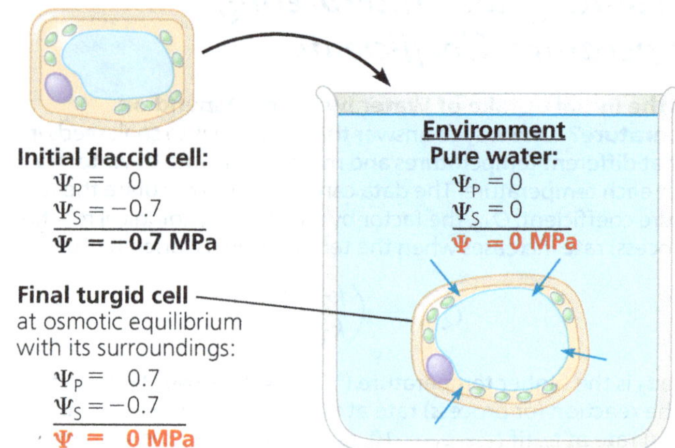

(b) Initial conditions: cellular $\Psi <$ environmental Ψ. Net uptake of water by osmosis makes the cell turgid. When this tendency for water to enter is offset by the back pressure of the elastic wall, water potentials are equal for the cell and its surroundings.

transport of water molecules across plant cell plasma membranes. Aquaporin channels, which can open and close, affect the *rate* at which water moves osmotically across the membrane. Their permeability is decreased by increases in cytosolic Ca^{2+} or decreases in cytosolic pH.

Long-Distance Transport: The Role of Bulk Flow

Diffusion is an effective transport mechanism over the spatial scales typically found at the cellular level. However, diffusion is much too slow to function in long-distance transport within a plant. Although diffusion from one end of a cell to the other takes just seconds, diffusion from the roots to the top of a giant mountain ash or New Zealand kauri would take several centuries. Instead, long-distance transport occurs through **bulk flow**, the movement of liquid in response to a pressure gradient. The bulk flow of material always occurs from higher to lower pressure. Unlike osmosis, bulk flow is independent of solute concentration.

Long-distance bulk flow occurs within specialised cells in the vascular tissue, namely, the tracheids and vessel elements of the xylem and the sieve-tube elements of the phloem. In leaves, the branching of veins ensures that no cell is more than a few cells away from the vascular tissue **(Figure 36.8)**.

The structures of the conducting cells of the xylem and phloem facilitate bulk flow. Mature tracheids and vessel elements are dead cells and therefore have no cytoplasm, and the cytoplasm of sieve-tube elements (also called sieve-tube members) is almost devoid of organelles (see Figure 35.10). If you have dealt with a partially clogged drain, you know that the volume of flow depends on the pipe's diameter. Clogs reduce the effective diameter of the drainpipe. Such experiences help us understand how the structures of plant cells specialised for bulk flow fit their function. Like unclogging a drain, the absence or reduction of cytoplasm in a plant's "plumbing" facilitates bulk flow through the xylem and phloem. Bulk flow is also enhanced by the perforation plates at the ends of vessel elements and the porous sieve plates connecting sieve-tube elements.

Diffusion, active transport, and bulk flow act in concert to transport resources throughout the whole plant. For example, bulk flow due to a pressure difference is the mechanism of long-distance transport of sugars in the phloem, but active transport of sugar at the cellular level maintains this pressure difference. In the next three sections, we'll examine in more detail the transport of water and minerals from roots to shoots, the control of evaporation, and the transport of sugars.

▼ **Figure 36.8 Venation in an aspen leaf.** The finer and finer branching of leaf veins in eudicot leaves ensures that no leaf cell is far removed from the vascular system.

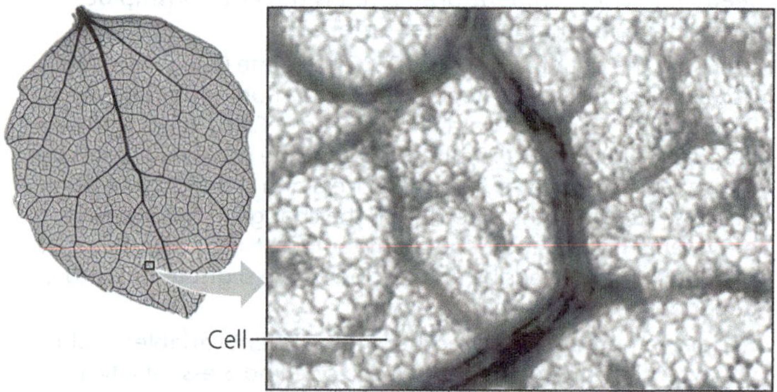

VISUAL SKILLS *In this leaf, what is the maximum number of cells any mesophyll cell is from a vein?*

CONCEPT CHECK 36.2

1. If a plant cell immersed in distilled water has a Ψ_S of −0.7 MPa and a Ψ of 0 MPa, what is the cell's Ψ_P? If you put it in an open beaker of solution that has a Ψ of −0.4 MPa, what would be its Ψ_P at equilibrium?
2. How would a reduction in the number of aquaporin channels affect a plant cell's ability to adjust to new osmotic conditions?
3. How would the long-distance transport of water be affected if tracheids and vessel elements were alive at maturity? Explain.
4. **WHAT IF?** What would happen if you put plant protoplasts in pure water? Explain.

For suggested answers, see Appendix A.

CONCEPT 36.3

Transpiration drives the transport of water and minerals from roots to shoots via the xylem

Picture yourself struggling to carry a 19-L container of water weighing 19 kg up several flights of stairs. Imagine doing this 40 times a day. Then consider the fact that an average-sized tree, despite having neither heart nor muscle, transports up to 800 kg of water effortlessly on a daily basis. How do trees accomplish this feat? To answer this question, we'll follow each step in the journey of water and minerals from roots to leaves.

Absorption of Water and Minerals by Root Cells

Although all living plant cells absorb nutrients across their plasma membranes, the cells near the tips of roots are particularly important because most of the absorption of water and minerals occurs there. In this region, the epidermal cells are permeable to water, and many are differentiated into root hairs, modified cells that account for much of the absorption of water by roots (see Figure 35.3). The root hairs absorb the soil solution, which consists of water molecules and dissolved mineral ions that are not bound tightly to soil particles. The soil solution is drawn into the hydrophilic walls of epidermal cells and passes freely along the cell walls and the extracellular spaces into the root cortex. This flow enhances the exposure of the cells of the cortex to the soil solution, providing a much greater membrane surface area for absorption than the surface area of the epidermis alone. Although the soil solution usually has a low mineral concentration, active transport enables roots to accumulate essential minerals, such as K^+, to concentrations hundreds of times greater than in the soil.

Transport of Water and Minerals into the Xylem

Water and minerals that pass from the soil into the root cortex cannot be transported to the rest of the plant until they enter the xylem of the vascular cylinder, or stele. The **endodermis**, the innermost layer of cells in the root cortex, functions as a last checkpoint for the selective passage of minerals from the cortex into the vascular cylinder (**Figure 36.9**). Minerals already in the symplast when they reach the endodermis continue through the plasmodesmata of endodermal cells and pass into the vascular cylinder. These minerals were already screened by the plasma membrane they had to cross to enter the symplast in the epidermis or cortex.

Minerals that reach the endodermis via the apoplast encounter a dead end that blocks their passage into the vascular cylinder. This barrier, located in the transverse and radial walls of each endodermal cell, is the **Casparian strip**, a belt made of suberin, a waxy material impervious to water and dissolved minerals (see Figure 36.9). Because of the Casparian strip, water and minerals cannot cross the endodermis and enter the vascular cylinder via the apoplast. Instead, water and minerals that are passively moving through the apoplast must cross the *selectively permeable* plasma membrane of an endodermal cell before they can enter the vascular cylinder. In this way, the endodermis transports needed minerals from the soil into the xylem and keeps many unneeded or toxic substances out. The endodermis also prevents solutes that have accumulated in the xylem from leaking back into the soil solution.

The last segment in the soil-to-xylem pathway is the passage of water and minerals into the tracheids and vessel elements of the xylem. These water-conducting cells lack protoplasts when mature and are therefore parts of the apoplast. Endodermal cells, as well as living cells within the vascular cylinder, discharge minerals from their protoplasts into their own cell walls. Both diffusion and active transport are involved in this transfer of solutes from the symplast to the apoplast, and the water and minerals can now enter the tracheids and vessel elements, where they are transported to the shoot system by bulk flow.

Bulk Flow Transport via the Xylem

Water and minerals from the soil enter the plant through the epidermis of roots, cross the root cortex, and pass into the vascular cylinder. From there the **xylem sap**, the water and dissolved minerals in the xylem, is transported long distances by bulk flow to the veins that branch throughout each leaf. As noted earlier, bulk flow is much faster than diffusion or active transport. Peak velocities in the transport of xylem sap can range from 15 to 45 m/hr for trees with wide vessel elements. The stems and leaves depend on this rapid delivery system for their supply of water and minerals.

The process of transporting xylem sap involves the loss of an astonishing amount of water by **transpiration**, the loss of water vapour from leaves and other aerial parts of the plant. A single corn plant, for example, transpires 60 L

▼ **Figure 36.9 Transport of water and minerals from root hairs to the xylem.**

VISUAL SKILLS *After studying the figure, explain how the Casparian strip forces water and minerals to pass through the plasma membranes of endodermal cells.*

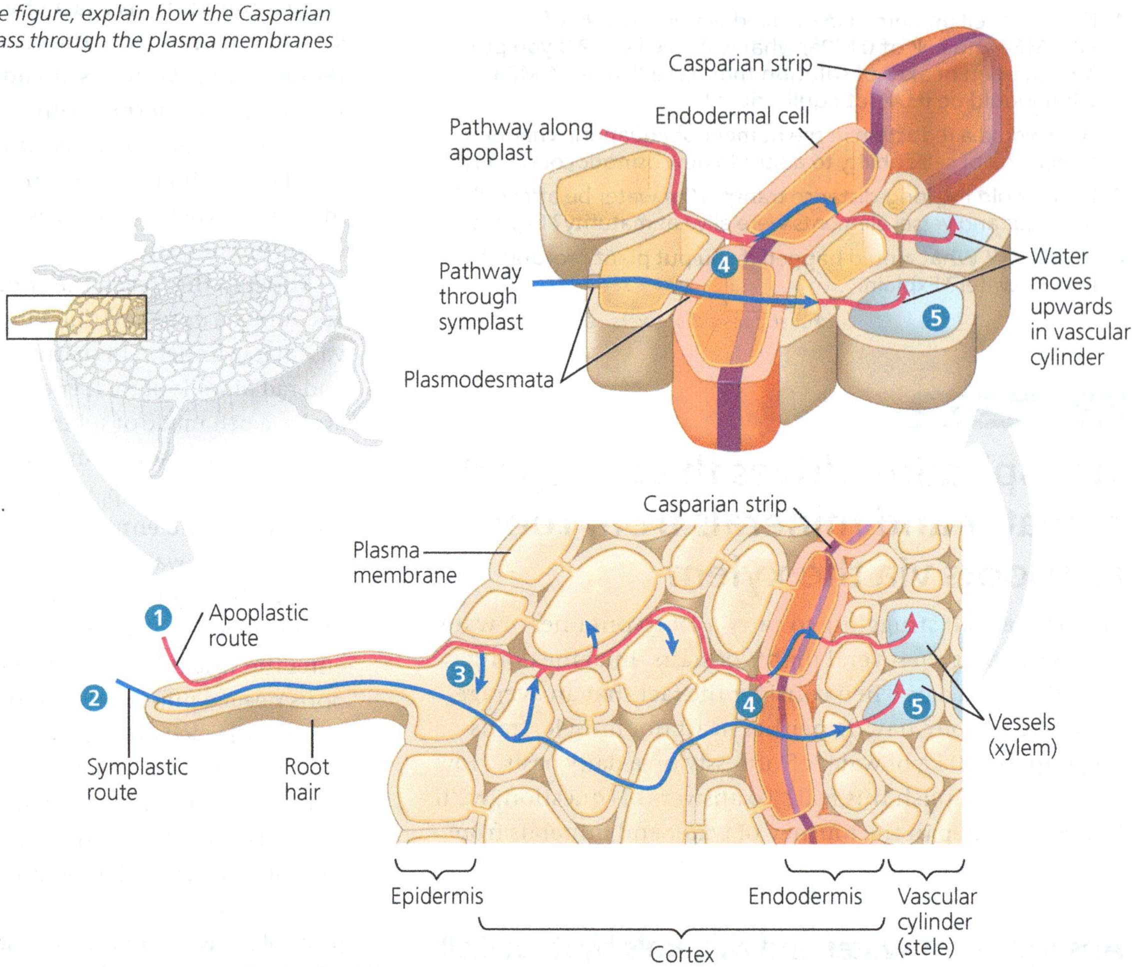

1 Apoplastic route. Uptake of soil solution by the hydrophilic walls of root hairs provides access to the apoplast. Water and minerals can then diffuse into the cortex along this matrix of walls and extracellular spaces.

2 Symplastic route. Minerals and water that cross the plasma membranes of root hairs can enter the symplast.

3 Transmembrane route. As soil solution moves along the apoplast, some water and minerals are transported into the protoplasts of cells of the epidermis and cortex and then move inwards via the symplast.

4 The endodermis: controlled entry to the vascular cylinder (stele). Within the transverse and radial walls of each endodermal cell is the Casparian strip, a belt of waxy material (purple band) that blocks the passage of water and dissolved minerals. Only minerals already in the symplast or entering that pathway by crossing the plasma membrane of an endodermal cell can detour around the Casparian strip and pass into the vascular cylinder (stele).

5 Transport in the xylem. Endodermal cells and also living cells within the vascular cylinder discharge water and minerals into their walls (apoplast). The xylem vessels then transport the water and minerals by bulk flow upwards into the shoot system.

of water during a growing season. A corn crop growing at a typical density of 60,000 plants per hectare (10,000 m^2) transpires almost 4 million litres of water per hectare every growing season. If the transpired water is not replaced with water transported from the roots, the leaves will wilt, and the plants will eventually die.

Xylem sap rises to heights of more than 120 m in the tallest trees. Is the sap mainly *pushed* upwards from the roots, or is it mainly *pulled* upwards? Let's evaluate the relative contributions of these two mechanisms.

Pushing Xylem Sap: Root Pressure

At night, when there is almost no transpiration, root cells continue actively pumping mineral ions into the xylem of the vascular cylinder. Meanwhile, the Casparian strip of the endodermis prevents the ions from leaking back out into the cortex and soil. The resulting accumulation of minerals lowers the water potential within the vascular cylinder. Water flows in from the root cortex, generating **root pressure**, a push of xylem sap. The root pressure sometimes causes more water to enter the leaves than is transpired, resulting in **guttation**, the exudation of water droplets that can be seen in the morning on the tips or edges of some plant leaves **(Figure 36.10)**. Guttation fluid should not be confused with dew, which is condensed atmospheric moisture.

▶ **Figure 36.10 Guttation.** Root pressure is forcing excess water from this strawberry leaf.

In most plants, root pressure is a minor mechanism driving the ascent of xylem sap, pushing water only a few metres at most. The positive pressures produced are simply too weak to overcome the gravitational force of the water column in the xylem, particularly in tall plants. Many plants do not generate any root pressure or do so only during part of the growing season. Even in plants that display guttation, root pressure cannot keep pace with transpiration after sunrise. For the most part, xylem sap is not pushed from below by root pressure but is pulled up.

Pulling Xylem Sap: The Cohesion-Tension Hypothesis

As we have seen, root pressure, which depends on the active transport of solutes by plants, is only a minor force in the ascent of xylem sap. Far from depending on the metabolic activity of cells, most of the xylem sap that rises through a tree does not even require living cells to do so. As demonstrated by Eduard Strasburger in 1891, leafy stems with their lower end immersed in toxic solutions of copper sulfate or acid will readily draw these poisons up if the stem is cut below the surface of the liquid. As the toxic solutions ascend, they kill all living cells in their path, eventually arriving in the transpiring leaves and killing the leaf cells as well. Nevertheless, as Strasburger noted, the uptake of the toxic solutions and the loss of water from the dead leaves can continue for weeks.

In 1894, a few years after Strasburger's findings, two Irish scientists, John Joly and Henry Dixon, put forward a hypothesis that remains the leading explanation of the ascent of xylem sap. According to their **cohesion-tension hypothesis**, transpiration provides the pull for the ascent of xylem sap, and the cohesion of water molecules transmits this pull along the entire length of the xylem from shoots to roots. Hence, xylem sap is normally under negative pressure, or tension. Since transpiration is a "pulling" process, our exploration of the rise of xylem sap by the cohesion-tension mechanism begins not with the roots but with the leaves, where the driving force for transpirational pull begins.

Transpirational Pull Stomata on a leaf's surface lead to a maze of internal air spaces that expose the mesophyll cells to the CO_2 they need for photosynthesis. The air in these spaces is saturated with water vapour because it is in contact with the moist walls of the cells. On most days, the air outside the leaf is drier; that is, it has lower water potential than the air inside the leaf. Therefore, water vapour in the air spaces of a leaf diffuses down its water potential gradient and exits the leaf via the stomata. It is this loss of water vapour by diffusion and evaporation that we call transpiration.

But how does loss of water vapour from the leaf translate into a pulling force for upward movement of water through a plant? The negative pressure potential that causes water to move up through the xylem develops at the surface of mesophyll cell walls in the leaf **(Figure 36.11)**. The cell wall acts like a very thin capillary network. Water adheres to the cellulose microfibrils and other hydrophilic components

▼ Figure 36.11 Generation of transpirational pull. Negative pressure (tension) at the air–water interface in the leaf is the basis of transpirational pull, which draws water out of the xylem.

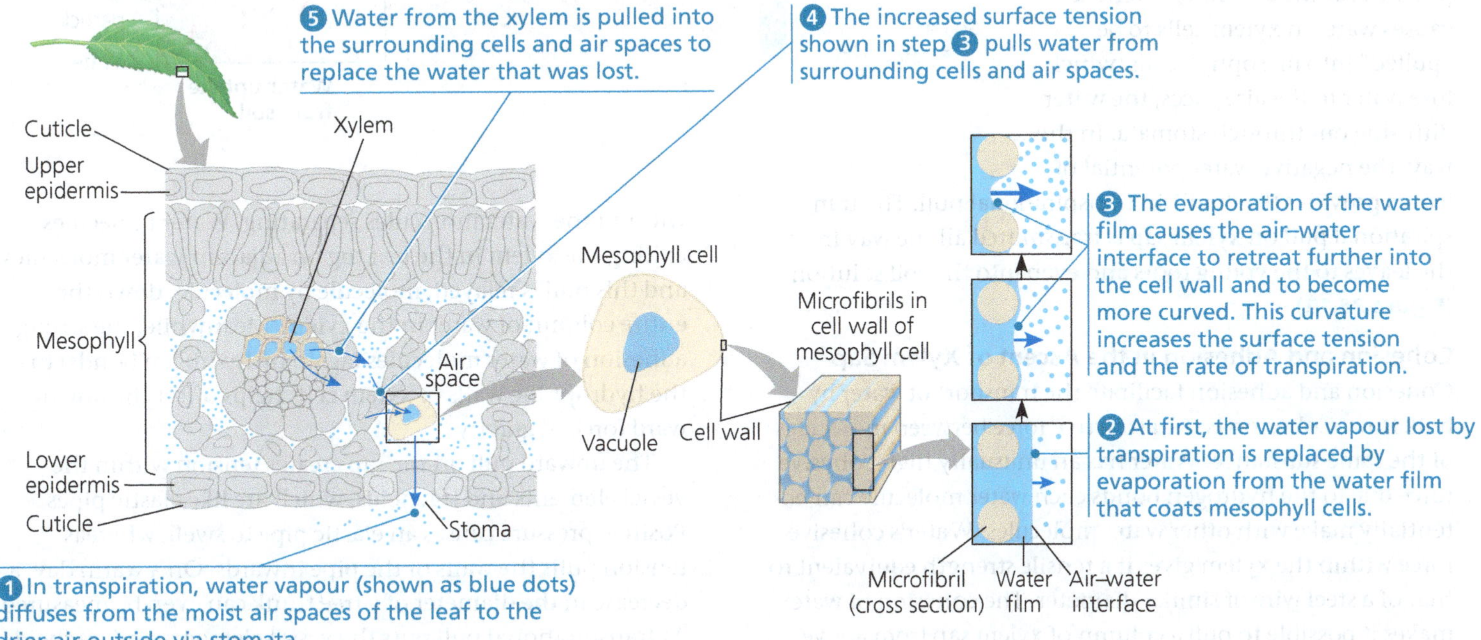

of the cell wall. As water evaporates from the water film that covers the cell walls of mesophyll cells, the air-water interface retreats further into the cell wall. Because of the high surface tension of water, the curvature of the interface induces a tension, or negative pressure potential, in the water. As more water evaporates from the cell wall, the curvature of the air–water interface increases and the pressure of the water becomes more negative. Water molecules from the more hydrated parts of the leaf are then pulled towards this area, reducing the tension. These pulling forces are transferred to the xylem because each water molecule is cohesively bound to the next by hydrogen bonds. Thus, transpirational pull depends on several of the properties of water discussed in Concept 3.2: adhesion, cohesion, and surface tension.

The role of negative pressure potential in transpiration is consistent with the water potential equation because negative pressure potential (tension) *lowers* water potential. Because water moves from areas of higher water potential to areas of lower water potential, the more negative pressure potential at the air–water interface causes water in xylem cells to be "pulled" into mesophyll cells, which lose water to the air spaces, the water diffusing out through stomata. In this way, the negative water potential of leaves provides the "pull" in transpirational pull. The transpirational pull on xylem sap is transmitted all the way from the leaves to the young roots and even into the soil solution **(Figure 36.12)**.

▼ Figure 36.12 Ascent of xylem sap. Hydrogen bonding forms an unbroken chain of water molecules extending from leaves to the soil. The force driving the ascent of xylem sap is a gradient of water potential (Ψ). For bulk flow over long distance, the Ψ gradient is due mainly to a gradient of the pressure potential (Ψ_P). Transpiration results in the Ψ_P at the leaf end of the xylem being lower than the Ψ_P at the root end. The Ψ values shown at the left are a "snapshot." They may vary during daylight, but the direction of the Ψ gradient remains the same.

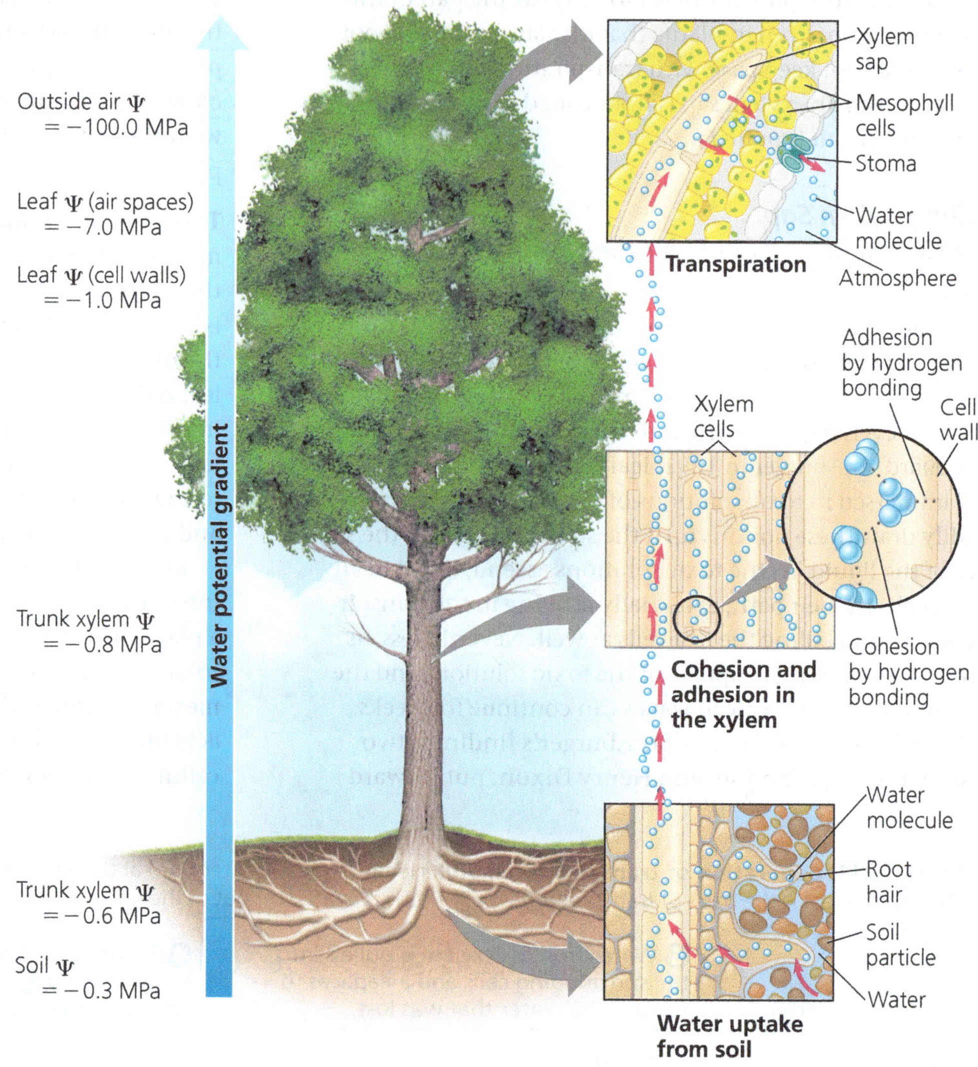

Cohesion and Adhesion in the Ascent of Xylem Sap

Cohesion and adhesion facilitate the transport of water by bulk flow. Cohesion is the attractive force between molecules of the same substance. Water has an unusually high cohesive force due to the hydrogen bonds each water molecule can potentially make with other water molecules. Water's cohesive force within the xylem gives it a tensile strength equivalent to that of a steel wire of similar diameter. The cohesion of water makes it possible to pull a column of xylem sap from above without the water molecules separating. Water molecules exiting the xylem in the leaf tug on adjacent water molecules, and this pull is relayed, molecule by molecule, down the entire column of water in the xylem. Meanwhile, the strong adhesion of water molecules (again by hydrogen bonds) to the hydrophilic walls of xylem cells helps offset the downward force of gravity.

The upward pull on the sap creates tension within the vessel elements and tracheids, which are like elastic pipes. Positive pressure causes an elastic pipe to swell, whereas tension pulls the walls of the pipe inwards. On a warm day, a decrease in the diameter of a tree trunk can even be measured. As transpirational pull puts the vessel elements and tracheids under tension, their thick secondary walls prevent them from collapsing, much as wire rings maintain the shape of a vacuum-cleaner hose. The tension produced by transpirational pull lowers water potential in the root xylem to such an extent that water flows passively from the soil, across the root cortex, and into the vascular cylinder.

Transpirational pull can extend down to the roots only through an unbroken chain of water molecules. Cavitation, the formation of a water vapour pocket, breaks the chain. It is more common in wide vessel elements than in tracheids and can occur during drought stress or when xylem sap freezes in winter. The air bubbles resulting from cavitation expand and block water channels of the xylem. The rapid expansion of air bubbles produces clicking noises that can be heard by placing sensitive microphones at the surface of the stem.

The interruption of xylem sap transport by cavitation is not always permanent. The chain of water molecules can detour around the air bubbles through pits between adjacent tracheids or vessel elements (see Figure 35.10). Moreover, root pressure enables small plants to refill blocked vessel elements. Recent evidence suggests that cavitation may even be repaired when the xylem sap is under negative pressure, although the mechanism by which this occurs is uncertain. In addition, secondary growth adds a layer of new xylem each year. Only the youngest, outermost secondary xylem layers transport water. Although the older secondary xylem no longer transports water, it does provide support for the tree (see Figure 35.22).

Xylem Sap Ascent by Bulk Flow: *A Review*

The cohesion-tension mechanism that transports xylem sap against gravity is an excellent example of how physical principles apply to biological processes. In the long-distance transport of water from roots to leaves by bulk flow, the movement of fluid is driven by a water potential difference at opposite ends of xylem tissue. The water potential difference is created at the leaf end of the xylem by the evaporation of water from leaf cells. Evaporation lowers the water potential at the air–water interface, thereby generating the negative pressure (tension) that pulls water through the xylem.

Bulk flow in the xylem differs from diffusion in some key ways. First, it is driven by differences in pressure potential (Ψ_P); solute potential (Ψ_S) is not a factor. Therefore, the water potential gradient within the xylem is essentially a pressure gradient. Also, the flow does not occur across plasma membranes of living cells, but instead within hollow, dead cells. Furthermore, it moves the entire solution together—not just water or solutes—and at much greater speed than diffusion.

The plant expends no energy to lift xylem sap by bulk flow. Instead, the absorption of sunlight drives most of transpiration by causing water to evaporate from the moist walls of mesophyll cells and by lowering the water potential in the air spaces within a leaf. Thus, the ascent of xylem sap, like the process of photosynthesis, is ultimately solar powered.

CONCEPT CHECK 36.3

1. A horticulturalist notices that when *Zinnia* flowers are cut at dawn, a small drop of water collects at the surface of the rooted stump. However, when the flowers are cut at noon, no drop is observed. Suggest an explanation.
2. **WHAT IF?** Suppose an *Arabidopsis* mutant lacking functional aquaporin proteins has a root mass three times greater than that of wild-type plants. Suggest an explanation.
3. **MAKE CONNECTIONS** How are the Casparian strip and tight junctions similar (see Figure 6.30)?

For suggested answers, see Appendix A.

CONCEPT 36.4

The rate of transpiration is regulated by stomata

Leaves generally have large surface areas and high surface-to-volume ratios. The large surface area enhances light absorption for photosynthesis. The high surface-to-volume ratio aids in CO_2 absorption during photosynthesis as well as in the release of O_2, a by-product of photosynthesis. Upon diffusing through the stomata, CO_2 enters a honeycomb of air spaces formed by the spongy mesophyll cells (see Figure 35.18). Because of the irregular shapes of these cells, the leaf's internal surface area may be 10 to 30 times greater than the external surface area.

Although large surface areas and high surface-to-volume ratios increase the rate of photosynthesis, they also increase water loss by way of the stomata. Thus, a plant's tremendous requirement for water is largely a consequence of the shoot system's need for ample exchange of CO_2 and O_2 for photosynthesis. By opening and closing the stomata, guard cells help balance the plant's requirement to conserve water with its requirement for photosynthesis.

Stomata: Major Pathways for Water Loss

About 95% of the water a plant loses escapes through stomata, although these pores account for only 1–2% of the external leaf surface. The waxy cuticle limits water loss through the remaining surface of the leaf. Each stoma is flanked by a pair of guard cells. Guard cells control the diameter of the stoma by changing shape, thereby widening or narrowing the gap between the guard cell pair. Under the same environmental conditions, the amount of water lost per leaf depends largely on stomatal density **(Figure 36.13)** and the average size of the stomatal pores.

▼ **Figure 36.13 Partially open stomata in the epidermis of a bean (*Phaseolus*) leaf.**

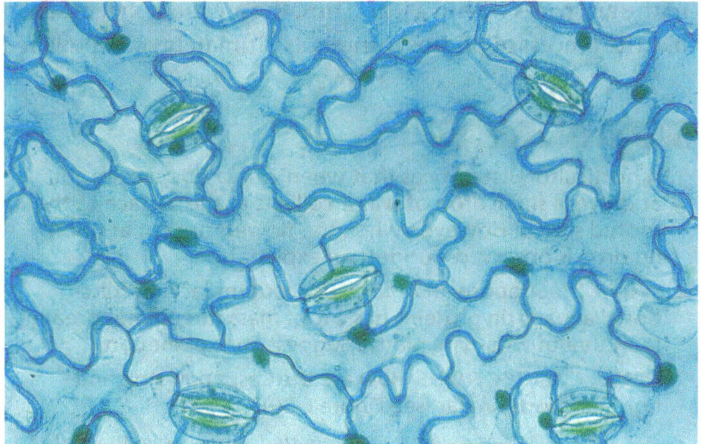

VISUAL SKILLS *If the micrograph above is of a section of epidermis 500 μm by 300 μm, what is the stomatal density of this bean leaf? Give your answer in number of stomata per square centimetre.*

Stomatal density can be under both genetic and environmental control. As a result of evolution by natural selection, shade-tolerant species tend to have lower stomatal densities than shade-intolerant species because CO_2 uptake doesn't limit photosynthesis under shady conditions. Stomatal density, however, can also be regulated by environmental changes. Low CO_2 levels during leaf development, for example, lead to increased stomatal densities in many species, an adaptation that facilitates CO_2 uptake under these conditions. By measuring the stomatal density of leaf fossils, scientists have gained insight into the levels of atmospheric CO_2 in past climates. A recent British survey found that stomatal density of many woodland species has decreased since 1927, when a similar survey was made. This observation is consistent with other findings that atmospheric CO_2 levels increased dramatically during the late 1900s.

Mechanisms of Stomatal Opening and Closing

When guard cells take in water from neighbouring cells by osmosis, they become more turgid. In most angiosperm guard cells, the cell walls facing the stomatal pore are thicker, and the cellulose microfibrils are oriented in a direction that causes the guard cells to bow outwards when turgid **(Figure 36.14a)**. This bowing outwards increases the size of the pore between the guard cells. When the cells lose water and become flaccid, they become less bowed, and the pore closes.

The changes in turgor pressure in guard cells result primarily from the reversible absorption and loss of K^+. Stomata open when guard cells actively accumulate K^+ from neighbouring epidermal cells **(Figure 36.14b)**. The flow of K^+ across the plasma membrane of the guard cell is coupled to the generation of a membrane potential by proton pumps (see Figure 36.6a). Stomatal opening correlates with active transport of H^+ out of the guard cell. The resulting voltage (membrane potential) drives K^+ into the cell through specific membrane channels. The absorption of K^+ causes the water potential to become more negative within the guard cells, and the cells become more turgid as water enters by osmosis. Because most of the K^+ and water are stored in the vacuole, the vacuolar membrane also plays a role in regulating guard cell dynamics. Stomatal closing results from a loss of K^+ from

▼ **Figure 36.14 Mechanisms of stomatal opening and closing.**

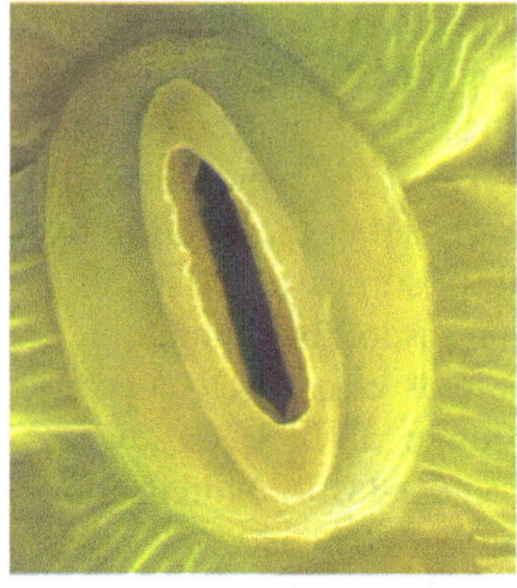

Guard cells turgid/Stoma open

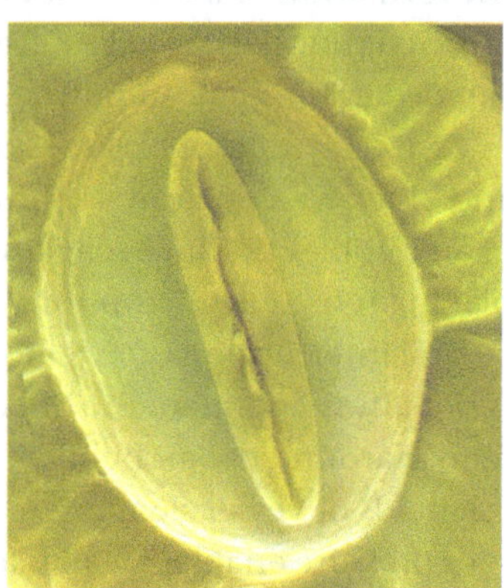

Guard cells flaccid/Stoma closed

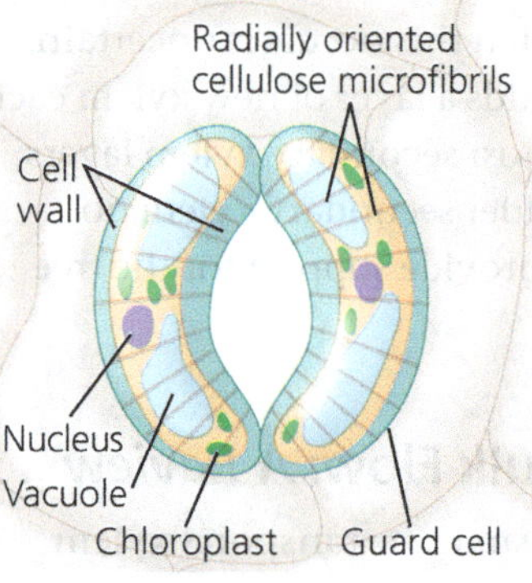

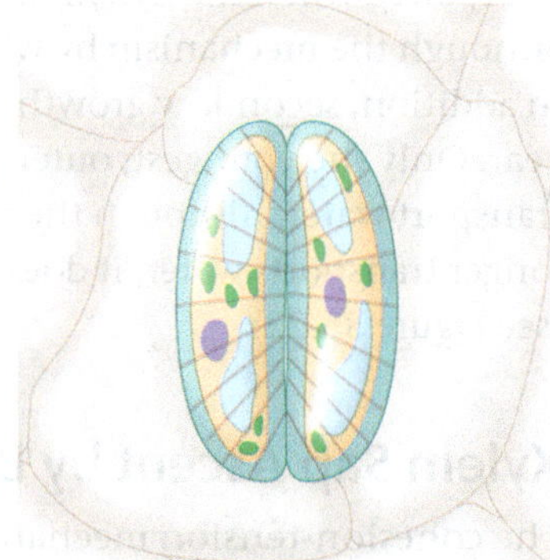

(a) Changes in guard cell shape and stomatal opening and closing (surface view). Guard cells of a typical angiosperm are illustrated in their turgid (stoma open) and flaccid (stoma closed) states. The radial orientation of cellulose microfibrils in the cell walls causes the guard cells to increase more in length than width when turgor increases. Since the two guard cells are tightly joined at their tips, they bow outwards when turgid, causing the stomatal pore to open.

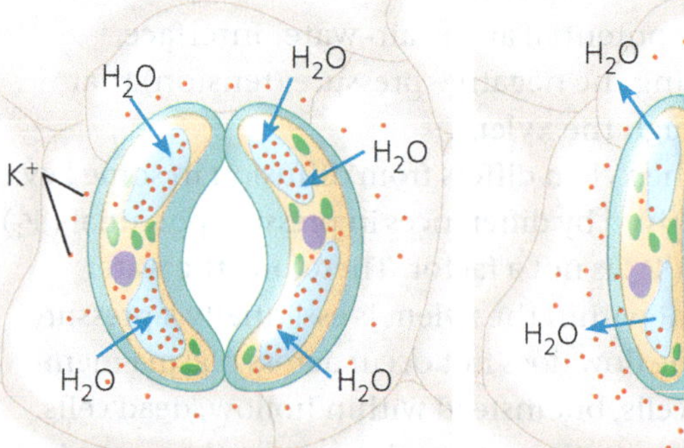

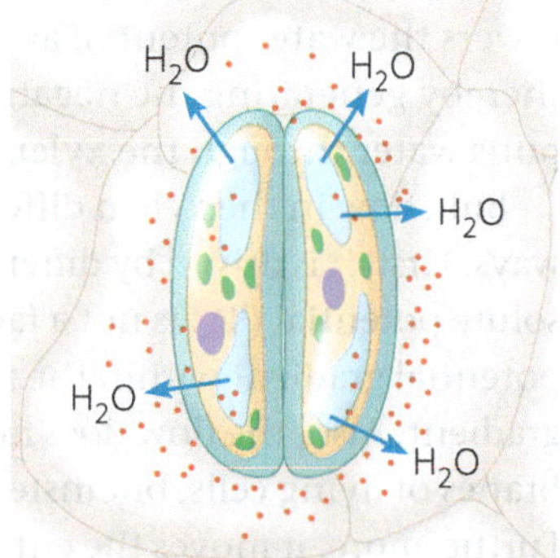

(b) Role of potassium ions (K^+) in stomatal opening and closing. The transport of K^+ (symbolised here as red dots) across the plasma membrane and vacuolar membrane causes the turgor changes of guard cells. The uptake of anions, such as malate and chloride ions (not shown), also contributes to guard cell swelling.

guard cells to neighbouring cells, which leads to an osmotic loss of water. Aquaporins also help regulate the osmotic swelling and shrinking of guard cells.

Stimuli for Stomatal Opening and Closing

In general, stomata are open during the day and mostly closed at night, preventing the plant from losing water under conditions when photosynthesis cannot occur. At least three cues contribute to stomatal opening at dawn: light, CO_2 depletion, and an internal "clock" in guard cells.

Light stimulates guard cells to accumulate K^+ and become turgid. This response is triggered by illumination of blue-light receptors in the plasma membrane of guard cells. Activation of these receptors stimulates the activity of proton pumps in the plasma membrane of the guard cells, in turn promoting absorption of K^+.

Stomata also open in response to depletion of CO_2 within the leaf's air spaces as a result of photosynthesis. As CO_2 concentrations decrease during the day, the stomata progressively open if sufficient water is supplied to the leaf.

An internal "clock" in the guard cells ensures that stomata continue their daily rhythm of opening and closing. This rhythm occurs even if a plant is kept in a dark location. All eukaryotic organisms have internal clocks that regulate cyclic processes. Cycles with intervals of approximately 24 hours are called **circadian rhythms** (which you'll learn more about in Concept 39.3).

Drought stress can also cause stomata to close. A hormone called **abscisic acid (ABA)**, which is produced in roots and leaves in response to water deficiency, signals guard cells to close stomata. This response reduces wilting but also restricts CO_2 absorption, thereby slowing photosynthesis. ABA also directly inhibits photosynthesis. Water availability is closely tied to plant productivity not because water is needed as a substrate in photosynthesis, but because freely available water allows plants to keep stomata open and take up more CO_2.

Guard cells control the photosynthesis-transpiration compromise on a moment-to-moment basis by integrating a variety of internal and external stimuli. Even the passage of a cloud or a transient shaft of sunlight through a forest can affect the rate of transpiration.

Effects of Transpiration on Wilting and Leaf Temperature

As long as most stomata remain open, transpiration is greatest on a day that is sunny, warm, dry, and windy because these environmental factors increase evaporation. If transpiration cannot pull sufficient water to the leaves, the shoot becomes slightly wilted as cells lose turgor pressure. Although plants respond to such mild drought stress by rapidly closing stomata, some evaporative water loss still occurs through the cuticle. Under prolonged drought conditions, leaves can become severely wilted and irreversibly injured.

Transpiration also results in evaporative cooling, which can lower a leaf's temperature by as much as 10°C compared with the surrounding air. This cooling prevents the leaf from reaching temperatures that could denature enzymes involved in photosynthesis and other metabolic processes.

Adaptations That Reduce Evaporative Water Loss

Water availability is a major determinant of plant productivity. The main reason water availability remains intimately connected to plant productivity is not related to photosynthesis's direct need for water as a substrate. Instead, freely available water allows plants to keep stomata open and take up more CO_2. The problem of reducing water loss is especially acute for desert plants. Plants adapted to arid environments are called **xerophytes** (from the Greek *xero*, dry). Figure 36.16 provides examples of xerophytic plants from Australia's and New Zealand's high and lowland environments, where water remains scarce.

Many species of desert plants avoid drying out by completing their short life cycles during the brief rainy seasons. Rain comes infrequently in deserts, but when it arrives, the vegetation is transformed as dormant seeds of annual species quickly germinate and bloom, completing their life cycle before dry conditions return.

Other xerophytes have unusual physiological or morphological adaptations that enable them to withstand harsh desert conditions. The stems of many xerophytes are fleshy because they store water for use during long dry periods. Cacti have highly reduced leaves that resist excessive water loss; photosynthesis is carried out mainly in their stems. Another adaptation common in arid habitats is crassulacean acid metabolism (CAM), a specialised form of photosynthesis found in succulents of the family Crassulaceae and several other families (see Figure 10.21). Because the leaves of CAM plants take in CO_2 at night, the stomata can remain closed during the day, when evaporative stresses are greatest. Other examples of xerophytic adaptations are discussed in **Figure 36.15**.

CONCEPT CHECK **36.4**

1. What are the stimuli that control the opening and closing of stomata?
2. The pathogenic fungus *Fusicoccum amygdali* secretes a toxin called fusicoccin that activates the plasma membrane proton pumps of plant cells and leads to uncontrolled water loss. Suggest a mechanism by which the activation of proton pumps could lead to severe wilting.
3. **WHAT IF?** If you buy cut flowers, why might the florist recommend cutting the stems underwater and then transferring the flowers to a vase while the cut ends are still wet?
4. **MAKE CONNECTIONS** Explain why the evaporation of water from leaves lowers their temperature (see Concept 3.2).

For suggested answers, see Appendix A.

▼ Figure 36.15 Some xerophytic adaptations.

▶ The leafless rock wattle (*Acacia aphylla*) is a rare species that is vulnerable to extinction. Only two isolated populations persist in Western Australia. The species evolved to survive and thrive in the dry and hot conditions. It has no true leaves; instead, it possesses thorn-like cylindrical stalks. The small surface-area-to-volume ratio of these stalks reduces heat absorption from the sun and surrounding environment. In this way, the plant reduces the amount of water required to evaporatively cool its body.

▲ New Zealand's tōtara (*Podocarpus totara*) possess needles with a similar cylindrical shape to the stalks of *Acacia aphylla*. Although, the tōtara's adaptation protects the plant from the cold, not the heat, the same xerophytic adaptations that minimise water loss in hot conditions reduce water loss from these relatives of pine trees throughout the year.

▼ *Banksia baxteri*, shown in the inset, grows commonly in arid climates. Leaves of this and other arid species possess a thick cuticle with multiple-layered epidermal tissue that reduces water loss. Stomata are recessed in cavities called "pits," an adaptation that reduces the rate of transpiration by protecting the stomata from hot, dry wind. Trichomes help minimise transpiration by breaking up the flow of air, allowing the chamber of the pit to have a higher humidity than the surrounding atmosphere (LM).

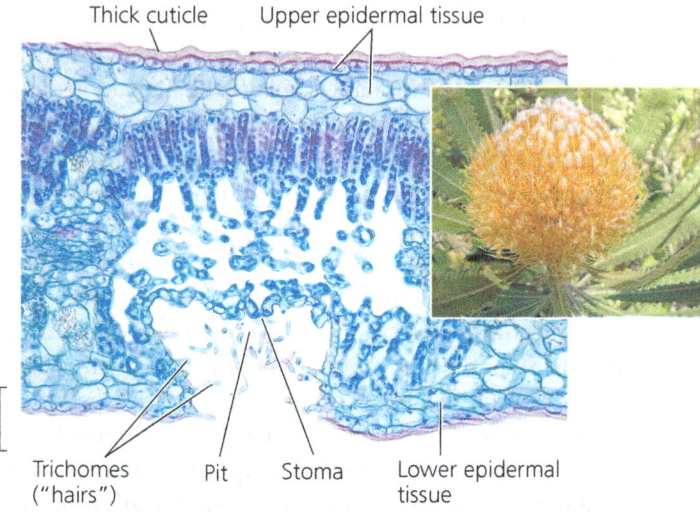

▶ Hairy white leaves at the ends of tightly interwoven branches lend New Zealand's *Raoulia* spp. the name and appearance of vegetable sheep. The compact growth form and xerophytic leaves minimise water loss in the dry mountain regions these plants inhabit.

CONCEPT 36.5

Sugars are transported from sources to sinks via the phloem

The unidirectional flow of water and minerals from soil to roots to leaves through the xylem is largely in an upward direction. In contrast, the movement of photosynthates often runs in the opposite direction, transporting sugars from mature leaves to lower parts of the plant, such as root tips that require large amounts of sugars for energy and growth. The transport of the products of photosynthesis, known as **translocation**, is carried out by another tissue, the phloem.

Movement from Sugar Sources to Sugar Sinks

Sieve-tube elements are specialised cells in angiosperms that serve as conduits for translocation. Arranged end to end, they form long sieve tubes (see Figure 35.10). Between these cells are sieve plates, structures that allow the flow of sap along the sieve tube. **Phloem sap**, the aqueous solution that flows through sieve tubes, differs markedly from the xylem sap that is transported by tracheids and vessel elements. By far the most prevalent solute in phloem sap is sugar, typically sucrose in most species. The sucrose concentration may be as high as 30% by weight, giving the sap a syrupy thickness. Phloem sap may also contain amino acids, hormones, and minerals.

In contrast to the unidirectional transport of xylem sap from roots to leaves, phloem sap moves from sites of sugar production to sites of sugar use or storage (see Figure 36.2). A **sugar source** is a plant organ that is a net producer of sugar, by photosynthesis or by breakdown of starch. In contrast, a **sugar sink** is an organ that is a net consumer or depository of sugar. Growing roots, buds, stems, and fruits are sugar sinks. Although expanding leaves are sugar sinks, mature leaves, if well illuminated, are sugar sources. A storage organ, such as a tuber or a bulb, may be a source or a sink, depending on the season. When stockpiling carbohydrates in the summer, it is a sugar sink. After breaking dormancy in the spring, it is a sugar source because its starch is broken down to sugar, which is carried to the growing shoot tips.

Sinks usually receive sugar from the nearest sugar sources. The upper leaves on a branch, for example, may export sugar to the growing shoot tip, whereas the lower leaves may export sugar to the roots. A growing fruit may monopolise the sugar sources that surround it. For each sieve tube, the direction of transport depends on the locations of the sugar source and sugar sink that are connected by that tube. Therefore, neighbouring sieve tubes may carry sap in opposite directions if they originate and end in different locations.

Sugar must be transported, or loaded, into sieve-tube elements before being exported to sugar sinks. In some species, it moves from mesophyll cells to sieve-tube elements via the symplast, passing through plasmodesmata. In other species, it moves by symplastic and apoplastic pathways. In corn leaves, for example, sucrose diffuses through the symplast from photosynthetic mesophyll cells into small veins. Much of it then moves into the apoplast and is accumulated by nearby sieve-tube elements, either directly or, as shown in **Figure 36.16a**, through companion cells. In some plants, the walls of the companion cells feature many ingrowths, enhancing solute transfer between apoplast and symplast.

In many plants, sugar movement into the phloem requires active transport because sucrose is more concentrated in sieve-tube elements and companion cells than in mesophyll. Proton pumping and H^+/sucrose cotransport enable sucrose to move from mesophyll cells to sieve-tube elements or companion cells **(Figure 36.16b)**.

Sucrose is unloaded at the sink end of a sieve tube. The process varies by species and organ. However, the concentration of free sugar in the sink is always lower than in the sieve tube because the unloaded sugar is consumed during growth and metabolism of the cells of the sink or converted to insoluble polymers such as starch. As a result of this sugar concentration gradient, sugar molecules diffuse from the phloem into the sink tissues, and water follows by osmosis.

Bulk Flow by Positive Pressure: The Mechanism of Translocation in Angiosperms

Phloem sap flows from source to sink at rates that are as great as 1 m/hr, which is much faster than diffusion or cytoplasmic streaming. Researchers concluded that it moves through the sieve tubes of angiosperms by bulk flow driven by positive pressure, known as *pressure flow* **(Figure 36.17)**. The building of pressure at the source and reduction of that pressure at the sink cause sap to flow from source to sink.

The pressure-flow hypothesis explains why phloem sap flows from source to sink, and experiments build a strong case for pressure flow as the mechanism of translocation in

▼ Figure 36.16 Loading of sucrose into phloem.

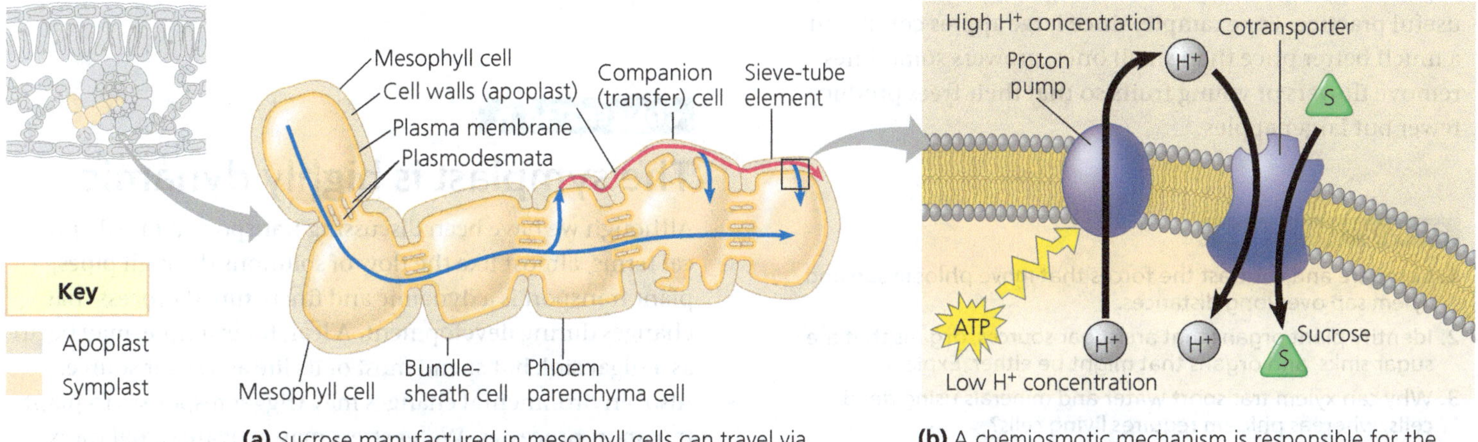

(a) Sucrose manufactured in mesophyll cells can travel via the symplast (blue arrows) to sieve-tube elements. In some species, sucrose exits the symplast near sieve tubes and travels through the apoplast (red arrow). It is then actively accumulated from the apoplast by sieve-tube elements and their companion cells.

(b) A chemiosmotic mechanism is responsible for the active transport of sucrose into companion cells and sieve-tube elements. Proton pumps generate an H^+ gradient, which drives sucrose accumulation with the help of a cotransport protein that couples sucrose transport to the diffusion of H^+ back into the cell.

▼ **Figure 36.17 Bulk flow by positive pressure (pressure flow) in a sieve tube.**

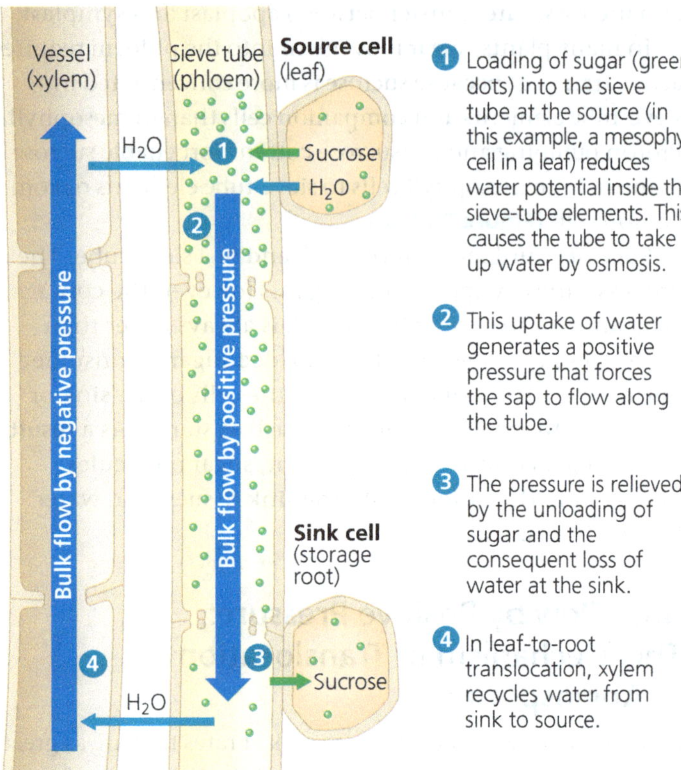

1. Loading of sugar (green dots) into the sieve tube at the source (in this example, a mesophyll cell in a leaf) reduces water potential inside the sieve-tube elements. This causes the tube to take up water by osmosis.

2. This uptake of water generates a positive pressure that forces the sap to flow along the tube.

3. The pressure is relieved by the unloading of sugar and the consequent loss of water at the sink.

4. In leaf-to-root translocation, xylem recycles water from sink to source.

angiosperms **(Figure 36.18)**. However, studies using electron microscopes suggest that in nonflowering vascular plants, the pores between phloem cells may be too small or obstructed to permit pressure flow.

Sinks vary in energy demands and capacity to unload sugars. Sometimes there are more sinks than can be supported by sources. In such cases, a plant might abort some flowers, seeds, or fruits—a phenomenon called *self-thinning*. Removing sinks can also be a horticulturally useful practice. For example, since large apples command a much better price than small ones, growers sometimes remove flowers or young fruits so that their trees produce fewer but larger apples.

CONCEPT CHECK 36.5

1. Compare and contrast the forces that move phloem sap and xylem sap over long distances.
2. Identify plant organs that are sugar sources, organs that are sugar sinks, and organs that might be either. Explain.
3. Why can xylem transport water and minerals using dead cells, whereas phloem requires living cells?
4. **WHAT IF?** Apple growers in Japan sometimes make a non-lethal spiral slash around the bark of trees that are destined for removal after the growing season. This practice makes the apples sweeter. Why?

For suggested answers, see Appendix A.

▼ **Figure 36.18 Inquiry**

Does phloem sap contain more sugar near sources than near sinks?

Experiment The pressure-flow hypothesis predicts that phloem sap near sources should have a higher sugar content than phloem sap near sinks. To test this idea, researchers used aphids that feed on phloem sap. An aphid probes with a hypodermic-like mouthpart called a stylet that penetrates a sieve-tube element. As sieve-tube pressure forced out phloem sap into the stylets, the researchers separated the aphids from the stylets, which then acted as taps exuding sap for hours. Researchers measured the sugar concentration of sap from stylets at different points between a source and sink.

Sap droplet

Aphid feeding

Stylet in sieve-tube element

Separated stylet exuding sap

Results The closer the stylet was to a sugar source, the higher its sugar concentration was.

Conclusion The results of such experiments support the pressure-flow hypothesis, which predicts that sugar concentrations should be higher in sieve tubes closer to sugar sources.

Data from S. Rogers and A. J. Peel, Some evidence for the existence of turgor pressure in the sieve tubes of willow (*Salix*), *Planta* 126:259–267 (1975).

WHAT IF? *Spittlebugs* (Clasirptora sp.) *are xylem sap feeders that use strong muscles to pump xylem sap through their guts. Could you isolate xylem sap from the excised stylets of spittlebugs?*

CONCEPT **36.6**

The symplast is highly dynamic

Although we have been discussing transport in mostly physical terms, almost like the flow of solutions through pipes, plant transport is a dynamic and finely tuned process that changes during development. A leaf, for example, may begin as a sugar sink but spend most of its life as a sugar source. Also, environmental changes may trigger responses in plant transport processes. Water stress may activate signal transduction pathways that greatly alter the membrane transport proteins governing the overall transport of water and minerals. Because the symplast is living tissue, it is largely responsible for the dynamic changes in plant transport processes.

We'll look now at some other examples: changes in plasmodesmata, chemical signalling, and electrical signalling.

Changes in Plasmodesmatal Number and Pore Size

Based mostly on the static images provided by electron microscopy, biologists formerly considered plasmodesmata to be unchanging, pore-like structures. More recent studies, however, have revealed that plasmodesmata are highly dynamic. They can open or close rapidly in response to changes in turgor pressure, cytosolic Ca^{2+} level, or cytosolic pH. Although some plasmodesmata form during cytokinesis, they can also form much later. Moreover, loss of function is common during differentiation. For example, as a leaf matures from a sink to a source, its plasmodesmata either close or are eliminated, causing phloem unloading to cease.

Early studies by plant physiologists and pathologists came to differing conclusions regarding pore sizes of plasmodesmata. Physiologists injected fluorescent probes of different molecular sizes into cells and recorded whether the molecules passed into adjacent cells. Based on these observations, they concluded that the pore sizes were approximately 2.5 nm—too small for macromolecules such as proteins to pass. In contrast, pathologists provided electron micrographs showing evidence of the passage of virus particles with diameters of 10 nm or greater **(Figure 36.19)**.

Subsequently, it was learned that plant viruses produce *viral movement proteins* that cause the plasmodesmata to dilate, enabling the viral RNA to pass between cells. More recent evidence shows that plant cells themselves regulate plasmodesmata as part of a communication network. The viruses can subvert this network by mimicking the cell's regulators of plasmodesmata.

A high degree of cytosolic interconnectedness exists only within certain groups of cells and tissues, which are known as *symplastic domains*. Informational molecules, such as proteins and RNAs, coordinate development between cells within each symplastic domain. If symplastic communication is disrupted, development can be grossly affected.

▼ Figure 36.19 Virus particles moving cell to cell through plasmodesma connecting turnip leaf cells. (TEM)

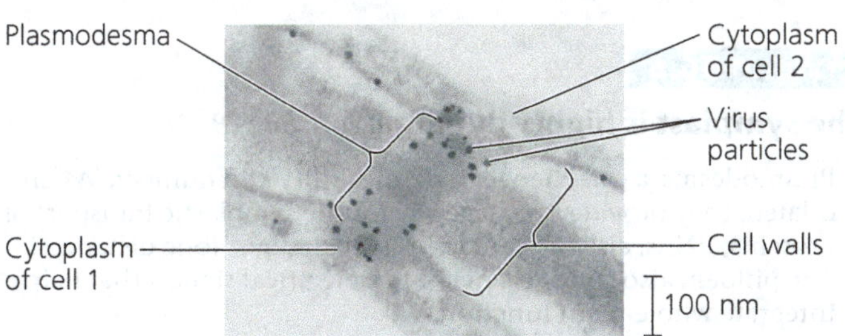

Phloem: An Information Superhighway

In addition to transporting sugars, the phloem is a "superhighway" for the transport of macromolecules and viruses. This transport is systemic (throughout the body), affecting many or all of the plant's systems or organs. Macromolecules translocated through the phloem include proteins and various types of RNA that enter the sieve tubes through plasmodesmata. Although they are often likened to the gap junctions between animal cells, plasmodesmata are unique in their ability to traffic proteins and RNA.

Systemic communication through the phloem helps integrate the functions of the whole plant. One classic example is the delivery of a flower-inducing chemical signal from leaves to vegetative meristems. Another is a defensive response to localised infection, in which chemical signals travelling through the phloem activate defence genes in noninfected tissues.

Electrical Signalling in the Phloem

Rapid, long-distance electrical signalling through the phloem is another dynamic feature of the symplast. Electrical signalling has been studied extensively in plants that have rapid leaf movements, such as the sensitive plant (*Mimosa pudica*) and Venus flytrap (*Dionaea muscipula*). However, its role in other species is less clear. Some studies have revealed that a stimulus in one part of a plant can trigger an electrical signal in the phloem that affects another part, where it may elicit a change in gene transcription, respiration, photosynthesis, phloem unloading, or hormonal levels. Thus, the phloem can serve a nerve-like function, allowing for swift electrical communication between widely separated organs.

The coordinated transport of materials and information is central to plant survival. Plants can acquire only so many resources in the course of their lifetimes. Ultimately, the successful acquisition of these resources and their optimal distribution are the most critical determinants of whether the plant will compete successfully.

CONCEPT CHECK 36.6

1. How do plasmodesmata differ from gap junctions?
2. Nerve-like signals in animals are thousands of times faster than their plant counterparts. Suggest a behavioural reason for the difference.
3. **WHAT IF?** Suppose plants were genetically modified to be unresponsive to viral movement proteins. Would this be a good way to prevent the spread of infection? Explain.

For suggested answers, see Appendix A.

36 Chapter Review

SUMMARY OF KEY CONCEPTS

CONCEPT 36.1

Adaptations for acquiring resources were key steps in the evolution of vascular plants *(pp. 826–828)*

- Leaves typically function in gathering sunlight and CO_2. Stems serve as supporting structures for leaves and as conduits for the long-distance transport of water and nutrients. Roots mine the soil for water and minerals and anchor the whole plant.
- Natural selection has produced plant architectures that optimise resource acquisition in the ecological niche in which the plant species naturally exists.

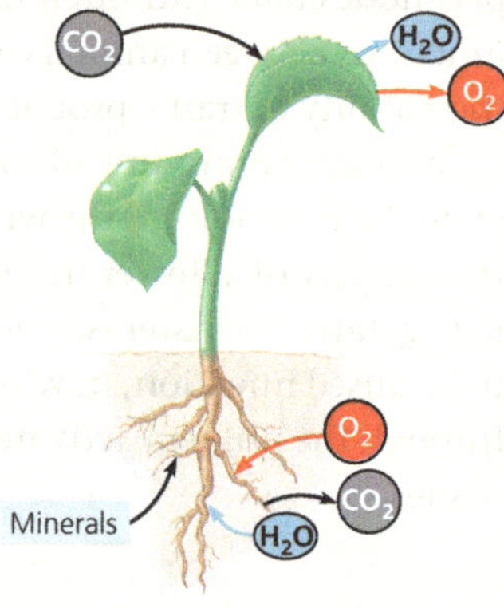

? *How did the evolution of xylem and phloem contribute to the successful colonisation of land by vascular plants?*

CONCEPT 36.2

Different mechanisms transport substances over short or long distances *(pp. 828–833)*

- The selective permeability of the plasma membrane controls the movement of substances into and out of cells. Both active and passive transport mechanisms occur in plants.
- Plant tissues have two major compartments: the **apoplast** (everything outside the cells' plasma membranes) and the **symplast** (the cytosol and connecting plasmodesmata).
- Direction of water movement depends on the **water potential**, a quantity that incorporates solute concentration and physical pressure. The **osmotic** uptake of water by plant cells and the resulting internal pressure that builds up make plant cells **turgid**.
- Long-distance transport occurs through **bulk flow**, the movement of liquid in response to a pressure gradient. Bulk flow occurs within the tracheids and vessel elements of the **xylem** and within the sieve-tube elements of the **phloem**.

? *Is xylem sap usually pulled or pushed up the plant?*

CONCEPT 36.3

Transpiration drives the transport of water and minerals from roots to shoots via the xylem *(pp. 833–837)*

- Water and minerals from the soil enter the plant through the epidermis of roots, cross the root cortex, and then pass into the vascular cylinder by way of the selectively permeable cells of the **endodermis**. From the vascular cylinder, the **xylem sap** is transported long distances by bulk flow to the veins that branch throughout each leaf.
- The **cohesion-tension hypothesis** proposes that the movement of xylem sap is driven by a water potential difference created at the leaf end of the xylem by the evaporation of water from leaf cells. Evaporation lowers the water potential at the air-water interface, thereby generating the negative pressure that pulls water through the xylem.

? *Why is the ability of water molecules to form hydrogen bonds important for the movement of xylem sap?*

CONCEPT 36.4

The rate of transpiration is regulated by stomata *(pp. 837–839)*

- **Transpiration** is the loss of water vapour from plants. **Wilting** occurs when the water lost by transpiration is not replaced by absorption from roots. Plants respond to water deficits by closing their stomata. Under prolonged drought conditions, plants can become irreversibly injured.
- Stomata are the major pathway for water loss from plants. A stoma opens when guard cells bordering the stomatal pore take up K^+. The opening and closing of stomata are controlled by light, CO_2, the drought hormone **abscisic acid**, and a **circadian rhythm**.
- **Xerophytes** are plants that are adapted to arid environments. Reduced leaves and CAM photosynthesis are examples of adaptations to arid environments.

? *Why are stomata necessary?*

CONCEPT 36.5

Sugars are transported from sources to sinks via the phloem *(pp. 840–841)*

- Mature leaves are the main **sugar sources**, although storage organs can be seasonal sources. Growing organs such as roots, stems, and fruits are the main **sugar sinks**. The direction of phloem transport is always from sugar source to sugar sink.
- Phloem loading depends on the active transport of sucrose. Sucrose is cotransported with H^+, which diffuses down a gradient generated by proton pumps. Loading of sugar at the source and unloading at the sink maintain a pressure difference that keeps **phloem sap** flowing through a sieve tube.

? *Why is phloem transport considered an active process?*

CONCEPT 36.6

The symplast is highly dynamic *(pp. 841–843)*

- Plasmodesmata can change in permeability and number. When dilated, they provide a passageway for the symplastic transport of proteins, RNAs, and other macromolecules over long distances. The phloem also conducts nerve-like electrical signals that help integrate whole-plant function.

? *By what mechanisms is symplastic communication regulated?*

TEST YOUR UNDERSTANDING

Levels 1-2: Remembering/Understanding

1. Which of the following is an adaptation that enhances the uptake of water and minerals by roots?
(A) mycorrhizae
(B) pumping through plasmodesmata
(C) active uptake by vessel elements
(D) rhythmic contractions by cells in the root cortex

2. Which structure or compartment is part of the symplast?
(A) the interior of a vessel element
(B) the interior of a sieve tube
(C) the cell wall of a mesophyll cell
(D) an extracellular air space

3. Movement of phloem sap from a source to a sink
(A) occurs through the apoplast of sieve-tube elements.
(B) depends ultimately on the activity of proton pumps.
(C) depends on tension, or negative pressure potential.
(D) results mainly from diffusion.

Levels 3-4: Applying/Analysing

4. Photosynthesis ceases when leaves wilt, mainly because
(A) the chlorophyll in wilting leaves is degraded.
(B) accumulation of CO_2 in the leaf inhibits enzymes.
(C) stomata close, preventing CO_2 from entering the leaf.
(D) photolysis, the water-splitting step of photosynthesis, cannot occur when there is a water deficiency.

5. What would enhance water uptake by a plant cell?
(A) decreasing the Ψ of the surrounding solution
(B) positive pressure on the surrounding solution
(C) the loss of solutes from the cell
(D) increasing the Ψ of the cytoplasm

6. A plant cell with a Ψ_S of −0.65 MPa maintains a constant volume when bathed in a solution that has a Ψ_S of −0.30 MPa and is in an open container. The cell has a
(A) Ψ_P of +0.65 MPa.
(B) Ψ of −0.65 MPa.
(C) Ψ_P of +0.35 MPa.
(D) Ψ_P of 0 MPa.

7. Compared with a cell with few aquaporin proteins in its membrane, a cell containing many aquaporin proteins will
(A) have a faster rate of osmosis.
(B) have a lower water potential.
(C) have a higher water potential.
(D) accumulate water by active transport.

8. Which of the following would tend to increase transpiration?
(A) spiny leaves
(B) sunken stomata
(C) a thicker cuticle
(D) higher stomatal density

Levels 5-6: Evaluating/Creating

9. EVOLUTION CONNECTION Large brown algae called kelps can grow as tall as 25 m. Kelps consist of a holdfast anchored to the ocean floor, blades that float at the surface and collect light, and a long stalk connecting the blades to the holdfast (see Figure 28.13). Specialised cells in the stalk, although nonvascular, can transport sugar. Suggest a reason why these structures analogous to sieve-tube elements might have evolved in kelps.

10. SCIENTIFIC INQUIRY • INTERPRET THE DATA A Minnesota gardener notes that the plants immediately bordering a walkway are stunted compared with those further away. Suspecting that the soil near the walkway may be contaminated from salt added to the walkway in winter, the gardener tests the soil. The composition of the soil near the walkway is identical to that further away except that it contains an additional 50 m*M* NaCl. Assuming that the NaCl is completely ionised, calculate how much it will lower the solute potential of the soil at 20°C using the *solute potential equation*:

$$\Psi_S = -iCRT$$

where i is the ionisation constant (2 for NaCl), C is the molar concentration (in mol/L), R is the pressure constant $[R = 0.00831\ (\text{L} \cdot \text{MPa})/(\text{mol} \cdot \text{K})]$, and T is the temperature in Kelvin $(273 + {}^\circ\text{C})$.

How would this change in the solute potential of the soil affect the water potential of the soil? In what way would the change in the water potential of the soil affect the movement of water in or out of the roots?

11. SCIENTIFIC INQUIRY Cotton plants wilt within a few hours of flooding of their roots. The flooding leads to low-oxygen conditions, increases in cytosolic Ca^{2+} concentration, and decreases in cytosolic pH. Suggest a hypothesis to explain how flooding leads to wilting.

12. WRITE ABOUT A THEME: ORGANISATION Natural selection has led to changes in the architecture of plants that enable them to photosynthesise more efficiently in the ecological niches they occupy. In a short essay (100–150 words), explain how shoot architecture enhances photosynthesis.

13. SYNTHESISE YOUR KNOWLEDGE

Imagine yourself as a water molecule in the soil solution of a forest. In a short essay (100–150 words), explain what pathways and what forces would be necessary to carry you to the leaves of these trees.

For selected answers, see Appendix A.

39 Plant Responses to Internal and External Signals

KEY CONCEPTS

Figure 39.1 Sunflowers track the sun from east to west each day. After sunset, they reverse direction, facing the direction of the next sunrise. By facing the hot sun during the day, the floral heads become warmer and release greater amounts of chemicals that attract pollinators. Light is just one of the many factors to which a plant responds.

Study Tip

Make a table: As you read the chapter, add specific examples for each of the general categories of responses shown in the diagram.

Factor	Example of plant response
Light	Seed germination in response to red light

What are some factors that plants sense and respond to?

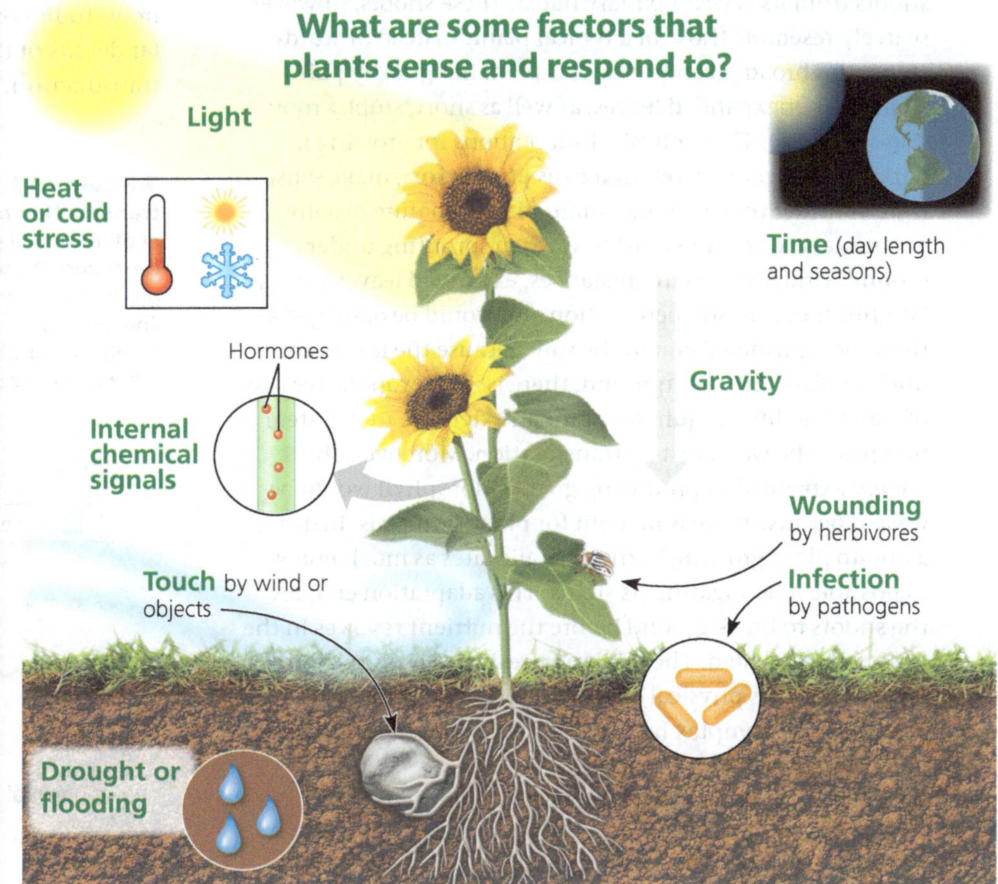

Go to Mastering Biology

to access Dynamic Study Modules for revision, 3D BioFlix® animations and high-quality videos, and your interactive Pearson eText.

CONCEPT 39.1

Signal transduction pathways link signal reception to response

The idea that plants are inert or passive is a common misconception. Nothing could be further from the truth. As shown in the examples of plant responses in **Figure 39.1**, plants must sense and integrate information about many facets of their environment. Although plant development is simpler than the development of multicellular animals, plant cells are just as complex as animal cells (see Figure 6.8). The molecular biology of plants is as complicated as that of animals: A Japanese plant (*Paris japonica*) currently has the largest genome on record, about fifty times greater than the human genome. Plants have among the largest genomes (see Table 21.1). At the levels of signal reception and signal transduction, your cells are not all that different from those of plants—the similarities far outweigh the differences. As an animal, however, your responses to environmental stimuli are generally quite different from those of plants. Animals commonly respond by movement; plants do so by altering growth and development.

As an example of a plant modifying its growth and development in response to environmental cues, consider a forgotten potato in the back corner of a kitchen cupboard. This modified underground stem, or tuber, has sprouted shoots from its "eyes" (axillary buds). These shoots, however, scarcely resemble those of a typical plant. Instead of sturdy stems and broad green leaves, this plant has ghostly pale stems and unexpanded leaves, as well as short, stubby roots **(Figure 39.2a)**. These physical adaptations for growing in darkness, collectively referred to as **etiolation**, make sense if we consider that a young potato plant in nature usually encounters continuous darkness when sprouting underground. Under these circumstances, expanded leaves would be a hindrance to soil penetration and would be damaged as the shoots pushed through the soil. Because the leaves are unexpanded and underground, there is little evaporative loss of water and little requirement for an extensive root system to replace the water lost by transpiration. Moreover, the energy expended in producing green chlorophyll would be wasted because there is no light for photosynthesis. Instead, a potato plant growing in the dark allocates as much energy as possible to elongating its stems. This adaptation enables the shoots to break ground before the nutrient reserves in the tuber are exhausted. The etiolation response is one example of how a plant's physical characteristics are tuned to its surroundings by complex interactions between environmental and internal signals.

When a shoot reaches light, the plant undergoes profound changes, collectively called **de-etiolation**

▼ **Figure 39.2 Light-induced de-etiolation (greening) of dark-grown potatoes.**

(a) Before exposure to light. A dark-grown potato has tall, spindly stems and nonexpanded leaves—morphological adaptations that enable the shoots to penetrate the soil. The roots are short, but there is little need for water absorption because little water is lost by the shoots.

(b) After a week's exposure to natural daylight. The potato plant begins to resemble a typical plant with broad green leaves, short sturdy stems, and long roots. This transformation begins with the reception of light by a specific pigment, phytochrome.

(informally known as greening). Stem elongation slows; leaves expand; roots elongate; and the shoot produces chlorophyll. In short, it begins to resemble a typical plant **(Figure 39.2b)**. In this section, we'll examine this de-etiolation response as an example of how a plant cell's reception of a signal—in this case, light—is transduced into a response (greening). Along the way, we will explore how studies of mutants provide insights into the molecular details of the stages of cell signal processing: reception, transduction, and response **(Figure 39.3)**.

▼ **Figure 39.3 Review of a general model for signal transduction pathways.** As discussed in Concept 11.1, a hormone or other kind of stimulus interacting with a specific receptor protein can trigger the sequential activation of relay proteins and also the production of second messengers that participate in the pathway. The signal is passed along, ultimately bringing about cellular responses. In this diagram, the receptor is on the surface of the target cell; in other cases, the stimulus interacts with receptors inside the cell.

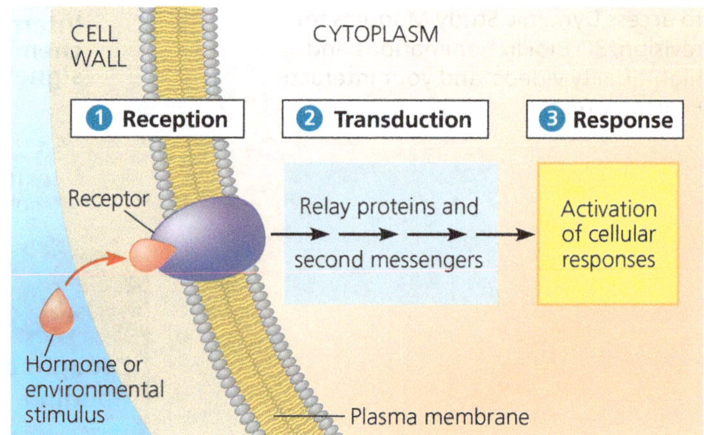

Reception

Signals are first detected by receptors, proteins that undergo changes in shape in response to a specific stimulus. The receptor involved in de-etiolation is a type of *phytochrome*, a member of a class of photoreceptors that we'll explore more fully later in the chapter. Unlike most receptors, which are built into the plasma membrane, the type of phytochrome that functions in de-etiolation is located in the cytoplasm. Researchers demonstrated the requirement for phytochrome in de-etiolation through studies of the tomato, a close relative of the potato. The *aurea* mutant of tomato, which has reduced levels of phytochrome, greens less than wild-type tomatoes when exposed to light. (*Aurea* is Latin for "gold." In the absence of chlorophyll, the yellow and orange accessory pigments called carotenoids are more obvious.) Researchers produced a normal de-etiolation response in individual *aurea* leaf cells by injecting phytochrome from other plants and then exposing the cells to light. Such experiments indicated that phytochrome functions in light detection during de-etiolation.

Transduction

Receptors can be sensitive to very weak environmental or chemical signals. Some de-etiolation responses are triggered by extremely low levels of light, in certain cases as little as the equivalent of a few seconds of moonlight. The transduction of these extremely weak signals involves **second messengers**—small molecules and ions in the cell that amplify the signal and transfer it from the receptor to other proteins that carry out the response **(Figure 39.4)**. Concept 11.3 discussed several kinds of second messengers (see Figures 11.12 and 11.14). Here, we examine the particular roles of two types of second messengers in de-etiolation: calcium ions (Ca^{2+}) and cyclic GMP (cGMP).

Changes in cytosolic Ca^{2+} levels play an important role in phytochrome signal transduction. The concentration of cytosolic Ca^{2+} is generally very low (about 10^{-7} *M*), but phytochrome activation leads to the opening of Ca^{2+} channels and a transient 100-fold increase in cytosolic Ca^{2+} levels. In response to light, phytochrome undergoes a change in shape that leads to the activation of guanylyl cyclase, an enzyme that produces the second messenger cyclic GMP. Both Ca^{2+}

▼ Figure 39.4 An example of signal transduction in plants: the role of phytochrome in the de-etiolation (greening) response.

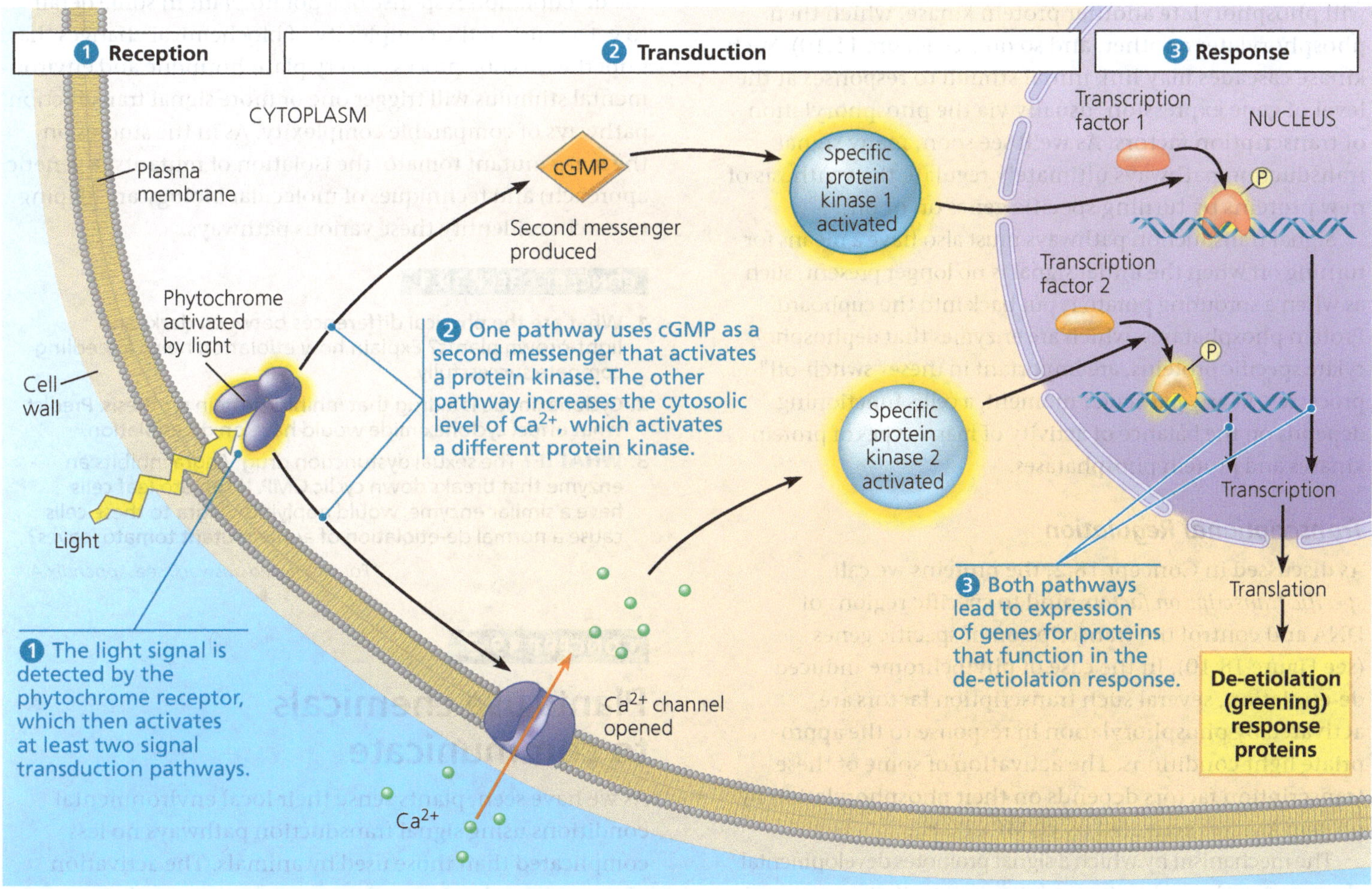

MAKE CONNECTIONS *Which panel in Figure 11.17 best exemplifies the phytochrome-dependent signal transduction pathway during de-etiolation? Explain.*

and cGMP must be produced for a complete de-etiolation response. The injection of cGMP into *aurea* tomato leaf cells, for example, induces only a partial de-etiolation response.

Response

Ultimately, second messengers regulate one or more cellular activities. In most cases, these responses involve the increased activity of particular enzymes. There are two main mechanisms by which a signalling pathway can enhance an enzymatic step in a biochemical pathway: transcriptional regulation and post-translational modification. Transcriptional regulation increases or decreases the synthesis of mRNA encoding a specific enzyme. Post-translational modification activates preexisting enzymes.

Post-translational Modification of Preexisting Proteins

In most signal transduction pathways, preexisting proteins are modified by the phosphorylation of specific amino acids, which alters the protein's hydrophobicity and activity. Many second messengers, including cGMP and Ca^{2+}, activate protein kinases directly. Often, one protein kinase will phosphorylate another protein kinase, which then phosphorylates another, and so on (see Figure 11.10). Such kinase cascades may link initial stimuli to responses at the level of gene expression, usually via the phosphorylation of transcription factors. As we'll see soon, many signal transduction pathways ultimately regulate the synthesis of new proteins by turning specific genes on or off.

Signal transduction pathways must also have a means for turning off when the initial signal is no longer present, such as when a sprouting potato is put back into the cupboard. Protein phosphatases, which are enzymes that dephosphorylate specific proteins, are important in these "switch-off" processes. At any particular moment, a cell's functioning depends on the balance of activity of many types of protein kinases and protein phosphatases.

Transcriptional Regulation

As discussed in Concept 18.2, the proteins we call *specific transcription factors* bind to specific regions of DNA and control the transcription of specific genes (see Figure 18.10). In the case of phytochrome-induced de-etiolation, several such transcription factors are activated by phosphorylation in response to the appropriate light conditions. The activation of some of these transcription factors depends on their phosphorylation by protein kinases activated by cGMP or Ca^{2+}.

The mechanism by which a signal promotes developmental changes may depend on transcription factors that are activators (which *increase* transcription of specific genes) or repressors (which *decrease* transcription) or both. For example, some *Arabidopsis* mutants, except for their pale colour, have a light-grown appearance when grown in the dark; they have expanded leaves and short, sturdy stems but are not green because the final step in chlorophyll production requires light directly. These mutants have defects in a repressor that normally inhibits the expression of other genes that are activated by light. When the repressor is eliminated by mutation, the pathway that is normally blocked proceeds. Thus, these mutants appear to have been grown in the light, except for their pale colour.

De-etiolation ("Greening") Proteins

What types of proteins are either activated by phosphorylation or newly transcribed during the de-etiolation process? Many are enzymes that function in photosynthesis directly; others are enzymes involved in supplying the chemical precursors necessary for chlorophyll production; still others affect the levels of plant hormones that regulate growth. For example, the levels of auxin and brassinosteroids, hormones that enhance stem elongation, decrease following the activation of phytochrome. That decrease explains the slowing of stem elongation that accompanies de-etiolation.

We have examined the signal transduction involved in the de-etiolation response of a potato plant in some detail to get a sense of the complexity of biochemical changes that underlie this one process. Every plant hormone and environmental stimulus will trigger one or more signal transduction pathways of comparable complexity. As in the studies on the *aurea* mutant tomato, the isolation of mutants (a genetic approach) and techniques of molecular biology are helping researchers identify these various pathways.

CONCEPT CHECK 39.1

1. What are the physical differences between dark- and light-grown plants? Explain how etiolation helps a seedling compete successfully.
2. Cycloheximide is a drug that inhibits protein synthesis. Predict what effect cycloheximide would have on de-etiolation.
3. **WHAT IF?** The sexual dysfunction drug Viagra inhibits an enzyme that breaks down cyclic GMP. If tomato leaf cells have a similar enzyme, would applying Viagra to these cells cause a normal de-etiolation of *aurea* mutant tomato leaves?

For suggested answers, see Appendix A.

CONCEPT 39.2

Plants use chemicals to communicate

As we have seen, plants sense their local environmental conditions using signal transduction pathways no less complicated than those used by animals. The activation of signal transduction pathways often changes the activity of enzymes and therefore the production of chemicals.

Plants use chemicals to communicate with the living world, as in the example of those released by the sunflowers in Figure 39.1. Plants also use chemicals for communicating between different parts of the plant, thereby optimising the response of the whole organism. Chemical communication within plants is an exploding field of research. The discovery of new mobile information molecules, such as hundreds of small RNAs and dozens of small peptides, will keep researchers busy for decades. It has also been discovered that plants have an attribute that animals do not: The cytoplasmic bridges between plant cells (plasmodesmata), unlike the analogous structures in animals (gap junctions), can dilate enough that macromolecules such as proteins can move from cell to cell. This recent research builds on a long history of classic experiments that provided the first clues that mobile signalling molecules called hormones are internal regulators of plant growth.

General Characteristics of Plant Hormones

A **hormone**, in the original meaning of the term, is a signalling molecule that is produced in low concentrations by one part of an organism's body and transported to other parts, where it binds to a specific receptor and triggers responses in target cells and tissues. In animals, hormones are usually transported through the circulatory system, a criterion often included in definitions of the term. Many modern plant biologists, however, argue that the hormone concept, which originated from studies of animals, is too limiting to describe plant physiological processes. For example, plants don't have circulating blood to transport hormone-like signalling molecules. Moreover, some signalling molecules that are considered plant hormones act only locally. Finally, there are some signalling molecules in plants, such as glucose, that typically occur in plants at concentrations thousands of times greater than a

Table 39.1 Overview of Plant Hormones

Hormone	Where Produced or Found in Plant	Major Functions
Auxin (IAA)	Shoot apical meristems and young leaves are the primary sites of auxin synthesis. Root apical meristems also produce auxin, although the root depends on the shoot for much of its auxin. Developing seeds and fruits contain high levels of auxin, but it is unclear whether it is newly synthesised or transported from maternal tissues.	Stimulates stem elongation (low concentration only); promotes the formation of lateral and adventitious roots; regulates development of fruit; enhances apical dominance; functions in phototropism and gravitropism; promotes vascular differentiation; retards leaf abscission
Cytokinins	These are synthesised primarily in roots and transported to other organs, although there are many minor sites of production as well.	Regulate cell division in shoots and roots; modify apical dominance and promote lateral bud growth; promote movement of nutrients into sink tissues; stimulate seed germination; delay leaf senescence
Gibberellins (GA)	Meristems of apical buds and roots, young leaves, and developing seeds are the primary sites of production.	Stimulate stem elongation, pollen development, pollen tube growth, fruit growth, and seed development and germination; regulate sex determination and the transition from juvenile to adult phases
Abscisic acid (ABA)	Almost all plant cells have the ability to synthesise abscisic acid, and its presence has been detected in every major organ and living tissue; it may be transported in the phloem or xylem.	Inhibits growth; promotes stomatal closure during drought stress; promotes seed dormancy and inhibits early germination; promotes leaf senescence; promotes desiccation tolerance
Ethylene	This gaseous hormone can be produced by most parts of the plant. It is produced in high concentrations during senescence, leaf abscission, and the ripening of some types of fruits. Synthesis is also stimulated by wounding and stress.	Promotes ripening of many types of fruit, leaf abscission, and the triple response in seedlings (inhibition of stem elongation, promotion of lateral expansion, and horizontal growth); enhances the rate of senescence; promotes root and root hair formation; promotes flowering in the pineapple family
Brassinosteroids	These compounds are present in all plant tissues, although different intermediates predominate in different organs. Internally produced brassinosteroids act near the site of synthesis.	Promote cell expansion and cell division in shoots; promote root growth at low concentrations; inhibit root growth at high concentrations; promote xylem differentiation and inhibit phloem differentiation; promote seed germination and pollen tube elongation
Jasmonates	These are a small group of related molecules derived from the fatty acid linolenic acid. They are produced in several parts of the plant and travel in the phloem to other parts of the plant.	Regulate a wide variety of functions, including fruit ripening, floral development, pollen production, tendril coiling, root growth, seed germination, and nectar secretion; also produced in response to herbivory and pathogen invasion
Strigolactones	These carotenoid-derived hormones and extracellular signals are produced in roots in response to low phosphate conditions or high auxin flow from the shoot.	Promote seed germination, control of apical dominance, and the attraction of mycorrhizal fungi to the root

typical hormone. Nevertheless, they activate signal transduction pathways that greatly alter the functioning of plants in a manner similar to a hormone. Thus, many plant biologists prefer the broader term *plant growth regulator* to describe organic compounds, either natural or synthetic, that modify or control one or more specific physiological processes within a plant. The terms *plant hormone* and *plant growth regulator* are used about equally, but for historical continuity this text will use the term *plant hormone* and adhere to the criterion that plant hormones are active at very low concentrations.

Although plant hormones are produced in very low concentrations, a tiny amount of hormone can have a profound effect on plant growth and development. Virtually every aspect of plant growth and development is under hormonal control to some degree. Each hormone has multiple effects, depending on its site of action, its concentration, and the developmental stage of the plant. Conversely, multiple hormones can influence a single process. Plant hormone responses commonly depend on both the amounts of the hormones involved and their relative concentrations. It is often the interactions between different hormones, rather than hormones acting in isolation, that control growth and development. These interactions will become apparent in the following survey of hormone function.

A Survey of Plant Hormones

Table 39.1 previews the major types and actions of plant hormones, including auxin, cytokinins, gibberellins, abscisic acid, ethylene, brassinosteroids, jasmonates, and strigolactones.

Auxin

The idea that chemical messengers exist in plants emerged from a series of classic experiments on how stems respond to light. As you know, the shoot of a houseplant on a windowsill grows towards light. Any growth response that results in plant organs curving towards or away from stimuli is called a **tropism** (from the Greek *tropos*, turn). The growth of a shoot towards light or away from it is called **phototropism**; the former is positive phototropism, and the latter is negative phototropism.

In natural ecosystems, where plants may be crowded, phototropism directs shoot growth towards the sunlight that powers photosynthesis. This response results from a differential growth of cells on opposite sides of the shoot; the cells on the darker side elongate faster than the cells on the brighter side.

Charles Darwin and his son Francis conducted some of the earliest experiments on phototropism in the late 1800s **(Figure 39.5)**. They observed that a grass seedling ensheathed in its coleoptile (see Figure 38.9b) could bend towards light only if the tip of the coleoptile was present. If the tip was removed, the coleoptile did not curve. The seedling also failed to grow towards light if the tip was covered with an opaque cap, but neither a transparent cap over the tip nor an opaque shield placed below the coleoptile tip prevented

▼ Figure 39.5 Inquiry

What part of a grass coleoptile senses light, and how is the signal transmitted?

Experiment In 1880, Charles and Francis Darwin removed and covered parts of grass coleoptiles to determine what part senses light. In 1913, Peter Boysen-Jensen separated coleoptiles with different materials to determine how the signal for phototropism is transmitted.

Results

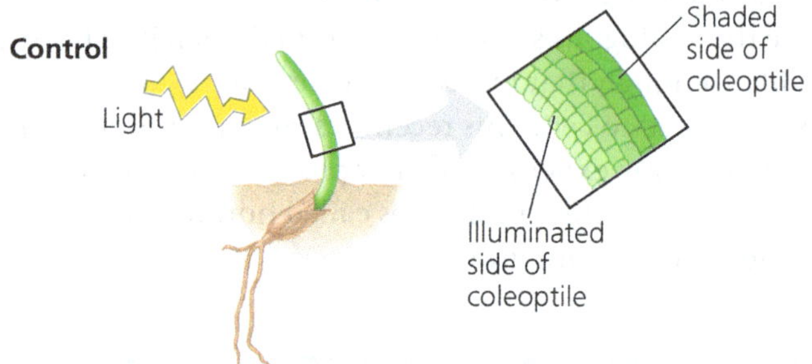

Darwin and Darwin: Phototropism occurs only when the tip is illuminated.

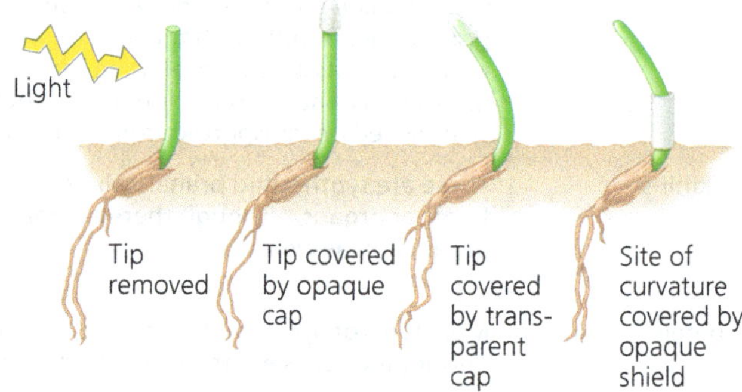

Boysen-Jensen: Phototropism occurs when the tip is separated by a permeable barrier but not an impermeable barrier.

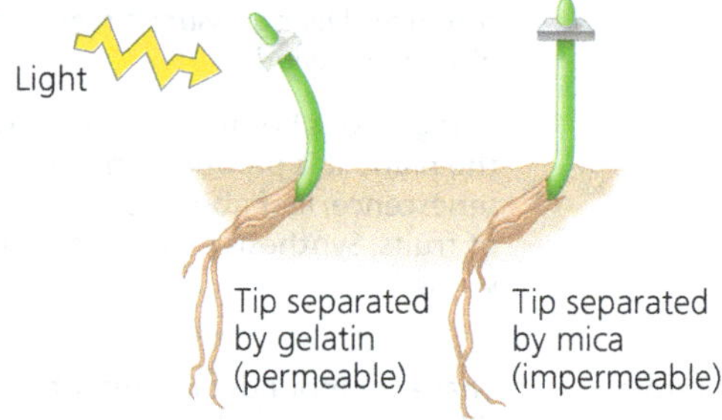

Data from C. R. Darwin, *The Power of Movement in Plants*, John Murray (1880). P. Boysen-Jensen, Concerning the performance of phototropic stimuli on the Avenacoleoptile, *Berichte der Deutschen Botanischen Gesellschaft* (*Reports of the German Botanical Society*) 31:559–566 (1913).

Conclusion The Darwins' experiment suggested that only the tip of the coleoptile senses light. The phototropic bending, however, occurred at a distance from the site of light perception (the tip). Boysen-Jensen's results suggested that the signal for the bending is a light-activated mobile chemical.

WHAT IF? *How could you experimentally determine which colours of light cause the most phototropic bending?*

the phototropic response. It was the tip of the coleoptile, the Darwins concluded, that was responsible for sensing light. However, they noted that the differential growth response that led to curvature of the coleoptile occurred some distance below the tip. The Darwins postulated that some signal was transmitted downwards from the tip to the elongating region of the coleoptile. A few decades later, the Danish scientist Peter Boysen-Jensen demonstrated that the signal was a mobile chemical substance. He separated the tip from the remainder of the coleoptile by a cube of gelatin, which prevented cellular contact but allowed chemicals to pass through. These seedlings responded normally, bending towards light. However, if the tip was experimentally separated from the lower coleoptile by an impermeable barrier, such as the mineral mica, no phototropic response occurred.

Subsequent research showed that a chemical was released from coleoptile tips and could be collected by means of diffusion into agar blocks. Little cubes of agar containing this chemical could induce "phototropic-like" curvatures even in complete darkness if the agar cubes were placed off-centre atop the cut surface of decapitated coleoptiles. Coleoptiles curve towards light because of a higher concentration of this growth-promoting chemical on the darker side of the coleoptile. Since this chemical stimulated growth as it passed down the coleoptile, it was dubbed "auxin" (from the Greek *auxein*, to increase). Auxin was later purified, and its chemical structure determined to be indoleacetic acid (IAA). The term **auxin** is used for any chemical, synthetic or not, that promotes coleoptile elongation. The major natural auxin in plants is IAA, which has many additional effects. Unless noted otherwise, the terms *auxin* and *IAA* will be used interchangeably.

Auxin is produced predominantly in shoot tips and is transported from cell to cell down the stem at a rate of about 1 cm/hr. It moves only from tip to base, not in the reverse direction. This unidirectional transport of auxin is called *polar transport*. Polar transport is unrelated to gravity; experiments have shown that auxin travels upwards when a stem or coleoptile segment is placed upside down. Rather, the polarity of auxin movement is attributable to the polar distribution of auxin transport protein in the cells. Concentrated at the basal end of a cell, the auxin transporters move the hormone out of the cell. The auxin can then enter the apical end of the neighbouring cell **(Figure 39.6)**. Auxin has a variety of effects, including stimulating cell elongation and regulating plant architecture.

The Role of Auxin in Cell Elongation One of auxin's chief functions is to stimulate elongation of cells within young developing shoots. As auxin from the shoot tip (see Figure 35.16) moves down to the region where cells are elongating, the hormone stimulates cell growth by binding to a receptor in the nucleus. Auxin stimulates growth only over a certain concentration range, from about 10^{-8} to 10^{-4} M. At higher concentrations, auxin may inhibit cell elongation by inducing production of ethylene, a hormone that generally hinders growth. We will return to this hormonal interaction when we look at ethylene.

According to a model called the *acid growth hypothesis*, proton pumps play a major role in the growth response of cells to auxin. In a shoot's region of elongation, auxin stimulates the plasma membrane's proton (H^+) pumps. This pumping of H^+ increases the voltage across the membrane (membrane potential) and lowers the pH in the cell wall within minutes. Acidification of the wall activates proteins called **expansins** that break the cross-links (hydrogen bonds) between cellulose microfibrils and other cell wall constituents, loosening the wall's fabric **(Figure 39.7)**. Increasing the membrane potential enhances ion uptake into the cell, which causes osmotic uptake of water and increased turgor. Increased turgor and increased cell wall plasticity enable the cell to elongate.

▼ Figure 39.6 Inquiry

What causes polar movement of auxin from shoot tip to base?

Experiment To investigate how auxin is transported unidirectionally, Leo Gälweiler and colleagues designed an experiment to identify the location of the auxin transport protein. They used a greenish yellow fluorescent molecule to label antibodies that bind to the auxin transport protein. Then they applied the antibodies to longitudinally sectioned *Arabidopsis* stems.

Results The light micrograph on the left shows that auxin transport proteins are not found in all stem tissues, but only in the xylem parenchyma. In the light micrograph on the right, a higher magnification reveals that these proteins are primarily localised at the basal ends of the cells.

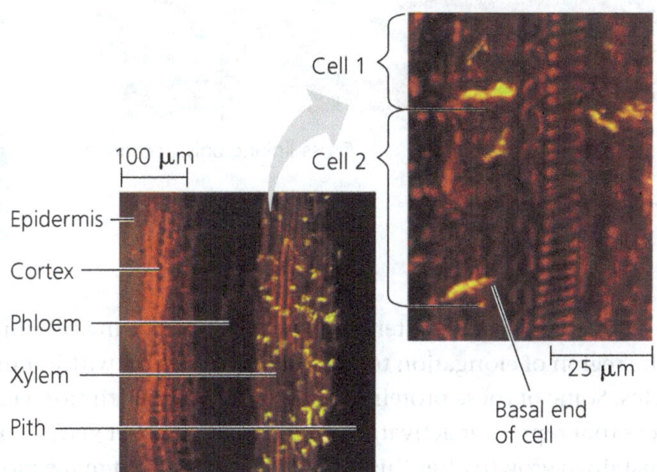

Data from L. Gälweiler et al., Regulation of polar auxin transport by AtPIN1 in *Arabidopsis* vascular tissue, *Science* 282:2226–2230 (1998).

Conclusion The results support the hypothesis that concentration of the auxin transport protein at the basal ends of cells mediates the polar transport of auxin.

WHAT IF? *If auxin transport proteins were equally distributed at both ends of the cells, would polar auxin transport still be possible? Explain.*

▼ **Figure 39.7 Cell elongation in response to auxin: the acid growth hypothesis.** The cell expands in a direction mainly perpendicular to the main orientation of the microfibrils in the cell wall (see Figure 35.28).

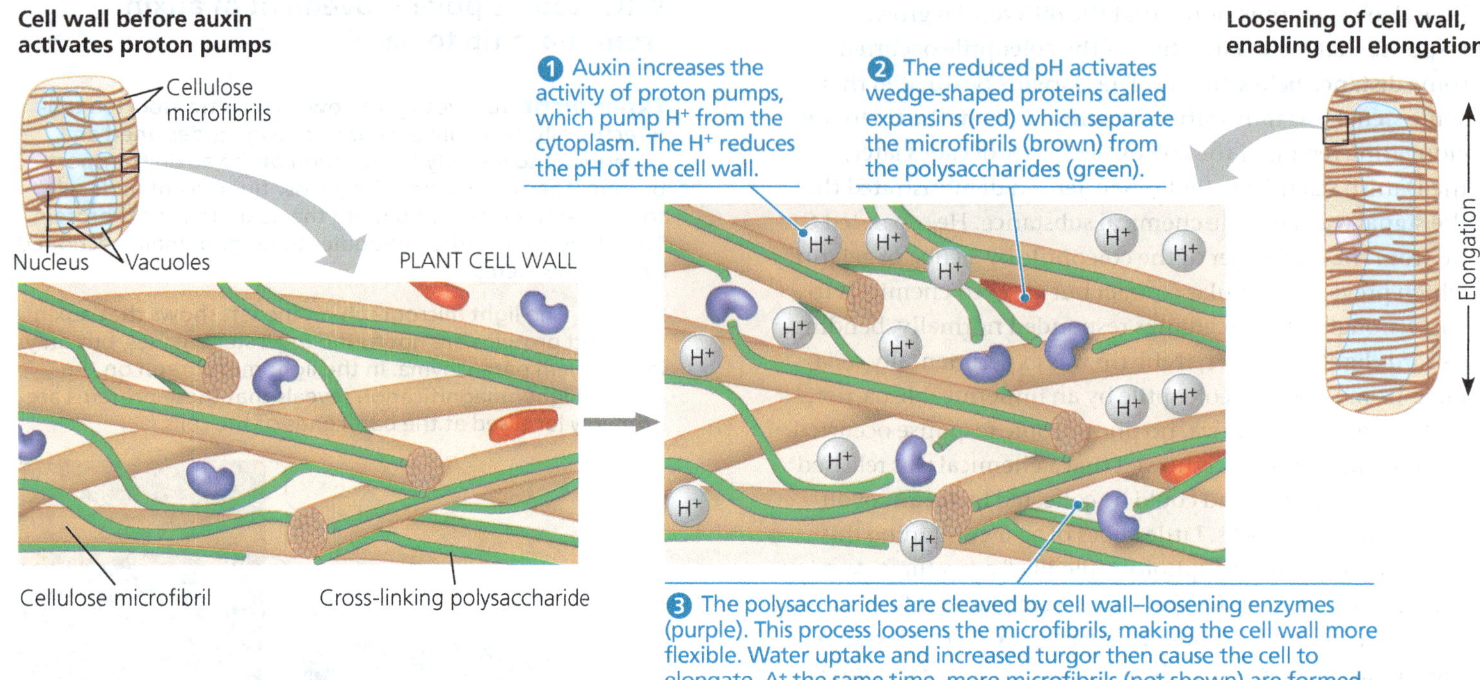

Auxin also rapidly alters gene expression, causing cells in the region of elongation to produce new proteins within minutes. Some of these proteins are short-lived transcription factors that repress or activate the expression of other genes. For sustained growth after this initial spurt, cells must make more cytoplasm and wall material. In addition, auxin stimulates this sustained growth response.

Auxin's Role in Plant Development The polar transport of auxin plays a major role in controlling the spatial organisation, or *pattern formation*, of a plant. The auxin synthesised in the shoot tips of a branch carries information about the growth potential of that branch. If a branch finds itself in an environment conducive for growth, it produces more auxin, and the plant diverts more resources to that branch. A reduced flow of auxin from a branch indicates that the branch is not being sufficiently productive: New branches are needed elsewhere. Thus, lateral buds below the branch are released from dormancy and begin to grow.

The transport of auxin also plays a key role in establishing the pattern of leaf emergence from the shoot apical meristem (see Figure 36.3). A leading model proposes that polar auxin transport in the shoot tip generates local peaks in auxin concentration that determine the site of leaf primordium formation and thereby the different leaf arrangements found in nature.

The polar transport of auxin from the leaf margin also directs the patterns of leaf veins. Inhibitors of polar auxin transport result in leaves that lack vascular continuity through the petiole and have broad, loosely organised main veins, an increased number of secondary veins, and a dense band of irregularly shaped vascular cells adjacent to the leaf margin.

The activity of the vascular cambium, the meristem that produces woody tissues, is also under the control of auxin transport. When a plant becomes dormant at the end of a growing season, there is a reduction in auxin transport capacity and the expression of genes encoding auxin transporters.

Auxin's effects on plant development are not limited to the familiar sporophyte plant that we see. Recent evidence suggests that the organisation of the microscopic angiosperm female gametophytes is regulated by an auxin gradient.

Practical Uses for Auxins Auxins, both natural and synthetic, have many commercial applications. For example, the natural auxin indolebutyric acid (IBA) is used in the vegetative propagation of plants by cuttings. Treating a detached leaf or stem with powder containing IBA often causes adventitious roots to form near the cut surface.

Certain synthetic auxins are widely used as herbicides, including 2,4-dichlorophenoxyacetic acid (2,4-D). Monocots, such as corn and turfgrass, can rapidly inactivate such synthetic auxins. However, eudicots cannot and therefore die from hormonal overdose. Spraying cereal fields or turf with 2,4-D eliminates eudicot (broadleaf) weeds.

Developing seeds produce auxin, which promotes fruit growth. In tomato plants grown in greenhouses, often fewer seeds are produced, resulting in poorly developed tomato fruits. However, spraying synthetic auxins on greenhouse-grown tomato vines induces normal fruit development, making the greenhouse-cultivated tomatoes commercially viable.

Cytokinins

Trial-and-error attempts to find chemical additives that would enhance the growth and development of plant cells in tissue culture led to the discovery of **cytokinins**. In the 1940s, researchers stimulated the growth of plant embryos in culture by adding coconut milk, the liquid endosperm of a coconut's giant seed. Subsequent researchers found that they could induce cultured tobacco cells to divide by adding degraded DNA samples. The active ingredients of both experimental additives were modified forms of adenine, a component of nucleic acids. These growth regulators were named cytokinins because they stimulate cytokinesis, or cell division. The most common natural cytokinin is zeatin, so named because it was discovered first in corn (*Zea mays*). Cytokinins influence cell division, cell differentiation, and apical dominance.

Control of Cell Division and Differentiation Cytokinins are produced in actively growing tissues, particularly in roots, embryos, and fruits. Cytokinins produced in roots reach their target tissues by moving up the plant in the xylem sap. Acting in concert with auxin, cytokinins stimulate cell division and influence the pathway of differentiation. The effects of cytokinins on cells growing in tissue culture provide clues about how this class of hormones may function in an intact plant. When a piece of parenchyma tissue from a stem is cultured in the absence of cytokinins, the cells grow very large but do not undergo mitosis. But if cytokinins are added along with auxin, the cells divide. Cytokinins alone have no effect. The ratio of cytokinins to auxin controls cell differentiation. When the concentrations of these two hormones are at certain levels, the mass of cells continues to grow, but it remains a cluster of undifferentiated cells called a callus (see Figure 38.15). If cytokinin levels increase, shoot buds develop from the callus. If auxin levels increase, roots form.

Control of Apical Dominance Apical dominance, the ability of the apical bud to suppress the development of axillary buds, is under the control of sugar and various plant hormones, including auxin, cytokinins, and strigolactones. The sugar demand of the shoot tip is critical for maintaining apical dominance. Cutting off the apical bud removes apical sugar demand and rapidly increases sugar (sucrose) availability to axillary buds. This increase of sugar is sufficient to initiate bud release. However, not all of the buds grow equally: Usually only one of the axillary buds closest to the cut surface will take over as the new apical bud.

Three plant hormones—auxin, cytokinins, and strigolactones—play a role in determining the extent to which specific axillary buds elongate **(Figure 39.8)**. In an intact plant, auxin transported down the shoot from the apical bud *indirectly* inhibits axillary buds from growing, causing a shoot to lengthen at the expense of lateral branching. The polar flow of auxin down the shoot triggers the synthesis of strigolactones, which *directly* repress bud growth. Meanwhile, cytokinins entering the shoot system from roots counter the action of auxin and strigolactones by signalling axillary buds to begin growing. Thus, in an intact plant, the cytokinin-rich axillary buds closer to the base of the plant tend to be longer than the auxin-rich axillary buds closer to the apical bud. Mutants that overproduce cytokinins or plants treated with cytokinins tend to be bushier than normal.

Removing the apical bud, a major site of auxin biosynthesis, causes the auxin and strigolactone levels in the stem to wane, particularly in those regions close to the cut surface (see Figure 39.8). This causes the axillary buds closest to the cut surface to grow most vigorously, and one of these axillary buds will eventually take over as the new apical bud. Applying auxin to the cut surface of the shoot tip resuppresses the growth of the lateral buds.

Anti-aging Effects Cytokinins slow the aging of certain plant organs by inhibiting protein breakdown, stimulating RNA and protein synthesis, and mobilising nutrients from surrounding tissues. If leaves removed from a plant are dipped in a cytokinin solution, they stay green much longer than otherwise.

Gibberellins

In the early 1900s, farmers in Asia noticed that some rice seedlings in their paddies grew so tall and spindly that they

▼ **Figure 39.8 Effects on apical dominance of removing the apical bud.** Apical dominance refers to the inhibition of the growth of axillary buds by the apical bud of a plant shoot. Removal of the apical bud enables lateral branches to grow. Multiple hormones play a role in this process, including auxin, cytokinin, and strigolactones.

toppled over before they could mature. In 1926, it was discovered that a fungus of the genus *Gibberella* causes this "foolish seedling disease." By the 1930s, it was determined that the fungus causes hyperelongation of rice stems by secreting a chemical, which was given the name **gibberellin**, or gibberellic acid. In the 1950s, researchers discovered that plants also produce gibberellins. Since that time, scientists have identified more than 100 different gibberellins that occur naturally in plants, although a much smaller number occur in each plant species. "Foolish rice" seedlings, it seems, suffer from too much gibberellin. Gibberellins have a variety of effects, such as stem elongation, fruit growth, and seed germination.

Stem Elongation The major sites of gibberellin production are young roots and leaves. Gibberellins are best known for stimulating stem and leaf growth by enhancing cell elongation *and* cell division. One hypothesis proposes that they activate enzymes that loosen cell walls, facilitating entry of expansin proteins. Thus, gibberellins act in concert with auxin to promote stem elongation.

The effects of gibberellins in enhancing stem elongation are evident when certain dwarf (mutant) varieties of plants are treated with gibberellins. For instance, some dwarf pea plants (including the variety Mendel studied; see Concept 14.1) grow tall if treated with gibberellins. But there is often no response if the gibberellins are applied to wild-type plants. Apparently, these plants already produce an optimal dose of the hormone. The most dramatic example of gibberellin-induced stem elongation is *bolting*, rapid growth of the floral stalk **(Figure 39.9a)**.

▼ Figure 39.9 Effects of gibberellins on stem elongation and fruit growth.

(a) Some plants develop in a rosette form, low to the ground with very short internodes, as in the *Arabidopsis* plant shown at the left. As the plant switches to reproductive growth, a surge of gibberellins induces bolting: Internodes elongate rapidly, elevating floral buds that develop at stem tips (right).

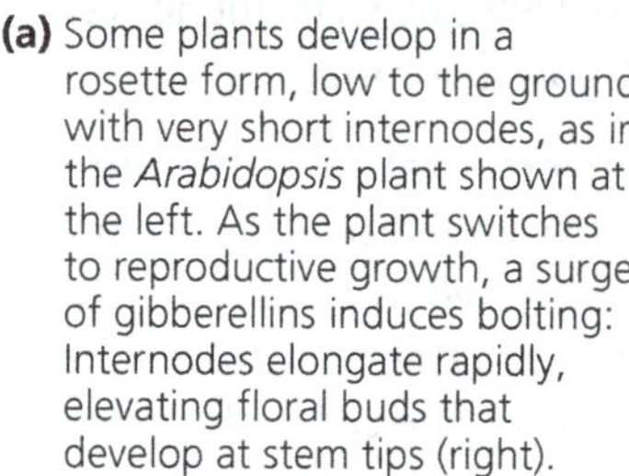

(b) The Thompson seedless grape bunch on the left is from an untreated control vine. The bunch on the right is growing from a vine that was sprayed with gibberellin during fruit development.

Fruit Growth In many plants, both auxin and gibberellins must be present for fruit to develop. The most important commercial application of gibberellins is in the spraying of Thompson seedless grapes **(Figure 39.9b)**. The hormone makes the individual grapes grow larger, a trait valued by the consumer. The gibberellin sprays also make the internodes of the grape bunch elongate, allowing more space for the individual grapes. By enhancing air circulation between the grapes, this increase in space also makes it harder for yeasts and other microorganisms to infect the fruit.

Germination The embryo of a seed is a rich source of gibberellins. After water is imbibed, the release of gibberellins from the embryo signals the seed to break dormancy and germinate. Some seeds that normally require particular environmental conditions to germinate, such as exposure to light or low temperatures, break dormancy if they are treated with gibberellins. Gibberellins support the growth of cereal seedlings by stimulating the synthesis of digestive enzymes such as α-amylase that mobilise stored nutrients **(Figure 39.10)**.

▼ Figure 39.10 Mobilisation of nutrients by gibberellins during the germination of grain seeds such as barley.

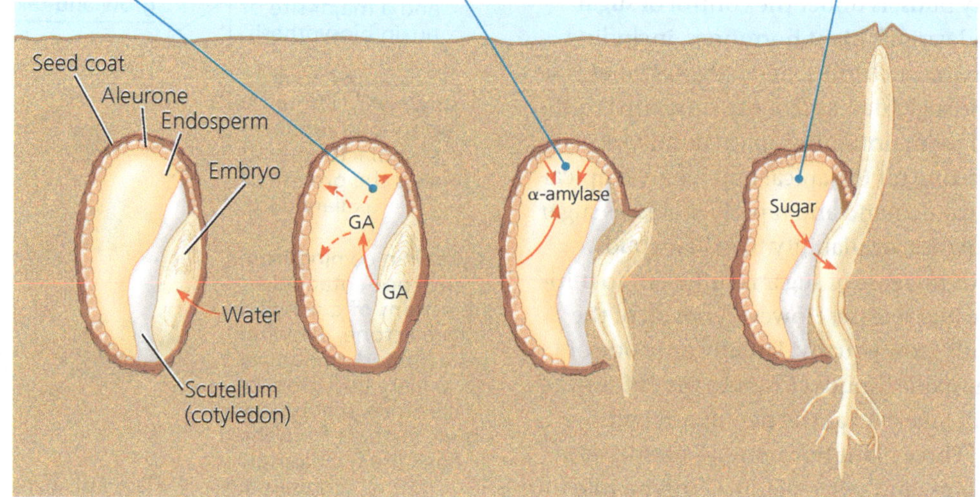

Abscisic Acid

In the 1960s, one research group studying the chemical changes that precede bud dormancy and leaf abscission in deciduous trees and another team investigating chemical changes preceding abscission of cotton fruits isolated the same compound, **abscisic acid (ABA)**. Ironically, ABA is no longer thought to play a primary role in leaf abscission, but it is very important in other functions. Unlike the growth-stimulating hormones discussed so far—auxin, cytokinins, gibberellins, and brassinosteroids—ABA *slows* growth. ABA often antagonises the actions of growth hormones, and the ratio of ABA to one or more growth hormones determines the final physiological outcome. We will consider here two of ABA's many effects: seed dormancy and drought tolerance.

Seed Dormancy Seed dormancy increases the likelihood that seeds will germinate only when there are sufficient amounts of light, temperature, and moisture for the seedlings to survive (see Concept 38.1). What prevents seeds dispersed in autumn from germinating immediately, only to die in the winter? What mechanisms ensure that such seeds do not germinate until spring? For that matter, what prevents seeds from germinating in the dark, moist interior of the fruit? The answer to these questions is ABA. The levels of ABA may increase 100-fold during seed maturation. The high levels of ABA in maturing seeds inhibit germination and induce the production of proteins that help the seeds withstand the extreme dehydration that accompanies maturation.

Many types of dormant seeds germinate when ABA is removed or inactivated. The seeds of some desert plants break dormancy only when heavy rains wash ABA out of them. Other seeds require light or prolonged exposure to cold to inactivate ABA. Often, the ratio of ABA to gibberellins determines whether seeds remain dormant or germinate, and adding ABA to seeds that are primed to germinate makes them dormant again. Inactivated ABA or low levels of ABA can lead to precocious (early) germination **(Figure 39.11)**. For example, a corn mutant with grains that germinate while still on the cob lacks a functional transcription factor required for ABA to induce expression of certain genes. Precocious germination of red mangrove seeds, due to low ABA levels, is actually an adaptation that helps the young seedlings to plant themselves like darts in the soft mud below the parent tree.

Drought Tolerance ABA plays a major role in drought signalling. When a plant begins to wilt, ABA accumulates in the leaves and causes stomata to close rapidly, reducing transpiration and preventing further water loss. By affecting second messengers such as calcium, ABA causes potassium channels in the plasma membrane of guard cells to open, leading to a massive loss of potassium ions from the cells. The accompanying osmotic loss of water reduces guard cell turgor and leads to closing of the stomatal pores (see Figure 36.14).

▼ Figure 39.11 Precocious germination of wild-type mangrove and mutant corn seeds.

◀ Red mangrove (*Rhizophora mangle*) seeds produce only low levels of ABA, and the seeds germinate while still on the tree. In this case, early germination is a useful adaptation. When released, the radicle of the dart-like seedling deeply penetrates the soft mudflats in which the mangroves grow.

▲ Precocious germination in this corn mutant is caused by lack of a functional transcription factor required for ABA action.

In some cases, water shortage stresses the root system before the shoot system, and ABA transported from roots to leaves may function as an "early warning system." Many mutants that are especially prone to wilting are deficient in ABA production.

Ethylene

During the 1800s, when coal gas was used as fuel for streetlights, leakage from gas pipes caused nearby trees to drop leaves prematurely. In 1901, the gas **ethylene** was demonstrated to be the active factor in coal gas. But the idea that it is a plant hormone was not widely accepted until the advent of a technique called gas chromatography simplified its identification.

Plants produce ethylene in response to stresses such as drought, flooding, mechanical pressure, injury, and infection. Ethylene is also produced during fruit ripening and programmed cell death and in response to high concentrations of externally applied auxin. Indeed, many effects previously ascribed to auxin, such as inhibition of root elongation, may be due to auxin-induced ethylene production. We focus here on four of ethylene's many effects: response to mechanical stress, senescence, leaf abscission, and fruit ripening.

The Triple Response to Mechanical Stress Imagine a pea seedling pushing upwards through the soil, only to come up against a stone. As it pushes against the obstacle, the stress in its delicate tip induces the seedling to produce ethylene. The hormone then instigates a growth manoeuvre known as the **triple response** that enables the shoot to avoid the obstacle. The three parts of this response are a slowing of stem elongation, a thickening of the stem (which makes it stronger), and a curvature that causes the stem to start growing horizontally. As the effects of the initial ethylene pulse lessen, the stem resumes vertical growth. If it again contacts a barrier, another burst of ethylene is released, and horizontal growth resumes. However, if the upward touch detects no solid object, then ethylene production decreases, and the stem, now clear of the obstacle, resumes its normal upward growth. It is ethylene that induces the stem to grow horizontally rather than the physical obstruction itself; when ethylene is applied to normal seedlings growing free of physical impediments, they still undergo the triple response **(Figure 39.12)**.

Studies of *Arabidopsis* mutants with abnormal triple responses are an example of how biologists identify a signal transduction pathway. Scientists isolated ethylene-insensitive (*ein*) mutants, which fail to undergo the triple response after exposure to ethylene **(Figure 39.13a)**. Some types of *ein* mutants are insensitive to ethylene because they lack a functional ethylene receptor. Mutants of a different sort undergo the triple response even out of soil, in the air, where there are no physical obstacles. Some of these mutants have a regulatory defect that causes them to produce ethylene

▼ Figure 39.12 The ethylene-induced triple response. In response to ethylene, a gaseous plant hormone, germinating pea seedlings grown in the dark undergo the triple response—slowing of stem elongation, stem thickening, and horizontal stem growth. The response is greater with increased ethylene concentration.

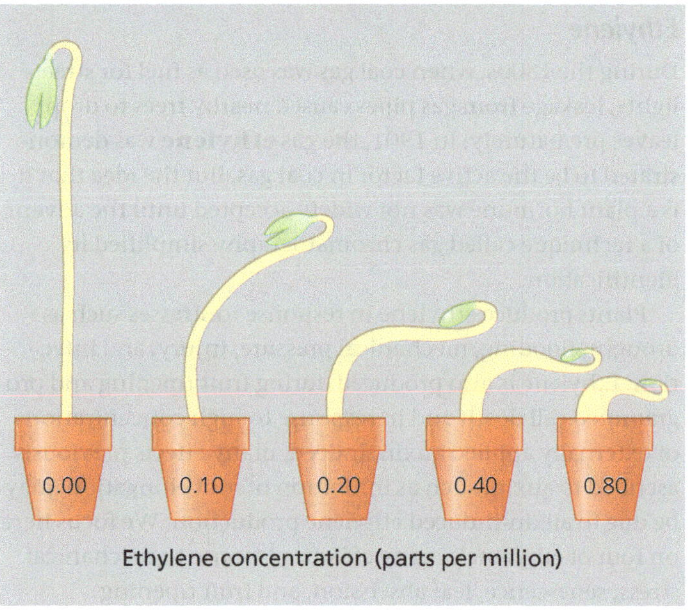

▼ Figure 39.13 Ethylene triple-response *Arabidopsis* mutants.

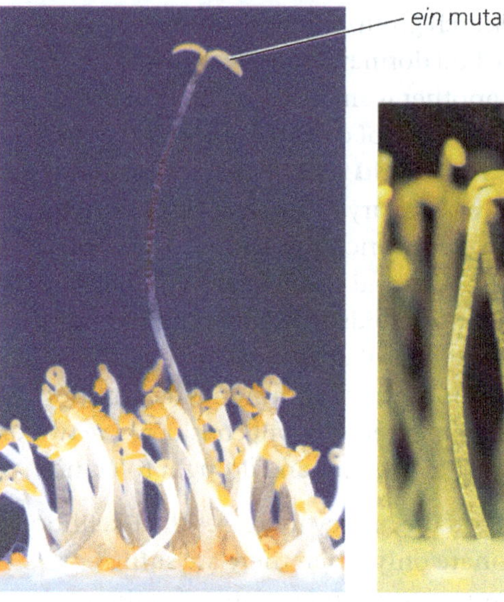

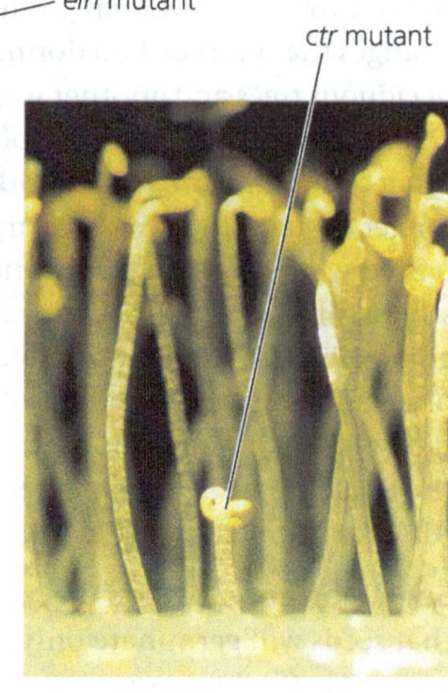

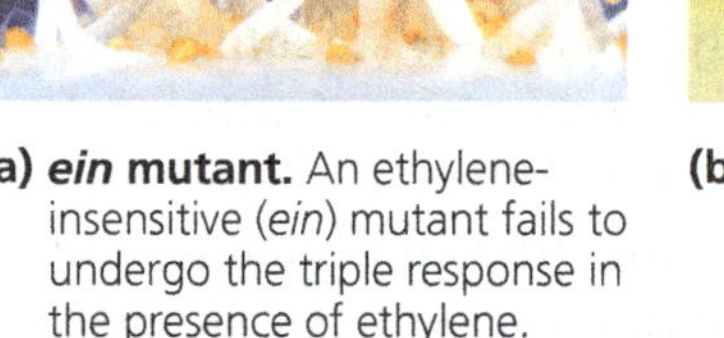

(a) *ein* mutant. An ethylene-insensitive (*ein*) mutant fails to undergo the triple response in the presence of ethylene.

(b) *ctr* mutant. A constitutive triple-response (*ctr*) mutant undergoes the triple response even in the absence of ethylene.

VISUAL SKILLS *If the* ein *single mutation is combined with an ethylene-overproducing* (eto) *mutation, would the phenotype of the double mutant be different from that of the single mutant? Explain.*

at rates 20 times normal. The phenotype of such ethylene-overproducing (*eto*) mutants can be restored to wild-type by treating the seedlings with inhibitors of ethylene synthesis. Other mutants, called constitutive triple-response (*ctr*) mutants, undergo the triple response in air but do not respond to inhibitors of ethylene synthesis **(Figure 39.13b)**. (Constitutive genes are genes that are continually expressed in all cells of an organism.) In *ctr* mutants, ethylene signal transduction is permanently turned on, even though ethylene is not present.

The affected gene in *ctr* mutants codes for a protein kinase. The fact that this mutation *activates* the ethylene response suggests that the normal kinase product of the wild-type allele is a *negative* regulator of ethylene signal transduction. Thus, binding of the hormone ethylene to the ethylene receptor normally leads to inactivation of the kinase, and the inactivation of this negative regulator allows synthesis of the proteins required for the triple response.

Senescence Consider the shedding of a leaf in autumn or the death of an annual after flowering. Or think about the final step in differentiation of a vessel element, when its living contents are destroyed, leaving a hollow tube behind. Such events involve **senescence**—the programmed death of certain cells or organs or the entire plant. Cells, organs, and plants genetically programmed to die on a schedule do not simply shut down cellular machinery and await death. Instead, at the molecular level, the onset of programmed cell death is a very busy time in a cell's life, requiring new gene expression. Newly formed enzymes break down many chemical components, including chlorophyll, DNA, RNA, proteins, and membrane lipids. The plant salvages many of the breakdown products. A burst of ethylene is almost always associated with the death of cells during senescence.

Leaf Abscission The loss of leaves from deciduous trees helps prevent desiccation during seasonal periods when the availability of water to the roots is severely limited. Before dying leaves abscise, many essential elements are salvaged from them and stored in stem parenchyma cells. These nutrients are recycled back to developing leaves during the following spring. The colours of autumn leaves are due to newly made red pigments as well as yellow and orange carotenoids (see Concept 10.3) that were already present in the leaves and are rendered visible by the breakdown of the dark green chlorophyll in autumn.

When an autumn leaf falls, it detaches from the stem at an abscission layer that develops near the base of the petiole **(Figure 39.14)**. The small parenchyma cells of this layer have very thin walls, and there are no fibre cells around the vascular tissue. The abscission layer is further weakened when enzymes hydrolyse polysaccharides in the cell walls. Finally, the weight of the leaf, with the help of the wind, causes a separation within the abscission layer. Even before the leaf falls, a layer of cork forms a protective scar on the twig side of the abscission layer, preventing pathogens from invading the plant.

▼ Figure 39.14 Abscission of a maple leaf. Abscission is controlled by a change in the ratio of ethylene to auxin. The abscission layer is seen in this longitudinal section as a vertical band at the base of the petiole. After the leaf falls, a protective layer of cork becomes the leaf scar that helps prevent pathogens from invading the plant (LM).

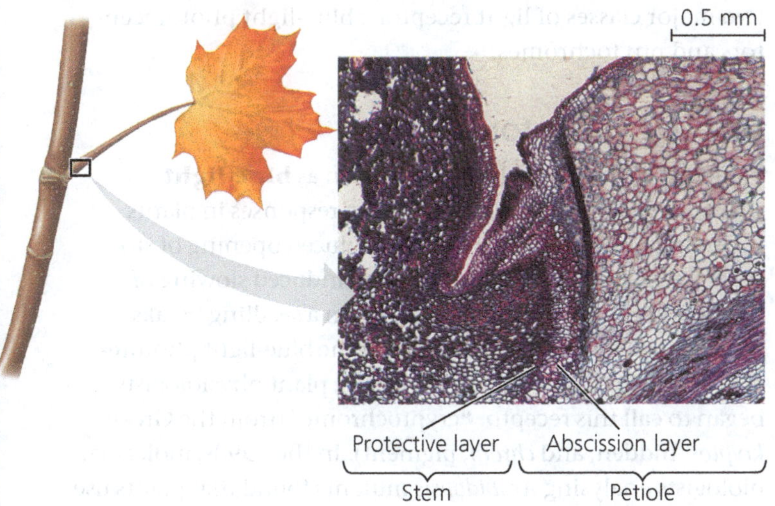

A change in the ratio of ethylene to auxin controls abscission. An aging leaf produces less and less auxin, rendering the cells of the abscission layer more sensitive to ethylene. As the influence of ethylene on the abscission layer prevails, the cells produce enzymes that digest the cellulose and other components of cell walls.

Fruit Ripening Immature fleshy fruits are generally tart, hard, and green—features that help protect the developing seeds from herbivores. After ripening, the mature fruits help *attract* animals that disperse the seeds (see Figures 30.11 and 30.12). In many cases, a burst of ethylene production in the fruit triggers the ripening process. The enzymatic breakdown of cell wall components softens the fruit, and the conversion of starches and acids to sugars makes the fruit sweet. The production of new scents and colours helps advertise ripeness to animals, which eat the fruits and disperse the seeds.

A chain reaction occurs during ripening: Ethylene triggers ripening, and ripening triggers more ethylene production. The result is a huge burst in ethylene production. Because ethylene is a gas, the signal to ripen spreads from fruit to fruit. If you pick or buy green fruit, you may be able to speed ripening by storing the fruit in a paper bag, allowing ethylene to accumulate. On a commercial scale, many kinds of fruits are ripened in huge storage containers in which ethylene levels are enhanced. In other cases, fruit producers take measures to slow ripening caused by natural ethylene. Apples, for instance, are stored in bins flushed with carbon dioxide. Circulating the air prevents ethylene from accumulating, and carbon dioxide inhibits synthesis of new ethylene. Stored in this way, apples picked in autumn can still be shipped to grocery stores the following summer.

Given the importance of ethylene in the postharvest physiology of fruits, the genetic engineering of ethylene signal transduction pathways has potential commercial applications. For example, by engineering a way to block the transcription of one of the genes required for ethylene synthesis, molecular biologists have created tomato fruits that ripen on demand. These fruits are picked while green and will not ripen unless ethylene gas is added. As such methods are refined, they will reduce spoilage of fruits and vegetables, a problem that ruins between one-third and half of the produce harvested in Australia and New Zealand.

More Recently Discovered Plant Hormones

Auxin, gibberellins, cytokinins, abscisic acid, and ethylene are often considered the five "classic" plant hormones. However, more recently discovered hormones have swelled the list of important plant growth regulators.

Brassinosteroids are steroids similar to cholesterol and the sex hormones of animals. They induce cell elongation and division in stem segments and seedlings at concentrations as low as 10^{-12} M. They also slow leaf abscission (leaf drop) and promote xylem differentiation. These effects are so qualitatively similar to those of auxin that it took years for plant physiologists to determine that brassinosteroids were not types of auxins.

The identification of brassinosteroids as plant hormones arose from studies of an *Arabidopsis* mutant that even when grown in the dark exhibited physical features similar to plants grown in the light. The researchers discovered that the mutation affects a gene that normally codes for an enzyme similar to one involved in steroid synthesis in mammals. They also found that this brassinosteroid-deficient mutant could be restored to the wild-type phenotype by applying brassinosteroids.

Jasmonates, including *jasmonate* (JA) and *methyl jasmonate* (MeJA), are fatty acid–derived molecules that play important roles both in plant defence (see Concept 39.5) and, as discussed here, in plant development. Chemists first isolated MeJA as a key ingredient producing the enchanting fragrance of jasmine (*Jasminum grandiflorum*) flowers. Interest in jasmonates exploded when it was realised that jasmonates are produced by wounded plants and play a key role in controlling plant defences against herbivores and pathogens. From studying jasmonate signal transduction mutants as well as the effects of applying jasmonates to plants, it soon became apparent that jasmonates and their derivatives regulate a wide variety of physiological processes in plants, including nectar secretion, fruit ripening, pollen production, flowering time, seed germination, root growth, tuber formation, mycorrhizal symbioses, and tendril coiling. In controlling plant processes, jasmonates also engage in cross-talk with phytochrome and various hormones, including GA, IAA, and ethylene.

Strigolactones are xylem-mobile chemicals that stimulate seed germination, suppress adventitious root formation, help establish mycorrhizal associations, and (as noted earlier) help control apical dominance. Their recent discovery relates back to studies of their namesake, *Striga*, a colourfully named genus of rootless parasitic plants that penetrate the roots of other plants, diverting essential nutrients from them and stunting their growth. (In Romanian legend, Striga is a vampire-like creature that lives for thousands of years, only needing to feed every 25 years or so.) Also known as witchweed, *Striga* may be the greatest obstacle to food production in Africa, infesting about two-thirds of the area devoted to cereal crops. Each *Striga* plant produces tens of thousands of tiny seeds that can remain dormant in the soil for many years until a suitable host begins to grow. Thus, *Striga* cannot be eradicated by growing non-grain crops for several years. Strigolactones, exuded by the host roots, were first identified as the chemical signals that stimulate the germination of *Striga* seeds.

CONCEPT CHECK 39.2

1. Fusicoccin is a fungal toxin that stimulates the plasma membrane H^+ pumps of plant cells. How may it affect the growth of isolated stem sections?
2. **WHAT IF?** If a plant has the double mutation *ctr* and *ein*, what is its triple-response phenotype? Explain your answer.
3. **MAKE CONNECTIONS** What type of feedback process is exemplified by the production of ethylene during fruit ripening? Explain. (See Figure 1.10.)

For suggested answers, see Appendix A.

CONCEPT 39.3

Responses to light are critical for plant success

Light is an especially important environmental factor in the lives of plants. In addition to being required for photosynthesis, light triggers many key events in plant growth and development, collectively known as **photomorphogenesis**. Light reception also allows plants to measure the passage of days and seasons.

Plants detect not only the presence of light signals but also their direction, intensity, and wavelength (colour). A graph called an **action spectrum** depicts the relative effectiveness of different wavelengths of radiation in driving a particular process, such as photosynthesis (see Figure 10.10b). Action spectra are useful in studying *any* process that depends on light. By comparing action spectra of various plant responses, researchers determine which responses are mediated by the same photoreceptor (pigment). They also compare action spectra with absorption spectra of pigments; a close correspondence for a given pigment suggests that the pigment is the photoreceptor mediating the response. Action spectra reveal that red and blue light are the most important colours in regulating a plant's photomorphogenesis. These observations led researchers to two major classes of light receptors: blue-light photoreceptors and phytochromes.

Blue-Light Photoreceptors

Pigments that absorb blue light, known as **blue-light photoreceptors**, initiate a variety of responses in plants, including phototropism, the light-induced opening of stomata (see Figure 36.14), and the light-induced slowing of hypocotyl elongation that occurs when a seedling breaks ground. The biochemical identity of the blue-light photoreceptor was so elusive that in the 1970s, plant physiologists began to call this receptor "cryptochrome" (from the Greek *kryptos*, hidden, and *chrom*, pigment). In the 1990s, molecular biologists analysing *Arabidopsis* mutants found that plants use different types of pigments to detect blue light. *Cryptochromes*, molecular relatives of DNA repair enzymes, are involved in the blue-light-induced inhibition of stem elongation that occurs, for example, when a seedling first emerges from the soil. *Phototropin* is a protein kinase involved in mediating blue-light-mediated stomatal opening, chloroplast movements in response to light, and phototropic curvatures **(Figure 39.15)**, such as those studied by the Darwins.

▼ Figure 39.15 Action spectrum for blue-light-stimulated phototropism in corn coleoptiles. Phototropic bending towards light is controlled by phototropin, a photoreceptor sensitive to blue and violet light, particularly blue light.

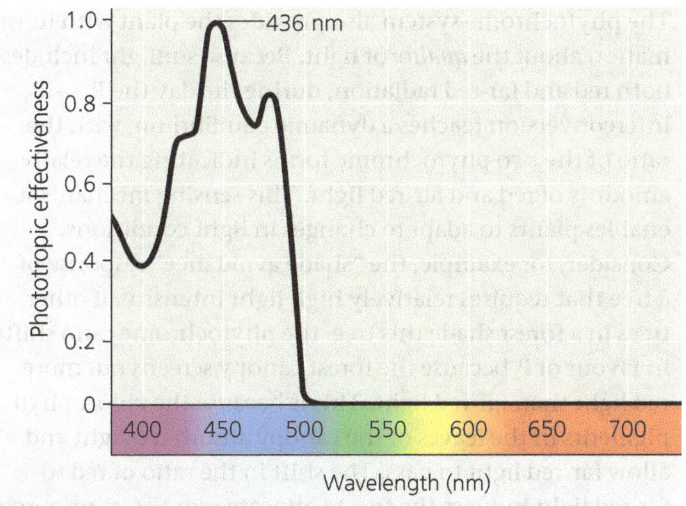

(a) This action spectrum illustrates that only light wavelengths below 500 nm (blue and violet light) induce curvature.

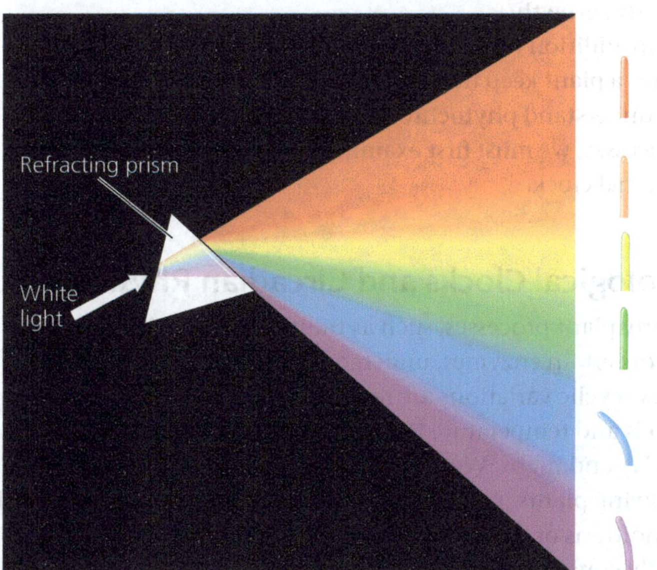

(b) When coleoptiles are exposed to light of various wavelengths as shown here, violet light induces slight curvature towards the light and blue light induces the most curvature. The other colours do not induce any curvature.

Phytochrome Photoreceptors

When exploring signal transduction in plants earlier in the chapter, we looked at the role of the plant pigments called phytochromes in the de-etiolation process. **Phytochromes**, pigments that absorb mostly red and far-red light, regulate many plant responses to light, including seed germination and shade avoidance.

Phytochromes and Seed Germination

Studies of seed germination led to the discovery of phytochromes. Because of limited nutrient reserves, many types of seeds, especially small ones, germinate only when the light environment and other conditions are near optimal. Such seeds often remain dormant for years until light conditions change. For example, the death of a shading tree or the plowing of a field may create a favourable light environment for germination.

In the 1930s, scientists determined the action spectrum for light-induced germination of lettuce seeds. They exposed water-swollen seeds to a few minutes of single-coloured light of various wavelengths and then stored the seeds in the dark. After two days, the researchers counted the number of seeds that had germinated under each light regimen. They found that red light of wavelength 660 nm increased the germination percentage of lettuce seeds maximally, whereas far-red light—that is, light of wavelengths near the upper edge of human visibility (730 nm)—*inhibited* germination compared with dark controls **(Figure 39.16)**. What happens when the lettuce seeds are subjected to a flash of red light followed by a flash of far-red light or, conversely, to far-red light followed by red light? The *last* flash of light determines the seeds' response: The effects of red and far-red light are reversible.

The photoreceptors responsible for the opposing effects of red and far-red light are phytochromes. So far, researchers have identified five phytochromes in *Arabidopsis*, each with a slightly different polypeptide component. In most phytochromes, the light-absorbing portion is photoreversible, converting back and forth between two forms, depending on the colour of light to which it is exposed. In its red-absorbing form (P_r), a phytochrome absorbs red (r) light maximally and is converted to its far-red-absorbing form (P_{fr}); in its P_{fr} form, it absorbs far-red (fr) light and is converted to its P_r form **(Figure 39.17)**. This $P_r \leftrightarrow P_{fr}$ interconversion is a switching mechanism that controls various light-induced events in the life of the plant. P_{fr} is the form of phytochrome that triggers many of a plant's developmental responses to light. For example, P_r in lettuce seeds exposed to red light is converted to P_{fr}, stimulating the cellular responses that lead to germination. When red-illuminated seeds are then exposed to far-red light, the P_{fr} is converted back to P_r, inhibiting the germination response.

▼ Figure 39.16 Inquiry

How does the order of red and far-red illumination affect seed germination?

Experiment Scientists at the US Department of Agriculture briefly exposed batches of lettuce seeds to red light or far-red light to test the effects on germination. After the light exposure, the seeds were placed in the dark, and the results were compared with control seeds that were not exposed to light.

Results The bar below each photo indicates the sequence of red light exposure, far-red light exposure, and darkness. The germination rate increased greatly in groups of seeds that were last exposed to red light (left). Germination was inhibited in groups of seeds that were last exposed to far-red light (right).

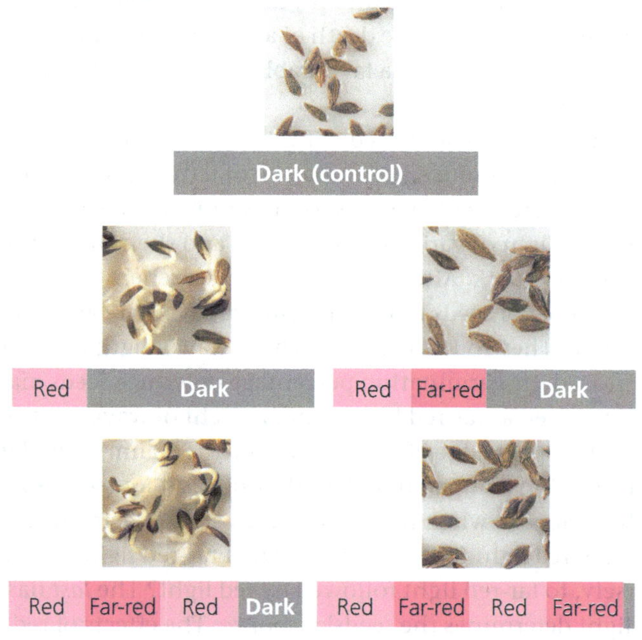

Data from H. Borthwick et al., A reversible photoreaction controlling seed germination, *Proceedings of the National Academy of Sciences USA* 38:662–666 (1952).

Conclusion Red light stimulates germination, and far-red light inhibits germination. The final light exposure is the determining factor. The effects of red and far-red light are reversible.

WHAT IF? *Phytochrome responds faster to red light than to far-red light. If the seeds had been placed in white light instead of the dark after their red light and far-red light treatments, would the results have been different?*

How does phytochrome switching explain light-induced germination in nature? Plants synthesise phytochrome as P_r, and if seeds are kept in the dark, the pigment remains almost entirely in the P_r form (see Figure 39.17). Sunlight contains both red light and far-red light, but the conversion to P_{fr} is faster than the conversion to P_r. Therefore, the ratio of P_{fr} to P_r increases in the sunlight. When seeds are exposed to adequate sunlight, the production and accumulation of P_{fr} trigger their germination.

▼ Figure 39.17 Phytochrome: a molecular switching mechanism. The absorption of red light causes P_r to change to P_{fr}. Far-red light reverses this conversion. In most cases, it is the P_{fr} form of the pigment that switches on physiological and developmental responses in the plant.

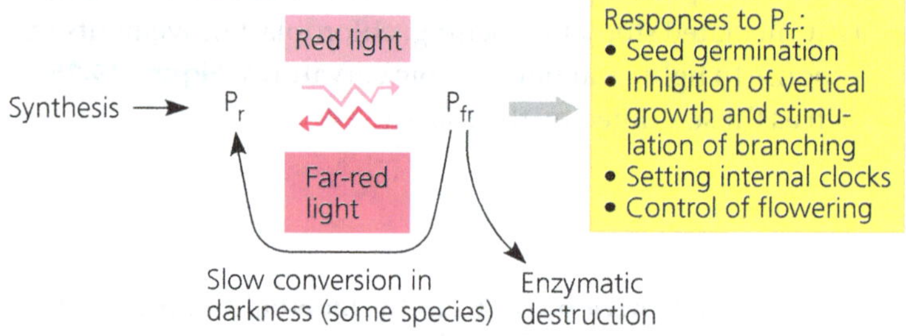

Phytochromes and Shade Avoidance

The phytochrome system also provides the plant with information about the *quality* of light. Because sunlight includes both red and far-red radiation, during the day the $P_r \leftrightarrow P_{fr}$ interconversion reaches a dynamic equilibrium, with the ratio of the two phytochrome forms indicating the relative amounts of red and far-red light. This sensing mechanism enables plants to adapt to changes in light conditions. Consider, for example, the "shade avoidance" response of a tree that requires relatively high light intensity. If other trees in a forest shade this tree, the phytochrome ratio shifts in favour of P_r because the forest canopy screens out more red light than far-red light. This is because the chlorophyll pigments in the leaves of the canopy absorb red light and allow far-red light to pass. The shift in the ratio of red to far-red light induces the tree to allocate more of its resources to growing taller. In contrast, direct sunlight increases the proportion of P_{fr}, which stimulates branching and inhibits vertical growth.

In addition to helping plants detect light, phytochrome helps a plant keep track of the passage of days and seasons. To understand phytochrome's role in these timekeeping processes, we must first examine the nature of the plant's internal clock.

Biological Clocks and Circadian Rhythms

Many plant processes, such as transpiration and the synthesis of certain enzymes, undergo a daily oscillation. Some of these cyclic variations are responses to the changes in light levels and temperature that accompany the 24-hour cycle of day and night. We can control these external factors by growing plants in growth chambers under rigidly maintained conditions of light and temperature. But even under artificially constant conditions, many physiological processes in plants, such as the opening and closing of stomata and the production of photosynthetic enzymes, continue to oscillate with a frequency of about 24 hours. For example, many

▼ Figure 39.18 Sleep movements of a bean plant (*Phaseolus vulgaris*). The movements are caused by reversible changes in the turgor pressure of cells on opposing sides of the pulvini, motor organs of the leaf.

legumes lower their leaves in the evening and raise them in the morning **(Figure 39.18)**. A bean plant continues these "sleep movements" even if kept in constant light or constant darkness; the leaves are not simply responding to sunrise and sunset. Such cycles, with a frequency of about 24 hours and not directly controlled by any known environmental variable, are called **circadian rhythms** (from the Latin *circa*, approximately, and *dies*, day).

Recent research supports the idea that the molecular "gears" of the circadian clock really are internal and not a daily response to some subtle but pervasive environmental cycle, such as geomagnetism or cosmic radiation. Organisms, including plants and humans, continue their rhythms even after being placed in deep mine shafts or when orbited in satellites, conditions that alter these subtle geophysical periodicities. However, daily signals from the environment can entrain (set) the circadian clock to a period of precisely 24 hours.

If an organism is kept in a constant environment, its circadian rhythms deviate from a 24-hour period (a period is the duration of one cycle). These free-running periods, as they are called, vary from about 21 to 27 hours, depending on the particular rhythmic response. The sleep movements of bean plants, for instance, have a period of 26 hours when the plants are kept in the free-running condition of constant darkness. Deviation of the free-running period from exactly 24 hours does not mean that biological clocks drift erratically. Free-running clocks are still keeping perfect time, but they are not synchronised with the outside world. To understand the mechanisms underlying circadian rhythms, we must distinguish between the clock and the rhythmic processes it controls. For example, the leaves of the bean plant in Figure 39.18 are the clock's "hands" but are not the essence of the clock itself. If bean leaves are restrained for several hours and then released, they will reestablish the position appropriate for the time of day. We can interfere with a biological rhythm, but the underlying clockwork continues to tick.

At the heart of the molecular mechanisms underlying circadian rhythms are oscillations in the transcription of certain genes. Mathematical models propose that the 24-hour period arises from negative-feedback loops involving the transcription of a few central "clock genes." Some clock genes may encode transcription factors that inhibit, after a time delay, the transcription of the gene that encodes the transcription factor itself. Such negative-feedback loops, together with a time delay, are enough to produce oscillations.

Researchers have used a novel technique to identify clock mutants of *Arabidopsis*. One prominent circadian rhythm in plants is the daily production of certain photosynthesis-related proteins. Molecular biologists traced the source of this rhythm to the promoter that initiates the transcription of the genes for these photosynthesis proteins. To identify clock mutants, scientists inserted the gene for an enzyme responsible for the bioluminescence of fireflies, called luciferase, to the promoter. When the biological clock turned on the promoter in the *Arabidopsis* genome, it also turned on the production of luciferase. The plants began to glow with a circadian periodicity. Clock mutants were then isolated by selecting specimens that glowed for a longer or shorter time than normal. The genes altered in some of these mutants affect proteins that normally bind photoreceptors. Perhaps these particular mutations disrupt a light-dependent mechanism that sets the biological clock.

The Effect of Light on the Biological Clock

As previously noted, the free-running period of the circadian rhythm of bean leaf movements is 26 hours. Consider a bean plant placed at dawn in a dark cabinet for 72 hours: Its leaves would not rise again until 2 hours after natural dawn on the second day, 4 hours after natural dawn on the third day, and so on. Shut off from environmental cues, the plant becomes desynchronised. Desynchronisation happens to humans when we fly across several time zones; when we reach our destination, the clocks on the wall are not synchronised with our internal clocks. Most organisms are probably prone to jet lag.

The factor that entrains the biological clock to precisely 24 hours every day is light. Both phytochromes and blue-light photoreceptors can entrain circadian rhythms in plants, but our understanding of how phytochromes do this is more complete. The mechanism involves turning cellular responses on and off by means of the $P_r \leftrightarrow P_{fr}$ switch.

Consider again the photoreversible system in Figure 39.17. In darkness, the phytochrome ratio shifts gradually in favour of the P_r form, partly as a result of turnover in the overall phytochrome pool. The pigment is synthesised in the P_r form, and enzymes destroy more P_{fr} than P_r. In some plant species, P_{fr} present at sundown slowly converts to P_r. In darkness, there is no means for the P_r to be reconverted to P_{fr}, but upon illumination, the P_{fr} level suddenly increases again as P_r is rapidly converted. This increase in P_{fr} each day at dawn resets the biological clock: Bean leaves reach their most extreme night position 16 hours after dawn.

In nature, interactions between phytochrome and the biological clock enable plants to measure the passage of night and day. The relative lengths of night and day, however, change over the course of the year (except at the equator). Plants use this change to adjust activities in timing with the seasons.

Photoperiodism and Responses to Seasons

Imagine the consequences if a plant produced flowers when pollinators were not present or if a deciduous tree produced leaves in the middle of winter. Seasonal events are of critical importance in the life cycles of most plants. Seed germination, flowering, and the onset and breaking of bud dormancy are all stages that usually occur at specific times of the year. The environmental cue that plants use to detect the time of year is the change in day length (*photoperiod*). A physiological response to specific night or day lengths, such as flowering, is called **photoperiodism**.

Photoperiodism and Control of Flowering

An early clue to how plants detect seasons came from a mutant variety of tobacco, Maryland Mammoth, that grew tall but failed to flower during summer. It finally bloomed in a greenhouse in December. After trying to induce earlier flowering by varying temperature, moisture, and mineral nutrition, researchers learned that the shortening days of winter stimulated this variety to flower. Experiments revealed that flowering occurred only if the photoperiod was 14 hours or shorter. This variety did not flower during summer because at Maryland's latitude the photoperiods were too long.

The researchers called Maryland Mammoth a **short-day plant** because it apparently required a light period *shorter* than a critical length to flower. Chrysanthemums, poinsettias, and some soybean varieties are also short-day plants, which generally flower in late summer, autumn, or winter. Another group of plants flower only when the light period is *longer* than a certain number of hours. These **long-day plants** generally flower in late spring or early summer. Spinach, for example, flowers when days are 14 hours or longer. Radishes, lettuce, irises, and many cereal varieties are also long-day plants. **Day-neutral plants**, such as tomatoes, rice, and dandelions, are unaffected by photoperiod and flower when they reach a certain stage of maturity, regardless of photoperiod.

Critical Night Length In the 1940s, researchers learned that flowering in short- and long-day plants is actually controlled by night length, not day length (photoperiod). Many of these scientists worked with cocklebur (*Xanthium strumarium*), a short-day plant that flowers only when days are 16 hours or shorter (and nights are at least 8 hours long). These researchers found that if the photoperiod is broken by a brief exposure to darkness, flowering proceeds. However, if the night length is interrupted by even a few minutes of dim light, cocklebur will not flower, and this turned out to be true for other short-day plants as well **(Figure 39.19a)**. Cocklebur is unresponsive to day length, but it requires at least 8 hours of continuous darkness to flower. Short-day plants are really long-night plants, but the older term is embedded firmly in the lexicon of plant physiology. Similarly, long-day plants are actually short-night plants. A long-day plant grown under long-night conditions that would not normally induce flowering will flower if the night length is interrupted by a few minutes of light **(Figure 39.19b)**.

Notice that long-day plants are *not* distinguished from short-day plants by an absolute night length. Instead, they are distinguished by whether the critical night length sets a *maximum* number of hours of darkness required for flowering (long-day plants) or a *minimum* number of hours of darkness required for flowering (short-day plants). In both cases, the actual number of hours in the critical night length is specific to each species of plant.

▼ Figure 39.19 Photoperiodic control of flowering.

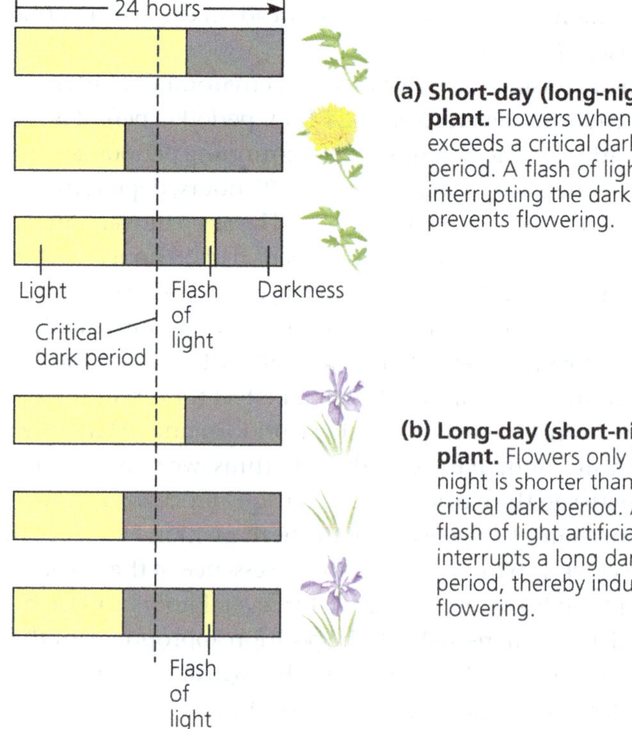

(a) Short-day (long-night) plant. Flowers when night exceeds a critical dark period. A flash of light interrupting the dark period prevents flowering.

(b) Long-day (short-night) plant. Flowers only if the night is shorter than a critical dark period. A brief flash of light artificially interrupts a long dark period, thereby inducing flowering.

▼ **Figure 39.20 Reversible effects of red and far-red light on photoperiodic response.** A flash of red (r) light shortens the dark period. A subsequent flash of far-red (fr) light cancels the red flash's effect.

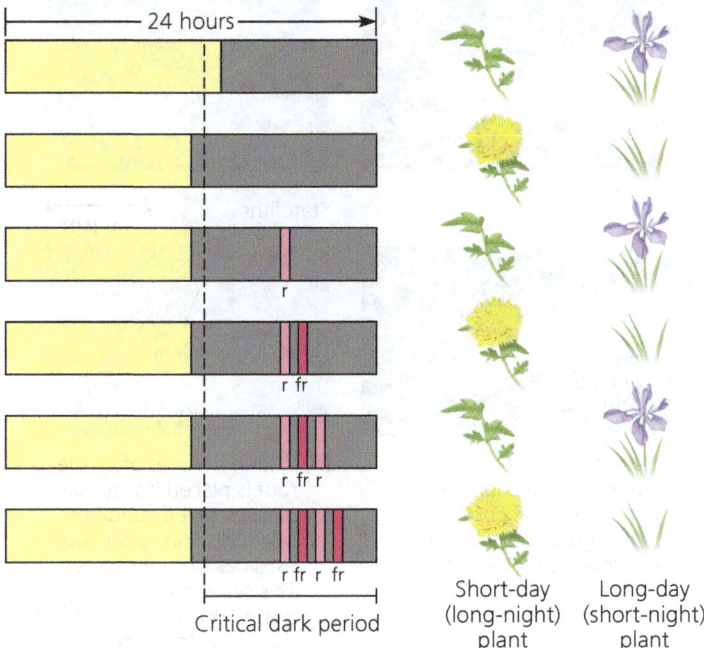

VISUAL SKILLS *Under long-day conditions (as in the top panel) or under short-day conditions (as in the second panel), how would a single flash of far-red light during the dark period affect flowering?*

Red light is the most effective colour in interrupting the night length. Action spectra and photoreversibility experiments show that phytochrome is the pigment that detects the red light **(Figure 39.20)**. For example, if a flash of red light during the night length is followed by a flash of far-red light, then the plant detects no interruption of night length. As in the case of phytochrome-mediated seed germination, red/far-red photoreversibility occurs.

Plants measure night lengths very precisely; some short-day plants will not flower if night is even 1 minute shorter than the critical length. Some plant species always flower on the same day each year. It appears that plants use their biological clock, entrained by night length with the help of phytochrome, to ascertain precisely the season of the year. The Australian and New Zealand floriculture (flower-growing) industry applies this knowledge to produce flowers out of season. Chrysanthemums, for instance, are short-day plants that normally bloom in autumn, but their blooming can be stalled until Mother's Day in May by punctuating each long night with a flash of light, thus turning one long night into two short nights.

Some plants bloom after a single exposure to the photoperiod required for flowering. Other species need several successive days of the appropriate photoperiod. Still others respond to a photoperiod only if they have been previously exposed to some other environmental stimulus, such as a period of cold. Winter wheat, for example, will not flower unless it has been exposed to several weeks of temperatures below 10°C. The use of pretreatment with cold to induce flowering is called **vernalisation** (from the Latin for "spring"). Several weeks after winter wheat is vernalised, a long photoperiod (short night) induces flowering.

A Flowering Hormone?

Although flowers form from apical or axillary bud meristems, it is leaves that detect changes in photoperiod and produce signalling molecules that cue buds to develop as flowers. In many short-day and long-day plants, exposing just one leaf to the appropriate photoperiod is enough to induce flowering. Indeed, as long as one leaf is left on the plant, photoperiod is detected and floral buds are induced. If all leaves are removed, the plant is insensitive to photoperiod.

Classic experiments revealed that the floral stimulus could move across a graft from an induced plant to a noninduced plant and trigger flowering in the latter. Moreover, the flowering stimulus appears to be the same for short-day and long-day plants, despite the different photoperiodic conditions required for leaves to send this signal **(Figure 39.21)**. The hypothetical signalling molecule for flowering, called **florigen**, remained unidentified for over 70 years as scientists focused on small hormone-like molecules. However, large macromolecules, such as mRNA and proteins, can move by the symplastic route via plasmodesmata and regulate plant development. It now appears that florigen is a protein. A gene called *FLOWERING LOCUS T* (*FT*) is activated in leaf cells during conditions favouring flowering, and the FT protein travels through the symplasm to the shoot apical meristem, initiating the transition of a bud's meristem from a vegetative to a flowering state.

▼ **Figure 39.21 Experimental evidence for a flowering hormone.** If grown individually under short-day conditions, a short-day plant will flower and a long-day plant will not. However, both will flower if grafted together and exposed to short days. This result indicates that a flower-inducing substance (florigen) is transmitted across grafts and induces flowering in both short-day and long-day plants.

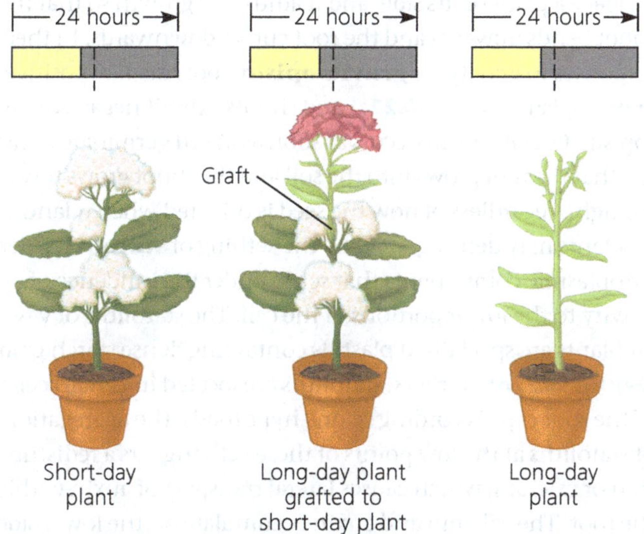

WHAT IF? *If flowering were inhibited in both parts of the grafted plants, what would you conclude?*

CONCEPT CHECK 39.3

1. If an enzyme in field-grown soybean leaves is most active at noon and least active at midnight, is its activity under circadian regulation?
2. **WHAT IF?** If a plant flowers in a controlled chamber with a daily cycle of 10 hours of light and 14 hours of darkness, is it a short-day plant? Explain.
3. **MAKE CONNECTIONS** Plants detect the quality of their light environment by using blue-light photoreceptors and red-light-absorbing phytochromes. After reviewing Figure 10.9, suggest a reason why plants are so sensitive to these colours of light.

For suggested answers, see Appendix A.

CONCEPT 39.4

Plants respond to a wide variety of stimuli other than light

Although plants are immobile, some mechanisms have evolved by natural selection that enable them to adjust to a wide range of environmental circumstances by developmental or physiological means. Light is so important in the life of a plant that we devoted the entire previous section to the topic of a plant's reception of and response to this particular environmental factor. In this section, we will examine responses to some of the other environmental stimuli that a plant commonly encounters.

Gravity

Because plants are photoautotrophs, it is not surprising that mechanisms for growing towards light have evolved. But what environmental cue does the shoot of a young seedling use to grow upwards when it is completely underground and there is no light for it to detect? Similarly, what environmental factor prompts the young root to grow downwards? The answer to both questions is gravity.

Place a plant on its side, and it adjusts its growth so that the shoot bends upwards and the root curves downwards. In their responses to gravity, or **gravitropism**, roots display positive gravitropism **(Figure 39.22a)** and shoots exhibit negative gravitropism. Gravitropism occurs as soon as a seed germinates, ensuring that the root grows into the soil and the shoot grows towards sunlight, regardless of how the seed is oriented when it lands.

Plants may detect gravity by the settling of **statoliths**, dense cytoplasmic components that settle under the influence of gravity to the lower portions of the cell. The statoliths of vascular plants are specialised plastids containing dense starch grains **(Figure 39.22b)**. In roots, statoliths are located in certain cells of the root cap. According to one hypothesis, the aggregation of statoliths at the low points of these cells triggers a redistribution of calcium, which causes lateral transport of auxin within the root. The calcium and auxin accumulate on the lower side of the root's zone of elongation. At high concentration, auxin

▼ Figure 39.22 Positive gravitropism in roots: the statolith hypothesis.

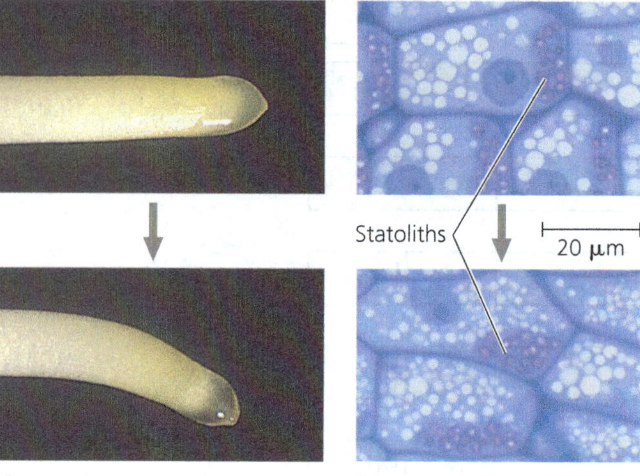

(a) Over the course of hours, a horizontally oriented primary root of corn bends gravitropically until its growing tip becomes vertically oriented (LMs).

(b) Within minutes after the root is placed horizontally, plastids called statoliths begin settling to the lowest sides of root cap cells. This settling may be the gravity-sensing mechanism that leads to redistribution of auxin and differing rates of elongation by cells on opposite sides of the root (LMs).

inhibits cell elongation, an effect that slows growth on the root's lower side. The more rapid elongation of cells on the upper side causes the root to grow straight downwards.

Falling statoliths, however, may not be necessary for gravitropism. For example, there are mutants of *Arabidopsis* and tobacco that lack statoliths but are still capable of gravitropism, though the response is slower than in wild-type plants. It could be that the entire cell helps the root sense gravity by mechanically pulling on proteins that tether the protoplast to the cell wall, stretching the proteins on the "up" side and compressing the proteins on the "down" side of the root cells. Dense organelles, in addition to starch granules, may also contribute by distorting the cytoskeleton as they are pulled by gravity. Statoliths, because of their density, may enhance gravitational sensing by a mechanism that simply works more slowly in their absence.

Mechanical Stimuli

Trees in windy environments usually have shorter, stockier trunks than a tree of the same species growing in more sheltered locations. This stunted shape enables the plant to hold its ground against strong gusts of wind. The term **thigmomorphogenesis** (from the Greek *thigma*, touch) refers to the changes in form that result from mechanical perturbation. Plants are very sensitive to mechanical stress: Even the act of measuring the length of a leaf with a ruler alters its subsequent growth. Rubbing the stems of a young

▼ Figure 39.23 Thigmomorphogenesis in *Arabidopsis*. The shorter plant on the left was rubbed twice a day. The untouched plant (right) grew much taller.

plant a couple of times daily results in plants that are shorter than controls **(Figure 39.23)**.

Some plant species have become, over the course of their evolution, "touch specialists." Acute responsiveness to mechanical stimuli is an integral part of these plants' "life strategies." Most vines and other climbing plants have tendrils that coil rapidly around supports (see Figure 35.7). These grasping organs usually grow straight until they touch something; the contact stimulates a coiling response caused by differential growth of cells on opposite sides of the tendril. This directional growth in response to touch is called **thigmotropism**, and it allows the vine to take advantage of whatever mechanical supports it comes across as it climbs upwards towards a forest canopy.

Other examples of touch specialists are plants that undergo rapid leaf movements in response to mechanical stimulation. For example, when the compound leaf of the sensitive plant *Mimosa pudica* is touched, it collapses and its leaflets fold together **(Figure 39.24)**. This response, which takes only a second or two, results from a rapid loss of turgor in cells within pulvini, specialised motor organs located at the joints of the leaf. The motor cells suddenly become flaccid after stimulation because they lose potassium ions, causing water to leave the cells by osmosis. It takes about 10 minutes for the cells to regain their turgor and restore the "unstimulated" form of the leaf. The function of the sensitive plant's behaviour invites speculation. Perhaps the plant appears less leafy and appetising to herbivores by folding its leaves and reducing its surface area when jostled.

A remarkable feature of rapid leaf movements is the mode of transmission of the stimulus through the plant. If one leaflet on a sensitive plant is touched, first that leaflet responds, then the adjacent leaflet responds, and so on, until all the leaflet pairs have folded together. From the point of stimulation, the signal that produces this response travels at a speed of about 1 cm/sec. An electrical impulse travelling at the same rate can be detected when electrodes are attached to the leaf. These impulses, called **action potentials**, resemble nerve impulses in animals, though the action potentials of plants are thousands of times slower. Action potentials have been discovered in many species of algae and plants and may be used as a form of internal communication. For example, in the Venus flytrap (*Dionaea muscipula*), action potentials are transmitted from sensory hairs in the trap to the cells that respond by closing the trap (see Figure 37.23). In the case of *Mimosa pudica*, a more violent stimulus, such as burning a leaf, causes *all* the leaves and leaflets to droop. This whole-plant response involves the spread of signalling molecules from the injured area to other parts of the shoot.

Environmental Stresses

Environmental stresses, such as flooding, drought, or extreme temperatures, can have devastating effects on plant survival, growth, and reproduction. In natural ecosystems, plants that cannot tolerate an environmental stress either die or are outcompeted by other plants. Thus, environmental stresses are an important factor in determining the geographic ranges of plants. In the last section of this chapter, we'll examine the defensive responses of plants to common **biotic** (living) stresses, such as herbivores and pathogens. Here we'll consider some of the more common **abiotic** (nonliving) stresses that plants encounter. Since these abiotic factors are major determinants of crop yields, there is currently much interest in trying to project how global climate change will impact crop production (see the **Problem-Solving Exercise**).

▼ Figure 39.24 Rapid turgor movements by the sensitive plant (*Mimosa pudica*).

(a) Unstimulated state (leaflets spread apart) **(b) Stimulated state (leaflets folded)**

PROBLEM-SOLVING EXERCISE

How will climate change affect crop productivity?

Plant growth is significantly limited by air temperature, water availability, and solar radiation. A useful parameter for estimating crop productivity is the number of days per year when these three climate variables are suitable for plant growth. Camilo Mora (University of Hawaii at Manoa) and his colleagues analysed global climate models to project the effect of climate change on suitable days for plant growth by the year 2100.

In this exercise, you will examine projected effects of climate change on crop productivity and identify the resulting human impacts.

Your Approach Analyse the map and table. Then answer the questions below.

Your Data The researchers projected the annual changes in suitable days for plant growth for all three climate variables: temperature, water availability, and solar radiation. They did so by subtracting recent averages (1996–2005) from projected future averages (2091–2100). The map shows the projected changes if no measures are taken to reduce climate change. The numbers identify locations of the 15 most populous nations. The table identifies their economies as either mainly industrial ([industrial icon]) or agricultural ([agricultural icon]) and their annual per capita income category.

Nation	Map location	Estimated population in 2014 (millions)	Type of economy	Income category*
China	1	1,350	[industrial icon]	$$$
India	2	1,221	[agricultural icon]	$$
United States	3	317	[industrial icon]	$$$$
Indonesia	4	251	[agricultural icon]	$
Brazil	5	201	[agricultural icon]	$$$
Pakistan	6	193	[agricultural icon]	$$
Nigeria	7	175	[agricultural icon]	$$
Bangladesh	8	164	[agricultural icon]	$
Russia	9	143	[industrial icon]	$$$$
Japan	10	127	[industrial icon]	$$$$
Mexico	11	116	[agricultural icon]	$$$
Philippines	12	106	[agricultural icon]	$$
Ethiopia	13	94	[agricultural icon]	$
Vietnam	14	92	[agricultural icon]	$
Egypt	15	85	[agricultural icon]	$$

Data from The World Bank
*Per capita income, based on World Bank categories: $ = low: < $1,035; $$ = lower middle: $1,036-$4,085; $$$ = upper middle: $4,086-$12,615; $$$$ = high: > $12,615.

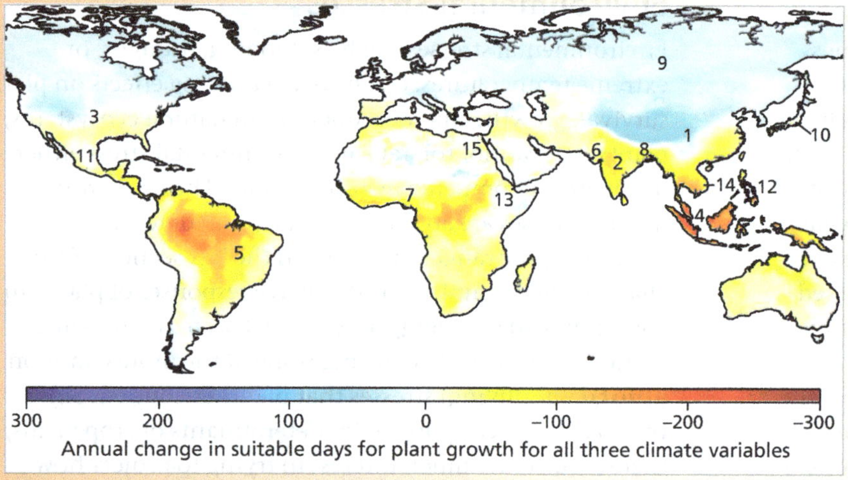

300 200 100 0 −100 −200 −300
Annual change in suitable days for plant growth for all three climate variables

Map data from Camilo Mora, et al. Days for Plant Growth Disappear under Projected Climate Change: Potential Human and Biotic Vulnerability. *PLoS Biol* 13(6): e1002167 (2015).

Your Analysis

1. Camilo Mora began the study as a result of talking with someone who claimed that climate change improves plant growth because it increases the number of days above freezing. Based on the map data, how would you respond to this claim?
2. What do the map and the table data indicate about the impact of the projected changes on humans?

Drought

On a sunny, dry day, a plant may wilt because its water loss by transpiration exceeds water absorption from the soil. Prolonged drought, of course, will kill a plant, but plants have control systems that enable them to cope with less extreme water deficits.

Many of a plant's responses to water deficit help the plant conserve water by reducing the rate of transpiration. Water deficit in a leaf causes stomata to close, thereby slowing transpiration dramatically (see Figure 36.14). Water deficit stimulates increased synthesis and release of abscisic acid in the leaves; this hormone helps keep stomata closed by acting on guard cell membranes. Leaves respond to water deficit in several other ways. For example, when the leaves of grasses wilt, they roll into a tubelike shape that reduces transpiration by exposing less leaf surface to dry air and wind. Other plants, such as the leafless rock wattle (see Figure 36.15), shed their leaves in response to seasonal drought. Although these leaf responses conserve water, they also reduce photosynthesis, which is one reason why a drought diminishes crop yield. Plants can even take advantage of early warnings in the form of chemical signals from wilting neighbours and prime themselves to respond more readily and intensely to impending drought stress (see the **Scientific Skills Exercise**).

Scientific Skills Exercise

Interpreting Experimental Results from a Bar Graph

Do Drought-Stressed Plants Communicate Their Condition to Their Neighbours? Researchers wanted to learn if plants can communicate drought-induced stress to neighbouring plants and, if so, whether they use aboveground or belowground signals. In this exercise, you will interpret a bar graph concerning widths of stomatal openings to investigate whether drought-induced stress can be communicated from plant to plant.

How the Experiment Was Done Eleven potted pea plants (*Pisum sativum*) were placed equidistantly in a row. The root systems of plants 6–11 were connected to those of their immediate neighbours by tubes, which allowed chemicals to move from the roots of one plant to the roots of the next plant without moving through the soil. The root systems of plants 1–6 were not connected. Osmotic shock was inflicted on plant 6 using a highly concentrated solution of mannitol, a natural sugar commonly used to mimic drought stress in vascular plants. Fifteen minutes following the osmotic shock to plant 6, researchers measured the width of stomatal openings in leaves from all the plants. A control experiment was also done in which water was added to plant 6 instead of mannitol.

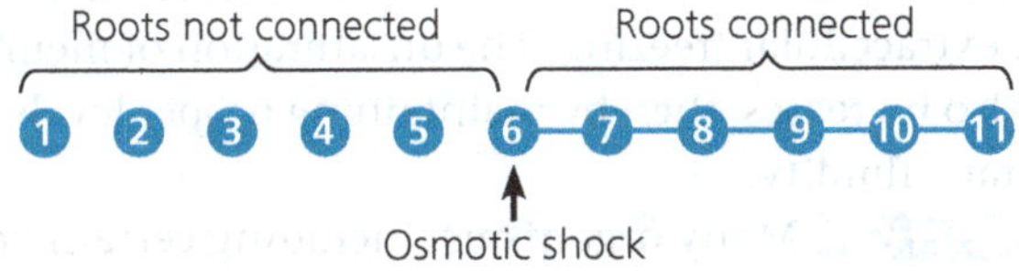

Data from the Experiment

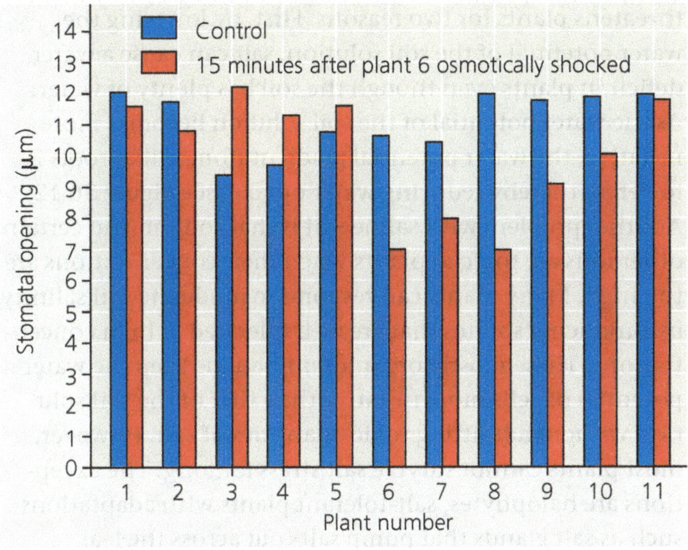

Data from O. Falik et al., Rumor has it...: Relay communication of stress cues in plants, *PLoS ONE* 6(11):e23625 (2011).

INTERPRET THE DATA

1. How do the widths of the stomatal openings of plants 6–8 and plants 9 and 10 compare with those of the other plants in the experiment? What does this indicate about the state of plants 6–8 and 9 and 10? (For information about reading graphs, see the Scientific Skills Review in Appendix D.)
2. Do the data support the idea that plants can communicate their drought-stressed condition to their neighbours? If so, do the data indicate that the communication is via the shoot system or the root system? Make specific reference to the data in answering both questions.
3. Why was it necessary to make sure that chemicals could not move through the soil from one plant to the next?
4. When the experiment was run for 1 hour rather than 15 minutes, the results were about the same except that the stomatal openings of plants 9–11 were comparable to those of plants 6–8. Suggest a reason why.
5. Why was water added to plant 6 instead of mannitol in the control experiment? What do the results of the control experiment indicate?

Flooding

Too much water is also a problem for a plant. An overwatered houseplant may suffocate because the soil lacks the air spaces that provide oxygen for cellular respiration in the roots. Some plants are structurally adapted to very wet habitats. For example, the submerged roots of mangroves, which inhabit coastal marshes, are continuous with aerial roots exposed to oxygen (see Figure 35.4). But how do less specialised plants cope with oxygen deprivation in waterlogged soils? Oxygen deprivation stimulates the production of ethylene, which causes some cells in the root cortex to die. The destruction of these cells creates air tubes that function as "snorkels," providing oxygen to the submerged roots **(Figure 39.25)**.

▼ **Figure 39.25 A developmental response of corn roots to flooding and oxygen deprivation. (a)** A cross section of a control root grown in an aerated hydroponic medium. **(b)** A root grown in a nonaerated hydroponic medium. Ethylene-stimulated programmed cell death creates the air tubes (SEMs).

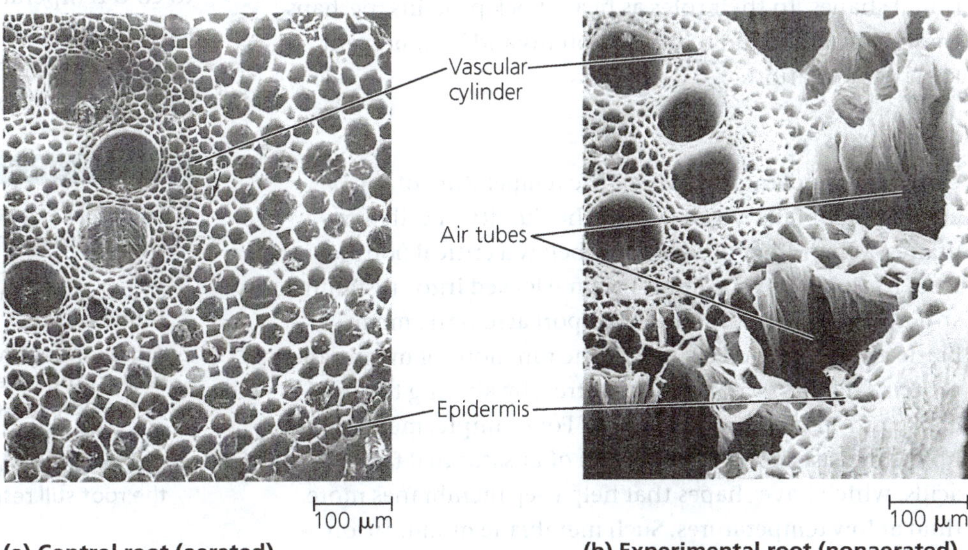

Salt Stress

An excess of sodium chloride or other salts in the soil threatens plants for two reasons. First, by lowering the water potential of the soil solution, salt can cause a water deficit in plants even though the soil has plenty of water. As the water potential of the soil solution becomes more negative, the water potential gradient from soil to roots is lowered, thereby reducing water uptake (see Figure 36.12). Another problem with saline soil is that sodium and certain other ions are toxic to plants when their concentrations are too high. Many plants can respond to moderate soil salinity by producing solutes that are well tolerated at high concentrations: These mostly organic compounds keep the water potential of cells more negative than that of the soil solution without admitting toxic quantities of salt. However, most plants cannot survive salt stress for long. The exceptions are halophytes, salt-tolerant plants with adaptations such as salt glands that pump salts out across the leaf epidermis.

Heat Stress

Excessive heat may harm and even kill a plant by denaturing its enzymes. Transpiration helps cool leaves by evaporative cooling. On a warm day, for example, the temperature of a leaf may be 3–10°C below the ambient air temperature. Hot, dry weather also tends to dehydrate many plants; the closing of stomata in response to this stress conserves water but then sacrifices evaporative cooling. This dilemma is one reason why very hot, dry days take a toll on most plants.

Most plants have a backup response that enables them to survive heat stress. Above a certain temperature—about 40°C for most plants in temperate regions—plant cells begin synthesising **heat-shock proteins**, which help protect other proteins from heat stress. This response also occurs in heat-stressed animals and microorganisms. Some heat-shock proteins function in unstressed cells as temporary scaffolds that help other proteins fold into their functional shapes. In their roles as heat-shock proteins, perhaps these molecules bind to other proteins and help prevent their denaturation.

Cold Stress

One problem plants face when the temperature of the environment falls is a change in the fluidity of cell membranes. When a membrane cools below a critical point, it loses its fluidity as the lipids become locked into crystalline structures. This alters solute transport across the membrane and also adversely affects the functions of membrane proteins. Plants respond to cold stress by altering the lipid composition of their membranes. For example, membrane lipids increase in their proportion of unsaturated fatty acids, which have shapes that help keep membranes more fluid at low temperatures. Such membrane modification requires from several hours to days, which is one reason why unseasonably cold temperatures are generally more stressful to plants than the more gradual seasonal drop in air temperature.

Freezing is another type of cold stress. At subfreezing temperatures, ice forms in the cell walls and intercellular spaces of most plants. The cytosol generally does not freeze at the cooling rates encountered in nature because it contains more solutes than the very dilute solution found in the cell wall, and solutes lower the freezing point of a solution. The reduction in liquid water in the cell wall caused by ice formation lowers the extracellular water potential, causing water to leave the cytoplasm. The resulting increase in the concentration of ions in the cytoplasm is harmful and can lead to cell death. Whether the cell survives depends largely on how well it resists dehydration. In regions with cold winters, native plants are adapted to cope with freezing stress. For example, before the onset of winter, the cells of many frost-tolerant species increase cytoplasmic levels of specific solutes, such as sugars, that are well tolerated at high concentrations and that help reduce the loss of water from the cell during extracellular freezing. The unsaturation of membrane lipids also increases, thereby maintaining proper levels of membrane fluidity.

EVOLUTION Many organisms, including certain vertebrates, fungi, bacteria, and many species of plants, have proteins that hinder ice crystals from growing, helping the organism escape freezing damage. First described in Arctic fish in the 1950s, these *antifreeze proteins* permit survival at temperatures below 0°C. They bind to small ice crystals and inhibit their growth or, in the case of plants, prevent the crystallisation of ice. The five major classes of antifreeze proteins differ markedly in their amino acid sequences but have a similar three-dimensional structure, suggesting convergent evolution. Surprisingly, antifreeze proteins from rye (*Secale cereale*) are homologous to antifungal defence proteins, but they are produced in response to cold temperatures and shorter days, not fungal pathogens. Progress is being made in increasing the freezing tolerance of crop plants by engineering antifreeze protein genes into their genomes.

CONCEPT CHECK 39.4

1. Thermal images are photographs of the heat emitted by an object. Researchers have used thermal imaging of plants to isolate mutants that overproduce abscisic acid. Suggest a reason why they are warmer than wild-type plants under conditions that are normally nonstressful.
2. A greenhouse worker finds that potted chrysanthemums nearest the aisles are often shorter than those in the middle of the bench. Explain this "edge effect," a common problem in horticulture.
3. **WHAT IF?** If you removed the root cap from a root, would the root still respond to gravity? Explain.

For suggested answers, see Appendix A.

CONCEPT 39.5

Plants respond to attacks by pathogens and herbivores

Through natural selection, plants have evolved many types of interactions with other species in their communities. Some interspecific interactions are mutually beneficial, such as the associations of plants with mycorrhizal fungi (see Figure 37.22) or with pollinators (see Figures 38.4 and 38.5). Many plant interactions with other organisms, however, do not benefit the plant. As primary producers, plants are at the base of most food webs and are subject to attack by a wide range of plant-eating (herbivorous) animals. A plant is also subject to infection by diverse viruses, bacteria, and fungi that can damage tissues or even kill the plant. Plants counter these threats with defence systems that deter animals and prevent infection or combat invading pathogens.

Defences Against Pathogens

A plant's first line of defence against infection is the physical barrier presented by the epidermis and periderm of the plant body (see Figure 35.19). This line of defence, however, is not impenetrable. The mechanical wounding of leaves by herbivores, for example, opens up portals for invasion by pathogens. Even when plant tissues are intact, viruses, bacteria, and the spores and hyphae of fungi can still enter the plant through natural openings in the epidermis, such as stomata. Once the physical lines of defence are breached, a plant's next lines of defence are two types of immune responses: PAMP-triggered immunity and effector-triggered immunity.

PAMP-Triggered Immunity

When a pathogen succeeds in invading a host plant, the plant mounts the first of two lines of immune defence, which ultimately results in a chemical attack that isolates the pathogen and prevents its spread from the site of infection. This first line of immune defence, called *PAMP-triggered immunity*, depends on the plant's ability to recognise **pathogen-associated molecular patterns** (**PAMPs**; formerly called *elicitors*), molecular sequences that are specific to certain pathogens. For example, bacterial *flagellin*, a major protein found in bacterial flagella, is a PAMP. Many soil bacteria, including some pathogenic varieties, get splashed onto the shoots of plants by raindrops. If these bacteria penetrate the plant, a specific amino acid sequence within flagellin is perceived by a Toll-like receptor, a type of receptor also found in animals, where it plays a key role in the innate immune system (see Concept 43.1). The innate immune system is an evolutionarily old defence strategy and is the dominant immune system in plants, fungi, insects, and primitive multicellular organisms. Unlike vertebrates, plants do not have an adaptive immune system: Plants neither generate antibody or T cell responses nor possess mobile cells that detect and attack pathogens.

PAMP recognition in plants leads to a chain of signalling events that lead ultimately to the local production of broad-spectrum, antimicrobial chemicals called *phytoalexins*, which have fungicidal and bactericidal properties. The plant cell wall is also toughened, hindering further progress of the pathogen. Similar but even stronger defences are initiated by the second plant immune response, effector-triggered immunity.

Effector-Triggered Immunity

EVOLUTION Over the course of evolution, plants and pathogens have engaged in an arms race. PAMP-triggered immunity can be overcome by the evolution of pathogens that can evade detection by the plant. These pathogens deliver **effectors**, pathogen-encoded proteins that cripple the plant's innate immune system, directly into plant cells. For example, some bacteria deliver effectors inside the plant cell that block the perception of flagellin. Thus, these effectors allow the pathogen to redirect the host's metabolism to the pathogen's advantage.

The suppression of PAMP-triggered immunity by pathogen effectors led to the evolution of *effector-triggered immunity*. Because there are thousands of effectors, this plant defence is typically made up of hundreds of disease resistance (*R*) genes. Each *R* gene codes for an R protein that can be activated by a specific effector. Signal transduction pathways then lead to an arsenal of defence responses, including a local defence called the *hypersensitive response* and a general defence called *systemic acquired resistance*. Local and systemic responses to pathogens require extensive genetic reprogramming and commitment of cellular resources. Therefore, a plant activates these defences only after detecting a pathogen.

The Hypersensitive Response An important mechanism that restricts the spread of a pathogen is the **hypersensitive response**, the formation of a ring of local cell death around the infection site. As indicated in Figure 39.26, the hypersensitive response, which results from effector-triggered immunity, involves the production of enzymes and chemicals that impair the pathogen's cell wall integrity, metabolism, or reproduction. Effector-triggered immunity also stimulates the formation of lignin and the cross-linking of molecules within the plant cell wall, responses that hinder the spread of the pathogen to other parts of the plant. As shown in step 2 of the figure, the hypersensitive response results in localised lesions on a leaf. As "sick" as such a leaf appears, it will still survive, and its defensive response will help protect the rest of the plant.

Systemic Acquired Resistance

The hypersensitive response is localised and specific. However, pathogen invasions can also produce signalling molecules that "sound the alarm" of infection to the whole plant. The resulting **systemic acquired resistance** arises from the plant-wide expression of defence genes. It is nonspecific, providing protection against a diversity

▼ **Figure 39.27**

MAKE CONNECTIONS

Levels of Plant Defences Against Herbivores

Herbivory, animals eating plants, is ubiquitous in nature. Plant defences against herbivores are examples of how biological processes can be observed at multiple levels of biological organisation: molecular, cellular, tissue, organ, organism, population, and community. (See Figure 1.3.)

Opium poppy fruit

Molecular-Level Defences

At the molecular level, plants produce chemical compounds that deter attackers. These compounds are typically terpenoids, phenolics, and alkaloids. Some terpenoids mimic insect hormones and cause insects to molt prematurely and die. Some examples of phenolics are tannins, which have an unpleasant taste and hinder the digestion of proteins. Their synthesis is often enhanced following attack. The opium poppy (*Papaver somniferum*) is the source of the narcotic alkaloids morphine, heroin, and codeine. These drugs accumulate in secretory cells called laticifers, which exude a milky-white latex (opium) when the plant is damaged.

Cellular-Level Defences

Some plant cells are specialised for deterring herbivores. Trichomes on leaves and stems hinder the access of chewing insects. Laticifers and, more generally, the central vacuoles of plant cells may serve as storage depots for chemicals that deter herbivores. *Idioblasts* are specialised cells found in the leaves and stems of many species, including taro (*Colocasia esculenta*). Some idioblasts contain needle-shaped crystals of calcium oxalate called *raphides*. They penetrate the soft tissues of the tongue and palate, making it easier for an irritant produced by the plant, possibly a protease, to enter animal tissues and cause temporary swelling of the lips, mouth, and throat. The crystals act as a carrier for the irritant, enabling it to seep deeper into the herbivore's tissues. The irritant is destroyed by cooking.

Raphide crystals from taro plant

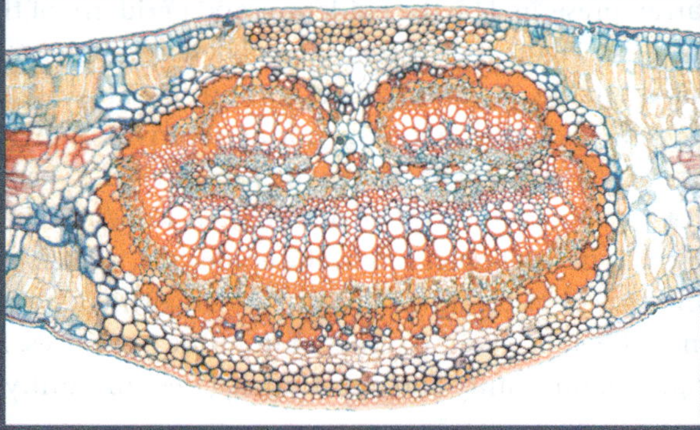

Tissue-Level Defences

Some leaves deter herbivores by being especially tough to chew as a result of extensive growth of thick, hardened sclerenchyma tissue. The bright orange cells with thick cell walls seen in this cross section through the major vein of an olive leaf (*Olea europaea*) are tough sclerenchyma fibres.

Organ-Level Defences

The shapes of plant organs may deter herbivores by causing pain or making the plant appear unappealing. Spines (modified leaves) and thorns (modified stems) provide mechanical defences against herbivores. Bristles on the spines of some cacti have fearsome barbs that tear flesh during removal. The leaf of the snowflake plant (*Trevesia palmata*) looks as if it has been partially eaten, perhaps making it less attractive. Some plants mimic the presence of insect eggs on their leaves, dissuading insects from laying eggs there. For example, the leaf glands of some species of *Passiflora* (passion flowers) closely imitate the bright yellow eggs of *Heliconius* butterflies.

Bristles on cactus spines

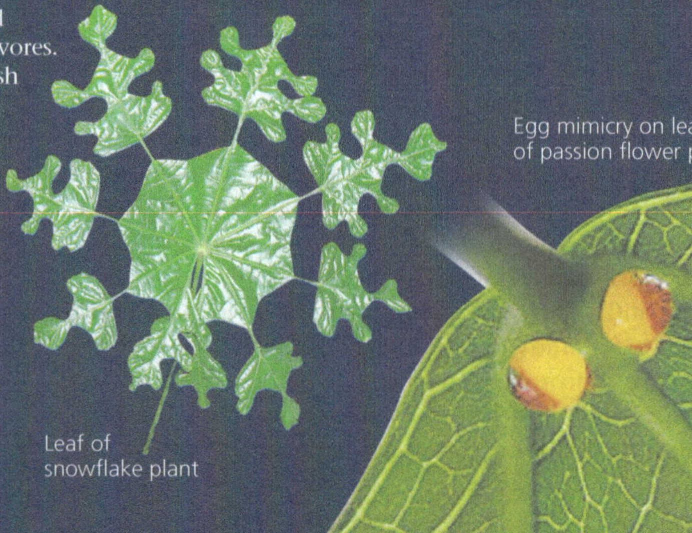

Leaf of snowflake plant

Egg mimicry on leaf of passion flower plant

Organismal-Level Defences

Mechanical damage by herbivores can greatly alter a plant's entire physiology, deterring further attack. For example, a species of wild tobacco called *Nicotiana attenuata* changes the timing of its flowering as a result of herbivory. It normally flowers at night, emitting the chemical benzyl acetone, which attracts hawk-moths as pollinators. Unfortunately for the plant, the moths often lay eggs on the leaves as they pollinate, and the larvae are herbivores. When the plants become too larvae-infested, they stop producing the chemical and instead open their flowers at dawn, when the moths are gone. They are then pollinated by hummingbirds. Research has shown that oral secretions from the munching larvae trigger the dramatic shift in the timing of flower opening.

Hummingbird pollinating wild tobacco plant

Population-Level Defences

In some species, a coordinated behaviour at the population level helps defend against herbivores. Some plants can communicate their distress from attack by releasing molecules that warn nearby plants of the same species. For example, lima bean (*Phaseolus lunatus*) plants infested with spider mites release a cocktail of chemicals that signal "news" of the attack to noninfested lima bean plants. In response, these neighbours instigate biochemical changes that make them less susceptible to attack. Another type of population-level defence is a phenomenon in some species called masting, in which a population synchronously produces a massive amount of seeds after a long interval. Regardless of environmental conditions, an internal clock signals each plant in the population that it is time to flower. Bamboo populations, for example, grow vegetatively for decades and suddenly flower en masse, set seed, and die. As much as 80,000 kg of bamboo seeds are released per hectare, much more than the local herbivores, mostly rodents, can eat. As a result, some seeds escape the herbivores' attention, germinate, and grow.

Flowering bamboo plants

Community-Level Defences

Some plant species "recruit" predatory animals that help defend the plant against specific herbivores. Parasitoid wasps, for example, inject their eggs into caterpillars feeding on plants. The eggs hatch within the caterpillars, and the larvae eat through their organic containers from the inside out. The larvae then form cocoons on the surface of the host before emerging as adult wasps. The plant has an active role in this drama. A leaf damaged by caterpillars releases compounds that attract parasitoid wasps. The stimulus for this response is a combination of physical damage to the leaf caused by the munching caterpillar and a specific compound in the caterpillar's saliva.

Parasitoid wasp cocoons on caterpillar host

Adult wasp emerging from a cocoon

MAKE CONNECTIONS

As with plant adaptations against herbivores, other biological processes can involve multiple levels of biological organisation (Figure 1.3). Discuss examples of specialised photosynthetic adaptations involving modifications at the molecular (Concept 10.5), tissue (Concept 36.4), and organismal (Concept 36.1) levels.

▼ Figure 39.26 Defence responses against pathogens. Plants can often prevent the systemic spread of infection by instigating a hypersensitive response. This response helps isolate the pathogen by producing lesions that form "rings of death" around the sites of infection.

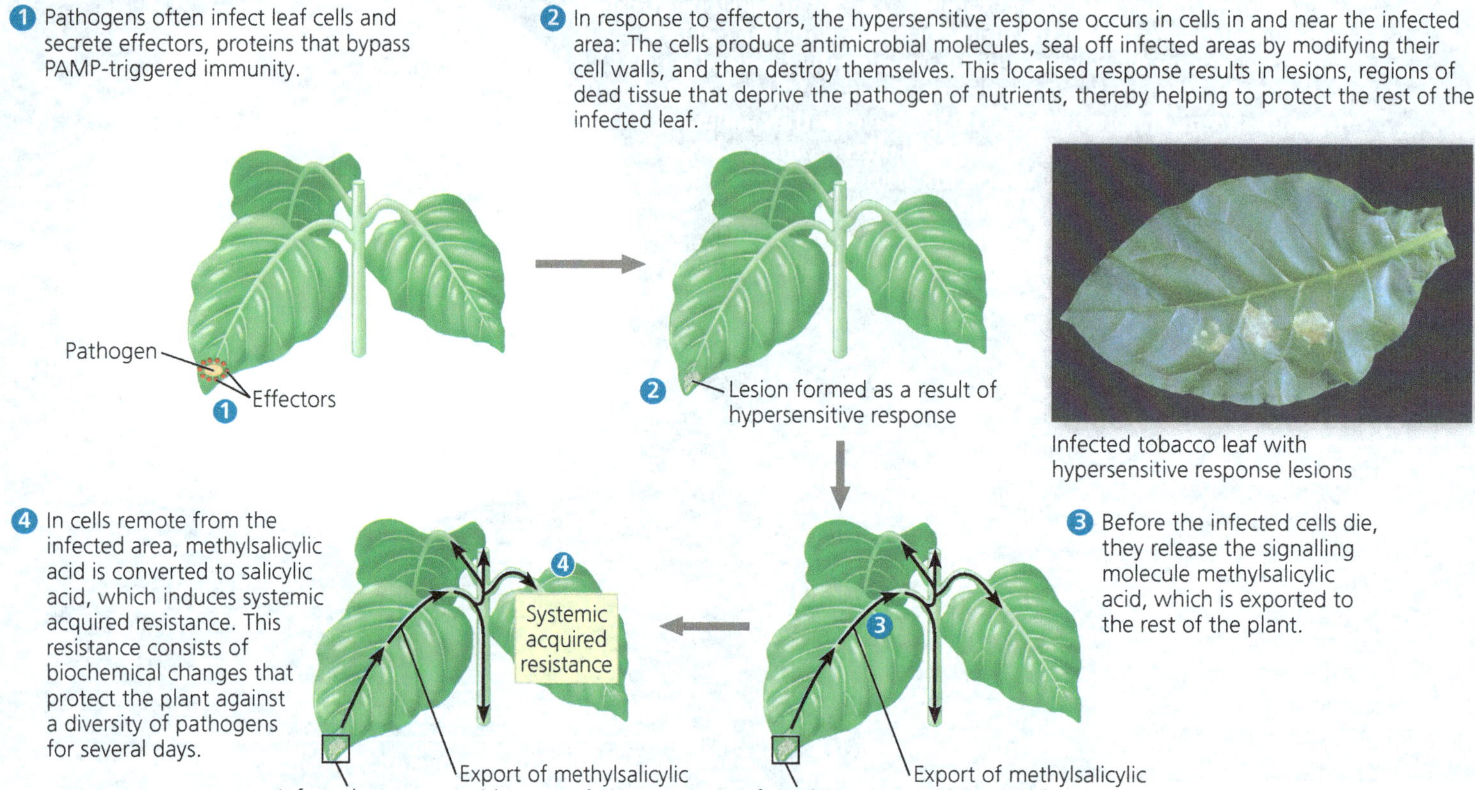

of pathogens that can last for days. A signalling molecule called methylsalicylic acid is produced around the infection site, carried by the phloem throughout the plant, and then converted to **salicylic acid** in areas remote from the sites of infection. Salicylic acid activates a signal transduction pathway that poises the defence system to respond rapidly to another infection (see step 4 of Figure 39.26).

Plant disease epidemics, such as the potato blight (see Concept 28.3) that caused the Irish potato famine of the 1840s, can lead to incalculable human misery. Other diseases, such as chestnut blight (see Concept 31.5) and sudden oak death (see Concept 54.5), can dramatically alter community structures. Plant epidemics are often the result of infected plants or timber being inadvertently transported around the world. As global commerce increases, such epidemics will become increasingly more common. To prepare for such outbreaks, plant biologists are stockpiling the seeds of wild relatives of crop plants in special storage facilities. Scientists hope that undomesticated relatives may have genes that will be able to curb the next plant epidemic.

Defences Against Herbivores

Herbivory, animals eating plants, is a stress that plants face in any ecosystem. The mechanical damage caused by herbivores reduces the size of plants, hindering ability to acquire resources. It can restrict growth because many species divert some energy to defend against herbivores. Also, it opens portals for infection by viruses, bacteria, and fungi. Plants prevent excessive herbivory through methods that span all levels of biological organisation **(Figure 39.27)**, including physical defences, such as thorns and trichomes (see Figure 35.9), and chemical defences, such as distasteful or toxic compounds.

CONCEPT CHECK 39.5

1. Why do pathogen-infected leaves often appear spotted?
2. Chewing insects mechanically damage plants and lessen the surface area of leaves for photosynthesis. In addition, these insects make plants more vulnerable to pathogen attack. Suggest a reason why.
3. Many fungal pathogens get food by causing plant cells to become leaky, releasing nutrients into the intercellular spaces. Would it benefit the fungus to kill the host plant in a way that results in all the nutrients leaking out? Explain.
4. **WHAT IF?** Suppose a scientist finds that a population of plants growing in a breezy location is more prone to herbivory by insects than a population of the same species growing in a sheltered area. Suggest an explanation.

For suggested answers, see Appendix A.

39 Chapter Review

SUMMARY OF KEY CONCEPTS

CONCEPT 39.1

Signal transduction pathways link signal reception to response *(pp. 892–894)*

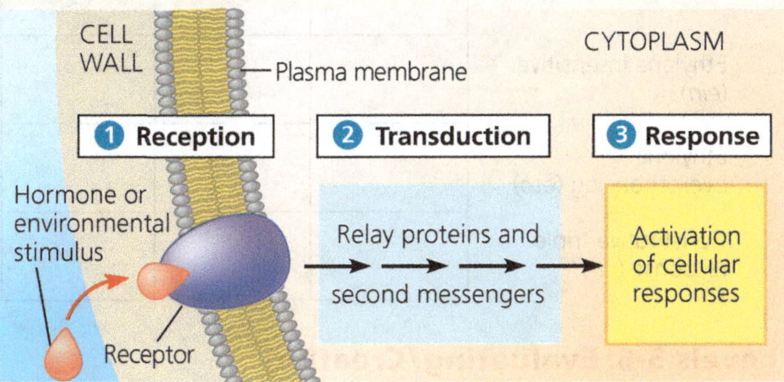

? *What are two common ways by which signal transduction pathways enhance the activity of specific enzymes?*

CONCEPT 39.2

Plants use chemicals to communicate *(pp. 894–904)*

- **Hormones** control plant growth and development by affecting the division, elongation, and differentiation of cells. Some also mediate the responses of plants to environmental stimuli.

Plant Hormone	Major Responses
Auxin (IAA)	Stimulates cell elongation; regulates branching and organ bending (phototropism and gravitropism)
Cytokinins	Stimulate plant cell division and differentiation; promote axillary bud growth
Gibberellins	Promote stem elongation; help seeds break dormancy and use stored reserves
Abscisic acid (ABA)	Promotes stomatal closure in response to drought; promotes seed dormancy
Ethylene	Mediates senescence, leaf abscission, fruit ripening, and obstacle avoidance by shoots (the triple response)
Brassinosteroids	Chemically similar to the sex hormones of animals; induce cell elongation and division
Jasmonates	Mediate plant defences against insect herbivores; regulate a wide range of physiological processes
Strigolactones	Regulate apical dominance, seed germination, and mycorrhizal associations

? *Is there any truth to the old saying "One bad apple spoils the whole bunch?" Explain.*

CONCEPT 39.3

Responses to light are critical for plant success *(pp. 904–910)*

- **Blue-light photoreceptors** control hypocotyl elongation, stomatal opening, and phototropism.
- **Phytochromes** act like molecular "on-off" switches that regulate shade avoidance and germination of many seed types. Red light turns phytochrome "on," and far-red light turns it "off."

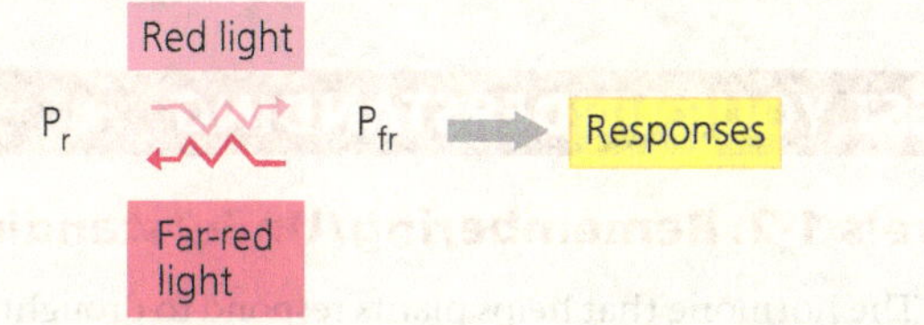

- Phytochrome conversion also provides information about the day length (photoperiod) and hence the time of year. **Photoperiodism** regulates the time of flowering in many species. **Short-day plants** require a night longer than a critical length to flower. **Long-day plants** need a night length shorter than a critical period to flower.
- Many daily rhythms in plant behaviour are controlled by an internal circadian clock. Free-running **circadian rhythms** are approximately 24 hours long but are entrained to exactly 24 hours by dawn and dusk effects on phytochrome form.

? *Why did plant physiologists propose the existence of a mobile molecule (florigen) that triggers flowering?*

CONCEPT 39.4

Plants respond to a wide variety of stimuli other than light *(pp. 910–914)*

- **Gravitropism** is bending in response to gravity. Roots show positive gravitropism, and stems show negative gravitropism. **Statoliths**, starch-filled plastids, enable roots to detect gravity.
- **Thigmotropism** is a growth response to touch. Rapid leaf movements involve transmission of electrical impulses.
- Plants are sensitive to environmental stresses, including drought, flooding, high salinity, and extremes of temperature.

Environmental Stress	Major Response
Drought	ABA production, reducing water loss by closing stomata
Flooding	Formation of air tubes that help roots survive oxygen deprivation
Salt	Avoiding osmotic water loss by producing solutes tolerated at high concentrations
Heat	Synthesis of heat-shock proteins, which reduce protein denaturation at high temperatures
Cold	Adjusting membrane fluidity; avoiding osmotic water loss; producing antifreeze proteins

? *Plants that have acclimated to drought stress are often more resistant to freezing stress as well. Suggest a reason why.*

CONCEPT 39.5

Plants respond to attacks by pathogens and herbivores *(pp. 915–918)*

- The **hypersensitive response** seals off an infection and destroys both pathogen and host cells in the region. **Systemic acquired resistance** is a generalised defence response in organs distant from the infection site.
- In addition to physical defences such as thorns and trichomes, plants produce distasteful or toxic chemicals, as well as attractants that recruit animals that destroy herbivores.

? *How can insects make plants more susceptible to pathogens?*

TEST YOUR UNDERSTANDING

Levels 1-2: Remembering/Understanding

1. The hormone that helps plants respond to drought is
(A) auxin.
(B) abscisic acid.
(C) cytokinin.
(D) ethylene.

2. A barley mutant lacking a gibberellic acid receptor would
(A) fail to make GA.
(B) catalyse starch more quickly.
(C) fail to make α-amylase.
(D) fail to take up water.

3. Charles and Francis Darwin discovered that
(A) auxin is responsible for phototropic curvature.
(B) red light is most effective in shoot phototropism.
(C) light destroys auxin.
(D) light is perceived by the tips of coleoptiles.

4. How may a plant respond to *severe* heat stress?
(A) by reorienting leaves to increase evaporative cooling
(B) by creating air tubes for ventilation
(C) by producing heat-shock proteins, which may protect the plant's proteins from denaturing
(D) by increasing the proportion of unsaturated fatty acids in cell membranes, reducing their fluidity

Levels 3-4: Applying/Analysing

5. The signalling molecule for flowering might be released earlier than usual in a long-day plant exposed to flashes of
(A) far-red light during the night.
(B) red light during the night.
(C) red light followed by far-red light during the night.
(D) far-red light during the day.

6. If a long-day plant has a critical night length of 9 hours, which 24-hour cycle would prevent flowering?
(A) 16 hours light/8 hours dark
(B) 14 hours light/10 hours dark
(C) 4 hours light/8 hours dark/4 hours light/8 hours dark
(D) 8 hours light/8 hours dark/light flash/8 hours dark

7. A plant mutant that shows normal gravitropic bending but does not store starch in its plastids would require a reevaluation of the role of ________ in gravitropism.
(A) auxin
(B) calcium
(C) statoliths
(D) differential growth

8. DRAW IT Indicate the response to each condition by drawing a straight seedling or one with the triple response.

	Control	Ethylene added	Ethylene synthesis inhibitor
Wild-type			
Ethylene insensitive (*ein*)			
Ethylene overproducing (*eto*)			
Constitutive triple response (*ctr*)			

Levels 5-6: Evaluating/Creating

9. EVOLUTION CONNECTION In general, light-sensitive germination is more pronounced in small seeds compared with germination of large seeds. Suggest a reason why.

10. SCIENTIFIC INQUIRY A plant biologist observed a peculiar pattern when a tropical shrub was attacked by caterpillars. After a caterpillar ate a leaf, it would skip over nearby leaves and attack a leaf some distance away. Simply removing a leaf did not deter caterpillars from eating nearby leaves. The biologist suspected that an insect-damaged leaf sent out a chemical that signalled nearby leaves. How could the researcher test this hypothesis?

11. SCIENCE, TECHNOLOGY, AND SOCIETY Describe how our knowledge about the control systems of plants is being applied to agriculture or horticulture.

12. WRITE ABOUT A THEME: INTERACTIONS In a short essay (100–150 words), summarise phytochrome's role in altering shoot growth for the enhancement of light capture.

13. SYNTHESISE YOUR KNOWLEDGE

In 1837, Australia's brushtail possum was introduced to New Zealand, where it favoured eating the shoot tips of native plants. Describe how this event would alter the physiology, biochemistry, structure, and health of the plant, and identify which hormones and other chemicals are involved in making these changes.

For selected answers, see Appendix A.

Unit 7 ANIMAL FORM AND FUNCTION

Meet the "phage wrangler," Steffanie Strathdee, Professor of Medicine in the Division of Infectious Diseases and Global Public Health at the University of California, San Diego, in the United States. Born in Canada, she completed her MSc and PhD degrees in epidemiology at the University of Toronto. Since joining the faculty at UC San Diego in 2004, she has focused on HIV research and prevention in the underserved populations of Tijuana, just across the Mexican border. In 2017, extreme personal circumstances prompted a radical shift in her research, as detailed in *The Perfect Predator: A Scientist's Race to Save Her Husband from a Deadly Superbug*, a book co-authored by Dr Strathdee and her husband. In 2018, Dr Strathdee became co-director of the newly established Center for Innovative Phage Applications and Therapeutics at UC San Diego.

AN INTERVIEW WITH

Steffanie Strathdee

Tell us about your start in science.

I always had a natural curiosity as a young girl. I did not have an innate ability at science or math, even though I was interested in them. I struggled in math. Even in university, I got D's in calculus. I still don't like math, but I've learned to surround myself with people that do really well in math. I think it's important for students to realise that just because they're not good at everything doesn't mean that they can't pursue their dreams.

How did you become interested in epidemiology?

After both my course instructor and my Masters and PhD advisor passed away from AIDS, I decided I wanted to focus on ending the HIV epidemic. I had started out in the laboratory and realised I was really lousy at tissue culture experiments. So I turned my attention to public health, of which epidemiology is one part. Epidemiology involves studying risk factors and patterns, not just at an individual level, like what people eat and how they behave, but also the social, political, and economic forces that drive those behaviours.

▼ Dr Steffanie Strathdee's husband, Dr Tom Patterson, holding an electron micrograph of his superbug (left), and Dr Strathdee, with an artist's rendition of the bacteriophage that defeated it (right).

What changed your research focus?

My husband became ill with a superbug infection—an infection caused by a bacterium that is resistant to many antibiotics. As an infectious disease epidemiologist, you would think that I would have understood the global crisis we are facing in multidrug-resistant organisms. But it wasn't until it hit me on a personal level that I fully understood the threat.

How did the story of your husband's illness unfold?

My husband and I were in Egypt, and he became very ill. The doctor gave him IV antibiotics and said, "He'll be fine," but he wasn't. It turned out that a gallstone had blocked his bile duct, causing an abscess (cavity) to form, and a superbug moved into that abscess and multiplied. Tom's particular bacterium had 51 different antibiotic-resistance genes. It was winning its battle against the immune system, none of the standard antibiotic treatments were working, and my husband was dying. I did a literature search and found a hundred-year-old, forgotten cure based on bacteriophages. These viruses, called phages for short, attack bacteria, but not human cells. Each phage is specific for a particular bacterium. Phages inject their DNA into bacteria, turning them into phage factories and killing the bacteria in the process. I emailed Tom's doctor, who is a colleague of Tom's and mine, and he thought the idea was worth a shot. So, I had to go out and find people who had phages that were active against Tom's bacterium, and that felt harder than looking for a needle in a haystack. Luckily, labs at the Naval Medical Research Center in Maryland and at Texas A&M University dedicated themselves to the search and found phages that matched. The phages were grown, purified, and injected into Tom's body, a billion phages per dose. Three days later, Tom woke up from his coma.

Tell us a bit more about your collaborators.

A PhD student at Texas A&M worked around the clock and found the phages that made up the first infusion. This student was at a really low point in her studies, thinking, "I don't know what I'm going to do with my career, I don't know if what I do really matters." And then she ended up helping save the life of a total stranger, and it's jump-started a whole new field—phage therapy.

> *"This student . . . ended up helping save the life of a total stranger, and it's jump-started a whole new field."*

What advice do you have for students starting to study biology?

Students should look for mentors with whom they share the same values. For me, that was what really turned my D in calculus and some of the troubles that I had in science into success. I found mentors who were encouraging and helped me identify my strengths. Knowing what your weaknesses and your strengths are is, I think, one of the big components to making a successful career.

40 Basic Principles of Animal Form and Function

KEY CONCEPTS

Study Tip

Draw a diagram: When you encounter an example in the chapter of how an animal maintains a steady internal state, draw a simple circuit diagram (see example—illustrations are optional!). Label the variable being controlled, a perturbation that affects the variable, the response, and its effect in restoring the normal state.

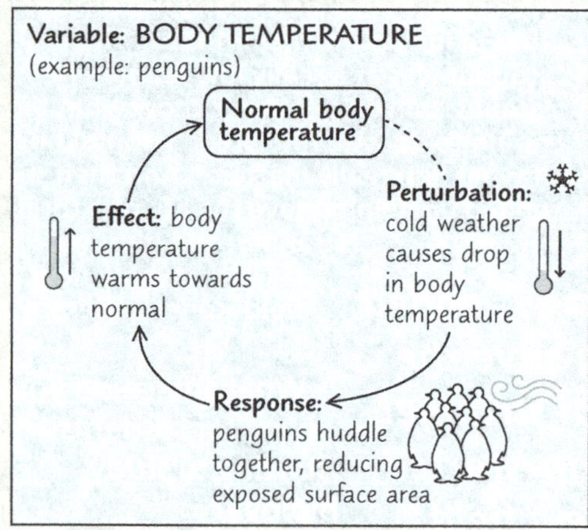

Go to Mastering Biology

to access Dynamic Study Modules for revision, 3D BioFlix® animations and high-quality videos, and your interactive Pearson eText.

Figure 40.1 Emperor penguins (*Aptenodytes forsteri*) live in Antarctica, Earth's coldest and windiest continent. In summer, these birds catch fish by diving down 500 m in water only 2°C above freezing. In winter, the females forage and the males incubate eggs while temperatures drop to –40°C and winds gust to 200 km per hour.

How do animals regulate their internal state even in changing or harsh environments?

Adaptations in **form, function, and behaviour** help maintain an animal's internal environment. Adaptations that limit variation in temperature and other internal variables are widespread and diverse. Consider, for example, three adaptations that help an Emperor penguin stay warm:

Form (anatomy): An insulating layer of fat (blubber) reduces heat loss from most of the penguin's body (blue body areas in this thermal image).

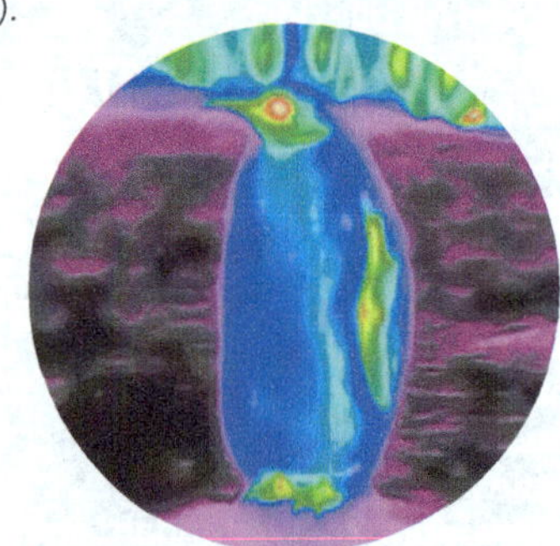

Function (physiology): Rapid cycles of muscle contraction and relaxation during shivering produce heat at a cellular level.

Behaviour: By packing together in groups of up to several thousand, Emperor penguins greatly reduce their exposure to wind and cold.

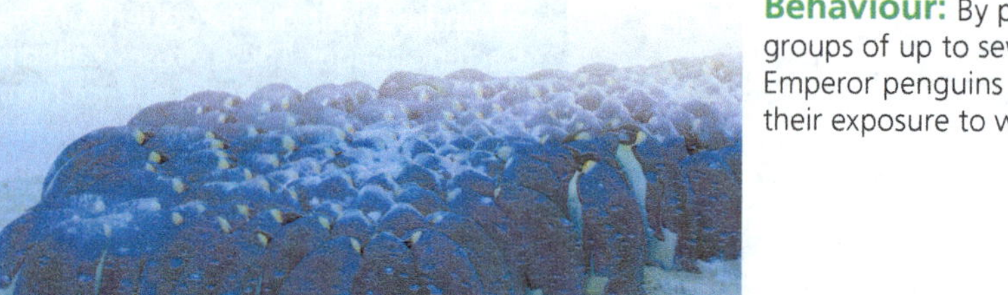

CONCEPT 40.1

Animal form and function are correlated at all levels of organisation

Over the course of its life, an Emperor penguin faces the same fundamental challenges as any other animal, whether bilby, kiwi, or human. All animals must obtain nutrients and oxygen, fight off infection, and survive to produce offspring. Given that all animal species share these and other basic requirements, why does their form, including **anatomy**—biological structure—vary so widely? The answer lies in natural selection and adaptation. Natural selection favours those variations in a population that increase relative fitness (see Concept 23.4). The evolutionary adaptations that enable survival vary among environments and species but frequently result in a close match of form to function, as illustrated for the Emperor penguins in Figure 40.1. Because structure and function are correlated, examining anatomy often provides clues to **physiology**—biological function.

An animal's size and shape are fundamental aspects of form that significantly affect the way the animal interacts with its environment. Although we may refer to size and shape as elements of a "body plan" or "design," this does not imply a process of conscious invention. The body plan of an animal is the result of a pattern of development programmed by the genome, itself the product of millions of years of evolution.

Evolution of Animal Size and Shape

EVOLUTION Many different body plans have arisen during the course of evolution, but these variations fall within certain bounds. Physical laws that govern strength, diffusion, movement, and heat exchange limit the range of animal forms.

As an example of how physical laws constrain evolution, let's consider how some properties of water limit the possible shapes for animals that are fast swimmers. Water is about 1,000 times denser than air and also far more viscous. Therefore, any bump on an animal's body surface that causes drag impedes a swimmer more than it would a runner or flyer. Tuna and other fast ray-finned fishes can swim at speeds up to 80 km per hour. Sharks, penguins, dolphins, and seals are also relatively fast swimmers. As illustrated by the three examples in **Figure 40.2**, these animals all have a shape that is fusiform, meaning tapered on both ends. The similar streamlined shape found in these speedy vertebrates is an example of convergent evolution (see Concept 22.3). Natural selection often results in similar adaptations when diverse organisms face the same environmental challenge, such as overcoming drag during swimming.

▼ Figure 40.2 Convergent evolution in fast swimmers.

Physical laws also influence animal body plans with regard to maximum size. As body dimensions increase, thicker skeletons are required to maintain adequate support. This limitation affects internal skeletons, such as those of vertebrates, as well as external skeletons, such as those of insects and other arthropods. In addition, as bodies increase in size, the muscles required for locomotion must represent an ever-larger fraction of the total body mass. At some point, mobility becomes limited. By considering the fraction of body mass in leg muscles and the effective force such muscles generate, scientists can estimate maximum speed for a wide range of body plans. In the case of the 6-metre-tall dinosaur *Tyrannosaurus rex*, there is controversy, with some scientists calculating a top running speed as fast as that of an Olympic sprinter—30 km per hour, but others inferring that *T. rex* was at best a fast walker.

Exchange with the Environment

Animals must exchange nutrients, waste products, and gases with their environment, and this requirement imposes an additional limitation on body plans. Exchange occurs as substances dissolved in an aqueous solution move across the plasma membrane of each cell. A single-celled organism, such as the amoeba in **Figure 40.3a**, has a sufficient membrane surface area in contact with its environment to carry out all necessary exchange. In contrast, an animal is composed of many cells, each with its own plasma membrane across which exchange must occur. The rate of exchange is proportional to the membrane surface area involved in exchange, whereas the amount of material that must be exchanged is proportional to the total body volume. A multicellular organisation therefore works only if every cell has access to a suitable aqueous environment, either inside or outside the animal's body.

Many animals with a simple internal organisation have body plans that enable direct exchange between almost all their cells and the external environment. For example, a pond-dwelling hydra has a saclike body plan and a body wall only two cell layers thick **(Figure 40.3b)**. Because its gastrovascular cavity opens to the external environment, both the outer and inner layers of cells are constantly bathed by pond water. Another common body plan that maximises exposure to the surrounding medium is a flat shape. Consider, for instance, a parasitic tapeworm, which can reach several metres in length (see Figure 33.11). A thin, flat shape places most cells of the worm in direct contact with its particular environment—the nutrient-rich intestinal fluid of a vertebrate host.

▼ **Figure 40.3 Direct exchange with the environment.**

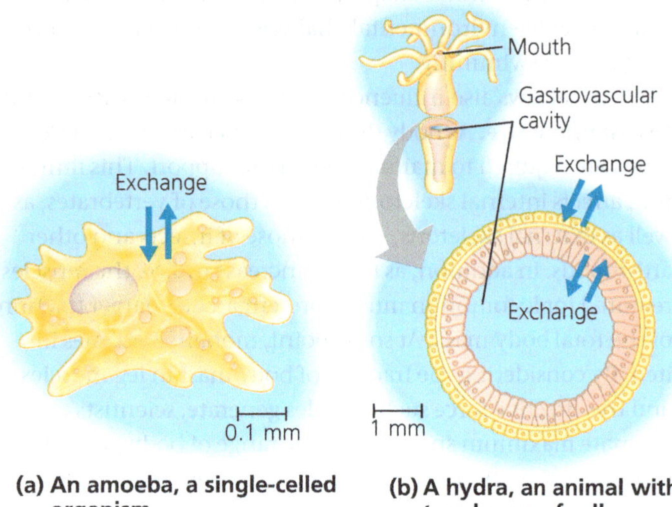

(a) An amoeba, a single-celled organism

(b) A hydra, an animal with two layers of cells

Our bodies and those of most other animals are composed of compact masses of cells, with an internal organisation much more complex than that of a hydra or a tapeworm. For such a body plan, increasing the number of cells decreases the ratio of outer surface area to total volume. As an extreme comparison, the ratio of outer surface area to volume for a whale is hundreds of thousands of times smaller than that for a water flea. Nevertheless, every cell in the whale has access to oxygen, nutrients, and other resources. How is this accomplished?

In whales and most other animals, the evolutionary adaptations that enable sufficient exchange with the environment are specialised surfaces that are extensively branched or folded **(Figure 40.4)**. In almost all cases, these exchange surfaces lie within the body, an arrangement that protects their delicate tissues from abrasion or dehydration and allows for streamlined body contours. The branching or folding greatly increases surface area (see Figure 33.8). In humans, for example, the exchange surfaces for digestion, respiration, and circulation each have an area more than 25 times larger than that of the skin.

Internal body fluids link exchange surfaces to body cells. The spaces between cells are filled with fluid, known in many animals as **interstitial fluid** (from the Latin for "stand between"). Complex body plans also include a circulatory fluid, such as blood. Exchange between the interstitial fluid and the circulatory fluid enables cells throughout the body to obtain nutrients and get rid of wastes (see Figure 40.4).

Complex body plans offer numerous benefits. For example, an external skeleton can protect against predators, and sensory organs can provide detailed information on the animal's surroundings. Internal digestive organs can break down

▶ **Figure 40.4 Internal exchange surfaces of complex animals.** Most animals have surfaces that are specialised for exchanging chemicals with the surroundings. These exchange surfaces are usually internal but are connected to the environment via openings on the body surface (the mouth, for example). The exchange surfaces are finely branched or folded, giving them a very large area. The digestive, respiratory, and excretory systems all have such exchange surfaces. Chemicals exchanged across these surfaces are transported throughout the body via the circulatory system.

VISUAL SKILLS *Using this diagram, explain how exchange carried out by animals can be described as both internal and external.*

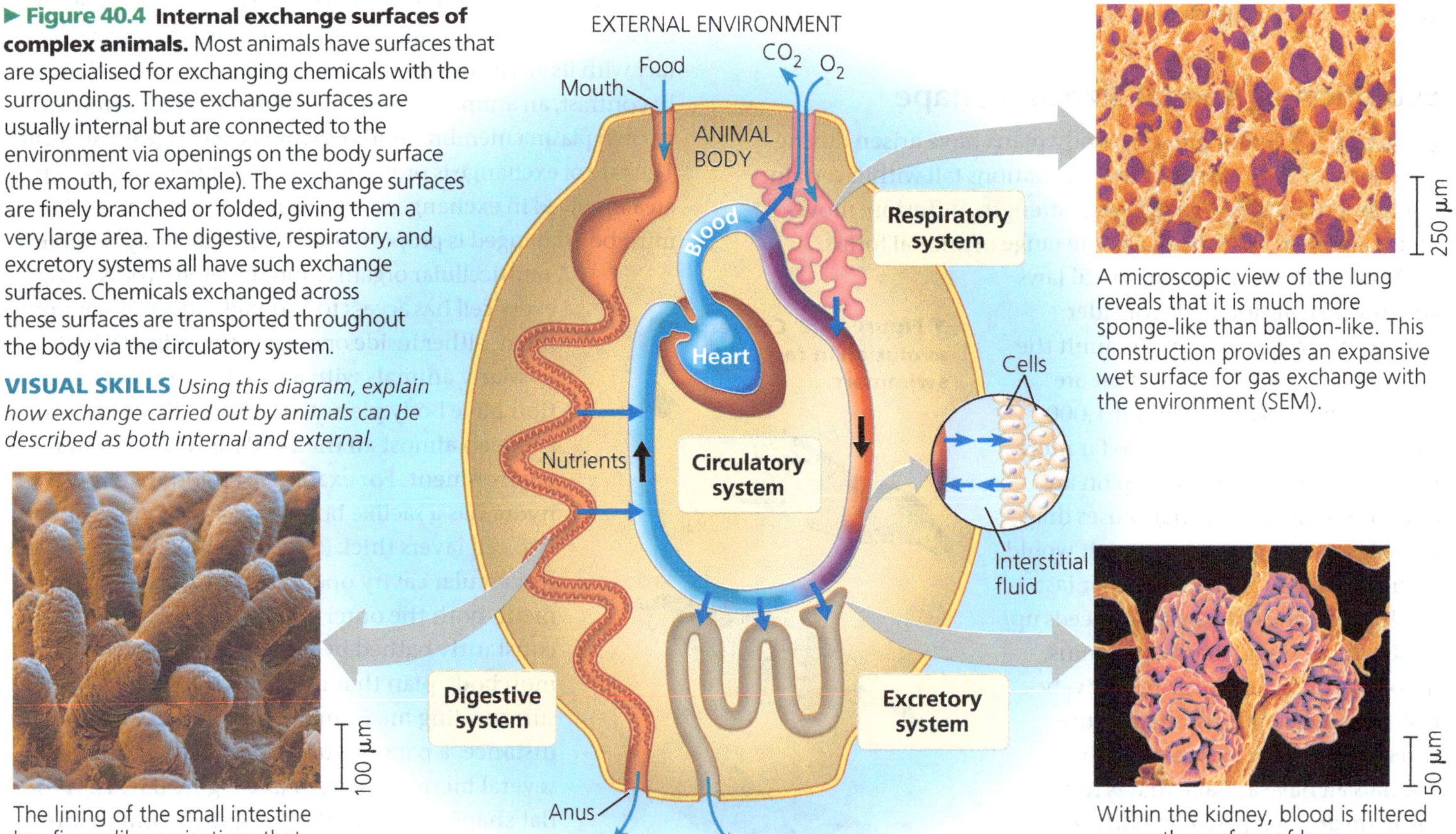

EXTERNAL ENVIRONMENT
Food
Mouth
CO_2 O_2
ANIMAL BODY
Blood
Respiratory system
Heart
Cells
Nutrients
Circulatory system
Interstitial fluid
Digestive system
Excretory system
Anus
Unabsorbed matter (faeces)
Metabolic waste products (nitrogenous waste)
250 µm
100 µm
50 µm

A microscopic view of the lung reveals that it is much more sponge-like than balloon-like. This construction provides an expansive wet surface for gas exchange with the environment (SEM).

The lining of the small intestine has finger-like projections that expand the surface area for nutrient absorption (SEM).

Within the kidney, blood is filtered across the surface of long, narrow blood vessels packed into ball-shaped structures (SEM).

food gradually, controlling the release of stored energy. In addition, specialised filtration systems can adjust the composition of the internal fluid that bathes the animal's body cells. In this way, an animal can maintain a relatively stable internal environment despite the fact that it is living in a changeable external environment. A complex body plan is especially advantageous for animals living on land, where the external environment may be highly variable.

Hierarchical Organisation of Body Plans

Cells form a working animal body through their *emergent properties*, which arise from successive levels of structural and functional organisation (see Concept 1.1). Cells are organised into **tissues**, groups of cells with a similar appearance and a common function. Different types of tissues are further organised into functional units called **organs**. (The simplest animals, such as sponges, lack organs or even true tissues.) Groups of organs that work together, providing an additional level of organisation and coordination, make up an **organ system** **(Table 40.1)**. Thus, for example, the skin is an organ of the integumentary system, which protects against infection and helps regulate body temperature.

Many organs have more than one physiological role. If the roles are distinct enough, we consider the organ to belong to more than one organ system. The pancreas, for instance, produces enzymes critical to the function of the digestive system but also regulates the level of sugar in the blood as a vital part of the endocrine system.

Just as viewing the body's organisation from the "bottom up" (from cells to organ systems) reveals emergent properties, a "top-down" view of the hierarchy reveals the multilayered basis of specialisation. Organ systems include specialised organs made up of specialised tissues and cells. Consider the human digestive system: Each organ has specific roles. In the case of the stomach, one role is to initiate protein breakdown. This process requires a churning motion powered by stomach muscles, as well as digestive juices secreted by the stomach lining. Producing digestive juices, in turn, requires highly specialised cell types: One cell type secretes a protein-digesting enzyme, a second generates concentrated hydrochloric acid, and a third produces mucus, which protects the stomach lining.

The specialised and complex organ systems of animals are built from a limited set of cell and tissue types. For example, lungs and blood vessels have different functions but are lined by tissues that are of the same basic type and therefore share many properties.

There are four main types of animal tissues: epithelial, connective, muscle, and nervous. **Figure 40.5** explores the structure and function of each type. In later chapters, we'll discuss how these tissue types contribute to the functions of particular organ systems.

Table 40.1 Organ Systems in Mammals

Organ System	Main Components	Main Functions
Digestive	Mouth, pharynx, esophagus, stomach, intestines, liver, pancreas, anus (See Figure 41.8.)	Food processing (ingestion, digestion, absorption, elimination)
Circulatory	Heart, blood vessels, blood (See Figure 42.5.)	Internal distribution of materials
Respiratory	Lungs, trachea, gills, other breathing tubes (See Figures 42.22–42.24.)	Gas exchange (uptake of oxygen; disposal of carbon dioxide)
Immune and lymphatic	Bone marrow, lymph nodes, thymus, spleen, lymph vessels (See Figure 43.6.)	Body defence (fighting infections and virally induced cancers)
Excretory	Kidneys, ureters, urinary bladder, urethra (See Figure 44.12.)	Disposal of metabolic wastes; regulation of osmotic balance of blood
Endocrine	Pituitary, thyroid, pancreas, adrenal, and other hormone-secreting glands (See Figure 45.8.)	Coordination of body activities (such as digestion and metabolism)
Reproductive	Ovaries or testes and associated organs (See Figures 46.12 and 46.13.)	Gamete production; promotion of fertilisation; support of developing embryo
Nervous	Brain, spinal cord, nerves, sensory organs (See Figure 49.6.)	Coordination of body activities; detection of stimuli and formulation of responses to them
Integumentary	Skin and its derivatives (such as hair, claws, sweat glands) (See Figure 50.5.)	Protection against mechanical injury, infection, dehydration; thermoregulation
Skeletal	Skeleton (bones, tendons, ligaments, cartilage) (See Figure 50.37.)	Body support, protection of internal organs, movement
Muscular	Skeletal muscles (See Figure 50.26.)	Locomotion and other movement

▼ Figure 40.5 Exploring Structure and Function in Animal Tissues

Epithelial Tissue

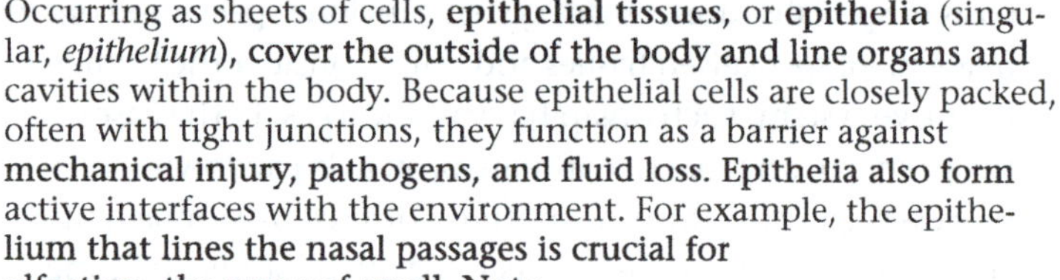

Occurring as sheets of cells, **epithelial tissues,** or **epithelia** (singular, *epithelium*), cover the outside of the body and line organs and cavities within the body. Because epithelial cells are closely packed, often with tight junctions, they function as a barrier against mechanical injury, pathogens, and fluid loss. Epithelia also form active interfaces with the environment. For example, the epithelium that lines the nasal passages is crucial for olfaction, the sense of smell. Note how different cell shapes and arrangements correlate with distinct functions.

Stratified squamous epithelium

Apical surface

Basal surface

A stratified squamous epithelium is multilayered and regenerates rapidly. New cells formed by division near the basal surface push outwards, replacing cells that are sloughed off. This epithelium is commonly found on surfaces subject to abrasion, such as the outer skin and the linings of the mouth, anus, and vagina.

Cuboidal epithelium

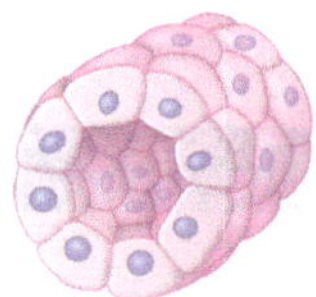

A cuboidal epithelium, with dice-shaped cells specialised for secretion, makes up the epithelium of kidney tubules and many glands, including the thyroid gland and salivary glands.

Simple columnar epithelium

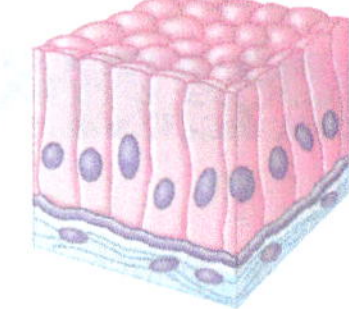

The large, brick-shaped cells of simple columnar epithelia are often found where secretion or active absorption is important. For example, a simple columnar epithelium lines the intestines, secreting digestive juices and absorbing nutrients.

Simple squamous epithelium

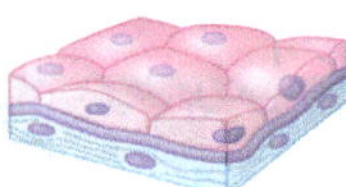

The single layer of platelike cells that form a simple squamous epithelium functions in the exchange of material by diffusion. This type of epithelium, which is thin and leaky, lines blood vessels and the air sacs of the lungs, where diffusion of nutrients and gases is essential.

Pseudostratified columnar epithelium

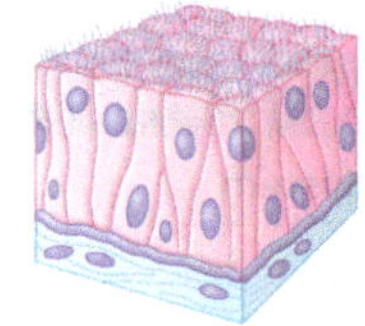

A pseudostratified epithelium consists of a single layer of cells varying in height and the position of their nuclei. In many vertebrates, a pseudostratified epithelium of ciliated cells forms a mucous membrane that lines portions of the respiratory tract. The beating cilia sweep the film of mucus along the surface.

Lumen

Apical surface

Basal surface

10 µm

Polarity of epithelia

All epithelia are polarised, meaning that they have two different sides. The *apical* surface faces the lumen (cavity) or outside of the organ and is therefore exposed to fluid or air. Specialised projections often cover this surface. For example, the apical surface of the epithelium lining the small intestine is covered with microvilli, projections that increase the surface area available for absorbing nutrients. Opposite the apical surface of each epithelium is the *basal* surface.

Connective Tissue

Connective tissue, consisting of a sparse population of cells scattered through an extracellular matrix, holds many tissues and organs together and in place. The matrix generally consists of a web of fibres embedded in a liquid, jellylike, or solid foundation. Within the matrix are numerous cells called **fibroblasts**, which secrete fibre proteins, and **macrophages**, which engulf foreign particles and any cell debris by phagocytosis.

Connective tissue fibres are of three kinds: *Collagenous fibres* provide strength and flexibility, *reticular fibres* join connective tissue to adjacent tissues, and *elastic fibres* make tissues elastic. If you pinch a fold of tissue on the back of your hand, the collagenous and reticular fibres prevent the skin from being pulled far from the bone, whereas the elastic fibres restore the skin to its original shape when you release your grip. Different mixtures of fibres and foundation form the major types of connective tissue shown below.

Loose connective tissue

The most widespread connective tissue in the vertebrate body is *loose connective tissue*, which binds epithelia to underlying tissues and holds organs in place. Loose connective tissue gets its name from the loose weave of its fibres, which include all three types. It is found in the skin and throughout the body.

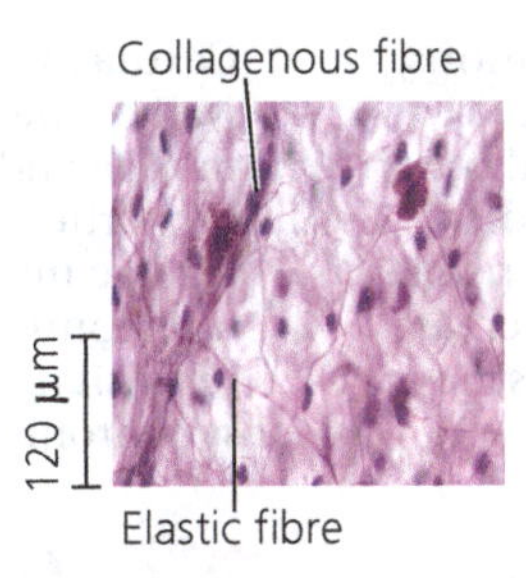

Fibrous connective tissue

Fibrous connective tissue is dense with collagenous fibres. It is found in **tendons**, which attach muscles to bones, and in **ligaments**, which connect bones at joints.

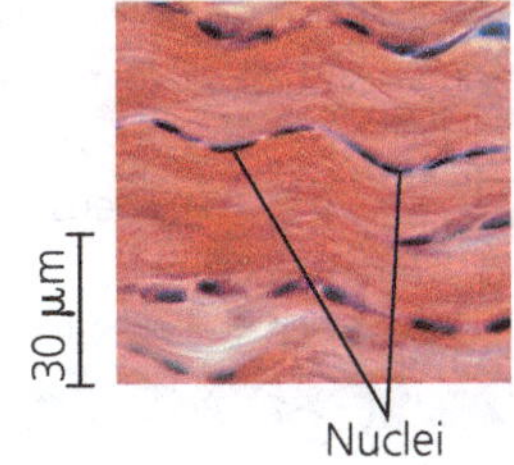

Bone

The skeleton of most vertebrates is made of **bone**, a mineralised connective tissue. Bone-forming cells called *osteoblasts* deposit a matrix of collagen. Calcium, magnesium, and phosphate ions combine into a hard mineral within the matrix. The microscopic structure of hard mammalian bone consists of repeating units called *osteons*. Each osteon has concentric layers of the mineralised matrix, which are deposited around a central canal containing blood vessels and nerves.

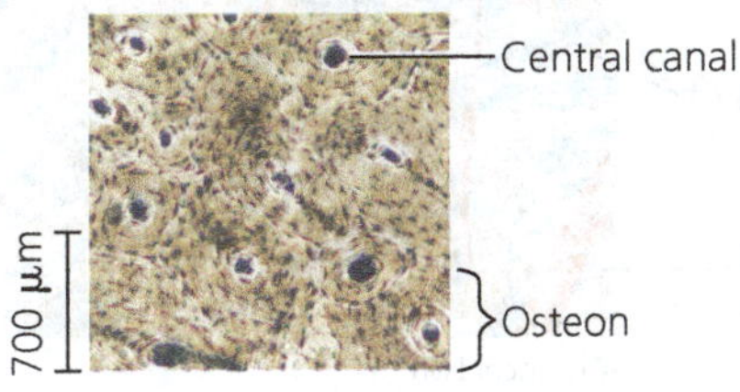

Adipose tissue

Adipose tissue is a specialised loose connective tissue that stores fat in adipose cells distributed throughout its matrix. Adipose tissue pads and insulates the body and stores fuel as fat molecules. Each adipose cell contains a large fat droplet that swells when fat is stored and shrinks when the body uses that fat as fuel.

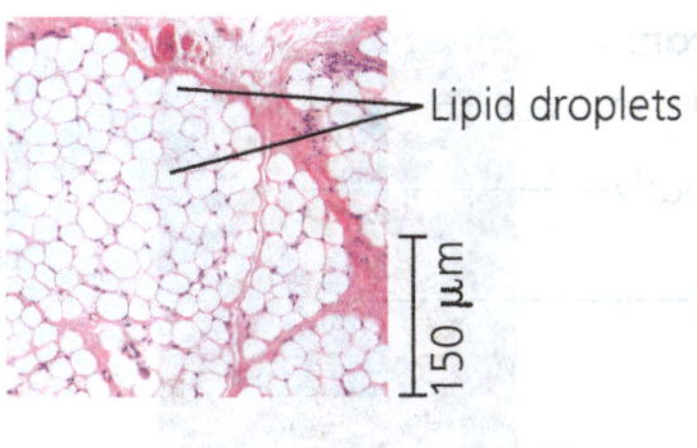

Blood

Blood has a liquid extracellular matrix called plasma, which consists of water, salts, and dissolved proteins. Suspended in plasma **are erythrocytes (red blood cells),** leucocytes (white blood cells), and cell fragments called platelets. Red cells carry oxygen, white cells function in defence, and platelets aid in blood clotting.

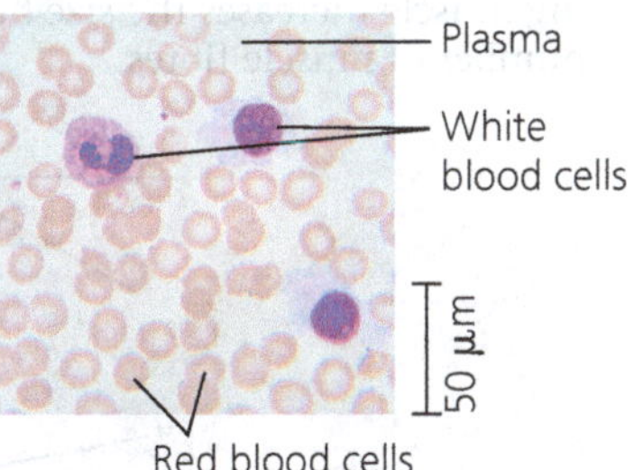

Cartilage

Cartilage contains collagenous fibres embedded in a rubbery protein-carbohydrate complex called chondroitin sulfate. Cells called *chondrocytes* secrete the collagen and chondroitin sulfate, which together make cartilage a strong yet flexible support material. The skeletons of many vertebrate embryos contain cartilage that is replaced by bone as the embryo matures. Cartilage remains in some locations, such as the disks that act as cushions between vertebrae.

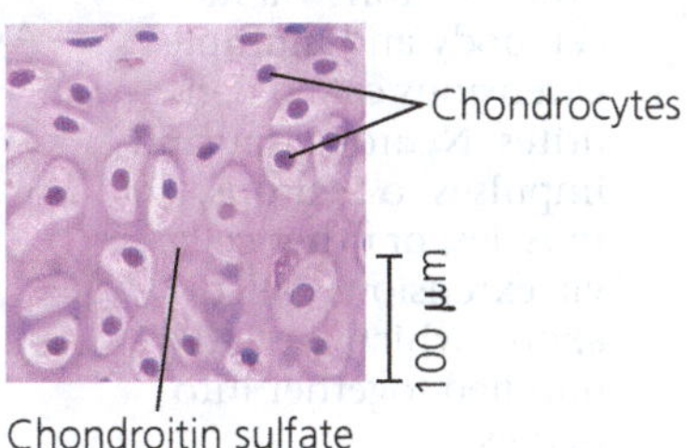

Muscle Tissue

The tissue responsible for nearly all types of body movement is **muscle tissue**. All muscle cells consist of filaments containing the proteins actin and myosin, which together enable muscles to contract. There are three types of muscle tissue in the vertebrate body: skeletal, smooth, and cardiac.

Skeletal muscle

Attached to bones by tendons, **skeletal muscle**, or *striated muscle*, is responsible for voluntary movements. Skeletal muscle consists of bundles of long cells that are called muscle fibres. During development, skeletal muscle fibres form by the fusion of many cells, resulting in multiple nuclei in each muscle fibre. The arrangement of contractile units, or sarcomeres, along the fibres gives the cells a striped (striated) appearance. In adult mammals, building muscle increases the size but not the number of muscle fibres.

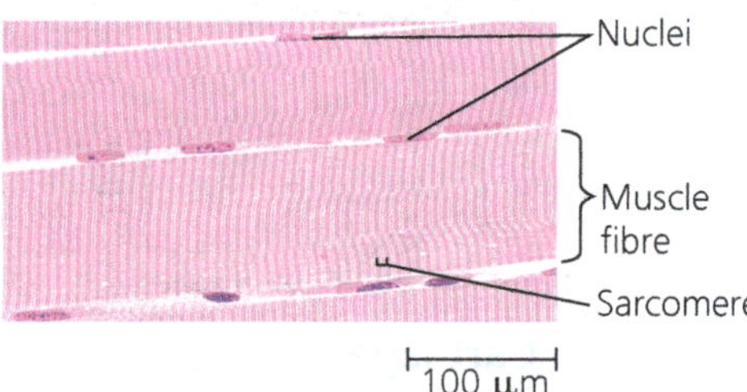

Smooth muscle

Smooth muscle, which lacks striations, is found in the walls of the digestive tract, urinary bladder, arteries, and other internal organs. The cells are spindle-shaped. Smooth muscles are responsible for involuntary body activities, such as churning of the stomach and constriction of arteries.

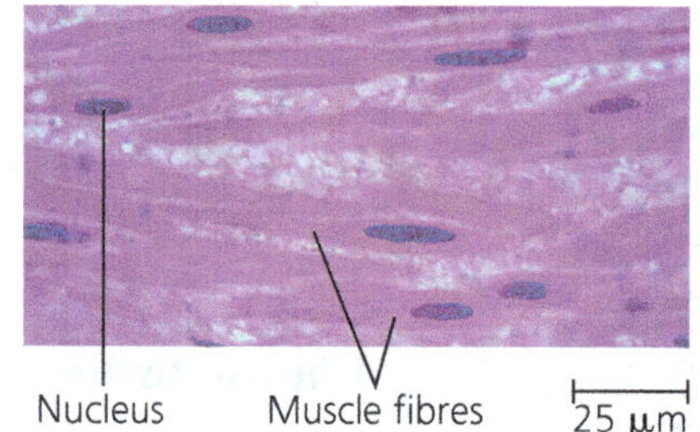

Cardiac muscle

Cardiac muscle forms the contractile wall of the heart. It is striated like skeletal muscle and has similar contractile properties. Unlike skeletal muscle, however, cardiac muscle has branched fibres that interconnect via intercalated disks, which relay signals from cell to cell and help synchronise heart contraction.

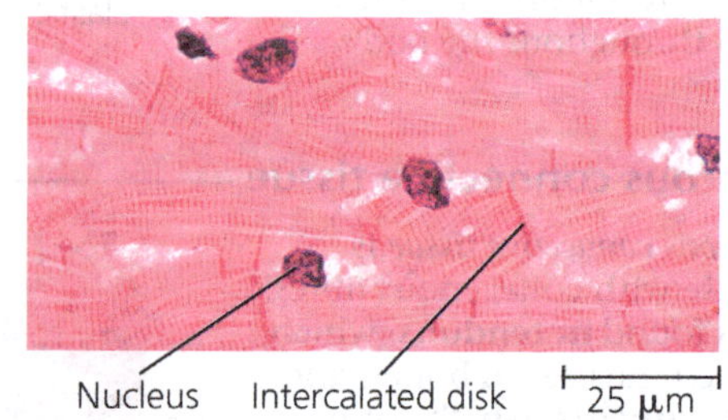

Nervous Tissue

Nervous tissue functions in the receipt, processing, and transmission of information. Nervous tissue contains **neurons**, or nerve cells, which transmit nerve impulses, as well as support cells called **glial cells**, or simply **glia**. In many animals, a concentration of nervous tissue forms a brain, an information-processing centre.

Neurons

Neurons are the basic units of the nervous system. A neuron receives nerve impulses from other neurons via its cell body and multiple extensions called dendrites. Neurons transmit impulses to neurons, muscles, or other cells via extensions called axons, which are often bundled together into nerves.

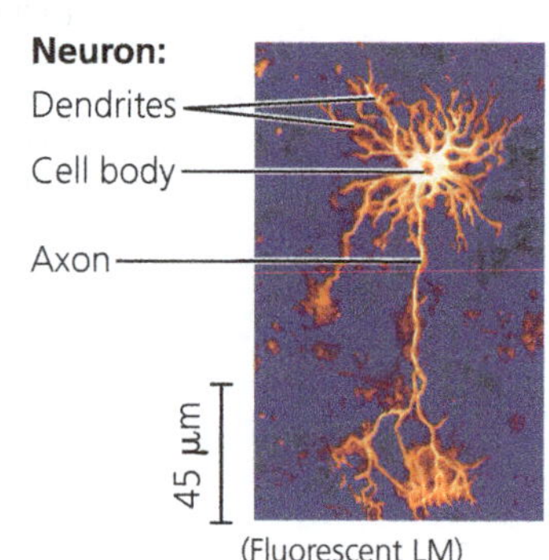

(Fluorescent LM)

Glia

The various types of glia help nourish, insulate, and replenish neurons, and in some cases, modulate neuron function.

Glia

15 μm

Axons of neurons

Blood vessel

(Confocal LM)

Coordination and Control

For an animal's tissues and organ systems to function effectively, they must act in concert with one another. For example, when the dingo shown in Figure 40.5 hunts, blood flow is regulated to bring adequate nutrients and gases to its leg muscles, which in turn are activated by the brain in response to cues detected by the nose. What signals coordinate activity? How do the signals move within the body?

Animals have two major systems for coordinating and controlling responses to stimuli: the endocrine and nervous systems **(Figure 40.6)**. In the **endocrine system**, signalling molecules released into the bloodstream by endocrine cells are carried to all locations in the body. In the **nervous system**, neurons transmit signals along dedicated routes connecting specific locations in the body. In each system, the type of pathway used is the same regardless of whether the signal's ultimate target is at the other end of the body or just a few cell diameters away.

The signalling molecules that are broadcast throughout the body by the endocrine system are called **hormones**. It takes seconds for hormones to be released into the bloodstream and carried throughout the body. The effects are often long-lasting, however, because hormones can remain in the bloodstream for minutes or even hours.

Different hormones cause distinct effects, and only cells that have receptors for a particular hormone respond (Figure 40.6a). Depending on which cells have receptors for that hormone, the hormone may have an effect in just a single location or in sites throughout the body. For example, thyroid-stimulating hormone (TSH) acts solely on cells in the thyroid gland. They in turn release thyroid hormone, which acts on nearly every body tissue to increase oxygen consumption and heat production.

In the nervous system, signals called nerve impulses travel to specific target cells along communication lines consisting mainly of axons (Figure 40.6b). Transmission in the nervous system is extremely fast; nerve impulses take only a fraction of a second to reach the target and last only a fraction of a second.

Nerve impulses can act on other neurons, on muscle cells, and on cells and glands that produce secretions. Unlike the endocrine system, the nervous system conveys information by the *pathway* the signal takes. For example, a person can distinguish different musical notes because each note's frequency activates neurons in the ear that connect to slightly different locations in the brain.

Communication in the nervous system usually involves more than one type of signal. Nerve impulses travel along axons, sometimes over long distances, as changes in voltage. In contrast, passing information from one neuron to another often involves very short-range chemical signals.

▼ Figure 40.6 Signalling in the endocrine and nervous systems.

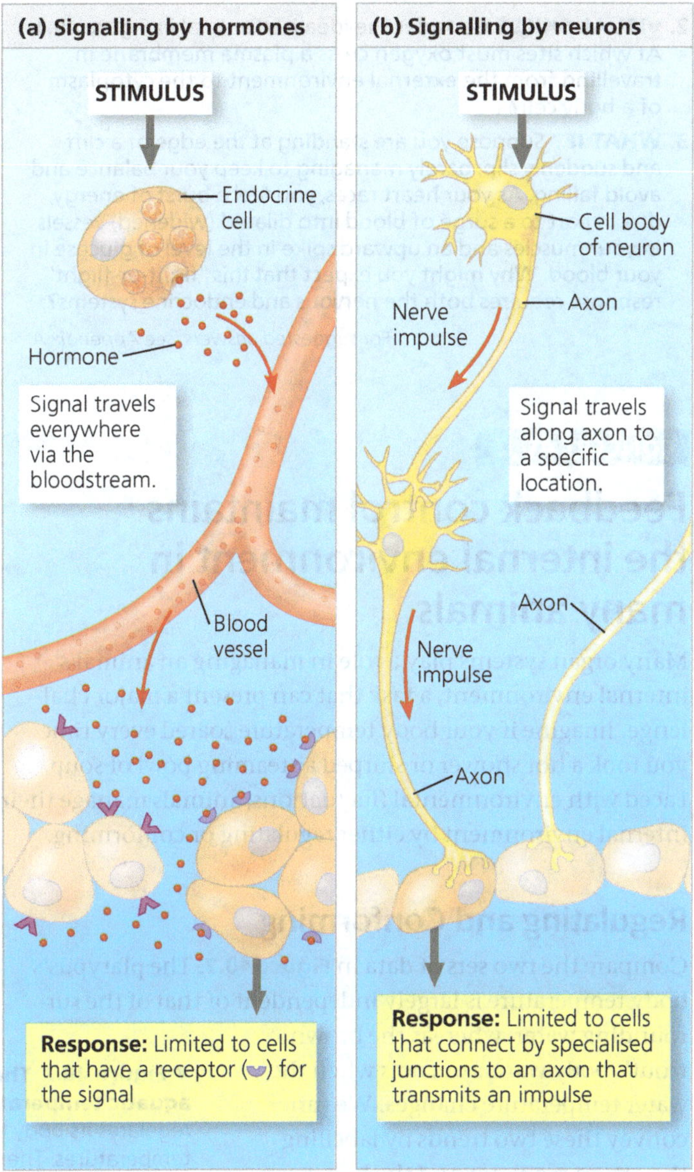

VISUAL SKILLS *After comparing the two diagrams, explain why a particular nerve impulse signal has only one physical pathway but a particular hormone molecule can have multiple physical pathways.*

Because the two major communication systems of the body differ in signal type, transmission, speed, and duration, it is not surprising that they are adapted to different functions. The endocrine system is especially well adapted for coordinating gradual changes that affect the entire body, such as growth, development, reproduction, metabolic processes, and digestion. The nervous system is well suited for directing immediate and rapid responses to the environment, such as reflexes and other rapid movements. Nevertheless, the two systems often work in close coordination. Both help maintain a stable internal environment, our next topic of discussion.

CONCEPT CHECK 40.1

1. What properties do all types of epithelia share?
2. **VISUAL SKILLS** Consider the idealised animal in Figure 40.4. At which sites must oxygen cross a plasma membrane in travelling from the external environment to the cytoplasm of a body cell?
3. **WHAT IF?** Suppose you are standing at the edge of a cliff and suddenly slip, barely managing to keep your balance and avoid falling. As your heart races, you feel a burst of energy, due in part to a surge of blood into dilated (widened) vessels in your muscles and an upward spike in the level of glucose in your blood. Why might you expect that this "fight-or-flight" response requires both the nervous and endocrine systems?

For suggested answers, see Appendix A.

CONCEPT 40.2

Feedback control maintains the internal environment in many animals

Many organ systems play a role in managing an animal's internal environment, a task that can present a major challenge. Imagine if your body temperature soared every time you took a hot shower or slurped a steaming bowl of soup. Faced with environmental fluctuations, animals manage their internal environment by either regulating or conforming.

Regulating and Conforming

Compare the two sets of data in **Figure 40.7**. The platypus's body temperature is largely independent of that of the surrounding water, whereas the brown trout's body warms or cools when the water temperature changes. We can convey these two trends by labelling the otter a regulator and the bass a conformer with regard to body temperature. An animal is a **regulator** for an environmental variable if it uses internal mechanisms to control internal change in the face of external fluctuation. In contrast, an animal is a **conformer** if it allows its internal condition to change in accordance with external changes in the particular variable.

An animal may allow some internal conditions to conform to the environment but regulate others. For instance, the bass conforms to the temperature of the water in which it lives, but regulates the solute concentration in its blood and interstitial fluid. In addition, conforming need not involve changes in an internal variable: Many marine invertebrates, such as spider crabs (genus *Libinia*), let their internal solute concentration conform to the relatively stable salinity of their ocean environment.

▼ **Figure 40.7 The relationship between body and environmental temperatures in an aquatic temperature regulator and an aquatic temperature conformer.** The platypus regulates its body temperature, keeping it stable across a wide range of environmental temperatures. The brown trout, meanwhile, allows its internal environment to conform to the water temperature.

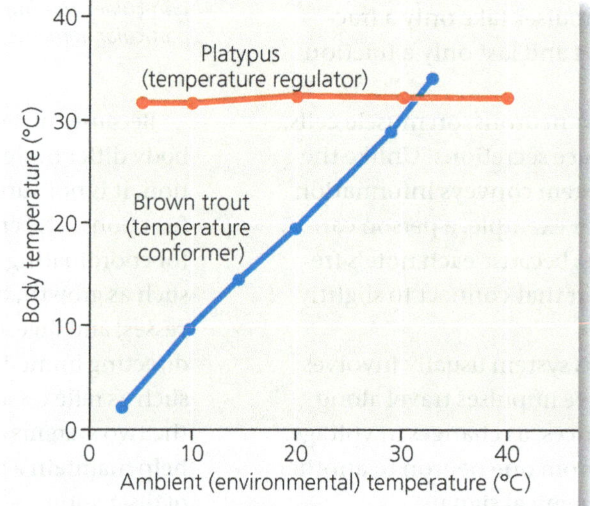

Homeostasis

The steady body temperature of a river otter and the stable concentration of solutes in a freshwater bass are examples of **homeostasis**, which means the maintenance of internal balance. In achieving homeostasis, animals maintain a "steady state"—a relatively constant internal environment—even when the external environment changes significantly.

Many animals exhibit homeostasis for a range of physical and chemical properties. For example, humans maintain a fairly constant body temperature of about 37°C, a blood pH within 0.1 pH unit of 7.4, and a blood glucose concentration that is predominantly in the range of 70–110 mg of glucose per 100 mL of blood.

Mechanisms of Homeostasis

Homeostasis requires a control system. Before exploring homeostasis in animals, let's get a basic picture of how a control system works by considering a nonliving example: the regulation of room temperature. Let's assume you want to keep a room at 20°C, a comfortable temperature for normal activity. You set a control device—the thermostat—to 20°C. A thermometer in the thermostat monitors the room temperature. If the temperature falls below 20°C, the thermostat turns on a radiator, furnace, or other heater

▼ **Figure 40.8 A nonliving example of temperature regulation: control of room temperature.** Regulating room temperature depends on a sensor/control centre (a thermostat) that detects temperature change and activates mechanisms that reverse that change.

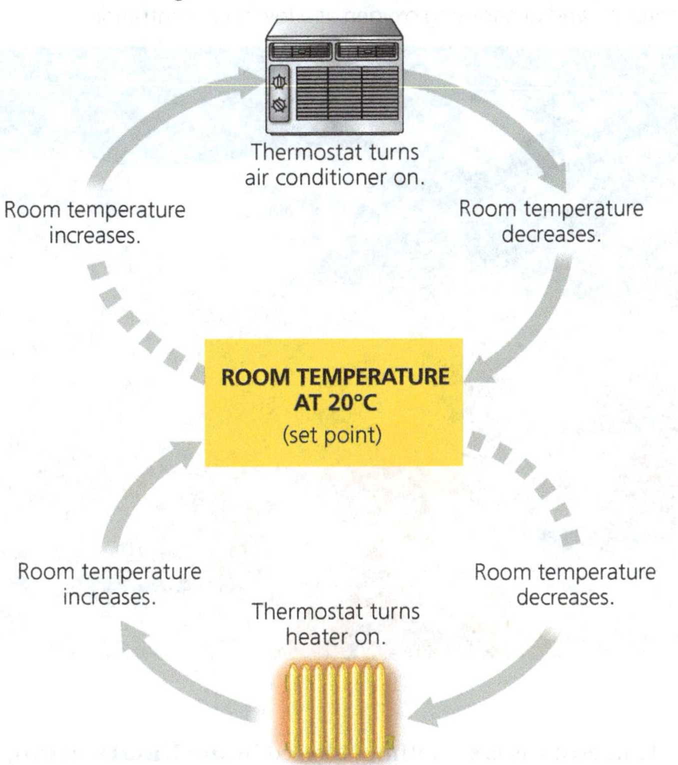

DRAW IT *Label at least one stimulus, response, and sensor/control centre in the above figure.*

(Figure 40.8). When the room exceeds 20°C, the thermostat switches off the heater. If the temperature then drifts below 20°C, the thermostat activates another heating cycle. If the temperature instead rises above 20°C, the thermostat activates a cooling mechanism, such as by turning on an air conditioner.

Like a home heating system, the homeostatic control system in animals maintains a variable, such as body temperature or solute concentration, at or near a particular value, or **set point**. A fluctuation in the variable above or below the set point serves as the **stimulus** detected by a **sensor**. The sensor signals a *control centre*, which triggers a **response**, a physiological activity that helps return the variable to the set point. In the home heating example, a drop in temperature below the set point acts as a stimulus, the thermostat serves as the sensor and control centre, and the heater produces the response.

Feedback Control in Homeostasis

If you examine the circuit in Figure 40.8, you can see that either response (heating or cooling) reduces the stimulus (the change in temperature) that triggered that response. The circuit thus displays **negative feedback**, a control mechanism that "damps" its stimulus (see Figure 1.10). This type of feedback regulation plays a major role in homeostasis in animals. For example, when you exercise vigorously, you produce heat, which increases your body temperature. Your nervous system detects this increase and triggers sweating. The evaporation of sweat from your skin then cools your body, helping return body temperature to its set point and eliminating the stimulus.

Homeostasis is a dynamic equilibrium, an interplay between external factors that tend to change the internal environment and internal control mechanisms that oppose such changes. Note that physiological responses to stimuli are not instantaneous, just as switching on a heater does not immediately warm a room. As a result, homeostasis moderates but doesn't eliminate changes in the internal environment. Fluctuation is greater if a variable has a *normal range*—an upper and lower limit—rather than a set point. This is equivalent to a heating system that is programmed to produce heat when the room temperature drops to 19°C and to stop heating when the temperature reaches 21°C. Regardless of whether there is a set point or a normal range, homeostasis is enhanced by adaptations that reduce fluctuations, such as insulation in the case of temperature and physiological buffers in the case of pH.

Unlike negative feedback, **positive feedback** is a control mechanism that amplifies the stimulus. In animals, positive-feedback loops do not play a major role in homeostasis, but instead help drive processes to completion. During childbirth, for instance, the pressure of the baby's head against sensors near the opening of the mother's uterus stimulates the uterus to contract. These contractions result in greater pressure against the opening of the uterus, heightening the contractions and thereby causing even greater pressure, ultimately causing the baby to be born.

Alterations in Homeostasis

The set points and normal ranges for homeostasis can change under various circumstances. In fact, *regulated changes* in the internal environment are essential to normal body functions. Some regulated changes occur during a particular stage in life, such as the radical shift in hormone balance that occurs during puberty. Other regulated changes are cyclic, such as the variation in hormone levels responsible for a woman's menstrual cycle (see Figure 46.17).

In all animals (and plants), certain cyclic alterations in metabolism reflect a **circadian rhythm**, a set of physiological changes that occur roughly every 24 hours **(Figure 40.9)**. One way to observe this rhythm is to monitor body temperature, which in humans typically undergoes a cyclic rise and fall of more than 0.6°C in every 24-hour period. Remarkably, a biological clock maintains this rhythm even when variations in human activity, room temperature, and light levels are minimised (see Figure 40.9a). A circadian rhythm is thus intrinsic to the body, although the biological clock is

▼ **Figure 40.9 Human circadian rhythm.**

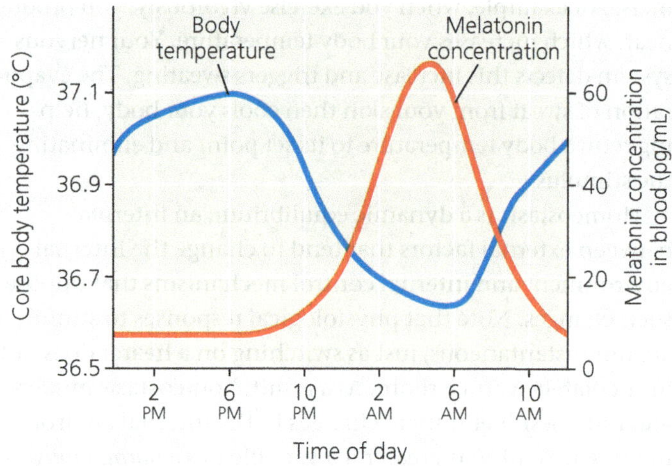

(a) Variation in core body temperature and melatonin concentration in blood. Researchers studied resting but awake volunteers in an isolation chamber with constant temperature and low light. (Melatonin is a hormone secreted by the pineal gland.)

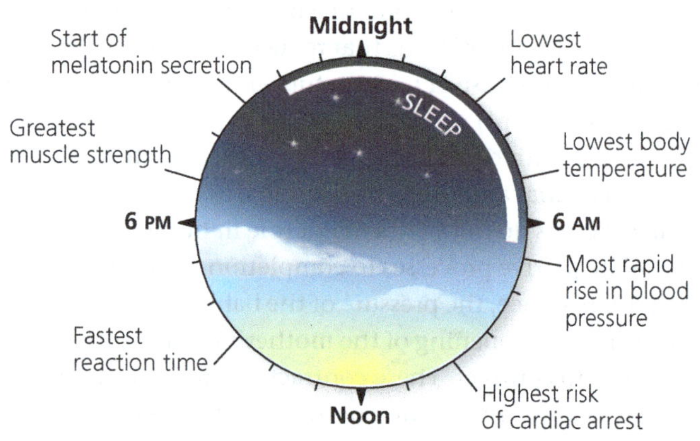

(b) The human circadian clock. Metabolic activities undergo daily cycles in response to the circadian clock. As illustrated for a typical individual who rises early in the morning, eats lunch around noon, and sleeps at night, these cyclic changes occur throughout a 24-hour day.

normally coordinated with the cycle of light and darkness in the environment (see Figure 40.9b). For example, the hormone melatonin is secreted at night, and more is released during the longer nights of winter. External stimuli can reset the biological clock, but the effect is not immediate. That is why flying across several time zones results in jet lag, a mismatch between the circadian rhythm and local environment that persists until the clock fully resets.

Noting the importance of biological clocks to human health and disease, the Nobel Prize Committee awarded the 2017 Nobel Prize in Physiology or Medicine to Americans Jeffrey Hall, Michael Rosbash, and Michael Young, who studied the fruit fly *Drosophila* to map out the molecular mechanisms that underlie circadian rhythms.

▼ **Figure 40.10 Acclimatisation by mountain climbers in the Himalayas.** To lessen the risk of altitude sickness when ascending a high peak, climbers acclimatise by camping partway up the mountain. Spending time at an intermediate altitude allows the circulatory and respiratory systems to become more efficient in capturing and distributing oxygen at a lower concentration.

Homeostasis is sometimes altered by **acclimatisation**, an animal's physiological adjustment to changes in its external environment. For instance, when an elk moves up into the mountains from sea level, the lower oxygen concentration in the high mountain air stimulates the animal to breathe more rapidly and deeply. As a result, more CO_2 is lost through exhalation, raising blood pH above its normal range. As the animal acclimatises over several days, changes in kidney function cause it to excrete urine that is more alkaline, returning blood pH to its normal range. Other mammals, including humans, are also capable of acclimatising to dramatic altitude changes **(Figure 40.10)**, although health risks remain.

CONCEPT CHECK 40.2

1. **MAKE CONNECTIONS** How does negative feedback in thermoregulation differ from feedback inhibition in an enzyme-catalysed biosynthetic process (see Figure 8.21)?
2. If you were deciding where to put the thermostat in a house, what factors would govern your decision? How do these factors relate to the fact that many homeostatic control sensors in humans are located in the brain?
3. **MAKE CONNECTIONS** Like animals, cyanobacteria have a circadian rhythm. By analysing the genes that maintain biological clocks, scientists concluded that the 24-hour rhythms of humans and cyanobacteria reflect convergent evolution (see Concept 26.2). What evidence would have supported this conclusion? Explain.

For suggested answers, see Appendix A.

CONCEPT 40.3

Homeostatic processes for thermoregulation involve form, function, and behaviour

In this section, we'll examine the regulation of body temperature as an example of how form and function work together in regulating an animal's internal environment. Later chapters in this unit will discuss other physiological systems involved in maintaining homeostasis.

Thermoregulation is the process by which animals maintain their body temperature within a normal range. Body temperatures outside the normal range can reduce the efficiency of enzymatic reactions, alter the fluidity of cellular membranes, and affect other temperature-sensitive biochemical processes, potentially with fatal results.

In talking about thermoregulation, we will need to talk about heat. Formally, heat is defined as thermal energy in transfer from one body of matter to another (see Concept 8.1). Here, however, we will use the term heat to refer simply to thermal energy.

Endothermy and Ectothermy

Heat for thermoregulation can come from either internal metabolism or the external environment. Humans and other mammals, as well as birds, are **endothermic**, meaning that they are warmed mostly by heat generated by metabolism. Some fishes and insect species and a few nonavian reptiles are also mainly endothermic. In contrast, amphibians, many nonavian reptiles and fishes, and most invertebrates are **ectothermic**, meaning that they gain most of their heat from external sources. Endothermy and ectothermy are not mutually exclusive, however. For example, a bird is mainly endothermic but may warm itself in the sun on a cold morning, much as an ectothermic lizard does.

Endotherms can maintain a stable body temperature even in the face of large fluctuations in the environmental temperature. In a cold environment, an endotherm generates enough heat to keep its body substantially warmer than its surroundings **(Figure 40.11a)**. In a hot environment, endothermic vertebrates have mechanisms for cooling their bodies, enabling them to withstand temperatures that are intolerable for most ectotherms.

Many ectotherms adjust their body temperature by behavioural means, such as seeking out shade or basking in the sun **(Figure 40.11b)**. Because their heat source is largely environmental, ectotherms generally need to consume much less food than endotherms of equivalent size—an advantage if food supplies are limited. Ectotherms also usually tolerate larger fluctuations in their internal temperature.

Variation in Body Temperature

Animals also differ in whether their body temperature is variable or constant. An animal whose body temperature varies with its environment is called a *poikilotherm* (from the Greek *poikilos*, varied). In contrast, a *homeotherm* has a relatively constant body temperature. For example, the brown trout is a poikilotherm, and the platypus is a homeotherm (see Figure 40.7).

From the descriptions of ectotherms and endotherms, it might seem that all ectotherms are poikilothermic and all endotherms are homeothermic. In fact, there is no fixed relationship between the source of heat and the stability of body temperature. Many ectothermic marine fishes and invertebrates inhabit waters with such stable temperatures that their body temperature varies less than that of mammals and other endotherms. Conversely, the body temperature of a few endotherms varies considerably. For example, the body temperature of some bats drops from 40°C to a few degrees above zero when they enter hibernation.

It is a common misconception that ectotherms are "cold-blooded" and endotherms are "warm-blooded." Ectotherms

▼ **Figure 40.11 Thermoregulation by internal or external sources of heat.** Endotherms obtain heat from their internal metabolism, whereas ectotherms rely on heat from their external environment.

(a) King penguins (*Aptenodytes patagonicus*), endotherms

(b) Eastern long-necked turtle (*Chelodina longicollis*), ectotherm

do not necessarily have low body temperatures. On the contrary, when sitting in the sun, many ectothermic lizards have higher body temperatures than mammals. Thus, the terms *cold-blooded* and *warm-blooded* are misleading and are avoided in scientific communication.

Balancing Heat Loss and Gain

Thermoregulation depends on an animal's ability to control the exchange of heat with its environment. That exchange can occur by any of four processes: radiation, evaporation, convection, and conduction **(Figure 40.12)**. In each, heat is transferred from an object of higher temperature to one of lower temperature.

The essence of thermoregulation is maintaining a rate of heat gain that equals the rate of heat loss. Animals do this through mechanisms that either reduce heat exchange overall or favour heat exchange in a particular direction. In mammals, several of these mechanisms involve the **integumentary system**, the outer covering of the body, consisting of the skin, hair, and nails (claws or hooves in some species).

Insulation

Insulation, which reduces the flow of heat between an animal's body and its environment, is a major adaptation for thermoregulation in both mammals and birds. Insulation is found both at the body surface—hair and feathers—and beneath—layers of fat formed by adipose tissue. In addition, some animals secrete oily substances that repel water, protecting the insulating capacity of feathers or fur. Birds, for example, secrete oils that they apply to their feathers during preening.

Often, animals can adjust their insulating layers to further regulate body temperature. Most land mammals and birds, for example, react to cold by raising their fur or feathers. This action traps a thicker layer of air, thereby increasing the effectiveness of the insulation. Lacking feathers or fur, humans must rely primarily on fat for insulation. We do, however, get "goose bumps," a vestige of hair raising inherited from our furry ancestors.

Insulation is particularly important for marine mammals, such as whales and walruses. These animals swim in water colder than their body core, and many species spend at least part of the year in nearly freezing polar seas. Furthermore, the transfer of heat to water occurs 50 to 100 times more rapidly than heat transfer to air. Survival under these conditions is made possible by an evolutionary adaptation called blubber, a very thick layer of insulating fat just under the skin. The insulation that blubber provides is so effective that marine mammals can maintain body core temperatures of about 36–38°C without requiring much more energy from food than land mammals of similar size.

Circulatory Adaptations

Circulatory systems provide a major route for heat flow between the interior and exterior of the body. Adaptations that regulate the extent of blood flow near the body surface or that trap heat within the body core play a significant role in thermoregulation.

In response to changes in the temperature of their surroundings, many animals alter the amount of blood (and hence heat) flowing between their body core and their skin. Nerve signals that relax the muscles of the vessel walls result in *vasodilation*, a widening of superficial blood vessels (those near the body surface). As a consequence of the increase in vessel diameter, blood flow in the skin increases. In endotherms, vasodilation usually increases the transfer of body heat to the environment by radiation, conduction, and convection (see Figure 40.12). The reverse process, *vasoconstriction*, reduces blood flow and heat transfer by decreasing the diameter of superficial vessels.

Like endotherms, some ectotherms control heat exchange by regulating blood flow. For example, when the

▼ Figure 40.12 Heat exchange between an organism and its environment.

Radiation is the emission of electromagnetic waves by all objects warmer than absolute zero. Here, a goanna absorbs heat radiating from the distant sun and radiates a smaller amount of energy to the surrounding air.

Evaporation is the removal of heat from the surface of a liquid that is losing some of its molecules as gas. Evaporation of water from a goanna's moist surfaces that are exposed to the environment has a strong cooling effect.

Convection is the transfer of heat by the movement of air or liquid past a surface, as when a breeze contributes to heat loss from a goanna's dry skin or when blood moves heat from the body core to the extremities.

Conduction is the direct transfer of thermal motion (heat) between molecules of objects in contact with each other, as when a goanna sits on a hot rock.

VISUAL SKILLS *If this figure showed a penguin (an endotherm) on an ice floe rather than a goanna (an ectotherm) on a rock, would any of the arrows point in a different direction? Explain.*

▼ Figure 40.13 Countercurrent heat exchangers. A countercurrent exchange system traps heat in the body core, thus reducing heat loss from the extremities, particularly when they are immersed in cold water or in contact with ice or snow. In essence, heat in the arterial blood emerging from the body core is transferred directly to the returning venous blood instead of being lost to the environment.

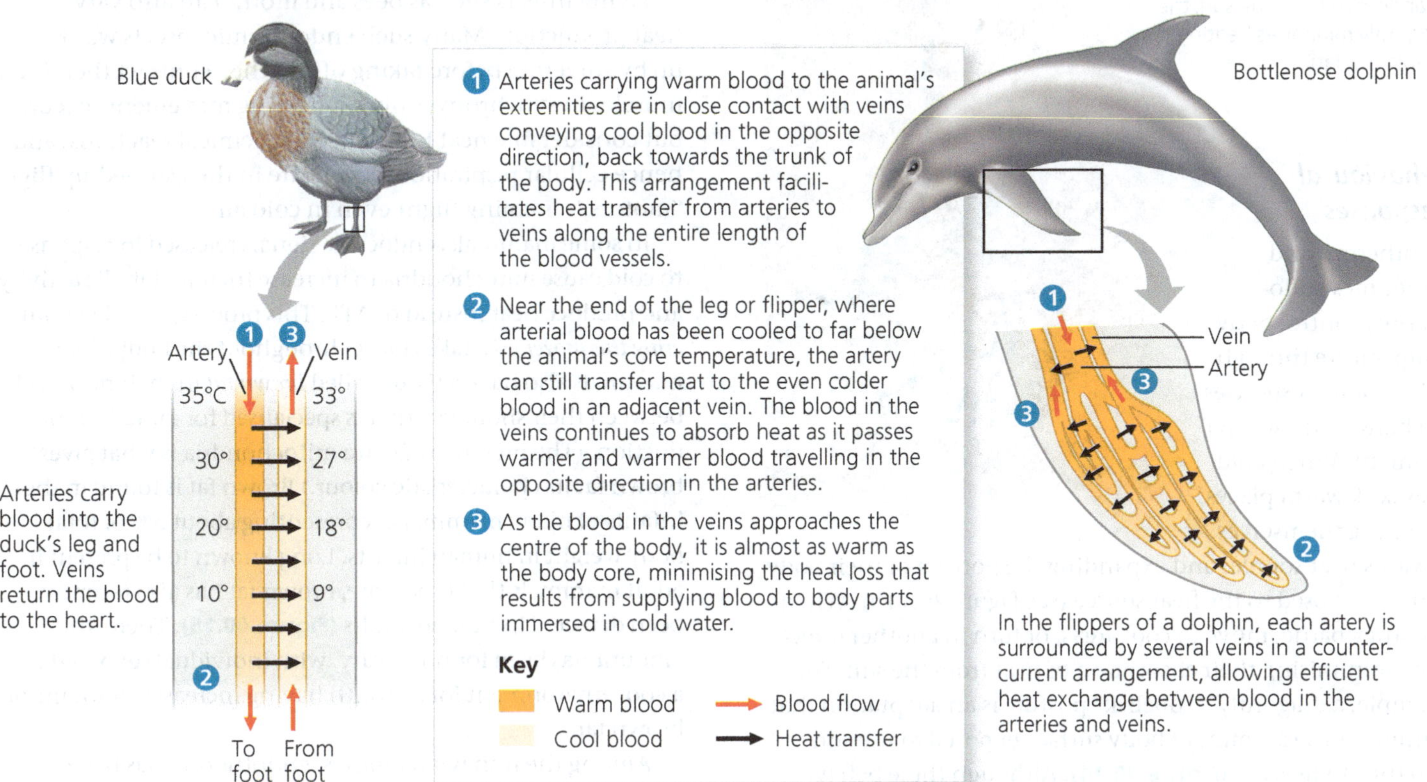

marine iguana of the Galápagos Islands swims in the cold ocean, its superficial blood vessels undergo vasoconstriction. This process routes more blood to the body core, conserving body heat.

In many birds and mammals, reducing heat loss from the body relies on **countercurrent exchange**, the transfer of heat (or solutes) between fluids that are flowing in opposite directions. In a countercurrent heat exchanger, arteries and veins are located adjacent to each other **(Figure 40.13)**. Because blood flows through the arteries and veins in opposite directions, this arrangement allows heat exchange to be remarkably efficient. As warm blood in the arteries moves outwards from the body core, it transfers heat to the colder blood in the veins returning from the extremities. Most importantly, heat is transferred along the entire length of the exchanger, maximising the rate of heat exchange and minimising heat loss to the environment.

Although most sharks and fishes are temperature conformers, countercurrent heat exchangers are found in some large, powerful swimmers, including great white sharks, bluefin tuna, and swordfish. By keeping the main swimming muscles warm, this adaptation enables vigorous, sustained activity. Similarly, many endothermic insects (carpenter bees, honeybees, and some moths) have a countercurrent exchanger that helps maintain a high temperature in their thorax, where flight muscles are located.

Cooling by Evaporative Heat Loss

Many mammals and birds live in places where regulating body temperature requires cooling at some times and warming at others. If environmental temperature is above body temperature, only evaporation can keep body temperature from rising. Water absorbs considerable heat when it evaporates (see Concept 3.2); this heat is carried away from the skin and respiratory surfaces with water vapour.

Some animals exhibit adaptations that greatly facilitate evaporative cooling. A few mammals, including horses and humans, have sweat glands. In many other mammals, as well as in birds, panting is important. Some birds have a pouch richly supplied with blood vessels in the floor of the mouth; fluttering the pouch increases evaporation. Pigeons can use this adaptation to keep their body temperature close to 40°C in air temperatures as high as 60°C, as long as they have sufficient water.

▶ **Figure 40.14**
Thermoregulatory behaviour in a dragonfly. By orienting its body so that the narrow tip of its abdomen faces the sun, the dragonfly minimises heating by solar radiation.

Behavioural Responses

Ectotherms, and sometimes endotherms, control body temperature through behavioural responses to changes in the environment. When cold, they seek warm places, orienting themselves towards heat sources and expanding the portion of their body surface exposed to the heat source (see Figure 40.11b). When hot, they bathe, move to cool areas, or turn in another direction, minimising their absorption of heat from the sun. For example, a dragonfly's "obelisk" posture is an adaptation that minimises the amount of body surface exposed to the sun and thus to heating **(Figure 40.14)**. Although these behaviours are relatively simple, they enable many ectotherms to maintain a nearly constant body temperature.

Social behaviour contributes to thermoregulation in both endotherms and ectotherms. Among endotherms, for example, behaviour contributes significantly to the winter survival of Emperor penguins (see Figure 40.1). Among ectotherms, honeybees are notable for their use of behaviour in achieving homeostasis for temperature. In cold weather, they increase heat production and huddle together, thereby retaining heat. Individuals move between the cooler outer edges of the huddle and the warmer centre, thus circulating and distributing the heat. In hot weather, honeybees cool the hive by transporting water to the hive and fanning with their wings, promoting evaporation and convection. Thus, a honeybee colony uses many of the mechanisms of thermoregulation characteristic of individual animals.

Adjusting Metabolic Heat Production

Because endotherms generally maintain a body temperature considerably higher than that of the environment, they must counteract continual heat loss. Endotherms can vary heat production—*thermogenesis*—to match changing rates of heat loss. Thermogenesis is increased by such muscle activity as moving or shivering. For example, shivering helps chickadees, birds with a body mass of only 20 g, remain active and hold their body temperature nearly constant at 40°C in environmental temperatures as low as −40°C.

Flying insects such as bees and moths can also vary heat production. Many such endothermic insects warm up by shivering before taking off. As they contract their flight muscles in synchrony, only slight wing movements occur, but considerable heat is produced. Chemical reactions, and hence cellular respiration, accelerate in the warmed-up flight "motors," enabling flight even in cold air.

In some mammals, endocrine signals released in response to cold cause mitochondria to increase their metabolic activity and produce heat instead of ATP. This process, called *nonshivering thermogenesis*, takes place throughout the body. Some mammals also have a tissue called *brown fat* in their neck and between their shoulders that is specialised for rapid heat production. (The presence of extra mitochondria is what gives brown fat its characteristic colour.) Brown fat is found in the infants of many mammals, representing about 5% of total body weight in human infants. Long known to be present in adult mammals that hibernate, brown fat has also recently been detected in human adults **(Figure 40.15)**. There, the amount has been found to vary, with individuals exposed to a cool environment for a month having increased amounts of brown fat.

Among the nonavian reptiles, endothermy has been observed in some large species in certain circumstances. For example, researchers found that a female Burmese python (*Python molurus bivittatus*) incubating eggs maintained a body temperature roughly 6°C above that of the surrounding air. Where did the heat come from? Further studies showed that such pythons, like birds, can raise their body temperature

▼ **Figure 40.15 Brown fat activity during cold stress.** This PET scan shows metabolically active brown fat deposits (indicated by the arrows) surrounding the neck.

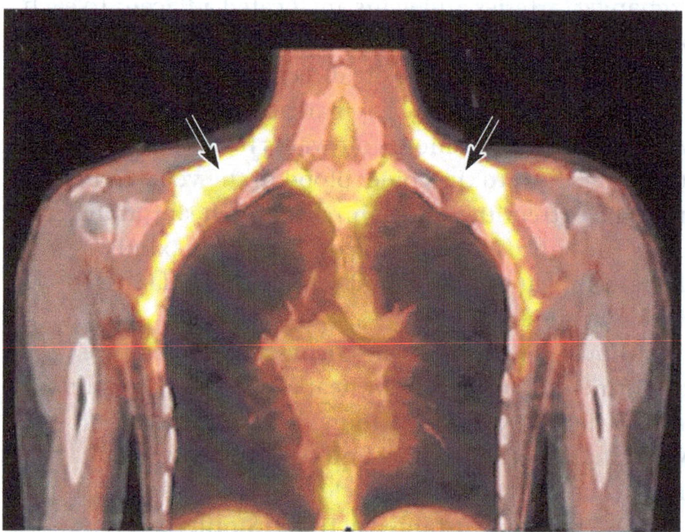

▼ **Figure 40.16** Inquiry

How does a Burmese python generate heat while incubating eggs?

Experiment Herndon Dowling and colleagues at the Bronx Zoo in New York observed that when a female Burmese python incubated eggs by wrapping her body around them, she raised her body temperature and frequently contracted the muscles in her coils. To learn if the contractions were elevating her body temperature, they placed the python and her eggs in a chamber. As they varied the chamber's temperature, they monitored the python's muscle contractions as well as her oxygen uptake, a measure of her rate of cellular respiration.

Results The python's oxygen consumption increased when the temperature in the chamber decreased. As shown in the graph, this increase in oxygen consumption paralleled an increase in the rate of muscle contraction.

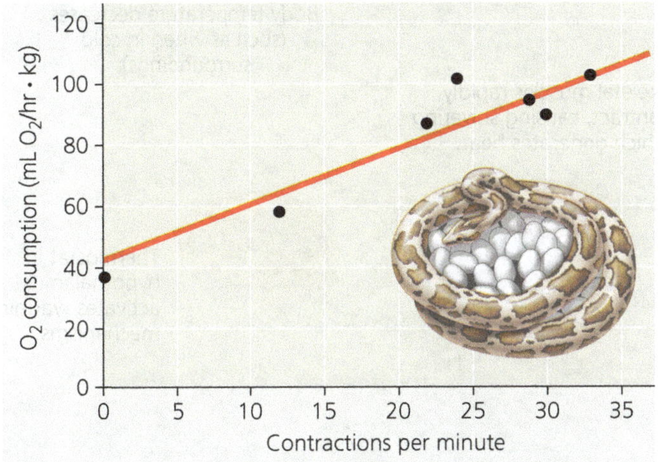

Conclusion Because oxygen consumption, which generates heat through cellular respiration, was correlated with the rate of muscle contraction, the researchers concluded that the muscle contractions, a form of shivering, were the source of the Burmese python's elevated body temperature.

Data from V. H. Hutchison, H. G. Dowling, and A. Vinegar, Thermoregulation in a brooding female Indian python, *Python molurus bivittatus*, *Science* 151:694–696 (1966).

WHAT IF? *Suppose you varied air temperature and measured oxygen consumption for a female Burmese python without a clutch of eggs. Since she would not show shivering behaviour, how would you expect the snake's oxygen consumption to vary with environmental temperature?*

through shivering **(Figure 40.16)**. Whether certain groups of Mesozoic dinosaurs were similarly endothermic is a matter of active debate.

Acclimatisation in Thermoregulation

Acclimatisation contributes to thermoregulation in many animal species. In birds and mammals, acclimatisation to seasonal temperature changes often includes adjusting insulation—growing a thicker coat of fur in the winter and shedding it in the summer, for example.

Acclimatisation in ectotherms often includes adjustments at the cellular level. Cells may produce variants of enzymes that have the same function but different optimal temperatures. Also, the proportions of saturated and unsaturated lipids in membranes may change; unsaturated lipids help keep membranes fluid at lower temperatures (see Figure 7.5).

Remarkably, some ectotherms can survive subzero temperatures, producing "antifreeze" proteins that prevent ice formation in their cells. In the Arctic and Southern (Antarctic) Oceans, these proteins enable certain fishes to survive in water as cold as −2°C, a full degree Celsius below the freezing point of body fluids in other species.

Physiological Thermostats and Fever

In humans and other mammals, the sensors responsible for thermoregulation are concentrated in the **hypothalamus**, the brain region that also controls the circadian clock. Within the hypothalamus, a group of nerve cells functions as a thermostat, responding to body temperatures outside the normal range by activating mechanisms that promote heat loss or gain **(Figure 40.17)**.

At body temperatures above the normal range, the hypothalamic thermostat promotes cooling of the body by dilation of vessels in the skin, sweating, or panting. When body temperatures instead drop below the normal range, the thermostat inhibits heat loss mechanisms and activates mechanisms that either save heat, such as constricting vessels in the skin, or generate heat, such as shivering.

In the course of certain bacterial and viral infections, mammals and birds develop *fever*, an elevated body temperature. A variety of experiments have shown that fever reflects an increase in the normal range for the biological thermostat. For example, artificially *raising* the temperature of the hypothalamus in an infected animal *reduces* fever in the rest of the body.

Among certain ectotherms, an increase in body temperature upon infection reflects what is called a behavioural fever. For example, the desert iguana (*Dipsosaurus dorsalis*) responds to infection with certain bacteria by seeking a warmer environment and then maintaining a body temperature that is elevated by 2–4°C. Similar observations in fishes, amphibians, and even cockroaches indicate that fever is common to both endotherms and ectotherms.

Now that we have explored thermoregulation, we'll conclude our introduction to animal form and function by considering the different ways that animals allocate, use, and conserve energy.

▶ **Figure 40.17 The thermostatic function of the hypothalamus in human thermoregulation.**

WHAT IF? *Suppose at the end of a hard run on a hot day you find that there are no drinks left in the cooler. If, out of desperation, you dunk your head into the cooler, how might the ice-cold water affect the rate at which your body temperature returns to normal?*

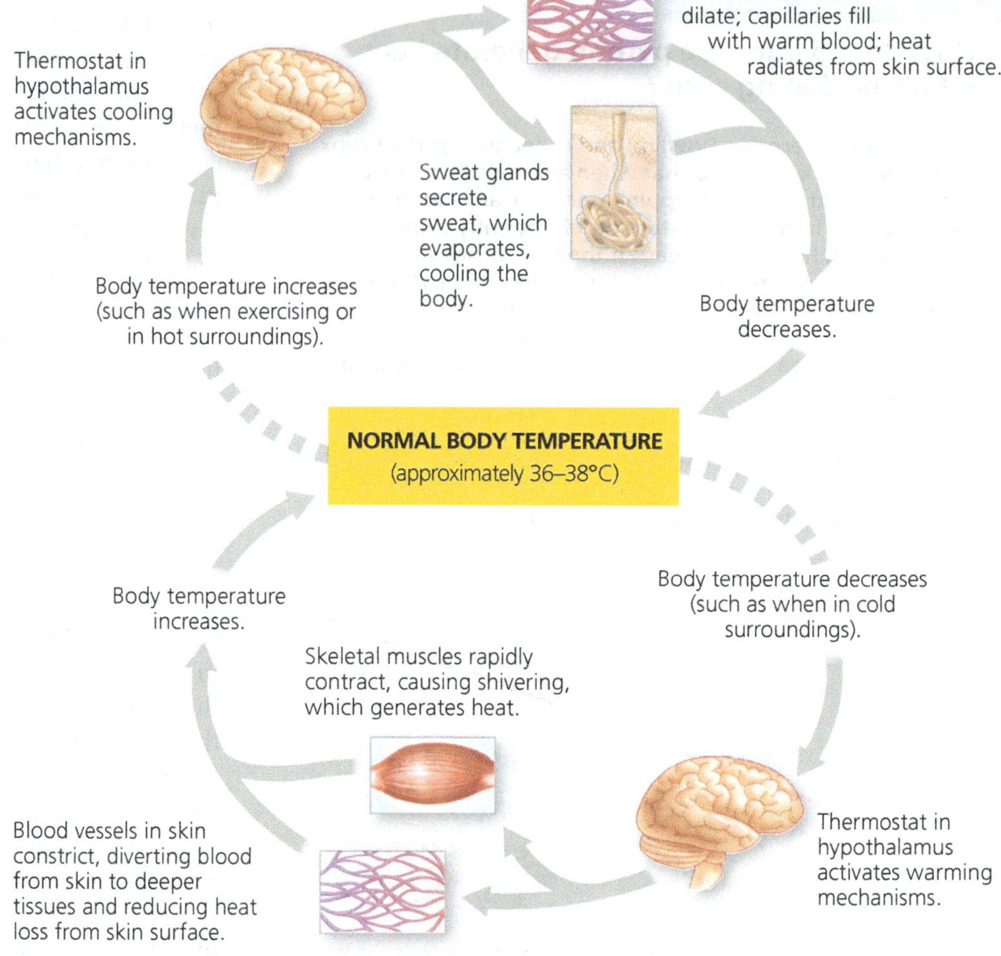

CONCEPT CHECK **40.3**

1. What mode of heat exchange is involved in "wind chill," when moving air feels colder than still air at the same temperature? Explain.
2. Flowers differ in how much sunlight they absorb. Why might this matter to a hummingbird seeking nectar on a cool morning?
3. **WHAT IF?** Why is shivering likely during the onset of a fever?

For suggested answers, see Appendix A.

CONCEPT **40.4**

Energy requirements are related to animal size, activity, and environment

One of the unifying themes of biology, introduced in Concept 1.1, is that life requires energy transfer and transformation. Like other organisms, animals use chemical energy for growth, repair, activity, and reproduction. The overall flow and transformation of energy in an animal—its **bioenergetics**—determines nutritional needs and is related to the animal's size, activity, and environment.

Energy Allocation and Use

Organisms can be classified by how they obtain chemical energy. Most *autotrophs*, such as plants, harness light energy to build energy-rich organic molecules and then use those molecules for fuel. Most *heterotrophs*, such as animals, obtain their chemical energy from food, which contains organic molecules synthesised by other organisms.

Animals use chemical energy harvested from the food they eat to fuel metabolism and activity. Food is digested by enzymatic hydrolysis (see Figure 5.2b), and nutrients are absorbed by body cells. The ATP (adenosine triphosphate) produced by cellular respiration and fermentation powers cellular work, enabling cells, organs, and organ systems to perform the functions that keep an animal alive. Other uses of energy in the form of ATP include biosynthesis, which is needed for

▼ **Figure 40.18 Bioenergetics of an animal: an overview.**

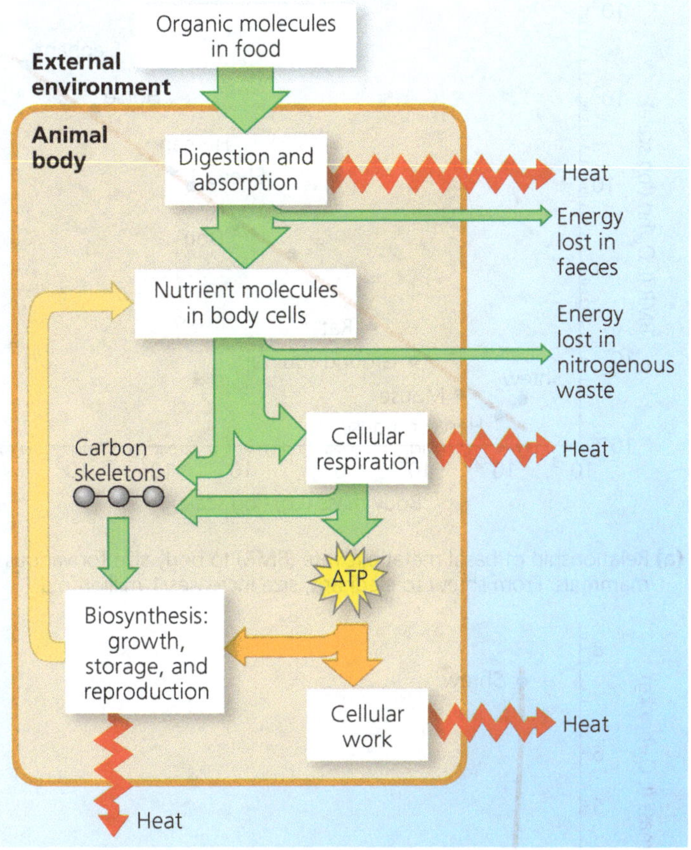

MAKE CONNECTIONS *Use the idea of energy coupling to explain why heat is produced in the absorption of nutrients, in cellular respiration, and in the synthesis of biopolymers (see Concept 8.3).*

body growth and repair, synthesis of storage material such as fat, and production of gametes. The production and use of ATP generate heat, which the animal eventually gives off to its surroundings **(Figure 40.18)**.

Quantifying Energy Use

How much of the total energy an animal obtains from food does it need just to stay alive? How much energy must be expended to walk, run, swim, or fly from one place to another? What fraction of the energy intake is used for reproduction? Physiologists answer such questions by measuring the rate at which an animal uses chemical energy and how this rate changes in different circumstances.

The sum of all the energy an animal uses in a given time interval is called its **metabolic rate**. Energy is measured in joules (J) and kilojoules (kJ), or in calories (cal) and kilocalories (kcal). A kilojoule equals 1,000 joules, or 0.239 kilocalories. (The unit Calorie, with a capital C, as used by many nutritionists, is actually a kilocalorie.)

Metabolic rate can be determined in several ways. Because nearly all of the chemical energy used in cellular respiration eventually appears as heat, metabolic rate can be measured by monitoring an animal's rate of heat loss. For this approach, researchers use a calorimeter, which is a closed, insulated chamber equipped with a device that records the heat an animal gives off to its environment. Metabolic rate can also be determined from the amount of oxygen consumed or carbon dioxide produced by an animal's cellular respiration **(Figure 40.19)**. To calculate metabolic rate over longer periods, researchers record the rate of food consumption, the energy content of the food (about 19–21 kJ per gram of protein or carbohydrate and about 38 kJ per gram of fat), and the chemical energy lost in waste products (faeces and urine or other nitrogenous wastes).

Minimum Metabolic Rate and Thermoregulation

Animals must maintain a minimum metabolic rate for basic functions such as cell maintenance, breathing, and circulation. Researchers measure this minimum metabolic rate differently for endotherms and ectotherms. The minimum metabolic rate of a nongrowing endotherm that is at rest, has an empty stomach, and is not experiencing stress is called the **basal metabolic rate (BMR)**. BMR is measured under a "comfortable" temperature range—a range that requires only the minimum generation or shedding of heat. The minimum metabolic rate of ectotherms is determined at a specific temperature because changes in the environmental temperature alter body temperature and therefore metabolic rate. The metabolic rate of a fasting, nonstressed

▼ **Figure 40.19 Measuring the rate of oxygen consumption by a swimming shark.** A researcher monitors the decrease in oxygen level over time in the recirculating water of a juvenile hammerhead's tank.

ectotherm at rest at a particular temperature is called its **standard metabolic rate (SMR)**.

Comparisons of minimum metabolic rates reveal the different energy costs of endothermy and ectothermy. The BMR for humans averages 6,695–7,930 kJ per day for adult males and 5,440–6,275 kJ per day for adult females. These BMRs are about equivalent to the rate of energy use by a 75-watt lightbulb. In contrast, the SMR of an American alligator is only about 250 kJ per day at 20°C. As this represents less than $\frac{1}{20}$ the energy used by a comparably sized adult human, it is clear that ectothermy has a markedly lower energetic requirement than endothermy.

Influences on Metabolic Rate

Metabolic rate is affected by many factors other than an animal being an endotherm or an ectotherm. Some key factors are age, sex, size, activity, temperature, and nutrition. Here we'll examine the effects of size and activity.

Size and Metabolic Rate

Larger animals have more body mass and therefore require more chemical energy. Remarkably, the relationship between overall metabolic rate and body mass is constant across a wide range of sizes and forms, as illustrated for various mammals in **Figure 40.20a**. In fact, for even more varied organisms ranging in size from bacteria to blue whales, metabolic rate remains roughly proportional to body mass to the three-quarter power ($m^{3/4}$). Scientists are still researching the basis of this relationship, which applies to ectotherms as well as endotherms.

The relationship of metabolic rate to size profoundly affects energy consumption by body cells and tissues. As shown in **Figure 40.20b**, the energy it takes to maintain each gram of body mass is inversely related to body size. Each gram of a mouse, for instance, requires about 20 times as many joules as a gram of an elephant, even though the whole elephant uses far more joules than the whole mouse. The smaller animal's higher metabolic rate per gram demands a higher rate of oxygen delivery. To meet this demand, the smaller animal must have a higher breathing rate, blood volume (relative to its size), and heart rate.

Thinking about body size in bioenergetic terms reveals how trade-offs shape the evolution of body plans. As body size decreases, each gram of tissue increases in energy cost. As body size increases, energy costs per gram of tissue decrease, but an ever-larger fraction of body tissue is required for exchange, support, and locomotion.

Activity and Metabolic Rate

For both ectotherms and endotherms, activity greatly affects metabolic rate. Even a person reading quietly at a desk or an insect twitching its wings consumes energy beyond the BMR or SMR. Maximum metabolic rates (the highest rates of ATP use) occur during peak activity, such as lifting a heavy object, sprinting, or swimming at high speed. In general, the maximum metabolic rate an animal can sustain is inversely related to the duration of activity.

For most terrestrial animals, the average daily rate of energy consumption is two to four times BMR (for endotherms) or SMR (for ectotherms). Humans in most developed countries have an unusually low average daily metabolic rate of about 1.5 times BMR—an indication of a relatively sedentary lifestyle.

The fraction of an animal's energy "budget" that is devoted to activity depends on many factors, including its environment, behaviour, size, and thermoregulation. In the

▼ Figure 40.20 The relationship of metabolic rate to body size.

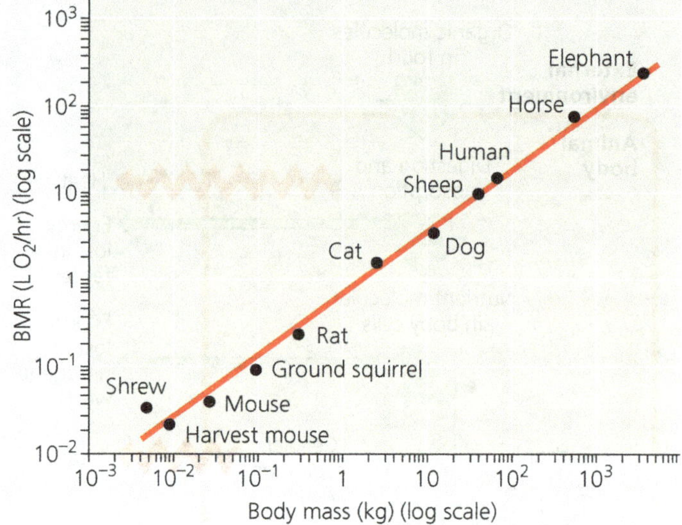

(a) Relationship of basal metabolic rate (BMR) to body size for various mammals. From shrew to elephant, size increases 1 millionfold.

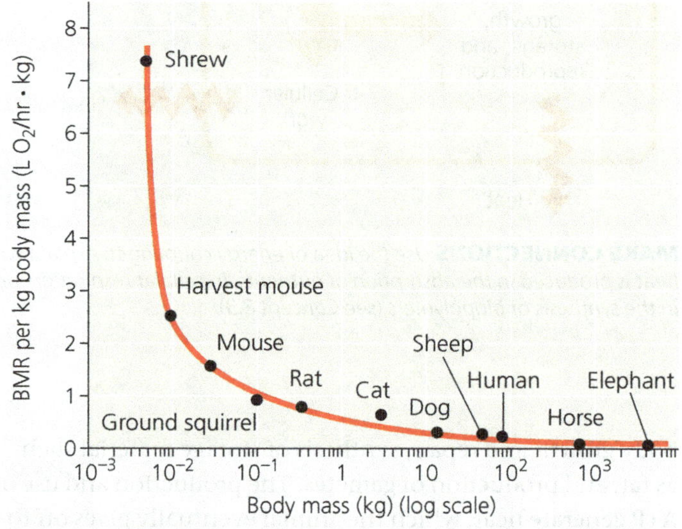

(b) Relationship of BMR per kilogram of body mass to body size for the same mammals as in (a).

INTERPRET THE DATA *Based on the graph in (a), one observer suggests that a group of 100 ground squirrels has the same basal metabolic rate as 1 dog. A second observer looking at the graph disagrees. Who is correct and why?*

Scientific Skills Exercise

Interpreting Pie Charts

How Do Energy Budgets Differ for Three Terrestrial Vertebrates? To explore bioenergetics in animal bodies, let's consider typical annual energy budgets for three terrestrial vertebrates that vary in size and thermoregulatory strategy: a 4-kg male Adélie penguin, a 25-g (0.025-kg) female deer mouse, and a 2-kg female ball python. The penguin is well-insulated against his Antarctic environment but must expend energy in swimming to catch food, incubating eggs laid by his partner, and bringing food to his chicks. The tiny deer mouse lives in a temperate environment where food may be readily available, but her small size causes rapid loss of body heat. Unlike the penguin and mouse, the python is ectothermic and keeps growing throughout her life. She produces eggs but does not incubate them. In this exercise, we'll compare the energy expenditures of these animals for five important functions: basal (standard) metabolism, reproduction, thermoregulation, activity, and growth.

How the Data Were Obtained Energy budgets were calculated for each of the animals based on measurements from field and laboratory studies.

Data from the Experiments Pie charts are a good way to compare *relative* differences in a set of variables. In the pie charts here, the sizes of the wedges represent the relative annual energy expenditures for the functions shown in the key. The total annual expenditure for each animal is given below its pie chart.

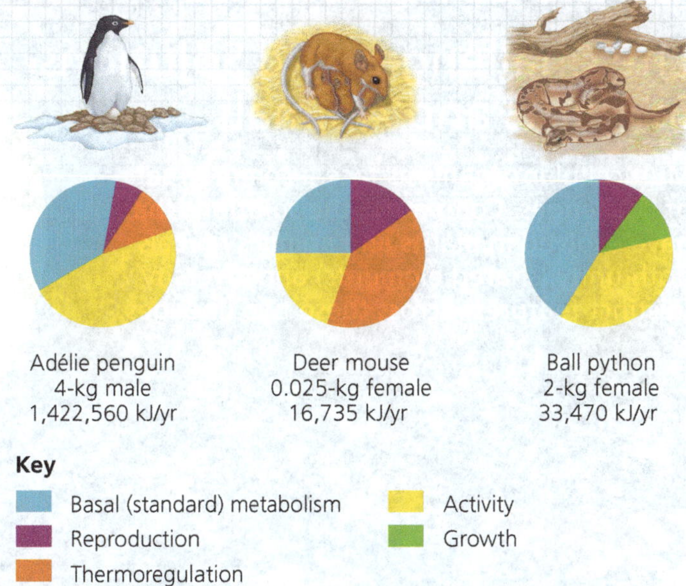

Data from M. A. Chappell et al., Energetics of foraging in breeding Adélie penguins, *Ecology* 74:2450–2461 (1993); M. A. Chappell et al., Voluntary running in deer mice: speed, distance, energy costs, and temperature effects, *Journal of Experimental Biology* 207:3839–3854 (2004); T. M. Ellis and M. A. Chappell, Metabolism, temperature relations, maternal behavior, and reproductive energetics in the ball python (*Python regius*), *Journal of Comparative Physiology B* 157:393–402 (1987).

INTERPRET THE DATA

1. You can estimate the contribution of each wedge in a pie chart by remembering that the entire circle represents 100%, half is 50%, and so on. What percent of the mouse's energy budget goes to basal metabolism? What percent of the penguin's budget is for activity?
2. Without considering the sizes of the wedges, how do the three pie charts differ in which functions they include? Explain these differences.
3. Does the penguin or the mouse expend a greater proportion of its energy budget on thermoregulation? Why?
4. Now look at the *total* annual energy expenditures for each animal. How much more energy does the penguin expend each year compared to the python?
5. Which animal expends the most kilojoules per year on thermoregulation?
6. If you monitored energy allocation in the penguin for just a few months instead of an entire year, you might find the growth category to be a significant part of the pie chart. Given that adult penguins don't grow from year to year, how would you explain this finding?

Scientific Skills Exercise, you'll interpret data on the annual energy budgets of three terrestrial vertebrates.

Torpor and Energy Conservation

Despite their many adaptations for homeostasis, animals may encounter conditions that severely challenge their abilities to balance their heat, energy, and materials budgets. For example, at certain times of the day or year, their surroundings may be extremely hot or cold, or food may be unavailable. A major adaptation that enables animals to save energy in the face of such difficult conditions is **torpor**, a physiological state of decreased activity and metabolism.

Many birds and small mammals (including many small Australian marsupials) exhibit a daily torpor that is well adapted to feeding patterns. Some bats (including some species in Australia and New Zealand) feed at night and go into torpor in daylight. Endothermic birds, such as the Australian owlet-nightjar and tawny frogmouths, also enter torpor, especially in winter months. The echidna *(Tachyglossus aculeatus)* has been shown to increase use of torpor in response to changes in habitat, such as after fire, due to reduced foraging opportunities.

All endotherms that exhibit daily torpor are relatively small; when active, they have high metabolic rates and thus very high rates of energy consumption. The changes in body temperature, and thus the energy savings, are often considerable: the body temperature of chickadees drops as much as 10°C at night, and the core body temperature of a hummingbird can fall 25°C or more.

Hibernation is long-term torpor that is an adaptation to winter cold and food scarcity. When a mammal enters hibernation, its body temperature declines as its body's thermostat is turned down **(Figure 40.21)**. Some hibernating mammals cool to as low as 1–2°C, and at least one,

▼ **Figure 40.23**

MAKE CONNECTIONS

Life Challenges and Solutions in Plants and Animals

Multicellular organisms face a common set of challenges. Comparing the solutions that have evolved in plants and animals reveals both unity (shared elements) and diversity (distinct features) across these two lineages.

Nutritional Mode

All living things must obtain energy and carbon from the environment to grow, survive, and reproduce. Plants are autotrophs, obtaining their energy through photosynthesis and their carbon from inorganic sources, whereas animals are heterotrophs, obtaining their energy and carbon from food. Evolutionary adaptations in plants and animals support these different nutritional modes. The broad surface of many leaves enhances light capture for photosynthesis. When hunting, a tuatara relies on stealth and speed to capture its preferred invertebrate prey. (See Figure 36.2 and Figure 41.16.)

Growth and Regulation

The growth and physiology of both plants and animals are regulated by hormones. In plants, hormones may act in a local area or be transported in the body. They control growth patterns, flowering, fruit development, and more. In animals, hormones circulate throughout the body and act in specific target tissues, controlling homeostatic processes and developmental events such as moulting. (See Figure 39.10 and Figure 45.12.)

Environmental Response

All forms of life must detect and respond appropriately to conditions in their environment. Specialised organs sense environmental signals. For example, the floral head of a sunflower and an insect's eyes both contain photoreceptors that detect light. Environmental signals activate specific receptor proteins, triggering signal transduction pathways that initiate cellular responses coordinated by chemical and electrical communication. (See Figure 39.19 and Figure 50.15.)

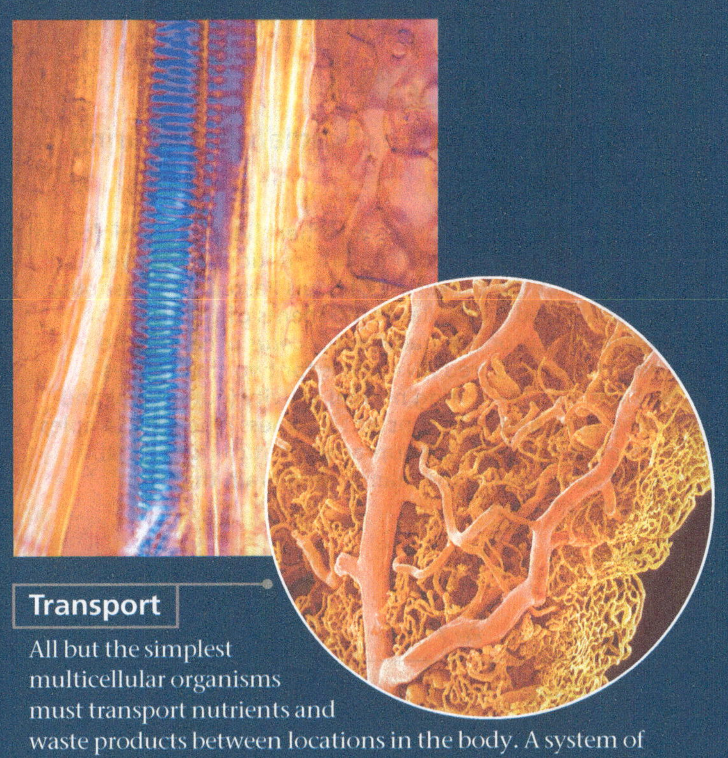

Transport

All but the simplest multicellular organisms must transport nutrients and waste products between locations in the body. A system of tubelike vessels is the common evolutionary solution, while the mechanism of circulation varies. Plants harness solar energy to transport water, minerals, and sugars through specialised tubes (left). In animals, a pump (heart) moves circulatory fluid through vessels (right). (See Figure 35.10 and Figure 42.9.)

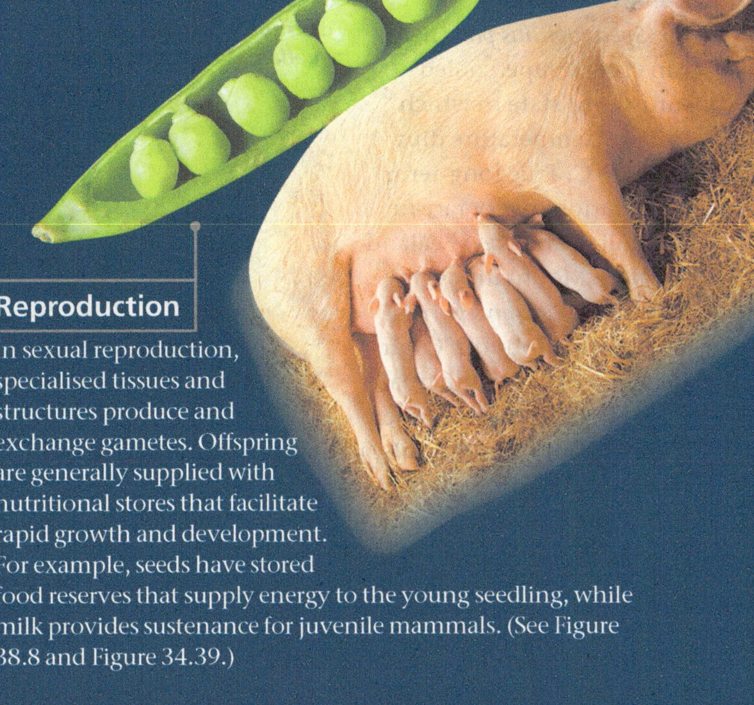

Reproduction

In sexual reproduction, specialised tissues and structures produce and exchange gametes. Offspring are generally supplied with nutritional stores that facilitate rapid growth and development. For example, seeds have stored food reserves that supply energy to the young seedling, while milk provides sustenance for juvenile mammals. (See Figure 38.8 and Figure 34.39.)

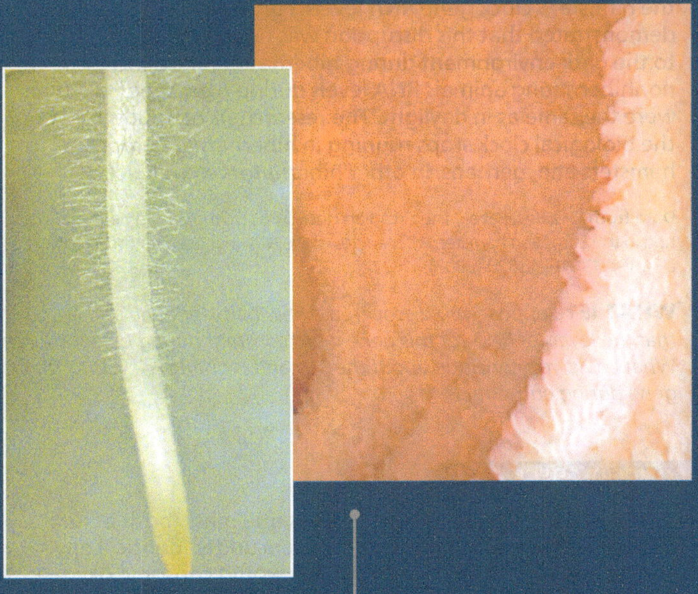

Absorption

Organisms need to absorb nutrients. The root hairs of plants (left) and the villi (projections) that line the intestines of vertebrates (right) increase the surface area available for absorption. (See Figure 35.3 and Figure 41.12.)

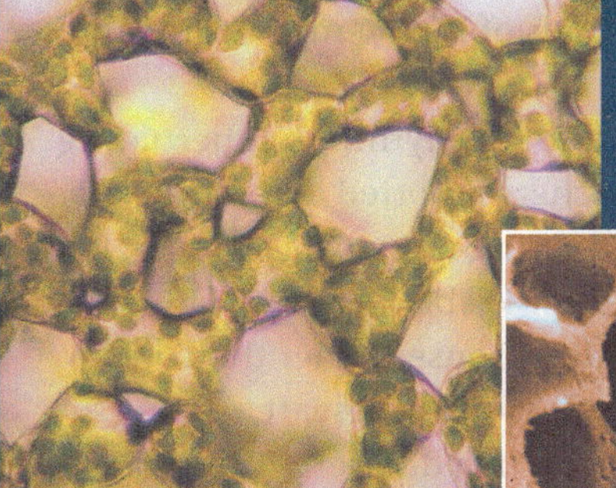

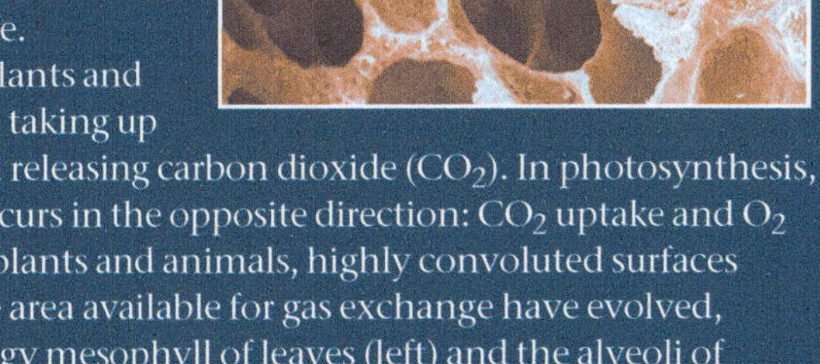

Gas Exchange

The exchange of certain gases with the environment is essential for life. Respiration by plants and animals requires taking up oxygen (O_2) and releasing carbon dioxide (CO_2). In photosynthesis, net exchange occurs in the opposite direction: CO_2 uptake and O_2 release. In both plants and animals, highly convoluted surfaces that increase the area available for gas exchange have evolved, such as the spongy mesophyll of leaves (left) and the alveoli of lungs (right). (See Figure 35.18 and Figure 42.24.)

MAKE CONNECTIONS

Compare the adaptations that enable plants and animals to respond to the challenges of living in hot and cold environments. See Concept 39.4 and Concept 40.3.

the Arctic ground squirrel (*Spermophilus parryii*), can enter a supercooled (unfrozen) state in which its body temperature dips below 0°C. True long-term hibernation is very rare in Australian and New Zealand animals. Mountain pygmy possums (*Burramys parvus*) can cool to around 2°C during winter as they hibernate in the boulder-fields of their alpine habitat. Periodically, perhaps every two weeks or so, hibernating animals undergo arousal, raising their body temperature and becoming active briefly before resuming hibernation. Metabolic rates during hibernation can be 20 times lower than if the animal attempted to maintain normal body temperatures of 36–38°C. As a result, hibernators such as the ground squirrel can survive through the winter on limited supplies of energy stored in the body tissues or as food cached in a burrow. Similarly, the slow metabolism and inactivity of *aestivation*, or summer torpor, enable animals to survive long periods of high temperatures and scarce water.

▼ Figure 40.21 A hazel dormouse (*Muscardinus avellanarius*) hibernating.

What happens to the circadian rhythm in hibernating animals? In the past, researchers reported detecting daily biological rhythms in hibernating animals. However, in some cases the animals were probably in a state of torpor from which they could readily arouse, rather than "deep" hibernation. More recently, a group of researchers in France addressed this question in a different way, examining the machinery of the biological clock rather than the rhythms it controls **(Figure 40.22)**. Working with the European hamster, they found that molecular components of the clock stopped oscillating during hibernation. These findings support the hypothesis that the circadian clock ceases operation during hibernation, at least in this species.

From tissue types to homeostasis, this chapter has focused on the whole animal. We also investigated how animals exchange materials with the environment and how size and activity affect metabolic rate. For much of the rest of this unit, we'll explore how specialised organs and organ systems enable animals to meet the basic challenges of life. In Unit 6, we investigated how plants meet the same challenges. **Figure 40.23**, on the next two pages, highlights some fundamental similarities and differences in the evolutionary adaptations of plants and animals. This figure is thus a review of Unit 6, an introduction to Unit 7, and, most importantly, an illustration of the connections that unify the myriad forms of life.

▼ Figure 40.22 Inquiry

What happens to the circadian clock during hibernation?

Experiment To determine whether the 24-hour biological clock continues to run during hibernation, Paul Pévet and colleagues at the University of Louis Pasteur in Strasbourg, France, studied molecular components of the circadian clock in the European hamster (*Cricetus cricetus*). The researchers measured RNA levels for two clock genes—*Per2* and *Bmal1*—during normal activity (euthermia) and during hibernation in constant darkness. The RNA samples were obtained from the suprachiasmatic nuclei (SCN), a pair of structures in the mammalian brain that control circadian rhythms.

Results

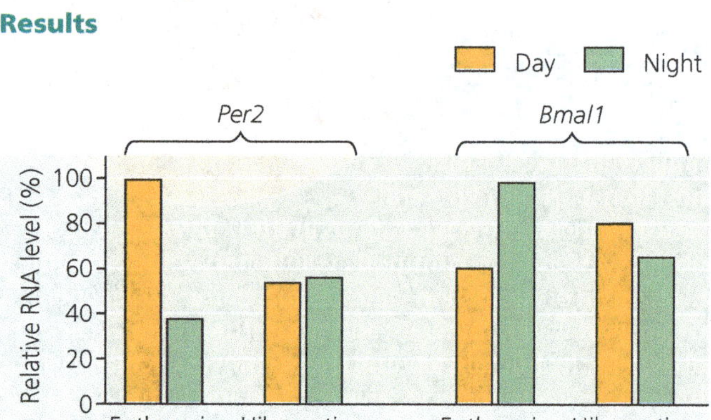

Conclusion Hibernation disrupted circadian variation in the hamster's clock gene RNA levels. Further experiments demonstrated that this disruption was not simply due to the dark environment during hibernation, since for nonhibernating animals RNA levels during a darkened daytime were the same as in daylight. The researchers concluded that the biological clock stops running in hibernating European hamsters and, perhaps, in other hibernators as well.

Data from F. G. Revel et al., The circadian clock stops ticking during deep hibernation in the European hamster, *Proceedings of the National Academy of Sciences USA* 104:13816–13820 (2007).

WHAT IF? *Suppose you discovered a new hamster gene and found that the levels of RNA for this gene were constant during hibernation. What could you conclude about the day and night RNA levels for this gene during euthermia?*

CONCEPT CHECK 40.4

1. If a mouse and a small lizard of the same mass (both at rest) were placed in experimental chambers under identical environmental conditions, which animal would consume oxygen at a higher rate? Explain.
2. Which animal must eat a larger proportion of its weight in food each day: a house cat or an African lion caged in a zoo? Explain.
3. **WHAT IF?** Suppose the animals at a zoo were resting comfortably and remained at rest while the nighttime air temperature dropped. If the temperature change were sufficient to cause a change in metabolic rate, what changes would you expect for an alligator and a lion?

For suggested answers, see Appendix A.

40 Chapter Review

SUMMARY OF KEY CONCEPTS

CONCEPT 40.1

Animal form and function are correlated at all levels of organisation *(pp. 923–930)*

- Physical laws constrain the evolution of an animal's size and shape. These constraints contribute to convergent evolution in animal body forms.
- Each animal cell must have access to an aqueous environment. Simple two-layered sacs and flat shapes maximise exposure to the surrounding medium. More complex body plans have highly folded internal surfaces specialised for exchanging materials.
- Animal bodies are based on a hierarchy of cells, **tissues, organs**, and **organ systems. Epithelial tissue** forms active interfaces on external and internal surfaces; **connective tissue** binds and supports other tissues; **muscle tissue** contracts, moving body parts; and **nervous tissue** transmits nerve impulses throughout the body.
- The **endocrine** and **nervous systems** are the two means of communication between different locations in the body. The endocrine system broadcasts signalling molecules called **hormones** everywhere via the bloodstream, but only certain cells are responsive to each hormone. The nervous system uses dedicated cellular circuits involving electrical and chemical signals to send information to specific locations.

? *For a large animal, what challenges would a spherical shape pose for carrying out exchange with the environment?*

CONCEPT 40.2

Feedback control maintains the internal environment in many animals *(pp. 930–932)*

- An animal is a **regulator** if it controls an internal variable and a **conformer** if it allows an internal variable to vary with external changes. **Homeostasis** is the maintenance of a steady state despite internal and external changes.
- Homeostatic mechanisms are usually based on **negative feedback**, in which the **response** reduces the **stimulus**. In contrast, **positive feedback** involves amplification of a stimulus by the response and often brings about a change in state, such as the transition from pregnancy to childbirth.

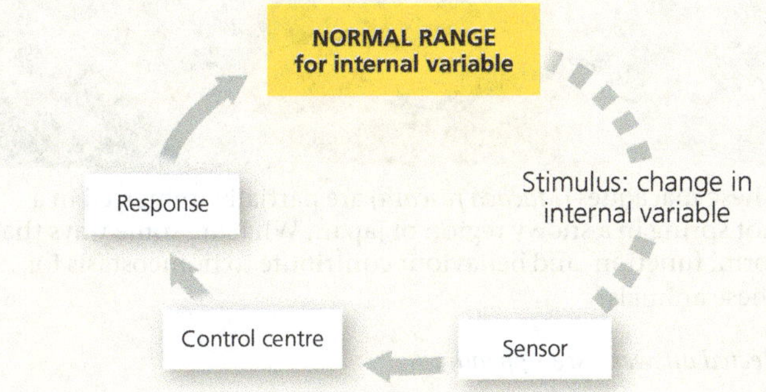

- Regulated change in the internal environment is essential to normal function. **Circadian rhythms** are daily fluctuations in metabolism and behaviour tuned to the cycles of light and dark in the environment. Other environmental changes may trigger **acclimatisation**, a temporary shift in the steady state.

? *Is it accurate to define homeostasis as a constant internal environment? Explain.*

CONCEPT 40.3

Homeostatic processes for thermoregulation involve form, function, and behaviour *(pp. 933–938)*

- An animal maintains its internal temperature within a tolerable range by **thermoregulation. Endothermic** animals are warmed mostly by heat generated by metabolism. **Ectothermic** animals get most of their heat from external sources. Endothermy requires a greater expenditure of energy. Body temperature may vary with environmental temperature, as in *poikilotherms*, or be relatively constant, as in *homeotherms*.
- In thermoregulation, physiological and behavioural adjustments balance heat gain and loss, which occur through radiation, evaporation, convection, and conduction. Insulation and **countercurrent exchange** reduce heat loss, whereas panting, sweating, and bathing increase evaporation, cooling the body. Many ectotherms and endotherms adjust their rate of heat exchange with their surroundings by vasodilation or vasoconstriction and by behavioural responses.
- Many mammals and birds adjust their amount of body insulation in response to changes in environmental temperature. Ectotherms undergo a variety of changes at the cellular level to acclimatise to shifts in temperature.
- The **hypothalamus** acts as the thermostat in mammalian regulation of body temperature. Fever reflects a resetting of this thermostat to a higher normal range in response to infection.

? *Given that humans thermoregulate, explain why your skin is cooler than your body core.*

CONCEPT 40.4

Energy requirements are related to animal size, activity, and environment *(pp. 938–944)*

- Animals obtain chemical energy from food, storing it for short-term use in ATP. The total amount of energy used in a unit of time defines an animal's **metabolic rate**.
- Under similar conditions and for animals of the same size, the **basal metabolic rate** of endotherms is substantially higher than the **standard metabolic rate** of ectotherms. Minimum metabolic rate per gram is inversely related to body size among similar animals. Animals allocate energy for basal (or standard) metabolism, activity, homeostasis, growth, and reproduction.
- **Torpor**, a state of decreased activity and metabolism, conserves energy during environmental extremes. Animals may enter torpor according to a circadian rhythm (daily torpor), in winter (**hibernation**), or in summer (aestivation).

? *Why do small animals breathe more rapidly than large animals?*

TEST YOUR UNDERSTANDING

Levels 1-2: Remembering/Understanding

1. The body tissue that consists largely of material located outside of cells is
(A) epithelial tissue.
(B) connective tissue.
(C) muscle tissue.
(D) nervous tissue.

2. Which of the following would increase the rate of heat exchange between an animal and its environment?
(A) feathers or fur
(B) vasoconstriction
(C) wind blowing across the body surface
(D) countercurrent heat exchanger

3. Consider the energy budgets for a human, an elephant, a penguin, a mouse, and a snake. The ________ would have the highest total annual energy expenditure, and the ________ would have the highest energy expenditure per unit mass.
(A) elephant; mouse
(B) elephant; human
(C) mouse; snake
(D) penguin; mouse

Levels 3-4: Applying/Analysing

4. Compared with a smaller cell, a larger cell of the same shape has
(A) less surface area.
(B) less surface area per unit of volume.
(C) the same surface-area-to-volume ratio.
(D) a smaller cytoplasm-to-nucleus ratio.

5. An animal's inputs of energy and materials would exceed its outputs
(A) if the animal is an endotherm, which must always take in more energy because of its high metabolic rate.
(B) if it is actively foraging for food.
(C) if it is growing and increasing its mass.
(D) never; due to homeostasis, these energy and material budgets always balance.

6. You are studying a large tropical reptile that has a high and relatively stable body temperature. How do you determine whether this animal is an endotherm or an ectotherm?
(A) You know from its high and stable body temperature that it must be an endotherm.
(B) You subject this reptile to various temperatures in the lab and find that its body temperature and metabolic rate change with the ambient temperature. You conclude that it is an ectotherm.
(C) You note that its environment has a high and stable temperature. Because its body temperature matches the environmental temperature, you conclude that it is an ectotherm.
(D) You measure the metabolic rate of the reptile, and because it is higher than that of a related species that lives in temperate forests, you conclude that this reptile is an endotherm and its relative is an ectotherm.

7. Which of the following animals uses the largest percentage of its energy budget for homeostatic regulation?
(A) marine jellyfish (an invertebrate)
(B) snake in a temperate forest
(C) desert insect
(D) desert bird

8. **DRAW IT** Draw a model of the control circuit(s) required for driving an automobile at a fairly constant speed over a hilly road. Indicate each feature that represents a sensor, stimulus, or response.

Levels 5-6: Evaluating/Creating

9. **EVOLUTION CONNECTION** In 1847, the German biologist Christian Bergmann noted that mammals and birds living at higher latitudes (further from the equator) are on average larger and bulkier than related species found at lower latitudes. Suggest an evolutionary hypothesis to explain this observation.

10. **SCIENTIFIC INQUIRY** Eastern tent caterpillars (*Malacosoma americanum*) live in large groups in silk nests resembling tents, which they build in trees. They are among the first insects to be active in early spring, when daily temperature fluctuates from freezing to very hot. Over the course of a day, they display striking differences in behaviour: Early in the morning, they rest in a tightly packed group on the tent's east-facing surface. In midafternoon, they are on its undersurface, each caterpillar hanging by a few of its legs. Propose a hypothesis to explain this behaviour. How could you test it?

11. **SCIENCE, TECHNOLOGY, AND SOCIETY** Medical researchers are investigating artificial substitutes for various human tissues. Why might artificial blood or skin be useful? What characteristics would these substitutes need in order to function well in the body? Why do real tissues work better? Why not use the real tissues if they work better? What other artificial tissues might be useful? What problems do you anticipate in developing and applying them?

12. **WRITE ABOUT A THEME: ENERGY AND MATTER** In a short essay (about 100–150 words) focusing on energy transfer and transformation, discuss the advantages and disadvantages of hibernation.

13. **SYNTHESISE YOUR KNOWLEDGE**

These macaques (*Macaca fuscata*) are partially immersed in a hot spring in a snowy region of Japan. What are some ways that form, function, and behaviour contribute to homeostasis for these animals?

For selected answers, see Appendix A.

45 Hormones and the Endocrine System

KEY CONCEPTS

45.1 **Hormones and other signalling molecules bind to target receptors, triggering specific response pathways** *p. 1050*

45.2 **Feedback regulation and coordination with the nervous system are common in hormone pathways** *p. 1054*

45.3 **Endocrine glands respond to diverse stimuli in regulating homeostasis, development, and behaviour** *p. 1061*

Study Tip

Make a flowchart: Many hormones, such as insulin, parathyroid hormone, and adrenaline, have multiple physiological effects in a single organism. To keep track of the action and function of each such hormone, make a flowchart like this example. Use arrows to indicate how the hormone's diverse effects contribute to an overall outcome for the organism.

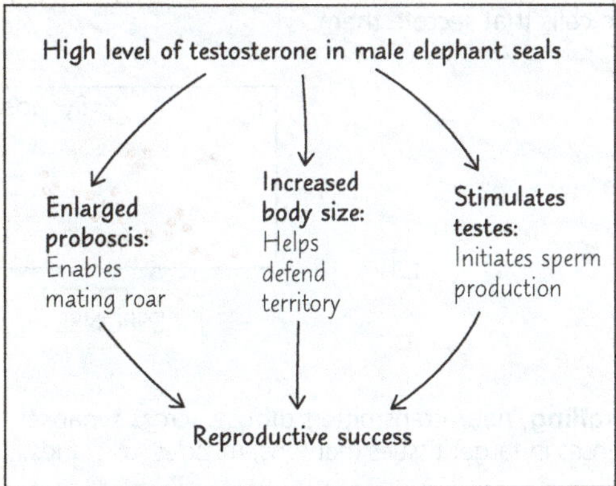

Go to Mastering Biology

to access Dynamic Study Modules for revision, 3D BioFlix® animations and high-quality videos, and your interactive Pearson eText.

Figure 45.1 Male and female elephant seals (*Mirounga angustirostris*) differ greatly in appearance and behaviour. The male is much larger, and only he has the prominent proboscis for which the species is named. The male is also far more territorial, using the proboscis to emit loud roars during mating season. Underlying each of these differences is a single hormone—testosterone. Like all hormones, testosterone is an endocrine signalling molecule that circulates in the blood throughout the body.

What variables shape a hormone's effect on an animal's body and behaviour?

Concentration of the hormone in the body:

Testosterone is present in both male and female mammals, but typically at a much higher concentration in males.

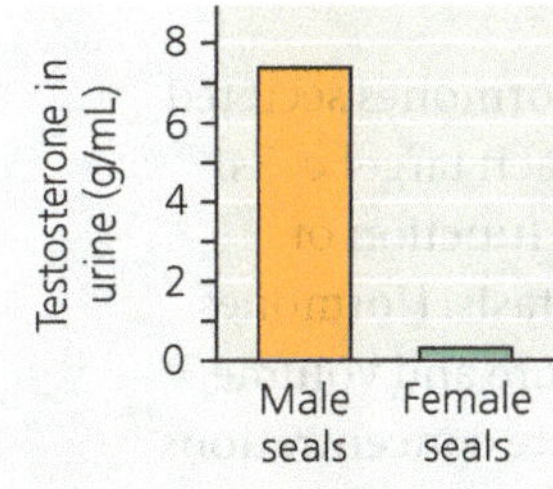

Presence of the hormone receptor in a cell:

A hormone circulates throughout the bloodstream, but cells only respond to a hormone if they have a receptor that binds that hormone specifically. The receptors in these cells can be present either on the surface of the cell (cell surface receptor) or in the inside of the cell (intracellular receptor).

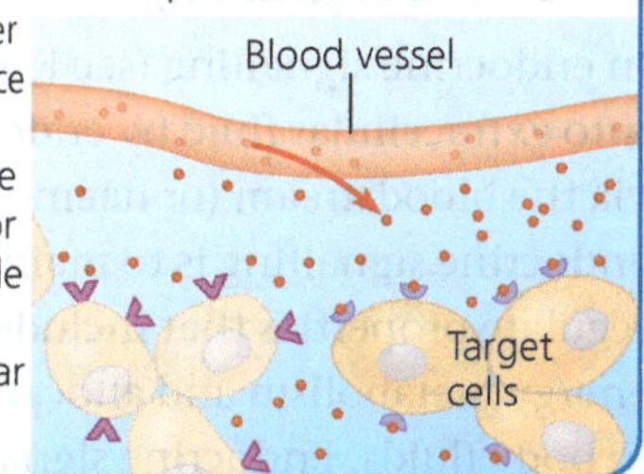

Response of the cell when the receptor binds the hormone:

Testosterone binding to receptor causes different responses in different cells.

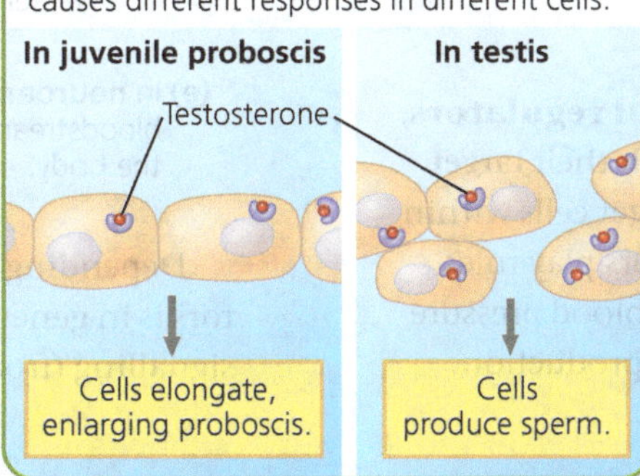

▼ **Male elephant seals sparring**

CONCEPT 45.1

Hormones and other signalling molecules bind to target receptors, triggering specific response pathways

A **hormone** (from the Greek *horman*, to excite) is a secreted molecule that circulates throughout the body and stimulates specific cells. Although a given hormone reaches all cells of the body, it only elicits a response—such as a change in metabolism—in specific *target cells*, those that have a receptor that binds the hormone specifically. Cells lacking a receptor for that hormone are unaffected.

Chemical signalling by hormones is the function of the **endocrine system**, one of the two basic systems for communication and regulation in the animal body. The other major communication and control system is the **nervous system**, a network of specialised cells—neurons—that transmit signals along dedicated pathways. These signals in turn regulate neurons, muscle cells, and endocrine cells. Because signalling by neurons can regulate the release of hormones, the nervous and endocrine systems often overlap in function.

As a background to our further exploration of the endocrine system, we'll begin with an overview of the diverse ways that animal cells use chemical signals to communicate.

Intercellular Information Flow

Communication between animal cells via secreted signals is often classified by two criteria: the type of secreting cell and the route taken by the signal in reaching its target. **Figure 45.2** illustrates five forms of signalling distinguished in this manner.

Endocrine Signalling

In endocrine signalling (see Figure 45.2a), hormones secreted into extracellular fluid by endocrine cells reach target cells via the bloodstream (or haemolymph). One function of endocrine signalling is to maintain homeostasis. Hormones regulate properties that include blood pressure and volume, energy metabolism and allocation, and solute concentrations in body fluids. Endocrine signalling also mediates responses to environmental stimuli, regulates growth and development, and triggers physical and behavioural changes underlying sexual maturity and reproduction (see Figure 45.1).

Paracrine and Autocrine Signalling

Many types of cells produce and secrete **local regulators**, molecules that act over short distances, reach their target cells solely by diffusion, and act on their target cells within seconds or even milliseconds. Local regulators play roles in many physiological processes, including blood pressure regulation, nervous system function, and reproduction.

▼ **Figure 45.2 Intercellular communication by secreted molecules.** In each type of signalling, secreted molecules (•) bind to a specific receptor protein (◡) expressed by target cells. Some receptors are located inside cells, but for simplicity, here all are drawn on the cell surface.

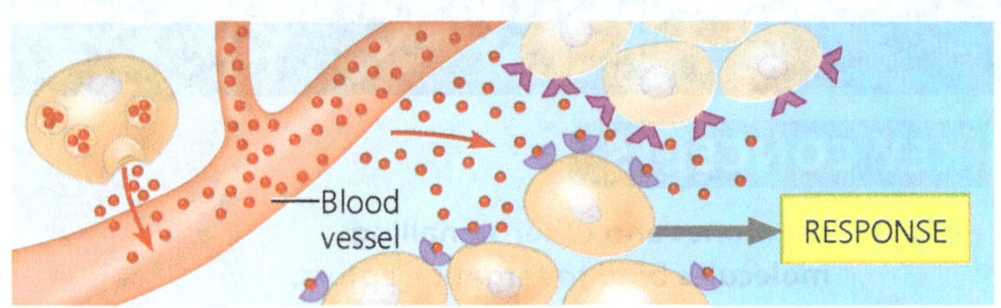

(a) In **endocrine signalling**, secreted molecules diffuse into the bloodstream and trigger responses in target cells anywhere in the body.

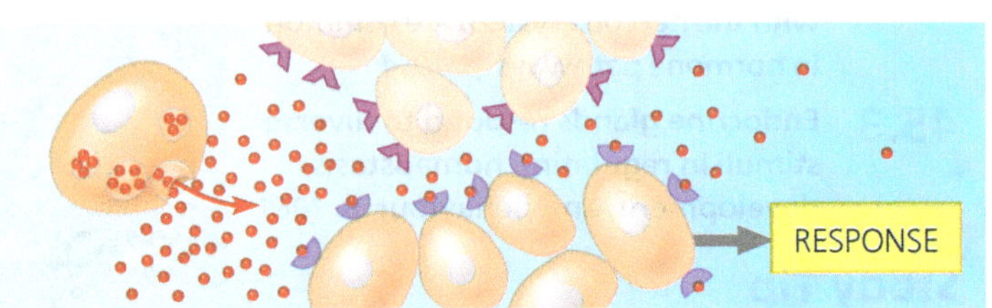

(b) In **paracrine signalling**, secreted molecules diffuse locally and trigger a response in neighbouring cells.

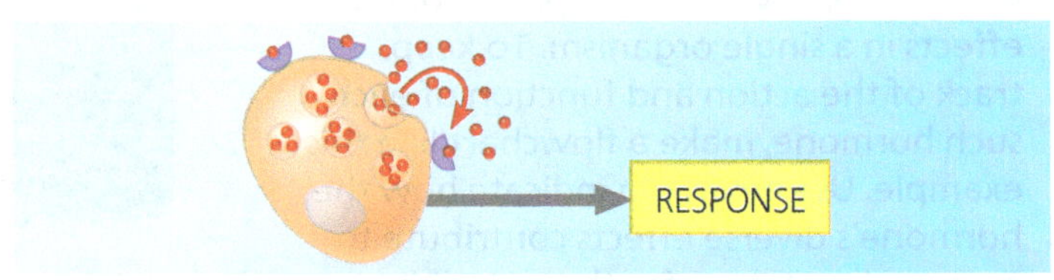

(c) In **autocrine signalling**, secreted molecules diffuse locally and trigger a response in the cells that secrete them.

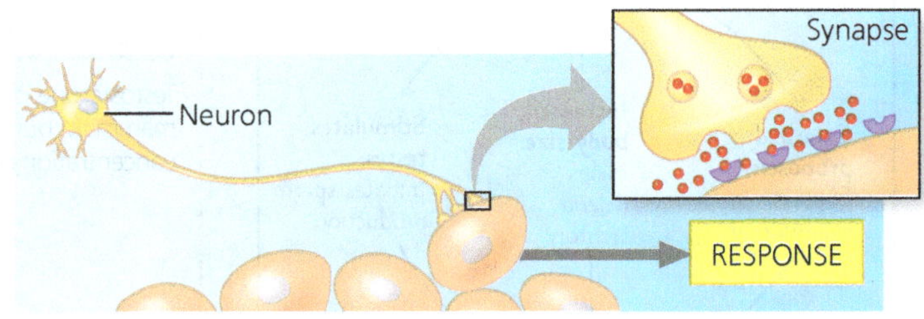

(d) In **synaptic signalling**, neurotransmitters diffuse across synapses and trigger responses in target tissues (neurons, muscles, or glands).

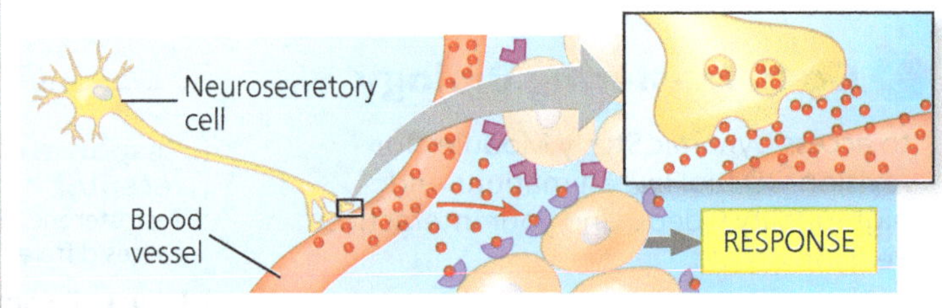

(e) In **neuroendocrine signalling**, neurohormones diffuse into the bloodstream and trigger responses in target cells anywhere in the body.

Depending on the target cell, signalling by local regulators is in general either paracrine or autocrine. In **paracrine** signalling (from the Greek *para*, to one side of), target cells lie near the secreting cell (see Figure 45.2b). In **autocrine** signalling (from the Greek *auto*, self), the secreting cells themselves are the target cells (see Figure 45.2c).

One group of local regulators are the **prostaglandins**, which are produced throughout the body and have diverse functions. In the immune system, for example, prostaglandins promote inflammation and the sensation of pain in response to injury. Drugs that block prostaglandin synthesis, such as aspirin and ibuprofen, prevent these activities, producing both anti-inflammatory and pain-relieving effects.

Prostaglandins are modified fatty acids. Many other local regulators are polypeptides, including cytokines, which enable immune cell communication (see Figure 43.16 and Figure 43.17), and growth factors, which promote cell growth, division, and development.

Some local regulators, such as **nitric oxide (NO)**, are gases. When the level of oxygen in the blood falls, endothelial cells in blood vessel walls synthesise and release NO. After diffusing into the surrounding smooth muscle cells, NO activates an enzyme that relaxes the cells. The result is vasodilation, which increases blood flow to tissues.

In human males, NO's ability to promote vasodilation enables sexual function by increasing blood flow into the penis, producing an erection. The drug Viagra (sildenafil citrate), a treatment for male erectile dysfunction, sustains an erection by prolonging activity of the NO response pathway.

Synaptic and Neuroendocrine Signalling

Secreted molecules are essential for the function of the nervous system. Neurons communicate with target cells, such as other neurons and muscle cells, via specialised junctions called synapses. At most synapses, neurons secrete molecules called **neurotransmitters** that diffuse a very short distance (a fraction of a cell diameter) and bind to receptors on the target cells (see Figure 45.2d). Such *synaptic signalling* is central to sensation, memory, cognition, and movement (as we'll explore in the chapters, "Neurons, Synapses, and Signalling," "Nervous Systems" and "Sensory and Motor Mechanisms").

In *neuroendocrine signalling*, neurons called neurosecretory cells secrete **neurohormones**, which diffuse from nerve cell endings into the bloodstream (see Figure 45.2e). One example of a neurohormone is antidiuretic hormone, which functions in kidney function and water balance as well as courtship behaviour. Many neurohormones regulate endocrine signalling, as we'll discuss later in this chapter.

▼ Figure 45.3 Signalling by pheromones. Using their lowered antennae, these Asian army ants (*Leptogenys distinguenda*) carry pupae and larvae along a pheromone-marked trail to a new nest site.

Signalling by Pheromones

Not all secreted signalling molecules act within the body. Members of a particular animal species sometimes communicate with each other via **pheromones**, chemicals that are released into the external environment. For example, when a foraging ant discovers a new food source, it marks its path back to the nest with a pheromone. Ants also use pheromones for guidance when a colony migrates to a new location **(Figure 45.3)**.

Pheromones serve a wide range of functions that include defining territories, warning of predators, and attracting potential mates. The polyphemus moth (*Antheraea polyphemus*) provides a noteworthy example: The sex pheromone released into the air by a female enables her to attract a male of the species from up to 4.5 km away. You'll read more about pheromone function when we take up the topic of animal behaviour in the chapter, "Animal Behaviour."

Chemical Classes of Hormones

Hormones fall into three major chemical classes: polypeptides, steroids, and amines **(Figure 45.4)**. The hormone insulin, for example, is a polypeptide that contains two chains in its active form. Steroid hormones, such as cortisol, are lipids

▼ Figure 45.4 Variation in hormone solubility and structure.

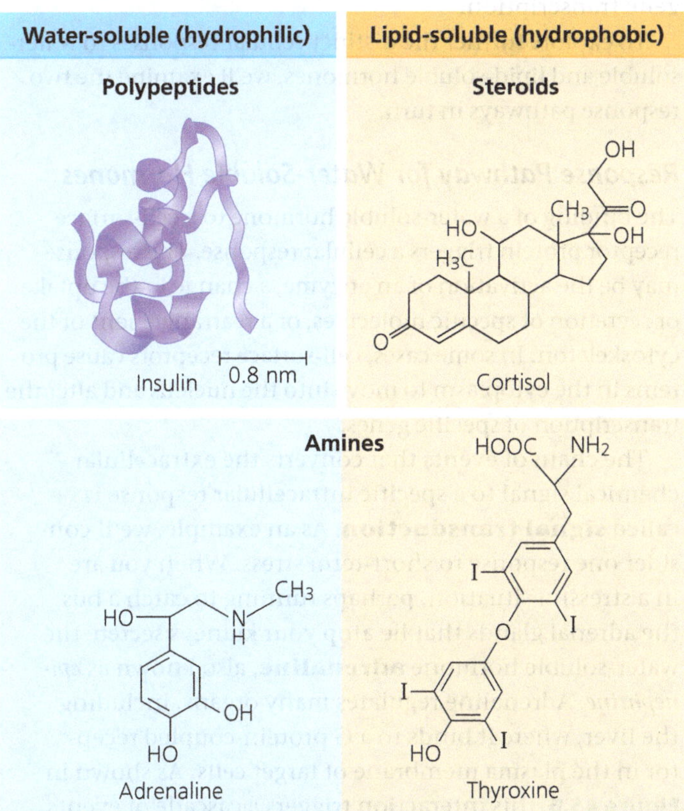

MAKE CONNECTIONS *Cells synthesise adrenaline from the amino acid tyrosine. On the structure of adrenaline shown above, draw a circle around the portion of the molecule corresponding to the R group of tyrosine (see Figure 5.14).*

that contain four fused carbon rings; all are derived from the steroid cholesterol (see Figure 5.12). Adrenaline and thyroxine are amine hormones, each synthesised from a single amino acid, either tyrosine or tryptophan.

As Figure 45.4 indicates, hormones vary in their solubility in aqueous and lipid-rich environments. Polypeptides and most amine hormones are water-soluble, whereas steroid hormones and other largely nonpolar (hydrophobic) hormones, such as thyroxine, are lipid-soluble.

Cellular Hormone Response Pathways

Water-soluble and lipid-soluble hormones differ in their response pathways. One key difference is the location of the receptor proteins in target cells. Water-soluble hormones are secreted by exocytosis and travel freely in the bloodstream. Being insoluble in lipids, they cannot diffuse through the plasma membranes of target cells. Instead, these hormones bind to cell-surface receptors, inducing changes in cytoplasmic molecules and sometimes altering gene transcription **(Figure 45.5a)**. In contrast, lipid-soluble hormones exit endocrine cells by diffusing out across the membranes. They then bind to transport proteins, which keep them soluble in blood. After circulating in the blood, they diffuse into target cells and typically bind to receptors in the cytoplasm or nucleus **(Figure 45.5b)**. The hormone-bound receptor then triggers changes in gene transcription.

To explore further the distinct cellular responses to water-soluble and lipid-soluble hormones, we'll examine the two response pathways in turn.

▼ **Figure 45.5 Variation in hormone receptor location.**

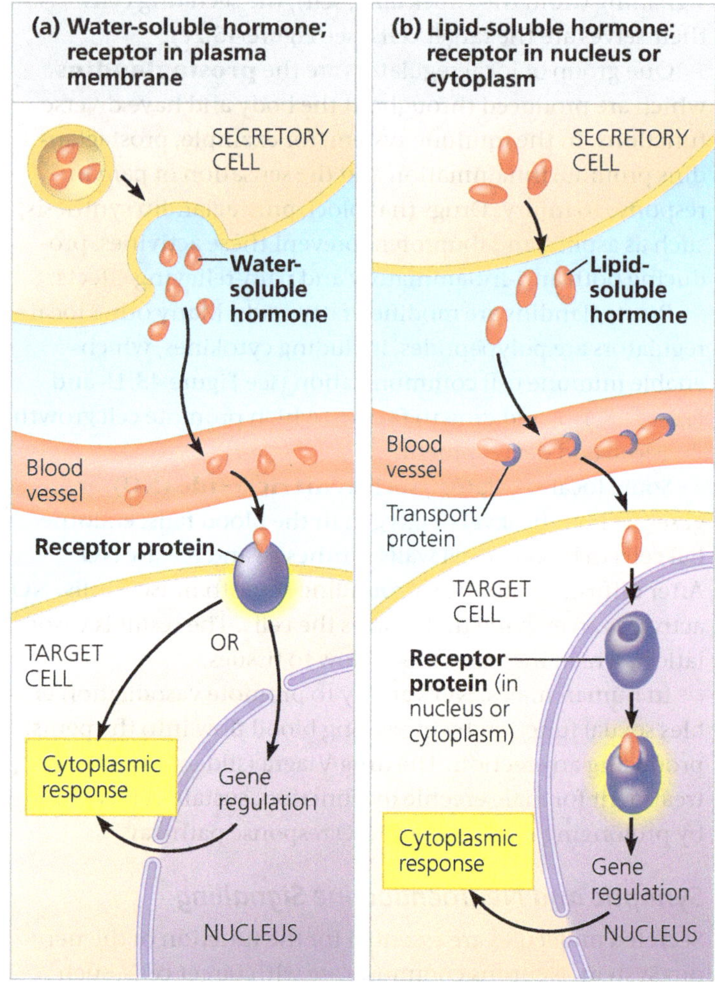

WHAT IF? *Suppose you are studying a cell's response to a particular hormone. You observe that the cell produces the same response to the hormone whether or not the cell is treated with a chemical that blocks transcription. What can you surmise about the hormone and its receptor?*

Response Pathway for Water-Soluble Hormones

The binding of a water-soluble hormone to a cell-surface receptor protein triggers a cellular response. The response may be the activation of an enzyme, a change in the uptake or secretion of specific molecules, or a rearrangement of the cytoskeleton. In some cases, cell-surface receptors cause proteins in the cytoplasm to move into the nucleus and alter the transcription of specific genes.

The chain of events that converts the extracellular chemical signal to a specific intracellular response is called **signal transduction**. As an example, we'll consider one response to short-term stress. When you are in a stressful situation, perhaps running to catch a bus, the adrenal glands that lie atop your kidneys secrete the water-soluble hormone **adrenaline**, also known as *epinephrine*. Adrenaline regulates many organs, including the liver, where it binds to a G protein-coupled receptor in the plasma membrane of target cells. As shown in **Figure 45.6**, this interaction triggers a cascade of events involving synthesis of cyclic AMP (cAMP) as a short-lived *second messenger*. Activation of protein kinase A by cAMP leads to activation of an enzyme required for breakdown of glycogen into glucose, as well as inactivation of an enzyme needed for glycogen synthesis.

Note that there are three enzymes in this signal transduction cascade—adenylyl cyclase (which converts AMP to its cyclic form), protein kinase A, and, for example, the enzyme that breaks down glycogen into glucose. Each enzyme-catalysed step in the cascade provides an opportunity for signal amplification: One enzyme molecule can catalyse many reactions, thereby generating multiple signals at that step in the cascade. Furthermore, because the three enzymes act at different steps in the same pathway, the net effect can be enormous. If, for instance, each enzyme carried out 1,000 reactions, the binding of one molecule of adrenaline to its receptor would trigger cleavage of a billion ($10^3 \times 10^3 \times 10^3$) glycogen molecules. The net result is that the liver releases a substantial

▼ **Figure 45.6 Signal transduction triggered by a cell-surface hormone receptor.**

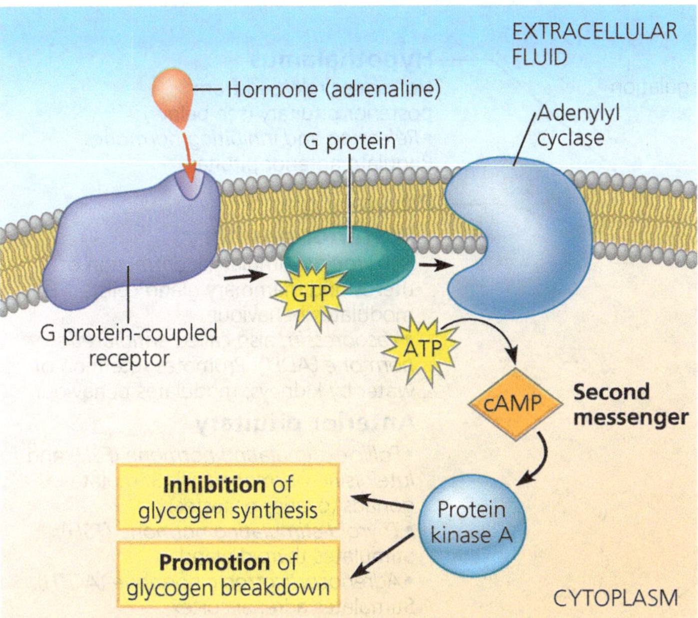

VISUAL SKILLS *A series of arrows represents the steps linking adrenaline to protein kinase A. How does the event represented by the arrow between ATP and cAMP differ from the other four?*

▼ **Figure 45.7 Direct regulation of gene expression by a steroid hormone receptor.**

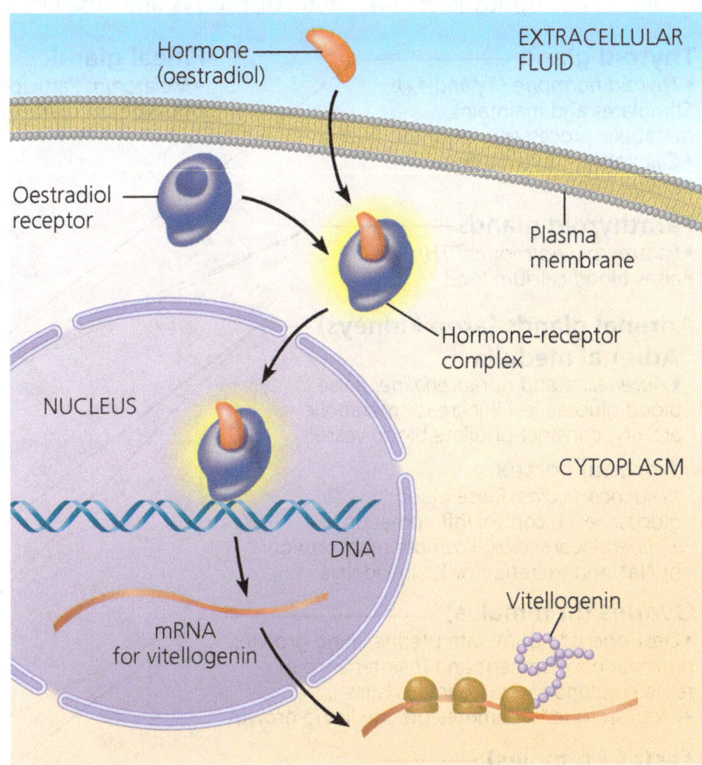

quantity of glucose into the bloodstream, quickly providing the body with extra fuel.

Response Pathway for Lipid-Soluble Hormones

Intracellular receptors for lipid-soluble hormones perform the entire task of transducing a signal within a target cell. The hormone activates the receptor, which then directly triggers the cell's response. In most cases, the response to a lipid-soluble hormone is a change in gene expression.

Most steroid hormone receptors are located in the cytosol prior to binding to a hormone. Binding of a steroid hormone to its cytosolic receptor forms a complex that moves into the nucleus (see Figure 11.9). There, the receptor portion of the complex interacts with a specific DNA-binding protein or response element in the DNA, altering transcription of particular genes. (In some cell types, steroid hormones trigger additional responses by interacting with other kinds of receptor proteins located at the cell surface).

Among the best-characterised steroid hormone receptors are those that bind to oestrogens, steroid hormones necessary for female reproductive function in vertebrates. For example, in female birds and frogs, oestradiol, a form of oestrogen, binds to a cytoplasmic receptor in liver cells. Binding of oestradiol to this receptor activates transcription of the vitellogenin gene **(Figure 45.7)**. Following translation of the messenger RNA, vitellogenin protein is secreted and transported in the blood to the reproductive system, where it is used to produce egg yolk.

Thyroxine, vitamin D, and other lipid-soluble hormones that are not steroids typically have receptors in the nucleus. These receptors bind to hormone molecules that diffuse from the bloodstream across both the plasma membrane and nuclear envelope. Once bound to a hormone, the receptor binds to specific sites in the cell's DNA and stimulates the transcription of specific genes.

Multiple Responses to a Single Hormone

Although hormones bind to specific receptors, a particular hormone can vary in its effects. A hormone can elicit distinct responses in particular target cells if those cells differ in receptor type or in the molecules that produce the response. In this way a single hormone can trigger a range of activities that together bring about a coordinated response to a stimulus. For example, the multiple effects of adrenaline form the basis for the "fight-or-flight" response, a rapid response to stress that you'll read about in Concept 45.3.

Endocrine Tissues and Organs

Some endocrine cells are found in organs that are part of other organ systems. For example, the stomach contains isolated endocrine cells that help regulate digestive processes by secreting the hormone gastrin. More often, endocrine cells are grouped in ductless organs called **endocrine glands**,

▼ **Figure 45.8 Human endocrine glands and their hormones.** This figure highlights the location and primary functions of the major human endocrine glands. Endocrine tissues and cells are also located in the thymus, heart, liver, stomach, kidneys, and small intestine.

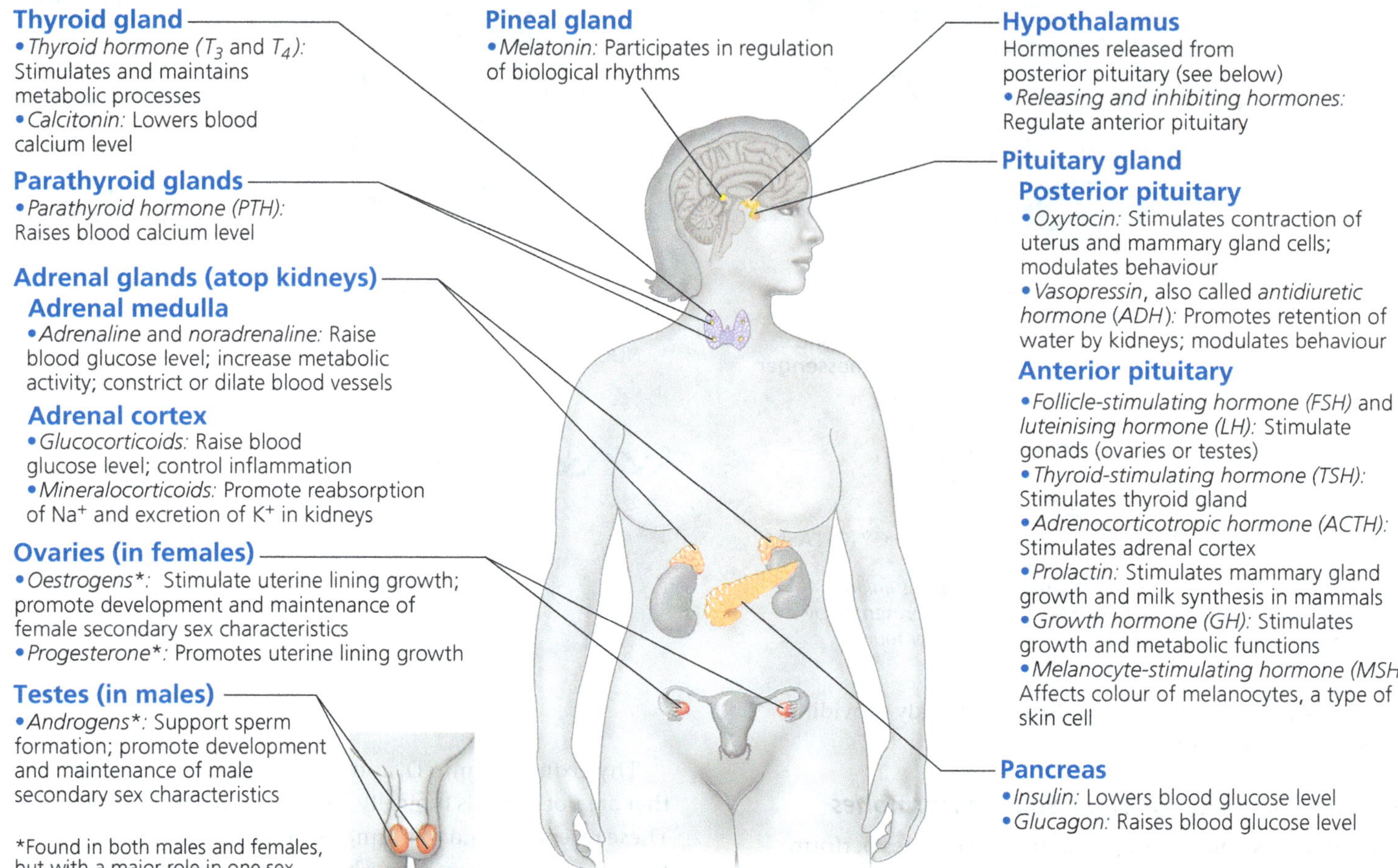

such as the thyroid and parathyroid glands and the gonads, either testes in males or ovaries in females **(Figure 45.8)**.

Note that endocrine glands secrete hormones directly into the surrounding fluid. In contrast, *exocrine glands* have ducts that carry secreted substances, such as sweat or saliva, onto body surfaces or into body cavities. This distinction is reflected in the glands' names: The Greek *endo* (within) and *exo* (out of) refer to secretion into or out of body fluids, while *crine* (from the Greek word meaning "separate") refers to movement away from the secreting cell. In the case of the pancreas, endocrine and exocrine tissues are found in the same gland: Ductless tissues secrete hormones, whereas tissues with ducts secrete enzymes and bicarbonate.

CONCEPT CHECK 45.1

1. How do response mechanisms in target cells differ for water-soluble and lipid-soluble hormones?
2. What type of gland would you expect to secrete pheromones? Explain.
3. **WHAT IF?** Predict what would happen if you injected a water-soluble hormone into the cytosol of a target cell.

For suggested answers, see Appendix A.

CONCEPT 45.2

Feedback regulation and coordination with the nervous system are common in hormone pathways

Having explored hormone structure, recognition, and response, we now consider how regulatory pathways controlling hormone secretion are organised.

Simple Endocrine Pathways

In a *simple endocrine pathway*, endocrine cells respond directly to an internal or environmental stimulus by secreting a particular hormone. The hormone travels in the bloodstream to target cells, where it interacts with its specific receptors. Signal transduction within target cells brings about a physiological response.

▼ Figure 45.9 A simple endocrine pathway. Endocrine cells respond to a change in some internal or external variable—the stimulus—by secreting hormone molecules that binds to a specific receptor protein expressed by target cells, triggering a particular response. In the case of secretin signalling, the simple endocrine pathway is self-limiting because the response to secretin (bicarbonate release) reduces the stimulus (low pH) through negative feedback.

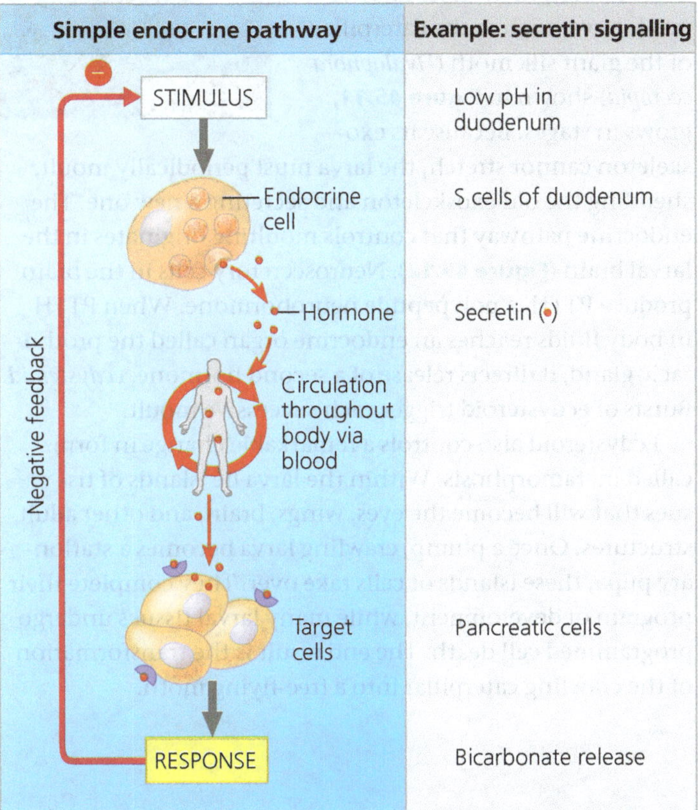

The activity of endocrine cells in the duodenum, the first part of the small intestine, provides a useful example of a simple endocrine pathway. During digestion, the partially processed food that enters the duodenum contains highly acidic digestive juices secreted by the stomach. Before further digestion can occur, this acidic mixture must be neutralised. **Figure 45.9** outlines the simple endocrine pathway that ensures neutralisation takes place.

The low pH of partially digested food entering the small intestine is detected by S cells, which are endocrine cells in the lining of the duodenum. In response, the S cells secrete the hormone *secretin*, which diffuses into the blood. Travelling throughout the circulatory system, secretin reaches the pancreas. Target exocrine cells in the pancreas have receptors for secretin and respond by releasing bicarbonate into ducts that lead to the duodenum. In the last step of the pathway, the bicarbonate released into the duodenum raises the pH, neutralising the stomach acid.

Simple Neuroendocrine Pathways

In a *simple neuroendocrine pathway*, the stimulus is received by a sensory neuron rather than endocrine tissue. The sensory

▼ Figure 45.10 A simple neuroendocrine pathway. Sensory neurons respond to a stimulus by sending nerve impulses to a neurosecretory cell, triggering secretion of a neurohormone. Upon reaching its target cells, the neurohormone binds to its receptor, triggering a specific response. In oxytocin signalling, the response increases the stimulus, forming a positive-feedback loop that amplifies signalling.

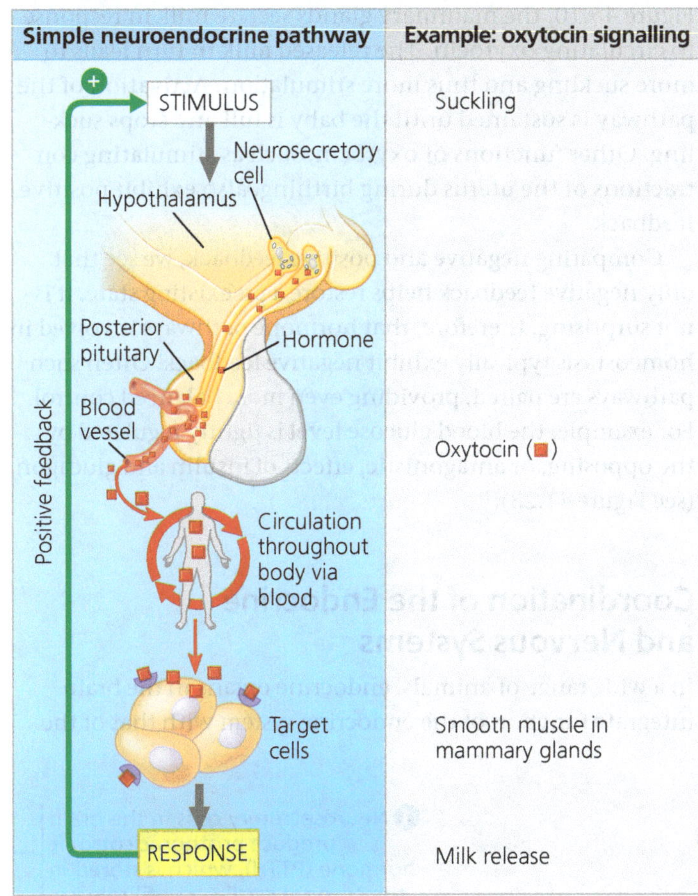

neuron in turn stimulates a neurosecretory cell. In response, the neurosecretory cell secretes a neurohormone. Like other hormones, the neurohormone diffuses into the bloodstream and travels in the circulation to target cells.

As an example of a simple neuroendocrine pathway, consider the regulation of milk release during nursing in mammals **(Figure 45.10)**. When an infant suckles, it stimulates sensory neurons in the nipples, generating nerve impulses that reach the hypothalamus. This input triggers the secretion of the neurohormone **oxytocin** from the posterior pituitary gland. Oxytocin then causes contraction of mammary gland cells, forcing milk from reservoirs in the gland.

Feedback Regulation

A feedback loop linking a response back to an initial stimulus is a feature of many control pathways. Often, this loop involves **negative feedback**, in which the response reduces the initial stimulus. For instance, bicarbonate released in response to secretin increases pH in the intestine, eliminating the stimulus and thereby shutting off secretin release (see Figure 45.9). By decreasing hormone signalling, negative-feedback regulation prevents excessive pathway activity.

Whereas negative feedback dampens a stimulus, **positive feedback** reinforces a stimulus, driving a process to completion. For example, in the pathway outlined in Figure 45.10, the mammary glands secrete milk in response to circulating oxytocin. The released milk in turn leads to more suckling and thus more stimulation. Activation of the pathway is sustained until the baby is full and stops suckling. Other functions of oxytocin, such as stimulating contractions of the uterus during birthing, also exhibit positive feedback.

Comparing negative and positive feedback, we see that only negative feedback helps restore a preexisting state. It is not surprising, therefore, that hormone pathways involved in homeostasis typically exhibit negative feedback. Often such pathways are paired, providing even more balanced control. For example, the blood glucose level is tightly regulated by the opposing, or antagonistic, effects of insulin and glucagon (see Figure 41.23).

Coordination of the Endocrine and Nervous Systems

In a wide range of animals, endocrine organs in the brain integrate function of the endocrine system with that of the nervous system. We'll explore the basic principles of such integration in invertebrates and vertebrates.

Invertebrates

▼ **Figure 45.11 Larva of the giant silk moth.**

The control of development in a moth illustrates neuroendocrine coordination in invertebrates. A moth larva, such as the caterpillar of the giant silk moth (*Hyalophora cecropia*) shown in **Figure 45.11**, grows in stages. Because its exoskeleton cannot stretch, the larva must periodically moult, shedding the old exoskeleton and secreting a new one. The endocrine pathway that controls moulting originates in the larval brain **(Figure 45.12)**. Neurosecretory cells in the brain produce PTTH, a polypeptide neurohormone. When PTTH in body fluids reaches an endocrine organ called the prothoracic gland, it directs release of a second hormone, *ecdysteroid*. Bursts of ecdysteroid trigger each successive moult.

Ecdysteroid also controls a remarkable change in form called metamorphosis. Within the larva lie islands of tissues that will become the eyes, wings, brain, and other adult structures. Once a plump, crawling larva becomes a stationary pupa, these islands of cells take over. They complete their program of development, while many larval tissues undergo programmed cell death. The end result is the transformation of the crawling caterpillar into a free-flying moth.

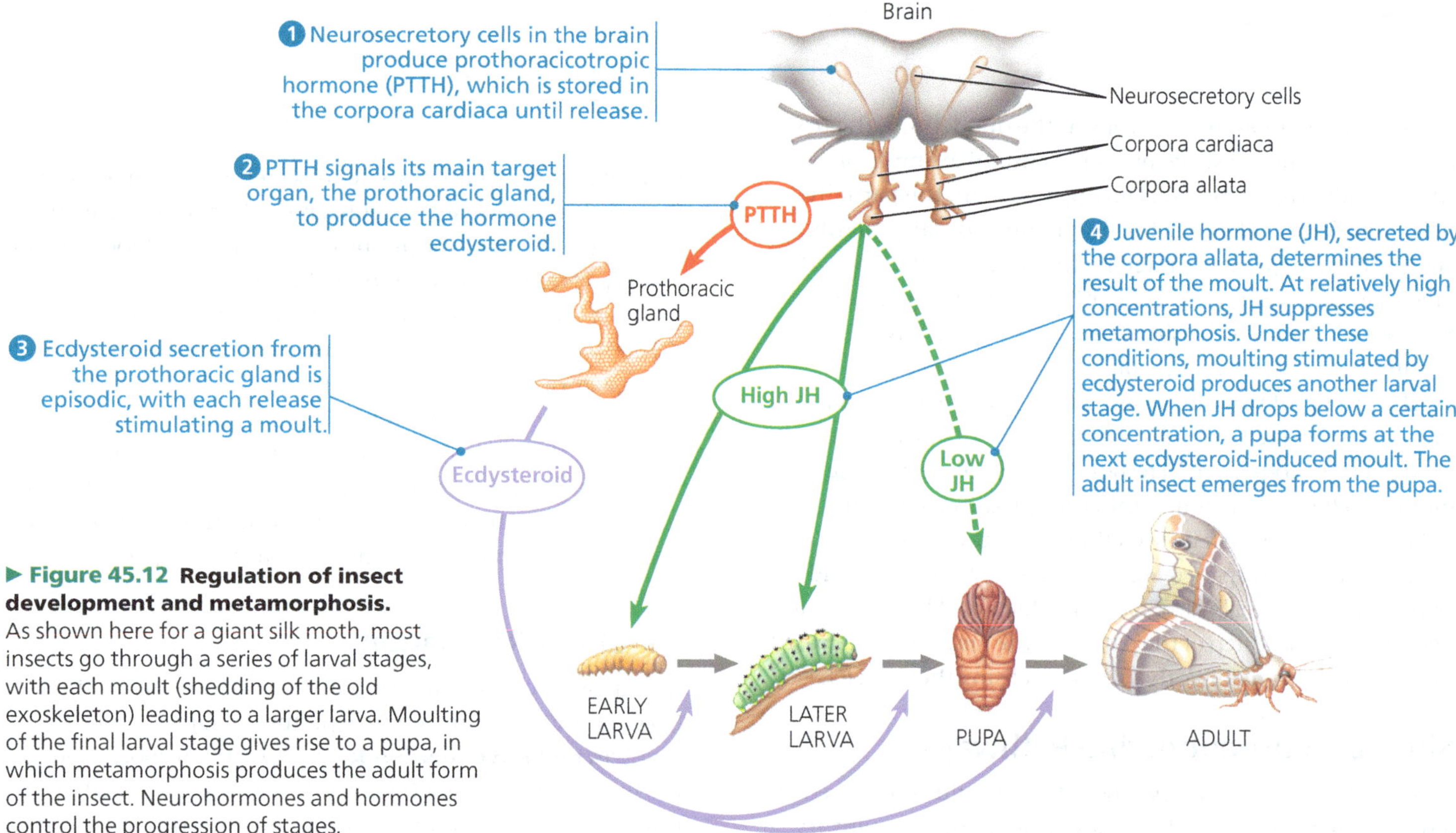

► **Figure 45.12 Regulation of insect development and metamorphosis.** As shown here for a giant silk moth, most insects go through a series of larval stages, with each moult (shedding of the old exoskeleton) leading to a larger larva. Moulting of the final larval stage gives rise to a pupa, in which metamorphosis produces the adult form of the insect. Neurohormones and hormones control the progression of stages.

Given that ecdysteroid can cause either moulting or metamorphosis, what determines which process takes place? The answer is another signal, juvenile hormone (JH), secreted by a pair of endocrine glands behind the brain. JH modulates ecdysteroid activity. When the level of JH in body fluids is high, ecdysteroid stimulates moulting (and thus maintains the "juvenile" larval state). When the JH level drops, ecdysteroid instead induces formation of a pupa, within which metamorphosis occurs.

Knowledge of the coordination between the nervous system and endocrine system in insects has provided a basis for novel methods of agricultural pest control. For example, one tool to control insect pests is a chemical that binds to the ecdysteroid receptor, causing insect larvae to moult prematurely and die.

Vertebrates

In vertebrates, coordination of endocrine signalling relies heavily on the **hypothalamus** (Figure 45.13). The hypothalamus receives information from nerves throughout the body and, in response, initiates neuroendocrine signalling appropriate to environmental conditions. In many vertebrates, for example, nerve signals from the brain pass sensory information to the hypothalamus about seasonal changes. The hypothalamus, in turn, regulates the release of reproductive hormones required during the breeding season.

Signals from the hypothalamus travel to the **pituitary gland**, a gland located at the base of the hypothalamus (see Figure 45.13). Roughly the size and shape of a lima bean, the pituitary is made up of two glands that fused during development but remain as discrete posterior and anterior parts, or lobes, that perform very different functions. The **posterior pituitary** is an extension of the neural tissue of the hypothalamus. Hypothalamic axons that reach into the posterior pituitary secrete neurohormones synthesised in the hypothalamus. In contrast, the **anterior pituitary** is an endocrine gland that synthesises and secretes hormones in response to hormones from the hypothalamus.

Posterior Pituitary Hormones Neurosecretory cells of the hypothalamus synthesise the two posterior pituitary hormones: antidiuretic hormone (ADH) and oxytocin. After travelling to the posterior pituitary within the long axons of the neurosecretory cells, these neurohormones are stored, to be released into the bloodstream in response to nerve impulses transmitted by the hypothalamus (Figure 45.14).

Antidiuretic hormone (ADH), or *vasopressin*, regulates kidney function. Circulating ADH increases water retention in the kidneys, helping maintain normal blood osmolarity (see Concept 44.5). ADH also has an important role in social behaviour (see Concept 51.4).

Oxytocin has multiple functions related to reproduction. As we have seen, in female mammals oxytocin controls milk secretion by the mammary glands and regulates uterine contractions during birthing. In addition, oxytocin has targets in

▼ Figure 45.13 Endocrine glands in the human brain. This side view of the brain indicates the position of the hypothalamus, the pituitary gland, and the pineal gland. (The pineal gland plays a role in regulating biological rhythms.)

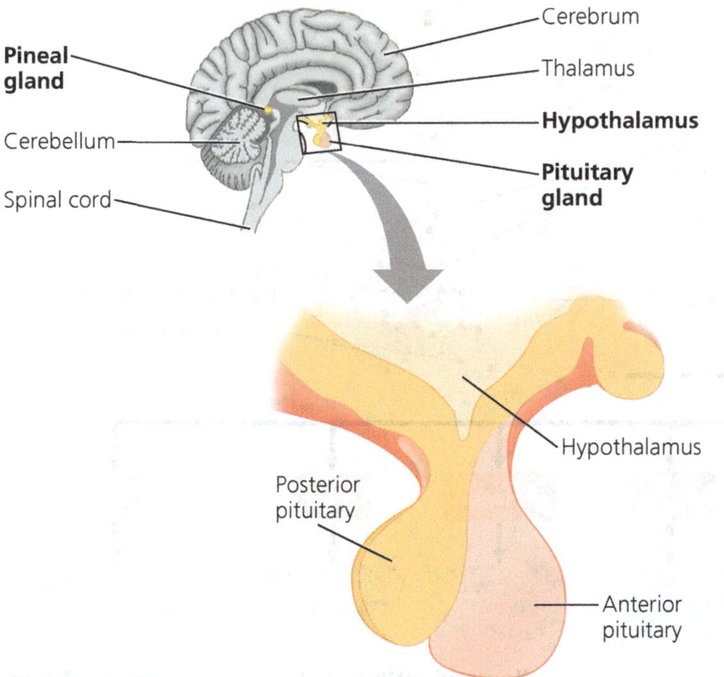

▼ Figure 45.14 Production and release of posterior pituitary hormones. The posterior pituitary gland is an extension of the hypothalamus. Certain neurosecretory cells in the hypothalamus make antidiuretic hormone (ADH) and oxytocin, which are transported to the posterior pituitary, where they are stored. Nerve signals from the brain trigger release of these neurohormones.

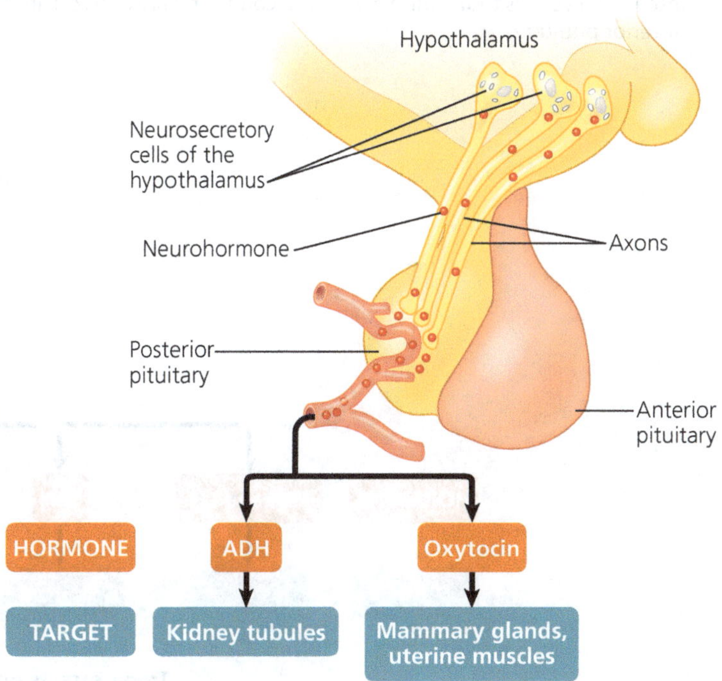

the brain, where it influences behaviours related to maternal care, pair bonding, and sexual activity.

Anterior Pituitary Hormones Hormones secreted by the anterior pituitary control diverse processes in the human body, including metabolism, osmoregulation, and reproduction. As illustrated in **Figure 45.15**, many anterior pituitary hormones, but not all, regulate endocrine glands or tissues.

Hormones secreted by the hypothalamus control the release of all anterior pituitary hormones. Each hypothalamic hormone that regulates release of one or more hormones by the anterior pituitary is called a *releasing* or *inhibiting* hormone. *Prolactin-releasing hormone*, for example, is a hypothalamic hormone that stimulates the anterior pituitary to secrete **prolactin**, which has activities that include stimulating milk production. Each anterior pituitary hormone is controlled by at least one releasing hormone. Some, such as prolactin, have both a releasing hormone and an inhibiting hormone.

The hypothalamic releasing and inhibiting hormones are secreted near capillaries at the base of the hypothalamus. The capillaries drain into short blood vessels, called portal vessels, which subdivide into a second capillary bed within the anterior pituitary. Releasing and inhibiting hormones thus have direct access to the gland they control.

In neuroendocrine pathways, sets of hormones from the hypothalamus, the anterior pituitary, and a target endocrine gland are often organised into a *hormone cascade*, a form of regulation in which multiple endocrine organs and signals act in series. Signals to the brain stimulate the hypothalamus to secrete a hormone that stimulates or inhibits release of a specific anterior pituitary hormone. The anterior pituitary hormone in turn stimulates another endocrine organ to secrete yet another hormone, which affects specific target tissues. In reproduction, for example, the hypothalamus signals the anterior pituitary to release the hormones FSH and LH, which in turn regulate hormone secretion by the gonads (ovaries or testes).

In a sense, hormone cascade pathways redirect signals from the hypothalamus to other endocrine glands. For this reason, the anterior pituitary hormones in such pathways are called *tropic* hormones, or tropins, and are said to have a *tropic* effect (from the Greek *trope*, to turn). Thus, FSH and LH are gonadotropins because they convey signals from the hypothalamus to the gonads. To learn more about tropic hormones and hormone cascade pathways, we'll turn next to thyroid gland function and regulation.

Thyroid Regulation: A Hormone Cascade Pathway

In mammals, **thyroid hormone** regulates bioenergetics; helps maintain normal blood pressure, heart rate, and muscle tone; and regulates digestive and reproductive functions. **Figure 45.16** provides an overview of the hormone cascade

▶ **Figure 45.15 Production and release of anterior pituitary hormones.** The release of hormones synthesised in the anterior pituitary gland is controlled by hypothalamic releasing and inhibiting hormones. The hypothalamic hormones are secreted by neurosecretory cells and enter a capillary network within the hypothalamus. These capillaries drain into portal vessels that connect with a second capillary network in the anterior pituitary.

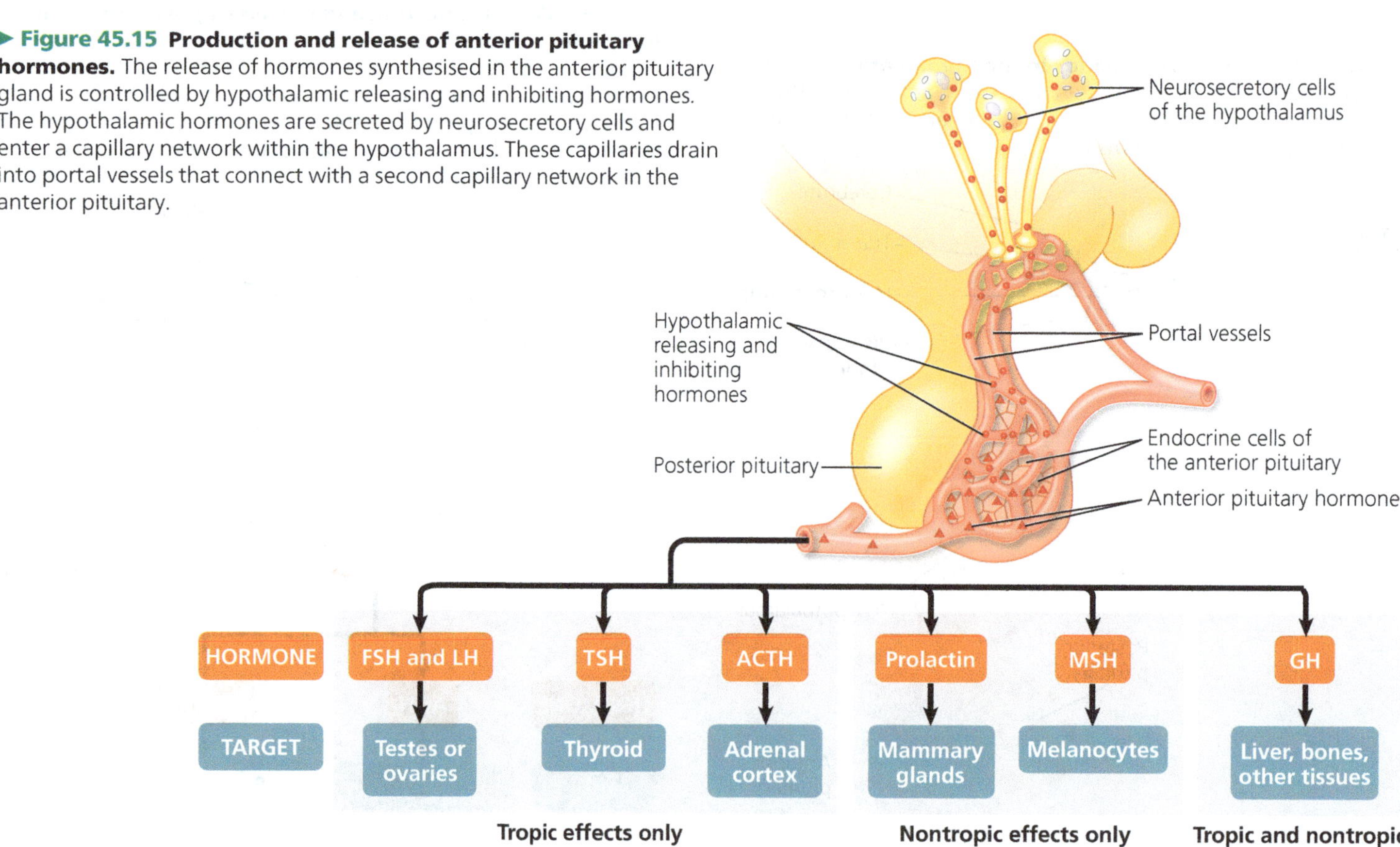

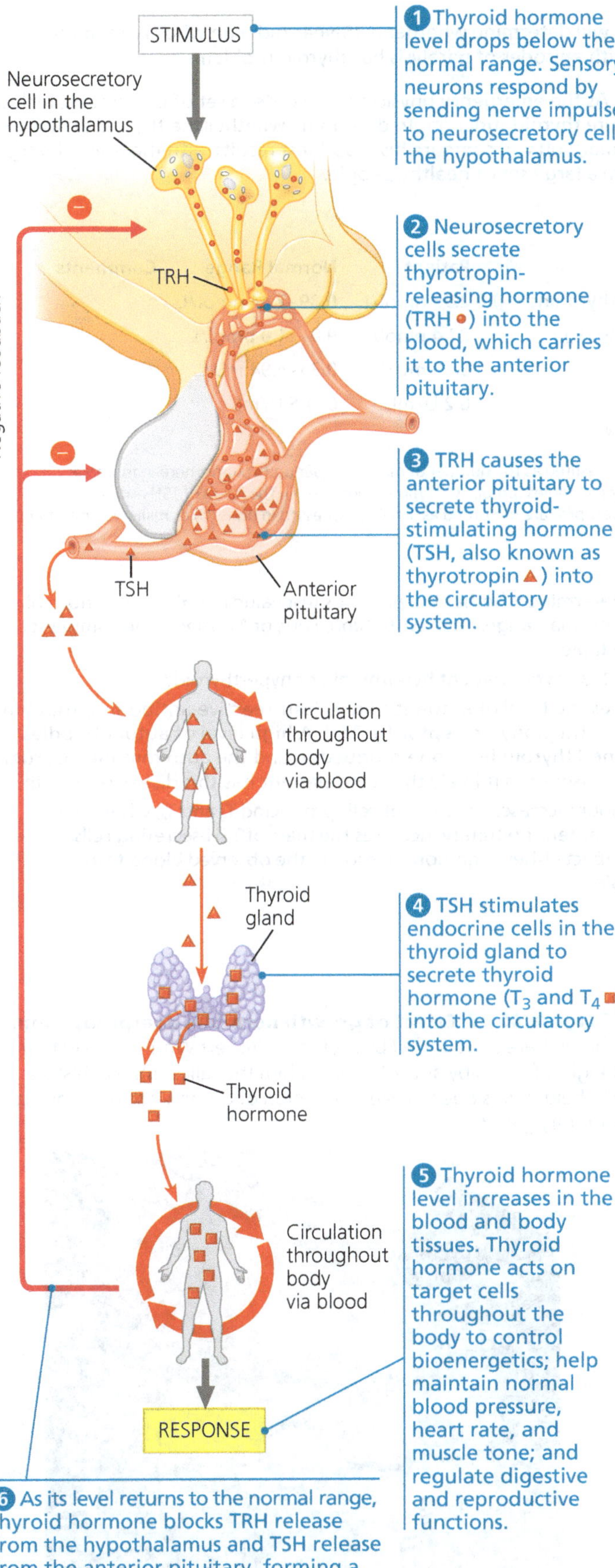

▼ **Figure 45.16 Regulation of thyroid hormone secretion: a hormone cascade pathway.**

pathway that regulates thyroid hormone release. If the level of thyroid hormone in the blood drops, the hypothalamus secretes thyrotropin-releasing hormone (TRH), causing the anterior pituitary to secrete thyrotropin, a tropic hormone also known as thyroid-stimulating hormone (TSH). TSH in turn stimulates the **thyroid gland**, an organ in the neck consisting of two lobes on the ventral surface of the trachea. The thyroid gland responds by secreting thyroid hormone, which increases metabolic rate.

As with other hormone cascade pathways, feedback regulation often occurs at multiple levels. For example, thyroid hormone exerts negative feedback on the hypothalamus and on the anterior pituitary, in each case blocking release of the hormone that promotes its production (see Figure 45.16).

Disorders of Thyroid Function and Regulation

Disruption of thyroid hormone production and regulation can result in serious disorders. One such disorder reflects the unusual chemical makeup of thyroid hormone, the only iodine-containing molecule synthesised in the body. *Thyroid hormone* is actually a pair of very similar molecules derived from the amino acid tyrosine. *Triiodothyronine* (T_3) contains three iodine atoms, whereas tetraiodothyronine, or *thyroxine* (T_4), contains four (see Figure 45.4).

Although iodine is readily obtained from seafood or iodised salt, people in many parts of the world lack enough iodine in their diet to synthesise adequate amounts of thyroid hormone. With only a low blood level of thyroid hormone, the pituitary receives no negative feedback and continues to secrete TSH. An elevated TSH level in turn causes the thyroid gland to enlarge, resulting in goitre, a marked swelling of the neck.

Hormonal Regulation of Growth

Growth hormone (GH), which is secreted by the anterior pituitary, stimulates growth through both tropic and nontropic effects. A major target, the liver, responds to GH by releasing *insulin-like growth factors* (*IGFs*), which circulate in the blood and directly stimulate bone and cartilage growth. (IGFs also appear to play a key role in aging in many animal species.) In the absence of GH, the skeleton of an immature animal stops growing. GH also exerts diverse metabolic effects that tend to raise the blood glucose level, thus opposing the effects of insulin.

Abnormal production of GH in humans can result in several disorders, depending on when the problem occurs and whether it involves hypersecretion (too much) or hyposecretion (too little). Hypersecretion of GH during childhood can lead to gigantism, in which the person grows unusually tall but retains relatively normal body

PROBLEM-SOLVING EXERCISE

Is thyroid regulation normal in this patient?

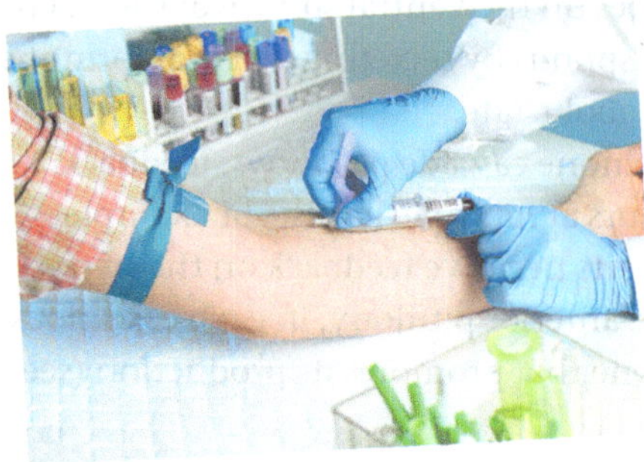

Normal health requires proper regulation of the thyroid gland. Hypothyroidism, the secretion of too little thyroid hormone (T_3 and T_4), can cause weight gain, lethargy, and intolerance to cold in adults. In contrast, excessive secretion of thyroid hormone, known as hyperthyroidism, can lead to high body temperature, profuse sweating, weight loss, muscle weakness, irritability, and high blood pressure. Thyroid-stimulating hormone (TSH) stimulates the thyroid to release thyroid hormone. Testing for levels of T_3, T_4, and TSH in the blood can help diagnose various medical conditions.

In this exercise, you will determine whether a 35-year-old man who came to the emergency room with episodes of paralysis has thyroid problems.

Your Approach As the emergency physician, you order a set of blood tests, including four that measure thyroid function. To determine whether the thyroid activity of your patient is normal, you will compare his blood test results with the normal range, as determined from a large set of healthy people.

Your Data

#	Test*	Patient	Normal Range	Comments
1	Total triiodothyronine (T_3)	2.93 nmol/L	0.89–2.44 nmol/L	
2	Free thyroxine (T_4)	27.4 pmol/L	9.0–21.0 pmol/L	
3	TSH	5.55 mU/L	0.35–4.94 mU/L	
4	TSH receptor autoantibody	0.2 U/mL	0–1.5 U/mL	

*T_3 and T_4 levels are measured as the number of molecules per unit volume: here, nanomoles (nmol, 10^{-9} moles) or picomoles (pmol, 10^{-12} moles) per litre (L). The levels of TSH and the autoantibody for its receptor are measured as activity, expressed in units (U) or milliunits (mU) per unit volume.

Your Analysis

1. For each test, determine whether the patient's test value is high, low, or normal relative to the normal range. Then write *High, Low*, or *Normal* in the comments column of the table.
2. Based on tests 1–3, is your patient hypothyroid or hyperthyroid?
3. Test 4 measures the level of autoantibodies (self-reactive antibodies) that bind to and activate the body's receptor for TSH. A high level of autoantibodies causes sustained thyroid hormone production and the autoimmune disorder called Graves' disease. Is it likely that your patient has this disease? Explain.
4. A thyroid tumour increases the mass of cells producing T_3 and T_4, whereas a tumour in the anterior pituitary increases the mass of TSH-secreting cells. Would you expect either condition to result in the observed blood test values? Explain.

proportions (**Figure 45.17**). Excessive GH production in adulthood stimulates bony growth in the few body parts that are still responsive to the hormone—predominantly the face, hands, and feet. The result is an overgrowth of the extremities called acromegaly (from the Greek *acros*, extreme, and *mega*, large).

Hyposecretion of GH in childhood retards long-bone growth and can lead to pituitary dwarfism. People with this disorder are for the most part properly proportioned but generally reach a height of only about 1.2 m. If diagnosed before puberty, pituitary dwarfism can be treated with human GH (also called HGH). Treatment with HGH produced by recombinant DNA technology is common.

Whereas the effects of altered growth hormone levels are readily related to a change in adult height, disrupting some endocrine pathways can have effects that appear unrelated to normal pathway function. The **Problem-Solving Exercise** explores one such example of a medical mystery difficult to diagnose based on symptoms alone.

▼ Figure 45.17 Effect of growth hormone overproduction. Shown here surrounded by his family, Robert Wadlow grew to a height of 2.7 m by age 22, making him the tallest man in history. His height was due to excess secretion of growth hormone by his pituitary gland.

CONCEPT CHECK 45.2

1. What are the roles of oxytocin and prolactin in regulating the mammary glands?
2. How do the two fused glands of the pituitary gland differ in function?
3. **WHAT IF?** Propose an explanation for why defects in a particular hormone cascade pathway observed in patients typically affect the final gland in the pathway rather than the hypothalamus or pituitary.
4. **WHAT IF?** Lab tests of two patients, each diagnosed with excessive thyroid hormone production, revealed an elevated level of TSH in one but not the other. Was the diagnosis of one patient necessarily incorrect? Explain.

For suggested answers, see Appendix A.

CONCEPT 45.3

Endocrine glands respond to diverse stimuli in regulating homeostasis, development, and behaviour

In the remainder of this chapter, we'll focus on endocrine function in homeostasis, development, and behaviour. We'll begin with another example of a simple hormone pathway, the regulation of calcium ion concentration in the circulatory system.

Parathyroid Hormone and Vitamin D: Control of Blood Calcium

Because calcium ions (Ca^{2+}) are essential to the normal functioning of all cells, homeostatic control of the level of calcium in the blood is vital. If the blood Ca^{2+} level falls substantially, skeletal muscles begin to contract convulsively, a potentially fatal condition. If the blood Ca^{2+} level rises substantially, calcium phosphate can form precipitates in body tissues, leading to widespread organ damage.

In mammals, the **parathyroid glands**, a set of four small structures embedded in the posterior surface of the thyroid (see Figure 45.8), play a major role in blood Ca^{2+} regulation. When the blood Ca^{2+} level falls below a set point of about 10 mg/100 mL, these glands release **parathyroid hormone (PTH)**.

PTH raises the level of blood Ca^{2+} through direct effects in bones and the kidneys and an indirect effect on the intestines **(Figure 45.18)**. In bones, PTH causes the mineralised matrix to break down, releasing Ca^{2+} into the blood. In the kidneys, PTH directly stimulates reabsorption of Ca^{2+} through the renal tubules. In addition, PTH indirectly raises the blood Ca^{2+} level by promoting production of vitamin D. A precursor form of vitamin D is obtained from food or synthesised by skin exposed to sunlight. Conversion of this precursor to active vitamin D begins in the liver. PTH acts in the kidney to stimulate completion of the conversion process. Vitamin D in turn acts on the intestines, stimulating the uptake of Ca^{2+} from food. As the blood Ca^{2+} level rises, a negative-feedback loop inhibits further release of PTH from the parathyroid glands (not shown in Figure 45.18).

The thyroid gland can also contribute to calcium homeostasis. If the blood Ca^{2+} level rises above the set point, the thyroid gland releases **calcitonin**, a hormone that inhibits bone breakdown and enhances Ca^{2+} excretion by the kidneys. In fishes, rodents, and some other animals, calcitonin is required for Ca^{2+} homeostasis. In humans, however, calcitonin is apparently needed only during the extensive bone growth of childhood.

▶ **Figure 45.18 The roles of parathyroid hormone (PTH) in regulating the blood calcium level in mammals.**

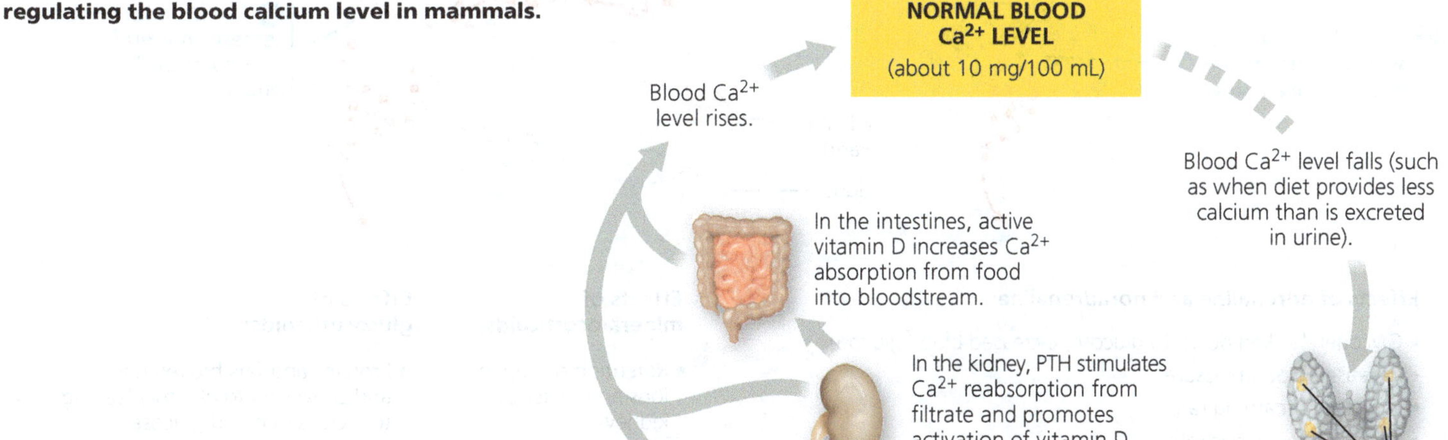

Adrenal Hormones: Response to Stress

The adrenal glands of vertebrates play a major role in the response to *stress*, a state of threatened homeostasis. Located atop the kidneys (the *renal* organs), each **adrenal gland** is actually made up of two glands with different cell types, functions, and embryonic origins: the adrenal *cortex*, the outer portion, and the adrenal *medulla*, the central portion **(Figure 45.19)**. The adrenal cortex consists of true endocrine cells, whereas the secretory cells of the adrenal medulla develop from neural tissue. Thus, like the pituitary gland, each adrenal gland is a fused endocrine and neuroendocrine gland.

The Role of the Adrenal Medulla

Imagine that while walking in the woods at night you hear a growling noise nearby. "A feral pig?" you wonder. Your heart beats faster, your breathing quickens, your muscles tense, and your thoughts speed up. These and other rapid responses to perceived danger comprise the "fight-or-flight" response. This coordinated set of physiological changes is triggered by two hormones of the adrenal medulla, adrenaline (epinephrine) and **noradrenaline** (also known as norepinephrine). Both are *catecholamines*, a class of amine hormones synthesised from the amino acid tyrosine. Both molecules also function as neurotransmitters, as you'll read in Concept 48.4.

As hormones, adrenaline and noradrenaline increase the amount of chemical energy available for immediate use (see Figure 45.19a). Both catecholamines increase the rate of glycogen breakdown in the liver and skeletal muscles and promote the release of glucose by liver cells and of fatty acids from fat cells. The released glucose and fatty acids circulate in the blood and can be used by body cells as fuel.

Catecholamines also exert profound effects on the cardiovascular and respiratory systems. For example, they increase heart rate and stroke volume and dilate the bronchioles in the lungs, actions that raise the rate of oxygen delivery to body cells. For this reason, doctors may prescribe adrenaline as a heart stimulant or to open the airways during an asthma attack. Catecholamines also alter blood flow, causing constriction of some blood vessels and dilation of others. The overall effect is to shunt blood away from the skin, digestive organs, and kidneys while increasing the blood supply to the heart, brain, and skeletal muscles.

▼ **Figure 45.19 Stress and the adrenal gland.**

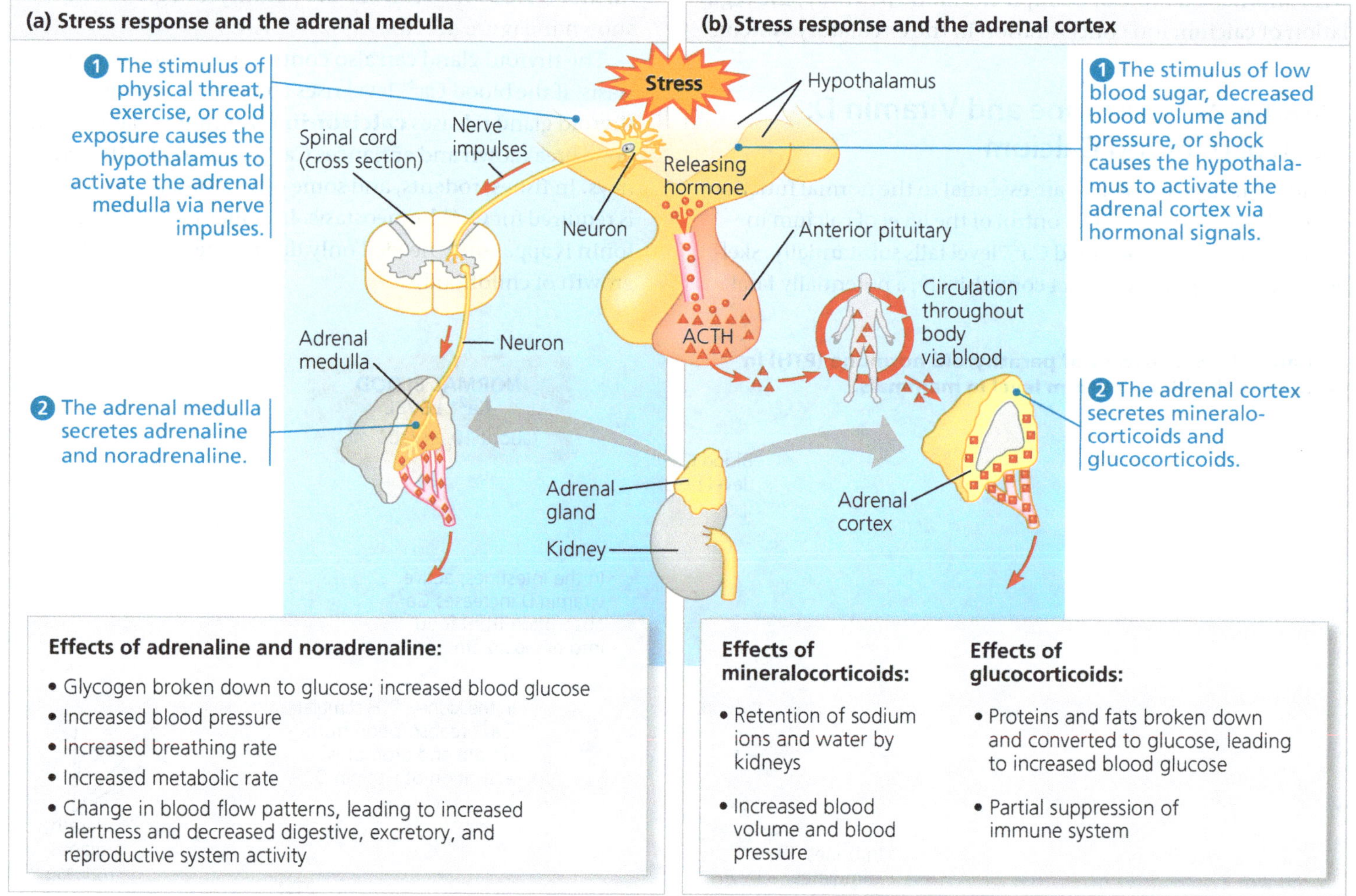

Adrenaline's Multiple Effects: *A Closer Look* How can adrenaline coordinate a response to stress that involves widely varying effects in individual tissues? We can answer that question by examining different response pathways **(Figure 45.20)** in a range of target cells:

- In liver cells, adrenaline binds to a β-type receptor in the plasma membrane. This receptor activates the enzyme protein kinase A, which in turn regulates enzymes of glycogen metabolism, causing release of glucose into the blood (see Figure 45.20a). Note that this is the signal transduction pathway illustrated in Figure 45.6.
- In the smooth muscle cells that line blood vessels supplying skeletal muscle, the same kinase activated by the same adrenaline receptor inactivates a muscle-specific enzyme. The result is smooth muscle relaxation, leading to vasodilation and hence increased blood flow to skeletal muscles (see Figure 45.20b).
- In the smooth muscle cells lining blood vessels of the intestines, adrenaline binds to an α-type receptor (see Figure 45.20c). This receptor triggers a signalling pathway that involves enzymes other than protein kinase A and that causes smooth muscle contraction rather than relaxation. The resulting vasoconstriction reduces blood flow to the intestines, facilitating the redirection of blood to active skeletal muscle.

Thus, adrenaline elicits multiple responses if its target cells differ in the receptor protein they express or in the molecules activated by the receptor upon hormone binding. As illustrated in these examples, such variation in response plays a key role in enabling adrenaline to trigger a range of activities that together bring about a coordinated rapid response to stressful stimuli.

▼ **Figure 45.20 One hormone, different effects.** Adrenaline, the primary "fight-or-flight" hormone, produces different responses in different target cells. Target cells with the same receptor exhibit different responses if they have different signal transduction pathways or effector proteins; compare **(a)** with **(b)**. Target cells with different receptors for the hormone often exhibit different responses; compare **(b)** with **(c)**.

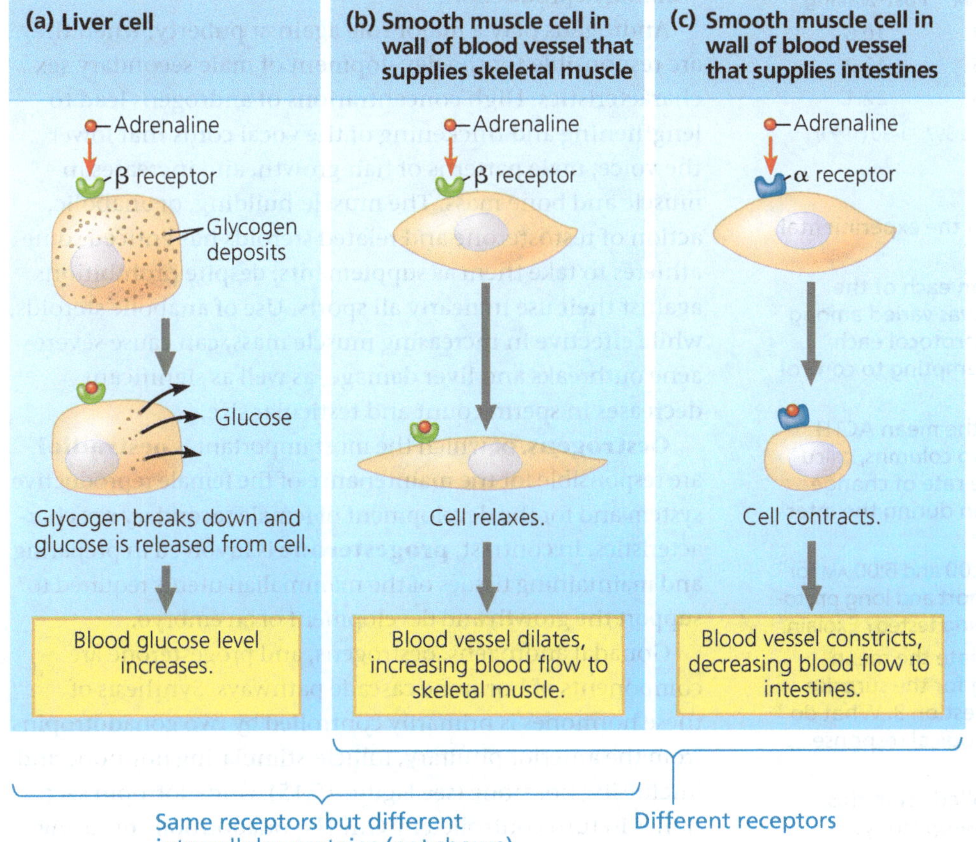

The Role of the Adrenal Cortex

Like the adrenal medulla, the adrenal cortex mediates an endocrine response to stress (see Figure 45.19b). The two portions of the adrenal gland differ, however, in both the types of stress that trigger a response and the targets of the hormones that are released.

The adrenal cortex becomes active under stressful conditions that include low blood sugar, decreased blood volume and pressure, and shock. Such stimuli cause the hypothalamus to secrete a releasing hormone that stimulates the anterior pituitary to release adrenocorticotropic hormone (ACTH), a tropic hormone. When ACTH reaches the adrenal cortex via the bloodstream, it stimulates the endocrine cells to synthesise and secrete a family of steroids called *corticosteroids*. The two main types of corticosteroids in humans are glucocorticoids and mineralocorticoids.

Glucocorticoids, such as cortisol (see Figure 45.4), make more glucose available as fuel by promoting glucose synthesis from noncarbohydrate sources, such as proteins. Glucocorticoids also act on skeletal muscle, causing the breakdown of muscle proteins into amino acids. These are transported to the liver and kidneys, converted to glucose, and released into the blood. The synthesis of glucose upon the breakdown of muscle proteins provides circulating fuel when the body requires more glucose than the liver can mobilise from its glycogen stores.

If glucocorticoids are introduced into the body at a level above that normally present, they suppress certain components of the body's immune system. For this reason, glucocorticoids are sometimes used to treat inflammatory diseases such as arthritis. However, their long-term use can have serious side effects on metabolism. Nonsteroidal anti-inflammatory drugs (NSAIDs), such as

Scientific Skills Exercise

Designing a Controlled Experiment

How Is Nighttime ACTH Secretion Related to Expected Sleep Duration? Humans secrete increasing amounts of adrenocorticotropic hormone (ACTH) during the late stages of normal sleep, with the peak secretion occurring at the time of spontaneous waking. Because ACTH is released in response to stressful stimuli, scientists hypothesised that ACTH secretion prior to waking might be an anticipatory response to the stress associated with transitioning from sleep to a more active state. If so, an individual's expectation of waking at a particular time might influence the timing of ACTH secretion. How can such a hypothesis be tested? In this exercise, you will examine how researchers designed a controlled experiment to study the role of expectation.

How the Experiment Was Done Researchers studied 15 healthy volunteers in their mid-20s over three nights. Each night, each subject was told when he or she would be awakened: 6:00 or 9:00 AM. The subjects went to sleep at midnight. Subjects in the "short" or "long" protocol group were awakened at the expected time (6:00 or 9:00 AM, respectively). Subjects in the "surprise" protocol group were told they would be awakened at 9:00 AM, but were actually awakened 3 hours early, at 6:00 AM. At set times, blood samples were drawn to determine plasma levels of ACTH. To determine the change (Δ) in ACTH concentration post-waking, the researchers compared samples drawn at waking and 30 minutes later.

Data from the Experiment

			Mean Plasma ACTH Level (pg/mL)		
Sleep Protocol	Expected Wake Time	Actual Wake Time	1:00 AM	6:00 AM	Δ in the 30 Minutes Post-waking
Short	6:00 AM	6:00 AM	9.9	37.3	10.6
Long	9:00 AM	9:00 AM	8.1	26.5	12.2
Surprise	9:00 AM	6:00 AM	8.0	25.5	22.1

Data from J. Born et al., Timing the end of nocturnal sleep, *Nature* 397:29–30 (1999).

INTERPRET THE DATA

1. Describe the role of the "surprise" protocol in the experimental design.
2. Each subject was given a different protocol on each of the three nights, and the order of the protocols was varied among the subjects that so that one-third had each protocol each night. What factors were the researchers attempting to control for with this approach?
3. For subjects in the short protocol, what was the mean ACTH level at waking? Using the data in the last two columns, calculate the mean level 30 minutes later. Was the rate of change faster or slower in that 30-minute period than during the interval from 1:00 to 6:00 AM?
4. How does the change in ACTH level between 1:00 and 6:00 AM for the surprise protocol compare to that for the short and long protocols? Does this result support the hypothesis being tested? Explain.
5. Using the data in the last two columns, calculate the mean ACTH concentration 30 minutes post-waking for the surprise protocol and compare to your answer for question 3. What do your results suggest about a person's physiological response immediately after waking?
6. What are some variables that weren't controlled for in this experiment that could be explored in a follow-up study?

aspirin and ibuprofen, are therefore generally preferred for treating chronic inflammatory conditions.

Mineralocorticoids act principally in maintaining salt and water balance. For example, the mineralocorticoid *aldosterone* functions in ion and water homeostasis of the blood (see Figure 44.22). Like glucocorticoids, mineralocorticoids not only mediate stress responses, but also participate in homeostatic regulation of metabolism. In the **Scientific Skills Exercise**, you can explore an experiment investigating changes in ACTH secretion as humans awaken from sleep.

Sex Hormones

Sex hormones affect growth, development, reproductive cycles, and sexual behaviour. Although the adrenal glands secrete small quantities of these hormones, the gonads (testes of males and ovaries of females) are their principal sources. The gonads produce and secrete three major types of steroid sex hormones: androgens, oestrogens, and progesterone. All three types are found in both males and females but in different proportions.

The testes primarily synthesise **androgens**, the main one being **testosterone**. In humans, testosterone first functions in male (XY) embryos, promoting development of male reproductive structures **(Figure 45.21)**. In female (XX) embryos, the absence of testosterone allows the development of female reproductive structures. You can learn more about this role of hormones in the development of an embryo as male or female in the Scientific Skills Exercise in the chapter, "Animal Reproduction."

Androgens play a major role again at puberty, when they are responsible for the development of male secondary sex characteristics. High concentrations of androgens lead to lengthening and thickening of the vocal cords that lower the voice, male patterns of hair growth, and increases in muscle and bone mass. The muscle-building, or anabolic, action of testosterone and related steroids has enticed some athletes to take them as supplements, despite prohibitions against their use in nearly all sports. Use of anabolic steroids, while effective in increasing muscle mass, can cause severe acne outbreaks and liver damage, as well as significant decreases in sperm count and testicular size.

Oestrogens, of which the most important is **oestradiol**, are responsible for the maintenance of the female reproductive system and for the development of female secondary sex characteristics. In contrast, **progesterone** is involved in preparing and maintaining tissues of the mammalian uterus required to support the growth and development of an embryo.

Gonadal androgens, oestrogens, and progesterone are components of hormone cascade pathways. Synthesis of these hormones is primarily controlled by two gonadotropins from the anterior pituitary, follicle-stimulating hormone and luteinising hormone (see Figure 45.15). Gonadotropin secretion is in turn controlled by GnRH (gonadotropin-releasing hormone) from the hypothalamus. We'll examine the

▼ Figure 45.21 Sex hormones regulate formation of internal reproductive structures in human development. In a male (XY) embryo, the bipotential gonads (gonads that can develop into either of two forms) become the testes, which secrete testosterone and anti-Müllerian hormone (AMH). Testosterone directs formation of sperm-carrying ducts (vas deferens and seminal vesicles), while AMH causes the female ducts to degenerate. In the absence of these testis hormones, the male ducts degenerate and female structures form, including the oviduct, uterus, and vagina.

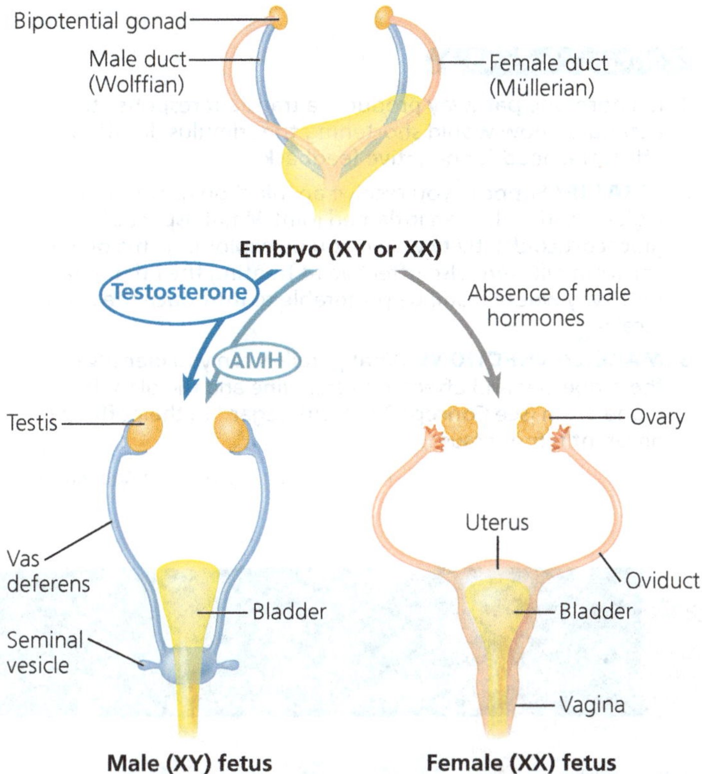

VISUAL SKILLS *Looking at this figure, explain why the adjective* bipotential *is only used to describe the gonad.*

feedback relationships that regulate gonadal hormone secretion in detail in the chapter, "Animal Reproduction."

Endocrine Disruptors

Between 1938 and 1971, some pregnant women at risk for pregnancy complications were prescribed a synthetic oestrogen called diethylstilboestrol (DES). What was not known until 1971 was that exposure to DES can alter reproductive system development in the fetus. Daughters of women who took DES more frequently developed certain reproductive abnormalities, including vaginal and cervical cancer, structural changes in the reproductive organs, and increased risk of miscarriage (spontaneous abortion). DES is now recognised as an *endocrine disruptor*, a foreign molecule that interrupts the normal function of a hormone pathway.

In recent years, some scientists have hypothesised that molecules in the environment also act as endocrine disruptors. For example, bisphenol A, a chemical used in making some plastics, has been studied for potential interference with normal reproduction and development. In addition, it has been suggested that some oestrogen-like molecules, such as those present in soybeans and other edible plant products, have the beneficial effect of lowering breast cancer risk. Sorting out such effects, whether harmful or beneficial, has proven quite difficult, in part because enzymes in the liver change the properties of any such molecules entering the body through the digestive system.

Hormones and Biological Rhythms

There is still much to be learned about the hormone **melatonin**, a modified amino acid that regulates functions related to light and the seasons. Melatonin is produced by the **pineal gland**, a small mass of tissue near the centre of the mammalian brain (see Figure 45.13).

Although melatonin affects skin pigmentation in many vertebrates, its primary effects relate to biological rhythms associated with reproduction and with the daily activity level (see Figure 40.9). Melatonin is secreted at night, and the amount released depends on the length of the night. In winter, for example, when days are short and nights are long, more melatonin is secreted. There is also good evidence that nightly increases in the level of melatonin play a significant role in promoting sleep.

The release of melatonin by the pineal gland is controlled by a group of neurons in the hypothalamus called the suprachiasmatic nucleus (SCN). The SCN functions as a biological clock and receives input from specialised light-sensitive neurons in the retina of the eye. Although the SCN regulates melatonin production during the 24-hour light/dark cycle, melatonin also influences SCN activity. We'll consider biological rhythms further in Concept 49.2, where we analyse experiments on SCN function.

Evolution of Hormone Function

EVOLUTION Over the course of evolution, the functions of a given hormone often diverge between species. An example is thyroid hormone, which across many evolutionary lineages plays a role in regulating metabolism (see Figure 45.16). In frogs, however, the thyroid hormone thyroxine (T_4) has taken on an apparently unique function: stimulating resorption of the tadpole's tail during metamorphosis **(Figure 45.22)**.

▼ Figure 45.22 Specialised role of a hormone in frog metamorphosis. The hormone thyroxine is responsible for the resorption of the tadpole's tail as the frog develops into its adult form.

▲ Tadpole

▲ Adult frog

The hormone *prolactin* has an especially broad range of activities. Prolactin stimulates mammary gland growth and milk synthesis in mammals, regulates fat metabolism and reproduction in birds, delays metamorphosis in amphibians, and regulates salt and water balance in freshwater fishes. These varied roles indicate that prolactin is an ancient hormone with functions that have diversified during the evolution of vertebrate groups.

Melanocyte-stimulating hormone (MSH), secreted by the anterior pituitary, provides another example of a hormone with distinct functions in different evolutionary lineages. In amphibians, fishes, and reptiles, MSH regulates skin colour by controlling pigment distribution in skin cells called melanocytes. In mammals, MSH functions in hunger and metabolism in addition to skin colouration.

The specialised action of MSH that has evolved in the mammalian brain may prove to be of particular medical importance. Many patients with late-stage cancer, AIDS, tuberculosis, and certain aging disorders develop a devastating wasting condition called cachexia. Characterised by weight loss, muscle atrophy, and loss of appetite, cachexia responds poorly to existing therapies. However, it turns out that activation of a brain receptor for MSH produces some of the same changes seen in cachexia. Moreover, in experiments on mice with mutations that cause cancer and consequently cachexia, treatment with drugs that blocked the brain receptor for MSH prevented cachexia. Whether such drugs can be used to treat cachexia in humans is an area of active study.

CONCEPT CHECK 45.3

1. If a hormone pathway produces a transient response to a stimulus, how would shortening the stimulus duration affect the need for negative feedback?
2. **WHAT IF?** Suppose you receive an injection of cortisone, a glucocorticoid, in an inflamed joint. What aspect of glucocorticoid activity would you be exploiting? If a glucocorticoid pill were also effective at treating the inflammation, why would it still be preferable to introduce the drug locally?
3. **MAKE CONNECTIONS** What parallels can you identify in the properties and effects of adrenaline and the plant hormone auxin (see Concept 39.2) with regard to their effects in different target tissues?

For suggested answers, see Appendix A.

45 Chapter Review

SUMMARY OF KEY CONCEPTS

CONCEPT 45.1

Hormones and other signalling molecules bind to target receptors, triggering specific response pathways *(pp. 1050–1054)*

- The forms of signalling between animal cells differ in the type of secreting cell and the route taken by the signal to its target. **Endocrine** signals, or **hormones**, are secreted into the extracellular fluid by endocrine cells or ductless glands and reach target cells via circulatory fluids. There the binding of a hormone to a receptor specific for that particular hormone triggers a cellular response. **Paracrine** signals act on neighbouring cells, whereas **autocrine** signals act on the secreting cell itself. **Neurotransmitters** also act locally, but **neurohormones** can act throughout the body. **Pheromones** are released into the environment for communication between animals of the same species.
- **Local regulators**, which carry out paracrine and autocrine signalling, include cytokines and growth factors (polypeptides), **prostaglandins** (modified fatty acids), and **nitric oxide** (a gas).
- Polypeptides, steroids, and amines comprise the major classes of animal hormones. Depending on whether they are water-soluble or lipid-soluble, hormones activate different response pathways. The endocrine cells that secrete hormones are often located in glands dedicated in part or in whole to endocrine signalling.

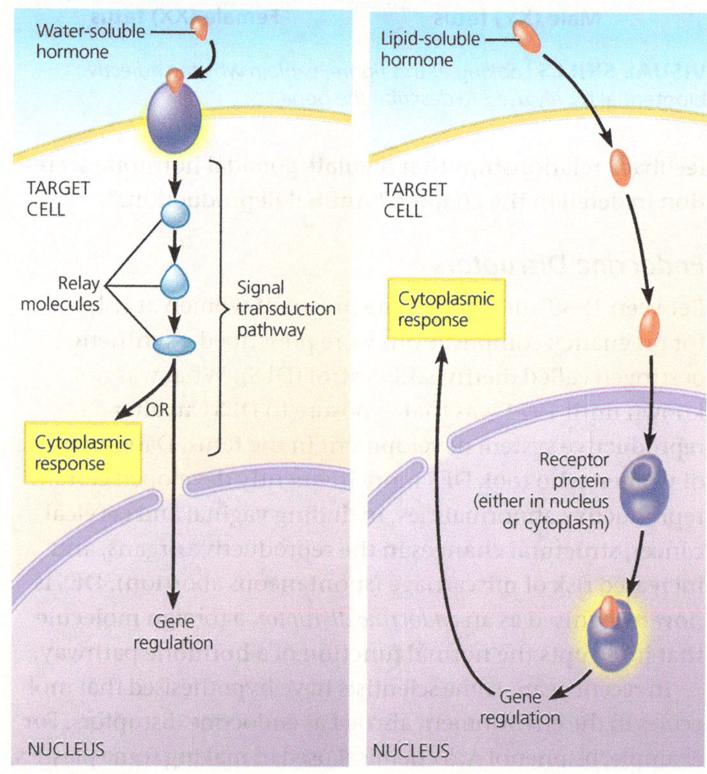

MAKE CONNECTIONS *What forms of signalling activate a helper T cell in immune responses (see Figure 43.18)?*

CONCEPT 45.2

Feedback regulation and coordination with the nervous system are common in hormone pathways *(pp. 1054–1061)*

- In a simple endocrine pathway, endocrine cells respond directly to a stimulus. By contrast, in a simple neuroendocrine pathway, a sensory neuron receives the stimulus.
- Hormone pathways may include **negative feedback**, which dampens the stimulus and thus limits the response, or **positive feedback**, which amplifies the stimulus and drives the response to completion.

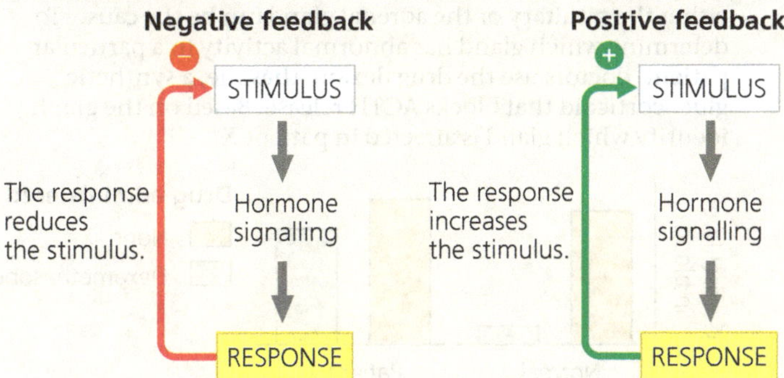

- In insects, moulting and development are controlled by three hormones: PTTH; ecdysteroid, whose release is triggered by PTTH; and juvenile hormone. Coordination of signals from the nervous and endocrine systems and modulation of one hormone activity by another bring about the sequence of developmental stages that lead to an adult form.
- In vertebrates, neurosecretory cells in the **hypothalamus** produce two hormones that are secreted by the **posterior pituitary** and that act directly on nonendocrine tissues: **oxytocin**, which induces uterine contractions and release of milk from mammary glands, and **antidiuretic hormone (ADH)**, which enhances water reabsorption in the kidneys.
- Other hypothalamic cells produce hormones that are transported to the **anterior pituitary**, where they stimulate or inhibit the release of particular hormones.
- Often, anterior pituitary hormones act in a cascade. For example, the secretion of thyroid-stimulating hormone (TSH) is regulated by thyrotropin-releasing hormone (TRH). TSH in turn induces the **thyroid gland** to secrete **thyroid hormone**, a combination of the iodine-containing hormones T_3 and T_4. Thyroid hormone stimulates metabolism and influences development and maturation.

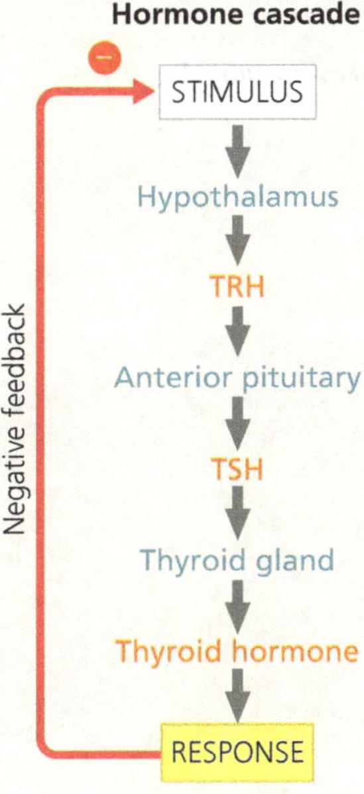

- Most anterior pituitary hormones are tropic hormones, acting on endocrine tissues or glands to regulate hormone secretion. Tropic hormones of the anterior pituitary include TSH, follicle-stimulating hormone (FSH), luteinising hormone (LH), and adrenocorticotropic hormone (ACTH). **Growth hormone (GH)** has both tropic and nontropic effects. It promotes growth directly, affects metabolism, and stimulates the production of growth factors by other tissues.

? *Which major endocrine organs described in Figure 45.8 are regulated independently of the hypothalamus and pituitary?*

CONCEPT 45.3

Endocrine glands respond to diverse stimuli in regulating homeostasis, development, and behaviour *(pp. 1061–1066)*

- **Parathyroid hormone (PTH)**, secreted by the **parathyroid glands**, causes bone to release Ca^{2+} into the blood and stimulates reabsorption of Ca^{2+} in the kidneys. PTH also stimulates the kidneys to activate vitamin D, which promotes intestinal uptake of Ca^{2+} from food. **Calcitonin**, secreted by the thyroid, has the opposite effects in bones and kidneys as PTH. Calcitonin is important for calcium homeostasis in adults of some vertebrates, but not humans.
- In response to stress, neurosecretory cells in the adrenal medulla release **adrenaline** and **noradrenaline**, which mediate various fight-or-flight responses. The adrenal cortex releases **glucocorticoids**, such as cortisol, which influence glucose metabolism and the immune system. It also releases **mineralocorticoids**, primarily aldosterone, which help regulate salt and water balance.
- Sex hormones regulate growth, development, reproduction, and sexual behaviour. Although the adrenal cortex produces small amounts of these hormones, the gonads (testes and ovaries) serve as the major source. All three types—**androgens, oestrogens**, and **progesterone**—are produced in males and females, but in different proportions.
- The **pineal gland**, located within the brain, secretes **melatonin**, which functions in biological rhythms related to reproduction and sleep. Release of melatonin is controlled by the SCN, the region of the brain that functions as a biological clock.
- Hormones have acquired distinct roles in different species over the course of evolution. **Prolactin** stimulates milk production in mammals but has diverse effects in other vertebrates. **Melanocyte-stimulating hormone (MSH)** influences fat metabolism in mammals and skin pigmentation in other vertebrates.

? *ADH and adrenaline act as hormones when released into the bloodstream and as neurotransmitters when released in synapses between neurons. What is similar about the endocrine glands that produce these two molecules?*

TEST YOUR UNDERSTANDING

Levels 1-2: Remembering/Understanding

1. Which statement is accurate?
   (A) Hormones that differ in effect reach their target cells by different routes through the body.
   (B) Pairs of hormones that have the same effect are said to have antagonistic functions.
   (C) Hormones are often regulated through feedback loops.
   (D) Hormones of the same chemical class usually have the same function.

2. The hypothalamus
   (A) synthesises all of the hormones produced by the pituitary gland.
   (B) influences the function of only one lobe of the pituitary gland.
   (C) produces only inhibitory hormones.
   (D) regulates both reproduction and body temperature.
3. Growth factors are local regulators that
   (A) are produced by the anterior pituitary.
   (B) are modified fatty acids that stimulate bone and cartilage growth.
   (C) are found on the surface of cancer cells and stimulate abnormal cell division.
   (D) bind to cell-surface receptors and stimulate growth and development of target cells.
4. Which hormone is *correctly* paired with its action?
   (A) oxytocin—stimulates uterine contractions during childbirth
   (B) thyroxine—inhibits metabolic processes
   (C) ACTH—inhibits the release of glucocorticoids by the adrenal cortex
   (D) melatonin—raises blood calcium level

Levels 3-4: Applying/Analysing

5. What do steroid and peptide hormones typically have in common?
   (A) their solubility in cell membranes
   (B) their requirement for travel through the bloodstream
   (C) the location of their receptors
   (D) their reliance on signal transduction in the cell
6. Which of the following is the most likely explanation for hypothyroidism in a patient whose iodine level is normal?
   (A) greater production of T_3 than of T_4
   (B) hyposecretion of TSH
   (C) hypersecretion of MSH
   (D) a decrease in the thyroid secretion of calcitonin
7. The relationship between the insect hormones ecdysteroid and PTTH is an example of
   (A) an interaction of the endocrine and nervous systems.
   (B) homeostasis achieved by positive feedback.
   (C) homeostasis maintained by antagonistic hormones.
   (D) competitive inhibition of a hormone receptor.
8. **DRAW IT** In mammals, milk production by mammary glands is controlled by prolactin and prolactin-releasing hormone. Draw a simple sketch of this pathway, including glands, tissues, hormones, routes for hormone movement, and effects.

Levels 5-6: Evaluating/Creating

9. **EVOLUTION CONNECTION** The intracellular receptors used by all the steroid and thyroid hormones are similar enough in structure that they are all considered members of one "superfamily" of proteins. Propose a hypothesis for how the genes encoding these receptors may have evolved. (Hint: See Figure 21.13.) Explain how you could test your hypothesis using DNA sequence data.
10. **SCIENTIFIC INQUIRY • INTERPRET THE DATA** A chronically high level of glucocorticoids can result in obesity, muscle weakness, and depression, a combination of symptoms called Cushing's syndrome. Excessive activity of either the pituitary or the adrenal gland can be the cause. To determine which gland has abnormal activity in a particular patient, doctors use the drug dexamethasone, a synthetic glucocorticoid that blocks ACTH release. Based on the graph, identify which gland is affected in patient X.

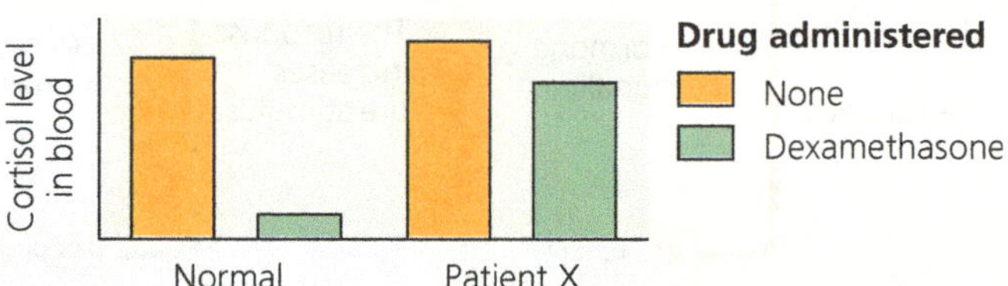

11. **WRITE ABOUT A THEME: INTERACTIONS** In a short essay (100–150 words), discuss the role of hormones in an animal's responses to changes in its environment. Use specific examples.
12. **SYNTHESISE YOUR KNOWLEDGE**

The frog on the left was injected with MSH, causing a change in skin colour within minutes due to a rapid redistribution of pigment granules in specialised skin cells. Using what you know about neuroendocrine signalling, explain how a frog could use MSH to match its skin colouration to that of its surroundings.

For selected answers, see Appendix A.

This page is intentionally blank.

Appendix A Answers

Chapter 1

Figure Questions

Figure 1.4 Dividing the length of the prokaryotic cell by the length of the scale bar, the length of the prokaryotic cell is about 1.4 scale bars. Each scale bar represents 1 μm, so the prokaryotic cell is about 1.4 μm long. Dividing the diameter of the eukaryotic cell by the length of the scale bar, the diameter of the eukaryotic cell is about 8.2 scale bars, which is 8.2 μm. **Figure 1.10** The response to insulin is glucose uptake by cells and glucose storage in liver cells. The initial stimulus is a high glucose level, which is reduced when glucose is taken up by cells. **Figure 1.18**

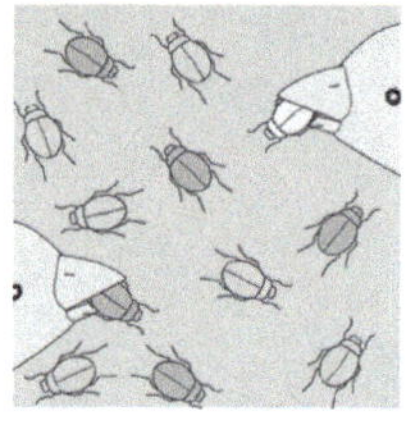

5 Environmental change resulting in survival of organisms with different traits

As the soil gradually becomes lighter grey, beetles that match the colour of the soil will not be seen by birds and therefore will not be eaten. For example, when the soil is medium-coloured, it will be easiest for birds to see and eat the darkest and lightest of the beetles in a population. (Most or all of the lighter beetles will have been eaten earlier, but new light beetles will arise due to variation in new generations of the population.) Thus, over time, the population will become lighter as the soil becomes lighter.

Concept Check 1.1

1. Examples: A molecule consists of *atoms* bonded together. Each organelle has an orderly arrangement of *molecules*. Photosynthetic plant cells contain *organelles* called chloroplasts. A tissue consists of a group of similar *cells*. Organs such as the heart are constructed from several *tissues*. A complex multicellular organism, such as a plant, has several types of *organs*, such as leaves and roots. A population is a set of *organisms* of the same species. A community consists of *populations* of the various species inhabiting a specific area. An ecosystem consists of a biological *community* along with the nonliving factors important to life, such as air, soil, and water. The biosphere is made up of all of Earth's *ecosystems*. **2.** (a) New properties emerge at successive levels of biological organisation: Structure and function are correlated. (b) Life's processes involve the expression and transmission of genetic information. (c) Life requires the transfer and transformation of energy and matter. **3.** Sample answers: *Organisation (Emergent properties)*: The ability of a human heart to pump blood requires an intact heart; it is not a capability of any of the heart's tissues or cells working alone. *Organisation (Structure and function)*: The strong, sharp teeth of a dingo are well suited to grasping and dismembering its prey. *Information*: Human eye colour is determined by the combination of genes inherited from the two parents. *Energy and Matter*: A plant, such as a grass, absorbs energy from the sun and transforms it into molecules that act as stored fuel. Animals can eat parts of the plant and use the food for energy to carry out their activities. *Interactions (Molecules)*: When your stomach is full, it signals your brain to decrease your appetite. *Interactions (Ecosystems)*: A mouse eats food, such as nuts or grasses, and deposits some of the food material as wastes (faeces and urine). Construction of a nest rearranges the physical environment and may hasten degradation of some of its components. The mouse may also act as food for a predator.

Concept Check 1.2

1. The naturally occurring heritable variation in a population is "edited" by natural selection because individuals with traits better suited to the environment survive and reproduce more successfully than others, and these traits are passed on to the next generation more often. Over time, better-suited individuals persist and their percentage in the population increases, while less well-suited individuals become less prevalent—a type of population editing. **2.** Here is one possible explanation: The ancestor species of the green warbler finch lived on an island where insects were a plentiful food source. Among individuals in the ancestor population, there was likely variation in beak shape and size. Individuals with slender, sharp beaks were likely more successful at picking up insects for food. Being well-nourished, they gave rise to more offspring than birds with thick, short beaks. Their many offspring inherited slender, sharp beaks (because of genetic information being passed from generation to generation, although Darwin didn't know this). In each generation, the offspring birds with the beaks of a shape best at picking up insects would eat more and have more offspring. Therefore, the green warbler finch of today has a slender beak that is very well matched (adapted) to its food source, insects.

3.

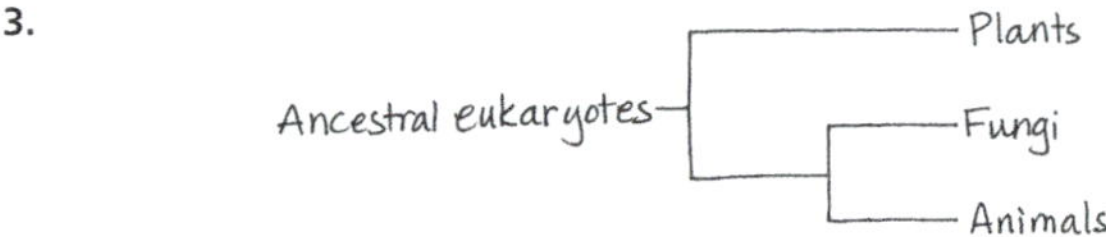

Concept Check 1.3

1. Mouse coat colour matches the environment for both beach and inland populations. **2.** Inductive reasoning derives generalisations from specific cases; deductive reasoning predicts specific outcomes from general premises. **3.** Compared to a hypothesis, a scientific theory is usually more general and substantiated by a much greater amount of evidence. Natural selection is an explanatory idea that applies to all kinds of organisms and is supported by vast amounts of evidence of various kinds. **4.** Based on the mouse colouration in Figure 1.25, you might expect that the mice that live on the sandy soil would be lighter in colour and those that live on the lava rock would be much darker. And in fact, that is what researchers have found. You would predict that each colour of mouse would be less preyed upon in its native habitat than it would be in the other habitat. (Research results also support this prediction.) You could repeat the Hoekstra experiment with coloured models, painted to resemble these two types of mouse. Or you could try transplanting some of each population to its non-native habitat and counting how many you can recapture over the next few days, then comparing the four samples as was done in Hoekstra's experiment. (The painted models are easier to recapture, of course!) In the live mouse transplantation experiment, you would have to do controls to eliminate the variable represented by the transplanted mice being in a new, unknown territory. You could control for the transplantation process by transplanting some dark mice from one area of lava rock to one far distant, and some light mice from one area of sandy soil to a distant area.

Concept Check 1.4

1. Science aims to understand natural phenomena and the underlying mechanisms that affect them, while technology involves application of scientific discoveries for a particular purpose or to solve a specific problem. **2.** Natural selection could be operating. Malaria is present in sub-Saharan Africa, so there might be an advantage to people with the sickle-cell disease form of the gene that makes them more able to survive and pass on their genes to offspring. Among those of African descent living in Australia or New Zealand, where malaria is absent, there would be no advantage, so they would be selected against more strongly, resulting in fewer individuals with the sickle-cell disease form of the gene.

Summary of Key Concepts Questions

1.1 Finger movements rely on the coordination of the many structural components of the hand (muscles, nerves, bones, etc.), each of which is composed of elements from lower levels of biological *organisation* (cells, molecules). The development of the hand relies on the genetic *information* encoded in chromosomes found in cells throughout the body. To power the finger movements that result in a text message, muscle and nerve cells require chemical *energy* that they transform in powering muscle contraction or in propagating nerve impulses. Texting is in essence communication, an *interaction* that conveys information between organisms, in this case of the same species. **1.2** Ancestors of the beach mouse may have exhibited variations in their coat colour. Because of the prevalence of visual predators, the better-camouflaged (lighter) mice in the beach habitat may have survived longer and been able to produce more offspring. Over time, a higher and higher proportion of individuals in the population would have had the adaptation of lighter fur that acted to camouflage the mouse in the beach habitat. **1.3** Gathering and interpreting data are core activities in the scientific process, and they are affected by, and affect in turn, three other arenas of the scientific process: exploration and discovery, community analysis and feedback, and societal benefits and outcomes. **1.4** Different approaches taken by scientists studying natural phenomena at different levels complement each other, so more is learned about each problem being studied. A diversity of backgrounds among scientists may lead to fruitful ideas in the same way that important innovations have often arisen where a mix of cultures coexist, due to multiple different viewpoints.

Test Your Understanding

1. B **2.** C **3.** C **4.** B **5.** C **6.** A **7.** D **8.** Your figure should show the following: (1) for the biosphere, the Earth with an arrow coming out of a tropical ocean; (2) for the ecosystem, a distant view of a coral reef; (3) for the community, a collection of reef animals and algae, with corals, fishes, some seaweed, and any other organisms you can think of; (4) for the population, a group of fish of the same species; (5) for the organism, one fish from your population; (6) for the organ, the fish's stomach; (7) for a tissue, a group of similar cells from the stomach; (8) for a cell, one cell from the tissue, showing its nucleus and a few other organelles; (9) for an organelle, the nucleus, where most of the cell's DNA is located; and (10) for a molecule, a DNA double helix. Your sketches can be very rough!

Chapter 2

Figure Questions

Figure 2.7 Atomic number = 12; 12 protons, 12 electrons; three electron shells; 2 valence electrons

Figure 2.14 One possible answer:

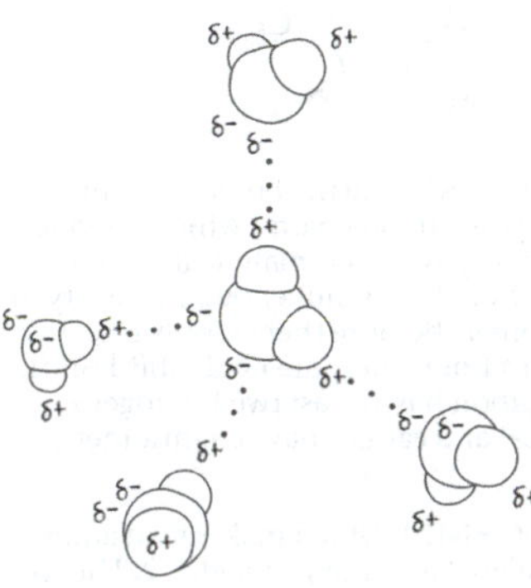

Figure 2.17

Sunlight

Leaf

Bubbles of O_2

$6\ O_2$

$6\ CO_2$

$C_6H_{12}O_6$

$6\ H_2O$

Concept Check 2.1

1. Table salt (sodium chloride) is made up of sodium and chlorine. We are able to eat the compound, showing that it has different properties from those of a metal (sodium) and a poisonous gas (chlorine). **2.** Yes. An organism requires trace elements, though only in small amounts. **3.** A person with an iron deficiency will probably show fatigue and other effects of a low oxygen level in the blood. (The condition is called anemia and can also result from too few red blood cells or abnormal haemoglobin.) **4.** Variant ancestral plants that could tolerate elevated levels of the elements in serpentine soils could grow and reproduce there. (Plants that were well adapted to nonserpentine soils would not be expected to survive in serpentine areas.) The offspring of the variants would also vary, with those most capable of thriving under serpentine conditions growing best and reproducing most. Over many generations, this probably led to the serpentine-adapted species we see today.

Concept Check 2.2

1. 7 **2.** $^{15}_{7}N$ **3.** 9 electrons; two electron shells; 1*s*, 2*s*, 2*p* (three orbitals); 1 electron is needed to fill the valence shell. **4.** The elements in a row all have the same number of electron shells. All the elements in a column have the same number of electrons in their valence shells.

Concept Check 2.3

1. In this structure, each carbon atom has only three covalent bonds instead of the required four. **2.** The attraction between oppositely charged ions, forming ionic bonds **3.** If you could synthesise molecules that mimic these shapes, you might be able to treat diseases or conditions caused by the inability of affected individuals to synthesise such molecules—or to block the function of such molecules if overproduction is the cause of the disorder.

Concept Check 2.4

1.

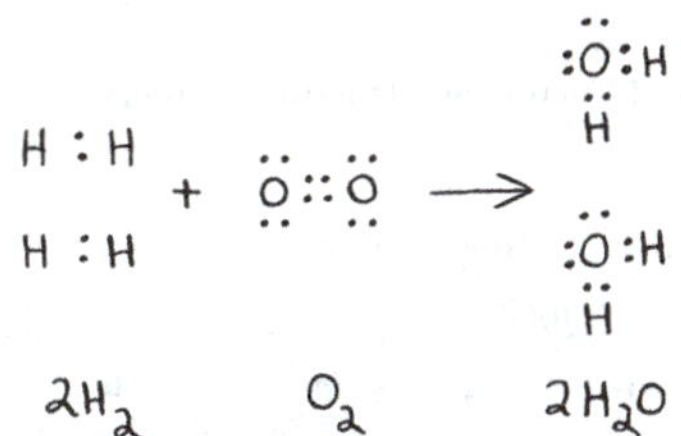

2. At equilibrium, the forward and reverse reactions occur at the same rate. **3.** $C_6H_{12}O_6 + 6\ O_2 \rightarrow 6\ CO_2 + 6\ H_2O$ + Energy. Glucose and oxygen react to form carbon dioxide and water, releasing energy. We breathe in oxygen because we need it for this reaction to occur, and we breathe out carbon dioxide because it is a by-product of this reaction. (This reaction is called cellular respiration, and you will learn more about it in Chapter 9.)

Summary of Key Concepts Questions

2.1 A compound is made up of two or more elements combined in a fixed ratio, while an element is a substance that cannot be broken down to other substances.
2.2

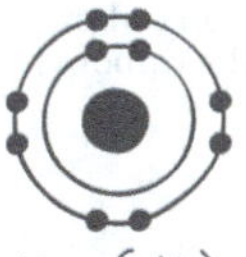

Both neon and argon have completed valence shells, containing 8 electrons. They do not have unpaired electrons that could participate in chemical bonds. **2.3** Electrons are shared equally between the two atoms in a nonpolar covalent bond. In a polar covalent bond, the electrons are drawn closer to the more electronegative atom. In the formation of ions, an electron is completely transferred from one atom to a much more electronegative atom. **2.4** The concentration of products would increase as the added reactants were converted to products. Eventually, an equilibrium would again be reached in which the forward and reverse reactions were proceeding at the same rate and the relative concentrations of reactants and products returned to where they were before the addition of more reactants.

Test Your Understanding

1. D **2.** A **3.** B **4.** A **5.** D **6.** B **7.** C **8.** D

9. a. This structure makes sense because all valence shells are complete, and all bonds have the correct number of electrons.

b. This structure doesn't make sense because H has only 1 electron to share, so it cannot form bonds with 2 atoms.

Chapter 3

Figure Questions

Figure 3.2 One possible answer:

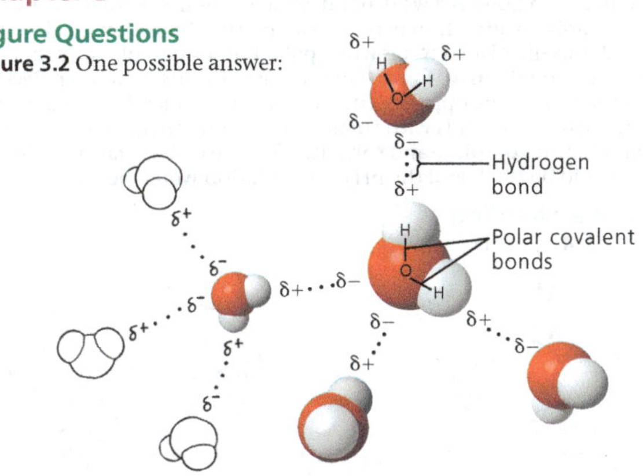

Figure 3.8 Heating the solution would cause the water to evaporate faster than it is evaporating at room temperature. At a certain point, there wouldn't be enough water molecules to dissolve the salt ions. The salt would start coming out of solution and re-forming crystals. Eventually, all the water would evaporate, leaving behind a pile of salt like the original pile. **Figure 3.12** Adding excess CO_2 to the oceans ultimately reduces the rate at which calcification (by organisms) can occur.

Concept Check 3.1

1. Electronegativity is the attraction of an atom for the electrons of a covalent bond. Because oxygen is more electronegative than hydrogen, the oxygen atom in H_2O pulls electrons towards itself, resulting in two partial negative charges on the oxygen atom and a partial positive charge on each hydrogen atom. Atoms in neighbouring water molecules with opposite partial charges are attracted to each other, forming a hydrogen bond. **2.** Due to its two polar covalent bonds, a water molecule has regions of partial negative charge on the O atom that allow it to form hydrogen bonds with hydrogen atoms on neighbouring water molecules, and regions of partial positive charge on the H atoms that allow them to form hydrogen bonds with oxygen atoms on neighbouring water molecules. **3.** The hydrogen atoms of one molecule, with their partial positive charges, would repel the hydrogen atoms of the adjacent molecule. **4.** The covalent bonds of water molecules would not be polar, so no regions of the molecule would carry partial charges and water molecules would not form hydrogen bonds with each other.

Concept Check 3.2

1. Hydrogen bonds hold neighbouring water molecules together. This cohesion helps chains of water molecules move upwards against gravity in water-conducting cells as water evaporates from the leaves. Adhesion between water molecules and the walls of the water-conducting cells also helps counter gravity. **2.** High humidity hampers cooling by suppressing the evaporation of sweat. **3.** As water freezes, it expands because water molecules move further apart in forming ice crystals. When there is water in a crevice of a boulder, expansion due to freezing may crack the boulder. **4.** The hydrophobic substance repels water, perhaps helping to keep the ends of the legs from becoming coated with water and breaking through the surface. If the legs were coated with a hydrophilic substance, water would be drawn up them, possibly making it more difficult for the water strider to walk on water.

Concept Check 3.3

1. 10^5, or 100,000 **2.** $[H^+] = 0.01\ M = 10^{-2}\ M$, so pH = 2. **3.** $CH_3COOH \rightarrow CH_3COO^- + H^+$. CH_3COOH is the acid (the H^+ donor), and CH_3COO^- is the base (the H^+ acceptor). **4.** The pH of the water should decrease from 7 to about 2 (as mentioned in the text); the pH of the acetic acid solution will decrease only a small amount, because as a weak acid, it acts (like carbonic acid) as a buffer. The reaction shown for question 3 will shift to the left, with CH_3COO^- accepting the influx of H^+ and becoming CH_3COOH molecules.

Summary of Key Concepts Questions

3.1

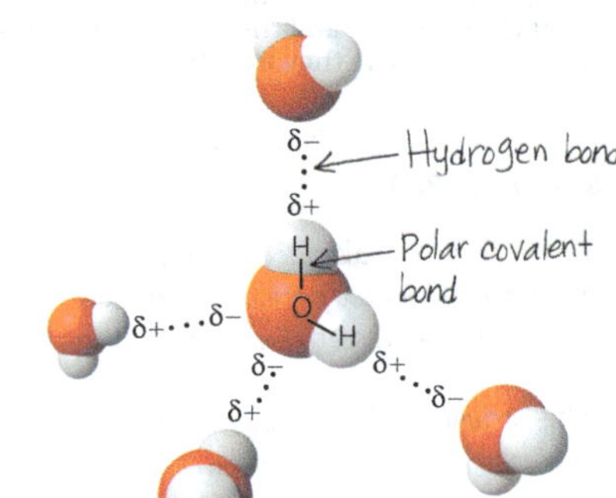

No. A covalent bond is a strong bond in which electrons are shared between two atoms. A hydrogen bond is a weak bond, which does not involve electron sharing, but is simply an attraction between two partial charges on neighbouring atoms. **3.2** Ions dissolve in water when polar water molecules form a hydration shell around them, with partially charged regions of water molecules being attracted to ions of the opposite charge. Polar molecules dissolve as water molecules form hydrogen bonds with them and surround them. Solutions are homogeneous mixtures of solute and solvent. **3.3** The concentration of hydrogen ions (H^+) would be 10^{-11}, and the pH of the solution would be 11.

Test Your Understanding

1. C **2.** D **3.** C **4.** A **5.** D

6.

7. Due to intermolecular hydrogen bonds, water has a high specific heat (the amount of heat required to increase the temperature of water by 1°C). When water is heated, much of the heat is absorbed in breaking hydrogen bonds before the water molecules increase their motion and the temperature increases. Conversely, when water is cooled, many H bonds are formed, which releases a significant amount of heat. This release of heat can provide some protection against freezing of the plants' leaves, thus protecting the cells from damage. **8.** Both global warming and ocean acidification are caused by increasing levels of carbon dioxide in the atmosphere, the result of burning fossil fuels.

Chapter 4

Figure Questions

Figure 4.2 Because the concentration of the reactants influences the equilibrium (as discussed in Concept 2.4), there might have been more HCN relative to CH_2O since there would have been a higher concentration of the reactant gas containing nitrogen.

Figure 4.4

Figure 4.6 The tails of fats contain only carbon-hydrogen bonds, which are relatively nonpolar. Because the tails occupy the bulk of a fat molecule, they make the molecule as a whole nonpolar and therefore incapable of forming hydrogen bonds with water.

Figure 4.7

Concept Check 4.1

1. The sparks provided energy needed for the inorganic molecules in the atmosphere to react with each other. (You'll learn more about energy and chemical reactions in Chapter 8.)

Concept Check 4.2

1.

2. The forms of C_4H_{10} in (b) are structural isomers, as are the butenes (forms of C_4H_8) in (c). **3.** Both consist largely of hydrocarbon chains, which provide fuel—petrol for engines and fats for plant embryos and animals. Reactions of both types of molecules release energy. **4.** No. There is not enough diversity in propane's atoms. It can't form structural isomers because there is only one way for three carbons to attach to each other (in a line). There are no double bonds, so *cis-trans* isomers are not possible. Each carbon has at least two hydrogens attached to it, so the molecule is symmetrical and cannot have enantiomers.

Concept Check 4.3

1. An amino acid has both an amino group (—NH_2), which makes it an amine, and a carboxyl group (—COOH), which makes it a carboxylic acid. **2.** The ATP molecule loses a phosphate, becoming ADP.

3. A chemical group that can act as a base has been replaced with a group that can act as an acid, increasing the acidic properties of the molecule. The shape of the molecule would also change, likely changing the molecules with which it can interact. The original cysteine molecule has an asymmetric carbon in the centre. After replacement of the amino group with a carboxyl group, this carbon is no longer asymmetric.

Summary of Key Concepts Questions

4.1 Miller showed that organic molecules could form under the physical and chemical conditions estimated to have been present on early Earth. This abiotic synthesis of organic molecules would have been a first step in the origin of life. **4.2** Acetone and propanal are structural isomers. Acetic acid and glycine have no asymmetric carbons, whereas glycerol phosphate has one. Therefore, glycerol phosphate can exist as forms that are enantiomers, but acetic acid and glycine cannot. **4.3** The methyl group is nonpolar and not reactive. The other six groups are called functional groups because they can participate in chemical reactions. Also, all except the sulfhydryl group are hydrophilic, increasing the solubility of organic compounds in water.

Test Your Understanding

1. B **2.** B **3.** C **4.** C **5.** A **6.** B **7.** A

8. The molecule on the right; the middle carbon is asymmetric.

9. Silicon has 4 valence electrons, the same number as carbon. Therefore, silicon would be able to form long chains, including branches, that could act as skeletons for large molecules. It would clearly do this much better than neon (with no valence electrons) or aluminum (with 3 valence electrons).

Chapter 5

Figure Questions

Figure 5.3 Glucose and fructose are structural isomers.

Figure 5.4

Four carbons are in the fructose ring, and two are not. (The latter two carbons are attached to carbons 2 and 5, which are in the ring.) The fructose ring differs from the glucose ring, which has five carbons in the ring and one that is not. (Note that the orientation of this fructose molecule is flipped horizontally relative to that of the one in Figure 5.5b; also, note that the oxygen on carbon 5 lost its proton and that the oxygen on carbon 2, which used to be the carbonyl oxygen, gained a proton.

Figure 5.5

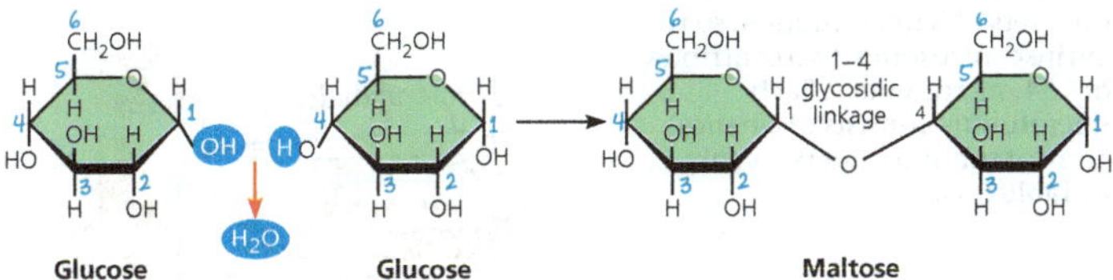

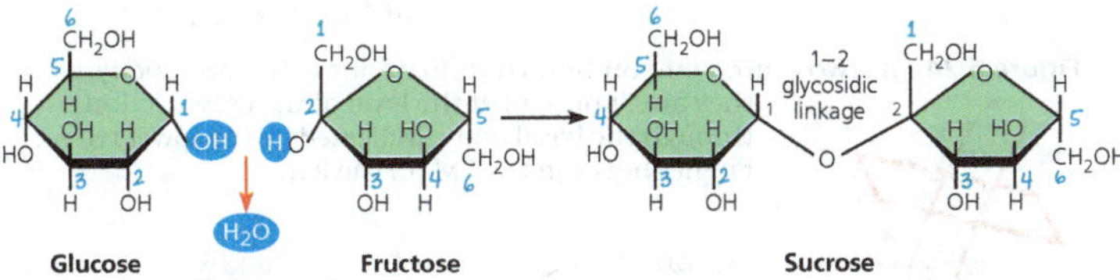

(a) In maltose, the linkage is called a 1–4 glycosidic linkage because the number 1 carbon in the left monosaccharide (glucose) is linked to the number 4 carbon in the right monosaccharide (also glucose). (b) In sucrose, the linkage is called a 1–2 glycosidic linkage because the number 1 carbon in the left monosaccharide (glucose) is linked to the number 2 carbon in the right monosaccharide (fructose). (Note that the fructose molecule is oriented differently from glucose in Figures 5.4 and 5.5b, where carbon 2 is on the right. In fructose in Figure 5.5b and here, carbon 2 of fructose is on the left.)

Figure 5.11 **Figure 5.12**

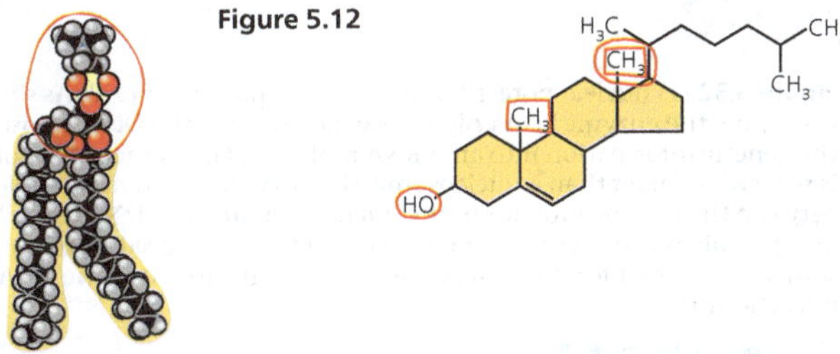

Figure 5.15

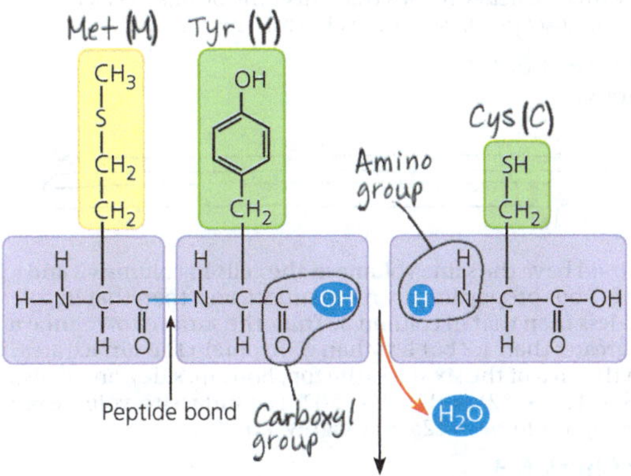

Figure 5.16 (1) The polypeptide backbone is most easily followed in the ribbon model.

(2)

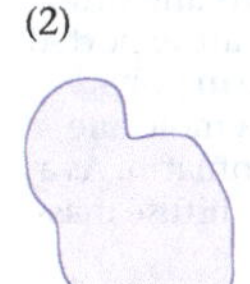

(3) The point of this diagram is to show that a pancreas cell secretes insulin proteins, so the shape is not important to the process being illustrated. **Figure 5.17** We can see that their complementary shapes allow the two proteins to fit together quite precisely. **Figure 5.19** The R group on glutamic acid is acidic and hydrophilic, whereas that on valine is nonpolar and hydrophobic. Therefore, it is unlikely that valine and glutamic acid participate in the same intramolecular interactions. A change in these interactions could (and does) cause a disruption of molecular structure. **Figure 5.26** Using a genomics approach allows us to use gene sequences to identify species and to learn about evolutionary relationships among any two species. This is because all species are related by their evolutionary history, and the evidence is in the DNA sequences. Proteomics—looking at proteins that are expressed—allows us to learn about how organisms or cells are functioning at a given time or in an association with another species.

Concept Check 5.1

1. The four main classes are proteins, carbohydrates, lipids, and nucleic acids. Lipids are not polymers. **2.** Nine, with one water molecule required to hydrolyse each connection between adjacent monomers **3.** The amino acids in the fish protein must be released in hydrolysis reactions and incorporated into other proteins in dehydration reactions.

Concept Check 5.2

1. $C_3H_6O_3$ **2.** $C_{12}H_{22}O_{11}$ **3.** The antibiotic treatment is likely to have killed the cellulose-digesting prokaryotes in the cow's gut. The absence of these prokaryotes would hamper the cow's ability to obtain energy from food and could lead to weight loss and possibly death. Thus, prokaryotic species are reintroduced, in appropriate combinations, in the gut culture given to treated cows.

Concept Check 5.3

1. Both have a glycerol molecule attached to fatty acids. The glycerol of a fat has three fatty acids attached, whereas the glycerol of a phospholipid is attached to two fatty acids and one phosphate group. **2.** Human sex hormones are steroids, a type of compound that is hydrophobic and thus classified as a lipid. **3.** The oil droplet membrane could consist of a single layer of phospholipids rather than a bilayer, because an arrangement in which the hydrophobic tails of the membrane phospholipids were in contact with the hydrocarbon regions of the oil molecules would be more stable.

Concept Check 5.4

1. Secondary structure involves hydrogen bonds between atoms of the polypeptide backbone. Tertiary structure involves interactions between atoms of the side chains of the amino acid subunits. **2.** The two ring forms of glucose are called α and β, depending on how the glycosidic bond dictates the position of a hydroxyl group. Proteins have α helices and β pleated sheets, two types of repeating structures found in polypeptides due to interactions between the repeating constituents of the chain (not the side chains). The haemoglobin molecule is made up of two types of polypeptides: It contains two molecules each of α-globin and β-globin. **3.** During formation of a polypeptide by polymerization of amino acids, the amino group of one amino acid reacts with the carboxyl group of the next, forming a peptide bond. Therefore, in a polypeptide, there is only one amino group on the N-terminus and one carboxyl group on the C-terminus, along with any carboxyl groups or amino groups located in the amino acid side chains (R groups). **4.** These are all nonpolar, hydrophobic amino acids, so you would expect this region to be located in the interior of the folded polypeptide, where it would not contact the aqueous environment inside the cell.

Concept Check 5.5

1.

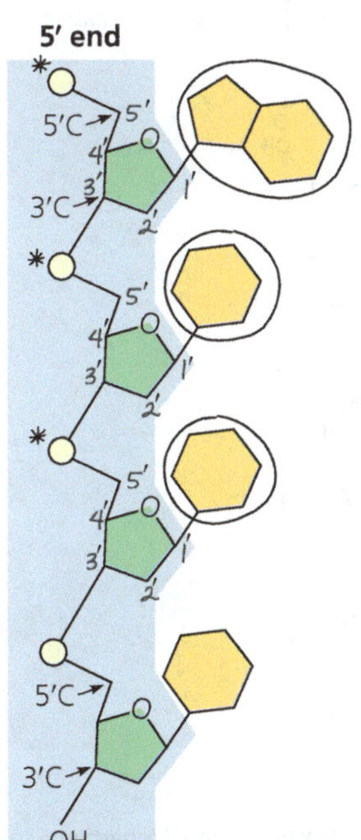

2.

5′–T A G G C C T–3′
3′–A T C C G G A–5′

Concept Check 5.6

1. The DNA of an organism encodes all of its proteins, and proteins are the molecules that carry out the work of cells, whether an organism is unicellular or multicellular. By knowing the DNA sequence of an organism, scientists would be able to catalog the protein sequences as well. **2.** Ultimately, the DNA sequence carries the information necessary to make the proteins that determine the traits of a particular species. Because the traits of the two species are similar, you would expect the proteins to be similar as well, and therefore the gene sequences should also have a high degree of similarity.

Summary of Key Concepts Questions

5.1 The polymers of large carbohydrates (polysaccharides), proteins, and nucleic acids are built from three different types of monomers (monosaccharides, amino acids, and nucleotides, respectively). **5.2** Both starch and cellulose are polymers of glucose, but the glucose monomers are in the α configuration in starch and the β configuration in cellulose. The glycosidic linkages thus have different geometries, giving the polymers different shapes and thus different properties. Starch is an energy-storage compound in plants; cellulose is a structural component of plant cell walls. Humans can hydrolyse starch to provide energy but cannot hydrolyse cellulose. Cellulose aids in the passage of food through the digestive tract. **5.3** Lipids are not polymers because they do not exist as a chain of linked monomers. They are not considered macromolecules because they do not reach the giant size of many polysaccharides, proteins, and nucleic acids. **5.4** A polypeptide, which may consist of hundreds of amino acids in a specific sequence (primary structure), has regions of coils and pleats (secondary structure), which are then folded into irregular contortions (tertiary structure) and may be noncovalently associated with other polypeptides (quaternary structure). The linear order of amino acids, with the varying properties of their side chains (R groups), determines what secondary and tertiary structures will form to produce a protein. The resulting unique three-dimensional shapes of proteins are key to their specific and diverse functions. **5.5** The complementary base pairing of the two strands of DNA makes possible the precise replication of DNA every time a cell divides, ensuring that genetic information is faithfully transmitted. In some types of RNA, complementary base pairing enables RNA molecules to assume specific three-dimensional shapes that facilitate diverse functions. **5.6** You would expect the human gene sequence to be most similar to that of the mouse (another mammal), then to that of the fish (another vertebrate), and least similar to that of the fruit fly (an invertebrate).

Test Your Understanding

1. D **2.** A **3.** B **4.** A **5.** B **6.** B **7.** C

8.

	Monomers or Components	Polymer or larger molecule	Type of linkage
Carbohydrates	Monosaccharides	Polysaccharides	Glycosidic linkages
Fats	Fatty acids	Triacylglycerols	Ester linkages
Proteins	Amino acids	Polypeptides	Peptide bonds
Nucleic acids	Nucleotides	Polynucleotides	Phosphodiester linkages

9.

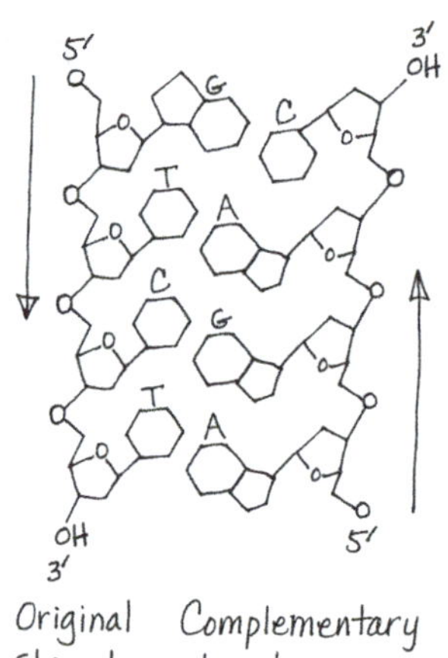

Chapter 6

Figure Questions

Figure 6.3 The cilia in the lower portion of the TEM were oriented lengthwise in the plane of the slice, while those in the upper portion of the TEM were oriented perpendicular to the plane of the slice. Therefore, the cilia in the lower portion were cut in longitudinal section, and the cilia in the upper portion were cut in cross section. **Figure 6.4** You would use the pellet from the final fraction, which is rich in ribosomes. These are the sites of protein translation. **Figure 6.6** The dark bands in the TEM correspond to the hydrophilic heads of the phospholipids, while the light band corresponds to the hydrophobic fatty acid tails of the phospholipids. **Figure 6.9** The DNA in a chromosome dictates synthesis of a messenger RNA (mRNA) molecule, which then moves out to the cytoplasm. There, the information is used for the production, on ribosomes, of proteins that carry out cellular functions. **Figure 6.10** Any of the bound ribosomes (attached to the endoplasmic reticulum) could be circled, because any could be making a protein that will be secreted. **Table 6.1** Three dimers **Figure 6.22** Each centriole has nine sets of 3 microtubules, so the entire centrosome (two centrioles) has 54 microtubules. Each microtubule consists of a helical array of tubulin dimers (as shown in Table 6.1).

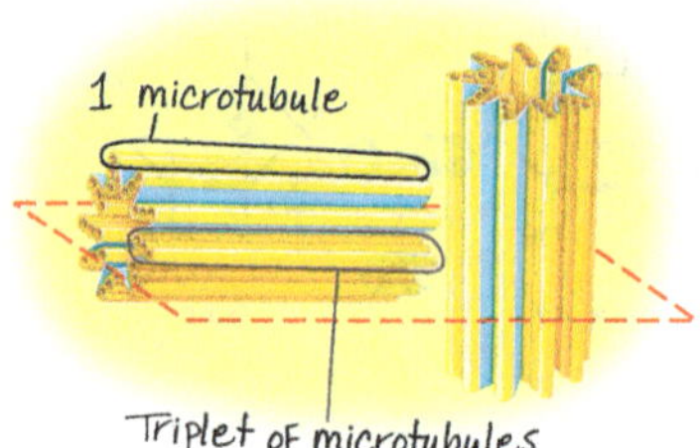

Figure 6.24 The two central microtubules terminate above the basal body, so they aren't present at the level of the cross section through the basal body, indicated by the lower red rectangle shown in the EM on the left.

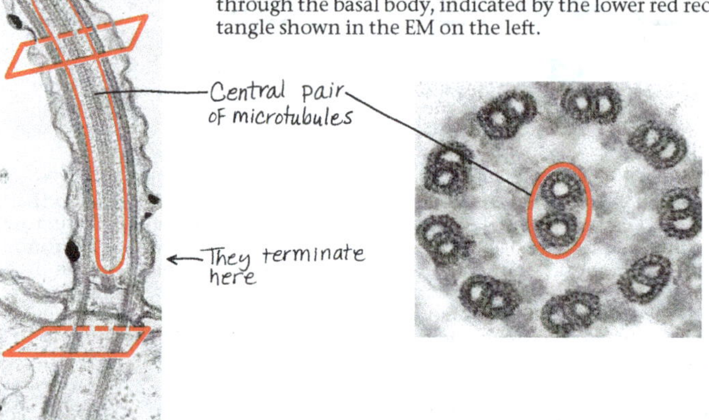

Figure 6.32 (1) nuclear pore, ribosome, proton pump, cyt *c*. (2) As shown in the figure, the enzyme RNA polymerase moves along the DNA, transcribing the genetic information into an mRNA molecule. Given that RNA polymerase is somewhat larger than a nucleosome, the enzyme would not be able to fit between the histone proteins of the nucleosome and the DNA itself. Thus, the group of histone proteins must be separated from or moved along the DNA somehow in order for the RNA polymerase enzyme to access the DNA. (3) A mitochondrion.

Concept Check 6.1

1. Stains used for light microscopy are coloured molecules that bind to cell components, affecting the light passing through, while stains used for electron microscopy involve heavy metals that affect the beams of electrons.
2. (a) Light microscope, (b) scanning electron microscope.

Concept Check 6.2

1. See Figure 6.8.
2.

This cell would have the same volume as the cells in columns 2 and 3 in Figure 6.7 but proportionally more surface area than that in column 2 and less than that in column 3. Thus, the surface-to-volume ratio should be greater than 1.2 but less than 6. To obtain the surface area, you would add the area of the six sides (the top, bottom, sides, and ends): $125 + 125 + 125 + 125 + 1 + 1 = 502$. The surface-to-volume ratio equals 502 divided by a volume of 125, or roughly 4.0.

Concept Check 6.3

1. Ribosomes in the cytoplasm translate the genetic message, carried from the DNA in the nucleus by mRNA, into a polypeptide chain. **2.** Nucleoli consist of DNA and the ribosomal RNAs (rRNAs) made according to its genes in the DNA, as well as proteins imported from the cytoplasm. Together, the rRNAs and proteins are assembled into large and small ribosomal subunits. (These are exported through nuclear pores to the cytoplasm, where they will participate in polypeptide synthesis.) **3.** Each chromosome consists of one long DNA molecule attached to numerous protein molecules, a combination called chromatin. As a cell begins division, each chromosome becomes "condensed" as its diffuse mass of chromatin coils up.

Concept Check 6.4

1. The primary distinction between rough and smooth ER is the presence of bound ribosomes on the rough ER. Both types of ER make phospholipids, but membrane proteins and secretory proteins are all produced by the ribosomes on the rough ER. The smooth ER also functions in detoxification, carbohydrate metabolism, and storage of calcium ions. **2.** Transport vesicles move membranes and the substances they enclose between other components of the endomembrane system. **3.** The mRNA is synthesised in the nucleus and then passes out through a nuclear pore to the cytoplasm, where it is translated on a bound ribosome, attached to the rough ER. The protein is synthesised into the lumen of the ER and may be modified there. A transport vesicle carries the protein to the Golgi apparatus. After further modification in the Golgi, another transport vesicle carries it back to the ER, where it will perform its cellular function.

Concept Check 6.5

1. Both organelles are involved in energy transformation, mitochondria in cellular respiration and chloroplasts in photosynthesis. They both have multiple membranes that separate their interiors into compartments. In both organelles, the innermost membranes—cristae, or infoldings of the inner membrane, in mitochondria and the thylakoid membranes in chloroplasts—have large surface areas with embedded enzymes that carry out their main functions. **2.** Yes. Plant cells are able to make their own sugar by photosynthesis, but mitochondria in plant cells (which are, of course, eukaryotic) are the organelles that are able to generate ATP molecules to be used for energy generation from sugars, a function required in all cells. **3.** Mitochondria and chloroplasts are not derived from the ER, nor are they connected physically or via transport vesicles to organelles of the endomembrane system. Mitochondria and chloroplasts are structurally quite different from vesicles derived from the ER, which are bounded by a single membrane.

Concept Check 6.6

1. Dynein arms, powered by ATP, move neighbouring doublets of microtubules relative to each other. Because they are anchored within the flagellum or cilium and with respect to one another, the doublets bend instead of sliding past each other. Synchronised bending of the nine microtubule doublets brings about bending of both cilia and flagella. **2.** Such individuals have defects in the microtubule-based movement of cilia and flagella. Thus, the sperm can't move because of malfunctioning or nonexistent flagella, and the airways are compromised because cilia that line the trachea malfunction or don't exist, and so mucus cannot be cleared from the lungs.

Concept Check 6.7

1. The most obvious difference is the presence of direct cytoplasmic connections between cells of plants (plasmodesmata) and animals (gap junctions). These connections result in the cytoplasm being continuous between adjacent cells.
2. The cell would not be able to function properly and would probably soon die, as the cell wall or ECM must be permeable to allow the exchange of matter between the cell and its external environment. Molecules involved in energy production and use must be allowed entry, as well as those that provide information about the cell's environment. Other molecules, such as products synthesised by the cell for export and the by-products of cellular respiration, must be allowed to exit.
3. The parts of the protein that face aqueous regions would be expected to have polar or charged (hydrophilic) amino acids, while the parts that go through the membrane would be expected to have nonpolar (hydrophobic) amino acids. You would predict polar or charged amino acids at each end (tail), in the region of the cytoplasmic loop, and in the regions of the two extracellular loops. You would predict nonpolar amino acids in the four regions that go through the membrane between the tails and loops.

Concept Check 6.8

1. *Colpidium colpoda* moves around in freshwater using cilia, projections from the plasma membrane that enclose microtubules in a "9 + 2" arrangement. The interactions between motor proteins and microtubules cause the cilia to bend synchronously, propelling the cell through the water. This is powered by ATP, obtained via breaking down sugars from food in a process that occurs in mitochondria. *C. colpoda* obtains bacteria as their food source, maybe via the same process (involving filopodia) the macrophage uses in Figure 6.31. This process uses actin filaments and other elements of the cytoskeleton to ingest the bacteria. Once ingested, the bacteria are broken down by enzymes in lysosomes. The proteins involved in all of these processes are encoded by genes on DNA in the nucleus of the *C. colpoda*.

Summary of Key Concepts Questions

6.1 Both light and electron microscopy allow cells to be studied visually, thus helping us understand internal cellular structure and the arrangement of cell components. Cell fractionation techniques separate out different groups of cell components, which can then be analysed biochemically to determine their function. Performing microscopy on the same cell fraction helps to correlate the biochemical function of the cell with the cell component responsible.
6.2 The separation of different functions in different organelles has several advantages. Reactants and enzymes can be concentrated in one area instead of spread throughout the cell. Reactions that require specific conditions, such as a lower pH, can be compartmentalised. And enzymes for specific reactions are often embedded in the membranes that enclose or partition an organelle.
6.3 The nucleus contains the genetic material of the cell in the form of DNA, which specifies messenger RNA, which in turn provides instructions for the synthesis of proteins (including the proteins that make up part of the ribosomes). DNA also codes for ribosomal RNAs, which are combined with proteins in the nucleolus into the subunits of ribosomes. Within the cytoplasm, ribosomes join with mRNA to build polypeptides, using the genetic information in the mRNA. **6.4** Transport vesicles move proteins and membranes synthesised by the rough ER to the Golgi for further processing and then to the plasma membrane, lysosomes, or other locations in the cell, including back to the ER.
6.5 According to the serial endosymbiont theory, mitochondria originated from an oxygen-using prokaryotic cell that was engulfed by a cell that was ancestral to eukaryotic cells. Over time, the host and endosymbiont evolved into a single unicellular organism containing a mitochondrion. Chloroplasts originated when at least one of these eukaryotic cells containing mitochondria engulfed and then retained a photosynthetic prokaryote, which eventually evolved into a chloroplast. **6.6** Inside the cell, motor proteins interact with components of the cytoskeleton to move cellular parts. Motor proteins "walk" vesicles along microtubules. The movement of cytoplasm within a cell involves interactions of the motor protein myosin and microfilaments (actin filaments). Whole cells can be moved by the rapid bending of flagella or cilia, which is caused by the motor protein–powered sliding of microtubules within these structures. Cell movement can also occur when pseudopodia form at one end of a cell (caused by actin polymerisation into a filamentous network), followed by contraction of the cell towards that end; this amoeboid movement is powered by interactions of microfilaments with myosin. Interactions of motor proteins and microfilaments in muscle cells causes muscle contraction that can propel whole organisms (for example, by walking or swimming). **6.7** A plant cell wall is primarily composed of microfibrils of cellulose embedded in other polysaccharides and proteins. The ECM of animal cells is primarily composed of collagen and other protein fibres, such as fibronectin and other glycoproteins. These fibres are embedded in a network of carbohydrate-rich proteoglycans. A plant cell wall provides structural support for the cell and, collectively, for the plant body. In addition to giving support, the ECM of an animal cell allows for communication of environmental changes into the cell. **6.8** The nucleus houses the chromosomes; each is made up of proteins and a single DNA molecule. The genes that exist along the DNA carry the genetic information necessary to make the proteins involved in ingesting a bacterial cell, such as the actin of microfilaments that form pseudopodia (filopodia), the proteins in the mitochondria responsible for providing the necessary ATP, and the enzymes present in the lysosomes that will digest the bacterial cell.

Test Your Understanding

1. B **2.** C **3.** B **4.** A **5.** D **6.** See Figure 6.8.

Chapter 7

Figure Questions

Figure 7.2

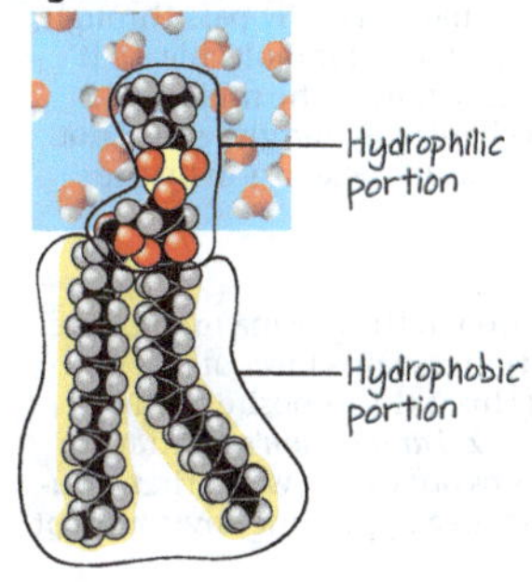

The hydrophilic portion is in contact with an aqueous environment (cytosol or extracellular fluid), and the hydrophobic portion is in contact with the hydrophobic portions of other phospholipids in the interior of the bilayer. **Figure 7.4** You couldn't rule out movement of proteins within membranes of the same species. You might propose that the membrane lipids and proteins from one species weren't able to mingle with those from the other species because of some incompatibility. **Figure 7.7** A transmembrane protein like the dimer in (f) might change its shape upon binding to a particular extracellular matrix (ECM) molecule. The new shape might enable the interior portion of the protein to bind to a second, cytoplasmic protein that would relay the message to the inside of the cell, as shown in (c). **Figure 7.8** The shape of a protein on the HIV surface is likely to be complementary to the shape of the receptor (CD4) and also to that of the co-receptor (CCR5). A molecule with a shape similar to that of the HIV surface protein could bind CCR5, blocking HIV binding. (Another answer would be a molecule that bound to CCR5 and changed the shape of CCR5 so it could no longer bind HIV; in fact, this is how maraviroc works.)

Figure 7.9

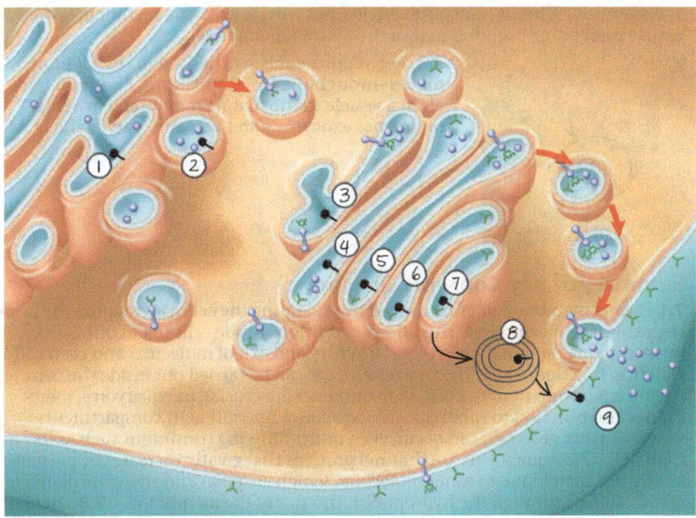

The protein would contact the extracellular fluid. (Because one end of the protein is in the ER membrane, no part of the protein extends into the cytoplasm.) The part of the protein not in the membrane extends into the ER lumen. Once the vesicle fuses with the plasma membrane, the "inside" of the ER membrane, facing the lumen, will become the "outside" of the plasma membrane, facing the extracellular fluid. **Figure 7.12** The orange dye would be evenly distributed throughout the solution on both sides of the membrane. The solution levels would not be affected because the orange dye can diffuse through the membrane and equalise its concentration. Thus, no additional osmosis would take place in either direction. In this experiment, the membrane is meant to represent the plasma membrane of a cell. **Figure 7.13** The cells take up water and become turgid, causing the stalk to lose its limpness and become crisp. **Figure 7.16** The sodium ion concentration ([Na^+]) is low inside the cell and high outside, while potassium ion concentration ([K^+]) is low outside the cell and high inside. Three Na^+ ions are moved out of the cell and two K^+ ions into the cell for each cycle. **Figure 7.17** The diamond solutes are moving into the cell (downwards), and the round solutes are moving out of the cell (upwards); each is moving against its concentration gradient. **Figure 7.21** (a) In the micrograph of the algal cell, the diameter of the algal cell is about 2.3 times longer than the scale bar, which represents 5 μm, so the diameter of the algal cell is about 11.5 μm. (b) In the micrograph of the coated vesicle, the diameter of the coated vesicle is about 1.2 times longer than the scale bar, which represents 0.25 μm, so the diameter of the coated vesicle is about 0.3 μm. (c) Therefore, the food vacuole around the algal cell will be about 40 × larger than the coated vesicle.

Concept Check 7.1

1. They are on the inside of the transport vesicle membrane. **2.** The grasses living in the cooler region would be expected to have more unsaturated fatty acids in their membranes because those fatty acids remain fluid at lower temperatures. The grasses living immediately adjacent to the hot springs would be expected to have more saturated fatty acids, which would allow the fatty acids to "stack" more closely, making the membranes less fluid and therefore helping them to stay intact at higher temperatures. (In plants, cholesterol is generally not used to moderate the effects of temperature on membrane fluidity because it is found at vastly lower levels in membranes of plant cells than in those of animal cells.)

Concept Check 7.2

1. O_2 and CO_2 are both small, nonpolar molecules that can easily pass through the hydrophobic interior of a membrane. **2.** Water is a polar molecule, so it cannot pass very rapidly through the hydrophobic region in the middle of a phospholipid bilayer. **3.** The hydronium ion is charged, while glycerol is not. Charge is probably more significant than size as a basis for exclusion by the aquaporin channel.

Concept Check 7.3

1. CO_2 is a nonpolar molecule that can diffuse through the plasma membrane. As long as it diffuses away so that the concentration remains low outside the cell, it will continue to exit the cell in this way. (This is the opposite of the case for O_2, described in this section of the text.) **2.** *Paramecium*'s contractile vacuole will become less active. The vacuole pumps out excess water that accumulates in the cell; this accumulation occurs only in a hypotonic environment.

Concept Check 7.4

1. These pumps use ATP. To establish a voltage, ions have to be pumped against their gradients, which requires energy. **2.** Each ion is being transported against its electrochemical gradient. If either ion were transported down its electrochemical gradient, this *would* be considered cotransport. **3.** The internal environment of a lysosome is acidic, so it has a higher concentration of H^+ than does the cytoplasm. Therefore, you might expect the membrane of the lysosome to have a proton pump such as that shown in Figure 7.18 to pump H^+ into the lysosome.

Concept Check 7.5

1. Exocytosis. When a transport vesicle fuses with the plasma membrane, the vesicle membrane becomes part of the plasma membrane.

2.

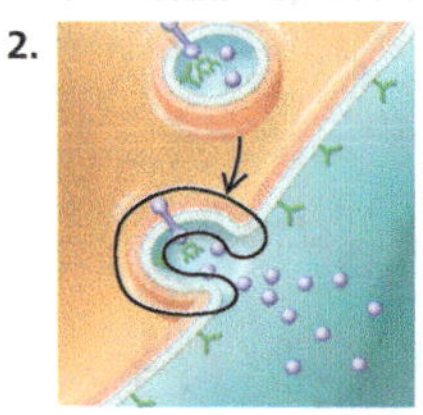

3. The glycoprotein is synthesised in the ER lumen, moves through the Golgi apparatus, and then travels in a vesicle to the plasma membrane, where it undergoes exocytosis and becomes part of the ECM.

Summary of Key Concepts Questions

7.1 Plasma membranes define the cell by separating the cellular components from the external environment. This allows conditions inside cells to be controlled by membrane proteins, which regulate entry and exit of molecules and even cell function (see Figure 7.7). The processes of life can be carried out inside the controlled environment of the cell, so membranes are crucial. In eukaryotes, membranes also function to subdivide the cytoplasm into different compartments where distinct processes can occur, even under differing conditions such as low or high pH. **7.2** Aquaporins are channel proteins that greatly increase the permeability of a membrane to water molecules, which are polar and therefore do not readily diffuse through the hydrophobic interior of the membrane. **7.3** There will be a net diffusion of water out of a cell into a hypertonic solution. The free water concentration is higher inside the cell than in the solution (where not as many water molecules are free, because many are clustered around the numerous solute particles). **7.4** One of the solutes moved by the cotransporter is actively transported against its concentration gradient. The energy for this transport comes from the concentration gradient of the other solute, which was established by an electrogenic pump that used energy to transport the other solute across the membrane. Because energy is required overall to drive this process (because ATP is used to establish the concentration gradient), it is considered active transport. **7.5** In receptor-mediated endocytosis, specific molecules bind to receptors on the plasma membrane in a region where a coated pit develops. The cell can acquire bulk quantities of those specific molecules when the coated pit forms a vesicle and carries the bound molecules into the cell.

Test Your Understanding

1. B **2.** C **3.** A **4.** C **5.** B

6. (a)

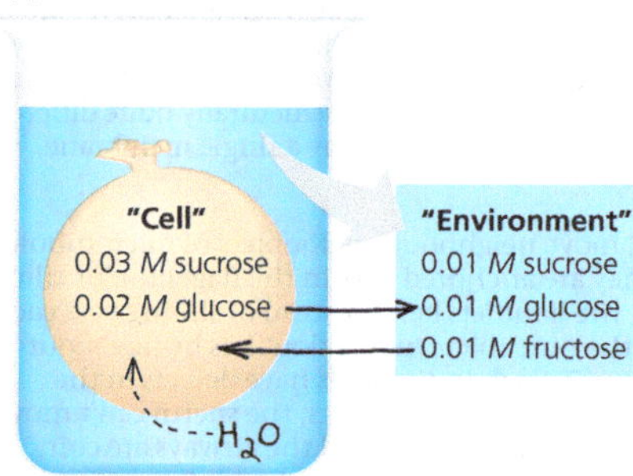

(b) The solution outside is hypotonic. It has less sucrose, which is a nonpenetrating solute. (c) See answer for (a). (d) The artificial cell will become more turgid. (e) Eventually, the two solutions will have the same solute concentrations. Even though sucrose can't move through the membrane, water flow (osmosis) will lead to isotonic conditions.

Chapter 8

Figure Questions

Figure 8.5 With a proton pump (Figure 7.18), the energy stored in ATP is used to pump protons across the membrane and build up a higher (nonrandom) concentration outside of the cell, so this process results in higher free energy. When solute molecules (analogous to hydrogen ions) are uniformly distributed, similar to the random distribution in the bottom of (b), the system has less free energy than it does in the top of (b). The system in the bottom can do no work. Because the concentration gradient created by a proton pump (Figure 7.18) represents higher free energy, this system has the potential to do work once there is a higher concentration of protons on one side of the membrane (as you will see in Figure 9.15). **Figure 8.10** Glutamic acid (Glu) has a carboxyl group at the end of its R group. Glutamine (Gln) has exactly the same structure as glutamic acid, except that there is an amino group in place of the $—O^-$ on the R group. (This O atom on the R group leaves during the synthesis reaction.) Thus, in this figure, Gln is drawn as a Glu with an attached NH_2.

Figure 8.13

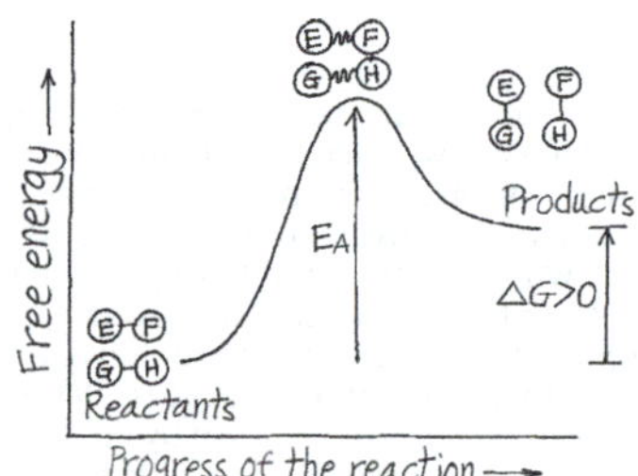

Figure 8.16

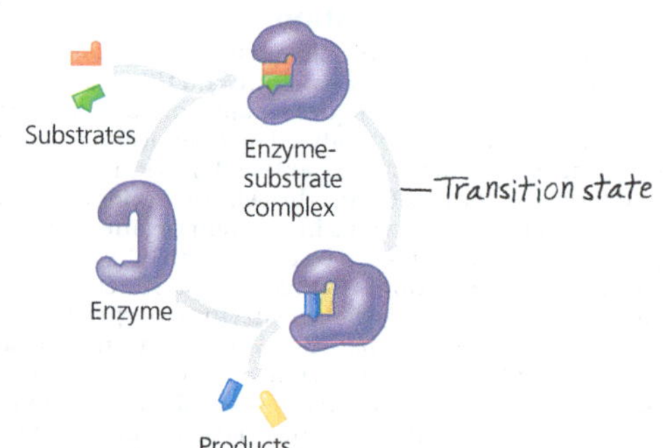

Concept Check 8.1

1. The second law is the trend towards randomisation, or increasing entropy. When the concentrations of a substance on both sides of a membrane are equal, the distribution is more random than when they are unequal. Diffusion of a substance to a region where it is initially less concentrated increases entropy, making it an energetically favourable (spontaneous) process as described by the second law. This explains the process seen in Figure 7.11. **2.** The apple has potential energy in its position hanging on the tree, and the sugars and other nutrients it contains have chemical energy. The apple has kinetic energy as it falls from the tree to the ground. Finally, when the apple is digested and its molecules broken down, some of the chemical energy is used to do work, and the rest is lost as thermal energy. **3.** The sugar crystals become less ordered (entropy increases) as they dissolve and become randomly spread out in the water. Over time, the water evaporates, and the crystals form again because the water volume is insufficient to keep them in solution. While the reappearance of sugar crystals may represent a "spontaneous" increase in order (decrease in entropy), it is balanced by the decrease in order (increase in entropy) of the water molecules, which changed from a relatively compact arrangement as liquid water to a much more dispersed and disordered form as water vapour.

Concept Check 8.2

1. Cellular respiration is a spontaneous and exergonic process. The energy released from glucose is used to do work in the cell or is lost as heat. **2.** Catabolism breaks down organic molecules, releasing their chemical energy and resulting in smaller products with more entropy, as when moving from the top to the bottom of Figure 8.5c. Anabolism consumes energy to synthesise larger molecules from simpler ones, as when moving from the bottom to the top of part (c). **3.** The reaction is exergonic because it releases energy—in this case, in the form of light. (This is a nonbiological version of the bioluminescence seen in Figure 8.1.)

Concept Check 8.3

1. ATP usually transfers energy to an endergonic process by phosphorylating (adding a phosphate group to) another molecule. (Exergonic processes, in turn, phosphorylate ADP to regenerate ATP.) **2.** A set of coupled reactions can transform the first combination into the second. Since this is an exergonic process overall, ΔG is negative and the first combination must have more free energy (see Figure 8.10). **3.** Active transport: The solute is being transported against its concentration gradient, which requires energy, provided by ATP hydrolysis.

Concept Check 8.4

1. A spontaneous reaction is a reaction that is exergonic. However, if it has a high activation energy that is rarely attained, the rate of the reaction may be low. **2.** O_2 is required as a substrate to react with the gas. There is O_2 in the lab air, which is why the Bunsen burner has a flame above the gas delivery opening, but there is no O_2 in the rubber tubing or the gas supply. **3.** In the presence of malonate, increase the concentration of the normal substrate (succinate) and see whether the rate of reaction increases. If it does, malonate is a competitive inhibitor.
4.

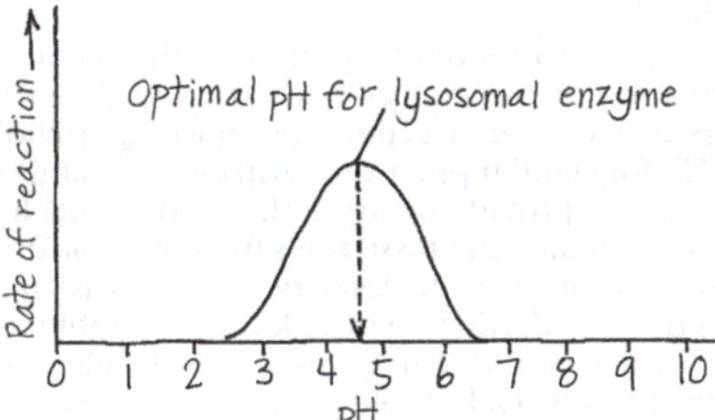

Concept Check 8.5

1. The activator binds in such a way that it stabilises the active form of an enzyme, whereas the inhibitor stabilises the inactive form. **2.** A catabolic pathway breaks down organic molecules, generating energy that is stored in ATP molecules. In feedback inhibition of such a pathway, ATP (one product) would act as an allosteric inhibitor of an enzyme catalysing an early step in the catabolic process. When ATP is plentiful, the pathway would be turned off and no more would be made.

Summary of Key Concepts Questions

8.1 The process of "ordering" a cell's structure is accompanied by an increase in the entropy (disorder) of the universe. For example, an animal cell takes in highly ordered organic molecules as the source of matter and energy used to build and maintain its structures. In the same process, however, the cell releases heat and the simple molecules of CO_2 and H_2O to the surroundings. The increase in entropy of the latter process offsets the entropy decrease in the former. **8.2** A spontaneous reaction has a negative ΔG and is exergonic. For a chemical reaction to proceed with a net release of free energy ($-\Delta G$), the enthalpy or total energy of the system must decrease ($-\Delta H$), and/or the entropy or disorder must increase (yielding a more negative term, $-T\Delta S$). Spontaneous reactions supply the energy to perform cellular work. **8.3** The free energy released from the hydrolysis of ATP may drive endergonic reactions through the transfer of a phosphate group to a reactant molecule, forming a more reactive phosphorylated intermediate. ATP hydrolysis also powers the mechanical and transport work of a cell, often by powering shape changes in the relevant motor proteins. Cellular respiration, the catabolic breakdown of glucose, provides the energy for the endergonic regeneration of ATP from ADP and Ⓟ$_i$.
8.4 Activation energy barriers prevent the complex molecules of the cell, which are rich in free energy, from spontaneously breaking down to less ordered, more stable molecules. Enzymes permit a regulated metabolism by binding to specific substrates and forming enzyme-substrate complexes that selectively lower the E_A for the chemical reactions in a cell. **8.5** A cell tightly regulates its metabolic pathways in response to fluctuating needs for energy and materials. The binding of activators or inhibitors to regulatory sites on allosteric enzymes stabilises either the active or the inactive form of the subunits. For example, the binding of ATP to a catabolic enzyme in a cell with excess ATP would inhibit that pathway. Such types of feedback inhibition preserve chemical resources within a cell. If ATP supplies are depleted, binding of ADP to the regulatory site of catabolic enzymes would activate that pathway, generating more ATP.

Test Your Understanding

1. B **2.** C **3.** B **4.** A **5.** C **6.** D **7.** C

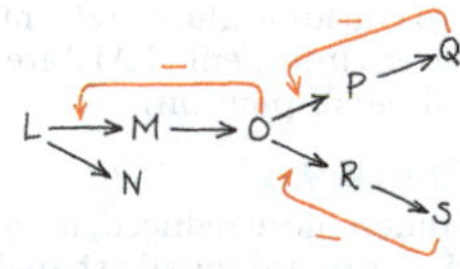

9.

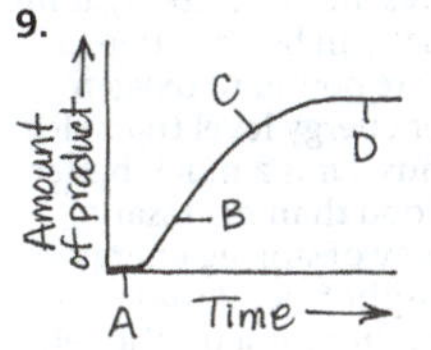

A. The substrate molecules are entering the pancreatic cells, so no product is made yet.
B. There is sufficient substrate, so the reaction is proceeding at a maximum rate.
C. As the substrate is used up, the rate decreases (the slope is less steep).
D. The line is flat because no new substrate remains and thus no new product appears.

Chapter 9

Figure Questions

Figure 9.2 The C atom is oxidised. The electrons that used to be equally shared with H atoms in methane are now much closer to the O atoms in CO_2 than they are to the C. **Figure 9.3** The reduced form has an extra hydrogen, along with 2 electrons, bound to the carbon shown at the top of the nicotinamide (opposite the N). There are different numbers and positions of double bonds in the two forms: The oxidised form has three double bonds in the ring, while the reduced form has only two. (In organic chemistry you may have learned, or will learn, that three double bonds in a ring are able to "resonate," or act as a ring of electrons. Having three resonant double bonds is more "oxidised" than having only two double bonds in the ring.) In the oxidised form there is a + charge on the N (because it is sharing 4 electron pairs), whereas in the reduced form it is only sharing 3 electron pairs (having a pair of electrons to itself). **Figure 9.6** Because there is no external source of energy for the reaction, it must be exergonic, and the reactants must be at a higher energy level than the products. **Figure 9.8** The removal would probably stop glycolysis, or at least slow it down, since it would push the equilibrium for step 5 towards DHAP (towards the bottom in this figure). If less (or no) glyceraldehyde 3-phosphate were available, step 6 would slow down (or be unable to occur). **Figure 9.12** The electrons in NADH are in a C—H bond (right side of Figure 9.3). In H_2O, the electrons are in an O—H bond. Because the electronegativities of C and H are similar, the electrons are equally shared in NADH. The electronegativity of O is much higher than that of C or H, so the electrons are much closer to O in H_2O and have "fallen" in potential energy. **Figure 9.14** At first, some ATP could be made, since electron transport could proceed as far as complex III, and a small H^+ gradient could be built up. Soon, however, no more electrons could be passed to complex III because it could not be reoxidised by passing its electrons to complex IV. **Figure 9.15** First, there are 2 NADH from the oxidation of pyruvate plus 6 NADH from the citric acid cycle (CAC); $8\,NADH \times 2.5\,ATP/NADH = 20\,ATP$. Second, there are 2 $FADH_2$ from the CAC; $2\,FADH_2 \times 1.5\,ATP/FADH_2 = 3\,ATP$. Third, the 2 NADH from glycolysis enter the mitochondrion through one of two types of shuttle. They pass their electrons either to 2 FAD, which become $FADH_2$ and result in 3 ATP, or to 2 NAD^+, which become NADH and result in 5 ATP. Thus, $20 + 3 + 3 = 26$ ATP, or $20 + 3 + 5 = 28$ ATP from all NADH and $FADH_2$.

Concept Check 9.1

1. Both processes include glycolysis, the citric acid cycle, and oxidative phosphorylation. In aerobic respiration, the final electron acceptor is molecular oxygen (O_2); in anaerobic respiration, the final electron acceptor is a different substance. **2.** $C_4H_6O_5$ would be oxidised and NAD^+ would be reduced.

Concept Check 9.2

1. NAD^+ acts as the oxidising agent in step 6, accepting electrons from glyceraldehyde 3-phosphate (G3P), which thus acts as the reducing agent.

Concept Check 9.3

1. NADH and $FADH_2$; one ATP is produced during substrate-level phosphorylation in step 5. **2.** The CO_2 that we exhale is produced by pyruvate oxidation and the citric acid cycle. **3.** In both cases, the precursor molecule loses a CO_2 molecule and then donates electrons to an electron carrier in an oxidation step. Also, the product has been activated due to the attachment of a CoA group by its S atom.

Concept Check 9.4

1. Oxidative phosphorylation would eventually stop entirely, resulting in no ATP production by this process. Without oxygen to "pull" electrons down the electron transport chain, H^+ would not be pumped into the mitochondrion's intermembrane space and chemiosmosis would not occur. **2.** Decreasing the pH means addition of H^+. This would establish a proton gradient even without the function of the electron transport chain, and we would expect ATP synthase to function and synthesise ATP. (In fact, it was experiments like this that

Appendix A Answers

provided support for chemiosmosis as an energy-coupling mechanism.) **3.** One of the components of the electron transport chain, ubiquinone (Q), must be able to diffuse within the membrane. It could not do so if the membrane components were locked rigidly into place.

Concept Check 9.5

1. A derivative of pyruvate, such as acetaldehyde during alcohol fermentation, or pyruvate itself during lactic acid fermentation; O_2; another electron acceptor at the end of an electron transport chain, such as sulfate (SO_4^{2-}) **2.** The cell would need to consume glucose at a rate about 16 times the consumption rate in the aerobic environment (2 ATP are generated by fermentation versus up to 32 ATP by cellular respiration).

Concept Check 9.6

1. The fat is much more reduced; it has many $-CH_2-$ units, and in all these bonds the electrons are equally shared. The electrons present in a carbohydrate molecule are already somewhat oxidised (shared unequally in bonds; there are more C—O and O—H bonds), as quite a few of them are bound to oxygen. Electrons that are equally shared, as in fat, have a higher energy level than electrons that are unequally shared, as in carbohydrates. Thus, fat is a much better fuel than carbohydrate. **2.** When you consume more food than necessary for metabolic processes, your body synthesises fat as a way of storing energy for later use. **3.** AMP will accumulate, stimulating phosphofructokinase, and thus increasing the rate of glycolysis. Since oxygen is not present, the cell will convert more pyruvate to lactate, providing a supply of ATP. **4.** When O_2 is present, the fatty acid chains containing most of the energy of a fat are oxidised and fed into the citric acid cycle and the electron transport chain. During intense exercise, however, O_2 is scarce in muscle cells, so ATP must be generated by glycolysis alone. A very small part of the fat molecule, the glycerol backbone, can be oxidised via glycolysis, but the amount of energy released by this portion is insignificant compared to that released by the fatty acid chains. (This is why moderate exercise, staying below 70% maximum heart rate, is better for burning fat—because enough O_2 remains available to the muscles.)

Summary of Key Concepts Questions

9.1 Most of the ATP produced in cellular respiration comes from oxidative phosphorylation, in which the energy released from redox reactions in an electron transport chain is used to produce ATP. In substrate-level phosphorylation, an enzyme directly transfers a phosphate group to ADP from an intermediate substrate. All ATP production in glycolysis occurs by substrate-level phosphorylation; this form of ATP production also occurs at one step in the citric acid cycle. **9.2** The oxidation of the three-carbon sugar, glyceraldehyde 3-phosphate, yields energy. In this oxidation, electrons and H^+ are transferred to NAD^+, forming NADH, and a phosphate group is attached to the oxidised substrate. ATP is then formed by substrate-level phosphorylation when this phosphate group is transferred to ADP. **9.3** The release of six molecules of CO_2 represents the complete oxidation of glucose. During the processing of two pyruvates to acetyl CoA, the fully oxidised carboxyl groups ($-COO^-$) are given off as 2 CO_2. The remaining four carbons are released as CO_2 in the citric acid cycle as citrate is oxidised back to oxaloacetate. **9.4** The flow of H^+ through the ATP synthase complex causes the rotor and attached rod to rotate, exposing catalytic sites in the knob portion that produce ATP from ADP and Ⓟ$_i$. ATP synthases are found in the inner mitochondrial membrane, the plasma membrane of prokaryotes, and membranes within chloroplasts. **9.5** Anaerobic respiration yields more ATP. The 2 ATP produced by substrate-level phosphorylation in glycolysis represent the total energy yield of fermentation. NADH passes its "high-energy" electrons to pyruvate or a derivative of pyruvate, recycling NAD^+ and allowing glycolysis to continue. In anaerobic respiration, the NADH produced during glycolysis, as well as additional molecules of NADH produced when pyruvate is oxidised, are used to generate ATP molecules. An electron transport chain captures the energy of the electrons in NADH via a series of redox reactions; ultimately, the electrons are transferred to an electronegative atom in a molecule other than oxygen. **9.6** The ATP produced by catabolic pathways is used to drive anabolic pathways. Also, many of the intermediates of glycolysis and the citric acid cycle are used in the biosynthesis of a cell's molecules.

Test Your Understanding

1. C **2.** C **3.** A **4.** B **5.** D **6.** A **7.** B

8. Since the overall process of glycolysis results in net production of ATP, it would make sense for the process to slow down when ATP levels have increased substantially. Thus, we would expect ATP to allosterically inhibit phosphofructokinase. **9.** The proton pump in Figures 7.18 and 7.19 is carrying out active transport, using ATP hydrolysis to pump protons against their concentration gradient. Because ATP is required, this is active transport of protons. The ATP synthase in Figure 9.13 is using the flow of protons down their concentration gradient to power ATP synthesis. Because the protons are moving down their concentration gradient, no energy is required, and this is passive transport.

10.

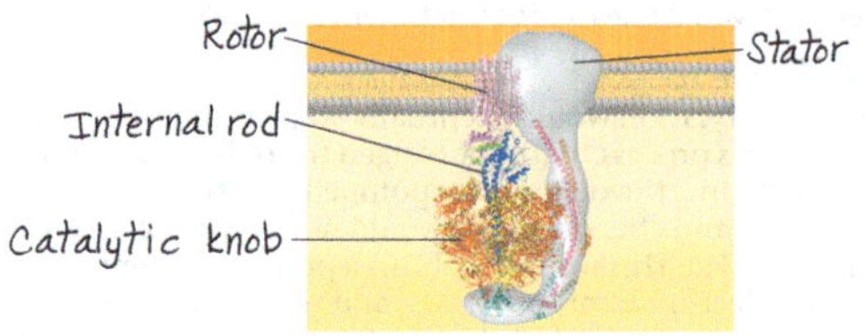

12.

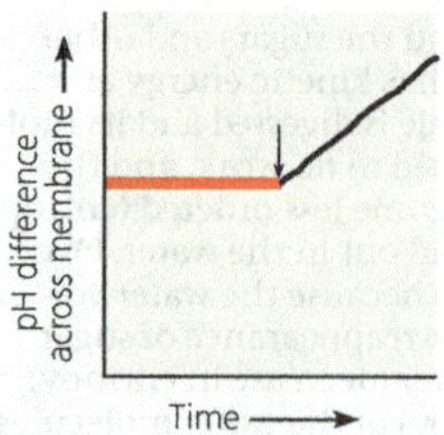

H^+ would continue to be pumped across the membrane into the intermembrane space, increasing the difference between the matrix pH and the intermembrane space pH. H^+ would not be able to flow back through ATP synthase, since the enzyme is inhibited by the poison, so rather than maintaining a constant difference across the membrane, the difference would continue to increase. (Over a longer period of time, the H^+ concentration in the intermembrane space would be so high that no more H^+ would be able to be pumped against the gradient. This wasn't asked for in the question, but if your graph levels off at the right end and this is your reasoning, your answer is correct.)

Chapter 10

Figure Questions

Figure 10.12 In the leaf, most of the chlorophyll electrons excited by photon absorption are used to power the reactions of photosynthesis. **Figure 10.16** The person at the top of the photosystem I tower would not turn to his left and throw his electron into the NADPH bucket. Instead, he would throw it onto the top of the ramp at his right, next to the photosystem II tower. The electron would then roll down the ramp, get energised by a photon, and return to him. This cycle would continue as long as light was available. (This is why it's called cyclic electron flow.) **Figure 10.17** You would (a) decrease the pH outside the mitochondrion (thus increasing the H^+ concentration) and (b) increase the pH in the chloroplast stroma (thus decreasing the H^+ concentration). In both cases, this would generate an H^+ gradient across the membrane that would cause ATP synthase to synthesise ATP. **Figure 10.18** Steps that increase [H^+] in the thylakoid space or decrease [H^+] in the stroma contribute to the [H^+] concentration gradient across the thylakoid membrane. In step 2, water is split in the thylakoid space, releasing 2 H^+ and increasing [H^+]. In step 3, as electrons travel down the electron transport chain, 4 H^+ are pumped into the thylakoid space, increasing [H^+]. In step 5, NADPH formation uses an H^+ from the stroma, decreasing [H^+] in the stroma. **Figure 10.23** The gene encoding hexokinase is part of the DNA of a chromosome in the nucleus. There, the gene is transcribed into mRNA, which is transported to the cytoplasm where it is translated on a free ribosome into a polypeptide. The polypeptide folds into a functional protein with secondary and tertiary structure. Once functional, it carries out the first reaction of glycolysis in the cytoplasm.

Concept Check 10.1

1. Because heterotrophs cannot photosynthesise, they cannot capture the energy of sunlight and produce energy-rich compounds like sugars, as autotrophs can. Sugars are oxidised by cellular respiration, providing energy (in the form of ATP) for cellular processes. Without this ability, heterotrophs depend on autotrophs to provide sugars as the food molecules that fuel their vital processes. **2.** Burning fossil fuels in power stations and internal combustion engines produces CO_2. Algae require CO_2 as one of the raw materials for the photosynthetic process. Placing photobioreactors filled with algae close to the sources of CO_2 emissions allows the algae to fix some of the CO_2 emissions into sugars. Photobioreactors can to some extent reduce atmospheric CO_2 concentrations around sites of high production. This process would reduce gases that would otherwise contribute to climate change; see Concept 1.1.

Concept Check 10.2

1. CO_2 enters the leaves via stomata, and, being a nonpolar molecule, can cross the leaf cell membrane and the chloroplast membranes to reach the stroma of the chloroplast. **2.** Using ^{18}O, a heavy isotope of oxygen, as a label, researchers were able to confirm van Niel's hypothesis that the oxygen atoms in O_2 produced during photosynthesis comes from H_2O, not from CO_2. **3.** The light reactions could *not* keep producing NADPH and ATP without the $NADP^+$, ADP, and Ⓟ$_i$ that the Calvin cycle generates. The two cycles are interdependent.

Concept Check 10.3

1. Green, because green light is mostly transmitted and reflected—not absorbed—by photosynthetic pigments **2.** Water (H_2O) is the initial electron donor; $NADP^+$ accepts electrons at the end of the electron transport chain, becoming reduced to NADPH. **3.** In this experiment, the rate of ATP synthesis would slow and eventually stop. Because the added compound would not allow a proton gradient to build up across the membrane, ATP synthase could not catalyse ATP production.

Concept Check 10.4

1. 6, 18, 12 **2.** The more potential energy and reducing power a molecule stores, the more energy and reducing power are required for the formation of that molecule. Glucose is a valuable energy source because it is highly reduced (has lots of C—H bonds), storing lots of potential energy in its electrons. To reduce CO_2 to glucose, a large amount of energy and a lot of reducing power are required in the form of large numbers of ATP and NADPH molecules, respectively. **3.** Yes, it would inhibit the dark reactions. The light reactions require ADP and $NADP^+$, which would not be formed in sufficient quantities from ATP and NADPH if the Calvin cycle stopped.

4.

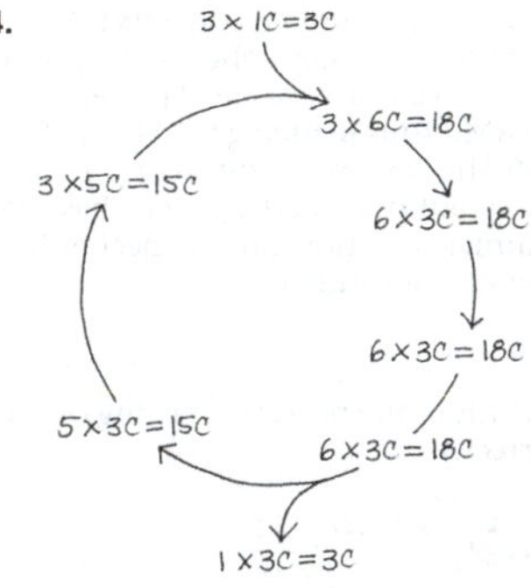

For every three turns of the Calvin cycle, the total number of carbon atoms remains constant because three carbons enter as part of CO_2 molecules, replacing the three that left as part of a G3P molecule. **5.** In glycolysis, G3P acts as an intermediate. The six-carbon sugar fructose 1,6-bisphosphate is cleaved into two three-carbon sugars, one of which is G3P. The other is an isomer called dihydroxyacetone phosphate (DHAP), which can be converted to G3P by an isomerase. Because G3P is the substrate for the next enzyme, it is constantly removed, and the reaction equilibrium is pulled in the direction of conversion of DHAP to more G3P. In the Calvin cycle, G3P acts as both an intermediate and a product. For every three CO_2 molecules that enter the cycle, six G3P molecules are formed, five of which must remain in the cycle and become rearranged to regenerate three five-carbon RuBP molecules. The one remaining G3P is a product, which can be thought of as the result of "reducing" the three CO_2 molecules that entered the cycle into a three-carbon sugar that can later be used to generate energy.

Concept Check 10.5

1. Photorespiration decreases photosynthetic output by adding O_2, instead of CO_2, to the Calvin cycle. As a result, no sugar is generated (no carbon is fixed), and O_2 is used rather than generated. **2.** Without PS II, no O_2 is generated in bundle-sheath cells. This avoids the problem of O_2 competing with CO_2 for binding to rubisco in these cells. **3.** Both problems are caused by a drastic change in Earth's atmosphere due to burning of fossil fuels. The increase in CO_2 concentration affects ocean chemistry by decreasing pH, thus affecting calcification by marine organisms. On land, CO_2 concentration and air temperature are conditions that plants have become adapted to, and changes in these characteristics have a strong effect on photosynthesis by plants. Thus, alteration of these two fundamental factors could have critical effects on organisms all around the planet, in all different habitats. **4.** You would expect that C_4 and CAM species would replace many of the C_3 species.

Concept Check 10.6

1. Plants can break down the sugar they make (in the form of glucose) by cellular respiration, producing ATPs for various cellular processes such as endergonic chemical reactions, transport of substances across membranes, and movement of molecules in the cell. ATPs are also used for the movement of chloroplasts during cellular streaming in some plant cells (see Figure 6.26).

Summary of Key Concepts Questions

10.1 Photosynthesising autotrophs are called producers because they use sunlight as energy to make their own organic molecules (including sugars) from CO_2 and H_2O. Heterotrophs are called consumers because they cannot make their own food and therefore must feed on other organisms, either autotrophs or other heterotrophs. Heterotrophs that consume the remains of dead organisms are called decomposers. Photosynthesis enables almost all organisms to survive because it makes food for photosynthesising autotrophs, which in turn are food for most heterotrophs. **10.2** CO_2 and H_2O are the products of cellular respiration; they are the reactants in photosynthesis. In respiration, glucose is oxidised to CO_2 and electrons are passed through an electron transfer chain from glucose to O_2, producing H_2O. In photosynthesis, H_2O is the source of electrons, which are energised by light, temporarily stored in NADPH, and used to reduce CO_2 to carbohydrate. **10.3** The action spectrum of photosynthesis shows that some wavelengths of light that are not absorbed by chlorophyll *a* are still effective at promoting photosynthesis. The light-harvesting complexes of photosystems contain accessory pigments such as chlorophyll *b* and carotenoids, which absorb different wavelengths and pass the energy to chlorophyll *a*, broadening the spectrum of light usable for photosynthesis.
10.4

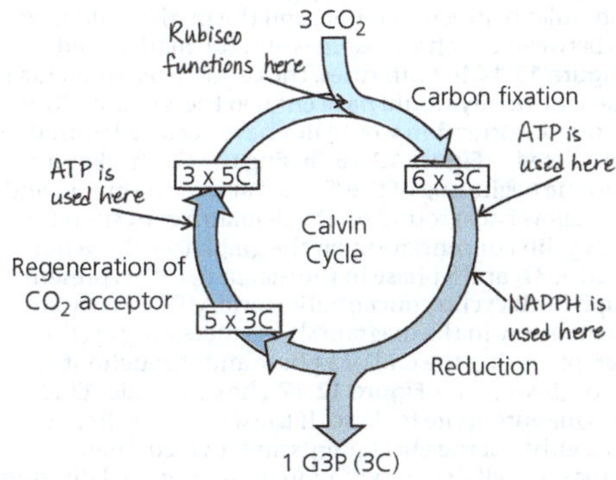

In the reduction phase of the Calvin cycle, ATP phosphorylates a three-carbon compound, and NADPH then reduces this compound to G3P. ATP is also used in the regeneration phase, when five molecules of G3P are converted to three molecules of the five-carbon compound RuBP. Rubisco catalyses the first step of carbon fixation—the addition of CO_2 to RuBP.

10.5 Both C_4 photosynthesis and CAM photosynthesis involve initial fixation of CO_2 to produce a four-carbon compound (in mesophyll cells in C_4 plants and at night in CAM plants). These compounds are then broken down to release CO_2 (in the bundle-sheath cells in C_4 plants and during the day in CAM plants). ATP is required for recycling the molecule that is used initially to combine with CO_2. These pathways avoid the photorespiration that consumes ATP and reduces the photosynthetic output of C_3 plants when they close stomata on hot, dry, bright days. Thus, hot, arid climates would favour C_4 and CAM plants. **10.6** Sucrose made in the leaves of plants is transported through veins to nonphotosynthetic parts of the plant, where some of it is oxidised by cellular respiration, producing ATP for cellular processes. Other sugar molecules enter anabolic pathways, where they are used for synthesis of proteins, lipids, and polysaccharides such as cellulose, the main component of cell walls. Excess sugar is stockpiled as glucose subunits of the polysaccharide starch.

Test Your Understanding

1. D **2.** B **3.** C **4.** A **5.** A **6.** B **7.** C

10.

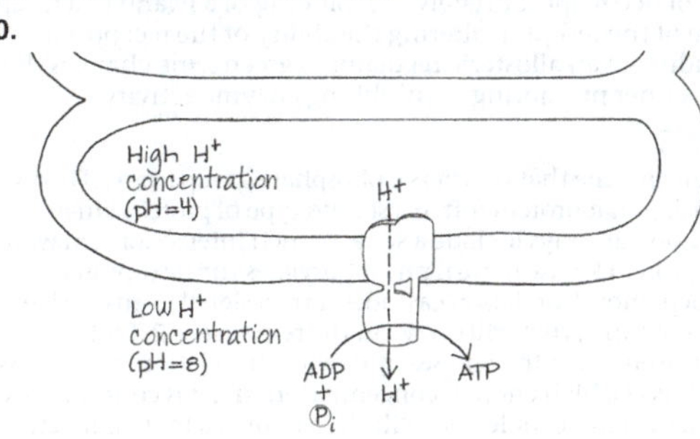

The ATP would end up outside the thylakoid. The thylakoids were able to make ATP in the dark because the researchers set up an artificial proton concentration gradient across the thylakoid membrane; thus, the light reactions were not necessary to establish the H^+ gradient required for ATP synthesis by ATP synthase.

Chapter 11

Figure Questions

Figure 11.6 Adrenaline is a signalling molecule outside the cell; presumably, it binds to a cell-surface receptor protein and thus is part of the signal reception step. **Figure 11.8** This is an example of passive transport. The ion is moving down its concentration gradient, and no energy is required. **Figure 11.9** The aldosterone molecule, a steroid, doesn't need a receptor protein—it is hydrophobic and can therefore pass directly through the hydrophobic lipid bilayer of the plasma membrane into the cell. (Hydrophilic molecules cannot do this.) **Figure 11.10** The entire phosphorylation cascade wouldn't operate. Regardless of whether or not the signalling molecule was bound, protein kinase 2 would always be inactive and would not be able to activate the purple-coloured protein leading to the cellular response. **Figure 11.11** The signalling molecule (cAMP) would remain in its active form and would continue to signal; the pathway would remain active even in the absence of ligand because the cAMP would persist.
Figure 11.12

GTP
ATP
cAMP
Protein Kinase A
Cellular responses

Figure 11.14 The Ca^{2+} pump shown in Figure 11.13 is carrying out active transport of Ca^{2+} ions against their concentration gradient, using ATP for energy. These pumps help maintain a significantly lower [Ca^{2+}] in the cytoplasm than outside the cell and inside the ER. The Ca^{2+} channel proteins seen in Figure 11.14 are facilitating diffusion of Ca^{2+} down its concentration gradient; here, the Ca^{2+} is moving from inside the ER, where it is more concentrated, to the cytoplasm. Because the ion is moving down its concentration gradient, no energy is required. **Figure 11.16** 100,000,000 (one hundred million, or 10^8) glucose molecules are released just from one adrenaline binding to the receptor. The first step results in 100× amplification (one adrenaline activates 100 G proteins); the next step does not amplify the response; the next step is a 100× amplification (10^2 active adenylyl cyclase molecules to 10^4 cyclic AMPs); the next step does not amplify; the next two steps are each 10× amplifications, and the final step is a 100× amplification. **Figure 11.17** The signalling pathway shown in Figure 11.14 leads to the splitting of PIP_2 into the second messengers DAG and IP_3, which produce different responses. (The response elicited by DAG is mentioned but not shown.) The pathway shown for cell B is similar in that it branches and leads to two responses.

Concept Check 11.1

1. The two cells of opposite mating type (**a** and **α**) each secrete a unique signalling molecule, which can only be bound by receptors carried on cells of the opposite mating type. Thus, the **a** mating factor cannot bind to another **a** cell and cause it to grow towards the first **a** cell. Only an **α** cell can "receive" the signalling molecule and respond by directed growth. **2.** Glycogen phosphorylase acts in the third stage, the cellular response to adrenaline signalling. **3.** Glucose 1-phosphate would not be generated because the activation of the enzyme requires an intact cell, with an intact receptor in the membrane and an intact signal transduction pathway. The enzyme cannot be activated directly by interaction with the signalling molecule in the cell-free mixture.

Concept Check 11.2

1. NGF is water-soluble (hydrophilic), so it cannot pass through the lipid membrane to reach intracellular receptors, as steroid hormones can. Therefore, you'd expect the NGF receptor to be in the plasma membrane—which is, in fact, the case. **2.** The cell with the faulty receptor would not be able to respond appropriately to the signalling molecule when it was present. This would most likely have dire consequences for the cell since regulation of the cell's activities by this receptor would not occur appropriately. **3.** Binding of a ligand to a receptor changes the shape of the receptor, altering the ability of the receptor to transmit a signal. Binding of an allosteric regulator to an enzyme changes the shape of the enzyme, either promoting or inhibiting enzyme activity.

Concept Check 11.3

1. A protein kinase is an enzyme that transfers a phosphate group from ATP to a protein, usually activating that protein (often a second type of protein kinase). Many signal transduction pathways include a series of such interactions, in which each phosphorylated protein kinase in turn phosphorylates the next protein kinase in the series. Such phosphorylation cascades carry a signal from outside the cell to the cellular protein(s) that will carry out the response. **2.** Protein phosphatases reverse the effects of the kinases by dephosphorylation, and unless the signalling molecule is at a high enough concentration that it is continuously rebinding the receptor, the kinase molecules will all be returned to their inactive states by phosphatases. **3.** The signal that is being transduced is the *information* that a signalling molecule is bound to the cell-surface receptor. Information is transduced by way of sequential protein-protein interactions that change protein shapes, causing them to function in a way that passes the signal (the information) along.
4. The IP_3-gated channel would open, allowing calcium ions to flow out of the ER and into the cytoplasm, which would raise the cytosolic Ca^{2+} concentration.

Concept Check 11.4

1. At each step in a cascade of sequential activations, one molecule or ion may activate numerous molecules functioning in the next step. This causes the response to be amplified at each such step and overall results in a large amplification of the original signal. **2.** Scaffolding proteins hold molecular components of signalling pathways in a complex with each other. Different scaffolding proteins would assemble different collections of proteins, facilitating different molecular interactions and leading to different cellular responses in the two cells.
3. A malfunctioning protein phosphatase would not be able to dephosphorylate a particular receptor or relay protein. As a result, the signalling pathway, once activated, would not be able to be terminated. (In fact, one study found altered protein phosphatases in cells from 25% of colorectal tumours.) **4.** The proteins in the two cells are different, so the cellular response is different. In heart muscle cells, the pathway shown in Figure 11.16 allows glucose to fuel faster muscle contractions and heart rate. In respiratory muscles, the relay proteins must be different, so that the effect is to block muscle contraction. (In fact, the steps are the same through protein kinase A (PKA), but in respiratory muscle cells, PKA phosphorylates a protein that is required for muscle contraction—and in this case, phosphorylation *inactivates* that protein. So muscle contraction cannot occur.)

Concept Check 11.5

1. In formation of the hand or paw in mammals, cells in the regions between the digits are programmed to undergo apoptosis. This serves to shape the digits of the hand or paw so that they are not webbed. (A lack of apoptosis in these regions in water birds results in webbed feet.) **2.** If a receptor protein for a death-signalling molecule were defective such that it was activated even in the absence of the death signal, this would lead to apoptosis when it wouldn't normally occur. Similar defects in any of the proteins in the signalling pathway would have the same effect if the defective proteins activated relay or response proteins in the absence of interaction with the previous protein or second messenger in the pathway. Conversely, if any protein in the pathway were defective in its ability to respond to an interaction with an early protein or other molecule or ion, apoptosis would not occur when it normally should. For example, a receptor protein for a death-signalling ligand might not be able to be activated, even when ligand was bound. This would stop the signal from being transduced into the cell.

Summary of Key Concepts Questions

11.1 A cell is able to respond to a hormone only if it has a receptor protein on the cell surface or inside the cell that can bind to the hormone. The response to a hormone depends on the specific signal transduction pathway within the cell, which will lead to the specific cellular response. The response can vary for different types of cells. **11.2** Both GPCRs and RTKs have an extracellular binding site for a signalling molecule (ligand) and one or more α-helical regions of the polypeptide that spans the membrane. A GPCR functions singly, while RTKs tend to dimerise or form larger groups of RTKs. GPCRs usually trigger a single transduction pathway, whereas the multiple activated tyrosines on an RTK dimer may trigger several different transduction pathways at the same time. **11.3** A protein kinase is an enzyme that adds a phosphate group to another protein. Protein kinases are often part of a phosphorylation cascade that transduces a signal. A second messenger is a small, nonprotein molecule or ion that rapidly diffuses and relays a signal throughout a cell. Both protein kinases and second messengers can operate in the same pathway. For example, the second messenger cAMP often activates protein kinase A, which then phosphorylates other proteins. **11.4** In G protein-coupled pathways, the GTPase portion of a G protein converts GTP to GDP, inactivating the G protein. Protein phosphatases remove phosphate groups from activated proteins, thus stopping a phosphorylation cascade of protein kinases. Phosphodiesterase converts cAMP to AMP, thus reducing the effect of cAMP in a signal transduction pathway. **11.5** The basic mechanism of controlled cell suicide evolved early in eukaryotic evolution, and the genetic basis for these pathways has been conserved during animal evolution. Such a mechanism is essential to the development and maintenance of all animals.

Test Your Understanding

1. D **2.** A **3.** B **4.** A **5.** C **6.** C **7.** C **8.** This is one possible drawing of the pathway. (Similar drawings would also be correct.)

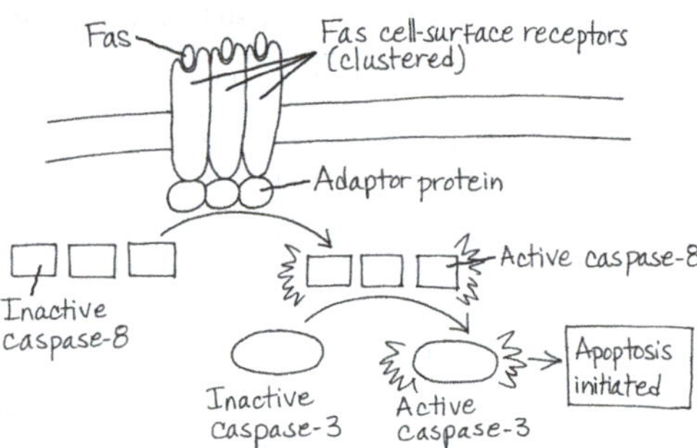

Chapter 12

Figure Questions

Figure 12.4

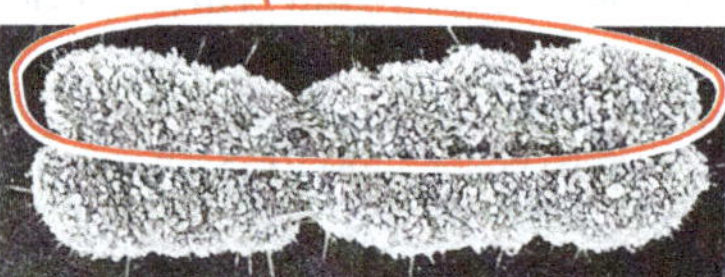

Circling the other chromatid instead would also be correct. **Figure 12.5** The chromosome has four arms. The single (duplicated) chromosome in ❷ becomes two (unduplicated) chromosomes in ❸. The duplicated chromosome in step 2 is considered one single chromosome.
Figure 12.7 12; 2; 2; 1
Figure 12.8

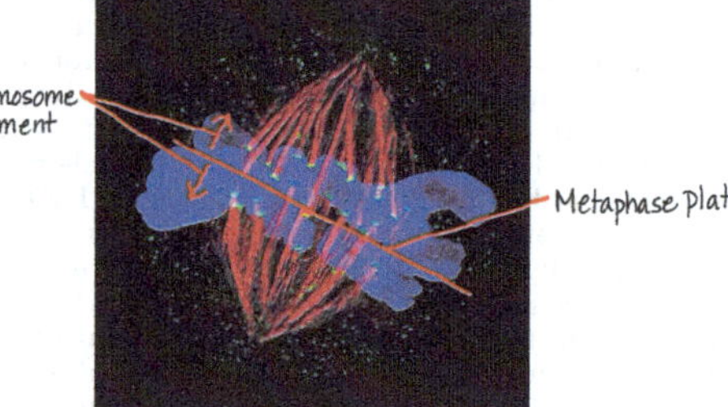

Figure 12.9 The mark would have moved towards the pole closer to it. The lengths of fluorescent microtubules between that pole and the mark would have decreased, while the lengths between the chromosomes and the mark would have remained the same. **Figure 12.14** In both cases, the G_1 nucleus would have remained in G_1 until the time it normally would have entered the S phase. Chromosome condensation and spindle formation would not have occurred until the S and G_2 phases had been completed. **Figure 12.16** Passing the G_2 checkpoint in the diagram corresponds to the beginning of the "Time" axis of the graph, and entry into the mitotic phase (yellow background on the diagram) corresponds to the peaks of MPF activity and cyclin concentration on the graph (see the yellow M banner over the peaks). During G_1 and S phase in the diagram, Cdk is present without cyclin, so on the graph both cyclin concentration and MPF activity are low. The curved purple gradient arrow in the diagram shows increasing cyclin concentration, seen on the graph during the end of S phase and throughout G_2 phase. Then the cell cycle begins again. **Figure 12.17** The cell would divide under conditions where it was inappropriate to do so. If the daughter cells and their descendants also ignored either of the checkpoints and divided, there would soon be an abnormal mass of cells. (This type of inappropriate cell division can contribute to the development of cancer.) **Figure 12.18** The cells in the vessel with PDGF would not be able to respond to the growth factor signal and thus would not divide. The culture would resemble that without the added PDGF.

Concept Check 12.1

1. 1; 1; 2 **2.** 39; 39; 78

Concept Check 12.2

1. 6 chromosomes; they are duplicated; 12 chromatids **2.** Following mitosis, cytokinesis results in two genetically identical daughter cells in both plant cells and animal cells. However, the mechanism of dividing the cytoplasm is different in animals and plants. In an animal cell, cytokinesis occurs by cleavage, which divides the parent cell in two with a contractile ring of actin filaments. In a plant cell, a cell plate forms in the middle of the cell and grows until its membrane fuses with the plasma membrane of the parent cell. A new cell wall grows inside the cell plate, thus eventually between the two new cells. **3.** From the end of S phase in interphase through the end of metaphase in mitosis **4.** During eukaryotic cell division, tubulin is involved in spindle formation and chromosome movement, while actin functions during cytokinesis. In bacterial binary fission, it's the opposite: Actin-like molecules are thought to move the daughter bacterial chromosomes to opposite ends of the cell, and tubulin-like molecules are thought to act in daughter cell separation. **5.** A kinetochore connects the spindle (a motor; note that it has motor proteins) to a chromosome (the cargo it will move). **6.** Microtubules made up of tubulin in the cell provide "rails" along which vesicles and other organelles can travel, based on interactions of motor proteins with tubulin in the microtubules. In muscle cells, actin in microfilaments interacts with myosin filaments to cause muscle contraction.

Concept Check 12.3

1. The nucleus on the right was originally in the G_1 phase; therefore, it had not yet duplicated its chromosomes. The nucleus on the left was in the M phase, so it had already duplicated its chromosomes. **2.** A sufficient amount of MPF has to exist for a cell to pass the G_2 checkpoint; this occurs through the accumulation of cyclin proteins, which combine with Cdk to form (active) MPF. MPF then phosphorylates other proteins, initiating mitosis. **3.** The intracellular receptor (for example, an oestrogen receptor), once activated, would be able to act as a transcription factor in the nucleus, turning on genes that may cause the cell to pass a checkpoint and divide. The RTK receptor, when activated by a ligand, would form a dimer, and each subunit of the dimer would phosphorylate the other. This would lead to a series of signal transduction steps, ultimately turning on genes in the nucleus. As in the case of the oestrogen receptor, the genes would code for proteins necessary to cause the cell to pass a checkpoint and divide.

Summary of Key Concepts Questions

12.1 The DNA of a eukaryotic cell is packaged into structures called *chromosomes*. Each chromosome is a long molecule of DNA, which carries hundreds to thousands of genes, with associated proteins that maintain chromosome structure and help control gene activity. This DNA-protein complex is called *chromatin*. The chromatin of each chromosome is long and thin when the cell is not dividing. Prior to cell division, each chromosome is duplicated, and the resulting sister *chromatids* are attached to each other by proteins at the centromeres and, for many species, all along their lengths (a phenomenon called sister chromatid cohesion). **12.2** A chromosome exists as a single DNA molecule in G_1 of interphase and in anaphase and telophase of mitosis. During S phase, DNA replication produces two sister chromatids per chromosome, which persist during G_2 of interphase and through prophase, prometaphase, and metaphase of mitosis.
12.3 Checkpoints allow cellular surveillance mechanisms to determine whether the cell is prepared to go to the next stage. Internal and external signals move a cell past these checkpoints. The G_1 checkpoint determines whether a cell will proceed forwards in the cell cycle or switch into the G_0 phase. The signals to pass this checkpoint often are external, such as growth factors. Passing the G_2 checkpoint requires sufficient numbers of active MPF complexes, which in turn orchestrate several mitotic events. MPF also initiates degradation of its cyclin component, terminating the M phase. The M phase will not begin again until sufficient cyclin is produced during the next S and G_2 phases. The signal to pass the M phase checkpoint is not activated until all chromosomes are attached to kinetochore fibres and are aligned at the metaphase plate. Only then will sister chromatid separation occur.

Test Your Understanding

1. B **2.** A **3.** C **4.** C **5.** A **6.** B **7.** A **8.** D **9.** See Figure 12.7 for a description of major events. Only one cell is indicated for each stage, but other correct answers are also present in this micrograph.

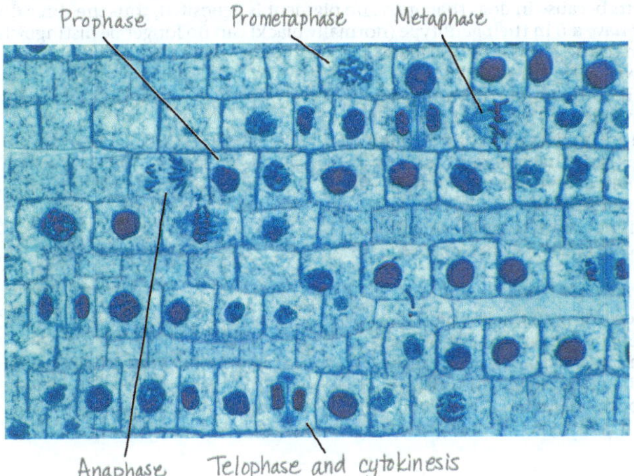

10.

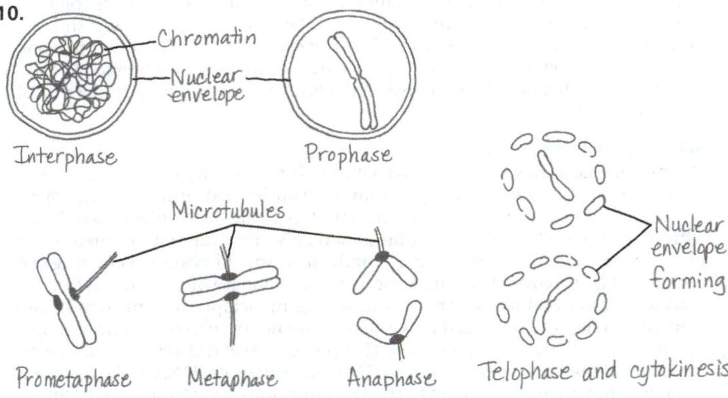

Chapter 13

Figure Questions

Figure 13.3 The chromosomes in a karyotype are photographed at their most condensed, at metaphase, which occurs during M phase of the cell cycle. At this point, the chromosomes have already been duplicated during S phase, so each chromosome exists as two sister chromatids. **Figure 13.4** Two sets of chromosomes are present. Three pairs of homologous chromosomes are present. One long chromosome would be red (maternal), and one would be blue (paternal); one medium chromosome would be red, and one would be blue; one short chromosome would be red, and one would be blue. **Figure 13.6** In (a), haploid cells do not undergo mitosis. In (b), haploid spores undergo mitosis to form the gametophyte, and haploid cells of the gametophyte undergo mitosis to form gametes. In (c), haploid cells undergo mitosis to form either a multicellular haploid organism or a new unicellular haploid organism, and these haploid cells undergo mitosis to form gametes.

Figure 13.7

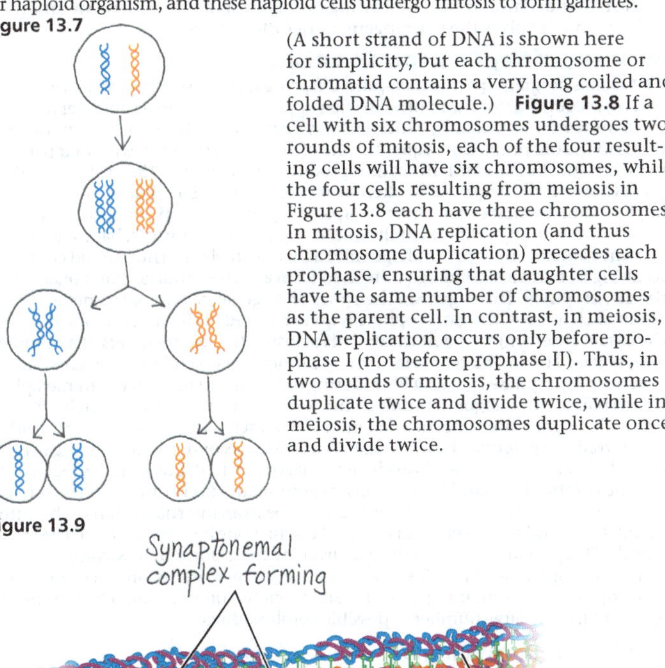

(A short strand of DNA is shown here for simplicity, but each chromosome or chromatid contains a very long coiled and folded DNA molecule.) **Figure 13.8** If a cell with six chromosomes undergoes two rounds of mitosis, each of the four resulting cells will have six chromosomes, while the four cells resulting from meiosis in Figure 13.8 each have three chromosomes. In mitosis, DNA replication (and thus chromosome duplication) precedes each prophase, ensuring that daughter cells have the same number of chromosomes as the parent cell. In contrast, in meiosis, DNA replication occurs only before prophase I (not before prophase II). Thus, in two rounds of mitosis, the chromosomes duplicate twice and divide twice, while in meiosis, the chromosomes duplicate once and divide twice.

Figure 13.9

Figure 13.10 Yes. Each of the six chromosomes (three per cell) shown in telophase I has one nonrecombinant chromatid and one recombinant chromatid. Therefore, eight (2 possibilities for the first chromosome × 2 for the second × 2 for the third) possible sets of chromosomes can be generated for the cell on the left and eight for the cell on the right.

Concept Check 13.1

1. Parents pass genes to their offspring; by dictating the production of messenger RNAs (mRNAs), the genes program cells to make specific enzymes and other proteins, whose cumulative action produces an individual's inherited traits.
2. Such organisms reproduce by mitosis, which generates offspring whose genomes are exact copies of the parent's genome (in the absence of mutation). **3.** She should clone (generate a clone of) it. Crossbreeding it with another plant would generate offspring that have additional variation, which she no longer desires now that she has obtained her ideal orchid.

Concept Check 13.2

1. Each of the six chromosomes is duplicated, so each contains two DNA molecules (double helices), so there are 12 DNA molecules in the cell. The haploid number, n, is 3. One set is always haploid. **2.** There are 23 pairs of chromosomes and two sets. **3.** A human diploid cell would have 23 "pairs of shoes." A haploid cell would have 23 "shoes," one of each pair. **4.** This organism has the life cycle shown in Figure 13.6c, since the zygote doesn't undergo mitosis but immediately undergoes meiosis. Therefore, it must be a fungus or a protist, perhaps an alga.

Concept Check 13.3

1. The chromosomes are similar in that each is composed of two sister chromatids, and the individual chromosomes are positioned similarly at the metaphase plate. The chromosomes differ in that in a mitotically dividing cell, sister chromatids of each chromosome are genetically identical, but in a meiotically dividing cell, sister chromatids are genetically distinct because of crossing over in meiosis I. Moreover, the chromosomes in metaphase of mitosis can be a diploid set or a haploid set, but the chromosomes in metaphase of meiosis II always consist of a haploid set. **2.** If crossing over did not occur, the two homologues would not be associated in any way. This is because each sister chromatid would be either all maternal or all paternal DNA, and the single DNA molecule would therefore not have been joined to the DNA of a nonsister chromatid, holding the complex together. The lack of association of homologues could easily result in the incorrect arrangement of homologues during metaphase I (both might go towards the same pole, for instance) and ultimately in formation of gametes with an abnormal number of chromosomes.

Concept Check 13.4

1. Mutations in a gene lead to the different versions (alleles) of that gene. **2.** Without crossing over, independent assortment of chromosomes during meiosis I theoretically can generate 2^n possible haploid gametes, and random fertilisation can produce $2^n \times 2^n$ possible diploid zygotes. Because the haploid number (n) of grasshoppers is 23 and that of fruit flies is 4, two grasshoppers would be expected to produce a greater variety of zygotes than would two fruit flies. **3.** If the segments of the maternal and paternal chromatids that undergo crossing over are genetically identical and thus have the same two alleles for every gene, then the recombinant chromosomes will be genetically equivalent to the parental chromosomes. Crossing over contributes to genetic variation only when it involves the rearrangement of different alleles.

Summary of Key Concepts Questions

13.1 Genes program specific traits, and offspring inherit their genes from each parent, accounting for similarities in their appearance to one or the other parent. Humans reproduce sexually, which ensures new combinations of genes (and thus traits) in the offspring. Consequently, the offspring are not clones of their parents (which would be the case if humans reproduced asexually). **13.2** Animals and plants both reproduce sexually, alternating meiosis with fertilisation. Both have haploid gametes that unite to form a diploid zygote, which then goes on to divide mitotically, forming a diploid multicellular organism. In animals, haploid cells become gametes and don't undergo mitosis, while in plants, the haploid cells resulting from meiosis undergo mitosis to form a haploid multicellular organism, the gametophyte. This organism then goes on to generate haploid gametes. (In plants such as trees, the gametophyte is quite reduced in size and not obvious to the casual observer.) **13.3** At the end of meiosis I, the two members of a homologous pair end up in different cells, so they cannot pair up and undergo crossing over during prophase II. **13.4** First, during independent assortment in metaphase I, each pair of homologous chromosomes lines up independently of each other pair at the metaphase plate, so a daughter cell of meiosis I randomly inherits either a maternal or a paternal chromosome of each pair. Second, due to crossing over, each chromosome is not exclusively maternal or paternal, but includes regions at the ends of the chromatid from a nonsister chromatid (a chromatid of the other homologue). (The nonsister segment can also be in an internal region of the chromatid if a second crossover occurs beyond the first one before the end of the chromatid.) This provides much additional diversity in the form of new combinations of alleles. Third, random fertilisation ensures even more variation since any sperm of a large number containing many possible genetic combinations can fertilise any egg of a similarly large number of possible combinations.

Test Your Understanding

1. A **2.** B **3.** A **4.** D **5.** C
6. (a) One possible answer:

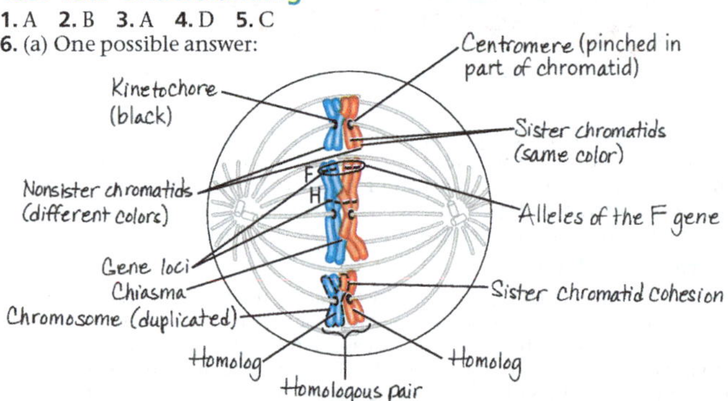

(b) A haploid set is made up of one long, one medium, and one short chromosome, no matter what combination of colours. For example, one red long, one blue medium, and one red short chromosome make up a haploid set. (In cases where crossovers have occurred, a haploid set of one colour may include segments of chromatids of the other colour.) All red and blue chromosomes together make up a diploid set. (c) Metaphase I **7.** This cell must be undergoing meiosis because the two homologues of a homologous pair are associated with each other at the metaphase plate; this does not occur in mitosis. Also, chiasmata are clearly present, meaning that crossing over has occurred, another process unique to meiosis.

Chapter 14

Figure Questions

Figure 14.3 All offspring would have purple flowers. (The ratio of purple to white flowers would be 1 purple: 0 white.) The P generation plants are true-breeding, so mating two purple-flowered plants produces the same result as self-pollination: All the offspring have the same trait. If Mendel had stopped after the F_1 generation, he could have concluded that the white factor had disappeared entirely and would not ever reappear.
Figure 14.8

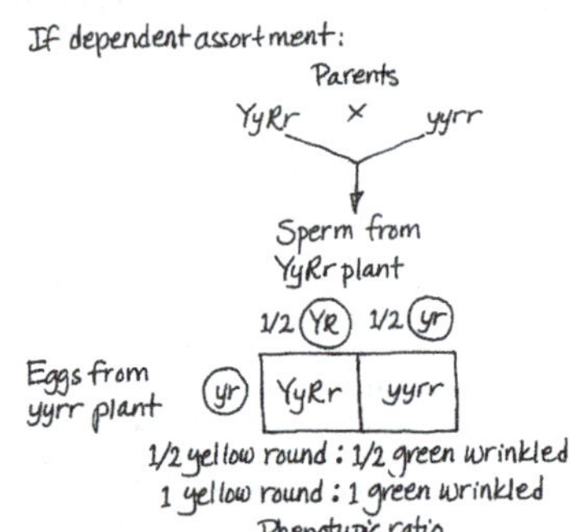

If independent assortment:
Parents
YyRr × yyrr
Sperm from YyRr plant

Eggs from yyrr plant	1/4 (YR)	1/4 (Yr)	1/4 (yR)	1/4 (yr)
(yr)	YyRr	Yyrr	yyRr	yyrr

1/4 yellow round : 1/4 yellow wrinkled: 1/4 green round : 1/4 green wrinkled
1 yellow round : 1 yellow wrinkled : 1 green round : 1 green wrinkled
Phenotypic ratio

Yes, this cross would also have allowed Mendel to make different predictions for the two hypotheses, thereby allowing him to distinguish the correct one. **Figure 14.10** Your classmate would probably point out that the F_1 generation hybrids show an intermediate phenotype between those of the homozygous parents, which supports the blending hypothesis. You could respond that crossing the F_1 hybrids results in the reappearance of the white phenotype, rather than identical pink offspring, which fails to support the idea of traits blending during inheritance. The blending hypothesis predicts that the white trait would have been lost after the F_1 generation. **Figure 14.11** Both the I^A and I^B alleles are dominant to the i allele because the i allele results in no attached carbohydrate. The I^A and I^B alleles are codominant; both are expressed in the phenotype of I^AI^B heterozygotes, who have type AB blood. **Figure 14.12** In this cross, the final "3" and "1" of a standard cross are lumped together as a single phenotype. This occurs because in dogs that are *ee*, no pigment is deposited, thus the three dogs that have a *B* in their genotype (normally black) can no longer be distinguished from the dog that is *bb* (normally brown). **Figure 14.16** In the Punnett square, two of the three individuals with normal colouration are carriers, so the probability is $\frac{2}{3}$. (Note that you must take into account everything you know when you calculate probability: You know she is not *aa*, so there are only three possible genotypes to consider.)

Parents
AaTt × AaTt
Sperm from AaTt plant

Eggs from AaTt plant	(AT)	(At)	(aT)	(at)
(AT)	AATT	AATt	AaTT	AaTt
(At)	AATt	AAtt	AaTt	Aatt
(aT)	AaTT	AaTt	aaTT	aaTt
(at)	AaTt	Aatt	aaTt	aatt

Concept Check 14.1

1. According to the law of independent assortment, 25 plants ($\frac{1}{16}$ of the offspring) are predicted to be *aatt*, or recessive for both characters. The actual result is likely to differ slightly from this value. **2.** The plant could make eight different gametes (*YRI, YRi, YrI, Yri, yRI, yRi, yrI,* and *yri*). To fit all the possible gametes in a self-pollination, a Punnett square would need 8 rows and 8 columns. It would have spaces for the 64 possible unions of gametes in the offspring. **3.** Self-pollination is sexual reproduction because meiosis is involved in forming gametes, which unite during fertilisation. As a result, the offspring in self-pollination are genetically different from the parent. (As mentioned in the footnote near the beginning of Concept 14.1, we have simplified the explanation in referring to the single pea plant as a parent. Technically, the gametophytes in the flower are the two "parents.")

Concept Check 14.2

1. $\frac{1}{2}$ homozygous dominant (*AA*), 0 homozygous recessive (*aa*), and $\frac{1}{2}$ heterozygous (*Aa*) **2.** $\frac{1}{4}$ *BBDD*; $\frac{1}{4}$ *BbDD*; $\frac{1}{4}$ *BBDd*; $\frac{1}{4}$ *BbDd* **3.** The genotypes that fulfill this condition are *ppyyII, ppyyIi, ppYYii, ppYyii, Ppyyii,* and *ppyyii.* Use the multiplication rule to find the probability of getting each genotype, and then use the addition rule to find the overall probability of meeting the conditions of this problem:

ppyyII	$\frac{1}{2}$ (probability of *pp*) × $\frac{1}{4}$ (*yy*) × $\frac{1}{4}$ (*II*)	$= \frac{1}{32}$
ppyyIi	$\frac{1}{2}$ (*pp*) × $\frac{1}{4}$ (*yy*) × $\frac{1}{2}$ (*Ii*)	$= \frac{1}{16} = \frac{2}{32}$
ppYYii	$\frac{1}{2}$ (*pp*) × $\frac{1}{4}$ (*YY*) × $\frac{1}{4}$ (*ii*)	$= \frac{1}{32}$
ppYyii	$\frac{1}{2}$ (*pp*) × $\frac{1}{2}$ (*Yy*) × $\frac{1}{4}$ (*ii*)	$= \frac{1}{16} = \frac{2}{32}$
Ppyyii	$\frac{1}{2}$ (*Pp*) × $\frac{1}{4}$ (*yy*) × $\frac{1}{4}$ (*ii*)	$= \frac{1}{32}$
ppyyii	$\frac{1}{2}$ (*pp*) × $\frac{1}{4}$ (*yy*) × $\frac{1}{4}$ (*ii*)	$= \frac{1}{32}$
Fraction predicted to be homozygous recessive for at least two of the three characters		$\frac{8}{32} = \frac{1}{4}$

Concept Check 14.3

1. Incomplete dominance describes the relationship between two alleles of a single gene, whereas epistasis relates to the genetic relationship between two genes (and the respective alleles of each). **2.** Half of the children would be expected to have type A blood and half type B blood. **3.** The black and white alleles are incompletely dominant, with heterozygotes being grey in colour. A cross between a grey rooster and a black hen should yield approximately equal numbers of grey and black offspring.

Concept Check 14.4

1. $\frac{1}{9}$ (Since cystic fibrosis is caused by a recessive allele, Lucia and Jared's siblings who have CF must be homozygous recessive. Therefore, each parent must be a carrier of the recessive allele. Since neither Lucia nor Jared has CF and are thus not *cc* (using *C*/*c* as the alleles for the CF gene), this means they each have a $\frac{2}{3}$ chance of being a carrier. If they are both carriers, there is a $\frac{1}{4}$ chance that they will have a child with CF; $\frac{2}{3} \times \frac{2}{3} \times \frac{1}{4} = \frac{1}{9}$); virtually 0 (Both Lucia and Jared would have to be carriers to produce a child with the disease, unless a very rare mutation (change) occurred in the DNA of cells making eggs or sperm in a noncarrier that resulted in the CF allele.) **2.** In normal haemoglobin, the sixth amino acid is glutamic acid (Glu), which is acidic (has a negative charge on its side chain). In sickle-cell haemoglobin, Glu is replaced by valine (Val), which is a nonpolar amino acid, very different from Glu. The primary structure of a protein (its amino acid sequence) ultimately determines the shape of the protein and thus its function. The substitution of Val for Glu enables the haemoglobin molecules to interact with each other and form long fibres, leading to the protein's deficient function and the deformation of the red blood cell. **3.** Juanita's genotype is *Dd*. Because the allele for polydactyly (*D*) is dominant to the allele for five digits per appendage (*d*), the trait is expressed in people with either the *DD* or *Dd* genotype. But because Juanita's father does not have polydactyly, his genotype must be *dd*, which means that Juanita inherited a *d* allele from him. Therefore, Juanita, who does have the trait, must be heterozygous. **4.** In the monohybrid cross involving flower colour, the ratio is 3.15 purple : 1 white, while in the human family in the pedigree, the ratio in the third generation is 1 taster of PTC : 1 nontaster of PTC. The difference is due to the small sample size (two offspring) in the human family. If the second-generation couple in this pedigree were able to have 929 offspring as in the pea plant cross, the ratio would likely be closer to 3:1. (Note that none of the pea plant crosses in Table 14.1 yielded *exactly* a 3:1 ratio.)

Summary of Key Concepts Questions

14.1 Alternative versions of genes, called alleles, are passed from parent to offspring during sexual reproduction. In a cross between purple- and white-flowered homozygous parents, the F_1 offspring are all heterozygous, each inheriting a purple allele from one parent and a white allele from the other. Because the purple allele is dominant, it determines the phenotype of the F_1 offspring to be purple, and the expression of the recessive white allele is masked. Only in the F_2 generation is it possible for some of the offspring to be homozygous recessive, which causes the white trait to be expressed.

14.2

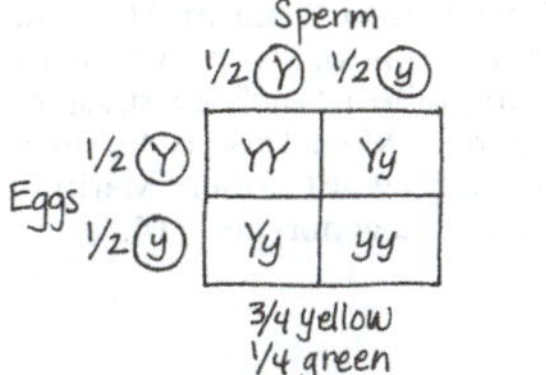

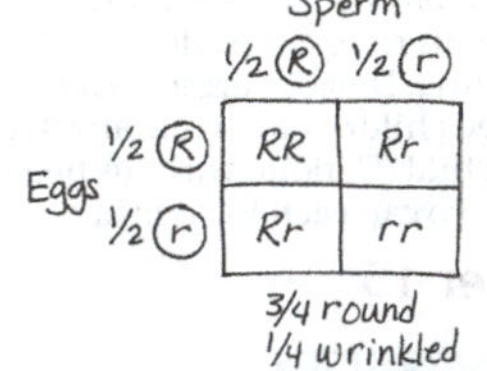

$\frac{3}{4}$ yellow × $\frac{3}{4}$ round $= \frac{9}{16}$ yellow round
$\frac{3}{4}$ yellow × $\frac{1}{4}$ wrinkled $= \frac{3}{16}$ yellow wrinkled
$\frac{1}{4}$ green × $\frac{3}{4}$ round $= \frac{3}{16}$ green round
$\frac{1}{4}$ green × $\frac{1}{4}$ wrinkled $= \frac{1}{16}$ green wrinkled
= 9 yellow round : 3 yellow wrinkled : 3 green round : 1 green wrinkled

14.3 The ABO blood group is an example of multiple alleles because this single gene has more than two alleles (I^A, I^B, and *i*). Two of the alleles, I^A and I^B, exhibit codominance since both carbohydrates (A and B) are present when these two alleles exist together in a genotype. I^A and I^B each exhibit complete dominance over the *i* allele. This situation is not an example of incomplete dominance because each allele affects the phenotype in a distinguishable way, so the result is not intermediate between the two phenotypes. Because this situation involves a single gene, it is not an example of epistasis or polygenic inheritance. **14.4** The chance of the fourth child having cystic fibrosis is $\frac{1}{4}$, as it was for each of the other children, because each birth is an independent event. We already know both parents are carriers, so whether their first three children are carriers or not has no bearing on the probability that their next child will have the disease. The parents' genotypes provide the only relevant information.

Test Your Understanding

1. A cross of *Ii* × *ii* would yield offspring with a genotypic ratio of 1 *Ii*: 1 *ii* (2:2 is an equivalent answer) and a phenotypic ratio of 1 inflated : 1 constricted (2:2 is equivalent).

2. Man I^Ai; woman I^Bi; child *ii*. Genotypes for future children are predicted to be $\frac{1}{4}$ I^AI^B, $\frac{1}{4}$ I^Ai, $\frac{1}{4}$ I^Bi, $\frac{1}{4}$ *ii*. **3.** $\frac{1}{2}$

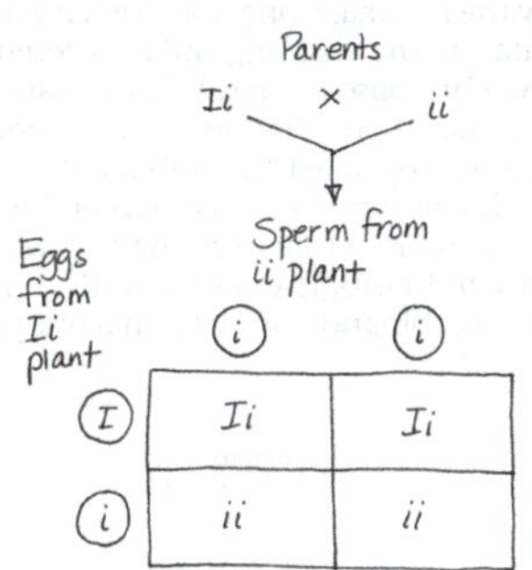

Genotypic ratio 1 *Ii* : 1 *ii*
(2:2 is equivalent)

Phenotypic ratio 1 inflated : 1 constricted
(2:2 is equivalent)

4. The green pod trait is dominant, so the green pod allele is *G* and the yellow pod allele is *g*; the inflated pod trait is dominant, so the inflated pod allele is *I* and the constricted pod allele is *i*. The cross described is *GgIi* × *GgIi*.

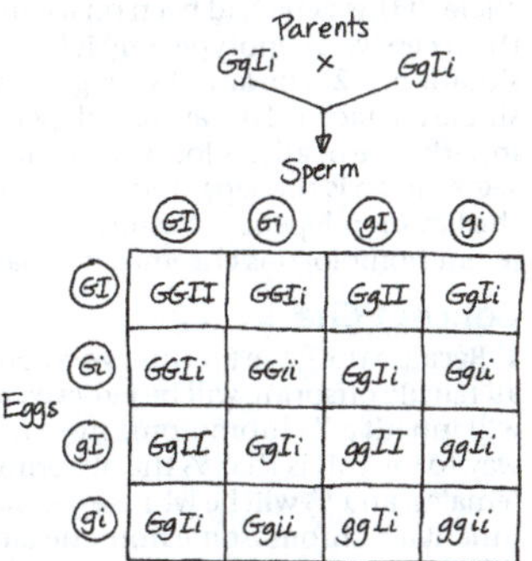

9 green inflated : 3 green constricted :
3 yellow inflated : 1 yellow constricted

5. (a) $\frac{1}{64}$; (b) $\frac{1}{64}$; (c) $\frac{1}{8}$; (d) $\frac{1}{32}$ **6.** (a) $\frac{3}{4} \times \frac{3}{4} \times \frac{3}{4} = \frac{27}{64}$; (b) $1 - \frac{27}{64} = \frac{37}{64}$; (c) $\frac{1}{4} \times \frac{1}{4} \times \frac{1}{4} = \frac{1}{64}$; (d) $1 - \frac{1}{64} = \frac{63}{64}$ **7.** (a) $\frac{1}{256}$; (b) $\frac{1}{16}$; (c) $\frac{1}{256}$; (d) $\frac{1}{64}$; (e) $\frac{1}{128}$ **8.** (a) 1; (b) $\frac{1}{32}$; (c) $\frac{1}{8}$; (d) $\frac{1}{2}$ **9.** $\frac{1}{9}$

10. Matings of the original mutant cat with true-breeding noncurl cats will produce both curl and noncurl F_1 offspring if the curl allele is dominant, but only noncurl offspring if the curl allele is recessive. Whether the curl trait is dominant or recessive, you would obtain some true-breeding offspring homozygous for the curl allele from matings between the F_1 cats resulting from the original curl × noncurl crosses. If dominant, you wouldn't be able to tell true-breeding, homozygous offspring from heterozygotes without further crosses. You know that cats are true-breeding when curl × curl matings produce only curl offspring. As it turns out, the allele that causes curled ears is dominant. **11.** 25%, or $\frac{1}{4}$, will be cross-eyed; all (100%) of the cross-eyed offspring will also be white. **12.** The dominant allele *I* is epistatic to the *P*/*p* locus, and thus the genotypic ratio for the F_1 generation will be 9 *I–P–* (colourless) : 3 *I–pp* (colourless) : 3 *iiP–* (purple) : 1 *iipp* (red). Overall, the phenotypic ratio is 12 colourless : 3 purple : 1 red. **13.** Recessive.

All individuals with alkaptonuria (Ariana, Benito, Elena, and Carlota) are homozygous recessive *aa*. Jorge is *Aa*, since some of his children with Ariana have alkaptonuria. Diego, Carmen, Hector, and Julio are each *Aa*, since they are all unaffected children with one affected parent. Miguel also is *Aa*, since he has an affected child (Carlota) with his heterozygous wife Carmen. Mariposa, Paloma, and Roberto can each have either the *AA* or *Aa* genotype. **14.** $\frac{1}{6}$

Chapter 15

Figure Questions

Figure 15.3 About $\frac{3}{4}$ of the F_2 offspring would have red eyes and about $\frac{1}{4}$ would have white eyes. Since the sex chromosomes are not involved in determining eye colour in this hypothetical cross, about half of the white-eyed flies would be female and half would be male; similarly, about half of the red-eyed flies would be female and half would be male. (Note that the autosomes with the eye colour alleles would be the same shape in the Punnett square—unlike the X and Y chromosomes—and each offspring would inherit two alleles. The sex of the flies would be determined separately by inheritance of the sex chromosomes. Thus your F_2 Punnett square would have four possible combinations in sperm and two in eggs; it would have eight squares altogether.) **Figure 15.4** The ratio would be 1 yellow round : 1 green round : 1 yellow wrinkled : 1 green wrinkled. (This is similar to a test cross.) **Figure 15.7** All the males would be colour-blind, and all the females would be carriers. (Another way to say this is that ½ the offspring would be colour-blind males, and ½ the offspring would be carrier females.) **Figure 15.9** The two largest classes would still be the offspring with the phenotypes of the true-breeding P generation flies, but now they would be grey vestigial and black normal, which is now the "parental type" because those were the specific allele combinations in the P generation (the linked alleles on their chromosomes). **Figure 15.10** The two chromosomes on the left side of the sketch below are like the two chromosomes inherited by the F_1 female, one from each P generation fly. They are passed by the F_1 female intact to the offspring and thus could be called "parental" chromosomes. The other two chromosomes result from crossing over during meiosis in the F_1 female. Because they have combinations of alleles not seen in either of the F_1 female's chromosomes, they can be called "recombinant" chromosomes. (Note that in this example, the allele combinations on the recombinant chromosomes, $b^+ \, vg^+$ and $b \, vg$, are the allele combinations that were on the parental chromosomes in the cross shown in Figures 15.9 and 15.10. The basis for calling them parental chromosomes is that they have the combination of alleles that was present on the P generation chromosomes.)

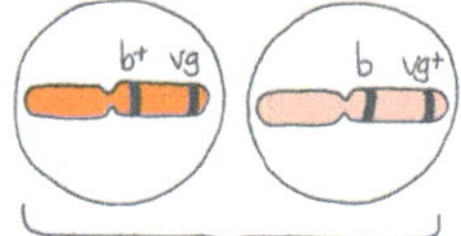

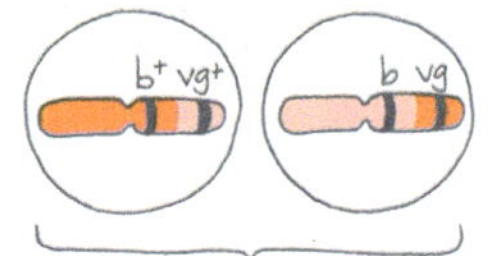

b+ vg+   b vg

Parental chromosomes   Recombinant chromosomes

Concept Check 15.1

1. To show the mutant phenotype, a male needs to possess only one mutant allele. If this gene had been on a pair of autosomes, an individual would show the recessive phenotype only if both alleles were mutant, a much less probable situation. **2.** The law of segregation describes the inheritance of alleles for a single character. The law of independent assortment of alleles describes the inheritance of alleles for two characters. **3.** The physical basis for the law of segregation is the separation of homologues in anaphase I. The physical basis for the law of independent assortment is the alternative arrangements of all the different homologous chromosome pairs in metaphase I.

Concept Check 15.2

1. Because the gene for this eye colour character is located on the X chromosome, all female offspring will be red-eyed and heterozygous ($X^{w^+}X^w$); all male offspring will inherit a Y chromosome from the father and be white-eyed (X^wY). (Another way to say this is that $\frac{1}{2}$ the offspring will be red-eyed heterozygous [carrier] females, and $\frac{1}{2}$ will be white-eyed males.) **2.** $\frac{1}{4}$ ($\frac{1}{2}$ chance that the child will inherit a Y chromosome from the father and be male $\times$ $\frac{1}{2}$ chance that he will inherit the X carrying the disease allele from his mother). If the child is a boy, there is a $\frac{1}{2}$ chance he will have the disease; a female would have zero chance (but $\frac{1}{2}$ chance of being a carrier). **3.** In a disorder caused by a dominant allele, there is no such thing as a "carrier," since those with the allele have the disorder. Because the allele is dominant, the females lose any "advantage" in having two X chromosomes, since one disorder-associated allele is sufficient to result in the disorder. All fathers who have the dominant allele will pass it along to *all* their daughters, who will also have the disorder. A mother who has the allele (and thus the disorder) will pass it to half of her sons and half of her daughters.

Concept Check 15.3

1. Crossing over during meiosis I in the heterozygous parent produces some gametes with recombinant genotypes for the two genes. Offspring with a recombinant phenotype arise from fertilisation of the recombinant gametes by homozygous recessive gametes from the double-mutant parent. **2.** In each case, the alleles contributed by the female parent (in the egg) determine the phenotype of the offspring because the male in this cross contributes only recessive alleles. Thus, identifying the phenotype of the offspring tells you what alleles were in the mother's (the dihybrid female's) egg. **3.** No. The order could be *A-C-B* or *C-A-B*. To determine which possibility is correct, you need to know the recombination frequency between *B* and *C*.

Concept Check 15.4

1. In meiosis, a combined 14-21 chromosome will behave as one chromosome. If a gamete receives the combined 14-21 chromosome and a normal copy of chromosome 21, trisomy 21 will result when this gamete combines with a normal gamete (with its own chromosome 21) during fertilisation. **2.** No. The child can be either I^AI^Ai or I^Aii. A sperm of genotype I^AI^A could result from nondisjunction in the father during meiosis II, while an egg with the genotype *ii* could result from nondisjunction in the mother during either meiosis I or meiosis II. **3.** Activation of this gene could lead to the production of too much of this kinase. If the kinase is involved in a signalling pathway that triggers cell division, too much of it could trigger unrestricted cell division, which in turn could contribute to the development of a cancer (in this case, a cancer of one type of white blood cell).

Concept Check 15.5

1. Inactivation of an X chromosome in females and genomic imprinting. Because of X inactivation, the effective dose of genes on the X chromosome is the same in males and females. As a result of genomic imprinting, only one allele of certain genes is phenotypically expressed. **2.** The genes for leaf colouration are located in plastids within the cytoplasm. Normally, only the maternal parent transmits plastid genes to offspring. Since variegated offspring are produced only when the female parent is of the B variety, we can conclude that variety B contains both the wild-type and mutant alleles of pigment genes, producing variegated leaves. (Variety A must contain only the wild-type allele of pigment genes.) **3.** Each cell contains numerous mitochondria, and in affected individuals, most cells contain a variable mixture of normal and mutant mitochondria. The normal mitochondria carry out enough cellular respiration for survival. (The situation is similar for chloroplasts.)

Summary of Key Concepts Questions

15.1 Because the sex chromosomes are different from each other and because they determine the sex of the offspring, Morgan could use the sex of the offspring as a phenotypic character to follow the parental chromosomes. (He could also have followed them under a microscope, as the X and Y chromosomes look different.) At the same time, he could record eye colour to follow the eye colour alleles. **15.2** Males have only one X chromosome, along with a Y chromosome, while females have two X chromosomes. The Y chromosome has very few genes on it, while the X has about 1,000. When a recessive X-linked allele that causes a disorder is inherited by a male on the X from his mother, there isn't a second allele present on the Y (males are hemizygous), so the male has the disorder. Because females have two X chromosomes, they must inherit two recessive alleles in order to have the disorder, a rarer occurrence. **15.3** Crossing over results in new combinations of alleles. Crossing over is a random occurrence, and the more distance there is between two genes, the more chances there are for crossing over to occur, leading to new allele combinations. **15.4** In inversions and reciprocal translocations, the same genetic material is present in the same relative amount but just organised differently. In aneuploidy, duplications, deletions, and nonreciprocal translocations, the balance of genetic material is upset, as large segments are either missing or present in more than one copy. Apparently, this type of imbalance is very damaging to the organism. (Although it isn't lethal in the developing embryo, the reciprocal translocation that produces the Philadelphia chromosome can lead to a serious condition, cancer, by altering the expression of important genes.) **15.5** In these cases, the sex of the parent contributing an allele affects the inheritance pattern. For imprinted genes, either the paternal or the maternal allele is expressed, depending on the imprint. For mitochondrial and chloroplast genes, only the maternal contribution will affect offspring phenotype because the offspring inherit these organelles from the mother, via the egg cytoplasm.

Test Your Understanding

1. 0; $\frac{1}{2}$; $\frac{1}{16}$ **2.** Recessive; if the disorder were dominant, it would affect at least one parent of a child born with the disorder. The disorder's inheritance is sex-linked because it is seen only in boys. For a girl to have the disorder, she would have to inherit recessive alleles from *both* parents. This would be very rare, since males with the recessive allele on their X chromosome die in their early teens. **3.** 17%; yes, it is consistent. In Figure 15.9, the recombination frequency was also 17%. (You'd expect this to be the case since these are the very same two genes, and their distance from each other wouldn't change from one experiment to another.)

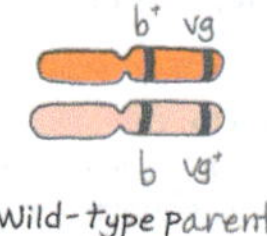

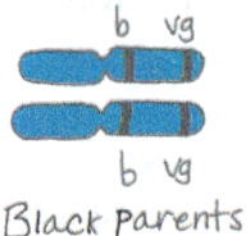

4. Between *T* and *A*, 12%; between *A* and *S*, 5% **5.** Between *T* and *S*, 16%; sequence of genes is *T-A-S* **6.** 6%; wild-type heterozygous for normal wings and red eyes $\times$ recessive homozygous for vestigial wings and purple eyes **7.** Fifty percent of the offspring will show phenotypes resulting from crossovers. These results would be the same as those from a cross where *A* and *B* were *not* on the same chromosome, and you would interpret the results to mean that the genes are unlinked. (Further crosses involving other genes between *A* and *B* on the same chromosome would reveal the genetic linkage and map distances.) **8.** 450 each of blue/oval and white/round (parentals) and 50 each of blue/round and white/oval (recombinants) **9.** About one-third of the distance along the chromosome from the vestigial wing locus towards the brown eye locus **10.** Because bananas are triploid, homologous pairs cannot line up during meiosis. Therefore, it is not possible to generate gametes that can fuse to produce a zygote with the triploid number of chromosomes. **12.** (a) For each pair of genes, you had to generate an F_1 dihybrid fly; let's use the *A* and *B* genes as an example. You obtained homozygous parental flies, either the first with dominant alleles of the two genes (*AABB*) and the second with recessive alleles (*aabb*), or the first with dominant alleles of gene *A* and recessive alleles of gene *B* (*AAbb*) and the second with recessive alleles of gene *A* and dominant alleles of gene *B* (*aaBB*). Breeding either of these pairs of P generation flies gave you an F_1 dihybrid, which you then testcrossed with a doubly homozygous recessive fly (*aabb*). You classed the offspring as parental or recombinant, based on the genotypes of the P generation parents (either of the two pairs described above). You added up the number of recombinant types and then divided by the total number of offspring. This gave you the recombination percentage (in this case, 8%), which you can translate into map units (8 map units) to construct your map.

(b)

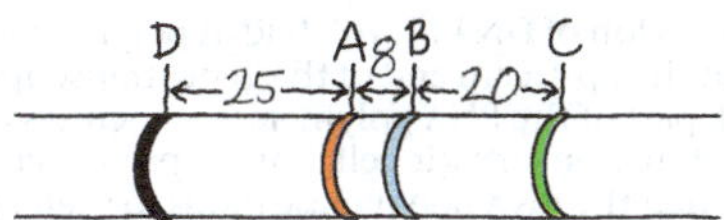

Chapter 16

Figure Questions

Figure 16.2 The living S cells found in the blood sample were able to reproduce to yield more S cells, indicating that the S trait is a permanent, heritable change, rather than just a one-time use of the dead S cells' capsules. **Figure 16.4** The radioactivity would have been found in the pellet when proteins were labelled (batch 1) because proteins would have had to enter the bacterial cells to program them with genetic instructions. It's hard for us to imagine now, but the DNA might have played a structural role that allowed some of the proteins to be injected while it remained outside the bacterial cell (thus no radioactivity would be found in the pellet in batch 2). **Figure 16.7** (1) The nucleotides in a single DNA strand are held together by covalent bonds between an oxygen on the 3′ carbon of one nucleotide and the phosphate group on the 5′ carbon of the next nucleotide in the chain. Instead of covalent bonds, the bonds that hold the two strands together are hydrogen bonds between a nitrogenous base on one strand and the complementary nitrogenous base on the other strand. (Hydrogen bonds are weaker than covalent bonds, but there are so many hydrogen bonds in a DNA double helix that, together, they are enough to hold the two strands together.) (2) One end, the 5′ end, has a phosphate group, which is attached to the 5′ carbon of the sugar, the one that is not in the ring. The other end, the 3′ end, has an —OH group attached to the 3′ carbon of the sugar; this carbon is in the ring. (3) The left diagram shows the most detail. It shows that each sugar-phosphate backbone is made up of sugars (blue pentagons) and phosphates (yellow circles) joined by covalent bonds (black lines). The middle diagram doesn't show any detail in the backbone. Both the left and middle diagrams label the bases and represent their complementarity by the complementary shapes at the ends of the bases (curves/indents for G/C or V's/notches for T/A). The diagram on the right is the least detailed, implying that the base pairs pair up, but showing all bases as the same shape so not including the information about specificity and complementarity visible in the other two diagrams. The left and right diagrams show that the strand on the left was synthesised most recently, as indicated by the light blue colour. All three diagrams show the 5′ and 3′ ends of the strands. **Figure 16.12** The tube from the first replication would look the same, with a middle band of hybrid ^{15}N–^{14}N DNA, but the second tube would not have the upper band of DNA molecules made up of two light blue strands. Instead, it would have a bottom band of DNA made up of two dark blue strands, like the bottom band in the result predicted after one replication in the conservative model. **Figure 16.13** In the bubble at the top of the micrograph in (b), arrows should be drawn pointing left and right to indicate the two replication forks. **Figure 16.15**

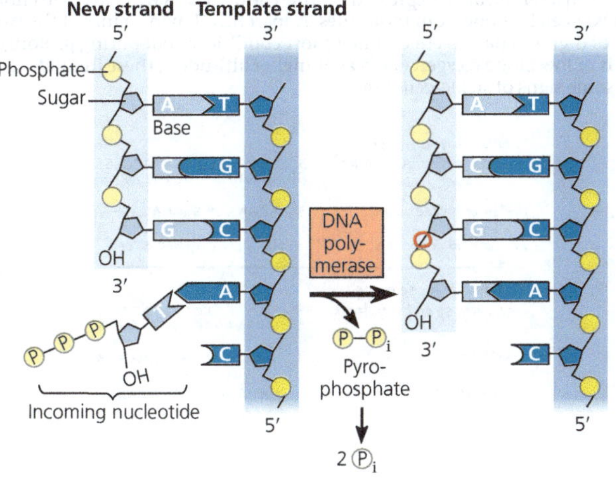

Figure 16.18

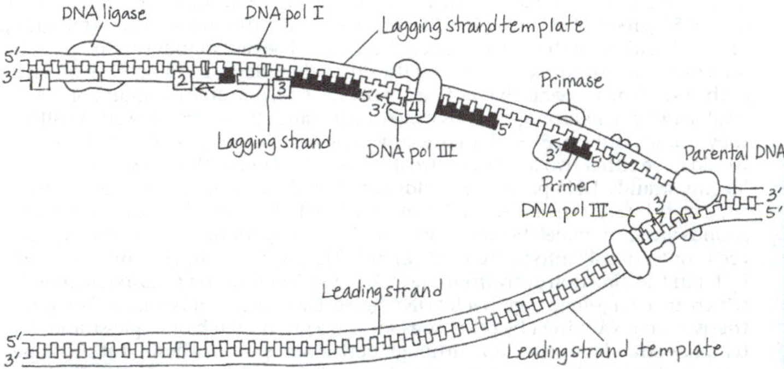

Figure 16.19

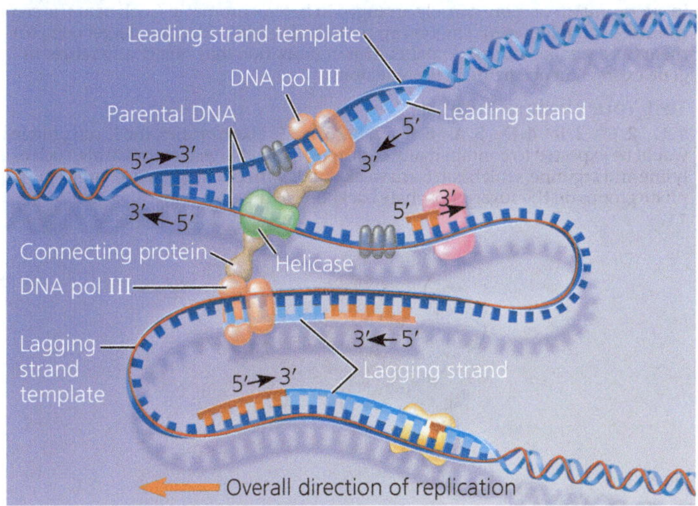

Figure 16.24 The two members of a homologous pair (which would be the same colour) would be associated tightly together at the metaphase plate during metaphase I of meiosis I. In metaphase of mitosis, however, each chromosome would be lined up individually, so the two chromosomes of the same colour would be in different places at the metaphase plate.

Concept Check 16.1

1. In order to tell which end is the 5′ end, you need to know which end has a phosphate group on the 5′ carbon (the 5′ end) and/or which end has an —OH group on the 3′ carbon (the 3′ end). **2.** Griffith expected that the mouse injected with the mixture of heat-killed S cells and living R cells would survive, since neither type of cell alone would kill the mouse.

Concept Check 16.2

1. Complementary base pairing ensures that the two daughter molecules are exact copies of the parental molecule. When the two strands of the parental molecule separate, each serves as a template on which nucleotides are arranged by the base-pairing rules; they will then be polymerised by enzymes into new complementary strands. **2.** DNA pol III covalently adds nucleotides to new DNA strands and proofreads each added nucleotide for correct base pairing. **3.** In the cell cycle, DNA synthesis occurs during the S phase, between the G_1 and G_2 phases of interphase. DNA replication is therefore complete before the mitotic phase begins. **4.** Synthesis of the leading strand is initiated by an RNA primer, which must be removed and replaced with DNA, a task that could not be performed if the cell's DNA pol I were nonfunctional. In the overview box in Figure 16.18, just to the left of the top origin of replication, a functional DNA pol I would replace the RNA primer of the leading strand (shown in red) with DNA nucleotides (blue). The nucleotides would be added on to the 3′ end of the first Okazaki fragment of the upper lagging strand (the right half of the replication bubble).

Concept Check 16.3

1. A nucleosome is made up of eight histone proteins, two each of four different types, around which DNA is wound. Linker DNA runs from one nucleosome to the next. **2.** The 10-nm fibre of euchromatin is less compacted during interphase than in mitosis and is accessible to the cellular proteins responsible for gene expression. In contrast, the 10-nm fibre of heterochromatin is relatively compacted (densely arranged) during interphase, and genes in heterochromatin are largely inaccessible to proteins necessary for gene expression. **3.** The nuclear lamina is a netlike array of protein filaments that provides mechanical support just inside the nuclear envelope and thus maintains the shape of the nucleus. Considerable evidence also supports the existence of a nuclear matrix, a framework of protein fibres extending throughout the nuclear interior.

Summary of Key Concepts Questions

16.1 Each strand in the double helix has polarity; the end with a phosphate group on the 5′ carbon of the sugar is called the 5′ end, and the end with an —OH group on the 3′ carbon of the sugar is called the 3′ end. The two strands run in opposite directions, one running 5′ → 3′ and the other alongside it running 3′ → 5′. Thus, each end of the molecule has both a 5′ and a 3′ end, one on each strand of the double helix. This arrangement is called antiparallel. If the strands were parallel, they would both run 5′ → 3′ in the same direction, so an end of the molecule would have either two 5′ ends or two 3′ ends. **16.2** On both the leading and lagging strands, DNA polymerase adds onto the 3′ end of an RNA primer synthesised by primase, synthesising DNA in the 5′ → 3′ direction. Because the parental strands are antiparallel, however, only on the leading strand does synthesis proceed continuously into the replication fork. The lagging strand is synthesised bit by bit in the direction away from the fork as a series of shorter Okazaki fragments, which are later joined together by DNA ligase. Each fragment is initiated by synthesis of an RNA primer by primase as soon as a given stretch of single-stranded template strand is opened up. Although both strands are synthesised at the same rate, synthesis of the lagging strand is delayed because initiation of each fragment begins only when sufficient template strand is available. **16.3** The chromatin in an interphase nucleus is present as the 10-nm fibre, either fairly loosely arranged in euchromatin or more densely arranged in heterochromatin (such as at the centromeres and telomeres). The euchromatin is also subdivided into larger compartments and smaller looped domains. This organisation may reflect differences in gene expression occurring in these regions.

Test Your Understanding

1. C **2.** C **3.** B **4.** D **5.** A **6.** D **7.** B **8.** A **9.** Like histones, the *E. coli* proteins would be expected to contain many basic (positively charged) amino acids, such as lysine and arginine, which can form weak bonds with the negatively charged phosphate groups on the sugar-phosphate backbone of the DNA molecule.
11.

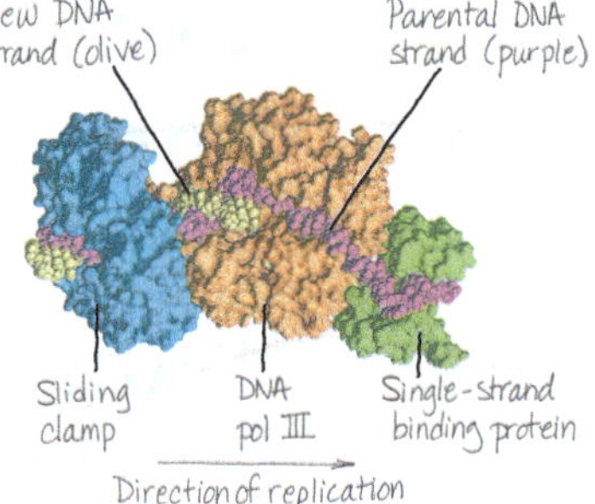

Chapter 17

Figure Questions

Figure 17.3 The previously presumed pathway would have been wrong. The new results would support this pathway: precursor → citrulline → ornithine → arginine. They would also indicate that class I mutants have a defect in the second step and class II mutants have a defect in the first step. **Figure 17.5** The mRNA sequence (5′-UGGUUUGGCUCA-3′) is the same as the nontemplate DNA strand sequence (5′-TGGTTTGGCTCA-3′), except there is a U in the mRNA wherever there is a T in the DNA. The nontemplate strand is probably used to represent a DNA sequence because it so closely resembles the mRNA sequence, containing codons. (This is why it's called the coding strand.) **Figure 17.6** Arg (or R)–Glu (or E)–Pro (or P)–Arg (or R) **Figure 17.8** The processes are similar in that polymerases form polynucleotides complementary to an antiparallel DNA template strand. In replication, however, both strands act as templates, whereas in transcription, only one DNA strand acts as a template. This reflects, of course, the fact that replication results in a double-stranded product (DNA) while transcription results in a single-stranded product (an RNA). **Figure 17.9** The RNA polymerase would bind directly to the promoter, rather than being dependent on the previous binding of transcription factors.
Figure 17.12

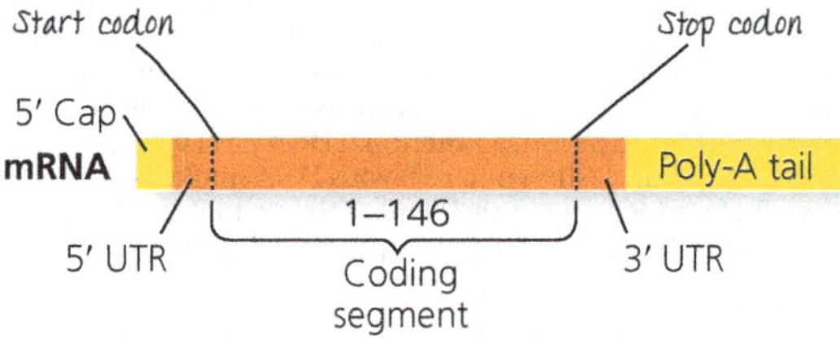

Figure 17.16 The anticodon on the tRNA is 3′-AAG-5′, so it would bind to the mRNA codon 5′-UUC-3′. This codon codes for phenylalanine (Phe, or F), which is the amino acid this tRNA would carry. **Figure 17.22** It would be packaged in a vesicle, transported to the Golgi apparatus for further processing, and then transported via a vesicle to the plasma membrane. The vesicle would fuse with the membrane, releasing the protein outside the cell. **Figure 17.24** The mRNA farthest to the right (the longest one) started transcription first. The ribosome at the top, closest to the DNA, started translating first and thus has the longest polypeptide.

Concept Check 17.1

1. Recessive **2.** A polypeptide made up of 10 Gly (glycine) amino acids
3.

Nontemplate sequence (from template sequence given in problem), written 3′ → 5′: 3′-ACGACTGAA-5′

If above used as template, mRNA sequence: 5′-UGCUGACUU-3′

Translated: Cys-STOP

If the nontemplate sequence could have been used as a template for transcribing the mRNA, the protein translated from the mRNA would have a completely different amino acid sequence, so it would not be able to function as the original protein (translated from an mRNA transcribed from the template strand). It would also be shorter because of the UGA stop signal shown in the mRNA sequence above—and possibly others earlier in the mRNA sequence.

Concept Check 17.2

1. A promoter is the region of DNA to which RNA polymerase binds to begin transcription. It is at the upstream end of the gene (transcription unit).
2. In a bacterial cell, part of the RNA polymerase recognises the gene's promoter and binds to it. In a eukaryotic cell, transcription factors must bind to the promoter first, then the RNA polymerase binds to them. In both cases, sequences in the promoter determine the precise binding of RNA polymerase so the enzyme is in the right location and orientation. **3.** The transcription factor that recognises the TATA sequence would be unable to bind, so RNA polymerase could not bind and transcription of that gene most likely would not occur.

Concept Check 17.3

1. Due to alternative splicing of exons, each gene can result in multiple different mRNAs and can thus direct synthesis of multiple different proteins. **2.** In watching a pre-recorded show, you watch segments of the show itself (exons) and fast-forward through the commercials, which are thus like introns. However, unlike introns, commercials remain in the recording, while the introns are cut out of the RNA transcript during RNA processing. **3.** Once the mRNA has exited the nucleus, the cap prevents it from being degraded by hydrolytic enzymes and facilitates its attachment to ribosomes. If the cap were removed from all mRNAs, the cell would no longer be able to synthesise any proteins and would probably die.

Concept Check 17.4

1. First, each aminoacyl-tRNA synthetase specifically recognises a single amino acid and attaches it only to an appropriate tRNA. Second, a tRNA charged with its specific amino acid binds only to an mRNA codon for that amino acid. **2.** A signal peptide on the leading end (amino end, or N-terminus) of the polypeptide being synthesised is recognised by a signal-recognition particle that brings the ribosome to the ER membrane. There, the ribosome attaches and continues to synthesise the polypeptide, depositing it in the ER lumen. **3.** Because of wobble, the tRNA could bind to either 5′-GCA-3′ or 5′-GCG-3′, both of which code for alanine (Ala, or A). Alanine would be attached to the tRNA (see diagram, upper right). **4.** When one ribosome terminates translation and dissociates, the two subunits would be very close to the cap. This could facilitate their rebinding and initiating synthesis of a new polypeptide, thus increasing the efficiency of translation.

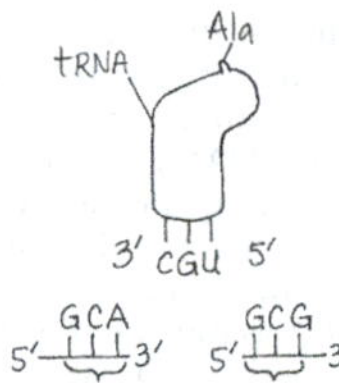

Concept Check 17.5

1. In the mRNA, the reading frame downstream from the deletion is shifted, leading to a long string of incorrect amino acids in the polypeptide, and in most cases, a stop codon will occur, leading to premature termination. The polypeptide will most likely be nonfunctional. **2.** Heterozygous individuals, said to have sickle-cell trait, have a copy each of the wild-type allele and the sickle-cell allele. Both alleles will be expressed, so these individuals will have both normal and sickle-cell haemoglobin molecules. Apparently, having a mix of the two forms of β-globin has no effect under most conditions, but during prolonged periods of low blood oxygen (such as at higher altitudes), these individuals can show some signs of sickle-cell disease.
3.

Normal DNA sequence (template strand is on top): 3′-TACTTGTCCGATATC-5′
5′-ATGAACAGGCTATAG-3′

mRNA sequence: 5′-AUGAACAGGCUAUAG-3′

Amino acid sequence: Met-Asn-Arg-Leu-STOP

Mutated DNA sequence (template strand is on top): 3′-TACTTGTCCAATATC-5′
5′-ATGAACAGGTTATAG-3′

mRNA sequence: 5′-AUGAACAGGUUAUAG-3′

Amino acid sequence: Met-Asn-Arg-Leu-STOP

No effect: The amino acid sequence is Met-Asn-Arg-Leu both before and after the mutation because the mRNA codons 5′-CUA-3′ and 5′-UUA-3′ both code for Leu. (The fifth codon is a stop codon.) **4.** A Cas9–guide RNA could be synthesised complementary to the mutated sequence, and the Cas9–guide RNA complex injected into a cell with the mutation. A region of double-stranded DNA with the correct sequence would also be provided. Cas9 would cut the mutated sequence, and the DNA repair system in the cell would repair the DNA, using the correct sequence as a template. Based on the answer to question 3, however, this would not be worthwhile because there is no amino acid change in the encoded protein, so the function of the protein encoded by the mutated gene is normal.

Summary of Key Concepts Questions

17.1 A gene contains genetic information in the form of a nucleotide sequence. The gene is first transcribed into an RNA molecule, and a messenger RNA molecule is ultimately translated into a polypeptide. The polypeptide makes up part or all of a protein, which performs a function in the cell and contributes to the phenotype of the organism. **17.2** Both bacterial and eukaryotic genes have promoters, regions where RNA polymerase ultimately binds and begins transcription. In bacteria, RNA polymerase binds directly to the promoter; in eukaryotes, transcription factors bind first to the promoter, and then RNA polymerase binds to the transcription factors and promoter together.
17.3 Both the 5′ cap and the 3′ poly-A tail help the mRNA exit from the nucleus and then, in the cytoplasm, help ensure mRNA stability and allow it to bind to ribosomes. **17.4** In the context of the ribosome, tRNAs function as translators between the nucleotide-based language of mRNA and the amino-acid-based language of polypeptides. A tRNA carries a specific amino acid, and the anticodon on the tRNA is complementary to the codon on the mRNA that codes for that amino acid. In the ribosome, the tRNA binds to the A site. Then the polypeptide being synthesised (currently on the tRNA in the P site) is joined to the new amino acid, which becomes the new (C-terminal) end of the polypeptide. Next, the tRNA in the A site moves to the P site. After the polypeptide is transferred to the new tRNA, thus adding the new amino acid, the now empty tRNA moves from the P site to the E site, where it exits the ribosome. **17.5** When a nucleotide base is altered chemically, its base-pairing characteristics may be changed. When that happens, an incorrect nucleotide is likely to be incorporated into the complementary strand during the next replication of the DNA, and successive rounds of replication will perpetuate the mutation. Once the gene is transcribed, the mutated codon may code for a different amino acid that inhibits or changes the function of a protein. If, however, the chemical change in the base is detected and repaired by the DNA repair system before the next replication, no mutation will result.

Test Your Understanding

1. B **2.** C **3.** A **4.** B **5.** B **6.** C **7.** D **8.** No. Transcription and translation are separated in space and time in a eukaryotic cell, as a result of the eukaryotic cell's nuclear membrane
9.

Type of RNA	Functions
Messenger RNA (mRNA)	Carries information specifying amino acid sequences of polypeptides from DNA to ribosomes
Transfer RNA (tRNA)	Serves as translator molecule in protein synthesis; translates mRNA codons into amino acids
Ribosomal RNA (rRNA)	In a ribosome, plays a structural role; as a ribozyme, plays a catalytic role (catalyzes peptide bond formation)
Primary transcript	Is a precursor to mRNA, rRNA, or tRNA, before being processed; some intron RNA acts as a ribozyme, catalyzing its own splicing
Small RNAs in spliceosome	Play structural and catalytic roles in spliceosomes, the complexes of protein and RNA that splice pre-mRNA

Chapter 18

Figure Questions

Figure 18.3 As the concentration of tryptophan in the cell falls, eventually there will be none bound to *trp* repressor molecules. These will then change into their inactive shapes and dissociate from the operator, allowing transcription of the operon to resume. The enzymes for tryptophan synthesis will be made, and they will again synthesise tryptophan in the cell.
Figure 18.10 Each of the two polypeptides has two regions—one that makes up part of MyoD's DNA-binding domain and one that makes up part of MyoD's activation domain. Each functional domain in the complete MyoD protein is made up of parts of both polypeptides. **Figure 18.12** In both types of cell, the albumin gene enhancer has the three control elements coloured yellow, grey, and red. The sequences in the liver and lens cells would be identical, since the cells are in the same organism. **Figure 18.18** Even if the mutant MyoD protein couldn't activate the *myoD* gene, it could still turn on genes for the other proteins in the pathway (other transcription factors, which would turn on the genes for muscle-specific proteins, for example). Therefore, some differentiation would occur. But unless there were other activators that could compensate for the loss of the MyoD protein's activation of the *myoD* gene, the cell would not be able to maintain its differentiated state. **Figure 18.22** Normal Bicoid protein would be made in the anterior end and compensate for the presence of mutant *bicoid* mRNA put into the egg by the mother. Development should be normal, with a head present. (This is what was observed.) **Figure 18.25** The mutation is likely to be recessive because it is more likely to have an effect if both copies of the gene are mutated and code for nonfunctional proteins. If one normal copy of the gene is present, its product could inhibit the cell cycle. (However, there are also known cases of dominant *p53* mutations, and the HNPCC gene, discussed later, is a dominant mutation in a tumour-suppressor pathway.) **Figure 18.27** Cancer is a disease in which cell division occurs without its usual regulation. Cell division can be stimulated by growth factors (see Figure 12.18), which bind to cell-surface receptors (see Figure 11.8). Cancer cells evade these normal controls and can often divide in the absence of growth factors (see Figure 12.19). This suggests that the receptor proteins or some other components in a signalling pathway are abnormal in some way (see, for example, the mutant Ras protein in Figure 18.24) or are expressed at abnormal levels, as seen for the receptors in this figure. Under some circumstances in the mammalian body, steroid hormones such as oestrogen and progesterone can also promote cell division. These molecules also use cell-signalling pathways, as described in Concept 11.2 (see Figure 11.9). Because signalling receptors are involved in triggering cells to undergo cell division, it is not surprising that altered genes encoding these proteins might play a significant role in the development of cancer. Genes might be altered through either a mutation that changes the function of the protein product or a mutation that causes the gene to be expressed at abnormal levels that disrupt the overall regulation of the signalling pathway.

Concept Check 18.1

1. Binding by the *trp* corepressor (tryptophan) activates the *trp* repressor, which binds to the *trp* operator, shutting off transcription of the *trp* operon. Binding by the *lac* inducer (allolactose) inactivates the *lac* repressor, so that it can no longer bind to the *lac* operator, leading to transcription of the *lac* operon. **2.** When glucose is scarce, cAMP is bound to CRP and CRP is bound to the *lac* promoter, favouring the binding of RNA polymerase. However, in the absence of lactose, the *lac* repressor is bound to the *lac* operator, blocking RNA polymerase from transcribing the *lac* operon genes. **3.** The cell would continuously produce β-galactosidase and the two other enzymes for using lactose, even in the absence of lactose, thus wasting cell resources.

Concept Check 18.2

1. Histone acetylation is generally associated with gene expression, while DNA methylation is generally associated with lack of expression. **2.** The same enzyme could not methylate both a histone and a DNA base. Enzymes are very specific in structure, and an enzyme that could methylate an amino acid of a protein would not be able to fit the base of a DNA nucleotide into the same active site. **3.** General transcription factors function in assembling the transcription initiation complex at the promoters for all genes. Specific transcription factors bind to control elements associated with a particular gene and, once bound, either increase (activators) or decrease (repressors) transcription of that gene. **4.** Regulation of translation initiation, degradation of the mRNA, activation of the protein (by chemical modification, for example), and protein degradation **5.** The three genes should have some similar or identical sequences in the control elements of their enhancers. Because of this similarity, the same specific transcription factors that are present in muscle cells could bind to the enhancers of all three genes and stimulate their expression coordinately.

Concept Check 18.3

1. Both miRNAs and siRNAs are small, single-stranded RNAs that associate with a complex of proteins and then can base-pair with mRNAs that have a complementary sequence. This base pairing leads to either degradation of the mRNA or blockage of its translation. In some yeasts, siRNAs associated with proteins in a different complex can bind back to centromeric chromatin, recruiting enzymes that cause condensation of that chromatin into heterochromatin. Both miRNAs and siRNAs are processed from double-stranded RNA precursors but have subtle variations in the structure of those precursors. **2.** The mRNA would persist and be translated into the cell division–promoting protein, and the cell would probably divide. If the intact miRNA is necessary for inhibition of cell division, then division of this cell might be inappropriate. Uncontrolled cell division could lead to formation of a mass of cells (tumour) that prevents proper functioning of the organism and could contribute to the development of cancer. **3.** The *XIST* RNA is transcribed from the *XIST* gene on the X chromosome that will be inactivated. It then binds to that chromosome and induces heterochromatin formation. A likely model is that *XIST* RNA somehow recruits chromatin modification enzymes that lead to formation of heterochromatin.

Concept Check 18.4

1. Cells undergo differentiation during embryonic development, becoming different from each other. Therefore, the adult organism is made up of many highly specialised cell types that are different from each other. **2.** By binding to a receptor on the receiving cell's surface and triggering a signal transduction pathway, involving intracellular molecules such as second messengers and transcription factors that affect gene expression **3.** The products of maternal effect genes, made and deposited into the egg by the mother, determine the head and tail ends, as well as the back and belly, of the egg and embryo (and eventually the adult fly). **4.** The lower cell is synthesising signalling molecules because the gene encoding them is activated, meaning that the appropriate specific transcription factors are binding to the gene's enhancer. The genes encoding these specific transcription factors are also being expressed in this cell because the transcription factor activators that can turn them on were expressed in the precursor to this cell. A similar explanation also applies to the cells expressing the receptor proteins. This scenario began with specific cytoplasmic determinants localised in specific regions of the egg. These cytoplasmic determinants were distributed unevenly to daughter cells, resulting in cells going down different developmental pathways.

Concept Check 18.5

1. A cancer-causing mutation in a proto-oncogene usually makes the gene product overactive, whereas a cancer-causing mutation in a tumour-suppressor gene usually makes the gene product nonfunctional. **2.** When an individual has inherited an oncogene or a mutant allele of a tumour-suppressor gene **3.** Apoptosis is signalled by the p53 protein when a cell has extensive DNA damage, so apoptosis plays a protective role in eliminating a cell that might contribute to cancer. If mutations in the genes in the apoptotic pathway blocked apoptosis, a cell with such damage could continue to divide and might lead to tumour formation.

Summary of Key Concepts Questions

18.1 A corepressor and an inducer are both small molecules that bind to the repressor protein in an operon, causing the repressor to change shape. In the case of a corepressor (like tryptophan), this shape change allows the repressor to bind to the operator, blocking transcription. In contrast, an inducer causes the repressor to dissociate from the operator, allowing transcription to begin.
18.2 In that specific type of cell, the chromatin must not be tightly condensed because it must be accessible to transcription factors. The appropriate specific transcription factors (activators), which are made in that type of cell, must bind to the control elements in the enhancer of the gene, while repressors must not be bound. The DNA must be bent by a bending protein so the activators can contact the mediator proteins and form a complex with general transcription factors at the promoter. Then RNA polymerase must bind and begin transcription.
18.3 miRNAs do not "code" for the amino acids of a protein—they are never translated. Each miRNA associates with a group of proteins to form a complex. Binding of the complex to an mRNA with a complementary sequence causes that mRNA to be degraded or blocks its translation. This is considered gene regulation because it controls the amount of a particular mRNA that can be translated into a functional protein. **18.4** The first process involves cytoplasmic determinants, including mRNAs and proteins, placed into specific locations in the egg by maternal cells. The embryonic cells that are formed from different regions of the egg during early cell divisions will have different proteins in them, which will direct different programs of gene expression. The second process involves the cell in question responding to signalling molecules secreted by neighbouring cells (induction). The signalling pathway in the responding cell also leads to a different pattern of gene expression. The coordination of these two processes results in each cell following a unique pathway in the developing embryo.
18.5 The protein product of a proto-oncogene is usually involved in a pathway that stimulates cell division. The protein product of a tumour-suppressor gene is usually involved in a pathway that inhibits cell division.

Test Your Understanding

1. C **2.** A **3.** B **4.** C **5.** C **6.** D **7.** A **8.** C **9.** B **10.** D
11. (a)

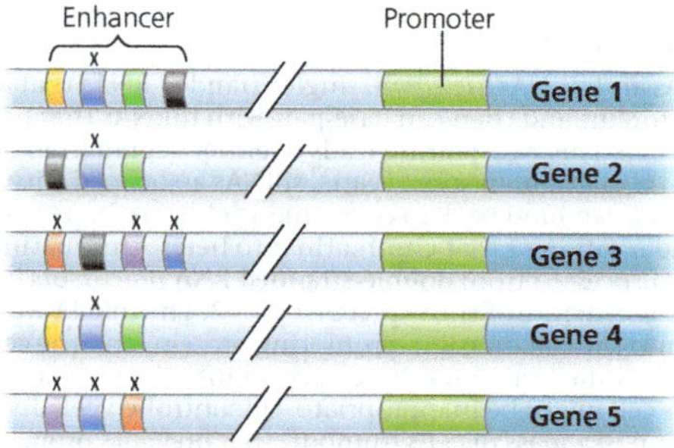

The purple, blue, and red activator proteins would be present.
(b)

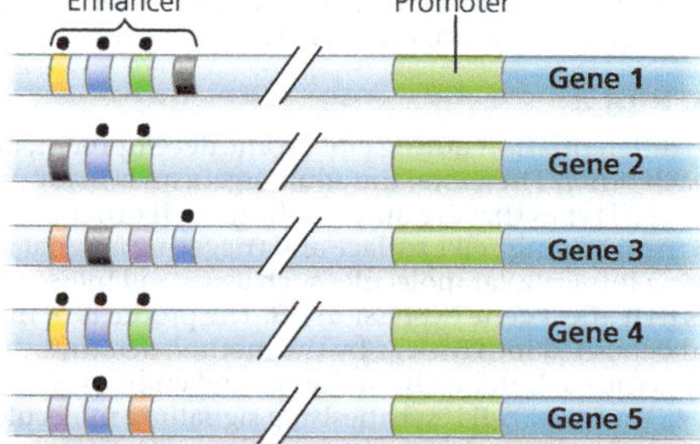

Only gene 4 would be transcribed.
(c) In nerve cells, the yellow, blue, green, and black activators would have to be present, thus activating transcription of genes 1, 2, and 4. In skin cells, the red, black, purple, and blue activators would have to be present, thus activating genes 3 and 5.

Chapter 19

Figure Questions

Figure 19.4 Beijerinck might have concluded that the agent was a toxin produced by the plant that was able to pass through a filter but that became more and more dilute. In this case, he would have concluded that the infectious agent could not replicate.
Figure 19.6

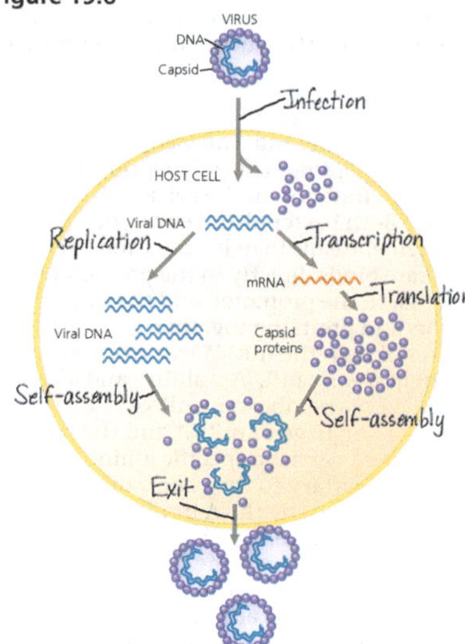

Figure 19.11 The main protein on the cell surface that HIV binds to is called CD4. However, HIV also requires a "co-receptor," which in many cases is a protein called CCR5. HIV binds to both of these proteins together and then is taken into the cell. Researchers discovered this requirement by studying individuals who seemed to be resistant to HIV infection despite multiple exposures. These individuals turned out to have mutations in the gene that encodes CCR5 such that the protein apparently cannot act as a co-receptor, and so HIV can't enter and infect cells.

Concept Check 19.1

1. Both viruses consist of RNA as the genetic material, associated with proteins. However, TMV consists of one molecule of RNA surrounded by a helical array of proteins, while the influenza virus has eight molecules of RNA, each associated with proteins and wound into a double helix. Another difference between the viruses is that the influenza virus has an outer envelope and TMV does not. **2.** The T2 phages were an excellent choice for use in the Hershey-Chase experiment because they consist of only DNA surrounded by a protein coat, and DNA and protein were the two candidates for macromolecules that carried genetic information. Hershey and Chase were able to radioactively label each type of molecule alone and follow it during separate infections of *E. coli* cells with T2. Only the DNA entered the bacterial cell during infection, and only labelled DNA showed up in some of the progeny phage. Hershey and Chase concluded that the DNA must carry the genetic information necessary for the phage to reprogram the cell and produce progeny phages.

Concept Check 19.2

1. Lytic phages can only carry out lysis of the host cell, whereas lysogenic phages may either lyse the host cell or integrate into the host chromosome. In the latter case, the viral DNA (prophage) is simply replicated along with the host chromosome. Under certain conditions, a prophage may exit the host chromosome and initiate a lytic cycle. **2.** Both the CRISPR-Cas system and miRNAs involve RNA molecules bound in a protein complex and acting as "homing devices" that enable the complex to bind a complementary sequence. However, miRNAs are involved in regulating gene expression (by affecting mRNAs) and the CRISPR-Cas system protects bacterial cells from foreign invaders—infecting phages. Thus the CRISPR-Cas system is more like an immune system than is the miRNA system. **3.** Both the viral RNA polymerase and the RNA polymerase in Figure 17.10 synthesise an RNA molecule complementary to a template strand. However, the RNA polymerase in Figure 17.10 uses one of the strands of the DNA double helix as a template, whereas the viral RNA polymerase uses the RNA of the viral genome as a template. **4.** HIV is called a retrovirus because it synthesises DNA using its RNA genome as a template. This is the reverse ("retro") of the usual DNA → RNA information flow. **5.** There are many steps that could be interfered with: binding of the virus to the cell, reverse transcriptase function, integration into the host cell chromosome, genome synthesis (in this case, transcription of RNA from the integrated provirus), assembly of the virus inside the cell, and budding of the virus. (Many of these, if not all, are targets of actual medical strategies to block progress of the infection in HIV-infected people.)

Concept Check 19.3

1. Mutations can lead to a new strain of a virus that can no longer be effectively fought by the immune system, even if an animal had been exposed to the original strain; a virus can jump from one species to a new host; and a rare virus can spread if a host population becomes less isolated. **2.** In horizontal transmission, a plant is infected from an external source of virus, which enters through a break in the plant's epidermis due to damage by herbivores or other agents. In vertical transmission, a plant inherits viruses from its parent either via infected seeds (sexual reproduction) or via an infected cutting (asexual reproduction).
3. Humans are not within the host range of TMV, so they can't be infected by the virus. (TMV can't bind to receptors on human cells and infect them.)

Summary of Key Concepts Questions

19.1 Viruses are generally considered nonliving because they are not capable of replicating outside of a host cell and are unable to carry out the energy-transforming reactions of metabolism. To replicate and carry out metabolism, they depend completely on host enzymes and resources. **19.2** Single-stranded RNA viruses require an RNA polymerase that can make RNA using an RNA template. (Cellular RNA polymerases make RNA using a DNA template.) Retroviruses require reverse transcriptases to make DNA using an RNA template. (Once the first DNA strand has been made, the same enzyme can promote synthesis of the second DNA strand.) **19.3** The mutation rate of RNA viruses is higher than that of DNA viruses because RNA polymerase has no proofreading function, so errors in replication are not corrected. Their higher mutation rate is one reason that RNA viruses change faster than DNA viruses, leading to their being able to have an altered host range and to evade immune defences in possible hosts.

Test Your Understanding

1. C **2.** D **3.** C **4.** D **5.** B

6. As shown below, the viral genome would be translated into capsid proteins and envelope glycoproteins directly, rather than after a complementary RNA copy was made. A complementary RNA strand would still be made, however, that could be used as a template for many new copies of the viral genome.

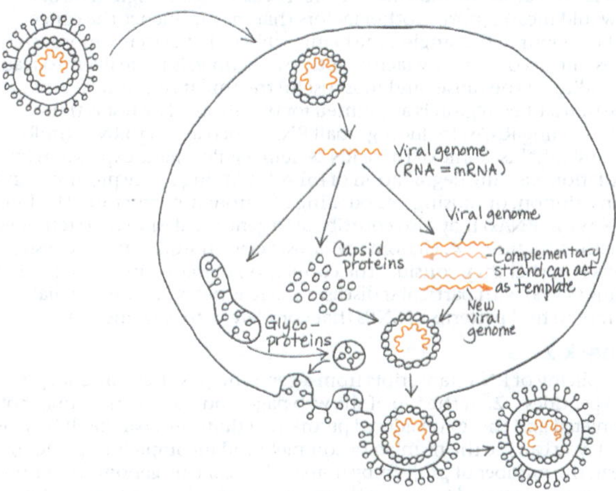

Chapter 20

Figure Questions

Figure 20.5

5′ AAGCTT 3′ / 3′ TTCGAA 5′ —HindIII→ 5′ A 3′ / 3′ TTCGA 5′ + 5′ AGCTT 3′ / 3′ A 5′

Figure 20.16 None of the eggs with the transplanted nuclei from the four-cell embryo at the upper left would have developed into a tadpole. Also, the result might include only some of the tissues of a tadpole, which might differ, depending on which nucleus was transplanted. (This assumes that there was some way to tell the four cells apart, as one can in some frog species.) **Figure 20.21** Using converted iPS cells would not carry the same risk, which is its major advantage. Because the donor cells would come from the patient, they would be perfectly matched. The patient's immune system would recognise them as "self" cells and would not mount an attack (which is what leads to rejection). On the other hand, cells that are rapidly dividing might carry a risk of inducing some type of tumour or contributing to development of cancer.

Concept Check 20.1

1. The covalent sugar-phosphate bonds of the DNA strands **2.** Yes, *PvuI* will cut the molecule (at the position indicated by the dashed red line).

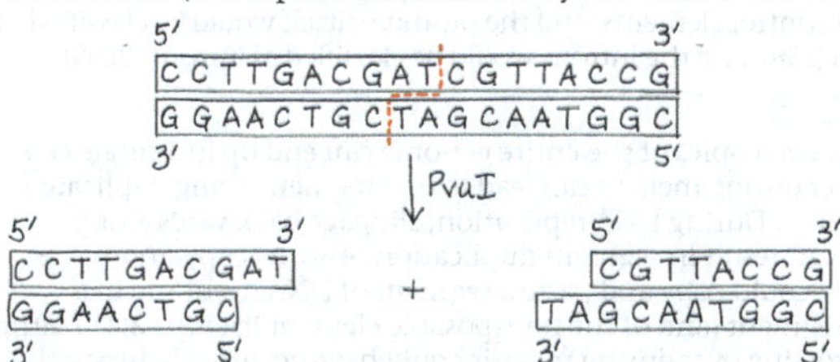

3. Some eukaryotic genes are too large to be incorporated into bacterial plasmids. Bacterial cells lack the means to process RNA transcripts into mRNA, and even if the need for RNA processing is avoided by using cDNA, bacteria lack enzymes to catalyse the post-translational processing that many eukaryotic proteins require to function properly. (This is often the case for human proteins, which are a focus of biotechnology.) **4.** During the replication of the ends of linear DNA molecules (see Figure 16.21), an RNA primer is used at the 5′ end of each new strand. The RNA must be replaced by DNA nucleotides, but DNA polymerase is incapable of starting from scratch at the 5′ end of a new DNA strand. During PCR, the primers are made of DNA nucleotides already, so they don't need to be replaced—they just remain as part of each new strand. Therefore, there is no problem with end replication during PCR, and the fragments don't shorten with each replication.

Concept Check 20.2

1. Complementary base pairing is involved in cDNA synthesis, which is required for all three techniques: RT-PCR, DNA micro-array analysis, and RNA sequencing. Reverse transcriptase uses mRNA as a template to synthesise the first strand of cDNA, adding nucleotides complementary to those on the mRNA. Complementary base pairing is also involved when DNA polymerase synthesises the second strand of the cDNA. Furthermore, in RT-PCR, the primers must base-pair with their target sequences in the DNA mixture, locating one specific region among many. Also, the DNA polymerase (for example, Taq polymerase) used in PCR relies on complementary base pairing to the template strand to add new nucleotides during synthesis of the fragments. In DNA micro-array analysis, the labelled cDNA probe binds only to the specific target sequence due to complementary nucleic acid hybridisation (DNA-DNA hybridisation). In RNA-seq, when sequencing the cDNAs, base complementarity plays a role in the sequencing process. **2.** As a researcher interested in how cancer develops, you would want to study genes represented by spots that are green or red because these are genes for which the expression level differs between the two types of tissue. Some of these genes may be expressed differently as a result of cancer, while others might play a role in causing cancer, so both would be of interest.

Concept Check 20.3

1. The state of chromatin modification in the nucleus from the intestinal cell was undoubtedly less similar to that of a nucleus from a fertilised egg, explaining why many fewer of these nuclei were able to be reprogrammed. In contrast, the chromatin in a nucleus from a cell at the four-cell stage would have been much more like that of a nucleus in a fertilised egg and therefore much more easily programmed to direct development. **2.** No, primarily because of subtle (and perhaps not so subtle) differences in the environment in which the clone develops and lives compared with that in which the original pet lived (see the differences noted in Figure 20.18). This does provoke ethical questions. To produce Dolly, also a mammal, several hundred embryos were cloned, but only one survived to adulthood. If any of the "reject" dog embryos survived to birth as defective dogs, would they be killed? Is it ethical to produce living animals that may be defective? You can probably think of other ethical issues as well. **3.** Given that muscle cell differentiation involves a master regulatory gene (*MyoD*), you might start by introducing either the MyoD protein or an expression vector carrying the *MyoD* gene into stem cells. (This is not likely to work, because the embryonic precursor cell in Figure 18.18 is more differentiated than the stem cells you are working with, and some other changes would have to be introduced as well. But it's a good way to start! And you may be able to think of others.)

Concept Check 20.4

1. Stem cells continue to reproduce themselves, ensuring that the corrective gene product will continue to be made. **2.** Herbicide resistance, pest resistance, disease resistance, drought resistance, and delayed ripening **3.** Because hepatitis A is an RNA virus, you could isolate RNA from the blood and try to detect copies of hepatitis A RNA by RT-PCR. You would first reverse-transcribe the blood mRNA into cDNA and then use PCR to amplify the cDNA, using primers specific to hepatitis A sequences. If you then ran the products on an electrophoretic gel, the presence of a band of the appropriate size would support your hypothesis. Alternatively, you could use RNA-seq to sequence all the RNAs in your patient's blood and see whether any of the sequences match up with that of hepatitis A. (Since you are only seeking one sequence, though, RT-PCR is probably a better choice.)

Summary of Key Concepts Questions

20.1 A plasmid vector and a source of foreign DNA to be cloned are both cut with the same restriction enzyme, generating restriction fragments with sticky ends. These fragments are mixed together, ligated, and reintroduced into bacterial cells. The plasmid has a gene for resistance to an antibiotic. That antibiotic is added to the host cells, and only cells that have taken up a plasmid will grow. (Another technique allows researchers to select only the cells that have a recombinant plasmid, rather than the original plasmid without an inserted gene.) **20.2** The genes that are expressed in a given tissue or cell type determine the proteins (and noncoding RNAs) that are the basis of the structure and functions of that tissue or cell type. Understanding which groups of interacting genes establish particular structures and carry out certain functions will help us learn how the parts of an organism work together. We will also be better able to treat diseases that occur when faulty gene expression leads to malfunctioning tissues. **20.3** (1) Cloning a mouse involves transplanting a nucleus from a differentiated mouse cell into a mouse egg cell that has had its own nucleus removed. Activating the egg cell and promoting its development into an embryo in a surrogate mother results in a mouse that is genetically identical to the mouse that donated the nucleus. In this case, the differentiated nucleus has been reprogrammed by factors in the egg cytoplasm. (2) Mouse ES cells are generated from inner cells in mouse blastocysts, so in this case the cells are "naturally" reprogrammed by the process of reproduction and development. (Cloned mouse embryos can also be used as a source of ES cells.) (3) iPS cells can be generated without the use of embryos from a differentiated adult mouse cell by adding certain transcription factors into the cell. In this case, the transcription factors are reprogramming the cells to become pluripotent. **20.4** First, the disease must be caused by a single gene, and the molecular basis of the problem must be understood. Second, the cells that are going to be introduced into the patient must be cells that will integrate into body tissues and continue to multiply (and provide the needed gene product). Third, the gene must be able to be introduced into the cells in question in a safe way, as there have been instances of cancer resulting from some gene therapy trials. (Note that this will require testing the procedure in mice; moreover, the factors that determine a safe vector are not yet well understood. Maybe one of you will go on to solve this problem!)

Test Your Understanding

1. D **2.** B **3.** C **4.** B **5.** C **6.** A **7.** B **8.** You would use PCR to amplify the gene. This could be done from genomic DNA. Alternatively, mRNA could be isolated from lens cells and reverse-transcribed by reverse transcriptase to make cDNA. This cDNA could then be used for PCR. In either case, the gene would then be inserted into an expression vector so you could produce the protein and study it. **9.** Crossing over, which causes recombination, is a random event. The chance of crossing over occurring between two loci increases as the distance between them increases. If a SNP is located very close to a disease-associated allele, it is said to be genetically linked. Crossing over will rarely occur between the SNP and the allele, so the SNP can be used as a genetic marker indicating the presence of the particular allele.
10.

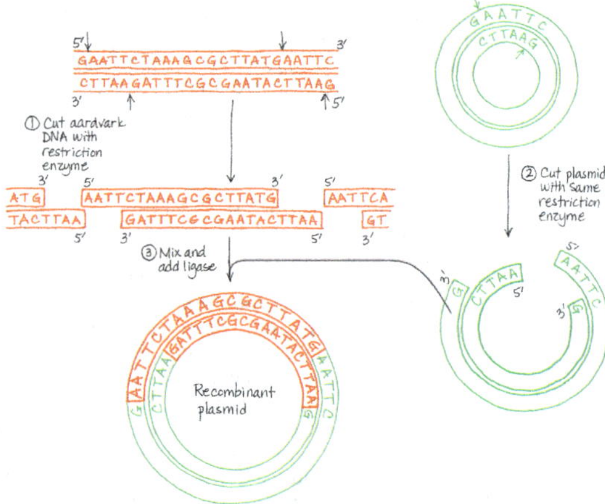

Chapter 21

Figure Questions

Figure 21.2 In step 2 of this figure, the order of the fragments relative to each other is not known and will be determined later by computer. The unordered nature of the fragments is reflected by their scattered arrangement in the diagram. **Figure 21.8** The transposon would be cut out of the DNA at the original site rather than copied, so the figure would show the original stretch of DNA without the transposon after the mobile transposon had been cut out. **Figure 21.10** The RNA transcripts extending from the DNA in each transcription unit are shorter on the left and longer on the right. This means that RNA polymerase must be starting on the left end of the unit and moving towards the right.
Figure 21.13

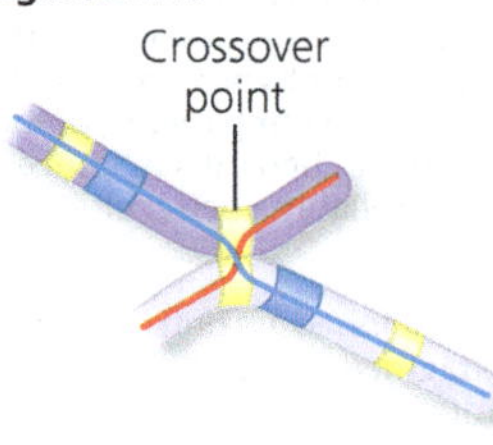

Figure 21.14 Pseudogenes are nonfunctional. They could have arisen by any mutations in the second copy that made the gene product unable to function. Examples are base changes that introduce stop codons in the sequence, alter amino acids, or change a region of the gene promoter so that the gene can no longer be expressed. **Figure 21.15** At position 5, there is an R (arginine) in lysozyme and a K (lysine) in α-lactalbumin; both of these are basic amino acids. **Figure 21.16** Let's say a transposable element (TE) existed in the intron to the left of the indicated EGF exon in the EGF gene, and the same TE was present in the intron to the right of the indicated F exon in the fibronectin gene. During meiotic recombination, these TEs could cause nonsister chromatids on homologous chromosomes to pair up incorrectly, as seen in Figure 21.13. One gene might end up with an F exon next to an EGF exon. Further mistakes in pairing over many generations might result in these two exons being separated from the rest of the gene and placed next to a single or duplicated K exon. In general, the presence of repeated sequences in introns and between genes facilitates these processes because it allows incorrect pairing of nonsister chromatids, leading to novel exon combinations. **Figure 21.18** Since you know that chimpanzees do not speak but humans do, you'd probably want to know how many amino acid differences there are between the human wild-type FOXP2 protein and that of the chimpanzee and whether these changes affect the function of the protein. (As we explain later in the text, there are two amino acid differences.) You know that humans with mutations in this gene have severe language impairment. You would want to learn more about the human mutations by checking whether they affect the same amino acids in the gene product that the chimpanzee sequence differences affect. If so, those amino acids might play an important role in the function of the protein in language. Going further, you could analyse the differences between the chimpanzee and mouse FOXP2 proteins. You might ask: Are they more similar than the chimpanzee and human proteins? (It turns out that the chimpanzee and mouse proteins have only one amino acid difference and thus are more similar than the chimpanzee and human proteins, which have two differences, and also are more similar than the human and mouse proteins, which have three differences.)

Concept Check 21.1

1. In the whole-genome shotgun approach, short fragments are generated by cutting the genome with multiple restriction enzymes. These fragments are cloned, sequenced, and then ordered by computer programs that identify overlapping regions.

Concept Check 21.2

1. The Internet allows centralisation of databases such as GenBank and software resources such as BLAST, making them freely accessible. Having all the data in a central database, easily accessible on the Internet, minimises the possibility of errors and of researchers working with different data. It streamlines the process of science, since all researchers are able to use the same software programs, rather than each having to obtain their own, possibly different, software. It speeds up dissemination of data and ensures as much as possible that errors are corrected in a timely fashion. These are just a few answers; you can probably think of more.
2. Cancer is a disease caused by multiple factors. Focusing on a single gene or a single defect would mean ignoring other factors that may influence the cancer and even the behaviour of the single gene being studied. The systems approach, because it takes into account many factors at the same time, is more likely to lead to an understanding of the causes and most useful treatments for cancer.
3. Some of the transcribed region is accounted for by introns. The rest is transcribed into noncoding RNAs, including small RNAs, such as microRNAs (miRNAs), siRNAs, and piRNAs. Some of these RNAs help regulate gene expression by blocking translation, causing degradation of mRNA, binding to the promoter and repressing transcription, or causing remodelling of chromatin structure. The long noncoding RNAs (lncRNAs) may also contribute to gene regulation or to remodelling of chromatin structure. **4.** Genome-wide association studies use the systems biology approach in that they consider the correlation of many single nucleotide polymorphisms (SNPs) with particular diseases, such as heart disease and diabetes, in an attempt to find patterns of SNPs that correlate with each disease.

Concept Check 21.3

1. Alternative splicing of RNA transcripts from a gene and post-translational processing of polypeptides **2.** At the top of the web page, you can see the number of genomes completed and those considered permanent drafts in a bar graph by year. Scrolling down, you can see the number of complete and incomplete sequencing projects by year, the number of projects by domain by year (the genomes of viruses and metagenomes are counted too, even though these are not "domains"), the phylogenetic distribution of bacterial genome projects, and projects by sequencing centre. Finally, near the bottom, you can see a pie chart of the "Project Relevance of Bacterial Genome Projects," which shows that about 58.6% have medical relevance. (This number may vary over time.) The web page ends with another pie chart showing the sequencing centres for archaeal and bacterial projects. **3.** A prokaryotic cell is generally smaller than a eukaryotic cell, and prokaryotes reproduce by binary fission. The evolutionary process involved is natural selection for more quickly reproducing cells: The faster they can replicate their DNA and divide, the more likely they will be able to dominate a population of prokaryotes. The less DNA they have to replicate, then, the faster they will reproduce.

Concept Check 21.4

1. The number of genes is higher in mammals, and the amount of noncoding DNA is greater. Also, the presence of introns in mammalian genes makes them larger, on average, than prokaryotic genes. **2.** The copy-and-paste transposon mechanism and retrotransposition **3.** In the rRNA gene family, identical transcription units, each containing genes for all three different RNA products, are present in long arrays, repeated one after the other. The large number of copies of the rRNA genes enable organisms to produce the rRNA for enough ribosomes to carry out active protein synthesis, and the single transcription unit for the three rRNAs ensures that the relative amounts of the different rRNA molecules produced are correct—every time one rRNA is made, a copy of each of the other two is made as well. Rather than numerous identical units, each globin gene family consists of a relatively small number of nonidentical genes. The differences in the globin proteins encoded by these genes result in production of haemoglobin molecules adapted to particular developmental stages of the organism. **4.** The exons would be classified as exons (1.5%); the enhancer region containing the distal control elements, the region closer to the promoter containing the proximal control elements, and the promoter itself would be classified as regulatory sequences (5%); and the introns would be classified as introns (20%).

Concept Check 21.5

1. If meiosis is faulty, two copies of the entire genome can end up in a single cell. Errors in crossing over during meiosis can lead to one segment being duplicated while another is deleted. During DNA replication, slippage backwards along the template strand can result in segment duplication. Also, a "copy-and-paste" transposable element could copy and paste a segment of DNA, resulting in a duplication of that segment (and of the transposable element itself.) **2.** For either gene, a mistake in crossing over during meiosis could have occurred between the two copies of that gene, such that one ended up with a duplicated exon. (The other copy would have ended up with a deleted exon.) This could have happened several times, resulting in the multiple copies of a particular exon in each gene.
3. Homologous transposable elements scattered throughout the genome provide sites where recombination can occur between different chromosomes. Movement of these elements into coding or regulatory sequences may change expression of genes, which can affect the phenotype in a way that is subject to natural selection.

Transposable elements also can carry genes with them, leading to dispersion of genes and in some cases different patterns of expression. Transport of an exon during transposition and its insertion into a gene may add a new functional domain to the originally encoded protein, a type of exon shuffling. (For any of these changes to be heritable, they must happen in germ cells, cells that will give rise to gametes.) **4.** Because more offspring are born to women who have this inversion, it must provide some advantage during the process of reproduction and development. Because proportionally more offspring have this inversion, we would expect it to persist and spread in the population. (In fact, evidence in the study allowed the researchers to conclude that it has been increasing in proportion in the population. You'll learn more about population genetics in the next unit.)

Concept Check 21.6

1. Because both humans and macaques are primates, their genomes are expected to be more similar than the macaque and mouse genomes are. The mouse lineage diverged from the primate lineage before the human and macaque lineages diverged.
2. Homeotic genes differ in their *non*homeobox sequences, which determine the interactions of homeotic gene products with other transcription factors and hence which genes are regulated by the homeotic genes. These nonhomeobox sequences differ in the two species, as do the expression patterns of the homeobox genes.
3. *Alu* elements underwent transposition more actively in the human genome for some reason. Their increased sites of insertion may have then allowed more recombination errors in the human genome, resulting in more or different duplications. The divergence of the organisation and content of the two genomes presumably made the chromosomes of each genome less similar to those of the other, thus accelerating divergence of the two species by making matings less and less likely to result in fertile offspring due to the mismatch of genetic information.

Summary of Key Concepts Questions

21.1 One focus of the Human Genome Project was to improve sequencing technology in order to speed up the process. During the project, many advances in sequencing technology allowed faster reactions and detection of products, which were therefore less expensive. **21.2** The most significant finding is that more than 75% of the human genome appears to be transcribed at some point in at least one of the cell types studied. Also, at least 80% of the genome contains an element that is functional, participating in gene regulation or maintaining chromatin structure in some way. The project was expanded to include other species to further investigate the functions of these transcribed DNA elements. It is necessary to carry out this type of analysis on the genomes of species that can be used in laboratory experiments. **21.3** (a) In general, bacteria and archaea have smaller genomes, lower numbers of genes, and higher gene density than eukaryotes. (b) Among eukaryotes, there is no apparent systematic relationship between genome size and phenotype. The number of genes is often lower than would be expected from the size of the genome—in other words, the gene density is often lower in larger genomes. (Humans are a good example.) **21.4** Transposable element–related sequences can move from place to place in the genome, and some of these sequences make a new copy of themselves when they do so. Thus, it is not surprising that they make up a significant percentage of the genome, and this percentage might be expected to increase over evolutionary time. **21.5** Chromosomal rearrangements within a species lead to some individuals having different chromosomal arrangements. Each of these individuals could still undergo meiosis and produce gametes, and fertilisation involving gametes with different chromosomal arrangements could result in viable offspring. However, during meiosis in the offspring, the maternal and paternal chromosomes might not be able to pair up, causing gametes with incomplete sets of chromosomes to form. Most often, when zygotes are produced from such gametes, they do not survive. Ultimately, a new species could form if two different chromosomal arrangements became prevalent within a population and individuals could mate successfully only with other individuals having the same arrangement. **21.6** Comparing the genomes of two closely related species can reveal information about more recent evolutionary events, perhaps events that resulted in the distinguishing characteristics of the two species. Comparing the genomes of very distantly related species can tell us about evolutionary events that occurred a very long time ago. For example, genes that are shared between two distantly related species must have arisen before the two species diverged.

Test Your Understanding

1. B **2.** A **3.** C **4.** Answers for (a) through (c):

Chimpanzee	PKSSD ... TSSTT ... NARRD
Mouse	PKSSE ... TSSTT ... NARRD
Gorilla	PKSSD ... TSSTT ... NARRD
Human	PKSSD ... TSSNT ... SARRD
Rhesus monkey	PKSSD ... TSSTT ... NARRD

(d) There is one difference between the sequence for the mouse and the sequence for the chimpanzee, gorilla, and rhesus monkey. There are two differences between the human sequence and the sequence for the chimpanzee, gorilla, and rhesus monkey. These facts might lead to the hypothesis that the *FOXP2* gene has been evolving faster in the human lineage than in other primates: Two differences between humans and other primates occurred during the 6 million years since they diverged, but only one difference occurred during the much longer period of 65 million years since rodents and primates diverged. However, as described in the text, later analysis that included more genomes that were more diverse failed to support this hypothesis.

Chapter 22

Figure Questions

Figure 22.6 You should have circled the branch located at the far left of Figure 1.20. Although three of the descendants (*Certhidea olivacea*, *Camarhynchus pallidus*, and *Camarhynchus parvulus*) of this common ancestor ate insects, the other three species that descended from this ancestor did not eat insects. **Figure 22.8** The common ancestor lived about 5.5 million years ago. **Figure 22.13** These results show that being reared from the egg stage on one plant species or the other did not result in the adult having a beak length appropriate for that host; instead, adult beak lengths were determined primarily by the population from which the eggs were obtained. Because an egg from a balloon vine population likely had long-beaked parents, while an egg from a golden rain tree population likely had short-beaked parents, these results indicate that beak length is an inherited trait.
Figure 22.15 Both strategies should increase the time that it takes *S. aureus* to become resistant to a new drug. If a drug that harms *S. aureus* does not harm other bacteria, natural selection will not favour resistance to that drug in the other species. This would decrease the chance that *S. aureus* would acquire resistance genes from other bacteria—thus slowing the evolution of resistance. Similarly, selection for resistance to a drug that slows the growth but does not kill *S. aureus* is much weaker than selection for resistance to a drug that kills *S. aureus*—again slowing the evolution of resistance. **Figure 22.18** Based on this evolutionary tree, crocodiles are more closely related to birds than to lizards because they share a more recent common ancestor with birds (ancestor 5) than with lizards (ancestor 4).
Figure 22.21 Hind limb structure changed first. *Rodhocetus* lacked flukes, but its pelvic bones and hind limbs had changed substantially from how those bones were shaped and arranged in *Pakicetus*. For example, in *Rodhocetus*, the pelvis and hind limbs appear to be oriented for paddling, whereas they were oriented for walking in *Pakicetus*.

Concept Check 22.1

1. Hutton and Lyell proposed that geologic events in the past were caused by the same processes operating today, at the same gradual rate. This principle suggested that Earth must be much older than a few thousand years, the age that was widely accepted in the early 19th century. Hutton's and Lyell's ideas also stimulated Darwin to reason that the slow accumulation of small changes could ultimately produce the profound changes documented in the fossil record. In this context, the age of Earth was important to Darwin, because unless Earth was very old, he could not envision how there would have been enough time for evolution to occur. **2.** By this criterion, Cuvier's explanation of the fossil record and Lamarck's hypothesis of evolution are both scientific. Cuvier thought that species did not evolve over time. He also suggested that sudden, catastrophic events caused extinctions in particular areas and that such regions were later repopulated by a different set of species that immigrated from other areas. These assertions can be tested against the fossil record. Lamarck's principle of use and disuse can be used to make testable predictions for fossils of groups such as whale ancestors as they adapted to a new habitat. Lamarck's principle of use and disuse and his associated principle of the inheritance of acquired characteristics can also be tested directly in living organisms.

Concept Check 22.2

1. Organisms share characteristics (the unity of life) because they share common ancestors; the great diversity of life occurs because new species have repeatedly formed when descendant organisms gradually adapted to different environments, thereby becoming different from their ancestors. **2.** The fossil reptile species (or its ancestors) would most likely have colonised the New Zealand Alps from within New Zealand, whereas ancestors of reptiles currently found in European mountains would most likely have colonised those mountains from other parts of Europe. As a result, the New Zealand fossil species would share a more recent common ancestor with New Zealand reptiles than with reptiles in Europe. Thus, for many of its traits, the fossil reptile species would probably more closely resemble reptiles that live in the lowland forests of New Zealand than reptiles that live on European mountains. It is also possible, however, that the New Zealand fossil reptile could resemble a reptile living in the European mountains because similar environments had selected for similar adaptations (even though the fossil and European species were only distantly related to one another).
3. As long as the white phenotype (encoded by the genotype *pp*) continues to be favoured by natural selection, the proportion of white individuals in the population should increase over time relative to the proportion of purple individuals (encoded by the genotypes *PP* and *Pp*). As a result, the frequency of the *p* allele in the population would likely increase over time.

Concept Check 22.3

1. An environmental factor such as a drug does not create new traits, such as drug resistance, but rather selects for traits among those that are already present in the population. **2.** (a) Despite their different functions, the forelimbs of different mammals are structurally similar because they all represent modifications of a structure found in the common ancestor; thus, they are homologous structures. (b) In this case, the similar features of these mammals represent analogous features that arose by convergent evolution. The similarities between the sugar glider and flying squirrel indicate that similar environments selected for similar adaptations despite different ancestry. **3.** At the time that dinosaurs originated, Earth's landmasses formed a single large continent, Pangaea. Because many dinosaurs were large and mobile, it is likely that early members of these groups lived on many different parts of Pangaea. When Pangaea broke apart, fossils of these organisms would have moved with the rocks in which they were deposited. As a result, we would predict that fossils of early dinosaurs would have a broad geographic distribution (this prediction has been upheld).

Summary of Key Concepts Questions

Concept 22.1 Darwin thought that descent with modification occurred as a gradual, steplike process. The age of Earth was important to him because if Earth were only a few thousand years old (as conventional wisdom suggested), there wouldn't have been sufficient time for major evolutionary change. **Concept 22.2** All species have the potential to overreproduce—that is, to produce more offspring than can be supported by the environment. This ensures that there will be what Darwin called a "struggle for existence" in which many of the offspring are eaten, starved, diseased, or unable to reproduce for a variety of other reasons. Members of a population exhibit a range of heritable variations, some of which make it likely that their bearers will leave more offspring than other individuals (for example, the bearer may escape predators more effectively or be more tolerant of the physical conditions of the environment). Over time, natural selection resulting from factors such as predators, lack of food, or the physical conditions of the environment can increase the proportion of individuals with favourable traits in a population (evolutionary adaptation). **Concept 22.3** The hypothesis that cetaceans originated from a terrestrial mammal and are closely related to even-toed ungulates is supported by several lines of evidence. For example, fossils document that early cetaceans had hind limbs, as expected for organisms that descended from a land mammal; these fossils also show that cetacean hind limbs became reduced over time. Other fossils show that early cetaceans had a type of ankle bone that is otherwise found only in even-toed ungulates, providing strong evidence that even-toed ungulates are the land mammals to which cetaceans are most closely related. DNA sequence data also indicate that even-toed ungulates are the land mammals to which cetaceans are most closely related.

Test Your Understanding

1. C **2.** D **3.** C **4.** A **5.** B
7. (a)

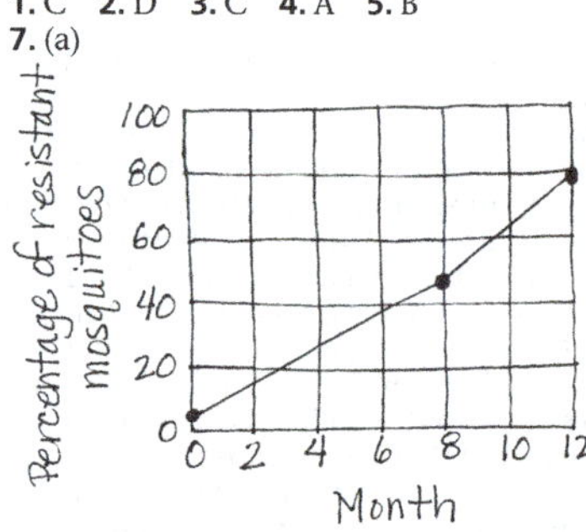

(b) The rapid rise in the percentage of mosquitoes resistant to DDT was most likely caused by natural selection in which mosquitoes resistant to DDT could survive and reproduce while other mosquitoes not resistant to DDT could not. (c) In India—where DDT resistance first appeared—natural selection would have caused the frequency of resistant mosquitoes to increase over time. If resistant mosquitoes then migrated from India (for example, transported by wind or in planes, trains, or ships) to other parts of the world where DDT was being used, the frequency of DDT resistance would increase there as well. In addition, if resistance to DDT were to arise independently in mosquito populations outside of India, those populations would also experience an increase in the frequency of DDT resistance.

Chapter 23

Figure Questions

Figure 23.4 The genetic code is redundant, meaning that more than one codon can specify the same amino acid. As a result, a substitution at a particular site in a coding region of the *Adh* gene might change the codon but not the translated amino acid, and thus not the resulting protein encoded by the gene. One way an insertion in an exon would not affect the gene produced is if it occurs in an untranslated region of the exon. (This is the case for the insertion at location 1,703.) **Figure 23.8** There should be 24 red balls. **Figure 23.9** The predicted frequencies are 36% C^RC^R, 48% C^RC^W, and 16% C^WC^W. **Figure 23.10** Overall, by chance the frequency of the C^W allele first increases in generation 2 and then falls to zero in generation 3—causing the C^R allele to become fixed (reach a frequency of 100%). **Figure 23.13** It would be expected that cold tolerance would increase in the southern population due to a barrier to gene flow from warm-adapted northern populations into southern populations. If cold tolerance is what is limiting the southern limit of the species range, it would be expected that, with an increase in cold tolerance in the southern population, there would be an southward expansion of the species range. **Figure 23.14** Directional selection. The seeds of golden rain tree are buried less deeply than are the seeds of the native host, balloon vine. Thus, in soapberry bug populations feeding on golden rain tree, bugs with shorter beaks had an advantage, resulting in directional selection for shorter beak length. **Figure 23.17** Crossing a single female's eggs with both an SC and an LC male's sperm allowed the researchers to directly compare the effects of the males' contribution to the next generation since both batches of offspring had the same maternal contribution. This isolation of the male's impact enabled researchers to draw conclusions about differences in genetic "quality" between the SC and LC males. **Figure 23.19** Under prolonged low-oxygen conditions, some of the red blood cells of a heterozygote may sickle, leading to harmful effects. This does not occur in individuals with two wild-type haemoglobin alleles, suggesting selection against heterozygotes in malaria-free regions (where heterozygote advantage does not occur). However, since heterozygotes are healthy under most conditions, selection against them is unlikely to be strong.

Concept Check 23.1

1. Within a population, genetic differences among individuals provide the raw material on which natural selection and other mechanisms can act. Without such differences, allele frequencies could not change over time—and hence the population could not evolve. **2.** Many mutations occur in somatic cells, which do not produce gametes and so are lost when the organism dies. Of the mutations that do occur in cell lines that produce gametes, many do not have a phenotypic effect on which natural selection can act. Others have a harmful effect and are thus unlikely to increase in frequency because they decrease the reproductive success of their bearers. **3.** Its genetic variation (whether measured at the level of the gene or at the level of nucleotide sequences) would probably drop over time. During meiosis, crossing over and the independent assortment of chromosomes produce many new combinations of alleles. In addition, a population contains a vast number of possible mating combinations, and fertilisation brings together the gametes of individuals with different genetic backgrounds. Thus, via crossing over, independent assortment of chromosomes, and fertilisation, sexual reproduction reshuffles alleles into fresh combinations each generation. Without sexual reproduction, the rate of forming new combinations of alleles would be vastly reduced, causing the overall amount of genetic variation to drop.

Concept Check 23.2

1. There are 700 individuals in the population: 85 of genotype *AA*, 320 of genotype *Aa*, and 295 of genotype *aa*. The genotype frequencies are thus 0.12 (85/700) for genotype *AA*, 0.46 (320/700) for genotype *Aa*, and 0.42 (295/700) for genotype *aa*. Each individual has two alleles, so the total number of alleles is 1,400. To calculate the frequency of allele *A*, note that each of the 85 individuals of genotype *AA* has two *A* alleles, each of the 320 individuals of genotype *Aa* has one *A* allele, and each of the 295 individuals of genotype *aa* has zero *A* alleles. Thus, the frequency (p) of allele *A* is

$$p = \frac{(2 \times 85) + (1 \times 320) + (0 \times 295)}{1{,}400} = 0.35$$

There are only two alleles (*A* and *a*) in our population, so the frequency of allele *a* must be $q = 1 - p = 0.65$. **2.** Because the frequency of allele *a* is 0.45, the frequency of allele *A* must be 0.55. Thus, the expected genotype frequencies are $p^2 = 0.3025$ for genotype *AA*, $2pq = 0.495$ for genotype *Aa*, and $q^2 = 0.2025$ for genotype *aa*. **3.** There are 120 individuals in the population, so there are 240 alleles. Of these, there are 124 *V* alleles—32 from the 16 *VV* individuals and 92 from the 92 *Vv* individuals. Thus, the frequency of the *V* allele is $p = 124/240 = 0.52$; hence, the frequency of the *v* allele is $q = 0.48$. Based on the Hardy-Weinberg equation, if the population were not evolving, the frequency of genotype *VV* should be $p^2 = 0.52 \times 0.52 = 0.27$; the frequency of genotype *Vv* should be $2pq = 2 \times 0.52 \times 0.48 = 0.5$; and the frequency of genotype *vv* should be $q^2 = 0.48 \times 0.48 = 0.23$. In a population of 120 individuals, these expected genotype frequencies lead us to predict that there would be 32 *VV* individuals (0.27×120), 60 *Vv* individuals (0.5×120), and 28 *vv* individuals (0.23×120). The actual numbers for the population (16 *VV*, 92 *Vv*, 12 *vv*) deviate from these expectations (fewer homozygotes and more heterozygotes than expected). This indicates that the population is not in Hardy-Weinberg equilibrium and hence may be evolving at this locus.

Concept Check 23.3

1. Natural selection is more "predictable" in that it alters allele frequencies in a nonrandom way: It tends to increase the frequency of alleles that increase the organism's reproductive success in its environment and decrease the frequency of alleles that decrease the organism's reproductive success. Alleles subject to genetic drift increase or decrease in frequency by chance alone, whether or not they are advantageous. **2.** Genetic drift results from chance events that cause allele frequencies to fluctuate at random from generation to generation; within a population, this process tends to decrease genetic variation over time. Gene flow is the transfer of alleles between populations, a process that can introduce new alleles to a population and hence may increase its genetic variation (albeit slightly, since rates of gene flow are often low). **3.** Selection is not important at this locus; furthermore, the populations are not small, and hence the effects of genetic drift should not be pronounced. Gene flow is occurring via the movement of pollen and seeds. Thus, allele and genotype frequencies in these populations should become more similar over time as a result of gene flow.

Concept Check 23.4

1. The relative fitness of a mule is zero, because fitness includes reproductive contribution to the next generation, and a sterile mule cannot produce offspring. **2.** Although both gene flow and genetic drift can increase the frequency of advantageous alleles in a population, they can also decrease the frequency of advantageous alleles or increase the frequency of harmful alleles. Only natural selection *consistently* results in an increase in the frequency of alleles that enhance survival or reproduction. Thus, natural selection is the only mechanism that consistently leads to adaptive evolution. **3.** The three modes of natural selection (directional, stabilising, and disruptive) are defined in terms of the selective advantage of different *phenotypes*, not different genotypes. Thus, the type of selection represented by heterozygote advantage depends on the phenotype of the heterozygotes. In this question, because heterozygous individuals have a more extreme phenotype than either homozygote, heterozygote advantage represents directional selection.

Summary of Key Concepts Questions

23.1 Much of the nucleotide variability at a genetic locus occurs within introns. Nucleotide variation at these sites typically does not affect the phenotype because introns do not code for the protein product of the gene. (Note: In certain circumstances, it is possible that a change in an intron could affect RNA splicing and ultimately have some phenotypic effect on the organism, but such mechanisms are not covered in this introductory text.) There are also many variable nucleotide sites within exons. However, most of the variable sites within exons reflect changes to the DNA sequence that do not change the sequence of amino acids encoded by the gene (and hence may not affect the phenotype). **23.2** No, this is not an example of circular reasoning. Calculating p and q from observed genotype frequencies does not imply that those genotype frequencies must be in Hardy-Weinberg equilibrium. For example, consider a population that has 195 individuals of genotype *AA*, 10 of genotype *Aa*, and 195 of genotype *aa*. Calculating p and q from these values yields $p = q = 0.5$. Using the Hardy-Weinberg equation, the predicted equilibrium frequencies are $p^2 = 0.25$ for genotype *AA*, $2pq = 0.5$ for genotype *Aa*, and $q^2 = 0.25$ for genotype *aa*. Since there are 400 individuals in the population, these predicted genotype frequencies indicate that there should be 100 *AA* individuals, 200 *Aa* individuals, and 100 *aa* individuals—numbers that differ greatly from the values that we used to calculate p and q. **23.3** It is unlikely that two such populations would evolve in similar ways. Since their environments are very different, the alleles favoured by natural selection would probably differ between the two populations. Although genetic drift may have important effects in each of these small populations, drift causes unpredictable changes in allele frequencies, so it is unlikely that drift would cause the populations to evolve in similar ways. Both populations are geographically isolated, suggesting that little gene flow would occur between them (again making it less likely that they would evolve in similar ways). **23.4** Compared to males, it is likely that the females of such species would be larger, more colourful, endowed with more elaborate ornamentation (for example, a large morphological feature such as the peacock's tail), and more apt to engage in behaviours intended to attract mates or prevent other members of their sex from obtaining mates.

Test Your Understanding

1. D **2.** C **3.** B **4.** A **5.** C

Chapter 24

Figure Questions

Figure 24.7 If this had not been done, the strong preference of "starch flies" and "maltose flies" to mate with like-adapted flies could have occurred simply because the flies could detect (for example, by sense of smell) what their potential mates had eaten as larvae—and preferred to mate with flies that had a similar smell to their own. **Figure 24.11** *Tragopogon dubius* and *T. pratenis* are the parent species of the polyploid species *T. miscellus*. *T. dubius* and *T. porrifolius* are the parent species of the other polyploid species, *T. mirus*. **Figure 24.12** In murky waters where females distinguish colours poorly, females of each species might mate often with males of the other species. Hence, since hybrids between these species are viable and fertile, the gene pools of the two species might become more similar over time.
Figure 24.13 The graph indicates that there has been gene flow of some fire-bellied toad alleles into the range of the yellow-bellied toad. Otherwise, all individuals located to the left of the hybrid zone portion of the graph would have allele frequencies equal to 1. **Figure 24.15** Because the populations had only just begun to diverge from one another at this point in the process, it is likely that any existing barriers to reproduction would weaken over time. **Figure 24.17** Over time, the chromosomes of the experimental hybrids came to resemble those of *H. anomalus*. This occurred even though conditions in the laboratory differed greatly from conditions in the field, where *H. anomalus* is found, suggesting that selection for laboratory conditions was not strong. Thus, it is unlikely that the observed rise in the fertility of the experimental hybrids was due to selection for life under laboratory conditions. **Figure 24.18** The presence of *M. cardinalis* plants that carry the *M. lewisii yup* allele would make it more likely that bumblebees would transfer pollen between the two monkey flower species. As a result, we would expect the number of hybrid offspring to increase.

Concept Check 24.1

1. (a) All except the biological species concept can be applied to both asexual and sexual species because they define species on the basis of characteristics other than the ability to reproduce. In contrast, the biological species concept can be applied only to sexual species. (b) The easiest species concept to apply in the field would be the morphological species concept because it is based only on the appearance of the organism. Additional information about its ecological habits or reproduction is not required. **2.** Because these birds live in fairly similar environments and can breed successfully in captivity, the reproductive barrier in nature is probably prezygotic; given the species' differences in habitat preference, this barrier could result from habitat isolation.

Concept Check 24.2

1. In allopatric speciation, a new species forms while in geographic isolation from its parent species; in sympatric speciation, a new species forms when both species are present in the absence of geographic isolation. Geographic isolation greatly reduces gene flow between populations, whereas ongoing gene flow is more likely in sympatric populations. As a result, allopatric speciation is more common than sympatric speciation. **2.** Gene flow between subsets of a population that live in the same area can be reduced in a variety of ways. In some species—especially plants—changes in chromosome number can block gene flow and establish reproductive isolation in a single generation. Gene flow can also be reduced in sympatric populations by habitat differentiation (as seen in the apple maggot fly, *Rhagoletis*) and sexual selection (as seen in Lake Victoria cichlids). **3.** Allopatric speciation would be less likely to occur on an island near a mainland than on a more isolated island of the same size. We expect this result because continued gene flow between mainland populations and those on a nearby island reduces the chance that enough genetic divergence will take place for allopatric speciation to occur. **4.** If all of the homologues failed to separate during anaphase I of meiosis, some gametes would end up with an extra set of chromosomes (and others would end up with no chromosomes). If a gamete with an extra set of chromosomes fused with a normal gamete, a triploid would result; if two gametes with an extra set of chromosomes fused with each other, a tetraploid would result.

Concept Check 24.3

1. Hybrid zones are regions in which members of different species meet and mate, producing some offspring of mixed ancestry. Such regions can be viewed as "natural laboratories" in which to study speciation because scientists can directly observe factors that cause (or fail to cause) reproductive isolation.
2. (a) If hybrids consistently survived and reproduced poorly compared with the offspring of intraspecific matings, reinforcement could occur. If it did, natural selection would cause prezygotic barriers to reproduction between the parent species to strengthen over time, decreasing the production of unfit hybrids and leading to a completion of the speciation process. If reinforcement did not occur, hybrids may continue to be produced even though they are selected against (as in the *Bombina* hybrid zone). (b) If hybrid offspring survived and reproduced as well as the offspring of intraspecific matings, indiscriminate mating between the parent species would lead to the production of large numbers of hybrid offspring. As these hybrids mated with each other and with members of both parent species, the gene pools of the parent species could fuse over time, reversing the speciation process.

Concept Check 24.4

1. The time between speciation events includes (1) the length of time that it takes for populations of a newly formed species to begin diverging reproductively from one another and (2) the time it takes for speciation to be complete once this divergence begins. Although speciation can occur rapidly once populations have begun to diverge from one another, it may take millions of years for that divergence to begin. **2.** Investigators transferred alleles at the *yup* locus (which influences flower colour) from each parent species to the other. *M. lewisii* plants with an *M. cardinalis yup* allele received many more visits from hummingbirds than usual; hummingbirds usually pollinate *M. cardinalis* but avoid *M. lewisii*. Similarly, *M. cardinalis* plants with an *M. lewisii yup* allele received many more visits from bumblebees than usual; bumblebees usually pollinate *M. lewisii* and avoid *M. cardinalis*. Thus, alleles at the *yup* locus can influence pollinator choice, which in these species provides the primary barrier to interspecific mating. Nevertheless, the experiment does not prove that the *yup* locus alone controls barriers to reproduction between *M. lewisii* and *M. cardinalis*; other genes might enhance the effect of the *yup* locus (by modifying flower colour) or cause entirely different barriers to reproduction (for example, gametic isolation or a postzygotic barrier). **3.** Crossing over. If crossing over did not occur, each chromosome in an experimental hybrid would remain as in the F_1 generation: composed entirely of DNA from one parent species or the other.

Summary of Key Concepts Questions

24.1 According to the biological species concept, a species is a group of populations whose members interbreed and produce viable, fertile offspring; thus, gene flow occurs between populations of a species. In contrast, members of different species do not interbreed and hence no gene flow occurs between their populations. Overall, then, in the biological species concept, species can be viewed as designated by the *absence* of gene flow—making gene flow of central importance to the biological species concept. **24.2** Sympatric speciation can be promoted by factors such as polyploidy, sexual selection, and habitat shifts, all of which can reduce gene flow between the subpopulations of a larger population. Of these factors, sexual selection and habitat shifts can also occur in allopatric populations and hence can also promote allopatric speciation. **24.3** If the hybrids are selected against, the hybrid zone could persist if individuals from the parent species regularly travel into the zone, where they mate to produce hybrid offspring. If hybrids are not selected against, there is no cost to the continued production of hybrids, and large numbers of hybrid offspring may be produced. However, natural selection for life in different environments may keep the gene pools of the two parent species distinct, thus preventing the loss (by fusion) of the parent species and once again causing the hybrid zone to be stable over time. **24.4** As the goatsbeard plant, Bahamas mosquitofish, and apple maggot fly illustrate, speciation continues to happen today. A new species can begin to form whenever gene flow is reduced between populations of the parent species. Such reductions in gene flow can occur in many ways: A new, geographically isolated population may be founded by a few colonists; some members of the parent species may begin to utilise a new habitat; or sexual selection may isolate formerly connected populations or subpopulations. These and many other such events are happening today.

Test Your Understanding

1. B **2.** C **3.** B **4.** A **5.** D **6.** C
8. Here is one possibility:

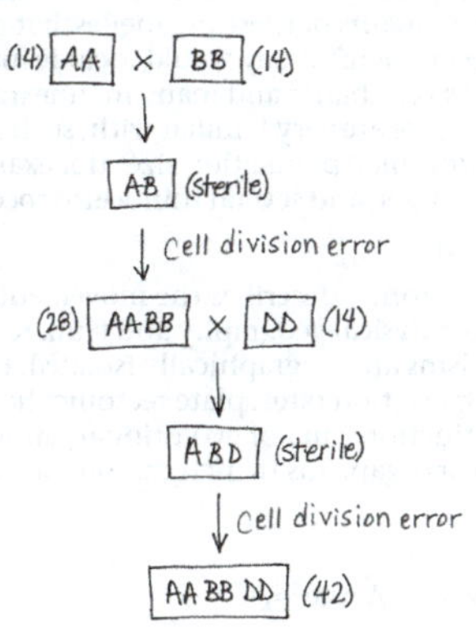

Chapter 25

Figure Questions

Figure 25.2 Proteins are almost always composed of the same 20 amino acids shown in Figure 5.14. However, many other amino acids could potentially form in this or any other experiment. For example, any molecule that had an R group that differed from those listed in Figure 5.14 would still be an amino acid as long as it also contained an α carbon, an amino group, and a carboxyl group—but that molecule would not be one of the 20 amino acids commonly found in nature. **Figure 25.4** The hydrophobic regions of such molecules are attracted to one another and excluded from water, whereas the hydrophilic regions have an affinity for water. As a result, the molecules can form a bilayer in which the hydrophilic regions are on the outside of the bilayer (facing water on each side of the bilayer) and the hydrophobic regions point towards each other (that is, towards the inside of the bilayer). **Figure 25.6** Because uranium-238 has a half-life of 4.5 billion years, the *x*-axis would be relabelled (in billions of years) as 4.5, 9, 13.5, and 18. **Figure 25.8** (1) The earliest direct evidence of life comes from fossils of prokaryotes that date to 3.5 billion years ago. Fossil evidence also shows that for the next 2 billion years (3.5 to 1.5 billion years ago), life on Earth consisted entirely of unicellular organisms. In fact, from 3.5 billion years ago to 1.8 billion years ago, all of Earth's organisms were prokaryotes; around 1.8 billion years ago, these unicellular prokaryotes were joined by unicellular eukaryotes (multicellular eukaryotes emerged about 1.3 billion years ago). (2) Sample answer: In Figure 25.12, there are two hatch marks on the *x*-axis. These hatch marks represent large time spans. Unless the graph was very wide, showing these entire time spans would obscure key details for the figure, such as when selected animal groups first appear in the fossil record. (3) The horizontal time scale indicates that prokaryotes originated 3.5 billion years ago and that the colonisation of land took place 500 million years ago. On a 1-hour time scale, this indicates that prokaryotes appeared about 46 minutes ago, while the colonisation of land took place less than 7 minutes ago. **Figure 25.12** You should have circled the node, shown in the tree diagram at approximately 635 million years ago (mya), that leads to the echinoderm/chordate lineage and to the lineage that gave rise to brachiopods, annelids, molluscs, and arthropods. To determine a minimum estimate of the age of the ancestor represented by this node, note that the most recent common ancestor of chordates and annelids must be at least as old as any of its descendants. Since fossil molluscs date to about 560 mya, the common ancestor represented by the circled branch point must be at least 560 million years old. **Figure 25.14** There are two speciation events and five extinctions in lineage A, while there are five speciation events and one extinction in lineage B during the last 2 million years. **Figure 25.18** The Australian plate's current direction of movement is roughly similar to the northeasterly direction the continent travelled over the past 66 million years. **Figure 25.45** The coding sequence of the *Pitx1* gene would differ between the marine and lake populations, but patterns of gene expression would not.

Concept Check 25.1

1. The hypothesis that conditions on early Earth could have permitted the synthesis of organic molecules from inorganic ingredients **2.** In contrast to random mingling of molecules in an open solution, segregation of molecular systems by membranes could concentrate organic molecules, assisting biochemical reactions. **3.** Today, genetic information usually flows from DNA to RNA, as when the DNA sequence of a gene is used as a template to synthesise the mRNA encoding a particular protein. However, the life cycle of retroviruses such as HIV shows that genetic information can flow in the reverse direction (from RNA to DNA). In these viruses, the enzyme reverse transcriptase uses RNA as a template for DNA synthesis, suggesting that a similar enzyme could have played a key role in the transition from an RNA world to a DNA world.

Concept Check 25.2

1. The fossil record shows that different groups of organisms dominated life on Earth at different points in time and that many organisms once alive are now extinct; specific examples of these points can be found in Figure 25.5. The fossil record also indicates that new groups of organisms can arise via the gradual modification of previously existing organisms, as illustrated by fossils that document the origin of mammals from their cynodont ancestors (see Figure 25.7). **2.** 22,920 years (four half-lives: $5{,}730 \times 4$)

Concept Check 25.3

1. Free oxygen attacks chemical bonds and can inhibit enzymes and damage cells. As a result, the appearance of oxygen in the atmosphere probably caused many prokaryotes that had thrived in anaerobic environments to survive and reproduce poorly, ultimately driving many of these species to extinction. **2.** All eukaryotes have mitochondria or remnants of these organelles, but not all eukaryotes have plastids. **3.** A fossil record of life today would include many organisms with hard body parts (such as vertebrates and many marine invertebrates), but might not include some species we are very familiar with, such as those that have small geographic ranges and/or small population sizes (for example, endangered species such as the giant panda, tiger, and several rhinoceros species).

Concept Check 25.4

1. The theory of plate tectonics describes the movement of Earth's continental plates, which alters the physical geography and climate of Earth, as well as the extent to which organisms are geographically isolated. Because these factors affect extinction and speciation rates, plate tectonics has a major impact on life on Earth. **2.** Mass extinctions; major evolutionary innovations; the diversification of another group of organisms (which can provide new sources of food); migration to new locations where few competitor species exist **3.** Evidence from previous mass extinctions indicates that the diversity of life on Earth would not recover, for millions of years, to what it had been before the mass extinction—a much greater period of time than our species has been in existence (about 200,000 years). Although new speciation events would eventually cause the total number of species on Earth to recover, the many species and evolutionary lineages driven to extinction would be gone forever, thus forever changing the course of evolution on our planet. In addition, previous evidence suggests that a sixth mass extinction would reduce thriving and complex ecological communities (such as forests and coral reefs) so greatly that they might hardly resemble what they are like now. A sixth mass extinction would also change the types of organisms that live in ecological communities and how those organisms interact with one another. Finally, a sixth mass extinction would pave the way for new adaptive radiations in some of the groups that survive the extinction.

Concept Check 25.5

1. Heterochrony can cause a variety of morphological changes. For example, if the timing of the onset of sexual maturity changes, retention of juvenile characteristics (paedomorphosis) may result. Paedomorphosis can be caused by small genetic changes that result in large changes in morphology, as seen in the axolotl salamander. **2.** In animal embryos, *Hox* genes influence the development of structures such as limbs and feeding appendages. As a result, changes in these genes—or in the regulation of these genes—are likely to have major effects on morphology. **3.** From genetics, we know that gene regulation is altered by how well transcription factors bind to noncoding DNA sequences called control elements. Thus, if changes in morphology are often caused by changes in gene regulation, portions of noncoding DNA that contain control elements are likely to be strongly affected by natural selection.

Concept Check 25.6

1. Complex structures do not evolve all at once, but in increments, with natural selection selecting for adaptive variants of the earlier versions. **2.** Although the myxoma virus is highly lethal, initially some of the rabbits are resistant (0.2% of infected rabbits are not killed). Thus, assuming resistance is an inherited trait, we would expect the rabbit population to show a trend for increased resistance to the virus. We would also expect the virus to show an evolutionary trend towards reduced lethality. We would expect this trend because a rabbit infected with a less lethal virus would be more likely to live long enough for a mosquito to bite it and hence potentially transmit the virus to another rabbit. (A virus that kills its rabbit host before a mosquito transmits the virus to another rabbit dies with its host.)

Summary of Key Concepts Questions

Concept 25.1 Particles of montmorillonite clay may have provided surfaces on which organic molecules became concentrated and hence were more likely to react with one another. Montmorillonite clay particles may also have facilitated the transport of key molecules, such as short strands of RNA, into vesicles. These vesicles can form spontaneously from simple precursor molecules, "reproduce" and "grow" on their own, and maintain internal concentrations of molecules that differ from those in the surrounding environment. These features of vesicles represent key steps in the emergence of protocells and (ultimately) the first living cells. **Concept 25.2** One challenge is that radioisotopes with very long half-lives are not used by organisms to build their bones or shells. As a result, fossils older than 75,000 years cannot be dated directly. Fossils are often found in sedimentary rock, but those rocks typically contain sediments of different ages, again posing a challenge when trying to date old fossils. To circumvent these challenges, geologists use radioisotopes with long half-lives to date layers of volcanic rock that surround old fossils. This approach provides minimum and maximum estimates for the ages of fossils sandwiched between two layers of volcanic rock. **Concept 25.3** The "Cambrian explosion" refers to a relatively short interval of time (535–525 million years ago) during which large forms of many present-day animal phyla first appear in the fossil record. The evolutionary changes that occurred during this time, such as the appearance of large predators and well-defended prey, were important because they set the stage for many of the key events in the history of life over the last 500 million years. **Concept 25.4** The broad evolutionary changes documented by the fossil record reflect the rise and fall of major groups of organisms. In turn, the rise or fall of any particular group results from a balance between speciation and extinction rates: A group increases in size when the rate at which its members produce new species is greater than the rate at which its member species are lost to extinction, while a group shrinks in size if extinction rates are greater than speciation rates.
Concept 25.5 A change in the sequence or regulation of a developmental gene can produce major morphological changes. In some cases, such morphological changes may enable organisms to perform new functions or live in new environments—thus potentially leading to an adaptive radiation and the formation of a new group of organisms. **Concept 25.6** Evolutionary change results from interactions between organisms and their current environments. No goal is involved in this process. As environments change over time, the features of organisms favoured by natural selection may also change. When this happens, what once may have seemed like a "goal" of evolution (for example, improvements in the function of a feature previously favoured by natural selection) may cease to be beneficial or may even be harmful.

Test Your Understanding

1. B **2.** A **3.** D **4.** B **5.** D **6.** C **7.** A

Chapter 26

Figure Questions

Figure 26.5 (1) In this tree, frogs are most closely related to a group consisting of lizards, chimps, and humans. (2) You should have circled the branch point splitting the frog lineage from the lineage leading to lizards, chimps, and humans. (3) Four: chimps–humans, lizards–chimps/humans; frogs–lizards/chimps/humans; and fishes–frogs/lizards/chimps/humans.

(4)

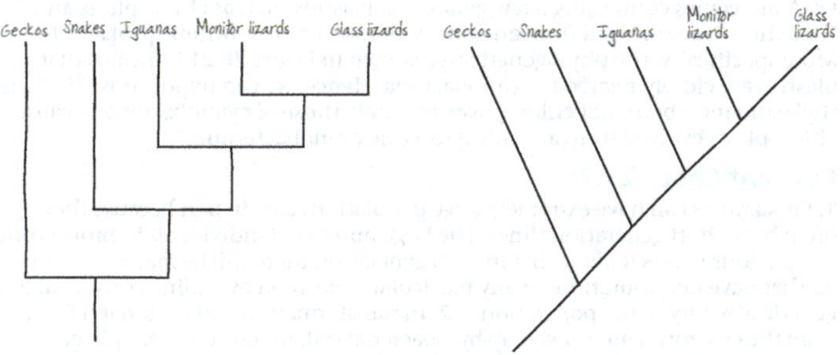

(5)

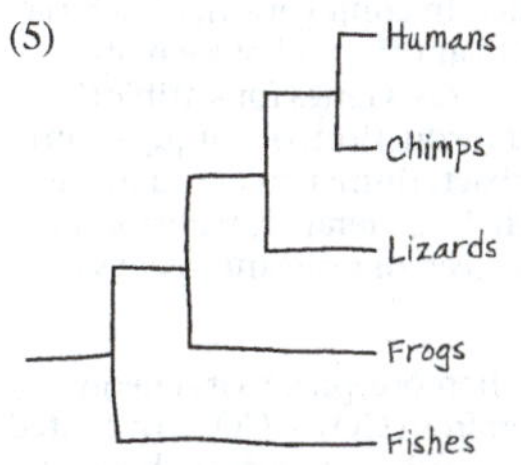

Each of the three trees identifies chimps and lizards as the two closest relatives of humans in these trees because they are the groups shown with whom we share the two most recent common ancestors. **Figure 26.6** Unknown 1b (a portion of sample 1) and unknowns 9–13 all would have to be located on the branch of the tree that currently leads to Minke (Southern Hemisphere) and unknowns 1a and 2–8. **Figure 26.9** There are four possible bases (A, C, G, T) at each nucleotide position. If the base at each position depends on chance, not common descent, we would expect roughly one out of four (25%) of them to be the same. **Figure 26.11** You should have circled the branch point that is drawn farthest to the left (the common ancestor of all taxa shown). Both cetaceans and seals descended from terrestrial lineages of mammals, indicating that the cetacean–seal common ancestor lacked a streamlined body form and hence would not be part of the cetacean–seal group. **Figure 26.12** Hinged jaws are a shared ancestral character for the group that includes frogs, turtles, and leopards. Thus, you should have circled the frog, turtle, and leopard lineages, along with their most recent common ancestor. **Figure 26.16** Crocodilians are the sister taxon to the dinosaur clade (which includes birds) because crocodilians and the dinosaur clade share an immediate common ancestor that is not shared by any other group. **Figure 26.21** This tree indicates that the sequences of rRNA and other genes in mitochondria are most closely related to those of proteobacteria, while the sequences of chloroplast genes are most closely related to those of cyanobacteria. These gene sequence relationships are what would be predicted from serial endosymbiont theory, which posits that both mitochondria and chloroplasts originated as engulfed prokaryotic cells.

Concept Check 26.1

1. In Figure 26.4, leopards are the sister taxon to a group consisting of the family Mustelidae (which includes badgers) and the family Canidae (which includes wolves). Since members of a sister group are each other's closest relatives, leopards are equally related to badgers and to wolves. **2.** The tree in (c) shows a different pattern of evolutionary relationships. In (c), C and B are sister taxa, whereas C and D are sister taxa in (a) and (b). **3.** The redrawn version of Figure 26.4 is shown below.

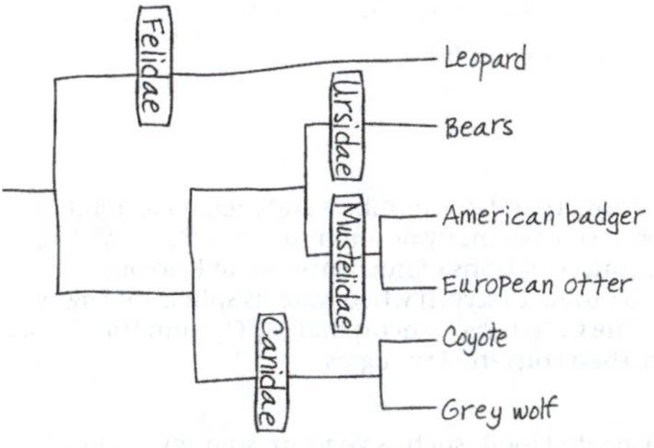

Concept Check 26.2

1. (a) Analogy, since porcupines and cacti are not closely related and since most other animals and plants do not have similar structures; (b) homology, since cats and humans are both mammals and have homologous forelimbs, of which the hand and paw are the lower part; (c) analogy, since owls and hornets are not closely related and since the structure of their wings is very different. **2.** Species B and C are more likely to be closely related. Small genetic changes (as between species B and C) can produce divergent physical appearances, but if many genes have diverged greatly (as in species A and B), then the lineages have probably been separate for a long time.

Concept Check 26.3

1. No; hair is a shared ancestral character common to all mammals and thus is not helpful in distinguishing different mammalian subgroups. **2.** The principle of maximum parsimony states that the hypothesis about nature we investigate first should be the simplest explanation found to be consistent with the facts. Actual evolutionary relationships may differ from those inferred by parsimony owing to complicating factors such as convergent evolution.
3.

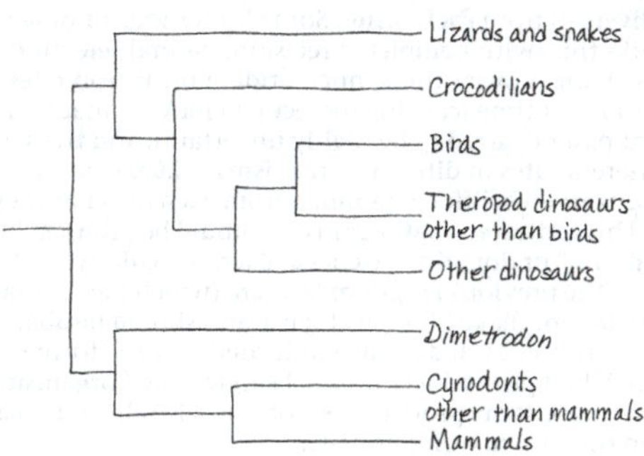

The traditional classification provides a poor match to evolutionary history, thus violating the basic principle of cladistics—that classification should be based on common descent. Both birds and mammals originated from groups traditionally designated as reptiles, making reptiles (as traditionally delineated) a paraphyletic group. These problems can be addressed by removing *Dimetrodon* and cynodonts from the reptiles and by regarding birds as a group of reptiles (specifically, as a group of dinosaurs).

Concept Check 26.4

1. Proteins are gene products. Their amino acid sequences are determined by the nucleotide sequences of the DNA that codes for them. Thus, differences between comparable proteins in two species reflect underlying genetic differences that have accumulated as the species diverged from one another. As a result, differences between the proteins can reflect the evolutionary history of the species.
2. Orthologous genes are homologous genes in which the homology results from a speciation event and hence occurs between genes found in different species. Paralogous genes are homologous genes in which the homology results from gene duplication. Orthologous genes should be used to infer phylogeny since differences among them reflect the history of speciation events. **3.** In RNA processing, the exons or coding regions of a gene can be spliced together in different ways, yielding different mRNAs and hence different protein products. As a result, different proteins could potentially be produced from the same gene in different tissues, thereby enabling the gene to perform different functions in these different tissues.

Concept Check 26.5

1. A molecular clock is a method of estimating the actual time of evolutionary events based on numbers of base changes in orthologous genes. It is based on the assumption that the regions of genomes being compared evolve at constant rates.
2. There are many portions of the genome that do not code for genes; mutations that alter the sequence of bases in such regions could accumulate through drift without affecting an organism's fitness. Even in coding regions of the genome, some mutations may not have a critical effect on genes or proteins.

Concept Check 26.6

1. The kingdom Monera included bacteria and archaea, but we now know that these organisms are in separate domains. Kingdoms are subsets of domains, so a single kingdom (like Monera) that includes taxa from different domains is not valid. **2.** Because of horizontal gene transfer, some genes in eukaryotes are more closely related to bacteria, while others are more closely related to archaea; thus, depending on which genes are used, phylogenetic trees constructed from DNA data can yield conflicting results. **3.** Eukaryotes are hypothesised to have originated when a heterotrophic prokaryote (an archaeal host cell) engulfed a bacterium that would later become an organelle found in all eukaryotes—the mitochondrion. Over time, a fusion of organisms occurred as the archaeal host cell and its bacterial endosymbiont evolved to become a single organism. As a result, we would expect the cell of a eukaryote to include both archaeal DNA and bacterial DNA, making the origin of eukaryotes an example of horizontal gene transfer.

Summary of Key Concepts Questions

26.1 The fact that humans and chimpanzees are sister species indicates that we share a more recent common ancestor with chimpanzees than we do with any other living primate species. But that does not mean that humans evolved from chimpanzees, or vice versa; instead, it indicates that both humans and chimpanzees are descendants of that common ancestor. **26.2** Homologous characters result from shared ancestry. As organisms diverge over time, some of their homologous characters will also diverge. The homologous characters of organisms that diverged long ago typically differ more than do the homologous characters of organisms that diverged more recently. As a result, differences in homologous characters can be used to infer phylogeny. In contrast, analogous characters result from convergent evolution, not shared ancestry, and hence can give misleading estimates of phylogeny. **26.3** All features of organisms arose at some point in the history of life. In the group in which a new feature first arose, that feature is a shared derived character that is unique to that clade. The group in which each shared derived character first appeared can be determined, and the resulting nested pattern can be used to infer evolutionary history. **26.4** Orthologous genes should be used; for such genes, the homology results from speciation and hence reflects evolutionary history. **26.5** A key assumption of molecular clocks is that nucleotide substitutions occur at fixed rates, and hence the number of nucleotide differences between two DNA sequences is proportional to the time since the sequences diverged from each other. Some limitations of molecular clocks: No gene marks time with complete precision; natural selection can favour certain DNA changes over others; nucleotide substitution rates can change over long periods of time (causing molecular clock estimates of when events in the distant past occurred to be highly uncertain); and the same gene can evolve at different rates in different organisms. **26.6** Genetic data indicated that many prokaryotes differed as much from each other as they did from eukaryotes. This indicated that organisms should be grouped into three "super-kingdoms," or domains (Archaea, Bacteria, Eukarya). These data also indicated that the previous kingdom Monera (which had contained all the prokaryotes) did not make biological sense and should be abandoned. Later genetic and morphological data also indicated that the former kingdom Protista (which had primarily contained single-celled organisms) should be abandoned because some protists are more closely related to plants, fungi, or animals than they are to other protists.

Test Your Understanding

1. A **2.** C **3.** B **4.** C **5.** B **6.** A **7.** D

9.

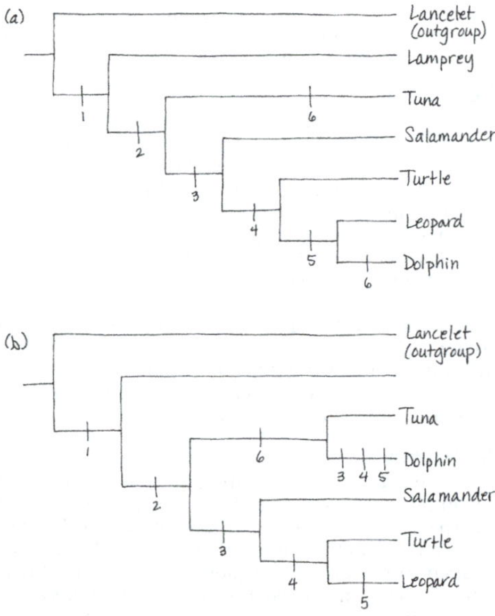

(c) The tree in (a) requires seven evolutionary changes, while the tree in (b) requires nine evolutionary changes. Thus, the tree in (a) is more parsimonious since it requires fewer evolutionary changes.

Chapter 27

Figure Questions

Figure 27.7 The top ring, to which the hook is attached, is embedded within the interior, hydrophobic portion of the lipid bilayer of the outer membrane, suggesting that the top ring is hydrophobic. Likewise, the third ring down is embedded within the hydrophobic portion of the plasma membrane's lipid bilayer, suggesting that this ring also is hydrophobic. **Figure 27.9** The third plasmid is the separate, small twisted loop located just above and to the left of the line pointing to the label "Chromosome." **Figure 27.10** It is likely that the expression or sequence of genes that affect glucose metabolism may have changed; genes for metabolic processes no longer needed by the cell also may have changed. **Figure 27.11** Transduction results in horizontal gene transfer when the host and recipient cells are members of different species. **Figure 27.16** Eukarya **Figure 27.18** Thermophiles live in very hot environments, so it is likely that their enzymes can continue to function normally at much higher temperatures than can the enzymes of other organisms. At low temperatures, however, the enzymes of thermophiles may not function as well as the enzymes of other organisms. **Figure 27.19** From the graph, plant uptake can be estimated as 0.72, 0.62, and 0.96 mg K^+ for strains 1, 2, and 3, respectively. These values average to 0.77 mg K^+. If bacteria had no effect, the average plant uptake of K^+ for strains 1, 2, and 3 should be close to 0.51 mg K^+, the value observed for plants grown in bacteria-free soil. **Figure 27.22** Penicillin.

Concept Check 27.1

1. Adaptations include the capsule (shields prokaryotes from the host's immune system) and endospores (enable cells to survive harsh conditions and to revive when the environment becomes favourable). **2.** Prokaryotic cells lack the complex compartmentalisation associated with the membrane-enclosed organelles of eukaryotic cells. Prokaryotic genomes have much less DNA than eukaryotic genomes, and most of this DNA is contained in a single ring-shaped chromosome located in the nucleoid rather than within a true membrane-enclosed nucleus. In addition, many prokaryotes also have plasmids, small ring-shaped DNA molecules containing a few genes. **3.** Plastids such as chloroplasts are thought to have evolved from an endosymbiotic photosynthetic prokaryote. More specifically, the phylogenetic tree shown in Figure 26.21 indicates that plastids are closely related to cyanobacteria. Hence, we can hypothesise that the thylakoid membranes of chloroplasts resemble those of cyanobacteria because chloroplasts evolved from an endosymbiotic cyanobacterium.

Concept Check 27.2

1. Prokaryotes can have extremely large population sizes, in part because they often have short generation times. The large number of individuals in prokaryotic populations makes it likely that in each generation there will be many individuals that have new mutations at any particular gene, thereby adding considerable genetic diversity to the population. **2.** In transformation, naked, foreign DNA from the environment is taken up by a bacterial cell. In transduction, phages carry bacterial genes from one bacterial cell to another. In conjugation, a bacterial cell directly transfers plasmid or chromosomal DNA to another cell via a mating bridge that temporarily connects the two cells. **3.** Yes. Genes for antibiotic resistance could be transferred (by transformation, transduction, or conjugation) from the nonpathogenic bacterium to a pathogenic bacterium; this could make the pathogen an even greater threat to human health. In general, transformation, transduction, and conjugation tend to increase the spread of resistance genes.

Concept Check 27.3

1. A phototroph derives its energy from light, while a chemotroph gets its energy from chemical sources. An autotroph derives its carbon from CO_2, HCO_3^-, or related compounds, while a heterotroph gets its carbon from organic nutrients such as glucose. Thus, there are four nutritional modes: photoautotrophic, photoheterotrophic (unique to prokaryotes), chemoautotrophic (unique to prokaryotes), and chemoheterotrophic. **2.** Chemoheterotrophy; the bacterium must rely on chemical sources of energy, since it is not exposed to light, and it must be a heterotroph if it requires a source of carbon other than CO_2 (or a related compound, such as HCO_3^-).
3. If humans could fix nitrogen, we could build proteins using atmospheric N_2 and hence would not need to eat high-protein foods such as meat, fish, or soy. Our diet would, however, need to include a source of carbon, along with minerals and water. Thus, a typical meal might consist of carbohydrates as a carbon source, along with fruits and vegetables to provide essential minerals (and additional carbon).

Concept Check 27.4

1. Molecular systematic studies indicate that some organisms once classified as bacteria are more closely related to eukaryotes and belong in a domain of their own: Archaea. Such studies have also shown that horizontal gene transfer is common and plays an important role in the evolution of prokaryotes. By not requiring that organisms be cultured in the laboratory, metagenomic studies have revealed an immense diversity of previously unknown prokaryotic species. Over time, the ongoing discovery of new species by metagenomic analyses may alter our understanding of prokaryotic phylogeny greatly. **2.** The three domains shown in Figure 27.16 are not valid under this assumption because domain Eukarya would be nested within domain Archaea. As such, Eukarya would be a subset of Archaea, not a separate domain of its own.

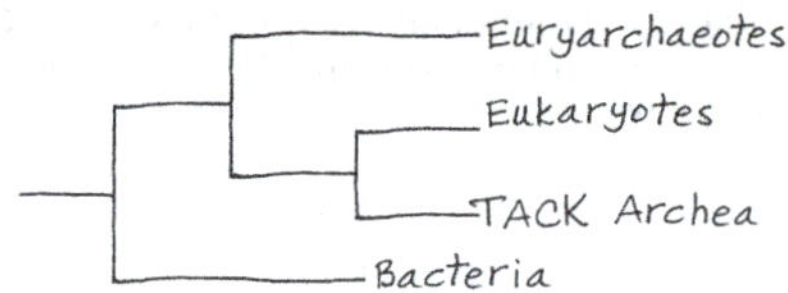

Concept Check 27.5

1. Although prokaryotes are small, their large numbers and metabolic abilities enable them to play key roles in ecosystems by decomposing wastes, recycling chemicals, and affecting the concentrations of nutrients available to other organisms. **2.** Cyanobacteria produce oxygen when water is split in the light reactions of photosynthesis. The Calvin cycle incorporates CO_2 from the air into organic molecules, which are then converted to sugars.

Concept Check 27.6

1. Sample answers: eating fermented foods such as yoghurt, sourdough bread, or cheese; receiving clean water from sewage treatment; taking medicines produced by bacteria **2.** No. If the poison is secreted as an exotoxin, live bacteria could be transmitted to another person. But the same is true if the poison is an endotoxin—only in this case, the live bacteria that are transmitted may be descendants of the (now-dead) bacteria that produced the poison. **3.** Some of the many different species of prokaryotes that live in the human gut compete with one another for resources (from the food that you eat). Because different prokaryotic species have different adaptations, a change in diet may alter which species can grow most rapidly, thus altering species abundance.

Summary of Key Concepts Questions

27.1 Specific structural features that enable prokaryotes to thrive in diverse environments include their cell walls (which provide shape and protection), flagella (which function in directed movement), and ability to form capsules or endospores (both of which can protect against harsh conditions). Prokaryotes also possess biochemical adaptations for growth in varied conditions, such as those that enable them to tolerate extremely hot or salty environments. **27.2** Many prokaryotic species can reproduce extremely rapidly, and their populations can number in the trillions. As a result, even though mutations are rare, every day many offspring are produced that have new mutations at particular gene loci. In addition, even though prokaryotes reproduce asexually and hence the vast majority of offspring are genetically identical to their parent, the genetic variation of their populations can be increased by transduction, transformation, and conjugation. Each of these (nonreproductive) processes can increase genetic variation by transferring DNA from one cell to another—even among cells that are of different species. **27.3** Prokaryotes have an exceptionally broad range of metabolic adaptations. As a group, prokaryotes perform all four modes of nutrition (photoautotrophy, chemoautotrophy, photoheterotrophy, and chemoheterotrophy), whereas eukaryotes perform only two of these (photoautotrophy and chemoheterotrophy). Prokaryotes are also able to metabolise nitrogen in a wide variety of forms (again unlike eukaryotes), and they frequently cooperate with other prokaryotic cells of the same or different species.
27.4 Phenotypic criteria such as shape, motility, and nutritional mode do not provide a clear picture of the evolutionary history of the prokaryotes. In contrast, molecular data have elucidated relationships among major groups of prokaryotes. Molecular data have also allowed researchers to sample genes directly from the environment; using such genes to construct phylogenies has led to the discovery of major new groups of prokaryotes **27.5** Prokaryotes play key roles in the chemical cycles on which life depends. For example, prokaryotes are important decomposers, breaking down corpses and waste materials, thereby releasing nutrients to the environment where they can be used by other organisms. Prokaryotes also convert inorganic compounds to forms that other organisms can use. With respect to their ecological interactions, many prokaryotes form life-sustaining mutualisms with other species. In some cases, such as hydrothermal vent communities, the metabolic activities of prokaryotes provide an energy source on which hundreds of other species depend; in the absence of the prokaryotes, the community would collapse. **27.6** Human well-being depends on our associations with mutualistic prokaryotes, such as the many species that live in our intestines and digest food that we cannot. Humans also can harness the remarkable metabolic capabilities of prokaryotes to produce a wide range of useful products and to perform key services such as bioremediation. Negative effects of prokaryotes result primarily from bacterial pathogens that cause disease.

Test Your Understanding

1. D **2.** A **3.** B **4.** C **5.** D **6.** A

Chapter 28

Figure Questions

Figure 28.3 The diagram shows that a single secondary endosymbiosis event gave rise to the stramenopiles and alveolates—thus, these groups can trace their ancestry back to a single heterotrophic protist (shown in yellow) that ingested a red alga. In contrast, euglenids and chlorarachniophytes each descended from a different heterotrophic protist (one of which is shown in grey, the other in brown). Hence, it is likely that stramenopiles and alveolates are more closely related than are euglenids and chlorarachniophytes.
Figure 28.5

Simplified tree that shows 4 supergroups:

Excavata
SAR
Archaeplastida
Unikonta

Simplified tree that shows Unikonta as sister group to all other eukaryotes:

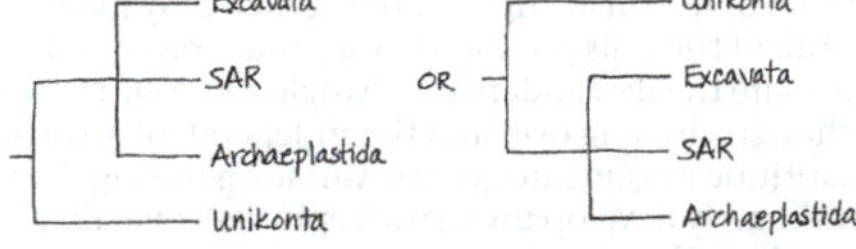

Figure 28.14 The sperm cells in the diagram are produced by the asexual (mitotic) division of cells in a single male gametophyte, which was itself produced by the asexual (mitotic) division of a single zoospore. Thus, the sperm cells are all derived from a single zoospore and so are genetically identical to one another. **Figure 28.18** Merozoites are produced by the asexual (mitotic) cell division of haploid sporozoites; similarly, gametocytes are produced by the asexual cell division of merozoites. Hence, it is likely that individuals in these three stages have the same complement of genes and that morphological differences between them result from changes in gene expression.
Figure 28.19 These events have a similar overall effect to fertilisation. In both cases, haploid nuclei that were originally from two genetically different cells fuse to form a diploid nucleus. **Figure 28.25** The following stage should be circled: step 6, where a mature cell undergoes mitosis and forms four or more daughter cells. In step 7, the zoospores eventually grow into mature haploid cells, but they do not produce new daughter cells. Likewise, in step 2, a mature cell develops into a gamete, but it does not produce new daughter cells.
Figure 28.26

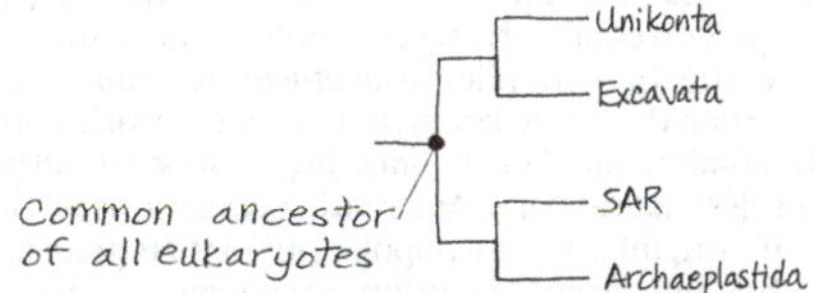

Figure 28.28 They would be haploid because originally each of these cells was a haploid, solitary amoeba.

Concept Check 28.1

1. Sample response: Protists include unicellular, colonial, and multicellular organisms; photoautotrophs, heterotrophs, and mixotrophs; species that reproduce asexually, sexually, or both ways; and organisms with diverse physical forms and adaptations. **2.** Strong evidence shows that eukaryotes acquired mitochondria after a host cell (either an archaean or a cell closely related to the archaea) first engulfed and then formed an endosymbiotic association with an alpha proteobacterium. Similarly, chloroplasts in red and green algae appear to have descended from a photosynthetic cyanobacterium that was engulfed by an ancient heterotrophic eukaryote. Secondary endosymbiosis also played an important role: Various protistan lineages acquired plastids by engulfing unicellular red or green algae. **3.** Four. The first (and primary) genome is the DNA located in the chlorarachniophyte nucleus. A chlorarachniophyte also contain remnants of a green alga's nuclear DNA, located in the nucleomorph. Finally, mitochondria and chloroplasts contain DNA from the (different) bacteria from which they evolved. These two prokaryotic genomes comprise the third and fourth genomes contained within a chlorarachniophyte.

Concept Check 28.2

1. Their mitochondria do not have an electron transport chain and so cannot function in aerobic respiration. **2.** Since the unknown protist is more closely related to diplomonads than to euglenids, it must have originated after the lineage leading to the diplomonads and parabasalids diverged from the euglenozoans. In addition, since the unknown species has fully functional mitochondria—yet both diplomonads and parabasalids do not—it is likely that the unknown species originated *before* the last common ancestor of the diplomonads and parabasalids.

Concept Check 28.3

1. Because foram tests are hardened with calcium carbonate, they form long-lasting fossils in marine sediments and sedimentary rocks. **2.** The plastid DNA would likely be more similar to the chromosomal DNA of cyanobacteria based on the well-supported hypothesis that eukaryotic plastids (such as those found in the eukaryotic groups listed) originated by an endosymbiosis event in which a eukaryote engulfed a cyanobacterium. If the plastid is derived from the cyanobacterium, its DNA would be derived from the bacterial DNA. **3.** Figure 13.6b. Algae and plants with alternation of generations have a multicellular haploid stage *and* a multicellular diploid stage. In the other two life cycles, either the haploid stage or the diploid stage is unicellular. **4.** During photosynthesis, aerobic algae produce O_2 and use CO_2. O_2 is produced as a by-product of the light reactions, while CO_2 is used as an input to the Calvin cycle (the end products of which are sugars). Aerobic algae also perform cellular respiration, which uses O_2 as an input and produces CO_2 as a waste product.

Concept Check 28.4

1. Many red algae contain a photosynthetic pigment called phycoerythrin, which gives them a reddish colour and allows them to carry out photosynthesis in relatively deep coastal water. Also unlike brown algae, red algae have no flagellated stages in their life cycle and must depend on water currents to bring gametes together for fertilisation. **2.** *Ulva* contains many cells and its body is differentiated into leaflike blades and a rootlike holdfast. *Caulerpa*'s body is composed of multinucleate filaments without cross-walls, so it is essentially one large cell. **3.** Red algae have no flagellated stages in their life cycle and hence must depend on water currents to bring their gametes together. This feature of their biology might increase the difficulty of reproducing on land. In contrast, the gametes of green algae are flagellated, making it possible for them to swim in thin films of water. In addition, a variety of green algae contain compounds in their cytoplasm, cell wall, or zygote coat that protect against intense sunlight and other terrestrial conditions. Such compounds may have increased the chance that descendants of green algae could survive on land.

Concept Check 28.5

1. Amoebozoans have lobe- or tube-shaped pseudopodia, whereas forams have threadlike pseudopodia. **2.** Slime molds are fungus-like in that they produce fruiting bodies that aid in the dispersal of spores, and they are animal-like in that they are motile and ingest food. However, slime molds are more closely related to tubulinids and entamoebas than to fungi or animals. **3.** The genes used to estimate the tree shown in Figure 28.26 were transferred from an alpha proteobacterium to an early eukaryote. Based on the sequences of these genes, the eukaryotes should be more closely related to alpha proteobacteria than they are to any other lineage of prokaryotes. Thus, alpha proteobacteria are well suited as an outgroup to the eukaryotes (the group of species whose relationships we are trying to determine).

Concept Check 28.6

1. Because photosynthetic protists constitute the base of aquatic food webs, many aquatic organisms depend on them for food, either directly or indirectly. (In addition, a substantial percentage of the oxygen produced by photosynthesis is made by photosynthetic protists.) **2.** Protists form mutualistic and parasitic associations with other organisms. Examples include photosynthetic dinoflagellates that form a mutualistic symbiosis with coral polyps; parabasalids that form a mutualistic symbiosis with termites; and the stramenopile *Phytophthora ramorum*, a parasite of oak trees. **3.** Corals depend on their dinoflagellate symbionts for nourishment, so coral bleaching could cause the corals to die. As the corals die, less food would be available for fishes and other species that eat coral. As a result, populations of these species might decline, and that, in turn, might cause populations of their predators to decline. **4.** The two approaches differ in the evolutionary changes they may bring about. A strain of *Wolbachia* that confers resistance to infection by *Plasmodium* and does not harm mosquitoes would spread rapidly through the mosquito population. In this case, natural selection would favour any *Plasmodium* individuals that could overcome the resistance to infection conferred by *Wolbachia*. If insecticides are used, mosquitoes that are resistant to the insecticide would be favoured by natural selection. Hence, use of *Wolbachia* could cause evolution in *Plasmodium* populations, while using insecticides could cause evolution in mosquito populations.

Summary of Key Concepts Questions

28.1 Sample response: Protists, plants, animals, and fungi are similar in that their cells have a nucleus and other membrane-enclosed organelles, unlike the cells of prokaryotes. These membrane-enclosed organelles make the cells of eukaryotes more complex than the cells of prokaryotes. Protists and other eukaryotes also differ from prokaryotes in having a well-developed cytoskeleton that enables them to have asymmetric forms and to change in shape as they feed, move, or grow. With respect to differences between protists and other eukaryotes, most protists are unicellular, unlike animals, plants, and most fungi. Protists also have greater nutritional diversity than other eukaryotes. **28.2** Unique cytoskeletal features are shared by many excavates. In addition, some members of Excavata have an "excavated" feeding groove for which the group was named. Moreover, recent genomic studies support the monophyly of the excavate supergroup. **28.3** Stramenopiles and alveolates are hypothesised to have originated by secondary endosymbiosis. Under this hypothesis, we can infer that the common ancestor of these two groups had a plastid, in this case of red algal origin. Thus, we would expect that apicomplexans (and alveolate or stramenopile protists) either would have plastids or would have lost their plastids over the course of evolution. **28.4** Red algae, green algae, and plants are placed in the same supergroup because considerable evidence indicates that these organisms all descended from the same ancestor, an ancient heterotrophic protist that acquired a cyanobacterial endosymbiont. **28.5** The unikonts are a diverse group of eukaryotes that includes many protists, along with animals and fungi. Most of the protists in Unikonta are amoebozoans, a clade of amoebas that have lobe- or tube-shaped pseudopodia (as opposed to the threadlike pseudopodia of rhizarians). Other protists in Unikonta include several groups that are closely related to fungi and several other groups that are closely related to animals. **28.6** Sample response: Ecologically important protists include photosynthetic dinoflagellates that provide essential sources of energy to their symbiotic partners, the corals that build coral reefs. Other important protistan symbionts include those that enable termites to digest wood and *Plasmodium*, the pathogen that causes malaria. Photosynthetic protists such as diatoms are among the most important producers in aquatic communities; as such, many other species in aquatic environments depend on them for food.

Test Your Understanding

1. D **2.** B **3.** B **4.** D **5.** D **6.** C

7.

Excavata
SAR
Ancestral prokaryote
Archaeplastida
Amoebozoans
Animals
Choanoflagellates
Unikonta
Fungi
Nucleariids

Pathogens that share a relatively recent common ancestor with humans will likely also share metabolic and structural characteristics with humans. Because drugs target the pathogen's metabolism or structure, developing drugs that harm the pathogen but not the patient should be most difficult for pathogens with whom we share the most recent evolutionary history. Working backwards in time, we can use the phylogenetic tree to determine the order in which humans shared a common ancestor with pathogens in different taxa. This process leads to the prediction that it should be hardest to develop drugs to combat animal pathogens, followed by choanoflagellate pathogens, fungal and nucleariid pathogens, amoebozoans, other protists, and finally prokaryotes.

Chapter 29

Figure Questions

Figure 29.3 The life cycles of plants and some algae, shown in Figure 13.6b, have alternation of generations; other life cycles do not. Unlike in the animal life cycle (Figure 13.6a), in the plant/algal life cycle, meiosis produces spores, not gametes. These haploid spores then divide repeatedly by mitosis, ultimately forming a multicellular haploid individual that produces gametes. There is no multicellular haploid stage in the animal life cycle. An alternation of generations life cycle also has a multicellular diploid stage, whereas the life cycle of most fungi and some protists shown in Figure 13.6c does not. **Figure 29.6** Plants, vascular plants, and seed plants are monophyletic because each of these groups includes the common ancestor of the group and all of the descendants of that common ancestor. The other two categories of plants, the nonvascular plants and the seedless vascular plants, are paraphyletic: These groups do not include all of the descendants of the group's most recent common ancestor. **Figure 29.7** Yes. As shown in the diagram, the sperm cell and the egg cell that fuse each resulted from the mitotic division of spores produced by the same sporophyte. However, these spores would differ genetically from one another because they were produced by meiosis, a cell division process that generates genetic variation among the offspring cells. **Figure 29.9** Because the moss reduces nitrogen loss from the ecosystem, species that typically colonise the soils after the moss probably experience higher soil nitrogen levels than they otherwise would. The resulting increased availability of nitrogen may benefit these species because nitrogen is an essential nutrient that often is in short supply. **Figure 29.12** A fern that had wind-dispersed sperm would not require water for fertilisation, thus removing a difficulty that ferns face when they live in arid environments. The fern would also be under strong selection to produce sperm above ground (as opposed to the current situation, where some fern gametophytes are located below ground).

Concept Check 29.1

1. Plants share some key traits only with charophytes: rings of cellulose-synthesising complexes and similarity in sperm structure. Comparisons of nuclear, chloroplast, and mitochondrial DNA sequences also indicate that certain groups of charophytes (such as *Zygnema*) are the closest living relatives of plants. **2.** Possible answers include walls toughened by sporopollenin (protects against harsh environmental conditions); multicellular, dependent embryos (provide nutrients and protection to the developing embryo); cuticle (reduces water loss); stomata (control gas exchange and reduce water loss) **3.** The multicellular diploid stage of the life cycle would not produce gametes. Instead, both males and females would produce haploid spores by meiosis. These spores would give rise to multicellular male and female haploid stages—a major change from the single-celled haploid stages (sperm and eggs) that we actually have. The multicellular haploid stages would produce gametes and reproduce sexually. An individual at the multicellular haploid stage of the human life cycle might look like us, or it might look completely different.

Concept Check 29.2

1. Most bryophytes do not have a vascular transport system, and their life cycle is dominated by gametophytes rather than sporophytes. **2.** Answers may include the following: Large surface area of protonema enhances absorption of water and minerals; the vase-shaped archegonia protect eggs during fertilisation and transport nutrients to the embryos via placental transfer cells; the stalk-like seta conducts nutrients from the gametophyte to the capsule, where spores are produced; the peristome enables gradual spore discharge; stomata enable CO_2/O_2 exchange while minimising water loss; lightweight spores are readily dispersed by wind. **3.** Effects of global warming on peatlands could result in positive feedback, which occurs when an end product of a process increases its own production. In this case, global warming is expected to lower the water levels of some peatlands. This would expose peat to air and cause it to decompose, thereby releasing stored CO_2 to the atmosphere. The release of more stored CO_2 to the atmosphere could cause additional global warming, which in turn could cause further drops in water levels, the release of still more CO_2 to the atmosphere, additional warming, and so on: an example of positive feedback.

Concept Check 29.3

1. Lycophytes have microphylls, whereas seed plants and monilophytes (ferns and their relatives) have megaphylls. Monilophytes and seed plants also share other traits not found in lycophytes, such as the initiation of new root branches at various points along the length of an existing root. **2.** Both seedless vascular plants and bryophytes have flagellated sperm that require moisture for fertilisation; this shared similarity poses challenges for these species in arid regions. With respect to key differences, seedless vascular plants have lignified, well-developed vascular tissue, a trait that enables the sporophyte to grow tall and that has transformed life on Earth (via the formation of forests). Seedless vascular plants also have true leaves and roots, which, when compared with bryophytes, provide increased surface area for photosynthesis and improve their ability to extract nutrients from soil. **3.** Three mechanisms contribute to the production of genetic variation in sexual reproduction: independent assortment of chromosomes, crossing over, and random fertilisation. If fertilisation were to occur between gametes from the same gametophyte, all of the offspring would be genetically identical. This would be the case because all of the cells produced by a gametophyte—including its sperm and egg cells—are the descendants of a single spore and hence are genetically identical. Although crossing over and the independent assortment of chromosomes would continue to generate genetic variation during the production of spores (which ultimately develop into gametophytes), overall the amount of genetic variation produced by sexual reproduction would drop.

Summary of Key Concepts Questions

29.1

Apical meristems
Liverworts
Mosses
COMMON ANCESTOR OF ALL PLANTS
Hornworts
Lycophytes
Monilophytes
Vascular tissue
Gymnosperms
Seeds
Angiosperms

29.2 Some mosses colonise bare, sandy soils, leading to the increased retention of nitrogen in these otherwise low-nitrogen environments. Other mosses harbour nitrogen-fixing cyanobacteria that increase the availability of nitrogen in the ecosystem. The moss *Sphagnum* is often a major component of deposits of peat (partially decayed organic material). Boggy regions with thick layers of peat, known as peatlands, cover broad geographic regions and contain large reservoirs of carbon. By storing large amounts of carbon—in effect, removing CO_2 from the atmosphere—peatlands affect the global climate, making them of considerable ecological importance. **29.3** Lignified vascular tissue provided the strength needed to support a tall plant against gravity, as well as a means to transport water and nutrients to plant parts located high above ground. Roots were another key trait, anchoring the plant to the ground and providing additional structural support for plants that grew tall. Tall plants could shade shorter plants, thereby outcompeting them for light. Because the spores of a tall plant disperse further than the spores of a short plant, it is also likely that tall plants could colonise new habitats more rapidly than short plants.

Test Your Understanding

1. B **2.** D **3.** C **4.** A **5.** C
6. (a) diploid; (b) haploid; (c) haploid; (d) diploid
7. Based on our current understanding of the evolution of major plant groups, the phylogeny has the four branch points shown here:

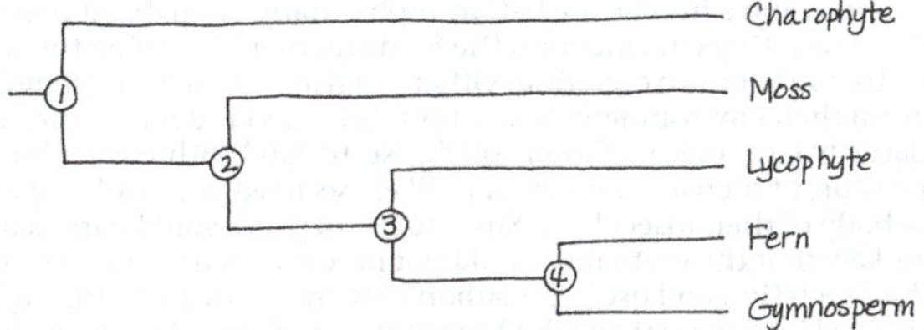

Derived characters unique to the charophyte and plant clade (indicated by branch point 1) include rings of cellulose-synthesising complexes and flagellated sperm structure. Derived characters unique to the plant clade (branch point 2) include alternation of generations; multicellular, dependent embryos; walled spores produced in sporangia; and apical meristems. Derived characters unique to the vascular plant clade (branch point 3) include life cycles with dominant sporophytes, complex vascular systems (xylem and phloem), and well-developed roots and leaves. Derived characters unique to the monilophyte and seed plant clade (branch point 4) include megaphylls and roots that can branch at various points along the length of an existing root.

Chapter 30

Figure Questions

Figure 30.2 Retaining the gametophyte within the sporophyte shields the egg-containing gametophyte from UV radiation. UV radiation is a mutagen. Hence, we would expect fewer mutations to occur in the egg cells produced by a gametophyte retained within the body of a sporophyte. Most mutations are harmful. Thus, the fitness of embryos should increase because fewer embryos would carry harmful mutations. **Figure 30.3** The seed contains cells from three generations: (1) the current sporophyte (cells of ploidy $2n$, found in the seed coat and in the megasporangium remnant that surrounds the spore wall), (2) the female gametophyte (cells of ploidy n, found in the food supply), and (3) the sporophyte of the next generation (cells of ploidy $2n$, found in the embryo). **Figure 30.4** Mitosis. A single haploid megaspore divides by mitosis to produce a multicellular, haploid female gametophyte. (Likewise, a single haploid microspore divides by mitosis to produce a multicellular male gametophyte.)

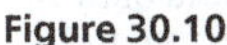
Figure 30.10

Figure 30.15 No. The branching order shown could still be correct if species on the lineages leading to basal angiosperms and magnoliids had originated prior to 150 million years ago, but fossils of that age from those lineages had not yet been discovered. In such a situation, the 140-million-year-old date for the origin of the angiosperms shown on the phylogeny would be incorrect.

Concept Check 30.1

1. To reach the eggs, the flagellated sperm of seedless plants must swim through a film of water, usually over a distance of no more than a few centimetres. In contrast, the sperm of seed plants do not require water because they are produced within pollen grains that can be transported long distances by wind or by animal pollinators. Although flagellated in some species, the sperm of seed plants do not require mobility because pollen tubes convey them from the point at which the pollen grain is deposited (near the ovules) directly to the eggs. **2.** The reduced gametophytes of seed plants are nurtured by sporophytes and protected from stress, such as drought conditions and UV radiation. Pollen grains, with walls containing sporopollenin, provide protection during transport by wind or animals. Seeds have one or two layers of protective tissue, the seed coat, that improve survival by providing more protection from environmental stresses than do the walls of spores. Seeds also contain a stored supply of food, which provides nourishment for growth after dormancy is broken and the embryo emerges as a seedling. **3.** If a seed could not enter dormancy, the embryo would continue to grow after it was fertilised. As a result, the embryo might rapidly become too large to be dispersed, thus limiting its transport. The embryo's chance of survival might also be reduced because it could not delay growth until conditions become favourable.

Concept Check 30.2

1. Although gymnosperms are similar in not having their seeds enclosed in ovaries and fruits, their seed-bearing structures vary greatly. For instance, cycads have large cones, whereas some gymnosperms, such as *Ginkgo* and *Gnetum*, have small cones that look somewhat like berries, even though they are not fruits. Leaf shape also varies greatly, from the needles of many conifers to the palmlike leaves of cycads to *Gnetum* leaves that look like those of flowering plants. **2.** The pine life cycle illustrates heterospory, as ovulate cones produce megaspores and pollen cones produce microspores. The reduced gametophytes are evident in the form of the microscopic pollen grains that develop from microspores and the microscopic female gametophyte that develops from the megaspore. The egg is shown developing within an ovule, and a pollen tube is shown conveying the sperm. The figure also shows the protective and nutritive features of a seed.
3.

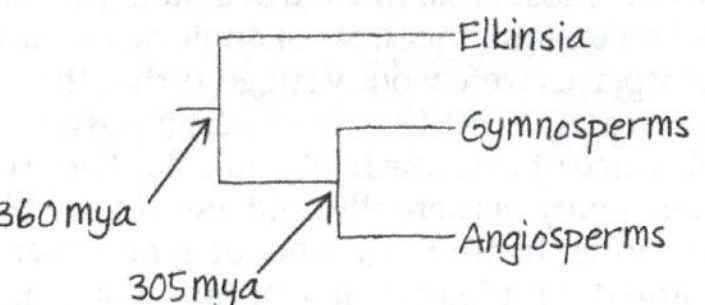

Concept Check 30.3

1. In the eucalypt's life cycle, the tree (the sporophyte) produces flowers, which contain gametophytes in pollen grains and ovules; the eggs in ovules are fertilised; the mature ovaries develop into dry fruits called acorns. We can view the eucalypt's life cycle as starting when the seeds germinate, resulting in embryos giving rise to seedlings and finally to mature trees, which produce flowers—and then more eucalypt seeds. **2.** Pine cones and flowers both have sporophylls, modified leaves that produce spores. Pine trees have separate pollen cones (with pollen grains) and ovulate cones (with ovules inside cone scales). In flowers, pollen grains are produced by the anthers of stamens, and ovules are within the ovaries of carpels. Unlike pine cones, many flowers produce both pollen and ovules. **3.** The fact that the clade with bilaterally symmetrical flowers had more species establishes a correlation between flower shape and the rate of plant speciation. Flower shape is not necessarily responsible for the result because the shape (that is, bilateral or radial symmetry) may have been correlated with another factor that was the actual cause of the observed result. Note, however, that flower shape was associated with increased speciation rates when averaged across 19 different pairs of plant lineages. Since these 19 lineage pairs were independent of one another, this association suggests—but does not establish—that differences in flower shape cause differences in speciation rates. In general, strong evidence for causation can come from controlled, manipulative experiments, but such experiments are usually not possible for studies of past evolutionary events.

Concept Check 30.4

1. While other continents and islands feature frequent natural immigration and emigration, Australia remains relatively isolated and New Zealand remains almost completely isolated from natural biological introductions. When Australia and New Zealand separated from Antarctica, they took with them a unique and limited suite of plant families. Over time, evolution of the plant families spawned many species with unique adaptations to the Australian and New Zealand environments. With rare exceptions, these species are endemic to our shores. **2.** Angiosperms possessed a reproductive system that allowed multiple generations to be produced in the time it took the gymnosperms to produce one. They possessed a more rapid reproductive cycle that produced seeds more quickly than gymnosperms. The shorter period to reach reproductive maturity among the angiosperms provided the raw materials on which natural selection could act that helped the angiosperms. **3.** Take pollinators, for example. In the absence of extensive New Zealand mammalian pollinators, birds, reptiles, and some insects filled niches occupied in other nations by mammals. New Zealand's fauna would probably have evolved different suites of relationships among plants and pollinators had a broader diversity of animals been taken into isolation aeons ago.

Concept Check 30.5

1. Plant diversity can be considered a resource because plants provide many important benefits to humans; as a resource, plant diversity is nonrenewable because if a species is lost to extinction, that loss is permanent. **2.** A detailed phylogeny of the seed plants would identify many different monophyletic groups of seed plants. Using this phylogeny, researchers could look for clades that contained species in which medicinally useful compounds had already been discovered. Identification of such clades would allow researchers to concentrate their search for new medicinal compounds among clade members—as opposed to searching for new compounds in species that were selected at random from the more than 290,000 existing species of seed plants.

Summary of Key Concepts Questions

30.1 The integument of an ovule develops into the protective coat of a seed. The ovule's megaspore develops into a haploid female gametophyte, and two parts of the seed are related to that gametophyte: The food supply of the seed is derived from haploid gametophyte cells, and the embryo of the seed develops after the female gametophyte's egg cell is fertilised by a sperm cell. A remnant of the ovule's megasporangium surrounds the spore wall that encloses the seed's food supply and embryo. **30.2** Gymnosperms arose about 305 million years ago, making them a successful group in terms of their evolutionary longevity. Gymnosperms have the five derived traits common to all seed plants (reduced gametophytes, heterospory, ovules, pollen, and seeds), making them well adapted for life on land. Finally, because gymnosperms dominate immense geographic regions today, the group is also highly successful in geographic distribution. **30.3** Based on fossils known during his lifetime, Darwin was troubled by the relatively sudden and geographically widespread appearance of angiosperms in the fossil record. Recent fossil evidence shows that angiosperms arose and began to diversify over a period of 20–30 million years, a less rapid event than was suggested by the fossils known during Darwin's lifetime. Fossil discoveries have also uncovered extinct lineages of woody seed plants thought to have been more closely related to angiosperms than to gymnosperms; one such group, the Bennettitales, had flowerlike structures that may have been pollinated by insects. In addition, phylogenetic analyses have identified a woody species, *Amborella*, as the most basal lineage of extant angiosperms. The fact that both the extinct seed plant ancestors of angiosperms and the most basal taxon of extant angiosperms were woody suggests that the common ancestor of angiosperms also was woody. **30.4** Gymnosperms possess a range of adaptations that confer tolerance of extreme heat and cold. These characteristics allow the group to dominate many perpetually cold areas of the planet. New Zealand and Australian alpine zones feature a number of gymnosperm species, as does the perpetually hot and dry mulga country of Australia. Angiosperms compete well in seasonally cold environments, through a suite of adaptations that confer a narrow range of cold tolerance, rapid growth, reproduction, and establishment during the warmer parts of the year. Angiosperms can complete their life cycle more quickly than gymnosperms. For example, some flowering plants can germinate, grow to maturity, and produce seed that germinates and grows to a mature plant in a matter of weeks. Mature gymnosperms may take—at one extreme—18 months to produce seed. Shorter generation times among the angiosperms provide multiple generations over which natural selection may favour particular traits. The longer generation times of the gymnosperms will take comparatively longer for unique, and favourable traits to become common among the population. Consequently, gymnosperms tend to perform well in consistent conditions, while angiosperms have the capacity to evolve rapidly in response to changing conditions. Each group lives, breeds, and survives in niches to which it is well adapted. **30.5** The loss of tropical forests could contribute to global warming (which would have negative effects on many human societies). People also depend on Earth's biodiversity for many products and services and hence would be harmed by the loss of species that would occur if the world's remaining tropical forests were cut down. With respect to a possible mass extinction, tropical forests harbour at least 50% of the species on Earth. If the remaining tropical forests were destroyed, large numbers of these species could be driven to extinction, thus rivaling the losses that occurred in the five mass extinction events documented in the fossil record.

Test Your Understanding

1. C **2.** B **3.** A **4.** D **5.** C

6.

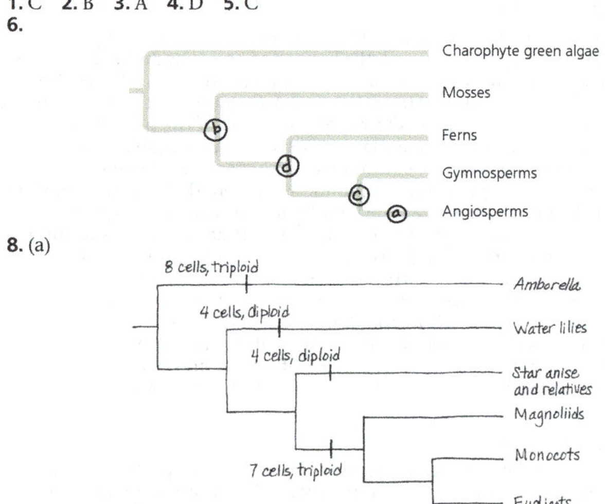

8. (a)

(b) The phylogeny indicates that basal angiosperms differed from other angiosperms in terms of the number of cells in female gametophytes and the ploidy of the endosperm. The ancestral state of the angiosperms cannot be determined from these data alone. It is possible that the common ancestor of angiosperms had seven-celled female gametophytes and triploid endosperm and hence that the eight-celled and four-celled conditions found in basal angiosperms represent derived traits for those lineages. Alternatively, either the eight-celled or four-celled condition may represent the ancestral state.

Chapter 31

Figure Questions

Figure 31.2 DNA from each of these mushrooms would be identical if each mushroom is part of a single hyphal network, as is likely. **Figure 31.5** The haploid spores produced in the sexual portion of the life cycle develop from haploid nuclei that were produced by meiosis; because genetic recombination occurs during meiosis, these spores will differ genetically from one another. In contrast, the haploid spores produced in the asexual portion of the life cycle develop from nuclei that were produced by mitosis; as a result, these spores are genetically identical to one another. **Figure 31.15** One or both of the following would apply to each species: DNA analyses would reveal that it is a member of the ascomycete clade, or aspects of its sexual life cycle would indicate that it is an ascomycete (for example, it would produce asci and ascospores). **Figure 31.16** The hypha is composed of cells that are haploid (n), as indicated by the teal-coloured arrow behind it. **Figure 31.18** The mushroom is a basidiocarp, or fruiting body, of the dikaryotic mycelium, and so a cell from its stalk would be dikaryotic ($n + n$). **Figure 31.20** Two possible controls would be E−P− and E+P−. Results from an E−P− control could be compared with results from the E−P+ experiment, and results from an E+P− control could be compared with results from the E+P+ experiment. Together, these two comparisons would indicate whether the addition of the pathogen causes an increase in leaf mortality. Results from an E−P− experiment could also be compared with results from the second control (E+P−) to determine whether adding the fungal endophytes has a negative effect on the plant.

Concept Check 31.1

1. Both a fungus and a human are heterotrophs. Many fungi digest their food externally by secreting enzymes into the food and then absorbing the small molecules that result from digestion. Other fungi absorb such small molecules directly from their environment. In contrast, humans (and most other animals) ingest relatively large pieces of food and digest the food within their bodies. **2.** The ancestors of such a mutualist most likely secreted powerful enzymes to digest the body of their insect host. Since such enzymes would harm a living host, it is likely that the mutualist would not produce such enzymes or would restrict their secretion and use. **3.** Carbon that enters the plant through stomata is fixed into sugar through photosynthesis. Some of these sugars are absorbed by the fungus that partners with the plant to form mycorrhizae; others are transported within the plant body and used in the plant. Thus, the carbon may be deposited in either the body of the plant or the body of the fungus.

Concept Check 31.2

1. The majority of the fungal life cycle is spent in the haploid stage, whereas the majority of the human life cycle is spent in the diploid stage. **2.** The two mushrooms might be reproductive structures of the same mycelium (the same organism). Or they might be parts of two separate organisms that have arisen from a single parent organism through asexual reproduction (for example, from two genetically identical asexual spores) and thus carry the same genetic information.

Concept Check 31.3

1. DNA evidence indicates that fungi, animals, and their protistan relatives form a clade, the opisthokonts. Furthermore, chytrids and other fungi thought to be members of basal lineages have posterior flagella, as do most other opisthokonts. This suggests that other fungal lineages lost their flagella after diverging from ancestors that had flagella. **2.** Mycorrhizae form extensive networks of hyphae through the soil, enabling nutrients to be absorbed more efficiently than a plant can do on its own; this is true today, and similar associations were probably very important for the earliest plants (which lacked roots). Evidence for the antiquity of mycorrhizal associations includes fossils showing arbuscular mycorrhizae in the early plant *Aglaophyton* and molecular results showing that genes required for the formation of mycorrhizae are present in liverworts and other basal plant lineages. **3.** Fungi are heterotrophs. Prior to the colonisation of land by plants, terrestrial fungi would have lived where other organisms (or their remains) were present and provided a source of food. Thus, if fungi colonised land before plants, they could have fed on prokaryotes or protists that lived on land or by the water's edge—but not on the plants or animals on which many fungi feed today.

Concept Check 31.4

1. Flagellated spores; molecular evidence also suggests that chytrids include species that belong to lineages that diverged from other fungi early in the history of the group. **2.** Possible answers include the following: In mucoromycetes, the sturdy, thick-walled zygosporangium can withstand harsh conditions and then undergo karyogamy and meiosis when the environment is favourable for reproduction. In one group of mucoromycetes, the glomeromycetes, the hyphae have a specialised morphology that enables the fungi to form arbuscular mycorrhizae with plant roots. In ascomycetes, the asexual spores (conidia) are often produced in chains or clusters at the tips of conidiophores, where they are easily dispersed by wind. The often cup-shaped ascocarps house the sexual spore-forming asci. In basidiomycetes, the basidiocarp supports and protects a large surface area of basidia, from which spores are dispersed. **3.** Such a change to the life cycle of an ascomycete would reduce the number and genetic diversity of ascospores that result from a mating event. Ascospore number would drop because a mating event would lead to the formation of only one ascus. Ascospore genetic diversity would also drop because in ascomycetes, one mating event leads to the formation of asci by many different dikaryotic cells. As a result, genetic recombination and meiosis occur independently many different times—which could not happen if only a single ascus was formed. It is also likely that if such an ascomycete formed an ascocarp, the shape of the ascocarp would differ considerably from that found in its close relatives.

Concept Check 31.5

1. A suitable environment for growth, retention of water and minerals, protection from intense sunlight, and protection from being eaten
2. A hardy spore stage enables dispersal to host organisms through a variety

of mechanisms; their ability to grow rapidly in a favourable new environment enables them to capitalise on the host's resources. **3.** Many different outcomes might have occurred. Organisms that currently form mutualisms with fungi might have gained the ability to perform the tasks currently done by their fungal partners, or they might have formed similar mutualisms with other organisms (such as bacteria). Alternatively, organisms that currently form mutualisms with fungi might be less effective at living in their present environments. For example, the colonisation of land by plants might have been more difficult. And if plants did eventually colonise land without fungal mutualists, natural selection might have favoured plants that formed more highly divided and extensive root systems (in part replacing mycorrhizae).

Summary of Key Concepts Questions

31.1 The body of a multicellular fungus typically consists of thin filaments called hyphae. These filaments form an interwoven mass (mycelium) that penetrates the substrate on which the fungus grows and feeds. Because the individual filaments are thin, the surface-to-volume ratio of the mycelium is maximised, making nutrient absorption highly efficient.

31.2

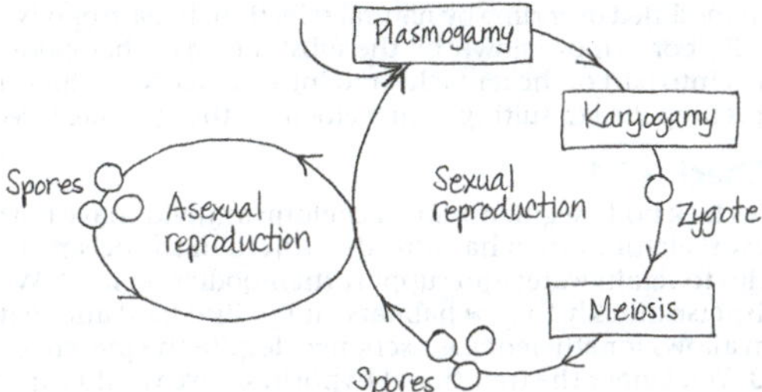

31.3 Phylogenetic analyses show that fungi and animals are more closely related to each other than either is to other multicellular eukaryotes (such as plants or multicellular algae). These analyses also show that fungi are more closely related to single-celled protists called nucleariids than they are to animals, whereas animals are more closely related to a different group of single-celled protists, the choanoflagellates, than they are to fungi. In combination, these results indicate that multicellularity evolved in fungi and animals independently, from different single-celled ancestors.

31.4

Cryptomycetes
Microsporidians
Chytrids
Zoopagomycetes
Mucoromycetes
Acomycetes
Basidiomycetes

31.5 As decomposers, fungi break down the bodies of dead organisms, thereby recycling elements between the living and nonliving environments. Without the activities of fungi and bacterial decomposers, essential nutrients would remain tied up in organic matter, and life as we know it would cease. As an example of their key role as mutualists, fungi form mycorrhizal associations with plants. These associations improve the growth and survival of plants, thereby indirectly affecting the many other species (humans included) that depend on plants. As pathogens, fungi harm other species. In some cases, fungal pathogens have caused their host populations to decline across broad geographic regions, as in the case of the American chestnut.

Test Your Understanding

1. B **2.** D **3.** A

4. Australian plants evolved in isolation from pests and diseases from other areas of the world. Without an evolved resistance to pests and diseases, plants are particularly susceptible to infection. Human activities that might contribute to the spread of plant diseases include unintended transport of fungal spores across the land by vehicles, animals, equipment, clothes and footwear, or when soil or vegetation carrying spores is transported to a new area. Movement of invasive diseases across the landscape will increase because of more extensive human-mediated transport mechanisms.

Chapter 32

Figure Questions

Figure 32.3 As described in 1 and 2, choanoflagellates and a broad range of animals have collar cells. Since collar cells have never been observed in plants, fungi, or non-choanoflagellate protists, this suggests that choanoflagellates may be more closely related to animals than to other eukaryotes. If choanoflagellates are more closely related to animals than to any other group of eukaryotes, choanoflagellates and animals should share other traits that are not found in other eukaryotes. The data described in 3 are consistent with this prediction. **Figure 32.10** The cells of an early embryo with deuterostome development typically are not committed to a particular developmental fate, whereas the cells of an early embryo with protostome development typically are committed to a particular developmental fate. As a result, an embryo with deuterostome development would be more likely to contain stem cells that could give rise to cells of any type.

Concept Check 32.1

1. In most animals, the zygote undergoes cleavage, which leads to the formation of a blastula. Next, in gastrulation, one end of the embryo folds inwards, producing layers of embryonic tissue. As the cells of these layers differentiate, a wide variety of animal forms are produced. Despite the diversity of animal forms, animal development is controlled by a similar set of *Hox* genes across a broad range of taxa. **2.** The imaginary plant would require tissues composed of cells that were analogous to the muscle and nerve cells found in animals: "Muscle" tissue would be necessary for the plant to chase prey, and "nerve" tissue would be required for the plant to coordinate its movements when chasing prey. To digest captured prey, the plant would need to either secrete enzymes into one or more digestive cavities (which could be modified leaves, as in a Venus flytrap) or secrete enzymes outside of its body and feed by absorption. To extract nutrients from the soil—yet be able to chase prey—the plant would need something other than fixed roots, perhaps retractable "roots" or a way to ingest soil. To conduct photosynthesis, the plant would require chloroplasts. Overall, such an imaginary plant would be very similar to an animal that had chloroplasts and retractable roots.

Concept Check 32.2

1. c, b, a, d **2.** The red-coloured portion of the tree represents ancestors of animals that lived between 1 billion years ago and 770 million years ago. Although these ancestors are more closely related to animals than to fungi, they would not be classified as animals. One example of an ancestor represented by the red-coloured portion of this tree is the most recent common ancestor shared by choanoflagellates and animals (see Figure 32.3). That common ancestor was not an animal (or a choanoflagellate), but it was a direct ancestor of the animals. **3.** In descent with modification, an organism shares characteristics with its ancestors (due to their shared ancestry), yet it also differs from its ancestors (because organisms accumulate differences over time as they adapt to their surroundings). As an example, consider the evolution of animal cadherin proteins, a key step in the origin of multicellular animals. These proteins illustrate both of these aspects of descent with modification: Animal cadherin proteins share many protein domains with a cadherin-like protein found in their choanoflagellate ancestors, yet they also have a unique "CCD" domain that is not found in choanoflagellates.

Concept Check 32.3

1. Grade-level characteristics are those that multiple lineages share regardless of evolutionary history. Some grade-level characteristics may have evolved multiple times independently. Features that unite clades are derived characteristics that originated in a common ancestor and were passed on to the various descendants. **2.** A snail has a spiral and determinate cleavage pattern; a human has radial, indeterminate cleavage. In a snail, the coelomic cavity is formed by splitting of mesoderm masses; in a human, the coelom forms from folds of archenteron. In a snail, the mouth forms from the blastopore; in a human, the anus develops from the blastopore. **3.** Most triploblasts have two openings to their digestive tract, a mouth and an anus. As such, their bodies have a structure that is analogous to that of a doughnut: The digestive tract (the hole of the doughnut) runs from the mouth to the anus and is surrounded by various tissues (the solid part of the doughnut). The doughnut analogy is most obvious at early stages of development (see Figure 32.10c).

Concept Check 32.4

1. Cnidarians possess tissues, while sponges do not. Also unlike sponges, cnidarians exhibit body symmetry, though it is radial and not bilateral as in most other animal phyla.

2.

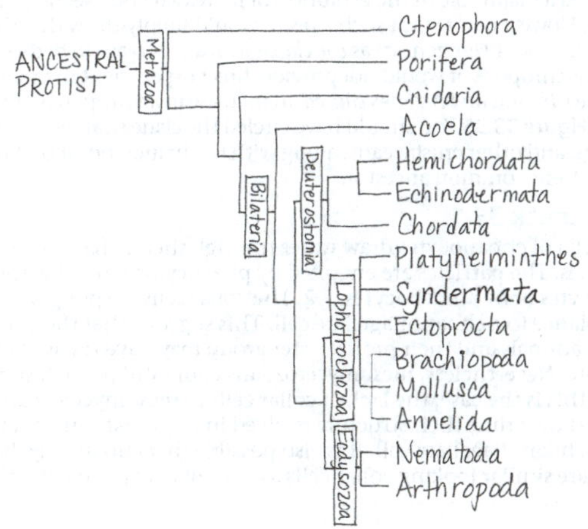

Under the hypothesis that ctenophores are basal metazoans, sponges (which lack tissues) would be nested within a clade whose other members all have tissues. As a result, a group composed of animals with tissues would not form a clade. **3.** The phylogeny in Figure 32.11 indicates that molluscs are members of Lophotrochozoa, one of the three main groups of bilaterians (the others being Deuterostomia and Ecdysozoa). As seen in Figure 25.11, the fossil record shows that molluscs were present tens of millions of years before the Cambrian explosion. Thus, long before the Cambrian explosion, the lophotrochozoan clade had formed and was evolving independently of the evolutionary lineages leading to Deuterostomia and Ecdysozoa. Based on the phylogeny in Figure 32.11, we can also conclude that the lineages leading to Deuterostomia and Ecdysozoa were independent of one another before the Cambrian explosion. Since the lineages leading to the three main clades of bilaterians were evolving independently of one another prior to the Cambrian explosion, that explosion could be viewed as consisting of three "explosions," not one.

Summary of Key Concepts Questions

32.1 Unlike animals, which are heterotrophs that ingest their food, plants are autotrophs, and fungi are heterotrophs that grow on their food and feed by absorption. Animals lack cell walls, which are found in both plants and fungi. Animals also have muscle tissue and nerve tissue, which are not found in either plants or fungi. In addition, the sperm and egg cells of animals are produced by meiotic division, unlike what occurs in plants and fungi (where reproductive cells such as sperm and eggs are produced by mitotic division). Finally, animals regulate the development of body form with *Hox* genes, a unique group of genes that is not found in either plants or fungi. **32.2** Current hypotheses about the cause of the Cambrian explosion include new predator-prey relationships, an increase in atmospheric oxygen, and an increase in developmental flexibility provided by the origin of *Hox* genes and other genetic changes. **32.3** Body plans provide a helpful way to compare and contrast key features of organisms. However, phylogenetic analyses show that similar body plans have arisen independently in different groups of organisms. As such, similar body plans may have arisen by convergent evolution and hence may not be informative about evolutionary relationships. **32.4** Listed in order from the most to the least inclusive clade, humans belong to Metazoa, Eumetazoa, Bilateria, Deuterostomia, and Chordata.

Test Your Understanding

1. A **2.** D **3.** C **4.** B

Chapter 33

Figure Questions

Figure 33.7 *Obelia* is an animal, and its life cycle is indeed most similar to the generalised life cycle of animals (Figure 13.6a). In *Obelia*, both the polyp and the medusa are diploid organisms. As in other animals, in *Obelia* only the single-celled gametes are haploid. By contrast, plants and some algae (Figure 13.6b) have a multicellular haploid generation and a multicellular diploid generation. *Obelia* also differs from fungi and some protists (Figure 13.6c) in that the diploid stage of those organisms is unicellular. **Figure 33.8** Possible examples include the Golgi apparatus (flattening; increases area for receiving and transporting proteins), the cristae of mitochondria (folding; increases the surface area available for cellular respiration), the cardiovascular system (branching; increase area for materials exchange in tissues), and root hairs (projections; increase area for absorption). **Figure 33.10** Adding fertiliser to the water supply would probably increase the abundance of algae, and that, in turn, would likely increase the abundance of snails (which eat algae). If the water was also contaminated with infected human faeces or urine, an increase in the number of snails would likely lead to an increase in the abundance of blood flukes (which require snails as an intermediate host). As a result, the occurrence of schistosomiasis might increase. **Figure 33.21** The extinction of freshwater bivalves might lead to an increase in the abundance of photosynthetic protists and bacteria. Because these organisms are at the base of aquatic food webs, increases in their abundance could have major effects on aquatic communities (including both increases and decreases in the abundance of other species). **Figure 33.29** Such a result would be consistent with the origin of the *Ubx* and *abd-A Hox* genes having played a major role in the evolution of increased body segment diversity in arthropods. However, note that such a result would simply show that the presence of the *Ubx* and *abd-A Hox* genes was *correlated with* an increase in body segment diversity in arthropods; it would not provide direct experimental evidence that the origin of the *Ubx* and *abd-A* genes *caused* an increase in arthropod body segment diversity. **Figure 33.35** You should have circled the clade that includes the insects, remipedians, and other crustaceans, along with the branch point that represents their most recent common ancestor.

Concept Check 33.1

1. The flagella of choanocytes draw water through their collars, which trap food particles. The particles are engulfed by phagocytosis and digested, either by choanocytes or by amoebocytes. **2.** The collar cells of sponges bear a striking resemblance to a choanoflagellate cell. This suggests that the last common ancestor of animals and their protist sister group may have resembled a choanoflagellate. Nevertheless, mesomycetozoans could still be the sister group of animals. If this is the case, the lack of collar cells in mesomycetozoans would indicate that over time their structure evolved in ways that caused it to no longer resemble a choanoflagellate cell. It is also possible that choanoflagellates and sponges share similar looking collar cells as a result of convergent evolution.

Concept Check 33.2

1. Both the polyp and the medusa are composed of an outer epidermis and an inner gastrodermis separated by a gelatinous layer, the mesoglea. The polyp is a cylindrical form that adheres to the substrate by its aboral end; the medusa is a flattened, mouth-down form that moves freely in the water. **2.** Both a feeding polyp and a medusa are diploid, as indicated by the pink arrow in the diagram. The medusa stage produces haploid gametes. **3.** Evolution is not goal oriented; hence, it would not be correct to argue that cnidarians are not "highly evolved" simply because their form had changed relatively little over the past 560 million years. Instead, the fact that cnidarians have persisted for hundreds of millions of years indicates that the cnidarian body plan is a highly successful one.

Concept Check 33.3

1. Tapeworms can absorb food from their environment and release ammonia into their environment through their body surface because their body is very flat, due in part to the lack of a body cavity. **2.** The inner tube is the alimentary canal, which runs the length of the body. The outer tube is the body wall. The two tubes are separated by the coelom. **3.** All molluscs have inherited a foot from their common ancestor. However, in different groups of molluscs, the structure of the foot has been modified over time by natural selection. In gastropods, the foot is used as a holdfast or to move slowly on the substrate. In cephalopods, the foot has been modified into part of the tentacles and into an excurrent siphon, through which water is propelled (resulting in movement in the opposite direction).

Concept Check 33.4

1. Nematodes lack body segments and a coelom; annelids have both. **2.** The arthropod exoskeleton, which had already evolved in the ocean, allows terrestrial species to retain water and support their bodies on land. Wings allow insects to disperse quickly to new habitats and to find food and mates. The tracheal system allows for efficient gas exchange despite the presence of an exoskeleton. **3.** Yes. Under the traditional hypothesis, we would expect body segmentation to be controlled by similar *Hox* genes in annelids and arthropods. However, if annelids are in Lophotrochozoa and arthropods are in Ecdysozoa (as current evidence suggests), body segmentation may have evolved independently in these two groups. In such a case, we might expect that different *Hox* genes would control the development of body segmentation in the two clades.

Concept Check 33.5

1. Each tube foot consists of an ampulla and a podium. When the ampulla squeezes, it forces water into the podium, which causes the podium to expand and contact the substrate. Adhesive chemicals are then secreted from the base of the podium, thereby attaching the podium to the substrate. **2.** Both insects and nematodes are members of Ecdysozoa, one of the three major clades of bilaterians. Therefore, a characteristic shared by *Drosophila* and *Caenorhabditis* may be informative for other members of their clade—but not necessarily for members of Deuterostomia. Instead, Figure 33.2 suggests that a species within Echinodermata or Chordata might be a more appropriate invertebrate model organism from which to draw inferences about humans and other vertebrates. **3.** Echinoderms include species with a wide range of body forms. However, even echinoderms that look very different from one another, such as sea stars and sea cucumbers, share characteristics unique to their phylum, including a water vascular system and tube feet. The differences between echinoderm species illustrate the diversity of life, while the characteristics they share illustrate the unity of life. The match between organisms and their environments can be seen in such echinoderm features as the eversible stomachs of sea stars (enabling them to digest prey that are larger than their mouth) and the complex, jaw-like structure that sea urchins use to eat seaweed.

Summary of Key Concepts Questions

33.1 The sponge body consists of two layers of cells, both of which are in contact with water. As a result, gas exchange and waste removal occur as substances diffuse into and out of the cells of the body. Choanocytes and amoebocytes ingest food particles from the surrounding water. Choanocytes also release food particles to amoebocytes, which then digest the food particles and deliver nutrients to other cells. **33.2** The cnidarian body plan consists of a sac with a central digestive compartment, the gastrovascular cavity. The single opening to this compartment serves as both a mouth and an anus. The two main variations on this body plan are sessile polyps (which adhere to the substrate at the end of the body opposite to the mouth/anus) and motile medusae (which move freely through the water and resemble flattened, mouth-down versions of polyps). **33.3** No. Some lophotrochozoans have a crown of ciliated tentacles that function in feeding (called a lophophore), while others go through a distinctive developmental stage known as trochophore larvae. Many other lophotrochozoans do not have either of these features. As a result, the clade is defined primarily by DNA similarities, not morphological similarities. **33.4** Many nematode species live in soil and in sediments on the bottom of bodies of water. These free-living species play important roles in decomposition and nutrient cycling. Other nematodes are parasites, including many species that attack the roots of plants and some that attack animals (including humans). Arthropods have profound effects on all aspects of ecology. In aquatic environments, crustaceans play key roles as grazers (of algae), scavengers, and predators and some species, such as krill, are important sources of food for whales and other vertebrates. On land, it is difficult to think of features of the natural world that are not affected in some way by insects and other arthropods, such as spiders and ticks. There are more than 1 million species of insects, many of which have enormous ecological effects as herbivores, predators, parasites, decomposers, and vectors of disease. Insects are also key sources of food for many organisms, including humans in some regions of the world. **33.5** Echinoderms and chordates are both members of Deuterostomia, one of the three main clades of bilaterian animals. As such, chordates (including humans) are more closely related to echinoderms than we are to animals in any of the other phyla covered in this chapter. Nevertheless, echinoderms and chordates have evolved independently for over 500 million years. This statement does not contradict the close relationship of echinoderms and chordates, but it does make clear that "close" is a relative term indicating that these two phyla are more closely related to each other than either is to animal phyla not in Deuterostomia.

Test Your Understanding

1. A **2.** C **3.** B **4.** C **5.** C **6.** D

Chapter 34

Figure Questions

Figure 34.2

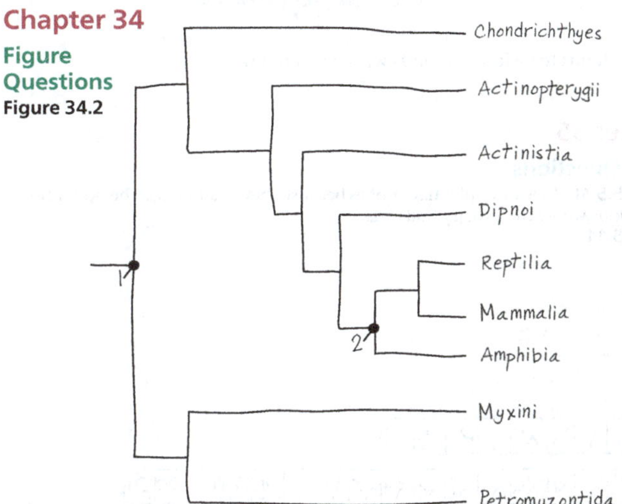

The redrawn tree shows mammals (including humans) as nested near the middle of the evolutionary tree of vertebrates. Showing the vertebrate tree in this way provides a visual illustration of the fact that the evolutionary history of vertebrates did not consist of a series of steps "leading to" humans. **Figure 34.6** The patterns in these figures suggest that specific *Hox* genes, as well as the order in which they are expressed, have been highly conserved over the course of evolution. **Figure 34.20** *Tiktaalik* was a lobe-fin fish that had both fish and tetrapod characters. Like a fish, *Tiktaalik* had fins, scales, and gills. As described by Darwin's concept of descent with modification, such shared characters can be attributed to descent from ancestral species—in this case, *Tiktaalik*'s descent from fish ancestors. *Tiktaalik* also had traits that were unlike a fish but like a tetrapod, including a flat skull, a neck, a full set of ribs, and the skeletal structure of its fin. These characters illustrate the second part of descent with modification, showing how ancestral features had become modified over time. **Figure 34.21** Sometime between 350 mya and 340 mya. We can infer this because amphibians must have originated after the most recent common ancestor of amphibians and amniotes (and that ancestor is shown as having lived 350 mya), but no later than the date of the earliest known fossils of amphibians (shown in the figure as 340 mya). **Figure 34.24** Pterosaurs did not descend from the common ancestor of all dinosaurs; hence, pterosaurs are not dinosaurs. However, birds are descendants of the common ancestor of the dinosaurs. As a result, a monophyletic clade of dinosaurs must include birds. In that sense, birds are dinosaurs. **Figure 34.36** In a catabolic pathway, like the aerobic processes of cellular respiration, water is released as a by-product when an organic compound such as glucose is mixed with oxygen. The kangaroo rat can retain and use that water, decreasing its need to drink water. **Figure 34.37** In general, the process of exaptation occurs as a structure that had one function acquires a different function via a series of intermediate stages. Each of these intermediate stages typically has some function in the organism in which it is found. The incorporation of articular and quadrate bones into the mammalian ear illustrates exaptation because these bones originally evolved as part of the jaw, where they functioned as the jaw hinge, but over time they became co-opted for another function, namely, the transmission of sound. **Figure 34.43** As shown in this phylogeny, chimpanzees and humans represent the tips of separate branches of evolution. As such, the human and chimpanzee lineages have evolved independently after they diverged from their common ancestor—an event that took place about 8 million years ago. Hence, it is incorrect to say that humans evolved from chimpanzees (or vice versa). If humans had descended from chimpanzees, for example, the human lineage would be nested within the chimpanzee lineage, much as birds are nested within the reptile clade.

Figure 34.50 Fossil evidence indicates that Neanderthals did not live in Africa; hence there would have been little opportunity for mating (gene flow) between Neanderthals and humans in Africa. However, as humans migrated from Africa, mating may have occurred between Neanderthals and humans in the first region where the two species encountered one another: the Middle East. Humans carrying Neanderthal genes may then have migrated to other locations, explaining why Neanderthals are equally related to humans from France, China, and Papua New Guinea.

Concept Check 34.1

1. The four characters are a notochord; a dorsal, hollow nerve chord; pharyngeal slits or clefts; and a muscular, post-anal tail. **2.** In humans, these characters are present only in the embryo. The notochord becomes discs between the vertebrae; the dorsal, hollow nerve cord develops into the brain and spinal cord; the pharyngeal clefts develop into various adult structures, and the tail is almost completely lost. **3.** You would expect the vertebrate groups Actinopterygii, Actinistia, Dipnoi, Amphibia, Reptilia, and Mammalia to have lungs or lung derivatives. All of these groups originate to the right of (evolved after) the hatch mark indicating the appearance of this derived character in their lineage.

Concept Check 34.2

1. Parasitic lampreys have a round, rasping mouth, which they use to attach to fish. Non-parasitic lampreys feed only as larvae; these larvae resemble lancelets and like them, are suspension feeders. Conodonts had two sets of mineralised dental elements, which may have been used to impale prey and cut it into smaller pieces. **2.** Such a finding suggests that early organisms with a head were favoured by natural selection in several different evolutionary lineages. However, while a logical argument can be made that having a head was advantageous, fossils alone do not constitute proof. **3.** In armored jawless vertebrates, bone served as external armor that may have provided protection from predators. Some species also had mineralised mouthparts, which could be used for either predation or scavenging.

Concept Check 34.3

1. Both are gnathostomes and have jaws, four clusters of *Hox* genes, enlarged forebrains, and lateral line systems. Shark skeletons consist mainly of cartilage, whereas tuna have bony skeletons. Sharks also have a spiral valve. Tuna have an operculum and a swim bladder, as well as flexible rays supporting their fins. **2.** Aquatic gnathostomes have jaws (an adaptation for feeding) and paired fins and a tail (adaptations for swimming). Aquatic gnathostomes also typically have streamlined bodies for efficient swimming and swim bladders or other mechanisms (such as oil storage in sharks) for buoyancy.
3.

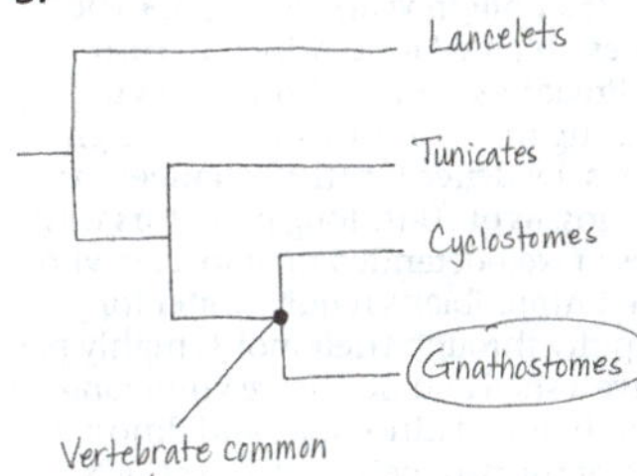

4. Yes, that could have happened. The paired appendages of aquatic gnathostomes other than the lobe-fins could have served as a starting point for the evolution of limbs. The colonisation of land by aquatic gnathostomes other than the lobe-fins might have been facilitated in lineages that possessed lungs, as that would have enabled those organisms to breathe air.

Concept Check 34.4

1. Tetrapods are thought to have originated about 365 million years ago when the fins of some lobe-fins evolved into the limbs of tetrapods. In addition to their four limbs with digits—a key derived trait for which the group is named—other derived traits of tetrapods include a neck (consisting of vertebrae that separate the head from the rest of the body) and a pelvic girdle that is fused to the backbone. **2.** Some fully aquatic species are paedomorphic, retaining larval features for life in water as adults. Species that live in dry environments may avoid dehydration by burrowing or living under moist leaves, and they protect their eggs with foam nests, viviparity, and other adaptations. **3.** Many amphibians spend part of their life cycle in aquatic environments and part on land. Thus, they may be exposed to a wide range of environmental problems, including water and air pollution and the loss or degradation of aquatic and/or terrestrial habitats. In addition, amphibians have highly permeable skin, providing relatively little protection from external conditions, and their eggs do not have a protective shell.

Concept Check 34.5

1. The amniotic egg provides protection to the embryo and allows the embryo to develop on land, eliminating the necessity of a watery environment for reproduction. Another key adaptation is rib cage ventilation, which improves the efficiency of air intake and may have allowed early amniotes to dispense with breathing through their skin. Finally, not breathing through their skin allowed amniotes to develop relatively impermeable skin, thereby conserving water. **2.** Yes. Although snakes lack limbs, they descended from lizards with legs. Some snakes retain vestigial pelvic and leg bones, providing evidence of their descent from an ancestor with legs. **3.** Birds have weight-saving modifications, including the absence of teeth, a urinary bladder, and a second ovary in females. The wings and feathers are adaptations that facilitate flight, as do efficient respiratory and circulatory systems that support a high metabolic rate. **4.** (a) synapsids; (b) tuataras; (c) turtles.

Concept Check 34.6

1. Monotremes lay eggs. Marsupials give birth to very small live young that attach to a nipple in the mother's pouch, where they complete development. Eutherians give birth to more developed live young. **2.** Hands and feet adapted for grasping, flat nails, large brain, forward-looking eyes on a flat face, parental care, and movable big toe and thumb **3.** Mammals are endothermic, enabling them to live in a wide range of habitats. Milk provides young with a balanced set of nutrients, and hair and a layer of fat under the skin help mammals retain heat. Mammals have differentiated teeth, enabling them to eat many different kinds of food. Mammals also have relatively large brains, and many species are capable learners. Following the mass extinction at the end of the Cretaceous period, the absence of large terrestrial dinosaurs may have opened many new ecological niches to mammals, promoting an adaptive radiation. Continental drift also isolated many groups of mammals from one another, promoting the formation of many new species.

Concept Check 34.7

1. Hominins are a clade within the ape clade that includes humans and all species more closely related to humans than to other apes. The derived characters of hominins include bipedal locomotion and relatively larger brains. **2.** In hominins, bipedal locomotion evolved long before large brain size. *Homo ergaster*, for example, was fully upright, bipedal, and as tall as modern humans, but its brain was significantly smaller than that of modern humans. **3.** Yes, both can be correct. *Homo sapiens* may have established populations outside of Africa as early as 180,00 years ago, as indicated by the fossil record. However, those populations may have left few or no descendants today. Instead, all living humans may have descended from Africans that spread from Africa roughly 50,000 years ago, as indicated by genetic data.

Summary of Key Concepts Questions

34.1 Lancelets are the most basal group of living chordates, and as adults they have key derived characters of chordates. This suggests that the chordate common ancestor may have resembled a lancelet in having an anterior end with a mouth along with the following four derived characters: a notochord; a dorsal, hollow nerve cord; pharyngeal slits or clefts; and a muscular, post-anal tail. **34.2** Conodonts, among the earliest vertebrates in the fossil record, were very abundant for 300 million years. While jawless, their well-developed teeth provide early signs of bone formation. Other species of jawless vertebrates developed armor on the outside of their bodies, which probably helped protect them from predators. Like lampreys, these species had paired fins for locomotion and an inner ear with semicircular canals that provided a sense of balance. There were many species of these armored jawless vertebrates, but they all became extinct by the close of the Devonian period, 359 million years ago. **34.3** The origin of jaws altered how fossil gnathostomes obtained food, which in turn had large effects on ecological interactions. Predators could use their jaws to grab prey or remove chunks of flesh, stimulating the evolution of increasingly sophisticated means of defence in prey species. Evidence for these changes can be found in the fossil record, which includes fossils of 10-m-long predators with remarkably powerful jaws, as well as lineages of well-defended prey species whose bodies were covered by armored plates. **34.4** Amphibians require water for reproduction; their bodies can lose water rapidly through their moist, highly permeable skin; and amphibian eggs do not have a shell and hence are vulnerable to desiccation. **34.5** Birds are descended from theropod dinosaurs, and dinosaurs are nested within the archosaur lineage, one of the two main reptile lineages. Thus, the other living archosaur reptiles, the crocodilians, are more closely related to birds than they are to non-archosaur reptiles such as lizards. As a result, birds are considered reptiles. (Note that if reptiles were defined as excluding birds, the reptiles would not form a clade; instead, the reptiles would be a paraphyletic group.) **34.6** Mammals are members of a group of amniotes called synapsids. Early (nonmammalian) synapsids laid eggs and had a sprawling gait. Fossil evidence shows that mammalian features arose gradually over a period of more than 100 million years. For example, the jaw was modified over time in nonmammalian synapsids, eventually coming to resemble that of a mammal. By 180 million years ago, the first mammals had appeared. There were many species of early mammals, but most of them were small, and they were not abundant or dominant members of their community. Mammals did not rise to ecological dominance until after the extinction of the dinosaurs. **34.7** The fossil record shows that from 4.5 to 2.5 million years ago, a wide range of hominin species walked upright but had relatively small brain sizes. About 2.5 million years ago, the first members of genus *Homo* emerged. These species used tools and had larger brains than those of earlier hominins. Fossil evidence indicates that multiple members of our genus were alive at any given point in time. Furthermore, until about 1.3 million years ago, these various *Homo* species also coexisted with members of earlier hominin lineages, such as *Paranthropus*. The different hominins alive at the same periods of time varied in body size, body shape, brain size, dental morphology, and the capacity for tool use. Ultimately, except for *Homo sapiens*, all of these species became extinct. Overall, human evolution can be viewed as an evolutionary tree with many branches—the only surviving lineage of which is our own.

Test Your Understanding

1. D **2.** C **3.** B **4.** C **5.** D **6.** A

8. (a) Because brain size tends to increase consistently in such lineages, we can conclude that natural selection favoured the evolution of larger brains and hence that the benefits outweighed the costs. (b) As long as the benefits of brains that are large relative to body size are greater than the costs, large brains can evolve. Natural selection might favour the evolution of brains that are large relative to body size because such brains confer an advantage in obtaining mates and/or an advantage in survival.

(c)

Mortality rate

Deviation of brain size from expected

Mortality tends to be lower in birds with larger brains.

Chapter 35

Figure Questions

Figure 35.5 All three examples have nodes because they're all stems. The nodes are the locations where the axillary buds arise.

Figure 35.11

(1)

Mature Tissues

Primary meristems

Root apical meristem

Root cap

(2)

X1 X2 X3 X4 X5 V P3 P2 P1

X1 X2 X3 X4 X5 X6 X7 X8 X9 X10 V P6 P5 P4 P3 P2 P1

As a result of the addition of secondary xylem cells, the vascular cambium is pushed further to the outside.

Figure 35.15

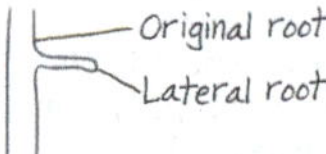

Figure 35.17 Pith and cortex are defined, respectively, as ground tissue that is internal and ground tissue that is external to vascular tissue. Since vascular bundles of monocot stems are scattered throughout the ground tissue, there is no clear distinction between internal and external relative to the vascular tissue. **Figure 35.19** The vascular cambium produces growth that increases the diameter of a stem or root. The tissues that are exterior to the vascular cambium cannot keep pace with the growth because their cells no longer divide. As a result, these tissues rupture. **Figure 35.23** Periderm (mainly cork and cork cambium), primary phloem, secondary phloem, vascular cambium, secondary xylem (sapwood and heartwood), primary xylem, and pith. At the base of ancient redwood that is many centuries old, the remnants of primary growth (primary phloem, primary xylem, and pith) would be quite insignificant. **Figure 35.30** Every root epidermal cell would develop a root hair. **Figure 35.32** Another example of homeotic gene mutation is the mutation in a *Hox* gene that causes legs to form in place of antennae in *Drosophila* (see Figure 18.20).

Figure 35.33

(a)

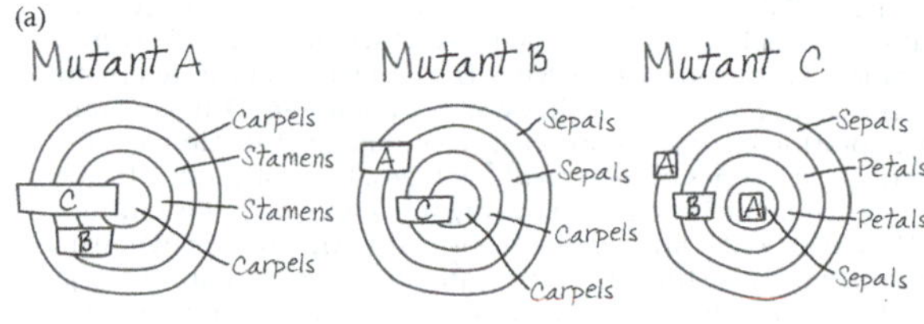

(b)

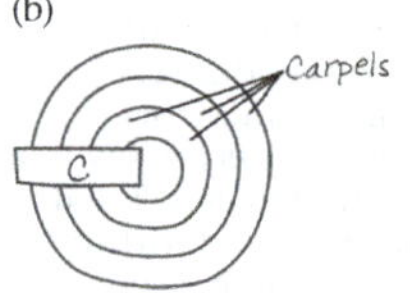

Concept Check 35.1

1. The vascular tissue system connects leaves and roots, allowing sugars to move from leaves to roots in the phloem and allowing water and minerals to move to the leaves in the xylem. **2.** To get sufficient energy from photosynthesis, we would need lots of surface area exposed to the sun. This large surface-to-volume ratio, however, would create a new problem—evaporative water loss. We would have to be permanently connected to a water source—the soil, also our source of minerals. In short, we would probably look and behave very much like plants. **3.** As plant cells enlarge, they typically form a huge central vacuole that contains a dilute, watery sap. Central vacuoles enable plant cells to become large with only a minimal investment of new cytoplasm. The orientation of the cellulose microfibrils in plant cell walls affects the growth pattern of cells.

Concept Check 35.2

1. Yes. In a woody plant, secondary growth is occurring in the older parts of the stem and root, while primary growth is occurring at the root and shoot tips.
2. The largest, oldest leaves would be lowest on the shoot. Since they would probably be heavily shaded, they would not photosynthesise much regardless of their size. Determinate growth benefits the plant by keeping it from investing an ever-increasing amount of resources into organs that provide little photosynthetic product. **3.** No. The carrot roots will probably be smaller at the end of the second year because the food stored in the roots will be used to produce flowers, fruits, and seeds.

Concept Check 35.3

1. In roots, primary growth occurs in three successive stages, moving away from the tip of the root: the zones of cell division, elongation, and differentiation. In shoots, it occurs at the tip of apical buds, with leaf primordia arising along the sides of an apical meristem. Most growth in length occurs in older internodes below the shoot tip.
2. The fossil probably came from a floating leaf because having stomata exclusively on the upper surface would be a poor adaptation for a desert plant since it would lead to water loss. A floating leaf, on the other hand, would benefit from having stomata on its upper surface since only that surface is in contact with the gaseous environment.
3. Root hairs are cellular extensions that increase the surface area of the root epidermis, thereby enhancing the absorption of minerals and water. Microvilli are extensions that increase the absorption of nutrients by increasing the surface area of the gut.

Concept Check 35.4

1. The sign will still be 2 m above the ground because this part of the tree is no longer growing in length (primary growth); it is now growing only in thickness (secondary growth). **2.** Stomata must be able to close because evaporation is much more intensive from leaves than from the trunks of woody trees as a result of the higher surface-to-volume ratio in leaves. **3.** Since there is little seasonal temperature variation in the tropics, the growth rings of a tree from the tropics would be difficult to discern unless the tree came from an area that had pronounced wet and dry seasons. **4.** The tree would die slowly. Girdling removes an entire ring of secondary phloem (part of the bark), completely preventing transport of sugars and starches from the shoots to the roots. After several weeks, the roots would have used all of their stored carbohydrate reserves and would die.

Concept Check 35.5

1. Although all the living vegetative cells of a plant have the same genome, they develop different forms and functions because of differential gene expression.
2. Plants show indeterminate growth; juvenile and mature phases are found on the same individual plant; and cell differentiation in plants is more dependent on final position than on lineage. **3.** One hypothesis is that tepals arise if *B* gene activity is present in all three of the outer whorls of the flower.

Summary of Key Concepts Questions

35.1 Here are a few examples: The cuticle of leaves and stems protects these structures from desiccation. Collenchyma and sclerenchyma cells have thick walls that provide support for plants. Strong, branching root systems help anchor plants in the soil. **35.2** Primary growth arises from apical meristems and involves production and elongation of organs. Secondary growth arises from lateral meristems and adds to the diameter of roots and stems. **35.3** Lateral roots emerge from the pericycle and destroy plant cells as they emerge. In stems, branches arise from axillary buds and do not destroy any cells. **35.4** With the evolution of secondary growth, plants were able to grow taller and shade competitors. **35.5** The orientation of cellulose microfibrils in the innermost layers of the cell wall causes growth along one axis. Microtubules in the cell's outermost cytoplasm play a key role in regulating the axis of cell expansion because it is their orientation that determines the orientation of cellulose microfibrils.

Test Your Understanding

1. D **2.** C **3.** C **4.** A **5.** D **6.** C **7.** D **8.** B **9.** D **10.** D
11.

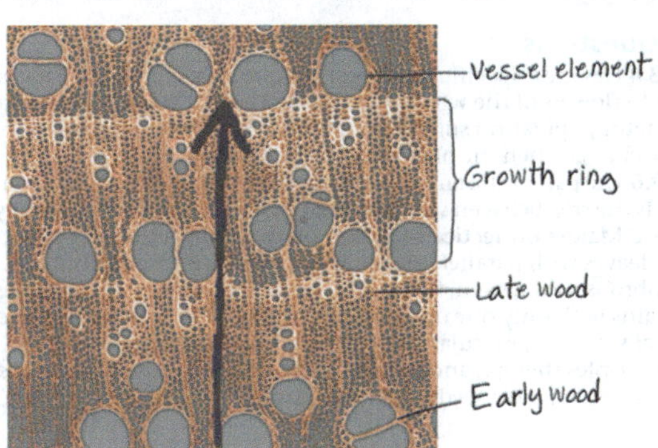

12. The tea and iris leaves differ in the arrangement of photosynthetic mesophyll cells. In the tea leaf, the mesophyll cells are divided into two layers, with the lower part of the leaf having a loosely arranged spongy layer with large air spaces, and the upper part having a palisade layer of tightly packed cells. In contrast, the mesophyll cells in the iris leaf are evenly distributed with no large air spaces. These differences are associated with the natural orientation of the leaves of these two species. Tea leaves are horizontally oriented and are more likely to receive light on the upper surface, so most of the mesophyll cells are concentrated in a palisade layer to absorb light efficiently. In contrast, iris leaves are vertically oriented with both sides illuminated about equally during the course of the day, so the mesophyll cells are evenly distributed. In both leaves, structure fits function because the arrangement of mesophyll cells maximises photosynthesis.
13. Pranksters must have been carried the bicycle high into the tree and threaded it over a young woody branch: Primary growth did not raise the bicycle off the ground. Over time, secondary growth in the region of the tree notch enveloped the bicycle.

Chapter 36

Figure Questions

Figure 36.2 There would be no reference to photosynthesis because it ceases at night. Also, the directions of the CO_2 and O_2 arrows associated with the leaves would be reversed because at night only the gas exchange related to cellular respiration is occurring. **Figure 36.3** The leaves are being produced in a counterclockwise spiral. The next leaf primordium will emerge approximately between and to the inside of leaves 8 and 13. **Figure 36.4** A higher leaf area index will not necessarily increase photosynthesis because of upper leaves shading lower leaves. **Figure 36.6** A proton pump inhibitor would depolarise (increase) the membrane potential because fewer hydrogen ions would be pumped out across the plasma membrane. The immediate effect of an inhibitor of the H^+/sucrose transporter would be to hyperpolarise (decrease) the membrane potential because fewer hydrogen ions would be leaking back into the cell through these cotransporters. An inhibitor of the H^+/NO_3^- cotransporter would have no effect on the membrane potential because the simultaneous cotransport of a positively charged ion and a negatively charged ion has no net effect on charge difference across the membrane. An inhibitor of the potassium ion channels would decrease the membrane potential because additional positively charged ions would not be accumulating outside the cell. **Figure 36.8** Few, if any, mesophyll cells are more than three cells from a vein. **Figure 36.9** The Casparian strip blocks water and minerals from moving between endodermal cells or moving around an endodermal cell via the cell's wall. Therefore, water and minerals must pass through an endodermal cell's plasma membrane. **Figure 36.13** A section 500 μm by 300 μm is equal to 150,000 μm^2 or 0.0015 cm^2. Since there are five stomata visible in 0.0015 cm^2, the stomatal density of this bean leaf is approximately 3,333 stomata per square centimetre of leaf surface. **Figure 36.18** Because the xylem is under negative pressure (tension), excising a stylet that had been inserted into a tracheid or vessel element would probably introduce air into the cell. No xylem sap would exude unless positive root pressure was predominant.

Concept Check 36.1

1. Vascular plants must transport minerals and water absorbed by the roots to all the other parts of the plant. They must also transport sugars from sites of production to sites of use. **2.** Increased stem elongation would raise the plant's upper leaves. Erect leaves and reduced lateral branching would make the plant less subject to shading by the encroaching neighbours. **3.** Pruning shoot tips removes apical dominance, resulting in lateral shoots (branches) growing from axillary buds (see Concept 35.3). This branching produces a bushier plant with a higher leaf area index.

Concept Check 36.2

1. The cell's Ψ_P is 0.7 MPa. In a solution with a Ψ of –0.4 MPa, the cell's Ψ_P at equilibrium would be 0.3 MPa. **2.** The cell would still adjust to changes in its osmotic environment, but its responses would be slower. Although aquaporins do not affect the water potential gradient across membranes, they allow for more rapid osmotic adjustments. **3.** If tracheids and vessel elements were alive at maturity, their cytoplasm would impede water movement, preventing rapid long-distance transport. **4.** The protoplasts would burst. Because the cytoplasm has many dissolved solutes, water would enter the protoplast continuously without reaching equilibrium. (When present, the cell wall prevents rupturing by limiting expansion of the protoplast.)

Concept Check 36.3

1. At dawn, a drop is exuded from the rooted stump because the xylem is under positive pressure due to root pressure. At noon, the xylem is under negative pressure (tension) when it is cut, and the xylem sap is pulled back into the rooted stump. Root pressure cannot keep pace with the increased rate of transpiration at noon. **2.** Perhaps greater root mass helps compensate for the lower water permeability of the plasma membranes. **3.** The Casparian strip and tight junctions both prevent movement of fluid between cells.

Concept Check 36.4

1. Stomatal opening at dawn is controlled mainly by light, CO_2 concentration, and a circadian rhythm. Environmental stresses such as drought, high temperature, and wind can stimulate stomata to close during the day. Water deficiency during the peak of the day can trigger release of the plant hormone abscisic acid, which signals guard cells to close stomata. **2.** The activation of the proton pumps of stomatal cells would cause the guard cells to take up K^+. The increased turgor of the guard cells would lock the stomata open and lead to extreme evaporation from the leaf. **3.** After the flowers are cut, transpiration from any leaves and from the petals (which are modified leaves) will continue to draw water up the xylem. If cut flowers are transferred directly to a vase, air pockets in xylem vessels prevent delivery of water from the vase to the flowers. Cutting stems again underwater, a few centimetres from the original cut, will sever the xylem above the air pocket. The water droplets prevent another air pocket from forming while the flowers are transferred to a vase. **4.** Water molecules are in constant motion, travelling at different speeds. If water molecules gain enough energy, the most energetic molecules near the liquid's surface will have sufficient speed, and therefore sufficient kinetic energy, to leave the liquid in the form of gaseous molecules (water vapour). As the molecules with the highest kinetic energy leave the liquid, the average kinetic energy of the remaining liquid decreases. Because a liquid's temperature is directly related to the average kinetic energy of its molecules, the temperature drops as evaporation proceeds.

Concept Check 36.5

1. In both cases, the long-distance transport is a bulk flow driven by a pressure difference at opposite ends of tubes. Pressure is generated at the source end of a sieve tube by the loading of sugar and resulting osmotic flow of water into the phloem, and this pressure *pushes* sap from the source end to the sink end of the tube. In contrast, transpiration generates a negative pressure potential (tension) that *pulls* the ascent of xylem sap. **2.** The main sources are fully grown leaves (producing sugar by photosynthesis) and fully developed storage organs (producing sugar by breakdown of starch). Roots, buds, stems, expanding leaves, and fruits are powerful sinks because they are actively growing. A storage organ may be a sink in the summer when accumulating carbohydrates but a source in the spring when breaking down starch into sugar for growing shoot tips. **3.** Positive pressure, whether it be in the xylem when root pressure predominates or in the sieve-tube elements of the phloem, requires active transport. Most long-distance transport in the xylem depends on bulk flow driven by the negative pressure potential generated ultimately by the evaporation of water from the leaf and does not require living cells. **4.** The spiral slash prevents optimal bulk flow of the phloem sap to the root sinks. Therefore, more phloem sap can move from the source leaves to the fruit sinks, making them sweeter.

Concept Check 36.6

1. Plasmodesmata, unlike gap junctions, have the ability to pass RNA, proteins, and viruses from cell to cell. **2.** Long-distance signalling is critical for the integrated functioning of all large organisms, but the speed of such signalling is much less critical to plants because their responses to the environment, unlike those of animals, do not typically involve rapid movements. **3.** Although this strategy would eliminate the systemic spread of viral infections, it would also severely impact the development of the plants.

Summary of Key Concepts Questions

36.1 Plants with tall shoots and elevated leaf canopies generally had an advantage over shorter competitors. A consequence of the selective pressure for tall shoots was the further separation of leaves from roots. This separation created problems for the transport of materials between root and shoot systems. Plants with xylem cells were more successful at supplying their shoot systems with soil resources (water and minerals). Similarly, those with phloem cells were more successful at supplying sugar sinks with carbohydrates. **36.2** Xylem sap is pulled up the plant by transpiration much more often than it is pushed up the plant by root pressure. **36.3** Hydrogen bonds are necessary for the cohesion of water molecules to each other and for the adhesion of water to other materials, such as cell walls. Both adhesion and cohesion of water molecules are involved in the ascent of xylem sap under conditions of negative pressure.
36.4 Although stomata account for most of the water lost from plants, they are necessary for exchange of gases—for example, for the uptake of carbon dioxide needed for photosynthesis. The loss of water through stomata also drives the long-distance transport of water that brings soil nutrients from roots to the rest of the plant. **36.5** Although the movement of phloem sap depends on bulk flow, the pressure gradient that drives phloem transport depends on the osmotic uptake of water in response to the loading of sugars into sieve-tube elements at sugar sources. Phloem loading depends on H^+ cotransport processes that ultimately depend on H^+ gradients established by active K^+ pumping.
36.6 Electrical signalling, cytoplasmic pH, cytoplasmic Ca^{2+} concentration, and viral movement proteins all affect symplastic communication, as do developmental changes in the number of plasmodesmata.

Test Your Understanding

1. A **2.** B **3.** B **4.** C **5.** B **6.** C **7.** A **8.** D

Chapter 37

Figure Questions

Figure 37.3 Cations. At low pH, there would be more protons (H^+) to displace mineral cations from negatively charged soil particles into the soil solution.
Table 37.1 During photosynthesis, CO_2 is fixed into carbohydrates, which contribute to the dry mass. In cellular respiration, O_2 is reduced to H_2O and does not contribute to the dry mass. **Figure 37.9** Some other examples of mutualism are the following relationships. *Flashlight fish and bioluminescent bacteria:* The bacteria gain nutrients and protection from the fish, while the bioluminescence attracts prey and mates for the fish. *Flowering plants and pollinators:* Animals distribute the pollen and are rewarded by a meal of nectar or pollen. *Vertebrate herbivores and some bacteria in the digestive system:* Microorganisms in the alimentary canal break down cellulose to glucose and, in some cases, provide the animal with vitamins or amino acids. Meanwhile, the microorganisms have a steady supply of food and a warm environment. *Humans and some bacteria in the digestive system:* Some bacteria provide humans with vitamins, while the bacteria get nutrients from the digested food. **Figure 37.19** Both ammonium and nitrate. A decomposing animal would release amino acids into the soil that would be converted into ammonium by ammonifying bacteria. Some of this ammonium could be used directly by the plant. A large part of the ammonium, however, would be converted by nitrifying bacteria to form nitrate ions that could also be absorbed by the plant root system. **Figure 37.20** The legume plants benefit because the bacteria fix nitrogen that is absorbed by their roots. The bacteria benefit because they acquire photosynthetic products from the plants. **Figure 37.21** All three plant tissue systems are affected. Root hairs (dermal tissue) are modified to allow *Rhizobium* penetration. The cortex (ground tissue) and pericycle (vascular tissue) proliferate during nodule formation. The vascular tissue of the nodule connects to the vascular cylinder of the root to allow for efficient nutrient exchange.

Concept Check 37.1

1. Overwatering deprives roots of oxygen. Overfertilising is wasteful and can lead to soil salinisation and water pollution. **2.** As lawn clippings decompose, they restore mineral nutrients to the soil. If they are removed, the minerals lost from the soil must be replaced by fertilisation. **3.** Because of their small size and negative charge, clay particles would increase the number of binding sites for cations and water molecules and would therefore increase cation exchange and water retention in the soil. **4.** Due to hydrogen bonding between water molecules, water expands when it freezes, and this causes mechanical fracturing of rocks. Water also coheres to many objects, and this cohesion combined with other forces, such as gravity, can help tug particles from rock. Finally, water, because it is polar, is an excellent solvent that allows many substances, including ions, to become dissolved in solution.

Concept Check 37.2

1. No. Even though macronutrients are required in greater amounts, all essential elements are necessary for a plant to complete its life cycle. **2.** No. The fact that the addition of an element results in an increase in the growth rate of a crop does not mean that the element is strictly required for the plant to complete its life cycle. **3.** Inadequate aeration of the roots of hydroponically grown plants would promote alcohol fermentation, which uses more energy and may lead to the accumulation of ethanol, a toxic by-product of fermentation.

Concept Check 37.3

1. The rhizosphere is the zone in the soil immediately adjacent to living roots. It harbours many rhizobacteria with which the root systems form beneficial mutualisms. Some rhizobacteria produce antibiotics that protect roots from disease. Others absorb toxic metals or make nutrients more available to roots. Still others convert gaseous nitrogen into forms usable by the plant or produce chemicals that stimulate plant growth. **2.** Soil bacteria and mycorrhizae enhance plant nutrition by making certain minerals more available to plants. For example, many types of soil bacteria are involved in the nitrogen cycle, and the hyphae of mycorrhizae provide a large surface area for the absorption of nutrients, particularly phosphate ions. **3.** Mixotrophy refers to the strategy of using photosynthesis and heterotrophy for nutrition. Euglenids are well-known mixotrophic protists. **4.** Saturating rainfall may deplete the soil of oxygen. A lack of soil oxygen would inhibit nitrogen fixation by the peanut root nodules and decrease the nitrogen available to the plants. Alternatively, heavy rain may leach nitrate from the soil. A symptom of nitrogen deficiency is yellowing of older leaves.

Summary of Key Concepts Questions

37.1 The term *ecosystem* refers to the communities of organisms within a given area and their interactions with the physical environment around them. Soil is teeming with many communities of organisms, including bacteria, fungi, animals, and the root systems of plants. The vigor of these individual communities depends on nonliving factors in the soil environment, such as minerals, oxygen, and water, as well as on interactions, both positive and negative, between different communities of organisms. **37.2** No. Plants can complete their life cycle when grown hydroponically, that is, in aerated salt solutions containing the proper ratios of all the minerals needed by plants. **37.3** No. Some parasitic plants obtain their energy by siphoning off carbon nutrients from other organisms.

Test Your Understanding

1. B **2.** B **3.** A **4.** D **5.** B **6.** B **7.** D **8.** C **9.** D
10.

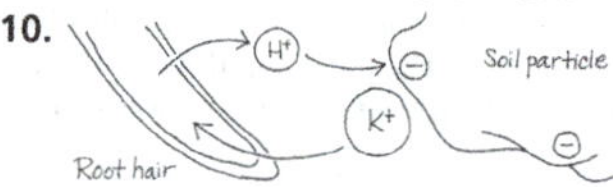

Chapter 38

Figure Questions

Figure 38.4 Having a specific pollinator is more efficient because less pollen gets delivered to flowers of the wrong species. However, it is also a risky strategy: If the pollinator population suffers to an unusual degree from predation, disease, or climate change, then the plant may not be able to produce seeds.
Figure 38.6 The part of the angiosperm life cycle characterised by the most mitotic divisions is the step between seed germination and the mature sporophyte.
Figure 38.8 Make Connections In addition to having a single cotyledon, monocots have leaves with parallel leaf venation, scattered vascular bundles in their stems, a fibrous root system, floral parts in threes or multiples of threes, and pollen grains with only one opening. In contrast, eudicots have two cotyledons, netlike leaf venation, vascular bundles in a ring, taproots, floral parts in fours or fives or multiples thereof, and pollen grains with three openings. **Visual Skills** The mature garden bean seed lacks an endosperm at maturity. Its endosperm

was consumed during seed development, and its nutrients were stored anew in the cotyledons. **Figure 38.9** Beans use a hypocotyl hook to push through the soil. The delicate leaves and shoot apical meristem are also protected by being sandwiched between two large cotyledons. The coleoptile of corn seedlings helps protect the emerging leaves.

Concept Check 38.1

1. In angiosperms, pollination is the transfer of pollen from an anther to a stigma. Fertilisation is the fusion of the egg and sperm to form the zygote; it cannot occur until after the growth of the pollen tube from the pollen grain.
2. Long styles help to weed out pollen grains that are genetically inferior and not capable of successfully growing long pollen tubes. **3.** No. The haploid (gametophyte) generation of plants is multicellular and arises from spores. The haploid phase of the animal life cycles is a single-celled gamete (egg or sperm) that arises directly from meiosis: There are no spores.

Concept Check 38.2

1. Flowering plants can avoid self-fertilisation by self-incompatibility, having male and female flowers on separate plants (dioecious species), or having stamens and styles of different heights on separate plants ("pin" and "thrum" flowers). **2.** Asexually propagated crops lack genetic diversity. Genetically diverse populations are less likely to become extinct in the face of an epidemic because there is a greater likelihood that a few individuals in the population are resistant. **3.** In the short term, selfing may be advantageous in a population that is so dispersed and sparse that pollen delivery is unreliable. In the long term, however, selfing is an evolutionary dead end because it leads to a loss of genetic diversity that may preclude adaptive evolution.

Concept Check 38.3

1. Traditional breeding and genetic engineering both involve artificial selection for desired traits. However, genetic engineering techniques facilitate faster gene transfer and are not limited to transferring genes between closely related varieties or species. **2.** *Bt* corn suffers less insect damage; therefore, *Bt* corn plants are less likely to be infected by fumonisin-producing fungi that infect plants through wounds. **3.** In such species, engineering the transgene into the chloroplast DNA would not prevent its escape in pollen; such a method requires that the chloroplast DNA be found only in the egg. An entirely different method of preventing transgene escape would therefore be needed, such as male sterility, apomixis, or self-pollinating closed flowers.

Summary of Key Concepts Questions

38.1 After pollination and fertilisation, a flower changes into a fruit. The petals, sepals, and stamens typically fall off the flower. The stigma of the pistil withers, and the ovary begins to swell. The ovules (embryonic seeds) inside the ovary begin to mature. **38.2** Asexual reproduction can be advantageous in a stable environment because individual plants that are well suited to that environment pass on all their genes to offspring. Also, asexual reproduction generally results in offspring that are less fragile than the seedlings produced by sexual reproduction. However, sexual reproduction offers the advantage of dispersal of tough seeds. Moreover, sexual reproduction produces genetic variety, which may be advantageous in an unstable environment. The likelihood is better that at least one offspring of sexual reproduction will survive in a changed environment. **38.3** "Golden Rice," although not yet in commercial production, has been engineered to produce more vitamin A, thereby raising the nutritional value of rice. A protoxin gene from a soil bacterium has been engineered into *Bt* corn. This protoxin is lethal to invertebrates but harmless to vertebrates. *Bt* crops require less pesticide spraying and have lower levels of fungal infection and fungal toxins. The nutritional value of cassava is being increased in many ways by genetic engineering. Enriched levels of iron and beta-carotene (a vitamin A precursor) have been achieved, and cyanide-producing chemicals have been almost eliminated from the roots.

Test Your Understanding

1. A **2.** C **3.** C **4.** C **5.** D **6.** D **7.** D
8.

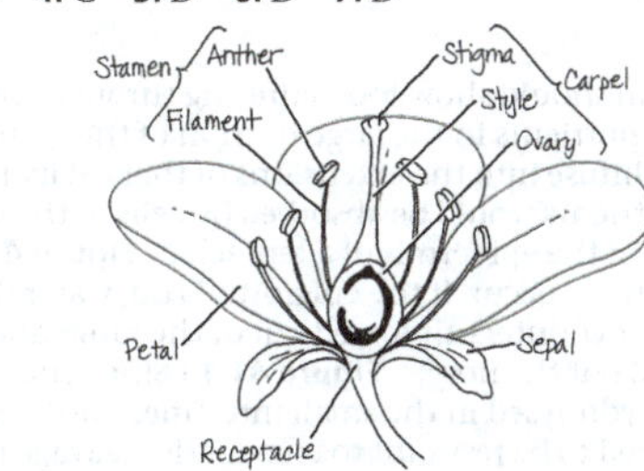

Chapter 39

Figure Questions

Figure 39.4 Panel B in Figure 11.17 shows a branching signal transduction pathway that resembles the branching phytochrome-dependent pathway involved in de-etiolation. **Figure 39.5** To determine which wavelengths of light are most effective in phototropism, you could use a glass prism to split white light into its component colours and see which colours cause the quickest bending (the answer is blue; see Figure 39.15). **Figure 39.6** No. Polar auxin transport depends on the distribution of auxin transport proteins at the basal ends of cells. **Figure 39.13** No. Since the *ein* mutation renders the seedling "blind" to ethylene, enhancing ethylene production by adding an *eto* mutation would have no effect on phenotype compared with the *ein* mutation alone. **Figure 39.16** Yes. The white light, which contains red light, would stimulate seed germination in all treatments. **Figure 39.20** Since far-red light, like darkness, causes an accumulation of the red-absorbing form (P_r) of phytochrome, single flashes of far-red light at night would have no effect on flowering beyond what the dark periods alone would have. **Figure 39.21** If this were true, then florigen would be an inhibitor of flowering, not an inducer. **Figure 39.27** Photosynthetic adaptations can occur at the molecular level, as is apparent in the fact that C_3 plants use rubisco to fix carbon dioxide initially, whereas C_4 and CAM plants use PEP carboxylase. An adaptation at the tissue level is that plants have different stomatal densities based on their genotype and environmental conditions. At the organismal level, plants alter their shoot architectures to make photosynthesis more efficient. For example, self-pruning removes branches and leaves that respire more than they photosynthesise.

Concept Check 39.1

1. Dark-grown seedlings are etiolated: They have long stems, underdeveloped root systems, and unexpanded leaves, and their shoots lack chlorophyll. Etiolated growth is beneficial to seeds sprouting under the dark conditions they would encounter underground. By devoting more energy to stem elongation and less to leaf expansion and root growth, a plant increases the likelihood that the shoot will reach the sunlight before its stored foods run out. **2.** Cycloheximide should inhibit de-etiolation by preventing the synthesis of new proteins necessary for de-etiolation. **3.** No. Applying Viagra, like injecting cyclic GMP as described in the text, should cause only a partial de-etiolation response. Full de-etiolation would require activation of the calcium branch of the signal transduction pathway.

Concept Check 39.2

1. Fusicoccin's ability to cause an increase in plasma H^+ pump activity has an auxin-like effect and promotes stem cell elongation. **2.** The plant will exhibit a constitutive triple response. Because the kinase that normally prevents the triple response is dysfunctional, the plant will undergo the triple response regardless of whether ethylene is present or the ethylene receptor is functional. **3.** Since ethylene often stimulates its own synthesis, it is under positive-feedback regulation.

Concept Check 39.3

1. Not necessarily. Many environmental factors, such as temperature and light, change over a 24-hour period in the field. To determine whether the enzyme is under circadian control, a scientist would have to demonstrate that its activity oscillates even when environmental conditions are held constant. **2.** It is impossible to say. To establish that this species is a short-day plant, it would be necessary to establish the critical night length for flowering and that this species only flowers when the night is longer than the critical night length.
3. According to the action spectrum of photosynthesis, red and blue light are the most effective in photosynthesis. Thus, it is not surprising that plants assess their light environment using blue- and red-light-absorbing photoreceptors.

Concept Check 39.4

1. A plant that overproduces ABA would undergo less evaporative cooling because its stomata would not open as widely. **2.** Plants close to the aisles may be more subject to mechanical stresses caused by passing workers and air currents. The plants nearer the centre of the bench may also be taller as a result of shading and less evaporative stress. **3.** No. Because root caps are involved in sensing gravity, roots that have their root caps removed are almost completely insensitive to gravity.

Concept Check 39.5

1. An infection can trigger the hypersensitive response, which causes a ring of cell death around the infection site, thereby isolating the pathogen from living host cells and preventing systemic infection. **2.** Mechanical damage breaches a plant's first line of defence against infection, its protective dermal tissue.
3. No. Pathogens that kill their hosts would soon run out of victims and might themselves go extinct. **4.** Perhaps the breeze dilutes the local concentration of a volatile defence compound that the plants produce.

Summary of Key Concepts Questions

39.1 Signal transduction pathways often activate protein kinases, enzymes that phosphorylate other proteins. Protein kinases can directly activate certain preexisting enzymes by phosphorylating them, or they can regulate gene transcription (and enzyme production) by phosphorylating specific transcription factors.
39.2 Yes, there is truth to the old adage that one bad apple spoils the whole bunch. Ethylene, a gaseous hormone that stimulates ripening, is produced by damaged, infected, or overripe fruits. Ethylene can diffuse to healthy fruit in the "bunch" and stimulate their rapid ripening. **39.3** Plant physiologists proposed the existence of a floral-promoting factor (florigen) based on the fact that a plant induced to flower could induce flowering in a second plant to which it was grafted, even though the second plant was not in an environment that would normally induce flowering in that species. **39.4** Plants subjected to drought stress are often more resistant to freezing stress because the two types of stress are quite similar. Freezing of water in the extracellular spaces causes free water concentrations outside the cell to decrease. This, in turn, causes free water to leave the cell by osmosis, leading to the dehydration of cytoplasm, much like what is seen in drought stress. **39.5** Chewing insects make plants more susceptible to pathogen invasion by disrupting the waxy cuticle of shoots, thereby creating an opening for infection. Moreover, substances released from damaged cells can serve as nutrients for the invading pathogens.

Test Your Understanding

1. B **2.** C **3.** D **4.** C **5.** B **6.** B **7.** C

8.

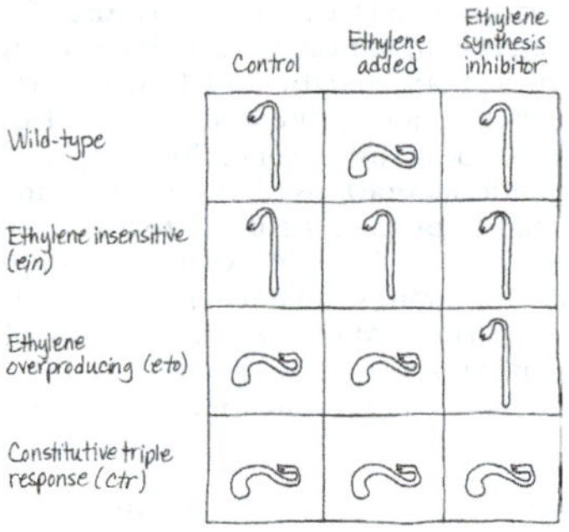

	Control	Ethylene added	Ethylene synthesis inhibitor
Wild-type			
Ethylene insensitive (ein)			
Ethylene overproducing (eto)			
Constitutive triple response (ctr)			

Chapter 40

Figure Questions

Figure 40.4 Such exchange surfaces are internal in the sense that they are inside the body. However, they are also continuous with openings on the external body surface that contact the environment. **Figure 40.6** Signals in the nervous system always travel on a direct route between the sending and receiving cell. In contrast, hormones that reach target cells can have an effect regardless of the path by which they arrive or how many times they travel through the circulatory system. **Figure 40.8** The stimuli are the room temperature increasing in the top loop and decreasing in the bottom loop. The responses could include the heater turning off and the temperature decreasing in the top loop and the heater turning on and the temperature increasing in the bottom loop. The sensor/control centre is the thermostat. The air conditioner would form a second control circuit, cooling the house when air temperature exceeded the set point. Such opposing, or antagonistic, pairs of control circuits increase the effectiveness of a homeostatic mechanism. **Figure 40.12** The conduction arrows would be in the opposite direction, transferring heat from the penguin to the ice because the penguin is warmer than the ice. **Figure 40.16** If a female Burmese python were not incubating eggs, her oxygen consumption would decrease with decreasing temperature, as for any other ectotherm. **Figure 40.17** The ice water would cool tissues in your head, including blood that would then circulate throughout your body. This effect would accelerate the return to a normal body temperature. If, however, the ice water reached the eardrum and cooled the blood vessel that supplies the hypothalamus, the hypothalamic thermostat would respond by inhibiting sweating and constricting blood vessels in the skin, slowing cooling elsewhere in the body. **Figure 40.18** The transport of nutrients across membranes and the synthesis of RNA and protein are coupled to ATP hydrolysis. These processes proceed spontaneously because there is an overall drop in free energy, with the excess energy given off as heat. Similarly, less than half of the free energy in glucose is captured in the coupled reactions of cellular respiration. The remainder of the energy is released as heat. **Figure 40.22** Nothing. Although genes that show a circadian variation in expression during euthermia exhibit constant RNA levels during hibernation, a gene that shows constant expression during hibernation might also show constant expression during euthermia. **Figure 40.23** In hot environments, both plants and animals experience evaporative cooling as a result of transpiration (in plants) or bathing, sweating, and panting (in animals); both plants and animals synthesise heat-shock proteins, which protect other proteins from heat stress; and animals also use various behavioural responses to minimise heat absorption. In cold environments, both plants and animals increase the proportion of unsaturated fatty acids in their membrane lipids and use antifreeze proteins that prevent or limit the formation of intracellular ice crystals; plants increase cytoplasmic levels of specific solutes that help reduce the loss of intracellular water during extracellular freezing; and animals increase metabolic heat production and use insulation, circulatory adaptations such as countercurrent exchange, and behavioural responses to minimise heat loss.

Concept Check 40.1

1. All types of epithelia consist of cells that line a surface, are tightly packed, are situated on top of a basal lamina, and form an active and protective interface with the external environment. **2.** An oxygen molecule must cross a plasma membrane when entering the body at an exchange surface in the respiratory system, in both entering and exiting the circulatory system, and in moving from the interstitial fluid to the cytoplasm of the body cell. **3.** You need the nervous system to perceive the danger and provoke a split-second muscular response to keep from falling. Nervous system stimulation of blood vessels is primarily via α_1-adrenergic receptors, which have a higher affinity to noradrenaline, causing generalised vasoconstriction. Blood vessels in skeletal muscle, however, have β_2-adrenergic receptors, which only have affinity to the hormone adrenaline released by the endocrine system, bringing about vasodilation and increased blood flow to the active muscle. This surge of adrenaline also stimulates liver stores to release glucose into the blood.

Concept Check 40.2

1. In thermoregulation, the product of the pathway (a change in temperature) decreases pathway activity by reducing the stimulus. In an enzyme-catalysed biosynthetic process, the product of the pathway (in this case, isoleucine) inhibits the pathway that generated it. **2.** You would want to put the thermostat close to where you would be spending time, where it would be protected from environmental perturbations, such as direct sunshine, and not right in the path of the output of the heating system. Similarly, the sensors for homeostasis located in the human brain are separated from environmental influences and can monitor conditions in a vital and sensitive tissue. **3.** In convergent evolution, the same biological trait arises independently in two or more species. Gene analysis can provide evidence for an independent origin. In particular, if the genes responsible for the trait in one species lack significant sequence similarity to the corresponding genes in another species, scientists conclude that there is a separate genetic basis for the trait in the two species and thus an independent origin. In the case of circadian rhythms, the clock genes in cyanobacteria appear unrelated to those in humans.

Concept Check 40.3

1. "Wind chill" involves heat loss through convection, as the moving air contributes to heat loss from the skin surface. **2.** The hummingbird, being a very small endotherm, has a very high metabolic rate. If by absorbing sunlight certain flowers warm their nectar, a hummingbird feeding on these flowers is saved the metabolic expense of warming the nectar to its body temperature. **3.** To raise body temperature to the higher range of fever, the hypothalamus triggers heat generation by muscular contractions, or shivering. The person with a fever may in fact say that they feel cold, even though their body temperature is above normal.

Concept Check 40.4

1. The mouse would consume oxygen at a higher rate because it is an endotherm, so its basal metabolic rate is higher than the ectothermic lizard's standard metabolic rate. **2.** The house cat; smaller animals have a higher metabolic rate per unit body mass and a greater demand for food per unit body mass. **3.** The alligator's body temperature would decrease along with the air temperature. Its metabolic rate would therefore also decrease as chemical reactions slowed. In contrast, the lion's body temperature would not change. Its metabolic rate would increase as it shivered and produced heat to keep its body temperature constant.

Summary of Key Concepts Questions

40.1 Animals exchange materials with their environment across their body surface, and a spherical shape has the minimum surface area per unit volume. As body size increases, the ratio of surface area to body volume decreases. **40.2** No; an animal's internal environment fluctuates slightly around set points or within normal ranges. Homeostasis is a dynamic state. Furthermore, there are sometimes programmed changes in set points, such as those resulting in radical increases in hormone levels at particular times in development. **40.3** Heat exchange across the skin is a primary mechanism for the regulation of body core temperature, with the result that the skin is cooler than the body core. **40.4** Small animals have a higher BMR per unit mass and therefore consume more oxygen per unit mass than large animals. A higher breathing rate is required to support this increased oxygen consumption.

Test Your Understanding

1. B **2.** C **3.** A **4.** B **5.** C **6.** B **7.** D

8.

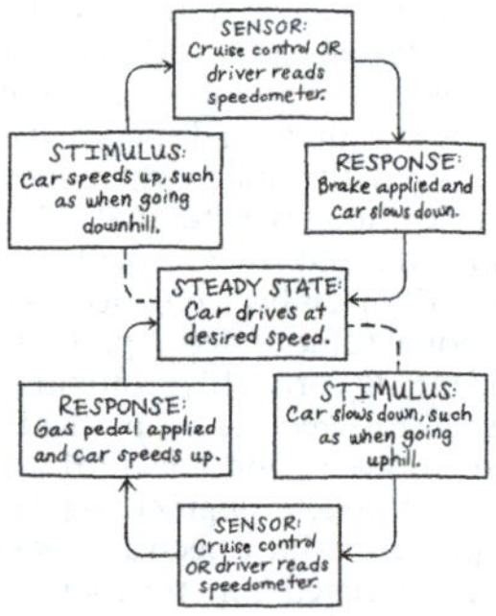

SENSOR: Cruise control OR driver reads speedometer.
RESPONSE: Brake applied and car slows down.
STEADY STATE: Car drives at desired speed.
STIMULUS: Car slows down, such as when going uphill.
SENSOR: Cruise control OR driver reads speedometer.
RESPONSE: Gas pedal applied and car speeds up.
STIMULUS: Car speeds up, such as when going downhill.

Chapter 41

Figure Questions

Figure 41.6 Your diagram should show food entering through the hydra's mouth and being digested into nutrients in the large portion of the gastrovascular cavity. The nutrients then diffuse into the extensions of that cavity that reach into the tentacles. There, nutrients would be absorbed by cells of the gastrodermis and transported to cells of the epidermis of a tentacle. **Figure 41.9** The airway must be open for exhaling to occur. If the epiglottis is up, water that entered the throat from the mouth encounters air forced out of the lungs and is carried along into the nasal cavity and out the nose. **Figure 41.11** Since enzymes are proteins, and proteins are hydrolysed in the small intestine, the digestive enzymes in that compartment need to be resistant to enzymatic cleavage other than the cleavage required to activate them. **Figure 41.12** None. Since digestion is completed in the small intestine, tapeworms simply absorb predigested nutrients through their large body surface. **Figure 41.13** Yes. The exit of the chylomicrons involves exocytosis, an active process that consumes energy in the form of ATP. In contrast, the entry of monoglycerides and fatty acids into the cell by diffusion is a passive process that does not consume energy. **Figure 41.23** Both insulin and glucagon are involved in negative feedback circuits.

Concept Check 41.1

1. The only essential amino acids are those that an animal cannot synthesise from other molecules. **2.** Many vitamins serve as enzyme cofactors, which, like enzymes themselves, are unchanged by the chemical reactions in which they participate. Therefore, only very small amounts of vitamins are needed. **3.** To identify the essential nutrient missing from an animal's diet, a researcher could supplement the diet with individual nutrients one at a time and determine which nutrient eliminates the signs of malnutrition.

Concept Check 41.2

1. A gastrovascular cavity is a digestive pouch with a single opening that functions in both ingestion and elimination; an alimentary canal is a digestive tube with a separate mouth and anus at opposite ends. **2.** As long as nutrients are within the cavity of the alimentary canal, they are in a compartment that is continuous with the outside environment via the mouth and anus and have not yet crossed a membrane to enter the body. **3.** In both cases, high-energy fuels are consumed, complex molecules are broken down into simpler ones, and waste products are eliminated. In addition, petrol, like food, is broken down in a specialised compartment, so that surrounding structures are protected from disassembly. Finally, just as food and wastes remain outside the body in a digestive tract, neither petrol nor its waste products enter the passenger compartment of the vehicle.

Concept Check 41.3

1. Because parietal cells in the stomach pump hydrogen ions into the stomach lumen where they combine with chloride ions to form HCl, a proton pump inhibitor reduces the acidity of chyme and thus the irritation that occurs when chyme enters the esophagus. **2.** Bile acids act as emulsifiers, dispersing large fat globules into smaller fat droplets. The lipid-soluble surfaces of the bile acids bind to the fat droplets and the water-soluble surfaces of the bile acids interact with the digestive fluids. As a result, surface tension is reduced and the droplets are stabilised, facilitating enzymatic digestion of fats by lipases. **3.** Proteins would be denatured and digested into peptides. Further digestion, to individual amino acids, would require enzymatic secretions found in the small intestine. No digestion of carbohydrates or lipids would occur.

Concept Check 41.4

1. The increased time for transit through the alimentary canal allows for more extensive processing, and the increased surface area of the canal provides greater opportunity for absorption. **2.** A mammal's digestive system provides mutualistic microorganisms with an environment that is protected against other microorganisms by saliva and gastric juice, that is held at a constant temperature conducive to enzyme action, and that provides a steady source of nutrients. **3.** For the yoghurt treatment to be effective, the bacteria from yoghurt would have to establish a mutualistic relationship with the small intestine, where disaccharides are broken down and sugars are absorbed. Conditions in the small intestine are likely to be very different from those in a yoghurt culture. The bacteria might be killed before they reach the small intestine, or they might not be able to grow there in sufficient numbers to aid in digestion.

Concept Check 41.5

1. Over the long term, the body stores excess calories in fat, whether those calories come from fat, carbohydrate, or protein in food. **2.** In most individuals, leptin levels decline during fasting. Individuals in the group with low levels of leptin are likely to be unable to produce leptin, so leptin levels would remain low regardless of food intake. Individuals in the group with high leptin levels are likely to be unable to respond to leptin, but they still should shut off leptin production as fat stores are used up. **3.** Excess insulin production will cause the blood glucose level to decrease below normal physiological levels. It will also trigger glycogen synthesis in the liver, further decreasing the blood glucose level. However, a low blood glucose level will stimulate the release of glucagon from alpha cells in the pancreas, which will trigger glycogen breakdown. Thus, there will be antagonistic effects in the liver.

Summary of Key Concepts Questions

41.1 Since the cofactor is necessary in all animals, those animals that do not require it in their diet must be able to synthesise it from other organic molecules. **41.2** A liquid diet containing glucose, amino acids, and other building blocks could be ingested and absorbed without the need for mechanical or chemical digestion. **41.3** The small intestine has a much larger surface area than the stomach. **41.4** The assortment of teeth in our mouth and the short length of our cecum suggest that our ancestors' digestive systems were not specialised for digesting plant material. **41.5** When mealtime arrives, nervous inputs from the brain signal the stomach to prepare to digest food through secretions and churning.

Test Your Understanding

1. B **2.** A **3.** B **4.** B **5.** B **6.** D **7.** B

8.

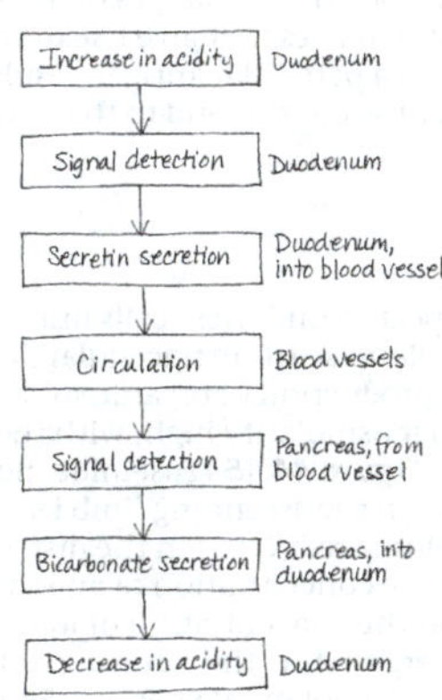

Chapter 42

Figure Questions

Figure 42.2 Although gas exchange might be improved by a steady, one-way flow of fluid, there would likely be inadequate time for food to be digested and nutrients absorbed if fluid flowed through the cavity in this manner. **Figure 42.5** Two capillary beds. The molecule of carbon dioxide would need to enter a capillary bed in the thumb before returning to the right atrium and ventricle, then travel to the lung and enter a capillary from which it could diffuse into an alveolus and be available to be exhaled. **Figure 42.8** Each feature of the ECG recording, such as the sharp upwards spike, occurs once per cardiac cycle. Using the *x*-axis to measure the time in seconds between successive spikes and dividing that number by 60 would yield the heart rate as the number of cycles per minute. **Figure 42.25** The reduction in surface tension results from the presence of surfactant. Therefore, for all the infants who had died of RDS, you would expect the amount of surfactant to be near zero. For infants who had died of other causes, you would expect the amount of surfactant to be near zero for body masses less than 1,200 g but much greater than zero for body masses above 1,200 g. **Figure 42.27** Since exhalation is largely passive, the recoil of the elastic fibres in alveoli helps force air out of the lungs. When alveoli lose their elasticity, as occurs in the disease emphysema, less air is exhaled. Because more air is left in the lungs, less fresh air can be inhaled. With a smaller volume of air exchanged, there is a decrease in the partial pressure gradient that drives gas exchange. **Figure 42.28** Breathing at a rate greater than that needed to meet metabolic demand (hyperventilation) would lower the blood CO_2 level. Sensors in major blood vessels and the medulla would signal the breathing control centre to decrease the rate of contraction of the diaphragm and rib muscles, decreasing the breathing rate and restoring normal CO_2 level in the blood and other tissues. **Figure 42.29** The resulting increase in tidal volume would enhance ventilation within the lungs, increasing P_{O_2} and decreasing P_{CO_2} in the alveoli.

Concept Check 42.1

1. In both an open circulatory system and a fountain, fluid is pumped through a tube and then returns to the pump after collecting in a pool. **2.** The ability to shut off blood supply to the lungs when the animal is submerged **3.** The O_2 content would be abnormally low because some oxygen-depleted blood returned to the right atrium from the systemic circuit would mix with the oxygen-rich blood in the left atrium.

Concept Check 42.2

1. The pulmonary veins carry blood that has just passed through capillary beds in the lungs, where it accumulated O_2. The venae cavae carry blood that has just passed through capillary beds in the rest of the body, where it lost O_2 to the tissues. **2.** The delay allows the atria to empty completely, filling ventricles fully before they contract. **3.** The heart, like any other muscle, becomes stronger through regular exercise. You would expect a stronger heart to have a greater stroke volume, which would allow for the decrease in heart rate.

Concept Check 42.3

1. The large total cross-sectional area of the capillaries **2.** An increase in blood pressure and cardiac output combined with the diversion of more blood to the skeletal muscles would increase the capacity for action by increasing the rate of blood circulation and delivering more O_2 and nutrients to the skeletal muscles. **3.** Additional hearts could be used to improve blood return from the legs. However, it might be difficult to coordinate the activity of multiple hearts and to maintain adequate blood flow to hearts far from the gas exchange organs.

Concept Check 42.4

1. An increase in the number of white blood cells (leucocytes) may indicate that the person is combating an infection. **2.** Clotting factors do not initiate clotting but are essential steps in the clotting process. **3.** The chest pain results from inadequate blood flow in coronary arteries. Vasodilation promoted by nitric oxide from nitroglycerin increases blood flow, providing the heart muscle with additional oxygen and thus relieving the pain. **4.** Embryonic stem cells are pluripotent rather than multipotent, meaning they can give rise to many rather than a few different cell types.

Concept Check 42.5

1. Their interior position helps gas exchange tissues stay moist. If the respiratory surfaces of lungs extended into the terrestrial environment, they would quickly dry out, and diffusion of O_2 and CO_2 across these surfaces would stop. **2.** Earthworms need to keep their skin moist for gas exchange, but they need air outside this moist layer. If they stay in their waterlogged tunnels after a heavy rain, they will suffocate because they cannot get as much O_2 from water as from air. **3.** In fish, water passes over the gills in the direction opposite to that of blood flowing through the gill capillaries, maximising the extraction of oxygen from the water along the length of the exchange surface. Similarly, in the extremities of some vertebrates, blood flows in opposite directions in neighbouring veins and arteries; this countercurrent arrangement maximises the recapture of heat from blood leaving the body core in arteries, which is important for thermoregulation in cold environments.

Concept Check 42.6

1. An increase in blood CO_2 concentration causes an increase in the rate of CO_2 diffusion into the cerebrospinal fluid, where the CO_2 combines with water to form carbonic acid. Dissociation of carbonic acid releases hydrogen ions, decreasing the pH of the cerebrospinal fluid. **2.** Increased heart rate increases the rate at which CO_2-rich blood is delivered to the lungs, where CO_2

is removed. **3.** A hole would allow air to enter the space between the inner and outer layers of the double membrane, resulting in a condition called a pneumothorax. The two layers would no longer stick together, and the lung on the side with the hole would collapse and cease functioning.

Concept Check 42.7

1. Differences in partial pressure between the capillaries and the surrounding tissues or medium; the net diffusion of a gas occurs from a region of higher partial pressure to a region of lower partial pressure. **2.** The Bohr shift causes haemoglobin to release more O_2 at a lower pH, such as is found in the vicinity of tissues with high rates of cellular respiration and CO_2 release. **3.** The doctor is assuming that the rapid breathing is the body's response to low blood pH. Metabolic acidosis, the lowering of blood pH as a result of metabolism, can have many causes, including complications of certain types of diabetes, shock (extremely low blood pressure), and poisoning.

Summary of Key Concepts Questions

42.1 In a closed circulatory system, an ATP-driven muscular pump generally moves fluids in one direction on a scale of millimetres to metres. Exchange between cells and their environment relies on diffusion, which involves random movements of molecules. Concentration gradients of molecules across exchange surfaces can drive rapid net diffusion on a scale of 1 mm or less. **42.2** Replacement of a defective valve should increase stroke volume. A lower heart rate would therefore be sufficient to maintain the same cardiac output. **42.3** Blood pressure in the arm would fall by 25–30 mm Hg, the same difference as is normally seen between your heart and your brain. **42.4** One microlitre of blood contains about 5 million erythrocytes and 5,000 leucocytes, so leucocytes make up only about 0.1% of the cells in the absence of infection. **42.5** Because CO_2 is such a small fraction of atmospheric gas (0.29 mm Hg/760 mm Hg, or less than 0.04%), the partial pressure gradient of CO_2 between the respiratory surface and the environment always strongly favours the release of CO_2 to the atmosphere. **42.6** Because the lungs do not empty completely with each breath, incoming and outgoing air mix. Lungs thus contain a mixture of fresh and stale air. **42.7** An enzyme speeds up a reaction without changing the equilibrium and without being consumed. Similarly, a respiratory pigment speeds up the exchange of gases between the body and the external environment without changing the equilibrium state and without being consumed.

Test Your Understanding

1. C **2.** A **3.** D **4.** C **5.** C **6.** A **7.** A
8.

Chapter 43

Figure Questions

Figure 43.3 Dicer-2 binds double-stranded RNA without regard to size or sequence and then cuts that RNA into fragments, each 21 base pairs long. The Argo complex binds to double-stranded RNA fragments that are each 21 base pairs long, displaces one strand, and then uses the remaining strand to match to a particular target sequence in a single-stranded mRNA. **Figure 43.4** Cell-surface TLRs recognise molecules on the surface of pathogens, whereas TLRs in vesicles recognise internal molecules of pathogens after the pathogens are broken down.
Figure 43.5 Because the pain of a splinter stops almost immediately when you remove it from the skin, you can correctly deduce that the signals that mediate the inflammatory response are quite short-lived. **Figure 43.10** Part of the enzyme or antigen receptor provides a structural "backbone" that maintains overall shape, while interaction occurs at a surface with a close fit to the substrate or antigen. The combined effect of multiple noncovalent interactions at the active site or binding site is a high-affinity interaction of tremendous specificity.
Figure 43.14 After gene rearrangement, a lymphocyte and its daughter cells make a single version of the antigen receptor. In contrast, alternative splicing is not heritable and can give rise to diverse gene products in a single cell. **Figure 43.16** A single B cell has more than 100,000 identical antigen receptors on its surface, not four, and there are more than 1 million B cells differing in their antigen specificity, not three. **Figure 43.19** These receptors enable memory cells to present antigen on their cell surface to a helper T cell. This presentation of antigen is required to activate memory cells in a secondary immune response. **Figure 43.23** Primary response: arrows extending from Antigen (1st exposure), Antigen-presenting cell, Helper T cell, B cell, Plasma cells, Cytotoxic T cell, and Active cytotoxic T cells; secondary response: arrows extending from Antigen (2nd exposure), Memory helper T cells, Memory B cells, Memory cytotoxic T cells, Plasma cells, and Active cytotoxic T cells **Figure 43.25** There would be no change in the results. Because the two antigen-binding sites of an antibody have identical specificity, the two bacteriophages bound would have to display the same viral peptide.

Concept Check 43.1

1. Because pus contains white blood cells, fluid, and cell debris, it indicates an active and at least partially successful inflammatory response against invading pathogens. **2.** Whereas the ligand for the TLR receptor is a foreign molecule, the ligand for many signal transduction pathways is a molecule produced by the organism itself. **3.** Mounting an immune response would require recognition of some molecular feature of the wasp egg not found in the host. It might be that only some potential hosts have a receptor with the necessary specificity.

Concept Check 43.2

1. See Figure 43.9. The transmembrane regions lie within the C regions, which also form the disulfide bridges. In contrast, the antigen-binding sites are in the V regions. **2.** Generating memory cells ensures both that a receptor specific for a particular epitope will be present and that there will be more lymphocytes with this specificity than in a host that had never encountered the antigen.
3. If each B cell produced two different light and heavy chains for its antigen receptor, different combinations would make four different receptors. If any one were self-reactive, the lymphocyte would be eliminated in the generation of self-tolerance. For this reason, many more B cells would be eliminated, and those that could respond to a foreign antigen would be less effective at doing so due to the variety of receptors (and antibodies) they express.

Concept Check 43.3

1. A child lacking a thymus would have no functional T cells. Without helper T cells to help activate B cells, the child would be unable to produce antibodies against extracellular bacteria. Furthermore, without cytotoxic T cells or helper T cells, the child's immune system would be unable to kill virus-infected cells.
2. Since the antigen-binding site is intact, the antibody fragments could neutralise viruses and opsonise bacteria. **3.** If the handler developed immunity to proteins in the antiveniom, another injection could provoke a severe immune response.

Concept Check 43.4

1. Myasthenia gravis is considered an autoimmune disease because the immune system produces antibodies against self molecules (certain receptors on muscle cells). **2.** A person with a cold is likely to produce oral and nasal secretions that facilitate viral transfer. In addition, since sickness can cause incapacitation or death, a virus that is programmed to exit the host when there is a physiological stress has the opportunity to find a new host at a time when the current host may cease to function. **3.** A person with a macrophage deficiency would have frequent infections. The causes would be poor innate responses, due to diminished phagocytosis and inflammation, and poor adaptive responses, due to the lack of macrophages to present antigens to helper T cells.

Summary of Key Concepts Questions

43.1 Lysozyme in saliva destroys bacterial cell walls; the viscosity of mucus helps trap bacteria; acidic pH in the stomach kills many bacteria; and the tight packing of cells lining the gut provides a physical barrier to infection. **43.2** Sufficient numbers of cells to mediate an innate immune response are always present, whereas an adaptive response requires selection and proliferation of an initially very small cell population specific for the infecting pathogen. **43.3** No. Immunological memory after a natural infection and that after vaccination are very similar. There may be minor differences in the particular antigens that can be recognised in a subsequent infection. **43.4** No. AIDS refers to a loss of immune function that can occur over time in an individual infected with HIV. However, certain multidrug combinations ("cocktails") or rare genetic variations usually prevent progression to AIDS in people infected with HIV.

Test Your Understanding

1. B **2.** C **3.** C **4.** B **5.** B **6.** B **7.** C
8. One possible answer:

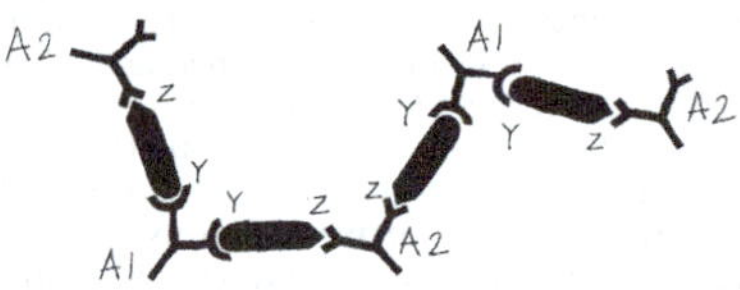

9. Lamarck's discredited idea was that organisms changed their form to fit challenges and then somehow passed those changes on to their descendants. In clonal selection, heritable differences that give rise to variation arise prior to any challenge. An encounter with a particular antigen results in proliferation of the variants best suited to recognise and respond to that challenge.

Chapter 44

Figure Questions

Figure 44.13 You would expect to find these cells lining tubules where they pass through the renal medulla. Because the extracellular fluid of the renal medulla has a very high osmolarity, production of organic solutes by tubule cells in this region keeps intracellular osmolarity high, with the result that these cells maintain normal volume. **Figure 44.15** Frusemide increases urine volume. The absence of ion transport in the ascending limb leaves the filtrate too concentrated for substantial volume reduction in the distal tubule and collecting duct. **Figure 44.18** When the concentration of an ion differs across a plasma membrane, the difference in the concentration of ions inside and outside represents chemical potential energy, while the resulting difference in charge inside and outside represents electrical potential energy. **Figure 44.21** The ADH levels would likely be elevated in both sets of patients with mutations because either defect prevents the recapture of water that restores blood osmolarity to normal levels. **Figure 44.22** Arrows that would be labelled "secretion" are the arrows indicating secretion of aldosterone, angiotensinogen, and renin.

Concept Check 44.1

1. Because the salt is moved against its concentration gradient, from low concentration (fresh water) to high concentration (blood) **2.** A freshwater osmoconformer would have body fluids too dilute to carry out life's processes. **3.** Without a layer of insulating fur, the camel must use the cooling effect of evaporative water loss to maintain body temperature, thus linking thermoregulation and osmoregulation.

Concept Check 44.2

1. Because uric acid is largely insoluble in water, it can be excreted as a semisolid paste, thereby reducing an animal's water loss. **2.** Humans produce uric acid from purine breakdown, and reducing purines in the diet often lessens the severity of gout. Birds, however, produce uric acid as a waste product of general nitrogen metabolism. They would therefore need a diet low in all nitrogen-containing compounds, not just purines.

Concept Check 44.3

1. In flatworms, ciliated cells draw interstitial fluids containing waste products into protonephridia. In earthworms, waste products pass from interstitial fluids into the coelom. From there, cilia move the wastes into metanephridia via a funnel surrounding an internal opening to the metanephridia. In insects, the Malpighian tubules pump fluids from the haemolymph, which receives waste products during exchange with cells in the course of circulation. **2.** Filtrate is formed when the glomerulus filters blood from the renal artery within Bowman's capsule. Some of the filtrate contents are recovered, enter capillaries, and exit in the renal vein; the rest remain in the filtrate and pass out of the kidney in the ureter. **3.** The presence of Na^+ and other ions (electrolytes) in the dialysate would limit the extent to which they would be removed from the filtrate during dialysis. Adjusting the electrolytes in the starting dialysate can thus lead to the restoration of proper electrolyte concentrations in the plasma. Similarly, the absence of urea and other waste products in the starting dialysate facilitates their removal from the filtrate.

Concept Check 44.4

1. The numerous nephrons and well-developed glomeruli of freshwater fishes produce urine at a high rate, while the small numbers of nephrons and smaller glomeruli of marine fishes produce urine at a low rate. **2.** The kidney medulla would absorb less water; thus, the drug would increase the amount of water lost in the urine. **3.** A decline in blood pressure in the afferent arteriole would reduce the rate of filtration by moving less material through the vessels.

Concept Check 44.5

1. Alcohol inhibits the release of ADH, causing an increase in urinary water loss and increasing the chance of dehydration. **2.** The consumption of a very large amount of water in a short period of time, coupled with an absence of solute intake, can reduce sodium levels in the blood below tolerable levels. This condition, called hyponatremia, leads to disorientation and, sometimes, respiratory distress. It has occurred in some marathon runners who drink water rather than sports drinks. (It has also caused the death of a fraternity pledge as a consequence of a water hazing ritual and the death of a contestant in a water-drinking competition.) **3.** High blood pressure

Summary of Key Concepts Questions

44.1 Water moves into a cell by osmosis when the fluid outside the cells is hypoosmotic (has a lower solute concentration than the cytosol). **44.2** As cofactors for the enzymes that catalyse metabolism, nitrogen-containing molecules such as NAD^+/NADH are "recycled" during cellular respiration. They thus are not broken down and their components are not absorbed or excreted. **44.3** Filtration produces a fluid for exchange processes that is free of cells and large molecules, which are of benefit to the animal and could not readily be reabsorbed. **44.4** Both types of nephrons have proximal tubules that can reabsorb nutrients, but only juxtamedullary nephrons have loops of Henle that extend deep into the renal medulla. Thus, only kidneys containing juxtamedullary nephrons can produce urine that is more concentrated than the blood. **44.5** Patients who don't produce ADH have symptoms relieved by treatment with the hormone, but many patients with diabetes insipidus lack functional receptors for ADH.

Test Your Understanding

1. C **2.** A **3.** C **4.** D **5.** C **6.** B

Chapter 45

Figure Questions

Figure 45.4

CH_3
HO
N
H
OH
HO
Adrenaline

Figure 45.5 The hormone is water-soluble and has a cell-surface receptor. Such receptors, unlike those for lipid-soluble hormones, can cause observable changes in cells without hormone-dependent gene transcription. **Figure 45.6** ATP is enzymatically converted to cAMP. The other steps represent binding reactions. **Figure 45.21** The embryonic gonad can become either a testis or an ovary. In contrast, the ducts either form a particular structure or degenerate, and the bladder forms in both males and females.

Concept Check 45.1

1. Water-soluble hormones, which cannot penetrate the plasma membrane, bind to cell-surface receptors. This interaction triggers an intracellular signal transduction pathway that ultimately alters the activity of a preexisting protein in the cytoplasm and/or changes transcription of specific genes in the nucleus. Steroid hormones are lipid-soluble and can cross the plasma membrane into the cell interior, where they bind to receptors located in the cytosol or nucleus. The hormone-receptor complex then functions directly as a transcription factor that changes transcription of specific genes. **2.** An exocrine gland, because pheromones are not secreted into interstitial fluid but instead are typically released onto a body surface or into the environment **3.** Because receptors for water-soluble hormones are located on the cell surface, facing the extracellular space, injecting the hormone into the cytosol would not trigger a response.

Concept Check 45.2

1. Prolactin regulates milk production, and oxytocin regulates milk release. **2.** The posterior pituitary, an extension of the hypothalamus that contains the axons of neurosecretory cells, is the storage and release site for two neurohormones, oxytocin and antidiuretic hormone (ADH). The anterior pituitary contains endocrine cells that make at least six different hormones. Secretion of anterior pituitary hormones is controlled by hypothalamic hormones that travel via blood vessels to the anterior pituitary. **3.** The hypothalamus and pituitary glands function in many different endocrine pathways. Many defects in these glands, such as those affecting growth or organisation, would therefore disrupt many hormone pathways. Only a very specific defect, such as a mutation affecting a particular hormone receptor, would alter just one endocrine pathway. The situation is quite different for the final gland in a pathway, such as the thyroid gland. In this case, a wide range of defects that disrupt gland function would disrupt only the one pathway or small set of pathways in which that gland functions. **4.** Both diagnoses could be correct. In one case, the thyroid gland may produce excess thyroid hormone despite normal hormonal input from the hypothalamus and anterior pituitary. In the other, abnormally elevated hormonal input (an elevated TSH level) may be the cause of the overactive thyroid gland.

Concept Check 45.3

1. If the function of the pathway is to provide a transient response, a short-lived stimulus would be less dependent on negative feedback. **2.** You would be exploiting the anti-inflammatory activity of glucocorticoids. Local injection avoids the effects on glucose metabolism that would occur if glucocorticoids were taken orally and transported throughout the body in the bloodstream. **3.** Both hormones produce opposite effects in different target tissues. In the fight-or-flight response, adrenaline increases blood flow to skeletal muscles and reduces blood flow to smooth muscles in the digestive system. In establishing apical dominance, auxin promotes the growth of apical buds and inhibits the growth of lateral buds.

Summary of Key Concepts Questions

45.1 As shown in Figure 43.18, helper T cell activation by cytokines acting as local regulators involves both autocrine and paracrine signalling. **45.2** The pancreas, parathyroid glands, and pineal gland **45.3** Both the pituitary and the adrenal glands are formed by fusion of neural and nonneural tissue. ADH is secreted by the neurosecretory portion of the pituitary gland, and adrenaline is secreted by the neurosecretory portion of the adrenal gland.

Test Your Understanding

1. C **2.** D **3.** D **4.** A **5.** B **6.** B **7.** A

8.

Prolactin-releasing hormone circulates in body via blood
↓
Anterior pituitary secretes prolactin (o)
Prolactin circulates in body via blood
↓
Mammary glands
↓
Milk production

Chapter 46

Figure Questions

Figure 46.10 Newly formed sperm enter the seminal vesicle from the testis and exit via the ejaculatory duct during intercourse. Sperm enter the spermatheca after intercourse and, after storage, are released into the oviduct to fertilise an egg moving into the uterus. **Figure 46.11** When successfully courted by a second male, regardless of his genotype, about one-third of the females rid themselves of all sperm from the first mating. Thus, two-thirds retained some sperm from the first mating. We would therefore predict that two-thirds of those females would have some offspring exhibiting the small-eye phenotype of the dominant mutation carried by the males with which the females mated first. **Figure 46.14** The

analysis would be informative because the polar bodies contain all of the maternal chromosomes that don't end up in the mature egg. For example, finding two copies of the disease gene in the polar bodies would indicate its absence in the egg. This method of genetic testing is sometimes carried out when oocytes collected from a female are fertilised with sperm in a laboratory dish.
Figure 46.18 The embryo normally implants about a week after conception, but it spends several days in the uterus before implanting, receiving nutrients from the endometrium. Therefore, the fertilised egg should be cultured for several days in liquid that is at normal body temperature and contains the same nutrients as those provided by the endometrium before implantation. **Figure 46.19** Testosterone can pass from fetal blood to maternal blood via the placental circulation, temporarily upsetting the hormonal balance in the mother. **Figure 46.21** Oxytocin would most likely induce labour, starting a positive-feedback loop that would direct labour to completion. Synthetic oxytocin is in fact frequently used to induce labour when prolonged pregnancy might endanger the mother or fetus.

Concept Check 46.1

1. The offspring of sexual reproduction are more genetically diverse. However, asexual reproduction can produce more offspring over multiple generations. **2.** Unlike other forms of asexual reproduction, parthenogenesis involves gamete production. By controlling whether or not haploid eggs are fertilised, species such as honeybees can readily switch between asexual and sexual reproduction. **3.** No. Owing to random assortment of chromosomes during meiosis, the offspring may receive the same copy or different copies of a particular parental chromosome from the sperm and the egg. Furthermore, genetic recombination during meiosis will result in reassortment of genes between pairs of parental chromosomes. **4.** Fragmentation occurs in both plants and animals. Also, budding in animals and the growth of adventitious plant roots both involve emergence of new individuals from outgrowths of the parent.

Concept Check 46.2

1. Internal fertilisation allows sperm to reach the egg without either gamete drying out. **2.** (a) Animals with external fertilisation tend to release many gametes at once, resulting in the production of enormous numbers of zygotes. This increases the chances that some will survive to adulthood. (b) Animals with internal fertilisation produce fewer offspring but generally exhibit greater care of the embryos and the young. **3.** Like the uterus of an insect, the ovary of a plant is the site of fertilisation. Unlike the plant ovary, the uterus is not the site of egg production, which occurs in the insect ovary. In addition, the fertilised insect egg is expelled from the uterus, whereas the plant embryo develops within a seed in the ovary.

Concept Check 46.3

1. Spermatogenesis occurs normally only when the testicles are cooler than normal body temperature. Extensive use of a hot tub (or of very tight-fitting underwear) can cause a decrease in sperm quality and number. **2.** In humans, the secondary oocyte combines with a sperm before it finishes the second meiotic division. Thus, oogenesis is completed after, not before, fertilisation.
3. The only effect of sealing off each vas deferens is an absence of sperm in the ejaculate. Sexual response and ejaculate volume are unchanged. The cutting and sealing off of these ducts, a *vasectomy*, is a common surgical procedure for men who do not wish to produce any (more) offspring.

Concept Check 46.4

1. In the testis, FSH stimulates the Sertoli cells, which nourish developing sperm. LH stimulates the production of androgens (mainly testosterone), which in turn stimulate sperm production. In both females and males, FSH encourages the growth of cells that support and nourish developing gametes (follicle cells in females and Sertoli cells in males), and LH stimulates the production of sex hormones that promote gametogenesis (oestrogens, primarily oestradiol, in females and androgens, especially testosterone, in males). **2.** In oestrous cycles, which occur in most female mammals, the endometrium is reabsorbed (rather than shed) if fertilisation does not occur. Oestrous cycles often occur just once or a few times a year, and the female is usually receptive to copulation only during the period around ovulation. Menstrual cycles are found only in humans and some other primates. They control the buildup and breakdown of the uterine lining, but not sexual receptivity. **3.** The combination of oestradiol and progesterone would have a negative-feedback effect on the hypothalamus, blocking release of GnRH. This would interfere with LH secretion by the pituitary, thus preventing ovulation. This is in fact one basis of action of the most common hormonal contraceptives. **4.** In the viral replicative cycle, the production of new viral genomes is coordinated with capsid protein expression and with the production of phospholipids for viral coats. In the reproductive cycle of a human female, there is hormonally based coordination of egg maturation with the development of support tissues of the uterus.

Concept Check 46.5

1. The secretion of hCG by the early embryo stimulates the corpus luteum to make progesterone, which helps maintain the pregnancy. During the second trimester, however, hCG production drops, the corpus luteum disintegrates, and the placenta completely takes over progesterone production. **2.** Both tubal ligation and vasectomy block the movement of gametes from the gonads to a site where fertilisation could take place. **3.** The introduction of a sperm nucleus directly into an oocyte bypasses the sperm's acquisition of motility in the epididymis, its swimming to meet the egg in the oviduct, and its fusion with the egg.

Summary of Key Concepts Questions

46.1 No. Because parthenogenesis involves meiosis, the mother would pass on to each offspring a random and therefore typically distinct combination of the chromosomes she inherited from her mother and father. **46.2** None. **46.3** The small size and lack of cytoplasm characteristic of a sperm are adaptations well suited to its function as a delivery vehicle for DNA. The large size and rich cytoplasmic contents of eggs support the growth and development of the embryo. **46.4** Circulating anabolic steroids mimic the feedback regulation of testosterone, turning off pituitary signalling to the testes and thereby blocking the release of signals required for spermatogenesis. **46.5** Oxygen in maternal blood diffuses from pools in the endometrium into fetal capillaries in the chorionic villi of the placenta and from there travels throughout the circulatory system of the fetus.

Test Your Understanding

1. D **2.** B **3.** B **4.** C **5.** A **6.** B **7.** C **8.** C
9.

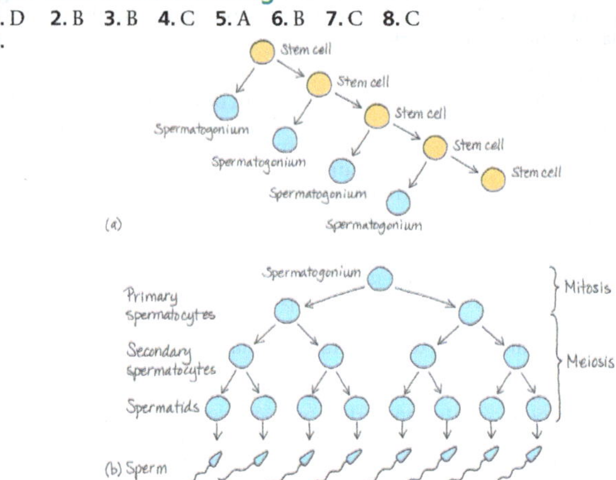

(c) The supply of stem cells would be used up, and spermatogenesis would not be able to continue.

Chapter 47

Figure Questions

Figure 47.3 You could inject the compound into an unfertilised egg, expose the egg to sperm, and see whether the fertilisation envelope forms.
Figure 47.6 There would be fewer cells, and they would be closer together.
Figure 47.8 (1) The blastocoel forms a single compartment that surrounds the gut, much like a doughnut surrounds a hole. (2) Ectoderm forms the outer covering of the animal, and endoderm lines the internal organs, such as the digestive tract. Mesoderm fills much of the space between these two layers.
Figure 47.19 Eight cell divisions are required to give rise to the intestinal cell closest to the mouth. **Figure 47.22** When the researchers allowed normal cortical rotation to occur, the "back-forming" determinants were activated. When they then forced the opposite rotation to occur, the back was established on the opposite side as well. Because the molecules on the normal side were already activated, forcing the opposite rotation apparently did not "cancel out" the establishment of the back side by the first rotation.
Figure 47.23 Draw It

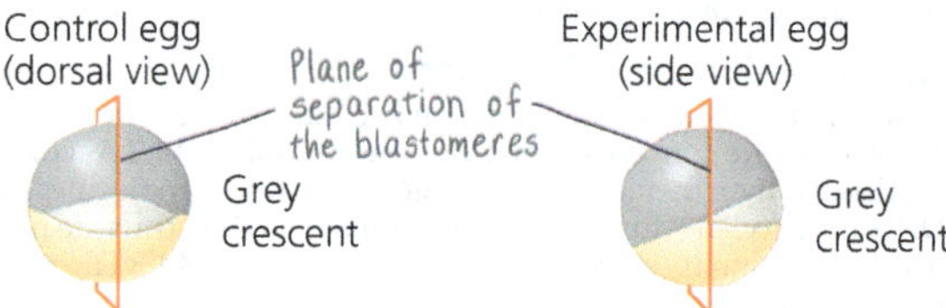

What If? In Spemann's control, the two blastomeres were physically separated, and each grew into a whole embryo. In Roux's experiment, remnants of the dead blastomere were still contacting the live blastomere, which developed into a half-embryo. Therefore, molecules present in the dead cell's remnants may have been signalling to the live cell, inhibiting it from making all the embryonic structures.
Figure 47.24 You could inject the isolated protein (or an mRNA encoding it) into ventral cells of an earlier gastrula. If dorsal structures form on the ventral side, that would support the idea that the protein is the signalling molecule secreted or presented by the dorsal lip. You should also do a control experiment to make sure the injection process alone did not cause dorsal structures to form. **Figure 47.26** Either Sonic hedgehog mRNA or protein can serve as a marker of the zone of polarising activity (ZPA). The absence of either one after removal of the apical ectodermal ridge would support your hypothesis. You could also block fibroblast growth factor function and see whether the ZPA formed (by looking for Sonic hedgehog).

Concept Check 47.1

1. The fertilisation envelope forms after cortical granules release their contents outside the egg, causing the vitelline membrane to rise and harden. The fertilisation envelope serves as a barrier to fertilisation by more than one sperm. **2.** The increased Ca^{2+} concentration in the egg would cause the cortical granules to fuse with the plasma membrane, releasing their contents and causing a fertilisation envelope to form, even though no sperm had entered. This would prevent fertilisation. **3.** You would expect it to fluctuate. The fluctuation of MPF drives the transition between DNA replication (S phase) and mitosis (M phase), which is still required in the abbreviated cleavage cell cycle.

Concept Check 47.2

1. The cells of the notochord migrate towards the midline of the embryo (converge), rearranging themselves so there are fewer cells across the notochord, which thus becomes longer overall (extends; see Figure 47.17). **2.** Because microfilaments would not be able to contract and decrease the size of one end of the cell, both the inwards bending in the middle of the neural tube and the outwards bending of the hinge regions at the edges would be blocked. Therefore, the neural tube probably would not form. **3.** Dietary intake of the vitamin folic acid dramatically reduces the frequency of neural tube defects.

Concept Check 47.3

1. Axis formation establishes the location and polarity of the three axes that provide the coordinates for development. Pattern formation positions particular tissues and organs in the three-dimensional space defined by those coordinates. **2.** Morphogen gradients act by specifying cell fates across a field of cells through variation in the level of a determinant. Morphogen gradients thus act more globally than cytoplasmic determinants or inductive interactions between pairs of cells. **3.** Yes, a second embryo could develop because inhibiting BMP-4 activity would have the same effect as transplanting an organiser. **4.** The limb that developed probably would have a mirror-image duplication, with the most posterior digits in the middle and the most anterior digits at either end.

Summary of Key Concepts Questions

47.1 The binding of a sperm to a receptor on the egg surface is very specific and likely would not occur if the two gametes were from different species. Without sperm binding, the sperm and egg membranes would not fuse. **47.2** Apoptosis functions to eliminate structures required only in an immature form, nonfunctional cells from a pool larger than the number required, and tissues formed by a developmental program that is not adaptive for the organism as it has evolved. **47.3** Mutations that affected both limb and kidney development would be more likely to alter the function of monocilia because these organelles are important in several signalling pathways. Mutations that affected limb development but not kidney development would more likely alter a single pathway, such as Hedgehog signalling.

Test Your Understanding

1. A **2.** B **3.** D **4.** A **5.** D **6.** C **7.** B

8.

Chapter 48

Figure Questions

Figure 48.7 Potassium and sodium channels must differ in the structure of the channel through which the ions pass. The channel could differ in the size of the opening, the distribution of charge, or other properties that would allow one type of ion but not others to diffuse through the channel. **Figure 48.8** Adding chloride channels would make the membrane potential less positive. Adding potassium channels would have no effect because there are no potassium ions present. **Figure 48.10** In the absence of other forces, chemical concentration gradients govern net diffusion. In this case, ions are more concentrated outside of the cell and move in when the channel opens.

Figure 48.11

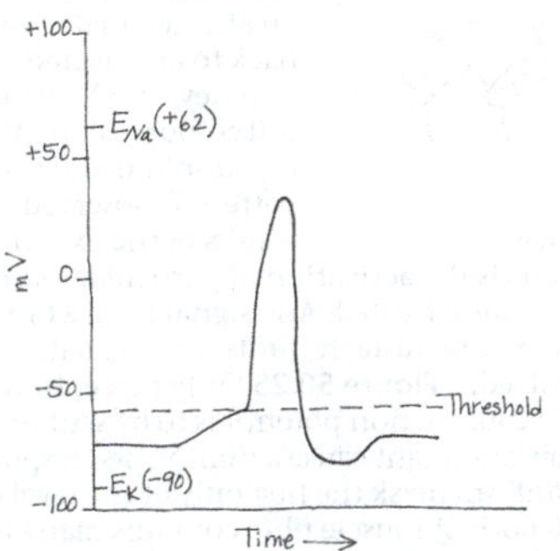

Figure 48.12

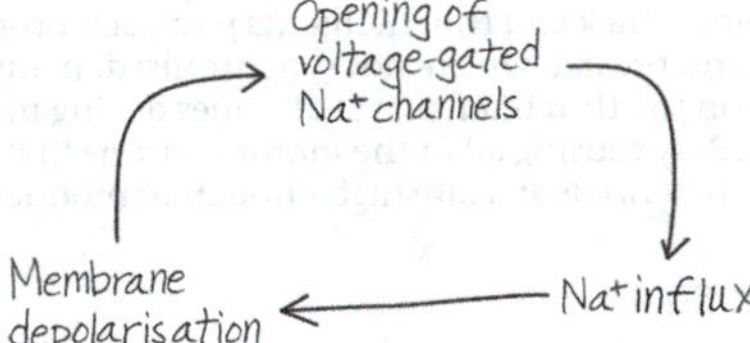

Figure 48.13

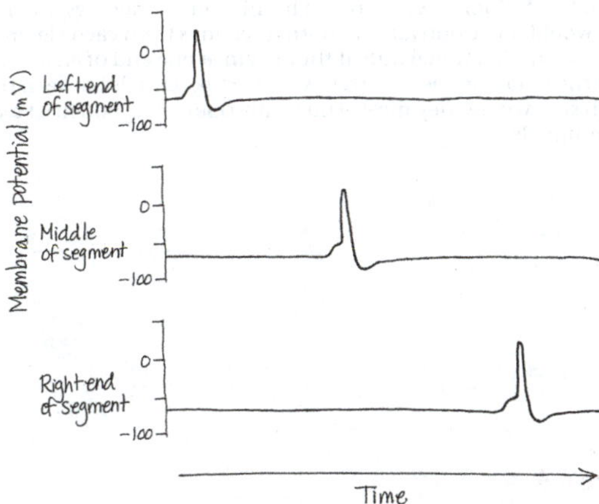

Figure 48.16 The production and transmission of action potentials would be unaffected. However, action potentials arriving at chemical synapses would be unable to trigger release of neurotransmitter. Signalling at such synapses would thus be blocked. **Figure 48.17** Summation only occurs if inputs occur simultaneously or nearly so. Thus, spatial summation, in which input is received from two different sources, is in effect also temporal summation.

Concept Check 48.1

1. A typical neuron has multiple dendrites and one axon. Dendrites transfer information to the cell body, whereas axons transmit information from the cell body. Both axons and dendrites extend from the cell body and function in information flow. **2.** Sensors in your ear transmit information to your brain. There, the activity of interneurons in processing centres enables you to recognise your name. In response, signals transmitted via motor neurons cause contraction of muscles that turn your neck. **3.** Increased branching would allow control of a greater number of postsynaptic cells, enhancing coordination of responses to nervous system signals.

Concept Check 48.2

1. Ions can flow against a chemical concentration gradient if there is an opposing electrical gradient of greater magnitude. **2.** A decrease in permeability to K^+, an increase in permeability to Na^+, or both **3.** Charged dye molecules could equilibrate only if other charged molecules could also cross the membrane. If not, a membrane potential would develop that would counterbalance the chemical gradient.

Concept Check 48.3

1. A graded potential has a magnitude that varies with stimulus strength, whereas an action potential has an all-or-none magnitude that is independent of stimulus strength. **2.** Loss of the insulation provided by myelin sheaths leads to a disruption of action potential propagation along axons. Voltage-gated sodium channels are restricted to the nodes of Ranvier, and without the insulating effect of myelin, the inward current produced at one node during an action potential cannot depolarise the membrane to the threshold at the next node. **3.** Positive feedback is responsible for the rapid opening of many voltage-gated sodium channels, causing the rapid outflow of sodium ions responsible for the rising phase of the action potential. As the membrane potential becomes positive, voltage-gated potassium channels open in a form of negative feedback that helps bring about the falling phase of the action potential. **4.** The maximum frequency would decrease because the refractory period would be extended.

Concept Check 48.4

1. It can bind to different types of receptors, each triggering a specific response in postsynaptic cells. **2.** These toxins would prolong the EPSPs that acetylcholine produces because the neurotransmitter would remain longer in the synaptic cleft. **3.** Membrane depolarisation, exocytosis, and membrane fusion each occur in fertilisation and in neurotransmission.

Summary of Key Concepts Questions

48.1 It would prevent information from being transmitted away from the cell body along the axon. **48.2** There are very few open sodium channels in a resting neuron, so the resting potential either would not change or would become slightly more negative (hyperpolarisation). **48.4** A given neurotransmitter can have many receptors that differ in their location and activity. Drugs that target receptor activity rather than neurotransmitter release or stability are therefore likely to exhibit greater specificity and potentially have fewer undesirable side effects.

Test Your Understanding

1. C **2.** C **3.** C **4.** B **5.** A **6.** D

7. The activity of the sodium-potassium pump is essential to maintain the resting potential. With the pump inactivated, the sodium and potassium concentration gradients would gradually disappear, resulting in a greatly reduced resting potential. **8.** Since GABA is an inhibitory neurotransmitter in the CNS, this drug would be expected to decrease brain activity. A decrease in brain activity might be expected to slow down or reduce behavioural activity. Many sedative drugs act in this fashion. **9.** As shown in this pair of drawings, a pair of action potentials would move outwards in both directions from each electrode. (Action potentials are unidirectional only if they begin at one end of an axon.) However, because of the refractory period, the two action potentials between the electrodes both stop where they meet. Thus, only one action potential reaches the synaptic terminals.

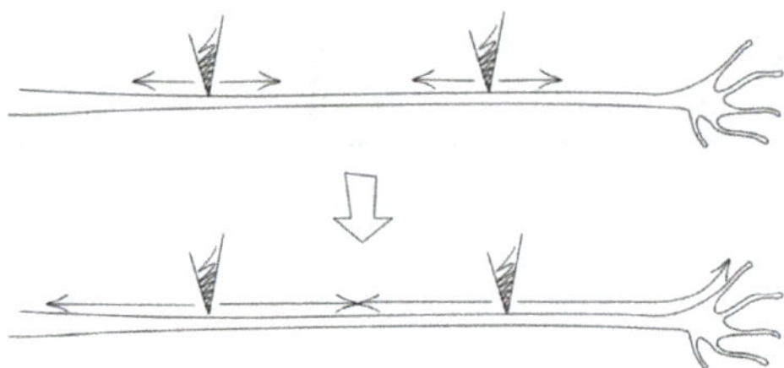

Chapter 49

Figure Questions

Figure 49.5 During swallowing, muscles along the oesophagus alternately contract and relax, resulting in peristalsis. One model to explain this alternation is that each section of muscle receives nerve impulses that alternate between excitation and inhibition, just as the quadriceps and hamstring receive opposing signals in the knee-jerk reflex. **Figure 49.15** The grey areas have a different shape and pattern, indicating different planes through the brain. This fact indicates that the nucleus accumbens and the amygdala are in different planes. **Figure 49.17** The hand is shown larger than the forearm because the hand receives more innervation than the forearm for sensory input to the brain and motor output from the brain. **Figure 49.24** If the depolarisation brings the membrane potential to or past threshold, it should initiate action potentials that cause dopamine release from the VTA neurons. This should mimic natural stimulation of the brain reward system, resulting in positive and perhaps pleasurable sensations.

Concept Check 49.1

1. The sympathetic division would likely be activated. It mediates the "fight-or-flight" response in stressful situations. **2.** Nerves contain bundles of axons, some that belong to motor neurons, which send signals outwards from the CNS, and some that belong to sensory neurons, which bring signals into the CNS. Therefore, you would expect effects on both motor control and sensation. **3.** Neurosecretory cells of the adrenal medulla secrete the hormones adrenaline and noradrenaline in response to preganglionic input from sympathetic neurons. These hormones travel in the circulation throughout the body, triggering responses in many tissues.

Concept Check 49.2

1. The cerebral cortex on the left side of the brain initiates voluntary movement of the right side of the body. **2.** Alcohol diminishes function of the cerebellum. **3.** A coma reflects a disruption in the cycles of sleep and arousal regulated by communication between the midbrain and pons (reticular formation) and the cerebrum. You would expect this group to have damage to the midbrain, the pons, the cerebrum, or any part of the brain between these structures. Paralysis reflects an inability to carry out motor commands transmitted from the cerebrum to the spinal cord. You would expect this group to have damage to the portion of the CNS extending from the spinal cord up to but not including the midbrain and pons.

Concept Check 49.3

1. Brain damage that disrupts behaviour, cognition, memory, or other functions provides evidence that the portion of the brain affected by the damage is important for the normal activity that is blocked or altered. **2.** Broca's area, which is active during the generation of speech, is located near the motor cortex, which controls skeletal muscles, including those in the face. Wernicke's area, which is active when speech is heard, is located in the posterior part of the temporal lobe, which is involved in hearing. **3.** Each cerebral hemisphere is specialised for different parts of this task—the right for face recognition and the left for language. Without an intact corpus callosum, neither hemisphere can take advantage of the other's processing abilities.

Concept Check 49.4

1. There can be an increase in the number of synapses between the neurons or an increase in the strength of existing synaptic connections. **2.** If consciousness is an emergent property resulting from the interaction of many different regions of the brain, then it is unlikely that localised brain damage will have a discrete effect on consciousness. **3.** The hippocampus is responsible for organising newly acquired information. Without hippocampal function, the links necessary to retrieve information from the cerebral cortex will be lacking, and no functional memory, short- or long-term, will be formed.

Concept Check 49.5

1. Both are progressive brain diseases whose risk increases with advancing age. Both result from the death of brain neurons and are associated with the accumulation of peptide or protein aggregates. **2.** The symptoms of schizophrenia can be mimicked by a drug that stimulates dopamine-releasing neurons. The brain's reward system, which is involved in drug addiction, is composed of dopamine-releasing neurons that connect the ventral tegmental area to regions in the cerebrum. Parkinson's disease results from the death of dopamine-releasing neurons. **3.** Not necessarily. It might be that the plaques, tangles, and missing regions of the brain seen at death reflect secondary effects, the consequence of other unseen changes that are actually responsible for the alterations in brain function.

Summary of Key Concepts Questions

49.1 Because reflex circuits involve only a few neurons—the simplest consist of a sensory neuron and a motor neuron—the path for information transfer is short and simple, increasing the speed of the response. **49.2** The midbrain coordinates visual reflexes; the cerebellum controls coordination of movement that depends on visual input; the thalamus serves as a routing centre for visual information; and the cerebrum is essential for converting visual input to a visual image. **49.3** You would expect the right side of the body to be paralysed because it is controlled by the left cerebral hemisphere, where language generation and interpretation are localised. **49.4** Learning a new language likely requires the maintenance of synapses that are formed during early development but are otherwise lost prior to adulthood. **49.5** Whereas amphetamine stimulates dopamine release, PCP blocks glutamate receptors, suggesting that schizophrenia does not reflect a defect in the function of just one neurotransmitter.

Test Your Understanding

1. B **2.** B **3.** D **4.** D **5.** C **6.** A
7.

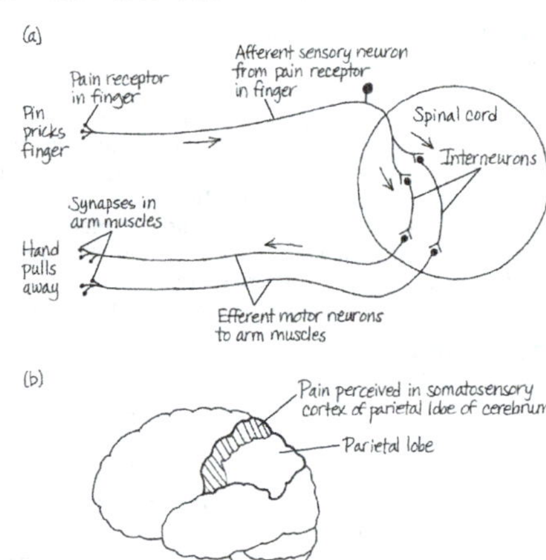

Chapter 50

Figure Questions

Figure 50.17

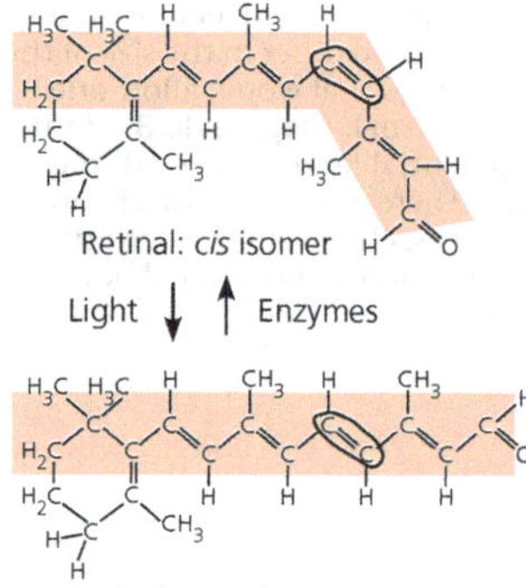

Figure 50.19 Each of the three types of cones is most sensitive to a different wavelength of light. A cone might be fully depolarised when there is light present if the light is of a wavelength far from its optimum. **Figure 50.21** In humans, an X chromosome with a defect in the red or green opsin gene is much less common than a wild-type X chromosome. Colour blindness therefore typically skips a generation as the defective allele passes from an affected male to a carrier daughter and back to an affected grandson. In squirrel monkeys, no X chromosome can confer full colour vision. As a result, all males are colour-blind and no unusual inheritance pattern is observed. **Figure 50.23** The results of the experiment would have been identical. What matters is the activation of particular sets of neurons, not the manner in which they are activated. Any signal from a bitter cell will be interpreted by the brain as a bitter taste, regardless of the nature of the compound and the receptor involved. **Figure 50.25** Only perception. Binding of an odourant to its receptor will cause action potentials to be sent to the brain. Although an excess of that odourant might cause a diminished response through adaptation, another odourant can mask the first only at the level of perception in the brain. **Figure 50.26** Both. A muscle fibre contains many myofibrils bundled together and divided lengthwise into many sarcomeres. A sarcomere is a contractile unit made up of portions of many myofibrils, and each myofibril is a part of many sarcomeres. **Figure 50.28** Hundreds of myosin heads participate in sliding each pair of thick and thin filaments past each other. Because cross-bridge formation and breakdown are not synchronised, many myosin heads are exerting force on the thin filaments at all times during muscle contraction. **Figure 50.33** By causing all of the motor neurons that control the muscle to generate action potentials at a rate high enough to produce tetanus in all of the muscle fibres.

Concept Check 50.1

1. Electromagnetic receptors in general detect only external stimuli. Nonelectromagnetic receptors, such as chemoreceptors or mechanoreceptors, can act as either internal or external sensors. **2.** The capsaicin present in the peppers activates the thermoreceptor for high temperatures. In response to the perceived high temperature, the nervous system triggers sweating to achieve evaporative cooling. **3.** You would perceive the electrical stimulus as if the sensory receptors that regulate that neuron had been activated. For example, electrical stimulation of the sensory neuron controlled by the thermoreceptor activated by menthol would likely be perceived as a local cooling.

Concept Check 50.2

1. Sonar-producing structures allow an animal to orient itself in its environment, providing information essential to its survival in places where light cues are absent. **2.** As a sound that changes gradually from a very low to a very high pitch **3.** The stapes and the other middle ear bones transmit vibrations from the tympanic membrane to the oval window. Fusion of these bones (as occurs in a disease called otosclerosis) would block this transmission and result in hearing loss. **4.** In animals, the statoliths are extracellular. In contrast, the statoliths of plants are found within an intracellular organelle. The methods for detecting their location also differ. In animals, detection is by means of mechanoreceptors on ciliated cells. In plants, the mechanism appears to involve calcium signalling.

Concept Check 50.3

1. Planarians have ocelli that cannot form images but can sense the intensity and direction of light, providing enough information to enable the animals to find protection in shaded places. Flies have compound eyes that form images and excel at detecting movement. **2.** The person can focus on distant objects but not close objects (without glasses) because close focusing requires the lens to become almost spherical. This problem is common after age 50. **3.** The signal produced by rod and cone cells is glutamate, and their release of glutamate decreases upon exposure to light. However, a decrease in glutamate production causes other retinal cells to increase the rate at which action potentials are sent to the brain, so that the brain receives more action potentials in light than in dark. **4.** Absorption of light by retinal converts retinal from its *cis* isomer to its *trans* isomer, initiating the process of light detection. In contrast, a photon absorbed by chlorophyll does not bring about isomerisation, but instead boosts an electron to a higher energy orbital, initiating the electron flow that generates ATP and NADPH.

Concept Check 50.4

1. Both taste cells and olfactory cells have receptor proteins in their plasma membrane that bind certain substances, leading to membrane depolarisation through a signal transduction pathway involving a G protein. However, olfactory cells are sensory neurons, whereas taste cells are not. **2.** Since animals rely on chemical signals for behaviours that include finding mates, marking territories, and avoiding dangerous substances, it is adaptive for the olfactory system to have a robust response to a very small number of molecules of a particular odourant. **3.** Because the sweet, bitter, and umami tastes involve GPCR proteins but the sour taste does not, you might predict that the mutation is in a molecule that acts in the signal transduction pathway common to the different GPCRs.

Concept Check 50.5

1. In a skeletal muscle fibre, Ca^{2+} binds to the troponin complex, which moves tropomyosin away from the myosin-binding sites on actin and allows cross-bridges to form. In a smooth muscle cell, Ca^{2+} binds to calmodulin, which activates an enzyme that phosphorylates the myosin head and thus enables cross-bridge formation. **2.** *Rigor mortis*, a Latin phrase meaning "stiffness of death," results from the complete depletion of ATP in skeletal muscle. Since ATP is required to release myosin from actin and to pump Ca^{2+} out of the cytosol, muscles become chronically contracted beginning about 3–4 hours after death. **3.** A competitive inhibitor binds to the same site as the substrate for the enzyme. In contrast, the troponin and tropomyosin complex masks, but does not bind to, the myosin-binding sites on actin.

Concept Check 50.6

1. The main problem in swimming is drag; a fusiform body minimises drag. The main problem in flying is overcoming gravity; wings shaped like airfoils provide lift, and adaptations such as air-filled bones reduce body mass. **2.** In modelling peristalsis you would constrict the toothpaste tube at different points along its length, using your hand to encircle the tube and squeeze concentrically. To demonstrate movement of food through the digestive tract you would want the cap off the toothpaste tube, whereas you would want the cap on to show how peristalsis contributes to worm locomotion. **3.** When you grasp the sides of the chair, you are using a contraction of the triceps to keep your arms extended against the pull of gravity on your body. As you lower yourself slowly into the chair, you gradually decrease the number of motor units in the triceps that are contracted. Contracting your biceps would jerk you down, since you would no longer be opposing gravity.

Summary of Key Concepts Questions

50.1 Nociceptors overlap with other classes of receptors in the type of stimulus they detect. They differ from other receptors only in how a particular stimulus is perceived. **50.2** Volume is encoded by the frequency of action potentials transmitted to the brain; pitch is encoded by which axons are transmitting action potentials. **50.3** The major difference is that neurons in the retina integrate information from multiple sensory receptors (photoreceptors) before transmitting information to the central nervous system. **50.4** Our olfactory sense is responsible for most of what we describe as distinct tastes. A head cold or other source of congestion blocks odourant access to receptors lining portions of the nasal cavity. **50.5** Hydrolysis of ATP is required to convert myosin to a high-energy configuration for binding to actin and to power the Ca^{2+} pump that removes cytosolic Ca^{2+} during muscle relaxation. **50.6** Human body movements rely on the contraction of muscles anchored to a rigid endoskeleton. Tendons attach muscles to bones, which in turn are composed of fibres built up from a basic organisational unit, the sarcomere. The thin and thick filaments have separate points of attachment within the sarcomere. In response to nervous system motor output, the formation and breakdown of cross-bridges between myosin heads and actin ratchet the thin and thick filaments past each other. Because the filaments are anchored, this sliding movement shortens the muscle fibres. Furthermore, because the fibres themselves are part of the muscles attached at each end to bones, muscle contraction moves bones of the body relative to each other. In this way, the structural anchoring of muscles and filaments enables muscle function, such as the bending of an elbow by contraction of the biceps.

Test Your Understanding

1. C **2.** A **3.** B **4.** C **5.** B **6.** D

7.

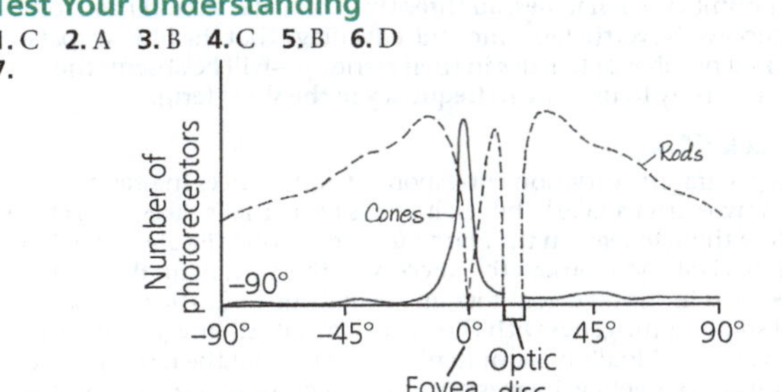

The answer shows the actual distribution of rods and cones in the human eye. Your graph may differ, but should have the following properties: only cones at the fovea; fewer cones and more rods at both ends of the *x*-axis; no photoreceptors in the optic disc.

Chapter 51

Figure Questions

Figure 51.2 The fixed action pattern based on the sign stimulus of a red belly ensures that the male will chase away any invading males of his species. By chasing away such males, the defender decreases the chance that another male will fertilise eggs laid in his nesting territory. **Figure 51.5** The straight-run portion conveys two pieces of information: direction, via the angle of that run relative to the wall of the hive, and distance, via the number of waggles performed during the straight run. At a minimum, the portions between the straight runs identify the activity as a waggle dance. Since they also provide contact with workers to one side and then the other, they may ensure transmission of information to a larger number of other bees. **Figure 51.7** There should be no effect. Imprinting is an innate behaviour that is carried out anew in each generation. Assuming the nest was not disturbed, the offspring of the geese imprinted on a human would imprint on the mother goose. **Figure 51.8** Perhaps the wasp doesn't use visual cues. It might also be that wasps recognise objects native to their environment, but not foreign objects, such as the pinecones. Tinbergen addressed these ideas before carrying out the pinecone study. When he swept away the pebbles and sticks around the nest, the wasps could no longer find their nests. If he shifted the natural objects in their natural arrangement, the shift in the landmarks caused a shift in the site to which the wasps returned. Finally, if natural objects around the nest site were replaced with pinecones while the wasp was in the burrow, the wasp nevertheless found her way back to the nest site. **Figure 51.28** It might be that the birds require stimuli during flight to exhibit their migratory preference. If this were true, the birds would show the same orientation in the funnel experiment despite their distinct genetic programming. **Figure 51.30** It holds true for some, but not all individuals. If a parent has more than one reproductive partner, the offspring of different partners will have a coefficient of relatedness less than 0.5.

Concept Check 51.1

1. The proximate explanation for this fixed action pattern might be that nudging and rolling are released by the sign stimulus of an object outside the nest, and the behaviour is carried to completion once initiated. The ultimate explanation might be that ensuring that eggs remain in the nest increases the chance of producing healthy offspring. **2.** There might be selective pressure for other prey fish to detect an injured fish because the source of the injury might threaten them as well. Among predators, there might be selection for those that are attracted to the alarm substance because they would be more likely to encounter crippled prey. Fish with adequate defences might show no change because they have a selective advantage if they do not waste energy responding to the alarm substance. **3.** In both cases, the detection of periodic variation in the environment results in a reproductive cycle timed to environmental conditions that optimise the opportunity for success.

Concept Check 51.2

1. Natural selection would tend to favour convergence in colour pattern because a predator learning to associate a pattern with a sting or bad taste would avoid all other individuals with that same colour pattern, regardless of species. **2.** You might move objects around to establish an abstract rule, such as "past landmark A, the same distance as A is from the starting point," while maintaining a minimum of fixed metric relationships, that is, avoiding having the food directly adjacent to or a set distance from a landmark. As you might surmise, designing an informative experiment of this kind is not easy. **3.** Learned behaviour, just like innate behaviour, can contribute to reproductive isolation and thus to speciation. For example, learned bird songs contribute to species recognition during courtship, thereby helping ensure that only members of the same species mate.

Concept Check 51.3

1. Certainty of paternity is higher with external fertilisation. **2.** Balancing selection could maintain the two alleles at the *forager* locus if population density fluctuated from one generation to another. At times of low population density, the energy-conserving sitter larvae (carrying the for^S allele) would be favoured, while at higher population density, the more mobile Rover larvae (for^R allele) would have a selective advantage. **3.** Because females would now be present in much larger numbers than males, all three types of males should have some reproductive success. Nevertheless, since the advantage that the blue-throats rely on—a limited number of females in their territory—will be absent, the yellow-throats are likely to increase in frequency in the short term.

Concept Check 51.4

1. Because this geographic variation corresponds to differences in prey availability between two garter snake habitats, it seems likely that snakes with characteristics enabling them to feed on the abundant prey in their locale would have had increased survival and reproductive success. In this way, natural selection would have resulted in the divergent foraging behaviours. **2.** The fact that the individual shares some genes with the offspring of its sibling (in the case of humans, with the individual's niece or nephew) means that the reproductive success of that niece or nephew increases the representation of those genes in the population (selects for them). **3.** The older individual cannot be the beneficiary because he or she cannot have extra offspring. However, the cost is low for an older individual performing the altruistic act because that individual has already reproduced (but perhaps is still caring for a child or grandchild). There can therefore be selection for an altruistic act by a postreproductive individual that benefits a young relative.

Summary of Key Concepts Questions

51.1 Circannual rhythms are typically based on the cycles of light and dark in the environment. As the global climate changes, animals that migrate in response to these rhythms may shift to a location before or after local environmental conditions are optimal for reproduction and survival. **51.2** For the goose, all that is acquired is an object at which the behaviour is directed. In the case of the sparrow, learning takes place that will give shape to the behaviour itself. **51.3** Because feeding the female is likely to improve her reproductive success, the genes from the sacrificed male are likely to appear in a greater number of progeny. **51.4** Studying the genetic basis of these behaviours reveals that changes in a single gene can have large-scale effects on even complex behaviours.

Test Your Understanding

1. C **2.** B **3.** B **4.** A **5.** C **6.** A
7.

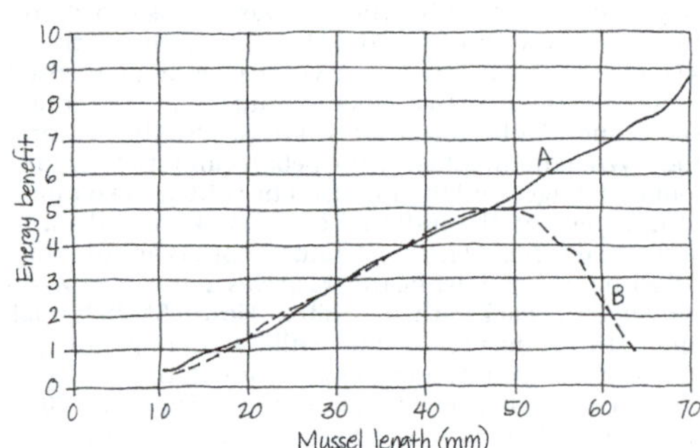

You could measure the size of mussels that oystercatchers successfully open and compare that with the size distribution in the habitat.

Chapter 52

Figure Questions

Figure 52.8 The species' distribution could be altered by dispersal limitations, the activities of people (such as a broad-scale conversion of forests to agriculture or selective harvesting), or many other factors, including those discussed later in the chapter (see Figure 52.19). **Figure 52.19** Some factors, such as fire, are relevant only for terrestrial systems. At first glance, water availability is primarily a terrestrial factor, too. However, species living along the intertidal zone of oceans or along the edge of lakes also suffer desiccation. Salinity stress is important for species in some aquatic and terrestrial systems. Oxygen availability is an important factor primarily for species in some aquatic systems and in soils and sediments.

Concept Check 52.1

1. In the tropics, high temperatures evaporate water and cause warm, moist air to rise. The rising air cools and releases much of its water as rain over the tropics. The remaining dry air descends at approximately 30° north and south, causing deserts to occur in those regions. **2.** The microclimate around the stream will be cooler, moister, and shadier than that around the unplanted agricultural field. **3.** Trees that require a long time to reach reproductive age are likely to evolve more slowly than annual plants in response to climate change, constraining the potential ability of such trees to respond to rapid climate change. **4.** Plants with C_4 photosynthesis are likely to expand their range globally as Earth's climate warms. C_4 photosynthesis minimises photorespiration and enhances sugar production, an advantage that is especially useful in warmer regions where C_4 plants are found today.

Concept Check 52.2

1. The biggest difference between the two biomes is the higher amounts of precipitation that the forest receives. **2.** Answers will vary by location but should be based on the information and maps in Figure 52.13. How much your local area has been altered from its natural state will influence how much it reflects the expected characteristics of your biome, particularly the expected plants and animals. **3.** Deserts will replace grasslands along the boundary between these biomes. With increased global temperatures, precipitation will likely decline. Increased temperatures and reduced precipitation will move conditions to the upper left-hand corner of the climagraph, where deserts will likely dominate. (see Figure 52.11).

Concept Check 52.3

1. In the oceanic pelagic zone, the ocean bottom lies below the photic zone, so there is too little light to support benthic algae or rooted plants. **2.** Aquatic organisms either gain or lose water by osmosis if the osmolarity of their environment differs from their internal osmolarity. Water gain can cause cells to swell, and water loss can cause them to shrink. To avoid excessive changes in cell volume, organisms that live in estuaries must be able to compensate for both water gain (under freshwater conditions) and water loss (under saltwater conditions). **3.** Oxygen serves as a reactant when decomposers break down the bodies of dead algae using aerobic respiration. Following an algal bloom, there are many dead algae; hence, decomposers may use a lot of oxygen to break down the bodies of dead algae, causing the lake's oxygen levels to drop.

Concept Check 52.4

1. (a) Humans might transplant a species to a new area that it could not previously reach because of a geographic barrier. (b) Humans might eliminate a predator or herbivore species, such as sea urchins, from an area. **2.** One test would be to build a fence around a plot of land in an area that has trees of that species, excluding all rabbits from the plot. You could then compare the abundance of tree seedlings inside and outside the fenced plot over time. **3.** Because the ancestor of the silverswords reached isolated Hawaii early in the islands' existence, it likely faced little competition and was able to occupy many unfilled niches. The cattle egret, in contrast, arrived in the Americas only recently and has to compete with a well-established group of species. Thus, its opportunities for adaptive radiation have probably been much more limited.

Concept Check 52.5

1. Changes in how organisms interact with one another and their environment can cause evolutionary change. In turn, an evolutionary change, such as an improvement in the ability of a predator to detect its prey, can alter ecological interactions. **2.** As cod adapt to the pressure of commercial fishing by reproducing at younger ages and smaller sizes, the number of offspring they produce each year will be lower. This may cause the population to decline as time goes on, thereby further reducing the population's ability to recover. If that happened, as the population becomes smaller over time, effects of genetic drift might become increasingly important. Drift could, for example, lead to the fixation of harmful alleles, which would further hinder the ability of the cod population to recover from overfishing.

Summary of Key Concepts Questions

52.1 Because dry air would descend at the equator instead of at 30° north and south latitude (where deserts exist today), deserts would be more likely to exist along the equator (see Figure 52.3). **52.2** The dominant plants in savanna ecosystems tend to be adapted to fire and tolerant of seasonal droughts. The savanna biome is maintained by periodic fires, both natural and set by humans, but humans are also clearing savannas for agriculture and other uses. **52.3** An aphotic zone is most likely to be found in the deep waters of a lake, the oceanic pelagic zone, or the marine benthic zone.
52.4 You might arrange a flowchart that begins with abiotic limitations—first determining the physical and chemical conditions under which a species could survive—and then moves through the other factors listed in the flowchart. **52.5** Because the introduced species had few predators or parasites, it might outcompete native species and thereby increase in number and expand its range in the new location. As the introduced species increased in abundance, natural selection might cause evolution in populations of competing species, favouring individuals with traits that made them more effective competitors with the introduced species. Selection could also cause evolution in populations of potential predator or parasite species, in this case favouring individuals with traits that enabled them to take advantage of this new potential source of food. Such evolutionary changes could modify the outcome of ecological interactions, potentially leading to further evolutionary changes, and so on.

Test Your Understanding

1. B **2.** B **3.** C **4.** D **5.** C **6.** A **7.** A **8.** B

Chapter 53

Figure Questions

Figure 53.3 The dispersion of the penguins would likely appear clumped as you flew over densely populated islands and sparsely populated ocean.
Figure 53.4 Ten percent (100/1,000) of the females survive to be 3 years old.
Figure 53.6 #109 **Figure 53.7** The population with $r = 1.0$ (blue curve) reaches 1,500 individuals in about 7.5 generations, whereas the population with $r = 0.5$ (red curve) reaches 1,500 individuals in about 14.5 generations.

Figure 53.22

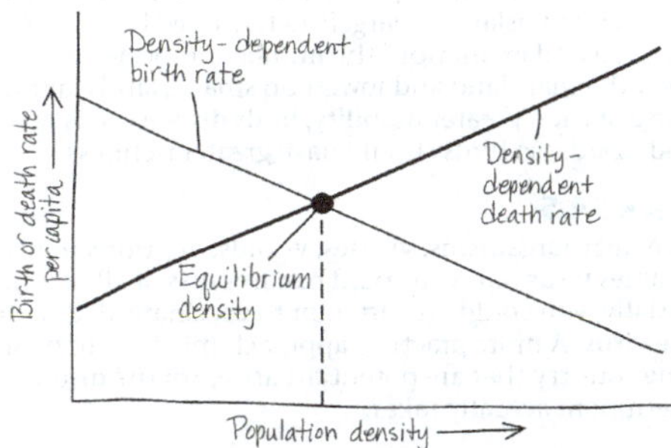

Figure 53.29 Based on Figure 53.28, which highlights the long-term, explosive growth of the human population, one might conclude (mistakenly) that the growth rate of the human population has not decreased in recent decades. However, the growth rate of the human population *has* decreased in recent decades—a slowdown that is evident in the blue curve shown in Figure 53.29. Both curves are accurate, but they convey different messages because they differ in the time scale over which human population size is represented. The time period covered by Figure 53.28 is so long (over 6,000 years in the "unbroken" portion of the *x*-axis lying to the right of the hatch mark) that the recent slowdown in how fast the human population is growing is not visually apparent. In contrast, Figure 53.29 covers only 100 years, a time period that is short enough to show the recent decrease in the growth rate of the human population. **Figure 53.31** If the average ecological footprint were 8 gha per person, Earth could support about 1.5 billion people in a sustainable fashion. This estimate is obtained by dividing the total amount of Earth's productive land (11.9 billion gha) by the number of global hectares used per person (8 gha/person), which yields 1.49 billion people.

Concept Check 53.1

1.

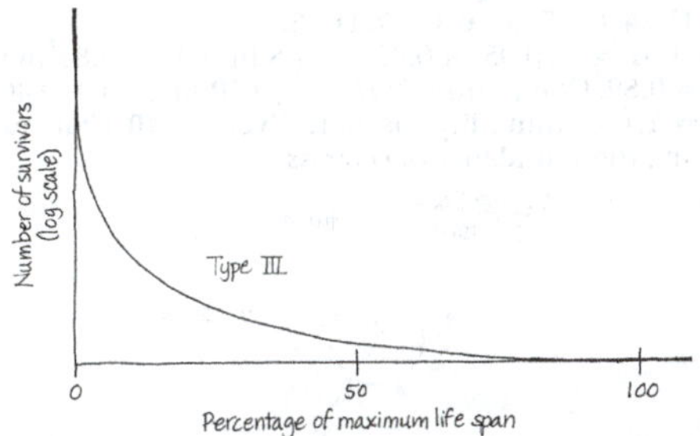

A Type III survivorship curve is most likely because very few of the young probably survive.
2. The proportion alive at the start of year 0–1 is 485/485 = 1.0. The proportion alive at the start of year 1–2 is 218/485 = 0.449. **3.** Male sticklebacks would likely have a uniform pattern of dispersion, with antagonistic interactions maintaining a relatively constant spacing between them.

Concept Check 53.2

1. Although *r* is constant, the population size (*N*) is increasing. As *r* is applied to an increasingly large *N*, population growth (*rN*) accelerates, producing the J-shaped curve. **2.** Exponential growth is more likely in the area where a forest was destroyed by fire. The first plants that found suitable habitat there would encounter an abundance of space, nutrients, and light. In the undisturbed forest, competition among plants for these resources would be intense. **3.** The equation for the number of people added to the population each year is $\Delta N/\Delta t = r_{\Delta t}N$. Therefore, the net population growth in 2020 was

$$\Delta N/\Delta t = 0.005 \times 26{,}000{,}000 = 128{,}000 \text{ (Australia)}$$

$$\Delta N/\Delta t = 0.005 \times 4{,}800{,}000 = 24{,}000 \text{ (New Zealand)}$$

152,000 people when aggregated across both nations. Whether you wish to determine if the Australian or New Zealand population, or the combined populations of both countries, is growing exponentially, you would need to determine whether $r > 0$ and if it is constant through time (across multiple years).

Concept Check 53.3

1. When *N* (population size) is small, there are relatively few individuals producing offspring. When *N* is large, near the carrying capacity, the per capita growth rate is relatively small because it is limited by available resources. The steepest part of the logistic growth curve corresponds to a population with a number of reproducing individuals that is substantial but not yet near carrying capacity.
2. All else being equal, you would expect a plant species to have a larger carrying capacity at the equator than at high latitudes because there is more incident sunlight near the equator. **3.** The sudden change in environmental conditions might alter the phenotypic traits favoured by natural selection. Assuming the newly favoured traits were encoded at least in part by genes, natural selection might alter gene frequencies in this population. In addition, a substantial drop in the carrying capacity of the population could cause the size of the population to drop considerably. If this occurred, effects of genetic drift could become more pronounced—and that in turn could lead to the fixation of harmful alleles, hindering the ability of the population to rebound in size.

Concept Check 53.4

1. Three key life history traits are when reproduction begins, how often reproduction occurs, and how many offspring are produced per reproductive episode. Organisms differ widely for each of these traits. For example, the age of first reproduction is typically 3–4 years in coho salmon compared to 30 years in loggerhead turtles. Similarly, an agave reproduces only once during its lifetime, whereas an oak tree reproduces many times. Finally, the white rhinoceros produces a single calf when it reproduces, while most insects produce many offspring each time they reproduce. **2.** By preferentially investing in the eggs it lays in the nest, the peacock wrasse increases the chance those eggs will survive. The eggs it disperses widely and does not provide care for are less likely to survive, at least some of the time, but require a lower investment by the adults. (In this sense, the adults avoid the risk of placing all their eggs in one basket.) **3.** If a parent's survival is compromised greatly by bearing young during times of stress, the animal's fitness may increase if it abandons its current young and survives to produce healthier young at a later time.

Concept Check 53.5

1. Three attributes are the size, quality, and isolation of patches. A patch that is larger or of higher quality is more likely to attract individuals and to be a source of individuals for other patches. A patch that is relatively isolated will undergo fewer exchanges of individuals with other patches. **2.** You should have circled the portion of the curve where it is close to *K* (after generation 10). **3.** You would need to study the population for more than one cycle (longer than 10 years and probably at least 20) before having sufficient data to examine changes through time. Otherwise, it would be impossible to know whether an observed decrease in the population size reflected a long-term trend or was part of the normal cycle. **4.** In negative feedback, the output, or product, of a process slows that process. In populations that have a density-dependent birth rate, such as dune fescue grass, an accumulation of product (more individuals, resulting in a higher population density) slows the process (population growth) by decreasing the birth rate.

Concept Check 53.6

1. A bottom-heavy age structure, with a disproportionate number of young people, portends continuing growth of the population as these young people begin reproducing. In contrast, a more evenly distributed age structure predicts a more stable population size, and a top-heavy age structure predicts a decrease in population size because relatively fewer young people are reproducing.
2. The growth rate of Earth's human population has dropped by half since the 1960s, from 2.2% in 1962 to 1.1% today. Nonetheless, the yearly increase in population size has not slowed as much because the smaller growth rate is counterbalanced by increased population size; hence, the number of additional people on Earth each year remains enormous—approximately 80 million. **3.** Each student will calculate his or her own ecological footprint. Each of us influences our ecological footprint by how we live—what we eat, how much energy we use, and the amount of waste we generate—as well as by how many children we have. Making choices that reduce our demand for resources makes our ecological footprint smaller.

Summary of Key Concepts Questions

53.1 Ecologists can potentially estimate birth rates by counting the number of young born each year, and they can estimate death rates by seeing how the number of adults changes each year. **53.2** Under the exponential model, both populations will continue to grow to infinite size, regardless of the specific value of *r* (see Figure 53.7). **53.3** There are many things you can do to increase the carrying capacity of the species, including increasing its food supply, protecting it from predators, and providing more sites for nesting or reproduction. **53.4** Ecological trade-offs are common because organisms do not have access to unlimited amounts of energy and resources. As a result, the use of energy or resources for one function (such as reproduction) can decrease the energy or resources available to support other functions (such as growth or survival). **53.5** An example of a biotic factor is disease caused by a pathogen; natural disasters, such as earthquakes and floods, are examples of abiotic factors. **53.6** Humans are unique in our potential ability to reduce global population through contraception and family planning. Humans also are capable of consciously choosing their diet and personal lifestyle, and these choices influence the number of people Earth can support.

Test Your Understanding

1. B **2.** A **3.** A **4.** D **5.** C **6.** B **7.** C **8.** A **9.** C

Chapter 54

Figure Questions

Figure 54.3 Its realised and fundamental niches would be similar, unlike those of *Chthamalus*. **Figure 54.5** Beak depths in the *G. fortis* population would likely decrease over time. With the extinction of *G. fuliginosa*, the small seeds eaten by that species would increase in abundance. As a result, natural selection would favour *G. fortis* individuals with smaller beaks because those individuals can eat small seeds more efficiently than could *G. fortis* individuals with larger beaks. **Figure 54.6** Individuals of a harmless species that resembled a distantly related harmful species might be attacked by predators less often than were other individuals that did not resemble the harmful species. As a result, individuals of the harmless species that resembled a harmful species would tend to contribute more offspring to the next generation than would other individuals of the harmless species. Over time, as natural selection by predators continued to favour those individuals of the harmless species that most closely resembled the harmful species, the resemblance of the harmless species to the harmful species would increase. However, selection is not the only process that could cause a harmless species to resemble a closely related harmful species. In this case, the two species could also resemble each other because they descended from a recent common ancestor and hence share many traits (including a resemblance to one another). **Figure 54.16** An increase in the abundance of carnivores that eat zooplankton might cause zooplankton abundance to drop, thereby causing phytoplankton abundance to increase. **Figure 54.17** The number of types of organisms eaten is zero for phytoplankton; one for copepods, crab-eater seals, and baleen whales; two for krill, carnivorous plankton, elephant seals, and sperm whales; three for squids, fishes, and leopard seals; and five for birds and smaller toothed whales. The two groups that both consume and are consumed by each other are fishes and squids. **Figure 54.18** Zooplankton are primary consumers of phytoplankton. Fish larvae are secondary consumers of zooplankton. Sea nettles function as secondary consumers when they eat zooplankton, but as tertiary consumers when they eat fish larvae. Juvenile striped bass are tertiary consumers of fish larvae. **Figure 54.20** The death of individuals of *Mytilus*, a competitively dominant species, should open up space for individuals of other species and thereby increase species richness even in the absence of *Pisaster*. **Figure 54.26** At the earliest stages of primary succession, free-living prokaryotes in the soil would reduce atmospheric N_2 to NH_3. Symbiotic nitrogen fixation could not occur until plants were present at the site. **Figure 54.30** We would expect that (a) population sizes would decrease because there would be fewer resources and less suitable habitat; (b) the extinction curve would rise more rapidly as the number of species on the island increased because small islands generally have fewer resources, less diverse habitats, and smaller population sizes; and (c) the predicted equilibrium species number would be smaller than shown in Figure 54.30. **Figure 54.33** Shrew populations in different locations and habitats might show substantial genetic variation in their susceptibility to the Lyme pathogen. As a result, there might be fewer infected ticks where shrew populations are less susceptible to the Lyme pathogen and more infected ticks where shrews are more susceptible.

Concept Check 54.1

1. Competition has negative effects on individuals of both species (−/−). In predation, members of the predator population benefit by killing and eating members of the prey population; this is an example of exploitation (+/−). Mutualism is an interaction in which individuals of both species benefit (+/+). **2.** One of the competing species will become locally extinct because of the greater reproductive success of the more efficient competitor. **3.** By specialising in eating seeds of different plant species, individuals of the two finch species may be less likely to come into contact in the separate habitats, thereby reinforcing a reproductive barrier to hybridisation.

Concept Check 54.2

1. Species richness, the number of species in the community, and relative abundance, the proportions of the community represented by the various species, both contribute to species diversity. Compared to a community with a very high proportion of one species, one with a more even proportion of species is considered more diverse. **2.** A food chain presents a set of one-way transfers of food energy up to successively higher trophic levels. A food web documents how food chains are linked together, with many species weaving into the web at more than one trophic level. **3.** In bottom-up control, adding extra predators would have little effect on lower trophic levels, particularly vegetation. If the top-down model applied, increased dingo numbers would decrease dunnart numbers, increase lizard numbers, decrease grasshopper numbers, and increase grass biomass. **4.** A decrease in krill abundance might increase the abundance of organisms that krill eat (phytoplankton and copepods), while decreasing the abundance of organisms that eat krill (baleen whales, crabeater seals, birds, fishes, and carnivorous plankton); baleen whales and crabeater seals might be particularly at risk because they eat only krill. However, many of these possible changes could lead to other changes as well, making the overall outcome hard to predict. For example, a decrease in krill abundance could cause an increase in copepod abundance—but an increase in copepod abundance could counteract some of the other effects of decreased krill abundance (since like krill, copepods eat phytoplankton and are eaten by carnivorous plankton and fishes).

Concept Check 54.3

1. High levels of disturbance are generally so disruptive that they eliminate many species from communities, leaving the community dominated by a few tolerant species. Low levels of disturbance permit competitively dominant species to exclude other species from the community. On the other hand, moderate levels of disturbance can facilitate coexistence of a greater number of species in a community by preventing competitively dominant species from becoming abundant enough to eliminate other species from the community. **2.** Early successional species can facilitate the arrival of other species in many ways, including increasing the fertility or water-holding capacity of soils or providing shelter to seedlings from wind and intense sunlight. **3.** The absence of fire for 100 years would represent a change to a low level of disturbance. According to the intermediate disturbance hypothesis, this change should cause diversity to decline as competitively dominant species gain sufficient time to exclude less competitive species.

Concept Check 54.4

1. Ecologists propose that the greater species richness of tropical regions is the result of their longer evolutionary history and the greater solar energy input and water availability in tropical regions. **2.** Immigration of species to islands declines with distance from the mainland and increases with island area. Extinction of species is lower on larger islands and on less isolated islands. Since the number of species on islands is largely determined by the difference between rates of immigration and extinction, the number of species will be highest on large islands near the mainland and lowest on small islands far from the mainland. **3.** Because of their greater mobility, birds disperse to islands more often than snakes and lizards, so birds should have greater richness.

Concept Check 54.5

1. Pathogens are microorganisms, viruses, viroids, or prions that cause disease. **2.** To keep the rabies virus out, you could ban imports of all mammals, including pets. Potentially, you could also attempt to vaccinate all dogs in the British Isles against the virus. A more practical approach might be to quarantine all pets brought into the country that are potential carriers of the disease, the approach the British government actually takes.

Summary of Key Concepts Questions

54.1 Note: Sample answers follow; other answers could also be correct. Competition: a fox and a bobcat competing for prey. Predation: an orca eating a sea otter. Herbivory: a bison eating grass. Parasitism: a parasitoid wasp that lays its eggs on a caterpillar. Mutualism: a fungus and an alga that make up a lichen. Commensalism: a wildflower that grows in a maple forest and a maple tree. **54.2** Not necessarily if the more species-rich community is dominated by only one or a few species **54.3** Similar to clear-cutting a forest or plowing a field, some species would be present initially. As a result, the disturbance would initiate secondary succession in spite of its severe appearance. **54.4** Glaciations are major disturbances that can completely destroy communities found in temperate and polar regions. As a result, tropical communities may be older than temperate or polar communities. This can cause species diversity to be high in the tropics simply because there has been more time for speciation to occur. **54.5** A keystone species is one with a pivotal ecological role. Hence, a pathogen that reduces the abundance of (or otherwise harms) a keystone species could greatly alter the structure of the community. For example, if a novel pathogen drove a keystone species to local extinction, drastic changes in species diversity could occur.

Test Your Understanding

1. D **2.** C **3.** C **4.** C **5.** B **6.** C **7.** D **8.** B
9. Community 1: $H = -(0.05 \ln 0.05 + 0.05 \ln 0.05 + 0.85 \ln 0.85 + 0.05 \ln 0.05) = 0.59$. Community 2: $H = -(0.30 \ln 0.30 + 0.40 \ln 0.40 + 0.30 \ln 0.30) = 1.1$. Community 2 is more diverse. **10.** Crab numbers should increase, reducing the abundance of eelgrass.

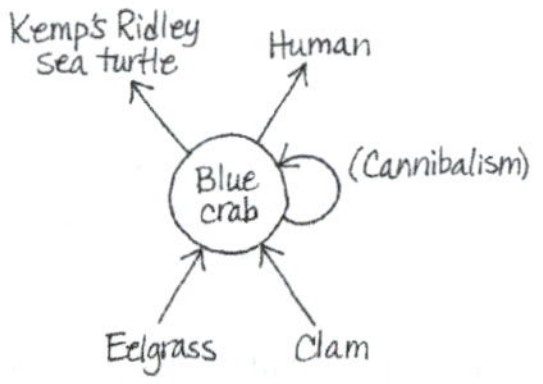

Chapter 55

Figure Questions

Figure 55.4 The blue arrow leading to *Primary consumers* could represent a grasshopper feeding on a plant. The blue arrow leading from *Primary consumers* to *Detritus* could represent the remains of a dead primary consumer (such as a grasshopper) becoming part of the detritus found in the ecosystem. The blue arrow leading from *Primary consumers* to *Secondary and tertiary consumers* could represent a bird (the secondary consumer) eating a grasshopper (the primary consumer). Finally, the blue arrow leading from *Primary consumers* to *Primary producers* could represent CO_2 released by a grasshopper in cellular respiration. **Figure 55.5** The map does not accurately reflect the productivity of wetlands, coral reefs, and coastal zones because these habitats cover areas that are too small to show up clearly on global maps. **Figure 55.11** The availability of water and exposure to light are other factors that may have varied across the sites. Factors such as these that are not included in the experimental design could make the results more difficult to interpret. Multiple factors can also be correlated to each other in nature, so ecologists must be careful that the factor they are studying is actually causing the observed response and is not just correlated with it. **Figure 55.12** (1) If the rate of decomposition slowed, more organic materials would be transferred from reservoir A to reservoir B; eventually, this might lead to more organic material becoming fossilised into fossil fuels. In addition, a decrease in decomposition rate would cause fewer inorganic materials to become available as nutrients in reservoir C, which would ultimately slow the rates of nutrient uptake and photosynthesis by living organisms. (2) Materials move into and out of reservoir A on a much shorter time scale than they move into reservoir B. Materials may remain in reservoir B for a very long time, or humans may remove them at a rapid pace by excavating and burning fossil fuels. **Figure 55.14** If the y-axis had a consistent scale with no break, the change in nitrate concentration in runoff from one tick mark to the next would remain constant throughout the entire axis. For example, if the y-axis were redrawn to have a consistent scale that ran from 0 to 80 mg/L with nine evenly spaced tick marks, the nitrate concentration would increase by 10 mg/L from each tick mark to the next. Drawing the graph with a consistent scale would emphasise the dramatic increase in nitrate concentration that occurred in 1966, but it would be harder to see other, comparatively small changes that occurred from 1965 to 1968 in both control and deforested areas. **Figure 55.24** Populations evolve as organisms interact with each other and with the physical and chemical conditions of their environment. As a result, any human action that alters the environment has the potential to cause evolutionary change. In particular, since climate change has greatly affected arctic ecosystems, we would expect that climate change will cause evolution in arctic tundra populations.

Concept Check 55.1

1. Energy passes through an ecosystem, entering as sunlight and leaving as heat. It is not recycled within the ecosystem. **2.** You would need to know how much biomass the wildebeests ate from your plot and how much nitrogen was contained in that biomass. You would also need to know how much nitrogen they deposited in urine or faeces. **3.** The second law states that in any energy transfer or transformation, some of the energy is dissipated to the surroundings as heat. For the ecosystem to remain intact, this "escape" of energy from the ecosystem must be offset by the continuous influx of solar radiation.

Concept Check 55.2

1. Only a fraction of solar radiation strikes plants or algae, only a portion of that fraction is of wavelengths suitable for photosynthesis, and much energy is reflected or lost as heat. **2.** By manipulating the level of the factors of interest, such as phosphorus availability or soil moisture, and measuring responses by primary producers **3.** It is likely that NEP would decline after the fire. To see why, recall that NEP = GPP − R_T, where GPP is gross primary production and R_T is the total amount of cellular respiration in the ecosystem. By killing trees and other plants, the fire would cause GPP to decline from its pre-fire level. In addition, as decomposers broke down the remains of trees killed by fire, the overall amount of cellular respiration (R_T) in the ecosystem could increase (because of increased cellular respiration by decomposers). **4.** The enzyme rubisco, which catalyses the first step in the Calvin cycle, is the most abundant protein on Earth. Like all proteins, rubisco contains nitrogen, and because photosynthetic organisms require so much rubisco, they also require considerable nitrogen to make it. Phosphorus is also needed as a component of several metabolites in the Calvin cycle and as a component of both ATP and NADPH (see Figure 10.19).

Concept Check 55.3

1. 20 J; 40% **2.** Nicotine protects the plant from herbivores. **3.** Total net primary production is 10,000 + 1,000 + 100 + 10 J = 11,110 J. This is the amount of energy theoretically available to decomposers.

Concept Check 55.4

1. For example, for the carbon cycle:

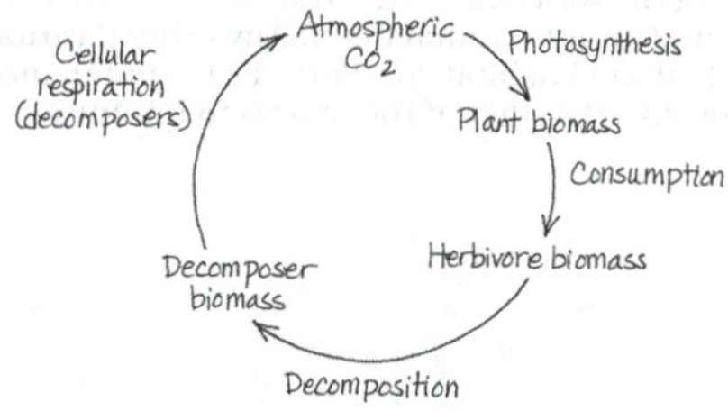

2. Removal of the trees stops nitrogen uptake from the soil, allowing nitrate to accumulate in the soil. The nitrate is washed away by precipitation and enters the streams. **3.** Most of the nutrients in a tropical rain forest are contained in the trees, so removing the trees by logging rapidly depletes nutrients from the ecosystem. The nutrients that remain in the soil are quickly carried away into streams and groundwater by the abundant precipitation.

Concept Check 55.5

1. The main goal is to restore degraded ecosystems to a more natural state. **2.** Bioremediation uses organisms—generally prokaryotes, fungi, or plants—to detoxify or remove pollutants from ecosystems. Biological augmentation uses organisms, such as nitrogen-fixing plants, to add essential materials to degraded ecosystems. **3.** The Atherton Crater Lakes re-establish canopy cover and seed production in the forest, and facilitate the bird dispersal of seeds from nearby rain forest patches, a self-sustaining outcome. Ecologists at the Maungatautari reserve will need to maintain the integrity of the fence indefinitely, an outcome that is not self-sustaining in the long term.

Summary of Key Concepts Questions

55.1 Because energy conversions are inefficient, with some energy inevitably lost as heat, you would expect that a given mass of primary producers would support a smaller biomass of consumers. **55.2** If you know NPP and want to estimate NEP, you must be able to determine how much of the total respiration (R_T) results from heterotrophs and how much results from autotrophs. In a sample of ocean water, primary producers and other organisms are usually mixed together, making their respective respirations hard to separate. **55.3** Runners use much more energy in respiration when they are running than when they are sedentary, reducing their production efficiency. **55.4** Factors other than temperature, including a shortage of water and nutrients, slow decomposition in hot deserts. **55.5** If the topsoil and deeper soil are kept separate, the engineers could return the deeper soil to the site first and then apply the more fertile topsoil to improve the success of revegetation and other restoration efforts.

Test Your Understanding

1. B **2.** B **3.** A **4.** C **5.** A **6.** D **7.** D **8.** D
9. (a)

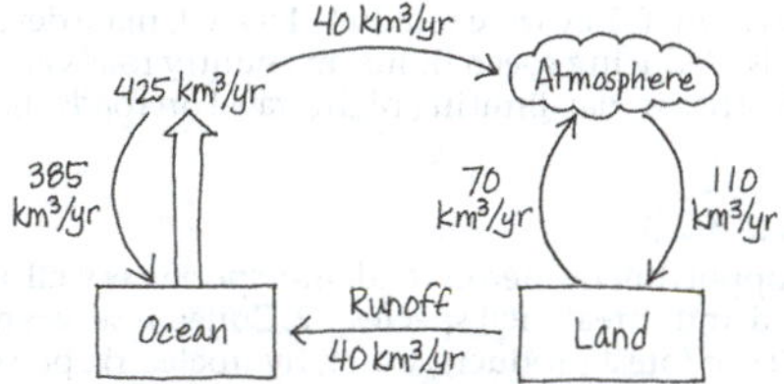

(b) On average, the ratio is 1, with equal amounts of water moving from the ocean to land as precipitation and moving from land to ocean in runoff. (c) During an ice age, the amount of ocean evaporation falling on land as precipitation would be greater than the amount returning to the oceans in runoff; thus, the ratio would be 71. The difference would build up on land as ice.

Chapter 56

Figure Questions

Figure 56.3 You would need to know the complete range of the species and that it is missing across all of that range. You would also need to be certain that the species isn't hidden, as might be the case for an animal that is hibernating underground or a plant that is present in the form of seeds or spores. **Figure 56.11** The two examples are similar in that segments of DNA from the harvested samples were analysed and compared with segments from specimens of known origin. One difference is that the whale researchers investigated relatedness at species and population levels to determine whether illegal activity had occurred, whereas the elephant investigators determined relatedness at the population level to determine the precise location of the poaching. Another difference is that mtDNA was used for the whale study, whereas nuclear DNA was used for the elephant study. The primary limitations of such approaches are the need to have (or generate) a reference database and the requirement that the organisms have sufficient variation in their DNA to reveal the relatedness of samples. **Figure 56.13** The higher the pH, the lower the acidity. Thus, the precipitation in this forest is becoming less acidic. **Figure 56.19** Removal of introduced predators will increase survivorship of nesting female kiwis, their eggs, chicks, young birds, and birds that grow to maturity. From a larger population of reproductively mature individuals, more birds will contribute to subsequent generations to maintain and expand the population. Increasing the area of kiwi habitat provides opportunities for the highly territorial kiwis to disperse into unoccupied territories, where birds can invest their energies in survival and breeding, instead of fighting over territories. **Figure 56.20** The photo shows edges between forest and grassland ecosystems, and grassland and river ecosystems. **Figure 56.29** The PCB concentration increased by a factor of 4.9 from phytoplankton to zooplankton, 41.6 from phytoplankton to smelt, 8.5 from zooplankton to smelt, 4.6 from smelt to lake trout, 119.2 from smelt to herring gull eggs, and 25.7 from lake trout to herring gull eggs. **Figure 56.36** Ocean acidification reduces the availability of carbonate ions (CO_3^{2-}). Corals and many other marine organisms require carbonate ions to build their shells. Since shell-building organisms depend upon their shells for survival, scientists have predicted that ocean acidification will cause many shell-building organisms to die. In turn, increased mortality rates of organisms that build shells would cause many other changes to ecological communities. For example, increased mortality rates of corals would harm the many other species that seek protection in coral reefs or that feed upon the species living there. **Figure 56.37** The model results in the blue curve (natural factors only) and the results in the purple curve (natural *and* human factors) both provide a good match to observed temperature changes until about 1960. After 1960, however, the results in the blue curve are a poor match to observed temperature changes, whereas the results in the purple curve continue to provide a good match. These results suggest that human activities such as burning fossil fuels have contributed to the observed rise in global temperatures, especially for the period 1960 to the present.

Concept Check 56.1

1. In addition to species loss, the biodiversity crisis includes the loss of genetic diversity within populations and species and the degradation of entire ecosystems. **2.** Habitat destruction, such as deforestation, channelising of rivers, or conversion of natural ecosystems to agriculture or cities, deprives species of places to live. Introduced species, which are transported by humans to regions outside their native range, often reduce the population sizes of native species through competition or by feeding on them (as predators, herbivores, or pathogens). Overharvesting has reduced populations of plants and animals or driven them to extinction. Finally, global change is altering the environment to the extent that it reduces the capacity of Earth to sustain life. **3.** If both populations breed separately, then gene flow between the populations would not occur and genetic differences between them would be greater. As a result, the loss of genetic diversity would be greater than if the populations interbreed.

Concept Check 56.2

1. Reduced genetic variation decreases the capacity of a population to evolve in the face of change. **2.** The effective population size, N_e, would be

$4(30 \times 10)/(30 + 10) = 30$ birds. **3.** Reducing numbers of feral animals in the core habitat of the Tasmanian devil provides one potential management strategy. Relocating devil populations away from roads may provide a temporary reduction in road deaths of devils. However, male devils travel up to 15 km in one evening so even if they were remotely located, male devils may travel to areas near roads. Reducing speed limits on country roads at night may limit the number of deaths, as may limiting night travel on roads that traverse devil habitats.

Concept Check 56.3

1. A small area supporting numerous endemic species as well as a large number of endangered and threatened species **2.** Zoned reserves may provide sustained supplies of forest products, water, hydroelectric power, educational opportunities, and income from tourism. **3.** Habitat corridors can increase the rate of movement or dispersal of organisms between habitat patches and thus the rate of gene flow between subpopulations. They thus help prevent a decrease in fitness attributable to inbreeding. They can also minimise interactions between organisms and humans as the organisms disperse; in cases involving potential predators, such as dingos or feral cats, minimising such interactions is desirable.

Concept Check 56.4

1. Adding nutrients causes population explosions of algae and the organisms that feed on them. Increased respiration by algae and consumers, including decomposers, depletes the lake's oxygen, which the fish require. **2.** Decomposers are consumers that use nonliving organic matter as fuel for cellular respiration, which releases CO_2 as a by-product. Because higher temperatures lead to faster decomposition, organic matter in these soils could be decomposed to CO_2 more rapidly, thereby speeding up global warming. **3.** Reduced concentrations of ozone in the atmosphere increase the amount of UV radiation that reaches Earth's surface and the organisms living there. UV radiation can cause mutations by producing disruptive thymine dimers in DNA.

Concept Check 56.5

1. Sustainable development is an approach to development that works towards the long-term prosperity of human societies and the ecosystems that support them, which requires linking the biological sciences with the social sciences, economics, and humanities. **2.** Biophilia, our sense of connection to nature and all forms of life, may act as a significant motivation for the development of an environmental ethic that resolves not to allow species to become extinct or ecosystems to be destroyed. Such an ethic is necessary if we are to become more attentive and effective custodians of the environment. **3.** At a minimum, you would want to know the size of the population and the average reproductive rate of the individuals in it. To develop the fishery sustainably, you would seek a harvest rate that maintains the population near its original size and maximises its harvest in the long term rather than the short term.

Summary of Key Concepts Questions

56.1 Nature provides us with many beneficial services, including a supply of reliable, clean water, the production of food and fibre, and the dilution and detoxification of our pollutants. **56.2** A more genetically diverse population is better able to withstand pressures from disease or environmental change, making it less likely to become extinct over a given period of time. **56.3** Habitat fragmentation can isolate populations, leading to inbreeding and genetic drift, and it can make populations more susceptible to local extinctions resulting from edge effects, including a change in physical conditions and an increase in competition or predation with edge-adapted species. **56.4** It's healthier to feed at a lower trophic level because biological magnification increases the concentration of toxins at higher levels. **56.5** One goal of conservation biology is to preserve as many species as possible. Sustainable approaches that maintain the quality of habitats are required for the long-term survival of organisms.

Test Your Understanding

1. C **2.** D **3.** B **4.** A **5.** B **6.** D

7.

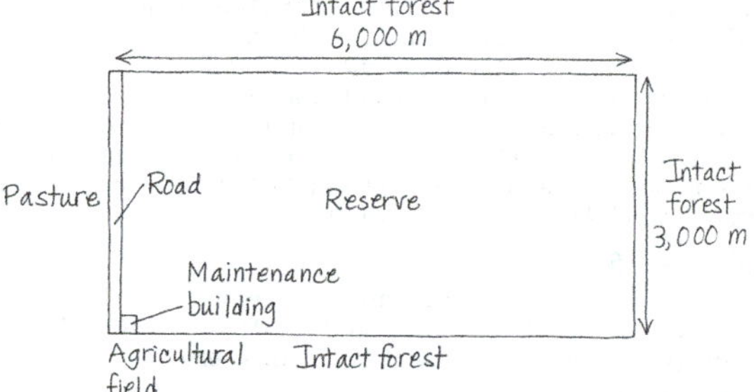

To minimise the area of forest into which the cowbirds penetrate, you should locate the road along the west edge of the reserve (since that edge abuts deforested pasture and an agricultural field). Any other location would increase the area of affected habitat. Similarly, the maintenance building should be in the southwest corner of the reserve to minimise the area susceptible to cowbirds.

Appendix B Classification of Life

This appendix presents a taxonomic classification for the major extant groups of organisms discussed in this text; not all phyla are included. The classification presented here is based on the three-domain system, which assigns the two major groups of prokaryotes, bacteria and archaea, to separate domains (with eukaryotes making up the third domain).

Various alternative classification schemes are discussed in Unit 5 of this text. The taxonomic turmoil includes debates about the number and boundaries of kingdoms and about the alignment of the Linnaean classification hierarchy with the findings of modern cladistic analysis.

DOMAIN BACTERIA

- **Proteobacteria**
- **Chlamydia**
- **Spirochetes**
- **Cyanobacteria**
- **Gram-positive Bacteria**

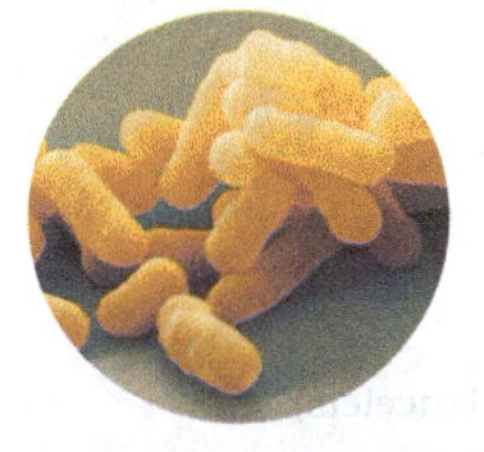

DOMAIN ARCHAEA

- **Euryarchaeota**
- **Thaumarchaeota**
- **Aigarchaeota**
- **Crenarchaeota**
- **Korarchaeota**

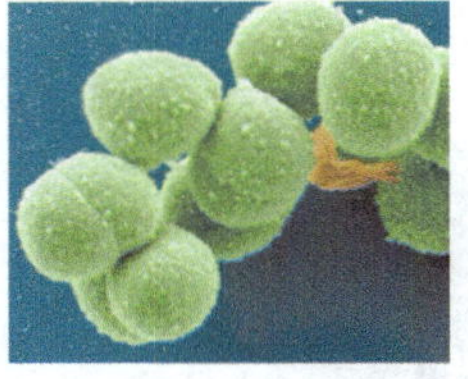

DOMAIN EUKARYA

In the phylogenetic hypothesis we present in the chapter, "Protists," major clades of eukaryotes are grouped together in the four "supergroups" listed in blue type. Formerly, all the eukaryotes generally called protists were assigned to a single kingdom, Protista. However, advances in systematics have made it clear that some protists are more closely related to plants, fungi, or animals than they are to other protists. As a result, the kingdom Protista has been abandoned.

Excavata
- Diplomonadida (diplomonads)
- Parabasala (parabasalids)
- Euglenozoa (euglenozoans)
  - Kinetoplastida (kinetoplastids)
  - Euglenophyta (euglenids)

SAR
- Stramenopila (stramenopiles)
  - Oomycota (oomycetes)
  - Phaeophyta (brown algae)
  - Bacillariophyta (diatoms)

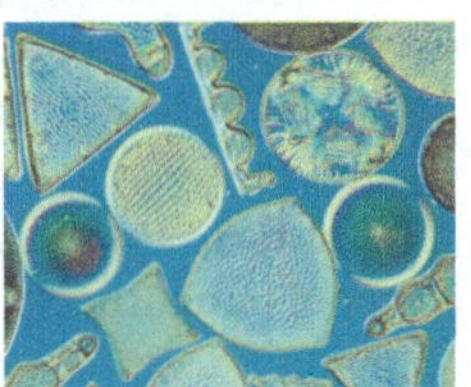

- Alveolata (alveolates)
  - Dinoflagellata (dinoflagellates)
  - Apicomplexa (apicomplexans)
  - Ciliophora (ciliates)
- Rhizaria (rhizarians)
  - Radiolaria (radiolarians)
  - Foraminifera (forams)
  - Cercozoa (cercozoans)

Archaeplastida
- Rhodophyta (red algae)
- Chlorophyta (green algae: chlorophytes)
- Charophyta (green algae: charophytes)
- Plantae
  - Nonvascular plants (bryophytes):
    - Phylum Hepatophyta (liverworts)
    - Phylum Bryophyta (mosses)
    - Phylum Anthocerophyta (hornworts)
  - Seedless vascular plants:
    - Phylum Lycophyta (lycophytes)
    - Phylum Monilophyta (ferns, horsetails, whisk ferns)
  - Seed plants:
    - Gymnosperms:
      - Phylum Ginkgophyta (ginkgo)
      - Phylum Cycadophyta (cycads)
      - Phylum Gnetophyta (gnetophytes)
      - Phylum Coniferophyta (conifers)
    - Angiosperms:
      - Phylum Anthophyta (flowering plants)

DOMAIN EUKARYA, continued

Unikonta (also called Amorphea)

- Amoebozoa (amoebozoans)
  - Tubulinea (tubulinids)
  - Myxogastrida (plasmodial slime moulds)
  - Dictyostelida (cellular slime moulds)
  - Entamoeba (entamoebas)
- Nucleariida (nucleariids)
- Fungi
  - Phylum Cryptomycota (cryptomycetes)
  - Phylum Microsporidia (microsporidians)
  - Phylum Chytridiomycota (chytrids)
  - Phylum Zoopagomycota (zoopagomycetes)
  - Phylum Mucoromycota (mucoromycetes)
  - Phylum Ascomycota (ascomycetes)
  - Phylum Basidiomycota (basidiomycetes)

- Choanoflagellata (choanoflagellates)
- Animalia
  - Phylum Porifera (sponges)
  - Phylum Ctenophora (comb jellyfish)
  - Phylum Cnidaria (cnidarians)
    - Medusozoa (hydrozoans, jellyfish, box jellyfish)
    - Anthozoa (sea anemones and most corals)
  - Phylum Acoela (acoel flatworms)
  - Phylum Placozoa (placozoans)
  - Lophotrochozoa (lophotrochozoans)
    - Phylum Platyhelminthes (flatworms)
      - Catenulida (chain worms)
      - Rhabditophora (planarians, flukes, tapeworms)
    - Phylum Nemertea (ribbon worms)
    - Phylum Ectoprocta (ectoprocts)
    - Phylum Brachiopoda (brachiopods)
    - Phylum Syndermata (rotifers and spiny-headed worms)
    - Phylum Gastrotricha (gastrotrichs)
    - Phylum Cycliophora (cycliophorans)
    - Phylum Mollusca (molluscs)
      - Polyplacophora (chitons)
      - Gastropoda (gastropods)
      - Bivalvia (bivalves)
      - Cephalopoda (cephalopods)
    - Phylum Annelida (segmented worms)
      - Errantia (errantians)
      - Sedentaria (sedentarians)
  - Ecdysozoa (ecdysozoans)
    - Phylum Loricifera (loriciferans)
    - Phylum Priapula (priapulans)
    - Phylum Nematoda (roundworms)
    - Phylum Arthropoda (This survey groups arthropods into a single phylum, but some zoologists now split the arthropods into multiple phyla.)
      - Chelicerata (horseshoe crabs, arachnids)
      - Myriapoda (millipedes, centipedes)
      - Pancrustacea (crustaceans, insects)
    - Phylum Tardigrada (tardigrades)
    - Phylum Onychophora (velvet worms)
  - Deuterostomia (deuterostomes)
    - Phylum Hemichordata (hemichordates)
    - Phylum Echinodermata (echinoderms)
      - Asteroidea (sea stars, sea daisies)
      - Ophiuroidea (brittle stars)
      - Echinoidea (sea urchins, sand dollars)
      - Crinoidea (sea lilies)
      - Holothuroidea (sea cucumbers)
    - Phylum Chordata (chordates)
      - Cephalochordata (cephalochordates: lancelets)
      - Urochordata (urochordates: tunicates)
      - Vertebrates:
        - Cyclostomata (cyclostomes)
          - Myxini (hagfishes)
          - Petromyzontida (lampreys)
        - Gnathostomata (gnathostomes)
          - Chondrichthyes (sharks, rays, chimaeras)
          - Actinopterygii (ray-finned fishes)
          - Actinistia (coelacanths)
          - Dipnoi (lungfishes)
          - Amphibia (amphibians: frogs, salamanders, caecilians)
          - Reptilia (reptiles: tuataras, lizards, snakes, turtles, crocodilians, birds)
          - Mammalia (mammals)

Appendix C A Comparison of the Light Microscope and the Electron Microscope

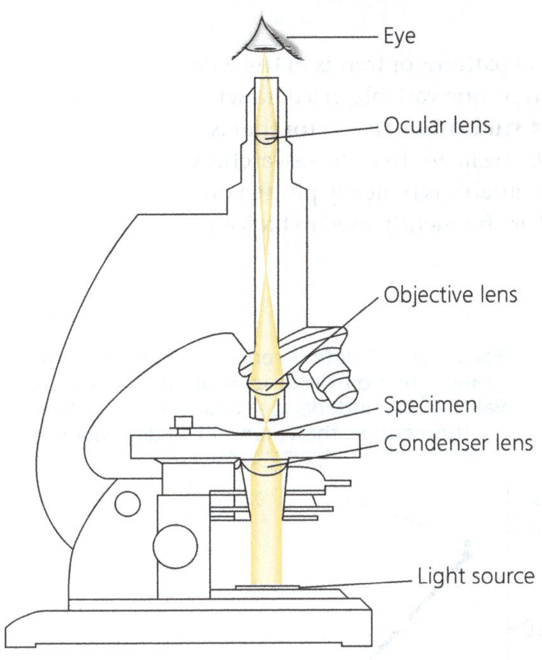

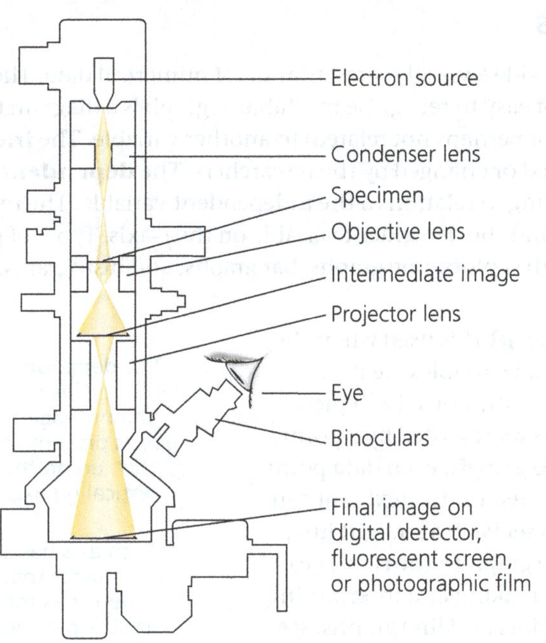

Light Microscope

In light microscopy, light is focused on a specimen by a glass condenser lens; the image is then magnified by an objective lens and an ocular lens for projection on the eye, digital camera, digital video camera, or photographic film.

Electron Microscope

In electron microscopy, a beam of electrons (top of the microscope) is used instead of light, and electromagnets are used instead of glass lenses. The electron beam is focused on the specimen by a condenser lens; the image is magnified by an objective lens and a projector lens for projection on a digital detector, fluorescent screen, or photographic film.

Appendix D Scientific Skills Review

Graphs

Graphs provide a visual representation of numerical data. They may reveal patterns or trends in the data that are not easy to recognise in a table. A graph is a diagram that shows how one variable in a data set is related (or perhaps not related) to another variable. The **independent variable** is the factor that is manipulated or changed by the researchers. The **dependent variable** is the factor that the researchers are measuring in relation to the independent variable. The independent variable is typically plotted on the *x*-axis and the dependent variable on the *y*-axis. Types of graphs that are frequently used in biology include scatter plots, line graphs, bar graphs, and histograms.

➤ A **scatter plot** is used when the data for all variables are numerical and continuous. Each piece of data is represented by a point. In a **line graph**, each data point is connected to the next point in the data set with a straight line, as in the graph to the right. (To practise making and interpreting scatter plots and line graphs, see the Scientific Skills Exercises in Chapters 2, 3, 7, 8, 10, 13, 19, 24, 34, 43, 47, 49, 50, 52, 54, and 56.)

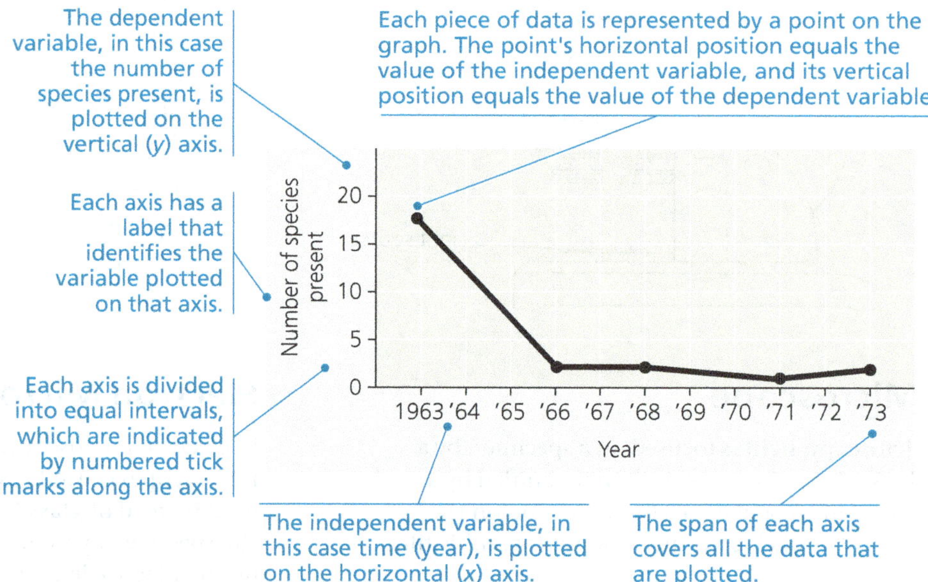

➤ Two or more data sets can be plotted on the same line graph to show how two dependent variables are related to the same independent variable. (To practise making and interpreting line graphs with two or more data sets, see the Scientific Skills Exercises in Chapters 7, 43, 47, 49, 50, 52, and 56.)

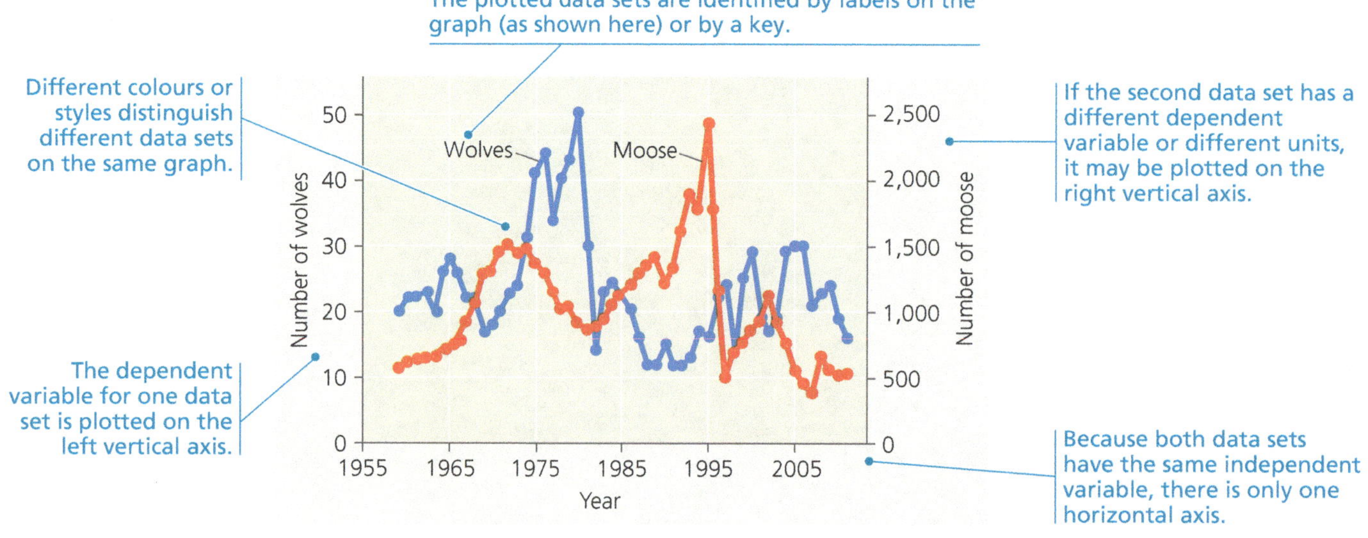

In some scatter plot graphs, a straight or curved line is drawn through the entire data set to show the general trend in the data. A straight line that mathematically best fits the data is called a *regression line*. Alternatively, a mathematical function that best fits the data may describe a curved line, often termed a *best-fit curve*. (To practise making and interpreting regression lines, see the Scientific Skills Exercises in Chapters 3, 10, and 34.)

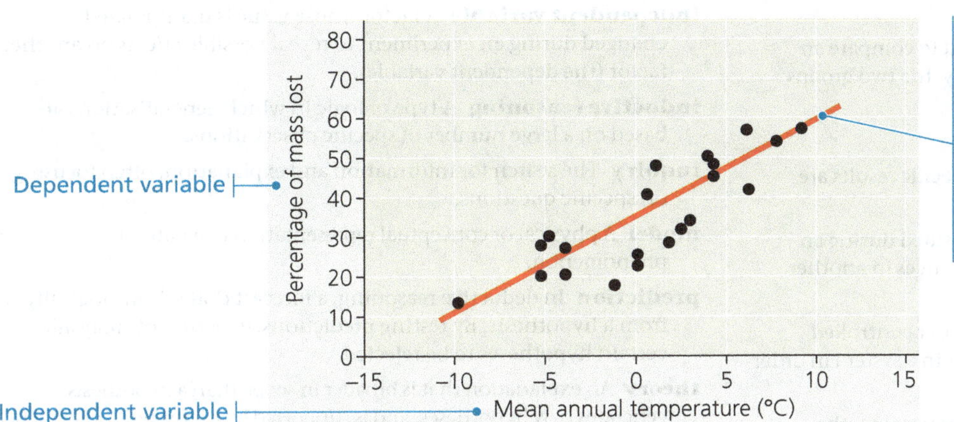

A **bar graph** is a kind of graph in which the independent variable represents groups or nonnumerical categories and the values of the dependent variable(s) are shown by bars. (To practise making and interpreting bar graphs, see the Scientific Skills Exercises in Chapters 1, 9, 18, 22, 25, 29, 33, 35, 39, 51, 52, and 54.)

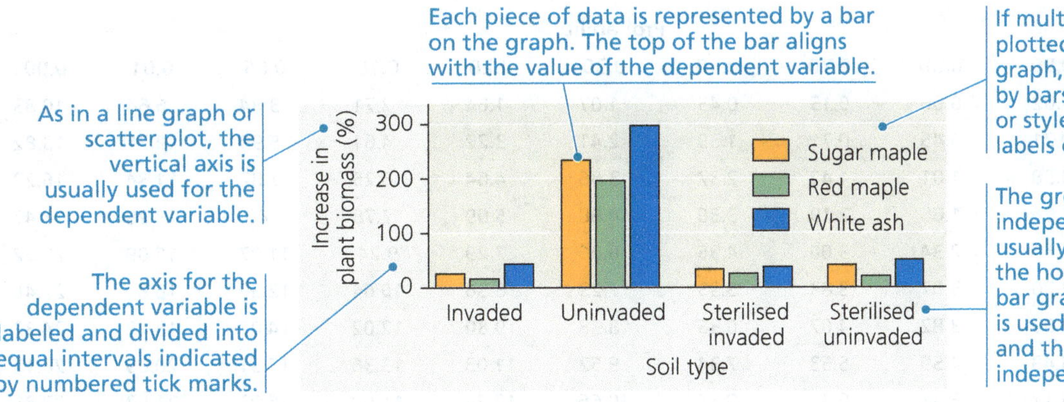

A variant of a bar graph called a **histogram** can be made for numeric data by first grouping, or "binning," the variable plotted on the *x*-axis into intervals of equal width. The "bins" may be integers or spans of numbers. In the histogram at right, the intervals are 25 mg/dL wide. The height of each bar shows the percentage (or, alternatively, the number) of experimental subjects whose characteristics can be described by one of the intervals plotted on the *x*-axis. (To practise making and interpreting histograms, see the Scientific Skills Exercises in Chapters 12, 14, and 42.)

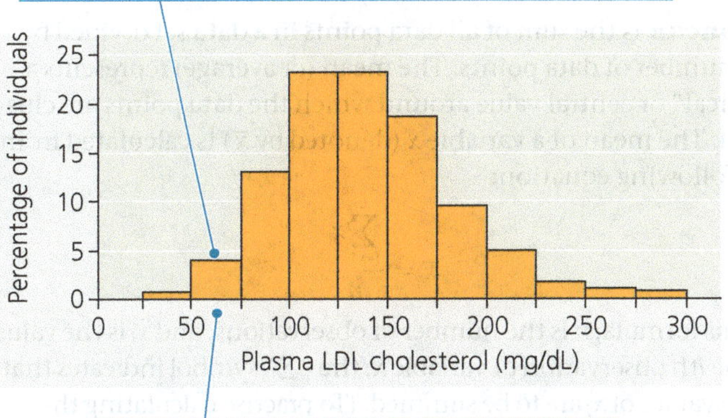

Glossary of Scientific Inquiry Terms

See Concept 1.3 for more discussion of the process of scientific inquiry.

control group In a controlled experiment, a set of subjects that lacks (or does not receive) the specific factor being tested. Ideally, the control group is identical to the experimental group in other respects.

controlled experiment An experiment designed to compare an experimental group with a control group; ideally, the two groups differ only in the factor being tested.

data Recorded observations.

deductive reasoning A type of logic in which specific results are predicted from a general premise.

dependent variable A factor whose value is measured during an experiment to see whether it is influenced by changes in another factor (the independent variable).

experiment A scientific test. Often carried out under controlled conditions that involve manipulating one factor in a system in order to see the effects of changing that factor.

experimental group A set of subjects that has (or receives) the specific factor being tested in a controlled experiment. Ideally, the experimental group is identical to the control group for all other factors.

hypothesis A testable explanation for a set of observations based on the available data and guided by inductive reasoning. A hypothesis is narrower in scope than a theory.

independent variable A factor whose value is manipulated or changed during an experiment to reveal possible effects on another factor (the dependent variable).

inductive reasoning A type of logic in which generalisations are based on a large number of specific observations.

inquiry The search for information and explanation, often focusing on specific questions.

model A physical or conceptual representation of a natural phenomenon.

prediction In deductive reasoning, a forecast that follows logically from a hypothesis. By testing predictions, experiments may allow certain hypotheses to be rejected.

theory An explanation that is broader in scope than a hypothesis, generates new hypotheses, and is supported by a large body of evidence.

variable A factor that varies during an experiment.

Chi-Square (χ^2) Distribution Table

To use the table, find the row that corresponds to the degrees of freedom in your data set. (The degrees of freedom is the number of categories of data minus 1.) Move along that row to the pair of values that your calculated χ^2 value lies between. Move up from those numbers to the probabilities at the top of the columns to find the probability range for your χ^2 value. A probability of 0.05 or less is generally considered significant. (To practise using the chi-square test, see the Scientific Skills Exercise in Chapter 15.)

Degrees of Freedom (df)	Probability										
	0.95	0.90	0.80	0.70	0.50	0.30	0.20	0.10	0.05	0.01	0.001
1	0.004	0.02	0.06	0.15	0.45	1.07	1.64	2.71	3.84	6.64	10.83
2	0.10	0.21	0.45	0.71	1.39	2.41	3.22	4.61	5.99	9.21	13.82
3	0.35	0.58	1.01	1.42	2.37	3.66	4.64	6.25	7.82	11.34	16.27
4	0.71	1.06	1.65	2.19	3.36	4.88	5.99	7.78	9.49	13.28	18.47
5	1.15	1.61	2.34	3.00	4.35	6.06	7.29	9.24	11.07	15.09	20.52
6	1.64	2.20	3.07	3.83	5.35	7.23	8.56	10.64	12.59	16.81	22.46
7	2.17	2.83	3.82	4.67	6.35	8.38	9.80	12.02	14.07	18.48	24.32
8	2.73	3.49	4.59	5.53	7.34	9.52	11.03	13.36	15.51	20.09	26.12
9	3.33	4.17	5.38	6.39	8.34	10.66	12.24	14.68	16.92	21.67	27.88
10	3.94	4.87	6.18	7.27	9.34	11.78	13.44	15.99	18.31	23.21	29.59

Mean and Standard Deviation

The **mean** is the sum of all data points in a data set divided by the number of data points. The mean (or average) represents a "typical" or central value around which the data points are clustered. The mean of a variable x (denoted by $\bar{x}$) is calculated from the following equation:

$$\bar{x} = \frac{\sum_{i=1}^{n} x_i}{n}$$

In this formula, n is the number of observations, and x_i is the value of the ith observation of variable x; the "Σ" symbol indicates that the n values of x_i are to be summed. (To practise calculating the mean, see the Scientific Skills Exercises in Chapters 27, 32, and 34.)

The **standard deviation** provides a measure of the variation found in a set of data points. The standard deviation (s) of a variable x is calculated from the following equation:

$$s = \sqrt{\frac{\sum_{i=1}^{n} (x_i - \bar{x})^2}{n - 1}}$$

In this formula, n is the number of observations, x_i is the value of the ith observation of variable x, and $\bar{x}$ is the mean of x; the "Σ" symbol indicates that the n values of $(x_i - \bar{x})^2$ are to be summed. (To practise calculating standard deviation, see the Scientific Skills Exercises in Chapters 27, 32, and 34.)

Performing a *t*-Test

One way to assess whether the results of an experiment are statistically significant is to perform a *t*-test. Consider an experiment in which one group of bean plants was treated with fertiliser, whereas a control group was not. Before beginning the experiment, the researchers hypothesised that fertiliser would not affect plant height (the fertiliser would neither increase plant height nor decrease it).

When the experiment was completed, the fertilised plants appeared overall to have grown taller than the unfertilised ones—that is, the mean height of the fertilised plants was greater than the mean height of the unfertilised ones. This result suggests that the fertiliser did, in fact, have an effect and, hence, that the two means are not equal. However, it is also possible that the different means in the two study groups resulted from natural variation of plant heights within the two groups, particularly if the total number of plants is small. How can we determine the likelihood that the observed differences are meaningful and therefore indicate that the fertiliser had an effect? The *t*-test provides a standardised way to decide whether the fertiliser had a significant effect on mean plant height.

To perform a *t*-test, the first step is to calculate the value T (named for "*t*-test"):

$$T = \frac{\bar{x}_1 - \bar{x}_2}{\sqrt{\dfrac{(s_1^2 + s_2^2)}{n}}}$$

In this equation, $\bar{x}_1$ is the mean for the experimental group (fertilised plants), $\bar{x}_2$ is the mean for the control group (unfertilised plants), s_1 is the standard deviation for the experimental group, and s_2 is the standard deviation for the control group. Finally, n is the number of observations in each of the groups. [Note: The formula shown here is valid when the experimental and control groups have the same number of observations (n). A different formula would be used if the two groups had different numbers of observations.]

To calculate T, plug in the values for $\bar{x}_1$, $\bar{x}_2$, s_1, s_2, and n. T will be close to zero when the means $\bar{x}_1$ and $\bar{x}_2$ are nearly equal, and T will be farther from zero (will differ more greatly from zero) when the means are considerably different.

Is the calculated value T different enough from zero to reject the hypothesis that the two means are equal? This decision is based on the probability (p) that an observed difference between two means could have occurred simply by chance (assuming the initial hypothesis was correct, that is, that the two means are equal). The value of p can be determined using a t distribution that has $2(n-1)$ degrees of freedom, where n is the number of observations. When p is small (typically, when it is less than 0.05), we reject the hypothesis that the means $\bar{x}_1$ and $\bar{x}_2$ are equal. The value of p can be obtained from an online calculator or looked up in tables for t distributions in a statistics textbook.

Credits

Photo Credits

Front Matter **p. ix bioluminescence** Biosphoto/Alamy Stock Photo; **mice** Permission for use granted by Randy Jirtle, Professor of Epigenetics, NC State University, Raleigh, NC; **Dutch Hunger Winter** History and Art Collection/Alamy Stock Photo; **p. x graph** Data from K. Kupferschmidt, Resistance Fighters. *Science* 352(6287):758-761. 13 May 2016; **p. xii koala** Francois Gohier/Science Source; **p. xvi human** lanych/Shutterstock; **monkey** David Bagnall/Alamy Stock Photo; **gibbon** Eric Isselee/Shutterstock; **frogs** Joel Sartore/Getty Images; **laptop** Lemberg Vector studio/Shutterstock; **p. xvii students** YanLev/Shutterstock; **p. xxi space-filling model, ribbon model** Data from PDB ID 2LYZ: R. Diamond. Real-Space Refinement of the Structure of Hen Egg-white Lysozyme. *Journal of Molecular Biology* 82(3):371–91 (Jan. 25, 1974); **wireframe model** Clive Freeman, The Royal Institution/Science Source; **p. xxiii Allwood** Queensland University of Technology; **Bautista** Mark Joseph Hanson; **Mojica** University of Alicante; **Hoffmann** Photographer: Peter Casamento; **Campbell** Courtesy of Hamish Campbell; **Gonsalves** Courtesy of Dennis Gonsalves; **Strathdee** UCSD Health; **Hoegh-Guldberg** Professor Ove Hoegh-Guldberg; **p. xxix seal** Hemis Morales/agefotostock; **p. xxx researchers** CDC; **p. xxxi impala** Nature Picture Library/Alamy Stock Photo; **pea blossom** John Swithinbank/Agefotostock; **p. xxxiii DNA** 4X-image/E+/Getty Images; **p. xxxiv cell infected with SARS-CoV-2** National Institute of Allergy and Infectious Diseases (NIAID); **sea horse** Rich Cary/Shutterstock; **p. xxxv hare** L. Scott Mills; **amber** Iolanda Astor/AGE fotostock/Alamy Stock Photo; **p. xxxvi dulse** Andrew J. Martinez/Science Source; **p. xxxxvii seeds of a dandelion** Martin Turner/Getty Images; ***Pilobolus*** G L Barron/Biological Photo Service; **p. xxxviii ladybug** André Skonieczny/F1online digitale Bildagentur GmbH/Alamy Stock Photo; **shark** Gino Santa Maria/shutterstock; **tree** Raimund Linke/Photodisc/Getty Images; **p. xxxix dust storm** BeyondImages/Getty Images; **pollen** Dartmouth College Electron Microscope Facility; **p. xl penguins** Paul Nicklen/National Geographic Image Collection/Getty Images; **p. xli virus** Kateryna Kon/Shutterstock; **p. xlii zebras** Mike Taylor/Alamy Stock Photo; **p. xliii plover** Joel Sartore/National Geographic Images; **p. xliv reef** Matthew Banks/Alamy Stock Photo Image; **possum** Edwin Giesbers/Nature Picture Library.

Chapter 1 **1.1 top** Kerri-Lee Harris; **Banksia serrata** KarenHBlack/Shutterstock; **middle left** Paul Whitington; **1.2 sunflower** John Foxx/ImageState Media Partners; **seahorse** R. Dirscherl/OceanPhoto/Frank Lane Picture Agency; **bilby** Martin Harvey/Alamy; **butterfly** Louise Docker Sydney Australia/Moment/Getty Images; **seedling** Frederic Didillon/Garden Picture Library/Getty Images; **venus flytrap** Maximilian Weinzierl/Alamy Stock Photo; **giraffe** Malcolm Schuyl/Frank Lane Picture Agency; **1.3 ecosystems** Tristan Schmurr/Wikimedia Commons; **communities** Nazera Salam/Shutterstock; **populations** Bernard Spragg/Wikimedia Commons; **organisms** Bob Gibbons/Alamy Stock Photo; **organs** Bob Hilscher/Getty Images; **tissues** Photo Researchers/Science Source; **cells** Andreas Holzenburg, University of Texas Rio Grande Valley; **organelles** Jeremy Burgess/Science Source; **p. 6 hummingbird** Jim Zipp/Science Source; **1.4 left** Steve Gschmeissner/Science Source; **1.4 right** A. Barry Dowsett/Science Source; **1.5** Conly L. Rieder, Wadsworth Center, Albany, NY; **1.6** Wallenrock/Shutterstock; **1.8a** Carol Yepes/Moment/Getty Images; **inset** Ralf Dahm/Max Planck Institute of Neurobiology; **1.11** James Balog/Aurora/Getty Images; **1.12** Kefca/Shutterstock; **1.13a, b** Eye of Science/Science Source; **1.13c plantae** John Delapp/Design Pics/Getty Images; **fungi** Carpeng/Shutterstock; **animalia** Anup Shah/Nature Picture Library; **protists** M. I. Walker/Science Source; **1.14 lake** Basel101658/Shutterstock; **paramecium** SPL/Science Source; **cilium** Dartmouth College Electron Microscope Facility; **cilia** Steve Gschmeissner/Science Source; **1.15** Robert Clark/National Geographic Image Collection; **1.16 left** Science Source; **1.16 right** Origin of Species/Charles Darwin, 1859. Murray edition; **1.17 hawk** jhayes44/E+/Getty Images; **robin** New Zealand Birds Online © David Boyle; **cassowary** David Watts/Alamy; **penguin** Volodymyr Goinyk/Shutterstock; **1.19** Dorling Kindersley ltd/Alamy Stock Photo; **1.19** © Steve Bourne; **1.21** Michael Nichols/National Geographic Image Collection; **1.23 top** Martin Shields/Alamy Stock Photo; **centre** xpacifica/Getty Images; **bottom left** Rolf Hicker Photography/All Canada Photos/Alamy Stock Photo; **bottom right** Maureen Spuhler/Pearson Education; **1.24 left** HildeAnna/Shutterstock; **inset** Courtesy of Hopi Hoekstra/Harvard University; **right** Sacha Vignieri; **inset** Shawn P. Carey, Migration Productions; **1.25** From: The selective advantage of cryptic coloration in mice. Vignieri, S. N., J. Larson, and H. E. Hoekstra. 2010. *Evolution* 64:2153–2158. Fig. 1; **Scientific Skills Exercise** Rolf Nussbaumer Photography/Alamy Stock Photo; **p. 26 gecko** Chris Mattison/Alamy Stock Photo.

Unit 1 Interview Queensland University of Technology/AAP.

Chapter 2 **2.1 top** www.pqpictures.co.uk/Alamy Stock Photo; **bottom** Nature Picture Library/Alamy Stock Photo; **2.2 left** sciencephotos/Alamy Stock Photo; **centre, right** Stephen Frisch/Pearson Education; **2.5** National Library of Medicine; **Scientific Skills Exercise** Pascal Goetgheluck/Science Source; **2.13** Stephen Frisch/Pearson Education; **p. 39 gecko** nico99/Shutterstock; **hairs** Andrew Syred/Science Source; **2.17** Nigel Cattlin/Science Source; **p. 43 top** Dr Jacinta Zalucki/Australia & Pacific Science Foundation; **bottom** From: Spray aiming in the bombardier beetle: photographic evidence. T. Eisner et al. *Proc Natl Acad Sci U S A* 1999 Aug 17;96(17):9705-9. Fig. 1.

Chapter 3 **3.1 top** Hemis Morales/agefotostock; **bottom** Paul Nicklen/National Geographic Image Collection; **3.3** Alasdair James/E+/Getty Images; **3.4** N.C Brown Center for Ultrastructure Studies, SUNY-ESF, Syracuse, NY; **3.6** Four Oaks/Shutterstock; **3.10a** Science Photo Library/Alamy Stock Photo; **3.10b** NASA/JPL-Caltech; **3.11** JPL/University of Arizona/NASA; **3.12 lemon** Paulista/Fotolia; **ant** Nature Picture Library/Alamy Stock Photo; **blood cells** SCIEPRO/SPL/AGE Fotostock; **bleach** Beth Van Trees/Shutterstock; **Scientific Skills Exercise** Vlad61/Shutterstock; **p. 57 cat** Eric Guilloret/Biosphoto/Science Source.

Chapter 4 **4.1** Florian Möllers/Nature Picture Library; **Scientific Skills Exercise** Mandeville Special Collections Library; **vials** Jeffrey Bada/Scripps Institution of Oceanography/University of California San Diego; **4.6a** David M. Phillips/Science Source; **p. 67 lions** George Sanker/Nature Picture Library.

Chapter 5 **5.1** Mark J. Winter/Science Source; **5.6a arm** Dougal Waters/Getty Images; **plastids** Omikron/Science Source; **5.6b** Mediscan/Alamy Stock Photo; **5.6c cell** John Durham/Science Source; **5.6c microfibrils** Biophoto Associates/Science Source; **5.8** blickwinkel/Alamy Stock Photo; **5.10a** Vincent Giordano Photo/Shutterstock; **5.10b** Bamorgan91/Shutterstock; **5.13 eggs** Andrey Stratilatov/Shutterstock; **muscle, collagen** Nina Zanetti/Pearson Education; **5.16 wireframe model** Clive Freeman, The Royal Institution/Science Source; **5.17** Peter M. Colman; **5.18 spider** Dieter Hopf/imageBROKER/AGE Fotostock; **blood cells** SCIEPRO/SPL/AGE Fotostock; **5.19** Eye of Science/Science Source; **5.21 top** David and Jane Richardson; **bottom** Laguna Design/Science Source; **5.25** Centers for Disease Control and Prevention (CDC); **5.26 DNA** Alfred Pasieka/Science Source; **neanderthal** Mark Thiessen/National Geographic/Alamy Stock Photo; **hippo** Frontline Photography/Alamy Stock Photo; **whale** WaterFrame_mus/Alamy Stock Photo; **5.26 doctor** Chassenet/BSIP/Alamy Stock Photo; **elephants** Villiers Steyn/Shutterstock; **roots** D.J. Read, Department of Animal and Plant Sciences, University of Sheffield; **Scientific Skills Exercise human** lanych/Shutterstock; **monkey** David Bagnall/Alamy Stock Photo; **gibbon** Eric Isselee/Shutterstock; **Problem-Solving Exercise** Cindy Hopkins/Alamy Stock Photo; **p. 92 butter** Vincent Giordano Photo/Shutterstock; **p. 92 olive oil** mamorgan91/Shutterstock; **p. 93 chick** Africa Studio/Shutterstock.

Unit 2 Interview **top** Mark Joseph Hansen; **bottom** Robin Heyden.

Chapter 6 **6.1** M. I. Walker/Science Source; **6.3 brightfield, phase-contrast, Nomarski** Elisabeth Pierson/Pearson Education; **fluorescence** Michael W. Davidson/The Florida State University Research Foundation; **confocal** Karl Garsha; **deconvolution** Hans van der Voort SVI; **super-resolution** Muthugapatti K. Kandasamy, Biomedical Microscopy Core, University of Georgia; **SEM, TEM** Steve Gschmeissner/Science Source; **cryo-EM** Veronica Falconieri and Siriam Subramaniam, National Cancer Institute; **6.5b** CNRI/Science Source; **6.6a** Don W. Fawcett/Science Source; **Scientific Skills Exercise** Kelly Tatchell; **6.8 human cells** S. Cinti/Science Source; **yeast SEM** SPL/Science Source; **yeast TEM** A. Barry Dowsett/Science Source; **duckweed** Biophoto Associates/Science Source; **alga SEM** SPL/Science Source; **alga TEM** From: Flagellar microtubule dynamics in *Chlamydomonas*: cytochalasin D induces periods of microtubule shortening and elongation; and colchicine induces disassembly of the distal, but not proximal, half of the flagellum. W. L. Dentler et al. *J Cell Biol.* 1992 Jun;117(6):1289-98. Fig. 10d; **p. 104 nucleus** Thomas Deerinck/Mark Ellisman/NCMIR; **6.9 nuclear envelope** Biophoto Associates/Science Source; **pore complex** Don W. Fawcett/Science Source; **nuclear lamina** Ueli Aebi; **6.10 left** Don W. Fawcett/Science Source; **right** Harry Noller; **6.11** R. W. Bolender; Don W. Fawcett/Science Source; **6.12** Don W. Fawcett/Science Source; **6.13a** Steve Gschmeissner/Science Source; **6.13b** Don W. Fawcett/Science Source; **6.14** Eldon H. Newcomb; **6.17a** Keith R. Porter/Science Source; **6.17c** From: The shape of mitochondria and the number of mitochondrial nucleoids during the cell cycle of *Euglena gracilis*. Y. Hayashi and K. Ueda. *Journal of Cell Science*, 93:565-570, fig. 3. Copyright © 1989 by Company of Biologists; **6.18b** Ed Reschke/Photolibrary/Getty Images; **6.18c** Jeremy Burgess/Mary Martin/Science Source; **6.19** Eldon H. Newcomb; **6.20** Albert Tousson; **6.21b** Bruce J. Schnapp; **Table 6.1 left to right** Gopal Murti/Science Source, Nikon MicroscopyU (www.microscopyu.com), Mark Ladinsky; **6.22** Kent L. McDonald; **6.23a** Biophoto Associates/Science Source; **6.23b** Oliver Meckes, Nicole Ottawa/Eye of Science/Science Source; **6.24a** Omikron/Science Source; **6.24b** Dartmouth College Electron Microscope Facility; **6.24c** Richard W. Linck; **6.25** From: Organization of actin, myosin, and intermediate filaments in the brush border of intestinal epithelial cells. Hirokawa et al. *J Cell Biol.* 1982 Aug;94(2):425-43. Fig. 1. The Rockefeller University Press; **6.26a** Clara Franzini-Armstrong/University of Pennsylvania; **6.26b** M. I. Walker/Science Source; **6.26c** Michael Clayton/University of Wisconsin; **6.27** G. F. Leedale/Science Source; **6.29** Eldon H. Newcomb, University of Wisconsin, Department of Botany; **6.30 top** Reproduced with permission from: *Freeze-Etch Histology*, by L. Orci and A. Perrelet, Springer-Verlag, Heidelberg, 1975. Plate 32. Page 68. Copyright 1975 by Springer-Verlag GmbH & Co KG; **centre** From: Fine structure of desmosomes, hemidesmosomes, and an adepidermal globular layer in developing newt epidermis. DE Kelly. *J Cell Biol.* 1966 Jan; 28(1):51-72. Fig. 7. Reproduced by permission of Rockefeller University Press; **bottom** From: Low resistance junctions in crayfish. Structural changes with functional uncoupling. C. Peracchia and A. F. Dulhunty, *The Journal of Cell Biology.* 1976 Aug; 70(2 pt 1):419-39. Fig. 6. Reproduced by permission of Rockefeller University Press; **6.31** Eye of Science/Science Source; **p. 127 epithelial cell** Susumu Nishinaga/Science Source.

Chapter 7 **7.1** David Goodsell; **p. 131 mosaic** camerawithlegs/Fotolia; **7.10** Used by permission of B. L. de Groot. Related to work done for: Water Permeation Across Biological Membranes: Mechanism and Dynamics of Aquaporin-1 and GlpF. B. L de Groot, H. Grubmüller. *Science* 294:2353-2357 (2001); **7.14** Michael Abbey/Science Source; **p. 137 ion channel** From: Crystal structure of a mammalian voltage-dependent Shaker family K^+ channel. S. B. Long et al. *Science.* 2005 Aug 5;309(5736):897-903. Epub 2005 Jul 7. Cover image; **Scientific Skills Exercise** Photo Fun/Shutterstock; **7.21 left** Biophoto Associates/Science Source; **centre** Don W. Fawcett/Science Source; **right** From: M.M. Perry and A.B. Gilbert, *Journal of Cell Science* 39: 257–272, Figs. 11 and 13 (1979). © 1979 The Company of Biologists Ltd.; **p. 144 spray** Kristoffer Tripplaar/Alamy Stock Photo.

Chapter 8 **8.1** Biosphoto/Alamy Stock Photo; **8.2** Stephen Simpson/Getty Images; **8.3a** Rdeb22/Shutterstock; **8.3b** blickwinkel/Alamy; **8.4 starfish** Image Quest Marine;

8.4 cactus asharkyu/Shutterstock; **8.15** Thomas Steitz; **Scientific Skills Exercise** Fer Gregory/Shutterstock; **8.17** Jack Dykinga/Nature Picture Library/Alamy Stock Photo; **8.22** Keith R. Porter/Science Source; **p. 165 penguins** Flickr/Getty Images.

Chapter 9 **9.1** Courtesy of Mike Bowie, Lincoln University; **top** David Wall/Alamy Stock Photo; **bottom** Courtesy of Mike Bowie, Lincoln University; **9.3** Dionisvera/Fotolia; **9.9 pyruvate dehydrogenase** Z. Hong Zhou, University of California, Los Angeles; **Scientific Skills Exercise** Thomas Kitchin & Victoria Hurst/Design Pics/Alamy Stock Photo; **p. 182 cacao** Aedka Studio/Shutterstock; **p. 188 ATP synthase model** Medical Research Council; **CoQ10** Stephen Rees/Shutterstock.

Chapter 10 **10.1 tree** Rolf Roeckl/mauritius images GmbH/Alamy Stock Photo; **caterpillar** Denis Crawford/Alamy Stock Photo; **10.2a** STILLFX/Shutterstock; **10.2b** NatalieJean/Shutterstock; **10.2c** M. I. Walker/Science Source; **10.2d** Michael Abbey/Science Source; **10.2e** Heide N. Schulz-Vogt, Leibniz Institute for Baltic Sea Research Warnemuende; **10.3** Qiang Hu; **10.4 mesophyll cell, 10.22** Andreas Holzenburg, University of Texas Rio Grande Valley; **10.4 chloroplast** Jeremy Burgess/Science Source; **p. 196 *C. thermalis*** Dennis Nürnberg; **10.12b** Christine L. Case; **Scientific Skills Exercise** The Ohio State University; **10.21a** Doukdouk/Alamy Stock Photo; **10.21b** Keysurfing/Shutterstock; **p. 213 snow** gary yim/Shutterstock.

Chapter 11 **11.1 top, 11.5c** Federico Veronesi/Gallo Images/Alamy Stock Photo; **bottom** Nature Picture Library/Alamy Stock Photo; **11.2a–c** A. Dale Kaiser/Stanford University; **11.2d** Michiel Vos; **Problem-Solving Exercise** Bruno Coignard and Jeff Hageman, CDC; **11.7** From: High-resolution crystal structure of an engineered human beta2-adrenergic G protein-coupled receptor. V. Cherezov, et al. *Science*. 2007 Nov 23;318(5854):1258-65. Epub 2007 Oct 25; **11.19** Gopal Murti/Science Source; **11.21** William Wood; **p. 235 chips** Maureen Spuhler/Seelevel.com.

Chapter 12 **12.1** George von Dassow; **12.2a, c** Biophoto/Science Source; **12.2b** Biology Pics/Science Source; **12.3** Andrew S. Bajer, University of Oregon, Eugene; **12.4, 12.5** Biophoto/Science Source; **12.7** Conly L. Rieder, Wadsworth Center, Albany, NY; **12.8 left** Jane Stout and Claire Walczak, Indiana University; **right** Matthew J. Schibler; **12.10a** Don W. Fawcett/Science Source; **12.10b** B. A. Palevitz and E. H. Newcomb, University of Wisconsin; **12.11** Elizabeth Pierson, Pearson Education; **12.18** Guenter Albrecht-Buehler; **12.19** Lan Bo Chen; **12.20** Nature's Geometry/Science Source; **Scientific Skills Exercise** Molecular Expressions; **p. 254 onion cells** Scenics & Science/Alamy Stock Photo; **HeLa cells** Steve Gschmeissner/Science Source.

Unit 3 Interview **Mojica, flask, petri dish** University of Alicante; **lake** Richard Brown/Alamy Stock Photo.

Chapter 13 **13.1** Attila Csaszar/Getty Images; **p. 257 sperm** Don W. Fawcett/Science Source; **13.2a** Roland Birke/Okapia/Science Source; **13.2b** Auscape International Pty Ltd/Alamy Stock Photo; **13.3 top** Ermakoff/Science Source; **bottom** CNRI/Science Source; **Scientific Skills Exercise** SciMAT/Science Source; **13.12** Mark Petronczki and Maria Siomos; **13.13** John Walsh, Micrographia.com; **p. 270 bananas** Randy Ploetz.

Chapter 14 **14.1** John Swithinbank/Agefotostock; **14.14a** Maximilian Weinzierl/Alamy Stock Photo; **14.14b** Paul Dymond/Alamy Stock Photo; **Scientific Skills Exercise** Apomares/E+/Getty Images; **14.15** Barbara Bowman, Pearson Education; **14.16** Patricia Willocq; **14.18** Michael Ciesielski Photography; **14.19** CNRI/Science Source; **p. 295 cat** Imagebroker/Alamy Stock Photo; **family** Rene MALTETE/Gamma-Rapho/Getty Images.

Chapter 15 **15.1** Courtesy of Peter Lichter; **15.2** Martin Shields/Alamy Stock Photo; **15.5** Andrew Syred/Science Source; **15.6b** Li Jingwang/E+/Getty Images; **15.6c** Kosam/Shutterstock; **15.6d** Creative images/Fotolia; **15.8** Jagodka/Shutterstock; **Scientific Skills Exercise** Oliver911119/Shutterstock; **15.15 left** CNRI/Science Source; **right** Denys_Kuvaiev/Fotolia; **15.18** Phomphan/Shutterstock; **p. 315 butterfly** James K Adams.

Chapter 16 **16.1** 4X-image/E+/Getty Images; **16.3** Oliver Meckes/Eye of Science/Science Source; **Scientific Skills Exercise** Marevision Agency/AGE Fotostock/Alamy Stock Photo; **16.6a** Pictorial Press/Alamy Stock Photo; **16.6b** Science Source; **16.13 left** From: Enrichment and visualization of small replication units from cultured mammalian cells. D. J. Burks et al. *J Cell Biol*. 1978 Jun;77(3):762-73. Fig. 6A; **right** Jerome Vinograd; **16.22** Peter Lansdorp; **16.23 DNA strand** Gopal Murti/Science Source; **nucleosome** Victoria E. Foe; **chromosome** Biophoto/Science Source; **16.24a** Thomas Reid, Genetics Branch/CCR/NCI/NIH; **16.24b** Michael R. Speicher/Medical University of Graz; **p. 336 models** Thomas A. Steitz/Yale University.

Chapter 17 **17.1** Francois Gohier/Science Source; **koala** Cloudrest Images/Shutterstock; **albino koala** Tom McHugh/Science Source; **17.7a** Keith V. Wood; **17.7b** Sinclair Stammers/Science Source; **17.18** Joachim Frank; **17.23b** Barbara Hamkalo; **17.24** Oscar Miller/Science Source; **17.26** Eye of Science/Science Source; **Problem-Solving Exercise** Duplass/Shutterstock; **p. 366 cat** Vasiliy Koval/Shutterstock.

Chapter 18 **18.1 top** gallimaufry/Shutterstock; **18.1 bottom** Andreas Werth; **18.8a** Permission for use granted by Randy Jirtle, Professor of Epigenetics, NC State University, Raleigh, NC; **18.8b** History and Art Collection/Alamy Stock Photo; **Scientific Skills Exercise** hidesy/E+/Getty Images; **18.13** Michael Speicher and Nigel Carter, Medical University of Graz; **18.16** Mike Wu; **18.20** F. Rudolf Turner, Indiana University; **18.21** Wolfgang Driever, University of Freiburg, Freiburg, Germany; **18.22** Ruth Lahmann, The Whitehead Institution; **18.27** Bloomberg/Getty Images; **p. 399 fish** Peter Herring/Image Quest Marine.

Chapter 19 **19.1** National Institute of Allergy and Infectious Diseases(NAID); **19.2** WHO COVID-19 Dashboard. Geneva: World Health Organization, 2020. Available online: http://covid19.who.int (last cited: 20 May 2021); **19.3a** © 2021 Commonwealth of Australia as represented by the Department of Health; **19.3b** New Zealand Ministry of Health; **19.4** Peter von Sengbusch, Botanik; **19.5a** Science Source; **19.5b** Linda M. Stannard, University of Cape Town/Science Source; **19.5c** Hazel Appleton, Health Protection Agency Centre for Infections/Science Source; **19.5d** Ami Images/Science Source; **19.9 computer model** molekuul.be/Fotolia; **19.11 top** Charles Dauguet/Science Source; **bottom** Petit Format/Science Source; **19.12a** CDC; **19.12b** Kuhn and Rossmann research groups, Purdue University; **19.12c** Cynthia Goldsmith, CDC; **Scientific Skills Exercise** Dong yanjun/Imaginechina/AP Images; **19.13** Olivier Asselin/Alamy Stock Photo; **inset** James Gathany, CDC; **19.14** Nigel Cattlin/Alamy Stock Photo; **p. 418 Tamiflu** Nelson Hale/Shutterstock.

Chapter 20 **20.1** Ian Derrington; **20.2** P. Morris, Garvan Institute of Medical Research; **20.6b** Scott Sinklier/Alamy Stock Photo; **20.9** Ethan Bier; **20.13** George S. Watts and Bernard W. Futscher, University of Arizona Cancer Center; **20.14** Stephen McNally, UC Berkeley; **20.17** Roslin Institute; **20.18** Pat Sullivan/AP Images; **20.19** Steve Gschmeissner/Science Photo Library/Alamy Stock Photo; **20.23** Robert MacColl/Newspix/News Ltd; **p. 445 hot spring** Galyna Andrushko/Shutterstock.

Chapter 21 **21.1 elephant shark** Image Quest Marine; **sea horse** Rich Cary/Shutterstock; **21.4** University of Toronto Lab; **21.5** Affymetrix; **21.7 left** AP Images; **right** Virginia Walbot; **21.10a** Oscar L. Miller Jr., Dept. of Biology, University of Virginia; **21.18 mice** Nicholas Bergkessel, Jr./Science Source; **brain cells** From: Altered ultrasonic vocalization in mice with a disruption in the *Foxp2* gene. W. Shu et al. *Proc Natl Acad Sci U S A*. 2005 Jul 5;102(27):9643-8. Epub 2005 Jun 27. Fig. 3; **p. 468 sea urchins** Waterframe/Alamy Stock Photo; **p. 470 treehopper** Patrick Landmann/Science Source.

Unit 4 Interview Photographer: Peter Casamento.

Chapter 22 **22.1** Lighthouse/UIG/AGE Fotostock; **22.2 rhino fossil** *Recherches sur les ossemens fossiles*. G. Cuvier. Atlas, pl. 17 (1836); **iguana** Wayne Lynch/All Canada Photos/AGE Fotostock; **frog** The Natural History Museum, London/Alamy Stock Photo; **Alfred Russel Wallace** The Natural History Museum/Alamy Stock Photo; ***The Origin of Species*** 1859 Murray edition of *The Origin of Species* by Charles Darwin; **22.4** Karen Moskowitz/Stone/Getty Images; **22.5 left** FineArt/Alamy Stock Photo; **right** Photo Researchers/Science History Images/Alamy Stock Photo; **22.6a** Michel Gunther/Science Source; **22.6b** David Hosking/Frank Lane Picture Agency; **22.6c** David Hosking/Alamy Stock Photo; **22.7** Darwin, C. R. Notebook B: Transmutation of species (1837-1838), p. 36. CUL-DAR121; **22.8** Artwork by Utako Kikutani (as appeared in "What Can Make a Four-Ton Mammal a Most Sensitive Beast?" by Jeheskel Shoshani, from Natural History, November 1997, Volume 106(1), 36–45). Copyright © 1997 by Utako Kikutani. Reprinted with Permission of the Artist; **22.9 Brussels sprouts** Arena Photo UK/Fotolia; **kale** Željko Radojko/Fotolia; **cabbage** Guy Shapira/Shutterstock; **wild mustard** Gerhard Schulz/Naturephoto; **broccoli** YinYang/E+/Getty Images; **kohlrabi** Motorolka/Shutterstock; **22.10** Karen Burke Da Silva; **22.11** Coral Brunner/Shutterstock; **22.12a** Caitlin Henderson (She's Got Legs Photography); **22.12b&c** © Alan Henderson – Minibeast Wildlife; **22.13** Scott P. Carroll; **22.14a&b** Photographs courtesy of Scott P. Carroll; **22.17 left** Keith Wheeler/Science Source; **right** Omikron/Science Source; **22.19 left** ANT Photo Library/Science Source; **right** Joe McDonald/Steve Bloom Images/Alamy Stock Photo; **22.20** Chris Linz, Thewissen lab, Northeastern Ohio Universities College of Medicine (NEOUCOM); **p. 490 honeypot ant** © Joel Sartore/National Geographic Image Collection.

Chapter 23 **23.1** Sylvain Cordier/Science Source; **23.3** Christopher R. Friesen; **23.5** Erick Greene; **23.6** © Noel Meyers; **23.7** © 2016 Gilmour JP, Underwood, JN, Howells, EJ; **Scientific Skills Exercise** DLeonis/Fotolia; **23.12a** Oliver Strewe/Lonely Planet Images/Getty Images; **23.12b** Pearson Education; **23.13a** Antoine Morin; **23.15 bottom** John Visser/Bruce Coleman/Photoshot; **23.16** Dave Blackey Agency/All Canada Photos/Alamy Stock Photo; **23.19 blood cells** Eye of Science/Science Source; **doctor with child** Caroline Penn/Alamy Stock Photo; **mosquito** Kletr/Shutterstock; **p. 512 lake** weltmainzer/Fotolia.

Chapter 24 **24.1** Joel Sartore/National Geographic Image Collection; **24.2a left** David Hosking/Alamy; **right** John Carnemolla/Shutterstock; **24.2b top left** Robert Kneschke/Kalium/AGE Fotostock; **top centre** Global Warming Images/Alamy; **top right** Ryan Mcvay/Getty Images; **bottom left** Dragon Images/Shutterstock; **bottom centre** Blaine Harrington III/Corbis; **bottom right** Jaki good photography - celebrating the art of life/Moment Open/Getty Images; **24.3a** Phil Huntley-Franck; **24.3b** Jerry A. Payne, USDA Agricultural Research Service, Bugwood.org; **24.3c** Hogle Zoo; **24.3d** USDA; **24.3e** Imagebroker/Alamy Stock Photo; **24.3f** Takahiro Asami; **24.3g** William E. Ferguson; **24.3h** Chuck Brown/Science Source; **24.3i** Eyewire Collection/Getty Images; **24.3j** Bagicat/Fotolia; **24.3k** FreeReinDesigns/Fotolia; **24.3l** Kazutoshi Okuno; **24.4 top** CLFProductions/Shutterstock; **right** Boris Karpinski/Alamy Stock Photo; **bottom** Troy Maben/AP Images; **24.6a** Courtesy of Brian Langerhans; **24.8 crayfish** Jason Sulda; **Scientific Skills Exercise** John Shaw/Avalon/Photoshot/Alamy Stock Photo; **24.11** Ole Seehausen; **24.12** Jeroen Speybroeck, Research Institute for Nature and Forest; **Problem-Solving Exercise** Philimon Bulawayo/Reuters; **24.14** Ole Seehausen; **24.16** Jason Rick and Loren Rieseberg; **24.18** Cell lineage analysis in ascidian embryos by intracellular injection of a tracer enzyme. III. Up to the tissue restricted stage. H. Nishida. *Develomental Biology*, 1987 June, 121(2):526-41; **p. 532 frog** Rolf Nussbaumer Photography/Alamy Stock Photo.

Chapter 25 **25.1** Juergen Ritterbach/Alamy Stock Photo; **25.2** Stringer/Chile/Reuters/Newscom; **25.3 left** NASA; **right** Deborah S. Kelley; **25.4b** From: Chemically-Induced Birthing and Foraging in Vesicle Systems. F. M. Menger, and Kurt Gabrielson. *J. Am. Chem. Soc.*, February 1994, 116 (4), pp 1567–1568. Fig. 1; **25.4c** Jack W. Szostak; **25.5a** The Royal Botanic Gardens, Sydney; **25.5b** Ron Erwin/Alamy Stock Photo; **25.5c** AuscapeUniversal Images Group via Getty Images; **25.5d** Iolanda Astor/AGE Fotostock/Alamy Stock Photo; **25.5e** Government of Yukon; **25.8 stromatolites** Biosphoto/Alamy Stock Photo; **stromatolite (inset)** Sinclair Stammers/Science Source; **microfossil** David Lamb; ***Coccosteus*** Roger Jones; ***Tiktaalik*** Ted Daeschler/Academy of Natural Sciences; ***Archaefructus*** David L. Dilcher and Ge Sun; **25.11a** Xunlai Yuan; **25.11b** From: The most probable Eumetazoa among late Precambrian macrofossils. A.Y. Ivantsov. *Invertebrate Zoology*. Vol.14. No.2: 127–133 [in English], 2017. Fig 2; **25.13a** From: Four hundred-million-year-old vesicular arbuscular mycorrhizae. W. Remy et al. *Proc Natl Acad Sci USA*. 1994 Dec 6;91(25):11841-3. Figure 1; **inset** From: Four hundred-million-year-old vesicular arbuscular mycorrhizae. Remy W1, Taylor TN, Hass H, Kerp H. *Proc Natl Acad Sci USA*. 1994 Dec 6;91(25):11841-3. Figure 4; **Scientific Skills Exercise** Biophoto Associates/Science Source; **25.19** Richard Jones/Science Photo Library; **25.20** Pearson Education; **25.21** Michael Rosskothen/Shutterstock; **25.22 inset** New Zealand satellite image from NASA's Earth Observatory; **25.23** titoOnz/Shutterstock Elements of this image furnished by NASA; **25.24** Bill Bachman/Alamy Stock Photo; **25.25** Auscape/

Contributor/Getty Images; **25.26a** Ian Beattie/Alamy Stock Photo; **25.26b** Artist: Dorothy Dunn. 'Riversleigh', Archer, M., Hand, S.J., Godthelp, H., 1991. Riversleigh. Reed Books: Sydney; **25.27** © 18831819/123RF; **25.29** Photograph by G. Chaloupka; **25.31** Roman Garcia Mora/Stocktrek Images/Alamy Stock Photo; **25.32** Pearson Education; **25.33** Frans Lanting/Corbis; **25.35a&b** © Rod Morris/www.rodmorris.co.nz; **25.41** ***Dubautia laxa, Argyroxiphium sandwicense, Dubautia waialealae, Dubautia scabra, Dubautia linearis*** Gerald D. Carr; ***Carlquistia muirii*** Bruce G. Baldwin; **25.42** Jean Kern; **25.43** Juniors Bildarchiv GmbH/Alamy Stock Photo; **25.45 top** David Horsley; **25.46** Sinclair Stammers/Science Source; **p. 576 volcano** Solent News/Splash News/Newscom.

Unit 5 Interview Courtesy of Hamish Campbell.

Chapter 26 **26.1** Reproduced courtesy of Museum Victoria/Photographer Ian R McCann. https://collections.museumsvictoria.com.au/species/8389; **26.17a** Mick Ellison; **26.17b** Julius T. Csotonyi/Science Source; **26.22** Gary Crabbe/Enlightened Images/Alamy Stock Photo; **inset** Gerald Schoenknecht; **Scientific Skills Exercise** Nigel Cattlin/Alamy Stock Photo; **p. 597 manatee** David Fleetham/Alamy Stock Photo.

Chapter 27 **27.1 top** © A Life Beneath Stars/Shutterstock; **centre right** Janice Haney Carr, CDC; **bottom left** Irina Sen/Shutterstock; **bottom right** Oliver Meckes/Eye of Science/Science Source; **27.2a** Janice Haney Carr, CDC; **27.2b** CDC; **27.2c** Stem Jems/Science Source; **27.3** L. Brent Selinger/Pearson; **27.4** Immo Rantala/SPL/Science Source; **27.5** Oliver Meckes/Eye of Science/Science Source; **27.6** Kwangshin Kim/Science Source; **27.7** David DeRosier; **27.8a** From: Taxonomic Considerations of the Family Nitrobacteraceae Buchanan: Requests for Opinions. Stanley W. Watson, *IJSEM (International Journal of Systematic and Evolutionary Microbiology* formerly (in 1971) *Intl. Journal of Systematic Bacteriology*), July 1971 vol. 21 no. 3, 254-270. FIg. 14; **27.8b** From: Light-dependent governance of cell shape dimensions in cyanobacteria. B. L. Montgomery. *Front Microbiol*. 2015 May 26;6:514. doi: 10.3389/fmicb.2015.00514. eCollection 2015. Fig. 1. CC BY 4.0; **27.9** Huntington Potter; **27.12** Charles C. Brinton, Jr; **27.14** John Walsh/Science Source; **27.15** Paul Gunning/Science Source; **27.17 spirochetes** Cnri/SPL/Science Source; **proteobacteria** Yuichi Suwa; **cyanobacteria** Michael Abbey/Science Source; **chlamydia** Moredon Animal Health/SPL/Science Source; **gram-positive bacteria** Paul Alan Hoskisson; **27.18** Filip Fuxa/123RF; **27.19** Pascale Frey-Klett; **27.20** WaterFrame/Alamy Stock Photo; **27.21 left** Steve Heap Agency/Zoonar GmbH/Alamy Stock Photo; **centre** David M. Phillips/Science Source; **right** James Gathany, CDC; **Scientific Skills Exercise** Slava Epstein; **27.24** From: RNA-directed gene editing specifically eradicates latent and prevents new HIV-1 infection. W. Hu et al. *Proc Natl Acad Sci U S A*. 2014 Aug 5;111(31):11461-6. Fig. 3D; **27.25** From: Synthesis of High-Molecular-Weight Polyhydroxyalkanoates by Marine Photosynthetic Purple Bacteria. M. Higuchi-Takeuchi et al. *PLoS One*. 2016 Aug 11;11(8):e0160981. doi: 10.1371/journal.pone.0160981. eCollection 2016. Fig. 2; **27.26** Accent Alaska/Alamy Stock Photo; **p. 617 pin** Biophoto Associates/Science Source.

Chapter 28 **28.1, 28.2** Brian S. Leander; **Scientific Skills Exercise** Marbury/Shutterstock; **28.4** Ken Ishida; **28.5** ***Giardia*** Tony Brain/Science Source; **diatom** M I Walker/NHPA/Photoshot/Newscom; ***Volvox*** Frank Fox/Science Source; ***Volvox* (inset)** David J. Patterson; ***Globigerina*** Howard Spero, University of California Davis; ***Globigerina* (inset)** National Oceanic and Atmospheric Administration (NOAA); **amoeba** Michael Abbey/Science Source; **28.6a** The Natural History Museum, London/Science Source; **28.6b** CSIRO; **28.7** David M. Phillips/Science Source; **28.8** David J. Patterson; **28.9** Oliver Meckes/Science Source; **28.10** David J. Patterson; **28.11** CDC; **28.12** © Ivo Grigorov; **28.13** © Leon Perrie; **28.14** Paul Kay/Oxford Scientific/Getty Images; **28.15a** Jennifer L. Matthews; **28.15b** Noble Proctor/Science Source; **28.16** Guy Brugerolle; **28.17a** David M. Phillips/Science Source; **28.17b** Science Source; **28.18** ©1979 Rockefeller University Press. *Journal of Experimental Medicine*. 149:172-184. doi:10.1084/jem.149.1.172; **28.19a** M. I. Walker/Science Source; **28.20** Perennou Nuridsany/Science Source; **28.21** Nature Picture Library/Alamy Stock Photo; **28.22** Eva Nowack; **28.23** ***Asparagopsis armata*** Damsea/Shutterstock; ***Palmaria palmata*** Andrew J. Martinez/Science Source; **nori** Biophoto Associates/Science Source; **sushi** Dorling Kindersley ltd/Alamy Stock Photo; **28.24a** Michael Abbey/Science Source; **28.24b** Laurie Campbell/Photoshot; **28.24c** David L. Ballantine; **28.25** William L. Dentler; **28.27** Ken Hickman; **28.28** Robert Kay; **28.29** Patrick Keeling; **28.30** Doug Houghton NZ/Alamy Stock Photo; **p. 642** ***Didinium*** Greg Antipa/Biophoto Associates/Science Source.

Chapter 29 **29.1** Ken Griffiths/Shutterstock; **rosette cellulose-synthesising proteins** Malcolm R. Brown Jr; **29.3 embryo** Linda Graham/University of Wisconsin-Madison; **placental transfer cell** Karen S. Renzaglia; **sporangia** Arterra Picture Library/Alamy Stock Photo; ***Mnium* sporangium** Mike Peres RBP SPAS/CMSP Biology/Newscom; **liverwort** David John Jones; **root tip** Ed Reschke/Getty Images; **shoot** Ed Reschke/Getty Images; **29.4** Charles H. Wellman; **29.5** From: The early evolution of land plants, from fossils to genomics: a commentary on Lang (1937) 'On the plant-remains from the Downtonian of England and Wales'. D. Edwards and P. Kenrick. *Philos Trans R Soc Lond B Biol Sci*. 2015 Apr 19;370(1666). pii: 20140343. doi: 10.1098/rstb.2014.0343; **29.7** Custom Life Science Images/Alamy Stock Photo; **brood body** Bill Malcolm & Nancy Malcolm; **29.8 thalloid liverwort** Alvin E. Staffan/Science Source; **sporophyte** Linda E. Graham; **leafy liverwort, hornwort** The Hidden Forest; **moss** Tony Wharton/Fundamental Photographs; **Scientific Skills Exercise** Richard Becker/Fundamental Photographs; **29.10a** © Noel Meyers; **29.10b** Thierry Lauzun/Iconotec/Alamy Stock Photo; **29.11** Hans Kerp; **29.13 top** Maureen Spuhler/Pearson Education; **bottom** FloralImages/Alamy Stock Photo; **29.14 spikemoss** Purdue University; **quillwort** Murray Fagg/Australian National Botanic Gardens; **mountain club moss** blickwinkel/Alamy Stock Photo; **rough tree fern** C T Johansson; **horsetail** Stephen P. Parker/Science Source; **whisk fern** Francisco Javier Yeste Garcia; **29.15** Michael Clayton; **p. 661** Ed Reschke/Getty Images; **p. 662 stomata** © W. Barthlott, lotus-salvinia.de.

Chapter 30 **30.1** Lyn Topinka, USGS; **inset** Marlin Harms; **Scientific Skills Exercise** Guy Eisner; **30.5** Rudolph Serbet, Natural History and Biodiversity Institute, University of Kansas; **30.6** Claus Habfast; **30.7** ***Macrozamia communis*** AYArktos/Wikimedia Commons; **ginkgo seeds** www.biolib.de; ***Ginkgo biloba*** Travis Amos/Pearson; ***Welwitschia*** Thomas Schoepke; ***Gnetum*** Michael Clayton; **kauri** Photograph by William Shipway, from http://shipway.net/fraser/kauripinetrunk.jpg; **Dwarf mountain pine** © M. Fagg, Australian National Botanic Gardens; **juniper** Svetlana Tikhonova/Shutterstock; **Pahautea** David Reed/Alamy; **sequoia** Daniel Acevedo/AGE Fotostock/Alamy Stock Photo; ***Wollemi* fossil** Jaime Plaza/Royal Botanic Gardens Sydney; ***Wollemi* forest** Wildlight Photo Agency/Alamy Stock Photo; **pencil pine** © Noel Meyers; **30.10 top** Jardin du ruisseau de léglise/Alamy Stock Photo; **bottom** Paul Atkinson/Shutterstock; **30.11 tomatoes** Dave King/Dorling Kindersley Limited; **grapefruit** Andy Crawford/Dorling Kindersley Limited; **nectarine** Dave King/Dorling Kindersley Limited; **macadamia nuts** sommai/Fotolia; **milkweed** Maria Dryfhout/123RF; **30.12 maple tree** Alessandro Zocchi/123RF; **kereru** Denis La Touche/Alamy; **banksia** KrystynaSzulecka/Alamy; **cobbler's peg** John McQueen/Alamy; **cobbler's peg attached to shoes** Forest & Kim Starr; **30.14a** David L. Dilcher; **30.16** © Esther Beaton Wild Pictures www.estherbeaton.com; **30.17** Cindy Hopkins/Alamy Stock Photo; **30.19 water lily** Howard Rice/Dorling Kindersley Limited; **star anise** Floridata.com; ***Amborella trichopoda*** Joel McNeal; ***Magnolia grandiflora*** Dorling Kindersley Limited/Alamy Stock Photo; **orchid** Eric Crichton/Dorling Kindersley Limited; **tussock grass** mundoview/Fotolia; **palm** John Dransfield **snow pea** Maria Dattola/Getty Images; **apple flower** Mirscho/Fotolia; **30.20a** © Noel Meyers; **30.20b** Radius Images/Alamy; **30.20c** Jack dykinga/Nature Picture Library; **30.21** ***Acacia unguicula*** Sheryl Caston/Alamy Stock Photo; ***Eremophila scaberula*, Beadle's grevillea & silver gum** © M. Fagg, Australian National Botanic Gardens; ***Melaleuca sciotostyla*** Peter Davies/WA Herbarium; **30.22** Pearson Education; **30.23a** Dave Watts/Nature Picture Library; **30.23b** Grant Dixon/Lonely Planet Images/Getty; **30.23c** Matt Smith/Alamy; **30.24** Rudolph 89/Wikimedia Commons; **30.25** LazingBee/Getty; **30.26** Science Photo Library/Alamy; **30.27** NASA; **p. 687 dandelion** Martin Turner/Getty Images.

Chapter 31 **31.1 top** © Bruce Fuhrer; **bottom** Ted M. Kinsman/Science Source; **p. 690** ***Cortinarius capratus*** Matthijs Wetterauw/Alamy Stock Photo; **31.2 top** Nata-Lia/Shutterstock; **bottom** Fred Rhoades; **bottom (inset)** George L. Barron; **31.4a** Biological Photo Service; **Scientific Skills Exercise** U.S. Department of Energy/DOE Photo; **31.6** Olga Popova/123RF; **inset** Biophoto Associates/Science Source; **31.7** Stephen J. Kron; **31.9** Dirk Redecker; **31.10 chytrids** John Taylor; **zygomycetes** Ray Watson; **glomeromycetes** Ecological Society of America; **ascomycetes** blickwinkel/Alamy Stock Photo; **basidiomycetes** Phil A Dotson/Science Source; **31.11** William E. Barstow; **31.12 bread** Antonio D'Albore/Getty Images; ***Rhizopus*** Culture Collection of Fungi (CCF); **Sporangia** George L. Barron; **zygosporangium** Ed Reschke/Getty Images; **31.13** G L Barron/Biological Photo Service; **31.14** Biological Photo Service; **31.15 left** Bryan Eastham/Fotolia; **right** Jacana/Science Source; **31.16 neurospora** Fred Spiegel; **31.17 top** Frank Paul/Alamy Stock Photo; **centre** kichigin19/Fotolia; **bottom** Fletcher and Baylis/Science Source; **31.18** Biophoto Associates/Science Source; **31.19** University of Tennessee, Entomology and Plant Pathology; **31.21** Mark Bowler/Science Source; **31.22 top** Benvie/Wild Wonders of Europe/Nature Picture Library; **centre** AGE fotostock/Alamy Stock Photo; **bottom** Ralph Lee Hopkins/National Geographic/Getty Images; **31.23** Eye of Science/Science Source; **31.24a** Scott Camazine/Alamy Stock Photo; **31.24b** Peter Chadwick/Dorling Kindersley; **31.24c** Blickwinkel/Alamy Stock Photo; **31.25** Vance T. Vredenburg; **31.26** Gary Strobel; **p. 709 wasp** Erich G Vallery/USDA Forest Service.

Chapter 32 **32.1 chameleon** Rolf Nussbaumer Photography/Alamy Stock Photo; **koala** Tom Brakefield/Stockbyte/Getty Images; **neuron** James Cavallini/Science Source; **muscle** Nina Zanetti/Pearson Education; **32.5a** Lisa-Ann Gershwin/Museum of Paleontology **32.5b** Smith609 at en.wikipedia; **32.6** From: Predatorial borings in late precambrian mineralized exoskeletons. S. Bengtson and Y. Zhao. *Science*. 1992 Jul 17;257(5068):367-9. Fig. 3. Reprinted with permission from AAAS; **32.7** The Natural History Museum Trading Company Ltd; **inset** Chip Clark; **32.12a** Blickwinkel/Alamy Stock Photo; **32.13** © Dr Vera Weisbecker/South Australian Museum Mammalogy Collection; **p. 724 organism** WaterFrame/Alamy Stock Photo.

Chapter 33 **33.1** Paul Anthony Stewart; **33.2 sponge** Andrew J. Martinez/Science Source; **jellyfish** Sea Tops/Alamy Stock Photo; **Acoela** Teresa Zuberbühler; **placozoan** From: Global diversity of the Placozoa. M. Eitel et al. *PLoS One*. 2013;8(4):e57131. doi: 10.1371/journal.pone.0057131. Epub 2013 Apr 2. Fig. 1; **ctenophore** Gregory G. Dimijian/Science Source; **marine flatworm** Robinson Ed/Perspectives/Getty Images; **rotifer** M. I. Walker/Science Source; **ectoprocts** blickwinkel/Alamy Stock Photo; **brachiopod** Image Quest Marine; **gastrotrich** Sinclair Stammers/Nature Picture Library; **ribbon worm** Sue Daly/Nature Picture Library; **cycliophor** Peter Funch; **annelid** cbimages/Alamy Stock Photo; **octopus** Photonimo/Shutterstock; **loriciferan** Reinhart Mobjerg Kristensen; **priapulan** Andreas Altenburger/Alamy Stock Photo; **onychophoran** Thomas Stromberg; **roundworm** London Scientific Films/Oxford Scientific/Getty Images; **tardigrades** Andrew Syred/Science Source; **spider** Reinhard Hölzl/ImageBROKER/AGE Fotostock; **acorn worm** Leslie Newman & Andrew Flowers/Science Source; **tunicate** Ethan Daniels/Stocktrek Images/Alamy Stock Photo; **sea urchin** Louise Murray/robertharding/Alamy Stock Photo; **33.3** Andrew J. Martinez/Science Source; **33.6a left** SeaTops/Alamy Stock Photo; **right** David Doubilet/National Geographic; **33.6b left** Neil G. McDaniel/Science Source; **right** Mark Conlin/V&W/Image Quest Marine; **33.7** Biophoto Associates/Science Source; **33.8 top left** blickwinkel/Alamy Stock Photo; **bottom left** Amar and Isabelle Guillen-Guillen Photo LLC/Alamy Stock Photo; **top right** Eldon H. Newcomb; **bottom right** Science Photo Library/Alamy Stock Photo; **33.10** CDC; **33.11** Eye of Science/Science Source; **33.11** Eye of Science/Science Source; **33.12** M. I. Walker/Science Source; **33.13** Holger Herlyn, University of Mainz, Germany; **33.14a** blickwinkel/Alamy Stock Photo; **33.14b** Image Quest Marine; **33.16** Image Quest Marine; **33.17a** Lubos Chlubny/Fotolia; **33.17b** Terry Moore/Stocktrek Images/Alamy Stock Photo; **Scientific Skills Exercise** Christophe Courteau/Water Rights/Alamy Stock Photo; **33.18** Andrew J. Martinez/Science Source; **33.20 top** Mark Conlin/VWPics/Alamy Stock Photo; **centre** Photonimo/Shutterstock; **bottom** SeaTops/Alamy Stock Photo; **33.21 left** Dave Clarke/Zoological Society of London; **right** The U.S. Bureau of Fisheries; **33.22** Fredrik Pleijel; **33.23** Wolcott Henry/National Geographic; **33.24** Astrid Michler, Hanns-Frieder Michler/Science Source; **33.25** Wayne Taylor/The AGE/Fairfax Media via Getty Images; **33.26** London Scientific Films/Oxford Scientific/Getty Images; **33.27** Power and Syred/Science Source; **33.28** Dan Cooper; **33.29b** Courtesy of Sean B. Carroll; **33.31** Mark Newman/Frank Lane Picture Agency; **33.32 top** Tim Flach/The Image Bank/Getty Images; **centre** Andrew Syred/Science Source; **bottom** Reinhard

Hölzl/ImageBROKER/AGE Fotostock; **33.34a** Premaphotos/Nature Picture Library; **33.34b** Tom McHugh/Science Source; **33.36** Maximilian Weinzierl/Alamy Stock Photo; **33.37** Peter Herring/Image Quest Marine; **33.38** Peter Parks/Image Quest Marine; **33.40** André Skonieczny/F1online digitale Bildagentur GmbH/Alamy Stock Photo; **33.41a, b, d, e** Cathy Keifer/Shutterstock; **33.41c** Jim Zipp/Science Source; **33.42 Archaeognatha** Kevin Murphy; **Zygentoma** Denis Crawford/Alamy Stock Photo; **Coleoptera** Premaphotos/Nature Picture Library; **Diptera** Bruce Marlin; **Hymenoptera** John Cancalosi/Nature Picture Library; **Lepidoptera** © FLPA/Alamy Stock Photo; **Hemiptera** Dante Fenolio/Science Source; **Orthoptera** Chris Mattison/Alamy Stock Photo; **33.43** Andrey Nekrasov/Image Quest Marine; **33.44** Daniel Janies; **33.45** Jeff Rotman/Science Source; **33.46** Louise Murray/robertharding/Alamy Stock Photo; **33.47** Jurgen Freund/Nature Picture Library; **33.48** Hal Beral/Getty Images; **p. 756 beetles** Lucy Arnold.

Chapter 34 **34.1 top to bottom** Derek Siveter, Tom McHugh/Science Source, Gino Santa Maria/Shutterstock; Digital Vision/Photodisc/Getty Images, Arnaz Mehta, Tom McHugh/Science Source, Rolf Nussbaumer Photography/Alamy Stock Photo, Visceralimage/Fotolia; **34.4** Natural Visions/Alamy Stock Photo; **34.5c** Stocktrek Images, Inc./Alamy Stock Photo; **34.8** Tom McHugh/Science Source; **34.9** Marevision/AGE Fotostock; **inset** A Hartl/AGE Fotostock; **34.10** Junyuan Chen/Nanjing Institute of Geology and Palaeontology, Chinese Academy of Sciences; **34.14** Field Museum Library/Premium Archive/Getty Images; **34.15a** Carlos Villoch/Image Quest Marine; **34.15b** Masa Ushioda/Image Quest Marine; **34.15c** Andy Murch/Image Quest Marine; **34.17 tuna** Michael Patrick O'Neill/Alamy; **lionfish** Jez Tryner/Image Quest Marine; **sea horse** George Grall/National Geographic; **eel** Fred McConnaughey/Science Source; **34.18** Reprinted by permission from Macmillan Publishers Ltd: From: The oldest articulated osteichthyan reveals mosaic gnathostome characters. M. Zhu. *Nature.* 2009 Mar 26;458(7237):469-74. doi: 10.1038/nature07855. Fig. 2; **34.19** Laurent Ballesta/www.blancpain-ocean-commitment.com/www.andromede-ocean.com and iSimangaliso Wetland Park Authority; **34.20 fossil, ribs, scales** Ted Daeschler/Academy of Natural Sciences/Vireo; **fin** Kalliopi Monoyios Studio; **34.22a** Alberto Fernández/AGE Fotostock; **34.22b** Paul A. Zahl/Science Source; **34.22c** Zeeshan Mirza/ephotocorp/Alamy Stock Photo; **34.23a** DP Wildlife Vertebrates/Alamy Stock Photo; **34.23b** FLPA/Alamy Stock Photo; **34.23c** John Cancalosi/Photolibrary/Getty Images; **Problem-Solving Exercise** Joel Sartore/National Geographic; **34.26** Nobumichi Tamura; **34.27** Chris Mattison/Alamy Stock Photo; **34.28a** Natural Visions/Alamy Stock Photo; **34.28b** Lee T. Matt; **34.28c** Doug Cheeseman/Photolibrary/Getty; **34.28d** Juniors Bildarchiv/AGE Fotostock; **34.28e** Mirko Zanni/WaterFrame/Getty; **34.29a** John Megahan/PLos Biology; **34.29b** The Natural History Museum/Alamy Stock Photo; **34.31** Boris Karpinski/Alamy Stock Photo; **34.32** DLILLC/Corbis/VCG/Getty Images; **34.33** Mariusz Blach/Fotolia; **34.34** The Africa Image Library/Alamy Stock Photo; **inset** mychicport/Shutterstock; **34.35** Gianpiero Ferrari/Frank Lane Picture Agency Limited; **34.38** Clearviewstock/Shutterstock; **inset** Commonwealth Scientific and Industrial Research Organization; **34.39a** John Cancalosi/Alamy Stock Photo; **34.39b** Martin Harvey/Alamy Stock Photo; **34.42** ImageBroker/Alamy Stock Photo; **34.44a** Kevin Schafer/AGE Fotostock; **34.45b** J & C Sohns/Picture Press/Getty Images; **34.45a** Morales/AGE Fotostock; **34.45b** Juniors Bildarchiv GmbH/Alamy Stock Photo; **34.45c** T.J. Rich/Nature Picture Library; **34.45d** E.A. Janes/AGE Fotostock; **34.45e** Martin Harvey/Photolibrary/Getty Images; **34.47** T. White/David L. Brill Photography; **34.48a** John Reader/Science Source; **34.48b** Mauricio Anton/Science Source; **Scientific Skills Exercise** Golfx/Shutterstock; **34.49** Alan Walker; **34.51a** Erik Trinkaus; **34.52** David L. Brill Photography; **34.53** From: *Homo naledi*, a new species of the genus *Homo* from the Dinaledi Chamber, South Africa. L. R. Berger et al. *eLife* 2015;4:e09560. Fig. 6; **34.54** C. Henshilwood; **p. 797 animal** Tony Heald/Nature Picture Library.

Unit 6 Interview Courtesy of Dennis Gonsalves.

Chapter 35 **35.1 tree** Raimund Linke/Photodisc/Getty Images; **seedling** Beata Becia/Shutterstock; **leaf cross section** P&R Fotos/AGE Fotostock/Alamy Stock Photo; **chloroplasts** John Durham/Science Source; **tube-shaped cells** Science Photo Library/Alamy Stock Photo; **root hairs** Scenics & Science/Alamy Stock Photo; **35.3** Jeremy Burgess/Science Source; **35.4 buttress roots** Karl Weidmann/Science Source; **prop roots** Natalie Bronstein; **beet** Rob Walls/Alamy Stock Photo; **pneumatophores** Bjorn Svensson/AGE Fotostock/Alamy Stock Photo; **strangling roots** Dana Tezarr/Photodisc/Getty Images; **35.5 top** Maureen Spuhler/Seelevel.com; **centre** Dorling Kindersley ltd/Alamy Stock Photo; **bottom** Toshihiko Watanabe/Aflo/Alamy Stock Photo; **35.7 tendrils** Neil Cooper/Alamy Stock Photo; **spines** Martin Ruegner/Photodisc/Getty Images; **storage leaves** Dmytro Skorobogatov/123RF; **carnivorous leaves** Gary Meszaros/Dembinsky Photo Associates; **Scientific Skills Exercise** Dorling Kindersley Ltd/Alamy Stock Photo; **35.9** Steve Gschmeissner/SPL/AGE Fotostock; **35.10 parenchyma** M I (Spike) Walker/Alamy Stock Photo; **35.10 collenchyma** Keith Wheeler/Science Source; **sclereid** Graham Kent/Pearson Education; **fibre** Graham Kent/Pearson Education; **tracheids and vessels** N.C Brown Center for Ultrastructure Studies; **sieve-tube element (TEM)** From: *Plant Cell Biology* on DVD: Information for students and a resource for teachers Springer-Verlag 2009, by B Gunning; **sieve-tube element (LM)** Ray F. Evert; **sieve plate** Graham Kent/Pearson Education; **35.13** From: ABA-mediated ROS in mitochondria regulate root meristem activity by controlling PLETHORA expression in *Arabidopsis*. Yang L. *PLoS Genet.* 2014 Dec 18;10(12):e1004791. doi: 10.1371/journal.pgen.1004791. eCollection 2014 Dec. Figure 6G; **35.14a** Ed Reschke; **35.14b top** Chuck Brown/Science Source; **34.14b bottom** Ed Reschke; **35.15, 35.16** Michael Clayton; **35.17, 35.18** Ed Reschke; **35.20 left** Michael Clayton; **right** Alison W. Roberts; **35.23** University of Southern California; **35.25 left** Bob Gibbons/Alamy; **right** David Wall/Alamy; **35.26** From: Natural variation in *Arabidopsis*: from molecular genetics to ecological genomics. D. Weigel. *Plant Physiol.* 2012 Jan;158(1):2-22. doi: 10.1104/pp.111.189845. Epub 2011 Dec 6. Fig. 1; **35.28** From: Microtubule plus-ends reveal essential links between intracellular polarization and localized modulation of endocytosis during division-plane establishment in plant cells. P. Dhonukshe. *BMC Biol.* 2005 Apr 14;3:11. Fig4B; **35.28 (inset)** B. Wells and K. Roberts; **35.29** From: The making of a compound leaf: genetic manipulation of leaf architecture in tomato. D. Hareven. *Cell.* 1996 Mar 8;84(5):735-44. Fig. 1; **35.30** From: A common position-dependent mechanism controls cell-type patterning and GLABRA2 regulation in the root and hypocotyl epidermis of *Arabidopsis*. C. Y. Hung et al. *Plant Physiol.* 1998 May;117(1):73-84. Fig. 2g; **35.31** Lawrence Jensen; **35.32** From: Genetic interactions among floral homeotic genes of *Arabidopsis*. JL Bowman, DR Smyth, EM Meyerowitz. *Development.* 1991 May;112(1):1-20; Fig. 1A; **p. 823 woody eudicot** From: Anatomy of the vessel network within and between tree rings of *Fraxinus lanuginosa* (Oleaceae). P. B. Kitin et al. *American Journal of Botany.* 2004;91:779-788. Fig. 1; **p. 824 tea leaves** Volodymyr Burdiak/Shutterstock; **tea leaf cross section** Keith Wheeler/Science Source; **iris leaves** Rob Stark/Shutterstock; **iris leaf cross section** www.willemsmicroscope.com; **bicycle** Janet Horton/Alamy Stock Photo; ***Hakea purpurea*** Biophoto Associates/Science Source.

Chapter 36 **36.1** velislava/Alamy Stock Photo; **36.3** Rolf Rutishauser and Evelin Pfeifer; **p. 831 plant** Nigel Cattlin/Alamy Stock Photo; **36.8** Benjamin Blonder and David Elliott; **36.10** Scott Camazine/Science Source; **36.13** AGE Fotostock/Alamy Stock Photo; **36.14** Power and Syred/Science Source; **36.15 leafless rock wattle** Doug Steley/Alamy; **totara** Edward Parker/Alamy; **inset** Nature Photographers Ltd/Alamy; **Banksia baxteri** Erika Buresch/Fotolia; **vegetable sheep** Patrik Stedrak/Fotolia; **36.18** M. H. Zimmerman/Harvard Forest; **36.19** From: A coiled-coil interaction mediates cauliflower mosaic virus cell-to-cell movement. L. Stavolone et al. *Proc Natl Acad Sci U S A.* 2005 Apr 26;102(17):6219-24. Epub 2005 Apr 18. Fig. 5c; **p. 845 forest** Catalin Petolea/Alamy Stock Photo.

Chapter 37 **37.1 top** BeyondImages/Getty Images; **bottom** Pearson Education; **37.2** ARS/USDA; **37.4** Emmanuel LATTES/Alamy Stock Photo; **37.7** Noel Meyers; **37.8** Noel Meyers; **37.9** 1xpert/123RF; **37.16 healthy** View Stock RF/AGE Fotostock; **nitrogen-deficient** Guillermo Roberto Pugliese/International Plant Nutrition Institute (IPNI); **phosphorus-deficient** C. Witt/IPNI; **potassium-deficient** M.K. Sharma and P. Kumar/IPNI; **Scientific Skills Exercise** Nigel Cattlin/Science Source; **37.17 lichen** David T. Webb, University of Montana; **lichen section** Courtesy of Ralf Wagner; **puffer** Andrey Nekrasov/Pixtal/AGE Fotostock; ***Azolla*** Daniel L Nickrent; **ant** Juan Carlos Vindas/Moment Open/Getty Images; **fungal garden** Martin Dohrn/Nature Picture Library; **root** Yoshihiro Kobae; **nectar** Oxford Scientific/Getty Images; **37.18** Sarah Lydia Lebeis; **37.20** Scimat/Science Source; **37.22 sheath** Hugues B. Massicotte/University of Northern British Columbia Ecosystem and Management Program, Prince George, BC, Canada; **cells, arbuscules** Mark Brundrett; **37.23 fern** David Wall/Alamy Stock Photo; **mistletoe** Peter Lane/Alamy Stock Photo; **dodder** Emilio Ereza/Alamy Stock Photo; **Indian pipe** Martin Shields/Alamy Stock Photo; **pitcher plants** Dorling Kindersley ltd/Alamy Stock Photo; **ant on pitcher plant** Paul Zahl/Science Source; **sundew** Fritz Polking/Frank Lane Picture Agency Limited W. Rolfes/Arco Images GmbH/Alamy Stock Photo; **Venus flytrap** Chris Mattison/Nature Picture Library; **p. 870 footprint** Mode Images/Alamy Stock Photo.

Chapter 38 **38.1** blickwinkel/Alamy Stock Photo; **inset** Nicolas J. Vereecken; **38.4 hazel carpellate** Friedhelm Adam/imageBROKER/Getty Images; **hazel staminate** Wildlife GmbH/Alamy Stock Photo; **dandelions** © Bjørn Rørslett/NN/Samfoto/Sipa USA; **moth** Doug Backlund/WildPhotosPhotography.com; **fruit flies** Dr Jacinta Zalucki/Australia & Pacific Science Foundation; **bat** © Hans & Judy Beste-Lochman Transparencies; **hummingbird** Rolf Nussbaumer/Nature Picture Library; **38.5** W. Barthlott and W.Rauh/Nees Institute for Biodiversity of Plants; **38.6 top** Michael Clayton/Botany Dept., University of Wisconsin; **bottom** Ed Reschke/Photolibrary/Getty Images; **38.10** Blickwinkel/Alamy Stock Photo; **38.12 coconut** Kevin Schafer/Alamy Stock Photo; ***Alsomitra macrocarpa*** Aquiya/Fotolia; **dandelion** Steve Bloom Images/Alamy Stock Photo; **booyong plant** Peter Woodard; **hairy panic grass** Mark Jesser/Fairfax; ***Tribulus terrestris*** California Department of Food and Agriculture's Plant Health and Pest Prevention Services; **rat** © Jiri Lochman-Lochman Transparencies; **seeds in faeces** Kim A. Cabrera; **ant** Benoit Guénard; **38.13** Noel Meyers; **Scientific Skills Exercise hummingbird** Dec Hogan/Shutterstock; **38.14a** Marcel Dorken; **38.14b** Nobumitsu Kawakubo; **38.15** Meriel G. Jones, University of Liverpool School of Biological Sciences; **38.16** Dorling Kindersley ltd/Alamy Stock Photo; **38.17** Gary P. Munkvold; **38.18** ton koene/Alamy Stock Photo; **p. 890 pollen** Dartmouth College Electron Microscope Facility.

Chapter 39 **39.1** Christopher Ison/Alamy Stock Photo; **39.2** Natalie Bronstein; **39.6** From: Regulation of polar auxin transport by AtPIN1 in *Arabidopsis* vascular tissue. L. Gälweiler et al. *Science.* 1998 Dec 18;282(5397):2226-30; Fig. 4; **39.9a** Richard Amasino; **39.9b** Fred Jensen, Kearney Agricultural Center; **39.11 left** Mia Molvray; **right** Karen E. Koch; **39.13a** Kurt Stepnitz; **39.13b** Joseph J. Kieber; **39.14** Ed Reschke; **39.16** Nigel Cattlin/Alamy Stock Photo; **39.18** Martin Shields/Alamy Stock Photo; **39.22** Michael L. Evans/Ohio State University; **39.23** From the cover of *Cell*, Volume 60, Issue 3, 9 February 1990. Janet Braam, Ronald W. Davis. Used by permission, Copyright ©1990 Cell Press. Image courtesy of Elsevier Sciences, Ltd; **39.24** Martin Shields/Alamy Stock Photo; **39.25** J. L. Basq/M. C. Drew; **39.26** New York State Agricultural Experiment Station/Cornell University College of Agriculture and Life Sciences; **39.27 poppy seed** De Meester Johan/Arterra Picture Library/Alamy Stock Photo; **taro plant** David T. Webb; **olive leaf** Science Photo Library/Alamy Stock Photo; **cactus spines** Susumu Nishinaga/Science Source; **snowflake plant** Giuseppe Mazza; **passion flower** Lawrence E. Gilbert/University of Texas-Austin; **hummingbird** Danny Kessler; **bamboo plants** Kim Jackson/Mode Images/Alamy Stock Photo; **wasp, cocoons** Custom Life Science Images/Alamy Stock Photo; **p. 920 possum** FLPA/Alamy Stock Photo.

Unit 7 Interview UCSD Health.

Chapter 40 **40.1 top** Paul Nicklen/National Geographic; **centre left** CNRS/IPEV/IPHC, France; **bottom** Nature Picture Library/Alamy Stock Photo; **40.2 seal** Dave Fleetham/Robert Harding World Imagery; **penguin** WILDLIFE GmbH/Alamy Stock Photo; **tuna** Andre Seale/Image Quest Marine; **40.4 intestine** Eye of Science/Science Source; **lung, kidney** Susumu Nishinaga/Science Source; **epithelia** Steve Downing/Pearson Education; **loose connective tissue, adipose tissue, bone, skeletal muscle** Nina Zanetti/Pearson Education; **40.5 blood** Jarun Ontakrai/Shutterstock; **cartilage** Chuck Brown/Science Source; **fibrous connective tissue, smooth muscle, cardiac muscle** Ed Reschke/Photolibrary/Getty Images; **neuron** James Cavallini/Science Source; **glia** Thomas Deerinck; **40.7 platypus** Zoos Victoria; **trout** Kletr/Shutterstock; **40.10** Meiqianbao/Shutterstock; **40.11a** Paul Souders/Danita Delimont Creative/Alamy Stock Photo; **40.11b** Ken Griffiths/Shutterstock; **40.12** William Robinson/Alamy Stock Photo; **40.14** Mirko Graul/Shutterstock; **40.15** From: Assessment of oxidative metabolism in brown fat using PET imaging. Otto Muzik, Thomas J. Mangner and

James G. Granneman. *Front. Endocrinol.*, 08 February 2012 | http://dx.doi.org/10.3389/fendo.2012.00015 Fig. 2; **40.19** Jeff Rotman/Alamy Stock Photo; **40.21** FLPA/Alamy Stock Photo; **40.23 plants** Irin-K/Shutterstock; **tuatara** Minden Pictures/imagefolk; **sunflowers** Phil_Good/Fotolia; **fly** WildPictures/Alamy Stock Photo; **sprouts** Bogdan Wankowicz/Shutterstock; **moulting** Nature's Images/Science Source; **plant vessels** Last Refuge/Robert Harding Picture Library Ltd/Alamy Stock Photo; **blood vessels** Susumu Nishinaga/Science Source; **peas** Scott Rothstein/Shutterstock; **pigs** steven goodier/Alamy Stock Photo; **intestinal lining** David M. Martin/Science Source; **root hairs, mesophyll** Rosanne Quinnell © The University of Sydney. eBot http://hdl.handle.net/102.100.100/1463, http://hdl.handle.net/102.100.100/2574; **alveoli** David M. Phillips/Science Source; **p. 946 macaques** Yoshiteru Takahashi/Sebun Photo/amana images/Getty Images.

Chapter 41 **41.1** Milo Burcham/First Light/Getty Images; **41.3** Stefan Huwiler/Rolf Nussbaumer Photography/Alamy Stock Photo; **41.5 baleen** Vicki Beaver/Alamy Stock Photo; **caterpillar** Stuart Wilson/Science Source; **fly** Peter Parks/Image Quest Marine; **python** Gunter Ziesler/Photolibrary/Getty Images; **41.16 left** Kitch Blain/Shutterstock; **right** Tom Brakefield/Stockbyte/Getty Images; **41.19** James Archer, CDC; **41.21** Peter Batson/Image Quest Marine; **Scientific Skills Exercise** ORNL/Science Source; **p. 969 owl** Stefan Huwiler/imageBROKER RF/AGE Fotostock.

Chapter 42 **42.1** John Cancalosi/Alamy Stock Photo; **42.2a** Reinhard Dirscherl/WaterFrame/Getty Images; **42.2b** Eric Grave/Science Source; **42.4 kangaroo** Galina Sinelnikova/Shutterstock; **42.9 top** Indigo Instruments; **bottom** Ed Reschke/Photolibrary/Getty Images; **42.18** Eye of Science/Science Source; **42.19** Image Source Plus/Alamy Stock Photo; **Scientific Skills Exercise** cassis/Fotolia; **42.21a** Peter Batson/Image Quest Marine; **42.21b** Olgysha/Shutterstock; **42.21c** Greg Amptman/Shutterstock; **42.23c** Prepared by Dr. Hong Y. Yan, University of Kentucky and Dr. Peng Chai, University of Texas; **42.24** Motta and Macchiarelli, Anatomy Dept., Univ. La Sapienza, Rome/Science Source; **42.26** Hans-Rainer Duncker, Institute of Anatomy and Cell Biology, Justus-Liebig-University Giessen; **42.32** Doug Allan/Nature Picture Library; **p. 1001 spider** CB2/ZOB/WENN.com/Newscom.

Chapter 43 **43.1 macrophage** SPL/Science Source; **virus** James Cavallini/BSIP SA/Alamy Stock Photo; **bacterium** Chris Bjornberg/Science Source; **fungus** Callista Images/Cultura Creative (RF)/Alamy Stock Photo; **influenza** Kateryna Kon/Shutterstock; **43.15** Steve Gschmeissner/Science Source; **43.27** CNRI/Science Source; **Scientific Skills Exercise** Eye of Science/Science Source; **43.29** Stephen C. Harrison/The Laboratory of Structural Cell Biology/Harvard Medical School; **p. 1026 vaccine** Tatan Yuflana/AP Images.

Chapter 44 **44.1** David Wall/Alamy Stock Photo; **44.2** Gusmonkeyboy/Wikimedia Commons; **44.3a** sdubrov/Fotolia; **44.3b** Gunther Schmida/Lochman Transparencies; **44.4** Eye of Science/Science Source; **Scientific Skills Exercise** Jiri Lochman/Lochman Transparencies; **44.6 left** GeorgePeters/E+/Getty Images; **centre** covenant/Shutterstock; **right** Ventura/Shutterstock; **44.7** Stephane Bidouze/Shutterstock; **44.12** Steve Gschmeissner/Science Source; **44.14** Jiri Lochman/Lochman Transparencies; **44.16** Michael Lynch/Shutterstock; **44.17** deb22/Shutterstock; **44.18 fish** Roger Steene/Image Quest Marine; **stomata** Eye of Science/Science Source; **frog** F1online digitale Bildagentur GmbH/Alamy Stock Photo; **bacterium** Power and Syred/Science Source; **p. 1048 iguana** Steven A. Wasserman.

Chapter 45 **45.1 top** Phillip Colla/Oceanlight.com; **bottom** Craig K. Lorenz/Science Source; **45.3** Volker Witte/Ludwig-Maximilians-Universitat Munchen; **45.11** Cathy Keifer/123RF; **Problem-Solving Exercise** angellodeco/Fotolia; **45.17** AP Images; **45.22 left** Blickwinkel/Alamy Stock Photo; **right** Jurgen and Christine Sohns/Frank Lane Picture Agency; **p. 1068 frogs** Eric Roubos.

Chapter 46 **46.1 coral** Auscape/UIG/Getty Images; **hydra** Roland Birke/Okapia/Science Source; **nudibranchs** Colin Marshall/Frank Lane Picture Agency; **frogs** Andy Sands/Nature Picture Library; **sperm** Don W. Fawcett/Science Source; **cardinals** William Leaman/Alamy Stock Photo; **zebras** Mike Taylor/Alamy Stock Photo; **46.2** Colin Marshall/Frank Lane Picture Agency; **46.3a** P. de Vries/Crews, David; **46.5** Andy Sands/Nature Picture Library; **46.6** John Cancalosi/Alamy Stock Photo; **Scientific Skills Exercise** Tierbild Okapia/Science Source; **46.15** Design Pics Inc/Alamy Stock Photo; **46.20** Tidningarnas Telelgrambyra AB; **46.24** Phanie/Superstock; **1094 Komodo dragon** Dave Thompson/AP Images.

Chapter 47 **47.1** Brad Smith/Stamps School of Art & Design, University of Michigan; **inset** Oxford Scientific/Getty Images; **47.3 top** Victor D. Vacquier; **bottom** From: Wave of free calcium at fertilization in the sea urchin egg visualized with fura-2. M. Hafner et al, *Cell Motil Cytoskeleton*. 1988;9(3):271-7. Fig. 1; **47.6** George von Dassow; **47.7 top** Jürgen Berger/Max Planck Institute for Developmental Biology, Tübingen Germany; **bottom** Andrew J. Ewald, Johns Hopkins Medical School; **47.13b** Alejandro Díaz Díez/AGE Fotostock/Alamy Stock Photo; **47.14a** P. Huw Williams and Jim Smith, The Wellcome Trust/Cancer Research UK Gurdon Institute; **47.14c** Thomas Poole, SUNY Health Science Center; **47.15b** Keith Wheeler/Science Source; **47.18b** From: Cell lineage analysis in ascidian embryos by intracellular injection of a tracer enzyme. III. Up to the tissue restricted stage. H. Nishida. *Dev Biol.* 1987 Jun;121(2):526-41. Fig. 1. Reprinted by permission of Academic Press; **47.19** From: Post-embryonic cell lineages of the nematode, *Caenorhabditis elegans*. E. Sulston et al. *Dev Biol.* 1977 Mar;56(1):110-56. Fig. 1; **47.20** Susan Strome; **47.21** Susan Strome; **47.24** From: Dorsal-ventral patterning and neural induction in *Xenopus* embryos. E. M. De Robertis and H. Kuroda. *Annu Rev Cell Dev Biol.* 2004;20:285-308. Fig. 1; **47.25a** Kathryn Tosney, University of Michigan; **47.26** Based on Honig and Summerbell, courtesy of Lawrence S. Honig; **p. 1118 turtle** James Gerholdt/Getty Images.

Chapter 48 **48.1** Franco Banfi/Science Source; **48.2** Edwin R. Lewis; **48.5** Thomas Deerinck/National Center for Microscopy and Imaging Research, University of California, San Diego; **48.14** Alan Peters; **p. 1136 rattlesnake** B.A.E./Alamy Stock Photo.

Chapter 49 **49.10** Tamily Weissman; **49.11** Larry Mulvehill/Corbis; **49.15** From: A functional MRI study of happy and sad affective states induced by classical music. M. T. Mitterschiffthaler et al. *Hum Brain Mapp.* 2007 Nov. 28(11):1150-62. Fig. 1; **49.18** Marcus E. Raichle, Washington University Medical Center. From research based on "Positron emission tomographic studies of the cortical anatomy of single-word processing". S.E. Petersen et al. *Nature* 331:585-589 (1988); **49.19** National Library of Medicine (NLM); **49.25** Martin M. Rotker/Science Source; **p. 1158 microphone** Eric Delmar/Getty Images.

Chapter 50 **50.1** Doug Perrine/Alamy Stock Photo; **50.2 whale** Doug Perrine/Alamy Stock Photo; **squid** Citron/Wikimedia Commons; **50.6a** R. A. Steinbrecht, Max Planck Institute; **50.6b** CSIRO Publishing; **50.7a** Michael Nolan/Robert Harding World Imagery; **50.7b** Grischa Georgiew/Panther Media/AGE Fotostock; **50.9** From: Richard Elzinga, *Fundamentals of Entomology*, 3rd ed. ©1987, p. 185. Reprinted by permission of Prentice-Hall, Upper Saddle River, NJ; **50.10** SPL/Science Source; **50.16a** APHIS Animal and Plant Health Inspection Service/USDA; **50.17** Steve Gschmeissner/Science Source; **50.21** Neitz Laboratories; **50.26** Clara Franzini-Armstrong; **50.27** H. E. Huxley; **50.34** YAY Media AS/Alamy Stock Photo; **50.39** Dave Watts/NHPA/Science Source; **Scientific Skills Exercise** Vance A. Tucker; **p. 1190 hound** Dogs/Fotolia.

Chapter 51 **51.1** Robert Koss/Shutterstock; **51.3** Manamana/Shutterstock; **51.5b** Scott Camazine/Alamy Stock Photo; **p. 1195 spider** Ian Fletcher/Shutterstock; **p. 1196 twins** Dustin Finkelstein/Getty Images; **51.7** Thomas D. McAvoy/The LIFE Picture Collection/Getty Images; **51.9** Lincoln Brower/Sweet Briar College; **51.10** Anacleto Rapping/Los Angeles Times/Getty Images; **51.11** Sebastian Micke/Paris Match/Getty Images; **51.12** Flinders University; **51.13** The Magpie Whisperer; **51.14** Amalia Bastos, used with permission; **51.15** Dr Clive Bromhall/Oxford Scientific/Getty Images; **51.16** Richard Wrangham; **inset** Mike Korostelev www.mkorostelev.com/Moment Open/Getty Images; **Scientific Skills Exercise** Matt Goff; **51.18a** Matt T. Lee; **51.18b** P. Barden/Wikimedia Commons; **51.18c** Rudie Kuiter; **51.19** Fotograferen.net/Alamy Stock Photo; **51.20** Gerald S. Wilkinson; **51.21** Juniors Bildarchiv/F300/Alamy Stock Photo; **51.24** Martin Harvey/Photolibrary/Getty Images; **51.25** Erik Svensson/Lund University, Sweden; **51.26** Lowell Getz; **51.27** Rory Doolin; **51.29** feathercollection/Shutterstock; **51.31** Fred van Wijk/Alamy Stock Photo; **51.32** Jupiterimages/Creatas/Thinkstock/Getty Images; **p. 1217 woodpecker** William Leaman/Alamy Stock Photo.

Unit 8 Interview Professor Ove Hoegh-Guldberg.

Chapter 52 **52.1 top** Christopher Austin; **bottom, left to right:** Siepmann/imageBROKER/Alamy Stock Photo, Janelle Lugge/Shutterstock, Digital Vision/Photodisc/Getty Images, NOAA Okeanos Explorer Program; **52.2 top to bottom:** Mel Shroder/NSW Office of Environment and Heritage, Entomology/CSIRO, Brook Mitchell/Getty Images, Ilya Genkin/Alamy Stock Photo, Shuang Li/Shutterstock, 1xpert/Fotolia; **52.9** Courtesy University of Wisconsin–Madison Arboretum. © University of Wisconsin Board of Regents.; **52.11 desert** Ted Mead/Getty Images; **grassland** Renclif Media/Shutterstock; **broadleaf forest** Jaaske M/Shutterstock; **tropical forest** Siepmann/ImageBroker/Alamy Stock Photo; **coniferous forest** Bent G. Nordeng/Shutterstock; **tundra** Photodisc/Getty Images; **52.13 left** Moment/Getty Images; **right** Science Photo Library/Alamy Stock Photo; **52.14 tropical forest** Siepmann/imageBROKER/Alamy Stock Photo; **desert** Ted Mead/Getty Images; **savanna** Robert Harding Picture Library/Alamy Stock Photo; **heathland** Steven David Miller/Nature Picture Library; **grassland** David Halbakken/AGE Fotostock; **coniferous forest** Bent Nordeng/Shutterstock; **broadleaf forest** Ashley Whitworth/Fotolia; **tundra** Juan Carlos Munoz/Nature Picture Library; **52.17 oligotrophic lake** Steven Brown/Fotolia; **eutrophic lake** AfriPics.com/Alamy Stock Photo; **wetland** Robert Harding World Imagery/Alamy Stock Photo; **headwater stream** scubaluna/Shutterstock; **Loire river** Photononstop/SuperStock; **estuaries** Juan Carlos Munoz/AGE Fotostock; **intertidal zone** Stuart Westmorland/Danita Delimont/Alamy Stock Photo; **ocean** Tatonka/Shutterstock; **coral reef** Digital Vision/Photodisc/Getty Images; **benthic zone** NOAA Okeanos Explorer Program; **52.18** Smileus/Fotolia; **52.20** Randy Bjorklund/Shutterstock; **52.21** Scott Ling; **52.24** Patrik Stedrak/Fotolia; **52.25** Auscape/UIG/AGE Fotostock; **52.26** Cephas Picture Library/Alamy Stock Photo; **52.27** Richard Kingsford 2009; **52.28** Auscape/UIG/AGE Fotostock; **Scientific Skills Exercise *Spartina*** John W. Bova/Science Source; ***Typha*** Dave Bevan/Alamy Stock Photo; **52.29** Dr Nick Fitzgerald; **p. 1249 giraffe** Daryl Balfour/The Image Bank/Getty Images.

Chapter 53 **53.1 top** Joel Sartore/National Geographic Image Collection; **bottom** Villiers Steyn/Shutterstock; **53.2** Todd Pusser/Nature Picture Library; **53.3a** Bernard Castelein/Nature Picture Library/Alamy Stock Photo; **53.3b** Michael S Nolan/AGE Fotostock; **53.3c** Alexander Chaikin/Shutterstock; **Table 53.1 top** Kevin Ebi/Alamy Stock Photo; **bottom** Jennifer A. Dever; **53.8** Villiers Steyn/Shutterstock; **53.11** Flip Nicklin/Minden Pictures/Corbis; **Scientific Skills Exercise** Lebendkulturen.de/Shutterstock; **53.12** Eucalyptus 99/Wikimedia Commons; **53.14** Jouan & Rius/Nature Picture Library; **53.15** jeanoz/Shutterstock; **53.16** James Wood/Tasmanian Seed Conservation Centre; **53.17** Burke's Backyard/Alamy Stock Photo; **53.18** VanderWolf Images/Fotolia; **53.19** Danita Delmont/Alamy Stock Photo; **53.20a** Stone Nature Photography/Alamy Stock Photo; **53.20b** Mikhail Mischenko/123RF; **inset** Forest Flora; **53.21** Dietmar Nill/Nature Picture Library; **53.22a** Steve Bloom Images/Alamy Stock Photo; **53.22b left** Science Photo Library/Alamy Stock Photo; **right** Universal Images Group North America LLC/DeAgostini/Alamy Stock Photo; **53.24** National Geographic/Alamy Stock Photo; **53.25 wheat** FotoVoyager/E+/Getty Images; **cheetah** Ian Cumming/Axiom/Design Pics/Alamy Stock Photo; **humans** Jorge Dan/Reuters; **mice** Nicholas Bergkessel Jr./Science Source; **yeast** Andrew Syred/Science Source; **53.27** Alan & Sandy Carey/Science Source; **53.28** Robert Pickett/Papilio/Alamy Stock Photo; **p. 1270 transponder** From: Tracking butterfly movements with harmonic radar reveals an effect of population age on movement distance. O. Ovaskainen et al. *Proc Natl Acad Sci U S A.* 2008 Dec 9;105(49):19090-5. doi: 10.1073/pnas.0802066105. Epub 2008 Dec 5. Fig. 1; **53.33** NASA; **p. 1277 locusts** Carlos Guevara/Reuters/Newscom.

Chapter 54 **54.1 eel** Jeremy Brown/123RF; **coral reef** Jan Wlodarczyk/Alamy Stock Photo; **triggerfish** imageBROKER/Alamy Stock Photo; **shark** Andrey Armyagov/Alamy Stock Photo; **coral bleaching** Reinhard Dirscherl/Alamy Stock Photo; **54.2 left** Joseph T. Collins/Science Source; **right** National Museum of Natural History/Smithsonian Institution; **54.4** Frank W Lane/Frank Lane Picture Agency Limited; **Scientific Skills Exercise** Johan Larson/Shutterstock; **54.6a** Tony Heald/Nature Picture Library; **54.6b** Tom Brakefield/Getty Images; **54.6c** Dirk Ercken/Shutterstock; **54.6d** Barry Mansell/Nature Picture Library; **54.6e left** Daniel Janzen/JANZEN.UPENN.EDU/Caters News; **right** Robert Pickett/Papilio/

Alamy Stock Photo; **54.6f left** David J Martin/Shutterstock; **right** Lightwriter1949/Alamy Stock Photo; **54.7** Roger Steene/Image Quest Marine; **54.8** Andrea Izzotti/Fotolia; **54.9a** Bazzano Photography/Alamy Stock Photo; **54.9b** Nicholas Smythe/Science Source; **54.10** Daryl Balfour/Gallo Images/Getty Images; **54.11a** Sally D. Hacker; **54.13** Gary W. Saunders; **54.14** Dung Vo Trung/Science Source; **54.15** Cedar Creek Ecosystem Science Reserve, University of Minnesota; **54.20a** Genny Anderson; **54.21** pierdest/Shuterstock; **54.25a** Charles D. Winters/Science Source; **54.25b** Keith Boggs; **54.25c** Terry Donnelly/Mary Liz Austin; **54.25d** Glacier Bay National Park and Preserve; **54.26 left to right** Charles D. Winters/Science Source, Keith Boggs, Terry Donnelly/Mary Liz Austin, Glacier Bay National Park/Preserve; **54.27 top** R. Grant Gilmore/NOAA; **bottom** Lance Horn/National Undersea Research Center/University of North Carolina-Wilmington/NOAA; **54.31** Tim Laman/National Geographic/Getty Images; **54.33** Nelish Pradhan/Bates College/Lewiston, ME; **p. 1301 flower** Jim Holden/Alamy Stock Photo.

Chapter 55 **55.1** All Canada Photos/Alamy Stock Photo; **55.2a** Avalon.red/Alamy Stock Photo; **55.2b** Noel Meyers; **55.3 left** Scimat/Science Source; **right** Justus de Cuveland/imageBROKER/AGE Fotostock; **55.5** MODIS Science Team/Earth Observatory/NASA; **55.7** A. T. Willett/Alamy Stock Photo; **Problem-Solving Exercise tree** Steven Katovich/USDA Forest Service; **mountain pine beetles** British Columbia Ministry of Forests, Lands and Natural Resource Operations; **55.8** Matt Meadows/Photolibrary/Getty Images; **Scientific Skills Exercise** David R. Frazier Photolibrary/Science Source; **55.14** Hubbard Brook Research Foundation/USDA Forest Service; **55.15** Nick Polanszky/Alamy Stock Photo; **55.16** Giedriius/Shutterstock; **55.17** Dr Nick Fitzgerald; **55.18** Modular Artificial Reef Structure (MARS)/Alex Goad & Reef Design Lab; **55.19** Alex Goad & Reef Design Lab; **55.20** Living Seawalls installation at Balmain © Leah Wood, Sydney Institute of Marine Science (SIMS); **55.21** Mark Gallagher/Princeton Hydro, LLC/Ringoes, NJ; **55.22 top to bottom** Jean-Paul Ferrero/Auscape International Pty Ltd/Alamy Stock Photo, Jean Hall/FLPA/Science Source, Tim Day/Xcluder Pest Proof Fencing Company, From: Species richness accelerates marine ecosystem restoration in the Coral Triangle. S. L. Williams et al. *Proc Natl Acad Sci U S A*. 2017 Nov 7;114(45):11986-11991. doi: 10.1073/pnas.1707962114. Epub 2017 Oct 24. Fig. 1a. Photos courtesy of D. Trockel, University of California, Davis, CA; **55.23** U.S. Department of Energy; **p. 1327 beetle** Dr Eckart Pott/NHPA/Photoshot.

Chapter 56 **56.1 gecko** Phung My Trung, vncreatures.net; **clearcutting** Mason Vranish/Alamy Stock Photo; **tusks, 56.8** Benezeth M. Mutayoba; **energy** Ververidis Vasilis/Shutterstock; **park** Edwin Giesbers/Nature Picture Library; **reef** Matthew Banks/Alamy Stock Photo Image; **56.3 top** Alexius Sutandio.123RF; **bottom** Rob Suisted/Nature's Pic Images; **56.4** Merlin D. Tuttle/Science Source; **56.5** Scott Camazine/Science Source; **56.7** Michael Storer; **56.8** Ken Griffiths/Shutterstock; **56.9** Michael Hammer; **56.10a** Janelle Lugge/Shutterstock; **56.10c** Gerry Pearce/Alamy Stock Photo; **56.10d** Noel Meyers; **56.11** National Academy of Sciences; **56.12** Travel Pictures/Alamy Stock Photo; **56.14** M.J. Tyler; **56.16** worldswildlifewonders/Shutterstock; **56.17a** Tui De Roy/Nature Picture Library/Alamy Stock Photo; **56.17b** Frans Lanting Studio/Alamy Stock Photo; **56.18** David Wall/Alamy Stock Photo; **56.19** mihailzhukov/Fotolia; **56.20** Vladimir Melnikov/Shutterstock; **56.21** Photo by Rob Beirregaard; **56.22** Frans Lemmens/Alamy Stock Photo; **56.24a** Tom Reeves Photo/Shutterstock; **56.24b** Photo: Menna Jones. 'To Lose Both Would Look Like Carelessness: Tasmanian Devil Facial Tumour Disease.' McCallum H, Jones M, PLoS Biology Vol. 4/10/2006, e342. doi:10.1371/journal.pbio.0040342; **56.26** Mark Chiappone; **56.27** Lyda Bergman, Green Teams of Canada; **56.30** Alfred Eisenstaedt/The LIFE Picture Collection/Getty Images; **56.32** Claire Fackler, NOAA National Marine Sanctuaries; **56.33** Courtesy of Bette Willis and Joleah Lamb; **Scientific Skills Exercise** Hank Morgan/Science Source; **56.36 resin canal** Biophoto Associates/Science Source; **tunnels** Ladd Livingston, Idaho Department of Lands, Bugwood.org; **dead trees** Dezene Huber; **melomys** © The State of Queensland; **Bramble Cay** Natalie Waller; **caribou** E.A. Janes/Robert Harding World Imagery; **chickweed** Gilles Delacroix/Garden World Images/AGE Fotostock; **urchin** Scott Ling; **56.39** NASA Ozone Watch; **56.41a** Serge de Sazo/Science Source; **56.41b** Javier Trueba/MSF/Science Source; **56.41c** Gabriel Rojo/Nature Picture Library; **56.41d** Titus Lacoste/The Image Bank/Getty Images; **p. 1361 possum** Edwin Giesbers/Nature Picture Library.

Appendix A **Figure 2.17** Nigel Cattlin/Science Source; **Figure 6.24 left** Omikron/Science Source; **6.24 right** Dartmouth College Electron Microscope Facility; **Ch. 9 Test Your Understanding 10** Dr Tanya Izzard, Medical Research Council; **Figure 12.4** Biophoto/Science Source; **Figure 12.8** J. Richard McIntosh; **Ch. 12 Test Your Understanding 9** Scenics & Science/Alamy Stock Photo; **Ch. 16 Test Your Understanding 11** Thomas A. Steitz, Yale University, New Haven; **Figure 30.9** Paul Atkinson/Shutterstock; **Ch. 35 Test Your Understanding 11** From: Anatomy of the vessel network within and between tree rings of *Fraxinus lanuginosa* (Oleaceae). Peter B. Kitin, Tomoyuki Fujii, Hisashi Abe and Ryo Funada. *American Journal of Botany*. 2004;91:779-788.

Appendix B **bacteria, archaea** Eye of Science/Science Source; **diatoms** M I Walker/NHPA/Photoshot/Newscom; **lily** Howard Rice/Dorling Kindersley, Ltd./Alamy Stock Photo; **fungus** Daksel/Fotolia; **chimpanzees** E.A. Janes/AGE Fotostock.

Illustration and Text Credits

Chapter 1 **1.23** Adapted from *The Real Process of Science* (2013), Understanding Science website. The University of California Museum of Paleontology, Berkeley, and the Regents of the University of California. Retrieved from http://undsci.berkeley.edu/article/howscienceworks_02; **1.25** Data from S. N. Vignieri, J. G. Larson, and H. E. Hoekstra, The Selective Advantage of Crypsis in Mice, *Evolution* 64:2153–2158 (2010); **Scientific Skills Exercise** Data from D. W. Kaufman, Adaptive Coloration in *Peromyscus polionotus*: Experimental Selection by Owls, *Journal of Mammalogy* 55:271–283 (1974).

Chapter 2 **Scientific Skills Exercise** Data from R. Pinhasi et al., Revised Age of late Neanderthal Occupation and the End of the Middle Paleolithic in the Northern Caucasus, *Proceedings of the National Academy of Sciences USA* 147:8611–8616 (2011). doi 10.1073/pnas.1018938108.

Chapter 3 **3.7 map** based on NOAA Fisheries, Bowhead Whale (*Balaena mysticetus*); sea ice extent from National Snow and Ice Data Center (https://nsidc.org/arcticseaicenews/); **3.9** Based on Simulating Water and the Molecules of Life by Mark Gerstein and Michael Levitt, from *Scientific American*, November 1998; **Scientific Skills Exercise** Data from C. Langdon et al., Effect of Calcium Carbonate Saturation State on the Calcification Rate of an Experimental Coral Reef, *Global Biogeochemical Cycles* 14:639–654 (2000).

Chapter 4 **4.2** Data from S. L. Miller, A Production of Amino Acids Under Possible Primitive Earth Conditions, *Science* 117:528–529 (1953); **Scientific Skills Exercise** Data from E. T. Parker et al., Primordial Synthesis of Amines and Amino Acids in a 1958 Miller H_2S-rich Spark Discharge Experiment, *Proceedings of the National Academy of Sciences USA* 108:5526–5531 (2011). www.pnas.org/cgi/doi/10.1073/pnas.1019191108; **4.7** Adapted from Becker, Wayne M.; Reece, Jane B.; Poenie, Martin F., *The World of the Cell*, 3rd Ed., ©1996. Reprinted and electronically reproduced by permission of Pearson Education, Inc., Upper Saddle River, New Jersey.

Chapter 5 **5.11** Adapted from Wallace/Sanders/Ferl, *Biology: The Science of Life*, 3rd Ed., ©1991. Reprinted and electronically reproduced by permission of Pearson Education, Inc., Upper Saddle River, New Jersey; **5.13** Collagen Data from Protein Data Bank ID 1CGD: "Hydration Structure of a Collagen Peptide" by Jordi Bella et al., from *Structure*, September 1995, Volume 3(9); **5.16 space-filling model, ribbon model** Data from PDB ID 2LYZ: R. Diamond. Real-Space Refinement of the Structure of Hen Egg-white Lysozyme. *Journal of Molecular Biology* 82(3):371–91 (Jan. 25, 1974); **5.18 transthyretin** Data from PDB ID 3GS0: S.K. Palaninathan, N.N. Mohamedmohaideen, E. Orlandini, G. Ortore, S. Nencetti, A. Lapucci, A. Rossello, J.S. Freundlich, J.C. Sacchettini. Novel Transthyretin Amyloid Fibril Formation Inhibitors: Synthesis, Biological Evaluation, and X-ray Structural Analysis. *Public Library of Science ONE* 4:e6290–e6290 (2009); **5.18 collagen** Data from PDB ID 1CGD: J. Bella, B. Brodsky, and H.M. Berman. Hydration Structure of a Collagen Peptide, *Structure* 3:893–906 (1995); **haemoglobin** Data from PDB ID 2HHB: G. Fermi, M.F. Perutz, B. Shaanan, R. Fourme. The Crystal Structure of Human Deoxyhaemoglobin at 1.74 Å resolution. *J. Mol. Biol.* 175:159–174 (1984).

Chapter 6 **6.6** Adapted from Becker, Wayne M.; Reece, Jane B.; Poenie, Martin F., *The World of the Cell*, 3rd Ed., ©1996. Reprinted and electronically reproduced by permission of Pearson Education, Inc. Upper Saddle River, New Jersey; **6.8 animal cell** Adapted from Marieb, Elaine N.; Hoehn, Katja, *Human Anatomy and Physiology*, 8th Ed., © 2010. Printed and electronically reproduced by permission of Pearson Education, Inc., Upper Saddle River, New Jersey; **6.9–6.13, 6.17, 6.22, 6.24 small cell** Adapted from Marieb, Elaine N.; Hoehn, Katja, *Human Anatomy and Physiology*, 8th Ed., ©2010. Printed and electronically reproduced by permission of Pearson Education, Inc., Upper Saddle River, New Jersey; **6.15** Adapted from Marieb, Elaine N.; Hoehn, Katja, *Human Anatomy and Physiology*, 8th Ed., ©2010. Printed and electronically reproduced by permission of Pearson Education, Inc., Upper Saddle River, New Jersey; **Table 6.1** Adapted from Hardin Jeff; Bertoni Gregory Paul, Kleinsmith, Lewis J., *Becker's World of the Cell*, 8th Edition, © 2012, p. 423. Reprinted and electronically reproduced by permission of Pearson Education, Inc. Upper Saddle River, New Jersey; **6.32** Data from: Proton pump: PDB ID 3B8C: Crystal Structure of the Plasma Membrane Proton Pump, Pedersen, B.P., Buch-Pedersen, M. J., Morth, J.P., Palmgren, M.G., Nissen, P. (2007) *Nature* 450: 1111–1114; calcium channel: PDB ID 5E1J: Structure of the Voltage-Gated Two-Pore Channel TPC1 from *Arabidopsis thaliana*, Guo, J., Zeng, W., Chen, Q., Lee, C., Chen, L., Yang, Y., Cang, C., Ren, D., Jiang, Y. (2016) *Nature* 531: 196–201; aquaporin: PDB ID 5I32: Crystal Structure of an Ammonia-Permeable Aquaporin, Kirscht, A., Kaptan, S.S., Bienert, G.P., Chaumont, F., Nissen, P., de Groot, B.L., Kjellbom, P., Gourdon, P., Johanson, U. (2016) *Plos Biol.* 14: e1002411–e1002411; BRI1 and SERK1 co-receptors: PDB ID 4LSX: Molecular mechanism for plant steroid receptor activation by somatic embryogenesis co-receptor kinases, Santiago, J., Henzler, C., Hothorn, M. (2013) *Science* 341: 889–892; BRI1 kinase domain: PDB ID 4OAC: Crystal structures of the phosphorylated BRI1 kinase domain and implications for brassinosteroid signal initiation, Bojar, D., Martinez, J., Santiago, J., Rybin, V., Bayliss, R., Hothorn, M. (2014) *Plant J.* 78: 31–43; BAK1 kinase domain: PDB ID 3UIM: Structural basis for the impact of phosphorylation on the activation of plant receptor-like kinase BAK1, Yan, L., Ma, Y.Y., Liu, D., Wei, X., Sun, Y., Chen, X., Zhao, H., Zhou, J., Wang, Z., Shui, W., Lou, Z.Y. (2012) *Cell Res.* 22: 1304–1308; BSK8 pseudokinase: PDB ID: 4I92 Structural Characterization of the RLCK Family Member BSK8: A Pseudokinase with an Unprecedented Architecture, Grutter, C., Sreeramulu, S., Sessa, G., Rauh, D. (2013) *J. Mol. Biol.* 425: 4455–4467; ATP synthase PDB ID 1E79: The Structure of the Central Stalk in Bovine F(1)-ATPase at 2.4 Å Resolution, Gibbons, C., Montgomery, M.G., Leslie, A.G.W., Walker, J.E. (2000) *Nat. Struct. Biol.* 7: 1055; ATP synthase PDB ID 1C17: Structural changes linked to proton translocation by subunit c of the ATP synthase, Rastogi, V.K., Girvin, M.E. (1999) *Nature* 402: 263–268; ATP synthase PDB ID 1L2P: The "Second Stalk" of *Escherichia coli* ATP Synthase: Structure of the Isolated Dimerization Domain, Del Rizzo, P.A., Bi, Y., Dunn, S.D., Shilton, B.H. (2002) *Biochemistry* 41: 6875–6884; ATP synthase PDB ID 2A7U: Structural Characterization of the Interaction of the Delta and Alpha Sub-units of the *Escherichia coli* F(1) F(0)-ATP Synthase by NMR Spectroscopy, Wilkens, S., Borchardt, D., Weber, J., Senior, A.E. (2005) *Biochemistry* 44: 11786–11794; Phosphofructokinase: PDB ID 1PFK: Crystal Structure of the Complex of Phosphofructokinase from *Escherichia coli* with Its Reaction Products, Shirakihara, Y., Evans, P.R. (1988) *J. Mol. Biol.* 204: 973–994; Hexokinase: PDB ID 4QS8: Biochemical and Structural Study of *Arabidopsis* Hexokinase 1, Feng, J., Zhao, S., Chen, X., Wang, W., Dong, W., Chen, J., Shen, J.-R., Liu, L., Kuang, T. (2015) *Acta Crystallogr.*, Sect. D 71: 367–375; Isocitrate dehydrogenase: PDB ID 3BLW: Allosteric Motions in Structures of Yeast NAD+-specific Isocitrate Dehydrogenase, Taylor, A.B., Hu, G., Hart, P.J., McAlister-Henn, L. (2008) *J. Biol. Chem.* 283:10872–10880; NADH-quinone oxidoreduc-tase: PDB ID 3M9S: The architecture of respiratory complex I, Efremov, R.G., Barada-ran, R., Sazanov, L.A. (2010) *Nature* 465: 441–445; NADH-quinone oxidore-ductase: PDB ID 3RKO: Structure of the membrane domain of respiratory complex I, Efremov, R.G., Sazanov, L.A. (2011) *Nature* 476: 414–420; Succinate dehydrogenase: PDB ID 1NEK: Architecture of Succinate Dehydrogenase and Reactive Oxygen Species Generation, Yankovskaya, V., Horsefield, R., Tornroth, S., Luna-Chavez, C., Miyoshi, H., Leger, C., Byrne, B., Cecchini, G., Iwata, S. (2003) *Science* 299: 700–704; Ubiquinone: http://www.proteopedia.org/wiki/index.php/Image: Coenzyme_Q10.pdb; Cytochrome bc1: PDB ID 1BGY: Complete structure of the 11-subunit bovine mitochondrial cytochrome bc1 complex, Iwata, S., Lee, J.W.,

Okada, K., Lee, J.K., Iwata, M., Rasmussen, B., Link, T.A., Ramaswamy, S., Jap, B.K. (1998) *Science* 281: 64–71; Cytochrome *c*: PDB ID 3CYT: Redox Conformation Changes in Refined Tuna Cytochrome c, Takano, T., Dickerson, R.E. (1980) *Proc. Natl. Acad. Sci. USA* 77: 6371–6375; Cytochrome *c* oxidase: PDB ID 1OCO: Redox-Coupled Crystal Structural Changes in Bovine Heart Cytochrome c Oxidase, Yoshikawa, S., Shinzawa-Itoh, K., Nakashima, R., Yaono, R., Yamashita, E., Inoue, N., Yao, M., Fei, M.J., Libeu, C.P., Mizushima, T., Yamaguchi, H., Tomizaki, T., Tsukihara, T. (1998) *Science* 280: 1723–1729; Rubisco: PDB ID 1RCX: The Structure of the Complex between Rubisco and its Natural Substrate Ribulose 1,5-Bisphosphate, Taylor, T.C., Andersson, I. (1997) *J. Mol. Biol.* 265: 432–444; Photosystem II: PDB ID 1S5L: Architecture of the Photosynthetic Oxygen-Evolving Center, Ferreira, K.N., Iverson, T.M., Maghlaoui, K., Barber, J., Iwata, S. (2004) *Science* 303: 1831–1838; Plastoquinone: http://www.rcsb.org/pdb/ligand/ligandsummary.do? hetId=PL9; Photosystem I: PDB ID 1JB0: Three-Dimensional Structure of Cyanobacterial Photosystem I at 2.5 Å Resolution, Jordan, P., Fromme, P., Witt, H.T., Klukas, O., Saenger, W., Krauss, N. (2001) *Nature* 411: 909–917; Ferredoxin-NADP+ reductase: PDB ID 3W5V: Concentration-Dependent Oligomerization of Cross-Linked Complexes between Ferredoxin and Ferredoxin-NADP(+) Reductase; DNA:PDB ID 1BNA: Structure of a B-DNA Dodecamer: Conformation and Dynamics, Drew, H.R., Wing, R.M., Takano, T., Broka, C., Tanaka, S., Itakura, K., Dickerson, R.E. (1981) *Proc. Natl. Acad. Sci. USA* 78: 2179–2183; RNA polymerase: PDB ID 2E2I: Structural basis of transcription: role of the trigger loop in substrate specificity and catalysis, Wang, D., Bushnell, D.A., Westover, K.D., Kaplan, C.D., Kornberg, R.D. (2006) *Cell* (Cambridge, Mass.) 127: 941–954; Nucleosome: PDB ID 1AOI: Crystal Structure of the Nucleosome Core Particle at 2.8 Å Resolution, Luger, K., Mader, A.W., Richmond, R.K., Sargent, D.F., Richmond, T.J. (1997) *Nature* 389: 251–260; tRNA: PDB ID 4TNA: Further refinement of the structure of yeast tRNAPhe, Hingerty, B., Brown, R.S., Jack, A. (1978) *J. Mol. Biol.* 124: 523–534; Ribosome: PDB ID 1FJF: Structure of the 30S Ribosomal Subunit, Wimberly, B.T., Brodersen, D.E., Clemons Jr., W.M., Morgan-Warren, R.J., Carter, A.P., Vonrhein, C., Hartsch, T., Ramakrishnan, V. (2000) *Nature* 407: 327–339; Ribosome: PDB ID 1JJ2: The Kink-Turn: A New RNA Secondary Structure Motif, Klein, D.J., Schmeing, T.M., Moore, P.B., Steitz, T.A. (2001) *EMBO J.* 20: 4214–4221; Microtubule: PDB ID 3J2U: Structural Model for Tubulin Recognition and Deformation by Kinesin-13 Microtubule Depolymerases, Asenjo, A.B., Chatterjee, C., Tan, D., Depaoli, V., Rice, W.J., Diaz-Avalos, R., Silvestry, M., Sosa, H. (2013) *Cell Rep.* 3: 759–768; Actin microfilament: PDB ID 1ATN: Atomic Structure of the Actin:DNase I Complex. Kabsch, W., Mannherz, H.G., Suck, D., Pai, E.F., Holmes, K.C. (1990) *Nature* 347: 37–44; Myosin: PDB ID 1M8Q: Molecular Modeling of Averaged Rigor Crossbridges from Tomograms of Insect Flight Muscle, Chen, L.F., Winkler, H., Reedy, M.K., Reedy, M.C., Taylor, K.A. (2002) *J. Struct. Biol.* 138: 92–104; Phosphoglucose Isomerase: PDB ID 1IAT: The Crystal Structure of Human Phosphoglucose Isomerase at 1.6 Å Resolution: Implications for Catalytic Mechanism, Cytokine Activity and Haemolytic Anaemia, Read, J., Pearce, J., Li, X., Muirhead, H., Chirgwin, J., Davies, C. (2001) *J. Mol. Biol.* 309: 447–463; Aldolase: PDB ID 1ALD: Activity and Specificity of Human Aldolases, Gamblin, S.J., Davies, G.J., Grimes, J.M., Jackson, R.M., Littlechild, J.A., Watson, H.C. (1991) *J. Mol. Biol.* 219: 573–576; Triosephosphate Isomerase: PDB ID 7TIM: Structure of the Triose-Phosphate Isomerase-Phosphoglycolohydroxamate Complex: An Analogue of the Intermediate on the Reaction Pathway, Davenport, R.C., Bash, P.A., Seaton, B.A., Karplus, M., Petsko, G.A., Ringe, D. (1991) *Biochemistry* 30: 5821–5826; Glyceraldehyde-3-Phosphate Dehydrogenase: PDB ID 3GPD: Twinning in Crystals of Human Skeletal Muscle D-Glyceraldehyde-3-Phosphate Dehydrogenase, Mercer, W.D., Winn, S.I., Watson, H.C. (1976) *J. Mol. Biol.* 104: 277–283; Phosphoglycerate Kinase: PDB ID 3PGK: Sequence and Structure of Yeast Phosphoglycerate Kinase, Watson, H.C., Walker, N.P., Shaw, P.J., Bryant, T.N., Wendell, P.L., Fothergill, L.A., Perkins, R.E., Conroy, S.C., Dobson, M.J., Tuite, M.F. (1982) *EMBO J.* 1: 1635–1640; Phosphoglycerate Mutase: PDB ID 3PGM: Structure and Activity of Phosphoglycerate Mutase, Winn, S.I., Watson, H.C., Harkins, R.N., Fothergill, L.A. (1981) *Philos. Trans. R. Soc. London, Ser. B* 293: 121–130; Enolase: PDB ID 5ENL: Inhibition of Enolase: The Crystal Structures of Enolase-Ca2(+)-2-Phosphoglycerate and Enolase-Zn2(+)-Phosphoglycolate Complexes at 2.2-Å Resolution, Lebioda, L., Stec, B., Brewer, J.M., Tykarska, E. (1991) *Biochemistry* 30: 2823–2827; Pyruvate Kinase: PDB ID 1A49: Structure of the Bis(Mg2+)-ATP-Oxalate Complex of the Rabbit Muscle Pyruvate Kinase at 2.1 Å Resolution: ATP Binding over a Barrel, Larsen, T.M., Benning, M.M., Rayment, I., Reed, G.H. (1998) *Biochemistry* 37: 6247–6255; Citrate Synthase: PDB ID 1CTS: Crystallographic Refinement and Atomic Models of Two Different Forms of Citrate Synthase at 2.7 and 1.7 Å Resolution, Remington, S., Wiegand, G., Huber, R. (1982) *J. Mol. Biol.* 158: 111–152; Succinyl-CoA Synthetase: PDB ID 2FP4: Interactions of GTP with the ATP-Grasp Domain of GTP-Specific Succinyl-CoA Synthetase, Fraser, M.E., Hayakawa, K., Hume, M.S., Ryan, D.G., Brownie, E.R. (2006) *J. Biol. Chem.* 281: 11058-11065; Malate Dehydrogenase: PDB ID 4WLE: Crystal Structure of Citrate Bound MDH2, Eo, Y.M., Han, B.G., Ahn, H.C. To Be Published; **Summary art nucleus, Golgi apparatus and endoplasmic reticulum** Adapted from Marieb, Elaine N.; Hoehn, Katja, *Human Anatomy and Physiology*, 8th Ed., ©2010. Printed and electronically reproduced by permission of Pearson Education, Inc., Upper Saddle River, New Jersey.

Chapter 7 **7.4** Data from L. D. Frye and M. Edidin, The Rapid Intermixing of Cell Surface Antigens after Formation of Mouse-human Heterokaryons, *Journal of Cell Science* 7:319 (1970); **7.6** Based on Similar Energetic Contributions of Packing in the Core of Membrane and Water-Soluble Proteins by Nathan H. Joh et al., from *Journal of the American Chemical Society*, Volume 131(31); **Scientific Skills Exercise** Data from Figure 1 in T. Kondo and E. Beutler, Developmental Changes in Glucose Transport of Guinea Pig Erythrocytes, *Journal of Clinical Investigation* 65:1–4 (1980).

Chapter 8 **Scientific Skills Exercise** Data from S. R. Commerford et al., Diets Enriched in Sucrose or Fat Increase Gluconeogenesis and G-6-pase but not Basal Glucose Production in Rats, *American Journal of Physiology—Endocrinology and Metabolism* 283:E545–E555 (2002); **8.19** Data from Protein Data Bank ID 3e1f: "Direct and Indirect Roles of His-418 in Metal Binding and in the Activity of Beta-Galactosidase (*E. coli*)" by Douglas H. Juers et al., from *Protein Science*, June 2009, Volume 18(6); **8.20** Data from Protein Data Bank ID 1MDYO: "Crystal Structure of MyoD bHLH Domain-DNA Complex: Perspectives on DNA Recognition and Implications for Transcriptional Activation" from *Cell*, May 1994, Volume 77(3); **8.22 small cell** Adapted from Marieb, Elaine N.; Hoehn, Katja, *Human Anatomy and Physiology*, 8th Ed., ©2010. Printed and electronically reproduced by permission of Pearson Education, Inc., Upper Saddle River, New Jersey.

Chapter 9 **9.4** Adaptation of Figure 2.69 from *Molecular Biology of the Cell*, 4th Edition, by Bruce Alberts et al. Garland Science/Taylor & Francis LLC; **9.8** Figure adapted from *Biochemistry*, 4th Edition, by Christopher K. Mathews et al. Pearson Education, Inc.; **Scientific Skills Exercise** Data from M. E. Harper and M. D. Brand, The Quantitative Contributions of Mitochondrial Proton Leak and ATP Turnover Reactions to the Changed Respiration Rates of Hepatocytes from Rats of Different Thyroid Status, *Journal of Biological Chemistry* 268:14850–14860 (1993).

Chapter 10 **10.10** Data from T. W. Engelmann, Bacterium Photometricum. Ein Beitrag zur Vergleichenden Physiologie des Lichtund Farbensinnes, *Archiv. für Physiologie* 30:95–124 (1883); **10.13b** Data from Architecture of the Photosynthetic Oxygen-Evolving Center by Kristina N. Ferreira et al., from *Science*, March 2004, Volume 303(5665); **10.15** Adaptation of Figure 4.1 from *Energy, Plants, and Man*, by Richard Walker and David Alan Walker. © 1992 by Richard Walker and David Alan Walker. Reprinted with permission of Richard Walker; **Scientific Skills Exercise** Data from D. T. Patterson and E. P. Flint, Potential Effects of Global Atmospheric CO_2 Enrichment on the Growth and Competitiveness of C3 and C4 Weed and Crop Plants, Weed *Science* 28(1):71–75 (1980).

Chapter 11 **Problem-Solving Exercise** Data from N. Balaban et al., Treatment of *Staphylococcus aureus* Biofilm Infection by the Quorum-Sensing Inhibitor RIP, *Antimicrobial Agents and Chemotherapy*, 51:2226–2229 (2007); **11.8, 11.12** Adapted from Becker, Wayne M.; Reece, Jane B.; Poenie, Martin F., *The World of the Cell*, 3rd Edition, © 1996. Reprinted and electronically reproduced by permission of Pearson Education, Inc., Upper Saddle River, New Jersey.

Chapter 12 **12.9** Data from G. J. Gorbsky, P. J. Sammak, and G. G. Borisy, Chromosomes Move Poleward in Anaphase along Stationary Microtubules that Coordinately Disassemble from their Kinetochore Ends, *Journal of Cell Biology* 104:9–18 (1987); **12.13** Adaptation of Figure 18.41 from *Molecular Biology of the Cell*, 4th Edition, by Bruce Alberts et al. Garland Science/Taylor & Francis LLC; **12.14** Data from R. T. Johnson and P. N. Rao, Mammalian Cell Fusion: Induction of Premature Chromosome Condensation in Interphase Nuclei, *Nature* 226:717–722 (1970); **Scientific Skills Exercise** Data from K. K. Velpula et al., Regulation of Glioblastoma Progression by Cord Blood Stem Cells is Mediated by Downregulation of Cyclin D1, *PLoS ONE* 6(3): e18017 (2011).

Chapter 14 **14.3, 14.8** Data from G. Mendel, Experiments in Plant Hybridization, *Proceedings of the Natural History Society of Brünn* 4:3–47 (1866).

Chapter 15 **15.3** Data from T. H. Morgan, Sex-limited inheritance in *Drosophila, Science* 32:120–122 (1910); **15.9** Based on the data from "The Linkage of Two Factors in *Drosophila* That Are Not Sex-Linked" by Thomas Hunt Morgan and Clara J. Lynch, from *Biological Bulletin*, August 1912, Volume 23(3).

Chapter 16 **16.2** Data from F. Griffith, The Significance of Pneumococcal Types, *Journal of Hygiene* 27:113–159 (1928); **16.4** Data from A. D. Hershey and M. Chase, Independent Functions of Viral Protein and Nucleic Acid in Growth of Bacteriophage, *Journal of General Physiology* 36:39–56 (1952); **Scientific Skills Exercise** Data from several papers by Chargaff: for example, E. Chargaff et al., Composition of the Desoxypentose Nucleic Acids of Four Genera of Sea-urchin, *Journal of Biological Chemistry* 195:155–160 (1952); **pp. 320–321 quote** J. D. Watson and F. H. C. Crick, Genetical Implications of the Structure of Deoxyribonucleic Acid, *Nature* 171:964–967 (1953); **16.11** Data from M. Meselson and F. W. Stahl, The Replication of DNA in *Escherichia coli, Proceedings of the National Academy of Sciences USA* 44:671–682 (1958).

Chapter 17 **17.3** Data from A. M. Srb and N. H. Horowitz, The Ornithine Cycle in Neurospora and Its Genetic Control, *Journal of Biological Chemistry* 154:129–139 (1944); **17.12** Adapted from Becker, Wayne M.; Reece, Jane B.; Poenie, Martin F., *The World of the Cell*, 3rd Edition, © 1996. Reprinted and electronically reproduced by permission of Pearson Education, Inc., Upper Saddle River, New Jersey; **17.14** Adapted from Klein-Smith, Lewis J., Kish, Valerie M.; *Principles of Cell and Molecular Biology*. Reprinted and electronically reproduced by permissions of Pearson Education, Inc., Upper Saddle River, New Jersey; **17.18** Adapted from Mathews, Christopher K.; Van Holde, Kensal E., *Biochemistry*, 2nd ed., ©1996. Reprinted and electronically reproduced by permission of Pearson Education, Inc. Upper Saddle River, New Jersey; **Scientific Skills Exercise** Material provided courtesy of Dr. Thomas Schneider, National Cancer Institute, National Institutes of Health, 2012; **Problem-Solving Exercise** Data from N. Nishi and K. Nanjo, Insulin Gene Mutations and Diabetes, *Journal of Diabetes Investigation* Vol. 2: 92–100 (2011).

Chapter 18 **18.10** Data from PDB ID 1MDY: P. C. Ma et al. Crystal structure of MyoD bHLH Domain-DNA Complex: Perspectives on DNA Recognition and Implications for Transcriptional Activation, *Cell* 77:451–459 (1994); **Scientific Skills Exercise** Data from J. N. Walters et al., Regulation of Human Microsomal Prostaglandin E Synthase-1 by IL-1b Requires a Distal Enhancer Element with a Unique Role for C/EBPb, *Biochemical Journal* 443:561–571 (2012); **18.26** Adapted from Becker, Wayne M.; Reece, Jane B.; Poenie, Martin F., *The World of the Cell*, 3rd Edition, © 1996. Reprinted and electronically reproduced by permission of Pearson Education, Inc., Upper Saddle River, New Jersey.

Chapter 19 **19.4** Data from M. J. Beijerinck, Concerning a Contagium Vivum Fluidum as Cause of the Spot Disease of Tobacco Leaves, *Verhandelingen der Koninkyke Akademie Wettenschappen te Amsterdam* 65:3–21 (1898). Translation published in English as Phytopathological Classics Number 7 (1942), American Phytopathological Society Press, St. Paul, MN; **Scientific Skills Exercise** Data from J.-R. Yang et al., New Variants and Age Shift to High Fatality Groups Contribute to Severe Successive Waves in the 2009 Influenza Pandemic in Taiwan, *PLoS ONE* 6(11): e28288 (2011).

Chapter 20 **20.7** Adapted from Becker, Wayne M.; Reece, Jane B.; Poenie, Martin F., *The World of the Cell*, 3rd Edition, © 1996. Reprinted and electronically reproduced by permission of Pearson Education, Inc., Upper Saddle River, New Jersey; **20.16** Data from J. B. Gurdon et al., The Developmental Capacity of Nuclei Transplanted from Keratinized Cells of Adult Frogs, *Journal of Embryology and Experimental Morphology* 34:93–112 (1975); **20.21** Data from K. Takahashi et al., Induction of pluripotent stem cells from adult human fibroblasts by defined factors, *Cell* 131:861–872 (2007).

Chapter 21 **21.3** Simulated screen shots based on Mac OS X and from data found at NCBI, U.S. National Library of Medicine using Conserved Domain Database, Sequence Alignment Viewer, and Cn3D; **21.8, 21.9** Adapted from Becker, Wayne M.; Reece, Jane B.; Poenie, Martin F., *The World of the Cell*, 3rd Edition, © 1996. Reprinted and electronically reproduced by permission of Pearson Education, Inc., Upper Saddle River, New Jersey; **21.10 haemoglobin** Data from PDB ID 2HHB: G. Fermi, M.F. Perutz, B. Shaanan, and R. Fourme. The Crystal Structure of Human Deoxyhaemoglobin at 1.74 Å resolution. *J. Mol. Biol.* 175:159–174 (1984); **21.15a** Drawn from data in Protein Data Bank ID 1LZ1: "Refinement of Human Lysozyme at 1.5 Å Resolution Analysis of Non-bonded and Hydrogen-bond Interactions" by P. J. Artymiuk and C. C. Blake, from *Journal of Molecular Biology*, 1981, 152:737–762; **21.15b** Drawn from data in Protein Data Bank ID 1A4V: "Structural Evidence for the Presence of a Secondary Calcium Binding Site in Human Alpha-Lactalbumin" by N. Chandra et al., from *Biochemistry*, 1998, 37:4767–4772; **haemoglobin in Scientific Skills Exercise** PDB ID 2HHB: G. Fermi, M.F. Perutz, B. Shaanan, and R. Fourme. The Crystal Structure of Human Deoxyhaemoglobin at 1.74 Å Resolution. *J. Mol. Biol.* 175:159–174 (1984); **Scientific Skills Exercise** Compiled using data from NCBI; **21.18** Data from W. Shu et al., Altered ultrasonic vocalization in mice with a disruption in the *Foxp2* gene, *Proceedings of the National Academy of Sciences USA* 102:9643–9648 (2005); **21.19** Adapted from *The Homeobox: Something Very Precious That We Share with Flies, From Egg to Adult* by Peter Radetsky, © 1992. Reprinted by permission from William McGinnis; **21.20** Adaptation from "Hox Genes and the Evolution of Diverse Body Plans" by Michael Akam, from *Philosophical Transactions of the Royal Society B: Biological Sciences*, September 29, 1995, Volume 349(1329): 313–319. Reprinted by permission from The Royal Society.

Chapter 22 **22.8** Artwork by Utako Kikutani (as appeared in "What Can Make a Four-Ton Mammal a Most Sensitive Beast?" by Jeheskel Shoshani, from *Natural History*, November 1997, Volume 106(1), 36–45). Copyright © 1997 by Utako Kikutani. Reprinted with permission of the artist; **22.13** Data from "Host Race Radiation in the Soapberry Bug: Natural History with the History" by Scott P. Carroll and Christin Boyd, from *Evolution*, 1992, Volume 46(4); **22.15** Figure created by Dr. Binh Diep on request of Michael Cain. Copyright © 2011 by Binh Diep. Reprinted with permission; **p. 488 quote** Darwin, C. (1859). On the Origin of Species by Means of Natural Selection: The Preservation of Favoured Races in the Struggle for Life. London: John Murray; **Scientific Skills Exercise** Data from J. A. Endler, Natural Selection on Color Patterns in *Poecilia reticulata*, *Evolution* 34:76–91 (1980); **p. 488 quote** Darwin, C. (1859). On the Origin of Species by Means of Natural Selection: The Preservation of Favoured Races in the Struggle for Life. London: John Murray; **Test Your Understanding Question 7** Data from C. F. Curtis et al., Selection for and Against Insecticide Resistance and Possible Methods of Inhibiting the Evolution of Resistance in Mosquitoes, *Ecological Entomology* 3:273–287 (1978).

Chapter 23 **23.4** Based on the data from *Evolution*, by Douglas J. Futuyma. Sinauer Associates, 2006; and Nucleotide Polymorphism at the Alcohol Dehydrogenase Locus of *Drosophila melanogaster* by Martin Kreitman, from *Nature*, August 1983, Volume 304(5925); **23.13 map** Modified from Schiffer, M. and Mcevey, S. F., Drosophila bunnanda—A new species from northern Australia with notes on other Australian members of the montium subgroup (Diptera: Drosophilidae), *Zootaxa* (2006) 1-23 10.5281/zenodo.174253; **23.13 graph** Simplified adaptation of Figure 3 from Magiafoglou, A., Carew, M.E. and Hoffmann, A.A., Shifting clinal patterns and microsatellite variation in Drosophila serrata populations: a comparison of populations near the southern border of the species range, *Journal of Evolutionary Biology* (2002), 15: 763-774; **23.15** Based on many sources: *Evolution* by Douglas J. Futuyma. Sinauer Associates 2005; and *Vertebrate Paleontology and Evolution* by Robert L. Carroll. W.H. Freeman & Co., 1988; **23.17** Data from A. M. Welch et al., Call Duration as an Indicator of Genetic Quality in Male Gray Tree Frogs, *Science* 280:1928–1930 (1998); **23.18** Adapted from Frequency-Dependent Natural Selection in the Handedness of Scale-Eating Cichlid Fish by Michio Hori, from *Science*, April 1993, Volume 260(5105); **Test Your Understanding Question 7** Data from R. K. Koehn and T. J. Hilbish, The Adaptive Importance of Genetic Variation, *American Scientist* 75:134–141 (1987).

Chapter 24 **24.6** Original unpublished graph created by Brian Langerhans; **24.7** Data from D. M. B. Dodd, Reproductive Isolation as a Consequence of Adaptive Divergence in *Drosophila pseudoobscura*, *Evolution* 43:1308–1311 (1989); **24.8** Shull, H. C., Pérez-Losada, M., Blair, D., Sewell, K., Sinclair, E. A., Lawler, S., Ponniah, M. and Crandall, K. A. (2005) Phylogeny and biogeography of the freshwater crayfish Euastacus (Decapoda: Parastacidae) based on nuclear and mitochondrial DNA, *Molecular Phylogenetics and Evolution*, 37: 249–263; Ponniah, M. and Hughes, J. M. (2006) The evolution of Queensland spiny mountain crayfish of the genus Euastacus. II. Investigating simultaneous vicariance with intraspecific genetic data, *Marine and Freshwater Research*, 57: 349–362; **24.8 map** Reprinted from Phylogeny and biogeography of the freshwater crayfish Euastacus (Decapoda: Parastacidae) based on nuclear and mitochondrial DNA, H. C. Shull, et al., Molecular Phylogenetics and Evolution, 37, p.251: Fig. 1. Collection sites and distributions of 43 E., Copyright 2005, with permission from Elsevier; **Scientific Skills Exercise** Data from S. G. Tilley, A. Verrell, and S. J. Arnold, Correspondence between Sexual Isolation and Allozyme Differentiation: A Test in the Salamander *Desmognathus ochrophaeus*, *Proceedings of the National Academy of Sciences USA* 87:2715–2719 (1990); **24.12** Based on *Hybrid Zone and the Evolutionary Process*, edited by Richard G. Harrison. Oxford University Press.

Chapter 25 **25.2** Based on data from The Miller Volcanic Spark Discharge Experiment by Adam P. Johnson et al., from *Science*, October 2008, Volume 322(5900); **25.4** Based on "Experimental Models of Primitive Cellular Compartments: Encapsulation, Growth, and Division" by Martin M. Hanczyc, Shelly M. Fujikawa, and Jack W. Szostak, from *Science*, October 2003, Volume 302(5645); **25.6** Eicher, D. L, *Geologic Time*, 2nd Ed., ©1976, p. 119. Adapted and electronically reproduced by permission of Pearson Education, Inc., Upper Saddle River, New Jersey; **25.7 first four skulls** Adapted from many sources including D.J. Futuyma, *Evolution*, Fig. 4.10, Sunderland, MA: Sinauer Associates, Sunderland, MA (2005) and from R.L. Carroll, *Vertebrate Paleontology and Evolution*. W.H. Freeman & Co. (1988); **last skull** Adapted from Z. Luo et al., A New Mammaliaform from the Early Jurassic and Evolution of Mammalian Characteristics, *Science* 292:1535 (2001); **25.8** Adapted from When Did Photosynthesis Emerge on Earth? by David J. Des Marais, from *Science*, September 2000, Volume 289(5485). **25.9** Adapted from The Rise of Atmospheric Oxygen by Lee R. Kump, from *Nature*, January 2008, Volume 451(7176); **Scientific Skills Exercise** Data from T. A. Hansen, Larval Dispersal and Species Longevity in Lower Tertiary Gastropods, *Science* 199:885–887 (1978); **25.16** Based on *Earthquake Information Bulletin*, December 1977, Volume 9(6), edited by Henry Spall; **25.17** Nick Mortimer, Hamish J. Campbell, Andy J. Tulloch, Peter R. King, Vaughan M. Stagpoole, Ray A. Wood, Mark S. Rattenbury, Rupert Sutherland, Chris J. Adams, Julien Collot, Maria Seton, Zealandia: Earth's Hidden Continent. *GSA Today*, 27(3); **25.28** Roman Uchytel; **25.30** Illustration by Peter Schouten; **25.34** Based on Trevor H. Worthy, Moa - Scientific classification, *Te Ara - the Encyclopedia of New Zealand*, http://www.TeAra.govt.nz/en/diagram/11361/ratite-birds; **25.36** Based on many sources: D.M. Raup and J. J. Sepkoski, Jr., Mass Extinctions in the Marine Fossil Record, *Science* 215:1501–1503 (1982); J. J. Sepkoski, Jr., A Kinetic Model of Phanerozoic Taxonomic Diversity. III. Post-Paleozoic Families and Mass Extinctions, *Paleobiology* 10:246–267 (1984); and D. J. Futuyma, *The Evolution of Biodiversity*, p. 143, Fig. 7.3a and p. 145, Fig. 7.6, Sinauer Associates, Sunderland, MA; **25.38** Based on data from A Long-Term Association between Global Temperature and Biodiversity, Origination and Extinction in the Fossil Record by P.J. Mayhew, G.B. Jenkins and T.G. Benton, *Proceedings of the Royal Society B: Biological Sciences* 275(1630):47–53. The Royal Society, 2008; **25.39** Adapted from Anatomical and Ecological Constraints on Phanerozoic Animal Diversity in the Marine Realm by Richard K. Bambach et al., from *Proceedings of the National Academy of Sciences USA*, May 2002, Volume 99(10); **25.44** Based on data from The Miller Volcanic Spark Discharge Experiment by Adam P. Johnson et al., from *Science*, October 2008, Volume 322(5900); **25.45** Data from Genetic and Developmental Basis of Evolutionary Pelvic Reduction in Threespine Sticklebacks by Michael D. Shapiro et al., from *Nature*, April 2004, Volume 428(6984); **25.47** Adaptations of Figure 3-1 (a–d, f) from *Evolution*, 3rd Edition, by Monroe W. Strickberger. Jones & Bartlett Learning, Burlington, MA.

Chapter 26 **26.6** Data from C. S. Baker and S. R. Palumbi, Which Whales Are Hunted? A Molecular Genetic Approach to Monitoring Whaling, *Science* 265:1538–1539 (1994); **26.13** Based on The Evolution of the Hedgehog Gene Family in Chordates: Insights from Amphioxus Hedgehog by Sebastian M. Shimeld, from *Developmental Genes and Evolution*, January 1999, Volume 209(1); **26.19** Based on *Molecular Markers, Natural History, and Evolution*, 2nd ed., by J.C. Advise. Sinauer Associates, 2004; **26.20** Adapted from Timing the Ancestor of the HIV-1 Pandemic Strains by B. Korber et al., *Science* 288(5472):1789–1796 (6/9/00); **Scientific Skills Exercise** Data from Nancy A. Moran, Yale University. See N. A. Moran and T. Jarvik, Lateral transfer of genes from fungi underlies carotenoid production in aphids, *Science* 328:624–627 (2010); **26.23** Adapted from Phylogenetic Classification and the Universal Tree by W.F. Doolittle, *Science* 284(5423):2124–2128 (6/25/99).

Chapter 27 **27.10** Graph Data from V. S. Cooper and R. E. Lenski, The Population Genetics of Ecological Specialization in Evolving *Escherichia coli* Populations, *Nature* 407:736–739 (2000); **27.19** Data from Root-Associated Bacteria Contribute to Mineral Weathering and to Mineral Nutrition in Trees: A Budgeting Analysis by Christophe Calvaruso et al., *Applied and Environmental Microbiology*, February 2006, Volume 72(2); **27.22** Data from K. Kupferschmidt, Resistance Fighters. *Science* 352(6287):758-761. 13 May 2016; **Scientific Skills Exercise** Data from L. Ling et al. A New Antibiotic Kills Pathogens without Detectable Resistance, *Nature* 517:455–459 (2015); **Test Your Understanding Question 8** Data from J. J. Burdon et al., Variation in the Effectiveness of Symbiotic Associations between Native Rhizobia and Temperate Australian Acacia: Within Species Interactions, *Journal of Applied Ecology* 36:398–408 (1999).

Chapter 28 **Scientific Skills Exercise** Data from D. Yang et al., Mitochondrial Origins, *Proceedings of the National Academy of Sciences USA* 82:4443–4447 (1985); **28.5** Adapted from Figure 2 from by P. Keeling, *Annual Reviews*, Inc.; **28.19** Adaptation of illustration by Kenneth X. Probst, from *Microbiology* by R.W. Bauman. Copyright © 2004 by Kenneth X. Probst; **28.26** Data from R. Derelle et al., Bacterial proteins pinpoint a single eukaryotic root, *Proceedings of the National Academy of Sciences USA* 112:E693–699 (2015); **28.32** Based on Global Phytoplankton Decline over the Past Century by Daniel G. Boyce et al., from *Nature*, July 29, 2010, Volume 466(7306); and authors' personal communications.

Chapter 29 **29.9** Modified from M. Delgado-Baquerizo, F. Maestre, D. Eldridge, M. Bowker, V. Ochoa, B. Gozalo, M. Berdugo, J. Val, B. Singh, Biocrust-forming mosses mitigate the negative impacts of increasing aridity on ecosystem multifunctionality in drylands, *New Phytologist* 209:1540-1552 (2015); **Scientific Skills Exercise** Data from T.M. Lenton et al, First Plants Cooled the Ordovician. *Nature Geoscience* 5:86–89 (2012); **Test Your Understanding Question 8** Data from O. Zackrisson et al., Nitrogen Fixation Increases with Successional Age in Boreal Forests, *Ecology* 85:3327–3334 (2006).

Chapter 30 **Scientific Skills Exercise** Data from S. Sallon et al, Germination, Genetics, and Growth of an Ancient Date Seed. *Science* 320:1464 (2008); **30.15a** Adapted from "A Revision of Williamsoniella" by T. M. Harris, from *Proceedings of the Royal Society B: Biological Sciences*, October 1944, Volume 231(583): 313–328; **30.15b** Adaptation of Figure 2.3, *Phylogeny and Evolution of Angiosperm*, 2nd Edition, by Douglas E. Soltis et al. (2005). Sinauer Associates, Inc.

Chapter 31 **Scientific Skills Exercise** Data from F. Martin et al., The genome of *Laccaria bicolor* provides insights into mycorrhizal symbiosis, *Nature* 452:88–93 (2008); **31.20** Data from A. E. Arnold et al., Fungal Endophytes Limit Pathogen Damage in a Tropical Tree, *Proceedings of the National Academy of Sciences USA* 100:15649–15654 (2003); **31.25** Adaption of Figure 1 from "Reversing Introduced Species Effects: Experimental Removal of Introduced Fish Leads to Rapid Recovery of a Declining Frog" by Vance T. Vredenburg, from *Proceedings of the National Academy of Sciences USA*, May 2004, Volume 101(20). Copyright (2004) National Academy of Sciences, U.S.A.

Test Your Understanding Question 5 Data from R. S. Redman et al., Thermotolerance Generated by Plant/Fungal Symbiosis, *Science* 298:1581 (2002).

Chapter 32 Scientific Skills Exercise Data from Bradley Deline, University of West Georgia, and Kevin Peterson, Dartmouth College, 2013.

Chapter 33 Scientific Skills Exercise Data from R. Rochette et al., Interaction between an Invasive Decapod and a Native Gastropod: Predator Foraging Tactics and Prey Architectural Defenses, *Marine Ecology Progress Series* 330:179–188 (2007); **33.21** Adaptation of Figure 3 from "The Global Decline of Nonmarine Mollusks" by Charles Lydeard et al., from *Bioscience*, April 2004, Volume 54(4). American Institute of Biological Sciences. Oxford University Press; **33.29** Tree Data from J. K. Grenier et al., Evolution of the Entire Arthropod Hox Gene Set Predated the Origin and Radiation of the Onychophoran/Arthropod Clade, *Current Biology* 7:547–553 (1997).

Chapter 34 34.10 Adaptation of Figure 1a from "Fossil Sister Group of Craniates: Predicted and Found" by Jon Mallatt and Jun-yuan Chen, from *Journal of Morphology*, May 15, 2003, Volume 258(1). John Wiley & Sons, Inc.; **34.12** Adapted from *Vertebrates: Comparative Anatomy, Function, Evolution* (2002) by Kenneth Kardong. The McGraw-Hill Companies, Inc.; **34.18** Adaptation of Figure 3 from "The Oldest Articulated Osteichthyan Reveals Mosaic Gnathostome Characters" by Min Zhu et al., from *Nature*, March 26, 2009, Volume 458(7237); **34.21** Adaptation of Figure 4 from "The Pectoral Fin of *Tiktaalik roseae* and the Origin of the Tetrapod Limb" by Neil H. Shubin et al., from *Nature*, April 6, 2006, Volume 440(7085). Macmillan Publishers Ltd.; **34.37a** Based on many sources including Figure 4.10 from *Evolution*, by Douglas J. Futuyma. Sinauer Associates, 2005; and *Vertebrate Paleontology and Evolution* by Robert L. Carroll. W.H. Freeman & Co., 1988; **34.46** Based on many photos of fossils. Some sources are *O. tugenensis* photo in "Early Hominid Sows Division" by Michael Balter, from *Science Now*, Feb. 22, 2001; *A. garhi* and *H. neanderthalensis* based on *The Human Evolution Coloring Book* by Adrienne L. Zihlman and Carla J. Simmons. Harper Collins, 2001; *K. platyops* based on photo in "New Hominin Genus from Eastern Africa Shows Diverse Middle Pliocene Lineages" by Meave Leakey et al., from *Nature*, March 2001, Volume 410(6827); *P. boisei* based on a photo by David Brill; *H. ergaster* based on a photo at www.museumsinhand.com; *S. tchadensis* based on Figure 1b from "A New Hominid from the Upper Miocene of Chad, Central Africa" by Michel Brunet et al., from *Nature*, July 2002, Volume 418(6894); **Scientific Skills Exercise** Data from Dean Falk, Florida State University, 2013; **Test Your Understanding Question 8** Data from D. Sol et al., Big-Brained Birds Survive Better in Nature, *Proceedings of the Royal Society B* 274:763–769 (2007).

Chapter 35 Scientific Skills Exercise Data from D. L. Royer et al., Phenotypic Plasticity of Leaf Shape Along a Temperature Gradient in Acer rubrum, *PLOS ONE* 4(10):e7653 (2009); **35.21** Data from "Mongolian Tree Rings and 20th-Century Warming" by Gordon C. Jacoby, et al., from *Science*, August 9, 1996, Volume 273(5276): 771–773.

Chapter 36 Scientific Skills Exercise Data from J. D. Murphy and D. L. Noland, Temperature Effects on Seed Imbibition and Leakage Mediated by Viscosity and Membranes, *Plant Physiology* 69:428–431 (1982); **36.18** Data from S. Rogers and A. J. Peel, Some Evidence for the Existence of Turgor Pressure in the Sieve Tubes of Willow (*Salix*), *Planta* 126:259–267 (1975).

Chapter 37 37.5 Modified from Figure 6.2, N. McKenzie et al. *Australian Soils and Landscapes: An Illustrated Compendium*, CSIRO Publishing, Collingwood, Australia (2004); **37.6** Adapted from Environment New Zealand 2007, Ministry for the Environment, Manatu Mo Te Taiao, New Zealand Government 9.1-9.2, 218-219; **37.12** Map based on data from 'Australian Dryland Salinity Assessment Spatial Data - NLWARA 2001', Australian Bureau of Agriculture and Resource Economics and Sciences (ABARES), © Commonwealth of Australia (2012); **37.13** Adapted from R. Fitzpatrick et al. *Atlas of Australian Acid Sulphate Soils v2*, CSIRO (2011), doi:10.4225/08/512E79A0BC589; **37.18b** Data from D.S. Lundberg et al., Defining the Core *Arabidopsis thaliana* Root Microbiome, *Nature* 488:86–94 (2012).

Chapter 38 Scientific Skills Exercise Data from S. Sutherland and R. K. Vickery, Jr. Trade-offs between Sexual and Asexual Reproduction in the Genus *Mimulus*. *Oecologia* 76:330–335 (1998).

Chapter 39 39.5 Data from C. R. Darwin, *The Power of Movement in Plants*, John Murray, London (1880). P. Boysen-Jensen, Concerning the Performance of Phototropic Stimuli on the Avenacoleoptile, *Berichte der Deutschen Botanischen Gesellschaft (Reports of the German Botanical Society)* 31:559–566 (1913); **39.6** Data from L. Gälweiler et al., Regulation of Polar Auxin Transport by AtPIN1 in *Arabidopsis* Vascular Tissue, *Science* 282:2226–2230 (1998); **39.15a** Based on *Plantwatching: How Plants Remember, Tell Time, Form Relationships and More* by Malcolm Wilkins. Facts on File, 1988; **39.16** Data from H. Borthwick et al., A Reversible Photo Reaction Controlling Seed Germination, *Proceedings of the National Academy of Sciences USA* 38:662–666 (1952); **Problem-Solving Exercise** Map data from Camilo Mora et al. Days for Plant Growth Disappear under Projected Climate Change: Potential Human and Biotic Vulnerability. *PLoS Biol.* 13(6): e1002167 (2015); **Scientific Skills Exercise** Data from O. Falik et al., Rumor Has It ...: Relay Communication of Stress Cues in Plants, *PLoS ONE* 6(11):e23625 (2011).

Chapter 40 40.16 Data from V. H. Hutchison, H. G. Dowling, and A. Vinegar, Thermoregulation in a Brooding Female Indian Python, *Python molurus bivittatus*, *Science* 151:694–696 (1966); **Scientific Skills Exercise** Based on the data from M. A. Chappell et al., Energetics of Foraging in Breeding Adélie Penguins, *Ecology* 74:2450–2461 (1993); M. A. Chappell et al., Voluntary Running in Deer Mice: Speed, Distance, Energy Costs, and Temperature Effects, *Journal of Experimental Biology* 207:3839–3854 (2004); T. M. Ellis and M. A. Chappell, Metabolism, Temperature Relations, Maternal Behavior, and Reproductive Energetics in the Ball Python (*Python regius*), *Journal of Comparative Physiology* B 157:393–402 (1987); **40.22** Data from F. G. Revel et al., The Circadian Clock Stops Ticking During Deep Hibernation in the European Hamster, *Proceedings of the National Academy of Sciences USA* 104:13816–13820 (2007).

Chapter 41 41.4 Data from R. W. Smithells et al., Possible Prevention of Neural-Tube Defects by Periconceptional Vitamin Supplementation, *Lancet* 315:339–340 (1980); **41.8** Adapted from Marieb, Elaine; Hoehn, Katja, *Human Anatomy and Physiology*, 8th Edition, 2010, p. 852, Reprinted and electronically reproduced by permission of Pearson Education, Upper Saddle River, New Jersey; **41.17** Adapted from Ottman N., Smidt H., de Vos W.M. and Belzer C. (2012) The function of our microbiota: who is out there and what do they do? *Front. Cell. Inf. Microbiol.* 2:104. doi: 10.3389/fcimb.2012.00104; **41.24** Republished with permission of American Association for the Advancement of Science, from Cellular Warriors at the Battle of the Bulge by Kathleen Sutliff and Jean Marx, from *Science*, February 2003, Volume 299(5608); **Scientific Skills Exercise** Based on the data from D. L. Coleman, Effects of Parabiosis of Obese Mice with Diabetes and Normal Mice, Diabetologia 9:294–298 (1973).

Chapter 42 Scientific Skills Exercise Data from J. C. Cohen et al., Sequence Variations in PCSK9, Low LDL, and Protection Against Coronary Heart Disease, *New England Journal of Medicine* 354:1264–1272 (2006); **42.25** Data from M. E. Avery and J. Mead, Surface Properties in Relation to Atelectasis and Hyaline Membrane Disease, *American Journal of Diseases of Children* 97:517–523 (1959).

Chapter 43 43.5 Adapted from *Microbiology: An Introduction*, 11th Edition, by Gerard J. Tortora, Berdell R. Funke, and Christine L. Case. Pearson Education, Inc.; **43.6** Adapted from Marieb, Elaine N.; Hoehn, Katja, *Human Anatomy and Physiology*, 8th Ed., © 2010. Reprinted and electronically reproduced by permission of Pearson Education, Inc., Upper Saddle River, New Jersey; **43.24** Based on multiple sources: WHO/UNICEF Coverage Estimates 2014 Revision. July 2015. Map Production: Immunization Vaccines and Biologicals (IVB). World Health Organization, 16 July 2015; Our Progress Against Polio, May 1, 2014. CDC; **Scientific Skills Exercise** Data from sources: L. J. Morrison et al., Probabilistic Order in Antigenic Variation of *Trypanosoma brucei*, *International Journal for Parasitology* 35:961-972 (2005); and L. J. Morrison et al., Antigenic Variation in the African Trypanosome: Molecular Mechanisms and Phenotypic Complexity, *Cellular Microbiology* 1: 1724–1734 (2009).

Chapter 44 Scientific Skills Exercise Data from R. E. MacMillen et al., Water Economy and Energy Metabolism of the Sandy Inland Mouse, *Leggadina hermannsburgensis*, *Journal of Mammalogy* 53:529–539 (1972); **44.6** Adapted from Mitchell, Lawrence G., *Zoology*, © 1998. Reprinted and electronically reproduced by permission of Pearson Education, Inc., Upper Saddle River, New Jersey; **44.13 & 44.15 kidney structure** Adapted from Marieb, Elaine N.; Hoehn, Katja, *Human Anatomy and Physiology*, 8th Ed., 2010. Reprinted and electronically reproduced by permission of Pearson Education, Inc., Upper Saddle River, New Jersey; **44.21** Data in tables from P. M. Deen et al., Requirement of Human Renal Water Channel Aquaporin-2 for Vasopressin-Dependent Concentration of Urine, *Science* 264:92–95 (1994); **Summary Figure** Adapted from Beck, *Life: An Introduction to Biology*, 3rd Ed., ©1991, p. 643. Reprinted and electronically reproduced by permission of Pearson Education, Inc., Upper Saddle River, New Jersey; **Test Your Understanding Question 7** Data for kangaroo rat from *Animal Physiology: Adaptation and Environment* by Knut Schmidt-Nielsen. Cambridge University Press, 1991.

Chapter 45 Scientific Skills Exercise Data from J. Born et al., Timing the End of Nocturnal Sleep, *Nature* 397:29–30 (1999).

Chapter 46 46.11 Data from R. R. Snook and D. J. Hosken, Sperm Death and Dumping in *Drosophila*, *Nature* 428:939–941 (2004); **Scientific Skills Exercise** Data from A. Jost, Recherches Sur la Differenciation Sexuelle de l'embryon de Lapin (Studies on the Sexual Differentiation of the Rabbit Embryo), *Archives d'Anatomie Microscopique et de Morphologie Experimentale* 36:271–316 (1947); **46.19** Adapted from Marieb, Elaine N., Hoehn, Katja, *Human Anatomy and Physiology*, 8th Ed., 2010. Reprinted and electronically reproduced by permission of Pearson Education, Inc., Upper Saddle River, New Jersey.

Chapter 47 47.3 Data from "Intracellular Calcium Release at Fertilization in the Sea Urchin Egg" by R. Steinhardt et al., from *Developmental Biology*, July 1977, Volume 58(1); **Scientific Skills Exercise** Data from J. Newport and M. Kirschner, A Major Developmental Transition in Early *Xenopus* Embryos: I. Characterization and Timing of Cellular Changes at the Midblastula Stage, *Cell* 30:675–686 (1982); **47.10** Adapted from Keller, R. E. 1986. The Cellular Basis of Amphibian Gastrulation. In L. Browder (ed.), *Developmental Biology: A Comprehensive Synthesis*, Vol. 2. Plenum, New York, pp. 241–327; **47.14** Based on "Cell Commitment and Gene Expression in the Axolotl Embryo" by T. J. Mohun et al., from *Cell*, November 1980, Volume 22(1); **47.17** *Principles of Development*, 2nd Edition by Wolpert (2002), Fig. 8.26, p. 275. By permission of Oxford University Press; **47.19** Republished with permission of Garland Science, Taylor & Francis Group, from *Molecular Biology of the Cell*, Bruce Alberts et al., 4th Edition, © 2002; permission conveyed through Copyright Clearance Center, Inc.; **47.23** Data from H. Spemann, *Embryonic Development and Induction*, Yale University Press, New Haven, CT (1938); **47.24** Data from H. Spemann and H. Mangold, Induction of Embryonic Primordia by Implantation of Organizers from a Different Species, Trans. V. Hamburger (1924). Reprinted in *International Journal of Developmental Biology* 45:13–38 (2001); **47.26** Data from L. S. Honig and D. Summerbell, Maps of strength of positional signaling activity in the developing chick wing bud, *Journal of Embryology and Experimental Morphology* 87:163–174 (1985); **47.27** Adapted from Marieb, Elaine N.; Hoehn, Katja, *Human Anatomy and Physiology*, 8th Edition, 2010. Reprinted and electronically reproduced by permission of Pearson Education, Inc., Upper Saddle River, New Jersey.

Chapter 48 48.12 Graph Based on Figure 6-2d from *Cellular Physiology of Nerve and Muscle*, 4th Edition, by Gary G. Matthews. Wiley-Blackwell, 2003; **Scientific Skills Exercise** Data from C. B. Pert and S. H. Snyder, Opiate Receptor: Demonstration in Nervous Tissue, *Science* 179:1011–1014 (1973).

Chapter 49 49.7 Adapted from Marieb, Elaine N.; Hoehn, Katja, *Human Anatomy and Physiology*, 8th Ed., © 2010. Reprinted and electronically reproduced by permission of Pearson Education, Inc., Upper Saddle River, New Jersey; **49.12** Based on "Sleep in Marine Mammals" by L. M. Mukhametov, from *Sleep Mechanisms*, edited by Alexander A. Borberly and J. L. Valatx. Springer; **Scientific Skills Exercise** Data from M. R. Ralph et al., Transplanted Suprachiasmatic Nucleus Determines Circadian Period, *Science* 247:975–978 (1990); **49.20** Adaptation of Figure 1c from "Avian Brains and a New Understanding of Vertebrate Brain Evolution" by Erich D. Jarvis et al., from *Nature Reviews Neuroscience*, February 2005, Volume 6(2); **49.23** Adaptation of Figure 10 from *Schizophrenia Genesis: The Origins of Madness* by Irving I. Gottesman. Worth Publishers.

Chapter 50 **50.12a, 50.13, 50.17 eye structure, 50.24a, 50.26, 50.31** Adapted from Marieb, Elaine N; Hoehn, Katja, *Human Anatomy and Physiology*, 8th Ed., © 2010 Reprinted and electronically reproduced by Permission of Pearson Education, Inc., Upper Saddle River, New Jersey; **50.23** Data from K. L. Mueller et al., The receptors and coding logic for bitter taste, *Nature* 434:225–229 (2005); **50.35 grasshopper** Based on Hickman et al., *Integrated Principles of Zoology*, 9th ed., p. 518, Fig. 22.6, McGraw-Hill Higher Education, NY (1993); **Scientific Skills Exercise** Data from K. Schmidt-Nielsen, Locomotion: Energy Cost of Swimming, Flying, and Running, *Science* 177:222–228 (1972).

Chapter 51 **51.4** Based on "*Drosophila*: Genetics Meets Behavior" by Marla B. Sokolowski, from *Nature Reviews: Genetics*, November 2001, Volume 2(11); **51.8** Data from *The Study of Instinct*, N. Tinbergen, Clarendon Press, Oxford (1951); **51.17** Adapted from Evolution of Foraging Behavior in *Drosophila* by Density Dependent Selection by Maria B. Sokolowski et al., from *Proceedings of the National Academy of Sciences USA*, July 8, 1997, Volume 94(14); **Scientific Skills Exercise** Data from Shell Dropping: Decision-Making and Optimal Foraging in Northwestern Crows by Reto Zach, from *Behaviour*, 1979, Volume 68(1–2); 51; **51.22** Reprinted by permission from Klaudia Witte; **51.28 illustration** Adaptations of photograph by Jonathan Blair, Figure/PhotoID: 3.14, as appeared in *Animal Behavior: An Evolutionary Approach*, 8th Edition, Editor: John Alcock, p. 88. Reprinted by permission; **51.28 map** Data from "Rapid Microevolution of Migratory Behaviour in a Wild Bird Species" by P. Berthold et al., from *Nature*, December 1992, Volume 360(6405); **art for Concept 51.2 Summary**: Data from *The Study of Instinct*, N. Tinbergen, Clarendon Press, Oxford (1951).

Chapter 52 **52.20** Based on data from the Department of Environment and Heritage (2005) © Commonwealth of Australia; **52.21** Based on the data from W.J. Fletcher, Interactions among Subtidal Australian Sea Urchins, Gastropods and Algae: Effects of Experimental Removals, *Ecological Monographs* 57:89–109 (1987); **52.23** Based on S. D. Ling et al. Climate-Driven Range Extension of a Sea Urchin: Inferring Future Trends by Analysis of Recent Population Dynamics, *Global Change Biology* (2009) 15, 719–731, doi: 10.1111/j.1365-2486.2008.01734.x; **Scientific Skills Exercise** Based on the data from C. M. Crain et al., Physical and Biotic Drivers of Plant Distribution Across Estuarine Salinity Gradients, *Ecology* 85:2539–2549 (2004); **52.30** Based on Rana W. El-Sabaawi et al, Assessing the Effects of Guppy Life History Evolution on Nutrient Recycling: From Experiments to the Field, Freshwater Biology (2015) 60, 590–601, doi:10.1111/fwb.12507; **graph for Test Your Understanding Question 11** Based on the data from J. Clausen et al., *Experimental Studies on the Nature of Species. III. Environmental Responses of Climatic Races of Achillea,* Carnegie Institution of Washington Publication No. 581 (1948).

Chapter 53 **53.2** Data from A. M. Gormley et al., Capture-Recapture Estimates of Hector's Dolphin Abundance at Banks Peninsula, New Zealand, *Marine Mammal Science* 21:204–216 (2005); **Table 53.1** Data from P. W. Sherman and M. L. Morton, Demography of Belding's Ground Squirrel, *Ecology* 65:1617–1628 (1984); **53.4** Based on Demography of Belding's Ground Squirrels by Paul W. Sherman and Martin L. Morton, from *Ecology*, October 1984, Volume 65(5); **53.13** Original cartographic map by Joan Blaeu, 1659. Reproduced with detail in *Hollandia Nova detecta 1644; Terre Australe decouuerte l'an 1644* by Melchisedech Thevenot, 1672; **53.21** Data from Brood Size Manipulations in the Kestrel (*Falco tinnunculus*): Effects on Offspring and Parent Survival by C. Dijkstra et al., from *Journal of Animal Ecology*, 1990, Volume 59(1); **53.23** Based on Climate and Population Regulation: The Biogeographer's Dilemma by J. T. Enright, from *Oecologia*, 1976, Volume 24(4); **53.24** Based on the data from Predator Responses, Prey Refuges, and Density-Dependent Mortality of a Marine Fish by T.W. Anderson, *Ecology* 82(1):245–257 (2001); **53.26** Based on the data provided by Dr. Rolf O. Peterson; **53.29** Based on U.S. Census Bureau International Data Base; **53.30** Based on U.S. Census Bureau International Data Base; **53.31** Based on U.S. Census Bureau International Data Base; **53.32** Based on Ewing B., D. Moore, S. Goldfinger, A. Oursler, A. Reed, and M. Wackernagel. 2010. *The Ecological Footprint Atlas 2010*. Oakland: Global Footprint Network, p. 33 (www.footprintnetwork.org).

Chapter 54 **54.1, 54.2** Based on A. Stanley Rand and Ernest E. Williams. The Anoles of La Palma: Aspects of Their Ecological Relationships, *Breviora*, Volume 327: 1–19. Museum of Comparative Zoology, Harvard University; **54.3** Data from J. H. Connell, The Influence of Interspecific Competition and Other Factors on the Distribution of the Barnacle *Chthamalus stellatus*, *Ecology* 42:710–723 (1961); **Scientific Skills Exercise** Based on the data from B. L. Phillips and R. Shine, An Invasive Species Induces Rapid Adaptive Change in a Native Predator: Cane Toads and Black Snakes in Australia, *Proceedings of the Royal Society B* 273:1545–1550 (2006); **54.11** Based on the data from Sally D. Hacker and Mark D. Bertness, Experimental Evidence for Factors Maintaining Plant Species Diversity in a New England Salt Marsh. *Ecology*, September 1999, Volume 80(6); **54.14 graph** Data from N. Fierer and R. B. Jackson, The Diversity and Biogeography of Soil Bacterial Communities, *Proceedings of the National Academy of Sciences USA* 103:626–631 (2006); **54.17** Based on George A. Knox. Antarctic Marine Ecosystems, from *Antarctic Ecology*, Volume 1, edited by Martin W. Holdgate. Academic Press, 1970; **54.18** Adapted from Denise L. Breitburg et al., Varying Effects of Low Dissolved Oxygen on Trophic Interactions in an Estuarine Food Web. *Ecological Monographs*, November 1997, Volume 67(4). Used by permission of the Ecological Society of America; **54.19** Based on B. Jenkins et al., Productivity, Disturbance and Food Web Structure at a Local Spatial Scale in Experimental Container Habitats. *OIKOS*, November 1992, Volume 65(2); **54.20 graph** Data from R. T. Paine, Food web complexity and species diversity, *American Naturalist* 100:65–75 (1966); **54.24** Based on the data from C.R. Townsend, M.R. Scarsbrook, and S. Doledec, The Intermediate Disturbance Hypothesis, Refugia, and Biodiversity in Streams, *Limnology and Oceanography* 42:938–949 (1997); **54.25** Based on Robert L. Crocker and Jack Major. Soil Development in Relation to Vegetation and Surface Age at Glacier Bay, Alaska. *Journal of Ecology*, July 1955, Volume 43(2); **54.26** Adapted from F. Stuart Chapin et al., Mechanisms of Primary Succession Following Deglaciation at Glacier Bay. *Ecological Monographs*, May 1994, Volume 64(2). Ecological Society of America; **54.28** Adapted from D. J. Currie. Energy and Large-Scale Patterns of Animal-and Plant-Species Richness. *American Naturalist*, January 1991, Volume 137(1): 27–49; **54.29** Adapted from Robert H. MacArthur and Edward O. Wilson, An Equilibrium Theory of Insular Zoogeography. *Evolution*, December 1963, Volume 17(4). Society for the Study of Evolution; **54.32** Based on Daniel S. Simberloff and Edward O. Wilson. 1969. Experimental Zoogeography of Islands: The Colonization of Empty Islands. *Ecology*, Vol. 50, No. 2 (Mar., 1969), pp. 278–296.

Chapter 55 **55.4** Based on Figure 1.2 from Donald L. DeAngelis (1992), *Dynamics of Nutrient Cycling and Food Webs*. Taylor & Francis; **Table 55.1** Data from D. W. Menzel and J. H. Ryther, Nutrients Limiting the Production of Phytoplankton in the Sargasso Sea, with Special Reference to Iron, *Deep Sea Research* 7:276–281 (1961); **55.6** Data from Ryther, J. H., & Dunstan,. M. (January 01, 1971). Nitrogen, Phosphorus, and Eutrophication in the Coastal Marine Environment. Science, 171, 3975, 1008; **55.7** Based on Fig. 3c and 3d from Temperate Forest Health in an Era of Emerging Megadisturbance, Constance I. Millar and Nathan L. Stephenson, *Science* 349, 823 (2015); doi: 10.1126/science.aaa9933; **Scientific Skills Exercise** Data from J. M. Teal, Energy Flow in the Salt Marsh Ecosystem of Georgia, *Ecology* 43:614–624 (1962); **55.11a** Data from J. A. Trofymow and the CIDET Working Group, The Canadian Intersite Decomposition Experiment: Project and Site Establishment Report (Information Report BC-X-378), Natural Resources Canada, Canadian Forest Service, Pacific Forestry Centre (1998) and T. R. Moore et al., Litter decomposition rates in Canadian forests, *Global Change Biology* 5:75–82 (1999); **55.11b** Data from Trofymow, J. A., Pacific Forestry Centre., & CIDET Working Group (Canada). (1998). The Canadian Intersite Decomposition Experiment (CIDET): Project and site establishment report. Victoria, B.C: Pacific Forestry Centre; **55.13** Adapted from Figure 7.4 from Robert E. Ricklefs (2001), *The Economy of Nature*, 5th edition. W.H. Freeman and Company; **55.23b** Based on the data from Wei-Min Wu et al. (2006), Pilot-Scale in Situ Bioremediation of Uranium in a Highly Contaminated Aquifer. 2. Reduction of U(VI) and Geochemical Control of U(VI) Bioavailability. *Environmental Science Technology* 40 (12):3986–3995 (5/13/06); **art for Concept 55.1 Summary** Based on Figure 1.2 from Donald L. DeAngelis (1992). *Dynamics of Nutrient Cycling and Food Webs*. Taylor & Francis.

Chapter 56 **56.6** From Metcalfe, D and Bul, E. State of the Environment. Australian Government Department of the Environment and Energy, Canberra (2016). Adapted from Tulloch, et al. Understanding the importance of small patches of habitat for conservation, *Journal of Applied Ecology* 53:418-429, using data from the National Vegetation Information System (NVIS); **56.13** Based on data from Gene Likens; **56.15** Krebs, Charles J., *Ecology: The Experimental Analysis of Distribution and Abundance*, 5th Ed., © 2001. Reprinted and electronically reproduced by permission of Pearson Education, Inc., Upper Saddle River, New Jersey; **56.23** Adapted from Norman Myers et al. (2000). Biodiversity Hotspots for Conservation Priorities, *Nature*, February 24, 2000, Volume 403(6772); **56.24c** Based on Hamede, R, Owen, R, Siddle, H, et al. The ecology and evolution of wildlife cancers: Applications for management and conservation, *Evolutionary Applications* 13: 1719–1732 (2020), doi:10.1111/eva.12948; **56.34** Based on CO_2 data from www.esrl.noaa.gov/gmd/ccgg/trends. Temperature data from www.giss.nasa.gov/gistemps/graphs/Fig. A.lrg.gif; **Scientific Skills Exercise** Based on data from National Oceanic & Atmospheric Administration, Earth System Research Laboratory, Global Monitoring Division; **56.36 map** © The State of Queensland; **56.38** Based on the data from "History of the Ozone Hole," from NASA website, February 26, 2013; and "Antarctic Ozone," from British Antarctic Society website, June 7, 2013; **56.40** Based on the data from Instituto Nacional de Estadistica y Censos de Costa Rica and Centro Centroamericano de Poblacion, Universidad de Costa Rica.

Appendix A **Figure 5.11** Wallace/Sanders/Ferl, *Biology: The Science of Life*, 3rd Ed., © 1991. Reprinted and electronically reproduced by permission of Pearson Education, Inc., Upper Saddle River, New Jersey.

Glossary

Pronunciation Key

ā	ace
a/ah	ash
ch	chose
ē	meet
e/eh	bet
g	game
ī	ice
i	hit
ks	box
kw	quick
ng	song
ō	robe
o	ox
oy	boy
s	say
sh	shell
th	thin
ū	boot
u/uh	up
z	zoo

′ = primary accent

′ = secondary accent

5′ cap A modified form of guanine nucleotide added onto the 5′ end of a pre-mRNA molecule.

ABC hypothesis A model of flower formation identifying three classes of organ identity genes that direct formation of the four types of floral organs.

abiotic (ā′-bī-ot′-ik) Nonliving; referring to the physical and chemical properties of an environment.

abortion The termination of a pregnancy in progress.

abscisic acid (ABA) (ab-sis′-ik) A plant hormone that slows growth, often antagonising the actions of growth hormones. Two of its many effects are to promote seed dormancy and facilitate drought tolerance.

absorption The third stage of food processing in animals: the uptake of small nutrient molecules by an organism's body.

absorption spectrum The range of a pigment's ability to absorb various wavelengths of light; also a graph of such a range.

abyssal zone (uh-bis′-ul) The part of the ocean's benthic zone between 2,000 and 6,000 m deep.

acanthodian (ak′-an-thō′-dē-un) Any of a group of ancient jawed aquatic vertebrates from the Silurian and Devonian periods.

accessory fruit A fruit, or assemblage of fruits, in which the fleshy parts are derived largely or entirely from tissues other than the ovary.

acclimatisation (uh-klī′-muh-tī -zā′-shun) Physiological adjustment to a change in an environmental factor.

acetyl CoA Acetyl coenzyme A; the entry compound for the citric acid cycle in cellular respiration, formed from a two-carbon fragment of pyruvate attached to a coenzyme.

acetylcholine (as′-uh-til-kō′-lēn) One of the most common neurotransmitters; functions by binding to receptors and altering the permeability of the postsynaptic membrane to specific ions, either depolarising or hyperpolarising the membrane.

acid A substance that increases the hydrogen ion concentration of a solution.

acoelomate (uh-se-lo-mat) A solid-bodied animal lacking a cavity between the gut and outer body wall.

acquired immunodeficiency syndrome (AIDS) The symptoms and signs present during the late stages of HIV infection, defined by a specified reduction in the number of T cells and the appearance of characteristic secondary infections.

acrosomal reaction (ak′-ruh-sōm′-ul) The discharge of hydrolytic enzymes from the acrosome, a vesicle in the tip of a sperm, when the sperm approaches or contacts an egg.

acrosome (ak′-ruh-sōm) A vesicle in the tip of a sperm containing hydrolytic enzymes and other proteins that help the sperm reach the egg.

actin (ak′-tin) A globular protein that links into chains, two of which twist helically about each other, forming microfilaments (actin filaments) in muscle and other kinds of cells.

action potential An electrical signal that propagates (travels) along the membrane of a neuron or other excitable cell as a nongraded (all-or-none) depolarisation.

action spectrum A graph that profiles the relative effectiveness of different wavelengths of radiation in driving a particular process.

activation energy The amount of energy that reactants must absorb before a chemical reaction will start; also called free energy of activation.

activator A protein that binds to DNA and stimulates gene transcription. In prokaryotes, activators bind in or near the promoter; in eukaryotes, activators generally bind to control elements in enhancers.

active immunity Long-lasting immunity conferred by the action of B cells and T cells and the resulting B and T memory cells specific for a pathogen. Active immunity can develop as a result of natural infection or immunisation.

active site The specific region of an enzyme that binds the substrate and that forms the pocket in which catalysis occurs.

active transport The movement of a substance across a cell membrane against its concentration or electrochemical gradient, mediated by specific transport proteins and requiring an expenditure of energy.

adaptation Inherited characteristic of an organism that enhances its survival and reproduction in a specific environment.

adaptive evolution A process in which traits that enhance survival or reproduction tend to increase in frequency over time, resulting in a better match between organisms and their environment.

adaptive immunity A vertebrate-specific defence that is mediated by B lymphocytes (B cells) and T lymphocytes (T cells) and that exhibits specificity, memory, and self-nonself recognition; also called acquired immunity.

adaptive radiation Period of evolutionary change in which groups of organisms form many new species whose adaptations allow them to fill different ecological roles in their communities.

addition rule A rule of probability stating that the probability of any one of two or more mutually exclusive events occurring can be determined by adding their individual probabilities.

adenosine triphosphate *See* ATP (adenosine triphosphate).

adenylyl cyclase (uh-den′-uh-lil) An enzyme that converts ATP to cyclical AMP in response to an extracellular signal.

adhesion The clinging of one substance to another, such as water to plant cell walls, in this case by means of hydrogen bonds.

adipose tissue A connective tissue that insulates the body and serves as a fuel reserve; contains fat-storing cells called adipose cells.

adrenaline A catecholamine that, when secreted as a hormone by the adrenal medulla, mediates the "fight, flight, or freeze" response to short-term stresses; also released by some neurons as a neurotransmitter.

adrenal gland (uh-drē′-nul) One of two endocrine glands located adjacent to the kidneys in mammals. Endocrine cells in the outer portion (cortex) respond to adrenocorticotropic hormone (ACTH) by secreting steroid hormones that help maintain homeostasis during long-term stress. Neurosecretory cells in the central portion (medulla) secrete adrenaline and nonadrenaline in response to nerve signals triggered by short-term stress.

aerobic respiration A catabolic pathway for organic molecules, using oxygen (O_2) as the final electron acceptor in an electron transport chain and ultimately producing ATP. This is the most efficient catabolic pathway and is carried out in most eukaryotic cells and many prokaryotic organisms.

age structure The relative number of individuals of each age in a population.

aggregate fruit A fruit derived from a single flower that has more than one carpel.

AIDS (acquired immunodeficiency syndrome) The symptoms and signs present during the late stages of HIV infection, defined by a specified reduction in the number of T cells and the appearance of characteristic secondary infections.

alcohol fermentation Glycolysis followed by the reduction of pyruvate to ethyl alcohol, regenerating NAD^+ and releasing carbon dioxide.

alga (plural, **algae**) A general term for any species of photosynthetic protist, including both unicellular and multicellular forms. Algal species are included in three eukaryote supergroups (Excavata, SAR, and Archaeplastida).

alimentary canal (al′-uh-men′-tuh-rē) A complete digestive tract, consisting of a tube running between a mouth and an anus.

alkaline vent A deep-sea hydrothermal vent that releases water that is warm (40–90°C) rather than hot and that has a high pH (is basic). These vents consist of tiny pores lined with iron and other catalytic minerals that some scientists hypothesise might have been the location of the earliest abiotic synthesis of organic compounds.

allele (uh-lē′-ul) Any of the alternative versions of a gene that may produce distinguishable phenotypic effects.

allopatric speciation (al′-uh-pat′-rik) The formation of new species in populations that are geographically isolated from one another.

allopolyploid (al′-ō-pol′-ē-ployd) A fertile individual that has more than two chromosome sets as a result of two different species interbreeding and combining their chromosomes.

allosteric regulation The binding of a regulatory molecule to a protein at one site that affects the function of the protein at a different site.

alpha (α) helix (al′-fuh hē′-liks) A coiled region constituting one form of the secondary structure of proteins, arising from a specific pattern of hydrogen bonding between atoms of the polypeptide backbone (not the side chains).

alternation of generations A life cycle in which there is both a multicellular diploid form, the sporophyte, and a multicellular haploid form, the gametophyte; characteristic of plants and some algae.

alternative RNA splicing A type of eukaryotic gene regulation at the RNA-processing level in which different mRNA molecules are produced from the same primary transcript, depending on which RNA segments are treated as exons and which as introns.

altruism (al′-trū-iz-um) Selflessness; behaviour that reduces an individual's fitness while increasing the fitness of another individual.

alveolates (al-vē′-uh-lets) One of the three major subgroups for which the SAR eukaryotic supergroup is named. This clade arose by secondary endosymbiosis, and its members have membrane-enclosed sacs (alveoli) located just under the plasma membrane.

alveolus (al-vē′-uh-lus) (plural, **alveoli**) One of the dead-end air sacs where gas exchange occurs in a mammalian lung.

Alzheimer's disease (alts′-hī-merz) An age-related dementia (mental deterioration) characterised by confusion and memory loss.

amino acid (uh-mēn′-ō) An organic molecule possessing both a carboxyl and an amino group. Amino acids serve as the monomers of polypeptides.

amino group (uh-mēn′-ō) A chemical group consisting of a nitrogen atom bonded to two hydrogen atoms; can act as a base in solution, accepting a hydrogen ion and acquiring a charge of 1+.

aminoacyl-tRNA synthetase An enzyme that joins each amino acid to the appropriate tRNA.

ammonia A small, toxic molecule (NH_3) produced by nitrogen fixation or as a metabolic waste product of protein and nucleic acid metabolism.

ammonite A member of a group of shelled cephalopods that were important marine predators for hundreds of millions of years until their extinction at the end of the Cretaceous period (65.5 million years ago).

amniocentesis (am′-nē-ō-sen-tē′-sis) A technique associated with prenatal diagnosis in which amniotic fluid is obtained by aspiration from a needle inserted into the uterus. The fluid and the fetal cells it contains are analysed to detect certain genetic and congenital defects in the fetus.

amniote (am′-nē-ōt) A member of a clade of tetrapods named for a key derived character, the amniotic egg, which contains specialised membranes, including the fluid-filled amnion, that protect the embryo. Amniotes include mammals as well as birds and other reptiles.

amniotic egg An egg that contains specialised membranes that function in protection, nourishment, and gas exchange. The amniotic egg was a major evolutionary innovation, allowing embryos to develop on land in a fluid-filled sac, thus reducing the dependence of tetrapods on water for reproduction.

amoeba (uh-mē′-buh) A protist characterised by the presence of pseudopodia.

amoebocyte (uh-mē′-buh-sīt′) An amoeba-like cell that moves by pseudopodia and is found in most animals. Depending on the species, it may digest and distribute food, dispose of wastes, form skeletal fibres, fight infections, or change into other cell types.

amoebozoan (uh-mē′-buh-zō′-an) A protist in a clade that includes many species with lobe- or tube-shaped pseudopodia.

amphibian A member of the clade of tetrapods that includes salamanders, frogs, and caecilians.

amphipathic (am′-fē-path′-ik) Having both a hydrophilic region and a hydrophobic region.

amplification The strengthening of stimulus energy during transduction.

amygdala (uh-mig′-duh-luh) A structure in the temporal lobe of the vertebrate brain that has a major role in the processing of emotions.

amylase (am′-uh-lās′) An enzyme that hydrolyses starch (a glucose polymer from plants) and glycogen (a glucose polymer from animals) into smaller polysaccharides and the disaccharide maltose.

anabolic pathway (an′-uh-bol′-ik) A metabolic pathway that consumes energy to synthesise a complex molecule from simpler molecules.

anaerobic respiration (an-er-ō′-bik) A catabolic pathway in which inorganic molecules other than oxygen accept electrons at the "downhill" end of electron transport chains.

analogous Having characteristics that are similar because of convergent evolution, not homology.

analogy (an-al′-uh-jē) Similarity between two species that is due to convergent evolution rather than to descent from a common ancestor with the same trait.

anaphase The fourth stage of mitosis, in which the chromatids of each chromosome have separated and the daughter chromosomes are moving to the poles of the cell.

anatomy The structure of an organism.

anchorage dependence The requirement that a cell must be attached to a substratum in order to initiate cell division.

androgen (an′-drō-jen) Any steroid hormone, such as testosterone, that stimulates the development and maintenance of the male reproductive system and secondary sex characteristics.

aneuploidy (an′-yū-ploy′-dē) A chromosomal aberration in which one or more chromosomes are present in extra copies or are deficient in number.

angiosperm (an′-jē-ō-sperm) A flowering plant, which forms seeds inside a protective chamber called an ovary.

anhydrobiosis (an-hī′-drō-bī-ō′-sis) A dormant state involving loss of almost all body water.

animal pole The point at the end of an egg in the hemisphere where the least yolk is concentrated; opposite of vegetal pole.

anion (an′-ī-on) A negatively charged ion.

anterior Pertaining to the front, or head, of a bilaterally symmetrical animal.

anterior pituitary A portion of the pituitary gland that develops from nonneural tissue; consists of endocrine cells that synthesise and secrete several tropic and nontropic hormones.

anther In an angiosperm, the terminal pollen sac of a stamen, where pollen grains containing sperm-producing male gametophytes form.

antheridium (an-thuh-rid′-ē-um) (plural, **antheridia**) In plants, the male gametangium, a moist chamber in which gametes develop.

anthropoid (an′-thruh-poyd) A member of a primate group made up of the monkeys and the apes (gibbons, orangutans, gorillas, chimpanzees, bonobos, and humans).

antibody A protein secreted by plasma cells (differentiated B cells) that binds to a particular antigen; also called immunoglobulin. All antibodies have the same Y-shaped structure and in their monomer form consist of two identical heavy chains and two identical light chains.

anticodon (an′-tī-kō′-don) A nucleotide triplet at one end of a tRNA molecule that base-pairs with a particular complementary codon on an mRNA molecule.

antidiuretic hormone (ADH) (an′-tī-dī-yū-ret′-ik) A peptide hormone, also called vasopressin, that promotes water retention by the kidneys. Produced in the hypothalamus and released from the posterior pituitary, ADH also functions in the brain.

antigen (an′-ti-jen) A substance that elicits an immune response by binding to receptors of B or T cells.

antigen presentation (an′-ti-jen) The process by which an MHC molecule binds to a fragment of an intracellular protein antigen and carries it to the cell surface, where it is displayed and can be recognised by a T cell.

antigen-presenting cell (an′-ti-jen) A cell that upon ingesting pathogens or internalising pathogen proteins generates peptide fragments that are bound by class II MHC molecules and subsequently displayed on the cell surface to T cells. Macrophages, dendritic cells, and B cells are the primary antigen-presenting cells.

antigen receptor (an′-ti-jen) The general term for a surface protein, located on B cells and T cells, that binds to antigens, initiating adaptive immune responses. The antigen receptors on B cells are called B cell receptors, and the antigen receptors on T cells are called T cell receptors.

antiparallel Referring to the arrangement of the sugar-phosphate backbones in a DNA double helix (they run in opposite $5' \rightarrow 3'$ directions).

aphotic zone (ā′-fō′-tik) The part of an ocean or lake beneath the photic zone, where light does not penetrate sufficiently for photosynthesis to occur.

apical bud (ā′-pik-ul) A bud at the tip of a plant stem; also called a terminal bud.

apical dominance (ā′-pik-ul) Tendency for growth to be concentrated at the tip of a plant shoot because the apical bud partially inhibits axillary bud growth.

apical ectodermal ridge (AER) (ā′-pik-ul) A thickened area of ectoderm at the tip of a limb bud that promotes outgrowth of the limb bud.

apical meristem (ā′-pik-ul mār′-uh-stem) A localised region at a growing tip of a plant body where one or more cells divide repeatedly. The dividing cells of an apical meristem enable the plant to grow in length.

apicomplexan (ap′-ē-kom-pleks′-un) A group of alveolate protists, this clade includes many species that parasitise animals. Some apicomplexans cause human disease.

apomixis (ap′-uh-mik′-sis) The ability of some plant species to reproduce asexually through seeds without fertilisation by a male gamete.

apoplast (ap′-ō-plast) Everything external to the plasma membrane of a plant cell, including cell walls, intercellular spaces, and the space within dead structures such as xylem vessels and tracheids.

apoptosis (ā-puh-tō′-sus) A type of programmed cell death, which is brought about by activation of enzymes that break down many chemical components in the cell.

aposematic colouration (ap′-ō-si-mat′-ik) The bright warning colouration of many animals with effective physical or chemical defences.

appendix A small, finger-like extension of the vertebrate cecum; contains a mass of white blood cells that contribute to immunity.

aquaporin A channel protein in a cellular membrane that specifically facilitates osmosis, the diffusion of free water across the membrane.

aqueous solution (ā′-kwē-us) A solution in which water is the solvent.

arachnid A member of a subgroup of the major arthropod clade Chelicerata. Arachnids have six pairs of appendages, including four pairs of walking legs, and include spiders, scorpions, ticks, and mites.

arbuscular mycorrhiza (ar-bus′-kyū-lur mī′-kō-rī′-zuh) Association of a fungus with a plant root system in which the fungus causes the invagination of the host (plant) cells' plasma membranes; also called endomycorrhiza.

arbuscular mycorrhizal fungus (ar-bus′-kyū-lur) A symbiotic fungus whose hyphae grow through the cell wall of plant roots and extend into the root cell (enclosed in tubes formed by invagination of the root cell plasma membrane).

arbuscules Specialised branching hyphae that are found in some mutualistic fungi and exchange nutrients with living plant cells.

Archaea (ar′-kē′-uh) One of two prokaryotic domains, the other being Bacteria.

Archaeplastida (ar′-kē-plas′-tid-uh) One of four supergroups of eukaryotes proposed in a current hypothesis of the evolutionary history of eukaryotes. This monophyletic group, which includes red algae, green algae, and plants, descended from an ancient protistan ancestor that engulfed a cyanobacterium. *See also* Excavata, SAR, and Unikonta.

archegonium (ar-ki-gō′-nē-um) (plural, **archegonia**) In plants, the female gametangium, a moist chamber in which gametes develop.

archenteron (ar-ken′-tuh-ron) The endoderm-lined cavity, formed during gastrulation, that develops into the digestive tract of an animal.

archosaur (ar′-kō-sōr) A member of the reptilian group that includes crocodiles, alligators and dinosaurs, including birds.

arteriole (ar-ter′-ē-ōl) A vessel that conveys blood between an artery and a capillary bed.

artery A vessel that carries blood away from the heart to organs throughout the body.

arthropod A segmented ecdysozoan with a hard exoskeleton and jointed appendages. Familiar examples include insects, spiders, millipedes, and crabs.

artificial selection The selective breeding of domesticated plants and animals to encourage the occurrence of desirable traits.

ascocarp The fruiting body of a sac fungus (ascomycete).

ascomycete (as′-kuh-mī′-sēt) A member of the fungal phylum Ascomycota, commonly called sac fungus. The name comes from the saclike structure in which the spores develop.

ascus (plural, **asci**) A saclike spore capsule located at the tip of a dikaryotic hypha of a sac fungus.

asexual reproduction The generation of offspring from a single parent that occurs without the fusion of gametes. In most cases, the offspring are genetically identical to the parent.

A site One of a ribosome's three binding sites for tRNA during translation. The A site holds the tRNA carrying the next amino acid to be added to the polypeptide chain. (A stands for aminoacyl tRNA.)

assisted migration The translocation of a species to a favourable habitat beyond its native range for the purpose of protecting the species from human-caused threats.

associative learning The acquired ability to associate one environmental feature (such as a colour) with another (such as danger).

atherosclerosis A cardiovascular disease in which fatty deposits called plaques develop in the inner walls of the arteries, obstructing the arteries and causing them to harden.

atom The smallest unit of matter that retains the properties of an element.

atomic mass The total mass of an atom, numerically equivalent to the mass in grams of 1 mole of the atom. (For an element with more than one isotope, the atomic mass is the average mass of the naturally occurring isotopes, weighted by their abundance.)

atomic nucleus An atom's dense central core, containing protons and neutrons.

atomic number The number of protons in the nucleus of an atom, unique for each element and designated by a subscript.

ATP (adenosine triphosphate) (a-den′-ō-sēn trī-fos′-fāt) An adenine-containing nucleoside triphosphate that releases free energy when its phosphate bonds are hydrolysed. This energy is used to drive endergonic reactions in cells.

ATP synthase A complex of several membrane proteins that functions in chemiosmosis with adjacent electron transport chains, using the energy of a hydrogen ion (proton) concentration gradient to make ATP. ATP synthases are found in the inner mitochondrial membranes of eukaryotic cells and in the plasma membranes of prokaryotes.

atrial natriuretic peptide (ANP) (ā′-trē-ul na′-trē-yū-ret′-ik) A peptide hormone secreted by cells of the atria of the heart in response to high blood pressure. ANP's effects on the kidney alter ion and water movement and reduce blood pressure.

atrioventricular (AV) node A region of specialised heart muscle tissue between the left and right atria where electrical impulses are delayed for about 0.1 second before spreading to both ventricles and causing them to contract.

atrioventricular (AV) valve A heart valve located between each atrium and ventricle that prevents a backflow of blood when the ventricle contracts.

atrium (ā′-trē-um) (plural, **atria**) A chamber of the vertebrate heart that receives blood from the veins and transfers blood to a ventricle.

autocrine Referring to a secreted molecule that acts on the cell that secreted it.

autoimmune disease An immunological disorder in which the immune system turns against self.

autonomic nervous system (ot′-ō-nom′-ik) An efferent branch of the vertebrate peripheral nervous system that regulates the internal environment; consists of the sympathetic and parasympathetic divisions and the enteric nervous system.

autopolyploid (ot′-ō-pol′-ē-ployd) An individual that has more than two chromosome sets that are all derived from a single species.

autosome (ot′-ō-sōm) A chromosome that is not directly involved in determining sex; not a sex chromosome.

autotroph (ot′-ō-trōf) An organism that obtains organic food molecules without eating other organisms or substances derived from other organisms. Autotrophs use energy from the sun or from oxidation of inorganic substances to make organic molecules from inorganic ones.

auxin (ok′-sin) A term that primarily refers to indoleacetic acid (IAA), a natural plant hormone that has a variety of effects, including cell elongation, root formation, secondary growth, and fruit growth.

axillary bud (ak′-sil-ār-ē) A structure that has the potential to form a lateral shoot, or branch. The bud appears in the angle formed between a leaf and a stem.

axon (ak′-son) A typically long extension, or process, of a neuron that carries nerve impulses away from the cell body towards target cells.

B cells The lymphocytes that complete their development in the bone marrow and become effector cells for the humoral immune response.

Bacteria One of two prokaryotic domains, the other being Archaea.

bacteriophage (bak-tēr′-ē-ō-fāj) A virus that infects bacteria; also called a phage.

bacteroid A form of the bacterium *Rhizobium* contained within the vesicles formed by the root cells of a root nodule.

balancing selection Natural selection that maintains two or more phenotypic forms in a population.

bar graph A graph in which the independent variable represents groups or nonnumerical categories and the values of the dependent variable(s) are shown by bars.

bark All tissues external to the vascular cambium, consisting mainly of the secondary phloem and layers of periderm.

Barr body A dense object lying along the inside of the nuclear envelope in cells of female mammals, representing a highly condensed, inactivated X chromosome.

basal angiosperm A member of one of three clades of early-diverging lineages of extant flowering plants. Examples are *Amborella*, water lilies, and star anise and its relatives.

basal body (bā′-sul) A eukaryotic cell structure consisting of a "9 + 0" arrangement of microtubule triplets. The basal body may organise the microtubule assembly of a cilium or flagellum and is structurally very similar to a centriole.

basal metabolic rate (BMR) The metabolic rate of a resting, fasting, and nonstressed endotherm at a comfortable temperature.

basal taxon In a specified group of organisms, a taxon whose evolutionary lineage diverged early in the history of the group.

base A substance that reduces the hydrogen ion concentration of a solution.

basidiocarp Elaborate fruiting body of a dikaryotic mycelium of a club fungus.

basidiomycete (buh-sid′-ē-ō-mī′-sēt) A member of the fungal phylum Basidiomycota, commonly called club fungus. The name comes from the club-like shape of the basidium.

basidium (plural, **basidia**) (buh-sid′-ē-um, buh-sid′-ē-ah) A reproductive appendage that produces sexual spores on the gills of mushrooms (club fungi).

Batesian mimicry (bāt′-zē-un mim′-uh-krē) A type of mimicry in which a harmless species resembles an unpalatable or harmful species to which it is not closely related.

behaviour Individually, an action carried out by muscles or glands under control of the nervous system in response to a stimulus; collectively, the sum of an animal's responses to external and internal stimuli.

behavioural ecology The study of behavioural interactions between individuals within populations and communities, usually in an evolutionary context.

benign tumour A mass of abnormal cells with specific genetic and cellular changes such that the cells are not capable of surviving at a new site and generally remain at the site of the tumour's origin.

benthic zone The bottom surface of an aquatic environment.

benthos (ben′-thōz) The communities of organisms living in the benthic zone of an aquatic biome.

beta oxidation A metabolic sequence that breaks fatty acids down to two-carbon fragments that enter the citric acid cycle as acetyl CoA.

beta (β) pleated sheet One form of the secondary structure of proteins in which the polypeptide chain folds back and forth. Two regions of the chain lie parallel to each other and are held together by hydrogen bonds between atoms of the polypeptide backbone (not the side chains).

bilateral symmetry Body symmetry in which a central longitudinal plane divides the body into two equal but opposite halves.

bilaterian (bī′-luh-ter′-ē-uhn) A member of a clade of animals with bilateral symmetry and three germ layers.

bile A mixture of substances that is produced in the liver and stored in the gallbladder; enables formation of fat droplets in water as an aid in the digestion and absorption of fats.

binary fission A method of asexual reproduction in single-celled organisms in which the cell grows to roughly double its size and then divides into two cells. In prokaryotes, binary fission does not involve mitosis, but in single-celled eukaryotes that undergo binary fission, mitosis is part of the process.

binomial A common term for the two-part, latinised format for naming a species, consisting of the genus and specific epithet; also called a binomen.

biodiversity hot spot A relatively small area with numerous endemic species and a large number of endangered and threatened species.

bioenergetics (1) The overall flow and transformation of energy in an organism. (2) The study of how energy flows through organisms.

biofilm A surface-coating colony of one or more species of unicellular organisms that engage in metabolic cooperation; most known biofilms are formed by prokaryotes.

biofuel A fuel produced from biomass.

biogeochemical cycle Any of the various chemical cycles, which involve both biotic and abiotic components of ecosystems.

biogeography The scientific study of the past and present geographical distributions of species.

bioinformatics The use of computers, software, and mathematical models to process and integrate biological information from large data sets.

biological augmentation An approach to restoration ecology that uses organisms to add essential materials to a degraded ecosystem.

biological clock An internal timekeeper that controls an organism's biological rhythms. The biological clock marks time with or without environmental cues but often requires signals from the environment to remain tuned to an appropriate period. *See also* circadian rhythm.

biological magnification A process in which retained substances become more concentrated at each higher trophic level in a food chain.

biological species concept Definition of a species as a group of populations whose members have the potential to interbreed in nature and produce viable, fertile offspring but do not produce viable, fertile offspring with members of other such groups.

biology The scientific study of life.

biomass The total mass of organic matter comprising a group of organisms in a particular habitat.

biome (bī′-ōm) Any of the world's major ecosystem types, often classified according to the predominant vegetation for terrestrial biomes and the physical environment for aquatic biomes and characterised by adaptations of organisms to that particular environment.

bioremediation The use of organisms to detoxify and restore polluted and degraded ecosystems.

biosphere The entire portion of Earth inhabited by life; the sum of all the planet's ecosystems.

biotechnology The manipulation of organisms or their components to produce useful products.

biotic (bī-ot′-ik) Pertaining to the living factors—the organisms—in an environment.

bipolar disorder A depressive mental illness characterised by swings of mood from high to low; also called manic-depressive disorder.

birth control pill A hormonal contraceptive that inhibits ovulation, retards follicular development, or alters a woman's cervical mucus to prevent sperm from entering the uterus.

blade (1) A leaflike structure of a seaweed that provides most of the surface area for photosynthesis. (2) The flattened portion of a typical leaf.

blastocoel (blas′-tuh-sēl) The fluid-filled cavity that forms in the centre of a blastula.

blastocyst (blas′-tuh-sist) The blastula stage of mammalian embryonic development, consisting of an inner cell mass, a cavity, and an outer layer, the trophoblast. In humans, the blastocyst forms 1 week after fertilisation.

blastomere An early embryonic cell arising during the cleavage stage of an early embryo.

blastopore (blas'-tō-pōr) In a gastrula, the opening of the archenteron that typically develops into the anus in deuterostomes and the mouth in protostomes.

blastula (blas'-tyū-luh) A hollow ball of cells that marks the end of the cleavage stage during early embryonic development in animals.

blood A connective tissue with a fluid matrix called plasma in which red blood cells, white blood cells, and cell fragments called platelets are suspended.

blue-light photoreceptor Any of several classes of light-absorbing molecules that have physiological effects when activated by blue light.

body cavity A fluid- or air-filled space between the digestive tract and the body wall.

body plan In multicellular eukaryotes, a set of morphological and developmental traits that are integrated into a functional whole—the living organism.

Bohr shift A lowering of the affinity of haemoglobin for oxygen, caused by a drop in pH. It facilitates the release of oxygen from haemoglobin in the vicinity of active tissues.

bolus A lubricated ball of chewed food.

bone A connective tissue consisting of living cells held in a rigid matrix of collagen fibres embedded in calcium salts.

book lung An organ of gas exchange in spiders, consisting of stacked plates contained in an internal chamber.

bottleneck effect Genetic drift that occurs when the size of a population is reduced, as by a natural disaster or human actions. Typically, the surviving population is no longer genetically representative of the original population.

bottom-up control A situation in which the abundance of organisms at each trophic level is limited by nutrient supply or the availability of food at lower trophic levels; thus, the supply of nutrients controls plant numbers, which control herbivore numbers, which in turn control predator numbers.

Bowman's capsule (bō'-munz) A cup-shaped receptacle in the vertebrate kidney that is the initial, expanded segment of the nephron, where filtrate enters from the blood.

brachiopod (bra'-kē-uh-pod') A marine lophotrochozoan with a shell divided into dorsal and ventral halves; also called lamp shells.

brain Organ of the central nervous system where information is processed and integrated.

brainstem A collection of structures in the vertebrate brain, including the midbrain, the pons, and the medulla oblongata; functions in homeostasis, coordination of movement, and conduction of information to higher brain centres.

branch point The representation on a phylogenetic tree of the divergence of two or more taxa from a common ancestor. A branch point is usually shown as a dichotomy in which a branch representing the ancestral lineage splits (at the branch point) into two branches, one for each of the two descendant lineages.

brassinosteroid A steroid hormone in plants that has a variety of effects, including inducing cell elongation, retarding leaf abscission, and promoting xylem differentiation.

breathing Ventilation of the lungs through alternating inhalation and exhalation.

bronchiole (brong'-kē-ōl') A fine branch of the bronchi that transports air to alveoli.

bronchus (brong'-kus) (plural, **bronchi**) One of a pair of breathing tubes that branch from the trachea into the lungs.

brown alga A multicellular, photosynthetic protist with a characteristic brown or olive colour that results from carotenoids in its plastids. Most brown algae are marine, and some have a plantlike body.

bryophyte (brī'-uh-fīt) An informal name for a moss, liverwort, or hornwort; a nonvascular plant that lives on land but lacks some of the terrestrial adaptations of vascular plants.

buffer A solution that contains a weak acid and its corresponding base. A buffer minimises changes in pH when acids or bases are added to the solution.

bulk feeder An animal that eats relatively large pieces of food.

bulk flow The movement of a fluid due to a difference in pressure between two locations.

bundle-sheath cell In C_4 plants, a type of photosynthetic cell arranged into tightly packed sheaths around the veins of a leaf.

C_3 plant A plant that uses the Calvin cycle for the initial steps that incorporate CO_2 into organic material, forming a three-carbon compound as the first stable intermediate.

C_4 plant A plant in which the Calvin cycle is preceded by reactions that incorporate CO_2 into a four-carbon compound, the end product of which supplies CO_2 for the Calvin cycle.

caecum (sē'-kum) (plural, **caeca**) The blind pouch forming one branch of the large intestine.

calcitonin (kal'-si-tō'-nin) A hormone secreted by the thyroid gland that lowers the blood calcium level by promoting calcium deposition in bone and calcium excretion from the kidneys; nonessential in adult humans.

callus A mass of dividing, undifferentiated cells growing at the site of a wound or in culture.

calorie (cal) The amount of heat energy required to raise the temperature of 1 g of water by 1°C; also the amount of heat energy that 1 g of water releases when it cools by 1°C. The Calorie (with a capital C), usually used to indicate the energy content of food, is a kilocalorie.

Calvin cycle The second of two major stages in photosynthesis (following the light reactions), involving fixation of atmospheric CO_2 and reduction of the fixed carbon into carbohydrate.

Cambrian explosion A relatively brief time in geological history when many present-day phyla of animals first appeared in the fossil record. This burst of evolutionary change occurred about 535–525 million years ago and saw the emergence of the first large, hard-bodied animals.

cAMP (cyclic AMP) Cyclical adenosine monophosphate, named because of its ring structure, is a common chemical signal that has a diversity of roles, including as a second messenger in many eukaryotic cells, and as a regulator of some bacterial operons.

CAM plant A plant that uses crassulacean acid metabolism, an adaptation for photosynthesis in arid conditions. In this process, CO_2 entering open stomata during the night is converted to organic acids, which release CO_2 for the Calvin cycle during the day, when stomata are closed.

canopy The uppermost layer of vegetation in a terrestrial biome.

capillary (kap'-il-ār'-ē) A microscopic blood vessel that penetrates the tissues and consists of a single layer of endothelial cells that allows exchange between the blood and interstitial fluid.

capillary bed (kap'-il-ār'-ē) A network of capillaries in a tissue or organ.

capsid The protein shell that encloses a viral genome. It may be rod-shaped, polyhedral, or more complex in shape.

capsule (1) In many prokaryotes, a dense and well-defined layer of polysaccharide or protein that surrounds the cell wall and is sticky, protecting the cell and enabling it to adhere to substrates or other cells. (2) The sporangium of a bryophyte (moss, liverwort, or hornwort).

carbohydrate (kar'-bō-hī'-drāt) A sugar (monosaccharide) or one of its dimers (disaccharides) or polymers (polysaccharides).

carbon fixation The initial incorporation of carbon from CO_2 into an organic compound by an autotrophic organism (a plant, another photosynthetic organism, or a chemoautotrophic prokaryote).

carbonyl group (kar'-buh-nil) A chemical group present in aldehydes and ketones and consisting of a carbon atom double-bonded to an oxygen atom.

carboxyl group (kar-bok'-sil) A chemical group present in organic acids and consisting of a single carbon atom double-bonded to an oxygen atom and also bonded to a hydroxyl group.

cardiac cycle (kar'-dē-ak) The alternating contractions and relaxations of the heart.

cardiac muscle (kar'-dē-ak) A type of striated muscle that forms the contractile wall of the heart. Its cells are joined by intercalated discs that relay the electrical signals underlying each heartbeat.

cardiac output (kar'-dē-ak) The volume of blood pumped per minute by each ventricle of the heart.

cardiovascular system A closed circulatory system with a heart and branching network of arteries, capillaries, and veins. The system is characteristic of vertebrates.

carnivore An organism that consumes animals for nutrition.

carotenoid (kuh-rot'-uh-noyd') An accessory pigment, either yellow or orange, in the chloroplasts of plants and in some prokaryotes. By absorbing wavelengths of light that chlorophyll cannot, carotenoids broaden the spectrum of colours that can drive photosynthesis.

carpel (kar'-pul) The ovule-producing reproductive organ of a flower, consisting of the stigma, style, and ovary.

carrier In genetics, an individual who is heterozygous at a given genetic locus for a recessively inherited disorder. The heterozygote is generally phenotypically normal for the disorder but can pass on the recessive allele to offspring.

carrying capacity The maximum population size that can be supported by the available resources, symbolised as *K*.

cartilage (kar′-til-ij) A flexible connective tissue with an abundance of collagenous fibres embedded in chondroitin sulfate.

Casparian strip (ka-spār′-ē-un) A water-impermeable ring of wax in the endodermal cells of plants that blocks the passive flow of water and solutes into the stele by way of cell walls.

catabolic pathway (kat′-uh-bol′-ik) A metabolic pathway that releases energy by breaking down complex molecules to simpler molecules.

catalysis (kuh-ta′-luh-sis) A process by which a chemical agent called a catalyst selectively increases the rate of a reaction without being consumed by the reaction.

catalyst (kat′-uh-list) A chemical agent that selectively increases the rate of a reaction without being consumed by the reaction.

cation (cat′-ī′-on) A positively charged ion.

cation exchange (cat′-ī′-on) A process in which positively charged minerals are made available to a plant when hydrogen ions in the soil displace mineral ions from the clay particles.

cell Life's fundamental unit of structure and function; the smallest unit of organisation that can perform all activities required for life.

cell body The part of a neuron that houses the nucleus and most other organelles.

cell cycle An ordered sequence of events in the life of a cell, from its origin in the division of a parent cell until its own division into two. The eukaryotic cell cycle is composed of interphase (including G_1, S, and G_2 phases) and M phase (including mitosis and cytokinesis).

cell cycle control system A cyclically operating set of molecules in the eukaryotic cell that both triggers and coordinates key events in the cell cycle.

cell division The reproduction of cells.

cell fractionation The disruption of a cell and separation of its parts by centrifugation at successively higher speeds.

cell-mediated immune response The branch of adaptive immunity that involves the activation of cytotoxic T cells, which defend against infected cells.

cell plate A membrane-bounded, flattened sac located at the midline of a dividing plant cell, inside which the new cell wall forms during cytokinesis.

cellular respiration The catabolic pathways of aerobic and anaerobic respiration, which break down organic molecules and use an electron transport chain for the production of ATP.

cellulose (sel′-yū-lōs) A structural polysaccharide of plant cell walls, consisting of glucose monomers joined by β glycosidic linkages.

cell wall A protective layer external to the plasma membrane in the cells of plants, prokaryotes, fungi, and some protists. Polysaccharides such as cellulose (in plants and some protists), chitin (in fungi), and peptidoglycan (in bacteria) are important structural components of cell walls.

central nervous system (CNS) The portion of the nervous system where signal integration occurs; in vertebrate animals, the brain and spinal cord.

central vacuole In a mature plant cell, a large membranous sac with diverse roles in growth, storage, and sequestration of toxic substances.

centriole (sen′-trē-ōl) A structure in the centrosome of an animal cell composed of a cylinder of microtubule triplets arranged in a "9 + 0" pattern. A centrosome has a pair of centrioles.

centromere (sen′-trō-mēr) In a duplicated chromosome, the region on each sister chromatid where it is most closely attached to its sister chromatid by proteins that bind to the centromeric DNA. Other proteins condense the chromatin in that region, so it appears as a narrow "waist" on the duplicated chromosome. (An unduplicated chromosome has a single centromere, identified by the proteins bound there.)

centrosome (sen′-trō-sōm) A structure present in the cytoplasm of animal cells that functions as a microtubule-organising centre and is important during cell division. A centrosome has two centrioles.

cercozoan An amoeboid or flagellated protist that feeds with threadlike pseudopodia.

cerebellum (sār′-ruh-bel′-um) Part of the vertebrate hindbrain located dorsally; functions in unconscious coordination of movement and balance.

cerebral cortex (suh-rē′-brul) The surface of the cerebrum; the largest and most complex part of the mammalian brain, containing nerve cell bodies of the cerebrum; the part of the vertebrate brain most changed through evolution.

cerebrum (suh-rē′-brum) The dorsal portion of the vertebrate forebrain, composed of right and left hemispheres; the integrating centre for memory, learning, emotions, and other highly complex functions of the central nervous system.

cervix (ser′-viks) The neck of the uterus, which opens into the vagina.

chaperonin (shap′-er-ō′-nin) A protein complex that assists in the proper folding of other proteins.

character An observable heritable feature that may vary among individuals.

character displacement The tendency for characteristics to be more divergent in sympatric populations of two species than in allopatric populations of the same two species.

checkpoint A control point in the cell cycle where stop and go-ahead signals can regulate the cycle.

chelicera (kē-lih′-suh-ruh) (plural, **chelicerae**) One of a pair of clawlike feeding appendages characteristic of chelicerates.

chelicerate (kē-lih-suh′-rāte) An arthropod that has chelicerae and a body divided into a cephalothorax and an abdomen. Living chelicerates include sea spiders, horseshoe crabs, scorpions, ticks, and spiders.

chemical bond An attraction between two atoms, resulting from a sharing of outer-shell electrons or the presence of opposite charges on the atoms. The bonded atoms gain complete outer electron shells.

chemical energy Energy available in molecules for release in a chemical reaction; a form of potential energy.

chemical equilibrium In a chemical reaction, the state in which the rate of the forward reaction equals the rate of the reverse reaction, so that the relative concentrations of the reactants and products do not change with time.

chemical reaction The making and breaking of chemical bonds, leading to changes in the composition of matter.

chemiosmosis (kem′-ē-oz-mō′-sis) An energy-coupling mechanism that uses energy stored in the form of a hydrogen ion gradient across a membrane to drive cellular work, such as the synthesis of ATP. Under aerobic conditions, most ATP synthesis in cells occurs by chemiosmosis.

chemoautotroph (kē′-mō-ot′-ō-trōf) An organism that obtains energy by oxidising inorganic substances and needs only carbon dioxide as a carbon source.

chemoheterotroph (kē′-mō-het′-er-ō-trōf) An organism that requires organic molecules for both energy and carbon.

chemoreceptor A sensory receptor that responds to a chemical stimulus, such as a solute or an odourant.

chiasma (plural, **chiasmata**) (kī-az′-muh, kī-az′-muh-tuh) The X-shaped, microscopically visible region where crossing over has occurred earlier in prophase I between homologous nonsister chromatids. Chiasmata become visible after synapsis ends, with the two homologues remaining associated due to sister chromatid cohesion.

chitin (kī′-tin) A structural polysaccharide, consisting of amino sugar monomers, found in many fungal cell walls and in the exoskeletons of all arthropods.

chlorophyll (klōr′-ō-fil) A green pigment located in membranes within the chloroplasts of plants and algae and in the membranes of certain prokaryotes. Chlorophyll *a* participates directly in the light reactions, which convert solar energy to chemical energy.

chlorophyll *a* (klōr′-ō-fil) A photosynthetic pigment that participates directly in the light reactions, which convert solar energy to chemical energy.

chlorophyll *b* (klōr′-ō-fil) An accessory photosynthetic pigment that transfers energy to chlorophyll *a*.

chloroplast (klōr′-ō-plast) An organelle found in plants and photosynthetic protists that absorbs sunlight and uses it to drive the synthesis of organic compounds from carbon dioxide and water.

choanocyte (kō-an′-uh-sīt) A flagellated feeding cell found in sponges. Also called a collar cell, it has a collar-like ring that traps food particles around the base of its flagellum.

cholesterol (kō-les′-tuh-rol) A steroid that forms an essential component of animal cell membranes and acts as a precursor molecule for the synthesis of other biologically important steroids, such as many hormones.

chondrichthyan (kon-drik′-thē-an) A member of the clade Chondrichthyes, vertebrates with skeletons made mostly of cartilage, such as sharks and rays.

chordate A member of the phylum Chordata, animals that at some point during their development have a notochord; a dorsal, hollow nerve cord; pharyngeal slits or clefts; and a muscular, post-anal tail.

chorionic villus sampling (CVS) (kōr′-ē-on′-ik vil′-us) A technique associated with prenatal diagnosis in which a small sample of the fetal portion of the placenta is removed for analysis to detect certain genetic and congenital defects in the fetus.

chromatin (krō′-muh-tin) The complex of DNA and proteins that makes up eukaryotic chromosomes. When the cell is not dividing, chromatin exists in its dispersed form, as a mass of very long, thin fibres that are not visible with a light microscope.

chromosome (krō′-muh-sōm) A cellular structure consisting of one DNA molecule and associated protein molecules. A duplicated chromosome has two DNA molecules. (In some contexts, such as genome sequencing, the term may refer to the DNA alone.) A eukaryotic cell typically has multiple, linear chromosomes, which are located in the nucleus. A prokaryotic cell often has a single, circular chromosome, which is found in the nucleoid, a region that is not enclosed by a membrane. *See also* chromatin.

chromosome theory of inheritance (krō′-muh-sōm) A basic principle in biology stating that genes are located at specific positions (loci) on chromosomes and that the behaviour of chromosomes during meiosis accounts for inheritance patterns.

chylomicron (kī′-lō-mī′-kron) A lipid transport globule composed of fats mixed with cholesterol and coated with proteins.

chyme (kīm) The mixture of partially digested food and digestive juices formed in the stomach.

chytrid (kī′-trid) A member of the fungal phylum Chytridiomycota, mostly aquatic fungi with flagellated zoospores that represent an early-diverging fungal lineage.

ciliate (sil′-ē-it) A type of protist that moves by means of cilia.

cilium (sil′-ē-um) (plural, **cilia**) A short appendage containing microtubules in eukaryotic cells. A motile cilium is specialised for locomotion or moving fluid past the cell; it is formed from a core of nine outer doublet microtubules and two inner single microtubules (the "9 + 2" arrangement) ensheathed in an extension of the plasma membrane. A primary cilium is usually nonmotile and plays a sensory and signalling role; it lacks the two inner microtubules (the "9 + 0" arrangement).

circadian rhythm (ser-kā′-dē-un) A physiological cycle of about 24 hours that persists even in the absence of external cues.

***cis-trans* isomer** One of several compounds that have the same molecular formula and covalent bonds between atoms but differ in the spatial arrangements of their atoms owing to the inflexibility of double bonds; also called a geometric isomer.

citric acid cycle A chemical cycle involving eight steps that completes the metabolic breakdown of glucose molecules begun in glycolysis by oxidising acetyl CoA (derived from pyruvate) to carbon dioxide; occurs within the mitochondrion in eukaryotic cells and in the cytosol of prokaryotes; together with pyruvate oxidation, the second major stage in cellular respiration.

clade (klād) A group of species that includes an ancestral species and all of its descendants. A clade is equivalent to a monophyletic group.

cladistics (kluh-dis′-tiks) An approach to systematics in which organisms are placed into groups called clades based primarily on common descent.

class In Linnaean classification, the taxonomic category above the level of order.

cleavage (1) The process of cytokinesis in animal cells, characterised by pinching of the plasma membrane. (2) The succession of rapid cell divisions without significant growth during early embryonic development that converts the zygote to a ball of cells.

cleavage furrow The first sign of cleavage in an animal cell; a shallow groove around the cell in the cell surface near the old metaphase plate.

climate The long-term prevailing weather conditions at a given place.

climate change A directional change in temperature, precipitation, or other aspect of the global climate that lasts for three decades or more.

climograph A plot of the temperature and precipitation in a particular region.

clitoris (klit′-uh-ris) An organ at the upper intersection of the labia minora that engorges with blood and becomes erect during sexual arousal.

cloaca (klō-ā′-kuh) A common opening for the digestive, urinary, and reproductive tracts found in many nonmammalian vertebrates but in few mammals.

clonal selection The process by which an antigen selectively binds to and activates only those lymphocytes bearing receptors specific for the antigen. The selected lymphocytes proliferate and differentiate into a clone of effector cells and a clone of memory cells specific for the stimulating antigen.

clone (1) A group of genetically identical individuals or cells. (2) In popular usage, an individual that is genetically identical to another individual. (3) As a verb, to make one or more genetic replicas of an individual or cell. *See also* gene cloning.

cloning vector In genetic engineering, a DNA molecule that can carry foreign DNA into a host cell and replicate there. Cloning vectors include plasmids and bacterial artificial chromosomes (BACs), which move recombinant DNA from a test tube back into a cell, and viruses that transfer recombinant DNA by infection.

closed circulatory system A circulatory system in which blood is confined to vessels and is kept separate from the interstitial fluid.

cnidocyte (nī′-duh-sīt) A specialised cell unique to the phylum Cnidaria; contains a capsule-like organelle housing a coiled thread that, when discharged, explodes outwards and functions in prey capture or defence.

cochlea (kok′-lē-uh) The complex, coiled organ of hearing that contains the organ of Corti.

coding strand Nontemplate strand of DNA, which has the same sequence as the mRNA except it has thymine (T) instead of uracil (U).

codominance The situation in which the phenotypes of both alleles are exhibited in the heterozygote because both alleles affect the phenotype in separate, distinguishable ways.

codon (kō′-don) A three-nucleotide sequence of DNA or mRNA that specifies a particular amino acid or termination signal; the basic unit of the genetic code.

coefficient of relatedness The fraction of genes that, on average, are shared by two individuals.

coelom (sē′-lōm) A body cavity lined by tissue derived only from mesoderm.

coelomate (sē-·lō-māt) An animal that possesses a true coelom.

coenocytic fungus (sē′-no-si′-tic) A fungus that lacks septa and hence whose body is made up of a continuous cytoplasmic mass that may contain hundreds or thousands of nuclei.

coenzyme (kō-en′-zīm) An organic molecule serving as a cofactor. Most vitamins function as coenzymes in metabolic reactions.

coevolution The joint evolution of two interacting species, each in response to selection imposed by the other.

cofactor Any nonprotein molecule or ion that is required for the proper functioning of an enzyme. Cofactors can be permanently bound to the active site or may bind loosely and reversibly, along with the substrate, during catalysis.

cognition The process of knowing that may include awareness, reasoning, recollection, and judgment.

cognitive map A neural representation of the abstract spatial relationships between objects in an animal's surroundings.

cohesion The linking together of like molecules, often by hydrogen bonds.

cohesion-tension hypothesis The leading explanation of the ascent of xylem sap. It states that transpiration exerts pull on xylem sap, putting the sap under negative pressure, or tension, and that the cohesion of water molecules transmits this pull along the entire length of the xylem from shoots to roots.

cohort A group of individuals of the same age in a population.

coleoptile (kō′-lē-op′-tul) The covering of the young shoot of the embryo of a grass seed.

coleorhiza (kō′-lē-uh-rī′-zuh) The covering of the young root of the embryo of a grass seed.

collagen A glycoprotein in the extracellular matrix of animal cells that forms strong fibres, found extensively in connective tissue and bone; the most abundant protein in the animal kingdom.

collecting duct The location in the kidney where processed filtrate, called urine, is collected from the renal tubules.

collenchyma cell (kō-len′-kim-uh) A flexible plant cell type that occurs in strands or cylinders that support young parts of the plant without restraining growth.

colon (kō′-len) The largest section of the vertebrate large intestine; functions in water absorption and formation of faeces.

commensalism (kuh-men′-suh-lizm) A +/0 ecological interaction that benefits the individuals of one species but neither harms nor helps the individuals of the other species.

communication (1) In behaviour, a process involving the transmission and reception of signals between organisms. (2) Transfer of information from one cell or molecule to another by means of chemical or physical signals.

community All the organisms that inhabit a particular area; an assemblage of populations of different species living close enough together for potential interaction.

community ecology The study of how interactions between species affect community structure and organisation.

community structure The number of species found in an ecological community, the particular species that are present, and the relative abundance of these species.

companion cell A type of plant cell that is connected to a sieve-tube element by many plasmodesmata and whose nucleus and ribosomes may serve one or more adjacent sieve-tube elements.

competition A −/− interaction that occurs when individuals of different species both use a resource that limits the survival and reproduction of each species.

competitive exclusion The concept that when populations of two similar species compete for the same limited resources, one population will use the resources more efficiently and have a reproductive advantage that will eventually lead to the elimination of the other population.

competitive inhibitor A substance that reduces the activity of an enzyme by entering the active site in place of the substrate, whose structure it mimics.

complement system A group of about 30 blood proteins that may amplify the inflammatory response, enhance phagocytosis, or directly lyse extracellular pathogens.

complementary DNA (cDNA) A double-stranded DNA molecule made *in vitro* using mRNA as a template and the enzymes reverse transcriptase and DNA polymerase. A cDNA molecule corresponds to the exons of a gene.

complete dominance The situation in which the phenotypes of the heterozygote and dominant homozygote are indistinguishable.

complete flower A flower that has all four basic floral organs: sepals, petals, stamens, and carpels.

complete metamorphosis The transformation of a larva into an adult that looks very different, and often functions very differently in its environment, than the larva.

compound A substance consisting of two or more different elements combined in a fixed ratio.

compound eye A type of multifaceted eye in insects and crustaceans consisting of up to several thousand light-detecting, focusing ommatidia.

concentration gradient A region along which the density of a chemical substance increases or decreases.

conception The fertilisation of an egg by a sperm in humans.

cone A cone-shaped cell in the retina of the vertebrate eye, sensitive to colour.

conformer An animal for which an internal condition conforms to (changes in accordance with) changes in an environmental variable.

conidium (plural, **conidia**) A haploid spore produced at the tip of a specialised hypha in ascomycetes during asexual reproduction.

conifer A member of the largest gymnosperm phylum. Most conifers are cone-bearing trees, such as pines and firs.

conjugation (kon′-jū-gā′-shun) (1) In prokaryotes, the direct transfer of DNA between two cells that are temporarily joined. When the two cells are members of different species, conjugation results in horizontal gene transfer. (2) In ciliates, a sexual process in which two cells exchange haploid micronuclei but do not reproduce.

connective tissue Animal tissue that functions mainly to bind and support other tissues, having a sparse population of cells scattered through an extracellular matrix.

conodont An early, soft-bodied vertebrate with prominent eyes and dental elements.

conservation biology The integrated study of ecology, evolutionary biology, physiology, molecular biology, and genetics to sustain biological diversity at all levels.

consumer An organism that feeds on producers, other consumers, or nonliving organic material.

contraception The deliberate prevention of pregnancy.

contractile vacuole A membranous sac that helps move excess water out of certain freshwater protists.

control element A segment of noncoding DNA that helps regulate transcription of a gene by serving as a binding site for a transcription factor. Multiple control elements are present in a eukaryotic gene's enhancer.

control group In a controlled experiment, a set of subjects that lacks (or does not receive) the specific factor being tested. Ideally, the control group is identical to the experimental group in other respects.

controlled experiment An experiment designed to compare an experimental group with a control group; ideally, the two groups differ only in the factor being tested.

convergent evolution The evolution of similar features in independent evolutionary lineages.

convergent extension A process in which the cells of a tissue layer rearrange themselves in such a way that the sheet of cells becomes narrower (converges) and longer (extends).

cooperativity A kind of allosteric regulation whereby a shape change in one subunit of a protein caused by substrate binding is transmitted to all the other subunits, facilitating binding of additional substrate molecules to those subunits.

coral reef Typically a warm-water, tropical ecosystem dominated by the hard skeletal structures secreted primarily by corals. Some coral reefs also exist in cold, deep waters.

corepressor A small molecule that binds to a bacterial repressor protein and changes the protein's shape, allowing it to bind to the operator and switch an operon off.

cork cambium (kam′-bē-um) A cylinder of meristematic tissue in woody plants that replaces the epidermis with thicker, tougher cork cells.

corpus callosum (kor′-pus kuh-lō′-sum) The thick band of nerve fibres that connects the right and left cerebral hemispheres in mammals, enabling the hemispheres to process information together.

corpus luteum (kor′-pus lū′-tē-um) A secreting tissue in the ovary that forms from the collapsed follicle after ovulation and produces progesterone.

cortex (1) The outer region of cytoplasm in a eukaryotic cell, lying just under the plasma membrane, that has a more gel-like consistency than the inner regions due to the presence of multiple microfilaments. (2) In plants, ground tissue that is between the vascular tissue and dermal tissue in a root or eudicot stem.

cortical nephron In mammals and birds, a nephron with a loop of Henle located almost entirely in the renal cortex.

cotransport The coupling of the "downhill" diffusion of one substance to the "uphill" transport of another against its own concentration gradient.

cotyledon (kot′-uh-lē′-dun) A seed leaf of an angiosperm embryo. Some species have one cotyledon, others two.

countercurrent exchange The exchange of a substance or heat between two fluids flowing in opposite directions. For example, blood in a fish gill flows in the opposite direction of water passing over the gill, maximising diffusion of oxygen into and carbon dioxide out of the blood.

countercurrent multiplier system A countercurrent system in which energy is expended in active transport to facilitate exchange of materials and generate concentration gradients.

covalent bond (kō-vā′-lent) A type of strong chemical bond in which two atoms share one or more pairs of valence electrons.

crassulacean acid metabolism (CAM) (crass-yū-lā′-shen) An adaptation for photosynthesis in arid conditions, first discovered in the family Crassulaceae. In this process, a plant takes up CO_2 and incorporates it into a variety of organic acids at night; during the day, CO_2 is released from organic acids for use in the Calvin cycle.

CRISPR-Cas9 system A technique for editing genes in living cells, involving a bacterial protein called Cas9 associated with a guide RNA complementary to a gene sequence of interest.

crista (plural, **cristae**) (kris′-tuh, kris′-tē) An infolding of the inner membrane of a mitochondrion. The inner membrane houses electron transport chains and molecules of the enzyme catalysing the synthesis of ATP (ATP synthase).

critical load The amount of added nutrient, usually nitrogen or phosphorus, that can be absorbed by plants without damaging ecosystem integrity.

crop rotation The practice of growing different crops in succession on the same land chiefly to preserve the productive capacity of the soil.

cross-fostering study A behavioural study in which the young of one species are placed in the care of adults from another species.

crossing over The reciprocal exchange of genetic material between nonsister chromatids during prophase I of meiosis.

cross-pollination In angiosperms, the transfer of pollen from an anther of a flower on one plant to the stigma of a flower on another plant of the same species.

cryptic colouration Camouflage that makes a potential prey difficult to spot against its background.

cryptomycete A member of the fungal phylum Cryptomycota, unicellular fungi that have flagellated spores; cryptomycetes and their sister taxon (microsporidians) are a basal fungal lineage.

culture A system of information transfer through social learning or teaching that influences the behaviour of individuals in a population.

cuticle (kyū′-tuh-kul) Any of a variety of tough but flexible, non-mineral outer coverings of an organism, or parts of an organism, which provide protection.

cyclic AMP (cAMP) Cyclical adenosine monophosphate, named because of its ring structure, is a common chemical signal that has a diversity of roles, including as a second messenger in many eukaryotic cells, and as a regulator of some bacterial operons.

cyclic electron flow A route of electron flow during the light reactions of photosynthesis that involves only one photosystem and that produces ATP but not NADPH or O_2.

cyclin (sī′-klin) A cellular protein that occurs in a cyclically fluctuating concentration and that plays an important role in regulating the cell cycle.

cyclin-dependent kinase (Cdk) (sī′-klin) A protein kinase that is active only when attached to a particular cyclin.

cyclostome (sī′-cluh-stōm) Member of one of the two main clades of vertebrates; cyclostomes lack jaws and include lampreys and hagfishes. *See also* gnathostome.

cystic fibrosis (sis′-tik fī-brō′-sis) A human genetic disorder caused by a recessive allele for a chloride channel protein; characterised by an excessive secretion of mucus and consequent vulnerability to infection; fatal if untreated.

cytochrome (sī′-tō-krōm) An iron-containing protein that is a component of electron transport chains in the mitochondria and chloroplasts of eukaryotic cells and the plasma membranes of prokaryotic cells.

cytokine (sī′-tō-kīn′) Any of a group of small proteins secreted by a number of cell types, including macrophages and helper T cells, that regulate the function of other cells.

cytokinesis (sī′-tō-kuh-nē′-sis) The division of the cytoplasm to form two separate daughter cells immediately after mitosis, meiosis I, or meiosis II.

cytokinin (sī′-tō-kī′-nin) Any of a class of related plant hormones that retard ageing and act in concert with auxin to stimulate cell division, influence the pathway of differentiation, and control apical dominance.

cytoplasm (sī′-tō-plaz-um) The contents of the cell bounded by the plasma membrane; in eukaryotes, the portion exclusive of the nucleus.

cytoplasmic determinant A maternal substance, such as a protein or RNA, that when placed into an egg influences the course of early development by regulating the expression of genes that affect the developmental fate of cells.

cytoplasmic streaming A circular flow of cytoplasm, involving interactions of myosin and actin filaments, that speeds the distribution of materials within cells.

cytoskeleton A network of microtubules, microfilaments, and intermediate filaments that extend throughout the cytoplasm and serve a variety of mechanical, transport, and signalling functions.

cytosol (sī′-tō-sol) The semifluid portion of the cytoplasm.

cytotoxic T cell A type of lymphocyte that, when activated, kills infected cells as well as certain cancer cells and transplanted cells.

dalton A measure of mass for atoms and subatomic particles; the same as the atomic mass unit, or amu.

data Recorded observations.

day-neutral plant A plant in which flower formation is not controlled by photoperiod or day length.

decomposer An organism that absorbs nutrients from nonliving organic material such as corpses, fallen plant material, and the wastes of living organisms and converts them to inorganic forms; a detritivore.

deductive reasoning A type of logic in which specific results are predicted from a general premise.

de-etiolation The changes a plant shoot undergoes in response to sunlight; also known informally as greening.

dehydration reaction A chemical reaction in which two molecules become covalently bonded to each other with the removal of a water molecule.

deletion (1) A deficiency in a chromosome resulting from the loss of a fragment through breakage. (2) A mutational loss of one or more nucleotide pairs from a gene.

demographic transition In a stable population, a shift from high birth and death rates to low birth and death rates.

demography The study of changes over time in the vital statistics of populations, especially birth rates and death rates.

denaturation (dē-nā′-chur-ā′-shun) In proteins, a process in which a protein loses its native shape due to the disruption of weak chemical bonds and interactions, thereby becoming biologically inactive; in DNA, the separation of the two strands of the double helix. Denaturation occurs under extreme (noncellular) conditions of pH, salt concentration, or temperature.

dendrite (den′-drīt) One of usually numerous, short, highly branched extensions of a neuron that receive signals from other neurons.

dendritic cell An antigen-presenting cell, located mainly in lymphatic tissues and skin, that is particularly efficient in presenting antigens to helper T cells, thereby initiating a primary immune response.

density The number of individuals per unit area or volume.

density dependent Referring to any characteristic that varies with population density.

density-dependent inhibition The phenomenon observed in normal animal cells that causes them to stop dividing when they come into contact with one another.

density independent Referring to any characteristic that is not affected by population density.

deoxyribonucleic acid (DNA) (dē-ok′-sē-rī′-bō-nū-klā′-ik) A nucleic acid molecule, usually a double-stranded helix, in which each polynucleotide strand consists of nucleotide monomers with a deoxyribose sugar and the nitrogenous bases adenine (A), cytosine (C), guanine (G), and thymine (T); capable of being replicated and determining the inherited structure of a cell's proteins.

deoxyribose (dē-ok′-si-rī′-bōs) The sugar component of DNA nucleotides, having one fewer hydroxyl group than ribose, the sugar component of RNA nucleotides.

dependent variable A factor whose value is measured during an experiment or other test to see whether it is influenced by changes in another factor (the independent variable).

depolarisation A change in a cell's membrane potential such that the inside of the membrane is made less negative relative to the outside. For example, a neuron membrane is depolarised if a stimulus decreases its voltage from the resting potential of −70 mV in the direction of zero voltage.

dermal tissue The outer protective covering of plants.

desert A terrestrial biome characterised by very low precipitation.

desmosome A type of intercellular junction in animal cells that functions as a rivet, fastening cells together.

determinate cleavage A type of embryonic development in protostomes that rigidly casts the developmental fate of each embryonic cell very early.

determinate growth A type of growth characteristic of most animals and some plant organs, in which growth stops after a certain size is reached.

determination The progressive restriction of developmental potential in which the possible fate of each cell becomes more limited as an embryo develops. At the end of determination, a cell is committed to its fate.

detritus (di-trī′-tus) Dead organic matter.

deuteromycete (dū′-tuh-rō-mī′-sēt) Traditional classification for a fungus with no known sexual stage.

deuterostome development (dū′-tuh-rō-stōm′) In animals, a developmental mode distinguished by the development of the anus from the blastopore; often also characterised by radial cleavage and by the body cavity forming as outpockets of mesodermal tissue.

Deuterostomia (dū′-tuh-rō-stōm′-ē-uh) One of the three main lineages of bilaterian animals. *See also* Ecdysozoa and Lophotrochozoa.

development The events involved in an organism's changing gradually from a simple to a more complex or specialised form.

diabetes mellitus (dī′-uh-bē′-tis mel′-uh-tus) An endocrine disorder marked by an inability to maintain glucose homeostasis. The type 1 form results from autoimmune destruction of insulin-secreting cells; treatment usually requires daily insulin injections. The type 2 form most commonly results from reduced responsiveness of target cells to insulin; obesity and lack of exercise are risk factors.

diacylglycerol (DAG) (dī-a′-sil-glis′-er-ol) A second messenger produced by the cleavage of the phospholipid PIP_2 in the plasma membrane.

diaphragm (dī′-uh-fram′) (1) A sheet of muscle that forms the bottom wall of the thoracic cavity in mammals. Contraction of the diaphragm pulls air into the lungs. (2) A dome-shaped rubber cup fitted into the upper portion of the vagina before sexual intercourse. It serves as a physical barrier to the passage of sperm into the uterus.

diapsid (dī-ap′-sid) A member of an amniote clade distinguished by a pair of holes on each side of the skull. Diapsids include the lepidosaurs and archosaurs.

diastole (dī-as′-tō-lē) The stage of the cardiac cycle in which a heart chamber is relaxed and fills with blood.

diastolic pressure Blood pressure in the arteries when the ventricles are relaxed.

diatom Photosynthetic protist in the stramenopile clade; diatoms have a unique glass-like wall made of silicon dioxide embedded in an organic matrix.

dicot A term traditionally used to refer to flowering plants that have two embryonic seed leaves, or cotyledons. Recent molecular evidence indicates that dicots do not form a clade; species once classified as dicots are now grouped into eudicots, magnoliids, and several lineages of basal angiosperms.

differential gene expression The expression of different sets of genes by cells with the same genome.

differentiation The process by which a cell or group of cells becomes specialised in structure and function.

diffusion The random thermal motion of particles of liquids, gases, or solids. In the presence of a concentration or electrochemical gradient, diffusion results in the net movement of a substance from a region where it is more concentrated to a region where it is less concentrated.

digestion The second stage of food processing in animals: the breaking down of food into molecules small enough for the body to absorb.

dihybrid (dī′-hī′-brid) An organism that is heterozygous with respect to two genes of interest. All the offspring from a cross between parents doubly homozygous for different alleles are dihybrids. For example, parents of genotypes *AABB* and *aabb* produce a dihybrid of genotype *AaBb*.

dihybrid cross (dī′-hī′-brid) A cross between two organisms that are each heterozygous for both of the characters being followed (or the self-pollination of a plant that is heterozygous for both characters).

dikaryotic (dī′-kār-ē-ot′-ik) Referring to a fungal mycelium with two haploid nuclei per cell, one from each parent.

dinoflagellate (dī′-nō-flaj′-uh-let) A member of a group of mostly unicellular photosynthetic algae with two flagella situated in perpendicular grooves in cellulose plates covering the cell.

dinosaur A member of an extremely diverse clade of reptiles varying in body shape, size, and habitat. Birds are the only extant dinosaurs.

dioecious (dī-ē′-shus) In plant biology, having the male and female reproductive parts on different individuals of the same species.

diploblastic Having two germ layers.

diploid cell (dip′-loyd) A cell containing two sets of chromosomes (2*n*), one set inherited from each parent.

diplomonad A protist that has modified mitochondria, two equal-sized nuclei, and multiple flagella.

directional selection Natural selection in which individuals at one end of the phenotypic range survive or reproduce more successfully than do other individuals.

disaccharide (dī-sak′-uh-rid) A double sugar, consisting of two monosaccharides joined by a glycosidic linkage formed by a dehydration reaction.

dispersal The movement of individuals or gametes away from their parent location. This movement sometimes expands the geographical range of a population or species.

dispersion The pattern of spacing among individuals within the boundaries of a population.

disruptive selection Natural selection in which individuals on both extremes of a phenotypic range survive or reproduce more successfully than do individuals with intermediate phenotypes.

distal tubule In the vertebrate kidney, the portion of a nephron that helps refine filtrate and empties it into a collecting duct.

disturbance A natural or human-caused event that changes a biological community and usually removes organisms from it. Disturbances, such as fires and storms, play a pivotal role in structuring many communities.

disulfide bridge A strong covalent bond formed when the sulfur of one cysteine monomer bonds to the sulfur of another cysteine monomer.

DNA (deoxyribonucleic acid) (dē-ok′-sē-rī′-bō-nū-klā′-ik) A nucleic acid molecule, usually a double-stranded helix, in which each polynucleotide strand consists of nucleotide monomers with a deoxyribose sugar and the nitrogenous bases adenine (A), cytosine (C), guanine (G), and thymine (T); capable of being replicated and determining the inherited structure of a cell's proteins.

DNA cloning The production of multiple copies of a specific DNA segment.

DNA ligase (lī′-gās) A linking enzyme essential for DNA replication; catalyses the covalent bonding of the 3′ end of one DNA fragment (such as an Okazaki fragment) to the 5′ end of another DNA fragment (such as a growing DNA chain).

DNA methylation The presence of methyl groups on the DNA bases (usually cytosine) of plants, animals, and fungi. (The term also refers to the process of adding methyl groups to DNA bases.)

DNA micro-array assay A method to detect and measure the expression of thousands of genes at one time. Tiny amounts of a large number of single-stranded DNA fragments representing different genes are fixed to a glass slide and tested for hybridisation with samples of labelled cDNA.

DNA polymerase (puh-lim′-er-ās) An enzyme that catalyses the elongation of new DNA (for example, at a replication fork) by the addition of nucleotides to the 3′ end of an existing chain. There are several different DNA polymerases; DNA polymerase III and DNA polymerase I play major roles in DNA replication in *E. coli*.

DNA replication The process by which a DNA molecule is copied; also called DNA synthesis.

DNA sequencing Determining the complete nucleotide sequence of a gene or DNA segment.

DNA technology Techniques for sequencing and manipulating DNA.

domain (1) A taxonomic category above the kingdom level. The three domains are Archaea, Bacteria, and Eukarya. (2) A discrete structural and functional region of a protein.

dominant allele An allele that is fully expressed in the phenotype of a heterozygote.

dormancy A condition typified by extremely low metabolic rate and a suspension of growth and development.

dorsal In an animal with bilateral symmetry, pertaining to the top (in most animals) or back (in animals with upright posture) of the body.

dorsal lip The region above the blastopore on the dorsal side of the amphibian embryo.

double bond A double covalent bond; the sharing of two pairs of valence electrons by two atoms.

double circulation A circulatory system consisting of separate pulmonary and systemic circuits, in which blood passes through the heart after completing each circuit.

double fertilisation A mechanism of fertilisation in angiosperms in which two sperm cells unite with two cells in the female gametophyte (embryo sac) to form the zygote and endosperm.

double helix The form of native DNA, referring to its two adjacent antiparallel polynucleotide strands wound around an imaginary axis into a spiral shape.

Down syndrome A human genetic disease usually caused by the presence of an extra chromosome 21; characterised by developmental delays and heart and other defects that are generally treatable or non-life-threatening.

Duchenne's muscular dystrophy (duh-shens′) A human genetic disease caused by a sex-linked recessive allele; characterised by progressive weakening and a loss of muscle tissue.

duodenum (dū′-uh-dēn′-um) The first section of the small intestine, where chyme from the stomach mixes with digestive juices from the pancreas, liver, and gallbladder as well as from gland cells of the intestinal wall.

duplication An aberration in chromosome structure due to fusion with a fragment from a homologous chromosome, such that a portion of a chromosome is duplicated.

dynein (dī′-nē-un) In cilia and flagella, a large motor protein extending from one microtubule doublet to the adjacent doublet. ATP hydrolysis drives changes in dynein shape that lead to bending of cilia and flagella.

E site One of a ribosome's three binding sites for tRNA during translation. The E site is the place where discharged tRNAs leave the ribosome. (E stands for exit.)

Ecdysozoa (ek'-dē-sō-zō'-uh) One of the three main lineages of bilaterian animals; many ecdysozoans are moulting animals. *See also* Deuterostomia and Lophotrochozoa.

echinoderm (i-kī'-nō-derm) A slow-moving or sessile marine deuterostome with a water vascular system and, in larvae, bilateral symmetry. Echinoderms include sea stars, brittle stars, sea urchins, feather stars, and sea cucumbers.

ecological footprint The aggregate land and water area required by a person, city, or nation to produce all of the resources it consumes and to absorb all of the waste it generates.

ecological niche (nich) The sum of a species' use of the biotic and abiotic resources in its environment.

ecological species concept Definition of a species in terms of ecological niche, the sum of how members of the species interact with the nonliving and living parts of their environment.

ecological succession Transition in the species composition of a community following a disturbance; establishment of a community in an area virtually barren of life.

ecology The study of how organisms interact with each other and their environment.

ecosystem All the organisms in a given area as well as the abiotic factors with which they interact; one or more communities and the physical environment around them.

ecosystem ecology The study of energy flow and the cycling of chemicals among the various biotic and abiotic components in an ecosystem.

ecosystem engineer An organism that influences community structure by causing physical changes in the environment.

ecosystem service A function performed by an ecosystem that directly or indirectly benefits humans.

ecotone The transition from one type of habitat or ecosystem to another, such as the transition from a forest to a grassland.

ectoderm (ek'-tō-durm) The outermost of the three primary germ layers in animal embryos; gives rise to the outer covering and, in some phyla, the nervous system, inner ear, and lens of the eye.

ectomycorrhiza (plural, **ectomycorrhizae**) (ek'-tō-mī'-kō-rī'-zuh, ek'-tō-mī'-kō-rī'-zē) Association of a fungus with a plant root system in which the fungus surrounds the roots but does not cause invagination of the host (plant) cell's plasma membrane.

ectomycorrhizal fungus A symbiotic fungus that forms sheaths of hyphae over the surface of plant roots and also grows into extracellular spaces of the root cortex.

ectoparasite A parasite that feeds on the external surface of a host.

ectopic Occurring in an abnormal location.

ectoproct A sessile, colonial lophotrochozoan; also called a bryozoan.

ectothermic Referring to organisms for which external sources provide most of the heat for temperature regulation.

Ediacaran biota (ē'-dē-uh-keh'-run bī-ō'-tuh) An early group of macroscopic, mostly soft-bodied, multicellular eukaryotes known from fossils that range in age from 635 million to 541 million years old.

effective population size An estimate of the size of a population based on the numbers of females and males that successfully breed; generally smaller than the total population.

effector A pathogen-encoded protein that cripples the host's innate immune system.

effector cell (1) A muscle cell or gland cell that carries out the body's response to stimuli as directed by signals from the brain or other processing centre of the nervous system. (2) A lymphocyte that has undergone clonal selection and is capable of mediating an adaptive immune response.

egg The female gamete.

ejaculation The propulsion of sperm from the epididymis through the muscular vas deferens, ejaculatory duct, and urethra.

electrocardiogram (ECG or EKG) A record of the electrical impulses that travel through heart muscle during the cardiac cycle.

electrochemical gradient The diffusion gradient of an ion, which is affected by both the concentration difference of an ion across a membrane (a chemical force) and the ion's tendency to move relative to the membrane potential (an electrical force).

electrogenic pump An active transport protein that generates voltage across a membrane while pumping ions.

electromagnetic receptor A receptor of electromagnetic energy, such as visible light, electricity, or magnetism.

electromagnetic spectrum The entire spectrum of electromagnetic radiation, ranging in wavelength from less than a nanometre to more than a kilometre.

electron A subatomic particle with a single negative electrical charge and a mass about 1/2,000 that of a neutron or proton. One or more electrons move around the nucleus of an atom.

electron microscope (EM) A microscope that uses magnets to focus an electron beam on or through a specimen, resulting in a practical resolution that is 100-fold greater than that of a light microscope using standard techniques. A transmission electron microscope (TEM) is used to study the internal structure of thin sections of cells. A scanning electron microscope (SEM) is used to study the fine details of cell surfaces.

electron shell An energy level of electrons at a characteristic average distance from the nucleus of an atom.

electron transport chain A sequence of electron carrier molecules (membrane proteins) that shuttle electrons down a series of redox reactions that release energy used to make ATP.

electronegativity The attraction of a given atom for the electrons of a covalent bond.

electroporation A technique to introduce recombinant DNA into cells by applying a brief electrical pulse to a solution containing the cells. The pulse creates temporary holes in the cells' plasma membranes, through which DNA can enter.

element Any substance that cannot be broken down to any other substance by chemical reactions.

elimination The fourth and final stage of food processing in animals: the passing of undigested material out of the body.

embryo sac (em'-brē-ō) The female gametophyte of angiosperms, formed from the growth and division of the megaspore into a multicellular structure that typically has eight haploid nuclei.

embryonic lethal A mutation with a phenotype leading to death of an embryo or larva.

embryophyte Alternate name for land plants that refers to their shared derived trait of multicellular, dependent embryos.

emergent properties New properties that arise with each step upwards in the hierarchy of life, owing to the arrangement and interactions of parts as complexity increases.

emigration The movement of individuals out of a population.

enantiomer (en-an'-tē-ō-mer) One of two compounds that are mirror images of each other and that differ in shape due to the presence of an asymmetric carbon.

endangered species A species that is in danger of extinction throughout all or a significant portion of its range.

endemic (en-dem'-ik) Referring to a species that is confined to a specific geographical area.

endergonic reaction (en'-der-gon'-ik) A non-spontaneous chemical reaction in which free energy is absorbed from the surroundings.

endocrine gland (en'-dō-krin) A ductless gland that secretes hormones directly into the interstitial fluid, from which they diffuse into the bloodstream.

endocrine system (en'-dō-krin) In animals, the internal system of communication involving hormones, the ductless glands that secrete hormones, and the molecular receptors on or in target cells that respond to hormones; functions in concert with the nervous system to effect internal regulation and maintain homeostasis.

endocytosis (en'-dō-sī-tō'-sis) Cellular uptake of biological molecules and particulate matter via formation of vesicles from the plasma membrane.

endoderm (en'-dō-durm) The innermost of the three primary germ layers in animal embryos; lines the archenteron and gives rise to the liver, pancreas, lungs, and the lining of the digestive tract in species that have these structures.

endodermis In plant roots, the innermost layer of the cortex that surrounds the vascular cylinder.

endomembrane system The collection of membranes inside and surrounding a eukaryotic cell, related either through direct physical contact or by the transfer of membranous vesicles; includes the plasma membrane, the nuclear envelope, the smooth and rough endoplasmic reticulum, the Golgi apparatus, lysosomes, vesicles, and vacuoles.

endometriosis (en'-dō-mē-trē-ō'-sis) The condition resulting from the presence of endometrial tissue outside of the uterus.

endometrium (en'-dō-mē'-trē-um) The inner lining of the uterus, which is richly supplied with blood vessels.

endoparasite A parasite that lives within a host.

endophyte A harmless fungus, or occasionally another organism, that lives between cells of a plant part or multicellular alga.

endoplasmic reticulum (ER) (en′-dō-plaz′-mik ruh-tik′-yū-lum) An extensive membranous network in eukaryotic cells, continuous with the outer nuclear membrane and composed of ribosome-studded (rough) and ribosome-free (smooth) regions.

endorphin (en-dōr′-fin) Any of several hormones produced in the brain and anterior pituitary that inhibit pain perception.

endoskeleton A hard skeleton buried within the soft tissues of an animal.

endosperm In angiosperms, a nutrient-rich tissue formed by the union of a sperm with two polar nuclei during double fertilisation. The endosperm provides nourishment to the developing embryo in angiosperm seeds.

endospore A thick-coated, resistant cell produced by some bacterial cells when they are exposed to harsh conditions.

endosymbiosis A relationship between two species in which one organism lives inside the cell or cells of another organism. *See also* serial endosymbiont theory.

endothelium (en′-dō-thē′-lē-um) The simple squamous layer of cells lining the lumen of blood vessels.

endothermic Referring to organisms that are warmed by heat generated by their own metabolism. This heat usually maintains a relatively stable body temperature higher than that of the external environment.

endotoxin A toxic component of the outer membrane of certain gram-negative bacteria that is released only when the bacteria die.

energetic hypothesis The concept that the length of a food chain is limited by the inefficiency of energy transfer along the chain.

energy The capacity to cause change, especially to do work (to move matter against an opposing force).

energy coupling In cellular metabolism, the use of energy released from an exergonic reaction to drive an endergonic reaction.

enhancer A segment of eukaryotic DNA containing multiple control elements, usually located far from the gene whose transcription it regulates.

enteric nervous system Within the autonomic nervous system, a distinct network of neurons that exerts partially independent control over the digestive tract, pancreas, and gallbladder.

entropy A measure of molecular disorder, or randomness.

enzyme (en′-zīm) A macromolecule serving as a catalyst, a chemical agent that increases the rate of a reaction without being consumed by the reaction. Most enzymes are proteins.

enzyme-substrate complex (en′-zīm) A temporary complex formed when an enzyme binds to its substrate molecule(s).

eosinophil Immune system cell that secretes destructive enzymes and helps defend against multicellular pathogens.

epicotyl (ep′-uh-kot′-ul) In an angiosperm embryo, the embryonic axis above the point of attachment of the cotyledon(s) and below the first pair of miniature leaves.

epidemic A widespread outbreak of a disease.

epidermis (1) The dermal tissue of nonwoody plants, usually consisting of a single layer of tightly packed cells. (2) The outermost layer of cells in an animal.

epididymis (ep′-uh-did′-uh-mus) A coiled tubule located adjacent to the mammalian testis where sperm are stored.

epigenetics The study of the inheritance of traits transmitted by mechanisms that do not involve the nucleotide sequence.

epiphyte (ep′-uh-fīt) A plant that nourishes itself but grows on the surface of another plant for support, usually on the branches or trunks of trees.

epistasis (ep′-i-stā′-sis) A type of gene interaction in which the phenotypic expression of one gene alters that of another independently inherited gene.

epithelial tissue (ep′-uh-thē′-lē-ul) Sheets of tightly packed cells that line organs and body cavities as well as external surfaces.

epithelium (plural, **epithelia**) An epithelial tissue.

epitope A small, accessible region of an antigen to which an antigen receptor or antibody binds.

equilibrium potential (E_{ion}) The magnitude of a cell's membrane voltage at equilibrium; calculated using the Nernst equation.

erythrocyte (eh-rith′-ruh-sīt) A blood cell that contains haemoglobin, which transports oxygen; also called a red blood cell.

erythropoietin (EPO) (eh-rith′-rō-poy′-uh-tin) A hormone that stimulates the production of erythrocytes. It is secreted by the kidney when body tissues do not receive enough oxygen.

essential amino acid An amino acid that an animal cannot synthesise itself and must be obtained from food in prefabricated form.

essential element A chemical element required for an organism to survive, grow, and reproduce.

essential fatty acid An unsaturated fatty acid that an animal needs but cannot make.

essential nutrient A substance that an organism cannot synthesise from any other material and therefore must absorb in preassembled form.

estuary The area where a freshwater stream or river merges with the ocean.

ethylene (eth′-uh-lēn) A gaseous plant hormone involved in responses to mechanical stress, programmed cell death, leaf abscission, and fruit ripening.

etiolation Plant morphological adaptations for growing in darkness.

euchromatin (yū-krō′-muh-tin) The less condensed form of eukaryotic chromatin that is available for transcription.

eudicot (yū-dī′-kot) A member of a clade that contains the vast majority of flowering plants that have two embryonic seed leaves, or cotyledons.

euglenid (yū′-glen-id) A protist, such as *Euglena* or its relatives, characterised by an anterior pocket from which one or two flagella emerge.

euglenozoan A member of a diverse clade of flagellated protists that includes predatory heterotrophs, photosynthetic autotrophs, and pathogenic parasites.

Eukarya (yū-kār′-ē-uh) The domain that includes all eukaryotic organisms.

eukaryote A single-celled or multicellular organism comprised of eukaryotic cells; eukaryotes include protists, plants, fungi, and animals.

eukaryotic cell (yū′-kār-ē-ot′-ik) A type of cell with a membrane-enclosed nucleus and membrane-enclosed organelles. Organisms with eukaryotic cells (protists, plants, fungi, and animals) are called eukaryotes.

eumetazoan (yū′-met-uh-zō′-un) A member of a clade of animals with true tissues. All animals except sponges and a few other groups are eumetazoans.

eurypterid (yur-ip′-tuh-rid) An extinct carnivorous chelicerate; also called a water scorpion.

Eustachian tube (yū-stā′-shun) The tube that connects the middle ear to the pharynx.

eutherian (yū-thēr′-ē-un) Placental mammal; mammal whose young complete their embryonic development within the uterus, joined to the mother by the placenta.

eutrophic lake (yū-trōf′-ik) A lake that has a high rate of biological productivity supported by a high rate of nutrient cycling.

eutrophication A process by which nutrients, particularly phosphorus and nitrogen, become highly concentrated in a body of water, leading to increased growth of organisms such as algae or cyanobacteria.

evaporative cooling The process in which the surface of an object becomes cooler during evaporation, a result of the molecules with the greatest kinetic energy changing from the liquid to the gaseous state.

evapotranspiration The total evaporation of water from an ecosystem, including water transpired by plants and evaporated from a landscape, usually measured in millimetres and estimated for a year.

evo-devo Evolutionary developmental biology; a field of biology that compares developmental processes of different multicellular organisms to understand how these processes have evolved and how changes can modify existing organismal features or lead to new ones.

evolution Descent with modification; the process by which species accumulate differences from their ancestors as they adapt to different environments over time; also defined as a change in the genetic composition of a population from generation to generation.

evolutionary lineage The sequence of ancestral organisms leading to a particular taxon; represented by a branch (line) in a phylogenetic tree.

evolutionary tree A branching diagram that reflects a hypothesis about evolutionary relationships among groups of organisms.

Excavata (ex′-kuh-vah′-tuh) One of four supergroups of eukaryotes proposed in a current hypothesis of the evolutionary history of eukaryotes. Excavates have unique cytoskeletal features, and some species have an "excavated" feeding groove on one side of the cell body. *See also* SAR, Archaeplastida, and Unikonta.

excitatory postsynaptic potential (EPSP) An electrical change (depolarisation) in the membrane of a postsynaptic cell caused by the binding of an excitatory neurotransmitter from a presynaptic cell to a postsynaptic receptor; makes it more likely for a postsynaptic cell to generate an action potential.

excretion The disposal of nitrogen-containing metabolites and other waste products.

exergonic reaction (ek′-ser-gon′-ik) A spontaneous chemical reaction in which there is a net release of free energy.

exocytosis (ek′-sō-sī-tō′-sis) The cellular secretion of biological molecules by the fusion of vesicles containing them with the plasma membrane.

exon A sequence within a primary transcript that remains in the RNA after RNA processing; also refers to the region of DNA from which this sequence was transcribed.

exoskeleton A hard encasement on the surface of an animal, such as the shell of a mollusc or the cuticle of an arthropod, that provides protection and points of attachment for muscles.

exotoxin (ek′-sō-tok′-sin) A toxic protein that is secreted by a prokaryote or other pathogen and that produces specific symptoms, even if the pathogen is no longer present.

expansin Plant enzyme that breaks the cross-links (hydrogen bonds) between cellulose microfibrils and other cell wall constituents, loosening the wall's fabric.

experiment A scientific test. Often carried out under controlled conditions that involve manipulating one factor in a system in order to see the effects of changing that factor.

experimental group A set of subjects that has (or receives) the specific factor being tested in a controlled experiment. Ideally, the experimental group is identical to the control group for all other factors.

exploitation A +/− ecological interaction in which individuals of one species benefit by feeding on (and thereby harming) individuals of the other species. Exploitative interactions include predation, herbivory, and parasitism.

exponential population growth Growth of a population in an ideal, unlimited environment, represented by a J-shaped curve when population size is plotted over time.

expression vector A cloning vector that contains a highly active bacterial promoter just upstream of a restriction site where a eukaryotic gene can be inserted, allowing the gene to be expressed in a bacterial cell. Expression vectors are also available that have been genetically engineered for use in specific types of eukaryotic cells.

extinction vortex A downward population spiral in which inbreeding and genetic drift combine to cause a small population to shrink and, unless the spiral is reversed, become extinct.

extracellular matrix (ECM) The meshwork surrounding animal cells, consisting of glycoproteins, polysaccharides, and proteoglycans synthesised and secreted by cells.

extraembryonic membrane One of four membranes (yolk sac, amnion, chorion, and allantois) located outside the embryo that support the developing embryo in reptiles and mammals.

extreme halophile An organism that lives in a highly saline environment, such as the Great Salt Lake or the Dead Sea.

extreme thermophile An organism that thrives in hot environments (often 60–80°C or hotter).

extremophile An organism that lives in environmental conditions so extreme that few other species can survive there. Extremophiles include extreme halophiles ("salt lovers") and extreme thermophiles ("heat lovers").

F_1 generation The first filial, hybrid (heterozygous) offspring arising from a parental (P generation) cross.

F_2 generation The offspring resulting from interbreeding (or self-pollination) of the hybrid F_1 generation.

facilitated diffusion The passage of molecules or ions down their electrochemical gradient across a biological membrane with the assistance of specific transmembrane transport proteins, requiring no energy expenditure.

facultative anaerobe (fak′-ul-tā′-tiv an′-uh-rōb) An organism that makes ATP by aerobic respiration if oxygen is present but that switches to anaerobic respiration or fermentation if oxygen is not present.

faeces (fē′-sēz) The wastes of the digestive tract.

family In Linnaean classification, the taxonomic category above genus.

fast-twitch fibre A muscle fibre used for rapid, powerful contractions.

fat A lipid consisting of three fatty acids linked to one glycerol molecule; also called a triacylglycerol or triglyceride.

fate map A territorial diagram of embryonic development that displays the future derivatives of individual cells and tissues.

fatty acid A carboxylic acid with a long carbon chain. Fatty acids vary in length and in the number and location of double bonds; three fatty acids linked to a glycerol molecule form a fat molecule, also called triacylglycerol or triglyceride.

feedback inhibition A method of metabolic control in which the end product of a metabolic pathway acts as an inhibitor of an enzyme within that pathway.

feedback regulation The regulation of a process by its output or end product.

fermentation A catabolic process that makes a limited amount of ATP from glucose (or other organic molecules) without an electron transport chain and that produces a characteristic end product, such as ethyl alcohol or lactic acid.

fertilisation (1) The union of haploid gametes to produce a diploid zygote. (2) The addition of mineral nutrients to the soil.

fetus (fē′-tus) A developing mammal that has all the major structures of an adult. In humans, the fetal stage lasts from the 9th week of gestation until birth.

F factor In bacteria, the DNA segment that confers the ability to form pili for conjugation and associated functions required for the transfer of DNA from donor to recipient. The F factor may exist as a plasmid or be integrated into the bacterial chromosome.

fibroblast (fī′-brō-blast) A type of cell in loose connective tissue that secretes the protein ingredients of the extracellular fibres.

fibronectin An extracellular glycoprotein secreted by animal cells that helps them attach to the extracellular matrix.

filament In an angiosperm, the stalk portion of the stamen, the pollen-producing reproductive organ of a flower.

filter feeder An animal that feeds by using a filtration mechanism to strain small organisms or food particles from its surroundings.

filtrate Cell-free fluid extracted from the body fluid by the excretory system.

filtration In excretory systems, the extraction of water and small solutes, including metabolic wastes, from the body fluid.

fimbria (plural, **fimbriae**) A short, hairlike appendage of a prokaryotic cell that helps it adhere to the substrate or to other cells.

first law of thermodynamics The principle of conservation of energy: Energy can be transferred and transformed, but it cannot be created or destroyed.

fission The separation of an organism into two or more individuals of approximately equal size.

fixed action pattern In animal behaviour, a sequence of unlearned acts that is essentially unchangeable and, once initiated, usually carried to completion.

flaccid (flas′-id) Limp. Lacking turgor (stiffness or firmness), as in a plant cell in surroundings where there is a tendency for water to leave the cell. (A walled cell becomes flaccid if it has a higher water potential than its surroundings, resulting in the loss of water.)

flagellum (fluh-jel′-um) (plural, **flagella**) A long cellular appendage specialised for locomotion. Like motile cilia, eukaryotic flagella have a core with nine outer doublet microtubules and two inner single microtubules (the "9 + 2" arrangement) ensheathed in an extension of the plasma membrane. Prokaryotic flagella have a different structure.

florigen A flowering signal, probably a protein, that is made in leaves under certain conditions and that travels to the shoot apical meristems, inducing them to switch from vegetative to reproductive growth.

flower In an angiosperm, a specialised shoot with up to four sets of modified leaves, bearing structures that function in sexual reproduction.

fluid feeder An animal that lives by sucking nutrient-rich fluids from another living organism.

fluid mosaic model The currently accepted model of cell membrane structure, which envisions the membrane as a mosaic of protein molecules drifting laterally in a fluid bilayer of phospholipids.

follicle (fol′-uh-kul) A microscopic structure in the ovary that contains the developing oocyte and secretes oestrogens.

follicle-stimulating hormone (FSH) (fol′-uh-kul) A tropic hormone that is produced and secreted by the anterior pituitary and that stimulates the production of eggs by the ovaries and sperm by the testes.

food chain The pathway along which food energy is transferred from trophic level to trophic level, beginning with producers.

food vacuole A membranous sac formed by phagocytosis of microorganisms or particles to be used as food by the cell.

food web The interconnected feeding relationships in an ecosystem.

foot (1) The portion of a bryophyte sporophyte that gathers sugars, amino acids, water, and minerals from the parent gametophyte via transfer cells. (2) One of the three main parts of a mollusc; a muscular structure usually used for movement. *See also* mantle and visceral mass.

foraging The seeking and obtaining of food.

foram (foraminiferan) An aquatic protist that secretes a hardened shell containing calcium carbonate and extends pseudopodia through pores in the shell.

forebrain One of three ancestral and embryonic regions of the vertebrate brain; develops into the thalamus, hypothalamus, and cerebrum.

fossil A preserved remnant or impression of an organism that lived in the past.

foundation species A species that has strong effects on its community as a result of its large size, high abundance, or pivotal role in community dynamics. Foundation species may provide significant habitat or food for other species; they may also be competitively dominant in exploiting key resources.

founder effect Genetic drift that occurs when a few individuals become isolated from a larger population and form a new population whose gene pool composition is not reflective of that of the original population.

fovea (fō′-vē-uh) The place on the retina at the eye's centre of focus, where cones are highly concentrated.

F plasmid The plasmid form of the F factor.

fragmentation A means of asexual reproduction whereby a single parent breaks into parts that regenerate into whole new individuals.

frameshift mutation A mutation occurring when nucleotides are inserted in or deleted from a gene and the number inserted or deleted is not a multiple of three, resulting in the improper grouping of the subsequent nucleotides into codons.

free energy The portion of a biological system's energy that can perform work when temperature and pressure are uniform throughout the system. The change in free energy of a system (ΔG) is calculated by the equation $\Delta G = \Delta H - T\Delta S$, where ΔH is the change in enthalpy (in biological systems, equivalent to total energy), ΔT is the absolute temperature, and ΔS is the change in entropy.

frequency-dependent selection Selection in which the fitness of a phenotype depends on how common the phenotype is in a population.

fruit A mature ovary of a flower. The fruit protects dormant seeds and often functions in their dispersal.

functional group A specific configuration of atoms commonly attached to the carbon skeletons of organic molecules and involved in chemical reactions.

fusion In evolutionary biology, a process in which gene flow between two species that can form hybrid offspring weakens barriers to reproduction between the species. This process causes their gene pools to become increasingly alike and can cause the two species to fuse into a single species.

G_0 phase A nondividing state occupied by cells that have left the cell cycle, sometimes reversibly.

G_1 phase The first gap, or growth phase, of the cell cycle, consisting of the portion of interphase before DNA synthesis begins.

G_2 phase The second gap, or growth phase, of the cell cycle, consisting of the portion of interphase after DNA synthesis occurs.

gallbladder An organ that stores bile and releases it as needed into the small intestine.

game theory An approach to evaluating alternative strategies in situations where the outcome of a particular strategy depends on the strategies used by other individuals.

gametangium (gam′-uh-tan′-jē-um) (plural, **gametangia**) Multicellular plant structure in which gametes are formed. Female gametangia are called archegonia, and male gametangia are called antheridia.

gamete (gam′-ēt) A haploid reproductive cell, such as an egg or sperm, that is formed by meiosis or is the descendant of cells formed by meiosis. Gametes unite during sexual reproduction to produce a diploid zygote.

gametogenesis (guh-mē′-tō-gen′-uh-sis) The process by which gametes are produced.

gametophore (guh-mē-‐tō-fōr) The mature gamete-producing structure of a moss gametophyte.

gametophyte (guh-mē′-tō-fīt) In organisms (plants and some algae) that have alternation of generations, the multicellular haploid form that produces haploid gametes by mitosis. The haploid gametes unite and develop into sporophytes.

ganglion (gan′-glē-uhn) (plural, **ganglia**) A cluster (functional group) of nerve cell bodies.

gap junction A type of intercellular junction in animal cells, consisting of proteins surrounding a pore that allows the passage of materials between cells.

gas exchange The uptake of molecular oxygen from the environment and the discharge of carbon dioxide to the environment.

gastric juice A digestive fluid secreted by the stomach.

gastrovascular cavity A central cavity with a single opening in the body of certain animals, including cnidarians and flatworms, that functions in both the digestion and distribution of nutrients.

gastrula (gas′-trū-luh) An embryonic stage in animal development encompassing the formation of three layers: ectoderm, mesoderm, and endoderm.

gastrulation (gas′-trū-lā′-shun) In animal development, a series of cell and tissue movements in which the blastula-stage embryo folds inwards, producing a three-layered embryo, the gastrula.

gated channel A transmembrane protein channel that opens or closes in response to a particular stimulus.

gated ion channel A gated channel for a specific ion. The opening or closing of such channels may alter a cell's membrane potential.

gel electrophoresis (ē-lek′-trō-fōr-ē′-sis) A technique for separating nucleic acids or proteins on the basis of their size and electrical charge, both of which affect their rate of movement through an electric field in a gel made of agarose or another polymer.

gene A discrete unit of hereditary information consisting of a specific nucleotide sequence in DNA (or RNA, in some viruses).

gene annotation Analysis of genomic sequences to identify protein-coding genes and determine the function of their products.

gene cloning The production of multiple copies of a gene.

gene drive A process that biases inheritance such that a particular allele is more likely to be inherited than are other alleles, causing the favoured allele to spread (be "driven") through the population.

gene editing Altering genes in a specific, predictable way.

gene expression The process by which information encoded in DNA directs the synthesis of proteins or, in some cases, RNAs that are not translated into proteins and instead function as RNAs.

gene flow The transfer of alleles from one population to another, resulting from the movement of fertile individuals or their gametes.

gene pool The aggregate of all copies of every type of allele at all loci in every individual in a population. The term is also used in a more restricted sense as the aggregate of alleles for just one or a few loci in a population.

gene therapy The introduction of genes into an afflicted individual for therapeutic purposes.

genetic drift A process in which chance events cause unpredictable fluctuations in allele frequencies from one generation to the next. Effects of genetic drift are most pronounced in small populations.

genetic engineering The direct manipulation of genes for practical purposes.

genetic map An ordered list of genetic loci (genes or other genetic markers) along a chromosome.

genetic profile An individual's unique set of genetic markers, detected most often today by PCR or, previously, by electrophoresis and nucleic acid probes.

genetic recombination General term for the production of offspring with combinations of traits that differ from those found in either parent.

genetic variation Differences among individuals in the composition of their genes or other DNA sequences.

genetically modified organism (GMO) An organism that has acquired one or more genes by artificial means.

genetics The scientific study of heredity and hereditary variation.

genome (jē′-nōm) The genetic material of an organism or virus; the complete complement of an organism's or virus's genes along with its noncoding nucleic acid sequences.

genome-wide association study (jē′-nōm) A large-scale analysis of the genomes of many people having a certain phenotype or disease, with the aim of finding genetic markers that correlate with that phenotype or disease.

genomic imprinting (juh-nō′-mik) A phenomenon in which expression of an allele in offspring depends on whether the allele is inherited from the male or female parent.

genomics (juh-nō′-miks) The systematic study of whole sets of genes (or other DNA) and their interactions within a species, as well as genome comparisons between species.

genotype (jē′-nō-tīp) The genetic makeup, or set of alleles, of an organism.

genus (jē′-nus) (plural, **genera**) A taxonomic category above the species level, designated by the first word of a species' two-part scientific name.

geological record A standard time scale dividing Earth's history into time periods, grouped into four eons—Hadean, Archaean, Proterozoic, and Phanerozoic—and further subdivided into eras, periods, and epochs.

germ layer One of the three main layers in a gastrula that will form the various tissues and organs of an animal body.

gestation (jes-tā′-shun) Pregnancy; the condition of carrying one or more embryos in the uterus.

gibberellin (jib′-uh-rel′-in) Any of a class of related plant hormones that stimulate growth in the stem and leaves, trigger the germination of seeds and breaking of bud dormancy, and (with auxin) stimulate fruit development.

glans In humans, the rounded head of the penis (in males) or of the clitoris (in females); the glans is highly sensitive to stimulation.

glia (glial cells) Cells of the nervous system that support, regulate, and augment the functions of neurons.

global ecology The study of the functioning and distribution of organisms across the biosphere and how the regional exchange of energy and materials affects them.

glomeromycete (glō′-mer-ō-mī′-sēt) A member of the fungal phylum Glomeromycota, characterised by a distinct branching form of mycorrhizae called arbuscular mycorrhizae.

glomerulus (glō-mār′-yū-lus) A ball of capillaries surrounded by Bowman's capsule in the nephron and serving as the site of filtration in the vertebrate kidney.

glucocorticoid A steroid hormone that is secreted by the adrenal cortex and that influences glucose metabolism and immune function.

glucagon (glū′-kuh-gon) A hormone secreted by the pancreas that raises blood glucose levels. It promotes glycogen breakdown and release of glucose by the liver.

glyceraldehyde 3-phosphate (G3P) (glis′-er-al′-de-hīd) A three-carbon carbohydrate that is the direct product of the Calvin cycle; it is also an intermediate in glycolysis.

glycogen (glī′-kō-jen) An extensively branched glucose storage polysaccharide found in the liver and muscle of animals; the animal equivalent of starch.

glycolipid A lipid with one or more covalently attached carbohydrates.

glycolysis (glī-kol′-uh-sis) A series of reactions that ultimately splits glucose into pyruvate. Glycolysis occurs in almost all living cells, serving as the starting point for fermentation or cellular respiration.

glycoprotein A protein with one or more covalently attached carbohydrates.

glycosidic linkage A covalent bond formed between two monosaccharides by a dehydration reaction.

gnathostome (na′-thu-stōm) Member of one of the two main clades of vertebrates; gnathostomes have jaws and include sharks and rays, ray-finned fishes, coelacanths, lungfishes, amphibians, reptiles, and mammals. *See also* cyclostome.

Golgi apparatus (gol′-jē) An organelle in eukaryotic cells consisting of stacks of flat membranous sacs that modify, store, and route products of the endoplasmic reticulum and synthesise some products, notably non-cellulose carbohydrates.

gonad (gō′-nad) A male or female gamete-producing organ.

G protein A GTP-binding protein that relays signals from a plasma membrane signal receptor, known as a G protein-coupled receptor, to other signal transduction proteins inside the cell.

G protein-coupled receptor (GPCR) A signal receptor protein in the plasma membrane that responds to the binding of a signalling molecule by activating a G protein. Also called a G protein-linked receptor.

graded potential An electrical response of a cell to a stimulus, consisting of a change in voltage across the membrane proportional to the stimulus strength.

Gram stain A staining method that distinguishes between two different kinds of bacterial cell walls; may be used to help determine medical response to an infection.

gram-negative Describing the group of bacteria that have a cell wall that is structurally more complex and contains less peptidoglycan than the cell wall of gram-positive bacteria. Gram-negative bacteria are often more toxic than gram-positive bacteria.

gram-positive Describing the group of bacteria that have a cell wall that is structurally less complex and contains more peptidoglycan than the cell wall of gram-negative bacteria. Gram-positive bacteria are usually less toxic than gram-negative bacteria.

granum (gran′-um) (plural, **grana**) A stack of membrane-bounded thylakoids in the chloroplast. Grana function in the light reactions of photosynthesis.

gravitropism (grav′-uh-trō′-pizm) A response of a plant or animal to gravity.

green alga A photosynthetic protist, named for green chloroplasts that are similar in structure and pigment composition to the chloroplasts of plants. Green algae are a paraphyletic group; some members are more closely related to plants than they are to other green algae.

greenhouse effect The warming of Earth due to the atmospheric accumulation of carbon dioxide and certain other gases, which absorb reflected infrared radiation and reradiate some of it back towards Earth.

grey matter Regions of clustered neuron cell bodies within the CNS.

gross primary production (GPP) The total primary production of an ecosystem.

ground tissue Plant tissue that is neither vascular nor dermal, fulfilling a variety of functions, such as storage, photosynthesis, and support.

growth factor (1) A protein that must be present in the extracellular environment (culture medium or animal body) for the growth and normal development of certain types of cells. (2) A local regulator that acts on nearby cells to stimulate cell proliferation and differentiation.

growth hormone (GH) A hormone that is produced and secreted by the anterior pituitary and that has both direct (nontropic) and tropic effects on a wide variety of tissues.

guard cells The two cells that flank the stomatal pore and regulate the opening and closing of the pore.

gustation The sense of taste.

guttation The exudation of water droplets from leaves, caused by root pressure in certain plants.

gymnosperm (jim′-nō-sperm) A vascular plant that bears naked seeds—seeds not enclosed in protective chambers.

haemocoel A body cavity lined by tissue derived from mesoderm and by tissue derived from endoderm.

haemoglobin (hē′-mō-glō′-bin) An iron-containing protein in red blood cells that reversibly binds oxygen.

haemolymph (hē′-mō-limf′) In invertebrates with an open circulatory system, the body fluid that bathes tissues.

haemophilia (hē′-muh-fil′-ē-uh) A human genetic disease caused by a sex-linked recessive allele resulting in the absence of one or more blood-clotting proteins; characterised by excessive bleeding following injury.

hagfish Marine jawless vertebrates that have highly reduced vertebrae and a skull made of cartilage; most hagfishes are bottom-dwelling scavengers.

hair cell A mechanosensory cell that alters output to the nervous system when hairlike projections on the cell surface are displaced.

half-life The amount of time it takes for 50% of a sample of a radioactive isotope to decay.

halophile *See* extreme halophile.

Hamilton's rule The principle that for natural selection to favour an altruistic act, the benefit to the recipient, devalued by the coefficient of relatedness, must exceed the cost to the altruist.

haploid cell (hap′-loyd) A cell containing only one set of chromosomes (n).

haplodiploidy A unique sex-determination system where males develop from unfertilised eggs and are consequently haploid, whereas females develop from fertilised eggs and are consequently diploid.

Hardy-Weinberg equilibrium The state of a population in which frequencies of alleles and genotypes remain constant from generation to generation, provided that only Mendelian segregation and recombination of alleles are at work.

heart A muscular pump that uses metabolic energy to elevate the hydrostatic pressure of the circulatory fluid (blood or haemolymph). The fluid then flows down a pressure gradient through the body and eventually returns to the heart.

heart attack The damage or death of cardiac muscle tissue resulting from prolonged blockage of one or more coronary arteries.

heart murmur A hissing sound that most often results from blood squirting backwards through a leaky valve in the heart.

heart rate The frequency of heart contraction (in beats per minute).

heat Thermal energy in transfer from one body of matter to another.

heathlands A biome of dense evergreen shrubs frequently less than 2 m tall. Heathlands grow in wind and salt-spray affected coastal areas to higher altitude sites, often above Australia's ephemeral snowline and below the permanent snowline in New Zealand. Typically, heathlands establish on substrates of sand, clay, or soils derived from the breakdown of sedimentary rock.

heat of vaporisation The quantity of heat a liquid must absorb for 1 g of it to be converted from the liquid to the gaseous state.

heat-shock protein A protein that helps protect other proteins during heat stress. Heat-shock proteins are found in plants, animals, and microorganisms.

heavy chain One of the two types of polypeptide chains that make up an antibody molecule and B cell receptor; consists of a variable region, which contributes to the antigen-binding site, and a constant region.

helicase An enzyme that untwists the double helix of DNA at replication forks, separating the two strands and making them available as template strands.

helper T cell A type of T cell that, when activated, secretes cytokines that promote the response of B cells (humoral response) and cytotoxic T cells (cell-mediated response) to antigens.

hepatic portal vein A large vessel that conveys nutrient-laden blood from the small intestine to the liver, which regulates the blood's nutrient content.

herbivore (hur′-bi-vōr′) An animal that mainly eats plants or algae.

herbivory A +/− ecological interaction in which an organism eats part of a plant or alga.

heredity The transmission of traits from one generation to the next.

hermaphrodite (hur-maf′-ruh-dīt′) An individual that functions as both male and female in sexual reproduction by producing both sperm and eggs.

hermaphroditism (hur-maf′-rō-dī-tizm) A condition in which an individual has both female and male gonads and functions as both a male and a female in sexual reproduction by producing both sperm and eggs.

heterochromatin (het′-er-ō-krō′-muh-tin) Eukaryotic chromatin that remains highly compacted during interphase and is generally not transcribed.

heterochrony (het′-uh-rok′-ruh-nē) Evolutionary change in the timing or rate of an organism's development.

heterocyst (het′-er-ō-sist) A specialised cell that engages in nitrogen fixation in some filamentous cyanobacteria; also called a heterocyte.

heterokaryon (het′-er-ō-kār′-ē-un) A fungal mycelium that contains two or more haploid nuclei per cell.

heteromorphic (het′-er-ō-mōr′-fik) Referring to a condition in the life cycle of plants and certain algae in which the sporophyte and gametophyte generations differ in morphology.

heterosporous (het-er-os′-pōr-us) Referring to a plant species that has two kinds of spores: microspores, which develop into male gametophytes, and megaspores, which develop into female gametophytes.

heterotroph (het′-er-ō-trōf) An organism that obtains organic food molecules by eating other organisms or substances derived from them.

heterozygote An organism that has two different alleles for a gene (encoding a character).

heterozygote advantage Greater reproductive success of heterozygous individuals compared with homozygotes; tends to preserve variation in a gene pool.

heterozygous (het′-er-ō-zī′-gus) Having two different alleles for a given gene.

hibernation A long-term physiological state in which metabolism decreases, the heart and respiratory system slow down, and body temperature is maintained at a lower level than normal.

high-density lipoprotein (HDL) A particle in the blood made up of thousands of cholesterol molecules and other lipids bound to a protein. HDL scavenges excess cholesterol.

hindbrain One of three ancestral and embryonic regions of the vertebrate brain; develops into the medulla oblongata, pons, and cerebellum.

histamine (his′-tuh-mēn) A substance released by mast cells that causes blood vessels to dilate and become more permeable in inflammatory and allergic responses.

histogram A variant of a bar graph that is made for numeric data by first grouping, or "binning," the variable plotted on the *x*-axis into intervals of equal width. The "bins" may be integers or ranges of numbers. The height of each bar shows the percent or number of experimental subjects whose characteristics can be described by one of the intervals plotted on the *x*-axis.

histone (his′-tōn) A small protein with a high proportion of positively charged amino acids that binds to the negatively charged DNA and plays a key role in chromatin structure.

histone acetylation (his′-tōn) The attachment of acetyl groups to certain amino acids of histone proteins.

HIV (human immunodeficiency virus) The infectious agent that causes AIDS. HIV is a retrovirus.

holdfast A rootlike structure that anchors a seaweed.

homeobox (hō′-mē-ō-boks′) A 180-nucleotide sequence within homeotic genes and some other developmental genes that is widely conserved in animals. Related sequences occur in plants and yeasts.

homeostasis (hō′-mē-ō-stā′-sis) The steady-state physiological condition of the body.

homeotic gene (hō-mē-o′-tik) Any of the master regulatory genes that control placement and spatial organisation of body parts in animals, plants, and fungi by controlling the developmental fate of groups of cells.

hominin (hō′-mi-nin) A group consisting of humans and the extinct species that are more closely related to us than to chimpanzees.

homologous chromosomes (or homologues) (hō-mol′-uh-gus) A pair of chromosomes of the same length, centromere position, and staining pattern that possess genes for the same characters at corresponding loci. One homologous chromosome is inherited from the organism's father, the other from the mother. Also called a homologous pair.

homologous pair *See* homologous chromosomes.

homologous structures (hō-mol′-uh-gus) Structures in different species that are similar because of common ancestry.

homologues *See* homologous chromosomes.

homology (hō-mol′-ō-jē) Similarity in characteristics resulting from a shared ancestry.

homoplasy (hō′-muh-play′-zē) A similar (analogous) structure or molecular sequence that has evolved independently in two species.

homosporous (hō-mos′-puh-rus) Referring to a plant species that has a single kind of spore, which typically develops into a bisexual gametophyte.

homozygote An organism that has a pair of identical alleles for a gene (encoding a character).

homozygous (hō′-mō-zī′-gus) Having two identical alleles for a given gene.

horizontal gene transfer The transfer of genes from one genome to another through mechanisms such as transposable elements, plasmid exchange, viral activity, and perhaps fusions of different organisms.

hormone In multicellular organisms, one of many types of secreted chemicals that are formed in specialised cells, travel in body fluids, and act on specific target cells in other parts of the organism, changing the target cells' functioning.

hornwort A small, herbaceous, nonvascular plant that is a member of the phylum Anthocerophyta.

host The larger participant in a symbiotic relationship, often providing a home and food source for the smaller symbiont.

host range The limited number of species whose cells can be infected by a particular virus.

Human Genome Project An international collaborative effort to map and sequence the DNA of the entire human genome.

human immunodeficiency virus (HIV) The infectious agent that causes AIDS (acquired immunodeficiency syndrome). HIV is a retrovirus.

humoral immune response (hyū′-mer-ul) The branch of adaptive immunity that involves the activation of B cells and that leads to the production of antibodies, which defend against bacteria and viruses in body fluids.

humus (hyū′-mus) Decomposing organic material that is a component of topsoil.

Huntington's disease A human genetic disease caused by a dominant allele; characterised by uncontrollable body movements and degeneration of the nervous system; usually fatal 10 to 20 years after the onset of symptoms.

hybrid Offspring that results from the mating of individuals from two different species or from two true-breeding varieties of the same species.

hybrid zone A geographical region in which members of different species meet and mate, producing at least some offspring of mixed ancestry.

hybridisation In genetics, the mating, or crossing, of two true-breeding varieties.

hydration shell The sphere of water molecules around a dissolved ion.

hydrocarbon An organic molecule consisting only of carbon and hydrogen.

hydrogen bond A type of weak chemical bond that is formed when the slightly positive hydrogen atom of a polar covalent bond in one molecule is attracted to the slightly negative atom of a polar covalent bond in another molecule or in another region of the same molecule.

hydrogen ion A single proton with a charge of 1+. The dissociation of a water molecule (H_2O) leads to the generation of a hydroxide ion (OH^-) and a hydrogen ion (H^+); in water, H^+ is not found alone but associates with a water molecule to form a hydronium ion.

hydrolysis (hī-drol′-uh-sis) A chemical reaction that breaks bonds between two molecules by the addition of water; functions in disassembly of polymers to monomers.

hydronium ion A water molecule that has an extra proton bound to it; H_3O^+, commonly represented as H^+.

hydrophilic (hī′-drō-fil′-ik) Having an affinity for water.

hydrophobic (hī′-drō-fō′-bik) Having no affinity for water; tending to coalesce and form droplets in water.

hydrophobic interaction (hī′-drō-fō′-bik) A type of weak chemical interaction caused when molecules that do not mix with water coalesce to exclude water.

hydroponic culture A method in which plants are grown in mineral solutions rather than in soil.

hydrostatic skeleton A skeletal system composed of fluid held under pressure in a closed body compartment; the main skeleton of most cnidarians, flatworms, nematodes, and annelids.

hydrothermal vent An area on the seafloor where heated water and minerals from Earth's interior gush into the seawater, producing a dark, hot, oxygen-deficient environment. The producers in a hydrothermal vent community are chemoautotrophic prokaryotes.

hydroxide ion A water molecule that has lost a proton; OH^-.

hydroxyl group (hī-drok′-sil) A chemical group consisting of an oxygen atom joined to a hydrogen atom. Molecules possessing this group are soluble in water and are called alcohols.

hyperpolarisation A change in a cell's membrane potential such that the inside of the membrane becomes more negative relative to the outside. Hyperpolarisation reduces the chance that a neuron will transmit a nerve impulse.

hypersensitive response A plant's localised defence response to a pathogen, involving the death of cells around the site of infection.

hypertension A disorder in which blood pressure remains abnormally high.

hypertonic Referring to a solution that, when surrounding a cell, will cause the cell to lose water.

hypha (plural, **hyphae**) (hī′-fuh, hī′-fē) One of many connected filaments that collectively make up the mycelium of a fungus.

hypocotyl (hī′-puh-cot′-ul) In an angiosperm embryo, the embryonic axis below the point of attachment of the cotyledon(s) and above the radicle.

hypothalamus (hī′-pō-thal′-uh-mus) The ventral part of the vertebrate forebrain; functions in maintaining homeostasis, especially in coordinating the endocrine and nervous systems; secretes hormones of the posterior pituitary and releasing factors that regulate the anterior pituitary.

hypothesis (hī-poth′-uh-sis) A testable explanation for a set of observations based on the available data and guided by inductive reasoning. A hypothesis is narrower in scope than a theory.

hypotonic Referring to a solution that, when surrounding a cell, will cause the cell to take up water.

imbibition The uptake of water by a seed or other structure, resulting in swelling.

immigration The influx of new individuals into a population from other areas.

immune system An organism's system of defences against agents that cause disease.

immunisation The process of generating a state of immunity by artificial means. In vaccination, an inactive or weakened form of a pathogen is administered, inducing B and T cell responses and immunological memory. In passive immunisation, antibodies specific for a particular pathogen are administered, conferring immediate but temporary protection.

immunoglobulin (Ig) (im′-yū-nō-glob′-yū-lin) *See* antibody.

imprinting In animal behaviour, the formation at a specific stage in life of a long-lasting behavioural response to a specific individual or object. *See also* genomic imprinting.

inclusive fitness The total effect an individual has on proliferating its genes by producing its own offspring and by providing aid that enables other close relatives to increase production of their offspring.

incomplete dominance The situation in which the phenotype of heterozygotes is intermediate between the phenotypes of individuals homozygous for either allele.

incomplete flower A flower in which one or more of the four basic floral organs (sepals, petals, stamens, or carpels) are either absent or nonfunctional.

incomplete metamorphosis A type of development in certain insects, such as grasshoppers, in which the young (called nymphs) resemble adults but are smaller and have different body proportions. The nymph goes through a series of molts, each time looking more like an adult, until it reaches full size.

independent variable A factor whose value is manipulated or changed during an experiment to reveal possible effects on another factor (the dependent variable).

indeterminate cleavage A type of embryonic development in deuterostomes in which each cell produced by early cleavage divisions retains the capacity to develop into a complete embryo.

indeterminate growth A type of growth characteristic of plants, in which the organism continues to grow as long as it lives.

induced fit Caused by entry of the substrate, the change in shape of the active site of an enzyme so that it binds more snugly to the substrate.

inducer A specific small molecule that binds to a bacterial repressor protein and changes the repressor's shape so that it cannot bind to an operator, thus switching an operon on.

induction A process in which a group of cells or tissues influences the development of another group through close-range interactions.

inductive reasoning A type of logic in which generalisations are based on a large number of specific observations.

inflammatory response An innate immune defence triggered by physical injury or infection of tissue involving the release of substances that promote swelling, enhance the infiltration of white blood cells, and aid in tissue repair and destruction of invading pathogens.

inflorescence A group of flowers tightly clustered together.

ingestion The first stage of food processing in animals: the act of eating.

ingroup A species or group of species whose evolutionary relationships are being examined in a given analysis.

inhibitory postsynaptic potential (IPSP) An electrical change (usually hyperpolarisation) in the membrane of a postsynaptic neuron caused by the binding of an inhibitory neurotransmitter from a presynaptic cell to a postsynaptic receptor; makes it more difficult for a postsynaptic neuron to generate an action potential.

innate behaviour Animal behaviour that is developmentally fixed and under strong genetic control. Innate behaviour is exhibited in virtually the same form by all individuals in a population despite internal and external environmental differences during development and throughout their lifetimes.

innate immunity A form of defence common to all animals that is active immediately upon exposure to a pathogen and that is the same whether or not the pathogen has been encountered previously.

inner cell mass An inner cluster of cells at one end of a mammalian blastocyst that subsequently develops into the embryo proper and some of the extraembryonic membranes.

inner ear One of the three main regions of the vertebrate ear; includes the cochlea (which in turn contains the organ of Corti) and the semicircular canals.

inositol trisphosphate (IP_3) (in-ō′-suh-tol) A second messenger that functions as an intermediate between certain signalling molecules and a subsequent second messenger, Ca^{2+}, by causing a rise in cytoplasmic Ca^{2+} concentration.

inquiry The search for information and explanation, often focusing on specific questions.

insertion A mutation involving the addition of one or more nucleotide pairs to a gene.

***in situ* hybridisation** A technique using nucleic acid hybridisation with a labelled probe to detect the location of a specific mRNA in an intact organism.

insulin (in′-suh-lin) A hormone secreted by pancreatic beta cells that lowers blood glucose levels. It promotes the uptake of glucose by most body cells and the synthesis and storage of glycogen in the liver and also stimulates protein and fat synthesis.

integral protein A transmembrane protein with hydrophobic regions that extend into and often completely span the hydrophobic interior of the membrane and with hydrophilic regions in contact with the aqueous solution on one or both sides of the membrane (or lining the channel in the case of a channel protein).

integrin (in′-tuh-grin) In animal cells, a transmembrane receptor protein with two subunits that interconnects the extracellular matrix and the cytoskeleton.

integument (in-teg′-yū-ment) Layer of sporophyte tissue that contributes to the structure of an ovule of a seed plant.

integumentary system The outer covering of a mammal's body, including skin, hair, and nails, claws, or hooves.

interferon (in′-ter-fēr′-on) A protein that has antiviral or immune regulatory functions. For example, interferons secreted by virus-infected cells help nearby cells resist viral infection.

intermediate disturbance hypothesis The concept that moderate levels of disturbance can foster greater species diversity than low or high levels of disturbance.

intermediate filament A component of the cytoskeleton that includes filaments intermediate in size between microtubules and microfilaments.

interneuron An association neuron; a nerve cell within the central nervous system that forms synapses with sensory and/or motor neurons and integrates sensory input and motor output.

internode A segment of a plant stem between the points where leaves are attached.

interphase The period in the cell cycle when the cell is not dividing. During interphase, cellular metabolic activity is high, chromosomes and organelles are duplicated, and cell size may increase. Interphase often accounts for about 90% of the cell cycle.

intersexual selection A form of natural selection in which individuals of one sex (usually the females) are choosy in selecting their mates from the other sex; also called mate choice.

interspecific interaction A relationship between individuals of two or more species in a community.

interstitial fluid The fluid filling the spaces between cells in most animals.

intertidal zone The shallow zone of the ocean adjacent to land and between the high- and low-tide lines.

intrasexual selection A form of natural selection in which there is direct competition among individuals of one sex for mates of the opposite sex.

intrinsic rate of increase (*r*) In population models, the per capita rate at which an exponentially growing population increases in size at each instant in time.

introduced species A species moved by humans, either intentionally or accidentally, from its native location to a new geographical region; sometimes called a non-native species, exotic species, or invasive species.

intron (in′-tron) A noncoding, intervening sequence within a primary transcript that is removed from the transcript during RNA processing; also refers to the region of DNA from which this sequence was transcribed.

inversion An aberration in chromosome structure resulting from reattachment of a chromosomal fragment in a reverse orientation to the chromosome from which it originated.

invertebrate An animal without a backbone. Invertebrates make up 95% of animal species.

***in vitro* fertilisation (IVF)** (vē′-trō) Fertilisation of oocytes in laboratory containers followed by artificial implantation of the early embryo in the mother's uterus.

***in vitro* mutagenesis** A technique used to discover the function of a gene by cloning it, introducing specific changes into the cloned gene's sequence, reinserting the mutated gene into a cell, and studying the phenotype of the mutant.

ion (ī′-on) An atom or group of atoms that has gained or lost one or more electrons, thus acquiring a charge.

ion channel (ī′-on) A transmembrane protein channel that allows a specific ion to diffuse across the membrane down its concentration or electrochemical gradient.

ionic bond (ī-on′-ik) A chemical bond resulting from the attraction between oppositely charged ions.

ionic compound (ī-on′-ik) A compound resulting from the formation of an ionic bond; also called a salt.

iris The coloured part of the vertebrate eye, formed by the anterior portion of the choroid.

isomer (ī′-sō-mer) One of two or more compounds that have the same numbers of atoms of the same elements but different structures and hence different properties.

isomorphic Referring to alternating generations in plants and certain algae in which the sporophytes and gametophytes look alike, although they differ in chromosome number.

isotonic (ī′-sō-ton′-ik) Referring to a solution that, when surrounding a cell, causes no net movement of water into or out of the cell.

isotope (ī′-sō-tōp′) One of several atomic forms of an element, each with the same number of protons but a different number of neutrons, thus differing in atomic mass.

iteroparity Reproduction in which adults produce offspring over many years; also called repeated reproduction.

jasmonate Any of a class of plant hormones that regulate a wide range of developmental processes in plants and play a key role in plant defence against herbivores.

joule (J) A unit of energy: 1 J = 0.239 cal; 1 cal = 4.184 J.

juxtaglomerular apparatus (JGA) (juks′-tuh-gluh-mār′-yū-ler) A specialised tissue in nephrons that releases the enzyme renin in response to a drop in blood pressure or volume.

juxtamedullary nephron In mammals and birds, a nephron with a loop of Henle that extends far into the renal medulla.

karyogamy (kār′-ē-og′-uh-mē) In fungi, the fusion of haploid nuclei contributed by the two parents; occurs as one stage of sexual reproduction, preceded by plasmogamy.

karyotype (kār′-ē-ō-tīp) A display of the chromosome pairs of a cell arranged by size and shape.

keystone species A species that is not necessarily abundant in a community yet exerts strong control on community structure by the nature of its ecological role or niche.

kidney In vertebrates, one of a pair of excretory organs where blood filtrate is formed and processed into urine.

kilocalorie (kcal) A thousand calories; the amount of heat energy required to raise the temperature of 1 kg of water by 1°C.

kinetic energy (kuh-net′-ik) The energy associated with the relative motion of objects. Moving matter can perform work by imparting motion to other matter.

kinetochore (kuh-net′-uh-kōr) A structure of proteins attached to the centromere that links each sister chromatid to the mitotic spindle.

kinetoplastid A protist, such as a trypanosome, that has a single large mitochondrion that houses an organised mass of DNA.

kingdom A taxonomic category, the second broadest after domain.

kin selection Natural selection that favours altruistic behaviour by enhancing the reproductive success of relatives.

***K*-selection** Selection for life history traits that are sensitive to population density.

labia majora A pair of thick, fatty ridges that enclose and protect the rest of the vulva.

labia minora A pair of slender skin folds that surround the openings of the vagina and urethra.

lacteal (lak′-tē-ul) A tiny lymph vessel extending into the core of an intestinal villus and serving as the destination for absorbed chylomicrons.

lactic acid fermentation Glycolysis followed by the reduction of pyruvate to lactate, regenerating NAD^+ with no release of carbon dioxide.

lagging strand A discontinuously synthesised DNA strand that elongates by means of Okazaki fragments, each synthesised in a $5' \rightarrow 3'$ direction away from the replication fork.

lamprey Any of the jawless vertebrates with highly reduced vertebrae that live in freshwater and marine environments. Almost half of extant lamprey species are parasites that feed by clamping their round, jawless mouth onto the flank of a live fish; nonparasitic lampreys are suspension feeders that feed only as larvae.

lancelet A member of the clade Cephalochordata, small blade-shaped marine chordates that lack a backbone.

landscape An area containing several different ecosystems linked by exchanges of energy, materials, and organisms.

landscape ecology The study of how the spatial arrangement of habitat types affects the distribution and abundance of organisms and ecosystem processes.

large intestine The portion of the vertebrate alimentary canal between the small intestine and the anus; functions mainly in water absorption and the formation of faeces.

larva (lar′-vuh) (plural, **larvae**) A free-living, sexually immature form in some animal life cycles that may differ from the adult animal in morphology, nutrition, and habitat.

larynx (lăr′-inks) The portion of the respiratory tract containing the vocal cords; also called the voice box.

lateralisation Segregation of functions in the cortex of the left and right cerebral hemispheres.

lateral line system A mechanoreceptor system consisting of a series of pores and receptor units along the sides of the body in fishes and aquatic amphibians; detects water movements made by the animal itself and by other moving objects.

lateral meristem (mār′-uh-stem) A meristem that thickens the roots and shoots of woody plants. The vascular cambium and cork cambium are lateral meristems.

lateral root A root that arises from the pericycle of an established root.

law of conservation of mass A physical law stating that matter can change form but cannot be created or destroyed. In a closed system, the mass of the system is constant.

law of independent assortment Mendel's second law, stating that each pair of alleles segregates, or assorts, independently of each other pair during gamete formation; applies when genes for two characters are located on different pairs of homologous chromosomes or when they are far enough apart on the same chromosome to behave as though they are on different chromosomes.

law of segregation Mendel's first law, stating that the two alleles in a pair segregate (separate from each other) into different gametes during gamete formation.

leading strand The new complementary DNA strand synthesised continuously along the template strand towards the replication fork in the mandatory $5' \rightarrow 3'$ direction.

leaf The main photosynthetic organ of vascular plants.

leaf primordium (plural, **primordia**) A finger-like projection along the flank of a shoot apical meristem, from which a leaf arises.

learning The modification of behaviour as a result of specific experiences.

lens The structure in an eye that focuses light rays onto the photoreceptors.

lenticel (len′-ti-sel) A small raised area in the bark of stems and roots that enables gas exchange between living cells and the outside air.

lepidosaur (leh-pid′-uh-sōr) A member of the reptilian group that includes lizards, snakes, and two species of New Zealand animals called tuataras.

leucocyte (lū′-kō-sīt′) A blood cell that functions in fighting infections; also called a white blood cell.

lichen The mutualistic association between a fungus and a photosynthetic alga or cyanobacterium.

life cycle The generation-to-generation sequence of stages in the reproductive history of an organism.

life history The traits that affect an organism's schedule of reproduction and survival.

life table A summary of the age-specific survival and reproductive rates of individuals in a population.

ligament A fibrous connective tissue that joins bones together at joints.

ligand (lig′-und) A molecule that binds specifically to another molecule, usually a larger one.

ligand-gated ion channel (lig′-und) A transmembrane protein containing a pore that opens or closes as it changes shape in response to a signalling molecule (ligand), allowing or blocking the flow of specific ions; also called an ionotropic receptor.

light chain One of the two types of polypeptide chains that make up an antibody molecule and B cell receptor; consists of a variable region, which contributes to the antigen-binding site, and a constant region.

light-harvesting complex A complex of proteins associated with pigment molecules (including chlorophyll *a*, chlorophyll *b*, and carotenoids) that captures light energy and transfers it to reaction-centre pigments in a photosystem.

light microscope (LM) An optical instrument with lenses that refract (bend) visible light to magnify images of specimens.

light reactions The first of two major stages in photosynthesis (preceding the Calvin cycle). These reactions, which occur on the thylakoid membranes of the chloroplast or on membranes of certain prokaryotes, convert solar energy to the chemical energy of ATP and NADPH, releasing oxygen in the process.

lignin (lig′-nin) A strong polymer embedded in the cellulose matrix of the secondary cell walls of vascular plants that provides structural support in terrestrial species.

limiting nutrient An element that must be added for production to increase in a particular area.

limnetic zone In a lake, the well-lit, open surface waters far from shore.

linear electron flow A route of electron flow during the light reactions of photosynthesis that involves both photosystems (I and II) and produces ATP, NADPH, and O_2. The net electron flow is from H_2O to $NADP^+$.

line graph A graph in which each data point is connected to the next point in the data set with a straight line.

linkage map A genetic map based on the frequencies of recombination between markers during crossing over of homologous chromosomes.

linked genes Genes located close enough together on a chromosome that they tend to be inherited together.

lipid (lip′-id) Any of a group of large biological molecules, including fats, phospholipids, and steroids, that mix poorly, if at all, with water.

littoral zone In a lake, the shallow, well-lit waters close to shore.

liver A large internal organ in vertebrates that performs diverse functions, such as producing bile, maintaining blood glucose level, and detoxifying poisonous chemicals in the blood.

liverwort A small, herbaceous, nonvascular plant that is a member of the phylum Hepatophyta.

loam The most fertile soil type, made up of roughly equal amounts of sand, silt, and clay.

lobe-fin Member of a clade of osteichthyans having rod-shaped muscular fins. The group includes coelacanths, lungfishes, and tetrapods.

local regulator A secreted molecule that influences cells near where it is secreted.

locomotion Active motion from place to place.

locus (plural, **loci**), (lō′-kus), (lō′-sī) A specific place along the length of a chromosome where a given gene is located.

logistic population growth Population growth that levels off as population size approaches carrying capacity.

long-day plant A plant that flowers (usually in late spring or early summer) only when the light period is longer than a critical length.

long noncoding RNA (lncRNA) An RNA between 200 and hundreds of thousands of nucleotides in length that does not code for protein but is expressed at significant levels.

long-term memory The ability to hold, associate, and recall information over one's lifetime.

long-term potentiation (LTP) An enhanced responsiveness to an action potential (nerve signal) by a receiving neuron.

loop of Henle (hen′-lē) The hairpin turn, with a descending and ascending limb, between the proximal and distal tubules of the vertebrate kidney; functions in water and salt reabsorption.

lophophore (lof′-uh-fōr) In some lophotrochozoan animals, including brachiopods, a crown of ciliated tentacles that surround the mouth and function in feeding.

Lophotrochozoa (lo-phah′-truh-kō-zō′-uh) One of the three main lineages of bilaterian animals; lophotrochozoans include organisms that have lophophores or trochophore larvae. *See also* Deuterostomia and Ecdysozoa.

low-density lipoprotein (LDL) A particle in the blood made up of thousands of cholesterol molecules and other lipids bound to a protein. LDL transports cholesterol from the liver for incorporation into cell membranes.

lung An infolded respiratory surface of a terrestrial vertebrate, land snail, or spider that connects to the atmosphere by narrow tubes.

luteinising hormone (LH) (lū′-tē-uh-nī′-zing) A tropic hormone that is produced and secreted by the anterior pituitary and that stimulates ovulation in females and androgen production in males.

lycophyte (lī′-kuh-fīt) An informal name for a member of the phylum Lycophyta, which includes club mosses, spike mosses, and quillworts.

lymph The colourless fluid, derived from interstitial fluid, in the lymphatic system of vertebrates.

lymph node An organ located along a lymph vessel. Lymph nodes filter lymph and contain cells that attack viruses and bacteria.

lymphatic system A system of vessels and nodes, separate from the circulatory system, that returns fluid, proteins, and cells to the blood.

lymphocyte A type of white blood cell that mediates immune responses. The two main classes are B cells and T cells.

lysogenic cycle (lī′-sō-jen′-ik) A type of phage replicative cycle in which the viral genome becomes incorporated into the bacterial host chromosome as a prophage, is replicated along with the chromosome, and does not kill the host.

lysosome (lī′-suh-sōm) A membrane-enclosed sac of hydrolytic enzymes found in the cytoplasm of animal cells and some protists.

lysozyme (lī′-sō-zīm) An enzyme that destroys bacterial cell walls; in mammals, it is found in sweat, tears, and saliva.

lytic cycle (lit′-ik) A type of phage replicative cycle resulting in the release of new phages by lysis (and death) of the host cell.

macroevolution Evolutionary change above the species level. Examples of macroevolutionary change include the origin of a new group of organisms through a series of speciation events and the impact of mass extinctions on the diversity of life and its subsequent recovery.

macromolecule A giant molecule formed by the joining of smaller molecules. Polysaccharides, proteins, and nucleic acids are macromolecules.

macronutrient An essential element that an organism must obtain in relatively large amounts. *See also* micronutrient.

macrophage (mak′-rō-fāj) A phagocytic cell present in many tissues that functions in innate immunity by destroying microorganisms and in acquired immunity as an antigen-presenting cell.

magnoliid A member of the angiosperm clade that is most closely related to the combined eudicot and monocot clades. Extant examples are magnolias, laurels, and black pepper plants.

major depressive disorder A mood disorder characterised by feelings of sadness, lack of self-worth, emptiness, or loss of interest in nearly all things.

major histocompatibility complex (MHC) molecule A host protein that functions in antigen presentation. Foreign MHC molecules on transplanted tissue can trigger T cell responses that may lead to rejection of the transplant.

malignant tumour A cancerous tumour containing cells that have significant genetic and cellular changes and are capable of invading and surviving in new sites. Malignant tumours can impair the functions of one or more organs.

Malpighian tubule (mal-pig′-ē-un) A unique excretory organ of insects that empties into the digestive tract, removes nitrogenous wastes from the haemolymph, and functions in osmoregulation.

mammal A member of the clade Mammalia, amniotes that have hair and mammary glands (glands that produce milk).

mammary gland An exocrine gland that secretes milk for nourishing the young. Mammary glands are characteristic of mammals.

mantle One of the three main parts of a mollusc; a fold of tissue that drapes over the mollusc's visceral mass and may secrete a shell. *See also* foot and visceral mass.

mantle cavity A water-filled chamber that houses the gills, anus, and excretory pores of a mollusc.

map unit A unit of measurement of the distance between genes. One map unit is equivalent to a 1% recombination frequency.

marine benthic zone The ocean floor.

mark-recapture method A sampling technique used to estimate the size of animal populations.

marsupial (mar-sū′-pē-ul) A mammal, such as a koala, kangaroo, or opossum, whose young complete their embryonic development inside a maternal pouch called the marsupium.

mass extinction The elimination of a large number of species throughout Earth, the result of global environmental changes.

mass number The total number of protons and neutrons in an atom's nucleus.

mast cell Immune system cell that secretes histamine; plays role in inflammatory response and allergies.

mate-choice copying Behaviour in which individuals in a population copy the mate choice of others, apparently due to social learning.

maternal effect gene A gene that, when mutant in the mother, results in a mutant phenotype in the offspring, regardless of the offspring's genotype. Maternal effect genes, also called egg-polarity genes, were first identified in *Drosophila melanogaster.*

matter Anything that takes up space and has mass.

maximum likelihood As applied to DNA sequence data, a principle that states that when considering multiple phylogenetic hypotheses, one should take into account the hypothesis that reflects the most likely sequence of evolutionary events, given certain rules about how DNA changes over time.

maximum parsimony The principle that when considering multiple explanations for an observation, one should first investigate the simplest explanation that is consistent with the facts.

mean The sum of all data points in a data set divided by the number of data points.

mechanoreceptor A sensory receptor that detects physical deformation in the body's environment associated with pressure, touch, stretch, motion, or sound.

medulla oblongata (meh-dul′-uh ob′-long-go′-tuh) The lowest part of the vertebrate brain, commonly called the medulla; a swelling of the hindbrain anterior to the spinal cord that controls autonomic, homeostatic functions, including breathing, heart and blood vessel activity, swallowing, digestion, and vomiting.

medusa (muh-dū′-suh) (plural, **medusae**) The floating, mouth-down form of the cnidarian body plan. The alternate form is the polyp.

megafauna Large animals that existed in Australia during the Pleistocene period (between 1.8 million and 10,000 years ago), including reptiles, birds, and marsupials.

megapascal (MPa) (meg′-uh-pas-kal′) A unit of pressure equivalent to about 10 atmospheres of pressure.

megaphyll (meh′-guh-fil) A leaf with a highly branched vascular system, found in almost all vascular plants other than lycophytes. *See also* microphyll.

megaspore A spore from a heterosporous plant species that develops into a female gametophyte.

meiosis (mī-ō′-sis) A modified type of cell division in sexually reproducing organisms consisting of two rounds of cell division but only one round of DNA replication. It results in cells with half the number of chromosome sets as the original cell.

meiosis I (mī-ō′-sis) The first division of a two-stage process of cell division in sexually reproducing organisms that results in cells with half the number of chromosome sets as the original cell.

meiosis II (mī-ō′-sis) The second division of a two-stage process of cell division in sexually reproducing organisms that results in cells with half the number of chromosome sets as the original cell.

melanocyte-stimulating hormone (MSH) A hormone produced and secreted by the anterior pituitary with multiple activities, including regulating the behaviour of pigment-containing cells in the skin of some vertebrates.

melatonin A hormone that is secreted by the pineal gland and that is involved in the regulation of biological rhythms and sleep.

membrane potential The difference in electrical charge (voltage) across a cell's plasma membrane due to the differential distribution of ions. Membrane potential affects the activity of excitable cells and the transmembrane movement of all charged substances.

memory cell One of a clone of long-lived lymphocytes, formed during the primary immune response, that remains in a lymphoid organ until activated by exposure to the same antigen that triggered its formation. Activated memory cells mount the secondary immune response.

menopause The cessation of ovulation and menstruation marking the end of a human female's reproductive years.

menstrual cycle (men′-strū-ul) In humans and certain other primates, the periodic growth and shedding of the uterine lining that occurs in the absence of pregnancy.

menstruation The shedding of portions of the endometrium during a uterine (menstrual) cycle.

meristem (mār′-uh-stem) Plant tissue that remains embryonic as long as the plant lives, allowing for indeterminate growth.

mesoderm (mez′-ō-derm) The middle primary germ layer in a triploblastic animal embryo; develops into the notochord, the lining of the coelom, muscles, skeleton, gonads, kidneys, and most of the circulatory system in species that have these structures.

mesohyl (mez′-ō-hīl) A gelatinous region between the two layers of cells of a sponge.

mesophyll (mez′-ō-fil) Leaf cells specialised for photosynthesis. In C_3 and CAM plants, mesophyll cells are located between the upper and lower epidermis; in C_4 plants, they are located between the bundle-sheath cells and the epidermis.

messenger RNA (mRNA) A type of RNA, synthesised using a DNA template, that attaches to ribosomes in the cytoplasm and specifies the primary structure of a protein. (In eukaryotes, the primary RNA transcript must undergo RNA processing to become mRNA.)

metabolic pathway A series of chemical reactions that either builds a complex molecule (anabolic pathway) or breaks down a complex molecule to simpler molecules (catabolic pathway).

metabolic rate The total amount of energy an animal uses in a unit of time.

metabolism (muh-tab'-uh-lizm) The totality of an organism's chemical reactions, consisting of catabolic and anabolic pathways, which manage the material and energy resources of the organism.

metagenomics The collection and sequencing of DNA from a group of species, usually an environmental sample of microorganisms. Computer software sorts partial sequences and assembles them into genome sequences of individual species making up the sample.

metamorphosis (met'-uh-mōr'-fuh-sis) A developmental transformation that turns an animal larva into either an adult or an adult-like stage that is not yet sexually mature.

metanephridium (met'-uh-nuh-frid'-ē-um) (plural, **metanephridia**) An excretory organ found in many invertebrates that typically consists of tubules connecting ciliated internal openings to external openings.

metaphase The third stage of mitosis, in which the spindle is complete and the chromosomes, attached to microtubules at their kinetochores, are all aligned at the metaphase plate.

metaphase plate An imaginary structure located at a plane midway between the two poles of a cell in metaphase on which the centromeres of all the duplicated chromosomes are located.

metapopulation A group of spatially separated populations of one species that interact through immigration and emigration.

metastasis (muh-tas'-tuh-sis) The spread of cancer cells to locations distant from their original site.

methanogen (meth-an'-ō-jen) An organism that produces methane as a waste product of the way it obtains energy. All known methanogens are in domain Archaea.

methyl group A chemical group consisting of a carbon bonded to three hydrogen atoms. The methyl group may be attached to a carbon or to a different atom.

microbiome The collection of microorganisms living in or on an organism's body, along with their genetic material.

microclimate Climate patterns on a very fine scale, such as the specific climatic conditions underneath a log.

microevolution Evolutionary change below the species level; change in the allele frequencies in a population over generations.

microfilament A cable composed of actin proteins in the cytoplasm of almost every eukaryotic cell, making up part of the cytoskeleton and acting alone or with myosin to cause cell contraction; also called an actin filament.

micronutrient An essential element that an organism needs in very small amounts. *See also* macronutrient.

microphyll (mī'-krō-fil) A small, usually spine-shaped leaf supported by a single strand of vascular tissue, found only in lycophytes.

microplastic A plastic particle less than 5 mm in size; microplastics have contaminated all of the world's oceans as well as freshwater and terrestrial ecosystems.

micropyle A pore in the integuments of an ovule.

microRNA (miRNA) A small, single-stranded RNA molecule, generated from a double-stranded RNA precursor. The miRNA associates with one or more proteins in a complex that can degrade or prevent translation of an mRNA with a complementary sequence.

microspore A spore from a heterosporous plant species that develops into a male gametophyte.

microsporidian A member of the fungal phylum Microsporidia, unicellular parasites of protists and animals; microsporidians and their sister taxon (cryptomycetes) are a basal fungal lineage.

microtubule A hollow rod composed of tubulin proteins that makes up part of the cytoskeleton in all eukaryotic cells and is found in cilia and flagella.

microvillus (plural, **microvilli**) One of many fine, finger-like projections of the epithelial cells in the lumen of the small intestine that increase its surface area.

midbrain One of three ancestral and embryonic regions of the vertebrate brain; develops into sensory integrating and relay centres that send sensory information to the cerebrum.

middle ear One of three main regions of the vertebrate ear; in mammals, a chamber containing three small bones (the malleus, incus, and stapes) that convey vibrations from the eardrum to the oval window.

middle lamella (luh-mel'-uh) In plants, a thin layer of adhesive extracellular material, primarily pectins, found between the primary walls of adjacent young cells.

migration A regular, long-distance change in location.

mineral In nutrition, a simple nutrient that is inorganic and therefore cannot be synthesised in the body.

mineralocorticoid A steroid hormone secreted by the adrenal cortex that regulates salt and water homeostasis.

minimum viable population (MVP) The smallest population size at which a species is able to sustain its numbers and survive.

mismatch repair The cellular process that uses specific enzymes to remove and replace incorrectly paired nucleotides.

missense mutation A nucleotide-pair substitution that results in a codon that codes for a different amino acid.

mitochondrial matrix The compartment of the mitochondrion enclosed by the inner membrane and containing enzymes and substrates for the citric acid cycle, as well as ribosomes and DNA.

mitochondrion (mī'-tō-kon'-drē-un) (plural, **mitochondria**) An organelle in eukaryotic cells that serves as the site of cellular respiration; uses oxygen to break down organic molecules and synthesise ATP.

mitosis (mī-tō'-sis) A process of nuclear division in eukaryotic cells conventionally divided into five stages: prophase, prometaphase, metaphase, anaphase, and telophase. Mitosis conserves chromosome number by allocating replicated chromosomes equally to each of the daughter nuclei.

mitotic (M) phase The phase of the cell cycle that includes mitosis and cytokinesis.

mitotic spindle An assemblage of microtubules and associated proteins that is involved in the movement of chromosomes during mitosis.

mixotroph An organism that is capable of both photosynthesis and heterotrophy.

model A physical or conceptual representation of a natural phenomenon.

model organism A particular species chosen for research into broad biological principles because it is representative of a larger group and usually easy to grow in a lab.

molarity A common measure of solute concentration, referring to the number of moles of solute per litre of solution.

mole (mol) The number of grams of a substance that equals its molecular or atomic mass in daltons; a mole contains Avogadro's number of the molecules or atoms in question.

molecular clock A method for estimating the time required for a given amount of evolutionary change, based on the observation that some regions of genomes evolve at constant rates.

molecular mass The sum of the masses of all the atoms in a molecule; sometimes called molecular weight.

molecule Two or more atoms held together by covalent bonds.

monilophyte An informal name for a member of the phylum Monilophyta, which includes ferns, horsetails, and whisk ferns and their relatives.

monoclonal antibody (mon'-ō-klōn'-ul) Any of a preparation of antibodies that have been produced by a single clone of cultured cells and thus are all specific for the same epitope.

monocot A member of a clade consisting of flowering plants that have one embryonic seed leaf, or cotyledon.

monogamous (muh-nog'-uh-mus) Referring to a type of relationship in which one male mates with just one female.

monohybrid An organism that is heterozygous with respect to a single gene of interest. All the offspring from a cross between parents homozygous for different alleles are monohybrids. For example, parents of genotypes *AA* and *aa* produce a monohybrid of genotype *Aa*.

monohybrid cross A cross between two organisms that are heterozygous for the character being followed (or the self-pollination of a heterozygous plant).

monomer (mon'-uh-mer) The subunit that serves as the building block of a polymer.

monophyletic (mon'-ō-fī-let'-ik) Pertaining to a group of taxa that consists of a common ancestor and all of its descendants. A monophyletic taxon is equivalent to a clade.

monosaccharide (mon'-ō-sak'-uh-rīd) The simplest carbohydrate, active alone or serving as a monomer for disaccharides and polysaccharides. Also called simple sugars, monosaccharides have molecular formulas that are generally some multiple of CH_2O.

monosomic Referring to a diploid cell that has only one copy of a particular chromosome instead of the normal two.

monotreme An egg-laying mammal, such as a platypus or echidna. Like all mammals, monotremes have hair and produce milk, but they lack nipples.

morphogen A substance, such as Bicoid protein in *Drosophila*, that provides positional information in the form of a concentration gradient along an embryonic axis.

morphogenesis (mōr′-fō-jen′-uh-sis) The development of the form of an organism and its structures.

morphological species concept Definition of a species in terms of measurable anatomical criteria.

moss A small, herbaceous, nonvascular plant that is a member of the phylum Bryophyta.

motor neuron A nerve cell that transmits signals from the brain or spinal cord to muscles or glands.

motor protein A protein that interacts with cytoskeletal elements and other cell components, producing movement of the whole cell or parts of the cell.

motor system An efferent branch of the vertebrate peripheral nervous system composed of motor neurons that carry signals to skeletal muscles in response to external stimuli.

motor unit A single motor neuron and all the muscle fibres it controls.

mould Informal term for a fungus that grows as a filamentous fungus, producing haploid spores by mitosis and forming a visible mycelium.

moulting A process in ecdysozoans in which the exoskeleton is shed at intervals, allowing growth by the production of a larger exoskeleton.

movement corridor A series of small clumps or a narrow strip of quality habitat (usable by organisms) that connects otherwise isolated patches of quality habitat.

MPF Maturation-promoting factor (or M-phase-promoting factor); a protein complex required for a cell to progress from late interphase to mitosis. The active form consists of cyclin and a protein kinase.

mucoromycete A member of the fungal phylum Mucoromycota, characterised by the formation of a sturdy structure called a zygosporangium during sexual reproduction.

mucus A viscous and slippery mixture of glycoproteins, cells, salts, and water that moistens and protects the membranes lining body cavities that open to the exterior.

Müllerian mimicry (myū-lār′-ē-un mim′-uh-krē) Reciprocal mimicry by two unpalatable species.

multifactorial Referring to a phenotypic character that is influenced by multiple genes and environmental factors.

multigene family A collection of genes with similar or identical sequences, presumably of common origin.

multiple fruit A fruit derived from an entire inflorescence.

multiplication rule A rule of probability stating that the probability of two or more independent events occurring together can be determined by multiplying their individual probabilities.

muscle tissue Tissue consisting of long muscle cells that can contract, either on its own or when stimulated by nerve impulses.

mutagen (myū′-tuh-jen) A chemical or physical agent that interacts with DNA and can cause a mutation.

mutation (myū-tā′-shun) A change in the nucleotide sequence of an organism's DNA or in the DNA or RNA of a virus.

mutualism (myū′-chū-ul-izm) A +/+ ecological interaction that benefits individuals of both interacting species.

mycelium (mī-sē′-lē-um) (plural, **mycelia**) The densely branched network of hyphae in a fungus.

mycorrhiza (plural, **mycorrhizae**) (mī′-kō-rī′-zuh, mī′-kō-rī′-zē) A mutualistic association of plant roots and fungus.

mycosis (mī-kō′-sis) General term for a fungal infection.

myelin sheath (mī′-uh-lin) Wrapped around the axon of a neuron, an insulating coat of cell membranes from Schwann cells or oligodendrocytes. It is interrupted by nodes of Ranvier, where action potentials are generated.

myofibril (mī′-ō-fī′-bril) A longitudinal bundle in a muscle cell (fibre) that contains thin filaments of actin and regulatory proteins and thick filaments of myosin.

myoglobin (mī′-uh-glō′-bin) An oxygen-storing, pigmented protein in muscle cells.

myosin (mī′-uh-sin) A type of motor protein that associates into filaments that interact with actin filaments to cause cell contraction.

myriapod (mir′-ē-uh-pod′) A terrestrial arthropod with many body segments and one or two pairs of legs per segment. Millipedes and centipedes are the two major groups of living myriapods.

NAD^+ The oxidised form of nicotinamide adenine dinucleotide, a coenzyme that can accept electrons, becoming NADH. NADH temporarily stores electrons during cellular respiration.

NADH The reduced form of nicotinamide adenine dinucleotide that temporarily stores electrons during cellular respiration. NADH acts as an electron donor to the electron transport chain.

$NADP^+$ The oxidised form of nicotinamide adenine dinucleotide phosphate, an electron carrier that can accept electrons, becoming NADPH. NADPH temporarily stores energised electrons produced during the light reactions.

NADPH The reduced form of nicotinamide adenine dinucleotide phosphate; temporarily stores energised electrons produced during the light reactions. NADPH acts as "reducing power" that can be passed along to an electron acceptor, reducing it.

natural killer cell A type of white blood cell that can kill tumour cells and virus-infected cells as part of innate immunity.

natural selection A process in which individuals that have certain inherited traits tend to survive and reproduce at higher rates than other individuals because of those traits.

negative feedback A form of regulation in which accumulation of an end product of a process slows the process; in physiology, a primary mechanism of homeostasis, whereby a change in a variable triggers a response that counteracts the initial change.

negative pressure breathing A breathing system in which air is pulled into the lungs.

nematocyst (nem′-uh-tuh-sist′) In a cnidocyte of a cnidarian, a capsule-like organelle containing a coiled thread that when discharged can penetrate the body wall of the prey.

nephron (nef′-ron) The tubular excretory unit of the vertebrate kidney.

neritic zone The shallow region of the ocean overlying the continental shelf.

nerve A fibre composed primarily of the bundled axons of neurons.

nervous system In animals, the fast-acting internal system of communication involving sensory receptors, networks of nerve cells, and connections to muscles and glands that respond to nerve signals; functions in concert with the endocrine system to effect internal regulation and maintain homeostasis.

nervous tissue Tissue made up of neurons and supportive cells.

net ecosystem production (NEP) The gross primary production of an ecosystem minus the energy used by all autotrophs and heterotrophs for respiration.

net primary production (NPP) The gross primary production of an ecosystem minus the energy used by the producers for respiration.

neural crest In vertebrates, a region located along the sides of the neural tube where it pinches off from the ectoderm. Neural crest cells migrate to various parts of the embryo and form pigment cells in the skin and parts of the skull, teeth, adrenal glands, and peripheral nervous system.

neural tube A tube of infolded ectodermal cells that runs along the anterior-posterior axis of a vertebrate, just dorsal to the notochord. It will give rise to the central nervous system.

neurohormone A molecule that is secreted by a neuron, travels in body fluids, and acts on specific target cells, changing their functioning.

neuron (nyūr′-on) A nerve cell; the fundamental unit of the nervous system, having structure and properties that allow it to conduct signals by taking advantage of the electrical charge across its plasma membrane.

neuronal plasticity The capacity of a nervous system to change with experience.

neuropeptide A relatively short chain of amino acids that serves as a neurotransmitter.

neurotransmitter A molecule that is released from the synaptic terminal of a neuron at a chemical synapse, diffuses across the synaptic cleft, and binds to the postsynaptic cell, triggering a response.

neutral variation Genetic variation that does not provide a selective advantage or disadvantage.

neutron A subatomic particle having no electrical charge (electrically neutral), with a mass of about 1.7×10^{-24} g, found in the nucleus of an atom.

neutrophil The most abundant type of white blood cell. Neutrophils are phagocytic and tend to self-destruct as they destroy foreign invaders, limiting their life span to a few days.

nitric oxide (NO) A gas produced by many types of cells that functions as a local regulator and as a neurotransmitter.

nitrogen cycle The natural process by which nitrogen, either from the atmosphere or from decomposed organic material, is converted by soil bacteria to compounds assimilated by plants. This incorporated nitrogen is then taken in by other organisms and subsequently released, acted on by bacteria, and made available again to the nonliving environment.

nitrogen fixation The conversion of atmospheric nitrogen (N_2) to ammonia (NH_3). Biological nitrogen fixation is carried out by certain prokaryotes, some of which have mutualistic relationships with plants.

nociceptor (nō′-si-sep′-tur) A sensory receptor that responds to noxious or painful stimuli; also called a pain receptor.

node A point along the stem of a plant at which leaves are attached.

node of Ranvier (ron′-vē-ā′) Gap in the myelin sheath of certain axons where an action potential may be generated. In saltatory conduction, an action potential is regenerated at each node, appearing to "jump" along the axon from node to node.

nodule A swelling on the root of a legume. Nodules are composed of plant cells that contain nitrogen-fixing bacteria of the genus *Rhizobium*.

noncompetitive inhibitor A substance that reduces the activity of an enzyme by binding to a location remote from the active site, changing the enzyme's shape so that the active site no longer effectively catalyses the conversion of substrate to product.

nondisjunction An error in meiosis or mitosis in which members of a pair of homologous chromosomes or a pair of sister chromatids fail to separate properly from each other.

nonequilibrium model A model that maintains that communities change constantly after being buffeted by disturbances.

nonpolar covalent bond A type of covalent bond in which electrons are shared equally between two atoms of similar electronegativity.

nonsense mutation A mutation that changes an amino acid codon to one of the three stop codons, resulting in a shorter and usually nonfunctional protein.

norepinephrine (nor-ep′-i-nef′-rin) A catecholamine that is chemically and functionally similar to adrenaline and acts as a hormone or neurotransmitter; also called noradrenaline.

northern coniferous forest A terrestrial biome characterised by long, cold winters and dominated by cone-bearing trees.

no-till agriculture A plowing technique that minimally disturbs the soil, thereby reducing soil loss.

notochord (nō′-tuh-kord′) A longitudinal, flexible rod made of tightly packed mesodermal cells that runs along the anterior-posterior axis of a chordate in the dorsal part of the body.

nuclear envelope In a eukaryotic cell, the double membrane that surrounds the nucleus, perforated with pores that regulate traffic with the cytoplasm. The outer membrane is continuous with the endoplasmic reticulum.

nuclear lamina A netlike array of protein filaments that lines the inner surface of the nuclear envelope and helps maintain the shape of the nucleus.

nucleariid A member of a group of unicellular, amoeboid protists that are more closely related to fungi than they are to other protists.

nuclease An enzyme that cuts DNA or RNA, either removing one or a few bases or hydrolysing the DNA or RNA completely into its component nucleotides.

nucleic acid (nū-klā′-ik) A polymer (polynucleotide) consisting of many nucleotide monomers; serves as a blueprint for proteins and, through the actions of proteins, for all cellular activities. The two types are DNA and RNA.

nucleic acid hybridisation (nū-klā′-ik) The base pairing of one strand of a nucleic acid to the complementary sequence on a strand from *another* nucleic acid molecule.

nucleic acid probe (nū′-klā′-ik) In DNA technology, a labelled single-stranded nucleic acid molecule used to locate a specific nucleotide sequence in a nucleic acid sample. Molecules of the probe hydrogen-bond to the complementary sequence wherever it occurs; radioactive, fluorescent, or other labelling of the probe allows its location to be detected.

nucleoid (nū′-klē-oyd) A non-membrane-enclosed region in a prokaryotic cell where its chromosome is located.

nucleolus (nū-klē′-ō-lus) (plural, **nucleoli**) A specialised structure in the nucleus, consisting of chromosomal regions containing ribosomal RNA (rRNA) genes along with ribosomal proteins imported from the cytoplasm; site of rRNA synthesis and ribosomal subunit assembly. *See also* ribosome.

nucleosome (nū′-klē-ō-sōm′) The basic, bead-like unit of DNA packing in eukaryotes, consisting of a segment of DNA wound around a protein core composed of two copies of each of four types of histone.

nucleotide (nū′-klē-ō-tīd′) The building block of a nucleic acid, consisting of a five-carbon sugar covalently bonded to a nitrogenous base and one to three phosphate groups.

nucleotide excision repair (nū′-klē-ō-tīd′) A repair system that removes and then correctly replaces a damaged segment of DNA using the undamaged strand as a guide.

nucleotide-pair substitution (nū′-klē-ō-tīd′) A type of point mutation in which one nucleotide in a DNA strand and its partner in the complementary strand are replaced by another pair of nucleotides.

nucleus (1) An atom's central core, containing protons and neutrons. (2) The organelle of a eukaryotic cell that contains the genetic material in the form of chromosomes, made up of chromatin. (3) A cluster of neurons.

nutrition The process by which an organism takes in and makes use of food substances.

obligate aerobe (ob′-lig-et ār′-ōb) An organism that requires oxygen for cellular respiration and cannot live without it.

obligate anaerobe (ob′-lig-et an′-uh-rōb) An organism that carries out only fermentation or anaerobic respiration. Such organisms cannot use oxygen and in fact may be poisoned by it.

ocean acidification The process by which the pH of the ocean is lowered (made more acidic) when excess CO_2 dissolves in seawater and forms carbonic acid (H_2CO_3).

oceanic pelagic zone Most of the ocean's waters far from shore, constantly mixed by ocean currents.

odourant A molecule that can be detected by sensory receptors of the olfactory system.

oesophagus (eh-sof′-uh-gus) A muscular tube that conducts food, by peristalsis, from the pharynx to the stomach.

oestradiol (es′-truh-dī′-ol) A steroid hormone that stimulates the development and maintenance of the female reproductive system and secondary sex characteristics; the major oestrogen in mammals.

oestrogen (es′-trō-jen) Any steroid hormone, such as oestradiol, that stimulates the development and maintenance of the female reproductive system and secondary sex characteristics.

oestrous cycle (es′-trus) A reproductive cycle characteristic of female mammals except humans and certain other primates, in which the endometrium is reabsorbed in the absence of pregnancy and sexual response occurs only during a mid-cycle point known as oestrus.

Okazaki fragment (ō′-kah-zah′-kē) A short segment of DNA synthesised away from the replication fork on a template strand during DNA replication. Many such segments are joined together to make up the lagging strand of newly synthesised DNA.

olfaction The sense of smell.

oligodendrocyte A type of glial cell that forms insulating myelin sheaths around the axons of neurons in the central nervous system.

oligotrophic lake A nutrient-poor, clear lake with few phytoplankton.

ommatidium (ōm′-uh-tid′-ē-um) (plural, **ommatidia**) One of the facets of the compound eye of arthropods and some polychaete worms.

omnivore An animal that regularly eats animals as well as plants or algae.

oncogene (on′-kō-jēn) A gene found in viral or cellular genomes that is involved in triggering molecular events that can lead to cancer.

oocyte (ō′-uh-sīt) A cell in the female reproductive system that differentiates to form an egg.

oogenesis (ō′-uh-jen′-uh-sis) The process in the ovary that results in the production of female gametes.

oogonium (ō′-uh-gō′-nē-em) (plural, **oogonia**) A cell that divides mitotically to form oocytes.

open circulatory system A circulatory system in which fluid called haemolymph bathes the tissues and organs directly and there is no distinction between the circulating fluid and the interstitial fluid.

operator In bacterial and phage DNA, a sequence of nucleotides near the start of an operon to which an active repressor can attach. The binding of the repressor prevents RNA polymerase from attaching to the promoter and transcribing the genes of the operon.

operculum (ō-per′-kyuh-lum) In aquatic osteichthyans, a protective bony flap that covers and protects the gills.

operon (op′-er-on) A unit of genetic function found in bacteria and phages, consisting of a promoter, an operator, and a coordinately regulated cluster of genes whose products function in a common pathway.

opisthokont (uh-pis′-thuh-kont′) A member of an extremely diverse clade of eukaryotes that includes fungi, animals, and several closely related groups of protists.

opposable thumb A thumb that can touch the ventral surface (fingerprint side) of the fingertip of all four fingers of the same hand with its own ventral surface.

opsin A membrane protein bound to a light-absorbing pigment molecule.

optimal foraging model The basis for analysing behaviour as a compromise between feeding costs and feeding benefits.

oral cavity The mouth of an animal.

orbital The three-dimensional space where an electron is found 90% of the time.

order In Linnaean classification, the taxonomic category above the level of family.

organ A specialised centre of body function composed of several different types of tissues.

organelle (ōr-guh-nel′) Any of several membrane-enclosed structures with specialised functions, suspended in the cytosol of eukaryotic cells.

organic chemistry The study of carbon compounds (organic compounds).

organism An individual living thing, consisting of one or more cells.

organismal ecology The branch of ecology concerned with the morphological, physiological, and behavioural ways in which individual organisms meet the challenges posed by their biotic and abiotic environments.

organ of Corti (kor′-tē) The actual hearing organ of the vertebrate ear, located in the floor of the cochlear duct in the inner ear; contains the receptor cells (hair cells) of the ear.

organogenesis (ōr-gan′-ō-jen′-uh-sis) The process in which organ rudiments develop from the three germ layers after gastrulation.

organ system A group of organs that work together in performing vital body functions.

origin of replication Site where the replication of a DNA molecule begins, consisting of a specific sequence of nucleotides.

orthologous genes Homologous genes that are found in different species because of speciation.

osculum (os′-kyuh-lum) A large opening in a sponge that connects the spongocoel to the environment.

osmoconformer An animal that is isoosmotic with its environment.

osmolarity (oz′-mō-lār′-uh-tē) Solute concentration expressed as molarity.

osmoregulation Regulation of solute concentrations and water balance by a cell or organism.

osmoregulator An animal that controls its internal osmolarity independent of the external environment.

osmosis (oz-mō′-sis) The diffusion of free water across a selectively permeable membrane.

osteichthyan (os′-tē-ik′-thē-an) A member of a vertebrate clade with jaws and mostly bony skeletons.

outer ear One of the three main regions of the ear in reptiles (including birds) and mammals; made up of the auditory canal and, in many birds and mammals, the pinna.

outgroup A species or group of species from an evolutionary lineage that is known to have diverged before the lineage that contains the group of species being studied. An outgroup is selected so that its members are closely related to the group of species being studied, but not as closely related as any study-group members are to each other.

oval window In the vertebrate ear, a membrane-covered gap in the skull bone, through which sound waves pass from the middle ear to the inner ear.

ovarian cycle (ō-vār′-ē-un) The cyclical recurrence of the follicular phase, ovulation, and the luteal phase in the mammalian ovary, regulated by hormones.

ovary (ō′-vuh-rē) (1) In flowers, the portion of a carpel in which the egg-containing ovules develop. (2) In animals, the structure that produces female gametes and reproductive hormones.

oviduct (ō′-vuh-duct) A tube passing from the ovary to the vagina in invertebrates or to the uterus in vertebrates, where it is also called a fallopian tube.

oviparous (ō-vip′-uh-rus) Referring to a type of development in which young hatch from eggs laid outside the mother's body.

ovoviviparous (ō′-vō-vī-vip′-uh-rus) Referring to a type of development in which young hatch from eggs that are retained in the mother's uterus.

ovulation The release of an egg from an ovary. In humans, an ovarian follicle releases an egg during each uterine (menstrual) cycle.

ovule (o′-vyūl) A structure that develops within the ovary of a seed plant and contains the female gametophyte.

oxidation The complete or partial loss of electrons from a substance involved in a redox reaction.

oxidative phosphorylation (fos′-fōr-uh-lā′-shun) The production of ATP using energy derived from the redox reactions of an electron transport chain; the third major stage of cellular respiration.

oxidising agent The electron acceptor in a redox reaction.

oxytocin (ok′-si-tō′-sen) A hormone produced by the hypothalamus and released from the posterior pituitary. It induces contractions of the uterine muscles during labour and causes the mammary glands to eject milk during nursing.

***p53* gene** A tumour-suppressor gene that codes for a specific transcription factor that promotes the synthesis of proteins that inhibit the cell cycle.

paedomorphosis (pē′-duh-mōr′-fuh-sis) The retention in an adult organism of the juvenile features of its evolutionary ancestors.

pain receptor A sensory receptor that responds to noxious or painful stimuli; also called a nociceptor.

paleoanthropology The study of human origins and evolution.

paleontology (pā′-lē-un-tol′-ō-jē) The scientific study of fossils.

pancreas (pan′-krē-us) A gland with exocrine and endocrine tissues. The exocrine portion functions in digestion, secreting enzymes and an alkaline solution into the small intestine via a duct; the ductless endocrine portion functions in homeostasis, secreting the hormones insulin and glucagon into the blood.

pancrustacean A member of a diverse arthropod clade that includes lobsters, crabs, barnacles and other crustaceans, as well as insects and their six-legged terrestrial relatives.

pandemic A global epidemic.

Pangaea (pan-jē′-uh) The supercontinent that formed near the end of the Paleozoic era, when plate movements brought all the landmasses of Earth together.

parabasalid A protist, such as a trichomonad, with modified mitochondria.

paracrine Referring to a secreted molecule that acts on a neighbouring cell.

paralogous genes Homologous genes that are found in the same genome as a result of gene duplication.

paraphyletic (pār′-uh-fī-let′-ik) Pertaining to a group of taxa that consists of a common ancestor and some, but not all, of its descendants.

parareptile A basal group of reptiles, consisting mostly of large, stocky quadrupedal herbivores. Parareptiles died out in the late Triassic period.

parasite (pār′-uh-sīt) An organism that feeds on the cell contents, tissues, or body fluids of another species (the host) while in or on the host organism. Parasites harm but usually do not kill their host.

parasitism (pār′-uh-sit-izm) A +/− ecological interaction in which one organism, the parasite, benefits by feeding upon another organism, the host, which is harmed; some parasites live within the host (feeding on its tissues), while others feed on the host's external surface.

parasympathetic division A division of the autonomic nervous system; generally enhances body activities that gain and conserve energy, such as digestion and reduced heart rate.

parathyroid gland One of four small endocrine glands, embedded in the surface of the thyroid gland, that secrete parathyroid hormone.

parathyroid hormone (PTH) A hormone secreted by the parathyroid glands that raises blood calcium level by promoting calcium release from bone and calcium retention by the kidneys.

parenchyma cell (puh-ren′-ki-muh) A relatively unspecialised plant cell type that carries out most of the metabolism, synthesises and stores organic products, and develops into a more differentiated cell type.

parental type An offspring with a phenotype that matches one of the true-breeding parental (P generation) phenotypes; also refers to the phenotype itself.

Parkinson's disease A progressive brain disease characterised by difficulty in initiating movements, slowness of movement, and rigidity.

parthenogenesis (par′-thuh-nō′-jen′-uh-sis) A form of asexual reproduction in which females produce offspring from unfertilised eggs.

partial pressure The pressure exerted by a particular gas in a mixture of gases (for instance, the pressure exerted by oxygen in air).

passive immunity Short-term immunity conferred by the transfer of antibodies, as occurs in the transfer of maternal antibodies to a fetus or nursing infant.

passive transport The diffusion of a substance across a biological membrane with no expenditure of energy.

pathogen An organism or virus that causes disease.

pathogen-associated molecular pattern (PAMP) A molecular sequence that is specific to a certain pathogen.

pattern formation The development of a multicellular organism's spatial organisation, the arrangement of organs and tissues in their characteristic places in three-dimensional space.

peat Extensive deposits of partially decayed organic material often formed primarily from the wetland moss *Sphagnum*.

pedigree A diagram of a family tree with conventional symbols, showing the occurrence of heritable characters in parents and offspring over multiple generations.

pelagic zone The open-water component of aquatic biomes.

penis The copulatory structure of male mammals.

PEP carboxylase An enzyme that adds CO_2 to phosphoenolpyruvate (PEP) to form oxaloacetate in mesophyll cells of C_4 plants. It acts prior to photosynthesis.

pepsin An enzyme present in gastric juice that begins the hydrolysis of proteins.

pepsinogen The inactive form of pepsin secreted by chief cells located in gastric pits of the stomach.

peptide bond The covalent bond between the carboxyl group on one amino acid and the amino group on another, formed by a dehydration reaction.

peptidoglycan (pep′-tid-ō-glī′-kan) A type of polymer in bacterial cell walls consisting of modified sugars cross-linked by short polypeptides.

perception The interpretation of sensory system input by the brain.

pericycle The outermost layer in the vascular cylinder, from which lateral roots arise.

periderm (pār′-uh-derm′) The protective coat that replaces the epidermis in woody plants during secondary growth, formed of the cork and cork cambium.

peripheral nervous system (PNS) The sensory and motor neurons that connect to the central nervous system.

peripheral protein A protein loosely bound to the surface of a membrane or to part of an integral protein and not embedded in the lipid bilayer.

peristalsis (pār′-uh-stal′-sis) (1) Alternating waves of contraction and relaxation in the smooth muscles lining the alimentary canal that push food along the canal. (2) A type of movement on land produced by rhythmic waves of muscle contractions passing from front to back, as in many annelids.

peristome (pār′-uh-stōme′) A ring of interlocking, tooth-like structures on the upper part of a moss capsule (sporangium), often specialised for gradual spore discharge.

peritubular capillary One of the tiny blood vessels that form a network surrounding the proximal and distal tubules in the kidney.

peroxisome (puh-rok′-suh-sōm′) An organelle containing enzymes that transfer hydrogen atoms from various substrates to oxygen (O_2), producing and then degrading hydrogen peroxide (H_2O_2).

personalised medicine A type of medical care in which each person's specific genetic profile can provide information about diseases or conditions for which the person is especially at risk and help make health-care decisions.

petal A modified leaf of a flowering plant. Petals are the often colourful parts of a flower that advertise it to insects and other pollinators.

petiole (pet′-ē-ōl) The stalk of a leaf, which joins the leaf to a node of the stem.

P generation The true-breeding (homozygous) parent individuals from which F_1 hybrid offspring are derived in studies of inheritance. (P stands for parental.)

pH A measure of hydrogen ion concentration equal to $-\log[H^+]$ and ranging in value from 0 to 14.

phage (fāj) A virus that infects bacteria; also called a bacteriophage.

phagocytosis (fag′-ō-sī-tō′-sis) A type of endocytosis in which large particulate substances or small organisms are taken up by a cell. It is carried out by some protists and by certain immune cells of animals (in mammals, mainly macrophages, neutrophils, and dendritic cells).

pharyngeal cleft (fuh-rin′-jē-ul) In chordate embryos, one of the grooves that separate a series of arches along the outer surface of the pharynx and may develop into a pharyngeal slit.

pharyngeal slit (fuh-rin′-jē-ul) In chordate embryos, one of the slits that form from the pharyngeal clefts and open into the pharynx, later developing into gill slits in many vertebrates.

pharynx (fār′-inks) (1) An area in the vertebrate throat where air and food passages cross. (2) In flatworms, the muscular tube that protrudes from the ventral side of the worm and ends in the mouth.

phase change (1) A shift from one developmental phase to another. (2) In plants, a morphological change that arises from a transition in shoot apical meristem activity.

phenotype (fē′-nō-tīp) The observable physical and physiological traits of an organism, which are determined by its genetic makeup.

pheromone (fār′-uh-mōn) In animals and fungi, a small molecule released into the environment that functions in communication between members of the same species. In animals, it acts much like a hormone in influencing physiology and behaviour.

phloem (flō′-em) Vascular plant tissue consisting of living cells arranged into elongated tubes that transport sugar and other organic nutrients throughout the plant.

phloem sap (flō′-em) The sugar-rich solution carried through a plant's sieve tubes.

phosphate group A chemical group consisting of a phosphorus atom bonded to four oxygen atoms; important in energy transfer.

phospholipid (fos′-fō-lip′-id) A lipid made up of glycerol joined to two fatty acids and a phosphate group. The hydrocarbon chains of the fatty acids act as nonpolar, hydrophobic tails, while the rest of the molecule acts as a polar, hydrophilic head. Phospholipids form bilayers that function as biological membranes.

phosphorylated intermediate (fos′-fōr-uh-lā′-ted) A molecule (often a reactant) with a phosphate group covalently bound to it, making it more reactive (less stable) than the unphosphorylated molecule.

phosphorylation cascade (fos′-fōr-uh-lā′-shun) A series of chemical reactions during cell signalling mediated by enzymes (kinases), in which each kinase in turn phosphorylates and activates another, ultimately leading to phosphorylation of many proteins.

photic zone (fō′-tic) The narrow top layer of an ocean or lake, where light penetrates sufficiently for photosynthesis to occur.

photoautotroph (fō′-tō-ot′-ō-trōf) An organism that harnesses light energy to drive the synthesis of organic compounds from carbon dioxide.

photoheterotroph (fō′-tō-het′-er-ō-trōf) An organism that uses light to generate ATP but must obtain carbon in organic form.

photomorphogenesis Effects of light on plant morphology.

photon (fō′-ton) A quantum, or discrete quantity, of light energy that behaves as if it were a particle.

photoperiodism (fō′-tō-pēr′-ē-ō-dizm) A physiological response to photoperiod, the interval in a 24-hour period during which an organism is exposed to light. An example of photoperiodism is flowering.

photophosphorylation (fō′-tō-fos′-fōr-uh-lā′-shun) The process of generating ATP from ADP and phosphate by means of chemiosmosis, using a proton-motive force generated across the thylakoid membrane of the chloroplast or the membrane of certain prokaryotes during the light reactions of photosynthesis.

photoreceptor An electromagnetic receptor that detects the radiation known as visible light.

photorespiration A metabolic pathway that consumes oxygen and ATP, releases carbon dioxide, and decreases photosynthetic output. Photorespiration generally occurs on hot, dry, bright days, when the stomata close and the O_2:CO_2 ratio in the leaf increases, favouring the binding of O_2 rather than CO_2 by rubisco.

photosynthesis (fō′-tō-sin′-thi-sis) The conversion of light energy to chemical energy that is stored in sugars or other organic compounds; occurs in plants, algae, and certain prokaryotes.

photosystem A light-capturing unit located in the thylakoid membrane of the chloroplast or in the membrane of some prokaryotes, consisting of a reaction-centre complex surrounded by numerous light-harvesting complexes. There are two types of photosystems, I and II; they absorb light best at different wavelengths.

photosystem I (PS I) A light-capturing unit in a chloroplast's thylakoid membrane or in the membrane of some prokaryotes; it has two molecules of P700 chlorophyll *a* at its reaction centre.

photosystem II (PS II) One of two light-capturing units in a chloroplast's thylakoid membrane or in the membrane of some prokaryotes; it has two molecules of P680 chlorophyll *a* at its reaction centre.

phototropism (fō'-tō-trō'-pizm) The bending of a plant or other organism in response to light, either towards the source of light (positive phototropism) or away from it (negative phototropism).

phylogenetic tree A branching diagram that represents a hypothesis about the evolutionary history of a group of organisms.

phylogeny (fī-loj'-uh-nē) The evolutionary history of a species or group of related species.

phylum (fī'-lum) (plural, **phyla**) In Linnaean classification, the taxonomic category above class.

physiology The processes and functions of an organism.

phytochromes (fī'-tuh-krōm) Plant pigments that absorb mostly red and far-red light and regulate many plant responses, such as seed germination and shade avoidance.

phytoremediation An emerging technology that seeks to reclaim contaminated areas by taking advantage of some plant species' ability to extract heavy metals and other pollutants from the soil and to concentrate them in easily harvested portions of the plant.

pilus (plural, **pili**) (pī'-lus, pī'-lī) In bacteria, a structure that links one cell to another at the start of conjugation; also called a sex pilus or conjugation pilus.

pineal gland (pī'-nē-ul) A small gland on the dorsal surface of the vertebrate forebrain that secretes the hormone melatonin.

pinocytosis (pī'-nō-sī-tō'-sis) A type of endocytosis in which the cell ingests extracellular fluid and its dissolved solutes.

pistil A single carpel (a simple pistil) or a group of fused carpels (a compound pistil).

pith Ground tissue that is internal to the vascular tissue in a stem; in many monocot roots, parenchyma cells that form the central core of the vascular cylinder.

pituitary gland (puh-tū'-uh-tār'-ē) An endocrine gland at the base of the hypothalamus; consists of a posterior lobe, which stores and releases two hormones produced by the hypothalamus, and an anterior lobe, which produces and secretes many hormones that regulate diverse body functions.

placenta (pluh-sen'-tuh) A structure in the uterus of a pregnant eutherian mammal that nourishes the fetus with the mother's blood supply; formed from the uterine lining and embryonic membranes.

placoderm A member of an extinct group of fishlike vertebrates that had jaws and were enclosed in a tough outer armour.

planarian A free-living flatworm found in ponds and streams.

plasma (plaz'-muh) The liquid matrix of blood in which the blood cells are suspended.

plasma membrane (plaz'-muh) The membrane at the boundary of every cell that acts as a selective barrier, regulating the cell's chemical composition.

plasmid (plaz'-mid) A small, circular, double-stranded DNA molecule that carries accessory genes separate from those of a bacterial chromosome; in DNA cloning, plasmids are used as vectors carrying up to about 10,000 base pairs (10 kb) of DNA. Plasmids are also found in some eukaryotes, such as yeasts.

plasmodesma (plaz'-mō-dez'-muh) (plural, **plasmodesmata**) An open channel through the cell wall that connects the cytoplasm of adjacent plant cells, allowing water, small solutes, and some larger molecules to pass between the cells.

plasmogamy (plaz-moh'-guh-mē) In fungi, the fusion of the cytoplasm of cells from two individuals; occurs as one stage of sexual reproduction, followed later by karyogamy.

plasmolysis (plaz-mol'-uh-sis) A phenomenon in walled cells in which the cytoplasm shrivels and the plasma membrane pulls away from the cell wall; occurs when the cell loses water to a hypertonic environment.

plastid One of a family of closely related organelles that includes chloroplasts, chromoplasts, and amyloplasts. Plastids are found in cells of photosynthetic eukaryotes.

plate tectonics The theory that the continents are part of great plates of Earth's crust that float on the hot, underlying portion of the mantle. Movements in the mantle cause the continents to move slowly over time.

platelet A pinched-off cytoplasmic fragment of a specialised bone marrow cell. Platelets circulate in the blood and are important in blood clotting.

pleiotropy (plī'-o-truh-pē) The ability of a single gene to have multiple effects.

pluripotent Describing a cell that can give rise to many, but not all, parts of an organism.

point mutation A change in a single nucleotide pair of a gene.

polar covalent bond A covalent bond between atoms that differ in electronegativity. The shared electrons are pulled closer to the more electronegative atom, making it slightly negative and the other atom slightly positive.

polar molecule A molecule (such as water) with an uneven distribution of charges in different regions of the molecule.

polarity A lack of symmetry; structural differences in opposite ends of an organism or structure, such as the root end and shoot end of a plant.

pollen grain In seed plants, a structure consisting of the male gametophyte enclosed within a pollen wall.

pollen tube A tube that forms after germination of the pollen grain and that functions in the delivery of sperm to the ovule.

pollination (pol'-uh-nā'-shun) The transfer of pollen to the part of a seed plant containing the ovules, a process required for fertilisation.

poly-A tail A sequence of 50–250 adenine nucleotides added onto the 3′ end of a pre-mRNA molecule.

polygamous Referring to a type of relationship in which an individual of one sex mates with several of the other.

polygenic inheritance (pol'-ē-jen'-ik) An additive effect of two or more genes on a single phenotypic character.

polymer (pol'-uh-mer) A long molecule consisting of many similar or identical monomers linked together by covalent bonds.

polymerase chain reaction (PCR) (puh-lim'-uh-rās) A technique for amplifying DNA *in vitro* by incubating it with specific primers, a heat-resistant DNA polymerase, and nucleotides.

polynucleotide (pol'-ē-nū'-klē-ō-tīd) A polymer consisting of many nucleotide monomers in a chain. The nucleotides can be those of DNA or RNA.

polyp The sessile variant of the cnidarian body plan. The alternate form is the medusa.

polypeptide (pol'-ē-pep'-tīd) A polymer of many amino acids linked together by peptide bonds.

polyphyletic (pol'-ē-fī-let'-ik) Pertaining to a group of taxa that includes distantly related organisms but does not include their most recent common ancestor.

polyploidy (pol'-ē-ploy'-dē) A chromosomal alteration in which the organism possesses more than two complete chromosome sets. It is the result of an accident of cell division.

polyribosome (polysome) (pol'-ē-rī'-buh-sōm') A group of several ribosomes attached to, and translating, the same messenger RNA molecule.

polysaccharide (pol'-ē-sak'-uh-rīd) A polymer of many monosaccharides, formed by dehydration reactions.

polyspermy The fertilisation of an egg by more than one sperm.

polytomy (puh-lit'-uh-mē) In a phylogenetic tree, a branch point from which more than two descendant taxa emerge. A polytomy indicates that the evolutionary relationships between the descendant taxa are not yet clear.

pons A portion of the brain that participates in certain automatic, homeostatic functions, such as regulating the breathing centres in the medulla.

population A group of individuals of the same species that live in the same area and interbreed, producing fertile offspring.

population dynamics The study of how complex interactions between biotic and abiotic factors influence variations in population size.

population ecology The study of populations in relation to their environment, including environmental influences on population density and distribution, age structure, and variations in population size.

positional information Molecular cues that control pattern formation in an animal or plant embryonic structure by indicating a cell's location relative to the organism's body axes. These cues elicit a response by genes that regulate development.

positive feedback A form of regulation in which an end product of a process speeds up that process; in physiology, a control mechanism in which a change in a variable triggers a response that reinforces or amplifies the change.

positive interaction A +/+ or +/0 ecological interaction between individuals of two species in which at least one individual benefits and neither is harmed; positive interactions include mutualism and commensalism.

positive pressure breathing A breathing system in which air is forced into the lungs.

posterior Pertaining to the rear, or tail end, of a bilaterally symmetrical animal.

posterior pituitary An extension of the hypothalamus composed of nervous tissue that secretes oxytocin and antidiuretic hormone made in the hypothalamus; a temporary storage site for these hormones.

postzygotic barrier (pōst′-zī-got′-ik) A reproductive barrier that prevents hybrid zygotes produced by two different species from developing into viable, fertile adults.

potential energy The energy that matter possesses as a result of its location or spatial arrangement (structure).

predation An interaction in which an individual of one species, the predator, kills and eats an individual of the other species, the prey.

prediction In deductive reasoning, a forecast that follows logically from a hypothesis. By testing predictions, experiments may allow certain hypotheses to be rejected.

pregnancy The condition of carrying one or more embryos in the uterus; also called gestation.

prepuce (prē-·pyūs) A fold of skin covering the head of the clitoris or penis.

pressure potential (Ψ_P) A component of water potential that consists of the physical pressure on a solution, which can be positive, zero, or negative.

prezygotic barrier (prē′-zī-got′-ik) A reproductive barrier that impedes mating between species or hinders fertilisation if interspecific mating is attempted.

primary cell wall In plants, a relatively thin and flexible layer that surrounds the plasma membrane of a young cell.

primary consumer A herbivore; an organism that eats plants or other autotrophs.

primary electron acceptor In the thylakoid membrane of a chloroplast or in the membrane of some prokaryotes, a specialised molecule that shares the reaction-centre complex with a pair of chlorophyll *a* molecules and that accepts an electron from them.

primary growth Growth produced by apical meristems, lengthening stems and roots.

primary immune response The initial adaptive immune response to an antigen, which appears after a lag of about 10–17 days.

primary meristems The three meristematic derivatives (protoderm, procambium, and ground meristem) of an apical meristem.

primary oocyte (ō′-uh-sīt) An oocyte prior to completion of meiosis I.

primary producer An autotroph, usually a photosynthetic organism. Collectively, autotrophs make up the trophic level of an ecosystem that ultimately supports all other levels.

primary production The amount of light energy converted to chemical energy (organic compounds) by the autotrophs in an ecosystem during a given time period.

primary structure The level of protein structure referring to the specific linear sequence of amino acids.

primary succession A type of ecological succession that occurs in an area where there were originally no organisms present and where soil has not yet formed.

primary transcript An initial RNA transcript from any gene; also called pre-mRNA when transcribed from a protein-coding gene.

primase An enzyme that joins RNA nucleotides to make a primer during DNA replication, using the parental DNA strand as a template.

primer A short polynucleotide with a free 3′ end, bound by complementary base pairing to the template strand and elongated with DNA nucleotides during DNA replication.

prion An infectious agent that is a misfolded version of a normal cellular protein. Prions appear to increase in number by converting correctly folded versions of the protein to more prions.

problem solving The cognitive activity of devising a method to proceed from one state to another in the face of real or apparent obstacles.

producer An organism that produces organic compounds from CO_2 by harnessing light energy (in photosynthesis) or by oxidising inorganic chemicals (in chemosynthetic reactions carried out by some prokaryotes).

product A material resulting from a chemical reaction.

production efficiency The percentage of energy stored in assimilated food that is not used for respiration or eliminated as waste.

progesterone A steroid hormone that contributes to the menstrual cycle and prepares the uterus for pregnancy; the major progestin in mammals.

progestin Any steroid hormone with progesterone-like activity.

prokaryote A single-celled organism of the domain Bacteria or Archaea.

prokaryotic cell (prō′-kār′-ē-ot′-ik) A type of cell lacking a membrane-enclosed nucleus and membrane-enclosed organelles. Organisms with prokaryotic cells (bacteria and archaea) are called prokaryotes.

prolactin A hormone produced and secreted by the anterior pituitary with a great diversity of effects in different vertebrate species. In mammals, it stimulates growth of and milk production by the mammary glands.

prometaphase The second stage of mitosis, in which the nuclear envelope fragments and the spindle microtubules attach to the kinetochores of the chromosomes.

promoter A specific nucleotide sequence in the DNA of a gene that binds RNA polymerase, positioning it to start transcribing RNA at the appropriate place.

prophage (prō′-fāj) A phage genome that has been inserted into a specific site on a bacterial chromosome.

prophase The first stage of mitosis, in which the chromatin condenses into discrete chromosomes visible with a light microscope, the mitotic spindle begins to form, and the nucleolus disappears but the nucleus remains intact.

prostaglandin (pros′-tuh-glan′-din) One of a group of modified fatty acids that are secreted by virtually all tissues and that perform a wide variety of functions as local regulators.

prostate gland (pros′-tāt) A gland in human males that secretes an acid-neutralising component of semen.

protease (prō′-tē-āz) An enzyme that digests proteins by hydrolysis.

protein (prō′-tēn) A biologically functional molecule consisting of one or more polypeptides folded and coiled into a specific three-dimensional structure.

protein kinase (prō′-tēn kī′-nās) An enzyme that transfers phosphate groups from ATP to a protein, thus phosphorylating the protein.

protein phosphatase (prō′-tēn fos′-fuh-tās) An enzyme that removes phosphate groups from (dephosphorylates) proteins, often functioning to reverse the effect of a protein kinase.

proteoglycan (prō′-tē-ō-glī′-kan) A large molecule consisting of a small core protein with many carbohydrate chains attached, found in the extracellular matrix of animal cells. A proteoglycan may consist of up to 95% carbohydrate.

proteome The entire set of proteins expressed by a given cell, tissue, or organism.

proteomics (prō′-tē-ō′-miks) The systematic study of sets of proteins and their properties, including their abundance, chemical modifications, and interactions.

protist An informal term applied to any eukaryote that is not a plant, animal, or fungus. Most protists are unicellular, though some are colonial or multicellular.

protocell An abiotic precursor of a living cell that had a membrane-like structure and that maintained an internal chemistry different from that of its surroundings.

proton (prō′-ton) A subatomic particle with a single positive electrical charge, with a mass of about 1.7×10^{-24} g, found in the nucleus of an atom.

protonema (prō′-tuh-nē′-muh) (plural, **protonemata**) A mass of green, branched, one-cell-thick filaments produced by germinating moss spores.

protonephridium (prō′-tō-nuh-frid′-ē-um) (plural, **protonephridia**) An excretory system, such as the flame bulb system of flatworms, consisting of a network of tubules lacking internal openings.

proton-motive force (prō′-ton) The potential energy stored in the form of a proton electrochemical gradient, generated by the pumping of hydrogen ions (H^+) across a biological membrane during chemiosmosis.

proton pump (prō′-ton) An active transport protein in a cell membrane that uses ATP to transport hydrogen ions out of a cell against their concentration gradient, generating a membrane potential in the process.

proto-oncogene (prō′-tō-on′-kō-jēn) A normal cellular gene that has the potential to become an oncogene.

protoplast The living part of a plant cell, which also includes the plasma membrane.

protostome development In animals, a developmental mode distinguished by the development of the mouth from the blastopore; often also characterised by spiral cleavage and by the body cavity forming when solid masses of mesoderm split.

provirus A viral genome that is permanently inserted into a host genome.

proximal tubule In the vertebrate kidney, the portion of a nephron immediately downstream from Bowman's capsule that conveys and helps refine filtrate.

pseudocoelomate (sū′-dō-sē-·lō-māt) An animal whose body cavity is lined by tissue derived from mesoderm and endoderm.

pseudogene (sū′-dō-jēn) A DNA segment that is very similar to a real gene but does not yield a functional product; a DNA segment that formerly functioned as a gene but has become inactivated in a particular species because of mutation.

pseudopodium (sū′-dō-pō′-dē-um) (plural, **pseudopodia**) A cellular extension of amoeboid cells used in moving and feeding.

P site One of a ribosome's three binding sites for tRNA during translation. The P site holds the tRNA carrying the growing polypeptide chain. (P stands for peptidyl tRNA.)

pterosaur Winged reptile that lived during the Mesozoic era.

pulse The rhythmic bulging of the artery walls with each heartbeat.

punctuated equilibria In the fossil record, long periods of apparent stasis, in which a species undergoes little or no morphological change, interrupted by relatively brief periods of sudden change.

Punnett square A diagram used in the study of inheritance to show the predicted genotypic results of random fertilisation in genetic crosses between individuals of known genotype.

pupil The opening in the iris, which admits light into the interior of the vertebrate eye. Muscles in the iris regulate its size.

purine (pyū′-rēn) One of two types of nitrogenous bases found in nucleotides, characterised by a six-membered ring fused to a five-membered ring. Adenine (A) and guanine (G) are purines.

pyrimidine (puh-rim′-uh-dēn) One of two types of nitrogenous bases found in nucleotides, characterised by a six-membered ring. Cytosine (C), thymine (T), and uracil (U) are pyrimidines.

quantitative character A heritable feature that varies continuously over a range rather than in an either-or fashion.

quaternary structure (kwot′-er-nār′-ē) The particular shape of a complex, aggregate protein, defined by the characteristic three-dimensional arrangement of its constituent subunits, each a polypeptide.

radial cleavage A type of embryonic development in deuterostomes in which the planes of cell division that transform the zygote into a ball of cells are either parallel or perpendicular to the vertical axis of the embryo, thereby aligning tiers of cells one above the other.

radial symmetry Symmetry in which the body is shaped like a pie or barrel (lacking a left side and a right side) and can be divided into mirror-imaged halves by any plane through its central axis.

radicle An embryonic root of a plant.

radioactive isotope An isotope (an atomic form of a chemical element) that is unstable; the nucleus decays spontaneously, giving off detectable particles and energy.

radiolarian A protist, usually marine, with a shell generally made of silica and pseudopodia that radiate from the central body.

radiometric dating A method for determining the absolute age of rocks and fossils, based on the half-life of radioactive isotopes.

radula A straplike scraping organ used by many molluscs during feeding.

***ras* gene** A gene that codes for Ras, a G protein that relays a growth signal from a growth factor receptor on the plasma membrane to a cascade of protein kinases, ultimately resulting in stimulation of the cell cycle.

ratite (rat′-īt) A member of the group of flightless birds.

ray-finned fish A member of the clade Actinopterygii, aquatic osteichthyans with fins supported by long, flexible rays, including tuna, bass, and herring.

reabsorption In excretory systems, the recovery of solutes and water from filtrate.

reactant A starting material in a chemical reaction.

reaction-centre complex A complex of proteins associated with a special pair of chlorophyll *a* molecules and a primary electron acceptor. Located centrally in a photosystem, this complex triggers the light reactions of photosynthesis. Excited by light energy, the pair of chlorophylls donates an electron to the primary electron acceptor, which passes an electron to an electron transport chain.

reading frame On an mRNA, the triplet grouping of ribonucleotides used by the translation machinery during polypeptide synthesis.

receptacle The base of a flower; the part of the stem that is the site of attachment of the floral organs.

reception In cellular communication, the first step of a signalling pathway in which a signalling molecule is detected by a receptor molecule on or in the cell.

receptor-mediated endocytosis (en′-dō-sī-tō′-sis) The movement of specific molecules into a cell by the infolding of vesicles containing proteins with receptor sites specific to the molecules being taken in; enables a cell to acquire bulk quantities of specific substances.

receptor potential A graded potential occurring in a receptor cell.

receptor tyrosine kinase (RTK) A receptor protein spanning the plasma membrane, the cytoplasmic (intracellular) part of which can catalyse the transfer of a phosphate group from ATP to a tyrosine on another protein. Receptor tyrosine kinases often respond to the binding of a signalling molecule by dimerising and then phosphorylating a tyrosine on the cytoplasmic portion of the other receptor in the dimer.

recessive allele An allele whose phenotypic effect is not observed in a heterozygote.

reciprocal altruism Altruistic behaviour between unrelated individuals, whereby the altruistic individual benefits in the future when the beneficiary reciprocates.

recombinant chromosome A chromosome created when crossing over combines DNA from two parents into a single chromosome.

recombinant DNA molecule A DNA molecule made *in vitro* with segments from different sources.

recombinant type (recombinant) An offspring whose phenotype differs from that of the true-breeding P generation parents; also refers to the phenotype itself.

rectum The terminal portion of the large intestine, where the faeces are stored prior to elimination.

red alga A photosynthetic protist, named for its colour, which results from a red pigment that masks the green of chlorophyll. Most red algae are multicellular and marine.

redox reaction (rē′-doks) A chemical reaction involving the complete or partial transfer of one or more electrons from one reactant to another; short for **red**uction-**ox**idation reaction.

reducing agent The electron donor in a redox reaction.

reduction The complete or partial addition of electrons to a substance involved in a redox reaction.

reference genome A complete sequence that researchers agree best represents the genome of a given species, arrived at by sequencing multiple individuals.

reflex An automatic reaction to a stimulus, mediated by the spinal cord or lower brain.

refractory period (rē-frakt′-ōr-ē) A period, immediately following a response to stimulation, during which a cell or organ is unresponsive to further stimulation.

regulator An animal for which mechanisms of homeostasis moderate internal changes in a particular variable in the face of external fluctuation of that variable.

regulatory gene A gene that codes for a protein, such as a repressor, that controls the transcription of another gene or group of genes.

reinforcement In evolutionary biology, a process in which natural selection strengthens prezygotic barriers to reproduction, thus reducing the chances of hybrid formation. Such a process is likely to occur only if hybrid offspring are less fit than members of the parent species.

relative abundance The proportional abundance of different species in a community.

relative fitness The contribution an individual makes to the gene pool of the next generation, relative to the contributions of other individuals in the population.

renal cortex The outer portion of the vertebrate kidney.

renal medulla The inner portion of the vertebrate kidney, beneath the renal cortex.

renal pelvis The funnel-shaped chamber that receives processed filtrate from the vertebrate kidney's collecting ducts and is drained by the ureter.

renin-angiotensin-aldosterone system (RAAS) A hormone cascade pathway that helps regulate blood pressure and blood volume.

repetitive DNA Nucleotide sequences, usually noncoding, that are present in many copies in a eukaryotic genome. The repeated units may be short and arranged tandemly (in series) or long and dispersed in the genome.

replication fork A Y-shaped region on a replicating DNA molecule where the parental strands are being unwound and new strands are being synthesised.

repressor A protein that inhibits gene transcription. In prokaryotes, repressors bind to the DNA in or near the promoter. In eukaryotes, repressors may bind to control elements within enhancers, to activators, or to other proteins in a way that blocks activators from binding to DNA.

reproductive isolation The existence of biological factors (barriers) that impede members of two species from producing viable, fertile offspring.

reptile A member of the clade of amniotes that includes tuataras, lizards and snakes, turtles, crocodilians, and birds.

reservoir In biogeochemical cycles, location of a chemical element, consisting of either organic or inorganic materials that are either available for direct use by organisms or unavailable as nutrients.

residual volume The amount of air that remains in the lungs after forceful exhalation.

resource partitioning The division of environmental resources by coexisting species such that the niche of each species differs by one or more significant factors from the niches of all coexisting species.

respiratory pigment A protein that transports oxygen in blood or haemolymph.

response (1) In cellular communication, the change in a specific cellular activity brought about by a transduced signal from outside the cell. (2) In feedback regulation, a physiological activity triggered by a change in a variable.

resting potential The membrane potential characteristic of a nonconducting excitable cell, with the inside of the cell more negative than the outside.

restriction enzyme An endonuclease (type of enzyme) that recognises and cuts DNA molecules foreign to a bacterium (such as phage genomes). The enzyme cuts at specific nucleotide sequences (restriction sites).

restriction fragment A DNA segment that results from the cutting of DNA by a restriction enzyme.

restriction site A specific sequence on a DNA strand that is recognised and cut by a restriction enzyme.

retina (ret′-i-nuh) The innermost layer of the vertebrate eye, containing photoreceptor cells (rods and cones) and neurons; transmits images formed by the lens to the brain via the optic nerve.

retinal The light-absorbing pigment in rods and cones of the vertebrate eye.

retrotransposon (re′-trō-trans-pō′-zon) A transposable element that moves within a genome by means of an RNA intermediate, a transcript of the retrotransposon DNA.

retrovirus (re′-trō-vī′-rus) An RNA virus that replicates by transcribing its RNA into DNA and then inserting the DNA into a cellular chromosome; an important class of cancer-causing viruses.

reverse transcriptase (tran-skrip′-tās) An enzyme encoded by certain viruses (retroviruses) that uses RNA as a template for DNA synthesis.

reverse transcriptase–polymerase chain reaction (RT-PCR) A technique for determining expression of a particular gene. It uses reverse transcriptase and DNA polymerase to synthesise cDNA from all the mRNA in a sample and then subjects the cDNA to PCR amplification using primers specific for the gene of interest.

rhizarians (rī-za′-rē-uhns) One of the three major subgroups for which the SAR eukaryotic supergroup is named. Many species in this clade are amoebas characterised by threadlike pseudopodia.

rhizobacterium A soil bacterium whose population size is much enhanced in the rhizosphere, the soil region close to a plant's roots.

rhizoid (rī′-zoyd) A long, tubular single cell or filament of cells that anchors bryophytes to the ground. Unlike roots, rhizoids are not composed of tissues, lack specialised conducting cells, and do not play a primary role in water and mineral absorption.

rhizosphere The soil region close to plant roots and characterised by a high level of microbiological activity.

rhodopsin (rō-dop′-sin) A visual pigment consisting of retinal and opsin. Upon absorbing light, the retinal changes shape and dissociates from the opsin.

ribonucleic acid (RNA) (rī′-bō-nū-klā′-ik) A type of nucleic acid consisting of a polynucleotide made up of nucleotide monomers with a ribose sugar and the nitrogenous bases adenine (A), cytosine (C), guanine (G), and uracil (U); usually single-stranded; functions in protein synthesis, in gene regulation, and as the genome of some viruses.

ribose The sugar component of RNA nucleotides.

ribosomal RNA (rRNA) (rī′-buh-sō′-mul) RNA molecules that, together with proteins, make up ribosomes; the most abundant type of RNA.

ribosome (rī′-buh-sōm) A complex of rRNA and protein molecules that functions as a site of protein synthesis in the cytoplasm; consists of a large and a small subunit. In eukaryotic cells, each subunit is assembled in the nucleolus. *See also* nucleolus.

ribozyme (rī′-buh-zīm) An RNA molecule that functions as an enzyme, such as an intron that catalyses its own removal during RNA splicing.

RNA interference (RNAi) A mechanism for silencing the expression of specific genes. In RNAi, double-stranded RNA molecules that match the sequence of a particular gene are processed into siRNAs that either block translation or trigger the degradation of the gene's messenger RNA. This happens naturally in some cells and can be carried out in laboratory experiments as well.

RNA polymerase An enzyme that links ribonucleotides into a growing RNA chain during transcription, based on complementary binding to nucleotides on a DNA template strand.

RNA processing Modification of RNA primary transcripts, including splicing out of introns, joining together of exons, and alteration of the 5′ and 3′ ends.

RNA sequencing (RNA-seq) (RNA-sēk) A method of analysing large sets of RNAs that involves making cDNAs and sequencing them.

RNA splicing After synthesis of a eukaryotic primary RNA transcript, the removal of portions of the transcript (introns) that will not be included in the mRNA and the joining together of the remaining portions (exons).

rod A rodlike cell in the retina of the vertebrate eye, sensitive to low light intensity.

root An organ in vascular plants that anchors the plant and enables it to absorb water and minerals from the soil.

root cap A cone of cells at the tip of a plant root that protects the apical meristem.

root hair A tiny extension of a root epidermal cell, growing just behind the root tip and increasing surface area for absorption of water and minerals.

root pressure Pressure exerted in the roots of plants as the result of osmosis, causing exudation from cut stems and guttation of water from leaves.

root system All of a plant's roots, which anchor it in the soil, absorb and transport minerals and water, and store food.

rooted Describing a phylogenetic tree that contains a branch point (often, the one farthest to the left) representing the most recent common ancestor of all taxa in the tree.

rough ER That portion of the endoplasmic reticulum with ribosomes attached.

round window In the mammalian ear, the point of contact where vibrations of the stapes create a travelling series of pressure waves in the fluid of the cochlea.

R plasmid A bacterial plasmid carrying genes that confer resistance to certain antibiotics.

***r*-selection** Selection for life history traits that maximise reproductive success in uncrowded environments.

rubisco (rū-bis′-kō) Ribulose bisphosphate (RuBP) carboxylase-oxygenase, the enzyme that normally catalyses the first step of the Calvin cycle (the addition of CO_2 to RuBP). When excess O_2 is present or CO_2 levels are low, rubisco can bind oxygen, resulting in photorespiration.

ruminant (rūh′-muh-nent) A cud-chewing animal, such as a cow or sheep, with multiple stomach compartments specialised for an herbivorous diet.

salicylic acid (sal′-i-sil′-ik) A signalling molecule in plants that may be partially responsible for activating systemic acquired resistance to pathogens.

salivary gland A gland associated with the oral cavity that secretes substances that lubricate food and begin the process of chemical digestion.

salt A compound resulting from the formation of an ionic bond; also called an ionic compound.

saltatory conduction (sol′-tuh-tōr′-ē) Rapid transmission of a nerve impulse along an axon, resulting from the action potential jumping from one node of Ranvier to another, skipping the myelin-sheathed regions of membrane.

SAR One of four supergroups of eukaryotes proposed in a current hypothesis of the evolutionary history of eukaryotes. This supergroup contains a large, extremely diverse collection of protists from three major subgroups: stramenopiles, alveolates, and rhizarians. *See also* Excavata, Archaeplastida, and Unikonta.

sarcomere (sar′-kō-mēr) The fundamental, repeating unit of striated muscle, delimited by the Z lines.

sarcoplasmic reticulum (SR) (sar′-kō-plaz′-mik ruh-tik′-yū-lum) A specialised endoplasmic reticulum that regulates the calcium concentration in the cytosol of muscle cells.

saturated fatty acid A fatty acid in which all carbons in the hydrocarbon tail are connected by single bonds, thus maximising the number of hydrogen atoms that are attached to the carbon skeleton.

savanna A tropical grassland biome with scattered individual trees and large herbivores and maintained by occasional fires and drought.

scaffolding protein A type of large relay protein to which several other relay proteins are simultaneously attached, increasing the efficiency of signal transduction.

scanning electron microscope (SEM) A microscope that uses an electron beam to scan the surface of a sample, coated with metal atoms, to study details of its topography.

scatter plot A graph in which each piece of data is represented by a point. A scatter plot is used when the data for all variables are numerical and continuous.

schizophrenia (skit′-suh-frē′-nē-uh) A severe mental disturbance characterised by psychotic episodes in which patients have a distorted perception of reality.

Schwann cell A type of glial cell that forms insulating myelin sheaths around the axons of neurons in the peripheral nervous system.

science An approach to understanding the natural world.

scion (sī′-un) The twig grafted onto the stock when making a graft.

sclerenchyma cell (skluh-ren′-kim-uh) A rigid, supportive plant cell type usually lacking a protoplast and possessing thick secondary walls strengthened by lignin at maturity.

sclerophyll (or **sclerophylly**) Plant traits of hard, leather-like leaves that grow close together and hang perpendicular to the ground. In addition to the leathery leaves mitigating damage from winds and insects, adaptations of these leaves minimise water loss. When leaves hang with their tips towards the ground, their flattened sides present the least surface area to the midday sun. Through minimising heat loads during the hottest part of the day, plants can avoid the need to cool their leaves evaporatively, which could rapidly deplete the water sources available to the plant. Typically, sclerophyll traits occur in New Zealand and Australian heathlands and open forests. Sclerophylly occurs commonly in environments throughout the world that feature hot, dry summer climates, and often seasonally wet winters, that vary between warm and cool temperatures.

scrotum A pouch of skin outside the abdomen that houses the testes; functions in maintaining the testes at the lower temperature required for spermatogenesis.

second law of thermodynamics The principle stating that every energy transfer or transformation increases the entropy of the universe. Usable forms of energy are at least partly converted to heat.

second messenger A molecule that relays messages in a cell from a receptor to a target where an action within the cell takes place.

secondary cell wall In plant cells, a strong and durable matrix that is often deposited in several laminated layers around the plasma membrane and provides protection and support.

secondary consumer A carnivore that eats herbivores.

secondary endosymbiosis A process in eukaryotic evolution in which a heterotrophic eukaryotic cell engulfed a photosynthetic eukaryotic cell, which survived in a symbiotic relationship inside the heterotrophic cell.

secondary growth Growth produced by lateral meristems, thickening the roots and shoots of woody plants.

secondary immune response The adaptive immune response elicited on second or subsequent exposures to a particular antigen. The secondary immune response is more rapid, of greater magnitude, and of longer duration than the primary immune response.

secondary oocyte (ō′-uh-sīt) An oocyte that has completed meiosis I.

secondary production The amount of chemical energy in consumers' food that is converted to their own new biomass during a given time period.

secondary structure Regions of repetitive coiling or folding of the polypeptide backbone of a protein due to hydrogen bonding between constituents of the backbone (not the side chains).

secondary succession A type of succession that occurs where an existing community has been cleared by some disturbance that leaves the soil or substrate intact.

secretion (1) The discharge of molecules synthesised by a cell. (2) The active transport of wastes and certain other solutes from the body fluid into the filtrate in an excretory system.

seed An adaptation of some terrestrial plants consisting of an embryo packaged along with a store of food within a protective coat.

seed coat A tough outer covering of a seed, formed from the outer coat of an ovule. In a flowering plant, the seed coat encloses and protects the embryo and endosperm.

seedless vascular plant An informal name for a plant that has vascular tissue but lacks seeds. Seedless vascular plants form a paraphyletic group that includes the phyla Lycophyta (club mosses and their relatives) and Monilophyta (ferns and their relatives).

selective permeability A property of biological membranes that allows them to regulate the passage of substances across them.

self-incompatibility The ability of a seed plant to reject its own pollen and sometimes the pollen of closely related individuals.

semelparity (seh′-mel-pār′-i-tē) Reproduction in which an organism produces all of its offspring in a single event; also called big-bang reproduction.

semen (sē′-mun) The fluid that is ejaculated by the male during orgasm; contains sperm and secretions from several glands of the male reproductive tract.

semicircular canals A three-part chamber of the inner ear that functions in maintaining equilibrium.

semiconservative model Type of DNA replication in which the replicated double helix consists of one old strand, derived from the parental molecule, and one newly made strand.

semilunar valve A valve located at each exit of the heart, where the aorta leaves the left ventricle and the pulmonary artery leaves the right ventricle.

seminal vesicle (sem′-i-nul ves′-i-kul) A gland in males that secretes a fluid component of semen that lubricates and nourishes sperm.

seminiferous tubule (sem′-i-nif′-er-us) A highly coiled tube in the testis in which sperm are produced.

senescence (se-nes′-ens) The programmed death of certain cells or organs or the entire organism.

sensitive period A limited phase in an animal's development when learning of particular behaviours can take place; also called a critical period.

sensor In homeostasis, a receptor that detects a stimulus.

sensory adaptation The tendency of sensory neurons to become less sensitive when they are stimulated repeatedly.

sensory neuron A nerve cell that receives information from the internal or external environment and transmits signals to the central nervous system.

sensory reception The detection of a stimulus by sensory cells.

sensory receptor A specialised structure or cell that responds to a stimulus from an animal's internal or external environment.

sensory transduction The conversion of stimulus energy to a change in the membrane potential of a sensory receptor cell.

sepal (sē′-pul) A modified leaf in angiosperms that helps enclose and protect a flower bud before it opens.

septum (plural, **septa**) One of the cross-walls that divide a fungal hypha into cells. Septa generally have pores large enough to allow ribosomes, mitochondria, and even nuclei to flow from cell to cell.

serial endosymbiont theory The theory that mitochondria and plastids originated as prokaryotic cells engulfed by a host cell. The engulfed cell and its host cell then evolved into a single organism. *See also* endosymbiosis.

serial endosymbiosis A hypothesis for the origin of eukaryotes consisting of a sequence of endosymbiotic events in which mitochondria, chloroplasts, and perhaps other cellular structures were derived from small prokaryotes that had been engulfed by larger cells.

set point In homeostasis in animals, a value maintained for a particular variable, such as body temperature or solute concentration.

seta (sē′-tuh) (plural, **setae**) The elongated stalk of a bryophyte sporophyte.

sex chromosome A chromosome responsible for determining the sex of an individual.

sex-linked gene A gene located on either sex chromosome. Most sex-linked genes are on the X chromosome and show distinctive patterns of inheritance; there are very few genes on the Y chromosome.

sexual dimorphism (dī-mōr′-fizm) Differences between the secondary sex characteristics of males and females of the same species.

sexual reproduction Reproduction arising from fusion of two gametes.

sexual selection A process in which individuals with certain inherited characteristics are more likely than other individuals of the same sex to obtain mates.

Shannon diversity index An index of community diversity symbolised by H and represented by the equation $H = -(p_A \ln p_A + p_B \ln p_B + p_C \ln p_C + \cdots)$, where A, B, C ... are species, p is the relative abundance of each species, and ln is the natural logarithm.

shared ancestral character A character, shared by members of a particular clade, that originated in an ancestor that is not a member of that clade.

shared derived character An evolutionary novelty that is unique to a particular clade.

shoot system The aerial portion of a plant body, consisting of stems, leaves, and (in angiosperms) flowers.

short tandem repeat (STR) Simple sequence DNA containing multiple tandemly repeated units of two to five nucleotides. Variations in STRs act as genetic markers in STR analysis, used to prepare genetic profiles.

short-day plant A plant that flowers (usually in late summer, fall, or winter) only when the light period is shorter than a critical length.

short-term memory The ability to hold information, anticipations, or goals for a time and then release them if they become irrelevant.

sickle-cell disease A recessively inherited human blood disorder in which a single nucleotide change in the α-globin gene causes haemoglobin to aggregate, changing red blood cell shape and causing multiple symptoms in afflicted individuals.

sieve plate An end wall in a sieve-tube element, which facilitates the flow of phloem sap in angiosperm sieve tubes.

sieve-tube element A living cell that conducts sugars and other organic nutrients in the phloem of angiosperms; also called a sieve-tube member. Connected end to end, they form sieve tubes.

sign stimulus An external sensory cue that triggers a fixed action pattern by an animal.

signal Any kind of information sent from one organism to another, or from one place in an organism to another place.

signal peptide A sequence of about 20 amino acids at or near the leading (amino) end of a polypeptide that targets it to the endoplasmic reticulum or other organelles in a eukaryotic cell.

signal-recognition particle (SRP) A protein-RNA complex that recognises a signal peptide as it emerges from a ribosome and helps direct the ribosome to the endoplasmic reticulum (ER) by binding to a receptor protein on the ER.

signal transduction The linkage of a mechanical, chemical, or electromagnetic stimulus to a specific cellular response.

signal transduction pathway A series of steps linking a mechanical, chemical, or electrical stimulus to a specific cellular response.

silent mutation A nucleotide-pair substitution that has no observable effect on the phenotype; for example, within a gene, a mutation that results in a codon that codes for the same amino acid.

simple fruit A fruit derived from a single carpel or several fused carpels.

simple sequence DNA A DNA sequence that contains many copies of tandemly repeated short sequences.

single bond A single covalent bond; the sharing of a pair of valence electrons by two atoms.

single circulation A circulatory system consisting of a single pump and circuit, in which blood passes from the sites of gas exchange to the rest of the body before returning to the heart.

single-lens eye The camera-like eye found in some jellyfish, polychaete worms, spiders, and many molluscs.

single nucleotide polymorphism (SNP) (snip) A single base-pair site in a genome where nucleotide variation is found in at least 1% of the population.

single-strand binding protein A protein that binds to the unpaired DNA strands during DNA replication, stabilising them and holding them apart while they serve as templates for the synthesis of complementary strands of DNA.

sinoatrial (SA) node (sī′-nō-ā′-trē-uhl) A region in the right atrium of the heart that sets the rate and timing at which all cardiac muscle cells contract; the pacemaker.

sister chromatids Two copies of a duplicated chromosome attached to each other by proteins at the centromere and, sometimes, along the arms. While joined, two sister chromatids make up one chromosome. Chromatids are eventually separated during mitosis or meiosis II.

sister taxa Groups of organisms that share an immediate common ancestor and hence are each other's closest relatives.

skeletal muscle A type of striated muscle that is generally responsible for the voluntary movements of the body.

sliding-filament model The idea that muscle contraction is based on the movement of thin (actin) filaments along thick (myosin) filaments, shortening the sarcomere, the basic unit of muscle organisation.

slow-twitch fibre A muscle fibre that can sustain long contractions.

small interfering RNA (siRNA) One of multiple small, single-stranded RNA molecules generated by cellular machinery from a long, linear, double-stranded RNA molecule. The siRNA associates with one or more proteins in a complex that can degrade or prevent translation of an mRNA with a complementary sequence.

small intestine The longest section of the alimentary canal, so named because of its small diameter compared with that of the large intestine; the principal site of the enzymatic hydrolysis of food macromolecules and the absorption of nutrients.

smooth ER That portion of the endoplasmic reticulum that is free of ribosomes.

smooth muscle A type of muscle lacking the striations of skeletal and cardiac muscle because of the uniform distribution of myosin filaments in the cells; responsible for involuntary body activities.

social learning Modification of behaviour through the observation of other individuals.

sociobiology The study of social behaviour based on evolutionary theory.

sodium-potassium pump A transport protein in the plasma membrane of animal cells that actively transports sodium out of the cell and potassium into the cell.

soil horizon A soil layer with physical characteristics that differ from those of the layers above or beneath.

solute (sol′-yūt) A substance that is dissolved in a solution.

solute potential (Ψ_S) A component of water potential that is proportional to the molarity of a solution and that measures the effect of solutes on the direction of water movement; also called osmotic potential, it can be either zero or negative.

solution A liquid that is a homogeneous mixture of two or more substances.

solvent The dissolving agent of a solution. Water is the most versatile solvent known.

somatic cell (sō-mat′-ik) Any cell in a multicellular organism except a sperm or egg or their precursors.

somite One of a series of blocks of mesoderm that exist in pairs just lateral to the notochord in a vertebrate embryo.

soredium (suh-rē′-dē-um) (plural, **soredia**) In lichens, a small cluster of fungal hyphae with embedded algae.

sorus (plural, **sori**) A cluster of sporangia on a fern sporophyll. Sori may be arranged in various patterns, such as parallel lines or dots, which are useful in fern identification.

spatial learning The establishment of a memory that reflects the environment's spatial structure.

speciation (spē′-sē-ā′-shun) An evolutionary process in which one species splits into two or more species.

species (spē′-sēz) A population or group of populations whose members have the potential to interbreed in nature and produce viable, fertile offspring but do not produce viable, fertile offspring with members of other such groups.

species-area curve (spē′-sēz) The biodiversity pattern that shows that the larger the geographical area of a community is, the more species it has.

species diversity (spē′-sēz) The number and relative abundance of species in a biological community.

species richness (spē′-sēz) The number of species in a biological community.

specific heat The amount of heat that must be absorbed or lost for 1 g of a substance to change its temperature by 1°C.

spectrophotometer An instrument that measures the proportions of light of different wavelengths absorbed and transmitted by a pigment solution.

sperm The male gamete.

spermatheca (sper′-muh-thē′-kuh) (plural, **spermathecae**) In many insects, a sac in the female reproductive system where sperm are stored.

spermatogenesis (sper-ma′-tō-gen′-uh-sis) The continuous and prolific production of mature sperm in the testis.

spermatogonium (sper-ma′-tō-gō′-nē-um) (plural, **spermatogonia**) A cell that divides mitotically to form spermatocytes.

S phase The synthesis phase of the cell cycle; the portion of interphase during which DNA is replicated.

sphincter (sfink′-ter) A ringlike band of muscle fibres that controls the size of an opening in the body, such as the passage between the oesophagus and the stomach.

spiral cleavage A type of embryonic development in protostomes in which the planes of cell division that transform the zygote into a ball of cells are diagonal to the vertical axis of the embryo. As a result, the cells of each tier sit in the grooves between cells of adjacent tiers.

spliceosome (splī′-sō-sōm) A large complex made up of proteins and RNA molecules that splices RNA by interacting with the ends of an RNA intron, releasing the intron and joining the two adjacent exons.

spongocoel (spon′-jō-sēl) The central cavity of a sponge.

spontaneous process A process that occurs without an overall input of energy; a process that is energetically favourable.

sporangium (spōr-an′-jē-um) (plural, **sporangia**) A multicellular organ in fungi and plants in which meiosis occurs and haploid cells develop.

spore (1) In the life cycle of a plant or alga undergoing alternation of generations, a haploid cell produced in the sporophyte by meiosis. A spore can divide by mitosis to develop into a multicellular haploid individual, the gametophyte, without fusing with another cell. (2) In fungi, a haploid cell, produced either sexually or asexually, that produces a mycelium after germination.

sporocyte (spō′-ruh-sīt) A diploid cell within a sporangium that undergoes meiosis and generates haploid spores; also called a spore mother cell.

sporophyll (spō′-ruh-fil) A modified leaf that bears sporangia and hence is specialised for reproduction.

sporophyte (spō′-ruh-fīt) In organisms (plants and some algae) that have alternation of generations, the multicellular diploid form that results from the union of gametes. Meiosis in the sporophyte produces haploid spores that develop into gametophytes.

sporopollenin (spōr-uh-pol′-eh-nin) A durable polymer that covers exposed zygotes of charophyte algae and forms the walls of plant spores, preventing them from drying out.

stability In evolutionary biology, a term referring to a hybrid zone in which hybrids continue to be produced; this causes the hybrid zone to be "stable" in the sense of persisting over time.

stabilising selection Natural selection in which intermediate phenotypes survive or reproduce more successfully than do extreme phenotypes.

stamen (stā′-men) The pollen-producing reproductive organ of a flower, consisting of an anther and a filament.

standard deviation A measure of the variation found in a set of data points.

standard metabolic rate (SMR) Metabolic rate of a resting, fasting, and nonstressed ectotherm at a particular temperature.

starch A storage polysaccharide in plants, consisting entirely of glucose monomers joined by glycosidic linkages.

start point In transcription, the nucleotide position on the promoter where RNA polymerase begins synthesis of RNA.

statocyst (stat′-uh-sist′) A type of mechanoreceptor that functions in equilibrium in invertebrates by use of statoliths, which stimulate hair cells in relation to gravity.

statolith (stat′-uh-lith′) (1) In plants, a specialised plastid that contains dense starch grains and may play a role in detecting gravity. (2) In invertebrates, a dense particle that settles in response to gravity and is found in sensory organs that function in equilibrium.

stele (stēl) The vascular tissue of a stem or root.

stem A vascular plant organ consisting of an alternating system of nodes and internodes that support the leaves and reproductive structures.

stem cell Any relatively unspecialised cell that can produce, during a single division, two identical daughter cells or two more specialised daughter cells that can undergo further differentiation, or one cell of each type.

steroid A type of lipid characterised by a carbon skeleton consisting of four fused rings with various chemical groups attached.

sticky end A single-stranded end of a double-stranded restriction fragment.

stigma (plural, **stigmata**) The sticky part of a flower's carpel, which receives pollen grains.

stimulus In feedback regulation, a fluctuation in a variable that triggers a response.

stipe A stemlike structure of a seaweed.

stock The plant that provides the root system when making a graft.

stoma (stō′-muh) (plural, **stomata**) A microscopic pore surrounded by guard cells in the epidermis of leaves and stems that allows gas exchange between the environment and the interior of the plant.

stomach An organ of the digestive system that stores food and performs preliminary steps of digestion.

stramenopiles (strah′-men-ō′-pē-lēs) One of the three major subgroups for which the SAR eukaryotic supergroup is named. This clade arose by secondary endosymbiosis and includes diatoms and brown algae.

stratum (strah′-tum) (plural, **strata**) A rock layer formed when new layers of sediment cover older ones and compress them.

strigolactone Any of a class of plant hormones that inhibit shoot branching, trigger the germination of parasitic plant seeds, and stimulate the association of plant roots with mycorrhizal fungi.

strobilus (strō-bī′-lus) (plural, **strobili**) The technical term for a cluster of sporophylls known commonly as a cone, found in most gymnosperms and some seedless vascular plants.

stroke The death of nervous tissue in the brain, usually resulting from rupture or blockage of arteries in the neck or head.

stroke volume The volume of blood pumped by a heart ventricle in a single contraction.

stroma (strō′-muh) The dense fluid within the chloroplast surrounding the thylakoid membrane and containing ribosomes and DNA; involved in the synthesis of organic molecules from carbon dioxide and water.

stromatolite Layered rock that results from the activities of prokaryotes that bind thin films of sediment together.

structural isomer One of two or more compounds that have the same molecular formula but differ in the covalent arrangements of their atoms.

style The stalk of a flower's carpel, with the ovary at the base and the stigma at the top.

substrate The reactant on which an enzyme works.

substrate feeder An animal that lives in or on its food source, eating its way through the food.

substrate-level phosphorylation (fos′-fōr-uh-lā′-shun) The enzyme-catalysed formation of ATP by direct transfer of a phosphate group to ADP from an intermediate substrate in catabolism.

sugar sink A plant organ that is a net consumer or storer of sugar. Growing roots, buds, stems, and fruits are examples of sugar sinks supplied by phloem.

sugar source A plant organ in which sugar is being produced by either photosynthesis or the breakdown of starch. Mature leaves are the primary sugar sources of plants.

sulfhydryl group A chemical group consisting of a sulfur atom bonded to a hydrogen atom.

summation A phenomenon of neural integration in which the membrane potential of the postsynaptic cell is determined by the combined effect of EPSPs or IPSPs produced in rapid succession at one synapse or simultaneously at different synapses.

suprachiasmatic nucleus (SCN) (sūp′-ruh-kē′-as-ma-tik) A group of neurons in the hypothalamus of mammals that functions as a biological clock.

surface tension A measure of how difficult it is to stretch or break the surface of a liquid. Water has a high surface tension because of the hydrogen bonding of surface molecules.

surfactant A substance secreted by alveoli that decreases surface tension in the fluid that coats the alveoli.

survivorship curve A plot of the number of members of a cohort that are still alive at each age; one way to represent age-specific mortality.

suspension feeder An animal that feeds by removing suspended food particles from the surrounding medium by a capture, trapping, or filtration mechanism.

sustainable agriculture Long-term productive farming methods that are environmentally safe.

sustainable development Development that meets the needs of people today without limiting the ability of future generations to meet their needs.

swim bladder In aquatic osteichthyans, an air sac that enables the animal to control its buoyancy in the water.

symbiont (sim′-bē-ont) The smaller participant in a symbiotic relationship, living in or on the host.

symbiosis An ecological relationship between organisms of two different species that live together in direct and intimate contact.

sympathetic division A division of the autonomic nervous system; generally increases energy expenditure and prepares the body for action.

sympatric speciation (sim-pat′-rik) The formation of new species in populations that live in the same geographical area.

symplast In plants, the continuum of cytosol connected by plasmodesmata between cells.

synapse (sin′-aps) The junction where a neuron communicates with another cell across a narrow gap via a neurotransmitter or an electrical coupling.

synapsid (si-nap′-sid) A member of an amniote clade distinguished by a single hole on each side of the skull. Synapsids include the mammals.

synapsis (si-nap′-sis) The pairing and physical connection of one duplicated chromosome to its homologue during prophase I of meiosis.

synaptonemal complex (si-nap′-tuh-nē′-muhl) A zipper-like structure composed of proteins, which connects a chromosome to its homologue tightly along their lengths during part of prophase I of meiosis.

systematics A scientific discipline focused on classifying organisms and determining their evolutionary relationships.

systemic acquired resistance A defensive response in infected plants that helps protect healthy tissue from pathogenic invasion.

systemic circuit The branch of the circulatory system that supplies oxygenated blood to and carries deoxygenated blood away from organs and tissues throughout the body.

systems biology An approach to studying biology that aims to model the dynamic behaviour of whole biological systems based on a study of the interactions among the system's parts.

systole (sis′-tō-lē) The stage of the cardiac cycle in which a heart chamber contracts and pumps blood.

systolic pressure Blood pressure in the arteries during contraction of the ventricles.

taproot A main vertical root that develops from an embryonic root and gives rise to lateral (branch) roots.

tastant Any chemical that stimulates the sensory receptors in a taste bud.

taste bud A collection of modified epithelial cells on the tongue or in the mouth that are receptors for taste in mammals.

TATA box A DNA sequence in eukaryotic promoters crucial in forming the transcription initiation complex.

taxis (tak′-sis) An oriented movement towards or away from a stimulus.

taxon (plural, **taxa**) A named taxonomic unit at any given level of classification.

taxonomy (tak-son′-uh-mē) A scientific discipline concerned with naming and classifying the diverse forms of life.

Tay-Sachs disease A human genetic disease caused by a recessive allele for a dysfunctional enzyme, leading to accumulation of certain lipids in the brain. Seizures, blindness, and degeneration of motor and mental performance usually become manifest a few months after birth, followed by death within a few years.

T cells The class of lymphocytes that mature in the thymus; they include both effector cells for the cell-mediated immune response and helper cells required for both branches of adaptive immunity.

technology The application of scientific knowledge for a specific purpose, often involving industry or commerce but also including uses in basic research.

telomere (tel′-uh-mēr) The tandemly repetitive DNA at the end of a eukaryotic chromosome's DNA molecule. Telomeres protect the organism's genes from being eroded during successive rounds of replication. *See also* repetitive DNA.

telophase The fifth and final stage of mitosis, in which daughter nuclei are forming and cytokinesis has typically begun.

temperate broadleaf forest A biome located throughout midlatitude regions where there is sufficient moisture to support the growth of large, broadleaf deciduous trees.

temperate grassland A terrestrial biome that exists at midlatitude regions and is dominated by grasses and forbs.

temperate phage A phage that is capable of replicating by either a lytic or lysogenic cycle.

temperature A measure in degrees of the average kinetic energy (thermal energy) of the atoms and molecules in a body of matter.

template strand The DNA strand that provides the pattern, or template, for ordering, by complementary base pairing, the sequence of nucleotides in an RNA transcript.

tendon A fibrous connective tissue that attaches muscle to bone.

terminator In bacteria, a sequence of nucleotides in DNA that marks the end of a gene and signals RNA polymerase to release the newly made RNA molecule and detach from the DNA.

territoriality A behaviour in which an animal defends a bounded physical space against encroachment by other individuals, usually of its own species.

tertiary consumer (ter′-shē-ār′-ē) A carnivore that eats other carnivores.

tertiary structure (ter′-shē-ār′-ē) The overall shape of a protein molecule due to interactions of amino acid side chains, including hydrophobic interactions, ionic bonds, hydrogen bonds, and disulfide bridges.

test In foram protists, a porous shell that consists of a single piece of organic material hardened with calcium carbonate.

testcross Breeding an organism of unknown genotype with a homozygous recessive individual to determine the unknown genotype. The ratio of phenotypes in the offspring reveals the unknown genotype.

testis (plural, **testes**) The male reproductive organ, or gonad, in which sperm and reproductive hormones are produced.

testosterone A steroid hormone required for development of the male reproductive system, spermatogenesis, and male secondary sex characteristics; the major androgen in mammals.

tetanus (tet′-uh-nus) The maximal, sustained contraction of a skeletal muscle, caused by a very high frequency of action potentials elicited by continual stimulation.

tetrapod A vertebrate clade whose members have limbs with digits. Tetrapods include mammals, amphibians, and birds and other reptiles.

thalamus (thal′-uh-mus) An integrating centre of the vertebrate forebrain. Neurons with cell bodies in the thalamus relay neural input to specific areas in the cerebral cortex and regulate what information goes to the cerebral cortex.

theory An explanation that is broader in scope than a hypothesis, generates new hypotheses, and is supported by a large body of evidence.

thermal energy Kinetic energy due to the random motion of atoms and molecules; energy in its most random form. *See also* heat.

thermocline A narrow stratum of abrupt temperature change in the ocean and in many temperate-zone lakes.

thermodynamics (ther′-mō-dī-nam′-iks) The study of energy transformations that occur in a collection of matter. *See also* first law of thermodynamics and second law of thermodynamics.

thermophile *See* extreme thermophile.

thermoreceptor A receptor stimulated by either heat or cold.

thermoregulation The maintenance of internal body temperature within a tolerable range.

theropod A member of a group of dinosaurs that were bipedal carnivores.

thick filament A filament composed of staggered arrays of myosin molecules; a component of myofibrils in muscle fibres.

thigmomorphogenesis (thig′-mō-mor′-phō-gen′-uh-sis) A response in plants to chronic mechanical stimulation, resulting from increased ethylene production. An example is thickening stems in response to strong winds.

thigmotropism (thig-mō′-truh-pizm) A directional growth of a plant in response to touch.

thin filament A filament consisting of two strands of actin and two strands of regulatory protein coiled around one another; a component of myofibrils in muscle fibres.

threatened species A species that is considered likely to become endangered in the foreseeable future.

threshold The potential that an excitable cell membrane must reach for an action potential to be initiated.

thrombus (plural, **thrombi**) A fibrin-containing clot that forms in a blood vessel and blocks the flow of blood.

thylakoid (thī′-luh-koyd) A flattened, membranous sac inside a chloroplast. Thylakoids often exist in stacks called grana that are interconnected; their membranes contain molecular "machinery" used to convert light energy to chemical energy.

thymus (thī′-mus) A small organ in the thoracic cavity of vertebrates where maturation of T cells is completed.

thyroid gland An endocrine gland, located on the ventral surface of the trachea, that secretes two iodine-containing hormones, triiodothyronine (T_3) and thyroxine (T_4), as well as calcitonin.

thyroid hormone Either of two iodine-containing hormones (triiodothyronine and thyroxine) that are secreted by the thyroid gland and that help regulate metabolism, development, and maturation in vertebrates.

thyroxine (T_4) One of two iodine-containing hormones that are secreted by the thyroid gland and that help regulate metabolism, development, and maturation in vertebrates.

tidal volume The volume of air a mammal inhales and exhales with each breath.

tight junction A type of intercellular junction between animal cells that prevents the leakage of material through the space between cells.

tissue An integrated group of cells with a common structure, function, or both.

tissue system One or more tissues organised into a functional unit connecting the organs of a plant.

Toll-like receptor (TLR) A membrane receptor on a phagocytic white blood cell that recognises fragments of molecules common to a set of pathogens.

tonicity The ability of a solution surrounding a cell to cause that cell to gain or lose water.

top-down control A situation in which the abundance of organisms at each trophic level is controlled by the abundance of consumers at higher trophic levels; thus, predators limit herbivores, and herbivores limit plants.

topoisomerase A protein that breaks, swivels, and rejoins DNA strands. During DNA replication, topoisomerase helps to relieve strain in the double helix ahead of the replication fork.

topsoil A mixture of particles derived from rock, living organisms, and decaying organic material (humus).

torpor A physiological state in which activity is low and metabolism decreases.

totipotent (tō′-tuh-pōt′-ent) Describing a cell that can give rise to all parts of the embryo and adult, as well as extraembryonic membranes in species that have them.

trace element An element indispensable for life but required in extremely minute amounts.

trachea (trā′-kē-uh) The portion of the respiratory tract that passes from the larynx to the bronchi; also called the windpipe.

tracheal system In insects, a system of branched, air-filled tubes that extends throughout the body and carries oxygen directly to cells.

tracheid (trā′-kē-id) A long, tapered water-conducting cell found in the xylem of nearly all vascular plants. Functioning tracheids are no longer living.

trait One of two or more detectable variants in a genetic character.

***trans* fat** An unsaturated fat, formed artificially during hydrogenation of oils, containing one or more *trans* double bonds.

transcription The synthesis of RNA using a DNA template.

transcription factor A regulatory protein that binds to DNA and affects transcription of specific genes.

transcription initiation complex The completed assembly of transcription factors and RNA polymerase bound to a promoter.

transcription unit A region of DNA that is transcribed into an RNA molecule.

transduction A process in which phages (viruses) carry bacterial DNA from one bacterial cell to another. When these two cells are members of different species, transduction results in horizontal gene transfer. *See also* signal transduction pathway.

transfer RNA (tRNA) An RNA molecule that functions as a translator between nucleic acid and protein languages by picking up a specific amino acid and carrying it to the ribosome, where the tRNA recognises the appropriate codon in the mRNA.

transformation (1) The process by which a cell in culture acquires the ability to divide indefinitely, similar to the division of cancer cells. (2) A change in genotype and phenotype due to the assimilation of external DNA by a cell. When the external DNA is from a member of a different species, transformation results in horizontal gene transfer.

transgene A gene that has been transferred naturally or by a genetic engineering technique from one organism to another.

transgenic Pertaining to an organism whose genome contains DNA introduced from another organism of the same or a different species.

translation The synthesis of a polypeptide using the genetic information encoded in an mRNA molecule. There is a change of "language" from nucleotides to amino acids.

translocation (1) An aberration in chromosome structure resulting from attachment of a chromosomal fragment to a nonhomologous chromosome. (2) During protein synthesis, the third stage in the elongation cycle, when the RNA carrying the growing polypeptide moves from the A site to the P site on the ribosome. (3) The transport of organic nutrients in the phloem of vascular plants.

transmembrane protein A type of integral protein that spans the entire membrane.

transmission electron microscope (TEM) A microscope that passes an electron beam through very thin sections stained with metal atoms and is primarily used to study the internal structure of cells.

transpiration The evaporative loss of water from a plant.

transport epithelium One or more layers of specialised epithelial cells that carry out and regulate solute movement.

transport protein A transmembrane protein that helps a certain substance or class of closely related substances to cross the membrane.

transport vesicle A small membranous sac in a eukaryotic cell's cytoplasm carrying molecules produced by the cell.

transposable element A segment of DNA that can move within the genome of a cell by means of a DNA or RNA intermediate; also called a transposable genetic element.

transposon A transposable element that moves within a genome by means of a DNA intermediate.

transverse (T) tubule An infolding of the plasma membrane of skeletal muscle cells.

triacylglycerol (trī-as′-ul-glis′-uh-rol) A lipid consisting of three fatty acids linked to one glycerol molecule; also called a fat or triglyceride.

trichome An epidermal cell that is a highly specialised, often hairlike outgrowth on a plant shoot.

triple response A plant growth manoeuvre in response to mechanical stress, involving slowing of stem elongation, thickening of the stem, and a curvature that causes the stem to start growing horizontally.

triplet code A genetic information system in which a series of three-nucleotide-long words specifies a sequence of amino acids for a polypeptide chain.

triploblastic Possessing three germ layers: the endoderm, mesoderm, and ectoderm. All bilaterian animals are triploblastic.

trisomic Referring to a diploid cell that has three copies of a particular chromosome instead of the normal two.

trochophore larva (trō′-kuh-fōr) Distinctive larval stage observed in some lophotrochozoan animals, including some annelids and molluscs.

trophic efficiency The percentage of production transferred from one trophic level to the next higher trophic level.

trophic level The position an organism occupies in a food chain.

trophic structure The different feeding relationships in an ecosystem, which determine the route of energy flow and the pattern of chemical cycling.

trophoblast The outer epithelium of a mammalian blastocyst. It forms the fetal part of the placenta, supporting embryonic development but not forming part of the embryo proper.

tropical dry forest A terrestrial biome characterised by relatively high temperatures and precipitation overall but with a pronounced dry season.

tropical rain forest A terrestrial biome characterised by relatively high precipitation and temperatures year-round.

tropics Latitudes between 23.5° north and south.

tropism A growth response that results in the curvature of whole plant organs towards or away from stimuli due to differential rates of cell elongation.

tropomyosin The regulatory protein that blocks the myosin-binding sites on actin molecules.

troponin complex The regulatory proteins that control the position of tropomyosin on the thin filament.

true-breeding Referring to organisms that produce offspring of the same variety over many generations of self-pollination.

tubal ligation A means of sterilisation in which a woman's two oviducts (fallopian tubes) are tied closed and a segment of each is removed to prevent eggs from reaching the uterus.

tube foot One of numerous extensions of an echinoderm's water vascular system. Tube feet function in locomotion and feeding.

tumour-suppressor gene A gene whose protein product inhibits cell division, thereby preventing the uncontrolled cell growth that contributes to cancer.

tundra A terrestrial biome at the extreme limits of plant growth. At the northernmost limits, it is called arctic tundra, and at high altitudes, where plant forms are limited to low shrubby or matlike vegetation, it is called alpine tundra.

tunicate A member of the clade Urochordata, sessile marine chordates that lack a backbone.

turgid (ter′-jid) Swollen or distended, as in plant cells. (A walled cell becomes turgid if it has a lower water potential than its surroundings, resulting in entry of water.)

turgor pressure The force directed against a plant cell wall after the influx of water and swelling of the cell due to osmosis.

turnover The mixing of waters as a result of changing water-temperature profiles in a lake.

tympanic membrane Another name for the eardrum, the membrane between the outer and middle ear.

Unikonta (yū′-ni-kon′-tuh) One of four supergroups of eukaryotes proposed in a current hypothesis of the evolutionary history of eukaryotes. This clade, which is supported by studies of myosin proteins and DNA, consists of amoebozoans and opisthokonts. *See also* Excavata, SAR, and Archaeplastida.

unsaturated fatty acid A fatty acid that has one or more double bonds between carbons in the hydrocarbon tail. Such bonding reduces the number of hydrogen atoms attached to the carbon skeleton.

urban ecology The study of organisms and their environment in urban and suburban settings.

urea A soluble nitrogenous waste produced in the liver by a metabolic cycle that combines ammonia with carbon dioxide.

ureter (yū-rē′-ter) A duct leading from the kidney to the urinary bladder.

urethra (yū-rē′-thruh) A tube that releases urine from the mammalian body near the vagina in females and through the penis in males; also serves in males as the exit tube for the reproductive system.

uric acid A product of protein and purine metabolism and the major nitrogenous waste product of insects, land snails, and many reptiles. Uric acid is relatively nontoxic and largely insoluble in water.

urinary bladder The pouch where urine is stored prior to elimination.

uterine cycle The cyclical changes in the endometrium (uterine lining) of mammals that occur in the absence of pregnancy. In certain primates, including humans, the uterine cycle is a menstrual cycle.

uterus A female organ where eggs are fertilised and/or development of the young occurs.

vaccine A harmless variant or derivative of a pathogen that stimulates a host's immune system to mount defences against the pathogen.

vacuole (vak′-yū-ōl′) A membrane-bounded vesicle whose specialised function varies in different kinds of cells.

vagina Part of the female reproductive system between the uterus and the outside opening; the birth canal in mammals. During copulation, the vagina accommodates the male's penis and receives sperm.

valence The bonding capacity of a given atom; the number of covalent bonds that an atom can form, which usually equals the number of unpaired electrons in its outermost (valence) shell.

valence electron An electron in the outermost electron shell.

valence shell The outermost energy shell of an atom, containing the valence electrons involved in the chemical reactions of that atom.

van der Waals interactions Weak attractions between molecules or parts of molecules that result from transient local partial charges.

variable A factor that varies in an experiment.

variation Differences between members of the same species.

vas deferens In mammals, the tube in the male reproductive system in which sperm travel from the epididymis to the urethra.

vasa recta The capillary system in the kidney that serves the loop of Henle.

vascular cambium A cylinder of meristematic tissue in woody plants that adds layers of secondary vascular tissue called secondary xylem (wood) and secondary phloem.

vascular plant A plant with vascular tissue. Vascular plants include all living plant species except liverworts, mosses, and hornworts.

vascular tissue Plant tissue consisting of cells joined into tubes that transport water and nutrients throughout the plant body.

vasectomy The cutting and sealing of each vas deferens to prevent sperm from entering the urethra.

vasoconstriction A decrease in the diameter of blood vessels caused by contraction of smooth muscles in the vessel walls.

vasodilation An increase in the diameter of blood vessels caused by relaxation of smooth muscles in the vessel walls.

vasopressin *See* antidiuretic hormone (ADH).

vector An organism that transmits pathogens from one host to another.

vegetal pole The point at the end of an egg in the hemisphere where most yolk is concentrated; opposite of animal pole.

vegetative propagation Asexual reproduction in plants that is facilitated or induced by humans.

vegetative reproduction Asexual reproduction in plants.

vein (1) In animals, a vessel that carries blood towards the heart. (2) In plants, a vascular bundle in a leaf.

ventilation The flow of air or water over a respiratory surface.

ventral In an animal with bilateral symmetry, pertaining to the underside (in most animals) or front (in animals with upright posture) of the body.

ventricle (ven′-tri-kul) (1) A heart chamber that pumps blood out of the heart. (2) A space in the vertebrate brain, filled with cerebrospinal fluid.

venule (ven′-yūl) A vessel that conveys blood between a capillary bed and a vein.

vernalisation The use of cold treatment to induce a plant to flower.

vertebrate A chordate animal with vertebrae, the series of bones that make up the backbone.

vesicle (ves′-i-kul) A membrane-bound sac in or outside a cell.

vessel A continuous water-conducting micropipe found in most angiosperms and a few nonflowering vascular plants.

vessel element A short, wide, water-conducting cell found in the xylem of most angiosperms and a few nonflowering vascular plants. Dead at maturity, vessel elements are aligned end to end to form micropipes called vessels.

vestigial structure A feature of an organism that is a historical remnant of a structure that served a function in the organism's ancestors.

villus (plural, **villi**) (1) A finger-like projection of the inner surface of the small intestine. (2) A finger-like projection of the chorion of the mammalian placenta. Large numbers of villi increase the surface areas of these organs.

viral envelope A membrane, derived from membranes of the host cell, that cloaks the capsid, which in turn encloses a viral genome.

virulent phage A phage that replicates only by a lytic cycle.

virus An infectious particle incapable of replicating outside of a cell, consisting of an RNA or DNA genome surrounded by a protein coat (capsid) and, for some viruses, a membranous envelope.

visceral mass One of the three main parts of a mollusc; the part containing most of the internal organs. *See also* foot and mantle.

visible light That portion of the electromagnetic spectrum that can be detected as various colours by the human eye, ranging in wavelength from about 380 nm to about 740 nm.

vital capacity The maximum volume of air that a mammal can inhale and exhale with each breath.

vitamin An organic molecule required in the diet in very small amounts. Many vitamins serve as coenzymes or parts of coenzymes.

viviparous (vī-vip′-uh-rus) Referring to a type of development in which the young are born alive after having been nourished in the uterus by blood from the placenta.

voltage-gated ion channel A specialised ion channel that opens or closes in response to changes in membrane potential.

vulva Collective term for the female external genitalia.

water potential (ψ) The physical property predicting the direction in which water will flow, governed by solute concentration and applied pressure.

water vascular system A network of hydraulic canals unique to echinoderms that branches into extensions called tube feet, which function in locomotion and feeding.

wavelength The distance between crests of waves, such as those of the electromagnetic spectrum.

wetland A habitat that is inundated by water at least some of the time and that supports plants adapted to water-saturated soil.

white matter Tracts of axons within the CNS.

whole-genome shotgun approach Procedure for genome sequencing in which the genome is randomly cut into many overlapping short segments that are sequenced; computer software then assembles the complete sequence.

wild type The phenotype most commonly observed in natural populations; also refers to the individual with that phenotype.

wilting The drooping of leaves and stems as a result of plant cells becoming flaccid.

wobble Flexibility in the base-pairing rules in which the nucleotide at the 5′ end of a tRNA anticodon can form hydrogen bonds with more than one kind of base in the third position (3′ end) of a codon.

xerophyte (zir′-ō-fīt′) A plant adapted to an arid climate.

X-linked gene A gene located on the X chromosome; such genes show a distinctive pattern of inheritance.

X-ray crystallography A technique used to study the three-dimensional structure of molecules. It depends on the diffraction of an X-ray beam by the individual atoms of a crystallised molecule.

xylem (zī′-lum) Vascular plant tissue consisting mainly of tubular dead cells that conduct most of the water and minerals upwards from the roots to the rest of the plant.

xylem sap (zī′-lum) The dilute solution of water and minerals carried through vessels and tracheids.

yeast Single-celled fungus. Yeasts reproduce asexually by binary fission or by the pinching of small buds off a parent cell. Many fungal species can grow both as yeasts and as a network of filaments; relatively few species grow only as yeasts.

yolk Nutrients stored in an egg.

zero population growth (ZPG) A period of stability in population size, when additions to the population through births and immigration are balanced by subtractions through deaths and emigration.

zona pellucida The extracellular matrix surrounding a mammalian egg.

zoned reserve An extensive region that includes areas relatively undisturbed by humans surrounded by areas that have been changed by human activity and are used for economic gain.

zone of polarising activity (ZPA) A block of mesoderm located just under the ectoderm where the posterior side of a limb bud is attached to the body; required for proper pattern formation along the anterior-posterior axis of the limb.

zoonotic pathogen A disease-causing agent that is transmitted to humans from other animals.

zoopagomycete A member of the fungal phylum Zoopagomycota, multicellular parasites or commensal symbionts of animals; sexual reproduction, where known, involves the formation of a sturdy structure called a zygosporangium.

zoospore Flagellated spore found in chytrid fungi and some protists.

zygomycete (zī′-guh-mī′-sēt) A member of the fungal phylum Zygomycota, characterised by the formation of a sturdy structure called a zygosporangium during sexual reproduction.

zygosporangium (zī′-guh-spōr-an′-jē-um) (plural, **zygosporangia**) In zygomycete fungi, a sturdy multinucleate structure in which karyogamy and meiosis occur.

zygote (zī′-gōt) The diploid cell produced by the union of haploid gametes during fertilisation; a fertilised egg.

Index

NOTE: Page numbers followed by *f* and *t* indicate figures and tables. Page numbers in **bold** indicate definitions of key terms.

A

B

C

Index

D

E

I

M

O

Index

T

U

V